DYNAMICS AND CONTROL OF PROCESS SYSTEMS 2004
(DYCOPS-7)

A Proceedings volume from the7th IFAC Symposium, Cambridge, Massachusetts, USA, 5 –7 July 2004

Edited by

SIRISH L. SHAH
Department of Chemical and Materials Engineering, University of Alberta, Edmonton, Canada

and

JOHN MacGREGOR
Department of Chemical Engineering McMaster University, Hamilton, Canada

(In two volumes)

Volume 2

Published for the

INTERNATIONAL FEDERATION OF AUTOMATIC CONTROL

by

ELSEVIER LIMITED

ELSEVIER Ltd
The Boulevard, Langford Lane
Kidlington, Oxford OX5 1GB, UK

Elsevier Internet Homepage
http://www.elsevier.com

Consult the Elsevier Homepage for full catalogue information on all books, journals and electronic products and services.

IFAC Publications Internet Homepage
http://www.elsevier.com/locate/ifac

Consult the IFAC Publications Homepage for full details on the preparation of IFAC meeting papers, published/forthcoming IFAC books, and information about the IFAC Journals and affiliated journals.

First edition 2005

Library of Congress Cataloging in Publication Data

A catalogue record for this book is available from the Library of Congress

British Library Cataloguing in Publication Data

A catalogue record for this book is available from the British Library

ISBN 0-08-044297 8
ISSN 1474-6670

Printed in Great Britain

To Contact the Publisher

Elsevier welcomes enquiries concerning publishing proposals: books, journal special issues, conference proceedings, etc. All formats and media can be considered. Should you have a publishing proposal you wish to discuss, please contact, without obligation, the publisher responsible for Elsevier's industrial and control engineering publishing programme:

Christopher Greenwell
Publishing Editor
Elsevier Ltd
The Boulevard, Langford Lane
Kidlington, Oxford
OX5 1GB, UK

Phone: +44 1865 843230
Fax: +44 1865 843920
E.mail: c.greenwell@elsevier.com

General enquiries, including placing orders, should be directed to Elsevier's Regional Sales Offices – please access the Elsevier homepage for full contact details (homepage details at the top of this page).

7th IFAC SYMPOSIUM ON DYNAMICS AND CONTROL OF PROCESS SYSTEMS 2004

Sponsored by
International Federation of Automatic Control (IFAC)
Technical Committee on Chemical Process Control

Organized by
International Federation of Automatic Control (IFAC)
American Automatic Control Council

***International Programme Committee* (IPC)**

Shah, S. L — Co-Chair
MacGregor, J. — Co-Chair
Marquardt, W. — IFAC Representative

Industrial Co-Chairs

Cott, B. (USA)
Duennebier, G. (Germany)

Technical Area Co-Chairs

Allgower, F. (Germany)
Bonvin, D. (Switzerland)
Braatz, R. (USA)
Doyle, F. (USA)
Dochain, D. (Belgium)
Foss, B. (Norway)
Hasebe, S. (Japan)
Huang, B. (Canada)
Jorgensen, S.B. (Denmark)
Marlin, T. (Canada)
Marquardt, W. (Germany)
Ricker, L. (USA)
Perkins, J. (UK)
Perrier, M. (Canada)
Pistikopoulos, S. (UK)
Qin, J. (USA)
Swartz, C. (Canada)
Thornhill, N. (UK)

IPC Members

Alvarez, J. (Mexico)
Arkun, Y. (Turkey)
Bakshi, B. (USA)
Biegler, L. (USA)
Boje, E. (South Africa)
Brambilla, A. (Italy)
Chen, B. (China)
Chu, J. (China)
Cinar, A. (USA)
Craig, I. (South Africa)
Daoutidis, P. (USA)
Edgar, T. (USA)
Engell, S. (Germany)
Forbes, F. (Canada)
Fujii, H. (Japan)
Jamsa-Jounela, S.-L. (Finland)
Gao, F. (HK China)
Guay, M. (Canada)
Gudi, R. (India)
Hagglund, T. (Sweden)
Hoo, K. (USA)
Huang, H.P. (Taiwan)
Isaksson, A. (Sweden)
Jacobsen, E. (Sweden)
Koivo, H. (Finland)
Kravaris, C. (Greece)
Lakshminarayanan, S. (Singapore)
Lee, J. (USA)
Lee, P. (Australia)
Lewin, D. (Israel)
Li Ping (China)
Mandler, J. (USA)
Morari, M. (Switzerland)
Narasimhan, S. (India)
Ohshima, M. (Japan)
Olivares, L.B. (Chile)
Palazoglu, A. (USA)
Park, S. (Korea)
Patwardhan, S. (India)
Preisig, H. (Norway)
Rawlings, J. (USA)
Romagnoli, J. (Australia)
Rossiter, A. (UK)
Scali, C. (Italy)
Secchi, A. (Brazil)
Tade, M.O. (Australia)
Williamson, C. (New Zealand)
Wongsri, M. (Thailand)
Yihui, J. (China)
Yoon, E.S. (Korea)
Yu, C.C. (Taiwan)

***National Organizing Committee* (NOC)**

Seider, W.D. — Chair

Beaverstock, M.
Barton, P.I.
Henson, M.A.
Kothare, M.V.
McAvoy, T.J.
Muske, K.R.
Ogunnaike, B.
Soroush, M.

FOREWORD

Dycops-2004 as held in Boston was an international event. It attracted a total of 170 participants from industry and academia around the world.

Over 200 papers were submitted from which 155 were selected for presentation at this conference including 3 plenary addresses and 8 keynote talks.

Editors:

Sirish L. Shah (University of Alberta, Canada)
John MacGregor (McMaster University, Canada)

CONTENTS

VOLUME 1

PLENARY PAPER 1

MODELING AND IDENTIFICATION 1

MONITORING AND FAULT DETECTION 1

ADVANCES IN CONTROL 1

KEYNOTE PAPERS 1-4

MODELING AND IDENTIFICATION 2

DESIGN AND CONTROL

ADVANCES IN CONTROL 2

PLENARY PAPER 2

MODELING AND IDENTIFICATION 3

MONITORING AND FAULT DETECTION 2

BIOPROCESS MODELING AND CONTROL 1

KEYNOTE PAPERS 5 and 6

CONTROL, OPTIMIZATION AND SCHEDULING 1

BATCH PROCESS MODELING AND CONTROL

IMPACT OF BIOLOGICAL ENIGNEERING ON PROCESS CONTROL EDUCATION

POSTER PAPERS

Bioprocess Modeling and Control

VOLUME 2

POSTER PAPERS (continued)

Monitoring and Fault Detection

Control, Optimization and Scheduling

Advances in Control

Modeling and Identification

Sensor Technologies

Industrial Applications

Batch and Semi-Batch Processes

KEYNOTE PAPERS 7 AND 8

CONTROL, OPTIMIZATION AND SCHEDULING 2

SENSOR TECHNOLOGY

BIOPROCESS MODELING AND CONTROL 2

ELSEVIER
IFAC
PUBLICATIONS
www.elsevier.com/locate/ifac

OBSERVER-BASED SUPERVISION AND FAULT DETECTION OF A FCC UNIT MODEL PREDICTIVE CONTROL SYSTEM

O.A.Z. Sotomayor[1], D. Odloak[1], E. Alcorta-Garcia[2], P. de León-Cantón[2]

[1] LSCP – Polytechnic School of the University of São Paulo
Av. Prof. Luciano Gualberto, trav.3, n.380, 05508-900 São Paulo-SP, BRAZIL
e-mail: oscar@lscp.pqi.ep.usp.br, odloak@usp.br
[2] FIME – Autonomous University of Nuevo León
Pedro de Alba S/N, 66450 San Nicolás de los Garza-NL, MEXICO
e-mail: ealcorta@ieee.org, plinin@hotmail.com

Abstract: This work is concerned with the design of a fault detection and isolation (FDI) system to monitor malfunctions in sensors and actuators of a fluid catalytic cracking (FCC) unit model predictive control (MPC) system. The control system is based on an infinite-horizon MPC algorithm. The fault detection scheme is developed based on two banks of robust observers, while the fault isolation task is done employing a structured residual approach. The models used in the control and in the FDI of the FCC unit are obtained by using subspace identification methods. The effectiveness of the proposed strategy is verified through numerical simulations carried out on a dynamical model of an industrial FCC unit under abrupt fault in its control devices. *Copyright © 2004 IFAC*

Keywords: Fluid catalytic cracking, Fault detection and isolation, Model predictive control, Unknown input observers, Subspace model identification.

1. INTRODUCTION

Market globalization has induced the high level of automatization in chemical and petrochemical plants. As an example, nowadays there is no petroleum refiner that does not use advanced process engineering, such as advanced control systems, to improve process operations (Moro, 2003). In these cases and under normal situations, the control system acts partly as operators to keep a satisfactory production process. Nevertheless, recent catastrophic accidents, namely the explosion at the Kuwait Petrochemical's Mina Al-Ahmedhi refinery in 2000, which resulted in about $400 million in damages, and the explosion at the offshore oil platform P-36 of Petrobras, Brazil, in 2001, with estimated loss about $5 billion (Venkatasubramanian, 2003), have shown that advanced control systems may not be sufficient to deal with abnormal situations.

Although these catastrophic accidents are infrequent and represent extreme cases within the spectrum of major abnormal situations, minor accidents are very common with significant economic, safety and environment effects. According to Cochran *et al.* (1996), the inability of automated control systems and plant operators to deal with abnormal situations cost the U.S. economy at least $20 billion annually, about half of which in direct losses related to petrochemical companies alone. Similar cases, cost the British economy up to $27 billion per year (Laser, 2002). Estimates indicate that elimination of all abnormal situations in the petrochemical industry could increase profits up to 5% (Harrold, 1998).

The performance of a control system depends on other control devices, namely sensors (to read the data that it receives) and actuators (to execute the commands that it issues). Obviously, any abnormal situation (or fault) in these devices or in the plant itself, can significantly affect the performance of the control system. For instance, a biased sensor lead the process to an operating point far from its optimal one and a stuck valve lead to a loss of control effectiveness. In any case, the control system can, eventually, hide a gradual fault incident till the failure of control is unavoidable and, in other cases, sudden faults can be amplified by the closed-loop control, inducing to premature plant shutdown, with loss of productivity, loss of expensive equipment and possible damage to human health and environment. Therefore, it is important to supervise the production process, as well as to detect faults while the plant is still operating in a controllable region, to support operators to deal with abnormal situations promptly. In this sense, fault diagnosis or fault detection and isolation (FDI) can be seen as part of a large scheme for optimal process operation.

There are several methods for FDI and they can be grouped into three general categories: quantitative model-based methods, qualitative model-based methods and process history based methods (Venkatasubramanian, 2003). Quantitative model-based FDI techniques, namely parity space approach, observer-based approach and parameter estimation approach, have received considerable attention in recent years. These approaches are based on the concept of *analytical redundancy* which makes use of a mathematical model of the process to obtain measurement estimates. The inconsistencies between the estimated and actual behaviors are *residuals* or fault indicators, which reflect the abnormal situation of the supervised process. Next, the residuals are analyzed aiming at the localization of the fault. Although there is a close relationship among the quantitative model-based FDI techniques, observer-based approach have become the most popular and important method for model-based FDI (Frank and Ding, 1997), especially within the automatic control community.

On the other hand, petroleum refining is one of the major industrial enterprices worldwide. It is a very competitive and specific market, where little improvements in the operation of the processes may lead to large economical benefits. The functional heart of a petroleum refinery, and the most economically critical component, is the FCC unit. It is operated to convert heavy petroleum fractions into light and more valuable hydrocarbon products, such as gasoline and LPG. It involves a highly nonlinear, slow and multivariable process, with strong interactions among its variables, subject to a number of operational constraints, and working at high temperatures and pressures, with a slate of fires and explosions arising from malfunctions. This complex characteristic together with its economic significance makes the FCC unit a potential candidate and challenging problem to the application of advanced process engineering tools. Although numerous papers have been written concerning different modelling approaches, advanced control (predominantly model predictive control, MPC) and optimization systems for the FCC unit, only few have treated about the application of FDI to the FCC system. Among these, Yang *et al.* (2000) use an approach based on neural network, Heim *et al.* (2003) use causal and knowledge based models, Huang *et al.* (2003) use an heuristic-based extended Kalman filter, Sundarraman and Srinivasan (2003) use a trend analysis-based approach and Wang *et al* (2003) use a recursive partial least squares (PLS) algorithm. A common feature in all these works is that the FDI system is mainly directed to detect process parameter faults and the FCC unit operates under conventional regulatory control. The only work available in the literature, in which the system is controlled by MPC was presented by Pranatyasto and Qin (2001). They propose an algorithm based on principal component analysis (PCA) for self-validating sensors in a FCC unit.

The present paper focuses on the design of a FDI system for the sensors and actuators of a FCC unit under MPC control. Closed-loop systems are vulnerable to faults and these are in general difficult to handle. In addition to the well-known damages, faults in control devices may cause structural changes of the process model as well. Two separated banks of robust observers are built for fault detection (FD) in the inputs and outputs of the FCC system. Both are based on the unknown input observers (UIO) theory. Fault isolation is obtained through a structured residual approach. The models used in the control and in the FDI system of the FCC unit are obtained by subspace identification methods. In the context of this study it is used the FCC RECOPE benchmark.

2. THE FCC UNIT DYNAMIC MODEL

The FCC RECOPE benchmark is basically the same as the one presented by Moro and Odloak (1995) for the FCC unit Kellog Orthoflow model F, and it has become a standard for validation of FCC control structures in the Petrobras's refineries. A simplified schematic diagram of this process is shown in figure 1. The process is constituted of three major parts, the riser, the reactor and the regenerator. At nominal steady-state conditions, the air flowrate to the regenerator is $u_1 = 221$ ton/h, the regenerated catalyst valve opening $u_2 = 82\%$, the total feed flowrate (gasoil + deasphalted heavy oil) introduced into the riser is $u_3 = 9700\,\text{m}^3/\text{d}$ and the temperature of the total feed flowrate is $u_4 = 235\,°\text{C}$. For the stable operation of the FCC unit, the benchmark includes necessarily a conventional regulatory control system, based on three PI controllers (not shown in figure 1), aiming at to keep the differential pressure between reactor and regenerator, the catalyst inventory in the reactor, and the suction pressure of the wet gas compressor at $0.65\,\text{kgf/cm}^2$, 90 ton. and $1.0\,\text{kgf/cm}^2$, respectively.

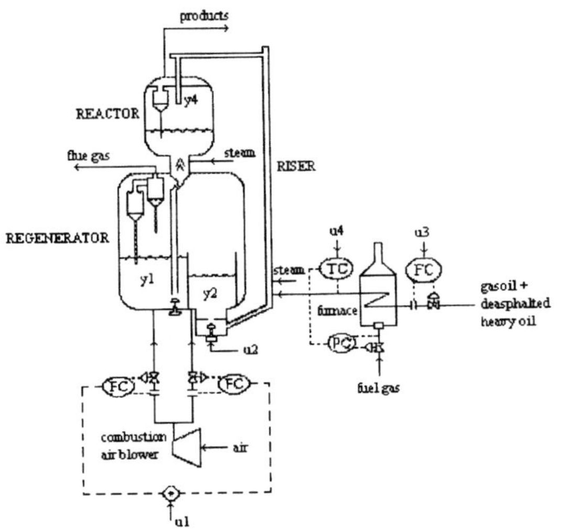

Fig. 1. Scheme of the FCC Kellog Orthoflow model F

The benchmark, represented by a 9x9 MIMO system, with 26 ODE's and 74 linear and nonlinear algebraic equations, was validated with industrial data from the FCC unit of the Petrobras's Henrique Lajes (REVAP) refinery in São José dos Campos. This simulator was written in language C and implemented in Simulink/Matlab™ platform by the RECOPE process control group, and it can be downloaded from *www.enq.ufrgs.br/recope/FCC/*. The model equations and the process parameters are here omitted but they can be found in Moro and Odloak (1995).

3. THE MPC CONTROL SYSTEM

MPC, or model-based predictive control (MBPC) or receding (moving) horizon control, is currently the most widely implemented advanced control system for process plants. It has several advantages over other control techniques because it incorporates an explicit process model into the control calculation and because of its close relationship with on-line optimization, which allows it to deal with multivariable process, coupling, inverse response, time-delays, process constraints, modelling uncertainties and measurement errors. MPC's are commonly found in the medium level of a plant-wide control scheme. A survey summarizing about 4500 successful MPC industrial applications is presented by Qin and Badgwell (2003).

Since a robust controller can desensitize fault effects and make the FDI very difficult (Chen and Patton, 1999), in this paper it is used the nominal version of the robust infinite-horizon MPC (IHMPC) controller proposed by Rodrigues and Odloak (2003).

3.1 The IHMPC algorithm

There are plenty of papers on MPC algorithms and their application to FCC units. When compared to other MPC algorithms, the IHMPC has the following advantages (Rodrigues and Odloak, 2003):

- The IHMPC is based on a class of output prediction oriented model (OPOM), where the state-space model of the process is modified to represent the output as a continuous function of time. With this model, for a finite control horizon, the infinite output horizon MPC cost function can be integrated explicity. The approach becomes more efficient as the computation of the control sequence does not include the solution of the Lyapunov equation to compute the terminal state penalty. The on-line solution of this equation may be time consuming for large systems.
- The cost function is modified by inclusion of slack variables in the output error. This is done to overcome the problem of an unbounded cost due to offset in the controlled outputs or unfeasibility of the infinite-horizon control problem. The cost function is minimized by using linear matrix inequalities (LMI) tools.
- It is not necessary to known the system steady state. Hence, the IHMPC can be applied for the servo problem as for the regulator problem.

3.2 Control structure for the FCC unit

The FCC control is by definition a multivariable one rather than a multiloop control. Nevertheless, the control performance of a FCC process highly depends on the control structure selection. This problem has been addressed by several authors, mainly focusing on possible combinations of manipulated and controlled variables for exact and partial control.

In this paper, the control structure is based on the original version of the MPC controller proposed for the FCC unit of the REVAP refinery. From this scheme we consider only a 4x4 MIMO control system, where the inputs or manipulated variables are as described in Section 2, and the outputs or controlled variables are: the temperature of the regenerator 1st stage dense phase (y_1), the temperature of the regenerator 2nd stage dense phase (y_2), the estimated cracking reaction severity (y_3) and the riser (reactor) temperature (y_4). The control objetive for the FCC unit is to stabilize the process despite undesired disturbances by keeping the outputs on their steady-state conditions, i.e. 670.14 °C, 700.88 °C, 77.5% and 542.2 °C, respectively. In the implementation of the IHMPC algorithm, $T = 3$ min and $m = 2$. For more information, see Sotomayor *et al.* (2004).

3.3 Controller model identification

The MPC requires a process model to represent the dynamical input-output behavior and it should be as simple as possible. In the industrial practice, a simple process model with two or three parameters are adjusted to produce the step response for the MPC controller. In this case, sixteen 2nd order SISO models are obtained by individually exciting the inputs with m-level pseudo-random signals (around $\pm 2\,\%$ their nominal values) and using the MOESP subspace identification method (Verhaegen and Dewilde, 1992) to identify initial models and to improve them by the PEM method of the System Identification Toolbox/Matlab™. These models are then converted to extract coefficients of the step response function, which are required by the IHMPC.

4. THE FDI SYSTEM

4.1 UIO and FD system based on UIO

Unknown input observers (UIO) are well known from the control theory. They constitute a class of robust observers that under certain conditions allow one to decouple the effect of unknown inputs (or disturbances) on the estimate of the true state. With slight modifications, they can be used to solve the disturbance decoupled residual problem in FD systems. Consider a dynamic system described by a discrete-time LTI model with an *additive* unknown disturbance as follows:

$$\begin{aligned} x_{k+1} &= Ax_k + Bu_k + Ed_k \\ y_k &= Cx_k \end{aligned} \tag{1}$$

where x is the state vector, u is the input vector, y is the output vector and d is the unknown input. A, B, C and E are known matrices with appropriate dimensions. The structure of a full-order UIO is described by:

$$\begin{aligned} z_{k+1} &= Fz_k + TBu_k + Ky_k \\ \hat{x}_k &= z_k + Hy_k \end{aligned} \tag{2}$$

where z is the state of the UIO, $\hat{x}$ is the estimated state vector, whilst F, T, K and H are matrices to be designed to achieve unknown input decoupling and other design requirements. When the observer (2) is applied to system (1), the estimation error ($\in_k = x_k - \hat{x}_k$) will approach to zero asymptotically if the following relationships hold true (Chen and Patton, 1999):

$$\left.\begin{aligned} T &= I - HC \\ TE &= 0 \\ F &= TA - K_1C \\ K_2 &= FH \\ K &= K_1 + K_2 \end{aligned}\right\} \tag{3}$$

It can be shown that if $rank(CE) = rank(E)$ then $H = E(CE)^+$, where ($^+$) denotes the Moore-Penrose pseudo-inverse. If the pair (C, TA) is observable, an UIO exists and F can be easily stabilized by a K_1 obtained using the pole placement routine in the Control System Toolbox/Matlab™. By employing the UIO structure for FD purposes, now called UIFDO, the residual (r) is generated as shown in figure 2.

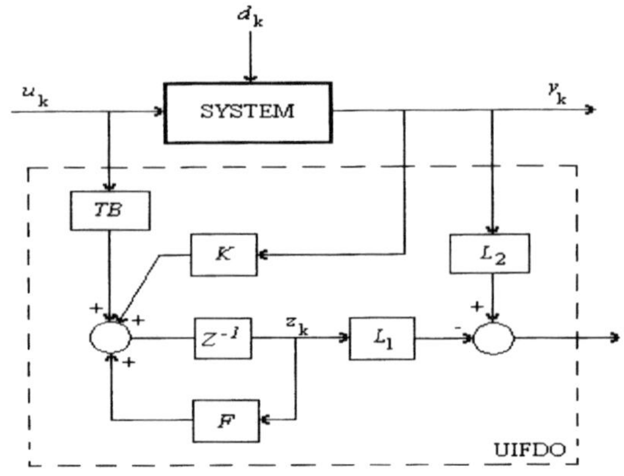

Fig. 2. Structure of the FD structure based on UIO

In figure 2, $L_1 = C$ and $L_2 = I - CH$. Note that by setting $T = I$, $H = 0$ and $E = 0$, the observer (2) is the classical Luenberger observer (LO).

4.2 FD model identification

The use of a simple process model for FDI applications may constitute a source of false alarms, which can corrupt the performance of the FD system to such an extent that it may even become totally useless. Therefore, it is necessary to obtain a highly precise process model (Frank *et al.*, 2000).
In this case, a 13th order MIMO linear state-space model, that reproduces very well the main dynamical characteristics of the process, is obtained by using the MOESP algorithm. The model order is chosen by applying the optimal statistical Akaike's information criterion (Akaike, 1973). The identification procedure was carried out in open-loop, by simultaneously exciting the control inputs with m-level pseudo-random signals (around $\pm 2\%$ their nominal values). To improve the model, the single-loop PI control signals are considered as inputs, and the temperature of the regenerator 1st stage dilute phase and the temperature of the regenerator general dilute phase are also considered as outputs. For more details, see Sotomayor *et al.* (2003).

4.3 FDI schemes and results

The main task of robust fault detection is to generate residual signals. To facilitate fault isolation, here it is used a structured residual approach, i.e. each residual is designed to be sensitive to a certain group of faults and insensitive to others. Taking into account these sensitivity and insensitivity properties, the fault isolation is possible. The ideal case is to make each residual only sensitive to a particular fault and insensitive to all other faults. However, this ideal situation is normally difficult to achieve. According to Chen and Patton (1999), an alternative solution is to make each residual to be sensitive to all the faults except a particular fault. Moreover, as full

insensitivity is difficult to achieve, the fault isolation task is obtained by choosing adequate thresholds.
Consider the open-loop system represented in (1) with possible sensor (output) and actuator (input) faults, which can be described as follows:

$$\begin{aligned} x_{k+1} &= Ax_k + Bu_k + Ef_{a,k} \\ y_k &= Cx_k + f_{s,k} \end{aligned} \tag{4}$$

where f_a and f_s denote actuator and sensor faults.

.1) Sensor FDI scheme

To uniquely detect and isolate a fault concerning one of the four sensors of the FCC control system, a bank of four LO, in a DOS (dedicated observer scheme) configuration (Wünnenberg, 1990), is used. It is assumed that the actuators are fault-free (i.e. $f_a = 0$). Figure 3 shows the absolute value of the residuals responses for the case of a single abrupt fault (bias) in each sensor, occurring at $t = 100$ min, with magnitude of +5%, +5%, +5% and +1% of the output value, respectively.

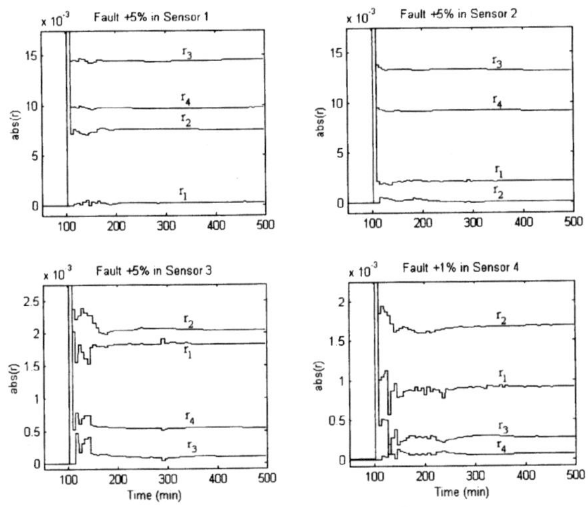

Fig. 3. Residual responses for faults in sensors.

Biased sensors lead to the saturation of controlled actuators. For example, a fault of +5% in sensor 2 is alarmed by the residuals r_1, r_3 and r_4 which are sensitive and not by the residual r_2 that is insensitive. The IHMPC tries to compensate this situation saturating input u_3. However, an acceptable control of the output variables is not achieved since bias appear in all the outputs.

.2) Actuator FDI scheme

To uniquely detect and isolate a fault concerning one of the four actuators of the FCC control system, a bank of four UIFDO, in a GOS (generalized observer scheme) configuration (Wünnenberg, 1990), is used considering that the sensors are fault-free (i.e. $f_s = 0$). Here, matrix E is one of the first four columns of the matrix B. Figure 4 shows the absolute value of the residuals responses for the case of a single abrupt fault (bias) in each actuator, occurring at $t = 100$ min, with magnitude of +5%, +10%, +5% and +5% of the input value, respectively.

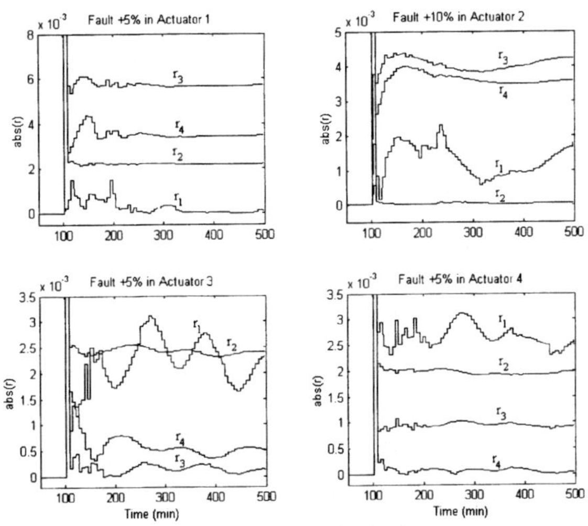

Fig. 4. Residual responses for faults in actuators.

For example, for a fault of +5% in actuator 1, the residuals r_2, r_3 and r_4 are sensitive and the residual r_1 is insensitive. In partial combustion mode, this fault scenario drives the process to the full combustion mode, severely affecting the quality of products. The IHMPC will try to revert this situation decreasing, mainly, the control signal of actuator 1. However, if the fault becomes more severe, the control fails and can cause a disastrous break down of the production.
It is necessary to state that in both presented cases, the fault is detected before the first 10 minutes after the fault occurrence. The instantaneous peaks in figures 3 and 4 are generated by the abrupt fault ocurrence and they can be considered as an incipient FD system.
To summarize the obtained results, table 1 shows the fault signatures in the structured residuals form.

Table 1. Structured residuals of the FDI system

	Sensor FDI scheme				Actuator FDI scheme			
	$f_{s,1}$	$f_{s,2}$	$f_{s,3}$	$f_{s,4}$	$f_{a,1}$	$f_{a,2}$	$f_{a,3}$	$f_{a,4}$
r_1	0	1	1	1	1	1	1	1
r_2	1	0	1	1	1	1	1	1
r_3	1	1	0	1	1	1	1	1
r_4	1	1	1	0	1	1	1	1
r_1	1	1	1	1	0	1	1	1
r_2	1	1	1	1	1	0	1	1
r_3	1	1	1	1	1	1	0	1
r_4	1	1	1	1	1	1	1	0

1=sensitive, 0=insensitive

5. CONCLUSIONS

Fault diagnosis in a FCC unit is an important task. Although, disturbances and parameter changes can be compensated by the advanced control system, the performance of the FCC control system is highly dependent of sensors and actuators. The timely detection of malfunctions in these control devices can help operators to avoid disastrous consequences. In this paper, we have presented a robust observer-based FDI system for a FCC unit under MPC control. It is shown that abrupt fault in a sensor or an actuator was accurate and rapidly detected. Later, the FDI information will be used for controller reconfiguration. The integration FDI-IHMPC provides a useful framework for an active fault tolerant MPC control. MPC seems a natural and promising basis for it.

Acknowledgements: The authors gratefully thank the finantial support from FAPESP/Brazil (grant N°02/08119-2) and CONACyT/ Mexico.

REFERENCES

Akaike, H. (1973). Information theory and an extension of the maximum likelihood principle, *2nd International Symposium on Information Theory*, Budapest, Hungary.

Chen, J., R.J. Patton (1999). *Robust model-based fault diagnosis for dynamic systems*, Kluwer Academic Publishers: Massachusetts.

Cochran, E., C. Miller, P. Bullemer (1996). Abnormal situation management in petrochemical plants: can a pilot's associate crack crude?, *1996 IEEE National Aerospace and Electronics Conference, NAECON 1996*, Dayton-KY, USA.

Frank, P.M., X. Ding (1997). Survey of robust residual generation and evaluation methods in observer-based fault detection systems, *Journal of Process Control*, 7(6), 403-424.

Frank, P.M., E. Alcorta-Garcia, B. Köppen-Seliger (2000). Modelling for fault detection and isolation versus modelling for control. *Mathematics and Computers in Simulation*, 53(4-6), 259-271.

Harrold, D. (1998). How to avoid abnormal situations, *Control Engineering*, 45(12), 68-76.

Heim, B., S. Gentil, B. Celse, S. Cauvin, L. Travé-Massuyès (2003). FCC diagnosis using several causal and knowledge based models, *5th IFAC Symposium on Fault Detection, Supervision and Safety of Technical Processes*, SAFEPROCESS 2003, Washington, D.C., USA.

Huang, Y., G.V. Reklaitis, V. Venkatasubramanian (2003). A heuristic extended Kalman filter based estimator for fault identification in a fluid catalytic cracking unit, *Industrial and Engineering Chemistry Research*, 42(14), 3361-3371.

Laser, M. (2002). Recent safety and environmental legislation, *Process Safety and Environmental Protection*, 80(B5), 277-281.

Moro, L.F.L. (2003). Process technology in the petroleum refining industry – current situation and future trends, *Computers and Chemical Engineering*, 27(8-9), 1303-1305.

Moro, L.F.L., D. Odloak (1995). Constrained multivariable control of fluid catalytic cracking converters, *Journal of Process Control*, 5(1), 29-39.

Pranatyasto, T.N., S.J. Qin (2001). Sensor validation and process fault diagnosis for FCC units under MPC feedback, *Control Engineering Practice*, 9(8), 877-888.

Qin, S.J., T.A. Badgwell (2003). A survey of industrial model predictive control technology, *Control Engineering Practice*, 11(7), 733-764.

Rodrigues, M.A., D. Odloak (2003). MPC for stable linear systems with model uncertaintly, *Automatica*, 39(4), 569-583.

Sotomayor, O.A.Z., D. Odloak, J.F. Tisza-Contreras (2003). Modelling of complex industrial processes using subspace identification methods, *II Encuentro Científico Internacional, ECI 2003i*, Lima-Peru. (In Spanish).

Sotomayor, O.A.Z., D. Odloak, E. Alcorta-Garcia, J.F. Tisza-Contreras (2004). Multivariable model predictive control of an industrial fluid catalytic cracking (FCC) unit, *XV Brazilian Automatic Control Conference, CBA 2004*, Gramado-RS, Brazil. (Submitted).

Sundarraman, A., R. Srinivasan (2003). Monitoring transitions in chemical plants using enhanced trend analysis, *Computers and Chemical Engineering*, 27(10), 1455-1472.

Venkatasubramanian, V. (2003). Abnormal events managements in complex process plants: challenges and opportunities in intelligent supervisory control, *4th International Conference on Foundation of Computer-Aided Process Operations, FOCAPO 2003*, Coral Springs-FL, USA.

Verhaegen, M., P. Dewilde (1992). Subspace model identification – Part I: The output-error state-space model identification class of algorithms, *International Journal of Control*, 56(5), 1187-1210.

Wang, X., U. Kruger, B. Lennox (2003). Recursive partial least squares algorithms for monitoring complex industrial processes, *Control Engineering Practice*, 11(6), 613-632.

Wünnenberg, J. (1990). *Observer-based fault detection in dynamic systems*, PhD thesis, Gerhard-Mercator-University of Duisburg, Germany.

Yang, S.H., B.H. Chen, X.Z. Wang (2000). Neural network based fault diagnosis using unmeasurable inputs, *Engineering Applications of Artificial Intelligence*, 13(3), 345-356.

ELSEVIER
IFAC
PUBLICATIONS
www.elsevier.com/locate/ifac

A STUDY ON THE SELECTION OF MODEL DIMENSIONS AND SENSITIVITY OF PCA-BASED FAULT DETECTION

Masayuki Tamura * and Shinsuke Tsujita *

* *Technical Research Institute, Tokyo Gas Co., Ltd.*
16-25 Shibaura 1-Chome, Minato-ku, Tokyo 105-0023, Japan
m-tamura@tokyo-gas.co.jp

Abstract: A method of determining model dimensions (number of principal components) which maximize the sensitivity of fault detection was studied. In this paper it is shown that the sensitivity of PCA-based fault detection generally depends on the number of principal components. Most of the existing methods give only one value as a recommended number of components, and so sometimes the sensitivity is poor for certain kinds of faults. Among existing methods, although the Variance of Reconstruction Error (VRE) criterion gives a recommended model dimension which depends on the kind of fault, it is not intended to maximize sensitivity. This paper presents a new method of determining the model dimension which maximizes the sensitivity of PCA-based fault detection.

Keywords: Fault detection, Principal component analysis, Model dimension, Number of components

1. INTRODUCTION

Principal component analysis (PCA) is one of the most intensively studied methods for fault detection of plants (Kresta *et al.*, 1991; MacGregor *et al.*, 1994; Kourti, 2002). The selection of model dimensions (number of components) is an ambiguous problem in PCA model building. There have been various comparative studies on methods for determining model dimensions for fault detection. Himes et al. (Himes *et al.*, 1994) compared four methods using the Tennessee Eastman simulation. Valle et al. (Valle *et al.*, 1999) tested 11 methods and concluded that the VRE criterion is preferred. Most of the previous studies use the concept that the best solution is the number of components which is the "true" dimension of the system, which is usually unknown. In this paper, we focus on the sensitivity to fault, which is one of the most important issues in fault detection. Both the Q-statistic (or SPE: Squared prediction error) and T^2-statistic which are used for fault detection, show different behaviors under different model dimensions, and the model dimensions that give the maximum sensitivity of fault detection depend on the kind of fault, as will be shown in this paper. We propose determining the model dimension based on the signal to noise ratio (SNR) of fault detection.

2. BACKGROUND

Data matrix $\mathbf{X}$, which consists of data under the normal operating conditions of a plant, is represented by a PCA model in a lower dimension a as follows.

$$\mathbf{X} = \mathbf{TP}' + \mathbf{E} = \sum_{h=1}^{a} \mathbf{t_h p'_h} + \mathbf{E} \qquad (1)$$

$\mathbf{T}$, $\mathbf{P}$ and $\mathbf{E}$ are score, loading and error matrices with appropriate dimensions respectively. The first term of the right-hand side represents a dimensional model subspace (or representative space), and the second term represents $m-a$ dimensional residual subspace which cannot be expressed by the PCA model. (m is the number of sensors.)

Operation data are analyzed by the constructed model. One of the indices used for fault detection is the Q-statistic:

$$Q = \sum_{j=1}^{m} (x_{newj} - \hat{x}_{newj})^2 \qquad (2)$$
$$= ||\mathbf{x}'_{new} - \mathbf{x}'_{new}\mathbf{P}_a\mathbf{P}'_a||^2 = ||\mathbf{x}'_{new}\tilde{\mathbf{P}}_a\tilde{\mathbf{P}}'_a||^2$$

$\mathbf{x}_{new}$ is a new measured vector to be diagnosed. $\mathbf{P}_a$ is an $m \times a$ matrix which consists of first a columns of the loading matrix $\mathbf{P}$, and $\tilde{\mathbf{P}}_a$ is an $m \times (m-a)$ matrix which consists of the remaining columns of $\mathbf{P}$. If this statistic shows an unexpectedly large value, it means that the data point went into the residual subspace. This means that correlations between sensors which were observed during normal operation conditions were lost. The other index used for fault detection is Hotelling's T^2-statistic:

$$T^2 = \mathbf{t}\mathbf{S}^{-1}\mathbf{t}' = \mathbf{x}'_{new}\mathbf{P}_a\mathbf{S}^{-1}\mathbf{P}'_a\mathbf{x}_{new} = \sum_{h=1}^{a} \frac{t_h^2}{s_{t_h}^2} \qquad (3)$$

$\mathbf{S}$ is a diagonal matrix, which is the estimated covariance matrix of the principal component scores, and $\mathbf{t}$ is the vector which contains the scores at the time of observation. This indicates deviation from normal values inside a dimensional representative space. That is, sensors show unexpected values but the linear relations among them do not necessarily change in this case.

3. DETERMINATION OF MODEL DIMENSION

3.1 Existing methods

Generally, each sensor output includes noise, and so the determination of model dimension a is not straightforward. There are known criteria used in existing methods such as the scree plot, eigenvalue limit, cumulative contribution limit, cross validation, variance of reconstruction error (VRE) criterion, and so on.

The scree plot and eigenvalue limit are based on the concept that components which have small eigenvalues are not important for modeling. In the cumulative contribution limit method, the minimum model dimension which can express, for example, 80% of the variance of the data is selected. Cross validation (Wold, 1978) is a method which uses a part of the training samples for model construction; the remaining samples are compared with the prediction by the model, and when PRESS (prediction residual sum of squares) becomes smaller, the new component is added to the model. All of these methods are popular and detailed descriptions can be found in the literature (Jackson, 1991; Zwick and Velicer, 1986). The VRE criterion was proposed for minimizing the error of fault "reconstruction"; fault reconstruction is a procedure for estimating the true state of the system without a faulty sensor (Dunia and Qin, 1998).

3.2 Proposed method

An observed vector during operation with a fault is written as the vector sum of the normal data and fault vector:

$$\mathbf{x}_f = \mathbf{x} + \mathbf{f} \qquad (4)$$

In order to determine the SNR of fault detection by monitoring the Q-statistic, it is reasonable to consider the norm of a projected vector of $\mathbf{x}_f$ onto residual subspace as a signal. It is also natural to regard the statistical control limit as SNR noise. The control limit of the Q-statistic which is shown in a reference (Jackson and Mudholkar, 1979) can be used, or simply a certain percentile of the Q-statistic of sample data (normal condition data) can be used. The SNR for Q-statistic monitoring is defined as:

$$SNR_Q = \frac{||\mathbf{x}'_f\tilde{\mathbf{P}}_a\tilde{\mathbf{P}}'_a||^2}{Q_a} \qquad (5)$$

The denominator Q_a is a statistical control limit of the Q-statistic. For monitoring using the T^2-statistic, the SNR can be defined in a similar way:

$$SNR_{T^2} = \frac{\mathbf{x}'_f\mathbf{P}_a\mathbf{S}^{-1}\mathbf{P}'_a\mathbf{x}_f}{\frac{a(n-1)}{n-a}F_{a,n-a;\alpha}} \qquad (6)$$

The denominator is an approximation of a control limit of the T^2-statistic. n is the number of samples. $F_{a,n-a;\alpha}$ is $100 \times (1-\alpha)$ percentage point (95% for example) of the F-distribution.

Note that the numerators of both SNR depend on the direction of the fault vector, whereas their denominators are independent of the kind of fault. We propose calculating the model dimension which maximizes these SNR. The recommended dimension will be different for Q and T^2, and it depends on the kind of fault, too.

In order to calculate these SNR, $\mathbf{x}_f$ is needed. In the case of simple sensor faults, fault vector $\mathbf{f}$ is easily determined. If the i-th sensor is faulty, the value of $\mathbf{f}$ is determined as follows:

$$\mathbf{f} = [0, \cdots, 0, I, 0, \cdots, 0] \qquad (7)$$
$$\uparrow ith$$

This can be added to normal condition data $\mathbf{x}$. In the case of complicated process faults, a reference dataset which includes the fault is needed to obtain $\mathbf{x}_f$.

This means that the proposed method is based on a priori knowledge of the fault if it is a process fault. This is a drawback because usually one of the advantages of PCA in fault detection is that the model can be constructed without information of abnormal situations. However, if a priori information of data of abnormal situations exists, this method is reasonably useful to increase the sensitivity. From another viewpoint, a speculation can be derived from this method. Namely, as far as the model dimension which maximizes the sensitivity of fault detection depends on the kinds of faults generally, models with various (desirably all) dimensions should be monitored if computational load allows.

If there are limitations such as computational load, more than one kind of fault may have to be detected by a single PCA model. In this case the following criterion is proposed. Based on many experiments and simulations, the authors found that there exists a model dimension a_c where SNR_Q takes its maximum and the sensitivity of the Q-statistic for fault detection becomes significantly worse when a model dimension larger than a_c is employed. In contrast, the T^2-statistic shows maximum sensitivity at higher order than $a = a_c$, and selection of fewer components makes the model insensitive to the fault. This is thought to be due to a principal component whose direction is close to $\mathbf{f}$, which is taken into the representative space at a dimension $a_c + 1$ when the dimension of the model is increased. A criterion can be considered based on this characteristic. A model dimension a_c is calculated as:

$$a_c = \underset{a}{\mathrm{argmax}} \left(\sum_i w_i SNR_Q \right) \tag{8}$$

where i is a suffix which denotes the kind of fault and w_i is a positive weight which expresses the significance of the fault. This expression is based on the same concept as the formula which had been proposed for the VRE criterion (Dunia and Qin, 1998). In the case of the SNR criterion proposed here, a pair of PCA models whose model dimension is a_c and $a_c + 1$ are employed and the former is used for Q-statistic monitoring and the latter for T^2-statistic monitoring. This is because the principal component whose direction is close to $\mathbf{f}$ is in the residual subspace at $a = a_c$, and is in the representative space at $a = a_c + 1$. As a result, the sensitivities of the $Q-$statistic and T^2-statistic are high at $a = a_c$ and $a_c + 1$, respectively.

4. RESULT: THE TENNESSEE EASTMAN PLANT SIMULATION

The Tennessee Eastman plant simulation is a plant-wide process control problem developed by Downs and Vogel. The process consists of five major units (a reactor, a product condenser, a vapor-liquid separator, a recycle compressor and a product stripper), as described in the reference (Downs and Vogel, 1993). In this study, the simulation conditions were as follows:

- Sampling interval: 3 minutes
- Data length: 500 points (for each dataset)
- Control: The cascade control system proposed by McAvoy and Ye (McAvoy and Ye, 1994)
- Sensors used for PCA modeling: 16 sensors proposed by Chen and McAvoy (Chen and McAvoy, 1998)
- Fault occurrence time: 1 hour (20 points) after the start of simulations
- Random seed: Changed for every simulation run

4.1 Definition of sensitivity and fault detection

In this paper, the sensitivity of fault detection is defined as follows. First, statistical control limits with 95% confidence limit for each PCA model are calculated. Then the Q-statistic and T^2-statistic of test data are calculated and divided by each control limit. Therefore, the value unity means that the statistic is equal to the control limit. In the following analysis we defined that the fault is detected when any of the PCA models has shown six succeeding values larger than one. (The value "six" was determined empirically to suppress false alarms, and has no significant meaning itself.)

4.2 Sensor faults

Three kinds of sensor faults are studied here: offset of reactor inlet flow sensor (XMEAS 6), drift of separator pressure sensor (XMEAS 13) and random noise of stripper pressure sensor (XMEAS 16), denoted by SF 1, 2 and 3, respectively. These three sensors are selected because they are not used in the control loop. Note that the use of faulty sensors for feedback control causes complicated dynamics of the system and the faults cannot then be treated as simple sensor faults.

First, a PCA loading matrix was calculated using training data under normal conditions. The scree plot is shown in Fig. 1. It implies that three components should be retained. The numbers of components retained by existing methods are listed in Table 1. The eigenvalue limit method suggests that five components should be retained and cumulative R^2 exceeds 80% with eight components. The cross validation method selected only two components in this case.

Subsequently, SNR were calculated using $\mathbf{f}$ vectors in the form of Eq. 7 for each fault. Model dimensions which maximize SNR_Q and SNR_{T2} are shown in Table 2, as well as the dimension of the VRE criterion selected.

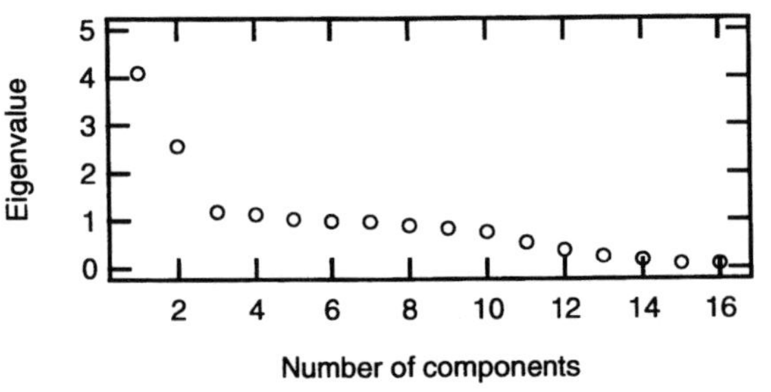

Fig. 1. Scree plot.

Table 1. Model dimensions selected by existing methods

Scree	Eigenvalue limit	Cumulative R^2	Cross validation
3	5	8	2

Table 2. Model dimensions selected by VRE and SNR criteria

Fault	VRE	SNR_Q	SNR_{T^2}
SF1	2	4	8
SF2	11	12	14
SF3	12	13	14

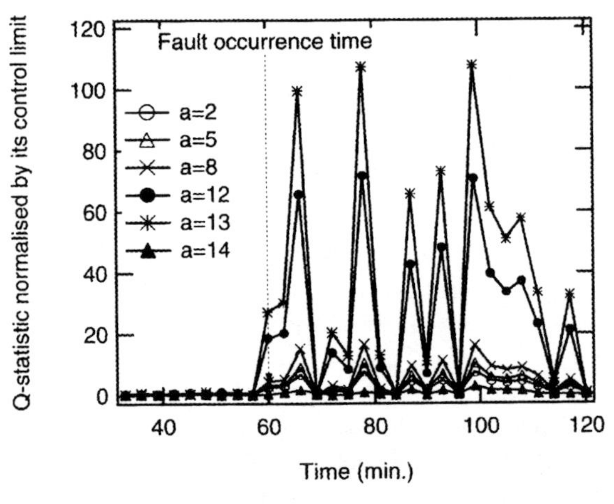

Fig. 2. Q-statistic normalized by its control limits for simulation case SF 3 (random noise of stripper pressure sensor (XMEAS 16)).

An example of the time series plot of sensitivities (Q-statistic and T^2-statistic normalized by their control limit) in the case of SF 3 are shown in Figures 2 and 3, respectively. These values are from a test by data with different random seed (independent data from the samples for modeling and dimension selection procedure). As shown in these figures, models with dimensions $a = 13$ for Q-statistic monitoring and $a = 14$ for T^2-statistic monitoring, which were determined by the SNR criterion, presented the highest sensitivities. The SNR criterion was also the best for the other two sensor faults in terms of the sensitivity.

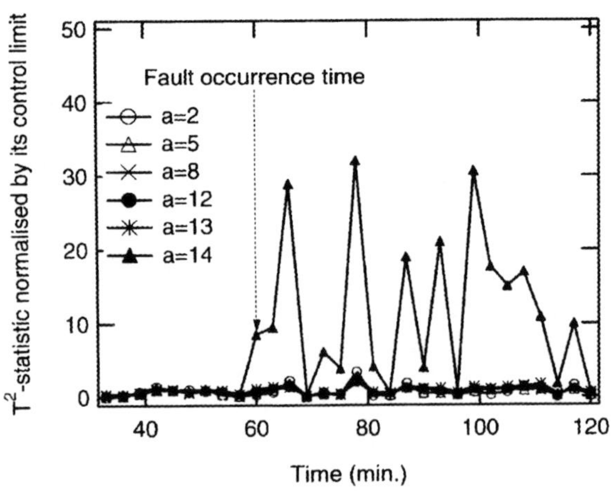

Fig. 3. T^2-statistic normalized by its control limits for simulation case SF 3 (random noise of stripper pressure sensor (XMEAS 16)).

4.3 Process faults

Disturbances prepared in the original Tennessee Eastman simulation code (the disturbances are denoted by "IDV" in their program) were used for process fault investigations. First, a PCA model was calculated by training data under normal conditions. The model dimension is tentatively determined by the eigenvalue limit method (here it was $a = 5$). Second, training data which contain each fault were used to obtain a reference for fault directions. Ten points from five points before the point where the fault was first detected were used as $\mathbf{x}_f$. SNR were calculated using these reference data. Note that the proposed method needs only a few samples to know the direction of the fault vector.

Using independent test data (with different random seeds), sensitivities were investigated. Figures 4 and 5 show time series plots of sensitivities for simulation case IDV 10 (random variation of feed temperature of reactant C). Sensitivity to the fault depends on the model dimensions and in this case $a = 14$ was maximum for Q-statistic monitoring and $a = 15$ was maximum for T^2-statistic monitoring, which coincides with the result of dimension selection by the SNR criterion.

Results of other investigated faults are not shown here due to lack of space, however, most of the cases showed that the SNR criterion is the best in terms of the sensitivity.

In case all the faults should be detected by a single PCA model, Eq. 8 is used for model selection. Assuming that all w_is are equal, Eq. 8 gave $a_c = 11$. On the other hand, the VRE criterion gave 2 as a recommended model dimension, which is because of the existence of fault SF 1. Figures 6 and 7 show time series plots of normalized Q-statistic and T^2-statistic. The fault vector of SF 1 is included in the representative space almost perfectly until $a = 10$, and the sensitivities for higher dimension models become poor. This makes u_i, which is an index to be minimized

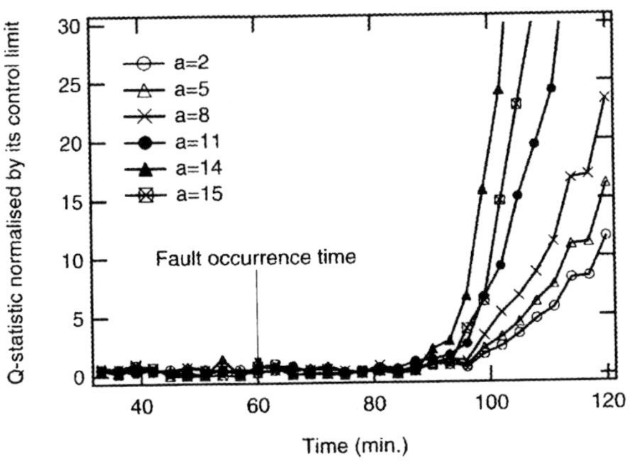

Fig. 4. Q-statistic normalized by its control limits for simulation case IDV 10 (random variation of feed temperature of reactant C).

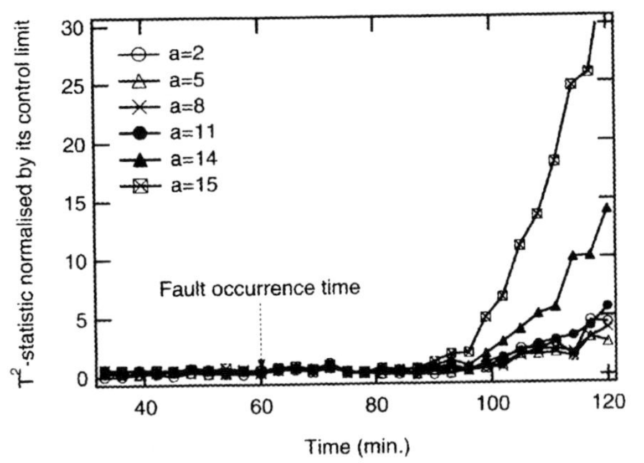

Fig. 5. T^2-statistic normalized by its control limits for simulation case IDV 10 (random variation of feed temperature of reactant C).

for dimension selection (Dunia and Qin, 1998), quite large for large model dimensions. This greatly affects the result and the VRE criterion recommends $a = 2$, which is quite insensitive to other faults.

If only Q-statistic monitoring is performed, the sensitivity for SF 1 is poor when $a = 11$, which is recommended by the method proposed in this paper, is selected. However, if T^2-statistic monitoring with the $a = 12$ PCA model is performed simultaneously as suggested, the sensitivity for SF 1 is also assured (see Figures 6 and 7). T^2-statistic monitoring and Q-statistic monitoring thus work cooperatively with each other.

It is interesting to examine the results of previous studies and compare the behavior of the Q-statistic and T^2-statistic. Himes et al. (Himes *et al.*, 1994) analyzed their data from the Tennessee Eastman plant. They analyzed in a different way and used all the 51 sensors in the simulation. According to their result, Q-control gave the fastest detection of a fault for models with 28–30 components, whereas the T^2-control gave the fastest detection for models with 34 or more components. For another fault, the fastest detection was given

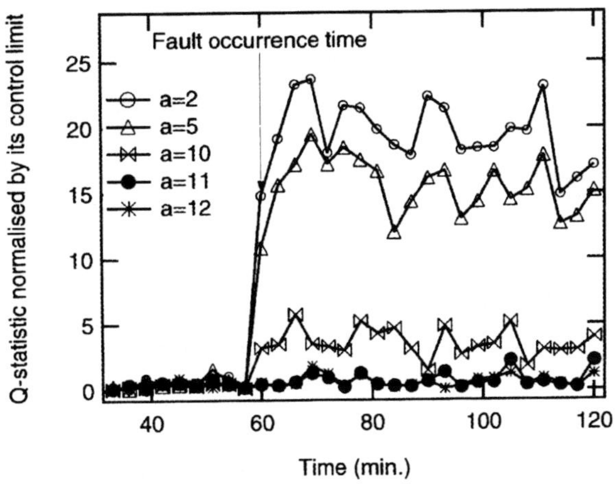

Fig. 6. Q-statistic normalized by its control limits for simulation case SF1 (offset of reactor inlet flow sensor (XMEAS 6)).

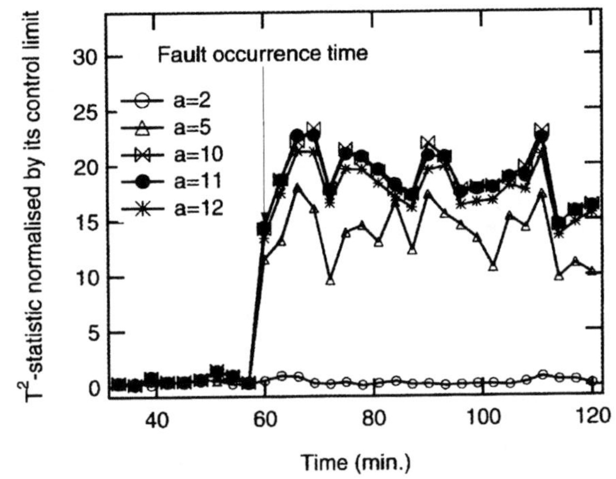

Fig. 7. T^2-statistic normalized by its control limits for simulation case SF1 (offset of reactor inlet flow sensor (XMEAS 6)).

for models with 28–30 components for Q-control and 32–42 components for T^2-control. Their results are consistent with the characteristics identified in this paper.

4.4 Consideration of false alarm

False alarms are another important point in fault detection, but were not considered in the analysis given above. Generally, false alarms are expected to increase when many components are retained because the model may include noise which yields large prediction errors. We analyzed the frequency of occurrence and magnitude of false alarms using independent normal data. The data used for this analysis consisted of 10000 points, which corresponds to 500 hours of normal operation of the plant. The result was as expected. The frequency of false alarms (type I error) becomes about 40% for several models with high dimensions although they were designed to show 5% type I error. However, care must be taken when assessing the method presented in this paper with this data. As far as the method is designed to maximize the sensitivity, the important point is the magnitude of false alarms. If the false alarms are not large, it is possible to avoid them by setting the control limit slightly higher. Actually, in our simulation result it was enough to set the control limit at only 2–3 times higher to suppress type I errors below 5%. If the prediction error shows a normal distribution, we can estimate how the set point of the control limit should be changed. If the observed false alarms are 40%, it can be expected from statistical tables of standard normal distributions that setting the control limit at 2–3 times higher makes type I errors about 5 %. In our case, the advantage of selecting the model dimension by the method based on SNR was not lost because the variation of the sensitivity in different model dimensions was much larger than the magnitude of false alarms. This point needs further investigations for generalization.

5. CONCLUSIONS

A new method for model dimension selection of the PCA model based on the sensitivity of fault detection was proposed. The proposed method is based on signal to noise ratios which are defined using the Q-statistic and T^2-statistic, respectively. The method was validated by the Tennessee Eastman plant simulation and the advantage of the new method in terms of sensitivity to faults was shown. On the other hand, the method does not maximize fault reconstruction reliability. It is recommended to use this method and the VRE criterion depending on the purpose of analysis because each method has respective strengths and limitations. An important difference with other existing methods is that existing methods except for the VRE criterion are based on the concept of estimation of the "true" dimension of the system. This is natural in view of the fact that PCA is a tool to explain the hidden mechanism of the system by computation of correlated data. However, our goal of using PCA for fault detection is slightly different and the sensitivity to faults is one of the most important points. Although it does not use a "true" system description, our method is effective in terms of sensitivity. The method uses a priori knowledge of faults to determine SNR. Although this is a drawback, this approach is useful to increase the sensitivity of the method when a priori information of data with faults exists.

REFERENCES

Chen, G. and T. J. McAvoy (1998). Predictive on-line monitoring of continuous processes. *J. Proc. Cont.* **8**, 409–420.

Downs, J. J. and E. F. Vogel (1993). A plant-wide industrial process control problem. *Computers. Chem. Engng.* **17**, 245–255.

Dunia, A. and S. J. Qin (1998). Joint diagnosis of process and sensor faults using principal componenent analysis. *Control Eng. Pract.* **6**, 457–469.

Himes, D. M., R. H. Storer and C. Georgakis (1994). Determination of the number of principal components for disturbance detection and isolation. *Proc. American Control Conference* pp. 1279–1283.

Jackson, J. E. (1991). *A User's Guide to Principal Components*. John Wiley & Sons, Inc.. Hoboken, New Jersey.

Jackson, J. E. and G. S. Mudholkar (1979). Control procedures for residuals associated with principal component analysis.. *Technometrocs* **21**, 341–349.

Kourti, T. (2002). Process analysis and abnormal situation detection: From theory to practice. *IEEE Control Systems Magazine* **22 (5)**, 10–25.

Kresta, J. V., J. F. MacGregor and T. E. Marlin (1991). Multivariative statistical monitoring of process operating performance. *Can. J. Chem. Eng.* **69**, 35–47.

MacGregor, J. F., C. Jaeckle, C. Kiparissides and M. Koutoudi (1994). Process monitoring and diagnosis by multiblock PLS methods. *AIChE J.* **40**, 826–838.

McAvoy, T. J. and N. Ye (1994). Base control for the tennessee eastman problem. *Computers Chem. Engng.* **18**, 383–413.

Valle, S., W. Li and S. J. Qin (1999). Selection of the number of principal components: The variance of the reconstruction eror criterion with a comparison to other methods. *Ind. Eng. Chem. Res.* **38**, 4389–4401.

Wold, S. (1978). Cross-validatory estimation of the number of components in factor and principal components models. *Technometrics* **20**, 397–405.

Zwick, W. R. and W. F. Velicer (1986). Comparison of five rules for determining the number of components to retain. *Psychological Bull.* **99**, 432–442.

ELSEVIER
IFAC
PUBLICATIONS
www.elsevier.com/locate/ifac

ON EVALUATING CONTROL PERFORMANCE ON LARGE DATA SETS

Alexander Horch[1], Friedrun Heiber[2]

[1]*ABB Corporate Research, Germany*
[2]*University of Stuttgart, Germany*

Abstract: The evaluation of control performance by means of performance indices from large amounts of measurement data is investigated. The focus is twofold: Firstly to assess information that can be deduced from *many* data sets and secondly to investigate the usefulness of *simple* performance measures. Established methods and some useful new ideas are evaluated on many industrial data sets and the results are discussed.

Keywords: Performance indices, control loops, signal processing, data collection.

1 INTRODUCTION

The evaluation of control performance mostly deals with the computation of performance indices. These indices range from very simple ones, e.g. the control error mean and variance (e.g. [Salsbury, 1999] up to rather complex ones involving e.g. subspace identification [Bezergianni and Georgakis, 2000].

Obviously, simple indices are very appealing since for performance monitoring, typically many control loops are assessed regularly. Unfortunately, complexity and information are usually correlated such that simple indices mostly imply simple answers. One exception is the minimum-variance control performance index by Harris [Harris, 1989]. It combines low computational effort with important information about the current loop performance. This fact contributed to its popularity over the last decade.

Traditionally, control loop performance assessment deals with the kind of information that can be deduced from the evaluation of a specific data set. This approach is sound and valid but carries some pitfalls: the data set may reflect unusual behaviour of the control loop in question. In the industrial practice, single data batches are often erroneous since plant shut downs or other unusual events might have been reflected in the data.
Such data sets will in general not be able to givea fair picture of the control loop performance. They should ideally be excluded from performance evaluation; however, a mechanism to automatically discard all such cases seems to be very difficult to achieve.

This paper attacks control performance assessment from a slightly different point of view: Given the fact that the time constant of good control is in the range of months rather than days, it would make sense to base a performance assessment on much more than only a few data sets.

The first issue discussed in this paper is: can useful information be found in simple control performance indices when evaluated on many data sets? Some indices (both simple and more advanced ones) will be evaluated for 20 control loops from a pulp mill. For each of the control loops more than 400 data batches were analysed.

A second aspect discussed is the continous collection of information that can be combined to new knowledge about a control loop. The availability of many data sets can be used to build a nonlinearity map of the process. Such information is of great use for tuning procedures.

The paper is organised as follows: In Section 2 the data used in this study is described and presented. Section 3 presents the evaluation of some performance indices for all data sets and conclusions thereof. Section 4 discusses the use of data in order to build a knowledge database that grows with each data set that was analysed.

Examples accompany both sections and the paper concludes with a summary and requirements for industrial control performance monitoring tools.

2 INDUSTRIAL DATA SETS

This study makes intensive use of more than 400 data sets containing control loop data from a stock preparation section in a pulp mill. Each data set contains about 20 minutes of data at a sampling rate of 1 second. The loop setpoint (SP), the process variable (PV) and the controller output (OP) were logged.

A typical data batch containing 20 control loops (11 flow loops, 3 level loops and 6 composition loops) is shown in Figure 1.

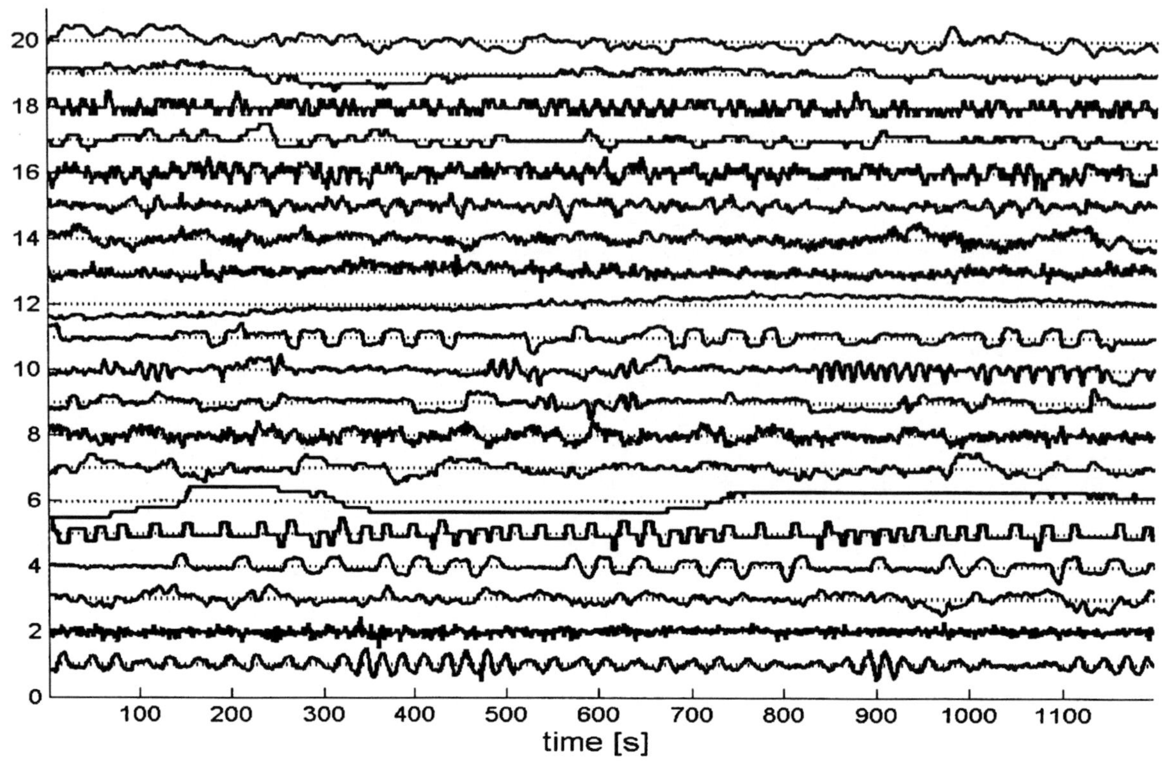

Figure 1: Example data (SP and PV respectively) for 20 loops from one data collection occasion. The signals are scaled such that they have equal standard deviation and are plotted on top of each other.

The data collection was done autmatically, once per day for more than a year resulting in 424 data sets.

3 PERFORMANCE INDICES

Quite a number of performance measures for assessing controller performance have been proposed in the literature, especially during the last decade. Most of them targeted to be computed from normal operating data only. It is the constraint of not allowing experiments that outperforms the computation of similar performance measures that are typically used in controller design (e.g. the loop overshoot or rise time).

3.1 Simple statistics

The term 'simple indices' refers to indices that can be evaluated with a modest amount of computations and that do not require any non-trivial a priori knowledge. Table 1 shows the simple performance indices that were evaluated in this study.

The control error mean should of course be centered around zero with no off-set and a sufficiently small standard deviation. Long or excessive deviation can easily be identified (see loops no. 2, 12, 15 and 20 in Figure 2.

index	description
CE mean [%]	mean of control error
CE std [%]	standard dev. of control err.
OP std [%]	st. dev. of controller ouptut
CE skewness	skewness of control error
CE kurtosis	kurtosis of control error
std ratio	ratio of std of control error and controller output
maximum bic	max. bicoherence
correlation coefficient	correlation coefficient between control error and controller output

Table 1: Simple performance indices that were evaluated for all data sets. The units [%] refer to the operating ranges of OP and PV.

The (normalised) standard deviation of the control error can also give considerable insight into loop behaviour, see Figure 3. It can be clearly seen that loop No. 8 has a problem with increased variability between logs no. 250 and 300. At this time, the loop had a strong oscillatory behaviour.

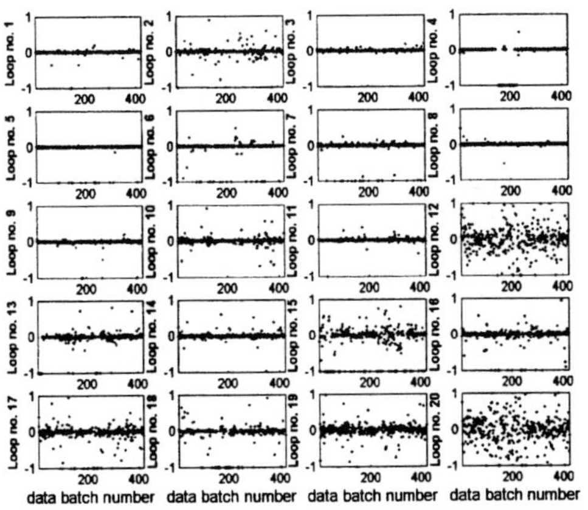

Figure 2: Trend of loop standard deviation.

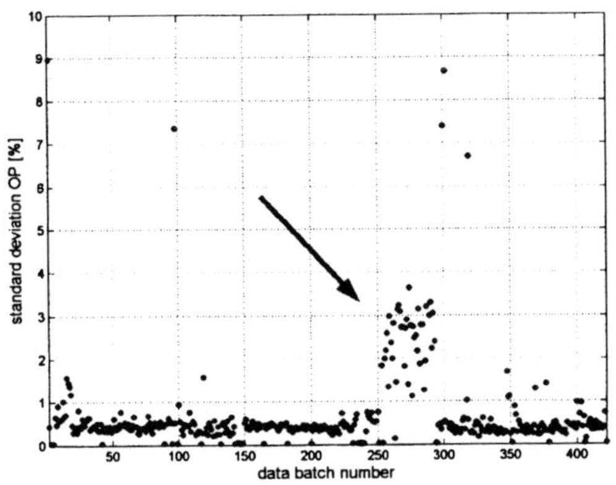

Figure 3: Standard deviation of control error (loop No. 8).

Another interesting statistics is the control error skewness. Skew data often indicate problems of nonlinear character. As an example see Figure 4 where it can be seen that for many data sets, the skewness is rather large, indicating regular problems of nonlinear source. As a matter of fact, this loop exhibits stiction regularly resulting in the presented non-symmetrical data distribution.

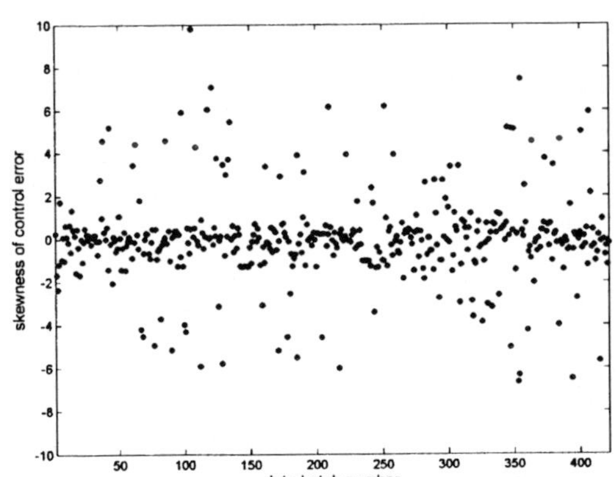

Figure 4: Skewness trend of control error for loop No. 2.

Consider another example in Figure 5 where the control error kurtosis for loop No. 12 is shown. For Gaussian signals, the kurtosis should be centered around zero. In this case, it is clearly around –1 instead, indicating rather non- Gaussian signals. The reason in this case is a slow periodic behaviour with a cycle time of around one hour. The data batches are too short for a detection algorithm to detect this oscillation.

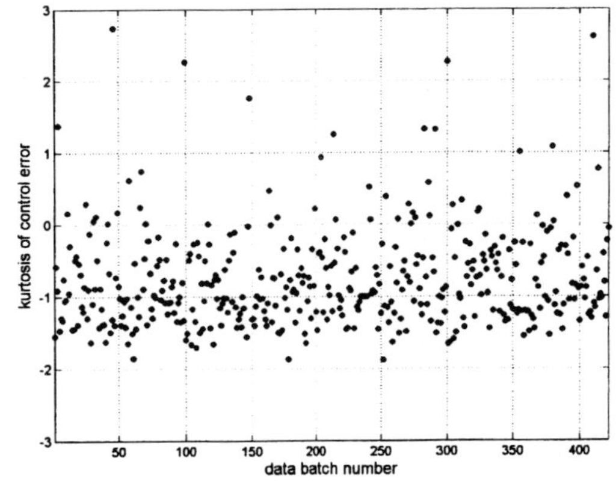

Figure 5: Control error kurtosis trend for loop 12.

In Figure 6, two trends of the ratio of the standard deviations of control error and controller output are shown

$$R = \frac{\sigma_{SP-PV}}{\sigma_{OP}}.$$

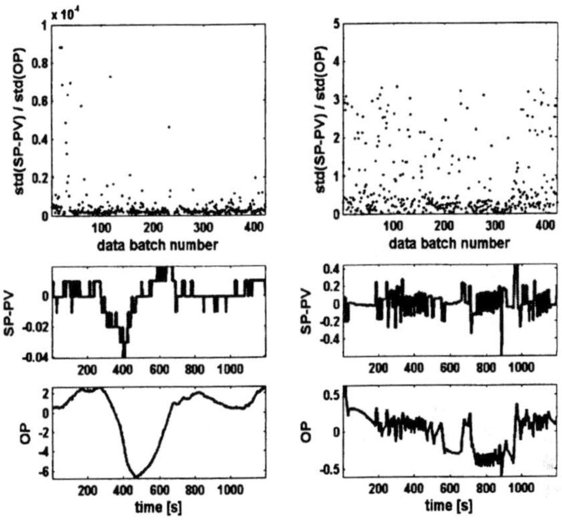

Figure 6: Example for ratio of standard deviations of control error and controller output. Left: case with not moving process output, loop No.6; right:loop with normal behaviour, loop No. 5

It can be seen that the value is 5 orders of magnitude smaller in the left example than in the right one. The reason is that the process variable in the left loop (No. 6) does not move much more than the quantisation level, whereas the loop on the right (No. 5) moves reasonably much. In loop No. 6, either the sensor signal is corrupt or the loop actuator does hardly move.

For a last example consider Figure 7 where the maximum bicoherence of the control error is plotted for all loops. In [Choudhury et al., 2004] it is shown that the bicoherence plot can be used to assess signal nonlinearity. Single evaluations may tend to contradict this hypothesis but when considering many data sets, it turns out that such a measure may be able to detect loops that exhibit nonlinearity problems (in this case loops No. 1, 2, 5, 6, 9, 10 and 11). This is in line with the knowledge about these loops.

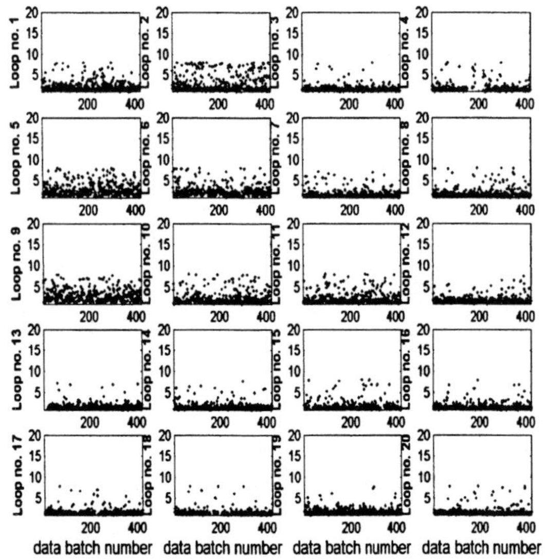

Figure 7: Trend of maximum bicoherence over all loops.

3.2 More advanced indices

Another group of indices involves more complex computations and eventually more prior knowledge. As already mentioned, the most prominent of these indices is the Harris index that compares actual loop variability to minimum-variance variability leading to an index between 0 and 1 where 1 equals minimum-variance performance,

$$I = \frac{\sigma^2_{\text{min var}}}{\sigma^2_{SP-PV}}.$$

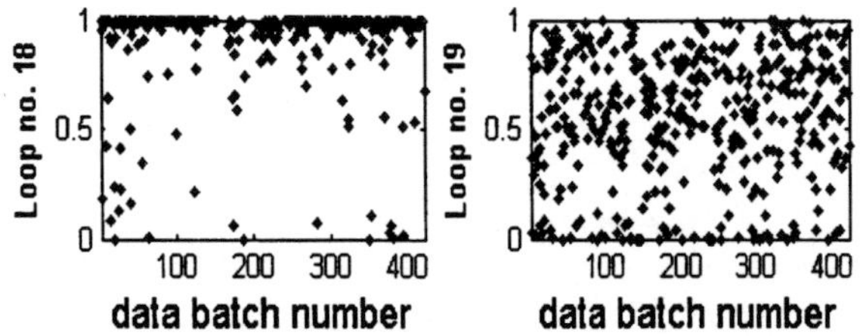

Figure 8: Example of trends for the Harris index.

The example in Figure 8 shows the need for interpreting the Harris index with a grain of salt. The left loop shows a good behaviour, however, for some batches the index is very low. The right loop on the other side offers the complete range of index values, indicating that the performance (or its assessment) is very different at different days. To rely on a few or a single data set only can be misleading when dealing with the Harris index.

Badly performing loops often exhibit oscillatory behaviour. Therefore, oscillation indices for oscillation detection and assessment are the most important quantity that should be monitored. More and more industrial applications start focussing on periodic disturbances in addition to the Harris index, which was mostly discussed in the mid-90s.

Oscillation detection can be done in various ways, see [Hägglund, 1995], [Forsman and Stattin, 1999] or [Seborg and Miao, 1999].

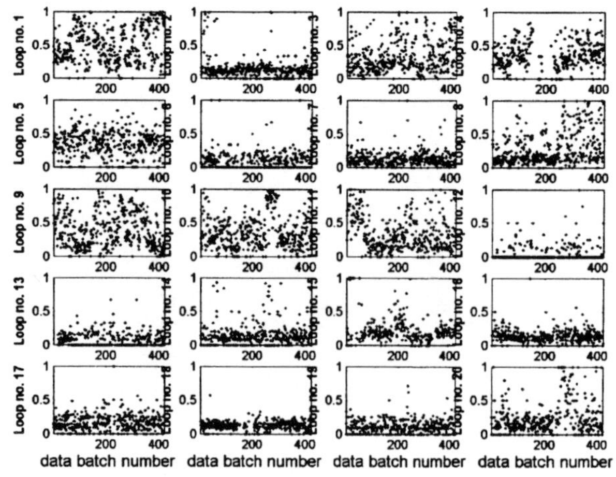

Figure 9: Oscillation index for all loops.

In Figure 9, trends of the oscillation index [Forsman and Stattin, 1999] are plotted. It can easily be seen that loops No. 1, 3, 4, 5, 8, 9, 10, and 11 exhibit regular oscillatory behaviour. The human eye recognises an oscillation when the index is larger than about 0.3.

4 Combination of indices

Performance indices provide a good means of analysing plant behaviour. However, it is often the *combination* of indices that gives significant insight into bad plant performance.

A very compact assessment would be the test if certain control loops exhibit bad behaviour simultaneously. Hence, a correlation of a specific index for different loops would be a valuable source of information.

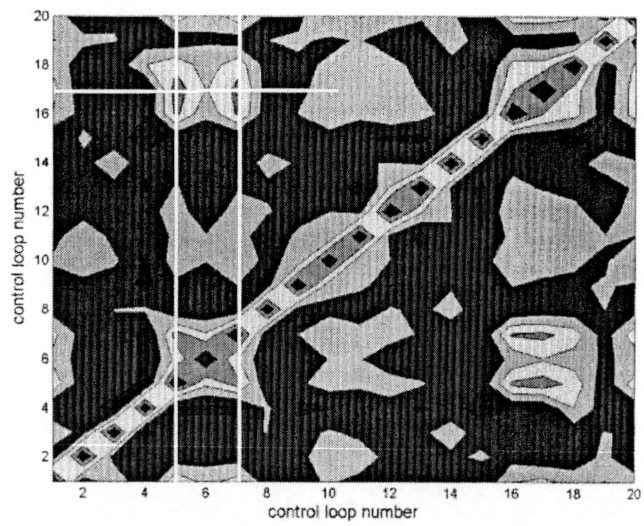

Figure 10: Correlation coefficients for oscillation indices over all loops.

Consider Figure 10 where oscillation indices are correlated for each loop. Such a plot indicates which loops typically oscillate simultaneously. The plot reveals a common oscillatory behaviour between loops no. 5, 7 and 17; see Figure 11. Note that the equality of frequency is of no importance for the correlation of the oscillation indices.

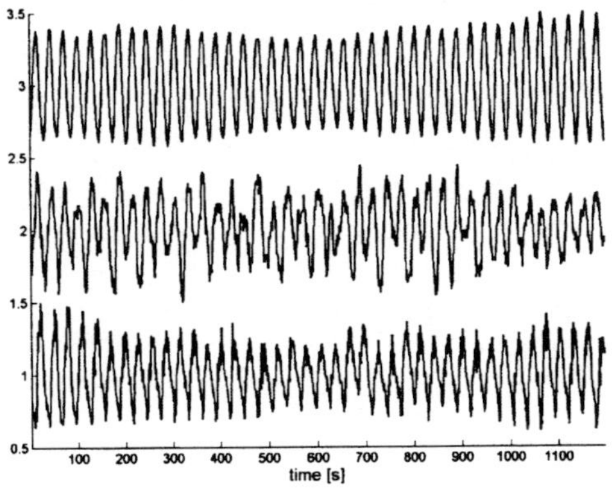

Figure 11: Three oscillating signals (with normalised variance) as indicated by correlation of their oscillation indices.

4.1 Increasing information

As mentioned in the introduction, the step beyond storing indices for single data batches is to collect and combine assessment information from single evaluations for future use. Examples for such applications are:

- Creation of static input – output maps
- Indication of data sets suitable for model identification

A static input output map is often of importance when control loop tuning is performed. Many commercial tuning tools offer ways to analyse experimental data where the process input is changed stepwise. The automatic generation of static y-u-maps can avoid costly experiments and thus enable faster controller tuning without disturbing current production.

If the data within a data batch is sufficiently stationary, then an algorithm can extract stationary values of OP and PV and store them. Figure 12 shows examples of static maps for all control loops analysed. For loops where sufficiently many different operating points have been found, a quadratic function has been fit to the data using a least-squares method.

For most data sets, a linear function would be sufficient to describe the static input output relationship well. Loops where the quadratic function fits the map better are No. 1, 2 and 8. However, the nonlinearity does not seem to be too severe, such that it could also be neglected.

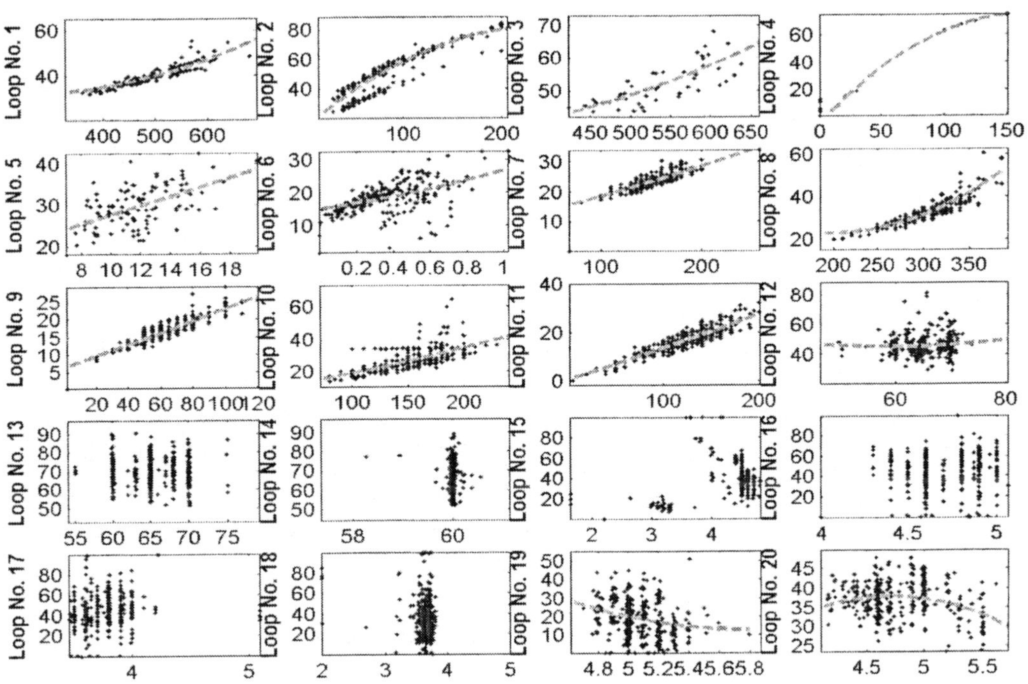

Figure 12: Static maps for all 20 control loops. The fitted curve is quadratic. Vertical axis is the controller output and horizontal axis is the process variable.

It was mentioned that static maps provide useful insight into the process model when dealing with controller tuning. A natural question is then: Could the regular analysis of normal operating data be used to detect data sets that are suitable for model identification? This would be data sets where (a) the setpoint is changed abruptly by a significant amount or, (b) the loop is in manual mode and the operator changes the process input stepwise. For both cases, a regular analysis of data could raise and store a flag if suitable identification data is available.

Since it may not be sufficient to flag for setpoint changes (or input changes in manual loop mode) only, it was chosen to use a flag that indicates if an estimated dynamic model for the process has sufficiently good quality. The quality is measured by goodness of fit test as they are used in standard system identification packages. Figure 13 shows an example of a data set that typically would flag for being suitable for system identification.

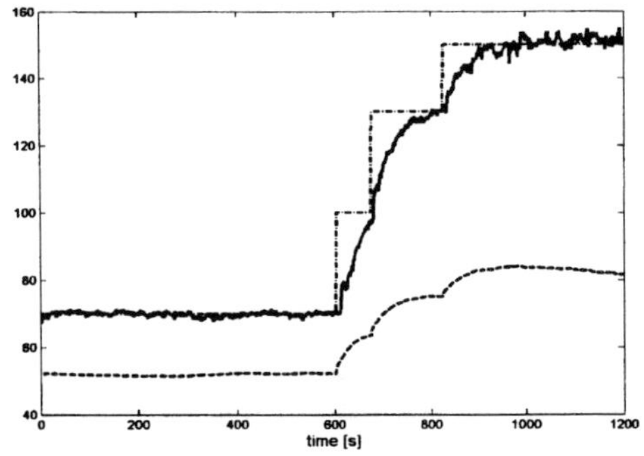

Figure 13: Data set suitable for model identification and controller tuning.

Clearly, some loops would never generate these flags since setpoints may never be changed or the loops are never taking into manual mode. Note that disturbances alone never qualify data to be useful for process model identification. In these cases, only the controller can be identified.

Using both static maps and the described model fit flags, it is hence possible – at least for some loops – to generate the information that is usually required for controller tuning without being forced to perform experiments.

Yet another flag that is useful to store for later use in controller tuning is whether the loop exhibited stiction behaviour [Horch, 1999]. Such information should be available when tuning loops.

5 IMPLICATIONS FOR INDUSTRIAL TOOLS

From the above results, some implications for the controller performance tools shall be drawn. A senseful tool should ...

- ... enable analysis of performance indices such as plot combinations, correlation, trend plots etc.,
- ... enable application-dependent selection /discarding of indices,
- ... offer an index database for search queries,
- ... help to retrieve data collection dates and – if possible – specific data sets.

6 CONCLUSIONS

The usefulness of performance indices for automatic controller performance assessment is an accepted fact in the process industry. A question that has received little attention so far is which indices to use and what kind of information can be deduced from each of them.

The focus in this paper was to show the strengths of some selected indices when a large amount of data batches is available. From the analyses, some general conclusions shall be drawn:

- Simple statistics are most useful for fast and overview-like scans of large amounts of data.
- More complex indices are very useful when averages over many data sets are available. Single evaluations may be misleading.
- Combination (e.g. correlation) of indices is useful and gives insight into the plant dynamics
- Storage of results for later usage are very helpful, especially for tuning (linearity, stiction).
- Trending of indices presents an extreme data and information compression for comfortable reporting.

REFERENCES

Bezergianni, S. and C.Georgakis (2000). 'Controller performance assessment based on minimum and open-loop output variance'. Control Engineering Practice, **8**, pp. 791-797.

Harris, T. (1989). 'Assessment of control loop performance'. The Canadian Journal of Chemical Engineering, **67**, pp 856-861.

Salsbury, T.I. (1999). 'A practical algorithm for diagnosing control loop problems'. Energy and Buildings, **29**, pp. 217-227.

Hägglund, T. (1995). 'A control loop performance monitor', Control Engineering Practice, **3**(11), pp 1543-1551.

Forsman, K and A. Stattin (1999). 'A new criterion for detecting oscillations in control loops.', In European Control Conference ECC, Karlsruhe, Germany. Session CP8-3.

Seborg, D. and T. Miao (1999). 'Automatic detection of excessively oscillatory feedback control loops'. In IEEE Int. Conf. on Control Applications, Hawaii, USA, pp 359-364.

Choudhury, M.A.A.S., S.L.Shah and N.F. Thornhill (2004). "Diagnosis of Poor Control Loop Performance using Higher Order Statistics", accepted for publication in Automatica.

Horch, A. (1999). 'A simple method for detection of stiction in process control loops.', In Control Eng. Practice, **7**(10), pp 1221-1231.

ELSEVIER
IFAC
PUBLICATIONS
www.elsevier.com/locate/ifac

A FILTER BASED APPROACH FOR ESTIMATION OF PI ACHIEVABLE PERFORMANCE

Mranal Jain and S. Lakshminarayanan *

Department of Chemical and Biomolecular Engineering
National University of Singapore, Singapore 117576
**Corresponding Author. Telephone: +65-68748484 Email: chels@nus.edu.sg*

Abstract: We present a new filter based method for the calculation of PI achievable performance for simple feedback control systems. We also extend this new approach to multiloop systems and address some key questions such as (a) what would be the performance with alternate loop pairing? (b) what performance improvement will accrue through the use of decouplers in multiloop control systems? *Copyright © 2004 IFAC*

Keywords: Control loop performance, PI achievable performance, Multiloop systems, Pairing, Decouplers

1. INTRODUCTION

In a typical process plant, there are thousands of control loops designed and implemented in order to achieve specific objectives. Continuous good performance of existing controllers is indispensable to meet the demands of the consumer in a cost effective way and to generate sustained benefits. This indicates the need for control loop performance benchmarks and measures. Astrom (1970), Harris (1989) and Stanfelj *et al.* (1993) used minimum variance control as the benchmark to assess control loop performance. Since then, there has been an interest in the academia and the industry for developing measures and benchmarks for control loop performance. In addition to the minimum variance benchmark, several alternate benchmarks also exist (Tyler and Morari, 1996; Kendra and Cinar, 1997; Swanda and Seborg, 1999). Applications of performance assessment techniques can be found in Thornhill *et al.* (1999). Huang *et al.* (1996) and Harris *et al.* (1996) extended the performance assessment concepts to MIMO feedback controllers. These works employ the multivariable minimum variance controller as the benchmark.

In the works described above, no restriction is placed on the form that the feedback controller Q can possibly take. If the structure of controller Q is complicated, it is not amenable for implementation on standard industrial DCS. Desborough and Miller (2001) surveyed the status of controllers employed in the chemical industry and concluded that a typical chemical plant has 98% PID type controllers and a vast majority of these controllers are PI controllers. Hence, it is essential to know to what extent can the given process be effectively regulated (for stochastic disturbances) with PI controllers. Therefore, the PI achievable performance is more realistic benchmark from the perspective of the chemical industry. The Approximate Stochastic Disturbance Realization (ASDR) technique developed by Ko and Edgar (1998) for estimating the PI achievable performance assumes that the process model is known. The ASDR method approximates the noise model from routine operating data. Recently, Agrawal and Lakshminarayanan (2003) described a method to determine the control loop performance achievable with PI type controllers, the optimal control settings that will yield the best performance and the expected robustness margins using closed loop transfer functions identified from closed loop experimental data.

The objective of this article is to describe a method to calculate the PI achievable performance of feedback control systems using the knowledge of process model alone. The key idea is the derivation of a filter that can provide routine closed loop operating data for any controller Q^* using information of the process model and original controller Q alone.

The paper is organized as follows. Section 2 provides a brief overview of existing methods that are available for the calculation of the PI achievable performance. In section 3, we propose a new method for the calculation of the PI achievable performance in single input single output systems. The proposed method is extended to multiloop systems in section 4. In section 4, we also outline a method to obtain PI achievable performance for alternate control loop pairing and for multiloop control systems with decouplers. Examples to demonstrate the proposed method are shown in section 5, followed by conclusions.

2. COMPUTATION OF PI ACHIEVABLE PERFORMANCE

We assume the reader to be familiar with the basics of the control loop performance assessment theory that employs minimum variance controller as the benchmark. The interested reader is referred to the exceptional coverage provided by Qin (1998), Huang and Shah (1999) and Harris *et al.* (1999).

The maximum control loop performance that can be attained by restricting the controller structure to PI type is called PI achievable performance. Assuming that the open loop model (including time delay) is known, Ko and Edgar (1998) used routine closed loop operating data (no set point change is made to excite the process) to estimate the PI achievable performance. Using an ARIMA (p,1,1) model with $2 \leq p \leq 5$, they approximate the disturbance (noise) transfer function N by matching the first few coefficients of the estimated closed loop disturbance impulse response model. Once, the process and noise models are known, Ko and Edgar (1998) employ a numerical optimization procedure to estimate the highest control performance index reachable by restricting the feedback controller Q to a PI or PID form.

Agrawal and Lakshminarayanan (2003) proposed an alternate way of determining the PI achievable performance from closed loop experimental data (set point excited data). Their method uses identified closed loop process as well as disturbance models. The relationship between the controlled variable and the set point under closed loop is:

$$Y = \frac{QT}{1+QT} Y_{sp} + \frac{N}{1+QT} a = G Y_{sp} + H a \qquad (1)$$

In equation (1), Y represents the controlled variable, Y_{sp} its set point; T is the open loop process model; Q is the feedback controller that is probably of the PID type; N is the open loop disturbance model; G is the closed loop servo response model and H is the closed loop disturbance model.

From equation (1), we can write

$$T = \frac{G}{(1-G)\,Q} \qquad (2)$$

and

$$N = \frac{H}{(1-G)} \qquad (3)$$

Assuming time invariant process (T) and noise dynamics (N), for a new controller Q^* the closed loop impulse response H^* is given by

$$H^* = \frac{N}{1+Q^* T} = \frac{\dfrac{H}{(1-G)}}{1+Q^* \dfrac{G}{(1-G)Q}} = \frac{H}{1+G\left(\dfrac{Q^*}{Q}-1\right)} \qquad (4)$$

We know that the control loop performance index could be obtained from the estimated closed loop impulse response H and the process delay d. Equation (4) implies that with the knowledge of the current closed loop impulse response H, closed loop servo transfer function G and the current controller Q, it is possible to estimate the closed loop impulse response H^* for any given controller Q^*. Given that the process delay d remains constant, it is possible to determine the optimal PI type controller Q^* that maximizes the performance. Hence, the PI achievable control loop performance can be computed from the knowledge of the current controller and current closed loop servo and disturbance transfer functions. Agrawal and Lakshminarayanan (2003) demonstrated the workability of the above scheme using several examples. They also ensured that deterministic control loop performance measures like the normalized integral absolute error, gain and phase margins are also within acceptable limits.

3. THE PROPOSED METHOD FOR PI ACHIEVABLE PERFORMANCE

Now, consider the process driven by white noise sequence a_t. The expression for the controlled variable Y_t is given by

$$Y_t = \left\{\frac{N}{1+TQ}\right\} a_t \qquad (5)$$

If the current controller Q is replaced by the new controller Q^*, then new output data series Y_t^* is given by

$$Y_t^* = \left\{\frac{N}{1+TQ^*}\right\} a_t \qquad (6)$$

Again, assuming a time invariant process T and noise dynamics N, equations (5) & (6) can be used to derive the following expression

$$\frac{Y_t^*}{Y_t} = \frac{1+TQ}{1+TQ^*} \qquad (7)$$

The right hand side of equation (7) represents the filter which gives the new "routine closed loop data series" Y_t^* when the current output data series Y_t^* passes through it. For any new controller Q^*, the filter is specified (if the open loop process model T and the original controller Q are known) and the original routine data Y_t can be used to "generate" the routine closed loop data Y_t^* that would be obtained with the controller Q^*. Using this Y_t^*, one can calculate the control loop performance index corresponding to the new controller Q^*. Incorporating this methodology in conjunction with an optimization routine, it is possible to determine the controller Q_{opt} that provides the maximum possible control loop performance. In this work, we are interested in having the optimal controller Q^* to be of PID type (more specifically PI type).

Also, substituting the expression for T from equation (2) into equation (7) will result in

$$\frac{Y_t^*}{Y_t} = \frac{1+\dfrac{G}{(1-G)}}{1+\dfrac{GQ^*}{(1-G)Q}} = \frac{Q}{(1-G)Q+GQ^*} \quad (8)$$

Equations (7) and (8) demonstrate that it is possible to obtain controllers that maximize the performance index with only the knowledge of the original controller and the open loop model T (equation 7) or the closed loop servo model G (equation 8). Notice that, in contrast to the earlier methods, we do not explicitly use the closed loop noise model H to determine the "best" controller. The noise component is implicit in the original closed loop data Y_t and will reflect in the generated Y_t^*. Y_t^* is then modelled using an ARMA structure to get the performance index for controller Q^*. Obviously, this method (like other methods which try to compute the PI achievable performance) requires fairly accurate open loop process model T or closed loop servo model G. This would inevitably require some experimentation on the process. The proposed method is therefore most useful when a first level control audit indicates poor performance of this loop and one desires to improve its performance.

4. EXTENSION TO MULTILOOP SYSTEMS

The above-described method can be easily extended to multiloop systems. Again, the central idea is to estimate the output data series for the new controller using known process model (open loop process model or closed loop servo model) and the present controller.

For the multiloop case, Equation (7) becomes

$$\frac{Y_t^*}{Y_t} = \frac{I+TQ}{I+TQ^*} \quad (9)$$

Equation (8) becomes

$$\frac{Y_t^*}{Y_t} = \frac{Q}{(I-G)Q+GQ^*} \quad (10)$$

where I is the identity matrix of appropriate dimension.

Consider the simplest multiloop example, i.e a 2 x 2 system. In this case, G, T, Q, Q^*, Y_t and Y_t^* take the following matrix form:

$$T = \begin{bmatrix} T_1 & T_3 \\ T_2 & T_4 \end{bmatrix}, \quad G = \begin{bmatrix} G_1 & G_3 \\ G_2 & G_4 \end{bmatrix},$$

$$Q = \begin{bmatrix} Q_1 & 0 \\ 0 & Q_2 \end{bmatrix}, \quad Q^* = \begin{bmatrix} Q_1^* & 0 \\ 0 & Q_2^* \end{bmatrix},$$

$$Y_t = \begin{bmatrix} Y_{1t} \\ Y_{2t} \end{bmatrix} \quad \text{and} \quad Y_t^* = \begin{bmatrix} Y_{1t}^* \\ Y_{2t}^* \end{bmatrix}$$

Again using equation (9) or equation (10), the output data series for any new controller Q^* can be predicted. This predicted series can be used to estimate the control loop performance index if the controller Q^* were in place. Using these equations within an optimization routine enables the calculation of the PI achievable control loop performance for the multiloop case.

The key element again is the knowledge of the process model T or the closed loop servo transfer function model G. Estimating either of these will involve some experimentation on the process. Some experiment is indispensable because of the need to compute the interactor matrix (a generalization of the time delay for multivariate systems) for performance assessment of multivariate systems (Huang *et al.*, 1997).

PI achievable with alternate pairing: In multiloop systems, the choice of input-output pairing is an important issue. Even after employing a certain pairing, one needs to ascertain if with alternate pairing the control loop performance can be improved. This question can be answered using the method proposed in this paper.

Continuing with the 2 x 2 example, the alternate pairing is chosen by changing the Q^* structure, i.e.

$$Q^* = \begin{bmatrix} 0 & Q_1^* \\ Q_2^* & 0 \end{bmatrix} \quad (11)$$

Equations (8) and (9) can still predict the data corresponding to any new controller with alternate pairing. Again, the use of an optimization procedure, will give us the optimal controller Q^* (corresponding to alternate loop pairing) that maximizes the CLPI for the process.

Optimal control loop performance with decouplers: Another well recognized problem in multiloop systems is that of interaction. Severe interaction among the loops deteriorates the performance. Decouplers can improve the performance in the interacting system by diagonalizing the plant effectively removing the interactions. The question of the extent of performance improvement possible with the use of decouplers is explored next.

Assuming that decoupled process is T^* is equivalent to the diagonalized T,i.e.

$$T = \begin{bmatrix} T_1 & T_3 \\ T_2 & T_4 \end{bmatrix} \quad \text{and} \quad T^* = \begin{bmatrix} T_1 & 0 \\ 0 & T_4 \end{bmatrix}$$

Now, for the process driven by white noise sequence, expression for Y_t is given by

$$Y_t = \left\{ \frac{N}{1 + T\,Q} \right\} a_t \tag{12}$$

For the decoupled process driven by white noise sequence with new controller Q^*, Y_t^* can be written as

$$Y_t^* = \left\{ \frac{N}{1 + T^*\,Q^*} \right\} a_t \tag{13}$$

Using equations (12) and (13) gives

$$\frac{Y_t^*}{Y_t} = \frac{1+TQ}{1+T^*Q^*} \tag{14}$$

Utilizing the filter obtained in equation (14), the PI achievable performance with decoupler and the corresponding controller parameters for Q^* can be found out by using an optimizer.

5. CASE STUDIES

The theory outlined thus far is validated with many examples in this section. For the SISO case, the process model is identified from the experimental data, obtained by set point changes in the plant. For the multiloop case, perfectly known process models are used just to demonstrate the efficacy of the developed theory. The developed method will be of greater practical relevance if identified models can be used. This will also demonstrate the robustness of the proposed method. Such investigations are currently in progress. The "fminsearch" routine available in the Optimization Toolbox (Matlab Version 6.5 Release 13) is employed in all the following examples.

Example of a simulated SISO system: This is a simulation of the closed loop system for a first order plus time delay process regulated by a PI controller. In particular, the process, noise and controller transfer functions are given by $T(z^{-1}) = \dfrac{0.2\,z^{-3}}{1-0.8\,z^{-1}}$, $N(z^{-1}) = \dfrac{1}{1-0.95\,z^{-1}}$ and $Q(z^{-1}) = \dfrac{0.14 - 0.12\,z^{-1}}{1-z^{-1}}$ respectively. With the present controller CLPI is calculated to be 0.3922.

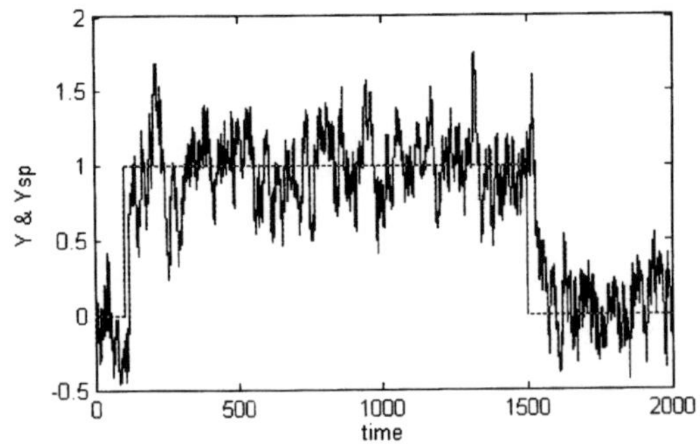

Fig. 1: Closed loop experimental data for example 1

With the experimental closed loop data shown in Figure 1, the closed loop servo transfer function G is identified. Standard tools from MATLAB's System Identification Toolbox are employed for this purpose. This identified model is used in Equation (8) to determine the optimal PI controller Q^* that maximizes the control loop performance. Using this approach, we predict the best possible performance for a PI controller (PI achievable performance) to be 0.78 and the optimal controller to be $Q^*(z^{-1}) = \dfrac{1.568 - 1.543\,z^{-1}}{1-z^{-1}}$. With complete knowledge of the open loop process model, we get the PI achievable performance to be 0.83 and the optimal PI controller is $Q^*(z^{-1}) = \dfrac{1.696 - 1.632\,z^{-1}}{1-z^{-1}}$. We therefore note that our method provides a very good estimate of what is theoretically achievable for this process with a PI controller.

A Multiloop example: Consider a 2 x 2 example with the following models for the process and noise. Note that we have integrated white noise affecting the process.

$$T = z^{-2} \begin{bmatrix} \left(\dfrac{-0.19}{1-0.905z^{-1}}\right) & \left(\dfrac{0.948}{1-0.368z^{-1}}\right) \\ \left(\dfrac{0.948}{1-0.368z^{-1}}\right) & \left(\dfrac{0.19}{1-0.905z^{-1}}\right) \end{bmatrix}$$

$$N = \begin{bmatrix} \left(\dfrac{1}{1-z^{-1}}\right) & \left(\dfrac{-0.6}{1-z^{-1}}\right) \\ \left(\dfrac{0.5}{1-z^{-1}}\right) & \left(\dfrac{1}{1-z^{-1}}\right) \end{bmatrix}$$

Initially, the process is controlled using diagonal pairing (1-1/2-2). The controller matrix Q is as follows

$$Q(1,1) = \frac{-0.01 + 0.005z^{-1}}{1 - z^{-1}},$$

$$Q(2,2) = \frac{0.01 - 0.005z^{-1}}{1 - z^{-1}} \text{ and}$$

$$Q(2,1) = Q(1,2) = 0$$

With the present PI controllers the obtained performance index is 0.0562.

Using the theory described in section 4 and taking the present controllers as the initial guess, the PI achievable performance for multiloop system is estimated to be 0.3453. The optimal PI controllers obtained in this case are:

$$Q^*(1,1) = \frac{-0.4761 + 0.4106z^{-1}}{1 - z^{-1}}$$

and

$$Q^*(2,2) = \frac{0.4437 - 0.3920z^{-1}}{1 - z^{-1}}.$$

Further, for finding out the PI achievable performance with the alternate (1-2/2-1) pairing, a stable initial guess for controllers Q(1,2) and Q(2,1) are provided based on the knowledge of process model. In this example, the initial guess used is

$$Q(1,2) = \frac{0.0654 - 0.0327z^{-1}}{1 - z^{-1}}$$

and

$$Q(2,1) = \frac{0.0654 - 0.0327z^{-1}}{1 - z^{-1}}.$$

With this initial guess, the performance index is 0.1021. Again, proceeding with the developed equations, the PI achievable with alternate pairing is predicted to be 0.7217. The optimal controllers with off-diagonal pairing are:

$$Q^*(1,2) = \frac{0.7038 - 0.4137z^{-1}}{1 - z^{-1}} \text{ and}$$

$$Q^*(2,1) = \frac{0.6859 - 0.4435z^{-1}}{1 - z^{-1}}.$$

It is quite evident from the results obtained that the off-diagonal pairing gives a much better regulatory performance for this process than does the diagonal pairing.

Finally, to estimate the effect of decouplers in improving the control performance of the process, the theory developed in section 4 is followed. Starting with the present controllers as the initial guess, the PI achievable performance obtained with decouplers is found to be 0.8472. The optimal controllers for this decoupled process comes out as

$$Q^*(1,1) = \frac{-2.3662 + 2.190z^{-1}}{1 - z^{-1}} \text{ and}$$

$$Q^*(2,2) = \frac{2.4389 - 2.2505z^{-1}}{1 - z^{-1}}.$$

We therefore conclude that, for this example, the off-diagonal control pairing can raise the PI achievable performance by more than 100%. Alternately, if the diagonal pairing is employed, the performance of the control system (for stochastic disturbances affecting this process) can be improved significantly (from 0.35 to 0.85) if perfect decouplers are employed. All of these predictions are being made with only the knowledge of the process model T and current controller Q. It is possible to estimate the enhancement in the controller performance without actually implementing the decouplers.

Another Multiloop example: We next consider the well-known Wood & Berry model of an experimental distillation column. The process and noise models are as follows

$$T = \begin{bmatrix} \left(\frac{12.8 * e^{-s}}{16.7s+1}\right) & \left(\frac{-18.9 * e^{-3s}}{21s+1}\right) \\ \left(\frac{6.6 * e^{-7s}}{10.9s+1}\right) & \left(\frac{-19.4 * e^{-3s}}{14.4s+1}\right) \end{bmatrix}$$

$$N = \begin{bmatrix} \left(\frac{3.8 * e^{-8.1s}}{14.9s+1}\right) & \left(\frac{0.22 * e^{-7.7s}}{22.8s+1}\right) \\ \left(\frac{4.9 * e^{-3.4s}}{13.2s+1}\right) & \left(\frac{0.14 * e^{-9.2s}}{12.1s+1}\right) \end{bmatrix}$$

The Relative Gain Array (RGA) (Bristol, 1966) for this process is given by

$$\Lambda = \begin{bmatrix} 2.0095 & -1.0095 \\ -1.0095 & 2.0095 \end{bmatrix}$$

The negative $(1,2)$ element of the RGA matrix suggests that no PI controller can achieve stability when off-diagonal pairing is used. Also, the Niederlinski index (Niederlinski, 1971) for the off-diagonal pairing turns out to be -0.9906, which confirms the instability.

The initial controller for the process is given by

$$Q(1,1) = \frac{0.5625 - 0.4946z^{-1}}{1 - z^{-1}}$$

$$Q(2,2) = \frac{-0.1125 + 0.1077z^{-1}}{1 - z^{-1}}$$

This controller gives a performance index of 0.5210. Further, using these present controllers as initial guess, the maximum performance index equal to 0.54. Hence, we can say that present controller (initial controller) is quite good and controller tuning is not contributing to the poor loop performance. Also, the alternate pairing need not be considered as it points to potential instability.

Interaction may be the factor for this low performance (0.54). Further analysis gives the PI achievable performance of 0.6627 with perfect decouplers in place. From the results obtained, it is quite evident that eliminating interaction does not significantly improve the loop performance. Further increase in performance may be possible with more complex controllers or by improved process design.

6. CONCLUSIONS

A novel method is developed to compute the PI achievable performance for SISO as well as multiloop systems. The effect of loop pairing and the implementation of decouplers are examined and the PI achievable performance is determined for these scenario *in silico*. The proposed method can therefore quantify the potential benefits in the performance that will result if alternate pairing or decouplers are employed.

REFERENCES

Agrawal P. and S. Lakshminarayanan, (2003). Tuning PID controllers using Achievable Performance Indices, *Industrial and Engineering Chemistry Research,* **42**(22), 5576-5582.

Åström, K.J. (1970). *Introduction to stochastic control theory*, Academic press, New York.

Bristol, E. H. (1966) On a new measure of interaction for multivariable process control, *IEEE Transactions on automatic control*, **AC-4**, 133-134.

Desborough, L. D.; Miller, R. M. (2001). Increasing Customer Value of Industrial Control Performance Monitoring – Honeywell's Experience, *Proceedings of CPC VI, Tucson,* USA.

Harris, T. J. (1989). Assessment of control loop performance. *Canadian Journal of Chemical Engineering*, **67**, 85.

Harris, T. J., F. Boudreau, and J. MacGregor. (1996). Performance assessment of multivariable feedback controllers *Automatica*, **32**(11), 1505-1518.

Harris, T.J., C. Seppala, and L. Desborough (1999). A review of performance monitoring and assessment techniques for univariate and multivariate control systems. *Journal of Process Control*, **9**, 1-17.

Huang, B., S. L. Shah, and K. Kwok (1996). How good is your controller? Application of control loop performance assessment techniques to MIMO processes. *Proceedings of 13th IFAC World Congress,* **Volume M**, San Franciso.

Huang, B., S. L. Shah, and H. Fujii (1997). The unitary interactor matrix and its estimation using closed-loop data. *Journal of Process Control*, **7**(3), 195-207.

Huang, B. and Shah, S.L. (1999). *Performance Assessment of Control Loops: Theory and Applications*, Springer-Verlag, London.

Ko, B.-S., & Edgar, T.F. (1998). Assessment of achievable PI control performance for linear processes with dead time. *Proceedings of the American Control Conference*, Philadelphia, USA, 1548-1552.

Kendra, S. and A. Cinar (1997).Controller performance assessment by frequency domain techniques. *Journal of Process Control*,**7**(3), 181-194.

Niederlinski, A. (1971).A heuristic approach to the design of linear multivariable interacting control systems. *Automatica*,**7**,691-701.

Optimization Toolbox, Version 2.2, *MATLAB* (Release 13), The MathWorks, Inc. USA.

Qin, S. J. (1998).Control performance monitoring – a review and assessment. *Computers and Chemical Engineering*, **23**, 173.

Stanfelj, N.; Marlin, T. E., MacGregor, J. F(1993). Monitoring and diagnosing control loop performance: The single loop case. *Industrial and Engineering Chemistry Research*, **32**, 301.

Swanda, A.P., & Seborg, D.E. (1999).Controller performance assessment based on setpoint response data. *Proceedings of the American Control Conference*, San Diego, USA, 3863-3867.

System Identification Toolbox, Version 2.2, *MATLAB* (Release 13), The MathWorks, Inc. USA.

Thornhill, N. F., Oettinger, M.; Fedenczuk, P. (1999).Refinery-wide control loop performance assessment. *Journal of Process Control*, **9**, 109.

Tyler, M. and M. Morari (1996). Performance monitoring of control systems using likelihood methods. *Automatica*, **32**, 1145- 1162.

ELSEVIER
IFAC
PUBLICATIONS
www.elsevier.com/locate/ifac

High-Pressure Polyethylene Process Monitoring Using PCA Based Bayesian Classification

Jialin Liu

Department of Information Management, Fortune Institute of Technology, 125-8, Chyi-Wen Rd., Chyi-Shan 842 Kaohsiung Country, Taiwan, Republic of China
E-Mail: jialin@center.fitc.edu.tw

Abstract: Using PCA based Bayesian Classification to monitor the real plant with different operated conditions is proposed. Since the process conditions are time-variant, as the PCA subspace cannot explain the data of new events, the PCA should be re-performed. In this work the method of updating Bayesian model is developed. Only the data of new events are trained in the newer subspace. The ability of PCA based Bayesian classification for monitoring different operating conditions is demonstrated using the real data from high-pressure polyethylene plant. *Copyright © 2004 IFAC*

Keywords: Process monitoring, principal component analysis, Bayesian classification

1. INTRODUCTION

A modern plant is well equipped with automation control system. Large amounts of operation data are generated and recorded. In plant history, the data of different operating conditions, normal and abnormal, are included. Data mining from historical data and finding out some of useful information provide to operators for decision-making, especially in abnormal situation that will effectively reduce plant accidents. Feature extraction from amounts of historical data is the first step of data mining. In complex chemical process, the measured variables are highly correlated with each other. Principal component analysis is one of popular approaches to compress highly dimensional variables to fewer principal components (Jackson, 1991). It is more efficient using extracted features for fault detection and isolation. Such as, Teppola et al. (1999) monitored the waste-water treatment plant in paper mill using PCA and fuzzy c-means clustering. Choi et al. (2003) predicted SOx and NOx from a power plant of steel mill through partial least square and credibilistic fuzzy c-means method.

Bayesian classification (Theodoridis and Koutroumbas, 1999) is a popular method of cluster analysis. Wang and McGreavy (1998) isolated the different operating conditions of a refinery fluid catalytic cracking process using automatic classification. Chen and Liu (1999) used mixture PCA to detect the abnormalities in different operating conditions. Generally, for clustering data, the expectation maximization (EM) algorithm is used for fitting the parameters of mixture model. However, EM algorithm suffers three significant difficulties. (1) Due to EM is a local maximum seeker; the different initial values will cause the different convergences. (2) The outliers easily affect the searching path that results in the lack of robustness of mixture model. (3) Predefined the number of clusters is required. Typically it needs multiple runs with various numbers of clusters for cluster validity. Ueda and Nakano (1998) proposed deterministic annealing EM (DAEM) algorithm in order to overcome the local maximum problem. They analogised the concept of maximum entropy of statistical mechanics, iteratively maximized entropy using a "temperature" parameter to achieve a stable solution. Medasani and Krishnapuram (2001) proposed robust mixture decomposition (RMD) algorithm retain "good" data for fitting parameters, reduced the effect of outliers. There are a lot of criteria for cluster validation, like Akaike's information criterion (AIC), Bayesian information criterion (BIC), normalized entropy criterion (NEC) and so no. McLachlan and Peel (2000) have compared the differences of these criteria. In this work, the BIC is used to determine the number of clusters.

This paper is organized as followed. The basic theory of PCA and Bayesian classification related algorithm is presented in section 2. Section 3 shows the method of PCA based Bayesian classification for process monitoring. Besides, the updating procedure of mixture model is proposed. Applied proposed process monitoring method to real plant data from high-pressure polyethylene plant is demonstrated in section 4. Finally the conclusions are given in section 5.

2. BASIC THEORY

2.1 *Principal Component Analysis*

The concept of PCA is linear combination of variables to perform a subspace. The dataset can be represented by less principal components or latent variables. Therefore, numbers of variables are reduced, and meanwhile data are explained properly.

Considered the data matrix $\mathbf{Y} \in R^{m \times n}$ with m rows (observations) and n columns (variables), and each column are normalized to zero mean and unit variance.

$$\mathbf{X} = \left(\mathbf{Y} - \overline{\mathbf{Y}}\right)\mathbf{S}^{-1} \tag{1}$$

where $\overline{\mathbf{Y}}$ is a mean vector and $\mathbf{S}$ is a diagonal matrix of standard deviations. The eigenvectors of covariance matrix and the score vectors are performed as followed.

$$\begin{aligned} &\mathbf{S} = \frac{1}{m-1}\mathbf{X}^T\mathbf{X} \\ &\mathbf{S}\mathbf{p}_i = \lambda_i \mathbf{p}_i, i=1\ldots n \\ &\mathbf{t}_i = \mathbf{X}\mathbf{p}_i, i=1\ldots n \end{aligned} \tag{2}$$

where λ_i are the eigenvalues associated with eigenvectors $\mathbf{p}_i$ and arranged in descending order. The data matrix $\mathbf{X}$ can be decomposed as:

$$\begin{aligned} \mathbf{X} &= \mathbf{t}_1\mathbf{p}_1^{\mathrm{T}} + \mathbf{t}_2\mathbf{p}_2^{\mathrm{T}} + \cdots + \mathbf{t}_K\mathbf{p}_K^{\mathrm{T}} + \mathbf{t}_{K+1}\mathbf{p}_{K+1}^{\mathrm{T}} + \cdots + \mathbf{t}_n\mathbf{p}_n^{\mathrm{T}} \\ &= \hat{\mathbf{X}} + \mathbf{E} \end{aligned} \tag{3}$$

The data matrix $\mathbf{X}$ is divided into two parts, systemic part $\hat{\mathbf{X}}$ and residual part $\mathbf{E}$. As the first K terms of components can explain data matrix properly, i.e. $\mathbf{X} \approx \hat{\mathbf{X}}$. The numbers of dimensions of data matrix are reduced from n to K.

There are amount of researches using PCA for fault detection (Russell *et al.*, 2001). The main concept is building PCA subspace via normal operating data from plant. In the on-line monitoring stage, the residuals of the new data $\mathbf{x}$ can be calculated from Eq. 3. The statistic Q is defined:

$$Q = \mathbf{e}\mathbf{e}^{\mathrm{T}} = \left(\mathbf{x} - \hat{\mathbf{x}}\right)\left(\mathbf{x} - \hat{\mathbf{x}}\right)^{\mathrm{T}} = \mathbf{x}\left(\mathbf{I} - \mathbf{P}_K\mathbf{P}_K^{\mathrm{T}}\right)\mathbf{x}^{\mathrm{T}} \tag{4}$$

The loading vectors $\mathbf{P}_K \in R^{m \times K}$ are the first K terms of eigenvectors of covariance matrix. Statistic Q is a measure of the approximated error. The confidence limit of Q is defined as followed:

$$\begin{aligned} &Q_\alpha = \Theta_1\left[\frac{c_\alpha\sqrt{2\Theta_2 h_0^2}}{\Theta_1} + 1 + \frac{\Theta_2 h_0\left(h_0 - 1\right)}{\Theta_1^2}\right]^{1/h_0} \\ &h_0 = 1 - \frac{2\Theta_1\Theta_3}{3\Theta_2^2}, \Theta_i = \sum_{j=K+1}^{n}\lambda_j^i \quad i = 1,2,3 \end{aligned} \tag{5}$$

The percentile α is the probability of Type I error in hypothesis testing and c_α is the normal deviate cutting off an area of the α upper tail of normal distribution. Another measuring difference between new data and the subspace of PCA is statistic T^2.

$$T^2 = \mathbf{x}\mathbf{P}_K ?^{-1}\mathbf{P}_K^{\mathrm{T}}\mathbf{x}^{\mathrm{T}} \tag{6}$$

The diagonal matrix ? is K terms of eigenvalues. The confidence limit is defined:

$$T_\alpha^2 = \frac{K(m-1)}{m-K}F_{K,m-1,\alpha} \tag{7}$$

where $F_{K,m-1,\alpha}$ is the F distribution with degrees of freedom K and $m-1$. In this paper, statistic Q and T^2 are used to verify the score vector can represent the new data or not. If not, using score vector for classifying will possibly result in misclassification.

2.2 *Bayesian Classification*

Given a classification task of c classes, the priori probabilities are $P_j, j = 1\ldots c$. The joint probabilities density functions of $\mathbf{x}_i$ and class j are $p(\mathbf{x}_i, j)$. Then, the conditional probability density functions of $\mathbf{x}_i$ in class j are:

$$p\left(\mathbf{x}_i \middle| j;?\right) = p\left(\mathbf{x}_i, j\right)/P_j \tag{8}$$

where ? are the parameter vectors of conditional probability density functions. According to total probability theorem, the probability density functions of $\mathbf{x}_i$ are $p(\mathbf{x}_i;?) = \sum_{k=1}^{c} p(\mathbf{x}_i|k;?)P_k$. The posteriori probabilities can be gotten from Bayes rule.

$$P\left(j\middle|\mathbf{x}_i;?,\mathbf{P}\right) = \frac{p\left(\mathbf{x}_i|j;?\right)P_j}{p\left(\mathbf{x}_i;?\right)} = \frac{p\left(\mathbf{x}_i|j;?\right)P_j}{\sum_{k=1}^{c} p\left(\mathbf{x}_i|k;?\right)P_k} \tag{9}$$

Here $\mathbf{P}$ are vector of priori probabilities, not the loading vector of PCA. If the parameters of conditional probability density functions and priori probabilities are known, the posteriori probabilities can be obtained from Eq. 9.

Expectation-Maximization Algorithm. The expectation maximization algorithm maximizes the expectation of loglikelihood function. The two steps of EM algorithm are:

E-Step:

$$\begin{aligned} Q(\mathbf{T};\mathbf{T}(t)) &\equiv E\left[\sum_{i=1}^{m} ln\left(p\left(\mathbf{x}_i|j;?\right)P_j\right)\right] \\ &= \sum_{j=1}^{c}\sum_{i=1}^{m} P\left(j|\mathbf{x}_i;\mathbf{T}(t)\right) ln\left(p\left(\mathbf{x}_i|j;?\right)P_j\right) \end{aligned} \tag{10}$$

where $\mathbf{T}(t)$ are the parameters at the tth step of iteration. The posteriori probabilities $P(j|\mathbf{x}_i;\mathbf{T}(t))$ can be obtained from Bayes rule.

M-Step:
Compute the next estimating parameters through maximizing the expectation of loglikelihood function.

$$\frac{\partial Q(\mathbf{T};\mathbf{T}(t))}{\partial \mathbf{T}}=0 \tag{11}$$

Take a multivariate Gaussian probability density function as an illustrated example:

$$p(\mathbf{x}_i|j;?)=\frac{exp\left(-\frac{1}{2}(\mathbf{x}_i-\boldsymbol{\mu}_j)\mathbf{S}_j^{-1}(\mathbf{x}_i-\boldsymbol{\mu}_j)^T\right)}{(2\pi)^{n/2}|\mathbf{S}_j|^{1/2}} \tag{12}$$

where $\boldsymbol{\mu}_j$, $\mathbf{S}_j$ are the mean vector and covariance matrix of the jth class. Using Lagrange multipliers the iterative steps can be derived:

$$P_j(t+1)=\frac{1}{m}\sum_{i=1}^{m}P(j|\mathbf{x}_i;\mathbf{T}(t)),j=1\ldots c \tag{13}$$

$$\boldsymbol{\mu}_j(t+1)=\frac{\sum_{k=1}^{m}P(j|\mathbf{x}_k;\mathbf{T}(t))\mathbf{x}_k}{\sum_{k=1}^{m}P(j|\mathbf{x}_k;\mathbf{T}(t))} \tag{14}$$

$$\mathbf{S}_j(t+1)=\frac{\sum_{k=1}^{m}P(j|\mathbf{x}_k;\mathbf{T}(t))(\mathbf{x}_k-\boldsymbol{\mu}_j(t))^T(\mathbf{x}_k-\boldsymbol{\mu}_j(t))}{\sum_{k=1}^{m}P(j|\mathbf{x}_k;\mathbf{T}(t))} \tag{15}$$

The fitting vector of parameters $\mathbf{T}$ can be obtained by repeating E and M steps until $\|\mathbf{T}(t+1)-\mathbf{T}(t)\|\le\varepsilon$, ε is the properly convergence criterion.

Deterministic Annealing EM Algorithm In general, the initial values of parameters $\mathbf{T}(0)$ are randomly selected, the posteriors are not reliable in early stage of the iteration. Relatively mislead the direction of seeking solution. However, the fitting parameters will be easily trapped into local maximum. Ueda and Nakano (1998) proposed deterministic annealing EM algorithm to resolve this problem. They introduced another posterior $F(j|\mathbf{x}_i;\mathbf{T}(t))$:

$$F(j|\mathbf{x}_i;\mathbf{T}(t))=\frac{\left[p(\mathbf{x}_i|j;?(t))P_j(t)\right]^{\beta}}{\sum_{k=1}^{c}\left[p(\mathbf{x}_i|k;?(t))P_k(t)\right]^{\beta}},$$
$$0\le\beta\le1 \tag{16}$$

In a special case $\beta=0$, the posterior is a uniform distribution unaffected by initial value of parameters. In another case $\beta=1$, the posterior reduces to the original posterior. The β increases from 0 to 1, that means the posterior changes from a uniform distribution to original posterior. This procedure analogises to simulated annealing, $1/\beta$ is like the "temperature" of annealing. They add a β-loop on the EM-steps and replace posterior by Eq. 16.

The DAEM algorithm can be derived.

1. Setting initial value of β, $\beta\approx0$.
2. The best solution of $\mathbf{T}_\beta$ can be obtained by iterating EM-steps until convergence.
3. If $\beta<1$, increasing β and setting initial value of parameters through $\mathbf{T}_\beta$ of step 2. Repeat from step 2 until $\beta=1$.

Robust Mixture Decomposition Algorithm Although the DAEM can overcome the local maximum problem, the outliers still affect the solution of DAEM. Unfortunately, outliers are unavoidable in real plant data. To resolve this difficulty, Medasani and Krishnapuram (2001) proposed the robust mixture decomposition (RMD) algorithm. They argued that the value of $p(\mathbf{x}_i|?)$ of the outlier $\mathbf{x}_i$ would have a low value when ? is correctly estimated. Therefore, they trimmed a fraction of the data by sorting probability functions for each $\mathbf{x}_i$, i.e. $p(\mathbf{x}_1|?)\ge p(\mathbf{x}_2|?)\ge\cdots\ge p(\mathbf{x}_m|?)$ before each EM iteration, retain the first r terms of data for fitting parameters. That will effectively reduce the influence of outliers. The retention ratio (r/m) can be estimated from the data quality, or be included as one of the parameters to be optimised. The smaller retention ratio may underestimate the covariances, but can stabilize the solution for fitting parameters.

Bayesian Information Criterion Using the fitting parameters $\mathbf{T}$ approximate the expectation of loglikelihood. The BIC is defined as:

$$BIC_c=-2\times Q(\mathbf{T})+v_c\,ln(m) \tag{17}$$

where v_c is number of fitting parameters. However, maximized the expectation of loglikelihood is to minimize BIC. Using multiple runs with various numbers of parameters can decide the number of cluster from the minimizing BIC.

3. PCA BASED BAYESIAN CLASSIFICATION

The PCA based Bayesian classification is to find out the clusters of different operating conditions using score vectors. Firstly, the subspace of PCA can be constructed from plant data. The highly dimensional variables can be effectively reduced that will simplify the Bayesian model and depress the computing time of fitting parameters. In order to obtain the reasonable clusters, the score vectors of historical data are clustered through combination of DAEM and RMD algorithms in this work. However, the operating conditions of real plant are time-variant. As new data cannot be explained by existed subspace, the subspace should be reconstructed from all of data. It is more consistent using one PCA subspace describes the whole dataset than multiple PCA subspaces. Besides, since the new data don't belong to the existed subspace that must be independent with existed classes. Therefore, only the data of new events are clustered on the newer subspace, the trained dataset don't need to be performed clustering again. The method of updating Bayesian model is proposed; the Bayesian model of the previous subspace can be updated to the newer subspace.

3.1. Fault Isolation

The statistic Q and T^2 of on-line data can be conducted from Eq. 4 and Eq. 6. As $Q > Q_\alpha$ or $T^2 > T_\alpha^2$, indicate the new event has occurred, existed PCA cannot explain the new data properly. The subspace of PCA should be rebuilt. On the other hand, $Q < Q_\alpha$ and $T^2 < T_\alpha^2$, the score vectors of on-line data are used for calculating the posteriors $P(j|\mathbf{t}) = p(\mathbf{t}|j;?)P_j \Big/ \sum_{k=1}^{c} p(\mathbf{t}|k;?)P_k$. That can be used to identify the on-line data belonging to which operated condition has happened. If the posteriors are evaluated using score vectors without verify the on-line date can be explained by existed subspace, that could result in misclassifying the operating condition, since the new data might not belong to the existed subspace.

3.2. Scheme of Updating Bayesian Model

Adding the data of new events to the original data matrix, the total number of data is added from m to m*. The data matrix $\mathbf{Y}^* \in R^{m^* n}$, $\mathbf{Y}^{*T} = \left[\mathbf{Y}^T \quad \mathbf{Y}_{new}^T\right]$, is normalized to $\mathbf{X}^* = \left(\mathbf{Y}^* - \overline{\mathbf{Y}}^*\right)\mathbf{S}^{*-1}$. The mean vector and diagonal matrix of standard deviation are $\overline{\mathbf{Y}}^*$ and $\mathbf{S}^*$. The first K* ($K^* \geq K$, the original number of PCs) terms of loading vectors $\mathbf{P}^*$ and correspondingly score vectors $\mathbf{T}^*$ can be determined by $\mathbf{E}^* \approx 0$. Since the original dataset can be explained by both of subspaces.

$$\mathbf{Y} \approx \overline{\mathbf{Y}} + \mathbf{t}\mathbf{P}^T\mathbf{S} \approx \overline{\mathbf{Y}}^* + \mathbf{t}^*\mathbf{P}^{*T}\mathbf{S}^* \tag{18}$$

Besides, the values of conditional probability density functions should be equalized on both subspaces. For example, the conditional probability density functions of score vectors in the jth class is:

$$p(\mathbf{t}|j) = exp\left(-\frac{1}{2}(\mathbf{t}-\boldsymbol{\mu}_j)\mathbf{S}_j^{-1}(\mathbf{t}-\boldsymbol{\mu}_j)^T\right)\Big/(2\pi)^{K/2}|\mathbf{S}_j|^{1/2}$$

On both of subspaces, the exponential parts have to be equalized.

$$(\mathbf{t}-\boldsymbol{\mu}_j)\mathbf{S}_j^{-1}(\mathbf{t}-\boldsymbol{\mu}_j)^T = (\mathbf{t}^*-\boldsymbol{\mu}_j^*)\mathbf{S}_j^{*-1}(\mathbf{t}^*-\boldsymbol{\mu}_j^*)^T \tag{19}$$

The new parameters of conditional probability density functions can be obtained.

$$\boldsymbol{\mu}_j^* = \left(\boldsymbol{\mu} - \Delta\overline{\mathbf{y}}\mathbf{S}^{-1}\mathbf{P}\right)\mathbf{A}^+ \tag{20}$$

$$\mathbf{S}_j^* = \mathbf{A}^{+T}\mathbf{S}_j\mathbf{A}^+ \tag{21}$$

$$\mathbf{A} = \mathbf{P}^{*T}\mathbf{S}^*\mathbf{S}^{-1}\mathbf{P} \tag{22}$$

where defines $\Delta\overline{\mathbf{y}} \equiv \overline{\mathbf{y}}^* - \mathbf{y}$, $\mathbf{A}^+$ is the pseudo inverse of **A**, and updated priors are $P_j^* = m \times P_j / m^*$. Through Eq. 20 to Eq. 22, the previous Bayesian model can be updated to the newer subspace.

4. EXAMPLE

This polyethylene plant is in Taiwan, there are five different operation lines; produce HDPE, LDPE, linear LDPE and so on. In this work the high-pressure process is studied, the operating conditions are more than 20 different grades in this process. The high purity ethylene (FI-001 in Figure 1) mixed with recycled gas before entering primary compressor, the amount of different modifiers (FIC-003 and FIC-004) will be added via the recipe. The pressure through secondary compressor (FI-002) is around 1500~2000 kg/cm^2-g. There are several reacted zones in reactor; the initiators maintain the temperature profile. The positions of temperature controllers (TIC-001, TIC-002 and TIC-003) are located closely on the top, middle and bottom of the reactor. The most important product quality is melt index (MI) that is significantly affected by temperature profile and pressure in reactor. The pressure is regulated by PIC-001 through the under flow of reactor. The mixture of polyethylene and unreacted gas will be flashed in two stages. Firstly, the higher-pressure gas is recycled to the entrance of secondary compressor from the top of high-pressure separator. Almost all of polymer and rest of unreacted gas will flow to extrusion hopper. PE will be pumped to extruder from bottom of extrusion hopper; the lower-pressure gas is recycled to the entrance of primary compressor from top of extrusion hopper. The reaction conversion is only around 10~15%, the most of unreacted gas continuously recycle in this process. The byproducts will be accumulated until the product quality is out of control limit and operator will be notified to purge recycled gas from FIC-006 by operator experience.

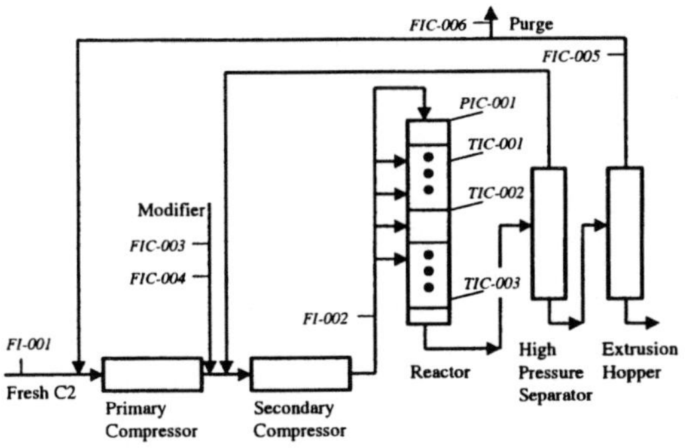

Fig. 1: Process flow diagram of high-pressure polyethylene plant

In this work, the reactor condition monitoring is focused. Not only the melt index is influenced by the reacted condition, but also this exothermic reaction is easily run away with slightly unstable temperature control. Besides, the pressure of reactor is higher than 1500 kg/cm^2-g, that is a most important issue to monitor the operat ing condition of reactor in this plant. The plant data have been collected every 5 minutes for a month. The data of shutdown has been trimmed, the training set contain 8928 data for fitting the parameters of Bayesian model. Thirteen process variables are collected, including three temperatures, respectively flow rates of initiators and the pressure in the reactor. Besides, the flow rates of two different modifiers, the top flow rate of extruder hopper and the other flow rates including purge, fresh ethylene, and secondary compressor. There exist seven different production grades in training dataset. The temperature profile and pressure for each grade are shown in Figure 2, the production grades are labelled from grade #1 to #7. For confidential concerned the process values are hidden, noteworthily the temperature difference can be greater than 70 degree C between different grades, such as grade #1 and #6. The variation of operating condition will be taken into account for production scheduling; in order to avoid the run away reaction will be caused as the transition of grades.

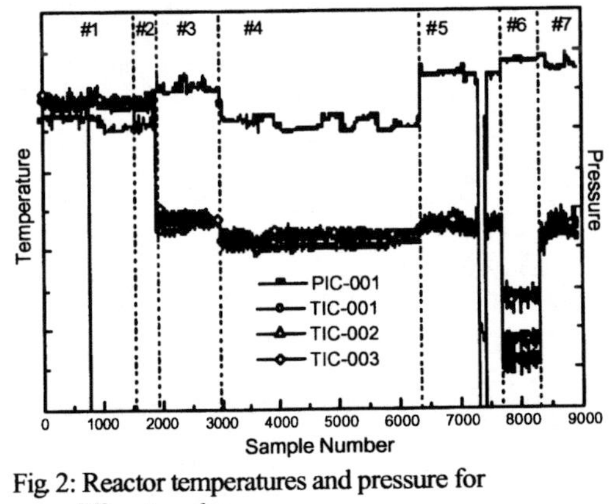

Fig. 2: Reactor temperatures and pressure for different grades.

Defined the subspace of PCA from plant data, the four principal components explain 90% of total variances. Due to the outliers of raw data, the retention ratio of RMD is set 0.8. In this example, the initial value of β in DAEM is set to 0.2 and the iterative procedure with $\beta_{new} = 1.5\times\beta_{old}$ until $\beta = 1$. The 9 clusters are determined with the minimizing BIC. The clustering result projected to the first two PCs is shown in Figure 3. Since the operat ing conditions of grade #1 and #2 are similar, those data share with the cluster C1&2. The data of grade #3 and #5 look like overlapped, but in the second and third PCs those data can be well separated. In the grade #4, there nearly exist two clusters C4_1 and C4_2. Consult with the expert of the plant; it is due to the different sources of ethylene feed. The purity of ethylene is slightly varied from different upstream vendors. That will cause the byproducts are easily accumulated as a little of impurity in feed ethylene. The operator will maintain the purity of recycled gas through purging the recycled gas. That is the reason of the different operated conditions in the same grade.

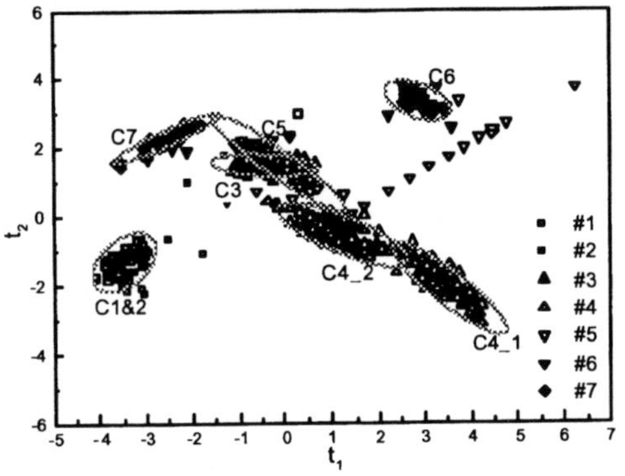

Fig. 3: First and second score vectors of trained data, and 9 clusters.

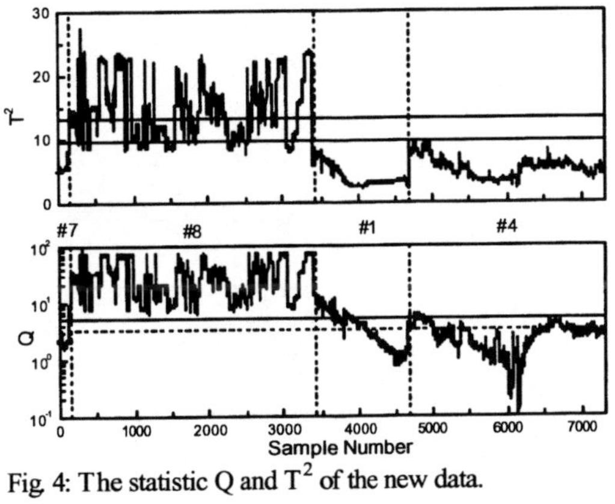

Fig. 4: The statistic Q and T^2 of the new data.

The new data are collected after training dataset the period is also one month. The dataset includes lasting data of grade #7, a new grade labelled #8, and grade #1, #4 have appeared in the training dataset. Before using the score vectors of PCA for classification, the statistic Q and T^2 of the new data have to be verified. In Figure 4, the statistic Q and T^2 of the new data for grade #7, #4 and part of #1 are within control limits. The data of grade #8 and leading part of grade #1 cannot be explained by existed subspace. Since grade #8 is a new event, it is reasonable for PCA cannot explain these data. The melt index of grade #8 is much lower than grade #1; the ratio of two melt indexes is around 0.01. For the lower MI the impurity of recycled gas must be as less as possible. Therefore, the higher MI can suffer more impurity. Since that operator will not purge recycled gas in the transition from the lower MI to higher MI until the impurity reach the tolerable limit. It is the reason that the data of leading part of grade #1 cannot be explained by existed subspace.

Project the data can be explained by PCA to the subspace, the first two of score vectors are shown in Figure 5. That conform the previous result of the cluster. In order to include the new events, the training data set is added the data of new events. The subspace of PCA is reconstructed; the first 5 PCs explain 94% of total variances. Only the dataof new events need to be clustered in the newer subspace. The clusters C8 and C1_2 can be gotten through DAEM and RMD algorithms. The previous clustering result can be updated to the new subspace using the proposed method described in section 3.2. In Figure 6, the clustering result in the new subspace is shown.

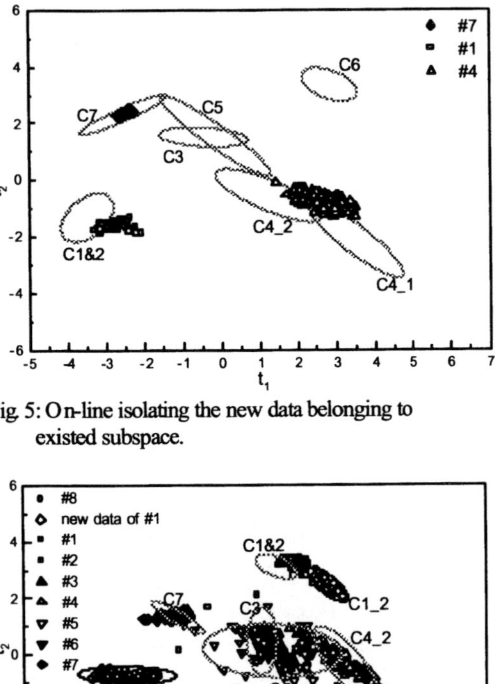

Fig. 5: On-line isolating the new data belonging to existed subspace.

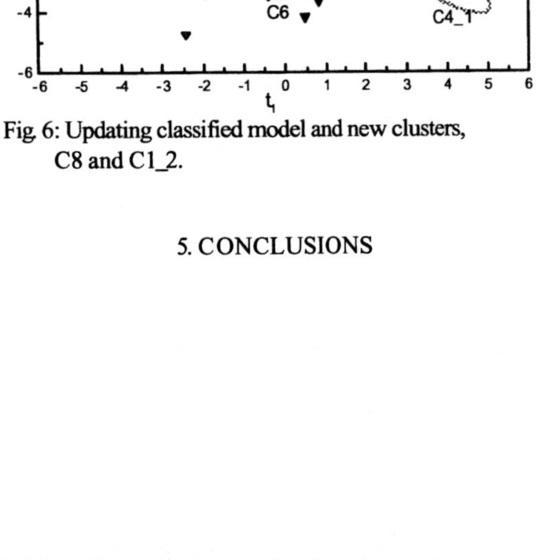

Fig. 6: Updating classified model and new clusters, C8 and C1_2.

5. CONCLUSIONS

The proposed method is successfully applied to the high-pressure polyethylene process monitoring. There exist several operating conditions for different grades in this process. Before using score vectors of PCA for on-line monitoring, the statistic Q and T^2 must be verified by hypothetically testing. As the data of new events cannot be explained by existed subspace, the newer subspace will be reconstructed by all of data for covering all of events. In the newer subspace, only the data of new events are clustered with DAEM and RMD algorithms. The number of clusters is determined through minimizing BIC. The previous clustering result can be directly updated to newer subspace with proposed method. Besides, the physical meaning of each cluster should be identified. Consult with the expert of plant or find out through operator log will be helpful to recognize the root causes.

ACKNOWLEDGMENT

This work is supported by the National Science Council, Republic of China, under Grant NSC-92-2213-E-268-001.

REFERENCES

Chen, J. and Liu, J. (1999) Mixture principal component analysis models for process monitoring. *Ind. Eng. Chem. Res.* **38**, 1478-1488.

Choi, S. W., Yoo, C. K. and Lee, I. (2003) Overall statistical monitoring of static and dynamic patterns. *Ind. Eng. Chem. Res.*, **42**, 108-117.

Jackson, J. E. (1991). *A User's Guide to Principal Components*, Wiley, New York.

McLachlan, G. J. and Peel, D. (2000) *Finite mixture models*. Wiley, New York.

Medasani S. and Krishnapuram R. (2001) Categorization of image databases for efficient retrieval using robust mixture decomposition. *Computer Vision and Image Understanding.*, **83**, 216-235.

Nimmo, I. (1995). Adequately address abnormal situation operations. *Chem. Eng. Prog.* **91**, 36-45.

Russell, E. L., Chiang, L. H. and Braatz, R. D., (2001). *Fault Detection and Diagnosis in Industrial Systems.* Springer, London.

Teppola, P., Mujunen, S and Minkkinen, P. (1999) Adaptive Fuzzy C-Means clustering in process monitoring. *Chemometrics and Intelligent Laboratory Systems*, **45**, 23-38.

Theodoridis, S. and Koutroumbas, K. (1999). *Pattern Recognition*. Academic Press.

Ueda, N. and Nakano, R. (1998) Deterministic annealing EM algorithm. *Neural Networks.*, **11**, 271-282.

Wang, X. Z. and McGreavy C. (1998) Automatic classification for mining process operational data. *Ind. Eng. Chem. Res.*, **37**, 2215-2222.

ELSEVIER
IFAC
PUBLICATIONS
www.elsevier.com/locate/ifac

MODELLING TEMPERATURE-INDUCED SPECTRAL VARIATIONS IN CHEMICAL PROCESS MONITORING

Zeng-Ping Chen, Julian Morris, Elaine Martin

Centre for Process Analytics and Control Technology,
School of Chemical Engineering and Advanced Materials,
University of Newcastle upon Tyne, NE1 7RU, UK

Abstract: With a view to ensuring the validity of multivariate calibration models in the presence of temperature variations, a new methodology, individual contribution standardization (ICS), is proposed to correct for temperature-induced spectral variations. The methodology was applied to shortwave NIR spectral data sets recorded at different temperatures. The results showed that ICS can almost eliminate temperature effects resulting in calibration models that exhibit good predictive performance. Compared with other methods, such as continuous piecewise direct standardization, ICS has the advantages of easy implementation, simple selection of the calibration model and less restrictions on the training samples. *Copyright© 2004 IFAC*

Keywords: Chemical process monitoring, Temperature-induced spectral variations, Multivariate calibration model.

1. INTRODUCTION

Spectroscopic techniques are increasingly being applied in process analysis (Blaser, et al., 1995; Lewiner, et al., 2001) as a result of their potential for on-line application and the short analysis time. Due to the lack of selectivity of the target species, in many chemical processes, multivariate methods such as Partial Least Squares (PLS) (Höskuldsson, 1988) are applied for the building of robust calibration models. Generally these models are valid only when the spectra of the future samples (test samples) are measured at the same external conditions as the calibration samples (training samples). However in industrial applications, on-line spectral process measurements are often affected by fluctuations in external variables such as temperature, thereby invalidating the multivariate calibration model if the effects of the external variations on the spectral measurements are not taken into account when developing the model.

One approach proposed for addressing external sources of variation has been to implicitly incorporate the effect of temperature (or other variables) into the calibration experimental design. Wülfert, et al. (1998) studied the effect of fluctuating temperature on the predictive ability of a multivariate calibration model for NIR spectra. They concluded that by implicitly including temperature in the model, satisfactory predictive performance was observed, although the complexity of the model was increased. As neither the temperatures of the calibration samples nor that of the future samples is included explicitly in the model as independent or dependent variables, the accuracy of the results will not be optimal.

An alternative approach to the handling of temperature influences is through the explicit inclusion of temperature into the calibration model by treating the temperature of the samples as an extra independent variable appended to the spectra or as another dependent variable, for example. Wülfert, et al., (2000a) investigated this approach. Their results showed that none of the approaches that explicitly include temperature in the calibration model results in an improvement in the performance of the calibration model when compared to the more basic idea of implicit inclusion.

It was proved that non-linearities such as temperature effects (Iwata, et al., 1997) cannot be filtered out or resolved through the application of an orthogonal basis transformation (Wülfert, et al., 2000a), such as the wavelet transformation. An alternative approach has been to use robust variable selection models to decrease the influence of temperature variations on a model's predictive performance (Swierenga, et al., 2000). It was observed that models based on robust variable selection may be better than, or similar to, implicit global models with respect to prediction errors. However, a disadvantage related to variable selection models is that there is a need for an understanding of the underlying chemical system.

Wang, et al. (1991, 1993) developed piecewise direct standardization (PDS) to correct for complex non-linear spectral variations to enable calibration transfer between measurements performed on two different instruments or under two different sets of conditions (say two different temperatures). However, a requirement is that measurements are made for a selected subset of samples for both the calibration and test temperatures, for example, thereby making the approach unsuitable for correcting temperature-induced spectral variations in process analysis, since it is not possible to measure the spectra of a selected subset of samples under all possible temperatures since temperature is a continuous variable.

Continuous piecewise direct standardization (CPDS) (Wülfert, et al., 2000b) is a generalization of PDS to continuous variables. To overcome the restriction of PDS, CPDS fits polynomials to the corresponding elements of the discrete PDS transformation matrices under different temperatures. The transformation matrix for a new temperature can then be estimated from the established polynomials. CPDS can provide results that are superior to the implicit or explicit inclusion of temperature into the calibration model.

Three parameters (the number of latent variables, the width of the window size and the degree of the polynomial) have to be estimated to build the CPDS correction model. The optimization procedures are complex and time consuming. The requirement that the spectral measurements are obtained for exactly the same subset of samples at all calibration temperatures cannot always be satisfied in process analytical applications. Moreover, the corresponding elements of the discrete PDS transformation matrices under different temperatures may not follow smooth non-linear models. The difference between the transformation matrix estimated for a new temperature by the polynomial models established and the actual one may not be trivial, hence the resulting prediction would be inaccurate.

In this paper, a new methodology is proposed, Individual Contribution Standardization (ICS), for the correction of temperature-induced spectral variations. The method is evaluated using a NIR data set and the results are compared with CPDS.

2. THEORY AND ALGORITHM

Suppose the rows of the spectral data matrices $\mathbf{X}(t_1)$, $\mathbf{X}(t_2)$,..., $\mathbf{X}(t_K)$ are the corresponding spectra of the training mixture samples measured at training temperature t_1, t_2, ..., and t_K, respectively with known concentration matrix $\mathbf{Y}$. The objective is to build multivariate calibration models that can predict the concentrations of the chemical components in future mixture samples from the spectral data measured at certain temperatures. If the temperature of future mixture sample, t_k, is equivalent to any of $t_1, t_2, \ldots, t_K$, a prediction can be made using the local model built between $\mathbf{X}(t_k)$ and $\mathbf{Y}$.

However temperature, as stated, is a continuous variable. It can therefore take any value. Due to the difference in temperature between the future samples and the training samples, the local model built on the training samples cannot give correct predictions for future samples if the temperature-induced spectral variations have not been appropriately corrected for.

To correct for the influence of temperature, the continuous piecewise direct standardization (CPDS) algorithm (Wülfert, et al., 2000b) adopts the following strategy. First a partial least square (PLS) calibration model between the spectral data $\mathbf{X}(t_{ref})$ and the concentration matrix $\mathbf{Y}$ is established for a selected reference temperature t_{ref} from the training temperatures. The next step is the calculation of the discrete calibration transformation matrices between the spectra measured at the remaining training temperatures and the reference temperature. These are found through piecewise direct standardization (PDS) (Wang, et al., 1991):

$$\mathbf{X}(t_{ref}) = \mathbf{X}(t_k)\mathbf{P}(t_k), \quad (k = 1,2,\cdots,K; t_k \neq t_{ref}) \tag{1}$$

The transformation matrix $\mathbf{P}(t_k)$ is a banded matrix that can correct for the spectral variations resulting from the temperature differences between two distinct measurement temperatures. The transformation matrix $\mathbf{P}(t_k)$ is obtained by regressing the elements, $\mathbf{x}_j(t_{ref})$ (wavelength j, reference temperature t_{ref}) on a window $\mathbf{x}_{j-w}(t_k)$ to $\mathbf{x}_{j+w}(t_k)$. The widow width is $(2\times w+1)$. The regression vectors $\mathbf{b}_j(t_k)$ are then calculated using PLS and the transformation matrix $\mathbf{P}(t_k)$ is constructed by placing $\mathbf{b}_j(t_k)$ in the jth column, that ranges from row $j-w$ to $j+w$. A polynomial is then fitted to each element along the bands of the main diagonal of the transformation matrices $\mathbf{P}(t_k)$ $(k=1,2,\cdots,K; t_k \neq t_{ref})$ against the temperature difference Δt_k $(\Delta t_k = t_k - t_{ref})$:

$$p_{i,j}(t_k) = a_{i,j} + b_{i,j}\Delta t_k + c_{i,j}\Delta t_k^2 + e_{i,j} \tag{2}$$

Once the parameters of the polynomials have been estimated, the transformation matrix $\mathbf{P}(t_{test})$ for the spectral matrix $\mathbf{X}(t_{test})$ of the test samples measured at temperature t_{test} can be calculated. The temperature effects can then be removed:

$$\mathbf{X}(t_{test} \rightarrow t_{ref}) = \mathbf{X}(t_{test})\mathbf{P}(t_{test}) \tag{3}$$

The concentrations can be predicted by using the local model at the reference temperature. To obtain satisfactory results, three parameters have to be optimized prior to building the final CPDS model, the number of latent variables for calculating the regression vector $\mathbf{b}_j(t_k)$, the width of the bands about the main diagonal (window size$(2\times w+1)$) and the degree of the polynomials to fit the elements of the transformation matrices. The best values for the three parameters are determined by leave-one-out cross validation. This approach is time consuming. Furthermore, the selection of the reference temperature is critical for accurate predictions.

In their original paper, Wülfert, et al. (2000b) proposed selecting a midpoint calibration temperature for better results. Intuitively, for test samples measured at different temperatures, the best reference temperatures should be different. The smaller the difference between the reference and test temperature, the better the results. However, building CPDS models for all calibration temperatures is challenging. The optimal values for the three parameters may not be the same when different reference temperatures are considered. Consequently considerable time has to be spent optimising these parameters for each reference temperature. Even if the optimal values for the three parameters are found, the elements on the bands of the transformation matrices may not follow smooth non-linear functions. Hence, the results of CPDS may not be reliable.

Generally, temperature has a different impact on different chemical species in mixture samples, and the absorbance of each chemical species at different wavelengths will follow simple monotonic smooth non-linear functions with respect to temperature. These facts clearly indicate that the temperature-induced spectral variations should be modelled separately for each chemical species in mixture samples. Consequently the individual contribution standardization (ICS) approach developed adopts the following strategy. The response patterns $\mathbf{s}_l(t_k)$ $(l=1,2,\cdots,m)$ of chemical species l under temperature t_k can be resolved using the following direct regression approach:

$$\mathbf{S}(t_k) = \mathbf{Y}^+\mathbf{X}(t_k)$$
$$\mathbf{S}(t_k) = \begin{bmatrix} \mathbf{s}_1'(t_k) \\ \mathbf{s}_2'(t_k) \\ \vdots \\ \mathbf{s}_m'(t_k) \end{bmatrix}, \quad (k=1,2,\cdots,K) \tag{4}$$

A polynomial is then fitted for each element of $\mathbf{S}(t_k)$ $(k=1,2,\cdots,K)$ against temperature t_k.

$$s_{i,j}(t_k) = a_{i,j} + b_{i,j}t_k + c_{i,j}t_k^2 + e_{i,j} \tag{5}$$

After $a_{i,j}$, $b_{i,j}$ and $c_{i,j}$ have been estimated, the response pattern matrix $\mathbf{S}(t_{test})$ of the chemical species under test temperature t_{test} can be calculated. Given the spectral matrix $\mathbf{X}(t_{test})$ of test samples measured under temperature t_{test}, the contribution of the chemical species can be approximately estimated:

$$\hat{\mathbf{Y}} = \mathbf{X}(t_{test})\mathbf{S}^+(t_{test}) \tag{6}$$

As a result of the existence of interferences such as background, $\hat{\mathbf{Y}}$ is not necessarily an acceptable estimate of the concentration matrix for the test samples. However, it can be used to transform $\mathbf{X}(t_{test})$ to $\mathbf{X}(t_{test} \rightarrow t_{ref})$, as though it were measured under reference temperature, t_{ref}:

$$\mathbf{X}(t_{test} \rightarrow t_{ref}) = \hat{\mathbf{Y}}(\mathbf{S}(t_{ref}) - \mathbf{S}(t_{test})) + \mathbf{X}(t_{test}) \tag{7}$$

Here, $\mathbf{S}(t_{ref})$ is the response pattern matrix of the chemical species for reference temperature t_{ref}. $\hat{\mathbf{Y}}(\mathbf{S}(t_{ref}) - \mathbf{S}(t_{test}))$ can be viewed as the difference introduced by the temperature variation between $\mathbf{X}(t_{test})$ and $\mathbf{X}(t_{test} \rightarrow t_{ref})$.

After the spectral variation introduced by the difference in temperatures has been corrected for, calibration models established under reference temperatures can be developed to predict the concentration of all the chemical species in the test sample. It should be observed that from a set of calibration models for the spectra measured for all training temperatures (t_1, t_2, ...,and t_K), the one with the temperature closest to that of the test sample is selected. Therefore, the most appropriate calibration model can be used to predict the concentration of the test sample. In addition ICS does not require the same subset of samples for all training temperatures. This characteristic of ICS offers some advantages over CPDS in process applications.

3. APPLICATION

An NIR data set was selected to compare the performance of CPDS and ICS. The data was collated from ninety five NIR spectra for 19 ternary mixtures of ethanol (C_1), water (C_2) and 2-propanol (C_3) recorded at five different temperature levels (30°C, 40°C, 50°C, 60°C and 70°C) in the range 580nm to 1049nm with a resolution of 1nm using an HP 8454 spectrophotometer equipped with a thermostable cell holder. The spectral region between 749nm and 849nm was used for slope and offset corrections and the 200 absorbances in the range 850nm to 1049nm were selected for the data analysis.

The 19 samples at each temperature were divided into 13 training and 6 test samples. Experimental details can be found in Wülfert, et al. (2000b), and the data is available at http://www-its.chem.uva.nl. The root-mean-squared error of prediction (RMSEP) was used as the performance criterion to assess the predictive power of ICS and CPDS. The RMSEP was calculated as follows:

$$RMSEP=\sqrt{\sum_{i=1}^{N_{test}}\frac{(y_{i,test}-\hat{y}_{i,test})^2}{N_{test}}} \quad (8)$$

Here, $y_{i,test}$ is the known concentration of the ith test sample, $\hat{y}_{i,test}$ is the corresponding value estimated by ICS or the CPDS model. N_{test} denotes the number of test samples. The data analysis was undertaken on a Pentium computer using Matlab version 6.5 (Mathworks, Inc). All the programmes including ICS, CPDS and PLS1 were written in house.

4. RESULTS AND DISCUSSIONS

Mean-centred spectra ($\mathbf{X}$) were used as the prediction variables with the concentrations of water, ethanol and 2-propanol defining the predicted variables ($\mathbf{y}$). For each chemical component, five local PLS1 calibration models with four latent variables (suggested by Wülfert, et al. 2000b) were built from the spectra of the training samples measured at five different temperatures (30 °C, 40°C, 50°C, 60°C and 70°C). The results of the local PLS1 models with 4 latent variables are provided for comparison in Table 1.

Table 1. RMSEP ($\times 10^{-2}$Mole fraction) of the test set for local models

Temp. (°C)		Components		
T_{pred}	T_{cal}	C_1	C_2	C_3
30	30	1.8	0.9	1.2
40	40	1.1	0.7	0.9
50	50	1.7	1.1	2.2
60	60	1.0	0.4	0.8
70	**70**	1.1	0.4	1.5

Before any detailed discussion on the performance of ICS is progressed, the validity of the assumption underpinning ICS is investigated. Fig. 1 shows the response patterns of all the three chemical species resolved by direct regression. It is evident that the variations in the response patterns with respect to temperature follow simple monotonic smooth non-linear functions.

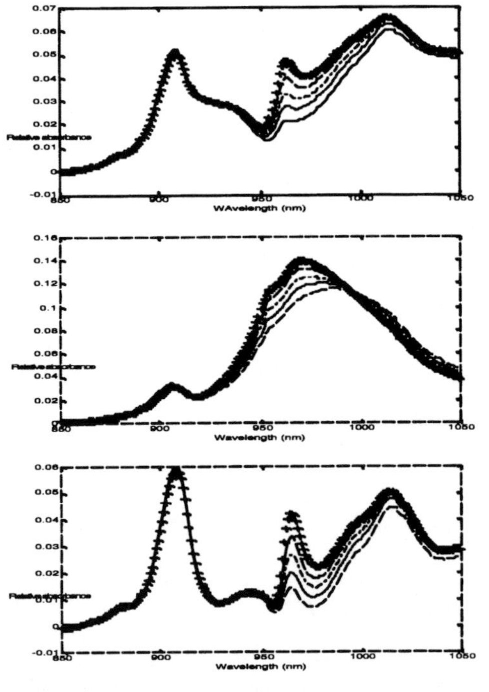

Fig.1. The resolved response patterns of ethanol (upper), water (middle) and 2-propanol (bottom) at 30°C (solid), 40°C (dotted), 50°C (dash dotted), 60°C (dash) and 70°C (plus).

The appropriateness of applying second order polynomials to model the temperature effects is also apparent from the agreement between the response patterns of the three chemical species at 50°C resolved by direct regression and the corresponding ones estimated by applying ICS using the spectra measured at temperatures other than 50°C (Fig. 2)

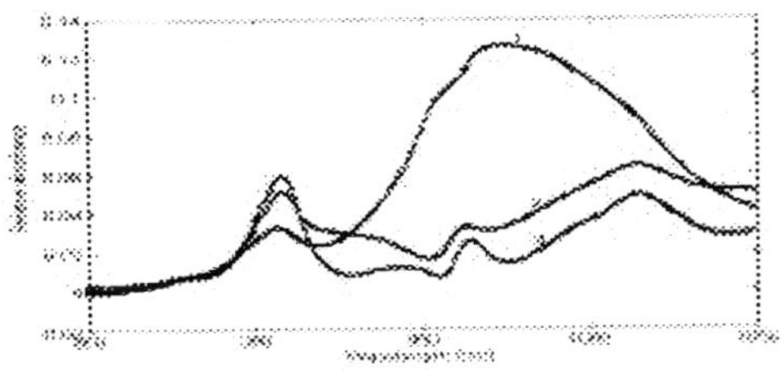

Fig.2. The response patterns of the three chemical species at 50 °C. Solid lines: resolved by direct regression; Dotted lines: estimated by ICS.

The previous results confirm the theoretical basis of ICS. The effectiveness of ICS for the correction of temperature induced spectral variations is illustrated through a typical set of results. Fig. 3 shows the 38 spectra for the 19 samples measured at 30°C and 50°C. Significant spectral differences introduced by temperature variations can be observed. When the ICS correction model built on the spectra of all the samples measured at training temperatures other than 50°C is applied, it can be observed that the temperature effects are effectively removed (Fig. 4).

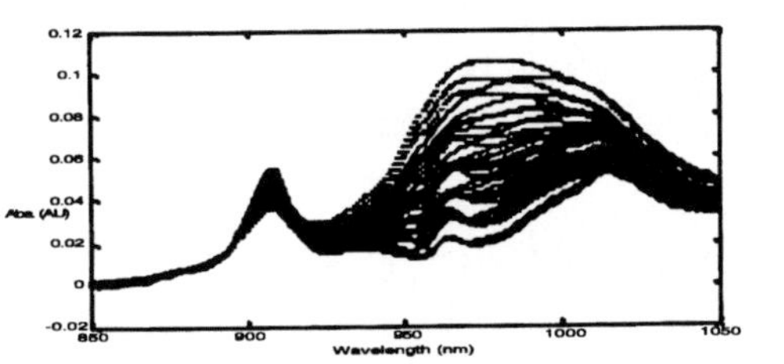

Fig.3. Spectra of all the 19 samples measured at 30°C (solid line) and 50°C (dotted line).

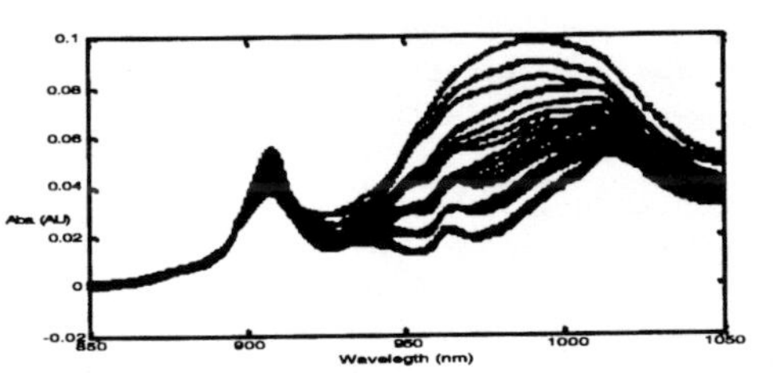

Fig.4. Spectra of all the 19 samples measured at 30°C (solid line) and those after corrections (dotted line) from 50°C to 30°C by ICS.

The predictive performance of the ICS correction model and those of CPDS are presented in Tables 2 and 3, respectively. The best results for the ICS and CPDS models for the test samples measured at 40°C, 50°C and 60°C are comparable to those for the local models. This demonstrates that both ICS and CPDS can remove, to some extent, the influence of the temperature-induced spectral variations on the predictions.

Table 2. RMSEP ($\times 10^{-2}$ Mole fraction) of test sets after application of ICS correction model

T_{pred} (°C)	Comp.	T_{cal} (°C)			
		40	50	60	70
30	C_1	1.2	1.3	1.2	1.3
	C_2	2.3	2.3	2.7	2.8
	C_3	2.4	2.9	3.1	3.7
		30	50	60	70
40	C_1	1.4	1.2	1.0	0.8
	C_2	0.7	0.6	0.8	1.0
	C_3	1.5	1.1	1.3	1.6
		30	40	60	70
50	C_1	2.1	0.9	1.5	1.6
	C_2	1.8	1.3	1.3	1.4
	C_3	1.7	1.4	2.3	2.9
		30	40	50	70
60	C_1	1.8	1.2	1.1	0.9
	C_2	0.9	0.7	0.6	0.9
	C_3	1.8	1.4	1.1	1.1
		30	40	50	60
70	C_1	3.6	2.0	2.0	1.5
	C_2	5.9	4.6	3.6	2.3
	C_3	4.2	4.2	3.2	2.6

Table 3. RMSEP ($\times 10^{-2}$ Mole fraction) of test sets after application of CPDS correction model

T_{pred} (°C)	Comp	T_{cal} (°C)			
		40	50	60	70
30	C_1	5.3	2.5	4.7	7.1
	C_2	2.6	1.2	1.1	1.2
	C_3	4.8	1.6	3.8	5.9
		30	50	60	70
40	C_1	1.2	1.3	1.5	1.7
	C_2	2.4	0.8	0.9	0.9
	C_3	2.5	1.6	1.8	2.1
		30	40	60	70
50	C_1	1.8	1.0	1.4	1.3
	C_2	1.8	1.4	1.0	0.7
	C_3	2.0	1.4	1.6	1.3
		30	40	50	70
60	C_1	1.7	1.1	1.0	3.4
	C_2	1.3	0.6	0.8	1.2
	C_3	1.4	1.2	0.7	4.0
		30	40	50	60
70	C_1	2.1	2.2	3.3	9.6
	C_2	3.3	2.6	1.7	3.6
	C_3	1.2	3.0	4.1	13

However the ICS and CPDS models demonstrate different abilities in terms of extrapolation. The average prediction errors of the CPDS models for the two extreme test temperatures, 30°C and 70°C, are significantly larger than those for ICS. In the CPDS models, with respect to different reference temperatures, variations in the prediction errors for water (C_2) for the test samples for the two extreme test temperatures are considerably smaller, compared with ethanol (C_1) and 2-propanol (C_3). This is a consequence of the higher absorption coefficients and larger temperature effects of water. The temperature effect of water dominates the variance in the spectra. CPDS tends to model the temperature effects of water. Hence when the CPDS models, with respect to different reference temperatures, are used for extrapolation, variations in the prediction errors of ethanol and 2-propanol will be larger than those for water. In contrast, ICS tries to model the temperature effects of each chemical species individually.

When ICS models, with respect to different reference temperatures, are used for extrapolation, the variations in the prediction errors of the three chemical species should be comparable. The results in Table 2 validate this hypothesis. Moreover, the results confirm that the calibration model built at the temperature closest to the test temperature provides better prediction results for the test samples after ICS correction, while CPDS does not possess such an advantage (Table 3).

5. CONCLUSION

It has been demonstrated that the influence of temperature-induced spectral variation on the predictive abilities of multivariate calibration models can effectively be removed through the proposed ICS approach. This is achieved by fitting polynomials for each element of the response patterns of the chemical species in mixtures against temperature. The estimated polynomials can then be used to predict the response patterns of the chemical species at the test temperature. The difference between the spectrum of the test sample measured at the test temperature and the corresponding spectrum theoretically measured at the reference temperature can then be calculated. After the temperature-induced spectral variations have been corrected, the calibration model can provide satisfactory predictions for test samples measured at temperatures other than the reference temperature.

Compared with CPDS, ICS has the following advantages: 1) Its implementation is straightforward. Unlike CPDS, only one parameter (the order of the polynomial) needs to be preset. In practice, the order of the polynomials can be set to a pre-defined vale of two. 2) ICS does not require the same training samples to be measured at different temperatures. 3) Since the calibration model built on the temperature closest to the test temperature can give more accurate predictions for test samples, it is straightforward to select the appropriate calibration model for the test samples.

A limiting factor in the application of ICS is that the system under study should be a white system (such as a crystallization process), i.e. there is no unknown chemical species in the system. This is not necessary for CPDS. Hence, CPDS is still the only chemometric method available for the correction of temperature fluctuations in grey systems, despite its limitations in terms of its application. Further research is focussing on standardization methods for the spectra of grey systems.

ACKNOWLEDGMENT

Dr Chen acknowledges the financial support of the EPSRC (GR/R19366/01 - KNOW-HOW and GR/R43853/01 – Chemicals Behaving Badly II).

REFERENCES

Blaser, W. W., R.A. Bredeweg, R.S. Harner, M.A. LaPack, A. Leugers, D.P. Martin, R.J. Pell, J. Workman Jr., and L.G. Wright (1995). Process analytical chemistry. *Anal. Chem.*, **67**, pp. 47R-70R.

Höskuldsson, A. (1988). PLS regression methods. *J. Chemom.*, **2**, pp. 211-228.

Iwata, T., J. Koshoubu, C. Jin and Y. Okubo (1997). Temperature dependence of the mid-Infrared OH spectral band in liquid water. *Appl. Spectrosc.*, **51**, pp. 1269-1275.

Lewiner, F., J.P. Klein, F. Puel and G. Févotte (2001). On-line ATR FTIR measurement of supersaturation during solution crystallization processes. calibration and applications on three solute/solvent. *Chemical Engineering Science*, **56**, pp. 2069-2084.

Swierenga, H., F. Wülfert, O.E. de Noord, A. P. Weijer, A.K. Smilde and L.M.C Buydens (2000). Development of robust calibration models in near infra-red spectrometric applications. *Anal. Chim. Acta*, **411**, pp. 121-135.

Wang, Y., D.J. Veltkamp and B.R. Kowalski (1991). Multivariate instrument standardization. *Anal. Chem.*, **63**, pp. 2750-2756.

Wang, Y. and B.R. Kowalski (1993). Temperature-compensating calibration transfer for near-infrared filter instruments. *Anal. Chem.*, **65**, pp. 1301-1303.

Wülfert, F., W. T. Kok, and A.K. Smilde (1998). Influence of temperature on vibrational spectra and consequences for the predictive ability of multivariate models. *Anal Chem.*, **70**, pp. 1761-1767.

Wülfert, F., W.T. Kok, O.E. de Noord and A.K. Smilde (2000a). Linear techniques to correct for temperature-induced spectral variation in multivariate calibration. *Chemom. Intell. Lab. Syst.*, **51**, pp. 189-200.

Wülfert, F., W.T. Kok, O.E. de Noord and A.K. Smilde (2000b). Correction of temperature-induced spectral variation by continuous piecewise direct standardization. *Anal. Chem.*, **72**, pp. 1639-1644.

ELSEVIER
IFAC
PUBLICATIONS
www.elsevier.com/locate/ifac

MONITORING OF DISTILLATION COLUMN OPERATION THROUGH SELF-ORGANIZING MAPS

Y.S. Ng and R. Srinivasan*

Laboratory for Intelligent Applications in Chemical Engineering,
Department of Chemical and Biomolecular Engineering,
National University of Singapore, 10 Kent Ridge Crescent, Singapore 119260
**Corresponding Author: Email: chergs@nus.edu.sg*

Abstract: Monitoring of chemical processes is becoming increasingly difficult as a result of growing complexity and larger scale of operation. In this paper, Kohonen self-organizing map (SOM) is used to monitor the operation of a lab-scale distillation column and to identify process states. SOM projects high-dimensional data to a lower two dimensional grid map while preserving the metric relations of the original data. The results from this paper show that using this property of SOM, process monitoring can be performed effectively through observing time series trajectory of process operations on SOM while fulfilling the objective of state identification at the same time. Occurrence of a fault will result in the deviation from the normal operating trajectory. Root cause identification can also be performed through simple visualization of component planes.

Keywords: monitoring, self-organizing systems, fault diagnosis, Neural-networks

1. INTRODUCTION

Incidents occurred in the past have shown the catastrophic consequences of improper monitoring of chemical processes; the resulting effect is not merely the loss of capital from shareholders but also injuries and loss of precious human lives. There is an increasing need for monitoring technique so that abnormal situations can be detected in the shortest amount of time. Over the past few decades, a lot of monitoring techniques have been introduced e.g. enhanced trend analysis, principal-component analysis, artificial neural-networks, etc; some have been implemented in industry successfully (Venkatasubramanian, *et al.*, 2003), but what is still lacking in the chemical plant is a powerful visualization technique that help plant personnel to visualize the evolution of a process and detect abnormal situations efficiently. These situations are worsening with the growing scale and complexity of chemical processes; at any instant, the number of variables that plant personnel need to track can easily range in the thousands. Due to the above mentioned reasons, monitoring through conventional control charts has become less effective and there is a need for new and effective methods to help plant personnel to monitor the progression of chemical processes.

In this paper, Self-Organizing Map (SOM) has been used for the purpose of monitoring of chemical processes. SOM is a powerful projection method which is capable of projecting high dimensional data onto a lower two dimensional grid while preserving the metric relationships of the original data. The effort to perform process monitoring is reduced many fold by SOM as compared to conventional methods since visualization is performed on a much lower two dimensional grid. The organization of this paper is as follows: A brief description of the underlying concepts behind SOM is presented in Section 2. Section 3 presents the review of SOM-based process analysis while Section 4 presents the method of using SOM for process monitoring and state identification. Monitoring and fault detection through SOM for a lab-scale distillation column is presented in Section 5.

2. THE SELF ORGANIZING-MAP

SOM was first proposed by Kohonen (Kohonen, 1981) and has been widely used since then. SOM algorithm employs nonparametric regression technique which involves the fitting of discrete, ordered reference vectors, to the distribution of input feature vectors. The dimensions of the prototype vectors are equal to the dimensions of input vectors. The hexagonal lattice type has been employed as it is more effective for visualization. Consider an input vector, $\vec{x}$, given by

$$\vec{x} = \{x_1, x_2, x_3, \ldots, x_n\} \quad (1)$$

$\vec{x}$ will be compared with all reference vectors, say $\vec{m}_i$,

$$\vec{m}_i = \{m_1, m_2, m_3, \ldots, m_n\} \quad (2)$$

Each $\vec{m}_i$ also represents a node on a hexagonal output grid. The map $\vec{m}_i$ which gives the smallest Euclidean distance with $\vec{x}$ is defined as the best-matching unit (BMU), represented here as c,

$$c = \arg\min_i \{\| \vec{x} - \vec{m}_i \|\} \quad (3)$$

During the training of the SOM, the reference weight vector $\vec{m}_i$ will be updated and the topological neighbors around the BMU are updated as well by moving them towards the training sample $\vec{x}$. The SOM learning rule is given by

$$\vec{m}_i(t+1) = \vec{m}_i(t) + \alpha(t)h_{ci}(t)[\vec{x}(t) - \vec{m}_i(t)] \quad (4)$$

where $h_{ci}(t)$ is the neighborhood function and $\alpha(t)$ is the learning rate factor. It is necessary that h_{ci} reduces to zero as the training proceeds to guarantee convergence. A large value of α is employed initially and usually decreases monotonically with time (Kohonen, 2000). During the training phase, SOM will fold onto the pattern formed by the training data and neighboring units are pulled nearer together because of the neighborhood relations. Thus neighboring units are in a sense, similar to each other. The SOM is usually visualized using the U-matrix (Ultsch, 1990), where distances of each map unit to its corresponding neighbors are calculated and displayed through gray or color-scales. A similar technique can also be applied to visualize the individual components (or variables) so that correlations among components can be investigated and abnormality can be detected at the variable level. The combinatory use of U-matrix and component planes can so fulfill the objectives of monitoring of chemical processes on both a plant wide and variable level.

3. SOM-BASED PROCESS ANALYSIS

SOM has been widely used as a visualization tool for unsupervised learning. (Vesanto, 2002). It has several powerful features which make it popular for the task of process monitoring. SOM implements an ordered dimensionality reduction through the mapping of the input feature vectors while preserving the most crucial topological and metric relationships of the original data, hence producing a similarity map of the input feature vectors The self-organizing maps is readily explainable, simple and easy to visualize (Vesanto, 1999). SOM has found successes in diverse fields of applications. Li Xiao, *et al.* (2003) used component plane presentation of SOM for microarray data analysis of yeast cells and human breast tumors, permitting the visualization of transcriptional changes of tumor sample at a genome scale. López-Rubio, *et al.* (2003) proposed a self-organizing neural model that performs principal components analysis. Kolehmainen, *et al.*(2003) used SOM and Sammon's mapping to identify the growth phases of brewer's yeast. In their work, the exhaust gas from the fermenter was measured with ion mobility spectrometry and a new growth phase for the fermentation process has been identified in addition to the four phases reported in the literature. While the previous applications of SOM are mainly on data clustering, some recent applications of SOM on process monitoring and fault diagnosis are presented in the next subsection.

Deventer, *et al.* (1996) demonstrated how disturbances in a plant for froth flotation of minerals can be visualized with SOM. They track the changes in operating conditions through an on-line computer system utilizing features extracted from froth images and visualize the degree of dispersion of the various input feature vectors through SOM. Chan, *et al.* (2001) presented a modified version of Kohonen SOM called constrained Kohonen networks (CKN) to overcome the problem of redundant sensors by constraining the weight vectors in the parity space. Srinivasan and Gopal (2002) showed how SOM can be used to extract operating information from operating data from a fluidized catalytic cracking unit. Jämsä-Jounela, *et al.* (2003) presented a SOM based fault diagnosis system for a smelter based on heuristic rules. SOM has been used to determine the coefficient for oxygen enrichment and detection of aggregations in various part of the plant after principal component analysis. Abonyi, *et al.*, (2003) applied SOM to a polyethylene process for product quality estimation. They developed multiple local linear model for the Philips polyethylene process through piecewise linear regression with SOM. More than 4000 scientific publication on SOM have been written to date (Kohonen, 2001) and the majority of them deal with the classification of input feature vectors and data mining. A comprehensive reference of SOM research has been compiled by Kaski, *et al.*, (1998). In this work, we propose a methodology for process monitoring and state identification through the use of SOM.

4. PROPOSED SOM-BASED MONITORING METHODOLOGY

The proposed methodology can be summarized as follows. Measurements of the normal operating process will first be collected and range normalized before being projected onto the U-matrix of SOM. The clusters formed on the U-matrix are interpreted and different states, which correspond to different operating phases is identified. Different clusters will be separated by large distances, modeled in the SOM's U-matrix by a border of darker color, predefined by the user. During the on-line monitoring phase, the trajectory being formed by the real time data will follow the trajectory of the normal operating data and significant deviation from the normal operating trajectory can be regarded as a fault. Duration of operation can be read from the size of the hexagon on the U-matrix, in which the size of the hexagon is directly proportional to the duration of operation. State identification of the process is done through the identification of the location of the projected operating data (data hit) on SOM. The proposed methodology is illustrated next through a case study.

5. MONITORING OF DISTILLATION-COLUMN OPERATION

The proposed method has been tested on an Armfield lab-scale distillation column. The schematic of the distillation column is shown in Figure 1. The distillation column is of 2 meters height and 20cm width and has 10 trays, where the feed enters at tray 4. The system is well integrated with a control console for controllers setting and data recording. Cold startup of the distillation column with ethanol-water at 30% v/v as bottoms is performed following the standard operating procedure (SOP) as shown in Table 1. The feed passes through a heat exchanger before being fed to the column. 19 variables --- all tray temperatures, reboiler and condenser temperature, reflux ratio, top and bottom column temperature, feed pump power, reboiler heat duty, cooling water inlet and outlet temperature - are measured at 10-second interval. The startup normally takes two hours and different faults such as sensor fault, failure to open pump, too high a reflux ratio etc., can be introduced at different states of operation.

The constructed SOM for a normal startup of distillation column is shown in Figure 2. The fully trained SOM consists of 30 x 8 map units and its corresponding bar-plane is shown in Figure 3. Each unit of the bar-plane is a representation of the magnitude of the 19 variables of the distillation column after normalization. Two observations can be seen from Figure 2. Firstly, the startup process has been observed to follow a trajectory on the U-matrix. Each label represents the corresponding operating time (x 10 seconds). Secondly, it has been observed that any significant deviation from the normal operating trajectory is an indication of a faulty operation.

By comparing Figure 2 and Figure 3, even without any prior knowledge on the process, one can deduce that during the normal run, most of the variables are increasing over time as represented by the trajectory on the U-matrix as indicated in Figure 2. Different clusters which correspond to different operating phases can be defined for each U-matrix or component plane constructed. As can be seen from the U-matrix on Figure 2, four clusters of operations can be identified for the startup of distillation column, each representing different phases of the start-up of distillation column. The border of each cluster is separated by map units of darker color

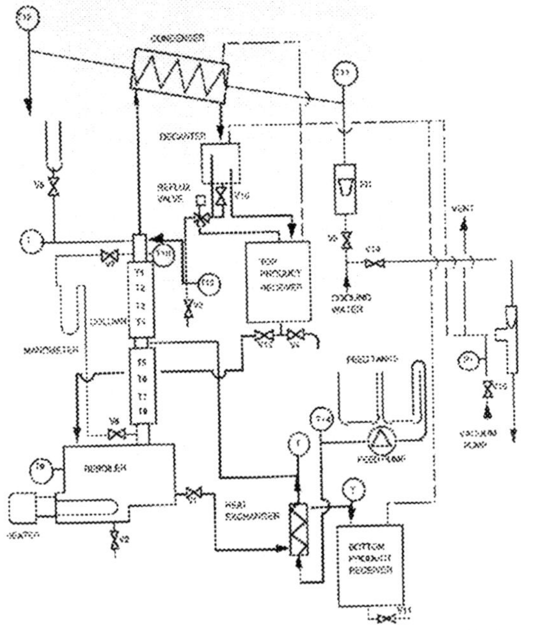

Figure 1. Schematic of the lab scale distillation column.

Table 1 Standard operating procedures (SOP) for the startup of distillation column

Distillation column startup SOP	
1.	Set all controllers to manual
2.	Fill reboiler with liquid bottom product
3.	Open reflux valve and operate the column on full reflux
4.	Establish cooling water flow to condenser
5.	Start the reboiler heating coil power
6.	Wait for all of the temperatures to stabilize
7.	Start feed pump
8.	Activate reflux control and set reflux ratio
9.	Open bottom valve to collect product
10.	Wait for all the temperatures to stabilize

which can be interpreted as a boundary marker between different clusters. The color code on the right tells the exact distance among neighboring units, which were trained using Gaussian neighborhood function during the batch training phase. Operating trajectory from the starting of the heating process till time 3630s can be regarded as a separate state from the rest of the data points – reboiler heating phase. Time series 3630s to 3820s is the boiling phase where evaporation of the bottom liquid takes place and there is an abrupt rise of temperature in the distillation tower. Time series 3840s to 4800s correspond to the full reflux heating phase before the feed was introduced and thus causing a switch in operating phase as shown from time trajectory from 4930s to 6340s, which is representing the steady state operating phase of the tower.

The use of SOM to monitor faulty operations is shown in Figure 4 and Figure 5. Scenario 1: Figure 4

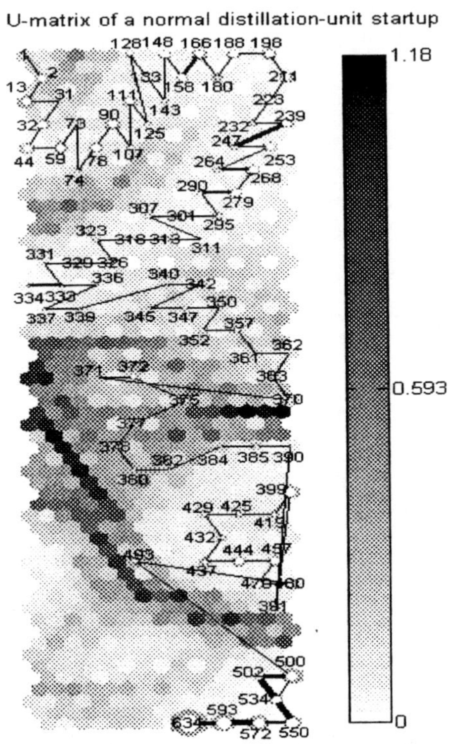

Figure 2. Trajectory of a normal run on U-matrix. Labels indicate the corresponding operating time (x10 seconds).

Figure 3. Bar-plane for the trained self-organizing maps.

shows the trajectory of a faulty run. Two trajectories can be observed from Figure 4, the light solid trajectory is the trajectory of a normal run whereas the dark solid line is the trajectory of the faulty run. The first fault occurred from 2650s (shown as 265 in Figure 4), where operator failed to turn on the feed pump and thus causing deviation from normal run. The situation is later rectified by the operator and the recovery can be seen on the U-matrix where the

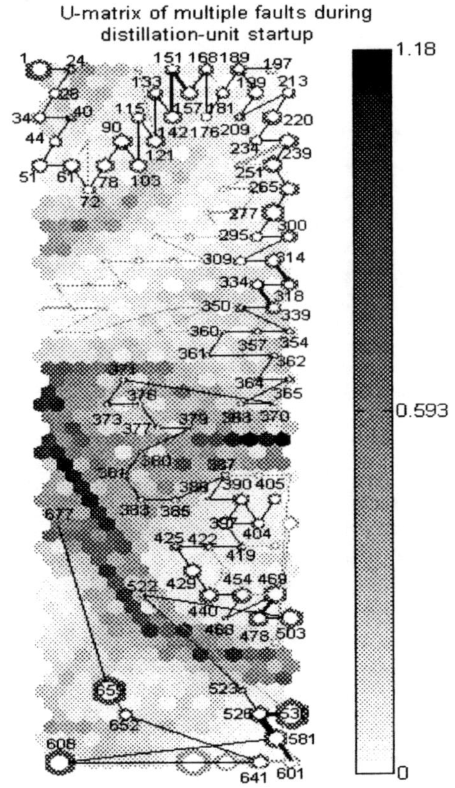

Figure 4. Trajectory of Scenario 1 (Dark solid line represents the faulty run while light solid line represents the normal run).

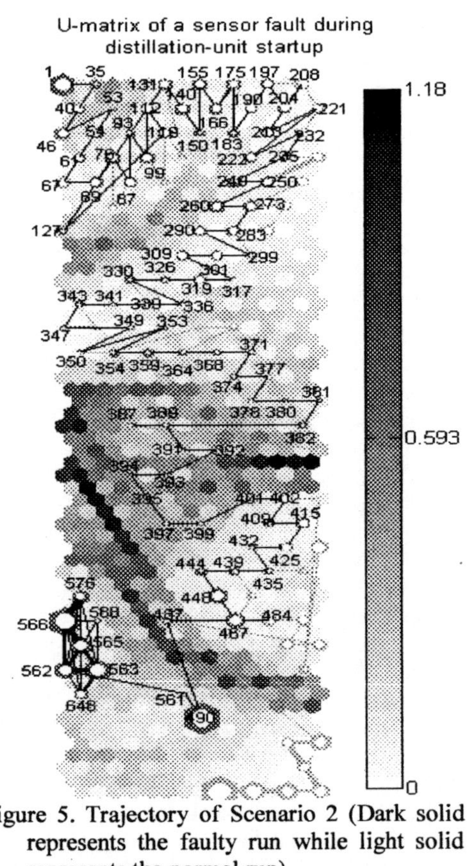

Figure 5. Trajectory of Scenario 2 (Dark solid line represents the faulty run while light solid line represents the normal run).

trajectory returns to follow the trajectory of a normal run at 3570s. A second fault occurred at time 6520s, when a feed flow of 1.5 times the normal value was introduced. The corresponding trajectory was found to deviate from normal operating trajectory from 6520s to 6770s. Secnario 2: The second scenario consists of high reflux ratio and temperature sensor fault as shown in Figure 5. At 4900s, the operator accidentally set the reflux ratio to 2 times higher as compared to the normal run and the column is operated in this condition for a long period of time (as seen from the size of the data hits), before a temperature sensor break down at 5600s.

While U-matrix is a simple 2-dimensional plot for operators to perform process monitoring, it does not contain sufficient information for diagnosing the fault. The component planes of the corresponding run have to be investigated in order to locate the fault. For the second scenario (Figure 5), variable 6 can be identified to be the main cause of deviation as it moves into a cluster of lower value (see Figure 6). The corresponding operating chart for variable 6 is given in Figure 7. Similar analysis can also be applied to other scenarios for fault identification (not shown in this paper).

The knowledge inferred from the normal operating run in Figure 2 can also be applied to other cases such as the one in Figure 4 and Figure 5. The operating state of a different run can be directly obtained from the location of the data point on the self-organizing maps. As an illustrative example, for the operation of distillation column shown in Figure 4, even though there are faults in the system, the process of state identification can still be done fairly successfully. Reboiler heating phase can be observed from time 0 – 3650s. Evaporation of bottoms has been observed to take place from 3680s – 3870s while full reflux operation prior to feed is from 3890s – 5220s, and lastly the steady state operation can be seen to take place from 5230s – 6770s. Similar interpretations can be applied to other case studies as well.

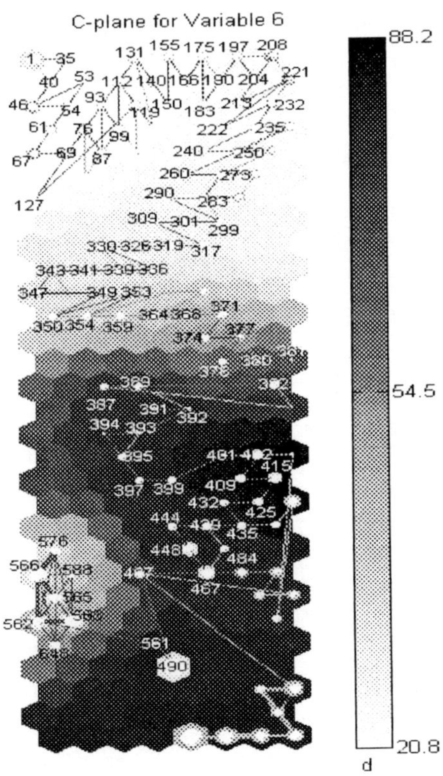

Figure 6. Component plane of variable 6 (5th tray temperature sensor).

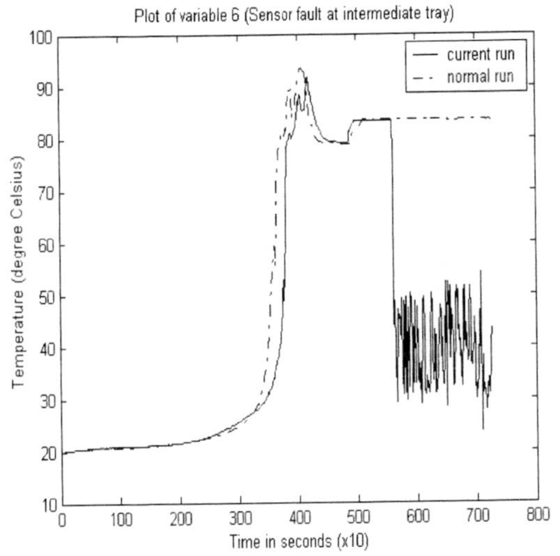

Figure 7. Time variations of variable 6 (5th tray temperature sensor).

6. CONCLUSIONS AND DISCUSSION

This paper illustrates how Kohonen SOM can be used for monitoring of process dynamic transitions, in this case the operation of a distillation column. High dimensional data has been projected onto a lower two dimensional grid maps, consequently the task of process monitoring has been simplified as compared to conventional control charts approach. One only needs to keep track of the trajectory in order to deduce whether the operation is normal or abnormal. A faulty run will result in deviation from normal trajectory, and component planes can then be analyzed to locate faulty variables for root cause identification. This paper has also shown that SOM can be used for state-identification. The exact phase of different run at any instant can be inferred from the location of the data hit on the self-organizing maps. Our current research is targeted at automating the SOM-based approach to improve the efficiency of monitoring. Two methods that use the results from SOM for automated abnormality detection and diagnosis have been formulated. Firstly, statistical analysis has been used to generate variable specific residual by using SOM as local model during the transition. Secondly, a pattern recognition approach is being explored for diagnosis and fault isolation purposes. By using a fault database and machine learning approaches, faults can be identified in a short amount of time. A robust formulation of normal trajectories is also being investigated to accommodate multiple normal operating conditions.

REFERENCES

Abonyi, J., Nemeth, S., Vincze., Arva, P., (2003). Process analysis and product quality estimation by self-organizing maps with an application to polyethylene production. *Computers in Industry 52.* 221-234.

Chan, C.W., Jin, H., Cheung, K.C., Zhang, H.Y., (2001). Fault detection of systems with redundant sensors using constrained Kohonen networks. *Auomatica 37.* 1671-1676.

Deventer, J.S.J.V., Moolman, D.W., Aldrich, C., (1996). Visualization of plant disturbances using self-organizing maps, *Computers Chem. Engng Vol.20*, pp.S1095-S1100.

Jämsä-Jounela, S.L., Vermasvuori, M., Endén, P., Haavisto,S., (2003). A process monitoring system based on the Kohonen self-organizing maps. *Control Engineering Practice 11.* 83-92.

Kaski, S., Kangas, J., Kohonen, T., (1998). Bibliography of self-organizing map (SOM) papers: 1981-1997, *Neural Computing Surveys*, 1:102-350.

Kohonen, T., (1981). Automatic formation of topological maps of patterns in a self-organizing system, Proc. 2SCIA, Scand. *Conf. on Image Aanlysis*, pp 214-220, Helsinki, Finland.

Kohonen, T., (2000). *Self-Organizing Maps, Springer Series in Information Sciences*, Springer, Berlin, Germany.

Kohonen, T., (2001). Seiffert, U., Jain, L. C., Editors, *Self-Organizing Neural Networks: Recent Advances and Applications*, Phisica-Verlag Heidelberg, Springer-Verlag, Germany.

Kolehmainen, M., Rönkkö, P., Raatikainen, O., (2003). Monitoring of yeast fermentation by ion mobility spectrometry measurement and data visualization with self-organizing maps. *Analytica Chimica Acta 484.* 93-100.

López-Rubio,E., Muñoz-Pérez, J., Gómez-Ruiz, J.A., (2003). A principle component analysis self-organizing map. *Neural Networks*. In-press.

Srinivasan, R., Gopal, S., (2002). Extracting information from high-dimensional operations data using visualization techniques, *AIChe meeting*, Indianapolis, #271c.

Ultsch,A., Siemon, H.P., (1990). Kohonen's self organizing feature maps for exploratory data analysis, *Proceedings of International Neural Network Conference (INNC'90)*, Kluwer academic Publishers, Dordrecht, pp. 305-308.

Venkatasubramanian, V., Rengaswamy, R., Yin, K. & Kavuri, S.N., (2003). A review of fault detection and diagnosis Part I: Quantitative model-based methods, *Computers and Chemical Engineering 27*, 293-311.

Vesanto, J., (1999). SOM-based data visualization methods, *Intelligent Data Analysis 3*, 111-126.

Vesanto, J., (2002). *Data Exploration Process Based on the Self-Organizing Map, phD thesis*, Helsinki University of Technology, Acta Polytechnica Scandincavica: Mathematics and Computing Series No. 115, Department of Computer Science and Engineering, FIN-02015 HUT, Finland.

Vesanto, J., Himberg., Alhoniemi, E., Parhankangas, J., (2000). *SOM Toolbox for Matlab 5.* SOM toolbox team, Helsinki University of Technology, Finland.

Xiao, L., Wang, K., Teng, Y., Zhang, J., (2003). Component plane presentation integrated self-organizing map for microarray data analysis. *FEBS Letters 538.* 117-124.

ELSEVIER
IFAC
PUBLICATIONS
www.elsevier.com/locate/ifac

DATA-DRIVEN QUALITY IMPROVEMENT: HANDLING QUALITATIVE VARIABLES

Manabu Kano * **Koichi Fujiwara** * **Shinji Hasebe** *
Hiromu Ohno **

* *Kyoto University, Kyoto 606-8501, Japan*
** *Kobe University, Kobe 657-0013, Japan*

Abstract: A data-based methodology for improving product quality is proposed. Referred to as Data-Driven Quality Improvement (DDQI), the proposed method can cope with qualitative as well as quantitative variables, determine the operating conditions that can achieve the desired product quality, optimize operating condition under constraints, and also evaluate the validity of the results. The desired yield is specified instead of the quality for a qualitative quality variable. This paper aims to formulate DDQI and demonstrate its usefulness with an illustrative example. In addition, possible extensions and remaining problems are discussed based on the authors' experience of suceeding in improving product quality by applying DDQI to several industrial processes. *Copyright ©2004 IFAC*

Keywords: Quality Improvement, Statistical Quality Control, Statistical Process Control, Principal Component Regression, Optimization

1. INTRODUCTION

How can we improve product quality and yield? More than ever, the answer to this question is vital as product life cycles are getting shorter and international competition is getting keener. Since this question arises repeatedly when a new product is developed, quality improvement should be achieved faster and in a more systematic way.

In the present work, a data-based methodology for improving product quality is proposed. The proposed method, referred to as Data-Driven Quality Improvement (DDQI), is based on a statistical model. DDQI can cope with qualitative as well as quantitative variables, determine the operating conditions that can achieve the desired product quality, optimize operating condition under constraints, and also evaluate the validity of the results. This paper aims to formulate DDQI, show its usefulness via a case study, and also point out possible extensions and remaining problems.

The most important contribution of this work is to provide a new method that can handle qualitative quality variables. By building a model that can relate operating condition to yield, the desired product quality can be specified, the operating condition can be optimized, and also the achievable quality at a certain operating condition can be estimated, even if the product quality is not measured quantitatively.

2. FORMULATION OF DATA-DRIVEN QUALITY IMPROVEMENT

In this section, Data-Driven Quality Improvement (DDQI) is formulated. Jaeckle and MacGregor (1998) proposed a product design method based on linear/nonlinear multivariate analysis. Their method can derive the operating conditions that can achieve the desired product quality, but it does not account for qualitative variables. DDQI can handle qualitative as well as quantitative variables in a unified framework. This characteristic expands the application area and the usefulness of DDQI.

2.1 Preprocessing Data

A quality data matrix $\boldsymbol{Y} \in \Re^{N\times Q}$ and an operating condition data matrix $\boldsymbol{X} \in \Re^{N\times P}$ are observed. N, Q, and P are the numbers of samples, quality variables, and operating condition variables, respectively. The nth measurements of the qth quality variable and the pth operating condition variable are denoted by y_{nq} and x_{np}, respectively. For simplicity, it is assumed that quality data $\boldsymbol{y}_q \in \Re^{N\times 1}$ and operating condition data $\boldsymbol{x}_p \in \Re^{N\times 1}$ are mean-centered. In addition, they are appropriately scaled if necessary.

If quality is given as qualitative information, such as good and bad, the qualitative variable should be quantified to build a model. For example, good and bad can be quantified and denoted by 1 and 0, respectively.

$$y_{nq} = \begin{cases} 1, & \text{for } y_{nq} \in C_1 \\ 0, & \text{for } y_{nq} \in C_2 \end{cases} \tag{1}$$

where C_1 and C_2 are classes consisting of good and bad samples, respectively. A quality variable can be quantified in a similar way even if there are more than two classes. This quantified variable is then mean-centered. This quantification is standard and widely used.

On the other hand, if an operator needs to decide whether to use particular equipment or to decide which equipment to use, such operating conditions should be quantified. For example, when one of K pieces of equipment is used for production, the information on which equipment is used can be quantified by introducing K 0-1 variables. The kth 0-1 variable e_{nk} is defined as:

$$e_{nk} = \begin{cases} 1, & \text{when } k\text{th equipment is used.} \\ 0, & \text{when } k\text{th equipment is not used.} \end{cases} \tag{2}$$

By simple extension, combinations of equipment to use can be quantified in a similar way. It should be noted here that only $K-1$ variables are necessary for distinguishing K operating status because the Kth equipment must be used if the first to the K–1th equipment is not used. However, if only $K-1$ variables are used, the Kth variable does not have any coefficient in the quality model developed, and thus it becomes difficult to understand the influence of the Kth equipment on the quality variables and the relationship between the Kth equipment and other operating conditions. In DDQI, therefore, K 0-1 variables are introduced for distinguishing K operating status and then they are mean-centered.

2.2 Modeling Quality and Operating Conditions

DDQI is based on a statistical quality model that relates operating conditions with quality. Multiple regression analysis (MRA) is the simplest method for building a quality model, but it cannot be used if a colinearity problem occurs. Principal component regression (PCR) can cope with colinearity, and it includes MRA as its special case. That is, PCR is more general than MRA. On the other hand, linear discriminant analysis (LDA) can be used for classification if a quality variable is qualitative, but it cannot cope with colinearity. The colinearity problem can be solved by applying principal component analysis (PCA) to reduce the dimensionality before LDA is used. This method is referred to as PCA-LDA. Since LDA for two classes is the same as MRA when qualitative variables are quantified, PCA-LDA is essentially the same as PCR. Therefore, the formulation of DDQI is based on PCR.

The singular value decomposition of an operating condition data matrix $\boldsymbol{X}$ is written as

$$\begin{aligned} \boldsymbol{X} &= \boldsymbol{U}\boldsymbol{S}\boldsymbol{V}^T \\ &= \left[\boldsymbol{U}_R\ \boldsymbol{U}_0\right] \begin{bmatrix} \boldsymbol{S}_R & \boldsymbol{0} \\ \boldsymbol{0} & \boldsymbol{S}_0 \end{bmatrix} \left[\boldsymbol{V}_R\ \boldsymbol{V}_0\right]^T \end{aligned} \tag{3}$$

where $\boldsymbol{U}$ and $\boldsymbol{V}$ are orthogonal matrices. R denotes the number of principal components to be retained in a PCA model. The diagonal matrix $\boldsymbol{S}$ has singular values s_r in its diagonal elements in decreasing order. The rth principal component is given as the rth column $\boldsymbol{v}_r$ of the loading matrix $\boldsymbol{V}_R$, and the rth score $\boldsymbol{t}_r$ is given by

$$\boldsymbol{t}_r = \boldsymbol{X}\boldsymbol{v}_r = s_r\boldsymbol{u}_r\,. \tag{4}$$

The score matrix is given by

$$\boldsymbol{T}_R = \boldsymbol{X}\boldsymbol{V}_R = \boldsymbol{U}_R\boldsymbol{S}_R\,. \tag{5}$$

The score $\boldsymbol{t}_r$ are uncorrelated to each other. In fact, the covariance matrix of $\boldsymbol{T}_R$ becomes diagonal.

$$\frac{1}{N-1}\boldsymbol{T}_R^T\boldsymbol{T}_R = \frac{1}{N-1}\boldsymbol{S}_R^2 \tag{6}$$

The dimensionality of $\boldsymbol{X}$ can be reduced from P to R via PCA, and $\boldsymbol{T}_R$ is obtained. Then, $\boldsymbol{X}$ can be reconstructed as $\hat{\boldsymbol{X}}$ by projecting $\boldsymbol{T}_R$ onto the original P dimensional space.

$$\hat{\boldsymbol{X}} = \boldsymbol{T}_R\boldsymbol{V}_R^T = \boldsymbol{X}\boldsymbol{V}_R\boldsymbol{V}_R^T \tag{7}$$

The size of $\hat{X}$ is the same as that of X, but the rank of $\hat{X}$ is only R.

In PCR, the scores are used as input variables of MRA. The colinearity problem does not occur any more, because the scores are uncorrelated with each other. The PCR model can be written as

$$Y = T_R K + E \tag{8}$$

where K is a regression coefficient matrix and E is an error matrix. By least squares, K is determined:

$$K = (T_R^T T_R)^{-1} T_R^T Y \quad . \tag{9}$$

The prediction of quality can be given by

$$\hat{Y} = T_R K = X V_R K = X K_{PCR} \tag{10}$$

where

$$K_{PCR} \equiv V_R K . \tag{11}$$

The coefficient matrix K_{PCR} shows the influence of operating conditions on quality. Basically, the operating condition variable with the larger coefficient has greater influence on the quality. It should be noted here, however, that the estimated coefficients are biased due to correlation among input variables..

2.3 Searching Operating Conditions That Can Achieve Desired Product Quality

A method to determine the operating conditions that can achieve desired product quality is described here. The method solves an inverse problem of the PCR model. In general, the number of quality variables Q is less than that of principal components R, and thus, the operating condition cannot be determined uniquely. In DDQI, a space where operating conditions can achieve the desired quality is searched within R dimensional space spanned by principal components.

Desired product quality cannot be specified when the quality variable is qualitative. For the qualitative quality variable, the yield, i.e., the percentage of good products to all products, or a similar quantitative measure can be specified instead of the quality itself on the basis of a histogram for each category.

Based on Eq. (10), the operating conditions $\tilde{x} \in \Re^{P\times1}$ that can achieve the desired quality $\tilde{y} \in \Re^{Q\times1}$ are determined. The operating conditions must exist in the space spanned by R principal components, because correlation among operating condition variables, which is theoretically defined by physical or chemical laws, is extracted from operation data by using PCA in DDQI. Therefore, at the first step to solving the inverse problem, the scores $\tilde{t} \in \Re^{R\times1}$ that can achieve the desired quality $\tilde{y}$ are determined. The scores $\tilde{t}$ are related to the desired quality $\tilde{y}$ via

$$\tilde{y} = K^T \tilde{t} . \tag{12}$$

If $\tilde{t}$ is found, then the operating conditions $\tilde{x}$ can be determined.

$$\tilde{x} = V_R \tilde{t} \tag{13}$$

When the number of quality variables Q is larger than that of principal components R, there is no score $\tilde{t}$ that can achieve the desired quality. In such a case, the score that achieves as desired quality as possible should be determined.

$$\tilde{t} = (KK^T)^{-1} K \tilde{y} \tag{14}$$

From Eq. (13), the operating condition recommended for quality improvement is

$$\tilde{x} = V_R (KK^T)^{-1} K \tilde{y} . \tag{15}$$

On the other hand, there is a unique score vector $\tilde{t}$ that can achieve the desired quality when $Q = R$. In this case, the operating condition recommended for quality improvement is also determined by Eq. (15) or

$$\tilde{x} = V_R (K^T)^{-1} \tilde{y} . \tag{16}$$

In many cases, however, Q is less than R. Therefore, the scores $\tilde{t}$ that can achieve the desired quality cannot be determined uniquely. In such a case, a solution set $\mathcal{S}_{\tilde{t}}$ where any score can achieve the desired quality is searched.

In general, a column space $\mathcal{S}_c(A)$, a row space $\mathcal{S}_r(A)$, and a null space (kernel) $\mathcal{S}_0(A)$ of a matrix A have the following relationships:

$$\mathcal{S}_r(A) = \mathcal{S}_0(A)^{\perp} \tag{17}$$

$$\mathcal{S}_r(A) = \mathcal{S}_c(A^T) \tag{18}$$

where $\perp$ denotes orthogonal complement. Therefore, any element $\tilde{t}$ in the solution set $\mathcal{S}_{\tilde{t}}$ is uniquely decomposed into the orthogonal projection, $\tilde{t}_{ps}$, of K^T onto its row space $\mathcal{S}_r(K^T)$ and the other orthogonal projection, $\tilde{t}_0$, of K^T onto its null space $\mathcal{S}_0(K^T)$.

$$\tilde{t} = \tilde{t}_{ps} + \tilde{t}_0 \tag{19}$$

Since $\tilde{t}_0$ is the orthogonal projection onto the null space $\mathcal{S}_0(K^T)$,

$$\boldsymbol{K}^T\tilde{\boldsymbol{t}}_0 = \boldsymbol{0}. \quad (20)$$

Substituting these equations for Eq. (12) gives

$$\tilde{\boldsymbol{y}} = \boldsymbol{K}^T(\tilde{\boldsymbol{t}}_{ps} + \tilde{\boldsymbol{t}}_0) = \boldsymbol{K}^T\tilde{\boldsymbol{t}}_{ps}. \quad (21)$$

That is, $\tilde{\boldsymbol{t}}_{ps}$ is a particular solution of the inverse problem, and it can be determined by using the minimum norm generalized inverse $(\boldsymbol{K}^T)^+$ of $\boldsymbol{K}^T$.

$$\tilde{\boldsymbol{t}}_{ps} = (\boldsymbol{K}^T)^+\tilde{\boldsymbol{y}} \quad (22)$$

Since a general solution $\tilde{\boldsymbol{t}}$ is given by Eq. (19), it is necessary to determine $\tilde{\boldsymbol{t}}_0$ that is an element of the null space $\mathcal{S}_0(\boldsymbol{K}^T)$. By defining a matrix $\tilde{\boldsymbol{T}}_0 \in \Re^{R\times(R-Q)}$ which has all bases of $\mathcal{S}_0(\boldsymbol{K}^T)$ as its columns, $\tilde{\boldsymbol{t}}_0$ can be described as

$$\tilde{\boldsymbol{t}}_0 = \tilde{\boldsymbol{T}}_0\boldsymbol{\phi} \quad (23)$$

by using any coefficient vector $\boldsymbol{\phi} \in \Re^{(R-Q)\times 1}$. Therefore,

$$\tilde{\boldsymbol{t}} = (\boldsymbol{K}^T)^+\tilde{\boldsymbol{y}} + \tilde{\boldsymbol{T}}_0\boldsymbol{\phi} \quad . \quad (24)$$

Finally the operating conditions that can achieve the desired quality $\tilde{\boldsymbol{y}}$ are given by

$$\tilde{\boldsymbol{x}} = \boldsymbol{V}_R\left\{(\boldsymbol{K}^T)^+\tilde{\boldsymbol{y}} + \tilde{\boldsymbol{T}}_0\boldsymbol{\phi}\right\}. \quad (25)$$

Any general solution $\tilde{\boldsymbol{x}}$ is an element of the column space of the loading matrix $\boldsymbol{V}_R$, and thus it is clear that $\tilde{\boldsymbol{x}}$ exists in the space spanned by R principal components.

Unfortunately, desired product quality cannot be specified when the quality variable is qualitative. Even in such a case, the quality variable can be quantified as described before, and the quality model can be built via PCR.

$$\hat{\boldsymbol{y}} = \boldsymbol{X}\boldsymbol{k}_{PCR} \quad (26)$$

For the qualitative quality variable, the yield, i.e., the percentage of good products to all products, or a similar quantitative measure can be specified instead of the quality itself on the basis of a histogram for each category. The histograms can be obtained from operation data, and they can be drawn against the axis defined by $\boldsymbol{k}_{PCR}$. For example, histograms of both good products and bad products can be calculated and drawn against the axis defined by $\boldsymbol{k}_{PCR}$, and then the desired yield can be specified. Once the desired yield is specified, operating conditions that can achieve the desired yield can be found by following the above-mentioned approach. An illustrative example is shown in the next section.

The searching algorithm cannot cope with 0-1 variables e_k, which is used for quantifying operating status when an operator selects particular equipment to use. To derive a feasible operating condition, 0-1 variables should be determined in advance. Then, operating conditions that can achieve the desired product quality can be calculated. It should be noted here that only a few combinations should be checked because the quality model helps one to judge which equipment is to be used.

2.4 Optimizing Operating Condition

The operating conditions that can achieve the desired product quality cannot be uniquely determined in general, but it can be optimized if an objective function is provided.

An objective function is assumed to be provided as a quadratic function. The objective function is optimized under the following three constraints: 1) the desired product quality is achieved, 2) the operating condition exists in the space spanned by principal components, and 3) all operating condition variables exist within their upper and lower bounds. Since the objective function is quadratic and the constraints are linear, the optimization problem can be solved by quadratic programming.

The operating conditions $\tilde{\boldsymbol{x}}$ that satisfy the constraints 1 and 2 are given by Eq. (25) as a function of the coefficient vector $\boldsymbol{\phi}$. Therefore, the first two constraints are automatically satisfied if Eq. (25) is substituted for the objective function $J(\tilde{\boldsymbol{x}})$. As a result, the objective function $J'(\boldsymbol{\phi})$ needs to be optimized under the third constraint. The upper and lower bound constraints on $\tilde{\boldsymbol{x}}$ can be transformed into the linear constraints on $\boldsymbol{\phi}$ via Eq. (25). Therefore, the optimization problem can be formulated as follows:

$$\max_{\boldsymbol{\phi}} J'(\boldsymbol{\phi}) \quad (27)$$

$$\text{s.t. } \boldsymbol{V}_R\tilde{\boldsymbol{T}}_0\boldsymbol{\phi} \geq \boldsymbol{x}_l - \boldsymbol{V}_R(\boldsymbol{K}^T)^+\tilde{\boldsymbol{y}} \quad (28)$$

$$\boldsymbol{V}_R\tilde{\boldsymbol{T}}_0\boldsymbol{\phi} \leq \boldsymbol{x}_u - \boldsymbol{V}_R(\boldsymbol{K}^T)^+\tilde{\boldsymbol{y}} \quad (29)$$

where $\boldsymbol{x}_u$ and $\boldsymbol{x}_l$ are the upper and lower constraints on $\tilde{\boldsymbol{x}}$. For example, the objective function is given as

$$J(\tilde{\boldsymbol{x}}) = \|\tilde{\boldsymbol{x}} - \boldsymbol{x}_0\|^2_{\boldsymbol{W}} + \tilde{\boldsymbol{x}}^T\boldsymbol{\theta} \quad (30)$$

where $\boldsymbol{x}_0$, $\boldsymbol{W}$, and $\boldsymbol{\theta}$ denote the reference operating condition, a weighting matrix, and a coefficient vector describing the operation cost.

If there is no solution that satisfies all constraints, i.e., the imposed specifications on quality are too severe, the operating condition that achieves as desired quality as possible should be determined under the second and the third constraints. This optimization problem can be solved by quadratic programming if its objective function is defined as a weighted squared norm of the difference between the desired quality and the achieved quality. Or, the problem can be solved by linear programming if its objective function is defined as a linear function of the achieved quality. The relationship between the quality and the scores is modeled by Eq. (12). Therefore, the second constraint is automatically satisfied if Eq. (12) is substituted for the objective function $J_{sub}(\boldsymbol{y})$. As a result, the objective function $J'_{sub}(\boldsymbol{t})$ needs to be optimized under the third constraint. The constraints on $\boldsymbol{x}$ can be transformed into the linear constraints on $\boldsymbol{t}$ via Eq. (13). Therefore, the optimization problem can be formulated as follows:

$$\max_{\boldsymbol{t}} \; J'_{sub}(\boldsymbol{t}) \tag{31}$$

$$\text{s.t.} \; \boldsymbol{V}_R \boldsymbol{t} \geq \boldsymbol{x}_l \tag{32}$$

$$\boldsymbol{V}_R \boldsymbol{t} \leq \boldsymbol{x}_u \, . \tag{33}$$

2.5 Evaluating Validity of Quality Model

DDQI provides a unified framework for improving product quality. An operating condition can be optimized so that the desired quality is achieved under constraints. Since DDQI is based on PCR, the results are valid inside the region where the PCR model can give a close approximation to the real process. The validity will deteriorate outside the operating conditions where process data used for modelling are obtained.

To check the validity of the optimization results, that is, to check the validity of the PCR model, SPC is integrated with DDQI. In PCA-based SPC, two statistics, T^2 and Q, are monitored (Jackson and Mudholkar, 1979).

$$T^2 = \sum_{r=1}^{R} \frac{t_r^2}{\sigma_{t_r}^2} \tag{34}$$

$$Q = \sum_{p=1}^{P} (x_p - \hat{x}_p)^2 \tag{35}$$

where t_r is the rth score and $\sigma_{t_r}^2$ is the variance of t_r. x_p and $\hat{x}_p$ are a measurement of the pth variable and its predicted (reconstructed) value, respectively. The T^2 statistic is a measure of the variation within the PCA model, and the Q statistic is a measure of the amount of variation not captured by the PCA model. The optimization results as well as the PCR model are judged to be valid when both statistics are below the predetermined thresholds.

Furthermore, the operating condition can be optimized within the space where the quality model is valid by imposing the constraints on scores

$$\boldsymbol{t}_l \leq \boldsymbol{t} \leq \boldsymbol{t}_u \, . \tag{36}$$

By using constraints on scores instead of T^2 statistic, the optimization problem can be solved by quadratic programming.

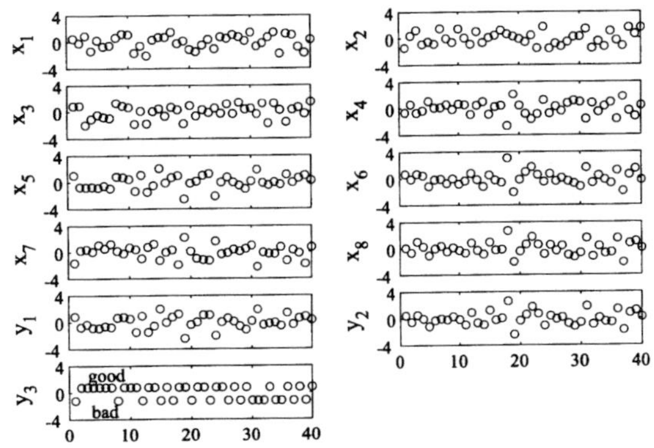

Fig. 1. Operating condition and quality data

3. NUMERICAL EXAMPLE

In this section, a case study shows how DDQI works, in particular, how it handles a qualitative quality variable. Simulation data of eight operating condition variables $\boldsymbol{x}$ and three quality variables $\boldsymbol{y}$ are generated. In this case study, $\boldsymbol{x}$ is linear combinations of random variables. The quality variable y_3 is qualitative, the judgment of which is given as good or bad, while y_1 and y_2 are quantitative variables. All variables are normalized and shown in Fig. 1.

PCR is used for modeling, and four principal components are retained. The operating condition $\tilde{\boldsymbol{x}}$ that can achieve the desired product quality $\tilde{\boldsymbol{y}}$ must exist in one dimensional space because there are three quality variables ($Q = 3$) and four principal components ($R = 4$). In this case study, four optimization problems are solved. The objective function is given as Eq. (30), where

$$\boldsymbol{x}_0 = \boldsymbol{0}, \; \boldsymbol{W} = \boldsymbol{I}, \; \boldsymbol{\theta} = [5\,1\,4\,0\,0\,0\,0\,0]^T \quad . \tag{37}$$

The desired product quality and the constraints are summarized in Table 1. The desired quality of y_3 cannot be specified because y_3 is a qualitative variable. Therefore, the desired yield is specified by using the expected yield function, which is determined on the basis of histograms of good products and bad products as shown in Fig. 2. The horizontal axis y_3 represents the dummy quality variable estimated via Eq. (26). The optimal operating conditions are summarized in Table 1 and Fig. 2 (right-bottom). Each solid line in Fig. 2 shows the projection of the solution set

Table 1. Constraints and optimal operating conditions

	y_1	y_2	y_3	x_1	x_2	x_3	x_4	x_5	x_6	x_7	x_8
upper bound				2.0	2.0	2.0	3.0	3.0	4.0	3.0	3.0
lower bound				-4.0	-2.0	-2.0	-3.0	-3.0	-3.0	-3.0	-3.0
optimization 1	1.0	0.5	70%	-1.59	-1.78	-1.03	-0.09	-0.77	0.22	-0.36	-0.35
optimization 2	1.5	0.7	75%	-2.68	-1.89	-1.16	0.02	-1.15	0.30	-0.37	-0.52
optimization 3	2.0	0.9	80%	-3.77	-2.00	-1.27	0.12	-1.53	0.38	-0.38	-0.69
optimization 4	3.0	2.0	100%	The desired quality cannot be achieved							
(achieved)	2.5	2.4	95%	2.00	2.00	0.42	-1.84	2.08	1.52	-1.12	2.29

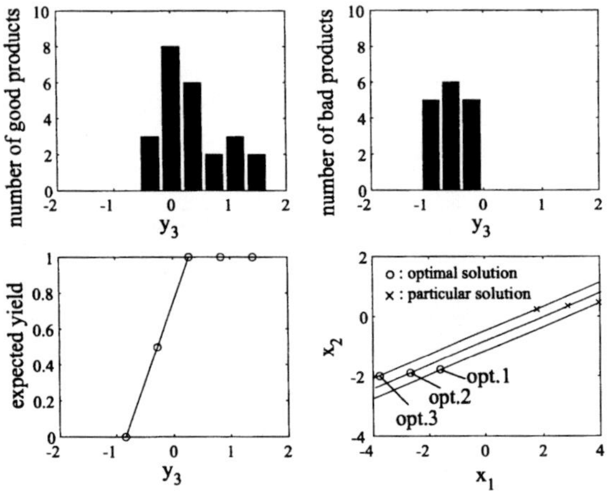

Fig. 2. Histogram of good products (left-top), histogram of bad products (right-top), expected yield (left-bottom), and derived operating conditions (right-bottom)

onto $x_1 - x_2$ space. The derived optimal operating conditions exist inside the space spanned by the principal components and limited by the constraints, and they can achieve the desired product quality. In the last case (optimization 4), however, the desired quality cannot be achieved under constraints. Therefore, operating condition is optimized to realize as desired quality as possible.

4. EXTENSIONS AND PROBLEMS

In this section, possible extensions and remaining problems are discussed briefly.

Since DDQI formulated here is based on a linear model, its applicability to nonlinear processes is limited. A simple extension can be made by using a nonlinear model. However, a serious problem to consider is the limited number of samples as pointed out by Jaeckle and MacGregor (1998). In many processes, where product life cycles are short, product quality needs to be improved by using a small number of samples. In such a case, nonlinear modeling is not practical and should be avoided. From the authors' experience of succeeding in improving yield and product quality by applying DDQI to the steel and the semiconductor industries, linear models are sufficiently practical and useful for real problems. In practice, an operating condition cannot be changed drastically at a time, it should be changed carefully and step by step. Therefore, a linear model is useful.

In practice, only a few manipulated variables can be changed simultaneously. It would be usually unacceptable to change all variables. Therefore, it is necessary to derive a new operating condition that can achieve the desired product quality by changing only a few operating condition variables. In DDQI, new operating conditions must exist in the space spanned by principal components. This constraint might need to be relaxed for searching the optimal operating condition in which only a few variables need to be changed. For this purpose, constraints on T^2 and Q may be used.

To date, DDQI has been tested in the steel and the semiconductor industries, and it has succeeded in finding new operating conditions to achieve higher product quality. However, model building and validation are time-consuming and require expertise. To realize an online DDQI system as automatic process control, an efficient and robust procedure needs to be developed.

5. CONCLUSIONS

In the present work, a data-based methodology for improving product quality – Data-Driven Quality Improvement (DDQI) – is proposed. DDQI can cope with qualitative as well as quantitative variables, determine the operating conditions that can achieve the desired product quality, optimize operating condition under constraints, and evaluate the validity of the results. In addition, possible extensions and remaining problems are discussed.

REFERENCES

Jackson, J.E. and G.S. Mudholkar (1979). Control Procedures for Residuals Associated with Principal Component Analysis. *Technometrics*, **21**, 341–349.

Jaeckle, C.M. and J.F. MacGregor (1998). Product Design through Multivariate Statistical Analysis of Process Data. *AIChE J.*, **44**, 1105–1118.

ELSEVIER
IFAC
PUBLICATIONS
www.elsevier.com/locate/ifac

NONLINEAR RESIDUAL FEEDBACK OBSERVER FOR PROCESS FAULT DIAGNOSIS

Pramod Vachhani, Sridharakumar Narasimhan, Raghunathan Rengaswamy [1]

Department of Chemical Engineering, Clarkson University, Potsdam, NY -13699, USA

Abstract: Faults lead to loss of productivity and can eventually lead to loss of human lives. Therefore, fault diagnosis is a critical procedure for increased reliability and safety. Diagnostic observers, especially Unknown Input Observers (UIO) (Frank, 1990), have been well studied in literature. In this paper a novel residual feedback structure is proposed for fault diagnosis of a class of nonlinear systems. Conditions under which such a feedback system converges are discussed. Simulation results of a residual feedback nonlinear observer show that exact fault magnitude estimates are achieved.

Keywords: Fault Diagnosis, Nonlinear systems, Observers

1. INTRODUCTION

Process fault diagnosis is the problem of identifying the causal origins of malfunctions in a process, given current sensor data and *a priori* knowledge about the process behaviour under normal and abnormal conditions. Fault diagnosis is an important procedure relevant to process safety and process economics (Nimmo, 1995). The reliability of fault tolerant schemes is dependent on the diagnostic scheme identifying the correct fault with reliable estimates for the magnitude of failure.

Various techniques have been used to address the problems in fault diagnosis (Isermann and Balle, 1997). Of the various model based techniques, the unknown input observer has been widely studied for residual generation. The structured residual approach integrated with a decision logic is used for fault isolation (Frank, 1990; Garcia and Frank, 1997). Observer theory can also be used to estimate fault magnitudes if each observer residual reflects the occurrence of a single fault. However, the class of nonlinear systems for which such a fault decoupled observer can be constructed is limited. Seliger and Frank (1991) demonstrate the use of such an observer for fault diagnosis in a tank system. The theory of input reconstruction can also be used for diagnosis to get reliable fault magnitude estimates. In this strategy, the faults are posed as unknown inputs to be reconstructed from the measurements and known inputs to the process. This theory is applicable for linear (Hou and Patton, 1998) and linear descriptor (Hou and Patton, 1999*a*; Hou and Patton, 1999*b*) systems.

In this paper, the design for an unknown input observer (Frank, 1990) in addition with fault decoupling is used to develop a nonlinear observer to track parametric and sensor faults. This design is applicable for fault affine processes. The heat exchanger case study used is nonlinear in faults and restricted linearization is used to cast the model in a fault affine form. As shown later, neglecting this critical information leads to biased magnitude estimates, making the construction of a decision logic for fault isolation a difficult task.

[1] Corresponding Author: raghu@clarkson.edu

A residual feedback strategy is proposed to compensate the nonlinearity in fault. The convergence criterion for the feedback strategy is also presented. The residual feedback observer gives unbiased fault magnitude estimates. The results for the residual feedback observer are presented for the heat exchanger case study.

2. DESIGN OF NONLINEAR OBSERVER (NLO)

The unknown input observer (Frank, 1990) utilizes a nonlinear model of the system given as,

$$\begin{aligned}\dot{\mathbf{x}} &= \mathbf{Ax} + \mathbf{B(u,y)} + \mathbf{Ed} + \mathbf{K(x,u)f} \\ \mathbf{y} &= \mathbf{Cx} + \mathbf{Gf}\end{aligned} \tag{1}$$

where, $\mathbf{x}$ is $(n \times 1)$ state vector, $\mathbf{u}$ is $(p \times 1)$ input vector, $\mathbf{y}$ is $(q \times 1)$ vector of measured outputs, $\mathbf{d}$ is the unknown inputs/disturbance vector, $\mathbf{f}$ are the faults and $\mathbf{A}, \mathbf{B}, \mathbf{C}, \mathbf{E}, \mathbf{K}, \mathbf{G}$ are known matrices of appropriate dimensions. The term $\mathbf{Ed}$ models the unknown inputs to the actuator and to the dynamic process, $\mathbf{K(x,u)f}$ models the process and actuator faults, and the term $\mathbf{Gf}$ models the sensor faults. From these equations, it can be seen that the dynamic equations may be nonlinear only in input and output variables and the input distribution matrix of the fault vector may be a function of the inputs and the states.

The nonlinear observer is given by the equations (Frank, 1990)

$$\begin{aligned}\dot{\mathbf{z}} &= \mathbf{Rz} + \mathbf{J(u,y)} + \mathbf{Sy} \\ \mathbf{r} &= \mathbf{L_1 z} + \mathbf{L_2 y} \\ \mathbf{z(t=0)} &= \mathbf{z_0}\end{aligned} \tag{2}$$

where $\mathbf{z} = \mathbf{Tx}$ **iff** $\mathbf{f} = \mathbf{0}$. This observer is valid for fault detection if the residual $\mathbf{r}$ has the following properties

$$\lim_{t \to \infty} \mathbf{r(t)} = \mathbf{0} \quad \textbf{if } \mathbf{f} = \mathbf{0} \quad \forall\, \mathbf{u}, \mathbf{d}, \mathbf{x_0}, \mathbf{z_0} \tag{3}$$

and

$$\mathbf{r} \neq \mathbf{0} \quad \textbf{if } \mathbf{f} \neq \mathbf{0} \tag{4}$$

The derivation of the conditions on the observer matrices follows from the linear equivalent (Frank, 1990), and the necessary/sufficient conditions for their existence (Frank and Wunnenberg, 1989) are available in literature . For sake of brevity, the matrix conditions are presented here.

$$\begin{aligned}\mathbf{TA} - \mathbf{RT} &= \mathbf{SC} \\ \mathbf{J(u,y)} &= \mathbf{TB(u,y)} \\ \mathbf{TE} &= \mathbf{0} \\ \mathbf{TK} &\neq \mathbf{0} \\ \mathbf{L_1T} + \mathbf{L_2C} &= \mathbf{0}\end{aligned} \tag{5}$$

For the residuals to go to zero, $\mathbf{R}$ is required to be stable i.e., have eigenvalues in the left half of the s-plane.

3. DESIGN OF NONLINEAR OBSERVER WITH RESIDUAL FEEDBACK (NLOFB)

In the previous section, a nonlinear observer design for fault affine processes was presented. To account for nonlinearity in fault, a modified observer structure based on the previous design is proposed. A nonlinear residual feedback is used for direct compensation of the nonlinearity in fault. The nonlinear model of the process is of the form,

$$\begin{aligned}\dot{\mathbf{x}} &= \mathbf{Ax} + \mathbf{B(u,y)} + \mathbf{K(y,u)f} + \phi\mathbf{(y,f)} \\ \mathbf{y} &= \mathbf{Cx} + \mathbf{Gf}\end{aligned} \tag{6}$$

where $\phi\mathbf{(y,f)}$ is Lipschitz in $\mathbf{f}$, i.e

$$\|\phi(\mathbf{y},\mathbf{f_1}) - \phi(\mathbf{y},\mathbf{f_2})\| \leq \gamma(\mathbf{y})\|\mathbf{f_1} - \mathbf{f_2}\|$$

and the input distribution matrix may be a function of the measurements and inputs.

The observer constructed is of the form,

$$\begin{aligned}\dot{\mathbf{z}} &= \mathbf{Rz} + \mathbf{J(u,y)} + \mathbf{Sy} + \mathbf{T}\phi\mathbf{(y,r)} \\ \mathbf{r} &= \mathbf{L_1 z} + \mathbf{L_2 y} \\ \mathbf{z(t=0)} &= \mathbf{z_0}\end{aligned} \tag{7}$$

Claim 1. For step faults in the dynamic system as defined in Equation 7, the residuals converge to the faults if matrices $\mathbf{R}, \mathbf{J}, \mathbf{S}, \mathbf{T}, \mathbf{L_1}, \mathbf{L_2}$ exist and the following conditions are satisfied.

$$\begin{aligned}\mathbf{TA} - \mathbf{RT} &= \mathbf{SC} \\ \mathbf{J(u,y)} &= \mathbf{TB(u,y)} \\ \mathbf{TK} &\neq \mathbf{0} \\ \mathbf{R} &= -c_0\mathbf{I}, \ c_0 > 0 \\ \mathbf{L_1T} + \mathbf{L_2C} &= \mathbf{0} \\ \mathbf{L_1(SG - TK)} &= \mathbf{(RL_2G - R)} \\ \|\mathbf{L_1T}(\phi(\mathbf{y},\mathbf{f_1}) - \phi(\mathbf{y},\mathbf{f_2}))\|_2 &< c_0\|\mathbf{f_1} - \mathbf{f_2}\|\end{aligned} \tag{8}$$

PROOF. Construct the following Lyapunov function,

$$v = \frac{1}{2}(\mathbf{r} - \mathbf{f})^{\mathbf{T}}(\mathbf{r} - \mathbf{f})$$

The derivative of the Lyapunov function is ($\dot{\mathbf{f}}$ is zero for step faults, and $\mathbf{e} = \mathbf{z} - \mathbf{Tx}$),

$$\begin{aligned}\dot{v} &= (\mathbf{r} - \mathbf{f})^{\mathbf{T}}\dot{\mathbf{r}} \\ &= (\mathbf{r} - \mathbf{f})^{\mathbf{T}}[\mathbf{L_1}(\dot{\mathbf{e}} + \mathbf{T}\dot{\mathbf{x}}) + \mathbf{L_2}\dot{\mathbf{y}}]\end{aligned} \tag{9}$$

Using $\mathbf{L_1T} + \mathbf{L_2C} = \mathbf{0}$, $\dot{\mathbf{f}} = \mathbf{0}$,

$$\dot{v} = (\mathbf{r}-\mathbf{f})^{\mathbf{T}}\mathbf{L_1}\dot{\mathbf{e}}$$

Given that the matrix conditions in the claim hold, the error dynamics of the feedback nonlinear observer in presence of fault nonlinearity can be written as,

$$\begin{aligned}\dot{\mathbf{e}} &= \dot{\mathbf{z}} - \mathbf{T}\dot{\mathbf{x}} \\ &= \mathbf{Re} + (\mathbf{SG} - \mathbf{TK}(\mathbf{y},\mathbf{u}))\mathbf{f} + \mathbf{T}(\phi(\mathbf{y},\mathbf{r}) - \phi(\mathbf{y},\mathbf{f}))\end{aligned}$$

This leads to,

$$\begin{aligned}\dot{v} &= (\mathbf{r}-\mathbf{f})^{\mathbf{T}}[\mathbf{L_1R}(\mathbf{z}-\mathbf{Tx}) + \mathbf{L_1}(\mathbf{SG}-\mathbf{TK})\mathbf{f}] \\ &\quad + (\mathbf{r}-\mathbf{f})^{\mathbf{T}}\mathbf{L_1T}(\phi(\mathbf{y},\mathbf{r}) - \phi(\mathbf{y},\mathbf{f}))\end{aligned} \tag{10}$$

Since $\mathbf{R} = -c_0\mathbf{I}$, $\mathbf{L_1Rz} = \mathbf{RL_1z} = \mathbf{R}(\mathbf{r}-\mathbf{L_2y})$, $\mathbf{RL_2C} - \mathbf{L_1RT} = \mathbf{R}(\mathbf{L_2C} - \mathbf{L_1T}) = \mathbf{0}$, the first term in the above expression can be simplified to

$$(\mathbf{r}-\mathbf{f})^{\mathbf{T}}[\mathbf{Rr} - (\mathbf{RL_2G} - \mathbf{L_1}(\mathbf{SG}-\mathbf{TK}))\mathbf{f}]$$

If $\mathbf{L_1}$ is chosen as in Equation 8, the above term further simplifies to $-c_0\|(\mathbf{r}-\mathbf{f})\|^2$. Therefore,

$$\begin{aligned}\dot{v} &= \; -c_0\|(\mathbf{r}-\mathbf{f})\|^2 \\ &\quad + (\mathbf{r}-\mathbf{f})^{\mathbf{T}}\mathbf{L_1T}(\phi(\mathbf{y},\mathbf{r}) - \phi(\mathbf{y},\mathbf{f})) \\ \dot{v} &\leq -c_0\|\mathbf{r}-\mathbf{f}\|^2 \\ &\quad + \|(\mathbf{r}-\mathbf{f})^{\mathbf{T}}\mathbf{L_1T}(\phi(\mathbf{y},\mathbf{r}) - \phi(\mathbf{y},\mathbf{f}))\|\end{aligned} \tag{11}$$

Using the Cauchy-Schwarz-Buniakowsky inequality,

$$\begin{aligned}\dot{v} &\leq -c_0\|\mathbf{r}-\mathbf{f}\|^2 \\ &\quad + \|(\mathbf{r}-\mathbf{f})\|\|\mathbf{L_1T}(\phi(\mathbf{y},\mathbf{r}) - \phi(\mathbf{y},\mathbf{f}))\|\end{aligned} \tag{12}$$

If the nonlinearity in fault (ϕ) is Lipschitz in $\mathbf{f}$, so is $\mathbf{L_1T}\phi(\mathbf{y},\mathbf{f})$ and by the condition in Equation 8

$$\|(\mathbf{L_1T}(\phi(\mathbf{y},\mathbf{r}) - \phi(\mathbf{y},\mathbf{f}))\| < c_0\|\mathbf{r}-\mathbf{f}\| \tag{13}$$

Therefore,

$$\dot{v} < 0 \tag{14}$$

The derivative of the Lyapunov function is negative definite and the residuals ($\mathbf{r}$) asymptotically converge to the actual fault magnitude ($\mathbf{f}$).

4. HEAT EXCHANGER CASE STUDY

The heat exchanger is an important process unit in any chemical engineering operation and the transients of the process are extremely nonlinear. The first principles model of a counter-current heat exchanger is as follows

$$\begin{aligned}V_c\dot{T}_c &= q_c(T_{ci} - T_c) + \frac{UA}{\rho_c C_{pc}}\Delta T \\ V_h\dot{T}_h &= q_h(T_{hi} - T_h) - \frac{UA}{\rho_h C_{ph}}\Delta T\end{aligned} \tag{15}$$

where, V is the volume and subscript 'c' and 'h' refer to cold and hot side respectively, q is the known flow rate, ρ and C_p are the density and specific heat of the two streams and ΔT is the log mean temperature difference,

$$\Delta T = \frac{(T_{hi} - T_c) - (T_h - T_{ci})}{\log(T_{hi} - T_c) - \log(T_h - T_{ci})} \tag{16}$$

Table 1. Parameters for the heat exchanger

Notation	Variable	Steady state value
V_c	Holdup on cold side	0.05 m^3
V_h	Holdup on hot side	0.05 m^3
UA	Heat transfer coeff. × area	10 $KJ/^oC\ min$
T_c	Cold stream temp.	53.03 ^oC
T_h	Hot stream temp.	62.63 ^oC
T_{ci}	Cold stream inlet temp.	25 ^oC
T_{hi}	Hot stream inlet temp.	100 ^oC
ρ_c, ρ_h	Fluid densities	500 kg/m^3
C_{ph}, C_{pc}	Specific heats	3 $KJ/^oCkg$
q_c	Cold stream flow rate	0.01 m^3/min
q_h	Hot stream flow rate	0.0075 m^3/min

The steady state and parameters for the heat exchanger operation are presented in Table 1. The inputs to the heat exchanger, i.e. flow rates of hot and cold stream are known. The outlet temperatures (T_c and T_h) are measured, while the inlet temperatures (T_{ci} and T_{hi}) are the simulated parametric faults.

5. PARAMETRIC FAULTS CASE STUDY

In the following case study, the previously discussed observers are designed to monitor the changes in the inlet temperatures. The heat exchanger problem is nonlinear in fault and restricted linearization is carried out to cast the process model in a fault affine form for design of the nonlinear observer. For the residual feedback observer, the actual nonlinearity in fault is compensated. Therefore, it is expected that the nonlinear observer would give biased fault magnitude estimates, while the residual feedback observer should give exact estimates of the fault severity.

5.1 Observer Design- NonLinear Observer (NLO)

For design of nonlinear observer, the heat exchanger model equations given in Equation 15 are cast in the fault affine (Ref Equation 1), by restricted linearization where,

$$\mathbf{y} = \mathbf{x} = \begin{bmatrix} T_c \\ T_h \end{bmatrix} \quad \text{and} \quad \mathbf{f} = \begin{bmatrix} \Delta T_{ci} \\ \Delta T_{hi} \end{bmatrix}$$

$$\mathbf{A} = \begin{bmatrix} -\frac{q_c}{V_c} & 0 \\ 0 & -\frac{q_h}{V_h} \end{bmatrix}$$

$$\mathbf{B}(\mathbf{u},\mathbf{y}) = \begin{bmatrix} K_c\frac{\theta_1 - \theta_2}{\log\frac{\theta_1}{\theta_2}} + \frac{q_c}{V_c}25 \\ -K_h\frac{\theta_1 - \theta_2}{\log\frac{\theta_1}{\theta_2}} + \frac{q_h}{V_h}100 \end{bmatrix}$$

$$\mathbf{K(y,u)} = \begin{bmatrix} \frac{q_c}{V_c} + \frac{\alpha K_c}{d} & \frac{\beta K_c}{d} \\ -\frac{\alpha K_h}{d} & \frac{q_h}{V_h} - \frac{\beta K_h}{d} \end{bmatrix}$$

where

$$K_c = \frac{UA}{\rho_c C_{pc} V_C} \text{ and } K_h = \frac{UA}{\rho_h C_{ph} V_h}$$

$$\theta_1 = 100 - T_c,\ \theta_2 = T_h - 25 \text{ and } d = \left(\log\frac{\theta_1}{\theta_2}\right)^2$$

$$\alpha = \left(\log\frac{\theta_1}{\theta_2} + 1 - \frac{\theta_1}{\theta_2}\right)$$

$$\beta = \left(\log\frac{\theta_1}{\theta_2} - 1 + \frac{\theta_2}{\theta_1}\right)$$

The nonlinear observer as described in Equation 2 is implemented with the following choice of the observer matrices that satisfies the conditions in Equation 5: $\mathbf{T} = \mathbf{I_2}$, $\mathbf{R} = -\mathbf{I_2}$, $\mathbf{S} = \mathbf{A} - \mathbf{R}$, $\mathbf{J} = \mathbf{TB(u,y)}$, $\mathbf{L_1} = -\mathbf{K(y,u)}^{-1}$ and $\mathbf{L_2} = -\mathbf{L_1}$.

5.2 Observer Design- NonLinear FeebBack Observer (NLOFB)

The system equations (Ref Equation 15) are conveniently written in the following form

$$\dot{\mathbf{x}} = \mathbf{Ax} + \mathbf{B(u,y)} + \mathbf{K(y,u)f} + \phi(\mathbf{y},\mathbf{f})$$

where

$$\mathbf{y} = \mathbf{x} = \begin{bmatrix} T_c \\ T_h \end{bmatrix} \text{ and } \mathbf{f} = \begin{bmatrix} \Delta T_{ci} \\ \Delta T_{hi} \end{bmatrix}$$

$$\mathbf{A} = \begin{bmatrix} -\frac{q_c}{V_c} & 0 \\ 0 & -\frac{q_h}{V_h} \end{bmatrix} \text{ and } \mathbf{B(u,y)} = \begin{bmatrix} \frac{q_c}{V_c}25 \\ \frac{q_h}{V_h}100 \end{bmatrix}$$

$$\mathbf{K(y,u)} = \begin{bmatrix} \frac{q_c}{V_c} & 0 \\ 0 & \frac{q_h}{V_h} \end{bmatrix} \text{ and } \phi = \begin{bmatrix} K_c \\ -K_h \end{bmatrix} \Delta T$$

where ΔT is as defined in Equation 16. The nonlinear observer as described in Equation 7 is implemented with the following choice of the observer matrices that satisfies the conditions in Equation 8: $\mathbf{T} = \mathbf{I_2}$, $\mathbf{R} = -\mathbf{I_2}$, $\mathbf{S} = \mathbf{A} - \mathbf{R}$, $\mathbf{J} = \mathbf{TB(u,y)}$, $\mathbf{L_1} = \mathbf{RK(y,u)}^{-1}$ and $\mathbf{L_2} = -\mathbf{L_1}$.

In addition to the matrix conditions, the following condition is a sufficient condition for the convergence of the observer.

$$\|\mathbf{L_1T}(\phi(\mathbf{y},\mathbf{f_1}) - \phi(\mathbf{y},\mathbf{f_2}))\|_2 < c_0\|\mathbf{f_1} - \mathbf{f_2}\| \quad (17)$$

For the particular choice of the design matrices, this implies

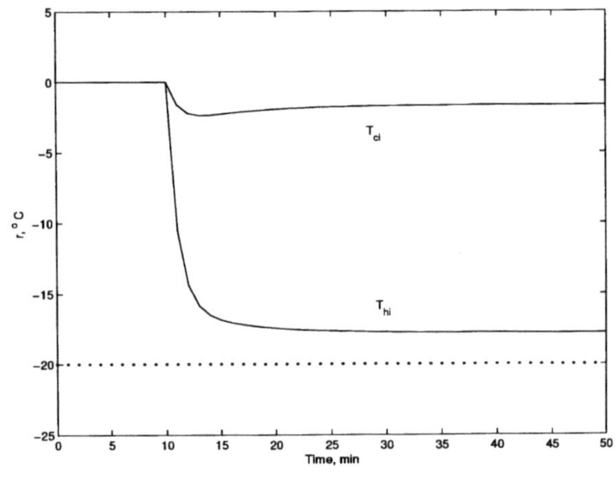

Fig. 1. NLO result: $\Delta T_{hi} = -20^oC$

$$\frac{|\Delta T(\mathbf{f}) - \Delta T(\mathbf{r})|}{\|\mathbf{r} - \mathbf{f}\|_2}\left(\left(\frac{K_h V_h}{q_h}\right)^2 + \left(\frac{K_c V_c}{q_c}\right)^2\right)^{0.5} < 1 \quad (18)$$

where

$$\Delta T(\mathbf{f}) = \frac{(T_{hi} + f_2 - T_c) - (T_h - T_{ci} - f_1)}{\log(T_{hi} + f_2 - T_c) - \log(T_h - T_{ci} - f_1)}$$
$$\mathbf{f} = [f_1, f_2]^t \quad (19)$$

It should be noted that the above is only a sufficiency condition and so, it is possible that the $\mathbf{r} \to \mathbf{f}$ as $t \to \infty$ even if the above condition is violated. This may occur, for example, when the fault magnitudes are large. The observer design is validated through simulation results involving different fault magnitudes and directions, which are presented in the subsequent section.

5.3 Case Study Results

Results of the NonLinear Observer (NLO) for two fault cases ($\Delta T_{hi} = -20^oC$ and $\Delta T_{ci} = 10^oC$) are presented in Figures 1 and 2 respectively. It is seen that, the estimates of the fault magnitudes are biased; this occurs because the nonlinearity in fault is not compensated for during restricted linearization. For example, when ($\Delta T_{hi} = -20^oC$), the residual for T_{ci} is non-zero and the magnitude of the residual for T_{hi} depends on the magnitude of the actual T_{hi} fault. Therefore, the use of absolute thresholding of residuals is not possible to achieve fault isolation.

Results of the NonLinear Observer with FeedBack (NLOFB) are presented in Figures 3, 4 and 5 for three different fault levels: $\Delta T_{hi} = -20^oC$, $\Delta T_{ci} = 10^oC$ and a simultaneous occurrence of both faults: $\Delta T_{hi} = 30^0C$ and $\Delta T_{ci} = 20^oC$. The terminal values of the residuals approach the corresponding fault magnitudes for all cases. Thus, this ensures complete decoupling of faults and correct estimation of the fault magnitudes. This validates the design of the NLOFB.

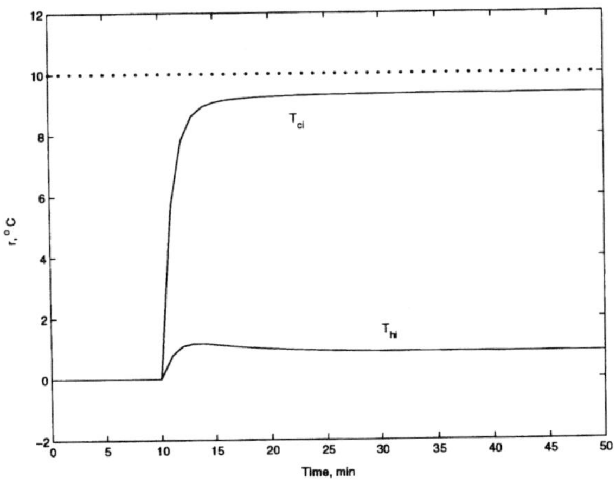

Fig. 2. NLO result: $\Delta T_{ci} = 10^oC$

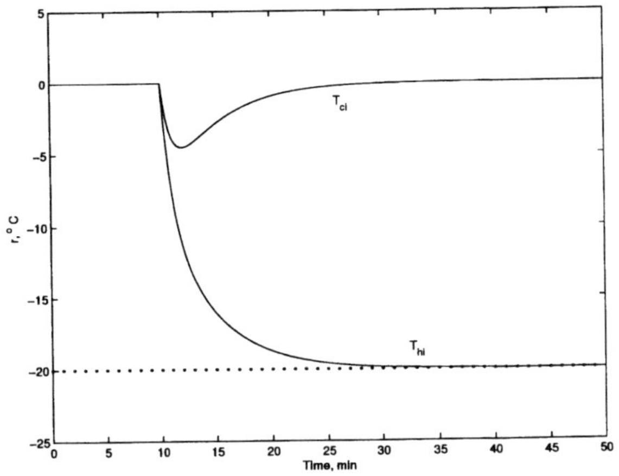

Fig. 3. NLOFB result: $\Delta T_{hi} = -20^oC$

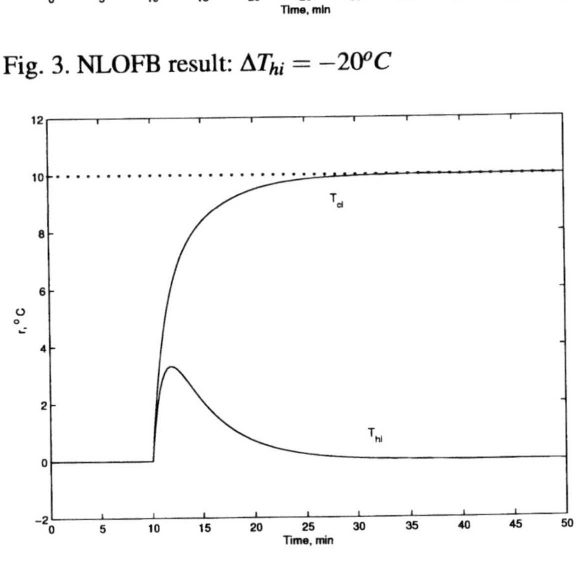

Fig. 4. NLOFB result: $\Delta T_{ci} = 10^oC$

6. CONCLUSIONS

A feedback structure for the unknown input observer for a system with a specified structure of nonlinearity has been presented. This feedback structure extends the capacity of the diagnostic observer towards fault identification by mimicking the nonlinearity in fault by aid of residuals. This has been demonstrated by achieving unbiased fault magnitude estimates in the

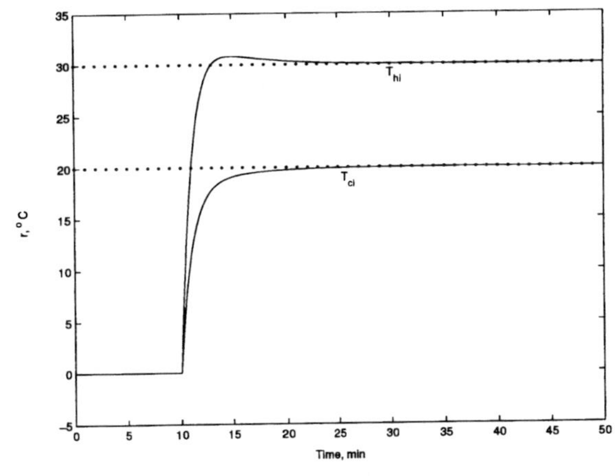

Fig. 5. NLOFB result: $\Delta T_{hi} = 30^oC, \Delta T_{ci} = 20^oC$

case of parametric faults in the heat exchanger case study.

REFERENCES

Frank, P. M. (1990). Fault diagnosis in dynamic systems using analytical and knowledge-based redundancy – a survey and some new results. *Automatica* **26**, 459–474.

Frank, P. M. and J. Wunnenberg (1989). Robust fault diagnosis using unknown input observer scheme. In: *Fault Diagnosis in Dynamic Systems: Theory and Application* (R. J. Patton, P. M. Frank and R. N. Clark, Eds.). Prentice Hall, NY.

Garcia, E. and P. Frank (1997). Deterministic nonlinear observer-based approaches to fault diagnosis: A survey. *Control Eng. Practice* **5**, 663–670.

Hou, M. and R. Patton (1998). Input observability and input reconstruction. *Automatica* **34**, 789–794.

Hou, M. and R. Patton (1999*a*). Causal observability of descriptor systems. *IEEE Transactions on Automatic Control* **44**, 158–163.

Hou, M. and R. Patton (1999*b*). Observer design for descriptor systems. *IEEE Transactions on Automatic Control* **44**, 164–168.

Isermann, R. and P. Balle (1997). Trends in the application of model-based fault detection and diagnosis of technical processes. *Control Eng. Practice* **5**, 709–719.

Nimmo, I. (1995). Adequately address abnormal situation operations. *Chem. Engng. Prog.* **91(9)**, 36–45.

Seliger, R. and P. Frank (1991). Fault-diagnosis by disturbance decoupled nonlinear observers. In: *Proceedings of the 30th Conference on Decision and Control*. Brighton, England. pp. 2248–2253.

ELSEVIER
IFAC
PUBLICATIONS
www.elsevier.com/locate/ifac

OBJECT-BASED DIAGNOSTIC NETWORK BASED ON STATISTICAL LEARNING

Sang Ok Song, En Sup Yoon

School of Chemical Engineering, Seoul National University

Abstract: This study proposes an effective diagnostic framework that can be constructed using historical data readily available from the process. The whole system is decomposed into its local diagnostic models based on the direct and local causalities of process variables and the statistical learning model for each local relation is developed using data available from the process. The decomposed local models and the underlying fault assumptions compose of an object-based diagnostic network to perform on-line fault diagnosis. The diagnostic performance of the proposed method has been successfully illustrated in CSTR process. *Copyright © 2004 IFAC*

Keywords: fault diagnosis, fault detection and isolation, statistical learning, residual generation and evaluation.

1. INTRODUCTION

As increasing the importance of early and accurate fault detection and diagnosis for modern industrial plants, various process monitoring techniques have been developed for the past decade. While techniques based on first-principles models have been around for more than two decades, their contribution to industrial practice has not been pervasive and data-driven approaches or process history based methods have been widely used for process monitoring due to the high cost and time required to develop sufficiently detailed physical models.

Most of conventional data-driven monitoring methods use process data collected during both normal operating conditions and specific faults to develop the measures for diagnosing faults based on supervised classification. However, the process data during abnormal situations are hardly available in real world industries and then conventional data-based fault diagnosis has a major disadvantage for their application to real processes (Lee, 2003).

2. CONVENTIONAL METHODS AND THEIR LIMITATIONS

The fault diagnosis procedure consists of two separate stages: residual generation and residual evaluation. In other words, the automatic fault diagnosis can be viewed as a sequential process involving the symptom extraction and the actual diagnostic task. As usual, the residual generation process is based on a comparison between the measured and predicted system outputs. As a result, the difference of the so-called residual is expected to be near zero under normal operation, but on the occurrence of fault a deviation from zero should appear. In the meantime, the residual evaluation module is dedicated to the analysis of the residual signal in order to determine whether a fault has occurred and to isolate the fault in a particular system device.

One of the most known approaches to residual generation is a model-based concept. In the general case, this concept can be realized using different kinds of models: analytical, knowledge-based and data-based ones. Unfortunately, the analytical model-based approach is usually restricted to simpler systems described by linear models. Industrial plants are often complex dynamic systems described by nonlinear high-order differential equations. For their quantitative modelling for residual generation, simplifications are inevitable. This usually concerns both the reduction of dynamics order and linearization. Another problem arises from unknown or time variant process parameters. Due to all these difficulties, conventional analytical models often turn out to be not accurate enough for an effective residual generation. Therefore, when there are no

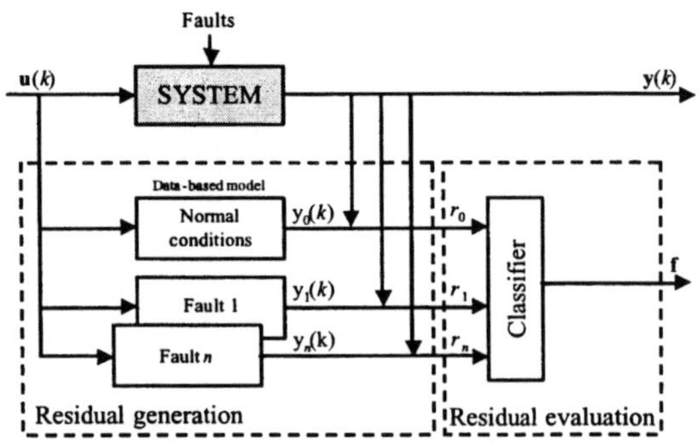

Fig.I. Conventional fault diagnostic system using data-based model

mathematical models of a system or the complexity of a system is so high that the task of modelling is difficult, analytical models cannot be applied or cannot give satisfactory results (Patan, 2000).
In these cases data-based models can be used instead of unavailable analytical models. For example, a data-based model replaces the analytical model that describes the process under normal operation. Fig. I illustrates how a fault diagnostic system can be designed using data-based technique. Residuals are generated based on a comparison between systems outputs and model outputs. And then a specific fault is identified by a supervised classifier.
However, this conventional approach has some inherent limitations on their ability to effectively identify and diagnose faults. As mentioned above, this approach requires data in every faulty situation for building a fault classifier but those data are not generally available at the real process. Hence this method cannot be used to diagnose faults that have not previously occurred. This is a weak point of conventional data-based approaches. To take advantage of data-based model approaches and overcome their limitations, a practical diagnostic method that uses data easily available from the process should be developed.

4. THE PROPOSED DIAGNOSTIC METHOD

In order to overcome the aforementioned limitation of data-driven model based diagnostic method, an object-based diagnostic network has been developed (Fig. II). The diagnostic network consists of a series of diagnostic data-based models and their qualitative knowledge base. The qualitative knowledge base is represented as a set of assumptions about underlying faults. These fault assumptions represent the assumptions that the associated model is based upon (Petti, 1992). Each diagnostic model is trained using normal operation data. The data-based models generate residuals based on a comparison between the measured and predicted system outputs. If all the assumptions of a model hold, the residual is close to zero. By examining the direction and extent to which each model is violated against its fault assumptions, the most likely failed assumption (fault) can be deduced. An assumption that is common to many violated models is strongly suspect, whereas satisfaction of models provides evidence that the associated assumptions are valid. How the models

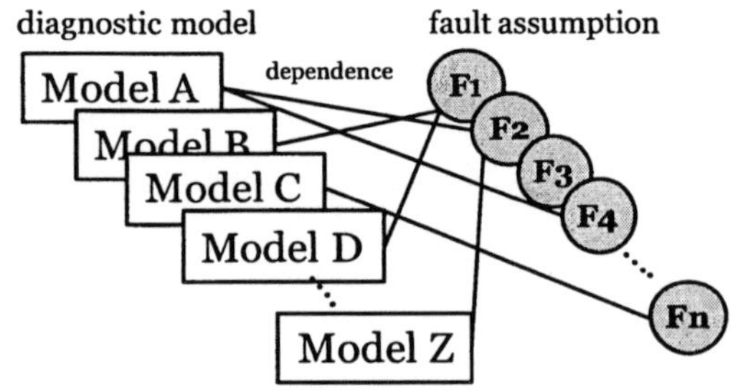

Fig. II. Object-based diagnostic network

can be formulated and how the fault assumptions can be associated with the models is explained in the next sub-sections.

4.1 *Design of local diagnostic models*

The formulation of diagnostic models is based on system decomposition by using local and direct causalities between process variables. The system decomposition approach narrows the diagnostic focus to a particular decomposed subsystem and performs diagnosis on this subsystem (Finch, 1988). System decomposition has the advantages of providing flexible diagnosis throughout operating condition changes, reducing the size of the knowledge base, and simplifying the understanding of complex process to process interactions.
The local subsystem models can be designed based on the information about local causalities between process variables. An example of subsystem models for a simple gravity flow tank is shown in Fig. III. The variables having a direct effect on L (target variable) are F_{in} and F_{out}. And F_{out} (target variable) is first affected by R_s and L. From these direct and local causalities, two models for L and F_{out} can be designed as shown in Fig. III. F_{in} and R_s are considered exogenous to a system because their values are determined outside of the system and controlled externally.
The availability of local subsystem models highly depends on the measurability of the process variables. Notice that unmeasured variables cannot participate in formulating diagnostic models due to their unavailability for training models. The outflow model contains an unmeasured variable R_s (the pipe resistance) and cannot be learned from history data. However, assuming that the unmeasured variable R_s is constant and does not affect the outflow model, it can be used for diagnostic task.

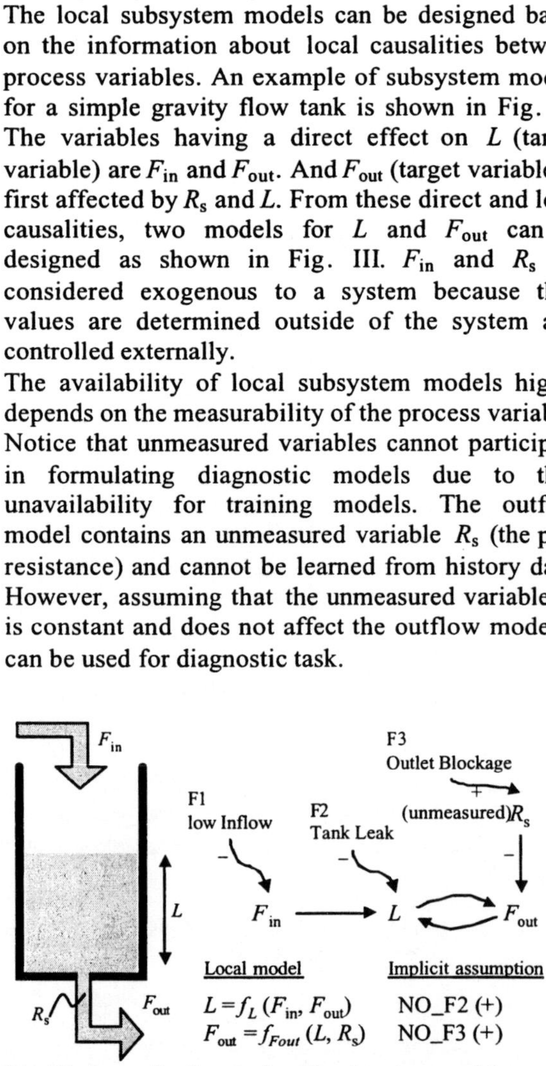

Fig. III. A gravity flow tank with subsystem models

4.2 Fault assumptions

A set of assumptions is associated with each diagnostic local model as shown in Fig. II. These assumptions guarantee the satisfaction of the model, if satisfied. If an assumption fails, the connected models do not hold anymore and the residuals become large. The failure of an assumption means the occurrence of a fault. Based on these symptoms, the most probable fault can be identified. Therefore, the task of selecting appropriate assumptions is crucial to the success of the proposed fault diagnosis system.

Some of the assumptions, such as correct sensor readings are explicit in the model and some are implicit, for example, the fact that there are no tank leaks. Consider the tank in Fig. III once again. The model L has following assumptions.

- Inflow sensor is OK (explicit).
- Outflow sensor is OK (explicit).
- No tank leaks (implicit).

Explicit fault assumptions are readily connected to the diagnostic models that include the same sensor variables. For example, the assumption that the outflow (F_{out}) sensor is ok is connected with the local model for L because the local model L contains the variable F_{out}. However, some reasoning procedures are required for implicit fault assumptions. Based on the following procedures, implicit fault assumptions are connected to each model.

Considering a fault (F_k) and the corresponding fault assumption (NO_F_k), there exist one or more primary deviation variables (x_{Fk}), i.e., variables included in the system which is directly affected by the root cause.

Case 1: When x_{Fk} is measured, the assumption, NO_F_k is connected with the model for x_{Fk}.

Case 2: When x_{Fk} is unmeasured, any measured variable which is first affected by x_{Fk} is considered as the primary deviation variable and then apply to *case 1*.

In the tank example, the tank level (L) is the primary deviation variable for tank leak (F2) and the assumption NO_F2 is associated with the level model. Outflow (F_{out}) is considered as the primary deviation variable for the outlet blockage (F3) because the pipe resistance (R_s) which is actually first affected by F3 is unmeasured and F_{out} is first affected by R_s. Therefore, NO_F3 is the implicit fault assumption of the outflow model.

In case of low inflow disturbance (F1), we do not have the local model for the primary deviation variable F_{in} because F_{in} is an exogenous variable. If F_{in} is steady state, F1 can be detected by checking some tolerance limit.

The fault assumptions associated with their local diagnostic models have their qualitative information represented the direction (+ or -) to which the associated local model will be violated when a fault assumption is not satisfied. If a fault assumption is connected with its associated local model positively (negatively), it implies that the predicted output of the local model will deviate from its measured value in the positive (negative) direction when the fault occurs. For example, NO_F2 is positively connected with the local model L. This means that the predicted value of L becomes larger than its measured one when the Tank Leaks.

4.3 Training local models

For residual generation, each local diagnostic model designed at previous step has to be trained using process data. In this step, the local relations between the input variables and the output variable of each model can be learned from data. The performance of data-driven methods highly depends on the quality of process data. So the pre-treatment procedures such as data reconciliation and gross error detection of raw data are necessary. But this issue is not covered in this study.

Training data sets are prepared from the historical database of the process. They need to be collected from various operating conditions, in order to learn more generalized and robust models. In addition to normal operating data, the process data which can be obtained in the presence of intentional step changes, external disturbances occurred frequently, or some faults happened previously can be good candidates for training data sets. However, it must be noted that the process data obtained in the presence of a particular fault should not be used to train the local model(s) with the same fault assumption. All training data sets are required to satisfy the underlying assumptions of each local model.

In the proposed method, the statistical learning techniques, specially focusing on regression are used for training local models. The values of target variable can be predicted using statistical learning machines for local models. Various statistical methods for regression such as splines, neural networks and Support Vector Machine can be a tool for building learning machines. The current prevailing view in the statistical and neural network community is that there is no single best method for all regression problems (Friedman, 1994) and the comparative work of various learning techniques is not the point of this research. Comparative studies for various regression methods are found in (Cherkassky, 1998). In our work, Support Vector Regression method was selected as a learning technique because it leads to a more tractable formulation of the optimization problem and the notion of model complexity can be separated from dimensionality. In addition, SVM can handle nonlinearities of the model in the easy manner by selecting appropriate kernel function. Other advantages of SVM can be found in many references (Cherkassky, 1998; Vapnik, 1995).

4.4 Residual generation and evaluation

After finishing the training via an appropriate statistical learning technique, the trained local models are ready for on-line residual generation. The residuals, between predicted values from local models and measured values are monitored and the CUSUM (cumulative summation) control chart is used to detect the changes from normal operating state of a process. The CUSUM test first introduced by Page (Page, 1954) has been shown to be efficient

in detecting changes in the mean of a process. As the name implies, the CUSUM chart cumulates deviations of the sample readings (x_i) from the target or desired value (μ). Once these cumulative summations reach either a high or low limit, an out-of-control signal is given. The parameters in the CUSUM chart are defined as follows:
k: the threshold for cumulative summation, which can be defined as the minimum difference between sample average and target that will cause the cumulative summation to begin.
SH_i and SL_i: the cumulative summation terms which are calculated as follows.

$$\begin{aligned} SH_i &= \max\left[0, SH_i' + SH_{i-1}\right] \\ SL_i &= \max\left[0, SL_i' + SL_{i-1}\right] \\ \text{where} & \\ SH_i' &= \text{highsidecumulativeterm} \\ &= (x_i - \mu) - k \\ SL_i' &= \text{lowsidecumulativeterm} \\ &= -(x_i - \mu) - k = (\mu - x_i) - k \end{aligned} \tag{2}$$

h: the control limit
Since the residuals are not uniform in magnitude, they are transformed into a metric between ? and 1 which indicates the degree to which the model equation is satisfied: 0 for perfectly satisfied, 1 for severely violated high, and ? for severely violated low. These values constitute the satisfaction vector, sf, which is calculated using the CUSUM results, SH_i and SL_i.
For the jth diagnostic model,

$$sf_j = \frac{sign(SH_j - SL_j)\left((SH_j - SL_j)/h_j\right)^n}{1 + \left((SH_j - SL_j)/h_j\right)^n} \tag{3}$$

The value of sf_j is given a positive value for a positive residual and a negative value for a negative residual. The curve is a general sigmoidal function with the steepness determined by the constant n.
A matrix of sensitivity values, S which describes the relations between local models and fault assumptions is assigned to weight the sf values as evidence. The ijth element of S, represents the sensitivity of the jth model to the ith assumption. Unless experience suggests otherwise, these values are usually given −1, 0, or 1; 0 for the local models independent of an assumption, −1 for the local models negatively connected with an assumption, and 1 for the local models positively connected. In some cases when the effect of a fault is uncertain if negative or positive, additional sensitivity matrix S' is created, whose values are given 1 or 0; 1 for the local models dependent of an assumption, 0 for the local models independent of an assumption.
In addition, a matrix of non-sensitivity values, NS which also describes the relations between local models and fault assumptions is assigned to give a penalty to non-sensitivity symptoms. These values are given 1 or 0; 1 for the local model independent of an assumption, 0 for the others.

The conclusion about the satisfaction of each assumption is made by combining the evidence from the local model, sf with consideration to the sensitivity matrix S, S', and the compensation matrix NS. This is done through the calculation of a vector of failure likelihoods, FL such that

$$FL_i = \frac{\sum_{j=1}^{N}(S_i \cdot sf_j)}{\sum_{j=1}^{N}|S_i|} + \frac{\sum_{j=1}^{N}(|S'_{ij} \cdot sf_j|)}{\sum_{j=1}^{N} S'_{ij}} - \frac{\sum_{j=1}^{N}(|NS_{ij} \cdot sf_j|)}{\sum_{j=1}^{N} NS_{ij}} \tag{4}$$

where N is the number of local model. It is evident that this method of combination allows the sf values of those local models which are most sensitive to ith fault assumption to be weighted the most heavily in the calculation of FL_i. The failure likelihood is interpreted as indicating a likely condition of ith fault assumption failing high as the value of FL_i approaches 1, while an FL_i tending toward −1 indicates a likely failure low.

5. ILLUSTRATIVE EXAMPLE

The example process is simple but displays the most common characteristics of industrial processes. This process is simulated by the model of Sorsa (Sorsa, 1991). The sampling interval is 5 s and faults are introduced at 100 s. Some selected faults of the CSTR process is described in Table 1.

5.1 *Design of local diagnostic models*

Based on the knowledge of the CSTR process, we obtained a direct causal relation between variables. The resulting local causalities of the CSTR process are illustrated in Fig V and the whole system could be decomposed into several subsystems from these direct and local causalities. For example, the variables having a direct effect on T_R are F_R, T, and F_W. From this causality, a local model for recycle flow rate (T_R) can be designed such as $T_R = f(F_W, F_R, T)$.

5.2 *Fault assumptions*

Each fault assumptions should be associated with corresponding local models before on-line diagnostic task. Explicit fault assumptions such as sensor faults are naturally associated with the local models that contain the corresponding sensor variables. For example, *FSBL* (flow sensor biased low) represents a failure for the recycle flow (F_R) sensor. The assumption (NO_*FSBL*) is connected with the local models for F_R, T and T_R that have the variable F_R.

Table 1 Selected faults of the CSTR process

NO.	name	description
1	*FSBL*	flow sensor biased low
2	*FVBL*	flow control valve biased low
3	*LEAK*	reactor leaking
4	*LSBL*	level sensor biased low
5	*LVBH*	level control valve biased high
6	*TSBH*	temperature sensor biased high
7	*TVBH*	temperature control valve biased high

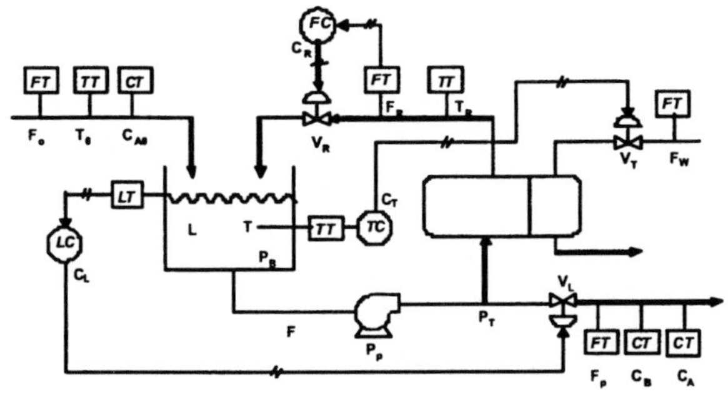

Fig. IV. Process flow diagram of the CSTR process

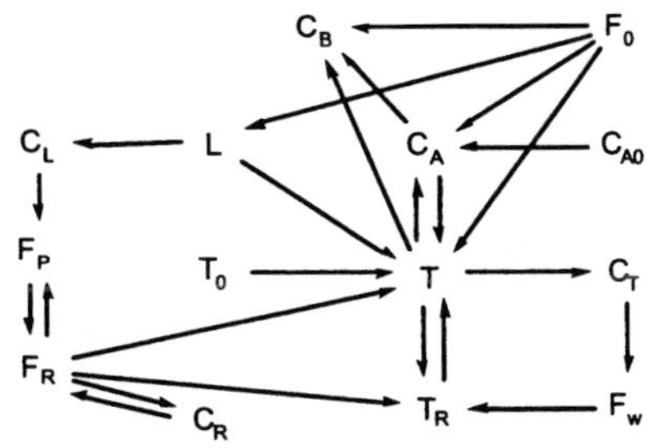

Fig. V. The direct causalities of the CSTR process

On the other hand, implicit fault assumptions are associated with their corresponding local model according to the procedures explained in section 4.2. For example, *LVBH* (level control valve biased high) represents a fault for valve V_L (Fig IV.). This fault affects the valve output first. Because the valve output is not measured, the output flow rate (F_P) that is first affected by the valve output (V_L) becomes the primary deviation variable for the fault *LVBH* and the local model for F_P is associated with the fault assumption, *LVBH*. In this manner, all faults in Table 1 are connected with decomposed local models and the object-based diagnostic network for the CSTR process is constructed as shown in Fig VI. The solid and dotted lines indicate the direction in which the associated model will deviate when the fault occurs. For example, *FSBL* is connected with the local model for F_R positively (solid line). It implies that the predicted output of the model F_R will deviate from its measured value in the positive direction when the fault *FSBL* occurs.

5.3 Training local models

Each diagnostic local model has been trained via support vector regression technique. Learning data for training the local models designed at the previous steps were collected under the condition where some external disturbances occurred or the set points of the controllers were changed.

<u>External disturbance data</u>
Cooling water temperature high
Feed composition change low
Feed flow rate change high
Feed temperature change low

<u>Controller set point change data</u>
Level controller set point high
Temperature controller set point high
Flow controller set point high

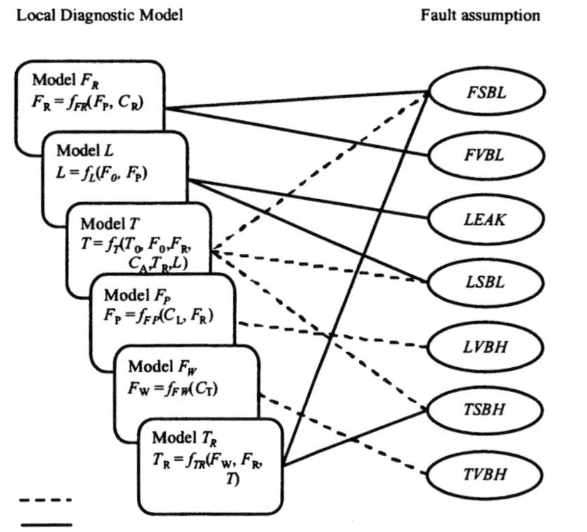

Fig. VI. The object-based diagnostic network for the CSTR process

The input matrix of each local model includes a part of its past observations to handle the time-dependency of the variables. The past values for all input variables always do not contribute to predict the output of a local model. For some input variables, both their past values and current values affect the current output of a local model. But in other cases, only the current values of the input mainly affect the output of a model. Whether or not the past values of the variables should be included in the input matrix is determined on the basis of the knowledge of the system and the comparison of the prediction performance of a local model via cross validation. Polynomial kernel was considered when applying Support Vector Machine regression technique. The determination of appropriate kernel parameters (e.g., polynomial degree of the polynomial kernel) was performed by cross validation. Table 2 summarizes the input-output structure of each local model and model parameters selected by cross validation.

5.4 On-line residual generation and evaluation

This study used 2σ of the residual distribution as the threshold for cumulative summation (*k*) and 6σ of the CUSUM distribution as the control limit (*h*).

Table 2 The input-output structure of diagnostic local models and selected parameters

Local Model	Inputs	Output	Kernel function in SVM
F_R	F_R(t-1), F_P(t-1) C_R(t), F_P(t)	F_R(t)	Polynomial (n=1)
L	L(t-1) F_0(t), F_P(t)	L(t)	Polynomial (n=1)
T	T(t-1) T_0(t), F_0(t), C_A(t) L(t), T_R(t), F_R(t)	L(t)	Polynomial (n=1)
F_P	F_P(t-1) C_L(t), F_R(t)	F_P(t)	Polynomial (n=1)
F_W	C_T(t)	F_W(t)	Polynomial (n=1)
T_R	T_R(t-1), F_W(t-1) T(t-1), F_R(t-1) F_W(t), T(t)	T_R(t)	Polynomial (n=2)

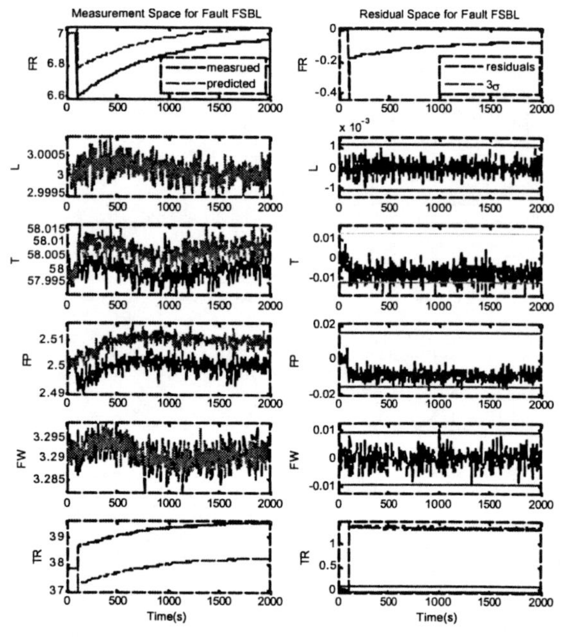

Fig. VII. The measured/predicted values and residuals for *FSBL*

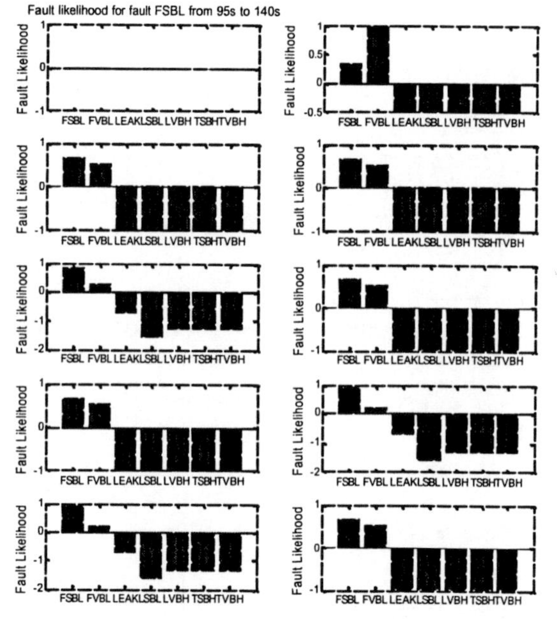

Fig. VIII. The failure likelihoods for *FSBL*

Fig. VII and VIII show the results for *FSBS (flow sensor biased low)*. The flow sensor for recycle flow is biased low (-0.7 kg/s). This fault affects the measurement F_R directly and the biased measurements break the internal relationships of the corresponding local models. The diagnostic network for the CSTR process (Fig. VI) shows that the fault *FSBL* is associated with the local models for F_R, T_R and T. The local model for F_R detected an abnormal behaviour at 100s and the local model for T_R gave a high signal at 105s and the local model for T gave a low signal at 110s (Fig. VII). Fig. VIII shows the fault likelihoods of considered faults for 10 sampling period. The fault likelihood of *FSBL* was largest after 105 sec and could be successfully diagnosed.
The diagnosis results of selected faults are summarized in Table 3.

5. CONCLUSION

This study proposes an object-based diagnostic network which can diagnose process faults effectively without every faulty data and an illustrative example showed the proficiency of the proposed method. The proposed diagnostic method is very practical in that data easily available from the process are used only and the modelling effort required for the development of diagnostic system is very cheap in comparison to analytical modelling.

Table 1 Selected faults of the CSTR process

Fault	Isolation Time (s)	Mismatch rate
FSBL	105	0.0026
FVBL	100	0
LEAK	110	0.2128
LSBL	105	0.2868
LVBH	100	0
TSBH	105	0.0026
TVBH	100	0

This approach provides a natural way of integrating qualitative information of the process into statistical learning structure

ACKNOWLEDGEMENT

We acknowledge the financial aid for this research provided by the Brain Korea 21 Program supported by the Ministry of Education. In addition, we would like to thank the Automation & Systems Research Institute and Research Institute of Engineering Science of Seoul National University.

REFERENCES

Lee, G and S.-O. Song (2003). Multiple-fault diagnosis based on system decomposition and dynamic PLS. *Ind. Eng. Chem. Res.*, **42**, 6145-6154.

Patan, K (2000). *Artificial dynamic neural networks and their applications in modelling of industrial processes*. Chapter 5, Ph. D. Dissertation, Warsaw University of Technology, Poland.

Petti, T. F. (1992). *Using mathematical models in knowledge-based control systems*. Ph. D. Thesis, University of Delaware, USA.

Finch, F. E. and M. A. Kramer (1988). Narrowing diagnostic focus using functional decomposition. *AIChE Journal*, **34**, 25-36.

Friedman, J. H. (1994). An overview of predictive learning and function approximation. In: *From Statistics to Neural Networks*, NATO ASI Series F 136, Springer Verlag, New York.

Cherkassky, V. and F. Muler (1998). *Learning from data: Concepts, Theory, and Methods*, John Wiley & Sons, New York.

Vapnik, V. N. (1995). *The Nature of Statistical Learning Theory*, Springer Verlag, New York.

Page, E. S. (1954). Continuous inspection scheme, *Biometrica*, **41**, 100-115.

Sorsa, T. and H. N. Koivo, Neural Networks in Process Fault Diagnosis, *Trans. Systems, Man and Cybernetics*, **21**, 815-825.

ELSEVIER
IFAC
PUBLICATIONS
www.elsevier.com/locate/ifac

IMPLICIT RELATIONS AND DISCRETE EVENTS IN PROCESS SIMULATION

David I. Wilson * **Christian Haag** **,[1]

* *Dept of Electrotechnology, Auckland University of Technology, New Zealand*
** *Dept. of Electrical Engineering, Karlstad University, Sweden*

Abstract: Simulation of hybrid systems is considerably more complex than pure continuous-time systems due to blocking states, exceptions due to discrete events, and Zeno-type behavior. Rather than esoteric, hybrid systems are actually the norm for industrial process simulation when including the possibility of tanks running dry, vessels overflowing etc. This paper compares two modelling tools; (one block-diagram, one causal) for a simple, but typical process engineering problem. Matlab/Simulink illustrates a traditional block diagram modelling tool, while MathModelica with its strong emphasis on objects and symbolic manipulation was used to demonstrate a novel approach to modelling. *Copyright © 2004 IFAC*

Keywords: Process simulation, symbolic manipulators, Modelica, DAEs, hybrid

1. INTRODUCTION

Process simulators have long been considered an essential tool in the chemical processing industry. Commercial simulators are now mature products with detailed libraries of unit operations, advanced control options, historians, and quality assurance programs. At their very core, chemical plants are often best described as interconnections of tanks and vessels, partially filled with liquids and it is essential that any simulator can robustly cope with this.

The typical chemical plant model is large, sparse, with a small number of important instances of highly nonlinear relations all with soft and hard constraints. While the importance of constraints and the nonlinearities does depend on the criteria of the simulation scenario under investigation, increasingly it is the unusual operating conditions such as start up and shut downs, recovery from major disturbances that effect the overall economics or where the payoff is to be found.

A good example of an industry that has modest demands for thermodynamics or esoteric unit operations is a paper mill. Here the key intermediate is dilute pulp which has rheological properties similar to water. Perhaps it is due to the low toxicity or the low relative value of the material, but in our experience in the pulp and paper industry, vessels regularly overflowed, or ran dry during operation. For this reason we are interested in simulation tools that reliably handle simple flow dynamics and events such as tanks running dry.

Our motivation to challenge the status quo of modelling tools is a direct result of the mediocre impact of a comprehensive pulp and paper simulation project reported in (Wilson and Balderud, 2000; Haag and Wilson, 2001; Balderud and Wilson, 2003). At the time, we favoured a block-diagram cause and effect modelling paradigm and used the dedicated simulator Simon's IDEAS specially tailored for the pulp and paper industry,

[1] Partially supported by the Skogsindustricentrum, Karlstad University, Sweden.

(McGarry *et al.*, 1997). In hindsight we have questioned that decision, and some comparisons and challenges have been raised in (Haag and Wilson, 2003). For example, Fig. 1 illustrates the small, but annoying simulation artifact of persistently oscillating levels in a scenario where we model the level of liquid in two interconnected tanks at the default tolerances and step sizes using a commercial process simulator.

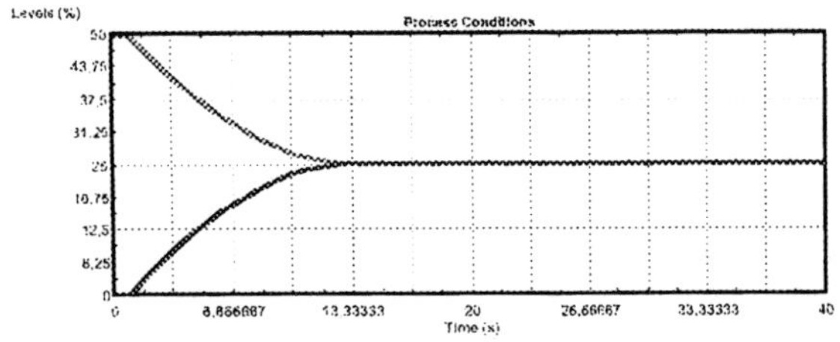

(a) Level trends of two connected tanks when valve is opened.

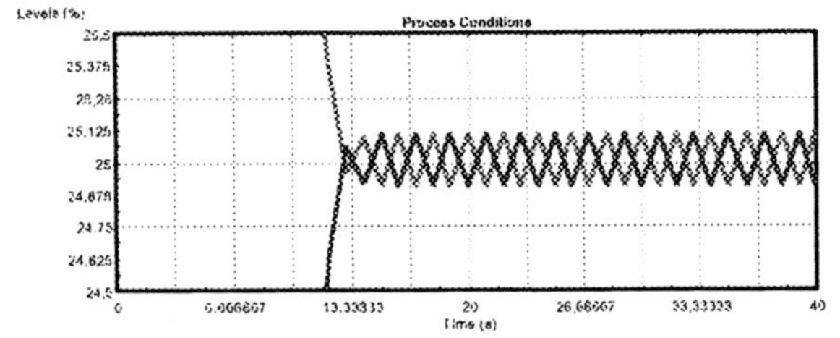

(b) Zoom version of Fig. 1(a) showing a persistent oscillation.

Fig. 1. Simulation results of levels in two interconnected tanks using IDEAS.

Consequently we were interested in the potential of 'acausal' modelling, perhaps best realised in the product DYMOLA, (Åström *et al.*, 1998). Causal/acausal tools represent an alternative classification for modelling tools from the block-oriented/equation-oriented division traditionally used by say (Marquardt, 1996). Acausal modelling tools combine object oriented thinking with a symbolic manipulator relieving the modeller from having to decide *apriori* information flow directions, i.e. what is an input, and what is an output. It is argued that this is a more natural way to build models, particularly for engineering staff whom may not be experienced modellers.

Our second interest was to explore the similarity between our immediate problem of interest (simulating levels and flows with possible underflow/overflow) with hybrid systems. Systems where continuous dynamics interact with discrete events are termed hybrid. The study of hybrid systems is an active research area due to the increasing importance of digital control, embedded systems and the modelling of phenomena that does not lend itself readily to differential equation type descriptions.

This has led some to take the approach that if one is not interested in logic circuits or bang-bang control policies, and avoids through good operating practice events such as tanks emptying, then process simulators that are primarily designed to solve large, sparse ODE systems are sufficient.

However we believe that rather than being esoteric or of only specialised interest, hybrid systems are actually far more prevalent in process control applications than seems generally appreciated. Certainly other process modelling texts such as (Hangos and Cameron, 2001) recognise the both the challenge and ubiquity of hybrid systems.

The outline of the remainder of the paper is as follows: section 2 introduces a simple, but representative process simulation model. Important hybrid considerations are summarised in section 2.1. Simulation results for a variety of schemes using standard tools are given in section 3. Finally some general observations relevant to process simulation are given in sections 4 and 5.

2. LEVEL IN A SPHERICAL TANK

The modelling of liquid level in a spherical vessel as shown in Fig. 2 is a minor modification of a classical exercise in process modelling. The

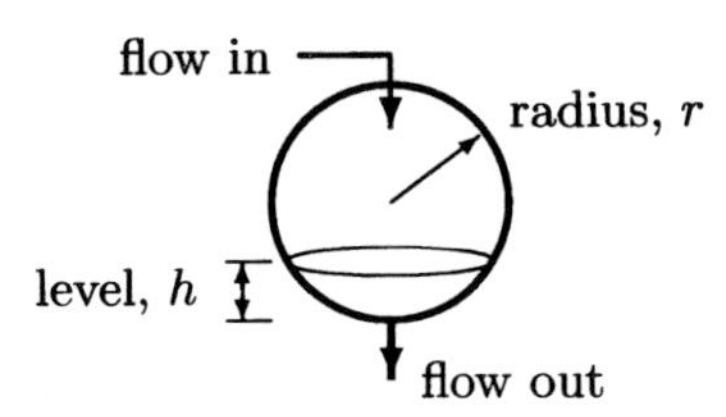

Fig. 2. Spherical tank/flow system

volume is given by a mass balance

$$\frac{dV}{dt} = F_{\text{in}} - F_{\text{out}}, \qquad V_0 = V(t=0) \tag{1}$$

where V_0 is the initial liquid volume in the tank, F_{in} is the flow rate into the tank, and the outflow, F_{out}, follows Torricelli's theorem

$$F_{\text{out}} = \beta\sqrt{h} \tag{2}$$

where β is a constant. The volume of liquid at height h in a spherical tank is

$$V = \frac{\pi}{3}\, h^2(3r - h) \tag{3}$$

and is physically meaningful in the range $0 \leq h \leq 2r$. Eqns 1–3 define the process to be simulated.

There are two implementation issues with this index-1 DAE system. One is the presence of multiple roots due to the inversion of Eqn. 3 as shown in Fig. 3, and the other is how to reliably simulate events such as the tank emptying.

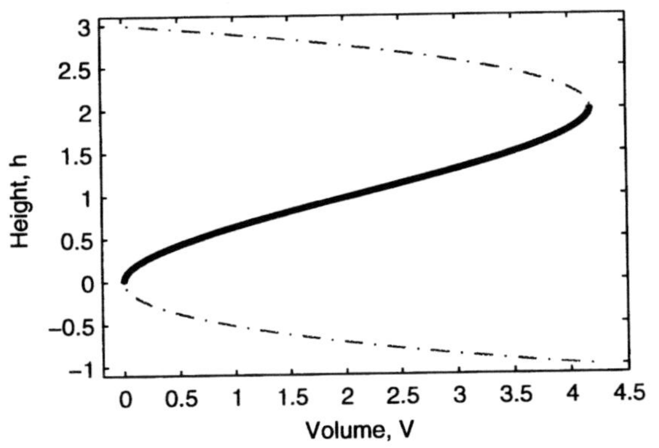

Fig. 3. The relationship between level, h, and volume, V for a spherical tank with radius $r = 1$, Eqn. 3.

2.1 Hybrid modelling

From a hybrid perspective, the process model is not simply Eqns 1–3 as stated above. The hybrid automaton given in Fig. 4 gives a more complete description by including the three discrete states, $Q \in$ {partially full, empty, overflowing}, that characterize the system.

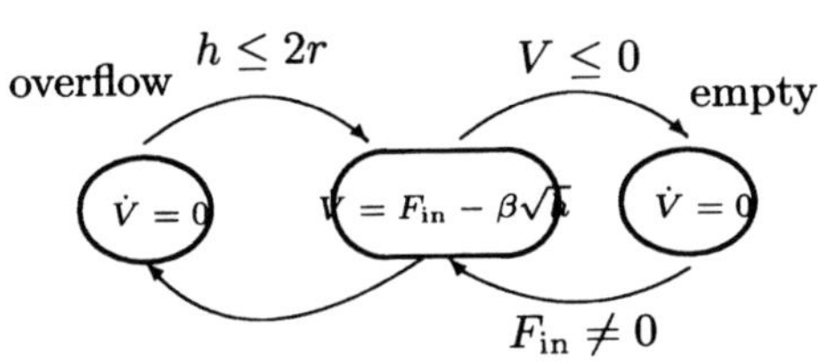

Fig. 4. Hybrid automaton for the tank level system

Under certain scenarios, the simulator rapidly switches between two discrete operating states, effectively stalling the simulation. This phenomena known as Zeno executions, (Zhang *et al.*, 2001), is unique to hybrid systems and is clearly of practical importance to simulators. In addition to accuracy and simulation time concerns, Zeno systems, or those sufficiently close to Zeno, can lead to a false sense of security when checking the hybrid system for criteria such as blocking states. This is because Zeno behaviour is not a true reflection of any physical system, rather it is an artifact of the modelling process.

There are two further challenging characteristics of the emptying tank simulation that are not shared by other common benchmark hybrid problems such as relays or bang/bang control problems. The first is that the continuous state (volume) becomes complex if the ODE system is applied outside the valid operating space. This poses problems for the numerical integrator routine that is attempting to accurately establish the emptying event location by using the standard zero crossing algorithms. The second problem is that the gradient of dV/dh at V or $h = 0$ is infinite causing further difficulties to the numerical root finder.

3. SIMULATION EXPERIENCES

For the purposes of illustration, we have chosen two general-purpose modelling environments. The first is MATLAB/Simulink which is a widely used directed block-diagram language clearly showing its classical control engineering heritage. MathModelica, (Jirstrand, 2000), is an acausal modelling product and, as the name suggests, is a hybrid of the computer algebra system Mathematica and the Modelica language compiler Dymola.

In the simulations that follow we model the emptying of a spherical tank with $r = 1, \beta = 0.09$ with no inflow, $F_{\text{in}} = 0$.

3.1 Retaining the implicit expression for h

From the modeller's perspective, the preferred option is to simply leave h as an implicit function of V and require that the simulation environment compute h when required. Both SIMULINK and MathModelica support DAE problems.

Fig. 5 illustrates the usage of the *Algebraic Constraint* option in SIMULINK to solve Eqn. 3. The advantage of such an approach is that the block diagram follows directly from the governing dynamic equations. The drawback is evident in the simulation results presented in Fig. 6. As the tank empties, h becomes negative, and SIMULINK quietly converts $\sqrt{h}$ to $\text{sign}(h)\sqrt{|h|}$ thereby retaining real, as opposed to complex, numbers. From this point on, *increasing* volume corresponds to *decreasing* level which is clearly nonsense.

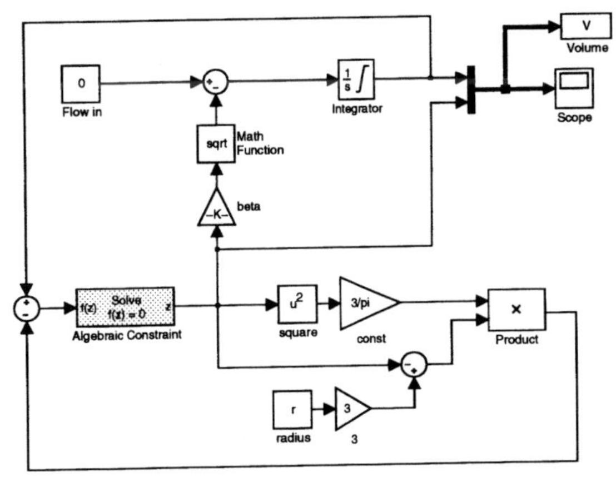

Fig. 5. Dynamic system in SIMULINK using the implicit Eqn. 3.

Initialising the simulation with the tank 100% full leads to a similar phenomenon. Here the solver again 'slips' into the wrong region and returns values of h higher than the top of the tank

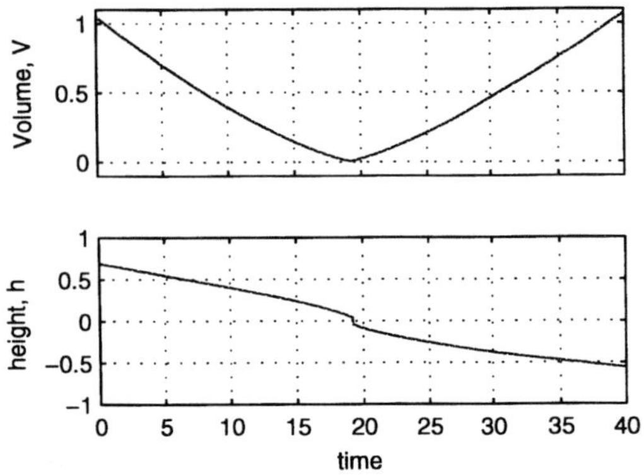

Fig. 6. Simulation result from Fig. 5 showing unrealisable levels.

indicating that *decreasing* volume corresponds to *increasing* level.

Both these phenomena are clearly artifacts due to how the model was implemented. The solver of the implicit algebraic equation is unaware of the physically valid limits of h. Adding a saturation block to constrain h to the interval $[0 \;\; 2r]$ avoids the case where the level becomes physically unrealisable, but at the expense of a spike at the emptying event, Fig. 7.

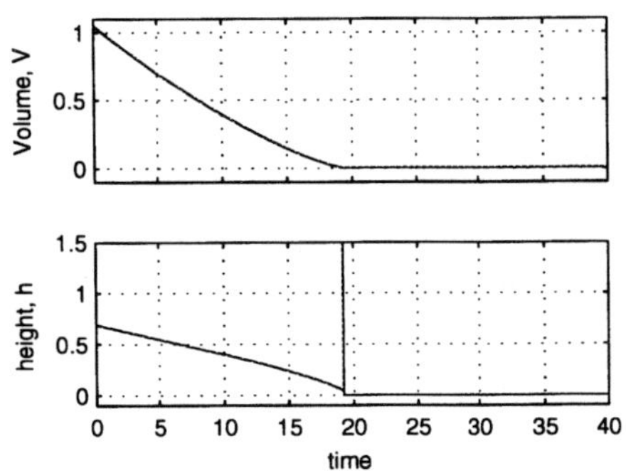

Fig. 7. Level and volume trends when saturating the level.

This crude method of trying to limit h is clearly suboptimal as it only cuts the algebraic equation solver's guess of h, $\hat{h}$, at the limits. It does not actually limit the solver's search region. Once the solver searches outside the limits, the input is effectively decoupled from the output, making any evaluation impossible.

3.2 MathModelica

One of MathModelica's design aims was to leverage Mathematica's symbolic capability within a simulation environment. In the MathModelica listing below, one simply lists the constituent equations leaving any necessary re-arrangement to the solver leading to a modelling approach many process engineers would be immediately comfortable with. To help ensure physically meaningful results, we specify the valid ranges for h and V.

```
model SphereTank "Spherical tank model"
// Constants
   constant Real pi=3.141592;
   constant Real g=9.81;

// Parameters
   parameter Real r=1.0 "Radius [m]";
   parameter Real beta = 0.02*sqrt(2*g);
   parameter Real Fin=0.0 "Inflow [m^3/s]";

// Variables
   Real V(min=0.0, max=4/3*pi*r^3,
        start=4/3*pi*r^3*0.5) "Volume of fluid [m^3]";
   Real h(min=0.0, max=2*r, start=2*r*0.5) "Level [m]";
   Real Fout "Outflow [m^3/s]";

equation
   V =  pi*r*h^2 - pi*h^3/3; // volume-level eqn
   Fout = beta*sqrt(abs(h)); // flow out
   der(V) = Fin - Fout;      // Conservation of mass
end SphereTank;
```

Fig. 8 indicates that MathModelica can easily manage the DAE, but as the level approached zero, the simulation stalled. Studying the diagnostics from the integrator routine revealed that it was still attempting to use negative numbers for h regardless of the specified valid range for h. Such Zeno-like behaviour is unsatisfactory in a process simulator.

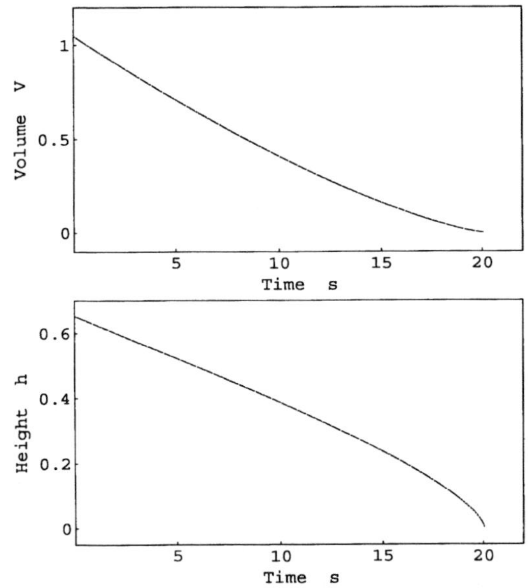

Fig. 8. MathModelica simulation stalling at the emptying event, $t \approx 20$s.

Subsequent communication with the developers confirmed that the Dymola kernel solver does not support the maximum/minimum attributes (as stated in the documentation), but rather only tests the valid range at $t = 0$.

3.3 Explicitly solving for h

The algebraic constraint, Eqn. 3, is a cubic equation which can be solved analytically, perhaps using a computer algebra tool, for h

$$h_1 = r + \frac{2^{1/3}\pi r^2}{M} + \frac{M}{2^{1/3}\pi} \quad (4)$$

$$h_2 = r - \frac{(1+i\sqrt{3})\pi r^2}{2^{2/3}M} - \frac{(1-i\sqrt{3})M}{\sqrt[3]{16\pi}} \quad (5)$$

$$h_3 = r - \frac{(1-i\sqrt{3})\pi r^2}{2^{2/3}M} - \frac{(1+i\sqrt{3})M}{\sqrt[3]{16\pi}} \quad (6)$$

where

$$M = \sqrt[3]{2\pi^3 r^3 - 3\pi^2 V + \pi^2\sqrt{3}\sqrt{3V^2 - 4\pi r^3 V}} \quad (7)$$

Not obvious from the above equations is that while all three roots for h are real in the region of interest $0 \le V \le 4\pi r^3/3$ as shown in Fig. 3, it is only h_3 that gives physically meaningful values. This can be proved from the observation that it is only dh_3/dV that is positive in the valid range. The SIMULINK implementation in Fig. 9 incorporates the explicit solution of Eqns 6 and 7.

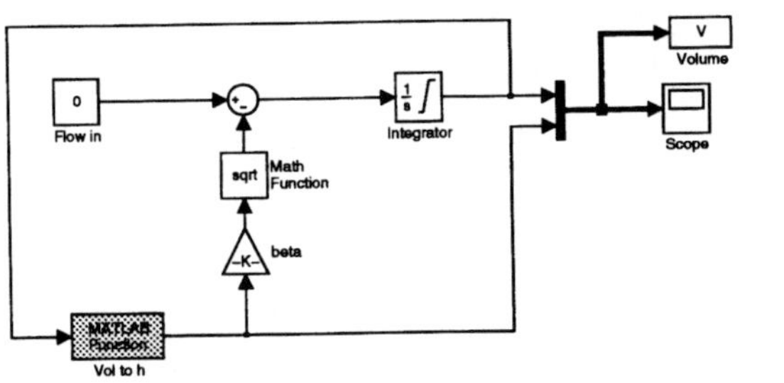

Fig. 9. Simulink model using the explicit expression for h given V.

Not surprisingly given the manual intervention required to construct the analytical inversion of Eqn. 3, the simulation performs as expected without the artifacts exhibited by the previous methods. However as the trends in Fig. 10 show, even this scheme is not completely robust. The 'hiccups' evident in the level, and to a lesser degree in the volume when the vessel is practically empty are almost two orders of magnitude above the requested numerical tolerance of 10^{-4}.

As Modelica has no intrinsic support for complex numbers, the implementation of equations 6 and 7 requires substantial re-working. While it is possible to manually define a complex number as a *struct* and develop the corresponding complex arithmetical libraries, this was considered too much work in general.

3.4 Casting level as the state

As this problem is an index-1 DAE, (refer (Ascher and Petzold, 1998, p232)), then a standard index reduction method is to differentiate the constraint equation, Eqn. 3, once and solving for dh/dt,

$$\frac{dh}{dt} = \frac{F_{\text{in}} - \beta\sqrt{h}}{\pi h(2r - h)} \quad (8)$$

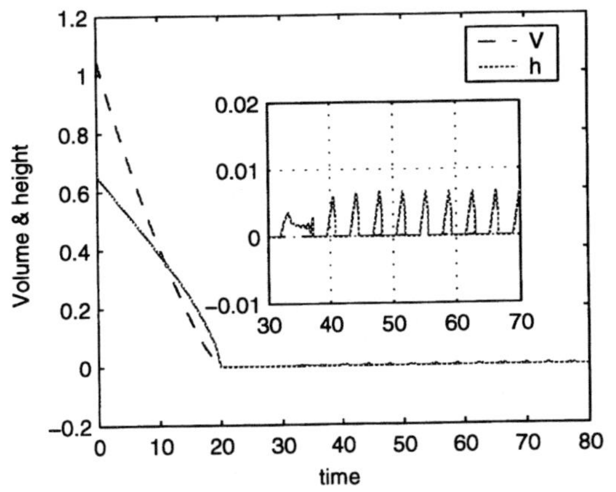

Fig. 10. Volume and level trends using the explicit expression from Fig. 9. The zoom portion in the insert shows evidence of chattering.

giving an index-0 problem. This is equivalent to considering level as the state variable as opposed to volume.

This eliminates the DAE problem, and by defining $F_{\text{out}} = \beta\text{sign}(h)\sqrt{|h|}$, as was done inadvertently in Fig. 6, then one can employ standard zero crossing detection techniques which is the commonly recommended solution approach to hybrid problems. One integrates up until the state-dependent event, stops and possibly adds or removes state dynamics, and then subsequently restarts until the next exception. The redefinition for F_{out} is even correct physically in the case where the height reflects the pressure differential between the bottom of the tank and atmospheric, and a reverse flow is possible. However changing the corresponding level dynamics from Eqn. 8 to $(F_{\text{in}} - \beta\sqrt{h})/(\pi|h|(2r-h))$ introduces a singularity at the event which cannot be integrated past using an adaptive scheme such as Runga-Kutta-Fehlberg.

4. DISCUSSION

The modelling of vessels emptying should be a trivial simulation exercise. However combining non-standard vessel shapes with the possibility of running dry, causes admittedly small, but still annoying, simulation hiccups. The emptying of a spherical tank combines implicit constraints with important exceptions. These problems are not just limited to tanks running dry, similar process scenarios are models involving implicit friction factor expressions operating across different flow regimes, reaction kinetics coupled with thermodynamics, and mechanical systems with friction.

While the emptying of a tank is clearly a toy problem, a full paper machine such as reported in (Balderud *et al.*, 2001) is not. In this application, adaptive error control was needed to complete the simulations in a reasonable time, but it should not stumble or exhibit Zeno-like behaviour when reaching underflows and overflows.

If approximate trends suffice, then crude integrators without error control typically suffice, oblivious of the numerical subtleties of the problem. To obtain clean results takes considerably more effort, effort that a modern process simulation environment could in principle do.

It is possible with manual intervention, employing analytical solutions where possible, judicious use of 'G-stops' and resetting the integrator, we can circumvent some of the simulation hiccups. However when using specific analytical solutions, one must take care that small changes to the problem formulation (such as plant rebuilds) do not necessitate major changes to the solution methodology.

Our SIMULINK simulations typically delivered a level trend, but not always the trend expected. When faced with complex levels,the automatic corrective action taken is justifiable. Adding saturation gives reasonable trends before and after the event, but spikes at the event causing severe problems to any derivative based level controller.

MathModelica, with the potential to combine symbolic manipulation and numerical calculation failed to simulate past the emptying event, and it lacked support for complex numbers making the alternative approach (solving the constraint equations off-line) unwieldy.

5. CONCLUSIONS

The realisation that the hybrid system is important in process simulation and not just appropriate in discrete logic circuits or discrete manufacturing is, we believe, important. The hybrid nature stems from both events that have often traditionally been sidestepped, if not outright ignored, in process simulation such as tanks emptying, in combination with the numerical integrator within the process simulator environment.

This paper investigates the potential of acausal modelling tools for process engineering applications that use symbolic manipulation as a means to ease the modeller's development task. Acausal tools have not yet quite reached the point where the modeller can simply specify the various dynamic and constraint equations, and confidently leave the resultant solution to the executive. Issues such as singularities and Zeno executions are still common in any non-trivial process simulation. Notwithstanding, we believe that process simulators that make use of symbolic manipulators are the way to achieve this goal.

REFERENCES

Ascher, Uri M. and Linda R. Petzold (1998). *Computer Methods for Ordinary Differential Equations and Differential-Algebraic Equations.* Society for Industrial and Applied Mathematics. Philadelphia, USA.

Åström, K.J., H. Elmqvist and S.E. Mattsson (1998). Evolution of Continuous-Time Modeling and Simulation. In: *The 12th European Simulation Multiconference, ESM'98.* Manchester, UK.

Balderud, Jonas and David Wilson (2003). The dynamic behaviour of the wet-end of a board machine. In: *CHEMECA 2003: 31st Annual Australasian Chemical Engineering Conference* (Dzuy Nguyen, Ed.). Adelaide, Australia.

Balderud, Jonas, Christian Haag and David Wilson (2001). Large-scale dynamic paper machine models. In: *6th World Congress of Chemical Engineering.* Melbourne, Australia. ISBN 0 7340 2201 8.

Haag, Christian and David Wilson (2001). Experiences of large-scale board machine modelling. In: *Proceedings of the IASTED Modelling, Identification and Control* (M.H. Hamza, Ed.). Innsbruck, Austria. pp. 745–750.

Haag, Christian and David Wilson (2003). Potential for symbolic-assisted process simulation. In: *IASTED conference on Modelling and Simulation 2003* (M. H. Hamza, Ed.). IASTED/ACTA Press. Palm Springs, Ca, USA. pp. 572–577.

Hangos, Katalin and Ian Cameron (2001). *Process Modelling and Model Analysis.* Academic Press.

Jirstrand, Mats (2000). Mathmodelica — a full system simulation tool. In: *Product Models 2000 - The Sixth Conference on Product Models, Global Product Development.* Linköping, Sweden.

Marquardt, W. (1996). Trends in Computer-Aided Process modelling. *Computers in Chemical Engineering* **20**(6/7), 591–609.

McGarry, M., B. Bickell and G. Pelkey (1997). A case study of the use of actual controls in simulation trainers. In: *83rd Annual meeting technical section — paper, people, progress.* Vol. 78. Canadian Pulp and Paper Association. Montreal, Que, Canada. pp. B357–B360.

Wilson, David I. and Jonas Balderud (2000). Experiences building and validating a large-scale dynamic paper machine model. In: *Model Validation for Plant Control and Condition Monitoring* (D.H. Owens, Ed.). IEE, Savoy Place London, WC2R 0BL, UK. pp. 2/1–2/7.

Zhang, Jun, Karl Henrik Johansson, John Lygeros and Shankar Sastry (2001). Zeno hybrid systems. *International Journal of Robust and Nonlinear Control* **11**, 435–451.

ELSEVIER
IFAC
PUBLICATIONS
www.elsevier.com/locate/ifac

PLANT-WIDE OPTIMAL CONTROL WITH DECENTRALIZED MPC

**Aswin N. Venkat * James B. Rawlings *,[1]
Stephen J. Wright ****

* *Dept. of Chemical and Biological Engineering, University of Wisconsin, Madison-53706*
** *Dept. of Computer Science, University of Wisconsin, Madison-53706*

Abstract: Most standard MPC implementations partition the plant into several units and apply MPC individually to these units. It is known that this strategy can lead to sub-optimal plant-wide control performance, especially if the units interact strongly. This paper tackles the problem of achieving optimal control performance in plants with such an MPC structure. A modeling framework, geared for use in MPC, that incorporates the interactions between the subsystems is employed. One may think that modeling the interactions and communicating the control actions between the controllers is sufficient to improve controller performance. We show that this idea is incorrect and can lead to closed-loop instability. A cooperation based MPC algorithm that converges to the plant-wide optimum is developed. In practical implementations, the cooperation based MPC scheme may have to be terminated before convergence is reached. To permit such flexibility, we propose a feasible cooperation based MPC algorithm. All cooperative iterates in this algorithm are feasible and the resulting MPC controller is closed-loop stable. Two examples comparing the performance of optimal and sub-optimal MPC controllers are presented.

Keywords: MPC, plant-wide control

1. INTRODUCTION

With ever increasing demands on improvements in productivity and efficiency of operation, the chemical industry today places significant importance on plant-wide automation. Improvements in practical control technology can potentially cut costs and raise profits. Over the last decade, Model Predictive Control (MPC) has established itself as one of the popular choices for advanced process control.

A number of articles have focused on improved plant-wide decentralized control. (Sandell-Jr. *et al.*, 1978) provide a survey of decentralized control methods for large scale systems. Some of these decentralized controller design approaches approximate or ignore the interactions between the various subsystems (Lunze, 1992; Siljak, 1991). Other approaches lead to a sub-optimal plant-wide control strategy (Acar and Ozguner, 1988; Samyudia and Kadiman, 2002). (Cui and Jacobsen, 2002) describe performance limitations arising due to the decentralized control framework. Most attention in the area of MPC has focussed on the centralized control framework. However, centralized control may not be practical, especially for large systems. In recent times, there has been some interest in studying plant-wide control within the MPC framework. (Lu, 2000; Kulhavy *et al.*, 2001) outlined the idea of cross-functional integration within the MPC framework and discussed requirements and potential benefits/impact of such technology. (Katebi and Johnson, 1997) proposed a two level de-

[1] Corresponding author, email:jbraw@bevo.che.wisc.edu

composition coordination strategy for Generalized Predictive Control(GPC) based on the master-slave paradigm. (Zhu *et al.*, 2000; Zhu and Henson, 2002) described a plant-wide control strategy based on the integration of linear and nonlinear MPC coupled with a plant decomposition procedure. A sub-optimal strategy for MPC of interconnected systems was proposed by (Antwerp and Braatz, 2000). While these methods have been demonstrated to work well for the cases considered, no safeguards against failure or closed-loop properties have been established. (Camponogara *et al.*, 2002) proposed a distributed model predictive control scheme to coordinate multiple agents. In their scheme, the MPC controllers exchange state and input trajectory information amongst themselves during a sampling interval. We shall show through examples in section 6 that a strategy based purely on the exchange of trajectory information (i.e. communication) is unreliable and may cause closed-loop instability.

The purpose of this work is to understand the idea of optimal plant-wide control from the MPC framework and provide algorithms that attain the best achievable plant-wide control performance through the suitable integration of the various MPC controllers. The inherent communication capability of the prediction horizons and an enhanced modeling framework are utilized in arriving at these algorithms. For the methods described in this work, we prove nominal properties like optimality, feasibility and closed-loop stability. These properties are central to the practical applicability of plant-wide control methods.

2. MODELING INTERACTIONS FOR PERFORMANCE IMPROVEMENT

Decentralized models Consider a plant comprising of M subsystems. Let the decentralized (local) model for each subsystem be represented by a discrete linear time invariant (LTI) model of the form

$$x_{ii}(k+1) = A_{ii}x_{ii}(k) + B_{ii}u_i(k), \quad i = 1, 2, \ldots, M \tag{1}$$

in which k is discrete time, and we assume (A_{ii}, B_{ii}, C_{ii}) is a minimal realization for each (u_i, y_i) input-output pair. Owing to material, energy and/or information flows there exists a level of interaction between the subsystems. In the decentralized modeling framework, it is assumed that the interactions have a negligible effect on local variables. This assumption is not reliable in many situations and can lead to deterioration in control performance. The centralized modeling framework, on the other hand, results in models that are larger than necessary. In many cases, centralized control is not feasible for implementation due to its sheer size, multiple time scales of operation, and limited operational flexibility.

Interaction models (IM): We employ IM to quantitatively assess the interactions between the subsystems. Modeling the interactions between subsystems provides a framework for improving plant-wide control performance while retaining most of the advantages of the decentralized control approach.

Consider a subsystem i $(i = 1, 2, \ldots, M)$. We represent the effect of an interacting subsystem j, $j \neq i$ on subsystem i through a discrete LTI model of the form

$$x_{ij}(k+1) = A_{ij}x_{ij}(k) + B_{ij}u_j(k) \quad j = 1, \ldots, M \tag{2}$$

in which B_{ij} represents the effect of the inputs of subsystem j on the states of subsystem i.

For each subsystem i, the decentralized state vector x_{ii} is augmented with states arising due to interactions with other subsystems. Let x_i denote the augmented set of states for subsystem i. The interaction model (IM) for the entire plant can therefore be expressed as a union of decentralized and interaction models. After identification of the significant interactions from closed-loop operating data, we expect that many of the interaction terms will be zero. In the decentralized model, all of the interaction terms are zero.

3. PROBLEM FORMULATION AND ASSUMPTIONS

In this work, we consider five formulations ($\mathcal{P}^1$ to $\mathcal{P}^5$) for unconstrained model predictive control (MPC), or LQR. In each case, a controller is defined by using the first input from the solution to the corresponding optimization problem.

$$\mathcal{P}^1: \quad \textbf{Centralized MPC} \tag{3}$$

$$\begin{aligned} \min_{\boldsymbol{x},\boldsymbol{u}} \quad & \Phi(\boldsymbol{x},\boldsymbol{u}) = \sum_i \Phi_i(\boldsymbol{x},\boldsymbol{u}) \\ \text{s.t.} \quad & x(t+1) = Ax(t) + Bu(t) \\ & x(k) = \widehat{x}(k) \\ & t = k, k+1, \ldots, k+N-1 \end{aligned}$$

in which

$$\begin{aligned} \boldsymbol{x} &= \{x(k), x(k+1), \ldots, x(k+N)\} \\ \boldsymbol{u} &= \{u(k), u(k+1), \ldots, u(k+N-1)\} \\ & x \in \mathbb{R}^n, u \in \mathbb{R}^m \end{aligned}$$

represent the centralized state and input trajectories through the control horizon. The horizon length is N. The cost function for subsystem i is Φ_i. The vector $\widehat{x}(k)$ represents the current estimate of the centralized model states at discrete time k.

$$\mathcal{P}_i^2: \quad \textbf{Decentralized MPC} \tag{4}$$

$$\begin{aligned} \min_{\boldsymbol{x}_{ii},\boldsymbol{u}_i} \quad & \Phi_i(\boldsymbol{x}_{ii},\boldsymbol{u}_i) \\ \text{s.t.} \quad & x_{ii}(t+1) = A_{ii}x_{ii}(t) + B_{ii}u_i(t) \\ & x_{ii}(k) = \widehat{x}_{ii}(k) \\ & t = k, k+1, \ldots, k+N-1 \end{aligned}$$

in which $\boldsymbol{x}_{ii}$ represents the decentralized state trajectory for subsystem i.

$$\boldsymbol{x}_{ii} = \{x_{ii}(k), x_{ii}(k+1), \ldots, x_{ii}(k+N)\}$$

with $x_{ii}(k) \in \mathbb{R}^{n_{ii}}$. $\widehat{x}_{ii}(k)$ denotes the estimate of the decentralized model states at discrete time k. For communication and cooperation based MPC, an iteration and exchange of variables between subsystems is performed during a sample time. We denote this iteration number as p.

$$\mathcal{P}_i^3: \quad \textbf{Communication based MPC} \qquad (5)$$

$$\begin{aligned} \min_{\boldsymbol{x}_i^p, \boldsymbol{u}_i^p} \quad & \Phi_i\left(\boldsymbol{x}_i^p, \boldsymbol{u}_i^p, \boldsymbol{u}_{j\neq i}^{p-1}\right) \\ \text{s.t.} \quad & x_i^p(t+1) = A_i x_i^p(t) + B_{ii} u_i^p(t) \\ & \quad + \sum_{j\neq i} B_{ij} u_j^{p-1}(t) \\ & x_i(k) = \widehat{x}_i(k) \\ & t = k, k+1, \ldots, k+N-1 \end{aligned}$$

in which

$$\begin{aligned} \boldsymbol{x}_i &= \{x_i(k), x_i(k+1), \ldots, x_i(k+N)\} \\ \boldsymbol{u}_i^p &= \{u_i^p(k), u_i^p(k+1), \ldots, u_i^p(k+N-1)\} \end{aligned}$$

and $\widehat{x}_i(k)$ represents the current estimate of the communication model states. Notice that the input sequence for subsystem i, $\boldsymbol{u}_i^p$, is optimized to produce its value at iteration p, but the other subsystem's inputs are not updated during this optimization; they remain at iterate $p-1$. The objective function is the one for subsystem i only.

We next modify the objective functions of the subsystem's controllers in order to provide a means for cooperative behavior among the controllers. We replace the objective Φ_i with an objective that measures the entire system performance. Many suitable objectives are possible. Here we choose the simplest case, the overall plant objective, which is the weighted sum of all the subsystems' objectives, $\Phi = \sum_i \mathrm{w}_i \Phi_i$. When all the cost functions are quadratic, this formulation exactly matches the centralized cost.

$$\mathcal{P}_i^4: \quad \textbf{Cooperation based MPC} \qquad (6)$$

$$\begin{aligned} \min_{\boldsymbol{x}_i^p, \boldsymbol{u}_i^p} \quad & \Phi\left(\boldsymbol{x}_i^p, \boldsymbol{u}_i^p, \boldsymbol{x}_{j\neq i}^{p-1}, \boldsymbol{u}_{j\neq i}^{p-1}\right) \\ \text{s.t.} \quad & x_i^p(t+1) = A_i x_i^p(t) + B_{ii} u_i^p(t) \\ & \quad + \sum_{j\neq i} B_{ij} u_j^{p-1}(t) \\ & x_i(k) = \widehat{x}_i(k) \\ & t = k, k+1, \ldots, k+N-1 \end{aligned}$$

4. COOPERATION BASED CONTROL

Sub-optimality (in the plant-wide sense) of the communication based MPC scheme provides the necessary motivation to seek an alternate approach–one that is plant-wide optimal in the nominal case. We note that while the communication based scheme accounts for the effect of the interacting subsystems, it fails to consider the effect of changes in local variables on other subsystems. Such an exchange scheme can give rise to conflicts between the various communicating controllers and lead to deterioration in control performance (see section 6.1). The cooperation based MPC scheme eliminates the possibility of such controller conflicts by utilizing a cost function that reflects the plant-wide effect of local variable changes. Each local regulator now charts a control path that not only minimizes the local objectives but also a projected cost at the plant-wide level. The algorithm and some useful properties of this method are outlined below.

Algorithm 1. Given $(\boldsymbol{x}_i^0, \boldsymbol{u}_i^0)$, $\quad i = 1, 2, \ldots, M$ and $\quad \epsilon > 0$
$p \leftarrow 1, \quad e_i \leftarrow \Gamma\epsilon \quad \Gamma \gg 1$
while $\quad e_i > \epsilon$
$(\boldsymbol{x}_i^p, \boldsymbol{u}_i^p) = \arg\{\mathcal{P}_i^4\}, \quad i = 1, 2, \ldots, M$
$e_i = \|(\boldsymbol{x}_i^p, \boldsymbol{u}_i^p) - (\boldsymbol{x}_i^{p-1}, \boldsymbol{u}_i^{p-1})\|$
$p \leftarrow p + 1$
end (while)

At discrete time k, define

$$\overline{\boldsymbol{x}}_i(k) = [x_i(k+1|k), \ldots, x_i(k+N|k)]^T, \qquad (7)$$

$$\overline{\boldsymbol{u}}_i(k) = [u_i(k|k), \ldots, u_i(k+N-1|k)]^T \qquad (8)$$

to be the vector of predicted states and inputs respectively, through the control horizon. For the particular case for which all the cost functions are quadratic, we can represent the cost function for the ith subsystem as

$$\Phi_i\left(\overline{\boldsymbol{x}}_i(k), \overline{\boldsymbol{u}}_i(k)\right) = \frac{1}{2}\overline{\boldsymbol{x}}_i^T(k)\mathbb{Q}_i\overline{\boldsymbol{x}}_i(k) + \frac{1}{2}\overline{\boldsymbol{u}}_i^T(k)\mathbb{R}_i\overline{\boldsymbol{u}}_i(k) \qquad (9)$$

in which

$$\begin{aligned} \mathbb{Q}_i &= \text{diag}\left(Q_i(1), Q_i(2), \ldots Q_i(N)\right) \quad \text{and} \\ \mathbb{R}_i &= \text{diag}\left(R_i(0), R_i(1), \ldots R_i(N-1)\right) \end{aligned}$$

denote the state and input penalties in the regulator through the control horizon. Eliminating the states from the model equations ((1), (2)) and propagating the inputs through the control horizon, we can rewrite the interaction model for subsystem i as

$$\overline{\boldsymbol{x}}_i(k) = E_{ii}\overline{\boldsymbol{u}}_i(k) + \sum_{j=1, j\neq i}^{M} E_{ij}\overline{\boldsymbol{u}}_j(k) + f_i x_i(k) \qquad (10)$$

in which

$$x_i(k) = \widehat{x}_i(k) \qquad (11)$$

$$E_{ij} = \begin{bmatrix} B_{ij} & 0 & \ldots & \ldots & 0 \\ A_i B_{ij} & B_{ij} & 0 & \ldots & 0 \\ \vdots & \vdots & \vdots & \vdots & \vdots \\ A_i^{N-1} B_{ij} & \ldots & \ldots & \ldots & B_{ij} \end{bmatrix} \qquad f_i = \begin{bmatrix} A_i \\ A_i^2 \\ \vdots \\ \vdots \\ A_i^N \end{bmatrix} \qquad (12)$$

Theorem 1. Given $Q_i(j) > 0$, $R_i(j) > 0$ and $Q_i(N) > 0 \quad j = 0, 1, 2, \ldots N-1, \quad i = 1, 2, \ldots, M$. Algorithm 1 converges to an optimal limit point.

Theorem 2. If the initial trajectory $z^0 = (x^0, u^0)$ satisfies $\|z^0 - z^*\| \leq \frac{1}{\epsilon^\delta}$ for some $\delta \geq 0$ and $\epsilon > 0$ then we have $\|z^p - z^*\| < \epsilon$ for all $p > \frac{(\delta+1)\log\epsilon}{\log\Omega}$, $0 < \lambda < \Omega < 1$ in which $0 < \lambda < 1$ is a system dependent property.

5. FEASIBLE COOPERATION BASED CONTROL

Theorem 2 provides a conservative lower bound on the number of cooperative iterates required for convergence. It is possible that the process sampling time is shorter than the time required for convergence of the cooperative iterates. To facilitate the practical use of the cooperative control methodology, we need the iterates generated by the cooperation based MPC scheme to be feasible and closed-loop stable. The above two properties permit us to terminate the cooperative scheme at the end of each sampling interval and inject the final iterate into the plant even if convergence has not been attained. In order to satisfy feasibility and closed-loop stability, we propose modifications to the existing cooperation based MPC framework. For the purpose of this study, we restrict ourselves to MPC controllers with quadratic cost functions.

To guarantee feasibility, we eliminate the model constraints and solve the cooperation based MPC problem as a collection of unconstrained optimization problems of the form

$\mathcal{P}_i^5$: **Feasible cooperation based MPC**

$$\min_{u_i^p} \ \Phi_i^p(x_i(k), \overline{u}_{j\neq i}^{p-1}) = \frac{1}{2}\overline{u}_i^{p^T}(k)\mathfrak{R}_i\overline{u}_i^p(k) + \left(r_i(k) + \sum_{j=1, j\neq i}^{M} H_{ij}\overline{u}_j^{p-1}(k)\right)^T \overline{u}_i^p(k) \tag{13}$$

in which

$$\mathfrak{R}_i = \mathbb{R}_i + E_{ii}^T\mathbb{Q}_i E_{ii} + \sum_{j\neq i}^{M} E_{ji}^T\mathbb{Q}_j E_{ji} \tag{14}$$

$$r_i(k) = E_{ii}^T\mathbb{Q}_i f_i x_i(k) + \sum_{j\neq i}^{M} E_{ji}^T\mathbb{Q}_j f_i x_j(k) \tag{15}$$

$$H_{ij} = E_{ii}^T\mathbb{Q}_i E_{ij} + E_{ji}^T\mathbb{Q}_j E_{jj} \tag{16}$$

$$x_j(k) = \widehat{x}_j(k) \tag{17}$$

We have the following results for feasible cooperation based MPC

Lemma 3. Given the MPC formulation $\mathcal{P}_i^5$, the sequence of cost functions

$$\{\Phi^p\} = \{\sum_{i=1}^{M} \Phi_i^p\left(x_i(k), \overline{u}_{j\neq i}^{p-1}\right)\}$$

is a non-increasing function of p.

At time k, let

$$\overline{u}_i^{p*}\left(x_i(k), \overline{u}_{j\neq i}^{p-1}\right) = \left[u_i^{p*}(k|k), \ldots, u_i^{p*}(k+N-1|k)\right]^T$$

represent the solution to the optimization problem $\mathcal{P}_i^5$ at iteration number p. The corresponding optimum value of the cost function is denoted as $\Phi_i^{p*}(x_i(k), \overline{u}_{j\neq i}^{p-1})$. The control law is obtained through a receding horizon implementation of optimal control whereby the input applied to subsystem i is $u_i^p(x_i(k), \overline{u}_{j\neq i}^{p-1}) = u_i^{p*}(k|k)$. Lemma 3 leads to the following theorem on closed-loop stability.

Theorem 4. Given the MPC formulation $\mathcal{P}_i^5$ $i = 1, 2, \ldots, M$. Suppose the following assumptions are satisfied

- (A_i, B_i) stabilizable
- Perfect knowledge of the states at each sampling instant k (state feedback).
- $Q_i(0) = Q_i(1) = \cdots = Q_i(N-1) = Q_i$
- $R_i(0) = R_i(1) = \cdots = R_i(N-1) = R_i$
- Stage cost $L_i(k+j|k) = \frac{1}{2}x_i(k+j|k)^T Q_i(j) x_i(k+j|k) + \frac{1}{2}u_i(k+j|k)^T R_i(j) u_i(k+j|k) > 0, \quad j = 0, 1, \ldots, N-1.$
- Terminal state constraint $x_i(k+N|k) = 0$
- $u_i(k+j|k) = 0, j \geq N$

then the origin is an asymptotically stable equilibrium point for the closed-loop system $x(k+1) = Ax(k) + Bu^p(x(k), \overline{u}^{p-1})$, in which $u^p(x(k), \overline{u}^{p-1}) = \left[u_1^p(x_1(k), \overline{u}_{j\neq 1}^{p-1}), \ldots, u_M^p(x_M(k), \overline{u}_{j\neq M}^{p-1})\right]^T$, for all $x(k)$ and all $p = 1, 2, \ldots$.

6. EXAMPLES

We present two examples to demonstrate and compare the control performance of the various MPC schemes described in this paper ($\mathcal{P}^1 - \mathcal{P}^5$). To provide a consistent platform for comparison, we ensure the following

- Perfect state feedback.
- Each regulator utilizes the optimal targets i.e. the centralized model targets, in evaluating its optimal control trajectory.
- No disturbances affect the plant.

To assess controller performance , we use the cumulative stage cost (CSC) as an index to quantify controller performance. The CSC index is given by $\sum_k \sum_{i=1}^{M} L_i(k|k)$. In addition, the iterative schemes for methods $\mathcal{P}_i^3, \mathcal{P}_i^4$ and $\mathcal{P}_i^5$ are initialized with the corresponding decentralized control solution.

6.1 *Example 1*

We consider a plant constituted by two subsystems. The first subsystem is represented by a LTI state space model comprising of 3 inputs, 3 outputs and 5 states. The LTI model for the second

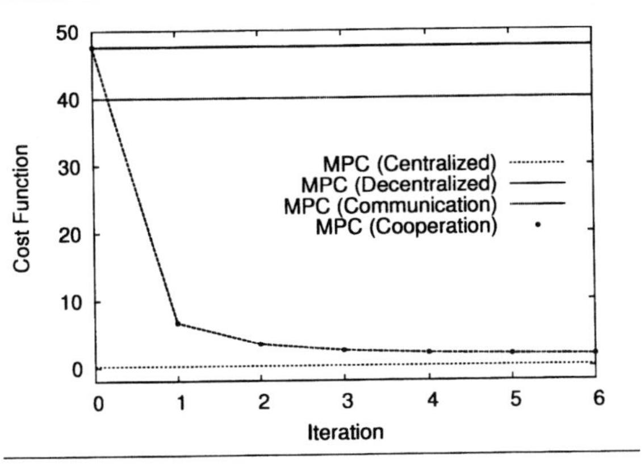

Fig. 1. Performance of the various schemes within a sampling interval

Table 1. Controller performance

MPC Configuration	Cumulative Stage Cost
Decentralized MPC	344.45
Communication based MPC	∞ (Unstable)
Cooperation based MPC (1 iterate)	5.581
Cooperation based MPC (10 iterates)	3.307
Cooperation based MPC (convergence)	2.654
Centralized MPC	2.654

subsystem consists of 4 inputs, 4 outputs and 8 states. The two units are such that the behavior of one subsystem affects the performance of the other.

We investigate the performance of the decentralized, centralized, communication and cooperation based MPC frameworks when a set point change is made to the outputs of subsystems 1 and 2. For scheme $\mathcal{P}_i^5$, the terminal state constraint (Theorem 4) is employed to guarantee closed-loop stability. Table 1 summarizes the controller performance for the various schemes considered ($\mathcal{P}^1 - \mathcal{P}_i^5$).

The decentralized MPC regulators attempt to track their respective set points with no knowledge of the effect of the interacting subsystem. Control performance consequently suffers and the deterioration in control performance is reflected as a relatively large CSC index. The centralized MPC regulator on the other hand is equipped with the complete model and all interactions are accounted for exactly. This controller therefore achieves the optimal nominal plant-wide control performance. The sequence of iterates generated by the communication based MPC scheme converge to a non-optimal limit point. In this case, the resulting communication based MPC controller leads to unstable closed-loop behavior. This example illustrates the unreliability of the communication based MPC scheme ($\mathcal{P}_i^3$). In accordance with Theorem 1, the cooperation based MPC scheme $\mathcal{P}_i^4$ converges and the performance is identical to the centralized MPC controller. For scheme $\mathcal{P}_i^5$, we examine closed loop performance when the cooperative scheme is terminated after 1 iterate and when the scheme is terminated after 10 iterates. As expected from lemma 3, the CSC index decreases with the iteration number until we reach the centralized CSC index. Figure 1 shows the performance of the various schemes within a sampling interval. It is verified that the controller resulting from scheme $\mathcal{P}_i^5$ stabilizes the plant in closed-loop for all values of p.

6.2 *Example 2*

Table 2. Controller performance

MPC Configuration	Cumulative Stage Cost
Decentralized MPC	76.953
Communication based MPC	52.773
Cooperation based MPC (1 iterate)	51.142
Cooperation based MPC (10 iterates)	50.529
Cooperation based MPC (convergence)	50.495
Centralized MPC	50.495

The second example is a reactor separator with recycle. The control challenges posed by the dynamics of the recycle stream has been studied by (Luyben, 1993*a*), (Luyben, 1993*b*) and (Luyben, 1994). (Wu *et al.*, 2002) and (Monroy-Loperena *et al.*, 2004) have addressed the problem of plant-wide control structure selection for reactor separator processes with recycle. For the process considered, fresh feed of reactant A enters a CSTR where an exothermic reaction A → B takes place. The outlet stream from the reactor is a mixture of product B and unreacted A. This stream is fed to a distillation column where the separation of the two occurs. The distillate, rich in reactant A, is recycled back to the reactor. The bottoms stream is almost pure B and is drawn out. It is desired to maintain the reactor temperature at 331.5°C. The flow rate to the reactor is manipulated to achieve this objective. The control objective for the distillation column is to maintain the top and bottom compositions of A at their desired specs by manipulating the reflux to the column and the vapor boil up flow rate.

The flow of the reactor outlet stream to the distillation column and the recycle of the distillate back to the reactor are sources of complex interaction. A change in the reactor operating conditions affects the performance of the distillation column and vice-versa. For the purpose of this study, a linearized model around the desired steady state is used to represent the plant.

Due to changes in operating conditions, it is desired to decrease the reactor temperature by 5°C and increase the composition of A in the distillate by 0.05. Decentralized MPC gives unsatisfactory performance as this mode of operation does not consider the two-way interaction between the reactor and the distillation column. In comparison, communication based MPC leads to an improvement in control performance. However, as observed in section 6.1, communication based MPC is an unreliable control strategy as it lacks well defined convergence and closed-loop properties.

With cooperation based MPC, we require a maximum of 28 cooperative iterates to achieve convergence. To permit real-time implementation, we employ the feasible cooperation based MPC controller in conjunction with Theorem 4. This controller is guaranteed to be feasible and closed-loop stable. In accordance with Lemma 3, the cost function is observed to be a non-increasing function of the iteration number p at each time step. The performance of the various MPC controllers are summarized in Table 2. For this example, we note that a single cooperative exchange leads to a 33.5 percent improvement in control performance in comparison with decentralized MPC.

7. CONCLUSIONS

In this work, the problem of integrating several MPC controllers to achieve optimal plant-wide control was addressed. The proposed cooperative MPC methodology is plant-wide optimal at convergence. To facilitate real time implementation, we propose a feasible cooperative MPC strategy. All iterates generated by this scheme are feasible and the resulting MPC controller is closed-loop stable. The theoretical aspects and performance of the proposed methods are illustrated through two examples.

REFERENCES

Acar, L. and U. Ozguner (1988). A completely decentralized suboptimal control strategy for moderately coupled interconnected systems. *Proceedings of the American Control Conference* **45**, 1521–1524.

Antwerp, J.G.V. and R.D. Braatz (2000). Model predictive control of large scale processes. *J. Proc. Control* **10**, 1–8.

Camponogara, Eduardo, Dong Jia, Bruce H. Krogh and Sarosh Talukdar (2002). Distributed model predictive control. pp. 44–52.

Cui, H. and E.W. Jacobsen (2002). Performance limitations in decentralized control. *J. Proc. Control* **12**, 485–494.

Katebi, M.R. and M.A. Johnson (1997). Predictive control design for large-scale systems. *Automatica* **33**, 421–425.

Kulhavy, R., J. Lu and T. Samad (2001). Emerging technologies for enterprise optimization in the process industries. In: *Chemical Process Control–CPC 6*. pp. 201–216. Tuscon, Arizona.

Lu, J. (2000). Multi-zone control under enterprise optimization: needs, challenges and requirements. In: *Nonlinear Model Predictive Control* (F. Allgower and A. Zheng, Eds.). Vol. 26. pp. 393–402. Birkhauser.

Lunze, J. (1992). *Feedback Control of Large Scale Systems*. Prentice-Hall. U.K.

Luyben, William L. (1993*a*). Dynamics and control of recycle systems. 1. simple open-loop and closed-loop systems. *Ind. Eng. Chem. Res.* **32**, 466–475.

Luyben, William L. (1993*b*). Dynamics and control of recycle systems. 2. comparison of alternative process designs. *Ind. Eng. Chem. Res.* **32**, 476–486.

Luyben, William L. (1994). Snowball effects in reactor/separator processes with recycle. *Ind. Eng. Chem. Res.* **33**, 299–305.

Monroy-Loperena, Rosendo, Rocio Solar and Jose Alvarez-Ramirez (2004). Balanced control scheme for reactor/separator processes with material recycle. *Ind. Eng. Chem. Res.* **43**, 1853–1862.

Samyudia, Y. and K. Kadiman (2002). Control design for recycled, multi unit processes. *J. Proc. Control* **13**, 1–14.

Sandell-Jr., Nils R., Pravin Varaiya, Michael Athans and Michael Safonov (1978). Survey of decentralized control methods for larger scale systems. **23**(2), 108–128.

Siljak, D.D. (1991). *Decentralized Control of Complex Systems*. Academic Press. London.

Wu, Kwo-Liang, Cheng-Ching Yu, William L. Luyben and Sigurd Skogestad (2002). Reactor/separator processes with recycles-2. design for composition control. **27**, 401–421.

Zhu, G. and M.A. Henson (2002). Model predictive control of interconnected linear and nonlinear processes. *Ind. Eng. Chem. Res.* **41**, 801–816.

Zhu, G., M.A. Henson and B.A. Ogunnaike (2000). A hybrid model predictive control strategy for nonlinear plant-wide control. *J. Proc. Control* **10**, 449–458.

ELSEVIER
IFAC
PUBLICATIONS
www.elsevier.com/locate/ifac

OPTIMAL SENSOR SELECTION FOR SUCCESSFUL REAL-TIME OPTIMIZATION

Weitao Zhong and Thomas E. Marlin*

Department of Chemical Engineering, McMaster University,
1280 Main Street West, Hamilton, Ontario, Canada L8S 4L7

Abstract: In Real-Time Optimization (RTO) systems, the performance of the RTO depends strongly on the selection of measured variables and of sensors used for their measurements. Many studies have investigated the selection of variables (e.g., Krishnan, *et al.*, 1992; Fraleigh, *et al.*, 1998). This study extends previous work by considering the error in the implementation of the RTO results due to imperfect sensors used in process control. A practical optimal sensor design strategy based on nonlinear closed-loop Monte-Carlo simulations is developed. Case study results demonstrate that the implementation error can be essential for determining the appropriate sensors. *Copyright © 2004 IFAC*

Keywords: Sensor design; Implementation error; Real-time optimization; Uncertainty; Closed-loop; Monte-Carlo simulation;

1. INTRODUCTION

Significant economic benefits can be obtained by applying operational optimization to chemical units or plants (Marlin and Hrymak, 1997). During the past ten years, with the development of numerical methods and computing power, many successful industrial applications of real-time optimization (RTO) have been reported (van Wijk and Pope, 1992; Bailey *et al.*, 1993; Brewer and Lopez, 1998).

The various existing process optimization techniques can be classified into two general categories: direct search and model-based optimization (Garcia and Morari, 1981). Direct search methods explore the plant response surface directly for the optimum operating conditions through a series of plant experiments, while model-based methods use process models to estimate the plant optimum. This paper concentrates on model-based process optimization systems.

A typical model-based RTO system is composed of several subsystems: process data validation, model update, economic optimization, and RTO results implementation. A general structure for the model-based RTO system is shown in Fig. 1.

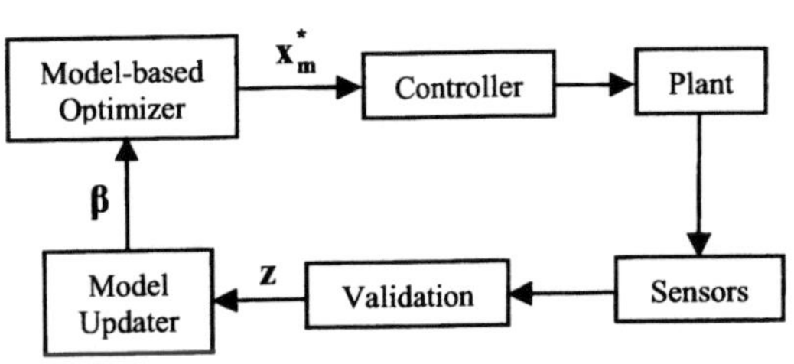

Fig. 1 General structure of RTO system

From Fig. 1, it can be seen that, in real-time optimization systems, feedback is applied to compensate for model errors and unmeasured disturbances. As a result, the performance of the RTO depends strongly on the selection of measured variables and of sensors used for their measurements. Several studies have investigated the selection of variables (Kage and Joseph, 1990; Krishnan *et al.*, 1992; Zhang, 1998). These studies, however, only consider the parameter updater as an isolated system rather than examining the effect of measurements on the RTO performance. Fraleigh *et al.* (1998) considered the whole closed-loop RTO system and developed a sensor system design cost criterion which allows the designer to compare various measurement combinations with respect to uncaptured profit due to setpoint variance and bias.

*Author to whom all correspondence should be addressed
Email: marlint@mcmaster.ca

This study extends previous work by considering the error in the implementation of the RTO results due to imperfect sensors used in process control. A practical optimal sensor design strategy based on nonlinear closed-loop Monte-Carlo simulations is developed. Sensor design candidates are evaluated in terms of true plant economic performance with consideration of various typical uncertainties experienced in industrial processes. The optimal sensor design strategy is illustrated with a distillation optimization example. Simulated case studies show that considering implementation error can be essential for determining the appropriate sensors.

In the next section, methods for optimal sensor design are briefly described. A design strategy based on closed-loop Monte-Carlo simulations is then presented with considering several critical issues in sensor selection problem. Section 3 contains the problem formulation and case study design. Case study results are presented in Section 4. The final section are the conclusions.

2. OPTIMAL SENSOR SELECTION FOR RTO SYSTEMS

For illustrative purpose in the following discussion, the closed-loop RTO system of Fig. 1 is simplified to include only three major subsystems shown in Fig. 2. Here it is assumed that the process measurements $\mathbf{z}$ are corrupted by random, zero-mean, stationary noise N and the process controllers are capable of implementing the calculated setpoints.

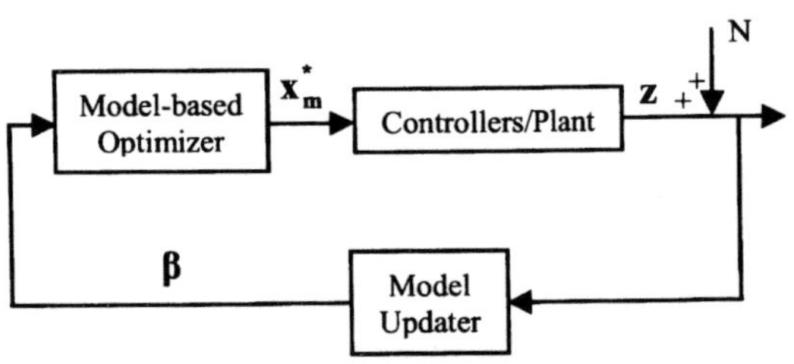

Fig. 2 Simplified RTO system

Generally, two approaches can be used for optimal sensor design: linear approximation and Monte-Carlo simulation. Linear approximation method was developed by Forbes and Marlin (1996). The central idea is based on the fact that each element of Fig. 2 can be considered as a nonlinear map. For arbitrarily small deviations from the plant optimum, the nonlinear maps of each element can be represented by their first-order approximations. The resulting system of linear equations can be reduced to an iterative relationship in perturbations to the predicted optimum setpoints ($\boldsymbol{\delta}\mathbf{x}^*_m$):

$$\boldsymbol{\delta}\mathbf{x}^*_m(k)=\left(\frac{d\mathbf{x}^*_m}{d\boldsymbol{\beta}}\frac{d\boldsymbol{\beta}}{dz}\frac{dz}{d\mathbf{x}}\right)_{\mathbf{x}^*_p,\boldsymbol{\beta}^*,\mathbf{z}^*}\boldsymbol{\delta}\mathbf{x}^*_m(k-1)+\left(\frac{d\mathbf{x}^*_m}{d\boldsymbol{\beta}}\frac{d\boldsymbol{\beta}}{dz}\right)_{\boldsymbol{\beta}^*,\mathbf{z}^*}N(k) \quad (1)$$

where $\mathbf{x}^*_p$ is the plant optimum setpoints; $\boldsymbol{\beta}^*$ and $\mathbf{z}^*$ are model parameters and process measurements at the plant optimum, respectively; k is the iteration counter.

For RTO system excited by process noise and with $\boldsymbol{\delta}\mathbf{x}^*_m(0)=\mathbf{0}$, the setpoint covariance matrix $Q_{\mathbf{x}^*_m}$ is:

$$Q_{\mathbf{x}^*_m}=\sum_{i=0}^{\infty}\left(\frac{d\mathbf{x}^*_m}{d\boldsymbol{\beta}}\frac{d\boldsymbol{\beta}}{dz}\frac{dz}{d\mathbf{x}}\right)^i_{\mathbf{x}^*_p,\boldsymbol{\beta}^*,\mathbf{z}^*}\left(\frac{d\mathbf{x}^*_m}{d\boldsymbol{\beta}}\frac{d\boldsymbol{\beta}}{dz}\right)_{\boldsymbol{\beta}^*,\mathbf{z}^*}Q_z\left[\left(\frac{d\mathbf{x}^*_m}{d\boldsymbol{\beta}}\frac{d\boldsymbol{\beta}}{dz}\frac{dz}{d\mathbf{x}}\right)^i_{\mathbf{x}^*_p,\boldsymbol{\beta}^*,\mathbf{z}^*}\left(\frac{d\mathbf{x}^*_m}{d\boldsymbol{\beta}}\frac{d\boldsymbol{\beta}}{dz}\right)_{\boldsymbol{\beta}^*,\mathbf{z}^*}\right]^T \quad (2)$$

where Q_z is the given covariance matrix of process measurements.

Thus, by using differential sensitivity analysis, a closed-loop RTO system shown in Fig. 2 can be represented by a linear approximation. The sensor design candidates can be evaluated by studying the setpoints covariance matrix $Q_{\mathbf{x}^*_m}$.

Although the linear approximation method has been demonstrated to be effective for small-scale process problems (Fraleigh, 1998), when applied to large-scale optimization systems, it may be impractical due to the potentially intense computational expense associated with the derivative calculations. Another shortcoming of this method is that the design procedure is based on linear approximation of the closed-loop RTO system, which means the analysis is only locally valid and the range of validity is unknown.

Monte-Carlo simulation, on the other hand, is a relatively straightforward method. Instead of performing linear sensitivity analysis, it evaluates design candidates by closed-loop RTO simulations. Given the sensor noise and typical uncertainties, the entire closed-loop RTO system can be simulated by considering the interaction of the three major subsystems shown in Fig. 2.

Compared to linear approximation approaches, Monte-Carlo simulation does not approximate the sensitivity of the loop elements and hence provides more accurate results for optimal sensor design problem. This advantage is significant especially when the systems considered are highly nonlinear. In addition, it can be applied in most of the commercial available optimization system products. In this paper, closed-loop Monte-Carlo simulation method is employed for the optimal sensor design with several important issues to be recognized.

The first is the impact of implementation error on the RTO performance when the process control system enforces the RTO results. No previous study has considered this factor. In industrial practice, high accuracy sensors may not be available or might not be economically justified. In this case, the implementation error has to be considered. It's a trade-off between sensor cost and plant profit: when high accuracy sensors are not used, RTO performance might be degraded by selecting relatively lower accuracy sensors instead.

Another important issue in optimal sensor design is the design strategy. To make the design decision quickly, a screening method is also proposed. The motivation for incorporating the screening method in design strategy is that in industrial practice, there are usually many, say, tens or even hundreds of design candidates to be evaluated. To make the important design decisions quickly, the screening method can be used to eliminate some of the design candidates before testing the best candidate on closed-loop Monte-Carlo simulations. The detailed strategy is described in the following steps:

1. Assume high accuracy sensors are available, perform closed-loop Monte-Carlo simulations without considering implementation error;
2. Assume high accuracy sensors are not available, set proper error level for each of the individual setpoints;
3. Use screening method to evaluate the design candidates. Detailed procedure is given below:
   a) For each design candidate, randomly select one or several sets of setpoints from previous closed-loop RTO iteration points;
   b) For the selected set of setpoints:
      i) Generate the implementation error according to the error level given in 2. and add it to the setpoints;
      ii) Run plant Monte-Carlo simulations;
   c) Eliminate some of the design candidates by comparing their RTO performance;
4. Test the design candidates that pass step 3 by closed-loop Monte-Carlo simulations with implementation error.

By following the design steps stated above, we can get both of the optimal sensor design results with/without implementation error, with reduced computing time.

It should be noted that the RTO performance should be calculated based on the plant model, not on the model used by the optimizer. This practice arises from the fact that due to model/plant mismatch, the optimum setpoints predicted by the optimizer don't necessarily guarantee an optimum performance in the true plant. Since we are more concerned about economic profit of the true plant, it is essential to evaluate the sensor design candidates in terms of the plant performance.

3. PROBLEM FORMULATION AND CASE STUDY DESIGN

3.1 Process description

The case study considered is a deisobutanizer from Sunoco Sarnia hydrocracker plant (Bailey *et al.*, 1993). It is a 65 bubble cap tray distillation column, operating at approximate 500kPa, which separates isobutane from normal butane. The nominal feed conditions are given in Table 1.

Table 1 Deisobutanizer feed conditions

Composition (mol %)	Propane (C_3)	3.83
	iButane (iC_4)	63.76
	Butane (nC_4)	28.22
	iPentane (C_5)	4.19
Flow (Mmols/day)		6.424
Temperature (°C)		94.8
Pressure (kPa)		1730.4

To match as closely as possible the behavior of the realistic distillation operation, the deisobutanizer model has been modified to include additional components, such as a total condenser model, flooding correlations, and pressure drop correlations, *etc.*

There are 32 ideal trays in the final model with one feed flow that enters the tower on tray 14. The model has 488 equations which are written in an equation-based format. All the variables are scaled by selecting appropriate engineering units. The case studies are solved using MINOS in GAMS (Brooke *et al.*, 1998).

3.2 Problem formulation

Model updater Mathematically, the model updater is a special type of nonlinear optimization problem, in which there is no distinction between independent and dependent variables. The process model $\mathbf{h}$ can be expressed in terms of measured variables $\mathbf{x}^m$, unmeasured variables $\mathbf{x}^u$, fixed parameters $\boldsymbol{\alpha}$ and estimated parameters $\boldsymbol{\beta}$ (the data in Table 1 except pressure), as given in (3):

$$\mathbf{h}(\mathbf{x}^m, \mathbf{x}^u, \boldsymbol{\alpha}, \boldsymbol{\beta}) = \mathbf{0} \tag{3}$$

Many formulations are available for the model updater. In this study, we consider both of the data reconciliation and parameter estimation in model updater. Naturally, we have considered the gross errors in process measurements that have been identified and removed prior to the updating phase.

$$\begin{aligned} \min_{\mathbf{x}^m, \mathbf{x}^u, \boldsymbol{\beta}} \quad & (\mathbf{x}^m - \mathbf{z})^T Q_z^{-1} (\mathbf{x}^m - \mathbf{z}) \\ s.t. \quad & \mathbf{h}(\mathbf{x}^m, \mathbf{x}^u, \boldsymbol{\alpha}, \boldsymbol{\beta}) = \mathbf{0} \end{aligned} \tag{4}$$

Model-based optimizer Model-based optimizer provides optimum operation policy based on the updated model and an economic objective function. The objective function accounts for the profit/cost of plant operation. Generally, it is a function of product profit and the costs of energy and raw materials. In this study, a nonlinear objective function based on plant cost is employed (Seferlis, 1995). It includes the cost of the utilities and the differential value of the components in the two product streams. The model-based optimizer can be stated as follows:

$$\begin{aligned} \min_{\mathbf{x}^i,\mathbf{x}^d} \Phi &= C_B Q_B + C_D Q_D + C_{hk} x_{D,hk} D + C_{lk} x_{B,lk} B \\ s.t. \quad & \mathbf{h}(\mathbf{x}^i,\mathbf{x}^d,\boldsymbol{\alpha},\boldsymbol{\beta}) = \mathbf{0} \\ & \mathbf{g}(\mathbf{x}^i,\mathbf{x}^d,\boldsymbol{\alpha},\boldsymbol{\beta}) \le \mathbf{0} \end{aligned} \tag{5}$$

where Φ is the plant cost (\$/d); $\mathbf{x}^i$ are the optimization variables; $\mathbf{x}^d$ are the dependent variables; $\boldsymbol{\alpha}$ are the fixed parameters; $\boldsymbol{\beta}$ are the updated parameters; $\mathbf{g}$ are the inequality constraints account for the product quality specifications; Q_B and Q_D are the reboiler and condenser duties (TJ/d); $x_{D,hk}$ is the heavy key component in the top product; $x_{B,lk}$ is the light key component in the bottom product; D, B are the top and bottom products (Mmol/d), respectively; C_B, C_D, C_{hk}, C_{lk} are the cost coefficients. The values of the cost coefficients are given in Table 2.

Table 2 Cost coefficients for objective function

C_B	C_D	C_{hk}	C_{lk}
2.40	0.40	15000	15000

There are three optimization variables in the model-based optimizer, one of which is the overhead pressure, the other two are either component compositions or tray temperatures, depending on the case scenarios (See Table 3).

Plant After the three optimization variables are determined, the degrees of freedom of the plant is zero, and the plant simulation can be performed by solving the following equations.

$$\mathbf{h}(\mathbf{x}^i,\mathbf{x}^d,\boldsymbol{\alpha},\boldsymbol{\beta}) = \mathbf{0} \tag{6}$$

3.3 Case study design

As we know, for a distillation tower, there are usually three possible positions for the composition analyzers: on the top product, on the bottom product, or on the feed, leading to 8 analyzer combinations available for the distillation RTO system. As on-line composition analyzers are expensive, it is necessary to determine which combination is the best for the distillation operation.

Setpoints are selected to match measurements available. For each case scenario, there are three setpoints, one of which is the overhead pressure, the other two are either component compositions or tray temperatures, depending on the corresponding analyzer and measurement sets. As previously stated, the updated parameters $\boldsymbol{\beta}$ in this study are the feed composition, flow rate, and temperature.

The sensor design candidates, their setpoints, and measurement sets are given in Table 3. The other measurements are the common measurements that are based on the on-line process data currently available in the deisobutanizer operation and are used by all the design candidates.

Table 3 Sensor design candidates

	Setpoints			Measurements					
				x_D		x_B		x_{fd}^*	others
1	x_{D,nC_4}	x_{B,iC_4}	P_{ov}	x_{D,C_3}	x_{D,nC_4}	x_{B,iC_4}	x_{B,C_5}	x_{fd}	
2	x_{D,nC_4}	x_{B,iC_4}	P_{ov}	x_{D,C_3}	x_{D,nC_4}	x_{B,iC_4}	x_{B,C_5}	---	F_{fd},
3	x_{D,nC_4}	T_{25}	P_{ov}	x_{D,C_3}	x_{D,nC_4}	---	---	x_{fd}	F_D, F_B,
4	x_{D,nC_4}	T_{25}	P_{ov}	x_{D,C_3}	x_{D,nC_4}	---	---	---	F_R, F_{st},
5	T_5	x_{B,iC_4}	P_{ov}	---	---	x_{B,iC_4}	x_{B,C_5}	x_{fd}	T_2, T_5,
6	T_5	x_{B,iC_4}	P_{ov}	---	---	x_{B,iC_4}	x_{B,C_5}	---	T_{18}, T_{25},
7	T_5	T_{25}	P_{ov}	---	---	---	---	x_{fd}	T_{31}, P_{ov}
8	T_5	T_{25}	P_{ov}	---	---	---	---	---	

*all components measured

In Table 3, x_D, x_B, x_{fd} are the compositions of top product, bottom product and feed; P_{ov} is the overhead pressure; F_{fd}, F_D, F_B, F_R, F_{st} are the flow rates of feed, top product, bottom product, reflux and steam to boiler, respectively; T_i are the tray temperature measurements.

In this case study, all the measurements have zero-mean white noise added to the true plant values; the noise on each of the individual measurements is not correlated, nor is the noise autocorrelated in time. The standard deviations for the measurements are 3% for flow rates, 0.5% for pressures, 5% for compositions, and 0.5°C for temperatures; the percentages are of the base case value.

Besides measurement noise, process disturbances are also considered. As feed composition variation is the major disturbance in distillation operation, four typical feed composition disturbance cases are investigated in this paper. Table 4 lists these disturbance cases.

Table 4 Disturbance cases

		Case1	Case2	Case3	Case4
Feed Comp. (mol %)	x_{fd,C_3}	5.06	3.83	10.03	9.03
	x_{fd,iC_4}	58.00	41.22	59.56	59.56
	x_{fd,nC_4}	14.07	50.76	26.22	23.22
	x_{fd,C_5}	22.87	4.19	4.19	4.19

Model/plant mismatch is another typical uncertainty experienced in optimization. In this study, structural model/plant mismatch is introduced in the number of ideal trays in the RTO system model. The feed tray number and tray temperature measurements are also changed proportionally to the total ideal tray number.

Table 5 Model/plant mismatch

	True Plant	RTO System
Tray No.	32	29
Feed Tray No.	14	13
Tray Temp. Measurements	2	2
	5	4
	18	16
	25	23
	31	28

4. CASE STUDY RESULTS

The 8 design candidates stated above are evaluated by using the strategy developed in this study. In each case scenario the simulated data are obtained by taking 300 closed-loop RTO iterations with each iteration involving model updater, model-based optimizer and plant, of which 100 iterations are for the base case, the other 200 are for the disturbance case. The disturbance is introduced at the 100th closed-loop RTO iteration.

RTO performance is calculated for each case. As a measure of the loss in economic performance of the RTO system due to imperfect optimization, the additional cost, that is, the offset between the mean value of the simulated data and the minimum value of the true plant cost is used to evaluate the design candidates. As any setpoints other than the plant optimum setpoints will lead to degraded plant performance, which in turn, yield higher plant cost, the offset is always positive, and the smaller the magnitude of the offset, the better the RTO performance. Fig. 3 illustrates the offset data of candidate 6 in disturbance case 1 without considering implementation error.

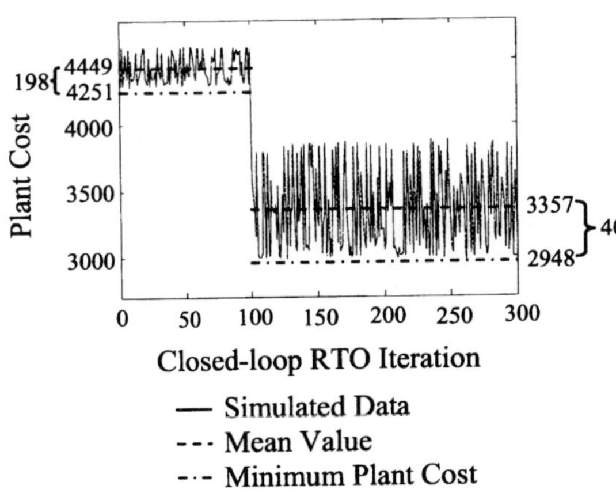

Fig. 3 RTO Performance Evaluation

To evaluate the design candidates, first, we perform closed-loop RTO simulations without considering the implementation error. In this case it is assumed that perfect sensors are available in the deisobutanizer operation. The design results are given in Table 6. It should be noted that the base case data in Table 6 are averaged from those of the base cases in the four disturbance cases.

Table 6 Sensor design offsets without implementation error

		Base	Case1	Case2	Case3	Case4
True Cost		4251	2948	4514	3982	3761
Cost Offset ($/d)	1	153	92	165	143	133
	2	151	90	163	141	131
	3	73	57	94	69	69
	4	40	27	53	37	42
	5	62	51	61	92	90
	6	198	409	136	381	397
	7	21	96	20	219	223
	8	116	459	77	333	385

From Table 6, it can be seen that candidates 1 - 5 consistently perform well in all the case studies. Candidates 6 - 8, on the other hand, yield significantly higher plant cost. Thus in the case where perfect sensors are available, design candidates 1 - 5 are recommended and 3 - 5 perform slightly better than 1 and 2.

In the subsequent studies, it is assumed that the perfect sensors are not available and hence the implementation error should be considered with appropriate implementation error. The implementation error level is the same as that of the sensor noise defined in Section 3.3. To reduce the computing time, instead of performing closed-loop RTO simulations directly, a screening method is first used to eliminate some of the design candidates. In this study, we select four sets of intermediate setpoints, say, 20th, 40th, 60th, 80th iteration points from the previous simulated data, and carry out 100 open-loop plant simulations for each set of setpoints. The offset data used to evaluate the design candidates are obtained by averaging the simulated data that are generated from the four sets of setpoints. Table 7 gives the screening results on base case. The intermediate setpoints come from the simulated data that produce the base case results in Table 6.

Table 7 Sensor design offsets by screening method

		20th	40th	60th	80th	Average
Cost Offset ($/d)	1	152	158	158	155	156
	2	148	148	165	157	155
	3	166	202	161	228	189
	4	190	178	212	153	183
	5	1692	2385	2667	2505	2312
	6	1319	2143	2058	2146	1917
	7	1498	1767	2068	1689	1756
	8	1623	2017	1727	1951	1830

Comparing Table 6 and 7, significant cost increases are observed in candidates 5 - 8, while the offset changes in 1 and 2 are negligible. Though RTO performance of 3 and 4 are worse, they are still acceptable. Thus by using the screening method, we can directly eliminate design candidates 5 - 8 without performing closed-loop RTO simulations.

To test and evaluate the design results in Table 7, further closed-loop RTO simulations are performed on design candidates 1 - 4 in order to give the design decisions with the consideration of implementation error. Table 8 gives the final sensor design results. From Table 8, it is clearly shown that design candidates 1 - 2 perform better than 3 - 4. Thus in the cases where perfect sensors are not available, design candidates 1 - 2 are recommended.

Table 8 Sensor design offsets with implementation error

		Base	Case1	Case2	Case3	Case4
True Cost		4251	2948	4514	3982	3761
	1	156	93	164	146	132
Cost Offset	2	154	92	165	143	134
($/d)	3	203	149	326	165	160
	4	176	128	226	185	150

Comparing the sensor design results with and without implementation error, it can be seen that the implementation error critically affects the design decisions. For example, candidate 5 would be the preferred design if implementation errors were not considered because it has a close approach to the best performance and only one costly analyzer. However, we see in Table 7 that this design gives very high costs with implementation errors. In addition, we see that the capital cost for three analyzers is not justified by the performance reported for candidate 1 in Table 8. Therefore, candidates 2 - 4, which all use two analyzers, remain as viable candidates. From Table 8, candidate 2 has the lowest cost and therefore, is the preferred design when considering implementation error.

CONCLUSIONS

Optimal sensor design provides a systematic way for selecting sensors of the RTO systems. In this paper, the importance of the implementation error to sensor selection problem is studied. A practical sensor design strategy based on nonlinear closed-loop Monte-Carlo simulations is developed. The strategy incorporates a screening method to eliminate some of the design candidates in a preprocessing phase, which helps to make the important design decisions in a timely manner.

A series of case studies based on a distillation column have been performed with typical uncertainties experienced in industrial processes. The case study results demonstrate that the explicit consideration of implementation error is especially important for the cases where high accuracy sensors are not available or are very costly. It is also shown that this factor can be essential for correctly determining the appropriate sensors.

REFERENCES

Bailey, J. K., Hrymak, A. N., Treiber, S. S., Hawkins, R. B. (1993). Nonlinear optimization of a hydrocracker fractionation plant. *Computers and Chemical Engineering, 17*(2), 123-138.

Brewer, W. M., & Lopez, S. F. (1998). Successful closed-loop olefins plant optimization. *Hydrocarbon Processing, 6*, 83-89.

Brooke, A., Kendrick, D., Meeraus, A., & Raman, R. (1998). *GAMS: A User's Guide*, GAMS Development Corporation.

Forbes, J. F., & Marlin, T. E. (1996). Design cost: a systematic approach to technology selection for model-based real-time optimization systems. *Computers and Chemical Engineering, 20*(6/7), 717-734.

Fraleigh L. M., Guay, M., & Forbes, J. F. (1998). Optimal sensor network design for real-time optimization. *American Institute of Chemical Engineering Annual Meeting*, Miami.

Garcia, C. E., & Morari, M. (1981). Optimal operation of integrated processing systems. *American Institute of Chemical Engineeing Journal, 27*(6), 960-968.

Kage, K., & Joseph, B. (1990). Measurement selection and detection of measurement bias in the context of model based control and optimization. *Industrial Engineering and Chemical Research, 29*, 2037-2044.

Krishnan, S., Barton, G. W., & Perkins, J. D. (1992). Robust parameter estimation in on-line optimization – part I: methodology and simulated case study. *Computers and Chemical Engineering, 16*(6), 545-562.

Marlin, T. E., & Hrymak, A. N. (1997). Real-time optimization of continuous processes. Fifth International Conference on Chemical Process Control. *American Institute of Chemical Engineering Symposium Series, 316*, 156-164.

Seferlis, P. (1995). *Collocation models for distillation units and sensitivity analysis studies in process optimization.* Ph.D. Thesis, McMaster University, Hamilton, Ontario, Canada.

van Wijk, R. A., & Pope, M. R. (1992). Advanced process control and on-line optimization in Shell refineries. *Computers and Chemical Engineering, 16*(Suppl.), S69-S80.

Zhang, C. (1998). *Nonlinear data reconciliation and parameter estimation for real-time optimization.* M. Eng. Thesis, McMaster University, Hamilton, Ontario, Canada.

ELSEVIER
IFAC
PUBLICATIONS
www.elsevier.com/locate/ifac

NONLINEAR OPTIMAL CONTROL USING DYNAMIC PROGRAMMING IN CELL SPACE — APPLICATION TO NONLINEAR CSTR

Sridhar Ungarala, [1] Zhongzhou Chen and Keyu Li

*Department of Chemical and Biomedical Engineering,
Cleveland State University, Cleveland, OH 44115*

Abstract: Optimal control problems for nonlinear systems are generally difficult to study. Closed-form solutions are often limited to linear systems and quadratic performance indices. This paper proposes a numerical approach to solve nonlinear optimal control problems using Bellman's principle of optimality in discrete time and space. The approach is based on Simple Cell Mapping (SCM), a numerical procedure to approximately describe nonlinear state-space dynamics. The paper discusses the construction of a general control database and performance index database along with a backward search algorithm to formulate optimal control policy. The cell space dynamic programming methodology is investigated on a nonlinear CSTR model. Numerical studies dealing with the influence of space and time discretization on computational feasibility are presented.

Keywords: Optimal control, nonlinear systems, cell mapping, dynamic programming.

1. INTRODUCTION

Optimal control of nonlinear dynamic systems has been an active research area for several decades. In general, the optimal control problem poses formidable difficulties, which are only compounded by nonlinearities and constraints placed on state and control variables. Since many chemical engineering systems fall under this category, they present challenging problems. For example, the Continuously Stirred Tank reactor (CSTR) is known to exhibit highly nonlinear behavior. Since process operations are under increasing demand to adapt and perform over a wide range of operating conditions, the limitations of traditional linear control theory are becoming more pronounced. As a result, research in nonlinear control theory has become one of the most important and relevant areas at present.

In theory, optimal control problems may be studied by analytical techniques such as calculus of variations, Pontryagin maximum principle, Bellman's principle of optimality and the Hamilton-Jacobi-Bellman equation (Lee, 1964; Lapidus and Luus, 1967). However, in the presence of strong nonlinearities and constraints, very few closed-form solutions are found. Alternatively, numerical techniques tend to be both cumbersome and computationally prohibitive (Kushner and Dupuis, 1992; Dupuis and James, 1998).

In this paper, we investigate the application of dynamic programming in discrete time and space to solve an optimal control problem for a CSTR. This numerical procedure is based on Bellman's principle of optimality in discrete domain. Discretization of the independent variable is commonly used for numerical procedures using nonlinear programming (Biegler, 1984; Logsdon and Biegler, 1989; Tieu *et al.*, 1995; Luus, 1994). This paper uses an approach based on the discretization of dependent variables also.

The foundation for the approach is provided by cell-to-cell mapping methods introduced by Hsu (1980). These methods have been primarily used for perform-

[1] Corresponding author: s.ungarala@csuohio.edu

ing global analysis of nonlinear systems. Several extensions to optimal control problems have been proposed since (Hsu, 1985; Crespo and Sun, 2002). These methods have yet to be investigated for process systems related to chemical engineering.

The idea of discretization of state and input variables is not new. Dynamic programming over discrete grids has been suggested by Bellman and Kalaba (1965) and Lapidus and Luus (1967). However, Bellman cited the "menace of expanding grid" and the "curse of dimensionality" as the causes of practical limitations, which continued to plague most numerical techniques. At present, the cell-mapping based methods offer a systematic and efficient methodology to explore nonlinear dynamics. Although the "menace" and the "curse" are still alive, they are rendered less potent due to the rapidly advancing computing power as well as cheap storage and fast data retrieval. At least for low dimensional systems, the cell mapping methodology offers a computationally feasible alternative.

The remainder of the paper is organized as follows. The concept of Simple Cell Mapping (SCP) is explained followed by a discussion of dynamic programming in cell space. The nonlinear CSTR is employed for a simulation example.

2. OPTIMAL CONTROL

A typical optimal control problem of dynamic systems is defined as follows: for a fixed time horizon $t \in [0,t_f]$, find the control signal policy $u(t)$ such that a performance index J, is maximized (or minimized),

$$J(x(0),t_f) = \phi(x(t_f)) + \int_0^{t_f} \psi(x(t),u(t))\, dt \qquad (1)$$

where ϕ is the arrival performance index and ψ is the incremental performance index. The optimization is typically subject to system equation in the form of a nonlinear ODE, initial condition and constraints as shown below,

$$\frac{dx}{dt} = f(x,u) \qquad (2)$$

$$x(0) = x_0 \qquad (3)$$

$$h(x,u) = 0 \qquad (4)$$

$$g(x,u) \geq 0 \qquad (5)$$

where x is the state vector and u is the control signal vector, both of which may be subjected to additional upper and lower bounds.

3. CELL-TO-CELL MAPPING

Consider the nonlinear dynamic system in Eq. (2). If the control signal is assumed constant, for a trajectory emanating from the initial condition $x(t)$, the arrival point after a time Δt is uniquely determined by

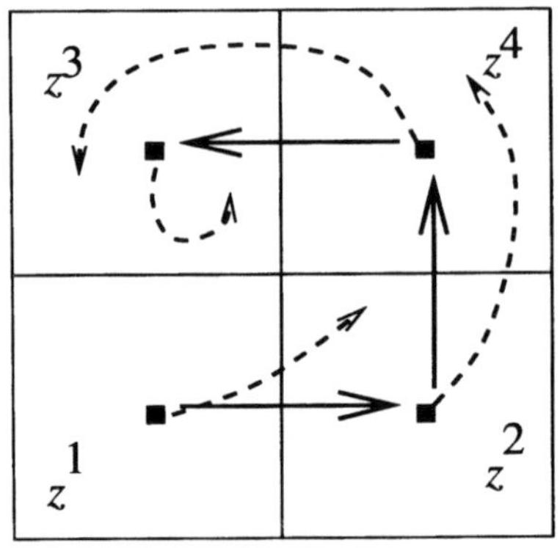

Fig. 1. Simple Cell Mapping (SCM)

$$x(t+\Delta t) = x(t) + \int_t^{t+\Delta t} f(x,u)\, d\tau \qquad (6)$$

In principle, the system equation may be integrated to define a point mapping in the state space for investigating the system dynamics parameterized by the mapping time step Δt. The discretization of time or the independent variable is commonly used for numerical procedures and also to reconcile with discretely observed systems. Eq. (6) may be written as a discrete system,

$$x_{k+1} = F(x_k, u_k, k) \qquad (7)$$

where $k = t/\Delta t$.

Consider a finite region $R \subset \mathbb{R}^n$, where the system dynamics, subject to constraints and bounds, are likely to be observed. Let R be partitioned into a collection of finite number of connected sets called cells, $\{z^i,\ i = 1,2,\ldots,N\}$. State space outside the region of interest is a single infinite sized cell called the sink cell z_0. The continuous state space $\mathbb{R}^n$ is approximated by the discrete cell space $Z = \{z^i\}_{i=0}^N$, and $Z \rightarrow \mathbb{R}^n$ as $N \rightarrow \infty$. Bounds automatically define R and constraint surfaces further limit the likely regions of R. The inclusion of constraints and bounds on state variables actually helps in defining a smaller cell space than otherwise is necessary.

Cell-to-cell mapping considers the system evolution in discrete time as a mapping among the cells. State transitions from point to point described by (7) possess analogous cell transitions in cell space. Transitions from cells $\{z^i,\ i = 1,2,\ldots,N\}$ into the sink cell z_0, are considered terminal. Simple Cell Mapping (SCM) uses a single point (usually the geometric center) of a cell to represent the entire volume of the cell (Hsu, 1980). The departure from a given cell z^j is evaluated by Eq. (7) to locate the arrival point in cell z^i, known as the image cell. The cell mapping function is defined as

$$z^i = C(z^j, \Delta t) \qquad (8)$$

which is Eq. (7) invoked at cell center.

Figure 1 illustrates the concept of SCM. The state space trajectories from four integrations from four cell centers are shown in dotted curves. The resulting cell trajectory is $z^1 \to z^2$, $z^2 \to z^4$, $z^4 \to z^3$ and $z^3 \to z^3$. In SCM terminology, a cell is classified as a *periodic* cell if it maps into itself in p mapping steps or $p \times \Delta t$ time period. For example, cell z^3 is a periodic cell with $p = 1$ containing a steady-state. Periodicity is used to identify steady-states and limit cycles in state space. The sink cell is a periodic cell with period equal to one.

Each group of periodic cells (called an attractor) has a set of *transient* cells, forming the domain of attraction, which map into the attractor within r mapping steps. For example cells z^1, z^2 and z^4 form the domain of attraction for cell z^3.

Consider the control vector u as a discrete representation $U = \{u^i\}_{i=1}^{M}$. Thus, the allowable control moves are limited to the set U and are executed as step functions at time intervals $k\Delta t$. A similar representation is used in Model Predictive Control (MPC).

For a constant u^i, the state space, partitioned into N cells, is perturbed for one mapping step by evaluating N integrations from N cell centers. The location of the end points of a trajectory assigns an image cell to each cell. The general control database Ω, consists of SCM implementations for all allowable controls in U, i.e., $N \times M$ mappings.

4. OPTIMAL CONTROL TABLE

The discrete version of the performance index is represented as follows,

$$J(x_0, H) = \phi(x_H) + \sum_{k=1}^{H} \psi(x_k, u_{k-1}) \qquad (9)$$

where $H = t_f/\Delta t$. The terminal performance index ϕ is evaluated at the center of the cell containing $x(t_f)$ and the incremental performance index ψ is similarly evaluated at the centers of the respective cells and the corresponding control vectors.

The incremental performance index database Λ, contains the values of the incremental performance index for all possible cells and the allowable controls, which is the general control database Ω. The optimal control policy is achieved by maximizing the sum of the terminal performance index and the accumulative performance index, which is simply the sum of the incremental performance index at discrete times.

According to the principle of optimality (Bellman and Kalaba, 1965): "an optimal policy has the property that whatever the initial state and the initial decision are, the remaining decisions must constitute an optimal policy with regard to the state resulting from the first decision."

The principle of optimality suggests that the search for minimizing or maximizing the performance index must proceed backwards in time from H to 1 (Lapidus and Luus, 1967). For instance, the optimal value of the performance index carried through H steps is

$$J^*(x_0, H) = \max\left\{\phi(x_H) + \sum_{k=1}^{H} \psi(x_k, u_{k-1})\right\} \qquad (10)$$

which may also be written as

$$J^*(x_0, H) = \max\{\psi(x_1, u_0)\} + \max\left\{\phi(x_H) + \sum_{k=2}^{H} \psi(x_k, u_{k-1})\right\} \qquad (11)$$

Thus, we reveal the fundamental recurrent nature of dynamic programming,

$$J^*(x_0, H) = \max\{\psi(x_1, u_0) + J^*(x_1, H-1)\} \qquad (12)$$

Going backwards in time, if $H - 1$ optimal controls have been determined, the next optimal control may be determined by considering the new step plus the already computed policy for the previous $H - 1$ steps.

The backward search algorithm in cell space over the general control database Ω and the incremental performance index database Λ can be summarized now. Let S^* denote the cells representing the optimal target value of the state vector and V^* denote the corresponding set of optimal control vectors. At the last stage H, the set S^* contains the cell representing the value x_H.

Starting at step H, initialize the cumulative performance index $J^* = \phi(x_H)$ and an intermediate target set $S = S^*$,

(1) Locate the cells in the general control database Ω, which possess the elements of S as image cells.
(2) Look up the value of the incremental performance index from the performance index database Λ, for all the maps found.
(3) Compute the accumulative performance index for each map and locate the maximum to update J^*.
(4) Expand the set S^* and V^* with the respective cells corresponding to J^*.
(5) Reinitialize S with the newest entry of S^*.

The backward search yields the optimal control table for all cells in the cell space. For a particular case of x_0, the initial cell z_0 is identified and the corresponding J^* is identified and thus the optimal control u_0 and optimal cell z_1 are located from the table. Continuing forward, we eventually recover the optimal control policy V^*.

5. SIMULATION EXAMPLE

We consider the optimal control of a nonlinear CSTR. The dimensionless model is the following set of ODEs (Lapidus and Luus, 1967),

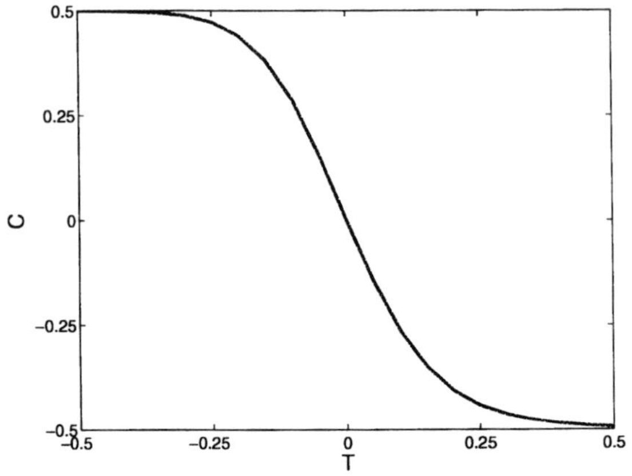

Fig. 2. Locus of steady-state points.

$$\frac{dT}{dt} = -(T+0.25) + (C+0.5)e^{\left(\frac{25T}{T+2}\right)} - u(T+0.25) \quad (13)$$

$$\frac{dC}{dt} = 0.5 - C - (C+0.5)e^{\left(\frac{25T}{T+2}\right)} \quad (14)$$

where u is the control signal with a maximum of 2 (fully cooled) and a minimum of 0 (adiabatic), and T and C are deviations from steady-state dimensionless temperature and concentration respectively. The model does not exhibit steady-state multiplicity, Figure 2 shows the locus of steady-state points.

Under proportional control, the control signal is given by $u = 1 + K_c T$, where K_c is the proportional gain. It is desired to find an optimal control policy so that when the reactor is started at $T = 0.05$ and $C = 0$, the origin $T = 0$ and $C = 0$ is reached in a fixed time horizon $t_f = 2$ at a minimal cost defined as follows,

$$J = \int_0^{t_f} T^2 + C^2 + u^2 \, dt \quad (15)$$

The state space $T \in [-0.1, 0.1]$ and $C \in [-0.1, 0.1]$ is discretized into 51 intervals on each dimension for a cell space Z, comprising 2,602 cells including the sink cell. A mapping time step of $\Delta t = 0.1$ is employed. The control signal is discretized into intervals of 0.5 over $u \in [0, 2]$ to define the allowable control set U comprising five elements.

The general control database Ω is computed using SCM. The mapping is performed by the fourth order Runge-Kutta integration scheme. The computation time required to perform 10,305 mappings is 2.5 secs on a Pentium 600 MHz machine using compiled code. The incremental performance index database is also computed in the same step. The storage requirement for the databases is approximately 256 KB. The backward search algorithm consumes 0.15 sec of CPU time. The computational demand and memory requirements are deemed fairly minimal for practical applications.

Figure 3 shows the results of the optimal control policy. The first two panels show the profiles of temperature and concentration respectively, while the third panel displays the control signal. The optimal trajectories of T and C in response to the control signal are shown in squares, which represent the optimal cell trajectories. Results of proportional control with $K_c = 17$ are also superimposed in solid curves. The optimal policy reaches the set point in nine steps.

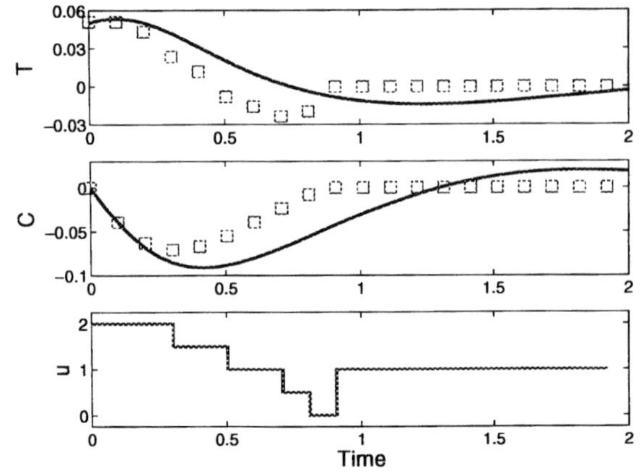

Fig. 3. Optimal control of CSTR along with proportional control.

6. CONCLUSIONS

A numerical procedure to solve optimal control problems for nonlinear systems using Bellman's principle of optimality in discrete time and space is presented. The point evolution of the system in state space is approximated as coarse grained dynamics of cell mappings in discretized cell space. The Simple Cell Mapping method is used to compile a general control database for all allowable controls in a discrete set. The approach bears close resemblance to state and control variable discretization schemes proposed in mid 1960s for dynamic programming. In contrast, the cell mapping approach avoids the difficulties of interpolations necessary therein. The accuracy of the solution in cell space is directly linked to the cell size and mapping time step size. Fine discretization is generally desired to minimize the approximation errors resulting from the center point method. However, it comes with added expense in computation and storage. Long mapping steps are detrimental since they may cause overshoots in the controlled response, which is a direct consequence of discrete control action. Application to a nonlinear CSTR shows promising results in terms of computational cost and controller performance.

ACKNOWLEDGEMENTS

Financial support from Established Full-Time Faculty Research Development Program and Doctoral Dissertation Research Expenses Award Program at Cleveland State University is gratefully acknowledged.

REFERENCES

Bellman, R. and R. Kalaba (1965). *Dynamic Programming and Modern control Theory*. Academic Press. New York.

Biegler, L. T. (1984). Solution of dynamic optimization problems by successive quadratic programing and orthogonal collocation. *Comp. Chem. Engng.* **8**, 243.

Crespo, L. G. and J. Q. Sun (2002). Stochastic optimal control of nonlinear systems via short-time Gaussian approximation andcell mapping. *Nonlin. Dyn.* **28**, 323–342.

Dupuis, P. and M. R. James (1998). Rates of conversion for approximation schemes in optimal control. *SIAM J. Control. Optim.* **36**(2), 719–741.

Hsu, C. S. (1980). A theory of cell-to-cell mapping dynamical systems. *J. App. Mech.* **47**, 931–939.

Hsu, C. S. (1985). A discrete method of optimal control based upon the cell state space concept. *J. Optim. Theo. Appl.* **18**, 199–221.

Kushner, H. J. and P. Dupuis (1992). *Numerical Methods for Stochastic Control Problems in Continuous Time*. New York. Springer Verlag.

Lapidus, L. and R. Luus (1967). *Optimal Control of Engineering Processes*. Blaisdell Pub. Co.. Waltham, MA.

Lee, R. C. K. (1964). *Optimal Estimation, Identification, and Control*. MIT Press. Cambridge, MA.

Logsdon, J. S. and L. T. Biegler (1989). Accurate solution of differential algebraic equations. *Ind. Eng. Chem. Res.* **28**, 1628–1639.

Luus, R. (1994). Optimal control of batch reactors by iterative dynamic programming. *J. proc. Control* **4**, 218–226.

Tieu, D., W. R. Cluett and A. Penlidis (1995). A comparrison of collocation methods for solving dynamic optimization problems. *Comp. Chem. Engng.* **19**, 375–381.

ELSEVIER
IFAC
PUBLICATIONS
www.elsevier.com/locate/ifac

OPTIMIZATION STRATEGY FOR MAXIMIZING PRODUCTION OF CYCLOHEXANOL

Delba N. C. Melo, Eduardo C. Vasco de Toledo,
Salah D. M. Hasan, Rubens M. Filho

Laboratory of Optimization, Design and Advanced Control (LOPCA). Faculty of Chemical Engineering. State University of Campinas (Unicamp), P.O. Box 6066, 13081-970, Campinas, SP, Brazil; email:delba@lopca.feq.unicamp.br,Tel.:+55-19-37883971; Fax:+55-19-37883965;

Abstract: The main issue objective of this work is to implement an optimization strategy through a two-layer structure where the search for optimal operational variables is made through of the algorithms optimization - Successive quadratic programming – SQP. The considered process is based on the hydrogenation of phenol for production of cyclo-hexanol in the presence of a nickel catalyst. The objective function to be considered for the optimization is to maximize production of cyclo-hexanol, through a steady-state model while maintaining the conversion of phenol at the exit of the reactor under environmental constraints. Factorial design and response surface analysis in a simulation study of a three-phase catalytic slurry reactor is used to identify the impact that each variable has on the system as well as the interaction among then. An improvement of 5.5% in cyclo-hexanol conversion was observed for the optimized variables (Q_H = 259 kg/h, Q_{Ni} = 53 kg/h).The resulted model for the response surface was used as a first objective function for the SQP optimization. The optimization results can be then used as set-point values to the advanced controller. *Copyright © 2004 IFAC*

Keywords: hydrogenation, dynamics modeling, response surface analysis, optimization, three-phase catalytic reactor; integration of control and optimization, algorithms, simulation, steady-state modeling.

1. INTRODUCTION

Hydrogenation reactions are widely applied industrially and reactors have been designed for this purpose. It is a highly non-linear process, multivariable, with exothermic reaction. Model assessment has mostly been reported for a single reaction, or for reactions obeying simplified kinetic laws under isothermal conditions. Nevertheless, exothermic reactions undergoing a multi-step reaction scheme with complex kinetics are industrially of a main interest, and rigorous comparisons of the performances of several multiphase reactors for such reactions have hardly ever been published (Bergault et al., 1997). With computational advances the process simulation has become an important tool for study and optimization of complex processes that normally involve a large number of variables. Hence, it is necessary to submit the process to an initial screening design prior to optimization (Kalil et al., 2000). The methodology of Placket & Burman (1946) is a tool for this initial screening, since it makes it possible to determine the influence of various factors with only a small of number of trials, instead of using more extensive factorial design, which would furnish more complete information, but which involves unfeasible complexity. If, however, the process is subjected to initial screening, response surface analysis can be limited to the most critical variables for the determination of their optimal bands of operation. Performing optimization encompasses the task of establishing process and economics models, choosing a predictive controller and modeling the optimization strategy. Yet no final answer has been given on how best to integrate the control and optimization of the operation of the reactor. Different optimization approaches have been attempted in the literature, for example, the two-layers approach where the economic optimization problem is solved separately from the control one. In this approach the economic optimization makes use of a rigorous steady-state process model and determines the set-point values for some controlled variables. Some approaches are just related to the selection of outputs and their appropriate reference values that will guarantee a probable optimal operation of the plant.

SQP algorithms (Sequential quadratic programming) are detached as non-linear (PNL) programming algorithms for the resolution of problems of no-large scale. More effective researchs have been done with the use of the SQP method, particularly for applications in complex systems. Robust algorithms have been considered and the difficulties that are found are detailed for solving the non-linear problems (Tvrzská de Gouvêa & Odloak, 1998; Ternet & Biegler, 1998; Lúcia et al., 1996). In order to reduce the convergence problems of some SQP algorithms, it is crucial to understand and identify the main difficulties of the considered optimization problem. The present work introduces response surface analysis in a simulation study of a three-phase catalytic slurry reactor to determine the operational conditions that lead to high conversion of the desired product. It is also proposed the use of statistical model from surface response analysis methodology to built of the objective function to be used in the SQP optimization algorithm.

2. MATHEMATICAL MODELING

2.1 Process modeling

The reactor considered in this work is constituted of a series of tubes, immersed in a boiler. In fact, they consist of concentrical tubes, where the reactants flow through the tubular as well as through the external annular region, while the coolant fluid flows inside the other regions. The model formulation was made focusing on the hydrogenation reaction of phenol to obtain the cyclo-hexanol (CHL), in the presence of a catalyst Ni/SiO_2 catalyst (Santana, 1999). The model consisted on mass and energy balance equations for the reactants and the coolant fluid. The scheme utilized to represent the reactor is shown in Figure 1 (Toledo et al., 2001).

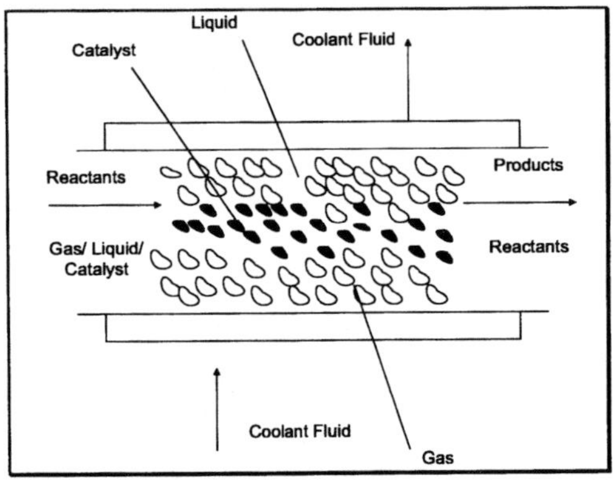

Figure 1. Three-phase reactor.

The following hypothesis were adopted when developing models that were studied in order to represent the steady-state behavior of the reactor by Santana (1999): a) plug-flow for the reactant and thermal fluid; b) homogeneous suspension (liquid + solid), considered as a pseudo-fluid; c) negligible pressure variations; d) reaction of the type: A (g) + νB (l) → νC (l), occurring at the catalyst and with a kinetic that is dependent of the concentrations of A and B; e) no phase change in the system; f) solid phase considered as pseudo-stationary. The radial dispersion has been neglected for the fluid phase, which is a very common assumption in multiphase reactor as normally found in literature. The kinetic expressions and the mass and balance equations were obtained from Gianetto & Silveston (1986), Coussemant & Jungers (1950) and Froment & Bischoff (1990). The operational parameters of the reactor, mass and energy balance coefficients, and physical properties have been considered constant. Some of these parameters are generated by correlations (Santana, 1999).

3. SEQUENTIAL QUADRATIC PROGRAMMING (SQP)

Sequential quadratic programming, or successive, or recursive, or iterative, or methods of the metric variable with restrictions, are the various form of SQP method representation, which is basically used to solve the equations of Karush-Kuhn-Tucker (KKT), or first-order conditions. The optimal result is obtained by the solution of the no-linear system of equations. The basic idea is the Newton method approach, generally used to decide problems with no restrictions. For this case study, it was necessary to establish an objective function, which is dependent upon the decision variables, and to include the restrictions related to these variables, so that the optimization problem may be previously defined. In order to reach well implemented optimization by the SQP method, an adequate choice of the decision variables must be done, with no convergence limitations. For this, it was previously used the experimental design and the response surface analysis as basis for a first optimization aiming to a definitive formulation of the SQP problem.

4. STATISTICAL ANALYSIS OF THE PROCESS

4.1Plackett-Burman Design

Packett-Burman (PB) designs are very useful for picking the most important factors from a list of candidate factors (Kalil et al., 2000), and require fewer runs than a comparable fractional design Haaland (1989). For screening purpose, ten variables on cyclo-hexanol conversion (X_{CHL}) were evaluated. The different variables were prepared in two levels, -1 for low level and +1 for high level, based on a Plackett-Burman statistical design. Dummy variables are used to estimate the standard error during analysis of data. Ten variables were considered in this study: Feed reactant temperature (T_F = 175.5 – 214.5°C), feed reactant pressure (P = 27 – 33 Kgf/cm^2), catalyst rate (Q_{Ni} = 225 – 275 kg/h), hydrogen feed rate (Q_H = 270 – 330 kg/h), coolant rate in tubes 1 to 6 (Q_{C1} = 657 – 803 kg/h, Q_{C2} = 3060 – 3740 kg/h, Q_{C3} = 5400 – 6600 kg/h, Q_{C4} = 1044 – 1276 kg/h, Q_{C5} = 405 – 495 kg/h, Q_{C6} = 234 – 286 kg/h).

4.2Response surface analysis

Once the relevant variables were selected by screening, simulations were planned to obtain a linear or a quadratic model, consisting of 2^n trials (n = number of relevant variables) plus a star configuration, with a central point. Surfaces were then built using the obtained model for the statistically significant variables. The software STATISTICA (Statsoft v. 5.0) was used to analyze the results.

5. THE OPTIMZATION STRATEGY AND THE INTEGRATION PROBLEM

The resulted model for the response surface was used as a first objective function for the SQP optimization. The optimization results can be then used as set-point values to the advanced controller. The real time process integration involving the optimization and control of the process, a step still being developed by the authors, is the final aim of the work. The process integration will be done with the two-layers approach, schematically shown in Figure 2.
In this approach, the control is set in a hierarchical structure, where an optimization layer calculates the set-points to the advanced controller, which is based on the Dynamic Matrix Control (DMC) procedure. The optimization layer is composed of an objective function and a steady-state model for the process. The control layer is composed of an economic objective function and a dynamic model for the process. Numerical results can, then, be used as comparative parameters for others strategies of the integration, such as one-layer approach for the cyclo-hexanol production.

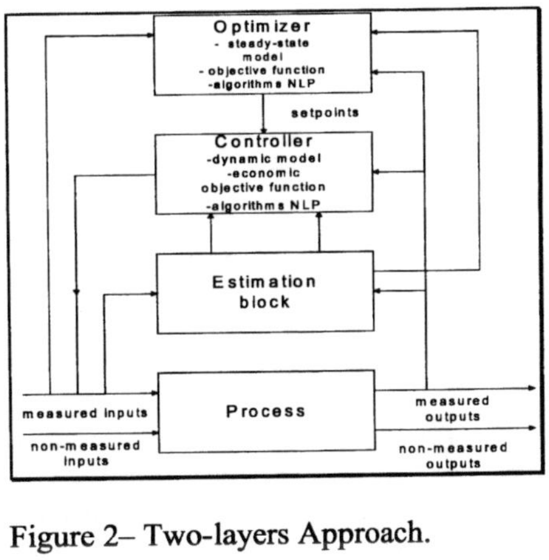

Figure 2– Two-layers Approach.

6.0 RESULTS AND DISCUSSION

6.1Evaluation of factors affecting cyclo-hexanol conversion (X_{CHL})

The experiments (simulations) were conducted according to the PB planning (which is composed by 20 trials) for the ten specified variables. The results of cyclo-hexanol conversion (X_{CHL}) were obtained for each trial. The Pareto chart was used for identifying which estimated effects are the most important. As it can be seen (Figure 3), only two parameters, catalyst and hydrogen rate (Q_{Ni} and Q_H, respectively) were statistically significant for (X_{CHL}). The negative values of effects indicate that higher values of cyclo-hexanol conversion can be obtained for lower values of Q_{Ni} and Q_H.

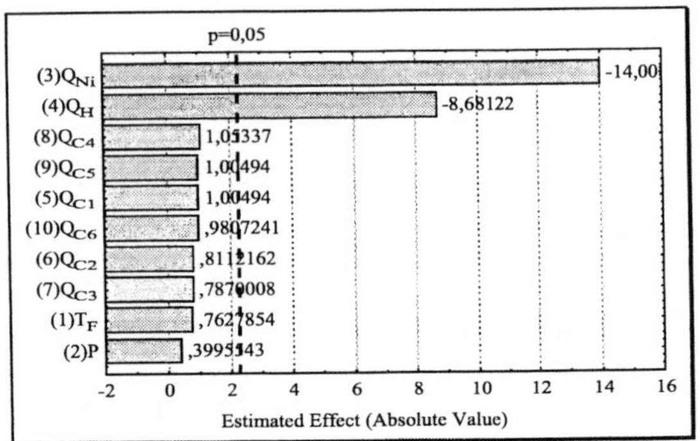

Figure 3. Pareto chart of effects for X_{CHL} (at 95% of confidence level), for the PB design.

6.2Process optimization - response surface analysis

A first central composite design (CCD) was carried out for the main variables Q_{Ni} and Q_H. It consisted on a complete factorial design 2^2 (levels −1 and +1) plus a star configuration (levels $-2^{1/2}$ and $+2^{1/2}$), with a central point (level 0). The central point is normally used with repetition for error estimation. In this case, however, only one point was used because there are no errors in computer simulations. The other eight variables were fixed at their central point of the PB design. Figure 4 shows the representation of the first central composite design results. Best values of (X_{CHL}) were obtained between simulations (trials) 1 and 6. Consequently, a new region, with lower values for Q_{Ni} and Q_H, must be explored. In such case, new simulations were carried out for values of Q_{Ni} and Q_H lower than 191.1 kg/h and 262.1 kg/h, respectively, and best value for (X_{CHL}) was obtained for Q_{Ni} = 53.0 kg/h and Q_H = 259.0 kg/h. Thus, a second CCD (similarly to the first CCD) was developed around this region for the determination of the optimal point. By the analysis of variance (ANOVA) for X_{CHL}, the F-test showed that the model was reliable since the calculated F-value was 4.29 times greater than the listed F-value, for 90% of confidence level. As a practical rule, a model has statistical significance when the calculated F-value is at least 3-5 times greater than the listed value (Kalil et al., 2000). The response has a correlation coefficient of 0.9705.

The model, expressed by equation (1), was generated by non-linear multiple regression of the data and is a function of the more significant variables,

$$X_{CHL}=0.9245-0.001Q_{Ni}-0.0028Q_{Ni}^2+0.0018Q_H-0.0024Q_H^2-0.0011Q_{Ni}\cdot Q_H \quad (1)$$

where Q_{Ni} and Q_H are in coded form. The response surface is represented in Figure 5. As it can be seen, X_{CHL} reaches a maximum value for Q_{Ni} = 53.0 kg/h and Q_H = 259.0 kg/h (corresponding to the central point of the second CCD).

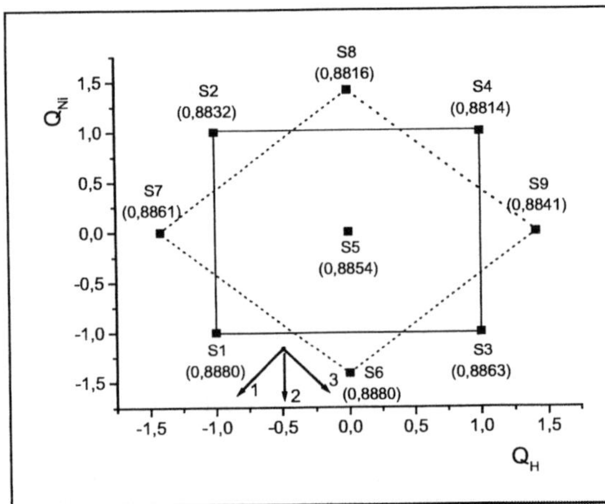

Figure 4. Representation of first central composite design. The values in brackets are the X_{CHL} results of simulations (S). The arrows 1, 2 and 3 indicate the trajectory at which Q_{Ni} and Q_H must be investigated (between S1 and S6) for the obtainment of higher X_{CHL} values.

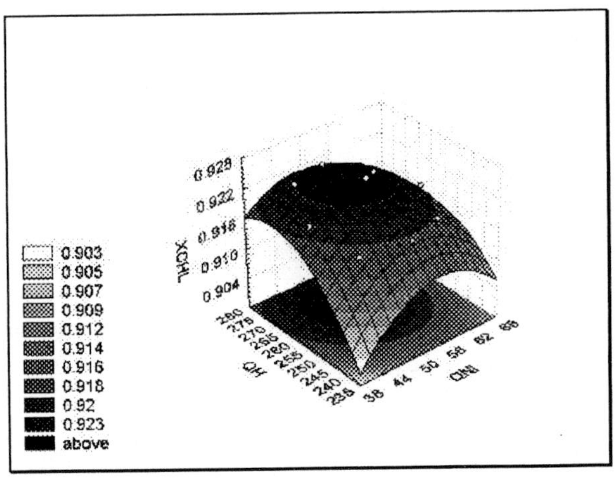

Figure 5. Response surface of conversion (X_{CHL}) expressed as a function of Q_{Ni} and Q_H, obtained for the second central composite design. Q_{Ni} e Q_H with real values.

6.3 Results of SQP Optimization

The following equation was used as objective function in the SQP implementation and represents cyclo-hexanol conversion dependency from the decision variables Q_{Ni} and Q_H. This equation was obtained from the response surface analysis where Q_{Ni} and Q_H are in real form and aims to cyclo-hexanol conversion maximization and, consequently, cycle-hexane minimization (undesired byproduct).

$$X_{CHL}=0.9245-0.001\cdot\left(\frac{(Q_{Ni}-53.0)}{(8.5)}\right)-0.0028\cdot\left(\frac{(Q_{Ni}-53.0)}{(8.5)}\right)^2+0.0018\cdot\left(\frac{Q_H-259.0}{12.4}\right)$$
$$-0.0024\cdot\left(\frac{Q_H-259.0}{12.4}\right)^2-0.0011\cdot\left(\frac{(Q_{Ni}-53.0)}{(8.5)}\right)\cdot\left(\frac{Q_H-259.0}{12.4}\right) \quad (2)$$

The advantage of using simplified process models in optimization problems is the reduction in convergence problems, so that the optimum point is reached easier.
The SQP code was implemented according to the algorithm developed by Schittowski (1995). Simultaneously, it was tested the routines of optimization of Martinez et al (2000), and the BDCONF routine (from IMSL library).
The optimal values of Q_{Ni} and Q_H were obtained from the routine EASY of Martinez et al. (2000) and were close to the statistical results (second CCD). The maximization of the objective function was reached at a CPU time of 1.5625 10-0.002 seconds. In a second stage, the original steady-state model of the process was incorporated in the routine following pre-adjusted parameters and convergence was obtained under a rational computational time. The convergence was possible through a strait band of nickel and hydrogen rates and the optimal point (X_{CHL} = 0.9245) was obtained for Q_{Ni} = 53.0 kg/h and Q_H = 259.0 kg/h.
The routines of the IMSL Fortran's library were tested and convergence problems were observed due to the high dimension of the problem. The DNCONF subroutine includes searching in unfeasible regions resulting in convergence limitations for the optimization problem.

7. CONCLUSIONS

In this paper a procedure for real time optimization was proposed through the coupling of SQP based algorithm and statistical analysis. The optimization results will be then used as set-point values to the advanced controller. Experimental design techniques was used to identify the main process variables with a significant impact on the process behavior, after a just screening by PB methods. Response surface analysis were implemented to find out a possible optimal region as well a to generate a reduced model statically validated.
Response surface methodology was found to be a powerful tool for the optimization of complex processes such as three-phase catalytic slurry reactor. In this study, an improvement of 5.5% was achieved for cyclo-hexanol conversion (0.9245 at the second CCD against 0.8763 at the previous PB design), with a decreasing in the by-product conversion. Plackett-Burman design offers a good and fast screening procedure and mathematically computes the significance of a large number of factors in one experiment, which is time saving and maintains convincing information on each component. The rate of hydrogen (Q_H) and catalyst (Q_{Ni}) showed strong effect on the cyclo-hexanol production and consequently should have special attention in the designing and operating process control. Surface response methodology is an interesting procedure to be used as alternative optimization or even to identify the feasible region for optimal operating conditions. In this case, the model obtained at the surface response methodology (eq. 1) was used as a first objective function for the SQP optimization..

ACKNOWLEDGEMENTS

The authors are grateful to the Fundação de Amparo à Pesquisa do Estado de São Paulo – FAPESP for the financial support.

REFERENCES

Bergault, I., M.V. Rajashekharam, R.V. Chaudhari, D. Schweich and H. Delmas. (1997). Modeling and comparison of acetophenone hydrogenation in trickle-bed and slurry airlift reactors. *Chem. Eng. Sci,.* **52**, 4033-4043.

Coussemant, F. and J.C. Jungers. (1950). La Cinétique de 1'Hydrogénation Catalytique des Phénols. *Bull. Soc. Chim. Belg.* **59**, 295-326.

Froment, G.F. and K. B. Bischoff. (1990). *Chemical reactor analysis and design.* New York: John Wiley & Sons, 2 nd. Ed, p. 664.

Gianetto, A. and P. L. Silveston. (1986). Multiphase chemical reactors: theory, design, scale-up. *Hemisphere Publishing Corporation, Washington, DC*, 682.

Haaland, P.D. (1989). Experimental design in biotechnology. *New York: Marcel Dekker.*

Kalil, S.J., F. Maugeri and M.I. Rodrigues. (2000). Response surface analysis and simulation as a tool for bioprocess design and optimization. *Process Biochem.*, **35**, 539-550.

Lucia, A., J. Xu, and K. M. Lauyn. (1996). Nonconvex process optimization. Comp. Chem. Engng., **20**, 1375-1398.

Martinez, J.M., N. Krejic, M.P. Mello and E. A. Pilotta. (2000).Validation of an augmented lagrangian algorithm with a Gauss-Newton Hessian approximation using a set of hard-spheres problems. *Computational Optimization and Applications*, **16**, 247-263.

Plackett R.L. and J. P. Burman. (1946). The design of optimum multifactorial experiments. *Biometrika*, **33**, 305-325.

Santana, P.L. (1999). Mathematical modeling for a three-phase reactor: deterministic, neural and hybrid models. *Ph.D. thesis*, Unicamp, Campinas, SP, Brazil.

Schittowski, K. (1985). NLQL: A FORTRAN-subroutine for solving constrained nonlinear programming problems. *Annals of Operations Research*, **5**, 485-500.

Ternet, D. J. and L.T Biegler (1998). Recent improvements to a multiplier-free reduced Hessian successive quadratic programming algorithm. *Computers and Chemical Engineering*, **22**, 963-978

Toledo, E.C.V., P.L. Santana, M.R.W. Maciel and R.M. Filho. (2001). Dynamic modeling of a three-phase catalytic slurry reactor. *Chem. Eng. Sci.*, **56**, 6055-6061.

Tvrzská de Gouvêa, M. and D. Odloak (1998). A new treatment of inconsistent quadratic programs in SQP-based algorithm. *Computers and Chemical engineering* ,**22**, 1623-1651.

ELSEVIER
IFAC
PUBLICATIONS
www.elsevier.com/locate/ifac

A MULTI-LEVEL, CONTROL-THEORETIC FRAMEWORK FOR INTEGRATION OF PLANNING, SCHEDULING AND RESCHEDULING

Munawar, S. A. and Gudi, R. D.

Department of Chemical Engineering, Indian Institute of Technology, Bombay, Mumbai – 400 076, India.

Abstract: In this paper, an integrated multi-level, control-theoretic framework has been proposed for effectively handling integration of planning, scheduling and rescheduling. The overall problem is decomposed into three levels with different horizons; each level has an abstraction of the lower level, with the philosophy of decentralized decision making and the flexibility and amenability to rescheduling. The main feature of the proposed framework is the integration of reactive scheduling that is motivated by some of the process control principles like cascade control and the concepts of receding horizon. An illustrative case study of a simple refinery flow sheet involving continuous lube production in a hybrid flowshop is presented.

Keywords: Integration, Planning, Scheduling, Rescheduling /Reactive scheduling, Multilevel, Control-theory.

1. INTRODUCTION

In the recent past, development of methods for efficient integration of planning and scheduling has received momentous attention in the industrial sector and in the research community, largely because of the challenges and the high economic incentives involved. Most large-scale multi-enterprise facilities which are distributed over distinct geographic locations, try to integrate the planning and scheduling activities as best as possible, from the stand point of a central planning at the corporate office for production and distribution, thus setting targets for single-site plants. Often, they employ some improvised techniques for this integration and are generally discontented with the resulting inconsistencies in decision-making.

Over the last few years, though some progress has been made in this direction for development of better frameworks for such integrations, still there is a large scope for further improvements. Traditionally, the decisions in an enterprise flow in a top-down manner leaving less degree of freedom at lower levels for rescheduling, leading to frequent revision of targets set by the top levels. At each level the horizons of interest are widely different. The upper levels must also reflect accurate abstractions of the lower levels and should be revised as infrequently as possible compared to the lower levels. Embedding contingency measures for integration of rescheduling has been ignored in most of the works.

In the literature there are several works on planning and scheduling. Shah (1998) gives a detailed review and current status on single and multi-site planning and scheduling. Grossmann et al. (2001) revised the classification of planning and scheduling models arising in process operations and the recent developments in their solution techniques. They proposed a general disjunctive model for integration of planning and scheduling. Shobrys and White (2000) examined the incentives and barriers in the integration of planning, scheduling and control functions in the process industries. They recounted some of the success stories in this direction and analyzed the reasons for failure of other companies that could not achieve integration despite multiple initiatives. They identified two non-techno challenges of coping with the human and organizational behavior and made some recommendations to overcome these barriers to integration. Recently, Van den Heever and Grossmann (2003) presented a two-level decomposition model for integration of planning and reactive scheduling in hydrogen supply networks. They used a simplified pipeline model at the planning level and a detailed pipeline description for the scheduling level.

The area of process control is well matured and recently there have been increasing applications of control-oriented frameworks for supply chain management and integration of planning and scheduling. Perea et al. (2000) proposed a dynamic approach to supply chain management with ideas from process dynamics and control. Vargas-Villamil and Rivera (2000) proposed a model predictive control (MPC) formulation for scheduling of reentrant semiconductor manufacturing lines. Bose and Pekny (2000) proposed MPC for integration of planning and scheduling for a multi-period operation of consumer goods supply chain. In each period, they used a forecasted model to calculate target inventories (control variable) for future periods and a scheduling model to achieve these targets by scheduling tasks (manipulated variable). Perea-Lopez et al. (2003) proposed an MPC strategy for supply chain optimization with a rolling horizon approach for updating the changes to the supply chain.

In this paper, we consider a hierarchical decomposition for integration of planning and scheduling and embed some decision-making ability at each level to provide the necessary degrees of freedom to make the model amenable to rescheduling. The principles of cascade-control and a receding horizon approach are used for integrating reactive scheduling to handle unexpected machine breakdowns in a local fashion. In section 2, we present a general multi-level, control-theoretic formulation for multi-site facilities. Due to space constraints, as proof of concept we demonstrate the proposed methodology for a single-site lube production in a hybrid flowshop facility for which the problem definition and solution approach are presented in section 3. A case study with model details is illustrated in section 4 followed by conclusions in the last section.

2. A GENERAL MULTI-LEVEL, CONTROL-THEORETIC FRAMEWORK

Consider a general multi-site, multiproduct planning and scheduling problem with several plants located in different geographic locations. At each plant a generic hybrid flowshop configuration of various machines is assumed that can be easily simplified to any problem specific topology of series and/or parallel configuration of different stages. We additionally consider resource constraints on the feed side of each plant where the feed could be a continuous stream with finite storage space. Again as a problem specific situation these resource constraints may be easily simplified or dropped. The overall problem can be traditionally decomposed into two major levels, a primary level for strategic (or long-term) planning across multiple sites and a secondary level for tactical (or mid-term) planning and scheduling at each site. The primary level has demands over a longer horizon (say 1 year) and has abstractions of each plant in terms of the average production and inventory capacities. The global objective here is meeting commitments made to a large number of customers. Accordingly, based on the production and transportation costs, the model at this level (LP/MILP) sets production targets for each plant. The primary level is revised on a less frequent basis (say monthly/quarterly).

At the secondary level, for mid-term planning and scheduling the horizon of interest is smaller (1-3 months) catering to less number of customers, and the model here may be revised on a frequent (say weekly/monthly) basis. To meet the global objectives of the overall problem, as in cascade-control, we propose that any local disturbances (machine breakdowns etc.) at the plant level have to be attenuated locally before they affect the global performance. If a machine breakdown occurs in a given period then we consider a receding horizon window from that time instant onwards and meet the slacks with increased production rates (the degrees of freedom provided here) and if necessary we may intrude into the next time period without sending a feedback until the next scheduled revision of the upper level. In contrast, if we do the normal feedback control option then in the same time horizon we may have to trigger the primary level thus forcing frequent revisions of the commitments made to customers. Only when these disturbances cannot be handled locally, a feedback to the primary level is sent thus seeking a target revision.

The multi-site formulation presented above is for the sake of generality. In the present work we consider the secondary level for single-site planning and scheduling with detailed model formulations and results as described in the following sections. Nevertheless, the model at primary level can be easily integrated vertically with the following models for the secondary level.

3. PROBLEM DEFINITION AND SOLUTION APPROACH

Consider a single-site plant with a general M-stage hybrid flowshop facility as shown in Fig. 1.

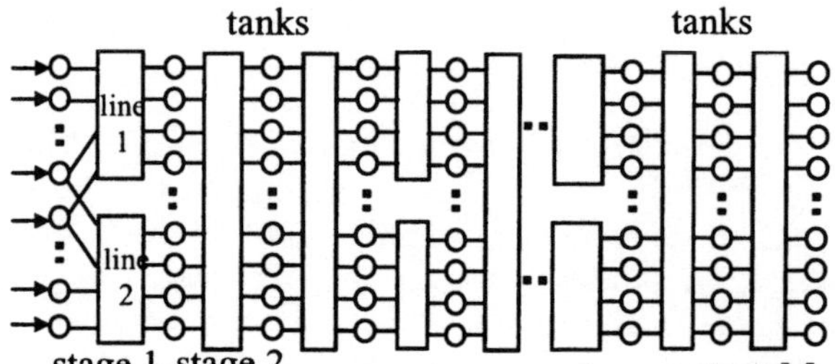

Fig. 1. General single-site hybrid flowshop facility

At each stage there could be either a single machine or multiple machines in parallel. The resource constraints are in terms of feed to the plant envelope being a continuous stream with finite storage space and hence the plant schedule is governed by both the feed side inventories and the demand side constraints. Earlier, we (Munawar et al., 2003a) proposed a generalized MINLP model for cyclic scheduling of this configuration with detailed inventory constraints. This model accounts for feed losses during grade changeovers (also known as slopping losses in continuous plants) through a modified time slot definition. It could handle special cases leading to empty slots (zero time duration) in the model resulting due to splitting of products across parallel lines. However, such model becomes intractable when we consider integration of planning and scheduling for a multi-period operation. Hence, at the secondary level for single-site plants, we propose a three level decomposition of the overall problem. Though, the case study in this paper is for cyclic scheduling of continuous multi-product plants, the proposed framework can however be appropriately extended to other cases of short-term scheduling.

At the single-site plant, the top level (level-1) for mid-term planning and scheduling has 1-3 month time horizon, the production targets for which are set from the primary level of the multi-site problem. At this level, we consider an abstract model with assumed slopping losses, and simplified inventory constraints in terms of upper bounds on processing times. In the next level (level-2) we consider detailed inventory constraints with the modified time slot definition to account for actual slopping losses. The horizon at level-2 is that of a single period and we do not consider detailed inventory management at this level, but consider an abstraction of the total inventory available for each product. The inventory management on an hourly/daily basis for individual product-to-tank assignments is done heuristically at the lower most level (level-3). For efficient usage of the available tank volumes an Inventory Slicing and Tank Reallocation (ISTR) algorithm is proposed at level-3. In the next section we present the model details for each level demonstrated for a simplified hybrid flowshop facility.

4. CASE STUDY AND MODEL DETAILS

Consider a single-site lube production in hybrid flowshop facility for producing 4 products on 3 stages as shown in Fig. 2, where stage 1 has two parallel machines line 1 and line 2 that relate to the same processing task (for e.g. extraction).

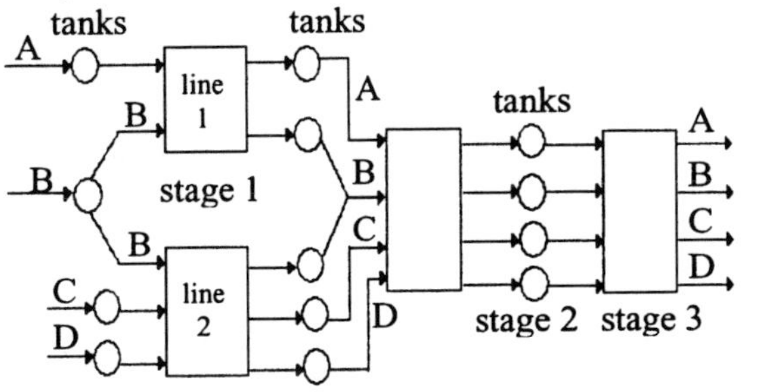

Fig. 2. Lube production in a hybrid flowshop facility

With the exception of component B, which can be processed in both the lines, all the products have to be processed one at a time in all the stages in the same sequence: line 1 or line 2 followed by stages 2 and 3.

The feed rates for base stocks of continuous streams A, B, C and D are assumed to be constant: 21, 28, 17 and 16 m^3/hr. There are some conversion losses in each stage and the processing rates and grade transitions are sequence and stage/line dependant, with the processing rates constrained between specified lower and upper bounds. The total inventory space available is finite and is about 3800 m^3 overall across all stages/lines, except for the product side where unlimited inventory space (UIS) is considered. Since the feed is a continuous stream received from upstream plants, the feed inventory tanks would almost always be busy. Nevertheless, if some inventory is unused then there is a potential for usage elsewhere in the plant from a reactive scheduling point of view. Hence, the objective at the operator level is to minimize the available inventory usage as best as possible.

4.1 Level-1 formulation:

Consider mid-term planning at level-1 for a multi-period operation over a small horizon, with demands specified for three periods each of 1000 hr, 900 hr and 800 hr duration respectively. As an abstraction of the inventories to be handled at level-2, upper bounds are specified on processing times for all products and are assumed to be 25 hr, 20 hr and 15 hr for stages 1, 2 and 3 respectively (based on past experience or some heuristics). The slopping losses are reflected by reducing the conversion rates of grades appropriately based on some heuristics. Here, we do not present the details of these heuristics for continuity purposes.

For a multi period operation, it is evident that an effective planning would require the processing tasks to be distributed uniformly, rather tightly in the early periods, so as to be optimal with respect to all the periods. For example, a high demand for a particular product in a given period would require its processing to be distributed over the entire time up to that period (including the previous periods), even though the demand for this product in the previous periods is low or zero. Even otherwise, from a reactive scheduling perspective we propose that the scheduler must push in tighter schedules in the early periods so that there is some leeway in the forthcoming periods to accommodate the unforeseen events. Hence, we allow over production in the first period of interest. As level-1 is anyway revised at the end of each period, the first period of interest rolls through, and finally we have the best possible production rates in each time period. The inventory costs for the overproduced quantities can always be weighed (trade-off) against the slacks in demands met otherwise if we do not produce in the early periods. We consider the objective function at level-1 to be maximization of production in the first period of interest subject to penalties for overall grade changeovers in all the time periods. Table 1(a) gives a comparison of the actual demands specified (set points) and the output from level-1 for the three periods considered.

Table 1. Summary of results for demand (m^3) predictions of level-1 and level-2

(a) Demands specified and projected by level-1 for all three periods

	set points			output		
	t=1	t=2	t=3	t=1	t=2	t=3
A	6237	7713	8190	7501.41	6751.27	6001.13
B	8414	7572	6731	8953.56	7032.44	6731
C	1470	2780	1975	3148.70	1101.30	0
D	1470	2780	1975	1817.14	3367.96	0

(b) Demands specified and projected by level-1 for last two periods

	set points		output	
	t=2	t=3	t=2	t=3
A	6751.27	7887.32	6751.27	6001.128
B	7032.44	6731	7032.44	6731
C	1101.30	3407.22	1101.30	0
D	3367.96	1039.9	3367.96	0

(c) Results of level-2 solved sequentially for each period

	t=1		t=2		t=3	
	set point	output	set point	output	set point	output
A	7501.41	7501.41	6751.27	6751.27	7887.32	6505.81
B	8953.56	8953.56	7032.44	7032.44	6731	6731
C	3148.70	1716.5	1101.30	1019.85	3488.67	1279.32
D	1817.14	1817.14	3367.96	1008.34	3399.52	2762.84

The model constraints at level-1 are similar to the model proposed by Munawar et al. (2003a) for level-2, except that we do not consider the detailed inventory constraints and the variables corresponding to slopping. Hence, we refer to Munawar et al. (2003a) for technical details.

For a demand limited scenario we may have to make provision for some idle time in the subsequent periods (except the first period because we are anyway allowing over production in the first period of interest) to avoid the case where machines are forced to run at their lower bound on processing rates to fill the gap in time period; instead an idle time may be preferred for the remaining time if the demands in some periods are low. However, we considered a capacity limited scenario and hence the total demands at the end of all time periods could not be met. As there is no incentive for production in the subsequent periods (except first period) the output predicted by level-1 shows some zero productions in Table 1(a). Here, the schedule is meant to be aggressive with respect to only the first period of interest and hence now we solve level-2 for the first time period (t=1) with the demands projected by level-1 as set points. Level-2 will predict the best possible demands in the presence of actual slopping losses and detailed inventory constraints. If all of the demands projected by level-1 cannot be met at level-2 (meaning heavy over production projected by level-1) then the slack (actually meaning less over production done at level-2) can be added on to the last period of the horizon. When level-1 is solved again for the remaining two periods (t=2 and 3), since we allow over production in the first period of interest (t=2), thus adding constraint of type ' $\geq$ 'on productions, it may lead to infeasibility if we add this slack to second time period. Hence, we always add the slacks to the last known period of interest. If it were feasible for this slack to be produced in t=2 itself, it is anyway feasible due to over production being allowed.

The output of level-1 for the last two periods is shown in Table 1(b). Now we again solve level-2 for t=2 with these demands as set points and check for feasibility. Finally for the last period (t=3) we need not solve level-1 as we can directly solve level-2. All the above runs are solved as an offline activity to find the best possible demands that can be met by level-2 in presence of real slopping losses and inventory constraints. Finally, the output of level-2 for each period is given in Table 1(c).

4.2 Level-2 formulation:

The model at level-2 involves detailed inventory constraints and actual slopping losses. The objective is maximizing profit, where the inventory and grade changeover costs are penalized subject to meeting maximum possible demands. In the continuous time domain representation the definition of a time slot is modified to account for the feed losses in slopping and additional slopping variables are defined. The model at this level is same as in Munawar et al., 2003a where we consider detailed inventory constraints in terms of the inventory breakpoints that define the total inventory profile for consumption and discharge of material.

The detailed product-to-tank assignments are not done here but are considered at level-3. At level-2 however, we consider abstractions of the available inventory volumes for each product. For the problem shown in Fig. 2 the inventory available is about 2500 m^3 for the feed tanks and 600 m^3 for tanks after line 1, 400 m^3 for tanks after line 2 and 300 m^3 for tanks between stage 2 and stage 3, totaling 3800 m^3. As we can see later in level-3 tanks are reused for storage of different grades over the time horizon and hence we consider an over estimation for the available inventory volumes based on some heuristics or past experience. Here, we use about 800 m^3 volume for each grade on the feed side, 600 m^3 for each grade for storage after line 1, 400 m^3 for each grade after line 2, and 300 m^3 for each grade after stage 2. The above volumes are used as upper bounds on the maximum inventory breakpoint for each grade at level-2. Though the feed tanks are almost always busy, some re-use of tanks is still observed at level-3, and hence we consider slightly higher upper bounds for inventories of the feed tanks as well. The problem at level-2 is solved sequentially as an offline activity for each period with the demand inputs taken from the first period of interest of level-1 as discussed earlier. The inventory profiles from level-2 in the first period are shown in Fig. 3.

Reactive scheduling formulation for machine breakdowns at level-2: Now we consider the model for reactive scheduling and analyze the interaction of level-1 and level-2 and simulate some scenarios of machine breakdowns leading to loss of available time for production in a given time period. We use a receding horizon time window as done by Munawar et al. (2003b) and find the amount of time required to be intruded into the next time period for compensating the slack in the current period. However, the proposed model in this paper is more generalized as it has provision for empty slots leading to zero time duration and simplified slopping representations at level-1.

In order to simulate the unforeseen breakdown of machines, a shut down time is introduced at the end of ten cycles (724.4 hrs) in the first time period. The current levels of inventory at the end of 724.4 hrs and the processing sequence are fixed as that of the nominal schedule. The objective function at level-2 for reactive scheduling, is posed as minimization of the amount of time required to be intruded into the next due date in order to meet the overall demand (slack in the current due date + demands in the next due date for the intruded period) at end of the receding horizon window. The shut down times are gradually increased from 1 hr to 110 hrs. For shut down times of 1, 10, 50, 75 and 99 hrs, the receding horizon approach predicted that we need to intrude 8.7, 87.8, 447.6, 673.9 and 891.5 hrs into the second period respectively. For shut down times beyond 100 hrs, even if all of the second period is included in the receding horizon the bulk demands could not be met, so, either the third period also has to be included in the receding horizon or the slack may be fed back to leve-1 for effective redistribution of demands.

4.3 Level-3 formulation:

We assume that we are given product sequencing and the total inventory profiles as input from level-2 and we need to figure out at level-3 if this volume can be met from the set of available tanks. We first focus on the triangular inventory breakpoints between stage 2 and stage 3 as shown in Fig. 3(d). We know that if these profiles do not overlap in the time frame, then we can use the same tanks repeatedly. For example the tanks that are used for storing either grade D or C can be used again for storing grades A and B. We exploit this feature in the proposed heuristic algorithm and make an efficient usage of the non-overlapping profiles. However, since the feed is a continuous stream received from upstream plants, the feed inventory tanks would almost always be busy as shown in Fig. 3(a), thus rendering less probability for reuse of these tanks elsewhere.

Note that the inverted triangular profiles for most of the grades in Fig. 3(a) continue to be in use as some feed is stored for use in next cycle, unlike the rest that are traditional triangular profiles which if freed can be used elsewhere. As already mentioned earlier, from the reactive scheduling point of view the objective for the operator at this level is to minimize the available inventory usage as best as possible.

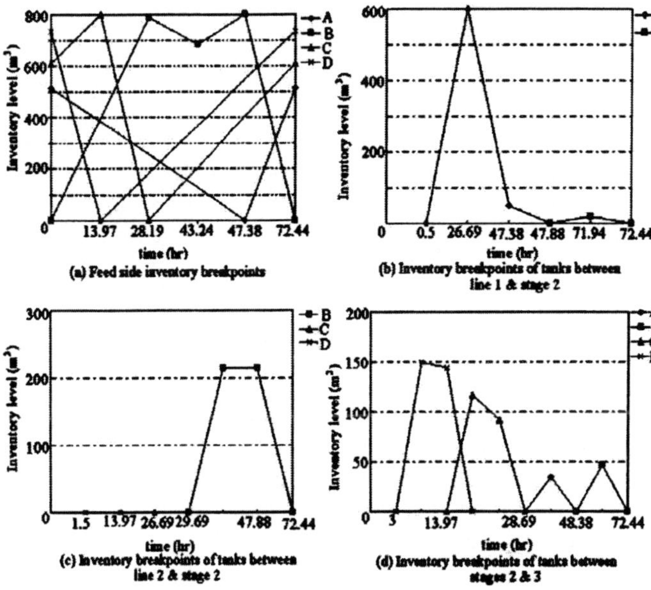

Fig. 3. Inventory profiles for level-2 in the first period of 1000 hr

Let us pose the problem here as one of finding the minimum number of tanks of each (say) 50 m^3 capacity required to store these grades. For each profile we first generate 'sub-profiles' by demarcating the 50 m^3 tank capacities as shown in Fig. 3(d) and find the corresponding timings on x-axis by linear interpolation. For example, for the profile of grade D, we get three sub-profiles of the shape 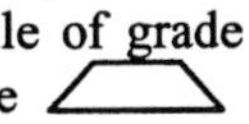 stacked on each other as seen Fig. 3(d). For each of these profiles suppose if we use three tanks T1-T3 each of 50 m^3 size. Then we know the times at which each of these tanks would be occupied and free to use again. For example T3 would be empty at the end of 16.18 hr, T2 at the end of 18.68 hr and T1 at the end of 21.18 hr. Now we consider the pool of all such sub-profiles across all grades and look out for which of these sub-profiles do not overlap in the time frame and try to reuse these tanks. Using this heuristic for the inventory profiles between stage 2 and stag 3, we can find that we need a minimum of four tanks of each 50 m^3 size. Without reuse of tanks we would need eight such tanks.

When we consider many stages with numerous inventory profiles, it is difficult to visualize and apply this heuristic manually. With this motive, a simple heuristic algorithm termed as Inventory Slicing and Tank Re-allocation (ISTR) is proposed here (in appendix) which automates the generation of sliced profiles, checks for the non-overlapping zones (sub-profiles) and finds the minimum number of tanks of a given capacity (say 50 m^3 or100 m^3) required to manage these inventories efficiently at level-3. Using the ISTR algorithm for the total thirteen profiles across all lines/stages of Fig. 3., the minimum number of tanks of 50 m^3 capacity was found to be 50 (totaling 2500m^3) in the first period and 57 and 50 tanks respectively in the second and third periods. If the number of tanks suggested by ISTR at level-3 is less than the actual number of tanks available, then level-2 is revised with modified upper bounds on inventories.

Amenability for reactive scheduling at level-3: Consider the horizon at level-3 as one cycle, 72.44 hr (approx. 3 days) and the tank assignments as suggested by ISTR algorithm. From the view point of reactive scheduling, if there are any tank breakdowns in the last two days of the horizon, then they can be mostly taken care locally in a cascade control fashion, without sending any feedback to level-2, as most of the tanks are free in the last two third of the horizon. If there are some tank breakdowns on the first day of the horizon itself, then we consider a receding horizon model from that time instance onwards and if required intrude into the next cycle without sending feedback up to level-2 until its scheduled revision. Otherwise, the only way to reject such disturbance locally would be to provide some resiliency margin or back off from the best schedule, i.e. instead of deploying minimum number of tanks as suggested by ISTR, we may deploy additional tanks thus making the problem amenable to reactive scheduling. The analogy in process control applications is clear here in terms of a compromise between aggressive but non-resilient control and robust but relatively less aggressive control.

5. CONCLUSION

An integrated multi-level, control-theoretic framework has been proposed in this work for integration of planning, scheduling and rescheduling. The proposed methodology has been demonstrated for single-site lube production in a hybrid flowshop.

Appendix: *ISTR algorithm for triangular profiles:* Consider the triangular inventory breakpoints between stage 2 and stage 3. In this algorithm for each profile we first generate sub-profiles of the given tank capacity i.e. we generate time vs. volume data (say V_{ik}) for each grade by slicing the inventory profiles as discussed earlier. This completes inventory slicing. In the V_{ik} data, all the inventory breakpoints would then be multiples of the given tank capacity (50 m^3), and the volumes get repeated grade-wise, except the maximum breakpoint. We exploit this feature in our algorithm. We know that the first occurrence of these entries correspond to the time at which a tank needs to be deployed and the repeated occurrence corresponds to the time at which such tank would be freed. At each time instance (k), for each grade (i), for each breakpoint, V_{ik}, we start deploying new tanks until we encounter a maximum breakpoint (say V_i^{max}). For the entry corresponding to V_i^{max} we assign the same tank as was used for the previous V_{ik} of the same grade (indicating the same tank has been still in use). Now after crossing the maximum breakpoint, before assigning a new tank at each time instance, we additionally check for each grade if same V_{ik} entry already exists in the previous time instances. After finding the first such occurrence (in the backward search from the current time instance, k), we assign the same tank number as that was used earlier (indicating the same tank has been in use till now). And we mark the status of this tank number as freed and available from this time instance onwards. Now, for the subsequent time instances, before assigning new tanks we also check if some freed tanks are available and if so we use them and remove the corresponding entry from the available tank list. Only when the list of freed tanks is empty we deploy a new tank. For inverted triangular profiles the algorithm had to be modified to account for their typical characteristic features, the details of which are beyond the scope of this paper.

REFERENCES

Bose, S., and J.F. Pekny (2000). A model predictive framework for planning and scheduling problems: a case study of consumer goods supply chain. *Comp. Chem. Eng.*, **24**, 329-335.

Grossmann, I.E., S.A. Van den Heever and I. Harjunkoski (2001). Discrete optimization methods and their role in the integration of planning and scheduling. *AIChE Symp. Ser.*, **98** *(326)*, 150-168.

Munawar, S.A., M. Bhushan, R.D. Gudi and A.M. Belliappa (2003a). Cyclic scheduling of continuous multiproduct plants in a hybrid flowshop facility. *Ind. Eng. Chem. Res.*, **42**, 5861-5882.

Munawar, S.A., M. Bhushan, R.D. Gudi and A.M. Belliappa (2003b). A multi-level, control-theoretic approach to reactive scheduling in continuous plants. In: Grossmann, I.E. and C.M. McDonald (Eds.), *Proceedings of the conference in foundations of computer aided process operations, CACHE Publications*, 397-400.

Perea E., I.E. Grossmann, E. Ydstie and T. Tahmassebi (2000). Dynamic modeling and classical control theory for supply chain management. *Comp. Chem. Eng.*, **24**, 1143-1149.

Perea-Lopez E., B.E. Ydstie and I.E. Grossmann (2003). A model predictive control strategy for supply chain optimization. *Comp. Chem. Eng.*, **27**, 1201-1218.

Shah, N. (1998). Single and multisite planning and scheduling: current status and future challenges. *AIChE Symp. Ser.*, **94** *(320)*, 75.

Shobrys, D. E., and D. C. White (2000). Planning, scheduling and control systems: why can they not work together. *Comp. Chem. Eng.*, **24**, 163-173.

Van den Heever, S.A., and I.E. Grossmann (2003). A strategy for the integration of production planning and reactive scheduling in the optimization of a hydrogen supply network. *Comp. Chem. Eng.*, **27**, 1813-1839.

Vargas-Villamil, F.D. and D.E. Rivera (2000). Multilayer optimization and scheduling using model predictive control: application to reentrant semiconductor manufacturing lines. *Comp. Chem. Eng.*, **24**, 2009-2021.

ELSEVIER
IFAC
PUBLICATIONS
www.elsevier.com/locate/ifac

FAULT-TOLERANT CONTROL OF MULTI-UNIT PROCESS SYSTEMS USING COMMUNICATION NETWORKS [1]

Nael H. El-Farra, Adiwinata Gani and Panagiotis D. Christofides [2]

Department of Chemical Engineering
University of California, Los Angeles, CA 90095-1592

Abstract: This work proposes a methodology for the design of fault-tolerant control systems for chemical plants with distributed interconnected processing units. Bringing together tools from Lyapunov-based nonlinear control and hybrid systems theory, the approach is based on a hierarchical architecture that integrates lower-level feedback control of the individual units with upper-level logic-based supervisory control over communication networks. The local control systems consist each of a family of control configurations connected, via a local communication network, to a local supervisor that orchestrates switching between them on the basis of the stability regions in the event of failures. The local supervisors communicate, through a plant-wide communication network, with a plant supervisor responsible for monitoring the different units and coordinating their responses in a way that minimizes the propagation of failure effects. The communication logic is designed to ensure efficient transmission of information between the units while respecting the inherent limitations in network resources. The proposed approach provides explicit guidelines for managing the various interplays between the tasks of feedback control, switching and communication. The efficacy of the proposed approach is demonstrated through a chemical process example.

Keywords: Hybrid control, Switching logic, Stability regions, Fault-tolerance, Supervisory control, Communication networks, Process systems.

1. INTRODUCTION

Safety and reliability are primary goals in the operation of industrial chemical plants. An important national need currently exists for enhancing the safety and reliability of chemical plants in ways that reduce their vulnerability to serious failures. Increasingly faced with these requirements and other economic drivers, plant operation is relying extensively on highly automated process control systems. Automation, however, tends to increase vulnerability of the plant to faults (e.g., defects/malfunctions in process equipment, sensors and actuators, failures in the controllers or in the control loops) which, if not appropriately handled in the control system design, can potentially cause a host of undesired economic, environmental, and safety problems that seriously degrade the operating efficiency of the plant. These considerations provide a strong motivation for the development of systematic methods and strategies for the design of fault-tolerant control systems and have motivated many research studies in this area (e.g., see (Willsky, 1998; Yang *et al.*, 1998; Bao *et al.*, 2002) and the references therein).

Given the complex dynamics of chemical processes (due, for example, to the presence of nonlinearities and constraints) and the geographically distributed, interconnected nature of plant units, as well as the large number of distributed sensors and actuators typically involved, the success of any fault-tolerant control strategy requires an integrated approach that brings together several essential elements, including: (1) the design of advanced feedback control algorithms that handle complex dynamics effectively, (2) the design of supervisory switching schemes that orchestrate the transition from the failed control configuration to available well-functioning fall-back configurations to ensure fault-tolerance, and (3) the efficient exchange of information and communication between the different plant units through a high-level supervisor that coordinates the overall plant response in failure situations and minimizes the effects of failure propagation. The realization of such an ap-

[1] Financial support from NSF, CTS-0129571, is gratefully acknowledged.
[2] Corresponding author (e-mail: pdc@seas.ucla.edu)

proach is increasingly aided by a confluence of recent, and ongoing, advances in several areas of process control research, including advances in nonlinear controller designs for constrained chemical processes (e.g., (El-Farra and Christofides, 2001; El-Farra and Christofides, 2003*a*)) and advances in the analysis and control of hybrid process systems (e.g., (Bemporad and Morari, 1999; El-Farra and Christofides, 2002; El-Farra and Christofides, 2003*b*)).

In addition to the above fundamental advances, recent innovations in actuator/sensor and communication technologies are increasingly enabling the integration of communication and control domains. For example, the use of communication networks as media to interconnect the different components in an industrial control system is rapidly increasing and expected to replace the more costly point-to-point connection schemes currently employed in distributed control systems. In addition to the advantage of reduced system wiring, the increased flexibility and ease of maintenance of a system using a network to transfer information is an appealing goal. In the context of fault-tolerant control, systems designed in this manner allow for easy modification of the control strategy by rerouting signals, having redundant systems that can be activated automatically when component failure occurs, and in general they allow having a high-level supervisor control over the entire plant. Currently, networked control systems is an active area of research within control engineering (e.g., see (Walsh *et al.*, 2002; Montestruque and Antsaklis, 2003; Tipsuwan and Chow, 2003; Patankar, 2003) and the references therein). The appealing features of communication networks motivate investigating ways for integrating them in the design of fault-tolerant control systems to ensure a timely and coordinated response that minimizes failure propagation effects between plant units.

In a previous work (El-Farra *et al.*, 2004), we presented an approach for fault-tolerant control of single-unit process systems using the idea of integrating feedback and supervisory control over networks. In this paper, we extend our previous work and develop a fault-tolerant control system design methodology, for plants with multiple (distributed) interconnected processing units, that accounts explicitly for the inherent complexities in supervisory control and communication tasks resulting from the distributed interconnected nature of plant units. The approach brings together tools from Lyapunov-based control and hybrid systems theory and is based on a hierarchical distributed architecture that integrates lower-level feedback control of the individual units with upper-level logic-based supervisory control over networks.

The local control systems consist each of a family of feedback control configurations together with a local supervisor that communicates with actuators and sensors, via a local communication network, to orchestrate the transition between the control configurations, on the basis of their fault-recovery regions in the event of failures. The local supervisors communicate, through a plant-wide communication network, with a plant supervisor responsible for monitoring the different units and coordinating their responses in a way that minimizes the propagation of failure effects. The communication logic is designed to ensure efficient transmission of information between the units while respecting the inherent limitations in network resources by minimizing unnecessary network usage and accounting explicitly for the effects of possible delays due to fault-detection, control computations, network communication and actuator activation.

2. PRELIMINARIES

2.1 *System description - problem formulation*

We consider the class of continuous-time, multivariable nonlinear process systems with constraints on the manipulated input, represented by the following state-space description:

$$\begin{aligned} \dot{x}(t) &= f_{k(t)}(x(t)) + \sum_{i=1}^{m} g^i_{k(t)}(x(t))u^i_{k(t)} \\ |u_{k(t)}| &\leq u^k_{max},\ k(t) \in \mathscr{K} = \{1,\cdots,N\} \end{aligned} \quad (1)$$

where $x(t) \in \mathbb{R}^n$ denotes the vector of process state variables and $u_k := [u^1_k \cdots u^m_k]^T$ denotes the vector of constrained manipulated inputs associated with the k-th control configuration. $k(t)$, which takes values in the finite index set, $\mathscr{K}$, represents a discrete state that indexes the vector fields $f_k(\cdot)$, $g^i_k(\cdot)$ as well as the manipulated inputs $u^i_k(\cdot)$. For each value that k assumes in $\mathscr{K}$, the process is controlled via a different set of manipulated inputs which define a given control configuration. Switching between the available N control configurations is controlled by a high-level supervisor that monitors the process and orchestrates, accordingly, the transition between the different control configurations in the event of control system failure. This in turn determines the temporal evolution of the discrete state, $k(t)$. The supervisor ensures that only one control configuration is active at any given time, and allows only a finite number of switches over any finite interval of time.

It is assumed that the origin is the equilibrium point of the nominal process and that the vector functions $f_k(\cdot)$ and $g^i_k(\cdot)$ are sufficiently smooth, for all k. The control objective is to stabilize the process of Eq.1 in the presence of actuator constraints and faults in the control system. The basic problem is how to coordinate switching between the different control configurations in a way that respects actuator constraints and guarantees closed-loop stability in the event of faults. Throughout the paper, the notation $|\cdot|$ is used to denote the Euclidean norm of a vector and the notation L_fV denotes the Lie derivative of a scalar function, V, with respect to the vector field, f. To simplify the presentation of our results, we will focus only on the state feedback problem where measurements of all process states are available for all times.

2.2 *Motivating example*

In this section, we introduce a simple benchmark example that will be used throughout the paper to illustrate the design and implementation of the fault-tolerant control design methodology to be proposed in section 3. While the discussion will center around this example, we note that the proposed framework can be applied to more complex plants involving more complex arrangements of processing units. To this end, consider two well-mixed, non-isothermal continuous stirred tank reactors in series, where three parallel irreversible elementary exothermic reactions of the form $A \xrightarrow{k_1} B$, $A \xrightarrow{k_2} U$ and $A \xrightarrow{k_3} R$ take place, where A is the reactant species, B is the desired product and U, R are undesired byproducts. The feed to CSTR 1 consists of pure A at flow rate F_0, molar concentration C_{A0} and temperature T_0, and the feed to CSTR 2 consists of the output of CSTR 1 and an additional fresh stream feeding pure A at flow rate F_3, molar concentration C_{A03} and temperature T_{03}. Due to the non-isothermal nature of the reactions, a jacket is used to remove/provide heat to both reactors. Under standard modeling assumptions, a mathematical model of the plant can be derived from material and energy balances and takes the following form:

$$\begin{aligned}
\frac{dT_1}{dt} &= \frac{F_0}{V_1}(T_0 - T_1) + \sum_{i=1}^{3} R_i(C_{A1}, T_1) + \frac{Q_1}{\rho c_p V_1} \\
\frac{dC_{A1}}{dt} &= \frac{F_0}{V_1}(C_{A0} - C_{A1}) - \sum_{i=1}^{3} k_{i0} e^{\frac{-E_i}{RT_1}} C_{A1} \\
\frac{dT_2}{dt} &= \frac{F_1}{V_2}(T_1 - T_2) + \frac{F_3}{V_2}(T_{03} - T_2) \\
&\quad + \sum_{i=1}^{3} R_i(C_{A2}, T_2) + \frac{Q_2}{\rho c_p V_2} \\
\frac{dC_{A2}}{dt} &= \frac{F_1}{V_2}(C_{A1} - C_{A2}) + \frac{F_3}{V_2}(C_{A03} - C_{A2}) \\
&\quad - \sum_{i=1}^{3} k_{i0} e^{\frac{-E_i}{RT_2}} C_{A2}
\end{aligned} \tag{2}$$

where $R_i(C_{Aj}, T_j) = \frac{(-\Delta H_i)}{\rho c_p} k_{i0} \exp\left(\frac{-E_i}{RT_j}\right) C_{Aj}$, for $j = 1, 2$. T, C_A, Q, and V denote the temperature of the reactor, concentrations of the species A, rate of heat input/removal from reactor, and volume of reactor, respectively, with subscript 1 denoting CSTR 1 and subscript 2 denoting CSTR 2, ΔH_i, k_i, E_i, $i = 1, 2, 3$, denote the enthalpies, pre-exponential constants and activation energies of the three reactions, respectively, c_p and ρ denote the heat capacity and density of the reactor. Using typical values of process parameters (values are omitted for brevity), CSTR 1, with $Q_1 = 0$, has three steady-states: two locally asymptotically stable and one unstable at $(T_{1s}, C_{A1s}) = (388.57\ K, 3.59\ kmol/m^3)$. The unstable steady-state of CSTR 1 corresponds to three steady-states for CSTR 2 (with $Q_2 = 0$), one of which is unstable at $(T_{2s}, C_{A2s}) = (429.24\ K, 2.55\ kmol/m^3)$.

The control objective is to stabilize both reactors at the (open-loop) unstable steady-states. To accomplish this objective under normal conditions (with no fail-

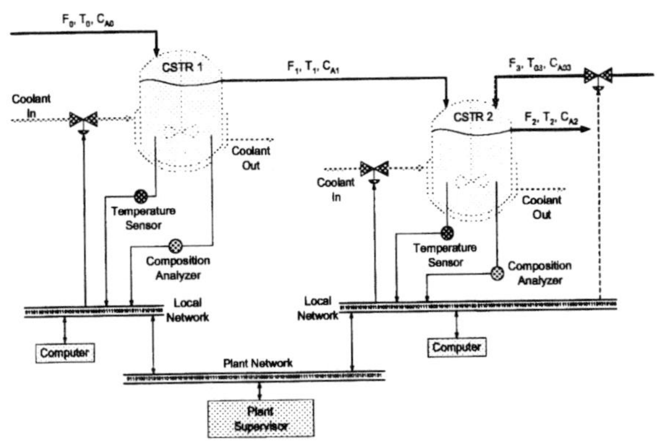

Fig. 1. Process flow diagram of two CSTRs in series.

ures), we choose as manipulated inputs the rate of heat inputs, $u_1^1 = Q_1$, subject to the constraint $|Q_1| \leq u_{max}^{Q_1} = 2.7 \times 10^6\ KJ/hr$ and $u_1^2 = Q_2$, subject to the constraint $|Q_2| \leq u_{max}^{Q_2} = 2.8 \times 10^6\ KJ/hr$. As shown in Fig.1, each unit has a local control system with its sensors and actuators connected through a communication network. The local control systems in turn communicate with the plant supervisor (and with each other) through a plant-wide communication network. Note that in designing each control system, only measurements of the local process variables are used (for example, the controller for the second unit uses only measurements of T_2 and C_{A2}).

The fault-tolerant control problem under consideration involves a catastrophic failure in both control systems after some time of startup, with the failure of the first unit being permanent. Our objective will be to preserve closed-loop stability of CSTR 2 by switching to an alternative control configuration involving, as manipulated variables, the rate of heat input, $u_2^1 = Q_2$, subject to the same constraint, and the inlet reactant concentration, $u_2^2 = C_{A03} - C_{A03s}$, subject to the constraint $|C_{A03} - C_{A03s}| \leq u_{max}^{C_{A03}} = 0.4\ kmol/m^3$ where $C_{A03s} = 3.0\ kmol/m^3$. The main question, which we address in the next section, is how to devise the switching and network communication logics in a way that ensures fault-tolerance in the second unit and, simultaneously, accounts for the inherent limitations in network resources and possible delays in fault-detection, communication and actuator activation.

3. FAULT-TOLERANT CONTROL SYSTEM DESIGN METHODOLOGY

In this section, we outline the main steps involved in the fault-tolerant control system design procedure, as applied to the reactors example of section 2.2. A major feature of the design methodology is the inherent coupling between the feedback, supervisory control and communication tasks.

(a) *Constrained feedback controller synthesis*

The main issue in this step is how to design the feedback control law for the fall-back configuration in CSTR 2 in a way that respects its actuators' constraints and guarantees closed-loop stability of this unit. The choice of the feedback law depends on our choice of the communication policy. To explain this connection, we first note that a total failure in the control system of CSTR 1 will cause the states (C_{A1}, T_1) to move away from the desired steady-state for this unit. Therefore, unless the feedback controller for CSTR 2 is redesigned to account for this incoming "disturbance", the evolution of C_{A2}, T_2 will be affected causing them not to converge to the desired steady-state. To account for this disturbance, one option could be to send available measurements of C_{A1} and T_1, through the network, to the second unit and redesign its controller accordingly. From a communications cost point of view, this option may be costly since it requires continuous usage of the network after failure, which can adversely affect the performance of other units sharing the same communication medium due to bandwidth limitations and overall delays.

To reduce unnecessary network usage, we propose an alternative approach where we view the failure in CSTR 1 as a bounded non-vanishing disturbance affecting CSTR 2 and use the available process model of CSTR 1 to capture, or estimate, the size of this disturbance (by comparing, for example, the evolution of the process variables under the failed and well-functioning control configurations through simulations). In this formulation, in lieu of measurements of C_{A1} and T_1, only bounds on the disturbance size are needed and transmitted from CSTR 1 to CSTR 2, which involves using the network only at the failure time and not continuously. The disturbance information can then be used to design an appropriate robust controller to attenuate the effect of disturbances and enforce robust closed-loop stability in the second unit. To this end, we can rewrite the model of CSTR 2 in the following general form:

$$\dot{\bar{x}} = \bar{f}(\bar{x}) + \bar{G}(\bar{x})u + \bar{W}(\bar{x})\theta \qquad (3)$$

where $\bar{x} = [T_2 \;\; C_{A2}]^T$, $u = [Q_2 \;\; (C_{A03} - C_{A03s})]^T$ are the inputs of the fall-back configuration, and $\theta = [T_1 \;\; C_{A1}]^T$ are the time-varying, but bounded, disturbances. For this system, one possible choice for the controller is a variation of the class of bounded robust Lyapunov-based control laws proposed in (El-Farra and Christofides, 2003*a*) (see also (Lin and Sontag, 1991)) and has the general form:

$$u = -r(\bar{x}, u_{max}, \theta_b)(L_{\bar{G}}V)^T \qquad (4)$$

where $r(x, u_{max}, \theta_b) =$

$$\frac{L_{\bar{f}}^*V + \sqrt{(L_{\bar{f}}^*V)^2 + (u_{max}|(L_{\bar{G}}V)^T|)^4}}{|(L_{\bar{G}}V)^T|^2\left[1 + \sqrt{1 + (u_{max}|(L_{\bar{G}}V)^T|)^2}\right]} \qquad (5)$$

$L_{\bar{f}}^*V = L_{\bar{f}}V + \chi\theta_b|(L_{\bar{W}}V)^T|$, θ_b is the disturbance bound (maximum size of the norm of the disturbance vector), V is a control Lyapunov function, $L_{\bar{G}}V = [L_{g^1}V \;\; L_{g^2}V]$, $L_{\bar{W}}V = [L_{w^1}V \;\; L_{w^2}V]$ are row vectors, g^i, w^i are column vectors of the matrices $\bar{G}$ and $\bar{W}$, respectively, and $\chi > 0$ is a tuning parameter. The nonlinear gain function, $r(\cdot)$ in Eqs.4-5, which depends on the size of actuator constraints, the disturbance size and the type of configuration used, is shaped in a way that guarantees constraint satisfaction and robust closed-loop stability, with an arbitrary degree of attenuation of the effect of the disturbances, within a well-characterized region in the state space. The characterization of this region is given next.

(b) *Characterization of fault-recovery regions*

Having designed the feedback control law, the next step is to explicitly characterize the set of admissible states starting from where the constrained fall-back control configuration is stabilizing (fault-recovery region). As discussed in step (c), this characterization is necessary for the design of the switching policy that ensures fault-recovery. For the controller of Eqs.4-5, using a Lyapunov argument, one can show that the set

$$\Pi(u_{max}, \theta_b) = \{\bar{x} : L_{\bar{f}}^*V(\bar{x}) \leq u_{max}|(L_{\bar{G}}V)^T(\bar{x})|\} \qquad (6)$$

describes a region in the state space where the control action satisfies the constraints and the Lyapunov function decays monotonically along the trajectories of the closed-loop system (see (El-Farra and Christofides, 2003*a*) for the detailed mathematical analysis). Note that the size of this set depends both on the magnitude of the constraints and the size of the disturbance (which in turn depends on the failure time). For a given control configuration, one can use the above inequality to estimate the fault-recovery region by constructing, for example, the largest invariant subset of Π, which we denote by $\Omega(u_{max}, \theta_b)$.

(c) *Supervisory switching logic design*

In the general case where more than one fall-back control configuration is available, an important question is how to decide which of the available configurations can be activated following the failure of the primary configuration in order to preserve closed-loop stability. The key idea here is that the supervisor can only activate the configuration for which the closed-loop state, at the time of failure, is within the stability region. For the reactor example of section 2.2, and since a single fall-back configuration, (Q_2,C_{A03}), is considered, this implies that if failure occurs at any time T for which $\bar{x}(T) \in \Omega(u_{max}, \theta_b)$, then switching to this configuration guarantees preservation of closed-loop stability. The implementation of this switching rule requires monitoring the closed-loop state trajectory with respect to the fault-recovery region. The idea of constructing the switching logic based on the stability regions was first proposed in (El-Farra and Christofides, 2002) for the control of switched nonlinear systems. If failure occurs at times when the states are outside the stability region, our analysis suggests that either the constraints should be relaxed to enlarge the fault-recovery region of the given configuration, or additional fall-back control loops must be introduced. The latter option is ultimately limited by the maximum allowable number of control loops that can be designed for the process.

(d) *Design of the communication logic*

An essential element in the design of the fault-tolerant control system is the use of a communication medium that ensures fast and efficient transmission of information during failure events. Communication networks offer such a medium which is both fast (relative to the typically slow dynamics of chemical processes) and inexpensive (relative to current point-to-point connection schemes which require extensive cabling and higher maintenance time and costs). The ability of the network to fulfill this role, however, requires that the communication policy be devised in a way that respects the inherent limitations in network resources, such as bandwidth constraints and overall delays, by minimizing unnecessary usage of the network.

In step (a), we have already discussed how bandwidth constraints can be handled by formulating the problem as a robust control problem, where the failure in the first unit is treated as a bounded non-vanishing disturbance to the second unit. This policy avoids unnecessary overloading of the network while also guaranteeing fault-tolerance. The idea of using knowledge of the plant dynamics to manage the tradeoff between bandwidth limitations (which favor reduced communication of measurements) and optimum control performance (which favors increased communication of measurements) is conceptually aligned with the notion of minimum attention control (e.g., see (Brockett, 1997; Montestruque and Antsaklis, 2003)). In our work, however, this idea is utilized in the context of fault-tolerant control.

The second consideration in devising the communication logic is the issue of time-delays which typically result from the time sharing of the communication medium as well as the computing time required for the physical signal coding and communication processing. The characteristics of these time delays depend on the network protocols adopted as well as the hardware chosen. For our purposes here, we consider an overall fixed time-delay (which we denote by τ_{max}) that combines the contribution of several delays, including: (1) delays in fault-detection, (2) the time the local supervisor of unit 1 takes to compute the effective disturbance bounds (through simulations comparing the open-loop and closed-loop state evolution), (3) the time the local supervisor of unit 1 takes to send the information to the plant supervisor, (4) the time it takes the plant supervisor to forward the information to the local supervisor of unit 2, (5) the time it takes the local supervisor for unit 2 to compute the fault-recovery region for the given fall-back configurations using the information arriving from unit 1 and the time it takes for the supervisor's decision to reach and activate the appropriate fall-back configuration, and (6) the inherent actuator/sensor dead-times.

Failure to take such delays into account can result in activating the wrong control configuration and subsequent instability. For example, even though failure of a given loop may take place at $t = T$, the backup configuration will not be switched in before $t = T + \tau_{max}$. If the delay is significant, then the switching rule in part (c) should be modified such that the supervisor activates the configuration for which $\bar{x}(T + \tau_{max}) \in \Omega(u_{max}, \theta_b)$. This modification is yet another manifestation of the inherent coupling between the switching and communication logics. The implementation of the modified switching rule (which accounts for delays) requires that the local supervisor of unit 2 be able to predict where the trajectory will be at $t = T + \tau_{max}$ (e.g., through simulations using the process model) and check whether the state at this time is within the fault-recovery region of the fall-back configuration.

4. SIMULATION RESULTS

In this section, we revisit the example of section 2.2 to illustrate the implementation of the proposed fault-tolerant control methodology. To this end, the reactors are initialized at $(T_1(0), C_{A1}(0)) = (300\,K, 4.0\,mol/L)$, and $(T_2(0), C_{A2}(0)) = (440\,K, 4.0\,mol/L)$. Under normal operation (with no failures), each reactor is controlled by manipulating the rate of heat input using bounded nonlinear Lyapunov-based control algorithms (e.g., see (El-Farra and Christofides, 2002); details of the nominal controller design are omitted for brevity). It was verified that, in the absence of failures, the controllers stabilize both reactors at the desired steady-states. Consider now a catastrophic failure in both controllers (Q_1 and Q_2) at $t_{failure} = 5\,min$. In this case, both reactors revert to their open-loop modes and, consequently, the states are expected to move away from the desired steady-state as shown by the dashed lines in Figs.3(a)-(b) for CSTR 2.

Recall that the objective is to preserve closed-loop stability of CSTR 2 despite this failure. Using the proposed methodology, the supervisor of CSTR 1, at the failure time, runs both open- and closed-loop simulations of the process model to estimate the size of the disturbance that will affect CSTR 2 and transmits this information to CSTR 2 through the network. The maximum disturbance size is proportional to the largest discrepancy between the values of C_{A1}, T_1 in the failed and in the well-functioning modes. Using this information, the local supervisor designs a robust control law of the form of Eqs.4-5 to stabilize CSTR 2, using the fall-back configuration with (Q_2, C_{A03}) as the manipulated inputs, and constructs the associated fault-recovery region (the shaded area in Fig.2) with the aid of Eq.6. From Fig.2, we observe that failure occurs when the states are within the fault-recovery region. Therefore, assuming no delays in the fault-detection, computations and communication processing, when the fall-back controllers are activated, closed-loop stability is preserved and the states converge to the desired steady-state as shown by the solid lines in Figs.3(a)-(b).

When delay effects are taken into account, the dotted line in Fig.2 shows that if an overall delay of 3 min elapses between the failure and the activation of the

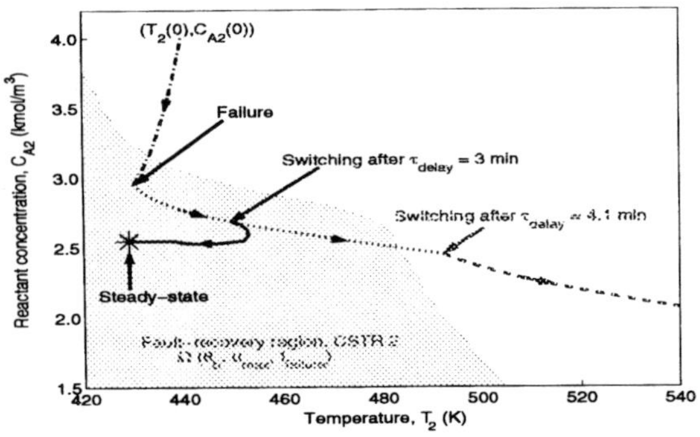

Fig. 2. Fault-recovery region of (Q_2, C_{A03}) config.

(Q_2, C_{A03}) configuration, the state at the end of the delay still resides within the fault-recovery region and, therefore, closed-loop stability can be preserved by switching as shown by the solid lines in Figs.3(c)-(d). By contrast, when the overall delay is 4.1 min, the state lies outside the fault-recovery region and the fall-back configuration cannot stabilize the system at the desired steady-state as can be seen from the dashed lines in Figs.3(c)-(d). Finally, examination of Fig.2 reveals how the interplay between the switching and communication tasks can be managed to ensure fault-tolerance. For example, the picture suggests that large communication delays can be tolerated by enlarging the fault-recovery region (e.g., by relaxing the constraints).

5. REFERENCES

Bao, J., W. Z. Zhang and P. L. Lee (2002). Passivity-based decentralized failure-tolerant control. *Ind. & Eng. Chem. Res.* **41**, 5702–5715.

Bemporad, A. and M. Morari (1999). Control of systems integrating logic, dynamics and constraints. *Automatica* **35**, 407–427.

Brockett, R. (1997). Minimum attention control. In: *Proceedings of 36th Conference on Decision and Control*. San Diego, CA. pp. 2628–2632.

El-Farra, N. H., A. Gani and P. D. Christofides (2004). Fault-tolerant control of process systems: Integrating supervisory and feedback control over networks. In: *Proceedings of 5th International Symposium on Advanced Control of Chemical Processes*. Hong Kong, P. R. China. pp. 784–789.

El-Farra, N. H. and P. D. Christofides (2001). Integrating robustness, optimality, and constraints in control of nonlinear processes. *Chem. Eng. Sci.* **56**, 1841–1868.

El-Farra, N. H. and P. D. Christofides (2002). Switching and feedback laws for control of constrained switched nonlinear systems. In: *Lecture Notes in Computer Science Series*. Vol. 2289. Tomlin, C. J. and M. R. Greenstreet (Eds.), Berlin, Germany: Springer-Verlag. pp. 164–178.

El-Farra, N. H. and P. D. Christofides (2003*a*). Bounded robust control of constrained multivariable nonlinear processes. *Chem. Eng. Sci.* **58**, 3025–3047.

El-Farra, N. H. and P. D. Christofides (2003*b*). Coordinating feedback and switching for control of

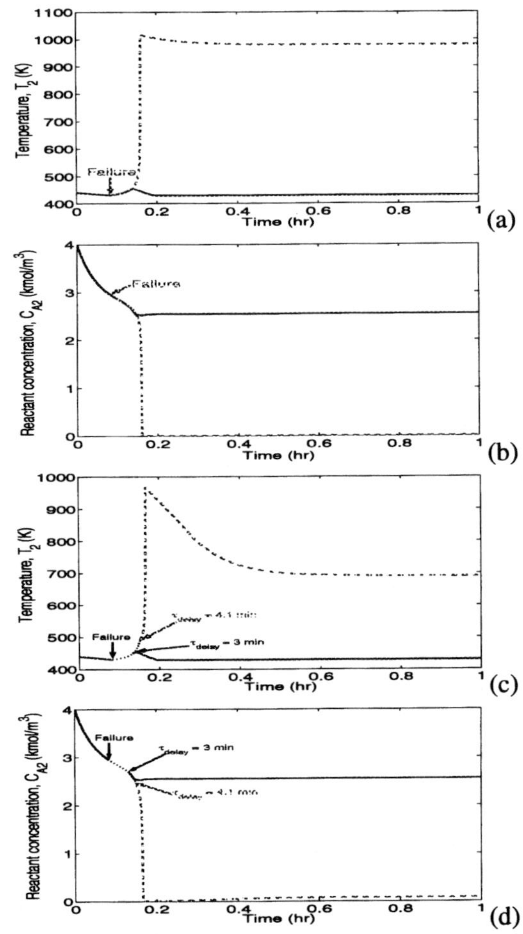

Fig. 3. Evolution of closed-loop state profiles when the fall-back controllers are not activated after failure (a–b, dashed), activated with no delays (a–b, solid), activated after total delays of 3 min (c–d, solid) and 4.1 min (c–d, dashed).

hybrid nonlinear processes. *AIChE J.* **49**, 2079–2098.

Lin, Y. and E. D. Sontag (1991). A universal formula for stabilization with bounded controls. *Systems & Control Letters* **16**, 393–397.

Montestruque, L. A. and P. J. Antsaklis (2003). On the model-based control of networked systems. *Automatica* **39**, 1837–1843.

Patankar, R. (2003). A model for fault-tolerant networked control system using TTP/C communication. In: *Proceedings of American Control Conference*. Denver, CO. pp. 533–537.

Tipsuwan, Y. and M.-Y. Chow (2003). Control methodologies in networked control systems. *Contr. Eng. Prac.* **11**, 1099–1111.

Walsh, G. C., H. Ye and L. G. Bushnell (2002). Stability analysis of networked control systems. *IEEE Trans. Contr. Syst. Tech.* **10**, 438–446.

Willsky, A. S. (1998). A survey of design methods for failure detection in dynamic systems. *Automatica* **12**, 601–611.

Yang, G. H., S. Y. Zhang, J. Lam and J. Wang (1998). Reliable control using redundant controllers. *IEEE Trans. Automat. Contr.* **43**, 1588–1593.

ELSEVIER
IFAC
PUBLICATIONS
www.elsevier.com/locate/ifac

FAULT DIAGNOSIS AND FAULT TOLERANT CONTROL USING REDUCED ORDER MODELS

Seema Manuja [*] **Sachin Patwardhan** [**]
Shankar Narasimhan [*,1]

[*] *Department of Chemical Engineering, Indian Institute of Technology, Madras, India.*
[**] *Department of Chemical Engineering, Indian Institute of Technology, Bombay, India.*

Abstract: The present work considers a model reduction approach for fault diagnosis with generalised likelihood ratio (GLR) method to improve upon its diagnostic performance and computational efficiency in large dimensional applications. Model reduction techniques are widely used in controller design. A similar concept with balanced truncation technique is employed to obtain a reduced order diagnostic model of the process for GLR implementation. The proposed method is incorporated in the fault tolerant control strategy developed by Prakash *et al.* (2002). Simulation results for a binary distillation column demonstrate the efficacy of the proposed fault detection and identification (FDI) method and the fault tolerant control strategy in comparison to the full-scale implementation for a variety of faults.

Keywords: fault detection and identification, GLR method, model reduction, balanced truncation, soft faults, fault tolerant control

1. INTRODUCTION

During the past two decades there has been an increasing trend towards the design and development of fault tolerant control system (FTCS), which can maintain an acceptable control performance despite the occurrence of faults. In an active FTCS design, the controller is designed for normal operation, and as and when a fault is deemed to have occurred, the control law is reformulated or restructured on-line.

An important component of an active FTCS design is the FDI method. Among the FDI approaches, model based methods employing analytical redundancy are found to be most suitable for control applications, since the information it provides facilitates explicit corrective measures. Recently, Prakash *et al.* (2002) developed an active FTCS scheme, which integrates a FDI methodology based on *generalised likelihood ratio* (GLR) method (Willsky and Jones, 1976); (Narasimhan and Mah, 1988) with a controller, through an on-line compensation mechanism. They demonstrated through simulation and experimental studies, the superior performance of the FTCS over the conventional control scheme in the presence of various types of soft faults such as sensor and actuator biases, abrupt changes in unmeasured disturbances and process parameters.

The key component in their FTCS design is the use of a linear model (and a Kalman filter) derived

[1] Author to whom correspondence should be addressed. Email: naras@che.iitm.ac.in

from first principles. The *first principles* modeling approach used in their work enables identification of various types of faults. However, the models of process units such as a distillation column or a subsystem consisting of several process units are of high order, involving hundreds of state variables. This can lead to excessive computational demand in real-time application of the proposed FTCS. In addition to this, system observability is a prime requirement in the Kalman filter design for obtaining unique and unbiased estimates of states. Typically, not all the states are observable from the relatively few measurements that are available in such systems. Thus, there is a need to develop an alternate formulation based on the first principles model, which will alleviate computational difficulties and improve the state estimation particularly when the system dimension is large.

In the proposed approach, we still make use of a linearised model of the process developed from first principles in the FDI method, in order to retain the ability to distinguish between different types of faults. However, instead of using the full high order model so developed, we propose to obtain a reduced order normal and fault models of the process (essentially a *diagnostic* model with fault dynamics incorporated) for FDI implementation. An ideal binary distillation column simulation is considered as a prototypical example of a large dimensional process to illustrate the proposed scheme. The objective of this study is to investigate the effectiveness of the proposed approach in comparison to the full-scale implementation.

2. REDUCED MODEL BASED FDI STRATEGY

Fault detection and identification (FDI) is concerned with detecting whether the fault has occurred, and if so, identify the cause and provide an estimate of the fault magnitude. We briefly describe the FDI strategy as developed by Narasimhan and Mah (1988) for identification of various types of faults.

2.1 FDI Strategy

The FDI strategy makes use of process models under normal and faulty operating conditions. A linear discrete stochastic state space equation is used to model the process as follows

$$x(k+1) = \Phi x(k) + \Gamma_u u(k) + w(k) \quad (1)$$

$$y(k) = Cx(k) + v(k) \quad (2)$$

where $x \in R^n$ represents the states, $u \in R^m$ represents the manipulated inputs, $y \in R^r$ represents the measured outputs, $w \in R^q$ and $v \in R^r$ are assumed to be independent zero mean Gaussian white noise sequences with covariance matrix R_1 and R_2 respectively. Under normal operation, the above model can be used to obtain the optimal estimates of the state variables using a Kalman filter (Astrom and Wittenmark, 1994). The innovations (or measurement residuals) generated by the Kalman filter are given as

$$\gamma(k) = y(k) - C\hat{x}(k/k-1) \quad (3)$$

where $\hat{x}(k/k-1)$ denote the state estimates predicted at time k using all measurements made up to time $(k-1)$. Simple *chi-square statistical tests* based on these innovations are employed for fault detection and subsequent fault confirmation over a time window. Fault identification is carried out with the GLR method, where for each hypothesised fault, the characteristic innovations trend termed as *fault signature* is determined from the corresponding fault model and the normal estimator model. For soft faults caused by biases in sensors and actuators which are assumed to occur as step changes, the corresponding fault models can be easily obtained as follows. In the presence of a sensor bias, the measurement model Eq. (2) is given by

$$y(k) = Cx(k) + v(k) + be_{y,i} \quad (4)$$

for $k \geq t_f$, where t_f is the time of occurrence of a fault. Likewise, in the presence of an actuator bias the state transition Eq. (1) is modeled as

$$x(k+1) = \Phi x(k) + \Gamma_u u(k) + w(k) + b\Gamma_u e_{u,i} \quad (5)$$

In the above equations, b represents the magnitude of the fault, $e_{y,i}$ and $e_{u,i}$ are unit vectors of appropriate dimensions, subscripts y, u signifies the fault type with i as the index of the measurement or actuator where the bias occurs.

The above fault models for sensor and actuator biases can be used whether the normal process model (defined by Eqs. 1 and 2) is derived from first principles, or obtained from input-output data using system identification methods. *However, if a step change in a process parameter or disturbance variable occurs, then the appropriate fault model can be derived, provided a first principles modeling approach is used (Prakash et al., 2002).* The fault models for these type of faults can be obtained as

$$x(k+1) = \Phi x(k) + \Gamma_u u(k) + w(k) + b\Gamma_{f'} e_{f',i} \quad (6)$$

where the matrix $\Gamma_{f'}$ depends on the type of fault that has occurred (disturbance or parameter) with $f' \in \{d, p\}$.

From the linearity of the system and filter, the effect of each fault on the expected values of the innovations at any time can be obtained as

$$E(\gamma(k)) = \left\{ \begin{array}{ll} 0 & k < t_f \\ bG_f(k;t_f)g_{f,i} & k \geq t_f \end{array} \right\} \quad (7)$$

where the subscript $f \in \{y, u, d, p\}$ denotes the fault type. Here, the matrix $G_f(k;t_f)$ is referred to as signature matrix and depends upon the time t_f at which the fault has occurred and time k at which the innovations are computed. The vector $g_{f,i}$ which we refer to as the fault signature vector, depends upon the fault type and location. The computational details of these signature matrices and signature vectors for different faults are given in Prakash et al. (2002). *The GLR method essentially identifies the fault whose signature best fits the observed innovations pattern.*

2.2 Reduced order FDI development

Model reduction techniques are widely used in controller design. Among the various approaches, the *balanced truncation* technique (Moore, 1981), removes after suitable transformations, the states that are difficult to control or observe. Since such states contribute little to the understanding of the process input-output behavior, their removal does not significantly alter the quality of the model predictions. The distinct advantage of this reduction approach is that one can obtain an *observable subsystem* for the Kalman filter design to be used in FDI development.

As explained in the preceding section, the GLR based FDI strategy makes use of process models under normal as well as faulty operating conditions. Therefore, the requirement in fault diagnosis is to adequately capture *fault-output dynamics* when compared to a controller design where the effect of a manipulated input on an output is important. The standard approach for obtaining reduced order models to design controllers needs to be modified.

A generic state space process model which describes the effect of manipulated inputs, parametric and disturbance faults, can be obtained by combining the normal and fault model (Eqs. 1, 5 and 6) as follows

$$x(k+1) = \Phi x(k) + \Gamma_u u_c(k) \quad (8)$$

$$y(k) = Cx(k) \quad (9)$$

where

$$u_c = \left[u^T \; d^T \; p^T \right]^T \quad (10)$$

where $x \in R^n$, $y \in R^r$ are variables as defined with Eqs 1 and 2. The input vector $u_c \in R^{m_c}$ denotes a complete input set with $\Gamma = \left[\Gamma_u \; \Gamma_d \; \Gamma_p \right]$ as the corresponding input coupling matrix with individual components as explained in Eqs 5 and 6. Let ζ be a vector of variables related to x by a balancing transformation $x = T\zeta$. The state space model Eqs. (8) and (9) in transformed domain is given as

$$\zeta(k+1) = \hat{\Phi}\zeta(k) + \hat{\Gamma}u_c(k) \quad (11)$$

$$y(k) = \hat{C}\zeta(k) \quad (12)$$

where $\hat{\Phi} = T^{-1}\Phi T$; $\hat{\Gamma} = T^{-1}\Gamma$ and $\hat{C} = CT$. In a balanced realisation, each state in the transformed domain ζ_i is just as controllable as it is observable. A measure of the state joint controllability and observability is given by the size of its associated Hankel singular value σ_i. The size of σ_i is a relative measure of the contribution that state ζ_i makes to the input-output behavior. The balanced truncation approach removes the states corresponding to the small Hankel singular values. Partition the state vector ζ of dimension n into ζ_1 and ζ_2 where ζ_2 is a vector of $(n - l)$ states whose effect is negligible compared to ζ_1 on the input-output response. With appropriate partitioning of $\hat{\Phi}$, $\hat{\Gamma}$ and $\hat{C}$ the state space equations becomes

$$\begin{bmatrix} \zeta_1(k+1) \\ \zeta_2(k+1) \end{bmatrix} = \begin{bmatrix} \hat{\Phi}_{11} & \hat{\Phi}_{12} \\ \hat{\Phi}_{21} & \hat{\Phi}_{22} \end{bmatrix} \begin{bmatrix} \zeta_1(k) \\ \zeta_2(k) \end{bmatrix} + \begin{bmatrix} \hat{\Gamma}_1 \\ \hat{\Gamma}_2 \end{bmatrix} u_c(k) \quad (13)$$

$$y(k) = \left[\hat{C}_1 \; \hat{C}_2 \right] \begin{bmatrix} \zeta_1(k) \\ \zeta_2(k) \end{bmatrix} \quad (14)$$

Setting $\zeta_2 = 0$ gives the following reduced order model

$$\zeta_1(k+1) = \hat{\Phi}_{11}\zeta_1(k) + \hat{\Gamma}_1 u_c(k) \quad (15)$$

$$\tilde{y}(k) = \hat{C}_1\zeta_1(k) \quad (16)$$

From the above equations, one can extract the reduced order model under normal operation, as well as the reduced order model for each type of fault to be used in FDI design.

3. BASIC FAULT TOLERANT CONTROL SCHEME

The efficacy of the proposed FDI method is assessed by incorporating it in the fault tolerant control scheme developed by Prakash *et al.* (2002). The basic idea of their FTCS is to compensate / adapt the controller and state estimator on-line once a fault is identified. As stated earlier, the soft faults caused by biases in sensors, actuators and abrupt changes in disturbance and operating parameters are considered in the design of FTCS.

In the BFTCS design (figure 1), the FDI module is like an external interface to the existing feedback

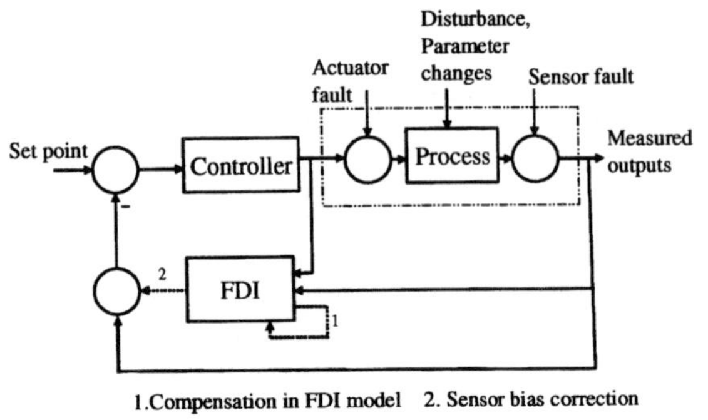

Fig. 1. Schematic of Basic Fault Tolerant Control Scheme

control system. Typically, a conventional feedback controller with integral action can satisfactorily deal with actuator biases, disturbances and process parameter changes in the absence of input constraints. However, in the presence of a sensor bias, it takes corrective action based on the biased value thereby introducing an offset in the true value of the plant output. The BFTCS was developed to give good control performance despite the occurrence of any type of soft fault, by appropriately integrating the FDI method with controller. In case a *sensor bias* is identified, the magnitude of the bias estimated by the FDI method is used to correct the faulty measurement, before being used in the controller. For all types of faults identified, the FDI model is suitably adapted so that any subsequent fault occurring in the process can be detected.

4. SIMULATION STUDIES

The performance of the proposed scheme is tested through simulation of a 20-tray single feed binary distillation column (Luyben, 1990). The following assumptions are made in deriving a nonlinear model from first principles: constant overflows, constant relative volatility, linear liquid flow dynamics, constant pressure, no vapor holdup, total condenser, perfect level control in condenser and reboiler. The only modification made to the column simulation by Luyben (1990) is that the Murphree vapor phase stage efficiency η is considered (assumed the same for all trays) in order to demonstrate parametric faults. The resulting model equations consists of two differential (material balance and component balance) and three algebraic equations (vapor liquid equilibrium, tray efficiency and liquid hydraulic relation) for each stage i. A full-order model of the process derived from first principles (with material and component balances) consists of 42 nonlinear ordinary differential equations. The reflux flow rate (R) and vapor boil-up rate (V) are manipulated to control the top and bottom product compositions (x_d, x_d).

Normal operating data is as reported in table 1. Seven different faults are hypothesized namely, bias in the two concentration measurements (x_d and x_b), bias in the two actuators corresponding to the manipulated variables, reflux R and vapor boil- up V, step change in the two disturbance variables (feed flow rate F and feed composition z_f) and a step change in the tray efficiency parameter (η). We have in total *five* inputs (R, V, F, z_f and η) to be considered for obtaining a reduced order model for fault diagnosis. We next describe the reduction procedure.

Table 1. Normal operating values

Distillate concentration x_d	0.904
Bottom concentration x_b	0.272
Reflux R	124.08 (mol/min)
Vapor boil-up V	178.01 (mol/min)
Feed flow rate F	100.00 (mol/min)
Feed composition z_f	0.5
Tray efficiency parameter η	0.7

4.1 Model reduction and FTCS development

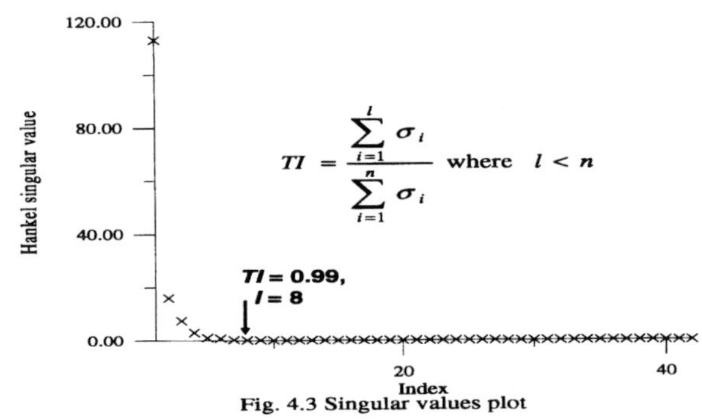

Fig. 2. Hankel singular values plot

A linearized continuous state space model is obtained by linearising the nonlinear equations around the normal operating point (table 1). The state space matrices and the fault coupling matrices are obtained by numerically evaluating the Jacobians of the differential equations with respect to the state, input and fault variables. The MATLAB function '*sysbal*' is used for computing the balancing transformation and '*strunc*' for obtaining the reduced dimensional state space subsystem by truncation. A truncation index (TI) of 0.99 was chosen to decide upon the model order of 8 as explained in the plot of Hankel singular values *vs* state indices (figure 2).

Figure 3 compares step responses of the distillate product composition obtained with respect to different normal and fault inputs for the full-scale and reduced order model. As can be seen from the Figures 3(a)-(e), the reduced order model is able to capture the system dynamics reasonably well for all the inputs considered. Similar kind of close match is obtained for responses of the bottom

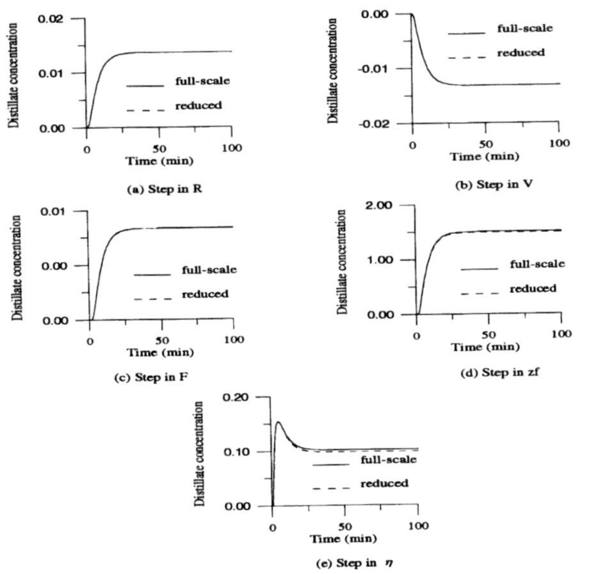

Fig. 3. Step responses of distillate concentration x_d with respect to different fault inputs

product composition. A point to note is that the MATLAB function for balancing is written for the continuous system description. Since the balancing theory is equivalent for continuous and discrete systems, the model reduction is carried out first in the continuous domain and discretised later. An 8th reduced order continuous model so obtained is discretised with a sampling time of 1 min for use in FDI and controller development.

One can develop a reduced order model for FDI and controller independently since the FDI is like an external interface in the developed FTC scheme. However, for the sake of consistency and ease in maintainability, the controller is designed using the reduced order normal model of the process obtained from Eq. (15). An unconstrained DMC control law is developed with tuning parameters as given in Table 2. The reduced model based FDI strategy is incorporated in the BFTCS design with the tuning parameters as given in Table 3. To decide upon the parameter values, certain guidelines as given by Prakash *et al.* (2002) are used. In all simulation runs, 50 simulation trials (N_t) were performed with each trial lasting for 1000 sampling instants. State noise is simulated with random fluctuations in disturbance inputs (feed flow rate F and feed composition z_f) around the mean value as Gaussian zero-mean white noise sequences with standard deviation as 5% and 1% of their nominal values respectively. Measurement noise is simulated as Gaussian zero-mean white noise sequence with standard deviation as 1% of their respective nominal values. The Kalman filter for FDI development is designed using reduced order normal description of the process with these noise characteristics. For all the simulation cases reported, the fault is introduced at the 1st sampling instant. The magnitude of the sensor and

Table 2. Parameter values for DMC

Parameter	value
Prediction horizon (p)	20
Control horizon (q)	1
Error weighting matrix W_e	diag[1 37]

Table 3. Parameter values for FDI

Parameter	value
Window length (N)	60
Level of significance for FDT (α_{FDT})	0.75
Level of significance for FCT (α_{FCT})	0.01

actuator bias introduced was in proportion to the standard deviation of the output or input, as the case may be under normal operation.

4.2 Performance measures

In order to assess the performance of the FTCS, the proposed FTCS as well as conventional control scheme are implemented, and the following performance measures are evaluated.

- *Performance index (PI)*

$$PI = \frac{ISE_i(\text{FTCS})}{ISE_i(\text{Conventional control})} \tag{17}$$

where ISE_i is the sum squared difference between the true value of a controlled variable and the corresponding setpoint over the simulation trial, averaged over all the trials in a run.

- *Percentage of successful trials (PST)*

$$PST = \frac{N_S}{N_T} \times 100 \tag{18}$$

where N_S is the number of simulation trials in which the fault is correctly identified at least once.

- *False alarm index (FAI)*

$$FAI = \frac{N_F}{N_T \times \frac{L}{N}} \tag{19}$$

where N_F is the total number of faults falsely identified in the simulation run.

5. RESULTS AND DISCUSSIONS

A comparative study of FDI and BFTCS control performance with different degrees of plant model mismatch is made for the following three *cases* namely,

- *FDI and process simulation using full-scale linear model (FS-L)*
- *FDI based on reduced order linear model, process simulated using full-scale linear model (RO-L)*

- *FDI based on reduced order linear model, process simulated using full-scale nonlinear model (RO-NL)*

As can be seen in tables 4 and 5, for sensor and parametric faults with linear process simulation $(FS - L, RO - L)$, the FDI and BFTCS control performance obtained in terms of number of successful detections, false alarms committed and the PI values is almost identical with the two different accuracy models used for FDI. Further, in the presence of a sensor bias, the conventional control scheme produces an offset between the true value and the setpoint whereas the fault tolerant scheme corrects for the bias and therefore improves upon the control performance as indicated by the lower PI values. With parametric faults, the BFTCS does not further improves the control performance as the DMC conventional controller can itself handle such small magnitude process parameter changes.

Table 4. Comparison of FDI and BFTCS control performance for a sensor bias in $x_d(3\sigma_{x_d} = 0.027)$

Model	$\hat{b}$	PST	FAI	$(PI)_{x_d}$	$(PI)_{x_b}$
FS-L	0.0265	100	0.008	0.08	1.007
RO-L	0.0266	100	0.006	0.076	1.006
RO-NL	0.0239	100	0.014	0.09	0.964

Table 5. Comparison of FDI and BFTCS control performance for a parametric fault in $\eta(\sim 10\% = -0.075)$

Model	$\hat{b}$	PST	FAI	$(PI)_{x_d}$	$(PI)_{x_b}$
FS-L	-0.076	100	0.014	1.028	1.001
RO-L	-0.077	100	0.007	1.004	1.0002
RO-NL	-0.10	100	0.002	1.013	1.0002

With full-scale model, for actuator and disturbance faults, there is a significant deterioration in FDI performance in terms of increasing false alarms and in control performance because of subsequent fault accommodation. In fact these are quite unacceptable results from control performance point of view. However, the reduced dimensional model gives acceptable performance for all the fault cases and the model approximation introduced via reduction does not deteriorates the FDI performance. On the contrary, Kalman filter design based on an observable system model improves state estimation which leads to a better diagnostic performance.

Table 6. Comparison of FDI and BFTCS control performance for an actuator bias in $R(-1\sigma = -5)$

Model	$\hat{b}$	PST	FAI	$(PI)_{x_d}$	$(PI)_{x_b}$
FS-L	-4.87	66	0.096	1.36	1.01
RO-L	-5.07	74	0.03	1.013	1.004
RO-NL	-5.33	96	0.03	1.029	1.006

Table 7. Comparison of FDI and BFTCS control performance for a disturbance step change in $F(10\% = 10)$

Model	$\hat{b}$	PST	FAI	$(PI)_{x_d}$	$(PI)_{x_b}$
FS-L	9.9	100	0.14	1.79	1.01
RO-L	10.001	100	0.07	1.007	1.06
RO-NL	10.18	100	0.056	1.032	1.001

In all the cases, the nonlinear process simulation deteriorates the FDI and BFTCS control performance when compared to the linear case because of the resulting plant model mismatch. However, the PI values indicate that the resulting FTCS control performance is superior to conventional control in the presence of a sensor bias and is as good as the conventional control for the other input faults that are considered.

6. CONCLUSIONS

The proposed reduced order model based FDI scheme enables efficient real-time FTCS implementation for large dimensional processes. Furthermore, state estimation using reduced order observable models improves the diagnostic performance considerably when compared to the full-scale implementation.

Acknowledgements Financial support for carrying out this work was provided by the Department of Science and Technology through the project CHE/00-01/051/DST/SACHX and Council of Scientific and Industrial Research (CSIR SRF No. 9/84(303)/98-EMR-I), INDIA.

REFERENCES

Astrom, K.J. and B. Wittenmark (1994). *Computer Controlled Systems.* Prentice-Hall.

Luyben, W. L. (1990). *Process Modeling, Simulation and Control for Chemical Engineers.* Mc Graw-Hill.

Moore, B. C. (1981). Principal component analysis in linear systems: Controllability, observability, and model reduction. *IEEE Trans. on Auto. Cont.* **AC-26**(1), 17–32.

Narasimhan, S. and R. S. H. Mah (1988). Generalised likelihood ratios for gross error identification in dynamic systems. *AIChE J.* **34**, 1321–1334.

Prakash, J., S.C. Patwardhan and S. Narasimhan (2002). A supervisory approach to fault tolerant control of linear multivariable systems. *Ind. Eng. Chem. Res.* **41**(9), 1181–2270.

Willsky, A.S. and H.L. Jones (1976). A generalised likelihood ratio approach to the detection and estimation of jumps in linear systems. *IEEE Trans. on Aut. Cont.* pp. 108–112.

ELSEVIER
IFAC
PUBLICATIONS
www.elsevier.com/locate/ifac

EKF AND ANFIS ESTIMATOR DESIGN IN MULTICOMPONENT BATCH DISTILLATION COLUMNS

Ugur Yildiz[1], Evren Güner[1], Canan Özgen[1] and Kemal Leblebicioglu[2]

Department of Chemical Engineering[1], Department of Electrical and Electronics Engineering[2]
Middle East Technical University,
Ankara, 06531, Turkey. Fax: +903122102636
uyildiz@metu.edu.tr, evrenguner@hotmail.com, cozgen@metu.edu.tr, lebleb@metu.edu.tr

Abstract: In the control of batch distillation columns, one of the problems is the difficulty of monitoring the compositions. This problem can be handled by estimating the compositions from readily available online temperature measurements using a state estimator. In this study, an extended Kalman Filter (EKF) and an Adaptive Neuro-Fuzzy Inference System (ANFIS) state estimators that infer the product composition in a multicomponent batch distillation column (MBDC) from the temperature measurements are designed and tested using a batch column simulation. The designed EKF and ANFIS estimators are successfully used in the composition – feedback inferential control of MBDC operated under variable reflux-ratio policy with an acceptable deviation from the desired purity level of the products. *Copyright © 2004 IFAC*

Keywords: Batch Distillation, Simulation, State Observer, Kalman Filter, Adaptive Neuro-Fuzzy Inference System.

1. INTRODUCTION

"Batch distillation is generally used as a separation unit in the fine speciality chemicals, pharmaceuticals, biochemical and food industries. The demand and the uncertainty in specifications for these chemicals have increased recently, which increased the popularity of the use of batch distillation" (Barolo, M. and Cengio, P.D., 2001). Instead of using many continuous columns in series, multiple products can be obtained from a single batch distillation column during a single batch run. Moreover, batch distillation processes can easily handle variations both in the product specifications and in the feed compositions. This flexibility of batch distillation processes provides the ability to cope with a market characterized by short product life times and strict specification requirements.

In batch distillation, the operation of the column with optimized operation scenario; including reflux ratio policy, switching times, and method of recycling, is required to be realized in a convenient control system. However, in order to employ the operation scenario; the designed controller will require continuous information flow from the column, including the compositions throughout the column or temperatures reflecting the composition knowledge. The reason for this requirement is that, the value of reflux ratio and switching between product and slop cut distillations are optimized which are subject to the composition profile along the column and obtained as a function of it. Therefore, the need for knowledge of current composition in the column becomes obvious.

This composition knowledge can be generated by means of direct composition analyzers. Although there is a great development in the technology of online composition analyzers, such as gas chromatography, they bring large measurement delays and high investment and maintenance costs (Mejdell, T. and Skogestad, S., 1991; Venkateswarlu, C. and Avantika, S., 2001). The most popular alternative to the composition controllers utilizing analyzers is standard temperature feedback controllers. Although, temperature measurements are inexpensive and have negligible measurement delays, they are not accurate indicators of composition (Mejdell, T. and Skogestad, S., 1991). Another alternative is inferential control systems incorporating state estimators which use secondary temperature measurements.

State estimation can be defined as the process of extracting information from data which contain valuable information about a system and state estimator is the tool responsible for gathering valuable measurements to infer the desired information. Modern estimators also use known relationships in computing the desired information; taking into account the measurement errors, the effects of disturbances and control actions on the system, and prior knowledge about the system and measuring devices. While gathering these elements, they make use of some error criteria and try to minimize errors in some respect. The criteria and the method of minimization characterize the method of estimation and the use of minimization makes the estimate (extracted information) "optimal". If this optimality is realized statistically, the estimator type becomes stochastic; if deterministically it becomes deterministic. One of the estimators used in this work falls in the stochastic category and it is named as Kalman Filter.

However, if the sufficient input-output information is generated from the process, artificial intelligence methods that use this collected information can also be used to design a state estimator. ANFIS is one of the examples of these methods in which a fuzzy inference system is implemented in the framework of adaptive networks. It constructs an input-output mapping based both on human knowledge (in the form of fuzzy if-then rules) and on generated input output data pairs by using a hybrid algorithm that is the combination of the gradient descent and least square estimates (Jang 1993).

In this study, the aim is to design state estimators that infer the component concentrations of the multicomponent batch distillation column from the measured tray temperatures. The designed estimator is further tested using a rigorous column simulation to find its performance. The extended Kalman Filter (EKF) and the Adaptive Neuro-Fuzzy Inference System (ANFIS) are selected as the state estimators. Because it is based on the linear dynamic model of the process, the rigorous model used in the simulation is adapted to the EKF estimator algorithm mainly by simplifying the equilibrium model and by means of linearization. ANFIS estimator is designed using the data sets including temperature values and corresponding composition values obtained from the rigorous model. The performances of the developed estimators are tested by using the rigorous column simulation and discrete measurements of the top product compositions.

2. PROCESS DESCRIPTION

The case column for simulation is the one which was simulated by Mujtaba, I.M. and Macchietto, S., 1993 in their study on the subject of optimal operation of MBDC. The column is used to separate the mixture of cyclo-hexane, n-heptane and toluene.

The batch distillation column is under the perfect control of reflux-drum level and has two degrees-of-freedom for manipulation which are reboiler heat load, Q_1 and reflux-ratio, R. In this study, the reboiler heat load, Q_1 is kept at its maximum value given by design while the reflux-ratio, R is used as manipulated variable in order to realize the optimal operation policy recommended by Mujtaba, I.M. and Macchietto, S., 1993. This optimal operation policy is used to yield two product-cuts with the desired purity levels of 0.9 and 0.8 from the mixture of cyclo-hexane, n-heptane and toluene with the composition of (0.407, 0.394, 0.199). In the simulations, this optimal reflux ratio profile is employed.

There are many different rigorous models of batch distillation columns. They use the same basic strategy in the simulation model development which was used initially by the first studies on rigorous modelling of distillation columns. In batch column modelling, this common strategy was initiated by Meadows, E.L., 1963 and Distefano, G.P., 1968 which were followed by Stewart et al., 1973. The rigorous model used in this study is based on the study of Distefano, G.P., 1968 and its details are given by Yildiz (2002).

3. EKF ESTIMATOR

The Extended Kalman Filter (EKF) is defined as "optimal recursive data processing algorithm" (Maybeck, P.S., 1979), handling the estimation issues in the nonlinear system theory. EKF uses the nonlinear model of the system given by Eq.(1)

$$\dot{\underline{x}}(t) = f(\underline{x}(t), \underline{u}(t), t) + G(t)\underline{w}(t) \tag{1}$$

where f is the vector of the nonlinear system functions and the noise process, w(t) is modelled as white Gaussian noise with statistics

$$E\{\underline{w}(t)\} = 0 \tag{2}$$

$$E\{\underline{w}(t)\underline{w}(t')^T\} = \begin{cases} \underline{Q}(t), & t = t' \\ 0, & t \neq t' \end{cases} \tag{3}$$

and the nonlinear measurement model written as

$$\underline{z}(t_k) = h(\underline{x}(t_k), t_k) + \underline{v}(t_k) \tag{4}$$

where h is the vector of the nonlinear measurement functions and noise process, $\underline{v}(t_k)$ is modelled as white Gaussian noise with statistics

$$E\{\underline{v}(t_k)\}=0 \tag{5}$$

$$E\{\underline{v}(t_k)\underline{v}(t_l)^T\}=\begin{cases}\underline{R}(t), t_k=t_l \\ 0 \quad , t_k\neq t_l\end{cases} \tag{6}$$

Moreover, EKF algorithm needs the linearized versions of these two models, specified by the Jacobian matrices and for the system it is given by Eq. (7) and for the measurement process by Eq. (8).

$$\underline{F}(\hat{\underline{x}}(t),\underline{u}(t),t)=\left.\frac{df(\underline{x}(t),\underline{u}(t),t)}{d\underline{x}(t)}\right|_{\underline{x}=\hat{\underline{x}}} \tag{7}$$

$$\underline{H}(\hat{\underline{x}}^-(t_k),t_k)=\left.\frac{dh(\underline{x}(t_k),t_k)}{d\underline{x}(t_k)}\right|_{\underline{x}(t_k)=\hat{\underline{x}}^-(t_k)} \tag{8}$$

The extended Kalman Filter has a two-step recursive calculation algorithm. The first named as *the propagation stage* is responsible to calculate the prediction of the state at the current time using the best state estimate at the previous time step. The second is named as *the update stage* and updates the prediction found in the first stage using the measurements taken from the actual process and calculates the best state estimate. The propagation stage integrates the state and error covariance derivatives from the previous time step t_{k-1} to the current time t_k and uses the best state estimate $\hat{\underline{x}}^+(t_{k-1})$ and its error covariance $\underline{P}^+(t_{k-1})$ at the previous time step t_{k-1}, in order to calculate the prediction of the state, $\hat{\underline{x}}^-(t_k)$ and its error covariance $\underline{P}^-(t_k)$ at the current time step t_k. The update stage inovates the prediction of the state $\hat{\underline{x}}^-(t_k)$ and its error covariance $\underline{P}^-(t_k)$ at the current timestep t_k. In order to initiate the Kalman Filter algorithm, the initial conditions incorporating the initial state, $\hat{\underline{x}}(t_0)=\hat{\underline{x}}_0$ and its error covariance, $\underline{P}(t_0)=\underline{P}_0$ are required. At the initialization time step, when the first measurement is taken, requirements of the best state estimate, $\hat{\underline{x}}^+(t_{k-1})=\hat{\underline{x}}^+(t_{-1})$ and it error covariance, $\underline{P}^+(t_{k-1})=\underline{P}^+(t_{-1})$ at the time step t_{-1} are supplied by replacing with the initial state $\hat{\underline{x}}_0$ and its error covariance $\underline{P}_0$.

The technique of EKF estimation will be applied to MBDC in order to infer the column compositions from the temperature measurements. At first, the observability of a multicomponent batch distillation column, which is a must to be able to estimate the system states, is to be analyzed. Employing a degree-of-freedom concept, Yu, c.C. and Luyben, W.L., 1987 found that a distillation column is observable if the number of measurements is at least (NC-1). The study of Quintero-Marmol et al., 1991, dealing with the design of an extended Luenberger Observer for MBDC, concluded that even though the linear observer in theory needs only (NC-1) temperature measurements to be observable, the nonlinear observer needed at least (NC) thermocouples to be effective. In addition, to improve the convergence without affecting the robustness, the use of (NC+2) measurements is recommended by Quintero-Marmol et al., 1991.

As a result, the nonlinear models for the system and for the temperature measurements are to be developed in the form required for EKF algorithm. However, the model developed for rigorous simulation of the batch column is not suitable for realistic situation in order to be implemented in EKF algorithm. The complexity of the simulation model requires high computational time and memory. Therefore, the rigorous column model for simulation is to be simplified and then the obtained nonlinear model is to be linearzed to achieve the Jacobian matrix both for the system and the measurement processes.

Some additional assumptions are needed for the simplification of the rigorous simulation model of MBDC. These assumptions are constant molar holdup on trays, neglecting the energy dynamics in the column, ideal trays, and use of Raoult's Law with Antoine's vapor pressure correlation for VLE description. As a result, the vapor flowrates throughout the column become equal as well as the liquid flowrates. The simplified model equations for MBDC are given by Yildiz (2002).

Lastly, the general forms of the linear system matrix

$$\underline{F}'(\underline{x}(t),\underline{u}(t),t)=\frac{d\underline{f}(\underline{x}(t),\underline{u}(t),t)}{d\underline{x}(t)} \tag{9}$$

and the linear measurement matrix

$$\underline{H}'(\underline{x}(t_k),t_k)=\frac{d\underline{h}(\underline{x}(t_k),t_k)}{d\underline{x}(t_k)} \tag{10}$$

are evaluated analytically. Their expanded forms and the details of the derivation are given by Yildiz (2002).

Consequently, all the information required for EKF estimator has been obtained. This information incorporates nonlinear and linearized models for the system of MBDC and the measurement process given respectively by $\underline{f},\underline{h},\underline{F}',\underline{H}'$.

4. ANFIS ESTIMATOR

In ANFIS, Takagi-Sugeno (TS) fuzzy model is used. A simple example of ANFIS with two TS type fuzzy rules is given in Figure 1. The output of each rule can be a linear combination of input variables plus a constant term or can be only a constant term. The final output is the weighted average of each rule's output.

Rule1: If x is A_1 and y is B_1, then $f_1 = p_1x + q_1y + r_1$

Rule2: If x is A_2 and y is B_2, then $f_2 = p_2x + q_2y + r_2$

The node functions in the same layer are the same as described below:

Layer 1: Every node i in this layer is an adaptive node with a node function as:

$$0_{1,i} = \mu_{A_i}(x), \; for \;\; i = 1,2$$
$$0_{1,i} = \mu_{B_i - 2}(y), \; for \;\; i = 3,4 \qquad (11)$$

where x is the input to node i, and A_i (or B_{i-2}) is a membership function (MF) associated with this node. Parameters in this layer are the MF's parameters.

Layer 2: Every node in this layer is a fixed node labeled as Π, whose output is the product of all incoming signals:

$$0_{2,i} = w_i = \mu_{A_i}(x)\mu_{B_i}(y), \quad i = 1,2 \qquad (12)$$

Each node output represents the *firing strength* of a fuzzy rule.

Layer 3: Every node in this layer is a fixed node labeled N. The ith node calculates the ratio of the rule's firing strength to the sum of all rules' firing strengths:

$$0_{3,i} = w_i = w_i / (w_1 + w_2), \quad i = 1,2 \qquad (13)$$

Outputs of this layer are called "normalized firing strengths".

Layer 4: Every node i in this layer is an adaptive node with a node function as:

$$0_{4,i} = \overline{w_i} f_i = \overline{w_i}(p_i x + q_i y + r_i) \qquad (14)$$

where $\overline{w_i}$ is a normalized firing strength from layer 3 and $\{p_i, q_i, r_i\}$ is the parameter set of this node.

Layer 5: The single node in this layer is a fixed node labeled Σ that computes the overall output as the summation of all incoming signals:

$$0_{5,i} = overall \;\; output = \sum_i \overline{w_i} f_i = \frac{\sum_i w_i f}{\sum_i w_i} \qquad (15)$$

The computation of parameters is facilitated by a gradient vector, which provides a measure of how well the ANFIS is modeling the input output data for a given parameter set. Once the gradient vector is obtained, backpropagation or hybrid learning algorithm, combination of the gradient descent and least square estimates, can be applied in order to adjust the parameters (Jang 1993). Detailed design procedure for the ANFIS estimator can be found in the study of Güner (2003).

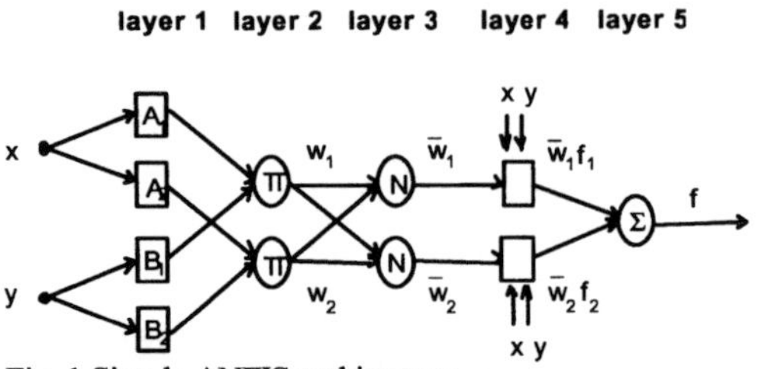

Fig. 1 Simple ANFIS architecture

5. RESULTS AND DISCUSSION

The EKF estimator implementation is performed in three phases. First, an extended Kalman Filter for the estimation of product compositions for a MBDC from temperature measurements is designed. Then the designed EKF is implemented on the case MBDC to check the performance of the EKF. In the third phase the designed EKF is utilized for control purposes in the MBDC.

After designing the EKF state estimator, several simulation test runs are done to obtain the optimum values of the tuning parameters by selecting the integral absolute error (IAE) as the performance criteria reflecting the fitness of the EKF design parameters. The tuning parameters for EKF are the diagonal terms of process noise covariance matrix, q, and the diagonal terms of measurement model noise covariance matrix, r. Also, the effect of number of measurement points and measurement period, Δt_m have been considered.

The optimum value of the diagonal terms of process noise covariance matrix, q and the diagonal terms of measurement model noise covariance matrix, r are searched in the range where the EKF estimator is stable. Performing some trial runs, the stability region of the estimator is found where the value of q and r are in the range of 50 - $1x10^{-7}$ and 0.5 - $5x10^8$, respectively. The best result (i.e. one having the lowest IAE sum) is obtained for the diagonal terms of process noise covariance matrix, q=5000 and for the diagonal terms of measurement model noise covariance matrix, r=0.00001. The best result is obtained for three measurement points and for the measurement location set with reboiler, 4th and top trays. The effect of measurement period, Δt_m on IAE scores is also investigated. Although, the measurement period selected in this simulation study is the integration time step (i.e. the minimum possible value), in real-time estimation problems, the value of Δt_m is to be chosen considering the limits of the computational power. The measurement periods of 3 min. even 5 min. can satisfactorily be used without much change in EKF performance. Given this optimum set of the tuning parameters, the actual and estimated reflux-drum composition profiles are shown in Figure 2 for the case column under the open-loop control actions.

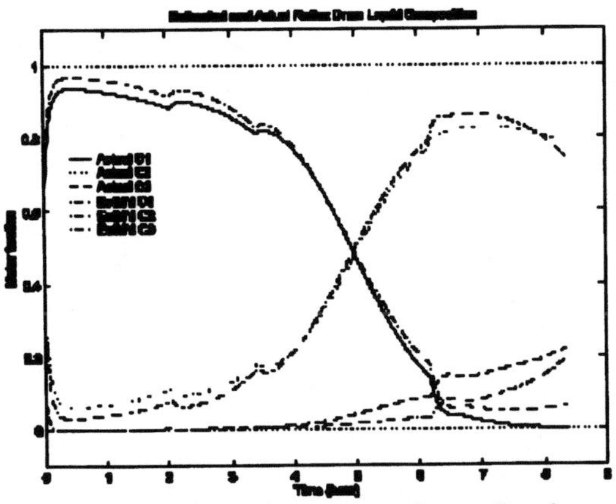

Fig.2. Actual and estimated reflux-drum composition profiles for the optimum values of the tuning parameters under open-loop control.

In last phase of the study, it is aimed to analyze the performance of the EKF estimator for a MBDC system in a composition-feedback inferential control structure which realizes an actual scheduling policy explained previously. In the simulation of this control structure, the compositions can be obtained directly from the process simulation or from the EKF estimator. Firstly, to create a reference point, a simulation is done, taking the composition knowledge directly from the column as the feedback information to the controller. The response of this reference run in terms of the liquid compositions, both in the reflux-drum and the reboiler are given in Fig.3.

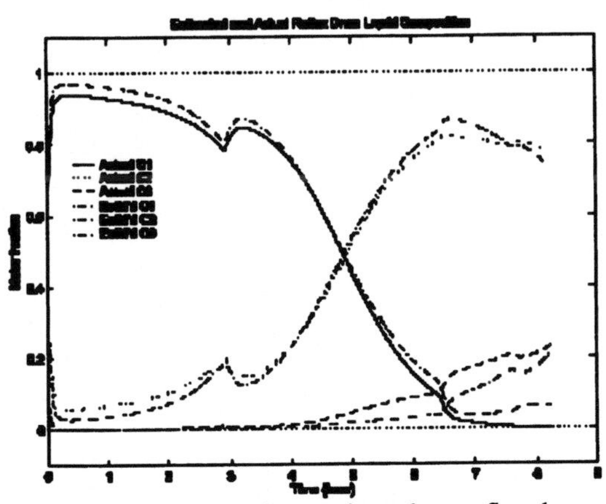

Fig.3. Actual and estimated reflux-drum composition profiles for the optimum values of the tuning parameters under open-loop control.

To see the performance of EKF, secondly the same simulation is realized using the estimated compositions as the feedback to the controller. For the design parameters of the EKF estimator, previously obtained optimum values are used. Actual and estimated compositions of the reflux-drum are shown in Fig.4. Thus, this batch distillation column can be controlled satisfactorily for variable reflux ratio policy by the use of EKF estimator utilizing a simplified model.

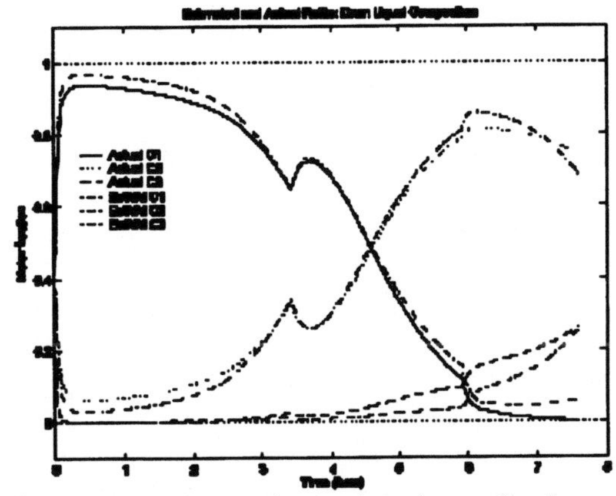

Fig.4. Actual and estimated reflux-drum composition profiles for the optimum values of the tuning parameters under closed-loop control.

The ANFIS estimator design is performed in three phases. First, input process variables which are in operational range are changed and output variables are obtained from rigorus model simulations. Then, training data sets are generated using these input-output data and different ANFIS architectures are trained with these data sets. Finally, trained architectures are implemented on the case MBDC to check the performance of the estimators. The structure that gives the minimum IAE score after the simulations is selected as an ANFIS estimator. Fig5 shows the performance of the selected ANFIS state estimator. It can be seen from the Fig. 5 that ANFIS estimator performance is very good. It can also be seen from the Fig.6. that performance of the ANFIS estimator is better than the EKF state estimator.

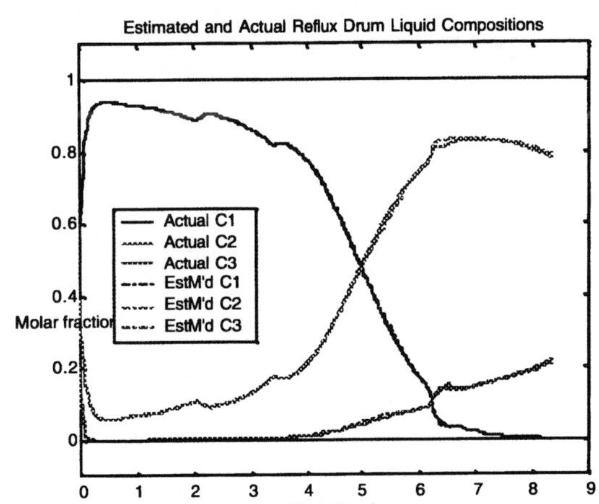

Fig.5. Actual and estimated reflux-drum composition profiles for the selected ANFIS estimator under open-loop control

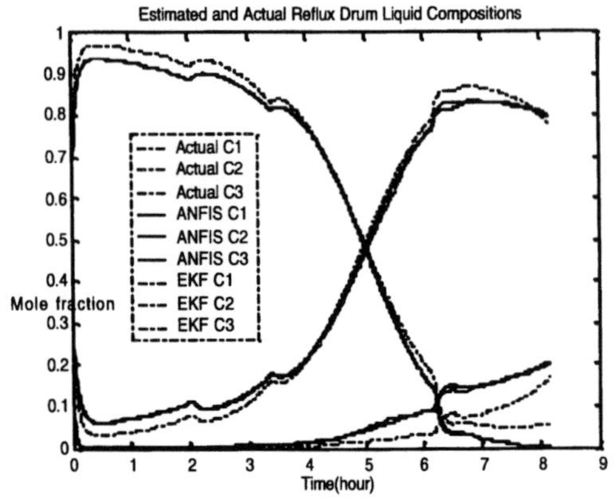

Fig.6. Comparison of the selected ANFIS estimator and EKF estimator under open-loop control

6. CONCLUSION

The study is aimed to estimate the compositions in the multicomponent batch distillation column from temperature measurements using EKF and ANFIS estimators, separately. It is found that, the most important part of the modelling affecting the performance of the EKF estimator is the selection of the VLE formulation. EKF parameters of the diagonal terms of process noise covariance matrix and the diagonal terms of measurement model noise covariance matrix are selected basing on the least sum of individual IAE scores for the reflux-drum and the reboiler composition estimates. From the input-output data of the rigorous plant simulation ANFIS estimator is trained and from several simulation runs for different architectures of ANFIS estimator the estimator performance is optimized. The designed EKF and ANFIS estimators are implemented in the open-loop control of the MBDC. It is seen that ANFIS estimator performs more accurate estimations of the reflux-drum compositions than the EKF does. The superior performance of the ANFIS is due to having the information directly taken from the input-output data of the simulated plant. However, the EKF estimator incorporates the simplified model and its linearized version of the plant causing plant-model mismatches.

REFERENCES

Barolo, M. and Cengio, P.D. (2001). Closed-loop Optimal Operation of Batch Distillation Columns. *Computers and Chemical Engineering*, **25**, 561-569.

Distefano, G.P. (1968). Mathematical Modelling and Numerical Integration of Multicomponent Batch Distillation Equations. *AIChE J.*, **14**, 190-199.

Furlonge, H.I., Pantelides, C.C., Sorensen, E. (1999). Optimal Operation of Multivessel Batch Distillation Columns. *AIChE J.*, **45**, 781-801.

Güner, E. (2003). Adaptive Neuro-Fuzzy Inference Applications in Chemical Processes, *M.Sc. Thesis*, Middle East Technical University, Ankara.

Jacobs, O.L.R. (1974). *Introduction to Control Theory*, Clarendon Press.

Jang, R.J. (1993). ANFIS: Adaptive-Network-Based Fuzzy Inference System. *IEEE Transections on Man, and Cybernetics*, **23**, No. 3, 665-683.

Luyben, W.L. (1988). Multicomponent Batch Distillation. 1. Ternary Systems with Slop Recycle. *Ind. Eng. Chem. Res.*, **27**, 642-647.

Maybeck, P.S. (1979), *Stochastic Models, Estimation, and Control*, Academic Press Inc.

Meadows, E.L. (1963). *Chem. Eng. Progr. Symp. Ser. 46*, **59**, 48.

Mejdell, T. and Skogestad, S. (1991). Estimation of Distillation Compositions from Multiple Temperature Measurements Using Partial-Least-Squares Regression. *Ind. Eng. Chem. Res.*, **30**, 2543-2555.

Mujtaba, I.M. and Macchietto, S. (1993). Optimal Operation of Multicomponent Batch Distillation---Multiperiod Formulation and Solution. *Computers chem. Engng.*, **17**, 1191-1207.

Perry, R.H., Green, D.W., Maloney, J.O. (1999). *Perry's Chemical Engineer's Handbook*. McGraw -Hill.

Quintero-Marmol, E., Luyben, W.L. and Georgakis, C. (1991). Application of an Extended Luenberger Observer to the Control of Multicomponent Batch Distillation. *Ind. Chem Eng. Res.*, **30**, 1870-1880.

Stewart, R.R., Weisman, E., Goodwin, B.M., Speight, C.E. (1973). Effect of Design Parameters in Multicomponent Batch Distillation. *Ind. Eng. Chem. Process Des. Develop.*, **12**, 130-136.

Venkateswarlu, C. and Avantika, S. (2001). Optimal State Estimation of Multicomponent Batch Distillation. *Chemical Engineering Science*, **56**, 5771-5786.

Yildiz U. (2002). Multicomponent Batch Distillation Column Simulation and State Observer Design. *M.Sc. Thesis*, Middle East Technical University, Ankara.

Young, S.C. and Luyben, W.L. (1987). Rigorous Dynamic Models of Distillation Columns. *Ind. Eng Chem. Res.*, **26**, 2158-2161.

Yu, c.C. and Luyben, W.L. (1987). Control of Multicomponent Distillation Columns Using Rigorous Composition Estimators, Distillation and Absorption. *Institute of Chemical Engineering of Symposium Series*, **104**, A29-A69.

ELSEVIER
IFAC
PUBLICATIONS
www.elsevier.com/locate/ifac

INTERNAL MODEL CONTROL OF A PHYSICAL VAPOR DEPOSITION EFFUSION SOURCE

S. Tobias Junker [*,***], **Robert W. Birkmire** [**,***], **Francis J. Doyle III** [****]

[*] *Department of Chemical Engineering*
[**] *Institute of Energy Conversion*
[***] *University of Delaware, Newark, DE 19716, USA*
[****] *Department of Chemical Engineering, University of California, Santa Barbara, CA 93106, USA*

Abstract: Internal model based temperature controllers are developed for effusion sources that are part of a continuous process for production of thin-film photovoltaic modules. Operation of this system in a research framework necessitates the ability to perform fast, overshoot-free reference changes. To address input constraints, controllers are implemented using an anti-windup scheme. Both empirical ARMAX models and a simplified fundamental model are utilized. Controllers were designed in simulation studies and tested in experiments and a very good match between simulation and experiment was obtained. *Copyright © 2004 IFAC*

Keywords: solar cells, ARMAX models, dynamic models, control oriented models, model based control

1. INTRODUCTION

The effusion sources are part of a continuous process for production of thin-film photovoltaic modules. The approach considered in the present work is the manufacture of Copper-Indium-Gallium-Diselenide (CIGS) films by deposition onto a flexible substrate using a roll-to-roll processing scheme (Birkmire and Eser, 1997). The film is deposited by thermal evaporation from a series of elemental sources located sequentially through the deposition zone.

A critical issue in commercializing this approach is to keep composition and thickness of the CIGS film at their setpoint over long deposition times. Due to long start-up and shutdown times, it is equally important to efficiently perform setpoint changes such that multiple recipes can be tested in a single experiment.

1.1 Process Model and Control Structure

The process can be divided into two subsystems: (1) the effusion subsystem, and (2) the deposition subsystem.

The effusion subsystem consists of the elemental sources where electric energy to resistive heaters is supplied as the input and the outputs are temperature and absorbance measured by thermocouples and atomic absorption spectroscopy sensors for

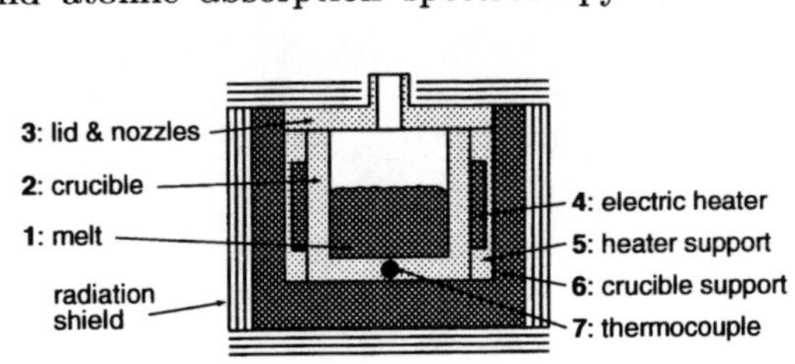

Fig. 1. Schematic cross section of effusion cell.

each source. Each effusion cell (see schematic in Figure 1) is resistively heated in order to first melt and then evaporate the metal. The crucible is covered by a lid with two nozzles. Junker *et al.* (2003) developed a fundamental mass- and heat transfer model of the effusion sources.

The deposition subsystem consists of the moving substrate to which the elemental vapor fluxes are delivered as the input and the outputs are the final film thickness and composition measured by X-ray fluorescence (Junker *et al.*, 2004).

A characteristic difference between these subsystems is their dynamic behavior. The effusion subsystem has a time delay on the order of seconds and thus reacts quickly to input changes. The deposition subsystem, on the other hand, has a long time delay on the order of several minutes.

Since the effusion subsystem's output is the deposition subsystem's input, disturbances of the former affect the latter. Based on this disturbance structure, the process should be controlled by a cascade structure where the outer control loop regulates the deposition process by providing flowrate/temperature setpoints to the inner control loop that regulates the effusion process.

For the outer loop, Junker *et al.* (2004) discuss the application of internal model control and model predictive control. For the inner loop, the present work discusses efficient internal model control for reference tracking of the copper source's temperature. The developed controllers are based on (1) a simplified form of the fundamental model, and (2) empirical models identified based on input/output data. Empirical models are developed to provide a basis of comparison for the fundamental model based controller.

1.2 Overview

The paper is organized in four main sections: development of (1) empirical ARMAX and (2) simplified fundamental models that are suitable for controller design, (3) introduction to the internal model control framework including the utilized anti-windup scheme, and (4) discussion of the obtained reference tracking results and comparison to the corresponding simulations.

2. EMPIRICAL ARMAX MODEL

A standard model structure for system description and control is the ARMAX model (Ljung, 1999, p. 83). It is a time-invariant parametric input-output model that describes the current output $y(k)$ as a function of past outputs, past inputs $u(k)$, and an estimate of the disturbance via modeling of the error $e(k)$.

2.1 Model Structure

The general model structure is given by the difference equation (Ljung, 1999, p. 83)

$$A(q)y(k) = B(q)u(k - n_k) + C(q)e(k) \quad (1)$$

where $e(k)$ is assumed to be white noise and the transfer operators A, B, and C are defined as

$$A(q) = 1 + a_1 q^{-1} + \ldots + a_{n_a} q^{-n_a} \quad (2)$$

$$B(q) = b_1 + \ldots + b_{n_b} q^{-(n_b - 1)} \quad (3)$$

$$C(q) = 1 + c_1 q^{-1} + \ldots + c_{n_c} q^{-n_c} \quad (4)$$

where q^{-1} is a backward shift operator, i.e.,

$$u(k-1) = u(k) \cdot q^{-1} \quad (5)$$

and the model order is defined by the integers (n_a, n_b, n_c, n_k). Here, n_a is the number of poles, $(n_b - 1)$ the number of zeros, and n_k the time delay. Different model structures are identified by the notation $\mathbf{ARMAX}_{\mathbf{n_a}, \mathbf{n_b}, \mathbf{n_c}, \mathbf{n_k}}$.

2.2 Experiment Design

Empirical system identification (ID) of ARMAX models is based on I/O data collected from the process to be modeled. Since the true system is nonlinear the experiment should be carried out at the nominal operating point and the input data should be "rich", i.e., excite the system and force it to show its properties (Ljung, 1999, p. 411).

A common input sequence is a random binary sequence that randomly switches the input between two values u_1 and u_2. The minimum switching frequency must neither be too fast to allow the process to react nor too slow to prevent getting a series of step responses. For the effusion source, a value of $n_{\min} = 4$ samples at a sampling time of $T_s = 2$ seconds gives good results.

A random binary sequence at the operating point is implemented in two stages: (1) the process is driven to the setpoint via PID control and the control move $u^\star$ required to sustain this setpoint is noted, (2) the upper and lower limits are chosen as a 10% variation of $u^\star$, i.e., $u_{1,2} = (1 \pm 0.1)u^\star$.

2.3 Model Estimation

Model estimation and analysis is carried via routines from MATLAB's system identification toolbox (Ljung, 2002). Since the models are used for controller design, proper prediction of the dynamic (high frequency) behavior is crucial. This is more important than the absolute values (low frequency) since incorrect absolute values are more easily compensated for by the controller.

In addition, the residuals should be independent of the input and of each other (Ljung, 2002, p. 2-32). This is tested via the cross-correlation and auto-correlation functions, respectively, that should lie entirely within their 99% confidence intervals for the validation data set.

2.4 Results

Based on an old data set, an $\text{ARMAX}_{2,1,6,2}$ model was identified (see Table 1, left); a comparison of experimental and predicted data is shown in Figure 2. The residuals (not shown) lie almost entirely within their 99% confidence intervals, however, the dynamic behavior is not very well predicted.

Table 1. ARMAX model parameters.

$\text{ARMAX}_{2,1,6,2}$ Coefficient	Value	$\text{ARMAX}_{4,1,1,10}$ Coefficient	Value
a_1	-1.9705	a_1	-1.9482
a_2	$9.7059 \cdot 10^{-1}$	a_2	$7.9386 \cdot 10^{-1}$
b_1	$1.3123 \cdot 10^{-3}$	a_3	$2.8465 \cdot 10^{-1}$
c_1	$-3.4350 \cdot 10^{-1}$	a_4	$-1.3019 \cdot 10^{-1}$
c_2	$-1.9255 \cdot 10^{-1}$	b_1	$9.3174 \cdot 10^{-4}$
c_3	$-3.7459 \cdot 10^{-1}$	c_1	$-9.6967 \cdot 10^{-1}$
c_4	$-1.3116 \cdot 10^{-1}$		
c_5	$7.9861 \cdot 10^{-2}$		
c_6	$4.0042 \cdot 10^{-2}$		

For comparison, an $\text{ARMAX}_{4,1,1,10}$ model was identified based on a more recent data set (see Table 1, right); experimental and predicted data are shown in Figure 3. Unlike the previous model, a separate validation data set was available and the shown results use this data set. Clearly, the results are very good since the dynamic behavior is very well captured and since the residuals (not shown) lie entirely within their 99% confidence interval.

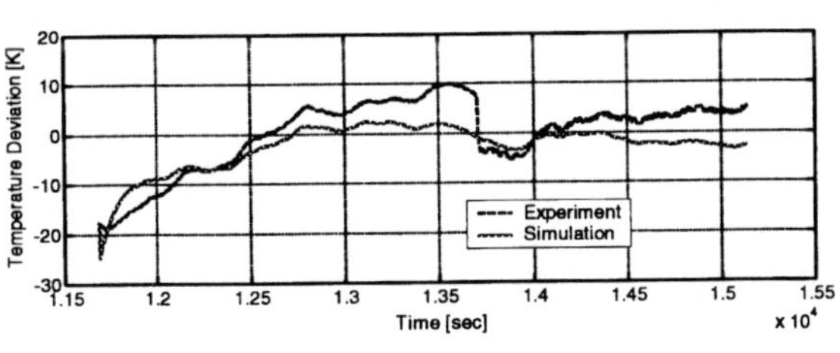

Fig. 2. Simulation of $\text{ARMAX}_{2,1,6,2}$ model.

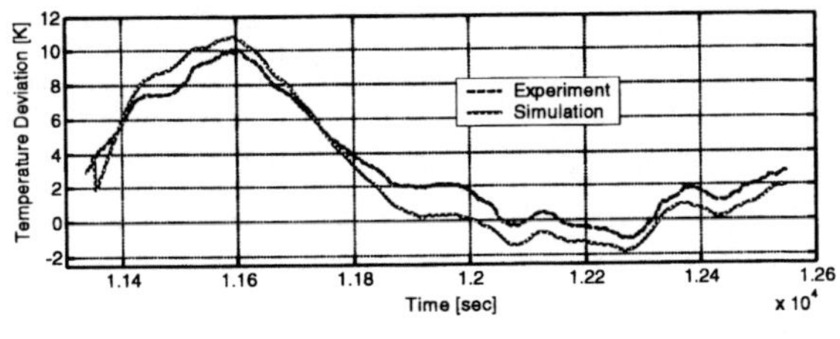

Fig. 3. Simulation of $\text{ARMAX}_{4,1,1,10}$ model.

3. SIMPLIFIED FUNDAMENTAL MODEL

The fundamental model developed by Junker *et al.* (2003) is a set of nine coupled nonlinear ordinary differential equations. Since this form is not suitable for internal model control design, simple first and second order plus time delay models are determined based on a simulated process reaction curve. This is preferred to a Taylor linearization since the low model orders simplify controller design. It is justified *a posteriori* since the controllers perform equally well as those based on the more detailed ARMAX models developed in the previous section.

The models' transfer functions are given by

$$M_1 = \frac{K_1 e^{-\alpha_1 s}}{\tau_{1,1} s + 1} \tag{6}$$

$$M_2 = \frac{K_2 e^{-\alpha_2 s}}{(\tau_{2,1} s + 1)(\tau_{2,2} s + 1)} \tag{7}$$

where the first index refers to the model order. The gain K is computed directly as the quotient of final process output and input step size while time constants τ and time delay α are computed via a least-squares optimization that minimizes the sum-squared error between step response and prediction (see Table 2).

Table 2. Parameters of first and second order plus time delay models.

K_1 $\left[\frac{°C}{\%power}\right]$	α_1 [s]	$\tau_{1,1}$ [s]	K_2 $\left[\frac{°C}{\%power}\right]$	α_2 [s]	$\tau_{2,1}$ [s]	$\tau_{2,2}$ [s]
8.19	78.16	539.96	8.19	17.64	527.23	69.88

The normalized residuals are shown in Figure 4. Consistent with intuition, the second order plus time delay (SOPTD) model is a better representation the process's high frequency character. Therefore, controllers are designed based on this model only.

4. INTERNAL MODEL CONTROL

4.1 Basic Structure

Internal model control (IMC) is characterized by two main features (see Figure 5): (1) the output

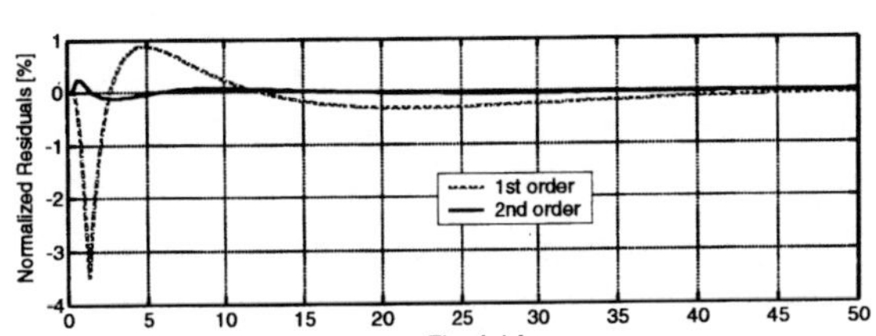

Fig. 4. Residuals of first and second order plus time delay models.

y of process P is compared to the output $\hat{y}$ of model M in order to obtain an estimate of the unmeasured disturbance d, and (2) the controller C is designed from an inverse of the process model (Garcia and Morari, 1982).

Model invertibility is limited by time delays and right half plane zeros. If contained in the model, it has to be factorized into an invertible part M_- and a noninvertible part M_+ (Garcia and Morari, 1982). The controller is then given by

$$C = \frac{F}{M_-} \tag{8}$$

where F is an appropriately chosen filter (Morari and Zafiriou, 1989; Ogunnaike and Ray, 1994). A typical n^{th} order filter is

$$F = \frac{1}{(\lambda s + 1)^n} \tag{9}$$

where λ is the tuning parameter — a larger λ leads to more sluggish control action.

4.2 Anti-windup Design

Control of the effusion source is limited by actuator constraints since the electric heater has a maximum power output of 2kW. Without proper compensation, this results in very sluggish control since the controller is unaware of the input saturation.

The situation can be remedied by using the anti-windup framework of Zheng *et al.* (1994). The basic idea is to minimize the difference between the unconstrained and constrained process outputs by introducing an additional feedback loop around the nonlinear saturation block (see Figure 6).

5. CONTROL RESULTS

Anti-windup IMC results are presented for empirical models based on two ID experiments and three

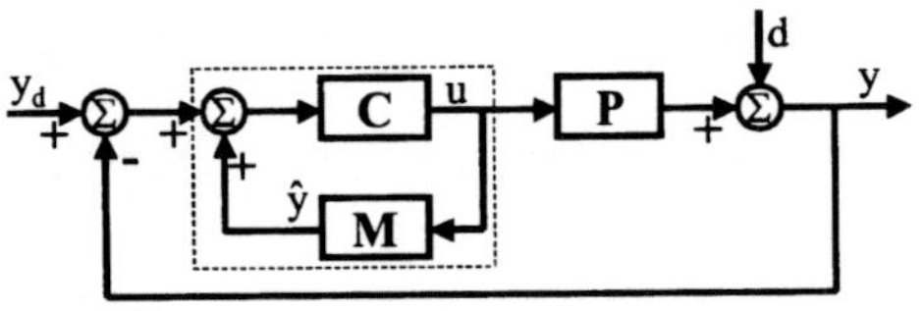

Fig. 5. Re-routed IMC structure.

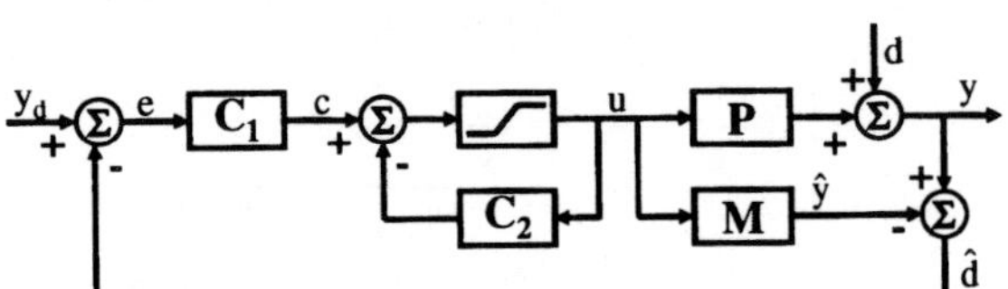

Fig. 6. Anti-windup IMC structure.

ARMAX models and for the simplified SOPTD model. For each model, a sluggish and an aggressive tuning are tested, in order to demonstrate the dependence of controller performance and noise sensitivity on the tuning. Proper controller tuning is tested in simulation studies before performing any experiments.

5.1 Empirical Model

$ARMAX_{2,1,6,2}$ Model The results for a sluggish ($\lambda = 50s$) and an aggressive ($\lambda = 10s$) tuning are shown in Figures 7 and 8, respectively. Clearly, the simulated results are not in very good agreement with the experimental data. For the sluggish tuning, the controller is more sluggish than predicted in the tuning experiment, for the aggressive tuning, it is more aggressive and very oscillatory. These results are contradicting each other. The controller should either be always too aggressive or always too sluggish.

Since the used model is based on old data, the observed behavior is most likely caused by changes made to the experimental system between the ID and control experiments. The achieved control

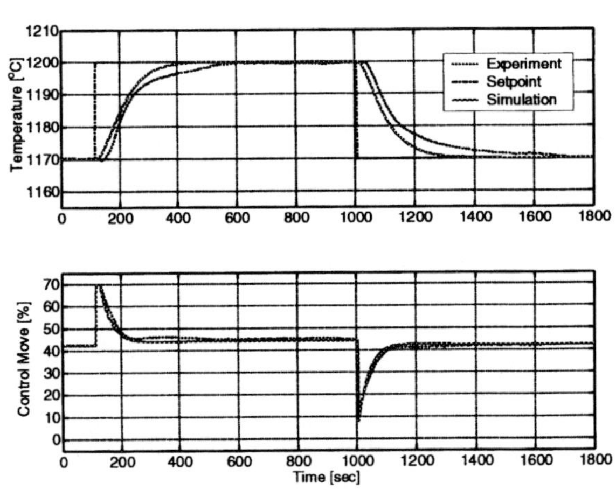

Fig. 7. $\text{ARMAX}_{2,1,6,2}$ model with $\lambda = 50s$.

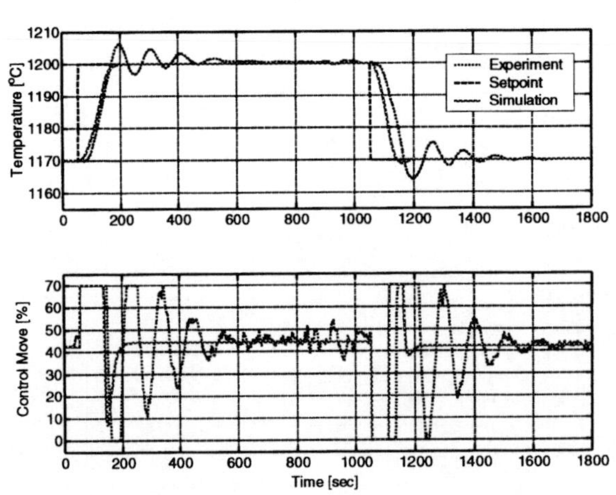

Fig. 8. $\text{ARMAX}_{2,1,6,2}$ model with $\lambda = 10s$.

action is unacceptable such that the model in its current form is of limited value for controller design.

ARMAX$_{4,1,1,10}$ *Model* Being based on a more recent ID experiment, the model's main difference besides higher n_a and n_c is a five fold increase of n_k from 1 to 5. At $T_s = 2s$, this closely matches the $\alpha = 25s$ delay of the fundamental model (Junker *et al.*, 2003).

The results for a sluggish ($\lambda = 100s$) and an aggressive ($\lambda = 22s$) tuning are shown in Figures 9 and 10, respectively. In the simulations, the experimentally observed noise was approximated by bandlimited white noise with a power of $5{\cdot}10^{-6}$ and a frequency of 50 Hz.

Unlike the ARMAX$_{2,1,6,2}$ model, the new model is very sensitive to measurement noise. This sensitivity increases as λ decreases. For $\lambda = 22s$ this clearly has a negative effect on the output which overshoots resulting in unacceptable performance. For $\lambda = 100s$ the noise amplification is acceptable and the simulation is in good agreement with the experiment.

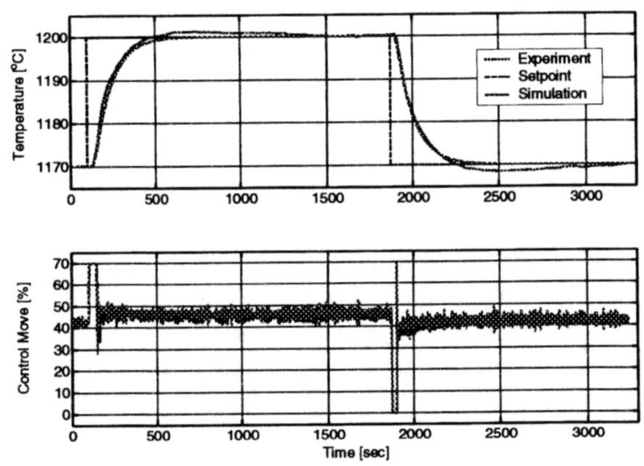

Fig. 9. ARMAX$_{4,1,1,10}$ model with $\lambda = 100s$. Simulation with bandlimited white noise (power $= 5{\cdot}10^{-6}$ @ 50 Hz).

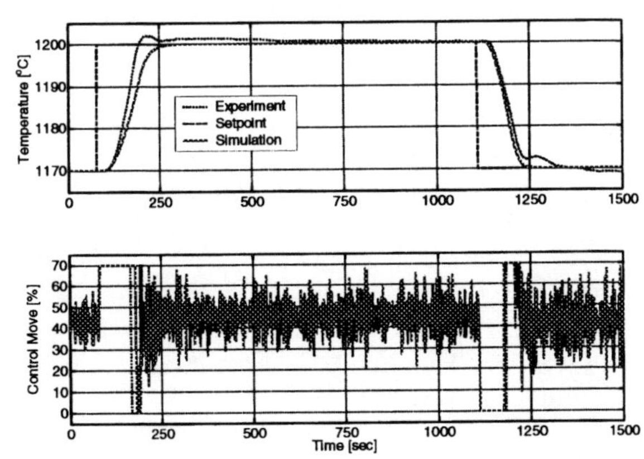

Fig. 10. ARMAX$_{4,1,1,10}$ model with $\lambda = 22s$. Simulation with bandlimited white noise (power $= 5{\cdot}10^{-6}$ @ 50 Hz).

Analysis The results obtained this far indicate that reference changes in less than three minutes are possible. Since neither of the two models suffices, a new model is devised by compounding the noise insensitive dynamic part (n_a, n_b, n_c) of ARMAX$_{2,1,6,2}$ and the time delay n_k of ARMAX$_{4,1,1,10}$ to a new ARMAX$_{2,1,6,2+8}$ model. Applicability of this new model was verified by reproducing the ARMAX$_{2,1,6,2}$ results in a simulation study.

5.1.1. ARMAX$_{2,1,6,2+8}$ *Model* Using the same tuning constants as for ARMAX$_{2,1,6,2}$ the results are shown in Figures 11 and 12. Both experiments show an improved performance and are closer to the simulation results.

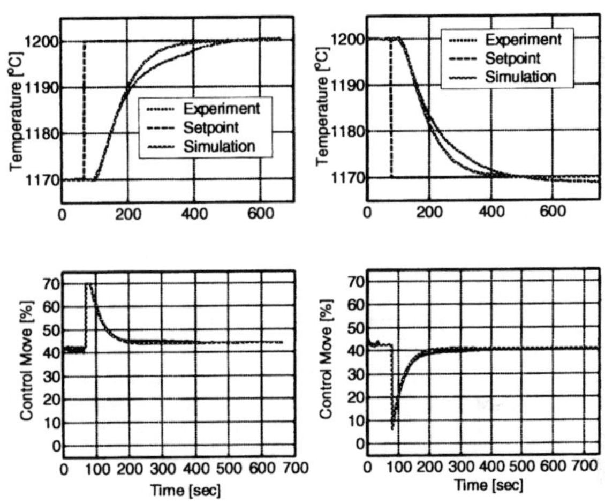

Fig. 11. ARMAX$_{2,1,6,2+8}$ model with $\lambda = 50s$. Simulation with bandlimited white noise (power $= 5{\cdot}10^{-6}$ @ 50 Hz).

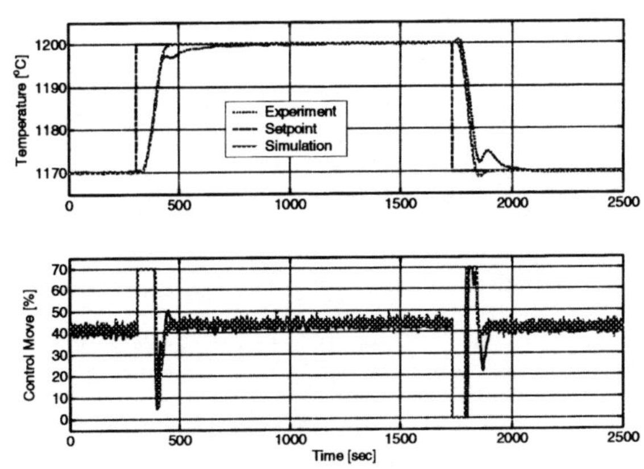

Fig. 12. ARMAX$_{2,1,6,2+8}$ model with $\lambda = 10s$. Simulation with bandlimited white noise (power $= 5{\cdot}10^{-6}$ @ 50 Hz).

5.2 SOPTD Model

For the simplified fundamental model, the tuning constant λ is defined with respect to the model time constants via the dimensionless tuning constant L, i.e.,

$$\lambda = L \cdot \frac{\tau_{2,1} + \tau_{2,2}}{2} \qquad (10)$$

The results for a sluggish ($L = 0.25$) and an aggressive ($L = 0.1$) tuning are shown in Figures 13 and 14, respectively. The experimental studies were performed in a single run, first for the sluggish, then for the aggressive tuning.

During the first reference change (Figure 13, left), the agreement between model and simulation is excellent. For following reference changes, there is a slight offset in the control move, however, the dynamic behavior of the control moves matches very closely.

The experimental reference changes exhibit a slight overshoot which the controller is not able to compensate for in a timely fashion. A similar overshoot can be reproduced when introducing plant model mismatch in the simulation studies, such that this is the most likely cause for the observed behavior.

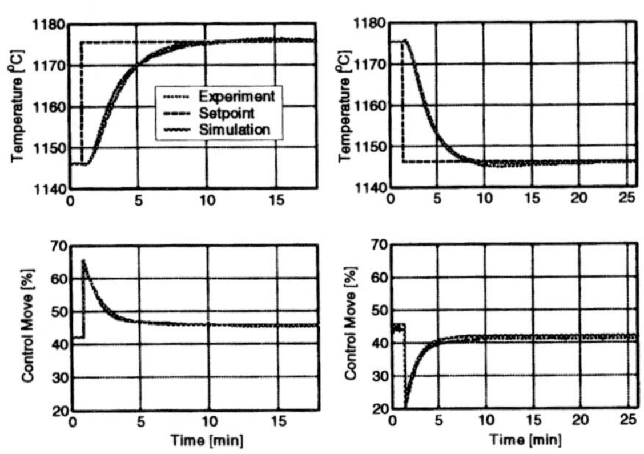

Fig. 13. SOPTD model with $\lambda = 74.64s$.

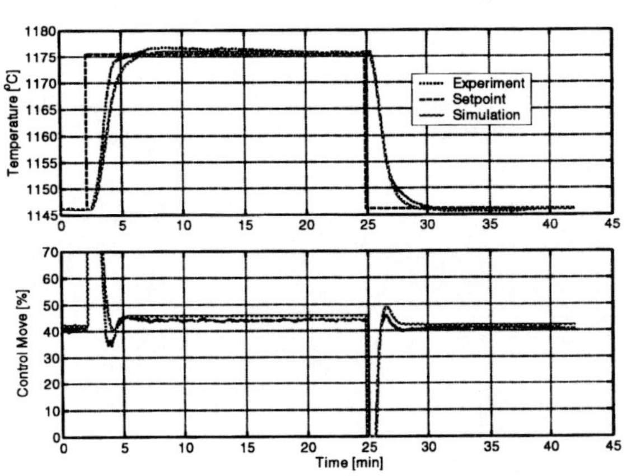

Fig. 14. SOPTD model with $\lambda = 29.86s$.

Overall, the experimental validation shows an excellent agreement with the simulation which proves the usefulness of the developed models for *a priori* controller tuning. Results obtained with the simple SOPTD model compare very favorably with the more complicated ARMAX models.

6. SUMMARY

In this paper, anti-windup internal model controllers for temperature control of a physical vapor deposition effusion source were presented. The controllers are based on both empirical ARMAX models and a simplified fundamental model. Controllers were designed in simulation studies and tested in experiments. A very good match between simulation and experiment was obtained which proves the usefulness of the developed models for *a priori* controller tuning. Controllers based on both models achieved good performance for fast, overshoot-free reference tracking.

ACKNOWLEDGEMENTS

Funding from *Global Solar Energy* and the *Delaware Research Partnership* is gratefully acknowledged.

REFERENCES

Birkmire, R.W. and E. Eser (1997). Polycrystalline thin film solar cells: Present status and future potential. *Annu. Rev. Mater. Sci.* **27**, 625–653.

Garcia, C.E. and M. Morari (1982). Internal model control - 1. Unifying review and some new results. *Ind. Eng. Chem. Proc. DD* **21**(2), 308–323.

Junker, S.T., R.W. Birkmire and F.J. Doyle III (2003). Mass- and heat transfer modeling of a physical vapor deposition effusion source. *AIChE J.*, **submitted**.

Junker, S.T., R.W. Birkmire and F.J. Doyle III (2004). Manufacture of thin-film solar cells: Modeling and control of $Cu(InGa)Se_2$ physical vapor deposition onto a moving substrate. *Ind. Eng. Chem. Res.* **43**(2), 566–576.

Ljung, L. (1999). *System Identification: Theory for the User.* 2nd ed.. Prentice-Hall, Engelwood Cliffs NJ.

Ljung, L. (2002). System Identification Toolbox, User's Guide Version 5. The MathWorks, Inc.

Morari, M. and E. Zafiriou (1989). *Robust Process Control.* Prentice-Hall, Englewood Cliffs NJ.

Ogunnaike, B.A. and W.H. Ray (1994). *Process Dynamics, Modeling, and Control.* Oxford University Press, New York.

Zheng, A., M.V. Kothare and M. Morari (1994). Anti-windup design for internal model control. *Int. J. Control* **60**(5), 1015–1024.

ELSEVIER
IFAC
PUBLICATIONS
www.elsevier.com/locate/ifac

MPC DESIGN FOR CONSTRAINED MULTIVARIABLE SYSTEMS UNDER ACTUATOR BACKLASH

H. Zabiri *,[1] **Y. Samyudia** *,[2]

* *Department of Chemical Engineering, McMaster University, Hamilton, ON, Canada.*

Abstract: A new approach to the design of Model Predictive Controller (MPC) that simultaneously addresses the actuator saturation and backlash is proposed in this paper. The discrete characteristics of the actuator backlash allows one to reformulate the input constraints as a set of mixed integer linear inequalities. As a result, the MPC is designed by solving a *Mixed-integer Quadratic Programming* problem. Simulation results are presented to show how this new approach performs as compared to the existing techniques of the backlash compensation when they are applied to an industrial case study. *Copyright © 2004 IFAC*

Keywords: actuator backlash, constrained systems, model predictive control, mixed integer quadratic programming

1. INTRODUCTION

Actuators, e.g. control valves in most industrial processes, play an important role in automatic control systems. As commonly the weakest link of the control loop, the control valve accounts for about 32% of "poor" or "fail" control loops [Miller and Desborough, 2001]. A recent survey in Yang and Clarke [1999] indicated that 30% of all control loops in Canadian paper mills were oscillating because of valve problems. Also, [Ruel, 2000] observed a similar problem in refinery plants. Poor control valve performance is therefore the single biggest source of controller-induced variability.

Control valves contain static (e.g. saturation) or dynamic (e.g. backlash) non-linearities as well as other non-linear characteristics such as stiction, hysteresis, etc [Corradini and Orlando, 2002]. Model Predictive Control (MPC) has had a significant and widespread impact on industrial process control because of its ability to explicitly handle actuator saturation. The performance of MPC, however, is sensitive to model uncertainty [Kothare et al., 1996]. Since the dynamics of actuator backlash is not accommodated in the MPC design formulation, the existence of backlash results in uncertainty in the plant inputs. The dynamics of actuator backlash introduces a limit cycle in the proximity of the steady state operating points even if the controlled system behaves linearly. As a result, the MPC performance would significantly degrade if the actuator backlash is present in the feedback loop. Various approaches have been adopted for compensating the backlash effect, including nonlinear inverse strategy [Corradini and Orlando, 2002; Tao and Kokotovic, 1993; Nordin and Gutman, 2002; Chow and Clarke, 1994], or re-tuning strategy Ghazzawi et al. [2001].

In this paper, we propose a tactical approach that simultaneously addresses the actuator backlash and saturation within the MPC design framework. This approach is adopted to alleviate the drawbacks of the re-tuning strategy (i.e. a sluggish

[1] Current affiliation is at Universiti Teknologi Petronas, Malaysia
[2] Corresponding author: samyudi@mcmaster.ca.

response as well as the trial-and-error nature), and the nonlinear-inverse strategy (i.e. in the case of encountering actuator saturation limits).

2. MPC DESIGN FRAMEWORK

Assume that the system is described by the following linear time-invariant model:

$$x(k+1) = Ax(k) + Bu(k) \tag{1}$$

$$y(k) = Cx(k) \tag{2}$$

with $x(k)$ is the state vector $\in \mathbb{R}^n$, $u(k)$ is the input vector $\in \mathbb{R}^m$, and $y(k)$ is the output vector $\in \mathbb{R}^m$. The matrices $(\mathbf{A}, \mathbf{B}, \mathbf{C})$ are associated with the state, input and output matrices, respectively, and are assumed to be stabilizable and detectable.

Consider the problem of tracking a constant set-point y_s and rejecting a time-varying output disturbance $\{d(k)\}$: it is wished to regulate the error $e(k)$ to zero: $e(k) = y(k) + (d(k) - y_s)$. Then, let us define d_e as the external signal: $d_e(k) = d(k) - y_s$. Given knowledge of the external signal d_e and the current state measurement $x(0)$ (or its on-line estimates), the MPC design is aimed at finding the M-move control sequence $\{u(0), u(1), \ldots, u(M-1)\}$ that minimizes the finite-horizon performance index:

$$\begin{aligned} J_0 &= [x(P) - x_s]^T \mathbf{\Psi}_f [x(P) - x_s] \\ &+ \sum_{k=0}^{P-1} e^T(k) \mathbf{\Psi} e(k) \\ &+ \sum_{k=0}^{M-1} [u(k) - u_s]^T \mathbf{\Phi} [u(k) - u_s] \end{aligned} \tag{3}$$

where $\mathbf{\Psi} \geq 0$, $\mathbf{\Phi} \geq 0$, $\mathbf{\Psi}_f \geq 0$. In (3), P is the prediction horizon, $M \leq P$ is the control horizon. This optimization can be translated into a convex QP problem as:

$$U^{OPT} = arg \min_{\substack{U \\ LU \leq K}} U^T \mathbf{W} U + 2U^T \mathbf{V} \tag{4}$$

where $\mathbf{W}$ and $\mathbf{V}$ are the dynamic matrices obtained from the linear model for the P prediction horizon of the outputs, and the M-move of input sequence.

3. MODEL OF BACKLASH AND ITS INVERSE

Fig. 1 describes an input-output map of the actuator backlash. Let us define two quantities to represent two situations; (i) if $u_I(t)$ is inside the backlash and on the positive side of the backlash,

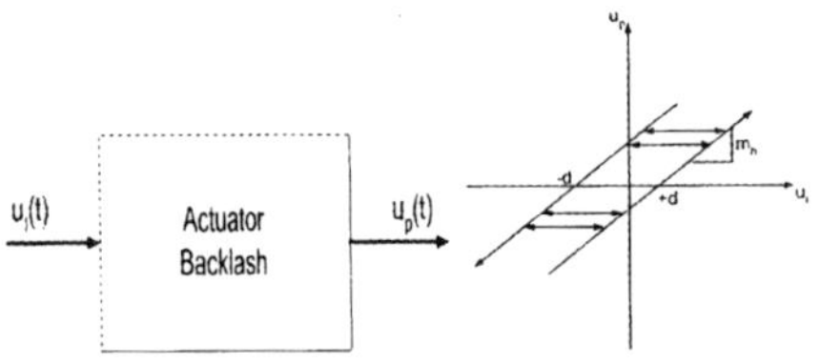

Fig. 1. An input-output map of backlash

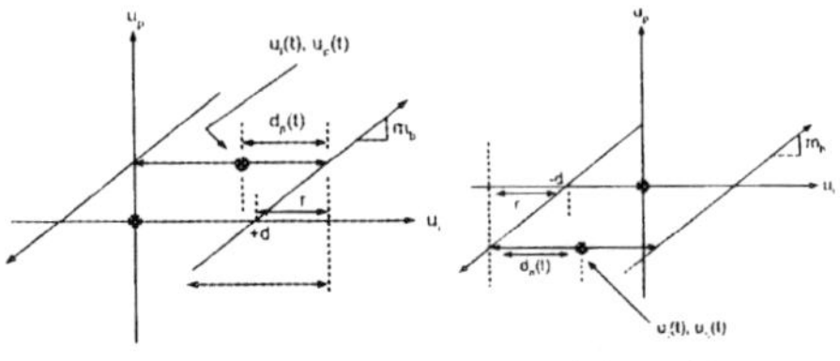

(a) $d_p(t)$ (b) $d_n(t)$

Fig. 2. Computation of $d_p(t)$ and $d_n(t)$

the distance from the positive boundary is defined as d_p (see Fig. 2(a)), and (ii) if $u_I(t)$ is inside the backlash and on negative side, the distance from its negative boundary is defined as d_n (see Fig. 2(b)). Without loss of generality, we assume that $m_b = 1$. Then, we have:

$$d_p(t) = [u_p(t) + d] - u_I(t)$$

$$d_n(t) = [u_p(t) + (-d)] - u_I(t)$$

Hence, we obtain the following backlash model:

$$u_p(t) = \begin{cases} u_I(t) - d & : \Delta u_I(t) > d_p(t-1) \\ u_I(t) - (-d) & : \Delta u_I(t) < d_n(t-1) \\ u_p(t-1) & : otherwise \end{cases}$$

The inverse of the backlash model is then given as:

$$u_I(t) = \begin{cases} u(t) + d & : \Delta u(t) > 0 \\ u(t) - d & : \Delta u(t) < 0 \\ u_I(t-1) & : \Delta u(t) = 0 \end{cases}$$

Clearly, the nonlinear inverse strategy of compensating the backlash effect is, in principle, to bump up the controller outputs without changing the input direction. Based on this property, we integrate the backlash inverse model within the MPC design framework.

4. MIQP-BASED DESIGN OF MPC

4.1 Reformulation of Input Constraints

In the QP-based MPC design, the input (saturation) constraint is given by:

$$u_{min} \leq u(k) \leq u_{max} \tag{5}$$

To incorporate the backlash dynamics within the MPC computational framework, we reformulate the input constraints so that the inverse model of the actuator backlash is integrated within the MPC design framework as in Fig. 3.

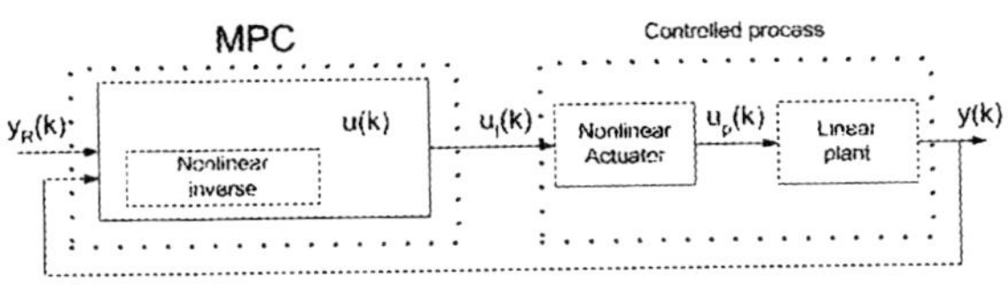

Fig. 3. An MPC scheme integrated with the backlash nonlinear inverse

By applying this integration strategy, $u_I(k)$ can be represented in terms of $u(k)$ with three different conditions depending upon the value of $\Delta u(k)$. The inverse model of the actuator backlash dictates that, for example, if the input change is positive (or $\Delta u(k) > 0$), the MPC should produce its output $u_I(t) = u(k) + d$. Similarly, for a negative input change (or $\Delta u(k) < 0$), the resulting MPC output is reduced by d. For a zero input change, the previous value of u_I should be used. Therefore, the inclusion of the backlash within the MPC computational framework is, in essence, to produce optimal input signals that are outside the dead-band region.

That basic idea of incorporating the actuator backlash is systematically implemented by reformulating the input constraints for the MPC design. Let us introduce a set of logical variables δ_{ij}, for $j = 1, 2, 3$, for representing the input change conditions (e.g. positive, negative, or zero). Then, we impose the following propositional logics on the inputs $u(k)$:

$$\delta_{i1} = 1 \longleftrightarrow u_i(k) \geq d_i \tag{6}$$

$$\delta_{i2} = 1 \longleftrightarrow u_i(k) \leq -d_i \tag{7}$$

$$\delta_{i3} = 1 \longleftrightarrow u_i(k) = 0 \tag{8}$$

$$\sum_{j=1}^{3} \delta_{ij} = 1 \tag{9}$$

for $i = 1, 2, ..., m$, with m is the number of manipulated inputs in the system. The propositional logics specified in (6), (7), and (8) can be transformed into a set of *mixed-integer linear inequalities*, i.e. linear inequalities involving both *continuous variables* $u \in \mathbb{R}^m$ and *logical variables* $\delta \in \{0, 1\}$, by adopting the framework described in Bemporad and Morari [1999]. The set of mixed-integer linear inequalities constraints for the ith input is then given by:

$$\begin{aligned} u_i(k) - d_i &\geq (u_{i_{min}} - d_i)(1 - \delta_{i1}) \\ u_i(k) - d_i &\leq (u_{i_{max}} - d_i)\delta_{i1} \\ u_i(k) + d_i &\geq (u_{i_{min}} + d_i)\delta_{i2} \\ u_i(k) + d_i &\leq (u_{i_{max}} + d_i)(1 - \delta_{i2}) \\ u_i(k) &\geq u_{i_{min}}(1 - \delta_{i3}) \\ u_i(k) &\leq u_{i_{max}}(1 - \delta_{i3}) \end{aligned} \tag{10}$$

for $i = 1, 2, ..., m$, with m is the number of manipulated inputs in the system.

Note that if $d_i = 0$, (10) reduces to $u_{i_{min}} \leq u_i(k) \leq u_{i_{max}}$. This implies that if the backlash is not active, we should assign $d_i = 0$. Otherwise, the MPC controller would have a sluggish response (or even infeasible solution) because of the shrinkage of feasible input constraints.

4.2 MIQP formulation

The MPC design is now aimed at determining its optimal $u(k)$ at each sampling time subject to the reconfigured input constraints in (10). At every time step, the dead-band region is completely avoided by suppressing a further movement of the input to avoid its direction change. This is done by activating *one and only one* logical variable at a time, and at the proximity of steady states where the backlash is active.

We show that this formulation belongs to an MIQP optimization problem. Let us consider, for illustration only, the case of m inputs and control horizon, $M = 1$. The linear inequality constraints in (10) is then re-arranged as:

$$\begin{bmatrix} -\underline{\mathbf{1}} & \theta & \mathbf{0} & \mathbf{0} \\ -\underline{\mathbf{1}} & \mathbf{0} & \mathbf{\Omega} & \mathbf{0} \\ -\underline{\mathbf{1}} & \mathbf{0} & \mathbf{0} & \mu \\ \underline{\mathbf{1}} & \alpha & \mathbf{0} & \mathbf{0} \\ \underline{\mathbf{1}} & \mathbf{0} & \beta & \mathbf{0} \\ \underline{\mathbf{1}} & \mathbf{0} & \mathbf{0} & \tau \\ \mathbf{0} & \underline{\mathbf{1}} & \underline{\mathbf{1}} & \underline{\mathbf{1}} \end{bmatrix} \begin{bmatrix} u_1 \\ \vdots \\ u_m \\ \delta_{11} \\ \vdots \\ \delta_{m1} \\ \delta_{12} \\ \vdots \\ \delta_{m2} \\ \delta_{13} \\ \vdots \\ \delta_{m3} \end{bmatrix} + \begin{bmatrix} \underline{\mathbf{U}}_{min} \\ -\underline{\mathbf{D}} \\ \underline{\mathbf{U}}_{min} \\ -\underline{\mathbf{D}} \\ -\underline{\mathbf{U}}_{max} \\ -\underline{\mathbf{U}}_{max} \\ -\underline{\mathbf{I}} \end{bmatrix} \leq \underline{\mathbf{0}} \tag{11}$$

and

$$\underline{\mathbf{1}} = diag[1_1, \ldots, 1_m]$$
$$\theta = diag[-(u_{1_{min}} - d_1), \ldots, -(u_{m_{min}} - d_m)]$$
$$\boldsymbol{\Omega} = diag[(u_{1_{min}} + d_1), \ldots, (u_{m_{min}} + d_m)]$$
$$\alpha = diag[-(u_{1_{max}} - d_1), \ldots, -(u_{m_{max}} - d_m)]$$
$$\beta = diag[(u_{1_{max}} + d_1), \ldots, (u_{m_{max}} + d_m)]$$
$$\mu = diag[u_{1_{min}}, \ldots, u_{m_{min}}]$$
$$\tau = diag[u_{1_{max}}, \ldots, u_{m_{max}}]$$
$$\underline{\mathbf{U}}_{min} = [u_{1_{min}}, \ldots, u_{m_{min}}]^T$$
$$\underline{\mathbf{U}}_{max} = [u_{1_{max}}, \ldots, u_{m_{max}}]^T$$
$$\underline{\mathbf{D}} = [d_1, \ldots, d_m]^T ; \underline{\mathbf{I}} = [1_1, \ldots, 1_m]^T$$

Define a matrix $\mathbf{Q}_{4m \times 4m}$ and a vector $\mathbf{b}_{4m \times 1}$ as

$$\mathbf{Q} = \begin{bmatrix} \mathbf{W} & 0 & \ldots & 0 \\ 0 & \vdots & \ddots & 0 \\ \vdots & \vdots & \ddots & 0 \\ 0 & 0 & \ldots & 0 \end{bmatrix} ; \mathbf{b} = \begin{bmatrix} 2\mathbf{V} \\ 0 \\ \vdots \\ 0 \end{bmatrix} \tag{12}$$

By defining the mixed linear inequalities of (11) as $Cz + \wp \leq 0$ with $z = [z_c, z_d]^T$, where z_c are the continuous variables u, and z_d are logical variables δ_{ij}, the MPC optimization problem can be expressed as:

$$\min_z \; J_{QP} \triangleq z^T \mathbf{Q} z + \mathbf{b}^T z \tag{13}$$
$$\text{s.t.}$$
$$Cz + \wp \leq 0$$
$$z = \begin{bmatrix} z_c \\ z_d \end{bmatrix}$$
$$z_c \in \mathbb{R}^{n_c}$$
$$z_d \in \{0,1\}^{n_d}$$

This is a *Mixed Integer Quadratic Programming* (MIQP) problem. Note that the matrix $\mathbf{Q}$ in (13) is positive semi-definite - an important criteria that is necessary if such an optimization needs to be solved *on-line* for a global optimality.

Since the proposed technique considers both the logical conditions and saturation simultaneously during the determination of the MPC optimal inputs, we avoid the problem of producing the infinite input for compensating the backlash due to the input saturation as is the case when the nonlinear inverse strategy is applied.

4.3 Stability Analysis of MIQP-based MPC

In this section, we address the stability condition for the MIQP-based MPC. For this purpose, we follow the stability analysis of Bemporad and Morari [1999] for the general system of Mixed Logical Dynamical. Define the states and inputs equilibrium as (x_s, u_s), and let the logical variables be definitely admissible, meaning that δ_s corresponds to the desired steady-state values for the logical variables. Let t be the current time t, and $x(t)$ the current state. The MPC optimization problem (3) can be equivalently expressed in the general form of Mixed Logical Dynamical system (MLD) as the following,

$$\min_u \; J \triangleq \sum_{k=0}^{M-1} \| u(k) - u_s \|^2_{\Phi} + \| \delta(k|t) - \delta_s \|^2_R + \| x(k|t) - x_s \|^2_{\Psi} \tag{14}$$
$$\text{s.t.}$$
$$x(M|t) = x_s$$
$$x(k+1|t) = Ax(k|t) + Bu(k)$$
$$y(k|t) = Cx(k|t)$$
$$E_2\delta(k|t) \leq E_1 u(k) + E_3 \tag{15}$$

where $\Phi > 0; R \geq 0; \Psi > 0$ and E_1, E_2 and E_3 can be straightforwardly defined from (11). Assume that the optimal solution is given by $U^* = \{u^*(0), u^*(1), \ldots, u^*(M-1)\}$, and the *receding horizon* strategy is applied as $u(t) = u^*(0)$ before repeating the whole optimization procedure at time $t+1$.

Theorem 1. Let (x_s, u_s) be an equilibrium pair and δ_s definitely admissible. Assume that the initial states $x(0)$ is such that a feasible solution of problem (14) exists at time $t = 0$. Then, the optimal inputs obtained by solving (14)-(15) stabilizes the system in the following sense:

$$lim_{t\to\infty} x(t) = x_s; \quad lim_{t\to\infty} u(t) = u_s;$$
$$lim_{t\to\infty} \| \delta(k|t) - \delta_s \|^2_R = 0$$

while fulfilling the constraints (15).

Proof. The proof follows from standard Lyapunov arguments. Let U^* denote the optimal control sequence $\{u^*(0), \ldots, u^*(M-1)\}$. Let

$$V(t) \triangleq J(U^*, x(t))$$

denote the corresponding value attained by the performance index. Also, let $U_1 = \{u^*(1), \ldots, u^*(M-2), u_s\}$. Then U_1 is feasible (but not optimal) at time $t+1$, along with the vectors $\delta(k|t+1) = \delta(k+1|t)$ for $k = 0, \ldots, M-1$. Also $\delta(M-1|t+1) = \delta_s$ since $x(M-1|t-1) = x(M|t) = x_s$ and δ_s definitely admissible. Hence,

$$V(t+1) \leq J(U_1, x(t+1)) = V(t) - \| u(k) - u_s \|^2_{\Phi} - \| \delta(k|t) - \delta_s \|^2_R - \| x(k|t) - x_s \|^2_{\Psi}$$

and $V(t)$ is decreasing. Since $V(t)$ is lower-bounded by zero, there exists $V_\infty = lim_{t\to\infty} V(t)$, which implies $V(t+1) - V(t) \to 0$. Therefore, each term of the sum

$$\| u(k) - u_s \|_\Phi^2 + \| \delta(k|t) - \delta_s \|_R^2 + \| x(k|t) - x_s \|_\Psi^2 \leq V(t) - V(t-1)$$

converges to zero as well, which proves the theorem.

Note that our result on the stability analysis of the MIQP-based MPC is a subset of the stability result of Bemporad and Morari [1999], where in their case, the integer variable may appear in the state and input variables.

5. APPLICATION TO AN INDUSTRIAL FCCU PROCESS

The FCCU case study was taken from [Grosdidier et al., 1993]. At the MPC level, we control seven variables, denoted as $y_1, ..., y_7$ by manipulating six variables $u_1, ..., u_6$, see [Grosdidier et al., 1993] for details. Three proportional-integral (PI) controllers were applied in the flows of the combustion air, the hot gas oils and the combined cold gas and recycle oils (u_1, u_2, u_3). Two PI controllers were to control the feed preheat (u_4) and the riser outlet temperature (u_5). Recycle flow (u_6) is regulated by adjusting the output of a hand controller. Each variable is constrained by their associated high and low limits. All manipulated variables are at their ideal resting values (IRV), which is assumed to be their steady-state, except for combustion air flow (u_1) which is allowed to move freely. In this study, we consider the following conditions: (1) Backlash is known to be active only in the hot feed flow (u_2) valve; (2) Prediction horizon, P is fixed at 20; Control horizon, M is fixed at 1; and (3) The proposed MIQP MPC is then compared to the Nonlinear Inverse strategy ([Chow and Clarke, 1994]) and to the strategy of re-tuning λ, where λ is the element of Φ.

5.1 *Effect of Backlash*

In response to a disturbance change of +2.5% in u_2, Fig. 4 shows the existence of backlash in u_2, where its effect propagates to other input channels because of the interacting nature of the FCCU process. This backlash result in a significant degradation in the performance of MPC designed using the standard approach, producing oscillations in the manipulated inputs and the controlled variables. In the presence of backlash of 2.5%, a small oscillation in the controller outputs makes the backlash active. Furthermore, we should note that the backlash in the u_2 valve is not active when the change in the controller outputs is large; however, as the movement nearing the steady-state and the controller outputs are still oscillating, the backlash is then active.

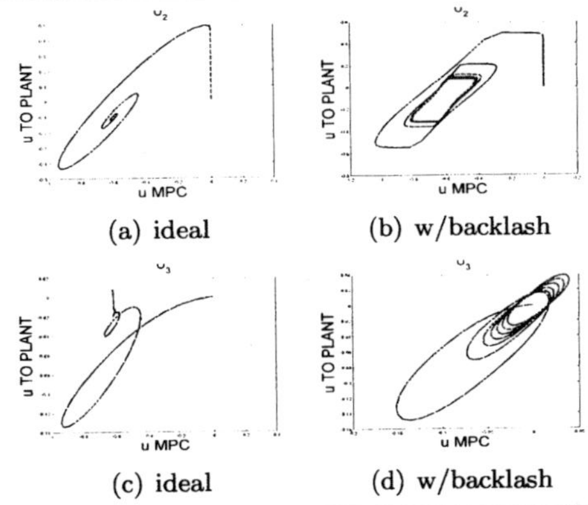

Fig. 4. u_2 and u_3 response with and without backlash.

5.2 *MIQP vs. Re-tuning Strategies*

Figs. 5 and 6 show the comparison of MIQP-based MPC with the strategy of re-tuning λ. The input moves u_2 for the different compensation methods are shown in Fig. 5. Fig. 6 shows the corresponding responses of the outputs for: the standard MPC with no backlash (i.e. *"ideal"*), MIQP-based MPC design (i.e. *"MIQP MPC"*) and the re-tuning of λ.

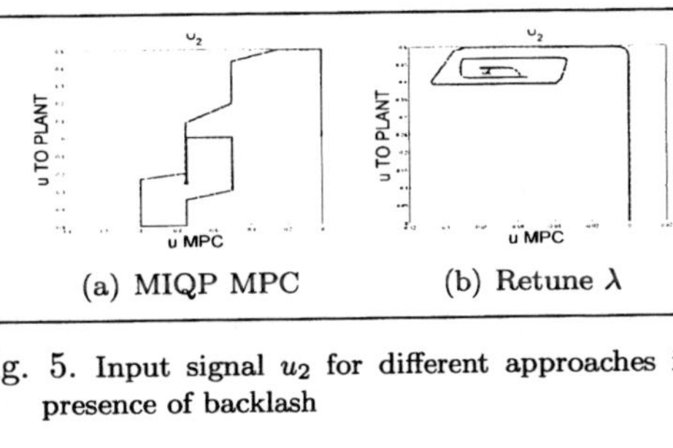

Fig. 5. Input signal u_2 for different approaches in the presence of backlash

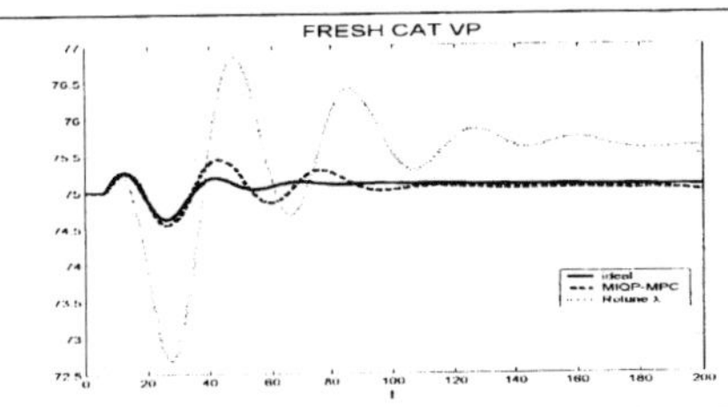

Fig. 6. Fresh cat VP in the presence of backlash

The re-tuning strategy, by increasing the move suppression factor λ, produces a more sluggish response in the manipulated and controlled variables (see Fig. 5(b)). This compensation effect is similar to the case when we apply the MIQP approach; however, note that the movement of u_2 in the transient is different. This is due to the fact that the MIQP technique suppresses the movement on u_2 only at the proximity of steady state in order to eliminate the oscillations caused by the backlash. On the other hand, the re-tuning strategy avoid the backlash by imposing a sluggish response of the controller from the beginning. This re-tuning strategy, however, results in a poor disturbance rejection as shown in Fig. 6.

5.3 MIQP vs. Nonlinear Inverse Strategies

We study the case when the valve of u_2 is operating near the saturation limit by considering the disturbance of -2.5%, with 5% backlash. The nonlinear inverse method fails to eliminate the backlash since the saturation limit eclipses the total amount of energy required to compensate the backlash (see Figs. 7 and 8(d)). The MIQP MPC method, on the other hand, produces a better performance than the nonlinear inverse strategy as it anticipates the saturation limits in u_2 when determining the optimal input moves.

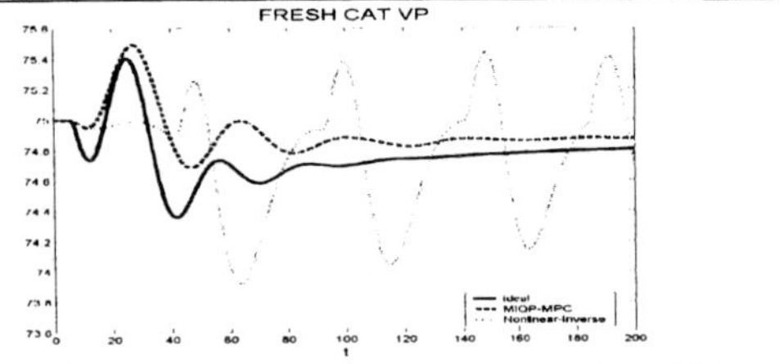

Fig. 7. Fresh cat VP; valve u_2 near saturation under disturbance change and 5% backlash

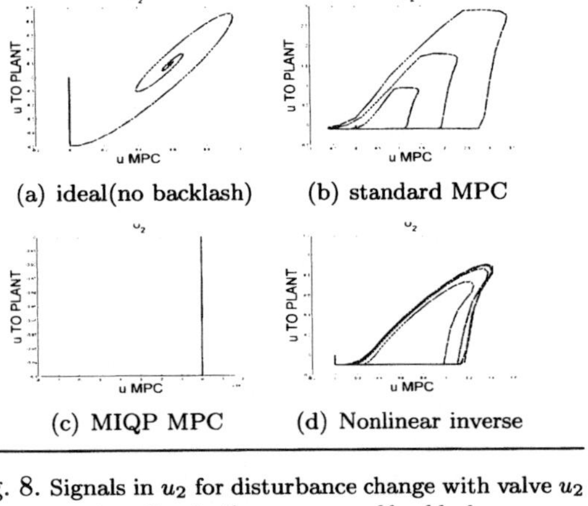

Fig. 8. Signals in u_2 for disturbance change with valve u_2 near saturation in the presence of backlash

6. CONCLUSIONS

In this paper, we have extended the MPC design methodology for addressing simultaneously the actuator backlash and saturation. We have shown that a set of logical constraints for the inputs has been produced so that the MPC design has been formulated as a mixed integer programming problem. Simulation studies using the FCCU process has been presented to compare the developed approach with other techniques i.e. the nonlinear inverse and re-tuning strategies. The MIQP-based MPC design has outperformed other compensation methods in the presence of both actuator saturation and backlash.

Acknowledgment: This work was partly supported by McMaster Advanced Control Consortium and NSERC Grant. Also, H. Zabiri would like to thank for a graduate scholarship received from University Technology Petronas, Malaysia.

REFERENCES

A. Bemporad and M. Morari. Control of Systems integrating logic, dynamics, and constraints. *Automatica*, 35(3):407–427, 1999.

C. M. Chow and D. W. Clarke. Actuator nonlinearities in predictive control. In D. Clarke, editor, *Advances in Model-Based Predictive Control*, pages 245–259. Oxford University Press, 1994.

M. L. Corradini and G. Orlando. Robust stabilization of nonlinear uncertain plants with backlash or dead zone in the actuator. *IEEE Transactions on Control Systems Technology*, 10(1):158–166, 2002.

A. Ghazzawi, E. Ali, A. Noud, and E. Zafiriou. Online tuning strategy for model predictive controllers. *Journal of Process Control*, 11:265–284, 2001.

P. Grosdidier, A. Mason, A. Aitolahti, P. Heinonen, and V. Vanhamaki. FCC Unit Reactor-Regenerator Control. *Computers and Chemical Engineering*, 17(2):165–179, 1993.

M. Kothare, V. Balakrishnan, and M. Morari. Robust Constrained Model Predictive Control using Linear Matrix Inequalities. *Automatica*, 32(10):1361–1379, 1996.

R. Miller and L. Desborough. Increasing customer value of industrial control performance - Honeywell's experience. *CPC-6*, 2001.

M. Nordin and P. Gutman. Controlling mechanical systems with backlash - a survey. *Automatica*, 38:1633–1649, 2002.

M. Ruel. Stiction the hidden menace. *Control Magazine (November Ed.)*, 2000.

G. Tao and P. V. Kokotovic. Adaptive Control of Systems with Backlash. *Automatica*, 29(2): 323–335, 1993.

J. C. Yang and D. W. Clarke. The self-validating actuator. *Control Engineering Practice*, 7:249–260, 1999.

ELSEVIER
IFAC
PUBLICATIONS
www.elsevier.com/locate/ifac

DYNAMIC OPTIMIZATION OF PROCESSING SYSTEMS WITH MIXED DEGREES OF FREEDOM

Olaf Stursberg

Process Control Laboratory (BCI-AST)
University of Dortmund, D-44221 Dortmund, Germany
Email: olaf.stursberg@uni-dortmund.de

Abstract: Procedures like start-up, product change-over, or shutdown of processing systems usually involve the manipulation of continuous and discrete controls. To optimize such procedures, the use of hybrid models is often appropriate to account for the change of dynamics during the transition. Some recently developed approaches for hybrid system optimization are based on mixed-integer linear programming. Applications reveal that the complexity inherent to these approaches considerably limits the applicability to industrial-size problems. This paper suggests as an alternative a tailor-made algorithm that optimizes the discrete and continuous degrees of freedom in a two-stage procedure. While the discrete controls are selected by a graph search algorithm, the continuous controls are obtained by embedded nonlinear programming. The method is illustrated for the start-up of a CSTR.

Keywords: Automata, Hybrid Systems, Integer Programming, Nonlinear Programming, Dynamic Optimization.

1. INTRODUCTION

It is certainly the majority of processing systems for which the operation is determined by the combination of continuous and discrete controls. While the continuous inputs are usually set by standard feedback control loops, discrete controls (like on-off actors occurring in form of valves, pumps, heaters, etc.) are determined predominantly by logic controllers. Traditionally, the control tasks that involve logical decisions and those which establish continuous regulation are separated and not designed in conjunction. For operations like start-up, shutdown, product change-over, or also the sequential operation of batch processes, the simultaneous consideration of both types of controls seems necessary. This is true in particular for the optimization of above-named procedures, since a separated design can not account for potentially opposing effects of both control components. This paper thus advocates a simultaneous optimization of both types of controls.

An important effect in model-based optimization of the considered procedures is that the process dynamics usually changes significantly. Is is often not possible to obtain one model that is sufficiently accurate for the complete procedure, but a set of models is required to represent the variation of the dynamics. Hybrid dynamic models have been identified in recent years as a suitable means to account for dynamic-dependent transitions between different models and for the combination of continuous as well as discrete (switching) inputs, see e. g. (Lynch *et al.*, 2003).

The task addressed in this paper is to model the process dynamics as a hybrid dynamic system, to formulate the transition procedure as an optimization problem, and – most importantly – to develop an algorithm to efficiently compute the (optimal) control inputs. The cost to be minimized can be the transition time in the simplest case, or more complex cost functions which depend on the state and input variables. Different approaches to the optimization of hybrid systems have been published in recent years, ranging from rather generic formulations to tailor-made methods for certain subtypes of hybrid systems, see e.g. (Branicky *et al.*, 1998; Avraam *et al.*, 1998; Broucke *et al.*, 2000; Gokbayrak and Cassandras, 2000; Hedlund and Rantzer, 2002; Barton and Lee, 2002; Shaikh and Caines, 2003; Oldenburg *et al.*, 2003). One branch of methods follows the idea of transforming the hybrid dynamics into a set of algebraic (in-)equalities that serve as constraints for a mixed-integer program (Bemporad and Morari, 1999; Stursberg *et al.*, 2002). If all constraints are written in linear form, mixed-integer linear (or quadratic) programming (MILP, MIQP) can be used for solution, i.e., standard solvers that employ branch-and-bound strategies, where bounds are obtained from linear relaxations, can be used. In (Till *et al.*, 2003) it has been shown exemplarily for the approach in (Stursberg *et al.*, 2002) that a drawback is the somewhat limited applicability when it comes to efficient solution for large-scale systems. The investigation allows the conclusion that one reason for the complexity is the fact that the MILP method does not restrict the search for the optimum to the true degrees of freedom (the model inputs) but has to cope with a large number of auxiliary integer variables and constraints introduced to map logical into algebraic constraints.

As an alternative, this paper suggest a method with the following characteristics: (a) the discrete degrees of freedom are determined by a tree search algorithm with tailor-made heuristics to determine the optimal discrete control sequence with low effort, (b) the continuous degrees of freedom are obtained from solving embedded (nonlinear) programming problems, (c) the cost function is evaluated by hybrid simulation which takes care of the state-dependent structural changes of the model.

2. HYBRID SYSTEM OPTIMIZATION

The type of model used to represent the dynamics to be considered for the transition procedure is first defined precisely. As argued in (Stursberg and Engell, 2002), the following type of hybrid automaton is suited for this purpose since it accounts for different nonlinear dynamics assigned to certain regions of the continuous state space (or different modes of operation), and for the combination of continuous and discrete inputs. The latter are useful to model, e.g., actuators with an 'on/off'-type of behavior.

Definition 2.1. Hybrid Automaton A
The hybrid automaton $A = (X, U, V, Z, inv, T, g, r, f)$ consists of the following components:

- the continuous state space $X \subseteq \mathbb{R}^{n_x}$ on which the *state vector* x is defined [1];
- the continuous input space $U = [u_1^-, u_1^+] \times \ldots [u_{n_u}^-, u_{n_u}^+]$ with $u_j^-, u_j^+ \in \mathbb{R}$ and the *continuous inputs* $u \in U$;
- the finite discrete input space $V = \{v_1, \ldots, v_{n_d}\}$ with *discrete inputs* $v \in V$ for which $v_j \in \mathbb{R}^{n_v}$;
- the finite set of *locations* $Z = \{z_1, \ldots, z_{n_z}\}$;
- a mapping $inv : Z \to 2^X$ which assigns a non-empty *invariant* of the form $inv(z_j) = \{x \mid \exists\, n_{p_j} \in \mathbb{N}, C_j \in \mathbb{R}^{n_{p_j} \times n_x}, d_j \in \mathbb{R}^{n_{p_j}}, x \in X : C_j \cdot x \leq d_j\}$ to each location $z_j \in Z$;
- the set of *transitions* $\Theta \subseteq Z \times Z$, where a transition from $z_1 \in Z$ into $z_2 \in Z$ is denoted by (z_1, z_2);
- a mapping $g : \Theta \to 2^X$ that associates a *guard* $g((z_1, z_2)) \subseteq X$ with each $(z_1, z_2) \in \Theta$ such that: $g((z_1, z_2)_j) = \{x \mid \exists\, n_{g_j} \in \mathbb{N}, C_j \in \mathbb{R}^{n_{g_j} \times n_x}, d_j \in \mathbb{R}^{n_{g_j}} : C_j \cdot x \leq d_j\}$. Given a state $z_1 \in Z$, it is required for all pairs of transitions originating from z_1 that the corresponding guards are disjoint;
- a *reset function* $r : \Theta \times X \to X$ which assigns an updated state $x' \in X$ to each $(z_1, z_2) \in \Theta$ and $x \in g((z_1, z_2))$;
- a *flow function* $f : Z \times X \times U \times V \to \mathbb{R}^n$ that defines a continuous vector field $\dot{x} = f(z, x, u, v)$ for each location $z \in Z$.

Semantics: Let $T = \{t_0, t_1, t_2, \ldots\}$ be an ordered set of time points $t_k \in \mathbb{R}^{\geq 0}$ which contains the initial time t_0, and all points of time at which an input change or a transition occurs. The states $x_k := x(t_k)$, $z_k := z(t_k)$ and the inputs $u_k = u(t_k)$, $v_k = v(t_k)$ are defined on T, i.e. their values are piecewise constant on intervals $[t_k, t_{k+1}[$, $k \in \mathbb{N} \cup \{0\}$. For given input sequences $u(t_k)$ and $v(t_k)$, a *feasible run* ϕ_σ of A is then defined as a sequence: $\phi_\sigma = (\sigma(t_0), \sigma(t_1), \sigma(t_2), \ldots)$ of hybrid states $\sigma_k := \sigma(t_k) = (z_k, x_k)$ such that:

- *Initialization*: $\sigma(t_0)$ is initialized to a given $z_0 = z(t_0) \in Z$ and $x_0 = x(t_0) \in X$ with $x_0 \in inv(z_0)$, $x_0 \notin g((z_0, \bullet))$ for any $(z_0, \bullet) \in \Theta$.
- *Progress*: $\sigma(t_{k+1})$ results from $\sigma(t_k)$ by:
  (i) continuous evolution: $\chi : [0, \tau] \to X$, $\tau \in \mathbb{R}^{>0}$ where $\chi(t_0) = x_k$, $\dot{\chi}(t) = f(z_k, \chi(t), u_k, v_k)$ with an existing unique solution for $t \in [0, \tau]$, and $\chi(t) \in inv(z_k)$ but for all $g(z_k, \bullet) \in \Theta$ and $t \in [0, \tau[$: $\chi(t) \notin g(z_k, \bullet)$;
  (ii) followed by a transition: $(z_k, z_{k+1}) \in \Theta$, $\chi(\tau) \in g(z_k, z_{k+1})$, and $x_{k+1} = r((z_k, z_{k+1}), \chi(\tau)) \in inv(z_{k+1})$. □

For the procedure developed in the next section it is important to note that a run of A is deterministic for given input trajectories. In particular, a transition is taken immediately if the correspond-

[1] For all parameters $n_\bullet$ in Def. 2.1 applies $n_\bullet \in \mathbb{N}$.

ing unique guard condition is met. Compared to the model used before in this context (Stursberg and Engell, 2002), the automaton in Def. 2.1 comprises two extensions: (a) the invariants $inv(z)$ do not necessarily establish a partition of the state space X, but can possibly overlap, such that different dynamics can be valid for the same region of X if different operating conditions apply; (b) if a transition is taken, the continuous state can be updated by the reset function r.

The objective of optimizing transition procedures of processing systems is now cast into a minimization problem employing the model according to Def. 2.1 as constraints. To be optimized is the procedure of driving the system from an initial state $\sigma(t_0)$ into a set of hybrid target states, defined as $\Sigma_{tar} = (z_{tar}, X_{tar})$ with $z_{tar} \in Z$ and $X_{tar} \subset inv(z_{tar})$. An additional requirement for the transition procedure is that a set of forbidden regions $F = \{F_1, \ldots, F_{n_F}\}$ is never entered, where each region $F_j = (z_{F,j}, X_{F,j})$ is a hybrid state set with $X_{F,j} \subset inv(z_{F,j})$. These regions are suitable to model the exclusion of undesired operating conditions, e.g., the concentration ranges of explosive mixtures.

The task is then to find the input trajectories that minimize a given cost functional Ω for the transition procedure. For simplicity of notation, $\phi_u = (u_0, u_1, \ldots, u_f)$ is written for a continuous input trajectory on a finite time set $T = \{t_0, \ldots, t_f\}$ and Φ_u for the set of all input trajectories of this length; likewise, $\phi_v = (v_0, v_1, \ldots, v_f)$ is a discrete input trajectory and Φ_v contains all ϕ_v. Given the sets Φ_u and Φ_v, the set of all corresponding feasible runs[2] is denoted by Φ_σ. The objective is to determine the input trajectories ϕ_u^*, ϕ_v^* that lead to a feasible run ϕ_σ^* which minimizes the cost function according to:

$$\min_{\phi_u \in \Phi_u, \phi_v \in \Phi_v} \Omega(t_f, \phi_\sigma, \phi_u, \phi_v) \tag{1}$$

$$\begin{aligned} s.t.\quad & \phi_\sigma = (\sigma_0, \ldots, \sigma_f) \text{ with: } \sigma_0 = (z_0, x_0), \\ & \sigma_f := (z(t_f), x(t_f)) \in \Sigma_{tar}, \text{ and for } \phi_\sigma \\ & \text{applies in each phase of cont. evolution} \\ & \text{(Def. 2.1)} : \chi(t) \notin F_j \ \forall\, F_j \in F,\ \forall\, t \in [0, \tau]. \end{aligned}$$

A slightly simplified version of this problem has been tackled in (Stursberg and Engell, 2002; Stursberg *et al.*, 2002) by an approach comprising the following steps: (a) the flow functions are approximated by discrete-time and linear models, (b) all logical parts of the model are reformulated in algebraic linear form such that a mixed-integer linear program results, (c) the optimization problem is solved in an MPC-like fashion where mixed-integer linear programming (MILP) is applied on a limited look-ahead horizon in each step.

[2] Of course, some combinations of ϕ_u and ϕ_v do not lead to feasible runs.

It has been identified as a drawback of this technique that the transformation into algebraic form requires a large set of auxiliary integer variables (and auxiliary constraints). When applying branch-and-bound techniques for MILP, the search tree encoding the discrete alternatives thus is considerably larger than one would expect from the true set of discrete degrees of freedom (i.e., the cardinality of V for each time point in T). As a consequence for larger systems, the method could be applied only for rather short prediction horizons. This is the motivation for an alternative procedure that does not use a completely algebraic formulation of the transition dynamics of A.

3. A GRAPH SEARCH ALGORITHM

The key idea is to separate the optimization of the continuous and discrete degrees of freedom in the following sense: The discrete choices (i. e., the input trajectories ϕ_v) are determined by a graph search algorithm resembling the well-known principle of shortest-path search. For each node contained in the search graph, an embedded optimization for the continuous degrees of freedom (and optionally for relaxed discrete degrees of freedom for future steps) is carried out. Within this embedded nonlinear programming, numerical simulation is employed to evaluate the hybrid dynamics of A, leading to a cost evaluation for the referring system evolution. These costs are used in the graph search to apply a branch-and-bound strategy, i.e., upper (and lower) bounds on the optimal costs for the transition procedure are iteratively computed to prune branches of the search tree as soon as possible.

In detail, the algorithm operates on a directed acyclic graph in which the nodes represent a system state at a time t_k for the case that a discrete input $v \in V$ has been applied in the previous step. As illustrated in Fig. 1, the graph can be understood as a representation for the hybrid states reachable over the time set T if all possible discrete input trajectories are applied. Clearly, the size of the graph grows exponentially with the number of discrete decisions, i.e. according to $|V|^{|T|-1}$ – hence, an objective of the optimization algorithm is to prune branches as quickly as possible. The information stored with each node is a structure $n = (\phi_\sigma, \phi_u, \phi_v, c_a, c_p)$, i.e. it contains the hybrid state trajectory ϕ_σ up to the current state (z_k, x_k), the input trajectories ϕ_u and ϕ_v that led to this state, as well as the costs c_a accumulated on the path into (z_k, x_k), and (optionally) a prediction c_p for the costs encountered on the remaining path into the target.

For a state trajectory ϕ_σ that leads from the initial state to the target, a *solution* is defined as a tuple

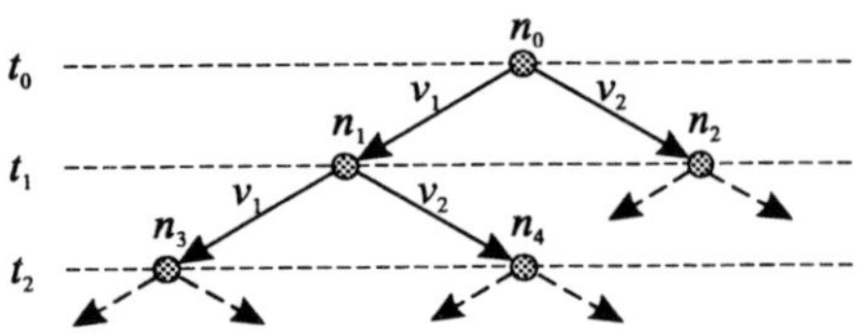

Fig. 1. Graph encoding the discrete choices for $V = \{v_1, v_2\}$.

$\phi_{sol} = (\phi_\sigma, \phi_u, \phi_v, c_a)$, and Φ_{sol} as the set of all found solutions. If, furthermore, the initial node is specified as $n_0 = (\sigma_0, -, -, 0, \infty)$ and n_{max} as the number of nodes to be explored maximally, the algorithm can be written as shown in Fig. 3: The set N denotes the set of all feasible nodes explored during the algorithm, L is the set of *live nodes* starting from which the system evolution still has to be investigated further, G is the set of nodes generated in the last iteration, and S are the nodes selected from L or G in the current iteration. This selection is realized by a function *select* which returns S depending on a chosen search mode. In an implementation of the algorithm the modes *breadth-first* ($S := L$), *depth-first* ($S := \{n_{best}\}$, where n_{best} is the node from G with minimum cost), and *best-first* ($S := \{n_{best}\}$ with n_{best} as the best node from L) can be chosen. In experiments

$N = L = G := \{n_0\};$
$S := \emptyset;\quad \Phi_{sol} := \emptyset;$
WHILE $(L \neq \emptyset) \wedge (|N| < n_{max})$ DO {
  $S := select(L, G);$
  $G := \emptyset;$
  FOR ALL $(n \in S)$ DO {
    IF $(n.c_a \leq ub)$:
      FOR ALL $(v \in V)$ DO {
        $(\sigma, u, \hat{\phi}_x, \hat{\phi}_u, \hat{\phi}_v) = extend(n, v);$
        $\phi_\sigma := (\phi_\sigma, \sigma);$
        $\phi_u := (\phi_u, u);$
        $\phi_v := (\phi_v, v);$
        $c_a = history(\phi_\sigma, \phi_u, \phi_v);$
        $c_p = future(\hat{\phi}_\sigma, \hat{\phi}_u, \hat{\phi}_v);$
        $n_{new} = (\phi_\sigma, \phi_u, \phi_v, c_a, c_p);$
        IF $(n_{new} \notin N) \wedge (feasible(\sigma))$:
          IF $(\sigma \in \Sigma_{tar})$:
            $N := N \cup \{n_{new}\};$
            $\Phi_{sol} := \Phi_{sol} \cup \{(\phi_\sigma, \phi_u, \phi_v, c_a)\};$
            IF $(c_a < ub)$: $ub := c_a$; END
          ELSE:
            IF $(c_a < ub)$:
              $N := N \cup \{n_{new}\}; G := G \cup \{n_{new}\};$
            END
          END
        END }
    END }
  $L := (L \cup G) \setminus S;$ }
$\phi^*_{sol} = \min_{\phi_{sol} \in \Phi_{sol}} \phi_{sol}.c_a$

Fig. 2. The optimization algorithm.

for different examples the following strategy has been found to be efficient: (a) the depth-first mode is applied first to find a feasible solution quickly (and thus start the pruning over upper bounds, see below), and then the procedure continues with the best-first or breadth-first mode; (b) the selection of n_{best} in both modes is based on the notion of a minimum distance of the current state σ_k to Σ_{tar}. These costs include, e.g., a Euclidean distance in X and a distance in Z defined as the number of transitions in Θ to be taken in order to reach z_{tar}.

The algorithm proceeds with computing successors for each $v \in V$ for all those nodes n in S, for which the costs $n.c_a$ accumulated along the path from n_0 to n do not exceed an upper bound ub for the costs for the complete path from n_0 to the target. Branches with $n.c_a > ub$ are pruned at this point. The computation of successors is accomplished by the function *extend*. It performs for the node n (reached at time t_k) and the corresponding input v_k the following optimization for an ordered time set $T' = \{t_k, \ldots, t_p\}$ with $p > k$:

$$\min_{\hat{\phi}_u, \hat{\phi}_v} \Omega(t_p, \hat{\phi}_\sigma, \hat{\phi}_u, \hat{\phi}_v) \tag{2}$$

where $\hat{\phi}_u = (u_k, u_{k+1}, \ldots, u_{p-1})$ and $\hat{\phi}_v = (v_{k+1}, \ldots, v_{p-1})$. It is important for the components of the trajectory $\hat{\phi}_v$ that they are not restricted to the set V, but each component of the vector v is relaxed to the range between the minimum and maximum value that occurs in V for the respective component. The optimization in Eq. 2 does then not contain any discrete degrees of freedom anymore, and can be solved by standard techniques for nonlinear programming. During this optimization, the solver computes the state trajectory $\hat{\phi}_x = (x_k, x_{k+1}, \ldots, x_p)$ that belongs to a guess for $\hat{\phi}_u$ and $\hat{\phi}_v$. This task can be solved by numerical simulation of the corresponding run of A and involves the following steps: (a) evaluating the continuous evolution, (b) detecting that a guard condition is satisfied, and (c) executing the transition with the associated reset function. The values of $\hat{\phi}_x$, $\hat{\phi}_u$, and $\hat{\phi}_v$ determine the costs $\Omega(t_p, \hat{\phi}_\sigma, \hat{\phi}_u, \hat{\phi}_v)$. As a result of the optimization, the function *extend* returns a hybrid successor state σ_{k+1}, the continuous input u_k to get there, and the predicted trajectories over the complete prediction horizon T'.

The function *history* determines the accumulated costs c_a from the initial state to the new hybrid state. A function *future*, on the other hand, computes future costs c_p over the predicted trajectories $\hat{\phi}_x$, $\hat{\phi}_u$, and $\hat{\phi}_v$. In the simplest case, c_p can just encode the distance of the last state in $\hat{\phi}_x$ to the target, i.e., it represents an estimation of how close the system can get to the target within the horizon T'. c_p is used by *select* to steer the exploration of the graph. The values c_p and

c_a together with the extended state and input trajectories determine the new node n_{new}.

If n_{new} was not contained in N before and if it is feasible, i.e., σ_{k+1} is within X but not in any forbidden region, it is considered for inclusion in N. If, in addition, σ_{k+1} is in the target, a new solution has been found and is included in Φ_{sol}. If the accumulated cost of n_{new} is below the current value of ub, this bound is updated. If on the other hand $\sigma_{k+1} \notin \Sigma_{tar}$ and the accumulated costs $n_{new}.c_a$ are still below ub, the node is included in N and G. If $n_{new}.c_a \geq ub$, this path is not further considered. At the end of an iteration the set L of live nodes is updated. The loop continues until L is empty, or the maximum number of nodes n_{max} is exceeded. If, at this stage, the set Φ_{sol} is non-empty, the element ϕ^*_{sol} with the smallest accumulated costs is the best solution found.

Two extensions of the algorithm are currently investigated: (i) A second important criterion for efficiently pruning the search graph is the use of lower bounds lb for the costs of the complete transition procedure: If the predicted state trajectory $\hat{\phi}_\sigma$ ends in Σ_{tar}, the associated costs c_p are a lower bound for the costs of the remaining path into Σ_{tar} (since the discrete degrees of freedom are relaxed to continuous ones). Hence, if the sum of c_a and c_p for a node is greater than the current value of ub, the node can not be an element of the optimal path, and the branch of the graph can be pruned. For certain cost functions (as, e. g., time-optimality of the transition procedure) this pruning can be applied even if $\hat{\phi}_\sigma$ does not end in Σ_{tar}. This means that the criterion is applicable also if the embedded nonlinear programming leads to solutions only for rather short prediction horizons T'. (ii) Another use of the result of the function *extend* is to take $\hat{\phi}_v$ as a heuristics for further exploring the subgraph emerging from the current node. As an obvious choice, one would first follow that sequence of discrete inputs ϕ_v in which the components are nearest to the relaxed input values of $\hat{\phi}_v$.

4. APPLICATION TO A CSTR START-UP

The method is illustrated by using the example of the start-up of a continuous stirred tank reactor (CSTR), as described in (Stursberg and Panek, 2002). The system consists of a tank equipped with two inlets, a heating coil, a cooling jacket, a stirrer, and one outlet (see Fig. 4). The inlets feed the reactor with two dissolved substances A and B which react exothermically to form a product D. The inlet flows F_1 and F_2 (with temperatures T_1 and T_2) can be switched discretely between two values each. The outlet flow F_3 is controlled continuously. In order to heat

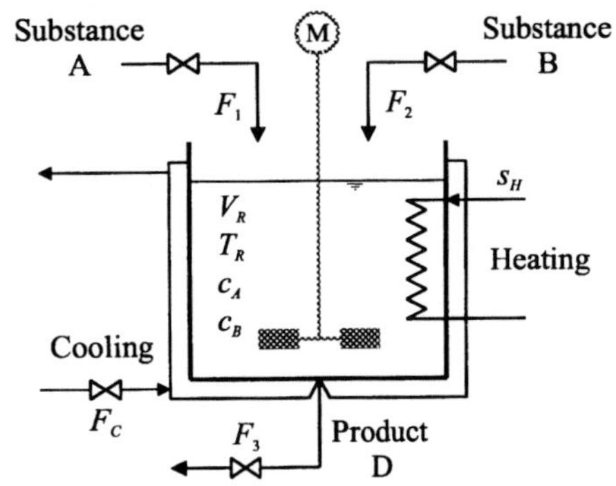

Fig. 3. Scheme of the CSTR.

up the reaction mixture to a desired temperature range with a high reaction rate, the heating can be switched on (denoted by a discrete variable $s_H \in \{0,1\}$). The continuously controlled cooling flow F_C serves as a means to remove an excess of heat once the reaction has started. It is the objective for this system to determine the input trajectories that drive the initially empty reactor into a status of desired operation in which the liquid volume V_R, the temperature T_R, and the concentrations c_A and c_B have reached nominal ranges. Additionally, regions with $T_R \geq 360$ and $V_R \geq 1.6$ are forbidden.

To model the system, the state vector is defined as $x := (V_R, T_R, c_A, c_B)^{\mathrm{T}}$, the continuous input vector as $u := (F_3, F_C)^{\mathrm{T}}$, and the discrete input vector as $v := (F_1, F_2, s_H)^{\mathrm{T}}$. Depending on the continuous state, the system dynamics can be written as $\dot{x} = f(z, x, u, v)$ where:

- for z_1 with $V_R \in [0.1, 0.8]$:

$$f^I = \begin{pmatrix} F_1 + F_2 - F_3 \\ (F_1(T_1 - T_R) + F_2(T_2 - T_R))/V_R \\ \quad + F_C k_1(T_C - T_R)(k_2/V_R + k_3) - k_4 q \\ (F_1 c_{A,1} - c_A(F_1 + F_2))/V_R + k_9 q \\ (F_2 c_{B,2} - c_B(F_1 + F_2))/V_R + k_{10} q \end{pmatrix}$$

- for z_2 with $V_R \in]0.8, 2.2]$:

$$f^{II} = \left(f_1^I,\ f_2^I + s_H k_6(T_H - T_R)(k_7 - \frac{k_8}{V_R}),\ f_3^I,\ f_4^I \right)^{\mathrm{T}}$$

and $q = c_A c_B^2 \exp(-k_5/T_R)$. The separation in two V_R-regions accounts for the fact that the heating is only effective above $V_R = 0.8$. The initial state is $x_0 = (0.1, 300, 0, 0)^{\mathrm{T}}$ and the target is given by z_2 and a hyper-ball with radius 0.1 around the continuous state $x_{tar} = (1.5, 345, 0.4, 0.2)^{\mathrm{T}}$. The optimization is run with the cost criterion that the transition time for the startup procedure is minimized. The strategy is chosen that depth-first search is used until a solution is found, then a breadth-first strategy is applied. For selecting the nodes from L and G, the criterion of minimum distance to the target is selected. Figure 4 shows the best solution obtained for a search with $n_{max} = 400$ nodes and with $|T'| = 2$. The result has been obtained within 125 seconds and is qualitatively comparable to the one obtained in 13 minutes with the method in (Stursberg *et al.*, 2002).

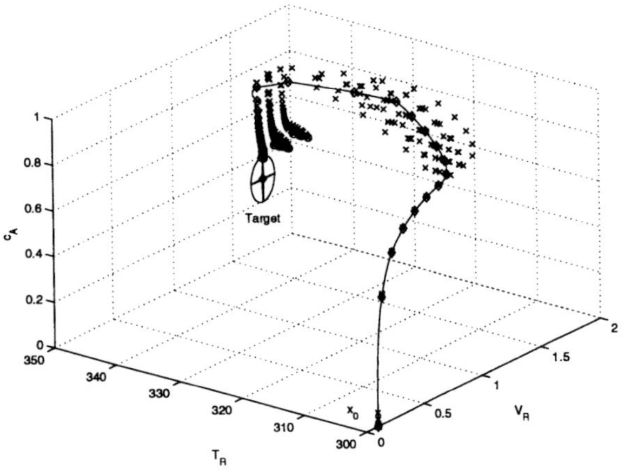

Fig. 4. CSTR: The optimal x-trajectory (solid line) is projected into the (V_R, T_R, c_a)-space. Explored nodes are marked by crosses.

5. CONCLUSIONS

Since space limitations exclude a detailed assessment of the proposed algorithm (including a comparison to other techniques), only the main differences to the approach in (Stursberg and Engell, 2002) are summarized briefly: (a) the method introduced in this paper evaluates directly the original hybrid model A and does not operate with linearized dynamics; (b) the hybrid model is extended with respect to the location invariants and resets (whereas not relevant for the example in Sec. 4); (c) the search is restricted to the true degrees of freedom and does not have to cope with auxiliary variables and constraints; (d) experiments also for a scalable example treated in (Till *et al.*, 2003) have shown that the computation times are by an order of magnitude smaller for the majority of tested configurations, (e) unlike the MPC-like approach in (Stursberg and Engell, 2002), which discards all but the best trajectory in each iteration, the graph search stores alternative paths in the set of live nodes for later exploration.

The focus of the current work is to introduce suitable heuristics to steer the graph search, to develop a scheme for adapting the time steps (rather than using constant sampling intervals), and to use tailor-made modifications for the embedded nonlinear programming to improve its performance.

6. REFERENCES

Avraam, M. P., R. Shah and C. C. Pantelides (1998). Modelling and optimisation of general hybrid systems in the cont. time domain. *Comp. Chem. Eng.* **22 (Suppl.)**, 221–228.

Barton, P.I. and C.K. Lee (2002). Modeling, simulation, sensitivity analysis and optimization of hybrid systems. *ACM Trans. Modeling and Comp. Simulation* **12**(4), 256–289.

Bemporad, A. and M. Morari (1999). Control of systems integrating logic, dynamics, and constraints. *Automatica* **35**(3), 407–427.

Branicky, M. S., V. S. Borkar and S. K. Mitter (1998). A unified framework for hybrid control: Model and optimal control theory. *IEEE Trans. Automatic Control* **43**(1), 31–45.

Broucke, M., M.D. Di Benedetto, S. Di Gennaro and A. Sangiovanni-Vincentelli (2000). Theory of optimal control using bisimulations. In: *Hybrid Systems: Comp. and Control.* Vol. 1790 of *LNCS.* Springer. pp. 89–102.

Gokbayrak, K. and C. G. Cassandras (2000). Hybrid controllers for hierarchically decomposed systems. In: *Hybrid Systems.* Vol. 1790 of *LNCS.* Springer. pp. 117–129.

Hedlund, S. and A. Rantzer (2002). Convex dynamic programming for hybrid systems. *IEEE TAC* **47**(9), 1539–1540.

Lynch, N., R. Segala and F. Vaandrager (2003). Hybrid i/o automata. *Information and Computation* **185**(1), 105–157.

Oldenburg, J., W. Marquardt, D. Heinz and D. Leineweber (2003). Mixed logic dynamic optimization applied to batch distillation process design. *AICHE Journ.* **49**, 2900–2917.

Shaikh, M.S. and P.E. Caines (2003). On the optimal control of hybrid systems. In: *Hybrid Systems: Comp. and Control.* Vol. 2623 of *LNCS.* Springer. pp. 466–481.

Stursberg, O. and S. Engell (2002). Optimal control of switched continuous systems using mixed-integer programming. In: *Proc. 15th IFAC World Congr. on Automatic Control.* Vol. Th-A06-4.

Stursberg, O. and S. Panek (2002). Control of switched hybrid systems based on disjunctive formulations. In: *Hybrid Systems.* Vol. 2289 of *LNCS.* Springer. pp. 421–435.

Stursberg, O., S. Panek, J. Till and S. Engell (2002). Generation of optimal control policies for systems with switched hybrid dynamics. In: *Modelling, Analysis, and Design of Hybrid Systems.* Vol. 279 of *LNCIS.* Springer. pp. 337–352.

Till, J., S. Engell, S. Panek and O. Stursberg (2003). Empirical complexity analysis of a milp-approach for optimization of hybrid systems. In: *IFAC Conf. Analysis and Design of Hybrid Systems.* pp. 159–164.

ELSEVIER
IFAC
PUBLICATIONS
www.elsevier.com/locate/ifac

Neural Predictive Control Design for Uncertain Nonlinear Systems

Wei Wu[1,*] and Jun-Xian Chang[2]

[1]*Department of Chemical Engineering, National Yunlin University of Science and Technology, Touliu, Yunlin 640, Taiwan, R.O.C.*
[2]*Department of Electrical Engineering, National Yunlin University of Science and Technology, Touliu, Yunlin 640, Taiwan, R.O.C.*

Abstract

Abstract : In this paper, adaptive neural-network predictive control strategies for general nonlinear systems are presented. The system is described by an unknown NARMAX model and neuro model is used to on-line learn the system. Despite state/parameter estimation, the neural predictive control scheme associated with the constrained optimization framework is implemented in a straightforward manner. Through the Lyapunov stability analysis, the network weight adaptation rule is derived, and guarantees the minimum error between the neuro output and plant output. An unstable reactor system is given to demonstrate the effectiveness of the proposed control schemes. *Copyright © 2004 IFAC*

Keywords: Predictive control; Neural network; Input constraints; Dynamic backpropagation algorithm; Reactor systems

1. Introduction

Research on neural-network-based control systems has received a considerable attention over the past several years [14]. It is because many control systems often exhibit strongly nonlinearities, and the implicit programming of neural network computing architectures is able to accurately approximate any nonlinear function. Since uncertain nonlinear systems without complete model information are usually called as "black boxes", the widely used structures of neural-network-based control designs lead to problems in nonlinear and adaptive control fashions. Ge et al. [5] provided a framework for structured dynamic modelling and adaptive control design for robots using neural networks, Kulawski and Brdyś [9] developed adaptive control technique for nonlinear stable systems using recurrent neural networks, and Calise et al. [2] presented an adaptive output feedback design procedure, where the design employed the feedback linearization coupled with an on-line neural network.

As it is mentioned above, the neuro identification process turns out to be of the central parts in constructing feasible adaptive controllers. However, the adaptive control combined with neuro-identifier would reduce the efficient and reliable computation due to the hybrid convergence manner. Recently, Poznyak et al. [17] indicated that 'dynamic' neural network architectures were successfully used to enhance tracking controllers, Poznyak et al. [18] proposed optimal adaptive controller based on online learning of the neural network parameters to realize the good tracking performance, Ku and Lee [10] developed a diagonal recurrent neural network to generalize the structure of the tracking controllers, and Gao et al. [3] provided a qualified real-time implementation to elaborate the merit of the diagonal recurrent neural network. Obviously, the adaptive neuro-based identifier can create robust training algorithms and reinforce the robustness of adaptive control techniques [23].

Recently, the study of neuro-optimizers has been mentioned in order to make the system identification more efficient, and the neural network is more expected to determine controller parameters. Whereas the predictive control technique is a dynamic optimization approach to control problems, and the flexible constraint-handling capability makes it most suitable for process control problems, Lazar and Pastravanu [11] indicated that the neural predictive control could reduce the significant obstacles encountered in conventional model predictive control (MPC) applications, Noriega and Wang [16] developed the adaptive neural-network control scheme within the unconstrained optimization framework, Vila and Wanger [20] used the feedforward neural network (FNN) to generate a constrained optimal control at each time step, and Wang and Wan [21] provided the high computational demand in solving the optimization problems associated with the MPC technique and structured neural network approach. In other respects, Hoo et al. [6] established a reliable estimate by incorporating directional information to improve the predictive capability of neural network models, and Ahmed and Tasadduq [1] developed neural servocontroller for a nonlinear multi-input multi-output (MIMO) system with the strong interaction of the input-output pairs.

*Author to whom correspondence should be addressed.
Email: weiwu@yuntech.edu.tw

In this paper, the flexible predictive control strategy using on-line neuro-based adaptation is developed. The control design procedure is implemented to highly nonlinear systems in the presence of parameter uncertainties and input constraints, that does not rely on state/parameter estimation. The resulting implementation of neural predictive controller is able to ensure a rapid and reliable solution to the control algorithm, so it is superior to 'deterministic' model-based predictive control applications.

2. Problem Statement

Consider the discrete-time uncertain system described by the NARMAX model [4] is shown as

$$y(k+1) = f_p(y(k), y(k-1), \cdots, y(k-n+1), \\ u(k), u(k-1), \cdots, u(k-m+1)) \tag{1}$$

where $u(k)$ and $y(k)$ are measurable scalar input and output, respectively. n is the number of past outputs used, m is the number of past inputs used, and $f_p(\bullet)$ is the unknown nonlinear difference equation representing the system dynamics. Since the neural network architecture can provide the accurate approximation, the off-line model identification is recommended that the training procedure for a class of FNN architectures corresponding to the NARMAX model associated with adequately delayed inputs and optimum weights is addressed [16].

2.1 Off-line neuro identification

The neural NARMAX model is given by

$$y_{NN}(k+1) = \hat{f}(Y(k-d_y), U(k-d_u)) \tag{2}$$

where $\hat{f}(\bullet)$ represents the input-output transfer function of the neural network, d_y and d_u are the transport delays of the input space, and the vector forms are shown as

$$U(k-d_u) = [u(k-d_u), \cdots, u(k-d_u-m+1)]^T \\ Y(k-d_y) = [y(k-d_y), \cdots, y(k-d_y-n+1)]^T \tag{3}$$

Remark 1. The model orders for the input and output and delays need to be determined for the non-recurrent neural modeling [22]. In this paper, a three-layer FNN with one net output and n+m+1 input is given by [13]

$$\hat{f}(z) = (W^O)^T \sigma[(W^I)^T z + W^B] \tag{4}$$

where $z = [U^T, Y^T, 1]^T$, and the activation function $\sigma(\bullet)$ is set as

$$\sigma(z_a) = \tanh(z_a), \quad z_a \in \Re \tag{5}$$

$W^I \in \Re^{m\times(m+n+1)}$ is the first-to-second layer interconnection weight, $W^O \in \Re^{1\times m}$ is the second-to-third layer interconnection weight, and $W^B \in \Re^{m\times 1}$ is the threshold offset. In general, the learning algorithm consists in sequentially adjusting the network weights for the minimal mean squared error between the nominal NARMAX model $f_p(\bullet)$ and the neural NARMAX model $\hat{f}(\bullet)$. Moreover, the time validation test of the FNN model is employed for new input signals required. Through the off-line training and validation algorithms, the nominal FNN model is established and provides a stable input-output representation of neural networks.

2.2 Neural predictive control algorithm

The objective of the predictive control strategy using FNN model is twofold: (i) to fulfill the output feedback control design and (ii) to minimize an objective function with a quadratic penalty function. Thus, the FNN model-based control problem is described by

$$\min_{u(k),\ldots,u(k+N_u-1)} \sum_{i=k}^{k+N_y} [y_{NN}(i) - r(i)]^2 + \lambda \sum_{j=k}^{k+N_u-1} [\Delta u(j)]^2 \tag{6}$$

subject to input constraint

$$u_{\min} \le u(j) \le u_{\max}, j=k, \ldots, k+N_u-1. \tag{7}$$

where Δu presents the manipulated variable increment, N_u is the control horizon, N_y is the prediction horizon, r is the reference trajectory, and λ is the weight factor. $u_{\min}$ and $u_{\max}$ represent lower and upper bounds, respectively. The output of FNN model is shown as

$$y_{NN}(i) = (\tilde{W}^O)^T \sigma[(\tilde{W}^I)^T z + \tilde{W}^B], \; i=k, \ldots, k+N_y \tag{8}$$

$(\tilde{W}^O, \tilde{W}^I, \tilde{W}^B)$ in (11) represent optimal values of the network weights. Note that the update of weights is corrected during the off-line training procedure.

Remark 3. Obviously, the above augmented objective function has been shown in [8, 13]. Because the quadratic problem (QP) framework is a nonlinear algebraic equation, the optimization toolbox in the Matlab environment provides the accurate computation. Inspired by the issue in [11], the one-step-ahead predictive control approach for the QP framework with N_y=1 and the single parameter λ in (9) would induce a uniformly convex function so that a rapid, reliable solution to the control algorithm can be achieved.

Remark 4. The tuning scheme of FNN model-based predictive control usually depends on the number of output prediction N_y and the weight factor λ. However, the tuning parameter N_y cannot effectively improve the control performance when the FNN model cannot be faithful representation of the uncertain system. Consider that the control horizon N_u is one and the control action is bounded by (10), Kambhampati et al. [7] indicated that the one-step-ahead predictive control design with the addition of the penalty function could be implemented to an open-loop unstable process.

3. Adaptive Neural-Network Predictive Controller

In fact, the above FNN model-based identification cannot instantly capture the real dynamic behavior. In this section, the on-line training procedure associated with network weight adaptation rule is utilized to establish an adaptive neural-network predictive control.

3.1 On-line neuro identification

An approach for on-line identification using the FNN-based predictive model in (11) is presented. The current square error between the plant output and the neural output is defined as

$$E_m(k) = \frac{1}{2}(y(k) - y_{NN}(k))^2 \tag{9}$$

and the learning law by a gradient method [17] for the minimization of $E_m(k)$ is written as

$$\pi(k+1) = \pi(k) + \eta(-\frac{\partial E_m}{\partial \pi}) \tag{10}$$

where η is a learning rate, and

$$\frac{\partial E_m(k)}{\partial \pi} = -(y(k) - y_{NN}(k))\frac{\partial y_{NN}(k)}{\partial \pi} \tag{11}$$

Obviously, from (13) and (14) it is a typically dynamic backpropagation (DBP) algorithm. According to the FNN architecture in (11), the learning rules of the weights are obtained as follows:

(1) Output weights:

$$w_i^O(k+1) = w_i^O(k) + \eta(y(k) - y_{NN}(k))\sigma(s_i(k)) \tag{12}$$

where $s_i(k) \equiv \sum_j w_{i,j}^I z_j(k) + w_i^B$.

(2) Input weights:

$$w_{i,j}^I(k+1) = w_{i,j}^I(k) + \eta(y(k) - y_{NN}(k))[w_i^O Q_{ij}(k)] \tag{13}$$

where $Q_{ij}(k) \equiv \sigma'(s_i(k))z_j(k)$ and $Q_{ij}(0) = 0$.

(3) Bias weights:

$$w_i^B(k+1) = w_i^B(k) + \eta(y(k) - y_{NN}(k))[w_i^O R_i(k)] \tag{14}$$

where $R_i(k) \equiv \sigma'(s_i(k))$ and $R_i(0) = 0$.

Remark 5. All weights are split up into the three-layer FNN architecture, Those tuning algorithms are used in the on-line fashion. The on-line update of the weights captures changes in the process dynamics. Apparently, the DBP algorithm can be treated as an adaptation approach. Referring the traditional adaptive mechanism in [19], the performance of those learning rules will rely critically on the choice of the learning rate η. However, the development of guideline in selecting the learning rate is omitted here due to the similar results which have been stated in [10]. For a small value of η, the convergence is guaranteed but its rate is slow. Oppositely, the too big η will induce the unstable updating algorithm. In the next demonstrated examples depend on the on-line tuning scheme for the parameter λ but the learning rate η is fixed.

3.2 Control algorithm

Inspired by the neural predictive control strategy in (9) and the DBP algorithm in (15)-(17), the one-step-ahead predictive controller using the on-line neuro identification is described by

$$\min_{u(k)} J_1(\lambda) = [\hat{y}_{NN}(k+1) - r(k+1)]^2 + \lambda \Delta u(k)^2 \tag{15}$$

$$\min_{\pi} J_2(\eta) = [\hat{y}_{NN}(k) - y(k)]^2 \tag{16}$$

subject to the current input constraint

$$u_{\min} \le u(k) \le u_{\max} \tag{17}$$

where the one-step-ahead FNN model is written as

$$\hat{y}_{NN}(k+1) = (W^O)^T \sigma[(W^I)^T z + W^B] \tag{18}$$

(W^O, W^I, W^B) in (18) represent updated network weights connected to optimal network weights in (11) as initials.

Remark 6. Obviously, the control scheme is dynamic and adaptable. Since the solution for minimization of J_2 is explicit, the DBP algorithm can provide the asymptotic convergence by adjusting a proper parameter η . Under the one-step-ahead predictive horizon and input constraint, the optimizer not only carries out the minimal tracking error, i.e., $\lim_{k\to\infty} \hat{y}_{NN}(k) \approx y(k) \approx r(k)$, but also the tuning parameter λ directly affects the closed-loop tracking performance and stability. Moreover, this adaptive neural-network predictive control (ANNPC) architecture is depicted in Fig. 1.

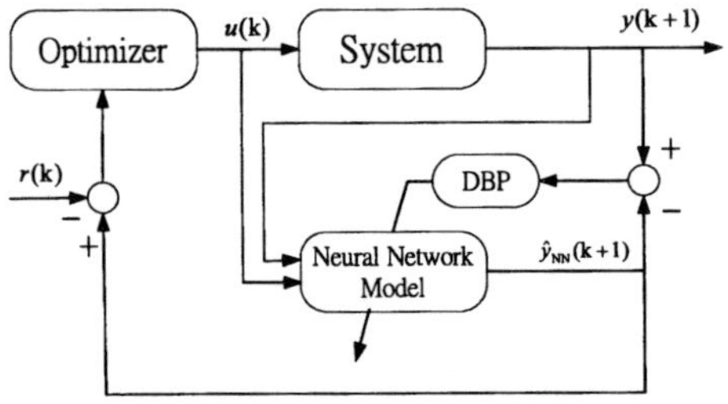

Fig. 1 Adaptive neural-network predictive control scheme.

4. Application to a Reactor System

A continuous stirred tank reactor (CSTR), in which an exothermic irreversible first-order reaction $(A \to B)$ takes place, is covered with a coolant stream flowing in a cocurrent fashion. The coolant flow rate is chosen to be manipulated input and the coolant temperature is allowed to vary along the length of the cooling coil [15]. Under the mass and energy balance, the dynamics of this CSTR model can be described as

$$\begin{aligned} \dot{C}_A &= \frac{q}{V}(C_{Af} - C_A) - k_0 C_A \exp(-\frac{E}{RT}) \\ \dot{T} &= \frac{q}{V}(T_f - T) + \frac{(-\Delta H)k_0 C_A}{\rho C_p}\exp(-\frac{E}{RT}) \\ &\quad + \frac{\rho_c C_{pc}}{\rho C_p V} q_c [1 - \exp(-\frac{hA}{q_c \rho_c C_{pc}})](T_{cf} - T) \end{aligned} \tag{19}$$

where C_A is the effluent concentration of component A, T is the reactor temperature, q is the feed flowrate, and q_c is the coolant flowrate. The nominal system parameters and operating point for the reactor are given in Table 1. Under input constraint by $80(\mathrm{l\,min^{-1}}) \le q_c \le 120(\mathrm{l\,min^{-1}})$, the control objective is to regulate C_A by manipulating q_c .

Table 1 Nominal CSTR operating conditions

$q = 100\ \mathrm{l\,min^{-1}}$	$E/R = 9.95\times10^3\ \mathrm{K}$
$C_{Af} = 1\ \mathrm{mol\,l^{-1}}$	$-\Delta H = 2\times10^5\ \mathrm{cal\,mol^{-1}}$
$T_f = 350\ \mathrm{K}$	$\rho, \rho_c = 1000\ \mathrm{g\,l^{-1}}$
$T_{cf} = 350\ \mathrm{K}$	$C_p, C_{pc} = 1\ \mathrm{cal\,g^{-1}\cdot K^{-1}}$
$V = 100\ \mathrm{l}$	$q_c = 103.41\ \mathrm{min^{-1}}$
$hA = 7\times10^5\ \mathrm{cal\,min^{-1}\cdot K^{-1}}$	$T = 440.2\ \mathrm{K}$
$k_0 = 7.2\times10^{10}\ \mathrm{min^{-1}}$	$C_A = 8.36\times10^{-2}\ \mathrm{mol\,l^{-1}}$

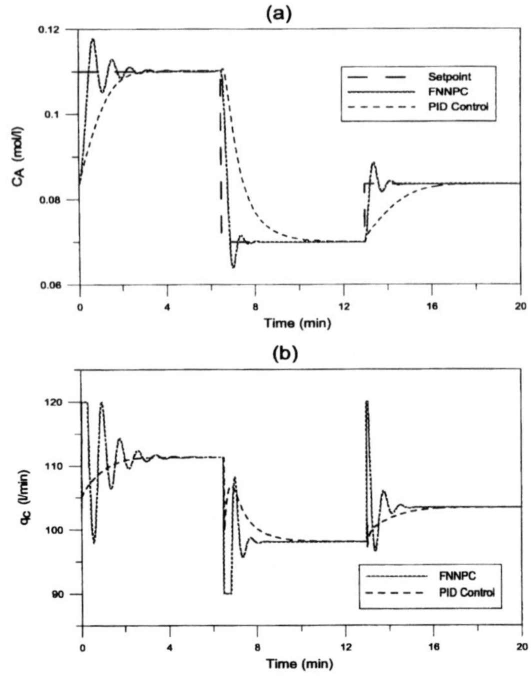

Fig. 2 Closed-loop output tracking responses for the reactor system, in the case of using the unconstrained FNNPC and PID controllers

According to the neural predictive control strategy in Section 2, Fig. 2 shows that the FNN-based predictive control (FNNPC) design can confirm better tracking performance than the conventional PID control approach. By the issue in [15], the PID controller parameters are $K_c = 190\ \mathrm{l^2/mol^{-1}\cdot min^{-1}}$, $\tau_I = 0.556$ min , and $\tau_D = 0.827$ min . For the output regulation of perturbed systems, the adaptive neural-network predictive control (ANNPC) in Section 3 is used. Under a small learning rate $\eta = 0.002$ for the on-line neuro adaptation, we find that the update of weights can keep a rapid, convergent manner. Fig. 3(a) depicts that the unconstrained ANNPC design can carry out the satisfactory disturbance attenuation for +20% change in the inlet concentration C_{Af} . However, Fig. 3(b) shows that the optimization-based control connected to the on-line adaptation procedure could cause the over-large control action. Within the constrained OP framework in (18), Fig. 4 shows that by adjusting λ the satisfactory output tracking subject to an unsaturated control action is superior to the PID control. Moreover, the same ANNPC design for unknown disturbance attenuation is depicted in Fig. 5. Obviously, the simulation demonstrates that the proposed ANNPC method is relatively robust against unknown disturbances.

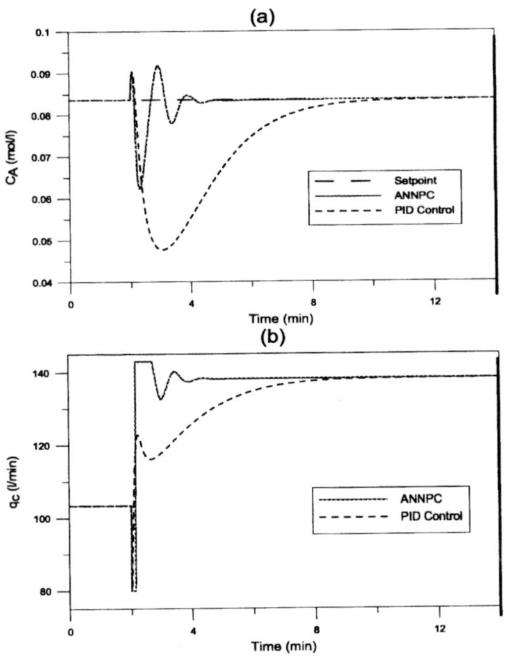

Fig. 3 Disturbance rejection for step change in feed concentration, in the case of using the unconstrained ANNPC and PID controllers

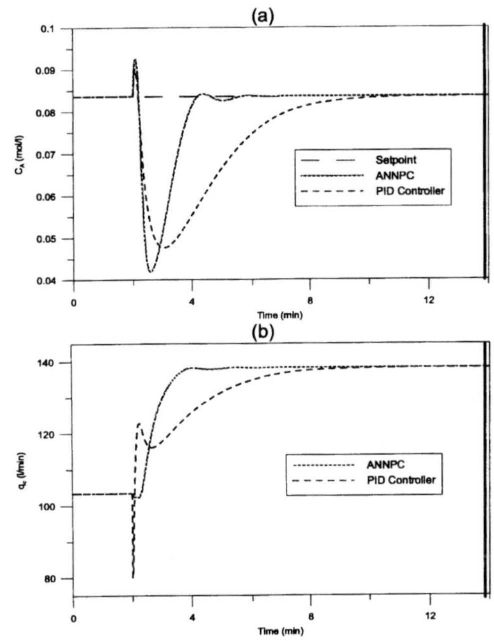

Fig. 5 Disturbance rejection for step change in feed concentration, in the case of using the constrained ANNPC and PID controllers:

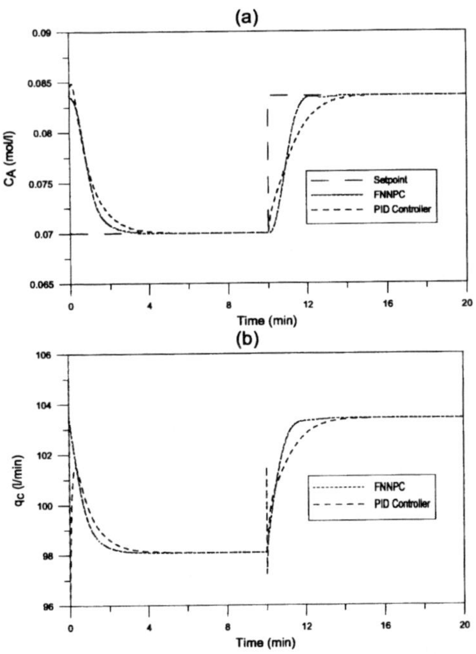

Fig. 4 Closed-loop output tracking responses for the reactor system, in the case of using the constrained FNNPC and PID controllers

5. Conclusion

In this work, predictive control strategies associated with FNN-based adaptation mechanism for constrained SISO and MIMO nonlinear systems are addressed. The discrete-time learning procedure enforce the minimal error between the FNN model output and plant output, and the one-step-ahead predictive control design can carry out the stable output regulation in spite of unknown disturbances. Definitely, the performance and robustness of the proposed control schemes have been successfully verified by a strongly nonlinear reactor system.

Acknowledgments: This work is supported by the National Science Council of the Republic of China under grant number NSC-92-2214-E-224-002.

REFERENCE

1. Ahmed, M. S., and Tasadduq, I. A.: 'Neural servocontroller for nonlinear MIMO plant', *IEE Proc.-Control Theory Appl.*, 1998, **145**, (3), pp. 277-290
2. Calise, A. J., Hovakimyan, N., and Idan, M., 'Adaptive output feedback control of nonlinear systems using neural networks', *Automatica*, 2001, **37**, (8), pp. 1201-1211
3. Gao, F., Wang, F., and Li, M.: 'A simple nonlinear controller with diagonal recurrent neural network', *Chem. Eng. Sci.*, 2001, **55**, (7), pp. 1283-1288
4. Chen, S., and Billings, S. A.: 'Representations of nonlinear systems: the NARMAX model', *Int. J. Control*, 1989, **49**, (10), pp. 1013-1032

5. Ge, S. S., Lee, T. H., and Harris, C. J.: 'Adaptive neural network control of robotic manipulators' (World Scientific Publishing Co., Singapore, 1998)
6. Hoo, K. A., Sinzinger E. D., and Piovoso, M. J.: "Improvements in the predictive capability of neural network', *Journal of Process Control*, 2002, **12**, (1), pp. 193-202
7. Jiang, Z. P., and Wang, Y.: 'Input-to-state stability for discrete-time nonlinear systems', *Automatica*, 2001, **37**, (2), pp. 857-869
8. Kambhampati, C., Mason, J. D., and Warwick, K.: 'A stable one-step-ahead predictive control of nonlinear systems', *Automatica*, 2000, **36**, (4), pp. 485-495
9. Kulawski, G. J., and Brdyś, M. A.: 'Stable adaptive control with recurrent networks', *Automatica*, 2000, **36**, (1), pp. 5-22
10. Ku, C. C., and Lee, K. Y.: 'Diagonal recurrent neural networks for dynamic systems control', *IEEE Transactions on Neural Networks*, 1995, **6**, (1), pp. 144-156
11. Lazar, M., and Pastravanu, O.: 'A neural predictive controller for nonlinear systems', *Mathematics and Computers in Simulation*, 2002, **60**, (2), pp. 315-324
12. Lewis, F. L., Jagannathan, S., and Yeşildirek, A.: 'Neural network control of robot manipulators and nonlinear systems' (Taylor & Francis, London, 1999)
13. Maciejowaki, J. M.: 'Predictive control with constraints' (Pearson Education Limited, England, 2002)
14. Narendra, K. S., and Lewis, F. L.: 'Introduction to the special issue on neural network feedback control', *Automatica*, 2001, **37**, (8), pp. 1147-1148
15. Nahas, E. P., Henson, M., and Seborg, D. E.: 'Nonlinear internal model control strategy for neural network models,' *Computers & Chem. Eng.*, 1992, **16**, (12), pp. 1039-1057
16. Noriega, J., and Wang, H.: 'A direct adaptive neural-network control for unknown nonlinear systems and its application', *IEEE Transactions on Neural Networks*, 1998, **9**, (1), pp. 27-33
17. Poznyak, A. S., Sanchez, E. N., and Yu, W.: 'Differential neural networks for robust nonlinear control' (World Scientific Publishing Co., Singapore, 2001)
18. Poznyak, A. S., Yu, W., Sanchez, E. N., and PEREZ, J. P.: 'Nonlinear adaptive trajectory tracking using dynamic neural networks', *IEEE Transactions on Neural Networks*, 1999, **10**, (6), pp. 1402-1411
19. Slotine, J.-J. E., and Li, W.: 'Applied nonlinear control' (Prentice Hall Englewood Cliffs, New Jersey, 1991)
20. Vila, J. P., and Wagner, V.: 'Predictive neurocontrol of uncertain systems: design and use of a neuro-optimizer', *Automatica*, 2003, **39**, (5), pp. 767-777
21. Wang, L. X., and Wan, F.: 'Structured neural networks for constrained model predictive control', *Automatica*, 2001, **37**, (8), pp. 1235-1243
22. Yu, D. L., Gomm, J. B., and Williams, D.: 'On-linear predictive control of a chemical process using neural network models', *14th World Congress of IFAC*, 1999, Beijing, P.R. China, pp. 121-126
23. Yu, W., and Li, X.: 'Discrete-time neuro identification without robust modification', IEE *Proc.-Control Theory Appl.*, 2003, **150**, (3), pp. 311-316

ELSEVIER
IFAC
PUBLICATIONS
www.elsevier.com/locate/ifac

A Smith Predictor Enhanced PID Controller

Shih-Bo Hung[1], Cheng-Ching Yu[2*] and Yu-Chang Cheng[1]

Dept. of Chemical Engineering[1]
National Taiwan University of Sci. Technol.
Taipei 106-07, TAIWAN
Dept. of Chemical Engineering[2]
National Taiwan University
Taipei 106-17, TAIWAN

Abstract: In this work, a Smith predictor enhanced PID controller, SP-PID, is proposed. A tuning parameter K_{sp} is devised which gradually transforms the controller from a PID controller to a Smith predictor as K_{sp} changes 0 to 1. Properties of the SP-PID are explored and design procedure is given to ensure a certain degree of robustness. Simulation results clearly indicate that the SP-PID takes advantage of the SP when small modeling error is encountered and it is gradually detuned to a PID controller, a user friendly controller, when the model quality degrades. Moreover, the proposed concept can be implemented to current process control computer with virtually no extra hardware cost.

Keywords: dead time compensation, Smith predictor, PID control, robust performance

1. Introduction

The Smith predictor (SP) introduced by Otto Smith (1957) provides a nice controller structure for deadtime compensation and it has received considerable attention over past 40 years (Meyer et al., 1976; Ogunnaike and Ray, 1979; Morari and Zafiriou, 1989; Huang et al., 1990; Palmer and Blau, 1994; Lee et al., 1996; Kwak et al., 1999; Majhi and Atherton, 2000; Kaya, 2001; Ingimundarson and Hägglund, 2002). However, the modeling requirement, non-trivial tuning, and unfamiliarity prevent wide-spread applications. A typical scenario is that, in order to maintain robust stability, the associated PI controller is detuned to such a degree that control performance is no better than a simple PI controller and this is especially true for systems with small deadtime to time constant ratio (D/τ) (Ingimundarson and Hägglund, 2002). Despite the clear advantage for deadtime dominant processes, the Smith predictor again finds limited applications throughout process industries. On the other hand, we have found PID controllers remain as a standard feature in process industries (Åström and Hägglund, 1995; Yu, 1999; Tan et al., 1999).

The modeling problem of the SP can be overcome using relay feedback tests which also becomes a standard feature in many process control computers (Åström and Hägglund, 1984; Luyben, 1987; Chang et al., 1992; Palmer and Blau, 1994; Yu, 1999; Majhi and Atherton, 2000; Wang et al., 2003). Based on the shape information, the model structure as well as model parameters can be identified in a single relay feedback test (Luyben, 2001; Thyagarajan and Yu, 2003). Once the process model is available, one can proceed with controller tuning. In this work, a new type of controller is proposed which includes the Smith predictor as an enhanced feature of a standard PID controller. Next, a new tuning parameter, K_{sp}, is devised which provides a gradual transition between the SP (K_{sp}=1) and PID (K_{sp}=0) controller. That is the Smith predictor is functioning at its full capacity when the model quality is good and the deadtime compensation feature is gradually turned off when plant-model mismatch develops, and, ultimately, the controller reduced to the familiar PID controller.

The remainder of this paper is organized as follows. The concept of SP-PID is depicted and design procedure is also given in section 2. Improved PID design is proposed and robust performance of SP-PID is explored in section 3 followed by the conclusion.

2. Smith predictor enhanced PID control

2.1 Concept

Ingimundarson and Hägglund (2002) present an interesting paper showing that the performance of the SP (Fig. 1b) is inferior than a PI (Fig. 1A) controller for first order plus deadtime systems (FOPDT) with the deadtime to time constant ratio (D/τ) less than 0.2. Moreover, for PID controller, the advantage of the SP can only be seen for FOPDT systems with D/τ greater than 10. It is troublesome because how can a modeled-based approach, with correct controller structure, fail to achieve improved performance. The reason is obvious that the SP is significantly detuned to achieve certain degree of robustness (or the PI settings were tightened to obtain better performance). In other words, if a model-based controller is loosely

* **corresponding author: fax: +886-2-2362-3040, e-mail: ccyu@ntu.edu.tw**

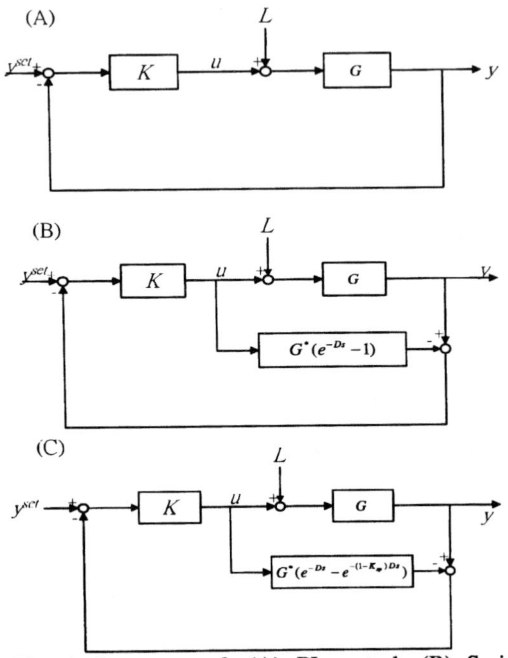

Fig. 1 Structures of: (A) PI control, (B) Smith predictor, and (C) Smith predictor enhanced PI control (SP-PID).

tuned, the performance can be no better than a PI controller. This scenario is often encountered in practice, especially when the users are not familiar with the controller and its tradeoff between the robustness and performance.

On the other hand, the progress in integrated circuits provided much increased computing ability for most controllers. From the hardware perspective, the implementation cost of a Smith predictor is virtually the same as a standard PID controller. How can we integrated the model-based SP into the standard PID controller while possesses some degree of user friendliness. Figure 1C shows the Smith predictor enhanced PID controller via a new tuning constant K_{sp}. Figure 1C indicates that when K_{sp} is set to one, it becomes the SP (Fig. 1B) and when K_{sp} is turned to 0, the feedback reduced to the conventional one, a PID type of controller. That is the amount of deadtime in the feedback (D_R, remaining deadtime) can be adjusted via K_{sp} from high performance/less robust/less familiarity (K_{sp}=1) to nominal performance/increased robustness/familiarity (K_{sp}=0). Certainly, this provides a gradual transition from a SP to a PID controller.

2.2 Analysis

Consider the block diagram in Fig.1 where the process G is expressed in terms of the delay-free part G^* and the deadtime portion e^{-Ds}.

$$G = G^* e^{-Ds} \tag{1}$$

The model also has a similar structure.

$$\tilde{G} = \tilde{G}^* e^{-\tilde{D}s} \tag{2}$$

when $\tilde{G}^*$ is the delay free part of the model and $\tilde{D}$ is the deadtime in the model. The closed-loop relationship of the SP can be expressed as

$$y = \frac{KG}{1+K\tilde{G}^*+KG-K\tilde{G}} y^{set} + \frac{(1+K\tilde{G}^*-K\tilde{G})G}{1+K\tilde{G}^*+KG-K\tilde{G}} L \tag{3}$$

where y^{set} is the set point, K is the feedback controller and L is the load variable. With the perfect model assumption (i.e., $G = \tilde{G}$), we have :

$$y = \frac{KG}{1+KG^*} y^{set} + \frac{(1+KG^*-KG)G}{1+KG^*} L \tag{4}$$

The advantage of the SP can clearly be seen from the characteristic equation (denominator of Eq. 4) that the controller K can be designed aggressively to achieve a large bandwidth. For the proposed SP-PID controller in Fig. 1c, the closed-loop relationship becomes:

$$y = \frac{KG}{1+K\tilde{G}^* e^{-(1-K_{sp})\tilde{D}s}+KG-K\tilde{G}} y^{set} + \frac{(1+K\tilde{G}^* e^{-(1-K_{sp})\tilde{D}s}-K\tilde{G})G}{1+K\tilde{G}^* e^{-(1-K_{sp})\tilde{D}s}+KG-K\tilde{G}} L \tag{5}$$

Similarly, when $G = \tilde{G}$, one obtains:

$$y = \frac{KG}{1+KG^* e^{-(1-K_{sp})Ds}} y^{set} + \frac{(1+KG^* e^{-(1-K_{sp})Ds}-KG)G}{1+KG^* e^{-(1-K_{sp})Ds}} L \tag{6}$$

Eq. 6 clearly contrasts the difference between the SP and the SP-PID where, for the later, the remaining deadtime (D_R=(1- K_{sp})D) in the feedback loop is adjustable via K_{sp}. It becomes a SP on one end (K_{sp} =1) and resumes the role of a PID controller on the other end (K_{sp} =0).

Let us use the first order plus deadtime (FOPDT) process to illustrate the performance and robustness characteristics of SP-PID control. Consider

$$G(s) = \frac{K_p e^{-Ds}}{\tau s+1} \tag{7}$$

Here, K_p is the steady-state gain, τ is the time constant, and D is the deadtime. Without loss of generality, we use IMC type of tuning rule for a PI controller (Morari and Zafiriou, 1989; Chien and Fruehauf, 1990). First, a closed-loop time constant λ is selected and the controller gain becomes:

$$K_c = \frac{\tau_I}{K_p \lambda} \tag{8}$$

and the rest time τ_I is simply set to:

$$\tau_I = \tau \tag{9}$$

Note that this assumption will be relaxed in a later section. In doing this, the closed-loop relationship for the SP-PID becomes:

$$y = \frac{e^{-Ds}}{\lambda s+e^{-(1-K_{sp})Ds}} y^{set} + \frac{\lambda s+e^{-Ds}\left(e^{K_{sp}Ds}-1\right)}{\lambda s+e^{-(1-K_{sp})Ds}} \frac{K_p e^{-Ds}}{\tau s+1} L \tag{10}$$

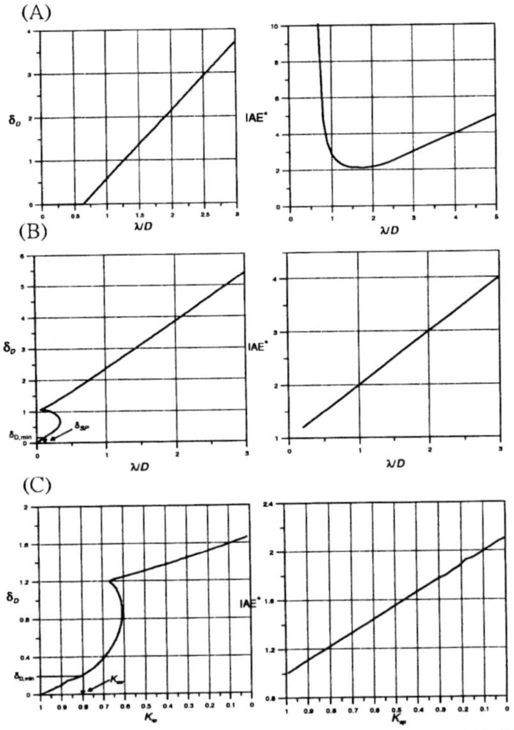

Fig. 2 Tolerable deadtime error and IAE for: (A) PI control, (B) Smith predictor, and (C) Smith predictor enhanced PI control.

where λ is the closed-loop time constant. The integrated error (IE) for a unit step setpoint change can be derived analytically using the final-value theorem.

$$(\mathrm{IE})_{\mathrm{setpoint}} = \lim_{s\to 0} s \frac{\lambda s + (e^{-(1-K_{sp})Ds} - e^{-Ds})}{\lambda s + e^{-(1-K_{sp})Ds}} \frac{1}{s}\cdot\frac{1}{s} \quad (11)$$

$$= \lambda + K_{sp} D$$

Eq. 10 shows that the IE is a function of the closed-loop time constant (λ) as well as K_{sp}. If λ is chosen to be a function of the remaining deadtime (D_R) in the feedback loop and following the IMC-PI tuning rule, we have:

$$\lambda = 1.7 D_{\mathrm{R}} = 1.7(1-K_{sp})D \quad (12)$$

and IE thus becomes:

$$(\mathrm{IE})_{\mathrm{setpoint}} = (1.7 - 0.7K_{sp})D$$

Despite the fact that the IE is a good performance measure only for monotonic step response, it provides a qualitative indication of the closed-loop performance for the SP-PID. Figure 2 show a more realistic performance measure, the normalized integrated absolute error (IAE*), for the proposed controller. For a FOPDT system, the normalized IAE is defined as:

$$\mathrm{IAE}^* = \frac{\mathrm{IAE}}{\mathrm{IAE}_{\min}} \quad (13)$$

where $\mathrm{IAE}_{\min}$ is the minimum IAE and for a FOPDT system it is simply $\mathrm{IAE}_{\min}$ =D for a unit step change. Figure 2c shows that the performance is improved as we turn on the deadtime compensation portion of the controller via an increased K_{sp}. Qualitative correct trend is also seen as compared to the simple performance measure in Eq. 12. Similarly, the performance of the SP can be conjectured from Eq. 10 by setting K_{sp} =1 and the normalized IAE (IAE*) is also shown in Fig. 2b. However, the performance degrades significantly for a PI controller at region of small λ (Fig. 2a) for an obvious reason: the stability constraint is almost violated and oscillatory responses result at small λ region. The performance part (IAE*) of Fig. 2 indicates that the SP-PID takes the advantage of the Smith predictor at the high performance region and it is reduced to a PID controller when performance requirement is not high.

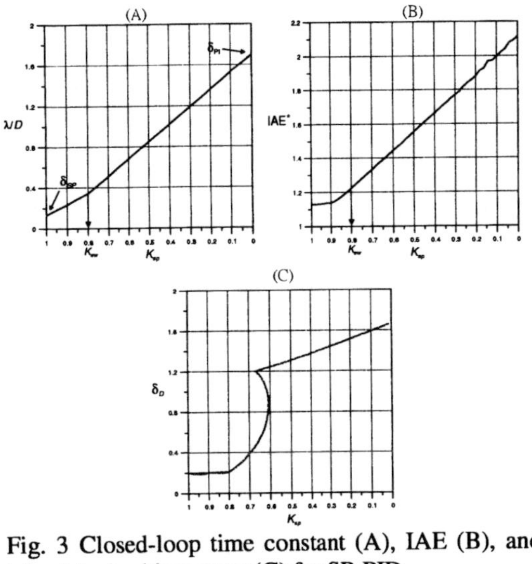

Fig. 3 Closed-loop time constant (A), IAE (B), and tolerable deadtime error (C) for SP-PID.

The characteristic equation in Eq. 5 can be used to evaluate robust stability with respect to deadtime error. Assume uncertainty occurs in the deadtime part (i.e., $D \neq \tilde{D}$) the closed-loop characteristic equation (CLCE) becomes:

$$\lambda s + e^{-(1-K_{sp})\tilde{D}s}(e^{-(K_{sp}+\delta_D)\tilde{D}s} - e^{-K_{sp}\tilde{D}s} + 1) = 0 \quad (14)$$

where δ is the multiplicative deadtime error, i.e. $\delta = (D-\tilde{D})/\tilde{D}$. Taking the extremes ($K_{sp}$ =0 and K_{sp} =1), the CLCE's for the PI and SP are:

$$\lambda s + e^{-(1+\delta)\tilde{D}s} = 0 \quad (15)$$

$$\lambda s + e^{-(1+\delta)\tilde{D}s} - e^{-\tilde{D}s} + 1 = 0 \quad (16)$$

Eqs. 14~16 can be used to compute tolerable deadtime errors (δ_D) for different closed-loop time constants (Cheng and Hwang, 1999). By tolerable deadtime error we mean the amount of deadtime error leads to the limit of stability. Interesting enough, conditional stability is observed for the SP as well as SP-PID over certain ranges of controller settings when is not observed for the PI controller. Figure 2 also reveals that extreme sensitivity for deadtime error when λ approaches zero or K_{sp} approaches one. Certainly, such settings are not acceptable in practice, despite almost perfect performance.

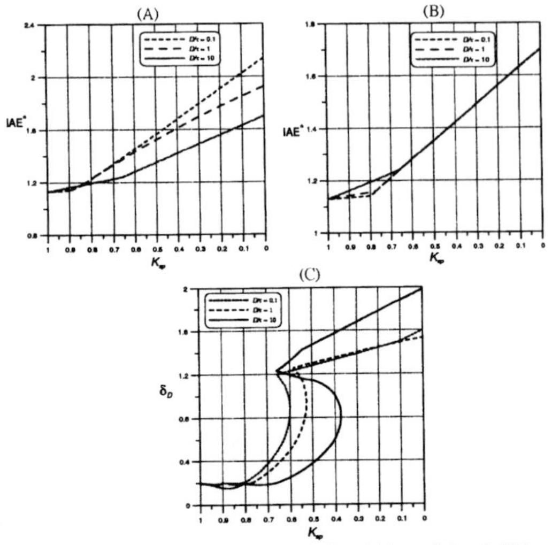

Fig. 4 IAE's for unit step setpoint (A) and load (B) changes and tolerable deadtime error (C) using SP-PID.

2.3 Design

The design principle is to turn the SP gradually toward the PID controller when the model quality degrades while maintain a certain of robustness at the SP end with respect to deadtime error. Let us consider controller tunings at two extremes, K_{sp} =0 and K_{sp} =1. At the PI control end (K_{sp} =0), the closed-loop time constant λ is set proportional to the deadtime. That is:

$$\lambda = \delta_{PI} \cdot D \qquad (17)$$

A typical value of $\delta_{PI} = 1.7$ is often employed according to the IMC-PI rule. At the SP side, λ is often selected to handle a pre-determined degree of deadtime error. Note that the setting does not have to be conservative because the tradeoff between robustness/performance is handled by the controller via K_{sp}. A typical value of 20% tolerable deadtime error ($\delta_{D,\min} = 0.2$) is recommended at the performance end. Therefore, we can read-off δ_{SP} directly from Fig. 2b given $\delta_{D,\min}$ (or computed δ_{SP} from Eq. 16). Thus, we have

$$\lambda = \delta_{SP} \cdot D \qquad (18)$$

In theory, Eqs 17 and 18 will be sufficient to formulate the tuning constant for the SP-PID. However, in order to ensure the tolerable deadtime error is larger than $\delta_{D,\min}$ over the entire K_{sp} range (0~1), a switching point of K_{sp} has to be located that gives the tolerable deadtime error to be $\delta_{D,\min}$. Then, a linear interpolation is employed to set the closed-loop time constant for K_{sp} greater than K_{sw} (Fig. 3). In other words, we have:

$$\begin{cases} \dfrac{\lambda}{D} = (1-K_{sp})\delta_{PI}, & K_{sp} \le K_{sw} \\ \dfrac{\lambda}{D} = (1-K_{sp})\delta_{PI} + \left(\dfrac{K_{sp}-K_{sw}}{1-K_{sw}}\right)\delta_{SP}, & K_{sp} > K_{sw} \end{cases} \qquad (19)$$

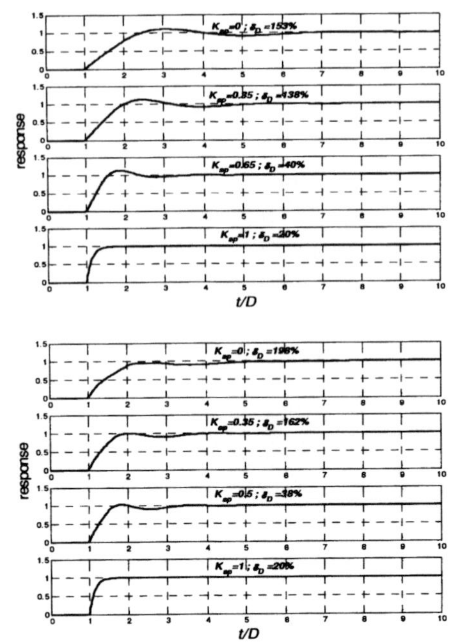

Fig. 5 Unit step setpoint responses usung different Ksp settings for systems with D/τ=1 (top) and 5 (bottom).

Therefore, the controller design procedure becomes:

S1. Select δ_{PI} (e.g., from IMC-PI tuning rule).

S2. Choose $\delta_{D,\min}$ (a typical value is $\delta_{D,\min}$ =0.2, 20% deadtime error).

S3. Find δ_{SP} from Eq.16 (or Fig. 2B) and K_{sw} from Eq. 15 (or Fig. 2C).

S4. Set the closed-loop time constant to

$$\frac{\lambda}{D} = \max\left[(1-K_{sp})\delta_{PI}\,,\,(1-K_{sp})\delta_{PI} + \left(\frac{K_{sp}-K_{sw}}{1-K_{sw}}\right)\delta_{SP}\right]$$

If IMC-PI tuning rule is employed (i.e., $\tau_I = \tau$ and δ_{PI} =1.7), for $\delta_{D,\min}$ =0.2, we have: K_{sw} =0.8 and $\delta_{SP} = \lambda / D$ =0.129. Figure 3 shows the relationships among closed-loop time constant λ / D, performance (IAE*), and robustness to deadtime error (δ_D). Note that Fig.3 remains the same for FOPDT systems with all possible D/τ ratios provided with the plain IMC-PI tuning, i.e., $\tau_I = \tau$ and $K_c K_p = \tau_I / \lambda$. Fig. 3B also shows that the IAE at the PI and (K_{sp} =0) is at 210% of the minimum value as opposed to 170% predicted by Eq. 9 using the integrated error (IE). The reason is obvious that the area above and below the set point cancelled out each other, a non-monotonic set point response. Control performance can be improved using improved IMC-PI tuning rules (Morari and Zafiriou, 1989; Chien and Fruehauf, 1990).

3. Performance

3.1 Improved PID tuning

The improved IMC-PI tuning rule takes the effect of

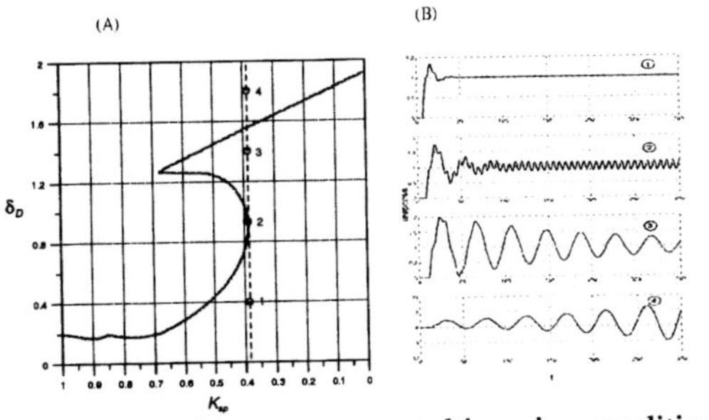

Fig. 6 Deadtime error resulting in conditional stability (A) and corresponding setpoint responses (B).errors

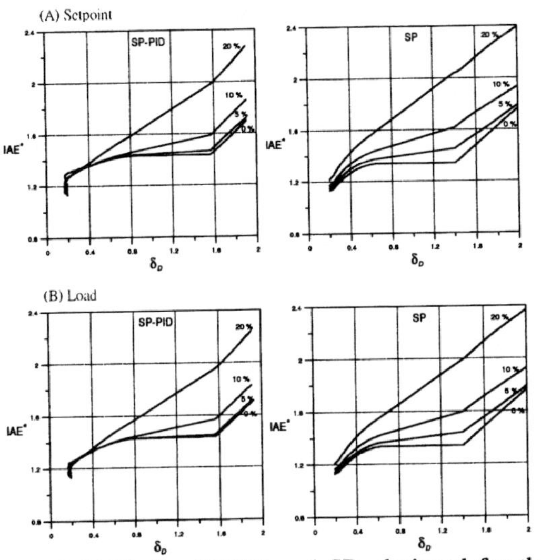

Fig. 7 IAE's for SP-PID and SP designed for the same amount of tolerable deadtime error under the same percentage (0, 5, 10, 20%) of deadtime errors.

deadtime into consideration and, thus, the reset time is adjusted according to $\tau_I = \tau + D_R/2$. Following the same line, the controller gain becomes:

$K_c K_p = \tau_I/(\delta_{PI}\cdot D_R) = \tau_I/(1.7\cdot D_R)$

Using this setting, the closed-loop characteristic equation for the SP-PID can be expressed as (cf. Eq. 14):

$$\lambda s+\frac{\{1+[\tau+(1-K_{sp})\tilde{D}/2]s\}}{\tau S+1}e^{-(1-K_{sp})\tilde{D}s}\left[e^{-(K_{sp}+\delta)\tilde{D}s}-e^{-K_{sp}\tilde{D}s}+1\right]=0 \tag{20}$$

Eq. 20 can be used to computed the switching value of K_{sp} (K_{sw}) and δ_{SP} given $\delta_{D,\min}$. Again, procedure in sec. 2.3 can be employed to design the SP-PID for systems with different D/τ values. Results reveal that the setpoint performance and robust stability are functions of D/τ while the regulatory performance is almost independent of the deadtime to time constant ratio as shown in Fig. 4. Furthermore, improved performance (e.g., IAE* =170% for D=10 with K_{sp} =0) actually can be achieved using a better tuning rule for the PI controller, especially at high D/τ region.

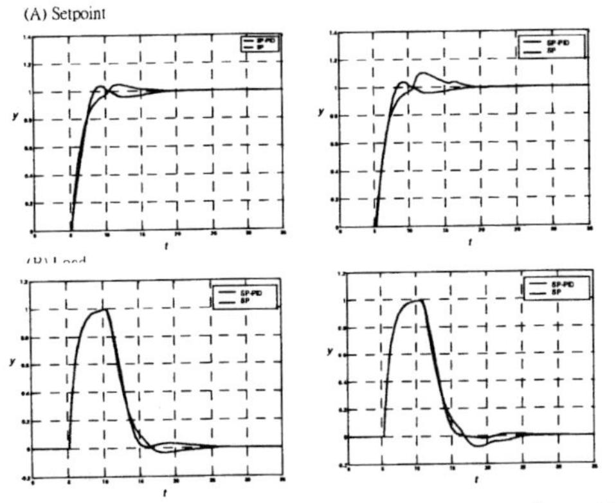

Fig. 8 Setpoint (A) and load (B) responses for +10% (left) and +20% (right) deadtime errors for D/τ=5 using SP and SP-PID (designed with the same amount of tolerable deadtime error)

Let us use a system with D=1 and D=5 to illustrate the performance of the SP-PID control. Figure 5 shows that better performance becomes evident as we gradually turn on the Smith predictor. But this is achieved with the expense of decreased robustness as also shown in Fig. 5,comparing the tolerable deadtime error (δ_D). Figure 5 clearly illustrates the advantage of the SP-PID control where aggressive tuning (provided with correct controller structure) can be applied when the model quality is good and the controller structure is detuned (via K_{sp}) as the model quality degrades.

3.2 Robust Performance

Up to this point, we have only studied the nominal performance of the SP-PID. One may wonder how the proposed one differs from a detuned Smith predictor? The SP-PID control has two clear advantages over a typical SP. First, it provides a transparent tuning constant K_{sp} in a finite range and the user can use the on-off concept to adjust the performance/robustness tradeoff while, for the SP, we really do not know to what degree the corresponding PI controller should be detuned when the model quality degrades. The second, but not an obvious one, is that the SP-PID gives a better robust performance. Let use design both the SP-PID (via K_{sp}) and the SP (via closed-loop time constant λ) with the same amount of tolerable deadtime error over a wide range of δ_D, 20%~200%. Again, let us use D=5 case as an example. Nominally, the SP gives better performance over the range of small δ_D as shown in Fig. 7. However, as the deadtime error increases to 5%, 10%, 20% of the nominal value, the SP-PID shows a slower rate of increase in the integrated absolute error (IAE*) as shown in Fig. 7. This is the case for both set point and load changes. Figure 8 compares the performance of the SP-PID and SP for 10 & 20% deadtime for the case of δ_D=40% (K_{sp}=0.52 for SP-PID and λ/D=0.264 for the SP). Similar behavior is observed for systems with different D/τ values.

Conclusion

In this work, a new type of controller, Smith predictor enhanced PID controller SP-PID, is proposed. A tuning parameter (K_{sp}) gradually transforms the controller from a PID controller to a Smith predictor as K_{sp} varies from 0 to 1, a controller with a well-defined range of settings. The PID parameters are adjusted according to the remaining deadtime in the feedback loop. The characteristics and stability of the SP-PID is explored and a design procedure is proposed to ensure a pre-specified degree of robustness. Following the design procedure with a minimum deadtime error of 20%, we have the following recommendations for 4 distinct settings:

(1) Set K_{sp}=1 right after autotuning. This correspond to tolerable deadtime error (δ_D) of 20% and normalized integrated absolute error (IAE*) of 1.13 for all D/τ.

(2) Set K_{sp}=0.65 for $D/\tau \le 1$ and K_{sp}=0.5 for $D/\tau > 1$ when oscillatory behavior occurs. This corresponds to δ_D = 40% & IAE* =1.38 for D/τ=1 and δ_D = 38% & IAE*=1.35 for D/τ=10.

(3) Set K_{sp}=0.35 when oscillatory behavior is again observed. This corresponds to δ_D = 138% & IAE*=1.65 for D/τ=1 and δ_D = 162% & IAE*=1.45 for D/τ=10.

(4) Turn the SP-PID to a PID controller, K_{sp} =0, when significant oscillation still exists. This gives δ_D = 153% & IAE* =1.92 for D/τ=1 and δ_D = 198% & IAE*=1.70 for D/τ=10.

Certainly, when the PID control (K_{sp}=0) still cannot provide satisfactory control performance, a further detuning and/or autotune identification should be initiated. Simulation results show that the SP-PID indeed takes advantage of the theoretically correct SP structure while maintaining the familiar PID structure, user friendliness. More importantly, it can be implemented on current process control computers with virtually no extra hardware cost.

Acknowledgement

This work is supported by the Ministry of Economic Affairs of Taiwan

REFERENCES

Åström, K. J. and Hägglund, T. (1984) Automatic tuning of simple regulators with specifications on phase and amplitude margins. *Automatica* , **20**,.645

Åström, K. J. and Hägglund, T.(1995) *PID controllers. 2nd ed.*, Instrumentation Society of America, Research Triangle Park, NC.

Chang, R. C., Shen, S. H. and Yu, C. C. (1992) Derivation of transfer function from relay feedback systems. *Ind. Eng. Chem. Res.*, **31(3)**, 855.

Cheng, S. L., Hwang, C. (1999) On stabilization of time-delay unstable systems using PID controllers. *J. Chin. Inst. Chem. Eng.*, **30 (2)**, 123.

Chien, I.-L.; Fruehauf, P. S.(1990) Consider IMC tuning to improve controller performance. *Chem. Eng. Prog.* , **86 (10)**, 33.

Huang, H. P.; Chen, C. L.; Chao, Y. C.; Chen, P. L.(1990) A modified Smith predictor with an approximate inverse of dead time. *AIChE J* , **36**, 1025.

Ingimundarson, A.; Hägglund, T. (2002) Performance comparison between PID and Dead-time compensating controllers, *J. Process Control* , **12**, 887.

Kaya, I. (2001) Tuning Smith predictors using simple formulas derived from optimal responses. *Ind. Eng. Chem. Res.* , **40 (12)**, 2654.

Kwak H. J.; Sung, S. W.; Lee, I. B.; Park, J. Y.(1999) A modified Smith predictor with a new structure for unstable processes. Ind. Eng. Chem. Res. , **38(2)**, 405.

Lee, T. H., Wang, Q. G., Tan, K. K. (1996) Robust Smith-predictor controller for uncertain delay systems. *AIChE J.* , **42 (4)**,1033.

Luyben, W.L. (1987) Derivation of transfer functions for highly nonlinear distillation columns. *Ind. Eng. Chem. Res.* , **26(12)**, 2490.

Luyben, W. L. (2001) Getting more information from relay feedback tests. *Ind. Eng. Chem. Res.*, **40(20)**, 4391.

Majhi, S.; Atherton, D. P.(2000) Obtaining controller parameters for a new Smith predictor using autotuning. *Automatica* , **36**,1651.

Meyer, C.; Seborg, D. E.; Wood, R. K.(1976) A comparison of the Smith predictor and conventional feedback control. *Chem. Eng. Sci.* , **31**,775.

Morari, M. and Zafiriou, E.(1989) *Robust process control*, Prentice Hall, Englewood Cliffs, NJ .

Ogunnaike, B. A.; Ray, W. H. (1979) Multivariable controller design for linear systems having multiple time-delays. *AIChE J.*, **25**, 1043.

Palmor, Z. J.; Blau, M. (1994) An auto-tuner for Smith dead time compensator. *International Journal of Control* , **60(1)**, 117.

Smith, O. J. M. (1957)Closed control of loops with dead time. *Chemical Engineering Progress* , **53 (5)**, 217.

Tan, K. K., Wang, Q. G. and Hang, C. C. (1999) *Advances in PID Control*, Springer-Verlag: London .

Thyagarajan, T.; Yu, C.C. (2003). Improved auto-tuning using shape factor from relay feed back. *Ind. Eng. Chem. Res*,. (ASAP article).

Wang, Q. G.; Lee, T. H.; Lin, C. (2003) *Relay Feedback*, Springer-Verlag: London .

Yu, C. C. (1999) *Autotuning of PID controllers*, Springer-Verlag: London .

ELSEVIER
IFAC
PUBLICATIONS
www.elsevier.com/locate/ifac

A RATIO CONTROL ARCHITECTURE FOR SET-POINT FOLLOWING AND LOAD DISTURBANCE REJECTION

Antonio Visioli * **Massimiliano Veronesi** **

* *Dipartimento di Elettronica per l'Automazione, University of Brescia (Italy), e-mail: antonio.visioli@ing.unibs.it*
** *Yokogawa Italia, Milan (Italy), e-mail: max.veronesi@it.yokogawa.com*

Abstract: In this paper a new ratio control architecture is proposed. Differently from the ratio controllers proposed in the literature, this one aims at achieving satisfactory performances with respect to both set-point following and load disturbances rejection specifications. Tuning rules are provided, so that no significant design effort is required to the user. Overall, the simplicity of the proposed control scheme makes it suitable to implement in the industrial context. Simulation results show the effectiveness of the methodology. *Copyright © 2004 IFAC*

Keywords: Ratio control, PID controllers, tuning, set-point following, load disturbance rejection.

1. INTRODUCTION

PID controllers are undoubtedly the most adopted controllers in industrial settings, because of the good cost/benefit ratio they are able to provide. This is due to the fact that they are able to achieve satisfactory performances for a wide variety of processes and, at the same time, they are simple to use. In order to help the operators to satisfy control specifications with a limited design effort, a large number of tuning rules have been devised (O'Dwyer, 2003), together with methodologies for the design of those additional functionalities (such as anti-windup, feedforward action, gain scheduling, adaptive control and so on) that make the adoption of the basic PID algorithm successful in practical cases (Aström and Hägglund, 1995).
Often, PID controllers are employed as a fundamental component in more complex control schemes, where couplings between simple control systems are exploited. A significant example in this context is ratio control, where the control specification is to keep a constant ratio between two process variables, despite set-point changes and load disturbances. A ratio control problem can be found in many industrial processes such as chemical dosing, water treatment, chlorination, mixing vessels, waste incinerators. For example, in combustion systems it is necessary to control accurately the air-to-fuel ratio in order to obtain a high efficiency, and in blending processes a selected ratio of different flows has to be maintained in order to achieve a constant product composition.
Design methodologies for ratio controllers have not been much investigated in the literature. A significant contribution is the work of Hagglund (Hägglund, 2001), where a ratio control architecture named Blend station is proposed to improve set-point responses with respect to classic schemes. A different method has been presented in (Visioli, 2003). Therein, in addition to provide a perfect set-point response in the absence of model uncertainties, load disturbances rejection performances are also taken into account.
Aiming at improving the load disturbances rejection performances without impairing the set-point response, a new design methodology is proposed in this paper. It has to be stressed that, in general, the

adoption of more complex control schemes should not imply an increase of the design effort for the operator. Thus, tuning rules have been devised to guarantee a simple implementation of the method and therefore its suitability for industrial applications.
The paper is organized as follows. In Section 2 ratio control is reviewed and techniques proposed in the past are briefly discussed. In Section 3 the new method is presented. Simulation results are shown in Section 4 and conclusions are drawn in Section 5.

2. RATIO CONTROL

A ratio controller aims at keeping a predetermined constant ratio a between two process variables y_2 and y_1, despite set-point changes and load disturbances that might occur on the plant. The most employed control scheme in this context is the one shown in Figure 1 where each variable is controlled by two separate controllers C_1 and C_2 (typically of PI type) and the output y_1 of the first process P_1 is multiplied by a and adopted as the set-point signal of the closed-loop control system of the second process P_2, i.e. it is $r_2(t) = ay_1(t)$ (Shinskey, 1996). It has to be noted that in this case the set-point response of the second closed-loop system is necessarily delayed with respect to the response of the first one. This drawback is somewhat limited by imposing that the dynamics of the second loop be much faster than that of the first loop.
Thus, this choice is usually preferred to that of imposing $r_2(t) = ar_1(t)$ because in this case the required ratio is no more followed when a load disturbance occurs in the first loop, although a perfect set-point response (i.e. the desired ratio a is achieved during the whole transient) can be potentially achieved by selecting the same dynamics for the two loops.
In order to improve the set-point response performances with respect to the classic architecture of Figure 1 a different control architecture named Blend station has been proposed in (Hägglund, 2001). It combines the previous approaches by selecting

$$r_2(t) = a(\gamma r_1(t) + (1-\gamma)y_1(t)) \tag{1}$$

where γ is a constant parameter that actually weights the relative influence of the set-point r_1 on r_2 with respect to y_1 (note that for $\gamma = 0$ the classic scheme of Figure 1 is obtained). This solution however yields to a decrement in the load disturbance rejection performances and therefore should not be used when these are likely to occur.
An alternative scheme (shown in Figure 2) has been proposed in (Visioli, 2003). By choosing

$$F(s) = \frac{C_1(s)P_1(s)}{C_2(s)P_2(s)} \tag{2}$$

a perfect set-point response is obtained and by suitably selecting the design parameters (note that tuning rules

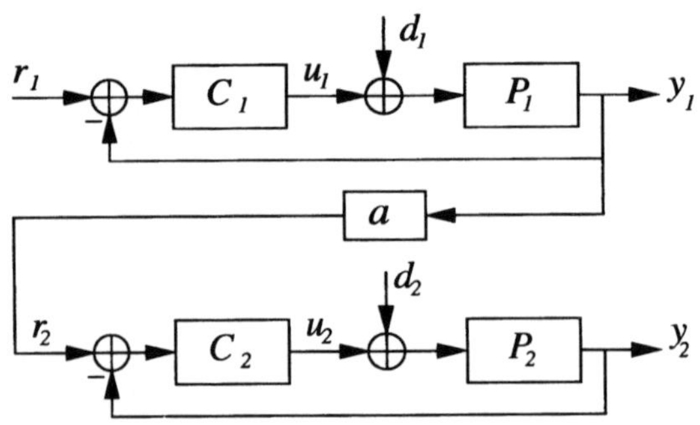

Fig. 1. The typical ratio control architecture.

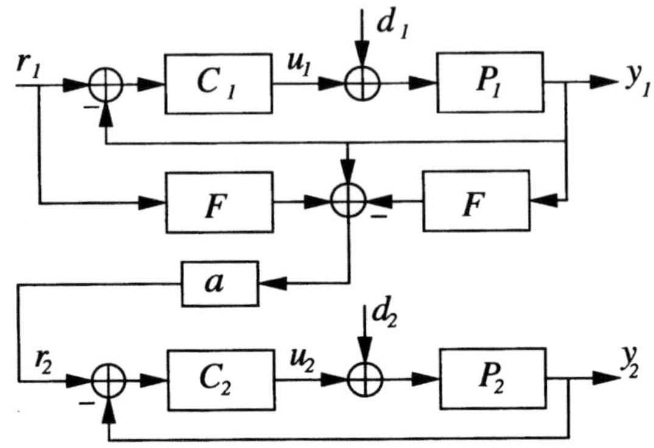

Fig. 2. An alternative ratio control architecture.

are explicitly provided) this is not paid by a decrement in the load disturbances rejection performances.
However, it appears that in all the considered methodologies the desired ratio is not followed when a load disturbance occurs in the second control loop (signal d_2 in Figures 1 and 2). In order to provide an effective solution for this problem a new architecture is proposed in the next section.

3. THE NEW ARCHITECTURE

The new control scheme is shown in Figure 3 and an explanation of it, together with tuning rules, is given in the following. First, a first order plus dead time (FOPDT) model for the two processes under control has to be obtained. This is what it is usually done in the industrial practice, and it can be performed with a variety of methods (e.g. the well-known area method (Aström and Hägglund, 1995)). Then, process P_1 has to be selected as the one with the largest estimated dead time, i.e. we have:

$$P_1(s) = \frac{K_1}{T_1 s+1}e^{-L_1 s} \qquad P_2(s) = \frac{K_2}{T_2 s+1}e^{-L_2 s} \qquad L_1 > L_2. \tag{3}$$

In case the two estimated dead times have the same value, the choice can be done arbitrarily.
By choosing

$$Q(s) := \frac{P_1(s)}{P_2(s)} = \frac{K_1(T_2 s+1)}{K_2(T_1 s+1)}e^{(L_1-L_2)s} \quad L_1 > L_2. \tag{4}$$

we have that the two loops have the same complementary sensitivity transfer function (whatever the controller C transfer function is) and therefore the desired ratio is maintained during the whole set-point response (assuming that no model uncertainties are present). In any case, in this context it is sensible to take C as a PI controller and the following tuning rule based on the pole-zero cancellation can be employed:

$$C(s) = K_c \frac{T_1 s + 1}{s}. \tag{5}$$

The choice of the design parameters K_c will be discussed in the next.

While filter Q is devoted to ensure good set-point following performances, filters H and G are devoted to improve the load disturbances response. The design of these transfer functions can be performed by taking into account the transfer functions from the load disturbance inputs d_1 and d_2 and the ratio error $e := ay_1 - y_2$, i.e. the input of the two filters. After somewhat trivial calculations, it results:

$$\begin{aligned} F_1(s) &:= \frac{e(s)}{d_1(s)} \\ &= -\frac{aP_1(s)}{1 + P_1(s)(aH(s) + C(s)) + P_2(s)G(s))} \end{aligned} \tag{6}$$

and

$$\begin{aligned} F_2(s) &:= \frac{e(s)}{d_2(s)} \\ &= \frac{P_2(s)}{1 + P_1(s)(aH(s) + C(s)) + P_2(s)G(s))}. \end{aligned} \tag{7}$$

Then, by choosing

$$H(s) = C(s) \tag{8}$$

and

$$G(s) = C(s) \frac{T_2 s + 1}{T_1 s + 1} = K_c \frac{T_2 s + 1}{s} \tag{9}$$

and by approximating the time delay term in the denominator by a first order power-series expansion $e^{-Ls} \approx 1 - Ls$ (Chen and Seborg, 2002), we obtain:

$$F_1(s) = -\frac{aK_1 s}{(T_1 s + 1)p(s)} e^{-L_1 s} \tag{10}$$

and

$$F_2(s) = \frac{K_2 s}{(T_2 s + 1)p(s)} e^{-L_2 s} \tag{11}$$

where

$$\begin{aligned} p(s) = \; & (1 - aK_1K_cL_1 - K_1K_cL_1 - K_2K_cL_2)s + \\ & aK_1K_c + K_1K_c + K_2K_c. \end{aligned} \tag{12}$$

Hence, it results that $F_1(s)$ and $F_2(s)$ are second order transfer functions where the location of a pole can be arbitrarily selected by suitably choosing the design parameter K_c. Indeed, by fixing

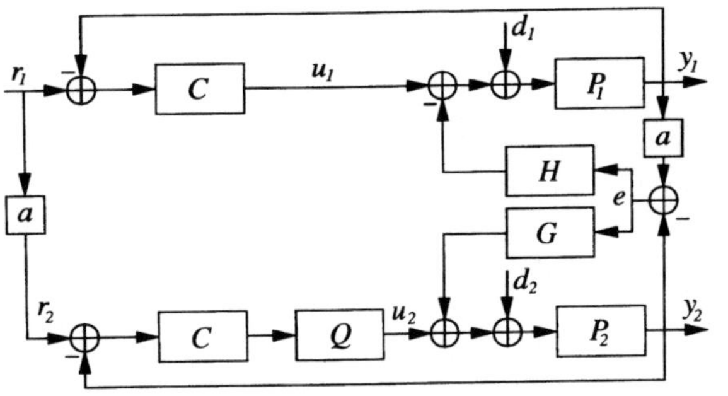

Fig. 3. The new ratio control architecture.

$$K_c = \frac{1}{aK_1L_1 + K_1L_1 + K_2L_2 + aT_fK_1 + T_fK_1 + T_fK_2} \tag{13}$$

we eventually have

$$F_1(s) = -\frac{aK_1/Ks}{(T_1 s + 1)(T_f s + 1)} e^{-L_1 s} \tag{14}$$

and

$$F_2(s) = \frac{K_2/Ks}{(T_2 s + 1)(T_f s + 1)} e^{-L_2 s} \tag{15}$$

where

$$K := aK_1K_c + K_1K_c + K_2K_c. \tag{16}$$

Summarizing, the devised control architecture has only one design parameter, namely the secondary time constant T_f of the load disturbances response, thus determining automatically the value of the proportional gain K_c of the PI controller.

In order to avoid even this design effort, a simple default choice can be imposed, for example by selecting

$$T_f = \frac{\min\{T_1, T_2\}}{10} \tag{17}$$

so that the load disturbances rejection performances are mainly limited by the processes time constants (note again that set-point following performances are addressed by equation (4)).

Finally, it is worth noting that the value of K is inversely proportional to the value of T_f, so that decreasing the value of T_f implies also that the peak value of the ratio error in the load disturbance response is decreased (see (14)-(15)).

Remark 1. It is worth stressing that Q results to be a simple lead/lag filter plus a dead time term and G and H are actually to PI controllers. Thus, no particular difficulty emerges in the implementation of the devised scheme.

4. SIMULATION RESULTS

Simulation results related to different kind of processes are presented in order to show the effectiveness of the devised approach. In each case, a unit set-point step is applied at time $t = 0$ s and then, a unit load disturbance step is applied to the first process P_1 at time $t = 40$ s and to the second process P_2 at time $t = 90$ s. For the sake of clarity the desired ratio a is fixed to one.

4.1 *Example 1 - FOPDT processes*

As a first example, two FOPDT processes have been considered:

$$P_1(s) = \frac{1}{4s+1}e^{-3s} \tag{18}$$

$$P_2(s) = \frac{1}{6s+1}e^{-2s}. \tag{19}$$

By applying the proposed method and the proposed tuning procedure, i.e. by applying formulas (4), (5), (8), (9), (13) and (17) with $K_1 = 1$, $T_1 = 4$, $L_1 = 3$, $K_2 = 1$, $T_2 = 6$, $L_2 = 2$, it results $T_f = 0.4$, $K_c = 0.11$, and

$$Q(s) = \frac{6s+1}{4s+1}e^{-s}, \tag{20}$$

$$C(s) = H(s) = 0.11\frac{4s+1}{s}, \tag{21}$$

$$G(s) = 0.11\frac{6s+1}{s}. \tag{22}$$

The two process outputs, together with the corresponding control variable signals are plotted in Figure 4. It appears that the desired ratio is maintained, as expected, during the whole set-point change transient response and that satisfactory performances are obtained also when a load disturbance occurs both on the first and on the second process. In order to provide a better understanding of the results, the ratio error $e(t)$ is plotted in Figure 5 and the outputs of the two filters H and G are plotted in Figure 6.

A comparison has been made with the method proposed in (Visioli, 2003) (see Figure 2). For this last technique, which outperforms the typical control scheme of Figure 1, results related to the process outputs and to the control variables are reported in Figure 7.

It turns out that the methodology proposed in this paper allows to address the occurrence of a load disturbance on the second process P_2 without decreasing significantly the performances obtained in the rejection of a load disturbance on the first process P_1.

4.2 *Example 2 - High order processes*

As a second example, a fourth order process and a second order plus dead time process have been consid-

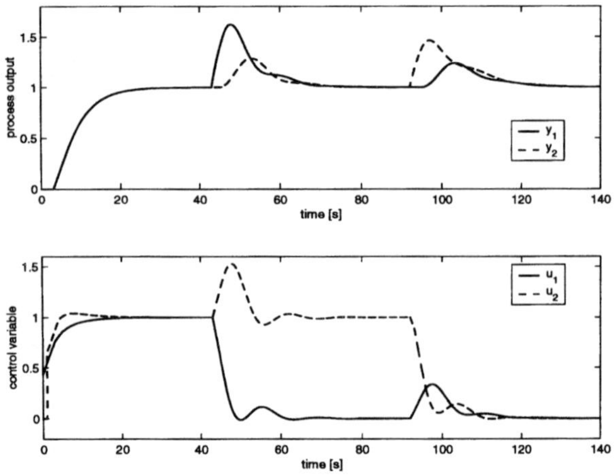

Fig. 4. Process outputs and control variables for example 1 with the new scheme.

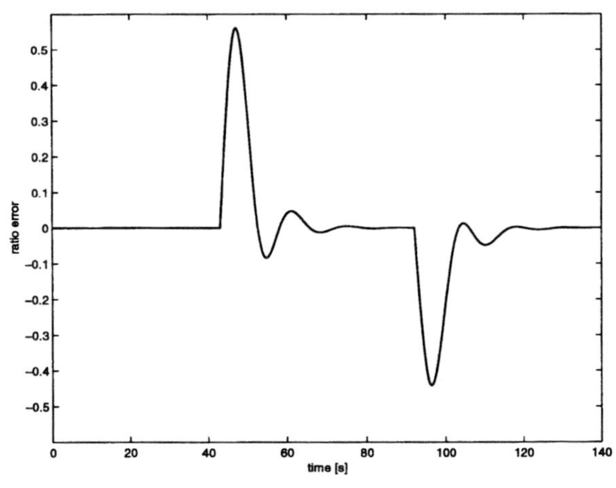

Fig. 5. Ratio error for example 1 with the new scheme.

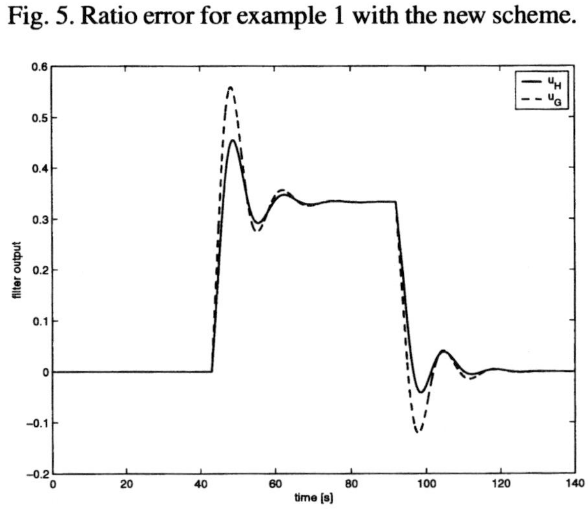

Fig. 6. Filters H and G outputs for example 1 with the new scheme.

ered, mainly in order to verify the effectiveness of the method in the presence of unstructured uncertainties:

$$P_1(s) = \frac{1}{(s+1)^4} \tag{23}$$

$$P_2(s) = \frac{1}{(s+1)^2}e^{-s}. \tag{24}$$

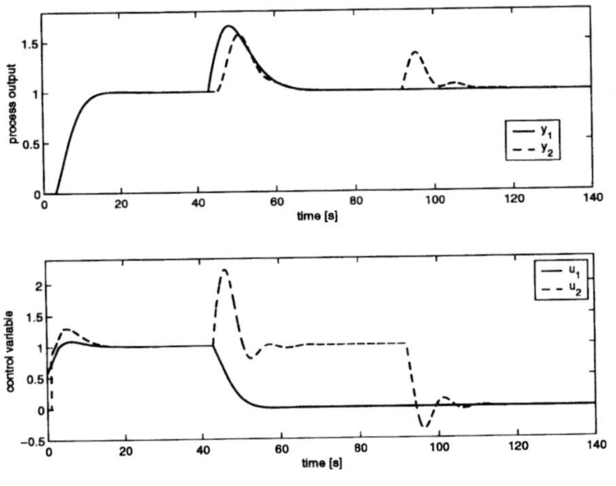

Fig. 7. Process outputs and control variables for example 1 with the scheme of Figure 2.

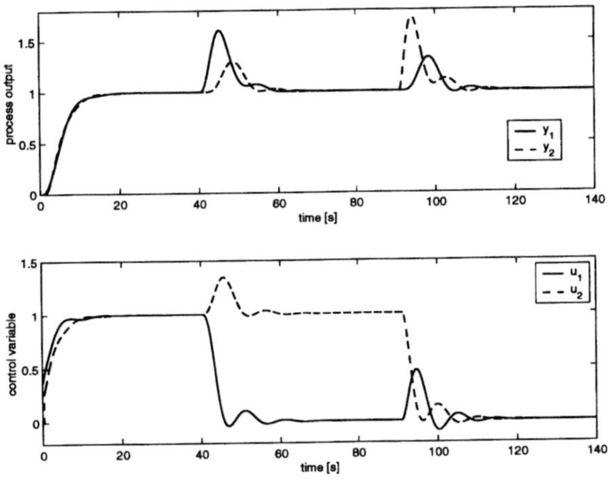

Fig. 8. Process outputs and control variables for example 2 with the new scheme.

Two FOPDT transfer functions have been estimated by applying the area method (which is based on the open-loop step response). It results $K_1 = 1$, $T_1 = 1.84$, $L_1 = 1.92$, $K_2 = 1$, $T_2 = 1.39$, $L_2 = 1.57$. Based on these values, it has been fixed $T_f = 0.14$, $K_c = 0.17$, and

$$Q(s) = \frac{1.39s+1}{1.84s+1}e^{-0.35s}, \tag{25}$$

$$C(s) = H(s) = 0.17\frac{1.84s+1}{s}, \tag{26}$$

$$G(s) = 0.17\frac{1.39s+1}{s}. \tag{27}$$

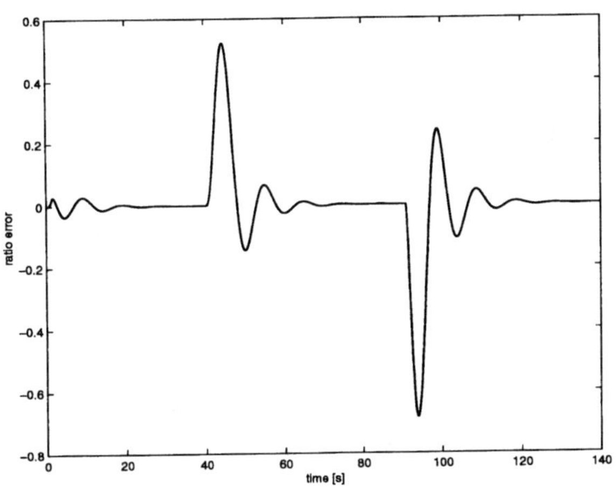

Fig. 9. Ratio error for example 2 with the new scheme.

Results obtained with the new method are reported in Figures 8-10, while those obtained with the architecture of Figure 2 are shown in Figure 11. Obviously, a perfect set-point response (in the sense that has been explained in Section 2) is no more achieved, but results are still satisfactory despite approximated models have been adopted. Comparing the two approaches, the same considerations done for example 1 can actually be done also for this example.

Remark 2. Note that results obtained with the typical ratio control architecture of Figure 1 have not been reported as they can be found in (Visioli, 2003). Further, results related to the Blend station (Hägglund, 2001) have also not been considered because in that approach the load disturbance rejection performances are not addressed.

Remark 3. It is worth stressing that better results can be possibly obtained with the new scheme if a different choice of the transfer functions of the two filters H and G is done. However, it has to be take into account that load disturbances rejection performances are in any case limited by the different (apparent) dead times of the two processes and most of all, the simplicity of the implementation (and therefore the avoidance of a significant design effort for the user) is a major requirement to obtain a satisfactory cost/benefit ratio in the industrial context.

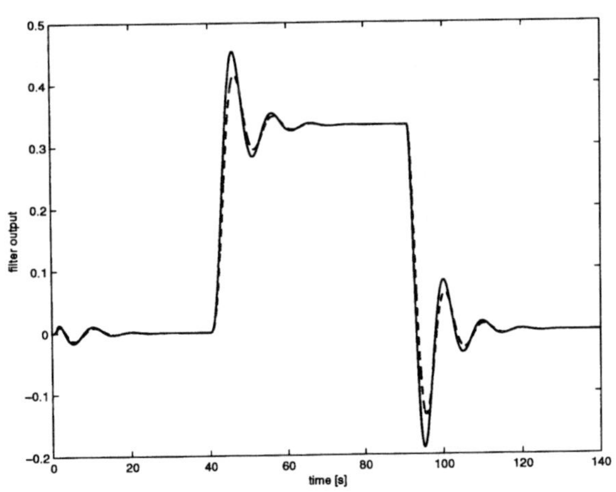

Fig. 10. Filters H and G outputs for example 2 with the new scheme.

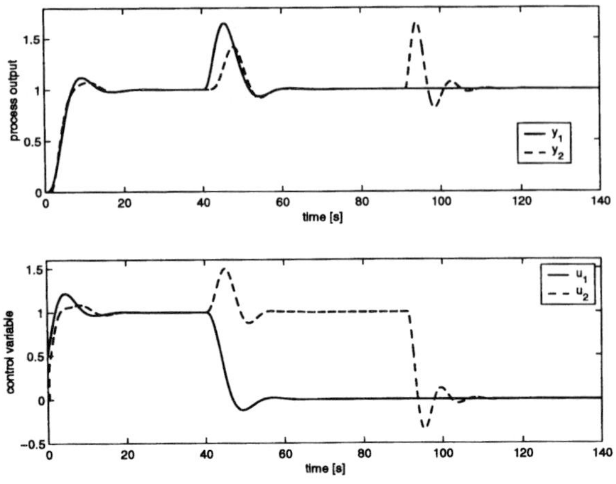

Fig. 11. Process outputs and control variables for example 2 with the scheme of Figure 2.

5. CONCLUSIONS

In this paper a new ratio control architecture has been proposed. The major feature of the devised architecture is that it addresses both set-point changes and load disturbances that might occur on the two processes. A tuning procedure has been proposed in order to avoid any significant tuning effort for the user and to make the proposed methodology suitable to be applied in the industrial context.
Results show that the technique represents a sound alternative choice with respect to the one presented in (Visioli, 2003) whereas disturbances occurring on the second process are of concern.

REFERENCES

Aström, K. and T. Hägglund (1995). *PID controllers: theory, design and tuning*. ISA Press. Research Triangle Park.

Chen, D. and D. E. Seborg (2002). Pi/pid controller design based on direct synthesis and disturbance rejection. *Industrial and Engineering Chemistry Research* **41**, 4807–4822.

Hägglund, T. (2001). The blend station - a new ratio control structure. *Control Engineering Practice* **9**, 1215–1220.

O'Dwyer, A. (2003). *Handbook of PI and PID controller tuning rules*. World Scientific Press.

Shinskey, F. G. (1996). *Process Control Systems - Application, Design, and Tuning*. McGraw-Hill. New York, USA.

Visioli, A. (2003). A new ratio control architecture. In: *Proceedings European Control Conference*.

ELSEVIER
IFAC
PUBLICATIONS
www.elsevier.com/locate/ifac

IMPROVING THE PERFORMANCE OF DUAL RATE CONTROL IN THE ABSENCE OF A FAST RATE MODEL

J.A. Rossiter*, T. Chen+, Sirish L. Shah+ *

* *Sheffield University, UK. S1 3JD,* + *University of Alberta, Canada. T6G 2V4*

Abstract: This paper discusses issues related to model predictive control for dual rate systems in which the input is updated faster than the output. When fast-rate models are not available, a simple and effective approach is proposed for receding horizon control which uses just a single prediction equation to generate fast control moves. *Copyright © 2004 IFAC*

Keywords: Multirate control, performance, predictive control, multirate systems.

1. INTRODUCTION

In many real systems the output can only be measured at a relatively slow sampling rate and certainly much slower than an achievable input sampling rate. In such cases it would seem wise to update the system input at a fast rate (FR) in order to improve performance. However, it is not obvious how a control law should be designed for such a case as most conventional design methods and models assume a single sampling rate. Moreover, as new information (that is measurements) come only at the slow rate (SR), one might wonder if there can be any benefits to updating control decisions at a FR as during intersample[1] periods. That is how can one improve on decisions without new information? This might suggest that one could just as well design a slow rate control law but that goes against the intuitive feel that surely one can get benefits by faster input updates if that is possible. The aim of this paper is to look at such issues in a given context, to be defined, and hence give some answers to what can be achieved.

[1] Intersample is used here to mean in between output measurements.

Many pieces of work on multi-rate systems assume that one has access to an underlying fast rate model (Lee *et al*, 1992). However such an assumption is simplistic as only SR output measurements are available and this places limitations on the identification. Recent work (Li *et al*, 2001) has shown that in some cases a FR model can be deduced but this is still an area of active research and the robustness and limitations of such models are still not well understood. Hence in this paper we take the more pragmatic approach that only a slow rate model is available. However such a model can nevertheless show the dependence on FR input sampling through the mechanism of lifting (Khargonekar *et al*, 1985; Kranc *et al*, 1957) whereby a multi-rate SISO system can be represented by a single rate MIMO system. This will be explained in more detail in section 2.

Given only a SR model there is no obvious mechanism for estimating outputs at intersample points. One common method is to use an internal model (Garcia *et al*, 1982) to estimate intersample outputs and use these in lieu of the actual measurements to allow the implementation of a FR control law. Here it is assumed that an internal model is not available so one might think that the best one can do is to use a SR control law based on lifting. The weakness of this is illustrated in (Rossiter *et al*, 2003).

This paper builds on (Rossiter *et al*, 2003) in that it proposes a simple solution without recourse to infinite horizons, that is it demonstrates how to use a slow rate lifted model and yet update the control law (optimisation) at the fast rate. We note that similar concepts were adopted in (Pan *et al*, 2003). One should also emphasise that a further aim of this paper is to propose only those solutions which gives significant improvements in performance for a fixed (small) computational load; hence the control horizon is restricted to small values.

The paper structured is: Section 2 gives background to multi-rate systems, lifting and predictive control; Section 3 discusses alternative algorithms and section 4 contains numerical comparisons.

2. BACKGROUND

2.1 Lifting for dual rate systems

Consider a system where the input u is updated every T seconds and the output y every $mT = T_m$ seconds. Use the index k for the slow sample rate so that y_k is the output at the kth slow sampling period and use the index l for intersample periods so that $u_{k,l}$ is the lth intersample value of the input during the kth period. One could consider this as a single rate system if the intersample inputs $u_{k,l}$ were grouped as follows:

$$U_k = \begin{bmatrix} u_{k,1} \\ u_{k,2} \\ \vdots \\ u_{k,m} \end{bmatrix} \tag{1}$$

where if $u_{k,1}$ is implemented synchronous with the measurement of y_k, then $u_{k,i}$ is implemented $(i-1)T$ seconds later[2]. Then, using the z-transform operator z to denote the time-delay operator at the slow sample rate T_m, a SISO system could be represented as a MISO system

$$y(z) = G(z)U(z) = [g_1(z)\ g_2(z)\ \dots\ g_m(z)]U(z) \tag{2}$$

where the output is still single dimensional but the input U is m dimensional. A state space model representation is:

$$x_{k+1} = Ax_k + BU_k; \quad y_k = Cx_k \tag{3}$$

[2] For convenience this paper will often use the equivalence $u_{k,m+i} = u_{k+1,i}$

The key point to remember here is that there is implied timing within the components of U_k. As a consequence, when designing for multi-rate systems, care has to be taken in the set up of the control law to respect the so called causality constraint (Chen *et al*, 1994; Meyer *et al*, 1990; Sheng *et al*, 2002). That is one cannot have current inputs dependent on future inputs. In this paper, because we restrict ourselves to dual rate systems, causality is not an issue.

2.2 Predictive control

For the purposes of this paper the illustrations will be based on a generalised pedictive control (GPC, (Clarke *et al*, 1987)) algorithm. Future work will consider algorithms with guaranteed stability (e.g.(Kouvaritakis *et al*, 1992; Scokaert *et al*, 1998)) but it should be noted that the means of performing such an extension is not necessarily obvious or straightforward for MR systems.

As predictive control (MPC) is by now well known, here only the key points are presented. First there is a need for prediction equations. For state space models (3) these can take the form:

$$\underset{\rightarrow k+1}{y} = P_x x_k + H\underset{\rightarrow k}{U}$$
$$\underset{\rightarrow k+1}{y} = \begin{bmatrix} y_{k+1} \\ y_{k+2} \\ \vdots \\ y_{k+n_y} \end{bmatrix} \quad \underset{\rightarrow k}{U} = \begin{bmatrix} U_k \\ U_{k+1} \\ \vdots \\ U_{k+n_y-1} \end{bmatrix} \tag{4}$$

where x_k is the state, y_k the output and U_k the input, all at sampling instant k and P_x, H are defined as:

$$P_x = \begin{bmatrix} A \\ A^2 \\ \vdots \\ A^{n_y} \end{bmatrix}; \ H = \begin{bmatrix} B & 0 & \dots & 0 \\ AB & B & \dots & 0 \\ \vdots & \vdots & \vdots & \vdots \\ A^{n_y-1}B & A^{n_y-2}B & \dots & B \end{bmatrix} \tag{5}$$

Remark 2.1. It is usual to limit the number of d.o.f. in the optimisation of J, hence typically one will minimise over the first n_u components of $\underset{\rightarrow k}{U}$ (that is $u_{k,1}, \dots, u_{k,n_u}$) and assume that the remaining components (that is $u_{k,n_u+1}, \dots, u_{k+n_y-1,m}$ are given by u_{ss}, the expected steady-state value of u. Hence one partitions the part of the prediction equation (4) dependent on future inputs as follows:

$$H\underset{\rightarrow k}{U} = \underbrace{[H_f|H_{ff}]}_{H} \left.\begin{bmatrix} \underset{\rightarrow f}{U} \\ \hline Lu_{ss} \end{bmatrix}\right\} \underset{\rightarrow k}{U}; \ \underset{\rightarrow f}{U} = \begin{bmatrix} u_{k,1} \\ \vdots \\ u_{k,n_u} \end{bmatrix} \tag{6}$$

where L is a vector of ones.

For convenience and to avoid unnecessary details here, one can assume $u_{ss} = 0$ in the initial design and then introduce offset free tracking in a later stage. Hence the term $H_{ff}Lu_{ss}$ is ignored.

In GPC it is usual to minimise a 2-norm performance index. A typical choice for the regulation problem is is

$$\min_{\underset{\rightarrow f}{U}} \quad J = \|\underset{\rightarrow}{y}\|_2^2 + \lambda\|\underset{\rightarrow f}{U}\|_2^2 \tag{7}$$

where λ is an input weight. Subsituting predictions (4) into cost (7) and minimising wrt to $\underset{\rightarrow f}{U}$ gives the optimal set of predicted future inputs as

$$\begin{aligned} \underset{\rightarrow f}{U} &= -\underbrace{[H_f^T H_f + \lambda I]^{-1} H_f^T P_x}_{K} x_k \\ &= -Kx_k \end{aligned} \tag{8}$$

In order to introduce tracking and to allow for offset free disturbance rejection (Muske *et al*, 1993) the control law is implemented as

$$\begin{aligned} \underset{\rightarrow f}{U} - L_{n_u} u_{ss} &= -K[x_k - x_{ss}] \\ &\Downarrow \\ \underset{\rightarrow f}{U} &= -Kx_k + f_k; \end{aligned} \tag{9}$$

where $f_k = Kx_{ss} + L_{n_u}u_{ss}$, L_{n_u} is an n_u vector of ones and u_{ss}, x_{ss} are the expected steady-state input and state giving no steady-state offset, that is $r = y$, $u = u_{ss}$, $x = x_{ss}$ are consistent (r is the set point). The reader is refered to (Muske *et al*, 1993) for details of how to estimate x_{ss}, u_{ss} and here we note only that it depends on state estimation and the set point but is otherwise straightforward.

In GPC it is usual to take only the first component $u_{k,1}$ of the optimal control trajectory $\underset{\rightarrow f}{U}$ and then recompute the optimum at each sampling instant. For a dual rate system one would implement the first component of $\underset{\rightarrow k}{U}$, that is U_k which the reader will recall actually has m seperate moves spread over for the intersample period. Should $n_u < m$, then, as mentioned in remark 2.1, $\underset{\rightarrow f}{U}$ is padded with u_{ss} $m - n_u$ times as this is consistent with the prediction assumption. Hence the first m control moves, those defined in U_k, are given from

$$U_k = \begin{bmatrix} -Kx_k + f_k \\ u_{ss} \\ u_{ss} \\ \vdots \end{bmatrix} \tag{10}$$

3. AN ALTERNATIVE APPROACH TO GPC FOR DUAL RATE SYSTEMS

3.1 The important property of predictions in GPC

Although very succesful in practice (DMC (Cutler *et al*, 1980) is essentially the same algorithm), GPC has one major weakness – there is no apriori stability guarantee for the general case. Although solutions to this exist (e.g. (Kouvaritakis *et al*, 1992; Scokaert *et al*, 1998)), it is pertinent to understand from an intuitive rather than mathematical viewpoint why problems can arise. The answer is very simple and given next.

First let us assume no model mismatch so that the predictions are exact. Nevertheless in GPC the closed-loop behaviour and the open-loop predictions used in the minimisation of J do not match. That is one is minimising predicted performance with a certain assumption (that the inputs become fixed after n_u steps) when what actually happens is something quite different. It is this mismatch that causes the problem, that is the optimisation is illposed as it is based on erroneous assumptions that do not match reality [3]. Whether this causes stability or performance problems depends upon how great the mismatch is. Fortunately for many processes the resulting closed-loop behaviour is quite close to the predictions and good performance results.

Secondly let us look in more detail at the prediction assumption that the input assume a fixed value after n_u steps. This assumption is taken to ensure computational tractability but clearly does not match the expected closed-loop behaviour where the input would be expected to change continuously, at least over the process rise time. MPC strategies deal with this mismatch by continuously updating the assumption, that is the so called receding horizon (RH). At each sampling instant a new d.o.f. (e.g. u_{k+n_u}) is introduced and the optimum input trajectory is recomputed.

So in summary, GPC does have a weakness in the prediction assumption adopted. However in practice, use of the receding horizon concept is sufficient to recover reasonable closed-loop performance. There are of course well documented exceptions where this mismatch can have catrastrophic consequences. The next section will demonstrate that multi-rate systems are one area where this weakness cannot be overlooked.

---

[3] The work of (Scokaert *et al*, 1998) provides an ideal solution when tractable as it removes this mismatch.

3.2 Does the receding horizon work well for dual rate systems ?

The problem with using the receding horizon to rectify what could be considered as a limited prediction class is that it can work well given a fast update of the optimisation, that is where the receding horizon concept is deployed every sampling instant. However, for dual rate systems the control law (10) is updated only every m^{th} sample (of the faster sample rate) so one has to live with the previously computed optimum for m samples before it can be improved. In this case if the prediction class is not close to the desired closed-loop dynamic, the RH update is too infrequent to correct the behaviour. This is particularly marked when $m > n_u$, an observation which the example section will illustrate.

Remark 3.1. This paper takes the assumption that n_u should be small, say 1 or 2 as that is common in practical implementations. Clearly for large n_u the issues are much less significant but the implied computational load may be considered intractable for some cases.

3.3 Improving the GPC algorithm for lifted systems

The normal expectation for GPC implementations is to use the receding horizon concept at the fast sampling rate, that is at the sample rate at which the input is updated. Here it will be shown how that can be achieved effectively in the lifted framework despite the lack of fast rate output measurements and the lack of a fast rate model. It is noted that the existence or not of new output measurements does not affect nominal performance, as for the certain case predictions are exact. In fact output measurements are needed primarily to give robust performance and disturbance rejection.

The key issue in the receding horizon is the introduction of a new d.o.f. It will now be shown that in the dual rate scenario it is straightforward to introduce new d.o.f. at intersample periods and hence to update the predictions at a fast rate. Define the future input trajectory $\underrightarrow{U}_k$ to have 3 parts, a past (that is inputs that have already been implemented), d.o.f. (that is available to the optimisation) and a steady state:

$$\underrightarrow{U}_k^T = [U_{k|\text{past}}^T, U_{k|\text{d.o.f.}}^T, U_{k|ss}] \tag{11}$$

The structure of the components of input trajectory $\underrightarrow{U}_k$ at the lth intersample period is given as

$$\begin{aligned} U_{k|\text{past}} &= U_{k,l|p} = \begin{bmatrix} u_{k,1} \\ \vdots \\ u_{k,l-1} \end{bmatrix} \\ U_{k|\text{d.o.f.}} &= U_{k,l|f} = \begin{bmatrix} u_{k,l} \\ \vdots \\ u_{k,l+n_u-1} \end{bmatrix} \\ U_{k|ss} &= U_{k,l|ss} = \begin{bmatrix} u_{ss} \\ u_{ss} \\ \vdots \end{bmatrix} \end{aligned} \tag{12}$$

Hence at the (k, l) sampling instant (that is the lth intersample period of the kth sample period), the optimisation should be set up as follows:

$$\min_{U_{k,l|f}} \quad J = \|\underrightarrow{y}\|_2^2 + \lambda \|\underrightarrow{U}\|_2^2 \tag{13}$$

where again for convenience it is assumed that $u_{ss} = 0$. Partition H analogously to the partition of $\underrightarrow{U}_k$ (which clearly depends upon l) and but ignore (ignore $U_{k,l|ss}$ (as $u_{ss} = 0$), so that

$$H\underrightarrow{U}_k = H_{p|l}U_{k,l|p} + H_{f|l}U_{k,l|f} \tag{14}$$

Hence substituting (14) into (13) and minimising gives the optimal control law as

$$\begin{aligned} U_{k,l|f} &= -[H_{f|l}^T H_{f|l} + \lambda I]^{-1} H_{f|l}^T \\ &\qquad [P_x x_k + H_{p|l} U_{k,l|p}] \\ &= -K_l x_k + M_l U_{k,l|p} \end{aligned} \tag{15}$$

where $P_L = [H_{f|l}^T H_{f|l} + \lambda I]^{-1} H_{f|l}^T$, $K_l = P_l P_x$, $M_l = P_l H_{p|l}$. One can introduce tracking and offset free disturbance rejection as in (9) so that the optimal control trajectory is given from

$$\begin{aligned} U_{k,l|f} - L_{n_u} u_{ss} &= -K_l(x_k - x_{ss}) \\ &\quad + M_l(U_{k,l|p} - L_l u_{ss}) \end{aligned} \tag{16}$$

where L_i is a i-dimensional vector of ones. As one is doing a fast RH update, only teh first element, that is $u_{k,l}$, is implemented.

Remark 3.2. Control law (16) is time varying within the intersample periods, that is the law depends upon l. This is an inevitable consequence of lifting (Khargonekar *et al*, 1985; Kranc *et al*, 1957; Sheng *et al*, 2002) but nevertheless it appears as a time invariant control law at the slow sampling rate. There can be problems with intersample ripple (Tangirala *et al*, 2001) due to this time varying nature but this should not occur in the presence of integral action which is deployed here.

Remark 3.3. The beauty of the approach proposed here is that it is based on a single prediction model

(4). The only change for intersample periods is how the matrix H is partitioned in (14).

4. NUMERICAL ILLUSTRATIONS

In this section it will be shown how the combination of lifting (which is necessary when a FR model does not exist) and small n_u render more typical GPC algorithms ineffective. By adopting the modification suggested in this paper one can recover the nominal performance to that achievable should the fast rate output measurements be available.

Consider the simple example with a fast rate state space model

$$x_{k,l+1} = \begin{bmatrix} 0.3 & 0.5 \\ 0.1 & 0.9 \end{bmatrix} x_{k|l} + \begin{bmatrix} 0.1 \\ 0.2 \end{bmatrix} u_{k|l}; \qquad (17)$$
$$y_k = [1 \;\; 0] x_k$$

Recall that this model is given for the readers benefit but the assumption is made that only the equivalent model of form (3) is known. Assume that the output is sampled 4 times slower than the input, i.e. $m = 4$. The GPC algorithms of (10) and (16) are implemented for $n_u = 1, 2, 3, 4$ with $n_y = 8$, $\lambda = 1$. The simulations are displayed in figures 1-4 for $n_u = 1, ..., 4$ respectively; circles and dotted lines are used for the fast RH algorithm of (16) and crosses and solid lines are used for the slow RH update algorithm of (10). The x-axis has units of the fast sample rate so new output measurements are given only every 4th sample. The corresponding closed-loop runtime costs are given in table 1.

RH Algorithm	$n_u = 1$	$n_u = 2$	$n_u = 3$	$n_u = 4$
Slow (eqn.(10))	3.18	2.65	2.33	2.14
Fast (eqn.(16))	2.21	2.17	2.14	2.11

Table 1: Closed-loop runtime costs J

It is clear both from the table and the figures that the proposed algorithm using a fast RH update based on an identical prediction model has vastly outperformed the nominal lifted algorithm when n_u is small, even though there has been no new output measurements. The limitation of the prediction assumption is very clear in figures 1-3 where it can be seen that the input moves to a poor default value, that is u_{ss}, during the later intersample periods. If one uses a fast RH law (as iin (16)) the negative effects of this poor assumption can be alleviated, but not if one uses a slow RH (as in 10). Naturally as n_u becomes larger (e.g. fig. 4) then the issue becomes less significant but there is a consequent increase in the online computational burden.

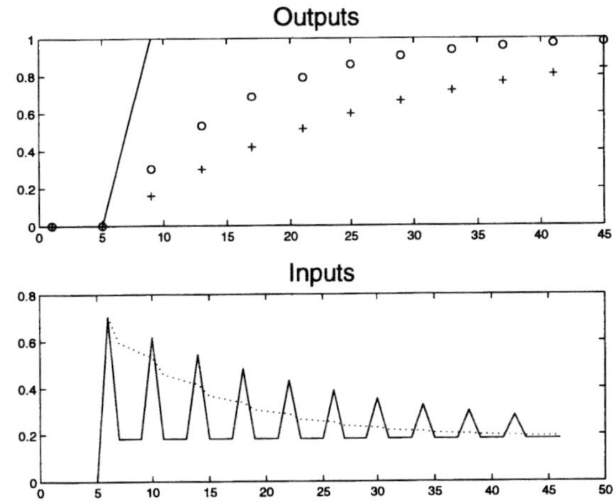

Fig. 1. Simulations with $n_u = 1$

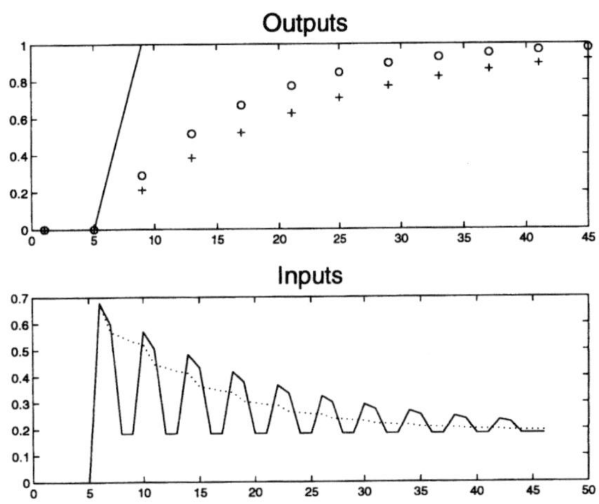

Fig. 2. Simulations with $n_u = 2$

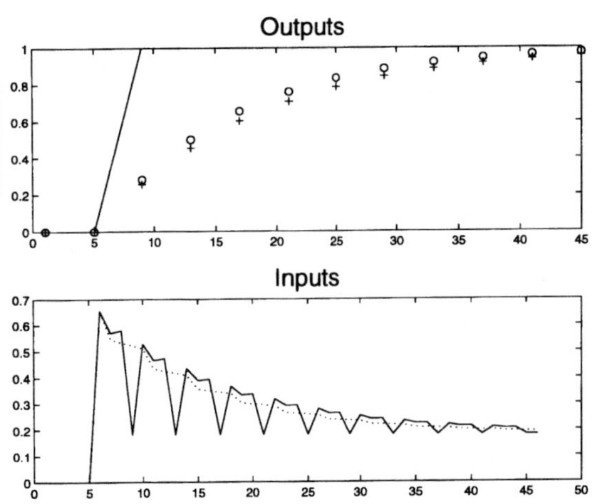

Fig. 3. Simulations with $n_u = 3$

5. CONCLUSIONS AND FURTHER WORK

This paper builds further on (Rossiter *et al*, 2003) which illustrated that in the context of dual rate systems, purely using lifting and a slow control law update, has severe and unnecessary limitations. It is shown here that even though no new obser-

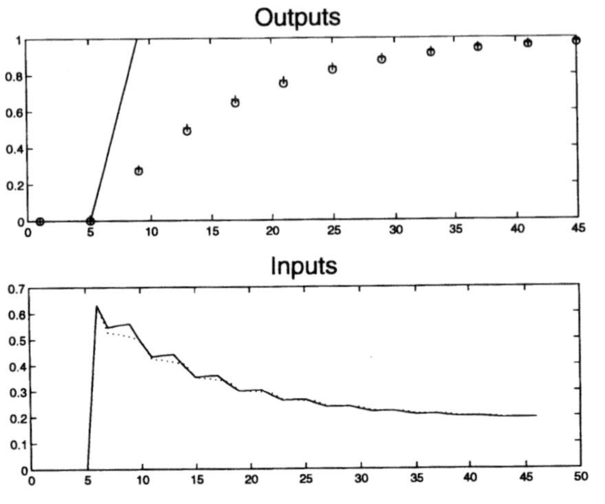

Fig. 4. Simulations with $n_u = 4$

vations are available during intersample periods, the predictive control algorithm is at its best only when used in a receding horizon sense. Hence this receding horizon property should be utilised to its maximum, even during intersample periods. Here a means of doing this has been presented which does not require more than one model (Pan*et al*, 2003) unlike other approaches (Sheng *et al*, 2002) which require different models for each intersample point. Rather it has been shown that one can simply use different partitions of a single prediction equation in order to find the m control laws most relevant to the m intersample periods. The benefits of this simple approach are evident from the examples.

Future work will look at the context of algorithms with guaranteed stability, e.g. (Scokaert *et al*, 1998), and investigate whether the same concept is equally applicable. Allied to this is the issue of constraint handling which was omitted here as secondary to the key contribution. Do constraints bring any new facets to the problem? Moreover there is a need to look at multi-rate systems which are not just dual rate as for these the partitioning of the intersample period is more complex. It should be re-emphasised here that the focus here has not been *what is the best one can do with arbitrary computing power?* but rather *what can one achieve with a relatively simple and undemanding algorithm?*

REFERENCES

Chen, T., and L. Qiu, $\mathcal{H}_\infty$ design of general multirate sampled-data control systems, *Automatica*, 30(7), pp. 1139-1152, 1994.

Cutler, C.R., and B.L. Ramaker, Dynamic matrix control - A computer control algorithm, *Proc. Joint Automatic Control Conference*, San Francisco, U.S.A., 1980.

Clarke, D.W., C. Mohtadi and P.S. Tuffs, Generalised predictive control, Parts 1 and 2, *Automatica*, 23, pp. 137-160, 1987.

Garcia, C.E. and M. Morari, Internal Model control 1. A unifying review and some new results, *I&EC Proc. Design and Dev.*, 21, pp. 308-323, 1982.

Khargonekar, P.P., K. Poolla, and A. Tannenbaum, Robust control of linear time-invariant plants using periodic compensation, *IEEE Trans. AC*, 30, pp. 1088-1096, 1985.

Kouvaritakis, B., J.A. Rossiter, and A.O.T. Chang, Stable generalized predictive control: an algorithm with guaranteed stability, *IEE Proc. Pt. D*, 140, pp. 364-372, 1992.

Kranc, G.M., Input output analysis of multi-rate feedback system, *IRE Trans, Automatic Control*, 3, pp. 21-28, 1957.

Lee, J.H., and M. Morari, Robust inferential control of multi-rate sampled data systems, *Chem. Eng Sci.*, 47, pp. 865-885, 1992.

Li, D., S.L. Shah, and T. Chen, Identification of fast-rate models from multirate data, *Int. Journal of Control*, 74(7), pp. 680-689, 2001.

Meyer, D.G., A parametrization of stabilizing controllers for multirate sampled-data systems, *IEEE Trans. AC*, 35, pp. 233-236, 1990.

Muske, K.R., and J. R. Rawlings, Model predictive control with linear models, *AIChE Journal*, 39(2), pp. 262-287, 1993.

Pan, Y., and J.H. Lee, Identification and control of processes with periodic operations or disturbances, INd Eng. Chem. Res., 42, 1938-1947, 2003.

Rossiter, J.A., T.Chen and S.L.Shah, Developments in Multi-rate Predictive Control, ADCHEM03, 2004

Scokaert, P.O.M., and J.B. Rawlings, Constrained linear quadratic regulation, *IEEE Trans AC*, 43(8), pp. 1163-1168, 1998.

Sheng, J., T. Chen, and S.L. Shah, Generalized predictive control for non-uniformly sampled systems, *Journal of Process Control*, (12), pp. 875-885, 2002.

Tangirala, A.K., D. Li, R.S. Patwardhan, S.L. Shah, and T. Chen, Ripple free conditions for lifted multi-rate control systems, *Automatica*, 37, pp. 1637-1645, 2001.

Copyright © IFAC Dynamics and Control of Process Systems,
Cambridge, Massachusetts, USA, 2004

ELSEVIER
IFAC
PUBLICATIONS
www.elsevier.com/locate/ifac

AN EXCHANGE LANGUAGE FOR PROCESS MODELLING AND MODEL MANAGEMENT

Huaizhong Li **C. Peng Lam**

School of Computer and Information Science
Edith Cowan University, Perth, WA 6050, Australia
email: {h.li,c.lam@ecu.edu.au}

Abstract: Many development tools and environments have been used in process design to develop models and to perform simulations for processes. These development tools and environments generally use proprietary format to represent the developed models. Frequently, a model developed using a particular tool can not be used in another without tedious modifications. It is also common in process design that the developed models are not systematically managed, hence it is often impossible to reuse an existing model due to difficulties to retrieve knowledge and design artifacts relevant to the model. These problems greatly limit the exchange of the available resources in process design. In this paper, we propose an exchange language for process modeling and model management to promote interoperability of the process models and to enable systematic management of the process models.

Keywords: Process modeling, model management, reuse, XML, data model

1. INTRODUCTION

Modeling and model-based simulation have been extensively used across all development phases of the industrial processes. Process modeling is a critical and enabling technology that conveys knowledge from research to the industry (Krieger, 1995).

Mathematical models are essential for process development, design and operation. Mathematical modeling is characterized by a number of experts using different modeling and simulation environments to build models for varying purposes in process design (Foss *et al.*, 1998). Many commercial and research modeling and simulation environments, for examples, gPROMS, Aspen+, Matlab, Model.LA (Bieszczad, 2000), ModKit (Bogusch *et al.*, 2001) and ModDev (Jensen and Gani, 1999), are widely used in both research society and industry to develop models for processes and alternatively to perform process simulations.

It is well-known that heterogeneous modeling and simulation tools for process modeling exclusively use proprietary formats to represent process models. These modeling and simulation tools are incompatible in general, therefore, the models developed using these tools are not inter-operable. Unfortunately, it is a common practice in process design to employ different tools for different purposes. For example, Aspen+ is commonly used for steady state simulation, gPROMS has advantage in dynamic simulation, while Matlab is perhaps the most used tool for control design and optimization. Thus, additional efforts may have to be devoted to convert a model developed in a tool into a format which is usable in another tool (Schopfer *et al.*, 2000). Therefore, it is advantageous that a tool-independent modeling representation can be developed to facilitate information exchange between the heterogeneous tools.

Model reuse refers to the practice that uses available design artifacts, such as model knowledge and tool specific implementations, in the development and maintenance of a process. Obviously, model reuse across different modeling and simulation tools can save time and resource for the process design. However, model reuse in the process design life cycle is also difficult. Due to the proprietary formats used to represent the models in the development tools, current model reuse practice is largely confined to within the same development tool only. Furthermore, besides the incompatibility of the modeling and simulation tools which hiders the reusability of the models, currently there is no well-established standard to develop process models. Hence individual engineers and organizations may develop and maintain models in their own ways which may be cryptical to others. A neutral exchange language for process modeling may help standardizing the model development procedure.

In the development life cycle for process design, model knowledge is frequently lost due to improper documentation or lack of systematic management mechanism. It is common that directories in a file system are adopted to serve the purpose of distinguishing different versions in the evolution of the process models. Existing models are either hard to be found, or hard to be used because of insufficient documentation for the models. As a result, repeated modeling is common in process research and industry (Schopfer *et al.*, 2000). From the authors' experience, a systematic model management strategy is helpful in model reuse for process modeling.

Model integration consists of vertical and horizontal integrations. Vertical integration means the integration of the models of comparable granularity but different functionality to form a complete process, while horizontal integration implies the integration of sub-models into a model to expand the functionality of the model. Model integration heavily involves model reuse in different ways. Model can be reused either *a-priori* where the reusable models are designed and implemented with specific consideration for integration, or *a-posteriori* where the models are reused and integrated almost arbitrarily (Marquardt *et al.*, 2000).

It is clear that there are two key issues related to model reuse and model integration:

Model Management Process models have to be managed in a systematic ways to promote reuse and evolution in the process development life cycle

Model Interoperability Process models have to be inter-operable for model reuse and integration between different development environments. This is especially important for *a-posteriori* model integration

One of the best known notions for reuse is 'design once, use many times'. A centralized model repository certainly reduces the risk of repeated design of the same model. A centralized model repository also eases the task of model management. These considerations clearly call for a framework for process modeling which has a centralized data management system, the associated external converters and integrators, and the model exporters. Such framework can act as an integrated part for a heterogeneous modeling and simulation environment to provide interoperability to the other heterogeneous modeling and simulation environments, and to offer the benefits of reusing different versions of the proprietary artifacts from the other heterogeneous modeling and simulation environments.

In the following, we present an exchange language for process modeling and model management. The exchange is defined by a XML schema.

2. AN EXCHANGE LANGUAGE FOR PROCESS MODELING AND MODEL MANAGEMENT

As discussed in the previous section, A centralized modeling framework can reduce the risk of repeated design of the same model and ease the task of model management. Hence, we propose a centralized framework for process modeling and model management. The conceptual architecture of the framework is illustrated in Fig. 1. The framework is currently under development, it has the following functionalities:

- A data exchange server to interact with the various modeling and simulation environments. The data exchange server accepts proprietary models derived from a modeling and simulation source tool, and exports client-specific models to a particular target tool
- An array of converters and integrators to convert the received proprietary models into a neutral exchange format used in the framework, and an array of exporters to transform the models in the exchange format into the proprietary formats which can readily be used in the client tools
- A modeling and management unit which is the core of the framework. This unit performs the following functions:
  - · Manage model repositories to store and to retrieve the transformed models from the converters and integrators
  - · Provide an on-line modeling tool as a virtual development environment for authenticated engineers to model processes and to store the derived models in the repositories

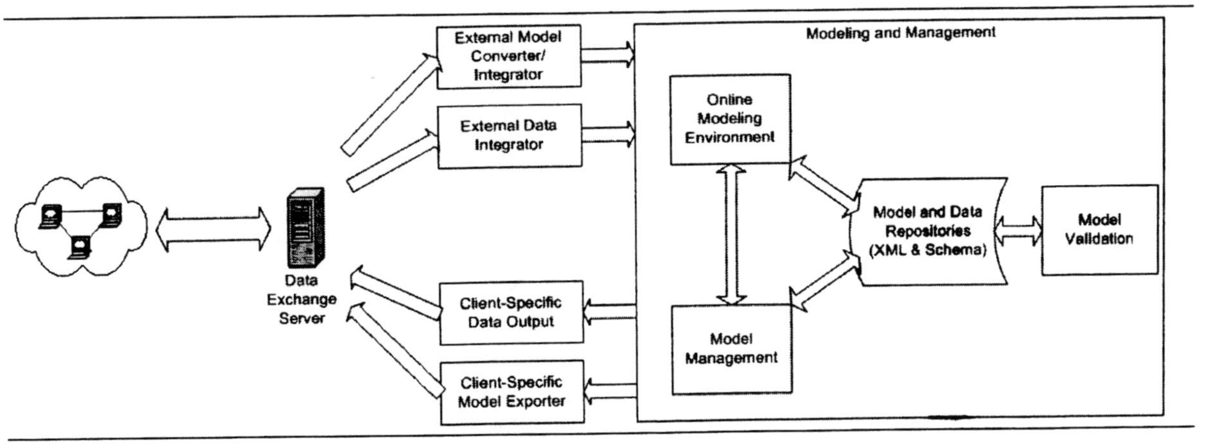

Fig. 1. Conceptual Process Modelling and Model Management Architecture

· Provide model management to the modeling artifacts.

To promote model interoperability and to enable model management, it is advantageous that the models used inside the proposed framework are represented using a neutral exchange language. One of the prominent possibilities for a neutral exchange language is XML. As stated by Wedel (Wedel, 2002), XML has great advantages to represent process models which can be easily represented using the XML flexible meta model. Many existing tools can be used and integrated in the process development environments to process XML data. XML representation of the models also enables systematic model management. Models represented in XML format can be managed and validated using a corresponding XML schema for process modeling.

In the following, we present an exchange language for process modeling and model management. The proposed exchange language is one of the key elements for the modeling and management unit.

We have developed a XML schema to define an exchange language for representation of the process models. As illustrated in Fig. 2, a process model (we call it 'process' in the schema) comprises three parts, namely, the model parts ('model') used to integrate the process, the connections between the models, and the elementary version control mechanism for the process. Due to space limitation, we only present the coarse skeleton of the exchange language in this paper. The actual schema, the relevant documentation, and some sample process XML instances can be obtained from the authors.

Connections between models refer to the coupling between the input ports and the output ports of the relevant models. As shown in Fig. 3, a connection has a source *connectFrom* from an output port of a model, and a target *connectTo* to an input port of a model. The nature of the connection is defined by *link* which contains a list

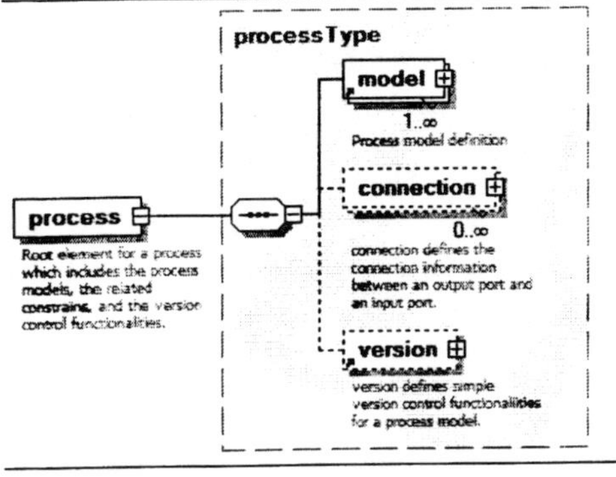

Fig. 2. Process Structure

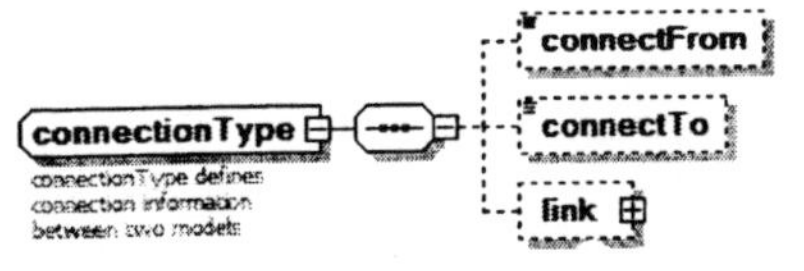

Fig. 3. Connection Structure

of the allowable connection types, for example, a *link* can be stream or liquid for chemical processes.

A process consists of one or more integrated model parts, referred to as horizontal model integration in the previous section. For example, a reaction/separation chemical process in (Lee *et al.*, 2000) is made up of two systems namely a reactor system and a separation system, the reactor system is a continuous stirred tank reactor (CSTR), and the separation system consists of an extractor unit, a flash unit and a distillation column. The CSTR, the extractor unit, the flash unit and the distillation column are all model parts in the developed schema. A model can be represented in a structure shown in Fig. 4. Specifically, a model has a name and a unique ID which can be used to assist in model management. The model can be a continuous, or a discrete, or a hybrid model indicated by *domain*. A model has one or many input ports and output ports, each port is described by a *portName*, a unique *PortID*, and a unit for the port. Such description eliminates any confusion in the connection of ports.

A model can also have sub-models, as illustrated in Fig. 4. The inclusion of the sub-models implements the vertical integration concept discussed in the previous section.

A process model has mathematical concepts like variables, parameters, and equations. The behavior of a model is determined by its variables which are constrained by mathematical equations. The variables, parameters and equations for a model are defined in our schema as *variables* and *equation*. Each variable or parameter has a name, a unique ID, the default value, its unit, and the constraint for the variable or parameter. the definition of *variables* enables the handling of general variables and parameters in various processes.

We have created an equation structure, shown in Fig. 5, to describe the equations for a model. A equation in the exchange language comprises the left-hand-side *equationLHS* and the right-hand-side *equationRHS*, *relation* is defined to describe the relationship between the two sides. Though the term 'equation' has been used in the exchange language, the equation type in the schema is capable to define interim variables, mathematic terms, equations or inequalities.

It should be mentioned that there are alternative ways to represent equations. In CapeML (Wedel, 2002), MathML is adopted for this purpose. However, as indicated in (Wedel, 2002), a MathML-based solution adds difficulties to the implementation of the tools, and the mathematical viewpoint of MathML is counterintuitive to many simulation tools. Unlike MathML, the equation definition in our schema originates from process viewpoint. Namely, the definition of an equation in the schema is designed to be as similar as possible to the one used in process research and industries. Through repeatable use of *operator*, *matrix* and *variableArray*, complex equations in the processes can be straightforwardly represented using the proposed schema.

To assist model management, the elementary version control mechanism has been implemented in the schema. A particular version of a process or a model part consists of the version ID, the version object and the version resource, as shown in Fig. 6. Creators of the model, checkin rules, checkout rules, version history and baselines for the version objects form the content of the resource. The version mechanism implemented in our schema is a simplified variant of many popular version control systems. However, we believe that this simplified version control mechanism is appropriate, and it should be also sufficient for the process model management.

Before we conclude the paper, we need to justify the exchange language proposed in this paper. To our knowledge, two exchange languages have been proposed for process modeling. One of the languages is Modelica (Modelica, 2002). Modelica can comprehensively represent models. However, as pointed out in (Wedel, 2002), Modelica is easy for human to read, but more difficult to be parsed by a software program. Unlike Modelica, our exchange language shares commonality with CapeML (Wedel, 2002) on the ground that the purpose of the language is to promote model reuse and integration for different modeling and simulation tools, hence it is more important if the model representation can be easily handled by computers.

Another exchange language is CapeML (Wedel, 2002) for CAPE-OPEN as previously discussed. CapeML is implemented using XML DTD, while our exchange language is defined using W3C schema. There is rich literature available about the differences between DTD and schema, or about the advantage of the W3C schema, hence we will not argue the benefits of using the W3C schema here. One thing that we have to mention is that it is possible to use reasoning in W3C schema. We plan to use reasoning in the next draft of the exchange language to further improve model exchange capability.

Both Modelica and CapeML do not concern about model management which is crucial for model reuse and model integration. As illustrated above, model management is part of the developed exchange language in this paper. Prototype research has also been carried out to manage the process models using the developed exchange language. The details will be reported elsewhere.

3. CONCLUSION

In this paper, we have proposed and implemented an exchange language for process modeling and model management. The exchange language is formulated in XML format and is defined by a XML schema to facilitate easy integration with existing tools and environments for process modeling. The developed exchanged language can be used to promote interoperability of process models and to enable systematic management of process models.

REFERENCES

Banares-Alcántara, R. (1995). Design support system for process engineering I. requirements and proposed solutions for a design process representation. *Computer & Chemical Engineering* **19**, 267–277.

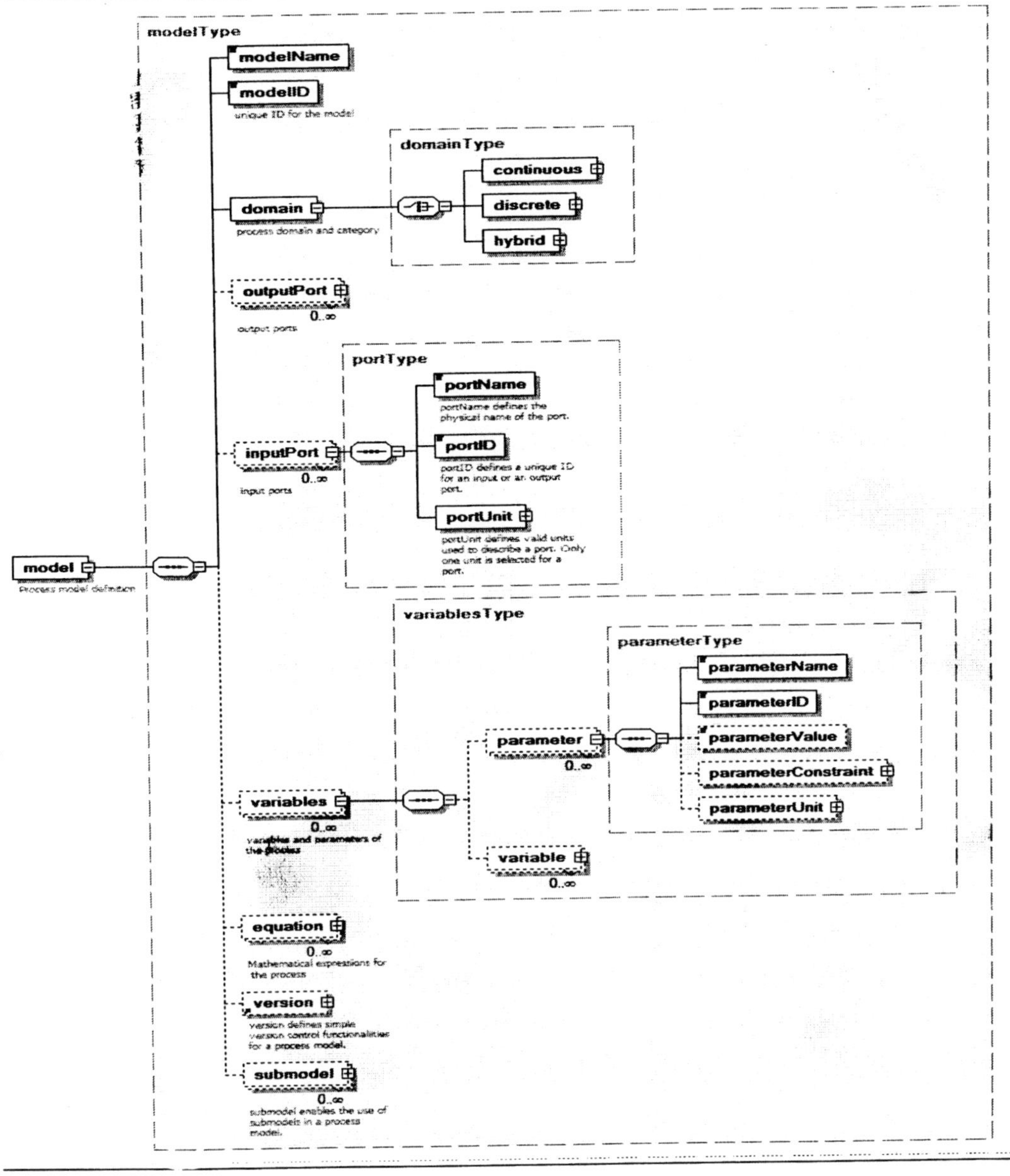

Fig. 4. Model Structure

Bieszczad, J. (2000). A framework for the language and logic of computer aided phenomena-based process modeling. PhD thesis. Massachusetts Institute of Technology.

Bogusch, R., B. Lohmann and W. Marquardt (2001). Computer-aided process modeling with modkit. *Computers and Chemical Engineering* **25**, 963–995.

Foss, B., B. Lohmann and W. Marquardt (1998). A field study of chemical process modeling. *J. Process Control* **8**, 325–337.

Jensen, A.K. and R. Gani (1999). A computer-aided modelling system. *Computers and Chemical Engineering* **23**, 673–678.

Krieger, J.H. (1995). Process simulation seen as pivotal in corporate information flow. *Chemical & Engineering News* **73**(13), 50–61.

Lee, P.L., H. Li and I. T. Cameron (2000). Decentralized control design for nonlinear multi-unit plants: a gap metric approach. *Chemical Engineering Science* **55**(18), 3743–3758.

Marquardt, W., L.v. Wedel and B. Bayer (2000). Perspectives on lifecycle process modeling. In: *Foundations of computer-aided process design* (M.F. Malone, J.A. Trainham and B. Carnahan, Eds.). Vol. 96. pp. 192–214. AIChE Symp. Series 323.

Modelica (2002). *Modelica: A unified object-oriented language for physical systems modeling*. http://www.Modelica.org/.

Schopfer, G., L.v. Wedel and W. Marquardt (2000). An environment architecture to support modeling and simulation in the process design lifecycle. In: *AIChe Annual Meeting*. Los Angeles.

Wedel, L.v. (2002). CapeML - A model exchange language for chemical process engineering. Technical report. RWTH Aachen.

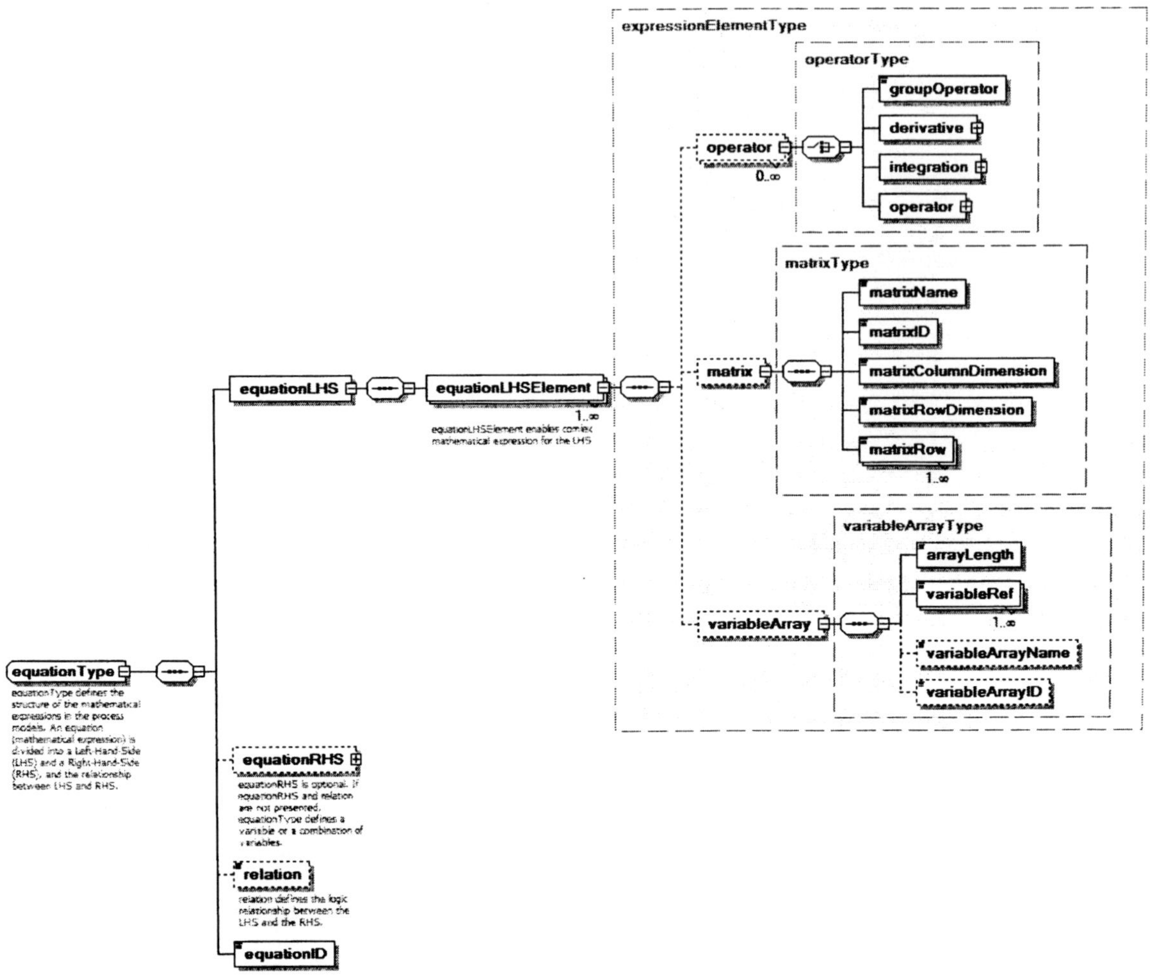

Fig. 5. Equation Structure

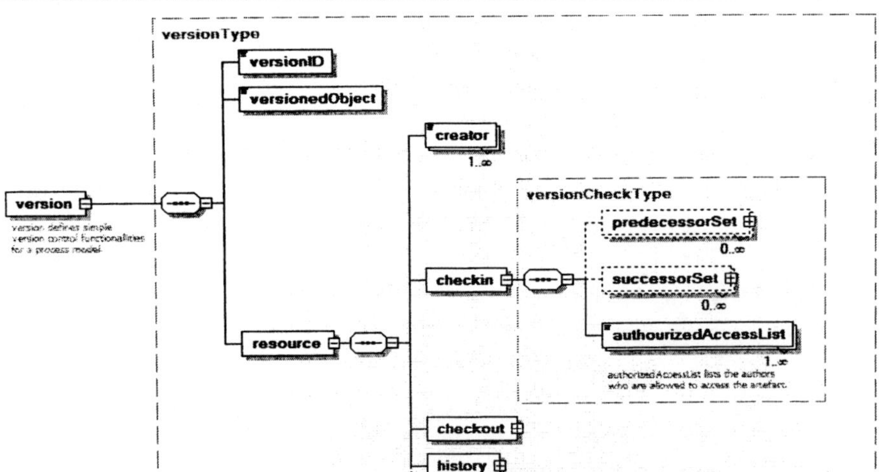

Fig. 6. Version Structure

ELSEVIER
IFAC
PUBLICATIONS
www.elsevier.com/locate/ifac

USE OF BP NEURAL NETWORK TO PREDICT HYDROGEN CONTENT IN COAL

Ju L.C.(Xian Jiaotong University, China), Tade M.O. (Curtin University, Australia) and Zhu J.N. (Curtin University, Australia)

Abstract: Neural network is a popular method for predicting unknown process variables from measured process data. Many learning algorithms have been proposed in the literature to improve model prediction. In this paper, we introduce the concept of sample study risk in neural network (NN) to improve the prediction of hydrogen content in coal using Back Propagation (BP) NN. Targeting the problem of training convergence quality impaired by the interfering information of some samples in BP NN, the validity of the concept of sample study in NN and the feasibility of analysing the chemical element in coal using NN are discussed. Copyright © 2004 IFAC

Keywords: neural network, BP neural network, sample study risk, hydrogen content in coal.

1. INTRODUCTION

The hydrogen content in coal is of great significance for indicating the quality of coal. It is especially important for a combustion process. However, direct determination of hydrogen content is not an easy task. Ultimate analysis, though can give individual elemental content including the hydrogen, the method is nevertheless rather complicated and in most cases it has to be conducted in an external laboratory. Therefore it is difficult to apply ultimate analysis frequently during coal fired power generation. On the contrary, proximate coal analysis is a simple process and is usually available everywhere. This analysis gives the most basic information on coal properties, including contents of moisture, fixed carbon, ash, volatile matter and lower heating value. Unfortunately, the information regarding elemental content is not provided. Nonetheless, it is speculated that the proximate analysis and ultimate analysis have some connections and these connections will be exploited using Artificial Neural Network (ANN) in this paper. Consequently, the ANN prediction model of hydrogen content in coal, the model training and the model checking by proximate analysis of coal, using the Back Propagation (BP) arithmetic of Neural Network (NN) are discussed.

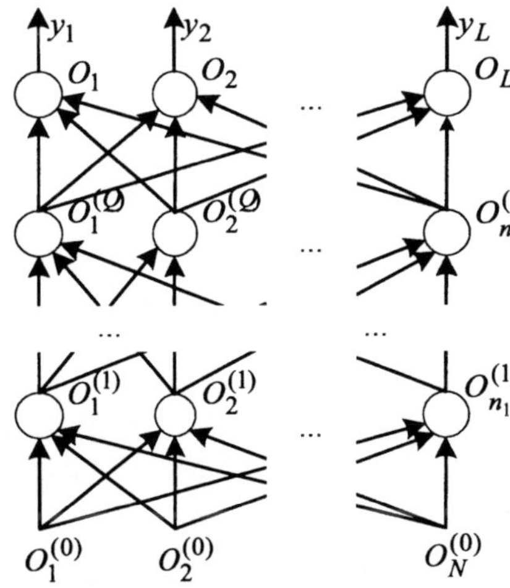

Fig.1 Structure of BP Neural Network

2. BP NEURAL NETWORK MODEL

The Back Propagation Neural Network (BP NN) structure adopted is shown in Figure 1. For the given volume of training samples: $(x_{p1}, x_{p2}, \cdots x_{pN}) \rightarrow (t_{p1}, t_{p2}, \cdots t_{pL}), p = 1,2\cdots$, (where P is the sample number, N, L are the input/output vector dimensions respectively), the average sum of square error between the results of the NN operation and the output of the training samples is:

$$E = \frac{1}{P} \sum_{p=1}^{P} E_p \qquad (1)$$

$$E_p = \frac{1}{2}\sum_{l=1}^{L}\left(t_{pl} - y_{pl}\right)^2 \quad (2)$$

where, t_{pl} is the expected output of the number l output cell in the number p sample; y_{pl} is the results of NN operation of the number l output cell in the number p sample.

The process of training study by the network includes the forward calculation in the inner network and the backward error propagation calculation. These calculations minimize the NN error by adjusting the inner connections of the NN. For the case of a single hidden layer BP NN and multi-layer BP NN connections between the implicit layer and the output layer is adjusted according to the rule of delta [Haykin, (2001)]. The connection between the hidden layers or between the output layer and hidden layer of the multi-layer BP NN is manipulated by the BP arithmetic [Sun et al, (1997), Mills, et al (1996)] (an advanced arithmetic) suggested by Rumelhart and others [Haykin, (2001)].

2.1 Formula for calculating network node

As shown in Figure 1, the forward calculation of the BP inner network is given by:

(1) The output of nodes in the input layer

For N dimensions, the input vector is

$$O_i^{(0)} = \left(x_i - \left(\max(x_i) + \min(x_i)\right)/2\right) \Big/ \left(\max(x_i) - \min(x_i)\right) + 0.5 \quad (3)$$

Where, max (x_i) and min (x_i) is the maximum and minimum value of the ith sample among all the samples.

(2) The input/output of nodes in hidden layers

For implicit layers the input of the number q is:

$$I_i^{(q)} = \sum_{j=1}^{n_{q-1}} w_{ij}^{(q)} O_j^{(q-1)} - \theta_i^{(q)} \quad \left(i = 1,2,\cdots,n_q\right) \quad (4)$$

Where, $w_{ij}^{(q)}$ is the connecting weight from the number j node of the number (q – 1) layer to the number i node of the number q layer; $\theta_i^{(q)}$ is the threshold value of the number i node in the number q layer; $O_j^{(q-1)}$ is the output of the number j node in the number (q – 1) layer; n_{q-1} is the number of nodes in the number (q – 1) layer.

The output of the nodes in the number q layer is:

$$O_i^{(q)} = f\left(I_i^{(q)}\right) \quad \left(i = 1,2,\cdots,n_q\right) \quad (5)$$

Where, f is the Sigmoid function: $f(x) = [1 + \exp(-x)]^{-1}$; $i = 1,2,\cdots, n_q$, n_q is the number of nodes in the number q layer.

(3) The input/output of the output layer

If the output vector has L dimensions, then the input of the nodes in the output layer is:

$$I_i = \sum_{j=1}^{n_Q} w_{ij} O_j^{(Q)} + \theta_i \quad \left(i = 1,2,\cdots,L\right) \quad (6)$$

Where, w_{ij} is the connecting intention from the number j node of the number q hidden layer to the number i node of the output layer; θ_i is the threshold value of the output layer; $O_j^{(Q)}$ is the output of the number j node in the number Q layer; n_Q is the number of nodes in the number Q layer.

The output of the nodes in the output layer is:

$$O_i = f(I_i) = [1 + \exp(-I_i)]^{-1} \quad (i = 1,2,\cdots L) \quad (7)$$

2.2 The iterative computation of connection weights

(1) The connection weight of the nodes between the output and the hidden layers is given by

$$w_{ij}(k+1) = w_{ij}(k) + \alpha\left[(1-\eta)D_{ij}(k) + \eta D_{ij}(k-1)\right] \quad (8)$$

$$D_{ij}(k) = \delta_i O_j^Q \quad (9)$$

$$\delta_i = (t_i - O_i)O_i\left(1 - O_i^Q\right) \quad (10)$$

Where, α is the study rate (step length), and $O < \alpha < 1$; η is the momentum factor, and $O \leq \eta < 1$; D_{ij} (k – 1) is the negative gradient at the (k – 1) time step; $i = 1,2,\ldots, L, j = 1,2,\ldots, n_Q$.

(2) The connection weight between the hidden layers is

$$w_{ij}^{(q)}(k+1)=w_{ij}^{(q)}(k)+\alpha\left[(1-\eta)D_{ij}^{(q)}(k)+\eta D_{ij}^{(q)}(k-1)\right] \tag{11}$$

$$D_{ij}^{(q)}=\delta_i^{(q)}O_j^{(q)} \tag{12}$$

$$\delta_i^{(q)}=\left[\sum_{k=1}^{n_{q+1}} w_{ki}^{(q+1)}\delta_k^{(q+1)}\right]\bullet O_i^{(q)}\left[1-O_i^{(q-1)}\right] \tag{13}$$

Where, q=0,1,...,Q, when q = 0, its input layer, and n_0 = N; $w_{ij}^{(q)}$ is the connecting weight from the number j node of the number (q - 1) hidden layer to the number i node of the number q layer; k is the iteration time.

To minimize the training error of the network (defined by equation (2)) and speed up the convergence of the training, the method of variable study rate is adopted. When $\Delta E_n = E_n - E_{n-1} < 0$, we choose a bigger study rate, and do the converse for the opposite situation. This can be expressed as follows:

$$\begin{cases}\text{if} \quad \Delta E_n < 0, \text{ then } \alpha=\alpha a, \quad a>1 \\ \qquad\qquad \text{else } \alpha=\alpha b, \quad 0<b<1\end{cases}$$

Where, "a" and "b" are constants; α is a variable study rate. This training method can automatically adjust the study rate according to the training error and solve the problem of poor convergence. One only needs to choose the study rate once by experience. The values of "a", "b" and α can be acquired by computer calculation according to the above principle. The optimised value of "a", "b" and α that was obtained from the literature [Yao et al, (1996)] is applied in this paper, α = 0.87, a = 1.05, b = 0.68.

For the general case the correct prediction model can be obtained using the method of variable step length. However, due to the complexity of coal property, even though the momentum item and the method of variable step length study are introduced, the quality of convergence in the training is not perfect. This is because the characteristic parameters of some coal sample are greatly different from the other samples. In other words, some samples can have greater incoherence than others. To describe such incoherence, the concept of study risk in the NN studies is adopted in this work.

3. THE RISK IN NN SAMPLE STUDY

During the course of BP network's online study, suppose the present network after P group samples $\mathbf{x}_p = [x_{p1}, x_{p2}, ..., x_{pn}]^T$, $\mathbf{d}_p = [d_{p1}, d_{p2}, ..., d_{pL}]^T$, (p = 1.2,....,P; **x** is the input and **d** is the output.) training, the matrix of connecting weights is $\mathbf{W}_p$ and the network's output is: $\mathbf{y}_p = [y_{pl}, y_{p2}, ..., y_{pL}]$. When the new sample (P + 1) is received, after one training, the matrix of connecting weights becomes $\mathbf{W}_{p+1}$ and the corresponding connecting weights of the previous P groups of samples' output is: $\mathbf{y}_{p+1} = [y_{(p+1)1}, y_{(p+1)2}, ..., y_{(p+1)L}]$, then the associated risk to the network due to the new study sample can be expressed as:

$$\sigma=\frac{1}{2}\sum_{p=1}^{P}\sum_{l=1}^{L}\left(y_{(p+1)l}-d_{pl}\right)^2 \Bigg/ \left[\frac{1}{2}\sum_{p=1}^{P}\sum_{l=1}^{L}\left(y_{pl}+d_{pl}\right)^2\right] \tag{14}$$

σ indicates the degree of consistency of the new sample P + 1 and the previous P groups of samples. Small σ shows that new sample's mapping relations is approaching the fixed BP NN's and the training is easier to reach convergence after receiving new sample with lower risk. On the contrary, a big σ shows that the new sample includes different information from the old ones and the training will not easily achieve convergence. Accordingly, the training is of a higher risk. Therefore, σ is called the study risk parameter for this sample.

Since the first P groups of samples are not based on a ready network, a hypothesis is necessary to confirm the study risk. Taking a network model, which has taken certain time to train with a comparatively stable error, as the basic network, then the study risk parameter for number p(p ≤ P) group samples' can be expressed as:

$$\sigma_i=\frac{1}{2}\sum_{p=1}^{P}\sum_{l=1}^{L}\left(y_{(p+1)l}-d_{pl}\right)^2 \Bigg/ \left[\frac{1}{2}\sum_{p=1}^{P}\sum_{l=1}^{L}\left(y_{pl}-d_{pl}\right)^2\right] \quad p\neq i, i\leq P \tag{15}$$

where $y_{(p+1)l}$, $(l \neq i)$ is the network's output after the basic network's first training with the number i sample, y_{pl} is the output of the basic network.

When the training error is relatively stable during the training process, the network can be said to have acquired certain optimisation. If further training can result in greater changes to other samples, it is certain that the connecting weights of the network have changed significantly relative to the basic network model. The consistency can be described by σ_i as shown in equation (15).

The study risk parameter of the sample indicates the degree of the dissimilarity of the sample with the other ones. The risk comes from two sources, namely: one is the samples contain significant interrupting information; the other is that the samples contain some new information that is not indicated by the other samples. It is necessary to control a large risk since it directly affects the convergence of BP NN. On the condition that the samples do not change, a large risk indicates that the sample contains significant interrupting information. In order to reduce the risk these interrupting information should be removed. However, this measure is on the other hand likely to sacrifice the opportunity of learning new object properties. Therefore, both the convergence of the network and the learning of new object properties have to be taken into account. If convergence is ensured, use of a large risk parameter is recommended. Controlling the study risk is sometimes very effective to screen out the influence of some large interrupting information and improve the convergence of the network. The following section illustrates how to use the concept of study risk on the prediction of hydrogen content in coal.

4. THE PREDICTION OF HYDROGEN CONTENT IN COAL USING NN

Since coal qualities usually vary from seam to seam, it is more difficult and impossible to predict the hydrogen content for any coal with a single NN model without compromising the precision of the prediction. In order to overcome this difficulty, we separate the coals into several categories according to the differences in ratios of volatile matter, ratios of ash content, ratios of dry air moisture and the ratios of lower heating values. NN is then applied to predict the hydrogen content of each type of the coal respectively.

(1) Selection of the structural parameters of the BP NN

The structural parameters of BP NN include layers and nodes. The numbers of input nodes and output nodes are decided based on the complexity of the process. The key problem is the number of hidden layers and nodes required. Unfortunately, there is no theory available so far to select this number. The ability of the network to approach non-linear function will increase with the enhancement of layers and nodes, but the amount of calculation increases rapidly. Therefore, the structural parameters of BP are obtained on the basis of large number of experiments. In this paper, we adopt four layers structure: one input layer, two hidden layers and one output layer. There are four nodes in the input layers, eight and four nodes in the hidden layers respectively, and one node in the output layer. The nodes of the input layers are the ratios of volatile matter, ratios of ash content, ratios of dry air moisture content and the ratios of lower heating values, respectively. According to the principle of section 2, the coefficient of study risk is confirmed to be 1.1 through repeated computation.

(2) Prediction results of hydrogen content in coal based on BP NN

We used the data of 904 groups of coals to gain the training of design and testing of the model. Considering the study risk of NN, we excluded 61 groups of sample data due to their larger study risk parameter and adopted 720 groups of data as the training sample. After the training of the parameters to the allowable range of error, the validity of the model is verified by the remaining 123 groups of data. The prediction result is shown in Figure 2 for the 123 samples with the model after training. In the figure, the abscissa is arranged in the increasing order of coal's ratio of volatilisation. The prediction error is shown in Figure 3. The biggest error is 8.7%, which is in the allowable range.

For a good model, the square difference of the estimated error should be equal or less than 10^{-2}. We can use the following formula as the square difference of the error [Bakirtzis et al, (1996) and Hou et al, (1996)].

$$\sigma^2 = \frac{1}{n}\sum_{k=1}^{n} e^2(k) \tag{16}$$

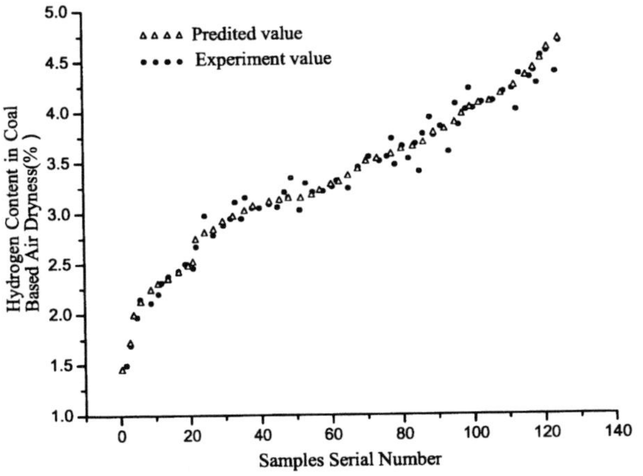

Fig.2 Prediction Results of Hydrogen Content in Coal based on Neural Network

where e(k) is the difference between the predicted value and the measured value. The square difference of the estimated error is shown in Table 1. From Table 1, we can see that the standard square difference of the estimated error for this model is less than 10^{-2}, which demonstrate that the NN can better predict the amount of hydrogen in the coal.

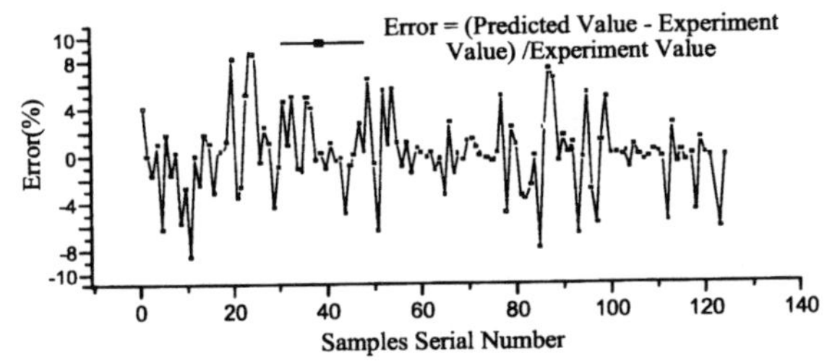

Fig.3 Prediction Errors of Hydrogen Content in Coal Based on Neural Network

Table 1: Validity Verification of Predicted Model

n	$\bar{y}$	squared difference	standardised squared difference
123	3.27	0.017	$0.0016\ \bar{y}^2$

In the above results only 123 samples are not training samples in all the samples, and hence they are the ones used to establish the validity and reliability of the model. The standard square difference of the estimated error for the model is also less than 10^{-2} when less samples were used for the checking (i.e. $n < 123$ in Table 1). Furthermore, the standardised squared difference of the estimated error for the model also improves substantially for $n > 123$ at the expense of increasing computational load.

5. CONCLUSIONS

NN has great potential for modelling non-linear systems and uncertain processes. In this paper, we proposed the concept of sample study risk for NN, and applied it to the prediction of the hydrogen content in the coal. The control of sample study risk can accelerate the convergence of training for the NN and accurate prediction of the hydrogen content in coal also demonstrates that the control of study risk is valid for the convergence of the trained NN.

6. REFERENCES

Hagkin, S. (2001). *Neural networks, a comprehensive foundation.* 2nd Ed. Tsinghua University Express, China.

Sun, Z., Zhang, Z and Deng, Z., (1997). *The theory and technology of aptitude control.* Tsinghua University Express and The Science and Technology Express. Guangxi.

Yao, R., Sun, G., and Tang. C., (1996). *An improved BP method in the NN for multi-layers.* The transactions of South-East University. Vol 26, No. 4. pgs 126-130.

Mills, P.M., Zomaya, A.Y. and Tadé, M.O. (1996). *Neuro-Adaptive Process Control – A Practical Approach.* John Wiley & Sons, New York.

Bakirtzis, A.G., Petridis, V., and Kiartzis, S. (1996). *A neural network short term load predicting model for the Greek power system. IEEE Trans on Power Systems.* Vol. 11, No. 2. pgs 858-863.

Hou, Z., Wu, J., and Zhang, Q., (1996). *Several improved means for the short term burthen in the electric power system.* The Automation of the Electric System. Vol. 20, No. 7. pgs 27-31.

ELSEVIER
IFAC
PUBLICATIONS
www.elsevier.com/locate/ifac

SIMULATION OF PULPING PROCESS USING NEURAL NETWORKS AND HYBRID MODEL

Polowski, N.V.; Aguiar, H.C.; Maciel Filho,R.

Laboratory of Optimization Design and Advanced Process Control. Faculty of Chemical Engineering- State University of Campinas ,P.O. 6066 – ZIP: 13081- 970 – Campus – SP- Brazil

ABSTRACT
A Neural Network Model is developed for a Kraft digester which is a fundamental stage in the pulp production. A deterministic model is used to describe the main features of the process and will provide the data to verify the performance of the artificial neural network modeling. The paper shows that the artificial neural network is a good way to represent the process since it gives equivalent results when compared to deterministic model, and it is easier and cheaper to be developed. *Copyright © 2004 IFAC*

keywords: neural networks, hybrid model, pulp, optimization

1. INTRODUCTION

Pulp and paper industry is one of the most important productive activities and, in fact, it is used as a way of measuring the development of the country. As it is the case for a number of processes, there is a large incentive for the development of cheaper and more environmentally, more friendly operation, but still enhancing product quality. This usually requires the use of a set of tools, as advanced control strategies and real time monitoring and sometimes optimization, which needs a suitable mathematical representation of the process.

Even in a net integrated environment, as real time optimization, in order to find out which are the best operating conditions and policy, a reliable and easy to use mathematical model is required. A deterministic representation, which is developed from the mass and energy balance equations, is a convenient way to obtain a model of the process. On the other hand, artificial neural networks model is an alternative approach which may be very general and easy to be developed since many industrial process have data acquisition system normally used to process monitoring. The objective of this paper is to show the potential of the neural network model to represent the kraft digester and to describe the most important steps to build up such kind of a model aimind the development of hybrid model.

The approach based on neural network is particularly interesting when the phenomena taking place in the process are very complex so that deterministic description will require knowledge and information on heat and mass transfer coefficients, besides kinetics, to allow a reasonable process description. This is the ease for many chemical processes but particulary for pulping process. In fact in such process there is a lack of data for kinetic and especially transport data related to diffusion through the wood chips by the chemicals. In this case, industrial process data may be not enough and laboratory data are usually expensive and time consuming to be obtained. This is typically the case for the wood digester. Wood is a complex structure of fibres glued together by lignin. In order to make pulp, it is necessary to separate them, which can be done mechanically, chemically or by a combination of the two processes. The focus here is on chemical pulp process, more specifically in kraft pulps since they are important industrial process. The kraft process consists of a chemical reaction between the lignin and a cooking liquor, composed mainly of sodium hydroxide and sodium sulfide. However, the liquor can also attack other wood components, such as carbohydrates (cellulose, hemicellulose and others), degrading them, which is not desirable since they are the final product. A high quality pulp should have, among other properties, fibres in good conditions, for that they will contribute significantly for the resistance of the final paper.

Although the composition of the wood used in the process will partially determine the properties of the paper, these latter are also altered during the manufacturing process. According to the operating conditions, the fibres can be more or less degraded, bleached or protected. It will depend on the way in which the reactions develop, attacking mostly the lignin, or the fibres, which is carried out in the digester (Aguiar and Maciel Filho, 1997).

2. KRAFT DIGESTER

The continuous digester is a large pressurized vessel, divided into zones (Figure 1). The first zone is responsible for the impregnation of the wood with liquor. It is very important that the wood receives liquor entirely and uniformly, so that, at the end of the cooking, the pulp is uniform and the amount of rejects is minimum. A heating zone follows the impregnation zone, leading the process to the suitable temperature for the lignin reaction which predominates over the carbohydrates reactions. The chips stay in the cooking zone for a period, and then proceed to the washing and discharge zones. The liquor is added and extracted from the digester in different areas, in order to keep the reaction happening properly, and its flow patterns vary from zone to zone (Assumpção et al.,1988).

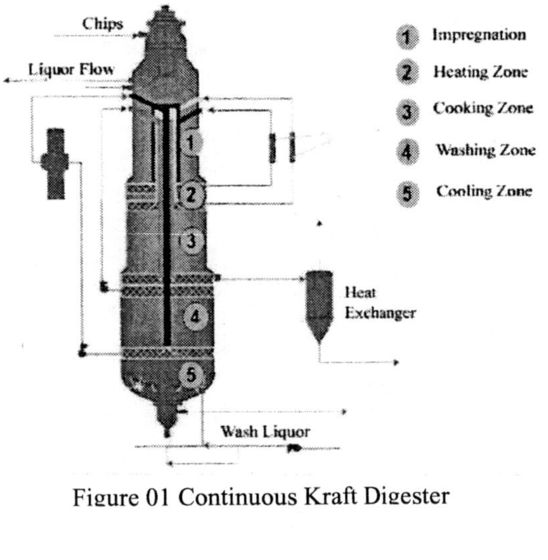

Figure 01 Continuous Kraft Digester

3. FIRST PRINCIPLES MODEL

The kraft pulping of wood is essentially based on the simultaneous occurrence of two phenomena: the first, of a physical nature, involves the mass transfer of the cooking chemicals from the bulk liquor surrounding the wood chips to the liquid filling the pores of the wood structure; and the second is related to the chemical reactions between these reactants, inside the chips, and the main wood components (i.e., lignin and carbohydrates). Because of them, it is of great importance for the understanding of the pulping process to quantify their behavior (Egas et. al., 2002).

Theoretically, first principles model should reproduce the modeled process in any situation, even in extreme conditions. For that reason, their application in simulators and control instruments are spread in various classes of processes. Many cooking models, with different approaches, can be found in literature. One of the earliest kinetic model was developed by Vroom (1957), who used an Arrhenius type expression for the reaction rate temperature dependence to derive the "H-factor". The H factor, which combines the cooking temperature and cooking time into a variable, is at the heart of many control schemes used for kraft pulping. H-factor models that are highly empirical in nature are very useful for batch pulping, but they are not suitable for continuous flow digesters such as the kamyr digester. The major disadvantage of using the H-factor is that it does not include mass transfer and the variation of chemical concentration during the process.

There are several different approaches to the chemistry of the process in the literature. Most of them use the same general form (Saltin, 1992), although some simplifications and modifications are usually applied (Aguiar, Maciel Filho, 1997):

$$-dW/dt=(k_1[OH^-]^a+k_2[OH^-]^b[HS^-]^c)(W-W_\infty) \quad (1)$$

where: W is the wood component concentration; k1and k2 are rate constants that are correlated as a function of temperature using the Arrhenius equation $\left(k = Ae^{-\Delta E/RT}\right)$; W_∞ is the wood component after infinite time; a,b and c are the kinetic orders; $[OH^-]$ and $[HS^-]$ are the OH^- and HS^- concentrations, respectively, and t is time.

Some approaches consider a three step reaction, with three differential equation of the same type above, one for each phase (Gustafson et al., 1983). Gustafson's model predicts the effect of inadequate diffusion on pulping properties but cannot predict the consequences of incomplete penetration on the pulping performance and final pulp properties. Another approach divide the lignin into high and low reactivity lignin, according to the kind of ethers they contain (Saltin, 1992; Mirams and Nguyen, 1994). The model developed by Mirams and Nguyen (1994), which is based on batch cooking process includes mass transfer and intrinsic reaction kinetics based on parallel reactions of lignin, hemicellulose and cellulose.

This study considers the simultaneous reactions of two lignins, one reacting faster and the other suffering a slow reaction. There is also a third type of lignin, in a very low concentration, that does not react with cooking liquor. It is assumed the presence of two types of cellulose and hemicellulose, which do not consume the sulfide ion. Unlike most models, this one considers a simple kinetic equation for the extractives. Wood influence is accounted for in the model through density and porosity (Mirams and Nguyen, 1994; Aguiar and Maciel Filho, 1998). A diffusion term is included to describe the transport of cooking chemicals, assuming the same diffusion coefficient for both, OH^- and HS^-.

The deterministic model equations can be written as follows:

- Lignin

$$\frac{dL_1}{dt} = k_{L_1} \exp\left(\frac{E_{L_1}}{RT}\right)[OH]^a [HS]^b L_1 \tag{2}$$

$$\frac{dL_2}{dt} = k_{L_2} \exp\left(\frac{E_{L_2}}{RT}\right)[OH]^a [HS]^b L_2 \tag{3}$$

- Hemicellulose

$$\frac{dH_1}{dt} = k_{H_1} \exp\left(\frac{E_{H_1}}{RT}\right)[OH]^a H_1 \tag{4}$$

$$\frac{dH_2}{dt} = k_{H_2} \exp\left(\frac{E_{H_2}}{RT}\right)[OH]^a H_2 \tag{5}$$

- Cellulose

$$\frac{dC_1}{dt} = k_{C_1} \exp\left(\frac{E_{C_1}}{RT}\right)[OH]^a C_1 \tag{6}$$

$$\frac{dC_2}{dt} = k_{C_2} \exp\left(\frac{E_{C_2}}{RT}\right)[OH]^a C_2 \tag{7}$$

- Extractives

$$\frac{dE}{dt} = -0.6E \tag{8}$$

- Hydroxide

$$\frac{\partial OH}{\partial t} = \frac{1}{X^2}\frac{\partial}{\partial x}\left(D\frac{\partial OH}{\partial x}\right) + \tag{9}$$
$$+\frac{\rho}{0.1\varepsilon}\left[3.2\times10^{-3}\frac{dL}{dt} + 6.1\times10^{-3}\left(\frac{dH}{dt}+\frac{dC}{dt}\right) + 7.0\times10^{-3}\frac{dE}{dt}\right]$$

$$\frac{\partial HS}{\partial t} = \frac{1}{X^2}\frac{\partial}{\partial x}\left(D\frac{\partial HS}{\partial x}\right) \tag{10}$$

The orthogonal collocation technique was used to solve the partial differential equations (the equations involving diffusion), and the complete system was solved by the LSODE integrator. One inconvenience when generating deterministic models is that parameter estimation methods require acquisition of data not readily available from mill DCS. For instance the concentrations of substances that are not usually measured at the mill. In order to overcome such difficulties, laboratory data is used to adjust mill process parameters (Aguiar, 1997).

4. NEURAL NETWORKS

The most common network used for process modeling is the feedforward neural net. Such net contains at least one input layer, one output layer and one hidden layer. The connections between neurons (or nodes) in this type of network are transmitted forward, from each neuron in a layer to all neurons in the following layer, as shown in Figure 2. Each process variable value is given to one node in the input layer. The neurons on the first hidden layer receive a weighted summation of the signals from input nodes, added to a bias term. Inasmuch as the weights are specific for each connection, and the bias terms are particular for each receiving node, all the neurons will receive a distinct value. The summed values are altered by a transfer function which transforms the signal to a value in a limited range (Figure 3). The sigmoidal function, which transfers numbers to the [0,1] range, is usually used, but other functions can be applied, such as sine, tangent, etc. The transformed value will be the output of the node and will be transmitted to the next layer in the same manner, and from one layer to the other until the output layer releases the net output.

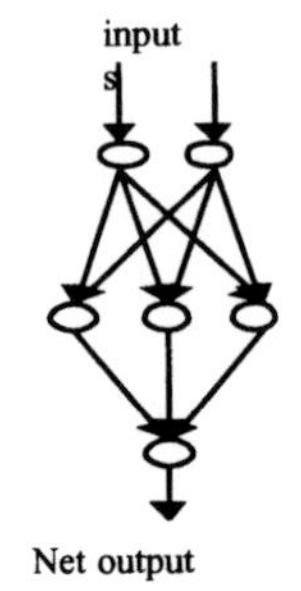

Figure 2 Feedfroward neural network information transference.

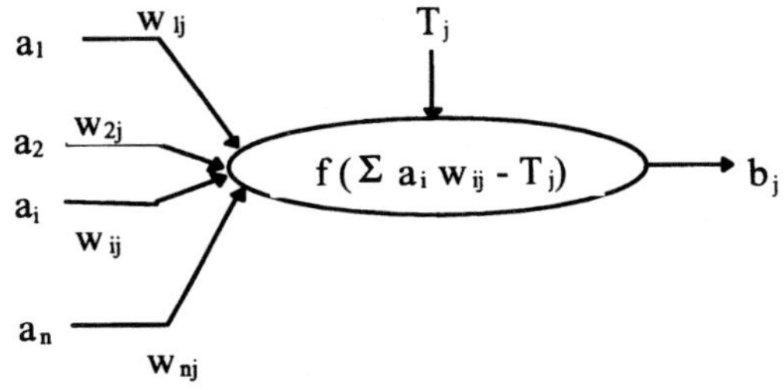

Figure 3 A single neuron

Many backpropagation algorithms can be found in literature. The algorithm used in this work is the Generalized Delta Rule (GDR) algorithm, a gradient descent method that minimizes the sum of squares of the residual (Savkovic-Stevanovic, 1994). The amount and quality of training sets are essential for the success of the artificial neural network simulations. The number of hidden layers and the neurons in each hidden layer can be tested in order to obtain better results. Also the learning rate, a coefficient equivalent to a step size and the momentum term, which keeps the direction of descent from changing too rapidly from step to step can be altered in order to improve the net efficiency. These parameters affect not only the accuracy of outputs, but also the training time, which can be considerably long in some cases especially for on-line trainny (Aguiar and Maciel Filho, 1997).

This ability to recognize and to reproduce cause-effect relationships through training, and filtering noise or irrelevant data, and yet the possibility of working with multiple input and output vectors (MIMO), make neural networks efficient in representing even nonlinear systems. That justifies the choice of neural nets to model the relationship between process operation and its product characteristics in complex processes. Furthermore, the models based on this method may be developed by using process variables that are familiar to mill personnel and easily accessed as inputs, in order to predict outputs that are usually difficult or expensive to obtain. Used in this fashion, the neural network can work as a process predictor. In this paper the neural network used to predict kappa number from process conditions will be described. The usual measurement of pulp properties and delignification extent is the kappa number, which indicates the amount of lignin left in the pulp.

Nevertheless, neural nets have their limitations. The training step is very time consuming, especially with industrial noisy data. They can not extrapolate far beyond their training range, and also do not allow interpretation of the phenomena they are simulating. An alternative approach to pick up advantage of both deterministic and Artificial Neural Network approach is to develop a hybrid model as follow.

5. HYBRID MODEL

Although mathematical models can describe relatively well some behavior of the process, they make a number of assumptions that turn them into approximated models, and often are not able to reproduce properly the nonlinearities of the systems. On the other hand, neural networks represent non linear patterns properly, but do not allow extrapolation. Table 1 describes the 9 inputs variables used on Hybrid Model..

Table 1 - Selected input variables

Variables	Unities
Total effective alkali	g/l NaOH
Temperature (top)	^oC
H Factor (top)	-
Temperature (CD1)	^oC
H Factor (CD1)	-
Effective alcali (CD1)	g/l NaOH
Temperature (CD2)	^oC
H Factor (CD2)	-
Effective alcali (CD2)	g/l NaOH

This work proposes a hybrid model, based on the known part of a first principle model, combined to a neural network to account for the phenomena that is not comprised in the deterministic equations, in order to overcome limitations of both methods.

Hybrid models may be developed in many different combinations. Knowledge of the process may be used to impose constraints (as inequalities) to the neural model. The network can also be used as default, but the deterministic model should be used when the variables are out of training range. Also, the mathematical model may be used in situations when data is not available. A common approach is having the first principles model as the basis, with the neural net calculating unknown parameters. Still, the network can be used to learn the deviation between the mathematical model output and the aimed output. In fact, this may be easier than learning the complex relationships between input variables and process results. Other possible approach is using the deterministic model as reinforcement for the function relationship between inputs and outputs.

The choice of how to combine both parts depends on the precedence of data or first principle system knowledge, and it is particular to each case. The approach considered in this work uses the mathematical model as a reinforcement for the relationship between inputs and outputs. The mathematical model receives information from process variables and generates the kappa number according to the kinetic equations. The result is then fed into the net along with other process variables, and the net yields a final kappa number. The scheme is described in Figure 4.

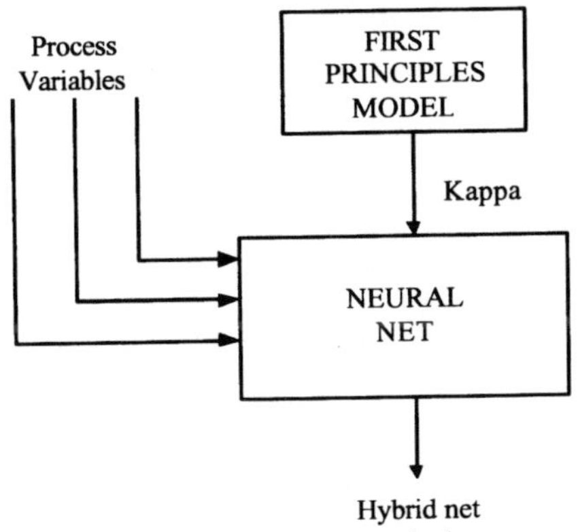

Figure 4 Hybrid model scheme

The mathematical model receives information from process variables and generates the kappa number according to the kinetic equations. The result is then fed into the net yields a final kappa number (Figure 4). Training is carried out as described for a pure network training section. However the mathematical model calculates an extra input variable, which is the calculated kappa number, for each training set. A hybrid network using the neural net to estimate the unknown parameters is other development not considered in this work.

6. RESULTS

Many adjustments in network configuration and analysis of the input variables set must be done in order to obtain the best network configuration for the modeled process. The net was initially set to be trained with one hidden layer and all 22 available variables. Table 2 describes all the 22 inputs variables. However, it did not converge to the predetermined minimum error. Changes in net configuration were done gradually until best results were obtained. The parameters to be optimized in a neural network are: number of layers and neurons in each la yer, momentum term, offset, learning rate, error tolerance, transfer function and random initial weights algorithm.

Table 2 - Initial input variables

Variable	Unities
Production rate	RPM/h
Total white liquor flow	m^3/h
Wood density	kg/m^3
Total Aakali charge	%
Total effective alkali	g/l NaOH
Sulfidity	%
Temperature (top, CD1, CD2)	°C
H Factor (top, CD1, CD2)	-
Aakali charge (top, CD1, CD2)	%
Effective alkali (top, CD1, CD2)	g/l NaOH
White liquor flow (top, CD1, CD2)	m^3/h

The predicted variable in the model was the kappa number, due to its importance to the next sections of the process and its significance for mill personnel. Regression and graphical analysis of the data were carried out in order to eliminate the variables that do not affect kappa number significantly. As a result, only 9 variables were used to feed the input neurons in the network. Figure 5 illustrates the improvement obtained for kappa number prediction with previous data analysis.

The number of industrial profiles was not the desired number of training sets for a neural network but it reflects the potential to be used even with relatively few data. In any case, it is expected that when more data is gathered, network prediction will improve considerably. The hybrid network was trained in the same way. The tendency dictated by the mathematical model result is responsible for a slight improvement in network prediction, as indicated in Figure 6. The training time, when measured as number of computational iterations, was shorter for the hybrid network than for the pure neural net. This is a relevant information when on-line applications either for advanced control or process optimization are desired.

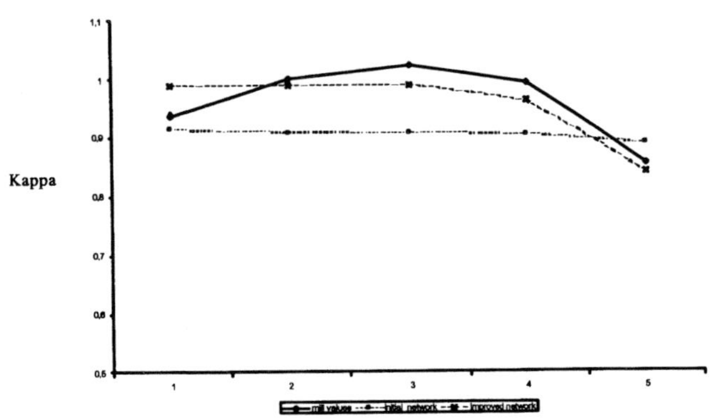

Figure 5 Kappa prediction

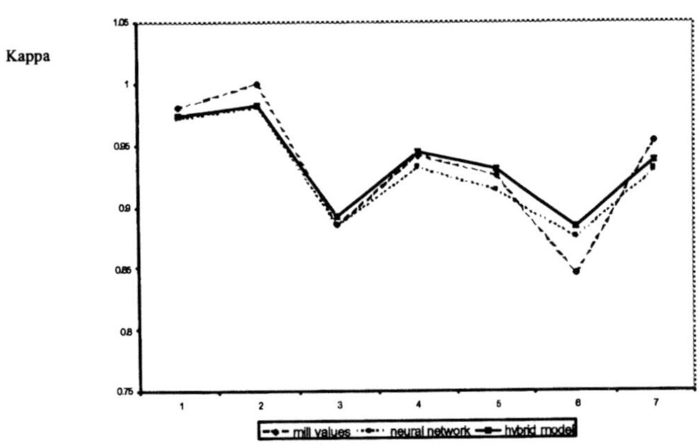

Figure 6 Comparison between hybrid model and neural model predictions

7. CONCLUDING REMARKS

Both, neural network and hybrid network models produced similar results, showing to be adequate tools to model an industrial cooking process, even with a reduced number of training sets. The simple approach used to combine first principles knowledge and neural network features produced improved results when compared to the pure net. However, it should be seen as the starting point for a more complex model, where process knowledge and pattern learning capacity can be better combined in order to improve prediction capability and to provide a more complete approach of the process (when all variables and relevant phenomena influencing results are accounted for).

The expected benefit is that simulation program assists in decision making on process policies according to the desired product characteristics, and enables mill personnel to correct the process before undesired pulp grades are produced. Furthermore the software can be used to design specific pulp characteristics that are requested by customers, and can be achieved by specific parameters controlled in the process. It can also be a useful tool for training operators and engineers. The software was written in Fortran 77 (public domain) with an easy t o use interface.

9. REFERENCES

Aguiar, H.C. and Maciel Filho, R., (1997). Modelling and Optimization of Pulp and Paper Processes Using Neural Networks, ESCAPE 8.

Assumpção, R.M.V. et al., (1998). Celulose e Papel – Tecnologia de fabricação da pasta celulósica, chapter VI, Vol. 1, IPT.

Egas, A. P. V., et al., (2002). Experimental Methodology for Heterogeneous Studies in Pulping of Wood, *Ind. Eng. Chem. Process Des. Dev.41.2529-2534.*

Gustafson, R.R. et al., (1983). Theoretical Model of Kraft Pulping Process, *Ind. Eng. Chem. Process Des. Dev.* 22(1):87.

Jiménez, G., Gustafson, R.R. and McKean, W.T., (1989). "Modeling Incomplete Penetration of Kraft Pulping Liquor", *Journal of pulp and paper science* 15(3):J110.

Mirams, S. and Nguyen, K. L., (1994). Application of Predictive Kinetic Models in Eucalypt Kraft Pulping, *International Pan Pacific Conference, Tappi Proceedings*, 73-82.

Saltin, J.F., (1992). A Predictive Dynamic Model for Continuous Digesters, *Pulping Conference, Tappi Proceedings*, 261-268.

Savkovic-Stevanovic, J, (1994). *Computers and Chem. Engng* 18(11/12):1149

Vroom, K. E., (1957). The 'H' Factor: A Means of Expressing Cooking Times and Temperature as a Single Variable, *Pulp Pap. Mag. Can.*, 58, n.3, 228.

ELSEVIER
IFAC
PUBLICATIONS
www.elsevier.com/locate/ifac

USE OF BIFURCATION ANALYSIS FOR MODEL IDENTIFICATION PURPOSES

Vega, M.P., Coimbra, K.B., Araújo, J.M. and Scheid, CM.

Departamento de Engenharia Química - Universidade Federal Rural do Rio de Janeiro, BR 465, km7 – CEP: 23890-000 – Seropédica – RJ – Brasil

Abstract: Nonlinear system identification poses challenging questions because a closed general theory is not available for this field. Particularly, models based on neural networks may present incompatible general process behavior, leading to improper closed loop responses, even when they allow a satisfactory one step ahead prediction of process dynamics, as required by traditional validation methods. It is shown here that performing detailed bifurcation and stability analysis may be very helpful for the adequate analysis of neural models. The study of the many parameters that are defined a priori during the training of the neural network is of paramount importance, as the spurious dynamic behavior is related mostly to the use of incomplete data sets during the learning process. Strategies to improve the quality of the identification procedure are provided and analyzed, using cyclones as a case study. *Copyright © 2004 IFAC*

Keywords: Neural networks, Stability analysis, Bifurcation diagram, Cyclones

1. INTRODUCTION

When all the important characteristics of the process are known, building a phenomenological model is normally an easy task. However, difficulties may emerge during the solution of the resulting system of equations. For this reason, a number of different model reduction techniques have been proposed in the literature, Benallou *et al.* (1986); Pinto *et al.* (1988); Levine *et al.* (1991).

Empirical linear modeling, Luyben (1990), is well studied but, as pointed out by Pearson *et al.* (1997), a well-developed theory for nonlinear system identification is not available. The neural network (NN) approach has proved to be a useful tool and is the most popular framework for empirical model development, although estimating the huge number of parameters frequently present in the model may be regarded as a major problem to be solved, Su *et al.* (1997) and Cybenko (1989).

Nonlinear system identification involves model parameters selection, determination of the forcing function which is introduced into the plant to generate the output response, estimation of model parameters and comparison of plant information and model predictions for data not used in model development. All steps represent very challenging theoretical and practical problems, for a general theory is not available. As a result, further investigation on systematic techniques for nonlinear model validation, characterization of the amount and type of process data required to build nonlinear empirical models with satisfactory predictive capability and the identification of nonlinear model structures which are capable of capturing a wide variety of process behaviors are future research issues that need to be explored.

The main objective of this paper is building a methodology for the analysis of NNs, which allow the development of confident model identification procedure for use in the laboratory and industrial environment. NNs were validated in terms of the traditional methods, Pollard *et al.* (1992) and Sriniwas *et al.* (1995), and in terms of their complex static and dynamic behavior, using bifurcation and stability analysis. As observed through many examples, Vega (2001), the use of traditional validation tests is not enough to guarantee successful use of NNs, as the complex dynamic behavior displayed by the model may be completely different from the one displayed by the plant, resulting in poor identification efficiency. Good identification performance can be detected using bifurcation diagrams, which can be computed with typical numerical packages, such as AUTO, Doedel (1986). It is proposed here that standard bifurcation analysis be used as an additional validation procedure for implementation of NN models. In order to illustrate this point, big and small cyclones, separating solid particles from gases, are taken as a case study. Bifurcation techniques are used to allow the development of confident NNs, based on experimental data. Experimental data are obtained elsewhere, Halasz *et al.* (2000). It is shown that the bifurcation analysis of NNs may be very helpful for the appropriate development and implementation of model identification. Therefore, bifurcation analysis should be included as an innovative validation criterion in the nonlinear system identification methodology.

2. THE PROCESS ANALYZED

A cyclone is a particle removal device without moving parts which spins a gas stream to collect entrained particles by centrifugal force. Figure 1 shows a typical cone-under-cylinder cyclone design. In this design, particle-laden gas enters the cyclone at the top of the cylinder and makes several revolutions due to the shape of the entry forming a vortex with a high tangential velocity which accelerates particles outward to the wall for collection. Below the bottom of the gas exit tube, the spinning gas gradually migrates inward, to a central core axially along the cylinder centerline, and from there up, finally out to the exit tube. Cyclones are used in the field of air pollution control with small cyclones for ambient and source sampling and large ones for industrial particulate control. Their simple design, low capital and maintenance costs, and adaptability to a wide range of operating conditions have made cyclones one of the most widely used industrial dust collectors.

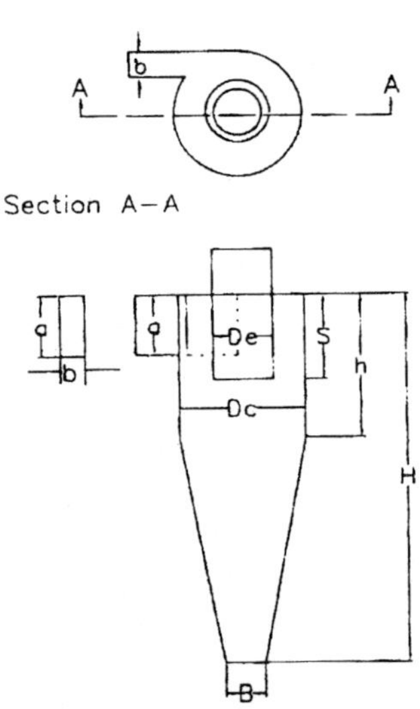

Figure 1 – Typical cyclone design

Due to the complex three-dimensional fluid flow in cyclones the exact mechanisms of removing particulate are still not fully understood. In addition, different operating conditions such as temperature, pressure and flow rate add even more difficulties to the already complicated problem. Therefore, most cyclone theories are based on a simplified model or depend upon empirical correlation equations. Although these theories are valid for certain cyclone operating conditions, none of them has been satisfactorily validated. Therefore, study on cyclones is still largely based on experiment methods and design of cyclones relies upon experience, trial and design guides.

Halasz *et al.*, 2000 employed a FLN (Functional Link Network) neural network for modeling the particle collection efficiency, according to Equations 1 and 2, using literature data for big and small diameter cyclones, respectively. The NN models were a function of operational conditions [flow (Q), diameter of the particle to be separated (dp) and ratio between fluid viscosity and particle density $P=(\mu/\rho_S)$], the equipment characteristic relations [(a/Dc), (b/Dc), (De/Dc), (H/Dc), (h/Dc), (B/Dc), (S/Dc)] and cylindrical section diameter, Dc. The lower and upper limit values for the parameters are reported in Table 1.

$$\begin{aligned}\eta(dp) &= 0.7152 + 20.048D_c - 164.69\frac{P}{D_c} \\ &- 0.0036\frac{1}{Q^2} - 25.40\left(D_c^{0.5}\right)^3 - 0.4738\left(\frac{D_c}{Q}\right) \\ &+ 1.4098\left(\frac{D_c^{0.5}P}{dp}\right) + 0.013\frac{D_c^{0.5}}{Q^2} \\ &- 4.9078D_c^{0.5}\left(\frac{b}{D_c}\right)\left(\frac{D_e}{D_c}\right)^2 \\ &+ 2.4907\times10^8\left(\frac{D_e}{D_c}\right)^2\left(\frac{b}{D_c}\right)P + 1.984\times10^{-7}\times \\ &\frac{\left(b/D_c\right)}{Qdp} - 12.0252\left(\frac{b}{D_c}\right)\left(\frac{a}{D_c}\right)\left(\frac{D_e}{D_c}\right)^2 \\ &+ 2.2\times10^{-18}\frac{1}{dp^3} - 3.4044\left(\frac{b}{D_c}\right)^2\left(\frac{S}{D_c}\right) \\ &+ 0.0025\left(\frac{B}{D_c}\right)^2\left(\frac{h}{D_c}\right)^4\end{aligned} \qquad (1)$$

3. STABILITY ANALYSIS

Bifurcation theory provides tools for a system stability analysis under its parametric changes. As the parameters undergo changes, the existence of multiple steady states, sustained oscillations and traveling waves might occur for highly nonlinear processes, Ray *et al.* (2000).

The quality of the different NNs was evaluated by analyzing their dynamic structure (attractors and respective stability characteristics). In order to do that, bifurcation and stability analyses were carried out to unveil the attractors of the NNs, employing well-known continuation methods. The computations presented in this paper were carried out with routines provided by AUTO, Doedel (1986). Steady states are stable if all eigenvalues of the Jacobian matrix are inside the unity circle. If any of the eigenvalues is outside the unity circle, the steady-state solution is unstable. At a Limit Point, an eigenvalue becomes identically equal to +1. At this point, multiple steady state solutions usually appear and a change in stability occurs. At a Hopf (Thorus) Bifurcation Point, a pair of complex eigenvalues crosses the unit circle with non-zero imaginary component and a branch of oscillatory solutions may appear. At a Period Doubling Bifurcation Point an eigenvalue becomes equal to -1 and branches of periodic solutions usually develops. It is assumed here that a good NN model should exhibit a steady-state bifurcation diagram that resembles the bifurcation diagram of the original process. This means that NNs should present the same bifurcation structure and bifurcation points of the original process. Therefore, in our particular case, a good NN should not display multiple steady-state solutions and steady-state unstable operation conditions.

$$
\begin{aligned}
\eta(dp) = {} & 1.26 - 2.49\times10^{-14}\left(\frac{D_e{}^2 S}{Qdp^2}\right) - 0.312h \\
& - 6.64\times10^{-19}\left(\frac{S}{Q^2dp^2}\right) - 0.039\left(D_c{}^{0.5}Hh^2\right) + \\
& 2.21\times10^{-48}\left(\frac{1}{P^4Qdp^2}\right) - 7.75\left(D_c ah\right) - 2.24 \\
& \times10^{-14}\left(\frac{B^4}{Qdp^2}\right) - 3.18\times10^{-32}\left(\frac{D_e{}^2}{P^2Qdp^2}\right) \\
& + 2.75\times10^{-26}\left(\frac{B^2}{Qdp^4}\right) - 19.1\left(D_c{}^{3/2}\right) \\
& - 6.97\times10^{-14}\left(\frac{D_c}{Qdp^2}\right) + 3.05\times10^{-15}\left(\frac{Sh}{Qdp^2}\right) \\
& + 1.45\times10^{-4}\left(\frac{D_e{}^2S^2}{Q}\right) + 3.12\times10^{-18}\left(\frac{D_e{}^2}{Q^2dp^2}\right) \\
& - 1.09\times10^{-7}\left(\frac{D_e{}^4}{Q^2}\right) + 2.31\left(D_e{}^8\right) + 5.4\left(D_c{}^{0.5}b^{0.5}h\right) \\
& - 7.86\times10^{-44}\left(\frac{1}{P^2Qdp^4}\right) - 2.14\times10^{-17}\left(\frac{b^{0.5}D_e{}^2S}{P^2}\right)
\end{aligned} \qquad (2)
$$

AUTO automatically detects bifurcation points and provides routines for computation of the multiple steady state solutions, oscillatory and periodic solutions that arise at these special points. Unstable behavior usually occurs in the vicinities of these bifurcation points, as at least one of the eigenvalues crosses the unity circle. Therefore, identification of the bifurcation points may allow the understanding of how and why the NNs models fail at certain process operation conditions.

Table 1 – Parameter range

	big diameter cyclone		small diameter cyclone	
parameter	lower limit	upper limit	lower limit	upper limit
D_c, m	0.18	0.4	0.01	0.05
$P\times10^9$, m^2/s	4	80	4	80
$Q\times10^3$, m^3/s	43	240	0.15	4.8
a/D_c	0.25	0.7	0.21	0.59
b/D_c	0.15	0.3	0.15	0.32
D_e/D_c	0.3	0.58	0.2	0.8
S/D_c	0.35	3.5	0.4	1.64
H/D_c	3	6	2.3	4.3
h/D_c	0.5	3.5	1.1	2.1
B/D_c	0.38	1	0.3	0.69
dp, m	1×10^{-6}	1×10^{-4}	1×10^{-6}	1×10^{-4}

4. RESULTS AND DISCUSSION

Literature information provides that as the parameter Dc improves, the collection efficiency decreases. The increase of P produces a decrease in the collection efficiency. As Q, S, h, B and dp increase the collection efficiency increases. In addition, as b and De increases, the collection efficiency diminishes. Finally, it is well-known that the collection efficiency varies from zero to unity.

The bifurcation diagrams presented in Figures 2 and 3 for big and small cyclones, respectively, were built using central parameter values (see Table 1). All NN models present incompatible behavior for the collection efficiency. Besides, as b increases the small cyclone NN model predicts a wrong increase in the collection efficiency. The small cyclone NN model also indicates that the collection efficiency is insensitive to parameters B, De, S and P. The big cyclone NN model displays an inverse behavior for the continuation parameter De. The big cyclone NN model presents an incompatible collection efficiency behavior as the continuation parameter dp varies. The small cyclone NN model presents opposite collection efficiency behavior as the continuation parameter h varies. Finally, the big and small cyclone NN models present spurious collection efficiency behavior as the continuation parameters P and S varies.

5. CONCLUSIONS

It was observed that NNs built to represent cyclones presented incompatible open loop steady state stable behavior. Also, the NNs generalization capacity was very poor in the central data region, producing spurious solutions, incompatible with the real system performance. Bifurcation diagrams built in the extreme parameter limits, not shown in the text, presented the same incompatible collection efficiency values. The collection efficiency is supposed to be severely affected by Q, dp, De, B and Dc. However, none of the bifurcation diagrams presented such behavior.

It may me concluded that although NNs may be used successfully for identification purposes, care must be taken regarding the number and range of data used, as the simple manipulation of the neuron activation functions, NN architecture and initial guesses used for NN training are not enough to guarantee the building of proper models. It is also shown that, detailed bifurcation analysis may be very useful for the proper design and identification. Bifurcation diagrams indicate whether spurious model responses are predicted by the model and therefore indicate whether additional effort is needed for proper model development. The result obtained with the bifurcation analysis validation criterion motivates further analysis for adequate NN cyclone model building. The authors are now working on the analysis of the training data set (500 points), used by Halasz *et al.* (2000), in order to investigate the reasons for the weak NN performance. Bifurcation analysis was used as an efficient tool for validating models, which were built in a supervisory fashion, using available experimental data. Unknown systems can be unveiled if the convergence of the bifurcation diagram to a final structure is used as a quality index in an iterative procedure. This sophisticated validation procedure is indicated for complex units operating in a large range of operating conditions and using nonlinear model based controllers.

6. REFERENCES

Benallou, A., Seborg, D.E. Dynamic compartmental models for separation processes. AIChE J., 32 (1986) 1067-1078.

Cybenko, G., Approximations by superpositions of a sigmoidal function. Math. Control Signal Systems, 2 (1989) 303-314.

Doedel, E., AUTO: Software for Continuation and Bifurcation Problems in Ordinary Differential Equations, California Institute of Technology, Pasadena, 1986.

Halasz, M.R.T. and Massarani, G., Performance analysis and design of small diameter cyclones. Braz. J. Chem. Eng., 17 (2000) 451-458.

Levine, J. and Rouchon, P., Quality control of binary distillation columns via nonlinear aggregated models. Automatica, 27 (1991) 463-480.

Luyben, W.L., Process Modeling, Simulation and Control for Chemical Engineers. McGraw-Hill, Singapore, 1990.

Pearson, R.K. and Ogunnaike, B.A., Nonlinear process control. Englewood Cliffs, NJ: Prentice-Hall, 1997.

Pinto, J.C. and Biscaia, E.C. Order reduction strategies for models of staged separation systems. Computers and Chemical Engineering, 12 (1988) 821-831.

Pollard, J.F., Broussard, M.R., Garrison, D.B. and San, K.Y., Process Identification using Neural Networks. Computers and Chemical Engineering, 16 (1992) 253-270.

Ray, W.H. and Villa, C.M., Nonlinear dynamics found in polymerization processes – a review. Chemical Engineering Science, 55 (2000) 275-290.

Sriniwas, G.R., Arkun, Y., Chien, I-L and Ogunnaike, B.A., Nonlinear identification and control of a high-purity distillation column: a case study. Journal of Process Control, 5 (1995) 149-162.

Su, H.T. and McAvoy, T.J., Nonlinear process control, eds M.A. Henson, D.E. Seborg, chapter 7, p.371, Englewood Cliffs, NJ: Prentice-Hall, 1997.

Vega, M.P., Identificação, monitoramento e controle em linha de reação de polimerização em reatores tubulares, DSc. Thesis, PEQ/COPPE/UFRJ, Rio de Janeiro (in Portuguese), 2001.

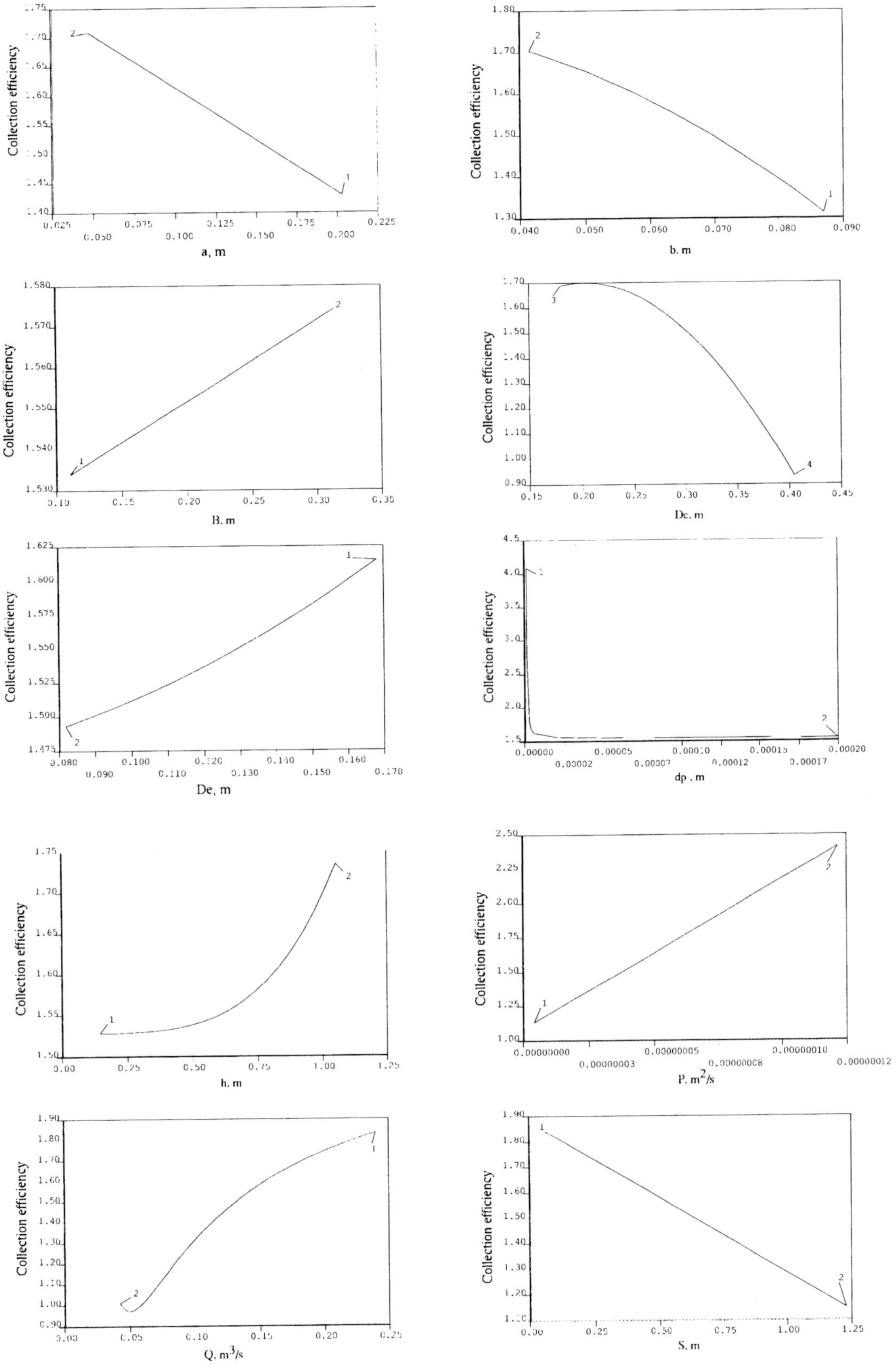

Figure 2 – Bifurcation diagrams for big diameter cyclones

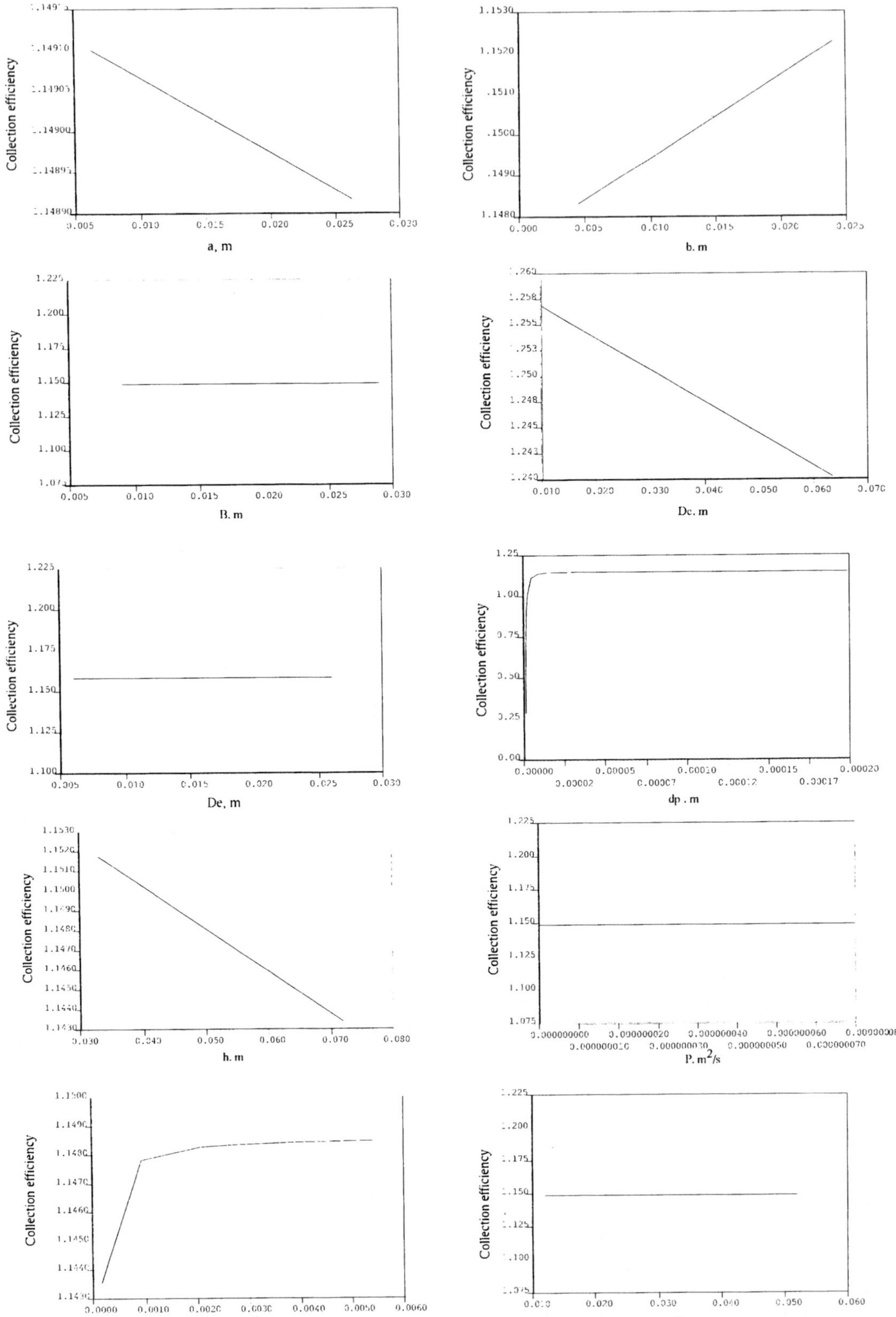

Figure 3 – Bifurcation diagrams for small diameter cyclones

ELSEVIER
IFAC
PUBLICATIONS
www.elsevier.com/locate/ifac

PARAMETER ESTIMATION FOR BATCH PROCESSES USING A BAYESIAN APPROACH

Zhen Lu, Elaine B. Martin and Julian Morris

Centre for Process Analytics and Control Technology
School of Chemical Engineering and Advanced Materials,
University of Newcastle, Newcastle upon Tyne, NE1 7RU, UK

Abstract: In a changing operational environment, a major challenge that still exits is the assured state and parameter estimation of dynamic processes. The values or expressions of important parameters can be difficult to determine and initial errors may be present in some parameters as a result of changes in the initial operating conditions. Furthermore as a result of variations in the environmental and operational conditions or the dynamic characteristics of the process, parameters may be time-varying. In this paper, an on-line Bayesian parameter estimator is developed and evaluated on a simulation of a batch methyl methacrylate process. *Copyright © 2004 IFAC*

Keywords: Bayesian Analysis; Parameter estimation; Polymerization.

1. INTRODUCTION

The operating objectives in many batch polymerisation processes must satisfy complex property requirements in terms of the final polymer whilst simultaneously achieving the best operational performance of the batch. Most properties of the polymer products are directly or indirectly linked with the molecular structural properties of polymer chains (e.g. molecular weight distribution, copolymer composition distribution, chain sequence length distribution, etc.), which are difficult (sometimes impossible) to measure on-line. Average polymer molecular weight properties (e.g. number and weight average molecular weight), which can be indirectly inferred from the on-line measurement of the solution viscosity or melt index of the polymer, are commonly selected as the control variables that need to be maintained within well-determined limits so that product quality criteria can be satisfied.

Control strategies require that pre-determined trajectories for key process variables (e.g. reactor temperature) are followed during batch operation (Thomas and Kiparissides, 1984). However, the operational performance of a batch polymerisation reactor is affected by process disturbances potentially resulting in changing operating conditions. A consequence of this is the introduction of process-model mismatch and hence the need for time-varying model parameter values and the regular updating of the optimal control trajectories during batch operation, otherwise the control strategy may fail to meet the product quality specifications and the operating requirements (Kiparissides, *et al,* 2002).

Despite the large number of papers discussing the control of polymer properties using an Extended Kalman Filter (EKF), very few tackle the issue of robust parameter updating (Gagnon and MacGregor, 1991; Kozub and MacGregor, 1992). MacGregor and co-workers (1991, 1992) analysed the parameter updating issue and emphasised the need to update as many parameters as possible in order to incorporate all possible changes that could affect the process.

The general problem of parameter estimation is associated with the fitting of a model to a set of measurements. Given a model with some unknown parameters, a parameter estimator calculates those parameter values that result in the model predicted values of the process outputs being closest to the corresponding measured values of the process outputs. In off-line parameter estimation, a model is fitted optimally to the process measurements from one or more completed process runs. In contrast, in on-line parameter estimation, a model is fitted to the past and present process measurements whilst the process is in operation (Narendra and Annaswamy; 1988, Bastin and Dochain, 1986; Yin, 1993).

Schuler and Schmidt (1992) discussed the effect of unmeasured and time varying parameters such as reactive heat flow and heat transfer coefficients, and presented a parameter adaptive EKF with additional stochastic states to approximate these parameters. Scali, *et al*, (1997) applied a parameter adaptive EKF to a methyl methacrylate (MMA) polymerisation in a continuous solution CSTR reactor. Sirohi and Choi (1996) applied the EKF and an optimisation based non-linear dynamic parameter estimator to a continuous olefin polymerisation reactor. They combined the two algorithms to estimate the uncertain kinetic parameters on-line.

Parameter estimation through on-line optimisation has also been widely applied. For this approach, parameter estimates are obtained by solving on-line a minimisation problem such as the sum of squared errors (Muske and Rawlings, 1995; Robertson, *et al*, 1996). As in state estimation, this parameter estimator can either have an increasing or constant horizon (Muske and Rawlings, 1995). Although this methodology is computationally expensive, it allows the inclusion of constraints in the estimation. With the advancement of computers in terms of speed, the computational problem will soon cease to be an issue.

An alternative approach to implementing parameter estimation is to make use of Bayesian technologies. Here prior information about the parameters is combined with the process observations.

2. BAYESIAN PARAMETER ESTIMATION

2.1 *Basic Framework of Bayesian Parameter Estimation*

Consider a non-linear process. It is assumed that the process has an $(n\times 1)$ state vector, $\mathbf{x}(t)$, that can be described by the continuous time process model:

$$\dot{\mathbf{x}}(t)=\mathbf{f}(\mathbf{x}(t),\boldsymbol{\theta},t) \qquad \mathbf{x}(0)=\mathbf{x}_0 \tag{1}$$

where the measurement model takes the form:

$$\mathbf{y}(t)=\mathbf{h}(\mathbf{x}(t),\boldsymbol{\theta},t)+\mathbf{v}_t \tag{2}$$

$\mathbf{f}(.)$ and $\mathbf{h}(.)$ are non-linear system and observation functions respectively. $\mathbf{y}(t)$ is an $(m\times 1)$ vector of observations over the time period $t=1,2,\ldots$. For convenience, the measurement information $\mathbf{y}(t)$ is represented by $\mathbf{y}_t$. Let $\mathbf{D}_t$ be the information set available prior to time t. Thus given the initial prior information $\mathbf{D}_0$ at time point $t=0$, at any future time t, the available information set is:

$$\mathbf{D}_t=\{\mathbf{y}_t,\mathbf{D}_{t-1}\} \tag{3}$$

$\boldsymbol{\theta}$ is a vector of unknown parameters about which there may be some prior beliefs that can be expressed as a probability density function, $p(\boldsymbol{\theta})$. This prior distribution can be defined using either previously monitored data, or an expert's knowledge. $\mathbf{v}_t$ is the vector of observational error, and is defined as a normal random variable with zero mean:

$$\mathbf{v}_t \sim N[0,\mathbf{V}_t] \tag{4}$$

Equations (1) and (2) define the model relating $\mathbf{y}_t$ to the parameter vector $\boldsymbol{\theta}_t$ at time t, and has a probability distribution that is a function of the unknown parameters. The dependence of $\mathbf{y}_t$ on $\boldsymbol{\theta}_t$ can be expressed as the conditional probability density function $p(\mathbf{y}_t\,|\,\boldsymbol{\theta}_t)$:

$$p(\mathbf{y}_t\,|\,\boldsymbol{\theta}_t)=p(\mathbf{h}(\mathbf{x}_t,\boldsymbol{\theta}_t)+\mathbf{v}_t\,|\,\boldsymbol{\theta}_t) \tag{5}$$

Once $\boldsymbol{\theta}_t$ has been defined, the following characteristics can be obtained:

$$\begin{aligned}
&E(\mathbf{y}_t\,|\,\boldsymbol{\theta}_t)=E(\mathbf{h}(\mathbf{x}_t,\boldsymbol{\theta}_t)\,|\,\boldsymbol{\theta}_t)+E(\mathbf{v}_t\,|\,\boldsymbol{\theta}_t)\\
&=\mathbf{h}(\mathbf{x}_t,\boldsymbol{\theta}_t)\\
&\mathrm{cov}(\mathbf{y}_t\,|\,\boldsymbol{\theta}_t)\\
&=\mathrm{cov}(\mathbf{h}(\mathbf{x}_t,\boldsymbol{\theta}_t)\,|\,\boldsymbol{\theta}_t)+\mathrm{cov}(\mathbf{v}_t\,|\,\boldsymbol{\theta}_t)=\mathbf{V}_t\\
&p(\mathbf{y}_t\,|\,\boldsymbol{\theta}_t)\propto\\
&\exp\{-(\mathbf{y}_t-\mathbf{h}(\mathbf{x}_t,\boldsymbol{\theta}_t))'\mathbf{V}_t^{-1}(\mathbf{y}_t-\mathbf{h}(\mathbf{x}_t,\boldsymbol{\theta}_t))/2)
\end{aligned} \tag{6}$$

Thus the observation equation can be expressed as a conditional distribution:

$$p(\mathbf{y}_t\,|\,\boldsymbol{\theta}_t)\sim N(\mathbf{h}(\mathbf{x}_t,\boldsymbol{\theta}_t),\mathbf{V}_t) \tag{7}$$

To update the probability density of the unknown $\boldsymbol{\theta}_t$ after new process data has been obtained, Bayes theorem is applied:

$$p(\boldsymbol{\theta}_t\,|\,\mathbf{y}_t)=\frac{p(\mathbf{y}_t\,|\,\boldsymbol{\theta}_t)\,p(\boldsymbol{\theta}_t)}{p(\mathbf{y}_t)} \tag{8}$$

The denominator, $p(\mathbf{y}_t)$ is the unconditional probability density of the observation data and from the law of total probability:

$$p(\mathbf{y}_t) = \int p(\mathbf{y}_t \mid \boldsymbol{\theta}_t) p(\boldsymbol{\theta}_t) d\boldsymbol{\theta}_t = c \quad (9)$$

This value acts as a normalising constant. To construct the Bayesian parameter estimator, the posterior density is expressed as:

$$p(\boldsymbol{\theta}_t \mid \mathbf{y}_t) = c^{-1} e^{\log p(\mathbf{y}_t \mid \boldsymbol{\theta}_t) + \log p(\boldsymbol{\theta}_t)} \quad (10)$$

A point estimate of the unknown parameters $\boldsymbol{\theta}_t$, $\hat{\boldsymbol{\theta}}_t$ is the value that maximises the a-posteriori density. That is, $\hat{\boldsymbol{\theta}}_t$ maximises the probability that the estimate is correct. To maximise the posterior probability density $p(\boldsymbol{\theta}_t \mid \mathbf{y}_t)$, the function $J(\boldsymbol{\theta}_t \mid \mathbf{y}_t)$ is minimised:

$$J(\boldsymbol{\theta}_t \mid \mathbf{y}_t) = -\log p(\mathbf{y}_t \mid \boldsymbol{\theta}_t) - \log p(\boldsymbol{\theta}_t) \quad (11)$$

This function is associated with the new measured process data and the prior information. The incorporation of measurement data and prior information provides a weighting based on knowledge of the statistical error. Consequently the parameters can be identified from measurements and constrained by the prior density. Thus, a working expression for $J(\boldsymbol{\theta}_t \mid \mathbf{y}_t)$ can be derived from equation (11):

$$J(\boldsymbol{\theta}_t \mid \mathbf{y}_t) = \quad (12)$$
$$\frac{1}{2}[\mathbf{y}_t - \mathbf{h}(\mathbf{x}_t, \boldsymbol{\theta}_t)]^T \mathbf{V}_t^{-1} [\mathbf{y}_t - \mathbf{h}(\mathbf{x}_t, \boldsymbol{\theta}_t)]$$
$$+ \frac{1}{2}\log(2\pi \mid \mathbf{V} \mid) - \log p(\boldsymbol{\theta}_t)$$

and the Bayesian parameter estimation problem is defined as a constrained minimization problem:

$$\min_{\boldsymbol{\theta}_t} J(\boldsymbol{\theta}_t \mid \mathbf{y}_t) \quad (13)$$
$$\text{s.t.} \quad \boldsymbol{\theta}_t \in \Omega_{\boldsymbol{\theta}}$$

where the constraints $\Omega_{\boldsymbol{\theta}}$, in the simplest case, is a set determined by the lower and upper bounds of the elements of $\boldsymbol{\theta}_t$.

2.2 *Bayesian Parameter Estimation for Dynamic Processes*

Uncertain or unknown parameters can result in model-plant mismatch that can lead to the failure of model based control and optimization strategies. For some of these parameters, it is difficult to determine the accurate value since operational conditions are constantly changing. Others may be time-varying and thus need to be estimated on-line at every sampling interval.

Performing discretisation of the dynamic model with respect to the state variables $\mathbf{x}_t$ gives:

$$\mathbf{x}_t = \mathbf{f}_t(\mathbf{x}_{t-1}, \boldsymbol{\theta}_t) + \mathbf{w}_t \quad \mathbf{w}_t \sim N[0, \mathbf{W}] \quad (14)$$
$$\mathbf{y}_t = \mathbf{h}_t(\mathbf{x}_t, \boldsymbol{\theta}_t) + \mathbf{v}_t \quad \mathbf{v}_t \sim N[0, \mathbf{V}]$$

The error sequences $\mathbf{v}_t$ and $\mathbf{w}_t$ are zero mean, white Gaussian random sequences with:

$$E[\mathbf{v}(t)\mathbf{w}^T(t)] = 0 \quad (15)$$
$$E[\mathbf{v}(t)\mathbf{v}^T(t+\tau)] = 0;$$
$$E[\mathbf{w}(t)\mathbf{w}^T(t+\tau)] = 0$$

2.3 *Estimation for Non-time-varying Parameters*

For those parameters that are not known accurately and that are not time-varying, such as the initiator concentration, or those parameters that change in magnitude only after a change in operational conditions occur, updated parameter estimates at every time instance is not necessary. After the process starts-up or a change in operational condition occurs, parameter estimation is necessary once sufficient observational data become available.

A Bayesian parameter estimator for non-time-varying parameters can be derived and has the following objective function:

$$J(\boldsymbol{\theta} \mid \sum_{k=1}^{n} \mathbf{y}_k) =$$
$$\frac{1}{2}[\sum_{k=1}^{n}(\mathbf{y}_k - \mathbf{h}_k(\mathbf{x}_k, \boldsymbol{\theta}_k))]^T (n\mathbf{V})^{-1} [\sum_{k=1}^{n}(\mathbf{y}_k - \mathbf{h}_k(\mathbf{x}_k, \boldsymbol{\theta}_k))]$$
$$+ \frac{1}{2}\log(2\pi \mid n\mathbf{V} \mid) - \log p(\boldsymbol{\theta}) \quad (16)$$

2.4 *On-line Estimation for Time-varying Parameters*

Time-varying parameters need to be estimated on-line to minimise model-plant mismatch and ensure acceptable performance of model based control and optimization strategies. The estimator structure derived in equation (16) is not suitable for estimating time-varying parameters. For time-varying parameters, at any time instant, *t*, the unknown parameters are estimated by:

$$\hat{\boldsymbol{\theta}}_t = \min J_t(\boldsymbol{\theta}_t \mid \mathbf{y}_t) \quad (17)$$
$$= \frac{1}{2}[\mathbf{y}_t - \mathbf{h}_t(\mathbf{x}_t, \boldsymbol{\theta}_t)]^T \mathbf{V}^{-1} [\mathbf{y}_t - \mathbf{h}_t(\mathbf{x}_t, \boldsymbol{\theta}_t)]$$
$$+ \frac{1}{2}\log(2\pi \mid \mathbf{V} \mid) - \log p(\boldsymbol{\theta}_t)$$

The Bayesian updating process is an iterative procedure. The prior distribution represents the analyst's state of knowledge prior to the last set of evidence, not prior to all evidence. In many cases therefore, when new observational data are obtained, the prior distributions are updated using Bayes theorem thereby becoming posterior distributions. The informative prior distribution is simply the posterior distribution from the last updating calculation. That is the prior distribution function $p(\boldsymbol{\theta}_t)$ can be the posterior distribution from the last updating calculation, $p(\boldsymbol{\theta}_{t-1} \mid \mathbf{y}_{t-1})$.

For a Bayesian normal model, the means of the posterior distribution of parameters at time *t-1* are the estimates of the parameters. At any time point *t*, the measurement vector is $\mathbf{y}_t$ and the estimation vector $\hat{\boldsymbol{\theta}}_{t-1}$ is the mean of the distribution $p(\boldsymbol{\theta}_t)$. Hence:

$$p(\boldsymbol{\theta}_t) = p(\boldsymbol{\theta}_{t-1} \mid \mathbf{y}_{t-1}) \text{ and } \overline{\boldsymbol{\theta}}_t = \hat{\boldsymbol{\theta}}_{t-1} \quad (18)$$

Thus the new estimates at time point *t* are $\hat{\boldsymbol{\theta}}_t$.

3. BAYESIAN PARAMETER ESTIMATION FOR A BATCH MMA POLYMERISATION PROCESS

The process studied is a pilot scale polymerisation reactor (Achilias and Kiparissides, 1992). The batch reaction is the free-radical batch polymerisation of methyl-methacrylate (MMA) with a water solvent and benzoyl peroxide initiator.

The jacketed reactor is provided with a stirrer for thorough mixing of the reactants. Heating and cooling of the reaction mixture is achieved by controlling the flows of two water streams (a hot and a cold water stream) at an appropriate temperature through the reactor jacket. The polymerisation temperature is controlled by a cascade control system consisting of a primary PID and two secondary PI controllers. The reactor temperature is fed back to the primary controller whose output is taken as the set-point of the two secondary controllers. The manipulated variables for the two secondary controllers are hot and cold water flow rates. The hot and cold water streams are mixed before entering the reactor jacket and provide heating or cooling for the reactor. The jacket outlet temperature is fed back to the two secondary controllers.

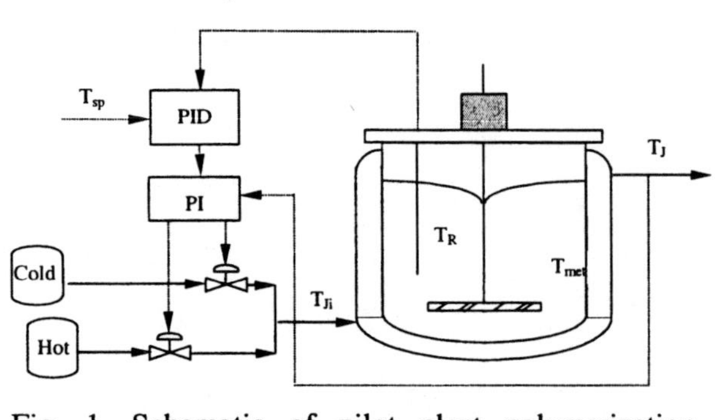

Fig. 1. Schematic of pilot plant polymerization reactor.

A detailed dynamic mathematical model of the reactor covering reaction kinetics, heat and mass balances has been described by Mourikas (1998). This simulation model serves as a realistic test bed for investigating parameter estimation.

The variables that potentially could be used for estimation include: reactor temperature, jacket inlet temperature, jacket outlet temperature, monomer conversion, coolant flow rate and reaction time. In this study it is assumed that monomer conversion and reactor and jacket temperatures are selected as the available measurement set for the estimation procedure. The measurements are corrupted by white Gaussian noise with standard deviations of 0.1%, 0.1K and 0.1K respectively. The sampling rate is 1 minute with batch duration being 120 minutes.

3.1 Erroneous Initial Initiator Concentration and Time-varying Initiator Decomposition Rate

Consider the case where two key parameters for the initiation reaction are incorrectly estimated, initiator concentration and initiator decomposition rate constant. The initiator concentration I is represented by $I = I_0 I_s$ with stochastic correction term I_s, while the initiator decomposition rate constant k_d is represented by $k_d = k_{d0} g^s_{d,corr}$ with the stochastic correction term $g^s_{d,corr}$. I_0 and k_{d0} are computed from kinetic equations (Mourikas, 1998). I_s and $g^s_{d,corr}$ describe the variation associated with the initiator concentration and initiator decomposition rate constant respectively. If there is no model mismatch then the calculated values I_s and $g^s_{d,corr}$ will be equal to the actual values I and k_d, that is the parameters I_s and $g^s_{d,corr}$ will be equal to unity. Uniform prior and normal prior distributions were formulated for the unknow parameters, $\boldsymbol{\theta} = [I_s, g^s_{d,corr}]$:

$$p(\theta_i) \sim U[0.01,10] \quad (19)$$
$$p(\theta_i) \sim N[1,0.25]; i = 1,2$$

Table 1 Parameter estimation results using different prior distributions and sizes of measurement data set.

Prior Distribution	Data set	Estimated Value	Error $(\hat{\theta}-\theta)/\theta$
$U[0.01,10]$	10	0.725; 0.640	44.9%;28.9%
$N[1,0.25]$	10	0.659; 0.686	31.8%;23.7%
$U[0.01,10]$	70	0.502; 0.896	0.34%;0.43%
$N[1,0.25]$	70	0.543; 0.844	8.5%; 6.17%
$U[0.01,10]$	120	0.499; 0.901	0.26%;0.12%
$N[1,0.25]$	120	0.500; 0.900	0.02%;0.02%

The results in Table 1 illustrate the poor performance of the estimator when only 10 observations are used to calculate the parameter estimates for both the uniform and normal prior distributions. Since two parameters are estimated simultaneously, the complexity of the estimation accordingly increases. The estimation errors for the unknown parameters using a uniform prior distribution decrease significantly (Table 1) when the size of the measurement set is increased to 70. In contrast, the errors of the estimates using the normal prior distributions are larger for the same data set size. When the size of the measurement data set is increased to 120, the errors for the unknown parameters using both the normal and uniform prior distribution are small thus indicating that both prior distributions are appropriate.

Comparison of the results in this section illustrate that when the prior information about the unknown parameters is described by a uniform prior distribution, the unknown parameters can be estimated accurately using a small number of samples. However for the normal prior distribution, a larger sample size is needed to obtain estimates with a level of accuracy comparable to that attained with the uniform distribution.

3.2 *Time-varying Termination Rate Constant*

In this second study, process model mismatch is introduced in the form of a time variation in a kinetic parameter. In practice, important kinetic parameters such as k_t, the termination rate constant, cannot be determined accurately and may vary during the polymerisation process. In this study, the termination rate constant is represented by $k_t = k_{t0} g_t g^s_{t,corr}$, where k_{t0} is the rate constant, g_t is a diffusion controlled function, which includes a number of parameters that are often unknown. The stochastic correction term $g^s_{t,corr}$ is used to account for the imprecise knowledge of g_t. In the model, a random walk is assumed for the behaviour of the stochastic state. In the process, the actual values of $g^s_{t,corr}$ are assumed to vary linearly from an initial value of 0.9 to 0.66. The time-varying parameter to be estimated is set as $\theta_t = g^s_{t,corr,t}$

It is assumed that the mean of the parameter, θ_0, is unity and the standard deviation is 0.05. Using maximum entropy, the prior distribution of the unknown parameter is $p_{\theta,0}(\theta_0) \sim N(1,0.05)$. According to the discussion in section 2.3, the prior distribution at time point t is:

$$p_{\theta,k}(\theta_k) \sim N(\mu_k, \sigma_k) \tag{20}$$
$$\theta_{k=0} = 1 ; \mu_k = \theta_{k-1} ; \sigma_k = 0.05$$

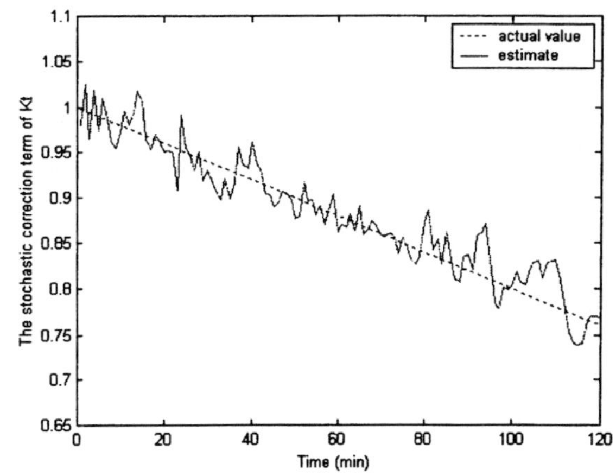

Fig. 2. Estimation of the time-varying correction term for the termination rate constant.

The parameter is assumed to decrease linearly with 10% initial error whilst the parameter in the model is fixed at unity. The parameter estimate is shown in Fig. 2. The Bayesian parameter estimator identifies the starting value of the time-varying parameter and follows the changes closely.

3.3 *Time-varying Reactor Heat Transfer Coefficient*

A common effect in polymerisation reactors is reactor fouling. It is well known that the heat transfer coefficient is strongly correlated with the viscosity of the reacting mixture, which varies during the polymerization. Thus the reactor heat transfer coefficient is an important parameter to monitor. However, it is difficult to identify the overall heat transfer coefficient of the reactor wall.

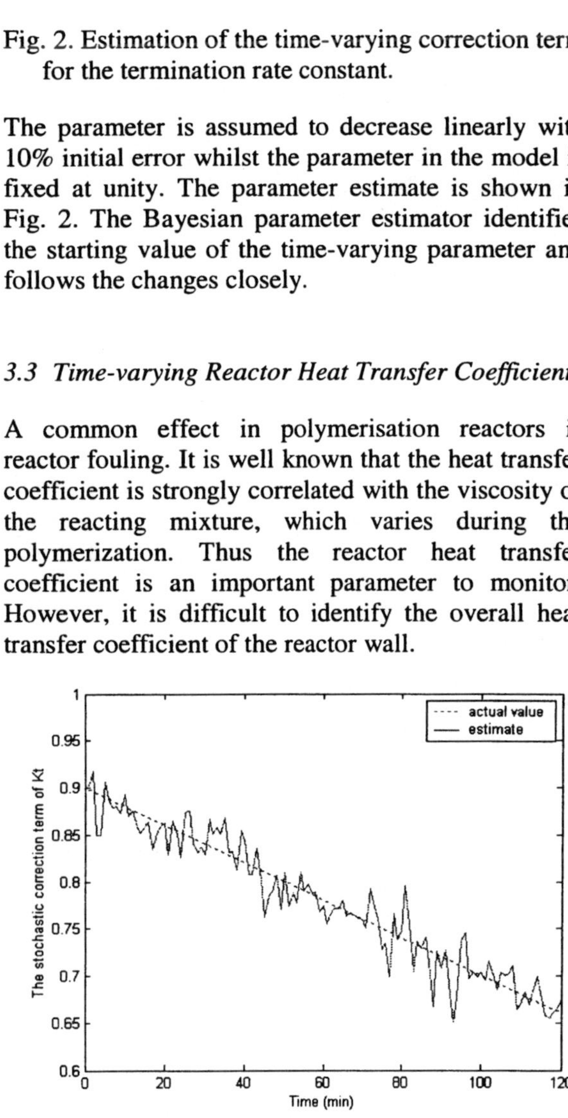

Fig. 3. Estimation of the time-varying correction term for the heat transfer coefficient.

In this study, the heat transfer coefficient is represented by $U = U_0 U_s$, where U_0 represents the experimental value of the heat transfer coefficient. U_s is introduced as a stochastic correction term for the heat transfer coefficient. If there is no reactor fouling, the stochastic correction term is equal to unity. In this study, the heat transfer coefficient is assumed to decrease. It is assumed that, at time point $t = 0$, the mean of the time-varying parameter, θ_0, is unity and the standard deviation is 0.05. The prior distribution of the time-varying parameter is:

$$p_{\theta,k}(\theta_k) \sim N(\mu_k, \sigma_k) \tag{21}$$
$$\theta_{k=0} = 1; \mu_k = \theta_{k-1}; \sigma_k = 0.05$$

Fig. 3 shows that the Bayesian parameter estimator can track the changes tightly for a time-varying heat transfer coefficient so that the impact of reactor fouling can be identified and the model accordingly updated.

4. CONCLUSIONS

This paper has presented a study on parameter estimation based on Bayesian theory. A basic Bayesian parameter estimation framework is derived from Bayesian Maximum a Posteriori. A Bayesian parameter estimator is constructed for unknown non time-varying parameters, which use historical measurements to update the model before implementing the model for model based optimisation and control. An on-line estimator was then derived to deal with unknown time-varying parameters.

Unlike conventional estimation methods, Bayesian parameter estimation allows expert knowledge and prior beliefs, in the form of a prior probability distribution, to be formally incorporated into the statistical analysis. The prior distribution summarises all available information and expert opinion relating to the parameter of interest before the data have been observed. All knowledge about the underlying parameter is then contained in the posterior probability distribution after observing the data. Thus Bayesian methods use the information to get the best estimate.

The construction of a prior distribution and the data set size affect the performance of the estimator. The impact of different prior distributions for the estimation of the parameters is evaluated by applying the method to a MMA process. It is shown that for a small data set, constructing an appropriate prior distribution is very important for parameter estimation.

The estimation of time-varying parameter was also evaluated. The simulation shows that the Bayesian parameter estimator can estimate the initial error of the parameter and track the time-varying parameter over the duration of the batch.

6. AKNOWLEDGMENTS

Zhen Lu acknowledges CPACT and the University of Newcastle for the financial support of his PhD and Prof. Kiparissides, CPERI, Thessaloniki, for the MMA reactor simulation. The authors acknowledge the EU Research Training Network, BatchPro, No. HPRN-CT-2000-00039.

REFERENCES

Achillas, D.S. and C. Kiparissides (1994). On the validity of the steady state approximations in high conversion diffusion-controlled free-radical copolymersiations reactions. *Polymer*, **35(8)**, pp. 1714-1721,

Bastin, G. and D. Dochain (1986). On-line estimation of microbial specific growth rates. *Automation*, **22**, pp. 705 – 709.

Gagnon, L. and J.F. MacGregor (1991). State estimation for continuous emulsion polymerization. *Can. J. Chem. Eng.*, **69**, pp. 648-656.

Kozub, D.J. and J.F. MacGregor (1992). State estimation for semi-batch polymerization reactors. *Chem. Eng. Sci.*, **47(5)**, pp. 1047-1061.

Kiparissides, C., G.P. Seferlis, G. Mourikas and A. J. Morris (2002). On-line optimizing control of molecular weight properties in a batch free radical polymerization reactors, *Ind. Eng. Chem. Res.*, **41**, pp. 61620-6131.

Mourikas, G. (1998). *Modelling, estimation and optimisation of polymerisation processes*. PhD Thesis, University of Newcastle, England.

Muske, K.R. and J.B. Rawlings (1995). Non-linear moving horizon state estimation. In: Berber R. (Ed.), *Methods of model based process control.* Netherlands: Kluwer Academic Publisher.

Narendra, K.S. and A.M. Annaswamy (1988). *State adaptive systems.* NJ: Englewood Cliffs, Prentice-Hall.

Robertson, D.G., Lee, J.H. and Rawlings, J.B. (1996). A moving horizon-based approach for least squares state estimation. *AIChE.* **42**, pp. 2209-2224.

Scali, C., M. Morretta and Semino, C. (1997). Control of the quality of polymer products in continuous reactors: Comparison of performance of state estimators with and without updating of parameters. *J. Process Control.* **7(5)** 357-369.

Schuler, H. and Ch-U Schmidt (1992). Calorimetric state estimators for chemical reactor diagnosis and control: review of methods and applications. *Chem. Engng Sci.*, **47**, pp. 899-915.

Sirohi, A. and K.Y. Choi (1996). On-line parameter estimation in a continuous polymerization process. *Ind. Engng Chem. Res.*, **35**, 1332-1343.

Thomas, I.M. and C. Kiparissides (1984). Computation of the near-optimal temperature and initiator policies for a batch polymerization reactor. *Can J Chem Eng* **62 (2)**, pp. 284-296.

Yin, K.W. (1994). Parameter estimation and related problems for external input-output models. *Chem. Engng Sci.*, **48**, pp. 3870-3875.

ELSEVIER
IFAC
PUBLICATIONS
www.elsevier.com/locate/ifac

USING A DITHERING SIGNAL IN THE REFERENCE TO IMPROVE THE ESTIMATES FROM SUBSPACE IDENTIFICATION METHODS ON CLOSED LOOP DATA

Geir Werner Nilsen* and David Di Ruscio**

**Department of Electrical Engineering, Information Technology and Cybernetics, Telemark University College, N-3914 Porsgrunn, Norway, E-mail: Geir.W.Nilsen@hit.no*
***Department of Electrical Engineering, Information Technology and Cybernetics, Telemark University College, N-3914 Porsgrunn, Norway, E-mail: David.Di.Ruscio@hit.no*

Abstract: It is shown that when an appropriate dithering signal is used in the reference of a specific process DSR gives eigenvalue estimates with smaller bias and variance than PEM on finite closed loop data. As dithering signal a sinusoid signal at different frequencies have been tried. Comparing the optimums of DSR and PEM we find that DSR provide eigenvalue estimates with smaller bias and variance at its optimum than PEM does at its. When using an external dithering signal in the reference the choice of the frequency of the dithering signal is important when using the input and output data for direct system identification in closed loop.

Keywords: Subspace methods; identification; closed loop; finite data sets; dithering signal.

1. INTRODUCTION

The reason for the problems that can occur when applying subspace identification (SID) algorithms for direct identification of closed loop data is the projection of the future outputs onto the future inputs. Future inputs and future outputs are assumed to be uncorrelated. In closed loop operation this assumption is not necessarily fulfilled. This is the cause of the bias. Di Ruscio (2003) shows that using a filter in the feedback loop is a method to reduce the bias problem. It is also stressed that biased estimates may be more reliable than estimates from an unbiased algorithm because the variance may be small. Van Overschee and De Moor (1996), (1997) are suggesting to use the Markov parameters of the controller in the algorithm to avoid the problem. The controller is assumed to be linear. Chou and Verhaegen (1997) have developed a SID algorithm for Errors In Variables (EIV) problems. They have shown that this algorithm will give consistent estimates on closed loop data, if a persistence of excitation requirement is satisfied. This leads to the use of signals with relatively high order of persistent excitation in the reference. Gustafsson (2001) has proposed an Instrument Variable (IV) approach SID algorithm as an improvement of the already existing EIV algorithm of Chou and Verhaegen (1997). The modified algorithm is named SIV (Subspace-based Identification using instrumental Variables). Our work is based on the already existing SID algorithm named DSR, Aoki and Havenner (1997), Di Ruscio (1996). We want to examine if an external dithering signal used in the reference of a system operating in closed loop can lead to improved parameter estimates. This is an idea presented in Di Ruscio (2003) together with the experience that an optimal experiment for the subspace methods is in general not a white noise input, but rather a colored input where the frequency spectrum is optimized to excite the parameters in the system as well as possible. The estimates from DSR will be compared to estimates from PEM (Prediction Error Method) in the system identification toolbox in Matlab 6.5. A description of prediction error methods in general and their properties can be found in amongst others Ljung (1999).

The rest of this paper is organized as follows. A model of a chemical reactor operating in closed loop is presented in Section 2.1. This model will be used throughout the paper. The effect of using a sinusoid signal as dithering signal in the reference of the chemical reactor operating in closed loop is investigated in Section 2.2. Closed loop eigenvalue estimates are presented i Section 2.2.1. In Section 2.2.2 the parameter estimates from DSR are considered as a function of the frequency of the sinusoid signal used as dithering signal and the prediction horizon used in DSR. In Section 2.2.3 the parameter estimates from DSR, with a constant prediction horizon, and the parameter estimates from PEM are presented as functions of the frequency of the sinusoid signal used as dithering signal. Alternative quality measures are presented in section 2.2.4. The results are compared. Some concluding remarks follows in Section 3.

2. A CASE STUDY IN CLOSED LOOP SYSTEM IDENTIFICATION

2.1 A model of a chemical reactor operating in closed loop

In order to generate closed loop data we choose to use a model of a chemical reactor operating in closed loop. The reaction mechanism of the reactor is given by:

$$A \xrightarrow{k_1} B \xrightarrow{k_2} C, \tag{1}$$

$$2A \xrightarrow{k_3} D. \tag{2}$$

The reaction from body A to body D is of order two, while the other reactions are of order 1. The manipulated variable is the feed flow (flow rate) u $[\frac{1}{hours}]$. The concentration of body A in the feed flow, u, is θ. The concentration of body A and body B in the tank is respectively x_1 and x_2. The connection from u to $y = x_2$ is then given by the model:

$$\dot{x}_1 = -k_1 x_1 - k_3 x_1^2 + (\theta - x_1)u, \tag{3}$$

$$\dot{x}_2 = k_1 x_1 - k_2 x_2^2 - x_2 u, \tag{4}$$

$$y = x_2, \tag{5}$$

where the reaction rate constants are given by $k_1 = 50$, $k_2 = 100$ and $k_3 = 10$. The stationary values of the states, disturbances, parameters and manipulated variable are given: $x_1^s = 2.5$, $x_2^s = 1$, $\theta^s = 10$ and $u^s = 25$. We assume that θ is constant and known.

Defining:

$$\delta u_k = u_k - u^s, \tag{6}$$

$$\delta x_k = x_k - x^s, \tag{7}$$

$$\delta y_k = y_k - y^s, \tag{8}$$

$$\delta \theta_k = \theta_k - \theta^s, \tag{9}$$

a linearized discrete model can be expressed on the form:

$$\delta x_{k+1} = A\delta x_k + B\delta u_k + C\delta\theta_k + v_k, \tag{10}$$

$$\delta y_k = D\delta x_k + E\delta u_k + w_k, \tag{11}$$

where we in addition have added additive process and measure noise, v_k and w_k, respectively.

After linearization and discretization the system is expressed by:

$$A = \begin{bmatrix} 0.8750 & 0 \\ 0.0500 & 0.8750 \end{bmatrix}, B = \begin{bmatrix} 0.0075 \\ -0.0010 \end{bmatrix}$$

$$C = \begin{bmatrix} 0.0250 \\ 0 \end{bmatrix}, D = \begin{bmatrix} 0 & 1 \end{bmatrix} \tag{12}$$

and the matrix E is the zero matrix. The linearization is done at the stationary point. The discretization is done by explicit Euler with uniform sampling interval $\delta t = 0.001$. We consider a time series from 0 to 1, with N=1001 discrete data points, $k = 0, 1, \ldots, N$.

The body B is controlled by a PI-controller with $k_p = 50$ and $T_i = \frac{1}{75}$. The PI-controller on discrete form is given by:

$$u_k = k_p e_k + z_k, \tag{13}$$

$$z_{k+1} = z_k + \delta t \frac{k_p}{T_i} e_k, \tag{14}$$

where e_k is given by:

$$e_k = r_k - y_k. \quad (15)$$

The linearized reactor model, (10) and (11), and the PI-controller, (13) and (14), are used in order to generate closed loop data which will be used throughout this paper for closed loop system identification. The reference will be constant $r_k = 1$ at each time instant k, superposed a dithering signal. The noise used is significant. The process noise variance used is $E(v_k v_k^T) = 0.1 \cdot I$ and the measuring noise variance used is $E(w_k^2) = 0.0001$. The reason for using a linearized discrete model of the chemical reactor is that we want to compare the estimated eigenvalues to the eigenvalues of the linearized discrete model. The closed loop system is illustrated in Figure 1.

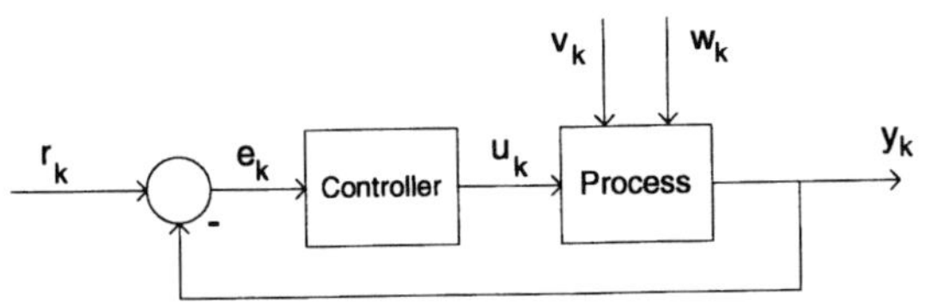

Fig. 1. System operating in closed loop

2.2 The effect of using a sinusoid signal as dithering signal in the reference of a chemical reactor operating in closed loop

2.2.1. Closed loop eigenvalue estimates A sinusoid signal with frequency $\omega = 600$ $[rad/s]$ and with magnitude ± 0.1 is chosen as dithering signal. The reference at time instant k is given by:

$$r_k = 1 + 0.1 \cdot \sin(\omega \delta t k). \quad (16)$$

We find it natural to compare DSR with PEM implemented in the system identification toolbox in Matlab 6.5. Throughout the paper the stationary values in both inputs and outputs will be removed prior to identification. Initial parameter estimates to PEM is provided by **canstart** in the Matlab system identification toolbox with the parameters chosen as: $orders = 2$, $D = 1$, $K = 1$ and $X = 1$.

The system was simulated 100 times. The same input was used each time but the noise realization was changed each time. Figure 2 shows the estimated eigenvalues from the Monte Carlo simulation. The linearized discrete system have multiple poles, actually two real eigenvalues at 0.875. The parameters in DSR are chosen as: $n = 2$, $g = 0$, $L = 12$ and $J = 13$. The default parameters are used in PEM. Figure 2 shows that by using an appropriate dithering signal DSR can provide an eigenvalue estimate with smaller bias and variance than PEM on data from a system operating in closed loop.

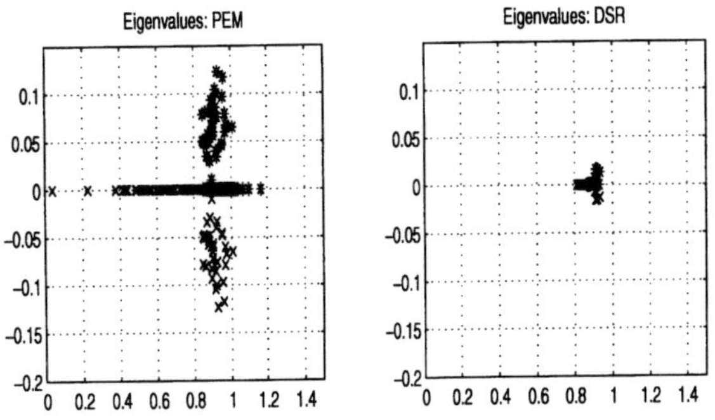

Fig. 2. Monte Carlo simulation using a constant reference superposed a sinusoid signal as dithering signal, $r_k = 1 + 0.1 \cdot \sin(\omega \delta t k)$ with $\omega = 600$, using DSR with $n = 2$, $g = 0$, $L = 12$ and $J = 13$ and PEM with default parameters for identification.

We have to make a remark that PEM in this case actually gives some estimates with unstable eigenvalues. To improve the estimates from PEM we have tried to reduce the tolerance (Tol) from the default value 0.01 and increase the maximum number of iterations (MaxIter) from the default value 20. Several values were tried with Tol as low as 0.0001 and MaxIter as high as 20000. This did not improve the estimates from PEM. By increasing the tolerance to 0.02 and reducing the maximum number of iterations to 15 we observed a marginal improvement of the estimates from PEM. Because of the insignificant improvement of the estimates from PEM when changing the algorithm parameters from the default values we have chosen to use the default values in PEM. An additional comment on the estimates from PEM with unstable eigenvalues is that when using Matlab 6.1 PEM does not only give estimates with eigenvalues larger than 1, but also eigenvalues with more negative values than -1. When using Matlab 6.0 we get approximately the same values as when using Matlab 6.5.

Table 1 contents the mean and the standard deviation (Std) of the eigenvalues of the estimated system matrices from the Monte Carlo simulation when using DSR. Table 2 contains corresponding data from PEM. The data in Table 1 and Table 2 support the conclusion from the visual inspection of Figure 2, that DSR gives estimates with smaller bias and variance than PEM in this case.

2.2.2. The parameter estimates from DSR as a function of the prediction horizon and the frequency of the sinusoid signal used as dithering signal in the reference In DSR there are four parameters g, n, L and J that can be chosen by the user. If the structure parameter g is 1, which is the default, the data matrix E in the state space model is identified. If g is put to zero the matrix

Table 1. Mean and standard deviation from the Monte Carlo simulation using a constant reference superposed a sinusoid signal as dithering signal, $r_k = 1 + 0.1 \cdot \sin(\omega \delta t k)$ with $\omega = 600$, and using DSR with $n = 2$, $g = 0$, $L = 12$ and $J = 13$ for identification.

DSR	Pole 1	Pole 2
	[*Re Im*]	[*Re Im*]
Mean	[0.9203 0.0006]	[0.8612 -0.0006]
Std	[0.0043 0.0027]	[0.0202 0.0027]

Table 2. Mean and standard deviation from the Monte Carlo simulation using a constant reference superposed a sinusoid signal as dithering signal, $r_k = 1 + 0.1 \cdot \sin(\omega \delta t k)$ with $\omega = 600$, and using PEM with default parameters for identification

PEM	Pole 1	Pole 2
	[*Re Im*]	[*Re Im*]
Mean	[0.9325 0.0241]	[0.7692 -0.0241]
Std	[0.0538 0.0349]	[0.1758 0.0349]

E is forced to be the zero matrix. The parameter n specify the model order. The parameter L is the number of block rows in the extended observability matrix. L can be interpreted as the identification horizon used to predict the number of states. This again limits the maximal system order which can be identified. The order must be chosen in the interval $1 \leq n \leq L \cdot m$, where m is the number of outputs. The parameter J is the number of time instants in the past horizon which is used defining the instrument variable matrix which are used to remove noise.

The experience so far in open loop cases is that the parameter L should be chosen as small as possible in order to reduce the variance of the estimates. This is especially important in case of poorly excitating input signals. The parameter J is usually chosen as $J = L$ or $J = L + 1$. It is normally not crucial which of these two alternatives which are chosen.

Related to this it is interesting to consider the quality of the parameter estimates from DSR as a function of the frequency, ω, of the dithering signal and the identification horizon, L, in the algorithm when data is collected from a system operating in closed loop. The parameter J is chosen as $J = L + 1$. The system used is still the chemical reactor operating in closed loop (10), (11), (13) and (14). Now we do not only consider a Monte Carlo simulation at one frequency but Monte Carlo simulations in 25 $[rad/s]$ steps from 25 $[rad/s]$ up to the Nyquist frequency which is the half of the sampling frequency. To evaluate the quality of the estimated model parameters we choose to use a quadratic criterion on the eigenvalues of the estimated model. The criterion used sums up the square of the deviation between the eigenvalues of the estimated model and the actual eigenvalues. The quality criterion (17) is presented using Matlab notation where $\lambda(A)\ R \in \mathbb{R}^n$ is the vector of the eigenvalues of the true state transition matrix and $\lambda(\overline{A}_i)\ R \in \mathbb{R}^N$ is the vector of the estimated eigenvalues of the state transition matrix at Monte Carlo run number i. M is the total number of Monte Carlo runs.

$$V = \sum_{i=1}^{M} \|\mathrm{Re}(\lambda(\overline{A}_i)) - \mathrm{Re}(\lambda(A))\|_F^2 + \sum_{i=1}^{M} \|\mathrm{Im}(\lambda(\overline{A}_i)) - \mathrm{Im}(\lambda(A))\|_F^2 \tag{17}$$

The squared Frobenius norm of a matrix $A \in \mathbb{R}^{m \times n}$ is equal to the trace of the product $A^T A$, and defined as follows:

$$\|A\|_F^2 = \mathrm{tr}(A^T A) = \sum_{i=1}^{m} \sum_{j=1}^{n} a_{ij}^2. \tag{18}$$

For a complex number $c = a + \mathrm{j} \cdot b$ we define:

$$\mathrm{Re}(c) = a \text{ and } \mathrm{Im}(c) = b. \tag{19}$$

Figure 3 shows the criterion (17) as a function of ω and L. Each point in the figure is the sum given by (17) from a Monte Carlo simulation at a specific ω with a specific L.

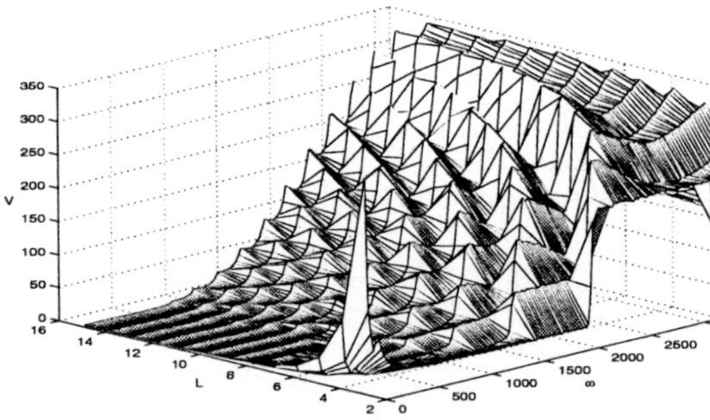

Fig. 3. The quality criterion (17) as a function of ω and L when using a constant reference superposed a sinusoid signal as dithering signal, $r_k = 1 + 0.1 \cdot \sin(\omega \delta t k)$, and DSR with $n = 2$, $g = 0$ and $J = L + 1$ for identification.

From Figure 3 it is obvious that the choice of ω is the most crucial parameter. When zooming in Figure 3 for ω in the interval up to 1000 $[rad/s]$ we get Figure 4. Figure 4 shows that in this closed loop data set it is not favorable to choose L as small as possible as in open loop cases.

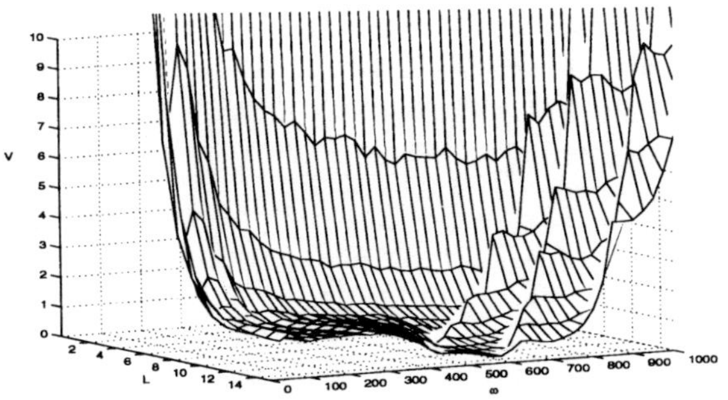

Fig. 4. The quality criterion (17) as a function of ω and L when using a constant reference superposed a sinusoid signal as dithering signal, $r_k = 1 + 0.1 \cdot \sin(\omega \delta t k)$, and DSR with $n = 2$, $g = 0$ and $J = L + 1$ for identification.

2.2.3. Comparing the parameter estimates from DSR and PEM as functions of the frequency of the sinusoid signal used as dithering signal in the reference In Figure 5 PEM and DSR are compared as functions of ω of the sinusoid signal used as dithering signal in the reference. The criterion used is given by (17). DSR is used with $L = 12$ and $J = L + 1$. From Figure 5 two important notes have to be done. The estimates from PEM and DSR have optimums for dithering signals in different frequency ranges. DSR can actually give estimates with smaller bias and variance than PEM by choosing an appropriate frequency of the dithering signal.

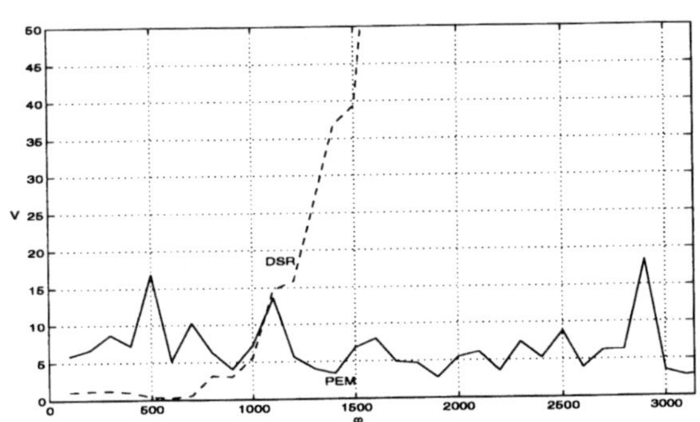

Fig. 5. The quality criterion (17) as a function of ω, when using a constant reference superposed a sinusoid signal as dithering signal $r_k = 1 + 0.1 \cdot \sin(\omega \delta t k)$, and the estimates from PEM with default parameters and DSR with $n = 2$, $g = 0$, $L = 12$ and $J = 13$. PEM is illustrated by the solid line and DSR is illustrated by the broken line

2.2.4. Alternative closed loop quality measures Transforming (12) to observable canonical form gives system matrices with the following structure:

$$A = \begin{bmatrix} 0 & 1 \\ a_{21} & a_{22} \end{bmatrix}, B = \begin{bmatrix} b_{11} \\ b_{21} \end{bmatrix}, D = \begin{bmatrix} 1 & 0 \end{bmatrix}, \quad (20)$$

and E is till the zero matrix. The parameters to be estimated in the observable canonical form are collected in a parameter vector:

$$\theta = [a_{21} \; a_{22} \; b_{11} \; b_{21}]. \quad (21)$$

The true parameter vector is:

$$\theta = [-0.7656 \; 1.7500 \; -0.0010 \; -0.0005]. \quad (22)$$

In figure 6 the estimated parameters in the parameter vector from DSR and PEM are compared as a function of the runs in the Monte Carlo simulation. The reference is given by (16) with $\omega = 600$. DSR is used with the parameters $n = 2$, $g = 0$, $L = 12$ and $J = 13$. The parameter estimates from DSR all have small variance. The parameters b_{11} and b_{21} have a bias, but considering the scale of the axes the bias is marginal. The parameter estimates from PEM are characterized by large variance compared to the estimates from DSR. The mean and standard deviation of the parameters in the parameter vector are listed in Table 3 and Table 4 for respectively DSR and PEM.

Table 3. Mean and standard deviation from the Monte Carlo simulation using a constant reference superposed a sinusoid signal as dithering signal, $r_k = 1 + 0.1 \cdot \sin(\omega \delta t k)$ with $\omega = 600$, and using DSR with $n = 2$, $g = 0$, $L = 12$ and $J = 13$ for identification

DSR	a_{21}	a_{22}	b_{11}	b_{21}
Mean	-0.7925	1.7815	-0.0057	-0.0037
Std	0.0185	0.0201	0.0003	0.0001

Table 4. Mean and standard deviation from the Monte Carlo simulation using a constant reference superposed a sinusoid signal as dithering signal, $r_k = 1 + 0.1 \cdot \sin(\omega \delta t k)$ with $\omega = 600$, and using PEM with default parameters for identification.

PEM	a_{21}	a_{22}	b_{11}	b_{21}
Mean	-0.7194	1.7024	-0.0014	-0.0004
Std	0.1732	0.1814	0.0027	0.0018

A common quality measure is the simulated error (SE) which is the deviation between the real outputs in Y and the simulated outputs in Y^d. Using the squared Frobenius norm (18) we introduce the Squared Simulated Error (SSE):

$$SSE = \|Y - Y^d\|_F^2. \quad (23)$$

The SSE is plotted for DSR and PEM in figure (7) for the data generated to evaluate the estimation of the parameters in the parameter vector.

The SSE from the parameter estimates from DSR have smaller bias and variance than the SSE from

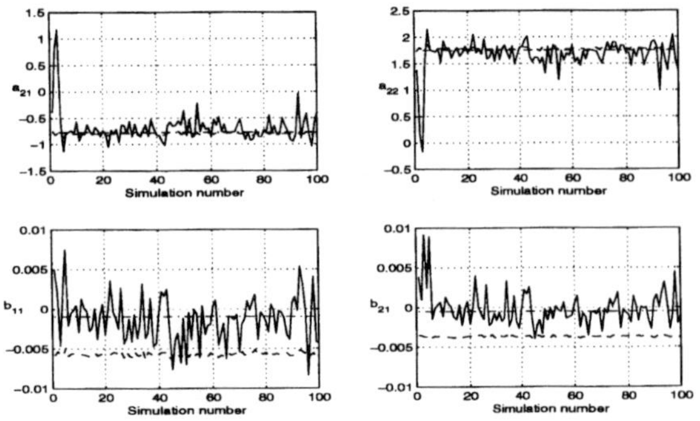

Fig. 6. The parameter vector (21) as a function of the runs in the Monte Carlo simulation using a constant reference superposed a sinusoid signal as dithering signal, $r_k = 1 + 0.1 \cdot \sin(\omega \delta t k)$ with $\omega = 600$, using DSR with $n = 2$, $g = 0$, $L = 12$ and $J = 13$ and PEM with default parameters for identification. PEM is illustrated by the solid line, DSR is illustrated by the broken line and the true value is illustrated by the dash dotted line.

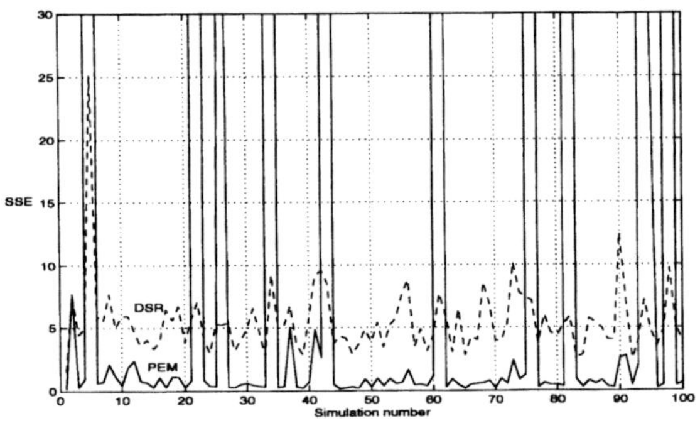

Fig. 7. Squared Simulated Error (SSE) for each of the runs in the Monte Carlo simulation using a constant reference superposed a sinusoid signal as dithering signal, $r_k = 1 + 0.1 \cdot \sin(\omega \delta t k)$ with $\omega = 600$, using DSR with $n = 2$, $g = 0$, $L = 12$ and $J = 13$ and PEM with default parameters for identification. PEM is illustrated by the solid line and DSR is illustrated by the broken line.

the parameter estimates from PEM. Actually the mean and standard deviation of the SSE is respectively approximately 5.5 and 2.8 for DSR. The corresponding values from PEM are both exceeding working precision. This because of the unstable models from PEM shown in Figure 2.

3. CONCLUDING REMARKS

Our simulations on finite data sets from a system operating in closed loop have shown that by using an appropriate dithering signal in the reference the DSR algorithm can provide eigenvalue estimates with smaller bias and variance than PEM, from the system identification toolbox in Matlab 6.5. As dithering signal we have tried sinusoid signals at different frequencies. We have observed that DSR and PEM have optimums in different frequency ranges. Comparing the optimums of DSR and PEM we find that DSR gives eigenvalue estimates with smaller bias and variance at its optimum than PEM does at its.

An other observation is that DSR provide eigenvalue estimates with smallest bias and variance when the dithering signal is low frequent.

When using an external dithering signal in the reference the choice of the frequency of the dithering signal is the most crucial parameter when using DSR on closed loop data, except the choice of model order which is not treated in this work. When choosing the prediction horizon L in DSR it is important to use an L that is large enough. This is opposite of what is used in identification of open loop data. Then keeping L as small as possible helps keeping the variance small, especially when input signals are poorly excited.

REFERENCES

Aoki, M. and Havenner, A. M. (1997). Lecture Notes in Statistics. Applications of Computer Aided Time Series Modeling. Springer-Verlag New York, Inc

Chou, C. T. and Verhaegen, M. (1997). Subspace Algorithms for the Identification of Multivariable Dynamic Errors-in-Variables Models. Automatica, Vol. 33, No. 10, pp. 1857-1997.

Di Ruscio, D. (1996). DSR Toolbox. Available on request.

Di Ruscio, D. (2003). Subspace System Identification of the Kalman Filter. Modeling, Identification and Control, Vol. 24, No. 3.

Gustafsson. T. (2001). Subspace identification using instrumental variable techniques. Automatica, Vol. 37, pp. 2005-2010.

Ljung, L. (1999). System Identification. Theory for the User. Second Edition. Prentice Hall.

Van Overschee, P. and De Moor, B. (1996). Closed loop subspace system identification. Internal report.Katholieke Universiteit Leuven. Departement Elektrotechniek. ESAT-SISTA/TR 1996-521

Van Overschee, P. andDe Moor, B. (1997). Closed loop subspace system identification. In the proceedings of the 36th Conference on Decision and Control 1997, San Diego, California, December 6-14

ELSEVIER
IFAC
PUBLICATIONS
www.elsevier.com/locate/ifac

DYNAMIC MODELLING OF A THREE-PHASE CATALYTIC SLURRY REACTOR CONSIDERING THE PHASE CHANGE PHENOMENON

Adriano Pinto Mariano, Eduardo Coselli Vasco de Toledo[1], José Marcos Francisco da Silva, Rubens Maciel Filho

Faculty of Chemical Engineering. State University of Campinas (UNICAMP)
CP 6066 - CEP 13081-970 - Campinas, SP - Brazil

Abstract: This work deals with the development of a model capable to simulate via deterministic modelling the phenomenon of phase change observed in three-phase catalytic slurry reactors, that occurs in the reacting medium and in the refrigerant fluid. A dynamic heterogeneous non-isothermal model was formulated to evaluate the impact of this phenomenon in the dynamic behaviour of the reactor. A multicomponent flash calculation was used to reproduce this effect in the reacting medium and an appropriate correction procedure of the global coefficient of heat transfer to analyse the effect in the refrigerant fluid.

Keywords: Reactor, Modelling, Three-Phase, Catalytic, Slurry, Phase Change.

1. INTRODUCTION

The development of an efficient and reliable model of three-phase catalytic reactors is still a difficult task, because it involves many aspects including hydrogenation, gas-liquid and liquid-solid mass transfers, heat transfer, pore diffusion, reaction kinetics and deactivation. Model assessment has mostly been reported for a single reaction, or reactions obeying simplified kinetic laws, under isothermal conditions. Nevertheless, exothermic reactions undergoing a multi-step reaction scheme and complex kinetics are industrially of a main interest and rigorous comparison of the performances of several multiphase reactors, for such reactions, has hardly ever been published (Vasco de Toledo et al., 2001). Therefore, despite the large amount of effort put into the modelling of three-phase catalytic reactors, many issues still remain to be solved, for instance, the phenomenon of phase change in both reacting medium and refrigerant fluid, which occurs, frequently, in three-phase reactors.

The impact of the phase change in the dynamic behaviour of the reactor can be very intensive, compromising seriously the efficiency of the controllers in their function of the maintenance of a stable operation of the reactor. The previous knowledge, obtained through simulations, of the operating conditions at which this phenomenon can occur is very important. So, the aim of the developed model presented in this work is to allow the user to simulate different aspects involving the phase change.

2. DESCRIPTION OF THE DYNAMIC MODEL

The main phenomena present in three-phase reactors were incorporated into the dynamic model. The model consists of mass and energy balance equations, and also an equation for the refrigerant fluid. Another aspect contemplated is the consideration of the variation of the physicochemical properties and of the mass and heat transfer coefficients, through correlations available in the literature. This work considers the hydrogenation of *o*-cresol on Ni/SiO_2 catalyst (Hichri et al., 1991).

[1]Author to whom correspondence should be addressed: urso@lopca.feq.unicamp.br

However, as there are many other hydrogenation reactions of industrial interest, the model can easily incorporate other reactions due to the generality adopted during its development.

The hypotheses adopted to the dynamic model of the tubular reactor are:

1) a pseudo-homogeneous media is assumed for the liquid and the solid with respect to solid movement. This means that there is no velocity difference between solid and liquid;
2) negligible pressure variations;
3) reaction of the type $A_{(g)} + \nu B_{(l)} \rightarrow \nu C_{(l)}$ taking place in the catalyst and having a kinetic that is dependent on the concentrations of A and B;
4) phase change happens in the system.

The scheme utilized to represent the tubular reactor is displayed in Figure 1.

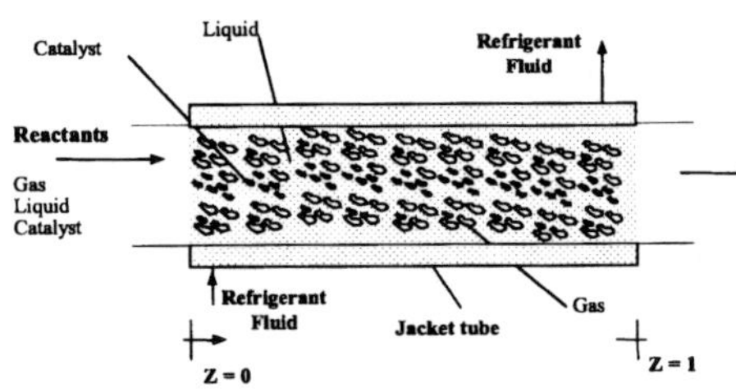

Fig. 1. Three-phase tubular reactor

The radial dispersion has been neglected for the fluid phase, which is a very common assumption in multi-phase reactors as normally found in literature (Gianetto and Silverston, 1986). The superficial velocities and the hold-ups are updated taking into account the changes in temperature, pressure and concentration. Some discussion on the formulation of dynamic models are presented elsewhere: Warna and Salmi (1996), Lange et al. (1999), Julcour et al. (1999), Salmi et al. (2000), Vasco de Toledo and Maciel Filho (2002) and Mariano (2003).

Aiming to adapt the equations of the model to the units (kmol/s) usually employed in multicomponent flash calculation, it was necessary to multiply the mass balance equations by the feed volumetric flow of the liquid phase (Q_l), that is a constant value and for this reason the results generated by the model are not changed.

The equations of the heterogeneous dynamic model developed are:

Fluid Phase:

Mass balance of reactant A in the gas phase:

$$\varepsilon_g \frac{\partial FA_g}{\partial t} = \frac{D_{eg}}{L^2}\frac{\partial^2 FA_g}{\partial z^2} - \frac{u_g}{L}\frac{\partial FA_g}{\partial z} - (K_{gl})_A a_{gl}(FA^* - FA_l) \qquad (1)$$

$$\left.\frac{D_{eg}}{L}\frac{\partial FA_g}{\partial z}\right|_{z=0} = u_g\left(FA_g(z=0) - FA_{gfo}\right) \qquad \left.\frac{\partial FA_g}{\partial z}\right|_{z=1} = 0 \qquad (2)$$

Mass balance of reactant A and B and product C in the liquid phase:

$$\varepsilon_l \frac{\partial FA_l}{\partial t} = \frac{D_{el}}{L^2}\frac{\partial^2 FA_l}{\partial z^2} - \frac{u_l}{L}\frac{\partial FA_l}{\partial z} + (K_{gl})_A a_{gl}(FA^* - FA_l) - (K_{ls})_A a_{ls}(FA_l - FA_s^s) \qquad (3)$$

$$\left.\frac{D_{el}}{L}\frac{\partial FA_l}{\partial z}\right|_{z=0} = u_l\left(FA_l(z=0) - FA_{lfo}\right) \qquad \left.\frac{\partial FA_l}{\partial z}\right|_{z=1} = 0 \qquad (4)$$

$$\varepsilon_l \frac{\partial FB_l}{\partial t} = \frac{D_{el}}{L^2}\frac{\partial^2 FB_l}{\partial z^2} - \frac{u_l}{L}\frac{\partial FB_l}{\partial z} - (K_{ls})_B a_{ls}(FB_l - FB_s^s) \qquad (5)$$

$$\left.\frac{D_{el}}{L}\frac{\partial FB_l}{\partial z}\right|_{z=0} = u_l\left(FB_l(z=0) - FB_{lfo}\right) \qquad \left.\frac{\partial FB_l}{\partial z}\right|_{z=1} = 0 \qquad (6)$$

$$\varepsilon_l \frac{\partial FC_l}{\partial t} = \frac{D_{el}}{L^2}\frac{\partial^2 FC_l}{\partial z^2} - \frac{u_l}{L}\frac{\partial FC_l}{\partial z} - (K_{ls})_C a_{ls}(FC_l - FC_s^s) \qquad (7)$$

$$\left.\frac{D_{el}}{L}\frac{\partial FC_l}{\partial z}\right|_{z=0} = u_l\left(FC_l(z=0) - FC_{lfo}\right) \qquad \left.\frac{\partial FC_l}{\partial z}\right|_{z=1} = 0 \qquad (8)$$

Energy balance in the fluid (gas + liquid) phase:

$$(\varepsilon_g \rho_g C_{pg} + \varepsilon_l \rho_l C_{pl})\frac{\partial T}{\partial t} = \frac{(\varepsilon_g \lambda_g + \varepsilon_l \lambda_l)}{L^2}\frac{\partial^2 T}{\partial z^2} - \frac{(\varepsilon_g \rho_g C_{pg} u_g + \varepsilon_l \rho_l C_{pl} u_l)}{L}\frac{\partial T}{\partial z} + h_s a_{ls}(T_s^s - T) - \frac{4U}{D_t}(T - T_r) - Q_{FLASH} \qquad (9)$$

$$\left.\frac{(\varepsilon_g \lambda_g + \varepsilon_l \lambda_l)}{L}\frac{\partial T}{\partial z}\right|_{z=0} = (\varepsilon_g \rho_g C_{pg} u_g + \varepsilon_l \rho_l C_{pl} u_l)\cdot (T(z=0) - T_{fo}) \qquad (10)$$

$$\left.\frac{\partial T}{\partial z}\right|_{z=1} = 0$$

Energy balance of the refrigerant fluid

$$\rho_r C_{pr}\frac{\partial T}{\partial t} = -\frac{\rho_r C_{pr} u_r}{L}\frac{\partial T}{\partial z} + \frac{4U}{D}(T - T_r) \qquad (11)$$

$$T_r = T_{ro}, \qquad z = 0$$

Solid Phase:

Mass balance of reactants A and B and product C in the solid phase:

$$\varepsilon_s \frac{\partial FA_s}{\partial t} = \frac{D_{ea}}{R_p^2} \frac{1}{r_p^2} \frac{\partial}{\partial r_p}\left(r_p^2 \frac{\partial FA_s}{\partial r_p}\right) - \rho_s Q_l R_W(FA_s, FB_s, T_s) \quad (12)$$

$$\frac{D_{ea}}{R_p} \frac{\partial FA_s}{\partial r_p}\bigg|_{r_p=1} = (K_{ls})_A \left(FA_l - FA_s^s(r_p=1)\right) \qquad \frac{\partial FA_s}{\partial r_p}\bigg|_{r_p=0} = 0 \quad (13)$$

$$\varepsilon_s \frac{\partial FB_s}{\partial t} = \frac{D_{eb}}{R_p^2} \frac{1}{r_p^2} \frac{\partial}{\partial r_p}\left(r_p^2 \frac{\partial FB_s}{\partial r_p}\right) - \nu \rho_s Q_l R_W(FA_s, FB_s, T_s) \quad (14)$$

$$\frac{D_{eb}}{R_p} \frac{\partial FB_s}{\partial r_p}\bigg|_{r_p=1} = (K_{ls})_B \left(FB_l - FB_s^s(r_p=1)\right) \qquad \frac{\partial FB_s}{\partial r_p}\bigg|_{r_p=0} = 0 \quad (15)$$

$$\varepsilon_s \frac{\partial FC_s}{\partial t} = \frac{D_{ec}}{R_p^2} \frac{1}{r_p^2} \frac{\partial}{\partial r_p}\left(r_p^2 \frac{\partial FC_s}{\partial r_p}\right) + \nu \rho_s Q_l R_W(FA_s, FB_s, T_s) \quad (16)$$

$$\frac{D_{ec}}{R_p} \frac{\partial FC_s}{\partial r}\bigg|_{r_p=1} = (K_{ls})_C \left(FC_l - FC_s^s(r_p=1)\right) \qquad \frac{\partial FC_s}{\partial r_p}\bigg|_{r_p=0} = 0 \quad (17)$$

Energy balance in the solid phase

$$\rho_s C_{ps} \frac{\partial T_s}{\partial t} = \frac{\lambda_s}{R_p^2} \frac{1}{r_p^2} \frac{\partial}{\partial r_p}\left(r_p^2 \frac{\partial T_s}{\partial r_p}\right) + \rho_s (-\Delta H_R) R_W(FA_s, FB_s, T_s) \quad (18)$$

$$\frac{\lambda_s}{R_p} \frac{\partial T_s}{\partial r_p}\bigg|_{r_p=1} = h_s \left(T - T_s^s(r_p=1)\right) \qquad \frac{\partial T_s}{\partial r_p}\bigg|_{r_p=0} = 0 \quad (19)$$

2.1 Solution of the dynamic model.

The partial differential equations (PDEs) were converted to ordinary differential equations (ODEs) by discretization using the global orthogonal collocation method (Villadsen and Michelsen, 1978). After the discretization of the spatial coordinates in the particle and in the reactor, a system of ODEs and LEs (linear equations) are obtained. An important feature of the system is its stiffness (Salmi et al., 2000). This implies that specific calculation methods must be used. In this work, the software DASSL was used, which is a specific software for algebraic-differential stiff systems.

Intending to consider the variations of the physical-chemical properties, transfer coefficients of mass and heat and velocities of the fluid phase, without making difficult the numerical solution procedure, it was utilized the strategy normally employed in literature of calculating at each instant of integration the values of this terms through correlations found in literature. In the case of the physical-chemical properties, polynomials generated by the HYSYS® simulator were converted to FORTRAN90® language, utilized in this work.

2.2 Phase Change of the Reacting Medium.

During simulations, a multicomponent flash calculation can be or not considered to evaluate the effect of the phase change in the reacting medium in the dynamic behaviour of the reactor.

Before applying the multicomponent flash, it is necessary to check the existence of two phases. Two methods were tested: the thermodynamic Bubble and Dew point calculation and another criteria proposed by Hanika et al. (1976), that has been used in works related to the study of the phenomenon of phase change in three-phase reactors (Kheshgi et al. (1992), Khadilkar et al. (1999) and Cheng et al. (2001)).

With respect to the non-ideality of the liquid and vapour phases, it was considered in the flash calculation through, respectively, models for activity (Ideal, NRTL, UNIQUAC) and for fugacity (Ideal, Virial) coefficients.

The numerical strategy used to couple the flash calculation to the dynamic model is: during pauses in the integration, the bubble and the dew points of the reacting medium or, as another option, the Hanika criteria (Hanika et al., 1976) are evaluated for each axial position of the reactor discretization grid. If the temperature of the reacting medium at each axial position is found between the bubble and dew points or the criteria of Hanika is satisfied, the multicomponent flash is calculated. Consequently, the concentration values of the components are updated and the heat involved in the phase change (Q_{flash}) is computed by the energy balance of the fluid phase (Equation 9).

An important detail introduced in the calculation of the multicomponent flash is the factor F. The results obtained with the flash procedure are multiplied by F, ($0 \leq F \leq 1$). This factor allows the user of the model to adjust the effect of the phase change in order to represent the dynamic behaviour of the reactor as it is observed in practice. This is accounted for each control volume element that has the fraction of m^3 dimensionality. Thus, F is considered a parameter that corrects the flash results since the fraction of the mass flow, in the reacting medium, that is suffering phase change at one specific axial point in the reactor, is not known (Vasco de Toledo and Maciel Filho, 2002).

2.3 Phase Change of the Refrigerant Fluid.

The great amount of energy released in hydrogenation reaction may increase the temperature of the refrigerant until evaporation. For this reason, the regime flow of the refrigerant fluid can present two situations: monophase (liquid or steam) and two-phase (liquid+steam).

The transfer of heat in monophase regime flow is reasonably susceptible to a predictive treatment, having a large number of correlations that predict the convective coefficient (Incropera and De Witt, 1990). The estimation of coefficients of heat transfer in ebullition (two-phase regime flow) was possible due to the methodology proposed by Chen and described by Stephan (1992).

The numerical strategy to incorporate this phenomenon in the model of the reactor is described next. During the integration of the dynamic model of the reactor, at each axial position of the grid of discretization of the reactor, the temperature of the refrigerant fluid (T_r) is compared to its saturation temperature (T_{sat}), that is dependent of the pressure. From this comparison, the convective heat transfer coefficient of the refrigerant fluid (*he*) may be calculated differently in three cases:

1) If $T_r < T_{sat}$, then the flow is monophase and liquid, thus *he* is calculated by the equation in Incropera and De Witt (1990);

2) If $T_r = T_{sat}$, then the flow is two-phase, thus *he* is calculated by the methodology proposed by Chen and described by Stephan (1992);

3) If $T_r > T_{sat}$, then the flow is monophase and vapour, thus *he* is calculated by the equation in Incropera and De Witt (1990) using the vapour physical properties.

In accordance with the methodology proposed by Chen, during the phase change is necessary to evaluate the amount of refrigerant fluid vaporized or condensed taking into account the latent heat of the refrigerant fluid. Thus, a possible variation of *he* due to the conditions of vaporization above mentioned may cause a variation of the global coefficient of heat transfer (*U*), whose value affects the energy balance of the fluid phase and of the refrigerant fluid (Eqs. 9 and 11). In the case of study of this work, the refrigerant fluid is water, which is observed in many industrial processes.

3. RESULTS

The simulations monitored the temperature of the reacting medium at the end of the tubular reactor (*T*). The studies focused on this variable because its knowledge is fundamental for elaboration of efficient control strategies, simulation studies and reactor design. For each simulation, a step perturbation related to one specific operating parameter was introduced in the system. The parameters that showed to have more influence on the reacting medium temperature (Vasco de Toledo and Maciel Filho, 2002) are the inlet temperatures of the fluid phase, T_{fo}, and of the refrigerant fluid, T_{ro}.

Figures 2 and 3 show the effect of the phase change of the reacting medium in the dynamic behaviour of the reactor after a step perturbation in T_{ro} and T_{fo}, respectively. This phenomenon gives rise to steep temperature gradients within the reactor due to the heat involved in the phase change according to the flash calculation. As a result, temperature increases when condensation occurs and it decreases as the reacting medium vaporizes, thus the reactor shows an oscillatory dynamic behaviour.

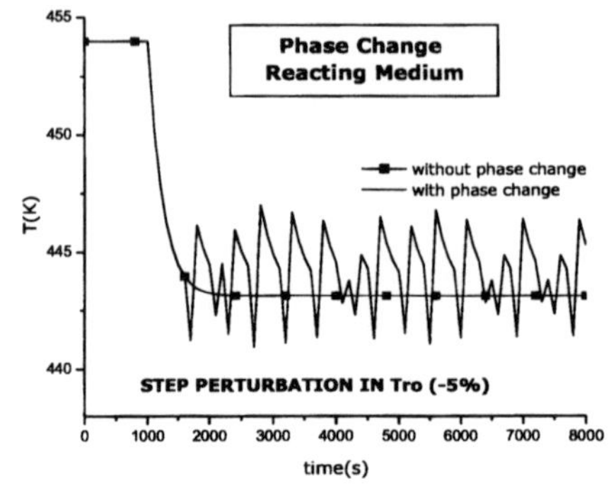

Fig. 2. Influence of the phase change of the reacting medium on the outlet temperature profile of the reactor. Step perturbation in T_{ro} (refrigerant fluid feeding temperature).

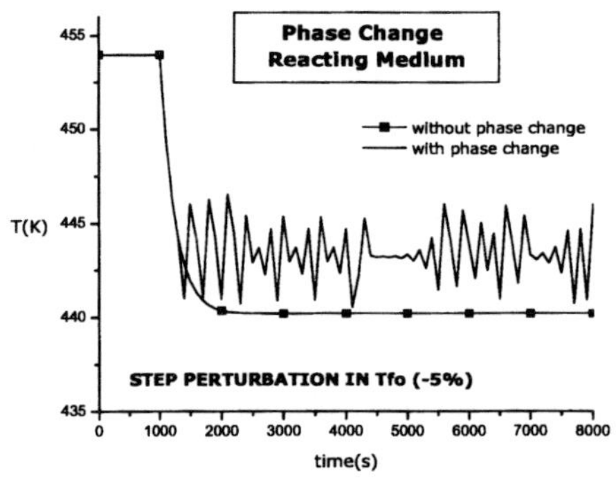

Fig. 3. Influence of the phase change of the reacting medium on the outlet temperature profile of the reactor. Step perturbation in T_{fo} (feeding temperature).

In Figures 4 and 5, the effects of considering the non-ideality of the, respectively, liquid and vapour phases, for the case displayed in Figure 2, can be seen. The differences between the results show that the hypotheses of ideal system may affect the dynamic behaviour of the reactor when, in fact, the system does not have these ideal characteristics. When the criteria of Hanika was used to check the occurrence of the phase change, in general, there was a conflict between the temperature pointed out by the criteria and the phase equilibrium equation (in flash calculation) that did not considered the values of temperature and pressure as conditions of existence for the two phases. Thereby, for this case of study, the flash calculation reached a better performance when the Bubble and Dew points option were utilized to check the temperature of the phase change.

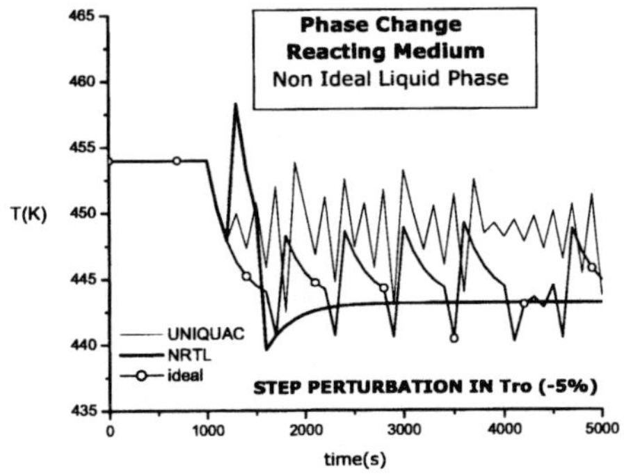

Fig. 4. Influence of the thermodynamic models (activity coefficients) on the outlet temperature profile of the reactor. Step perturbation in T_{ro}.

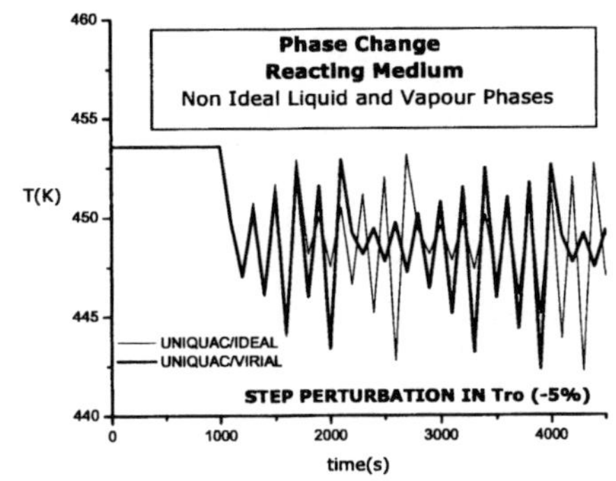

Fig. 5. Influence of the thermodynamic models (fugacity coefficients) on the outlet temperature profile of the reactor. Step perturbation in T_{ro}.

The simulations represented in Figure 6 show the influence of the factor F in the representation of the phenomenon of phase change of the reacting medium. According to the results, small values of F must be preferred, otherwise the variations in the temperature become wider than is observed in practice (~10°C). Thus, the factor F allows the user of the model to adjust the effect of the phase change as it is observed in practice.

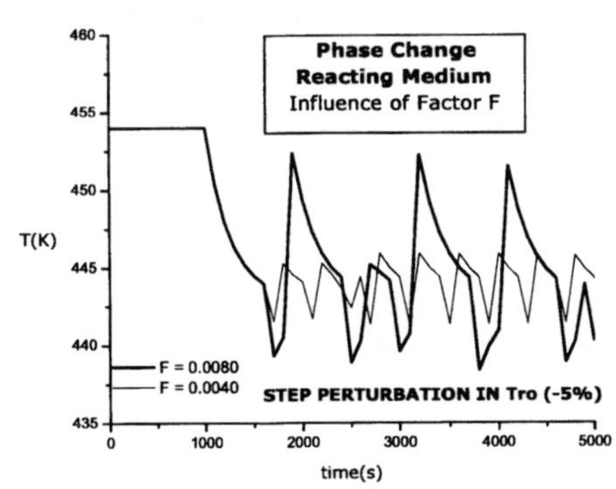

Fig. 6. Influence of the factor F on the outlet temperature profile of the reactor. Step perturbation in T_{ro}.

The effect of the phase change in the refrigerant fluid is represented in Figure 7 (the phase change of the reacting medium is not considered). The reactor temperature (*T*) falls during the vaporization of the refrigerant fluid because of an intensification of the heat exchange caused by a sharp rise in the convective heat transfer coefficient of the refrigerant fluid (*he*) (Figure 8) that affects the value of the global heat transfer coefficient (*U*). Otherwise, while the vaporization does not happen, *T* comes back to increase. Thus, an oscillatory dynamic behaviour is observed in the reactor.

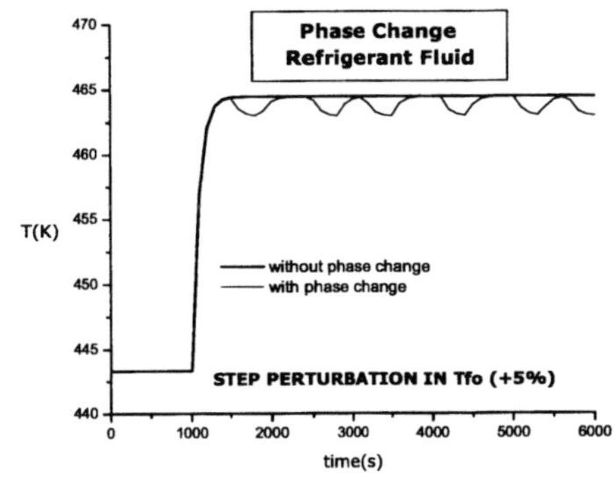

Fig. 7. Influence of the phase change of the refrigerant fluid on the outlet temperature profile of the reactor. Step perturbation in T_{fo}.

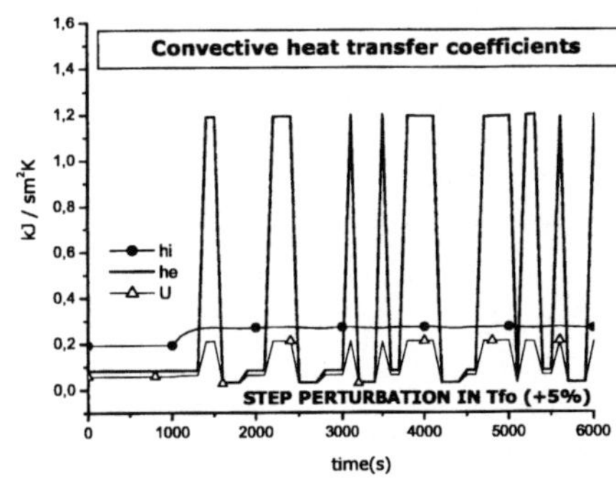

Fig. 8. Convective coefficients of heat transfer during simulation of Figure 7.

4. CONCLUDING REMARKS

The model developed and presented in this work allowed to simulate some aspects of the dynamic behaviour of three-phase slurry reactor. The model was capable to foresee the phenomenon of phase change of the reacting medium as well as of the refrigerant fluid that introduced perturbations on the temperature of the reactor that surely affect the product specifications. Therefore the knowledge obtained through the model is essential for a good design and control of three-phase catalytic slurry reactors.

NOTATION

a	interfacial area, m^{-1}
C_p	heat capacity, kj/kg.K
D_e	effective diffusivity, m^2/s
D_t	reactor intern diameter, m
D_{te}	reactor extern diameter, m
D_{ts}	diameter of the jacket tube, m
E	activity energy, J/mol
FA	molar flow of the component A, kmol/s
FB	molar flow of the component B, kmol/s
FC	molar flow of the component C, kmol/s
he	external convective heat transfer coefficient, $kj/m^2.s.K$
hi	internal convective heat transfer coefficient, $kj/m^2.s.K$
k	kinetic constant, kmol/ kg cat . s
K	mass transfer coefficients, cm/s
L	reactor length, m
Q_{flash}	heat of phase change of the reacting medium, kj/s
Q_l	liquid feed volumetric flow, m^3/s
r_p	dimensionless particle radial position
R_p	radius particle, m
R_w	rate of hydrogenation of *o*-cresol, kmol/kg-cat.s
T	temperature, K
t	time, s
u	linear velocity, m/s
U	global heat transfer coefficient, $kj/m^2.s.K$
w	catalyst concentration, $kgcat/m^3$
z	dimensionless reactor axial position

Greek letters

ΔH_R	heat of reaction, kj/kmol
λ	thermal conductivity, kj/m.s.K
ν	stoichiometric coefficient
ρ	density, kg/m^3
ε_g	gas phase hold-up
ε_l	liquid phase hold-up
ε_s	solid porosity
τ	tortuosity

Subscripts

g	gas phase
fo	feeding
l	liquid phase
i	initial value (reactor inlet)
p	particle
r	refrigerant fluid
s	solid

Superscripts

s	catalyst surface

REFERENCES

Cheng, Z. M., Abdulhakeim, M. A., Yuan, W. K. (2001). Intensification of Phase Transition on Multiphase Reactors, *AIChe Journal*, **Vol. 47**, n. 5.

Gianetto, A., Silveston, P. L. (1986). *Multiphase Chemical Reactors: Theory, Design, Scale-up*, Hemisphere Publishing Corporation, Washington.

Hanika, J., Sporka, K, Ruzicka, V, Hrstka, J. (1976). Measurement of axial temperature profiles in an adiabatic trickle-bed reactor. *Chem Eng. J.*, **Vol. 12**, 193.

Hichri, H., Armand, A. and Andrieu, J. (1991). Kinetics and Slurry-type Reactor Modelling during Catalytic Hydrogenation of o-Cresol on Ni/SiO_2. *Chem. Eng. Proc.*, **Vol. 30**, 133-140.

Incropera, F. P. and D. P. de Witt. (1990). *Fundamentals of Heat and Mass Transfer*. John Wiley & Sons, 3ed, 919pp, New York.

Julcour, C., Stüber, F., Le Lann J. M., Wilhelm, A. M., Delmas, H. (1999). Dynamics of a Three-Phase Upflow Fixed-Bed Catalytic Reactor, *Chem. Eng. Sci.*, **Vol. 54**, 2391-2400.

Khadilkar, M. R., Mills, P. L., Dudokovic, M. P. (1999). Trickle-bed reactor models for systems with a volatile liquid phase. *Chem. Eng. Sci.*, **Vol. 54**, 2421-2431.

Kheshgi, H. S., Reyes, S. C., Hu, R., Ho, T.C. (1992). Phase transition and steady state multiplicity in a trickle-bed reactor, *Chem. Eng. Sci.*, **Vol. 47**, 1771.

Lange, R., Gutsche, R., Hanika, J. (1999). Forced Periodic of a Trickle-Bed Reactor, *Chem. Eng. Sci.*, **Vol. 54**, 2569-2573.

Mariano, A. P. (2003). *Modelagem Determinística de Reatores de Lama Catalíticos Trifásicos: Aplicação para Reações de Hidrogenação*. MSc. Thesis, UNICAMP, Campinas, SP, Brazil.

Salmi, T., Wärna, J., Toppinen, S., Rönnholm, M. and Mikkola, J. P. (2000). Dynamic Modelling of Catalytic Three-Phase Reactors for Hydrogenation and Oxidation Processes. *Brazilian Journal of Chem. Eng.*, **Vol. 17**, No. 04-07, pp. 1023-1034.

Stephan, K. (1992). *Heat Transfer in Condensation and Boiling*, Springer, New York.

Vasco de Toledo, E. C., Santana, P. L., Wolf Maciel, M. R. and Maciel Filho, R. (2001). Dynamic Modelling of a Three-Phase Catalytic Slurry Reactor, *Chem. Eng. Sci.*, **Vol 56**, 6055-6061.

Vasco de Toledo, E. C., Maciel Filho, R. (2002). Development of a Dynamic Model for a Three-Phase Catalytic Slurry Reactor with Change of Phase of the Reacting Medium and the Cooling Fluid, *17th International Symposium on Chemical Reaction Engineering-ISCRE 17*, August 25-28, Hong Kong, China.

Villadsen, J., Michelsen, M. L. (1978). *Solution of Differential Equation Models by Polynomial Approximation*. Prentice-Hall, New Jersey.

Warna, J., Salmi, T. (1996). Dynamic Modeling of Catalytic Three Phase Reactors, *Comp. Chem. Eng.*, **Vol. 20(1)**, 39-47.

ELSEVIER
IFAC
PUBLICATIONS
www.elsevier.com/locate/ifac

NEURAL NETWORK-BASED IDENTIFICATION OF NONLINEAR ADSORPTION ISOTHERMS

Weihua Gao * Sebastian Engell *

* *Process Control Laboratory, Department of Biochemical and Chemical Engineering, University of Dortmund, Emil-Figge-Str. 70, D-44221 Dortmund, Germany*

Abstract: The adsorption isotherm is the most important parameter in chromatographic separation process. In this paper, a numerical isotherm estimation method based on neural networks is proposed. As not the absolute values but the slopes of the isotherms are most important in process simulation and optimization, the slopes of the isotherm instead of the absolute values are taken as the outputs of the neural network. A tailored method is used to initialize the neural network. The parameters of the neural network are adjusted to minimize the difference between the simulated and the measured profiles. The issue of the design of the experiments, which decides the amount of the isotherm information contained in the profiles, is discussed. Simulation and experimental studies illustrate the potential of the proposed method. *Copyright © 2004 IFAC*

Keywords: Chromatography, neural networks, nonlinear identification, adsorption isotherm.

1. INTRODUCTION

The adsorption isotherm describes the distribution of the solute between the two phases of a chromatographic system. Different components can be separated because they have different adsorption isotherms. In analytical applications of chromatography the concentration of the analyte in the mobile phase usually is very small. Therefore the assumption of linear isotherms is justified for these processes. In most preparative applications, however, the concentration range of interest is wide and therefore the linearity assumption is no longer valid. The functional relationship between the equilibrium concentrations in the two phases can become very complex, depending on the concentration range, the retention mechanism, and the phase system. Furthermore, in the case of multicomponent mixtures, an additional complexity results from the competition between the different components for interaction with the stationary phase.

The identification of the adsorption isotherm is very important for applying model-based methods to the optimization and the operation of chromatographic separation processes. Adsorption isotherms can be determined by static and dynamic experimental methods (Ruthven, 1984; Guiochon *et al.*, 1994). Static methods consist of near-equilibrium measurements of the extent of adsorption at specified, constant adsorbate concentrations. The extent of adsorption has been measured by methods including microbalance gravimetry, infrared absorption, and thermogravimetry. Static methods are tedious and time-consuming and require a large amount of material. Dynamic methods consist of frontal analysis (FA) (James and Phillips, 1954; Schay and Szekely, 1954; Glückauf, 1955), frontal analysis by characteristic points (FACP) (Glückauf, 1947), peak-maxima methods (Kipping and Winter, 1965), elution by characteristic point (ECP) (Cremer and Huber, 1961), and step-and-pulse (minor disturbance) methods (Helfferich and Peterson, 1963). Among them, the frontal analysis (FA) and step-and-pulse can be extended to measure multicomponent isotherms. The simple wave method can also be used for the determination of binary isotherms in cases in which the deviation of isotherm from the Langmuir model is moderate (Ma *et al.*, 1990). These experimental methods need large volumes of concentrated solutions. This is expensive for valuable substances. Moreover, the long experimental period is another drawback. An alternative approach is to approximate the adsorption isotherm numerically. The isotherm functional form is chosen at the outset. Then by minimizing the difference between calculated and experimental profiles, the parameter values are adjusted until a best fit to the experimental data is found. Dose *et al.* determined the equilibrium isotherms of N-benzoyl-(D,L)-alanine and N-benzoyl-(D,L)-phenylalanine on immobilized bovine serum albumin from the peak shapes of single component (Dose *et al.*, 1991). They used a modified simplex algorithm to find the best parameters of Bi-Langmuir isotherms and found a good agreement with the isotherms determined with frontal analysis. James and Sepúlveda developed a more sophisticated algorithm for the estimation of the isotherm parameters by the inverse method (James and Sepúlveda, 1994; James *et al.*, 1999). They used the conjugate gradient algorithm for the minimization of the objective function that - besides the conventional least squares - took into account the difference between the first moments of the peaks, as well. This algorithm was applied to the estimation of the competitive Bi-Langmuir and Moreau isotherms of Ketoprofen enantiomers on a Chiracel OJ column, as well as of the competitive Langmuir isotherm of benzyl alcohol and 2-phenylethanol on a C_{18} column from individual band profiles. The competitive isotherms estimated with the inverse method agreed well with the data obtained from conventional methods. Felinger *et al.* derived the parameters of competitive isotherms from the band profiles of a racemic mixture of the 1-phenyl-1-propanol enantiomers by means of a nonlinear least-squares method (Felinger *et al.*, 2003). The isotherms determined from the overloaded band profiles matched extremely well the isotherms determined by frontal analysis.

The numerical methods transfer the experimental effort to the computational effort. Compared with the experimental methods, they are more economical because only a small amount of concentrated solution is used. At the same time, the computational time is shortened by using high-speed computers. However the functional form of the isotherm must be assumed prior to the numerical approximation. If there is a structural mismatch between the assumed functional form and the actual isotherm, the numerical methods cannot guarantee a good result. In practice, several isotherm forms should be tried in order to estimate the most suitable model. On the other hand, a universal form of the isotherm should be used when there is no knowledge about the isotherm. This leads to the idea of using a neural network to represent the isotherm model and of estimating the neural network parameters from the experimental band profiles. This idea was first described in (Daniel *et al.*, 2002). In this paper, a systematic way to use neural network to identify isotherm is introduced.

2. THEORY

2.1 Chromatographic separation process

In a batch chromatographic separation process, one charge of the mixture to be separated is injected into the column together with a suitable solvent. The charge is carried through the column by continuously injecting additional desorbent. Due to the different adsorption affinities, the different components have different migration velocities and are gradually separated while moving along the column. The concentrations of the components are measured at the outlet of the column (Fig. 1).

The mathematical modelling of the chromatographic separation process has been extensively described in the literature by several authors, and is in most cases based on a differential mass balance. The most general 1-dimensional model, which takes into consideration all important effects, is the general rate model:

$$\frac{\partial c_i}{\partial t} + u\frac{\partial c_i}{\partial x} + \frac{3(1-\varepsilon)k_{l,i}}{\varepsilon r_p}(c_i - c_{p,i}(r_p)) = D_{ax,i}\frac{\partial^2 c_i}{\partial x^2}$$
$$(1-\varepsilon_p)\frac{\partial q_i}{\partial t} + \varepsilon_p\frac{\partial c_{p,i}}{\partial t} = \varepsilon_p D_{p,i}\left[\frac{1}{r^2}\frac{\partial}{\partial r}(r^2\frac{\partial c_{p,i}}{\partial r})\right] \tag{1}$$

These two partial differential equations describe the concentrations in the mobile phase (c_i) and in the stationary phase (q_i and $c_{p,i}$). Adsorption isotherms are used to relate the concentrations q_i (in the solid of the stationary phase) and $c_{p,i}$ (in the liquid of the stationary phase):

$$q_i = f_i(c_{p,1}, c_{p,2}, \cdots, c_{p,i}, \cdots, c_{p,n}). \tag{2}$$

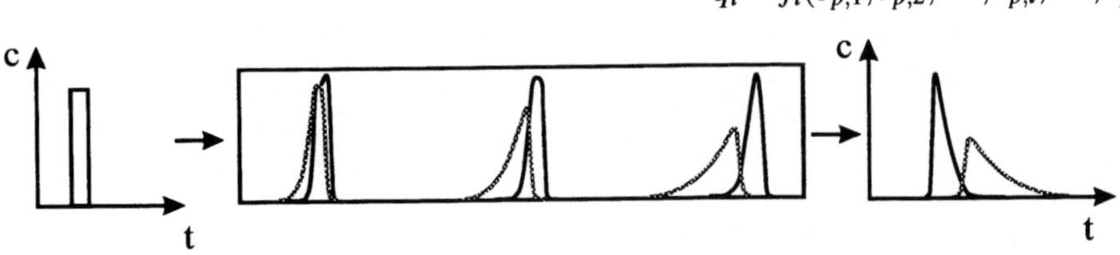

Fig. 1. Illustration of a chromatographic separation process.

It should be noted that in the isotherm equations, the concentration $c_{p,i}$ is usually denoted by c_i. This does not mean that they are the same. An efficient numerical solution for the general rate model incorporating arbitrary nonlinear isotherms was proposed by Gu (Gu, 1995). The mobil phase and the stationary phase are first discretized using the finite element and the orthogonal collocation methods respectively. The resulting ODE system is solved using an ODE solver that is based on the Gear's method. The numerical solution yields the concentrations of the components in the column at different locations and times. The concentrations at the outlet of the column are used to generate the band profiles. From the model (1), it is obvious that the derivatives $\partial q_i/\partial c_j$ and not the absolute values of the isotherm are critical.

2.2 *Neural network isotherm*

Two-layer neural networks with sigmoidal neurons in the hidden layer and linear neurons with bias in the output layer are used to describe the isotherms. The outputs of the neural networks are the slopes of the isotherms $\partial q_i/\partial c_j$, instead of the values q_i. Fig. 2 depicts the structure of a simple neural network for binary chromatography.

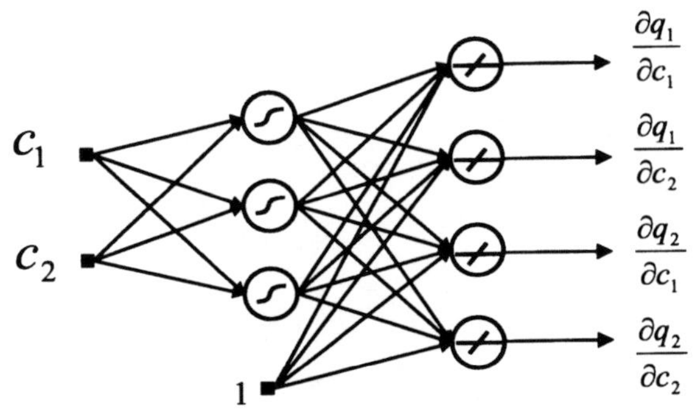

Fig. 2. Structure of neural network isotherm model.

2.3 *Identification of the neural network isotherm*

The identification principle is shown in Fig. 3. The neural network isotherm is used in the general rate model to simulate the band profiles. The experimental band profiles are compared with the simulated band profiles by evaluating the distance function

$$\sum_i \sum_j \left(c_{i,j}^{sim} - c_{i,j}^{meas}\right)^2, \tag{3}$$

where i is the index of component, j is the index of the sampling points. The parameters of the neural network isotherm are adjusted to minimize the distance, using the Levenberg-Marquardt method with line-search.

A simple method is proposed to initialize the neural network isotherm. The weights of the input layer are set to random numbers in $(0,1)$. The weights of the

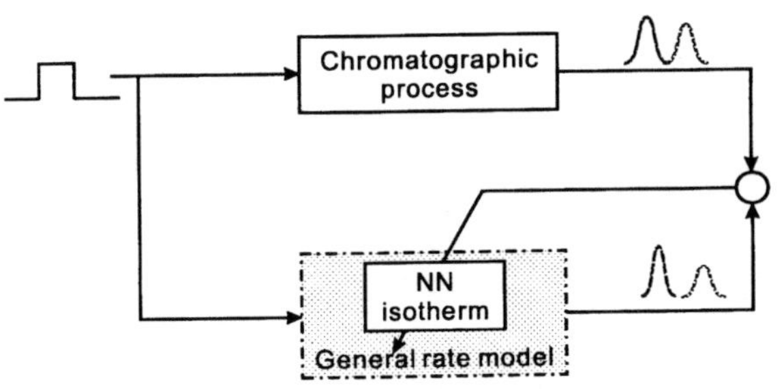

Fig. 3. Principle of neural network-based isotherm identification.

output layer are set to 0. The biases of $\partial q_i/\partial c_j$ $(i = j)$ are set by a trial and error procedure. They are adjusted until the simulated band profiles have roughly the same position as the experimental band profiles. The biases of $\partial q_i/\partial c_j$ $(i \neq j)$ are set to arbitrary negative numbers in $(-1,0)$. The steps in the solution of the identification problem are as follows:

(1) Analysis of the chromatographic process and design of proper experimental conditions, i.e. the operating parameters, the input concentration, and the choice of multi injection or single injection. Running the experiment and measuring the band profiles.
(2) Selection of the neural network structure. Besides the above structure, other more complex structures are also possible.
(3) Choice of the number of neurons in the hidden layer. The number of the neurons in the hidden layer decides the ability of the neural network to approximate a nonlinear function. An iterative procedure is used to determine this number.
(4) Estimation of the parameters using the Levenberg-Marquardt method with line-search.
(5) Validation of the neural network isotherm by comparing simulations of eq. (1) to measured data.

3. RESULTS AND DISCUSSION

The neural network isotherm was first checked in a simulation study, and then it was applied to estimate isotherms from experimental data.

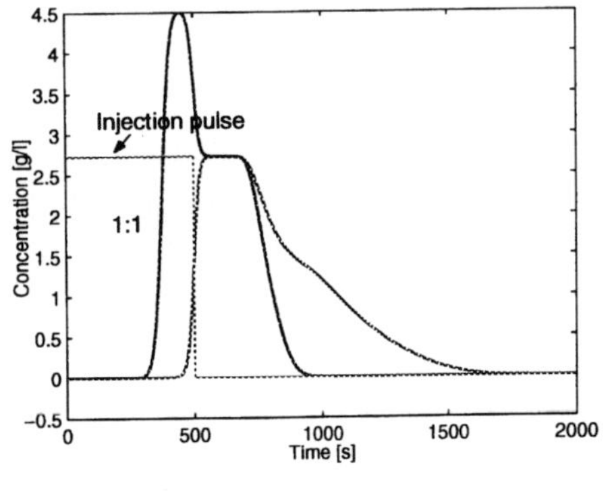

Fig. 4. Overloaded band profiles of the single injection, 1:1 feed mixture.

3.1 Simulation study

In the simulation study, an separation process which exhibits a strongly nonlinear adsorption isotherm was considered (Hanisch, 2002). The competitive adsorption to the solid phase of the two components is described by a Bi-Langmuir isotherm:

$$q_1 = \frac{0.0127c_1}{1+1000c_1+1008c_2} + \frac{23.412c_1}{1+333.1c_1+94.5c_2}$$
$$q_2 = \frac{8.979c_2}{1+211.4c_1+43.46c_2} \quad . \tag{4}$$

By simulating the general rate model with the Bi-Langmuir isotherm band profiles were acquired. They are regarded as the "experimental" band profiles in the identification procedure. Fig. 4 shows the band profiles for an overloaded injection. Different numbers of neurons in the hidden layer were tried. The results are listed in table 1. From this, the number of the neurons in the hidden layer is set to three.

Table 1. Estimation results for different numbers of neurons in the hidden layer

Number of neurons in the hidden layer	Residual ×500	Iterations	Function count
1	9.2085	29	438
2	0.1886	33	714
3	0.1184	40	1087
4	0.0409	49	1625

The estimated neural network isotherms were compared with the original Bi-Langmuir isotherm in Fig. 5. The neural network isotherm approaches the Bi-Langmuir isotherm in all four derivatives. The difference between them can be explained as insufficient isotherm information contained in the "experimental" band profiles. To show this, all the combinations of concentrations which are involved in the numerical solution of the model (1) are depicted in Fig. 6. The "experimental" band profiles are only the concentrations at the outlet of the column. Most of the points are along a closed curve in the concentration space. To improve the neural network isotherm, the experiment should be specially designed so that enough information is contained in the band profiles. As a first attempt, a multi injection was simulated. The injection consisted of three pulses, as shown in Fig. 7. For every pulse, there is different feed concentration ratio (1:1, 3:1 and 1:3). The concentration combinations in the numerical solution are shown in Fig. 8. Compared

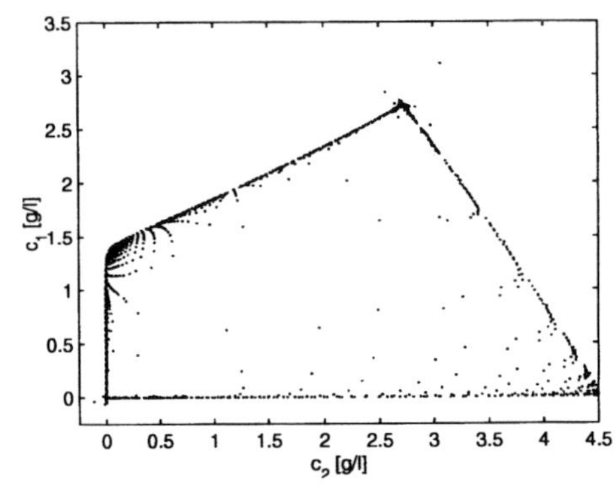

Fig. 6. Concentration combinations in the column, single injection, 1:1 feed mixture.

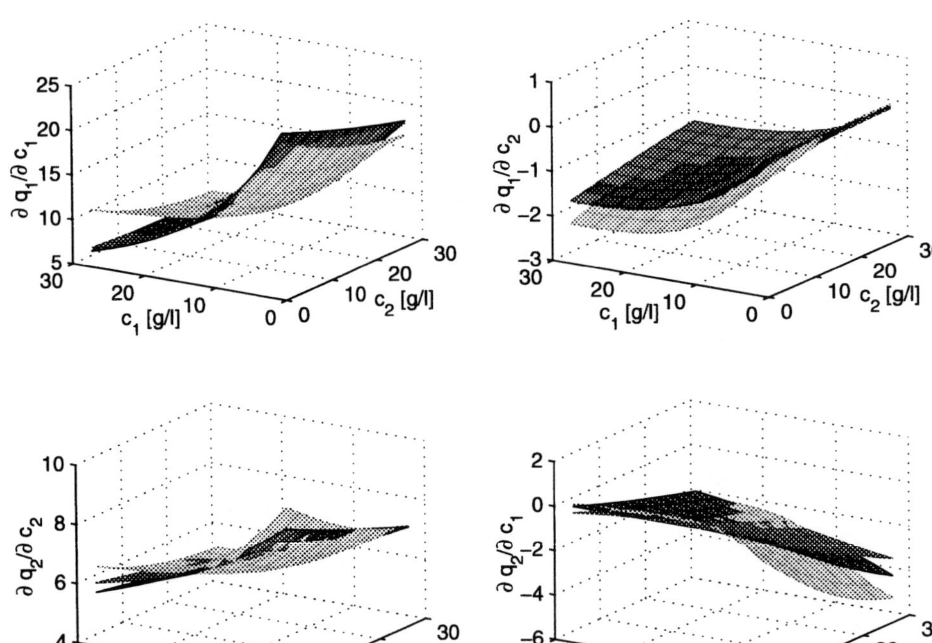

Fig. 5. Comparison of the Bi-Langmuir isotherm and the estimated neural network isotherms, black surface: Bi-Langmuir isotherm, light grey surface: neural network isotherm (single injection), dark grey surface: neural network isotherm (multi-injection).

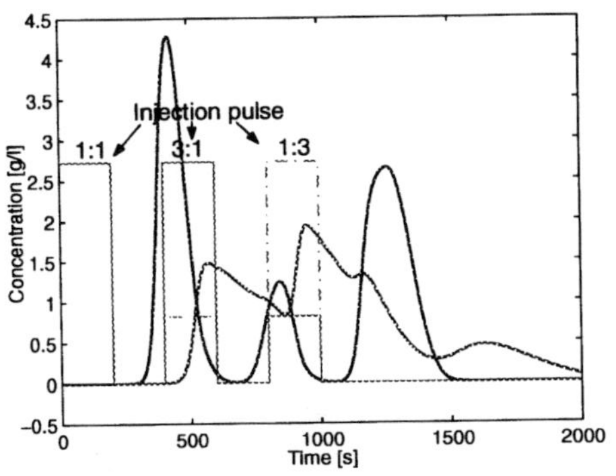

Fig. 7. Band profiles of the multi injection, (1:1), (3:1) and (1:3) feed mixture.

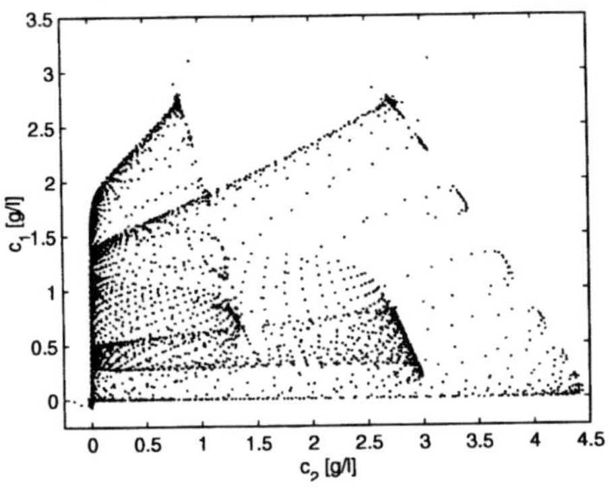

Fig. 8. Concentration combinations in the column, multi-injection.

with the single overloaded injection, a much larger area is covered. The same neural network isotherm model (3 neurons in the hidden layer) was used in the identification procedure. As Fig. 5 shows, the neural network isotherm model was remarkably improved.

3.2 *Estimating from experimental data*

The experimental data was supplied by the chair of process and plant design (Lehrstuhl für Anlagentechnik, Universität Dortmund, Germany). The separation of EMD53986 on a chiral stationary phase is considered. For the measurement of the EMD53986 a preparative UV detector cell was used. The band profiles were calculated from the UV signals.

The experimental data consists of five sets of band profiles for different injection volumes. The neural network isotherm was identified from the most overloaded band profiles (Fig. 9) and was tested using the remaining band profiles. The results are shown in Fig. 9 and Fig. 10, and compared with an optimized Bi-Langmuir isotherm. In Fig. 9, the simulated band profiles using the neural network isotherm match the experimental band profiles well, while with the Bi-Langmuir isotherm large errors are seen because of its fixed structure. In Fig. 10, the neural network isotherms give a better prediction in the overlapping

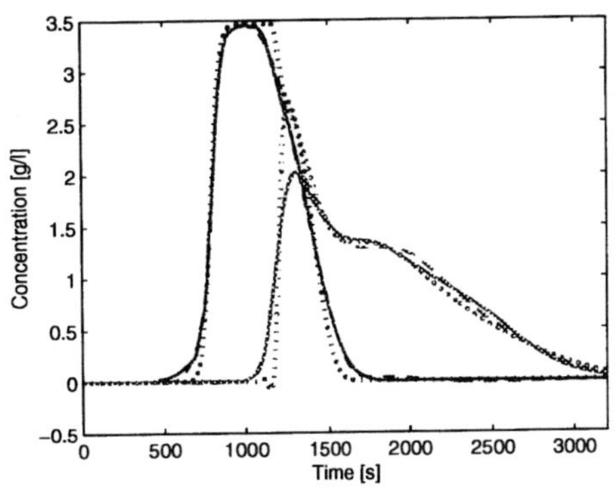

Fig. 9. Comparison of measurement and estimation for injection volume of 120 ml, solid lines: measurement, dashed lines: estimation of neural network isotherm (4 neurons in the hidden layer), dotted lines: estimation of Bi-Langmuir isotherm.

region than the Bi-Langmuir isotherm. This is important for model-based control and optimization because the purities of the products are determined by the overlap of the profiles.

4. CONCLUSIONS

Numerical isotherm identification methods have attracted considerable attention in preparative chromatography. This paper studies the possibility of using neural networks as the isotherm functional form in the numerical methods. The isotherm slopes are generated as the outputs of a neural network. By increasing the number of hidden layer neurons the neural network can represent any adsorption isotherm. The simulation and experimental studies prove that adsorption isotherms can be estimated well from band profiles by using neural networks as the isotherm functional form. The key issue is to design the experiment to generate enough information for the identification. The optimization of the experiment will be the focus of future research.

ACKNOWLEDGEMENTS

The work of the first author was supported by a scholarship of the NRW Graduate School of Production Engineering and Logistics at Universtität Dortmund. This support is very gratefully acknowledged.

REFERENCES

Cremer, E. and H. F. Huber (1961). Measurement of adsorption isotherms at high temperature with the help of gas-solid elution chromatography. *Angew. Chem.*, **73**, 461-465.

Daniel, G., J. Dienstuhl, S. Engell, S. Felske, K. Goser, R. Klinkenberg, K. Morik, O. Ritthoff and H. Schmidt-Traub (2002) Novel learning tasks, optimization, and their application. In: *Advances*

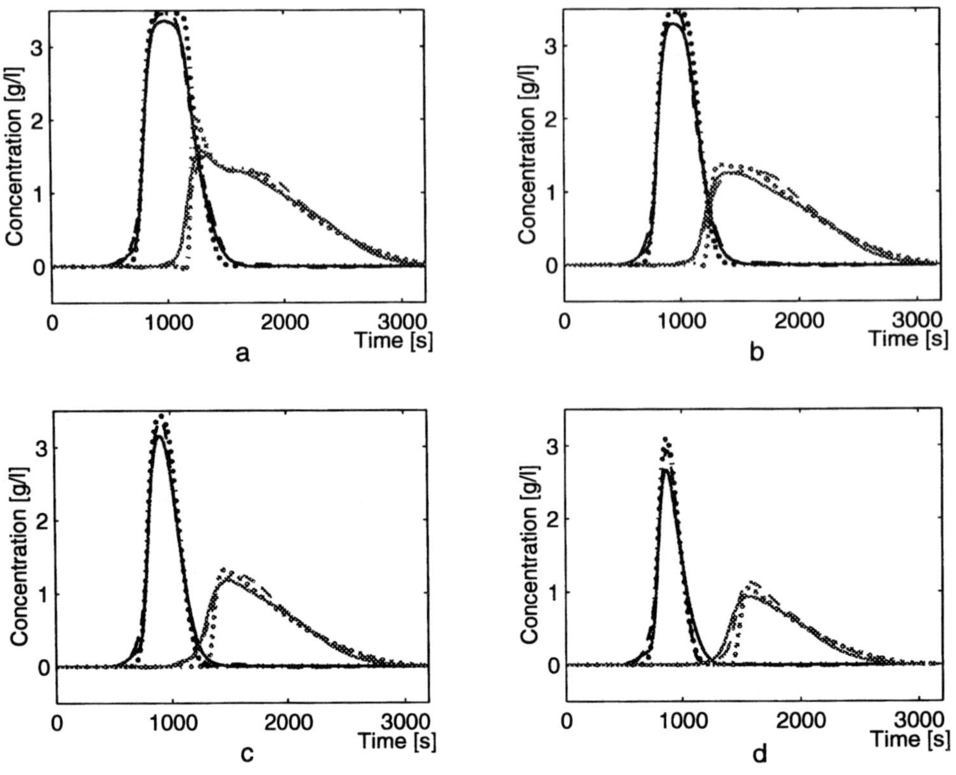

Fig. 10. Model validation for injection volumes of 100 ml, 80 ml, 60 ml and 40 ml, solid lines: measurements, dashed lines: simulation of neural network isotherm, dotted lines: simulation of a Bi-Langmuir isotherm.

in Computational Intelligence: Theory and Practice (H.-P. Schwefel, I. Wegener and K. Weinert, Ed.). Chap. 8, pp. 278-294. Springer Verlag, Berlin.

Dose, E. V., S. Jacobson and G. Guiochon (1991). Determination of isotherms from chromatographic peak shapes. *Anal. Chem.*, **63**, 833-839.

Felinger, A., A. Cavazzini and G. Guiochon (2003). Numerical determination of the competivite isotherm of enantiomers. *Journal of Chromatography A*, **986**, 207-225.

Glückauf, E. (1947). Theroy of chromatography: Part II. Chromatography of a single solute. *J. Chem. Soc.*, 1302–1308.

Glückauf, E. (1955). Theory of chromatography: Formulas for diffusion into spheres and their application to chromatography. *Trans. Faraday Soc.*, **51**, 1540.

Gu, T. (1995). *Mathematical Modelling and Scale Up of Liquid Chromatography*. Springer Verlag, New York.

Guiochon, G., S. Golshan-Shirazi and A. Katti (1994). *Fundamentals of Preparative and Nonlinear Chromatography*. Academic Press, Boston.

Hanisch, F. (2002). Prozessführung Präparativer Chromatographieverfahren (Operation of preparative chromatographic processes). Dr.-Ing. Dissertation. Shaker Verlag, Aachen. (in German)

Helfferich, F. G. and D. L. Peterson (1963). Chromatographic methods for sorption isotherms and phase equilibria. *Science*, **142**, 661-662.

James, D. H. and C. S. G. Phillips (1954). The chromatography of gases and vapours: Part III. the determination of adsorption isotherms. *J. Chem. Soc.*, 1066-1070.

James, F. and M. Sepúlveda (1994). Parameter identification for a model of chromatographic column. *Inverse Problems*, **10**, 1299-1314.

James, F., M. Sepúlveda, F. Scharton and I. Quinones (1999). Determination of binary competitive equilibrium isotherms from the individual chromatographic band profiles. *Chemical Engineering Science*, **54**, 1677-1696.

Kipping, P. J. and D. G. Winter (1965). The measurement of adsorption isotherms by a gas chromatographic technique. *Nature*, **205**, 1002-1003.

Ma, Z., A. Katti, B. Lin and G. Guiochon (1990). Simple wave effects in two-component nonlinear liquid chromatography. application to the measurement of competitive adsorption isotherms. *J. Phys. Chem.*, **94**, 6911-6922.

Ruthven, D. M. (1984). *Principles of Adsorption and Adsorption Processes*. Wiley, New York.

Schay, G. and G. Szekely (1954). Gas adsorption measurements in flow systems. *Acta Chim. Hung.*, **5**, 167-182.

ELSEVIER
IFAC
PUBLICATIONS
www.elsevier.com/locate/ifac

MODELING AND CONTROL OF A O_2/CO_2 GAS TURBINE CYCLE FOR CO_2 CAPTURE

Lars Imsland [*] Dagfinn Snarheim [*] Ragnhild Ulfsnes [**] Olav Bolland [**] Bjarne A. Foss [*,1]

[*] *Department of Engineering Cybernetics,*
{Lars.Imsland,Dagfinn.Snarheim,Bjarne.A.Foss}@itk.ntnu.no
[**] *Department of Energy and Process Engineering,*
{Ragnhild.E.Ulfsnes,Olav.Bolland}@ntnu.no
Norwegian Univ. of Science and Technology, 7491 Trondheim

Abstract: This article is concerned with the dynamics and control of a semi-closed O_2/CO_2 gas turbine cycle for CO_2 capture. In the first part the process is described and a model is developed. Thereafter, a control system structure is proposed, and an output feedback model predictive control algorithm is implemented and simulated on the process. *Copyright © 2004 IFAC*

Keywords: Model predictive control, gas turbines, semi-closed cycle, CO_2 capture

1. INTRODUCTION

Gas turbines are widely used for power production from gaseous fossil fuels. Although gas turbine engines are relatively clean burning, there is inevitably a production of CO_2 from combustion of fossil fuels. Thus, with todays increasing concern about global warming and climate change, there is an incentive to investigate gas turbine processes with CO_2 capture.

Focusing on gas turbines, it is generally acknowledged (see e.g. Bolland and Undrum (2003)) that there are three main concepts for CO_2 capture: a) Conventional power cycles where CO_2 is removed from the exhaust (post-combustion removal), b) Removal of carbon from fuel (pre-combustion removal), and c) Combustion with pure oxygen (instead of air), which leaves the exhaust consisting of CO_2 and water (easily condensed to obtain pure CO_2). While all these concepts have their pros and cons, we will in this paper concentrate on a process based on concept c).

The process we study (described in more detail in Section 2) recycles the exhaust gas, consisting mainly of CO_2 after water is removed, as working fluid in the gas turbine. We will investigate dynamics and control of this semi-closed O_2/CO_2 gas turbine cycle, where CO_2 capture is achieved since some CO_2 must be removed from the cycle to avoid accumulation. In particular, we look at the design of a predictive controller that aims at achieving close-to-optimal load control operation, despite disturbances that inevitably will excite the system, while handling important process constraints explicitly.

The literature on this specific process is scarce, at least as far as dynamics and control are concerned. On conventional (open) gas turbine processes, there are considerably more, for instance Rowen (1983) and Ordys *et al.* (1994). Predictive control of conventional gas turbines is suggested in Vroemen *et al.* (1999) with experiments in van Essen and de Lange (2001). The modeling in this work is based on Ulfsnes *et al.* (2003).

In the first part of the paper, the process is described and a model is developed. Thereafter, the main challenges for a control system are discussed, and closed-loop simulations using a model predictive control algorithm is compared to a "conventional" approach using PI controllers. A brief discussion ends the paper.

[1] Corresponding author.

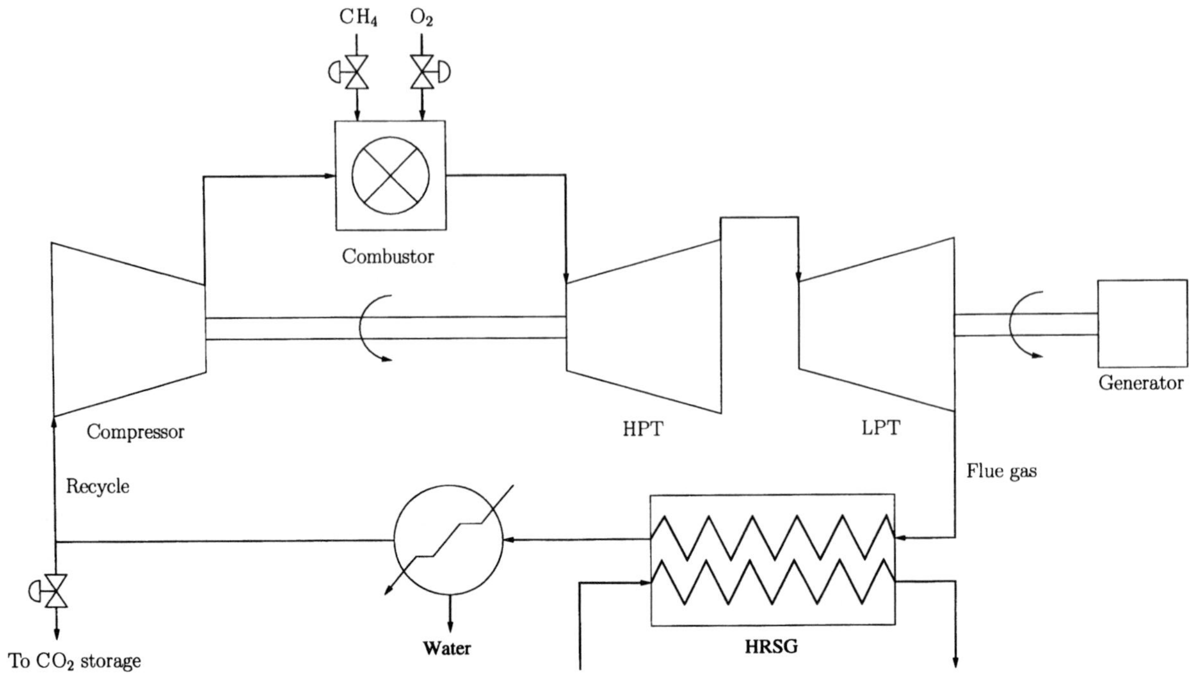

Fig. 1. Process layout

2. PROCESS DESCRIPTION

A sketch of the process is shown in Figure 1. In the combustion chamber, methane (CH_4) and oxygen (O_2) react at a ratio slightly above the stoichiometric ratio. Recycled gas, mainly consisting of CO_2, is compressed and used as an inert in the combustion to limit temperatures in the combustion chamber and turbine inlet. The gas leaving the combustor is expanded in two turbines. The high pressure turbine (HPT) drives the compressor, while the low pressure turbine (LPT) is connected to a generator. The exhaust gas leaves the power turbine with a temperature well suited to deliver heat to a steam bottoming cycle. After the heat recovery steam generator (HRSG) the gas has to be cooled in a condenser, and condensed water is removed from the cycle. The exhaust gas, now mainly consisting of CO_2 is split into two streams, one stream is recycled to the compressor; the other stream is removed from the cycle for storage.

For space reasons, we will not give the values for all parameters, or provide a complete nomenclature. We have tried to keep the notation standard, see also the nomenclature of Ulfsnes *et al.* (2003). Some typical ("design") values for key variables are given in the table below:

Variable	Symbol	Typical value
LPT power output	$\dot{W}_{LPT}$	100MW
Turbine inlet temperature	TIT	1597K
Compressor mass flow	$\dot{m}_c$	173kg/s
Exhaust gas temperature	TET	1095K
Mass flow CO_2 to storage	$\dot{m}_{CO_2}$	16kg/s
Fuel mass flow	$\dot{m}_{CH_4}$	6.1kg/s
O_2 mass flow	$\dot{m}_{O_2}$	25kg/s
Compressor inlet temp.	T_{in}	290K
Compressor pressure ratio		19.3

3. MODELING

The dynamic process model is based on Ulfsnes *et al.* (2003). Some simplifications are made, mainly for computational efficiency reasons. The modeling is performed using the modeling environment gPROMS (gPROMS, 2003). Thermodynamic properties have been determined with Multiflash, a physical property package.

3.1 Compressor

The power required for compression is equal to the increase in enthalpy,

$$\dot{W}_c = \dot{m}_c \cdot \Delta h_c.$$

The increase in specific enthalpy will be calculated by assuming it being somewhat larger (given by the efficiency) than the isentropic enthalpy increase $\Delta h_{c,s}$,

$$\Delta h_c \cdot \eta_{c,s} = \Delta h_{c,s}.$$

We have assumed a constant isentropic efficiency $\eta_{c,s}$.

For a given compressor, the static relation between (dimensionless) compressor speed, compressor mass flow and compressor pressure ratio is usually called the *compressor map.* The "reduced" quantities are the standard quantities used for compressors with air as the working fluid (Saravanamuttoo *et al.*, 2001):

$$N_{\text{dim}} = \frac{N_{\text{red}}}{N_{\text{red,design}}}, \quad N_{\text{red}} = \frac{N}{\sqrt{T_1}}$$

$$\dot{m}_{\text{dim}} = \frac{\dot{m}_{\text{red}}}{\dot{m}_{\text{red,design}}}, \quad \dot{m}_{\text{red}} = \frac{\dot{m}_c\sqrt{T_1}}{p_1}\sqrt{\frac{R}{\gamma_1}}.$$

The gas constant R is dependent on the molar weight M_c of the working fluid, $R = \bar{R}/M_c$.

In our case, we will assume that the compressor map given by that dimensionless "reduced speed" is proportional to dimensionless "reduced mass flow":

$$N_{\rm dim} = K \cdot \dot{m}_{\rm dim}.$$

This corresponds to having vertical lines in the compressor map. This can be a good approximation in the normal operating range of a gas turbine cycle.

3.2 Combustion

Due to the rapid response of the combustion process, we have assumed an instantaneous mass balance, which gives the following mass flow leaving the combustion chamber,

$$\dot{m}_{\rm out} = \dot{m}_{\rm c} + \dot{m}_{\rm CH_4} + \dot{m}_{\rm O_2}.$$

Similarly, the energy balance is given by[2]

$$\dot{m}_{\rm CH_4}\Delta h_{\rm CH_4} + \dot{m}_{\rm O_2}\Delta h_{\rm O_2} + \dot{m}_{\rm c}\Delta h_{\rm cc} + \dot{m}_{\rm out}\Delta h_{\rm rx} = 0,$$

where $\Delta h_{\rm rx}$ is the enthalpy of reaction, assuming all fuel reacts according to

$$CH_4 + 2O_2 \rightarrow CO_2 + 2H_2O.$$

Compared to using an equilibrium reactor as in Ulfsnes *et al.* (2003), this is a good approximation assuming the oxygen excess ratio

$$\lambda_{\rm O_2} = \frac{\dot{n}_{\rm O_2}}{2\dot{n}_{\rm CH_4}}$$

is larger than, say, 1.02. Furthermore, we assume a fixed percentage pressure drop over the combustion chamber.

The fuel (CH_4) and O_2 streams enter the combustion chamber through two valves. We assume both of these are controlled with flow controllers, and we assume that a perfect ratio controller controls the inflow of O_2, such that a constant oxygen excess ratio is maintained. The setpoint $\dot{m}_{\rm CH_4,ref}$ to the flow controller for the CH_4 stream is the manipulated variable for the controller to be designed. If we assume that this flow controller is well tuned, then we can write

$$\frac{d\dot{m}_{\rm CH_4}}{dt} = \frac{1}{\tau_{\rm CH_4}}(\dot{m}_{\rm CH_4,ref} - \dot{m}_{\rm CH_4}),$$

where $\tau_{\rm CH_4}$ is given by the bandwidth of the flow controller. Further, $\dot{m}_{\rm O_2}$ is a fixed ratio of $\dot{m}_{\rm CH_4}$ given by $\lambda_{\rm O_2}$ and the molar masses.

3.3 Turbine

The power generated by the high pressure turbine is

$$\dot{W}_{\rm HPT} = \dot{m}_{\rm HPT} \cdot \Delta h_{\rm HPT}$$

[2] The symbol $\Delta h_{\rm cc}$ is used to differentiate this enthalpy from the enthalpy increase in the compressor.

where the enthalpy drop is less than the isentropic enthalpy drop,

$$\Delta h_{\rm HPT} = \eta_{\rm HPT,s} \cdot \Delta h_{\rm HPT,s}$$

given by the (assumed constant) isentropic turbine efficiency $\eta_{\rm HPT,s}$.

The same relations are used for the low pressure turbine (exchange HPT with LPT).

Moreover, we assume that both turbines can be regarded as "choked nozzles", which is used to calculate the relationship between pressure drop, temperature and mass flow, when these differ from the design values ("off-design calculations"). The choked nozzle equation used here is given as

$$\tilde{p}_{\rm in} = \tilde{\dot{m}}\sqrt{\frac{\tilde{T}_{\rm in}}{\tilde{M}}},$$

where $\tilde{\cdot}$ denotes the ratio to the design value, e.g. for the molar weight, $\tilde{M} = M/M_{\rm design}$.

3.4 Rotating shaft

The high pressure turbine drives the compressor via a rotating shaft. Newton's second law gives

$$I\frac{d\omega}{dt} = \frac{\dot{W}_{\rm HPT} - \dot{W}_{\rm c}}{\omega}$$

where $\omega = \pi N/30$.

The low pressure turbine drives the generator via another rotating shaft. We assume that the generator delivers its power to an infinite bus, thus the rotating speed of the low pressure turbine will be fixed.

3.5 Heat recovery steam generator and condenser

In this work, we look at the heat recovery steam generator and condenser as a single counter flow heat exchanger. We do not model in any detail anything on the cold side of the heat exchanger. However, as the load of the plant varies, the amount of removed heat varies. For instance, a load increase will give a larger inlet mass flow, at about equal temperature. If the additional heat is not removed, then the compressor inlet temperature will inevitably increase, which will have a severe effect on the overall efficiency of the cycle. Thus, in a real plant, the steam bottoming cycle and condenser must be operated such that the changes in the compressor inlet temperature are suppressed. We have chosen to model this by letting a PI-controller decide the flow on the cold side of the heat exchanger such that the outlet temperature is kept constant. A suitable tuning of this controller represents the dynamics of the change in operating point for the steam cycle and condenser.

The heat transferred in the heat exchanger is modeled as proportional to the difference in average temperature between cold and hot side,

$$\dot{Q} = U_{\text{wall}} A_{\text{wall}} (T_{\text{cold,avg}} - T_{\text{hot,avg}}), \qquad (1)$$

where $U_{\text{wall}} A_{\text{wall}}$ is the heat transfer coefficient for the whole wall. We have used the arithmetic mean when calculating average temperature, since the (more correct) logarithmic mean proved to have a significant impact on computational performance.

The heat exchanger will not react instantly to changes in the inflow. We thus model the "real" outlet temperature as a first order lag of the outlet temperature given from (1). For the hot side, this is

$$\frac{dT_{\text{hot,out}}}{dt} = \frac{1}{\tau_{\text{HX}}} \left(\frac{\dot{Q}}{\dot{m}_{\text{hot}} c_{\text{p,hot}}} + T_{\text{hot,in}} - T_{\text{hot,out}} \right),$$

and accordingly on the cold side.

In order to model pressure variations, a mass balance together with the ideal gas law is used.

3.6 Valve and splitter

After most of the water is removed in the condenser, some of the CO_2 leaves the cycle through a valve. The flow through this valve is mainly determined by the pressure difference, using the valve equation

$$\dot{m}_{CO_2} = K_{\text{v}} \sqrt{\Delta p} u_{\text{v}},$$

where $0 \leq u_{\text{v}} \leq 1$ is the (rate constrained) valve opening, a control input.

4. CONTROL AND CLOSED LOOP SIMULATIONS

4.1 Control structure

The control problem we consider, is that of load control: Operate the process so it supplies a specified load to the grid. As the process is open loop stable, the control objective is to operate the process as efficiently as possible, under varying disturbances. The major disturbances that affect the operation and are considered herein, are load changes and disturbances affecting the heat transfer in the HRSG. This study does not include start-up and shutdown of the system.

Manipulated variables: Possible manipulated variables are fuel valve, O_2 valve, CO_2 valve, compressor variable inlet guiding vanes (VIGV), and a number of variables affecting the operation of the HRSG and the condenser.

As explained above, we assume a perfect ratio controller to manipulate the O_2 valve to obtain a constant ratio of inflow of CH_4 and O_2. We also assume a well-tuned controller controlling the fuel valve, leaving us with the reference value as a manipulated variable.

Furthermore, we have chosen to disregard any compressor VIGVs. This is a limitation that will be discussed in Section 5.

We have not developed a detailed model of the cold side of the HRSG and the condenser, thus any manipulated variables related to these systems are not available to us. However, according to Kehlhofer *et al.* (1999), these are not normally used for load control in a conventional combined cycle. Thus, for the steam bottoming cycle, these manipulated variables should be used to operate the steam cycle as efficiently as possible for varying loads, removing as much heat as possible from the turbine exhaust. The impact of this is modeled by the PI-controller controlling the mass flow on the cold side of the heat exchanger.

This leaves us with opening of CO_2 valve and fuel inflow controller reference as manipulated variables (u) for this study.

Controlled variables: For the semi-closed gas turbine cycle alone, the Carnot efficiency is maximized by keeping the turbine inlet temperature (TIT) as high as possible (limited by turbine material constraints), and keeping the compressor inlet temperature as low as possible. Combined cycle considerations clutter the picture slightly, since the efficiency of the steam cycle must also be considered. However, Ordys *et al.* (1994) recommends for conventional combined cycles to keep the turbine exhaust temperature (TET) as high as possible (subject to constraints) to maximize energy flow to the HRSG, and we have adopted this philosophy and choose to control TET. In a real application, the optimum TET setpoint will vary with load changes, but for simplicity we have chosen to keep it fixed. Note that for a given load, maximizing TET is close to maximizing TIT.

Since this study considers the load control problem, the controlled variables (z) will be LPT power output, $\dot{W}_{\text{LPT}}$ in addition to TET.

Measured variables: Although TIT imposes an important constraint, is not possible to measure this variable, and TIT must be inferred from other measurements. In our case, this is done through a Kalman filter, which is also needed to obtain the states for MPC prediction. We have used TET, N, $\dot{W}_{\text{LPT}}$ and the state of the steam cycle (the integral error of the PI controller controlling mass flow on cold side of HRSG) as measured variables (y).

4.2 Predictive control

Linear MPC refers to an online optimization where, at each sample instant, the control is determined by optimizing future behavior as predicted by a linear process model, subject to constraints on states (or controlled variables) and inputs, then applying the first part of the computed control on the process (Maciejowski, 2002).

The linear discrete-time process model used for prediction is on standard state-space form,

$$x_{k+1} = Ax_k + Bu_k + Ed_k,$$
$$z_k = C_z x_k + D_z u_k + F_z d_k,$$
$$y_k = C_y x_k + D_y u_k + F_y d_k$$

where z_k are the controlled outputs, and y_k are the measured outputs. A standard linear Kalman filter is used to estimate the state (x_k) from the measured variables and the inputs (u_k). The d_k is a disturbance state used in the Kalman-filter to obtain integral control. The linear model is obtained using the LINEARIZE-function of gPROMS. We assume linear constraints on states (or controlled outputs), input and input rate,

$$Ex_k \leq 0, \quad Fu_k \leq 0, \quad G(u_k - u_{k-1}) \leq 0.$$

We choose to minimize a quadratic objective function of the following form[3],

$$V(k) = \sum_{i=1}^{H_p} \|\hat{z}(k+i|k) - r(k+i)\|_Q^2 + \sum_{i=0}^{H_u} \|\hat{u}(k+i|k) - \hat{u}(k+i-1|k)\|_R^2$$

where $\hat{z}(k+i|k)$ and $\hat{u}(k+i|k)$ are predicted variables at time k (with $\hat{u}(k-1|k) = u(k-1)$), and $r(k)$ is a reference trajectory for the controlled variables.

The most important constraints that are imposed here, are the upper limit on turbine inlet temperature (1598K) and the constraint on valve operation (opening between 0 and 1, stroke time 15s). We used $H_p = H_u = 50$, with sample time 0.5s.

4.3 Closed loop simulations

The simulations are performed in gPROMS, while the controller calculations are done in Matlab. gPROMS communicates with Matlab via gPROMS' Foreign Process Interface. The QP-problem is solved using quadprog from the Optimization Toolbox in Matlab. At each sample instant, the measurements are transferred from gPROMS to Matlab, where an optimal control trajectory is computed, and the manipulated variables for the next sample interval are returned to gPROMS.

The MPC closed loop trajectories are compared to trajectories from a well-tuned PI control structure (with anti-windup) where the turbine exhaust temperature is controlled by the CO_2 valve controller, while the flow of fuel controls the power output[4]. In a conventional gas turbine, the power loop would incorporate logic to avoid too high TIT, but this is not implemented here. A conventional process would also reduce mass flow by using VIGV to keep high TIT/TET at part-load, but this is obtained in this process since the

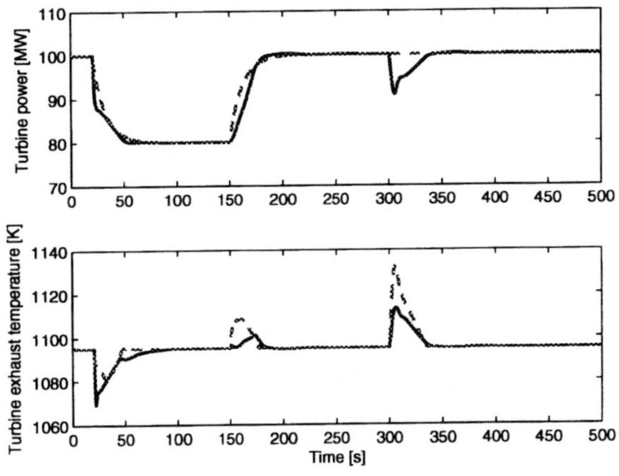

Fig. 2. Controlled var., MPC (-) and PI (- -).

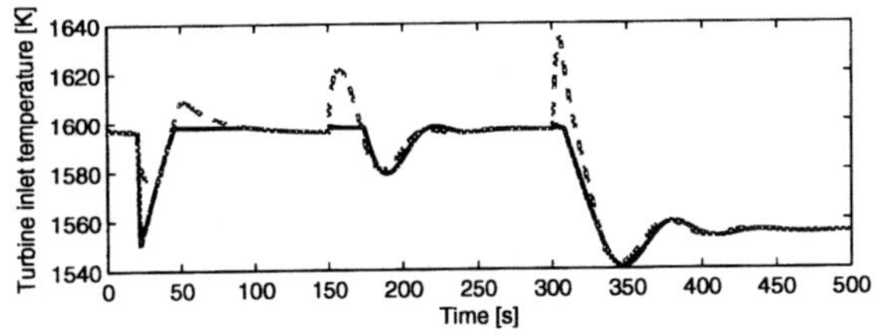

Fig. 3. TIT, MPC (-) and PI (- -).

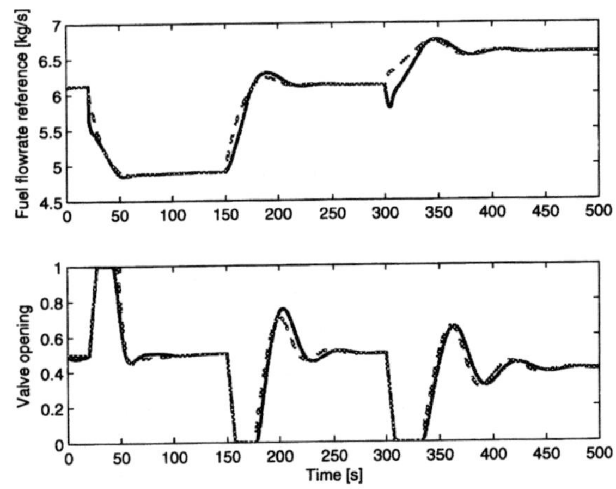

Fig. 4. Manipulated var., MPC (-) and PI (- -).

controller changes the total mass in the loop when operating the CO_2 valve.

A simulation of the closed loop is shown in Figures 2-4. The first disturbance (at 20s) is a change in power setpoint from 100MW to 80MW, and at 150s the setpoint goes back to 100MW. At 300s, an "event" in the HRSG/condenser causes an increase in compressor inlet temperature from 290K to 310K in less than 10s (the PI controller controlling the compressor inlet temperature has a 20K setpoint change). Note that both these disturbances are rather large considering how fast they happen.

We see that the PI controller obtains good control of power, at the cost of TIT constraint violations. Less deviations can be obtained by detuning the power controller. The MPC-controller obtain

[3] The norm $\|\cdot\|_H$ is defined by $\|z\|_H = \sqrt{z^T H z}$, $H > 0$.
[4] The setpoint to this controller is filtered to allow tighter control of the power subject to other disturbances.

much better control of temperature, at the cost of having a dip in power output at the last disturbance. How large this dip must be, is a matter of tuning – if a higher temperature limit is used (or the hard constraint is replaced with a soft constraint), then better power control can be achieved.

5. DISCUSSION

Modeling: The developed model is mainly based on first principles and thermodynamics. However, as no such process exists today, there is considerable uncertainty related to several dimensions and characteristics. As we see it, the main uncertainty factor related to the dynamics is the modeling of the HRSG/condenser. Other issues that will be looked upon are a more realistic compressor map, and using a single shaft gas turbine. Introducing isentropic efficiency maps for the compressor and the turbines will also increase model confidence, but we believe that this will not have a significant influence on the dynamic properties of the process.

Trade-off: As can be suspected, there is a trade-off between good control of temperature and good power control. The PI-controller can have considerably smaller maximum turbine inlet temperature if the power controller is detuned. As we can see in Figure 4, the PI power controller actually contributes to the temperature rise after the last disturbance. For the MPC controller, we see that the fuel input is used to keep the temperature below the limit, but we must pay with a dip in power output. If we can allow a higher temperature limit (or allow the temperature limit to be a soft constraint), this dip can be considerably reduced.

Limitations: The control setup considered in the previous section is rather simple. Most importantly, compressor VIGV are not used as an input for limiting mass flows at part-load. However, the simulations showed that since this process is operated in semi-closed cycle, it is possible to obtain this effect using the CO_2-valve, and hence (to a certain extent) it is viable to operate the cycle at part-load without using VIGV. Nevertheless, in future work, we aim at including VIGV as control input, to obtain more flexibility.

There are also some important constraints that are not taken into account in this paper:

- Compressor surge constraints (operate the compressor to avoid surge).
- Pressure constraints, especially related to the HRSG, due to closed cycle operation.
- Constraints related to processing (compressing) downstream the CO_2-valve.

Including VIGV as input should facilitate handling these constraints.

Obtaining models: In this work, we obtained linear models at the operating point from gPROMS, using the `LINEARIZE`-function. Due to the nonlinearities (mainly the CO_2-valve), the prediction became bad when far from the operating point. It was experimented with obtaining linear models from several operating points, and use them in the prediction in a gain-scheduled manner. The models were "scheduled" over the prediction horizon using a process variable (in our case, a manipulated variable) from the previous time-step. In doing this, the prediction was significantly improved, resulting in better control (especially, better constraint handling)

6. CONCLUSION

A modular dynamic model of a semi-closed O_2/CO_2 gas turbine cycle for CO_2 capture was developed. A predictive controller, designed based on a linearization of the model, was shown to control the process satisfactorily. In particular, the controller keeps temperature constraints without resorting to the logic normally used by gas turbine load controllers, and is able to maintain reasonably high efficiency at part-load without using variable inlet guiding vanes. In future work, we will study how efficiency at part-load should be optimized using the increased flexibility of variable inlet guiding vanes, subject to more realistic constraints.

ACKNOWLEDGEMENTS

The Gas Technology Center NTNU-SINTEF and NFR are acknowledged for financial support.

REFERENCES

Bolland, O. and H. Undrum (2003). A novel methodology for comparing CO_2 capture options for natural gas-fired combined cycle plants. *Advances in Environmental Research* **7**, 901–911.

gPROMS (2003). gPROMS Introductory User Guide. Process Systems Enterprise Ltd.

Kehlhofer, R. H., J. Warner, H. Nielsen and R. Bachmann (1999). *Combined-Cycle Gas Steam Turbine Power Plants.* PennWell.

Maciejowski, J. M. (2002). *Predictive Control with Constraints.* Prentice Hall.

Ordys, A. W., A. E. Pike, M. A. Johnson, R. M. Katebi and M. J. Grimble (1994). *Modelling and Simulation of Power Generation Plants.* Springer-Verlag.

Rowen, W. I. (1983). Simplified mathematical representations of heavy-duty gas turbines. *ASME J. Eng. Power* **105**, 865–869.

Saravanamuttoo, H. I. H., G. F. C. Rogers and H. Cohen (2001). *Gas Turbine Theory.* 5th ed.. Prentice Hall.

Ulfsnes, R. E., O. Bolland and K. Jordal (2003). Modelling and simulation of transient performance of the semi-closed O_2/CO_2 gas turbine cycle for CO_2-capture. In: *Proceedings of ASME TURBO EXPO 2003.* Atlanta, Georgia, USA. GT2003-38068.

van Essen, H. A. and R. de Lange (2001). Nonlinear model predictive control experiments on a laboratory gas turbine installation. *ASME J. Eng. Gas Turbines Power* **123**, 347–352.

Vroemen, B. G., H. A. van Essen, A. A. van Steenhoven and J. J. Kok (1999). Nonlinear model predictive control of a laboratory gas turbine installation. *ASME J. Eng. Gas Turbines Power* **121**, 629–634.

ELSEVIER IFAC PUBLICATIONS
www.elsevier.com/locate/ifac

WHICH IS THE BEST CRITERION FOR IDENTIFICATION OF DYNAMIC MODELS ?

J. O. Trierweiler* and V. C. Machado

*Group of Integration, Modelling, Simulation, Control and Optimization of Processes (GIMSCOP)
Department of Chemical Engineering, Federal University of Rio Grande do Sul (UFRGS)
Rua Duque de Caxias, 1303 apt. 603 CEP: 90010-283 - Porto Alegre - RS - BRAZIL,
Fax: +55 51 3316 3277, Phone: +55 51 3316 4072
E-MAIL: {vinicius, jorge}@enq.ufrgs.br*

Abstract: A novel criterion for system identification is proposed. It is based on calculations of the total quadratic error between the derivative of the plant data and the derivative of the model output. This criterion produces much better models when the experimental data are corrupted by unmeasured external disturbance. *Copyright © 2004 IFAC*

Keywords: - system identification, dynamic models, optimization criteria, dynamic models for process control

1 INTRODUCTION

In almost all identification procedure, the optimization problem is formulated as the minimization of a square error criterion, such as: $J = min \sum [e(t)]^2$, where the error, e, is the difference between the measured (y) and predicted (y_P) output, i.e., $e(t) = y(t)-y_P(t)$. But, this kind of criterion does not give good results, if it is used directly without a filter to identify dynamic models from the plant data corrupted by unmeasured external disturbances. Since unmeasured external disturbances are always present in the industrial scenarios, it would be of the great practical impact if a simple solution for this kind of problem could be found. To go one step ahead in this direction, in this paper, it is analyzed how a criterion based on the square error of derivative signal (i.e., $min \sum [de(t)/dt]^2$) can improve the quality of dynamic model.

The paper is structured as follows. In section 2, a motivation example to compare both criteria is presented. In section 3, it is shown how to calculate the derivatives from plant data. In section 4, both criteria are applied to the motivation example. Finally, section 5 compares the quality of the dynamic models identified using the novel optimization criterion.

2 MOTIVATION EXAMPLE

2.1 *Experimental Plant*

To compare and analyze both criteria for identification, it is used plant data from a heating experimental unit consisting of two tanks with constant volumes in series. Figure 1 shows the system schematically. The first tank is fed with a cold water stream (F) at temperature T_0 and has an adjustable electrical heating power, which is used as manipulated variable. The controlled variable is the outlet temperature of second tank, T_2. In this work, the inlet temperature T_0 is considered as unmeasured external disturbance and the inlet flowrate is kept constant during the experiments. The real plant data used in this work are from a unit available at Process Control Group of Dortmund University (Saecker, 1995).

---

* The corresponding author.

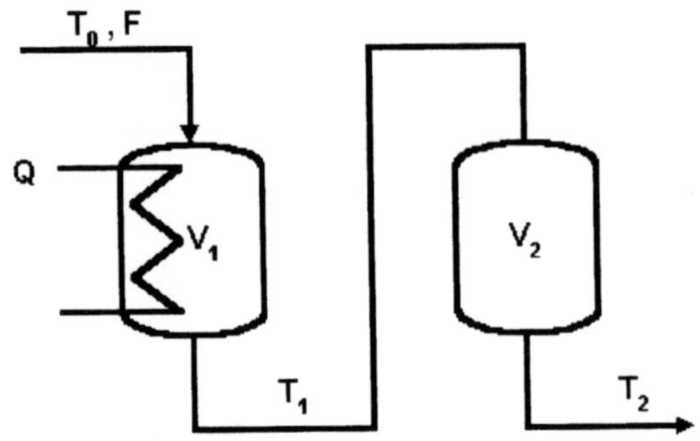

Figure 1. Schematic representation of the heating plant of two tanks.

2.2 Experimental data

An experimental data set with decreasing inlet temperature T_0 was selected and is shown in Figure 2. Although the temperature T_0 is measured, we will assume that this disturbance is not measured. To design a controller for this unit, it is necessary to have a good dynamic model between the manipulated variable (heating power, Q) and the controlled variable T_2. But the measured output temperature is influenced by both Q and T_0. Therefore, the output temperature T_2 is corrupted by the inlet temperature. It is possible to isolate the real effect of the heating duty into the outlet temperature making the temperature difference between inlet and outlet temperatures (i.e., $\Delta T = T_2 - T_0$), since the velocity as T_0 varies it is much slower than the dominant system's dynamic. Figure 3 compares the two cases: with and without inlet temperature compensation.

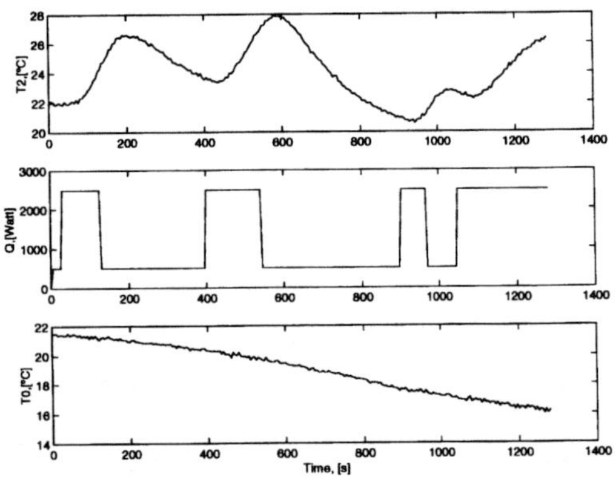

Figure 2. Experimental data set with decreasing inlet temperature T_0 (disturbance), heating duty Q (manipulated variable), and outlet temperature T_2.

Next section shows the model identification for these two data sets, applying Box-Jenkins and Output Error structures.

2.3 Black-box Identification

Now the two data sets shown in Figure 3 are identified using the Box-Jenkins and Output Error model structures available in the MATLAB System Identification Toolbox (Ljung, 1997).

Box-Jenkins models (BJ) give good results when unmeasured disturbance like T_0 are present, while the Output Error (OE) models can only succeed when the discrepancy between the model and experimental data is produced by normal distributed measurement noise.

Figure 4 shows the comparison between BJ, OE, and plant data using identified models from the data set $\Delta\Delta T$ (cf. Figure 3). Since this data set the inlet temperature effect T_0 is compensate, both model structures produce very good results.

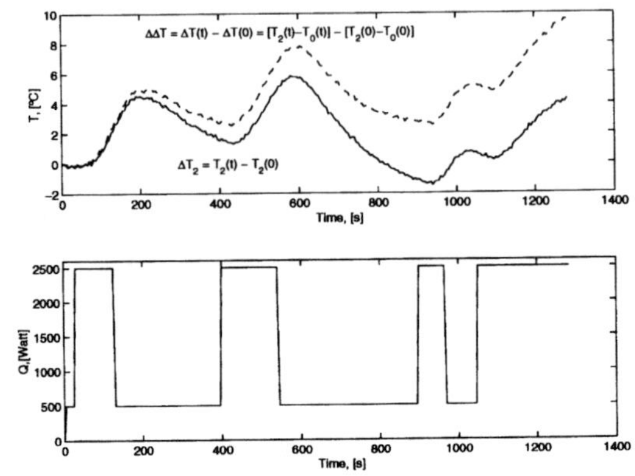

Figure 3. Data sets that are used in the identification procedure: (1) $\Delta T_2 = T_2(t) - T_2(0)$ (solid line) data without compensation of the inlet temperature effect and (2) $\Delta\Delta T = \Delta T(t) - \Delta T(0) = [T_2(t) - T_0(t)] - [T_2(0) - T_0(0)]$ (dashed line) data with compensation of the inlet temperature effect.

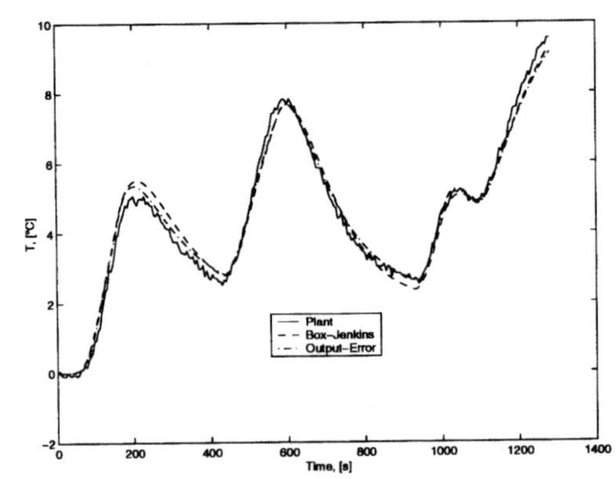

Figure 4. Comparison between plant data without effect of the variations in T_0 (i.e., $\Delta\Delta T$ of Figure 3) and the identified models OE and BJ.

The situation is completely different when the data set ΔT_2 is used. Figure 5 clearly shows that the model BJ has captured the tendency of the model, although has the larger square error if compared to the OE model. On the other hand, the OE model cannot capture the tendency as well as the BJ model. At this point an important question arises: "which is the best model?" If we compare with the results shown in

Figure 4, we can conclude that the BJ model is the correct one. Therefore, it is much better to follow the tendency (the derivative) than the output signal, in other words, the discrepancy between the model and experimental data is better described as a function of $[de(t)/dt]^2$ instead of $[e(t)]^2$. In the next section, we will see how we can calculate $de(t)/dt$ from experimental data.

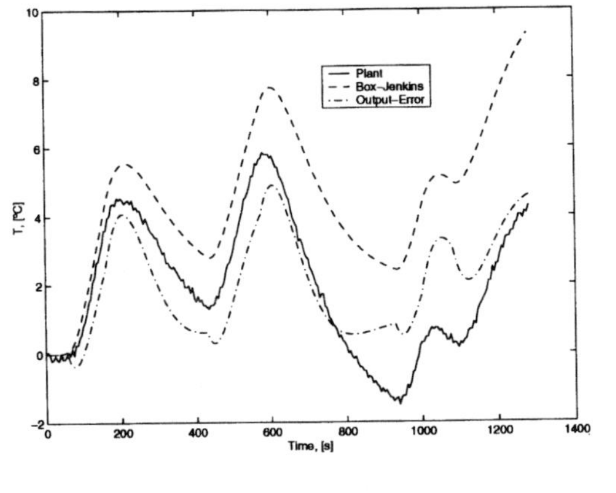

Figure 5. Comparison between plant data without compensation of T_0 (i.e., ΔT_2) and the identified models OE and BJ.

3 METHODS FOR DERIVATIVE CALCULATION

In most of the industrial cases, plant data are corrupted by noise, which increases the difficult to yield the correct derivative signal. An example of this is showed in Figure 6.

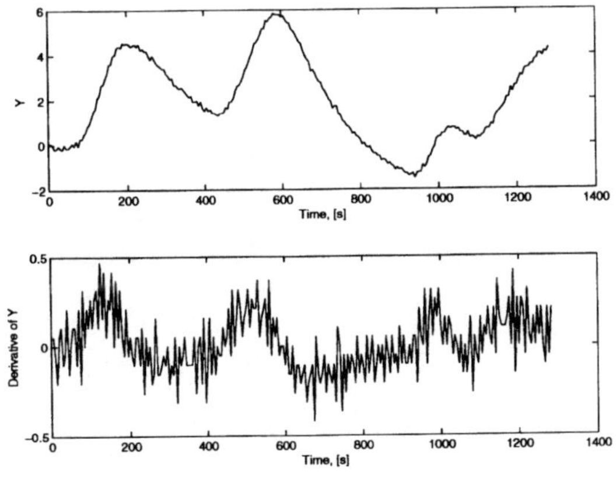

Figure 6. Plant data and its derivative.

To avoid this problem, it is necessary to filter the data before to apply the derivative. Here two different forms to filter the plant data are compared: low pass and Wavelets filters. These methodologies are implemented in MATLAB. After filtering the data, the derivative can be easily estimated, by finite differences method, for instance.

3.1 Low pass filter (Idfilt (Ljung,1997))

Filtering data with classical low pass filters as Butterworth filters usually given good results. These filters are implemented in (Ljung, 1997) and (Signal TB). In this paper the version of Butterworth filter implemented in the matlab function idfilt (Ljung,1997) was used.

3.2 Wavelets (Misra et al., 1999)

Wavelets provide a joint time-scale representation of a signal through its projection onto nested orthogonal subspaces; finer subspaces containing coarser subspaces, in a multi-resolution framework. Signal at each level of resolution is projected on an approximation subspace and a detail subspace. Differently of Fourier analysis, the Wavelets analysis is based on infinite sets of base functions, not only sines and cosines, which allow representing a noisy signal as a sum of other signals with different frequencies. If the objective is to remove noise, high frequencies component of the signal must be removed, in general.

To compare two methods to filter the data, Figure 7 shows the derivative of the filtered data by Idfilt and Wavelets algorithms. The data set is the same as in the Figure 6, but to make clearer the comparison, only the first 200s are shown.

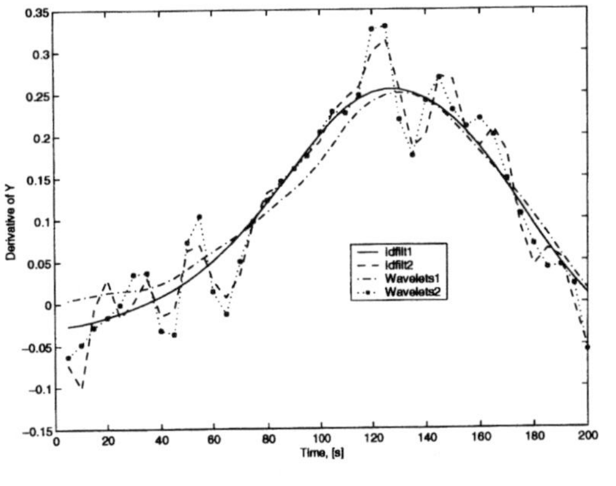

Figure 7. Calculated derivatives from filtered data by Idfilt and Wavelets algorithms. Only the first 200 s was plotted to increase the visual quality. Idfilt1 (filter order = 2, cut-off frequency = 0.1), Wavelets1 (decomposition level = 50), Idfilt2 (filter order = 2, cut-off frequency = 0.5), and Wavelets2 (decomposition level = 1).

Due to good results and simplicity, the Idfilt with filter order=2 and cut-off frequency=0.1 was selected to be used in the rest of the paper. For data collected at a corrected sampling rate (i.e., sampling time based on the process dynamic), these parameters for the filter can be in general applied and will produce good results. Therefore, we recommend using a Butterworth's second order filter with 0.1 cut-off frequency as a default and starting point selection.

4 APPLICATION IN EXAMPLE

Figure 5 shows the quality of the identified models OE and BJ when the effect of the unmeasured disturbance is not compensated. For BJ mode, as a disturbance model is simultaneously identified with the process model, can be expected better model than for OE model structure. Unfortunately, the better quality of the BJ model is masked if it is directly compared with the original data (cf. Figure 5). On the other hand, if the results are compared in a derivative basis, the BJ becomes much closer to the derivative data, as it is shown in Figure 8. Note that the OE model has a bigger discrepancy here than in the original data. To check this affirmation, the sum of the quadratic errors between the derivative of the model and the derivative of the plant data (i.e.,Σ [$de(t)/dt]^2$) and the sum of the quadratic errors (i.e.,Σ [$e(t)]^2$) were calculated. The results are summarized in Table 1 and confirm our visual impression, that the OE models fit better the original data and the BJ model the derivative data.

Table 1. Sum of the quadratic errors obtained from the Figures 7 and 8 for the OE and BJ models.

Sum of quadratic error $\Sigma [e(t)]^2$		Sum of quadratic derivative error ($\Sigma [de(t)/dt]^2$)	
OE	BJ	OE	BJ
0.75×10^5	4.25×10^5	1.73×10^2	0.32×10^2

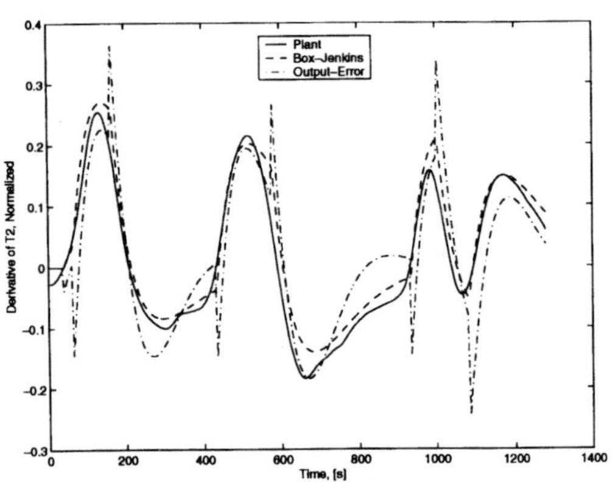

Figure 8: Comparison between the derivative of the plant data filtered by Idfilt and the derivatives of the identified models OE and BJ.

When data are affected by load disturbances, it is possible to affirm that quadratic error criterion brings wrong conclusions about the quality of the identified models. In the studied example, BJ model was better than OE model, and not the opposite, as quadratic error criterion would indicate. On the other hand, the quadratic derivative error criterion has reflected better the quality of the identified model. This makes us to think about to use a different criterion for system identification based on *min* Σ [$de(t)/dt]^2$. The next section explores this idea.

5 SQUARE DERIVATIVE ERROR AS CRITERION FOR IDENTIFICATION OF DYNAMIC MODELS

The most common cost function for system identification is to minimize the square error, i.e.

$$S1 = \min \sum_{k=1}^{n} \left(y_k - \hat{y}_{k|k-1}\right)^2 \quad (1)$$

where n is the number of experimental data points, y_k is the k value of the controlled variable, and $\hat{y}_{k|k-1}$ is the predicted value of the output variable (model). Here, we propose to use the square derivative error as a new criterion for system identification, i.e.

$$S2 = \min \sum_{k=1}^{n} \left(\frac{d}{dt} y_k - \frac{d}{dt} \hat{y}_{k|k-1}\right)^2 . \quad (2)$$

Just to show the benefices of the new criterion, we apply it to identify the parameter $a0$, $a1$, $a2$, $a3$, and $a4$ of the following second order transfer function with pure time delay:

$$\hat{G}(s) = \frac{(a0 \cdot s + a1) \cdot e^{-a4 \cdot s}}{a2 \cdot s^2 + a3 \cdot s + 1} \quad (3)$$

To illustrate the potentialities of the new criterion, it will be applied to two examples.

5.1 The two tanks heating system

This system was already as introduced as motivation example in section 2. Here, the criterion $S1$ (eq. 1) and $S2$ (eq. 2) are applied to the data set with and without inlet temperature compensation.

The $\Delta\Delta T$ data set (with inlet temperature compensation). Table 2 summarizes the parameters of the identified models. The transfer function $G1$ was identified using a discrete OE model structure. The discrete model was converted to continuos and the corresponding parameters of (3) are listed in the table. The solution of the nonlinear optimization problems $S1$ and $S2$ for the second order transfer function (3) gave the models $G2$ and $G3$ of Table 2, respectively. The optimization problems were solved using standard and generic optimization methods implemented in the *lsqnonlin* function of the Matlab Optimization Toolbox (Coleman et al., 1997).

Figure 9 shows that all three models are very similar what is also seen in Figure 10 where the corresponding step responses are presented. These results confirm that for experimental data without significant unmeasured external disturbance the optimization criteria $S1$ and $S2$ are similar.

Table 2. Transfer function coefficients were obtained from the $\Delta\Delta T$ experimental data

TF	Parameters of equation (3)				
	$a0$	$a1$	$a2$	$a3$	$a4$
$G1$ / OE	-0,01	$4.4.10^{-3}$	4997.5	154.1	30.0
$G2$ / S1	0.0	$4.3.10^{-3}$	5541.0	148.9	28.0
$G3$ / S2	0.0	$4.3.10^{-3}$	5663.0	150.5	28.0

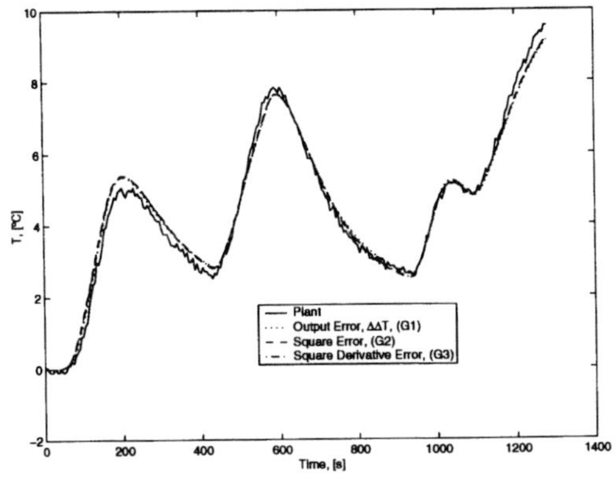

Figure 9. Models were identified using the $\Delta\Delta T$ experimental data set: $G1$ identified with OE (is assumed as true model, see fig. 4), $G2$ identified with criterion $S1$ and $G3$ with $S2$.

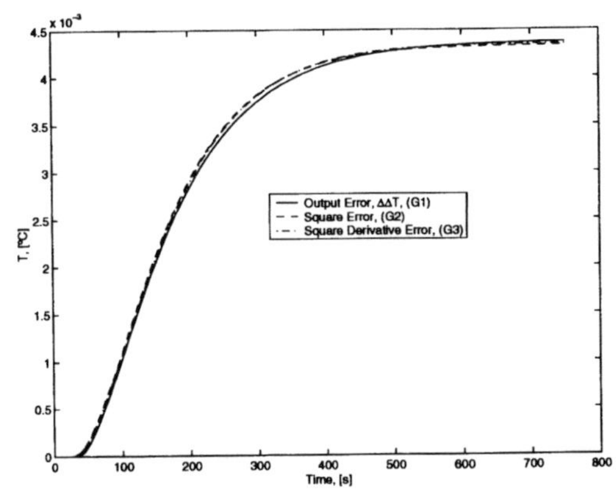

Figure 10. Step responses of $G1$, $G2$, and $G3$.

The ΔT_2 data set (without inlet temperature compensation). For this experimental data, the situation is completely different to the case before. Here, due to the external unmeasured/uncompensated external disturbance, the optimization criteria $S1$ and $S2$ converge to different parameters as shown in Table 3. In this table, $G4$ was identified using a discrete BJ model structure. The discrete model was converted to continuos and the corresponding parameters of (3) are listed. The solution of the nonlinear optimization problems $S1$ and $S2$ are $G5$ and $G6$, respectively.

Figure 11 shows clearly that the square derivative error criterion, i.e., $S2$, is the best criterion for identification, since it is closed to the Box-Jenkins model (assumed as the true model, for the case with load disturbances). The step responses shown in Figure 12 make clearer the difference between the models. Note that the model based on the square derivative error ($G6/S2$) has almost the same parameters and response as BJ and is almost unaffected by the presence of disturbance as we can see with the comparison between the models $G3$ and $G6$.

Table 3: Transfer function coefficients were obtained from the ΔT_2 experimental data

TF	Parameters of equation (3)				
	$a0$	$a1$	$a2$	$a3$	$a4$
$G4$ / BJ	-0,006	$4.4.10^{-3}$	6013.2	142.6	30.0
$G5$ / S1	0.0	$1.9.10^{-3}$	3158.0	112.4	30.0
$G6$ / S2	0.0	$4.0.10^{-3}$	5156.0	143.6	30.0

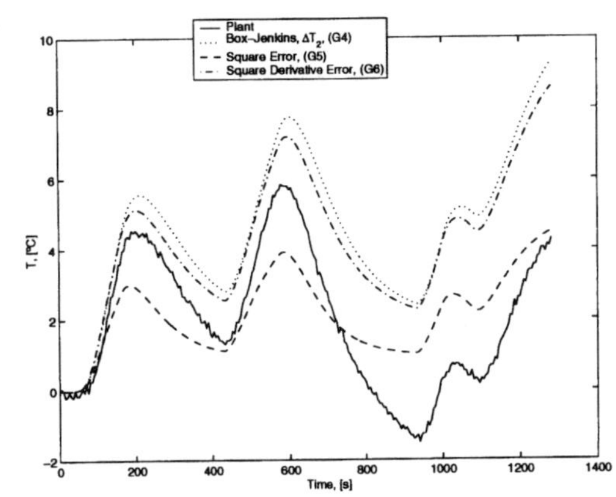

Figure 11: Models were identified using the ΔT_2 experimental data set: $G4$ identified with BJ (is assumed as true model, see fig. 5), $G5$ (dashed line) identified with criterion $S1$ and $G6$ (dashdot line) with $S2$.

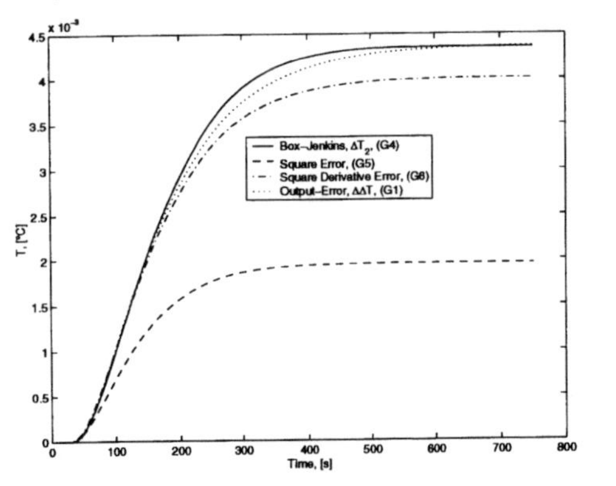

Figure 12. Step responses of $G4$, $G5$, and $G6$.

5.2 Additional example

The following LTI model is used as the true plant in this example:

$$G(s) = \frac{1 \cdot e^{-5 \cdot s}}{s^2 + 2s + 1} \tag{4}$$

Figure 13 shows the block diagram used to generate the data to be used in the identification procedure. The plant output y, the PRBS input u, and the output disturbance d used as experimental data are shown in Figure 14. It is assumed that the output disturbance is not measured, i.e., in the identification procedure only the signal u and y are used.

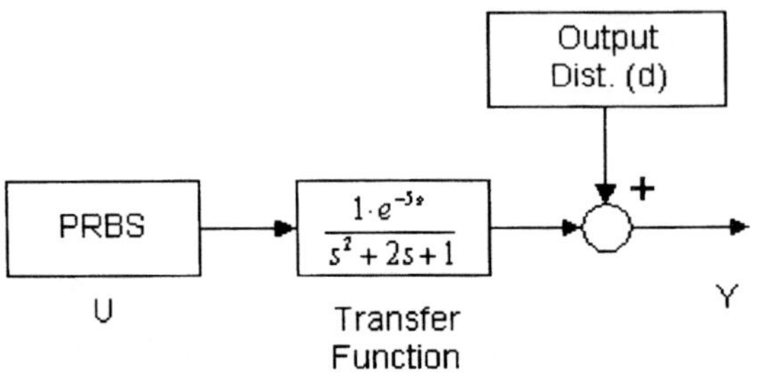

Figure 13. Block diagram used to generate the identification data

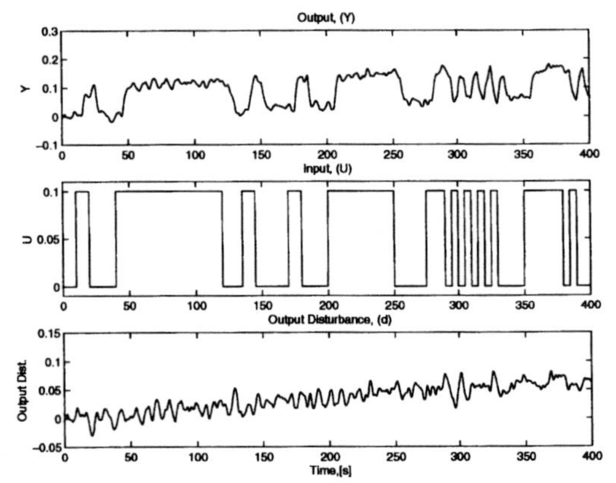

Figure 14. Open-loop data used in the identification test

Table 4 summarizes the parameters of the identified models. The transfer function $G7$ was identified using BJ model. The solution of the nonlinear optimization problems $S1$ and $S2$ gave the models $G8$ and $G9$, respectively.

Table 4: Transfer function coefficients for the additional example

TF	Parameters of equation (3)				
	a0	a1	a2	a3	a4
Plant	0	1	1	2	5
G7 / BJ	0,063	1.265	1.304	2.152	5
G8 / S1	0.0	1.436	4.871	4.414	3,5
G9 / S2	0.0	0.991	1.059	2.058	5

Figure 15 shows the step responses of the models. Note that the model $G9$ agrees very well with the true plant. Even the BJ model does not produce good results. This result clearly shows that the proposed square derivative error criterion (S2) is much better than the traditional square error criterion (S1).

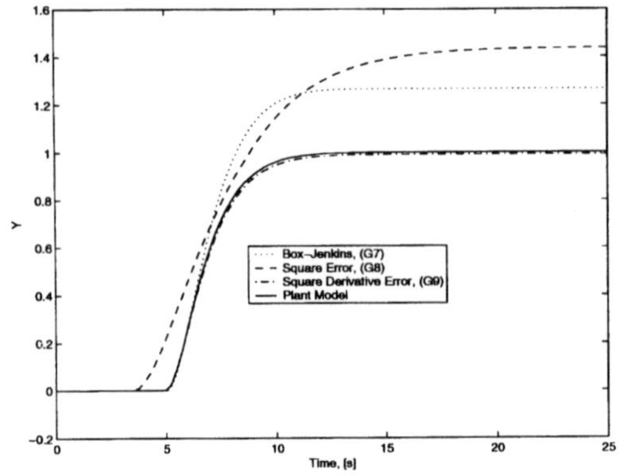

Figure 15. Step responses of the identified models: true plant (solid line), *G7*/BJ (dotted line), *G8*/S1 (dashdot line), and *G9*/S2 (dashed line).

6 CONCLUSIONS

The paper proposed a new criterion for system identification based on the square derivative error (i.e., $min \sum [de(t)/dt]^2$). It was shown that, special when the plant data is corrupted by unmeasured external disturbances, the $min \sum [de(t)/dt]^2$ criterion produces much better results than the traditional square error criterion (i.e., $J = min \sum [e(t)]^2$) . The reason why the new criterion produces better results is its capacity to capture the process tendency instead of trying to pass through all experimental points. Since unmeasured external disturbances are always present in the industrial scenarios, the new criterion can have significant impact in real process identification problems.

ACKNOWLEDGMENT

The authors thank to Prof. Dr. S. Engell and to the Process Control Group of Dortmund University for the plant data used in this work. The authors also thank the financial support given by PETROBRAS/FINEP.

REFERENCES

Coleman, T.; Branch, M. A.; Grace A. Optimization Toolbox for Matlab, Version 2.0, 1997.

Ljung, L., MATLAB System Identification Toolbox, User's Guide, Mathworks, Version 4.0.5. , (1997)

MATLAB Signal Processing Toolbox, User's Guide, Mathworks, Version 4.3., (1997)

Misra, M.; Kumar, S.; Joe Qin, S; Seemann, D.; "Error based criterion for on-line wavelet data compression.", *Journal of Process Control, **11***, pp.717-731, (1999).

Saecker, D., Studienarbeit, Universität Dortmund, (1995)

ELSEVIER
IFAC
PUBLICATIONS
www.elsevier.com/locate/ifac

ANALYSIS OF NONLINEAR PARTIAL LEAST SQUARES ALGORITHMS

S. Kumar * U. Kruger **,[1] E. B. Martin, and * A. J. Morris *

* *Centre of Process Analytics and Process Technology, University of Newcastle, NE1 7RU, U.K.*
** *Intelligent Systems and Control Group, Queen's University Belfast, BT9 5AH, U.K.*

Abstract: This paper presents an analysis of nonlinear extensions to Partial Least Squares (PLS) using error-based minimization techniques. The analysis revealed that such algorithms are maximizing the accuracy with which the response variables are predicted. Therefore, such algorithms are nonlinear reduced rank regression algorithms rather than nonlinear PLS algorithms.

Keywords: Nonlinear Systems, Identification Algorithms, Models, Gradient Methods, Prediction

1. INTRODUCTION

A number of extensions to conventional PLS have been proposed for identifying nonlinear relationship between two variable sets, which are usually referred to as the predictor set that is predictive and the response set that is predicted. For these extensions, the iterative nature of PLS is taken advantage of in terms of reducing the modelling task to the identification of score models in the latent variable space.

The score models rely on projecting recorded observations of the predictor and response variables on respective one-dimensional subspaces, defined by weight vectors. The projections of the response variables, further referred to as u-scores, are then predicted by nonlinear functions of the projections of the predictor variable, denoted as t-scores. (Wold *et al.*, 1989) suggested to identify the score models using second order polynomials. More flexible nonlinear models were proposed by (Frank, 1990) and (Wold, 1992). Whilst the former work included the use of a smoothing procedure, the technique by (Wold, 1992) incorporated spline functions.

(Qin and McAvoy, 1992; Holcomb and Morari, 1992; Wilson *et al.*, 1997) used artificial neural networks and more recently, (Patwardhan and Lakshminarayan, 1998) exploited the use of Hammerstein and Wiener models, (Hiden *et al.*, 1998) employed Genetic Programming and (Li *et al.*, 2001) utilized the Box-Tidwell transformation. However, (Berglund and Wold, 1997) argued that artificial neural networks and spline functions (i) inherent a considerable degree of flexibility and (ii) may lead to overfitting. A revision of the work by (Wold *et al.*, 1989), incorporating low order polynomials, was given by (Baffi *et al.*, 1999*a*), where a number of modifications were introduced to produce a simpler algorithm.

In this article, the work by (Wold *et al.*, 1989) and the modifications that (Baffi *et al.*, 1999*a*) introduced are investigated. This analysis revealed the following contributions, summarized in the

[1] Corresponding Author: uwe.kruger@ee.qub.ac.uk; Tel: +44(0)2890 274059; Fax: +44(0)2890 667023

presented work. Firstly, if a linear structure for each score model is considered, the technique by (Baffi *et al.*, 1999*a*) is equivalent to reduced rank regression (RRR). Therefore, each score model maximally reduces the variation of the response variables. Secondly, if nonlinear score models are identified, the algorithm by (Baffi *et al.*, 1999*a*) is consequently a nonlinear RRR (NLRRR) technique. Thus, it should not be considered as a nonlinear PLS technique, since minimizing the variation of the prediction error of the response variables does not guarantee maximum variance between pairs of score variables. A maximum covariance, however, is the criterion with which a conventional PLS model is identified. Thirdly, using NLRRR, the direction on which the response variables are projected is constrained to be equivalent to the direction for predicting the response variables by the predicted u-scores.

2. PARTIAL LEAST SQUARES PRELIMINARIES

PLS is designed to analyze the relationships between the predictor and the response variables. The observations of each variable are stored in matrices, i.e. the predictor matrix $\mathbf{X} \in \mathbb{R}^{K \times N}$ and response matrix $\mathbf{Y} \in \mathbb{R}^{K \times M}$. Each column refers to a particular predictor or response variable and each row refers to a particular observation, K represents the number of observations, N is the number of predictor variables and M is the number of response variables. Typically, $\mathbf{X}$ and $\mathbf{Y}$ are mean centered and appropriately scaled prior to the applications of PLS. It is assumed throughout this article that $K \gg N, M$ and that the rank of $\mathbf{X}$ and $\mathbf{Y}$ is N and M, respectively. Note, however, that these assumptions are for the convenience of the presentation rather than imposing restrictions onto the presented methods.

2.1 PLS Algorithms

PLS determines the kth pair of weight vectors, $\mathbf{v}_k$ and $\mathbf{w}_k$, and score vectors, $\mathbf{t}_k$ and $\mathbf{u}_k$, i.e. the vectors in which the t- and u-scores are stored, by maximizing $J_k^{(\mathbf{v},\mathbf{w})}$:

$$J_k^{(\mathbf{v},\mathbf{w})} = \mathbf{t}_k^T \mathbf{u}_k = \mathbf{w}_k^T \mathbf{X}_k^T \mathbf{Y}_k \mathbf{v}_k, \tag{1}$$

which is subject to the constraints $G_k^{(\mathbf{v})}$ and $G_k^{(\mathbf{w})}$ (Höskuldsson, 1988):

$$G_k^{(\mathbf{v})} = \|\mathbf{v}_k\|_2^2 - 1 = 0 \quad G_k^{(\mathbf{w})} = \|\mathbf{w}_k\|_2^2 - 1 = 0, \tag{2}$$

with $\|\circ\|_2^2$ being the squared norm of a vector. For determining subsequent weight and score vectors, PLS uses a deflation procedure, which involves the contribution of the kth pair of score vectors to be subtracted or *deflated* from the predictor and response matrices, (Geladi and Kowalski, 1986):

$$\mathbf{X}_{k+1} = \mathbf{X}_k - \mathbf{t}_k \mathbf{p}_k^T \qquad \mathbf{Y}_{k+1} = \mathbf{Y}_k - \mathbf{t}_k \mathbf{q}_k^T, \tag{3}$$

where $\mathbf{X}_k$, $\mathbf{Y}_k$ and $\mathbf{X}_{k+1}$, $\mathbf{Y}_{k+1}$ are the predictor and response matrices after $k+1$ and k deflation steps, respectively, and $\mathbf{p}_k$ and $\mathbf{q}_k$ are regression or loading vectors that represent the contribution of the t-score vector to the predictor and response matrices, respectively. The loading vectors are determined as follows:

$$\mathbf{p}_k = \frac{\mathbf{X}_k^T \mathbf{t}_k}{\mathbf{t}_k^T \mathbf{t}_k} \qquad \mathbf{q}_k = \frac{\mathbf{Y}_k^T \mathbf{t}_k}{\mathbf{t}_k^T \mathbf{t}_k}. \tag{4}$$

After computing n pairs of weight, score and loading vectors, often denoted as latent variables (LVs), a parametric regression matrix, $\mathbf{B}_{PLS}^{(n)}$, can be calculated:

$$\mathbf{B}_{PLS}^{(n)} = \mathbf{W}_n \left[\mathbf{P}_n^T \mathbf{W}_n\right]^{-1} \mathbf{Q}_n^T, \tag{5}$$

where $\mathbf{W}_n$, $\mathbf{P}_n$ and $\mathbf{Q}_n$ are matrices in which the n w-weight and p- and q-loading vectors are stored as column vectors in successive order.

The following properties of PLS are of importance in subsequent sections.

Remark 1. The vectors $\mathbf{v}_k$, $\mathbf{q}_k$ and $\mathbf{p}_k$ are functions of $\mathbf{w}_k$

Remark 2. The vectors $\mathbf{v}_k$ and $\mathbf{q}_k$ point to the same direction in the response space, i.e. $\mathbf{v}_k \propto \mathbf{q}_k$

Remark 3. The elements of each pair of score vectors, $\mathbf{t}_k$ and $\mathbf{u}_k$, are determined to have maximum covariance.

It follows from Remark 1 that the w-weight vector completely characterizes PLS, i.e. using $\mathbf{w}_k$, the score, loading and the v-weight vector can be computed.

3. NONLINEAR PARTIAL LEAST SQUARES

Various nonlinear extensions of PLS have been proposed that rely on a nonlinear mapping, where the u-scores are predicted using a nonlinear function of the t-scores:

$$\mathbf{u}_k = f(\mathbf{t}_k) + \mathbf{e}_k, \tag{6}$$

where $\mathbf{e}_k$ is the residual vector of $\mathbf{u}_k$. The following functions, $f(\circ)$, were introduced: second order polynomials (Wold *et al.*, 1989; Baffi *et al.*, 1999*a*), smoothing procedures (Frank, 1990), spline functions (Wold, 1992), artificial neural networks (Qin and McAvoy, 1992; Holcomb and Morari, 1992; Wilson *et al.*, 1997; Baffi *et al.*, 1999*b*) and Box-Tidwell transformations (Li *et al.*, 2001).

It follows from Remark 1 that the algorithms by (Wold *et al.*, 1989; Baffi *et al.*, 1999*a*; Baffi *et* *al.*, 1999*b*) exhibit a close relation to PLS. More precisely, $\mathbf{t}_k$, $\mathbf{v}_k$, $\mathbf{u}_k$, $\mathbf{p}_k$, $\mathbf{q}_k$ are linear or nonlinear functions of $\mathbf{w}_k$. The update procedure of the ith iteration step for the w-weight vector, $\mathbf{w}_{i,k}$, is given by (Baffi *et al.*, 1999*a*):

$$\Delta \mathbf{w}_{i+1,k} = \left[\mathbf{Z}_{i,k}^T \mathbf{Z}_{i,k}\right]^\dagger \mathbf{Z}_{i,k}^T \mathbf{e}_{i,k}$$
$$\mathbf{w}_{i+1,k} = \frac{\mathbf{w}_{i,k} + \Delta \mathbf{w}_{i+1,k}}{\|\mathbf{w}_{i,k} + \Delta \mathbf{w}_{i+1,k}\|_2}, \qquad (7)$$

where $\dagger$ represents the generalized inverse of a matrix and $\mathbf{Z}_{i,k}$ is a matrix in which the nonlinear function of the kth t-scores are stored. For example, the jth row of this matrix has the following elements for a second order polynomial, $u_{i,jk} = a_{i,1} + a_{i,2} \cdot t_{i,jk} + a_{i,3} \cdot t_{i,jk}^2 + e_{i,jk}$, $\mathbf{z}_{i,jk}^T = \left(a_{i,2} \cdot \mathbf{x}_{jk}^T \; 2 \cdot a_{i,3} \cdot \mathbf{x}_{jk}^T \right)$, where $a_{i,1}$, $a_{i,2}$ and $a_{i,3}$ are coefficients that are determined using a standards least squares procedure for instance and $\mathbf{x}_{jk}^T$ is the jth row vector of $\mathbf{X}_k$.

The algorithm by (Baffi *et al.*, 1999*a*) is based on the work by (Wold *et al.*, 1989). More precisely, (Baffi *et al.*, 1999*a*) introduced some simplifications to enhance the algorithm by (Wold *et al.*, 1989). The simplified algorithm was later extended to include the application of neural networks (Baffi *et al.*, 1999*b*). The error-based algorithm by (Baffi *et al.*, 1999*a*; Baffi *et al.*, 1999*b*) is analyzed in the next sections.

4. ANALYSIS OF ERROR-BASED NONLINEAR PLS

Since updating $\mathbf{w}_{i+1,k} = \frac{\mathbf{w}_{i,k}+\Delta\mathbf{w}_{i+1,k}}{\|\mathbf{w}_{i,k}+\Delta\mathbf{w}_{i+1,k}\|_2}$ is based on minimizing the residuals of the kth score model, $\mathbf{e}_{i,k}$, it is not guaranteed that the elements in each pair of score vectors are determined to have maximum covariance after converged. This can be seen by analyzing the cost function with which the w-weight vector is determined:

$$J_k^{(\mathbf{w})} = \min_{\mathbf{w}_k} \|\mathbf{e}_k\|_2^2 = \min_{\mathbf{w}_k} \|\mathbf{u}_k - f(\mathbf{t}_k)\|_2^2$$
$$= \min_{\mathbf{w}_k} \|\mathbf{Y}_k \mathbf{v}_k - f(\mathbf{X}_k \mathbf{w}_k)\|_2^2, \qquad (8)$$

which is subject to the following constraints:

$$\mathbf{G}_k^{(\mathbf{v})} = \mathbf{v}_k - \frac{\mathbf{Y}_k^T f(\mathbf{X}_k \mathbf{w}_k)}{\left\|\mathbf{Y}_k^T f(\mathbf{X}_k \mathbf{w}_k)\right\|_2} = \mathbf{0}$$
$$G_k^{(\mathbf{w})} = \|\mathbf{w}_k\|_2^2 - 1 = 0. \qquad (9)$$

To simplify the analysis of the above error-based algorithm, the function $f(\mathbf{t}_k)$ is assumed to be linear. This allows a strict comparison between conventional PLS and "linear" error-based PLS.

Using a linear function gives rise to the following Theorem, proven in Appendix A.

Theorem 1. If a linear function of the t-scores is used to predict the u-scores, i.e. $\mathbf{u}_k = a_k \mathbf{t}_k + \mathbf{e}_k$ with a_k being a regression coefficient, the weight vector $\mathbf{w}_k$ is the most dominant eigenvector of the matrix expression $\left[\mathbf{X}_k^T \mathbf{X}_k\right]^\dagger \mathbf{X}_k^T \mathbf{Y}_k \mathbf{Y}_k^T \mathbf{X}_k$.

The solution for obtaining $\mathbf{w}_k$, however, is equivalent to the solution of the RRR cost function, which gives rise to the following Conclusion.

Conclusion 1. If a linear relationship between the score variables is considered, the error-based iteration procedure by (Baffi *et al.*, 1999*a*; Baffi *et al.*, 1999*b*) produces, in fact, a RRR solution.

For completeness, the cost function of RRR and its solution are provided in Appendix B. Conclusion 1 can be reformulated as follows.

Conclusion 2. Since the error-based iteration procedure converges to produce a RRR solution, it is not guaranteed that the elements in the score vectors, $\mathbf{t}_k$ and $\mathbf{u}_k$ describe a maximum covariance.

An examples to support Conclusion 2 is given in Section 6. The equivalence between a linear version of error-based PLS and RRR follows from the v-weight vector being a function of the w-weight vector.

Theorem 2. By incorporating the constraint $\mathbf{G}_k^{(\mathbf{v})} = \mathbf{v}_k - \frac{\mathbf{Y}_k \widehat{\mathbf{u}}_k}{\|\mathbf{Y}_k \widehat{\mathbf{u}}_k\|_2} = \mathbf{v}_k - \frac{\mathbf{Y}_k \mathbf{t}_k}{\|\mathbf{Y}_k \mathbf{t}_k\|_2} = \mathbf{0}$, the error-based cost function minimizes the residuals of the response variables instead of minimizing the residuals of the kth score variables.

The proof of Theorem 1 can also be applied to prove Theorem 2. Note that the error-based procedure complies with Bellman's principle of optimality (Bellman, 1957) for predicting the response variables.

By applying a nonlinear mapping between the score variables, as shown in Equation (6), the kth u-score variable is predicted more accurately in case the underlying process exhibits nonlinear relationships between the predictor and response variables. On the basis of Theorem 2, this increased accuracy translates into an increased accuracy for predicting the response variables. This is, again, a result of the constraint applied to the v-weight vector, which gives rise to the following conclusion.

Conclusion 3. For nonlinear score models, the error-based iteration procedure converges to minimize the residual variance of the response variables.

Conclusion 3 can be argued on the basis of the following Lemmas.

Lemma 1. The constraint of the v-weight vector, used to calculated the u-score vector, is determined to maximize the contribution of the predicted u-score vector to the response matrix.

Lemma 2. After the iteration procedure has converged, the q-loading vector is obtained to maximize the contribution of the predicted u-score vector to the response variables and points in the same direction as the v-weight vector.

Lemma 3. The error-based iteration procedure converges to minimize the variance of the score model prediction and after conversion under the constraint of the v-weight vector being a function of the w-weight vector.

Lemma 1 and 2 are proven in Appendix C and Lemma 3 follows from Equation (9). With respect to PLS, Conclusions 2 and 3 give rise to postulate the following conclusion.

Conclusion 4. The error-based iteration procedure by (Baffi *et al.*, 1999*a*; Baffi *et al.*, 1999*b*) is a nonlinear extension to RRR rather than a nonlinear extension to conventional PLS.

The iterative error-based procedure is, in fact, a gradient descent approach for minimizing Equation (8) by incorporating the constraints of Equation (9). It should therefore be noted that the solution that is computed may be suboptimal, i.e. the solution may present a local minimum rather than a global minima. This might be circumvented by applying a Genetic Algorithm. Although computationally expensive this technique showed to provide more accurate NLRRR models, as shown by (Sharma *et al.*, 2003).

Another problem is that the series of iteration steps may not converge for each set of LVs. This, however, has not yet been reported and the applications of NLRRR in Section 6 also showed convergence. In contrast, the "linear" error-based iteration procedure converges to compute the most dominant eigenvector of a matrix expression. This, in turn, implies that (i) the results are optimal and (ii) convergence problems will usually not arise (Golub and van Loan, 1996).

5. APPLICATION STUDIES

In this section, the NLRRR algorithm is applied to a simulation study to illustrate the above findings. In addition, the covariance of the elements of each pair of t- and u-scores are determined for PLS and NLRRR.

The simulated process contained 4 predictor variables, denoted as x_1, x_2, x_3 and x_4, and 3 output variables, referred to as y_1, y_2 and y_3. The predictor variables were generated as follows:

$$\begin{aligned} x_1 &= 0.821 \cdot t_1 + 0.444 \cdot t_2 \\ x_2 &= 0.615 \cdot t_1 + 0.744 \cdot t_2 \\ x_3 &= 0.921 \cdot t_1 + 0.738 \cdot t_2 \\ x_4 &= 0.176 \cdot t_1 + 0.405 \cdot t_2, \end{aligned} \tag{10}$$

where t_1 and t_2 were normally distributed sequences of zero mean and unit variance, i.e. $t_1, t_2 \in \mathcal{N}\{0,1\}$. Using the 4 predictor variables, the response variables were then computed to be:

$$\begin{aligned} y_1 &= \exp\left(2 \cdot x_1 \cdot \sin\left(\pi \cdot x_4\right)\right) + \sin\left(x_2 \cdot x_3\right) \\ y_2 &= x_1^2 + \sin\left(\pi \cdot x_2 \cdot x_3\right) + x_4^4 \\ y_3 &= \exp\left(x_2 \cdot \cos\left(\pi \cdot x_4\right)\right) + x_1 \cdot x_4. \end{aligned} \tag{11}$$

A data set of 1000 samples was obtained as described above and an identically and independently distributed sequence of zero mean and variance 0.01, was superimposed on each variable to represent measurement noise. A division of the recorded data set was then carried out as follows. The first 800 samples were used as reference data to identify a NLRRR model and the remaining 200 samples were used to test the identified model.

Prior to the identification of the NLRRR model, the variables of the reference data set were normalized, i.e. mean centered and scaled to unit variance. The nonlinear function was selected to be second order polynomials. To highlight that the NLRRR algorithm may not calculate pairs of t- and u-scores that have a maximum covariance, each pair of score vectors was evaluated for PLS and NLRRR. The results of this comparison are summarized in Table (1).

Table 1. Variance captured by score models for PLS and NLRRR

#LV	PLS Model	NLRRR Model
1	2.1083	0.1614
2	0.2180	0.0319
3	0.0003	0.0002
4	0.0000	0.0000

It could clearly be seen from Table (1), that the PLS model determined pairs of t- and u-scores that described a significantly larger covariance than those computed by the NLRRR model. The model performance of the NLRRR model was evaluated on the reference data using the percentage contribution of each set of LVs and their cumulative percentage contribution captured in the predictor and response matrices. Furthermore, the mean squared prediction error (MSPE) were also analyzed. Table (2) summarize the performance of the identified NLRRR model.

Table 2. Performance of NLRRR model

#LV	Contr to $\mathbf{X}$	Tot Contr to $\mathbf{X}$	Contr to $\mathbf{Y}$	Tot Contr to $\mathbf{Y}$	MSPE
1	88.38	88.38	42.86	42.86	1.7121
2	11.50	99.88	8.09	50.95	1.4698
3	0.07	99.95	0.17	51.12	1.4648
4	0.05	100.00	0.08	51.20	1.4626

Given the construction of the predictor variables, the predictor matrix without the introduction of measurement noise has rank 2. Thus, only two sets of LVs were required to represent the structure between the predictor and response variables. This could also be revealed by the NLRRR models as their variance contribution of the 3rd and 4th set of LVs was insignificant for (i) predicting the response matrix and (ii) reconstructing the predictor matrix. Hence, only two score models were required to represented the variation of the predictor and response variables, respectively.

6. CONCLUSIONS

In this paper, nonlinear extensions of PLS, introduced by (Wold *et al.*, 1989) and (Baffi *et al.*, 1999*a*), were analyzed. This analysis revealed, firstly, that these algorithms, in fact, determine a linear RRR model if the nonlinear functions, incorporated in the score models, are replaced by linear functions. Secondly, these algorithms should therefore be seen as nonlinear extensions of RRR, instead of nonlinear extensions to PLS. Thirdly, the weight and loading vectors of the response variables point in the same direction.

The analysis yielded further that the algorithm by (Baffi *et al.*, 1999*a*) are based on a gradient descent technique and may therefore produce a sub-optimal solution. In addition, the series of iteration steps may not converge for each set of LVs. Convergence problems, however, have not yet been reported and were also not experienced in this work. To alleviate the determination of an optimal solution for the associated cost functions and their constraints, (Sharma *et al.*, 2003) introduced the application of a Genetic Algorithm strategies.

The above findings were demonstrated using an application studies to a synthetic example.

Appendix A. LINEAR ERROR-BASED PARTIAL LEAST SQUARES

For simplicity, the subscripts on matrices and vectors, that indicate which LV is currently being determined, are omitted. For the ith iteration step, the dependency of the weight vector $\mathbf{v}_i$, the regression coefficient b_i and the residual vector $\mathbf{e}_i$ upon the weight vector $\mathbf{w}_i$ is as follows:

$$\mathbf{v}_i = \frac{\mathbf{Y}^T\mathbf{X}\mathbf{w}_i}{\|\mathbf{Y}^T\mathbf{X}\mathbf{w}_i\|_2} \tag{A.1}$$

$$b_i = \frac{\mathbf{w}_i^T\mathbf{X}^T\mathbf{Y}\mathbf{Y}^T\mathbf{X}\mathbf{w}_i}{\mathbf{w}_i^T\mathbf{X}^T\mathbf{X}\mathbf{w}_i\,\|\mathbf{Y}^T\mathbf{X}\mathbf{w}_i\|_2}$$

$$\mathbf{e}_i = \frac{\mathbf{Y}\mathbf{Y}^T\mathbf{X}\mathbf{w}_i}{\|\mathbf{Y}^T\mathbf{X}\mathbf{w}_i\|_2} - \mathbf{X}\mathbf{w}_i\frac{\mathbf{w}_i^T\mathbf{X}^T\mathbf{Y}\mathbf{Y}^T\mathbf{X}\mathbf{w}_i}{\mathbf{w}_i^T\mathbf{X}^T\mathbf{X}\mathbf{w}_i\,\|\mathbf{Y}^T\mathbf{X}\mathbf{w}_i\|_2},$$

which implies that the update of the weight vector $\mathbf{w}_i$ is given by:

$$\Delta\mathbf{w}_{i+1} = \left[\mathbf{X}^T\mathbf{X}\right]^{\dagger}\frac{\mathbf{X}^T\mathbf{Y}\mathbf{Y}^T\mathbf{X}\mathbf{w}_i\,\|\mathbf{X}\mathbf{w}_i\|_2^2}{\|\mathbf{Y}^T\mathbf{X}\mathbf{w}_i\|_2^2} - \mathbf{w}_i. \tag{A.2}$$

Consequently, the w-weight vector for the $(i+1)^{th}$ iteration step is calculated as:

$$\mathbf{w}_{i+1} = \frac{\left[\mathbf{X}^T\mathbf{X}\right]^{\dagger}\mathbf{X}^T\mathbf{Y}\mathbf{Y}^T\mathbf{X}\mathbf{w}_i}{\left\|\left[\mathbf{X}^T\mathbf{X}\right]^{\dagger}\mathbf{X}^T\mathbf{Y}\mathbf{Y}^T\mathbf{X}\mathbf{w}_i\right\|_2}. \tag{A.3}$$

The above iteration is equivalent to the Power method (Golub and van Loan, 1996) for determining the most dominant eigenvector of the positive definite or semi-definite matrix $\left[\mathbf{X}^T\mathbf{X}\right]^{\dagger}\mathbf{X}^T\mathbf{Y}\mathbf{Y}^T\mathbf{X}$. On convergence, the most dominant eigenvalue is equal to $\mathbf{w}^T\left[\mathbf{X}^T\mathbf{X}\right]^{\dagger}\mathbf{X}^T\mathbf{Y}\mathbf{Y}^T\mathbf{X}\mathbf{w}$.

Appendix B. COST FUNCTION AND SOLUTION FOR REDUCED RANK REGRESSION

The residuals of the response matrix are given by:

$$\mathbf{E} = \mathbf{Y} - \mathbf{t}\mathbf{t}^T\mathbf{Y}, \tag{B.1}$$

which is based on the following constraints:

$$\mathbf{G}^{(\mathbf{t})} = \mathbf{t} - \mathbf{X}\mathbf{w} = \mathbf{0} \qquad G^{(\mathbf{w})} = \mathbf{w}^T\mathbf{X}^T\mathbf{X}\mathbf{w} - 1 = 0 \tag{B.2}$$

Using the above equations, the RRR cost function for $\mathbf{w}$ is:

$$\begin{aligned} J^{(\mathbf{w})} &= \min_{\mathbf{w}} \text{trace}\left\{\mathbf{E}^T\mathbf{E}\right\} \\ &= \min_{\mathbf{w}} \text{trace}\left\{\mathbf{Y}^T\mathbf{Y}\right\} - \text{trace}\left\{\mathbf{Y}^T\mathbf{t}\mathbf{t}^T\mathbf{Y}\right\}. \end{aligned} \tag{B.3}$$

The solution of the above cost function can be obtained as follows:

$$\frac{\partial J^{(\mathbf{w})}}{\partial \mathbf{t}}\frac{\partial \mathbf{t}}{\partial \mathbf{w}} - \lambda\frac{\partial G^{(\mathbf{w})}}{\partial \mathbf{w}} = \mathbf{0} \tag{B.4}$$

and is given by:

$$\begin{aligned} 2\mathbf{w}^T\mathbf{X}^T\mathbf{Y}\mathbf{Y}^T\mathbf{X} - \lambda 2\mathbf{w}^T\mathbf{X}^T\mathbf{X} &= \mathbf{0} \\ \lambda\mathbf{w} &= \left[\mathbf{X}^T\mathbf{X}\right]^{\dagger}\mathbf{X}^T\mathbf{Y}\mathbf{Y}^T\mathbf{X}\mathbf{w}. \end{aligned} \tag{B.5}$$

Consequently, $\mathbf{w}$ is the eigenvector associated with the largest eigenvalue of $\left[\mathbf{X}^T\mathbf{X}\right]^{\dagger}\mathbf{X}^T\mathbf{Y}\mathbf{Y}^T\mathbf{X}$.

Appendix C. DETERMINATION OF THE V-WEIGHT AND Q-LOADING VECTOR

The response matrix can be predicted using the prediction of the u-score vector, $f(\mathbf{t})$, and the loading vector $\mathbf{q}$:

$$\mathbf{Y} = f(\mathbf{t})\,\mathbf{q}^T + \mathbf{E}, \tag{C.1}$$

where $\mathbf{E}$ is a residual matrix. Given the case that $f(\mathbf{t})$ is predetermined, $\mathbf{q}^T$ can be regarded as an independent vector and the minimum variance of $\mathbf{E}$ is given by:

$$\mathbf{q}^T = \left(f(\mathbf{t})^T f(\mathbf{t})\right)^{-1} f(\mathbf{t})^T \mathbf{Y}, \tag{C.2}$$

which is equal to:

$$\mathbf{q} = \frac{\mathbf{Y}^T f(\mathbf{t})}{f(\mathbf{t})^T f(\mathbf{t})} \propto \mathbf{Y}^T f(\mathbf{t}), \tag{C.3}$$

as $f(\mathbf{t})^T f(\mathbf{t})$ is a scalar.The v-weight vector is also proportional to $\mathbf{Y}^T f(\mathbf{t})$, which implies that (i) the v-weight and the q-loading vector lie in the same direction and (ii) both are obtained to minimize the variance of the residual matrix $\mathbf{E}$.

REFERENCES

Baffi, G., E. B. Martin and A. J. Morris (1999*a*). Non-linear projection to latent structures revisited: the quadratic pls algorithm. *Computers and Chemical Engineering* **23**, 395–411.

Baffi, G., E. B. Martin and A. J. Morris (1999*b*). Non-linear projection to latent structures revisited: the neural network pls algorithm. *Computers and Chemical Engineering* **23**(9), 1293–1307.

Bellman, R. (1957). *Dynamic Programming.* Princeton University Press. Priceton, N.J., U.S.A.

Berglund, A. and S. Wold (1997). Inlr, implicit non-linear latent variable regression. *Journal of Chemometrics* **11**, 141–156.

Frank, I. E. (1990). A nonlinear pls model. *Chemometrics and Intelligent Laboratory Systems* **8**, 109–119.

Geladi, P. and B. R. Kowalski (1986). Partial least squares regression: A tutorial. *Analytica Chimica Acta* **185**, 231–246.

Golub, G. H. and C. F. van Loan (1996). *Matrix Computation.* 3 ed.. John Hopkins. Baltimore.

Hiden, H., B. McKay, M. Willis and G. Montague (1998). Non-linear partial least squares using genetic programming. In: *Proceedings of the 2nd Annual Conference on Genetic Programming (GP'98).* Madison, Wisconsin, U.S.A.

Holcomb, T. R. and M. Morari (1992). Pls/neural networks. *Computers and Chemical Engineering* **16**(4), 393–411.

Höskuldsson, A. (1988). Pls regression models. *Journal of Chemometrics* **2**, 211–228.

Li, B., E. B. Martin and A. J. Morris (2001). Box-tidwell transformation based partial least squares regression. *Computers and Chemical Engineering* **25**, 1219–1233.

Patwardhan, R. and S. Lakshminarayan (1998). Constraint nonlinear mpc using hammerstein and wiener models. *AIChE Journal* **44**(7), 1611–1622.

Qin, S. J. and T. J. McAvoy (1992). Nonlinear pls modelling using neural networks. *Computers and Chemical Engineering* **16**(4), 379–391.

Sharma, S., U. Kruger and G. W. Irwin (2003). Genetic learning methods for enhanced nonlinear partial least squares. In: *Proceedings of the IFAC International Conference on Intelligent Control Systems and Signal Processing (ICONS2003).* pp. 37–42. Faro, Portugal.

Wilson, D. J. H., G. W. Irwin and G. Lightbody (1997). Nonlinear pls modelling using radial basis functions. In: *Proceedings of the American Control Conference.* Albuquerque, New Mexico.

Wold, S. (1992). Non-linear partial least squares modelling. ii spline inner functions. *Chemometrics and Intelligent Laboratory Systems* **14**, 71–84.

Wold, S., N. Kettaneh-Wold and B Skagerberg (1989). Non-linear pls modelling. *Chemometrics and Intelligent Laboratory Systems* **7**, 53–65.

ELSEVIER
IFAC
PUBLICATIONS
www.elsevier.com/locate/ifac

STATE AND PARAMETER ESTIMATION IN CEMENT GRINDING CIRCUITS - PRACTICAL ASPECTS

R. Lepore *,[1] **A. Vande Wouwer** * **M. Remy** *
Ph. Bogaerts **

* *Laboratoire d'Automatique*
Faculté Polytechnique de Mons
Boulevard Dolez 31, B-7000 Mons, Belgium
** *Service d'Automatique et d'Analyse des Systèmes*
Université Libre de Bruxelles

Abstract: Due to the lack of reliable and/or inexpensive hardware sensors in cement grinding, development of software sensors is particularly significant for control and monitoring purposes. In this study, a nonlinear distributed-parameter, full-horizon observer is designed, which allows the contents of the mill to be described in terms of hold-up and particle size distribution. When measurements are available at relatively high sampling rates and at, at least, two spatial locations along the mill, fast observer convergence is obtained. However, in practical situations where measurements can be collected at the mill outlet only and with a relatively low sampling rate, the observer convergence deteriorates and , as the sampling rate decreases, performance becomes similar to an asymptotic (simulation) observer. The robustness of the software sensor can be improved by on-line identification of some time-varying parameters, such as the material grindability. These several concepts are discussed and tested in simulation based on a realistic process model.

Keywords: distributed-parameter systems; nonlinear models; softsensing; observers; process control; industrial production systems.

1. INTRODUCTION

Control and monitoring of cement grinding circuits are notoriously delicate tasks. Together with the strong nonlinearity of the process, the general lack of (reliable) measurement sensors of relevant process variables are among the major difficulties to be faced. When available, measurement signals are usually corrupted by noise (e.g., the mechanical vibrations affecting the elevator power used as a mill flow rate measurement), and collected at relatively low sampling rates (e.g., sample analysis in laboratory facilities to determine particle size distribution, color, etc.). On the other hand, the cost of on-line analyzers is still prohibitive in most industrial applications (Hodouin and Del Villar, 1994).

Therefore, the development of software sensors for reconstructing the process state variables on line is particularly significant. These software sensors are based on a dynamic model of the process and some available measurement signals.

With regard to the process nonlinearity and stochastic nature of the measurement signals, a natural approach is given by full-horizon observers

[1] Corresponding author:
e-mail: Renato.Lepore@fpms.ac.be
phone: +32 (0)65 374140 fax: +32 (0)65 374136

which estimate the most-likely initial conditions for the process model. The algorithm minimizes a maximum-likelihood criterion which measures the deviation between the model prediction and the real system outputs (Bogaerts and Hanus, 2001). In this particularly versatile approach, signals of different nature (e.g., mass flow rates, particle size distributions, etc.) can be incorporated in the criterion, measurement errors can be taken into account, initial conditions can be parameterized, and equality and/or inequality constraints can be easily included. In addition, periodic model updates can be computed as well, based on parameter estimation techniques. Finally, this optimization-based state estimation technique is consistent with a nonlinear model predictive control (NMPC) approach (Lepore *et al.*, 2003) in terms of concepts, model and software tools.

The contribution of this paper is therefore:

- to develop a distributed-parameter, full-horizon observer, based on a process model consisting of partial differential equations (Lepore *et al.*, 2002), which allows the particle size distribution inside the mill to be described. This feature has to be contrasted from results published in the mineral processing literature, where attention is mostly focused on global variables, such as sieve residue at the mill and/or classifier outlet or material mass content of the mill (Hodouin and Del Villar, 1994). The particle size distribution inside the mill is, however, of primary interest for the operator to evaluate the mill operation, and for model-based control, such as the nonlinear model predictive control schemes presented in (Lepore *et al.*, 2002; Lepore *et al.*, 2003);
- to analyze, through a simulation case study, the properties of the observer in terms of convergence and reliability. This study considers two situations: a) the ideal case, for which several sensors are available inside the mill and measurements are collected at high sampling rate, b) the nonideal (or real) case, for which measurements are available at low sampling rate at the mill outlet only;
- to discuss the potentiality of on-line parameter estimation to periodically update the process model.

The continuation of this article is divided into five sections. Section 2 describes the process and the model equations. The design of full-horizon observers in the context of cement grinding is discussed in Section 3. In Section 4, numerical results are presented, whereas Section 5 is devoted to future extensions, including on-line parameter updates. Finally, conclusions and perspectives can be found in Section 6.

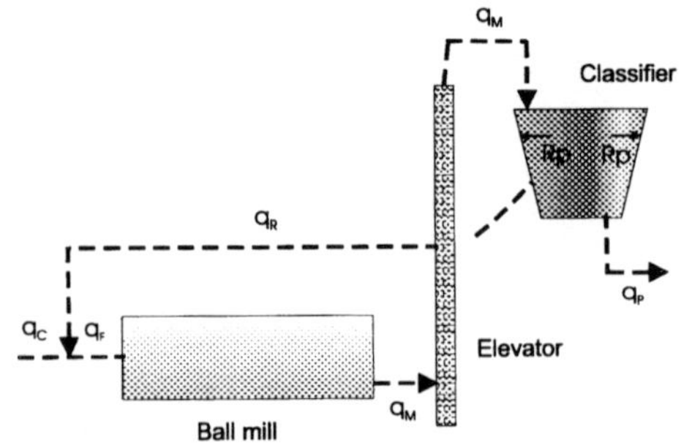

Fig. 1. Closed-loop grinding circuit

2. PROCESS DESCRIPTION AND MODELLING

2.1 Process description

A typical cement grinding circuit is represented in figure 1, which consists of a single-compartment ball mill in closed-loop with an air classifier. The raw material (usually clinker) flow q_C is fed to the rotating mill, in which balls perform the breakage of the material particles by fracture and/or attrition. At the other end, the output or mill flow q_M is lifted up by a bucket elevator onto the classifier which separates the material into two parts: the product flow q_P and the rejected flow q_R, which is recirculated to the mill inlet. The selectivity of the classifier and, in turn, the product fineness, can be modified by acting on special registers *Reg*. The sum of q_C and q_R is the total feed flow, denoted by q_F.

2.2 Process model

The size continuum is divided into three size intervals (noted *s.i.* in the following) numbered 1, 2 and 3 for the coarse, intermediate-size and fine particles, respectively. Mass balances lead to:

$$\frac{\partial X_i}{\partial t} = -u_i \frac{\partial X_i}{\partial x} + D_i \frac{\partial^2 X_i}{\partial x^2} + \sum_{j=1}^{2} k_{ij} \varphi_j \quad ; \text{i=1,2,3}$$

$$(k_{ij}) = \begin{pmatrix} -1 & 0 \\ +k & -1 \\ 1-k & +1 \end{pmatrix} \tag{1}$$

where:

- X_i is the mass per unit of length of the particles in s.i. i;
- k is the yield fraction of the particles in size interval 2 appearing from the breakage of the particles in s.i. 1; φ_j is the breakage rate of the material in s.i. j;
- u_i and D_i are the convection velocity and the diffusion coefficient, respectively, of the particles in s.i. i;

The partial differential equations (1) are supplemented by initial (2) and boundary (3) conditions:

$$X_i(0,x) = H_0(x)w_{0;i}(x) \quad \forall \text{ x; i=1,2,3} \qquad (2)$$

$$0 = u_i X_i - D_i \frac{\partial X_i}{\partial x} - q_F w_{F;i} \text{ x=0; i=1,2,3}$$
$$0 = \frac{\partial X_i}{\partial x} \qquad \text{x=L; i=1,2,3} \quad (3)$$

where:

- $H_0(x)$ is the initial material content per unit of length (e.g., in tons per meter), $w_{0;i}(x)$ is the corresponding mass fraction in s.i. i;
- q_F is the total feed flow rate, $w_{F;i}$ is the corresponding mass fraction in s.i. i.

The breakage rates are formulated as follows:

$$\varphi_j = \alpha_j X_j e^{-\beta H} \qquad ;\text{ j=1,2} \qquad (4)$$

where α_j is the specific rate of breakage for s.i. j, H is the hold-up, i.e., $(X_1 + X_2 + X_3)$, β is an inhibition coefficient.

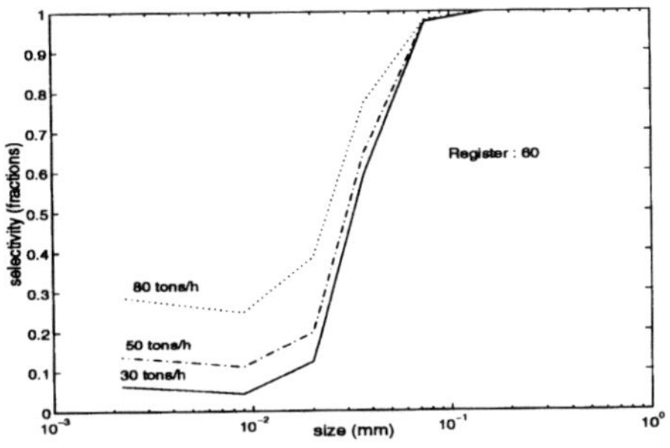

Fig. 2. Classifier selectivity (for a single register position and several mill flow rates)

The classifier has very fast dynamics compared to the mill and is therefore described by a nonlinear steady-state model. Selectivity is the fraction of material in each size interval which is recirculated (see the "fish-hook" curves in figure 2).

3. FULL-HORIZON OBSERVER

3.1 Basic principles

Consider the general nonlinear model:

$$\frac{dx}{dt} = f(x(t), u(t)), \qquad x(t_0) = x_0 \qquad (5)$$
$$y(t_k) = h(x(t_k)) + \epsilon(t_k) \qquad (6)$$

where $x(t)$, $u(t)$, $y(t)$ are the vectors of states, inputs and measurements, respectively, t_k are the sampling times and $\epsilon(t_k)$ is a white noise sequence with zero mean and variance $Q(t_k)$.

Following (Allgöwer *et al.*, 1999; Bogaerts and Hanus, 2001) and the references therein, the observer is defined as:

$$\hat{x}(t) = g(t, u(t), \hat{x}_{0;k}) \qquad \forall t \in [t_0 \; t_k[\qquad (7)$$

where $g(.)$ is the state trajectory of model (5) and $\hat{x}_{0;k}$ is the estimated initial-condition vector at time t_k.

$\hat{x}_{0;k}$ is obtained by minimizing the maximum-likelihood criterion $J_k(x_{0;k})$:

$$J_k(x_{0;k}) = \sum_{j \in HO_k} \epsilon(t_j)^T Q(t_j)^{-1} \epsilon(t_j) \qquad (8)$$
$$\epsilon(t_j) = y(t_j) - h(g(t_j; u(t_j); x_{0;k})) \qquad (9)$$

where HO_k is the observation horizon, of which $x_{0;k}$ is the initial condition, $y(t_j)$ are the measured values.

When all the measurements, from t_0 up to t_k, are used to estimate the initial condition, $HO_k = 0, 1, \ldots k$ and the observer is called a *full-horizon observer*.

3.2 Implementation

Spatial discretization of the PDE model (1-4) leads to a large number of variables to estimate (for instance, in a finite difference scheme, the number of state variables increases proportionnally with the number of spatial grid points). However, experimental and simulation results demonstrate that the spatial profiles of the hold-up and the weight fractions inside the mill can be reasonably approximated by simple, linear or quadratic functions. Hence, five parameters describe the state vector
$x_0 = H_0(x).[w_{1;0}(x) \; w_{2;0}(x) \; w_{3;0}(x)]^T$,
based on the following assumptions:

- hold-up: $H_0(x)$ is constant (requires one parameter);
- coarse particles: $w_{1;0}(x)$ is a concave quadratic law, the minimum value being reached at the outlet of the mill (requires two parameters);
- fine particles: $w_{3;0}(x)$ is a convex quadratic law, the maximum value being reached at the outlet of the mill (requires two parameters);
- intermediate-size particles: $w_{2;0}(x)$ requires no parameter since it is equal to $1 - w_{1;0}(x) - w_{3;0}(x)$.

On the other hand, several box and linear inequality constraints can be derived from the fact that:

- $H_0(x) \geq 0, \forall x$;
- $0 \leq w_{i;0}(x) \leq 1, \forall i, x$.

To keep the things clear, let us consider, say, that $w_{1;0}(x) = \theta_{w_1;1} x(x - 2L) + \theta_{w_1;2}$ is the quadratic law defined on $[0\ L]$ (L being the mill length). Convexity and $0 \leq w_{1;0}(0) \leq 1$ result in $\theta_{w_1;1} \geq 0$ and $0 \leq \theta_{w_1;2} \leq 1$, respectively (box contraints) whereas $0 \leq w_{1;0}(L) \leq 1$ results in $0 \leq -\theta_{w_1;1}L^2 + \theta_{w_1;2} \leq 1$ (linear inequality constraints). Similar arguments are applied on $H_0(x)$ and $w_{3;0}(x)$ leading to box and linear inequality constraints on the three other parameters.

In addition, in order to evaluate quantitatively and qualitatively the information available to determine the vector θ, the sensitivity functions, $S_{i;\theta_j}(t) = (\frac{\partial h_i}{\partial \theta_j})(t)$, are analyzed and, in order to estimate the confidence in the parameters, the covariance matrix is reasonably approximated by the inverse of the Fisher information matrix defined as follows:

$$F(\hat{\theta}) = \sum_{j \in HO_k} (\frac{\partial h}{\partial \theta})^T_{t_j} Q(t_j) (\frac{\partial h}{\partial \theta})_{t_j} \qquad (10)$$

The minimization of (8) is performed using the "Optimization toolbox 2.2", Matlab 6.5.0. The solution of the partial differential equations is achieved using (a) a "method of lines" Matlab procedure for spatial differentiation (b) standard solvers from Matlab 6.5.0 for the integration in time of the differential equations.

4. NUMERICAL RESULTS

A mathematical model, previously calibrated to mimic an industrial grinding circuit (Lepore *et al.*, 2002; Boulvin *et al.*, 2002), is used in this study as a reference system to test the performance of the observer. Figure 3, which represents the evolution of the relevant operating variables, illustrates a change in the operating point of the cement grinding circuit: from a steady state, the manipulated variables are moved at time 40 so as to produce more cement (see the increase in the product flow rate q_P) of lower quality (see the decrease in the weight fraction of the fine particles $w_{P;3}$) while maintaining the same grinding efficiency (in fact, one holds the mill flow rate and the weight fraction of the intermediate-size particles $w_{M;2}$ at constant values).
From time 0 on, several state estimations are performed. In all the observer designs considered in this study, the flow rate and the particle size distribution of the input material are supposed to be measured. In order to assess the quality of the state estimates, a global indicator,
$e_r(X) = \|\hat{X} - X\|/\|X\|$ is used, where $\hat{X}$ and X are the predicted and the actual state vector, respectively, and $\|.\|$ is the Euclidean norm.

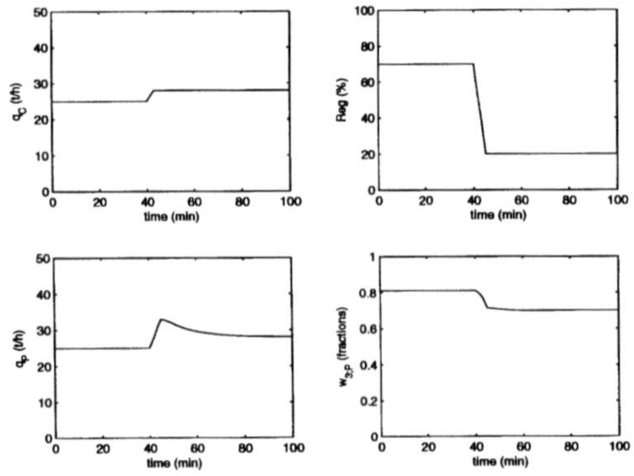

Fig. 3. Change in the cement production: decrease of the cement fineness

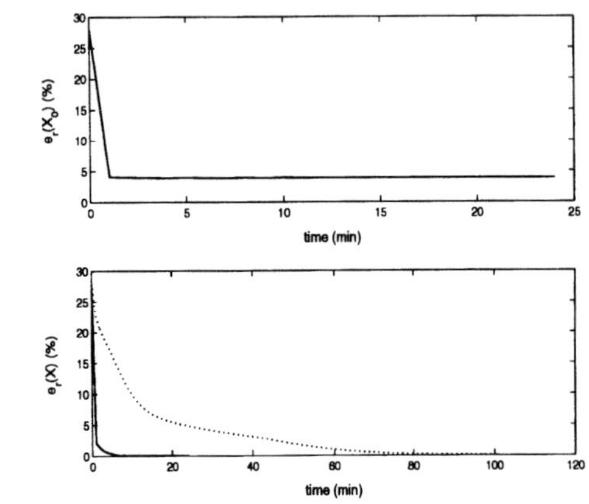

Fig. 4. Deviation between estimated and real values in the ideal case; upper: initial state vector; lower: state vector; (dotted: deviation when plain simulator is used)

4.1 Ideal case

In the following, we consider that the material hold-up and particle size distribution are available at two locations inside the mill (18% and 100% of the mill length), at sufficiently high sampling rate ($T_s = 1\ min$). The maximum error on the measurements is 0.005 t/m (overall mean value of the measurements is 0.25 t/m). Figure 4, which represents the temporal evolution of the deviation between the real and estimated states (at the initial time and at the current time), shows that the estimation error of the initial conditions quickly converges to a minimum, yet nonzero, value. This nonzero minimum results from the approximation of the real spatial profiles of the initial conditions by polynomials. Despite this residual error, the state estimates converge rapidly towards the system states, as depicted in figure 4. For comparison purposes, the convergence of a plain simulator (which can be viewed as an asymptotic observer) is also shown in figure 4. In addition, Table 1 confirms the high reliability of the procedure since the individual error on the initial condition estimates is very small (globally, a few percents).

θ_H	$\theta_{w1;1}$	$\theta_{w1;2}$	$\theta_{w3;1}$	$\theta_{w3;2}$
0.7473	0.2084	0.2264	−0.3425	0.1680
0.0006	0.0042	0.0031	0.0034	0.0024

Table 1. Confidence in the parameter estimates in the ideal case; row 1: parameter value; row 2: standard deviation

4.2 Nonideal (real) case

In the following, we consider only on-line measurements that can be achieved at reasonable costs in practice. Particularly, flow rates and particle size distributions obtained by a two-sieve technique (i.e., weight fractions of coarse, intermediate-size and fine particles) can be measured in real-life operation, provided that the sampling period is large enough. This way, at each time t_j in the horizon HO_k, the following measurements are available at the mill output: the flow rate q_M and the weight fractions $w_{M;i}$ $(i = 1 \ldots 3)$.

Figure 5 depicts the deviation between real and estimated states for three different sampling periods (20 *min*: solid; 10 *min*: dashed; 5 *min*: dash-dotted). The following conclusions can be drawn:

- $T_s = 20\,min$ appears to be a limiting period, for which the full-horizon observer and the asymptotic observer exhibit similar convergence speeds;
- the asymptotic values of the estimation error on the initial conditions are much larger than in the ideal case;
- however, despite this observation, it appears that a small correction of the initial conditions is responsible for a significant improvement in the estimation of the actual state.

However, very large uncertainties are associated with the parameters of the polynomial approximations of the initial condition profiles. From the evolution of the sensitivity functions $S_{M;i\theta}(t) = \frac{\partial X_{M;i}}{\partial \theta}(t)$ ($X_{M;i}$ being the material weight in size interval i at the mill output) (see figure 6), the following conclusions can be drawn:

- valuable information with respect to the initial hold-up is available until $t = 80\,min$;
- valuable information with respect to size interval 3 and size interval 1 are available until $t = 20\,min$ and $t = 10\,min$, respectively.

So, unless a sufficiently small sampling period is used (appropriately, $1 \sim 2\ min$), which is unfeasible in practice, the particle weight fractions cannot be determined with enough accuracy when using a single measurement at the mill output. On the other hand, since the hold-up can be reconstructed accurately in all cases, an additional sensor (such as an electronic ear) to measure the hold-up is probably useless (due to the low level of accuracy and reliability of these equipments).

5. EXTENSIONS

Whatever the solution adopted (full-horizon observer or asymptotic observer), an accurate model is a prerequisite to the design of a state estimator. However, it is by no means true that all the parameters are constant during normal operation. For instance, due to changing storage conditions, the grindability of the input material is subject to variations. In the following, we will focus on the estimation of the grindability factor, which is the variation of the fragmentation rates α_1 and α_2 from their reference values.
The estimate of the grindability factor $\hat{\theta}_k$ at time t_k is obtained by minimizing a maximum-likelihood criterion $J_k(\theta_k)$:

$$J_k(\theta_k) = \sum_{j \in HO_k} \epsilon(t_j)^T Q(t_j)^{-1} \epsilon(t_j) \quad (11)$$

$$\epsilon(t_j) = y(t_j) - h(g(t_j; u(t_j); x_{0;k}; \theta_k)) \quad (12)$$

As a test example, consider that the process is in steady state until time 30, at which a change in the process grindability factor of −20% occurs. This process variation is supposed to be detected, so that an identification procedure can start. Figure 7 represents 1) the real and estimated grindability factors, 2) the standard deviation of the estimates, for three horizon lengths, i.e., 1, 5 and 10 sampling periods (with $T_s = 10\,min$). The following conclusions can be drawn:

- the longer the horizon, the smaller is the deviation from the real value;
- whatever the horizon length, the results in terms of standard deviations of the estimates are very satisfactory.

At this stage, one should investigate other issues, such as estimating additional model parameters (e.g., transport velocity), selection of the horizon length (e.g., grindability could be considered as an

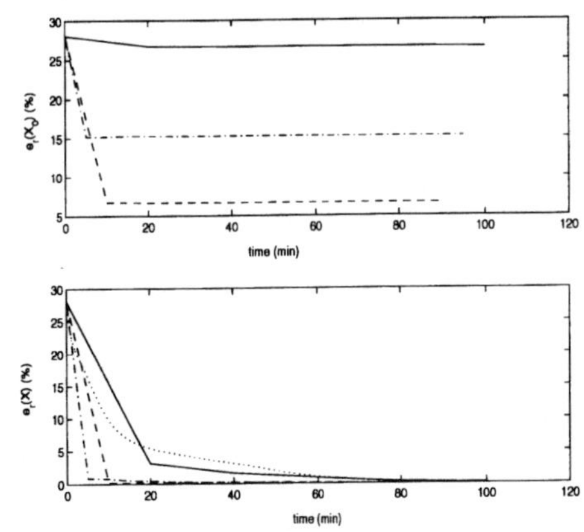

Fig. 5. Deviation between estimated and real values in the nonideal (real) case; upper: initial state vector; lower: state vector; (dotted: deviation when plain simulator is used);

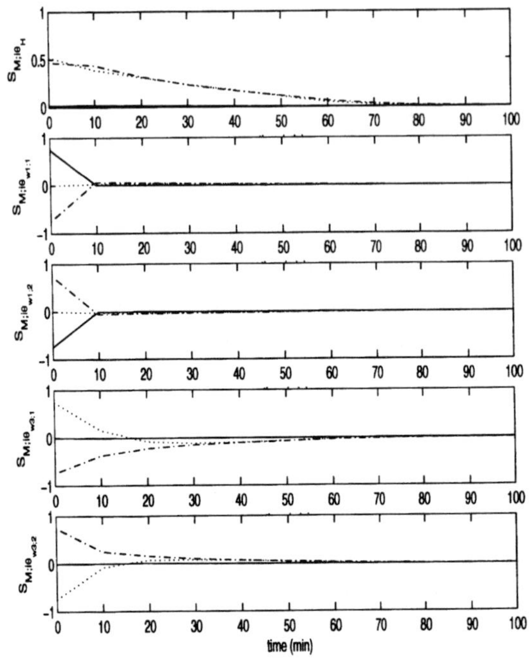

Fig. 6. Sensitivity functions at the mill output w.r.t. the parameters in the nonideal case; $i = 1$: solid, $i = 2$: dash-dotted, $i = 3$: dotted

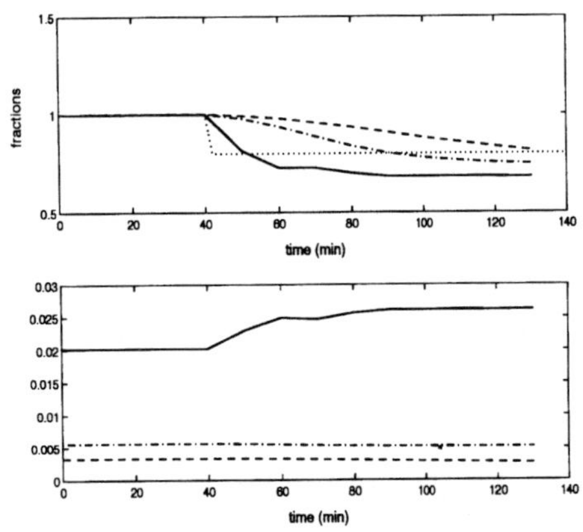

Fig. 7. Grindability factor - upper: real (dotted) and estimated values, lower: standard deviation; for three horizon lengths (in periods of $10\,min$) (1: solid; 5: dash-dotted; 10: dashed)

additional model state, estimated together with the initial conditions), stability issues, etc.

6. CONCLUSION

A distributed-parameter, full-horizon observer is implemented in order to predict the material contents of the mill (i.e., hold-up and particle size distribution). The observer is based on the computation of the most-likely initial conditions, expressed via a small number of parameters.

With a reference model calibrated to mimic an industrial plant, the observer has been evaluated in two different situations. In the ideal situation, for which several sensors are available along the mill and measurements are collected at high sampling rates, the algorithm exhibits good performance, especially when compared to the asymptotic observer, and reliability (or confidence) with respect to the estimates. On the other hand, in practical situations, for which only a few low-cost measurements (low sampling rate) are available at the mill output only, an important degradation of the performance is noticed, which is confirmed by the analysis of the sensitivity functions. This observation tends to designate the asymptotic observer as the only feasible solution in this real situation. Interestingly, hold-up is estimated accurately in all cases, which tends to demonstrate that additional equipements, such as electronic ear, could be of little use, considering the low accuracy and reliability of these hardware sensors.

Finally, in order to ensure robust estimation, on-line parametric adaptation of the observer is tested. Preliminary results are obtained for the estimation of the material grindability using a receding-horizon parameter estimation procedure. Future studies should focus on the estimation of more than one parameter (for example, grindability and transport velocity), considering the correlations between parameter estimates and stability issues. Work should also be devoted to the automatic detection of such parameter variations.

REFERENCES

Allgöwer, F., T.A. Badgwell, J.S. Qin, J.B. Rawlings and S.J. Wright (1999). Nonlinear predictive control and moving horizon estimation - an introduction overview. In: *Advances in Control (Highlights of ECC '99)* (P.M. Frank, Ed.). pp. 391–449. Springer-Verlag.

Bogaerts, Ph. and R. Hanus (2001). On-line state estimation of bioprocesses with full horizon observers. *Mathematics and computers in simulation* **56**, 425–441.

Boulvin, M., A. Vande Wouwer, R. Lepore, Ch. Renotte and M. Remy (2002). Modeling and control of cement grinding processes. *IEEE transactions on control systems technology.*

Hodouin, D. and R. Del Villar (1994). Conduite des unités de broyage. *Techniques de l'ingénieur* **JP**(J 3 110), 1–26.

Lepore, R., A. Vande Wouwer and M. Remy (2002). Modeling and predictive control of cement grinding circuits. IFAC '02, Barcelona, Spain.

Lepore, R., A. Vande Wouwer and M. Remy (2003). Nonlinear model predictive control of cement grinding circuits. ADCHEM 2003, Hong-Kong, China.

ELSEVIER
IFAC
PUBLICATIONS
www.elsevier.com/locate/ifac

AUTOMATIC DETECTION OF STICTION IN ACTUATORS: A TECHNIQUE TO REDUCE THE NUMBER OF UNCERTAIN CASES

M. Rossi * **C. Scali** *,[1]

* *Department of Chemical Engineering - CPCLab - University of Pisa (Italy)*

Abstract: The presence of stiction in actuators is responsible for deterioration of performance in control loops: the possibility of automatic detection of this phenomenon from the analysis of recorded data is certainly very important and should be included in a monitoring system. In the paper it is shown that the application on plant data of a largely diffused technique (Horch, 1999), may not be able to detect the presence of stiction in some cases. More in general, application of the technique in simulation on different stiction models, shows that, by varying process parameters, stiction can be hidden and the uncertainty zone (gray zone: where is not possible to take any decision) may become very large. A new technique, based on the shape of loop signals originated by stiction, is then proposed. Even though it is not able to guarantee detection of friction phenomena in all cases, (impossible for every technique, without any further knowledge on the process), the proposed technique allows a significant reduction of the gray zone; this is illustrated by intensive simulation and by application on plant data.

Keywords: Process Monitoring, Stiction, Actuators, Stiction Detection.

1. INTRODUCTION

Control loop performance monitoring is certainly a topics of large interest in research activity and in industrial applications. The possibility of assessing how the plant is behaving is of great importance, both for off-line analysis and for on-line application; the last is the the more challenging task because analysis technique must cope with constraints deriving from real time optimization tools. State of the art, with indications of most advanced fields of application, as well as of several aspects still unresolved in a systematic way, can be found in (Thornhill and Seborg, 2002).

Among different causes which can bring the controlled system to a non optimal behavior can be mentioned: wrong design or tuning of controllers (Ender, 1993), anomalies and failures of sensors, presence of friction in actuators, external perturbations. No doubts that the most popular problems addressed up to date is the uncorrect tuning of controllers, caused by initial errors (too prudential tuning) or by changes in process parameters (due to variations of operating conditions). Automatic techniques for rapid process identification and retuning of industrial controllers have been presented and nowadays find large applications in industry (see (Yu, 1999) for a review of autotuning methods).

One of most common cause of low performance of controlled loops can be found in the presence of stiction in actuators; this fact causes a delayed and sluggish actuation on the process of changes in manipulated variables, required by the control system (Hägglund, 2002). Usually the presence of friction in actuators shows up as oscillations in loop variables (output, manipulated and control signals). Oscillations can be also associated to aggressive tuning (close to marginal stability conditions) or to the inlet of periodic perturbations in the control loop.

[1] scali@ing.unipi.it

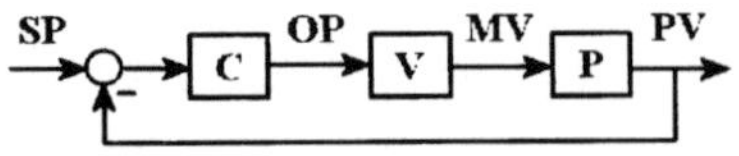

Figure 1. The reference scheme for the controlled loop

A desirable features of a monitoring system is certainly the detection of causes of oscillations, because actions to be taken will be different in the three cases: retuning of controllers, maintainance or substitution of valves, operations in upstream equipments in the plant. Therefore an automatic procedure to detect friction in actuators is called for; several techniques have been presented in literature and find applications in process industry: some of them (Hägglund, 1995), (Taha *et al.*, 1996), are not suitable for automatic implementation, while others can be (Horch, 1999), (Horch, 2000) and have been considered in this study.

After this synthetic introduction to the general problem, the paper reviews basic aspects of stiction, models to describe the phenomenon and techniques for its detection (section 2); applications of the (Horch, 1999) technique to plant data and simulation to different cases are presented in section 3; a new technique is proposed in section 4; a comparison of results from the application to the same cases is reported in section 5; conclusions and further work in section 6.

2. THE PROBLEM OF STICTION

2.1 Monitoring the presence of stiction

The basic control system is depicted in Figure 1, where SP, OP, MV, PV, represent set point, control signal, manipulated and controlled variables. Usually SP, OP, PV are recorded by the DCS, while the manipulated variable MV, which would allow a direct test for the presence of stiction by comparing MV and OP, is not available in the general industrial situation. In our case, the technique to detect the presence of stiction in the actuator is included in a general monitoring system (Rossi *et al.*, 2003), which performs a continuous analysis of data recorded by the DCS; the presence of significant perturbations is detected by the application of Hägglund indexes (Hägglund, 1995), (Hägglund, 1999). The cause of scarce performance must be determined (stiction, inadequate tuning or external perturbation), in order to decide the right operation to be performed on the plant. It is important that stiction is detected automatically, with very low margin of error or uncertainty, reducing to a minimum the need of interaction with operator.

2.2 Stiction Models

The presence of increasing stiction in the actuator introduces a delay and a non linear behavior between the control signal OP and the manipulated variable MV. Referring to Fig. 2: the valve do not move until the active force (F_a) is less than the static friction force (F_s); the integral action of the controller increases F_a over F_s: at this point the valve move and jumps to

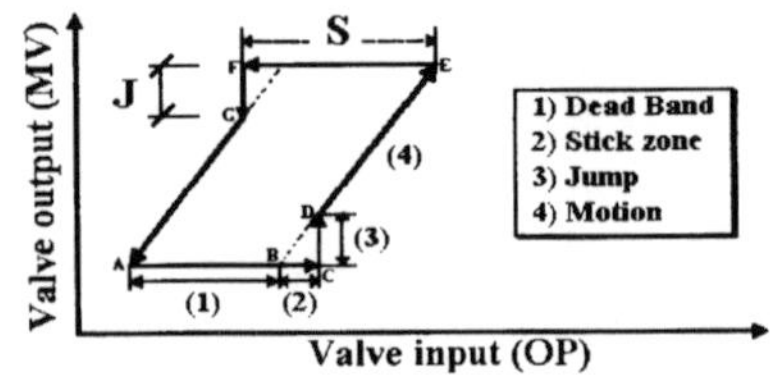

Figure 2. Characteristic Curve of the valve in the presence of stiction

a new position; the net force acting on the stem is always less than the nominal value corresponding to the control action, owing to the presence of dynamic friction (F_d) which depends on the velocity; finally the valve stops again when F_a becomes less than F_s. As a consequence of this phenomenon in the valve (stick-slip), characteristic oscillations appear in the output variable.

Many models have been proposed in literature to describe the presence of friction in the actuators; a survey is reported in (Armstrong-Hélouvry *et al.*, 1994). One of simplest and universally accepted models, is the Karnopp model:

$$m \cdot \frac{d^2x}{dt^2} = F_a - F - F_k \tag{1}$$

$$F = \begin{cases} -F_d sign(v) + F_v v & if \;\; v < -\Delta V \\ max[-F_s, F_a] & if \;\; -\Delta V < v < 0 \\ min[-F_s, F_a] & if \;\; 0 > v > \Delta V \\ F_d sign(v) + F_v & if \;\; v > \Delta V \end{cases} \tag{2}$$

The active force F_a generated by the controller is balanced by the friction force F and by the spring force F_k. The friction force is the sum of static and dynamic friction. The presence of static friction (that is not a function of v) is showed when valve velocity is $|v| < \Delta V$. Out of this range there is only the dynamic friction force (function of v).

Even though the number of parameters is not large, compared with other models, the lack of knowledge on values of key variables makes the model difficult to implement in a real plant. An alternative approach has been presented recently by (Choudhury *et al.*, 2003). By means of only two parameters (S and J in fig. 2) the model is able to describe characteristic curves of valves in the presence of friction phenomena: S is the range of OP values in which the valve is blocked owing to static friction, J represents the jump obtained when the valve starts to move.

2.3 Techniques for automatic detection of stiction

Several different techniques for detection of stiction have been proposed in literature. Unfortunately, some of them are manual or require strong interaction with the operator (Hägglund, 1995); other techniques require specific knowledge on process and actuators, usually not available (Taha *et al.*, 1996). These have not been taken into consideration in this study oriented towards automatic techniques, without any a priori specific knowledge about the system. The technique proposed by (Horch, 1999), is suitable for automatic

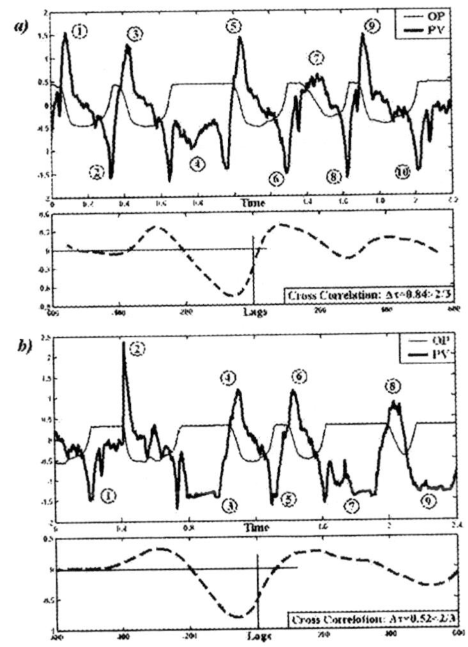

Figure 3. Application of the Horch technique on plant data. a) First set: detection of stiction, b) Second set: uncertainty

implementation and has been considered for adoption in the monitoring system. From the analysis of different plant data Horch observed that, in the presence of stiction, the control signal OP and the controlled variable PV, oscillate with a phase shift $\phi = \pi/2$, while in the presence of a sinusoidal perturbation the phase shift $\phi = \pi$. This observation has been theoretically explained and a first method based on the Cross-Correlation (C_{XY}) between *OP* and *PV* proposed: in the presence of stiction C_{XY} is an odd function ($\phi = \pi/2$), while in the presence of oscillating disturbances C_{XY} is an even function ($\phi = \pi$). This allows an automatic detection of the two cases, through the computation of two parameters:

$$\Delta\tau = \frac{|\tau_I - \tau S|}{\tau_I + \tau S}; \qquad \Delta\rho = \frac{|r_0 - r_{Max}|}{r_0 + r_{Max}} \tag{3}$$

τ_I e τ_S can be obtained as the two abscissas, closest to zero, where $C_{XY} = 0$; r_0 is the value of C_{XY} in zero and r_{Max} is the maximum value of C_{XY}. According to the values of these two parameters, conditions corresponding to the presence of stiction or sinusoidal perturbations can be indicated; also an uncertainty range, where no decision can be taken, is encountered (gray zone). In details:

$$\left.\begin{array}{l} 0 < \Delta\rho \leq 0.072 \\ 0 < \Delta\tau \leq 1/3 \end{array}\right\} \Rightarrow No \quad Stiction \tag{4}$$

$$\left.\begin{array}{l} 0.072 < \Delta\rho \leq 1/3 \\ 0 < \Delta\tau \leq 2/3 \end{array}\right\} \Rightarrow No \quad Decision \tag{5}$$

$$\left.\begin{array}{l} 1/3 < \Delta\rho \leq 1 \\ 2/3 < \Delta\tau \leq 1 \end{array}\right\} \Rightarrow Stiction \tag{6}$$

A second technique proposed by (Horch, 2000) should also be mentioned; the second derivative of the variable PV is analyzed and its sampling histogram is computed. In the presence of stiction the histogram has a *gaussian* trend; with an oscillating disturbance it has a *bimodal* trend. The technique may be difficult to implement in an industrial environment, because the second derivative can be too sensitive to the typology of noise; for this reason has not been taken into consideration in the present study.

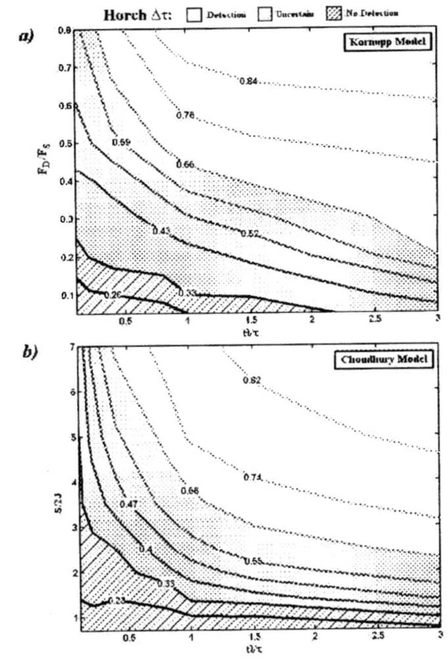

Figure 4. Application of Horch technique; a): Karnopp model, b): Choudhury model

3. APPLICATION OF THE DETECTION TECHNIQUE

3.1 Plant Data

The previously described technique has been applied on plant data obtained from one of the loops of a refinery industry. Examined data refer to a flow control loop, with evident presence of valve stiction and oscillation in controlled output PV. In Fig. 3, sets of recorded data in two different time windows are shown; in the upper part of the figure PV (thick line) and OP (thin line) are reported; in the lower part C_{XY} (dash line) between OP and PV is presented.

In the examined control loop Horch technique was applied choosing a variable number of oscillation with different time windows. The number of analysed oscillations seems not to influence the result and in most cases stiction is correctly detected ($\Delta\tau = 0.7 \div 0.85$); (recorded data and cross correlation showed in Fig. 3a). Using a different time window, again with a variable number of oscillations, results change and lie in the gray-zone ($\Delta\tau = 0.5 \div 0.6$), where no decision can be taken; (recorded data and cross correlation showed in Fig. 3b). Therefore it is important to realise why the presence of stiction can be masked and not detected by the technique. This objective has been addressed by analysing what happens for different extent of stiction and different process dynamics.

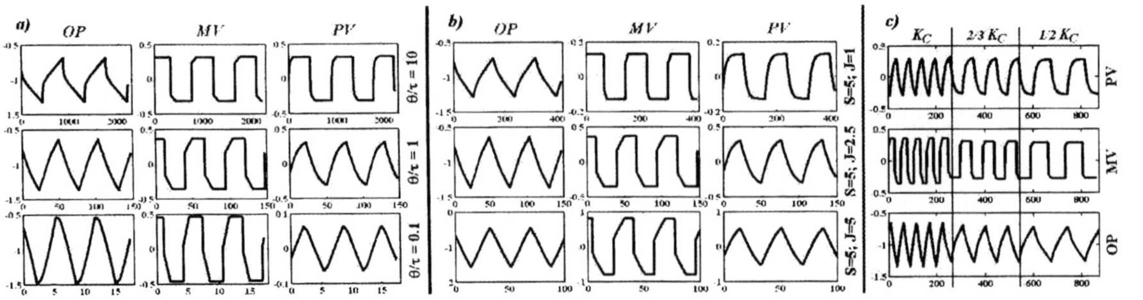

Figure 5. Wave shapes of OP, MV, PV in case of stiction by varying the system parameter: a) variations of θ/τ; b) variations of S/J; c) variations of the controller gain K_C

3.2 *Simulation of different cases*

To investigate previous point, Horch technique has been applied to situations of presence of stiction generated by two different stiction models (Karnopp and Choudury) for a first order plus time delay process dynamics. Results are reported in Figure 4a and 4b, expressed in terms of the index $\Delta\tau$; (analogous results can be obtained using $\Delta\rho$). Stiction parameters are: the ratio between dynamic and static friction (F_D/F_S, for the Karnopp model, Figure 4a) and the ratio between the dead zone and the jump (S/2J, for the Choudury model, Figure 4b). It can be seen that the presence of stiction is clearly indicated only for processes with large values of the ratio θ / τ and for stiction situations represented by large values of F_D/F_S (or S/2J). On the contrary, for smaller values, no decision can be taken or stiction is not indicated, even though it is present; this result does not depend on the adopted stiction model.

In particular, it is important to point out the effect of process dynamics: for low values of θ/τ (lag dominant processes), the presence of stiction can be hidden (or no decision can be taken), whatever the value of the stiction parameters. Therefore a further investigation of the phenomenon is worth, in order to realise how relevant process variables change with stiction and process dynamics and which of their characteristics do not change.

4. NARROWING THE GRAY ZONE

The analysis of stiction characteristics is based on the observation of different oscillations shape of OP, MV and PV variables. The effect of stiction, process and controller parameters is illustrated in Fig. 5. In Fig. 5a) parameters of stiction are fixed (S=5, J=2.5) and the ratio θ/τ is varied from 0.1 to 10; in Fig. 5b) process parameters are fixed (θ/τ=1) and the ratio S/J is varied ($J < S/2$; $J = S/2$; and $J > S/2$); in Fig. 5c) there is only a variation on the controller gain K_C ($\theta/\tau = 1$,$S/2J = 1$). Choudhury model has been adopted for all the simulations.

Main features can be pointed out as follows (further details in (Rossi and Scali, 2004):

(1) The manipulated variable MV maintains always typical square wave components; the almost perfect square wave shape, shown for θ / $\tau >> 1$, can be slightly modified to saw teeth shape, but the discontinuity on the derivative is maintained.
(2) The controlled variable PV, for decreasing values of S/J and θ/τ, changes its shape from square wave to triangular, much closer to a sinusoidal form.
(3) This effect is even more pronounced when higher order processes with underdamped elements are analyzed; (results reported in Figure 5 are for FOPTD dynamics).
(4) The effect of a change in controller parameters has the peculiarity of not influencing the oscillation shape, but affects the oscillation frequency; in Fig. 5c) a decrease of Kc from 1 to 2/3 to 1/2, causes a decrease of oscillation frequency from 0.116, to 0.067 to 0.044.

From points 2 and 3 it is evident that, basing the analysis on values of PV, it will not be always possible to obtain a sharp distinction between different behaviors, especially in the presence of process noise which tends to cover difference in wave shapes.

In a slight different scenario (relay identification for autotuning purposes (Luyben, 2001) and (Thyagarajan and Yu, 2003), wave shape similar to the ones reported in Figure 5 have been observed for output variables from a FOPTD process. The shape of signals is modified by process dynamics: this suggests the basic idea of the technique proposed here. Every significant recorded oscillation is fitted by using three different models: the output response of a first order plus time delay under relay control, a triangular wave and a sine wave.

The comparison of the error between real and fitted data can be used as an index of accuracy of approximation and then as an indication of the behavior observed in the analyzed variable: relay and triangular waves are associated with the presence of stiction, while sinusoidal shape with the presence of external perturbations.

The three approximating curves (respectively C_S for the sinusoidal wave, C_R for the relay wave and C_T for the triangular wave) are:

$$C_S = A_S \cdot sin(\omega_S t + \phi_S) \tag{7}$$

$$C_R = \begin{cases} A_R(1 - e^{-\frac{t-\phi_R}{\tau_R}}) & t \leqslant t_n \\ C_R^n - (A_R - C_R^n)(1 - e^{-\frac{t-t_n}{\tau_R}}) & t > t_n \end{cases} \tag{8}$$

$$C_T = \begin{cases} A_T(m_1 t + q_1) & t \leqslant t_n \\ A_T(m_2 t + (m_1 - m_2)t_n + q_2) & t > t_n \end{cases} \quad (9)$$

where C_R^n is C_R evaluated at $t = t_n$. There are three typologies of parameters: a parameter $n \in \mathbb{N}$; the parameters $\rho_{NL} = (\omega_S; \phi_S; \tau_R; \phi_R) \in \mathbb{R}$ used in *non linear* functions; the parameters $\rho_L = (A_S; A_R; A_T; m_1; m_2; q_1; q_2) \in \mathbb{R}$ used in a *linear* relation.

Consequentially the parameters are varied on three different levels for each curve: firstly, when requested by the approximating function, n is varied. Secondly, for each fixed value $\overline{n}$, ρ_{NL} parameters are varied with a non-linear optimization (using the *"fminsearch"* function in Matlab). Thirdly, for each fixed set $\overline{\rho_{NL}}$ the parameters ρ_L that generate the best-fit are calculated with linear regression techniques. From all the obtained curves the procedure chooses the three ones ($\overline{C_S}$,$\overline{C_R}$,$\overline{C_T}$) that generate the best-fit with smallest error.

5. RESULTS

5.1 Application on Simulations

The proposed technique has been tested on simulation in closed loop processes (PI-controllers, with Ziegler-Nichols tuning) and in the presence of stiction on the valve, by adopting Choudhury model. Process parameters (θ, τ) and friction parameters (S, J) are varied in the same range chosen for previous analysis (Fig. 4b), for a faster comparison of the old and new gray zones. Indicating with E_Sin, E_Rel and E_Tri respectively the mean square error between the curves C_S, C_R, C_T and the recorded data, two errors can be defined:

$$E_1 \triangleq min[E\_Rel, E\_Tri] \quad (10)$$

$$E_2 \triangleq E\_Sin \quad (11)$$

In the nominal case (no noise) it would be sufficient to impose $E_1 < E_2$. The presence of noise modifies the shape of oscillations and it can be possible to approximate sinusoidal waves with triangular waves (comparable errors); that would generate a false alarm, indicating stiction when it is not present. To avoid this problem, the proposed criterion becomes:

$$E_1 < E_2 \cdot K, \qquad with\, K < 1 \quad (12)$$

A suitable choice for the parameter K, in analogy with the parameter $\Delta\tau$ proposed by Horch, can be: $K = 2/3$. When this criterion is not satisfied, no decision can be taken on the basis of a simple analysis of recorded data, thus generating, unavoidably, a new gray zone. The study about the effect of the noise and overlapped disturbances is still in progress.

The application of the proposed technique, with the suggested criterion, brought results represented in Fig. 6. A logarithmic scale has been adopted for the x axis, in order to zoom in the low values range. It can be observed that a small zone gray zone remains for lag dominant processes ($\theta/\tau < 0.2$): in this range it is not possible to take any decision, also with the new technique; by a comparison with Fig. 4b, the extent of the new gray zone is strongly reduced. Therefore, without any additional knowledge about plant dynamics, it is not possible to detect the presence of stiction in all possible cases and to distinguish it from the presence of external sinusoidal perturbations.

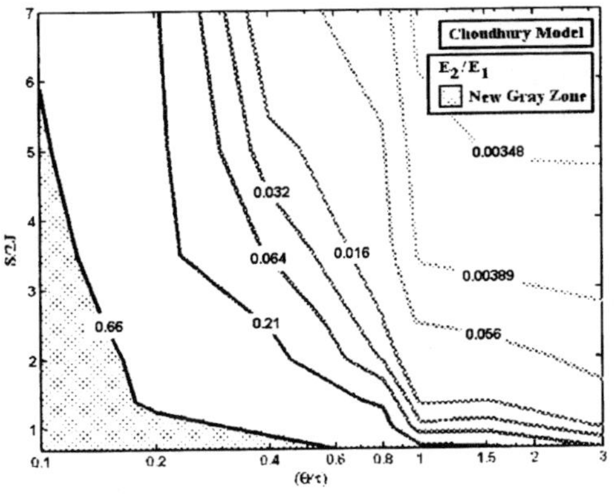

Figure 6. New gray zone with the application of the proposed technique (log x axis)

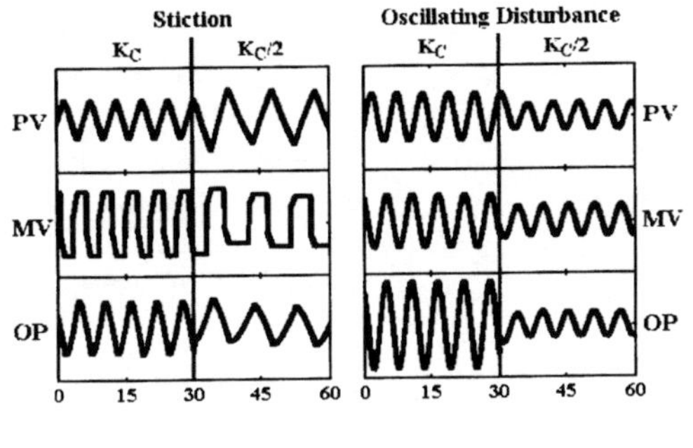

Figure 7. The effect of variations in K_C

Two possibilities are indicated here for a complete detection of stiction.

(1) By recalling that MV in the presence of stiction maintains the typical square wave shape, in the cases that MV is available (recorded by DCS), the gray zone is practically reduced to zero. So, the application of the proposed technique on this variable is able to distinguish correctly the sinusoidal wave (oscillating disturbances) from the square wave or the saw teeth wave (stiction).

(2) When MV is not available (and this is the most frequent situation), cases lying in the gray zone ($E_1 > K \cdot E2$), can be resolved by means of a plant test which decreases the controller gain K_C. In the case of oscillating disturbances, the frequency of the oscillation does not vary (the controller parameters influence only amplitude and phase). In the case of stiction, decreasing K_C causes a significant decrease in the frequency of the recorded oscillation. An example is reported in Fig. 7; this process has $\theta/\tau = 0.1$, so both the technique of Horch and the proposed techniques are in the gray zone. In case of stiction, when the controller gain decreases, also the oscillation frequency decreases ($\omega_1 = 1.25 \rightarrow \omega_2 = 0.68$), on the contrary, with oscillating disturbances it remains constant.

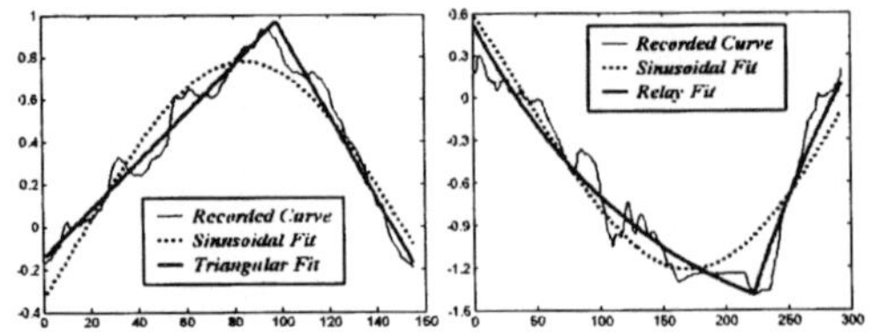

Figure 8. Example of applications on plant data

5.2 *Application on plant data*

The proposed technique has been applied on the same set of plant data, where the presence of stiction was known. Due to the presence of plant noise, oscillations can show some differences each other, so the technique has been applied on a minimum number of oscillations N_{OSC} equal to 5. The mean error obtained by the three approximating curves for the examined oscillations has been used to discriminate. The total number of oscillations (18) has been examined: values of the ratio $\overline{E_1}/\overline{E_2}$ are reported in Tab.1. It can be seen that for all the examined cases and chosen time windows, this ratio is always less than 2/3, to indicate the presence of stiction correctly. An example of the obtained best-fit for two different oscillations is showed in Fig. 8.

Table 1. Analysis of Plant Data

	$N_{Osc} = 6$	$N_{Osc} = 12$	$N_{Osc} = 18$
I set	0.46	0.50	0.46
II set	0.54	0.49	—
III set	0.39	0.46	—

6. CONCLUSIONS

The analysis carried out in this paper, about the influence of process dynamics and stiction parameters on the characteristics of the output variable (PV), puts into clear evidence that there is a continuous variations of basic properties of oscillating signals. Therefore the presence of stiction can be hidden according to process dynamics and no techniques for automatic detection of stiction can be successful in all possible cases, without any additional knowledge on the process.

These conclusions can be drawn from the application of Horch technique to plant data: stiction is not detected in some cases and simulation results show that there is a large gray zone, where no decision can be made.

The new proposed technique is originated by observing shapes of oscillation in the presence of stiction and compares errors obtained by means of different fitting curves on real data. Even though it is not able to detect stiction in all the cases, by the analysis of PV data only, it allows a significant reduction of gray zones. This is confirmed by the analysis of simulation results and application on plant data. The possibility of analyzing MV data would result in completely successful application; in the cases where no decision can be taken, it is suggest to perform a test on the plant, by decreasing the controller gain.

REFERENCES

Armstrong-Hélouvry, B., P. Dupont and C.C. de Wit (1994). A survey of models, analysis tools and compensation methods for the control of machine with friction. *Automatica* **30**, 1083–1138.

Choudhury, M.A.A.S., S.L. Shah and N.F. Thornhill (2003). A data-driven model for valve stiction. *IFAC-Adchem'03 Int. Conf., Hong Kong (HK), Preprints Volume I, session 2.2* pp. 261–266.

Ender, D.B. (1993). Process control performance: Not as good as you think. *Control Engineering* **40(10)**, 180–190.

Hägglund, T. (1995). A control loop performance monitor. *CEP* **3**, 1543–1551.

Hägglund, T. (1999). Automatic detection of sluggish control loop. *CEP* **7**, 1505–1511.

Hägglund, T. (2002). A friction compensator for pneumatic control valves. *Journal of Process Control* **12**, 897–904.

Horch, A. (1999). A simple method for detection of stiction in control valves. *CEP* **7**, 1221–1231.

Horch, A. (2000). Condition monitoring of control loops. *PhD. Thesis; Process Control Dept. (University of Stockholm- Sweden).*

Luyben, W.L. (2001). Getting more information from relay feedback tests. *Ind. Eng. Chem. Res.* **40(20)**, 4391–4402.

Rossi, M. and C. Scali (2004). Una tecnica automatica per individuare lattrito negli attuatori. *Automazione e Strumentazione, February* pp. 91–97.

Rossi, M., C. Scali and M. Amadei (2003). Development of a technique for performance evaluation of industrial controllers. *IFAC-Adchem'03 Int. Conf., Hong Kong (HK), Preprints Volume I, session 1.3* pp. 183–188.

Taha, Q., G.A. Dumont and M.S. Davies (1996). Detection and diagnosis of oscillation in control loops. *Proceedings of 35th Conference on Decision and Control, Kobe, (Japan).*

Thornhill, N.F. and D.E. Seborg (2002). New directions in control system performance monitoring. *T-Mo-A11, Session of 15th IFAC World Congress, Barcelona (Spain).*

Thyagarajan, T. and C.C. Yu (2003). Improved autotuning using the shape factor from relay feedback. *Ind. Eng. Chem. Res.* **40(10)**, 4425–4440.

Yu, C.C. (1999). Autotuning of pid controllers. *Springer Verlag.*

ELSEVIER
IFAC
PUBLICATIONS
www.elsevier.com/locate/ifac

SIMPLIFIED FIRST-PRINCIPLES MODELLING OF GLASS FURNACES FOR CONTROL PURPOSE

Auchet O.[1], Riedinger P.[2], Iung C.[3], Malassé O.[4]

*[1,2,3] CRAN - CNRS UMR 7039. Institut National Polytechnique de Lorraine.
2, avenue de la forêt de Haye 54116 Vandoeuvre-lès-Nancy, France. olivier.auchet@ensem.inpl-nancy.fr
[4] CER ENSAM de Metz –Technopôle 2000- 4, rue Augustin Fresnel, 57078 Metz CEDEX 3, France*

Abstract : Control of glass furnaces has known a great expansion in the last decades, particularly with model-based operation optimisation methods. The challenge is to get models covering the whole operating range and at the same time presenting a low computational load. The most widespread models today are data-based or issued from CFD techniques, and the need for accurate fast models is important. Our model based on first-principles and using the zonal approach shows a good compromise, and we will investigate its applicability to control purposes in further studies. *Copyright © 2004 IFAC*

Keywords: Model-based control, optimal control, physical models, multivariable control, Non-linear models.

1. INTRODUCTION

Operation of glass furnaces is becoming more and more delicate due to ever increasing social and environmental constraints coupled with classical energy costs minimizing and optimal operating constraints. Glass industry is an intensive energy consumer and is therefore severely controlled by the environment experts. To help the operators in controlling the furnace, performing control structures are developed (Carvahlo, *et al.*, 1997, Chmelar, *et al.*, 2001).

Among them, model-based methods have shown a great success due to the increasing computational power of computers. Knowledge of the system behaviour thanks to mathematical relations allows to optimise the control decisions (Carvahlo, *et al.*, 1996). One of the most widespread method is the Model Predictive Method (MPC), which consists in determining the input sequence that optimises the system evolution on a future horizon. Models used in control structures are widely disparate. From first principles computational fluid dynamic (CFD) models to empirical fuzzy logic learning systems, the robustness and performance of the control will depends on the accuracy of the model. On the other hand, the computational cost of simulating the model is also important to consider. More generally, two things are required for models for control purpose :

- To cover the furnace operating range to be controlled
- To present a low computational load

Therefore, when developing such a model, the compromise between accuracy and CPU time leads the choices. Time to build the model and return on investment period is also an important criterion to consider.

Empirical models are obtained thanks to heavy identification tests on actual plants, and the slow dynamics of glass furnaces make the tests campaigns very long (several weeks). Moreover, identification models refer to particular operating points and present therefore important deviation from reality when furnace operation changes. However, the advantage of empirical models is to allow very fast simulation. CFD models cover the whole range of furnace operation with great accuracy at the cost of heavy computational load. Their greatest advantage is to predict non measurable phenomenon and their principal drawback is long development time.

Today, research efforts are made to provide fast models for control purpose that are still accurate enough to represent relevant dynamics (Backx, *et al.*, 2002, Huisman, *et al.*, 2001). We are working on the simplified modelling of a glass furnace, based on first principles applied to a rather coarse mesh in the three parts of the furnace (combustion chamber, bath and walls). The advantages of such a model is to avoid long identification tests and CFD models heavy computational load. We will investigate in further studies whether or not this kind of model can be used for control purposes.

We present first the glass furnace and some control features associated to it, and we describe then our model.

2. THE GLASS FURNACE

2.1 *System description* (Tooley, 1961)

The glass furnace is the kernel of industrial glass producing lines. Depending on the type of glass produced, the technology varies but the principle is the same. Raw materials are loaded to the bath and thanks to heat release from the heating system, glass forming reactions occur. The molten glass is then mixed and homogenised due to natural convection streams. At the end of its residence time in the bath, the molten material can finally be used for further processing (blowing or floating).

The melting temperature of raw materials (69% SiO_2) is round 1500°K, and to achieve such temperatures in the bath, heat is obtained by oil or gas combustion in the combustion chamber and/or by electric Joule effect. The losses through walls make the furnaces overall efficiency low (50%). Glass furnaces differ mainly by their heating system (energy used) and their shape, both influenced by the type of product considered. We focus on regenerative or recuperative oil and gas fired furnaces. Regenerative systems stores the heat of combustion gases in huge refractory towers (regenerators) and use it back to preheat combustion air. The principle is to alternate the currents of cold air and hot gases in two regenerators. Recuperative systems are based on heat exchanger principle. The shape of the furnace depends on the type of glass produced, with following rule : the higher quality required, the longer the residence time needed in the bath, the longer the bath. Of course, the burners number and location will be influenced by the size of the melting zone.

2.2 *Control tasks* (Backx, *et al.,* 2002, Chmelar, *et al.,* 2001)

The control tasks inside a glass furnace are distributed on three different hierarchical levels (cf. table 1). The primary goal is of course the glass quality, which corresponds to purity from defects (blisters, cords, stones) in the molten glass ready for further processing. This control task is characteristic from glass industry, where as the other first level goals are classical industrial process control constraints. The overall control task is therefore optimisation of glass quality under optimal operating constraints.

Quality of molten glass depends on melting process and environment conditions. One distinguishes primary defects from secondary defects, where the firsts are consequence of a bad melting process due to bad reaction conditions, and the seconds come from perturbations of the reaction by the environment (reaction with flame, refractories exudation etc). So, quality control requires to create the optimal conditions in the bath for the melting reaction. Of course, every manipulated and controlled variable has to remain within specific limits.

Table 1 Hierarchical levels of glass furnaces control tasks

Level 1 : **Principal goals**	Glass quality maximisation	Pollutants production and emission minimisation. Thermal efficiency Maximisation	
	Furnace and refractory lifetime maximisation	Energetic consumption minimisation	
Level 2 : **physical control variables**	bath level, residence time, residence time distribution	Air-to-gas ratio, exhaust composition	Crown and bottom temperature profiles
	Atmosphere pressure and temperature		
Level 3 : **process control variables**	Batch charging system, bubbling, boosting	Fuel and air flows at each port. Preheating temperatures	Cooling air flows

The bath is composed of different zones (cf. fig. a). The batch zone (1) is where the raw materials containing all the ingredients needed for glass forming are introduced as wet packets into the bath. They float on the melt surface until the chemical reaction of melting occurs thanks to heat transfers. During melting, the newly formed glass sinks in the bath within a second zone (2) characterized by a great amount of remaining sand grains and bubbles. At this stage, the molten glass is not yet usable for forming and it needs to be homogenised and refined. Temperature gradients and additional systems in the bath create laminar natural convection streams that allow this homogenising to occur. During its residence time in the bath strongly depending on the convection streams pattern, molten glass will purify from sand grains and bubbles (3). The tank is designed such that molten glass in the last section, called buffer zone (4), is ready for end-product forming. We don't study the conditioning and forming sections, but great control challenges are also involved in them.

The physical phenomenon just described are distinguished as melting, mixing and homogenising, and fining/refining. These phenomenon involve other processes, as for example fining which is constituted of diffusion of gases and chemical reactions with the fining agents. All these phenomenon strongly depend on temperature and temperature gradients in the melt, and they need time and particular conditions of mixing to perform well. Their control is allowed by manipulating heat input to the bath (control task of level 2 in table 1), which is often related to the crown and bottom temperature profiles. Bubbling, cooling and raw materials pull rate also influences the different processes.

Control of combustion remains the greatest 2nd level control task in a glass furnace, because it is the major heat production source for the melting processes and for the driving of the flow. It involves maximum

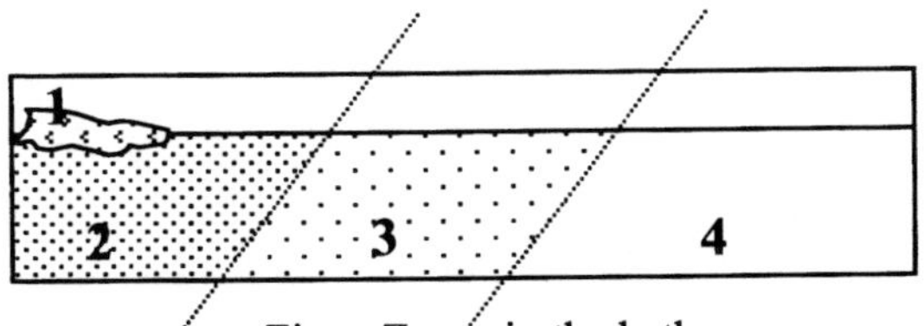

Fig a. Zones in the bath

flame temperature, total heat power and flame pattern. The manipulated variables are the air and fuel flows and preheating temperatures at each burner port. Many constraints directly linked on combustion process exist on pollutants emission, fuel efficiency, etc. Other technological constraints as refractories maximum temperature, direction of the flame are to be considered.

2.3 Control strategies (Chmelar, *et al.*, 2001)

The control of the above mentioned physical phenomenon in the bath is very complex for several reasons. There are strong couplings between them and also with the combustion process in the combustion chamber and the heat diffusion in the walls. This problem is typically a multi-objective optimisation problem of a highly-coupled non-linear multiple-input/multiple-output (MIMO) system.

To each level in the different control tasks table correspond adequate control strategies. Actually, a control task of one particular level will generate set-points for some control tasks located one level below. Level 1 control tasks use much operator knowledge and off-line simulations to determine the optimal operation conditions. Today tools include expert knowledge to allow automatic control at this level (at least help the operator in decision), and propose integrated control of the furnace. Level 2 control tasks are automated and state-of-the art control techniques propose very interesting strategies as will be shown below. At the bottom level, process variables are controlled in classical control loops as PID controllers.

PID controllers are still widely used today in 2nd level control tasks, as for example temperatures control in the furnace different sections. But the complex coupled dynamics, the long dead times as the multi-objective control tasks are hardly handled by PID strategies, better suited to single-input/single-output dynamics. Therefore, advanced process control seems to be the only solution, and multiple-input/multiple-output (MIMO) model-based operation optimisation methods have the greatest success. Model predictive algorithms optimise the future evolution of the plant according to multi-objective criterions by determining the adequate input sequence using the model of the system. The algorithm used is therefore known as prediction-optimisation. In the optimisation algorithm, the criterions of table 1 have to be translated into costs functions that will be minimized on a future horizon. To predict the system evolution, models are required and we later describe a dynamic model presenting an interesting alternative to today used models. There are namely first principles based CFD models and data-based models. The compromise that has to be considered with models for control lies between model accuracy and computational load.

CFD models cover the whole operating range with a great accuracy at the cost of very heavy computational load (Carvahlo, *et al.*, 1997). They often include many physical phenomenon, but cannot be simulated in real-time and are thus unsuited for control. Reducing techniques like Proper Orthogonal Decomposition allow to get fast models from CFD ones that can be used for control purpose ((Backx, *et al.*, 2002). Data-based models use actual plant measures to identify a model behaviour. They result in fast models but are hard to obtain due to long and painful tests campaigns. Moreover, they are only valid for particular operating points. The model we present here is based on first principles applied to a coarse decomposition of the three parts of the furnace (combustion chamber, bath and walls). It covers the whole operating range at a low computational cost, and hence is typically what is required. We will investigate its use for control purposes in further studies.

3. MODELLING

The model has the manipulated variables as inputs, and the controlled variables as outputs.

Table 2 : model inputs and outputs

Inputs	Outputs
• Fuel & oxidant flow at each burner port • Preheating temperatures and pressures of air & fuel flows at each burner port	• Combustion chamber temperatures points • Combustion chamber pressure points • Refractories temperatures points
• Raw material pull rate	• Bath temperatures points

The structure of glass furnace models follows the spatial location of the different physical phenomenon (Carvahlo, *et al.*, 1997). We therefore first decompose the furnace into three parts : the combustion space, the bath and the walls. This decomposition is done to handle the very different physical phenomenon occurring in these parts, in term of characteristic times principally. We thereby separately consider the turbulent reactive flow in the combustion space, the convection streams in the bath, and the heat transfers at and in the walls (see fig. b), and the coupling between these phenomenon.

The classical CFD approach (Ungan, 1996) is based on first-principles modelling, where conservation laws on energy, momentum, mass and mass fraction constitute the backbone of the model, and where closure models for turbulence, combustion etc. are used. Our model uses exactly the same concept, but to reduce the computational load, we have chosen to play on accuracy. The zonal method initially developed by Hottel (Hottel, *et al.*, 1958) for combustion chamber modelling considers volume

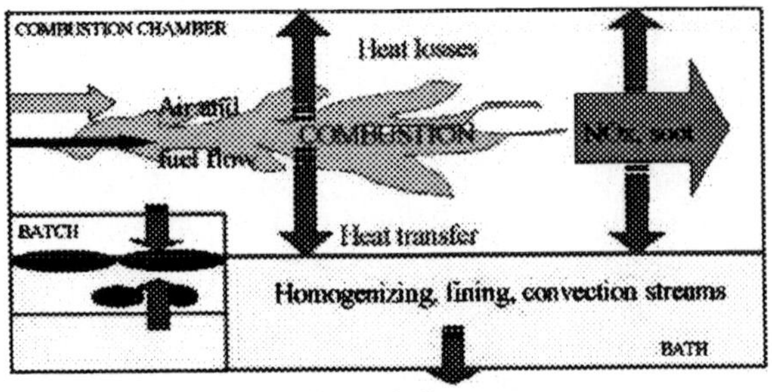

fig.b: Physical phenomenon

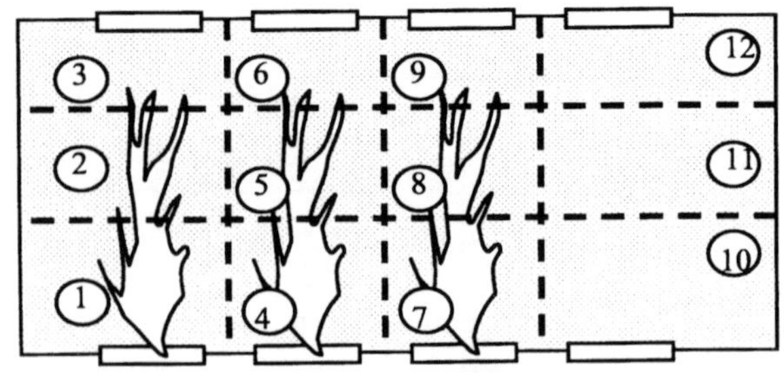

fig. c: 3 burners combustion chamber

and surface cells in which the classical conservation laws with closure models are applied. By choosing a rather big size for the cells, the number of equations decreases accordingly. Compared to several thousands nodes in classical CFD computations, our model treats between few tens and few hundreds cells and considers only the main physical phenomenon. The computation load is greatly reduced. Of course, the model has to show as little as possible deviation from reality. The challenge is to choose the zones in order to correctly represent the relevant phenomenon.

We motivate our research on positive experiences of leading glass modelling research centres using this approach in the combustion chamber and in the whole furnace (Goodson, *et al.*, 1979, Liu, *et al.*, 2001, Huisman, *et al.*, 2001).

3.1 Modelling of the combustion chamber

We apply the approach of TNO with its Rapid Combustion Model (Paarhuis *et al.*, 2000). The heat input profile from combustion chamber to the bath is of major influence on the melting, mixing and fining processes. This profile is manipulated through the fuel and air flows. It mainly depends on the flames, and this leads the spatial decomposition process.

The phenomenon to take into account are the turbulent flow, the heat transfers by radiation and convection. The flow is complex, with recirculation of gases above the flames, but we simplify considerably by considering it unidirectional in the chamber height. The direction is given by the flames and the exhaust ports location. We do a 2D spatial decomposition in the horizontal plan with burner sections divided into cells (typically between 3 and 10 cells per section). See the example of a three burners combustion chamber in fig. c.

The goal is to determine the heat release of the combustion gases in each cell. We have first to determine the *flow* in the cell, the *compositions* of the atmosphere and the associated *combustion reaction*. When this is done, we have to compute the *heat transfers* by radiation and convection to the load and to the surrounding (losses).

Modelling of the reactive flow

In each cell k of a combustion section, we have to compute the density ρ_k, the mass flow rates Q_k, the mass fractions Y_k^j of each specie j considered (specified later), the enthalpy h_k, and the temperature T_k, all assumed uniform. We assume the pressure known and uniform, as the mass flow rates at the entrance of the furnace. To compute these (4+s)*n unknowns for a n zones section in purely convective regime and a mixture composed of *s* chemical species, we use the following set of equations discretized with the one dimensional finite volume approach in the n volumes.

ρ : ideal gas law : (n equations)

$$\rho = \frac{P}{\sum_j \frac{Y_j}{M_j}.R.T} \tag{1}$$

Q_k : mass conservation : (n equations)

$$\frac{\partial \rho}{\partial t} + \frac{\partial \rho u}{\partial x} = 0 \tag{2}$$

Y^j : mass fraction transport (s*n eq.)

$$\frac{\partial \rho Y^j}{\partial t} + \frac{\partial \rho u Y^j}{\partial x} = \dot{\omega}_j \tag{3}$$

h : energy conservation (n equations)

$$\frac{\partial \rho h}{\partial t} + \frac{\partial \rho u h}{\partial x} = \dot{S} \tag{4}$$

T : iterative computation using enthalpy equation : (n equations)

$$\sum_{i \equiv all\ species} Y_i \left(\Delta h_{fi}^0 + \int_{T_{ref}}^{T} C_{pi} dT \right) = h \tag{5}$$

Where M_j is the molar mass of specie j, C_{pj} the specific heat of specie j, and R the gas constant. $\dot{S}$ is the energy sink/source term. Δh_{fi}^0 is the standard formation enthalpy of specie i.

Let us explicit the combustion related source term appearing in (3), which is the production/destruction term of specie j ($\dot{\omega}_j$). We consider combustion of hydrocarbon (C_mH_n in general, but we focus on methane) with oxidant (oxygen and nitrogen). We use the Single Chemically Reacting System (SCRS) assumption (irreversible forward single-step reaction kinetic) together with the *mixed-is-burned* assumption (infinitely fast chemistry). Due to the SCRS hypothesis, we only consider CO_2, H_2O, unburned and inerts in the combustion products. The regime of the combustion has to be determined by studying the richness of the mixture in each volume and the reaction rate is proportional to the fuel flow according to the mixture richness.

Due to the mixed-is-burned assumption, combustion would occur in the first cell of a section only, inducing short flames. It doesn't well represent the reality, where the flows of fuel and oxidant are mixed in the turbulent eddies, creating long flames. To represent this behaviour, we introduce in each cell a weight $(\alpha_k < 1)$ on the reaction speed $(\alpha_k \dot{\omega}_k)$. The coefficients α_k represent the influence of the mixing flow on the combustion. We tune the coefficients to best fit reference temperature fields.

Modelling of convective and radiative heat transfers

The source term $\dot{S}$ in energy equation corresponds to heat transfers by convection and radiation in the reactive flow.

Convection

Convective heat transfers between two adjacent elements i and j at temperatures T_i and T_j (flow cell i with flow cell j or with wall element j) presenting a common area A_{ij} is very simply computed by :

$$q_{ij} = h_{ij} \cdot A_{ij} \cdot (T_i - T_j) \tag{6}$$

At the current time, we estimate the forced convection coefficient h_{ij} (around 15 W/m²/K for flow/wall exchanges f.e.) but we will include an empirical method.

Radiation

Due to high temperatures, radiation is preponderant and particular attention is therefore paid for its modelling. Each element of the decomposition emits, reflects and absorbs radiation, and we have to take into account the presence of strongly emitting and absorbing particulates such as soot in the chamber atmosphere. The methods best suited to zonal models are those based on exchanges areas (Hottel, *et al.*, 1958) which use influence factors (called view factors) between all elements of a spatial decomposition (volumes and surfaces). The influence factors quantify the distribution of the power emitted by each element to all other elements of the decomposition. The greatest drawback that makes this method unsuitable for applications requiring precision is the grey body assumption. This leads to such great errors that one often prefers the computationally intensive resolution of radiative transfer equation. However, for our purpose, the exchange areas method is the more adequate.
We use the Gebhart formalism (Gebhart, 1971) which considers total absorption influence factors. In the computation of influence factors between two elements i and j, the method takes into account the multiple path of a ray emitted by i and travelling in the enclosure by mean of successive reflexions at walls until it reaches element j.

The complexity of exchange areas method is to compute the view factors because there are volume and surfaces integrals. For surface-surface view factors, we found simple analytical relations based on contour integration. For volume-volume and volume-surface view factors, we had to compute them by using the method described in (Emery, *et al.*, 1987). This method extends a scan line algorithm, based upon surface-surface radiation, to the computation of surface-gas an gas-gas radiation transmittances.

Finally, the radiant heat fluxes (W) of all elements are obtained thanks to the net radiative balance in a very simple vector form :

$$q_{rad} = BW - W \tag{7}$$

Where the second term on the right hand side is the emitted powers vector ($W \equiv \varepsilon T^4$), and the first term is the distribution of emitted powers between all elements taking gas absorption into account (B is the so called Gebhart matrix containing total absorption factors, taking multiple reflection into account).

The preceding relations allow to compute the reactive flow, the heat release by combustion and the preponderant heat transfers by radiation. This constitutes a boundary limit for the bath and walls models, that we describe now.

3.2 Modelling of the bath

The phenomenon occurring in the bath are numerous and complex. We have described the principal in the control tasks description part : Batch melting, mixing/homogenizing, fining/refining are strongly dependent on temperatures and temperatures gradients fields. Other processes important for the dynamic of the bath, as bubbling and gas releasing, are also depending on temperature, but they are not taken into account in the modelling,. The convection streams are driven mainly by temperature gradients. The temperatures are therefore the controlled variables in the bath. We use again a zonal decomposition in the bath, and we get a temperature point per cell. The model is articulated around mass and energy balances in each cell, considered as a well-stirred reactor. The flow pattern corresponds to mass exchanges between cells. We assume it known, and take typically two circulation rolls due to natural convection (cf. fig. d). It is possible to modify it during simulation or control. The source terms in the energy balance are diffusion and radiation heat transfers. Radiation is taken into account thanks to the Rosseland method, adequate because the medium is optically thick. An additive term is included in the heat conductivity for the radiation inside the bath and in the convection coefficient for the radiation to the walls. We neglect the effects of chemical reaction

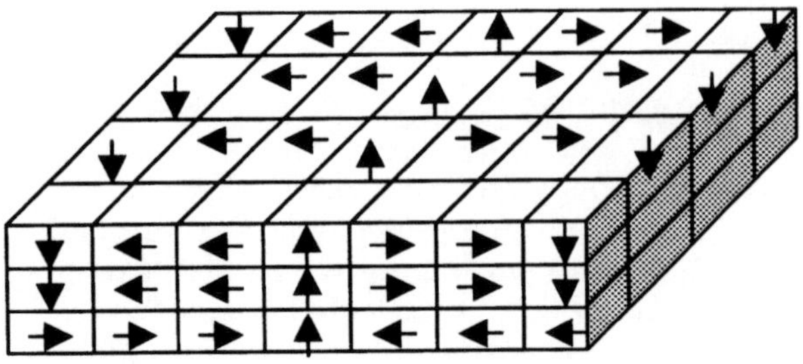

fig. d : coarse spatial decomposition in the bath

and refining in the glass. For simplifying, we consider the glass level constant.

3.3 Modelling of the walls

Refractories surface temperatures are very important to monitor, because they influence greatly the heat transfers to the bath. Crown and bottom temperature profiles are manipulated variables for bath temperature control. These temperatures have to remain in acceptable limits to avoid glass secondary defects.

The phenomenon to consider are the diffusion of heat in the walls, and convective and radiative transfers at the sufaces. We use the method of thermal quadrupoles (Maillet, *et al.*, 2000), which is based on time-Laplace transforming of diffusion equation and writing in a quadrupole form of the transfers of the temperature/thermal-flux vector in a flux tube. This formalism coupled with numerical inversion of Laplace transforms yields interesting results in term of rapidity, and the method is simple to implement.

In our case, the decomposition into volumes of the combustion chamber or bath induces a spatial decomposition of the walls into wall elements. We consider each wall element as a two layers slab (refractory and insulation) in 1D-diffusion and we compute the correponding transfer matrix of temperature-flux between internal and external sides. Knowing the temperatures inside and outside the furnace, as the radiative and convective flux at each surface of the wall element, we can determine the temperature at every point of the wall.

4. CONCLUSION

Control of glass furnaces is a so complicated task that advanced methods have to be used. Model-based methods are the more successful and the challenge is to get models showing short computation times for the best accuracy. We have proposed a model based on first principles applied to a coarse zonal decomposition.

In future studies, we will improve the modelling of the bath by testing different simple models, and we will test the applicability of the model to control. Its low computational load makes it anyway interesting for other applications like simulation for operator training.

REFERENCES

Backx, T.C., Huisman, L., Astrid, P., Beerkens, R. (2002) Model-based control of glass melting furnaces and forehearths : First principles-based model of predictive control system design. In : *63rd conference on Glass Problems.* Fawett Center for Tomorrow. Ohio State Univ.

Carvahlo, M.D.G., Xeira, A.P., Nogueira, M. (1996) Model-based operation optimisation of glass melting furnaces. *Int. Glass Journal.* **Vol 89.** p 31-37

Carvahlo, M.G, Speranskaia, N., Wang, J., Nogueira, M. (1997) Modeling of glass melting furnaces : applications to control, design, and operation optimization. In : *Advances in Fusion and Processing of Glass II.* vol. 82 (*Clare, Jones (Ed)).* p.109-135. Ceramic Transactions.

Chmelar, J., Bodi, R., Muysenberg, E. (2001). Supervisory advanced control of glass melters and forehearths by expert system. In : *Proc. Int. Congr. Glass.*, **Vol. 1**, p. 247-254.

Gebhart, B. (1971). *Heat Transfer.* 2nd ed. McGraw-Hill.

Goodson, R.E., McConnell, R.R. (1979). Modelling of glass furnaces design for improved energy efficiency. *Glass technology.* **Vol. 20**, p. 100-106.

Emery, A.F., Johansson O., Abrous A. (1987). Radiation heat transfer shapefactors for combustion systems. *Fundamentals and applications of radiation heat transfer.* **HTD-Vol. 72**, p. 119-126.

Hottel, H. C. and Cohen, E. S. (1958). Radiant heat exchange in a gas-filled enclosure: allowance for non-uniformity of gas temperature. *AIChE Journal.* **Vol. 4, No.1**, p. .3-14.

Huisman, L., Beerkens, R., Backx, A. (2001). Model based predictive control of glass melting furnaces Department of chemical engineering, Eindhoven University of Technology.

Liu M.S., Choi C.K., Leung C.W. (2001). Startup of oil-fired furnace – the smoothing Monte Carlo model approach. *Heat and Mass Transfer.* **Vol 37**, p. 449-457.

Maillet, D., Andre, S., Batsale, J-C., Degiovanni, A., Moyne, C. (2000). *Thermal quadrupoles.* Ed. John Wiley & Sons Ltd.

Paarhuis, B., Lankhorst, A., Riepen, M., Velthuis, J-F-M (2000). A rapid simulation model of the combustion space of glass melting furnaces. Technical report, TNO TPD.

Tooley, Fay V. (1961). *The handbook of glass manufacture.* Ashlee Publishing Co., Inc.

Ungan, A. (1996) Numerical simulation of glass melting furnaces : a review. In : *Proc. of Int. Symp. on Glass Problems. Vol. 1.* Istambul, Turkey. p. 351-366.

ELSEVIER
IFAC
PUBLICATIONS
www.elsevier.com/locate/ifac

INTEGRATING PROCESS INDICATORS WITH MONITORING METHOD HYBRIDS

Sirkka-Liisa Jämsä-Jounela, Tiina Komulainen

Laboratory of Process Control and Automation, Helsinki University of Technology, P.O.BOX 6100, FIN-02015 HUT, Finland
Phone: +358-9-451 3859, Fax: +358-9-451 3854, email: tiina.komulainen@hut.fi

Abstract: In this article the benefits of process monitoring are discussed. The different aspects of data-based monitoring methods and their combinations are reviewed. The role of process indicators is discussed, and the concept of combining a monitoring method library with the process indicators is presented. Finally, the benefits of process indicators and monitoring method hybrids are demonstrated with five industrial process monitoring applications created by the Laboratory of Process Control and Automation, Helsinki University of Technology. *Copyright © 2004 IFAC*

Keywords: monitoring, process indicators, industrial applications.

1. INTRODUCTION

Increasing international competition in the process industries is emphasizing the importance of product quality management; the quality of production is increased by better monitoring and control of the processes. According to several studies, inadequate management of abnormal situations causes annual losses of 20 billion dollars for the petrochemical industry in the USA. This, together with many other similar estimates, has led to extension of the field of diagnostic methods during the last decade. Since then, hundreds of successful applications of different monitoring methods have been reported. Diagnostic methods have proved to be useful and effective in industrial use, for instance in the chemical, mineral and metal, pulp and paper industries. Online process monitoring with fault detection can provide stability and efficiency for a wide range of processes. The current challenge is the automation of abnormal event management using intelligent systems, thereby providing operators with assistance in the most pressing area of need. This has also been viewed as the next major milestone in control systems research by the people working in the process industries. (Lennox and Sandoz 2002)

The diagnostic methods can be divided into model-based and data history-based methods. In the model-based approaches a priori knowledge about the process is needed, whereas in process history-based methods only the availability of a large amount of historical process data is the limiting factor.

The problem in model-based monitoring methods is that it is not always possible to construct dynamical process models that describe the process to an adequate degree of precision. Therefore most of the industrial applications are process history-based methods. The process history-based approach often lacks the ability to capture the process features and, in order to achieve good results, complex methods are required when using direct process measurements.

There are two ways to complement the weaknesses of the data-based monitoring methods. First, different monitoring methods possessing the desired properties can be combined, and secondly, the data-based and model-based approaches can be combined using process indicators as the variables of the monitoring applications.

In the following chapters the different monitoring methods and process indicators are discussed. A concept in which these two approaches are combined in advance in order to develop an operator support system is then presented, and finally some industrial cases using this concept are briefly reviewed.

2. COMBINING MONITORING METHODS

According to Dash and Venkatasubramanian (2000), diagnostic methods can be divided into model-based and process history-based methods, as shown in Figure 1. A fundamental understanding of the functionality of the studied process is necessary for model development in model-based methods. The process history-based approach, which is especially suitable for process monitoring purposes, requires a large amount of data in order to capture and model the features of the process. The history-based models can be subdivided into qualitative and quantitative models. The basis of qualitative models consists of rule-based and trend modeling methodologies, whereas the quantitative methods are divided into statistical and non-statistical, neural networks based on pattern recognition models (Venkatasubramanian et al. 2003).

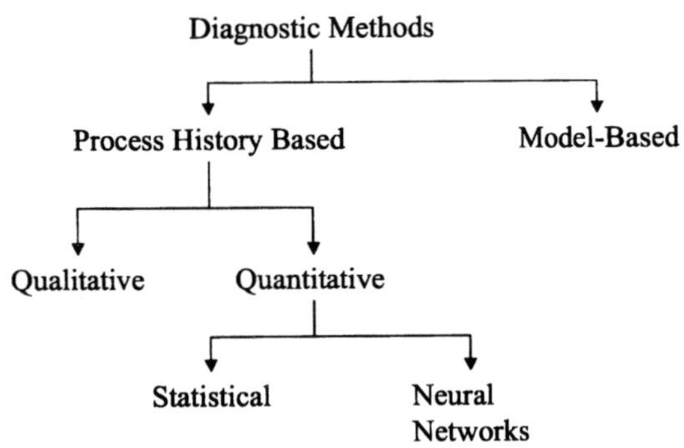

Fig. 1. Categories of the diagnostic methods. (Dash and Vankatasubramanian 2000)

Common features of the statistical methods used are their ability to reduce correlations between variables, compress data, and reduce the dimensionality of the data. These characteristics enable efficient extraction of the relevant information and analysis of the data. The most important statistical monitoring methods are based on principal component analysis (PCA) and partial least squares regression (PLS) (Wold et. al. 2001). Dynamic methods of PCA and PLS consider the dynamic nature of the monitored process and analyze both cross-correlation and auto-correlation. The dynamic methods are especially suitable for continuous processes with long time delays and varying throughputs on process variables (Ku et al. 1995, Chen et. al. 1998). Recursive methods for PCA and PLS have been proposed by Li et al. (2000) and Qin (1998). The recursive methods are especially suitable for time-dependent processes with slow changes. Multi-scale principal component analysis (MSPCA), a combination of PCA and wavelet analysis, removes the autocorrelations of variables by means of wavelet analysis, and eliminates cross-correlations between variables with PCA (Misra et al. 2002). The method is suitable for processes with auto-correlated measurements and time-varying characteristics. Nonlinear principal component analysis (NLPCA) is a combination of neural networks and PCA. Dong and McAvoy (1996) proposed an NLPCA method, which integrates a principal curve algorithm and neural networks. The idea of this method is to fit curves instead of lines to the data with the help of a feedforward network.

Neural network architectures are usually divided into three categories: feedforward, feedback and self-organizing networks. Neural networks are the most applicable to classification and regression problems, which do not need perfect precision. The availability of large amounts of data is especially important.
The self-organizing map, introduced by Kohonen (2001), is an unsupervised neural network that has been compared to NLPCA, because it adapts to the structure of the data, and the weight of the neurons tends to set the densest regions of the data and form an approximation of a curve fitted to the data. A neural net based on adaptive resonance theory differs fundamentally from a self-organizing map in the fact that the size and shape of the map are not determined beforehand. The ART map has many modifications, including combinations of ART maps, like ART3 and ARTnet, and hybrids of ART maps and fuzzy logic in fuzzyARTMAP (Wienke et al. 1996, Rallo et al. 2002). Radial basis function networks, introduced by Leonard and Kramer (1991), fit input data to radial basis functions, whereas traditional feed forward networks usually compare input signals to data vectors. RBFN is suitable for fault diagnosis.
Wavenets are combination of wavelets and neural networks with hierarchical multiresolution learning (Bakshi and Stephanopoulos 1993). This type of neural network is especially suitable for low-dimension dynamic fault diagnostic problems. (Zhao et. a. 1998)
Since no single method has all the desirable features, combinations of the methods have become more common in monitoring applications. The wealth of different monitoring methods is stunning, and selecting suitable methods for industrial problems requires plenty of time and considerable resources. These facts have lead to a situation in which the industrial applications far too often rely on mature methods instead of more sophisticated new approaches. Therefore, a systematic approach for the method choosing process would be necessary. In Table 1 some of the statistical, neural network and hybrid methods are compiled and their theoretical and practical aspects are compared. Compiling an open source extended table of monitoring methods might increase the applications of new monitoring methods in the industry. The theoretical aspects listed are fault detection and diagnostic abilities, robustness to missing measurements etc., immunity to signal noise, accuracy of the results, small amount of modeling data required, capability to handle multivariable data, amount of pre-processing required, and the dynamic, trend separation, nonlinear and adaptive properties, and finally an ability to detect novel states. The practical aspects of usability in industrial applications are: ease of implementation, ease of maintenance, visuality and simplicity of computation.

Table1: The qualities of different monitoring methods.

		Theorethical aspects													Usability			
		FAULT DETECTION	FAULT DIAGNOSIS	ROBUSTNESS (MISSING MEASUREMENTS)	NOISE IMMUNITY	ACCURACY	MODELLING DATA LOW	CAPABILITY WITH MULTIVARIABLE DATA	PREPROCESSING NECESSARY	DYNAMIC	TREND SEPARATION	NONLINEAR	ADAPTIVE	NOVELTY DETECTION	EASE OF IMPLEMENTATION	EASE OF MAINTENANCE	VISUALITY	SIMPLICITY OF COMPUTATION
STATISTICAL METHODS																		
PCA	principal component analysis	x	-	x	x	x	x	x	x	-	-	-	-	-	x	x	x	x
PLS	partial least squares	x	-	x	x	x	x	x	x	-	-	-	-	-	-	x	x	x
DPCA/PLS	dynamic	x	-	x	x	x	x	x	x	x	-	-	-	-	-	-	-	x
RPCA/PLS	recursive	x	-	x	x	x	x	x	x	-	x	-	x	x	x	x	x	x
NEURAL NETWORK METHODS																		
SOM	self-organizing map	-	x	x	x	-	-	-	x	-	-	x	-	-	x	-	x	-
RBFN	radial basis function	-	x	x	x	-	-	-		x	-	x	-	-	x	x	x	x
Wavenet	wavelet+neural network	-	x	-	x	x	-	-	x	x	x	x	-		-	-	x	-
ART-NN	adaptive resonance theory	-	x	-	x	-	-	-	x	-	-	x	x	x	x	x	x	x
HYBRID METHODS																		
NLPCA/PLS	nonlinear	x	-	x	x	x	-	-	x	-	-	x	-	-	-	-	x	-
MSPCA/PLS	multi-scale (wavelet)	x	-	-	x	x	-	-	x	x	x	x	-	x	-	-	x	-
DNNPCA/PLS	dynamic, nonlinear	x	-	x	x	x	-	-	x	x	-	x	-	-	-	-	-	-
FuzzyARTMAP	fuzzy logic + ART	-	x	x	x	-	-	-	x	-	-	x	x	x	-	x	x	-

3. COMBINING PROCESS INDICATORS WITH MONITORING METHOD LIBRARY

3.1. Process indicators

Model-based approaches are more suitable for capturing the basic characteristics of industrial processes than pure data-based methods using direct process measurements. Since it is often impossible to construct dynamical or mathematical models to the desired precision, data-based (process history based) methods are usually used. The nonlinearities and dynamical properties of the processes lead to the development of methods of ever-increasing complexity, while a combination of model-based and data-based methods would result in simpler computing and better understanding of the method. Relatively little research is being carried out on how to identify the right variables for process history-based methods and how to capture the essence of process knowledge.

The approach used here is based on process indicators, which are modeling process characteristics. The process characteristics can be captured by means of model-based computed variables that approximate, for example, enthalpy or the mass balance, and variable relations such as differences between temperatures or flow measurement ratios. The use of variables that are suitable combinations of measurements are better able to capture process non-linearities than complex monitoring methods. This ability obviously improves the identification capabilities of most data-based monitoring methods.

These process indicators have to be designed separately for every process, but the idea is universally applicable. Modeling the process nonlinearities with simple relationships and combining these with production goal specific indexes provides better understanding of the overall process.

3.2. The combination concept

The idea is to construct a general scheme of a combination of the monitoring method library and the process indicators, and to use these modules when constructing a monitoring application for a specific process, as shown in Fig. 2.

The monitoring method library module consists of the data-based monitoring methods and the table of theoretical and practical aspects of the methods. The process indicators module consists of higher level modeling rules for developing process characteristic indicators.

For the target process, the suitable monitoring methods and their hybrids are first chosen on the basis of the table of method properties. Then, the process characteristics are specified and modeled. Finally, these are combined together in advance to develop a monitoring application that is a part of the advanced operator support system of the plant.

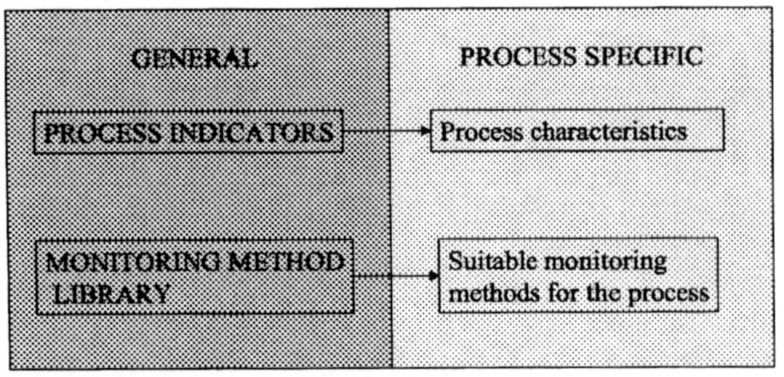

Fig. 2. Combination of process indicators and monitoring method library.

4. INDUSTRIAL EXPERIENCES

In this chapter the industrial applications, which have led to the development of the concept described above, are presented briefly. In the applications, monitoring methods with computed variables describing the process characteristics, have been used. The applications were done by the Laboratory of Process Control and Automation, Helsinki University of Technology in collaboration with industry. The application areas are refining, metals and minerals, and the pulp and paper industries. First, the online monitoring system developed for dearomatization process, is described and discussed in more detail. Then, the other applications are briefly described.

4.1. An online monitoring system for dearomatization process

Komulainen et. al. (2003) have described an online monitoring system for a dearomatization process. The aim of the monitoring system was first, to classify the process state, second, to detect whether the flash point and distillation curve analyzers were working correctly, and third, to provide a reliable prediction of the analyzer measurements. The application used dynamic PLS with the input combination of direct process measurements and process indicators.

The purpose of the dearomatization process is to remove aromatic compounds from the feedstock by hydrogenating them in a continuous process. The process consists of two trickle-bed reactors with packed beds of catalyst, a distillation column, several heat exchangers and separation drums and other unit operations.

Due to the complex, strongly nonlinear, nature of the process, a nine-stage systematic approach was introduced for the development of the online monitoring system. First, the direct process variables that affected the flash point and distillation curve of the product were determined, and the selected variables were then time-lagged. Next, process indicators were identified and the corresponding computed variables were created on the basis of the time-lagged, direct process variables. The combination of direct process variables and computed variables with the strongest influence on the flash point and distillation curve, was then selected. After the combination of variables had been created, the most suitable method for monitoring proved to be dynamic PLS. Different models for monitoring were developed and tested. The offline test was performed with the most suitable models. When the results of the offline tests were satisfactory, an online-monitoring system was developed and tested. Finally, the results of the online test were analyzed.

The process variables affecting the analyzer variables were selected on the basis of process knowledge and correlation analysis. The selected variables included: temperature measurements from reactors and distillation column, direct and recycle flow measurements from several points of the process, pressures and their set points, level measurements, and temperature difference and enthalpy from the reboiler of the column.

The process indicators were identified from the basic equations of the dearomatization process. The characteristics of the distillation especially were investigated, because distillation has a strong effect on the flash point and the initial point of the distillation curve. Based on the selected process indicators, the computed variables were constructed from 36 time-lagged process measurements as follows:

- The heat generated in the reactors, and the heat divided by the flow of the fresh feedstock to the first reactor.
- Several temperature differences in the reactors and the distillation column.
- Flow ratios were determined between feed to the process, feed to the distillation column, distillate, reflux, and product.
- Variables describing enthalpy were represented by simple products of the flow rate and temperature. Instead of direct enthalpies, the enthalpy ratios, i.e. one enthalpy variable divided by another, were used.
- The enthalpy ratios for the distillation column were computed between: feed, distillate, reflux and bottom product
- The process measurement describing the enthalpy of the reboiler, divided by the flow of the feed to the column, was also calculated.

A total of 23 computed variables were created. The combination of direct process measurements and computed variables was formed on the basis of the correlations between the variables and the analyzer variables. The final combination contained 21 direct, time-lagged, process measurements and 23 computed variables, which were constructed from the time-lagged process measurements.

Choosing the most suitable monitoring method the following requirements were specified; The method should be able to handle several dozen variables and the output of the method should give accurate predictions of the analyzer values. The method should be able to distinguish between process transitions at feedstock changes, and the process and analyzer faults. The method should also be able to distinguish between malfunctions of the analyzers and process faults as early as possible, and to give the operator an

alarm. A method that could be applied online and based on process history was preferred.
The need for precise prediction ruled out neural nets. The clear relationship between process and computed variables, and the quality variables of the product, led to the use of PLS-based methods. Owing to the dynamic nature of the process, dynamic partial least squares regression (DPLS) was selected. DPLS fulfilled all the requirements, and the relationship between the process variables and the analyzer variables made the structure of the method very simple.
The time-lagged direct process variables and the computed variables formed the input block, and the output block consisted of the analyzer variables. The predictions of the analyzer variables were made using the input block. The residuals between the predicted and the real analyzer values indicated whether the analyzers were functioning correctly. A possible process fault was detected by measuring the Hotelling T2 value of the input block.
The effectiveness of the computed variables was tested by constructing two DPLS models with the original 35 process variables and comparing the results of these models to the results of the DPLS model with computed variables. The teaching data set was the same for all the DPLS models. Testing data set included only data from normal process state. The DPLS model with computed variables had 5 latent variables, and the DPLS models without computed variables had 3 and 5 latent variables. The results with computed variables are excellent for all the analyzer variables where as the models without computed variables fail to classify the normal states correctly for distillation curve variables (Table 2).
During normal process states, the DPLS model with computed variables was used to predict the quality of the final product. The results of the offline test were encouraging; 96 – 99 % of the normal states and 67 – 97 % of the fault states of the analyzers were classified correctly. An online monitoring system was developed and tested in the Naantali Oil Refinery, Finland, for a time period of 144 hours. The monitoring system classified the two feed type changes correctly as normal states, and gave an alarm for an abnormal process state during the disturbance.

Table 2: Comparison of the results with and without computed variables

	With CVs	Without CVs	Without CVs
Latent variables	5	5	3
Captured variance			
Process var.	88.3 %	89.7 %	77.6%
Analyzer var	99.7 %	99.8 %	98.0 %
Correctly detected normal states			
Flash point	100 %	96 %	100 %
Distillation 0%	99 %	0 %	1 %
Distillation 5%	100 %	0 %	59 %
Distillation 10%	100 %	3 %	95 %

4.2. Other industrial applications

Kämpjärvi et al. (2003) developed an online monitoring system, which used a combination of PCA, SOM and RBFN to detect and identify faults. The inputs of the system consisted of direct process measurements and process indicators. The process indicators included relationships between the process variables and model-based computed variables. The main task of the system was to determine when the composition measurements of the online analyzers could be relied on for automatic control purposes. The system was successfully tested online at the Borealis ethylene plant in Porvoo, Finland. The system classified correctly 99,6 % normal states and 77,0 % fault states.
An application of SOM for monitoring the Outokumpu Harjavalta flash smelter was described by Jämsä-Jounela et al. (2003). The system detected equipment malfunctions and monitored process states using SOM in conjunction with heuristic rules. Model-based computed variables had a decisive influence on the success of the application.
Laine et al. (2000) reported an operator support system for ore type identification in the Hitura Mine, Finland. The monitoring system used SOM for on-line identification of the feed ore type and a knowledge database that contained information about how to handle a determined ore type. The key for the successful implementation was the right selection of variables for the ore type determination. The results of the project were remarkable, a commercial product and several industrial implementations.
Rantala (1999) developed a supervisory process monitoring system for the electrolytic refining of copper. This process is characterized by exceptionally large time constants and strong correlations between the variables. The monitoring application was based on a combination of principal component analysis and a self-organizing map. The classification ability of the monitoring system was excellent. The conditions producing high and poor quality copper cathodes were clearly separated in the SOM.
A combination of PLS and computed variables has also been applied successfully to pulp quality control.

5. CONCLUSIONS

Choosing the most suitable monitoring methods and the correct combination of direct process measurements and process indicators play an essential role in the successful development of an industrial process monitoring system. The challenge is to develop a comprehensive monitoring method library and a tool with the ability to choose the most suitable methods given the process description. Having the right method is, however, not enough; the process characteristics also have to be taken into account.
There are already many applications that employ the approach of computed variables modeling the process characteristics, but the research in this field is

minimal. The process indicators as a part of the supervisory process monitoring system is now under research in the Laboratory of Process Control and Automation, Helsinki University of Technology.

REFERENCES

Bakshi, B.R., Stephanopoulos, G. (1993), Wavenet: A multiresolution hierarchical neural network with localised learning, *AIChE Journal*, **39**, pp. 57 - 81.

Bergman, S., Sourander, M., Jämsä-Jounela, S.-L., Monitoring of an industrial dearomatisation process, *In IFAC' 02 15th Triennal World Congress*, Barcelona.

Dash, S., Venkatasubramanian, V. (2000), Challenges In the Industrial Applications of Fault Diagnosis Systems, *Computers & Chemical Engineering*, **24**, pp. 785 – 791.

Chen, G., McAvoy, T.J., Piovoso, M. J. (1998), A multivariate statistical controller for online quality improvement, *Journal of Process Control*, **8**, 139 – 149.

Dong, D. and McAvoy, T.J. (1996), Nonlinear principal component analysis--based on principal curves and neural networks, *Computers & Chemical Engineering*, **20**, pp. 65-78.

Jämsä-Jounela, S.-L., Vermasvuori, M., Endén, P., Haavisto, S. (2003), A process monitoring system based on the Kohonen self-organizing maps, *Control Engineering Practice*, **11**, pp.. 83 – 92.

Kämpjärvi, P., Sourander, M., Jämsä-Jounela, S.-L. (2003), Online analyzer validation and process fault diagnosis for ethylene cracking process under MPC feedback, Industrial Engineering & Chemistry Research, submitted.

Kohonen, T. (2001), *Self-Organizing Maps*, Springer, Berlin, 3rd Ed., 501.

Komulainen, T., Sourander, M., Jämsä-Jounela, S.-L. (2003), An online application of dynamic PLS to dearomatization process, *Computers & Chemical Engineering*, submitted.

Ku, W., Storer, R. H., Georgakis, C. (1995), Disturbance detection and isolation by dynamic principal component analysis, *Chemometrics and Intelligent Laboratory Systems*, **30**, 179 – 196.

Laine, S., Pulkkinen, K., S. -L. Jämsä-Jounela (2000), On-line determination of the concentrator feed type at Outokumpu Hitura mine, Minerals Engineering, **13**, pp. 881-895

Lennox, B. and Sandoz, D. (2002), Recent experiences in the industrial exploitation of principle component based fault detection methods, Proceedings of the IEEE International Symposium on Computer Aided Control System Design, Glasgow 2002, pp xxxv – xliv.

Leonard, J.A. and Kramer, M.A. (1991), Radial basis function networks for classifying process faults, *IEEE Control Systems*, **11**, pp. 31 - 38.

Li, W., Yue, H.H., Valle-Cervantes, S. and Qin S.J. (2000), Recursive PCA for adaptive process monitoring, *Journal of Process Control*, **10**, pp. 471-486.

Misra, M., Yue, H.H., Qin, S.J. and Ling, C. (2002), Multivariate process monitoring and fault diagnosis by multi-scale PCA, *Computers & Chemical Engineering*, **26**, 1281 - 1293.

Rallo, R., Ferre-Giné, J., Arenas, A. and Giralt, F. (2002), Neural virtual sensor for the inferential prediction of product quality from process variables, *Computers & Chemical Engineering*, **26**, 1735 – 1754.

Rantala, A. (1999), Identification of Process Disturbances in Electrolytic Refining of Copper, Master's Thesis, Helsinki University of Technology, Espoo, 103.

Venkatasubramanian, V., Rengaswamy, R., Kavuri, S. N. and Yin, K. (2003), A review of process fault detection and diagnosis Part III: Process history based methods, *Computers & Chemical Engineering*, **27**, pp. 327 – 346.

Wienke, D., van den Broek, W., Buydens, L., Huth-Fehre, T., Feldhoff, R., Kantimm, T. and Cammann, K. (1996), Adaptive resonance theory based neural network for supervised chemical pattern recognition (FuzzyARTMAP) Part 2: Classification of post-consumer plastics by remote NIR spectroscopy using an InGaAs diode array, *Chemometrics and Intelligent Laboratory Systems*, **32** (2) (1996), pp. 165-176.

Wold, S., Sjöström, M. and Eriksson, L. (2001), PLS-regression: a basic tool of chemometrics, *Chemometrics and Intelligent Laboratory Systems*, **58**, pp. 109 – 130.

Qin, S.J. (1998), Recursive PLS algorithms for adaptive data modeling, *Computers & Chemical Engineering*, **22**, 503-514.

Zhao, J., Chen, B., and Shen, J. (1998) Multidimensional non-orthogonal wavelet-sigmoid basis function neural network for dynamic process fault diagnosis, *Computers & Chemical Engineering*, **23**, pp. 83-92

ELSEVIER
IFAC
PUBLICATIONS
www.elsevier.com/locate/ifac

NONLINEAR FRICTION COMPENSATION DESIGN FOR SUPPRESSING STICK SLIP OSCILLATIONS IN OIL WELL DRILLSTRINGS

F. ABDULGALIL and H. SIGUERDIDJANE

École Supérieure D'Électricité,
Automatic Control Department
3, rue Joliot-Curie
91192 Gif-sur-Yvette, France
e-mail :Farag.abdulgalil@supelec.fr
e-mail :houria.siguerdidjane@supelec.fr

Abstract: Friction appears in all mechanical systems and has a significant impact on control. This paper presents simple but effective nonlinear friction compensation method for a drillstring system, by considering the stick- slip motion of the bit on the bottom hole assembly (BHA) in oil wells. Due to nonlinear problem at near- zero velocity where the torque exerted on bit is a function of the bit speed, the drillstring controlling is not an easy task. The control strategy proposed in this work involves a combination of a feedback model based compensation of friction and linear controller which enables to reduce the stick slip oscillations. The effectiveness of this approach is demonstrated through the resultants simulations. *Copyright © 2004 IFAC*

Keywords: Control systems, nonlinear systems, stick-slip oscillations, drilling system, friction compensation, feedback.

1. INTRODUCTION

In the drilling processes there are different types of vibrations, one of them and for which we are interested in this work is torsional oscillations (Stick-Slip). These oscillations are induced by nonlinear frictional torques between the drill bit at the rock surface. The amplitude of this torsional excitation can be two to four times the target or average angular speed (typically between 3.14 and 15.7rad/sec) set by the top-drive. The torsional excitation can give rise to enormously destructive fluctuating torques in the drill-string that, once out of control; invariably cause damage to the bit or drill-string. Even small amplitude slip-stick vibrations are thought to be a major cause of bit wear. Different control solutions have been presented in the literature to combat this instability as by using the H_∞ technique on the local linearized model about suppressing stick-slip in oil well drillstrings (Serrarens *et al.*, 1998), by applying optimal control by (Smit, 1995) and by using classical controller as PI (Tucker and Wang, 1999).
A wide variety of studies have been reported, which formulate nonlinear friction models, identifying it's parameters, and suggest compensation techniques for the friction. The friction model has been widely studied by numerous researchers (Candus *et al.* 1993).

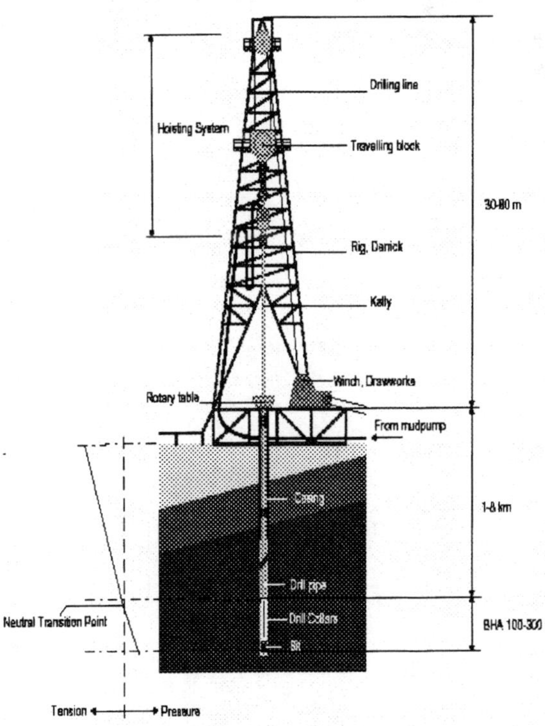

Fig. 1. Drilling equipment

The control strategy proposed in this paper consists of using compensation for the friction in drillstring system and together with a nonlinear controller.
The approach is based on the principle that it only feds back the friction in the control loop. The PI controller is used for reducing the steady state and the tracking error in the system.
Nevertheless, it has never been applied to drillstring system, at least to our knowledge. The effectiveness of this proposed approach is demonstrated by simulations on a drilling system.
An idealized system of a rotary drilling installation is shown in Figure 1, which one may be find in (Serrarens, 2002). Starting at the top end of the complete system, a heavy weight disc-shaped mass called the *rotary table* is driven by an electric motor. The rotary table acts as flywheel to approximately maintaining a constant speed of the subsequent system elements. These elements can be identified as the drillstring, the BHA and finally the very bottom end of the structure, a cutting tool called a *bit*. The medium to transport the energy from the motor to the bit is formed by a drillstring, mainly consisting of a number of relatively thin walled pipes.
The remainder of this paper is organized as follows. Section 2 presents a brief description of the elements of the rotary drilling model. In Section 3, nonlinear friction technique with PI controller is proposed. Friction method is applied on rotary drilling system for designing a nonlinear controller in Section 3. Simulation results are shown in Section 4 for the case of step input. Finally, Section 5 ends with some conclusions.

2. ROTARY DRILLING MODELLING

Figure 2 shows the main components of the model which contains two damped inertias mechanically coupled by an elastic intertialess (drillstring). The drillstring is modelled as a continuous mechanical system with infinity natural resonance frequencies. It is assumed that the drillstring is homogenous along its entire length and simply modelled as a single linear torsional spring with stiffness k.
The lowest part of the drillstring, BHA is composed of thick-walled tubulars loaded in compression, and, without buckling; it provides weight on the bit (WOB) required to generate sufficient cutting force. The BHA can be several hundreds of meters in length, and often contains specialized downhole tools. The connection of the drillstring to the rotary table is made by an interface element called the *Kelly*. This square or hexagonally shaped rod connected to the top of the drillstring, which fits into a square or hexagonal hole in the rotary table, allowing vertical motion of the drillstring.
The rotary drilling system consists of two main components:

- The drillstring and BHA
- The driven system

2.1 The drillstring and BHA

A simple model of torsional drillstring vibrations is obtained by assuming that the drillstring behaves as a torsional pendulum, i.e. the drill pipes are represented as a torsional spring, the drillcollars behave as a rigid body and the rotary table rotates at a constant speed. The corresponding equation of motion may be found in (Serrarens *et al.*, 1998) and is

$$J_1\dot{\Omega}_1 + C_1\Omega_1 - k\phi = -T_{tob}(\Omega_1) \tag{1}$$

Furthermore, the quantities
$\phi = \varphi_2 - \varphi_1$, $\Omega_1 = \dot{\varphi}_1$ and $\Omega_2 = \dot{\varphi}_2$ are defined, where φ_1 is the angular displacement of the bit and φ_2 is the angular displacement of the rotary table, J_1 is the equivalent of mass moment of inertia of the collars and the drillpipes, C_1 is the equivalent viscous damping coefficient of BHA, and T_{tob} is a nonlinear function which will be referred to be the torque on-bit.

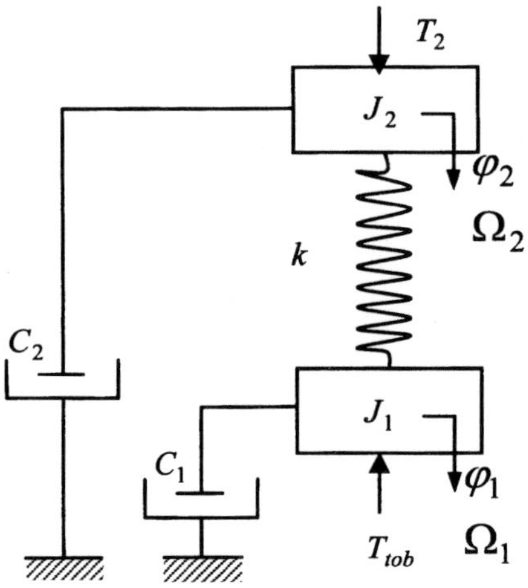

Fig. 2. Drilling rotary model

2.2 The driven system

The mechanical behaviour of the drive system is dominated by three components: the rotary table, a bevel gearbox with combined gear ratio of $1:n$, and an electric motor. The equations of this system are given in (Serrarens *et al.*, 1998) as

$$J_2 = J_{rot} + n^2 J_m$$

$$J_2\dot{\Omega}_2 + C_2\Omega_2 + k\phi = T_2 \tag{2}$$

$$C_2\Omega_{ref} + u = T_2 \tag{3}$$

where J_2 represents the inertia of the rotary table (J_{rot}) augmented with inertias of the electric motor (J_m) and the transmission gear box ratio of the real system, C_2 is the viscous damping of the rotary table, Ω_{ref} is the desired velocity of the bit and finally T_2 is the torque delivered by the motor to the system.

In general, the electrical behaviour of the motor has to be described in terms of nonlinear differential equations that depend on the type of the motor. Many drilling rigs are equipped with a separately excited DC motor, and in that case the equations reduce to the linear relationships (Serrarens *et al.*, 1998):

$$L\dot{I} + RI + V_{emf} = V_m$$
$$V_{emf} = K\Omega_2$$
$$T_2 = KI$$

where $I, R, L,$ and V_m are respectively defined as motor current, motor resistance, motor induction and motor input voltage. The back-electromotive force (back-emf) V_{emf} is linearly related to the rotary table speed, K is the motor constant multiplied by the transmission ratio such as $K = nK_m$.

2.3 The stick-slip oscillations

The modelling of exact friction characteristic is a quite difficult problem, because the friction characteristic can be changed easily due to the environment's changes, for example, the variation of the load, lubrication, and the surface roughness (nature of the rocks).
In the oil drillstring, these oscillations are driven by nonlinear friction (T_{tob}) at near-zero bit velocities. T_{tob} represents the combined effects of reactive torque on the bit and nonlinear frictional forces along the BHA. Little is known about these effects, and several functions have been proposed for analyzing stick-slip vibrations (Serrarens, 2002; Jansen, 1993; Cull, and Tucker 1999). The connection between the friction torque (T_{tob}) as a function of the bit speed is given in (Serrarens, 2002) by the following nonlinear function:

$$T_{tob}(\Omega_1) = T_{tobdyn}\,\frac{2}{\pi}\left(\alpha_1\Omega_1 e^{-\alpha_2|\Omega_1|} + a\tan(\alpha_3\Omega_1)\right) \quad (4)$$

where $T_{tobyn} = 0.5kNm$, $\alpha_1 = 9.5$, $\alpha_2 = 2.2$ and $\alpha_3 = 35.0$.

3. NONLINEAR FRICTION COMPENSATION METHOD

Many schemes form model based friction compensation have been proposed. If a good friction model is available it is possible to use a model based compensation scheme as shown in Figure 3. The system is provided with friction compensation and friction model. The linear controller is presented by PI (Proportional Integral) control, which is used to achieve the following performances: Increased damping, reduced rise time for step or rapid inputs, friction compensation (without much stick slip oscillation). PI may be any control structure of the form:

$$u(t) = k_p e(t) + k_i \int e(\tau)d\tau$$

where k_p, k_i are control parameters, $u(t)$ and $e(t)$ are the system input and error, respectively.

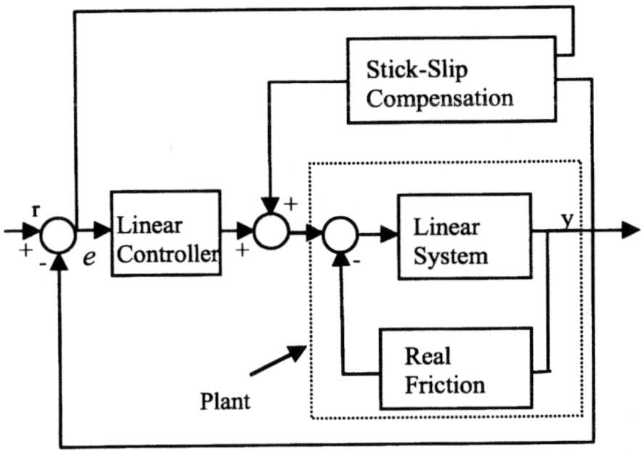

Fig.3. Block diagram of the model-based friction compensation scheme

4. CONTROLLER DESIGN FOR THE ROTARY DRILLING SYSTEM BASED ON FRICTION COMPENSATION

Figure 4 shows the block diagram of the system to be controlled. This system involves a plant which is driven by a PI controller. The plant has inherent driven system, drillstring, BHA and finally sticks slip friction.

Consider the nonlinear equation of the rotary drilling system given in (1)-(4), which may be written in the form:

$$\dot{x}(t) = f(x(t)) + g(x)u(t) \quad (8)$$
$$y(t) = h(x(t))$$

where $x(t) = [\Omega_1 \quad \phi \quad \Omega_2]^T \in R^3$ is the state vector, $y(t) \in R$ is the measured output variable,

$$f(x) = \begin{bmatrix} -\frac{C_1}{J_1}\Omega_1 + \frac{k}{J_1}\phi - \frac{1}{J_1}T_{tob}(\Omega\ ) \\ \Omega_2 - \Omega_1 \\ -\frac{C_2}{J_2}\Omega_2 - \frac{k}{J_2}\phi + \frac{C_2}{J_2}\Omega_{ref} \end{bmatrix}$$

$$g(x) = \begin{bmatrix} 0 & 0 & \frac{1}{J_2} \end{bmatrix}^T$$

and $h(x) = \Omega_1$

The control signal u is composed of a linear and a nonlinear part:

$$u = u_{lin} + \hat{T}_{tob}$$

where u_{lin} is the linear part and $\hat{T}ob$ is the estimated torque force.

$$u_{lin} = k_p e + k_i \int edt$$

where $e = \Omega_1 - \Omega_{ref}$

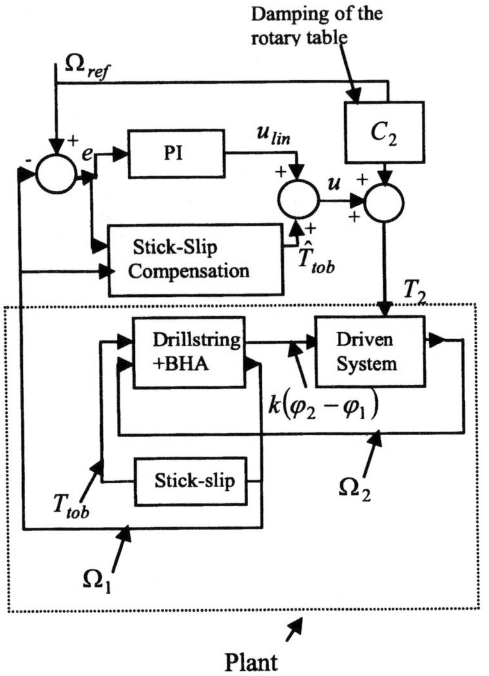

Fig.4. The system to be controlled

5. SIMULATION RESULTS

Simulations have been performed to investigate the efficiency level of the proposed nonlinear friction compensation method. The numerical values of the drilling system are listed in Table 1 and are taken from (Serrarens *et al.*, 1998).

Table 1: Numerical values of drilling system

Parameter	Description	Value	Unit
J_1	BHA + 1/3 drillstring inertia	374	$[kgm^2]$
J_2	Rotary table + drive interia	2122	$[kgm^2]$
C_1	BHA damping	0-50	$[Nms/rad]$
C_2	Rotary table damping	425	$[Nms/rad]$
k	Drillstring stiffness	473	$[Nm/rad]$
w_r	Eigen frequency	1.125	$[rad/\sec]$

Figure 5 shows the velocity of bit without the controller, one may observe the oscillations that are undesirable because they are often regarded as one of the most damaging modes of vibration when drilling with low speed.

Furthermore, one may notice that the reference velocity is involved in the system through equation (3).

The drilling system, whose schematic in figure 1, is commanded by a step set point with final value $10 rad/\sec$. This step signal presents the desired velocity of the bit (Ω_{ref}).The PI parameters have been found based on the local linearized system as $k_p = 2$, $k_i = 9$.

The angular velocity of the rotary table, shown in Figure 6, reaches the desired velocity with time response of about 25s. Figure 7 shows that the angular velocity of the bit reaches the desired rotary drilling velocity (10 rad/sec) without much oscillation and with the same settling time like the angular velocity of the rotary table.

However, at the stick slip vibrations, as the control system attempts to track a desired behavior, without any more oscillations. Thus, the importance of nonlinear compensation is demonstrated.

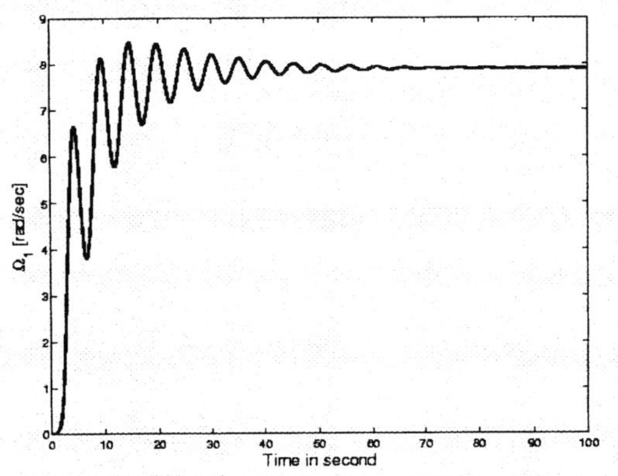

Fig.5. Angular velocity of the bit without the controller

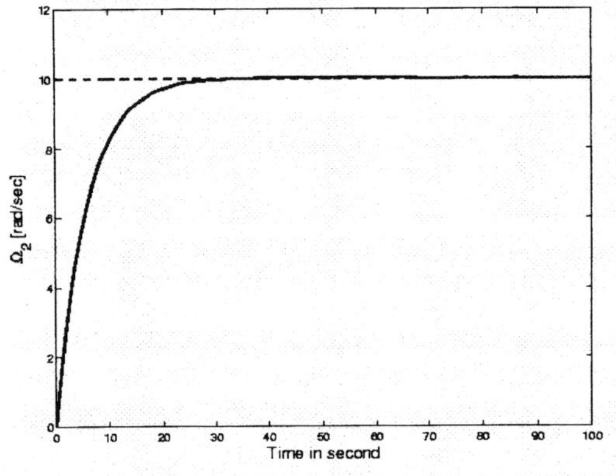

Fig.6. step response of the angular velocity of the rotary table (rad/sec)

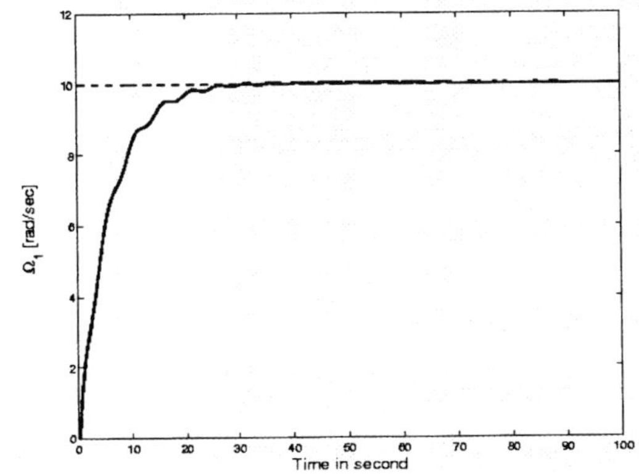

Fig. 7. Step response of the angular velocity of the bit (rad/sec) vs time (in sec) in closed loop

6. CONCLUISIONS

In this work, an effective method has been proposed to compensate the vibrations in oil well drillstrings, which has not seen applied up to now, at least to our knowledge. The nonlinear controller combines stick-slip compensation and a proportional integral controller.
In the controller development in this paper, it was assumed that all of the system parameters are known.

By the way, the modelling of exact friction characteristic is not an easy problem, because the friction characteristic can be changed easily due to the environment's changes, for example, the variation of the load, lubrication, and the surface roughness (nature of the rocks). However, this assumption is not always true. As for example the perturbations occur in the drillstring length during the process of drilling. The method has the potential to be extended by using an adaptive nonlinear controller. This will be reported in some near future work.

REFERENCES

Candus, C, H. Olsson, K. J. Astrom, and P. Lischinsky. (1993). Dynamic friction models and control design. *In Proc. Amer. Contol. Conf,* pp 10-12.

Cull S.J and R.W Tucker. (1999). On the modelling of coulomb friction. *J. Phys.A: Math.* **Gen.32,** pp. 2103-2113.

Jansen, J.D. (1993). Nonlinear dynamics of oilwell drillstrings. *PhD thesis, Delft University Press.*

Serrarens, A.F.A, M.J.G. van de Molengraft, J.J. Kok and L. Van den Steen (1998). H_∞ Control for suppressing stick slip in oil well drillstrings *IEEE Control systems,* **Vol. 18, No. 2,** pp. 19-30.

Serrarens, A.F.A (2002). H_∞ control as applied to torsional drillstring dynamics. *Master's thesis, Eindhoven, University of Technology.*

Tucker, R.W. and C. Wang (1999). On the effective control of torsional vibrations in drilling systems. *J. of Sound and Vibration,* **224(1),** 101-122.

Van den Steen, L (1997). Suppressing stick slip induced drill string oscillation: hyperstability approach. *PhD thesis, University of Twente.*

ELSEVIER
IFAC
PUBLICATIONS
www.elsevier.com/locate/ifac

MODEL PREDICTIVE CONTROL FOR BATCH PROCESSES USING LATENT VARIABLE METHODS

Jesus Flores-Cerrillo and John F. MacGregor

Department of Chemical Engineering
McMaster University, Hamilton, On. Canada

Abstract: A novel multivariate model predictive control strategy for trajectory tracking and disturbance rejection for batch processes, based on multiway PCA models, is presented. It directly computes the manipulated variable trajectory adjustments over a future horizon using the structure of the PCA model. The advantages and the modest data requirements are illustrated using an emulsion polymerization process for temperature tracking. *Copyright © 2004 IFAC*

Keywords: Model predictive control, Principal component analysis, Batch processes, Missing data.

1. INTRODUCTION

Batch/semi batch processes are commonly used because their flexibility to manage many different grades and types of products. In these processes, one of the requirements to achieve consistent final quality specifications and adequate operation is that the low level controllers can adequate follow the set-points determined by master controllers or optimizers. Proportional-integral (PI) and Proportional-integral-derivative (PID) controllers are by far the most common approach used in industry. However, batch processes usually exhibit long time delays, large time constants, and manage different type of products. Under this situation the standard PID controller might not achieve adequate control performance. Enhancements to conventional PID controllers have proven to lessen some of these deficiencies (Clarke-Pringle and MacGregor, 1997).

Several advanced control approaches have been presented in order to further improve PID performance. Cott and Macchietto (1989), and Aziz *et al.*, (2000) used generic model control to track reactor temperature set points. A considerable number of studies using nonlinear model predictive control have also been presented. Özkan *et al.*, (2001), for example, used nonlinear dynamic model control to track optimal reactor temperatures in the solution polymerization of styrene.

Differential geometry methods have also been applied for the control of batch and continuous processes (Kravaris *et al.*, 1989; Clarke-Pringle and MacGregor, 1997).

Several control approaches using PLS models have been presented by Flores-Cerrillo and MacGregor (2004). However, their approaches are mainly for controlling end product quality. Lee *et al.*, (1999) proposed a model predictive control technique for trajectory tracking aided with iterative learning. The methodology is illustrated for the temperature control of a laboratory batch reactor.

Statistical controllers for continuous processes based on Principal Component Analysis (PCA) have also been proposed (Chen and McAvoy, 1996). A related algorithm called covariance control can be found in Lu and Skelton (1997), and within the PCA framework in Shah *et al.*, (1998). The approach is illustrated in the control of reactor temperature and conversion for an industrial tubular reactor.

The purpose of this paper is to introduce a novel multivariate control strategy based on latent variable models for set-point trajectory tracking and disturbance rejection. The control strategy and manipulated variable adjustments, over a future horizon, are formulated in the space of a dynamic PCA model. The outline of the paper is as follows: in section 2 the methodology is introduced; in section 3, the control approach is illustrated using an emulsion polymerization process. Conclusions and future work are stated in section 4.

2. CONTROL METHODOLOGY

2.1 Basic Methodology

Modeling

The proposed algorithm uses historical databases and a few complementary identification experiments for model building. The empirical model is obtained using PCA.

The database used for building the PCA model, denoted as $\mathbf{\Omega}$, consists of K matrices, $\mathbf{X}_k$ (k=1,...,K), where K is the number of batches used for model building.

$$\mathbf{\Omega} = \begin{bmatrix} \mathbf{X}_1 \\ \vdots \\ \mathbf{X}_k \\ \vdots \\ \mathbf{X}_K \end{bmatrix}$$

Each matrix $\mathbf{X}_k$ contains the information of k-th batch. For the sake of simplicity, the subscription k is omitted in the following explanation of the structure of X_k.

Consider a batch run. Denote N as the total number of samples collected along the time. For sampling time i, define a vector ζ_i^T such that

$$\zeta_i^T = [\mathbf{x}_{me}^T \ \mathbf{y}_{cv}^T \ \mathbf{y}_{sp}^T \ \mathbf{u}_c^T]_i$$

where $\mathbf{x}_{me}$ is a vector of measurements on S on-line (and potentially off-line) process variables such as stirring, flows, etc. that can be incorporated to give information on disturbances and process conditions changes; $\mathbf{y}_{cv}$ is a vector of R process variables needed to be tracked; $\mathbf{y}_{sp}$ is the vector of its corresponding set-points; and $\mathbf{u}_c$ is a vector of A manipulated variables. The length of ζ_i^T is then equal to (S+2R+A). Notice that if time delays are important, these can be easily incorporated in ζ_i by simply shifting the delayed variables.

Every row vector of $\mathbf{X}$ is composed of vectors ζ^T at different sampling times, including the M-step future data, current data and the L-step past data.

$$\mathbf{X} = \begin{bmatrix} \zeta_1^T & \cdots & & \cdots & & \cdots & \zeta_{1+M+L}^T \\ \vdots & \ddots & & & & \cdot^{\cdot^{\cdot}} & \vdots \\ \zeta_{i-L}^T & \cdots & \zeta_{i-1}^T & \zeta_i^T & \zeta_{i+1}^T & \cdots & \zeta_{i+M}^T \\ \vdots & \cdot^{\cdot^{\cdot}} & & & & \ddots & \vdots \\ \zeta_{N+M-L}^T & \cdots & & \cdots & & \cdots & \zeta_N^T \end{bmatrix}$$

Past data   Current data   Future data

Therefore, the total dimension of matrix $\mathbf{\Omega}$ is [(N+M-L)K] × [(M+L+1) (S+2R+A)].

The data-set used for model building consists of representative operating data from past batches to capture information on most of the disturbances and operating policies normally encountered in the batch operation. In addition, data in which pseudo random binary sequences (PRBS) have been added on the output of the controller currently used (i.e. added on the manipulated variables) is needed to establish causal relationship between the manipulated variable changes and the tracked trajectories. Rebuilding the model by adding new batch data collected after implementing the proposed control scheme may also be done in order to further improve the causal relationship. The data requirements are further discussed in the examples.

PCA is then performed by projecting the scaled and mean centered matrix $\mathbf{\Omega}$ onto the score space $\mathbf{T}$:

$$\begin{aligned} \hat{\mathbf{\Omega}} &= \mathbf{T}\mathbf{P}^{\mathrm{T}} \\ \mathbf{T} &= \mathbf{\Omega}\mathbf{P} \end{aligned} \qquad (1)$$

where $\mathbf{P}$ is the loading matrix and $\mathbf{P}^T\mathbf{P} = \mathbf{I}$. Non-linear PCA methods can also be used. However, for simplicity, in the following discussion linear modeling is assumed.

Control

Consider sampling time i. We can rearrange the vector $\left[\zeta_{i-L}^T \ \cdots \ \zeta_i^T \ \cdots \ \zeta_{i+M}^T\right]$ such that

$$\left[\zeta_{i-L}^T \ \cdots \ \zeta_i^T \ \cdots \ \zeta_{i+M}^T\right] \rightarrow [\mathbf{x}_1^T \ \ \mathbf{x}_2^T]$$

where $\mathbf{x}_1^T$ contains the *known* information, including all past information (ζ_j^T, j=i-L, ..., i-1) and future controlled variable set point trajectories ($\mathbf{y}_{sp,j}^T$, j=i,...,i+M), while $\mathbf{x}_2^T$ contains the ***unknown*** information, including future measurements ($\mathbf{x}_{me,j}^T$ j=i,...,i+M), future controlled variables ($\mathbf{y}_{cv,j}^T$, j=i,...,i+M) and future control actions adjustments ($\mathbf{u}_{c,j}^T$,j=i,...,i+M).

Loading matrix $\mathbf{P}$ can also be decomposed into two corresponding parts:

$$\mathbf{P} = \begin{bmatrix} \mathbf{P}_1 \\ \mathbf{P}_2 \end{bmatrix}$$

Two conditions need to be considered to compute $\mathbf{x}_2$. First, the unknown information $\mathbf{x}_2$ should keep the correlation structure as in the training dataset. Therefore, from the PCA model,

$$\mathbf{x}_2^T = \hat{\mathbf{t}}^T \mathbf{P}_2^T \qquad (2)$$

where $\hat{\mathbf{t}}^T$ is the estimated score vector for the current batch. Second, $\hat{\mathbf{t}}^T$ should be the projection in the score space of $[\mathbf{x}_1^T \quad \mathbf{x}_2^T]$. Therefore, $\hat{\mathbf{t}}^T$ can be obtained by solving:

$$\min_{\hat{\mathbf{t}}} (\hat{\mathbf{t}}^T - \mathbf{x}_1^T \mathbf{P}_1 - \mathbf{x}_2^T \mathbf{P}_2)\mathbf{Q}(\hat{\mathbf{t}}^T - \mathbf{x}_1^T \mathbf{P}_1 - \mathbf{x}_2^T \mathbf{P}_2)^T$$

$$s.t. \quad \mathbf{x}_2^T = \hat{\mathbf{t}}^T \mathbf{P}_2^T \qquad (3)$$

The solution for (3) can be easily obtained (**Q**=**I**) as:

$$\hat{\mathbf{t}}^T = \mathbf{x}_1^T \mathbf{P}_1 (\mathbf{I} - \mathbf{P}_2^T \mathbf{P}_2)^{-1} \qquad (4)$$

Then, $\mathbf{x}_2^T$ can be obtained by substituting (4) in (2):

$$\mathbf{x}_2^T = \mathbf{x}_1^T \mathbf{P}_1 (\mathbf{I} - \mathbf{P}_2^T \mathbf{P}_2)^{-1} \mathbf{P}_2^T \qquad (5)$$

Equation (5) can be further restated as $\mathbf{x}_2^T = \mathbf{x}_1^T \mathbf{P}_1 (\mathbf{P}_1^T \mathbf{P}_1)^{-1} \mathbf{P}_2^T$ since $\mathbf{P}_1^T \mathbf{P}_1 + \mathbf{P}_2^T \mathbf{P}_2 = \mathbf{I}$. This equation is the same as the missing data algorithm known as Projection to the Plane (Nelson *et al.*, 1996)

This control algorithm is recomputed at every sampling time *i* until completion of the batch. Only the current control actions ($\mathbf{u}_{c,j}^T$, *j*=*i*) are implemented.

In practice it is possible that the matrix $\mathbf{P}_1^T \mathbf{P}_1$ becomes ill conditioned. In this case it is recommended to obtain $\hat{\mathbf{t}}$ by using either a pseudo-inverse procedure based on single value decomposition or numerically until convergence of $\hat{\mathbf{t}}$ is achieved.

In the objective function (Equation 3), if desired, hard constraints to the scores, manipulated variables, and movement suppression factors can also be included. In this case a quadratic programming problem (for linear models) will have to be solved at every time.

An alternative modeling and control strategy is to simply build a PCA model by defining ζ_i^T as

$$\zeta_i^T = [\mathbf{x}_{me}^T \ \mathbf{y}_{cv}^T \ \mathbf{u}_c^T]_i$$

which does not contain controlled variable set point information. In the controlling step, however, at sampling time *i*, the future controlled variable trajectories can be specified as their desired set-point trajectories, $\mathbf{y}_{cv,j}^T = \mathbf{y}_{sp,j}^T$, *j*=*i*,...,*i*+*M* and then included in $\mathbf{x}_1^T$. In this case, $\mathbf{x}_2^T$ will only include future measurements and future control actions that can be computed using (3).

Enforced Feedback

While the control algorithm (eq. 3) has induced feedback from the implemented control actions, measured controlled variables and measured process variables, it remains basically as an open-loop strategy. One possible way to more directly enforce feedback is by using a reference trajectory ($\mathbf{y}_{ref}$) instead of the original set points. $\mathbf{y}_{ref}$, at sampling time *i*, can be specified to follow a first order response:

$$y_{ref,j} = \lambda^{j-i} y^* + (1 - \lambda^{j-i}) y_{sp,j} \qquad j = i, i+1 \ldots, i+M$$

where $y^* = y_{sp} - \Delta y$, $\Delta y = (y_{cv} - y_{sp})_{i-1}$ and $0 \leq \lambda \leq 1$ is a forgetting factor. Other ways to enforce feedback are possible such as the introduction of prediction error used in MPC.

2.2 Adaptive methods

By using the structure in the last section, a single set of **P** coefficients is determined for the whole duration of the batch. Depending on the process, improved control may be obtained by using an adaptive modeling.

In the adaptive modeling scheme, a PCA model is obtained at each sampling time *i*, with proper weights added to the observations

$$\mathbf{X}^{(i)} = \begin{bmatrix} w(L+1-i) \times [\zeta_1^T & \cdots & \cdots & \cdots & \zeta_{1+M+L}^T] \\ \vdots \\ w(0) \times [\zeta_{i-L}^T & \cdots & \zeta_i^T & \cdots & \zeta_{i+M}^T] \\ \vdots \\ w(N+M-i) \times [\zeta_{N+M-L}^T & \cdots & \cdots & \cdots & \zeta_N^T] \end{bmatrix}$$

where $\mathbf{X}^{(i)}$ is the matrix **X** for sampling time *i*. Then a training dataset for sampling time *i* is obtained by

$$\boldsymbol{\Omega}^{(i)} = \begin{bmatrix} \mathbf{X}_1^{(i)} \\ \vdots \\ \mathbf{X}_k^{(i)} \\ \vdots \\ \mathbf{X}_K^{(i)} \end{bmatrix}$$

Loading matrix for sampling time *i*, $\mathbf{P}^{(i)}$, is computed by performing PCA on $\boldsymbol{\Omega}^{(i)}$. In the control algorithm, one only need to replace **P** in Equation (3) by $\mathbf{P}^{(i)}$ to compute the control action at sampling time *i*.

Different weighting vectors can be used. Here we use a weighting vector as shown in Figure 1. This weighting vector consists of two mirrored sigmoid functions, which can be calculated using:

$$w(\theta) = \begin{cases} \dfrac{1}{1 + \exp(6.9068 \dfrac{\theta - \mu_1}{\mu_2})} & \theta \geq 0 \\ \dfrac{1}{1 + \exp(-6.9068 \dfrac{\theta + \mu_1}{\mu_2})} & \theta < 0 \end{cases}$$

where $\mu_1 = (a+b)/2$, $\mu_2 = (b-a)/2$ (*a* and *b* are two adjusting factors, which can be used to change the shape of the weighting vector). For more sophisticated adaptive PCA methods, please refer to Rännar *et al.*, (1998).

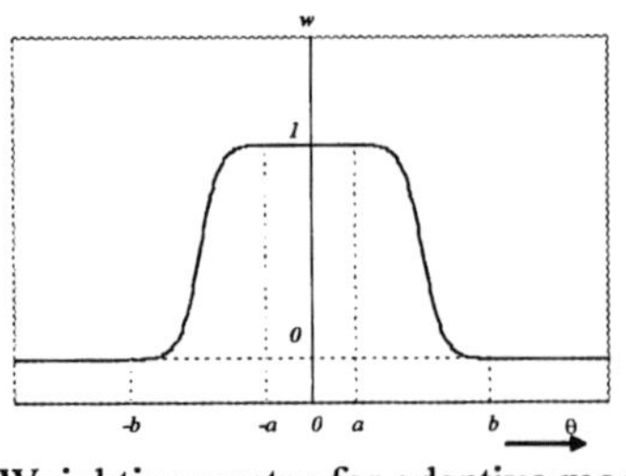

Figure 1. Weighting vector for adaptive modeling

3. SIMULATION EXAMPLES

The control algorithm was tested using two non-linear simulators: the emulsion polymerization of styrene (Lynch and Kiparissides, 1981) and the solution polymerization of vinyl acetate (Teymor, 1997). However, for brevity, only results on the emulsion polymerization process are presented.

Lynch and Kiparissides (1981) developed a non-linear model, with simple kinetics, for the styrene emulsion polymerization. This model, originally developed for tubular reactors with full recycle, has been adapted for use in batch and semi-batch processes. For a complete description of the model and model parameters the reader is referred to the original publication. The non-linear simulator is used for data generation and controller performance evaluation. The control objective is to perform reactor temperature (T_r) trajectory tracking by adjusting the inlet jacket temperature (T_j). On-line T_r measurements, considered to be available every 30sec, are corrupted by normally distributed random error with standard deviation σ=0.15°K. Simulation time is 300min. Control action is taken every 30sec.

3.1 Control studies

In order to evaluate the control algorithm, several studies involving the effect of information content of data available for model building, the type of model (adaptive versus fix), and the effect of inclusion of explicit feedback were performed for different set points and disturbances. In what follows some of these studies are presented.

As shown in section 2, PCA modeling allows having different species for model building. For example, one can use $\zeta_i^T = [T_{r,sp}\ T_j\ T_{r,cv}]_i^T$ or simply $\zeta_i^T = [T_j\ T_{r,cv}]_i^T$ ($\zeta_i^T = [\mathbf{u}_c^T\ \mathbf{y}_{cv}^T]_i$). Both situations were studied. For brevity only results in which $\zeta_i^T = [T_j\ T_{r,cv}]_i^T$ are presented.

The first step in model building is to obtain a data set. In most of the situations (at least for temperature control of batch reactors), the base controller is PI or PID. Therefore, it is feasible to assume that closed loop data containing some of the effects that T_j and normal operating disturbances have on T_r is available. However, in order to further establish a casual relationship between u_c= T_j and y_{cv}= T_r, a few complementary experiments, in which PRBS are added on the top of the output of the PI controller (u_c) are needed. Two scenarios are illustrated here: in the first one a dataset is obtained from a system in which a fine tuned PI (denoted as fine PI) is employed together with PRBS, while in the second one a slightly sluggish PI (denoted as sluggish PI) is used together with smaller PRBS. Comparison between the proposed algorithm and PI is given (PID performance is similar to those obtained from PI and therefore not shown).

Figure 2a shows the performance of the control algorithm (no feedback used) for the T_r trajectory tracking of scenario 1, Model 1, set-point 1 of Table 1 (set-point 1 is the set-point from which the model was identified, Figure 3) versus that obtained from a fine tuned PI. Figure 2b shows their corresponding manipulated variable. In Figure 2a (···) represents the reactor temperature set point (T_{sp}), while in Fig. 2a/2b (– –) T_r/T_j from Model 1 and (—) from the fine-tuned PI. It is evident that the trajectory tracking of T_r from the proposed control algorithm (Model 1) is better than that of PI, while having a much smother manipulated variable behavior.

The good control performance can be further observed by inspecting the average integral of the absolute error $IAE = \sum_{i=1}^{N} |y_{cv,i} - y_{sp,i}| / N$ for the CV, and the differentiator operator $\nabla = \sum_{i=1}^{N} |u_{c,i} - u_{c,i-1}| / N$ for the MV in Table 1. It can be seen that smaller values of IAE and ∇ are obtained by using the database algorithm. Slight improvement is still possible by enforcing feedback (Model 1a, Table 1).

The data requirements to build Model 1 are small considering that only 3 batches were used and that the PRBS do not affect substantially the batch operation as can bee seen in Figure 3. In this Figure (···) indicates the control obtained when in the MV (T_j) has been added PRBS, and (—) that under normal operation (no PRBS in the PI output). The IAE of the CV for this data set is IAE=0.85 (only 30% higher than when no PRBS is added).

One of the advantages in using dynamic PCA is that the usefulness of the model is not limited to only the set point from which the model was identified (as long as the correlation structure of the model does not change largely). This is illustrated in Figure 4a for scenario 1, Model 1 (identified using the data shown in Figure 3) for set point 2 of Table 1 (no feedback used), and compared to that obtained used a fine tuned PI. Figure 4b shows their corresponding manipulated variables (caption follows that of Figure 2a-b).

In spite of that the data requirements for Model 1 are not very demanding, it is not uncommon to find processes in which the PI performance is sluggish. Scenario 2 (Model 2 and 3 of Table 1) was designed in order to evaluate the performance of the algorithm when the *data* used for model building is obtained with a more sluggish PI, and when the PRBS is smaller than those used in Figure 3 (scenario 1).

Table 1 Control Results for the emulsion polymerization of styrene

Control	Scenario	Model	Model type	Data from PI tuned	k	Set point IAE[1]	∇^1	IAE[2]	∇^2
PCA	1	1	Fix	fine	3	0.49	4.5	0.32	2.8
	2	2	Fix	sluggish	3	0.72	4.7	0.48	2.9
	2	3	Adaptive	sluggish	3	0.34	4.3	0.37	4.3
PCA*	1	1a	Fix	fine	3	0.38	4.8	0.23	3.4
	2	2a	Fix	sluggish	3	0.47	5.8	0.32	3.4
PI	1			fine		0.66	13.4	0.35	14.1
	2			sluggish		1.09	5.9	0.54	6.1

*Enforced feedback, [1] set point change 1, [2] set point change 2.

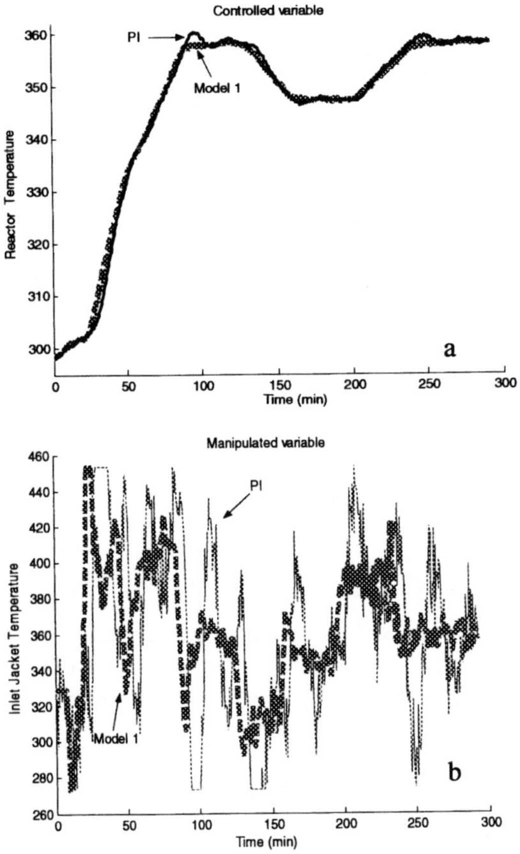

Fig. 2a-b. Control performance for Model 1 (Table 1) vs. fine-tuned PI. In Figure 2a (···) represents T_{sp}, while in Figure 2a/2b (– –) T_r/T_j from Model 1 and (—) from the fine-tuned PI.

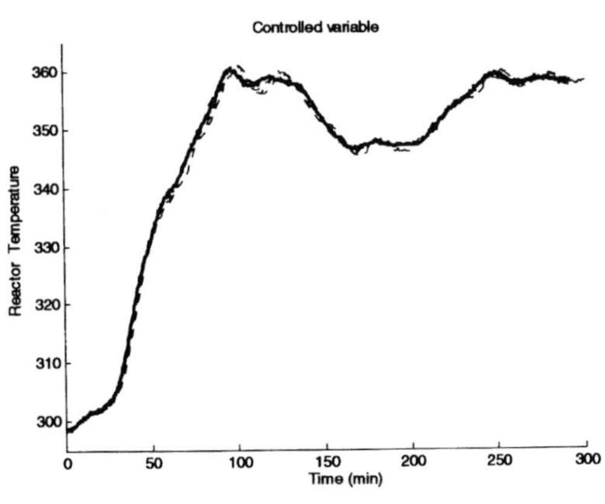

Fig. 3. Dataset used for building Model 1 (Table 1). (···) indicates the control obtained when in the MV (T_j) has been added PRBS, and (—) that under normal operation for the fine-tuned PI.

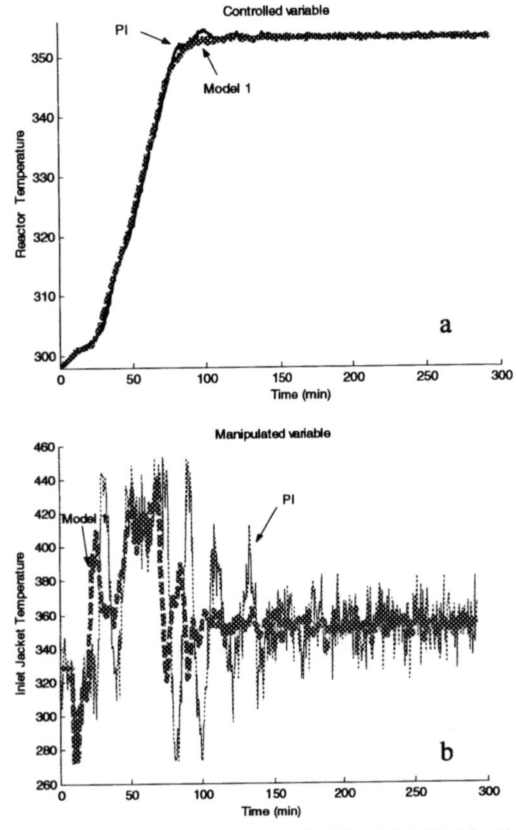

Fig. 4a-b. Control performance for Model 1 (Table 1) vs. fine-tuned PI for set point 2. In Figure 4a (···) represents T_{sp}, while in Figure 4a/4b (– –) T_r/T_j from Model 1 and (—) from the fine-tuned PI.

Figure 5a shows the performance of the control algorithm (no feedback used) for the T_r tracking of scenario 2, Model 2, set-point 1 of Table 1 versus that obtained from a sluggish PI. Figure 5b shows their corresponding manipulated variable. It is clear (see also Table 1) that also in this situation the performance of the proposed control algorithm (Model 2, IAE=0.72) is better than that of PI (IAE=1.09), while also having a smother manipulated variable behavior. However, it seems that at certain instances, the control from the data based algorithm is a little sluggish. This can be further corrected by using adaptive modeling (Model 3), which as can be seen in Figure 5a and Table 1 (IAE=0.34), it practically overlaps on the set points, or by including feedback. Notice that the advantage of the predictive part of the algorithm is evidently in Figure 5b, where the manipulated variable obtained from the algorithm forecast the changes in the set-points, while the MV of the PI responds much slower.

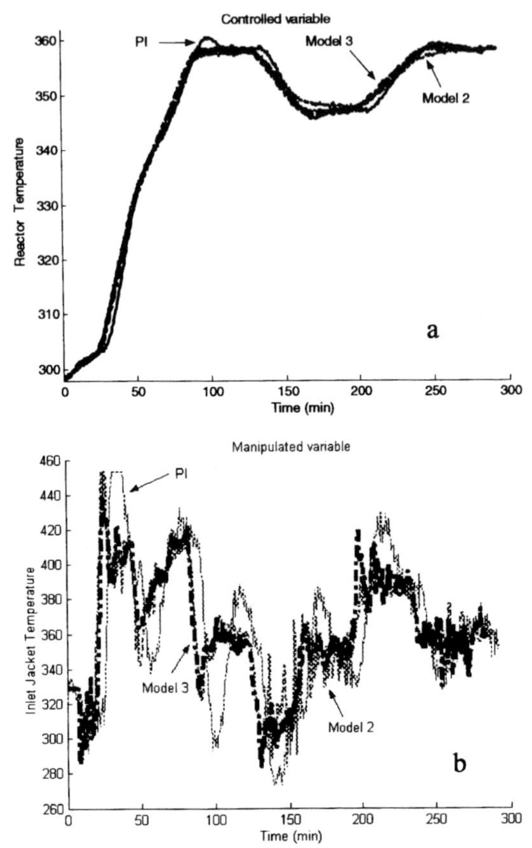

Fig. 5a-b. Control performance for Model 2 and 3 (Table 1) vs. sluggish PI. In Figure 5a (···) represents T_{sp}, while in Figure 5a/5b (— —) T_r/T_j from Model 2, (—·—) form Model 3 (Adaptive) and (—) from the PI.

From Table 1 and Figures 2,4 and 5, it can be seen that the control performance of the proposed control strategy is generally superior to that obtained using PI even in cases that the data set used for model building contains no information on new set-points.

The effect of the information content on the data-set is evident from the results: if a model is obtained using an initially better tuned PI the resulting data based controller (eq. 3) would perform better than when the model is obtained using a sluggish PI (This situation can be corrected, at considerable extent, by using adaptive modeling as can be seen in Table 1 and Figure 5a or by including feed-back). Moreover, independently of the PI and PRBS used to generate the model, the performance obtained from the data based method generally outperforms that obtained from PI (Similar conclusion were observed when using other recipes and when using a solution polymerization system (Teymor, 1997). However, these results are not shown here for brevity.)

4. CONCLUSIONS AND FUTURE WORK

A novel and simple multivariate model predictive control strategy (BMPC) for trajectory tracking based on PCA models is presented. It computes M steps ahead the trajectory manipulated variable adjustments using the structure of the PCA model. The advantages of the control algorithm and data requirements are illustrated using an emulsion polymerization process for reactor temperature tracking.

Future work includes more efficient ways to enforce feedback as well as testing of the methodology in a MIMO system. Comparison with other multivariate techniques is also foreseen.

REFERENCES

Aziz, N., M. A. Hussain and I. M. Mujtaba (2000). Performance of different types of controllers in tracking optimal temperature profiles in batch reactors. *Computers chem. Eng.*, **24**, 1069.

Chen, G. and T. J. McAvoy (1996). Process control utilizing data based multivariate statistical models. *The Canadian Journal of Chemical Engineering*. **74**, 1010.

Clarke-Pringle, T. and J. F. MacGregor (1997). Nonlinear adaptive temperature control of multi-product, semi-batch polymerization reactors. *Computers chem. Eng.*, **21** (12), 1395.

Cott, B. J. and I. M. Mujtaba (1989). Temperature control of exothermic batch reactors using generic model control. *Ind. Eng. Chem. Res.*, **28**, 1177.

Flores-Cerrillo, J. and J. F. MacGregor (in press). Control of batch product quality by trajectory manipulation using latent variable models. *JPC.*

Kravaris, C., R. A. Wright and J.F. Carrier (1989). Nonlinear controllers for trajectory tracking in batch processes. *Computers Chem. Eng.*, **13** 73.

Lee, K. S., I. Chin, H. J. Lee and J. H. Lee (1999). Model Predictive Control Technique Combined with Iterative Learning for Batch Processes. *AIChE J.* **45** (10) 2175.

Lu, J. and R. E. Skelton (1997). Covariance Control using Closed Loop Modeling for Structures. *Proceedings of the ASCE Structure Congress.*

Lynch, D. and C. Kiparissides (1981). Numerical Simulation of a Tubular Polymerization Reactor. *J. of Applied polymer Science*, **26**, 1283.

Nelson, P., J. F. MacGregor and P.A. Taylor (1996). Missing Data Methods in PCA and PLS: Score Calculations with Incomplete Observations. *Chem. & Intel. Lab. Sys.*, **35**, 45.

Özkan, G., S. Özen, S. Erdoğan, H. Hapoğlu and M. Alpbaz (2001). Nonlinear control of polymerization reactor. *Computers chem. Eng.*, **25**, 757.

Rännar, S., J. F. MacGregor and S. Wold (1998). Adaptive batch monitoring using hierarchical PCA. *Chem. & Intel. Lab. Sys.*, **41**, 73.

Shah, S., M. Randy, H. Takada, K. Morinaga and T. Satou (1998). Modeling and Control of a Tubular Reactor: A PCA Based Approach. *Proceedings of the 5th IFAC Symposium on Dynamics and Control of Process Systems*, 17.

Teymour, F. (1997). Dynamics of Semibatch Polymerization Reactors: I. Theoretical Analysis. *AIChE J.* **43** (1) 145.

ELSEVIER
IFAC
PUBLICATIONS
www.elsevier.com/locate/ifac

TIME OPTIMAL CONTROL OF THE MOLECULAR WEIGHT IN A SEMI BATCH EMULSION POLYMERIZATION

R. Gesthuisen [*,1], **K. Dadhe** [*], **S. Krämer** [*], **S. Engell** [*]

[*] *Process Control Laboratory,*
Department of Biochemical and Chemical Engineering,
University of Dortmund, D-44221 Dortmund, Germany,
Tel: +49/231/755-5127, Fax: +49/231/755-5129

Abstract: In this contribution a hierarchical approach to the control of the molecular weight in a semi-batch emulsion polymerization is derived, which ensures minimum batch time. Rather than performing a numerical optimization to determine an optimal trajectory for the process at any point in time, the existing boundary conditions are examined and used to determine the time optimal operation at each time step. This procedure provides the set points for the molar holdups of monomer and chain transfer agent (CTA) in the system. Either a decentralized PI controller or a nonlinear model predictive controller (NMPC) can be applied to track the calculated trajectory. As the set point calculation is a solution of a system of algebraic equations it can easily be performed on-line. It is shown that this concept can be used to produce multi modal molecular weight distributions and can also handle disturbances in the process, such as an unexpected nucleation of particles.

Keywords: optimization, emulsion polymerization, control, NMPC, molecular weight

1. INTRODUCTION

Semi-continuous emulsion polymerization (SEP) is one of the major processes in the production of polymer latices. Control of the molecular weight distribution is one major issue when running the process. Usually a seed, a pre-defined polymer dispersion with a known number of particles, particle size, and monomer concentration, is used to ensure well-known starting conditions. In order to produce a certain molecular weight distribution or a special co-polymer composition, control of the concentrations of the monomers and possibly of the chain transfer agent is necessary. Vicente *et al.* (2001) calculate an optimal trajectory by minimizing the inverse of the reaction rate to achieve an optimal process time. The equality constraints of this optimization problem are given by a rigorous dynamic model of the polymerization of styrene with butyl maleate (BuM), which acts as a CTA. The resulting set points for the molar holdups of styrene and BuM are tracked by controller based on model inversion. Similar control structures are reported by Gentric *et al.* (1999), Kravaris *et al.* (1989) or Seferlis and Kiparissides (2001). They use closed loop control concepts, i. e. trajectory tracking is applied. In each of these publications the numerical optimization is performed off-line as the computational effort is too demanding for an on-line application. Krämer *et al.* (2002) applied an on-line re-optimization to a semi-bath emulsion co-polymerization. In order to be able to re-calculate the optimal trajectory on-line, a simplified model for the co-polymer composition was used. Due to the simplifications, the model error may be problematic. In this paper, we propose a different strategy in which time optimality is guaranteed by operating the process at its constraints, which is an approach towards realizing an optimal operation and was proposed by Srinivasan *et al.* (2003). As a feedback controller has to be used to track the optimal set points, the proposed control scheme is of hierarchical structure.

The process investigated here is the emulsion polymer-

[1] Corresponding author
E-mail: r.gesthuisen@bci.uni-dortmund.de

ization of styrene using butyl maleate as a CTA (cf. (Vicente *et al.*, 2001)). The goal is to produce a pre-specified multi modal molecular weight distribution in minimum time.

2. PROCESS AND MODEL DESCRIPTION

The process investigated here is the seeded semi-batch emulsion polymerization of styrene (Vicente *et al.*, 2001). In order to control the molecular weight, a chain transfer agent (BuM) is used. The reactor is precharged with a well defined seed and is operated at 60°C. Temperature control is realized by a PI-controller. The process is simulated using a rigorous model that describes the volumes of water, polymer, the total volume in the reactor and the molar holdups of monomer and CTA, the concentration of initiator, the average number of radicals in the particle phase and the discretized molecular weight distribution. In contrast to many other publications we consider the jacket as as a plug flow reactor (PFR). This modification is necessary when pilot plants or industrial scale reactors are considered as the assumption of an ideally stirred jacket is not correct for reactors of larger volume. The resulting partial differential equation is approximated by the method of orthogonal collocation (the interested reader is referred to (Georgakis *et al.*, 1977). It is assumed that the heat of reaction is only produced by the polymerization reaction, i. e. there is no contribution by the reaction of CTA. Furthermore, we assume that the heat transfer coefficient decreases with increasing conversion, i. e. with increasing solids content.

3. HIERARCHICAL CONTROL APPROACH

Figure 1 depicts the hierarchical control concept on which the control algorithm is based. It consists of four basic steps:

(1) Calculation of the average molecular weights which have to be produced in certain conversion intervals (computed off-line).
(2) Calculation of the set point of the molar monomer holdup in the reactor.
(3) Calculation of the ratio between the total molar holdups of monomer and of CTA.
(4) Calculation of the control actions by a controller (decentralized PI-controller or NMPC).

Each of these steps is described briefly in the subsequent paragraphs.

Calculation of the average molecular weights

For linear polymers formed in an ideal polymerization it is possible to describe an arbitrary MWD by the combination of ideal MWDs with certain average molecular weights (Vicente *et al.*, 2001). Therefore, the free parameters in this optimization step are the average molecular weights of the different modes and the fractions of conversion for which these molecular weights are produced. Vicente *et al.* (2001) derive conditions for which the termination of radical chains in an emulsion polymerization with CTA is dominated by the chain transfer reaction and therefore by the ratio between the monomer and CTA concentration in the particle phase. For such a process the assumption that the instantaneous MWD is of Schulz-Zimm type holds. In this case the instantaneous average molecular weight is calculated by:

$$\bar{M}_{n,ins} = \frac{k_p[M]^p}{k_{tr,CTA}[CTA]^p} P_M. \quad (1)$$

where P_M denotes the molecular weight of the repeating unit. This equation holds for the following conditions:

$$\frac{[CTA]^p}{[M]^p} > 5.2 \times 10^{-6} \qquad [CTA]^p > 3.5 \times 10^{-6} \left[\frac{mol}{l}\right]. \quad (2)$$

The calculation is performed off-line, while all other calculations discussed in the sequel have to be performed on-line.

Calculation of the set point of the molar monomer holdup in the reactor

To determine the largest possible reaction rate two criteria have to be evaluated in order to find the highest possible reaction rate, i. e. the maximum admissible concentration of monomer in the particle phase. These two criteria are:

- A droplet phase has to be avoided.
- The heat generated by reaction must not exceed the maximum heat removal capability.

As a result of both calculations a maximum admissible number of moles is calculated, the minimum of the two is used as the set point of a lower level controller.

Avoiding a droplet phase When the maximum swelling of the polymer particles with monomer is reached, a droplet phase will form. As the reaction rate in the polymer particles depends solely on the amount of monomer in the particles, an additional droplet phase will not increase the reaction rate, but it will downgrade the controllability of the process as the droplets act as a reservoir and enable the direct influence on the concentrations in the particle phase. Thus, droplet phase is unnecessary and potentially dangerous and has to be avoided.

For the calculation of the maximum amount of monomer we will assume that the minimum amount of CTA has no strong influence on the phase distribution of the monomer, especially, when a safety margin for the maximum amount has been set. Thus, the necessary calculation is similar to the case of a homopolymerization.

In an earlier paper, we calculated a condition for the non-existence of a droplet phase (Krämer *et al.*, 2001):

$$V_W \geq (k^d - 1)V_M^w. \quad (3)$$

If the maximum total amount of monomer allowed in

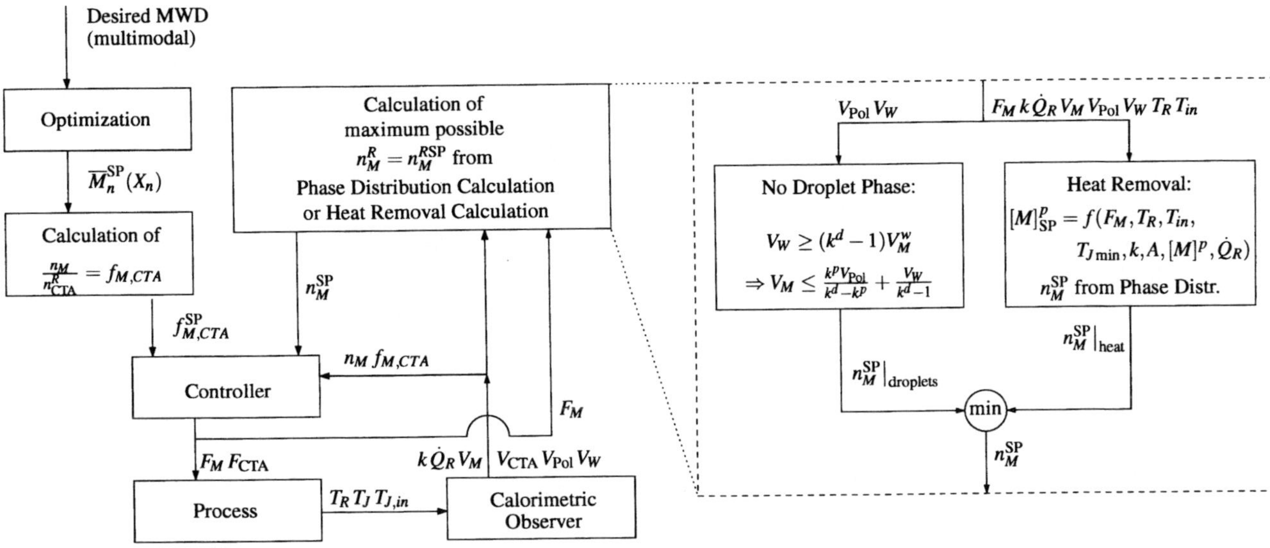

Fig. 1. Description and explanation of the hierarchical control concept applied

the system to avoid the formation of a droplet phase is to be calculated, the amount of water and the amount of polymer in the particles need to be known. Here, V_{Pol} and V_W are estimated and calculated from the mass balance, respectively. Thus, using the following equations and inequality (3) as an additional equation, the maximum amount of monomer can be calculated:

$$k^p \frac{V_M^w}{V^w} = \frac{V_M^p}{V^p} \tag{4}$$

$$V^w = V_M^w + V_W \tag{5}$$

$$V^p = V^p + V_{Pol} \tag{6}$$

$$V_M = V_M^p + V_M^w. \tag{7}$$

The result of this system of equations is the following condition for the total molar holdup n_M^{SP}:

$$n_M^{SP} = \varphi \left(\frac{k^p V_{Pol}}{k^d - k^p} + \frac{V_W}{k^d - 1} \right) \bar{V}_M. \tag{8}$$

Here, $\varphi < 1$ denotes a safety margin. With increasing conversion, e. g. with an increasing amount of polymer, the maximum amount of monomer will increase. On the other hand, the heat transfer coefficient will decrease significantly and will limit the heat removal and therefore the acceptable amount of free monomer.

Maximum Heat Removal To calculate the maximum heat removal, a number of internal states and the manipulated variables are needed. It has been shown that the inner states can be calculated by applying open loop observers as shown in (Vicente *et al.*, 2001). The scope of this contribution is the derivation of the control concept. Thus, we assume that all required states in the following calculations are available.

Assume that a minimum jacket inlet temperature $T_{J,\min}$ is given. Furthermore it is assumed that the process should be operated at isothermal conditions. Hence, from the stationary heat balance for the reactor and the assumption that the change of the heat capacity C_R in time depends mainly on the feed to the reactor, the following expression for the maximum heat of reaction which can be removed by the jacket cooling is derived:

$$\dot{Q}_{R,\max} = \dot{V}_{in}\rho_{in}c_{p,in}(T_R - T_{in}) - kA(\bar{T}_{J\min} - T_R).. \tag{9}$$

Here, $\bar{T}_{J\min}$ the specified minimum average temperature allowed in the jacket, i.e. in case of a CSTR-jacket it is the specified outlet temperature and in case of a PFR-jacket it is the mean temperature over the length of the pipe. This temperature is obviously depended on the minimum jacket inlet temperature.

The heat of reaction currently produced in the reactor is defined by:

$$\dot{Q}_R = k_p[M]^p \frac{\bar{n}N_T}{N_A}(-\Delta H). \tag{10}$$

From the ratio between the currently produced heat of reaction $\dot{Q}_R$ (10) and the maximum removable heat of reaction $\dot{Q}_{R,\max}$ (9) an equation for the maximum allowed concentration of monomer in the particle phase can be derived:

$$[M]_{\max}^p = (\dot{V}_{in}\rho c_p(T_R - T_{in}) - kA(\bar{T}_{J\min} - T_R)) \frac{[M]^p}{\dot{Q}_R}. \tag{11}$$

By adjusting $T_{J\min}$ a safety margin can be defined. The desired set-point n_M^{SP} can now be calculated from the inverse of the phase distribution calculation. Alternatively and less complex it can be assumed that no droplet phase exists and the solubility of the monomer in water is negligible:

$$n_M^{SP} = [M]_{\max}^p \frac{V_{Pol}}{1 - [M]_{\max}^p \bar{V}_M}. \tag{12}$$

This simplification is valid as the non-existence of a

droplet phase is checked by the foregoing criterion.

Calculation of the total molar holdup ratio

The off-line calculation of step 1 gives the ratio $f^p_{M,CTA}$ between monomer and CTA in the polymer particle phase. As the controlled variables are the total amounts of the reactants, the desired ratio of these values must be determined. Hence, the phase distribution algorithm (Arzamendi *et al.*, 1998) has to be reformulated and extended by the following equation:

$$V_{CTA} = \frac{\bar{V}_{CTA} V_M}{\bar{V}_M f^p_{M,CTA}} \frac{1 + \frac{V^w}{V^p k^p_{CTA}} + \frac{V^d k^d_{CTA}}{V^p k^p_{CTA}}}{1 + \frac{V^w}{V^p * k^p_M} + \frac{V^d k^d_M}{V^p k^p_M}}. \quad (13)$$

By introducing this equation the amount of CTA in the particle phase is fixed, but the total amount of CTA in the reactor becomes a free variable. From the solution of the modified phase distribution algorithm the desired ratio is given by

$$f_{M,CTA} = \frac{V_M \bar{V}_{CTA}}{V_{CTA} \bar{V}_M}. \quad (14)$$

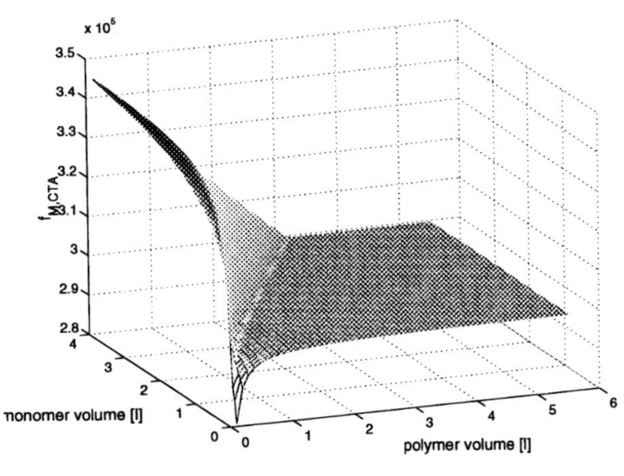

Fig. 2. Dependency of $f_{M,CTA}$ on the polymer volume and the monomer volume

Figure 2 shows the dependency of the ratio $f_{M,CTA}$ on the polymer and monomer volume. The desired ratio in the polymer is $f^p_{M,CTA} = 3 \times 10^5$. It follows that for a wide range of polymer and monomer volumes the two ratios (particle and reactor) are nearly identical. If a droplet phase exists, the necessary ratio of the molar holdup of monomer and CTA in the reactor differs significantly from the desired ratio in the particle phase.

$\bar{V}_{CTA}$ and $\bar{V}_M$ denotes the molar volumes of the CTA and the monomer, respectively. The set point of the total molar holdup reads as:

$$n^{SP}_{CTA} = \frac{n^{SP}_M}{f_{M,CTA}}. \quad (15)$$

In this manner, the process is always driven under time optimal conditions as it is performed at its upper constraints (Srinivasan *et al.*, 2003). Finally, controllers have to be designed that track the given set point trajectory.

Controller design

We investigated two types of controllers, simple decentralized PI-controllers for each of the molar holdups and an NMPC. The controlled variables are the total amounts of monomer and CTA in the reactor and the manipulated variables are defined by the molar flow rates of the to reactants.

NMPC-design The dynamic model used in the NMPC scheme is based on the rigorous model described in section 2. The state variables are:

- total reactor volume
- overall amount of monomer and CTA
- volume of water
- concentration of initator
- polymer volume
- concentration of radicals in the water phase
- average number of radicals per particle.

The NMPC is realized in a sequential approach using the SQP-routine `fmincon.m` provided by MATLAB. In order to make the computation more effective, the phase distribution equations were linearized around the initial states of the NMPC in each step what enables the formulation of an analytic solution. As a linear controller is used to keep the reactor temperature at the desired set point, it is unnecessary to take the heat balance equations for the reactor and the jacket into account for the NMPC. Different prediction and control horizon lengths were checked and tuning of the weighting matrices was performed by simulation studies. The cost function is chosen as:

$$J = \left| n^{SP} - n \right|^2_{Q_x} + \left| \Delta u \right|^2_{Q_{\Delta u}}. \quad (16)$$

The first term assesses the deviation from the desired set points of the molar holdups of monomer and CTA and the second term weights the changes of the manipulated variables to avoid too fast control actions. The upper and lower bounds for the flow rates are given by the equipment of the pilot plant.

Decentralized PI-controller design In general, emulsion polymerization processes are highly nonlinear in nature. Due to the avoidance of a droplet phase in the operation of the process and due to the fact, that monomer and CTA do not interact, the nonlinearities are not strong. Therefore, it is also possible to design linear decentralized controllers to track the set points of the molar holdups of monomer and CTA. The gains of the controllers were adjusted by simulation studies.

The proposed hierarchical control approach can also be applied to controlling the composition in a copolymerization. In this case the coupling between the different monomers is strong and the nonlinearities are significant. It is expected that in this situation an NMPC outperforms a linear MIMO-controller.

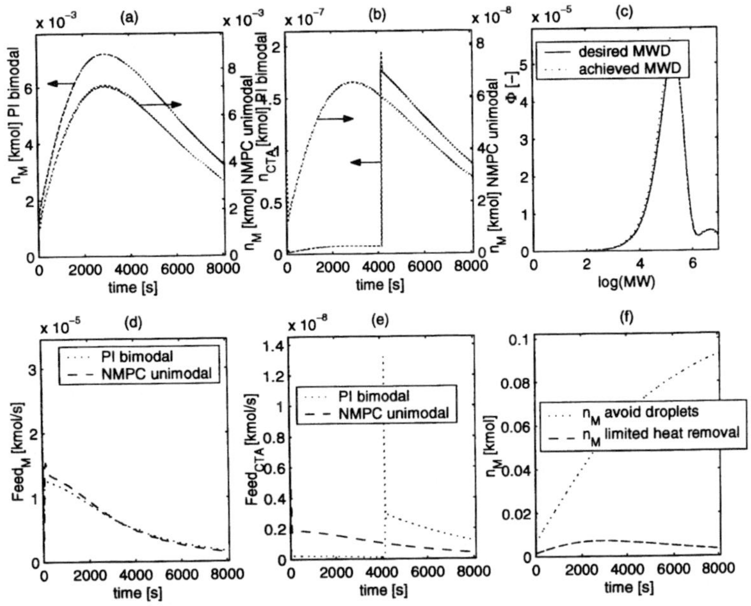

Fig. 3. Performance of the NMPC and the decentralized PI-controller scheme - PI controlled process: production of a bimodal MWD - NMPC controlled process: production of a uni-modal MWD - (a) Tracking of the molar holdup of monomer n_M, (b) Tracking of the molar holdup of CTA n_{CTA}, (c) comparison between desired and achieved MWD for PI-controlled process, (d) molar feed monomer Feed_M, (e) molar feed CTA Feed_{CTA}, (f) development of the constraints over batch time

4. SIMULATION STUDIES

The simulation results are obtained for a polymerization in a 10l stainless steel reactor with geometric parameters taken from a pilot plant at the Process Control Laboratory at the University of Dortmund. Approximately 7% of the total polymer to be produced is pre-charged as a seed. The production of uni-modal as well as multi-modal molecular weight distributions was investigated.
Figure 3(c) shows the simulation results for the production of a bi-modal molecular weight distribution using decentralized PI-controllers. It follows that the control strategy is able to produce the desired MWD. Figures 3(a-b) and (d-e) depict the set point tracking and the manipulated variables, e.g. the molar feed rates of monomer and CTA, for either an NMPC- (uni-modal) and PI-controlled (bi-modal) process. The NMPC was investigated as it is most often stated in the open literature that nonlinear control is required for this type of process. The simulation results in this contribution clearly demonstrate that this is not necessary. Figures 3 (a-b) demonstrate that both approaches are able to track the desired set points very quickly. Certainly, there is a difference in the performance between the NMPC-scheme and the decentralized PI-controllers as the NMPC is operated with a sample time of 30 seconds whereas the time discrete PI controllers are performed with a sample time of 3 seconds. The different sampling intervals result from the different computational effort to calculate the manipulated variables.
Due to the pre-charged seed, the upper constraints of the process are completely determined by the possible heat removal, see figure 3 (f). The set point of the molar holdup in the reactor increases for the first third of the process time. The reason for this behaviour is the increase of the polymer volume. As the reaction rate depends only on the concentration of the monomer in the polymer phase, it is possible to increase the total amount of monomer in the reactor without increasing the reaction rate. With increasing conversion/process time the decreasing heat removal capacity becomes more and more important and leads to a decrease of the allowed total amount of monomer in the reactor. Hence, the simulation results show that the hierarchical control approach is a possible solution to time optimal operation of the emulsion polymerization process at hand.
Figure 4 depicts the trajectory of the desired ratio of monomer and CTA and the output of the controlled process versus the total conversion. The control algorithm is able to track the desired ratio quickly. It has to be considered that in the beginning of the process the reaction rate is high and the conversion increases fast. It takes the PI-controllers less than 260 seconds to bring the process to a tolerance band of $\pm 1\%$ of the set point.
Figure 5 depicts the temperatures in the reactor and the jacket inlet for a disturbed process. It is assumed that an additional nucleation occurs at $t = 5000$ seconds. If an off-line calculated trajectory for the amounts of monomer and CTA is used, it is not possible to keep the temperature of the reactor at the desired value as the reaction rate increases significantly due to the larger number of reaction loci and the too large amount of monomer in the system. The proposed hierarchical control approach performs well and adjusts the trajectory of the monomer holdup to the new reaction conditions. The set point of the total amount of monomer is therefore an

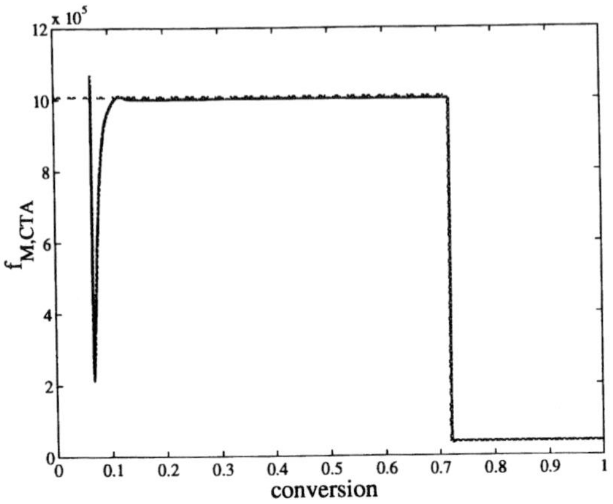

Fig. 4. Performance of the decentralized PI-controller scheme - tracking of the desired ratio $f_{M,CTA}$ (dashed: set point, solid: achieved ratio)

additional degree of freedom in the process which cannot be used if an off-line calculated trajectory is applied.

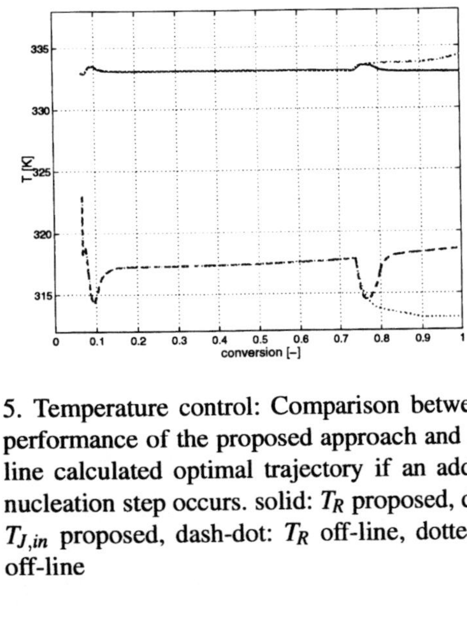

Fig. 5. Temperature control: Comparison between the performance of the proposed approach and an off-line calculated optimal trajectory if an additional nucleation step occurs. solid: T_R proposed, dashed: $T_{J,in}$ proposed, dash-dot: T_R off-line, dotted: $T_{J,in}$ off-line

5. CONCLUSIONS

A new hierarchical control approach for a time optimal production of a polymer with a prescribed molecular weight distribution in an emulsion polymerization using CTA has been developed. This approach avoids an off-line optimization to calculate trajectories for monomer and CTA. Instead, the constraints of the process have been investigated and are used to calculate at each point in time set points for the molar holdups of monomer and CTA. The desired molecular weight distribution is approximated by a combination of ideal distributions. The average molecular weights of these ideal distribution are produced for a certain fraction of conversion. The maximum possible amount of monomer is calculated by the avoidance of a droplet phase or the maximum heat removal capacity, respectively. Appropriate conditions were derived. Applying a modified phase distribution algorithm, the total amount of CTA can be calculated and finally it can be used as set point in a nonlinear model predictive controller or a decentralized PI-controller. It was shown by simulation that the derived approach performs well and is able to produce arbitrary multi modal molecular weight distributions. Another demonstrated feature of the proposed approach is the possibility to reject disturbances in the process, which cannot be achieved with off-line calculated trajectories.

ACKNOWLEDGEMENTS

Parts of this work were founded by the German Research Foundation (DFG-Deutsche Forschungsgemeinschaft, DFG-EN 152/31-1). The authors gratefully acknowledge this financial support.

6. REFERENCES

Arzamendi, Gurutze, Claudia Sayer, Nuria Zoco and José M. Asua (1998). Modeling of MWD in emulsion polymerization: Partial distinction approach. *Polymer Reaction Engineering* **6**(3-4), 193–223.

Gentric, C., F. Pla, M. A. Latifi and J. P. Corriou (1999). Optimization and nonlinear control of a batch emulsion polymerization reactor. *Chemical Engineering Journal* **75**, 31–46.

Georgakis, C., R. Aris and N. R. Amundson (1977). Studies in the control of tubular reactors. *Chemical Engineering Science* **32**, 1359–1369.

Krämer, Stefan, Ralf Gesthuisen and Sebastian Engell (2002). Re-optimization based control of copolymer quality in semi continuous emulsion polymerization. In: *Proceedings of the American Control Conference* (R. Russell Rhinehart, Ed.). American Automatic Control Council. Anchorage, USA. pp. 3870–3875.

Krämer, Stefan, Ralf Gesthuisen, Michael Wieland and Sebastian Engell (2001). An alternative model formulation for emulsion polymerization processes. In: *Conference Proceedings of the European Congress of Chemical Engineering* (VDI, Ed.). VDI-GVC. VDI-GVC. Frankfurt, Germany.

Kravaris, C., R. A. Wright and J. F. Carrier (1989). Nonlinear controllers for trajectory tracking in batch processes. *Computers and Chemical Engineering* **13**(1/2), 73–82.

Seferlis, P. and C. Kiparissides (2001). Optimizing control of molecular weight properties for batch free-radical polymerization reactors. *IFAC* pp. 251–256.

Srinivasan, B., S. Palanki and D. Bonvin (2003). Dynamic optimization of batch processes. i. characterization of the nominal solution. *Computers and Chemical Engineering* **27**, 1–26.

Vicente, Matias, Selwa Ben Amor, Luius M. Gugliotta, Jose R. Leiza and Jose M. Asua (2001). Control of molecular weight distribution in emulsion polymerization using on-line reaction calorimetry. *Industrial and Engineering Chemistry Research* **40**(1), 218–227.

ELSEVIER
IFAC
PUBLICATIONS
www.elsevier.com/locate/ifac

A STAGE-BASED MONITORING METHOD FOR BATCH PROCESSES WITH LIMITED REFERENCE DATA

Ningyun Lu [1,2], Yi Yang[1], Fuli Wang[2], Furong Gao[1*]

[1]*Department of Chemical Engineering*
The Hong Kong University of Science & Technology
Clearwater Bay, Kowloon, Hong Kong

[2]*School of Information Science and Engineering*
Northeastern University
Shenyang, Liaoning, P. R. China

Abstract: A method is proposed for batch process monitoring and fault diagnosis, starting from a single batch reference data and updating with accumulation of successive batches. A moving data window method is adopted for exploring local covariance structure, stage division, and the development of monitoring models. The application to an injection molding process shows the effectiveness and feasibility of the proposed method for batch industry. *Copyright © 2004 IFAC*

Keywords: Batch process monitoring, Principal component analysis, Multistage modeling, Injection molding.

1. INTRODUCTION

Batch processes are widely used in chemical, semiconductor, food and biology industry for producing high-value-added products to meet today's rapidly changing market. Batch processes are characterized by prescribed sequential operations in an infinite duration; they are subject to various disturbances that may affect the final product quality and the degree of reproducibility. Proper monitoring and diagnosis of the process is important for safety and quality improvement (Kourti and MacGregor, 1995; Kosanovic et al., 1996; Wold et al., 1998; Louwerse and Smilde, 2000; Sprange et al., 2002; Ündey and Cinar, 2002; Kourti, 2003).

Most batch process monitoring methods are based on such multivariate statistical projection techniques as multi-way principle component analysis (MPCA) and partial least squares (MPLS) (Nomikos and MacGregor, 1994, 1995a, 1995b). In those methods, all correlation information in process variables over the whole batch duration are extracted to develop a statistical model for the monitoring of the processes. This type of models is effective in determining whether or not a batch operation is normal as a post process analysis tool. The requirement of many cycle reference data and the drawbacks associated with on-line process monitoring may hinder its wider application to the industry.

With the existing methods, the reference data collected for modeling are expected covering statistically all normal batch-to-batch variation. This is relatively easy for batch processes with short operation duration and processes that are inexpensive to conduct many trial runs. Some slow batch processes, such as bio-related processes, however, may take exceptionally long time to complete a batch run. In this case, a method will be needed for developing a monitoring model with a minimal batch cycle, for example, one successful batch run. The model can be improved with the

* Corresponding author: E-mail: kefgao@ust.hk; Tel: +852-2358-7139; Fax: +852-2358-0054

newly available batches, without having to wait for sufficient successful batches.

This paper proposes to develop such a monitoring method starting with minimum successful batches. The initial database may contain only one successful batch run. The basic unit for modeling is a moving data window. In each window, data are arranged as a two-way matrix, where each row contains process measurements at each sampling interval, and rows are arranged according to time. Data windows are properly normalized before being used to derive a PCA model for each window to extract local covariance information. This method is termed the moving time-window PCA-based batch modeling in this paper.

The proposed moving time-window PCA is similar to the moving principal component analysis by Kano et al. (2001), which was developed for continuous processes for detecting changes of process operation condition. Lennox et al. (2001) had also presented a method called moving window PCA for batch processes. Their method, however, also requires sufficient successful batches for modeling.

As mentioned before, the PCA model of each time window contains local covariance information. Although batch process variables are time-varying, the local covariance structure in process measurements will be similar when the data of two windows have the similar underlying characteristics. This idea is similar to the method of the authors (Lu et al., 2004a), in which a batch process is divided into "stages" according to the changes of process correlation, and then a sub PCA model is built for each stage for monitoring. The differences between this proposed method and previous method lie in: (1) this paper focuses on the change of the covariance structure in time direction, not in batch direction; and (2) it requires only one reference batch for initial modeling, and the model is updating with the accumulation of new successful batch data. The details of the proposed method are given in section 2. An illustrative example is shown in section 3. Conclusions are given in the last section.

2. METHODOLOGY

2.1. Moving time-window PCA-based batch modeling

As mentioned before, it is attractive to develop a data-based model for process monitoring using limited batches. In this proposed modeling method, the reference data can be an arbitrary successful batch of history, represented as $X(N \times m)$, where N is the total sampling points in the batch run, and m is the number of process variables. As illustrated in Figure 1, a moving time window of the batch is proposed to extract the local covariance information.

The data in a window are arranged as a two-way matrix, noted as $\tilde{X}(n \times m)$, where n is the data length in a window. Obviously, a large window will result in stable local PCA models, reliable but slow fault detection; a small window can rapidly detect process abnormalities but may have excessive false alarms. In this paper, n is approximately of two or three times of the number of process variables, as recommended in the field of multivariable regression to ensure a reliable statistical model (Johnson and Wichern, 2002). Moving step can be set as small as 1 for prompt fault detection.

With this arrangement, $(N-n)$ number of windows can be resulted for each batch, designated as, $\tilde{X}^k$, $(k=1,\cdots,N-n)$. Each window should be mean-centered to provide the reference trajectories of process variables for normalizing the new batch for online process monitoring. At the same time, the measurements should be also properly scaled to eliminate the influence of different measuring units. After normalization, PCA can be applied to each moving time-window, $\tilde{X}^k = \tilde{T}^k \cdot (\tilde{P}^k)^T$. The resulted PCA loading matrices, $\tilde{P}^k(m \times m)$, contain process covariance information in the kth data window.

Although the batch process variables are time varying, the covariance structure of two neighboring windows will not change much, provided that the process characteristic does not change rapidly. During the time when the process is driven by the

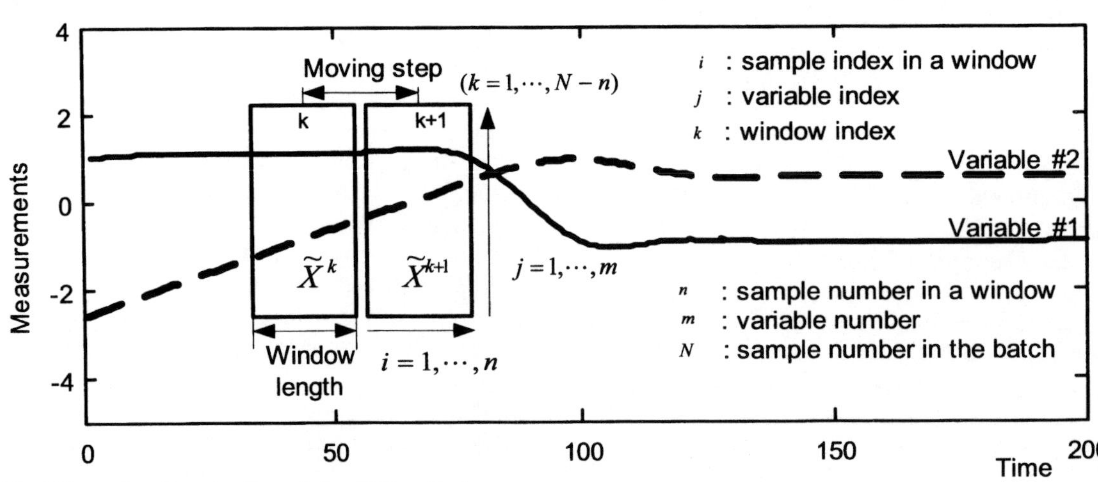

Fig. 1 Illustration of moving time-window PCA method

similar underlying characteristic, those windows will have similar covariance structures, consequently, the similar PCA loading matrices. In this way, a batch process can be divided into several stages or phases, of different process characteristics. Covariance structure changes, in corresponding to the changes of process characteristics, can be used to divide the process into stages.

There are several methods for defining the similarity of two PCA loading matrices. For instance, Krzanowski (1979) presented a PCA similarity factor as a measure of the similarity between two data spaces; and Johannesmeyer et al. (2002) discussed a statistical method for comparing two PCA models by calculating the T^2 and/or Q similarity factor. The existing methods are, however, ill-suited for partitioning simultaneously a large number of PCA models into groups according to their similarities. The PCA loading clustering algorithm (Lu et al., 2004a) is adopted here to divide the moving time-window PCA models into groups to reflect the change of process covariance structure. By the clustering algorithm, $(N-n)$ number of moving time-window PCA loading matrix $\tilde{P}^k$ are partitioned into C number of sub groups, where the loading matrices remain similar within the same group, but showing significant difference between groups. The clustering results, associated with process operation time, can be used to define process sub stages. The scheme of stage division is illustrated in Figure 2.

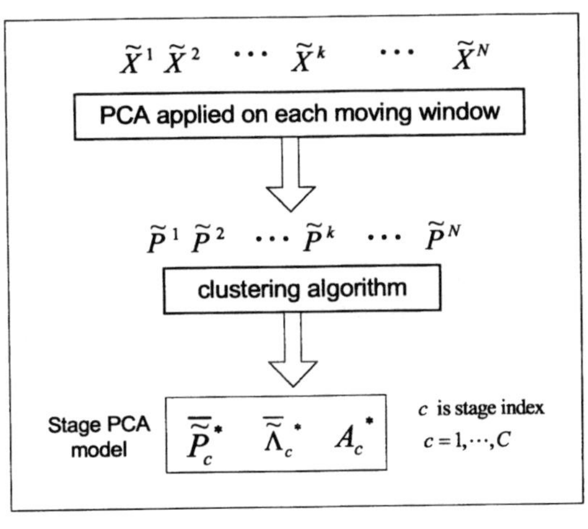

Fig. 2. Scheme of stage-division based on window's PCA loading matrices.

A stage PCA model can be developed by finding an optimal PCA loading matrix as the representative one for that stage, mathematically,

$$\tilde{P}_c^* = \underset{k}{Min}\left\|\tilde{P}^k - \tilde{P}_c^*\right\| = \frac{1}{n_c}\sum_k \tilde{P}^k (c=1,\cdots,C; k=1,\cdots,n_c), \tag{1}$$

where n_c is the number of loading matrices belonging to stage c $(c=1,\cdots,C)$. The stage representative loading matrix, $\tilde{P}_c^*$, is divided into two parts, $\bar{P}_c^*(m\times A_c^*)$ and $\bar{\bar{P}}_c^*(m\times(m-A_c^*))$, for principal component and residual subspaces, where the number of retained principal components, A_c^*, can be selected also by the method (Lu et al., 2004a). Similarly, the stage representative singular value matrices, $\bar{\Lambda}_c^*$, are also defined for monitoring.

For data $\mathbf{x}(1\times m)$, the principal component score and SPE can be calculated by the following stage PCA model,

$$\begin{aligned} \mathbf{t} &= \mathbf{x}\bar{P}_c^* \\ \mathbf{e} &= \mathbf{x} - \mathbf{x}\bar{P}_c^*(\bar{P}_c^*)^T . \\ SPE &= \mathbf{e}^T\mathbf{e} \end{aligned} \tag{2}$$

The initial control limit trajectories for the two statistics, Hotelling- T^2 and Q (or SPE), are estimated from the first successful reference batch. With new batches available, the control limits can be gradually updated to focus more on the batch-to-batch variation. The determination of the initial control limits and the updating procedures are given in the next.

2.2. On-line monitoring and model updating

In on-line monitoring, one should first determine which stage new data of the evolving batch belong to before calling the corresponding stage PCA model to calculate the two statistics. Since process operation time can be associated with the stage division results, one can determine the stage that new data belong to by simply checking which time span the current sampling falls to. Process monitoring is conducted by comparing the two statistics with the predetermined control limits. When the statistics go beyond the control limit, responding to an abnormality, the contribution plot (Miller et al., 1998), a commonly used diagnosis tool, can be used to show the variables impacted by the occurred abnormality.

The proposed stage monitoring method has been derived from a single reference batch, considering only within-batch information. Batch-to-batch variation information can also be explored with accumulation of batches, *e.g.*, by adjusting the stage representative PCA loading matrix and the control limit of SPE.

Initial Control limits of the two statistics

The control limits of Hotelling- T^2 are estimated for each stage, while the SPE limits are estimated within each moving window.

$$T^2 = \mathbf{t}(\bar{\Lambda}_c^*)^{-1}\mathbf{t}^T \sim \frac{A_c^*(n_c-1)}{(n_c-A_c^*-1)}F_{A_c^*,(n_c-A_c^*-1),\alpha} . \tag{3}$$

Adopting the works of Box (1954) and Jackson and Mudholkar (1979), the SPE statistic can be approximated by a weighted Chi-squared distribution, $g\chi_h^2$, where the weight g and the freedom degree h can be obtained following the same approach of

Nomikos and MacGregor (1995a). The parameters of the $g\chi^2_h$ distribution at time k are estimated from the SPE values in the moving window $\tilde{X}^k(n\times m)$, that is, $g^k = v^k/2m^k$ and $h^k = 2(m^k)^2/v^k$, where m^k is the average of the window's SPE values and v^k is the corresponding variance. Thus, SPE control limit at time k can be approximated by,

$$SPE^k_\alpha = (v^k/2m^k)\chi^2_{x(m^k)^2/v^k,\alpha}\,. \tag{4}$$

Stage model and the SPE control limits updating

When a new batch run follows the similar operation sequence and variable trajectories of the reference, it can be considered as a normal batch. The stage PCA models, reference trajectories and control limits, should be adjusted to include the normal batch-to-batch variation.

The stage representative PCA loading matrix as defined in Equation (1), *i.e.*, the center of the window's PCA loading matrices that belongs to stage *c*, may move with the accumulation of new successful batch data. The new center can be readily obtained by recalculating Equation (1) with the window's PCA loading matrices for the new normal batch.

The control limits detailed in the last section are derived from one reference batch, they may be too tight (or loose) in certain periods for monitoring a new batch. It is desirable to adjust the control limits by taking the batch-to-batch variation into account. For the initial SPE control limits, SPE^k_α are estimated from the SPE values of the data in the window, $\tilde{X}^k(n\times m)$. In the updating procedure, the data used for estimating the parameters in Equation (4) can be augmented to include the data of the new batch,

$$\tilde{X}^k_{new} = \begin{pmatrix} \tilde{X}^k \\ \mathbf{x}^k_{new} \end{pmatrix}, \tag{5}$$

where $\tilde{X}^k_{new}$ is gradually filled by the new data at time k of different batches. By doing so, the moving window will contain more and more batch-to-batch variation, and the SPE control limits estimated from this new data window will be correspondingly adjusted.

3. ILLUSTRATIVE EXAMPLE

The proposed modeling and monitoring method is tested on a batch process, injection molding process. Although the proposed method is developed for those long batches, the method works equally well with a relatively short batch such as the injection molding process.

3.1. Injection molding process

Injection molding (Rubin, 1972; Yang and Gao, 1999), a key process in polymer processing, transforms polymer materials into various shapes and types of products. A typical injection molding process consists of three physical stages, injection of molten plastic into the mold, packing-holding of the material into the mold under pressure, and cooling of the plastic in the mold until the part becomes sufficiently rigid for ejection. Plastication takes place in the barrel in the early cooling phase, where polymer is melted and conveyed to the front of barrel by screw rotation, preparing for next cycle.

Figure 3 shows a simplified diagram of a typical reciprocating-screw injection molding machine with instrumentations. Several normal batches are collected under the same operation condition to the previous work (Lu et al., 2004b), among which an arbitrary batch is used for modeling, and the other cycles are used for model updating. An abnormal batch with check-ring problem is also conducted to test the proposed process monitoring and diagnosis scheme. Ten process variables are selected for modeling, that is, Nozzle Pressure (No.1), Stroke (No.2), Injection Velocity (No.3), Hydraulic Pressure (No.4), Plastication Pressure (No.5), Cavity Pressure (No.6), Screw Rotation Speed (No.7), SV1 valve opening (No.8), SV2 valve opening (No.9), and Mold Temperature (No.10), respectively. Window length is set to be 30 and moving step is 1.

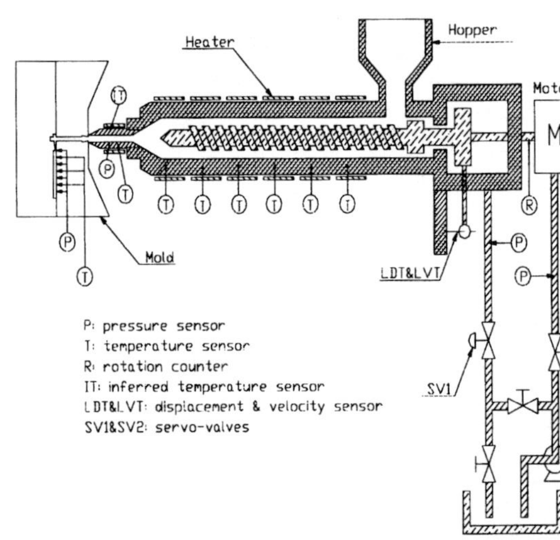

Fig. 3. Simplified illustration of injection molding machine and measuring points

3.2. Experimental results

Without using any prior process knowledge, the proposed algorithm automatically divides the trajectories of the injection molding batch into eight stages according to the change of local covariance structure, among which four long stages correspond to the four physical operation stages, *i.e.*, injection, packing-holding, plastication and cooling stages. Short new stages emerge between the four main stages, corresponding to the transitions from one major stage to the next. Dividing a batch process into "steady" and transient stages can not only benefit process monitoring and diagnosis, but enhance process understanding.

On-line process monitoring and fault diagnosis are conducted by judging whether the scores and SPE value of the coming measurements in a running batch are below the control limits. In the on-line monitoring of the batch with check-ring problem, both the T^2 and SPE charts can detect the fault, as shown in Figure 4. Hotelling-T^2 statistic is more sensitive than SPE because the failure of check-ring valve will not significantly deteriorate the covariance structure of time direction, however, the deviations from the reference variable trajectories result in large principal component scores, hence, have significant T^2 values.

By process knowledge, the check-ring problem occurs in the injection stage, also impacts the packing-holding and plastication stages, but have little influence to the cooling stage. The monitoring results shown in the T^2 chart of Figure 4 agree well with the aforementioned. As the check-ring problem has no influence on the cooling stage, the plot for this long stage is not given to allow more detailed presentation of the results in the affected stages. The SPE chart of Figure 5 shows, the monitoring of the same fault with control limits updated with ten successful cycles. A comparison of Figures 4 and 5 show the adjustment of the control limits, by considering the batch-to-batch variation. The results are similar, indicating that it is viable to monitor such a process starting with just one cycle data. The diagnosis result is shown in the contribution plot in Figure 5, where Nozzle Pressure (No.2), Stroke (No.3), Injection Velocity (No.4), Hydraulic Pressure (No.4), Cavity Pressure (No.6) and Mold Temperature (No.10) are seriously impacted by the detected fault. This fault pattern is well agreed with the check-ring problem. Detailed description on the check-ring problem can be found in reference (Lu et al., 2003).

4. CONCLUSION

A new batch monitoring method has been proposed for those cases where there exist difficulties in collecting sufficient cycles in limited time. The proposed method uses a single batch reference data to extract local covariance information by applying

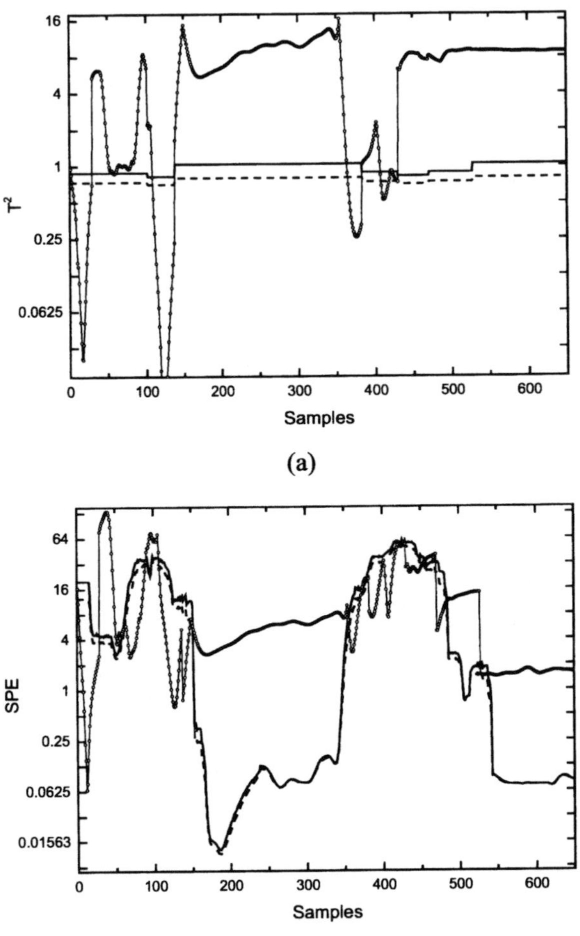

(a)

(b)

Fig. 4. Monitoring charts of a faulty batch with check-ring problem.
(a). Hotelling T^2 monitoring chart;
(b). SPE monitoring chart.
(Solid line, 99% control limit; Dash line, 95% control limit)

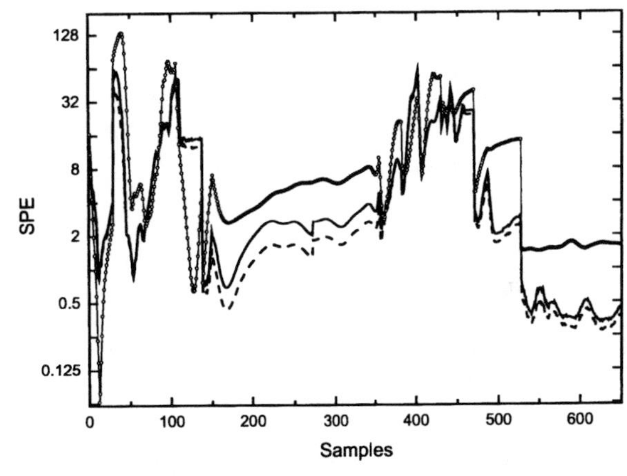

Fig. 5. SPE monitoring chart with updated control limits for the batch with check-ring problem.
(Solid line, 99% control limit; Dash line, 95% control limit)

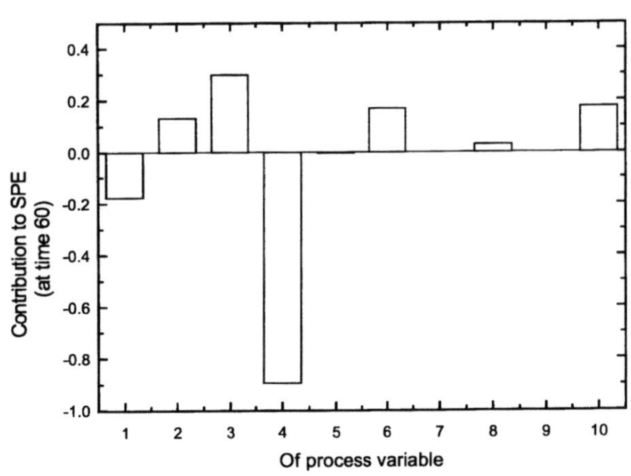

Fig. 6. Contribution plots for the batch with check-ring problem.

PCA on a moving time-window that scans process trajectory over the batch duration. Process stages are determined by analyzing the change of process covariance structure, by partitioning the window's PCA loading matrices using a clustering algorithm. A sub PCA model has been developed for each stage according to the clustering results. Model updating has also been discussed with newly available batches. The proposed modeling method can not only give a valid monitoring scheme with a minimal reference batch, but also it can divide the batch into stages. The division of a batch process without the requirement of prior process knowledge can improve process monitoring efficiency in terms of reduction of computation resources, also enhance process understanding, and consequently benefit fault diagnosis.

ACKNOWLEDGEMENT

This work was supported in part by Hong Kong RGC (DAG01/02 EG11) and the National Science Foundation of China (No. 60374003).

REFERENCES

Box G.E.P. (1954). Some theorems on quadratic forms applied in the study of analysis of variance problems: Effect of inequality of variance in one-way classification. *The annals of Mathematical Statistics*, **25**, 290.

Jackson J.E. (1991). A user's guide to principal components. *John Wiley & Sons, Inc.*

Jackson J.E. and Mudholkar, G.S. (1979). Control procedures for residuals associated with principal components analysis, *Technometrics*, **21**, 341.

Johannesmeyer M.C., Singhal A. and Seborg D.E. (2002). Pattern matching in historical data. *J. AIChE*, **48**, 2022.

Johnson R.A. and Wichern D.W. (2002). Applied multivariate statistical analysis. *Prentice Hall, N.J.*

Kosanovic K.A., Dahl K.S. and Piovoso M.J. (1996). Improved process understanding using multiway principal component analysis. *Ind. Eng. Chem. Res.*, **35**, 138.

Kourti T. and MacGregor J.F. (1995). Process analysis, monitoring and diagnosis, using multivariate projection methods. *Chemometrics Intell. Lab. Syst.*, **28**, 3.

Kourti T. (2003). Multivariate dynamic data modeling for analysis and statistical process control of batch processes, start-ups and grade transitions. *J. Chemometrics*, **17**, 93.

Krzanowski W.J. (1979). Between-groups comparison of principal components. *Journal of the American Statistical Association,* **74**, 703.

Louwerse D.J. and Smilde A.K. (2000). Multivariate statistical process control of batch processes based on three-way models. *Chem. Eng. Sci.*, **55**, 1225.

Lu N., Yang Y., Gao F. and Wang F. (2003). Stage-based multivariate statistical analysis for injection molding. *Proceedings of International symposium on advanced control of chemical processes (ADCHEM)*, Hong Kong.

Lu N., Gao F., and Wang F. (2004a). A sub-PCA modeling and on-line monitoring strategy for batch processes. *AIChE J.*, **50**, 255.

Lu N., Gao F., Yang Y., and Wang F. (2004b). A PCA-based Modeling and On-line Monitoring Strategy for Uneven-length Batch processes. *Ind. Eng. Chem. Res.*, In Presss

Miller P., Swanson R.E. and Heckler C.E. (1998). Contribution plots: A missing link in multivariate quality control. *Appl. Math. And Comp. Sci.*, **8**, 775.

Nomikos P. and MacGregor J.F. (1994). Monitoring of batch processes using multi-way principal component analysis. *AIChE J.*, **40**, 1361.

Nomikos P. and MacGregor J.F. (1995a). Multivariate SPC charts for monitoring batch processes. *Technometrics*, **37**, 41.

Nomikos P. and MacGregor J.F. (1995b). Multiway partial least squares in monitoring batch processes. *Chemometrics Intell. Lab. Syst.*, **30**, 97.

Rubin I.I. (1972). Injection molding theory and practice. *John Wiley & Sons, Inc.*

Sprange E.N.M., Ramaker, H.J., Westerhuis J., Smilde A. (2002). Critical evaluation of approaches for on-line batch process monitoring. *Chem. Eng. Sci.*, **57**, 3979.

Ündey C. and Cinar A. (2002). Statistical monitoring of multistage, multiphase batch processes. *IEEE Control Systems Magazine*, **22**, 40.

Wold S., Kettaneh N., Friden H. and Holmberg A. (1998). Modeling and diagnosis of batch processes and analogous kinetic experiments. *Chemometrics Intell. Lab. Syst.*, **44**, 331.

Yang Y. and Gao F. (1999). Cycle-to-cycle and within-cycle adaptive control of nozzle pressures during packing-holding for thermoplastic injection molding. *Polym. Eng. Sci.*, **39**, 2042.

ELSEVIER
IFAC
PUBLICATIONS
www.elsevier.com/locate/ifac

CONTROLLER DESIGN FOR INTEGRATING PROCESSES IN SISO OR MIMO SYSTEMS

Hsiao-Ping Huang, Feng-Yi Lin, and Jyh-Cheng Jeng

Department of Chemical Engineering, National Taiwan University, Taipei Taiwan 10617, R.O.C.

Abstract: A two-step design to control integrator processes in an SISO or MIMO system is proposed. For an SISO system, the first step of the design is to stabilize the integrator process with P&D modes in a secondary loop to achieve good stability robustness and modest performance. In the second step of design, a P&I controller in the primary loop is derived to improve the overall performance. By this way, an equivalent two-element control system comprising of a single PID control loop and a pre-filter applying on the set-point command is derived. The same design approach is then applied to design multi-loop SISO controllers for MIMO processes, where integrating behavior exists. By employing the same two-step design and the direct method of Huang et al. (2003), 2-df multi-loop PI/PID controllers can be obtained. The Simulation results show that this presented method is effective for both SISO and MIMO process with integrators. *Copyright © 2004 IFAC*

Keywords: Integrating process, two-element control, RGA, Effective open-loop process

1. INTRODUCTION

The integrating processes, serving as material and energy inventory, or, as retention of material for processing and heat transfer, are found in many chemical plants. For years, PID controllers have been used to control such processes. For inventory purposes, usually, proportional controllers with wide proportional bands will be sufficient, but, for tight control purposes, the controller design encounters difficulties due to the existence of integrator and process dead-time. In the past, study of controller design has been mostly focused on open-loop stable processes, until lately, the control of integrating processes starts to attract increasing attentions from the researchers in this field. Many PI or PID controllers have been proposed (Chien and Fruehauf 1990; Tyreus and Luyben, 1992). But, Kwak et al. (2000) showed that a conventional simple feedback system with PI/PID controller will not be able to give the prescribed and desired response unless they use advanced control configuration. Most popular way of design is to use a Smith-predictor type of controller (Kwak, et al., 2001; Chien, et al., 2002; Hang et al., 2003) or use a time-optimal controller (Visioli, 2003). Nevertheless, they are either sensitive to modeling errors or difficult for practical uses. Lately, the control structure incorporating a cascaded loops is also adopted to design control for integrating processes, for examples: Kwak et al. (2000), etc. By the use of the cascaded loops, a two-element control system is resulted. Examples include the PID-P controller of Wang and Cai (2002), and the PI-PD controller of Kaya (2003). Although the resulting performances are slightly poorer than those using predictors, the 2-df controllers are simpler and can be implemented with commercial controller products.

For MIMO processes to be controlled with multiple SISO controllers, the problem becomes more challenging then they are standing alone as SISO loop systems, when the process has integrator(s). Due to the interaction among the loops, the SISO designed controllers are difficult to be stabilized in the MIMO environment. A late work of Huang et al. (2003), decomposes the design of multi-loop controlled MIMO system into that of independent simple SISO loops based on the assumption of open-loop stable dynamics. In order to follow the same design approach for MIMO process with integrator(s), a two-step design approach is thus proposed. In this paper, a design method of multi-loop PI/PID controller to control MIMO processes that have integrators and input dead time is proposed. For SISO controllers in the multi-loop system, a simple two-element controller will be derived. The two-element configuration for the SISO control system is similar to the one of Kaya (2003). The cascaded system is implemented as an equivalent single loop system with a pre-filter at the set-point input (see Figure 1). In the following text, in the first place, a design method for the two-element SISO controller will be derived. Incorporation of this two-element controller design into the design of a multi-loop system will then follow. Guidelines for loop pairing using the RGA analysis are also included.

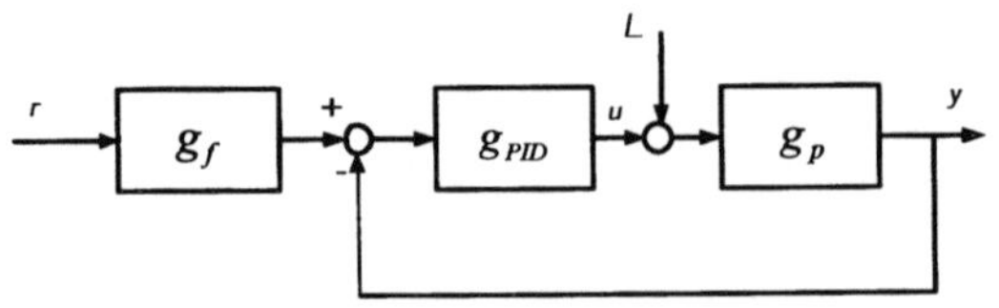

Fig. 1. Two-element control structure

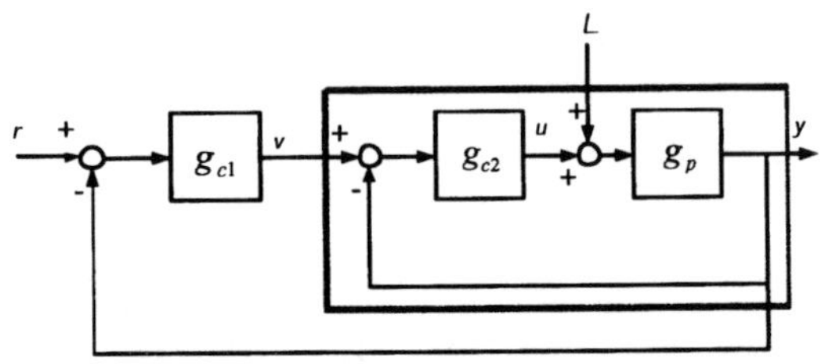

Fig. 2. Equivalent cascaded feedback loop for design.

2. DESIGN FOR TWO-ELEMENT CONTROL SYSTEM

The two-element control system of Fig.1 will be derived as an equivalence to the control configuration in Fig. 2. The design of the internal loop in Fig. 2 is aimed to stabilize the integrating process whose dynamics can be represented by a g_p of the following:

- Process 1

$$g_p(s)=\frac{ke^{-\theta s}}{s} \qquad (1)$$

- Process 2

$$g_p(s)=\frac{ke^{-\theta s}}{s(\tau s+1)} \qquad (2)$$

In fact, not only process having integrators, the above transfer functions can represent many slow dynamic processes also. In the cascaded system, a PD controller, $g_{c2}(s)$, is used in the internal loop which stabilizes the integrating processes with sufficient robustness. A PI-type controller, $g_{c1}(s)$, is used to control the resulting closed internal loop (the part boxed by dotted line). The PD and PI controllers used are considered to have the following forms:

$$g_{c1}=k_\beta\left(1+\frac{1}{\tau_R s}\right) \qquad (3)$$

$$g_{c2}(s)=k_\alpha(1+\tau_D s) \qquad (4)$$

Notice that, the product of $g_{c2}(s)g_p(s)$ has an integration mode due to the integrator in $g_p(s)$, and can be written in the following form:

$$g_{lp}(s)=g_{c2}(s)g_p(s)=\kappa(s)\frac{e^{-\theta s}}{s} \qquad (5)$$

The PD controllers are selected to compensate for the open-loop transfer functions to become the following standard forms individually, that is:

- For Model 1

$$g_{lp1}(s)=\frac{k_0(1+0.4\theta s)e^{-\theta s}}{\theta s} \qquad (6)$$

- For Model 2

$$g_{lp2}(s)=\frac{k_0 e^{-\theta s}}{\theta s} \qquad (7)$$

Where, the value of k_0 is chosen as 0.76 in Eq.(6) and 0.6 in Eq.(7) so that each $g_{lp}(s)$ has sufficient gain margin and phase margin. According to the $g_{lp}(s)$ chosen, the parameters of PD controllers are determined. The complementary sensitivity function, $h(s)$, can thus be written as:

$$h(s)=\frac{\kappa(s)\frac{e^{-\theta s}}{s}}{1+\kappa(s)\frac{e^{-\theta s}}{s}} \qquad (8)$$

Since k_0 is chosen to give modest performance for the internal loop, $h(s)$ in Eq.(8) can be approximated by a first-order plus time delay model, that is:

$$h(s)\cong\frac{e^{-\theta_m s}}{\tau_m s+1} \qquad (9)$$

As θ and k_0 are the only parameters, τ_m and θ_m can be written as functions of these two parameters. As has been studied in the range of θ = 0.2 to 10, the values of τ_m and θ_m for different values of k_0 are shown in Fig. 3 and Fig. 4. The data are also correlated and given in Table 1.

Since the dynamics of the internal loop is represented by an FOPDT model, a PI controller can be used to compensate for the internal loop to become the standard form of Eq. (7).

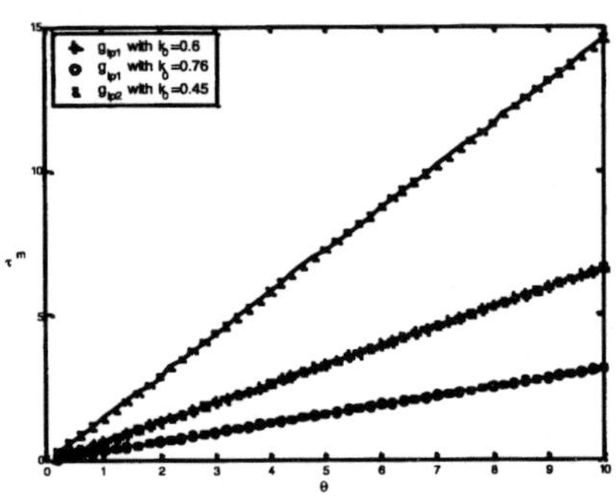

Fig. 3. Comparison of actual values and fitted equations (solid line) for τ_m.

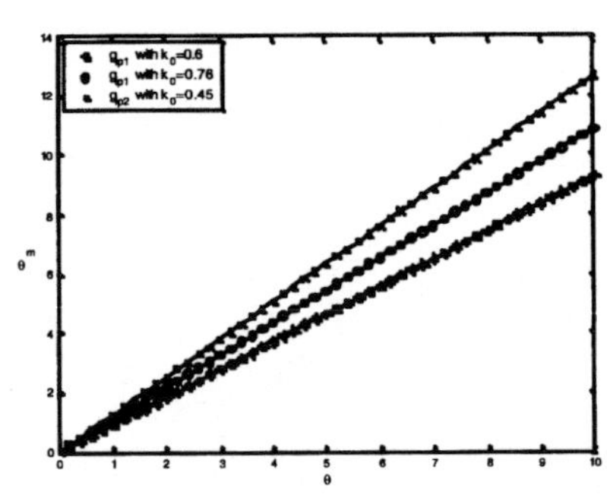

Fig. 4. Comparison of actual values and fitted equations (solid line) for θ_m.

Table 1.FOPDT model parameters of internal closed-loop for different loop transfer function.

Loop transfer function	Loop gain	Time constant	Time delay
			$\theta = 0.2\sim 10$
$g_{lp1}(s)$	0.76	$\tau_m = 0.3139\theta$	$\theta_m = 1.084\theta$
$g_{lp1}(s)$	0.6	$\tau_m = 0.663\theta$	$\theta_m = 0.9264\theta$
$g_{lp2}(s)$	0.45	$\tau_m = 1.452\theta$	$\theta_m = 1.263\theta$

With the cascaded system thus obtained, an equivalent two-element control system consisting of a single loop and a pre-filter at the command input, as shown in Fig. 2, can be derived. They are:

$$g_{PID}(s) = k_c'\left(1+\frac{1}{\tau_R' s}+\tau_D' s\right) \quad (10)$$

$$g_f(s) = \frac{\tau_R'\tau_D' s^2+\tau_R' s+1}{\tau_R'\tau_D'\left(1+\frac{1}{k_\beta}\right)s^2+\tau_R'\left(1+\frac{\tau_R}{k_\beta(\tau_R+\tau_D)}\right)s+1} \approx \frac{1}{\tau_f s+1} \quad (11)$$

where

$$k_c' = k_\alpha + k_\alpha k_\beta\left(\frac{\tau_R+\tau_D}{\tau_R}\right) \quad (12)$$

$$\tau_R' = \tau_R + \tau_D \quad (13)$$

$$\tau_D' = \frac{\tau_R\tau_D}{\tau_R+\tau_D} \quad (14)$$

3 MULTI-LOOP CONTROLLER DESIGN

Consider a n×n system of the following:

$$Y(s) = G(s)U(s)+D(s) \quad (15)$$

where $Y(s)$, $U(s)$ and $D(s)$ designate the output, input, and disturbance vectors, respectively. $G(s)$ is an open-loop transfer function matrix with some of its elements having integrators. Similar to the design approach for SISO systems, an internal controller is used to stabilize the integrating processes first. As show in Fig. 5, $V(s)$ is the input to the secondary loop, and $U(s)$ is the actual control input to the process, and, the resulting internal loop is designated as $\bar{G}(s)$. Assume that $g_{mn}(s)$ (i.e. the element of the (m,n)-th entry of $G_p(s)$) has an integrator, then $G_{c2}(s)$ and $E(s)$ in the internal loop are given as:

$$G_{c2} = \begin{cases} G_{c2}(n,n) = k_{c2,n}(1+\tau_{D2,n}) \\ G_{c2}(i,j) = \delta(i,j), \forall i,j \neq n \end{cases} \quad (16)$$

$$E(i,j) = \delta(i,n)\delta(j\ m) \quad (17)$$

where, $\delta(i,j)$ is a kronecker delta function. Substituting $U(s)$ with $V(s)$, Eq. (15) can be written as:

$$Y(s) = G(s)\left(I+G_{c2}(s)EG(s)\right)^{-1}G_{c2}(s)V(s) \quad (18)$$

$$Y(s) = \bar{G}(s)\,V(s) \quad (19)$$

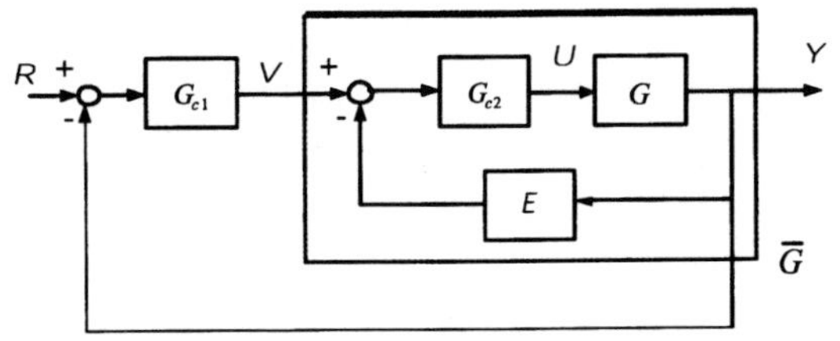

Fig. 5. Multi-loop control structure with inner feedback loops

Because of $G_{c2}(s)$, $\bar{G}(s)$ becomes an open-loop transfer function matrix. To control this $\bar{G}(s)$, the direct method (Huang, et al., 2003) for multi-loop PI/PID controller design can be applied. Before deriving the controllers, first, the pairing of variables has to be addressed.

3.1 Loop pairing

Relative gain analysis has become an accepted method for establish pairs of manipulated and controlled variables to minimize interactions in multivariable systems. However, this RGA computation can't be directly applied to the non-self-regulation system which has integrator in the transfer function matrix. Besides, the allocation of integrators in $G(s)$ will also complicate the pairing problem. As has been mentioned, since integrator is critical to system stability, internal loop(s) is to be used to stabilize the output(s) that has integrator(s) associated with inputs. The choice of pairing for the secondary loops will be clear when each of the output has only one integrator associated with only one manipulating input. But, it becomes more complicated when there are more then one integrator associated with more than one outputs. In the following, the above mentioned different pairing problems will be illustrated using a 3×3 systems. The results can be extended to the n×n systems.

The pairing problem is formulated in the following way:

(1) Assume a candidate set of pairs for constructing the secondary loops. When the secondary loops are closed, the resulting open-loop stable process (i.e. $\bar{G}(s)$) and its steady-state gain, $\bar{\mathbf{K}}$, will be computed, that is:

$$\bar{\mathbf{K}} = \lim_{s\to 0}\bar{G}(s) \quad (20)$$

Then, RGA, $\bar{\Lambda}$, is computed from the elements of the resulting $\bar{\mathbf{K}}$ according to:

$$\bar{\Lambda} = \bar{\mathbf{K}} \otimes \left(\bar{\mathbf{K}}^{-1}\right)^T \quad (21)$$

(2) Try the alternative pairing for secondary loops and compute the corresponding RGA, $\bar{\Lambda}$.

(3) Among the candidates of RGAs, find the one which has most comfortable relative gains for pairings. The pairing procedures will thus be discussed for the three occasions as follows.

[Case 1] ***One integrator to one output***

In this case, there is only one possible pair (say, y_a-

u_b) to stabilize the integrator with internal loop. The resulting $\bar{K}$, after closing the internal loop, will have the same $(n-1)\times(n-1)$ sub-matrix as that of K, that is:

$$\bar{K}=\begin{cases}\bar{k}(i,j)=1, & i=a \text{ and } j=b\\ \bar{k}(i,j)=k_{ij}, & i\neq a \text{ and } j\neq b\\ \bar{k}(i,j)=0, & \text{otherwise}\end{cases} \tag{22}$$

In other words, the remaining $(n-1)\times(n-1)$ subsystem can be paired according to the RGA calculated from the corresponding sub-matrix of K.

[Case 2] ***More than one integrators to one output.*** In this case, more integrators are associated with one output. No matter which integrator element is stabilized with a given secondary internal loop, the resulting RGA computing from the resulting $\bar{\mathbf{K}}$ will be the same. In other words, although the choice for pairing the secondary loops may differ, the interaction measure for the remaining sub-system in terms of the RGA will be the same. The proof is given in the following.

Proof. Assume that y_1 is engaged with integrating inputs, and g_{1k} is to be stabilized with an inner loop, we have:

$$u_k=c_{p1,k}(v_k-y_1) \tag{23}$$

$$u_i=v_i;\quad i\neq k \tag{24}$$

$$y_1=\frac{c_{p1,k}g_{1k}}{1+c_{p1,k}g_{1k}}v_k+\sum_{\substack{m=1\\m\neq k}}^{n}\frac{g_{1m}}{1+c_{p1,k}g_{1k}}v_m \tag{25}$$

$$y_i=\frac{c_{p1,k}g_{ik}}{1+c_{p1,k}g_{1k}}v_k+\sum_{\substack{m=1\\m\neq k}}^{n}\left(g_{im}-\frac{g_{ik}c_{p1,k}g_{m}}{1+c_{p1,k}g_{1k}}\right)v_m;\quad i\neq 1 \tag{26}$$

According to Eq. (25), the steady-state gains (designate as $\bar{K}^k$) at (1,k) entry, (i.e. $\bar{k}_{1k}$) becomes one. The other gains in the same row (i.e. $\bar{k}_{1j}$, $j\neq k$) becomes $k_{1j}/(k_{c,k}k_{1,k})$ when $g_{1j}(s)$ has integrator, or zero when $g_{1j}(s)$ has no integrator. By Eq. (26), the steady-state gains can be given for i=2~n as $\bar{k}_{ij}=k_{ij}-k_{1j}k_{ik}/k_{1k}$ when $g_{1,j}(s)$ has an integrator, and $\bar{k}_{ij}=k_{ij}$ when $g_{1,j}(s)$ does not have integrator. Similarly, if the g_{1h} is considered to be stabilized with an internal loop, its steady-state gain array $\bar{K}^h$ can be obtained in the same way. It can be shown that $\bar{K}^k$ and $\bar{K}^h$ has the following relation, that is:

$$\bar{K}^k=B\bar{K}^hA \tag{27}$$

where B and A are diagonal matrices:

$$A=\left[a_{ij}\right]=\begin{cases}a_{ii}=-k_{1h}/k_{1k}; & g_{1i}(s)\text{ has an integrator}\\ a_{ij}=0; & i\neq j\\ a_{ii}=1; & \text{otherwise}\end{cases} \tag{28}$$

$$B=\left[b_{ij}\right]=\begin{cases}b_{ij}=k_{ch}/k_{ck}; & i=j=1\\ b_{ij}=1; & i=j\neq 1\\ b_{ij}=0; & i\neq j\end{cases} \tag{29}$$

Then, it is obvious to have:

$$\Lambda^h=\Lambda^kP(h,k) \tag{30}$$

Where, $P(h,k)$ is a permutation matrix which makes hth and kth columns exchanged. As a result, to stabilize the integrating process, there will be no preference to the selection of pairing for secondary loops, as far as the interaction measures for the remaining subsystem is concerned, as the interaction measures for the remaining subsystem will be the same. The guideline for pairing the secondary loops should then be focused on the stability robustness of each individual secondary loop (e.g. choose larger $k_{i,j}$ or faster dynamics).

[Case 3] ***More integrators to more outputs.*** In this case, it can also be shown that stabilizing the integrator element with different secondary loop pairings has the same interaction measures for their sub-systems. Thus, there will be also no preference for one set of candidate pairs over the others, unless the concern of stability in each individual loop is considered. The guideline to pair the secondary loop for integrator, in the first step of design, is thus to focus on each secondary loop to have better stability robustness. In the second step, $\bar{\mathbf{K}}$ and the related RGA of $\bar{G}(s)$ is computed, and, input/output pairs for multi-loop control are determined accordingly.

3.2 Multi-loop controller design

By incorporating the secondary loops, the original MIMO process can be considered as an open-loop stable process. Regarding this resulted open-loop stable process (i.e. $\bar{G}(s)$), multi-loop PI/PID controllers can be designed by employing the direct method proposed by Huang et al. (2003). The transfer function that describes the effective transmission in each equivalent loop is considered as the effective open-loop process (designate as EOP) of that loop. Let $\bar{G}(s)$ and $G_{c1}(s)$ be partitioned into 2×2 form, that is:

$$\bar{G}(s)=\begin{bmatrix}\bar{g}_{1,1} & \bar{G}_{1,2}\\ \bar{G}_{2,1} & \bar{G}_{2,2}\end{bmatrix};\quad G_{c1}(s)=\begin{bmatrix}g_{c,1} & 0\\ 0 & \bar{G}_{c,1}\end{bmatrix} \tag{31}$$

As mentioned, the first equivalent loop (i.e. $\bar{\mathbf{g}}_1$) can be derived as following:

$$\bar{\mathbf{g}}_1=\bar{g}_{1,1}-\bar{G}_{1,2}\bar{G}_{2,2}^{-1}\left\{I-\left(I+\bar{G}_{2,2}\bar{G}_{c,1}\right)^{-1}\right\}\bar{G}_{2,1} \tag{32}$$

All other EOPs can thus be derived in the same way. Huang et al. (2003) provides a approximation formulation to include the effect of $\bar{G}_{c1}(s)$, without knowing exactly what they are, in the formulation of EOPs. As a result, the design of controllers for the EOPs can be carried out independently. Notice that, for each EOP that has integrator in its internal loop, only PI-type controller will be assigned. Other than these loops, PID-type controller will be used to improve each single loop control for the EOPs. The EOPs in Eq. (32), in general, are represented by either the FOPDT or the SOPDT models of the following:

- FOPDT dynamics:

$$\bar{\mathbf{g}}^*=\frac{k_pe^{-\theta s}}{\tau s+1} \tag{33}$$

- SOPDT dynamics (overdamped):

$$\bar{\mathbf{g}}^*=\frac{k_p(\tau_3s+1)e^{-\theta s}}{(\tau_1s+1)(\tau_2s+1)} \tag{34}$$

- SOPDT dynamics (underdamped):

$$\overline{\mathbf{g}}^{*} = \frac{k_p(\tau_3 s+1)e^{-\theta s}}{\tau^2 s^2 + 2\tau\zeta s + 1} \quad (35)$$

Methods to determine FOPDT models can be derived from ultimate gains and ultimate frequency, or, by formulating the following optimization problem. That is:

$$Arg\{\mathrm{P}\} = \min_{\mathrm{P}} \int_0^{w_f} \left\{ \left| \mathrm{Re}\left(\overline{\mathbf{g}}(w;\mathrm{P}) - \overline{\mathbf{g}}^{*}(w;\mathrm{P})\right) \right|^2 + \left| \mathrm{Im}\left(\overline{\mathbf{g}}(w;\mathrm{P}) - \overline{\mathbf{g}}^{*}(w;\mathrm{P})\right) \right|^2 \right\} dw \quad (36)$$

where P consists of parameters in $\overline{\mathbf{g}}^{*}$ and w_f is the frequency bandwidth concerned. Upon finishing the modeling for EOPs, PI and PID controllers can be selected to compensate for the loop transfer functions to become the desired loop transfer functions. Notice each loop gain can be assigned to weight the importance of each loop. For equal weight consideration, the value for each loop is defaulted to be taken as 0.6 for $g_{lp1}(s)$ and 0.45 for $g_{lp2}(s)$. Notice that there is no need to go through an iteration process to determine the final multi-loop controller, the one obtained for the EOP of Eq.(32) will be the one to use.

4. SIMULATION RESULTS

4.1 Example for SISO system

Example 1 Consider a process with the transfer function $0.0506e^{-6s}/s$, which was also given in the paper of Kaya (2003). The PI-PD controller parameters in that paper are $K_c = 1.6184$, $T_i = 17.599$, $K_f = 1.0956$ and $T_d = 2.4528$. Then for the proposed methods, the required parameters for PI-PD and PID-filter are found: $k_\alpha = 2.5033$, $\tau_D = 2.4$, $k_\beta = 0.1737$, $\tau_R = 1.8834$, $k_c' = 3.4922$, $\tau_R' = 15.1262$, $\tau_D' = 2.0192$ and $\tau_f = 11.2810$. Simulation results are given in Fig. 6. A unit step set point is introduced at time t=0, and a load disturbance d=1 is introduced at time t=100. The proposed method gives superior results than the other, especially for the disturbance rejection. Although the approximate filter is slightly degraded in system performance, it is pretty easy for implementation in the factory.

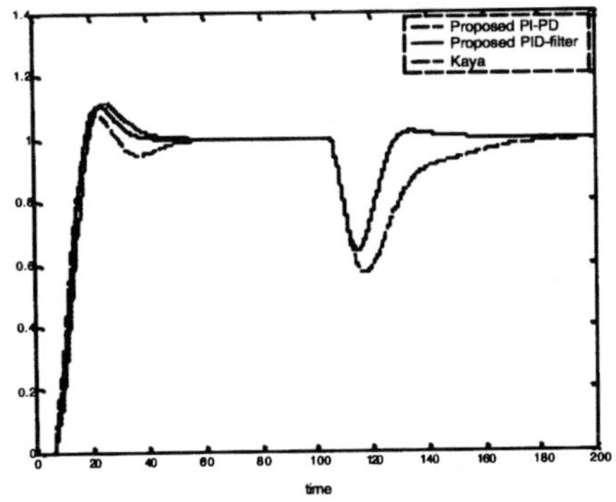

Fig. 6. Set point and disturbance responses for example 1

4.2 Multi-loop controllers design

Example 2. The proposed method is then applied to a system that has three loops. Considered a 3×3 system:

$$G_p = \begin{bmatrix} \frac{-1.986e^{-0.71s}}{s(66.67s+1)} & \frac{5.984e^{-2.24s}}{14.29s+1} & \frac{5.24e^{-60s}}{400s+1} \\ \frac{0.0204e^{-0.59s}}{(7.14s+1)^2} & \frac{2.38e^{-0.42s}}{(1.43s+1)^2} & \frac{-0.33e^{-0.68s}}{(2.38s+1)^2} \\ \frac{0.374e^{-7.75s}}{22.22s+1} & \frac{-9.811e^{-1.59s}}{11.36s+1} & \frac{-11.3e^{-3.79s}}{(21.74s+1)^2} \end{bmatrix} \quad (37)$$

where $g_{1,1}(s)$ isn't a pure integrating process which has an extra first order lag. Since only $g_{11}(s)$ has an integrator, the internal loop pairing should direct to the (1,1)-pair. The inner controller then is designed according to $g_{lp2}(s)$. The pairing for the remaining subsystem is then conducted by the computed RGA of $\overline{\mathbf{K}}$. After pairing the loops, the resulting dynamic models of the EOPs are given in Table 2. Based on these models, the PI and PID controllers are synthesized and are given in the same table. Simulation results for step changes are given in Fig. 7. The results show the performance of these loops is satisfactory.

5. CONCLUSION

In this paper, a two-element control system is applied to design the control for integrating process in either SISO, or MIMO system. For a single-loop system, a two-element control having a PID controller in the feedback loop and a pre-filter at the commend input is resulted. In a MIMO system, the same approach is applied. As a first step, constructing secondary loops to stabilize the integrators is required. Pairing issue for such secondary loops is addressed. It is found that stabilizing an integrator via pairing different input to the same output will result in the same interaction measures for the remaining loops, if the RGA analysis is used. It is also found that loop pairing in the proposed method will have the same result as that of Woolverton (1980). In design, selection of pairing for secondary loops is thus guided by focusing on obtaining the best stability robustness to individual secondary loops, then, the modified steady-state gain matrix and its RGA are computed for the selection of pairs for the remaining loops. A direct method by making use of effective open-loop process (EOP) to design the multi-loop controllers is then used. As a result, controllers in each loop can be designed independently. The effectiveness of this proposed method in designing single-loop and multi-loop controllers is illustrated with two simulated examples, one for a SISO system and one for a 3×3 system. The simulation results from the proposed design give satisfactory control performance for in both cases.

ACKNOWLEDGEMENT

This research is supported by the Ministry of Economic Affair, Taiwan, under grant No. 92-EC-17-A-09-S1-019.

Table 2. Model of EOP and controller tuning for example 2

	Loop 1	Loop 2	Loop 3
Ultimate frequency	2.2179	1.9263	0.2698
Ultimate gain	2.3494	10.2263	26.8778
Model of EOP	$\frac{e^{-0.9065s}}{0.9585s+1}$	$\frac{2.6665e^{-0.8663s}}{5.2834s+1}$	$\frac{-12.6603e^{-5.96s}}{99.5531s+1}$
Controller	$-11.03\left(1+\frac{1}{69.64s}+2.85s\right)$	$1.37\left(1+\frac{1}{5.28s}\right)(1+0.35s)$	$-0.79\left(1+\frac{1}{99.55s}\right)(1+2.38s)$
Pre-filter	$\frac{1}{1.86s+1}$		

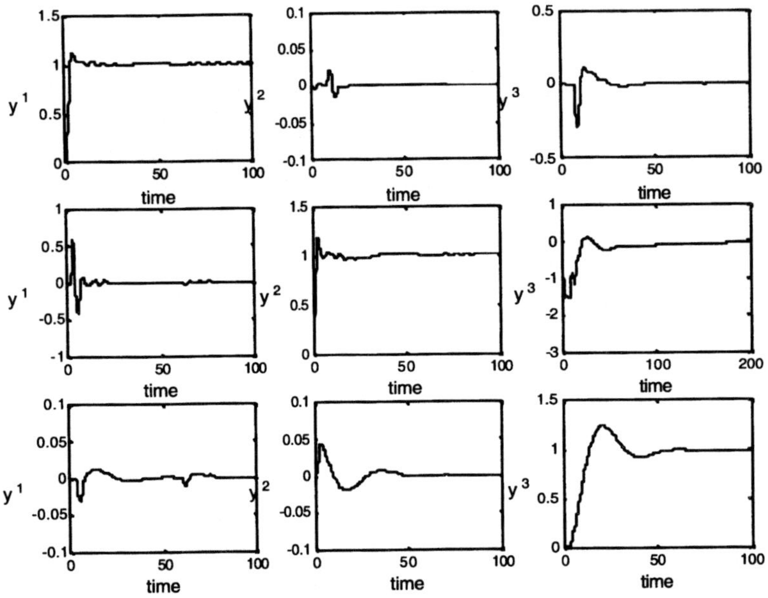

Fig. 7. Responses of multi-loop control for example 2.

REFERENCE

Chien, I. L. and P. S. Fruehauf (1990). *Consider IMC tuning to improve performance.* Chem. Eng. Prog., 86 (10), pp.33-41.

Chien, I. L., S. C. Peng and J. H. Liu (2002). *Simple control method for integrating processes with long deadtime* Journal of Process Control, 12, pp. 391-404.

Huang, H. P., M. W. Lee and C. L. Chen (2000). *Inverse-based design for a modified PID controller*. J. Chin. Inst. Chem. Engrs., 31, pp. 225-236.

Huang, H. P. and J. C. Jeng (2002). *Monitoring and assessment of control performance for single loop systems*. Ind. Eng. Chem. Res., 41, pp. 1297-1309

Huang, H. P., J. C. Jeng, C. H. Chiang and W. Pan (2003). *A direct method for multi-loop PI/PID controller design.* Journal of Process Control, 13, pp. 769-786.

Kaya, I. (2003). *A PI-PD controller design for control of unstable and integrating processes*. ISA Transactions, 42, pp. 111-121.

Kwak, H. J., S. W. Sung and I. B. Lee (1997). *On-line process identification and autotuning for integrating process*. Ind. Eng. Chem. Res., 36, pp. 5329-5338.

Kwak, H. J., S. W. Sung and I. B. Lee (2001). *Modified Smith predictors for integrating processes: Comparisons and Proposition.* Ind. Eng. Chem. Res., 40, pp. 1500-1506.

Kwak, H. J., S. W. Sung and I. B. Lee (2000). *On-line process identification and autotuning for unstable processes*. Chem. Eng. Res. Des., 78, pp. 549-556.

Tyreus, B. D. and W. L. Luyben (1992). *Tuning PI controllers for integrator/dead time processes.* Ind. Eng. Chem. Res., 31, pp. 2625-2628.

Wang, Y. G. and W. J. Cai (2002). *Advanced proportional-integral-derivative tuning for integrating and unstable processes with gain and phase margin specifications*. Ind. Eng. Chem. Res., 41, pp. 2910-2914.

Woolverton, P. F. (1980). *How to use relative gain analysis in systems with Integrating Variables*. InTech, 27, pp. 63-65

ELSEVIER
IFAC
PUBLICATIONS
www.elsevier.com/locate/ifac

IDENTIFICATION FOR CONTROL: ACHIEVEMENTS AND OPEN PROBLEMS [1]

Michel Gevers *

* *Center for Systems Engineering and Applied Mechanics (CESAME)*
Université Catholique de Louvain
Bâtiment Euler, 4 Av. Georges Lemaître,
B-1348 Louvain-la-Neuve, Belgium
Phone: +32-10-472590
Email: gevers@csam.ucl.ac.be

Abstract: This paper presents the author's views on the development of identification for control. The paper reviews the emergence of this subject as a specific topic within the larger discipline of system identification, the major progress accomplished over the last 15 years and its practical impact, as well as the remaining open problems.

Keywords: Identification, identification for control, robust control, optimal design.

1. INTRODUCTION

Identification for control! Identification and control have initially emerged as two separate disciplines, developed by two distinct communities.

Identification has for a long time been the territory of mathematicians, statisticians, time-series analysts and econometricians. The history of system identification goes as far back as the work of Gauss and Legendre in the late 18th and early 19th century. An excellent presentation of the history of system identification can be found in (Deistler, 2002). Control theory has always been the territory of engineers. Until about 1960, most of control design was based on model-free methods, using Bode, Nyquist, or Ziegler-Nichols plots, combined with graphical design techniques. The introduction of state-space models in 1960, together with the solution of optimal control and optimal filtering problems in a Linear Quadratic Gaussian framework (Kalman, 1960*a*; Kalman, 1960*b*), gave birth to a tremendous development of model-based control design methods. Successful applications abounded, particularly in aerospace, where accurate models were readily available.

1965 saw the start of identification activity in the control community. The paper (Ho and Kalman, 1965) set the stage for state space realization theory which, 25 years later, became the founding block for what is now called subspace identification. The paper (Åström and Bohlin, 1965) introduced into the control community the Maximum Likelihood framework for the identification of input-output models. This gave rise to the celebrated Prediction Error (PE) framework (Ljung, 1999) that has since proven so successful. Undoubtedly the advent of identification theory was spurred by a desire to extend the applicability of model-based control design to broader and broader fields of applications, for which no reliable models could be obtained from first principles.

From 1965 to the late eighties, model-based control was applied to ever growing classes of dynamical systems and processes, with models obtained via the newly emerging identification techniques.

[1] This paper presents research results of the Belgian Programme on Interuniversity Attraction Poles, initiated by the Belgian Federal Science Policy Office. The scientific responsibility rests with its author.

The prevailing habit at that time was to separate the identification step and the control design step. A model was identified first using the best available techniques; subsequently, a model-based control design was performed based on the "certainty equivalence principle", i.e. the model was treated as if it represented the true system. Dual control and adaptive control were two early attempts to address the issue of parametric uncertainty and model-based control design in a synergistic way. In dual control the parameter estimation and the control design mechanism are obtained as the result of a single but complex optimization problem. In adaptive control, the parameter adjustment scheme is subsidiary to the control objective. Both schemes were developed for the case where the structure of the true system is assumed to be known, which severely limits their practical applicability. The solution of the dual control problem proved to be computationally intractable, even in the simplest cases. As for adaptive control, the major difficulty is that the parameters of the feedback control system change at every sampling instant, making the closed-loop dynamics nonlinear and their stability analysis extremely complex. After convergence mechanisms had been devised for the ideal case where the system is in the model set, attempts were made to robustify the adaptive control algorithms in order to take account of some modest degree of uncertainty. These attempts essentially consisted of introducing cautionary safeguards in the computation of the gain of the parameter adjustment scheme; see e.g. (Anderson *et al.*, 1986).

One of the main contributions of the control community to system identification theory was to consider identification as an exercise in estimating the *best possible approximate model* within some model set, rather than as a search for the *true system*. Together with this effort came the characterization of the approximate model in terms of bias error and variance error on the estimated transfer functions.

If the model of a system is exact, it is optimal for all applications. However, if the model is only an approximation of the "true system", then the quality of the model should be dependent on the intended application. It thus makes sense to tune the identification towards the objective for which the model is to be used, i.e. to ensure that the distribution of model error is such that it does not deteriorate the objective too much. This gave rise to the paradigm of *goal-oriented identification* and it led one to view identification as a *design problem*. Identification for control has been the major outlet for this new paradigm. The reasons for this are many: (i) control is very often the main motivation for model building; (ii) high performance control can often be achieved with very simple models, provided some basic dynamical features of the system are accurately reflected; (iii) a powerful robust control theory, based on nominal models and uncertainty sets, had been developed all through the eighties, but these models and uncertainty sets were not data-based for lack of a proper theory; (iv) identification for control research led to iterative model and controller tuning tools that were intuitive, practical and easy to implement by the process engineers.

One early approach to optimal identification design for control, that established a direct link between experimental conditions and controller performance, was obtained in (Gevers and Ljung, 1986) by considering variance errors only, i.e. the system was assumed to be in the model set. This approach consists, for a given certainty equivalence control design procedure, of computing the experimental conditions of the identification that minimize the average performance degradation that results from the fact that the controller is computed from an estimated (and hence random) model rather than from the exact true system. Like with all optimal experiment design methods, the optimal experiment depends on the unknown system: see Section 4. Hence, even though such results give useful insights, they do not provide an operational design method for the identification for control problem. In addition, they are based on a certainty equivalence controller design mechanism, rather than on a robust control design.

Except for the methods already mentioned (dual control, adaptive control, control-oriented optimal design), which are restricted to the case where the true system is in the model set, the first contributions in which identification and control design with restricted complexity models were looked upon as a combined design problem appeared only around 1990. The plenary (Gevers, 1991) at the 1991 IFAC Symposium on System Identification addressed many of the key issues; however, it was more an agenda for research than a presentation of solutions. Indeed, there was very little understanding at that time about the interplay between system identification and robust control. The two theories had been developed by separate communities with very little interaction. In the nineties, the activity in identification for control surged: in the plenary (Hjalmarsson, 2003), the author estimated that about 1,500 papers have appeared on the topic of identification for control after the plenary (Gevers, 1991).

When the application of a model is the design of a controller, then what really matters is the performance achieved by this model-based controller on the "true system", not the intrinsic quality of the model. We illustrate this idea with a very simple example inspired by (Skelton, 1989). Let a 'true system' be represented by $G_0(s) = \frac{1}{s+1}$. Then the 'model' $\hat{G}(s) = \frac{1}{s}$ would clearly be deemed to be absurd as a model for G_0. However, if the objective is to design a high gain static output feedback controller, then $\hat{G}$ would be a perfectly acceptable model. Indeed, with a high gain static output feedback $u = -Ky$, the closed loop transfer functions $\frac{K}{s+1+K}$ and $\frac{K}{s+K}$ become indistinguishable. Thus, whether or not a model is appropriate for control design depends as much on the controller that will be implemented as it depends on the plant/model mismatch.

In practice, the true system is unknown, the model is unknown at the identification design stage, and the controller that will be implemented is unknown because it depends on that model. What is typically known in control-oriented identification is the control performance objective. Some prior knowledge about the true system may also be available. Ideally, the design of a control-oriented identification procedure could then be formulated as follows: *Given a control performance objective, design the identification in such a way that the performance achieved by the model-based controller on the true system is as high as possible.* Identification design includes many choices: input data, feedback configuration (possibly), data length, model structure, identification criterion, validation criterion, etc. The identification design problem is impossible to solve in such generality. In order to get a handle on the problem, it has been customary to fix some of the choices; the number of data is usually taken to be fixed, and so is the model structure.

In the first half of the nineties, the research focused on the bias error distribution, assuming that low complexity models were being used for the design of the controller. It produced a string of results on the design of control-oriented nominal models (Schrama and Bosgra, 1993; Lee *et al.*, 1993; Åström and Nilsson, 1994; Zang *et al.*, 1995; de Callafon and Van den Hof, 1997). The first and rather obvious result was to establish that a model is good for control design if the closed loop system obtained by the feedback connection of that model with the designed controller is close to the system obtained by the feedback connection of the true system with that same controller. Since the 'to be designed controller' is not available at the identification stage, this led to the necessity of using an iterative scheme of model updates and controller updates.

A second important result that emerged from the understanding of control-oriented nominal models was that the experimental conditions of the identification should resemble as much as possible the experimental conditions obtained when the designed controller is applied to the true system (Anderson and Gevers, 1998). This led to the observation that, for most control performance objectives, identification should be performed in closed loop (Hjalmarsson *et al.*, 1996; Forssell and Ljung, 2000). This observation triggered a revival of interest for closed loop identification, and the emergence of new identification methods specifically designed for this situation: see e.g. (Hansen *et al.*, 1989; Van den Hof and Schrama, 1993).

The iterative identification and control design schemes do not necessarily converge to a stationary point, corresponding to a stable closed loop system. Thus, they must be applied with caution, and a lot of work has gone into developing tools for safe model and controller updates. Even so, these iterative schemes have had a remarkably fast transfer into the world of applications. There are two main reasons for this:

- whereas much of the industrial world was still living with the belief that one should 'open the loop' to perform a valid identification experiment, here was a theory that showed the benefits of closed-loop identification; this came as welcome news to process control engineers who had never really liked the idea of opening the loop;
- in the process industry, thousands of measurements flow into the computer; here was a theory that showed how these data could be used for the design of a better controller.

As stated above, the work on control-oriented nominal models focused on the bias error distribution of the identified model, with the controller computed from the model in a certainty equivalence framework. Thus, that work did not incorporate the robust control concepts developed during the eighties. It focused on the design of identification criteria that minimize a (control-oriented) measure of the model error.

The second half of the nineties saw a shift towards the definition and estimation of *control-oriented uncertainty sets* (Kosut and Anderson, 1994; Mäkilä *et al.*, 1995; de Vries and Van den Hof, 1995; Bombois *et al.*, 2000) in order to put the control design into the framework of robust control design. The focus turned to shaping the distribution of the variance error of the identified models, i.e. on manipulating the shape of the uncertainty set. Indeed, the paradigm of robust control design is to compute a controller that achieves the best possible worst case performance, i.e. the best possible performance over all models in an uncertainty set. Such best worst case performance depends as much on the controller as it depends on the uncertainty set, and this set is directly dependent on the experimental conditions under which the identification is performed. The study of the interplay between experiment conditions of the identification and properties of the robust controller has been split up into two questions:

(1) what is the connection between a model uncertainty set and the properties of robust controllers computed from that set and, consequently, how should one define a control-oriented uncertainty set?
(2) how should one design the identification experiment in such a way that the uncertainty set around the identified model has such 'control-oriented' property?

Even though many new insights have been gained on the first question, there is at this point no clear view as to the most operational definition of a 'control-oriented uncertainty set'. We shall come back to this in Section 7.

To summarize this brief historical account, the work on identification of the last 15 years has been essentially developed in three directions: optimal control-oriented experiment design for identification, the definition and computation of control-oriented nominal models, and the connection between data-based uncertainty sets and the properties of robust controllers resulting from such sets. In addition, the demands of this highly visible field have generated lots of parallel work on estimation and validation of uncertainty sets from data, and on closed-loop identification. Some of the concepts that have emerged from these 15 years of work, such as the idea of improving the performance of an existing controller on the basis of closed loop data collected with the presently operating controller, have immediately found their way into practice. However, many questions remain unsolved, particularly on the optimal tuning of uncertainty sets, and we are still far from an automatic procedure that would go from the collection of experimental data to the design of a robust controller via a model and its uncertainty set.

In the rest of this paper, we explain in some more detail the specific achievements and the remaining unsolved questions in the different subtopics that we have sketched: what is optimal control-oriented experiment design? what is a control-oriented nominal model? why iterative design? what is a control-oriented uncertainty set? how can we match experiment design and uncertainty set? We have chosen to do the presentation of these ideas in a Prediction Error (PE) identification framework, because PE identification is by far the most successful and widely used identification method. In the next section, we first present the bare essentials of PE identifcation that are necessary to understand the remaining issues.

2. THE BARE ESSENTIALS OF PREDICTION ERROR IDENTIFICATION

The purpose of this paper is to present concepts rather than technicalities. Thus, to simplify the presentation, we assume that the unknown true system can be represented by a single-input single-output linear time-invariant system:

$$\mathcal{S} : y_t = G_0(z)u_t + v_t = G_0(z)u_t + H_0(z)e_t, \quad (1)$$

where $G_0(z)$ is a linear time-invariant causal operator, y is the measured output, u is the control input, and v is noise, assumed to be quasistationary, modelled as the output of a model $v_t = H_0(z)e_t$, where e is white noise. One considers a parametrized model set: $\mathcal{M} = \{(G(z,\theta), H(z,\theta), \quad \theta \in D_\theta \subset \mathbb{R}^d\}$ where $G(z,\theta)$ and $H(z,\theta)$ are typically rational transfer functions, and D_θ is a subset of admissible values for the parameter vector θ. To every θ corresponds a one-step ahead predictor:

$$\begin{aligned}\hat{y}_{t|t-1}(\theta) = & H^{-1}(z,\theta)G(z,\theta)u_t \\ & +[1 - H^{-1}(z,\theta)]y_t, \end{aligned} \quad (2)$$

and hence a one-step ahead prediction error:

$$\begin{aligned}\varepsilon_t(\theta) &\triangleq y_t - \hat{y}_{t|t-1}(\theta) \quad (3)\\ &= H^{-1}(z,\theta)[(G_0(z) - G(z,\theta))u_t + v_t]\end{aligned}$$

These prediction errors can, possibly, be filtered by a data filter $D(z)$, thus defining the filtered prediction errors $\varepsilon_t^f(\theta) = D(z)\varepsilon_t(\theta)$. The least squares PE estimate $\hat{\theta}_N$ based on N input-output data is then defined as

$$\hat{\theta}_N = \arg\min_{\theta \in D_\theta} V_N(\theta), \quad (4)$$

where the PE criterion is defined as

$$V_N(\theta) = \frac{1}{N}\sum_{t=1}^{N}[\varepsilon_t^f(\theta)]^2. \quad (5)$$

The estimate $\hat{\theta}_N$ defines the model $\hat{G} = G(z,\hat{\theta}_N)$, $\hat{H} = H(z,\hat{\theta}_N)$. Under reasonable conditions, $\hat{\theta}_N \overset{N\to\infty}{\longrightarrow} \theta^*$, where $\theta^* \triangleq \arg\min_{\theta\in D_\theta} \bar{V}(\theta)$, with $\bar{V}(\theta) \triangleq E[\varepsilon_t^f(\theta)]^2$ (Ljung, 1999) .

Two different situations need to be considered. The first one is when the model structure $\mathcal{M}$ has been chosen sufficiently complex that the true system belongs to the model set. This is denoted $\mathcal{S} \in \mathcal{M}$, and means that there exists a value $\theta_0 \in D_\theta$ such that $G(z,\theta_0) = G_0(z)$ and $H(z,\theta_0) = H_0(z)$. In such case, under reasonable conditions, $\theta^* = \theta_0$, which means that the PE estimates of the transfer functions converge to the true transfer functions: $G(z,\hat{\theta}_N) \overset{N\to\infty}{\longrightarrow} G_0(z)$, $H(z,\hat{\theta}_N) \overset{N\to\infty}{\longrightarrow} H_0(z)$. When $\mathcal{S} \in \mathcal{M}$, the parameter error converges to a Gaussian random variable:

$$\sqrt{N}(\hat{\theta}_N - \theta_0) \stackrel{N\to\infty}{\longrightarrow} N(0, P_\theta). \tag{6}$$

The asymptotic parameter covariance P_θ can be estimated from the data, and the true parameter vector θ_0 belongs to an ellipsoid:

$$U_\theta = \{\theta | N(\theta - \hat{\theta}_N)^T P_\theta^{-1}(\theta - \hat{\theta}_N) < \chi^2\} \tag{7}$$

with probability $\alpha(d, \chi^2) = Pr(\chi^2(d) \leq \chi^2)$, where $\chi^2(d)$ denotes the χ^2 distribution with d degrees of freedom. Thus, when the system is in the model set, PE identification delivers a nominal model $G(z, \hat{\theta}_N), H(z, \hat{\theta}_N)$, together with an ellipsoidal confidence region in parameter space. This, in turn, defines an uncertainty region in the space of transfer functions:

$$\mathcal{D} = \{G(z, \theta) \mid \theta \in U_\theta\}. \tag{8}$$

In the more general situation where the system is not in the model set, $\lim_{N\to\infty} \hat{\theta}_N = \theta^* \neq \theta_0$. In such case, the transfer function error, $G_0(e^{j\omega}) - G(e^{j\omega}, \hat{\theta}_N)$, at a given frequency ω, can be decomposed as:

$$G_0(e^{j\omega}) - G(e^{j\omega}, \hat{\theta}_N) = \underbrace{G_0(e^{j\omega}) - G(e^{j\omega}, \theta^*)}_{\text{bias error}} + \underbrace{G(e^{j\omega}, \theta^*) - G(e^{j\omega}, \hat{\theta}_N)}_{\text{variance error}}. \tag{9}$$

The bias error arises when the model structure is unable to represent the true system. The variance error is caused by the noise and the finiteness of the data set; by definition of θ^*, it goes to zero asymptotically. In the mid-eighties, Ljung produced some important formulas for the characterization of bias and variance errors of identified transfer functions (Ljung, 1985; Wahlberg and Ljung, 1986). The bias was characterized implicitly by representing θ^* as the minimizing argument of a frequency integral. A variance error estimate for the estimated transfer functions was obtained under an assumption of model order going to infinity. More recent work has produced formulas for the estimation of an uncertainty set $\mathcal{D}$ around $\hat{G}$, with the property that $G_0 \in \mathcal{D}$ with probability α, where α is any desired level close to 1 (e.g. $\alpha = 0.95$) even in the case where the system is not in the model set, and for finite model orders: see Section 6. Finally, we note that the results described in this section are valid for both open loop and closed loop identification.

3. THE GAME AND THE PLAYERS

In identification for control, a typical situation is that we can perform experiments on the true system (1) with the purpose of designing a feedback controller. The system may already be under feedback control, in which case the task is to replace the present controller by one that achieves better performance. This situation is representative of many practical industrial situations. We then denote the present controller by C_{id} and the reference signal, if any, by r_t:

$$u_t = C_{id}(z)[r_t - y_t]. \tag{10}$$

Using N data collected on the system, in open loop or in closed loop, we can compute a model $\hat{G}$ of the unknown G_0, and possibly also a noise model $\hat{H}$ of H_0 by PE identification. Since the complexity of a model-based controller is of the same order as that of the model, one often performs the identification with a low order model.

The traditional scenario in model-based robust control design was: *First estimate a model $\hat{G}$ and an uncertainty set $\mathcal{D}$, then design a new controller $C(z)$ that achieves closed-loop stability and meets the required performance with all models in $\mathcal{D}$, and hence with the unknown true system G_0.* For this scenario to be successful, a very accurate model $\hat{G}$ was typically required.

The objective in *identification for control* is to replace that traditional scenario by the following. *On the basis of the required performance, and of any knowledge of the unknown system, design a control-oriented identification experiment that produces a model $\hat{G}$ and an uncertainty set $\mathcal{D}$; then design a new controller C that achieves closed-loop stability and meets the required performance with all models in $\mathcal{D}$, and hence with the unknown true system G_0.* If necessary, repeat this design procedure, possibly with a more demanding performance criterion. In some scenarios, one first computes a class $\mathcal{C}(\hat{G}, \mathcal{D})$ of controllers, each of which achieves the required performance with all models in $\mathcal{D}$; the controller C is then chosen within this class in such a way as to have some additional nice features (e.g. low complexity).

The goal of the new scenario is to achieve the same or better performance based on models (and hence controllers) of lower complexity. The class $\mathcal{C}$ of controllers that achieve the required performance is larger if the model uncertainty set $\mathcal{D}$ can be tuned towards that aim.

The players within this (iterative) identification and robust control design scenario are therefore:

- the unknown plant G_0
- the optimal controller C^{opt} for G_0
- the present controller C_{id} (if any)
- the present model $\hat{G}_{init}$ (if any)
- the identified model $\hat{G}$
- the uncertainty set of models $\mathcal{D}$ around $\hat{G}$
- the set $\mathcal{C}$ of controllers that achieve the prescribed performance
- the new model-based controller $C \in \mathcal{C}$

Except for the unknown plant G_0 and its corresponding optimal controller C_0, the designer has a handle on all other players. It is the complexity of the interplay between these players that makes the problem challenging and interesting. To illustrate the interplay, it is important to understand that one deals with five different feedback loops, which impact on one another: see Figures 1 to 5.

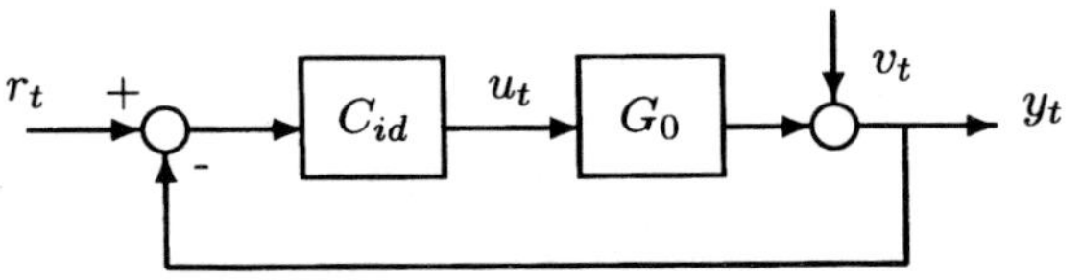

Fig. 1 : Experimental loop

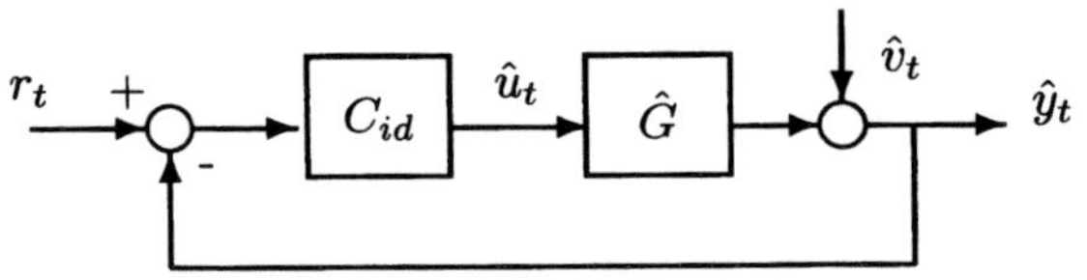

Fig. 2 : Identified system

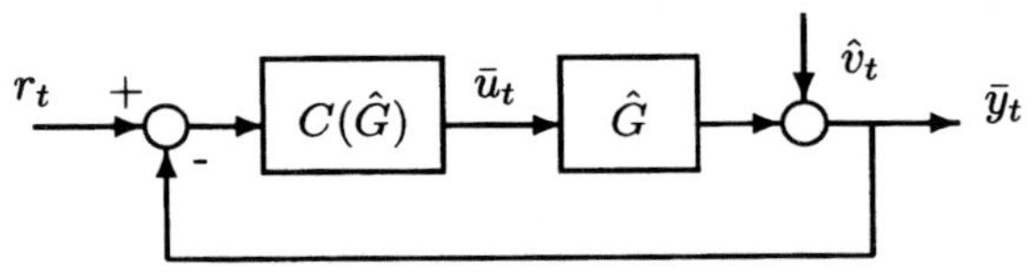

Fig. 3 : Design loop

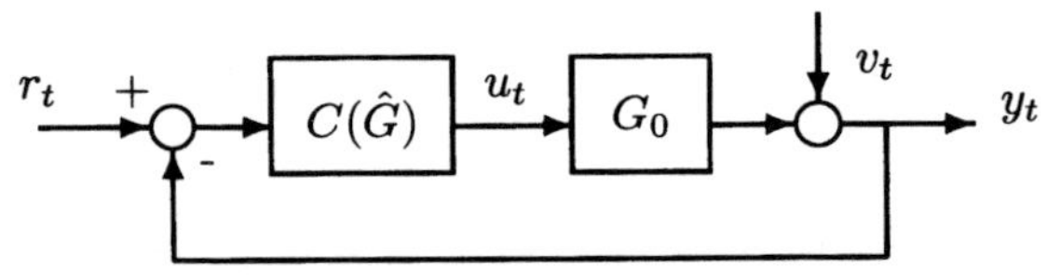

Fig. 4 : Achieved loop

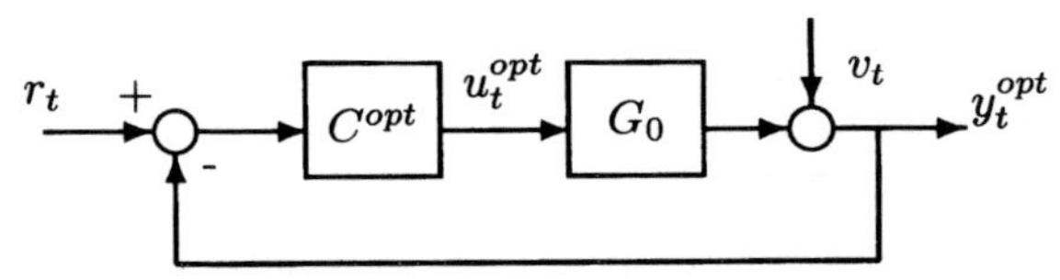

Fig. 5 : Optimal loop

In identification for control, the designer collects data on the experimental loop of Figure 1, and estimates a model $\hat{G}$ such that the closed loop system of Figure 2 is "as close as possible" to the actual system of Figure 1. Sometimes, one of his design choices is the choice of a controller C_{id} and of a reference signal r_t in the experimental loop. On the basis of the identified model $\hat{G}$, possibly with an estimated uncertainty set $\mathcal{D}$, the designer then computes the new controller $C(\hat{G})$, or $C(\hat{G}, \mathcal{D})$: this generates the designed loop of Figure 3. However, what really matters is the performance achieved by this controller with the real system, i.e. the performance of the achieved loop of Figure 4. Thus, if the identifier performs a good job, then the identified loop will be close to the experimental loop; but what is really desired is that the achieved performance on the loop of Figure 4 is close to the designed performance on the loop of Figure 3. If the experimental setup was such that C_{id} was identical (or at least close to) the "to-be-designed controller" C, then closeness of the loops of Figures 1 and 2 would entail closeness of the loops of Figures 3 and 4. This explains why it pays to have experimental conditions that closely match the conditions in which the "to-be-designed controller" will operate.

The first identification for control results, mentioned earlier, were based on a different approach. The aim was to find the optimal experiment conditions such that the output of the achieved loop of Figure 4 is as close as possible to the output of the optimal loop of Figure 5 that would be obtained if the true system were known exactly. Because of its historical precedence, and its intuitive appeal, we first present this approach in the next section.

4. OPTIMAL CONTROL-ORIENTED IDENTIFICATION DESIGN

In this approach an identification experiment design is called "optimal" if the controller computed from the estimated model is one that minimizes the average performance degradation vis-à-vis the performance that would be achieved with the ideal controller. The ideal controller is the controller that would be computed if the true system were known. We now explain this in some more detail.

We denote by $J(G, H, C)$ the control design criterion, and by $C = c(G, H)$ the certainty equivalence mapping that maps a model (G, H) into the corresponding optimal controller. In particular,

$$C^{opt} = c(G_0, H_0) = \arg\min_C J(G_0, H_0, C). \quad (11)$$

We consider PE identification of a parametric model from N data, and assume that $\mathcal{S} \in \mathcal{M}$. The control design mapping then defines a controller $\hat{C}_N = c(\hat{G}_N, \hat{H}_N)$ for each model $(\hat{G}_N, \hat{H}_N)$.[2] The controller $\hat{C}_N$ is a random variable, because the estimated parameter vector $\hat{\theta}_N$ is random, and hence also the model. Applying $\hat{C}_N$ (rather than C^{opt}) to the true system results in an achieved cost $J(G_0, H_0, \hat{C}_N) \geq J(G_0, H_0, C^{opt})$. This results in a "performance degradation" $J^{deg} = J(G_0, H_0, \hat{C}_N) - J(G_0, H_0, C^{opt})$, which is again a random variable.

The problem statement of optimal identification design for control is then phrased as follows: "Find the experimental conditions $\mathcal{X}$ that minimize the

[2] Here $\hat{G}_N$ is a shorthand notation for $G(z, \hat{\theta}_N)$, and similarly for $\hat{H}_N$.

average performance degradation" [3]. In view of what precedes, this can be formulated as:

$$\min_{\mathcal{X}} EJ(G_0, H_0, c(\hat{G}_N, \hat{H}_N)). \quad (12)$$

The expected value is taken with respect to the noise, which affects the model estimate, and hence the controller estimate.

In the context of certainty equivalence control design, this is probably the most logical (and ideal) problem formulation for an *optimal identification for control design.* However, there are several difficulties with this formulation (see Chapter 9 in (Albertos and Sala, 2002)), the main one being that the optimal experiment $\mathcal{X}$ defined by (12) necessarily depends on the unknown system (G_0, H_0). This is specific to all experiment design problems. It does not mean that such results are meaningless: they give useful guidelines for the identification design, and they may lead to iterative schemes that converge to the optimal experiment design: see e.g. (Hjalmarsson *et al.*, 1996).

The first application of this optimal experiment design concept to control-oriented identification was in (Gevers and Ljung, 1986), where an application to Minimum Variance control was treated. The results showed that the optimal experiment consists of performing closed-loop identification with the unknown optimal Minimum Variance controller in the loop. These results were later extended to other control performance criteria in (Hjalmarsson *et al.*, 1996; Forssell and Ljung, 2000).

5. ITERATIVE DESIGN FOR THE NOMINAL MODEL

In this section we discuss the control-oriented identification design of the nominal model, and we show why the pursuit of a control-oriented objective leads to iterative model and controller updates. This observation was made independently in the early nineties by several research teams, who were using different combinations of identification method and control design criterion. To understand the need for iterative design, consider the closed loop systems of Figures 3 and 4, and assume for simplicity that there is no noise ($v_t = 0$ and $\hat{v}_t = 0$), i.e. the control objective is a tracking performance objective.

[3] $\mathcal{X}$ denotes the set of all admissible experimental conditions that have an effect on the quality of the model estimates $(\hat{G}_N, \hat{H}_N)$, such as use of open-loop or closed-loop data, choice of input spectrum distribution, of regulator in the case of closed-loop identification, etc. By 'admissible' experimental conditions, we refer to conditions that obey possible constraints on signal powers or signal energies.

The controller is designed on the basis of the model $\hat{G}$ and then applied to the system G_0. The achieved performance will therefore be close to the desired performance if the two closed loop transfer functions, or a weighted version of these, are close to one another. Thus, we want the following error to be small [4]:

$$\frac{G_0 C}{1 + G_0 C} - \frac{\hat{G} C}{1 + \hat{G} C} = (G_0 - \hat{G}) C S_0 \hat{S} \quad (13)$$

where $S_0 = \frac{1}{1+G_0 C}$ and $\hat{S} = \frac{1}{1+\hat{G}C}$. Now, closed loop PE identification with a given model structure $\mathcal{M} = \{G(z,\theta) \mid \theta \in D_\theta\}$, and with a controller C_{id} in the loop, will asymptotically deliver a model $G(z, \theta^*)$, where $\theta^* = \arg\min_{\theta \in D_\theta} V(\theta)$, with $V(\theta)$ given by

$$V(\theta) = \int_{\pi}^{\pi} |G_0 - \hat{G}(\theta)|^2 |C_{id} S_0|^2 |D|^2 \Phi_r d\omega. (14)$$

Here $D(z)$ is the data filter that can be freely chosen by the user. Observe that, if $C_{id} = C$, and if the data filter is chosen such that $|D(e^{j\omega})|^2 \Phi_r(\omega) = |\hat{S}(e^{j\omega})|^2$, then the model $G(z, \theta^*)$ obtained asymptotically by such closed loop PE identification will make the error (13) small in an H_2 sense. By such design, the identification criterion would be matched to the control performance criterion. However, there are two difficulties with such design: (i) the controller C in the design loop of Figure 3 is a function of the identified model, $C = C(\hat{G})$, and it is therefore impossible to choose $C_{id} = C(\hat{G})$ at the identification design stage; (ii) the sensitivity function $\hat{S}$ is also a function of the estimated model, $\hat{S} = \hat{S}(\hat{G})$. These observations have led to the concept of iterative design, where successive steps of closed loop identification and model-based controller design are performed. Thus, at iteration k, where a model $\hat{G}_k$ has been obtained, select $C_{id,k} = C(\hat{G}_k)$, and $|D_k|^2 \Phi_r = |\hat{S}_k|^2$, where $\hat{S}_k = \frac{1}{1+\hat{G}_k C(\hat{G}_k)}$.

The research work of the early nineties on the definition and computation of control-oriented nominal models led to several important conclusions, that can be summarized as follows.

- The identification criterion for the nominal model should aim at minimizing the distance between the achieved and the designed loop, where this distance is measured in a norm determined by the control performance criterion. An application of this principle to LQG control can be found in (Zang *et al.*, 1995), to H_∞ control in (Schrama and Bosgra, 1993) and to GPC in (Shook *et al.*, 1992), the latter result being based on dual control ideas.

[4] For simplicity of notation, we omit all ω-dependent arguments whenever there is no risk of confusion.

- The identification must be performed in closed loop, with a specific data filter.
- The data filter is model-dependent (i.e. θ-dependent). Thus, one has to resort to iterative model/controller updates for the practical implementation of this design.

In summary, in identification for control, the control performance objective shapes the bias error distribution of the nominal model. This means that the nominal model has a bias error that is small in the frequency areas where it needs to be small for the design of a better controller, typically around the present cross-over frequency.

Iterative identification and control schemes flourished in the nineties, with various combinations of control criteria and identification criteria. The reader is referred to (Gevers, 1993; Bitmead, 1993; Van den Hof and Schrama, 1995) for details and for a survey on such iterative schemes. Unfortunately, it was found (Hjalmarsson *et al.*, 1995) that these iterative schemes do not generically converge to the achievable minimum (within the model/controller set) of the control performance cost.

Despite this, the concept of iterative identification and control design was rapidly adopted in process control applications: see e.g. (Partanen and Bitmead, 1995; Schrama and Bosgra, 1993; de Callafon *et al.*, 1993; Holmberg *et al.*, 2000; Cooley and Lee, 2001). One reason is that it is typical in such applications that large numbers of closed loop data are flowing into the control computer, and it then makes sense to use these data to replace the existing controller by one that achieves better performance. The practical impact of iterative model and controller redesign has been assessed in (Landau, 1999), where some interesting observations are made on the distinction between this batch-like mode of operation and the more classical methods of adaptive control.

6. MODEL UNCERTAINTY SETS AND THE ROBUST CONTROL PARADIGM

We have shown how control-oriented nominal models are obtained by minimization of an identification criterion that is determined by the overall control performance criterion. In establishing this result, one has assumed that the controller is computed from the estimated model $\hat{G}$ using the certainty equivalence paradigm. At no point has the model uncertainty due to the noise been taken into account. Conversely, the control design step that has led to the iterative design schemes developed above is not based on robust control principles. The paradigm of modern[5] robust control design can be briefly summarized as follows.

One wants to design a controller for an unknown system G_0. Some prior knowledge about G_0 allows one to assume that G_0 belongs to some *model uncertainty set* $\mathcal{D}$. Most often, a nominal model $\hat{G}$ of G_0 is available, typically the centre of $\mathcal{D}$. Some performance objective is often given in the form of a criterion $J(G,C)$, to be minimized. If the true system were perfectly known, the optimal controller would then be defined as $C^{opt} = \arg\min J(G_0, C)$, where the minimization is performed over some predefined set of admissible controllers $\mathcal{C}$. In the robust control paradigm, the true system is unknown, but the information $(\hat{G}, \mathcal{D})$ is available. One then seeks a *robust controller* C with the following properties:

(1) C must stabilize all models in $\mathcal{D}$;
(2) the worst performance achieved by C on any model in $\mathcal{D}$ must be as high as possible.

The robust controller C is often selected as one that stablizes all models in $\mathcal{D}$ and achieves the best *worst-case performance* over $\mathcal{D}$, i.e. C is computed as

$$C = \arg\min_{C \in \mathcal{C}} \sup_{G \in \mathcal{D}} J(G, C). \tag{15}$$

The uncertainty model set $\mathcal{D}$ plays a central role in this design strategy. In the robust control theory developed through the eighties and nineties, this set is god-given in that it is based on so-called "prior assumptions" about the model and its uncertainty. Many different descriptions of model uncertainty have been considered; most often they are expressed as frequency domain sets, containing both structured and unstructured components: see e.g. (Zhou *et al.*, 1995; Skogestad and Postlethwaite, 1996; Morari and Zafiriou, 1989). The following are representative examples of commonly used uncertainty sets; for simplicity, we consider scalar transfer functions only.

Additive uncertainty set:

$$\mathcal{D}_A = \{G_\Delta(z) \mid G_\Delta(z) = G(z) + \Delta(z), \quad \text{with } |\Delta(e^{j\omega})| < W(e^{j\omega}) \ \forall \omega\},$$

where $G(z)$ is a "nominal model" and $W(e^{j\omega})$ is a frequency weighting function.

Youla-Kucera uncertainty set:

$$\mathcal{D}_{YK} = \{G_\Delta(z) \mid G_\Delta(z) = \frac{N_x + D_c\Delta}{D_x - N_c\Delta}, \quad \text{with } |\Delta(e^{j\omega})| < W(e^{j\omega}) \ \forall \omega\},$$

[5] We refer to modern robust control design as the theory developed in the eighties, that is based on model uncertainty sets; in contrast, the classical robust design theory relies on robustness margins expressed in Bode, Nyquist or Nichols plots.

where $N_x, D_x, N_c, D_c, \Delta$ are stable, rational, proper transfer functions, $\hat{G} = \frac{N_x}{D_x}$ is a "nominal model", and $C = \frac{N_c}{D_c}$ is any stabilizing controller of G_0. This uncertainty set, dual of the Youla-Kucera set of all stabilizing controllers of a given system, was introduced for the description of model sets in (Hansen *et al.*, 1989).

The new approach, initiated around 1990, was to consider the estimation of uncertainty sets from data. Unfortunately, the available PE identification theory had rather little to offer to the existing robust control theory, for two reasons: (1) there were no adequate expressions for the estimation of the total error on an identified transfer function, or some upper bound on this error; the main difficulty was the estimation of the bias error, which could only be characterized by implicit integral expressions; (2) the available uncertainty descriptions were not given as frequency domain sets. As a result, a wide range of new identification techniques were developed. New model assumptions, noise assumptions and identification criteria were introduced whose main merit was to deliver computable error bounds on the estimated models. The survey paper (Ninness and Goodwin, 1995) is probably still one of the best presentations of these alternative methods. Some of these methods have been shown to lead to very conservative upper bounds on the uncertainty set. But clearly, this work has led to important advances in the characterization of bounds on transfer function error estimates.

One uncertainty set that was available in PE identification theory was the set (8) defined via the ellipsoidal set (7) in parameter space. For open loop identification, this set can be described as follows.

Prediction Error uncertainty set:

$$\mathcal{D}_{PE} = \{G(z,\theta) \mid G(z,\theta) = \frac{N(z,\theta)}{D(z,\theta)}$$
$$\text{with } (\theta - \hat{\theta}_N)^T R(\theta - \hat{\theta}_N) < 1\}$$

Here N and D are polynomials parametrized by θ, and R is proportional to the inverse of the estimated covariance matrix of $\hat{\theta}_N$. The difficulty with this PE uncertainty set is that it could be computed easily only in the case where the system is in the model set, and it did not connect with the robust control theory and design tools that were available in the early nineties. Thus, all through the nineties, a lot of research was produced by the identification community to

(1) extend the use of uncertainty sets in order to also include the bias error; this was achieved by either embedding the bias error in a stochastic framework (Goodwin *et al.*, 1992; Hakvoort and Van den Hof, 1997), or by estimating the bias error through a validation step that uses a full order model (Ljung, 2000; Gevers *et al.*, 2003).
(2) develop a robust control stability and performance theory for PE uncertainty sets characterized by ellipsoids in parameter space (Bombois *et al.*, 2001).

Many important advances have thus been made in the characterization of bounds on transfer function error estimates, but this subject will undoubtedly remain an object of intense debate and activity for years to come.

7. TOWARDS CONTROL-ORIENTED UNCERTAINTY SETS?

In the second half of the nineties, one began to seriously study the interplay between model uncertainty sets and robust control objectives, in order to address the question of building *control-oriented uncertainty sets.* The motivation for this is based on the following two observations.

- It follows from the properties that define a robust controller (see the previous section) that satisfaction of these two properties hinges as much on the choice of the controller C as it does on the uncertainty set $\mathcal{D}$.
- The shape of a data-based uncertainty set depends very much on the experimental conditions under which it is estimated.

Combining these two observations leads to the idea of constructing control-oriented uncertainty sets by proper choice of experimental conditions.

To illustrate the connection between experiment conditions and model-based control properties, we consider a very simple gedankenexample, presented in (Gevers *et al.*, 1998), where we focus attention only on robust stability. Consider a "true system" G_0 described by the following simple ARX model:

$$(1 - 1.4z^{-1} + 0.45z^{-2})y_t = z^{-1}(1 + 0.25z^{-1})u_t + e_t,$$

where e is a unit variance white noise. With a constant gain feedback law $u_t = r_t - Cy_t$, the closed loop system is stable for $C < C_{max} = 2.2$. Consider now that we estimate the parameters of this ARX model by PE identification, using the exact structure. We can thus estimate an ellipsoid in parameter space, to which the true parameters belong with probability 95%, say. We can then compute the proportional output feedback controller with the highest gain, $\hat{C}_{max}$, that stabilizes all models $\hat{G}$ whose parameters lie in that 95% uncertainty set. We call this controller the "optimal robust controller". In order to show the effect of the experimental setup of the identification on the optimal robust controller, we have compared, by Monte Carlo simulations, two experimental conditions:

(1) open loop identification with a unit variance white noise as input signal u;
(2) closed loop identification with a controller $u_t = r_t - y_t$ in the loop during data collection, and a white noise reference signal r with variance 18.38; this choice yields the same output variance as in the open loop experiment.

Each of these two experiments was run 100 times, each time with 1000 input-output data. For each run, the 95% confidence ellipsoid was computed in parameter space, and the gain $\hat{C}^i_{max}$ of the i-th run was computed as the largest gain that would stabilize all models in the corresponding model set. On the basis of 100 runs, we then computed the average $\hat{C}_{max}$ and its variance for each of the two experimental conditions. The following results were obtained for open loop (O.L.), respectively closed loop (C.L.), identification:

$$\text{O.L.}: \hat{C}_{max} = 1.36,\ \sigma^2_{\hat{C}max} = 0.12$$

$$\text{C.L.}: \hat{C}_{max} = 2.04,\ \sigma^2_{\hat{C}max} = 0.02$$

Remember that, for the true system G_0, we have $C_{max} = 2.2$. This example clearly shows that the experimental conditions used in the feedback experiment are more control-oriented than those in the open loop experiment, since they lead to a much less conservative estimate of the limit gain for the robust controller. This example serves to illustrate the effect of the identification experiment on the set of admissible controllers, via the model uncertainty set. In a more realistic setup, the interplay between the design of the identification experiment, the corresponding model uncertainty set, and the set of admissible controllers, is a lot harder to understand and analyze.

In order to relate the identification design with the properties of a robust controller, one has resorted to splitting up the problem into its two components:

- understanding the effect of experimental conditions on model quality, or more precisely on the properties of the uncertainty set;
- understanding the interplay between the uncertainty set and the properties of the ensuing robust controller.

While the first problem is reasonably easy and has already yielded many results, the second problem is much harder and there is, as of today, no consensus on the best way to address it. We now elaborate on these two subproblems.

In PE identification, there has been a continuing activity aimed at characterizing the bias, variance, and total mean square error on the estimation of parametric transfer functions, dating back to the mid-eighties. The variance formulas of (Ljung, 1985) were based on an assumption of model order going to infinity; more accurate approximations have recently been obtained for the variance formulas of finite order transfer function estimates (Ninness and Hjalmarsson, 2003). The bias formulas of (Wahlberg and Ljung, 1986) have given way to explicit expressions based on a stochastic embedding of the bias error, or on its estimation based on validation with full order models, as mentioned in the previous section.

All these formulas for the estimation of the bias, the variance or the total mean square error of identified transfer functions explicitly contain the effect of the experimental conditions (e.g. number of data, input spectrum, noise spectrum, feedback configuration, etc) on the error measure. This yields the possibility of optimizing over some relevant experimental design variable in order to minimize a particular measure of this error. In the 1970's, optimal input design for system identification was an active area of research, with different quality measures of the identified model being used for this optimal design (Zarrop, 1979; Goodwin and Payne, 1977). Thus, given a clear view as to what constitutes a *control-oriented uncertainty set*, one could certainly extend these optimal experiment design ideas to the construction of such sets.

So, what constitutes a *control-oriented uncertainty set*? There is no clear consensus yet on a good definition. One possible view is to say that a model uncertainty set $\mathcal{D}$ is "control-oriented" if the corresponding set of admissible controllers, $\mathcal{C}$, is large. Such track has been pursued in e.g. (Douma *et al.*, 2003) where different uncertainty structures are compared, and in (Gevers *et al.*, 2003) where the worst-case ν-gap has been proposed as a control-oriented measure of size of $\mathcal{D}$, because it is related to the size of the corresponding set $\mathcal{C}$ of stabilizing controllers. In (Hildebrand and Gevers, 2003) the corresponding optimal input design problem for the minimization of this worst-case ν-gap measure has been solved.

However, one could also define a control-oriented uncertainty set in a very different way as follows. Consider an uncertainty set $\mathcal{D}$ with center $\hat{G}$ containing the true G_0, and a controller $C = C(\hat{G}, \mathcal{D})$ with nominal stability margin $b(\hat{G}, C)$ and nominal performance $J(\hat{G}, C)$. Then $\mathcal{D}$ could be called control oriented if C stabilizes all models in $\mathcal{D}$, if the worst case stability margin $\sup_{G\in\mathcal{D}} b(G, C)$ of C with all models in $\mathcal{D}$ is close to the nominal margin $b(\hat{G}, C)$, and if the worst case performance $\sup_{G\in\mathcal{D}} J(G, C)$ is close to the nominal performance $J(\hat{G}, C)$.

Thus, much work remains to be done on the definition of control-oriented uncertainty sets, and on the computation of the corresponding quality measures. This is certainly one area where one can expect a lot of research activity in the coming years.

8. CONCLUSIONS

We have attempted to explain the major issues related to the problem of identification for control, to present the successive solutions that have been brought to the overall problem, and to display the major remaining open problems. Most certainly, the major impact so far in terms of transfer of technology to the industrial world has been the iterative schemes of model and controller updates, which have provided a methodological background for iterative data-based controller performance enhancement. As for the synergy of robust control concepts and identification design concepts, we believe that most challenges are still ahead of us. The main reason is that a full understanding of the interplay between identification design and robust control analysis and synthesis cannot bypass the role of the uncertainty set, and that the analysis of this problem is difficult and involves techniques from several different subdisciplines.

9. REFERENCES

Albertos, P. and A. Sala (2002). *Editors, Iterative Identification and Control.* Springer Verlag.

Anderson, B.D.O. and M. Gevers (1998). Fundamental problems in adaptive control. *Perspectives in Control, a Tribute to Ioan Doré Landau* pp. 9–21.

Anderson, B.D.O., R.R. Bitmead, C.R. Johnson Jr, P.V. Kokotovic, R.L. Kosut, I.M.Y. Mareels, L. Praly and B.D. Riedle (1986). *Stability of Adaptive Systems: Passivity and Averaging Analysis.* MIT Press. USA.

Åström, K.J. and J. Nilsson (1994). Analysis of a scheme for iterated identification and control. In: *Proc. IFAC Symp. on Identification.* Copenhagen, Denmark. pp. 171–176.

Åström, K.J. and T. Bohlin (1965). Numerical identification of linear dynamic systems from normal operating records. In: *Proc. IFAC Symposium on Self-Adaptive Systems.* Teddington, UK.

Bitmead, R.R. (1993). Iterative control design approaches. In: *Proc. 12th IFAC World Congress.* Vol. 9. Sydney, Australia. pp. 381–384.

Bombois, X., M. Gevers and G. Scorletti (2000). A measure of robust stability for an identified set of parametrized transfer functions. *IEEE Transactions on Automatic Control* **45**(11), 2141–2145.

Bombois, X., M. Gevers, G. Scorletti and B.D.O. Anderson (2001). Robustness analysis tools for an uncertainty set obtained by prediction error identification. *Automatica* **37**(10), 1629–1636.

Cooley, Brian L. and Jay H. Lee (2001). Control-relevant experiment design for multivariable systems described by expansions in othonormal bases. *Automatica* **37**(2), 273–281.

de Callafon, R.A. and P.M.J. Van den Hof (1997). Suboptimal feedback control by a scheme of iterative identification and control design. *Mathematical Modelling of Systems* **3**(1), 77–101.

de Callafon, R.A., P.M.J. Van den Hof and M. Steinbuch (1993). Control-relevant identification of a compact disc pick-up mechanism. In: *Proc. 1993 IEEE Conf. on Decision and Control.* San Antonio, Texas. pp. 2050–2055.

de Vries, D.K. and P.M.J. Van den Hof (1995). Quantification of uncertainty in transfer function estimation: a mixed probabilistic - worst-case approach. *Automatica* **31**, 543–558.

Deistler, M. (2002). System identification and time series analysis: Past, present, and future. In: *'Stochastic Theory and Control', Festschrift for Tyrone Duncan.* Springer. Kansas, USA. pp. 97–108.

Douma, S.G., P.M.J. Van den Hof and O.H. Bosgra (2003). Controller tuning freedom under plant identification uncertainty: double youla beats gap in robust stability. *Automatica* **39**, 325–333.

Forssell, U. and L. Ljung (2000). Some results on optimal experiment design. *Automatica* **36**, 749–756.

Gevers, M. (1991). Connecting identification and robust control: A new challenge. In: *Proc. IFAC/IFORS Symposium on Identification and System Parameter Estimation.* Budapest, Hungary. pp. 1–10.

Gevers, M. (1993). Towards a joint design of identification and control?. In: *Essays on Control: Perspectives in the Theory and its Applications, Birkhauser, New York* (H.L. Trentelman and J.C. Willems, Eds.). Birkhauser. New York. pp. 111–151.

Gevers, M. and L. Ljung (1986). Optimal experiment designs with respect to the intended model application. *Automatica* **22**, 543–554.

Gevers, M., B. D.O. Anderson and B. Codrons (1998). Issues in modeling for control. In: *Proc. American Control Conference.* Philadelphia, USA. pp. 1615–1619.

Gevers, M., X. Bombois, B. Codrons, G. Scorletti and B.D.O. Anderson (2003). Model validation for control and controller validation in a prediction error identification framework - Part I: theory. *Automatica* **39**(3), 403–415.

Goodwin, G. and R.L. Payne (1977). *Dynamic System Identification: Experiment Design and Data Analysis.* Academic Press. New York.

Goodwin, G.C., M. Gevers and B. Ninness (1992). Quantifying the error in estimated transfer functions with application to model order selection. *IEEE Trans. Automatic Control* **37**, 913–928.

Hakvoort, R.G. and P.M.J. Van den Hof (1997). Identification of probabilistic system uncertainty regions by explicit evaluation of bias and variance errors. *IEEE Trans. Automatic Control* **42**(11), 1516–1528.

Hansen, F., G. Franklin and R. Kosut (1989). Closed-loop identification via the fractional representation: Experiment design. *Proc. American Control Conference* pp. 1422–1427.

Hildebrand, R. and M. Gevers (2003). Identification for control: optimal input design with respect to a worst-case ν-gap cost function. *SIAM Journal on Control and Optimization* **41**(5), 1586–1608.

Hjalmarsson, H. (2003). From experiments to closed-loop control (plenary lecture). In: *CD-ROM of 13th IFAC Symposium on System Identification.* Rotterdam, The Netherlands.

Hjalmarsson, H., M. Gevers and F. De Bruyne (1996). For model-based control design, closed-loop identification gives better performance. *Automatica* **32**, 1659–1673.

Hjalmarsson, H., S. Gunnarsson and M. Gevers (1995). Optimality and sub-optimality of iterative identification and control design schemes. In: *Proc. American Control Conference 95.* Vol. 4. Seattle, Washington. pp. 2559–2563.

Ho, B.L. and R.E. Kalman (1965). Effective construction of linear state-variable models from input-output functions. *Regelungstechnik* **12**, 545–548.

Holmberg, U., S. Valentinotti and D. Bonvin (2000). An identification-for-control procedure with robust performance. *Control Engineering Practice* **8**, 1107–1117.

Kalman, R.E. (1960*a*). Contributions to the theory of optimal control. *Boletin de la Sociedad Matematica Mexicana* **5**, 102–119.

Kalman, R.E. (1960*b*). A new approach to linear filtering and prediction problems. *Transactions ASME, Series D, J. Basic Eng.* **82**, 34–45.

Kosut, R.L. and B.D.O. Anderson (1994). Least-squares parameter set estimation for robust control design. In: *Proc. American Control Conference.* Baltimore,Maryland. pp. 3002–3006.

Landau, I.D. (1999). From robust control to adaptive control. *Control Engineering Practice* **7**(9), 1113–1124.

Lee, W.S., B.D.O. Anderson, R.L. Kosut and I.M.Y Mareels (1993). A new approach to adaptive robust control. *Int. Journal of Adaptive Control and Signal Processing* **7**, 183–211.

Ljung, L. (1985). Asymptotic variance expressions for identified black-box transfer function models. *IEEE Trans. Automatic Control* **AC-30**, 834–844.

Ljung, L. (1999). *System Identification: Theory for the User, 2nd Edition.* Prentice-Hall. Englewood Cliffs, NJ.

Ljung, L. (2000). Model error modeling and control design. In: *CD-Rom Proc. of IFAC Symposium on System Identification, paper WeAM1-3.* Santa Barbara, California.

Mäkilä, P.M, J.R. Partington and T.K. Gustafsson (1995). Worst-case control-relevant identification. *Automatica* **31**(12), 1799–1819.

Morari, M. and E. Zafiriou (1989). *Robust Process Control.* Prentice-Hall. Englewood Cliffs, NJ.

Ninness, B. and G.C. Goodwin (1995). Estimation of model quality. *Automatica* **31**(12), 32–74.

Ninness, B. and H. Hjalmarsson (2003). Variance error quantifications that are exact for finite model order. *submitted for publication.*

Partanen, A.G. and R.R. Bitmead (1995). The application of an iterative identification and controller design to a sugar cane crushing mill. *Automatica* **31**, 1547–1563.

Schrama, R.J.P. and O.H. Bosgra (1993). Adaptive performance enhancement by iterative identification and control design. *Int. Journal of Adaptive Control and Signal Processing* **7**(5), 475–487.

Shook, D.S., C. Mohtadi and S.L. Shah (1992). A control-relevant identification strategy for gpc. *IEEE Trans. Automatic Control* **AC-37**(7), 975–980.

Skelton, R.E. (1989). Model error concepts in control design. *Int. Journal of Control* **49**(5), 1725–1753.

Skogestad, S. and I. Postlethwaite (1996). *Multivariabel Feedback Control: Analysis and Design.* J. Wiley. Chichester, U.K.

Van den Hof, P.M.J. and R.J.P. Schrama (1993). An indirect method for transfer function estimation from closed loop data. *Automatica* **29**, 1523–1527.

Van den Hof, P.M.J. and R.J.P. Schrama (1995). Identification and control - closed-loop issues. *Automatica* **31**, 1751–1770.

Wahlberg, B. and L. Ljung (1986). Design variables for bias distribution in transfer function estimation. *IEEE Trans. Automatic Control* **AC-31**, 134–144.

Zang, Z., R.R. Bitmead and M. Gevers (1995). Iterative weighted least-squares identification and weighted LQG control design. *Automatica* **31**(11), 1577–1594.

Zarrop, M. (1979). *Optimal experiment design for dynamic system identification.* Lecture Notes in Control and Information Sciences, Vol. 21, Springer Verlag. Berlin, New York.

Zhou, K., J.C. Doyle and K. Glover (1995). *Robust and Optimal Control.* Prentice Hall, New Jersey.

ELSEVIER
IFAC
PUBLICATIONS
www.elsevier.com/locate/ifac

ON MODEL SELECTION FOR STATE ESTIMATION FOR NONLINEAR SYSTEMS

Robert Bos *,[1] **Xavier Bombois** * **Paul M. J. Van den Hof** *

* *Delft Center for Systems and Control, Delft University of Technology, Mekelweg 2, 2628 CD Delft, The Netherlands, email: r.bos@dcsc.tudelft.nl*

Abstract: State filters can be used to produce online estimates of the state of a process. If an exact model for the true process is not known, but multiple candidate models are available to describe the current behavior of the true system, it is necessary to select that model that leads to the optimal state estimates. This paper describes a novel approach for model selection for state estimation by comparing the expected weighted prediction error using estimated states of different candidate models. The expected prediction error can not be computed exactly, but can be estimated using a newly derived generalized version of the FPE selection criterion. A simulation example of a time varying system is used to illustrate the performance of the selection method. *Copyright © 2004 IFAC*

Keywords: State estimation, FPE, model selection, fault detection

1. INTRODUCTION

The goal of monitoring can be to provide an operator with accurate online information on critical process variables. In this paper these physical variables will be referred to as the state of the system. Since the complete state of a complex process often cannot be measured directly, it has to be estimated using measurements related to the state. To obtain these estimates a model of the system is required that describes the dynamic behavior of the state and relates the states to the available measurements.

Assume that the model can be written in the following state-space form:

$$x_{k+1} = f(x_k, u_k, \theta) + w_k \tag{1}$$

$$y_k = h(x_k, \theta) + v_k, \tag{2}$$

[1] The work of Robert Bos is supported by Dutch research organization TNO-TPD.

in which x_k is the system state at time index k, u_k are the known inputs to the system, y_k are the measured system outputs, θ is a known parameter vector and w_k and v_k are zero mean independent gaussian noises that operate on the states and measurements respectively. In process industry,$f(\cdot)$ and $h(\cdot)$ will generally be complex nonlinear functions. If a model in this form is given, it is usually possible to construct a state filter, that produces an online estimate of a true state x_k using all measurements $y_0, \dots, y_k$. We will denote this estimate as $\hat{x}_{k|k}$. Examples of such filters are the Extended Kalman Filter (EKF) (Anderson and Moore, 1979), Unscented Kalman Filter (UKF) (Julier *et al.*, 2000), Moving Horizon estimators (MHE) (Robertson and Lee, 1995) and Particle Filters (Doucet *et al.*, 2001). Since a state filter estimates states based on measurements y_k and the model (1)-(2), the accuracy of the estimated states not only depends on the accuracy of the measurements, but also on the quality of the model. If, for instance, the model is incorrect, the state estimates could be severely biased.

In some situations multiple candidate models are available for describing the current behavior of the true system. An example of such a situation is in model based fault detection and isolation (FDI). In model based FDI, at least one model is given to describe the nominal process behavior, while a second set of models describe the behavior of the process after corresponding faults have occurred. These fault models could be the same as the nominal models, but with time varying parameters. In this paper we will focus on the problem of selecting the best possible model for state estimation.

Model selection for filtering problems is often described using Bayes conditional probability theory (Gustafsson, 2000). Suppose n models of the form (1)-(2) are available, denote these models as $\mathcal{M}_i$, with $i = 1, 2, \ldots, n$. Then the filtering procedure is carried out for each model, on data y. Afterwards, using Bayes conditional probability theory, the conditional probability $p(\mathcal{M}_i|y)$ is computed. The model with the highest conditional probability is then selected, and the state estimates based on this model are used. The conditional probability $p(\mathcal{M}_i|y)$ can be computed via: $p(\mathcal{M}_i|y) = p(y|\mathcal{M}_i)p(\mathcal{M}_i)/p(y)$. Using this equation in practice for complex process models is generally difficult, because the term $p(y|\mathcal{M}_i)$ is not trivial to compute for non-linear systems and knowledge of the a-priori probability of each model $p(\mathcal{M}_i)$ is rarely available.

An alternative approach to the model selection problem is given in (Tyler *et al.*, 2000). If a moving horizon state estimator (MHE) is used for state estimation, the state estimation problem is written as a weighted and regularized least squares problem. The problem of model selection is therefore approximately similar to model selection in system identification theory. Given this similarity, the model selection is done using the Akaike Information Criterion (AIC). Advantages of this approach are that exact probability distributions are no longer required, and the technique can also be easily adapted for non-linear models. Drawbacks of this approach are that the technique can only be used in conjunction with moving horizon estimators. Another drawback is that the AIC criterion may not be the best criterion, since it was derived only for least squares problems without weighting and regularization, while states are estimated with weighting and regularization.

In this paper a model selection procedure closely related to (Tyler *et al.*, 2000) will be considered. Instead of using AIC, a specialized criterion for weighted and regularized least squares problems is derived. Secondly we will show that the model selection can also be used with estimators different from the MHE. Finally a simulation example is presented in which model selection is used to accurately estimate the state of a time varying system.

2. MODEL SELECTION FOR STATE ESTIMATION

As mentioned in the introduction, our objective is to estimate the state vector of a true process using the output vector y_k, the input signal u_k and a model of the true process. Assume that this model of the true system is given by the linear state space equations:

$$x_{k+1} = Ax_k + Bu_k + w_k \tag{3}$$

$$y_k = Cx_k + v_k. \tag{4}$$

where w_k and v_k are gaussian noises with

$$\mathbb{E}\begin{bmatrix} w_k \\ v_k \end{bmatrix} = 0 \quad \mathbb{E}\begin{bmatrix} w_k \\ v_k \end{bmatrix}[w_l^T \; v_l^T] = \begin{bmatrix} Q & 0 \\ 0 & R \end{bmatrix}\delta_{k,l} \tag{5}$$

If the model (3)-(4) perfectly describes the true system, an optimal estimate of the state vector x_k at each time k can be obtained using a Kalman filtering procedure implemented using (3)-(4). Such a filtering procedure consists of two steps. In the prediction step, a prediction $\hat{x}_{k+1|k}$ of the state vector is given, along with its covariance matrix $\hat{P}_{k+1|k}$. In the measurement update, the measurement y_{k+1} is used to refine the prediction into estimate $\hat{x}_{k+1|k+1}$ and to produce its covariance matrix $\hat{P}_{k+1|k+1}$. An important result for the sequel is that the measurement update in a Kalman filtering procedure can be seen as the solution of a Weighted and Regularized Least Squares (WRLS) problem (Kailath *et al.*, 1999):

$$\hat{x}_{k|k} = \arg\min_x \|y_k - Cx\|^2_{R^{-1}} + \|x - \hat{x}_{k|k-1}\|^2_{\hat{P}^{-1}_{k|k-1}}, \tag{6}$$

with $\|z\|^2_W = z^T W z$ for $z \in \mathbb{R}^n$.

If the model of the true system that is assumed to perfectly describe the true system is nonlinear, such as e.g. in (1)-(2), an estimate of the state vector x_k can again be obtained using very similar procedures (e.g. using the EKF or UKF instead of the normal Kalman Filter) which also consist of two steps: a prediction step and a measurement update step. Moreover, it can be proven that, if the output function (2) can be well approximated by a linear function around $\hat{x}_{k|k-1}$, the measurement update in this procedure is still approximately a WRLS problem:

$$\hat{x}_{k|k} = \arg\min_{x} \|y_k - \hat{y}(x)\|^2_{R^{-1}} + \|x - \hat{x}_{k|k-1}\|^2_{\hat{P}^{-1}_{k|k-1}}, \quad (7)$$

where $\hat{y}(x)$ denotes the predictor of the output vector y_k using the state vector x and the available model (1)-(2) i.e. $\hat{y}(x) = h(x)$. By comparing (6) and (7), we see that (7) is also valid in the linear case with $\hat{y}(x) = Cx$.

The state estimation procedure delivers an estimate $\hat{x}_k = \hat{x}_{k|k}$ for x_k using the assumption that the available model perfectly describes the true system. In practice, the model is only an approximation of the true system and the quality of the estimate of x_k will depend on the quality of the chosen model. Consequently, the model will have to be chosen in such a way that the state estimation procedure based on this model delivers a good estimate of x_k. In this paper, we will consider the particular situation where we have several candidate-models and we have to select, among these models, the model which will deliver the best estimate of the state vector. To make this selection, only known inputs and measured outputs are available.

Our selection procedure will be based on a measure of the quality of the model for the estimation of the state vector. Different measures could be considered for this purpose. In this paper, we will define a measure of quality which is very similar to the quality measures used in system identification. Given a model of the type (1)-(2) and given the estimate $\hat{x}_k$ of the state vector at time k obtained using the WRLS problem (7), the measure of quality $\mathcal{V}(k)$ at time k is defined as follows:

$$\mathcal{V}(k) = \mathbb{E}_{\hat{x}_k} \overline{V}(\hat{x}_k, k). \quad (8)$$

$$\text{with } \overline{V}(x, k) = \mathbb{E}_{y_k} \|y_k - \hat{y}(x)\|^2_{R^{-1}}. \quad (9)$$

Using (9), we see that $\overline{V}(\hat{x}_k, k)$ represents the ability of the model and the estimate $\hat{x}_k$ to predict not only the particular realization of the output vector y_k that we used to estimate $\hat{x}_k$, but also all other possible realizations of y_k. In (9) the prediction error $y_k - \hat{y}(x)$ is weighted with R^{-1}, to take into account the variability of the measurements y_k. The quantity $\overline{V}(\hat{x}_k, k)$ is still a random variable since $\hat{x}_k$ is determined using noisy data. Therefore, it is safer to consider its mean as measure of quality for the model such as we have done in (8). From the definition of $\mathcal{V}(k)$, we see that the smaller the time function $\mathcal{V}(k)$ is, the better is the model.

We have thus defined a measure $\mathcal{V}(k)$ of the quality of a model. This quantity can only be used in a quality assessment procedure if it is possible to compute (or to approximate) $\mathcal{V}(k)$ using the available data. In order to find a method for computing (or approximating) $\mathcal{V}(k)$, we first notice the strong analogy between $\mathcal{V}(k)$ and the Final Prediction Error (FPE) that is used to assess the quality of a model in system identification theory (see (Ljung, 1999)). We can indeed see the FPE as a special case of $\mathcal{V}(k)$ where the matrix R in (9) has been replaced by the identity matrix and where $\hat{x}_k$ has been estimated using the criterion (7) with $R = I$ and with $\hat{P}^{-1}_{k+1|k} = 0$. In (Ljung, 1999) an approximation for FPE is also given for the case where $\hat{x}_k$ has been estimated using the problem (7) with $R = I$ and with $\hat{P}^{-1}_{k+1|k} = \delta I$ with δ a positive real constant. However, in practice $\hat{P}^{-1}_{k+1|k}$ cannot be written as δI, which implies that the approximations in (Ljung, 1999) can not be used directly in order to approximate $\mathcal{V}(k)$ if $\hat{x}_k$ has been estimated using the general WRLS problem (7). We have therefore extended the results of (Ljung, 1999) in order to be able to approximate $\mathcal{V}(k)$.

Proposition 1. Let us consider the time instant k and the output vector y_k collected from the true system at that instant. Let us also consider the measure of quality $\mathcal{V}(k)$ defined in (8)-(9). Furthermore assume that the estimate $\hat{x}_k$ of the state vector in (8)-(9) is obtained via the following weighted and regularized least squares problem which is equivalent to (7):

$$\hat{x}_k = \arg\min_{x} \Big(V(x, k) + (x - x_k^{\#})^T P^{-1} (x - x_k^{\#})\Big) \quad (10)$$

in which $x_k^{\#}$ is a given state vector, P^{-1} is a positive semi-definite regularization matrix, and $V(x, k)$ is a weighted least squares criterium:

$$V(x, k) = \|y_k - \hat{y}(x)\|^2_{R^{-1}}. \quad (11)$$

Then, if we can further assume that $x_k^{\#} \approx x_k^* = \arg\min_x \overline{V}(x, k)$, that $y_k - \hat{y}(x_k^*)$ is approximately a white noise and that the dimension of the vector y_k is sufficiently large, we have that

$$\mathcal{V}(k) \approx V(\hat{x}_k, k) + 4\mathrm{tr}\left\{[(\psi^T R^{-1} L R^{-T} \psi)] \times \left[\overline{V}''(x_k^*, k) + 2P^{-1}\right]^{-1}\right\}. \quad (12)$$

where $L = \mathbb{E}(ee^T)$ with $e = y_k - \hat{y}(x_k^*)$ and $\psi = \frac{\partial}{\partial x}\hat{y}(x)$ evaluated at $x = x_k^*$.

Proof. See Appendix A. □

As mentioned in the statement of Proposition 1, the derivation of the approximation (12) requires some additional assumptions. These assumptions are also present in the less general version of (Ljung, 1999) and in the same reference the consequences of these assumptions are discussed.

The estimated asymptotic fit $\mathcal{V}(k)$ in (12) is the sum of the achieved fit $V(\hat{x}_k, k)$ on the measurement data and a complex term. This term contains the second derivative of $\overline{V}(x_k^*, k)$. For general models this quantity can be approximated by the second derivative of $V(\hat{x}_k, k)$. Similarly, ψ can be approximated using the derivative of $\hat{y}(x)$ evaluated in $\hat{x}_k$ instead of x_k^*. If the output model is linear, for instance as is the case in (4), such approximations are not necessary, because in this case it is easy to show that $\overline{V}''(x_k^*, k) = 2CR^{-1}C^T$ and $\psi = C$. L represents the covariance of the minimal asymptotic prediction error. For additive measurement noise, $L = R$.

Let us now summarize and define our model selection procedure. We wanted to select, among a set of candidate-models, the model which delivers the best estimate of the state vector. We have defined for this purpose a measure $\mathcal{V}(k)$ of the quality of a model. This measure is a time function and can be approximated using (12). In order to make the selection, let us compute, with each model, the estimate of the state vector at time instants $k = 1...M$. Let us then approximate $\mathcal{V}_i(k)$ for each model and for $k = 1...M$ ($\mathcal{V}_i(k)$ denotes the quality measure of the ith model). Then, let us determine for each model i, the average quality measure over $k = 1...M$ i.e. $\mathcal{V}_{av,i} = (1/M)\sum_{k=1}^{M} \mathcal{V}_i(k)$. The best model is then defined as the model for which $\mathcal{V}_{av,i}$ is minimal.

Using the average $\mathcal{V}_{av,i}$ of the quality measure over $k = 1...M$ as selection criterion has two advantages: we replace a time function $\mathcal{V}_i(k)$ by a single number and we generally decrease the influence of the approximation errors in $\mathcal{V}_i(k)$.

3. MODEL SELECTION AND TIME VARYING SYSTEMS

State estimation for time varying systems can be improved by treating the estimation problem as a model selection problem which can be solved using the model selection procedure of section 2. This particular estimation setup is explained in the sequel of this paper.

Time varying systems are very common in the industrial world: the true behavior of a complex process will generally change over time. Changes can occur slowly, for example due to wear, but can also occur abruptly, e.g. due to a sudden fault in the system. If the system behavior changes, but the model used for state estimation remains the same, state estimates will most likely become biased.

If we assume that the dynamics of the true system can be described by (1)-(2) and that any change in the dynamics of this true system can be modelled as changes in the model parameters θ, then an extra equation can be added to the model, such that the model parameters are allowed to change:

$$\theta_{k+1} = f_\theta(x_k, \theta_k, u_k) + w_k^\theta, \tag{13}$$

where w_k^θ is a independent gaussian noise. If we define a new vector $\tilde{x} = [x_k^T \; \theta_k^T]^T$, it is always possible to combine (1)-(2) with (13) into a new model:

$$\tilde{x}_{k+1} = \tilde{f}(\tilde{x}_k, u_k) + \tilde{w}_k \tag{14}$$

$$y_k = \tilde{h}(\tilde{x}_k, u_k) + v_k, \tag{15}$$

with $\tilde{w}_k = [w_k^T \; {w_k^\theta}^T]^T$. Since the form of (14)-(15) is the same as that of the original model (1)-(2), a state filter can usually be designed to simultaneously estimate states x_k and to update parameters θ_k. The price for simultaneously estimating states and parameters is that the variance of the estimated states will increase, even if the true model parameters are constant. If we assume that parameters only change very slowly over time, this may be wasteful, because the estimation bias caused by slightly incorrect parameters may be less than the extra variance that is caused by simultaneously estimating states and parameters.

Let us summarize. We cannot use a model with a fixed parameter vector to estimate the state vector of time varying systems, because the estimate will become biased if the true parameter vector has changed. An estimator based on a flexible parameter model such as in (14)-(15) is also a suboptimal solution in this case since the estimate will have a large variance in the periods where the true parameter vector is constant.

Based on this, it is obvious that a better estimate of the state vector of a time varying true system would be obtained if the fixed parameter model (with the correct parameter values) is used at times where the true parameter remains constant and if the flexible parameter model is used when the true parameter vector is actually changing. In order to decide at a given time instant k, whether we have to use the fixed or the flexible parameter vector model, we apply our model selection procedure of section 2. The details of this model selection procedure are illustrated in an example in the next section.

4. SIMULATION EXAMPLE

In the simulation example we will use model selection to estimate states of a time varying system. Consider the following simple system:

$$x_{k+1} = \begin{bmatrix} \theta_k & 0.7 \\ 0 & 0.9 \end{bmatrix} x_k + \begin{bmatrix} 1 \\ 1.5 \end{bmatrix} u_k + w_k \tag{16}$$

$$y_k = [1 \; 1] x_k + v_k. \tag{17}$$

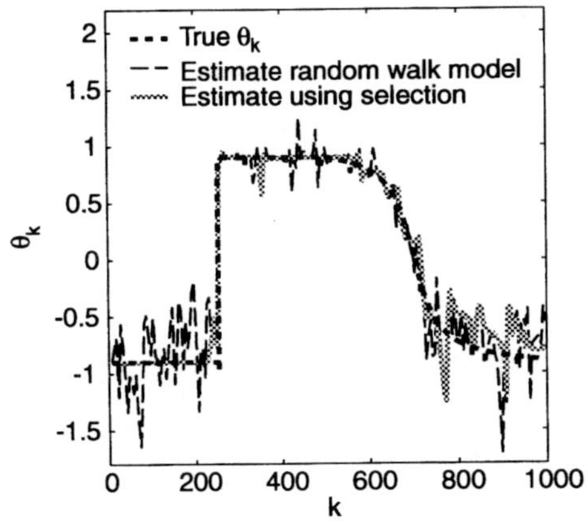

Fig. 1. True and estimated parameter θ_k in the simulation example.

The input signal u_k was chosen as a random binary signal with a switching probability of 0.05, the noises w_k and v_k were chosen as gaussian with:

$$\mathbb{E}\begin{bmatrix} w_k \\ v_k \end{bmatrix} = 0 \quad \mathbb{E}\begin{bmatrix} w_k \\ v_k \end{bmatrix}[w_k^T v_k^T] = \begin{bmatrix} Q & 0 \\ 0 & R \end{bmatrix} \quad (18)$$

$$Q = I \quad R = 10. \quad (19)$$

If the parameter θ_k is constant, these system equations describe a linear system for which the standard Kalman filter is the optimal state estimator. In this example however the parameter θ_k is chosen to be time varying. The parameter θ_k jumps from -0.90 to 0.90 at $k = 255$ and starting from $k = 500$, the parameter gradually returns to its original value, see Figure 1. This way, both a sudden jump (fault) and slowly changing behavior are modelled.

Using this system $N = 1000$ measurements y_k are generated. The true states x_k and parameters θ_k are also stored. Afterwards, using only the recorded measurements and inputs, the state of the system is estimated using the procedure based upon model selection as presented in the previous section. The estimation error is expressed in the mean squared error: $\frac{1}{N}\sum_{k=0}^{N-1} \|x_k - \hat{x}_{k|k}\|^2$.

For the model selection procedure two candidate models of the system are available. The first model is equal to the true model (16)-(17), but uses constant parameters, i.e.:

$$\theta_{k+1} = \theta_k. \quad (20)$$

The initial value for the constant parameter is -0.90, which is equal to the true starting value of θ_k. As proposed in section 3, the second candidate model is a flexible parameter model, i.e. the model (16)-(17) augmented by:

$$\theta_{k+1} = \theta_k + w_k^\theta \quad (21)$$

with w_k^θ chosen as a gaussian white noise with variance 0.1.

Table 1. Mean squares errors of estimated states for a time varying system. States were estimated using a fixed parameter model, a random walk parameter model, and using model selection. Results are averaged over 200 simulation runs.

Fixed parameter	Flexible par. filter	Selection filter
$1.1 \cdot 10^4$	15.1	11.8

In this example, the optimal model for state estimation will be selected using the model selection procedure of section 2. Two UKF filters are designed using the given models. At time instant $k = k_0$, both filters are initialized with the same estimate $\tilde{x}_{k_0|k_0}$ and $\tilde{P}_{k_0|k_0}$. Then both filters are run for $M = 15$ time steps and $\mathcal{V}_{av,i}$ of each filter model is computed. The results of the model with the lowest $\mathcal{V}_{av,i}$ is then chosen as the final selected result. At time $k_0 + M$ all filters are again initialized with the selected result and all filters are again run for M time steps, etcetera.

For illustration purposes, we have also estimated the state vector using only the fixed parameter model and using only the flexible parameter model. The results of all the simulations are given in Table 1. As expected, the fixed parameter model alone produces very poor results, due to the bias caused by using an inaccurate model. The results obtained using the flexible parameter model alone are much better. The extra degree of freedom in this model allows the filter to correct the parameter in the measurement update of the filtering procedure. The results using both models and our new model selection procedure are, as expected, much better than the results using the fixed model alone and better than the results of using the flexible parameter model alone. By looking at Figure 1, we indeed observe that the estimate using our new procedure is made using the fixed parameter model in the time periods where the true parameter remains constant, so that there is no extra variance in the estimated state.

5. CONCLUSIONS

This paper discusses a new algorithm for model selection for state estimation problems. The presented algorithm uses the expected weighted prediction error of estimated states, denoted as $\mathcal{V}(k)$, as a selection criterion. The expected weighted prediction error cannot be estimated using the standard FPE criterion, since parameters are estimated using a weighted and regularized least squares problem. To overcome this problem a new estimator for $\mathcal{V}(k)$ has been derived, that is also valid if states are estimated using weighting and regularization. Using $\mathcal{V}(k)$ as the selection criterion has the advantage that the exact conditional probability of a model structure given the data does not need to be computed. The new algorithm was also demonstrated in a time varying system simulation example.

REFERENCES

Anderson, B. D. O. and J. B. Moore (1979). *Optimal Filtering.* Prentice-Hall. Englewood Cliffs, N.J.

Doucet, A., N. De Freitas and N. Gordon (2001). *Sequential Monte Carlo Methods in Practice.* Springer. New York.

Gustafsson, F. (2000). *Adaptive Filtering and Change Detection.* Wiley & Sons. Chichester, United Kingdom.

Julier, S., J. Uhlmann and H.F. Durrant-Whyte (2000). A new method for the non-linear transformation of means and covariances in filters and estimators. *IEEE Trans. Autom. Control* **45**(3), 477–482.

Kailath, T., A.H. Sayed and B. Hassibi (1999). *Linear Estimation.* Prentice Hall. Upper Saddle River.

Ljung, L. (1999). *System Identification-Theory for the User.* second ed.. Prentice Hall. Upper Saddle River.

Robertson, D. G. and J. H. Lee (1995). A least squares formulation for state estimation. *J. Process Control* **5**(4), 291–299.

Tyler, M. L., K. Asano and M. Morari (2000). Application of moving horizon estimation based fault detection to cold tandem steel mill. *Int. J. Control* **73**(5), 427–438.

Appendix A. PROOF OF PROPOSITION 1

Define $W(x,k)$ as:

$$W(x,k) = V(x,k) + \|x - x_k^{\#}\|^2_{P^{-1}}, \quad \text{(A.1)}$$

then

$$\hat{x}_k = \arg\min_x W(x,k). \quad \text{(A.2)}$$

Define also $\overline{W}(x,k)$ as:

$$\overline{W}(x,k) = \overline{V}(x,k) + \|x - x_k^{\#}\|^2_{P^{-1}}. \quad \text{(A.3)}$$

In order to prove the proposition, let us first expand $\overline{V}(\hat{x}_k,k)$ around $x_k^* = \arg\min_x \overline{V}(x,k)$:

$$\overline{V}(\hat{x}_k,k) = \overline{V}(x_k^*,k) + \frac{1}{2}(\hat{x}_k - x_k^*)^T\overline{V}''(\zeta_k,k)(\hat{x}_k - x_k^*) \quad \text{(A.4)}$$

In the equation above, a $'$ means taking the partial derivative with respect to x. Similarly since $W'(\hat{x}_k,k) = 0$:

$$W(x_k^*,k) = W(\hat{x}_k,k) + \frac{1}{2}(\hat{x}_k - x_k^*)W''(\overline{\zeta}_k,k)(\hat{x}_k - x_k^*), \quad \text{(A.5)}$$

which is easily rewritten into:

$$W(\hat{x}_k,k) = W(x_k^*,k) - \frac{1}{2}(\hat{x}_k - x_k^*)W''(\overline{\zeta}_k,k)(\hat{x}_k - x_k^*). \quad \text{(A.6)}$$

The second derivative of $W(x,k)$ is:

$$W''(x,k) = V''(x,k) + 2P^{-1}. \quad \text{(A.7)}$$

Inserting (A.3) and (A.7) into (A.6) and using the assumption that $x_k^{\#} = x_k^*$, we obtain:

$$V(\hat{x}_k,k) = V(x_k^*,k) - \frac{1}{2}(\hat{x}_k - x_k^*)(V''(\overline{\zeta}_k,k) + 4P^{-1})(\hat{x}_k - x_k^*) \quad \text{(A.8)}$$

Take the expectations of (A.4) and (A.8) and use the following asymptotical relations:

$$\begin{aligned}\mathbb{E}_{\hat{x}_k}&(\hat{x}_k - x_k^*)(\overline{V}''(\zeta_k,k))(\hat{x}_k - x_k^*) \\ &= \mathbb{E}_{\hat{x}_k} \operatorname{tr}\left\{\overline{V}''(\zeta_k,k)(\hat{x}_k - x_k^*)(\hat{x}_k - x_k^*)^T\right\} \\ &\approx \operatorname{tr}\overline{V}''(x_k^*,k)P_x \quad \text{(A.9)}\end{aligned}$$

in which P_x is the asymptotic covariance matrix of $\hat{x}_k$. Also,

$$\mathbb{E}_{\hat{x}_k}(\hat{x}_k - x_k^*)(V''(\overline{\zeta}_k,k))(\hat{x}_k - x_k^*) \approx \operatorname{tr}\overline{V}''(x_k^*,k)P_x \quad \text{and} \quad V(x^*,k) \approx \overline{V}(x^*,k).$$

Using the last three relations together with (A.4) and (A.8) gives:

$$\mathbb{E}_{\hat{x}_k}\overline{V}(\hat{x}_k,k) \approx \overline{V}(x_k^*,k) + \frac{1}{2}\operatorname{tr}\overline{V}''(x_k^*,k)P_x$$

$$\mathbb{E}_{\hat{x}_k}V(\hat{x}_k,k) \approx \overline{V}(x_k^*,k) - \frac{1}{2}\operatorname{tr}(\overline{V}''(x_k^*,k) + 4P^{-1})P_x$$

Combining the two last expressions yields:

$$\mathbb{E}_{\hat{x}_k}\overline{V}(\hat{x}_k,k) \approx \mathbb{E}_{\hat{x}_k}V(\hat{x}_k,k) + \operatorname{tr}(\overline{V}''(x_k^*,k) + 2P^{-1})P_x \quad \text{(A.10)}$$

Using the theory of chapter 9, pages 281-282 in (Ljung, 1999), it can be shown that the covariance matrix P_x equals:

$$\begin{aligned}P_x &= 4\left[\overline{W}''(x_k^*,k)\right]^{-1}\left[\psi^T RLR^T\psi\right]\left[\overline{W}(x_k^*,k)\right]^{-1} \\ &= 4\left[\overline{V}''(x_k^*,k) + 2P^{-1}\right]^{-1}\left[\psi^T RLR^T\psi\right] \\ &\quad \times \left[\overline{V}''(x_k^*,k) + 2P^{-1}\right]^{-1} \quad \text{(A.11)}\end{aligned}$$

with ψ and L as defined in the statement of the theorem. To derive this last expression, we have used the assumption that $y_k - \hat{y}(x_k^*)$ is approximately a white noise and that $\overline{W}''(x_k^*,k)$ exists and is regular.

If we take result (A.10) (after replacing $\mathbb{E}_{\hat{x}_k}V(\hat{x}_k)$ with $V(\hat{x}_K)$, the only observation we have of it) and combine it with (A.11), we obtain the approximation (12). □

ELSEVIER
IFAC
PUBLICATIONS
www.elsevier.com/locate/ifac

GAUSSIAN PROCESS REGRESSION FOR BATCH PROCESS MODELLING

Xiaoling Ou, Julian Morris and Elaine Martin

*Centre for Process Analytics and Control Technology,
School of Chemical Engineering and Advanced Materials,
University of Newcastle, Newcastle upon Tyne, NE1 7RU, UK*

Abstract: This paper proposes the application of Gaussian process regression for the empirical modelling of batch processes to provide long range predictions. Gaussian processes are flexible, non-parametric Bayesian regression techniques. In the training stage, hyper-parameters that define the covariance structure of the Gaussian process can be obtained using Markov Chain Monte Carlo sampling. Model predictions can then be achieved by taking the average of the Monte Carlo samples. The proposed technique is evaluated by application to a benchmark simulation of a fed-batch bioreactor. The results show that comparable results can be achieved with other non-parametric modelling approaches such as recurrent neural networks. *Copyright © 2004 IFAC*

Keywords: Batch process, dynamic modelling, Gaussian process, Monte Carlo methods.

1. INTRODUCTION

It can be argued that the core task in process control and process optimisation is the establishment of an accurate model of the process. One approach has been to build a mechanistic model of the process. However, the development of mechanistic models for complex batch processes is challenging and time-consuming. In addition, the mismatch between the model and the real process is generally not negligible and hence needs to be incorporated within the analysis. Consequently a number of data-based parametric/non-parametric modelling methods have been applied in process manufacturing applications, including the linear latent variable techniques of Principal Component Analysis (PCA), Partial Least Squares (PLS) (Kourti, 2003) along with the non-linear approach of neural networks (e.g. Tian, *et al.*, 2002).

More recently, Bayesian statistical tools have emerged as promising solutions to a number of modelling and model related problems in process engineering. Bayesian inference is a procedure whereby a probability model is fitted to a set of data and the results are summarised in terms of the posterior probability distribution (Gelman, *et al.*, 1995; West and Harrison, 1997). The Bayesian approach provides a natural way to incorporate prior information with observed data from a real process, leading to generally better performance than non-Bayesian approaches in many applications, e.g. Nounou, *et al.*, (2002).

Gaussian processes are non-parametric Bayesian regression models formulated from Gaussian prior distributions defined over the infinite-dimensional space of all possible regression functions. The behaviour of a Gaussian process model can be controlled through the covariance function in terms of the underlying hyper-parameters. A wide variety of covariance functions can be applied in Gaussian process modelling subject to the requirement that the function must result in a non-negative definite covariance matrix (Neal, 1997).

For a Bayesian approach, the hyper-parameters have an associated prior distribution with their posterior distribution being identified using Markov Chain Monte Carlo (MCMC) approaches as opposed to analytical Bayesian inference. This is because of the difficulty of integrating analytically over a posterior distribution that is of complicated form (Brooks 1998). Predictions can then be achieved by averaging over the Monte Carlo samples.

In Rasmussen's (1996) work, it was demonstrated that Gaussian process models have comparable, if not better, predictive performance than other nonparametric regression methods such as neural networks. Their good predictive performance and analytical properties have made them increasingly attractive to users and has resulted in their widespread application in statistics, engineering and other fields (Neal, 1997; Shi, *et al.*; 2003Williams and Rasmussen, 1996).

In this paper a Gaussian process model is developed for a fed-batch bioreactor for long range predictions. The model is implemented using one of the more popular MCMC methods, the Metropolis-Hastings (MH) algorithm (Robert and Casella, 1999). Given the initial state of the fed-batch reactor, multi-step-ahead predictions can be calculated using an iterative approach. The results are sufficiently accurate to provide the basis of model based control algorithms.

2. GAUSSIAN PROCESS MODELLING WITH MCMC IMPLEMENTATION

The problem to be addressed is first described. Consider a noisy data set D comprising N pairs of d-dimensional input vectors and scalar outputs, i.e. $D=\{\mathbf{x}^{(k)}, y^{(k)}\}_{k=1}^{N}$. The primary objective is to establish a Gaussian process model based on this data set and use it to predict, y, given any new input $\mathbf{x}\notin D$. Adopting an iterative approach to the multi-step ahead case, the new input will include the previous predicted values of y.

2.1 *Prediction using Gaussian Process Model*

As for all regression approaches, a Gaussian process model can be defined as:

$$y = f(\mathbf{x}) + \varepsilon \tag{1}$$

where $f(\cdot)$ is a non-linear function mapping the input vector $\mathbf{x}$ to a scalar output y. ε is Gaussian noise with zero mean i.e. $\varepsilon \sim G(0,\sigma_v^2)$. The prior for $f(\cdot)$ is assumed to be a Gaussian process, $\{f(\cdot)\} \sim G(\mu,\mathbf{C})$, where μ is usually set to 0 and $\mathbf{C}$ is the covariance matrix. In practice normalisation is preferred for numerical stability, although it is not strictly necessary. Thus, $\{y^{(k)}\}_{k=1}^{N}$ is Gaussian with the mean and variance given by:

$$\hat{y}(\mathbf{x}) = \mathbf{k}^T(\mathbf{x})\boldsymbol{\Sigma}^{-1}\mathbf{y} \tag{2}$$

$$\hat{\sigma}_{\hat{y}}^2(\mathbf{x}) = \Sigma(\mathbf{x},\mathbf{x}) - \mathbf{k}^T(\mathbf{x})\boldsymbol{\Sigma}^{-1}\mathbf{k}(\mathbf{x}) \tag{3}$$

where $\boldsymbol{\Sigma}$ is the covariance matrix for the training data set:

$$\boldsymbol{\Sigma} = \mathbf{C} + \sigma_v^2\mathbf{I}, \quad \mathbf{I} \text{ is the identity matrix} \tag{4}$$

The entries are defined as $\Sigma_{ij} = \Sigma(\mathbf{x}^{(i)},\mathbf{x}^{(j)})$, $\mathbf{k}(\mathbf{x}) = \left(\Sigma(\mathbf{x},\mathbf{x}^{(1)}),\ldots,\Sigma(\mathbf{x},\mathbf{x}^{(N)})\right)^T$, $\mathbf{y} = (y^{(1)},\ldots,y^{(N)})^T$.

The only constraint on the choice of covariance function $\Sigma(\mathbf{x}^{(i)},\mathbf{x}^{(j)})$ is that it must generate a non-negative definite covariance matrix $\boldsymbol{\Sigma}$ for any set $\{\mathbf{x}^{(k)}\}_{k=1}^{N}$. A widely used form of $\Sigma(\mathbf{x}^{(i)},\mathbf{x}^{(j)})$ is:

$$\Sigma(\mathbf{x}^{(i)},\mathbf{x}^{(j)}) = v_0 \exp\left\{-\frac{1}{2}\sum_{l=1}^{d} w_l\left(x_l^{(i)} - x_l^{(j)}\right)^2\right\} + a_0 + a_1\sum_{l=1}^{d} x_l^{(i)}x_l^{(j)} + \sigma_v^2\delta(i,j) \tag{5}$$

where

$$\delta(i,j) = \begin{cases} 1, & \text{if } i=j \\ 0, & \text{otherwise} \end{cases}.$$

$\boldsymbol{\Theta} = \left(v_0, \{w_l\}_{l=1}^{d}, a_0, a_1, \sigma_v^2\right)$ defines the hyper-parameter vector. The covariance function consists of the following components, an exponential term ($v_0 \exp\left\{-\frac{1}{2}\cdot \sum_{l=1}^{d} w_l\left(x_l^{(i)} - x_l^{(j)}\right)^2\right\}$), a bias term ($a_0$), a linear term ($a_1\sum_{l=1}^{d} x_l^{(i)}x_l^{(j)}$) and a jitter term ($\sigma_v^2\delta(i,j)$). The exponential element expresses the hypothesis that cases with nearby inputs will have highly correlated outputs. The length scale, w_l, according to each input characterises the distance in that particular direction over which y is expected to significantly vary; and v_0 defines the vertical scale of variations of a typical function. The bias term adds a constant component to the regression function. The linear term expresses the linear trend of the function and finally the jitter term accounts for the noise in the data, where σ_v^2 is the variance of the noise. In practice, the covariance function can be defined as any combination of these four terms, according to the different properties of the function $f(\cdot)$.

2.2 *MCMC Implementation*

According to the Bayesian approach, the training process for the Gaussian process model commences with the definition of a prior distribution $P(\boldsymbol{\Theta})$ over the hyper-parameter space that is updated using the training data, D, to produce a posterior distribution $P(\mathbf{\Theta} \mid D)$:

$$p(\mathbf{\Theta} \mid D) \propto p(\mathbf{\Theta}) p(D \mid \mathbf{\Theta}) \tag{6}$$

where $p(D \mid \mathbf{\Theta})$ is the likelihood of the training data for a given $\mathbf{\Theta}$. Here the log-likelihood is used to avoid large rounding errors:

$$L = -\frac{1}{2}\log|\mathbf{\Sigma}| - \frac{1}{2}\mathbf{y}^T \mathbf{\Sigma}^{-1}\mathbf{y} - \frac{N}{2}\log 2\pi \tag{7}$$

This procedure requires the computation of the inversion of the covariance matrix $\mathbf{\Sigma}$ of order $N \times N$, which takes time, $O(N^3)$. This means that the implementation time increases cubically as the size of the training data set increases, and hence problems may result when dealing with large data sets.

To make predictions, integration over the posterior distribution of the hyper-parameters, $\mathbf{\Theta}$, is performed:

$$\bar{y}(\mathbf{x}) = \int \hat{y}_{\mathbf{\Theta}}(\mathbf{x}) P(\mathbf{\Theta} \mid D) d\mathbf{\Theta} \tag{8}$$

$$\bar{\sigma}_{\hat{y}}^2(\mathbf{x}) = \int \hat{\sigma}_{\hat{y}_{\mathbf{\Theta}}}^2(\mathbf{x}) P(\mathbf{\Theta} \mid D) d\mathbf{\Theta} \tag{9}$$

Because of the difficulty of performing this integration analytically, Markov Chain Monte Carlo (MCMC) is used to generate a series of samples for $\mathbf{\Theta}$ according to the posterior distribution, equation (6). The predictions are then calculated by averaging over the MCMC samples of $\mathbf{\Theta}$. Details of MCMC are given in Robert and Casell, (1999).

Priors for $\mathbf{\Theta}$. The priors for $\mathbf{\Theta}$ are defined as given by Rasmussen (1996) whilst the priors for $\{w_l\}$ are given by inverse Gamma distributions:

$$w_l^{-1} \sim Gamma\left(\frac{\alpha}{2}, \frac{\alpha}{2u}\right) \text{ with } u = u_0 d^{2/\alpha} \tag{10}$$

where u is the mean of w_l^{-1}. A small value of α results in a vague prior. $\alpha = 1$ and $u_0 = 1$ were selected. The priors for $\log(v_0)$, $\log(a_0)$ and $\log(\sigma_v^2)$ are all Gaussian distributions with large variances corresponding to fairly vague priors, i.e. $\log(v_0) \sim G(-3, 3^2)$, $\log(a_0) \sim G(-3, 3^2)$ and $\log(\sigma_v^2 v_0) \sim G(-1, 1^2)$.

The MCMC method. The Metropolis-Hastings algorithm is used to address the relationship in equation (6), which uses a candidate distribution $q(\mathbf{\Theta}, \mathbf{\Theta}')$ from which to sample, $p(\mathbf{\Theta} \mid D)$. The algorithm can be described as follows:

1. Initialization Step:
   Set $i = 0$ and generate a random value for $\mathbf{\Theta}^{(0)}$;
2. Iteration:
   For $i = 1 : K$

   Sample a candidate $\mathbf{\Theta} \sim q(\mathbf{\Theta}^{(i-1)}, \cdot)$;

   Evaluate the acceptance probability $\alpha\left(\mathbf{\Theta}^{(i-1)}, \mathbf{\Theta}\right)$:

$$\min\left\{\frac{p(\mathbf{\Theta} \mid D)/q(\mathbf{\Theta}^{(i-1)}, \mathbf{\Theta})}{p(\mathbf{\Theta}^{(i-1)} \mid D)/q(\mathbf{\Theta}, \mathbf{\Theta}^{(i-1)})}, 1\right\} \tag{11}$$

   Sample a uniformly distributed u as $u \sim U[0,1]$, if $u \le \alpha\left(\mathbf{\Theta}^{(i-1)}, \mathbf{\Theta}\right)$, then $\mathbf{\Theta}^{(i)} = \mathbf{\Theta}$, otherwise, $\mathbf{\Theta}^{(i)} = \mathbf{\Theta}^{(i-1)}$.

   End

In this study, $q(\mathbf{\Theta}, \mathbf{\Theta}')$ is set as the transition distribution of a random walk with a small perturbation, $\mathbf{z}$. A normally distributed $\mathbf{z}$ with a zero mean vector implies that $q(\mathbf{\Theta}, \mathbf{\Theta}') = q(\mathbf{\Theta}', \mathbf{\Theta})$. Then the acceptance probability becomes:

$$\begin{aligned} \alpha(\mathbf{\Theta}, \mathbf{\Theta}') &= \min\left\{\frac{p(\mathbf{\Theta}' \mid D)}{p(\mathbf{\Theta} \mid D)}, 1\right\} \\ &= \min\left\{\frac{p(D \mid \mathbf{\Theta}') p(\mathbf{\Theta}')}{p(D \mid \mathbf{\Theta}) p(\mathbf{\Theta})}, 1\right\} \end{aligned} \tag{12}$$

A simple slowly decreasing variance for $\mathbf{z}$ is selected although more elaborate strategies, such as simulated annealing, could be applied (e.g. Robert and Casella, 1999). Assuming the initial variance for $\mathbf{z}$ is $\sigma_z^2(0)$, at step i, the variance $\sigma_z^2(i)$ can be defined as:

$$\sigma_z^2(i) = \begin{cases} \sigma_z^2(0) & if \quad i \le T \\ \sigma_z^2(0)/(i-T)^{0.1} & otherwize \end{cases} \tag{13}$$

where $T < K$ is the start point for σ_z^2 to decrease.

After generating K samples of the hyper-parameters $\mathbf{\Theta}$, the predictions can be made by taking an average over the samples:

$$\bar{y}(\mathbf{x}) = \frac{1}{K}\sum_{i=1}^{K} \hat{y}_{\mathbf{\Theta}}^i(\mathbf{x}) \tag{14}$$

where $\hat{y}_{\mathbf{\Theta}}^i(\mathbf{x})$ is given by equation (2) for a particular sample $\mathbf{\Theta}^{(i)}$. The variance of the prediction can then be calculated in a similar way:

$$\bar{\sigma}_{\hat{y}}^2(\mathbf{x}) = \frac{1}{K}\sum_{i=1}^{K}\left(\hat{\sigma}_{\hat{y}}^{i\,2}(\mathbf{x})\right) \tag{15}$$

where $\hat{\sigma}_{\hat{y}}^{i\,2}(\mathbf{x})$ is given by equation (3).

2.3 Model for Processes with Multi-outputs.

The theory described holds for scalar outputs. For the modelling of processes with multiple outputs, although it is possible to define Gaussian processes with multi-outputs, it is not clear how the covariance function should be defined. Therefore, a separate Gaussian process model is established for each output to simplify the problem. However, since multiple outputs can share some common features, it may be advantageous to build this correlation structure into the analysis. This new approach is currently under investigation.

3. EXPERIMENTS AND RESULTS

3.1 The Fed-batch Bioreactor

The process used in the study is a benchmark fed-batch bio-reactor for the production of secreted protein (Park and Ramirez, 1988). They considered the secretion of foreign protein using Baker's yeast as the host organism. The mechanistic model that describes the dynamics of the process can be expressed by the following differential equations:

$$\dot{P}_M = A(S)(P_T - P_M) - \frac{q}{V} P_M \tag{16}$$

$$\dot{P}_T = B(S)X - \frac{q}{V} P_T \tag{17}$$

$$\dot{X} = C(S)X - \frac{q}{V} X \tag{18}$$

$$\dot{S} = -YC(S)X + \frac{q}{V}(m - S) \tag{19}$$

$$\dot{V} = q \tag{20}$$

where

$$A(S) = \Phi(\mu_x) = \frac{4.75C(S)}{0.12 + C(S)} \tag{21}$$

$$B(S) = f_P(S) = \frac{Se^{-5.0S}}{0.1 + S} \tag{22}$$

$$C(S) = \mu_x(S) = \frac{21.78S}{(S + 0.4)(S + 62.5)} \tag{23}$$

The state variables are the amount of secreted protein on a unit culture volume basis (P_M), the total protein amount on a unit volume basis (P_T), the culture cell density (X), the culture glucose concentration (S), and the culture volume (V). q is the feed flow rate which is used as the control variable, m is the glucose concentration of the feed stream, Y is the yield of glucose per cell mass and Φ, f_P, and μ_x are the protein secretion rate, the protein expression rate and the specific growth rate of the host cell, respectively.

Based on the above model, the performance of Gaussian process modelling of the bioreactor was investigated. The initial conditions of the state variables and parameters in the model were given by:

$P_M(t_0) = 0.0$ $V(t_0) = 1.0$ L
$P_T(t_0) = 0.0$ $m = 20.0$ g/L
$X(t_0) = 1.0$ g/L $Y = 7.3$
$S(t_0) = 5.0$ g/L

Data for 10 batches, each of 15 hours duration were generated. The data comprised the following simulated measurements, P_M, S, V, and q. It is assumed that the measurements of V and q are obtained on-line while the measurements of P_M and S are recorded off-line. This reflects current industrial practice where many bio-product quality variables are measured off-line through laboratory analysis. The data were generated by adding normally distributed random noise with zero mean to the nominal control policies (q) as reported in Tian, *et al.* (2002). Zero mean normal random noise was also added to all the measurements to simulate the effects of measurement noise. The standard deviations of the noise for q, P_M, S and V, were 0.02L/hr, 0.01, 0.02g/L and 0.05L respectively. The sampling time is 6 min, i.e. 150 samples per batch.

3.2 Simulation Study.

There are two states of interest to be predicted, the amount of secreted protein P_M and the culture glucose concentration S, which cannot be measured on-line. As described in Section 2.3, separate Gaussian models were constructed for each state, *GP_S* and *GP_ P_M*. The inputs for each model comprised the feed flow rate *q*, and the predicted value of V ($\hat{V}$) using:

$$\hat{V}(t) = \hat{V}(t-1) + q(t-1)\Delta t, \qquad \Delta t = 6\,\text{min} \tag{24}$$

and the previous predictions from the model (denoted $\hat{S}$ and $\hat{P}_M$), so that the model can make long-range predictions for the whole batch given the initial measurement of *V* and the initial values for S and P_M. As the number of inputs is related to the dimension of the hyper-parameter $\boldsymbol{\Theta}$, it is preferred that fewer input variables are included in the model to reduce computation time. Also it was observed that prediction performance was not compromised. Thus the model inputs are selected as follows. For *GP_S*, the inputs are:

$$q(t-1), q(t-2), \hat{V}(t-1), \hat{S}(t-1).$$

The culture glucose concentration S may have an impact on P_M, consequently two models were developed to predict P_M. $GP\_P_M\_1$ does not include information on S:

$$q(t-1), q(t-2), \hat{V}(t-1), \hat{P}_M(t-1),$$

whilst $GP\_P_M\_2$ includes the previous predictions of S from GP_S:

$$q(t-1), q(t-2), \hat{V}(t-1), \hat{S}(t-1), \hat{P}_M(t-1).$$

For the fed-batch process, the regression model is based on a smooth, continuous function, thus the covariance function is chosen as a simple form without the linear term:

$$\Sigma(\mathbf{x}^{(i)}, \mathbf{x}^{(j)}) = v_0 \exp\left\{-\frac{1}{2}\sum_{l=1}^{d} w_l \left(x_l^{(i)} - x_l^{(j)}\right)^2\right\} + a_0 + \sigma_v^2 \delta(i, j). \quad (25)$$

and thus $\mathbf{\Theta}$ is $\left(v_0, \{w_l\}_{l=1}^{d}, a_0, \sigma_v^2\right)$. During the training stage, 1200 iterations were run for the MCMC sampling of $\mathbf{\Theta}$. The algorithm converged quickly and tended to stabilise after about 200 iterations according to the values of the log-likelihood. These initial 200 values were discarded. One sample of $\mathbf{\Theta}$ was then selected from each consecutive set of 20 iterations resulting in a total of 50 samples being selected to perform the prediction. These samples can be approximately treated as independent and identically distributed, according to the relevant posterior distribution.

3.3 Results and Discussions.

Five batches were used to train the model with a further 5 (batches 6-10) being used for model validation. The initial state values S and P_M were set to the same values as in the simulation, i.e., $S_0 = 5.0$ g/L and $P_{M0} = 0.0$. Using the samples of $\mathbf{\Theta}$ obtained from the model training and the given initial states, the predictions for the whole batch were calculated using equations (2), (3), (14) and (15). The performance of the models was evaluated using the average Root Mean Square Error (RMSE) over the five validation batches, Table 1.

Table 1 Results for the Prediction of S and P_M

Model Type	Input variables	RMSE
GP_S	$q(t-1), q(t-2), \hat{V}(t-1), \hat{S}(t-1)$	0.1218
$GP\_P_M\_1$	$q(t-1), q(t-2), \hat{V}(t-1), \hat{P}_M(t-1)$	0.0435
$GP\_P_M\_2$	$q(t-1), q(t-2), \hat{V}(t-1), \hat{S}(t-1), \hat{P}_M(t-1)$	0.0384

The results show that the models exhibit quite good performance with the RMSEs being comparable with those obtained by the augmented recurrent neural network models proposed in Tian, *et al.* (2002). It is also observed that the RMSE of model $GP\_P_M\_2$ is smaller than for model $GP\_P_M\_1$, which implies that S has a direct effect on P_M and more accurate predictions of P_M can be obtained by including S as an additional model input.

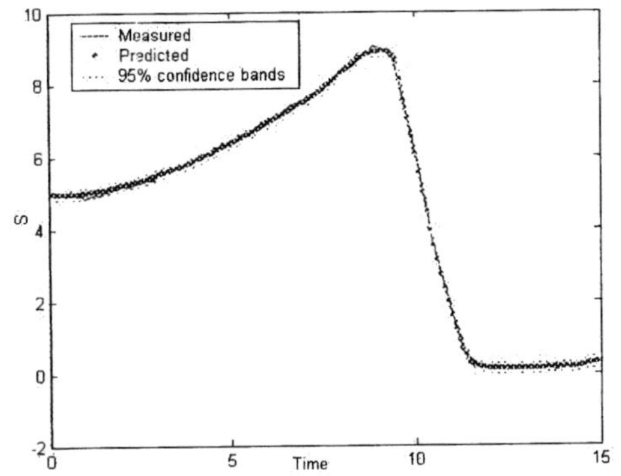

Fig. 1. Multi-step-ahead predictions for S on validation batch 9 (GP_S)

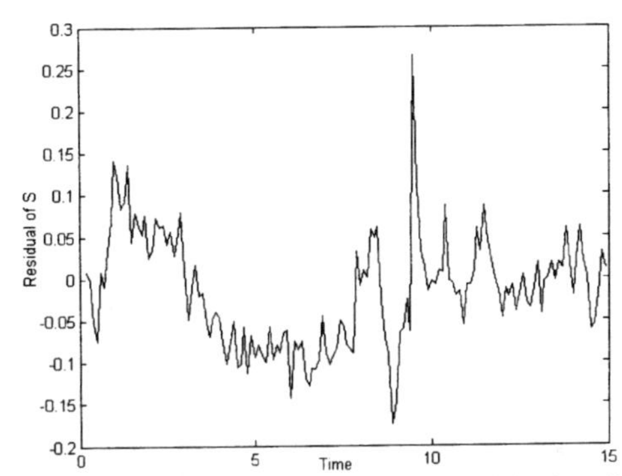

Fig. 2. Multi-step-ahead prediction residuals of S on validation batch 9 (GP_S)

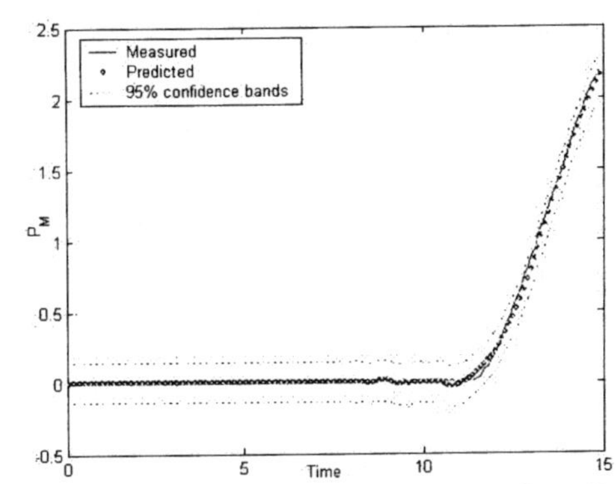

Fig. 3. Multi-step-ahead predictions for P_M on validation batch 9 ($GP\_P_M\_1$)

To further assess the performance of the model, the predictions with 95% confidence bands of each model (without considering the uncertainty of $\mathbf{x}$) are compared with the actual output values. Long range predictions of glucose concentration (S) for one of the validation data set (batch 9) and the corresponding residuals are plotted in Fig. 1 and Fig. 2. Fig. 3 - Fig. 5 show the long range predictions and the corresponding residuals of secreted protein (P_M) for the cases where the models do not have any information on S and where information on S is included.

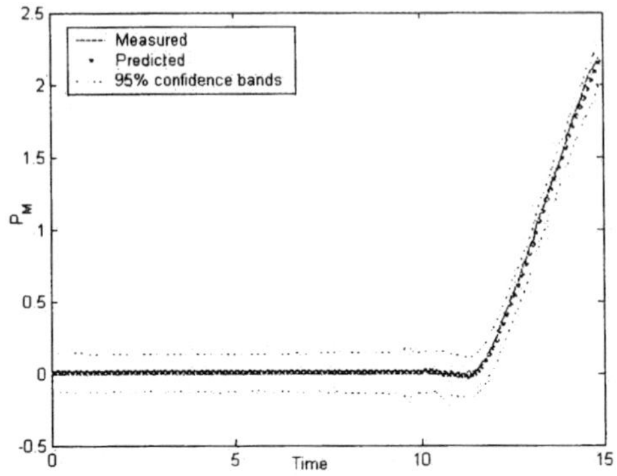

Fig. 4. Multi-step-ahead predictions for P_M on validation batch 9 (*GP_P_M_2*)

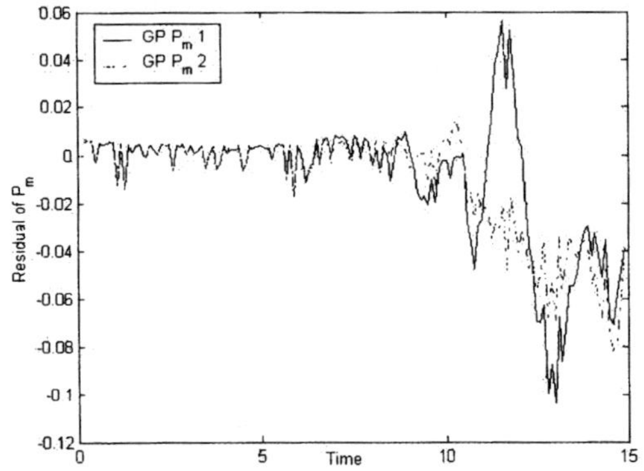

Fig. 5. Multi-step-ahead prediction residuals of P_M on validation batch 9 (*GP_P_M_1* and *GP_P_M_2*)

It can be seen that the predictions are satisfactory. It is also apparent that the predictions of *GP_P_M_2* are more accurate than those of *GP_P_M_1*, especially around 10hr30min to 12hr where the secreted protein is beginning to increase sharply.

4. CONLUSIONS AND DISCUSSION

This paper has proposed the use of Gaussian process regression for batch process modelling. Gaussian processes are non-parametric Bayesian regression models which are sufficiently flexible to represent a wide variety of data by defining an appropriate covariance matrix. Hyper-parameters, which determine the covariance functions can be estimated by Markov chain Monte Carlo simulation. The model can be implemented using matrix computations that are feasible for datasets comprising over a thousand samples. The effectiveness of the Gaussian process approach has been demonstrated by its application to the modelling of a fed-batch bioreactor.

Currently the research is aimed at extending Gaussian process approaches to mixtures of Gaussians (Shi, *et al.*, 2003) to account for heterogeneity between batches. A useful aspect of Gaussian process mixtures is that they can dramatically reduce the computational cost by splitting the training data. Conventional mixtures are defined according to different batches. The extension to Gaussian process mixtures for data monitored across batches is also on-going.

5. ACKNOWLEDGEMENTS

X. Ou acknowledges the financial support from CPACT and Newcastle University, for her PhD. The authors would also like to thank Dr. Jianqing Shi for discussions relating to Gaussian processes.

REFERENCES

Brooks, S. (1998). Markov Chain Monte Carlo Method and Its Application, *The Statistician*, **47**, pp. 69-100.

Gelman, A., J.B. Carlin, H.S. Stern and D.B. Rubin (2000). *Bayesian Data Analysis*, Chapman & Hall/CRC.

Neal, R.M. (1997). *Monte Carlo Implementation of Gaussian Process Models for Bayesian Regression and Classification*, Technical Report No. 9702, Dept. of Statistics, University of Toronto, available from http://www.cs.toronto.edu/~radford/publications.html.

Nounou M.N., B.R. Bakshi, P.K. Goel and X.T. Shen (2002). Bayesian Principal Component Analysis, *Journal of Chemometrics*, **16**, pp. 576-595.

Kourti T. (2003). Multivariate Dynamic Data Modelling for Analysis and Statistical Process Control of Batch Processes, Start-ups and Grade Transitions, *Journal of Chemometrics*, **17**, pp. 93-109

Park, S. and W.F. Ramirez (1988). Optimal Production of Secreted Protein Fed-Batch Reactors, *A.I.Ch.E Journal*, **39**, pp. 1550-1558.

Robert, C. and G. Casella (1999). *Monte Carlo Statistical Methods*, Springer, New York

Rasmussen, C.E. (1996). Evaluation of Gaussian Process and Other Methods for Non-linear Regression, PhD Thesis, University of Toronto, available from http://www.cs.toronto.edu/0~carl/gp.html.

Shi, J.Q., R. Murray-Smith and D.M. Titterington (2003), Bayesian Regression and Classification Using Mixtures of Gaussian Processes, *International Journal of Adaptive Control and Signal Processing*, **17**, pp.1-16.

Tian, Y., J. Zhang and A.J. Morris (2002). Optimal Control of A Fed-Batch Bioreactor Based upon An Augmented Recurrent Neural Network Model, *Neurocomputing*, **48**, pp. 919-936.

West, M. and J.F. Harrison (1997). *Bayesian Forecasting and Dynamic Models*, Springer Series in Statistics, 2nd edition.

Williams, C.K.I. and C.E. Rasmussen (1996). Gaussian processes for regression, *Neural Information Processing Systems*, **8**, pp. 514-520.

ELSEVIER
IFAC
PUBLICATIONS
www.elsevier.com/locate/ifac

DESIGN OF AN ON-LINE TITRATOR FOR NONLINEAR pH CONTROL

Alex D. Kalafatis * Liuping Wang ** William R. Cluett ***

* *AspenTech, Toronto, Canada*
** *School of Electrical & Computer Engineering, RMIT University, Melbourne, Australia*
*** *Department of Chemical Engineering & Applied Chemistry, University of Toronto, Toronto, Canada*

Abstract: A new on-line titrator for pH control has been developed. The on-line identification method generates an estimate of the inverse titration curve as a continuous smooth function eliminating the need for selection of the important break points of the titration curve. The time required to obtain the titration curve in the spanned pH range is approximately twice the settling time of the titrator and thus no significant identification lag is introduced. The results of the titrator are used directly to linearize the pH process output enabling the application of a simple and yet effective linearizing feedforward-feedback control strategy. *Copyright © 2004 IFAC*

Keywords: pH control; nonlinear control; recursive estimation; feedforward control

1. INTRODUCTION

The time-varying characteristics of pH processes make them difficult to control with fixed-gain controllers and motivate the use of adaptive control. The standard adaptive control approach to the pH problem is shown schematically in Figure 1. Linear adaptive pH control can provide satisfactory results for well-buffered processes, as long as the static gain (titration curve) does not change rapidly. Nonlinear adaptive control provides a better alternative, since some knowledge of the origin of the underlying nonlinearity is incorporated into the control strategy (Gustafsson and Waller, 1992; Henson and Seborg, 1994). However, many of these earlier approaches assume that the chemical composition of the process feed stream (i.e. all the species present and the dissociation constants of the weak species) is known. In reality, this is often not the case and, as a result, the application of such approaches is limited. When

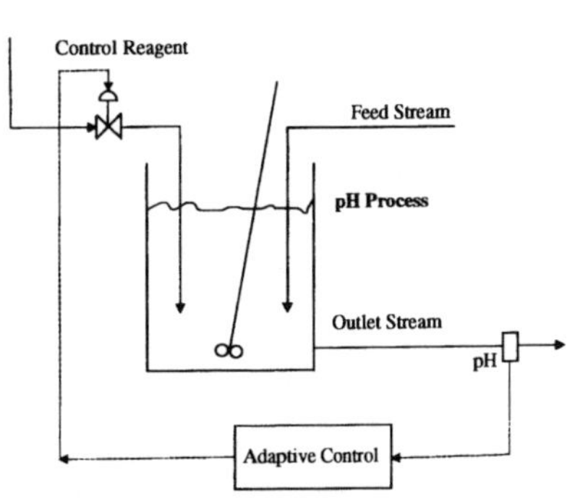

Fig. 1. Adaptive pH control approach.

the location of the buffer region of the titration curve does not change with time, the composition of the process feed stream can be approximated by

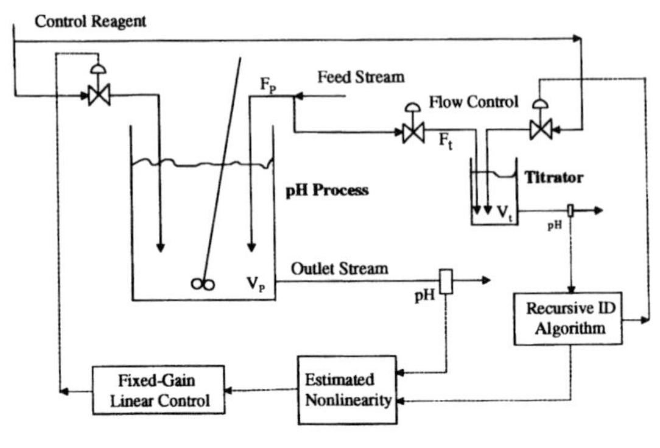

Fig. 2. On-line-titrator based pH control approach.

hypothetical species with dissociation constants that closely represent the nonlinear characteristics of the process. These dissociation constants could be measured off-line or estimated on-line to extend the applicability of these adaptive strategies. However, in the general case, when changes in the titration curve result from changes in the concentrations and composition (dissociation constants) of the feed stream, the increased number of parameters that need to be estimated may lead to deterioration of the controller performance.

In more recent years, attempts have been made to reduce the number of parameters required to represent the nonlinear behaviour of the pH process. The adaptive control strategy used by Sung *et al.* (1998) estimates on-line the total ion concentration and the dissociation constant of a fictitious weak acid. This idea, which has its roots in earlier work by Lee *et al.* (1994), works well as long as no new chemical species enter the feed stream that would drastically change the shape of the titration curve. In Wright *et al.* (1998), a system representation consisting of a combination of basis functions is proposed that relies on on-line identification of a small number of adjustable parameters. With this latter approach, the user must make a suitable choice of basis functions to ensure a reasonable fit to the titration curve as it changes over time.

In parallel with this work in adaptive control, there is continued interest in the development of on-line titration devices for identifying the process titration curve. This alternative approach to the pH control problem is schematically shown in Figure 2. Here, the device consists of a small neutralization tank that accepts a small side-stream of the process feed and performs separate identification of the titration curve using a small portion of the control reagent. The main advantage of these titrators is that they can be used to characterize titration curves with complex shapes over a wide pH region without relying on data generated by the pH process itself. This notion of physically separating the identification and control functions by introducing a titrator can be traced back to Mellichamp *et al.* (1966), Gupta and Coughanowr (1978) and, more recently, Sung *et al.* (1995).

Most commercial titrators mimic the traditional off-line approach of obtaining the titration curve. As a result, a substantial amount of time is required to obtain the titration curve, since for each steady-state point, a time equal to the settling time of the titrator vessel is needed. This introduces a lag in the identification, unless the dynamics of the titrator are very fast compared to the actual process and/or only a few points on the titration curve are identified. Limiting the number of identified points may deteriorate the quality of the nonlinear fit, especially for steep titration curves. Furthermore, selecting the location of these points, i.e. the break points of the titration curve, may be very crucial for the quality of the identification. Also, since the information obtained consists of only a set of discrete points, some kind of interpolation must still take place in order to represent the entire titration curve.

A new on-line titrator is proposed here for the control of pH processes that builds on earlier work described by the authors in Kalafatis *et al.* (2003*a*) and Kalafatis *et al.* (2003*b*). Both the identification in the titrator and the control in the main pH process are based on the idea of using a Wiener model structure to represent the pH process. Section 2 of this paper describes the design of the on-line titrator and Section 3 presents simulated pH control results that test the performance of the on-line titrator and associated control strategies under both increasing and decreasing buffering conditions.

2. DESIGN OF THE ON-LINE TITRATOR

As shown in Figure 2, the titrator consists of a small, well-mixed neutralization tank that has as its feed a small portion of the process feed stream and an adjustable flow rate of control reagent, and a pH sensor in its outlet stream. As a general rule, the portion of the feed that enters the titrator should be small enough to avoid the need for a large recycle stream of the off-specification titrator effluent but large enough to provide fast titrator dynamics.

A simple method has been developed in order to estimate in an on-line fashion the time-varying changes in the titration curve associated with the feed stream (Kalafatis *et al.*, 2003*b*). The method is based on the use of continuous sinusoidal adjustments to the flow rate of control reagent to the titrator and a recursive least squares algorithm. The method also incorporates automatic adjustment of the mean level of the control reagent flow rate in order to ensure that the titration curve is identified in the pH operating region of interest. The estimated inverse of the titration curve (or what is referred to as the estimated nonlinearity in Figure 2) is updated on a user-specified estimation cycle for use by the control strategy. Linearization by output transformation of the pH signal from the process outlet stream is then used to make the relationship between the control reagent process flow rate and the process pH reading appear linear, enabling the use of a linear, fixed-gain feedback controller. In addition to linearizing feedback control, a simple linearizing feedforward control scheme, based on the same estimate of the inverse titration curve in the vicinity of the target pH and a measurement of the process feed flow rate, is included to further enhance the pH control performance (Kalafatis *et al.*, 2003*a*).

The flow rate of the titrator feed (F_t) is controlled by a valve or a variable-speed pump and is maintained at a constant value. As a result, the titrator does not respond to variations in the process feed flow rate (F_p), which do not on their own alter the shape of the titration curve.

If concentration and/or composition changes occur in the process feed stream, the output pH of the titrator will move away from the desired operating point and the estimation method will provide the inverse of the titration curve over the spanned pH region. Automatic adjustments of the mean level of the titrator input signal eventually move this region around the pH setpoint of the process (usually after one or two estimation cycles).

In order to obtain fast titrator dynamics, and thus achieve fast identification cycles, the volume of the titrator vessel (V_t) should be small. The smaller the volume of the titrator, the faster the dynamic response, and hence, a faster data sampling rate must be used in order to acquire enough input-output data.

3. SIMULATED CLOSED-LOOP RESULTS USING THE ON-LINE TITRATOR

The model used for simulating the pH process is outlined in Kalafatis *et al.* (2003*a*) and consists of a continuous stirred tank reactor (CSTR) where a strong base ($NaOH$) reacts with two separate feed streams, a strong acid (HCl) and a buffering solution ($NaHCO_3$), while an overflow exit line maintains a constant liquid volume in the reactor. An identical model is used for the titrator, except with different parameter values. The pH process being controlled in Figure 2 has a mixing tank

Table 1. Nominal operating conditions of the simulated pH process and titrator.

Variable	Process	Titrator
Base stream flow rate (l/min)	11.64	1.164
Acid stream flow rate (l/min)	13.62	1.362
Buffer stream flow rate (l/min)	3.60	0.360
Base concentration (M)	0.00286	0.00286
Acid concentration (M)	0.00250	0.00250
Buffer concentration (M)	0.004	0.004
pH inside the tank	7.00	7.00
Volume of the tank (l)	158	2.1
Mixing tank time constant (min)	5.5	0.75
pH measurement time delay (sec)	30	10
pH probe time constant (sec)	25	10
Flow loop time constants (sec)	15	6
Sampling interval (sec)	15	1

volume of $V_p = 158\ l$ and a nominal feed flow rate (acid plus buffer) of $F_p = 17.22\ l$/min. The titrator volume is $V_t = 2.1\ l$ and the flow rate of the titrator feed stream is $F_t = 1.722\ l$/min. The nominal conditions for both the process and the titrator are summarized in Table 1 along with all other key model parameters. As indicated by the mixing tank time constants, the dynamics of the titrator are 7.5 times faster than those of the pH process considered. For all the simulations that follow, the estimation cycle in the titrator for updating the inverse nonlinearity is 10 min (approximately twice its settling time). The sampling interval used in the identification algorithm is 1 sec and the value of the fixed forgetting factor used in the recursive least squares algorithm is $\lambda = 0.97$. The sampling interval for the pH control algorithm is 15 sec.

3.1 *Decreased buffering conditions*

With both the process and the titrator initially at steady-state under the nominal conditions outlined in Table 1, at $t = 0$ min, a -83% change in the buffer stream flow rate is introduced. Although the change in the feed flow rate is small (-17.4%) the resulting new values of the reaction invariants (c_1, c_2) are significantly different as shown in Table 2. While the process remains under the decreased buffering conditions, additional changes of -25% and $+50\%$ are introduced into the feed flow rate at $t = 40$ min and at $t = 60$ min, respectively, while maintaining the ratio of acid to buffer stream flow rate constant (see Table 2). As expected, these latter flow rate changes do not affect the reaction invariant values and do not alter the titration curve. However, they still represent severe disturbances that force the output pH away from the setpoint, and thus form a good benchmark for evaluating the controller performance.

The titrator feed stream flow rate remains constant at its nominal value ($F_t = 1.722\ l$/min)

Table 2. Simulated process conditions under decreased buffering.

t (min)	acid flow (l/min)	buffer flow (l/min)	c_1 (mol/ml)	c_2 (mol/ml)
0-	13.620	3.60	-1.141	0.836
0	13.620	0.60	-2.226	0.169
40	10.215	0.45	-2.226	0.169
60	20.430	0.90	-2.226	0.169

and the composition/concentration of the titrator feed stream is always the same as the process feed stream. The base flow rate (control reagent) into the titrator is a single sinusoid with a frequency of $\bar{\omega} = 1.2566$ rad/min (corresponding to a period of 5 min, i.e. slightly greater than the titrator settling time) and an amplitude of 0.36 l/min around its mean value (which is adjusted at the end of each estimation cycle). As shown in Figure 3, at $t = 0$ min the titrator reacts to the severe

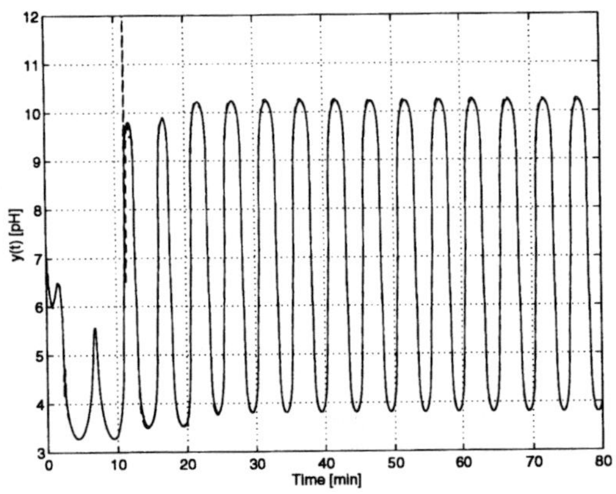

Fig. 3. Comparison of the simulated (solid line) and predicted (dashed line) titrator outputs. At $t = 0$, the process buffer stream flow rate is decreased by 83%.

buffer decrease and its spanned pH range is shifted into the region of pH 3.2 – 5.6, below the desired operating point of pH= 7. At the end of the first estimation cycle ($t = 10$ min), the inverse titration curve is identified across this pH region, and the mean level of the input signal is adjusted. As a result of this adjustment, the spanned pH range is shifted around the desired operating point during the second estimation cycle. The inverse titration curve is again obtained at the end of this cycle ($t = 20$ min) and is shown in Figure 4. Even better identification results are obtained in the third cycle where the mean input signal level converges to its final value. Since the titrator operates on a constant feed flow rate, it does not react to the process flow rate changes introduced at $t = 40$ min and $t = 60$ min.

Using the pH process parameters listed in Table 1 and a desired closed-loop time constant $\tau_c = 1.4$ min, the parameters of the linearizing feedback (PI) controller are calculated to be $K_C = 83.4$ l/min and $\tau_I = 5.5$ min. Figure 5 demonstrates the controller performance in response to

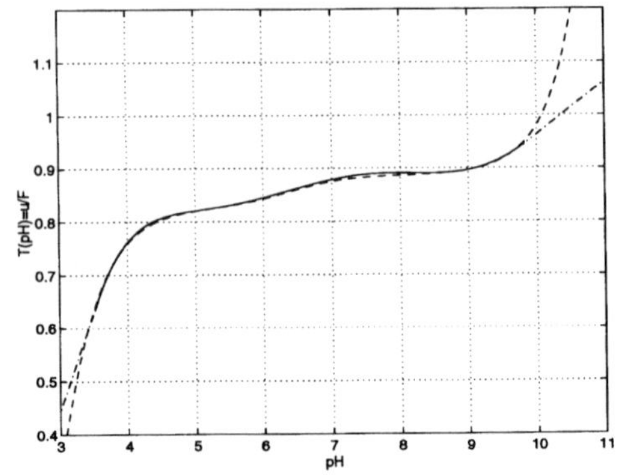

Fig. 4. Comparison of the identified (solid line) and theoretical (dashed line) inverse titration curves for an 83% decrease in the process buffer stream flow rate. The identified inverse nonlinearity is extrapolated with straight lines (dash-dotted) outside the excited input-output range.

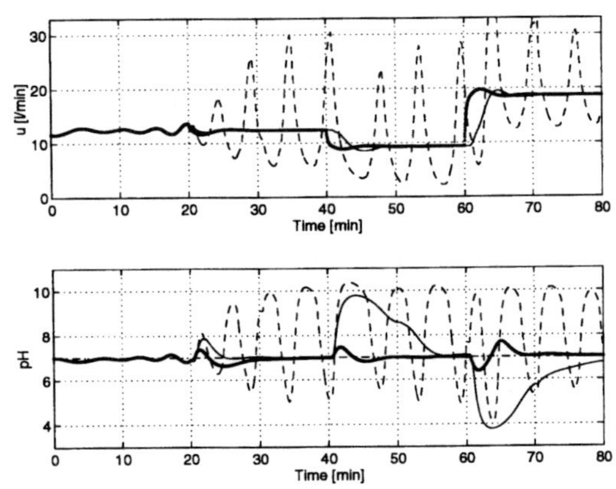

Fig. 5. Controller performance (top: manipulated control reagent flow rate; bottom: pH in process outlet stream) in the case of decreased buffering and in response to the disturbances listed in Table 2. (Dashed line: linearizing feedback control based on the initial titration curve; solid line: linearizing feedback control based on the identified inverse titration curve starting at $t = 20$ min; bold line: linearizing feedforward-feedback control based on the identified inverse titration curve starting at $t = 20$ min; dash-dotted line in the lower graph: pH setpoint.)

the decrease in buffering and the flow rate disturbances listed in Table 2, if the titrator identification results are not utilized and linearization by output transformation continues to be based on the initial titration curve. The increased process gain due to low buffering conditions makes the closed-loop system unstable, and the pH output experiences large oscillations at $t > 25$ min. The addition of linearizing feedforward control would not help the situation and would even worsen the closed-loop response, since the control action would be based on an incorrect titration curve.

By utilizing the new inverse titration curve identified at $t = 20$ min (shown in Figure 4) and basing the linearizing feedback controller on this new curve starting at $t = 20$ min and updating it every 10 min, the closed-loop system is stabilized. However, large deviations from the setpoint occur after the introduction of the disturbances in the feed stream flow rate at $t = 40$ min and $t = 60$ min. Due to the steep titration curve, resulting from the low buffering conditions, the feedback linearizing controller is not able to respond quickly enough to these disturbances. Even if the linearizing feedback controller is based on the exact theoretical titration curve, in which case the ideal feedback control performance would be obtained, the closed-loop results are very similar.

The addition of linearizing feedforward control starting at $t = 20$ min, based on the newly identified inverse titration curve, dramatically improves the control performance, as shown in Figure 5. The advantage of this combined feedforward-feedback controller scheme is that the feedforward control action does not rely on the output pH measurement. As a result, its response is very fast, while the feedback control action compensates for inaccuracies in estimating the inverse titration curve and in measuring the feed flow rate.

3.2 *Increased buffering conditions*

The previous set of simulations showed the effectiveness of the on-line titrator in maintaining the stability and performance of the control system under decreased buffering conditions. In the case where the amount of buffering increases (if the concentration of the weak species increases or new weak species enter the process feed) the stability of the system is not affected. However, the closed-loop response may be very sluggish and the controller performance unacceptable if the linearization by output transformation is based on the low buffering conditions. This occurs in practice if the tuning of the pH controller is based on the worst-case scenario, i.e. the lowest expected buffering conditions.

In the following simulations, the process is assumed to be initially under the decreased buffering conditions encountered in the previous simulations. At $t = 0$ min, the buffer stream flow rate is increased to its original nominal value. Then, at $t = 20$ min and at $t = 50$ min, a decrease of 50% and an increase of 100% in the feed flow rate are introduced, respectively, while maintaining the ratio of the acid to the buffer stream flow rate constant (see Table 3). The input signal to

Table 3. Simulated process conditions under increased buffering.

t (min)	acid flow (l/min)	buffer flow (l/min)	c_1 (mol/ml)	c_2 (mol/ml)
0-	13.62	0.60	-2.226	0.169
0	13.62	3.60	-1.141	0.836
20	6.81	1.80	-1.141	0.836
50	27.24	7.20	-1.141	0.836

the titrator is, as before, a single sinusoid with an amplitude of 0.36 l/min around its mean value (which is adjusted at the end of each estimation cycle). As shown in Figure 6, at $t = 0$ min the titrator responds to the buffer increase and its spanned pH range is now shifted in the region of pH 7.4–10.3, which is above the desired operating point of pH= 7. At the end of the first estimation cycle ($t = 10$ min), the inverse titration curve is identified across this pH region, and the mean level of the input signal is adjusted. At the end of the second estimation cycle ($t = 20$ min), the titration curve, shown in Figure 7, is obtained across the new spanned range that includes the

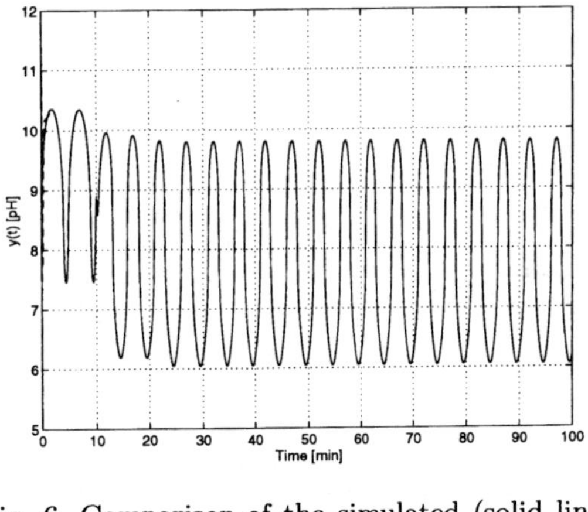

Fig. 6. Comparison of the simulated (solid line) and predicted (dashed line) titrator outputs. At $t = 0$, the process buffer stream flow rate is increased to its original nominal value.

desired operating point. The mean input level converges to its final value in the third identification cycle.

As shown in Figure 8, without utilizing the on-line titrator identification results, the closed-loop response to the flow rate disturbances is very sluggish. Linearization by output transformation based on the initial titration curve, which corresponds to low buffering conditions, produces a small loop gain when the amount of buffering increases. Thus control system performance is severely affected. Specifically in rejecting the second flow rate disturbance at $t = 50$ min, it takes more than one hour for the process output to return to its setpoint ($pH_{sp} = 7$). By basing the linearization on the identified inverse nonlinearity shown in Figure 7, starting at t = 20 min, the disturbance rejection is significantly improved as seen Figure 8. The addition of linearizing feedforward control to the feedback loop, substantially reduces the output deviations from setpoint even further (see Figure 8).

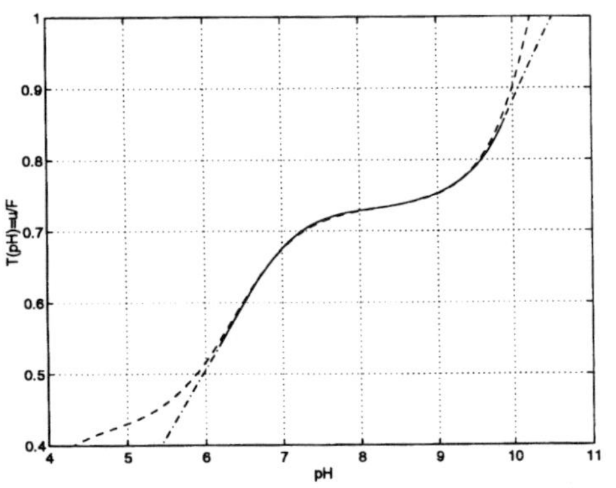

Fig. 7. Comparison of the identified (solid line) and theoretical (dashed line) inverse titration curves for an increase in the process buffer stream flow rate to its original nominal value. The identified inverse nonlinearity is extrapolated with straight lines (dash-dotted) outside the excited input-output range.

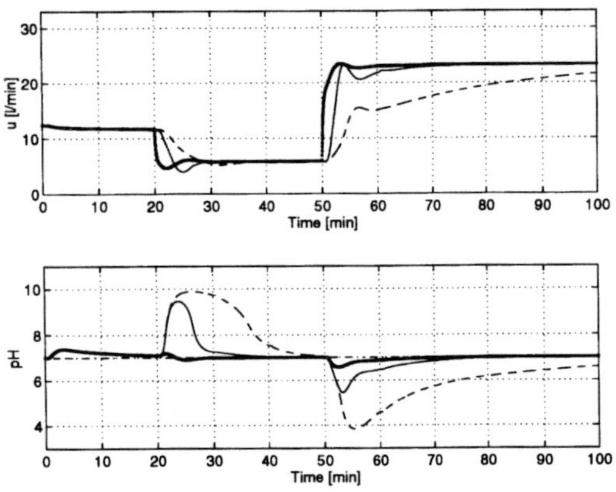

Fig. 8. Controller performance (top: manipulated control reagent flow rate; bottom: pH in process outlet stream) in the case of increased buffering and in response to the disturbances listed in Table 3. (Dashed line: linearizing feedback control based on the initial titration curve; solid line: linearizing feedback control based on the identified inverse titration curve starting at t = 20 min; bold line: linearizing feedforward-feedback control based on the identified inverse titration curve starting at t = 20 min; dash-dotted line in the lower graph: pH setpoint.)

4. CONCLUSIONS

An on-line titration device for pH control has been designed and tested using simulations. The titrator is placed in parallel with the pH process and accepts a portion of the process feed stream. Since the identification of the inverse titration curve takes place in the titrator, the actual pH process is not disrupted. The on-line identified titration curve is used directly to linearize the pH process output and thus allows for the use of a simple fixed-gain, linear controller (e.g. a PI-type). The effectiveness of a titrator-based linearizing feedforward-feedback control scheme has been evaluated under a variety of conditions and has demonstrated excellent performance.

5. REFERENCES

Gupta, S. R. and D. R. Coughanowr (1978). On-line gain identification of flow processes with application to adaptive pH control. *AIChE Journal* **24**(4), 654–664.

Gustafsson, T. K. and K. V. Waller (1992). Non-linear and adaptive control of pH. *Ind. Eng. Chem. Res.* **31**(12), 2681–2693.

Henson, M. A. and D. E. Seborg (1994). Adaptive nonlinear control of a pH neutralization process. *IEEE Trans. Control Systems Technology* **2**(3), 169–182.

Kalafatis, A. D., L. Wang and W. R. Cluett (2003*a*). Linearizing feedforward-feedback control of pH processes based on the Wiener model. *Journal of Process Control (accepted for publication).*

Kalafatis, A. D., L. Wang and W. R. Cluett (2003*b*). Recursive identification of pH processes using sinusoidal signals. *Automatica (under review).*

Lee, S. D., J. Lee and S. Park (1994). Nonlinear self-tuning regulator for pH systems. *Automatica* **30**(10), 1579–1586.

Mellichamp, D. A., D. R. Coughanowr and L. B. Koppel (1966). Identification and adaptation in control loops with time-varying gain. *AIChE Journal* **12**(1), 75–83.

Sung, S. W., I.-B. Lee and D. R. Yang (1995). pH control using an identification reactor. *Ind. Eng. Chem. Res.* **34**(7), 2418–2426.

Sung, S. W., I.-B. Lee, J. Y. Choi and J. Lee (1998). Adaptive control for pH systems. *Chem. Eng. Sci.* **53**(10), 2446–2461.

Wright, R. A., B. E. Smith and C. Kravaris (1998). On-line identification and nonlinear control of pH processes. *Ind. Eng. Chem. Res.* **37**(6), 2446–2461.

ELSEVIER
IFAC
PUBLICATIONS
www.elsevier.com/locate/ifac

ESTIMATING THE PREDICTION UNCERTAINTY OF DYNAMIC NEURAL NETWORK PROCESS MODELS

K. Dadhe [*,1], **R. Gesthuisen** [*], **S. Engell** [*]

[*] *Process Control Laboratory,*
Department of Biochemical and Chemical Engineering,
University of Dortmund, D-44221 Dortmund, Germany,
Tel: +49/231/755-5127, Fax: +49/231/755-5129

Abstract: In this paper, the explicit calculation of prediction intervals for multi-step-ahead predictions from dynamic neural network models is described. Usually, asymptotic methods that are based on linearized descriptions of neural networks are applied with the potential problem of large coverage errors and too optimistic prediction intervals. The potential sources of these problems are the negligence of the network parameter uncertainties and the non-normality of error distributions. To overcome these restrictions, the bootstrap method is used here. The bootstrap is a tool from computational statistics. New formulations are introduced to apply the bootstrap to nonlinear time series models with exogenous inputs. The proposed method is illustrated by an analysis of a neural network model of a bioreactor benchmark problem. *Copyright © 2004 IFAC*

Keywords: Prediction interval, neural networks, bootstrap, bioreactor.

1. INTRODUCTION

Chemical processes exhibit frequently strong nonlinearities and especially when operated in batch or semibatch mode or over a wide range cannot be controlled well by a linear controller which is designed around a nominal operation point. Hence, the application of nonlinear model predictive control (NMPC) schemes for chemical processes seems advantageous. A major drawback is the size and the complexity of rigoros dynamic models as the computational burden even with state of the art computers is prohibitive.

Nonlinear black box models may overcome this restriction and neural networks have gained considerably attention in both academic research as well as industrial applications. Their universal approximation abilities and the access to a wide range of software tools qualify them for the building of nonlinear dynamic black-box models which can be applied as prediction models in an NMPC scheme (see e.g. (Wang *et al.*, 2003)). In the application of this idea it turns out that a key issue is to estimate the prediction accuracy of the neural network models over medium to large prediction horizons.

However, a comparatively small number of articles in the literature deals with the assessment of neural network predictions. In the case of stationary mappings, several different approaches were reported. In (Chryssolouris *et al.*, 1996) a method based on linearization of the neural network is presented and in (Tibshirani, 1995) it is compared with two bootstrap methods.

The specific problem of the reliability or uncertainty of the neural network prediction within

[1] Corresponding author
E-mail: k.dadhe@bci.uni-dortmund.de

dynamic models has hardly been addressed systematically. On the other hand, neural network based predictive controller may show undesired behaviors when the process leaves the regime in which the prediction model had been trained. One possible solution is to divide the input space up into subspaces and to calculate the density of the training data in these subspaces. If the density is too low the neural network based NMPC switches to a conventional PID controller. This basic idea as well as technical considerations can be found in (Roßmann, 2002). A similar approach was described in (Tsai *et al.*, 2002) where a regional knowledge index is calculated. The index reflects the mean distance to all training patterns, and a supervising controller interpolates between the NMPC and a neural adaptive controller. Both approaches try to evaluate whether the current neural network input lies in a region with a high training data density or not. However, even if enough training data was used, the neural network predictions may be inaccurate, especially for growing prediction horizons.

In this paper, prediction intervals of multi-step-ahead forecasts of neural networks are calculated by bootstrap methods. A prediction interval is an upper and lower limit of the predicted value together with a probability α that future realizations of the process lie in between these bounds. For this purpose the residuals from the neural network training are taken as estimates of the true error distribution. The nominal neural network is then simulated many times with random draws from the error distribution. The newly generated data sets are used to estimate neural networks for each simulation run. The estimated error distribution represents the uncertainty of the network predictions with respect to noise or unmodeled process behavior, and the set of neural networks represents the uncertainty of the network parameters considering them as stochastic quantities as they are determined from noisy and limited data. The m-step-ahead prediction and its distribution are calculated from all neural network predictions and by adding an error term from the empirical error distribution during each prediction step.

The paper is structured as follows. Section 2 describes the basic methods used. In Section 3, extensions are made for dynamic models, Section 4 presents a linear method which is used in Section 5 where the results from a bioreactor simulation example are shown. The concluding Section 6 summarizes the presented results and gives an outlook to future work.

2. THE BOOTSTRAP

The bootstrap is a computer-based statistical method to estimate unknown distributions or parameters of distributions and can be separated into nonparametric or parametric techniques. The standard textbook which covers a wide range of applications is (Efron and Tibshirani, 1993). Basically, the bootstrap generates replicate pseudo-data sets by resampling from empirical distribution functions. The union of all bootstrap resamples is most likely to approximate the underlying true distribution. [2]

The standard nonparametric bootstrap method cannot be applied to dynamic models as it is based on the *iid*-assumption (independently and identically distributed) of the training data. As time series usually have serial correlations, the *iid*-assumption is obviously violated. In this case the bootstrap must be applied to the residuals of the training data rather than to the original data assuming that the residuals fulfill the *iid*-assumption.

A step-by-step guide for the calculation of prediction intervals without considerations of network uncertainties is as follows. The description is similar to that in (Clements and Taylor, 2001) and references therein even though only linear AR(p) models are considered there. The method will be denoted *Boot* in the sequel.

Step 1 The data is generated by the NAR(p) process

$$Y_k = f(\theta, Y_{k-1}, \ldots, Y_{k-p}) + E_k \qquad (1)$$

where E_k is a white noise process with distribution F_e, zero mean and finite second moment. θ is the parameter vector of the process. Simulate the process and get the time series $\mathcal{Y} = \{y_1, \ldots, y_{n+p}\}$ [3].

Step 2 Estimate the parameters of the neural network $\hat{\theta}$ and obtain the empirical distribution $F_{\hat{e}}$ [4] of the residuals

$$\hat{e}_k = y_k - f(\hat{\theta}, y_{k-1}, \ldots). \qquad (2)$$

If necessary center and rescale the residuals.

Step 3 Simulate B bootstrap continuations recursively for the m-step-ahead prediction

[2] This is the bootstrap paradigm. Statistical inference usually tries to find a description between a sample and the population the sample was drawn from. As the population and its parameters are generally unknown, the distribution of the (unknown) population is replaced by the (known) sample population and the original sample is replaced by the bootstrap one. A more formal definition is given in (Hall, 1988).

[3] Note the difference between the process Y_k and its realizations y_k.

[4] An empirical distribution function $F_{\hat{x}}$ assigns to each value of the sequence $\mathcal{X} = \{x_1, \ldots, x_n\}$ equal probabilities of n^{-1}. Sometimes n^{-1} is called mass.

$$y_{k+j}^{*b} = f\left(\hat{\theta}, y_{k+j-1}^{*b}, \ldots, y_{k+j-p}^{*b}\right) + e_{k+j}^{*b}, \quad (3)$$

$$j = 1, \ldots, m \quad b = 1, \ldots, B$$

and set $y_{k+s}^{*b} = y_{k+s}$ for $s \leq 0$ to condition the prediction interval on the last p observations which are supposed to be known. The e_{k+j}^{*b} are random draws from $F_{\hat{e}}$.

Step 4 Estimate the prediction intervals from the empirical distribution function of y_{k+j}^{*b}. Several methods may be applied, see (Hall, 1988) for a comprehensive treatment. Here, Efron's percentile method is used. Define the cumulative empirical distribution as

$$\alpha = G_B^*(h) = \frac{1}{B}\sum_{b=1}^{B} \mathcal{I}(y < h)$$

with the indicator function

$$\mathcal{I}(y < h) = \begin{cases} 1 & \text{if } y < h \\ 0 & \text{else} \end{cases},$$

the symmetric $100\alpha\%$-prediction interval is

$$\left[Q_B^*\left(\frac{1-\alpha}{2}\right), Q_B^*\left(\frac{1+\alpha}{2}\right)\right] \quad (4)$$

with $Q_B^* = G_B^{*-1}$.

This procedure does not account for the uncertainty in the neural network parameters. Extensions are presented to cope with this fact. The extended method is denoted *Boot-PU*.

Step 1-2 As above.

Step 3a Simulate B bootstrap time series $\mathcal{Y}^{*b}$ of length $n+p$ with the nominal model $f(\hat{\theta})$. Choose a random block of length p as initial values or fix them to $y_1, \ldots, y_p$.

Step 3b Estimate B neural networks with parameters $\hat{\theta}^{*b}$ for each $\mathcal{Y}^{*b}$.

Step 3c Simulate B bootstrap continuations recursively for the m-step-ahead prediction

$$y_{k+j}^{*b} = f\left(\hat{\theta}^{*b}, y_{k+j-1}^{*b}, \ldots, y_{k+j-p}^{*b}\right) + e_{k+j}^{*b}, \quad (5)$$

$$j = 1, \ldots, m \quad b = 1, \ldots, B$$

with the same settings as in eq. (3).

Step 4 As above.

Note that all B bootstrap models are trained in advance and that the potentially high number of neural networks does not rule out online applications as the evaluation of the network is comparatively fast in contrast to its training.

3. EXTENSIONS TO NARX MODELS

In the former section, the model class was restricted to nonlinear AR(p) models. The evaluation of the prediction uncertainty has gained considerable attention in the field of econometrics where exogenous inputs play a minor role. Control applications on the other hand demand for extensions to incorporate external inputs. Basically, there are three possible approaches. Assume that the model is of type

$$Y_k = f(\theta, Y_{k-1}, \ldots, Y_{k-p}, U_{k-1}, \ldots, U_{k-p}) + E_k. \quad (6)$$

Depending of the class of input signals, different strategies for resampling in Step 3a of the *Boot-PU* method can be applied. If the input signal is a deterministic one like sinusodial, chirp or rectangular, bootstrapping the empirical distribution function of the input sequence will produce signals with completely different spectral densities. As a consequence different process dynamics are excited and the estimated model most likely does not coincide with the nominal one. In this case it seems best to keep the input signal fixed or just shift it in phase. The resulting models and prediction intervals will thus be conditioned on the specific realization $\{u_k\}$ rather than the distribution F_u of the input signal.

For random signals with unknown distribution F_u, resampling from $\{u_1, \ldots, u_{n+p-1}\}$ will also generate bogus signals. A promising solution is the use of the empirical difference distribution $F_{\Delta\hat{u}}$. A bootstrap sample is most likely to have the same properties (e.g. spectral densities) as the original one and only rescaling of the bootstrap input signal might be necessary.

If the distribution of the input signals is known, it is obviously best to take random draws from the distribution function F_u. Step 3a is therefore extended as follows:

Step 3a, revised Simulate B bootstrap time series with the nominal model $f(\hat{\theta})$ to obtain $\mathcal{Y}^{*b}$ of length $n+p$.

$$y_k^{*b} = f\left(\hat{\theta}, y_{k-1}^{*b}, \ldots, y_{k-p}^{*b}, u_{k-1}^{*b}, \ldots, u_{k-p}^{*b}\right) + e_k^{*b}, \quad (7)$$

$$k = p+1, \ldots, p+n.$$

Choose a random block of length p as initial value or fix it to $\{y_1, \ldots, y_p, u_1, \ldots, u_p\}$. The e_k^{*b} is a random draw from $F_{\hat{e}}$. The input signals may be obtained by:

(1) Set $u_k^{*b} = u_{k-q^{*b}}$ to use the same input sequence as in $\mathcal{Y}$ with a potential (random) phase shift q^{*b}.

(2) Generate a difference data set $\Delta\mathcal{U} = \{u_2 - u_1, \ldots, u_{n+p-1} - u_{n+p-2}\}$ with corresponding distribution function $F_{\Delta\hat{u}}$. The input sequence will be $u_k^{*b} = u_{k-1}^{*b} + \Delta u_k^{*b}$ where Δu_k^{*b} is a random draw from $F_{\Delta\hat{u}}$. u_1^{*b} can be zero or any value from $F_{\hat{u}}$. The bootstrap input sequence $\mathcal{U}^{*b}$ should be rescaled to have the same range as $\mathcal{U}$.

(3) If F_u is explicitly known, the input sequence $\{u_k^{*b}\}$ is just a realization of U.

4. LINEARIZATION-BASED APPROACH

The nominal neural network is linearized around an input vector $[y_{k-1}, \ldots, y_{k-p}, u_{k-1}, \ldots, u_{k-p}]$. This yields

$$\begin{aligned} Y_{k|k-1} &= \delta + \phi_1 Y_{k-1} + \cdots + \phi_p Y_{k-p} \\ &\quad + \psi_1 U_{k-1} + \cdots + \psi_p U_{k-p} + E_k \end{aligned} \tag{8}$$

where $Y_{k|k-1}$ denotes the prediction obtained from a linearized model at instance $k-1$. As both input process U and parameters $\psi_{1\ldots p}$ and $\phi_{1\ldots p}$ are assumed to be known exactly, the approximated distribution of the m-step-ahead prediction is given by

$$Y_{k+j} \dot{\sim} \mathcal{N}(\hat{y}_{k+j} | \hat{\sigma}_e^2 + \sum_{i=1}^{p} \phi_i^2 \hat{\text{var}}(Y_{k+j-i})) \tag{9}$$

with $\hat{\text{var}}(Y_{k+j-i}) = 0$ for $j \leq i$ and variance $\hat{\sigma}_e^2$ estimated from the residuals between training data and neural network output. The resulting prediction interval is then

$$[\hat{y}_{k+j} - z_\alpha \hat{\text{se}}(Y_{k+j}); \hat{y}_{k+j} + z_\alpha \hat{\text{se}}(Y_{k+j})] \tag{10}$$

where z_α is the quantile of the normal distribution with $\Phi(z_\alpha) = \alpha$. The interval is symmetric about $\hat{y}_{k+j}$. The linearized method does not account for the uncertainty in the parameter estimates and for the possibly skewed distribution of Y_{k+j}.

5. SIMULATION EXAMPLE

5.1 *Model Description*

In this example, a continuous bioreactor benchmark is considered as described in (Puskorius and Feldkamp, 1994). The nonlinear coupled differential equations are given by

$$\dot{y}_1 = -y_1 u + y_1(1 - y_2)e^{y_2/\Gamma} \tag{11}$$

$$\dot{y}_2 = -y_2 u + y_1(1 - y_2)e^{y_2/\Gamma} \frac{1+\beta}{1+\beta-y_2} \tag{12}$$

where the state variables y_1 and y_2 represent the cell mass and nutrients conversion in dimensionless form. The exogenous variable u is the flow rate of nutrients into the constant volume tank and the nominal values of the parameters are $\Gamma = 0.48$ and $\beta = 0.02$. The process exhibits strong nonlinearities and has a tendency for instability and collapsing behavior ($y_{1,2} \to 0$ for constant u) for increasing flow rates. Fig. 1 shows the nonlinear behavior and reveals the necessity of a nonlinear model for e.g. model predictive control.

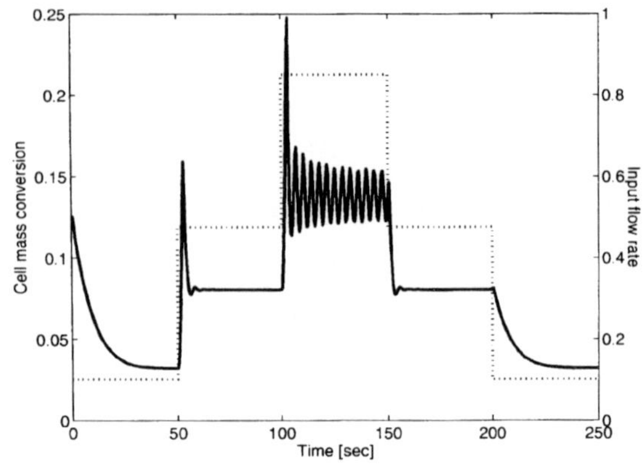

Fig. 1. Step responses of the bioreactor with initial values $y_{1,0} = 0.12$ and $y_{2,0} = 0.88$

5.2 *Neural Network Training*

The scope of this paper is not to find the best neural network structure and parameters for the problem of dynamic modeling of the bioreactor but to assess a neural network that has been obtained by standard methods.

(1) The rigoros process model is implemented in Matlab and simulated with the `ode23s`-integrator.
(2) The input signal for the training data set is an amplitude modulated pseudo random binary signal (APRBS) with a maximum flow of 0.85, a minimum of 0.1 and a switching time that is equally distributed over the interval $t_s \in [12.5s; 25s]$. The sampling interval is $0.5s$. For validation, the same type of signal is used with similar settings but different realizations.
(3) The feedforward neural network has one hidden layer with three neurons and sigmoidal activation functions. The order of the process (p) is four and the overall number of network parameters is 31. The process is simulated for 500 seconds which results in approximately 1000 training patterns. This is a relatively small number, in comparison (Puskorius and Feldkamp, 1994) use 10.000 points. An increase of generalization was obtained by the application of a ridge regression training procedure.
(4) Several networks with random initialization of the network parameters were trained and the one with the lowest root mean squared error on the training data set was selected.

Table 1 shows results for the validation data set. In the first column, the desired coverage probability α from eq. (4) is given, the second column denotes the method and the third one shows the actual coverage. An actual coverage of e. g. 88% means that 88% of the neural network five-step-ahead predictions for the validation data set lie within the calculated prediction intervals. If the methods for calculating the prediction intervals are (asymptotically) correct, then for large data sets and large bootstrap samples the actual coverage tends to the nominal one.

$$\lim_{n\to\infty} \lim_{B\to\infty} \text{Prob}\left[y_{\text{true}} \in \text{eq. (4)}\right] = \alpha \qquad (13)$$

The fourth column shows the width of the prediction intervals averaged over all validation data points. The standard deviation of the width is given in brackets ($\hat{\sigma}$).

Table 1. Coverage for five-step-ahead prediction for Validation Data Set.

Nominal Coverage	Method	Actual Coverage	Average Width ($\hat{\sigma}$)
50%	Linear	51.05%	0.52 (3.89)
	Boot	51.65%	0.01 (0.01)
	Boot-PU	58.54%	0.01 (0.03)
80%	Linear	64.34%	0.99 (7.38)
	Boot	81.12%	0.02 (0.04)
	Boot-PU	85.41%	0.02 (0.06)
90%	Linear	65.63%	1.27 (9.48)
	Boot	88.11%	0.03 (0.05)
	Boot-PU	90.61%	0.04 (0.08)
95%	Linear	67.13%	1.51 (11.30)
	Boot	92.11%	0.04 (0.07)
	Boot-PU	94.61%	0.05 (0.10)
99%	Linear	70.83%	1.98 (14.84)
	Boot	96.70%	0.07 (0.10)
	Boot-PU	97.70%	0.08 (0.17)

The method based on linearization constantly underestimates the prediction interval leading to far too optimistic results. The poor coverage and the relatively wide interval width originates from the fact that only symmetric intervals can be calculated due to the normality assumption. The large standard error $\hat{\sigma}$ is a consequence of the linearization as the estimated variance is rather high when step changes on the input variable occur and the nonlinearity of the process has a significant effect on the prediction. The bootstrap method which only uses random draws form the empirical residual distribution $F_{\hat{\epsilon}}$ for the prediction (denoted *Boot*) and therefore conditions them on the nominal network parameters $\hat{\theta}$ shows much better coverage than the linear method with the advantage of small average interval widths. It is only outperformed by the bootstrap method which accounts for the parameter uncertainty (hence denoted *Boot-PU*) which shows very good coverage with a slight tendency to overestimate the prediction intervals for low nominal coverage.

To illustrate the empirical distribution functions and the variation of the prediction intervals with time, the five-step-ahead prediction is compared with the plant behavior. Fig. 2 shows about 15% of the validation data set. The neural network shows good approximation abilities of the true process behavior. As the order of the NARX process is four, the five-step-ahead prediction $y_{k+5|k}$ is calculated completely from former neural network predictions because the values $y_{k+4|k}, \ldots, y_{k+1|k}$ are not known at time instance k. Only the large step at $t = 10$ sec causes a large error of the neural network prediction.

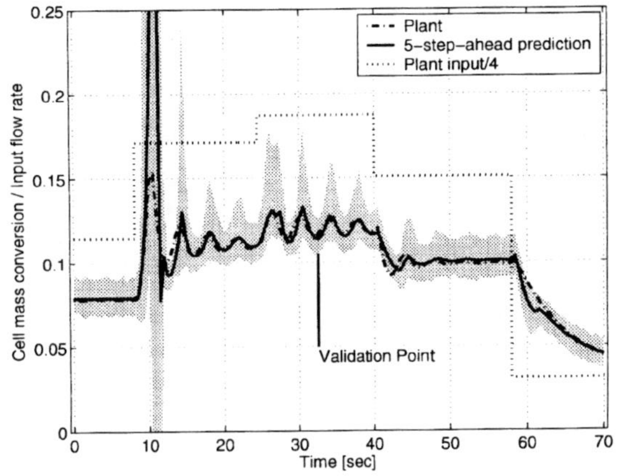

Fig. 2. Plant step response and neural network five-step-ahead prediction. The area shaded in gray is the 95% prediction interval of the *Boot-PU* method.

The distribution of the bootstrap predictions, the corresponding 95% confidence intervals as well as the normal approximations for the point indicated in Fig. 2 as validation point are shown in Fig. 3. The non-stationarity of the process at this point causes the linear method to estimate a large prediction uncertainty, whereas the neural network exhibits a sufficient performance there. The bootstrap distribution of the five-step-ahead prediction is slightly skewed to larger values of the predicted variable indicating that the nominal neural network may underestimate the true value. In comparison to the linear method, the width of the prediction interval is rather small. Both bootstrap methods yield almost equal intervals here.

Another important aspect is the asymptotic behavior and the rate of convergence to the limiting distribution. Table 2 shows the actual coverage for different nominal coverage with respect to an increasing number of bootstrap predictions B. The use of more than 400 bootstrap predictions does not increase the coverage accuracy significantly.

6. CONCLUSION

In this paper, a method has been introduced to evaluate the prediction uncertainty of neural net-

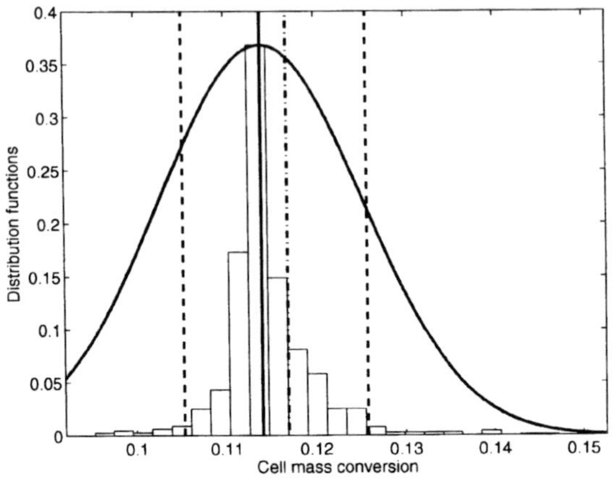

Fig. 3. Distribution functions at the point indicated in Fig. 2. The bars show the empirical distribution of the bootstrap predictions, the solid vertical line is the neural network prediction, the dash-dotted line is the real process output, the dashed lines are the 95% bootstrap intervals and the bell-shaped curve (normalized to the maximal bootstrap value) is the normal approximation of the error distribution that results from the linearization based method.

Table 2. Influence of the number of bootstrap samples

	Validation Data Set			
	Nominal coverage 80%		Nominal coverage 95%	
B	Boot	Boot-PU	Boot	Boot-PU
20	78.3%	82.1%	85.3%	89.0%
100	80.2%	83.5%	90.7%	92.8%
400	81.0%	85.1%	92.0%	94.6%
1000	81.1%	85.4%	92.1%	94.6%

works for modeling of nonlinear dynamic systems. This aspect has not yet attracted the necessary attention even though unreliable predictions can lead to erratic behavior of e.g. nonlinear model predictive controllers.

The bootstrap as a tool from computational statistics is used to circumvent the limitations that are imposed by the nonlinearity of the applied neural network models and by the normality assumption of the error distributions. The innovation improves the coverage largely and narrows the width of the prediction intervals as non-symmetric intervals are possible.

Although the proposed method gives additional insight into the prediction uncertainty of neural networks, further research is necessary to improve the estimation of prediction intervals. Using eq. (1) tacitly assumes that the data generating process lies in the class of neural networks used for modeling. If this assumption is not justified, the effects on the prediction intervals have to be examined as the empirical error distribution is potentially biased. Another aspect is the unconditionality of the empirical error distribution which implies that the variance of the error process does not depend on the current neural network input. If the neural network operates in ranges with low training data, sampling from the empirical error distribution might be too optimistic and if it operates in a range with high training data density sampling from the same distribution might be too pessimistic. Conditioning the empirical error distribution on the current point of operation will improve both the actual coverage of a complete data set and reduce the conservativeness of the empirical distribution of the multi-step-ahead prediction errors.

7. ACKNOWLEDGEMENTS

Part of this work was sponsored by the DFG (Deutsche Forschungsgemeinschaft).

8. REFERENCES

Chryssolouris, G., M. Lee and A. Ramsey (1996). Confidence interval prediction for neural network models. *IEEE Trans. Neural Networks* **7**(1), 229–232.

Clements, M. P. and N. Taylor (2001). Bootstrapping prediction intervals for autoregressive models. *International Journal of Forecasting* **17**, 247–267.

Efron, B and R. J. Tibshirani (1993). *An Introduction to the Bootstrap.* Chapman & Hall.

Hall, P. (1988). Theoretical comparison of bootstrap confidence intervals. *The Annals of Statistics* **16**(3), 927–953.

Puskorius, G. V. and L. A. Feldkamp (1994). Neurocontrol of nonlinear dynamical systems with kalman filter trained recurrent networks. *IEEE Trans. Neural Networks* **5**(2), 279–297.

Roßmann, V. (2002). Prädiktive Regelung mit Neuronalen Netzen (Predictive control with neural networks). PhD thesis. University of Dortmund, Department of Biochemical and Chemical Engineering.

Tibshirani, R. (1995). A comparison of some error estimates for neural networks. *Neural Computation* **8**, 152–163.

Tsai, P. F., J. Z. Chu, S. S. Jang and S. S. Shieh (2002). Developing a robust model predictive control architecture through regional knowledge analysis of artificial neural networks. *J. Proc. Contr.* **13**, 423–435.

Wang, C., K.-U. Klatt, G. Dünnebier, S. Engell and F. Hanisch (2003). Neural network-based identification of SMB chromatographic processes. *Control Engineering Practice* **11**(8), 949–959.

ELSEVIER
IFAC
PUBLICATIONS
www.elsevier.com/locate/ifac

IDENTIFICATION OF NONLINEAR OBSERVERS FOR MULTIVARIABLE SYSTEMS SUBJECTED TO UNMEASURED DISTURBANCES

Meka Srinivasarao, Raja Venkatasubramanian, Sachin C. Patwardhan

Department of Chemical Engineering, Indian Institute of Technology, Bombay, Mumbai , India.-400 076

Abstract: This work aims at the development of NARX type nonlinear multivariable stochastic models, which can explicitly account for the dynamics of unmeasured disturbances using input-output data. At the identification stage, the model is arranged to have Weiner like structure and the linear dynamic component of the Weiner model is parameterized using generalized orthonormal basis filters. The identified stochastic model can be further used to develop an extended Kalman filter. The efficacy of the proposed modeling scheme is demonstrated by simulation studies on the benchmark CSTR system, which exhibits input multiplicity and change in the sign of steady state gain in the operating region. The analysis of the simulation results reveals that the proposed model is able to capture the dynamics and the steady state behavior of the reactor reasonably accurately over a wide operating range. *Copyright © 2004 IFAC*

Keywords: Extended Kalman filters, Nonlinear Observers, Stochastic Modeling, NARX Models, Orthonormal Basis Filters.

1. INTRODUCTION

Linear model predictive control (MPC) schemes are now widely used in process industry for control of key unit operations. However, the use of linear dynamic models for prediction, limits their applicability to a narrow range of operation or to systems which exhibit mildly nonlinear dynamics. The need to achieve a tighter control of strongly nonlinear processes has led to a more general MPC formulation whereby nonlinear dynamic models are used for prediction(Meadows and Rawlings, 1997, Henson, 1998) Selection of a suitable form of nonlinear model to represent system dynamics is a crucial step in the development of a nonlinear MPC (NMPC) scheme. The NMPC schemes proposed in literature either use models developed from first principles or models identified from input-output data (i.e. nonlinear black box models) (Brengel and Seider, 1989; Hernandez and Arkun, 1993). The first principle models are valid globally and can predict system dynamics over the entire operating range. However, development of a reliable first principle model is, in general, a difficult and a time consuming task. The nonlinear black box models, on the other hand, have certain advantages over the mechanistic models in terms of development time and efforts. Thus, from a practical viewpoint, development of a NMPC scheme based on a nonlinear black box model appears to be an attractive option.

Given a model with known state and measurement noise characteristics, there are well-established procedures to design a nonlinear predictive control scheme based on an optimal state observer (extended Kalman filter) (Lee and Ricker, 1994). However, in most of the situations such stochastic models with known state and measurement noise characteristics are seldom available. One remedy to this problem is to develop stochastic models for unmeasured disturbances (or the state observers) directly from the input-output data. This can considerably simplify the implementation of control schemes based on stochastic models.

One of the linear black-box modelling techniques that has been receiving increasing attention in the past decade is the use of generalized orthonormal basis filter (GOBF) networks (Ninness and Gustovson,1997; Van den Hof et. al., 2000). The orthonormal basis filter approximations provide a simple and elegant method of representing stable systems. Advantages of using such orthonormal filters for process modelling can be summarized as follows:

- A good approximation can be obtained with a small number of terms due to the orthogonal property.
- Development of orthonormal filter models do not need any explicit knowledge about system time constant and time delays

The use of such orthonormal filters in combination with a memory less nonlinear map (referred to as OBF-Wiener model in the rest of the text) for modelling of nonlinear dynamical systems was originally proposed by Wiener (Billings, 1980). Boyd and Chuha(1985) have shown that a nonlinear moving average operator of this type is adequate for capturing dynamics of any time invariant nonlinear systems with fading memory. GOBF-Wiener model can be looked upon as a representation of Volterra series model which is parsimonious in parameters. Recently, Sentoni et al.(1998) and Saha (1999) have used deterministic models of this type for developing nonlinear model predictive control schemes. Their modeling schemes, however, do not attempt to capture effects of unmeasured disturbances.

If unmeasured disturbances have a significant effect on the process outputs, then NMPC formulation based on a deterministic model can lead to poor closed loop performance. Thus, this work aims at the development of NARX type nonlinear multivariable stochastic models, which can explicitly account for the dynamics of unmeasured disturbances. At the identification stage, the model is arranged to have Weiner like structure and the linear dynamic component of the Weiner model is parameterized using GOBF. The identified stochastic model can be further used to develop an extended Kalman filter. The efficacy of the proposed modeling scheme is demonstrated by simulation studies on a benchmark CSTR system, which exhibits input multiplicity and change in the sign of steady state gain in the operating region .

This paper is organized in five sections. The next section deals with the proposed model structure and formulation of parameter estimation problem. In the third section we present a simulation case study while the conclusions are presented in the final section.

2. GOBF-WEINER MODEL

In this work, it is desired to develop an NARX type state space model using input-output data. At the identification stage the proposed model can be arranged to have Wiener model like structure. Weiner model is a block oriented model where in a dynamic linear block is followed by a static nonlinear block as follows

$$\mathbf{x}(k+1) = \Phi\mathbf{x}(k) + \Gamma\boldsymbol{v}(k) \tag{1}$$

$$y(k) = \Omega(\mathbf{x}(k)) + \varepsilon(k) \tag{2}$$

where $\varepsilon(k)$ represents output error. Here input vector $\boldsymbol{v}(k)$ to the system (1-2) is not limited to manipulated inputs but can be any measured quantity. In the present work, the linear dynamic component in the Weiner model is parameterized using GOBF while the nonlinear state output map is constructed using quadratic polynomials.

2.1 Parameterizations of Linear Dynamics

Consider a SISO system represented by a strictly proper stable transfer function

$$\widehat{y}(z) = G(z)\,v(z) \tag{3}$$

where v represents input and $\widehat{y}$ represents the model output. Let $\{F_k(z) : k = 0, 1, 2, ...\}$ be an orthonormal basis for $\mathcal{H}_2$ (set of strictly proper stable transfer functions) such that

$$\langle F_j(z), F_i(z)\rangle = \left\{\begin{matrix} 1 & (i=j) \\ 0 & (i \neq j) \end{matrix}\right\}$$

Then, there exists a unique generalized Fourier series expansion $G(z)$ such that

$$G(z) = \sum_{i=1}^{\infty} c_i F_i(z) \tag{4}$$

where $\{c_i\}$ represent Fourier coefficients defined as $c_i = \langle G(z), F_i(z)\rangle$. Thus, given a linear time invariant system $G(z)$ the n'th order finite expansion model that approximates $G(z)$ best in an $\mathcal{H}_2$ sense is given by (Van den Hof, 2000)

$$G_n(z) = \sum_{i=1}^{n} c_i F_i(z) \tag{5}$$

Ninness and Gustafsson (1997) have shown that

$$F_l(z,\xi) = \frac{\sqrt{(1-|\xi_l|^2)}}{(z-\xi_l)} \prod_{i=1}^{l-1} \frac{(1-\xi_i^* z)}{(z-\xi_i)} \quad (6)$$

forms a complete orthogonal set in $\mathcal{H}_2$, where $\{\xi_p : l = 1, 2, ...\}$ is an arbitrary sequence of poles inside the unit circle appearing in complex conjugate pairs.

Defining a state vector $\mathbf{x}(k) \in R^n$ as

$$\mathbf{x}(k) = \left[\, F_1(z,\boldsymbol{\xi})v(k) \;......\; F_n(z,,\boldsymbol{\xi})v(k) \,\right]^T \quad (7)$$

where $\boldsymbol{\xi} = \left[\, \xi_1 \; \xi_2 \; .... \; \xi_l \,\right]^T$ represents the vector of GOBF poles, a single rate state realization of the form

$$\mathbf{x}(k+1) = \phi(\boldsymbol{\xi})\,\mathbf{x}(k) + \psi(\boldsymbol{\xi})\,v(k) \quad (8)$$

$$\widehat{y}(k) = C\,\mathbf{x}(k) \quad (9)$$

can be developed for the model given by equation (3). Here, ϕ represents $n \times n$ state transition matrix, ψ represents $n \times 1$ input coupling matrix and the elements of the C vector are the Fourier coefficients, i.e.

$$C = [c_1 \; c_2 \; ... \; ... \; c_n]^T \quad (10)$$

If the GOBF are used to parameterize the linear dynamic part of the Weiner type state space model (1-2), then

$$\mathbf{x}(k+1) = \phi(\boldsymbol{\xi})\,\mathbf{x}(k) + \psi(\boldsymbol{\xi})\,v(k) \quad (11)$$

$$y(k) = \Omega(\,\mathbf{x}(k)) + \varepsilon(k) \quad (12)$$

In developing a GOBF-Wiener type model to capture dynamics of MIMO nonlinear systems, one possibility is to model an $m \times q$ MIMO non-linear system as m MISO nonlinear systems. In order, to see how this can be achieved, consider a MISO GOBF-Wiener model relating the input vector with the ith output. Let $\mathbf{x}^{(ij)} \in R^{n_{ij}}$ represent the n_{ij} dimensional state vector associated with a realization of GOBF filter network consisting of n_{ij} number of filters associated with the jth input. The corresponding state dynamics can be represented as

$$\mathbf{x}^{(ij)}(k+1) = \phi^{(ij)}\,\mathbf{x}^{(ij)}(k) + \psi^{(ij)}\,v_j(k)$$

Defining an augmented state vector $\mathbf{X}^{(i)}(k)$ and input vector $\boldsymbol{\Upsilon}(k)$ as :-

$$\mathbf{X}^{(i)}(k) = \left[\left(\mathbf{x}^{(i1)}(k)\right)^T \; .... \; \left(\mathbf{x}^{(iq)}(k)\right)^T\right]^T$$

the state dynamics of the MISO model can be represented as

$$\mathbf{X}^{(i)}(k+1) = \boldsymbol{\Phi}^{(i)}\mathbf{X}^{(i)}(k) + \boldsymbol{\Gamma}^{(i)}\boldsymbol{v}(k) \;\; ;$$

for $i = 1,\, 2\ldots,\, m$ where,

$$\boldsymbol{\Phi}^{(i)} = block\ diag\left[\phi^{(i1)} \; \phi^{(i2)} \; .... \; \phi^{(iq)}\right]_{N_i \times N_i}$$

$$\boldsymbol{\Gamma}^{(i)} = block\ diag\left[\psi^{(i1)} \; \psi^{(i2)} \; .... \; \psi^{(iq)}\right]_{N_i \times m}$$

$N_i = \sum_{j=1}^{l} n_{ij}$,and $\boldsymbol{v} \in R^q$ denotes input vector. The output of MISO GOBF-Wiener model can be expressed as :-

$$y_i(k) = \Omega\left[\mathbf{X}^{(i)}(k)\right] + \varepsilon_i(k)$$

In the present work, the static non-linearity of the Weiner models is chosen to be a quadratic polynomial because the quadratic perturbation models are known to give good local approximations of local dynamics (Patwardhan and Madhavan, 1993). Thus, the state-output can be expressed as

$$\Omega\,[.] = \mathbf{C}^{(i)}\,\mathbf{X}^{(i)}(k) + \left(\mathbf{X}^{(i)}(k)\right)\,\mathbf{D}^{(i)}\,\left(\mathbf{X}^{(i)}(k)\right)^T \quad (13)$$

for $i = 1, ..., m$, where D_i is a $N_i \times N_i$ symmetrical matrix. Note that above state-output map is linear in parameters and can be expressed as

$$y_i(k) = \left(\boldsymbol{\Theta}^{(i)}\right)^T \mathbf{Z}^{(i)}(k) + \varepsilon_i(k) \;\; ; \quad (\text{for } i = 1, ..., m) \quad (14)$$

where

$$\mathbf{X}_q^{(i)}(k) = \left[\left(\mathbf{X}_1^{(i)}(k)\right)^2 \; 2\mathbf{X}_1^{(i)}(k)\mathbf{X}_2^{(i)}(k) \; ....\right]^T$$

$$\mathbf{Z}^{(i)}(k) = \left[\left(\mathbf{X}^{(i)}(k)\right)^T \; \left(\mathbf{X}_q^{(i)}(k)\right)^T\right]^T$$

and

$$\boldsymbol{\Theta}^{(i)}(k) = \left[\mathbf{C}^{(i)} \; \mathbf{D}_{11}^{(i)} \; \mathbf{D}_{12}^{(i)} \; .... \; \mathbf{D}_{N_i,N_i}^{(i)}\right]^T$$

Here, $\mathbf{D}_{jl}^{(i)}$ represents (j,l)'th element of matrix $\mathbf{D}^{(i)}$.

2.2 Parameter Estimation

Given input sequence $\{\boldsymbol{v}(0), \boldsymbol{v}(1), ...\boldsymbol{v}(N_s)\}$ and output $\{y_i(0), y_i(1), ...y_i(N_s)\}$ data sets, the least square estimate of the parameters can be obtained by solving the following minimization problem

$$(\widehat{\Theta}^{(i)}, \widehat{\boldsymbol{\xi}}^{(i)}) = \underset{\Theta^{(i)}, \boldsymbol{\xi}^{(i)}}{\text{arg min}} \frac{1}{N_s} \sum_{k=1}^{N_s} \left[\widehat{\varepsilon}_i(k, \Theta^{(i)}, \boldsymbol{\xi}^{(i)})\right]^2 \quad (15)$$

subject to

$$|\boldsymbol{\xi}_i| \leq 1; \;\; i = 1, 2....L_i \quad (16)$$

where $\boldsymbol{\xi}_i$ represents vector of GOBF poles. Now, given a set of GOBF poles $\boldsymbol{\xi}$, the above minimization problem can bc solved analytically using simple linear regression scheme as follows

$$\widehat{\Theta}^{(i)}(\boldsymbol{\xi}) = \left[\overline{E}(\mathbf{Z}^{(i)}(k)\mathbf{Z}^{(i)}(k))^T\right]^{-1} \overline{E}(\mathbf{Z}^{(i)}(k)\mathbf{Y}^{(i)})$$

$$= \left(\Xi^T \Xi\right)^{-1} \Xi_u \mathbf{Y}^{(i)} \quad (17)$$

where matrix Ξ_u and vector Y are defined as,

$$\Xi_u = \begin{bmatrix} \mathbf{Z}^{(i)}(1)^T \\ \ldots\ldots \\ \mathbf{Z}^{(i)}(k)^T \end{bmatrix}$$

$$\mathbf{Y}^{(i)} = \begin{bmatrix} y_i(1) \; y_i(2) \; \cdots \; y_i(N_s) \end{bmatrix}^T$$

and $\overline{E}(.)$ represents expected value operator. Thus, the parameter estimation problem can be reformulated as a constrained optimization problem

$$(\widehat{\Theta}^{(i)}, \widehat{\boldsymbol{\xi}}^{(i)}) = \underset{\boldsymbol{\xi}^{(i)}}{\arg\min} \; \frac{1}{N_s}\sum_{k=1}^{N_s}\left[\widehat{\varepsilon}_i(k, \widehat{\Theta}^{(i)}, \boldsymbol{\xi}^{(i)})\right]^2 \tag{18}$$

subject to the constraints (16) and (17). This approach can considerably simplify the original nonlinear optimization problem (15).

3. DEVELOPMENT OF NARX MODEL AND STATE ESTIMATION

In the present work we develop a NARX type model where by vector $\boldsymbol{v}$ consists of

$$\boldsymbol{v}(k) = \begin{bmatrix} \mathbf{u}(k)^T \; y_i(k) \end{bmatrix}^T$$

where $\mathbf{u}(k) \in R^m$ represents manipulated inputs. The resulting model can be expressed as

$$\mathbf{X}^{(i)}(k+1) = \Phi^{(i)}\mathbf{X}^{(i)}(k) + \Gamma_u^{(i)}\mathbf{u}(k) + \Gamma_y^{(i)}y_i(k)$$
$$y_i(k) = \Omega\left[\mathbf{X}^{(i)}(k)\right] + v_i(k)$$

for $i = 1, 2\ldots, m$.If the sequence$\{v_i(k)\}$ is found to be auto-correlated, then a liner filter is developed for whitening the sequence as follows

$$\mathbf{x}_e^{(i)}(k+1) = \Phi_e^{(i)}\mathbf{x}_{i,e}(k) + \Gamma_e^{(i)}v_i(k)$$
$$v_i(k) = C_e^{(i)}\mathbf{x}_{i,e}(k) + e_i(k) \tag{19}$$

Thus, the resulting NARX model given above can be rearranged as follows

$$\mathbf{X}_a^{(i)}(k+1) = \mathbf{F}\left[\mathbf{X}_a^{(i)}(k)\right] + \boldsymbol{\Gamma}^{(i)}\mathbf{u}(k) + \boldsymbol{\Lambda}^{(i)}e_i(k) \tag{20}$$
$$y_i(k) = G\left[\mathbf{X}_a^{(i)}(k)\right] + e_i(k) \tag{21}$$

for $i = 1....m$, where

$$\mathbf{X}_a^{(i)}(k) = \left[\left(\mathbf{X}^{(i)}(k)\right)^T \; \left(\mathbf{x}_e^{(i)}(k)\right)^T\right]^T$$

$$\boldsymbol{\Gamma}^{(i)} = \begin{bmatrix} \Gamma_u^{(i)} \\ [0] \end{bmatrix}; \quad \boldsymbol{\Lambda}^{(i)} = \begin{bmatrix} \Gamma_{yi} \\ \Gamma_{ie} \end{bmatrix}$$

$$\mathbf{F}[.] = \begin{bmatrix} (\Phi^{(i)} + \Gamma_y^{(i)}\Omega\left[\mathbf{X}^{(i)}(k)\right] + \Gamma_y^{(i)}C_e^{(i)}\mathbf{x}_{i,e}(k) \\ (\Phi_e^{(i)} + \Gamma_y^{(i)}C_e^{(i)})\mathbf{x}_{i,e}(k) \end{bmatrix}$$

$$G[.] = \begin{bmatrix} C^{(i)} \; C_e^{(i)} \end{bmatrix}\mathbf{X}_a^{(i)}(k) + \left[\mathbf{X}_a^{(i)}(k)\right]^T\begin{bmatrix} \mathbf{D}^{(i)} & [0] \\ [0] & [0] \end{bmatrix}\left[\mathbf{X}_a^{(i)}(k)\right]$$

It should be noted that the model structure in its final form is not Weiner like model and both state updation equation and state output map are nonlinear equations. Also it should be noted that the end result of the open loop identification exercise is an observer. The parameter estimation procedure discussed above generates NARX models of form (20-21) and an estimate of variance $\mathbf{v}_e^{(i)}$of white noise sequence $\{e_i(k)\}$.

The above MISO models can be combined to form a MIMO NARX model and expressed in standard form

$$\mathbf{X}(k+1) = \mathcal{F}\left[\mathbf{X}(k), \, \mathbf{u}(k)\right] + \mathbf{w}(k) \tag{22}$$
$$\mathbf{y}(k) = \mathbf{G}\left[\mathbf{X}(k)\right] + \mathbf{e}(k) \tag{23}$$

where $\mathbf{w}(k)$ and $\mathbf{e}(k)$ are white noise sequences with covariances $\mathbf{R}_1 = \boldsymbol{\Lambda}\mathbf{V}_e\boldsymbol{\Lambda}^T$, $\mathbf{R}_{12} = \boldsymbol{\Lambda}\mathbf{V}_e$and $\mathbf{R}_2 = \mathbf{V}_e$, respectively. The state space model together with state covariance information can be further used to estimate states using Extended Kalman filter as given by Muske and Edgar (1997) as follows

$$\widehat{\mathbf{X}}(k{+}1/k) = \mathcal{F}\left[\widehat{\mathbf{X}}(k-1/k-1), \, \mathbf{u}(k-1)\right] \tag{24}$$
$$\widehat{\mathbf{P}}_{(k/k-1)} = \Lambda_{(k-1)}\widehat{\mathbf{P}}_{(k-1/k-1)}\Lambda(k-1) + \mathbf{R}_1 \tag{25}$$
$$\Lambda_{(k-1)} = \left[\frac{\partial \mathcal{F}}{\partial \mathbf{X}}\right]_{(k-1)} \tag{26}$$
$$\mathbf{L}_{(k)} = \left(\widehat{\mathbf{P}}_{(k/k-1)}\Psi_{(k)}^T + \mathbf{R}_{12}\right)\left(\Psi_{(k)}\widehat{\mathbf{P}}_{(k/k-1)}\Psi_{(k)}^T + \mathbf{R}_2\right)^{-1} \tag{27}$$
$$\Psi_{(k)} = \left[\frac{\partial \mathbf{G}}{\partial \mathbf{X}}\right]_{(k/k-1)} \tag{28}$$
$$\widehat{\mathbf{X}}(k/k) = \widehat{\mathbf{X}}(k/k-1) + \mathbf{L}_{(k)}\gamma(k) \tag{29}$$
$$\gamma(k) = \mathbf{y}(k) - \mathbf{G}\left[\widehat{\mathbf{X}}(k/k-1)\right] \tag{30}$$
$$\widehat{\mathbf{P}}_{(k/k)} = \left(\mathbf{I} - \mathbf{L}_{(k)}\Psi_{(k)}\right)\widehat{\mathbf{P}}_{(k/k-1)} \tag{31}$$

4. SIMULATION CASE STUDY

The CSTR system under consideration consists of a reversible exothermic reaction $A \rightleftharpoons B$. The dynamic model for simulating the CSTR system is as follows

$$\frac{dC_a}{dt} = \frac{F_i}{h\,A_c}(C_{ai} - C_a) + K_1\,C_a - K_2\,C_b$$
$$\frac{dC_b}{dt} = -\frac{F_i}{h\,A_c}C_b + K_1\,C_a - K_2\,C_b$$
$$\frac{dT}{dt} = \frac{1}{h\,A_c}F_i(T_i - T) + \frac{-H_r}{\rho\,C_p}(K_1\,C_a - K_2\,C_b)$$
$$\frac{dh}{dt} = \frac{1}{A_c}(F_i - k\sqrt{h})$$
$$K_1 = k_f\,\exp(-E_f/T)\,; K_2 = k_b exp(-E_b/T)$$

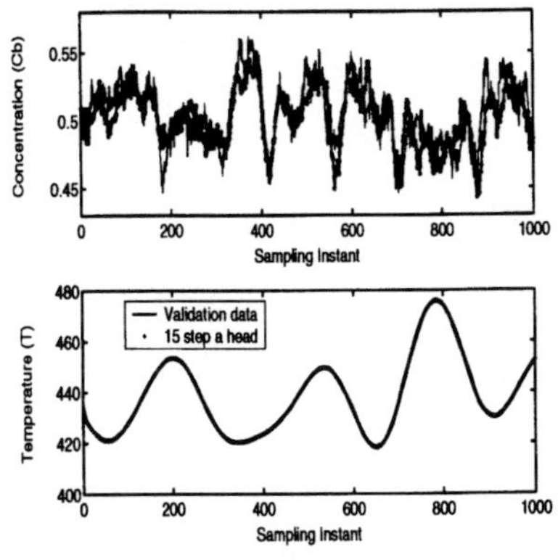

Fig. 1. Comparison of 15 step a head predictions and validation data for Cb and T.

The nominal parameters and the operating steady state used in the simulation studies can be found in Li and Biegler (1989). In the present work, the output concentration (C_b) and reactor temperature (T) in the CSTR are the two measured outputs of the system. The inlet flow rate and inlet temperature are used as manipulated variables and inlet concentration C_{ai} is treated as unmeasured disturbance. This system exhibits input multiplicity and change in the sign of steady state gain in the operating region. The difficulties associated with controlling such systems at the optimum operating point have been discussed in detail by Patwardhan and Madhawan (1993).

In the present study, sampling rate of 0.4 min was used. The inlet feed stream contains only A and its concentration fluctuates according to following stochastic process

$$\delta H_v(k) = \frac{0.05}{1 - 0.95z^{-1}} e(k) \qquad (32)$$

where $e(k)$ is a white noise sequence with standard deviation 0.2. Low frequency random Gaussian signals, with standard deviations of 0.15 m^3/s and 15 K and frequencies 0.05 (π/T_s) and 0.01 (π/T_s), respectively, were generated using the *'idinput'* function in the *System Identification Toolbox of MATLAB.* These signals were used to introduce simultaneous perturbations in both the inlet flowrate (F_{in}) and inlet temperature (T_i), respectively. An independent set of data, similar to the identification data set, was used to validate the models. The optimum set of poles obtained from the identification exercise for NARX model has been reported in Table 1. The comparison

Table 1. Optimum values of GOBF poles for NARX models for concentration and temperature

Variable	u_1	u_2	y
Cb	[0.42 0.42]	[2.18 3.68]	[4.02 3.99]
T	[3.89 0.47]	[4.03 0.59]	[2.94 5.99]

of the results for 15 step a head predictions with

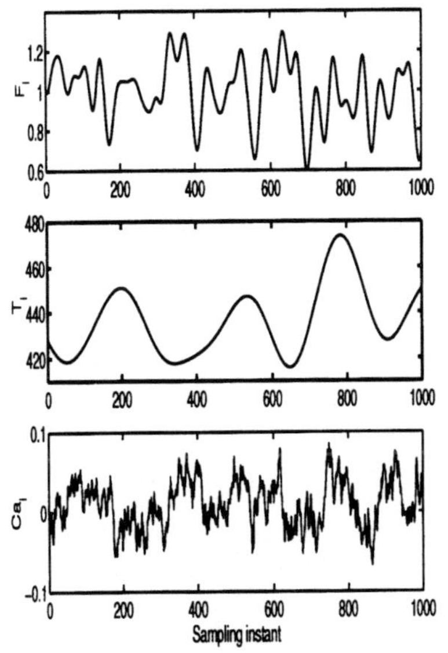

Fig. 2. Input signals Fi, Ti and unmeasured disturbance in Cai used to generate validation data.

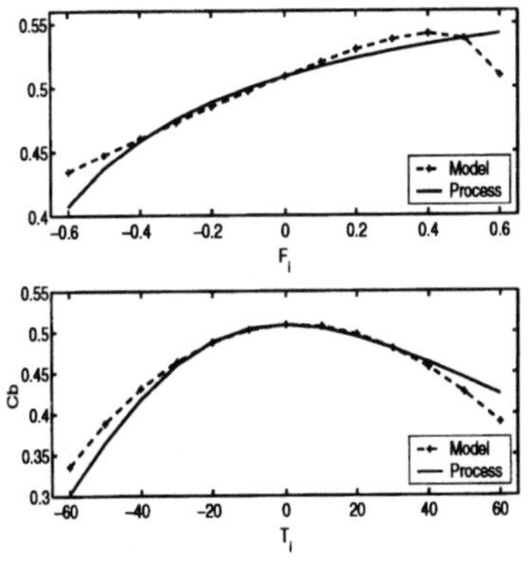

Fig. 3. Comparison of measured and predicted steady state concentration Cb in the reactor

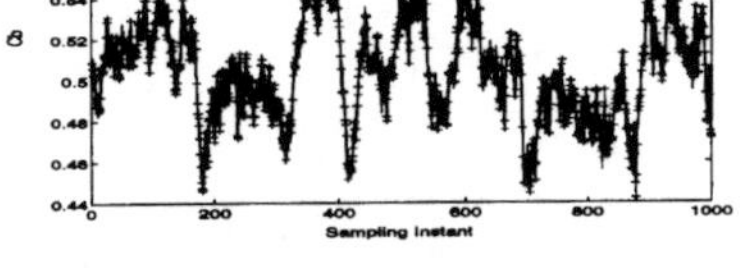

Fig. 4. Comparison of validation data with Extended Kalman filter predictions

the validation data set are presented in Figure 1. Figure 2 presents the corresponding manipulated inputs and unmeasured disturbance variations. In Table 2, integral squared estimation error (ISEE) and Percentage Prediction Error(PPE) values

$$(PPE)_i = \frac{\sum_{k=1}^{N_s}[y_i(k) - \hat{y}_i(k)]^2}{\sum_{k=1}^{N_s}[y_i(k) - \bar{y}_i]^2} \times 100 \qquad (33)$$

have been presented for one step a head and 15 step a head predictions for the validation data set. (Note that, $\bar{y}_i$ in Eq. (33) represents the mean value of the plant output data). The validation

Table 2. ISEE and PPE values for Cb and T based on the NARX model

Variable	Predictions	ISEE	PPE
cb	1 Step	0.0537	9.760
	15 steps	0.2398	43.630
	ekf	0.0768	——
T	1 Step	54.6630	0.024
	15 steps	57.8690	0.025
	ekf	54.7510	——

based on one step ahead predictions exhibits excellent match between the observed and predicted outputs. This also reflects in the small ISE and PPE values for one step a head predictions. The 15 step a head predictions, on the other hand, result in much larger PPE and ISE values. However, as evident from Fig. 1, the 15 step a head predictions capture the output variability of concentration (C_b) caused by the manipulated input variations reasonably well. In case of temperature both one step and 15 step a head predictions are showing excellent match. This can be attributed to the fact that the temperature is not affected significantly by the unmeasured disturbances in the input concentration.

Figure 3 presents the comparison of the steady state behavior of output concentration with respect to input flowrate and inlet temperature. As can be seen in Fig. 3, the identified NARX model captures the steady state behavior of the system over a wide operating range around extremum operating point. In particular the NARX model is able to capture the change in the sign of the steady state gain reasonably well. Figure 4 compares one step ahead predictions generated by the extended Kalman filter developed using the identified NARX model with the measured outputs in the validation data set. The EKF predictions show an excellent match with the measured outputs and this also reflects in the small ISEE values reported in the Table 2.

5. CONCLUSIONS

This work aims at the development of NARX type nonlinear multivariable stochastic models, which can explicitly account for the dynamics of unmeasured disturbances. At the identification stage, the model is arranged to have Weiner like structure and the linear dynamic component of the Weiner model is parameterized using GOBF. The identified stochastic model can be further used to develop an extended Kalman filter. The efficacy of the proposed modelling schemes is demonstrated by simulation studies on the benchmark CSTR system, which exhibits input multiplicity and change in the sign of steady state gain in the operating region. Analysis of the simulation results reveals that the proposed model is able to capture the dynamics and the steady state behavior of the reactor reasonably accurately over a wide operating range.

REFERENCES

Billings,S. A (1980). Identification of nonlinear systems: A survey. *IEE Proc.*, **127**, Pt. D., 6, 272- 285.

Boyd, S. and L. Chua (1985). Fading memory and the problem of approximating nonlinear operators with Volterra series. *IEEE Transactions on Circuits and Systems*, **CAS-32**, 11, 1150-1161.

Brengel,D. D. and W.D. Seider (1989). Multistep nonlinear predictive control. *Ind. Eng. Chem. Res.*, **28**, 1812-1822.

Henson, M.A. (1998). Nonlinear Model Predictive Control : Current status and future directions. *Computers and Chemical Engg,***23** , 187-202.

Hernandez, E. and Y. Arkun (1993). Control of nonlinear systems using polynomial ARMA models. *AIChE J*, **39**(3), 446-460.

Li, W. C. and L. T. Biegler (1988). Process Control Strategies for Constrained Nonlinear Systems, *Ind. Eng. Chem. Res.*, 27, 142.

Meadows, E.S.and J. B. Rawlings (1997). *Nonlinear Process Control*, (M.A. Henson and D.E. Seborg (eds.), New Jersey: Prentice Hall, Chapter 5.Rawlings and Meadows

Muske,K.R. and T.F.Edgar (1997). *Nonlinear Process Control*,(M.A. Henson and D.E. Seborg (eds.), New Jersey: Prentice Hall,Chapter 6.Muske and Edgar,

Ninness, B. M. and F. Gustafson (1997). A Unifying Construction of Orthonormal Basis for System Identification. *IEEE Transactions on Automatic Control*, **42**(4), 515-521.

Patwardhan,S.C.and K.P. Madhavan (1993).Nonlinear predictive control using approximate second order perturbation models. *Ind. Eng. Chem. Res.*, **32**, 334-344.

Lee, J.H. and N.L. Ricker (1994)."Extended Kalman Filter based Nonlinear Model Predictive Control",*AIChE J.*,33, 1530.

Saha,P. (1999), Ph.D thesis, IIT Madras, Chennai, India.

Sentoni G. B., J. P. Guiver, H. Zhao and L. T. Biegler (1998). A state space approach to nonlinear process modeling: Identification and universality. *AIChE J.* **44**,**10**,2229.

Van den Hof, P., B. Walhberg, P. Heuberger, B. Ninness, J. Bokor, T. Olivera e Silva (2000), "Modeling and Identification with Rational Orthonormal Basis Functions", *In Proc. of IFAC SYSID 2000,* Santa Barbara, California, June 21-23, 2000.

ELSEVIER
IFAC
PUBLICATIONS
www.elsevier.com/locate/ifac

ERROR DETECTION AND CONTROL IN GREY-BOX IDENTIFICATION OF DISTRIBUTED PARAMETER PROCESSES

Yi Liu and Elling W. Jacobsen

Automatic Control
Dept. of Signals, Sensors and Systems
Royal Institute of Technology
SE-100 44 Stockholm, Sweden
Email: yi.liu@s3.kth.se, jacobsen@s3.kth.se

Abstract: A systematic approach to grey-box identification of distributed parameter processes is proposed. The principle idea is to integrate a novel moving finite element method for model reduction, with the toolbox MoCaVa for grey-box modelling of lumped parameter systems. A method is proposed for separating the model reduction error from discrepancies between model structure and measurement data, based on estimating the model reduction error in combination with simple hypothesis testing. This method can be used for validating the PDE structure as well as the order of the reduced model. To keep the model order at a minimum, the reduced model is augmented with a mesh controller, which uses error feedback to determine the size and location of the finite elements.

1. INTRODUCTION

Grey-box identification concerns calibration and validation of dynamic process models with partially known but uncertain structure, and in the presence of disturbances from the environment. A number of grey-box identification methods and toolboxes have been developed for modelling lumped parameter systems governed by ordinary differential equations (ODE). However, they do in principle not cover distributed parameter system, which are nevertheless very common in the process industries. In order to extend the methods to distributed parameter systems, some form of model reduction, i.e., approximating PDEs with a set of ODEs, is required.

In principle, model reduction can be performed a priori using any discretization method, e.g., finite differences or finite elements, on any given discretization mesh. This is the typical approach, e.g., in software for parameter estimation (gPROMS, 2004). However, as model reduction will introduce a model error, it is in general difficult to a priori determine an appropriate discretization method and model order that will provide an acceptable accuracy. Moreover, it will in general be impossible to decide whether discrepancies between the model structure and the available data are due to the original model structure or the model reduction. Also, by employing an a priori fixed order discretization, valuable information about the original model structure is lost. Since a key-idea in grey-box identification is to exploit information about the underlying model structure, the knowledge that a model structure indeed involves PDEs should therefore somehow be conserved.

Our aim is to propose a systematic method for grey-box identification of distributed parameter processes. The ultimate goal is to integrate a flexible method for model reduction, based on moving finite elements (Liu and Jacobsen, 2001), with a method for grey-box identification as proposed in (Bohlin, 1991) and implemented in the software MoCaVa (Bohlin, 2001). The principle idea is to use the tools available in MoCaVa for calibration and validation also of the model reduction. See Figure 1. In this paper, we consider two specific problems. 1) Separating model reduction error from model-data discrepancies, by combining model reduction error estimation with simple hypothesis testing; 2) Model order selection, i.e., determining the required order of the discretized PDE model for identification purposes. The proposed method is demonstrated on a chromatography process used in bio-separation.

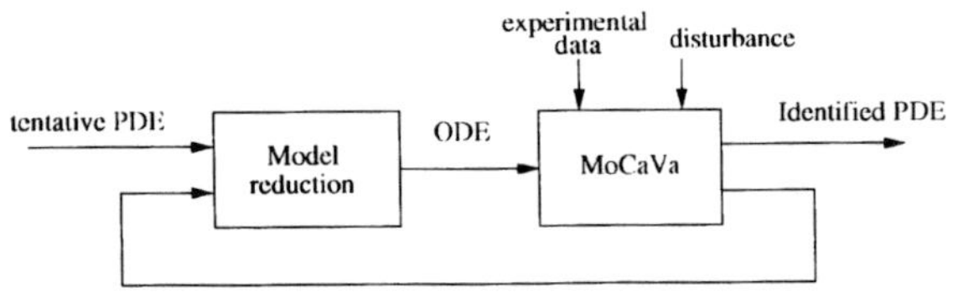

Figure 1. Integration of model reduction with MoCaVa

2. PROBLEM FORMULATION

Model reduction introduces a model error, and an important issue in grey-box identification of PDEs is to therefore distinguish between error sources, as illustrated in Figure 2. There are discrepancies between the tentative PDE model and measurement data (e_1), discretization errors from model reduction of the tentative PDE model (e_2), and discrepancies between the discretized model and the experimental data (e_3). Identification of the PDE models aims at making e_1 small, while the size of e_3 can be obtained when applying identification methods on the discretized model. In principle, a significant error e_3 can be caused by a poor model reduction rather than a poor structure of the tentative PDE model. Contrary, the ODE model may fit the experimental data well although e_1 is large. This may happen if also e_2 is large, and will thus lead to a falsely identified PDE model. Therefore, to ensure a small model error e_1, it is essential to establish whether the error e_3, as estimated by available identification methods, is due to e_1 or e_2.

In general, the problems discussed above cannot be solved by a priori discretization. Moreover, the required accuracy of the model reduction depends not only on the model structure, but also on the experimental data as well as the intended use of the model. For identification purpose, it is usually crucial to use a low order ODE model as the model order directly influences the complexity of the succeeding identification procedure.

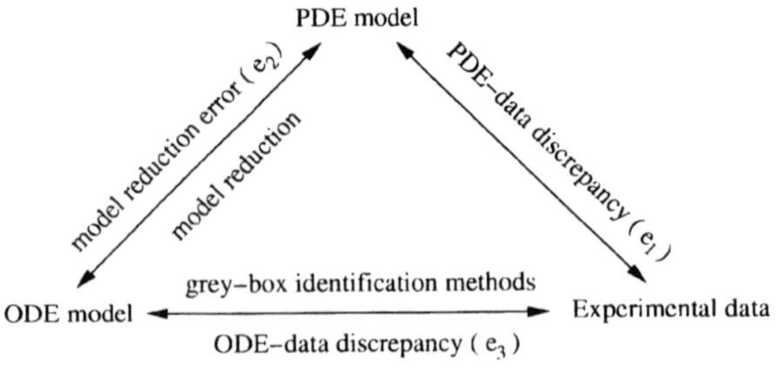

Figure 2. Error sources in grey-box identification of PDEs

2.1 Example: parameter estimation of a convection-diffusion process

To illustrate the potential problems with a priori discretization, consider a simple process involving convection and diffusion governed by

$$v_t = -v_x + 0.1v_{xx}, \quad x \in [0,1], \ t > 0 \qquad (1)$$

where x represents space and t time, with initial and boundary conditions given by

$$v(x,0) = 0, \ \ v(t)|_{x=0} = \sin(bt), \ \ v_x(t)|_{x=1} = 0$$

Treat $u = v(0,t)$ as the input and $y = v(1,t)$ as the output of this system. For simplicity, we assume no process disturbances nor measurement noises. The problem is to identify the system based on measured discrete-time outputs $y(t_k)$. Two tentative PDE models are considered, one in correct form, one neglecting the diffusion term:

$$\mathcal{M}_A: \quad v_t = -av_x + \varepsilon v_{xx} \qquad (2)$$

$$\mathcal{M}_B: \quad v_t = -av_x \qquad (3)$$

First, model reduction is applied to the tentative models. For stability reasons, an upwinding scheme is employed for discretizing the convection term v_x while central difference is used for the diffusion term v_{xx}, both on a fixed uniform mesh. Then, a weighted least square method is employed, i.e., parameter estimates are determined as the minimizer of some cost function V depending on the prediction error, $\hat{\theta} = \arg\min_\theta V(\theta)$. To directly relate the prediction error to the goodness of the fit, we employ a relative error. The cost function is defined as variance of the prediction error, weighted by variance of the simulated outputs (with output mean removed).

$$V(\theta) = \frac{\frac{1}{N_s}\sum_{k=1}^{N_s}[y(t_k) - \bar{y}(t_k,\theta)]^2}{var(\bar{y}) - [E(\bar{y})]^2} \qquad (4)$$

where $\bar{y}(t_k,\theta)$ are simulated outputs of the discretized PDE with the parameter set θ at time t_k, N_s is the number of measured outputs. $var(\bar{y})$ and $E(\bar{y})$ are variance and mean of $\bar{y}$, respectively. For

Test set	Model	Input	N	$\hat{a}$	$\hat{\varepsilon}$	$D(\hat{a})$	$D(\hat{\varepsilon})$	V_{min}	$V_{real}(\hat{a}, \hat{\varepsilon})$
(a)	$\mathcal{M}_{\mathcal{A}}$	$sin(t)$	5	1.0908	0.0162	1.2423e-4	5.9197e-5	1.7840e-6	0.0017
(b)	$\mathcal{M}_{\mathcal{A}}$	$sin(t)$	20	1.0064	0.0815	4.9710e-5	1.1088e-5	3.0927e-7	1.5668e-4
(c)	$\mathcal{M}_{\mathcal{B}}$	$sin(t)$	5	1.1237	-	8.7238e-4	-	8.8801e-5	0.0024
(d)	$\mathcal{M}_{\mathcal{B}}$	$sin(t)$	20	1.1316	-	0.0050	-	0.0030	0.0024
(e)	$\mathcal{M}_{\mathcal{A}}$	$sin(15t)$	20	1.0245	0.0748	0.0011	1.9025e-4	0.0014	1.6480e-4

Table 1. Estimation results for convection-diffusion process. True parameters: a=1, ε=0.1; N: order of the reduced model; $\hat{a}$ and $\hat{\varepsilon}$: parameter estimates; D: standard deviation; V_{min}: cost function with the estimates set, obtained from the discretized model; V_{real}: real cost function with tentative PDE.

the tentative model $\mathcal{M}_{\mathcal{A}}$, $\theta = [a \ \varepsilon]^T$, while for the tentative model $\mathcal{M}_{\mathcal{B}}$, $\theta = a$.

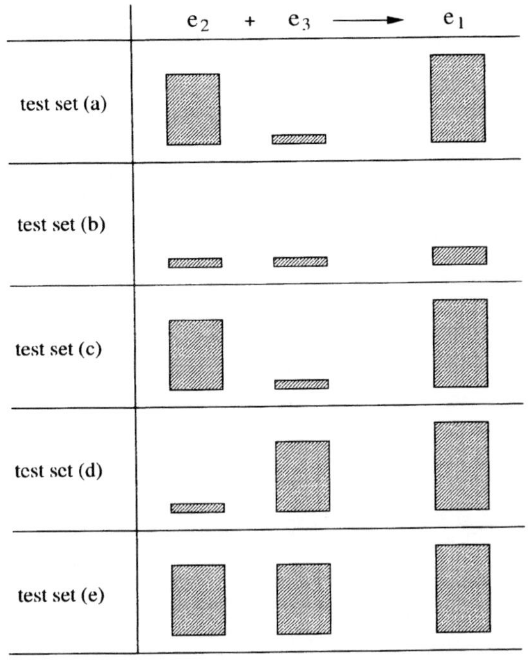

Figure 3. Error sources in different test sets for the illustrating example. The heights of the blocks represent size of the corresponding error.

In all test sets designed below, the input is a sinusoidal wave applied for 5 periods, and the output is recorded with 200 sampling points. The reason for using a simple sinusoid is to enable a simple illustration of the impact of the frequency content in the data. Tests (a) and (b) estimate parameters based on the model structure $\mathcal{M}_{\mathcal{A}}$, with different model orders. Tests (c) and (d) are similar, but for the erroneous model structure $\mathcal{M}_{\mathcal{B}}$. Test (e) is similar to test (b), except the input frequency is increased. A graphical illustration of the size of different error sources for all test sets are presented in Figure 3. The results are summarized in Table 1, from which three essential observations can be obtained. First, when the tentative model has a correct structure, significant biases in the estimates can be caused by a poor model reduction, which can then be improved by increasing the model order, as illustrated by (a) and (b). Note that in (a), although the minimized cost function V_{min} appears small, the *real* cost function, obtained from substituting the estimates into the tentative PDE model and discretizing it on a fine grid, is indeed large. This shows that the obtained parameter estimates involve significant biases even though the data-fitting seems fine. Second, (c) and (d) reveal that when a wrong model structure is used, e_3 may appear small, but it is due to a large e_1 counteracted by a large e_2. In such cases, increasing the model order decreases e_2 but increases e_3. The reason, in this particular example, is that the model reduction introduces an artificial diffusion which plays the role of the diffusion in the true system. Note, however, that this is not possible to detect by examining the standard deviations and the minimum cost function. Third, the required accuracy of the model reduction also depends on the frequency content of the data, as shown by (b) and (e). Due to the space limit, we do not include detailed discussion here but only point out that, for the purpose of identification, the discretized model should primarily approximate the tentative PDE model in the frequency range where the data are concentrated.

2.2 Separation of model reduction errors from model-data discrepancies

From the above, it is clear that in order to properly fit and validate PDE models it is essential to be able to have some knowledge about the model reduction error. If the model reduction error e_2 is guaranteed small, methods for evaluating estimation results applied on ODE systems can be more or less directly applied to PDE systems. In principle, e_2 decays to zero as the model order increases to infinity. However, higher order of the ODE model increases the complexity of the identification. Hence, it is critical to choose a suitable model order for accuracy and efficiency purposes. Below, two approaches to separate e_2 from model-data discrepancies, based on simple hypothesis postulation and testing, are considered.

The first approach starts by discretizing the tentative PDE model using the maximum model order N_{max} that one can practically accept, followed by parameter estimation based on this high order model. Assuming a negligible model reduction error e_2, a minimum cost function V_{min} large than a pre-defined threshold $(V_{min})_{tol}$ indicates the need to improve the model structure. Otherwise, reduce the model order gradually and repeat the parameter estimation, so as to make $V_{min}(N)$ approach the given threshold. The resulting model order is then considered as *"optimal"*, in the sense it provides both acceptable accuracy and efficiency. The main drawback of this approach is that it typically will require excessive computational loads. Further, it is difficult to determine whether the starting maximum model order yields a negligible e_2.

Second, start with a low model order and apply parameter estimation on the discretized model. Then, increase the order N gradually, and perform estimation on each discretized N^{th} order model, so as to find an optimal N such that $V_{min}(N)$ is (slightly) below the given threshold. This approach seems simple but indeed involves some problems. Since e_2 may compensate for the model structure error e_1, V_{min} calculated from the reduced model may not always decrease as N increases. For example, $V_{min}(N)$ may first decrease then increase, for an increasing N. If this happens, it implies that counteraction of significant e_2 and e_1 produces a false minimum V_{min}, and hence the structure of the tentative PDE model may need improvement. To deal with the problem one may, after obtaining the corresponding estimate $\hat{\theta}$ for a given N, re-discretize the PDE model with a high model order N_{max}. Substituting the previously obtained estimate $\hat{\theta}$ into this model, one can compare the resulting loss function $V(\hat{\theta}, N_{max})$ with that obtained for the given N, $V(\hat{\theta}, N)$. If the difference between the two loss functions is large, one can assume that the model reduction error is significant with model order N. This approach is more computationally efficient than the above approach, but still demanding due to the need to consider a large number of parameter estimations as well as high order models for validation of the results.

To summarize, a suitable model order is essential for achieving both accuracy and efficiency in grey-box identification of PDE processes. Since, in principle, the model reduction error asymptotically goes to zero as the model order goes to infinity, it is possible to separate e_2 from the model-data discrepancies by simple hypothesis testing, as discussed above. However, these approaches are computationally inefficient and sometimes involve numerical problems with the use of high model orders. To reduce the complexity, we discuss in the following how the problem can be handled by estimating the model reduction error and combining it with hypothesis testing.

3. A SYSTEMATIC APPROACH TO GREY-BOX MODELLING OF PDES

To enable grey-box modelling of PDEs in a more systematic fashion, we propose to include estimation of the error e_2, introduced by model reduction. The estimated error $\hat{e}_2$ can be used in combination with simple hypothesis testing to separate e_2 from model-data discrepancies. In order to keep the reduced model order close to the minimum required, a moving finite element method named OCMFE (*Orthogonal collocation on moving finite elements*), proposed in (Liu and Jacobsen, 2001), is employed. Available grey-box modelling methods for ODE models can then be applied to the reduced model.

The underlying discretization method employed in OCMFE is *orthogonal collocation on finite elements* (OCFE), proposed in (Carey and Finlayson, 1975). In OCFE, the spatial domain is divided into elements, and the solution is approximated by low order polynomials within each element.

In general, it is not possible to obtain the true model reduction error. For finite elements methods, the equation residuals are often used as measures of the solution accuracy. With the use of OCFE, for instance, residuals are forced to zero on the collocation points, but are in general nonzero at non-collocation points. The integral size of the residuals can hence be used to provide some estimate of the model reduction error e_2. In particular, large residuals will in general imply significant model reduction errors. However, the opposite does not necessarily hold, that is, small residuals does in the general case not guarantee small model errors. In (Liu and Jacobsen, 2001), it is shown that the residuals for linear problems, discretized with OCFE, always will be identically zero while the solution may be arbitrarily in error. To get a more reliable error estimate we consider combining the residual with an error measure often used in polynomial interpolation, based on Nevilles interpolation algorithm (Press *et al.*, 1988). This error measure is based on removing one interpolation point and then performing a new interpolation based on the remaining points. The difference in predicted profile then serves as a measure of the solution error. We combine the residual based and interpolation based error estimates, and denote it Q. By comparing the size of Q to the solution profile, an estimate of the model reduction error is proposed as

$$\hat{e}_2 = \int_t \frac{\int_x Q dx}{\int_x u dx} dt \qquad (5)$$

where the integration in time and space easily can be approximated by a simple quadrature.

Note that the error estimator as defined is data dependent.

Quantitatively, one can not draw any exact conclusion regarding the true error e_2 from the size of $\hat{e}_2$. However, it is reasonable to assume that a decrease in the size of the estimated error implies a decrease also in the true error. This makes the error estimation useful when combining it with simple hypothesis testing.

The method we propose here consist of starting with parameter estimation based on a low-order discretized model, and estimate the resulting model reduction error $\hat{e}_2(N, \hat{\theta})$. Then, increase the model order slightly to N^+ and re-discretize the PDE with the previously obtained $\hat{\theta}$. In general, this should decrease the estimated error i.e., $\hat{e}_2(N^+, \hat{\theta}) < \hat{e}_2(N, \hat{\theta})$. If the reduction is significant, N should be increased and the above procedure repeated until the difference becomes negligible. When so, check if V_{min} is below the pre-defined threshold, and if the standard deviations of the parameter estimates are reasonably low. If both are satisfied, it indicates acceptable estimation results and a suitable model order. Otherwise, we may consider to improve the PDE model structure and/or re-design experiments.

In addition to use the above proposed error estimate for validation of the model reduction, we here also use it to improve the efficiency of the model reduction itself. This is done by augmenting the discretized model with a feedback control law using the element sizes and positions as manipulated variables with the aim of spatially equidistributing the instantaneous discretization error, i.e., the time derivative of $\hat{e}_2$. Equidistribution of error measures is common in numerical methods, and is based on the inherent assumption that it leads to a reduced total error. The method amounts to a novel *moving mesh method*, and is described in (Liu and Jacobsen, 2001).

4. APPLICATION ON A CHROMATOGRAPHY PROCESS

We here apply the systematic approach discussed above on a chromatography process used for bioseparation. Chromatography processes typically possess a sharp and moving composition front and hence represent a challenging problem for finite dimensional modelling. Consider a real process studied by (Borgquist, 1999), and described by

$$\frac{\partial C}{\partial t} = \frac{1}{B_o}\frac{\partial^2 C}{\partial z^2} - \frac{\partial C}{\partial z}, \quad B_o = 1039.5 \qquad (6)$$

with initial and boundary conditions

$$C(z,0) = 0, \quad \left.\frac{\partial C}{\partial z}\right|_{z=0} = B_o(C-1), \quad \left.\frac{\partial C}{\partial z}\right|_{z=1} = 0$$

where C is the adsorbant concentration, z space and t time. $z \in [0,1]$, $t \geq 0$. Consider C at the right boundary, $C(z=1,t)$, as the output of this system, and assume that the tentative model is in correct form. For simplicity we assume noise free measurement data. We consider estimating the parameter B_o using the weighted lest square method presented in (4) above.

Below, four tests are performed using different discretization methods and model orders. The estimation results are summarized in Table 2 while the simulated outputs are plotted in Figure 4. Assume that the threshold for measuring the goodness of data-fitting is chosen as $(V_{min})_{tol} = 10^{-4}$, indicating that a relative prediction error below 0.01% is acceptable. In test (i) and (ii) we discretize the tentative model using central difference, resulting in 30 and 100 ODEs, respectively. Clearly, the parameter estimate is improved by increasing the model order. Both the minimum cost function and the standard deviation of $\hat{B}_o$ are decreased in (ii), compared to (i). However, even with the model order as high as 100, V_{min} from the discretized model is still larger than the threshold $(V_{min})_{tol}$, which hence requires further increase of N. Moreover, it is not trivial to determine a suitable model order unless using involved hypothesis testing.

Test (iii) employs OCFE on 5 elements with a 6th-order polynomial for each element, i.e., a total number of 29 ODEs. Thanks to this more efficient model reduction method, with a similar model order as in (i), the estimation results are apparently improved. Moreover, the proposed error estimation method can now be utilized. As can be seen, compared to the error estimate based on 29 ODEs, an error estimate based on a slightly higher order model is decreased about 30%, indicating that the model reduction needs to be improved.

In test (iv) we employ the moving mesh method OCMFE. The spatial discretization mesh again consists of 5 elements with 6th-order polynomials on each element. As the mesh control contributes 4 ODEs, the total model order now becomes 33. In this case, the difference between the estimated model reduction errors based on the selected model order and a slightly higher model order is much smaller. Compared to (iii), the estimated error $\hat{e}_2$ is reduced by a factor 8, and better estimation results are obtained, as indicated by the decrease of standard deviation of $\hat{B}_o$ and V_{min}. An excellent output fitting is also obtained, and V_{min} is now below the specified threshold. Indeed, the estimated parameter is very close to its true value.

Test set	Model reduction	N	$\hat{B}_o$	$D(\hat{B}_o)$	$V_{min}(N)$	$\hat{e}_2(N, \hat{B}_o)$	$\hat{e}_2(N^+, \hat{B}_o)$
(i)	Central difference	30	600.0	1.5992	0.0101	-	-
(ii)	Central difference	100	1028.1	0.8535	4.3332e-4	-	-
(iii)	OCFE	29	896.2	0.0598	5.4650e-4	0.0126	0.0091
(iv)	OCMFE	33	1043.7	0.0387	9.5380e-5	0.0018	0.0016

Table 2. Parameter estimation of the chromatography process, based on a tentative model $C_t = C_z/B_o - C_{zz}$. True parameter $B_o = 1039.5$; N: order of the discretized model; $\hat{B}_o$: estimate of B_o; D: standard deviation; V_{min}: cost function obtained from the discretized model; $\hat{e}_2(N, \hat{B}_o)$: estimated model reduction error based on N^{th} order discretized model with the parameter estimate $\hat{B}_o$.

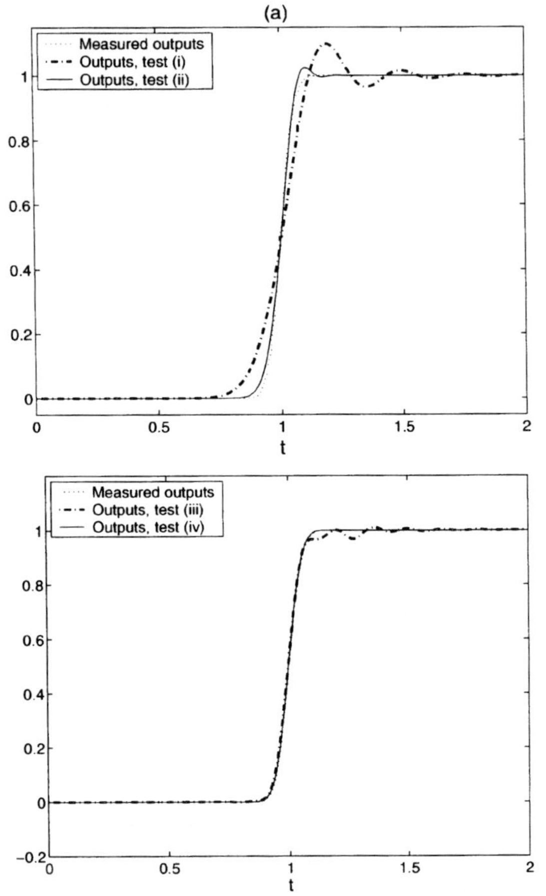

Figure 4. Data fitting for the chromatography process. (a) Measurements and modelled outputs in test (i) and (ii); (b) measurements and modelled outputs in test (iii) and (iv). Note that measured outputs and modelled outputs from test (iv) almost overlap.

5. CONCLUSIONS AND FUTURE WORK

We have considered the problem of grey-box identification of PDE systems in a systematic fashion. We showed that by estimating the model reduction error, the problem of separating this error from model-data discrepancies can be significantly simplified, which hence provides an effective complement to pure hypothesis postulation and testing. By employing the method of OCMFE for mesh control, a low order ODE model is achieved from the discretization of the original PDE model while preserving the model accuracy. This significantly reduces the complexity of the identification procedure. In the future, further problems involved in the integration of OCMFE with MoCaVa will be considered, e.g., the initialization and calibration of model order and control parameter in the model reduction, by utilizing statistical tools provided in MoCaVa.

REFERENCES

Bohlin, T. (1991). Interactive system identification: Prospects and pitfalls. *Springer-Verlag.*

Bohlin, T. (2001). A grey-box process identification tool: Theory and practice. *Technical report, IR-S3-REG-0103, Royal Institute of Technology, Sweden.*

Borgquist, P. (1999). Simulering av kromatografi processer. *Master thesis, TKKA-5005, Lund Institute of Technology, Sweden.*

Carey, G.F. and B.A. Finlayson (1975). Orthogonal collocation on finite elements.. *Chemical Engineering Science,* **30**, 587–596.

gPROMS (2004). *Process Systems Enterprise, London, UK.*

Liu, Y. and E.W. Jacobsen (2001). Effective model reduction for analysis of distributed parameter systems. *Proceedings EFChE ESCAPE11.* Kolding, Denmark.

Press, W.H., B.P. Flannery, S.A. Teukolsky and W.T. Vetterling (1988). Numerical recipes in C: The art of scientific computing. *Cambridge university Press* p. Ch3.1.

ELSEVIER
IFAC
PUBLICATIONS
www.elsevier.com/locate/ifac

ON-BOARD DIAGNOSTIC AND FAULT DETECTION STRATEGIES FOR AN AUTOMOTIVE THREE-WAY CATALYST

Kenneth R. Muske * **James C. Peyton Jones** **

* *Department of Chemical Engineering, Villanova University, Villanova, PA 19085*
** *Department of Electrical and Computer Engineering, Villanova University, Villanova, PA 19085*

Abstract: This work presents model-based strategies for on-board catalyst monitoring and fault detection using a simplified catalyst oxygen storage and reversible deactivation model. The performance of the catalyst is inferred from the error between the post-catalyst exhaust gas oxygen sensor measurement and the model predicted value. Each of these strategies is based on the use of some test statistic that is computed from a window of post-catalyst prediction error data updated in real-time. A fault is reported when the value of this statistic exceeds a threshold that is determined from some specified confidence level. We conclude with an evaluation of these strategies. *Copyright © 2004 IFAC*

Keywords: Automotive Catalyst Monitoring, Automotive On-board Diagnostic, Moving Horizon Monitoring

1. INTRODUCTION

In order to ensure that the components of a catalytic automotive emission control system continue to operate properly with age, effective and robust on-board system monitoring is essential. Because emissions from a small fraction of malfunctioning vehicles are believed to account for a large fraction of the total automotive-related emissions, on-board monitoring has the potential to significantly reduce hydrocarbon and nitrogen oxide air pollutants. This observation, along with recent legislation such as SULEV and OBD-II that mandate on-board diagnostic (OBD) systems to monitor the health and performance of the catalyst system, has led to interest in monitoring strategies that are able to reliably indicate when the emission control system is no longer meeting specification or when a fault is present.

On-board monitoring of the three-way catalyst component of catalytic automotive emission control systems is considered in this work. The specifications for catalyst operation are set based on the tailpipe hydrocarbon emissions. Under European legislation, these emissions must remain less than 0.4g/km during a specified test cycle. In the United States, OBD-II legislation requires hydrocarbon emissions to remain less than 1.5 times the 4000 mile emissions baseline. These requirements are particularly challenging due to the lack of a cost-effective and reliable automotive hydrocarbon sensor. This limitation has led to monitoring approaches that consider the oxygen storage dynamics of the catalyst rather than the hydrocarbon conversion efficiency. Although the oxygen storage capacity of the catalyst can not be measured directly, it can be inferred using pre- and post-catalyst exhaust gas oxygen (EGO) sensors. These sensors are often available as part of the engine air fuel ratio (AFR) control system making these monitoring approaches attractive.

2. CATALYST OXYGEN STORAGE

A key process in the reduction of tailpipe emissions by the three-way catalyst is the ability to store and release oxygen in response to the pre-catalyst exhaust air fuel ratio. Under rich (excess fuel) engine operation, the catalyst oxidizes the hydrocarbons and carbon monoxide present in the incoming engine exhaust by releasing previously stored oxygen. This oxygen release maintains stoichiometric combustion with commensurately low levels of hydrocarbon and carbon monoxide tailpipe emissions. Because of the finite storage capacity of the catalyst, however, this process cannot continue indefinitely. When the oxygen release rate of the depleted catalyst can no longer satisfy the demand, the post-catalyst air fuel ratio will decrease below stoichiometric and hydrocarbon breakthrough will eventually occur. A typical catalyst control system will therefore attempt to switch to lean (excess air) engine operation before this rich breakthrough condition is encountered. Under lean engine operation, the excess oxygen in the engine exhaust gas is now adsorbed onto the catalyst resulting in near-stoichiometric post-catalyst conditions and low tail-pipe emissions. As the oxygen storage capacity of the catalyst approaches its saturation condition, however, the post-catalyst oxygen concentration increases above stoichiometric and emission of nitrogen oxides will eventually occur. A typical catalyst control system will then attempt to switch back to rich engine operation before lean breakthrough. Through this cycling of the pre-catalyst air fuel ratio, the oxygen storage capacity of the catalyst is a buffer against tailpipe emissions by compensating for transient oxygen excess or deficiency.

Because of the nature of the three-way catalyst operation, a measure of the catalyst oxygen storage capacity would be expected to correlate well with the hydrocarbon conversion efficiency. In practice, however, the correlation is nonstationary and highly nonlinear resulting from differences in both the short-term and long-term dynamics between oxygen storage and hydrocarbon conversion. A further difficulty in the use of oxygen storage capacity for catalyst monitoring lies in the use of the post-catalyst EGO sensor. The sensor signal is not only a function of the exhaust gas air fuel ratio but also a function of the exhaust gas composition. In pre-catalyst applications, the dependency on gas composition can be largely removed by calibration. The composition of the post-catalyst exhaust, however, varies dynamically according to the reactions taking place on the catalyst. Static calibration in this case is insufficient to correct for composition dependent distortion. The resulting error is difficult to distinguish from oxygen storage and release effects and, unless treated explicitly, will propagate to the catalyst monitoring strategy.

Despite these difficulties, oxygen storage capacity (OSC) metrics for on-board monitoring have been proposed based on the assumption that the oxygen storage rate can be estimated from the difference between the pre-catalyst and post-catalyst EGO sensor measurements of the air fuel ratio. In order to measure catalyst OSC, the oxygen storage rate must be integrated over an appropriate time interval during which the catalyst oxygen storage state changes from an initial to a final state. It is the choice of the initial and final state of the integration that distinguishes these methods. The repeatability of the metric also depends on the accuracy with which the catalyst can be driven to the initial state and the determination of the point at which the catalyst has reached the final state. Measurement of these conditions is not straightforward because the level of stored oxygen cannot be observed directly.

A summary of these OSC metrics, along with the requirements on the engine and catalyst operation necessary to determine their value, is presented by Peyton Jones and Muske (2004). An inherent difficulty with diagnostic strategies based on these metrics is that they seek to encapsulate catalyst behavior using a single oxygen storage capacity determination method that is assumed indicative of catalyst health. Because catalyst dynamics are sufficiently complex, it is unlikely that such a metric can adequately describe hydrocarbon conversion efficiency for a variety of operating conditions. For this reason, model-based strategies are likely to be more accurate and discriminating.

A simple model-based monitoring strategy is to use the error between a model predicted and measured OSC metric as the catalyst diagnostic. When the catalyst is healthy, the model prediction and actual measurement should be in close agreement resulting in small prediction errors. As the catalyst performance degrades, the magnitude of these errors would be expected to increase. The effectiveness of this approach, however, depends on the model and the accuracy of both the measurement and the model prediction of the metric. A second consideration is the engine and catalyst operation that is necessary in order to provide sufficient information on the catalyst performance.

3. OXYGEN STORAGE MODEL

The oxygen storage model of Peyton Jones et al. (2000) is used in the monitoring strategies presented in this work. The catalyst oxygen storage is represented by the nonlinear integrating model

$$\dot{\phi} = K_{\lambda}\dot{m}_f\,(\Delta\lambda^{\triangleleft} - \mathcal{N}(\phi)) \qquad (1)$$

in which ϕ is the oxygen storage state of the catalyst, $\Delta\lambda^{\triangleleft}$ is the pre-catalyst air fuel ratio deviation from stoichiometric, $\dot{m}_f$ is the fuel mass flow rate to the engine, and $\mathcal{N}(\phi)$ is the catalyst oxygen capacity function. Here oxygen storage and release rates depend on the difference between the forcing function $\Delta\lambda^{\triangleleft}$, which promotes adsorption under conditions of oxygen excess, and the catalyst capacity $\mathcal{N}(\phi)$, which represents the desorption under similar conditions. This model describes stored oxygen relative to the equilibrium level when the pre-catalyst exhaust gas is stoichiometric, $\Delta\lambda^{\triangleleft} = 0$. We note that the pre-catalyst air fuel ratio deviation, $\Delta\lambda^{\triangleleft}$, is denoted using a left-facing triangle and that the post-catalyst air fuel ratio deviation, $\Delta\lambda^{\triangleright}$, is denoted using a right-facing triangle in this work.

The capacity function $\mathcal{N}(\phi)$ has a nonlinear spring characteristic. As ϕ increases from zero, it becomes progressively harder to store oxygen on the catalyst. As ϕ decreases from zero, it becomes progressively harder to remove oxygen from the catalyst. In this work, $\mathcal{N}(\phi)$ is parameterized using a polynomial expansion of ϕ.

$$\mathcal{N}(\phi) = a_1\phi + a_2\phi^2 + a_3\phi^3 + a_4\phi^4 + a_5\phi^5 \quad (2)$$

We note that the stored oxygen level does not hit hard saturation/depletion limits as in limited integrator models but instead attains an operating condition that is dependent on the steady-state pre-catalyst air fuel ratio. The oxygen saturation/depletion limits are approached asymptotically as would be expected from theory.

The post-catalyst air fuel ratio deviation from stoichiometric, $\Delta\lambda^{\triangleright}$, is determined from the capacity function by the following relationship.

$$\Delta\lambda^{\triangleright} = \begin{cases} 0, & (\Delta\lambda^{\triangleleft} < 0) \text{ and } (\phi > 0) \\ \mathcal{N}(\phi), & \text{otherwise} \end{cases} \quad (3)$$

In the case of rich pre-catalyst exhaust gas, $\Delta\lambda^{\triangleleft} < 0$, and an oxidized catalyst, $\phi > 0$, where stored oxygen is available for reducing the rich incoming exhaust gas, oxygen release is limited by the exhaust gas oxygen demand resulting in stoichiometric post-catalyst exhaust, $\Delta\lambda^{\triangleright} = 0$. This effect is clearly shown in Figure 1 where the post-catalyst air fuel ratio remains at stoichiometric for several seconds after the pre-catalyst air fuel ratio makes a lean to rich step change. As the stored oxygen becomes depleted from the catalyst, the post-catalyst air fuel ratio becomes rich until it eventually matches the pre-catalyst ratio. Oxygen storage on the catalyst is clearly shown after the pre-catalyst air fuel ratio rich to lean step change made at approximately seventy seconds.

The separation between the actual and model predicted air fuel ratios appearing after twenty seconds in Figure 1 is due to sensor distortion in the post-catalyst air fuel ratio sensor. As shown by Peyton Jones (2003), post-catalyst EGO sensor distortion under rich conditions can be related to

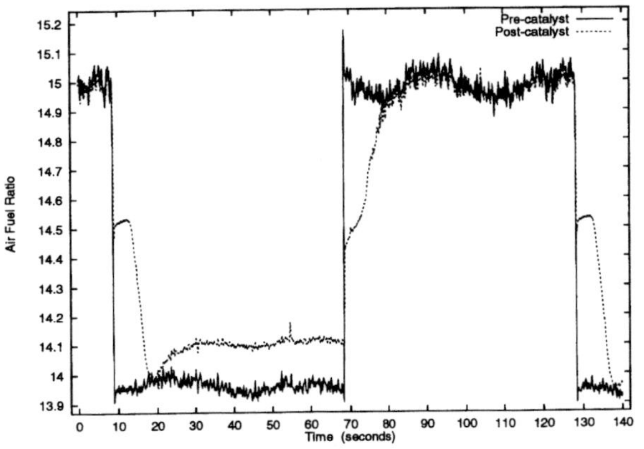

Fig. 1. Pre- and post-catalyst air fuel ratio.

the hydrogen generated by catalytic promotion of the water gas shift reaction.

$$CO + H_2O \rightleftharpoons CO_2 + H_2$$

If the reaction proceeds in the forward direction, the ability of hydrogen to diffuse faster than hydrocarbons and carbon monoxide results in a sensor output that appears richer than the true value. Reduced levels of post-catalyst hydrogen due to a progressive inhibition of the water gas shift reaction result in a sensor output that appears leaner that the true value. The rise toward leaner values of the measured post-catalyst air fuel ratio in Figure 1 is due to this inhibition effect.

The reversible short-term catalyst deactivation effects that inhibit hydrogen production through the water gas shift reaction can also have an effect on conversion efficiency particularly under rich or rich-biased operation. This behavior suggests that the gas-composition dependent error in the post-catalyst air fuel ratio sensor measurement can be used as a measure of this reversible catalyst deactivation effect. If the degree of water gas shift reaction inhibition is assumed proportional to the deactivated fraction of the catalyst surface, ψ, then the apparent post-catalyst AFR, $\Delta\lambda_a^{\triangleright}$, can be related to the true post-catalyst AFR, $\Delta\lambda^{\triangleright}$, as

$$\Delta\lambda_a^{\triangleright} = \Delta\lambda^{\triangleright} + K_H\psi \quad (4)$$

where the constant K_H represents the sensor sensitivity to hydrogen concentration changes and the inhibition of the water gas shift reaction due to reversible catalyst deactivation.

Assuming that the rate of deactivation is proportional to the post-catalyst oxygen deficiency, $-\Delta\lambda^{\triangleright}$, and the fraction of the surface that is already deactivated, ψ, results in the following model for the deactivated fraction of the catalyst surface

$$\dot{\psi}_d = \dot{m}_f K_d(-\Delta\lambda^{\triangleright} - \psi) \quad (5)$$

where K_d is the deactivation constant of proportionality. The presence of pre-catalyst free oxygen, $\Delta\lambda^{\triangleleft} > 0$, reverses the deactivation process at a rate proportional to the supply of pre-catalyst oxygen until the catalyst is reactivated, $\psi = 0$,

$$\dot{\psi}_r = -\dot{m}_f K_r \Delta\lambda^{\triangleleft} \quad (6)$$

where K_r represents the reactivation constant. The resulting reversible deactivation model is

$$\dot{\psi} = \begin{cases} \dot{\psi}_d, & (\Delta\lambda^{\triangleleft} < 0) \text{ and } (\Delta\lambda^{\triangleright} < 0) \\ \dot{\psi}_r, & (\Delta\lambda^{\triangleleft} > 0) \text{ and } (\psi > 0) \\ 0, & \text{otherwise} \end{cases} \quad (7)$$

where $\Delta\lambda^{\triangleright}$ is the true post-catalyst AFR determined from the oxygen storage model in Eq. 3 and the sensor distortion of the measured post-catalyst AFR is determined from Eq. 4.

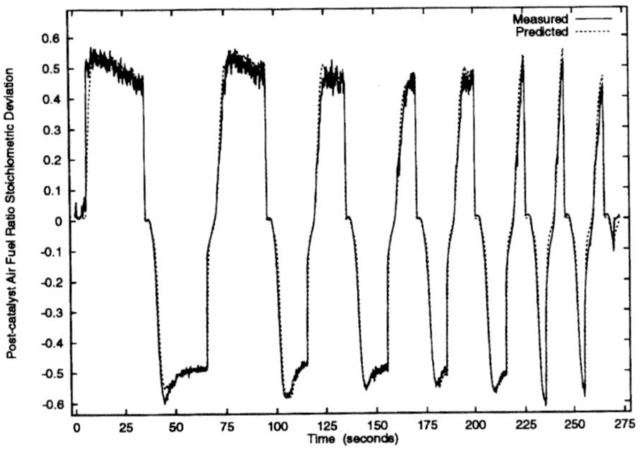

Fig. 2. Post-catalyst AFR model prediction.

Figure 2 presents a comparison between the measured post-catalyst air fuel ratio stoichiometric deviation and the combined oxygen storage-reversible catalyst deactivation rate model prediction for a series of step changes to the pre-catalyst air fuel ratio. As shown in this figure, the deviations at the end of each rich cycle are accounted for in the combined model. In particular, the combined model now characterizes the distortion that occurs in the post-catalyst sensor since there are no longer significant unmodeled dynamics. The close agreement between the predicted and actual measurements, especially during the periods of rich operation, demonstrates that the reversible catalyst deactivation dynamics are well described by this model.

4. STATISTICAL MODEL-BASED MONITORING STRATEGIES

In order to account for the uncertainty in both the measured and model predicted OSC metric, analysis of the prediction error is best carried out within a statistical framework. Previous work in statistical model-based monitoring strategies considered the Neyman-Pearson criterion to obtain a diagnosis at a given level of confidence. Analysis of the difference between the pre- and post-catalyst AFR measurements is considered by Arsie et al. (2000). The student t-test applied to multiple samples of the time to empty/fill the catalyst is presented by Brandt and Grizzle (2001). The disadvantage of these methods is the Gaussian distribution assumption for the metric which is not necessarily the case, especially given the biasing effects of sensor distortion. Catalyst deactivation effects may also change the distribution over time.

In this work, we consider the prediction error of the post-catalyst AFR as the basis for OSC metrics. The prediction error is easily computed and sensitive to both long-term catalyst deactivation effects and short-term emission control system device failures. It does not require catalyst operation that can result in increased tailpipe emissions or disruption in driveability as is possible during the direct determination of oxygen storage capacity. The monitoring strategies are based on a sample of the post-catalyst AFR prediction errors

$$\boldsymbol{e} = [e_1, e_2, \ldots, e_N]^T \quad (8)$$

where N is the sample population size, $e_k = \Delta\lambda^{\triangleright}_{m,k} - \Delta\lambda^{\triangleright}_{a,k}$ is the prediction error, $\Delta\lambda^{\triangleright}_{m,k}$ is the measured post-catalyst AFR, and $\Delta\lambda^{\triangleright}_{a,k}$ is the predicted post-catalyst AFR computed from Eq. 4 for samples $k = 1, \ldots, N$. The sample may be constructed using either a moving window or a batch approach.

We begin our discussion of monitoring strategies by analyzing the distribution of the post-catalyst AFR prediction errors for the engine operation in Figure 2. This distribution, shown in Figure 3, is nonsymmetric with a greater proportion of rich prediction errors but with larger lean prediction errors on average.

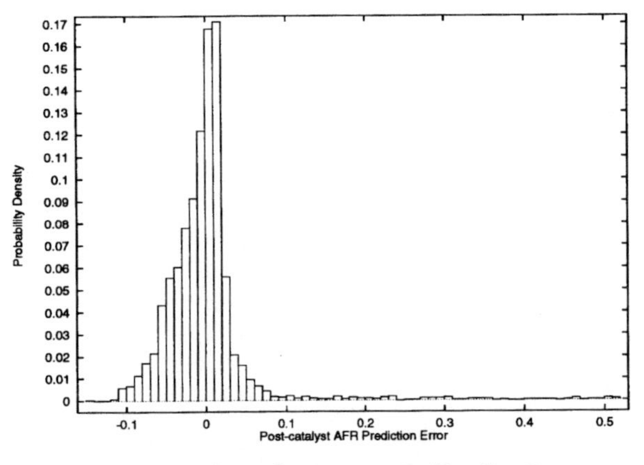

Fig. 3. Nominal prediction error distribution.

In order to estimate catalyst deactivation effects, the capacity function in Eq. 2 is compressed as

$$\tilde{N}(\phi) = \frac{a_1}{\beta}\phi + \frac{a_2}{\beta^2}\phi^2 + \frac{a_3}{\beta^3}\phi^3 + \frac{a_4}{\beta^4}\phi^4 + \frac{a_5}{\beta^5}\phi^5 \quad (9)$$

where $0 < \beta \leq 1$ is the fraction of oxygen storage capacity remaining. Simulated post-catalyst AFR operating data for a deactivated catalyst is constructed by computing the model predicted post-catalyst AFR using the modified catalyst capacity

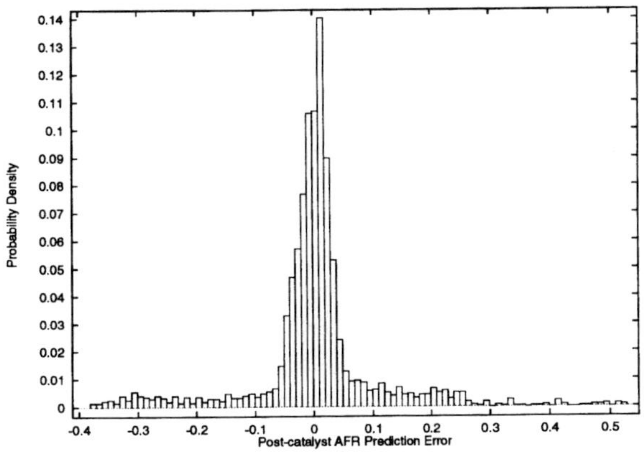

Fig. 4. Deactivated catalyst error distribution.

function in Eq. 9 with the same pre-catalyst conditions as Figure 2. The model prediction errors from Figure 2 are then inserted into this simulated data set. We choose these errors, as opposed to the addition of white noise, because the model prediction errors do show some correlation with the pre-catalyst AFR and we wish to retain this property. Figure 4 presents the prediction error distribution between the simulated deactivated operation with $\beta = 0.8$ and the model prediction with $\beta = 1$. As shown in this figure, there is a significant deviation from the nominal operation.

4.1 Mean Prediction Error Metric

In this section, we consider a statistical test on the mean of the population of prediction errors in the current sample as discussed by Christensen (1996). Provided that the sample population is large, a standard normal Gaussian distribution can be reasonably assumed for the test statistic

$$\mathcal{Z} = \frac{\bar{e}_{\mathcal{Y}} - \bar{e}_{\mathcal{X}}}{s_{\mathcal{Y}}^2/N_{\mathcal{Y}} + s_{\mathcal{X}}^2/N_{\mathcal{X}}} \tag{10}$$

in which $e_{\mathcal{Y}}$ is the prediction error sample drawn from the current operation with mean $\bar{e}_{\mathcal{Y}}$, variance $s_{\mathcal{Y}}^2$, and sample size $N_{\mathcal{Y}}$ and $e_{\mathcal{X}}$ is the nominal operation prediction error sample with mean $\bar{e}_{\mathcal{X}}$, variance $s_{\mathcal{X}}^2$, and sample size $N_{\mathcal{X}}$. The null hypothesis is $\bar{e}_{\mathcal{Y}} = \bar{e}_{\mathcal{X}}$ which indicates normal catalyst operation and is accepted if the test statistic $\mathcal{Z}$ is less than the threshold $\mathcal{T}_{\mathcal{Z}}(\alpha)$ chosen based on a significance level α. The alternative hypothesis is $\bar{e}_{\mathcal{Y}} > \bar{e}_{\mathcal{X}}$ which indicates improper catalyst operation. Because recursive estimates of the prediction error and its variance are easily computed, large sample populations are feasible which both reduces the sensitivity to outliers and improves the Gaussian distribution assumption made for $\mathcal{Z}$.

The symmetry of the deactivated distribution in Figure 4 suggests that the mean prediction error may not be an effective metric and, in this application, the mean absolute or squared error would be a better choice. Table 1 presents the value of the test statistic $\mathcal{Z}$ using each of these metrics for varying levels of catalyst oxygen storage capacity. A sample size of $N = 1000$ was used for each population where the samples for $e_{\mathcal{X}}$ were obtained by random sampling of the prediction errors from the operation in Figure 2 and the samples for $e_{\mathcal{Y}}$ were obtained by random sampling of the simulated deactivated operating data with varying values of β. As shown in this table, the metric based on mean absolute error is quite sensitive to small reductions in the oxygen storage capacity. Note that the threshold value is $\mathcal{T}_{\mathcal{Z}}(0.025) = 1.96$ for a 0.025 level of significance.

Storage	*Test Statistic*		$\mathcal{Z}$
Capacity	$\bar{e}$	$\lvert e \rvert$	e^2
95%	0.55	2.42	1.21
90%	1.00	6.38	3.92
85%	1.33	10.6	7.64
80%	1.56	14.3	11.4
75%	1.69	17.2	14.5
70%	1.78	19.6	17.0
65%	1.83	21.7	19.0
60%	1.86	23.6	20.6
55%	1.90	25.3	22.2
50%	1.94	27.0	23.7

Table 1. Test statistic for mean error.

4.2 Prediction Error Distribution Metric

The monitoring strategy employed in this section considers the distribution of the prediction errors in the samples. The use of the error distribution, as opposed to the mean, can increase sensitivity in the presence of model inadequacy. Because of the simplicity of the catalyst model, large model errors may occur even when the catalyst system is functioning properly. The error magnitude may also not change significantly when a fault is present. In this case, changes in the distribution of the errors can provide additional information on the catalyst operation. The Kolmogorov–Smirnov test, a nonparametric test for comparing cumulative distribution functions discussed by Conover (1980), is used. Although no assumptions about the form of the underlying distribution are necessary, this test does assume that the sampling is random from a continuously distributed population with sufficient measurement precision to avoid tied observations. These assumptions are well satisfied for catalyst system monitoring.

The cumulative distribution function (*cdf*) is the mapping $F_{\mathcal{X}}(z)$ that assigns the probability that a random variable $\mathcal{X}$ is less than or equal to a specific value z for all possible z values.

$$F_{\mathcal{X}}(z) = \mathcal{P}\{\mathcal{X} \leq z\}, \quad -\infty < z < \infty$$

The nominal catalyst operation prediction error is denoted by the random variable $\mathcal{X}$ and the current operation prediction error by the random variable

$\mathcal{Y}$. There are three test statistics that can be used with the Kolmogorov–Smirnov test

$$\mathcal{K} = \sup_{-\infty<z<\infty} |F_{\mathcal{Y}}(z) - F_{\mathcal{X}}(z)| \quad (11)$$

$$\mathcal{K}^+ = \sup_{-\infty<z<\infty} \left(F_{\mathcal{Y}}(z) - F_{\mathcal{X}}(z)\right) \quad (12)$$

$$\mathcal{K}^- = \sup_{-\infty<z<\infty} -\left(F_{\mathcal{Y}}(z) - F_{\mathcal{X}}(z)\right) \quad (13)$$

where $F_{\mathcal{Y}}(z)$ is the *cdf* for the current operation prediction error and $F_{\mathcal{X}}(z)$ is the *cdf* for the nominal operation prediction error. The $\mathcal{K}$ statistic in Eq. 11 is the maximum difference between the two distributions. The null hypothesis is $F_{\mathcal{Y}}(z) = F_{\mathcal{X}}(z)$ for all z indicating normal operation and is accepted if the test statistic $\mathcal{K}$ is less than the threshold $\mathcal{T}_{\mathcal{K}}(\alpha)$ chosen based on a significance level α. The alternative hypothesis is $F_{\mathcal{Y}}(z) \neq F_{\mathcal{X}}(z)$ for at least one value of z. The null hypothesis for the $\mathcal{K}^+$ statistic in Eq. 12 is $F_{\mathcal{Y}}(z) \leq F_{\mathcal{X}}(z)$ with the alternative hypothesis $F_{\mathcal{Y}}(z) > F_{\mathcal{X}}(z)$ for at least one value of z. The null hypothesis for the $\mathcal{K}^-$ statistic in Eq. 13 is $F_{\mathcal{Y}}(z) \geq F_{\mathcal{X}}(z)$ with the alternative hypothesis $F_{\mathcal{Y}}(z) < F_{\mathcal{X}}(z)$ for at least one value of z. The threshold values for each statistic are available as a function of the significance level and number of samples in each population. The *cdf* for both the nominal and current operation is approximated by a normalized histogram of the prediction errors in the corresponding sample. The normalized histogram is computed by taking the proportion of the sample population less than or equal to a specified prediction error. These approximate distribution functions approach the true distribution only in the asymptotic limit as the number of samples goes to infinity when the true underlying distribution is continuous. For large sample size, however, the density function based on a finite sample should be a reasonable approximation to the asymptotic density function. Because the calculations required to perform the test are relatively trivial, each of the three statistics in Eqs. 11 through 13 can be tested. The requirement of computing normalized histograms, however, will limit the sample population sizes in practice.

Table 2 presents average values of the test statistics $\mathcal{K}$, $\mathcal{K}^+$, and $\mathcal{K}^-$ for varying levels of catalyst oxygen storage capacity. The nominal error sample $e_{\mathcal{X}}$ was obtained by random sampling of the prediction errors from the operation in Figure 2. For each value of β, several current operation samples $e_{\mathcal{Y}}$ were obtained by random sampling of simulated deactivated operating data. A sample size of $N = 100$ was used for each population and the average test statistic values were determined using 1000 tests. The threshold value for $\mathcal{K}$ is $\mathcal{T}_{\mathcal{K}}(0.025) = 0.121$ for a 0.025 level of significance. Because of the smaller sample size, this test is not as sensitive as the mean absolute error in this example. When $\beta = 0.9$, the test statistic average is just below the threshold with approximately half of the samples above the threshold value.

Storage	*Test Statistic*		
Capacity	$\mathcal{K}$	$\mathcal{K}^+$	$\mathcal{K}^-$
100%	0.058	0.041	0.050
95%	0.093	0.091	0.080
90%	0.115	0.108	0.081
85%	0.138	0.131	0.093
80%	0.164	0.159	0.094
75%	0.172	0.169	0.096
70%	0.208	0.208	0.099
65%	0.228	0.228	0.115
60%	0.247	0.247	0.117
55%	0.256	0.256	0.122
50%	0.274	0.274	0.134

Table 2. Kolmogorov–Smirnov statistic.

5. CONCLUSIONS

A model-based automotive catalyst diagnostic and fault detection methodology based on the error between the measured and model predicted post-catalyst air fuel ratio is developed. Hypothesis tests using the mean absolute error and the Kolmogorov–Smirnov nonparametric test for cumulative distribution functions are used to determine statistically significant changes between the current and nominal catalyst operation. This methodology is demonstrated using engine data with simulated oxygen storage capacity reduction and shows promise for on-board application.

ACKNOWLEDGMENTS

National Science Foundation support from grant CTS-0215920 is gratefully acknowledged.

REFERENCES

E. Brandt and J. Grizzle. Three-way catalyst diagnostics for advanced emissions control systems. In *Proc. of the 2001 American Control Conference*, pages 3305–3311, 2001.

R. Christensen. *Analysis of Variance, Design and Regression.* Chapman & Hall, London, 1996.

W. Conover. *Practical Nonparametric Statistics.* Wiley, New York, 2nd edition, 1980.

I. Arsie, et al. Confidence level analysis for on-board estimation of SI engine catalytic converter efficiency. In *Proc. of the SYSID 2000 IFAC Symposium on System Identification*, pages 137–142, 2000.

J. Peyton Jones, et al. A simplified model for the dynamics of three-way catalytic converter. Paper 2000-01-0652, SAE World Congress, 2000.

J. Peyton Jones. Modeling combined oxygen storage and reversible deactivation dynamics for improved emissions predictions. Paper 2003-01-0999, SAE World Congress, 2003.

J. Peyton Jones and K. Muske. Model-based OBD for three-way catalyst systems. Paper 2004-01-0639, SAE World Congress, 2004.

ELSEVIER
IFAC
PUBLICATIONS
www.elsevier.com/locate/ifac

INDUSTRIAL EXPERIENCE ON PROCESS TRANSITION MONITORING FOR CONTINUOUS STEEL CASTING OPERATION

Yale Zhang* and Michael S. Dudzic

Process Automation, Dofasco Inc.
Box 2460, Hamilton, Ontario, Canada, L8N 3J5

Abstract: Process transitions are common in the iron and steel industry. Our investigation shows that more than 50% of catastrophic process failures in continuous steel casting operation are related to abnormal operations during the process transition period. As a multivariate statistical (MVS) analysis tool, Multi-way PCA (MPCA) is applied in this paper to monitor one important process transition in the continuous casting process: submerged entry nozzle (SEN) change. A novel scheme is proposed for synchronizing process trajectories over the SEN change and the missing data existing in the synchronized trajectories are handled subsequently in both modeling and monitoring parts. The monitoring results are demonstrated by an industrial example. It is shown to provide good detectability of various process abnormalities. The proposed scheme can be further extended to monitor other process transitions in continuous casting process such as flying tundish change and product grade change. *Copyright © 2004 IFAC*

Keywords: process transition; multivariate statistics; process monitoring; continuous caster; breakout.

1. INTRODUCTION

There are many process transitions in the iron and steel industry. Some common examples include process start-up, shutdown, product grade change and so forth. Since process transitions are not operating in a steady state, they are extremely difficult to control. During these transition periods, some process sensors or automatic control systems may be disabled which require operators taking a large amount of manual control actions. They are required to follow a series of predetermined standard operating procedures (SOP) to ensure the process stays within the safe operating zone. Any lapse from such procedures may lead to accidents or other process abnormalities. Our plant experiences show that catastrophic process failures often occur in the transition period, which not only create substantial profit loss due to non-productivity but also pose a significant safety risk to operation personnel. Therefore it is crucial to monitor process transitions in real-time to avert such abnormalities.

With the rapid development in advanced instrumentation technology and data acquisition systems, steel industries are currently storing massive volumes of plant data. To uncover useful process knowledge, MVS technology, including principal component analysis (PCA) and projection to latent structures (PLS), has been widely used for modeling and monitoring of complex industrial processes. This technology was introduced to the iron and steel industry in the mid 1990's and several online applications have been successfully implemented in various process units (Zhang, *et al.*, 2003a and Quinn, *et al.*, 2003).

Although there have been considerable efforts for applying MVS technology for process modeling and monitoring, most applications are limited to continuous, steady-state processes. Modeling and monitoring of process transitions have received very little attention. Some recent work in this area include: (1) Duchesne *et al.* (2002) reported a PLS-based transition monitoring scheme applied to re-start process of an industrial polymer reactor. (2) Zhang *et al.* (2003b) developed a real-time monitoring system for caster start-up operation in order to prevent start-cast breakouts. This system has been implemented at Dofasco's No.2 continuous caster and demonstrated very good predictability to start-cast breakouts. In this paper, our previous work has been further extended to monitor another important process transition in the continuous casting operation: SEN change where a novel scheme is proposed for synchronizing process trajectories over the SEN change and the missing data existing in the synchronized trajectories are handled subsequently in both modeling and monitoring parts. The monitoring results are demonstrated by one industrial example from Dofasco's No.2 continuous caster.

* Author to whom correspondence should be addressed. Tel.: +1(905)548-7200 ext. 3888. E-mail: yale_zhang@dofasco.ca.

2. MOTIVATION AND PROBLEM DESCRIPTION

Continuous casting is a key process in the steel-making industry whereby molten steel is solidified to produce solid steel slabs for subsequent rolling operations. A typical slab casting process is illustrated in Fig. 1. In this process, molten steel is poured into a tundish and then into a water-cooled copper mold through a device at the bottom of tundish to allow the flow of steel into the mold. This device is a submerged entry nozzle (SEN) with a stopper-rod that opens and closes to regulate the flow of steel into the mould to provide precise control of the steel level in the mould. Within the mold, the outer shell of the steel becomes solidified to form a steel strand with the liquid steel inside. The solidifying strand is continuously withdrawn from the mold into additional cooling chambers to promote solidification. Once the strand is fully solidified, it is cut to the steel slabs in various lengths to meet downstream product specifications.

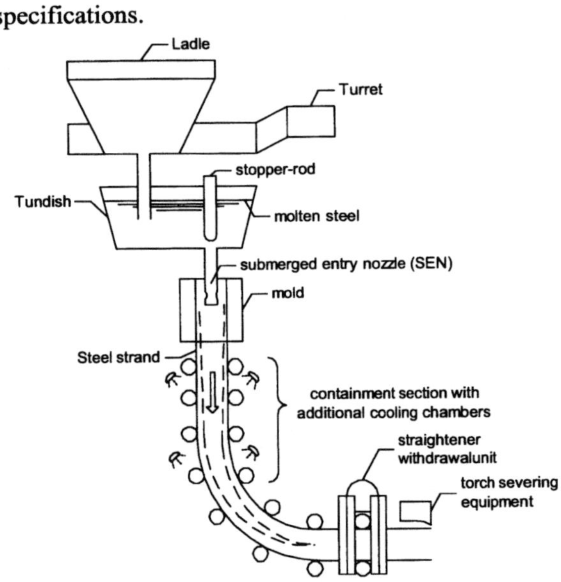

Fig. 1. Schematic diagram of continuous caster.

A typical sequence of a casting operation consists of a brief start-up operation (this is also referred to as start-cast), followed by a prolonged continuous, run-time production operation, and then a shutdown operation. Three types of process transitions in continuous casting operation are defined as follows: (1) start-up operation; (2) shutdown operation; and (3) special transitions during the continuous, run-time production operation, including SEN change and flying tundish change (FTC). These special transitions are critical in ensuring a long continuous, run-time operation. However, the nature of these transitions makes them particularly sensitive to process failures.

One of the most catastrophic process failures in the continuous casting operation is a *breakout*, which occurs when the solidifying strand shell ruptures or tears, leading to the molten steel pouring out beneath the mold. The direct economic loss of one breakout can be significant due to a large amount of process downtime, and the substantial cost associated with equipment damage. More seriously, breakouts may pose significant safety hazards to plant operators. Breakouts can be avoided by reducing the casting speed thereby providing more residence time in the mold for the steel to solidify. To avoid the occurrence of a breakout, it is critical to detect improper solidification of the steel shell in advance with enough lead-time to appropriately slow down the caster. A number of approaches have been proposed to predict breakouts in the continuous casting process. For example, the sticker detection method develops comprehensive rules to characterize patterns in the mold temperatures prior to the incidence of a sticker type breakout. If such patterns have been recognized in the current casting operation, then there is a high likelihood that a breakout will occur. MVS method has also been applied to develop a Caster Stable Operation Supervisory (SOS) system at Dofasco for breakout detection (Dudzic *et al.*, 2002). In this system, a PCA model is built using selected process measurements to model the normal operation of casting processes; certain hypothesis tests are then performed to detect abnormal operations including breakouts. With the aid of the Caster SOS system and continuously process improvements, the number of breakouts at Dofasco has been significantly reduced. Our plant experiences show that, in the current small number of breakouts, more than 50% of them are related to the abnormal operation of start-cast, SEN change, FTC and other process transitions (Rasmussen and Strobl, 2003). Such breakouts in process transitions cannot be detected by either a sticker detection system or the Caster SOS system. Our previous work proposed a novel method that is able to successfully detect the start-cast breakouts using MPCA algorithm (Zhang *et al.*, 2003b). In this paper, this work has been further extended to monitor the caster SEN change.

3. BATCH PROCESS MONITORING USING MULTI-WAY PCA

Prior to discussing SEN change monitoring in continuous casting operation, the fundamental concepts of MPCA is introduced in this section. More details can be found in Nomikos and MacGregor, (1994, 1995) and the recent published paper (Kourti, 2003).

MPCA is an extension of the standard PCA algorithm, which was developed for online monitoring of batch processes. Essentially, every batch process is a dynamic process. The process dynamics are shown in the process trajectories over the batch duration. In addition, batch process measurements are highly auto-correlated, which will violate the assumption for standard PCA that measurement samples are independent and process statistics are stationary over time. Therefore, standard PCA cannot be directly used to monitor a batch process. As shown in Fig. 2, batch process data can be organized into a 3-dimensional data block: there are a number of batch operations; for each operation, various process variables are measured at each sampling interval over time (these measurements are also referred to as observations). Batch processes often exhibit variable time duration. As illustrated in Fig. 2, they may end sooner or later depending on operating conditions.

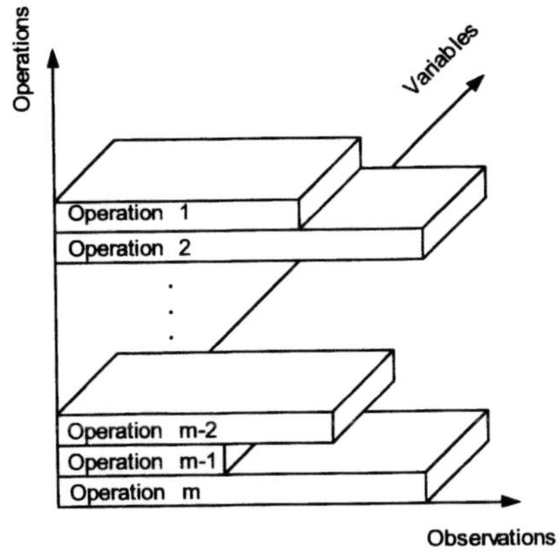

Fig. 2. 3-dimensional batch process data over various batch durations

The core concept of MPCA is to handle this 3-dimensional, dynamic data to monitor batch operation through the following four steps. (1) *Process trajectory synchronizing:* synchronization is performed to align all process trajectories such that the similarity of process dynamics of batch operations can be revealed. Our experiences show that the indicator variable method (Nomikos and MacGregor, 1995) works well for this purpose. In this method, a monotonically increasing variable that has the same value at the beginning and end of each batch is selected as the indicator variable. All process trajectories are then interpolated themselves upon the constant intervals in the indicator variable. After synchronization, each batch operation will have the same number of observations, as illustrated by the data block (**W**) in Fig. 3. (2) *Batch-wise unfolding*: the synchronized, 3-dimensional data block (**W**) was unfolded to a 2-dimensional data matrix (**X**) in Fig. 3. **W** is sliced vertically along the observation direction. The obtained slices are then juxtaposed to build the data **X** with a large column dimension such that each row corresponds to a batch operation. (3) *Modeling batch-to-batch variance*: a standard PCA algorithm is applied to **X** to build a statistical model to benchmark the normal batch operation. The data in **X** are projected to a new latent variable space defined by a loading matrix **P**, where most of the batch-to-batch variance contained in **X** is captured by only a few latent variables (*i.e.*, principal components). The values of principal components for each batch operation are called scores (denoted by **T**). *Squared Prediction Error* (SPE) and *Hotelling* T^2 statistic (HT) are defined at each observation such that they are able to describe how the process trajectories in a new batch operation are coincided with the normal operation. (4) *Monitoring a new batch*: as a new batch operation evolves, the real-time measurements of process variables are collected online. The process trajectories of these measurements from the beginning of the batch operation to the current time become known. The future trajectories are predicted based on the assumption that the current deviation from the average trajectory remains constant over the rest of the batch duration (Nomikos and MacGregor, 1995). The complete trajectories are then synchronized and unfolded in the same way in step (1) and (2). Their deviation from the normal batch operation is examined by SPE and HT statistic. Any difference that cannot be statistically attributed to common process variation indicates that the new batch operation is abnormal and an alarm will be generated.

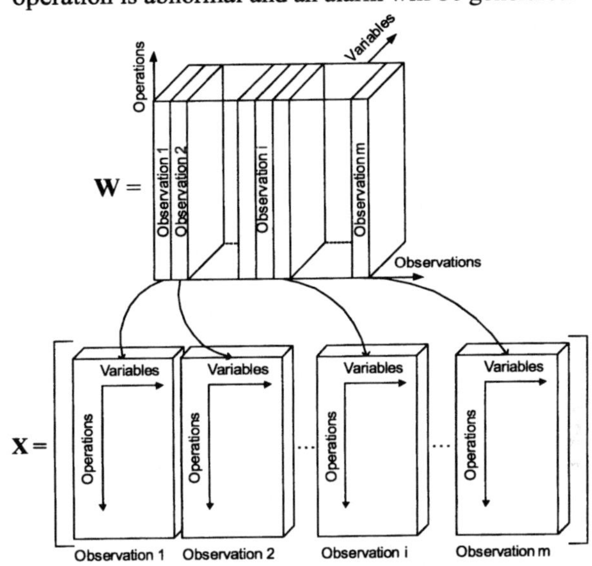

Fig. 3. 3-dimensional synchronized batch process data being unfolded to a 2-dimensoinal matrix to preserve the direction of batch operation.

4. ONLINE MONITORING OF PROCESS TRANSITION IN CASTER OPERATION

MPCA can be used to monitor a SEN change transition in the casting process by clearly defining the start and the end point of each transition. A SEN change is usually executed every 3 or 4 hours during the continuous casting production. Simply speaking, a SEN change consists of three operational steps as shown in Fig. 4: (1) Point A to B: slowing down the caster from the initial casting speed (at Point A) to approximately 0.6 meters per minute. (2) Point B to C: manually replacing the old SEN with the new one while the casting speed remains unchanged (the casting speed between B and C is also referred to as holding speed). (3) Point C to E: ramping up the casting speed to the desired normal operating condition (at Point E). Due to the changing operating conditions, the casting speed profile for each SEN change is very different. For example, the initial speed at Point A can be varying from 1.0 to 1.7 meters per minute, and the time duration between Point B and C can also be varying depending on how long it takes for operators to manually change a SEN.

Synchronization of SEN change Data. In order to monitor a SEN change using MPCA, all process trajectories during SEN change transition must be synchronized in such way that they progress similarly and have the same number of observations. As previously described, the indicator variable method is well-known for this purpose. To be the indicator variable, the selected process variable must satisfy two conditions simultaneously: it increase (or decrease) monotonically and have same values at the beginning and the end of each process transition. Unfortunately, there is no such a variable that can act as the indicator variable in a SEN change because of the varied operating conditions and inconsistent operators' behavior in a SEN change. The following procedures are applied to solve this problem.

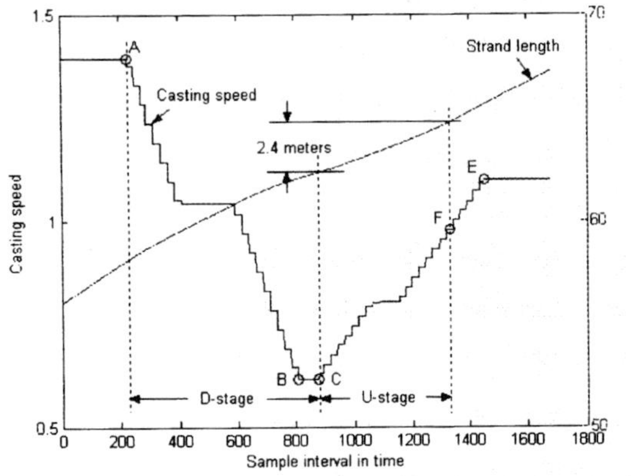

Fig. 4. Typical SEN change operation.

The entire SEN change operation is divided into two stages. The first stage (denoted by D-stage) starts from Point A and ends at Point C, in which the casting speed is continuously decreasing and therefore is chosen as the indicator variable. The second stage (denoted by U-stage) is from point C to F, where F is determined in such way that the strand length from C to F is equal to 2.4 meters[1]. Although the strand length is not measured, it can be calculated by the integral of the casting speed over time. In the U-stage, the strand length is chosen as the indicator variable. As a result, all process trajectories in U-stage are aligned by interpolating themselves based on the pre-defined synchronization scales in the strand length, which is illustrated in Fig. 5. It is worth noting that the non-uniform synchronization scales are used so that it possesses finer resolution of monitoring the beginning of the U-stage, where the process is changing fast and typical process failures often occur at that moment.

One issue of the casting speed being the indicator variable in D-stage is that the casting speed at the beginning and the end of D-stage are very different for each SEN change, which implies that the second condition for indicator variable is not satisfied. One method proposed by Garcia-Munoz (2003) can be used to deal with this situation: assume D-stage is 0% completed at Point A and it is 100% completed at Point C; then process trajectories can be synchronized based on the percentage of completeness of D-stage. Unfortunately, synchronization in this method can only be performed at the end of the stage when the casting speed at Point C becomes known. Obviously, this method is only good for offline process analysis and not suitable for online monitoring. In this paper, an alternative method is proposed for D-stage synchronization. Through process knowledge and all available historical process data, we know the maximum and minimum casting speed within D-stage, denoted by S_{max} and S_{min}, respectively. It is assumed that each D-stage starts at S_{max} and ends at S_{min}. Thus, all process trajectories can be resampled on m synchronization scales pre-defined by:

$$s(i) = S_{\max} - i \times \left(\frac{S_{\max} - S_{\min}}{m-1} \right) \quad i = 0,\ldots,m \quad \text{(eq.1)}$$

As shown in Fig. 6-a, the casting speed profile during a SEN change is very different due to various operating conditions. Compared with the raw process trajectories over time (Fig. 6-b), the resulting *synchronized* trajectories of a certain thermocouple temperature in both D-stage and U-stage are demonstrated in Fig. 6-c. Note that, in U-stage, all trajectories are synchronized with the same number of observations along the increasing strand length; whereas, the trajectories in D-stage show the different start and end value of casting speed for each SEN change. Apparently, some data are missing in D-stage due to the applied synchronization method. These missing data will be handled subsequently for both modeling and monitoring.

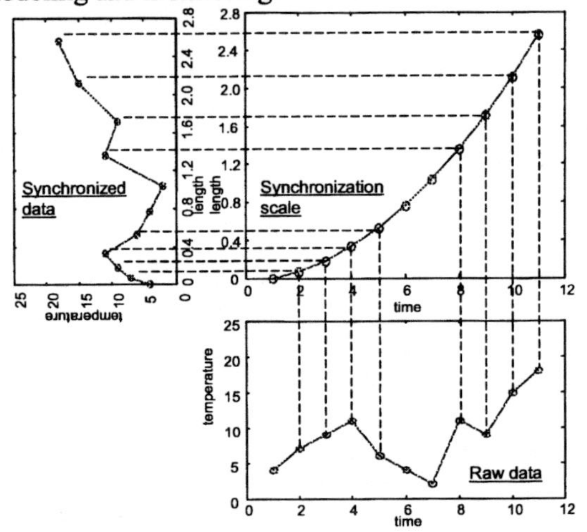

Fig. 5. Trajectory synchronization in U-stage.

Missing data handling. In a real industrial environment, process data can be missing due to various reasons, for instance, sensor failures, digital communication errors, *etc.* Missing data handling is a very important feature in an MVS-based online monitoring system. In this particular application of SEN change monitoring, a large portion of missing data is actually created by the synchronization method applied in D-stage. Fig. 7 gives an example of missing data pattern in one modeling set. In this synchronized, unfolded 2-dimensional matrix, any SEN change with the initial casting speed less than S_{max} or the holding casting speed greater than S_{min} may contain missing data blocks (illustrated by the

[1] This value is initially determined by prior process knowledge and then verified by the steady-state detection to ensure the casting operation reaches a pseudo steady-state at the end of the SEN change.

light grey color). Since the casting speed profile during SEN change is largely depending on the changing operating conditions and operators' preferences, we can assume that, in a large modeling set that includes sufficient number of SEN change operations to cover the entire operating region, these data are *randomly* missing.

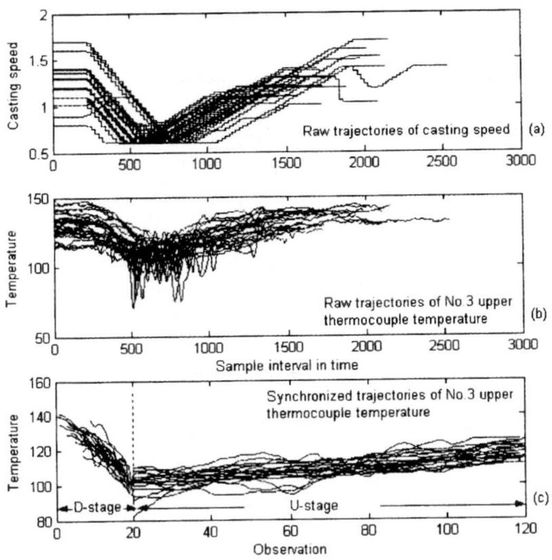

Fig. 6. Example of trajectory synchronization during SEN change.

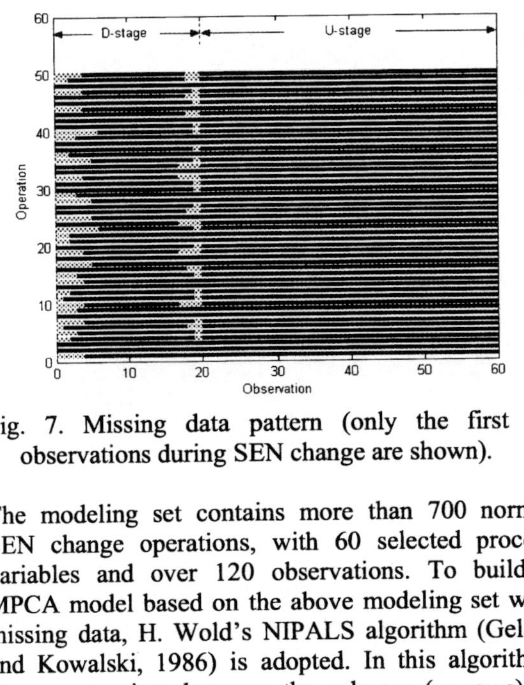

Fig. 7. Missing data pattern (only the first 60 observations during SEN change are shown).

The modeling set contains more than 700 normal SEN change operations, with 60 selected process variables and over 120 observations. To build a MPCA model based on the above modeling set with missing data, H. Wold's NIPALS algorithm (Geladi and Kowalski, 1986) is adopted. In this algorithm, linear regressions between the columns (or rows) of modeling data and the score (or loading) vectors are performed iteratively to obtain the converged scores **T** and loading matrix **P**. When data in any column or row of the modeling set are missing, the linear regressions are performed simply by ignoring the missing points. This algorithm has been proven very efficient to handle the missing data in PCA/MPCA model development. Moreover, missing data may also occur in observations of a new SEN change being monitored. If this is the case, the *Project to Model Plane* method proposed by Nelson *et al.* (1996) is used to estimate the scores from an existing MPCA model such that the available observations can be best explained by the scores. Thus, SPE and HT statistic can be calculated to determine if the new SEN change coincides with the normal operation.

Contribution plot of current observations. When an alarm is generated, it is important to quickly inform operators what may be causing the alarm. Such process diagnosis information is normally derived from contribution plots. The traditional PCA contribution plot includes all process variables involved in model calculation and the most likely process variables causing the alarms are identified by their largest contributions. However, as shown in Fig. 8-a, such a plot may suffer from a huge number of process variables and their observations over time in the MPCA model and won't provide helpful operating guidance to operators in a quick and clear manner. In the monitoring scheme of SEN change, a modified contribution plot is used, where only the observations that describe the current operating conditions are presented to operators. It is expected that, at current observation, a certain process variable with a high contribution to SPE or HT statistic in all normal SEN change within the modeling set should also have a high contribution in a new SEN change. Thus, under an alarming situation, if one process variable has a higher contribution at the current observation than what it usually has in the normal operation, it is the most likely process variable that caused the alarm. Under the above assumption, the contribution control limits at each observation can be determined based on the historical data in the modeling set. The process variables that most likely caused the alarms during a SEN change are then identified by the highest ratio of contributions at the current observation to their corresponding control limit, as illustrated in Fig. 8-b.

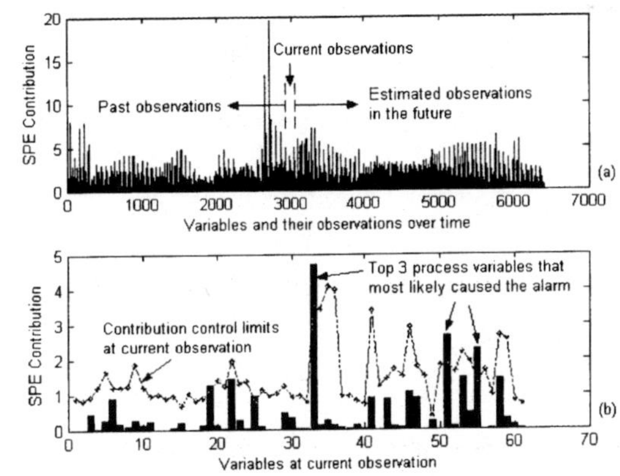

Fig. 8. (a) Traditional contribution plot in MPCA; (b) Contribution plot of current observations.

The scheme described in this section has been applied to monitor SEN changes at Dofasco's No.2 Caster. Industrial results show that the proposed methods provide continuous indication of process normality of a SEN change and some selected process failures such as sensor failures and equipment problems can be correctly detected.

5. INDUSTRIAL EXAMPLES

In this section, an industrial example is described, focusing on abnormal operation detection during SEN change. Since this monitoring scheme is still in a preliminary testing stage, all monitoring information is not made available to operators. There are no actions taken by operators to respond to the generated alarms.

This particular example occurred in April 2003. After a SEN change, operators found that there was a hole developing in the north-west side of the SEN. This ruptured SEN may have lead to some quality issues, even a breakout. Fig. 9-a shows the monitoring results calculated based on the historical data of this SEN change. Significant HT alarms were generated in D-stage, where the ruptured SEN was used in operation. The HT contribution plot of current observations is shown in Fig. 9-b, where most mold thermocouple temperatures and their differences on both north and south face exceeded their contribution control limits. It implied that the process fault was not isolated to a specific mold location but was globally distributed across the entire mold. Such observations are very consistent with our previous experience of SEN failures. After a new SEN was changed, the process was back to its normal operation and HT alarm was not observed in the U-stage.

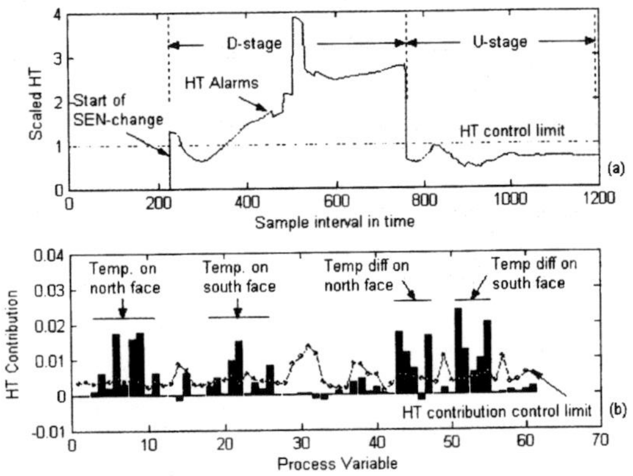

Fig. 9. HT alarms caused by a ruptured SEN in D-stage.

A background test was performed to monitor 101 SEN changes operated in July 2003 where 6% of SEN changes showed significant alarms, indicating certain abnormal situation occurred due to either process upsets or inappropriate operations. It is expected that, with the aid of an online monitoring scheme being commissioned to operators, the percentage number of abnormal operations can be significantly reduced.

6. SUMMARY AND CONCLUSIONS

In continuous steel casting processing, stable process transitions are very crucial for achieving excellent casting performance. The work presented in this paper focuses on developing an online monitoring scheme for one important process transition in the continuous casting process: SEN change. Some benefits of the monitoring scheme can be summarized as follows. (1) Severe process failures including breakouts can occur during SEN change. The online monitoring scheme is able to detect these process upsets such that the correct actions can be taken in advance by operators to avoid the occurrence of a catastrophic event. (2) Online monitoring scheme can provide a consistent indication on how SEN changes are executed. Some plant experiences have shown that stricter adherence to the SOP during SEN change or other process transitions may reduce the number of breakouts or other process failures. In summary, the proposed monitoring scheme for SEN change in continuous casting process has shown good detectability of various process abnormalities. Future work may include online commissioning this monitoring scheme to operation and extending it to monitor other process transitions in casting process.

REFERENCES

Duchesne C, T. Kourti and J.F. MacGregor (2002), Multivariate SPC for start-ups, and grade transitions, *AICHE J.* **48**, pp.2890-2901

Dudzic M. and I. Miletic (2002), Multivariate statistical monitoring of a continuous caster. In: *Proceeding of American Control Conference 2002*, pp. 600-601

Garcia-Munoz S., T. Kourti, J.F. MacGregor (2003), Troubleshooting of an industrial batch process using multivariate methods, *Ind. Eng. Chem. Res.*, **42**, pp.3592-3601

Geladi P. and B.R. Kowalski (1986), Partial least squares regression: a tutorial, *Analtica Chemica Acta*, **85**, pp. 1-17

Kourti T. (2003), Multivariate dynamic data modeling for analysis and statistical process control of batch processes, start-ups and grade transitions, *J. Chemometrics*, **17**, pp.93-109

Nelson P.R.C., P.A. Taylor and J.F. MacGregor (1996), Missing data methods in PCA and PLS: score calculations with incomplete observations, *Chemometric Intell. Lab. Syst.*, **35**, pp.45-65

Nomikos P and J.F. MacGregor (1994), Monitoring of batch processes using multi-way principal component analysis, *AICHE J.*, **40**, pp.1361-1375

Nomikos P and J.F. MacGregor (1995), Multivariate SPC charts for monitoring batch processes, *Technometrics*, **37**, pp.41-59

Quinn S. and V. Vaculik (2002), Improving the desulfurization process using adaptive multivariate statistical modeling, *AISE Steel Technology*, **10**, pp.37-41

Rasmussen P. and N.A. Strobl (2003), personal communication

Zhang Y. M. Dudzic and V. Vaculik (2003), Integrated monitoring solution to start-up and run-time operations for continuous casting, *Annual Reviews in Control,* **27**, pp.141-149

Zhang Y., V. Vaculik, M. Dudzic, I. Miletic, A. Smyth, and T. Holek (2003), Start cast breakouts preventative prediction using multi-way PCA technology, In: *IFAC workshop on new technologies for automation of metallurgical industry*, Shanghai, China, October 11-13, pp.97-102

ELSEVIER
IFAC
PUBLICATIONS
www.elsevier.com/locate/ifac

PRACTICAL MODEL AND DETECTION ALGORITHM FOR VALVE STICTION

Manabu Kano * Hiroshi Maruta *
Hidekazu Kugemoto ** Keiko Shimizu ***

* *Kyoto University, Kyoto 606-8501, Japan*
** *Sumitomo Chemical Co. Ltd., Niihama 792-8521, Japan*
*** *TOSHIBA Corporation, Yokohama 230-0051, Japan*

Abstract: Stiction is the most common problem in pneumatic control valves, which are widely used in the process industry. Stiction causes fluctuation of process variables, which lowers productivity. Therefore, developing a method to detect stiction and distinguish it from other causes is crucial to help operators take an appropriate action for improving control performance. The present work proposes a valve stiction model and new stiction detection algorithms. Using only two parameters, the proposed model can describe the stiction phenomenon with sufficient accuracy. The usefulness of the proposed detection methods are demonstrated by comparing them with a conventional method. The proposed methods are shown to successfully detect valve stiction, distinguish it from bad tuning or disturbances, and quantify the degree of stiction, by using simulation data sets and real operation data sets of several chemical processes. *Copyright ©2004 IFAC*

Keywords: Control Valve, Stiction, Control Performance Monitoring, Fault Detection, Fault Diagnosis

1. INTRODUCTION

The variability of process variables makes it difficult to keep operating conditions close to their bound and thus causes excessive energy consumption. To maximize productivity, control systems should achieve their best performance. The poor control performance is caused not only by bad controller tuning but also by undesired characteristics of control valves. Among the many types of undesired characteristics of control valves, stiction is the most common problem in the process industry. Therefore, it is important to develop a practical method that can detect stiction and distinguish it from other causes including inadequate controller tuning.

In the present work, a valve stiction model is built to understand the stiction phenomenon and simulate it. There are two types of models: 1) a detailed physical model that formulates the stiction phenomenon as precisely as possible (Muller, 1994), and 2) a data-driven model that describes the relationship between a controller output and a valve position (Choudhury et al., 2004). Since a detailed physical model has numbers of unknown physical parameters, it is not only time-consuming but not practical to simulate an actual control valve by using such a model. A data-driven model, on the other hand, is useful because it has only a few parameters that are easy to identify and simple to understand. However, the data-driven model proposed by Choudhury et al. (2004) has several disadvantages. The most serious problem would be their assumption that the input signal is deterministic; throughout their paper they only investigate sinusoidal inputs. In fact, when the input signal is stochastic the model does not function well. Although it has been pointed out that valve stiction causes persistent fluctuation of a process variable, stiction does not always cause persistent fluctuation. A valve stiction model should therefore be able to cope with both deterministic and stochastic signals. The model developed here is an improved and extended version of the model proposed by Choudhury et al. (2004). The accuracy of the model is validated by comparing simulation results with real operation data of a chemical process.

Furthermore, new methods for detecting valve stiction are proposed. Although several methods have been proposed (e.g., Horch, 2000), they are based on the assumption that a controlled variable such as flow rate follows a predefined probability distribution, and thus their judgment is not reliable when the assumption is not satisfied. In addition, conventional methods cannot quantify the degree of stiction even if they can detect it. The usefulness of new valve stiction detection methods is demonstrated by applying them to simulation data generated by using the developed model and real operation data obtained from several chemical processes.

2. VALVE STICTION MODEL

In this section, the stiction phenomenon of a pneumatic control valve, which is widely used in the process industry, is modeled.

2.1 Pneumatic Control Valve and Stiction

The general structure of a pneumatic control valve is shown in Fig. 1. This valve is closed by elastic force and opened by air pressure. Flow rate is changed according to the plug position, which is determined by the balance between elastic force and air pressure. The plug is connected to the valve stem. The stem is moved against static or kinetic frictional force caused by packing, which is a sealing device to prevent leakage of process fluid. Smooth movement of the stem is impeded by excessive static friction. The valve position cannot be changed until the controller output overcomes static friction, and it is suddenly and considerably changed when the difference between elastic force and air pressure exceeds the maximum static frictional force.

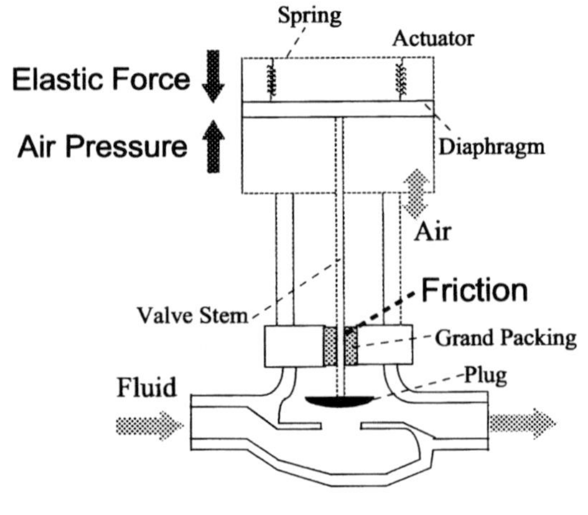

Fig. 1. Structure of pneumatic control valve

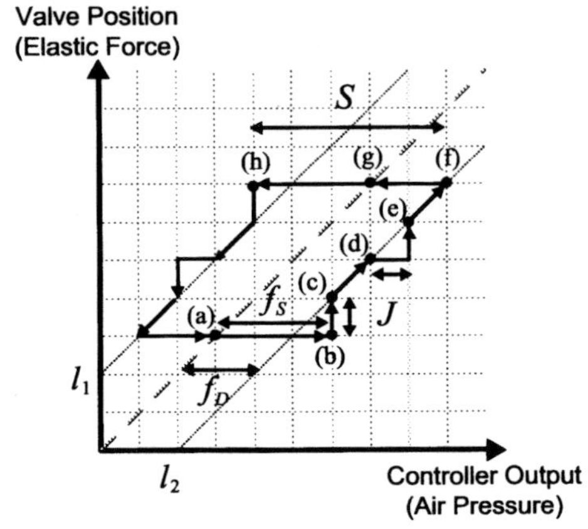

Fig. 2. Relation between controller output and valve position under valve stiction

2.2 Stiction Model

To model the relationship between the controller output and the valve position of a pneumatic control valve, the balance among elastic force, air pressure, and frictional force needs to be taken into account. The relationship can be described as shown in Fig. 2. The dashed line denotes the states where elastic force and air pressure are balanced. The controller output and the valve position change along this line in an ideal situation without any friction.

The ideal relationship is disturbed when friction arises. For example, the valve is resting at (a) where elastic force and air pressure are balanced. The valve position cannot be changed due to static friction even if the controller output, i.e., air pressure, is increased. The valve begins to open at (b) where the difference between air pressure and elastic force exceeds the maximum static frictional force. Since the frictional force changes from static f_S to kinetic f_D when the valve starts to move at (b), a slip-jump of the size

$$J = f_S - f_D \quad (1)$$

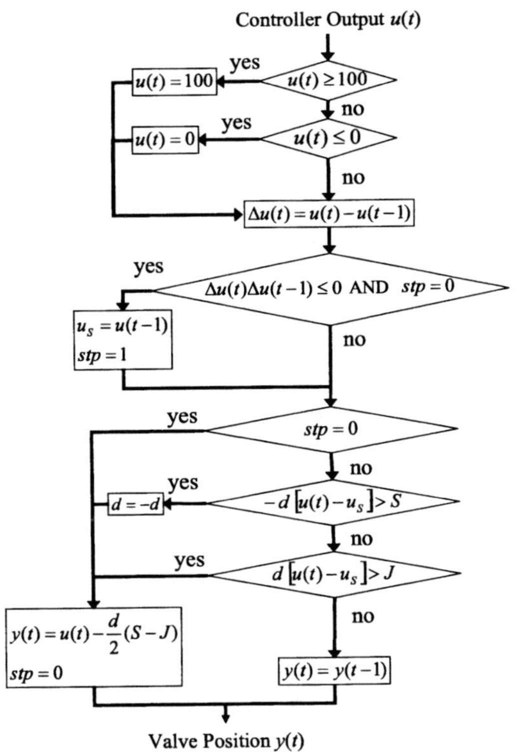

Fig. 3. Flowchart of valve stiction model

happens and the valve state changes from (b) to (c). Thereafter, the valve state changes along the line l_2 which deviates from the ideal line by f_D because the difference between air pressure and elastic force is equal to f_D. When the valve stops at (d), the difference between air pressure and elastic force needs to exceed f_S again for the valve to open further. Since the difference between them is f_D at (d), air pressure must increase by J to open the valve. Once air pressure exceeds elastic force by f_D, the valve state changes to (e) and then follows l_2.

Air pressure begins to decrease when the controller orders the valve to close at (f). At this moment, the valve changes its direction and comes to rest momentarily. The valve position does not change until the difference between elastic force and air pressure exceeds the maximum static frictional force f_S. The valve state (h) is just point-symmetric to (b). The difference of air pressure between (f) and (h) is given by

$$S = f_S + f_D. \tag{2}$$

The valve state follows the line l_1 while the valve position decreases.

The above-mentioned phenomena can be modeled as a flowchart shown in Fig. 3. The input and output of this valve stiction model are the controller output u and the valve position y, respectively. Here, the controller output is

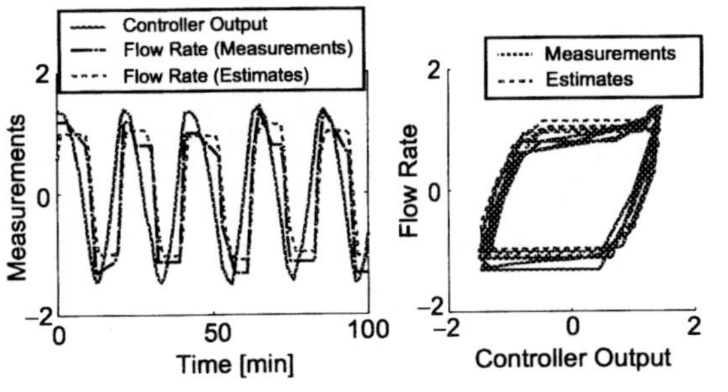

Fig. 4. Comparison between measurements and estimates of flow rate

transformed to the range corresponding to the valve position in advance.

The first two branches check if the upper and the lower bounds of the controller output are satisfied. In this model, two states of the valve are explicitly distinguished: 1) a moving state ($stp = 0$), and 2) a resting state ($stp = 1$). In addition, the controller output at the moment the valve state changes from moving to resting is defined as u_S. u_S is updated and the state is changed to the resting state ($stp = 1$) only when the valve stops or changes its direction ($\Delta u(t)\Delta u(t-1) \leq 0$) while its state is moving ($stp = 0$). Then, the following two conditions concerning the difference between $u(t)$ and u_S are checked unless the valve is in a moving state. The first condition judges whether the valve changes its direction and overcomes the maximum static friction (corresponding to (b) and (h) in Fig. 2). Here, $d = \pm 1$ denotes the direction of frictional force. The second condition judges whether the valve moves in the same direction and overcomes friction. If one of these two conditions is satisfied or the valve is in a moving state, the valve position is updated via the following equation.

$$y(t) = u(t) - df_D = u(t) - \frac{d(S-J)}{2} \tag{3}$$

On the other hand, the valve position is unchanged if the valve remains in a resting state.

To demonstrate the validity of this stiction model, simulation results are compared with operation data of a chemical process that suffers from valve stiction. The flow rate is estimated by calculating the valve position from the controller output with the stiction model and by assuming dynamics from the valve position to the flow rate is given by a first order model.

$$P_F(t) = \frac{1}{0.2s + 1} \tag{4}$$

Measurements and estimates of the flow rate are shown together with the controller output in Fig. 4. The flow rate measurements proved coincident with that estimated by the stiction model. That is, the proposed model can describe stiction successfully by using only two parameters S and J. By changing these two parameters, the stiction model can describe a variety of situations, for example, an ideal situation when $S=J=0$ and no slip-jump when $J=0$.

The valve stiction model developed here has several advantages compared with the model proposed by Choudhury et al. (2004). First, it can cope with stochastic input as well as deterministic input. Second, u_S can be updated at appropriate timings by introducing the valve state stp. Third, it can change the degree of stiction according to the direction of the valve movement.

3. STICTION DETECTION METHODS

In this section, two new methods for detecting valve stiction are proposed.

As shown in Fig. 2, the following characteristics are observed when stiction occurs.

(1) There are sections where the valve position does not change even though the controller output changes. Stiction is stronger as such sections are longer.
(2) The relationship between the controller output and the valve position takes the shape of a parallelogram if slip-jump J is neglected. Stiction is stronger as the distance between l_1 and l_2 is longer.

On the basis of these characteristics, new methods for detecting valve stiction are proposed. The first method, referred to as method A, is based on characteristic (1); and the second, method B, is based on characteristic (2).

The algorithm of method A is summarized as follows:

(1) Calculate the difference of valve position y.

$$\Delta y(t) = y(t) - y(t-1) \quad (5)$$

(2) Find time intervals when the following condition is satisfied.

$$|\Delta y(t)| < \varepsilon \quad (6)$$

where ε is a threshold.
(3) During each time interval found, calculate the difference between the maximum and the minimum of the controller output u and define it as $\tilde{u}$. Similarly, calculate the difference between the maximum and the minimum of the valve position y and define it as $\tilde{y}$. In addition, determine thresholds: ε_u for $\tilde{u}$ and ε_y for $\tilde{y}$.
(4) Conclude that stiction occurs when $\tilde{u} \geq \varepsilon_u$ and $\tilde{y} \leq \varepsilon_y$. Otherwise, conclude that stiction does not occur.
(5) Calculate the ratio ρ of total length of intervals when stiction occurs to total length of all intervals. In addition, calculate σ that is the mean of $\tilde{u}$ when stiction occurs.

There is higher possibility of stiction as the normalized measure ρ is closer to one. On the contrary, it is confirmed that no stiction occurs when ρ is zero. Furthermore, the degree of stiction can be quantified by using σ.

Method B is based on the fact that the relationship between the controller output and the valve position takes the shape of a parallelogram and the distance between l_1 and l_2 increases as stiction becomes stronger. To capture the difference of shapes, a function F is introduced.

$$F(t) = \max\left\{\min\left\{F(t-1)+\Delta u(t), F_{max}\right\}, 0\right\} \quad (7)$$

The function F indicates the difference between the controller output u and the value on the line l_1 at the same valve position. Therefore, $u-F$ will have a strong correlation with the valve position y. In addition, the maximum value F_{max} of F corresponds to $S-J$. Thus, the degree of stiction or deadband can be quantified by F_{max} because slip-jump J is much smaller than deadband S and F_{max} is usually close to S. This is the most important advantage of using method B. To date, no stiction detection method that enables one to quantify stiction has been proposed. F_{max} and its initial value $F(0)$ can be identified from operation data by solving an optimization problem that aims to maximize a correlation coefficient r between $u-F$ and y. As a result, the possibility of stiction gets larger as F_{max} becomes larger. However, it should be noted here that results of method B are reliable only when the correlation coefficient r is close to one.

4. APPLICATIONS

In this section, the proposed valve stiction detection methods, A and B, as well as a conventional method are applied to simulation data generated by using the stiction model and operation data obtained from several chemical processes; the usefulness of the proposed methods is also demonstrated.

4.1 Simulation

A flow control system and a level control system are investigated here. Block diagrams of both control systems are shown in Fig. 5. Process transfer functions of flow $P_F(s)$ and level $P_L(s)$ are given by

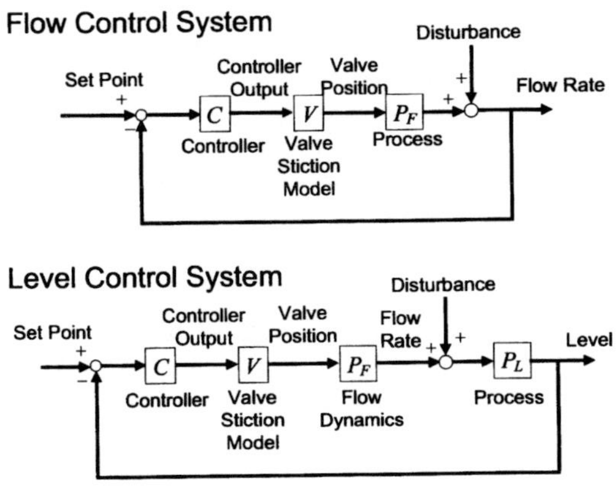

Fig. 5. Block diagram of control systems

Table 1. Controller tuning

	Propotional Gain	Integral Time [min]
Flow Control	0.5	0.3
Level Control	3	30

Table 2. Parameters of stiction model

	S[%]	J[%]
Case 1 (No Stiction)	0	0
Case 2 (Weak Stiction)	1	0.3
Case 3 (Strong Stiction)	5	1

Table 3. Application results

	A		B		H
	ρ	σ	F_{max}	r	Stiction
Flow Control (FC)					
Case 1	0.00	0.00	0.00	0.03	Yes
Case 2	0.77	0.60	0.00	0.18	Yes
Case 3	0.83	3.50	0.00	0.11	Yes
Level Control - F (LC-F)					
Case 1	0.00	0.00	0.00	1.00	Yes
Case 2	0.56	0.83	0.74	1.00	Yes
Case 3	0.79	4.54	4.20	0.99	Yes
Level Control - L (LC-L)					
Case 1	0.05	0.54	0.00	0.53	Yes
Case 2	0.02	0.68	0.00	0.65	Yes
Case 3	0.00	0.00	0.00	0.82	Yes

$$P_F(s) = \frac{1}{0.2s+1} \quad , \quad P_L(s) = \frac{1}{15s}e^{-s} \qquad (8)$$

where the unit of time is min. PI controllers are used for both control systems. Control parameters and valve stiction model parameters are summarized in Tables 1 and 2, respectively. Simulation results are shown in Figs. 6 and 7. Three methods are applied to normalized data of 300 samples. The results are summarized in Table 3. Here, H denotes the method proposed by Horch (2000). Method H assumes persistent fluctuation of the controlled variables, and its judgment is based on the probability distribution of the controlled variables. LC-F is the case where flow rate is used for detection. On the other hand, LC-L is the case where level is used. Average computational times of methods A, B, and H are 0.07 sec, 1.12 sec, and 79 sec, respectively. The computational load of method

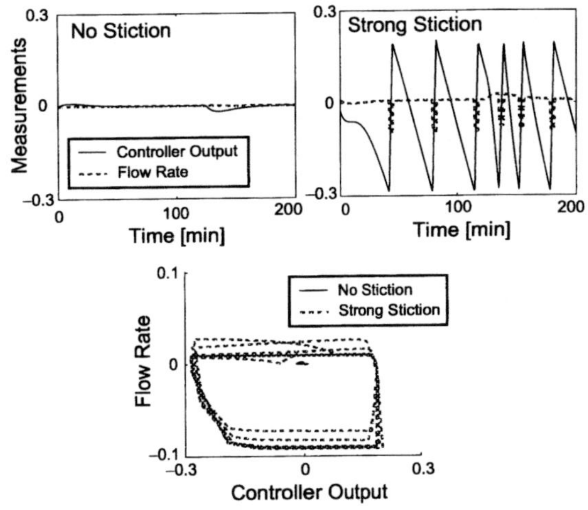

Fig. 6. Simulation results of flow control system

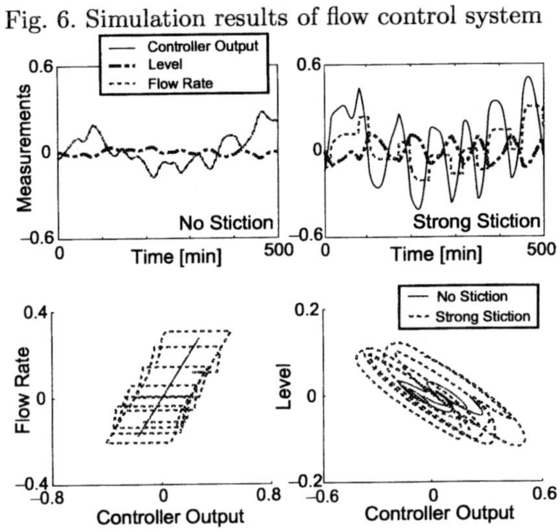

Fig. 7. Simulation results of level control system

H is excessively heavy, because it requires solving two optimization problems to fit data into two predefined probability distributions.

In case FC, only method A can detect stiction successfully. Method B fails because the relationship between the controller output and the flow rate does not take the shape of a parallelogram. Method H concludes that stiction occurs in all cases. Of course, this is not true. Method H does not function well in case 1 because flow rate does not fluctuate persistently. In case LC-F, methods A and B can detect stiction successfully. In addition, the degree of stiction was estimated from σ or F_{max} in each case. These values proved coincident with those in Table 2. That is, the proposed methods can detect valve stiction and also quantify the stiction, but method H does not function well. In case LC-L, no method can detect stiction successfully. Methods A and B do not function well in such a case where the controlled variable is delayed. Therefore, these methods should be used only when flow rate or valve position is measured. Method H is applicable

Table 4. Application results

Case		A ρ	σ	B F_{max}	r	Horch Stiction
1	LC (LC-F)	0.50	1.72	2.12	0.98	Yes
2	FC	0.31	1.21	1.92	0.97	Yes
3	LC (LC-F)	0.13	0.89	0.17	0.95	Yes
4	FC	0.03	0.80	0.00	0.13	Yes

only when persistent fluctuation is found in a controlled variable.

4.2 Operation Data from Chemical Processes

Normalized operation data obtained from four chemical processes are shown in Fig. 8. The results of applying three methods to the data are summarized in Table 4. In cases 1 and 2, all methods can detect stiction successfully. In case 3, where there is no stiction but bad controller tuning causes fluctuation, method H incorrectly concludes that there is stiction. However, methods A and B give very small indexes ρ and F_{max}, respectively, that is, methods A and B correctly conclude that there is no stiction. In case 4 where no stiction occurs, method H reaches a wrong conclusion again. Method A does, however, reach the right conclusion. Although method B reaches the right conclusion, the correlation coefficient r is too small and thus the result is not reliable.

5. CONCLUSIONS

In the present work, a valve stiction model is developed and new stiction detection algorithms are proposed. The validity of this model is demonstrated by comparing simulation results generated by using the model with operation data of a chemical process that suffers from valve stiction. Using only two parameters, the model can describe stiction phenomena with sufficient accuracy. In addition, the usefulness of the proposed detection methods is demonstrated by comparing them with a conventional method. The proposed methods are shown to successfully detect valve stiction, distinguish it from bad tuning or disturbances, and quantify the degree of stiction, with applications to simulation data sets and real operation data sets obtained from several chemical processes.

ACKNOWLEDGMENT

This research is partially supported by the consortium No.25 organized in the PSE 143th comittee of JSPS.

REFERENCES

Choudhury, M.A.A.S., N.F. Thornhill, and S.L. Shah (2004). A Data-Driven Model for Valve Stiction. *IFAC Symposium on Advanced Control of Chemical Processes (ADCHEM)*, Hong-Kong, Jan. 11–14.

Horch, A. (2000). Condition Monitoring of Control Loops. *PhD thesis*, Royal Institute of Technology, Stockholm, Sweden.

Muller, F. (1994). Simulation of an Air Operated Sticky Flow Control Valve. *Proc. of the 1994 Summer Computer Simulation Conference*, 742–745.

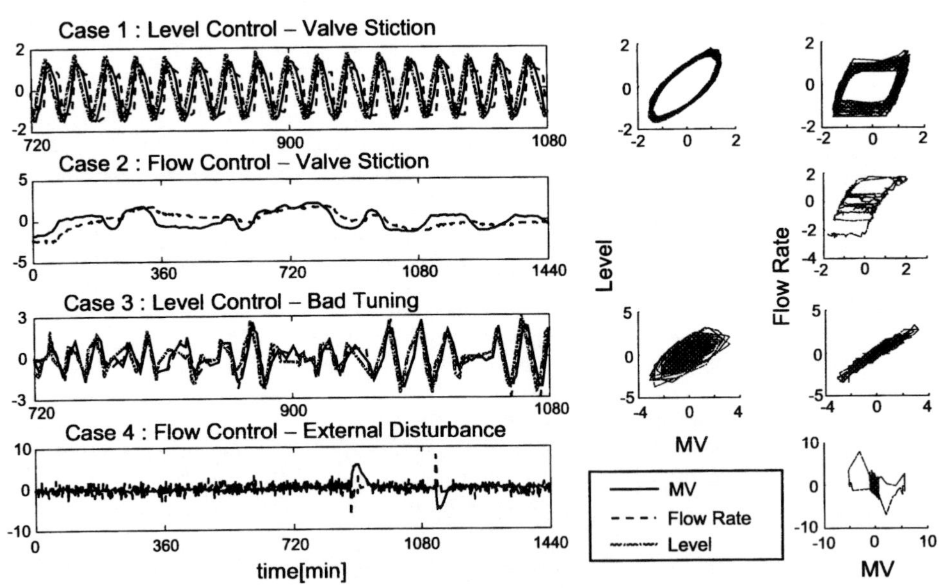

Fig. 8. Operation data of four control systems

ELSEVIER
IFAC
PUBLICATIONS
www.elsevier.com/locate/ifac

DETECTION AND QUANTIFICATION OF CONTROL VALVE STICTION

M. A. A. S. Choudhury * S. L. Shah *,[1]
N. F. Thornhill **

* Department of Chemical and Materials Engineering
University of Alberta, Edmonton AB, Canada, T6G 2G6
** Department of Electronic and Electrical Engineering
University College London, UK, WC1E 7JE

Abstract: Stiction is one of the most common problems in the spring-diaphragm type control valves, which are widely used in the process industry. Although there have been many attempts to understand and detect stiction in control valves, none of the current methods can simultaneously detect and quantify stiction. There is thus a clear need in the process industry for a non-invasive method that can not only detect but also quantify stiction so that the valves that need repair or maintenance can be identified, isolated and repaired. This work describes a method that can detect and quantify stiction that may present in control valves using routine closed loop operating data obtained from the process. No additional excitation or experimentation with the plant is required. Over a dozen industrial case studies have demonstrated the wide applicability and practicality of this method as an useful diagnostic aid in control-loop performance monitoring. *Copyright © 2004 IFAC*

Keywords: stiction, stickband, deadband, hysteresis, actuator, control valve, process control, nonlinearity, slip jump, clustering, ellipse, control loop, condition monitoring

1. INTRODUCTION

A typical chemical plant has hundreds of control loops. The presence of oscillation in a control loop increases the variability of the process variables thus causing inferior quality products, larger rejection rates, increased energy consumption, reduced average throughput and profitability. Bialkowski (1992) reported that about 30% of the loops are oscillatory due to control valve problems. In a recent study, Desborough and Miller (2002) reported that control valve problems account for about a third of the 32% of controllers classified as "poor" or "fair" in an industrial survey. If the control valve contains nonlinearities such as stiction, backlash, and deadband, the valve output may be oscillatory which in turn can cause oscillations in the process output. Among the many types of nonlinearities in control valves, stiction is one of the most common and long-standing problems in process industry. Stiction can easily be detected using invasive methods such as the valve travel or bump test. But to apply an invasive method across an entire plant site is neither feasible nor cost-effective because of the manpower, cost and time intensive nature of the method. Although there have been many studies (Taha *et al.*, 1996; Wallén, 1997; Sharif and Grosvenor, 1998; Gerry and Ruel, 2001) carried out for invasive analysis of control valve performance, only a few non-invasive methods ((Horch, 1999; Rengaswamy *et al.*, 2001; Stenman *et al.*, 2003) have appeared in literature. Horch's method (Horch, 1999; Horch, 2000) detects stic-

[1] author to whom all correspondence should be addressed.
E-mail: sirish.shah@ualberta.ca

tion with the use of the cross-correlation function between *pv* and *op*. Horch's method is successful mainly in detecting valve stiction in flow control loops. It can not be applied for loops involving an integrator or those carrying compressible fluids. The method described in Rengaswamy *et. al.* (2001) depends on the qualitative shape of the time trends of the data which is often distorted by the presence of noise and disturbance. Also, in real life the shape of the time trends of the data is heavily affected by the process and controller dynamics. Stenman *et. al.* (2003) described a model based segmentation method to detect stiction in control valves. This method requires the model of the process with a lot of tuning parameters. To obtain the closed loop model of the process from the routine operation data is very difficult. Moreover, all these methods can only detect stiction but can not quantify it. As pointed out rightly in (Desborough and Miller, 2002), a passive or non-invasive method that can reliably and automatically classify valve performance in closed loop is "*desperately needed in process industry*". An effective non-intrusive data-based monitoring method could reduce the cost of control loop performance maintenance by screening and short-listing those loops and/or valves that need maintenance. This work describes a data-based non-invasive method that can detect and quantify stiction present in control valves.

2. WHAT IS STICTION?

Different people or organizations have defined stiction in different ways. Some of these definitions have been presented in Choudhury *et al.* (2003*a*) . Based on a careful investigation of real process data a new definition of stiction was also proposed by the authors (Choudhury *et al.*, 2004) and is summarized as following. It is observed that

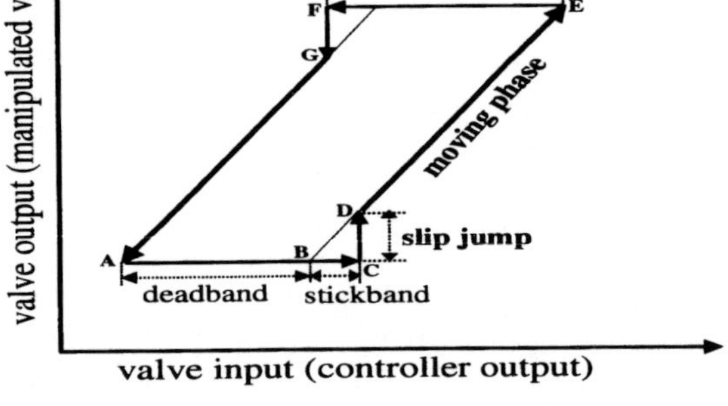

Fig. 1. *Typical input-output behavior of a sticky valve*

the phase plot of the input-output behavior of a valve "suffering from stiction" can be described by figure 1. It consists of four components: deadband, stickband, slip jump and the moving phase. When the valve comes to a rest or changes the direction at point A in figure 1, the valve sticks. After the controller output overcomes the deadband (AB) plus the stickband (BC) of the valve, the valve jumps to a new position (point D) and continues to move. Due to very low or zero velocity, the valve may stick again in between points D and E in figure 1 while travelling in the same direction (EnTech, 1998). In such a case the magnitude of deadband is zero and only stickband is present. This can be overcome if the controller output signal is larger than the stickband only. It is usually uncommon in industrial practice. The deadband and stickband represent the behavior of the valve when it is not moving though the input to the valve keeps changing. Slip jump represents the abrupt release of potential energy stored in the actuator chamber due to high static friction into kinetic energy as the valve starts to move. The magnitude of the slip jump is very crucial in determining the limit cyclic behavior introduced by stiction (Piipponen, 1996). Once the valve slips, it continues to move until it sticks again (point E in figure 1). In this moving phase dynamic friction is present which may be much lower than the static friction. Therefore, *"stiction is a property of an element such that its smooth movement in response to a varying input is preceded by a static part followed by a sudden abrupt jump called the slip-jump. Slip-jump is expressed as a percentage of the output span. Its origin in a mechanical system is static friction which exceeds the dynamic friction during smooth movement"*. This definition has been exploited in the next and subsequent sections for quantifying stiction in control valves. In the process industry, stiction is generally measured as a % of the valve travel or the span of the control signal (Gerry and Ruel, 2001). For example, a 2 % stiction means that when the valve gets stuck it will start moving only after the cumulative change of its control signal is greater than or equal to 2%. If the range of the control signal is 4 to 20 mA then a 2% stiction means that a change of the control signal less than 0.32 mA in magnitude will not be able to move the valve.

Note that it is difficult to estimate slip jump ('j') from the controlled output (*pv*) and the controller output (*op*) data because the slip jump in the valve output is destroyed by the process dynamics. This work will only quantify the parameter 's' (deadband plus stickband) which will be termed as the amount of stiction from this point forward.

3. DETECTION OF VALVE STICTION

In a control loop, nonlinearity may be present either in the process itself or in the control valve.

For our current analysis, we are assuming that the process nonlinearity is negligible in the steady state operating region during which the data has been collected. This is a reasonable assumption because the method works with the routine operating data of a control loop under regulatory control. For a particular operating region of a chemical plant, the plant is assumed to be locally linear under its feedback configuration. This is what makes linear controllers perform so well in the chemical process industries.

This method first examines the presence of a nonlinearity in a control loop. If a nonlinearity is detected then the process variable (*pv*), set point (*sp*) and controller output (*op*) signals are used to diagnose the possible causes of nonlinearity. The following section describes the method in brief.

3.1 *Detection of Loop Nonlinearity*

A control loop containing valve nonlinearities often produces non-Gaussian (e.g., a signal with asymmetric distribution) and nonlinear time series, namely process output (*pv*) and controller output (*op*). Higher order statistics based nonlinearity assessment can be used as a diagnostic tool for troubleshooting of hardware faults that may be present in control loop (Choudhury *et al.*, 2002; Choudhury *et al.*, 2003). As described in (Choudhury *et al.*, 2003), the test of Gaussianity and nonlinearity of the control error signal (*sp-pv*) is a useful diagnostic aid towards determining the poor performance of a control loop. The test described in Choudhury *et al.* (2003c) uses the sensitivity of the normalized bispectrum or bicoherence to detect the presence of nonlinear interactions in a time series signal. A distinctive characteristic of a non-linear time series is the presence of phase coupling such that the phase of one frequency component is determined by the phases of others. Phase coupling leads to higher order spectral features which can be detected in the bicoherence of a signal. Bicoherence is defined as:

$$bic^2(f_1, f_2) \triangleq \frac{|B(f_1, f_2)|^2}{E[|X(f_1)X(f_2)|^2]E[|X(f_1+f_2)|^2]} \quad (1)$$

where $B(f_1, f_2)$ is the bispectrum at frequencies (f_1,f_2) and is given by

$$B(f_1, f_2) \triangleq E[X(f_1)X(f_2)X^*(f_1+f_2)], \quad (2)$$

$X(f_1)$ is the discrete Fourier transform of the time series $x(k)$ at the frequency f_1, $X^*(f_1)$ is the complex conjugate and E is the expectation operator. A key feature of the bispectrum is that it has a non-zero value if there is significant phase coupling in the signal x between frequency components at f_1 and f_2. The bicoherence gives the same information but is normalized as a value between 0 and 1.

In (Choudhury *et al.*, 2003), two indices - the Non-Gaussianity Index (NGI) and the Non-Linearity Index (NLI) - have been defined as

$$NGI \triangleq \overline{\hat{bic^2}} - \overline{bic^2}_{crit} \quad (3)$$

$$NLI \triangleq |\, \hat{bic^2}_{max} - (\overline{\hat{bic^2}} + 2\sigma_{\hat{bic^2}}) \,| \quad (4)$$

where, $\overline{\hat{bic^2}}$ is the average squared bicoherence and $\hat{bic^2}_{max}$ is the maximum squared bicoherence, $\sigma_{\hat{bic^2}}$ is the standard deviation of the squared bicoherence and $\overline{bic^2}_{crit}$ is the statistical threshold/critical value obtained from the central chi-square distribution of squared bicoherence. As outlined in (Choudhury *et al.*, 2003), if both indices, NGI and NLI, are greater than zero, the signal is described as non-Gaussian and nonlinear. The details of the procedure are shown in figure 2. The test can be applied on any time series to check its non-Gaussianity and non-linearity. For a control loop, this test is applied on the error signal (*sp-pv*) to the controller because the error signal is more stationary than *pv* or *op* signal. If the error signal is found to be non-Gaussian and nonlinear, it is inferred that the loop in question exhibits significant non-linearity. The nonlinearity can be attributed to the control valve under the following assumptions:

- The process is assumed to be locally linear.
- No nonlinear disturbance is entering the loop.

3.2 *Use of pv-op Plot*

The long time practice in industrial studies has been the use of *pv-op* plots for the detection of valve problems, especially stiction. But experience shows that this type of method is successful only for a handful cases of flow control loops. The use of *pv-op* plot for detecting valve problems was not successful because it only takes into account the qualitative trend information of the time series which can be destroyed due to the presence of process dynamics, noise dynamics, disturbances and tightly tuned controllers. In our method, the *pv-op* plot is used as a second step to diagnose the valve nonlinearity problem. The detection of valve or process nonlinearity is first carried out using higher statistical method-based NGI and NLI indices. Once a nonlinearity is detected, only then the *pv-op* plot is used to isolate its cause. Because of the contamination of real life data with noise/disturbance, a *pv- op* plot is often unclear and ambiguous, and it is difficult to find any information from it. This necessitates the use of a filter to clean the data.

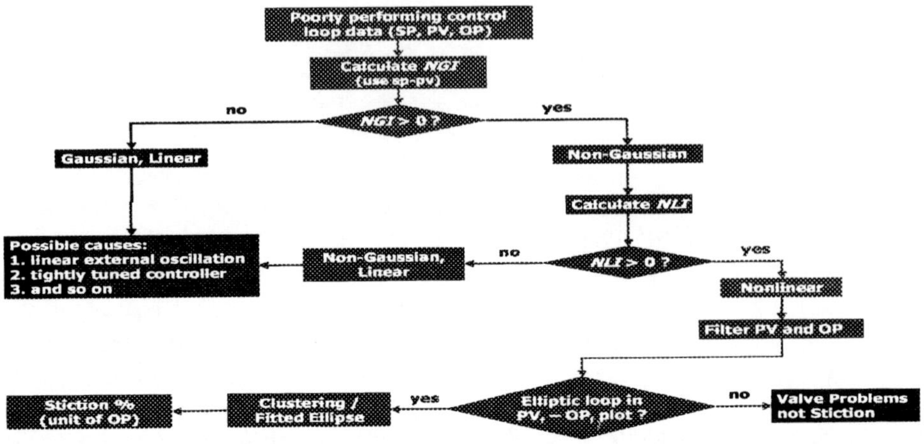

Fig. 2. *Decision Flow Diagram of the methodology for the detection and diagnosis of valve problems*

3.2.1. *Data Filtering or Pre-processing* Upon detection of a nonlinearity, the frequencies responsible for the significant nonlinear interactions can be determined from the significant peaks in the squared bicoherence plot. Then, a frequency domain Wiener filter is used to obtain the part(s) of the signal which contributes significantly to the signal nonlinearity. Both *pv* and *op* are filtered using a frequency domain Wiener filter. The detailed filter design algorithm is given in Thornhill *et. al.* (2003) . The frequency ranges for the filters are selected from the inspection of the peaks in the bicoherence plot. A segment of the data that has regular oscillations is then used for the construction of the pv_f-op_f plot (the pv_f and op_f are the filtered *pv* and *op*). If the pv_f-op_f plot shows an elliptical pattern, it is diagnosed as a signature of valve stiction, otherwise there is a valve problem which is not due to stiction. In this work, stiction is estimated as the maximum width of the cycles of the pv_f-op_f plot along the op_f axis. The quantified stiction is termed as "apparent stiction" because the actual amount of stiction to be obtained from the *mv-op* plot may differ from the estimated quantity because of the role of the controller in attempting to regulate the process variable.

4. QUANTIFYING STICTION

It is important to be able to quantify stiction so that a list of sticky valves in order of their maintenance priority can be prepared. The pv_f-op_f plot along with any of the following two methods can be used to quantify stiction in the unit of *op* signal. Note that there is no need to scale the data.

4.1 *Using a fitted ellipse*

An ellipse in the least square sense can be fitted to the pv_f-op_f plot and can be used for quantifying stiction. Since apparent stiction is defined as the maximum width of the ellipse along the *op* axis, the distance between two points lying on the intersections of the ellipse and a line parallel to the *op* axis and passing through the center of the ellipse will be the amount of stiction present in the loop. If m and n are the length of the major and minor axes of the fitted ellipse respectively, and α is the angle of rotation of the ellipse, the amount of stiction (length of AP in figure 3(e)) can be obtained using the following expression

$$stiction = AP = \frac{2mn}{\sqrt{(m^2\ sin^2\alpha + n^2\ cos^2\alpha)}} \quad (5)$$

4.2 *Clustering Technique*

Clustering is a method for dividing scattered groups of data into several groups. Since the *pv-op* plot for a control loop with a sticky valve exhibits elliptic patterns, the data corresponding to a narrow strip along the mean of *pv* and parallel to the *op* axis can be collected (see figure 3(d)) and used for quantifying stiction with the help of c-means clustering technique (Johnson and Wichern, 1998). The amount of stiction can be estimated from the absolute value of the difference between x co-ordinates of the centers of the two clusters. If the final centers of the clusters are (op_1, pv_1) and (op_2, pv_2), then the amount of stiction is obtained using the following expression:

$$stiction = |op_1 - op_2| \quad (6)$$

5. AN ILLUSTRATIVE EXAMPLE

The objective of this section is to explain the various steps of the proposed method with a detailed

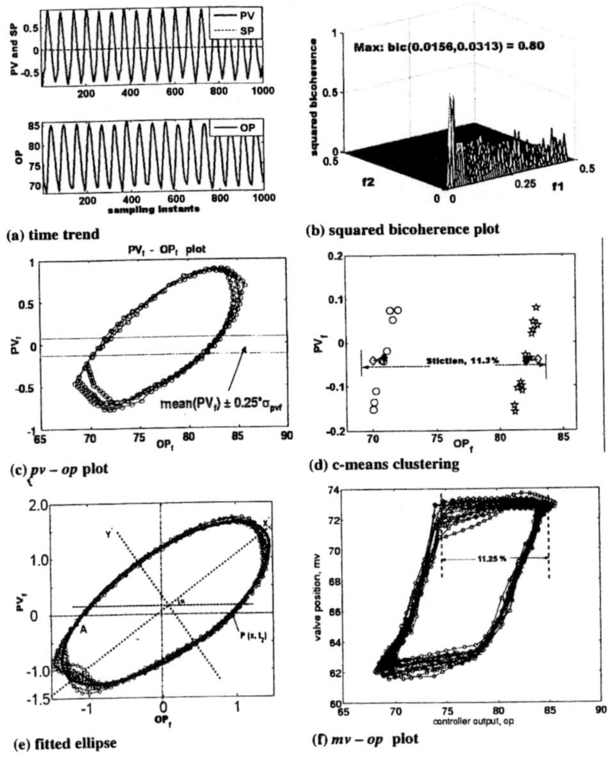

Fig. 3. *Results of a level control loop data analysis*

presentation of an industrial example. This example represents a level control loop in a power plant, which controls the level in a condenser located at the outlet of a turbine by manipulating the flow rate of the liquid condensate from the condenser. Figure 3(a) shows the time trends for level (*pv*), set point (*sp*) and the controller output (*op*). The loop shows oscillatory behavior. 4096 data points were used for the bicoherence calculation and figure 3(b) shows the squared bicoherence plot corresponding to the controller error signal (*sp-pv*). The values of NGI and NLI were found to be 0.04 and 0.61, respectively, indicating the presence of significant loop nonlinearity. To isolate the nonlinearity, the *pv-op* plot is found to be useful. Figure 3(c) demonstrates the c-means clustering technique used in quantification of the stiction. The points denoted with empty and filled diamonds are the initial and final centers of the clusters, respectively. This method quantifies the amount of stiction in this loop as 11.3%. Figure 3(e) shows the ellipse fitting technique and the amount of stiction estimated using this method is 11.40%. Both methods have produced identical results with practically tolerable limits of deviation from each other.

Validation of the Results of the Illustrative Example: Once the results of our analysis was sent to the plant people, plant engineers confirmed that this loop was suffering from stiction. For this loop, the valve positioner data was made available. Figure 3(f) shows the actual valve position (*mv*) vs. controller output (*op*) plot. This plot clearly shows that the valve was sticking during the change of its direction. From this plot, the amount of stiction can be estimated as 11.25% which is in agreement with the results obtained from the proposed methods.

Table 1. Results for Industrial Loops

Loop No.	Loop Type	NGI	NLI	Apparent Stiction %	
				c-means	ellipse
1	Level	0.08	0.44	9.3	8.7
2	Level	0.10	0.40	4.2	4.3
3	Level	-0.02	—	—	—
4	Flow	0.01	0.55	0.35	0.33
5	Flow	0.05	0.59	0.42	0.42
6	Temp	0.003	0.19	1.00	1.14

6. INDUSTRIAL CASE STUDIES

The objective of this section is to evaluate the proposed method on a number of selected control loop data obtained from different types of process industries. For each loop, the set point (*sp*), controlled output (*pv*) and controller output (*op*) data were available. Unless otherwise stated, a data length of 4096 was used for the squared bicoherence calculation for each case. The numerical results for all loops are provided in table 1. Due to space limitations, it is not possible to include figures for these examples. These data were analyzed before the prior knowledge of the control valve problems and the results of the analysis were confirmed later by the plant personnel.

- **Loops 1 and 2:** These loops are also level control loops in the same power plant described in the illustrative example. They also control the level of condensers located at the outlet of two different turbines by manipulating the flow rate of the liquid condensate. Both c-means clustering and fitted ellipse techniques provide the amount of stiction approximately as 9% for loop 1 and 4% for loop 2.
- **Loop 3:** This is another level control loop in the same power plant described in the illustrative example. It also controls the level of a condenser located at the outlet of a different turbine by manipulating the flow rate of the liquid condensate. The magnitude of NGI was -0.02, clearly indicating that nonlinearity is not a problem for this loop. From the valve positioner (*mv*) vs. controller output (*op*) plot (figure 4) it is obvious that the valve shows a linear response.
- **Loops 4 and 5:** These are flow control loops obtained from a refinery. The results of the analysis of these loops are given in

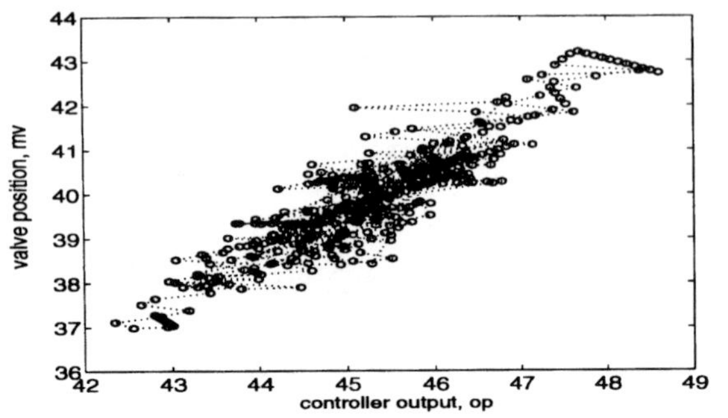

Fig. 4. *Valve positioner vs. controller output plot for loop 3*

the fourth and fifth rows of the table 1. The presence of small amount of stiction (0.35% for loop 3 and 0.4% for loop4) was causing large oscillations in these loops.

- **Loop 6:** It describes a temperature control loop on a furnace feed dryer system at the Tek-Cominco mine plant, BC, Canada. The temperature of the dryer combustion chamber is controlled by manipulating the flow rate of natural gas to the combustion chamber. The results are presented in the sixth row of table 1. The amount of stiction found in this loop was approximately 1%.

7. CONCLUSIONS

A non-invasive method for detecting and quantifying stiction in control valve has been presented in this paper. The method first detects nonlinearity in a control loop by the use of the sensitivity of the normalized bispectrum or bicoherence to the nonlinear interactions among various frequency components of the control error signal. If nonlinearity is detected, *pv* and *op* signals are filtered using a frequency domain Wiener filter to obtain filtered pv_f and op_f signals. If an ellipse can be fitted suitably to the pv_f-op_f plot, then it indicates a signature of valve stiction. C-means clustering or fitted ellipse techniques can be used for automatic quantification of stiction. The method has been extensively evaluated on simulated as well as industrial data sets.

8. REFERENCES

Bialkowski, W. L. (1992). Dreams vs. reality: A view from both sides of the gap. In: *Control Systems.* Whistler, BC, Canada. pp. 283–294.

Choudhury, M. A. A. S., Nina F. Thornhill and Sirish L. Shah (2004). A data-driven model for valve stiction. In: *to appear in ADCHEM 2004, Jan 11-14.* Hong Kong.

Choudhury, M. A. A. S., Sirish L. Shah and Nina F. Thornhill (2002). Detection and diagnosis of system nonlinearities using higher order statistics. In: *15th IFAC World Congress.* Barcelona, Spain.

Choudhury, M. A. A. S., Sirish L. Shah and Nina F. Thornhill (2003). Diagnosis of poor control loop performance using higher order statistics. *accepted for publication in Automatica, June 2003.*

Desborough, L. and R. Miller (2002). Increasing customer value of industrial control performance monitoring - honeywell's experience. In: *AIChE Symposium Series 2001.* number 326. pp. 172–192.

EnTech (1998). *EnTech Control Valve Dynamic Specification (version 3.0).*

Gerry, John and Michel Ruel (2001). How to measure and combat valve stiction online. ISA Society. Houston, Texas, USA. http://www.expertune.com/articles/isa2001/StictionMR.htm.

Horch, Alexander (1999). A simple method for detection of stiction in control valves. *Control Engineering Practice* **7**, 1221–1231.

Horch, Alexander (2000). Condition Monitoring of Control Loops. PhD thesis. Royal Institute of Technology. Stockholm, Sweden.

Johnson, R. A. and D. W. Wichern (1998). *Applied Multivariate Statistical Analysis.* Prentice-Hall. New Jersey.

Piipponen, Juha (1996). Controlling processes with nonideal valves: Tuning of loops and selection of valves. In: *Control Systems.* Chateau, Halifax, Nova Scotia, Canada. pp. 179–186.

Rengaswamy, R., T. Hagglund and V. Venkatasubramanian (2001). A qualitative shape analysis formalism for monitoring control loop performance.. *Engng. Appl. Artificial Intell.* **14**, 23–33.

Sharif, M. A. and R. I. Grosvenor (1998). Process plant condition monitoring and fault diagnosis. *Proc Instn Mec Engrs* **212**(Part E), 13–30.

Stenman, A., F. Gustafsson and K. Forsman (2003). A segmentation-based method for detection of stiction in control valves. *International Journal of Adaptive Control and Signal Processing* **17**, 625–634.

Taha, Othman, Guy A. Dumont and Michael S. Davies (1996). Detection and diagnosis of oscillations in control loops. In: *Proceedings of the 35th conference on Decision and Control.* Kobe, Japan.

Thornhill, N. F., B. Huang and H. Zhang (2003). Detection of multiple oscillations in control loops. *Journal of Process Control* **13**, 91–100.

Wallén, Anders (1997). Valve diagnostics and automatic tuning. In: *Proceedings of the American Control Conference.* Albuquerque, New Mexico. pp. 2930–2934.

ELSEVIER
IFAC
PUBLICATIONS
www.elsevier.com/locate/ifac

Multivariable Control of Multi-Zone Chemical Mechanical Polishing

Sheng-Jyh Shiu[1], Cheng-Ching Yu*[1], Shih-Haur Shen[2], and An-Jhih Su [1]

[1]*Dept. of Chem. Eng., National Taiwan UniversitTaipei 106-17, TAIWAN*
[2]*Applied Materials Taiwan Ltd., Hsin-Chu 300, TAIWAN*

Abstract: The modeling and multivariable control of the multi-zone chemical mechanical polishing (CMP) is studied in this work. For a three-zone CMP, the copper thickness across the radial position is measured with an in-situ sensor and then the measurements are converted to 60 data points across the radial position. In the process control notation, these are the controlled variables and the manipulated variables are the three pressures applied to each zone. Therefore, this is a 60x3 non-square multivariable control problem. Thus, a 60x3 steady-state gain matrix is obtained followed by finding dynamic elements according to the gas holdup in each chamber. As a result of the non-square system, it is not possible to keep all 60 outputs at their set points using only 3 inputs. The singular value decomposition (SVD) is used to design a non-square feedback controller. The proposed control system is test on incoming wafers with different surface profiles. Results show that achievable performance can be maintained using the proposed SVD controller. *Copyright © 2004 IFAC*

Keywords: CMP, multi-zone CMP, multivariable control, non-square system.

1. INTRODUCTION

In Cu CMP, the copper is removed using a two-step procedure. First, the overburden Cu is removed with a high removal rate and the objective is to reduce the step height to a given specification with a certain amount of copper remains. The reason for stopping short from the oxide surface is that within-wafer non-uniformity (WIWNU), as the results of incoming profile and CMP process, is often encountered in Cu CMP and the remaining Cu layer prevents unnecessary dishing in the thin spots on the wafer. This is followed by the overpolishing step where overburden metal and some barrier are removed and this step is generally carried out using a much smaller removal rate [Kao et al., 2002, Edgar et al., 2000, and Yao et al., 2000]. As implied by the name, CMP removes overburden metal or unnecessary materials by the combinative effects of mechanical abrasion and chemical reactions.

Despite recent advances in CMP, some manufacturing concerns associated with successful implementation of CMP remain to be overcome. In theory, CMP can achieve global planarity, but there is still a problem that different operating conditions will result in non-uniformity in thickness of wafer surface. The WIWNU indicates the variation in surface thickness across the wafer radial position, especially on the edge. Besides, the surface profile of wafers produced from electrochemical plating (ECP) process appears that the metallic layer is thicker on edge area. Thus, a new type of CMP, multi-zone CMP, offers an attractive alternative. Multi-zone CMP is expected to reduce WIWNU and to provide a wider processing window. Unlike the typical single zone configuration, the wafer carrier is divided into multiple zones in the radial position and different pressure can be applied to each zone (Fig. 1). The objective of this work is to devise a systematic approach to the modeling and control of such CMP processes.

2. MULTI-ZONE CMP

2.1. Process Description.

Consider a CMP system where the wafer carrier is divided into three zones in the radial position and different pressure can be applied to each zone (Fig. 1). For a wafer carrier with the radius of 150mm (i.e., a 300 mm wafer), zone #1 covers 0-130mm, the second zone ranges from 130-140mm, and third zone covers 140mm and beyond. Typically, the well-known Preston's equation [Preston, 1927] is used to model the polishing process. It describes the material removal as a linear function of pressure and rotation speed.

$$RR = K_p \cdot p \cdot v \qquad (1)$$

where K_p is the Preston constant, p is pressure, and v is rotation speed. The Preston equation can be extended to multi-zone CMP in a straightforward manner. For the multi-zone system, the relationship between removal rate at the ith radial position and input variables can be expressed as:

corresponding author; e-mail: ccyu@ntu.edu.tw; fax: +886-2-2362-3040

$$RR_i = \sum_j K_{p,i,j} p_j \cdot v \qquad (2)$$

where $K_{p,i,j}$ is the local Preston constant describing the effect of pressure from jth zone on the ith radial position, pj denotes the pressure of the jth zone, and v is again the rotation speed. Without loss of generality, assume that the rotation speed is fixed throughout all runs and v is absorbed into $K_{p,i,j}$ for the subsequent development.

2.2. Design of Experiments (DOE)

Copper (Cu) CMP is carried out on an Applied Materials' Reflexion™ polisher using Titan Profiler™ polishing heads. Polishing pad (Rodel IC1010) is used and the experimental copper CMP slurry is provided by Cabot Corporation. Electrochemical plating (ECP) process is employed for Cu plating. The 300mm wafers without pattern are used in all tests. All copper thickness measurements are performed with an *in situ* i-Scan sensor.

All the experiments were carried out according to a standard design of experiment procedure (DOE). The factors in this work are the three pressures applied to each zone and a pressure applied to the retaining ring (i.e., keeping the wafer in proper position). From center of the wafer to the edge, they are defined as p_1, p_2, p_3, and p_{rr}, respectively. We devise two levels for each factor such that a four-factor and two-level design of experiment (DOE) is carried out.

Before the test proceeds, the surfaces of 20 blanket wafers are measured. There are 119 measurement points from one edge to the other, so the pre-CMP thickness is defined as:

$$\mathbf{z_0} = \left[z_{0,1}, z_{0,2}, \ldots, z_{0,119}\right]^T \qquad (3)$$

The subscript 0 denotes the initial condition and i represents the radial position across the wafer. The post-CMP thickness is also recorded after polishing for 60 seconds.

$$\mathbf{z} = \left[z_1, z_2, \ldots, z_{119}\right]^T \qquad (4)$$

Because of the symmetry of wafers and the applied pressures, an averaged value of the thicknesses on the opposite of the center taken and it is defined as the output variables. Thus, the pre-CMP and post-CMP thicknesses are defined as follows:

$$\mathbf{y_0} = \left[y_{0,1}, y_{0,2}, \ldots, y_{0,60}\right]^T \qquad (5)$$

$$\mathbf{y} = \left[y_1, y_2, \ldots, y_{60}\right]^T \qquad (6)$$

where the subscript i (i=1-60) denotes the position on wafer from center to edge, and

$$y_i = \frac{z_{61-i} + z_{59+i}}{2} \qquad (7)$$

With the definition of the output variables and polish time, the local removal rate can be calculated by Eq. (8):

$$RR_i = \frac{\Delta y_i}{\Delta t} \qquad (8)$$

where Δy_i denotes the amount removed ($\Delta y_i = y_{0,i} - y_i$), and Δt denotes the total polish time. In modeling, the local removal rate is taken as the state variable (x) and it is also a 60x1 vector:

$$\mathbf{x} = \left[x_1, x_2, \ldots, x_{60}\right]^T \qquad (9)$$

For this multi-zone CMP, the manipulated variables are the three pressures applied:

$$\mathbf{p} = \left[p_1, p_2, p_3\right]^T \qquad (10)$$

Note that p_{rr} is discarded here because of negligible effect on wafer surface profile. The reason is that the pressure applied to the retaining ring is to keep the wafer in an acceptable range and it has little influence on the removal rate across the radial position. After discarding p_{rr}, the experimental conditions of some sets seem to be identical. To maintain the correctness, the raw data of those sets are averaged. Therefore, there will be 14 sets of experimental data left and used in the regression.

2.3. Steady-State Analysis

From the definitions of state and input variables, the steady-state behavior for the multivariable system can be written as follows:

$$\mathbf{x} = \mathbf{K_p} \cdot \mathbf{p} \qquad (11)$$

where $\mathbf{K_p}$ is the steady-state gain matrix. Given experimental data, $\mathbf{K_p}$ matrix ($\in \Re^{60x3}$) can be determined from the least square regression. The regression result versus position on wafer is given in Fig. 2a. The effects of input pressures to the removal rate on each zone appear to be quite reasonable. The pressure applied on zone #1, p_1, shows significant effect on removal rate on the range of 0-110mm and the influence degrades gradually toward the wafer edge as can be seen from the numerical values of the first column of $\mathbf{K_p}$ (i.e., $\mathbf{K_{p1}}$). The pressure applied on the second zone, p_2, shows a large value of Preston constant around 135mm and its influence on removal rate diminishes toward both ends as shown in $\mathbf{K_{p2}}$. As expected the input pressure of the third zone shows little effect on the removal rate until 135mm and then the influence grows linearly toward to edge. However, the unusual phenomena happened on edge area, for example, the reverse trend of $\mathbf{K_{p1}}$ and the sharp decrease of $\mathbf{K_{p2}}$, are detected. Those are attributed to the effect of noise. Therefore, the final data point is excluded from now on, and a new $\mathbf{K_p}$ matrix ($\in \Re^{60x3}$) is obtained.

Let us use the singular value decomposition (SVD) to analyze the system. The SVD decomposes the $\mathbf{K_p}$ matrix of model into three matrices:

$$\mathbf{K_p} = \mathbf{U \Sigma V}^T \qquad (12)$$

where $\mathbf{U}$ is a orthonormal matrix ($\in \Re^{60x3}$) in the output side, $\mathbf{V}$ is the input orthonormal matrix ($\in \Re^{3x3}$), and $\mathbf{\Sigma}$ is a 3x3 diagonal matrix with the singular value (σ_i) as the diagonal element. The ratio of the largest singular value to the smallest is the condition number ($\kappa=\sigma_{max}/\sigma_{min}$) which is a quantitative indicator of closeness to singularity. After proceeding SVD, the $\mathbf{V}$ matrix is found similar to identity. Then each column of $\mathbf{U}$ matrix is plotted against position on wafer (Fig. 2b). As can be seen in the figure, the trends of the three columns in **U** matrix are quite similar to those of $\mathbf{K_p}$ matrix. It indicates that the three directions are just enough to describe the characteristic of multi-zone system. The singular values sorted by number are 264, 31, and 16, respectively, and condition number (κ) is 16, i.e., quite far away from a singular system. Thus, this is a well-defined multivariable system with 60 states and 3 inputs.

Because of the characteristic of a non-square property of system, it is not possible to keep all 60 outputs at their set points using only 3 inputs. Thus, non-uniformity is an inherent property of this multi-zone CMP. Given the initial surface profile (y^0) (measured) and the desired thickness y^d, the amount needs to be removed (Δy^d) can easily calculated by

$$\Delta \mathbf{y}^d = \mathbf{y}^0 - \mathbf{y}^d \tag{13}$$

After the CMP with a polish time of Δt, the error between the desired and actual thicknesses can be expressed as

$$\begin{aligned}\mathbf{E} &= \mathbf{y} - \mathbf{y}^d = (\mathbf{y}^0 - \mathbf{y}^d) - (\mathbf{y}^0 - \mathbf{y}) \\ &= (\mathbf{x}^d - \mathbf{x}) \cdot \mathbf{\Delta t}\end{aligned} \tag{14}$$

Eq. (14) also indicates that the error is due to difference of actual removal rate (**x**) and desired removal rate ($\mathbf{x}^d$). In terms of the removal rate, we have:

$$\mathbf{x}^d = \mathbf{K_p} \cdot \mathbf{p} \tag{15}$$

Solving Eq. (15) by the least-square method, the input pressures are computed:

$$\mathbf{p} = (\mathbf{K}_\mathbf{p}^T \mathbf{K_p})^{-1} \mathbf{K}_\mathbf{p}^T \cdot \mathbf{x}^d = \mathbf{K}_\mathbf{p}^\dagger \cdot \mathbf{x}^d \tag{16}$$

where the superscript † denotes the pseudo-inverse. Substituting **p** into Eq. (11), one finally obtains:

$$\begin{aligned}\mathbf{E}_{\min} &= (\mathbf{I} - \mathbf{K_p}\mathbf{K}_\mathbf{p}^\dagger)\mathbf{x}^d \cdot \mathbf{\Delta t} \\ &= (\mathbf{I} - \mathbf{K_p}\mathbf{K}_\mathbf{p}^\dagger)\Delta \mathbf{y}^d\end{aligned} \tag{17}$$

The above equation indicates the achievable performance for this multi-zone CMP and it is characterized by the deviation of the projection matrix ($\mathbf{K_p}\mathbf{K}_\mathbf{p}^\dagger$) from an identity matrix. Consider the case when every entry of the vector Δy^d takes the value of 1. Fig. 3 shows the shape of the surface profile in the radial position. Since the 2-norm of E_{min} is equal to 0.0735, it implies an averaged error of 73.5Å for 1000Å removed.

2.4. Dynamics Analysis

After obtaining the 60x3 steady-state gain matrix, the dynamic elements which correspond to the gas holdup in each chamber are computed from the residence times. They are associated with the inputs (pressures). Assume that the flow rate of air is fixed so the time constant is proportional to the volume of the chamber in the carrier. The largest one is set to 3 s, and the others two are 0.479 s and 0.502 s, respectively. Therefore, we have a 60x3 process transfer function matrix with the same first order dynamics in each column. Moreover, from the definition of the removal rate, the removed thickness is defined as:

$$\dot{y}(t) = x(t) \tag{18}$$

Taking Laplace transformation, Eq. (18) becomes:

$$Y(s) = \frac{X(s)}{s} \tag{19}$$

where Y(s) denotes the Laplace transformed remaining thickness, X(s) is the Laplace transformed removal rate. Therefore, the polishing process can be expressed as:

$$\begin{aligned}Y(s) &= G_p(s)p(s) \\ &= \frac{1}{s} \cdot K_p \cdot diag[\frac{1}{\tau_1 s + 1}, \frac{1}{\tau_2 s + 1}, \frac{1}{\tau_3 s + 1}] \cdot p \\ &= \frac{1}{s} K_p D \cdot p\end{aligned} \tag{20}$$

where Gp is the process transfer function relating the input p(s) to the output Y(s), diag[.] denotes diagonal matrix, τ_i stands for the time constant associated with the ith input, and D represents the diagonal matrix. Note that Gp is a 60x3 transfer function matrix with the same dynamic element in each column and the dynamics in each column is represented by a second order integrating system.

3. CONTROL

In the process control notation, the measured data are the controlled variables and the manipulated variables are the three pressures applied to each zone. Therefore, this is a 60x3 multivariable control problem. The control objective is to maintain uniform surface profile using all three manipulated variables. Because this is a non-square system, it is not possible to keep all the outputs at their set points using only 3 inputs. To the best, a least square solution, minimizing the sum of square of errors between the set point (desired thickness) and controlled variables (measured thickness), can be obtained. A 2-norm-based objective function is then considered:

$$J = \left\| \mathbf{y}^f - \mathbf{y}^d \right\|_2 \tag{21}$$

where J is the objective function, $\mathbf{y}^f$ and $\mathbf{y}^d$ are vectors ($\in \Re^{60x1}$) of measured thicknesses (at end of each run, i.e., $y^f = y(t_f)$) and the desired thicknesses, respectively.

Some performance indices are introduced here.

i) Standard deviation (*SD*):

$$SD = \sqrt{\sum_{i=1}^{n} (y_i^f - \bar{y}^f)^2 / n} \tag{22}$$

where the overbar stands for the mean.

ii) ∞-norm of the deviation between measured thickness and the mean:

$$\infty\text{-norm} = \left\| y^f - \bar{y}^f \right\|_\infty \tag{23}$$

iii) The range of measured thickness:

$$\text{range} = y_{\max}^f - y_{\min}^f \tag{24}$$

where $y_{\max}^f = \max(y_i^f)$ and $y_{\min}^f = \min(y_i^f)$.

iv) Non-uniformity (*NU*):

$$NU = \frac{SD}{\bar{y}^f} \times 100\% \quad (25)$$

Those provide simple measures of quality after polishing. Obviously, the achievable performance defined by Eq. (17) also provides an absolute basis to evaluate control performance.

3.1. Ratio Control

Since the process model, at least Kp, is perfectly know, a simple control strategy can be implemented for the multi-zone CMP process. It is a feedforward control where the ratios of the pressures are computed from the model inverse and ratios are maintained throughout the run..

The final NU in flat case is about 5% which is acceptable in the fabrication. In other cases, the first two surface profiles with small SD value, IC1 and IC2, are regarded as finer initial conditions among the three. The indices of final status of those cases are slightly more than those of achievable status. However, their NUs are both kept within 5%. The SD of final status of IC3 is reduced while the other indices are not changing much. Basically, this kind of initial surface profile is hard to handle for the system. Thus, that indicates that ECP is supported to prevent production of that profile.

3.2. Multivariable Feedback Control

It is based on concept of feedback control. The controller adjusts the pressure input simultaneously according to the deviation between measured thickness by sensor and set points (desired thickness). To monitor the change of surface in a polish run, therefore, an *in situ* sensor is required to measure copper thickness on-line and a non-square feedback controller necessary to generate input pressures. Fig. 4a shows the non-square system with the tall process (more output than inputs) and a fat controller (more input to the controller than the control output). The SVD-based approach is employed to design the inversed-based controller. First, the steady-state gain matrix is decomposed into three matrices and the multivariable system can be expressed as (Fig. 4b):

$$\mathbf{Y} = \frac{1}{s}\mathbf{K_p} \cdot \mathbf{D} \cdot \mathbf{p} = \frac{1}{s}(\mathbf{U}\Sigma\mathbf{V}^T)\mathbf{D} \cdot \mathbf{p} = \mathbf{U} \cdot diag[\frac{\sigma_i}{s}] \cdot \mathbf{V}^T \cdot \mathbf{D} \cdot \mathbf{p} \quad (26)$$

Next, the derivative part of the controller is taken as the inverse of **D** and the output of the PI part of the controller becomes:

$$\mathbf{p} = \mathbf{D}^{-1}\mathbf{p}_{PI} \quad (27)$$

Thus, the relationship between the p_{PI} and Y becomes:

$$\mathbf{Y} = \mathbf{U} \cdot diag[\frac{\sigma_i}{s}] \cdot \mathbf{V}^T \mathbf{p}_{PI} \quad (28)$$

Multiply the both sides with $\mathbf{U}^T$ matrix (recall that both **U** and **V** are orthonormal matrices, i.e., $\mathbf{U}\mathbf{U}^T = I$ and $\mathbf{V}^T\mathbf{V} = I$), one obtains:

$$\mathbf{U}^T\mathbf{Y} = diag[\frac{\sigma_i}{s}] \cdot \mathbf{V}^T \cdot \mathbf{p}_{PI} \quad (29)$$

$\mathbf{U}^T\mathbf{Y}$ and $\mathbf{V}^T\mathbf{p}_{PI}$ correspond to the output and input in the principle directions and they are defined as $\mathbf{Y}^*$ ($\in \Re^{3\times1}$) and $\mathbf{p}^*(\in \Re^{3\times1})$, respectively. Therefore, we are left with a decoupled square system with simple diagonal elements, σ_i / s. Define the diagonal matrix as $\mathbf{G_p}^*$, the multivariable system becomes:

$$\mathbf{Y}^* = G_p^*\mathbf{p}^* \quad (30)$$

Fig. 4c also shows the result of transformation. Note the diagonalizing effort is absorbed into the controller as shown in Fig. 4b. Therefore, standard multivariable control method can be applied to this decoupled square system.

The internal model control (IMC) principle is employed to design the PID type of multivariable controller. In the IMC design, the relationship between the IMC controller (C_{IMC}) and the diagonal PI controller (C_{PI}) is:

$$C_{IMC} = [I + C_{PI}\hat{G}_p^*]^{-1}C_{PI} \quad (31)$$

where $\hat{G}_p^*$ is the model of the process. The IMC design consists of the following steps:

i. Decompose the diagonal process model $\hat{G}_p^*$ into:

$$\hat{G}_p^*(s) = \hat{G}_p^+(s) \cdot \hat{G}_p^-(s) \quad (32)$$

where $\hat{G}_p^+$ represents the terms of time delay and all the right-half plane zeros, $\hat{G}_p^-$ represents the invertible part. Since the system doesn't contain any non-minimum phase element, $\hat{G}_p^*(s)$ is equal to $\hat{G}_p^-(s)$.

ii. Because the trajectory for the removed thickness is a ramp function (increase linearly with time), the type-2 input is assumed for the design of the controller. Thus, the IMC filter (**F**) is chosen as:

$$\mathbf{F} = diag[f_i] \quad (33)$$

, and

$$f_i = \frac{2\tau_{CL,i}s + 1}{(\tau_{CL,i}s + 1)^2} \quad (34)$$

where $\tau_{CL,i}$ is the closed-loop (filter) time constant in each loop. In this work, the filter time constant is set to 30% of the open-loop time constant.

iii. The IMC controller is thus obtained.

$$C_{IMC} = (\hat{G}_p^*)^{-1}\mathbf{F} \quad (35)$$

Substituting C_{IMC} into Eq. (34), and the diagonal PI controller (C_{PI}) becomes:

$$C_{PI} = C_{IMC}[I - \hat{G}_p^*C_{IMC}]^{-1} = (\hat{G}_p^*(s))^{-1}\frac{f_i}{1 - f_i} \quad (36)$$

In a matrix form, the diagonal multivariable PI controller becomes:

$$C_{PI} = \Sigma^{-1} \cdot \begin{bmatrix} \frac{2\tau_{CL,1}s+1}{\tau_{CL,1}^2 s} & 0 & 0 \\ 0 & \frac{2\tau_{CL,2}s+1}{\tau_{CL,2}^2 s} & 0 \\ 0 & 0 & \frac{2\tau_{CL,3}s+1}{\tau_{CL,3}^2 s} \end{bmatrix} \quad (37)$$

Once the diagonal controllers become available, the full multivariable PID controller can be obtained by pre-multiplying the inverse of dynamic element ($\mathbf{D}^{-1}$) and the $\mathbf{V}_c$ matrix (from the SVD) (Fig. 4b). The resultant controller C_{PID} becomes:

$$C_{PID} = \begin{bmatrix} v_{11}C_{PI,1}(\tau_1 s+1) & v_{12}C_{PI,2}(\tau_1 s+1) & v_{13}C_{PI,3}(\tau_1 s+1) \\ v_{21}C_{PI,1}(\tau_2 s+1) & v_{22}C_{PI,2}(\tau_2 s+1) & v_{23}C_{PI,3}(\tau_2 s+1) \\ v_{31}C_{PI,1}(\tau_3 s+1) & v_{32}C_{PI,2}(\tau_3 s+1) & v_{33}C_{PI,3}(\tau_3 s+1) \end{bmatrix} \quad (38)$$

This multivariable controller consists of a diagonal PI type of controller, a $\mathbf{V}_c$ matrix which is generated from the SVD of system, and an inverse matrix of **D** (containing the derivative term). Thus, it becomes a full multivariable PID controller. Each element of this multivariable controller can be expressed analytically:

$$c_{PID,ij} = \sigma_i^{-1} v_{ij} \cdot \left(\left(\frac{\tau_i + 2\tau_{CL,i}}{\tau_{CL,i}^2}\right) + \left(\frac{1}{\tau_{CL,i}}\right)\frac{1}{s} + \left(\frac{2\tau_i}{\tau_{CL,i}}\right)s \right) \quad (39)$$

In terms of PID settings, the controller parameters, proportional gain, reset time and derivative time, thus become:

$$K_{c,ij} = \frac{v_{ij}(\tau_i + 2\tau_{CL,i})}{\sigma_i \tau_{CL,i}^2} \quad (40)$$

$$\tau_{I,ij} = \frac{v_{ij}}{\sigma_i \tau_{CL,i}} \quad (41)$$

$$\tau_{D,ij} = \frac{2v_{ij}\tau_i}{\sigma_i \tau_{CL,i}} \quad (42)$$

Taking the flat profile as an example, Fig. 5a shows the snapshots of the surface profiles throughout the polishing process. The control performance can be seen from the responses of the manipulated variables (Fig. 5 b & c) and the tracking error (Fig. 5d). In terms of the principle output (Fig. 5d) and input (Fig. 5c), the output settles to the set point in less than 20 seconds and acceptable control performance is obtained using the SVD-based multivariable controller. For CMP process, the ultimate performance measure in the uniformity. Table 1 summarizes all the indices using the ratio control as well as the multivariable feedback control. Comparison is made with respect to the achievable performance (Table 1). Results clearly indicate that the multivariable feedback control leads to the performance close to the achievable performance while the ratio control shows 4-15% deviations from the achievable one. More importantly, the seemingly complex control system can be design systematic manner with rather standard feedback design methodology.

4. CONCLUSION

In this work, a systematic modeling and control system design approaches are proposed for a multi-zone CMP system. The multi-zone CMP is intended to reduce the within-wafer non-uniformity (WIWNU) by manipulating different pressures across the radial position. This leads to a non-square multivariable system, a 60x3 system for the example studied. In the modeling phase, two 2^3 full factorial experimental design is carried out and steady-state gain matrix is obtained from the lease square regression followed by including the dynamic element associated with each input. For multivariable control, two control strategies with different degrees of complexity are proposed. First, a simple ratio control is designed based on the pseudo-inverse of the process model and, provided with the initial surface profile, the input pressures are computed and the ratios are maintained throughout the polishing run. The results show that, while giving reasonable uniformity, the performance is a little short from the achievable performance. Next, an on-line multivariable control system is designed. The singular value decomposition (SVD) is used to project the input and output in the principle directions and, therefore, the controllers can be designed in a reduced dimension (3x3) for a decoupled system. Then, the diagonal controllers are transformed back to the true input/output spaces. This significantly reduces the engineering effort in control system design. Results show that achievable control performance can be maintained using the SVD-based multivariable controllers.

ACKNOWLEDGEMENT

SJS and CCY thank National Science Council of Taiwan for the financial support under the grant NSC 92-2214-E002-032.

REFERENCES

Edgar, T. F.; S. W. Butler; W. J. Campbell; C. Pfeiffer; C. Bode; S. B. Hwang; K. S. Balakrishnan; J. Hahn, "Automatic control of microelectronics manufacturing: practices, challenges and possibilities", *Automatica, 36, 1567-1603 (2000).*

Kao, Y. C.; C. C. Yu; S. H. Shen "Robust Operation of Copper Chemical Mechanical Polishing", *Microelectronic Eng., **65**, 61-75 (2002).*

Preston, F. W., "The theory and design of plate glass polishing machines", *J. Soc. Glass Technology*, 11, 214-256 (1927).

Yao, C. H.; D. L. Feke; K. M. Robinson; S. Meikleb, "The influence of feature-scale surface geometry on CMP processes", *J. Electrochem. Soc.*, **147**, 3094-3099 (2000).

Table 1. Different uniformity measures of wafer surfaces before and after CMP using ratio control, multivariable feedback control and achievable performance (from Eq. 17) also given for comparison

Surface profile		Flat	IC1	IC2	IC3
Initial	SD	0	62.21	88.47	159.4
	∞-norm	0	378	636	603
	range	0	444	714	734
	NU (%)	0	0.71	1	1.15
Achievable	$\left\|E_{\min}\right\|_2$	727	535	698	1170
	SD	95.46	70.3	91.61	153.6
	∞-norm	189	142	321	457
	range	363	272	541	802
	NU (%)	4.77	3.51	4.58	7.68
Final (Top: ratio control Bottom: *multivariable feedback control*)	$\left\|E\right\|_2$	786 *751*	618 *579*	740 *703*	1218 *1201*
	SD	101 *95.23*	77.5 *70.04*	97 *91.52*	157 *153.33*
	∞-norm	225 *189*	223 *143*	330 *322*	444 *455*
	range	379 *362*	345 *273*	528 *541*	778 *800*
	NU (%)	5.01 *4.7*	3.83 *3.45*	4.83 *4.55*	7.72 *7.53*
Relative error*		8.12 *3.3*	15.51 *8.22*	6.02 *0.72*	4.1 *2.65*

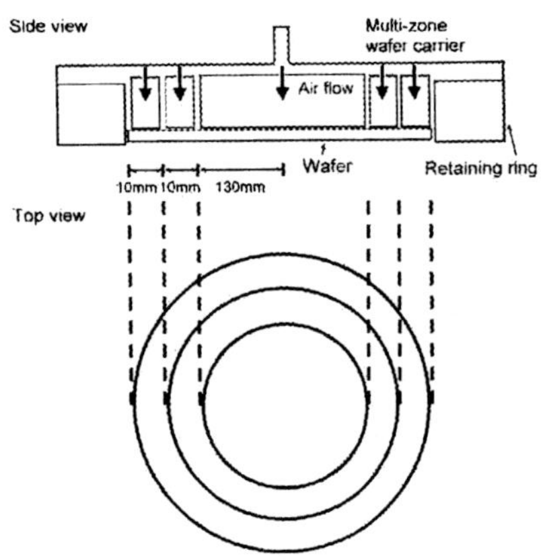

Fig. 1. Configuration of multi-zone CMP.

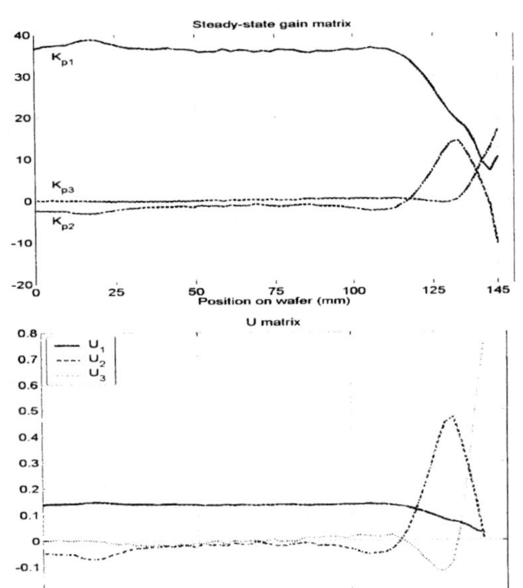

Fig. 2. Steady-state gain vectors and vectors of the U matrix from SVD

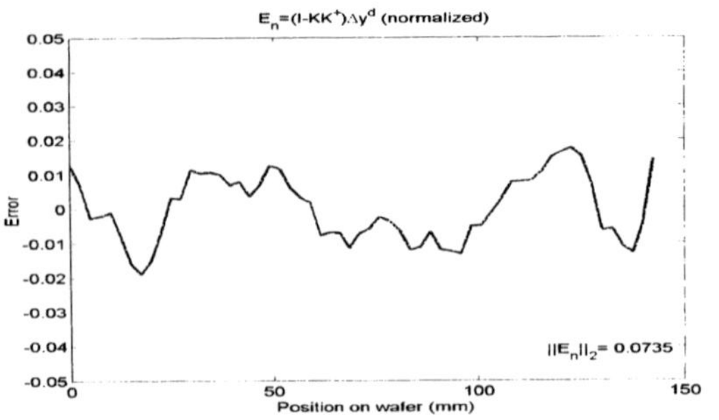

Fig. 3. The error vector of achievable performance by assuming input vector of unity.

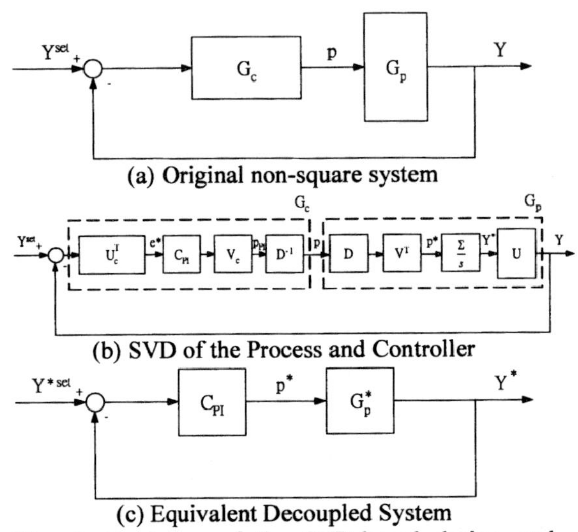

Fig. 4. Evolution of the SVD-based design and corresponding block diagrams.

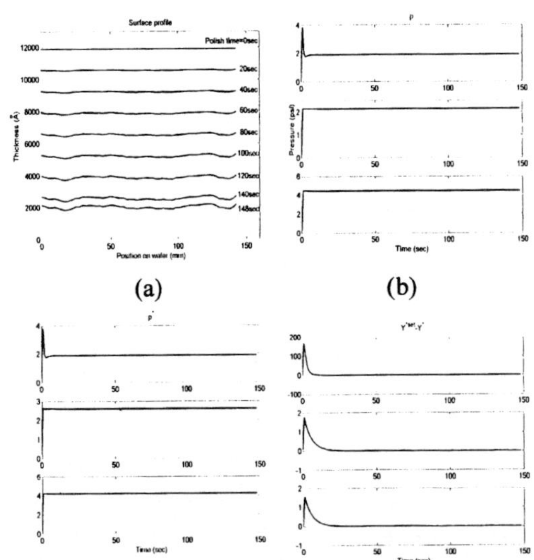

Fig. 5. Results of multivariable feedback control for incoming wafer with flat profile: (a) snap shots of surface profile, (b) pressures, (c) principal input, and (d) error in the principle direction.

ELSEVIER
IFAC
PUBLICATIONS
www.elsevier.com/locate/ifac

SYSTEM ANALYSIS OF COMPLEX REACTOR BEHAVIOR - A CASE STUDY.

Berit Floor Lund *,[1] **Bjarne A. Foss** *
Kjell Ragnar Lovasen ** **Birger Erik Ydstie** ***

* *Norwegian University of Science and Technology, NTNU*
** *Elkem ASA*
*** *Carnegie Mellon University*

Abstract: The paper examines the nonlinear behaviour between carbon coverage (input) and silicon production (output) in a submerged arc silicon furnace. A silicon furnace is a highly endothermic chemical reactor. Carbon coverage is the main input for control of the silicon yield and is determined manually. This is a difficult task since the dynamic response varies with the carbon level in the furnace. Internal interconnections between the reactions in the process cause the dominant zeros and poles in the response to move. For high input values and high silicon yield, inverse responses and finally unstability result. The paper describes how the dominant poles and zeros move and identifies the internal interconnections in the reactor that cause the change in dynamics. *Copyright © 2004 IFAC*

Keywords: Reactor dynamic behaviour, zero and pole drift. Submerged arc silicon furnace.

1. INTRODUCTION

In this paper we take a reactor perspective on the silicon furnace, and try to explain and untangle its highly complex and changing dynamic behavior. The focus is the dynamic response between carbon coverage and the tap-rate of silicon. Carbon coverage is defined as a percentage value of the stoichiometric ratio between the silicon in the quartz feed and the carbon in the carbon materials fed to the furnace. The silicon tap-rate is the amount of silicon (measured in kg) tapped during one hour.

Many interconnected phenomena inside the silicon furnace make the behavior change dramatically from low to high carbon coverage values. By applying the views often used to analyze the dynamic behavior of integrated process systems such as distillation columns and/or chemical reactors, this paper contributes with new insight into the dynamic behavior of the silicon furnace.

The production of silicon metal is a large industry world wide. Elkem has approximately 20% of the market share for silicon metal world wide with net sales of 2.2 BNOK (210 BTons) in 2002, (*Elkem Annual Report 2002*, 2002). Elkem has many plants in Norway, Iceland, Canada and the United States, each containing several furnaces. Silicon metal is used for the production of silicones in the chemical industry, and as an important additive to other metals in the casting industry. Silicon metal is also further refined for production of solar cells and electronics. A valuable byproduct of silicon metal production is microsilica which is used as an additive to concrete, among other things.

All data in this paper are obtained using "Simod" a dynamic simulator for submerged arc silicon furnaces developed by Elkem, NTNU and Sintef.

[1] Supported by NTNU and Elkem

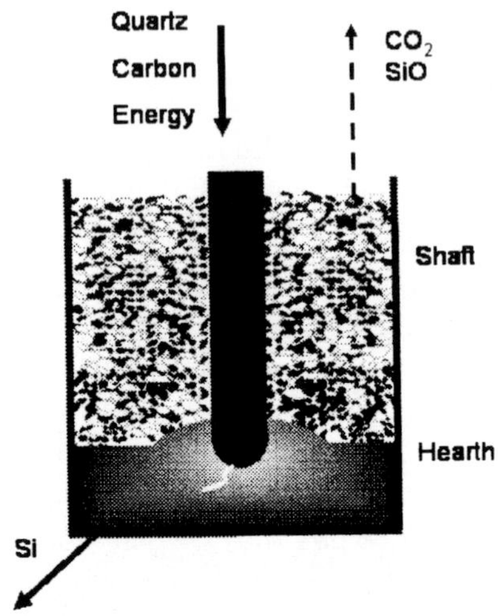

Fig. 1. Silicon furnace.

Simod is especially detailed with respect reaction kinetics and the thermodynamics of the furnace. The model is a nonlinear differential algebraic equation system, see (Foss and Wasbø, 2001). Simod was originally developed for teaching and training purposes of plant operation personnel in Elkem. It has also provided a valuable means to collect and consolidate Elkem's knowledge about the silicon furnace behavior. Simod has been used over several years, and its behavior has been verified by metallurgists and other process personnel.

The contents of the paper are as follows. An introduction to the fundamentals of the silicon production and chemistry is given in section 2. Section 3 describes and quantifies the properties of the steady state and dynamic response between carbon coverage and silicon tap-rate. The changing dynamic behavior of the process has been shown using small perturbations around a steady state for different carbon coverage levels. This can be considered a linear approximation to the nonlinear behavior. The time constants and gain of the linear model approximation have been estimated in order to quantify the changing dynamic behavior of the process. An analysis of the results is given, in section 4. A discussion and conclusion is given in section 5.

2. SILICON PRODUCTION

A silicon furnace is basically a large pot (3-4m high, 3-5m in diameter) in which carbon materials and quartz (SiO_2) are fed at the top and silicon metal (Si) is tapped at the bottom, see figure 1. Some of the SiO gas produced in the furnace escapes over the furnace top, oxidizes and forms microsilica which are captured in the off-gas filter.

The silicon furnace is highly endothermic. The main energy source to the process is electricity supplied through carbon electrodes. Depending on the size of the furnace, the electric power supply typically lies between 10 and 25MW. Most of this energy is released through an electric arc in the in the lower part of the furnace ("hearth") where silicon is formed and tapped (Valderhaug, 1992). The upper, cooler part of the furnace is called the "shaft".

The measurements of the mass transport of carbon, quartz into the furnace and amount of tapped silicon out of the furnace are registered as aggregated values on an hourly, shift (8 Hours) or day (24 Hours) basis. Microsilica production is in some plants measured for individual furnaces.

2.1 Silicon furnace chemistry

In essence, silicon production implies removing the oxygen atoms from the quartz molecule, SiO_2. The overall, gross reaction for the whole furnace can be written:

$$SiO_2+(1+x)C=xSi+(1-x)SiO+(1+x)CO \quad (1)$$

where x is the ratio of silicon metal recovered from the silicon in the quartz feed.

In reality a number of reactions take place in order for Si to be formed. The "engine" of the reactions in the furnace is silicon oxide gas, SiO. This gas is formed in the hearth in a highly endothermic reaction between silicon metal and melted SiO_2:

$$Si+SiO_2=2SiO \quad (2)$$

Some of the SiO is consumed within the hearth to form silicon metal according to the reaction:

$$SiC+SiO=2Si+CO \quad (3)$$

This reaction requires a sufficiently high partial pressure of SiO as well as a temperature of approximately 2000°C, (Schei *et al.*, 1998) which is obtained by the electric arc. Most of the silicon carbide required in reaction (3) is formed in the shaft in a reaction between SiO rising up from the hearth and the carbon feed in the shaft:

$$SiO+2C=SiC+CO \quad (4)$$

SiC is a solid, and travels down to the hearth. Reaction (4) may also take place in the hearth if unreacted carbon reaches the hearth.

Some of the SiO that rises up through the shaft condenses, and travels down with the other solids to the hearth or lower shaft as Si/SiO_2 condensate. The SiO condensation is a significant energy supply to materials in the shaft.

The component flows between the reactions and the hearth and shaft are shown in figure 2.

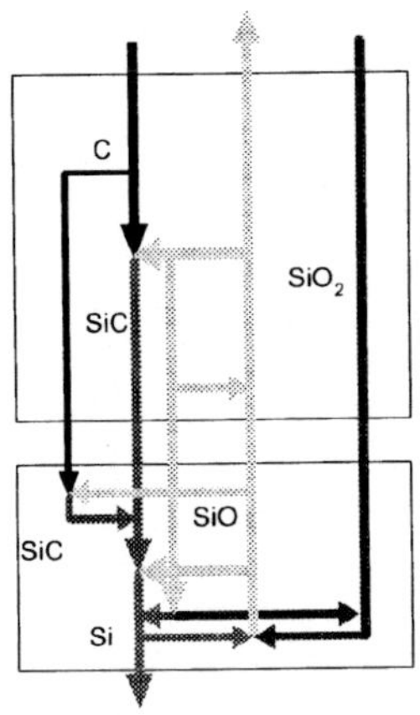

Fig. 2. Component flows in the silicon furnace shaft (upper) and hearth (lower).

3. CARBON COVERAGE TO SILICON TAP-RATE RESPONSE

This section gives a description of the dynamic behavior of the silicon furnace with an emphasis of the response with respect to changes in carbon coverage to the silicon tap-rate. The reasons for focusing on this particular input output pair is that carbon coverage is the main variable used for controlling the silicon yield. Also, silicon is the main product of the process is one of the key quantities for process operation personnel to control the process and evaluate its performance. We first will take a look at the steady state gain and dynamic responses at various carbon coverage levels. Next the changes in the dynamics are quantified.

3.1 Dynamic behavior and steady state gain

The steady state gain for carbon coverage to silicon tap-rate is plotted in figure 3. A carbon coverage of 100% would be consistent with 100% silicon recovery. Due to some loss of silicon over the furnace top, the average carbon coverage value should lie below 100%. In the curve below, the carbon coverage has been varied from 93% to 99%. These values include a 3% loss of carbon at the furnace top, and 3% should therefore be subtracted to get the net carbon coverage value entering the reactions of the furnace.

The simulations have been made with a net 22MW power supply and a reactivity parameter for the carbon of 0.56. A feed controller in the simulator ensures a constant material level of the furnace, and a tapping controller maintains a constant Si level in the hearth.

We see that the steady state gain is relatively constant all up to 99%. Above 99.5% there is a buildup of silicon carbide in the hearth. This

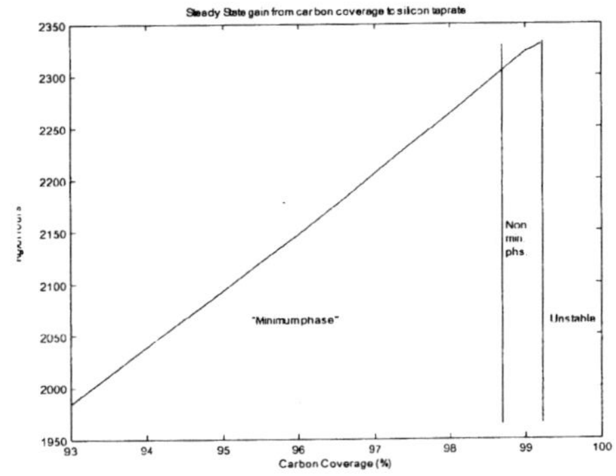

Fig. 3. Steady state gain between carbon coverage and taprate silicon.

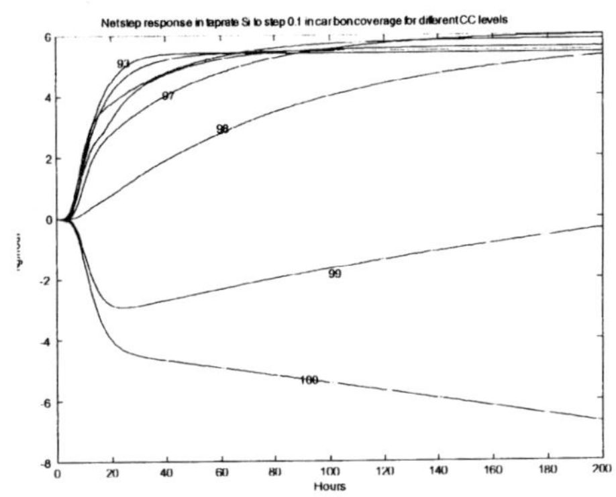

Fig. 4. Net step responses in taprate Si to a 0.1 step in carbon coverage (CC) at different carbon coverage values.

creates an integrating effect, and no steady state gain can therefore be defined for such high carbon coverage values.

The following method has been chosen to see how the dynamic response evolves for different carbon coverage values. For the carbon coverage values (including a 3% loss): 93, 94, 95, 96, 97, 98, 99,100 we have run the simulator to reach a steady state value (not obtainable at 100%). At each of these steady state situations, we have made a small positive perturbation (step) by adding 0.1 to the carbon coverage value at time 1 Hours and registered the silicon tap-rate response. In order to compare the dynamic response more easily, we have for each carbon coverage level, subtracted the steady state silicon tap-rate value so that each response starts from zero. As we can see, the responses are much faster for small carbon coverage values than for the higher values. The dominant time constant at a carbon coverage value of 93% (net 90%) lies around 8 Hours whereas for 98% it is approximately 100 Hours. We also can observe some nonlinear effects in the responses corresponding to 92% and 93%. At 99% carbon coverage we observe an inverse response in the tap-rate, thus we have indicated in figure 3

Carbon coverage	K	T1	T2	T3	T4
93	54,4	*	*	4,9	4,9
94	55,5	31,2	34,6	2,7	7,8
95	56,2	28,5	34,2	0,5	14,1
96	59,3	17,1	33,3	2,7	2,7
97	61,7	16,9	46,7	3,7	3,7
98	58,2	5,1	78,4	5,0	5,0
99	56,0	-306,7	688,8	3,1	3,1

Fig. 5. Identified poles and zeros (in Hours).

that this is a non-minimum region. Since we have included only 200 Hours, the figure does not reveal the fact that this response has approximately the same steady state gain as the other curves. For 100% (net 97%) we see that there is a net negative integrating effect, and no steady state gain can be defined for this input value. The negative gain makes the process is unstable in this region.

3.2 Quantification of the change in dynamic behavior

The following approximation to the linearized behavior of the process has been used

$$H(s) = \frac{K_1 e^{-\tau s}(1 + T_1 s)}{(1 + T_2 s)(1 + T_3 s)(1 + T_4 s)} \quad (5)$$

Even though we have said the process is nearly minimum-phase for small carbon coverage values, we still have included a small transport delay to capture some of the propagation time for the materials down through the furnace. This delay is estimated to approximately 2.5 Hours for all responses, and is not included in the table of figure 5. A third order model was necessary to achieve sufficient accuracy for the whole input range. The parameters have been estimated using a least squares method ("lsqcurvefit" in Matlab). All the time constants are in hours. As we can see from the table, the gain is relatively constant. The two right-most columns (T_3 and T_4) are the fastest poles of the process. The process seems to have a double pole which in the range 2.5-5.0 Hours. T_1 represents a zero and T_2 a pole. For 93% the response is better identified with an approximation model with only two poles. For higher carbon coverage values the time constant of the pole (T_2) increases. The time constant of the the zero (T_1) decreases, and finally goes negative. For a carbon coverage value of 99%, the zero is faster than the pole, and since the time constant is negative, the result is an inverse response. The negative integrating effect observed for carbon coverage equal to 100% can be interpreted as the time constant for the pole T_2 having become infinitely large. The development of the dominant time constant in the pole (T_2) and zero (T_1) have been plotted against the carbon coverage level in figure 6.

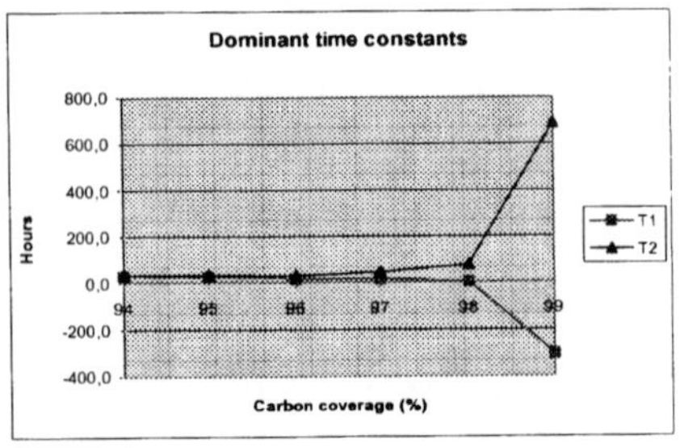

Fig. 6. Development of the dominant time constants.

4. ANALYSIS OF THE CHANGE IN DYNAMIC BEHAVIOR

The changing dynamic behaviour of the carbon coverage to silicon tap-rate response seen in figure 4 and identified in figure 5 has a dramatic effect on process operation, especially since carbon coverage is determined manually. At low silicon yield it is possible to have a "fast" process, and the effects can be observed within a shift. With high carbon coverage the response is extremely slow and difficult to relate to in manual operation. Our aim in this section is to explain the development of the dominant zero and pole by looking at the internal interconnections and component flows in the process.

4.1 Process interconnections and plant dynamics

The basis for a system theoretic analysis, is the literature dealing with the changing dynamics of reactors due to interconnections in process plants, see for instance (Morud and Skogestad, 1996). The paper classifies different process interconnections and gives an overview over the effect each type has on the plant dynamics. On the overall level they distinguish between external and internal interconnections. By external interconnections are meant interconnections between subsystems associated with different processing equipment. Internal interconnections mean interconnections between phenomena within one process vessel. Within the internal interconnection "class" they subdivide into recycle, parallel paths and series interconnections. Series interconnection is the simplest kind of interconnection and implies that there is a one-way flow between subsystems of material and/or energy. In a system with recycle, mass and/or energy flow is fed back in the process. Recycle generally moves the poles of the process. Which way, generally depends on whether the feedback is positive or negative. In most systems, recycle leads to positive feedback, according to (Morud and Skogestad, 1996). The third form of internal interconnections is parallel paths, or process "feedforward". The existence of parallel paths creates or moves the zeros of the process. If the effects of the parallel paths have opposite signs, they are competing effects which may give unstable zero dynamics, or inverse responses for linear systems. An example of this can be found in (Kuhlmann and Bogle, 1997) who describes a van de Vusse process in which there are competing effects causing the zeros of the process to move. In the van de Vusse process the unstable zero dynamics also cause a sign change in the steady state gain (input multiplicity).

To apply the ideas presented in (Morud and Skogestad, 1996) on the silicon furnace, we take a look back at figure 2 and try to identify the types of internal interconnections that we might have in the component flows in the silicon furnace from carbon feed to tapped silicon in order to explain the changing dynamics. There are many types of interconnections in this process, and the different types cannot be easily distinguished.

4.2 Steady state conversion levels of the reactions

In order to determine which are the dominant phenomena within the silicon furnace, we need to quantify the steady state conversion rates in the different reactions at different carbon coverage levels. In figure 7 we have plotted the steady state conversion rate of the Si producing reaction (4) and the SiO producing reaction (2). The difference in generated kmol/h of SiO is approximately 4 (twice the increase in the reaction) from the highest and to the lowest carbon coverage level, whereas the difference in consumed SiO in the Si producing reaction is almost 10 kmol/h. Since

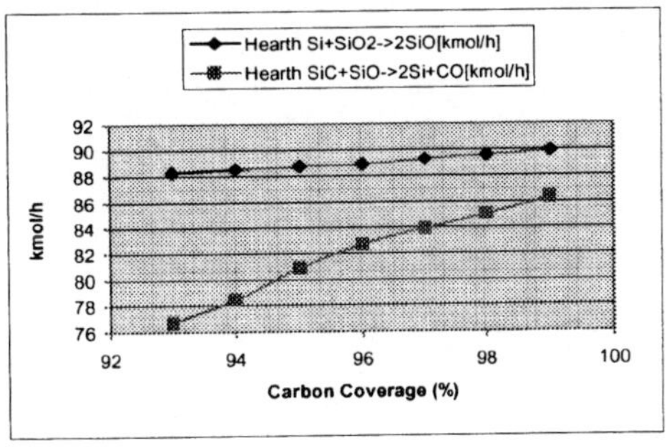

Fig. 7. Stationary conversion rates for SiO and Si producing reactions in hearth.

the SiO consumption in the hearth increases more than the production, the SiO level in the shaft will be lower for higher carbon coverage levels.

Next we look at the SiC generating reaction (4) which also consumes SiO. In figure 8 we have plotted the conversion rates for this reaction in the shaft (upper) and hearth (lower). For low carbon coverage rates, most of the SiC is formed in the shaft and little in the hearth. For higher carbon coverage rates less SiC is formed in the shaft and more in the hearth. The explanation lies the increased SiO consumption in the hearth due to increased production of Si, as seen in figure 7. In addition, the increased production of SiC in the hearth reduces the SiO in the shaft further. The last reaction we need to quantify is

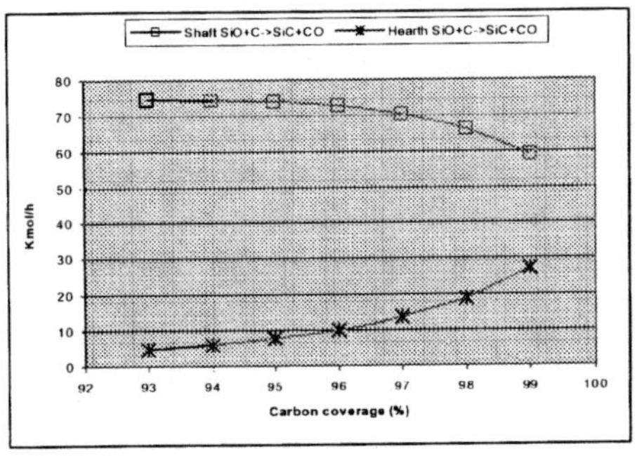

Fig. 8. SiC generation in the shaft (upper) and hearth (lower).

the recycling through condensation of SiO. The simulations show that most of this recycling takes place within the shaft itself, and that very little condensate reaches the hearth (0.01-0.02 kmol/h). This is a very small rate compared to the rates of the reactions in the figures above, and the "material recycle" of Si through SiO/SiC therefore dominates. We therefore ignore the effect of recycle through condensation of SiO in the shaft in the following analysis.

4.3 Analysis of the dominant mechanisms

By ignoring the recycling of SiO through condensation in the shaft, we can now redraw figure 2 emphasizing the response from carbon to tapped silicon. The boxes in figure 9 indicate reactions.

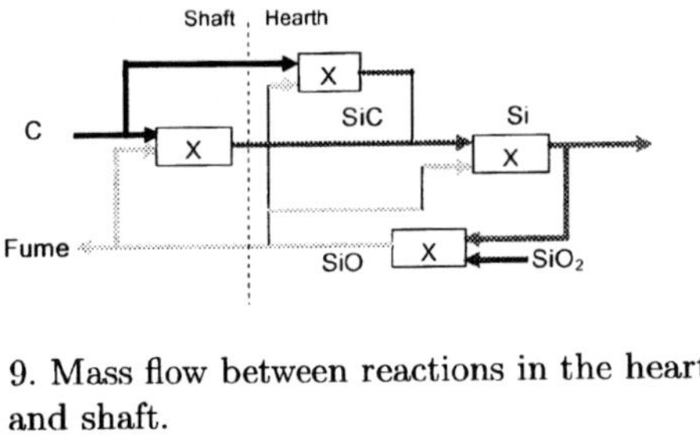

Fig. 9. Mass flow between reactions in the hearth and shaft.

The Si level in the hearth is kept constant by a level controller, and any surplus of Si is tapped. Si is also recycled through as SiO in the lower right box. SiO is consumed by the SiC generating and Si generating reactions in the hearth. What is not consumed in the hearth rises up through the shaft and is consumed by the SiC generating reaction there. The rest of the SiO gas escapes as fume over the furnace top.

One reasom for the slower response for high carbon coverage levels, is that the carbon in the hearth needs time to react to SiC, whereas for low carbon coverage values, the SiC is formed in the shaft and comes down to the hearth "ready" to enter the Si producing reaction. Another reason for the slower response at high carbon coverage rates is the decreased amount of SiO gas in the furnace, see the steady state profiles for SiO in figure 10. Elements 1 to 10 on the x-axis correspond to the top to bottom elements of the shaft, element 11 corresponds to the hearth. Since the partial pressure of SiO (above an equilibrium pressure) drives the SiC and Si generating reactions, less free SiO in the furnace will clearly give a slower response in all reactions. When looking for a source

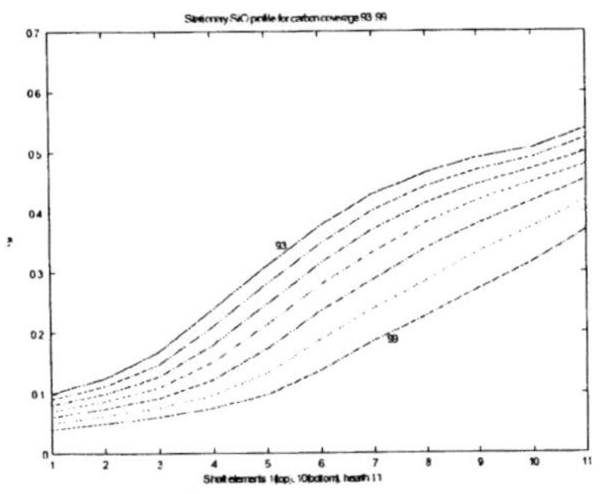

Fig. 10. The steady state SiO profile for carbon coverage values from 93% to 99%.

of the moving zero, (Morud and Skogestad, 1996) indicate that we should look parallel and possibly competing phenomena. The two reactions in the hearth producing SiC and Si are obvious candidates since they are exposed to the same partial pressure of SiO. SiC production in the hearth is insignificant for low carbon coverage values according to figure 8, but amounts to almost half of the SiC production for 99% carbon coverage. In the low carbon situation there is a relatively large surplus of SiO in all parts of the furnace. For a high carbon coverage level, the general level of SiO in the furnace goes down, more unreacted carbon will enter the hearth, and the SiC generating reaction in the hearth will take a more and more significant part of the SiO in the hearth.

5. DISCUSSION AND CONCLUSION

This paper has shown that speed of the response in silicon tap-rate to a change in carbon coverage is an indication of distance from maximal silicon production. A slow response indicates that the furnace operates very close to maximal production, whereas a fast response indicates that it is possible to increase the carbon coverage level. The carbon coverage level is in most furnaces determined manually. The results of the paper can used to help identify the operational margins wrt. carbon coverage level.

There are however other factors and inputs influencing the silicon furnace behaviour which have not been treated in this paper. Firstly, the reactivity of the carbon material is an obvious candidate for further studies since it enters linearly in the SiC generating reaction. A higher reactivity would cause more of the carbon to react in the shaft and less in the hearth.

Another effect not included in the simulations of this paper is the possibility for formation of a particularly hard and non reactive form of SiC. This effect may occur when there is a sufficiently high amount of carbon in the hearth (at maximal production), and may cause build up and severe tapping problems in the furnace. Inclusion of this effect would lower the optimal carbon coverage level , in the range of 0.5-1%.

Other inputs not considered in the paper are the electrical energy supply level, and vertical energy distribution in the furnace. In (Schei *et al.*, 1998) and (Lund *et al.*, 2004) it was shown that the energy supplied to the furnace affects the production level of silicon. Investigation of the effect of the reactivity and the power level on the dynamic behavior of the furnace should therefore be considered for future work and before an automatic controller structure is suggested.

REFERENCES

Elkem Annual Report 2002 (2002).

Foss, B. A. and S. O. Wasbø (2001). An integration scheme for stiff solid-gas reactor models. *Computer Methods in Applied Mechanics and Engineering* **190/45**, 6009–6021.

Kuhlmann, A. and D. Bogle (1997). Study on nonminimum phase behavious and optimal operation. *Computers Chem. Engineering* **21**, 397–402.

Lund, B. F., B. A. Foss, K. R. Lovasen and B. E. Ydstie (2004). Sensitivity analysis of a dynamic model for submerged arc silicon furnaces. *Proceedings of the Infacon X Congress.*

Morud, J. and S. Skogestad (1996). Dynamic behaviour of integrated plants. *Journal of Process Control* **6**(2/3), 145–156.

Schei, A., J. K. Tuset and H. Tveit (1998). *Production of High Silicon Alloys.* Tapir forlag. Trondheim, Norway.

Valderhaug, Aa. M. (1992). Modelling and Control of Submerged-Arc Ferrosilicon Furnaces. PhD thesis. Norwegian Institute of Technology.

ELSEVIER
IFAC
PUBLICATIONS
www.elsevier.com/locate/ifac

A FAST SUBOPTIMAL MULTI PARAMETRIC QUADRATIC PROGRAMMING ALGORITHM FOR PREDICTIVE CONTROL

J.A. Rossiter *

* *University of Sheffield, Dept. Automatic Control and Systems Engineering, S1 3JD, UK*

Abstract: Multi parameteric quadratic programming gives a full offline solution to a time varying quadratic programming (QP) problem arising during constrained predictive control. However, coding and implementation of this solution may be more burdensome than simply solving the original QP. This paper proposes an algorithm, which by accepting a small amount of suboptimality in the predicted control trajectories, achieves a large decrease in both the online computation and data storage requirements. *Copyright © 2004 IFAC*

Keywords: Efficient predictive control, constraints, quadratic programming

1. INTRODUCTION

Obstacles to more widespread use of predictive control (Mayne *et al.*, 2000) are the computational demand and the lack of transparency due to the use of an online optimisation. Typical MPC algorithms are based on the minimisation of a quadratic performance index subject to linear constraints, that is a quadratic programming (QP) problem. In many cases (Rossiter *et al.*, 2002), the control law must have a simpler implementation: (i) coding limitations may make a QP optimiser inpractical and (ii) computational time restricts implementation on fast systems.

A recent contribution (Bemporad *et al.*, 2002), denoted multi parametric quadratic programming (MPQP), solves offline, all possible QP problems that can arise on line and shows that the solution is piecewise affine, that is, within computable regions of the state space, the the input trajectory has a known affine dependence on the current state. Hence solving QP is equivalent to set membership tests; if the state is inside region 'k', then use the associated law. A major strength of MPQP is visibility; implementation is akin to gain scheduling in that the control law is modified according to the current state. Unfortunately, MPQP may not be efficient due to the potentially large number of regions/laws to be stored and hence it could be harder to implement than the original QP optimiser.

This paper proposes one means of reducing the data storage requirements and implementation time of MPQP by allowing a small degree of suboptimality. Other authors are also considering this problem (e.g.(Bemporad *et al.*, 2001; Grieder *et al.*, 2003; Tondel *et al.*, 2003)) but from a different perspective. The paper gives a brief overview of MPC and MPQP, shows how a suboptimal algorithm can reduce complexity with little detriment to performance.

2. BACKGROUND

2.1 Modelling and MPC

This paper makes use of state-space models,

$$\mathbf{x}_{k+1} = A\mathbf{x}_k + B\mathbf{u}_k; \quad \mathbf{y}_k = C\mathbf{x}_k \tag{1}$$

where $\mathbf{x}, \mathbf{u}, \mathbf{y}$ are the state, input and output respectively. Associated to the model are constraints:

$$\underline{\mathbf{u}} \leq \mathbf{u}_k \leq \overline{\mathbf{u}}; \quad \underline{\mathbf{x}} \leq \mathbf{x}_k \leq \overline{\mathbf{x}} \tag{2}$$

Define the d.o.f. via a dummy variable $\mathbf{c}$:

$$\begin{aligned} \mathbf{u}_k &= -K\mathbf{x}_k + \mathbf{c}_k; \quad & k = 0, ..., n_c - 1 \\ \mathbf{u}_k &= -K\mathbf{x}_k; & k \geq n_c \end{aligned} \tag{3}$$

where K is the optimal unconstrained feedback. Let $\mathbf{C}^T = [\mathbf{c}_0^T, \mathbf{c}_1^T, ..., \mathbf{c}_{n_c-1}^T]$. Then an MPC optimisation (Scokaert *et al.*, 1998; Rossiter *et al.*, 1998) of performance subject to constraint satisfaction can be written as:

$$\min_{\mathbf{C}} \; J = \mathbf{C}^T S \mathbf{C} \quad \text{s.t.} \quad M_c\mathbf{C} + M_x\mathbf{x}_0 - \mathbf{v} \leq 0 \tag{4}$$

Details of matrices S, M_x, M_c and $\mathbf{v}$ are omitted for brevity. The first component of $\mathbf{C}$, that is $\mathbf{c}_0$ is used to formulate the control law of (3).

2.2 Multi parametric quadratic programming

Consider the minimisation (4). The optimising $\mathbf{C}$ depends upon $\mathbf{x}$. Define regions $\mathcal{S}_i, \;\; i = 0, 1, ...$:

$$\mathcal{S}_i = \{\mathbf{x} : M_i\mathbf{x} - \mathbf{d}_i \leq 0\} \tag{5}$$

such that within each region the active set is the same and hence the $\mathbf{C}$ optimising (4) has a known affine dependence on $\mathbf{x}$. The following algorithm gives the optimal control:

Algorithm 2.1. Find i s.t. $\mathbf{x} \in \mathcal{S}_i$, then

$$\mathbf{x} \in \mathcal{S}_i \;\Rightarrow\; \mathbf{c} = \mathbf{e}_1^T(-\hat{K}_i\mathbf{x} + \mathbf{t}_i) \;\Rightarrow\; \mathbf{u} = -K_i\mathbf{x} + \mathbf{p}_i \tag{6}$$

where $\mathbf{e}_1^T = [I, 0, 0, ...]$, and K_i, $\mathbf{t}_i$, $\mathbf{p}_i$ have obvious definitions (Bemporad *et al.*, 2002).

If the number of sets $\mathcal{S}_i$ is small, then MPQP is very efficient. However, if the number of sets is large then the implied search and data storage requirements can become unmanageable.

2.3 ONEDOF algorithms

ONEDOF algorithms (Rossiter *et al.*, 2001) reduce the computational burden of constraint handling by: (i) allowing some suboptimality and (ii) making good use of offline information. In ONEDOF rather than minimising w.r.t. the control values $\mathbf{u}_k$, one uses a mixture of input trajectories associated to different control laws. For instance, with just two laws, the optimisation reduces to a single set membership test. The main weakness is possible restrictions to the feasible region as sometimes non-linear predictions are needed in transients to enlarge feasible regions.

The simplest algorithm makes use of a law $\mathbf{u} = -K_f\mathbf{x}$ with a large MAS and also the optimal control law $\mathbf{u} = -K\mathbf{x}$. One then takes a linear mix of the predicted control trajectories arising from each law and minimises the cost J over a scalar mixing variable, subject to constraint satisfaction. In summary the procedure reduces to:

Algorithm 2.2. **(ONEDOF)**

(1) Define the predicted future state and input trajectories (subscripts o, f denote optimal and feasible predictions) and α is the d.o.f. .

$$\mathbf{X} = (1-\alpha)\mathbf{X}_o + \alpha\mathbf{X}_f; \quad \mathbf{U} = (1-\alpha)\mathbf{U}_o + \alpha\mathbf{U}_f \tag{7}$$

(2) Substituting predictions (7) into the constraints of (2) gives the inequalities $\mathbf{M}\alpha - \mathbf{d} \leq 0$.

(3) Minimising J w.r.t. α is equivalent to:

$$\min_{\alpha} \;\; \alpha \quad \text{s.t.} \quad \begin{cases} \mathbf{M}\alpha - \mathbf{d} \leq 0 \\ 0 \leq \alpha \leq 1 \end{cases} \tag{8}$$

(4) Compute the current control action from

$$\mathbf{u}_k = \mathbf{e}_1^T[(1-\alpha)\mathbf{U}_o + \alpha\mathbf{U}_f]. \tag{9}$$

The minimisation (8) is trivial and hence efficient. Moreover, examples (Rossiter *et al.*, 2001) show that, where feasible, suboptimality is often negligible.

2.4 Summary

(1) MPQP suffers no loss in the size of the feasible region or optimality but may be inefficient.
(2) ONEDOF is very efficient, but may have a smaller feasible region and some suboptimality.

This paper combines the large feasible region of MPQP with the efficiency of ONEDOF.

3. IMPROVING THE EFFICIENCY OF MPQP

We can use inferences from control strategies associated to the MPQP feasibility boundary to gain a significant reduction in complexity and surprisingly little deterioration in performance.

3.1 MPQP and its region of attraction

The following lemmata are obvious and, due to space restrictions, stated without proof.

Lemma 3.1. The MPQP control law is defined and stabilising only in the region $\mathcal{S}_{max}$ where

$$\mathcal{S}_{max} = \bigcup_i \mathcal{S}_i = \{\mathbf{x} : M_{max}\mathbf{x} - \mathbf{d}_{max} \leq 0\} \quad (10)$$

Assume hereafter that $\mathcal{S}_{max}$ is in minimal form so that redundant constraints are removed. For a given pair K, n_c, $\mathcal{S}_{max}$ is the maximal volume feasible region.

Lemma 3.2. Each row of the linear inequalities describing $\mathcal{S}_{max}$ can be determined by forcing equality of a subset of the inequalities in $M_c\mathbf{C}+M_x\mathbf{x}-\mathbf{v} \leq 0$.

Theorem 3.1. If $\mathbf{x}$ lies on the facet of $\mathcal{S}_{max}$, then one can identify the associated region $\mathcal{S}_i$ and hence the underlying optimal control law far more efficiently than via a search over all the MPQP regions.

Proof: If $\mathbf{x}$ lies on a facet, then for some integer j:

$$\mathbf{e}_j^T[M_{max}\mathbf{x} - \mathbf{d}_{max}] = 0 \quad (11)$$

Finding j requires an equivalent computation of just one set membership test. One can now identify the region $\mathcal{S}_i$ in which $\mathbf{x}$ lies by doing a search over only those regions contributing to the ith facet. In general, the number of regions on a facet would be far fewer than the total number of regions. □

Hence an efficient procedure to find the optimal control law (for $\mathbf{x}$ on a facet) is:

Algorithm 3.1. **MPQP for boundary states**

(1) Identify the active facet from (11).

(2) From those regions which contribute to the jth facet, identify in which one $\mathbf{x}$ lies and implement the appropriate control law from (6).

Algorithm 3.1 is far more efficient that MPQP. First the number of facets will in general be less that the number of regions and moreover identifying the active facet requires only the computation of one inequality for each facet. Second, in general there will be only a few active sets on each facet, so the region search will be, relative to MPQP, simple.

3.2 Control law for states not on a facet

Here we extend algorithm 3.1 for interior points. Assume that $0 \in \mathcal{S}_{max}$. The following lemma can be used to establish the 'nearest' facet to a non-boundary state.

Lemma 3.3. Assume that $\mathbf{d}_{max}^T = [1, 1, ...]$ and the rows of M_{max} are m_j^T. Define P_j as the minimal volume polytopes containing the jth facet of $\mathcal{S}_{max}$ and the origin. Compute the values $\gamma_j = \mathbf{m}_j^T\mathbf{x}$ and compute j for which γ_j is a maximum. Then $\mathbf{x}$ lies in polytope P_j.

Corollary 3.1. Given $\mathbf{x} \in P_j$ and $\gamma_j = \mathbf{m}_j^T\mathbf{x}$ find the region $\mathcal{S}_i$ such that $\mathbf{x}/\gamma_j \in \mathcal{S}_i$. Then the control move

$$\mathbf{u} = -K_i\mathbf{x} + \gamma_j\mathbf{p}_i; \quad (12)$$

is the first move of a predicted control law with feasible and convergent predictions.

The control law implied by Corollary 3.1 requires only information about the boundary and therefore is far more efficient than MPQP: (i) only boundary regions and associated laws need be stored; (ii) set membership searchs are simplified to checking only those on a known facet. However, the cost of this simplification is suboptimality.

4. REDUCING SUBOPTIMALITY WITH ONEDOF

This section proposes a novel implementation of the ONEDOF algorithm to gain significant improvements in performance on algorithm 3.1 and control (12).

Algorithm 4.1. **ONEDOF:**

(1) Find the feasible $\mathbf{C}$ implied by Corollary (3.1). (γ_j times the $\mathbf{C}$ for $\mathbf{x}/\gamma_j$.)
(2) Solve the minimisation

$$\min_{\alpha} \ \alpha \ \text{s.t.} \ \alpha M_c \mathbf{C} + M_x \mathbf{x} - \mathbf{d} \leq 0 \quad (13)$$

(3) Implement the control law: $\mathbf{u} = -K\mathbf{x} + \mathbf{e}_1^T \alpha \mathbf{C}$.

This algorithm requires only a a trivial linear program, that is, in just one variable and therefore equivalent in complexity to a single set membership test. Despite this simplicity, one can often gain significantly by of way optimality compared to the control law of (12).

5. EXAMPLES

This section will illustrate two things. First the efficacy of control law (12) (denote as SCALING) and algorithm ONEDOF in giving near optimal performance and vast improvements in online efficiency.

5.1 Example 1

Take the discrete model and input and state limits

$$\mathbf{x}_{k+1} = \begin{bmatrix} 1 & 0.05 \\ 0 & 1 \end{bmatrix} + \begin{bmatrix} 0.0025 \\ 0.05 \end{bmatrix}; \quad y_k = [1 \ \ 0]\mathbf{x}_k$$

$$-1 \leq u_k \leq 1; \quad -0.5 \leq [0 \ \ 1]\mathbf{x}_k \leq 0.5 \qquad (14)$$

Compute the optimal control law with $Q = C^T C$, $R = 1$ as $K = [0.9653 \ \ 1.3655]$.

For $n_c = 2$ with state constraints and for $n_c = 6$ without state constraints, table 1 (and figures 1-3) gives the number of MPQP regions and facets. Comparing worst case scenarios, algorithm 4.1 gives a reduction in complexity over MPQP of a factor of at least 5. Set membership search complexity is at most 18 as opposed to 85 regions for $n_c = 2$ and at most 2 compared to 385 for $n_c = 6$! In both cases there are far fewer facets than regions; moreover locating the active facet from Lemma (3.3) is trivial.

Closed-loop simulations were performed for initial states near the feasibility boundary. The runtime cost was computed, based on J, for each algorithm and is summarised in table 2. Clearly ONEDOF has negligible suboptimality. Closed-loop simulations are given in figure 4 ((a,b) for one initial point and (c,d) for the other; the solid line for MPQP, dotted line for algorithm 4.1 and dashed line for the algorithm of eqn.(12)) and these show only small differences in behaviour.

MPQP	ONEDOF/SCALING	
Regions	Facets	Maximum number regions per facet
$n_c = 2$ with state constraints		
85	30	18
$n_c = 6$ without state constraints		
385	62	2

Table 1. Maximum search complexity

Initial state	MPQP	ONEDOF	SCALING
$n_c = 2$ with state constraints			
$\begin{bmatrix} -1.02 \\ 0.465 \end{bmatrix}$	17.91	18.08	19.87
$\begin{bmatrix} 1.1 \\ -0.13 \end{bmatrix}$	28.57	28.53	30.16
$n_c = 6$ without state constraints			
$\begin{bmatrix} -3.8 \\ 2.47 \end{bmatrix}$	200.98	202.29	206.30
$\begin{bmatrix} 3.2 \\ -1 \end{bmatrix}$	184.49	184.49	186.15

Table 2. Run time costs for example 1

5.2 Example 2

This model represents a simplified boiler for electricity generation:

$$A = \begin{bmatrix} 0.98 & 0 & 0.019 \\ 0.075 & 0.607 & 0.001 \\ 0 & 0 & 0.607 \end{bmatrix}; \quad B = \begin{bmatrix} 0.005 & -0.021 \\ 0.39 & 00.05 \end{bmatrix}; \quad C = \begin{bmatrix} 1.69 & 13.22 & 0 \\ 0.84 & 0 & 0 \end{bmatrix}$$

$$\begin{bmatrix} -1 \\ -1 \end{bmatrix} \leq \mathbf{u} \leq \begin{bmatrix} 1 \\ 1 \end{bmatrix}; \quad Q = C^T C, \quad R = I \quad (15)$$

A summary of the number of regions, facets and maximum regions per facet is given in table 3 for $n_c = 3$. Simulations are performed for initial points close to the boundary of $\mathcal{S}_{max}$ and the closed-loop run time costs are given in table 4. Once again it is evident that the ONEDOF algorithm has given only a small deterioration in performance relative to the huge increase in efficiency - search complexity down from 677 regions to 5!

MPQP	ONEDOF/SCALING	
Regions	Facets	Maximum number regions per facet
677	126	5

Table 3. Search complexity.

Initial state	MPQP	ONEDOF	SCALING
$\begin{bmatrix} 0.5 \\ 0.5 \\ 0.5 \end{bmatrix}$	56.51	68.61	149.6
$\begin{bmatrix} 0.6 \\ -0.6 \\ 0.6 \end{bmatrix}$	32.22	32.66	37.07
$\begin{bmatrix} 0.65 \\ -0.65 \\ -0.65 \end{bmatrix}$	32.50	32.61	36.68
$\begin{bmatrix} 0.55 \\ 0.55 \\ -0.55 \end{bmatrix}$	63.51	73.05	151.30

Table 4. Run time costs for example 2.

6. CONCLUSION

This paper has shown how the MPQP algorithm can be combined with the ONEDOF algorithm to obtain an algorithm with the same size of feasibility region, near optimal performance and yet a comparatively small computational load and far smaller data storage requirement. This algorithm has now been tested (and shown to be effective) on a large number of systems and the results are reported in (Rossiter *et al.*, 2004).

REFERENCES

Bemporad, A., M. Morari, V. Dua, E.N. Pistokopoulos, The explicit linear quadractic regulator for constrained systems, *Automatica*, 2002, 38, 1, pp 3-20

Bemporad, A. F. Borrelli and M. Morari, The explicit solution of constrained LP-based receding horizon control, Proc. European Control Conference, 2001.

Grieder, P., F. Borrelli, F. Torrisi and M. Morari, Computation of the constrained infinite time linear quadratic regulator, ACC, 2003

Mayne, D.Q., J.B. Rawlings, C.V. Rao and P.O.M. Scokaert, Constrained model predictive control: stability and optimality, *Automatica*, 2000, 36, pp789-814

Rossiter, J.A., M.J. Rice and B.Kouvaritakis, A numerically robust state-space approach to stable predictive control strategies, *Automatica*, 1998, 38, 1, 65-73

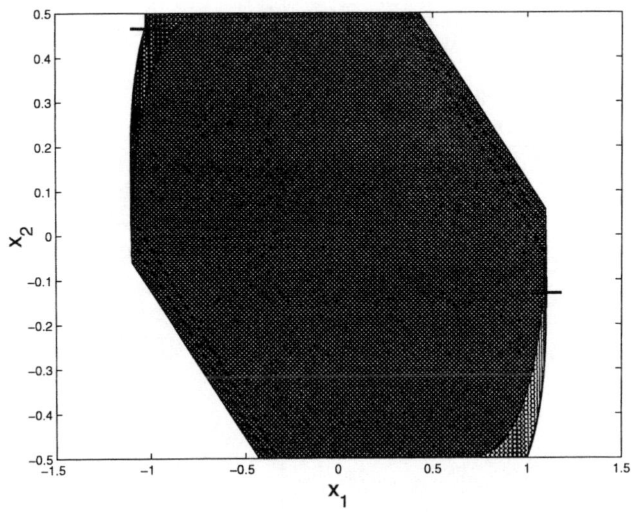

Fig. 1. Regions for example 1 with $n_c = 2$

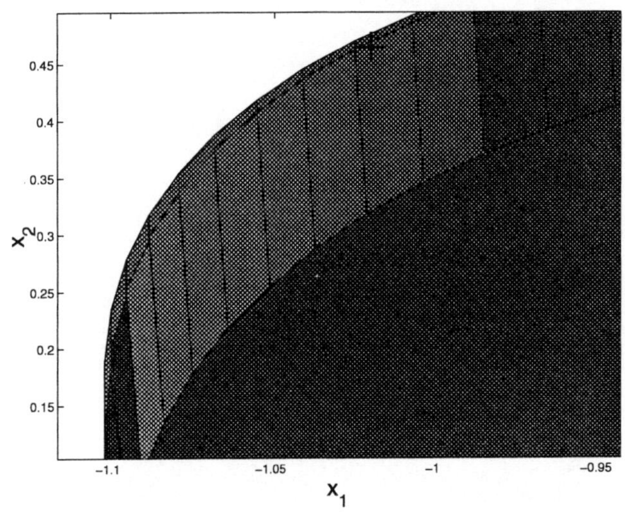

Fig. 2. Blown up view of regions for example 1 with $n_c = 2$

Rossiter, J.A., P.W. Neal and L. Yao, Applying predictive control to fossil fired power station, Proceedings Inst. MC, 2002, 24(3), 177-194.

Rossiter, J.A., B. Kouvaritakis and M. Cannon, Computationally efficient algorithms for constraint handling with guaranteed stability and near optimality, IJC, 2001, 74, 17, 1678-1689

Rossiter, J.A. and P. Grieder, Using Interpolation to Simplify Explicit Model Predictive Control, to appear ACC04

Scokaert, P.O.M. and J. B. Rawlings, Constrained linear quadratic regulation, *IEEE Trans AC*, 1998, 43, 8, pp1163-1168

Tøndel, P., T.A. Johansen, and A. Bemporad. Evaluation of piecewise affine control via binary search tree. *Automatica*, 39(5):945–950, 2003.

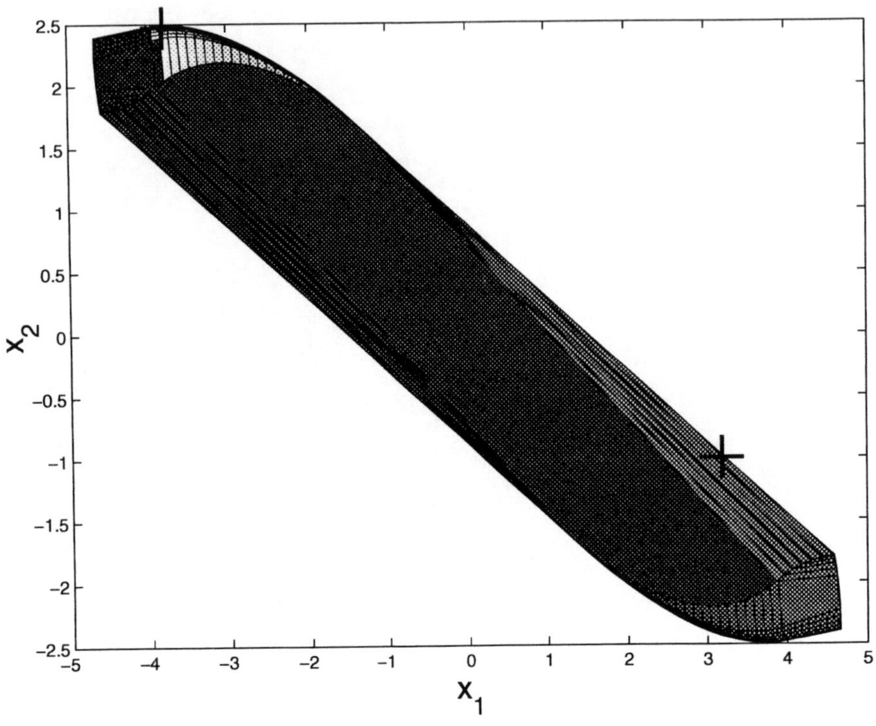

Fig. 3. Regions for example 1 with $n_c = 2$ and no state constraints

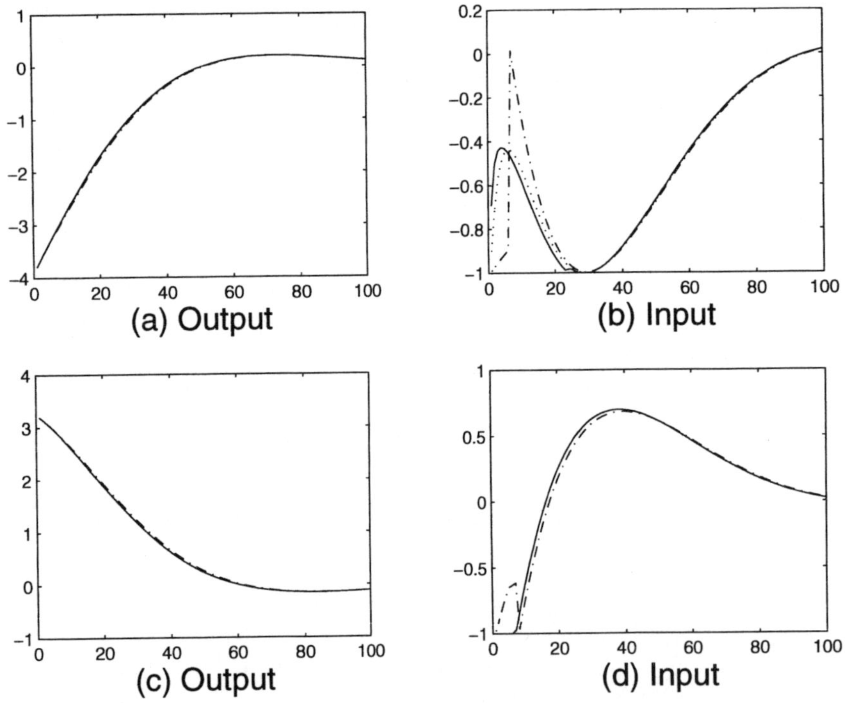

Fig. 4. Simulations for example 1 with $n_c = 6$ and no state constraints

ELSEVIER
IFAC
PUBLICATIONS
www.elsevier.com/locate/ifac

EMBEDDED MODEL PREDICTIVE CONTROL FOR SYSTEM-ON-A-CHIP APPLICATIONS

Leonidas G. Bleris * and Mayuresh V. Kothare **,[1]
Jesus Garcia *** and Mark G. Arnold ***

* *Electrical and Computer Engineering Department*
** *Chemical Engineering Department*
*** *Computer Science and Engineering Department*
Lehigh University, Bethlehem, PA 18015, U.S.A.

Abstract: We propose a framework for embedding model predictive control for Systems-on-a-Chip applications. In order to allow the implementation of such a computationally expensive controller on chip, we propose reducing the precision of the operations coupled with using logarithmic number system arithmetic. Two particular control problems are examined. We provide the methodology for choosing the design parameters; we emulate the performance of the embedded controller for the examined cases and we give the microprocessor architecture details. *Copyright © 2004 IFAC*

Keywords: Embedded Systems, Model Predictive Control, Systems-on-a-Chip.

1. INTRODUCTION

The realization of autonomous Systems-on-a-Chip (SoC) applications has been a very significant research area during the past decade. Moving on from simple prototype applications, new generations of more complex SoC are currently being developed for a variety of applications. The market for SoC is expanding rapidly to areas like medicine and bioengineering (DNA and genetic code analysis and synthesis, drug delivery, diagnostics and imaging), information technologies, avionics and aerospace (nano and microscale actuators and sensors, smart reconfigurable geometry wings and blades, microgyroscopes), automotive systems and transportation (accelerometers), microreactors for in-situ and on-demand chemical production. The functionality and performance of such a versatile SoC is directly related to the reliability and quality of the control logic used. For example processing DNA molecules in the microscale requires the fabrication of microfluidic devices capable of handling, mixing, thermal cycling and separating liquid micro-samples. Therefore efficient custom-made controllers have to be designed and integrated on chip. Furthermore the performance of the overall system has to be examined and verified through advanced simulation techniques.

With this paper we provide the framework for applying Model Predictive Control (MPC) and designing the microcontroller architecture that will allow optimal sensing-control-actuation performance for microchemical system applications. Microchemical systems are a new generation of miniature chemical systems that carry out chemical reactions and separations, in precisely fabricated three dimensional microreactor configurations. Our research effort is centered on eventually developing and embedding a controller to a catalytic reformer and separator microchemical system that can operate as a sustained source of hydrogen fuel for Proton Exchange Membrane (PEM) fuel cells. This SoC will be used poten-

[1] Corresponding author: phone (610) 758-6654, fax (610) 758-5057, e-mail: mayuresh.kothare@lehigh.edu

tially as an alternative to conventional portable sources of electricity such as batteries for laptop computers and mobile phones due to its ability to provide an uninterrupted supply of electricity as long as a supply of methanol, water and heat can be provided.

The application of real-time embedded model predictive control for microscale devices presents new technological challenges. We propose reducing the precision of the microprocessor to the minimum while maintaining stable control performance. Taking advantage of the low precision, a Logarithmic Number System (LNS) based microprocessor architecture is used, providing further energy and computational cost savings. In order to choose the optimal design parameters for this high-performance embedded controller we utilize computational tools to simulate the plant and we emulate the microcontroller arithmetic operations.

Recently a multi-parametric programming method was proposed (Bemporad *et al.*, 2002) to solve off-line the quadratic optimization problem associated with MPC. The constrained quadratic optimization is shown to be piecewise affine, using partitions of the state space determined by the constraints. The feedback laws are pre-computed and online calculations are greatly simplified, consisting of a table-lookup in memory and an affine transformation. However, the upper bound on the number of required regions is $N_r \leq \sum_{k=0}^{2^q-1} k!q^k$, where q is the number of constraints. For an input constrained problem, increasing the number of controlled variables (thus the number of constraints), yields prohibitive memory requirements for Systems-on-Chip.

As an alternative, we propose a framework for embedding MPC by reducing the precision, thereby increasing the optimization speed while reducing the power consumption and the overall chip area. The power consumption in the arithmetic units is related to the number of switching transistors. Therefore by reducing the transistor count we expect not only a lower cost for the embedded microcontroller but also a significant reduction in the consumed power. Additionally by integrating all the memory on-chip and reducing its size, the power dissipation can be further reduced. This provides a solution for embedding MPC for complex large applications.

With the rapid development of computers and the software tool capabilities, novel approaches can be used for the analysis of SoC. In this paper we use FEMLAB (COM, 2001), a Partial Differential Equation (PDE) solver. We use the ability to export complex micro-geometries into MATLAB environment and to analyze rigorously their dynamic behavior by solving problems such as heat transfer, convection-conduction and Navier-Stokes flows. Two control problems that are critical for the target microchemical system are provided. The control of the temperature distribution across a wafer geometry is initially examined. Secondly, we look into the problem of controlling the temperature of a fluid flowing in a microchannel and the flow velocity at the outlet of the microchannel by adjusting the inlet velocity and regulating the provided distributed energy supply.

2. THEORETICAL BACKGROUND

2.1 Model Predictive Control Theory

Model predictive control is also known as receding horizon control or moving horizon control. Generally controllers belonging to the MPC family are characterized by the following steps. Initially the future outputs are calculated at each sample interval over a predetermined horizon N, the prediction horizon, using the process model. These outputs $y(t+k|t)$ for $k = 1...N$ depend up to the time t on the past inputs and on the future signals $u(t+k|t)$, $k = 0...N-1$ which are those to be sent to the system. The next step is to calculate the set of future control moves by optimizing a determined criterion in order to keep the process as close as possible to a predefined reference trajectory. This criterion is usually a quadratic function of the difference between the predicted output signal and the reference trajectory. In some cases the control moves $u(t+k|t)$ are included in the objective function in order to minimize the control effort. Finally, the first control move $u(t|t)$ is sent to the system while the rest are rejected. This is because at the next sampling instant the output $y(t+1)$ is already known from the optimization results and the procedure is repeated with the new values so that we get an updated control sequence.

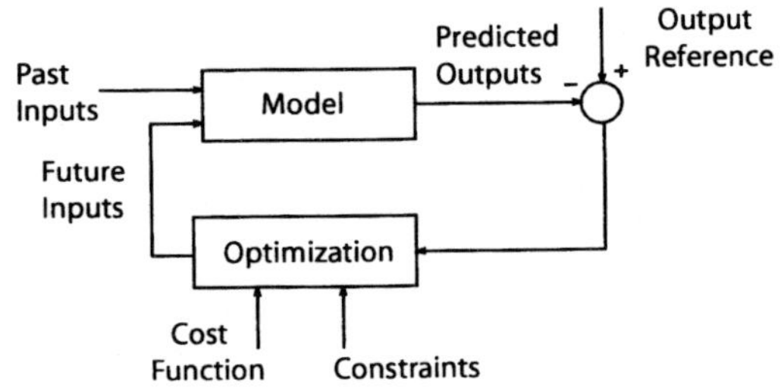

Fig. 1. Model Predictive Control block diagram.

Figure 1 shows how to implement the basic controller structure. A model is used to predict the future plant outputs based on past and current values and on the calculated optimal future control moves. These actions are calculated by the optimization that also takes into account the constraints. There are many models used in different formulations of the model predictive control.

These include step response models, impulse response models, transfer function models and the state space models. In this paper we build the geometry, define the PDE models and the boundary conditions in FEMLAB and we use the impulse responses as a model for the predictive control problem formulation.

2.2 Logarithmic Number System

One of the main advantages of floating point (FP) is its standardization, but for an architecture oriented towards embedded systems and therefore susceptible to being tailored for each particular application the need for novel approaches arises. The logarithmic number system (LNS) represents the value of the real number X using a sign bit and the base-b logarithm of $|X|$, x, so that

$$x = \text{round}(\log_b |X|), \tag{1}$$

where x is a fixed-point signed binary number, consisting of K integer bits and F fraction bits (Swartzlander and Alexopoulos, Dec. 1975). The round() operation approximates the value of $\log_b |X|$ so that it is representable in $N = K + F$ bits. The values for K and F define respectively the dynamic range (largest and closest to zero representable values), and the precision (distance between consecutive representable values) that are available. LNS provides a representation with numeric characteristics similar to those offered by FP (Arnold *et al.*, Aug. 1992). The first reason for preferring LNS is the constant relative error characteristic (as opposed to FP) that gives a superior worst-case error in LNS for the same word size. FP architectures can be scaled in a more efficient way for high precisions, but for low-to-moderate precisions (below 32 bits), LNS implementations offer an advantage in term of cost, power consumption and speed, that increases as the word size decreases.

3. CONTROL OF TEMPERATURE DISTRIBUTION ACROSS A WAFER

Consider the geometry of Figure 2 that illustrates a wafer with a distributed array of resistive heaters placed on top. We measure the temperature of this geometry by placing sensors on the bottom of the wafer at the exact opposite location of the actuators. We consider the 2D cross section of this wafer with nine distributed heaters and nine sensors. We are interested in heating and maintaining the temperature of the wafer at a predefined set-point, while constraining the heat supply. This is a significant control objective for our examined SoC application since the catalytic reformer and separator need to be heated to a specific temperature to ensure catalytic activation. Additionally the optimal working point for these two reactions is different; therefore, we need to maintain a temperature gradient within the wafer. The temperature distribution in the wafer is described by the time-dependent two-dimensional heat equation

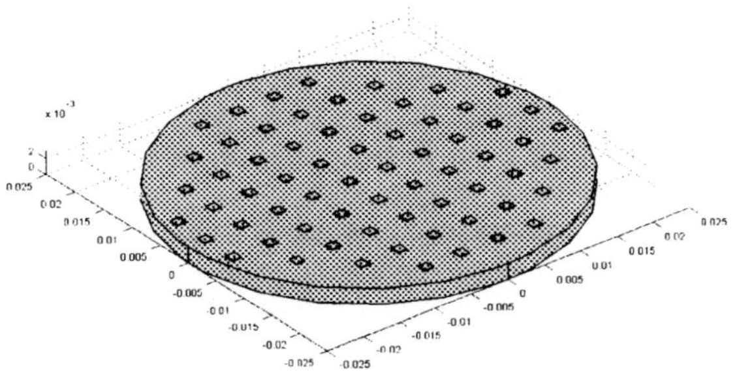

Fig. 2. Wafer geometry (dimensions in meters).

$$\frac{\partial T(t,x,y)}{\partial t} = \frac{1}{a}\nabla^2 T(t,x,y) \tag{2}$$

with initial conditions $T(0,x,y) = 300\text{K}$. Here $a = \rho C/\kappa$, κ is the thermal conductivity, ρ is the density and C is the heat capacity. The resistive actuator heat losses depend on the exposed surface $Q = hA(T - T_{air})$ where $T_{air} = 300\text{K}$, the natural convection heat-transfer coefficient h=25W/m^2K and A is the exposed heater area. With appropriate use of materials we have the ability (Bleris and Kothare, 2003) to control and maintain temperature gradients in the proposed geometry. The material used for the following FEM simulations is pyrex. We initially set the temperature set-point (Tsp) at the sensors placed at the boundaries at 370K and for the rest at 390K. The heat supply is constrained for each actuator at 5W/m^2. We apply this problem on a 2D geometry but the proposed approach can be easily extended in 3D, using an array of sensors and actuators.

The optimization is a fundamental part of MPC since it computes the optimal control input for the process. The computational effort required in MPC derives almost entirely from the optimization algorithm. Logically the choice of the optimization technique (Fletcher, 1987) is decisive to the performance of the controller. The algorithm that is implemented for the minimization of the cost function is a direct application of Newton's method (incorporating the constraints in the cost function using barrier functions). It has to be noted that from the architecture perspective the operations that are responsible for the majority of instructions that are executed in the processor are matrix inversions and abundant dot products. To estimate the performance of MPC using reduced precision, the basic arithmetic operations required by the optimization procedure were emulated using LNS. The design parameters of MPC as well as the number representation K and F (integer and fractional part of the fixed point exponents) were varied, allowing the selection of the desired compromise between precision reduction and response quality of the system.

The control problem of reaching the set-point is solved initially using $K = 7$ (integer bits) and $F = 20$ (fraction bits) for different control horizons (ch), Figure 3. The prediction horizon is fixed at 100. We define the error of the controller performance using

$$Error = \frac{\sum_{i=1}^{9} |Tsp(i) - T(i)|}{\sum_{i=1}^{9} (Tsp(i) - 300)} \times 100 \quad (3)$$

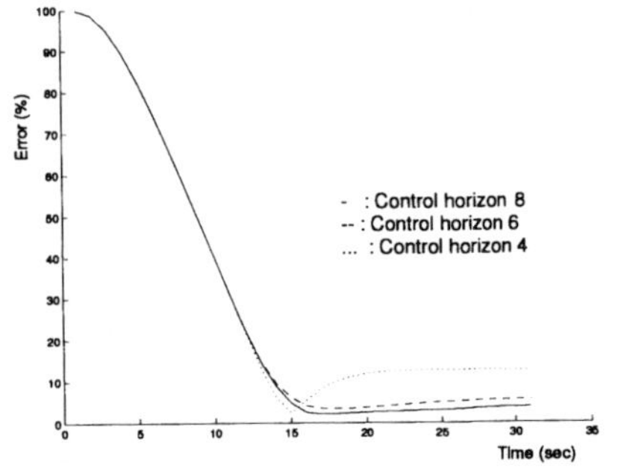

Fig. 3. Percentage of error using different control horizons.

From Figure 3 it is obvious that the performance of MPC is not satisfactory for $ch = 4$ and it is closest to the set-point using $ch = 8$. The total energy consumed using $ch = 8$ was $567\text{J}/m^2$ while using $ch = 6$ yields a 3.4% reduction. Additionally, since we are interested in minimizing the architectural complexity of the embedded controller we choose to use $ch = 6$. In Figure 4 we provide the simulation results. The objective now is to reduce further the implementation complexity, while maintaining a performance close to that of $K = 7$, $F = 20$ and $ch = 6$.

The next step for the proposed framework is to reduce the size of the integer part K. Emulation results of the coupled system of the MPC microcontroller and the FEMLAB models show exactly the same performance for $K = 7, 6, 5$ and significantly worse for $K = 4$. Therefore $K = 5$ was chosen. Subsequently the fractional part F was gradually reduced and in Figure 5 we provide the performance for different cases. From Figure 5 we notice that for $K = 5$, $F = 10$ and $ch = 6$ the error increases significantly. This can be explained by the fact that the subtraction operation at low precisions can be inaccurate when the result is close to zero. As the plant output

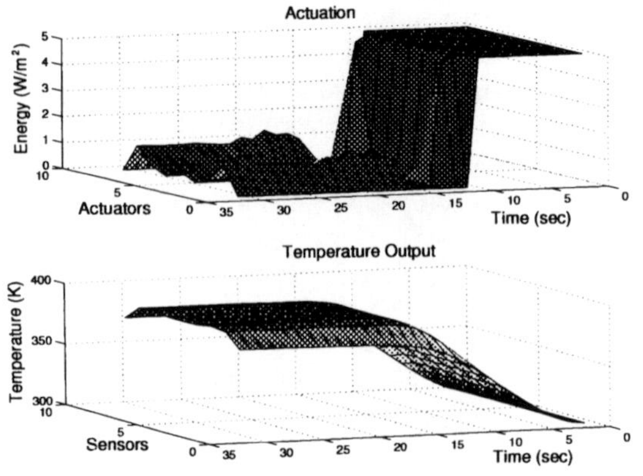

Fig. 4. Control variables and resulting temperature profile for K=7, F=20 and CH=6.

approaches the set-point, the values of the manipulated variables start oscillating (Figure 6) to keep the output close to the set-point. Furthermore this phenomenon creates added undesirable power consumption. To overcome this problem a hybrid MPC approach is under investigation, where the MPC switches to a Proportional Integral (PI) controller close to the set-point.

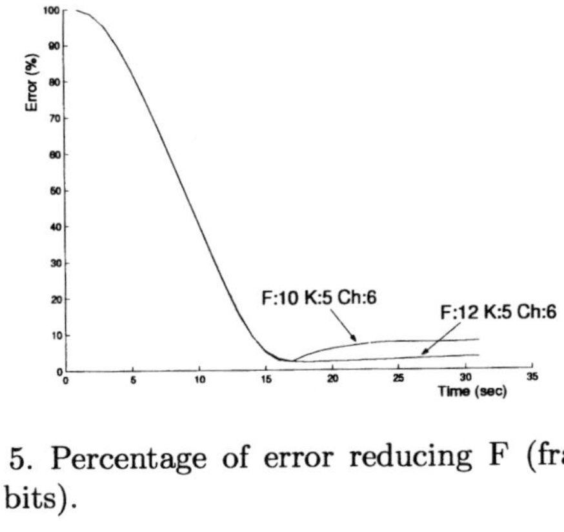

Fig. 5. Percentage of error reducing F (fraction bits).

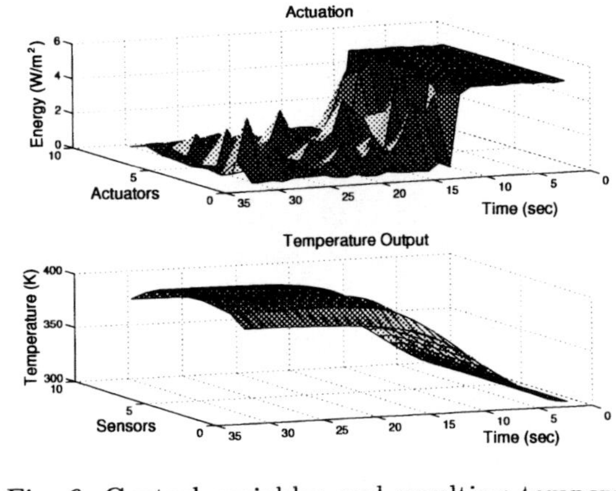

Fig. 6. Control variables and resulting temperature profile using K=5, F=10 and CH=6.

The preceding simulations show that for the examined problem, the minimum values for K and F were 5 and 12, respectively, which requires a 19-bit LNS representation (adding two bits for the signs

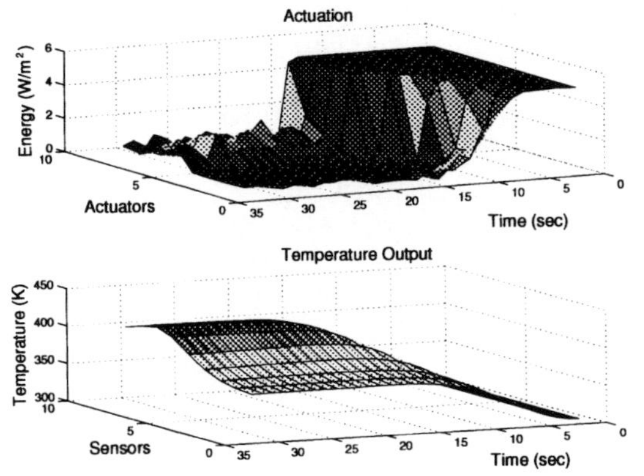

Fig. 7. Control variables and resulting temperature profile using K=5, F=12 and CH=6.

of K and F). We test this configuration using now a different set-point for the output temperature. The set-point is at 360K for the first 4 sensors, the 5th sensor is free to take any value and the final 4 sensors at 400K. From the simulation results of Figure 7, we notice the successful performance of the proposed reduced-precision controller.

4. NON-ISOTHERMAL FLOW

In this section we examine the coupled problem of microfluidic flow and convection-conduction in the 2D geometry of Figure 8. The microchannel has depth of 3mm and the arrows indicate the inlet, the outlet and the location of one of the ten distributed resistive heaters. We measure the temperature by placing ten sensors on the bottom of the geometry at the exact opposite location of the actuators. The dynamic model of the flow

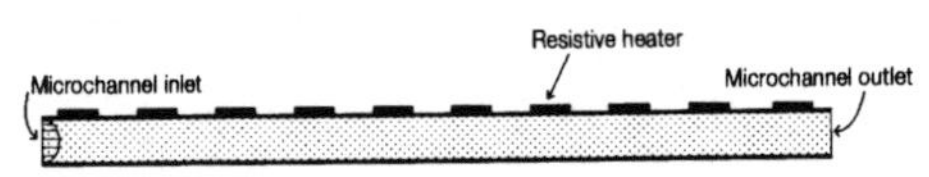

Fig. 8. Microchannel geometry.

can be described by the Navier-Stokes and the convection-conduction equations

$$\rho\frac{\partial v}{\partial t} - \nabla\cdot\eta(\nabla v + (\nabla v)^T) + \rho(v\cdot\nabla)v + \nabla p = 0 \quad (4)$$

$$\nabla\cdot(\rho v) = 0 \quad (5)$$

$$\rho C_p\frac{\partial T}{\partial t} + \nabla\cdot(-k\nabla T + \rho C_p T v) = 0 \quad (6)$$

where k is the thermal conductivity and C_p the heat capacity. Furthermore we use the ideal gas law to relate the changes of temperature to density $\rho = \frac{pM}{RT}$, where R is the gas constant and M is the molar mass. For the Navier-Stokes equations we define the boundary conditions to be no-slip at the walls and straight out at the outlet. For the heat balance the temperature at the inlet is 300K, at the top wall it is the energy provided by the resistive heaters, it is insulated at the bottom and at the outlet we assume that the transport is dominated by convection (the gradient of T perpendicular to the outlet is zero). We are interested in heating and maintaining the temperature of the fluid to a predefined set-point, while constraining the heat supply. Moreover due to the changes of density, the velocity of the fluid increases upon heating. We are therefore interested in controlling the velocity of the fluid at the outlet. Because of the fact that we have laminar flows in microscale (due to the low Reynolds number), we measure and control the maximum velocity which we expect to have at the middle of the channel. We solve the proposed control problem by decentralizing the system. Using proportional control the inlet velocity is adjusted to control the outlet velocity and MPC is applied for the regulation of the heat supply. Because of the fact that the fluid enters the microchannel at 300K, controlling the temperature profile becomes more important close to the inlet, especially for high velocities and elevated temperature set-point. We initially set the desired temperature at 350K throughout the microchannel and the maximum velocity at the outlet at 2mm/sec. The simulation results of applying MPC for controlling the temperature, with K=7, F=20, control horizon of 6 and prediction horizon 21, are illustrated in Figure 9. We

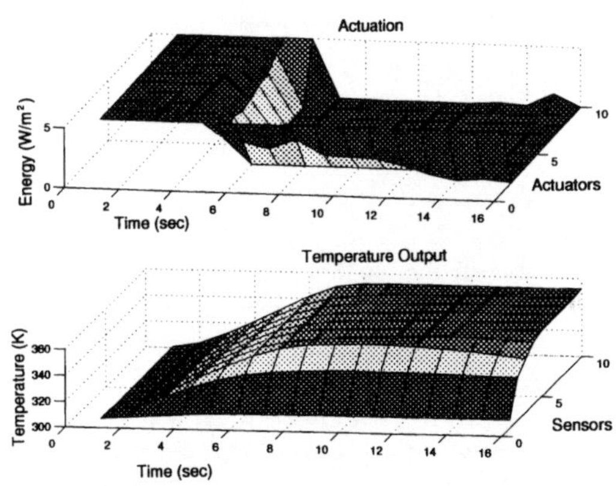

Fig. 9. Control variables and resulting temperature profile using K=7, F=20 and CH=6.

now simulate the system by gradually reducing the integer bits (K) and then the fraction bits (F) (Figure 10). We are interested in calculating the minimum precision while maintaining the same control performance. From the simulation results it becomes evident that we can reduce the controller size down to 17-bits without compromising the performance. As the LNS sign bit is treated separately, this requires only 16-bit addition to implement multiplication.

5. PROPOSED ARCHITECTURE

The requisites that an embedded MPC has to satisfy are: high speed, low power consumption

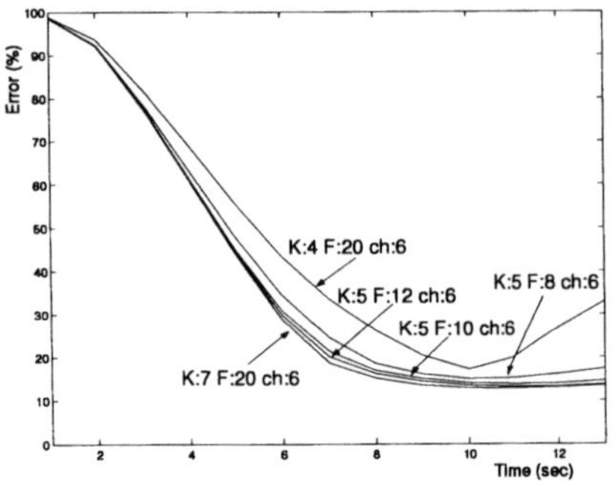

Fig. 10. Percentage of error using different design parameters.

and low cost. The last two are related, since power consumption is related to the transistor count and therefore the cost. The proposed architecture shown in Figure 11 utilizes macrocell technology. Commercially available designs of macrocells such as microprocessor cores and embedded DRAM exist. By using LNS custom arithmetic units, realistic MPC problems can be solved online with a system that integrates every component on the same chip. High speed in the embedded system

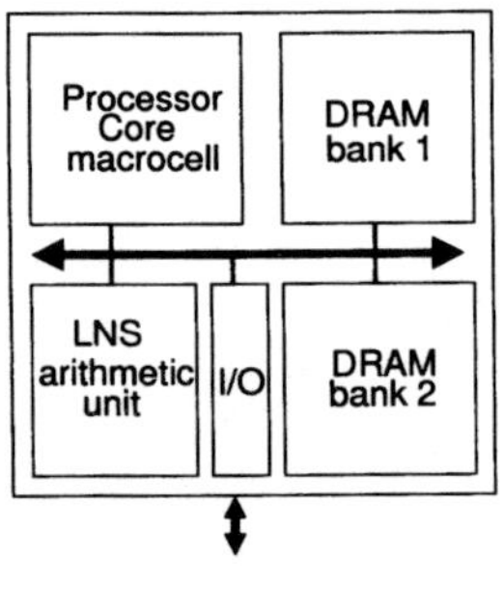

Fig. 11. Architecture block.

will depend both on fast arithmetic computations, and fast access to the data stored in memory. The main computing complexity comes from the abundant dot products and the matrix inversion. With double-precision floating point data the required ALU is expensive and operations would be slow and power consuming. Using a pipelined 16-bit LNS adder, neither scalar-vector multiplication nor vector addition represent a problem; these operations can be computed with a throughput of 1 operation/cycle. Long dot products can be unrolled in the adder pipeline, resulting in 1 op/cycle with a small overhead. To quantify the advantage of reducing the precision, estimations for both 64-bit FP and 16-bit LNS circuits show that for an arithmetic unit that computes addition, subtraction, multiplication and division, the size required is about 17 times larger for 64-bit FP, while the clock cycle is at least 3.23 times faster in 16-bit LNS. Having functional units that allow computing the majority of the operations with a throughput close to 1 operation/cycle, the other main obstacle to obtain high speed is memory access. Fast access to the RAM is a major problem for most microprocessors (Hennessy and Patterson, 1996). However, current technology allows embedding DRAM macrocells of several Mb, combining the lower DRAM cost (compared with SRAM as used in caches), with an access time of just one clock cycle. MPC algorithms use vectors and matrices of a fixed length (within the optimization loop, which represents the majority of the operations to be computed). Very regular loops of a fixed size, known *a priori*, allow customizing the memory system for these accesses. Page-access to the DRAM will maximize the data rate, by reading and writing vectors by rows. The data size reduction not only increases speed, but also reduces the memory and the size of the arithmetic units.

6. CONCLUSIONS

A framework for embedding model predictive control logic on chip has been provided. Reducing the precision of the operations coupled with the use of a logarithmic-number-system arithmetic allows a very efficient implementation. We have emulated the performance of the reduced precision LNS-based MPC for two examined cases and we have provided brief microprocessor architecture details.

Future work includes obtaining theoretical bounds for convergence as a function of precision and formulating a detailed embedded system architecture. Also incorporating a hybrid logic to the controller to allow an early termination of the optimization algorithm so that further computational and energy savings will be investigated.

REFERENCES

Arnold, M. G., T. A. Bailey, J. R. Cowles and M. D. Winkel (Aug. 1992). Applying Features of IEEE 754 to Sign/Logarithm Arithmetic. *IEEE Transactions on Computers* **41**, 1040–1050.

Bemporad, A., M. Morari, V. Dua and E. N. Pistikopoulos (2002). The explicit linear quadratic regulator for constrained systems. *Automatica* **38**(1), 3–20.

Bleris, L. G. and M. V. Kothare (2003). Model based control of temperature distribution in integrated microchemical systems. In: *Proceedings of the 2003 American Control Conference.* Denver, CO. pp. 1308–1313.

COM (2001). *FEMLAB Reference Manual.*

Fletcher, R. (1987). *Practical methods of optimization.* New York, Wiley Interscience.

Hennessy, J. L. and D. A. Patterson (1996). *Computer Architecture: A Quantitative Approach.* Morgan Kaufmann. San Francisco, CA.

Swartzlander, E. E. and A. G. Alexopoulos (Dec. 1975). The Sign/Logarithm Number System. *IEEE Transactions on Computers* **24**(12), 1238–1242.

ELSEVIER
IFAC
PUBLICATIONS
www.elsevier.com/locate/ifac

NUMERICAL METHODS FOR LARGE SCALE MOVING HORIZON ESTIMATION AND CONTROL

John Bagterp Jørgensen [*,1] **James B. Rawlings** [**]
Sten Bay Jørgensen [***]

[*] *2-control ApS, Høffdingsvej 34, DK-2500 Valby, Denmark*
[**] *Department of Chemical and Biological Engineering, University of Wisconsin-Madison, Madison, WI 53706 Wisconsin, USA*
[***] *CAPEC, Department of Chemical Engineering, Technical University of Denmark, DK-2800 Lyngby, Denmark*

Abstract: Both the linear moving horizon estimator and controller may be solved by solving a linear quadratic optimal control problem. A primal active set, a dual active set, and an interior point algorithm for solution of the linear quadratic optimal control problem are presented. The major computational effort in all these algorithms reduces to solution of certain unconstrained linear quadratic optimal control problems. A Riccati recursion procedure for efficient solution of such unconstrained problems is stated.

Keywords: Linear Model Predictive Control, Quadratic Programming, Numerical Methods

1. INTRODUCTION

Model predictive control has been established as the preferred advanced control technology in the process industries. Its success is attributed to its ability to handle hard constraints and its ability to use plant models identified by standard techniques. The implementation of model predictive control *systems* requires repeated on-line solution of a state estimation problem and an optimal control problem. Both the state estimator and the optimal controller can be formulated as constrained optimization problems (Allgöwer *et al.*, 1999; Binder *et al.*, 2001). Formulated as optimization problems and implemented in a moving horizon manner, the estimator is called a moving horizon estimator (MHE) and the controller is called a model predictive controller (MPC) or a moving horizon controller. Due to the requirement of real-time solution of both the estimation and control problem, the range and scale of processes that can be controlled by model predictive control systems depends critically on the ability to solve the constrained optimization problems efficiently and reliably.

In this paper we outline numerical methods for efficient solution of constrained moving horizon estimation and control problems for systems described by affine models and with quadratic objective functions. Both the moving horizon estimator and controller of such systems are formulated as sparse convex quadratic programs with equality and inequality constraints. This formulation is in contrast to the formulation in which the states are eliminated and the optimization problem is formulated as a dense quadratic program with inequalities only. Previously, both interior-point and active set algorithms for the solution of the quadratic program corresponding to a linear-quadratic optimal control based on a banded-matrix factorization have been proposed for solution of the model predictive control problem (Wright, 1996). The active set algorithm suggested in Wright (1996) is of the

[1] Corresponding author. E-mail: jbj@2-control.com. Phone: +45 70 222 404

primal type and the linear algebra is based on updating an LU-factorization each time a constraint is added to or removed from the current working set of active constraints. The quadratic programs corresponding to the moving horizon controller and estimator, respectively, have also been solved by interior-point algorithms using Riccati iterations for the matrix factorization (Rao *et al.*, 1998; Tenny and Rawlings, 2002). The Riccati facotization is very efficient for solution of the KKT system as it exploits the specific block-diagonal structure arising in the moving horizon control and estimation problem. Alternatively, a dual active set algorithm using the Schur-complement and a general sparse matrix factorization has been suggested for solution of the model predictive control problem (Bartlett *et al.*, 2000).

We show that in the primal active-set, the dual active-set, and the interior-point algorithm for general convex quadratic programming problems, the search direction may be computed by solving a Karush-Kuhn-Tucker (KKT) system with the same structure as an equality constrained quadratic program. In quadratic programs of constrained optimal control problems, the KKT-system corresponding to the unconstrained control problem has a special structure and may be solved efficiently by a Riccati recursion. Both the constrained moving horizon estimation problem and the moving horizon control problem may be formulated as an optimal control problem and solved efficiently using the Riccati recursion.

The key contribution of this paper is a dual algorithm based on Riccati based factorization for solution of the quadratic program constituting a constrained linear-quadratic optimal control problem. It is demonstrated how the constrained linear-quadratic optimal control problem may represent a moving horizon controller and a moving horizon estimator. Furthermore, the importance and key role of the KKT-system in numerical solution of the constrained-linear quadratic optimal control problem by the primal active set algorithm as well as the dual active set algorithm and the interior-point algorithm is emphasized and explained.

2. ALGORITHMS FOR SOLUTION OF CONVEX QPS

Consider the general convex quadratic program

$$\min_{x\in\mathbb{R}^n} \quad f(x) = \frac{1}{2}x'Gx + g'x \tag{1a}$$

$$s.t. \quad A'x = b \tag{1b}$$

$$C'x \geq d \tag{1c}$$

in which the Hessian matrix $G \in \mathbb{R}^{n\times n}$ is symmetric and positive semi-definite. Assume further that the matrix $A \in \mathbb{R}^{n\times m}$ has full column rank and that the KKT-matrix

$$\begin{bmatrix} G & -A \\ -A' & 0 \end{bmatrix} \tag{2}$$

is non-singular. Further let the columns of the matrix $C \in \mathbb{R}^{n\times m_I}$ be denoted $c_i \in \mathbb{R}^n$ for $i \in \mathcal{I} = \{1,2,\ldots,m_I\}$. The Lagrangian of (1) is

$$\mathcal{L}(x,\pi,\lambda) = \frac{1}{2}x'Gx+g'x-\pi'(A'x-b)-\lambda'(C'x-d) \tag{3}$$

and the necessary and sufficient conditions for optimality of (1) are

$$Gx + g - A\pi - \sum_{i\in\mathcal{A}(x)} c_i\lambda_i = 0 \tag{4a}$$

$$A'x = b \tag{4b}$$

$$C'x \geq d \tag{4c}$$

$$\lambda_i \geq 0 \qquad i \in \mathcal{A}(x) = \{i \in \mathcal{I} : c_i'x = d_i\} \tag{4d}$$

$$\lambda_i = 0 \qquad i \in \mathcal{I} \setminus \mathcal{A}(x) \tag{4e}$$

Active set algorithms for solution of (1) apply the conditions (4) in the search for the minimizer of (1). A current working set $\mathcal{W} \subset \mathcal{A}(x)$ of active constraints is recurred and the search direction is constructed by utilizing $\lambda_i = 0$ for $i \in \mathcal{I}\setminus\mathcal{W}$ and $c_i'x = d_i$ for $i \in \mathcal{W}$ and partly ignoring the remaining inequalities.

At a primal feasible point, primal active set algorithms compute the primal search direction p along which the objective function decreases (Gill and Murray, 1978). Along with the search direction p, the equality constraint Lagrange multipliers π, and the Lagrange multipliers λ associated with the current working set of active constraints are computed. (p, π, λ) are computed as the solution of

$$\begin{bmatrix} G & -A & -F \\ -A' & 0 & 0 \\ -F' & 0 & 0 \end{bmatrix} \begin{bmatrix} p \\ \pi \\ \lambda \end{bmatrix} = -\begin{bmatrix} Gx+g \\ 0 \\ 0 \end{bmatrix} \qquad F = [c_i]_{i\in\mathcal{W}} \tag{5}$$

The next iterate is $\bar{x} = x + \alpha p$ in which the step length α is selected to maintain primal feasibility, i.e. $c_i'\bar{x} \geq d_i$ for all $i \in \mathcal{I} \setminus \mathcal{W}$. Optimality of the current iterate is determined by the sign of λ. The current iterate is optimal if $\lambda \geq 0$ and $p = 0$. At each iteration either an inequality constraint is appended to or removed from the current working set of active constraints. This implies that a column is appended to the matrix F when an inequality constraint is appended to the current working set of active constraints. A column is removed from the matrix F when a constraint is removed from the current working set. Accordingly, the KKT-matrix at each iteration changes by a single column and row. Efficient active set algorithms exploit this simple modification of the KKT-matrix such that it is not refactorized at each iteration, but rather its factorization is updated. Initially, a primal feasible point must be provided as primal active set algorithms proceed from primal feasible points and maintains primal feasibility while improving the value of the program. Determination of the feasible point, for instance by solving a phase-I linear program, may be a substantial part of the overall computations in determining an optimal point.

Primal active set algorithms proceed by solving a sequence of equality constrained quadratic programs in which some of the inequality constraints are treated as equalities. Simultaneously, these algorithms generate the iterates such that primal feasibility is maintained. Dual active set algorithms proceed by generating iterates such that the objective function of the dual program is improved while maintaining dual feasibility (Goldfarb and Idnanini, 1983). The dual program of (1) is

$$\max_{x,\pi,\lambda} \quad \mathcal{L}(x,\pi,\lambda) \tag{6a}$$

$$s.t. \quad Gx - A\pi - C\lambda = -g \tag{6b}$$

$$\lambda \geq 0 \tag{6c}$$

Let inequality constraint r, i.e. $c_r'x \geq d_r$, be violated. In the dual active set algorithm, the search direction (p, w, v) along which the dual objective function increases is computed as the solution to the KKT system

$$\begin{bmatrix} G & -A & -F \\ -A' & 0 & 0 \\ -F' & 0 & 0 \end{bmatrix} \begin{bmatrix} p \\ w \\ v \end{bmatrix} = \begin{bmatrix} c_r \\ 0 \\ 0 \end{bmatrix} \qquad F = [c_i]_{i \in \mathcal{W}} \tag{7}$$

The next iterate $(\bar{x}, \bar{\pi}, \bar{\lambda})$ is computed by $\bar{x} = x + \alpha p$, $\bar{\pi} = \pi + \alpha w$, $\bar{\lambda}_{\mathcal{W}} = \lambda_{\mathcal{W}} + \alpha w$, $\bar{\lambda}_r = \lambda_r + \alpha$. $\lambda_{\mathcal{W}}$ is the Lagrange multipliers associated to the inequality constraints in the current working set of primal active constraints (and dual inactive constraints). The step length α is selected such that dual feasibility, i.e. $\lambda \geq 0$, is maintained. By construction of the search direction, the constraint (6b) is satisfied as $\lambda_i = 0$ for $i \in \mathcal{I} \setminus \mathcal{W}$. A particular advantage of the dual active set algorithm, is that an initial dual feasible point is readily available. It may be obtained as the solution to KKT-system of the equality constrained primal program (1a)-(1b), i.e. as the solution of

$$\begin{bmatrix} G & -A \\ -A' & 0 \end{bmatrix} \begin{bmatrix} x \\ \pi \end{bmatrix} = - \begin{bmatrix} g \\ b \end{bmatrix} \tag{8}$$

Interior-point algorithms for solution of (1) are not based on the optimality conditions (4) but on necessary and sufficient optimality conditions in the following form

$$Gx + g - A\pi - C\lambda = 0 \tag{9a}$$

$$A'x = b \tag{9b}$$

$$C'x \geq d \tag{9c}$$

$$\lambda \geq 0 \tag{9d}$$

$$\lambda_i(c_i'x - d_i) = 0 \qquad i \in \mathcal{I} \tag{9e}$$

Instead of the active set condition, $\lambda_i \geq 0$ for $i \in \mathcal{A}(x)$ and $\lambda_i = 0$ for $i \in \mathcal{I} \setminus \mathcal{A}(x)$, the formulation (9) is based on the complementarity condition (9e). Introduce slack variables, $t = C'x - d \geq 0$, and the notation

$$T = \begin{bmatrix} t_1 & & \\ & \ddots & \\ & & t_{m_I} \end{bmatrix} \quad \Lambda = \begin{bmatrix} \lambda_1 & & \\ & \ddots & \\ & & \lambda_{m_I} \end{bmatrix} \quad e = \begin{bmatrix} 1 \\ \vdots \\ 1 \end{bmatrix} \tag{10}$$

such that the conditions (9) may be expressed as

$$Gx + g - A\pi - C\lambda = 0 \tag{11a}$$

$$A'x - b = 0 \tag{11b}$$

$$C'x - d - t = 0 \tag{11c}$$

$$T\Lambda e = 0 \tag{11d}$$

$$(t, \lambda) \geq 0 \tag{11e}$$

These conditions may be regarded as a system of nonlinear equations represented as $F(x, \pi, \lambda, t) = 0$ with the requirement $(t, \lambda) \geq 0$. The Mehrotra predictor-corrector algorithm is an interior-point method which computes the search direction as a combination of a predictor and a corrector step (Rao *et al.*, 1998). At both the predictor and corrector step t and λ are maintained in the interior, i.e. $(t, \lambda) > 0$. The predictor step is a pure Newton step for $F(x, \pi, \lambda, t)$, i.e. (11a)-(11d), while the corrector step is a modified Newton step. In both cases, the structure of the equations solved in computing the search direction is

$$\begin{bmatrix} G & -A & -C & 0 \\ -A' & 0 & 0 & 0 \\ -C' & 0 & 0 & I \\ 0 & 0 & T_k & \Lambda_k \end{bmatrix} \begin{bmatrix} \Delta x \\ \Delta \pi \\ \Delta \lambda \\ \Delta t \end{bmatrix} = - \begin{bmatrix} r_G \\ r_A \\ r_C \\ r_\Lambda \end{bmatrix} \tag{12}$$

3. KKT-SYSTEMS

The major computational effort in algorithms for general convex quadratic programs (1) is computation of the search direction. This corresponds to solution of (5) in the primal active set algorithm, (7) in the dual active set algorithm, and (12) in the interior point algorithm, respectively.

Consider, the equality constrained quadratic subproblem of (1)

$$\min_{x \in \mathbb{R}^n} \quad f(x) = \frac{1}{2}x'Gx + g'x \tag{13a}$$

$$s.t. \quad A'x = b \tag{13b}$$

The solution of this program is computed as the solution of the following KKT system

$$\begin{bmatrix} G & -A \\ -A' & 0 \end{bmatrix} \begin{bmatrix} x \\ \pi \end{bmatrix} = - \begin{bmatrix} g \\ b \end{bmatrix} \tag{14}$$

The solution of (5), (7), and (12) may essentially be reduced to solution of systems with the structure (14). This is advantageous if (14) has a special structure that facilitates its efficient computation.

Proposition 1. (Schur-complement solution). Let $G \in \mathbb{R}^{n \times n}$ be symmetric positive semi-definite. Let $[A \; F] \in \mathbb{R}^{n \times (m+m_F)}$ have full column rank. Let the KKT-matrix (2) be non-singular. Then

$$S = [F' \; 0] \begin{bmatrix} G & -A \\ -A' & 0 \end{bmatrix}^{-1} \begin{bmatrix} F \\ 0 \end{bmatrix} \tag{15}$$

is symmetric positive definite and has the Cholesky factorization $S = LL'$. Furthermore, the unique solution (p, s, u) of

$$\begin{bmatrix} G & -A & -F \\ -A' & 0 & 0 \\ -F' & 0 & 0 \end{bmatrix} \begin{bmatrix} p \\ s \\ u \end{bmatrix} = - \begin{bmatrix} h \\ 0 \\ 0 \end{bmatrix} \tag{16}$$

may be obtained by solving the following sequence of equations

$$\begin{bmatrix} G & -A \\ -A' & 0 \end{bmatrix} \begin{bmatrix} p_0 \\ s_0 \end{bmatrix} = - \begin{bmatrix} h \\ 0 \end{bmatrix} \tag{17a}$$

$$LL'u = -F'p_0 \tag{17b}$$

$$\begin{bmatrix} G & -A \\ -A' & 0 \end{bmatrix} \begin{bmatrix} \Delta p \\ \Delta s \end{bmatrix} = - \begin{bmatrix} -Fu \\ 0 \end{bmatrix} \tag{17c}$$

$$\begin{bmatrix} p \\ s \end{bmatrix} = \begin{bmatrix} p_0 \\ s_0 \end{bmatrix} + \begin{bmatrix} \Delta p \\ \Delta s \end{bmatrix} \tag{17d}$$

Proof. See Ouellette (1981). □

Proposition 1 may be used for solution of (5) in the primal active set algorithm and solution of (7) in the dual active set algorithm. In the first case $h = Gx + g$ and in the latter case $h = -c_r$. The matrix S is not computed from scratch at each iteration. Rather, its Cholesky factorization is updated utilizing that the matrix F changes by a single column at each iteration in the active set algorithms. The Cholesky factor, L, is treated as a dense matrix. Hence, the method is most efficient when only a few inequality constraints are active at the optimal solution. The method is only efficient, when the KKT-matrix (2) has a sparse structure that can be utilized. The KKT matrix (2) used in proposition 1 remains constant and need only to be factorized once.

Proposition 2. (Interior-Point KKT System). Let $G \in \mathbb{R}^{n \times n}$ be symmetric positive semi-definite. Let (2) be non-singular. Let $(t_k, \lambda_k) > 0$. Then the unique solution $(\Delta x, \Delta\pi, \Delta\lambda, \Delta t)$ of (12) may be obtained by computation of the unique solution of

$$\begin{bmatrix} G + CT_k^{-1}\Lambda_k C' & -A \\ -A' & 0 \end{bmatrix} \begin{bmatrix} \Delta x \\ \Delta \pi \end{bmatrix} = - \begin{bmatrix} \tilde{r}_G \\ r_A \end{bmatrix} \tag{18a}$$

in which

$$\tilde{r}_G = -r_G + CT_k^{-1}(-r_\Lambda + \Lambda_k r_C) \tag{18b}$$

and subsequent computation of

$$\Delta t = -r_C + C'\Delta x \tag{18c}$$

$$\Delta\lambda = T_k^{-1}(-r_\Lambda - \Lambda_k \Delta t) \tag{18d}$$

Proof. Follows by simple rearrangement of (12). See Rao *et al.* (1998). □

As is evident from proposition 2, the computational burden in finding the search direction of the interior-point algorithm is solution of (18a). The structure of (18a) is identical to the structure of (14), and therefore the computation of the search direction in the interior-point algorithm is facilitated by efficient solution of (14). The KKT-matrix in (18a) changes as (λ_k, t_k) changes and must therefore be refactorized at each iteration.

4. LINEAR QUADRATIC OPTIMAL CONTROL

Proposition 1 and 2 provide methodologies for solution of inequality and equality constrained convex quadratic programming. The efficiency of these methods depends on the efficiency of the solution method for the corresponding equality constrained convex quadratic program.

The unconstrained linear quadratic optimal control problem is the equality constrained convex quadratic program

$$\min_{\{x_{k+1}, u_k\}_{k=0}^{N-1}} \quad \phi = \sum_{k=0}^{N-1} l_k(x_k, u_k) + l_N(x_N) \tag{19a}$$

$$s.t. \qquad x_{k+1} = A_k' x_k + B_k' u_k + b_k \tag{19b}$$

in which the stage costs of the objective function are

$$l_k(x_k, u_k) = \frac{1}{2}\left(x_k' Q_k x_k + 2x_k' M_k u_k + u_k' R_k u_k\right) + q_k' x_k + r_k' u_k + f_k \tag{20a}$$

$$l_N(x_N) = \frac{1}{2} x_N' P_N x_N + p_N' x_N + \gamma_N \tag{20b}$$

The matrices $\begin{bmatrix} Q_k & M_k \\ M_k' & R_k \end{bmatrix}$ and P_N are assumed to be symmetric positive semi-definite and the KKT-matrix of the problem is assumed to be non-singular.

The necessary and sufficient optimality conditions for (19) are

$$Q_k x_k + M_k u_k + q_k - \pi_{k-1} + A_k \pi_k = 0 \tag{21a}$$

$$M_k' x_k + R_k u_k + r_k + B_k \pi_k = 0 \tag{21b}$$

$$P_N x_N + p_N - \pi_{N-1} = 0 \tag{21c}$$

$$x_{k+1} = A_k' x_k + B_k' u_k + b_k \tag{21d}$$

In the case $N = 2$, the necessary and sufficient optimality conditions may be expressed as the KKT-system

$$\left[\begin{array}{cccc|cc} R_0 & & & & B_0 & \\ & Q_1 & M_1 & & -I & A_1 \\ & M_1' & R_1 & & & B_1 \\ & & & P_2 & & -I \\ \hline B_0' & -I & & & & \\ & A_1' & B_1' & -I & & \end{array}\right] \begin{bmatrix} u_0 \\ x_1 \\ u_1 \\ x_2 \\ \pi_0 \\ \pi_1 \end{bmatrix} = - \left[\begin{array}{c} r_0 + M_0' x_0 \\ q_1 \\ r_1 \\ p_2 \\ \hline b_0 + A_0' x_0 \\ b_1 \end{array}\right]$$

which may be rearranged to

$$\begin{bmatrix} R_0 & B_0 & & & & \\ B_0' & 0 & -I & & & \\ & -I & Q_1 & M_1 & A_1 & \\ & & M_1' & R_1 & B_1 & \\ & & A_1' & B_1' & 0 & -I \\ & & & & -I & P_2 \end{bmatrix} \begin{bmatrix} u_0 \\ \pi_0 \\ x_1 \\ u_1 \\ \pi_1 \\ x_2 \end{bmatrix} = - \begin{bmatrix} r_0 + M_0' x_0 \\ b_0 + A_0' x_0 \\ q_1 \\ r_1 \\ b_1 \\ p_2 \end{bmatrix}$$

The necessary and sufficient conditions (21) for optimality of (19) are sparse and highly structured. The following proposition prescribes a Riccati iteration procedure for solution of (21).

Proposition 3. (LQ Optimal Control Solution). Let x_0, $\{A_k, B_k, b_k, Q_k, M_k, R_k, q_k, r_k\}_{k=0}^{N-1}$, and $\{P_N, p_N\}$ be given. Then the solution $\{u_k, \pi_k, x_{k+1}\}_{k=0}^{N-1}$ of (21) may be obtained by the following procedure

(1) Compute

$$R_{e,k} = R_k + B_k P_{k+1} B_k' \tag{22a}$$

$$K_k = -R_{e,k}^{-1}(M_k' + B_k P_{k+1} A_k') \tag{22b}$$

$$a_k = -R_{e,k}^{-1}(r_k + B_k(P_{k+1}b_k + p_{k+1})) \tag{22c}$$

$$P_k = Q_k + A_k P_{k+1} A_k' - K_k' R_{e,k} K_k \tag{22d}$$

$$p_k = (A_k + K_k' B_k)(P_{k+1}b_k + p_{k+1}) + q_k + K_k' r_k \tag{22e}$$

for $k = N-1, N-2, \dots, 0$.

(2) Compute the primal solution $\{u_k, x_{k+1}\}_{k=0}^{N-1}$ for $k = 0, 1, \dots, N-1$ by

$$u_k = K_k x_k + a_k \tag{23a}$$

$$x_{k+1} = A_k' x_k + B_k' u_k + b_k \tag{23b}$$

(3) Obtain the dual solution $\{\pi_k\}_{k=0}^{N-1}$ by computing

$$\pi_{N-1} = P_N x_N + p_N \tag{24a}$$

$$\pi_{k-1} = A_k \pi_k + Q_k x_k + M_k u_k + q_k \tag{24b}$$

for $k = N-1, N-2, \dots, 1$.

Proof. See Rao *et al.* (1998) or Ravn (1999). □

Let $x \in \mathbb{R}^n$ and $u \in \mathbb{R}^m$. Then this method has complexity $O(N(n^3 + m^3))$ while a dense method on the same KKT-system has complexity $O(N^3(n+m)^3)$. Corresponding dense quadratic programs obtained by elimination of the states has complexity $O(N^3 m^3)$. The Riccati based factorization is thus two order of magnitudes faster than the dense based approach when the horizon N length is much larger than the state dimension n.

The constrained linear quadratic optimal control problem is (19) with the additional constraints

$$C_k' x_k + D_k' u_k + c_k \geq d_k \quad k = 0, 1, \dots, N-1 \tag{25a}$$

$$C_N' x_N + c_N \geq d_N \tag{25b}$$

The search direction in algorithms for solution of this problem is computed efficiently by combination of proposition 3 and either proposition 1 or 2.

The model predictive controller and the moving horizon estimator are particular instances of the optimal control problem, i.e. (19) and (25). For simplicity and to focus on the essentials, this is demonstrated for the model predictive controller and moving horizon estimator without inequality constraints. These formulations are easily extended to the inequality constrained cases.

5. MODEL PREDICTIVE CONTROL

The unconstrained model predictive controller for linear systems may be expressed as

$$\min_{\{y_k, w_{k+1}, u_k\}_{k=0}^{\infty}} \quad \phi = \sum_{k=0}^{\infty} \tilde{l}_k(y_k, \Delta u_k) \tag{26a}$$

$$s.t. \quad w_{k+1} = \tilde{A}_k' w_k + \tilde{B}_k' u_k + \tilde{b}_k \tag{26b}$$

$$y_k = \tilde{C}_k' w_k + \tilde{c}_k \tag{26c}$$

in which the stage cost is

$$\tilde{l}_k(y_k, \Delta u_k) = \frac{1}{2}||y_k - z_k||_{\tilde{Q}_k}^2 + \frac{1}{2}||\Delta u_k||_{\tilde{S}_k}^2 \tag{27}$$

The goal stated by this cost, is to keep the systems output $\{y_k\}_{k=0}^{\infty}$ close to some prescribed trajectory $\{z_k\}_{k=0}^{\infty}$ while simultenously limiting the actuator variation $\Delta u_k = u_k - u_{k-1}$. Let the data at stage k of this controller be $\mathcal{M}_k = \left\{\tilde{A}_k, \tilde{B}_k, \tilde{b}_k, \tilde{C}_k, \tilde{c}_k, \tilde{Q}_k, \tilde{S}_k\right\}$ and let $\mathcal{M} = \left\{\tilde{A}, \tilde{B}, \tilde{b}, \tilde{C}, \tilde{c}, \tilde{Q}, \tilde{S}\right\}$. Assume that the controller (26) is parameterized as

$$\mathcal{C} = \{\mathcal{M}_k\}_{k=0}^{\infty} = \{\mathcal{M}_0, \mathcal{M}_1, \dots, \mathcal{M}_{N-1}, \mathcal{M}, \mathcal{M}, \dots\} \tag{28}$$

and that the reference trajectory has the parameterization

$$\{z_k\}_{k=0}^{\infty} = \{z_0, z_1, \dots, z_{N-1}, z, z, \dots\} \tag{29}$$

The optimal steady state consistent with the controller model (26) is obtained by solution of the quadratic program

$$\min_{u,w,y} \quad \frac{1}{2}||y - z||_{\tilde{Q}}^2 + \frac{1}{2}||u - u_s||_{R_s}^2 \tag{30a}$$

$$s.t. \quad w = \tilde{A}' w + \tilde{B}' u + \tilde{b} \tag{30b}$$

$$y = \tilde{C}' w + \tilde{c} \tag{30c}$$

in which u_s is a target of the input if there are degrees of freedom in excess and R_s is computed by the procedure (Muske, 1997)

$$\tilde{N} = \text{Null}\left(\begin{bmatrix} I - \tilde{A}' & -\tilde{B}' \end{bmatrix}\right) = \begin{bmatrix} \tilde{N}_x \\ \tilde{N}_u \end{bmatrix} \tag{31a}$$

$$\alpha = \text{Null}\left(\tilde{N}_x' \tilde{C} \tilde{C}' \tilde{N}_x\right) \tag{31b}$$

$$R_s = \tilde{R} \tilde{N}_u \alpha \alpha' \tilde{N}_u' \tilde{R} \tag{31c}$$

When a target (u, w) has been computed the model predictive controller algorithm solves the dynamic quadratic program

$$\min \quad \phi = \sum_{k=0}^{N-1} l_k(x_k, u_k) + l_N(x_N) \tag{32a}$$

$$s.t. \quad x_{k+1} = A_k' x_k + B_k' u_k + b_k \tag{32b}$$

in which

$$x_k = \begin{bmatrix} w_k \\ u_{k-1} \end{bmatrix} \; A_k' = \begin{bmatrix} \tilde{A}_k' & 0 \\ 0 & 0 \end{bmatrix} \; B_k' = \begin{bmatrix} \tilde{B}_k' \\ I \end{bmatrix} \; b_k = \begin{bmatrix} \tilde{b}_k \\ 0 \end{bmatrix} \tag{33}$$

and $l_k(x_k, u_k)$ is of the form (20a) with the parameters

$$Q_k = \begin{bmatrix} \tilde{C}_k\tilde{Q}_k\tilde{C}_k' & 0 \\ 0 & \tilde{S}_k \end{bmatrix} \quad M_k = \begin{bmatrix} 0 \\ -\tilde{S}_k \end{bmatrix} \quad R_k = \tilde{S}_k \tag{34a}$$

$$q_k = \begin{bmatrix} \tilde{C}_k\tilde{Q}_k(\tilde{c}_k - z_k) \\ 0 \end{bmatrix} \qquad r_k = 0 \tag{34b}$$

$$f_k = \frac{1}{2}(\tilde{c}_k - z_k)'\tilde{Q}_k(\tilde{c}_k - z_k) \tag{34c}$$

Let

$$x = \begin{bmatrix} w \\ u \end{bmatrix} \quad Q = \begin{bmatrix} \tilde{C}\tilde{Q}\tilde{C}' & 0 \\ 0 & \tilde{S} \end{bmatrix} \quad M = \begin{bmatrix} 0 \\ -\tilde{S} \end{bmatrix} \quad R = \tilde{S} \tag{35}$$

and compute P from the Riccati equation

$$P = Q + APA' - (M + APB')(R + BPB')^{-1}(M + APB')' \tag{36}$$

Then the selected cost-to-go function in (32)

$$l_N(x_N) = \frac{1}{2}(x_N - x)'P(x_N - x) \tag{37}$$

is identical to (20b) with the parameters

$$P_N = P \quad p_N = -Px \quad \gamma_N = \frac{1}{2}x'Px \tag{38}$$

Consequently, the dynamic quadratic program of model predictive control is an instance of a linear quadratic optimal control problem.

6. CONCLUSION

Model predictive control and moving horizon estimation with affine models are instances of the linear quadratic optimal control problem. By utilizing the structure of the linear quadratic optimal control problem, efficient methods for solution of the linear quadratic optimal control problem have been outlined. These methods are linear in the control horizon. They are therefore particularly attractive for predictive control with long horizons. Long horizons in predictive control are often preferable, because nominal stability can be guaranteed provided the control horizon is sufficiently large.

The search-direction of active-set algorithms for solution of the constrained linear quadratic optimal control problems, (19) and the additional constraints (25), may be efficiently computed using the decomposition stated in proposition 1 and the Riccati iteration stated in proposition 3. When the model parameterization is fixed a priori, as for instance for linear time invariant systems, the factorization of the unconstrained linear quadratic optimal control problem may be done offline, and the method increases even further in efficiency.

The search direction of the Mehrotra predictor-corrector interior-point algorithm for solution of the constrained linear quadratic optimal control problem may be efficiently computed by combination of proposition 2 specialized to (19,25) and proposition 3.

The algorithms outlined for solution of the linear quadratic optimal control problem can be applied as the QP-solver for solution of the nonlinear optimal control problem by SQP algorithms.

REFERENCES

Allgöwer, F., T. A. Badgwell, J. S. Qin, J. B. Rawlings and S. J. Wright (1999). Nonlinear predictive control and moving horizon estimation - an introductory overview. In: *Advances in Control. Highlights of ECC'99* (Paul M. Frank, Ed.). Springer. pp. 391–449.

Bartlett, R. A., A. Wächter and L. T. Biegler (2000). Active set vs. interior point strategies for model predictive control. In: *Americal Control Conference 2000*. Chicago, Illinois. pp. 4229–4233.

Binder, Thomas, Luise Blank, H. Georg Bock, Roland Burlisch, Wolfgang Dahmen, Moritz Diehl, Thomas Kronseder, Wolfgang Marquardt, Johannes P. Schlöder and Oskar von Stryk (2001). Introduction to model based optimization of chemical processes on moving horizons. In: *Online Optimization of Large Scale Systems* (Martin Grötschel, Sven O. Krumke and Jörg Rambau, Eds.). pp. 295–339. Springer. Berlin.

Gill, Phillip E. and Walter Murray (1978). Numerically stable methods for quadratic programming. *Mathematical Programming* **14**, 349–372.

Goldfarb, D. and A. Idnanini (1983). A numerically stable dual method for solving strictly convex quadratic programs. *Mathematical Programming* **27**, 1–33.

Muske, Kenneth R. (1997). Steady-state target optimization in linear model predictive control. In: *Proceedings of the American Control Conference*. Albuquerque, New Mexico. pp. 3597–3601.

Ouellette, D. V. (1981). Schur complements and statistics. *Linear Algebra and Its Applications* **36**, 187–295.

Rao, C. V., S. J. Wright and J. B. Rawlings (1998). Application of interior-point methods to model predictive control. *Journal of Optimization Theory and Applications* **99**, 723–757.

Ravn, Hans F. (1999). *Discrete Time Optimal Control*. Technical University of Denmark. Lyngby, Denmark.

Tenny, Matthew J. and James B. Rawlings (2002). Efficient moving horizon estimation and nonlinear model predictive control. In: *Proocedings of the American Control Conference*. Ancorage. pp. 4475–4480.

Wright, Stephen J. (1996). Applying new optimization algorithms to model predictive control. In: *Fifth International Conference on Chemical Process Control* (Jeffrey C. Kantor, Carlos E. Garcia and Brice Carnahan, Eds.). CACHE, American Institute of Chemical Engineers. Tahoe City, California. pp. 147–155.

ELSEVIER
IFAC
PUBLICATIONS
www.elsevier.com/locate/ifac

ROBUST PREDICTIVE CONTROL BASED ON NEIGHBORING EXTREMALS

S. Gros, B. Srinivasan, and D. Bonvin

Laboratoire d'Automatique
École Polytechnique Fédérale de Lausanne
CH-1015 Lausanne, Switzerland

Abstract: The performance of predictive controller is typically poor when the true plant evolution deviates significantly from that predicted by the model. A robust approach that considers model uncertainty explicitly is then needed. However, it is often difficult to find a single input profile that works for the range of uncertainty considered. Thus, multiple input profiles, i.e. one for each realization of the uncertainty, need to be determined, which is computationally extremely expensive. This paper proposes an alternative approach, based on neighboring extremals, where the multiple input profiles are computed using a simple feedback law, thereby reducing considerably the computational burden. The idea is illustrated via the simulation of an inverted pendulum on a cart. *Copyright © 2004 IFAC*

Keywords: Predictive control, Robust control, Optimal control, Neighboring extremals.

1. INTRODUCTION

Predictive control is an effective approach for tackling problems with constraints and nonlinear dynamics, especially when analytical computation of the control law is difficult (Mayne *et al.*, 2000; Scokaert and Mayne, 1998). Classical predictive control involves recalculating at every sampling instant the input[1] that minimizes a criterion defined over a horizon window in the future, taking into account the current state of the system. Only the first part of the computed optimal input is applied to the system.

A crucial point in predictive control is the extensive use of the dynamic system model that, unfortunately, may not always correspond to the reality. Thus, the predicted state evolution may differ from the actual plant evolution. When the difference between the predicted and the true plant evolutions is significant (which occurs, for example, when the system is unstable), standard predictive control will not be able to provide the desired performance (Mayne *et al.*, 2000). One solution to this problem consists of pre-stabilizing the system with a simple feedback loop. However, there are no systematic ways of designing such a feedback (Ronco *et al.*, 2001; Morari and Lee, 1999; Bemporad, 1998). Another approach is to cast the problem into a robust framework, where optimization is performed by taking the uncertainty into account explicitly.

Standard robust predictive control computes an input that represent a compromise solution for the range of uncertainty considered (Bemporad and Morari, 1999; Lee and Yu, 1997; Kouvaritakis *et al.*, 2000). Furthermore, to prove robust stability, it is important to guarantee that the final state is within some bounded set. When

[1] The word input is considered here to be *singular* independent upon whether there are one or several inputs. This choice is made in order to be able later to distinguish between cases with a single or multiple input profiles corresponding to either a single nominal model or several models (one for each realization of the uncertainty).

the dispersion of the open-loop predicted state is large, especially in the case of unstable systems, it may not be possible to find a feasible solution to the robust optimization problem. However, due to the feedback introduced by the re-optimizations performed at subsequent sampling instants, the true state dispersion at the end of the prediction horizon will in fact be much smaller than the values given by open-loop prediction. Hence, this feedback needs to be incorporated in the robust predictive control formulation in order to reduce the conservatism and lead to feasible solutions.

There are two ways of expressing this inherent feedback in a robust optimization framework: (i) use multiple input profiles, i.e. one for each realization of the uncertainy, starting with the next sampling interval (Mayne *et al.*, 2000; Scokaert and Mayne, 1998), and (ii) approximate the inherent feedback by a control law (Bemporad, 1998). The former is computationally expensive, while an ad hoc method has been used for the approximation in the latter. This paper suggests using the neighboring extremal approach for approximating the feedback inherent to predictive control.

For small deviations from the optimal solution, a linear approximation of the system and a quadratic approximation of the cost are quite reasonable. In such a case, the theory of neighboring extremals (NE) provides a closed-form solution to the optimization problem (Bryson, 1999). Thus, the optimal input can be obtained using state feedback that approximates the feedback provided by explicit numerical optimization.

This approach can also be viewed as a novel way of performing robust optimization with multiple input profiles. Indeed, the proposed scheme optimizes the multiple profiles, one exactly via explicit optimization and the others approximately via the NE-approach. Since only one input profile is optimized explicitly, the computational complexity of the problem reduces considerably, while keeping the advantages of the robust optimization scheme with multiple input profiles. Note that the main emphasis is not in reducing the computational complexity of a general MPC problem as in (Wan and Kothare, 2003), but to really exploit the structure present in the robust optimization problem.

The paper is organized as follows. Background material regarding predictive control and neighboring extremals is presented in Section 2. The classical robust predictive control scheme and robust predictive control with multiple input profiles are presented in Section 3. The robust predictive control based on the NE-approach is presented in Section 4. In Section 5, the various schemes are compared on an illustrative example. Finally, concluding remarks are given in Section 6.

2. PRELIMINARIES

2.1 Classical Predictive Control

Consider the nonlinear system represented as:

$$\dot{x} = F(x, u, \theta), \qquad x(0) = x_0 \tag{1}$$

where the state x and the input u are vectors of dimension n and m, respectively. x_0 represents the initial conditions, $\theta \in \Theta$ the vector of uncertain parameters, assumed to lie in the admissible region Θ, and F the system dynamics.

In predictive control, the following optimization problem is solved repeatedly at discrete time instants:

$$\min_{u([t_k, t_k+T_f])} J(t_k) = \frac{1}{2} x(t_k+T_f)^T P\, x(t_k+T_f) \tag{2}$$

$$+ \frac{1}{2} \int_{t_k}^{t_k+T_f} \left(x(\tau)^T Q\, x(\tau) + u(\tau)^T R\, u(\tau) \right) d\tau$$

$$s.t. \quad \dot{x} = F(x, u, \theta), \qquad x(t_k) = x_k$$

$$x(t_k + T_f) \in \mathcal{X}$$

where P, Q, and R are positive-definite weighting matrices of appropriate dimensions, $\mathcal{X}$ the bounded region of state space where the final state should be, t_k the present time for which the optimization is performed, T_f the prediction horizon, and x_k the state measured or estimated at the time instant t_k. The optimal input computed by solving (2) is represented by $u^\star([t_k, t_k+T_f])$. The importance of having a terminal cost, and also a bounded region for the final state for the sake of stability, is discused in (Mayne *et al.*, 2000).

Let δ be the sampling period which, in general, is constant. The first part of the optimal input, $u^\star([t_k, t_k + \delta])$, is applied open loop, and the optimization problem is repeated at the time instant t_{k+1}. For implementation purposes, the infinite-dimensional input $u([t_k, t_k+T_f])$ needs to be parameterized using a finite number of decision variables, typically piecewise constant.

2.2 Neighboring Extremals

Including the constraints of the optimization problem (2) in the cost function, the augmented cost function, $\bar{J}$, can be written as (Bryson, 1999):

$$\bar{J} = \Phi(x(t_k+T_f)) + \int_{t_k}^{t_k+T_f} \left(H - \lambda^T \dot{x} \right) dt \tag{3}$$

where $\Phi = \frac{1}{2} x^T P x$, $H = \frac{1}{2}(x^T Q\, x + u^T R\, u) + \lambda^T F(x, u)$, and $\lambda(t) \neq 0$ is the n-dimensional vector of adjoint states (Lagrange multipliers for the system equations), whose dynamics are given by $\dot{\lambda}^T = -H_x$, $\lambda^T(t_k + T_f) = \Phi_x(t_k + T_f)$.

The notation $a_b = \frac{\partial a}{\partial b}$ is used. The necessary conditions of optimality read:

$$H_u = u^T R + \lambda^T F_u = 0 \tag{4}$$

At the optimal solution, the first variation of $\bar{J}$ is given by:

$$\Delta \bar{J} = \left(\Phi_x - \lambda^T\right) \Delta x\big|_{t_k+T_f} + \int_{t_k}^{t_k+T_f} \left[\left(H_x + \dot{\lambda}^T\right) \Delta x + H_u \Delta u\right] d\tau \tag{5}$$

where $\Delta x(t) = x(t) - x^\star(t)$ and $\Delta u(t) = u(t) - u^\star(t)$, with $x^\star$ and $u^\star$ being the optimal state and input, respectively. The conditions of optimality are derived from $\Delta \bar{J} = 0$. The second-order variation of $\bar{J}$ is given by:

$$\Delta^2 \bar{J} = \frac{1}{2} \Delta x(t_k + T_f)^T P \, \Delta x(t_k + T_f) + \frac{1}{2} \int_{t_k}^{t_k+T_f} \left[\Delta x^T \; \Delta u^T\right] \begin{bmatrix} H_{xx} & H_{xu} \\ H_{ux} & H_{uu} \end{bmatrix} \begin{bmatrix} \Delta x \\ \Delta u \end{bmatrix} d\tau \tag{6}$$

Choosing Δu to minimize $\Delta^2 \bar{J}$ under the linear dynamic constraint:

$$\Delta \dot{x} = F_x \Delta x + F_u \Delta u \tag{7}$$

represents a time-varying Linear Quadratic Regulator (LQR) problem, for which a closed-form solution is available:

$$\Delta u(t) = -K(t) \Delta x(t) \tag{8}$$

$$K = H_{uu}^{-1} \left(H_{ux} + F_u^T S\right) \tag{9}$$

$$\dot{S} = -H_{xx} - S F_x - F_x^T S + H_{xu} K + S F_u K$$

$$S(t_k + T_f) = P \tag{10}$$

The details of the development leading to this formulation can be found in (Bryson, 1999). The above controller, termed the neighboring extremal controller, will be used extensively in this paper.

3. EXISTING APPROACHES TO ROBUST PREDICTIVE CONTROL

3.1 Standard Robust Predictive Control

The state x and hence the cost function J are functions of the vector of uncertain parameters θ. For a given value of θ, let x_θ be the state and J_θ the cost function. In robust predictive control, the uncertainty is handled as part of the optimization problem, which is solved repeatedly at discrete time instants:

$$\min_{u([t_k, t_k+T_f])} E_{\theta \in \Theta}[J_\theta(t_k)] \tag{11}$$
$$\text{s.t.} \quad \dot{x}_\theta = F(x_\theta, u, \theta) \qquad x_\theta(t_k) = x_k$$
$$x_\theta(t_k + T_f) \in \mathcal{X}, \qquad \forall \theta \in \Theta$$

$$J_\theta(t_k) = \frac{1}{2} x_\theta(t_k + T_f)^T P \, x_\theta(t_k + T_f) + \frac{1}{2} \int_{t_k}^{t_k+T_f} \left(x_\theta(\tau)^T Q \, x_\theta(\tau) + u(\tau)^T R \, u(\tau)\right) d\tau \tag{12}$$

(where $E_{\theta \in \Theta}(x)$ denotes the expectation of x when the random variable θ is in the set Θ). The major difficulty with this formulation is that there may not be a solution, especially when the system is open-loop unstable. The state dispersion might be so large that it is not possible to find a single input $u([t_k, t_k+T_f])$ that satisfies $x_\theta(t_k+T_f) \in \mathcal{X}$, $\forall \theta \in \Theta$.

It should be noted here that a robust control scheme based on the expectation of the cost is a standard approach that, however leads to only limited robustness improvement (Nagy and Braatz, 2003).

3.2 Robust Predictive Control with Multiple Input Profiles

Though all computations in predictive control are performed open loop, there is inherent feedback due to the state measurement and the re-optimization. This feedback reduces the sensitivity to uncertainty, and thus the state dispersion is often much smaller than what is predicted from an open-loop perspective.

The difficulty with the classical formulation (11) is that it does not take into account the fact that the optimization will be repeated at subsequent time instants. Thus, the idea is to reformulate the robust predictive control problem and include re-optimization in the problem formulation (Mayne *et al.*, 2000):

$$\min_{u([t_k, t_{k+1}]), \bar{u}_\theta([t_{k+1}, t_k+T_f])} E_{\theta \in \Theta}[J_\theta(t_k)] \tag{13}$$
$$\text{s.t.} \quad \dot{x}_\theta = F(x_\theta, u_\theta, \theta) \qquad x_\theta(t_k) = x_k$$
$$x_\theta(t_k + T_f) \in \mathcal{X}, \qquad \forall \theta \in \Theta$$

$$u_\theta = \begin{cases} u & \text{if } t_k \le t < t_{k+1} \\ \bar{u}_\theta & \text{if } t_{k+1} \le t \le t_k + T_f \end{cases} \tag{14}$$

$$J_\theta(t_k) = \frac{1}{2} x_\theta(t_k + T_f)^T P \, x_\theta(t_k + T_f) + \frac{1}{2} \int_{t_k}^{t_k+T_f} \left(x_\theta(\tau)^T Q \, x_\theta(\tau) + u_\theta(\tau)^T R \, u_\theta(\tau)\right) d\tau \tag{15}$$

where $\bar{u}_\theta$ is the input for the realization θ during the prescribed interval. This means that the manipulated variables consist of one set of inputs for all realizations of θ between t_k and t_{k+1}. However, between t_{k+1} and $t_k + T_f$, different sets of inputs are required for different realizations of θ. This problem is computationally expensive to solve since $\bar{u}_\theta([t_{k+1}, t_k + T_f])$ needs to be optimized for every realization of θ. Note that, though $\bar{u}_\theta([t_{k+1}, t_k + T_f])$ is important for calculating $u([t_k, t_{k+1}])$, it will never be implemented on the real system. What will be implemented is the part $u([t_k, t_{k+1}])$, which represents the compromise input for all the uncertainty realizations considering all future possibilities.

4. ROBUST PREDICTIVE CONTROL BASED ON NEIGHBORING EXTREMALS

In order to avoid having to compute the optimal input for many different realizations of θ, a relationship between the uncertain parameters θ and the optimal input is needed. The idea proposed in this paper is to use the NE-approach. As seen in Section 2, the NE-approach provides the following relationship:

$$u_\theta([t_{k+1}, t_k + T_f]) = u_{\theta_0}([t_{k+1}, t_k + T_f]) - K(t)\Delta x([t_{k+1}, t_k + T_f]) \quad (16)$$

where θ_0 is the nominal parameter vector. The optimization problem then becomes:

$$\min_{u([t_k, t_k + T_f])} E_{\theta \in \Theta}[J_\theta(t_k)] \quad (17)$$

$$\begin{aligned} s.t.\quad & \dot{x}_{\theta_0} = F(x_{\theta_0}, u, \theta_0), \quad x_{\theta_0}(t_k) = x_k \\ & \dot{x}_\theta = F(x_\theta, u_\theta, \theta), \quad x_\theta(t_k) = x_k, \quad \forall \theta \neq \theta_0 \\ & x_\theta(t_k + T_f) \in \mathcal{X}, \quad \forall \theta \in \Theta \end{aligned}$$

$$u_\theta = \begin{cases} u & \text{if } t_k \le t < t_{k+1} \\ u - K(x_\theta - x_{\theta_0}) & \text{if } t_{k+1} \le t \le t_k + T_f \end{cases}$$

$$J_\theta(t_k) = \frac{1}{2} x_\theta(t_k + T_f)^T P\, x_\theta(t_k + T_f) + \frac{1}{2} \int_{t_k}^{t_k + T_f} \left(x_\theta(\tau)^T Q\, x_\theta(\tau) + u_\theta(\tau)^T R\, u_\theta(\tau) \right) d\tau$$

Note that the decision variables in (17) are only $u([t_k, t_k + T_f])$ for the nominal plant and not $u_\theta([t_k, t_k + T_f])$ for all realizations. The input profiles for realizations other than the nominal one are computed using the NE-controller. The parameterization of the nominal input can be chosen conveniently.

The NE-approach can be interpreted from two viewpoints: (i) From the view point of feedback, it is an approximation of the inherent feedback provided by the predictive control itself; (ii) From the point of view of robust predictive control with multiple input profiles, and given the nominal optimal input profile, the NE-approach computes to a first-order approximation the optimal input profiles that correspond to the various realization of θ. Note that, as in Subsection 3.2, the input computed via the NE-approach will never be implemented on the true system since it is computed for the time interval $[t_{k+1}; t_k + T_f]$. In that sense, the NE-feedback is a fictitious one that serves only a computational purpose.

The proposed NE-approach is only an approximation of robust control with multiple input profiles and thus should have an inferior performance. However, in the examples that have been worked out, the proposed NE-approach often led to a slight improvement in performance over robust control with multiple input profiles. This can be attributed to the fact that robust control with multiple input profiles requires the solution of an optimization problem with a large number of decision variables and, thus, often gets stuck in a local minimum due to sensitivity issues. Thus it might be advantageous to use a well-posed feedback law instead of a poorly posed open-loop problem.

Though this paper presents no stability proof for the proposed robust predictive controller based on the NE-approach, many pointers indicate that robust stability can indeed be established. The steps to be followed are: (i) Stability of the robust predictive control with multiple input profiles (Mayne *et al.*, 2000), (ii) Proof that robust predictive control based on NE is a first-order approximation of that with multiple input profiles (see preliminary work in this direction by (Ronco *et al.*, 2001)), and (iii) Effect of the approximation error on stability.

5. ILLUSTRATION: CONTROL OF AN INVERTED PENDULUM

5.1 System Model

This section illustrates the application of both classical and robust predictive control schemes to an inverted pendulum on a cart, ignoring the cart dynamics (Ronco *et al.*, 2001). The model equations are:

$$\dot{x}_1 = x_2 \quad (18)$$

$$\dot{x}_2 = \frac{m\,l}{J}[sin(x_1)\, g - cos(x_1) u] \quad (19)$$

where x_1 is the pendulum angle, x_2 its rotational velocity, and u the control torque. The control objective consists of regulating the pendulum around the upright position, starting from the downward position $x_0 = [\pi \quad 0]$. The following numerical values are used: $m = 1\,[kg]$, $g = 9.81\,\left[\frac{m}{s^2}\right]$, $l = 1\,[m]$ and $J = 1\,[kg \cdot m^2]$. In addition, the control is constrained, $-5\,\left[\frac{m}{s^2}\right] < u < 5\,\left[\frac{m}{s^2}\right]$. For all techniques, the same sampling period of $\delta = 0.25\,[s]$ is used.

5.2 Control Parameters

All schemes share the following features: (i) re-optimization of the input at each sampling instant, (ii) prediction horizon, $T_f = 1s$, (iii) control horizon = sampling period, i.e. the control sequence for $t_k \leq t \leq t_k + \delta$ is applied, leaving the rest of the sequence unused, (iv) $x(t_k + T_f) \in \mathcal{X}$ was not enforced, (v) $P = 10\ I$, $Q = 10^{-2}$, $R = 0.02$, where I is the identity matrix.

The input parameterization considers $u(t)$ constant over the time interval $[t_k, t_k + \delta]$. The rest of the input trajectory is obtained using the shooting method (Lewis, 1986), with the initial conditions of the adjoint variables serving as parameters. In the following simulations, it is assumed that the mass of the pendulum is unknown but lies somewhere between 0.5 $[kg]$ and 1.5 $[kg]$, with a uniform probability distribution. The expectation $E_{\theta\in\Theta}(J_\theta)$ is approximated as $\sum_{k=1}^{3} \frac{1}{3} J_{m_k}$, with $m_1 = 0.5\,[kg]$, $m_2 = 1\,[kg]$ and $m_3 = 1.5\,[kg]$. The predictive control schemes are applied to a simulated reality with a mass $m = 1.32\,[kg]$, thus making the real system slower than the nominal model ($m = 1\,[kg]$).

5.3 Classical Predictive Control

The classical non-robust re-optimization scheme (2) is applied first. The control is computed from a single nominal model with $m = 1\,[kg]$ and applied to the simulated reality with $m = 1.32\,[kg]$. The simulation results are displayed in Figure 1. They show that the approach does not work since the sampling period is too large for the control scheme to converge.

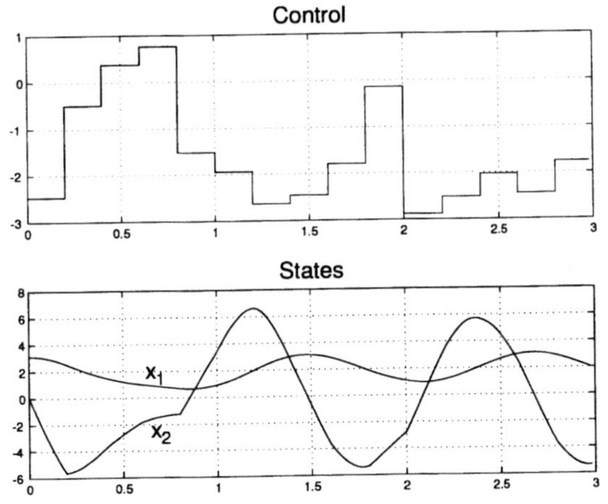

Fig. 1. Classical Predictive Control

5.4 Standard Robust Predictive Control

The robust predictive control scheme considered here uses three models (values of the mass: $m_1 = 0.5\,[kg]$, $m_2 = 1\,[kg]$, $m_3 = 1.5\,[kg]$). The optimization scheme (11) computes a single input profile for all three models. The simulation results are displayed in Figure 2. They show that this approach does not work either. The optimization is unable to find a single input profile that works well with all three models. Hence, the predicted value of the cost function is high and does not decrease with the number of re-optimizations. No final constraint is imposed since this optimization is unable to provide a feasible solution.

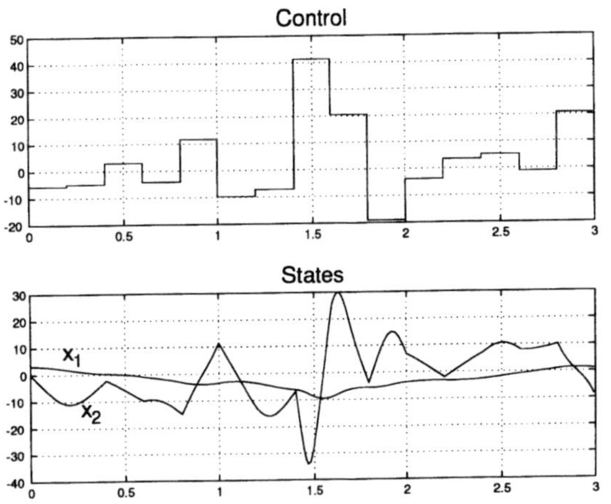

Fig. 2. Standard Robust Predictive Control

5.5 Robust Control with Multiple Input Profiles

The robust control scheme with multiple input profiles (13) is studied next. The same three models are used but, at each re-optimization, three different control sequences are computed, i.e. one for each model. The input for the first sampling interval is the same for all three models, but distinct for each model thereafter. The results of the simulation are displayed in Figure 3. The computational burden is very heavy, the duration of one re-optimization (when the system is not close to the reference) is approximately 5200 seconds. Also, since the number of decision variables and the number of simulations to be performed increase with the number of realizations considered, the computational time increases *quadratically* with the number of realizations.

5.6 Robust Control based on Neighboring Extremals

The proposed approach (17) is used, where the input consists of two parts. The first part, for $t = [t_k, t_k + \delta]$, is a piecewise constant input that is common to the three models. The second part, for $t = [t_k + \delta, t_k + T_f]$, is generated by a NE-controller and differs for each model. At each re-optimization, an optimal input is computed for the nominal model ($m = 1\,[kg]$), using a

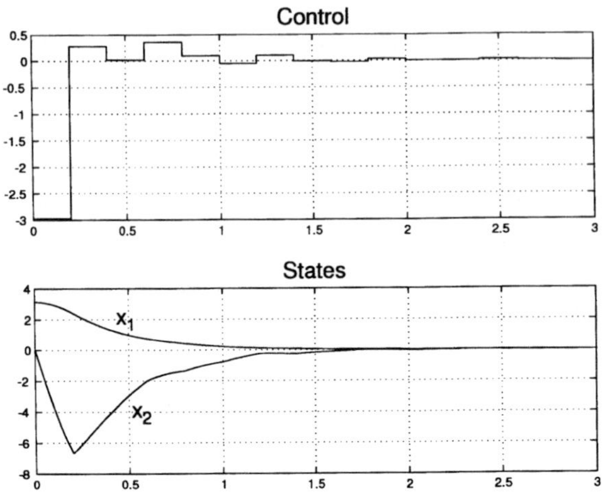

Fig. 3. Robust Control with Multiple Input Profiles

shooting method. Based on this optimal input, a NE-controller is designed to generate the optimal trajectory for each of the three models in the interval $t = [t_k + \delta, t_k + T_f]$.

The simulation results, displayed in Figure 4, show that this approach works well on this system. The computational burden is much lower than with multiple input profiles: the duration of a re-optimization (when the system is not close to the reference point) is of the order of 700 seconds, a considerable reduction compared to scheme (13). Moreover, with the NE-approach, the computational time increases *linearly* with the number of models used to represent the uncertainty.

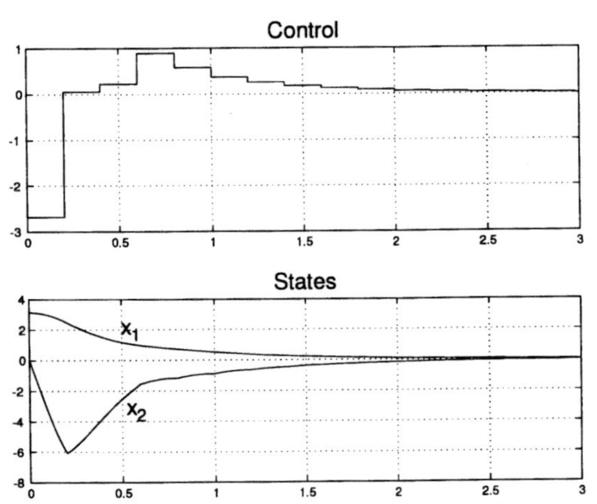

Fig. 4. Robust Control with Neighboring Extremals

6. CONCLUSION

This paper has addressed the problem of robust predictive control of systems for which the open-loop prediction of the future state evolution leads to very conservative results. The fact that there will be re-optimizations needs to be incorporated in the predictions, and this was done in this paper using the NE-approach. The proposed approach was illustrated and compared to other approaches on a simple unstable mechanical example.

The stability and performance of the NE-based robust predictive control scheme have not been addressed in this paper. These issues will form the subject of future research. The main issues involved therein will be how good the approximation is and how the approximation error influences stability. In this paper, the feedback computed using the NE-approach was only used as a fictitious one for computational purposes. Its use for implementation is another promising research direction.

REFERENCES

Bemporad, A. (1998). Reducing conservatism in predictive control of constrained systems with disturbances. In: *37th IEEE Control and Decision Conference*. Tampa, FL. pp. 1384–89.

Bemporad, A. and M. Morari (1999). *Robust Model Predictive Control: a Survey.* Springer Verlag.

Bryson, A. E. (1999). *Dynamic Optimization.* Addison-Wesley, Menlo Park, California.

Kouvaritakis, B., J. A. Rossiter and J. Schuurmans (2000). Efficient robust predictive control. *IEEE Trans. Automat. Contr.* **45**(8), 1545–49.

Lee, J. H. and Z. Yu (1997). Worst-case formulations of model-predictive control for systems with bounded parameters. *Automatica* **33**(5), 763–781.

Lewis, F. L. (1986). *Optimal Control.* John Wiley, New York.

Mayne, D. Q., J. B. Rawlings, C. V Rao and P. O. M. Scokaert (2000). Constrained model predictive control: Stability and optimality. *Automatica* **36**(6), 789–814.

Morari, M. and J. H. Lee (1999). Model predictive control: Past, present, and future. *Comp. Chem. Eng.* **23**, 667–682.

Nagy, Z. K. and R. D. Braatz (2003). Robust nonlinear model predictive control of batch processes. *AIChE Journal* **49**(7), 1776–1786.

Ronco, E., B. Srinivasan, J. Y. Favez and D. Bonvin (2001). Predictive control with added feedback for fast nonlinear systems. In: *European Control Conference.* Porto, Portugal. pp. 3167–3172.

Scokaert, P. O. and D. Q. Mayne (1998). Min-max feedback model predictive control for constrained linear systems. *IEEE Trans. Automat. Contr.* **43**, 1136–1142.

Wan, Z. and M. V. Kothare (2003). An efficient off-line formulation of robust model predictive control using linear matrix inequalities. *Automatica* **39**, 837–846.

ELSEVIER
IFAC
PUBLICATIONS
www.elsevier.com/locate/ifac

A FAST, EASILY TUNED, SISO, MODEL PREDICTIVE CONTROLLER

Gabriele Pannocchia [*,1] **Nabil Laachi** [**]
James B. Rawlings [**]

[*] *Department of Chemical Engineering – Univ. of Pisa Via Diotisalvi 2, 56126 Pisa (Italy)*
[**] *Department of Chemical and Biological Engineering – Univ. of Wisconsin 53706 Madison, WI (USA)*

Abstract: It is commonly believed that for SISO systems, well tuned PID controllers work as well as model-based controllers and that PID controllers are more robust to model errors. In this paper we present a novel offset-free constrained LQ controller for SISO systems, which is implemented in an efficient way so that the total controller execution time is similar to that of a PID. The proposed controller has three modules: a state and disturbance estimator, a target calculation, and a constrained dynamic optimization. It is shown that the proposed controller outperforms PID both in setpoint changes and disturbance rejection, it is robust to model errors, it is insensitive to measurement noise, it handles constraints much better than common anti-windup PID. Tuning the proposed controller is simple. In principle, there are three tuning parameters to choose, but in all examples presented only one was actually varied, obtaining a clear and intuitive effect on the closed-loop performance. *Copyright © 2004 IFAC.*

Keywords: Model predictive control (MPC), linear quadratic regulation (LQR), PID control, tuning, constraints.

1. INTRODUCTION

Digital PID control for single-input single-output (SISO) systems shows up everywhere in chemical process applications and process control education. Tuning rules are presented in numerous texts and, surprisingly, remain a topic of current control research (Chen and Seborg, 2001; Skogestad, 2003). In this paper we would like to raise the issue of whether this popularity is due to any concrete, technological advantage of digital PID controllers, or whether PID's continuing popularity is simply a historical accident stemming from the success of analog PID controllers. The main technical advantages ascribed to PID control are: PID is simple, fast, and easy to implement in hardware and software; it is easy to tune; it provides good nominal control performance; and it is robust to model errors.

Model-based control methods, such as model predictive control (MPC) of constrained systems, on the other hand, are regarded by many in process control as complex to implement and tune. MPC has become the advanced controller of choice by industry mainly for the economically important, large-scale, multivariable processes in the plant. The rationale for MPC in these applications is that the complexity of implementing MPC is justified only for the important loops with large payoffs.

To address this perception of complexity, we propose a constrained, SISO linear quadratic controller (CLQ) with the following features: it is essentially as fast to execute as PID (within a factor of five regardless of system order), it is easy to implement in software and hardware, and it displays both higher performance and better robustness than PID controllers.

Other researchers have explored the following, related topics. A SISO model predictive controller based on a first-order plus time delay model, with input horizon of one, is proposed by Mukati and Ogunnaike (2004). Soroush and Muske (2000) show that a particular MPC algorithm with input horizon of one, has PI or PID form when the system is first order or second order (without delay), respectively.

[1] Corresponding author. Email: g.pannocchia@ing.unipi.it, Fax: +39 050 511266.

2. OFFSET-FREE CONSTRAINED LQ CONTROLLER FOR SISO SYSTEMS

As most tuning rules for PID controllers require simple transfer function process models, we assume such a model is available. The offset-free constrained linear quadratic (CLQ) control algorithm has three main modules that use a state-space model of the system: a state and disturbance estimator, a constrained target calculation, and a constrained dynamic optimization.

2.1 Model and estimator

We assume that a state-space discrete-time model of the system to be regulated is known:

$$\begin{aligned} x_{k+1} &= Ax_k + Bu_{k-m} \\ y_k &= Cx_k \,, \end{aligned} \tag{1}$$

in which $x \in \mathbb{R}^n$ is the state vector, $u \in \mathbb{R}$ is the input, $y \in \mathbb{R}$ is the output and m is a non-negative integer, the time delay.

Assumption 1. (General). The pair (A,B) is controllable, the pair (C,A) is observable and

$$\operatorname{rank}\begin{bmatrix} I-A & -B \\ C & 0 \end{bmatrix} = n+1 \,. \tag{2}$$

The input u is assumed to be constrained as follows:

$$u_{\min} \leq u \leq u_{\max} \,, \tag{3}$$

in which $u_{\min} < u_{\max}$.

In order to guarantee offset-free control of y in the presence of plant/model mismatch and/or unmeasured nonzero disturbances, the system model (1) is augmented with an integrating disturbance according to the general methodology proposed in (Pannocchia and Rawlings, 2003). In this work we choose the so-called "input disturbance model", i.e. we add an integrating state d that enters the system at the same place as the input u. The resulting augmented system is:

$$\begin{aligned} \begin{bmatrix} x \\ d \end{bmatrix}_{k+1} &= \begin{bmatrix} A & B \\ 0 & 1 \end{bmatrix} \begin{bmatrix} x \\ d \end{bmatrix}_k + \begin{bmatrix} B \\ 0 \end{bmatrix} u_{k-m} \\ y_k &= [C \ 0] \begin{bmatrix} x \\ d \end{bmatrix}_k \,. \end{aligned} \tag{4}$$

Several studies have pointed out that such a disturbance model is an appropriate choice for rejecting unmeasured disturbances efficiently (Muske and Badgwell, 2002), and it provides good robustness to plant/model mismatch (Pannocchia, 2003).
The state x and the disturbance d are estimated from the plant measurement y by means of a steady-state Kalman filter. At each sampling time, an estimate of the state $\hat{x}_{k|k-1}$ and of the disturbance $\hat{d}_{k|k-1}$ based on previous measurements and inputs are available. Thus, the current filtered state and disturbance are:

$$\begin{aligned} \hat{x}_{k|k} &= \hat{x}_{k|k-1} + L_x(y_k - C\hat{x}_{k|k-1}) \\ \hat{d}_{k|k} &= \hat{d}_{k|k-1} + L_d(y_k - C\hat{x}_{k|k-1}) \,, \end{aligned} \tag{5}$$

in which the filter gains $L_x \in \mathbb{R}^n$ and $L_d \in \mathbb{R}$ are computed offline as described later in this paragraph. Given the input u_{k-m} (stored if $m > 0$, or computed as described in the next paragraph, if $m = 0$), the state and disturbance estimates for the next sampling time are:

$$\begin{aligned} \hat{x}_{k+1|k} &= A\hat{x}_{k|k} + Bu_{k-m} + B\hat{d}_{k|k} \\ \hat{d}_{k+1|k} &= \hat{d}_{k|k} \,. \end{aligned} \tag{6}$$

In order to compute the filter gains L_x and L_d, let

$$\hat{A} = \begin{bmatrix} A & B \\ 0 & 1 \end{bmatrix}, \quad \hat{Q} = \begin{bmatrix} q_x I_n & 0 \\ 0 & 1 \end{bmatrix}, \quad \hat{C} = [C \ 0] \,,$$

in which q_x is a non-negative scalar. Also let R_v be a positive scalar that represents the output noise variance. Then, the estimator steady-state Riccati equation is

$$\Pi = \hat{Q} + \hat{A}\Pi\hat{A}^T - \hat{A}\Pi\hat{C}^T \left(\hat{C}\Pi\hat{C}^T + R_v\right)^{-1} \hat{C}\Pi\hat{A}^T \,,$$

in which $\Pi \in \mathbb{R}^{(n+1)\times(n+1)}$ is symmetric semi-definite. Finally, the filter gain is:

$$L = \begin{bmatrix} L_x \\ L_d \end{bmatrix} = \Pi\hat{C}^T \left(\hat{C}\Pi\hat{C}^T + R_v\right)^{-1} \,. \tag{7}$$

Strictly speaking, q_x and R_v should be regarded as the estimator tuning parameters and more details about their effect are discussed in (Pannocchia *et al.*, 2003). In this paper, fixed values for q_x (0.05) and R_v (0.01) are used for all examples.

2.2 Constrained target calculation

At each sampling time, given the current disturbance estimate $\hat{d}_{k|k}$, we compute the steady-state targets for input and state such that the output ultimately reaches the setpoint. If the input were unconstrained, these targets would simply be the solution to the following square system:

$$\begin{bmatrix} I-A & -B \\ C & 0 \end{bmatrix} \begin{bmatrix} \bar{x}_k \\ \bar{u}_k \end{bmatrix} = \begin{bmatrix} B\hat{d}_{k|k} \\ \bar{y} \end{bmatrix} , \tag{8}$$

which exists and is unique because of Assumption 1. However, the solution to (8) may be such that the input target $\bar{u}_k$ violates (3). Moreover, for integrating processes it is possible that a steady state does not exist because of (3). For these reasons, we compute $(\bar{x}_k, \bar{u}_k)$ from the following QP:

$$(\bar{x}_k, \bar{u}_k) = \arg\min_{(\bar{x},\bar{u})} (C\bar{x} - \bar{y})^2 + \eta \left\| (I-A)\bar{x} - B(\bar{u} + \hat{d}_{k|k}) \right\|_2^2 \tag{9a}$$

subject to:

$$u_{\min} \leq \bar{u} \leq u_{\max} \,, \tag{9b}$$

in which η is a large positive number.

2.3 Constrained dynamic optimization problem: case of reachable setpoint

Given the current state and disturbance estimates, and given the current steady-state targets $(\bar{x}_k, \bar{u}_k)$, for the cases in which the desired setpoint is reachable, i.e. $C\bar{x}_k = \bar{y}$, we compute the control input by means of the following constrained dynamic optimization problem:

$$\min_{\{v_j\}_{j=0}^{N-1}} \left\{ \sum_{j=0}^{N-1} w_j^T Q w_j + s(v_j - v_{j-1})^2 \right\} + \begin{bmatrix} w_N \\ v_{N-1} \end{bmatrix}^T P \begin{bmatrix} w_N \\ v_{N-1} \end{bmatrix}, \quad (10a)$$

subject to:

$$w_0 = \hat{x}_{k+m|k} - \bar{x}_k, \qquad v_{-1} = u_{k-1} - \bar{u}_k, \quad (10b)$$

$$w_{j+1} = A w_j + B v_j, \quad (10c)$$

$$u_{\min} - \bar{u}_k \le v_j \le u_{\max} - \bar{u}_k, \quad (10d)$$

in which N is a positive integer, s a positive scalar and $Q = C^T C$. The matrix $P \in \mathbb{R}^{(n+1)\times(n+1)}$ is chosen as the positive semidefinite solution of the Riccati equation:

$$P = \tilde{Q} + \tilde{A}^T P \tilde{A} - \tilde{A}^T P \tilde{B}(\tilde{B}^T P \tilde{B} + s)^{-1} \tilde{B} P \tilde{A}, \quad (11)$$

in which

$$\tilde{A} = \begin{bmatrix} A & B \\ 0 & 1 \end{bmatrix}, \qquad \tilde{B} = \begin{bmatrix} B \\ 1 \end{bmatrix}, \qquad \tilde{Q} = \begin{bmatrix} Q & 0 \\ 0 & 0 \end{bmatrix}. \quad (12)$$

Let $v := (v_0, v_1, \ldots, v_{N-1})$ be a column vector of length N. We can write (10) as a strictly convex QP:

$$\min_v \frac{1}{2} v^T H v + v^T \tilde{c} \quad (13a)$$

subject to:

$$\mathbf{1} \cdot (u_{\min} - \bar{u}_k) \le v \le \mathbf{1} \cdot (u_{\max} - \bar{u}_k), \quad (13b)$$

in which

$$H = \mathscr{B}^T \mathscr{Q} \mathscr{B} + \mathscr{D}^T \mathscr{R} \mathscr{D}, \; \tilde{c} = \mathscr{B}^T \mathscr{Q} \mathscr{A} w_0 + \mathscr{D}^T \mathscr{R} \mathscr{C} v_{-1} \quad (14)$$

and the constant matrices $\mathscr{A}$, $\mathscr{B}$, $\mathscr{C}$, $\mathscr{D}$, $\mathscr{Q}$, $\mathscr{R}$ are not shown the sake of space (Pannocchia *et al.*, 2003). Let $v^* = (v_0^*, \ldots, v_{N-1}^*)$ denote the optimal solution to (13). Then, the current control input is defined by using a receding horizon implementation, that is:

$$u_k = \bar{u}_k + v_0^*. \quad (15)$$

2.4 *Constrained dynamic optimization problem: case of unreachable setpoint*

When the setpoint is not reachable, i.e. when

$$\bar{y}_k = C\bar{x}_k \ne \bar{y}, \quad (16)$$

the optimization problem (10) needs to be modified because the corresponding optimal input would drive the controlled variable to the reachable target $\bar{y}_k$ as quickly as possible. There are important cases in which this behavior is undesirable. These situations occur when a "large" disturbance enters the system, and the input constraints are such that the input will asymptotically saturate without rejecting completely the disturbance, and hence offset will occur. It is clear that, if the disturbance keeps affecting the system, steady-state offset is unavoidable but even in such cases it is desirable to keep the controlled variable close to the desired setpoint $\bar{y}$ as long as possible. This goal can be achieved by modifying the optimization problem (10) with a linear penalty (Bonné *et al.*, 2003)

$$\min_{\{v_j\}_{j=0}^{N-1}} \left\{ \sum_{j=0}^{N-1} w_j^T (Q w_j + 2q) + s(v_j - v_{j-1})^2 \right\} + \begin{bmatrix} w_N \\ v_{N-1} \end{bmatrix}^T \left(P \begin{bmatrix} w_N \\ v_{N-1} \end{bmatrix} + 2p \right), \quad (17a)$$

$$\text{subject to (10b)–(10d)}, \quad (17b)$$

in which P is given in (11) and

$$q = C^T(\bar{y}_k - \bar{y}), \quad p = \left(I - (\tilde{A} + \tilde{B}\tilde{K})^T\right)^{-1} \begin{bmatrix} q \\ 0 \end{bmatrix}, \quad (18)$$

and $\tilde{K} = -(s + \tilde{B}^T P \tilde{B})^{-1} \tilde{B}^T P \tilde{A}$. It is important to remark that if $\bar{y}_k = \bar{y}$ (i.e. the setpoint is reachable), we obtain the same formulation as in (10).

Let $v := (v_0, v_1, \ldots, v_{N-1})$ be a column vector of length N. We can write (17) as the same strictly convex quadratic program (QP) in (13) in which H is still given in (14) while $\tilde{c}$ is:

$$\tilde{c} = \mathscr{B}^T \mathscr{Q} \mathscr{A} w_0 + \mathscr{B}^T \mathscr{P} + \mathscr{D}^T \mathscr{R} \mathscr{C} v_{-1}, \quad (19)$$

in which $\mathscr{P} = \begin{bmatrix} q^T & \cdots & q^T & p^T \end{bmatrix}^T$. Let $v^* = (v_0^*, \ldots, v_{N-1}^*)$ denote the optimal solution to (13) with $\tilde{c}$ given in (19). Then, the current control input is still defined by (15).

3. PROPERTIES AND IMPLEMENTATION

If constraints are not present, the proposed controller reduces to an infinite horizon LQ controller with target calculation and origin "shifting". Hence, it is easy to show that it is nominally stable for any choice of the tuning parameters. It is also possible to derive a simple sufficiency test for nominal constrained stability using ellipsoid invariant set theory. The details are omitted for the sake of brevity (Pannocchia *et al.*, 2003). This test can be used online to detect if the terminal state is not in the output admissible set and to flag a warning for the operator. If one wishes to have a guarantee of nominal stability, we can easily formulate the regulator with the terminal constraint $w_N = 0$. But since N is chosen fairly small for computational speed, we find the terminal state constraint controller not as robust as the one presented here, and therefore do not recommend it for industrial practice.

Another important property of CLQ is that it guarantees offset-free control whenever the closed-loop system reaches a steady state in which the input is not saturated. This property holds independently of the plant dynamics, and is due to the presence of the integrating state d in (4) (Pannocchia and Rawlings, 2003, Th.1). Unlike PID, the proposed controller does not integrate the tracking error (i.e. $\bar{y} - y_k$). In fact from (5) and (6), one can write

$$\hat{d}_{k+1|k} = \hat{d}_{k|k-1} + L_d(y_k - C\hat{x}_{k|k-1}),$$

from which it is clear that there is integration of the prediction error (i.e. $y_k - C\hat{x}_{k|k-1}$). This approach is significantly different from integration of the tracking error as in PID control, and it does not require any anti-windup strategy when the input saturates.

Furthermore, unlike PID control, CLQ is a "two-degree of freedom" controller and it can provide both efficient setpoint tracking and disturbance rejection simultaneously. This feature is due to the use of the state and disturbance estimator, which can also be designed to be insensitive to measurement noise.

Two modules of CLQ – the target calculation and the constrained dynamic optimization – require one to solve a quadratic program at each sampling time. In order for the proposed method to be applicable to simple hardware and programming languages, we have developed an efficient method for solving these QPs (Pannocchia *et al.*, 2003). For space limitations details are omitted, but it is worth remarking that the proposed method for solving the constrained dynamic optimization problem (13) differs from the well known explicit LQR solutions (Bemporad *et al.*, 2002), and it is specifically tailored to SISO systems with input bound constraints only. This new method allows one to efficiently solve (13) on-line without using a "proper" QP solver, and also without using LP and QP solvers in the offline table generation as in (Bemporad *et al.*, 2002). Unlike previous research, the dimension of the state vector has no influence on the size of the stored solution table in this new method. Only the control horizon has an effect. This new method has two basic steps:

(1) The offline generation of a solution table using H, $u_{\min}$ and $u_{\max}$: this step involves some matrix inversions, multiplication and sums.
(2) The online table scanning given the current value of c: this step only involves multiplications and sums and checking conditionals. These same operations are required in PID control.

4. ILLUSTRATIVE EXAMPLES

In this section we present a number of examples of common processes to show that the proposed CLQ controller is simpler to tune than a PID controller, is robust to model errors, insensitive to noise measurements and guarantees superior performance both for setpoint changes and disturbance rejections. Since constraints are present, the common anti-windup "velocity" algorithm for PID is used (Ogunnaike and Ray, 1994). Also notice that if $T_d \neq 0$, the derivative action is suppressed when a setpoint change occurs, which is the common industrial practice to avoid the "derivative kick".

4.1 FOPTD system

The first example is a first order plus time delay (FOPTD) system:

$$G_1(s) = \frac{e^{-2s}}{10s+1},$$

sampled with $T_s = 0.25$. The input is constrained: $|u| \leq 1.5$, a horizon of $N = 4$ is used, and in all simulations the setpoint is changed from 0 to 1 at time zero. At time 25 a load disturbance of magnitude -0.25 enters the system; then at time 50 the disturbance magnitude becomes -1 (which makes the setpoint 1 unreachable); finally at time 75 the disturbance magnitude becomes -0.25 again. Fig. 1 shows the simulation results in the nominal case for two CLQ controllers

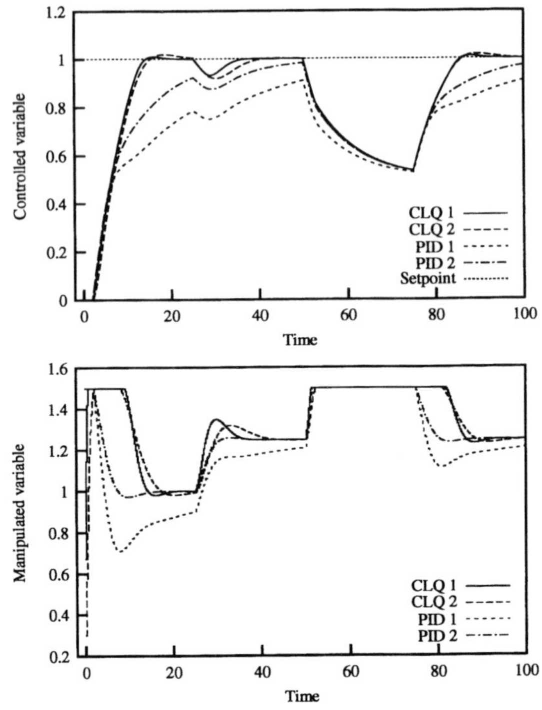

Fig. 1. FOPTD system: nominal case.

and two PID controllers. CLQ 1 uses a regulator input penalty of $s = 1$, while CLQ 2 uses $s = 10$. The tuning parameters for PID 1 are chosen according to Luyben's rules (Luyben and Luyben, 1997, p. 97): $K_c = 2.51$, $T_i = 17.3$, $T_d = 0$. The tuning parameters for PID 2 are chosen according to Skogestad's IMC rules (Skogestad, 2003): $K_c = 2.35$, $T_i = 10$, $T_d = 0$. Fig. 2 shows the simulation results for CLQ 1 and

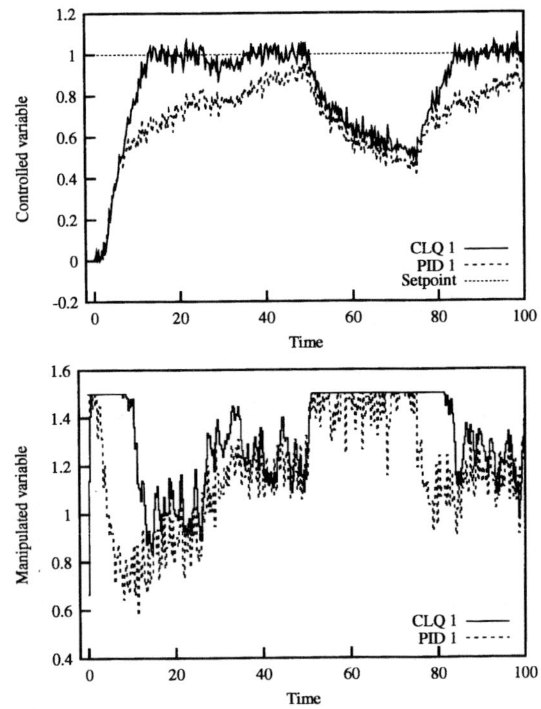

Fig. 2. FOPTD system: noisy case.

PID 1 in the presence of random output noise (with variance $\sigma^2 = 0.001$). Fig. 3 shows a comparison of

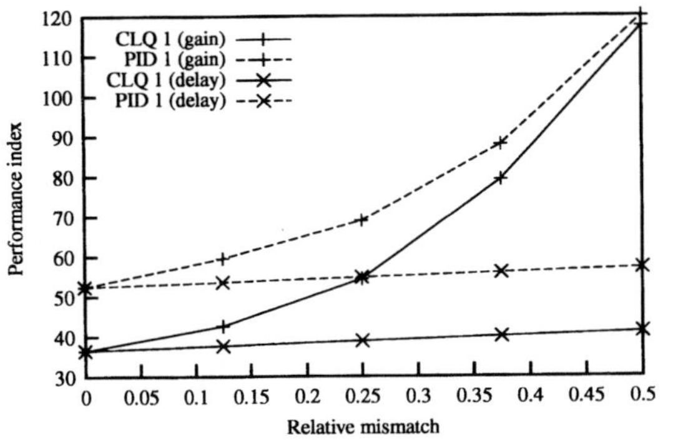

Fig. 3. FOPTD system: effect of plant/model mismatch.

the performance index $\Phi = \sum_{k=0}^{\infty}(y_k - \bar{y})^2 + (u_k - u_{k-1})^2$ *vs* the gain and delay relative plant/model mismatch, respectively, for CLQ 1 and PID 1.

4.2 Integrating system

The second example is an integrating system:

$$G_2(s) = \frac{e^{-2s}}{s},$$

sampled with $T_s = 0.25$. The same input constraints, horizon, setpoint change and disturbances as in the first example are considered. CLQ 1 uses a regulator input penalty of $s = 500$, while CLQ 2 uses $s = 5000$. The tuning parameters for PID 1 are chosen according to Luyben's rules: $K_c = 0.23$, $T_i = 18.7$, $T_d = 0$. The tuning parameters for PID 2 are chosen according to Skogestad's IMC rules: $K_c = 0.23$, $T_i = 17$, $T_d = 0$. Simulation results in the nominal case are reported

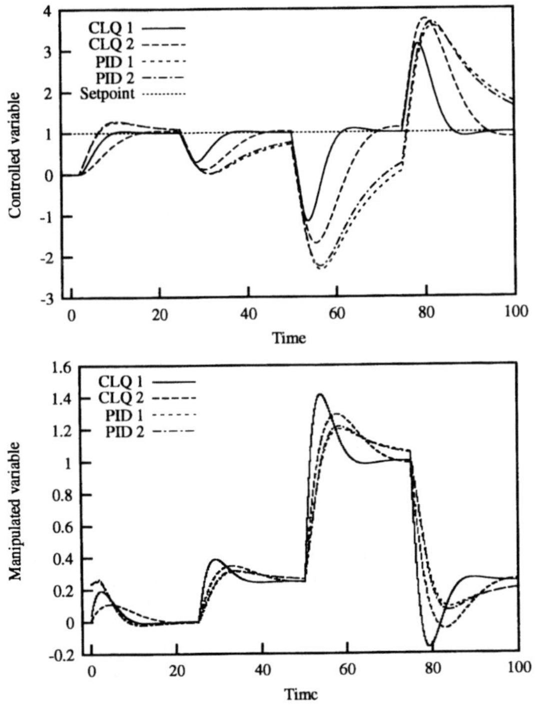

Fig. 4. Integrating system: nominal case.

in Fig. 4, while Φ *vs* the gain and delay relative plant/model mismatch is reported in Fig. 5.

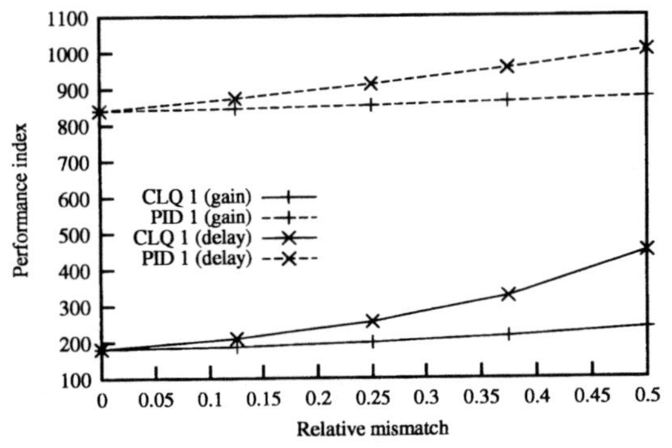

Fig. 5. Integrating system: effect of plant/model mismatch.

4.3 Under-damped system

The last example is a second order under-damped system:

$$G_3(s) = \frac{K}{\tau^2 s^2 + 2\tau\xi s + 1},$$

sampled with $T_s = 0.25$, and with nominal parameters of $K = 1$, $\tau = 5$ and $\xi = 0.2$. The same input constraints, horizon, setpoint change and disturbances as in the first example are considered. CLQ 1 uses a regulator input penalty of $s = 1$, while CLQ 2 uses $s = 10$. The tuning parameters for PID 1 are chosen according to Luyben's rules: $K_c = 5.0$, $T_i = 16.8$, $T_d = 0$. The tuning parameters for PID 2 are chosen using the same IMC approach as in (Skogestad, 2003): $K_c = 0.4$, $T_i = 2$, $T_d = 12.5$. Simulation results for the

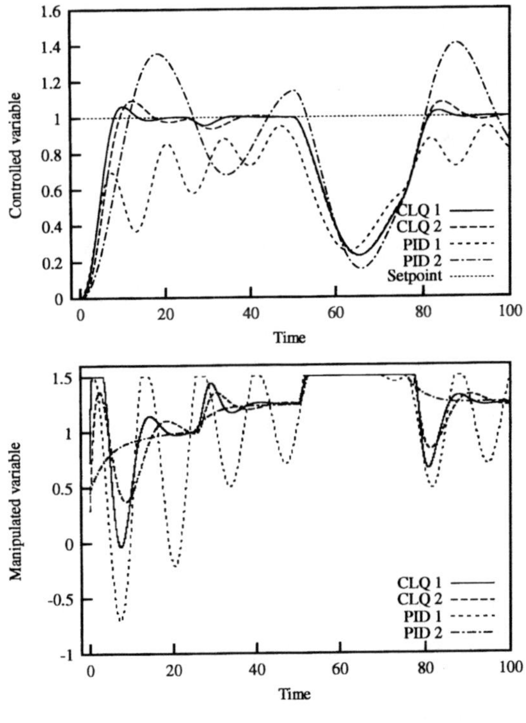

Fig. 6. Under-damped system: nominal case.

nominal case are reported in Fig. 6, while Φ *vs* the gain and damping factor relative uncertainties is reported in Fig. 7.

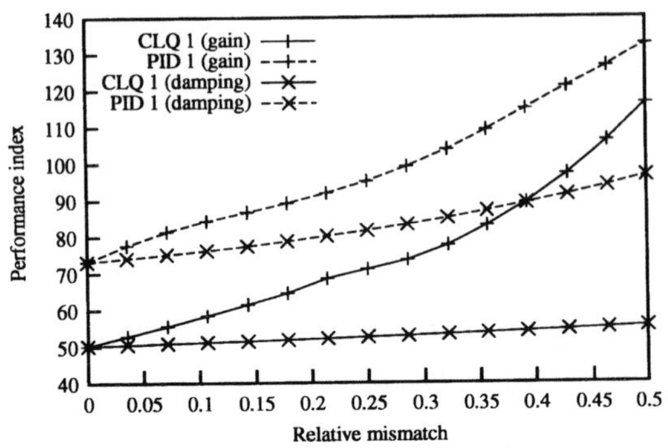

Fig. 7. Under-damped system: effect of plant/model mismatch.

4.4 *Discussion*

The results presented in the previous paragraphs clearly show that in all examples (as well as in several others not shown for the sake of space) CLQ outperforms PID both for setpoint changes and disturbance rejections. In the presence of constraints CLQ understands much better than PID "when" and "for how long" to saturate the input. Notice that CLQ does not require any anti-windup strategy.

Tuning CLQ is simple: basically one has to choose only the input penalty s, which trades off between tracking error and input usage. The effect of this tuning parameter is intuitive: the lower s the more aggressive the controller. CLQ is robust to plant/model mismatch: one can obtain high performance closed-loop response in the nominal case and still have robust performance and large stability margins. CLQ can be easily designed to be insensitive to measurement noise by adjusting the output noise variance R_v. If one detects high frequency oscillations in the manipulated variable it is sufficient to increase R_v to suppress this undesirable behavior. In this way, it is not necessary to "slow down" the controller's setpoint response (i.e. to increase s) when the measurement is noisy.

Finally, it is important to remark that CLQ is efficient and the computational burden is comparable to that of PID. The average CPU time required to compute the control input has been 0.22 ms for CLQ and 0.05 ms for PID (on a 1.7 GHz Athlon PC running Octave[2]). The maximum CPU time has been 0.55 ms for CLQ and 0.10 ms for PID. The computational efficiency comes about because only a small number of simple operations (addition, multiplication and comparison) are required at each sample time.

5. CONCLUSIONS

In this paper, a novel, offset-free, constrained, linear quadratic (CLQ) controller for SISO system was presented. The purpose of this work was to propose an alternative to digital PID controllers that are commonly available on the DCS. CLQ has three main modules based on a state-space model of the system: a state and disturbance estimator, a target calculation, a constrained dynamic optimization. Each module is implemented in an efficient way so that the overall CLQ algorithm has little computational cost and can be applied using simple hardware and software. As shown, the proposed controller outperforms PID controllers in all situations (setpoint changes or disturbance rejections, nominal case or in the presence of relevant model errors, noise-free or noisy measurements). Moreover, CLQ is much simpler to tune than PID. Strictly speaking, there are two parameters to choose for the estimator and one parameter for the dynamic optimization module. However, as shown in the examples, CLQ can achieve excellent results by varying only this last parameter, whose effect on the closed-loop performance is clear and intuitive. The proposed controller is "scalable", in the sense that it can be extended to larger multivariable systems in a straightforward fashion. Other possible extensions are:

- the use of feed-forward to reject measured disturbances even more efficiently,
- the coupling of several SISO CLQ controllers by appropriate exchange of information.

ACKNOWLEDGMENT

The authors would like to thank Drs. Badgwell, Brambilla, Downs and Ogunnaike for helpful discussion of this paper. The authors gratefully acknowledge the financial support of the industrial members of the Texas-Wisconsin Modeling and Control Consortium and NSF through grant #CTS-0105360.

REFERENCES

Bemporad, A., M. Morari, V. Dua and E. N. Pistikopoulos (2002). The explicit linear quadratic regulator for constrained systems. *Automatica* **38**, 3–20.

Bonné, D., J.B. Jørgensen, J.B. Rawlings and S.B. Jørgensen (2003). Model predictive control with infeasible targets. In preparation.

Chen, D. and D. E. Seborg (2001). PID controller design based on direct synthesis and disturbance rejection. In: *AIChE Annual Meeting*. Reno (NV), USA. p. paper 276i.

Luyben, W. L. and M. L. Luyben (1997). *Essentials of Process Control*. McGraw-Hill Int. Editions.

Mukati, K. and B. Ogunnaike (2004). Stability analysis and tuning strategies for a novel next generation regulatory controller. In: *Amer. Cont. Conf.*. Boston (MA), USA.

Muske, K. R. and T. A. Badgwell (2002). Disturbance modeling for offset-free linear model predictive control. *J. Proc. Cont.* **12**, 617–632.

Ogunnaike, B. A. and W. H. Ray (1994). *Process Dynamics, Modeling, and Control*. Oxford University Press.

Pannocchia, G. (2003). Robust disturbance modeling for model predictive control with application to multivariable ill-conditioned processes. *J. Proc. Cont.* **13**, 693–701.

Pannocchia, G. and J. B. Rawlings (2003). Disturbance models for offset-free model predictive control. *AIChE J.* **49**, 426–437.

Pannocchia, G., N. Laachi and J. B. Rawlings (2003). A candidate to replace PID control: SISO constrained LQ control. Submitted for publication in *AIChE J.*

Skogestad, S. (2003). Simple analytic rules for model reduction and PID controller tuning. *J. Proc. Cont.* **13**, 291–309.

Soroush, M. and K. R. Muske (2000). Analytical model predictive control. In: *Progress in Systems and Control Theory* (F. Allgower and A. Zheng, Eds.). pp. 163–179.

[2] Octave (`http://www.octave.org`) is freely distributed under the terms of the GNU General Public License.

ELSEVIER
IFAC
PUBLICATIONS
www.elsevier.com/locate/ifac

CONTROL SYSTEM SELECTION: A MEASURE OF CONTROL QUALITY LOSS IN ANALYTICAL CONTROL

Masoud Soroush *,[1] **Yiannis Dimitratos** **

* *Department of Chemical Engineering, Drexel University, Philadelphia, PA 19104, USA*
** *DuPont Engineering, Process Dynamics and Control, 1007 Market Street, Wilmington, DE 19898, USA*

Abstract: A question that often process control engineers face is for what class of processes one should use model predictive control that requires solving numerically a constrained optimization problem repeatedly on-line. The alternative is to use analytical control, such as P, PI, PID and differential geometric control, which do not require the on-line optimization. In other words, for what class of processes can analytical control provide control quality close to the optimal control quality that model predictive control (MPC) can provide? Here an analytical controller is defined as the one whose implementation does not require solving a constrained optimization problem numerically. This work presents a measure that allows one to quantify the degradation in closed-loop performance when one implements analytical control instead of MPC. A special case of the measure is used to derive a simple test that can be used to check if a given process with active input constraints can be controlled satisfactorily by analytical control. It is shown that processes with active input constraints and directionality benefit greatly from MPC. In other words, for input-constrained processes whose nonsingular characteristic (decoupling) matrix is independent of manipulated inputs and can be made diagonal by row or column rearrangements, control quality provided by analytical control can be adequate. The measure is used to see if four input-constrained process examples can be controlled satisfactorily by analytical control. Closed-loop responses are shown to confirm the usefulness of the measure.

Keywords: control system selection, model predictive control, analytical control, actuator saturation, directionality, decoupling matrix

1. INTRODUCTION

The degrading effect of constraints on control quality was recognized very early in practice. To minimize the degrading effect, a great number of strategies such as anti-windup schemes, directionality compensators, and optimization-based control methodologies (e.g., model predictive control) have been proposed and studied.

For a given process design, after identifying disturbances and selecting controlled outputs and manipulated inputs, an effective control system is chosen. The control system can be conventional, model-based, analytical (such as P, PI and PID), and/or numerical (model predictive). A question that often process control engineers face is for

[1] Corresponding author

what class of processes one should use model predictive control that requires solving numerically a constrained optimization problem repeatedly on-line. The alternative is to use analytical control, such as P, PI, PID and differential geometric control, which do not require the on-line optimization. In other words, control of what class of processes can significantly benefit from model predictive control?

In current industrial practice, this question is addressed in several ways, which generally depend on:

- The sector of the industry and the type of manufacturing operations (e.g. petrochemical, chemical, or food);
- The acceptance of the technology in operations (typically rooted in the history of introduction of the technology, experience with previous implementations, employee process control skills, and success in maintaining performance over time); and
- How strongly the financial benefits from a migration to MPC manifest themselves.

It is usually an obvious value preposition or a leap of faith in the benefits of the technology that would lead to the development and commissioning of an MPC application and not a detailed analysis.

Few companies have an in-house engineering capability to carry out an operability analysis, which allows for performance benchmarking and assessment of the financial benefits of the MPC migration. This analysis involves the development of a dynamic model of the process in question and the comparative assessment of an MPC versus an analytical implementation. In the best of the cases, a high-fidelity nonlinear simulator of the entire process is interfaced to the commercial MPC controller of preference to conduct the evaluation. The lack of a simple analysis tool with solid theoretical foundations to check whether an input-constrained process can be controlled adequately well by using analytical control is the best motivation for this study.

This work presents a measure that allows one to quantify the degradation in closed-loop performance when one implements analytical control instead of MPC for processes with active input constraints. The measure quantifies the deviation of (a) the response of the process under analytical control from (b) the response of the same process under model predictive control. A special case of the measure is used to derive a simple test that can be used to check if a given process with active input constraints can be controlled satisfactorily by analytical control. It is shown that processes with active input-constraints and directionality benefit greatly from MPC. The measure is used to check if four process examples can be controlled satisfactorily by analytical control. Closed-loop responses are shown to confirm the usefulness of the measure.

Section 2 describes the scope of the study and some preliminaries. An index of the control quality loss is presented in Section 3. The application of the index is illustrated by four examples in Section 4.

2. SCOPE AND PRELIMINARIES

Consider nonlinear multivariable processes with a model in the form

$$\begin{aligned} x(k+1) &= \Phi[x(k), u(k)], \qquad x(0) = x_0 \\ y(k) &= h[x(k)] \end{aligned} \tag{1}$$

with the input constraints

$$u_{l_j} \le u_j(k) \le u_{h_j}, \quad j = 1, \cdots, m \tag{2}$$

where $x = [x_1 \cdots x_n]^T$, $u = [u_1 \cdots u_m]^T$, and $y = [y_1 \cdots y_m]^T$ denote the vectors of state variables, manipulated inputs, and controlled outputs, respectively; $\Phi(x,u)$ and $h(x)$ are smooth vector functions; and u_{l_j}, u_{h_j}, Δu_{l_j}, and Δu_{h_j}, $j = 1, \cdots, m$, are constant scalar quantities.

For a process in the form of (1),

- Relative order (degree) of an output y_i with respect to the vector of manipulated inputs is denoted by r_i, where r_i is the smallest integer for which $y_i(k+r_i)$ depends explicitly on the present value of a manipulated input.
- Characteristic (decoupling) matrix

$$\mathcal{C}(x,u) \triangleq \begin{bmatrix} \dfrac{\partial}{\partial u} h_1^{r_1}(x,u) \\ \vdots \\ \dfrac{\partial}{\partial u} h_m^{r_m}(x,u) \end{bmatrix}$$

where

$$\begin{aligned} h_i^0(x) &\triangleq h_i(x), \quad i = 1, \cdots, m \\ h_i^\ell(x) &\triangleq h_i^{\ell-1}[\Phi(x,u)], \quad \ell = 1, \cdots, r_i - 1, \\ &\qquad\qquad i = 1, \cdots, m \end{aligned}$$

$$h_i^{r_i}(x,u) \triangleq h_i^{r_i-1}[\Phi(x,u)], \quad i = 1, \cdots, m$$

The decoupling (characteristic) matrix has been used to check the degree of dynamic decoupling in processes and specify the class of processes that do not exhibit directionality (Soroush and Valluri, 1999). It has also been called instantaneous process gain, because the strength of a process response over a very short horizon strongly depends on the decoupling matrix of the process. For processes with a diagonal and independent-of-manipulated-inputs decoupling matrix, completely decentralized control is more suitable. The matrix has also been used as basis for input-output pairing.

Let the time-varying static state feedback

$$u^*(k) = \Psi^*[x(k), k, y_{sp}(k)] \quad (3)$$

represent a general model predictive control law for the process of (1) with the input constraints of (2) [that is, the numerical solution to a general, constrained, moving horizon, optimization problem], and the time-invariant static state feedback

$$w(k) = \Psi[x(k), y_{sp}(k)] \quad (4)$$

denote the same general model predictive control law for the process of (1) without the input constraints of (2) [that is, the numerical solution to a general, unconstrained, moving horizon, optimization problem].

2.1 Directionality

Directionality has been defined as a controller performance degradation that is associated with actuator saturation (Campo and Morari, 1990) and as a systems property (Soroush and Valluri, 1999; Soroush and Mehranbod, 2002). As stated in (Soroush and Valluri, 1999; Soroush and Mehranbod, 2002), a process exhibits directionality when for every command signal $\hat{w}$, the output response of the process to sat($\hat{w}$) is not "closest" (in the controlled output space) to the output response of the process to $\hat{w}$, where

$$\text{sat}_\ell[w] \triangleq \begin{cases} u_{l_\ell}, & w_\ell < u_{l_\ell} \\ w_\ell, & u_{l_\ell} \le w_\ell \le u_{h_\ell}, \quad \ell = 1, \cdots, m \\ u_{h_\ell}, & w_\ell > u_{h_\ell} \end{cases}$$

A precise statement of this definition is given below.

Definition 1 (Soroush and Valluri, 1999; Soroush and Mehranbod, 2002): A process in the form of (1) does not exhibit directionality over time horizons $N_1, \cdots, N_m \ge 0$, if and only if for every sequence $\hat{w}(k) \in \Re^m$, $k = 0, 1, \cdots, N$, and for every initial condition $\hat{x}_0 \in \Re^n$, $u(k) = \text{sat}\{\hat{w}(k)\}$ minimizes

$$\sum_{i=1}^{m} \sum_{k=0}^{N_i} [\hat{y}_i(k) - \hat{y}_i^*(k)]^2$$

subject to $u_{l_\ell} \le u_\ell(k) \le u_{h_\ell}$, $\ell = 1, \cdots, m$, where $N = \max(N_1, \cdots, N_m)$,

$$\left.\begin{aligned} \hat{x}^*(k+1) &= \Phi[\hat{x}^*(k), \hat{w}(k)], \quad \hat{x}^*(0) = \hat{x}_0 \\ \hat{y}^*(k) &= h[\hat{x}^*(k)] \end{aligned}\right\}$$

$$\left.\begin{aligned} \hat{x}(k+1) &= \Phi[\hat{x}(k), u(k)], \quad \hat{x}(0) = \hat{x}_0 \\ \hat{y}(k) &= h[\hat{x}(k)] \end{aligned}\right\}$$

According to this definition, a SISO process that is not affine in its manipulated input can have the directionality.

Remark 1: The class of processes whose nonsingular characteristic (decoupling) matrix is independent of u and can be made diagonal by row or column rearrangements does not exhibit the directionality over the very small time horizons $r_1, \cdots, r_m$.

Remark 2: For linear processes in the form:

$$\left.\begin{aligned} \hat{x}(k+1) &= A\hat{x}(k) + Bu(k), \quad \hat{x}(0) = \hat{x}_0 \\ \hat{y}(k) &= C\hat{x}(k) \end{aligned}\right\} \quad (5)$$

where A, B and C are $n\times n$, $n\times m$ and $m\times n$ matrices respectively, the characteristic (decoupling) matrix

$$\mathcal{C} = \begin{bmatrix} c_1 A^{r_1-1} B \\ \vdots \\ c_m A^{r_m-1} B \end{bmatrix}$$

where c_i is the ith row of the matrix C. Linear processes whose characteristic matrix can be made diagonal by row or column rearrangements do not exhibit the directionality over the horizons $r_1, \cdots, r_m$.

Remark 3: The characteristic matrix of a linear process can be identified easily by subjecting the process at steady state $x = 0$ to step changes in its inputs. For example, the characteristic matrix entry $\mathcal{C}_{ij}$ can be calculated by subjecting the process at steady state $x = 0$ to a step change in u_j only and measuring the first nonzero value of y_i. In this case, $\mathcal{C}_{ij}$= [first nonzero value of y_i]/[size of the step change in u_j], $i = 1, \cdots, m$, $j = 1, \cdots, m$. In practice, the characteristic-matrix identification step may need to be repeated, if the process moves significantly away from the nominal steady state.

3. CONTROL QUALITY LOSS IN ANALYTICAL CONTROL

For linear processes, the general unconstrained MPC law of (4) is a linear state feedback that can be written in an analytical form. In the case of nonlinear processes, one may be able to find an analytical state feedback equivalent to the general unconstrained MPC law of (4). Thus, the general unconstrained MPC law of (4), in general, can be viewed as an analytical time-invariant state feedback. A question that one may ask is when the analytical state feedback of (4) together with clipping (the output of the state feedback) is adequate for constrained processes of the form (1). In other words, for what class of processes can this analytical control scheme provide a satisfactory control quality [which is not much less than the quality provided by the numerical state feedback of (3)]? To address this question, let us define the following index of control quality loss.

Definition 2: The index of control quality loss in analytical control over time horizons $N_1, \cdots, N_m$, is defined as:

$$I_N = \max_k \sum_{i=1}^{m} \sum_{\ell=0}^{N_i} \left[\frac{y_i(k+\ell)}{y_i^*(k+\ell)} - 1 \right]^2, \quad k = 0, \cdots, \infty$$

where $N_1 \geq r_1, \cdots, N_m \geq r_m$,

$$\left. \begin{aligned} x^*(k+1) &= \Phi[x^*(k), u^*(k)], \quad x^*(0) = x_0 \\ y^*(k) &= h[x^*(k)] \end{aligned} \right\}$$

$$\left. \begin{aligned} x(k+1) &= \Phi[x(k), sat\{w(k)\}], \quad x(0) = x_0 \\ y(k) &= h[x(k)] \end{aligned} \right\}$$

Special Case: The index of control quality loss in analytical control over time horizons $r_1, \cdots, r_m$, takes the form:

$$I_r = \max_k \sum_{i=1}^{m} \sum_{\ell=0}^{r_i} \left[\frac{y_i(k+\ell)}{y_i^*(k+\ell)} - 1 \right]^2, \quad k = 0, \cdots, \infty$$

The definition of the directionality over short horizons of $r_1, \cdots, r_m$, can be used to identify the class of processes that have a control-quality loss index of 0.

Remark 4: For processes that do not exhibit the directionality over the time horizons $r_1, \cdots, r_m$, the index of control quality loss, $I_r = 0$. The class of processes whose nonsingular characteristic matrix is independent of u and can be made diagonal by row or column rearrangements does not exhibit directionality over the time horizons $r_1, \cdots, r_m$. For processes that exhibit the directionality over the time horizons $r_1, \cdots, r_m$, $I_r \neq 0$ provided that there is an active input constraint.

4. EXAMPLES

4.1 Example 1

Consider the two-input two-output, time-invariant, linear process:

$$G(z) = \begin{bmatrix} \dfrac{1}{z-0.5} & \dfrac{1000(z-1)}{(z-0.5)^2} \\ \dfrac{z-1}{(z-0.5)^2} & \dfrac{2}{z-0.5} \end{bmatrix}$$

The steady state gain matrix (K_p) and the characteristic matrix ($\mathcal{C}$) of the process are:

$$K_p = \begin{bmatrix} 2 & 0 \\ 0 & 4 \end{bmatrix}, \quad \mathcal{C} = \begin{bmatrix} 1 & 1000 \\ 1 & 2 \end{bmatrix}$$

The steady state gain matrix is diagonal, implying that the process is "statically", completely decoupled. The process, however, is dynamically, strongly coupled, since its characteristic matrix is not diagonal; this process exhibits the directionality. Thus, for this process $I_r \neq 0$, implying that if analytical control is used and an input constraint becomes active, the loss of control quality will be significant.

4.2 Example 2

Consider the two-input two-output, time-invariant, linear process:

$$G(z) = \begin{bmatrix} \dfrac{1}{z-0.5} & \dfrac{1000}{(z-0.5)^2} \\ \dfrac{1}{(z-0.5)^2} & \dfrac{2}{z-0.5} \end{bmatrix}$$

The steady state gain matrix (K_p) and the characteristic matrix ($\mathcal{C}$) of the process are:

$$K_p = \begin{bmatrix} 2 & 4000 \\ 4 & 4 \end{bmatrix}, \quad \mathcal{C} = \begin{bmatrix} 1 & 0 \\ 0 & 2 \end{bmatrix}$$

implying that the process is "statically", highly coupled but is dynamically, weakly coupled (characteristic matrix is diagonal); the process does not exhibit the directionality. Thus, for this process $I_r = 0$, implying that if analytical control is used, the loss of control quality will not be significant.

Examples 1 and 2 confirmed that control system selection should be on the basis of dynamic rather than steady-state characteristics of a process. In particular, it is the nature of the characteristic matrix not that of the steady state gain matrix that determines when analytical control is adequate in the presence of active constraints.

4.3 Example 3: SISO Nonlinear Chemical Reactor

Consider an isothermal continuous stirred-tank reactor in which a non-elementary chemical reaction $A \rightarrow B$ takes place. The rate of production of B is given by:

$$R_B[C_A] = 0.036C_A^3 - 0.78C_A^2 + 4.4C_A - 1.0$$

where C_A denotes the concentration of the reactant. The reactor dynamics are described by

$$\frac{dC_B}{dt} = 0.036C_A^3 - 0.78C_A^2 + 4.4C_A - 1.0 - 0.625C_B \quad (6)$$

with $C_B(0) = 0$ and $0 \leq C_A \leq 10$, where C_B denotes the concentration of the product. Exact time-discretization (with a sampling period of 0.1 *hour*) of the reactor model leads to

$$C_B(k+1) = 0.9394C_B(k) + 0.0969R_B[C_A(k)] \quad (7)$$

The control objective is to maintain the concentration of the product, C_B, at a desired level ($C_{B_{sp}} = 8.0$ $kmol.m^{-3}$) by manipulating the reactant concentration, C_A. Here, the characteristic matrix

$$\mathcal{C} = 0.108u^2 - 1.56u + 4.4$$

Because $\mathcal{C}$ depends on u, the process has the directionality; For this process, the loss of control quality will be significant, if analytical control is used and one of the input constraints become active.

For this process, we use the two model predictive controllers (Grantz et al., 1998):

$$\begin{array}{l}\eta(k+1) = 0.9\eta(k) + 0.0394C_B(k) + 0.0969R_B[u(k)] \\ \min\limits_{u(k)} \{R_B[u(k)] - 1.0320[e(k)+\eta(k)] - 0.4066C_B(k)\}^2 \\ \text{subject to} \qquad 0 \le u(k) \le 10\end{array} \tag{8}$$

and

$$\begin{array}{l}\eta(k+1) = 0.9\eta(k) + 0.0394C_B(k) + 0.09694R_B[u(k)] \\ \min\limits_{w(k)} \{R_B[w(k)] - 1.0320[e(k)+\eta(k)] - 0.4066C_B(k)\}^2 \\ u(k) = \text{sat}(w(k))\end{array} \tag{9}$$

with $\eta(0) = C_B(0)$ and $e = 8 - C_B$. Numerical simulations are carried out for the following three cases:

- Case C1: the process of (7) without the input constraints, under the controller of (8) or (9).
- Case C2: the process of (7) with the input constraints, under the controller of (8).
- Case C3: the process of (7) with the input constraints, under the controller of (9).

In the absence of the input constraints, the controllers of (8) and (9) induce the first-order linear input-output response

$$C_B(k+1) - 0.9C_B(k) = 0.8 \tag{10}$$

As shown in Figure 1, the closed-loop responses are considerably different for the three cases. The response in case C1 has, of course, the lowest possible integral of squared error [ISE] (ISE=3.47×10^2). The advantage of constrained MPC (case C2 with an ISE = 3.90×10^2) over unconstrained MPC with clipping (case C3 with an ISE = 1.45×10^3) is quite obvious.

4.4 Example 4: Multivariable Linear Example

Consider the two-input two-output, time-invariant, linear process:

$$\left.\begin{array}{rl}x_1(k+1) &= 0.99x_1(k) + 0.40u_1(k) - 3.00u_2(k) \\ x_2(k+1) &= 0.99x_2(k) - 0.01u_1(k) + 0.40u_2(k) \\ y_1(k) &= x_1(k) \\ y_2(k) &= x_2(k)\end{array}\right\} \tag{11}$$

with $x_1(0) = 0$, $x_2(0) = 0$, $|u_1(k)| \le 1$, and $|u_2(k)| \le 1$.

For this linear example,

$$\mathcal{C} = \begin{bmatrix} 0.40 & -3.00 \\ -0.01 & 0.40 \end{bmatrix}$$

which is not diagonal, and thus the process cannot be regulated effectively by analytical control when one of the input constraints becomes active.

The performance of a model predictive controller is compared with that of two completely-decentralized proportional-integral (PI) controllers with conditional integration (to prevent integral windup) with the set point values $y_{sp_1} = 8$ and $y_{sp_2} = 3$. Figure 2 depicts the closed-loop process input and output responses under the two PI controllers (dashed line) and the model predictive controller (dotted line). In the absence of the input constraints, the closed-loop responses (solid line) under the decentralized PI controllers and the model predictive controller are the same. In the presence of the input constraints, however, while the PI controllers show poor responses in y_1, the model predictive controller provides a significantly better closed-loop performance. In this example, it happens that in the presence of the constraints the y_2 response is better when the PI controllers are used. With the model predictive controller, a y_2 response closer to the unconstrained one can be obtained by placing a higher weight on y_2.

5. ACKNOWLEDGEMENT

This material is based upon work supported by the National Science Foundation under Grant No. 0101133. Any opinions, findings, and conclusions or recommendations expressed in this material are those of the authors and do not necessarily reflect the views of the National Science Foundation.

REFERENCES

[1] Campo, P. J., and M. Morari, "Robust Control of Processes Subject to Saturation Nonlinearities," *Comput. & Chem. Eng.*, **14**, 343 (1990).

[2] Grantz, J., S. Valluri, and M. Soroush, "Discrete-Time Nonlinear Control of Processes with Actuator Saturation Nonlinearities," *AIChE J.*, **44(7)**, 1701 (1998).

[3] Soroush, M., and N. Mehranbod, "Optimal Compensation for Directionality in Processes with a Saturating Actuator," *Comput.& Chem. Engin.*, **26(11)**, 1633 (2002).

[4] Soroush, M., and S. Valluri, "Optimal Directionality Compensation in Processes with Input Saturation Nonlinearities," *International J. of Contr.*, **72(17)**, 1555 (1999).

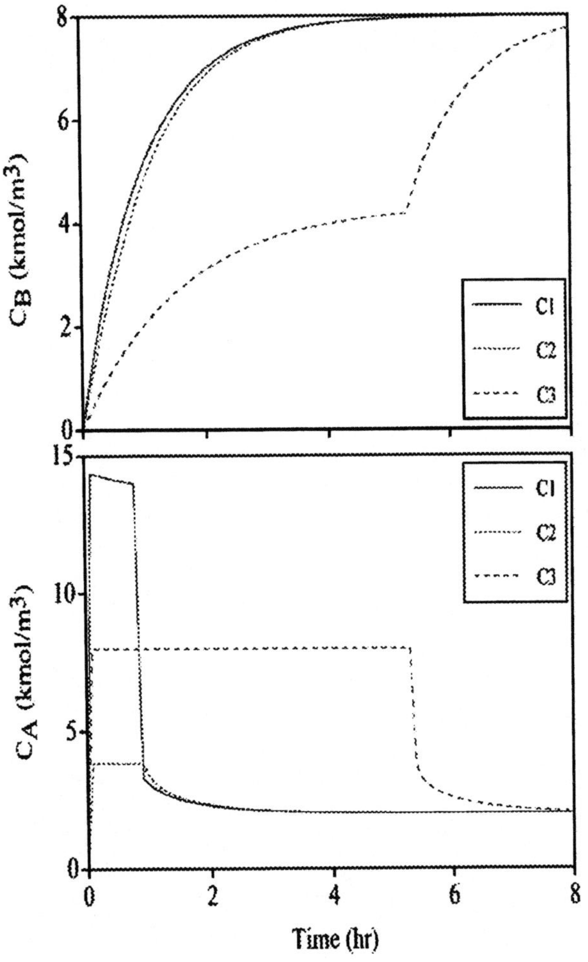

Fig. 1. Controlled output and manipulated input of Example 3.

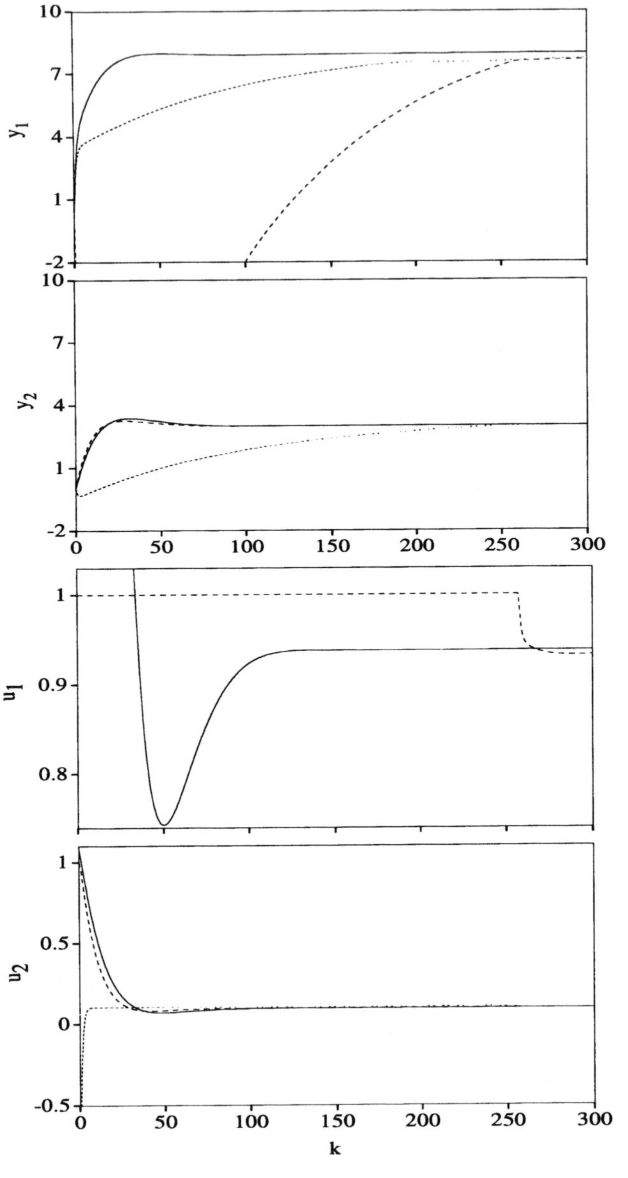

Fig. 2. Controlled outputs and manipulated inputs of Example 4.

ELSEVIER
IFAC
PUBLICATIONS
www.elsevier.com/locate/ifac

PROCESS ENGINEERING IS CHANGING YOU CAN $ENSE IT

Frank K. Schweighardt
Dave Zatko

Air Products and Chemicals, Inc.
7201 Hamilton Blvd Allentown, PA 18195

Abstract: The role and responsibilities of the Process Engineer are rapidly evolving. This is in part because economic drivers to automate control of batch and continuous operations for consistent molecular compositions also apply to maintaining consistent process performance to meet more demanding product specifications. Both of these criteria are showing a greater return on investment (ROI) initiated by traditional process control strategies, but enhanced with specific (on-line) composition analysis. Incorporation of state-of-the-art point-of-use physical property measurements (temperature/pressure/flow) combined with chemical specific (molecular species) sensors provides unique feedback enabling the process engineer to better apply "model predictive control (MPC) " techniques. Such sensor technologies may allow the engineer to know the process composition at each tray in a distillation tower, or at purification tower positions in temperature or pressure swing absorbers (TSP or PSA) in "real-time" to alert for break-thru. Chinese Taoist philosopher Lao Tzu circa 600 BCE said, "To Know and not to do is not to know". It is our role as chemists and engineers to know and to do to gain true process control. Knowing the capabilities and limitations of the new sensor technology and how such sensors are evolving, can provide unique insights for the process engineer, who in turn will challenge the process analytical chemist to improve sensor specificity and performance.

Keywords: Chemical microsensors; monitoring; chemical sensor; analytical approximations; active control

1. INTRODUCTION

The role and responsibilities of the Process Engineer are rapidly evolving. The process-manufacturing paradigm has changed from analytical testing, solely to document product quality and rejecting/recalling products of unacceptable quality, to a new paradigm of "make it right the first time"-using continuous quality assurance. This is in part because economic drivers to automate control of batch and continuous operations for consistent molecular compositions also apply to maintaining consistent process performance to meet more demanding product specifications. Both of these drivers give a greater return on investment (ROI) initiated by traditional process control strategies, but enhanced by specific composition analysis. Incorporation of state-of-the-art point-of-use physical property measurements (temperature /pressure /flow) combined with chemical (molecular species) sensors are providing unique feedback enabling the process engineer to better apply "model predictive control (MPC) " techniques (Smilde, et al 2002). Such sensor technologies may allow the engineer to know the process composition at each tray in a distillation tower, or at purification positions in temperature or pressure swing absorbers (TSP or PSA) in "real-time" to define break-thru. Knowing the capabilities and limitations of the new sensor technology and how such sensors are evolving, can provide unique insights for the process engineer, who in turn will challenge the process analytical chemist to improve sensor specificity and performance (Chauvel, et al 2002; Gunnell and van Vuuren, 2003).

This review will provide a brief historical perspective of process analysis, a view of current analytical technologies, a summary of the major sensor technologies being applied today, and an example of sensor technology applied to hydrogen production.

1.1 Historical Perspective.

Some one hundred years ago our sense of touch (temperature), taste (acid/sweet), smell (H_2S/NH_3), sight (color of smoke or product) and feeling/hearing (compressor sounds/motor torque-vibration) were our process analytical tools, usually associated with a few well-experienced operators/engineers. These skills were hard to find, difficult to train in others, and hard to quantify /document.

Figure I . Chemical Plant –1800's

Potyrail (Potyrail, 2001) reports "process chemicals analysis originates from the development of IR absorption (K.F. Luft at I.G. Farbenindustries-1938) and magnetic oxygen recorders in the late 1930s". Twenty-five years ago Exxon (ExxonMobil) led the way with Dow (Henslee, 1996), DuPont, Air Products, and others, with process engineering farsightedness to tie process control to the output of on-line/at-line analytical systems that sensed the molecular composition changes in chemical processing, refinery operations, bulk gas production, pharmaceutical and specialty chemical manufacturing. In some cases this has led to fully automated "lights out" manufacturing, and unattended delivery of liquefied gas products via the "auto-load" process at Air Products.

The development of the new electronic chip-based technologies over the past 20+ years has had major impact on the process analytical chemist's ability to provide the process engineer with "molecular thermocouples". The well-developed review by Harris *(Harris, 2002)* expanded upon by Workman, et al. (Workman, et al., 2001; Workman et al, 2003) provides insight to how such new analytical technologies have added to the tool kit of the process engineer and process analytical chemist. A good measure of such technology grew out of federally sponsored programs to detect biological species as weapons, chemical warfare agents, and new medical diagnostic systems. Workman, et al. (Workman et al, 2003) have stated "Measurement science has benefited recently from advances in technology in the computing and communication industries as well as from sensor-related research conducted in the US Government laboratories and agencies. Of particular mention are those developments in miniaturization, electrooptics, new materials (interfaces) and information/data handling."

The term applied to this advancing disciple that links the engineer and the chemist is Process Analytical Technology (PAT). Most recently, the US Food and Drug Agency (FDA) under Deputy Director Dr. Ajaz S Hussain (CDER Branch) put forth an initiative to foster the pharmaceutical application of PAT in the US (Mendricks, 2004).

PAT is defined by Hussain (Hussain, 2003) as a "system for analysis and control of manufacturing processes based on timely measurement during processing of critical quality parameters and performance attributes of raw and in-process materials and processes to assure acceptable product quality at the completion of the process".

1.2 Current Process Analytical Tools

Within the past thirty years (Creasy, 2003; Smilde et al., 2002) we have seen the movement of laboratory based (off-line) analytical systems into the plant environment (at-line or on-line) (Trevathan, 2003). For example, the HP/Agilent gas chromatograph (GC) became an ABB or Siemens or Daniels Process GC. For more complex feed/process streams requiring fast response time the ABB or Thermo Process Mass Spectrometer (MS) has moved from the lab (off-line) to an analyser shelter (at-line). These MS systems can have 1-10 sample streams being fed to them and carry out analyses every 30 sec or less from 500 feet away in the plant using continuous by-pass flow sampling technology. Such systems require 2-10 calibration/validation gases for routine operation. All actions feed back to a DCS computer to control the plant and manage the instrument remotely.

Fig II Agilent Lab Gas Chromatograph (GC)

Fig. III ABB Process Gas Chromatograph

Fig. IV ABB Questor IV Process Mass Spectrometer

The most recent advance in GC design is by both Sandia National Labs (Martin, 2003) and Honeywell Corporation (Bonne, 2003). Sandia made a GC on a chip "The Micro GC". The unit was most recently used by the Department of Defence in Iraq. The unit can detect, at the trace level, chemical warfare agents—or be used in a chemical plant to monitor specific off gases, for which it has been programmed, and do the analysis in < 60 seconds!

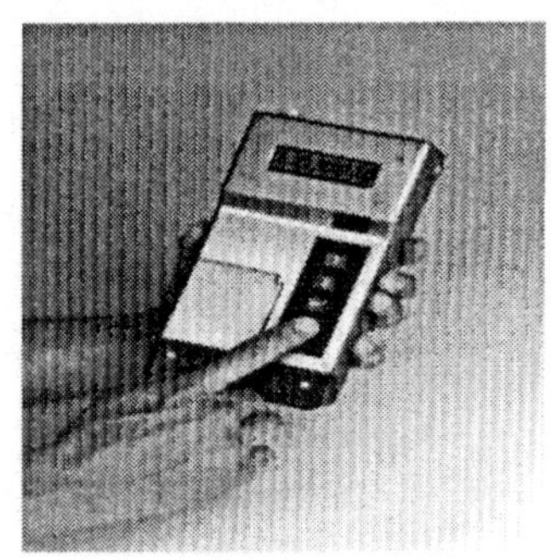

Fig. V Sandia National Labs –Micro GC

2 *Sensors*

New chemical specific sensors have benefited from micro machining to provide cost-effective processing (Goode, 2001; Bonne, et al 2004). Silicon-based metal-oxide chemical sensors were the first step. They were fabricated using standard semiconductor manufacturing and micromachining techniques (80-s/90-s). Such products deliver the key sensor benefits currently demanded by process analytical/engineers: low cost; lower power drain; robustness; small size; and ease of use.

2.1 Metal Oxide Sensor

Trautweiler, et al. (Trautweiler, et al, 1999) have summarized the metal oxide sensor technology. The figure below shows a MicroChemical Systems silicon-based metal-oxide chemical sensor. The sensor chip contains a thin-film sensing element fabricated from tin oxide and includes a buried polysilicon micro heater. The gas-sensitive layer measures concentration by electrical conductivity changes when exposed to air. The thin-film heater controls the operating temperature of the sensor to optimize selectivity and sensitivity.

Fig VI Metal Oxide Chemical sensors

Trautweiler et al. have summarized: "The heater and sensing element are built on a micro machined diaphragm precision formed from a silicon oxy-nitride ceramic with thermal properties that minimize sensor power consumption and response time. Heater power consumption levels are <40 mW with drive currents <40 mA, meaning that the chemical sensor is fully compatible with standard digital electronics components. This compatibility allows product engineers to design simpler, lower cost gas-detection products for consumer, industrial, and automotive applications. The underlying silicon micro-machined substrate is a standard sensor platform containing electrical interconnects to the sensor and device heater, as well as electrostatic discharge (ESD) protection diodes".

Trautweiler et al. go on to explain "The sensing film is constructed from an n-type tin oxide and ionic semiconducting material. During sensor operation, ambient O_2 molecules chemisorb onto the sensing film surface and dissociate into oxygen ions. Heating the tin-oxide film to ~450°C using the buried heating element facilitates this process. Formation of the oxygen ions removes free electrons from the tin-oxide grain surfaces, producing an increase in the sensing film's resistance. The presence of a reducing target gas such as CO reacts with the O_2 to produce CO_2 and releases electrons back into the sensing film with a corresponding increase in conductance or a decrease in film resistance. As CO concentration increases, the sensing film resistance decreases further. If the target gas were an oxidizing agent, such as NO_2, the opposite effect on sensing film resistance would occur."

2.2 Electrochemical Sensors

Electrochemical sensors are prevalent in the detection of oxygen and toxic environmental gases. "Each sensor is designed to be specific to the gas it is intended to detect. Electrochemical sensors are essentially fuel cells composed of noble metal electrodes in an electrolyte. The electrolyte can be an aqueous solution of strong inorganic acids or base. When a gas is detected, the cell generates a small current proportional to the concentration of the gas" (Anderson & Hadden 1999).

The RKI Instruments (R Pellissier, 2004) carbon monoxide (CO) and H_2S SAFETY Sensor Watches are examples of commercial electrochemical sensors; the O_2 sensor is of a Galvanic type. The Galvanic sensor is a special type of electrochemical cell. The galvanic oxygen sensor is essentially a battery that produces energy when exposed to oxygen and, is therefore consumed with exposure to oxygen. It is rugged and insensitive to shock and vibration.

Fig VII RKI Instruments GasWatch 2

"In its simplest form, an electrochemical sensor consists of a diffusion barrier, a sensing-electrode (sometimes called the working-electrode, measuring-electrode, or anode), a counter-electrode (sometimes called the cathode) and an electrolyte. In an environment free of chemically reactive gases, oxygen diffuses into the cell and adsorbs on both electrodes. The result is a stable potential between the two in which little, theoretically zero, current flows. The cell's chemical process at this point is" (Anderson & Hadden 1999).

$$O2 + 4H+ + 4e \rightarrow 2H2O \qquad (1)$$

"When a chemically reactive gas passes through the diffusion barrier it is either oxidized (accepts oxygen and/or gives up electrons) or reduced (gives up oxygen and/or accepts electrons), depending upon the gas. The resulting potential difference between the two electrodes causes a current to flow" (Anderson & Hadden 1999).

For instance, when carbon monoxide, a reducing gas, diffuses to the sensing electrode, it is oxidized causing the potential of the sensing electrode to shift in a negative (cathodic) direction. The cell's chemical process is:

Sensing-electrode

$$2CO + 2H_2O \rightarrow 2CO_2 + 4H^+ + 4e^- \quad (2)$$

Counter-electrode

$$O_2 + 4H^+ + 4e \rightarrow 2H_2O \quad (3)$$

Cell reaction

$$2CO + O_2 \rightarrow 2CO_2 \quad (4)$$

The newest form of toxic gas cell utilizes a third electrode called a reference electrode. This electrode has a stable potential from which no current is drawn. It is used to eliminate interference from side reactions with the counter-electrode. "In addition it allows the sensing-electrode potential to be biased with respect to its rest potential. Biasing is one method of controlling sensitivity to a particular gas. To provide for extended storage, a shorting clip is connected across the sensing and reference terminals. This short maintains the electrodes at the same potential and keeps current from flowing through the cell" (Anderson & Hadden 1999).

Electrochemical reactions are temperature dependent; therefore, electrochemical sensors incorporate a sensitive temperature sensor to compensate for temperature variations.

Electrochemical sensors are widely used and have the following advantages and disadvantages (Anderson & Hadden 1999).

- Advantages
  a) very linear
  b) good selectivity
  c) excellent repeatability and accuracy

- Disadvantages
  a) limited temperature range and sensitive to changes in temperature
  c) destabilized by humidity extremes
  d) sensitive to EMF/RFI
  e) limited storage life
  f) slow start-up if depolarized
  g) safety concerns-electrolyte has a strong acid.

2.3 *Interfering Gases*

Interfering gases, producing cross-sensitivities, are gases other than the target gas, which will cause the cell to respond. Like virtually any sensor, electrochemical cells are not completely specific. Despite the use of: carefully selected and engineered electrodes and electrolytes, changes in the operating potential of the sensing electrode, and selected chemical filters, it is difficult to develop a cell chemistry that will not respond to a more active gas than the target gas. Attempts to increase selectivity are not without side effects; for example, filters often slow the response of the cell.

2. 4 *Blocking Mechanisms*

Blocking is a condition that causes the cell to function poorly, or not at all until the condition is removed. Normally the block does not damage the sensor permanently as a poison would. The most common blocking mechanism for electrochemical cells is oxygen deprivation (Anderson & Hadden 1999).

2. 5 *Oxygen Deprivation*

Oxygen is an essential ingredient in the cell reaction with the target gas. If the oxygen supplied to the counter-electrode is cut off, the current cannot be sustained. Under normal conditions, when detecting small concentrations of a target gas in ambient air, an adequate supply of oxygen can easily be achieved (Anderson & Hadden 1999).

2. 6 *Catalytic Bead*

The catalytic bead sensor is made with two separate beads. One bead (active) is made of a small coil of wire, sealed in a ceramic substance, and coated with a catalyst. The second bead (reference) is identical to the active bead except a passivating substance is used in place of the catalyst. The reference bead compensates for changes in ambient temperature, humidity, and pressure variations.

Catalytic bead sensors operate above a threshold or "turn-on" voltage corresponding to the bead temperature that can, in the presence of the catalyst and oxygen, ignite the gas. Consequently, a flame arrestor is required around the sensors. (Anderson & Hadden 1999).

When a combustible gas diffuses through the flame arrestor, it oxidizes on the catalytically treated active bead. This oxidation reaction is exothermic, causes an increase in temperature relative to the temperature of the reference bead, and increases the electrical resistance of the small platinum coil embedded in the active bead. The change in resistance is proportional to the amount of chemical energy released by the oxidation reaction. Electronic circuitry immediately detects this increase in resistance and reduces electrical power to the bead until the active bead resistance is restored. The decrease in electrical power is linearly proportional to the combustible gas concentration present.

As the sensor ages, the catalyst slowly deactivates on the bead. The threshold voltage gradually increases, and the sensor sensitivity decreases. These changes cause increased zero drift and noise. The result is the sensor must be replaced.

2.7 *Field Effect Transistor (FET)*

The FET as summarized by Wroblewski, (Wroblewski, et al, 1996; Hughes 2003) measures the conductance of a semiconductor as a function of an electrical field perpendicular to the gate oxide surface. In a simple version, (i.e. a metal oxide semiconductor field effect transistor, n-channel MOSFET), a silicon substrate (bulk (B)) contains two n-type diffusion regions (source (S) and drain (D)). The structure is covered with a gate oxide layer (2) on top of which a metal gate electrode (1) is deposited.

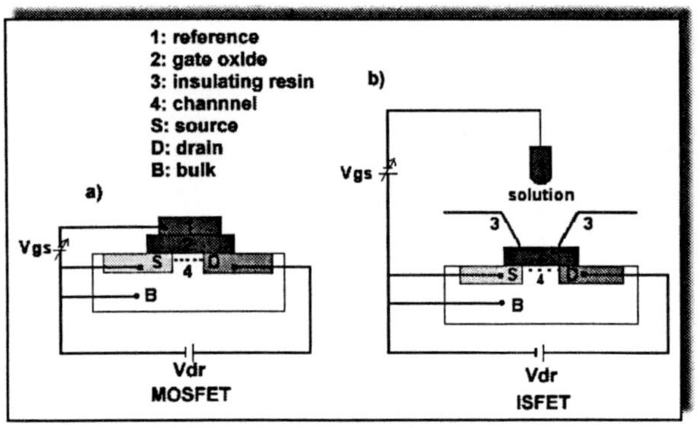

Fig VIII MOSFET & ISFET Diagram

When a positive voltage (with respect to the silicon) is applied to the gate electrode, electrons (which are the minority carriers in the substrate) are attracted to the surface of the semiconductor. Consequently, a conducting channel is created between the source and the drain, near the silicon dioxide interface. The conductivity of this channel can be modulated by adjusting the strength of electrical field between the gate electrode and the silicon, perpendicular to the substrate surface. At the same time a voltage can be applied between the drain and the source (Vds), which results in a drain current (Id) between the n-regions.

In the case of the ISFET, the gate metal electrode of the MOSFET is replaced by an electrolyte solution, which is contacted by a reference electrode. The metal part of the reference electrode can be considered as the gate electrode of the MOSFET.

When SiO2 is used as the insulator, the chemical nature of the interface oxide is reflected in the measured source-drain current. The surface of the gate oxide contains OH-functionalities, which are in electrochemical equilibrium with ions in the sample solutions (H+ and OH-). The hydroxyl groups at the gate oxide surface can be protonated and deprotonated and thus, when the gate oxide contacts an aqueous solution, a change of pH will change the SiO2 surface potential. A site-dissociation model describes the signal transduction as a function of the state of ionization of the amphoteric surface SiOH groups. Typical pH sensitivities measured with SiO2 ISFETs are 37-40 mV/ pH unit.

The selectivity and chemical sensitivity of the ISFET are completely controlled by the properties of the electrolyte/gate-oxide interface. Other inorganic gate materials for pH sensors like Al2O3, Si3N4 and Ta2O5 have better properties than SiO2 with respect to pH response, hysteresis and drift.

2.8 Microcantilver Sensors

Microcantilevers, as described by Thundat (Thundat et al., 2003; Dravid, 2003) appear to be a universal platform for real-time, in-situ measurement of physical and chemical properties. Based upon micro-electro-mechanical (MEMS) fabrication of silicon, combined with the sensitivity utilized in atomic force microscopy, the technique promises to revolutionize applications where multiple properties need to be economically monitored. A robust multielement sensor array can be constructed for pressure, temperature, viscosity and other physical measurements as well as the speciation of organic and inorganic constituents in harsh environments. By monitoring the deflection or resonance vibration of the cantilever, mass and stress changes induced by environmental conditions can be precisely and accurately recorded. For example, temperature changes can be measured to microdegrees by coating the cantilever with two dissimilar materials (bimetallic effect), and the viscosity of liquids and gases has been measured over four orders of magnitude using a single cantilever.

Microcantilever chemical sensors have also been made by coating the cantilevers with selectively absorbent films. For example, vapor detection of mercury, natural gas, and water vapor has been demonstrated with picogram sensitivity. Additional chemical coatings are being developed for inorganic, organic, and biological species (Hardy, 2004).

3. PAT Application – Hydrogen Sensor

A recent addition to process analytical technology is the H2scan hydrogen sensor system. This solid-state sensor detects concentrations of hydrogen gas, with or without oxygen present, in concentrations (actually partial pressures) from 10 parts per million to 100%.

H2Scan uses the Robust Hydrogen Sensor™, a thin film palladium-nickel (Pd/Ni) resistor deposited on a silicon substrate that changes resistance in the presence of hydrogen. The resistance of the Pd/Ni increases in direct correlation to the absorption of hydrogen molecules into the Pd/Ni lattice. No other gasses or environmental controls are necessary for these measurements; no oxygen is required for a reading; essentially the sensor counts hydrogen molecules.

In another configuration the Pd/Ni resistor becomes a Palladium Field Effect Transistor (FET) sensor that enables the unit to detect concentrations as low as 10 parts per million.

The sensor-module features a temperature sensing diode and on-board heaters. Using these, the system can maintain the sensor substrate at a constant temperature. This controlled elevated temperature assures that sensor operation is unaffected by ambient temperature.

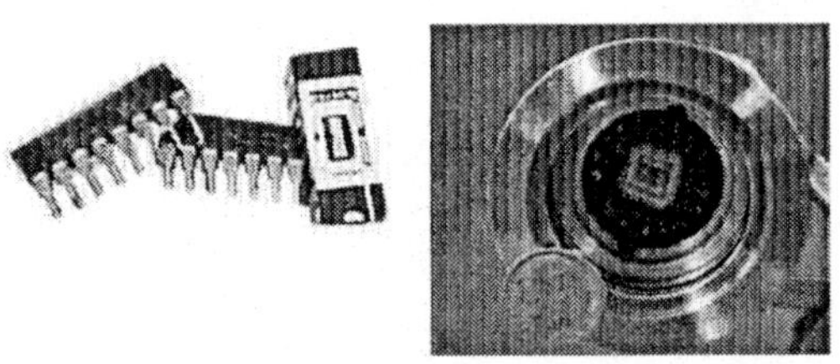

FIG IX H2Scan Hydrogen Sensors

The H2Scan sensor system can be provided in a Class I/Div II Group B housing with "Z" purge to permit its operation in a flammable gas plant environment.

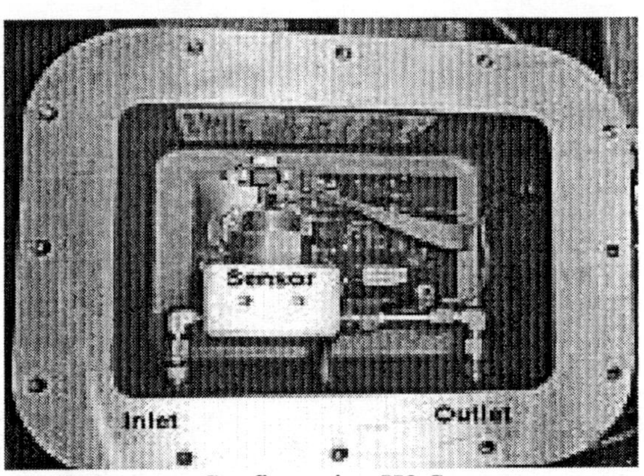

Fig X Process Configuration H2 Sensor

The H2scan system, with modifications by Air Products, was tested at an APCI plant for on-line monitoring of the H2 content (70-90 vol. %) in a mixed gas (Ar, N2, CO, CH4, H2O & CO2) process stream. The data were compared to an off-line GC over a 30 day period.

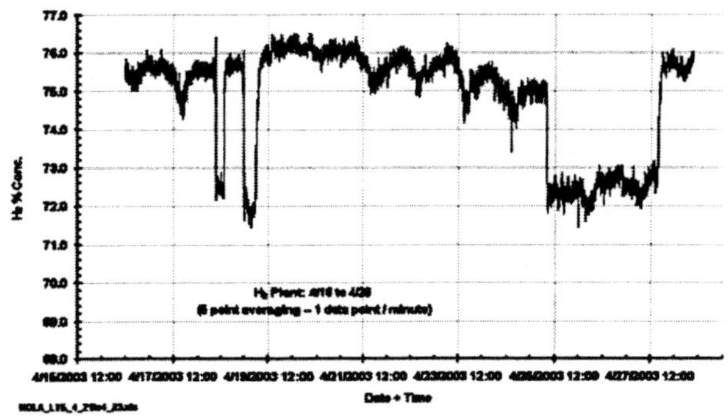

Fig XI On-Line Hydrogen Measurement

The H2 sensor system provided a partial pressure hydrogen content analytical result every 2 seconds, compared to the GC operation, which was routinely taken every four hours. During the trial, we observed periodic oscillations (dips) in the H2 content (+/- 3% H2). The dips correlated to maintenance times when modifications were being done to the Shift Reactor. The dips had not been quantified before in terms of H2 content. The concentration of H2 quantified correlated to the GC data (matched by time) +/- 0.5% H2.

4. *Sensor Technologies*

Tables (I, II, III) provide a summary for reference of the current sensor technologies on the market (Table I) or being developed, along with a means to contact the vendor. Because hydrogen is the current centerpiece to our Nation's economy for energy, fuel cells, and automotive application (Table III), as well as industrial applications (refinery/hydrogenation-edible oil-Table II) we have provided more details in this market segment.

5. *Summary*

The evolution of micro-sensor technology to replace our "five senses", and to replace sophisticated analytical instrumentation for on-line process control and final product certification is well under way. We are seeing the changes take place, slowly for sure, but taking place because we need to quantify the process variable (composition or T/P/F) to initiate closed loop process control in the time required to "Deliver the Difference" for safety, consistent process control and final product properties & performance at the lowest cost.

Table I Current Sensor Technology Vendors

Firm	Contact	Analyte	Technology
Alphasense	www.alphasense.com	O2, CO, H2S, SO2,Cl, NO2	Fuel cell, Electrochemica
Analytical Indust	www.aii1.com	O2	Galvanic cell
Applied sensors	www.appliedsensor.com	Various	MOS, (metal oxide) FE (field effect) QCM (quartz crystal microbalance)
Avalon Photonics	www.avap.ch	O2 and various	Vertical cavity surface emitting laser
Axsun	www.axsun.com	In development	NIR- Mems
ConductTech	www.conductivetech.com		Electrochemical-thin film
City Tech	www.citytech.com	O2, CO, H2S, SO2, Cl, NOX combustible	Electrochmical, Pellistor, IR, MOS
Cyranose Science	www.cyranosciences.com	Organic vapors	Sorption on carbon film
Detcon	www.detcon.com	H2S, CO2, O2,	MOS, catalytic bead, IR,electrochemical
Figaro	www.figaro.com	Combustible, toxic, CFCs , indoor, O2 auto	MOS
Hohner	www.surfacesensor.com	color, refraction, reflection	Optical encoder for Visable, UV
IR Micro	www.ir-microsystems.com	various	Pyroelectric array detector
IT		O2, CO, H2S, SO2, Cl, NOx	Electrochemical, galvanic
e2V	www.e2vtechnologies.com	Combustibles, CO2,methane	Cataytic, TCD,NIR
Ion Optics	www.ion-optics.com	aroius, CO2	NDIR
	www.sixth-sense.com	O2, CO, H2S, SO2, Cl, NOx	Electrochemical, galvanic
Microsensor	www.microsensorsystems.com	Toxic, EO, hydrocarbon	SAW
Sensicore	www.sensicore.com	Waste water analysis	Micro-electrochemical
TAOS	www.taosinc.com	Light, reflection	Micro-IR
Vaisala	www.iprocessmart.com/vaisala	CO2	Micro-R
Sensors, Inc	www.sensors-inc.com/	Emissions-auto	Portable IR,UV

Table II H_2 Sensor Technologies Process

Type	Property	Vendor	Range
Pd alloys (Thin film)	Resistance	H2SCAN	0.1 – 100%
Nanotube	Resistance	Nanomaterials Research	1000 ppm – 2%
Arrays (Thin film + FET)		H2SCAN	50 ppb – 100%
Electrochemical	Voltage	Unisense	0.02% to 25% H2 in H2O

Table III H_2 Sensor- Leak Detection

Type	Property	Vendor	Range
Pt/Ni alloys (Thin film)	Resistance	Applied Sensors	50 – 5000 ppm
PdO/Si	Resistance	Thermodyne Systems	10-1000 ppm
Pd/AlN/Si	MIS-capacitor	Wayne State Univ	1 ppm
Pd alloy	Resistance	Microchemical Systems	
FET	Resistance (iV)	H2SCAN	50-1000 ppm
Catalytic bead	Resistance	RKI Instruments	0-5%
Catalytic bead	Resistance	Fuel Cell Safety Inc	1-5%
Catalytic bead	Resistance	Analytical Technology Inc	0-100%LEL
Semicond oxides	Resistance	Dupont	0-1000 ppm
Semicond. oxide	Resistance	RKI Instruments	0-2000 ppm 0-10,000 ppm
Semicond. Oxide (SnO)	Resistance	Figaro Engineering, Inc	50-1000 ppm
Mixed metal oxide	Resistance	City Technologies	0-1000 ppm
Microheater and thin film	Heat of reaction	NGK Spark Plug, Ltd.	0-80%
Electro-chemical	Current	Control Instruments	100-1000 ppm
Electro-chemical	Current	Draeger Safety Inc	100-1000 ppm
Electro-chemical	Current	Bionics Instrument Co., Ltd	0-4%
Electro-chemical	Current	Toxalert, Inc	0-1000 ppm
Fiber optics	Optical absorption/ relection	Optisense	
Fiber Optics	Optical absorption/ relection	H2SCAN	
Fiber optics	Optical absorption	Davidson Instruments	
Nanotube	Resistance	Nanomix	
Nanotube	Resistance	Univ of Calif	1-10%
Nanotube	Resistance	NanoMaterials Research	10-1000 ppm
Nanotube	Resistance	Applied Nanotech Inc	

REFERENCES

Anderson G.L. and D. M. Hollen, (1999)n Electrochemical Sensors *–The Gas Monitoring Handbook* Published by Avocet Press Inc,1999

Bonne U. Private (2004) Private Communication

Bonne, U , D. Kubisiak , K. Creasy, and T. Francisco, (2002) Rugged Microsensor for Process Stream Monitoring. *Journal of Process Analytical Chemistry*, 7(2), 64-72.

Chauvel J.P., W.W. Henslee, and L. A. Melton (2002) Teaching Process Analytical Chemistry, *Analytical Chemistry* (2002) **74(13)** 381-384A

Creasy K, (2002) Process Analytical Chemistry- Smarter, Faster Stronger, *Journal of Process Analytical Chemistry*, 7(1), 1-7.

Dravid V.P. (2004) Private Communication, Northwestern University

Goode, B. G. (2001) Best of Sensors Expo Spring 2001, *Sensor Magazine* **18** (8).

Gunnell, P. and Van Vuuren, (2001) Process Analytical Systems: A Vision for the Future. *Journal of Process Analytical Chemistry*, **6**(1), 1-5.

Hardy, J. E., (2004) Private Communication, *Oak Ridge National Laboratory*

Harris, C. M., (2002) Analytical Chemistry **74**, 127A-133A

Henslee W. (1996) Effective Utilization of Process Analytical Technology, *Journal of Process Analytical Chemistry*, **2**(2) 189-234

Hughes, R. (2003) Private Communication.

Hussian A. S. (2003) Process analytical technologies (PAT) initiatives. US Food and Drug Administration. Center for Drug Evaluation and Research. http://www/fda.gov/cder/OPS/PAT.htm

Martin, S., Sandia National Labs (2003) Private Communication

Mendricks, J. Vanroeyen and H.Y. Wang, (2004) The Future of Pharmaceutical Engineering, *J. Pharm. Sci.*, **93**/2, 235-238

Pellissier, R. (2004) Private communication.

Potyrailo R.A. (2001) Process Chemical Analysis, *GE R&D Center-Technical Information Series* (**2001CRD138**) 1-19.

Smilde, A.K., F. W.J. van den Berg, and H. C. J. Hoefsloot, (2002) How to Choose the Right Process Analyzer, *Analytical Chemistry* (2002) **74(13)** 369-372A

Thundat T, G, Brown, and Hai-Feng Ji (2003) Environmental Monitoring Using Micro cantilever Sensors, *US DOE WORKSHOP-Third Session*, New Orleans, LA 1-5.

Trautweiler, S., N. Mosier, E S. Zdankiewicz, (1999) New Silicon-Based Metal –Oxide Chemical Sensors, *Sensor Magazine* **16**(9).

Trevathan V. L. (2003) What's So Special About Process Control Projects?, *Control Magazine*, **March**./2003 43-48

Workman Jr. J, K. Creasy, S. Doherty, L. Bond, M. Koch, A. Ullman, and D. J. Veltkamp, (2001) Process Analytical Chemistry, Analytical Chemistry **73** (120 2705-2718.

Workman Jr. J, M Koch and D J. Veltkamp (2003) , Process Analytical Chemistry, *Anal. Chem* 2003, **75** 2859-2876.

Wroblewski W, B. Mirzynska and Z Brzoka (1996) Field effect transistors (FETs) as transducers in electrochemical sensors, *Chemia Analityczna* (Warsaw, Poland) **41**, 697-706

ELSEVIER
IFAC
PUBLICATIONS
www.elsevier.com/locate/ifac

A NONLINEAR MODEL PREDICTIVE CONTROL ALGORITHM FOR BREAST CANCER TREATMENT

Jeffry A. Florian, Jr. * **Julie L. Eiseman** **,***
Robert S. Parker *,***,[1]

* *Department of Chemical and Petroleum Engineering, University of Pittsburgh School of Engineering, Pittsburgh, PA*
** *Department of Pharmacology, University of Pittsburgh School of Medicine, Pittsburgh, PA*
*** *Molecular Therapeutics and Drug Discovery Program, University of Pittsburgh Cancer Institute, Pittsburgh, PA*

Abstract: A nonlinear model predictive control (NMPC) algorithm was developed to dose the chemotherapeutic tamoxifen to mice bearing breast cancer xenografts. A novel saturating rate cell-cycle model (SCM) was developed to capture unperturbed tumor growth dynamics, and a bilinear tumor kill term was included in the G-phase to account for the cycle-specific nature of tamoxifen and its active metabolite. Drug pharmacokinetics were modeled using a three-compartment linear model, which successfully approximated parent compound and metabolite (4-hydroxytamoxifen) plasma concentrations as a function of time. Using daily tumor measurements, the model predictive control algorithm successfully reduced tumor volume along a specified reference trajectory over a period of 4 months. A more clinically-relevant implementation using weekly or biweekly tumor measurements, and a prediction horizon seven days beyond the measurement interval, also led to reduced tumor volumes. In the mismatch case, a controller based on the simpler linear cell-cycle model (LCM) was unable to track desired reductions in tumor volume. Controllers based on a lumped-parameter saturating Gompertz model (GM), however, can yield similar performance to those using the more complex saturating rate cell-cycle model. This performance was dependent on the cell-cycle phase of drug effect, with poorer results for M-phase targeted drugs. Overall, NMPC is a suitable algorithm for the class of chemotherapy problems with daily drug dosing, and the algorithm developed here may be adaptable to the clinical setting for the treatment of human breast cancer patients.

Keywords: biomedical systems, nonlinear models, nonlinear model predictive control

1. INTRODUCTION

Cancer is the second-leading cause of death in the United States and is responsible for $189.5 billion in health care and research costs (The American Cancer Society, 2004). It is estimated that more than 1.4 million new cancer cases will be diagnosed in the year 2004 alone (The American Cancer Society, 2004). Given the economic and societal impact of the disease, improved treatment algorithms could lead to dramatic reductions in morbidity and economic burden by increasing treatment efficacy and efficiency. Mathematical models of untreated or treated tumor growth have been developed with varying levels of complexity from the cellular to the macroscopic scale (Norton, 1988; Asachenkov *et al.*, 1994; Bajzer *et al.*, 1996). Furthermore, the medical community focuses significant energy on

[1] To whom correspondence should be addressed: rparker@pitt.edu; +1-412-624-7364; 1249 Benedum Hall, Pittsburgh, PA 15261 USA

the development of mathematical models of drug pharmacokinetics using available tools (D'Argenio and Schumitzky, 1997). Using case study specific models, the cancer chemotherapy treatment problem can be posed in a model-based framework.

In fact, for more than a decade, various model-based cancer treatment algorithms have been synthesized with a focus on batch-like problems (Martin, 1992; Panetta and Adam, 1995; Harrold *et al.*, 2003). The main focus of these approaches is to calculate a treatment profile over a fixed duration that satisfies a performance objective including, among other terms, minimizing tumor size at a specified future time. The durations may be as short as a cycle of therapy (28 days in (Harrold *et al.*, 2003)) or as long as a full course of therapy (52 weeks or more in (Martin, 1992)). The latter is a generic optimal treatment problem using a "drug" and "tumor" with assumed dynamic behavior; these problems can be solved using a variety of methods including control vector parameterization (Martin and Teo, 1994) and optimal control theory (Ledzewicz and Schättler, 2002). However, clinically-unrealistic assumptions about the efficacy of compounds or treatment modalities (such as continuous infusion over *days*) limit the impact of generic formulations. To improve the potential significance, case-study specific information from human or animal studies, which includes tumor growth, drug effect, toxicity, etc., are incorporated into the mathematical models of pharmacokinetics (PK) and pharmacodynamics (PD). Given these models, one can pose cyclical cancer treatment as a mixed-integer optimization problem (Harrold *et al.*, 2003).

Not all cancer treatment problems have the pseudo-batch structure outlined above. One such example, and the case study evaluated in this work, is the treatment of estrogen receptor positive (ER+) breast cancer xenografts by the chemotherapeutic tamoxifen (TM). The parent compound and its more active metabolite 4-hydroxytamoxifen (HTM) function through competitive inhibition with estrogen. Drug-binding prevents receptor activation and causes an accumulation of cells in late G_1 phase due to growth signal truncation. While neither the parent nor active metabolite directly induce cell apoptosis (programmed cell death), the cell stress resulting from accumulation causes increased apoptosis as a function of the total tumor size and the amount of drug administered. TM treatment is the method of choice against ER+ breast cancer, not only due to its effectiveness but also because of its low toxicity that enables long term dosing with minimal side effects (Ellis and Swain, 2001). A typical TM regimen might involve the patient taking daily TM doses for up to 5 years. Hence, it should be possible to develop a NMPC algorithm to dose breast cancer patients with TM. The present work addresses the synthesis of this algorithm in the context of a preclinical mouse study using mice bearing human breast cancer xenografts.

2. METHODS

Pharmacokinetic data were obtained from Robinson *et al.* (1991). A single large dose of TM (200 $\frac{\text{mg}}{\text{kg}}$) was administered orally (in 0.1 mL peanut oil) to non-tumor-bearing, ovariectomized, athymic mice, and plasma samples were taken at 0, 3, 6, 12, 24, 48, 72, and 96 hours after dosing. Plasma samples were purified using chromatography; concentrations of TM, HTM, and N-desmethyltamoxifen were determined via comparison to a standard curve and the resulting plasma concentrations represent the averaged values from three mice.

Tumor volume data for pharmacodynamic modeling were obtained from Conley and coauthors (1999). The study protocol involved athymic mice implanted subcutaneously with ER+ xenografts and estradiol pellets at day 0. Stratified groups were dosed via oral gavage at 7.5, 15, 30, of 60 $\frac{\text{mg}}{\text{kg}}$ tamoxifen per day dissolved in 0.1 mL of purified water (vehicle had no significant effect on tumor growth). Doses were given once a day for five consecutive days followed by two days off, beginning on day 12 of the study. This seven-day pattern was repeated twice. Twice each week, tumor volumes were recorded using digital calipers assuming an ellipsoidal tumor of volume $\frac{\ell w^2}{2}$ (ℓ = longest tumor axis, w = shortest tumor axis perpendicular to ℓ).

In this paper, different models with similar names and structures are presented. Tables 1 and 2 provide a summary of the model abbreviations employed throughout.

3. TUMOR GROWTH MODELING

Common models for describing untreated tumor growth include the exponential and Gompertz, among others (Martin and Teo, 1994). Biologically motivated

Table 1. Open-loop model nomenclature

SCM	saturating rate cell-cycle model (unperturbed)
SCM-G	saturating rate cell-cycle model with G-phase kill
SCM-M	saturating rate cell-cycle model with M-phase kill
GM	Gompertz model (with kill term)
LCM	linear cell-cycle model
LCM-G	linear cell-cycle model with G-phase kill

Table 2. Closed-loop model nomenclature

NC-G	nominal case; controller designed using SCM-G
NC-M	nominal case; controller designed using SCM-M
MC-Gu	mismatch case; SCM-G as the simulated mouse; controller designed using GM at equivalent input to SCM-G
MC-Mu	mismatch case; SCM-M as the simulated mouse; controller designed using GM at equivalent input to SCM-M
MC-My	mismatch case; SCM-M as the simulated mouse; controller designed using GM at equivalent output to SCM-M

cell-cycle models provide more insight into cell behavior (Panetta and Adam, 1995), but these linear state space models display exponential growth and are unable to capture saturating tumor growth dynamics. The Gompertz model (GM) is capable of capturing clinically observed tumor dynamics (Norton, 1988), but it does not capture information regarding the progression of cells through the individual phases of the cell-cycle. Cell cycle information is vital from a treatment perspective because cycle-specific anti-cancer compounds such as curacin A, Taxol, and tamoxifen can be modeled in a biologically meaningful manner (Lopes *et al.*, 1993; Panetta and Adam, 1995; Kozusko *et al.*, 2001). To accurately model both declining tumor growth rate and cell phase distribution, a saturating rate cell-cycle model (SCM) was constructed (Florian Jr. *et al.*, 2003):

$$\frac{dX_G}{dt} = -k_G X_G \ln\left(\frac{\theta}{N}\right) + 2k_M X_M \ln\left(\frac{\theta}{N}\right) \quad (1)$$

$$\frac{dX_S}{dt} = -k_S X_S + k_G X_G \ln\left(\frac{\theta}{N}\right) \quad (2)$$

$$\frac{dX_M}{dt} = -k_M X_M \ln\left(\frac{\theta}{N}\right) + k_S X_S \quad (3)$$

$$N = X_G + X_S + X_M \quad (4)$$

Here X_i corresponds to the volume (mm^3) of cells (assuming constant cell volume and 10^6 cells = 1 mm^3) in the G (growing), S (DNA synthesis), or M (mitosis) phases; θ is the plateau population of the tumor (10^5 mm^3); and k_i's denote transfer rates between cell phases ($\frac{1}{\text{hr}}$). The three states correspond to cells in G_0/G_1 (G), S, or G_2/M (M) phases, and groupings were selected based on experimental limitations. Representations of the five-compartment cell-cycle model and the reduced three-compartment cell-cycle model are shown in Figure 1. Flow cytometry separates cells based on total DNA content and a fluorescent DNA label, such as propodium iodine (Lee *et al.*, 2002). Cell phases possessing equivalent amounts of DNA emit the same fluorescence. The G_0 and G_1 phases contain the same amount of DNA, hence they are indistinguishable and are lumped together into the single phase, G. The same scenario applies for cells in the G_2 and M phases, which contain twice the DNA as the cell prepares to divide (Riesberg *et al.*, 2001).

Two alternative tumor growth models, the three compartment linear cell-cycle model (LCM) and the GM were simulated for comparison with the above model. Removing the logarithmic total population dependence from individual transition rates in Equations (1)-(3) results in the following governing equations for the LCM:

$$\frac{dX_G}{dt} = -k_G X_G + 2k_M X_M \quad (5)$$

$$\frac{dX_S}{dt} = -k_S X_S + k_G X_G \quad (6)$$

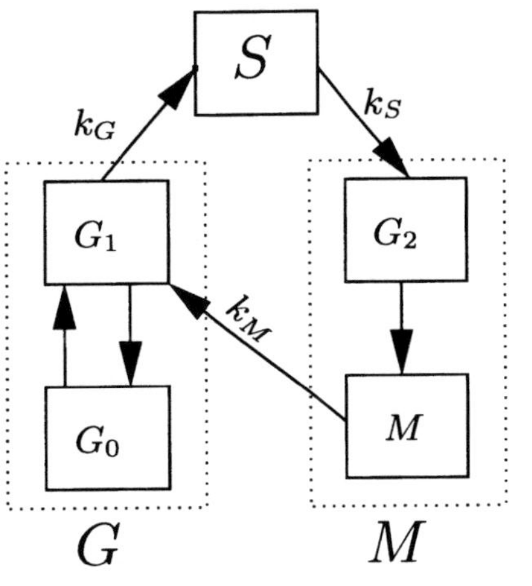

Fig. 1. The five phases of the cell-cycle: G_0 (senescence), G_1 (growth), S (DNA replication), G_2 (mitotic preparation), and M (mitosis). Evaluating fractional distributions via fluorescent DNA labeling and flow cytometry prevents unique quantification of G_0 from G_1 (lumped as G) and G_2 from M (lumped as M).

$$\frac{dX_M}{dt} = -k_M X_M + k_S X_S \quad (7)$$

$$N = X_G + X_S + X_M \quad (8)$$

The Gompertz model lumps cell phases into one homogeneous population, N, with saturating growth dynamics as $N \to \theta$:

$$\frac{dN}{dt} = \tau N \ln\left(\frac{\theta}{N}\right) \quad (9)$$

All constants and variables are equivalent to previous definitions; τ is a pseudo-doubling time ($\frac{1}{\text{hr}}$).

Values for the transfer rates (k_i) and the pseudo-doubling time (τ) were determined by minimizing the squared difference between model predictions and animal data (Conley *et al.*, 1999). The model parameters are summarized in Table 3. The tumor data and model predictions are plotted in Figure 2. Both the SCM and GM provide an accurate prediction of tumor growth over the duration of the experiment. In contrast, the LCM is unable to account for both rapid initial tumor proliferation and the decreasing growth rate observed with increasing tumor size. Rather, it predicts an average fit of the growth data.

Table 3. Parameter values for the tumor growth models: linear cell-cycle (LCM), Gompertz (GM), and saturating rate cell-cycle (SCM).

	LCM	GM	SCM
k_g	0.0168	—	0.0013
k_s	0.0088	—	0.0390
k_M	0.0429	—	0.0169
τ	—	.0010	—
θ	—	10^5	10^5

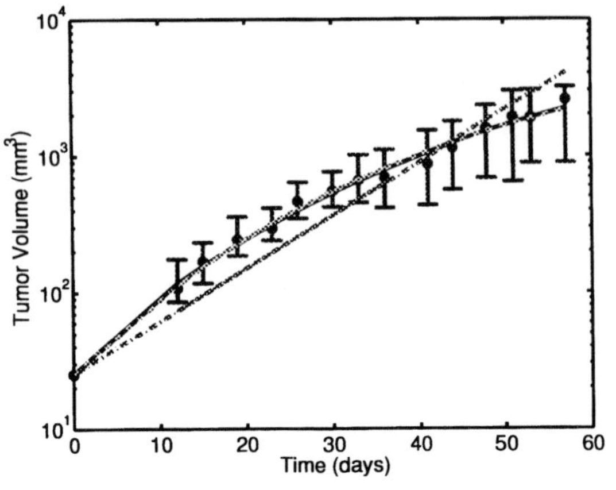

Fig. 2. Measured mean tumor volume data (circles; N = 20 control mice) and simulated tumor size (lines) as a function of time for LCM (dash-dot), SCM (solid), and GM (dash, overlay on solid) models. Error bars represent a 95% confidence interval.

Akaikie's Information Criteria, traditionally used by the medical community to assess whether increases in model complexity are justified by improvements in model accuracy, is given by (Akaike, 1979):

$$\mathrm{AIC} = n \ln\left(\frac{\mathrm{SSE}}{n}\right) + 2 * p \qquad (10)$$

Here, SSE is the sum squared error between the model prediction and the data at the measurement points, n is the number of data points, and p is the number of parameters in the model. Since the number of data points (17) and the SSE for both models were equivalent (13.98 for each) while the number of parameters increased from 1 to 3, AIC analysis clearly favored the simpler Gompertz description. However, this result was not unexpected and should result if the SCM accurately portrays Gompertz growth while maintaining cell phase information. The benefit of including additional complexity in this case allowed for an accurately targeted drug-effect when a cycle-specific chemotherapeutic was employed. It was anticipated that this improvement in model quality will lead to improved performance by controllers designed from this model (Morari and Zafiriou, 1989).

4. PHARMACOKINETICS

The pharmacokinetics of the oral agent TM and its active metabolite HTM were modeled using the first-order parent/metabolite with oral dosing compartmental structure in Figure 3. A TM dose is administered orally, and the parent compound is absorbed from the gut. After absorption, circulating TM can be cleared from the system (including conversion to metabolites such as tamoxifen N-oxide, N-desmethyltamoxifen, and others) or converted to HTM. Circulating HTM is also cleared from the

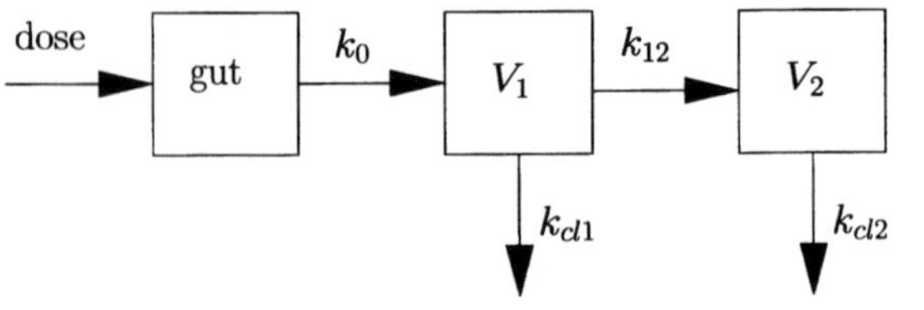

Fig. 3. 3-compartment model for plasma concentrations of tamoxifen and 4-hydroxytamoxifen after oral dosing

system. Of the metabolites, only HTM and N-desmethyltamoxifen appear in plasma at detectable concentrations (Ellis and Swain, 2001). Affinity studies have shown N-desmethyltamoxifen binds with estrogen receptors at $\frac{1}{50}$ the affinity of TM and $\frac{1}{1250}$ the affinity of HTM (Ellis and Swain, 2001). As such, only TM and HTM are postulated to contribute a significant anti-tumor effect.

The following equations describe the TM and HTM pharmacokinetics:

$$\begin{aligned}
\frac{dX_0}{dt} &= -k_0 X_0 + u(t) \\
\frac{dX_1}{dt} &= -k_{cl1} X_1 - k_{12} X_1 + k_0 X_0 \\
\frac{dX_2}{dt} &= -k_{cl2} X_2 + k_{12} X_1 \\
Y_1 &= \frac{X_1}{V_1} \\
Y_2 &= \frac{X_2}{V_2}
\end{aligned} \qquad (11)$$

Here X_0, X_1, and X_2 correspond to TM mass in the gut, TM mass in plasma, and HTM mass in plasma, respectively (μg), V_1 and V_2 are central compartment volumes for the parent compound and metabolite (mL), and k_i denote transfer rates ($\frac{1}{\mathrm{hr}}$). Variables Y_1 and Y_2 represent the measured plasma concentrations of TM and HTM ($\frac{\mu\mathrm{g}}{\mathrm{mL}}$).

The parameters in equations (11) were fit by minimizing the squared difference between model predictions and data obtained from non-tumor bearing athymic mice (Robinson *et al.*, 1991) and the resulting parameter values are summarized in Table 4.

Model predictions for both compounds can be seen in Figure 4. The pharmacokinetic model is able to capture the dynamic profile of the data with only a slight over-prediction of TM concentration

Table 4. PK parameter values for the oral dosing of tamoxifen.

Parameter	Value
k_0	0.052 $\frac{1}{\mathrm{hr}}$
k_{12}	0.340 $\frac{1}{\mathrm{hr}}$
k_{cl1}	0.349 $\frac{1}{\mathrm{hr}}$
k_{cl2}	14.866 $\frac{1}{\mathrm{hr}}$
V_1	2.354 mL
V_2	0.060 mL

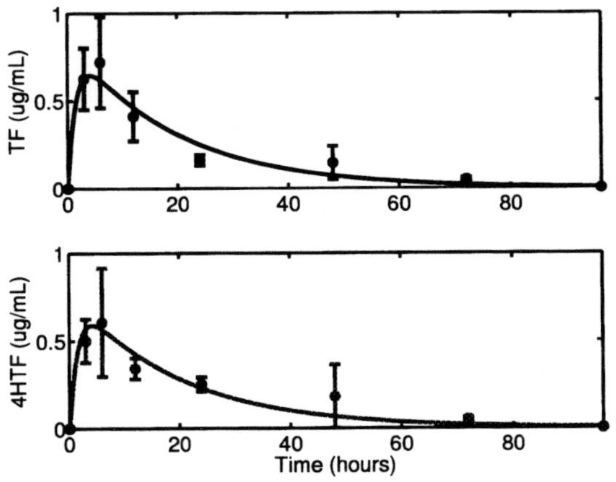

Fig. 4. Actual (circles) vs. simulated (lines) PK for TM (top) and HTM (bottom) after a 200 mg/kg oral dose of TM. Each data point represents mean plasma concentrations from three mice; error bars are ± 1 standard deviation.

at $t = 24$ hours. Additional compartments or delays could be incorporated into the TM model to more accurately fit the data at the cost of increased number of parameters, decreased parameter confidence, and increased mathematical complexity.

5. PHARMACODYNAMICS

The cycle-specific drug effects of TM and HTM were coupled with the tumor growth model by including a bilinear kill term in the differential equation for X_G in (1), termed the saturating rate cell-cycle model with G-phase kill (SCM-G). Cycle-specific drug effects were also incorporated for X_G in (5) and for N in (9), respectively. Conceptually, the bilinear structure means that increases in drug concentration have an increasing anti-tumor effect and a larger tumor will have more cells susceptible to a given dose. Adding the bilinear term yields the following modifications to equations (1), (5), and (9), respectively:

$$\frac{dX_G}{dt} = -k_G X_G \ln\left(\frac{\theta}{N}\right) + 2k_M X_M \ln\left(\frac{\theta}{N}\right) - k_D X_G \left(\frac{X_1}{V_1} + c\frac{X_2}{V_2}\right) \quad (12)$$

$$\frac{dX_G}{dt} = -k_G X_G + 2k_M X_M - k_D X_G \left(\frac{X_1}{V_1} + c\frac{X_2}{V_2}\right) \quad (13)$$

$$\frac{dN}{dt} = \tau N \ln\left(\frac{\theta}{N}\right) - k_D X_G \left(\frac{X_1}{V_1} + c\frac{X_2}{V_2}\right) \quad (14)$$

Here k_D is the drug-induced tumor death rate ($\frac{\text{mL}}{\mu\text{g hr}}$) and c represents the ratio between the binding affinities of TM and HTM. Combining the kill rates in this manner is justified based on the inhibition mechanism of the drug, which involves the blockage of estrogen receptor sites. A drug which binds more tightly to the receptor acts as a more potent blocker, and the relative strength of that bond serves as a basis of comparison for its effectiveness versus other drugs. As noted above, HTM has been shown experimentally to possess a binding affinity 25-50 times that of TM (Ellis and Swain, 2001), so c was set to 25 (possibly underestimating HTM anti-tumor effect).

Unperturbed tumor growth rates and PK parameters were fixed at identified values, and k_D was estimated by minimizing the squared difference between the data and SCM-G predictions at each of four dosing levels (60, 30, 15, and 7.5 $\frac{\text{mg}}{\text{kg}}$ per dose). The k_D rates corresponding to the three lower dosing levels were found to be in close agreement ($0.00251 \pm 0.00025\frac{\text{mL}}{\mu\text{g hr}}$), while the rate for 60 $\frac{\text{mg}}{\text{kg}}$ was significantly lower ($0.00192\frac{\text{mL}}{\mu\text{g hr}}$). This discrepancy is possibly due to a saturation of the estrogen receptor pathway at the high dose level. By limiting the dosing levels to $0 \le u(t) \le 30\frac{\text{mg}}{\text{kg}}$ per day (single daily dose), the value of k_D can be fixed to the mean value of $0.00251\frac{\text{mL}}{\mu\text{g hr}}$. Although Conley *et al.* (1999) did not demonstrate long term survival of mice on daily TM dosing, other authors have shown mouse longevity for up to 2 years at $40\frac{\text{mg}}{\text{kg}}$ daily dosing, thereby validating the dosing range used here as within acceptable toxicity limits (Martin *et al.*, 1997).

Estimating k_D for both the LCM-G and GM yielded similar rate trends for k_D. At the three lower dosing levels (30, 15, and $7.5\frac{\text{mg}}{\text{kg}}$), k_D values were in close agreement (LCM-G: $0.00531 \pm 0.00047\frac{\text{mL}}{\mu\text{g hr}}$; GM: $0.00325 \pm 0.00024\frac{\text{mL}}{\mu\text{g hr}}$) while rates at $60\frac{\text{mg}}{\text{kg}}$ (LCM-G: $0.00384\frac{\text{mL}}{\mu\text{g hr}}$; GM: $0.00192\frac{\text{mL}}{\mu\text{g hr}}$) were quite different. Estimated k_D rates between the GM and SCM-G were also similar while rates determined for the LCM-G were quite different. It is likely this change results from the linear vs. nonlinear dynamic structure of the models.

Figure 5 shows the ability of the pharmacodynamic models to fit tumor growth during treatment (dose $\le 30\frac{\text{mg}}{\text{kg}}$). The model over-predicts drug effect for doses of 15 and 30 $\frac{\text{mg}}{\text{kg}}$, however these over-predictions do not seriously impact overall tumor growth predictions over the 60 day period. A portion of this mismatch results from fixing the unperturbed tumor growth kinetics, which were different even among stratified groups. For example, at day twelve for $30\frac{\text{mg}}{\text{kg}}$ dosing (Figure 5, bottom) the predicted tumor size is greater than the measured tumor size on day 12 before drug is administered. Still, the model predicts the qualitative shape of drug-altered tumor growth and model predictions remain within the 95% confidence intervals.

6. CONTROLLER SYNTHESIS: SATURATING RATE MODEL

A daily dose of 12.0 mg/kg TM applied to the "simulated mouse" having PK governed by Equation (11) and SCM-G pharmacodynamics given by Equations (12), (2), (3), and (4) resulted in a steady state xenograft tumor burden of 100 mm^3. Note that the term "steady state" describes a tumor volume measurement taken at the same time each day. The bolus administration of TM resulted in periodic drug and tumor volume profiles, but measurements taken at 24-hour intervals were steady. A nonlinear model predictive control (NMPC) algorithm was developed to regulate tumor volume (measured variable) by manipulating TM dose on a daily schedule. In the nominal case, this algorithm uses the previously developed PK model in equation (11) and the PD model from equations (2) to (4) and (12) to explicitly predict the future tumor size based on the current and past dosing profile. The delivery problem formulation facilitated using the "$\Delta\mathscr{U}$" structure for the NMPC problem. The standard two-norm squared control objective was employed (Morari and Ricker, 1994):

$$\min_{\Delta\mathscr{U}(k|k)} \|\Gamma_y[\mathscr{R}(k+1) - \mathscr{Y}(k+1|k)]\|_2^2 + \|\Gamma_u \Delta\mathscr{U}(k|k)\|_2^2 \qquad (15)$$

Here $\mathscr{Y}(k+1|k)$ is the predicted future tumor volume over the prediction horizon of length p, $\mathscr{R}(k+1)$ is the desired tumor volume over the same horizon, and $\Delta\mathscr{U}(k|k)$ is the (move horizon, m, length) trajectory of future TM dosing changes calculated to minimize the objective (15). The weighting matrices Γ_y and Γ_u establish the relative importance of setpoint tracking error and manipulated variable movement, respectively; in the present work, these were set to $\Gamma_y = 1$ and $\Gamma_u = 0$ for all simulations. Magnitude

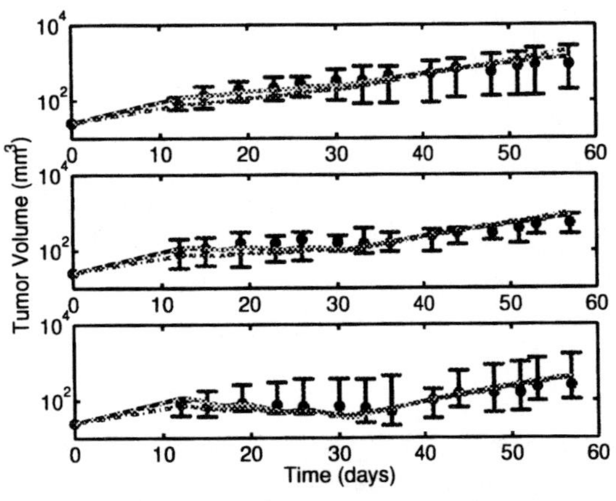

Fig. 5. Measured MCF-7 tumor xenograft data from tumor-bearing athymic mice (circles) vs. LCM-G (dash-dot), SCM-G (solid), and GM (dashed) model predictions at daily oral dose levels of 7.5 (top), 15 (middle) or 30 mg/kg (bottom) TM. Error bars represent a 95% confidence interval.

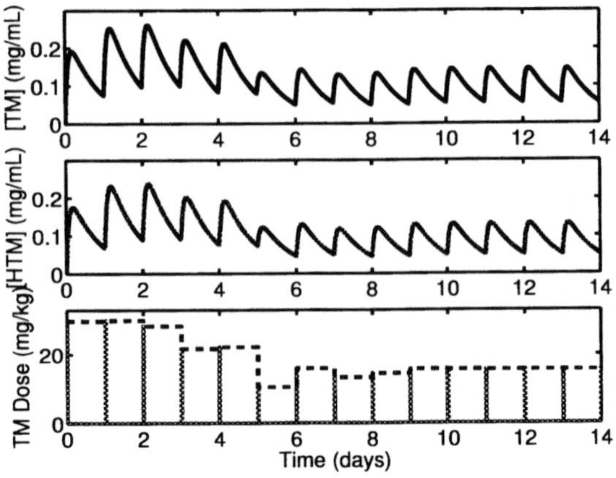

Fig. 6. TM plasma concentration (top), HTM plasma concentration (middle) and delivered TM oral dose (bottom) as a function of time. The dashed line in the bottom figure is the representation of NMPC results for easy viewing in longer simulations.

constraints were imposed to keep the daily TM dose within the bounds prescribed by the linearity in the PD effect with dose: $0 \leq \mathscr{U}(k|k) \leq 30\frac{\text{mg}}{\text{kg}}$. The resulting nonlinear programming problem was solved using the *fmincon* function in MATLAB (©2004, The MathWorks, Natick, MA). The studies evaluated in the remainder of this section are nominal control studies, with plant and model equivalent. The goal was to evaluate the performance of an NMPC controller formulation on a representative cancer chemotherapy problem.

Chemotherapy dosing differs from traditional engineering control systems in the manner of drug delivery. TM was delivered as a bolus each day by oral gavage in animal studies; this is conceptually similar to administering the chemotherapeutic as a pill. Hence, the "stepwise-constant" manipulated variable was replaced by a daily bolus. TM plasma concentration, HTM plasma concentration, and TM dose are shown in Figure 6. Each day a bolus was given (bottom, solid), and the corresponding TM (top) and HTM (middle) concentrations showed the characteristic rapid concentration rise and then decreasing plasma concentration over the course of the day. Of note is the accumulation of TM and HTM in the plasma; a pseudo-steady state was reached after about 10 days. The manipulated variable plot (bottom) in Figure 6 provides the actual dosing profile (solid) and this dose plotted in the traditional stepwise-constant shape (dashed). For ease of viewing in longer simulations, the stepwise-constant shape is plotted in future figures; *this stepwise shape is representative of the TM dose and is not the exact dosing profile.*

Controller tuning, *i.e.* setting the move and prediction horizons, is often accomplished on an *ad hoc* basis. Three candidate tuning sets for this case study are shown in Figure 7. The "simulated mouse" initially

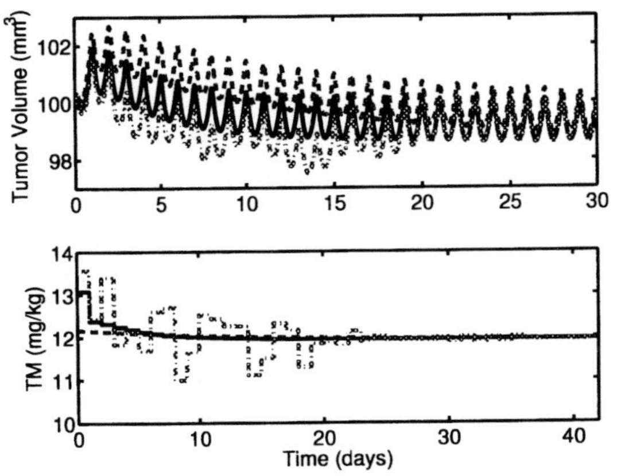

Fig. 7. Tumor volume (top) and TM dose representation (bottom) versus time for $m = 1$ with $p = 8$ (solid) and $p = 22$ (dashed) as well as $m = 2$, $p = 8$ (dash-dot).

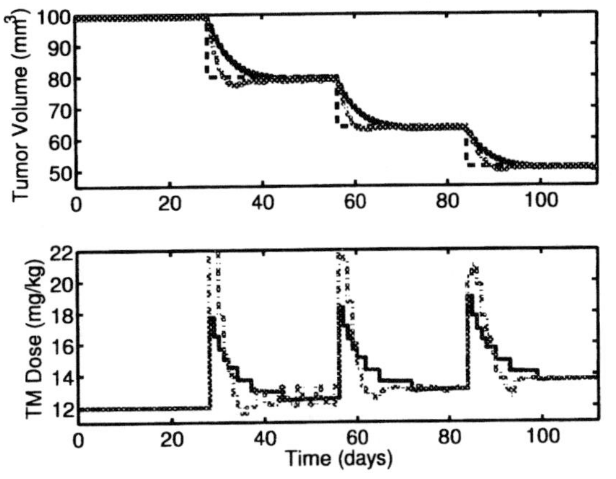

Fig. 8. Tumor volume (top), reference volume (dashed, top) and TM dose representation (bottom) as a function of time. Controller tunings are $m = 1$ (solid) and $m = 2$ (dash-dot) with $p = 8$.

had a 100 mm^3 tumor burden and TM and HTM plasma concentrations of 0; the NMPC controller was started with a reference value of 100 mm^3. Tumor measurements were taken at 24-hour intervals. All three tunings resulted in increased TM doses to offset the initial rise in tumor volume due to unperturbed tumor growth. As drug plasma concentrations rose over the first few days, tumor volumes started to decrease. After 56 days, the controller had achieved a pseudo-steady tumor burden of 100 mm^3. Settling to within 1% of the desired trajectory ($t_{99\%}$) for the $m = 1$, $p = 22$ case required markedly more time than the other two tests; this was rejected as a suitable controller tuning. From the bottom panel of Figure 7, the aggressiveness of the $m = 2$ controller can be observed; dose levels were changed about the mean value for approximately 3 weeks. The $m = 1$, $p = 8$ profile offered a smooth dosing profile and comparatively rapid return to the 100 mm^3 reference.

To determine which of the other two settings ($m = 1, p = 8$ or $m = 2, p = 8$) provided superior performance, a second trajectory was simulated. The ability of the different NMPC controller tunings to reduce tumor volume is shown in Figure 8. At four-week intervals (corresponding approximately to a "cycle" of traditional cancer chemotherapy), the tumor volume reference was decreased by 20% of the then-current value. The ability of the two controller tunings to track the reference changes from 100 to 80, 64, and 51.2 mm^3, respectively, was quite good. As expected, the more aggressive $m = 2$ tuning reduced tumor volume more quickly, but this led to increased undershoot with each step change and oscillatory dosing profiles. While the aggressive $m = 2$ trajectory reduced tumor size faster, the smoother $m = 1$ dosing profile is more acceptable from a clinical perspective.

An inherent assumption to this point is the ability to collect a tumor measurement on a daily basis. While this might be possible in animal studies at increased expense (using straightforward digital caliper measurements), the cost and impact on patient quality of life are prohibitive in human trials. The most frequent rate at which a tumor size measurement may be possible in humans is weekly without either the development of a new measurement technology or a reduction in time and cost associated with current measurements. In the nominal case, the following problem was posed. On day 0, and at 7-day (days 7, 14, etc.) or 14-day (days 14, 28, etc.) intervals, a tumor volume measurement was taken. The controller was formulated with $m = 1$ and $p = 14$ or $p = 21$ for weekly and biweekly measurement, respectively. The updated dose level calculated by the NMPC algorithm was delivered daily until the next measurement was obtained, and the controller computation was repeated. The results are shown in Figure 9. The weekly measurement had a $t_{99\%}$ value of 2 weeks, demonstrating excellent reference tracking performance. The slower measurement rate resulted in undershoot with a $t_{99\%}$ of 4 weeks which, as expected, was worse than the settling time from weekly updates. In the case that biweekly measurement is all that is available, the control algorithm can successfully reduce tumor volume.

7. CONTROLLER MODEL MISMATCH

The ability of the NMPC algorithm to perform despite mismatch between the actual and controller models was evaluated. The "simulated mouse" was taken as the actual system, and either the LCM-G or GM served as the controller model. As a tumor possesses a population of cells in different phases of the cell cycle and the overall growth dynamics saturate, these choices represent a fair depiction of overall tumor behavior. Furthermore, the robustness of developed control systems based on models failing to capture either nonlinear dynamic or cell-cycle information was assessed. Tracking tumor volume deviations with

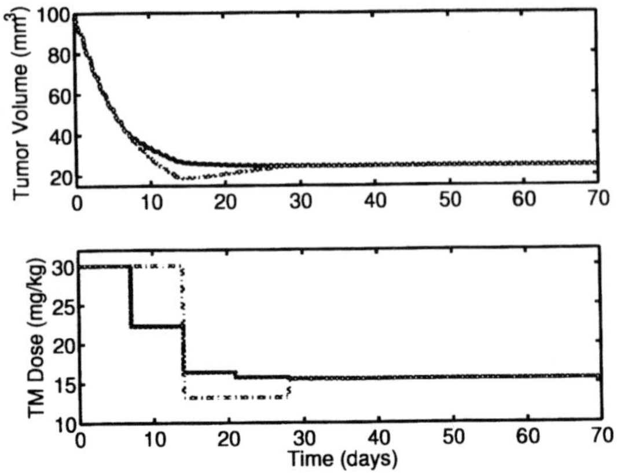

Fig. 9. Tumor volume (top) and TM dose representation (bottom) as a function of time. Tumor volumes are measured weekly (solid) or biweekly (dash-dot) under NMPC with $m = 1$, $p = 14$ (solid) or $p = 21$ (dash-dot). The tumor volume reference value is 25 mm^3.

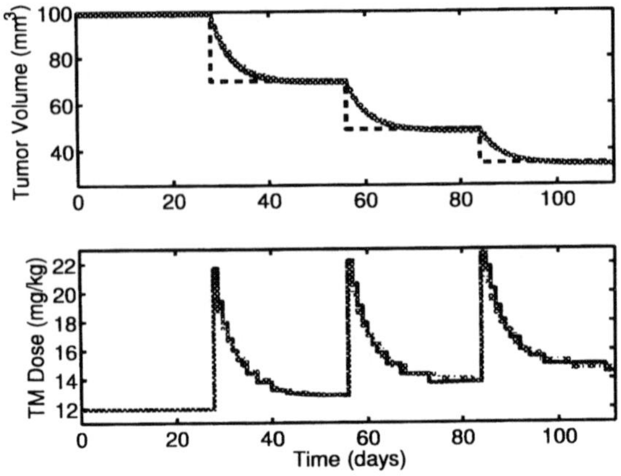

Fig. 10. Tumor volume (top), reference volume (dashed, top) and TM dose representation (bottom) as a function of time. Tuning parameters were chosen as m = 1 and p = 8 for both the nominal (NC-G, solid) and mismatch (MC-Gu, dash-dot) cases.

the LCM-G as the controller model was not possible, however, due to the differences in dynamic behavior between the LCM-G and SCM-G models over the dose range of interest. Further analysis with the linear cell-cycle model was discontinued as stability could not be guaranteed for progressive reductions in tumor volume.

Using the steady state input for the "simulated mouse" (12.0 $\frac{mg}{kg\ day}$ TM) resulted in a GM tumor burden of 104 mm^3. The controller model and the "simulated mouse" were initialized with the same steady state dose to allow manipulation of the daily dose over the full [0 $\frac{mg}{kg}$, 30 $\frac{mg}{kg}$] range. To assess controller performance under mismatch and to determine if the additional complexity of the SCM-G was necessary for high-performance control, closed-loop simulations were compared for two cases: a four-stage trajectory for tumor volume reduction with daily updates ($m = 1, p = 8$); and biweekly tumor volume updates ($m = 1; p = 21$) after a single reduction in desired tumor volume.

NMPC controller simulations for tracking stepwise tumor volume reductions under mismatched conditions employed desired tumor volume decrements of 30% between stages. This corresponded to reference trajectories of 100, 70, 51, and 34.3 mm^3 over each four-week cycle. The motivation for this alteration was to better utilize the allowable dosing range toward the maximum region (30 $\frac{mg}{kg}$); requiring greater tumor reduction during each step accomplished that task. The results are shown in Figure 10. The NMPC with the mismatched GM as the controller model initialized with the same input (MC-Gu) did not cause noticeable deviations from the nominal case with the SCM-G as the controller model (NC-G). The sum squared error increased by 10% (2603 vs. 2853 mm^6); this was primarily caused by the undershoot in the MC-Gu response profile. Differences in total drug delivered were negligible, though NC-G delivery profiles employed greater initial drug delivery in response to reference step changes. This resulted in a $t_{99\%}$ that was typically one day faster, occurring on day 14 (NC-G) vs. day 15 (MC-Gu) of each reference step. The MC-Gu also resulted in reference trajectory undershoot ranging from 1-2%. Both controllers led to tumor volumes that setttled to within 1% of the reference within the 4 weeks except the MC-Gu on the third stage (51 mm^3); 99% settling in the output occurred if the step was extended for longer than 28 days. Furthermore, the undershoot did not affect the ability of the MC-Gu to track the next tumor decrement.

Both controllers were also compared assuming tumor updates occurred every two weeks to better simulate clinical practice, and the results are shown in Figure 11. The input and output profiles were similar with reference deviation and total drug administered within 1% for the two systems. The $t_{99\%}$ settling time using the MC-Gu was markedly different, however, taking 111 days versus 29 for the NC-G.

Overall controller performance with the MC-Gu as the controller model did not differ significantly from controller performance using the NC-G, though the MC-Gu system displayed increased settling time in most cases. This agreement in performance was not unexpected. For the case of a G-phase-specific drug like TM, the SCM-G can be closely approximated as a Gompertz description at large tumor volumes. As the tumor size increases, a greater fraction of cells reside in G-phase; the SCM-G predicts 90% of the total cells in G-phase at a size of 100 mm^3 (increasing to 95% by 1000 mm^3). With the bulk of the population in G-

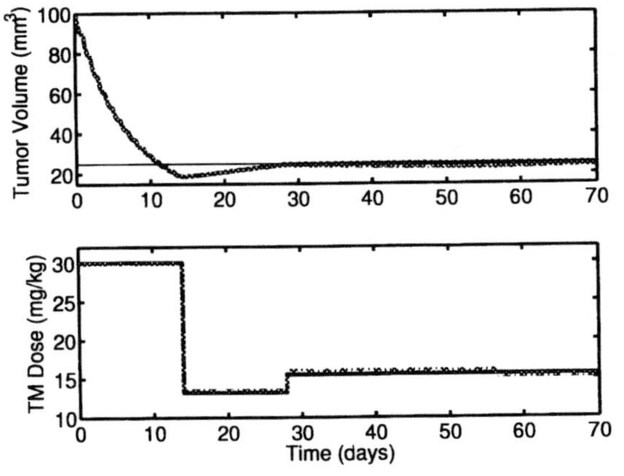

Fig. 11. Tumor volume (top) and TM dose representation (bottom) as a function of time. Tumor volumes are measured biweekly under NMPC with $m = 1$, $p = 21$ for the nominal (NC-G, solid) and mismatch (MC-Gu, dash-dot) cases. The tumor volume reference value is 25 mm^3.

phase along with the kill term, the dynamic behavior of the GM and SCM-G should be similar, as observed.

8. THEORETICAL STUDY: M-PHASE KILL

To test this hypothesis, a theoretical study was conducted using the same pharmacokinetic and pharmacodynamic data as the earlier studies within this paper. However, the PD effect was moved from the G-phase to the M-phase (where approximately 10% of the tumor cells reside), resulting in the following PD model (SCM-M):

$$\frac{dX_G}{dt} = -k_G X_G \ln\left(\frac{\theta}{N}\right) + 2k_M X_M \ln\left(\frac{\theta}{N}\right) \quad (16)$$

$$\frac{dX_S}{dt} = -k_S X_S + k_G X_G \ln\left(\frac{\theta}{N}\right)$$

$$\frac{dX_M}{dt} = -k_M X_M \ln\left(\frac{\theta}{N}\right) + k_S X_S - k_D X_G (X_1/V_1 + cX_2/V_2) \quad (17)$$

here equations (12) and (3) have been changed to equations (16) and (17), respectively, to reflect a drug acting primarily in the M-phase. The cell kill rate, k_D was determined as before for the four dosing levels (60, 30, 15, and 7.5 $\frac{mg}{kg}$). The rates for the three lower dosing levels were still in close agreement ($0.00161 \pm 0.00015 \frac{mL}{\mu g\,hr}$) while the kill rate corresponding to doses of 60 $\frac{mg}{kg}$ was again much lower (0.00090 $\frac{mL}{\mu g\,hr}$). This allowed the daily dosing range to be fixed as before, ($0 \leq u(t) \leq 30 \frac{mg}{kg}$) and the cell kill rate was set at 0.00161 $\frac{mL}{\mu g\,hr}$. A change in cell-phase specificity does not alter the GM PD effect, so k_D remained at 0.00325 $\frac{mL}{\mu g\,hr}$.

Figure 12 shows the predicted PD fit using the SCM-M, GM (including drug effect), and actual data. The SCM-M was able to capture the tumor response profile and coincides with the Gompertz predictions, demonstrating similar behavior for short dosing periods. In addition, it is interesting to note that the SCM-G also fit this data. All three models display similar tumor volume responses for the dosing regimen and time periods in question, despite each having a different location for kill term effect.

Similar to the previous mismatch studies, the SCM-M was set as the "simulated mouse" and tested using either the mismatch case Gompertz model as the controller model initialized with equivalent input (MC-Mu) or the nominal case saturating rate cell-cycle model with M-phase kill (NC-M) as the controller model. To maintain a steady 100 mm^3 xenograft tumor burden in the SCM-M, a daily dose of 9.5 $\frac{mg}{kg}$ was necessary, corresponding to a 262 mm^3 tumor volume for the GM. Initialization employed the same steady state dose to enable variable manipulation over the entire dosing range. The two test cases comparing stepwise reduction in tumor volume and biweekly dosing updates were simulated to determine whether performance similarity between the NC-G and MC-Gu controllers was a result of a G-phase chemotherapeutic acting on a predominantly G-phase tumor.

The results of the four-stage reduction in tumor volume simulation are shown in Figure 13. The MC-Mu controller was unable to follow the desired trajectory for the entire simulation, never reaching a stable tumor volume within 1% of desired by the time of the next decrement. The total drug delivered by the MC-Mu was 5% less than the NC-M system. Furthermore, the GM-based controller was less aggressive, delivering less drug initially, but

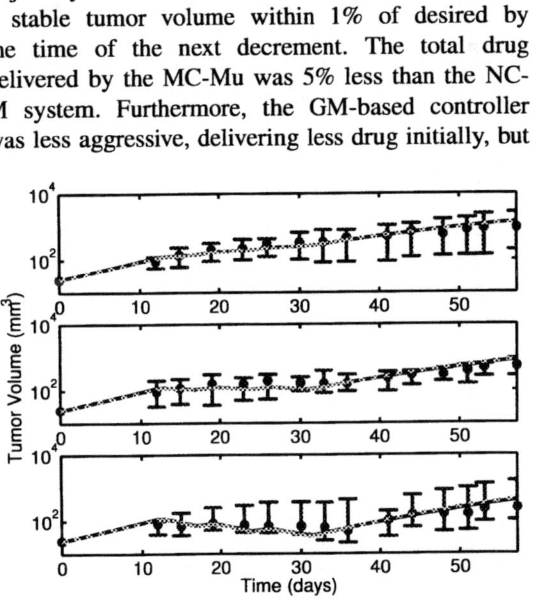

Fig. 12. Measured MCF-7 tumor xenograft data for tumor-bearing athymic mice (circles) vs. model predictions for the SCM-M (solid), and the GM (dashed) dosed orally at 7.5 (top), 15 (middle) or 30 mg/kg (bottom) TM per day. Error bars represent a 95% confidence interval.

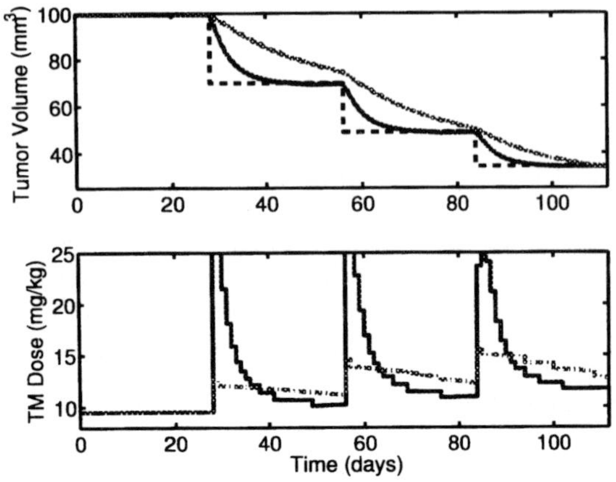

Fig. 13. Tumor volume (top), reference volume (dashed, top) and TM dose representation (bottom) as a function of time. Tuning parameters were chosen as m = 1 and p = 8 for both the nominal (NC-M, solid) and mismatch (MC-Mu, dash-dot) cases.

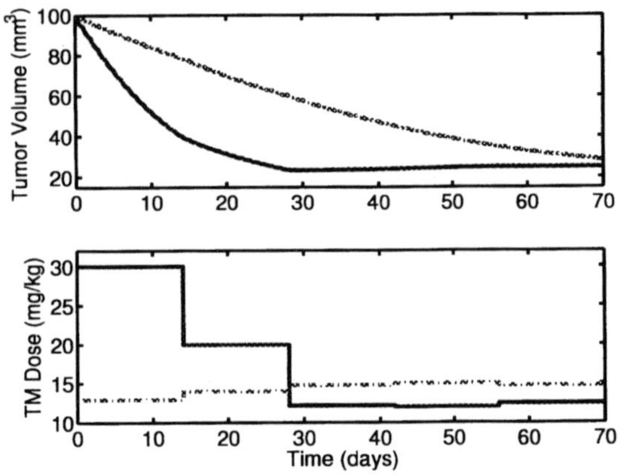

Fig. 14. Tumor volume (top) and TM dose representation (bottom) as a function of time. Tumor volumes are measured biweekly under NMPC with $m = 1$, $p = 21$ for the nominal (NC-M, solid) and mismatch (MC-Mu, dash-dot) cases. The tumor volume reference value is 25 mm^3.

maintaining an elevated dose level for an extended period of time. This significantly different dosing profile contributed to the MC-Mu system falling short of the desired trajectory by 1-5% at the end of each stage. A final point of concern is the trajectory taken by the MC-Mu controller following the step from 100 to 70 mm^3. The response was more sluggish, likely a result of the initial conditions for the model's operating state. While the NC-M maintains a steady tumor burden of 100 mm^3 at a dose of 9.5 $\frac{mg}{kg}$, the same dosing resulted in a GM predicted volume of 262 mm^3. At this operating state, small changes in dose were predicted to cause large alterations in the tumor volume, and the controller took a less aggressive approach toward the desired reference trajectory.

This point is further illustrated in Figure 14 where tumor dosing is updated biweekly. The NC-M controller rapidly drove the tumor volume toward the desired trajectory of 25 mm^3, settling to within 1% of the desired setpoint by 28 days. The MC-Mu controller, however, resulted in a trajectory that had not reached the reference criteria by 72 days. The dosing profile proves that the MC-Mu controller yielded a less aggressive response, a side effect of the large steady state tumor volume at the starting dose.

To improve controller performance, the GM was initialized with identical tumor volume (MC-My) to the "simulated mouse" (100 mm^3) instead of equivalent drug input (9.5 $\frac{mg}{kg}$). For this tumor burden, the GM required a steady state drug delivery of 12.0 $\frac{mg}{kg}$. Maintaining both dosing inputs within the constraints, $0 \leq \mathcal{U}(k|k) \leq 30\frac{mg}{kg}$, required redefining the input constraints as $2.5 \leq \mathcal{U}(k|k) \leq 27.5\frac{mg}{kg}$. The reduction in range for lower dosing levels did not significantly impact tumor treatment as the manipulated variable did not enter this range for all investigated simulations. Dose predictions within the excluded upper range, however, were implemented immediately following negative steps in the reference trajectory and contribute to the initial, rapid tumor size reduction.

Revisiting the constraint development, it was observed that doses within the range of 7.5 to $30\frac{mg}{kg}$ resulted in similar tumor-kill rates while k_D at doses of $60\frac{mg}{kg}$ were markedly reduced, possibly as a result of receptor saturation. The dose at which saturation begins (between $30 - 60\frac{mg}{kg}$) was not explicitly known. If one assumed that doses slightly above $30\frac{mg}{kg}$ would not achieve saturation, the following magnitude constraints can be enforced: $2.5 \leq \mathcal{U}(k|k) \leq 30\frac{mg}{kg}$. A performance comparison between the NC-M, MC-Mu, and GM controllers initialized at equivalent tumor volumes (MC-My), can be found in Figure 15. The NC-M response is identical to that for the first step in Figure 13 (analyzed above). In response to a decrease in desired tumor volume ($100 \rightarrow 70$ mm^3), the MC-My controller was more aggressive than the MC-Mu controller. Both had trajectories which undershoot and eventually settle to within 1% of the desired tumor volume, however the MC-My controller reached the desired tumor volume faster and had the smaller $t_{99\%}$.

The stepwise tumor volume reduction and biweekly update test cases were simulated for the NC-M and the MC-My controllers; results are shown in Figures 16 and 17, respectively. The MC-My controller dosed more aggressively than the MC-Mu controller for the four-stage trajectory, and the MC-My system reached the desired reference on the same day as the NC-M system. Undershoot occurred for each step and settling to within 1% of the desired value was not achieved

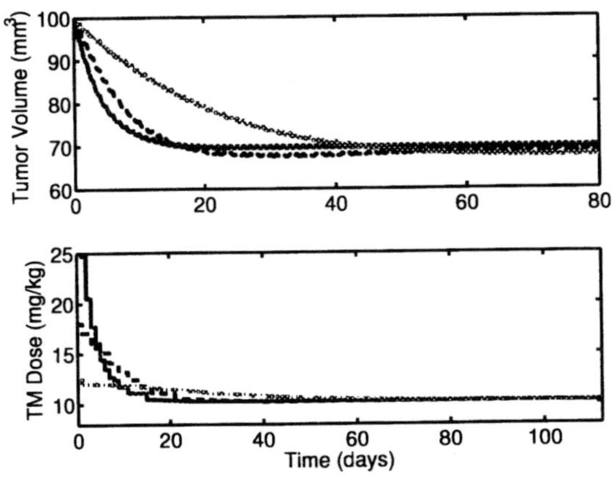

Fig. 15. Tumor volume (top) and TM dose representation (bottom) as a function of time. For tuning parameters of $m = 1$ and $p = 8$, the predicted trajectory is shown for the nominal (NC-M, solid) and the mismatch (MC-My, dashed; MC-Mu, dash-dot) cases. Reference tumor volume was 70 mm^3.

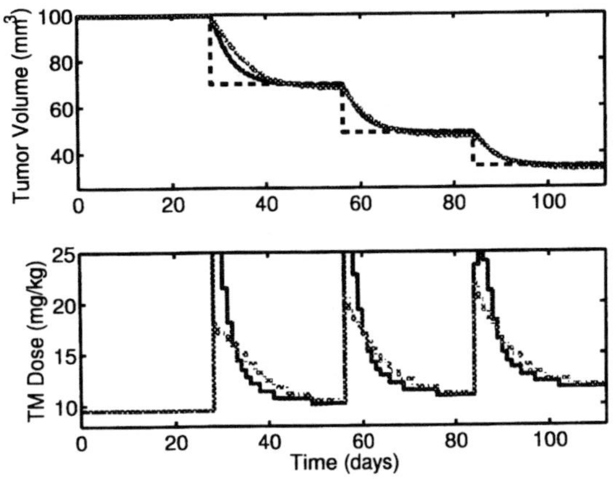

Fig. 16. Tumor volume (top) and TM dose representation (bottom) as a function of time. For tuning parameters of $m = 1$ and $p = 8$, the predicted trajectory is shown for the NC-M (solid) and MC-My (dash-dot) systems. Reference tumor volume decreased every 4 weeks, starting at 100 mm^3.

during any of the four-week stages. Similarly, a more aggressive trajectory was taken by the MC-My system in the biweekly update study. Undershoot resulted during weeks 3-4 with oscillations around the desired trajectory ceasing by week 16 (not shown). In both cases superior controller performance was achieved with the MC-My vs. the MC-Mu (results in Figures 13 and 14, provided that undershoot in tumor volume can be tolerated). These new controller settings were not able to outperform the NC-M, taking more time to settle in all studies. Furthermore, these results demonstrate the importance of incorporating cell-cycle information, especially if the fraction of the susceptible tumor population is small compared to the total tumor population (as is the case for an S- or M-phase targeted drug).

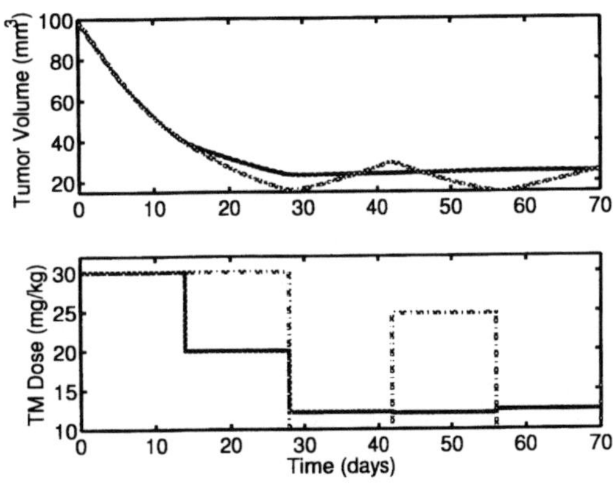

Fig. 17. Tumor volume (top) and TM dose representation (bottom) as a function of time. Tumor volumes are measured biweekly under NMPC with $m = 1$, $p = 21$ for the NC-M (solid) and the MC-My (dash-dot) cases. Reference tumor volume was 25 mm^3.

9. SUMMARY

A NMPC controller was designed to dose TM, a cycle-specific chemotherapeutic, to mice implanted with ER+ breast cancer xenografts. This controller is based on novel PK and PD models of TM dose and MCF-7 tumor growth, and the algorithm is ready for testing in a preclinical setting on mice. Adaptation to actual breast cancer patients would require TM PK and PD data from humans, but the model predictive control framework provides the groundwork for a patient-tailored dosing algorithm capable of constraining drug delivery within safety-induced bounds and responding to changes in patient response.

The closed loop system designed with the linear cell-cycle model in the controller was unable to track the four-stage trajectory for tumor volume reduction. Employing the less detailed nonlinear GM in the controller design resulted in similar trajectory tracking results when compared to controllers designed with the more complex SCM. Controller performance was dependent on the cycle phase of drug effect, with more significant performance deviation between the nominal and mismatch systems when the drug effect targeted a smaller portion of the the overall tumor population (M vs. G). Furthermore, given the rate at which tumor volume measurements are taken in the clinic, failure to track cell phase for cycle-specific chemotherapeutics can lead to gross misestimates in tumor volume reductions. Future foci include toxicity modeling/monitoring (*e.g.* white blood cell count) and development of a multi-rate NMPC algorithm with state estimation to allow toxicity- and efficacy-driven changes to the dosing profile in real-time.

10. ACKNOWLEDGMENTS

The authors would like to acknowledge funding in the form of a Whitaker Foundation Graduate Student Fellowship (JAF) and the Fulton C. Noss faculty fellowship (RSP). The animal studies from (Conley *et al.*, 1999) were funded by the NIH/NCI (N01-CM27752).

11. REFERENCES

Akaike, A. (1979). A Basian Extension of the Minimal AIC Procedures of Autoregressive Model Fitting. *Biometrika* **66**, 237.

Asachenkov, A., G. Marchuk, R. Mohler and S. Zuev (1994). *Disease Dynamics*. Birkhäuser. Boston.

Bajzer, Ž., M. Marušić and S. Vuk-Pavlović (1996). Conceptual frameworks for mathematical modeling of tumor growth dynamics. *Math. Comput. Model.* **23**(6), 31–46.

Conley, B. A., T. S. Ramsland, D. L. Sentz, S. Wu, D. M. Rosen, M. Wollman and J. L. Eiseman (1999). Antitumor activity, distribution, and metabolism of 13-*cis*-retinoic acid as a single agent or in combination with tamoxifen in established human MCF-7 xenografts in mice. *Cancer Chemother. Pharmacol.* **43**, 183–197.

D'Argenio, D. Z. and A. Schumitzky (1997). *ADAPT II Users Guide: Pharmacokinetic and Pharmacodynamic Systems Analysis Software*. Biomedical Simulations Resource, University of Southern California. Los Angeles, CA.

Ellis, M. J. and S. M. Swain (2001). Steroid hormone therapies for cancer. In: *Cancer Chemotherapy and Biotherapy: Principles and Practice* (B. A. Chabner and D. L. Longo, Eds.). 3rd ed.. pp. 103–13. Lippincott Williams and Wilkins.

Florian Jr., J. A., J. L. Eiseman and R. S. Parker (2003). Approximating cancer tumor growth dynamics using cell-cycle models in series. In: *AIChE Annual Meeting*. paper 439ab. San Francisco, CA.

Harrold, J. M., J. L. Eiseman, W. C. Zamboni and R. S. Parker (2003). A clinically relevant mixed integer approach to cancer chemotherapy treatment design. In: *AIChE Annual Meeting*. paper 478a. San Francisco, CA.

Kozusko, F., P. Chen, S. G. Grant, B. W. Day and J. C. Panetta (2001). A mathematical model of in vitro cancer cell growth and treatment with the antimitotic agent curacin A. *Mathimatical Biosciences* **170**, 1–16.

Ledzewicz, U. and H. Schättler (2002). Optimal bang-bang controls for a two-compartment model in cancer chemotherapy. *J. Optimization Theory Appl.* **114**(3), 609–637.

Lee, G. S., K. S. Ryu, S. P. Kim, S. E. Namkoong and K. T. Han (2002). Multiparametric flow cytometric analysis in a breast cancer cell line (MCF-7). *J. Obstet. Gynaecol. Res.* **28**(3), 141–8.

Lopes, N. M., E. G. Adams and B. K. Bhuyan (1993). Cell kill kinetics and cell cycle effects of taxol on human and hamster ovarian cell lines. *Cancer Chemoth. Pharm.* **32**(3), 235–242.

Martin, E. A., P. Carthew, I. N. White, R. T. Heydon, M. Gaskell, R. J. Mauthe, K. W. Turteltaub and L. L. Smith (1997). Invesitgation of the formation and accumulation of liver DNA adducts in mice chronically exposed to tamoxifen. *Carcinogenesis* **18**(11), 2209–15.

Martin, R. and K. L. Teo (1994). *Optimal Control of Drug Administration in Cancer Chemotherapy*. World Scientific. River Edge, NJ.

Martin, R. B. (1992). Optimal control drug scheduling of cancer chemotherapy. *Automatica* **28**, 1113–1123.

Morari, M. and E. Zafiriou (1989). *Robust Process Control*. Prentice-Hall, Englewood Cliffs, NJ.

Morari, M. and N. L. Ricker (1994). *Model Predictive Control Toolbox*. The MathWorks, Inc.. Natick, MA.

Norton, L. (1988). A Gompertzian model of human breast cancer growth. *Cancer Res.* **48**, 7067–7071.

Panetta, J. C. and J. Adam (1995). A mathematical model of cycle–specific chemotherapy. *Math. Comput. Model.* **22**(2), 67–82.

Riesberg, M., C. Kasper, K. F. Reardon and T. Scheper (2001). Flow cytometry in biotechnology. *Appl. Microbiol. Biotechnol.* **56**, 350–60.

Robinson, S. P., S. M. Langan-Fahey, A. J. Delinda and V. C. Jordan (1991). Metabolites, pharmacodynamics, and pharmacokinetics of tamoxifen in rats and mice compared to the breast cancer patient. *Drug Metabolism and Disposition* **19**(1), 36–43.

The American Cancer Society (2004). Cancer facts & figures: 2004. URL: http://www.cancer.org/downloads/STT/CAFF_finalPWSecured.pdf.

ELSEVIER
IFAC
PUBLICATIONS
www.elsevier.com/locate/ifac

Using Optimization to Detect Snowball Effects

by

Thomas J. Mc Avoy
Institute for Systems Research/Department of Chemical Engineering
University of Maryland
College Park, MD 20742

Abstract: In process plants with recycle streams some level control architectures are inoperable. Their use leads to large excursions in manipulated and/or controlled variables and this behavior has been termed the snowball effect. The snowball effect is a steady state phenomenon and it can be analyzed using steady state process models. In this paper a steady state mixed integer nonlinear programming (MINLP) approach is used to analyze for large excursions in process variables. This MINLP approach can be used to detect the likely occurrence of a snowball effect in a plant and to develop control architectures that can avoid this problem. The optimization-based approach is illustrated on a three-reactor three-distillation column plant taken from the literature. *Copyright © 2004 IFAC*

Introduction

The design of plantwide control systems has received increasing attention in the last few years. Tighter plant designs, due to such considerations as energy integration and reduced emissions, are leading to more challenging control problems. Several authors have proposed approaches to the problem of configuring plantwide control systems. These approaches range from those based on optimization[1-3] to those that are heuristic in nature[4]. In all plantwide control designs, it is necessary to control variables that are non-self regulating in nature. The most common process example of such variables is liquid level in a vessel. Buckley[5] was the first to address the problem of controlling inventory in a plant. If one considers only controlling liquid inventories, then controlling liquid levels can result in problems with other variables in the process. These problems have been labeled as snowball effects by Luyben and co-workers[4,6].

Luyben and Luyben[6] presented a control study of a three-reactor/three-distillation tower process that is shown in Fig. 1. Most control schemes for this plant were found to be inoperable, and these schemes exhibited the snowball effect. As discussed by Luyben et.al.[4] the snowball effect arises in systems with recycle flows when small changes in one variable result in extremely large changes in other process variables. Their results indicate that in designing a plantwide control system, liquid level loops need to be configured carefully so that a snowball effect does not occur. Luyben et.al.[4] point out that the snowball effect is essentially a steady state phenomenon. Thus, whether or not it is likely to occur can be analyzed using a steady state model that includes integrating variables. This paper proposes an approach, using mixed integer nonlinear programming (MINLP) on steady state process models coupled with a model for the integrating variables, that can detect whether the snowball effect is likely to occur in a process. Once a potential snowball problem is detected, then a second MINLP optimization can be solved to determine which additional variables need to be controlled to avoid the snowball problem. Using the results of the optimization one can systematically select level control systems together with any additional variables that need to be controlled to avoid snowball effects.

This paper is organized as follows. First the development of the steady state model used is discussed. Next the MINLP optimization approach is discussed and applied to the three-reactor/three-distillation tower process[6]. It is shown that snowball effects will occur in this plant. Then, a second optimization problem is proposed to determine if additional variables can be controlled to avoid snowball effects. Finally conclusions are presented.

Development of Steady State Model

The steady state model used for the snowball analysis consists of two parts. The measurement vector, y, is separated into y_N, for non-integrating measurements, and y_I for the integrating measurements. For y_N a typical nonlinear steady state process model is used:

$$0 = f(x,u,d_j) \qquad (1)$$

$$y_N = g(x,u,d_j) \qquad (2)$$

where x is the state vector, u is the vector of manipulated variables and d_j is the disturbance being considered. The y_I measurements are not included in a steady state model. Here they are included and it is assumed that a linearized model for their rate of change is acceptable for analyzing snowball effects. This model can be developed from a linearized dynamic model using the approach published by Arkun and Downs[7]. Alternatively it can be approximated as done by Mc Avoy and Miller[8]. The model for y_I has the form:

$$\dot{y}_I = K_I u + d_{j,I} \qquad (3)$$

where $d_{j,I}$ is the effect of the disturbance on the rate of change of the integrating measurements.

In order to use Eqns. 1 to 3 to analyze for snowball effects, the plant under consideration is assumed to have only integrating poles, and no right half plane poles. If the plant has right half plane poles then it must be stabilized to use the approach given below. The problem that is solved involves controlling at least all of the integrating variables in the plant. The approach taken is similar to that presented by Mc Avoy[3] for synthesizing plantwide control systems. Control of y_I requires that:

$$\dot{y}_I = K_I u + d_{j,I} = 0 \qquad (4)$$

Equation 4 assumes steady state control in which $\dot{y}_I = 0$, and in the optimization approach discussed below u is assumed to be a valve movement. In analyzing for potential snowball effects all potential disturbances need to be considered, one at a time.

Optimization-Based Analysis of the Snowball Effect

The optimization problem solved to detect the snowball effect involves determining which manipulated u's should be used to control the y_I's for a given disturbance, and whether any product measurements exceed reasonable bounds due to the control effort applied. One measure of how good a particular level control architecture is can be taken to be how much the manipulated valves it uses have to move from their original steady state positions when d_j occurs. A system that requires large changes in valve positions is deemed inferior to one requiring smaller changes. In formulating an objective function for detecting the snowball effect, both positive and negative changes in valve movement are penalized. To account for both changes, u is re-written in terms of v^+ and v^- as:

$$u = v^+ - v^- \tag{5}$$

where the elements of v^+ and v^- are all positive. To emphasize that valve movements are being considered, the symbol v is used in place of u. It is assumed that the v's all have the same units, e.g. between 0 and 100%. Then, an optimization problem to detect the snowball effect can be formulated as shown in Eqn. 6. In Eqn. 6 $v^+_{k,\max}$ and $v^-_{k,\max}$ are the maximum changes permitted for the valve movements away from their steady state values, Δy_{min} and Δy_{max} are the minimum and maximum allowable changes in the product measurements, κ is equal to the number of valves in the plant, and M is equal to the number of integrating measurements. The objective function is the sum of the absolute values of the movements of all of the valves used for controlling the integrating variables. The first equality constraint involves control of the integrating variables, Eqn. 4. The second equality constraint enforces steady state. The first inequality constraints involve product measurements. These constraints enforce reasonable limits on the excursions of these measurements after an upset. How to choose these limits is discussed below. The next 2 constraints deal with saturation of the manipulated variables. If a valve is selected $(z_k^+, z_k^- \neq 0)$, then it is constrained not to saturate either open or closed.

$$\begin{aligned}
&\min_{v^+, v^-, z^+, z^-} \quad \sum_{k=1}^{\kappa} (v_k^+ + v_k^-) \\
&s.t. \\
&K_I(v^+ - v^-) + d_{j,I} = 0 \\
&0 = f(x, v^+ - v^-, d_j) \\
&\Delta y_{\min} \le \Delta g(x, v^+ - v^-, d_j) \le \Delta y_{\max} \\
&v_k^+ \le z_k^+ v_{k,\max}^+ \quad (k = 1:\kappa) \\
&v_k^- \le z_k^- v_{k,\max}^- \quad (k = 1:\kappa) \\
&\sum_{k=1}^{\kappa} (z_k^+ + z_k^-) = M \\
&v_k^+, v_k^- \ge 0 \\
&z_k^+, z_k^- \in \{0,1\}
\end{aligned} \tag{6}$$

It can be noted that in Eqn. 6 the integer variables appear in a linear manner. The solution of Eqn. 6 gives the M manipulated variables that have to move the least in order to keep the y_I measurements at their setpoints. If Eqn. 6 has a feasible solution then one can conclude that there is a control structure for the integrating variables that does not have a snowball problem for the plant under consideration. Integer cuts can be used to generate additional feasible control architectures for the integrating variables by ruling out earlier solutions. The application of Eqn. 6 to a plant shown in Fig. 1 is discussed next.

Application of Optimization Methodology

For th e process shown in Fig. 1, there are 9 integrating liquid level measurements, 3 each for the reactors, reflux accumulators, and reboilers. The steady state considered here is similar to that given by Luyben and Luyben[6] in their Table 12. However, their flows do not material balance exactly, and their plant model is not exactly at steady state. Flows were changed slightly from their results so that the plant started at steady state. Since stream densities are independent of composition, it is straightforward to develop Eqn. 3 for the nine integrating levels in this process. For example for reactor 1 the following model can be used:

$$A_1 c\dot{l}_1 = F_{OA} + D_1 + D_3 - F_1 \qquad (7)$$

where A_1 is the cross sectional area, c is the molal density and the right hand side consists of the various flows into and out of the reactor. Since $\dot{l}$ is set to 0, one can use ±1 for the gains in the first row of Eqn. 3, i.e. the $1/A_1c$ in each gain can be factored out. Then these gains can be scaled in terms of % flow. The product measurements consist of the mole fraction of C in the distillate from tower 2, y_{2C}, and the mole fraction of F in the bottoms from tower 3, x_{3F}. There are potentially 19 manipulated flows in the plant, but depending on the disturbance not all can be used. Changes in the fresh feed of A were used as a disturbance by Luyben and Luyben[6], and its elimination results in 18 manipulated variables. Three disturbances, given in reference[6], are considered here, and they are: 1. a 20% step change in feed composition of A; 2. a 50% step change in the specific reaction rate in the first reactor; and 3. a 20% step change in the feed flow of A. Since disturbances 1 and 2 do not involve flow changes, they are simpler to analyze than disturbance 3. For disturbances 1 and 2 control of y_I does not require any changes in u. All the manipulated variables, except the makeup of B and D (fod and fob), are constrained to be within 0 and twice their steady state values. The two makeup streams are constrained to be within ±50. Product mole fractions are constrained to be within ±.02. The reason for the somewhat loose specification on these bounds is that level control only cannot be expected to yield tight control of product composition. Rather what one is looking for is very large excursions in these variables which in turn will cause a snowball effect when one attempts to tightly control them. Note that one could add other variables to be constrained in the formulation such as reactor compositions. The second optimization problem discussed below does consider tighter control of the product variables.

For disturbances 1 and 3 no feasible solution to Eqn. 6 is found. Table 1 gives results for all three disturbances where it can be seen that extremely large changes in both the distillate composition in tower 2 and the bottoms composition in tower 3 occur for disturbances 1 and 3. The results for disturbance 3 are for a particular level configuration, but they are typical of what is calculated for other level schemes. By contrast disturbance 2 yields a feasible solution to Eqn. 6. The compositions of A and B in reactor 1 change significantly for disturbance 2 since it involves a 50% increase in the kinetic constant for reactor 1, but the product compositions are hardly affected. Actually it is not necessary to analyze all disturbances, since lack of feasibility for ***any*** disturbance indicates a potential snowball effect. For the Luyben and Luyben plant one can conclude that a snowball problem is likely. The optimization results indicate that one could not operate the plant by controlling only the 9 liquid levels when disturbance 1 or 3 occurs. If the plant can be operated in these cases, additional variables need to be controlled. To determine whether the snowball problem can be avoided a second optimization problem, discussed next, can be solved.

Additional Optimization Problem

To determine if the snowball problem can be avoided, additional integer variables, ζ, can be added to Eqn. 6 as follows:

$$\min_{v^+,v^-,z^+,z^-,\zeta} \sum_{k=1}^{\kappa}(v_k^+ + v_k^-) + m\sum_{j=1}^{J}\zeta_j$$

$$\begin{aligned}
&s.t.\\
&K_I(v^+ - v^-) + d_{j,I} = 0\\
&0 = f(x, v^+ - v^-, d_j)\\
&\Delta y^*_{\min}(1-\zeta) \le \Delta g(x, v^+ - v^-, d_j) \le \Delta y^*_{\max}(1-\zeta)\\
&v_k^+ \le z_k^+ v_{k,\max}^+ \ (k = 1:\kappa)\\
&v_k^- \le z_k^- v_{k,\max}^- \ (k = 1:\kappa)\\
&\sum_{k=1}^{\kappa}(z_k^+ + z_k^-) = M + \sum_{j=1}^{J}\zeta_j\\
&v_k^+, v_k^- \ge 0\\
&z_k^+, z_k^-, \zeta_j \in \{0,1\}\\
&m = scalar
\end{aligned} \tag{8}$$

Eqn. 8 allows for additional non-integrating measurements, y*, to be controlled in order to achieve feasibility. The y* measurements include product measurements as well as other key measurements such as reactor compositions. If an additional measurement is chosen, then its ζ_j is 1. If v^+ and v^- are scaled in the range 0-100, then choosing m to be large, e.g. 1000, forces the optimization to limit the number of additional measurements that are controlled for feasibility. For measurements of product variables very tight constraints can be used for $\Delta y^*_{\min}$ and $\Delta y^*_{\max}$ since these constraints reflect how one would like the plant to operate. One could even specify that certain measurements need to be controlled exactly to steady state. Some engineering judgement is involved in the specification of bounds on the process measurements. Eqn. 8 is a difficult MINLP problem whose solution is currently under investigation. Since Eqn. 8 involves a larger number of integer variables than Eqn. 6 and some of these integer variables appear in the objective function its is more difficult to solve than Eqn. 6 is. The solution of Eqn. 8 yields plantwide architectures that avoid snowball problems.

Some preliminary results on the solution of Eqn. 8 have been calculated. If the fresh feed of A, F_{0A}, is a disturbance then it appears that no feasible solution of Eqn. 8 can be found. When F_{0A} changes too many light components escape in the bottom of tower 1 and these components exit in the distillate from tower 2, lowering the composition of C in this stream. Two feasible solutions of Eqn. 8 have been found and one is the same as scheme 1c presented by Luyben and Luyben. The second solution is similar, but it does not adhere to the rule of fixing a flow in every recycle loop. Results for the application of Eqn. 8 to the Luyben and Luyben process will be presented at the conference.

Conclusions

An optimization-based approach to detecting the snowball effect in plantwide control designs has been presented. The approach makes use of a steady state model and it involves solving a MINLP problem. The results of this problem indicated whether snowball effects are likely to occur in a particular plant. If such problems are likely, then one can solve a second problem to determine which additional variables need to be controlled to avoid the snowball effect. Preliminary results on a three-reactor/three-distillation tower process have been presented. The optimization-based approach holds promise as a systematic method for determining plantwide control architectures that avoid snowball problems.

Literature Cited

[1] Narraway, L. T.; Perkins, J. *Ind. Eng. Chem. Res.* **1993**, 32, 2681-2692.

[2] Kookos, I., and Perkins, J., *J. Process Control*, **2002**, 12, 85-100.

[3] McAvoy, T., *Ind. Eng. Chem. Res.* **1999**, 38, 2985-2994.

[4] Luyben, W. L.; Tyreus, B. D.; Luyben M. L. *Plantwide Process Control*; Mc Graw-Hill, New York, N.Y., 1998.

[5] Buckley, P., *Techniques of Process Control*; John Wiley and Sons, New York, N.Y., 1964.

[6] Luyben, M. L.; Luyben W. L. *Ind. Eng. Chem. Res.*, **1995**, 34, 3885-3898.

[7] Arkun, Y.; Downs,J. *Computers Chem. Eng.*, **1990**, 14, 1101-1110.

[8] Mc Avoy, T.; Miller, R. *Ind. Eng. Chem. Res.*, **1999**, 38, 412-420.

Table 1 Results for Solving Equation 6

Case	Mol. Fract. C Tower-2	Mol. Fract. F Tower-3
Base	0.9713	0.9980
disturbance 1	0.7762	0.8115
disturbance 2	0.9716	0.9980
disturbance 3	0.2398	0.3798

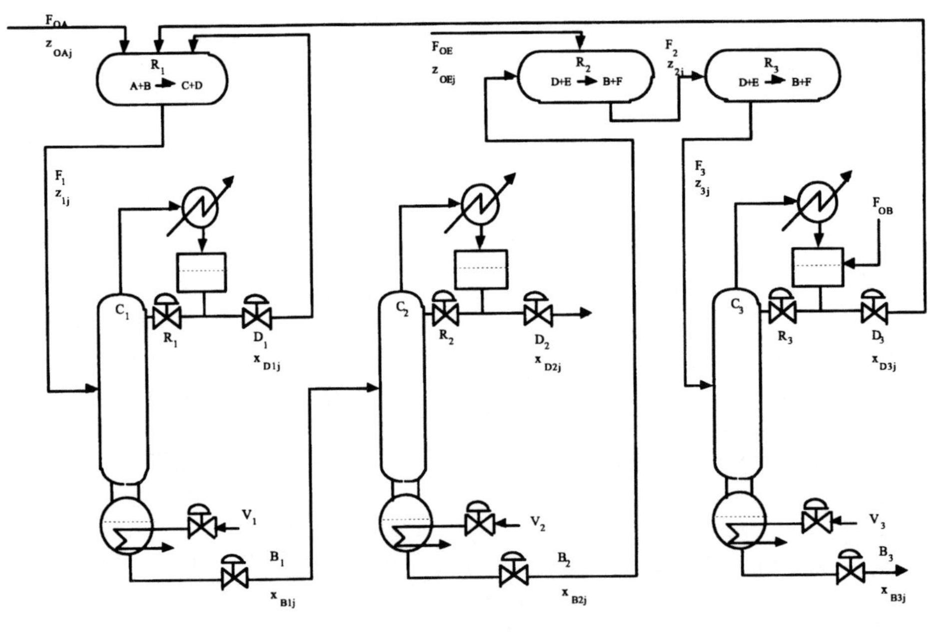

Figure 1 Luyben and Luyben Plant

ELSEVIER
IFAC
PUBLICATIONS
www.elsevier.com/locate/ifac

SENSITIVITY-BASED SOLUTION UPDATES IN CLOSED-LOOP DYNAMIC OPTIMIZATION

Jitendra V. Kadam* and Wolfgang Marquardt *,[1]

* *Lehrstuhl für Prozesstechnik, RWTH Aachen University*
Turmstr. 46, D-52064 Aachen, Germany
{kadam,marquardt}@lpt.rwth-aachen.de

Abstract: The challenge for real-time optimization-based control systems is to efficiently handle uncertainty. A novel approach for closed-loop optimization is presented that systematically combines a fast update strategy with rigorous optimization when necessary. A parametric sensitivity-based technique is used to calculate optimal first-order updates to a nominal reference solution. The technique does not assume that the active constraint set remains the same after changes in uncertain parameters. In closed-loop, the approach is very effective to handle uncertainty while requiring only a minimum number of full real-time optimizations reducing the on-line computational expense. The approach is illustrated by simulations of closed-loop real-time optimizations of a semi-batch reactor described by a model with different kinds and ranges of parametric uncertainty. It is observed that the closed-loop updated solution is almost identical to the true optimal solution corresponding to uncertainty. *Copyright © 2004 IFAC*

Keywords: dynamic real-time optimization, input adaptation, parametric sensitivity analysis, uncertainty, unknown active set

1. INTRODUCTION

Increasing competition requires a more agile chemical plant operation in order to increase productivity under flexible operating conditions while decreasing the overall production cost (Backx *et al.*, 1998). This demands an integration of economic optimization and control. However, existing applications use either stationary real-time optimization or off-line dynamic optimization together with linear model-based control. These techniques are limited with respect to the achievable flexibility and economic performance, especially when considering intentionally dynamic processes such as continuous processes during transitions or batch processes. Off-line dynamic optimization and trajectory tracking is not fully satisfactory for on-line application in the plant due to model uncertainty, process disturbances and changes in the external market conditions (e.g. future product demand). A basic schematic of dynamic real-time optimization (D-RTO) is shown in Figure 1 (left); its structure is identical to nonlinear model predictive control (NMPC) with output feedback. It is referred to as *single level D-RTO*. Controls are determined at the jth sampling time t_j by solving a dynamic optimization problem that minimizes an economical objective using a nonlinear process model on the prediction horizon $[t_j, t_{f_j}]$. The prediction horizon can be fixed (for continuous processes) or shrinking (for optimal transitions with fixed final time, e.g. batch operations). Estimates of the current process states ($\hat{x}^j$) and disturbance predictions ($\hat{d}^j(t)$) are provided by the solution

[1] Corresponding author. Tel: +49-241-8096712

of an estimation problem e.g. by means of an extended Kalman filter. Only the control (u) residing on the first sampling interval $[t_j, t_j+\Delta\tilde{t}]$ is implemented in the plant. The problem is thus repetitively solved to use the feedback through measurements. For large-scale industrial applications, the D-RTO problem is computationally expensive to solve even if the solution on a previous horizon is used to initialize the algorithm on the current horizon (Binder *et al.*, 2002). Due to the large computational requirements larger sampling intervals are demanded which may not be acceptable due to uncertainty. On the other hand, the current solution may not significantly differ from the previous one. Due to the complexity of D-RTO its acceptance in industry is limited. Hence, an integrated economic optimization and control strategy is required, which is less complex and hence computationally tractable in real-time, and which provides approximate control profiles of sufficient quality. We consider the *two-level D-RTO* strategy (Kadam *et al.*, 2002) shown in Figure 1 (right) to address these requirements. A

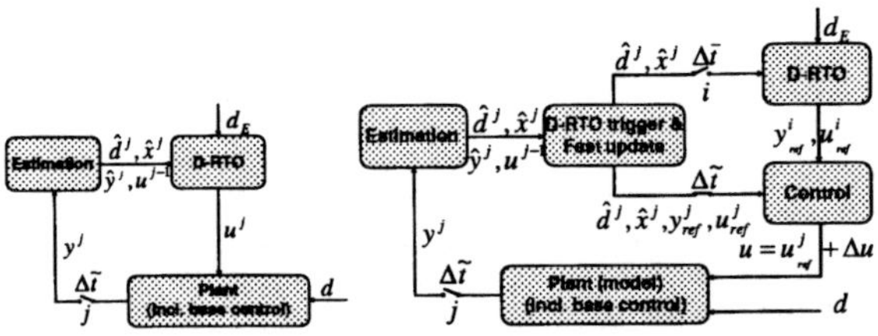

Fig. 1. Single level and two-level D-RTO

simple controller such as a PID controller or a predictive controller using a linear, possibly time-variant, model is used at the lower level to track the reference trajectories of the outputs y_{ref} and the controls u_{ref} provided by D-RTO at the upper level. To reject uncertainty, the reference solution is updated by a *fast update* technique at every sampling time t_j with the corresponding sample interval $\Delta\tilde{t}$. A so-called *D-RTO trigger* analyzes the updates in real-time for optimality and validity. When an estimate of an uncertain parameter differs from its nominal value, a re-optimization, denoted by the counter i, is triggered at non-equidistant sampling intervals $\Delta\bar{t} \geq \Delta\tilde{t}$ if the optimality criteria is not met. Moreover, the closed-loop strategy trades systematically off optimal solution accuracy for computational efficiency.

Parametric sensitivity analysis (Fiacco, 1983) is a strong tool to analyze an optimal solution for perturbations at a set of parameter values. Consequently, this analysis has been extensively used in steady-state and dynamic optimization for calculating updates due to parametric perturbations (cf. Büskens and Maurer (2002)) because it demands only negligible computational time. Diehl *et al.* (2002) use such an approach to provide updates as part of an NMPC algorithm tailored to large-scale nonlinear processes. The applicability of parametric sensitivity techniques, also referred to as neighboring extremal control, depends upon the strong assumption that the active constraint set does not change with perturbations, which is often quite restrictive. The assumption is only valid for sufficiently small perturbations entering the optimization problem. In this paper we focus on a *sensitivity-based fast solution update* in case of a *changing set of active constraints* in combination with a *D-RTO trigger to initiate rigorous re-optimization*. The paper is organized as follows: Section 2 presents a D-RTO problem formulation and discusses its numerical solution. The sensitivity analysis is given in Section 3. It is followed by the presentation of an algorithm integrating the techniques into the two-level approach. An illustrative example of dynamic real-time optimization of a semi-batch reactor is presented in Section 5.

2. PROBLEM FORMULATION AND SOLUTION

The main objectives of a D-RTO are minimization of operating cost and flexible and feasible operation in the presence of uncertainties. At t_j, the closed-loop single-level or two-level D-RTO problem with a fixed final time t_f, denoted by the counter j and i respectively, reads as

$$\min_{u(t)} \Phi(x(t), u(t), \hat{d}^j(t)) \qquad \text{(P)}$$

$$\text{s.t. } \bar{f}(\dot{x}(t), x(t), y(t), u(t), \hat{d}^j(t)) = 0;\ x(t_j) = \hat{x}^j,$$
$$\bar{g}(y(t), u(t), \hat{d}^j(t)) \geq 0;\ t \in [t_j, t_{f_j}].$$

For simplicity of notation, the superscript j of all the quantities in the this problem formulation is omitted. In closed-loop, the problem is solved repetitively at every t_j after a set of measurements (as feedback after the implementation of the control $u(t_{j-1})$) and a subsequent estimation of the initial state $\hat{x}^j$ and uncertainty $\hat{d}^j(t)$ are available. Due to the fixed final time considered here, at $t_{j+1}=t_j+\Delta\tilde{t}$, the previous time horizon $[t_j, t_{f_j}]$ is reduced by one sample interval $\Delta\tilde{t}$, i.e. $t_{f_{j+1}}=t_{f_j}$. In this formulation, Φ is a scalar *economic objective function* to be minimized over the time horizon $[t_j, t_{f_j}]$. $x(t) \in \mathbb{R}^{n_x}$ denote the system states with consistent initial conditions $\hat{x}^j$ at time t_j; and $y(t) \in \mathbb{R}^{n_y}$ are the algebraic variables. While, $u(t) \in \mathbb{R}^{n_u}$ denote the control variables. $\bar{f}(\cdot) : \mathbb{R}^{n_x} \times \mathbb{R}^{n_x} \times \mathbb{R}^{n_y} \times \mathbb{R}^{n_u} \times \mathbb{R}^{n_d} \to \mathbb{R}^{n_x+n_y}$ contains a differential-algebraic (DAE) process model. Any operational (path and endpoint) constraints are collected in $\bar{g}(\cdot) : \mathbb{R}^{n_y} \times \mathbb{R}^{n_u} \times \mathbb{R}^{n_d} \to \mathbb{R}^{n_{\bar{g}}}$. Furthermore, uncertainties (e.g. uncertain model parameters, disturbances) are collected in the vector $d(t) \in \mathbb{R}^{n_d}$. The elements of $d(t)$ can be either dependent on time or not. We consider here that a time-dependent uncertainty $d_i(t)$ (e.g. a drift in a reaction kinetic parameter) is parameterized by some expansion $d_i(t)=\sum_{k\in\mathcal{K}} c_{d_{i,k}}\phi_k(t)$, $i=1,\ldots,n_d$. The coefficients $[c_{d_{i,k}}]$ and other time-independent uncertain parameters are collected in a vector of uncertain parameters $p \in \mathbb{R}^{n_p}$ which is estimated along with the initial state x^j from past measurements.

The D-RTO problem (P) is solved numerically by a control vector parametrization (CVP) approach (Kraft, 1985). The dynamic optimization problem is converted into a nonlinear programming problem (NLP) by time-discretizing the controls u on the time horizon $[t_j, t_{f_j}]$ and using piecewise polynomial approximations. For the sake of simplicity, we consider a piecewise constant approximation $u_i(t_k)=c_{u_{i,k}}$, $k=1,\ldots,N$, $i=1,\ldots,n_u$, where N is the number of discretization intervals. Choosing the discretized controls $z := [c_{u_{i,k}}] \in \mathbb{R}^{n_z}, n_z = n_u \cdot N$, as the optimization variables, problem (P) can be transformed into the NLP

$$\min_z f(z,p) := \Phi(z,p) \qquad (\mathrm{P_D})$$
$$\text{s.t. } g(z,p) \geq 0.$$

The NLP problem is solved by employing an SQP algorithm. The objective function $f : \mathbb{R}^{n_z} \times \mathbb{R}^{n_p} \rightarrow \mathbb{R}$, constraints $g(\cdot) : \mathbb{R}^{n_z} \times \mathbb{R}^{n_p} \rightarrow \mathbb{R}^{n_g}$, $n_g=n_{\bar{g}} \cdot N$, and their gradients are evaluated by a simultaneous integration of the DAE model and the sensitivity equation system for given z and p. We employ the dynamic optimization solver DyOS (2002) in this work.

3. SENSITIVITY ANALYSIS

3.1 Basic concepts

Let us consider the NLP ($\mathrm{P_D}$) for sensitivity analysis. The objective function $f(\cdot)$ and constraints $g(\cdot)$ are assumed to be at least twice continuously differentiable in z. The Lagrangian function of the constrained optimization problem ($\mathrm{P_D}$) is defined as $L(z,p,\lambda) = f(z,p) - \lambda^T g(z,p)$, where $\lambda \in \mathbb{R}^{n_g}$ is the vector of Lagrange multipliers. Let us consider z_0^*, λ_0^* as the *nominal*[2] *optimal solution* of problem ($\mathrm{P_D}$) corresponding to p_0. To simplify notation, the superscript * is omitted in the sequel. At the optimal solution, the constraints g are divided into the active constraints g^a and the inactive constraints g^{ina} of dimension n_g^a and $n_g - n_g^a$, respectively. The corresponding Lagrange multipliers λ_0 are divided accordingly into λ_0^a and λ_0^{ina}. The active constraint set is denoted by $\mathcal{G}_0$. The first order necessary conditions of optimality (NCO) are[3]:

$$L_z(z_0,p_0,\lambda_0) = 0, \qquad (\mathrm{NCO})$$
$$g_i^a(z_0,p_0) = 0;\ \lambda_{0,i}^a > 0;\ i = 1,\ldots n_g^a \in \mathcal{G}_0,$$
$$g_j^{ina}(z_0,p_0) > 0;\ \lambda_{0,j}^{ina} = 0;\ j = 1,\ldots n_g^{ina} \notin \mathcal{G}_0.$$

Let the NCO and the strong second order sufficient conditions (SSC) be satisfied at z_0 and λ_0. Further, the active constraint gradients g_z^a are linearly independent. Also, assume that the active constraint set $\mathcal{G}$ at the changed parameter value $p = p_0+\Delta p$ is the same as $\mathcal{G}_0$. With these assumptions, Fiacco (1983) proved that the functions $z = z(p)$ and $\lambda^a = \lambda^a(p)$ are at least once differentiable in p. In the vicinity of p_0, the *parametric sensitivities* $z_p := \frac{dz}{dp}$ and $\lambda_p^a := \frac{d\lambda^a}{dp}$ of the optimal solution can be calculated by differentiating the NCO with respect to z, λ^a and p. The resultant linear equation system (Fiacco, 1983) to be solved is

$$\begin{bmatrix} L_{zz}(\cdot) & -g_z^{a,T}(\cdot) \\ g_z^a(\cdot) & 0 \end{bmatrix} \begin{bmatrix} z_p \\ \lambda_p^a \end{bmatrix} = -\begin{bmatrix} L_{zp}(\cdot) \\ g_p^a(\cdot) \end{bmatrix} \qquad (1)$$

where all the functions are evaluated at z_0, p_0, λ_0. A first order update due to a perturbation Δp can thus be calculated from a Taylor expansion as

$$\Delta z := z(p) - z_0 = z_p(p_0)\Delta p, \qquad (2a)$$
$$\Delta\lambda^a := \lambda^a(p) - \lambda_0^a = \lambda_p^a(p_0)\Delta p, \qquad (2b)$$
$$\Delta\lambda^{ina} := \lambda^{ina}(p) - \lambda_0^{ina} = 0. \qquad (2c)$$

In order to solve equation (1), we need first order partial derivatives of the constraint functions g_z, g_p and second order partial derivatives of the Lagrange function L_{zz}, L_{zp}. This computation requires a significant effort which can be prohibitive for large-scale problems. Finite difference formulae or automatic differentiation (Nocedal and Wright, 1999) can be used to provide sufficiently accurate second order derivatives.

3.2 The case of a changing active set

For a moderate perturbation Δp, the active constraint set $\mathcal{G}(p)$ can change. It is unknown a-priori whether such a change actually occurs. At the new optimal solution $z(p_0+\Delta p)$, some of the nominally active constraints $g^a(z_0,p_0)$ or the nominally inactive constraints $g^{ina}(z_0,p_0)$ can become inactive or active. Therefore, equation (1) is reformulated as the QP problem (Ganesh and Biegler, 1987)

$$\min_{\Delta z} \ 0.5\Delta z^T L_{zz}(\cdot)\Delta z \qquad (3a)$$
$$+\Delta p^T L_{zp}^T(\cdot)\Delta z + f_z^T(\cdot)\Delta z$$
$$\text{s.t. } g_z(\cdot)\Delta z \geq -g_p(\cdot)\Delta p - g(\cdot), \qquad (3b)$$

where all the functions are evaluated at z_0, p_0, λ_0. The solution of this QP problem, the updates Δz and new Lagrange multipliers $\bar{\lambda}=\lambda(p_0 + \Delta p)$, corresponds to taking a Newton step (not necessarily full) of Δz from z_0 of problem ($\mathrm{P_D}$) while respecting all inequality constraints to first order. The solution also detects a new active set

---

[2] A nominal quantity $(\cdot)_0$ is evaluated at p_0.

[3] $(\cdot)_z = \frac{\partial(\cdot)}{\partial z}$, $(\cdot)_{zz} = \frac{\partial^2(\cdot)}{\partial z^2}$, $(\cdot)_{zp} = \frac{\partial^2(\cdot)}{\partial z \partial p}$

$\bar{\mathcal{G}} := \mathcal{G}(p_0 + \Delta p)$ which is a better estimate of the true active set. Note that the updates always satisfy all constraints to first order, which cannot be guaranteed for the solution of the sensitivity equation system (1) due to the potential changes in the active set. If the active set does not change, the solutions of problems (3) and (1) are the same.

For computing L_{zz} and L_{zp}, an estimate of the true active set $\mathcal{G}$ and the corresponding Lagrange multipliers λ have to be known a-priori. One can use the nominal active set $\mathcal{G}_0$ as an initial guess and iterate as detailed below. Moreover, the Lagrange function is non-differentiable in z, p at a point where the active set changes. We assume strict complementary, e.g. $\bar{\lambda}(g_0(z_0, p_0) + g_z(z_0, p_0)\Delta z + g_p(z_0, p_0)\Delta p) = 0$, at any solution of QP (3). With this assumption, non-differentiability does not pose any difficulty in computing solutions of the QP problem, since the second-order derivatives are computed using a fixed active set. Again, note that Δz, $\bar{\lambda}$ and $\bar{\mathcal{G}}$ are feasible with respect to the linearized constraints (3), albeit the true active set may be different. Better optimal solution updates can be calculated by solving the QP problem iteratively to account for a change of the active set. The iterative strategy is as follows:

(1) By using a finite difference formula, compute $f_{zz}, g_{zz}, f_{zp}, g_{zp}$ at the nominal optimal solution z_0 corresponding to p_0.
(2) If the NLP (P_D) is nonlinear in p calculate f_z, g_z and g_p at z_0 and the changed uncertain parameter value $p{=}p_0{+}\Delta p$ by doing a sensitivity integration to account for the perturbation Δp; and assign $\Delta p{:=}0$.
(3) Initialize iteration $k{:=}0$, $z_k{:=}z_0$ $\lambda_k{:=}[\lambda_0^a, 0]$, $\mathcal{G}_k{:=}\mathcal{G}_0$.
(4) Assemble L_{zz} and L_{zp} from the pre-computed $f_{zz}, g_{zz}, f_{zp}, g_{zp}$ using λ_k and $\mathcal{G}_k$.
(5) Solve QP (3) to obtain Δz_k, z_k, $\bar{\lambda}_k$ and $\bar{\mathcal{G}}_k$.
(6) Set $\lambda_{k+1} = \bar{\lambda}_k$, $\mathcal{G}_{k+1} := \bar{\mathcal{G}}_k$, and $k := k+1$. If the active set is changed go back to step (4).
(7) Calculate g, f_z, g_z and g_p by doing a sensitivity integration for the updated solution z_k from steps (4)-(6) and p from step (2). If still $g(z_k, p) < 0$ go back to step (3).
(8) Declare that optimal updates $z = z_k$, $\lambda = \lambda_k$ and $\mathcal{G} := \mathcal{G}_k$ have been calculated.

The reduced Hessian strategy of Ganesh and Biegler (1987) can be employed to efficiently solve QP (3). For a given range of uncertainty, the accuracy of the updates is usually acceptable.

4. FAST UPDATE AND D-RTO TRIGGER

It is assumed that the perturbations Δp are measured or estimated without time delay, and a perfect controller is available to reject disturbances that are not considered in p, and which affect during the current sample interval. *Fast updates* of the controls are calculated as

$$z(p) = z_0 + \Delta z \text{ i.e. } u(t) = u_0 + \Delta u(t) \quad (4)$$

with Δz being the solution of QP (3). Output and state trajectories are updated by integrating the DAE model in p for the updated control.

The *optimality error of the updated inputs* $z(p)$ is defined here as residuals

$$\epsilon_{opt} = \frac{\|L_z(z, p, \lambda)\|_\infty}{\|\lambda\|_2}; \epsilon_{infs} = \frac{\|g(z, p)\|_\infty}{\|z\|_2}. \quad (5)$$

It is comprised of the error in the Lagrange sensitivity (ϵ_{opt}) and the nonlinear constraint infeasibility (ϵ_{infs}), which quantify optimality and feasibility of the updated controls with respect to uncertainty. Note that any changed active constraint set is taken into account in this calculation. We define τ_{opt}, τ_{infs} as a maximum allowable optimality error given by a controller. If the *D-RTO trigger criteria*, $\epsilon_{opt}{>}\tau_{opt}$ and $\epsilon_{infs}{>}\tau_{infs}$, are met a re-optimization is triggered to get a new set of optimal control and output trajectories.

The fast update (counter j) and D-RTO trigger (counter i) are embedded in the two-level strategy shown in Figure 1(right). The iterative algorithm is as follows: [4]

(1) Set counter $i{:=}1$ and counter $j{:=}1$;
(2) Solve the upper level D-RTO (P) on $t{\in}[0, t_{f_j}]$ to obtain the **reference solution** $z^i_{ref}, \lambda^i_{ref}$, $y^i_{ref}(t)$ computed at p_0. Assign $z^j_{ref}{=}z^i_{ref}$, $y^j_{ref} = y^i_{ref}$ as the reference trajectories for the lower level control and fast update. Compute L^j_{zz}, L^j_{zp} by finite differences.
(3) **for** $j = 2, N$ (number of sample intervals) **do**
    (a) At t_j, implement the control $u^{j-1}(t_{j-1})$ and get the **measurements** y^j and **estimates** $\hat{x}^j, \hat{d}^j$ to update p^j.
    (b) Reduce the time horizon for a batch operation by $\Delta\tilde{t}$ i.e. $t_j{=}t_{j-1}{+}\Delta\tilde{t}$, $t_{f_j}{=}t_{f_{j-1}}$. On the reduced horizon, **assemble** the shifted reference discretized controls $\bar{z}^j_{ref}$, $\bar{\lambda}^j_{ref}$ L^j_{zz}, L^j_{zp} from the corresponding quantities $z^{j-1}_{ref}, \lambda^{j-1}_{ref}, L^{j-1}_{zz}, L^{j-1}_{zp}$ on the previous horizon $[t_{j-1}, t_{f_{j-1}}]$. Evaluate the constraints g^j and sensitivities f^j_z, g^j_z for the controls $\bar{z}^j_{ref}$, estimated parameters p^j and states $\hat{x}^j$ by doing a sensitivity integration.
    (c) **Fast update:** Solve QP (3) to obtain Δz^j,λ^j and $\mathcal{G}^j$. Calculate the updated controls $z^j_{ref}{=}\bar{z}^j_{ref}{+}\Delta z^j$ and the Lagrange multipliers $\lambda^j_{ref}{=}\lambda^j$. Compute

[4] A reference quantity $(\cdot)_{ref}$ is any currently available quantity to be updated, and which is evolved through the previous fast updates and any rigorous re-optimizations.

y^j_{ref} by doing one integration of the DAE model for the updated control.

(d) Compute the optimality error (ϵ_{opt}, ϵ_{infs}) of the updated controls and check the **D-RTO trigger** criteria:
if ($\epsilon_{opt} > \tau_{opt}$) and ($\epsilon_{infs} > \tau_{infs}$) trigger a re-optimization and solve D-RTO (P) (set $i := i+1$) to get new optimal references $z^j_{ref} = u^j_{ref}, y^j_{ref}, \lambda^j$.

(4) **end for**; batch operation completed.

5. ILLUSTRATIVE EXAMPLE

A semi-batch reactor is considered here, which is derived from the continuous Williams-Otto benchmark reactor described by Forbes (1994). The following reactions are taking place in the reactor: $A+B \xrightarrow{k_1} C$, $C+B \xrightarrow{k_2} P+E$, $P+C \xrightarrow{k_3} G$. The reactor is fed initially with a fixed amount of reactant A; reactant B is fed continuously. The first-order reactions produce the desired products P and E. Product G is a waste. As the reactions are exothermic, the heat produced has to be removed from the cooling jacket by manipulating the cooling water temperature. A detailed dynamic process model and its parameters are given elsewhere (Kadam and Marquardt, 2003). During reactor operation, path constraints on the feed rate of reactant B ($F_{B_{in}}$), reactor temperature (T_r), hold-up (V) and cooling water temperature (T_w) should be respected. $F_{B_{in}}$ and T_w are the control variables. For this illustration the reactor is run for a fixed batch time of 1000 sec.

The operational objective is to maximize the yield of the main products at the end of batch. The dynamic optimization problem reads as

$$\max_{F_{B_{in}}(t), T_w(t)} \quad \Phi(t_f) = c_p n_p(t_f) + c_e n_e(t_f) \qquad (6)$$

s.t. *process model*, and

$$0 \le F_{B_{in}}(t) \le 5.784 \frac{\text{kg}}{\text{sec}}, \; V(t_f) \le 5 \text{ m}^3,$$
$$20 \le T_w(t) \le 100 \text{ °C}, \; 60 \le T_r(t) \le 90 \text{ °C}.$$

The initial reactor temperature $T_{r,0}$ and feed temperature T_{in} are fixed at their nominally optimal values of 60 °C and 35 °C, respectively, calculated by off-line optimization. A disturbance ΔT_{in} affects the feed temperature at, say, t=250 sec during batch operation. It is assumed to be measured directly. Further, the parameter b_1 in the reaction kinetic equation $k_1 = a_1 \exp(\frac{b_1}{T_r + 273.15})$ is assumed to vary about $\pm 25\%$ from its nominal value $b_1 = 6666.7$ sec^{-1}. This parameter is assumed to be estimated on-line. We collect T_{in} and b_1 in the vector of uncertain parameters p. The two-level D-RTO strategy is applied in closed-loop in the presence of the aforementioned uncertainty and disturbance.

Problem (6) is solved by employing DyOS (2002) to obtain a reference solution. Each control profile is time-discretized and approximated as piecewise constant function on 32 equidistant intervals to transform problem (6) into the NLP (P_D). The nominal optimal control and constraint profiles are depicted in Figure 2 by bold lines. These

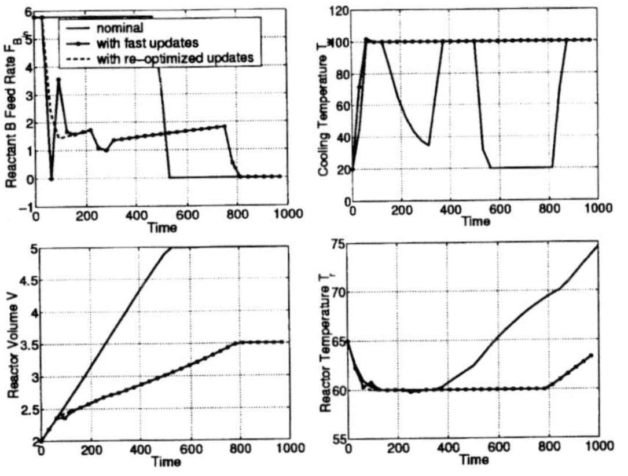

Fig. 2. Nominal and closed-loop simulation profiles of control and constraint variables: $\Delta b_1 = +10\%$ and $\Delta T_{in} = -10$ °C at t=250 sec

profiles have different arcs corresponding to active and inactive parts of the path constraints, which are characterized as follows: $F_{B_{in}}$ is initially kept at its *upper bound* and then switched to its *lower bound* when the *reactor hold-up* (V) reaches its *upper bound*. The second control variable T_w is manipulated to move the *reactor temperature* (T_r) to its lower bound at t=140 sec and keep it there. At the switching time t=360 sec, T_r is moved away from its lower bound by *manipulating* T_w in a bang-bang profile with the switching times computed implicitly by optimization. Note that T_w is at its lower bound at t=0 sec and quickly switched to its upper bound.

Two cases of uncertainty and disturbances are considered: a) $\Delta b_1 = +10\%$ and $\Delta T_{in} = -10$ °C at t=250 sec, and b) $\Delta b_1 = -25\%$ and $\Delta T_{in} = -10$ °C at t=250 sec. The algorithm of the two-level strategy with the fast update and D-RTO trigger described in Section 4 and Figure 1 (right) is applied in closed-loop. The closed-loop control and constraint profiles are shown in Figure 2 and 3 for the cases a) and b), respectively. The profiles shown by dash-dotted lines depict the response of the fast update and D-RTO trigger strategy. Moreover, the profiles shown by dashed lines depict the response of repetitive re-optimizations i.e. the single level D-RTO strategy (Figure (1)). When the algorithm is employed, only once a re-optimization was triggered at t=125 sec in case a) and none in case b). It can be observed in the figures that the closed-loop updated solution is almost identical to the true optimal solution (depicted by dashed lines) using the fast update strategy with only a minimum number of re-

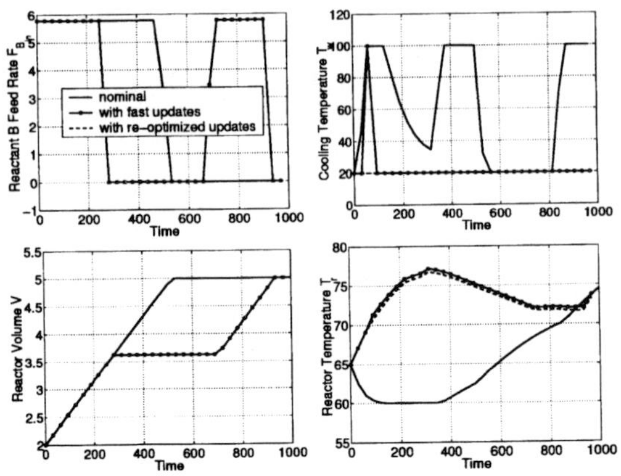

Fig. 3. Nominal and closed-loop simulation profiles of control and constraint variables: Δb_1=−25% and ΔT_{in}=−10 °C at t=250 sec

optimizations triggered. Note that the structure of the true optimal solution is drastically different from that of the nominal solution. Most interestingly, in case b), $F_{B_{in}}$ is stopped at t=282 sec, and again switched back to its upper bound at t=656 sec until the reactor hold-up reaches its upper bound. Furthermore, the reactor temperature is never at either of its bounds, while T_w is at its lower bound throughout the operation. These changed active sets are correctly and timely detected, and the batch operation is optimized in real-time. The computational time of one iterate of the D-RTO trigger integrated two-level strategy is approximately 15 sec. It is significantly smaller than approximately 60 sec required for a re-optimization which is warm-started with the nominal solution.

6. CONCLUSIONS

A novel solution update and D-RTO trigger strategy are developed for providing real-times updates and initiating a rigorous re-optimization when necessary. These techniques are integrated into a closed-loop two-level D-RTO strategy. With a case-study, it is pointed out that the nominally active constraints set can drastically change with perturbations in uncertain parameters. Hence, in such cases, parametric sensitivity based input update techniques that assume a constant active set (Büskens and Maurer, 2002) are not applicable. Changed active sets due to perturbations are detected, and first order updates are calculated by solution of a D-RTO problem consistent QP subproblem. Furthermore, the accuracy of closed-loop updates is acceptable. The two-level approach with the D-RTO trigger is shown to be quite effective in rejecting a large range of uncertainty in model parameters, and process disturbances. The closed-loop updated solution is almost identical to the true optimal solution corresponding to a perturbation in uncertain parameters. Moreover, when the accuracy of fast updates is not acceptable, the D-RTO trigger handles uncertainty by triggering a rigorous re-optimization.

REFERENCES

Backx, T., O. Bosgra and W. Marquardt (1998). Towards intentional dynamics in supply chain conscious process operation. In: *FOCAPO 98, Snowbird, Utah.* www.lpt.rwth-aachen.de/Publication/Techreport/1998/LPT-1998-25.html.

Binder, T., L. Blank, W. Dahmen and W. Marquardt (2002). *Online Optimization of Large Scale Systems.* Chap. Multiscale concepts for moving horizon optimization, pp. 341–361. Springer.

Büskens, C. and H. Maurer (2002). *Online Optimization of Large Scale Systems.* Chap. Sensitivity analysis and real-time optimization of parametric nonlinear-programming problems, pp. 3–16. Springer.

Diehl, M., H.G. Bock, J.P. Schloder, R. Findeisen, Z. Nagy and F. Allgower (2002). Real-time optimization and nonlinear model predictive control of processes governed by differential-algebraic equations. *Journal of Process Control* **12**(4), 577–585.

DyOS (2002). *DyOS User Manual, Release 2.1.* Lehrstuhl für Prozesstechnik, RWTH Aachen University. Aachen, Germany.

Fiacco, A. V. (1983). *Introduction to Sensitivity and Stability Analysis in Nonlinear Programming.* Academic Press, New York.

Forbes, J. Fraser (1994). Model structure and adjustable parameter selection for operations optimizations. PhD thesis. McMaster University.

Ganesh, N. and L. T. Biegler (1987). A reduced Hessian strategy for sensitivity analysis of optimal flowsheets. *AICHE J.* **33**(2), 282–296.

Kadam, J. V. and W. Marquardt (2003). A strategy for update of dynamic optimization solution in the presence of uncertainty. Technical report. Lehrstuhl für Prozesstechnik, RWTH Aachen University.

Kadam, J. V., W. Marquardt, M. Schlegel, R.L. Tousain, D.H. van Hessem, J. van den Berg and O. H. Bosgra (2002). A two-level strategy of integrated optimization and control of industrial processes: a case study. In: *European Symposium on Computer Aided Process Engineering - 12* (J. Grievink and J. v. Schijndel, Eds.). Elsevier. pp. 511–516.

Kraft, D. (1985). On converting optimal control problems into nonlinear programming problems. *Comput. Math. Prog.* **15**, 261–280.

Nocedal, J. and S.J. Wright (1999). *Numerical Optimization.* Springer-Verlag.

ELSEVIER
IFAC
PUBLICATIONS
www.elsevier.com/locate/ifac

DANTZIG-WOLFE DECOMPOSITION AND LARGE-SCALE CONSTRAINED MPC PROBLEMS

Ruoyu Cheng, J. Fraser Forbes*
Department of Chemical and Materials Engineering
University of Alberta
Edmonton, AB T6G 2G6, Canada

W. San Yip
Alberta Research Council
Edmonton, AB T6N 1E4, Canada

Abstract

Model Predictive Control (MPC) strategies are typically implemented in two levels: a steady-state target calculation and a control calculation. The steady-state target calculation consumes excess degrees of freedom within the control problem to provide optimal steady-state performance with respect to some specified objective. In some MPC approaches, the target calculation is formulated as a Linear Program (LP) with a pre-specified objective function and a linear or linearized steady-state model derived from that used in the control calculation. In large-scale problems, centralized MPC schemes find the optimal solution for the plant-wide optimization problem, but may not provide sufficient redundancy or reliability and can require substantial computation. On the other hand, in a decentralized MPC scheme, the target calculations are performed independently by ignoring interactions among units, and as a result will not usually find the optimal operation. In contrast to the centralized MPC approach, a decentralized MPC provides a high degree of redundancy with respect to the failure of an individual MPC. For large-scale process control problems, the desired characteristics for an MPC implementation include: accurate and quick tracking of the changing optimal steady-state operation, a high degree of reliability with respect to failure within the MPC application (i.e., failure of a portion of the control system), and low computational requirements. Fully centralized or monolithic MPC and independent block-wise decentralized MPC represent the two extremes in the "trade-off" among the desired characteristics of an implemented MPC system. In this paper, we propose a coordinated, decentralized approach to the steady-state target calculation problem. Our approach is based on the Dantzig-Wolfe decomposition principle and has been found to be effective at finding the optimal plant operation while providing a high degree of reliability at a reasonable computational load. *Copyright © 2004 IFAC*

Keywords

Model predictive control, Steady-state target calculation, Dantzig-Wolfe decomposition, Coordination, Linear programming.

Introduction

Model predictive control (MPC) strategies have been successful in a wide range of industrial applications. In the current control hierarchy, MPC plays an important role as a connection between local steady-state optimization and basic dynamic control (Kassmann *et al.* 2000). The MPC framework can be further divided into a steady-state calculation and a control calculation. The goal of the steady-state calculation is to calculate the desired targets for both output and input variables at a frequency much higher than those computed from local economic optimizers. The target calculation provides optimal achievable set-points that are passed to the control calculation.

Two approaches to using MPC in large-scale applications are centralized and decentralized MPC. By considering interactions among operating units,

* Email: fraser.forbes@ualberta.ca , phone: 780 492-0873, fax: 780 492-2881.

centralized MPC can offer plant-wide optimal control, including plant-wide target calculations. Unfortunately, the centralized MPC scheme contains a large-scale control calculation with little or no redundancy, implemented within relatively centralized computational environment. Alternatively, in the decentralized MPC scheme, the large-scale control problem is decomposed into a set of separate MPC subproblems being solved independently. Though the completely decentralized MPC target calculation may fail to yield plant-wide optimal solutions, it has higher degree of reliability to subsystem failure and may require less computation. In our work, reliability refers to the possibility that some control subsystems or portions thereof are able to function when other subsystems fail.

For large-scale process control problems, the essential trade-off is between: accurate and quick tracking of the changing optimal operation, reliability with respect to failure within the MPC application (i.e., failure of a portion of the control system), and computational requirements. Therefore, fully centralized or monolithic MPC and independent block-wise decentralized MPC represent the two extremes in the "trade-off" among these desired characteristics. In this paper, we propose an approach to coordinating decentralized steady-state target calculation that can yield the accuracy of centralization and the reliability of decentralization.

In a decentralized MPC scheme, the flowsheet is separated into sections that have separate control systems. In the decentralized approach, target calculation takes the strategy that the upstream units pass their decisions to the downstream units without considering interactions. The downstream units treat the upstream decisions as disturbances. Thus, the decentralized approach usually produces sub-optimal results. If the target calculation could be coordinated as shown in Fig. 1, it may be possible to produce an optimal solution similar to that of

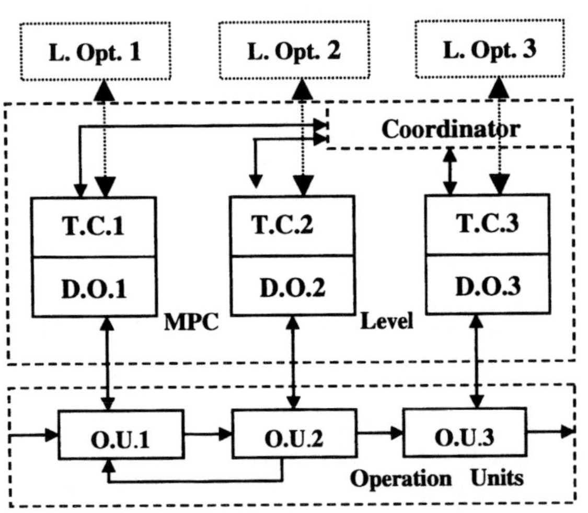

Figure 1. Coordinated, Decentralized MPC Target Calculation

the centralized approach. While Venkat and Rawlings (2003) are developing a communication-based scheme for the control calculations within a decentralized MPC scheme, the focus of our current study is steady-state targets optimization.

This paper discusses one possible approach, by taking advantage of the Dantzig-Wolfe decomposition principle, to coordinating the interactions among decentralized MPC units and provides insight into the information that must be communicated between MPC subsystems and the coordinator to ensure high performance levels. The outline is as follows. Firstly, we introduce the Dantzig-Wolfe decomposition principle. Secondly, we formulate a coordinated LP target calculation problem by introducing a coordination mechanism. Then, we implement the decomposition algorithm to solve the large-scale constrained MPC target calculation problem. An illustrative case study demonstrates the efficacy of the proposed approach.

Dantzig-Wolfe Decomposition Principle

The Dantzig-Wolfe decomposition principle (Dantzig and Thapa, 2002) is illustrated in Fig. 2.

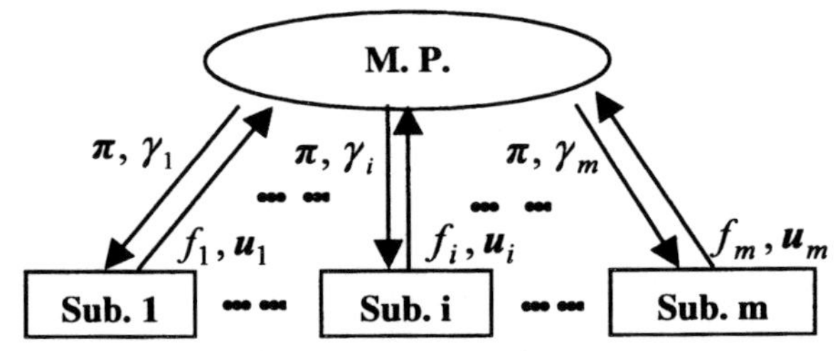

Figure 2. Mechanism of Dantzig-Wolfe Decomposition

A large-scale linear programming problem can be decomposed into independent subproblems, which are coordinated by a *master problem* (MP). The solution to the original large-scale problem can be shown to be equivalent to solving those subproblems and the MP through a finite number of iterations. During each iteration, the MP handles the linking constraints that connect the subproblems and the convexity constraints (which will be discussed later) for each subproblem, using the information supplied by the subproblems [f_i, $\boldsymbol{u}_i$], where f_i and $\boldsymbol{u}_i$ are the objective function value and solution of the ith subproblem, respectively. Then the MP sends sensitivity information [$\boldsymbol{\pi}$, γ_i], corresponding to the linking constraints and convexity constraints respectively, to all the subproblems for updating their objective functions. Consequently, the subproblems with updated objective functions are re-optimized independently. The above procedure is repeated until convergence and the solution is the optimum of the original large-scale problem.

Dantzig-Wolfe decomposition hinges on the theorem of convex combination and the column generation technique. To help understand the Dantzig-Wolfe decomposition principle, the following subsections briefly introduce these two fundamental components of the decomposition principle.

Theorem of Convex Combination

The theorem of convex combination (Lasdon, 2002), or D-W transformation (Dantzig and Thapa, 2002), states that an arbitrary point $\boldsymbol{x}$ in a convex polyhedral set $\boldsymbol{X}=\{\boldsymbol{x} \mid \boldsymbol{Ax}=\boldsymbol{b}, \boldsymbol{x}\geq \boldsymbol{0}\}$ can be written as a convex combination of the extreme points of $\boldsymbol{X}$ plus a nonnegative linear combination of the extreme rays (normalized homogeneous solutions) of $\boldsymbol{X}$, or:

$$\boldsymbol{x}=\sum_{i=1}^{L}\alpha_i \boldsymbol{u}^i+\sum_{j=1}^{M}\beta_j \boldsymbol{v}^j \tag{1}$$

$$\sum_{i=1}^{L}\alpha_i=1 \qquad \alpha_i,\beta_j\geq 0 \tag{2}$$

where $\{\boldsymbol{u}^i\}$ and $\{\boldsymbol{v}^j\}$ are the finite set of all extreme points and the finite set of all extreme rays, respectively. If the feasible region is bounded, we can reformulate the problem by using the extreme points only.

The Dantzig-Wolfe decomposition principle is particularly powerful for solving structured linear programs (Chvátal, 1983). Consider a block-wise linear programming problem:

$$\min \quad \sum_{i=1}^{m}\boldsymbol{c}_i{}^T\boldsymbol{x}_i$$

$$\text{s.t.} \quad \sum_{j=1}^{n}\boldsymbol{A}^j{}_{eq}\boldsymbol{x}_j=\boldsymbol{b}^0{}_{eq} \qquad n\leq m \tag{3a}$$

$$\boldsymbol{B}^i{}_{eq}\boldsymbol{x}_i=\boldsymbol{b}^i{}_{eq} \tag{3b}$$

$$\boldsymbol{G}^i\boldsymbol{x}_i\leq \boldsymbol{b}^i \qquad \mathrm{i}=1,2,......,\mathrm{m} \tag{3c}$$

where the equality constraints in (3a) represent the linking constraints associated with n interacting subproblems. The equality constraints in (3b) and the inequality constraints in (3c) are the local constraints of independent subproblems. The master problem (MP) can be formulated as follows using the linking constraints in (3a) and the convex combination as stated in (1), assuming that the feasible regions of the subproblems are bounded.

$$\min \quad \sum_{i=1}^{m}\sum_{j=1}^{p(i)} f_{ij}\lambda_{ij}$$

$$\text{s.t.} \quad \sum_{i=1}^{n}\sum_{j=1}^{p(i)}\boldsymbol{P}_{ij}\lambda_{ij}=\boldsymbol{b}^0{}_{eq} \tag{4}$$

$$\sum_{j=1}^{p(i)}\lambda_{ij}=1, \quad \lambda_{ij}\geq 0, i=1,2,......,m \tag{5}$$

where $p(i)$ represents the number of extreme points of the feasible region of the ith LP subproblem, and

$$\boldsymbol{x}_i=\sum_{j=1}^{p(i)}\lambda_{ij}\boldsymbol{u}_i^j \tag{6}$$

$$f_{ij}=\boldsymbol{c}_i{}^T\boldsymbol{u}_i^j \tag{7}$$

$$\boldsymbol{P}_{ij}=\boldsymbol{A}^i\boldsymbol{u}_i^j \tag{8}$$

The resulting MP problem has fewer rows in the coefficient matrix than the original problem. However, the number of columns in the MP is larger due to the increase in the number of variables associated with the extreme points of all the subproblems. The column generation method discussed in next section provides an approach for dealing with the increase in number of columns in the MP.

Column Generation Technique

To solve a large-scale LP problem practically, a column generation technique is applied in solving the MP (Gilmore and Gomory, 1961; Dantzig and Thapa, 2002). If the MP is solved via the Simplex method, we only need a basic set that has the same number of basic variables as the number of rows. Thus we do not need to explicitly know all the extreme points of subproblems. This leads to solving an equivalent problem, the *restricted master problem* (RMP).

Assume that we have a starting basic feasible solution to the RMP and it has a unique optimum, the optimal solution provides us with multiplier vectors [$\boldsymbol{\pi}$, $\boldsymbol{\gamma}$] for the linking constraints in Eq.(4) and convexity constraints in Eq.(5), respectively. Then, subproblems are formulated and solved to find the priced-out column associated with λ_{ij}:

$$f_{ij}=(\boldsymbol{c}_i-\boldsymbol{\pi}\boldsymbol{A}^i)\boldsymbol{x}_i^j-\gamma_i \tag{9}$$

and the ith subproblem is:

$$\min \quad z_i^0=(\boldsymbol{c}_i-\boldsymbol{\pi}\boldsymbol{A}^i)\boldsymbol{x}_i^j$$

$$\text{s.t.} \quad \boldsymbol{B}^i{}_{eq}\boldsymbol{x}_i=\boldsymbol{b}^i{}_{eq} \tag{10}$$

$$\boldsymbol{G}^i\boldsymbol{x}_i\leq \boldsymbol{b}^i \qquad \mathrm{i}=1,2,......,\mathrm{m}$$

Therefore, we reach an optimal solution when the following condition is satisfied:

$$\min_i \min_j f_{ij}=\min_i(z_i^0-\gamma_i)\geq 0, \quad i=1, 2, ..., m \tag{11}$$

Optimality and finite convergence have been proved by Dantzig and Thapa (2002).

Dantzig-Wolfe Decomposition and Coordinated, Decentralized MPC Target Calculations

Typically, in a chemical plant, the process model used for centralized MPC scheme has a block-wise structure. Therefore, most chemical plants are potential candidates for the implementation of Dantzig-Wolfe decomposition to coordinate independent MPC calculations. The coordinator can be designed by considering the interactions among operation units that can be modeled as the linking constraints.

In this work, we focus on the design of coordination mechanism for steady-state target calculations in MPC systems. Target calculation problems can be formulated either as an LP or a QP problem (Kassmann *et al.* 2000; Rao and Rawlings, 1999). In this work, we assume an LP formulation for the steady-state target optimization, with a linear or linearized plant model and a linear objective function.

LP Formulation for Decentralized Target Calculations

For each MPC subsystem, which includes a target calculation and a control calculation, we can formulate an LP problem shown in Eq.(12) for the steady-state target calculation at time k.

$$\begin{aligned} &\min \quad c^T x(k) \\ &\text{s.t.} \quad A_{eq} x(k) = b_{eq}(k) \qquad (12) \\ &\qquad G x(k) \le b(k) \end{aligned}$$

where $x(k) = [u_s(k), y_s(k)]$ or $[\Delta u(k), \Delta y(k)]$ is a vector of steady-state input and output variables for the subsystem. The latter corresponds to the velocity form, where:

$$\Delta u(k) = u_s(k) - u_p(k \mid k-1) \qquad (13)$$

$$\Delta y(k) = y_s(k) - y_p(k \mid k-1) \qquad (14)$$

in which u_s and y_s are the steady-state set-points computed at the kth target calculation, and u_p and y_p are the predicted inputs and outputs at the previous target calculation. Here, the equality constraints are taken from the linear dynamic model:

$$Y(s) = G(s)U(s) + G_d(s)D(s) + E(s) \qquad (15)$$

with the steady-state model:

$$y = Ku + K_d d + e \qquad (16)$$

where d stands for disturbances and e for unmeasured noise. The inequality constraints result from physical limitations on the inputs and outputs, such as actuator limits or production quality. Other MPC target calculation formulations are possible such as including model bias and soft output constraints (Kassmann *et al.* 2000; Lestage, 2002).

Coordinated, Decentralized Target Calculations

The solution from the decentralized MPC target calculation may not be optimal because of plant/model mismatch resulting from ignoring the interactions between operation units. Since the interactions can be modeled as the linking constraints, we can find the plant-wide optimum by designing a coordinator to coordinate independent MPC target calculations in the decentralized MPC scheme. Here, we propose a coordinated, decentralized approach to find the plant-wide optimum in MPC target calculations by applying the Dantzig-Wolfe decomposition.

The interactions between two operating units can always be modeled by equating the output variables from the upstream unit and the input variables to the downstream unit. In formulating the subproblems, those streams connecting different operating units are torn and consistency relationships can be used to model the interactions between different units. Assume that we have m separate operating units. Each of them is controlled by one MPC subsystem. If n units have interactions, by introducing interprocess stream consistency as the linking constraints in Eq.(3a), we can formulate a large-scale LP problem that has a form of Eq.(3). This problem has a block-wise structure which can be solved efficiently by Dantzig-Wolfe decomposition.

At each sampling instant of the plant-wide MPC target calculation, a coordinated LP problem will be formulated and solved. Then all the calculated steady-state targets both for inputs and outputs, including interprocess stream variables, are passed to MPC control calculation.

Illustrative Case-Study

The system under consideration in this paper is shown in Fig. 3. The normalized gains for the system are given in Eq.(17a) through Eq.(17c). An identity matrix is chosen for K_d in Eq.(16) assuming that the disturbances influence the outputs directly. The locations where the disturbances entering the plant are shown as dashed lines in Fig. 3.

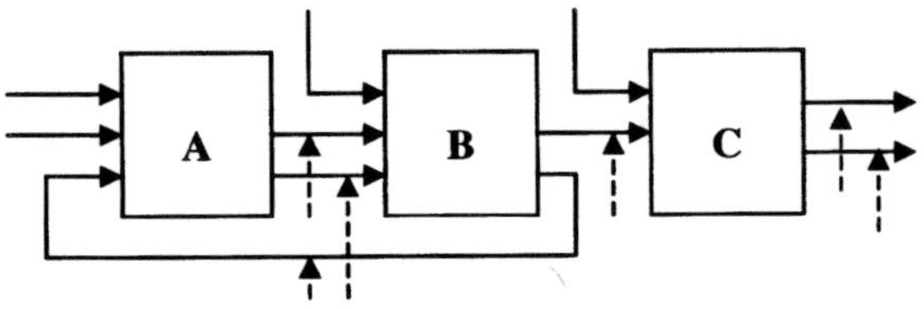

Figure 3. Interacting MIMO Operating Units

$$K_A = G_A(0) = \begin{bmatrix} 0.4 & 0.6 & 0.1 \\ 0.5 & 0.4 & 0.1 \end{bmatrix} \qquad (17a)$$

$$K_B = G_B(0) = \begin{bmatrix} 0.3 & 0.4 & 0.3 \\ 0.1 & 0.2 & 0.1 \end{bmatrix} \qquad (17b)$$

$$K_C = G_C(0) = \begin{bmatrix} 0.7 & 0.3 \\ 0.6 & 0.5 \end{bmatrix} \qquad (17c)$$

Each operating unit has its own objective, which is a subset of information used by plant-wide optimizers, and the profit function cost coefficients are:

$$c_A^T = [-1 \quad -1 \quad -1 \quad 3 \quad 3] \qquad (18a)$$

$$c_B^T = [-2 \quad -2 \quad -0.5 \quad 1 \quad 1] \qquad (18b)$$

$$c_C^T = [-1 \quad -2 \quad 5 \quad 5] \qquad (18c)$$

So for each operating unit, by tearing the interprocess stream, a linear program at the kth target calculation is:

$$\begin{aligned} &\min \quad c_j^T x_j(k) \\ &\text{s.t.} \quad K_j x_j(k) = b^j{}_{eq}(k) \qquad (19) \\ &\qquad G_j x_j(k) \le b^j(k) \quad j = A, B, C \end{aligned}$$

where G_j stands for the coefficient matrix associated with all the inequality constraints. The R.H.S. of the equality constraints $b^j{}_{eq}(k)$ represents the updated model bias at each target calculation execution. The R.H.S. of the inequality constraints $b^j(k)$ contains the lower bounds lb and upper bounds ub of the variables in the operation units. The bounds on the variables in this case study are shown in Eq.(20).

$$lb = [0.3\ \ 0.3\ \ 0.15\ \ 0.45\ \ 0.4\ \ 0.45\ \ 0.4\ \ 0.3\ \ 0.45\ \ 0.15\ \ 0.45\ \ 0.3\ \ 0.45\ \ 0.5] \quad (20a)$$

$$ub = [0.5\ \ 0.5\ \ 0.25\ \ 0.55\ \ 0.5\ \ 0.55\ \ 0.5\ \ 0.5\ \ 0.55\ \ 0.25\ \ 0.55\ \ 0.5\ \ 0.55\ \ 0.6] \quad (20b)$$

Three different MPC strategies, centralized MPC, decentralized MPC and coordinated, decentralized MPC, are implemented to evaluate their abilities to track the changing optimum in steady-state target calculations. For the centralized MPC target calculation, a direct LP problem is formulated treating all the inputs and outputs, including interprocess interactions, as decision variables. For the decentralized MPC scheme, separate LP problems are formulated by passing the upstream decisions to downstream units as disturbances. Finally, the coordinated MPC target calculation incorporates the linking constraints in modeling the interactions and solves the RMP and independent subproblems iteratively.

In our case study, unknown disturbances are generated by filtering random series of uniformly distributed variates in order to restrict these disturbances within the interval ±0.05. These unknown disturbances are directly imposed on the outputs when the optimized targets are implemented in our simulation. By using the autoregressive models in Eq.(21) as simplified disturbance models, we predict one-step ahead disturbances based on the past information of estimated disturbances. The estimated disturbances used to update the disturbance model in (21) are calculated by comparing the measured outputs and model predictions at every control execution. At the current control calculation, the parameters, c_1 and c_2, in the disturbance model are estimated using the estimated disturbances in the past 10 control execution periods. The one-step ahead disturbances are predicted using the estimated c_1 and c_2. The process models are updated using the predicted disturbances. The steady-state targets are then calculated by using the updated process models.

$$d(t) = \frac{1}{1 + c_1 q^{-1} + c_2 q^{-2}} e(t) \quad (21)$$

The following accumulated profit function is defined for performance comparison:

$$P = \sum Z(k) * T_s - \sum V(k) * T_s \quad (22)$$

where $Z(k)$ is the actual profit per unit time from the kth target calculation; $V(k)$ represents the penalty for constraint violations when we implement the calculated targets; and T_s is the sampling period between two target calculations.

A benchmark has to be defined for comparing the performances of different MPC steady-state target calculation strategies. The benchmark used for comparison is defined as the maximum profit achieved when the plant is operated at the true optimum, which is calculated using the perfect process model and exact knowledge of disturbances. Although this maximum profit is not achievable, it is a useful basis for performance comparison.

Fig. 4 shows the profit achieved as a function of control execution using different MPC steady-state target calculation strategies. The coordinated target calculation gives the same achievable optimum as the centralized MPC scheme does (because the same interactions are considered in both approaches), while the decentralized scheme yields lower performance as interactions are ignored in the calculation. Note that, in the figure, two curves for the centralized scheme and coordinated scheme overlap each other and are very close to the true optimum.

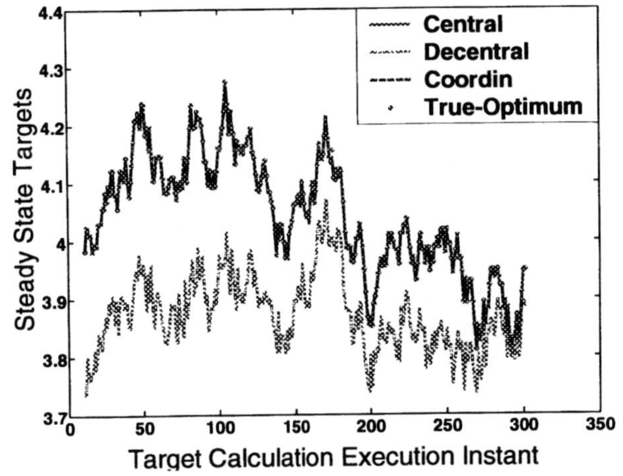

Figure 4. Calculated Targets Using Different Approaches

Table 1 compares the performance of different MPC strategies for a simulation of 290 target calculation executions. From Table 1, we can see that the centralized and the coordinated, decentralized target calculation give the same best achievable profit and achievable ratio to the true optimum, while the fully decentralized target calculation only captures around 95.7 % of the maximum profit.

Table 1. Performance Comparison

Strategy	Profit	Achiev. Ratio (%)	Comp.* Effort (s)	Problem Dimension
Central	1176.8	99.98	19.37	46×42
Decentral.	1126.8	95.73	34.14/3	15×15×3
Coordin.	1176.8	99.98	184.9/4	7×7 + 15×15×3
True-Opt.	1177.1	100	NA	NA

* All the simulations were performed in Matlab 6.5 on a Pentium III 1.0G Hz, 512M RAM machine.

Table 1 also reports the problem sizes for different steady-state target calculation strategies. The problem size is defined as the size of the coefficient matrix in the LP standard form used in the Simplex method. Therefore, slack and excess variables are added to convert the inequality constraints to equality constraints, and the columns of the coefficient matrices correspond to the process variables as well as the slack and excess variables. We can see that the centralized scheme has the largest problem size, which will grow significantly when the dimension of separate problems and the number of operation units in the flowsheet increase. The problem size for the decentralized MPC is reported as the dimension of the largest subproblem multiplied by the number of units. For the coordinated scheme, the problem size is expressed as the addition of two components, the dimension of RMP (the coordinator) and the problem dimension in the decentralized scheme (the subproblems).

The computational effort is also reported in Table 1. All the simulations were performed using one single-processor computer. Therefore, the computation times for the decentralized and coordinated, decentralized strategies reported are divided by the number of processors required for parallel computation, assuming a decentralized computational environment. As shown in Table 1, the decentralized scheme requires the least computational effort, because the large plant-wide optimization problem is broken down into smaller independent optimization problems being solved using multiple processors. Table 1 also shows that the coordinated approach requires approximately twice the computation time as compared to the centralized approach, even after accounting for the distributed computing environment that would normally be used for coordinated target calculations. To our best knowledge, little is known regarding the complexity or scaling of the Dantzig-Wolfe decomposition technique and since our example represents a single problem instance, no general conclusions can be drawn as to which approach is computationally "faster". What must be noted is that the coordinated approach does provide a level of subsystem failure tolerance, which the centralized approach cannot. Such failure tolerance is an important feature for industrial automation systems.

Further, in our example, no attempts were made to develop efficient codes for the decomposition algorithm such as those reported by Martinson and Tind (1999) and Dantzig and Thapa, (2002) in our simulations. Further studies are required to investigate the potential reduction in computation time using the coordinated, decentralized strategy. A key point that should be drawn from our study is that the proposed approach does yield acceptable computational effort when implemented in parallel. As such, it provides an approach to plant-wide control that does not require a centralized computing environment.

Conclusions and Future Work

In this paper, a new approach to solving plant-wide MPC target calculation by coordinating independent subproblems has been proposed. The coordinator is designed by considering the interactions between operation units, which are not considered in the existing decentralized MPC calculation scheme. In our study, the coordinated, decentralized MPC target calculation can provide the same performance as does the centralized scheme, while it can utilize parallel computation to ensure acceptable real-time calculation speeds. This scheme offers a higher reliability than the centralized scheme. In fact, the proposed scheme can provide the same reliability as the decentralized control scheme. The proposed approach only requires a minor modification in the existing decentralized MPC structure by applying Dantzig-Wolfe decomposition.

A number of challenges remain in the decomposition of the MPC target calculations. Since subproblems are solved independently, this property ought to be naturally employed for parallel computing, which may have strong potential to enhance the solution efficiency. Furthermore, a full understanding of the complexity and scaling behavior of Dantzig-Wolfe decomposition algorithm is required to determine the feasibility of the proposed approach on industrial-scale problems.

Acknowledgments

This study is part of the project on decomposition strategies for large-scale operations optimization problems. The authors would like to express their gratitude to NSERC for the financial support.

References

Chvátal, V. (1983). Linear Programming, W. H. Freeman and Company. New York.

Dantzig, G. B. and Wolfe, P. (1960). The Decomposition Algorithm for Linear Programming. Operations Res. 8.

Dantzig, G. B. and Thapa, M. N. (2002). Linear Programming 2: Theory and Extensions. Springer Verlag.

Gilmore, P. C. and Gomory, R. E. (1961). A Linear Programming Approach to the Cutting Stock Problem. Operation Res. 9.

Kassmann, D. E., Badgewell, T. A., Hawkins, R. B. (2000). Robust Steady-State Target Calculation for Model Predictive Control. AIChE Journal, Vol.46, No.5, 1007-1024.

Lestage, R., Pomerleau, A., Hodouin, D. (2002). Constrained Real-time Optimization of a Grinding Circuit Using Steady-state Linear Programming Supervisory Control. Power Tech., 124, 254-263.

Lasdon, L. S. (2002). Optimization Theory for Large Scale Systems, Dover Publications, INC. Mieola, New York.

Maciejowski, J. M. (2001). Model Predictive Control with Constraints.

Martinson, R. K. and Tind, J. (1999). An Interior Point Method in Dantzig-Wolfe Decomposition. Computers and Operation Research, 26, 1195-1216.

Rao, C. V., Rawlings, J. B. (1999). Steady States and Constraints in Model Predictive Control. AIChE Journal, Vol.45, No.6, 1266-1278.

Skogestad, S., Larsson, T. (1998). A Review of Plantwide Control. Technical Report, Norwegian University of Science and Technology.

Venkat, A. and Rawlings, J.B. (2003). Communication Based Decentralized MPC, Current projects of Rawlings Research Group, Chem. Eng. Depart. of Univ. of Wisconsin – Madison.

ELSEVIER
IFAC
PUBLICATIONS
www.elsevier.com/locate/ifac

CONTROL STRUCTURE DESIGN TO ACHIEVE MULTIPLE PERFORMANCE CRITERIA

Yongsong Cai and Thomas E. Marlin*

Department of Chemical Engineering, McMaster University
1280 Main Street West, Hamilton, Ontario, Canada, L8S 4L7

Abstract: The control structure has the dominant effect on control performance. Developing systematic methods for selecting the structure is challenging because of the multi-criteria nature of the problem. Some criteria include dynamic behavior of the controlled and manipulated variables, equipment capacities, and loop integrity. In this paper, a method is presented to address the complex nature of the problem, while using a formulation that can be solved in a reasonable computing time. The primary criterion is controlled variable performance, but additional criteria such as integrity are considered using meaningful metrics, such as Relative Gain Array. The crucial importance of using realistic scenarios and not relying on heuristics is demonstrated using two process examples. *Copyright © 2004 IFAC*

Keywords: control structure selection, decentralized control, plantwide control, complementarity formulation

1. INTRODUCTION

Control structure selection is an important decision at the beginning of the control system design. Available control design methods range from heuristic rules to approximate metrics to mathematical programming. Luyben, *et al.* (1998) presented a nine-step heuristic design procedure for plantwide control design. Bristol (1966) developed the most widely used metric, the Relative Gain Array (RGA), which only needs steady state gain and gives very useful information for integrity and robustness. A more complete analysis of control performance requires the transient response of key variables. Narraway and Perkins (1994) formulated the control structure selection and controller tuning together as a dynamic optimization using MINLP. Wang and McAvoy (2001) chose the control structure by optimizing dynamic performance for several subsets of controlled variables sequentially. Kookos and Perkins (2002) used linear dynamic model and integer variable for both structure and tuning selection, which makes a MILP problem with a very large number of integer variables.

This paper presents a control structure design method that extends the optimization of the time-domain transient response and gives the best overall behavior in industrial situations, even though the design might violate one or more common heuristics. Since control design is naturally a multiple criteria decision-making procedure, our method is flexible enough to consider several criteria simultaneously. In addition, the design must be performed for a set of conditions, i.e., a scenario, which represents the realistic situation. For example, the scenario should include realistic disturbances, noise, uncertainty and equipment bounds. Finally, since good controller tuning is essential when comparing alternative designs, our method includes a tuning optimization. To achieve these formidable results within reasonable computing time, we restrict ourselves to linear dynamic models and employ an efficient new optimization algorithm.

The paper is presented as follows; Section 2 describes the approach used for multiple criteria optimization, Section 3 defines the formulation for realistic scenarios, Section 4 discusses special

*Author to whom all correspondence should be addressed
Email: marlint@mcmaster.ca

aspects of the solution method, and Section 5 presents results from two cases studies.

2. MULTIPLE CRITERIA

There are many control design objectives (Marlin, 2000) that address a range of requirements including safety, smooth operation, product quality, and production rate. In addition, most systems involve inequality constraints on variables; some of these are hard (e.g., equipment performance limits) and others are soft (e.g., deviation from set point). The most often used method to formulate multiple criteria optimization is to penalize the weighted summation of all criteria. In spite of its mathematical simplicity, this method has a major drawback. The weighting should reflect the relative importance among criteria, but in most cases we do not know this sensitivity information before we solve the problem.

For multi-criteria optimization, we choose the e-constraint method (Steuer, 1986). It optimizes one criterion with the other criteria bounded as constraints.

$$\begin{aligned} &\min_{\mathbf{x}} c_j(\mathbf{x}) \\ &s.t.\ \ c_i(\mathbf{x}) \le e_i \quad \forall i \ne j \\ &\qquad \mathbf{h}(\mathbf{x}) = 0 \\ &\qquad \mathbf{g}(\mathbf{x}) \le 0 \end{aligned} \tag{1}$$

where $c_j(\mathbf{x})$ is the criterion as objective function and $c_i(\mathbf{x})$ are the criteria bounded by e_i. This method can be solved by standard optimization software and the result is easy to understand. Steuer (1986) points out that by solving the problem several times with different e_i, we can get a very useful byproduct, sensitivity among different criteria.

$$\frac{\partial c_j(\mathbf{x})}{\partial e_i} \tag{2}$$

Our major focus in this paper is dynamic performance; therefore, a measure of performance of the controlled variable deviation from set point is very natural choice for the objective function, c_j. Many other criteria could be included as the ancillary objectives, c_i. In this paper, we will demonstrate the approach by including integrity as an optional bound. Integrity has many definitions; here, we select "a system has integrity if the closed-loop dynamic system is stable when one or more loops are placed in manual (off) without changing the sign of the active feedback controller gains". One widely used metric is relative gain array (RGA), which is a necessary condition for good integrity.

3. FORMULATION WITH REALISTIC SCENARIO

Control performance is strongly affected by the scenario. Certainly, large, fast disturbances are worse than small, slow disturbances. However, many other characteristics are important in control design and many of these have been ignored in previously published methods. To achieve the flexibility required, we propose an optimization formulation that involves a linear dynamic model including a saturation model of the manipulated variable. The decision variables are the integer variables that specify the control structure and the controller tuning. The objective function is the integral error of the controlled variable deviation with a penalty for the control actions. The resulting formulation is given in the following equations.

$$\min_{\delta_{ij}, K_c, K_I, s_k^U(t), s_k^L(t)} \sum_k \sum_{t=1}^{tf} [\mathbf{Q}\cdot \mathbf{ae}_k(t) + \mathbf{R}\cdot \mathbf{au}_k(t)] \tag{3}$$

s.t.

$$\begin{aligned} &\mathbf{x}_k(t+1) = \mathbf{A}_k\mathbf{x}_k(t) + \mathbf{B}_k\mathbf{u}_k(t) + \mathbf{W}_k\mathbf{d}(t) \\ &\mathbf{y}_k(t) = \mathbf{C}_k\mathbf{x}_k(t) + \mathbf{V}_k\mathbf{d}(t) + N(t) \end{aligned} \tag{4}$$

$$\begin{aligned} &\Delta\mathbf{u}_k(t) = \mathbf{K}_c\left[\mathbf{e}_k(t) - \mathbf{e}_k(t-1) + \mathbf{K}_I\mathbf{e}_k(t)\right] \\ &\mathbf{e}_k(t) = \mathbf{sp}(t) - \mathbf{y}_k(t) \end{aligned} \tag{5}$$

$$\begin{aligned} &\mathbf{u}_k(t) = \mathbf{u}_k(t-1) + \Delta\mathbf{u}_k(t) - \mathbf{s}_k^U(t) + \mathbf{s}_k^L(t) \\ &s_{k,i}^U(t) \ge 0 \quad complements \quad u_{k,i}(t) \le u_{U,i} \\ &s_{k,i}^L(t) \ge 0 \quad complements \quad u_{k,i}(t) \ge u_{L,i} \end{aligned} \tag{6}$$

$$\begin{aligned} &-\mathbf{ae}_k(t) \le \mathbf{e}_k(t) \le \mathbf{ae}_k(t) \\ &-\mathbf{au}_k(t) \le \Delta\mathbf{u}_k(t) \le \mathbf{au}_k(t) \\ &\mathbf{ae}_k(t) \ge 0, \quad \mathbf{ae}_k(t) \ge 0 \end{aligned} \tag{7}$$

$$\begin{aligned} &-M_{ij}\delta_{ij} \le K_{c,ij} \le M_{ij}\delta_{ij} \\ &-M_{ij}\delta_{ij} \le K_{I,ij} \le M_{ij}\delta_{ij} \end{aligned} \tag{8}$$

$$\sum_i \delta_{ij} = 1, \quad \sum_j \delta_{ij} = 1 \tag{9}$$

$$\sum_j \lambda_{ji}\delta_{ij} \ge 0 \qquad \forall i \tag{10}$$

We use the summation of absolute error $\mathbf{ae_k}(t)$ and absolute manipulated variable movement $\mathbf{au_k}(t)$ as objective function in (3). Equation (7) calculates the absolute variable of error and manipulated variable movement that are used in objective function. The plant is described by a linear time-invariant state space model in. (4), and we use more than one model to represent model mismatch. In (4), subscription k describes different models, N(t) is measurement noise, $\mathbf{d}$(t) is disturbance. Multi-loop PI controllers are used, (5). The tuning parameters are controller gain, $\mathbf{K}_c$; integral gain, $\mathbf{K}_I$, which equals $\Delta T/\mathbf{T}_I$. Equation (6) expresses the capacity of the manipulated variable with upper and lower bounds, $u_{U,i}$ and $u_{L,i}$.

The slack variables, s^U_k(t) and s^L_k(t), express the amount that the controller output violates the bound on the manipulated variable. For instant, if the calculated manipulated variable output $u_{k,i}$(t-1)+$\Delta u_{k,i}$(t) is greater than its upper bound $u_{U,i}$, the actual manipulated variable output $u_{k,i}$(t) is clamped to its upper bound by allowing slack variable $s^U_{k,i}$(t) to be nonzero. The complementarity constraints guarantee that slack variables become nonzero only when the manipulated variables reach their limits. For instant, if manipulated variable $u_{k,i}$(t) is less than its upper bound $u_{U,i}$, that means inequality $u_{k,i}(t) \le u_{U,i}$ is not active, and by the complementarity constraints, the slack variable $s^U_{k,i}(t)$ must be zero. We note that this formulation also prevents reset-windup.

δ is a binary matrix that has the same size as the controller and defines loop pairing. We define $\delta_{ij} = 1$ if the controller pairs u_i with y_j and $\delta_{ij} = 0$ otherwise. Equation (8) states that the controller tuning parameters, $K_{c,ij}$ and $K_{I,ij}$ can be nonzero only when the corresponding δ_{ij} is one, which means the pairing $u_i - y_j$ is chosen. Equation (9) limits the possible structure to diagonal structure, which means that there is one and only one controller in each row and column.

Equation (10) forces the pairing we choose to have a positive RGA value, which is a necessary condition for integrity. As we will see in the case studies, enforcing integrity has a major effect on the control performance. Therefore, we will solve the problem with and without (10) to determine whether the good integrity property costs excessive degradation in control performance.

4. SOLUTION METHOD

Equations (3)-(10) define a mixed integer, non-convex, non-linear programming problem, which is challenging to solve. We take several steps to simplify the solution. First, we evaluate the relative gain at each node of the integer problem, i.e., at each pairing before optimizing the continuous problem. This procedure is appropriate because the (steady-state) relative gain depends on the process and the pairing, but not the disturbance or controller. To address the non-convexity of the continuous problem, we take two actions.

4.1 *Good Initial Points*

The optimal tuning problem is non-convex, so a good starting point is very important to solve the problem. Since we can choose a pairing with a negative RGA, some of the controller signs might need to be different from their SISO loop tuning. If a specific pairing has a negative relative gain, this does not ensure that the controller for this pairing should have its gain inverted. Clearly, the non-convex problem is complicated and covers a very wide range of tuning values. We propose a two-step procedure: a region elimination followed by an NLP in the identified region. The regions are defined by the controller gain (sign and magnitude) and the integral constant (magnitude). We have observed that good multiloop control can be achieved with an aggressive controller with a large $K_{c,ij}$ or a large $K_{I,ij}$ (small $T_{I,ij}$), depending on the particular problem. The regions are summarized in table 1.

We evaluate the control performance within each of these regions for the design scenario by simulation. While the table can contain many cases, each case is solved quickly. Then, we bound the tuning within the best region for the control design optimization problem.

Table 1 Possible tuning combinations

K_{c1}	K_{I1}	K_{c2}	K_{I2}	...
+low	low	+low	low	...
+high	low	-low	low	...
-low	high	+low	low	...
...	...	...	...	

4.2 *IPM Solver – IPOPT-C*

A major challenge remains because the valve saturation formulation (6) introduces the complementary constraints. In our experience, standard NLP solvers will not reliably solve this problem. Therefore, we have selected to use IPOPT-C, which is an Interior Point Method (IPM) solver solves general NLP. It has been extended to handle complementary constraints (Raghunathan and Biegler, 2003). The central-path following feature of this solver offers the potential for improved performance for this non-convex optimization. Good results have been obtained, but global optimization cannot be guaranteed.

5. CASE STUDIES

5.1 *Fluidized Catalytic Cracker (FCC)*

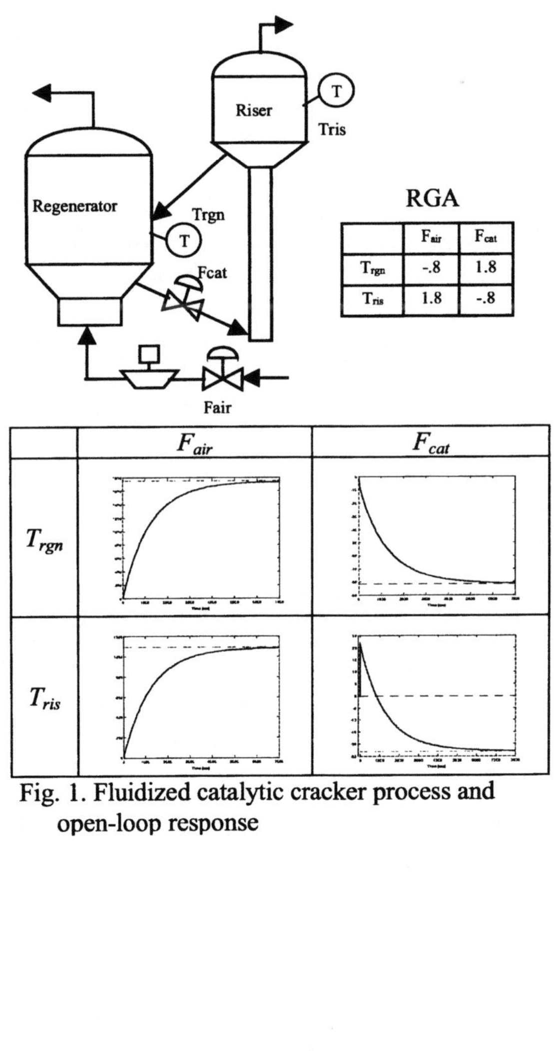

Fig. 1. Fluidized catalytic cracker process and open-loop response

We will apply the method to the loop pairing for the fluidized catalytic cracker unit shown in Fig. 1. It converts heavy oil into lighter and more valuable products, such as gasoline and fuel oil. The reaction section has two major units, one is a plug flow (transportation) reactor called the riser, which has a residence time of only a few seconds. The other unit is a fluidized bed vessel called regenerator, in which the catalyst is regenerated by burning the coke produced in the reactor using air. The inventory of catalyst in a typical regenerator is on the order of 60 tons. Two temperatures must be controlled to achieve smooth operation. The temperature of the riser (T_{ris}) is directly related to product quality and yield and to temperature limits of key equipment; therefore, it should be tightly controlled. The temperature of the regenerator (T_{rgn}) is not as crucial as riser, but it should be controlled in a range about its set point. Two available manipulated variables are flow of catalyst (F_{cat}) and flow of air (F_{air}).

The linearized model used in this study is based on Arbel *et al.* (1996). The open-loop response (Fig. 1) shows that the system has slow dynamics except loop T_{ris}–F_{cat}, which has a fast inverse response. The RGA for this 2x2 system is shown in Fig. 1. If we want the control system with integrity, we have only one choice, pairing on T_{rgn}–F_{cat}, T_{ris}–F_{air}. The other pairing will have poor integrity. Many heuristic design approaches would dictate that the relative gain must be positive. However, we will proceed with a systematic evaluation of the dynamic performance. The evaluation will begin with the simplest (unrealistic) scenario and add additional factors in the scenario until a realistic definition is attained (Table 2). Using this procedure, we will be able to follow the effects of the scenario on the control design.

Table 2 Cases for different scenarios

Case	A	B	C	D	E
RGA	+	+	+	+	-
Mismatch		×	×	×	×
Noise			×	×	×
Capacity				×	×
$K_p*K_c\approx$	10^6	100	1	1	1
IAE(T_{ris})	21.2	54.3	32.7	72.2	31.1

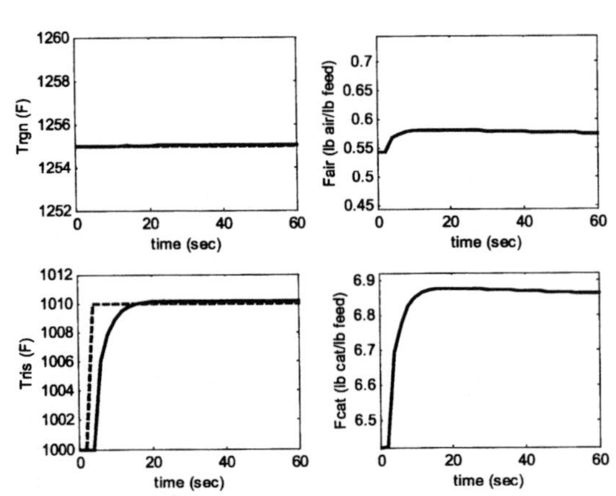

Fig. 2. Case A

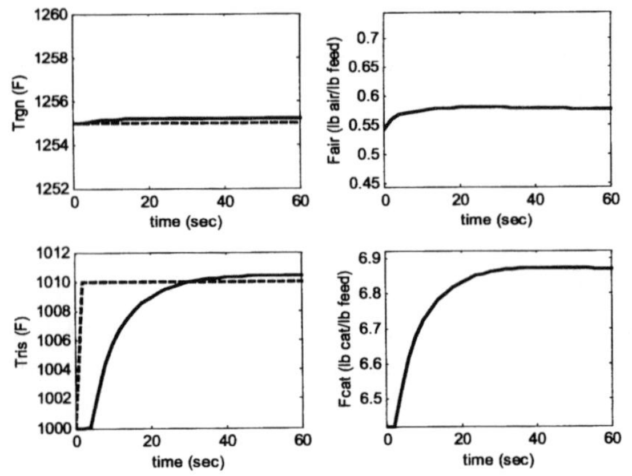

Fig. 3. Case B

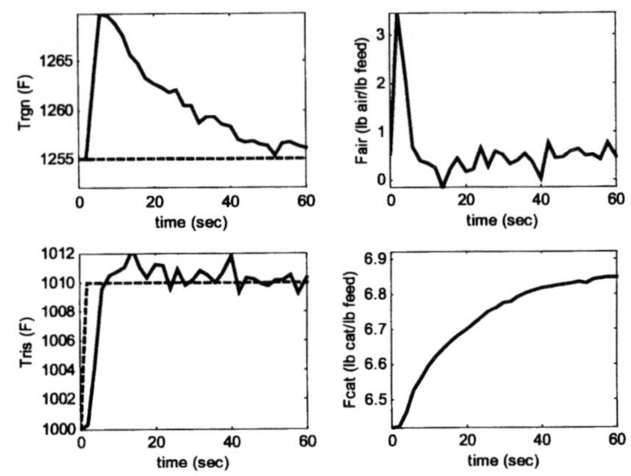

Fig. 4. Case C

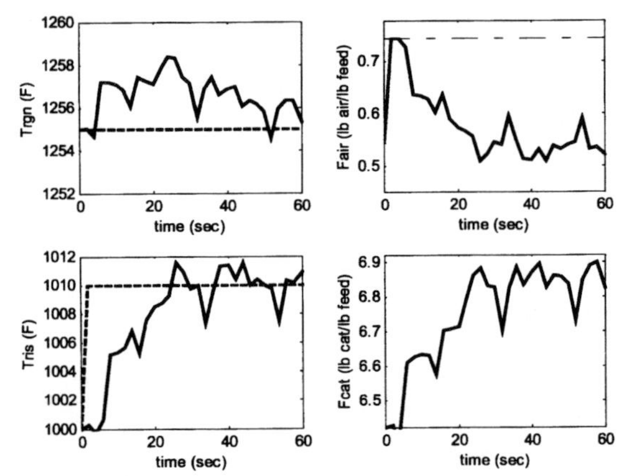

Fig. 5. Case D

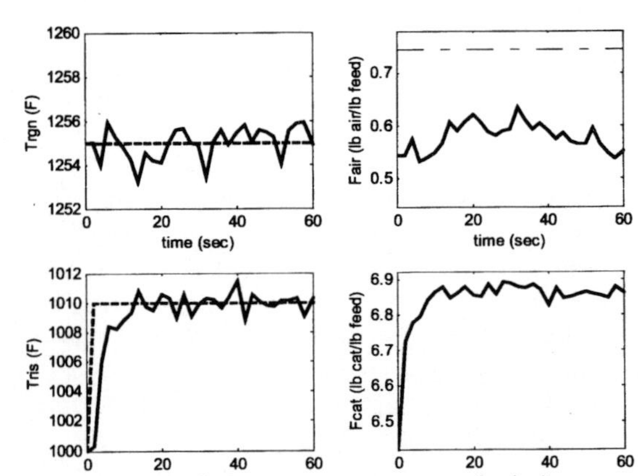

Fig. 6. Case E

We begin with the simplest scenario, without mismatch model, noise and valve constraints. The performance looks very good (Fig. 2). One guideline for tuning is that the product of the process gain K_p times the controller gain K_c should be around one. But, one of the controllers has a value of about 10^6! If we don't consider model mismatch, this tuning gives good performance, but a very small model mismatch will make the closed-loop system unstable. Fig. 3 shows the result for optimizing the base case and mismatch models simultaneously, with one tuning. The controller gain is significantly reduced but still very big. T_{rgn} only moves about 0.5°F, which will be overwhelmed by noise in real plant. The big controller gain will amplify the noise and make the performance much worse than its prediction. Fig. 4 shows the result with noise added in the formulation. The tuning now agrees with our guideline, and the performance is good. But in order to achieve this good performance, F_{air} has to make a big initial increase, which would require an air blower with about 600% extra capacity. This extra capacity would involve a two million dollar increase in plant capital investment. Fig. 5 shows the performance with a realistic extra plant capacity, which is the formulation we propose, which involves equations (3) – (10). The tuning and transient becomes reasonable, and the dynamic performance is realistic. In Fig. 5 the trajectory of F_{air} stays at its upper bound for several steps, which is saturation handled properly with the formulation. We note that the performance of the riser temperature has degraded in Fig. 5 and is likely not adequate for this critical variable.

Therefore, we chose to relax the requirement that the relative gain be positive. We report the results for the most demanding scenario in Fig 6, which shows the excellent performance for the riser temperature. The IAE with the negative RGA pairing is only half of the value achieved with the positive RGA pairing.

The final decision on the control structure is made by the engineer based on the rigorous optimization results. For FCC process, the temperature of riser is crucial, since it is directly related to profit and equipment protection. We want to keep it as close to its set point as possible. If the regenerator controller is placed in manual, the control system would be unstable because lack of integrity. However, the regenerator temperature would drift away slowly because of its huge volume, so the operators have time to respond. Based on this analysis we would choose dynamic performance and give up integrity; we choose negative RGA pairing, which is T_{rgn}–F_{air}, T_{ris}–F_{cat}. This is the industry standard control design for the FCC process considered in this example (Arbel, *et al.*, 1996). Note that it violates a commonly cited control heuristic, but it is widely applied.

5.2 Fired Heater

This case involves the control of the fired heater in Fig. 7, which has four burners and four coils in one firebox. The control goal is to manipulate the fuel to the four burners to control the four coil outlet temperatures. The transfer function model can be found in Manousiouthakis *et al.* (1986). The dynamics of each input-output is first-order, and the system has strong interaction. Equation (11) reports the relative gain array.

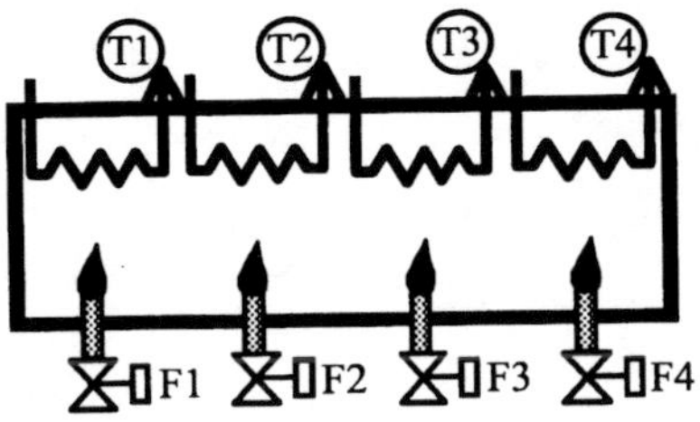

Fig. 7. Fired heater

$$RGA = \begin{bmatrix} \mathbf{1.748} & -0.686 & -0.096 & 0.034 \\ -0.727 & \mathbf{1.874} & -0.092 & -0.055 \\ -0.055 & -0.092 & \mathbf{1.874} & -0.727 \\ 0.034 & -0.096 & -0.686 & \mathbf{1.748} \end{bmatrix} \quad (11)$$

We see that only two pairings have positive RGA and good integrity. However, the previous case demonstrated the danger of basing a design on a heuristic. Therefore, we proceed to apply additional short-cut estimates of control performance that can be obtained from steady-state gain information. Here, we use the relative disturbance gain, RDG (Stanley et al, 1985). When the value of the RDG is less than 1.0, the multiloop performance, as measured by the integral error, is better than the single-loop performance.

$$RDG = \begin{bmatrix} 0.5479 & 0.2397 & 0.1027 & \mathbf{0.1096} \\ 0.3288 & 0.3425 & \mathbf{0.1370} & 0.1918 \\ 0.1918 & \mathbf{0.1370} & 0.3425 & 0.3288 \\ \mathbf{0.1096} & 0.1027 & 0.2397 & 0.5479 \end{bmatrix} \quad (12)$$

We see that all pairings are predicted to be better than single-loop, which indicates that the individual loops interact in a favorable manner. The highlighted pairings have the best (lowest) integral error. However, it is well known that the integral error can be small because of positive and negative cancellation (this is not the squared error).

Therefore, we proceed to the proposed design technique. We can evaluate the best performance with integrity (positive RGA pairings) and without the requirement for positive RGA pairings. In this case, the method finds the best dynamic control performance for a realistic scenario is achieved by the diagonal design highlighted in (11), see Fig. 8.

These rigorous results do not conform to the prediction from the RDG; in fact, the design indicated by (12) gives one of the poorest dynamic performances (Fig. 9). The reason is the complex feedback dynamics that occur with this pairing, which results in poor performance. This poor performance is not reflected in the heuristic, which is not affected by large positive and negative oscillations in the controlled variable that cancel. Again, we see that heuristics and approximate metrics can be misleading. We note that the RDG has proven useful in eliminating candidates in other studies, so that this case study does not generally invalidate the RDG metric, which should always be combined with a transient analysis before a design is finally selected.

The choice for the engineer in this case is easy, because good dynamic and integrity are achieved simultaneously. The diagonal controller highlighted in (11) is selected.

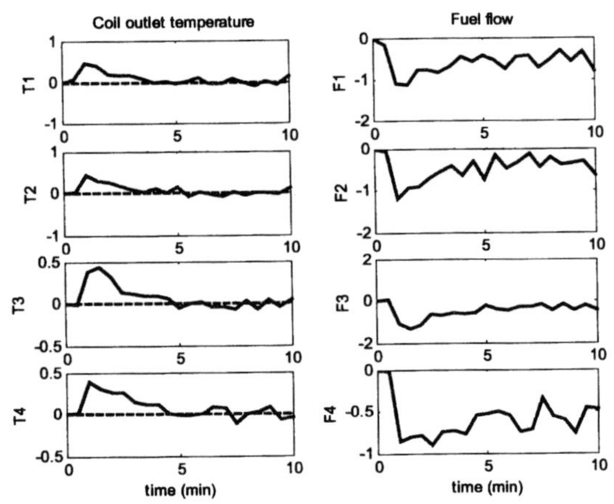

Fig. 8. Disturbance response of the pairing highlighted in (11).

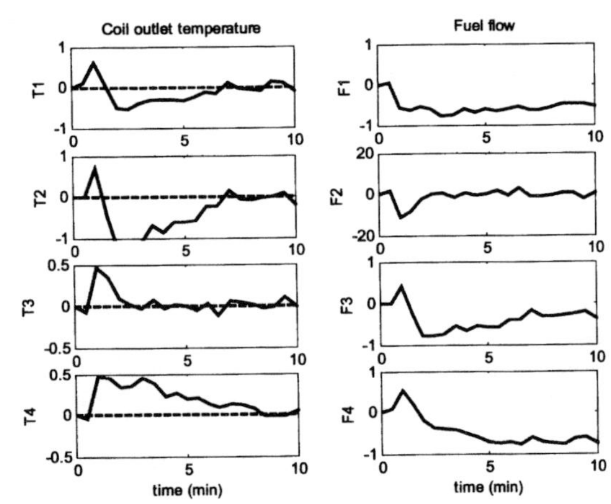

Fig. 9. Disturbance response of the pairing highlighted in (12).

6. CONCLUSION

This paper has discussed an optimization-based control structure design method. We demonstrate that a realistic scenario formulation is essential; this method includes important factors such as model mismatch, measurement noise and equipment capacity limits, to makes the resulting tuning and dynamic performance prediction close to the achievable performance in real. The multiple criteria framework enables us to tradeoff different design objectives, such as dynamic performance and integrity. We anticipate its application as an interactive procedure. Engineers can use their process knowledge to make the final decision based on results with different limits on the inequality constraints in (1).

Future work involves extending this methodology to handle more complex control structures, such as such a block-centralized structure.

REFERENCES

Arbel, A., R.H. Rinard and R. Shinnar (1996). Dynamics and control of fluidized catalytic crackers 3. Designing the control system: Choice of manipulated and measured variables for partial control. *Ind. Eng. Chem. Res.*, **35**, 2215-2233.

Bristol, E.H. (1966). On a new measure of interaction for multivariable process control. *IEEE Trans. Autom. Control*, **11**, 133-134.

Kookos, I.K. and J.D. Perkins (2002). Regulatory control structure selection of linear systems. *Compu. & Chem. Eng.*, **26**, 875-887.

Luyben, M.L., B.D. Tyreus and M.L. Luyben (1998). *Plantwide Process Control*. McGraw-Hill, New York.

Manousiouthakis,V., R.Savage and Y. Arkun (1986). Synthesis of decentralized process control structures using the concept of block relative gain. *AIChE Journal*, **32**, 991-1003.

Marlin, T.E. (2000). *Process control – designing process and control system for dynamic performance*, 2nd Ed, Chap 24. McGraw-Hill, New York.

Narraway, L. and J.D. Perkins (1994). Selection of process control structure based on economics. *Compu. & Chem. Eng.*, **18**, S511-S515

Raghunathan, A.U. and L.T. Biegler (2003). Mathematical programs with equilibrium constraints (MPECs) in process engineering. *Compu. & Chem. Eng.*, **27**, 1381-1392.

Stanley, G., M. Marino-Galarraga and T.J. McAvoy (1985). Short-cut operability analysis: 1. The relative disturbance gain. *Ind. & Eng. Chem. Proc. Des. Dev.* **24**, 1181-1188.

Steuer, R.E. (1986). *Multiple Criteria Optimization: Theory, Computation, and Application*. John Wiley & Sons, New York.

Wang, P. and T. McAvoy (2001). Synthesis of plantwide control systems using a dynamic model and optimization. *Ind. & Eng. Chem. Res.* **40**, 5732-5742.

ELSEVIER
IFAC
PUBLICATIONS
www.elsevier.com/locate/ifac

OPTIMAL SENSOR LOCATION FOR NONLINEAR DYNAMIC SYSTEMS VIA EMPIRICAL GRAMIANS

Abhay K. Singh and Juergen Hahn

Department of Chemical Engineering
Texas A&M University
3122 TAMU
College Station, Texas, 77845-3122, U.S.A.

Abstract: This paper presents a new approach for determining sensor locations for nonlinear dynamic systems. The method uses empirical observability gramians for observability analysis and combines the information from this investigation with observability measures which have been previously proposed in the literature. This approach offers the advantage over other methods in that it is directly applicable to nonlinear systems without resorting to linearization of the model. The presented procedure has been applied to a binary distillation column model. Additionally, the effect of scaling of a model for sensor placement has been examined as well as the conclusions which can be drawn from different observability measures. *Copyright © 2004 IFAC*

Keywords: Nonlinear systems, Observability, Sensor

1.0 INTRODUCTION

The chemical industry has gone through significant changes over the last few decades. Today, the emphasis is on running plants in the most optimal manner with due consideration to the safety aspects, rather than just achieving the design throughput. To meet these goals, information about a number of process variables and parameters is required. While it is possible to install additional sensors to monitor many of the aspects of a plant's operation, this can be expensive both in terms of the initial as well as the maintenance costs. In order to overcome this problem, model-based state estimation techniques can be used. By strategically measuring some key variables of the process, it is possible to compute many other variables and process parameters by using an observer. However, in order to gain the maximum benefit from this technique, the sensors have to be placed at "optimal" locations. This paper presents a method that allows computing such optimal sensor locations for plants described by nonlinear dynamic systems. Past research related to optimal sensor location has mainly been confined to linear or linearized systems (Dochain et al., 1997; Waldruff et al., 1998; Van der Berg et al., 2000, Muske, 2002). However, most chemical, petrochemical, or biochemical processes are accurately described by nonlinear dynamic systems and a linearized model may not represent the actual dynamics of the process over the entire region of operation. Due to this limitation, it is possible that different conclusions could be reached depending upon the operating point chosen for linearization. Therefore, it is desirable to use techniques, which will not have to resort to linearizing the model. The empirical gramians introduced by Lall et al. (2002) can form one piece of this investigation, because they can be used for determining observability/controllability of nonlinear systems over a specified operating region. Hahn and Edgar (2001) have shown that this new technique does provide a better representation of the input to state and state to output behavior of a nonlinear system over an operating region than the observability gramian of the linearized system. In this paper an approach for sensor placement based upon empirical observability gramians is presented. This is achieved by combining the results obtained from observability analysis via empirical gramians with measures, representing the degree of observability of a system. However, instead of proposing new measures for quantifying the degree of observability, existing measures have been used and adapted for the case where the observability analysis is based on empirical gramians. The advantage that such a technique has over other methods is that it does not resort to linearizing the system, while it can still be applied to nonlinear systems of significant complexity. Additionally, the results for different proposed measures are compared and the effect of scaling on this investigation is discussed. The methodology is applied to a distillation column separating a binary mixture.

2.0 PRELIMINARIES

2.1 Observability

Observability refers to the property of a system that allows the reconstruction of the values of the state variables given the outputs. For linear time invariant systems of the form

$$\dot{x}(t) = Ax(t) + Bu(t) \quad (1a)$$

$$y(t) = Cx(t) + Du(t) \quad (1b)$$

the observability gramian

$$W_o = \int_0^{\infty} e^{A^T t} C^T C e^{A t} dt \quad (2)$$

can be computed in order to determine observability of the system. If the observability gramian is a matrix of full rank then the system is observable. However, if the gramian is rank deficient then the system will not be observable and some of the states (or directions in state space) cannot be reconstructed from the output data.

2.2 Empirical observability gramian

While the observability gramian can be used for determining observability of linear systems, it may not result in sufficient information if the system is nonlinear. Extensive efforts have been undertaken in the last two decades to derive conditions for observability of nonlinear systems (Isidori, 1995). However, the results derived from these conditions are usually too complex to be interpreted for all but very simple systems. One alternative is to use the relatively new concept of empirical gramians that have been introduced by Lall et al. (2002). In order to present the definition of the empirical observability gramian the following quantities need to be defined:

$$T^n = \{T_1, \ldots, T_r; T_i \in R^{n\times n}, T_i^T T_i = I, i = 1, \ldots, r\}$$

$$M = \{c_1, \ldots, c_s; c_i \in R, c_i > 0, i = 1, \ldots, s\}$$

$$E^n = \{e_1, \ldots, e_n;\ \text{standard unit vectors in } R^n\ \}$$

where r is the number of matrices for perturbation directions, s the number of different perturbation sizes for each direction, and n the number of states of the system. The empirical observability gramian can be computed for general nonlinear systems

$$\dot{x}(t) = f(x(t), u(t)) \quad (3a)$$

$$y(t) = h(x(t)) \quad (3b)$$

and its definition is given by

$$W_O = \sum_{l=1}^{r} \sum_{m=1}^{s} \frac{1}{(rsc_m^2)} \int_0^{\infty} T_l \Psi^{lm}(t) T_l^T dt \quad (4)$$

where $\Psi^{lm}(t) \in R^{n\times n}$ corresponds to $\Psi_{ij}^{lm}(t) = (y^{ilm}(t) - y_{ss})^T (y^{jlm}(t) - y_{ss})$, $y^{ilm}(t)$ is the output of the system corresponding to the initial condition $x(0) = c_m T_l e_i + x_{ss}$ and y_{ss} is the steady state output of the system.

Empirical gramian can be computed for linear as well as nonlinear systems and reduces to the linear gramian (2) if the system is linear (Lall et al., 2002).

2.3 Measures for the degree of observability

While the rank of the observability gramian can be used in order to determine if a system is observable, this information is not sufficient for determining optimal sensor locations for a process. In order to address this problem, measures have been defined that quantify the degree of observability either of the entire system or of specific parts of the system. A wide variety of different measures for linear systems have been introduced over the last few decades and an overview over several of these is presented below.

Muller and Weber (1972) have outlined three candidates for measuring the degree of observability of a system based upon the linear observability gramian.

$$\mu_1 = \lambda_{\min}(W_o) \quad (5a)$$

$$\mu_2 = \frac{n}{trace(W_o^{-1})} \quad (5b)$$

$$\mu_3 = [\det(W_o)]^{1/n} \quad (5c)$$

Larger values of each of these measures imply an increased degree of observability of the system. However, these measures are mainly influenced by the smallest singular values of the observability gramian.

Dochain (1997) made use of the condition number

$$C.N. = \frac{\sigma_{\max}}{\sigma_{\min}} \quad (6)$$

for the observability analysis of a fixed bed bioreactor. The s refer to singular values of the observability gramian matrix. A smaller value of the condition number generally implies increased observability of a system. Waldruff (1998) made use of the minimum singular value of the observability matrix

$$N.S.(W_o) = \sigma_{\min}(W_o) \quad (7)$$

as one of the measures for determining sensor locations in tubular reactors. This method is similar to the measures based on the smallest eigenvalue given by Muller and Weber (1972). It serves as an indicator of how far the system is from being unobservable. Higher values of this criterion imply an increased degree of observability. All the above measures are strongly influenced by the minimum singular value of the system. Other approaches are based upon the maximal response which can be observed in sensor readings or upon the sum of the observability of all states. Van der Berg (2000) suggested two measures based upon these ideas. The first measure is the spectral radius

$$\rho(W_O) = \sigma_{\max}(W_O) \quad (8)$$

which corresponds to the largest singular value. A large value of the measure indicates that the dominant direction in the observability gramian can be easily observed. The second measure is based on the trace of the observability gramian:

$$trace(W_O) = \sum_{i=1}^{n} \sigma_i(W_O) \quad (9)$$

The trace can be interpreted as the sum of the singular values of the matrix. Larger values of the trace correspond to an increase in the overall observability of a system.

Summarizing, the presented measures can be put into two categories: measures which are mainly (or even exclusively) based upon the least observable direction in state space and measures which are predominantly influenced by the largest eigenvalue of the observability gramian. These findings are summarized in Table 1:

Table1: Measures for degree of observability

Smallest eigenvalue		*Largest eigenvalue*	
Measure	*Equation*	*Measure*	*Equation*
μ_1	(5a)	$\rho(W_O)$	(8)
μ_2	(5b)	$trace(W_O)$	(9)
μ_3	(5c)		
$C.N.$	(6)		
$N.S.(W_O)$	(7)		

2.4 Observability analysis and sensor placement

Sensor locations are often determined in order to be able to directly measure certain states to reconstruct required information of a system. As a minimum requirement, the controlled variables need to be measured for feedback control. However, other considerations, e.g. safety or product quality may result in a need for additional measurement locations. Observability analysis can be performed in order to maximize the amount of information gained from the available measurements. Thereby, it is possible to determine sensor location in order to either

- get the most information from a certain number of measurements
- use as small a number of measurements as possible in order to obtain a required amount of information about the system.

The following procedure is usually applied for determining optimal sensor locations for linear systems: observability gramians are computed for a variety of combinations of sensors at different locations. Scalar measures are computed from the gramians in order to compare the degree of observability of various locations for the sensor placement. A sensor configuration corresponding to the largest value of an observability measure indicates a good candidate for optimal sensor location.

3.0 OPTIMAL SENSOR LOCATION FOR NONLINEAR SYSTEMS

Since all systems in nature are nonlinear to a certain degree, a gramian of the linearized system will not always result in a good description of the state to output behavior if the operating conditions of a process can vary significantly. Due to this, it can happen that observability analysis based upon linear gramians may result in contradictory information depending upon the point of linearization of a system. In order to address these deficiencies, this work determines sensor locations by performing observability analysis of the nonlinear system via empirical gramians. This has the advantage that it can result in a more accurate description of the state to output behavior than if the gramian of the linearized system is used (Hahn and Edgar, 2002) and at the same time, it is computationally much less expensive than if Lie-algebra-based approaches (Isidori, 1995) would be used for observability analysis of nonlinear systems. While the empirical observability gramian is used for observability analysis, some of the measures that have originally been proposed for quantification of the degree of observability of linear/linearized systems can still be incorporated into this procedure: the measures need to be computed for the empirical observability gramian instead of the observability gramian of the linearized system. This is possible because the empirical gramian is an n-by-n, symmetric, and positive semi-definite matrix, similar to the gramian of a linear system. However, unlike the measures based upon linear gramians, these measures will make direct use of the nonlinear model without resorting to linearization. The procedure for computing optimal sensor locations by this approach is summarized in Figure 1.

3.1 Measuring observability of specific states

During process operations it is often required to have precise information about a particular state for process control or quality control of the product. In these cases, the values of specific states are of greater importance than the overall observability of a process. Many plants have requirements for both of these cases: certain states need to be exactly known and measured, whereas other measurements are used as an indicator of overall plant performance or safety. However, it is possible to take both of these situations into account with the proposed sensor placement framework. The observability of a specific state can be determined by analyzing the diagonal entry of the empirical observability gramian matrix corresponding to this state. The larger the magnitude of this entry, the easier it will be to reconstruct the value of this state from the available measurements. It should be pointed out that such an analysis is not always trivial. For example, if one is interested in the temperature at a specific location then this temperature should be directly measured. However, if the concentration of a product needs to be exactly known, then it is not always feasible to directly measure its value. Oftentimes, the value needs to be computed from measurements of other properties closely related to the state of interest. The presented sensor location framework can serve as an indicator which properties should be measured as well as where the measurements should be taken.

3.2 Effect of scaling on the optimal sensor location.

Most methods for determining optimal sensor locations are influenced by the magnitude of the singular values of the observability gramian. However, singular value decomposition is dependent upon scaling of the model, thereby possibly influencing sensor location decisions. A process model is generally scaled so that it reflects the sensitivity of the problem. That is, if a particular state is nearly unobservable then the scaled model should reflect poor response in the output for the perturbations in that state. The other reason to scale a model is to minimize the effect of round-off errors. Therefore, it would seem appropriate to perform scaling on all variables prior to the computation of the empirical gramian. However, scaling, though beneficial in most situations, can be a problem if the scaling factors are poorly chosen. In the past some authors have expressed their concern about bad-scaling choices. According to Paige (1981), a model should be well scaled so that it represents the actual physical picture. Waller and coworkers (1995) have stated that a bad choice of variable scaling may provide results that are not representative of the characteristics of the plant. Scaling of variables, which is essential in many cases for accurate physical interpretation and numerical stability of process model, represents an important part of

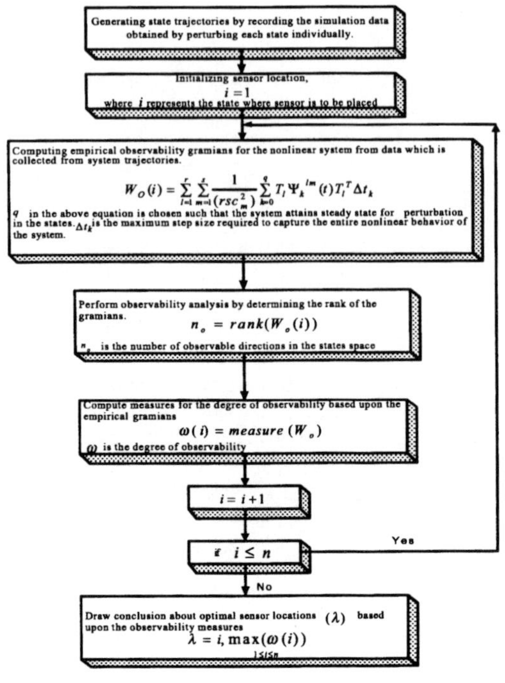

Fig. 1: Flowsheet for optimal sensor location.

computing measures for observability analysis. In order to address the scaling issue, the results from sensor location for the original model and for a scaled version of the model are compared in this work. Each variable, in the scaled model, is normalized by dividing it by its steady state value.

4.0 CASE STUDY

The presented sensor selection procedure will be applied to a specific example in this section. Since several different aspects will be investigated, the case study is limited to determining the optimal location if a single sensor is used. However, if the influence of placing a sensor on one part of the model has only a minor effect on the observability of other parts of the model, then conclusions can be drawn about subsequent placement of several sensors. The reader is referred to Hahn and Edgar (2002) for details on the calculation of the empirical gramians.

4.1 Distillation column model

For the analysis, a distillation column model with 30 trays for the separation of a binary mixture is considered (Hahn and Edgar, 2002). The column has 32 states and is assumed to have a constant relative volatility of 1.6. The feed is introduced in the middle compartment (17th tray) as a saturated liquid. The feed stream has a composition of x_f=*0.5*, distillate and bottom product purities are x_d=*0.935* and x_b=*0.065*, respectively. The reflux ratio is held at a constant value of 3.0.

4.2 *Measuring product observability*

In practice, composition sensors are generally placed at the top or bottom of the distillation column. The rationale behind this is to place the sensor close to the product for optimal product observability (Luyben, 1992). For the distillation column, the top and bottom products are given by the first and the last state of the model. Similarly, the first and the last diagonal entries of the empirical observability gramian correspond to the variance of the measurements that is caused by changes in these states. In order to gain as much information as possible about the product concentrations, it is desirable to determine sensor locations that maximize these two diagonal entries. Figure 2 shows a plot of the diagonal entries as the sensor location is varied along the height of the column for the original model and for the scaled system. It can be concluded from the figure that the most information about the bottom product can be obtained by directly measuring this product. However, contradictory conclusions could be drawn for the top product when comparing the results obtained for the scaled and the original model. The original model indicates that measuring the state corresponding to the top product will give the most information about this product. Additionally, the further the measurement is moved away from the top the harder it becomes to observe the top product. While the scaled model also indicates that measuring the top product is the best method for determining its value, the trend of decreasing "observability" of the top product as the measurement is moved away from this location does not hold for the scaled model. Instead, the scaled model could lead to the conclusion that measuring the bottom product may give more information about the top product than some states near the top product. These results are surprising at first glance, because they clearly indicate that scaling a model can have a negative effect on sensor selection by leading to conclusions that have no physical interpretation. The reason for this behavior is that all states in this model correspond to concentrations of the same component.

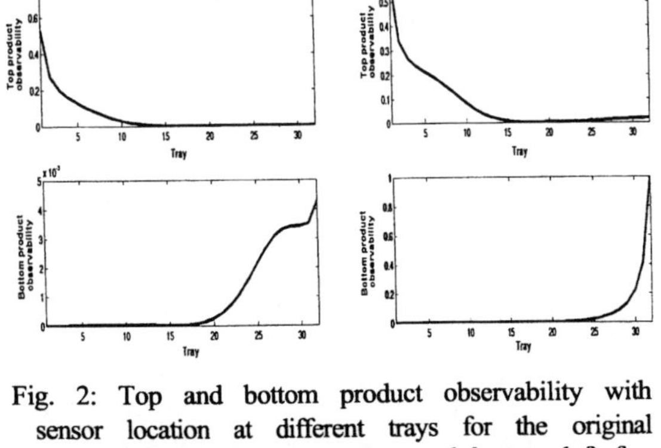

Fig. 2: Top and bottom product observability with sensor location at different trays for the original model (top left for top product and bottom left for bottom product) and scaled model (top right for top product and bottom right for bottom product)

Thereby, if a scaled model is used then a slight increase in the absolute value of the molar flow in one of the product streams can lead to a significant change of this value on a percentage basis. Due to this, the observability of the scaled model is skewed towards the states with lower concentrations (in this case towards the bottom of the column). While scaling in general may be an important part of this analysis, it can be concluded that at least for this specific case, it would result in misleading information.

4.3 *Determining optimal sensor locations*

Product measurement sensitivity is not the only factor when sensor locations are investigated. It is just as important to place sensors at locations where the most information about the state of the process can be gained. When analyzing the results obtained for sensor location by using the spectral norm and the trace of the empirical observability gramian, both measures return the same optimal location. However, when comparing the results for the scaled model and the original model (Figure 3), major difference between these two approaches can again be seen. For the original model the 6th state seems to be the best location and the 25th state would be the second best option, if one is interested in placing an additional sensor in the stripping section. For the scaled model, the bottom-most state in the stripping section (32nd state) would seem to be the best option. However, the sensitivity at the very top and bottom is at a minimum for most columns as the driving force for mass transfer is small compared to rest of the column (Luyben, 1992). Hence, the best sensor location should be at a certain distance from either end of the column. The results for the original model are in line with this physical knowledge, but the results for the scaled model would lead to contradicting conclusions. This investigation can serve as another indicator that special attention has to be paid when considering the use of scaling for sensor location. At least for this type of investigation, scaling should be avoided. Another conclusion that can be drawn from the results is that the feed tray is not a good choice for placing a sensor. This conclusion is based upon observations using both types of measures for the scaled and the original model. This seems physically

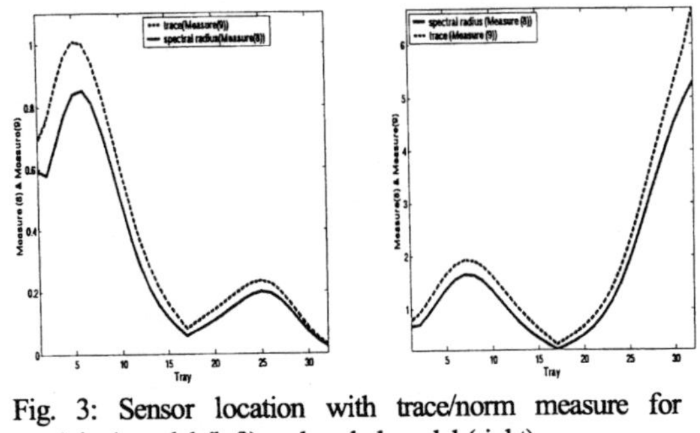

Fig. 3: Sensor location with trace/norm measure for original model (left) and scaled model (right)

justified as the feed tray is affected by the feed composition and is least sensitive to perturbations around its nominal point. In order to corroborate these findings, the location of the feed tray has been varied. Figure 4 shows the measures for a feed located at the 10th and the 25th tray. In both cases, the feed tray seems to be the worst location for placing a sensor. However, it should be pointed out that measuring the feed concentration, but not the concentration on the feed tray, can have beneficial effects if feedforward control strategies are to be applied. Another observation from this case study is that the measures based on the smallest singular value do not provide any relevant results for sensor placement.

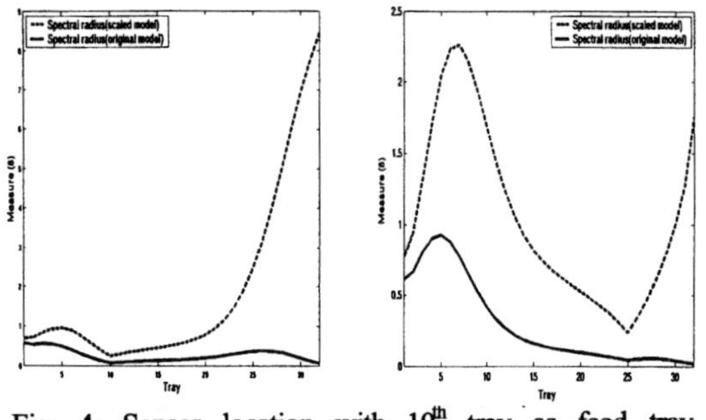

Fig. 4: Sensor location with 10th tray as feed tray (left) and 25th tray as feed tray (right)

5. CONCLUSION

This paper investigated a technique for computing optimal sensor locations for nonlinear dynamic systems. This has been achieved by first computing the empirical observability gramian for the nonlinear system over a pre-specified operating region and then determining observability measures based upon the empirical gramian. This method has the advantage over other techniques in that it does not resort to linearization of the model while at the same time it is computationally tractable. In addition it is possible to automate the presented procedure enabling computation of optimal sensor locations by formulating and solving an optimization problem.

When analyzing the various observability measures used for sensor location it can be concluded that for systems where the number of states far exceeds the number of measurements, as in the presented case study, methods based on maximal response energy provide better results than methods based upon maximizing observability of the least observable direction as the magnitude of the smallest singular value is often very close to zero. Spectral norm and trace of the empirical gramian are measures that can result in more accurate information, especially if some of the singular values are close to zero. In the presented case study, the measures indicate that there is one location in the stripping and one location in the rectifying section which are optimal for sensor placement.

Another important aspect of this work was investigation of the effect of scaling on the sensor placement. While scaling would seem to be a natural choice when computing gramians, it can lead to false conclusions in this investigation. The results indicate that scaling may not result in appropriate information if all the variables are representing the same type of physical property, i.e. concentrations for this case. The effect that scaling has on the computation of gramians when different types of variables are present, e.g. temperature and concentration variables, will be investigated in future work.

REFERENCES

Dochain, D., N. Tali-Mammar, and J.P. Babary (1997). On modeling, monitoring and control of fixed bed bioreactors. *Comp. Chem. Eng.*, 21, pp. 1255-1266.

Hahn, J. and T.F. Edgar (2002). An improved method for non-linear model reduction using balancing of empirical gramians. *Comp. Chem. Eng.*, 26, pp. 1379-1397.

Hahn, J. and T.F. Edgar (2001). A gramian based approach to non-linear quantification and model reduction. *Ind. & Eng. Chem. Res.*, 40, pp. 5724-5751.

Isidori, A. (1995, 3rd ed.). *Nonlinear Control Systems*; Springer Verlag, N.Y.

Lall, S., J.E. Marsden and S. Glavaski (2002). A subspace approach to balanced truncation for model reduction of non-linear control system. *Int. J. Robust and Non. Cont.*, 12, pp. 519-535.

Luyben, W.L. (1992). *Practical Distillation Control*; Van Nostrand Reinhold, N.Y.

Muller, P.C. and H.I. Weber (1972). Analysis and optimization of certain quantities of controllability and observability for linear dynamic system. *Automatica*, 8, pp. 237-246.

Muske, K.R. (2002). A methodology for optimal sensor selection in chemical processes. *Proc. Am. Control Conf.*, Anchorage, Alaska.

Paige, C.C. (1981). Properties of numerical algorithms related to computing controllability. *IEEE Trans. Aut. Control*, 26, pp. 130-138.

Waldruff, W., D. Dochain, S. Bourrel, and A. Magnus (1998). On the use of observability measures for sensor location in tubular reactor. *J. Process Control*, 8, pp. 497-505.

Waller, J.B. and K.V. Waller (1995). Defining directionality: Use of directionality measures with respect to scaling. *Ind. & Eng. Chem. Res.*, 34, pp. 1244-1252.

Van der Berg, F.W.J., H.C.J. Hoefsloot, H.F.M. Boelens, and A.K. Smilde (2000). Selection of optimal sensor position in a tubular reactor using robust degree of observability criteria. *Chem. Eng. Science*, 55, pp. 827-837.

ELSEVIER
IFAC
PUBLICATIONS
www.elsevier.com/locate/ifac

ROBUST DYNAMIC STATE ESTIMATION OF A BINARY DISTILLATION COLUMN

Roberto Baratti[(1)], Massimiliano Barolo[(2)], Fabrizio Bezzo[(2)] and Stefania Tronci[(1)]

[(1)]Dipartimento di Ingegneria Chimica e Materiali, Università degli Studi di Cagliari, Piazza D'Armi I-09123 Cagliari, Italy (baratti;tronci@dicm.unica.it)
[(2)]Dipartimento di Principi e Impianti di Ingegneria Chimica, Università di Padova, via Marzolo 9 I-35131 Padova, Italy (max.barolo;fabrizio.bezzo@unipd.it)

Abstract: A nonlinear geometric observer is developed, which infers the distillate and residue compositions in a highly-nonlinear binary distillation column from temperature measurements. The estimator performance is evaluated by imposing severe step changes to the input variables. In particular, the capability of the observer reconstruction is assessed with regard to ill-conditioning of the observability matrix, initialisation of the composition estimates, and noise in the temperature measurements. *Copyright © 2004 IFAC*

Keywords: Distillation column; State estimation; Geometric observer

1. INTRODUCTION

The product composition estimation problem in distillation columns stems from the fact that composition measurements are often not available on-line, either because an on-line composition analyzer is itself not available, or because the delay associated with the composition measurements makes it impossible to use these measurements within a composition control loop.

Estimating the dynamic evolution of product composition profiles in continuous distillation is an issue that has attracted a number of researchers. Basically, all approaches consider using temperature (and possibly flow) measurements to reconstruct the product composition dynamics. A pioneering work in this field is that of Joseph and Brosilow (1978), who used a linear combination of temperature and flow measurements to estimate product composition in a multicomponent column. More recently, Yu and Luyben (1989) proposed a nonlinear estimator based on the classical Wang-Henke computation algorithm; the selection of the optimal temperature measurements to be fed to the estimator was made through singular value decomposition of the steady state matrix of gains between tray temperatures and manipulated variables. Mejdell and Skogestad (1991) developed a static partial-least-square (PLS) estimator using steady-state temperature data only; because PLS regression is a linear technique, nonlinearities between temperatures and compositions were counteracted by using logarithmic transforms of these variables. Kano *et al.* (2000) proposed a dynamic PLS model where temperature, flow, heat duty and pressure data were regressed to provide product composition estimates; process dynamics was accounted for by including both current and past measurements into the input data set. Baratti *et al.* (1995) developed an extended Kalman filter using two tray temperature measurements to reconstruct the product composition dynamics in a pilot-plant column; they found that a critical issue in the reliability and robustness of the estimator is the tuning of the model error and measurement error covariance matrices.

Recently, Alvarez and Lopez (1999) proposed a nonlinear estimator based on the differential geometry theories. This method is principally characterized by its applicability to either observable or detectable systems, encompassing a broad class of plants in process system engineering. Furthermore, the proposed estimator design includes a robust local convergence and a systematic construction-tuning procedure. It was successfully applied to solve the local nonlinear estimation problem of a free-radical homopolymer reactor (Alvarez and Lopez, 1999; Alvarez, 2000), to infer the concentration in a catalytic reactor (Lopez *et al.*, 2002) and to monitor the pollutant concentration in an activated sludge process for wastewater treatment (Lopez *et al.*, 2004).

In this simulation work, a nonlinear geometric estimator is developed to provide on-line estimates of the dynamics of product compositions in a pilot-scale nonlinear binary distillation column using temperature and flow measurements. The estimator performance is discussed with respect to such issues as capability to reconstruct the system dynamics in response to severe inputs changes, and robustness with respect to composition profile initialization and noise in the temperature measurements.

2. THE PLANT MODEL

To avoid lengthy and expensive test runs on our pilot-scale distillation unit, the estimator performance is evaluated by using a detailed plant model as a surrogate of the real plant. The separation of a binary ethanol-water system at the atmospheric pressure is considered. It should be noted that this mixture exhibits a very marked nonideal thermodynamic behavior. As will be shown later, this provides a further challenge for the development of the estimator and may strongly impact its performance.

The model is a very reliable representation of a pilot-scale distillation unit equipped with thirty 300-mm sieve trays (tray spacing: 200 mm), a vertical thermosiphon reboiler, and a total shell-and-tube condenser. The average reflux drum and bottom holdups during normal operation are about 15 kg and 70 kg, respectively. The column is equipped with several Pt-100 probes to monitor on line the temperature profile on selected trays and on the bottom sump. Mass flow meters provide on-line measurements for the reflux rate, the distillate rate, the bottoms rate, and the feed rate. A vortex meter is used to measure the steam volumetric flowrate; the steam pressure and temperature are also measured on-line, so that a mass measurement is indirectly available for this stream. On-line composition measurements are not available, but liquid can be sampled and analyzed off-line by means of a gas chromatograph. All the measurements that are available on-line in the plant can be regarded as outputs that can be possibly made available to the estimator.

The plant dynamic model is based on the usual collection of material and energy balances, vapor-liquid equilibrium relationships, Murphree tray efficiency equations, and tray hydraulics correlations. In particular, the hydraulics of the liquid phase on the column trays is described by means of the nonlinear version of the Francis weir formula based on the actual tray geometry. The adjustable model parameters were kept to a minimum and were tuned by using the procedure outlined by Barolo **et al.** (2003). A very satisfactory match between simulated and experimental data was observed, both at steady state and in dynamic mode. In particular, temperature and composition profiles were reproduced very accurately, which is a crucial issue if the model has to be used to assess the performance of a composition estimator. In the following, this detailed model will be referred to as "the plant", against which the estimator performance is evaluated.

3. THE NONLINEAR ESTIMATOR

3.1 **The simplified model**

The nonlinear estimator is based on a simplified model of the plant, where the states simulated by the model are corrected by the available measurements. The main feature of the estimator is that it should be suited for on-line applications, hence the use of a simple model is to be preferred.

The simplified model (Baratti **et al.**, 1995) considers only the mass balance for one of the components of the binary system, while the energy balance on the trays is neglected by assuming that the vapor and liquid molar flows are constant in each section of the column. Following this assumption, the dynamics of the molar tray hold-ups is neglected. Tray holdup values are updated after an input variation by considering the steady-state values.

Vapor-liquid equilibria (VLE) at every stage are calculated by using the NRTL model for the estimation of activity coefficients, and assuming an ideal vapor phase. The Antoine equation is used to describe the vapor pressure as a function of temperature. The parameter values are reported in Gmehling and Onken (1977). Note that the parameters for the NRTL model and the vapor pressure equations are different from those used in the detailed model previously described. This mimics the fact that VLE cannot be described perfectly, an issue that may severely impact the estimator performance.

The composition y_j of the vapor phase leaving the generic j-th stage is calculated from the corresponding equilibrium value y_j^* by using the Murphree efficiency equation:

$$E_j = \frac{y_j - y_{j-1}}{y_j^* - y_{j-1}} \tag{1}$$

The efficiency was kept constant for all trays and was used as a tuning parameter to calibrate the steady state model, obtaining a value equal to 0.54 (this value is different from the one used in the detailed model). Finally, a linear pressure profile is assumed along the column, taking an overall pressure drop of 42 mmHg.

3.2 **Estimator design**

The nonlinear estimator developed to reconstruct the dynamics of the distillate and residue compositions is based on the differential geometry theory. A detailed discussion on the construction of the estimator can be found in Alvarez and Lopez (1999). Only a brief summary of the procedure will be given here.

Let consider the nonlinear dynamics of the distillation unit, according to the following model representation:

$$\dot{x} = f(x,u(t),r),\ \ x(t_0) = x_0,\ \ x \in \Re^n \qquad (2)$$
$$y = h(x,r),\ \ y \in \Re^m$$

where x are the states, u are the inputs, r are the parameters and y are the measured outputs, that in this case are the two pilot tray temperatures.

According to Alvarez and Lopez (1999), the motion $x(t)$ is detectable (i.e. partially observable) by means of the observability indexes $k1$ and $k2$ (where $k1$ and $k2$ are such as $k1+k2=k \leq n$) and the state partitions x_I and x_{II} (where x_I are the observable states and x_{II} are the unobservable states), if the following conditions are verified along the motion:

a) the map $\Phi(x,u(t),r)$ is invertible;
b) the unobservable dynamics x_{II}, is stable.

The map Φ is given by:

$$\Phi(x,u(t),r) = \left[h_1,...,L_f^{k1-1}h_1,h_2,...,L_f^{k2-1}h_2\right]^T \quad (3)$$

where $L^i_f h_j$ indicates the recursive directional derivative of the time-varying scalar field $h_j(\mathbf{x},t)$, with $j=1,2$ along with the time-varying vector-field $\mathbf{f}(\mathbf{x},t)$.

Condition b) is satisfied, since the column is stable. Condition a) can be verified noting that the observability matrix Q is equal to the Jacobian of Φ, that is:

$$Q = \frac{\partial \Phi}{\partial \mathbf{x}_I} \qquad (4)$$

Therefore condition a) is verified if the observability matrix has rank equal to $k=k1+k2$. If the above conditions are verified the following estimator can be constructed:

$$\dot{\hat{x}}_I = f_I(\hat{x}_I,\hat{x}_{II},u(t),r) + Q^{-1}K_0\left[y - h(\hat{x}_I,r)\right]$$
$$\dot{\hat{x}}_{II} = f_{II}(\hat{x}_I,\hat{x}_{II},u(t),r) \qquad (5)$$
$$\hat{y} = h(\hat{x}_I,r)$$

where Q^{-1} is the inverse of the observability matrix, and K_0 is the constant gain matrix , whose entries are selected such as the error dynamics is stable (Alvarez and Lopez, 1999).

The first step to consider is the selection of the estimator structure, that is the dimension of the observable and unobservable state vectors. The selection is strictly related to the robustness of the estimator. Here, only two temperature measurements are assumed to be available to the estimator; augmenting the dimension of the observable states leads to an increase of the Lie derivative order (see equation 3). This fact is detrimental to the robustness of the estimator and makes the correction algorithm highly complex from the algebraic point of view. It is worth noting that the VLE relationship between temperature and composition of the liquid phase is described by an implicit function, in which the NRTL model and the Antoine expression appear. The complexity of the map Φ, and then of the matrix Q, grows as the order of the Lie derivative is increased. As result, the computational effort grows significantly.

In this work, several estimator structures were considered, each one obtained by changing the dimension of the observed state vector and the location of the two measured outputs. The results that have been obtained show that observing only two variables leads to the best compromise among efficiency, robustness and simplicity of the nonlinear estimator. In this case, the invertibility of the matrix Φ is satisfied if the observable state vector entries are the liquid compositions on the tray where the sensors are located. These states depend on temperature through the VLE relationships. Therefore:

$$Q = \begin{bmatrix} q_{11} & q_{12} \\ q_{21} & q_{22} \end{bmatrix},$$
$$q_{11} = \frac{dy_1}{dx_{I,1}},\ \ q_{12} = q_{21} = 0,\ \ q_{22} = \frac{dy_2}{dx_{I,2}} \qquad (6)$$
$$Det[Q] \neq 0$$

where $x_{I,1}$ and $x_{I,2}$ represent the entries of the observable state vector $\mathbf{x}_I$, i.e. the liquid composition on the trays where the temperatures y_1 and y_2 are measured.

No particular strategy was used to select the measurement sensor locations, because this aspect was not of interest to the present work, but it will be considered in the future. A heuristic approach based on consideration about the dynamic behavior of the system, as well as on some test runs carried out using different measurements locations, showed that the best choice is to use the temperature of the bottom and of the 22nd tray (the 23rd stage of the column numbering from the reboiler). In particular, the bottom of the column was selected because it has the largest inertia of the system, i.e. the slowest dynamics. On the other hand the 22th tray is a compromise between closeness to the top of the column, and robust observability.

The choices described above lead to the following estimator structure:

$$k = 2,\ \ k1 = 1 \text{ and } k2 = 1$$

$$\mathbf{x}_I = \left[\mathbf{x}_{I,1}, \mathbf{x}_{I,2}\right]^T = \left[\mathbf{x}_1, \mathbf{x}_{23}\right]^T$$

$$\mathbf{x}_{II} = \left[\mathbf{x}_2, \ldots, \mathbf{x}_{22}, \mathbf{x}_{24}, \ldots, \mathbf{x}_{32}\right]^T \qquad (7)$$

$$y = \left[y_1, y_2\right]^T = \left[T_1, T_{23}\right]^T$$

where x_1 represents the ethanol mole fraction in the bottom, and x_{32} corresponds to the ethanol mole fraction in the distillate.

According to structure (6) it results:

$$\Phi = \left[T_1, T_{23}\right]^T$$

$$Q^{-1} = \begin{bmatrix} \left(\dfrac{dT_1}{dx_1}\right)^{-1} & 0 \\ 0 & \left(\dfrac{dT_{23}}{dx_{23}}\right)^{-1} \end{bmatrix} \qquad (8)$$

The entries of the gain matrix K_0 are the estimator tuning parameters. As proposed by Alvarez and Lopez (1999), they were selected as follows:

$$K_0 = \begin{bmatrix} k_{11} & 0 \\ 0 & k_{22} \end{bmatrix} \quad \begin{matrix} k_{11} = \omega_1 \\ k_{22} = \omega_2 \end{matrix} \qquad (9)$$

where ω_1 and ω_2 are the characteristic frequencies, which can be selected by considering the inherent dynamics of the column, that is by allowing the estimator response to be faster than the column one. It is worth noting that the gain matrix K_0 is multiplied by the inverse of the observability matrix, whose entries are the inverse of the derivative of temperature with respect to the observed states. This derivative is almost zero when the ethanol concentration is close to the azeotrope point, leading to a difficult calibration of the estimator. This situation can occur on the 22nd tray, when a high purity in the distillate is required. The best compromise between efficiency and robustness of the estimator was obtained when ω_1=0.003 and ω_2 =0.01.

4. RESULTS AND DISCUSSION

The estimator performance was evaluated by considering two different tests. The first one was a somewhat idealized situation, where the estimator was initialized correctly and the temperature measurements were noise-free. The second one was designed to assess the robustness of the estimator when the initial composition profile is not known exactly and temperature measurements are affected by noise.

The first experiment (Run 1) was mainly designed to evaluate the estimator structure, that is the dimension of the observability state vector and the location of measurement sensors. In this case the filter was initialized correctly by calculating the initial composition profile through the simplified model, and assuming that all the input variables were known exactly. Furthermore, the temperature measurements were not affected by noise. This experiment was conducted in several steps, which are summarized in Table 1.

In all the step tests, the feed temperature was kept constant at 295.3 K and the bottom pressure was equal to 801 mmHg.

Table 1. Operating conditions of the first test (Run 1)

Time (min)	Feed flowrate (kg/h)	Feed ethanol mole fraction	Reflux flowrate (kg/h)	Steam flowrate (kg/h)
0	125	0.143	107.4	90.0
100	125	0.143	161.0	90.0
300	125	0.143	107.4	90.0
500	125	0.143	107.4	67.5
700	125	0.143	107.4	90.0
900	162.5	0.143	107.4	90.0
1100	125	0.143	107.4	90.0
1300	125	0.186	107.4	90.0

The results related to Run 1 are shown in Figures 1a and 1b, where the ethanol mole fraction in the distillate stream and in the residue as estimated through the nonlinear geometric observer are compared to the relevant composition profiles in the plant. In the Figures, also the ethanol mole fraction calculated through the simplified model is reported, in order to show how the nonlinear geometric observer improves the estimation results. In particular, for both the top and the bottom products the geometric observer corrects the simplified model dynamics by improving the speed of response to the input changes. In fact, the results show that the simplified model exhibits a slower dynamics with respect to the actual one.

The nonlinear estimator reconstructs the bottom composition profile with a great accuracy (Figure 1b), except when a step change in the feed composition is imposed to the system (t = 1300 min). In this case the correction is less efficient and there is an offset at the new steady state condition. It is worth noting that in this situation the simplified model is not able to reconstruct the bottom composition dynamics at all. Note that the feed composition is not a measured input; hence its updated value was not fed either to the estimator or the simplified model.

Different results are obtained for the reconstruction of the distillate composition (Figure 1a). Although the nonlinear observer properly corrects the dynamics of the simplified model, offsets at the new steady states occur for all the step changes imposed, except when the feed flowrate is varied (t = 900 min). This behavior is related to the value of the derivative of the temperature with respect to the liquid composition on the 22nd tray. This derivative decreases by two orders of magnitude when the liquid composition on this tray varies from 0.135 to 0.85 mole fraction, leading to an increase of the correction term, that results in a excessive contribute of the latter with respect to the model term (see equations 5). A tuning parameter ω_2 smaller by two orders of magnitude would improve results at steady state, but with an unfavorable effect on the speed of response, making the estimator less prompt. Anyway, it is worth noting that the relative estimation error is always less than 1.3 %, and this value is within the experimental error of the liquid composition analysis. Furthermore, the ethanol mole fraction in the liquid phase at equilibrium is 0.88, when temperature is equal to 78.009 °C and pressure is kept constant at 750 mmHg (nominal condition on the 30th tray). When temperature is varied to 78.015 °C, the ethanol mole fraction at equilibrium becomes equal to 0.87. The temperature difference is much lower than the precision of the experimental measurement sensors; hence, requiring a higher precision in the distillate composition estimation is meaningless. Nevertheless, it is important to stress that the largest offset appears when the feed composition is changed (t = 1300 min), because the new feed composition is not updated to the estimator. Also in this case, the estimator significantly improves the simplified model composition profiles reconstruction.

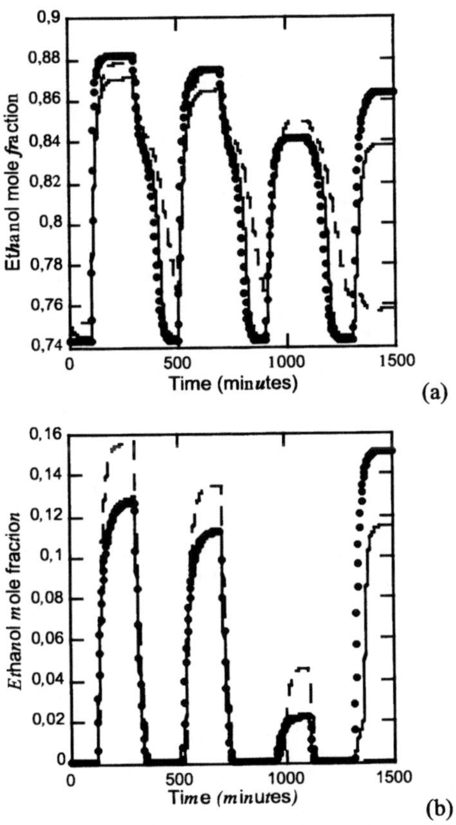

Figure 1. Comparison between plant (•), simplified model (dashed line) and estimated (solid line) ethanol mole fraction in the distillate (a) and residue (b) for Run 1.

The experiment described above deals with a somewhat idealized situation, because the temperature measurements were supposed to be noise-free and the estimator was initialized with a correct composition profile. In order to evaluate the robustness of the proposed nonlinear observer, a more demanding test was considered. In particular, the initial composition profile was calculated supposing that an incorrect feed composition (equal to 0.16 ethanol mole fraction) was available. Furthermore random numbers with zero mean and a standard deviation of 0.2 °C were superimposed to the temperature measurements in order to simulate noise.

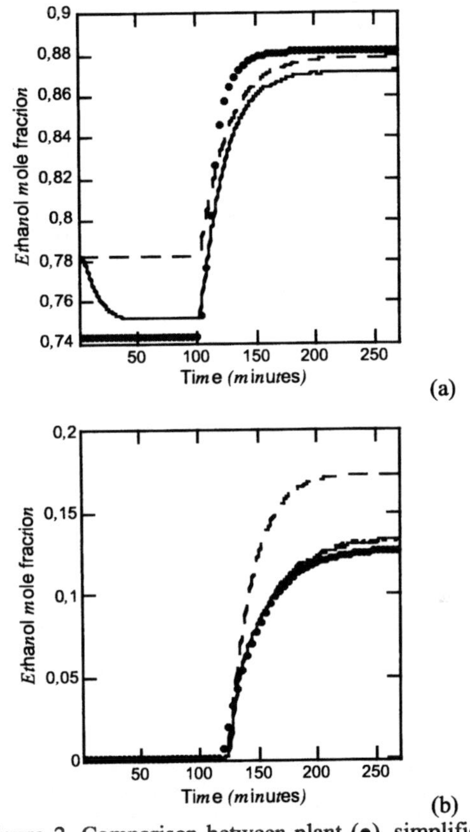

Figure 2. Comparison between plant (•), simplified model (dashed line) and estimated (solid line) ethanol mole fraction in the distillate (a) and residue (b) for Run 2.

The comparison between plant, simplified model and estimated ethanol mole fractions in the products are represented in Figures 2a and 2b, when a step change in the reflux flowrate from 107.4 to 161.0 kg/h is imposed to the system (Run 2). The results show that noise does not affect the estimator performance. This is mainly due to the low values of the constant gains k_{11} and k_{22}, which prevent that detrimental oscillations appear in the estimated composition profiles. On the other hand, the impact of the incorrect initialization is more evident. The estimator slowly corrects the initial states, reaching (in about 30 min) a new steady state value that differs from the exact one by ~0.003 (mole fraction). In this case the small values of the tuning parameters have a negative effect on the initial speed of response; however, it is important to note that the offset is significantly reduced with respect to the simplified model predictions.

5. CONCLUSIONS

A nonlinear geometric observer was developed to estimate the distillate and bottom composition profiles in a highly nonlinear binary distillation column. The results were quite promising, although the structure of the correction algorithm led to adjust the estimator tuning parameters in a conservative way. This is mainly due to the structure of the correction term, based on the inverse of the observability matrix. This structure implies that the strength of the estimator correction term becomes excessive when the matrix determinant value is close to zero, that is when the ethanol concentration value is close to the azeotrope point. With a proper location of the measurement sensors, the estimator was tuned in such a way as to obtain a good compromise between efficiency and robustness. The estimator correction term was somewhat ineffective when the ethanol concentration in the feed was changed. However, it was evident that the estimator greatly improved the results when compared to the simplified model.

Nevertheless, the conservative tuning also implied positive effects; in fact the estimator performance was not affected by measurement noise. Furthermore, the selected calibration allowed to reduce the offset between the actual and the estimated ethanol concentration in the distillate in response to the input changes, allowing the estimator to provide composition estimates within the expected experimental error.

The presented results are the outcomes of a preliminary study on the development of a geometric observer for a simulated distillation process, and further investigations have been already planned in order to complete the survey. In particular, future work will be addressed on developing a strategy to select measurement location as well as on the compensation of estimator bias, when it exists. Furthermore, the estimator performance will be tested on the real pilot plant.

ACKNOWLEDGMENT. The authors acknowledge MIUR (grant 2002095147, PRIN 2002) for the financial support, and Prof. Jesus Alvarez, who contributed to the development of the present paper by sharing his knowledge and ideas on the geometric approach.

REFERENCES

Alvarez, J. and T. Lopez (1999). Robust Nonlinear State Estimation of Nonlinear Plants. **AIChE** *J.*, *45*, 107-123

Alvarez, J. (2000). Nonlinear State Estimation with Robust Convergence. *J.* ***Process* Control**, *10*, 59–71.

Baratti, R., A. Bertucco, A. Da Rold, and M. Morbidelli (1995). Development of a Composition Estimator for Binary Distillation Columns. Application to a Pilot Plant. **Chem. Eng. Sci.**, *50*, 1541-1550.

Barolo, M., G. B. Guarise, C. Marchi, S. A. Rienzi and A. Trotta (2003). Rapid and Reliable Tuning of Dynamic Distillation Column Models on Experimental Data. **Chem. Eng. Commun.**, *190*, 1583-1600.

Gmehling, J., and U. Onken, 1977, Vapour-Liquid Equilibriun **Data Collection. Acqueos Organic Systems**. DECHEMA, Chemistry Data Series, Vol. 1, Part. 1 p.162

Joseph, B. and C. B. Brosilow (1978). Inferential Control of Processes. Part I: Steady State Analysis and Design. **AIChE** *J.*, *24*, 485-492.

Kano M., K. Miyazaki, S. Hasebe, and I. Hashimoto (2000). Inferential Control System of Distillation Compositions Using Dynamic Partial Least Squares Regression. *J.* **Process Control**, *10*, 157-166.

Lopez, T., S. Tronci, R. Baratti and J. Alvarez (2002). State Estimation in a Catalytic Reactor via a Contructive Approach. **IFAC 15th World Congress**, Barcelona (Spain).

Lopez, T., A. Pulis, M. Mulas, and R. Baratti (2004). A Software Sensor for a Wastewater Treatment Plant. **ADCHEM** Hong Kong.

Mejdell, T., and S. Skogestad (1991). Estimation of Distillation Compositions from Multiple Temperature Measurements Using Partial-Least-Squares Regression. **Ind. Eng. Chem. Res**, *30*, 2543-2555.

Yu C. C., and W. L. Luyben (1989). Control of Multicomponent Distillation Columns Using Rigorous Composition Estimators. In: **Distillation and Adsorption** *1997*, IChemE Symp. Ser.No.104, IChemE, London (U.K.), A29-A69.

ELSEVIER
IFAC
PUBLICATIONS
www.elsevier.com/locate/ifac

A CRITICAL COMPARISON OF LINEAR AND NONLINEAR PROPERTY ESTIMATORS IN INFERENTIAL CONTROL

Gabriele Pannocchia [*,1] **Paolo Leoni** [*]
Alessandro Brambilla [*]

[*] *Department of Chemical Engineering – University of Pisa*
Via Diotisalvi 2, 56126 Pisa (Italy)

Abstract: In this paper, a comparison of linear and nonlinear estimators with particular emphasis to the closed-loop properties of the resulting inferential control scheme is presented. The concept of closed-loop "consistency" is introduced as an effective criterion for choosing the auxiliary variables. An estimator is consistent if it guarantees low closed-loop steady-state offset in the true unmeasured controlled variables. By means of a case study of a high purity distillation column, a number of issues that can arise in inferential control are emphasized, and their implications on the closed-loop stability are discussed. It is shown that the use of some nonlinear estimators, which in general guarantee a superior precision, may be inappropriate because of the presence of zero gains and gain inversions that can lead the closed-loop system to instability. Moreover, in multi-input multi-output (MIMO) systems it is possible that the estimator requires the auxiliary variables to reach values that are not reachable by the actual plant. *Copyright © 2004 IFAC*

Keywords: inferential control, nonlinear estimators, partial least squares, consistency, steady-state offset.

1. INTRODUCTION

In the process industries it is common to adopt property estimators as a replacement to on-line analyzers, which are very expensive (or not available for some properties) and require significant maintenance work. From a control point a view, on-line analyzers suffer from relatively large time delays which can make the control task difficult. Indeed, a common alternative is to use a number of auxiliary measurements (such as temperatures, pressures, etc.) to infer the product properties, thus building an inferential control scheme. The issue of measurement selection is of crucial importance for the effectiveness of an estimator, and it has been the subject of extensive research in the chemical engineering community (Joseph and Brosilow, 1978; Mejdell and Skogestad, 1991; Morari and Stephanopoulos, 1980; Yu and Luyben, 1987). In particular when multiple auxiliary variables are used, problems related to potential collinearity of these variables need to be addressed. Linear multivariate regression techniques, like Partial Least-Squares regression (PLS) can be used to overcome these problems and improve the robustness of the estimators as suggested by Mejdell and Skogestad (1991).

Often property estimators are based on linear relations between the auxiliary variables and the property to be estimated. However, several nonlinear regression methods like Neural Networks and nonlinear PLS (Baffi *et al.*, 1999; Qin and McAvoy, 1992; Wold *et al.*, 1989) have been proposed, and there is therefore a natural interest in evaluating their applicability for building nonlinear property estimators.

The common approach is to evaluate the estimator effectiveness in terms of precision in fitting (training and validation) data, without addressing the implica-

[1] Corresponding author. Email: g.pannocchia@ing.unipi.it, Fax: +39 050 511266.

tions of the estimator characteristics on the resulting inferential control scheme. To address this problem, Pannocchia and Brambilla (2003) introduced a new concept of closed-loop "consistency", which allows one to choose the most appropriate auxiliary variables for designing an effective estimator to be used in an inferential control scheme. The estimator consistency is not necessarily related to the estimator precision, that is an estimator that is precise in fitting the data may be inappropriate for inferential control because it may lead to undesired steady-state offset. In the present paper this concept is revisited in a context of nonlinear estimators, and a critical evaluation of linear and nonlinear estimators in terms of closed-loop properties is presented.

2. REVIEW OF REGRESSION METHODS

PLS has been recognized as a powerful linear regression technique in many areas of process system engineering (MacGregor *et al.*, 1994; Mejdell and Skogestad, 1991; Qin and Dunia, 2000; Wise and Gallagher, 1996). Among the large number of nonlinear regression methods, the attention of this paper is focused on nonlinear modifications of the PLS algorithm. In particular, the quadratic PLS algorithm (Baffi *et al.*, 1999) is briefly reviewed in this section.

Given a vector $Y \in \mathbb{R}^n$ of measured values of the property of interest and the corresponding values of the m auxiliary variables arranged row-wise in a matrix $X \in \mathbb{R}^{n\times m}$ (both centered around reference values y^s and x^s, respectively, and scaled to unity variance), the objective of PLS is to extract a pairs of vectors called "latent variables" $t_i \in \mathbb{R}^n$ and $u_i \in \mathbb{R}^n$ related to each other as follows:

$$u_i = f(t_i) + e_i, \qquad i = 1, 2, \ldots, a, \tag{1}$$

where $f(t_i)$ depends on the particular algorithm chosen, and e_i is the approximation error. The auxiliary variables are related to the corresponding latent variables by a linear relation, i.e. :

$$t_i = X r_i, \qquad i = 1, 2, \ldots, a, \tag{2}$$

in which $r_i \in \mathbb{R}^m$ is a vector of appropriate coefficients. Linear PLS (Wold *et al.*, 2001) uses a linear "inner" relation $f(t_i) = b_i t_i$, while quadratic PLS (Wold *et al.*, 1989) uses the following relation:

$$f(t_i) = c_{1,i} t_i + c_{2,i} t_i^2, \qquad i = 1, 2, \ldots, a, \tag{3}$$

in which each element of $t_i^2 \in \mathbb{R}^n$ is the squared value of the corresponding element of t_i. In the original quadratic PLS (Wold *et al.*, 1989) the coefficients $c_i = [c_{1,i}, c_{2,i}]$ are computed by least-square regression between t_i and u_i. Baffi *et al.* (1999) proposed a different choice for c_i aimed at minimizing the regression error e_i. This is achieved by iteratively modifying the regression coefficients of r_i used to generate t_i from X. This approach is shown to be superior to the original quadratic PLS, and is chosen in the present paper.

For prediction purposes, given a vector of the m auxiliary variables (centered and scaled) $x \in \mathbb{R}^m$, one can compute the corresponding vector of x-latent variables, $t \in \mathbb{R}^{1\times a}$:

$$t = x^T R, \tag{4}$$

in which $R \in \mathbb{R}^{m\times a} = [r_1, r_2, \ldots, r_a]$. This step is followed by a quadratic "inner" relation to obtain the vector of y-latent variables, $\hat{u} \in \mathbb{R}^{1\times a}$:

$$\hat{u} = f(t) = f(x^T R). \tag{5}$$

Finally, one obtains the estimate of y as:

$$\hat{y} = \hat{u} Q^T = f(x^T R) Q^T = \mathscr{H}(x), \tag{6}$$

in which $Q \in \mathbb{R}^{1\times a}$ is an appropriate row vector, and $\mathscr{H}$ denotes the nonlinear relation between x and $\hat{y}$.

The estimator precision in fitting a set of data is typically expressed in terms of explained variance:

$$\mathrm{EV}(a) = 100\left(1 - \frac{\mathrm{MSE}(a)}{\mathrm{MSE}(0)}\right), \tag{7}$$

in which MSE is the mean square error, defined as

$$\mathrm{MSE}(a) = \frac{1}{n}\sum_{i=1}^{n}(y_i - \hat{y}_i(a))^2, \tag{8}$$

where $\hat{y}_i(a)$ is the predicted value of y_i obtained by an estimator model using a latent variables. The explained variance (often computed on a set of data not used in training) can be used to choose the number of latent variables, a (Wold *et al.*, 2001).

3. CONSISTENCY OF LINEAR AND NONLINEAR ESTIMATORS

Consider the inferential control scheme reported in Fig. 1 in which $u \in \mathbb{R}^p$ is the manipulated variable, $y \in \mathbb{R}^p$ is controlled variable (unmeasurable), $x \in \mathbb{R}^m$ is the auxiliary variable (measurable), $\hat{y} \in \mathbb{R}^p$ is the estimate of the controlled variable, and $d \in \mathbb{R}^q$ is the disturbance. Notice that the feedback controller C operates on the estimate of the controlled variable, $\hat{y}$, and therefore it is not possible (in general) to remove steady-state offset in the unmeasured controlled variable, y. Assuming that all variables are centered (and

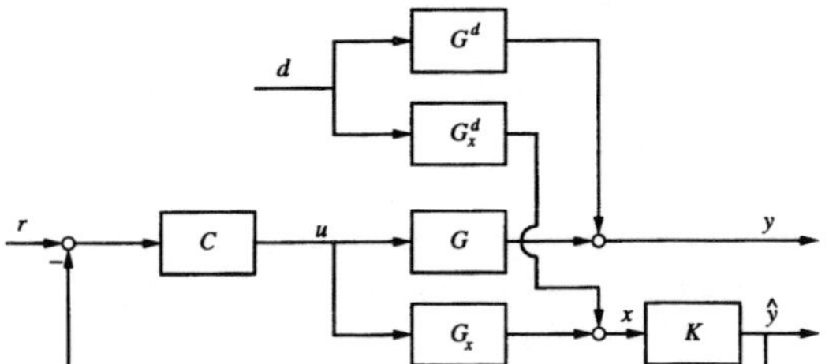

Fig. 1. Inferential control scheme

scaled) around their reference value, in a neighborhood of the origin, one can write:

$$y = Gu + G^d d \quad (9a)$$

$$x = G_x u + G_x^d d \quad (9b)$$

$$\hat{y} = \mathscr{H}(0) + Kx = Kx\,, \quad (9c)$$

in which

- $G \in \mathbb{R}^{p\times p}$ and $G_x \in \mathbb{R}^{m\times p}$ are the gain matrices from the manipulated variable to the controlled and the auxiliary variables, respectively;
- $G^d \in \mathbb{R}^{p\times q}$ and $G_x^d \in \mathbb{R}^{m\times q}$ are the gain matrices from the disturbance to the controlled and the auxiliary variables, respectively;
- $K \in \mathbb{R}^{p\times m}$ is the Jacobian of $\mathscr{H}$ evaluated at the origin, i.e.

$$K_{i,j} = \left.\frac{\partial \mathscr{H}_i}{\partial x_j}\right|_{x=0}, \quad i = 1,\ldots,p, \quad j = 1,\ldots,m\,. \quad (10)$$

Given the setpoint r, and assuming that the controller C has integral action, at steady state one can write

$$r = \hat{y} = Kx = KG_x u + KG_x^d d\,,$$

from which the corresponding input is

$$u = (KG_x)^{-1}(r - KG_x^d d)\,.$$

The steady-state closed-loop offset can be written as

$$e_{\text{CL}} = r - y = \left[I - G(KG_x)^{-1}\right] r + \left[G(KG_x)^{-1}KG_x^d - G^d\right] d\,. \quad (11)$$

For convenience of notation, let ε_r and ε_d be the following matrices

$$\varepsilon_r = I - G(KG_x)^{-1} \quad (12a)$$

$$\varepsilon_d = G(KG_x)^{-1}KG_x^d - G^d\,. \quad (12b)$$

The steady-state offset is given by $e_{\text{CL}} = \varepsilon_r r + \varepsilon_d d$. Thus, ε_r and ε_d can be regarded as the gain matrices from the setpoint reference and the disturbance, respectively, to the steady-state offset. Therefore, in order for an inferential control scheme to guarantee low closed-loop offset, it is desirable to keep some norm of ε_r and ε_d as low as possible. Such a property of an estimator is referred to as "consistency". It is important to recognize that consistency is not necessarily related to the estimator precision. As an example, consider the case in which the number of auxiliary variables equals the number of controlled variables, i.e. $m = p$. In such case, the matrix K is square and, hence, ε_d is independent of K, which implies that consistency for disturbance rejection is independent of the estimator model. In fact, in the presence of a disturbance the control system adjusts the p manipulated variables to bring the p auxiliary variables to the corresponding nominal reference value. Hence, in such cases the estimator consistency is only dependent on the choice of the auxiliary variables, and it is often the case that auxiliary variables that are not the most precise are indeed the most appropriate for guaranteeing low closed-loop offset.

4. CASE STUDY

As a case study we choose a high-purity two-product six-component distillation column, simulated by means of Aspen Plus 11.1. For space limitations, only a brief description of the case study is presented in this paper. More details can be found in (Leoni, 2003). Nominal compositions of feed and products are reported in Table 1, and main operating parameters in nominal conditions are reported in Table 2. The

Table 1. Nominal composition of feed and products (mole basis)

ID	Name	Feed	Top	Bottom
LLK_1	i-butane	0.05	0.111	0.000
LLK_2	2-methylpentane	0.10	0.223	0.000
LK	n-hexane	0.30	0.661	0.005
HK	2,2-dimethylpentane	0.30	0.005	0.541
HHK_1	2,3-dimethylpentane	0.15	0.000	0.272
HHK_2	2,2-dimethyleptane	0.10	0.000	0.182

column consists of 60 ideal stages, total condenser (stage no. 1) and Kettle reboiler, and the saturated liquid feed enters at the 32nd stage. In order to build

Table 2. Nominal operating parameters

Parameter	ID	Unit	Value
Feed rate	F	kmol/hr	100.0
Distillate rate	D	kmol/hr	44.95
Reboiler duty	Q	MMkcal/hr	2.035
Top reflux ratio	R	–	4.99
Condenser pressure	p_{top}	atm	1.1
Reboiler pressure	p_{bot}	atm	1.4

and compare several property estimators, a training database of 61 runs is obtained by varying the main operating parameters one-by-one in a wide range, as briefly described in Table 3. The data used are from open-loop tests. The possibility of using closed-loop data (Macgregor *et al.*, 1991), that is with a temperature controller on a tray, is appropriate only if the tray temperature represents a "consistent" auxiliary variable. (Moreover, data should be also collected by varying the setpoint of this temperature controller, otherwise all data correspond to approximately the same temperature, because of the feedback controller). If this is not the case, and it is not possible to move the temperature controller to a different tray (Semino and Brambilla, 1996), it may be convenient to remove the temperature controller and collect only open-loop data.

Table 3. Training set synopsis

Case	Runs	Run ID	Range
Base	1	1	–
Varying D	10	$2 \div 11$	$(0.7 \div 1.3)D_{\text{nom}}$
Varying Q	10	$12 \div 21$	$(0.65 \div 1.3)Q_{\text{nom}}$
Varying F	10	$22 \div 31$	$(0.7 \div 1.3)F_{\text{nom}}$
Varying feed comp.	20	$32 \div 51$	–
Varying p	10	$52 \div 61$	$(0.8 \div 1.32)p_{\text{nom}}$

4.1 SISO case

First, a SISO case of top product HK mole fraction control by manipulating the distillate rate is considered. Several estimators are designed by choosing the condenser pressure and one tray temperature as auxiliary variables. One latent variable is used by each estimator. In Fig. 2 the explained variance for linear and quadratic PLS estimators is plotted *vs* the tray number whose temperature is auxiliary variable. From these

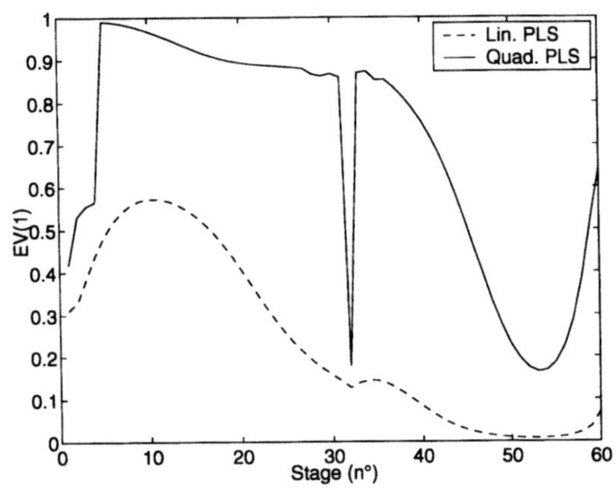

Fig. 2. Explained variance for top product estimators

results it appears that the most precise tray location is between the 8th and the 12th stage for the linear estimator, while it is between the 5th and the 10th stage for the nonlinear estimator. In Fig. 3 the most precise linear and nonlinear estimators (i.e. those with the highest explained variance) are compared in fitting the data (see also Table 3 for run IDs).

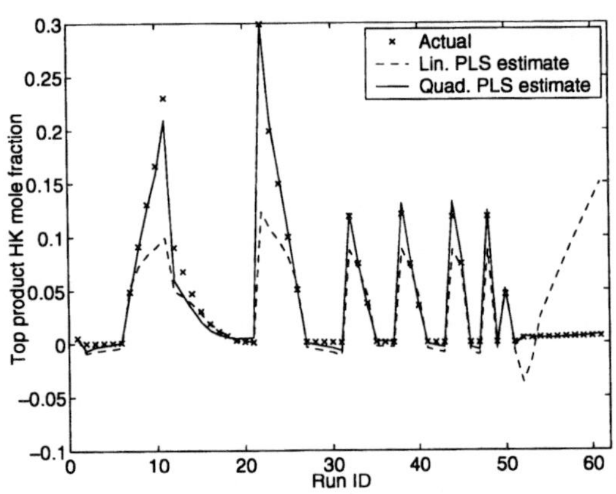

Fig. 3. Data fitting for top product estimators

From a consistency point of view, four normalized disturbances are considered:

$$d_1 = \frac{\Delta F}{F_{\text{nom}}}, \quad d_2 = \frac{\Delta\left(\frac{\text{LK}}{\text{HK}}\right)}{\left(\frac{\text{LK}}{\text{HK}}\right)_{\text{nom}}},$$

$$d_3 = \frac{\Delta\left(\frac{\text{LLK}_1+\text{LLK}_2}{\text{HHK}_1+\text{HHK}_2}\right)}{\left(\frac{\text{LLK}_1+\text{LLK}_2}{\text{HHK}_1+\text{HHK}_2}\right)_{\text{nom}}}, \quad d_4 = \frac{\Delta P}{P_{\text{nom}}}.$$

Using the training data to compute the gain matrices in (9) and numerical differentiation for computing the partial derivatives that define $K_{i,j}$ in (10), the disturbance consistency matrix ε_d is computed from (12b). Given the significant nonlinear behavior of the process, positive and negative variations of the disturbances are treated separately, thus building two consistency matrices ε_d^+ and ε_d^-, respectively. Then, the following scalar parameter is defined as a measure of closed-loop consistency for all disturbances:

$$\Phi_d = \frac{\left\|[\varepsilon_d^+,\ \varepsilon_d^-]\right\|_2}{2\dim d}. \tag{13}$$

In Fig. 4 the value of Φ_d for linear and nonlinear estimators is reported *vs* the tray number whose temperature is auxiliary variable. From these results, it appears

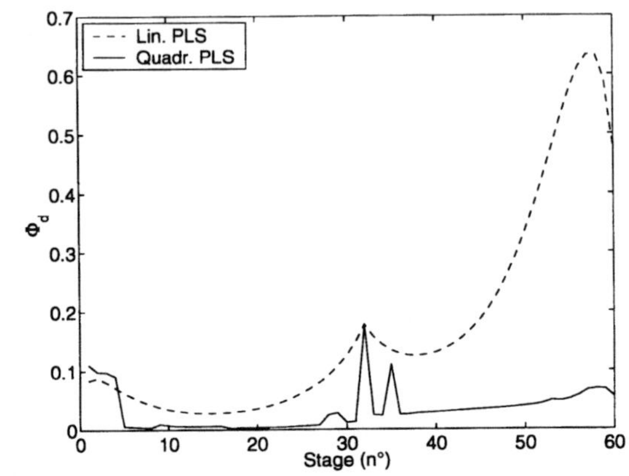

Fig. 4. Consistency parameter Φ_d for top product estimators

that the most consistent tray location is between the 12th and 18th stage for the linear estimator, while it is between the 6th to the 25th stage for the nonlinear estimator.

After this preliminary analysis, the following four estimators are compared in closed-loop rejection of the disturbances (using the Aspen Plus rigorous model):

- L_0: most precise linear estimator.
- NL_0: most precise nonlinear estimator.
- L_d: most consistent linear estimator.
- NL_d: most consistent nonlinear estimator.

Main characteristics of these estimators are reported in Table 4. In Table 5 the closed-loop offsets obtained

Table 4. Top product estimators

Est. ID	Tray	EV(1)%	$\Phi_d \cdot 10^3$
L_0	10	57.3	34.4
NL_0	5	99.1	5.6
L_d	15	53.2	27.9
NL_d	7	98.6	4.0

with each estimator in the presence of disturbances are reported. The first row (denoted with OL) refers to the open-loop offset. From these results, it appears that the estimators designed for consistency guarantee a significantly lower steady-state offset with respect to those designed for precision. On average the nonlinear estimators guarantee a lower offset than the linear estimators, even though for some disturbances (like

Table 5. Top product open-loop and closed-loop offset (multiplied by 10^3)

Est. ID	d_1		d_2		d_3		d_4		Mean
	+0.1	−0.1	+0.1	−0.1	+0.1	−0.1	+0.1	−0.08	
OL	3.9	-94.8	3.8	-32.7	3.6	-19.0	-0.3	0.5	19.9
L_0	-1.5	1.3	-0.1	0.1	-0.7	0.7	4.9	-33.6	5.4
NL_0	-2.2	2.0	0.6	-0.7	-2.0	2.1	-0.6	2.0	1.5
L_d	-1.1	1.2	-0.3	0.4	-0.4	0.4	4.	-20.7	3.7
NL_d	-1.8	1.6	0.2	-0.2	-1.2	1.3	-0.3	1.3	1.0

d_1 and d_3) the linear estimator L_d works better than the nonlinear one NL_d.

Despite these better precision and consistency properties, nonlinear estimators can exhibit a potentially harmful behavior, which arises from a non-monotonic relation between the auxiliary variables (in this case the tray temperature) and the controlled variable. As an example, Fig. 5 shows the top product HK mole fraction estimate obtained by the nonlinear estimator using the 16th stage temperature *vs* the corresponding temperature. It is clear that in such case if the setpoint

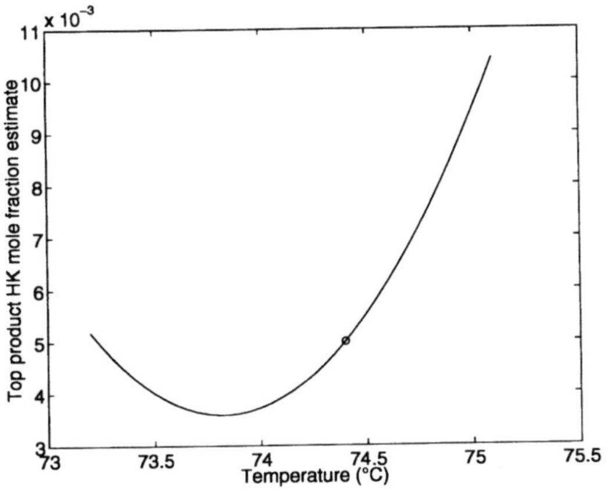

Fig. 5. Non-monotonic behavior of top product HK mole fraction estimate

is changed from 0.005 to 0.003, the control system will try to reduce the HK mole fraction estimate to the new setpoint by decreasing the distillate rate (and hence reducing the tray temperature). However, after the tray temperature reaches approximately 73.8°C, the HK mole fraction estimate increases again and the control system will further reduce the distillate rate ultimately leading to a potential closed-loop instability. Notice that this phenomenon can also occur in the presence of disturbances on the operating pressure, which move the curve in Fig. 5 vertically (up or down). This severe problem can occur with any nonlinear estimator which may have "zero gain" points and "gain inversions", while it cannot occur with linear estimators. In particular, this is likely to occurs with Neural Network models as criticized by Turner and Guiver (2002), who suggested an alternative "Bounded Derivative Network" model that is guaranteed to have monotonic relations between the auxiliary variables and the controlled variable.

Table 6. Top and bottom product estimators

Est. ID	Top product		Bottom product		Φ_d
	Tray	EV(1)%	Tray	EV(1)%	
L_0	10	57.3	50	56.6	0.126
NL_0	5	99.1	55	97.7	0.611
L_d	42	5.6	15	3.3	0.099
NL_d	13	94.2	43	83.8	0.018

4.2 MIMO case

Next, the MIMO case of top product HK and bottom product LK mole fraction control by manipulating the distillate rate and the reboiler duty is considered. For each property several estimators are designed by using the condenser pressure and one tray temperature as auxiliary variables. When addressing the problem from a consistency point of view one needs to recognize the following important observations:

- For obvious control reasons, top and bottom product estimators are not allowed to use the same tray temperature as auxiliary variable.
- Closed-loop consistency depends on both inferential control loops, so that all possible pairs of top and bottom product single temperature estimators need to be tested.
- Any estimator will be consistent for disturbances on the feed flow rate. In fact, if the feed flow rate changes and both product estimates are controlled, the steady-state temperature profile of the column will be the same as in the nominal case because both distillate rate and reboiler duty change proportionally to the feed-rate change (actually only pressure drop effects remain).

Hence, the disturbance d_1 is not considered when computing the consistency matrix ε_d and the corresponding scalar parameter Φ_d defined in (13).

Using the same rationale (best precision and best consistency) as for the SISO case, four (linear and nonlinear) estimators are designed and compared. The estimator characteristics are reported in Table 6. In Table 7 the closed-loop offsets of each product obtained by using different estimators in the presence of disturbances (not involving the operating pressure) are reported. From these results, similarly to the SISO case, it appears that the design of estimators based on consistency guarantees a lower closed-loop offset in the presence of disturbances. It is also interesting to notice that in this case the use of a nonlinear estimator model does not appear to be particularly appropriate. Furthermore, when disturbances in the operating pres-

Table 7. Top and bottom product open-loop and closed-loop offsets (multiplied by 10^3)

Est. ID	$d_2 = +0.1$		$d_2 = -0.1$		$d_3 = +0.1$		$d_3 = -0.1$		Mean	
	Top	Bottom	Top	Bottom	Top	Bottom	Top	Bottom	Top	Bottom
OL	3.8	-22.3	-37.7	2.3	3.7	-13.4	-19.0	2.1	16.1	10.0
L_0	-0.2	0.4	0.2	-0.5	-0.6	0.7	0.7	-0.6	0.4	0.6
NL_0	3.0	2.2	-0.1	1.2	-0.6	2.3	2.4	-0.8	1.5	1.6
L_d	-0.3	0.4	0.2	-0.6	-0.3	0.3	0.3	-0.3	0.3	0.4
NL_d	0.1	0.5	0.3	-0.6	-0.2	0.4	0.6	-0.3	0.3	0.4

Table 8. Estimator behavior for pressure disturbance ($P = 1.012$ atm)

Est. ID	Top			Bottom		
	T_{OL}	T_S	offset	T_{OL}	T_S	offset
L_0	68.9	72.5	-	88.7	91.1	-
NL_0	68.4	70.2	-	90.1	90.7	-
L_d	85.9	88.3	-	71.4	73.9	-
NL_d	70.7	70.3	$3.8 \cdot 10^{-3}$	86.3	86.8	$3.2 \cdot 10^{-3}$

sure are considered, both linear and nonlinear estimators may lead the closed-loop system to instability. In fact, Table 8 shows results obtained for the case of -8 % pressure change. For each estimator the open-loop tray temperature and the corresponding temperature required by the estimator to remove offset in the controlled variable estimates are reported. For all estimators, with the exception of NL_d the new "target" temperatures of the two trays cannot be reached simultaneously, and hence the control system will keep increasing the distillate rate and decreasing the reboiler duty, leading the closed-loop system to potential instability.

5. CONCLUSIONS

In this paper, a critical comparison of linear and nonlinear estimators in inferential control has been presented. The emphasis of the present work has been devoted to investigate the implications of the estimator design on inferential closed-loop systems, by revisiting the concept and the definition of "consistency" (Pannocchia and Brambilla, 2003) in the context of nonlinear estimators. It has been shown that the use of the most "consistent" auxiliary variables guarantees a lower steady-state offset than the use of the most "precise" ones, for both linear and nonlinear estimators. Moreover, severe problems that can occur in closed-loop have been discussed.

- When using some nonlinear estimators it is possible to have zero gains and gain inversions, which can make the closed loop unstable. In particular, this possibility is likely to occur when using Neural Networks (Turner and Guiver, 2002).
- In MIMO systems it is possible that, in order to remove offset, the estimator requires the auxiliary variables to reach values that cannot be reached by the actual plant. That may generate closed-loop instability. In particular, for high purity distillation columns this phenomenon can occur when disturbances in the operating pressure affect the system. Additional auxiliary variables, as the mole ratio L/V, can be used to improve the closed-loop response to such disturbances (Pastore *et al.*, 2004).

REFERENCES

Baffi, G., E. B. Martin and A. J. Morris (1999). Non-linear projection to latent structures revisited: the quadratic PLS algorithm. *Comput. Chem. Eng.* **23**, 395–411.

Joseph, B. and C. B. Brosilow (1978). Inferential control of processes. *AIChE J.* **24**, 485–509.

Leoni, P. (2003). Consistency of linear and nonlinear property estimators for distillation columns (in italian). MS Thesis in Chemical Engineering, University of Pisa.

MacGregor, J. F., C. Jaeckle, C. Kiparissides and M. Koutoudi (1994). Process monitoring and diagnosis by multiblock PLS methods. *AIChE J.* **40**, 826–838.

Macgregor, J. F., T. E. Marlin and J. V. Kresta (1991). Some comments on neural networks and other empirical modelling methods. In: *Fourth International Conference on Chemical Process Control* (Y. Arkun and W. H. Ray, Eds.). pp. 665–672.

Mejdell, T. and S. Skogestad (1991). Estimation of distillation composition from multiple temperature measurements using partial-least squares regression. *Ind. Eng. Chem. Res.* **30**, 2543–2555.

Morari, M. and G. Stephanopoulos (1980). Optimal selection of secondary measurements with the framework of state estimation in the presence of persistent unknown disturbances. *AIChE J.* **26**, 247–258.

Pannocchia, G. and A. Brambilla (2003). Consistency of property estimators in multicomponent distillation control. *Ind. Eng. Chem. Res.* **42**, 4452–4460.

Pastore, D., A. Brambilla and G. Pannocchia (2004). Estimator design with pls model for consistent control of refinery main fractionators. In: *Proceedings of IFAC AdChem*. Hong Kong.

Qin, S. J. and R. Dunia (2000). Determining the number of principal components for best reconstruction. *J. Proc. Cont.* **10**, 245–250.

Qin, S. J. and T. J. McAvoy (1992). Non-linear PLS modelling using neural networks. *Comput. Chem. Eng.* **16**, 379–391.

Semino, D. and A. Brambilla (1996). An efficient structure for parallel cascade control. *Ind. Eng. Chem. Res.* **35**, 1845–1852.

Turner, P. and J. Guiver (2002). Neural networks – a comprehensibely unsuitable technology for control. In: *AspenWorld 2003*. Orlando, FL.

Wise, B. M. and N. B. Gallagher (1996). Process chemometrics approach to process monitoring and fault detection. *J. Proc. Cont.* **6**, 329–348.

Wold, S., M. Sjöström and L. Eriksson (2001). PLS-regression: a basic tool of chemometrics. *Chemometrics and Int. Lab. Systems* **58**, 109–130.

Wold, S., N. Kettaneh-Wold and B. Skagerberg (1989). Non-linear PLS modelling. *Chemometrics and Int. Lab. Systems* **7**, 53–65.

Yu, C.-C. and W. L. Luyben (1987). Control of multicomponent distillation columns using rigorous composition estimators. In: *Distillation and Adsorption*. IChemE Symposium Series. The Institution of Chemical Engineers. Brighton, UK. pp. A29–A69.

ELSEVIER
IFAC
PUBLICATIONS
www.elsevier.com/locate/ifac

MULTIVARIATE IMAGE ANALYSIS FOR INFERENTIAL SENSING: A FRAMEWORK

Honglu Yu , John F. MacGregor

McMaster Advanced Control Consortium
Department of Chemical Engineering
McMaster University, Hamilton, On. Canada

Abstract: This paper presents a framework for developing vision-based inferential sensors. This framework not only gives a summary of existing methodologies, but also combines the methods used in other areas, such as traditional machine vision, multivariate image analysis and multivariate data analysis, and gives a broad vision for future developments of the area. *Copyright © 2004 IFAC*

Keywords: Inferential Sensor, Multivariate Image Analysis, Principal Component Analysis, Partial Least Squares, Machine Vision

1. INTRODUCTION

One of the most important elements in the successful monitoring and control of systems is fast, reliable and inexpensive on-line sensors. However, not all the important variables can be measured on-line. In this situation, a strategy often used is to estimate the unknown variable from other easily measured variables through a model, known as inferential control.

In this paper, attention is focused on a new generation of inferential sensing techniques, where process or product property variables difficult to measure are estimated using information extracted from online acquired images. This type of sensor can greatly improve the quality control performance in at least two areas: i) industries that produce solid products such as food, polymers, pulp and paper, where product properties can only be measured periodically in a laboratory; and ii) processes whose performance is related to their visual appearance, such as flame in a combustion process.

The petrochemical industry made rapid advances in multivariable model predictive control largely because they had the availability and abundance of inexpensive and informative sensors such as thermocouples, pressure transducers, flow meters, pH and ion-specific meters and gas chromatographs. This is a direct result of the fact that the major streams in petrochemical processes consist of well mixed gases and liquids which made the use of such sensors very easy. On the other hand, the solids processing industry has had much less success at implementing advanced control precisely because of the lack of such sensors. In industries that produce solid products, the product properties are generally measured periodically by manually collecting samples and then analyzing them in the laboratory. The analysis procedure may require that the samples be destroyed and the procedures are time consuming and manpower intensive. Therefore, it is usually impossible to obtain on-line quality measurements and even the simplest automatic control algorithm can hardly be implemented. In this situation, an inferential sensor based on an imaging system is very attractive. By taking images of the products, the samples remain untouched and the sampling rate can be very high. The product properties can then be predicted in real time from image data.

In fact, using imaging techniques for industrial production is not a new concept. Machine vision has been developed for more than 30 years. However, there are three major differences between conventional machine vision and vision-based inferential sensor. First, the objectives and the types of the problems dealing with are different. A large amount of the problems in machine vision area involve locating and identifying objects, measuring the size of products, sorting different types of products and detecting defects. These imaging systems are designed as a replacement of human vision and decision system. On the other hand, the objective of an inferential sensor is to quantitatively predict uneasily measured property variables for online monitoring and feedback control, which is actually beyond the capability of human vision system. Second, most images used in machine vision are greyscale or binary images while most images used for vision-based inferential sensors are multivariate images, such as multispectral images and color images. This is because multivariate images, compared with greyscale or binary images contain much more information. Third, since the problem type and image type are different, so do the methodologies developed. In machine vision area, most methods are used for greyscale images and to compute geometric characteristic of the objects. In the case of building inferential sensors, one is interested in extracting the subtle information (may or may not be seen) from multivariate images and relating the extracted information with some chemical or physical properties.

Research on multivariate image analysis started in 1970s mainly in remote sensing area. Afterwards, it has been developed in many other scientific areas, such as medical imaging in medical science and microscope multispectral imaging in chemistry. Most of these studies focus on how to efficiently display a multivariate image (e.g. visualisation of a hyperspectral image with 250 bands), on how to segment an image into different areas (e.g. segmentation of a satellite map with different features: road, river, land and grass) and on how to detect certain features (e.g. detection of cancer cells in an image obtained from Magnetic Resonance Imaging (MRI)). In other words, the purpose of these studies is to develop a human assistance system rather than an automatic system because the output is generally an image with enhanced appearance.

Recently, more and more attention has been paid to the development of vision-based inferential sensor. A series of papers have been published to solve several different problems, including seasoning level prediction on snack food (Yu and MacGregor, 2003a), pulp properties estimation using NIR imaging (Bharati et al., 2002) and environmental monitoring system for combustion process using flame color images (Yu and MacGregor, 2003b).

In this paper, a general framework for developing vision-based inferential sensor is presented. This framework not only gives a summary of the approaches published in literature, but also gives other possible options by combining the methods developed in traditional machine vision and multivariate image analysis areas. This framework can be used as a guideline for future applications.

2. MULTIVARIATE IMAGES

A multivariate image is a set of congruent images. The definition for congruence in imaging is given by Geladi and Grahn (1996): two or more images are congruent if they can be stacked so that for each pixel in one image there is a corresponding pixel in the other image(s) that can be referred to the same position in the object or scene depicted.

From the data storage point of view, a multivariate image is a three-way matrix. Two of the ways in this three-way matrix are the geometrical image coordinates which describe the image scene plane and usually treated as a pair. The third way is the 'variable' way. Hence one could also view a multivariate image as a two-way array of pixel intensity vectors. Multivariate image data often contain highly correlated information among the variables because the individual images within one multivariate image are congruent.

A multivariate image can be obtained from varied sources. A common way is to stack the images of an object at different radiation energies or wavelengths. This is a multispectral image. Magnetic Resonance Imaging (MRI) is another imaging technique to obtain multivariate images. By varying the parameters (most common parameters are the spin-echo time and the relaxation delay time), a set of images can be obtained. A multivariate image can also be constructed by combining images obtained from different instruments. A multi-temporal image is a type of multivariate image as well, which consists of the images of the same scene but taken at different times. A color image can be also considered as a 3-variable multivariate image because a color image consists of 3 color channels: red, green and blue. Another technique of creating multivariate images is to combine a grayscale image with the copies of it that are derived from different spatial filtering operations.

3. VISION-BASED INFERENTIAL SENSOR

A vision-based inferential sensor consists of three basic elements: an imaging system (e.g. camera), a computer and a model which predicts the property variables using information from the images. In this paper, we mainly consider the model building problem.

Though technically it is possible to build a vision-based inferential sensor based on fundamental model, most sensors are data-based. Like building any data-based model, the first thing is to collect a good training dataset, since the quality of the model is largely dependent on the quality of the training dataset. This training dataset should cover a large range of the operation range and the imaging condition should be kept as constant as possible. Both the images and the corresponding product properties need to be collected. If the dataset is collected in an online production mode, it is very crucial to ensure they have consistent time stamps in both images and property measurements.

Obviously, no regression methods can directly relate a set of three-way image matrices to a set of property variables. In most applications, a two-step procedure is adopted. In this two-step procedure, first a feature vector is extracted from each image, and then the extracted feature vectors are used as predictor to be regressed against properties variables. The feature extraction step is the key to build a successful model, not only because it largely reduces the problem dimension, but also, more importantly, it concentrates the related information. There are many different ways to extract features. In the framework proposed in the next section, we summarized the feature extraction methods into four categories. However, the choice of features is very case dependent and in many situations an iteration procedure is needed.

To predict the property variable for a new image, one needs to first compute its feature vectors and then calculate the estimation through the regressed model. Figure 1 shows a basic scheme for model building and for new image prediction.

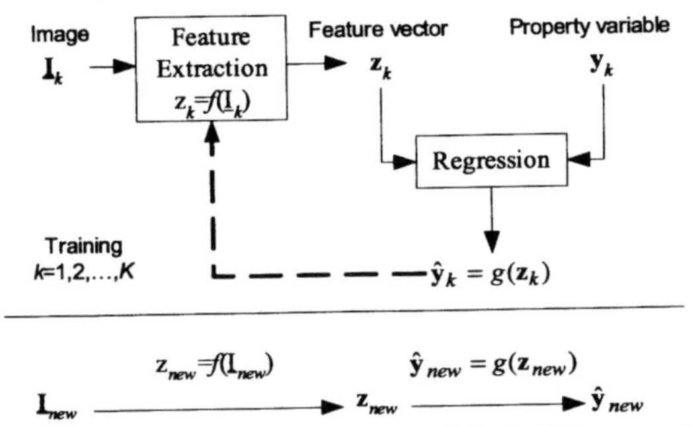

Figure 1: A scheme for model building and prediction

4. A FRAMEWORK FOR DEVELOPING VISION-BASED INFERENTIAL SENSOR

In this section, a framework is proposed to give a guideline for building an inferential sensor using multivariate images. This framework is originally a summary of several applications. However, it has been extended to give more options by combining many other approaches used in the traditional machine vision, multivariate image analysis and multivariate data analysis.

Figure 2 shows the proposed framework. In this framework, there are three steps: preprocessing, feature extraction and regression.

4.1 Preprocessing

Though other preprocessing techniques, such as removing noise and improving contrast may be used, here the preprocessing step mainly refers to variable dimension reduction.

Variable dimension reduction could be a significant preprocessing procedure in dealing with multivariate images, especially for images with high variable dimension. Because the individual bands of a multivariate image are often highly correlated, the proper variable dimension reduction can not only remove or reduce this variable redundancy and increase the computation efficiency, but can also often result in a better signal to noise ratio and concentrate on the useful information.

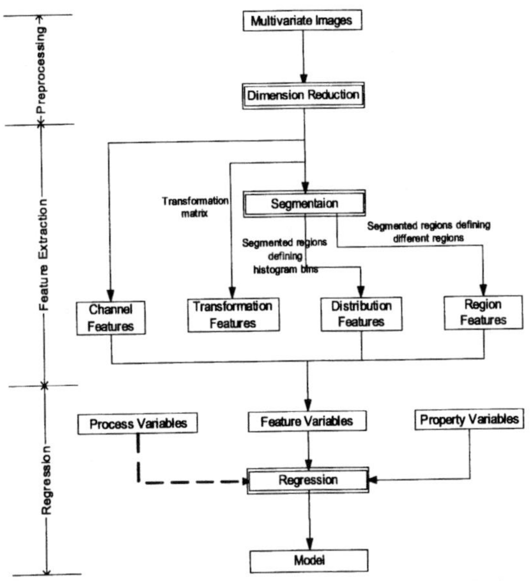

Figure 2 Proposed framework for predicting process/product properties using multivariate images

In both the snack food application (Yu and MacGregor , 2003a) and the flame application (Yu and MacGregor, 2003b), Principal Component Analysis (PCA) is used to reduce variable dimensions.

To perform PCA on an M-band multivariate image (often referred to as multi-way PCA), one first unfolds the image matrix $\underline{\mathbf{I}}$, where each pixel is considered as an observation. Normal PCA can be then performed on the unfolded matrix **I**.

$$\mathbf{I} = \sum_{a=1}^{A} \mathbf{t}_a \mathbf{p}_a^T + \mathbf{E}$$

where A is the number of principal components, the $\mathbf{t}_a$'s are score vectors and the corresponding $\mathbf{p}_a$'s are loading vectors and **E** is the residual.

Score $\mathbf{t}_a$'s can be refolded back into the original image size and form a new multivariate image with dimension A (the number of principal components), which could be much smaller than M (the original dimension). This new lower-dimensional multivariate image is sent to the feature extraction steps.

There are many other options for variable dimension reduction. These optional methods have been used in the traditional multivariate image analysis and data analysis.

PCA belongs to a family of linear projection approaches. In these linear projection approaches, a low dimensional set of projection directions are defined by using certain criteria and the multivariate image data is then projected to this low dimensional subspace. The differences among the methods are different criteria. PCA, finds the subspace that explains the largest percentage of the variation, Discriminant Analysis (DA, including Fisher's Discriminant Analysis, Canonical Component Transform and PLS Discriminant Analysis), finds the subspace that has the largest discriminating power (Maxwell, 1976; Lied and Esbensen, 2001), Independent Component Analysis (ICA) finds the subspace from which the projections of the data onto each of the basis vectors are independent (Kaarna et al., 2000), Projection Pursuit (PP) finds the subspace that captures the non-Gaussian distributed information (Ifarraguerri and Chang, 2000) and Decision Boundary Feature Extraction (DBFE) finds the subspace defined by the decision boundary (Lee and Landgrebe, 1993). Among these projection methods, PCA is the simplest one in computation and the one that has been most widely used.

Other than linear projection methods, several non-linear transformation methods, such as Self-Organizing Map (Kohonen, 1984) and Sammon's mapping (Sammon, 1969) can also be considered. However, the main drawback of these nonlinear mapping methods is that they often require long computation time (Bonnet et al., 1997).

4.2 Feature extraction

Feature extraction methods are summarized into four categories: channel features, transformation features, distribution features and region features. These features can be used independently or used in a combination manner.

Channel features

Channel features can be statistical measurements for each individual channel image, such as average values, standard deviations, moments and histogram. One example is found in Yu and MacGregor (2003a) where average colors of the snack food images were used as feature variables to predict seasoning level on the snack food product. Another example is given in Hätönen et al. (1999) where mean, standard deviation, skewness and kurtosis values of different RGB channels were used as feature variables to predict mineral concentrations in the flotation froth.

When computing channel features, each channel image is processed one at a time as a grayscale image. Therefore, many image processing methods for grayscale images can be used to extract certain information, such as the use of wavelet transforms on each channel image to extract texture information.

Transformation features

Transformation features refer to the transformation matrix obtained in variable dimension reduction step, such as the loading vectors $\mathbf{p}_a$'s when use PCA. In Bharati et al. (2002), where several properties of pulp were predicted using NIR multispectral images, the first principal component loading vectors were used as feature vectors. It has been pointed out in Yu and MacGregor (2003a) that if one performs PCA on an image data without mean-centering the first PCA loading vector is similar to the normalized average spectral (variable) response.

Distribution features

Pixel intensity histogram is an important tool and has been used extensively in grayscale image analysis. If in the dimension reduction step the variable dimension has been reduced to one, a histogram can be easily calculated and used as features.

However, if reduced variable dimension still larger than one, distribution of the spectral (variable) response needs to be described using a multi-dimensional histogram. Evidently, to obtain such a multi-dimensional histogram could be a very computation intensive work since the amount of calculation increases in an exponential manner as the number of dimensions increases. Furthermore, even if one does obtain such a high dimensional histogram as feature variables, he could encounter problems to obtain a good model in the regression step. In one of the feature extraction methods presented in Yu and MacGregor (2003a), all the snack food color images are first projected to a two-dimensional $\mathbf{t}_1$-$\mathbf{t}_2$ subspace and then an unfolded 32×32 histogram was used as feature variables to represent each image. The results showed that the model obtained using these feature variables had good prediction ability for the images having the same size with the training images but could have large errors when predicting from images with smaller size..

Yu and MacGregor (2003a) proposed another idea to further obtain a one-dimensional histogram. In their approach, $\mathbf{t}_1$-$\mathbf{t}_2$ score space was divided into several bins. Each bin was defined in a way that pixels falling into each bin were assumed having similar seasoning level. The counts of the pixels falling into each bin then formed a one-dimensional histogram.

Follow up this idea; one can obtain a one-dimensional histogram for a multivariate image by first segmenting the image into different parts and then counting the number of pixels falling into each part. Pixels falling into each part should have somehow common properties. In this way, the computation difficulties caused by high dimension are bypassed. However, how to segment the images becomes a new problem. More multivariate image segmentation approaches will be discussed in the 'region features' section.

The histogram is not the only distribution descriptor. It has been shown in Yu and MacGregor (2003a) that using cumulative histogram instead of histogram often can have better signal-to-noise ratio and result in more reasonable parameter estimates.

Region features

Sometimes, the imaging scene consists of several distinguished regions with different characteristics. In this situation, a natural thinking is to segment these regions first and then compute the features for each region, which can be channel features, transformation features and/or distribution features.

One example is the feature extraction method developed in Yu and MacGregor (2003b) for the flame application. In this application the objective is to predict several process properties, including heat of combustion of the waste feed and concentration of NO_x and SO_2 in the off-gas, using RGB flame images. To extract feature information from rapidly changing flame images, a flame luminous region is first separated from non-luminous region and nine feature variables are then computed for these two regions respectively. The separation of the flame luminous region and the non-luminous region was done by first projecting pixels onto the $\mathbf{t}_1$-$\mathbf{t}_2$ PCA score space and then defining a polygon mask in score space that indicates the flame luminous region. The boundary of this mask was obtained by a trial and error process, whereby one selects a mask area in the score plot, selects the pixels lying under it and highlights them in the image space, and iterates until one obtains a mask that segments the feature of interest. This segmentation approach was first introduced in 1989 by Esbensen and Geladi, and has become a very effective method. In Yu and MacGregor (2003a), a product mask was defined to separate snack food product pixels and conveyor belt pixels. However, in their work, the mask is not defined by a full manual trial-and-error procedure, but based on the computation of some covariance property of the score combinations.

As mentioned in the introduction section, segmentation is one of the major topics in traditional multivariate image analysis studies. Therefore, many other multivariate image segmentation techniques have been presented.

Basically, the segmentation approaches for multivariate images can be divided into two main categories. In the first category, only spectral (variable) information of each pixel in the image plane is considered. In another words, pixels are treated independently of one another. Therefore, a pixel can be treated as an observation in multivariate data analysis. The classical classification methods, such as K-means (Duda and Hart, 1973), fuzzy C-means (Boudraa et al., 2000), neural networks (Reddick et al. 1997), maximum likelihood (Liang et al. 1994) and discriminant analysis, can be directly used. Under the same category, segmentation can also be obtained by thresholding the pixel intensity values based on histogram information. This has been extensively used in the segmentation of grayscale images. For multivariate images, one option is to find a proper thresholding value for each individual band (variable) image and combine the results (Raya, 1990). However, in this way, the nature of the multivariate image has been ignored. Another option is to find a thresholding range in the M-dimensional space for an M-band image. For M=2, the thresholding range can be defined by the methods mentioned above (choosing a polygon mask). However, the computation of the M-dimensional histogram and choosing a proper range of the thresholding area will be very difficult for the case where there are more than two dimensions (M>2). A compromise solution is to choose polygon masks for each pair of the variable combinations.

In the second category, spatial information as well as spectral (variable) information is used for classification. This is also known as multivariate image texture analysis. There are three general approaches to segment an image based on textural information. In the first approach, spatial and spectral information are considered simultaneously. The approaches presented are often extended from grayscale image texture analysis. Several segmentation methods based on Markov Random Fields (MRF) models and maximizing a posteriori distribution probability (MAP), which utilize both spectral and spatial information to model the local correlation structure of an image, have been presented (Kartikeyan et al., 2002). Kovalev et al. (2001) proposed a texture analysis method for multivariate image that is based on extended co-occurrence matrices. Extraction and classification of homogeneous objects (ECHO) (Kettig and Landgrebe, 1976), is another texture segmentation method which is based on the spatial and spectral homogeneity of the regions. The listed methods above are only a few examples. However, methods considering both spatial and spectral information at the same time generally involve iterative computation and/or optimization that require intensive computation time. In the second approach, one or several representative images are generated by variable reduction or some other operation (e.g. using ratios). Then the texture methods for grayscale images can be applied on them individually. In the third approach, the images of one or several channels in a multivariate image are filtered or transformed. The newly filtered or transformed images are expected to contain the spatial information. A new multivariate image is then constructed by stacking the new images (and the original image) together. The segmentation methods in the first catalogue can then be used for classification. Liu and MacGregor (2002) studied both of the latter two approaches by using wavelets to extract the spatial information.

4.3 Regression

After extracting feature variables, the regression step is performed as in normal data analysis to build a model. Technically, any data based regression methods can be used in this step. The methods can be linear regression approaches, such as Multivariate Linear Regression (MLR), Partial Least Squares (PLS), Ridge Regression (RR). PLS is a method which can well handle highly correlated feature variables and has shown to be a simple and effective approach. Although PLS is a linear regression methods, transformation or non-linear terms can be used to handle the possible non-linearity. For example, in Bharati et al. (2002), one of the pulp properties, DCM Resin, was transformed into ln (DCM Resin). The regression methods can be also nonlinear regression approaches, such as neural networks. In Wang et al. (2002), a neural network was used to predict the NO_x emissive concentration of a boiler using features extracted from color flame images.

In some situation, besides the feature variables extracted from images, other process measurements can also be used to build the model. In Yu and MacGregor (2003b), a PLS model was built to predict the product of heat of combustion of waste stream and the fuel flow rate. The heat of combustion was then obtained by dividing the PLS model prediction by the liquid fuel flow rate. Other than the feature variables, the flow rates of liquid fuel and natural gas were also used as the predictors. This is shown in equation:

$$\hat{Q} = [F_{lf}\ F_{ng}\ \mathbf{v}] \cdot \hat{\gamma}, \text{ where } Q = H_{Lf} \cdot F_{Lf}$$
$$\hat{H}_{lf} = \hat{Q} / F_{lf}$$

where $\mathbf{v}$ is the feature vector extracted from the image data, F_{ng} is the natural gas flow rate, F_{lf} is the flow rate of liquid fuel, $\hat{\gamma}$ is the model regression coefficient vector and H_{Lf} is the heat of combustion.

5. SUMMARY AND CONCLUSIONS

A framework is proposed for building inferential sensors using multivariate image analysis for process monitoring and control. This framework not only gives a summary of existing methodologies, but also combines the methods used in other areas, such as traditional machine vision, multivariate image analysis and multivariate data analysis, and therefore gives a broad vision for the future development.

REFERENCES

Bharati, M. H., J. F. MacGregor, Marc Champagne and Marc Barrette (2002), Using NIR Multivariate Image Regression Techniques to Predict Pulp Properties, Control Systems 2002, Stockholm, Sweden, June 3-5

Bonnet, N., M. Herbin, and P. Vautrot (1997), "Multivariate Image Analysis and Segmentation in Microanalysis", *Scanning Microscopy*, **11**, 1

Boudraa, A., S. M. R. Dehak, Y. Zhu, C. Pachai, Y. Bao, and J. Grimaud (2000), "Automated Segmentation of Multiple Sclerosis Lesions in Multispectral MR Imaging Using Fuzzy Clustering," *Computers in Biology and Medicine*, **30**, 23

Duda, R. D., and P. E. Hart (1973), "Pattern Classification and Scene Analysis," New York, John Wiley & Sons

Esbensen, K., and P. Geladi (1989), "Strategy of Multivariate Image Analysis (MIA)," *Chemometrics and Intelligent Laboratory Systems*, **7**, 67

Geladi, P., and H. Grahn (1996), Multivariate Image Analysis, John Wiley & Sons, Chichester, UK

Hätönen, J., H. Hyötyniemi, J. Miettunen, and L. E. Carlsson (1999), "Using Image Information and PLS for Predicting Mineral Concentrations in the Flotation Froth," Proceedings of he Second International Conference on Intelligent Processing and Manufacturing of Materials (IPMM'99), July 10–15, Hawaii, USA

Ifarraguerri, A. and C. Chang (2000), "Unsupervised Hyperspectral Image Analysis with Projection Pursuit", *IEEE Trans. on Geoscience and Remote Sensing*, **38**, 2529

Kaarna, A., P. Zemcik, H. Kälviäinen, and J. Parkkinen (2000), "Compression of Multispectral Remote Sensing Images Using Clustering and Spectral Reduction," *IEEE Transactions on Geoscience and Remote Sensing*, **38**, 1073

Kartikeyan, B., K. L. Majumder, D. K. Pal, A. Sarkar, M. K. Biswas, and V. Kumar (2002), "A MRF Model-based Segmentation Approach to Classification for Multispectral Imagery," *IEEE Transactions on Geoscience and Remote Sensing*, **40**, 1102

Kettig, R. L. and D. A. Landgrebe (1976), "Classification of Multispectral Image Data by Extraction and Classification of Homogeneous Objects," *IEEE Transactions on GeoScience Electronics*, **GE4**, 19

Kohonen, T. (1984), Self-organization and Associative Memory, Springer, Berlin

Kovalev, V. A., F. Kruggel, H. Gertz, and D. Y. Von Gramon (2001), "Three-Dimensional Texture Analysis of MRI Brain Datasets," *IEEE Transactions on Medical Imaging*, **20**, 424

Lee, C. and D. A. Landgrebe (1993), "Decision Boundary Feature Extraction for Nonparametric Classification," *IEEE Transactions on Systems, Man, and Cybernetics*, **23**, 433

Liang, Z., J. R. MacFall, and D. P. Harrington (1994), "Parameter Estimation and Tissue Segmentation from Multispectral MR Images," *IEEE Transactions on Medical Imaging*, **13**, 441

Lied, T. T. and K. H. Esbensen (2001), "Principles of MIR, Multivariate Image Regression I: Regression Typology and Representative Application Studies," *Chemom. Intell. Lab. Syst.*, **58**, 213

Liu, J. and J. F. MacGregor (2002), "Multiresolutional Multivariate Image Analysis and its Application to Color Texture Analysis," *submitted to* Pattern Recognition

Maxwell, E. L. (1976), "Multivariate Systems Analysis of Multispectral Imagery", *Photogrammetric Engineering and Remote Sensing*, **42**, 189

Raya, S. P. (1990),"Low-Level Segmentation of 3-D magnetic Resonance Brain Images—A Rule-Based System," IEEE Trans. on Medical Imaging, **9**, 327

Reddick, W. E., J. O. Glass, E. N. Cook, T. D. Elkin and R. J. Deaton (1997), "Automated Segmentation and Classification of Multispectral Magnetic Resonance Images of Brain Using Artificial Neural Networks," *IEEE Transactions on Medical Imaging*, **16**, 911

Sammon, J. W. (1969), "A Nonlinear Mapping for Data Structure Analysis", *IEEE Trans. Comput.* **C18**, 401

Wang, F., X. J. Wang, Z. Y. Ma, J. H. Yan, Y. Chi, C. Y. Wei, M. J. Ni, and K. F. Cen (2002), "The Research on the Estimation for the NOx Emissive Concentration of the Pulverized Coal Boiler by the Flame Image Processing Technique," *Fuel*, **81**, 2113

Yu, H. and J. F. MacGregor (2003a). Multivariate Image Analysis and Regression for Prediction of Coating Content and Distribution in the Production of Snack Foods. *Chemom. Intel. Lab. Syst.* **67**, 125

Yu, H. and J. F. MacGregor (2003b). Monitoring Turbulent Nonpremixed Flames in an Industrial Boiler Using Multivariate Image Analysis (MIA). to appear in *AIChE J.*

ELSEVIER
IFAC
PUBLICATIONS
www.elsevier.com/locate/ifac

OPTIMAL PERIODIC CONTROL OF A DRUG DELIVERY SYSTEM

Subbarao Varigonda[1], **Tryphon T. Georgiou**[2],
Ronald A. Siegel[3], **Prodromos Daoutidis**[4]

Abstract: Administration of certain drugs at a steady rate results in a deterioration of the drug effect due to a phenomenon known as tolerance. Periodic drug delivery is an attractive option for maximizing the effect of drugs exhibiting tolerance. In this paper, periodic drug infusion strategies for maximizing an averaged measure of the drug effect are investigated. A simple pharmacokinetic-pharmacodynamic model of a system exhibiting tolerance is considered and optimal periodic control theory is employed. The regions in the parametric space where periodic infusion gives better drug effect than steady infusion are characterized using the so-called π test. The optimal drug delivery strategy obtained using two different computational approaches are presented for a representative set of parameter values and insight is provided into the results. The first method, proposed by the authors, is based on the notion of differential flatness and the second, is based on the standard shooting method for dynamic optimization problems.

1. INTRODUCTION

Periodic delivery of drugs has been observed to be more efficient than a steady delivery in the case of hormonal therapy or when the drug exhibits *tolerance* (Siegel 1997). In many studies on hormonal therapy, it has been observed that if a hormone or a drug that promotes the production of a hormone is administered periodically to match the natural rhythm of the hormone release in the body, it has a better effect. A drug is effective when its concentration is between the levels of a minimum effective concentration (MEC) and a mean toxic concentration (MTC). These limits do not change with time of exposure to the drug for normal drugs. However, for drugs that exhibit tolerance, MEC increases with increasing exposure and may even increase beyond MTC thus making the drug totally ineffective. Thus a periodic infusion of the drug is more suitable for drugs with tolerance. Some recent work on the design of periodic drug delivery systems can be found in (Siegel 1997, Zou and Siegel 1999, Leroux and Siegel 1999).

In this paper, a simple pharmaco-kinetic, pharmacodynamic model of a drug is considered and the question whether periodic drug infusion is more effective in a time-averaged sense is investigated. The problem is formulated as an optimal periodic control (OPC) problem so that the powerful analytical and computational techniques developed for OPC problems can be employed (Bittanti and Guardabassi 1986, Bernstein and Gilbert 1980, Maurer *et al.* 1998, Varigonda *et al.* 2004). In particular, the so-called π-test is used to assess when small periodic variations around the optimal steady state (OSS) can improve the OSS performance, a situation known as *local properness*. The region in a two dimensional parameter space

[1] Currently with United Technologies Research Center, East Hartford, CT.
[2] Department of Electrical and Computer Engineering, University of Minnesota
[3] Department of Pharmaceutics, University of Minnesota
[4] Corresponding author. Department of Chemical Engineering and Materials Science, University of Minnesota, Minneapolis, MN 55455, Ph (612) 625–8818, Email: daoutidi@cems.umn.edu

where the problem is locally proper is computed. Since the π-test is local and only concerns small sinusoidal inputs, more elaborate computations using dynamic optimization methods are required to obtain a solution to the OPC problem. Due to the periodicity of the constraints on the state and the unknown period, the OPC problem is computationally harder than other dynamic optimization problems. The solution to the OPC problem is obtained using two computational methods, both of which produce a consistent solution. The first method is based on flatness as described in (Varigonda *et al.* 2004) and the other, a shooting method provided by a commercially available software tool, gPROMS (PSE n.d.).

2. DRUG DELIVERY SYSTEM MODEL AND FORMULATION OF THE OPC PROBLEM

Here, a variation of the model for tolerance proposed in (Porchet *et al.* 1988) is considered for the cardio-accelerating effect of nicotine. It is demonstrated that periodic delivery gives better performance than steady delivery. The model accounts for tolerance of the drug through the action of a hypothetical metabolite produced by the drug. The metabolite acts as an antagonist and reduces the drug effect. The system consists of a two compartment pharmacokinetic model that describes the time evolution of the drug concentration, c and the antagonist concentration, a and a pharmacodynamic model that describes the effect of the drug in terms of c and a.

The non-dimensionalized dynamics are linear and are given by

$$\dot{c} = -c + u \tag{1}$$

$$\dot{a} = K_a(c - a) \tag{2}$$

where u is the drug infusion rate and K_a is the rate constant for antagonist elimination. The drug effect is described by

$$E(c, a) = \frac{c}{(1 + c)(1 + a/a^*)} \tag{3}$$

where a^* is a measure of the relative potency of the antagonist. The variables c, a and u are all constrained to be positive. In addition there is an upper bound on u reflecting the maximal rate at which the drug can be delivered.

The objective considered is to keep E in a prescribed interval $[E_1, E_2]$ which can be interpreted for the nicotine system as the desired interval for the heart rate. To represent this objective as a maximization, a smooth indicator function is employed. The indicator is close to unity when $E \in [E_1, E_2]$ and close to zero otherwise. The following indicator function is used:

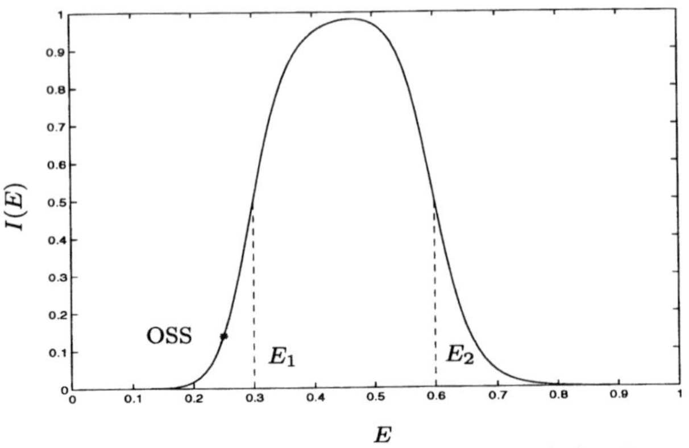

Fig. 1. Plot of the indicator function $I(E)$. The OSS value and the desired interval of the effect E are marked.

$$I(E) = \frac{(E/E_1)^\gamma}{[1 + (E/E_1)^\gamma][1 + (E/E_2)^{2\gamma}]}.$$

A plot of $I(E)$ is shown in Fig 1.

The objective is to optimize the drug effect indicator, I with respect to the drug infusion rate, u. Both steady and periodic profiles for u are considered. When u is steady, equilibrium solutions of the system in Eqs 1–2 are sought where the objective $J = I(E)$ is maximized. This results in an optimum steady state (OSS) problem. When u is periodic, the time average of the indicator over one time period, given by

$$J = \frac{1}{T}\int_0^T I(E)dt$$

is taken as the objective function for maximization. The period T is unknown and needs to be determined from optimization. In order to be able to repeat the periodic delivery strategy, the states of the system are required to be periodic as well, with the same period. This results in an optimal periodic control (OPC) problem.

3. ANALYSIS USING THE π TEST

A periodically varying u may provide better average drug effect than the OSS solution and in that case, the OPC problem is said to be *proper*. If performance improvements can be obtained by small sinusoidal perturbations of the input around the OSS, then the OPC problem is said to be *locally proper*. Local properness implies properness but the converse is not true. The so-called π-test can be used to determine local properness of an OPC problem (Bittanti *et al.* 1973, Bernstein and Gilbert 1980).

Consider the OPC problem of minimizing $J = g(y)$ over the inputs, $u(\cdot)$ and the period, T for dynamical system described by $\dot{x} = f(x, u)$ with averaged outputs $y = \frac{1}{T}\int_0^T \phi(x, u)\, dt$. Define the Hamiltonian of the system as

$$H(x, u, y, \lambda, \mu) = g(y) + \lambda' f(x, u) + \mu'(\phi(x, u) - y)$$

where $\lambda(\cdot), \mu$ are the Lagrange multipliers. Notice that the sign of μ is different from that of (Bernstein and Gilbert 1980). This definition ensures consistency of the OSS and OPC Hamiltonians. Let prime denote the transpose and an overbar denote the OSS value. Let $G(s)$ be the transfer function corresponding to the linearized dynamics at the OSS *i.e.*, $G(s) = (sI - \bar{f}_x)^{-1}\bar{f}_u$. The π-test for the local properness of OSS requires that the self-adjoint matrix $\pi(\omega)$ defined by

$$\pi(\omega) = G'(-j\omega)\bar{H}_{xx}G(j\omega) + \bar{H}_{ux}G(j\omega) + G'(-j\omega)\bar{H}_{xu} + \bar{H}_{uu}$$

be partially negative for some frequency, $\omega > 0$ (Bernstein and Gilbert 1980). The π-test can be used to readily determine if small sinusoidal perturbations of the input u around the OSS value $\bar{u}$ give any improvement in the performance.

The steady state of the system for any given u is given by $c = a = u$ and the optimum steady state (OSS) can be computed by maximizing $I(E)$ which gives $c = a = u = \sqrt{a^*}$ and $E = \frac{\sqrt{a^*}}{(1+\sqrt{a^*})^2}$. The nominal parameter values for the problem are $K_a = 0.1$, $a^* = 1$, $E_1 = 0.3$, $E_2 = 0.6$ and $\gamma = 10$. The optimum steady state (OSS) is at $c = a = u = 1$ and the OSS drug effect is $E_s = 0.25$. The indicator function at OSS is $I(E_s) = 0.1390$. The bounds on the input are taken as $u_{min} = 0$ and $u_{max} = 10$. The optimum steady state (OSS) is at $c = a = u = 1$ and the OSS drug effect is $E_s = 0.25$. The value of the indicator function at OSS is $I(E_s) = 0.1390$.

Since the drug effect E at the OSS is below E_1, steady state operation does not adequately fulfill the requirement that E be in the interval $[E_1, E_2]$. The π test is applied to determine the local properness. A plot of $\pi(\omega)$ *vs.* the frequency, ω is shown in Fig 2. Negative values of the π function over some frequency range indicates that the system is proper and periodic operation can improve the drug effect.

In order to understand the influence of the parameter values K_a and a^* on the local properness,

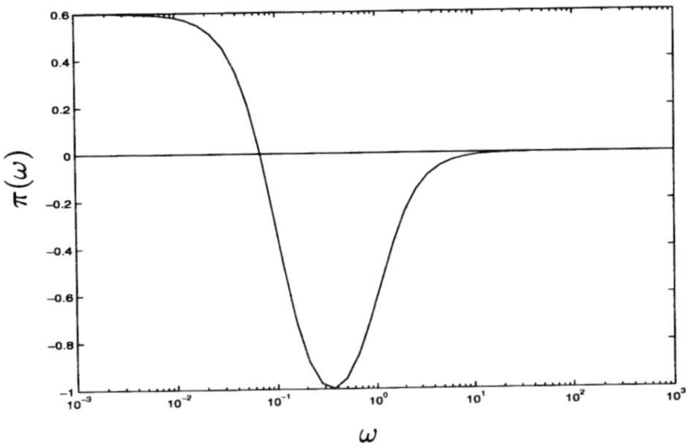

Fig. 2. Result of the π test for the drug effect model with $K_a = 0.1$ and $a^* = 1$

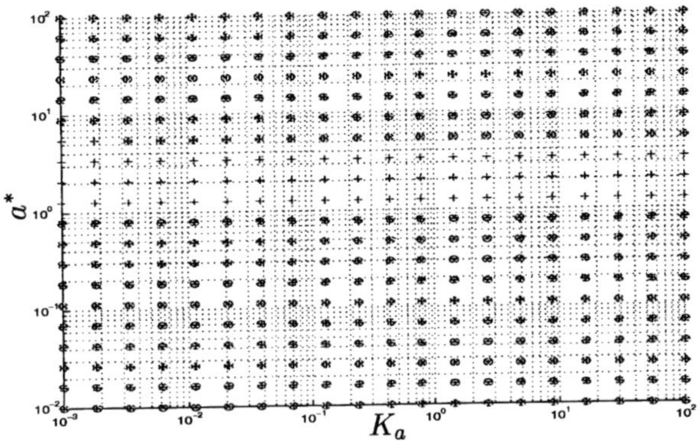

Fig. 3. Grid points in the K_a-a^* space (log-log scale) where the π test is applied (marked by +). Circled points indicate local properness, that is, the superiority of small amplitude periodic delivery over steady delivery.

the π-test is applied for several combinations of K_a and a^*. Fig 3 shows the region in the K_a-a^* space (in log-log scale) where the system is locally proper. It can be seen that the region over which the system is locally proper and small amplitude periodic drug delivery is superior to steady delivery is rather large. The surface plot of $\min_\omega \pi(\omega)$ in Fig 4 shows the extent of improvement that can be obtained with small amplitude periodic inputs in the K_a-a^* parameter space.

4. COMPUTATION OF THE OPTIMAL PERIODIC DRUG DELIVERY STRATEGY

Since the π-test is only local in nature, one cannot determine the exact shape of the periodic control signal that maximizes the drug effect from this test. Computational methods for dynamic optimization, with some modifications, need to be employed for this purpose (Bryson 1999, Maurer *et al.* 1998). It should be emphasized that the computation for the OPC problem is, in general, much harder than a regular optimal control problem due

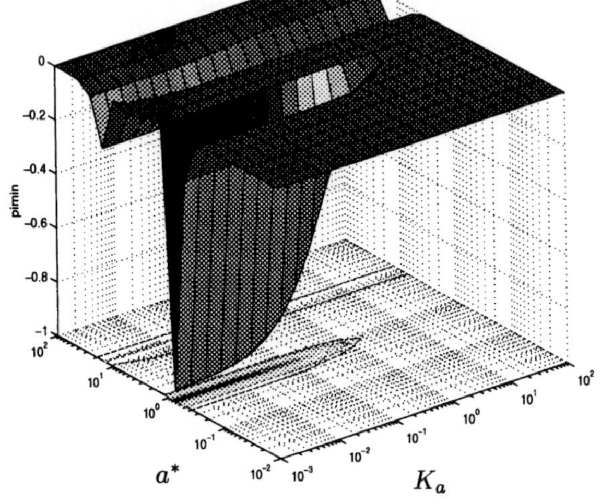

Fig. 4. Surface plot of $\min_{\omega} \pi(\omega)$ as a function of K_a and a^*. Large negative values of $\min \pi$ indicate stronger gains from periodic delivery.

to the fact that the constraints on the state are periodic and the time period is also unknown.

First, the method based on differential flatness described in (Varigonda *et al.* 2004) to obtain the optimal periodic drug infusion strategy is applied. Differential flatness (or simply, flatness), is the property of a dynamical system that is related to the concepts of absolute equivalence and dynamic feedback linearizability (van Nieuwstadt *et al.* 1998, Martin *et al.* 2003). The system described by $\dot{x} = f(x, u)$ is *differentially flat* if there exist outputs $\xi \in R^m$ depending on x, u and a finite number of time derivatives of u (*i.e.*, $\xi = h(x, u, \dot{u}, \ldots, u^{(\rho)})$) such that x, u can be expressed solely as functions of ξ and its derivatives upto order κ (Fliess and *et al.* 1995). The vector ξ must be of the same dimension as u and is called the flat output of the system. An optimal control problem on a flat system can be reformulated as a static optimization problem for efficient computation (Kansal *et al.* 2000, Oldenburg and Marquardt 2000, Mahadevan *et al.* 2000). The need for integration of differential equations is removed by restating the problem in terms of the so-called flat outputs. A computational method for OPC problems using flatness has been presented in (Varigonda *et al.* 2004) and this approach has been used for the current drug delivery problem to compute the OPC solution.

The objective to be maximized in the OPC problem is the time average of the indicator given by

$$J(u, x(0), T) = \frac{1}{T} \int_0^T I(E(c, a))dt.$$

The dynamics in (1)–(2) are linear and flat with the output $\xi = a$. The states and input can be determined from $\Xi := [\xi, \dot{\xi}, \ddot{\xi}]$ using the relations

$$a = \xi$$
$$c = \frac{\dot{\xi}}{K_a} + \xi$$
$$u = \frac{\ddot{\xi}}{K_a} + (1 + \frac{1}{K_a})\dot{\xi} + \xi.$$

The highest order derivative of the flat output is parametrized using $2N + 1$ Fourier basis functions as:

$$\xi^{(\kappa)}(t, \alpha) = \alpha_1 + \sum_{i=1}^{N} [\alpha_{2i} \sin(i\omega t) \quad (4)$$
$$+\alpha_{2i+1} \cos(i\omega t)] \quad (5)$$

where $\omega = 2\pi/T$ and $\alpha = (\alpha_1, \alpha_2, \ldots, \alpha_N)$. Then, the lower order derivatives $\xi^{(\kappa-1)}, \ldots, \xi$ are obtained by successively integrating Eq. 5 (*e.g.* using quadrature or Simpson's rule), κ times, with an integration constant β_j introduced during the jth integration. Imposing the periodicity constraints on the states one obtains $\alpha_1 = 0$ and $\beta_j = 0$ for all $j = 1, \ldots, \kappa - 1$ since these coefficients lead to polynomial terms in $\xi(t)$. Thus there are only $2N + 2$ unknown parameters, namely, T, α_2, ..., α_{2N+1} and β_κ to be determined by optimization. Let

$$\theta := [T, \alpha_2, \ldots, \alpha_{2N+1}, \beta_\kappa].$$

The constraints placed on the time period T and the drug infusion rate u are $10^{-2} \leq T \leq 10^2$ and $0 \leq u \leq 10$ respectively. In the approximation of $\ddot{\xi}$, 20 Fourier harmonics are employed (*i.e.*, $N = 20$). The solution obtained is shown in Fig 5. It should be noted that there are other locally optimal solutions also for this problem and the one selected is the one that gives the maximum improvement in the objective. The initial guess used and the final solution obtained for the optimization parameter θ are

$$\theta_0 = \begin{bmatrix} 18 \\ 0.01 \\ 0.03 \\ 0 \\ \vdots \\ 0 \\ 0.76 \end{bmatrix}, \quad \theta = \begin{bmatrix} 17.8138 \\ 0.0051 \\ 0.0295 \\ -0.0058 \\ \vdots \\ 0.0023 \\ 0.7444 \end{bmatrix}$$

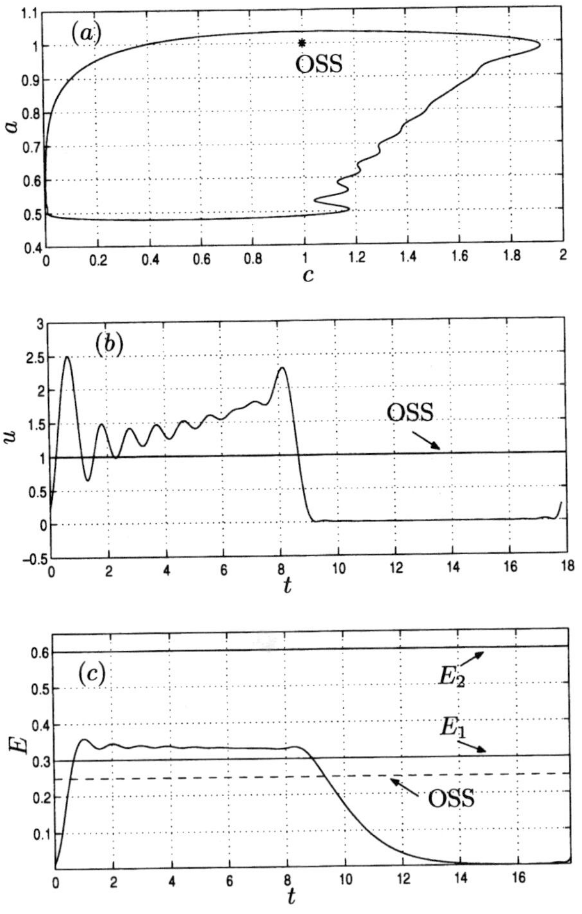

Fig. 5. Optimal periodic drug infusion strategy computed using the flatness method: (a) state space, (b) infusion rate and (c) the effect

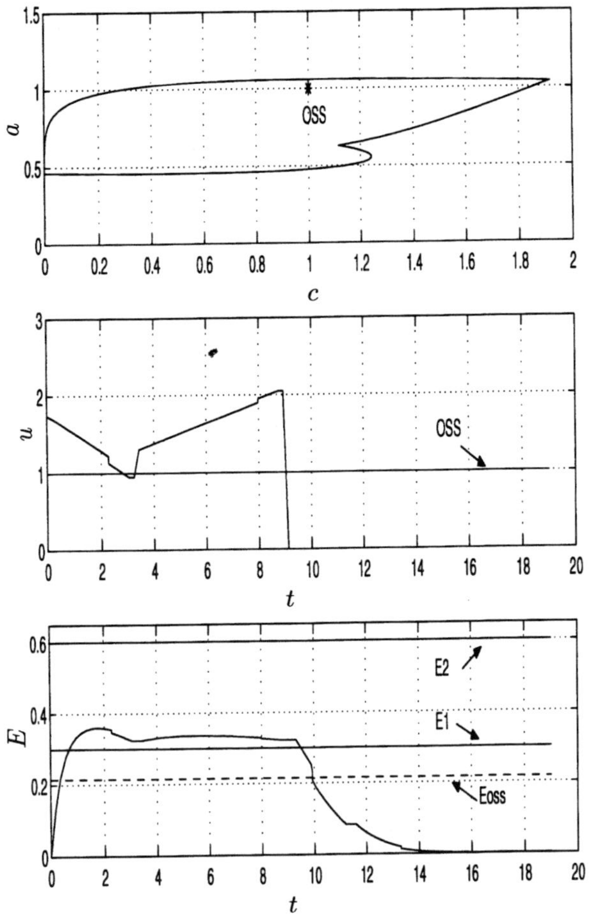

Fig. 6. Optimal periodic drug infusion strategy computed using gPROMS: (a) state space, (b) infusion rate and (c) the effect

where the first element of θ_0 is the time period T and the subsequent terms are the coefficients of the Fourier expansion of $\ddot{\xi}$. The final term is the integration constant β_2 that appears in ξ. The average of the indicator under periodic operation is approximately 0.3537 and the improvement over OSS is 0.2147.

Secondly, the same OPC problem is solved using the dynamic optimization feature of the commercial modeling software, gPROMS (PSE n.d.). Two solvers based on control vector parametrization (CVP) are provided in gPROMS for dynamic optimization. These are CVP_SS (single shooting) and CVP_MS (multiple shooting). CVP refers to the fact that the control signals u are assumed to be of certain shape (*e.g.*, piecewise linear and piecewise constant) and are represented in terms of a few parameters. In order to be able to solve the OPC problem using gPROMS, the problem was reformulated as a fixed time optimal control problem by rescaling time. Thus, the simulation time was fixed at unity while the actual time period T became an optimization parameter appearing on the right hand side of the differential equations. This enabled the objective function J to be correctly evaluated. The solution obtained using the CVP_SS method with 5 piecewise linear segments for u is shown in Fig 6. The similarities in the shape of the trajectory in state space, the input profile and the shape of the drug effect between Fig 5 and Fig 6 are obvious. The differences near the discontinuities are mainly due to the difference in the choice of basis functions.

The numerical improvement in the indicator $I(E)$ itself is not a significant factor since a higher magnitude of improvement can be obtained by using an indicator function that is steeper around the OSS. However, the qualitative features of the solution merit attention. The solution is not of bang-bang type but conforms to the intuitive idea that in presence of tolerance, the best strategy to administer the drug is to give a dose initially that will pump up the effect, while the antagonist is still building up. Once the effect reaches the desirable value, there is no more incentive in giving more drug. Hence the goal is to only maintain E at the current value for as long as possible. When the antagonist concentration a rises and starts reducing the drug effect, then u needs to be increased again to compensate for the decreasing effect. Ultimately, the antagonist concentration becomes dominating and then u needs to be shut down until a also goes down sufficiently. The cycle then repeats itself.

5. CONCLUSION

A simple pharmacokinetic, pharmacodynamic model of a drug that exhibits tolerance is proposed and an optimal periodic control problem to determine a periodic drug infusion strategy that improves the drug effect is formulated. The local properness of the OPC problem was established using the π-test. The region in the parameter space where the problem is locally proper was also determined using the π-test. The optimal periodic drug infusion rate as a function of time was computed using two very different methods and a consistent solution was obtained. Future work will focus on a better understanding of the modeling aspects that capture the observed superiority of periodic drug administration.

REFERENCES

Bernstein, D. S. and E. G. Gilbert (1980). Optimal periodic control: The π test revisited. *IEEE Trans. Automat. Contr.* **AC-25**(4), 673–684.

Bittanti, S. and G. Guardabassi (1986). Optimal periodic control and periodic systems analysis: An overview. In: *Proc. of the IEEE CDC*. Athens, Greece. pp. 1417–1423.

Bittanti, S., G. Fronza and G. Guardabassi (1973). Periodic control: A frequency domain approach. *IEEE Trans. Automat. Contr.* **AC-18**(1), 33–38.

Bryson, A. E. (1999). *Dynamic Optimization*. Addison Wesley.

Fliess, M. and *et al.* (1995). Flatness and defect of nonlinear systems: introductory theory and examples. *Int. J. Control* **61**(6), 1327–1361.

Kansal, S., J. F. Forbes and M. Guay (2000). Constrained optimization of nonlinear, dynamic chemical processes–a normalized form approach. In: *Preprints of IFAC ADCHEM*. Vol. 2. Pisa, Italy. pp. 791–796.

Leroux, J-C. and R. A. Siegel (1999). Autonomous gel/enzyme oscillator fueled by glucose: Preliminary evidence for oscillations. *Chaos* **9**(2), 267–275.

Mahadevan, R., S. K. Agrawal and F. J. Doyle III (2000). A flatness based approach to optimization in fed-batch bioreactors. In: *Preprints of IFAC ADCHEM*. Vol. 1. Pisa, Italy. pp. 111–116.

Martin, Ph., R. M. Murray and P. Rouchon (2003). Flat systems, equivalence and trajectory generation. Technical report. CDS Technical report.

Maurer, H., Ch. Büskens and G. Feichtinger (1998). Solution techniques for periodic control problems: A case study in production planning. *Optimal Control Appl. and Methods* **19**(3), 185–203.

Oldenburg, J. and W. Marquardt (2000). Dynamic optimization based on higher order differential model representations. In: *Preprints of IFAC ADCHEM*. Vol. 2. Pisa, Italy. pp. 809–814.

Porchet, H. C., N. L. Benowitz and L. Sheiner (1988). Pharmacodynamic model of tolerance: application to nicotine. *J. Pharmacol. Exp. Therap.* **244**, 231–236.

PSE (n.d.). gPROMS: General process modeling system. `http://www.psenterprise.com`.

Siegel, R. A. (1997). *Controlled Drug Delivery, Challenges and Strategies*. Chap. 25 (Modeling of Self-Regulating Oscillatory Drug Delivery). American Chemical Society. Washington D.C.. edited by Park, K.

van Nieuwstadt, M., M. Rathinam and R. M. Murray (1998). Differential flatness and absolute equivalence of nonlinear control systems. *SIAM J. Control Optim.* **36**(4), 1225–1239.

Varigonda, S., T. T. Georgiou and P. Daoutidis (2004). Numerical solution of the optimal periodic control problem using differential flatness. *IEEE Trans. Automat. Contr.* **49**(2), 271–5.

Zou, X. and R. A. Siegel (1999). Modeling of oscillatory dynamics of a simple enzyme-diffusion system with hysteresis. the case of lumped permeabilities. *J. Chem. Phys.* **110**(4), 2267–79.

ELSEVIER
IFAC
PUBLICATIONS
www.elsevier.com/locate/ifac

ROBUST GLOBAL STABILIZATION OF CONTINUOUS BIOREACTORS

L. Syrou*, I. Karafyllis, K. Stamatelatou*, G. Lyberatos* and C. Kravaris***

** Department of Chemical Engineering*
University of Patras
26500 Patras, Greece

***Department of Economics*
University of Athens
10559 Athens, Greece

Abstract: This paper studies the problem of designing controllers for enlarging the stability region of continuous stirred microbial bioreactors. A specific application is in anaerobic digestion, where the stability region can be very small if the operating steady state is selected to maximize the methane production rate. A control Lyapunov function approach is followed to construct a globally stabilizing state feedback control law. This turns out to be proportional output feedback in the case of anaerobic digestion, where the measurement is the methane production rate. The robustness properties of the feedback controller are also investigated. For situations of large, measurable changes in the organic load of the bioreactor, a feedforward measurement is incorporated in the control law, leading to improved robustness margin.

Keywords: Stabilization, Lyapunov methods, Robust stability, Bio control, Nonlinear analysis, Process control.

1. INTRODUCTION

Continuous stirred microbial bioreactors, often called chemostats, cover a wide range of applications; specialised "pure culture" biotechnological processes for the production of specialty chemicals (proteins, antibiotics etc.) as well as large-scale environmental technology processes of mixed cultures such as wastewater treatment. The dynamics of the chemostat is often adequately represented by a simple dynamic model involving two state variables, the microbial biomass x and the limiting organic substrate s. For control purposes, two operating variables are usually considered, the dilution rate D which is the manipulated input, and the feed substrate concentration S_0 which is a load variable. A general model for microbial growth on a limiting substrate in a chemostat is of the form:

$$\begin{aligned} \frac{dx}{dt} &= -Dx + \mu(s)x \\ \frac{ds}{dt} &= D(S_0 - s) - \frac{1}{Y_{x/s}}\mu(s)x \end{aligned} \tag{1}$$

where $\mu(s)$ is the specific growth rate, and $Y_{x/s}$ is a biomass yield factor. One important example is anaerobic digestion, which finds many applications e.g. in wastewater treatment, sludge management, energy from biomass, etc.

The purpose of this work is to study the problem of robust global stabilization of a bioreactor whose dynamics follows (1), the motivation coming from control problems in anaerobic digestion processes. Section 2 provides some background on anaerobic digestion and explains the nature of the control problem. In Section 3, a simple feedback controller is derived via a control Lyapunov function approach. The robustness properties of the feedback controller are examined in Section 4. Finally, Section 5 develops a feedforward / feedback controller with improved robustness margin.

2. EXAMPLE: ANAEROBIC DIGESTION

Anaerobic digestion is a complex biochemical process, in which organic compounds are mineralised to biogas (a useful energy product), consisting primarily of methane and carbon dioxide, through a series of reactions mediated by several groups of microorganisms. Under normal (or balanced) operation, the rate of production of the intermediates is matched by their consumption rate; hence there is very little accumulation of these compounds.

However, disturbances such as an increase in the concentration of organic compounds in the feed (organic overload), an increase in feed flow rate (hydraulic overload), presence of toxins in the feed, and temperature fluctuations, can cause an imbalance in the process (Switzenbaum et al., 1990), which results in accumulation of volatile organic acids. These acids cause a drop in the pH, inhibiting methanogenesis and the reactor fails. Such a failure has major consequences in the process economics since digester recovery can be a very cumbersome and costly process. For this reason, the development of appropriate control schemes for anaerobic digesters has received significant attention in recent years (Perrier and Dochain, 1993; Pind et al., 2003).

2.1 Mathematical model of anaerobic digestion.

For the description of anaerobic digestion, the mathematical model (1) can be used. This system of equations describes methanogenesis, the ultimate step in anaerobic digestion, which is rate limiting and is usually the most sensitive step. In other words, it is assumed that the bioconversion of organics into fatty acids (hydrolysis and acidification) has fast kinetics. Then x and s in (1) represent the methanogen and volatile fatty acid concentrations respectively. The specific growth rate is assumed to follow the Andrews kinetics (substrate inhibition) (Graef and Andrews, 1974):

$$\mu(s) = \frac{\mu_{\max} s}{K_S + s + \dfrac{s^2}{K_I}} \qquad (2)$$

The output of the system is the methane production rate,

$$Q = Y\mu(s)x \qquad (3)$$

where Y is a yield factor for methane production.

2.2 Optimal steady state for methane production

Apart from the washout steady state ($x = 0$, $s = S_0$), the bioreactor's steady states are calculated from the equations:

$$\begin{Bmatrix} \mu(s_s) = D_s \\ x_s = Y_{x/s}(S_0 - s_s) \end{Bmatrix} \qquad (4)$$

For a given feed, there is a value of the dilution rate that maximises the methane production rate. The steady state that corresponds to the maximization of methane production rate, i.e. $Q = YY_{x/s}\mu(s_s)(S_0 - s_s)$, draws technical interest. The methane production rate is maximized when:

$$\frac{dQ}{ds_s} = 0 \Leftrightarrow \frac{d\mu(s_s)}{ds}(S_0 - s_s) = \mu(s_s)$$

Solving the above equations for $\mu(s)$ given by (2), it is found that the optimal steady state is:

$$s_s^{opt} = \frac{S_0}{1 + \sqrt{1 + \dfrac{S_0}{K_S}\left(1 + \dfrac{S_0}{K_I}\right)}} \qquad (5)$$

For the following values of the parameters:

$S_0 = 10000\ mg/l, K_S = 100\ mg/l, \mu_{\max} = 0.5\ d^{-1}$, $K_I = 4000\ mg/l, Y_{x/s} = 0.05\ mg/mg$, $Y = 0.00746\ ll_{reactor}^{-1} d^{-1}$ the optimal steady state from equation (5) is $s_s = 506.714\ mg/l$. This corresponds to $x_s = 474.664\ mg/l$, $D_s = 0.377635\ d^{-1}$ and $Q_s = 1.337205\ ll_{reactor}^{-1} d^{-1}$.

The above numerical values of the parameters and the resulting optimal steady state conditions will be used in the numerical calculations throughout this paper.

2.3 Local asymptotic stability

The eigenvalues of the linearization of (1) are: $\lambda_1 = -\mu(s_s)$, $\lambda_2 = -\dfrac{d\mu(s_s)}{ds}(S_0 - s_s)$. Since $\mu(s) > 0$ for all s, local asymptotical stability is guaranteed as long as $\dfrac{d\mu(s_s)}{ds}(S_0 - s_s) > 0$. This condition is satisfied for the optimal steady state given by equation (5).

2.4 The need for control

Figure 1 depicts the phase portrait of the system dynamics under constant dilution rate *D*, in particular for $D = D_s = 0.377635\ d^{-1}$, which is the optimal steady state value. In the diagram, the points S and U represent the corresponding stable and the unstable steady states of the reactor, which are the solutions of equations (4). Notice that the optimal steady state S is locally stable but the stability region is very small. This makes the optimal operation of the biochemical reactor very sensitive to disturbances.

The goal of control is the stabilization of the system in the sense of enlargement of the stability region.

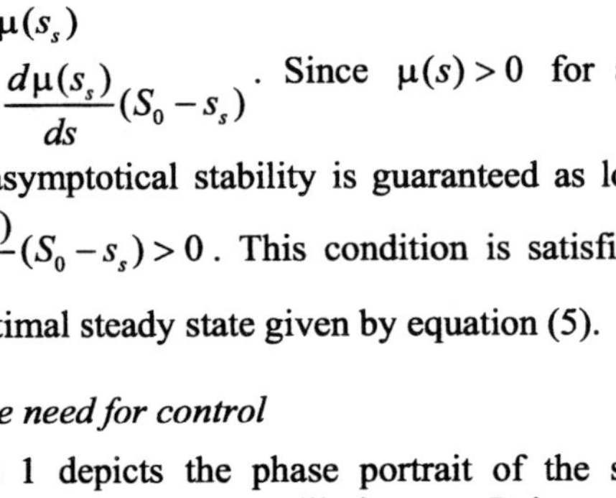

Fig.1. Phase portrait of the open-loop dynamics

3. FEEDBACK CONTROLLER SYNTHESIS: A CONTROL LYAPUNOV FUNCTION APPROACH

Consider the dynamic system (1) with $D \in [0, +\infty)$, $x \in (0, +\infty)$, $s \in (0, +\infty)$ where $\mu : R^+ \to R^+$ is a smooth function with $\mu(0) = 0, \mu(s) > 0$ for all $s > 0$. Also, consider the coordinate transformation:

$$x_1 = \ln\left(\frac{x}{Y_{x/s}(S_0 - s_s^{des})}\right)$$
$$x_2 = \ln\left(\frac{s}{s_s^{des}}\right) \tag{6}$$

where s_s^{des} is a design steady state value for s (e.g. from equation (5)). Transformation (6) maps the open first quadrant onto R^2. In particular, it maps the design steady state $\left\{\begin{matrix} x_s = Y_{x/s}(S_0 - s_s^{des}) \\ s_s = s_s^{des} \end{matrix}\right\}$ to the origin. Moreover, to facilitate the derivations, consider the input transformation:

$$D = \mu(s)u \tag{7}$$

Under the transformations (6) and (7), the system (1) becomes:

$$\frac{dx_1}{dt} = (1-u)\mu(s_s^{des}e^{x_2})$$
$$\frac{dx_2}{dt} = \left[u\left(\frac{S_0}{s_s^{des}}e^{-x_2} - 1\right) - \left(\frac{S_0}{s_s^{des}} - 1\right)e^{x_1-x_2}\right]\mu(s_s^{des}e^{x_2}) \tag{8}$$

Consider now the control Lyapunov function:

$$V = \frac{1}{2}(x_1^2 + x_2^2) \tag{9}$$

The time derivative of V along the trajectories of (8) is given by:

$$\dot{V} = \left\{ \left[x_1 - \left(\frac{S_0}{s_s^{des}} - 1\right)x_2 e^{x_1-x_2}\right](1-ue^{-x_1}) + u[x_1(e^{-x_1}-1) + x_2(e^{-x_2}-1)] \right\} \mu(s_s^{des}x_2) \tag{10}$$

Since $x(\exp(-x)-1) < 0$ for all $x \neq 0$, it is concluded from (10) that the smooth feedback law:

$$u = e^{x_1} \tag{11}$$

will globally stabilize the origin for system (8). Transforming the feedback law (11) back to the original coordinates (via (6) and (7)) results in

$$D = \frac{\mu(s)x}{Y_{x/s}(S_0 - s_s^{des})} \tag{12}$$

Notice that from equation (12), it is guaranteed that $D \geq 0$ for all times.

The feedback law (12) is a nonlinear state feedback law and, in general, requires measurement of both biomass and substrate. However, in the case of anaerobic digestion (see Section 2) where the measured biogas production rate is proportional to the product $\mu(s)x$, it becomes a linear output feedback law:

$$D = \frac{1}{YY_{x/s}(S_0 - s_s^{des})}Q \tag{13}$$

It is important to point out that the control law of equation (13) coincides with the constant yield control law (CYCL) of Pullammannappallil et al (1998), where the methane yield (defined as the ratio between the methane production rate and the feed flow rate) was maintained at a constant set-point value. Mailleret and Bernard (2001), using a different approach, arrived at the same control law and provided a rigorous justification of global closed-loop stability. The control law was implemented experimentally in Pullammannappallil et al (1998) and in Mailleret et al. (2003).

The closed-loop system under the feedback law (12) or (13) is:

$$\frac{dx}{dt} = \left(1 - \frac{x}{Y_{x/s}(S_0 - s_s^{des})}\right)\mu(s)x$$
$$\frac{ds}{dt} = \frac{(s_s^{des} - s)\mu(s)x}{Y_{x/s}(S_0 - s_s^{des})} \tag{14}$$

The form of the closed-loop system shows that:

i) the closed-loop system's equilibrium is at $x_s = Y_{x/s}(S_0 - s_s^{des})$ and $s_s = s_s^{des}$

ii) $\frac{d(s - s_s^{des})}{d(x - Y_{x/s}(S_0 - s_s^{des}))} = \frac{s - s_s^{des}}{x - Y_{x/s}(S_0 - s_s^{des})}$, which proves that the system's trajectories represent straight lines on the $s-x$ plane.

Figure 4 depicts a phase portrait of the closed-loop dynamics (14) for the particular parameter values and design conditions of Section 2.

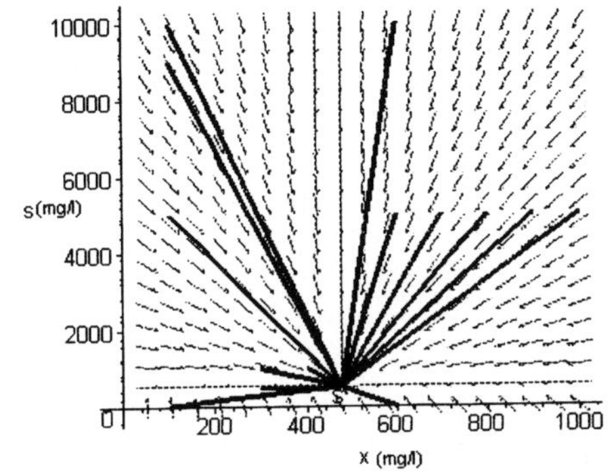

Fig.2. Phase portrait of the closed-loop system (14)

4. ROBUSTNESS PROPERTIES OF THE FEEDBACK CONTROL LAW

In the case of direct measurement of the methane production rate $Q = Y\mu(s)x$, the control law derived in the previous section is completely insensitive to errors in $\mu(s)$, in terms of achieving global stability. Of course, the computed s_s^{des} (e.g. from equation (5)) will depend on the parameters in $\mu(s)$ (like K_s and K_I) and will no longer be optimal in the presence of parameter errors. For the particular parameter values given in Section 2, which are representative of a typical anaerobic digestion process, it is found that a 50% error below the nominal values of K_s and K_I corresponds to 1.24% and 1.17% biogas loss respectively and that a 50% error above the nominal values corresponds to 0.63% and 0.22% biogas loss.

The situation is different in the presence of errors in S_0. Suppose that a design value S_0^{des} is used in the control law (13), i.e.

$$D = \frac{1}{YY_{x/s}(S_0^{des} - s_s^{des})} Q \tag{15}$$

whereas the true substrate concentration in the process is S_0. Then the resulting closed-loop dynamics will follow:

$$\begin{aligned} \frac{dx}{dt} &= \left(1 - \frac{x}{Y_{x/s}(S_0^{des} - s_s^{des})}\right)\mu(s)x \\ \frac{ds}{dt} &= \frac{\left[(S_0 - S_0^{des}) + s_s^{des} - s\right]}{Y_{x/s}(S_0^{des} - s_s^{des})}\mu(s)x \end{aligned} \tag{16}$$

From (16) it is seen that the closed-loop system's equilibrium will be at $\left\{\begin{matrix} x_s = Y_{x/s}(S_0^{des} - s_s^{des}) \\ s_s = s_s^{des} + (S_0 - S_0^{des}) \end{matrix}\right\}$

Notice that x_s equals the design value of biomass concentration, but there will be an offset of the substrate from the design conditions by an amount equal to the error in S_0. The problem is that, if S_0 is significantly lower than its design value S_0^{des}, s_s may be negative and therefore physically unrealistic.

Case 1: If $s_s^{des} + (S_0 - S_0^{des}) > 0$, then the system (16) can be transformed via the coordinate transformation:

$$\begin{aligned} x_1 &= \ln\left(\frac{x}{Y_{x/s}(S_0^{des} - s_s^{des})}\right) \\ x_2 &= \ln\left(\frac{s}{S_0 - S_0^{des} + s_s^{des}}\right) \end{aligned} \tag{17}$$

resulting in:

$$\begin{aligned} \frac{dx_1}{dt} &= (1 - e^{x_1})\mu((S_0 - S_0^{des} + s_s^{des})e^{x_2}) \\ \frac{dx_2}{dt} &= -(1 - e^{-x_2})e^{x_1}\mu((S_0 - S_0^{des} + s_s^{des})e^{x_2}) \end{aligned} \tag{18}$$

which is globally asymptotically stable, as can be easily seen through the Lyapunov function (9).

Case 2: If $s_s^{des} + (S_0 - S_0^{des}) < 0$, the trajectories of the system (16) will still be straight lines. However, they will not reach the physically unrealistic equilibrium point. Instead, they will terminate on the x- axis, which corresponds to process shut down (x=*const.*, s=0, D = 0).

Figures 2 and 3 depict representative phase portraits for S_0^{des}=10000 *mg/l* and the rest of the parameters as in Section 2. In Figure 3, $S_0 = 11000\ mg/l$ and this corresponds to Case 1, where the effect of the error in S_0 is just a shift in the steady-state value of s. In Figure 4, for $S_0 = 8000\ mg/l$ and this corresponds to Case 2, where the error in S_0 brings the system to shut down.

The conclusion from the foregoing analysis is that the feedback control law is robust for errors in S_0 such that: $s_s^{des} > (S_0^{des} - S_0)$. Otherwise, it brings the system to shut down, which is completely undesirable.

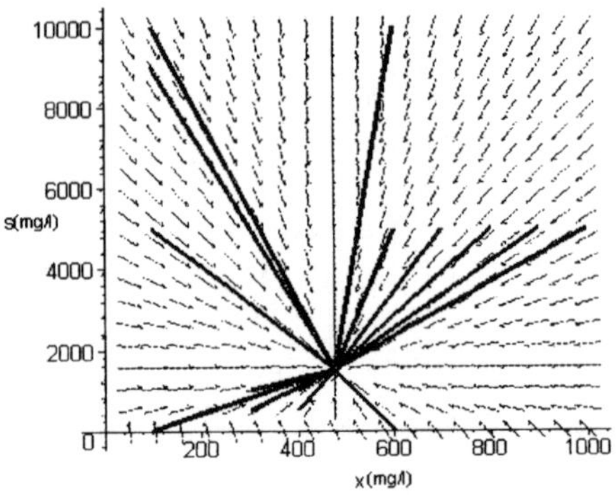

Fig.3. Phase portrait for $S_0 = 11000\ mg/l$

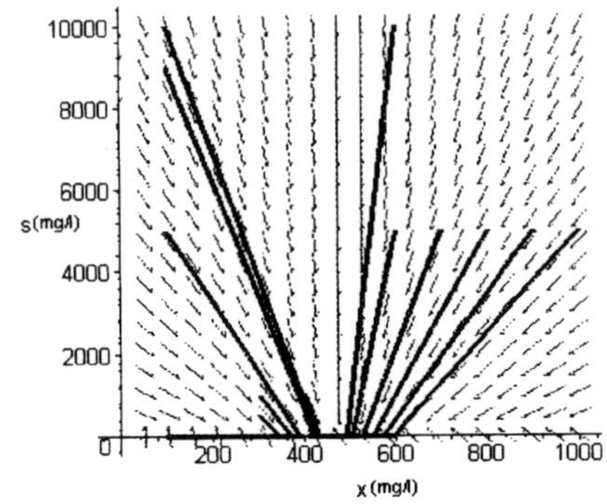

Fig.4. Phase portrait for $S_0 = 8000\ mg/l$

5. FEEDFORWARD /FEEDBACK CONTROL

If it is anticipated that the error in S_0 could be large enough to violate the robustness condition of the previous section, the control law will need to be modified to prevent shut down. One possibility is to measure the feed substrate concentration on-line and use it as a feedforward measurement in the control law. The measurement of S_0 is feasible using a Total Organic Carbon (TOC) analyzer.

Consider the control law (12) or (13), where S_0 is now the on-line measurement of the feed substrate concentration and, therefore, the control law now involves both feedback and feedforward action. Then, the resulting closed-loop system is still given by (14), even though now S_0 is a function of time, with $S_0(t) > 0$ for all $t > 0$. In the event that S_0 remains constant, the closed-loop system's equilibrium will be at: $\left\{\begin{matrix} x_s = Y_{x/s}(S_0 - s_s^{des}) \\ s_s = s_s^{des} \end{matrix}\right\}$. As long as $S_0 > s_s^{des}$, the equilibrium will be in the first quadrant and the closed-loop system will be globally stable as a result of the analysis of section 3.

The situation where $S_0(t)$ is time-varying but bounded requires further attention. Suppose that S_0^{des} is the value of S_0 for which the reactor has been designed to operate. Then the difference

$$z(t) = S_0(t) - S_0^{des} \tag{19}$$

represents a bounded but, in general, non-vanishing perturbation to the system. To be able to analyze the effect of this perturbation, consider the coordinate transformation:

$$x_1 = \ln\left(\frac{x}{Y_{x/s}(S_0^{des} - s_s^{des})}\right)$$
$$x_2 = \ln\left(\frac{s}{s_s^{des}}\right) \tag{20}$$

Under this transformation, (14) transforms into:

$$\frac{dx_1}{dt} = \left(1 - \frac{e^{x_1}}{1 + \dfrac{S_0(t) - S_0^{des}}{S_0^{des} - s_s^{des}}}\right)\mu(s_s^{des} e^{x_2})$$
$$\frac{dx_2}{dt} = (e^{-x_2} - 1)\frac{e^{x_1}}{1 + \dfrac{S_0(t) - S_0^{des}}{S_0^{des} - s_s^{des}}}\mu(s_s^{des} e^{x_2}) \tag{21}$$

Now assume that:

$$\left|\frac{S_0(t) - S_0^{des}}{S_0^{des} - s_s^{des}}\right| \le L, \quad \text{for all } t > 0 \tag{22}$$

where $L < \frac{1}{3}$. Then

$$\frac{d}{dt}\left(\frac{1}{2}x_2^2\right) = x_2(e^{-x_2} - 1)\frac{e^{x_1}}{1 + \dfrac{S_0(t) - S_0^{des}}{S_0^{des} - s_s^{des}}}\mu(s_s^{des} e^{x_2}) \le 0 \tag{23}$$

whereas the quantity

$$\frac{d}{dt}\left(\frac{1}{2}x_1^2\right) = \left\{x_1(1 - e^{x_1}) + \frac{x_1 e^{x_1}\dfrac{S_0(t) - S_0^{des}}{S_0^{des} - s_s^{des}}}{1 + \dfrac{S_0(t) - S_0^{des}}{S_0^{des} - s_s^{des}}}\right\}\mu(s_s^{des} e^{x_2}) \tag{24}$$

will be ≤ 0 as long as

$$x_1 \le \ln\left(1 - \frac{2L}{1+L}\right) \tag{25}$$

or

$$x_1 \ge \ln\left(1 + \frac{2L}{1-3L}\right) \tag{26}$$

From the above, using standard arguments of the theory of non-vanishing perturbations (Khalil, 1996, Chapter 5), it follows that for any initial conditions,

$$\lim_{t\to\infty} x_2(t) = 0$$

and there exists a time $T \ge 0$ such that

$$|x_1(t)| \le b, \forall t \ge T \tag{27}$$

where

$$b = \max\left\{-\ln\left(1 - \frac{2L}{1+L}\right), \ln\left(1 + \frac{2L}{1-3L}\right)\right\} \tag{28}$$

In other words, the solutions of (21) are uniformly ultimately bounded, with x_2 asymptotically approaching zero.

Consequently, the final value of the substrate concentration will equal its design value, $\lim_{t\to\infty} s(t) = s_s^{des}$, whereas the ratio of the biomas divided by its design value will be ultimately bounded as follows:

$$e^{-b} \le \frac{x(t)}{Y_{x/s}(S_0^{des} - s_s^{des})} \le e^{b} \tag{29}$$

for every initial condition within the open first quadrant.

Numerical simulations were performed under pure feedback action (FB) and combined feedforward and feedback action (FF/FB), when the system is initially at steady state with $S_0 = S_0^{des} = 10000\, mg/l$ but S_0 undergoes a step change to a new value.

Figures 5, 6 and 7 compare the responses of x, s and Q for some representative final values of S_0.

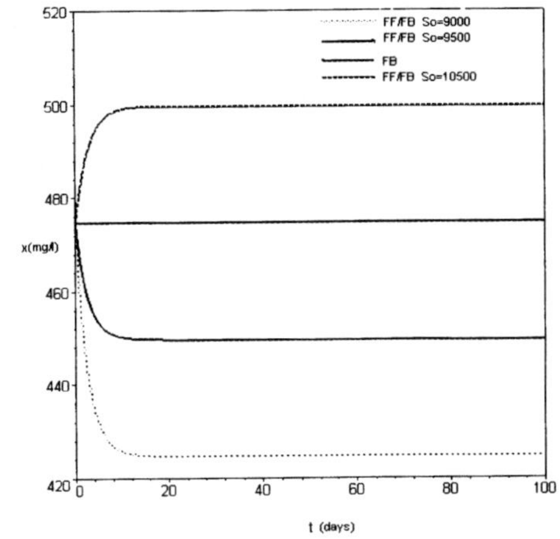

Fig.5.Responses of x for representative values of S_0

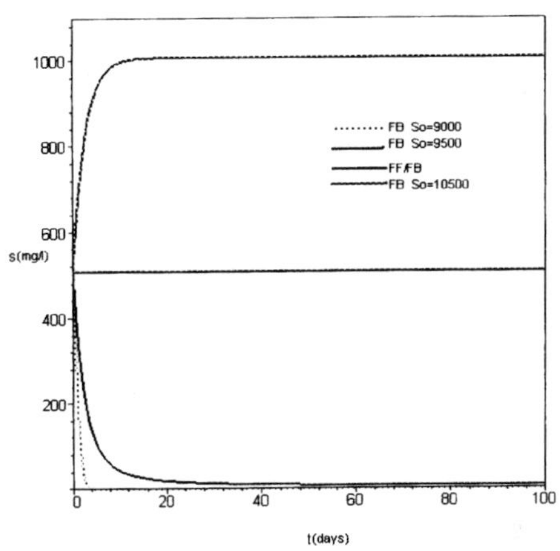

Fig.6.Responses of s for representative values of S_0

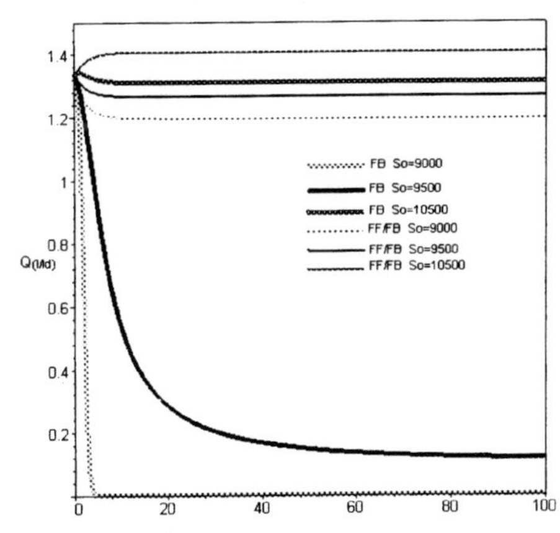

Fig.7. Response of Q for representative values of S_0

It is seen from Figures 5 and 6 that, as predicted by the analysis, in a step change in S_0, the steady state of x is unaffected under FB control, whereas the steady state of s is unaffected under FF/FB control.

The problem with FB control is when the robustness condition $s_s^{des} > (S_0^{des} - S_0)$ is violated, like in the case $S_0 = 9000\ mg/l$, where the system is brought to shut down. For $S_0 = 9500\ mg/l$, the reactor is not shut down because the robustness condition is satisfied, but the final value of the substrate is very small ($s \to 6.714\ mg/l$), leading to small biogas production ($Q \to 0.111\ ll_{reactor}^{-1} d^{-1}$). On the other hand, FF/FB control can tolerate much larger deviations.

In all the step changes considered, FF/FB has better performance than FB in the sense that the biogas production Q is larger (see Fig. 7).

Figures 8, 9 and 10 depict the responses of x, s and Q under a sinusoidal variation in S_0 of amplitude 500 *mg/l* and period 60 *d*. The system is initially at steady state with $S_0 = S_0^{des} = 10000\ mg/l$ and the average value of S_0 remains the same (10000 *mg/l*) for $t > 0$:

$$S_0(t) = 10000 + 500 \sin(\frac{2\pi}{60} t).$$

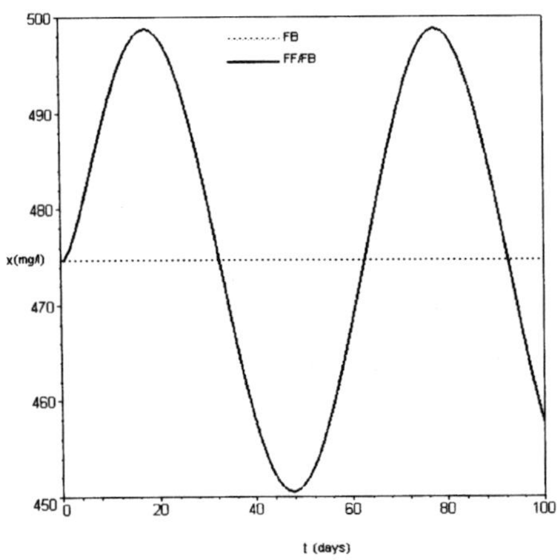

Fig.8. Response of x for a sinusoidal change in S_0

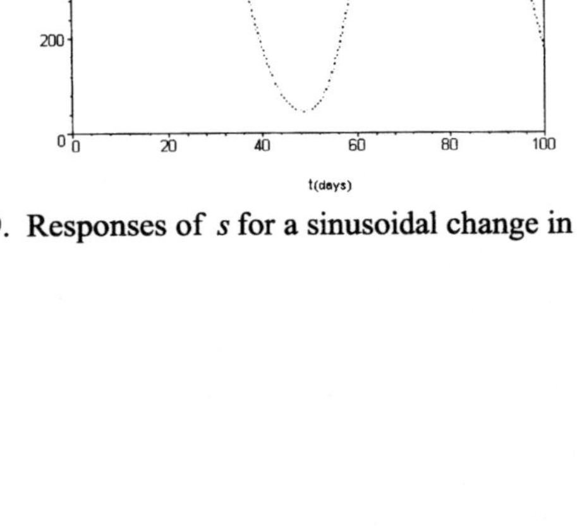

Fig.9. Responses of s for a sinusoidal change in S_0

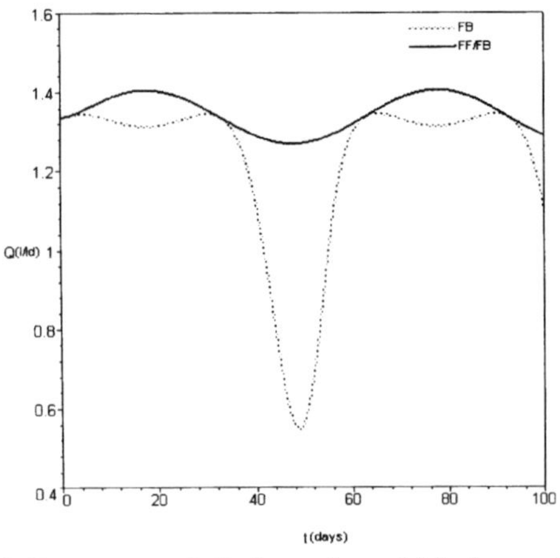

Fig.10. Response of Q for a sinusoidal change in S_0

Even though FF/FB cannot completely eliminate the effect of the discrepancy $S_0(t) - S_0^{des}$, the response of the system is bounded. Also, FF/FB leads to higher biogas production, as shown in Figure 10.

REFERENCES

Graef, S. P. and J. F. Andrews (1974). Stability and Control of Anaerobic Digestion. *WPCF Journal*, **46**, 667-682.

Khalil H. K. (1996). *Nonlinear Systems*, 2nd edition, Macmillan, New York.

Mailleret, L. and O. Bernard (2001). A Simple Robust Controller to Stabilize an Anaerobic Digestion Process. *Proc. 8th Conference on Computer Applications in Biotechnology*, 213-218.

Mailleret, L., O. Bernard and J.P. Steyer (2003). Robust Regulation of Anaerobic Digestion Processes. *Water Science and Technology*, **48**, 87-94.

Perrier, M. and D. Dochain (1993). Evaluation of Control Strategies for Anaerobic Digestion Processes. *International Journal on Adaptive Control and Signal Processing*, **7**, 309-321.

Pind P.F., I. Angelidaki, B.K. Ahring, K. Stamatelatou and G. Lyberatos (2003). Monitoring and Control of Anaerobic Reactors. In: *Biomethanation II* (Ahring, B.K. Ed.), *Adv. Biochem. Eng. Biotechnol., Vol. 82*, 135-182, Springer-Verlag, Heidelberg.

Pullammanappallil, P., S. A. Svoronos, D. P. Chynoweth and G. Lyberatos (1998). Expert system for control of anaerobic digesters. *Biotechnology and Bioengineering*, **58**, 13-22.

Switzenbaum, M. S., E. Giraldo-Gomez and R. F. Hickey (1990). Monitoring of the Anaerobic Methane Fermentation Process. *Enzyme Microb. Technol.*, **12**, 722-730.

ELSEVIER
IFAC
PUBLICATIONS
www.elsevier.com/locate/ifac

PUTTING THE "CONTROL" IN METABOLIC CONTROL ANALYSIS

Christopher V. Rao * **Herbert M. Sauro** ** **Adam P. Arkin** *

* *Howard Hughes Medical Institute*
Lawrence Berkeley National Laboratory
Department of Bioengineering, University of California
Berkeley, CA 94720
** *Keck Graduate Institute, Claremont, CA 91711*

Abstract: Metabolic control analysis is a framework for characterizing the parametric sensitivity of metabolic pathways and genetic networks. We establish a connection between metabolic control analysis and control theory. The main result is that we can use classical control theory and the associated signal-flow (block) diagrams to analyze biochemical reaction networks and that many results in metabolic control analysis have direct counterparts in control. In the process, we illustrate how some problems in biochemical network analysis can be reformulated in a control-theoretic framework. *Copyright © 2004 IFAC*

Keywords: Metabolic control analysis (MCA), sensitivity, biology, networks

1. INTRODUCTION

Metabolic control analysis (MCA) is a framework for analyzing metabolic pathways and genetic networks. The theory aims to link steady-state changes in the individual pathway components to steady-state changes in the systematic behavior of the network. The basic framework was proposed over the thirty years ago in a variety of forms by a number of biologists (Kacser and Burns, 1973; Burns, 1971; Heinrich and Rapoport, 1974*b*; Heinrich and Rapoport, 1974*a*; Savageau, 1971; Savageau, 1972). The theory was motivated by the realization that metabolic flux is not controlled by one rate-limiting enzyme. Rather, the control is shared by all of the enzymes in the pathway. This distributed control is particularly striking in regulated and branched pathways. The basic theory is now mature (Heinrich and Schuster, 1996; Fell, 1997), though many opportunities exist to extend MCA.

In addition to its name, there is striking similarity between the equations in MCA and those found in classical control theory. Intrigued by the cursory resemblance, we develop this connection and demonstrate that MCA can be entirely reformulated in control-theoretic terms. In the process we demonstrate that the MCA formulation, which is really just specialized sensitivity analysis, offers a promising foothold at the interface of biology and control theory.

2. BACKGROUND

We assume that the dynamics of the biochemical reaction network are described by the following differential equation

$$\dot{s}(t) = \mathbf{N}v(s(t), p), \tag{1}$$

where the vector $s \in \mathbb{R}^n$ contains the concentrations for n chemical species, the matrix $\mathbf{N} \in \mathbb{I}^{n \times m}$

defines the stoichiometry of the reaction network, the function $v(\cdot,\cdot) : \mathbb{R}^n \times \mathbb{R}^p \mapsto \mathbb{R}^m$ describes the rates for m chemical reactions, and the vector $p \in \mathbb{R}^p$ contains the relevant kinetic parameters for these rates such as enzyme concentrations and catalytic efficiencies.

As the inclusion of some species in the model (1) leads to redundant equations because of mass conservation constraints, one commonly removes these species to create a reduced stoichiometric matrix $\mathbf{N_r}$ with linearly independent rows (Reder, 1988). We can recover the full matrix by introducing the link matrix $\mathbf{L}$ where

$$\mathbf{N} = \mathbf{L}\mathbf{N_r}.$$

Let $s_i(t)$ denote the concentration of the reduced set of species. Then, we can recover $s(t)$ from $s_i(t)$ using the relation

$$s(t) = \mathbf{L}s_i(t) + T$$

with the appropriate choice for the constant vector T. The vector T accounts typically for mass conservation. For example, the mass constraint $a(t)+b(t) = 1$ yields the link matrix $\mathbf{L} = \begin{bmatrix} 1 & -1 \end{bmatrix}^T$ and constant vector $T = \begin{bmatrix} 0 & 1 \end{bmatrix}^T$ with $s(t) = \begin{bmatrix} a(t) & b(t) \end{bmatrix}^T$ and $s_i(t) = a(t)$.

MCA is concerned with how the properties of the network change when the parameters p are perturbed. The sensitivity function or **control coefficient** is defined as

$$x(t) \triangleq \frac{ds_i(t)}{dp}.$$

Using the chain rule for differentiation, one can show that the control coefficient $x(t)$ satisfies the following differential equation

$$\dot{x}(t) = \left(\mathbf{N_r}\frac{\partial v(s(t),p)}{\partial s(t)}\mathbf{L}\right) x(t) + \mathbf{N_r}\frac{\partial v(s(t),p)}{\partial p}. \quad (2)$$

Note that equation (2) is linear: it is identical to the linearization of equation (1). If the reaction network (1) is at a steady state s_{ss}, then the sensitivity equation (2) for infinitesimal perturbations about the steady state satisfies the linear, time-independent, differential equation

$$\dot{x}(t) = (\mathbf{N_r}\varepsilon_s\mathbf{L})\, x(t) + \mathbf{N_r}\varepsilon_p(t), \quad (3)$$

where

$$\varepsilon_x \triangleq \frac{\partial v(s_{ss},p)}{\partial s}, \quad \varepsilon_p \triangleq \frac{\partial v(s_{ss},p)}{\partial p}.$$

The matrices $\varepsilon_x \in \mathbb{R}^{m\times n}$ and $\varepsilon_p \in \mathbb{R}^{m\times p}$ are called the **elasticity coefficients**. They provide a measure for how strongly a single reaction in isolation is changed by infinitesimal perturbations either to the concentrations s or parameters p. It is possible to directly measure the elasticity coefficients from experiments without needing to know the rate laws $v(\cdot)$ explicitly. In fact, it is easier to determine experimentally the elasticity coefficients than the actual rate laws (Fell, 1997).

As equation (3) is linear and time invariant, we can take the Laplace transform and obtain the frequency response for the control coefficients:

$$X(j\omega) = (j\omega I - \mathbf{N_r}\varepsilon_x\mathbf{L})^{-1}\mathbf{N_r}\varepsilon_p, \quad (4)$$

where $X(j\omega)$ is the Laplace transform of $x(t)$ and j is the complex number $\sqrt{-1}$. While the perturbations are typically constant, it is also possible to consider time-varying perturbations $\Delta(t)$. To explore frequency variations, we replace ε_p with $\varepsilon_p\Delta(j\omega)$ in equation (4). At steady state ($\omega = 0$), the sensitivity equation (4) become

$$X(0) = -(\mathbf{N_r}\varepsilon_x\mathbf{L})^{-1}\mathbf{N_r}\varepsilon_p,$$

assuming the inverse exists. This equation is known as the **control equation**, because it relates the control coefficients $X(0)$ to the elasticity coefficients ε_p at steady state. It is a cornerstone of MCA (Burns, 1971; Heinrich and Rapoport, 1974*b*; Heinrich *et al.*, 1977; Reder, 1988).

In addition to the concentrations $s(t)$, MCA is concerned with perturbations to the flux through the network. We define the flux control coefficient as

$$y(t) \triangleq \frac{dv}{dp}.$$

The flux control coefficient $y(t)$ is related to the control coefficient $x(t)$ by the affine relation

$$\begin{aligned} y(t) &= \frac{\partial v}{\partial p} + \left(\frac{\partial v}{\partial s}\mathbf{L}\right) x(t), \\ &= \varepsilon_p + (\varepsilon_x\mathbf{L})\, x(t), \end{aligned}$$

where the second relation holds for perturbations about the steady state s_{ss}. Note the difference between y and ε_p.

If $\mathbf{A} \triangleq (\mathbf{N_r}\varepsilon_x\mathbf{L})$, $\mathbf{B} \triangleq \mathbf{N_r}$, $\mathbf{C} \triangleq \varepsilon_x\mathbf{L}$ and $\mathbf{D} \triangleq \mathbf{I}$, then we represent the control equations using the following state-space form:

$$\dot{x} = \mathbf{A}x + \mathbf{B}\varepsilon_p, \quad (5a)$$
$$y = \mathbf{C}x + \mathbf{D}\varepsilon_p, \quad (5b)$$

where ε_p is the input.

In the remainder of this article, we explore the control equation (5) and demonstrate how some elementary tools from control analysis can be applied to this equation to explore regulation in biochemical networks.

3. SIGNAL-FLOW GRAPHS

Molecular biologists commonly draw complex diagrams describing intracellular pathways in *cartoon* format. While these diagrams successfully enumerate the players and interactions within the pathway, they are unable to convey any information regarding dynamic or regulatory properties.

When analyzing and designing control systems, engineers often employ signal-flow graphs. Signal-flow graphs diagram the feedback structure using unambiguous notation. Signal-flow diagrams are also able to convey dynamic information. While signal-flow graphs are less relevant for design now with numerous CAD procedures, they are still indispensable for analysis, especially for linear systems.

As we demonstrate, it is possible to diagram the (linear) regulatory structure of pathways characterized by the control and elasticity coefficients using signal-flow graphs. While we would be too optimistic if we expect molecular biologists will adopt a similar convention, our pragmatic goal is to bridge biological network analysis with control theory, and in the process leverage the tools, theories, and intuition from control for biology. What better way than to start with a control diagram!

3.1 Linear Chain

Consider the following linear chain of reactions

$$() \underset{}{\overset{v_1}{\rightleftharpoons}} s_1 \underset{}{\overset{v_2}{\rightleftharpoons}} s_2 \underset{}{\overset{v_3}{\rightleftharpoons}} () \quad (6)$$

We assume that the reaction mechanisms are elementary to the degree that

$$v_1 = v_1(s_1, p), \;\; v_2 = v_2(s_1, s_2, p), \;\; v_3 = v_3(s_2, p).$$

The kinetic equations for this mechanism are of the form

$$\dot{s}_1 = v_1(s_1, p) - v_2(s_1, s_2, p),$$
$$\dot{s}_2 = v_2(s_1, s_2, p) - v_3(s_2, p).$$

Let

$$\varepsilon^i_j = \frac{\partial v_i(\cdot)}{\partial s_j},$$

where s_j denotes the jth element of the n-dimensional vector s.

The control equations for this reaction mechanism are

$$\dot{x}_1(t) = \left(\varepsilon^1_1 - \varepsilon^2_1\right) x_1(t) - \varepsilon^2_2 x_2(t) + \varepsilon^1_p - \varepsilon^2_p \quad (7a)$$
$$\dot{x}_2(t) = \left(\varepsilon^2_2 - \varepsilon^3_2\right) x_2(t) + \varepsilon^2_1 x_1(t) + \varepsilon^2_p - \varepsilon^3_p \quad (7b)$$

If we take the Laplace transform, then we can recast the control equations (7) in the frequency domain

$$j\omega X_1(j\omega) = \left(\varepsilon^1_1 - \varepsilon^2_1\right) X_1(j\omega) - \varepsilon^2_2 X_2(j\omega) + \varepsilon^1_p(j\omega) - \varepsilon^2_p(j\omega),$$
$$j\omega X_2(j\omega) = \left(\varepsilon^2_2 - \varepsilon^3_2\right) X_2(j\omega) + \varepsilon^2_1 X_1(j\omega) + \varepsilon^2_p(j\omega) - \varepsilon^3_p(j\omega).$$

We can represent the control equations for the single perturbation ε^1_p using a signal-flow diagram (Figure 1). For aesthetic reasons, we prefer this graphical representation to the block-diagram representation.

Evident from the diagram (Figure 1), simple reactions are equivalent to local feedback loops. These loops result from the reaction dynamics. For example, a degradation reaction is equivalent to a negative feedback loop. Consider the kinetic equation

$$\dot{s} = p - ks$$

where p is the rate of production and k is the degradation rate constant. In this mechanism, the degradation rate ks functions as a negative feedback loop on the concentration s. In the linear chain example, the reactions v_2 and v_3 act as negative feedback loops on s_1 and s_2 respectively ($-\varepsilon^2_1$ and $-\varepsilon^3_2$). For reversible reactions, the product negatively regulates the reactant ($-\varepsilon^2_2$). Likewise, if the reactions v_1 and v_2 are reversible, then the product inhibits its synthesis. This inhibition is equivalent to a negative feedback loop on the substrates s_1 and s_2 respectively ($\varepsilon^1_1 < 0$ and $e^2_2 < 0$). If these reactions are autocatalytic, then the product enhances its synthesis and the process is equivalent to a positive feedback loop ($\varepsilon^1_1 > 0$ and $e^2_2 > 0$). In terms of control, the perturbations ε_p are equivalent to additive disturbances.

We do not need to know the specific rate equations in order to construct the signal flow diagram. All we need to know is the stoichiometry of the reaction network and variable dependencies for the rate laws. If we wish to numerically evaluate the network, then we need to know the elasticity coefficients or the kinetic parameters.

For a signal-flow graph, the transfer function T_{ij} between i^{th} and j^{th} node is given by Mason's rule (Mason, 1953; Mason, 1956):

$$T_{ij} = \frac{\sum_k P_{ijk}\Delta_{ijk}}{\Delta}$$

where the summation is taken over all possible paths from node i to j and

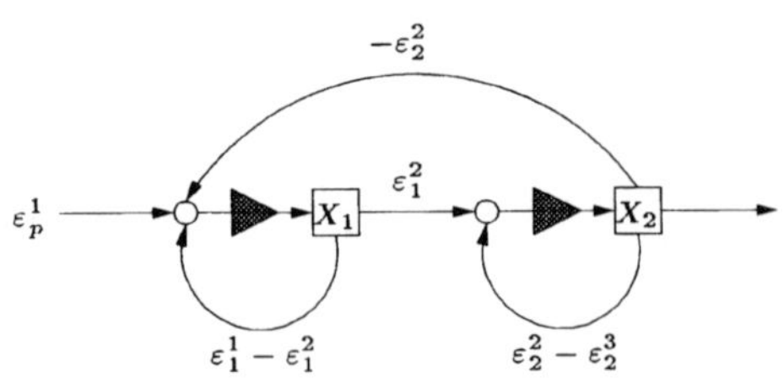

Fig. 1. A signal-flow representation of the dynamic control equations. The open circle denotes a summer and the shaded triangle denotes an integrator: $\frac{1}{j\omega}$. Evident from the diagram, reversible reactions introduce feedback loops as the concentration of the downstream species regulates the concentration of the upstream species.

$$
\begin{aligned}
P_{ijk} &= k^{\text{th}} \text{ path from node } i \text{ to } j, \\
\Delta &= \text{ determinant of flow graph}, \\
\Delta_{ijk} &= \text{ cofactor of the path } P_{ijk}.
\end{aligned}
$$

Numerous algorithms exist for applying Mason's rule to generic signal-flow networks.

Applying Mason's rule we obtain the following transfer function relating ε_p^1 to the control coefficients $X_1(j\omega)$ and $X_2(j\omega)$:

$$X_1(j\omega) = \frac{\left(j\omega - (\varepsilon_2^2 - \varepsilon_2^3)\right)\varepsilon_p^1 + \varepsilon_2^2\varepsilon_p^2}{-\omega^2 - \left((\varepsilon_1^1 - \varepsilon_1^2) + (\varepsilon_2^2 - \varepsilon_2^3)\right)j\omega + \varepsilon_1^1\varepsilon_2^2 - \varepsilon_1^1\varepsilon_2^3 + \varepsilon_1^2\varepsilon_2^3}$$

$$X_2(j\omega) = \frac{\varepsilon_2^1\varepsilon_p^1 + \left(j\omega + (\varepsilon_1^1 - \varepsilon_1^2)\right)\varepsilon_p^2}{-\omega^2 - \left((\varepsilon_1^1 - \varepsilon_1^2) + (\varepsilon_2^2 - \varepsilon_2^3)\right)j\omega + \varepsilon_1^1\varepsilon_2^2 - \varepsilon_1^1\varepsilon_2^3 + \varepsilon_1^2\varepsilon_2^3}$$

Note that both equations have the same denominator and the perturbations ε_p arise only in the numerator. This fact is well known in control: the denominator is the characteristic equation for the loop.

At steady state, we have the control equations

$$
\begin{aligned}
X_1(0) &= \frac{-(\varepsilon_2^2 - \varepsilon_2^3)\varepsilon_p^1 + \varepsilon_2^2\varepsilon_p^2}{\varepsilon_1^1\varepsilon_2^2 - \varepsilon_1^1\varepsilon_2^3 + \varepsilon_1^2\varepsilon_2^3}, \\
X_2(0) &= \frac{\varepsilon_2^1\varepsilon_p^1 + (\varepsilon_1^1 + \varepsilon_1^2)\varepsilon_p^2}{\varepsilon_1^1\varepsilon_2^2 - \varepsilon_1^1\varepsilon_2^3 + \varepsilon_1^2\varepsilon_2^3}.
\end{aligned}
$$

Both these equations describe how perturbing a single reaction effects the entire network.

3.2 Linear Chain With Feedback Inhibition

Consider the following set of reactions with end-product inhibition

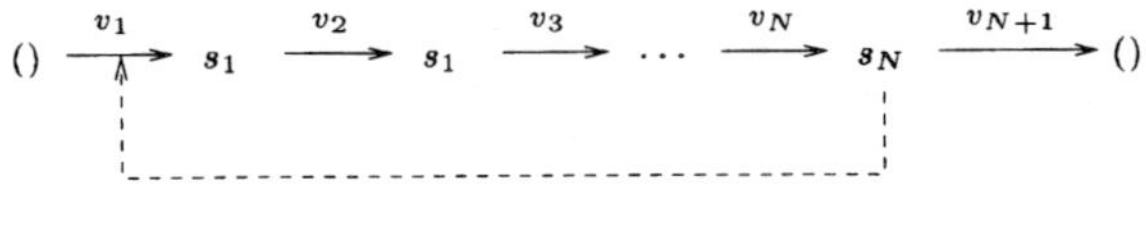

The kinetic equations for this mechanism are

$$
\begin{aligned}
\dot{s}_1 &= v_1(s_1, s_N, p) - v_2(s_1, s_2, p), \\
\dot{s}_2 &= v_2(s_1, s_2, p) - v_3(s_2, s_3, p), \\
&\vdots \\
\dot{s}_N &= v_N(s_{N-1}, s_N) - v_{N+1}(s_N, p).
\end{aligned}
$$

For this example, the control equations become

$$
\begin{aligned}
\dot{x}_1(t) &= \left(\varepsilon_1^1 - \varepsilon_1^2\right)x_1(t) - \varepsilon_2^2 x_2(t) + \varepsilon_N^1 x_N + \varepsilon_p^1 \\
\dot{x}_2(t) &= \left(\varepsilon_2^2 - \varepsilon_2^3\right)x_2(t) + \varepsilon_1^2 x_1(t) - \varepsilon_3^3 x_3(t) \\
&\vdots \\
x_N(t) &= (\varepsilon_N^N - \varepsilon_{N+1}^N)x_N + \varepsilon_{N-1}^N X_{N-1}
\end{aligned}
$$

Figure 2 diagrams the control equations as a signal-flow graph. As with the previous example, the reactions result in local feedback loops. However, there is also the global negative feedback loop resulting from endproduct inhibition ($-\varepsilon_5^1$). This example illustrates how it is often difficult to identify regulatory genes as the control may be embedded directly in the process (Savageau, 1972; Kacser and Burns, 1973).

3.3 Tryptophan Biosynthesis

The *trp* operon encodes five genes that synthesize tryptophan from chorismate, the common precursor for the aromatic amino acids phenylalanine, tyrosine, and tryptophan. As with many biosynthetic pathways, the process is subject to a hierarchy of regulatory feedback loops. At the level of the gene, expression of the operon is negatively regulated by tryptophan: tryptophan binds the *trp* repressor and inhibits transcription. Tryptophan, when bound with tRNAs, inhibits transcription through the process of translation attenuation. At the metabolic level, tryptophan allosterically inhibits the enzymes in the pathway.

A simple model (Xiu *et al.*, 1997) for tryptophan biosynthesis is described by the following differential equations

$$
\begin{aligned}
\dot{m} &= \underbrace{v_1(p)}_{\text{transcription}} - \underbrace{v_2(m)}_{\text{mRNA deg.}} \\
\dot{e} &= \underbrace{v_3(m)}_{\text{translation}} - \underbrace{v_4(e)}_{\text{prot. deg.}} \\
\dot{p} &= \underbrace{v_5(e,p)}_{\text{trp synthesis}} - \underbrace{v_6(p)}_{\text{trp deg.}} - \underbrace{v_7(p)}_{\text{trp consump.}}
\end{aligned}
$$

where m is the concentration of mRNA, e is the lumped concentration of the *trp* enzymes, and p is the concentration of tryptophan.

The control equations for this model take the following form

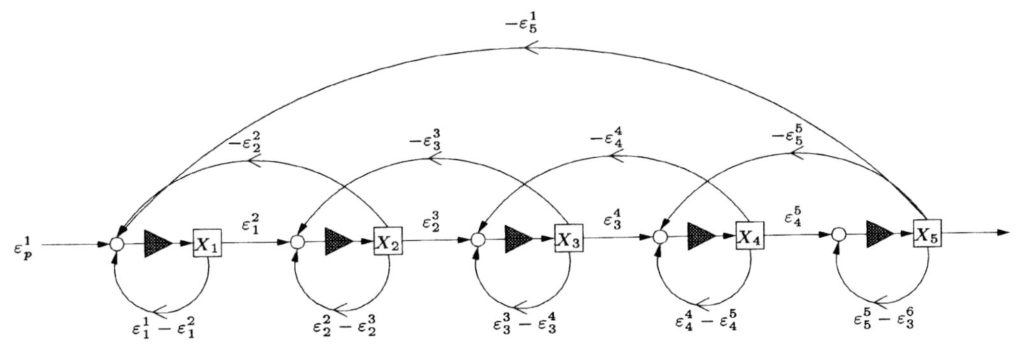

Fig. 2. Signal-flow diagram for a linear cascade with feedback inhibition ($N = 5$). The open circle denotes a summer and shaded triangle denotes the integrator $\frac{1}{j\omega}$.

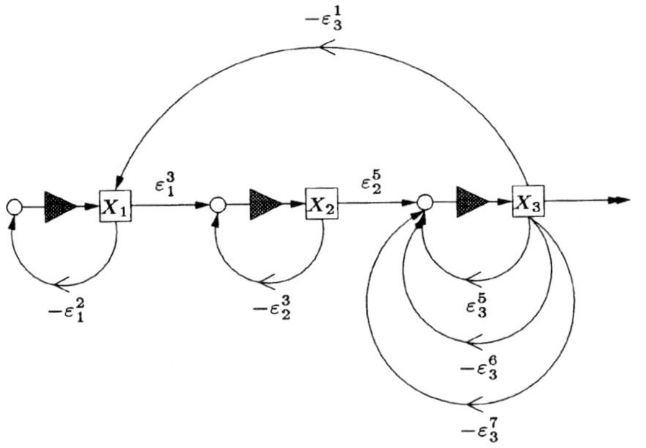

Fig. 3. Tryptophan Example: Signal-flow diagram for the tryptophan pathway. The open circle denotes a summer and the shaded triangle denotes the integrator $\frac{1}{j\omega}$.

$$\dot{x}_1 = \varepsilon_3^1 x_3 - \varepsilon_1^2 x_1$$
$$\dot{x}_2 = \varepsilon_1^3 x_1 + \varepsilon_3^3 x_3 - \varepsilon_2^4 x_2$$
$$\dot{x}_3 = \varepsilon_2^5 x_2 + (\varepsilon_3^5 - \varepsilon_3^6 - \varepsilon_3^7) x_3$$

where x_1 is the control coefficient for m, x_2 for e, and x_3 for p. The signal diagram (Figure 3) illustrates the signaling hierarchy in the tryptophan example. The global loop ($-\varepsilon_3^1$) results from the genetic regulatory mechanisms. The loops ($\varepsilon_3^5 < 0$, $-\varepsilon_3^6$, and $-\varepsilon_3^7$) result from allosteric inhibition by tryptophan, degradation, and consumption, respectively. In terms of the control equations and local dynamics, they are equivalent, though obviously they each have a unique role. One limitation of the MCA representation is that we see only a snapshot of the regulatory dynamics around the steady state. In the case of tryptophan, there are two feedback loops controlling gene expression. In terms of the control equations, these loops are equivalent. However, their relative strength depends on the concentration of tryptophan as they likely form a split-range controller. The control equations cannot capture this behavior as they are in reference to a single steady state.

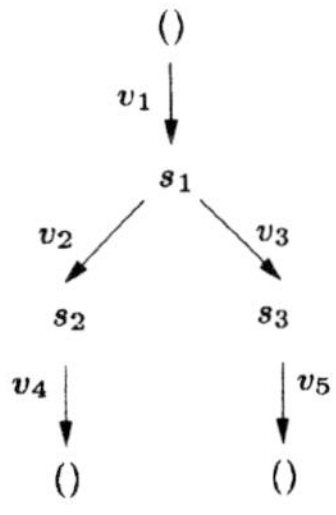

Fig. 4. Branched Pathway Example

3.4 Branched Pathway

Consider the branched pathway illustrated in Figure 4 where the goal is to produce substrate s_2 without effecting a key metabolic substrate s_3. The kinetic equations are

$$\dot{s}_1 = v_1(p) - v_2(s_1, p) - v_3(s_1, p),$$
$$\dot{s}_2 = v_2(s_1, p) - v_4(s_2, p),$$
$$\dot{s}_3 = v_3(s_1, p) - v_5(s_2, p).$$

The steady-state control equations are

$$x_1 = \frac{\varepsilon_p^1 - \varepsilon_p^2 - \varepsilon_p^3}{\varepsilon_p^2 + \varepsilon_p^3}$$
$$x_2 = \frac{1}{\varepsilon_2^4}\left[\frac{\varepsilon_1^2}{\varepsilon_1^2 + \varepsilon_1^3}\left(\varepsilon_p^1 - \varepsilon_p^2\right) + \varepsilon_p^2\right]$$
$$x_3 = \frac{1}{\varepsilon_3^5}\left[\frac{\varepsilon_1^2}{\varepsilon_1^2 + \varepsilon_1^3}\left(\varepsilon_p^1 - \varepsilon_p^2\right)\right]$$

The control equations are diagrammed in Figure 5. Simple inspection of these equations indicates that we need to simultaneously perturb reactions v_1 and v_2 in order to satisfy the design objective. This is also evident from the signal-flow diagram (Figure 5). While perturbing reaction v_2 increases the concentration of s_2, it depletes the concentration of s_1 and s_3. To compensate for the substrate depletion, we need to increase the rate for reaction v_1 to satisfy the design objective.

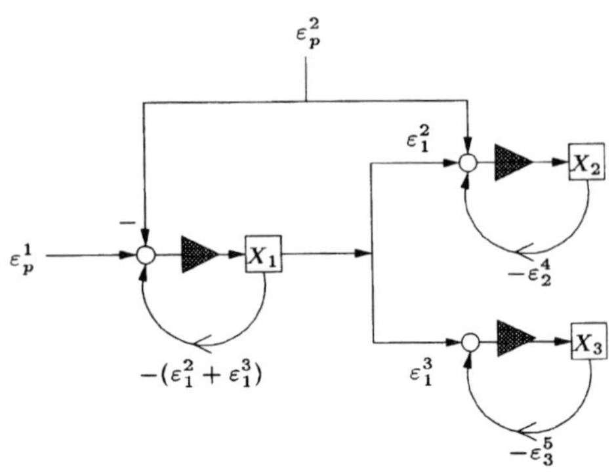

Fig. 5. Branched Pathway Example: Signal Flow Diagram. The open circle denotes a summer and the shaded triangle denotes the integrator $\frac{1}{j\omega}$.

3.5 Comments

The aim in the preceding sections was to demonstrate that MCA is subsumed by control theory. In fact, the previous discussion extends MCA which until recently was limited to steady-state analysis (Ingalls, 2004). For brevity, we have avoided explicit rate expressions and numerical values. However, one can imagine how tools from control theory can be applied to the analysis of biochemical networks (at least to first order approximation). Possible questions include controllability, robustness, and the influence of noise using frequency domain analysis.

4. CONCLUDING REMARKS

While MCA is a popular and powerful tool for analyzing the regulatory properties of intracellular networks, there is still much to be desired. The analysis is linear and consequently can only characterize infinitesimal perturbations to the dynamic behavior in a small neighborhood about the steady state. Despite its limitations, the MCA formulation provides a convenient gateway between biology and control. As much of the complexity in biology results from the regulatory interactions between individual genes, there are many interesting and important problems at the interface between control and biology. Currently, there is no satisfactory theory or set of tools (MCA included) for unraveling these complex regulatory interactions. By illustrating the similarity between the problem formulations in MCA and control, our goal was to show that this interface does exist and suggest MCA as one possible route for exploration. In some respects MCA is a mature theory, but in other respects it is still young and uncharted. Many important problems such as multivariable interactions, nonlinearity, noise, robustness, and discrete events have not been addressed yet in MCA. As each of these topics has a rich history in control, perhaps control can be useful for the analysis of biological networks.

REFERENCES

Burns, J. A. (1971). Studies on Complex Enzyme Systems. PhD thesis. University of Edinburgh. URL: http://www.cds.caltech.edu/~hsauro/Burns/jimBurns.pdf.

Fell, D.A. (1997). *Understanding the Control of Metabolism.* Portland Press.. London.

Heinrich, R. and S Schuster (1996). *The Regulation of Cellular Systems.* Chapman and Hall.

Heinrich, R. and T. A. Rapoport (1974*a*). A linear steady state treatment of enzymatic chains. critique of the crossover theorem and a general procedure to identify interaction sites with an effector.. *Eur. J. Biochem.* **42**, 97–105.

Heinrich, R. and T. A. Rapoport (1974*b*). A linear steady-state treatment of enzymatic chains; general properties, control and effector strength. *Eur. J. Biochem.* **42**, 89–95.

Heinrich, R., S. M. Rapoport and T. A. Rapoport (1977). Metabolic regulation and mathematical models.. *Prog. Biophys. Molec. Biol.* **32**, 1–82.

Ingalls, B. P. (2004). A frequency domain approach to sensitivity analysis of biochemical networks. *Journal of Physical Chemistry B* **108**, 1143–1152.

Kacser, H. and J. A. Burns (1973). The control of flux. In: *Rate Control of Biological Processes* (D. D. Davies, Ed.). Vol. 27 of *Symp. Soc. Exp. Biol.*. pp. 65–104. Cambridge University Press.

Mason, S. J. (1953). Feedback theory – some properties of signal flow graphs. *Proc. IRE* **41**, 1144–1156.

Mason, S. J. (1956). Feedback theory – further properties of signal flow graphs. *Proc. IRE* **44**, 920–926.

Reder, C. (1988). Metabolic control theory: A structural approach.. *J. Theor. Biol.* **135**, 175–201.

Savageau, M. A. (1971). Concepts relating the behaviour of biochemical systems to their underlying molecular properties.. *Arch. Biochem. Biophys.* **145**, 612–621.

Savageau, M. A. (1972). The behaviour of intact biochemical control systems. *Curr. Topics Cell. Reg.* **6**, 63–130.

Xiu, Z., A. Zeng and W. Deckwer (1997). Model analysis concerning the effects of growth rate and intracellular trptophan level on the stability and dynamics of tryptophan biosynthesis in bacteria. *J. Biotechnol.* **58**, 125–140.

ELSEVIER
IFAC
PUBLICATIONS
www.elsevier.com/locate/ifac

EVENT SOFTWARE SENSOR AND ADAPTIVE EXTREMUM SEEKING ALTERNATIVES FOR OPTIMIZING A CLASS OF FED-BATCH BIOREACTORS

Manuel J. BETANCUR[£+•]
Mariana TITICA[§]
Jaime A. MORENO[£#]
Denis DOCHAIN[§]
M. GUAY[&]

[£]*Institute of Engineering, Universidad Nacional Autónoma de México*
[§]*CESAME, Université Catholique de Louvain, 4-6 av. G. Lemaître, 1348, Louvain-la-Neuve, Belgium*
[&]*Department of Chemical Engineering, Queen's University, Kingston, Ontario, Canada, K7L 3N6*
[+]*Corresponding author: mbetancurb@iingen.unam.mx, on leave at CESAME-UCL*
[•]*A+D, Universidad Pontificia Bolivariana, Cq.1 #70-01, Medellín, Colombia*
[#]*On leave at the ISR, University Stuttgart, Germany*

Abstract: Optimization and control in spite of plant uncertainties is always a challenge, especially if additional constraints about measurements are present. Here, two different approaches to solve this problem are compared using simulation. They aim at reducing the operation time of bioreactors with inhibitory behavior where measuring the reaction rate is not feasible. The "Adaptive Extremum Seeking" proposed version relies on the structure information of the kinetic model and requires the measurements of the substrate and one other related variable. The "Event Driven Time Optimal Controller" strategy avoids the substrate measurement and, using an event software sensor, provides a nearly optimal solution without requiring a complete model. *Copyright © 2004 IFAC*

Keywords: Extremum seeking, software sensor, optimal control, fed-batch bioreactor, Haldane kinetics, control applications, biomass production

1. INTRODUCTION

The performance of fed-batch fermentation processes could typically be increased by the use of model-based optimization techniques. Using a fairly general model structure, many useful results have been derived in (Modak et al., 1986; Van Impe et al., 1994). One key result is that, in many cases, the final product yield (ratio of the amount of product formed and the amount of substrate consumed) can be maximized simply by maximizing the instantaneous yield. This can be achieved by maintaining the substrate concentration at an optimal level called S^*.

Over the past years, significant research effort has been done in the real time optimization for fed-batch bioreactors. The primary objective has been to overcome the performance limitations associated with the large uncertainty related to models of these processes (e.g. Bastin and Dochain, 1990).

In this paper, two approaches that are aimed at handling the uncertainties on the process kinetics are comparatively investigated. The version of the Adaptive Extremum Seeking (AES) proposed here utilizes explicit structure information of the objective function that depends on system states and unknown plant parameters. A Lyapunov-based adaptive control technique is used to estimate the unknown kinetic parameters and to drive the system to its unknown extremum. The second approach, the Event Driven Time Optimal Control (ED-TOC) uses bang-bang techniques based on the optimal solution approach, solving the problem without measuring the substrate nor needing the exact structure of the model.

The paper is organized as follows: in Section 2 the problem is stated. Sections 3 and 4, respectively, present the ED-TOC and the AES. A comparative simulation study is done in Section 5. Some conclusions and perspectives close the paper.

2. PROBLEM FORMULATION

In this work, we consider the optimization problem for a class of simple microbial growth processes with one gaseous product in a fed-batch bioreactor, described by the following dynamical model:

$$\frac{dX}{dt} = \mu X - DX \quad (1)$$

$$\frac{dS}{dt} = -k_1 \mu X + D(S_i - S) \quad (2)$$

$$y = k_2 \mu X \quad (3)$$

$$\frac{dV}{dt} = DV \quad (4)$$

where states X *(g/L)* and S *(g/L)* hold for biomass and substrate concentrations, respectively, μ (h^{-1}) is the specific growth rate, $D = \frac{F}{V}$ (h^{-1}) is the dilution rate, y *(g/L/h)* is the production rate of the reaction product, S_i *(g/L)* denotes the concentration of the substrate in the feed, k_1 and k_2 are yield coefficients, F *(L/h)* is the inflow and V *(L)* is the volume of the liquid medium in the tank.

A Haldane law approximates the specific growth rate:

$$\mu = \frac{\mu_0 S}{K_S + S + S^2 / K_I} \quad (5)$$

Where μ_0 is a parameter related to the maximum value of the specific growth rate μ^*, K_S denotes the saturation constant and K_I the inhibition constant. Such a model is typically used to describe the substrate inhibition effect in fed-batch bioreactors, as shown in Figure 1.

The control objective is to produce the maximum amount of biomass in the minimum possible time. It is well known (Moreno, 1999) that this will be achieved if μ is kept at its maximum value, μ^*, as shown in Figure 1.

3. EVENT DRIVEN TIME OPTIMAL CONTROL

The ED-TOC is based on the time optimal control solution to the problem (Moreno, J. 1999), which consists of bang-bang arcs and a singular arc. The singular arc maintains $\mu=\mu^*$ using a continuous control function, for which a good knowledge of the model would be required. The ED-TOC does the bang-bang arcs exactly but it approximates the singular arc by a switching strategy. This allows operating as near to the optimum as desired without the necessity of measuring many state variables and with a high degree of robustness against model and parameter uncertainties and changes. It considers also the physical restrictions of the actuator.

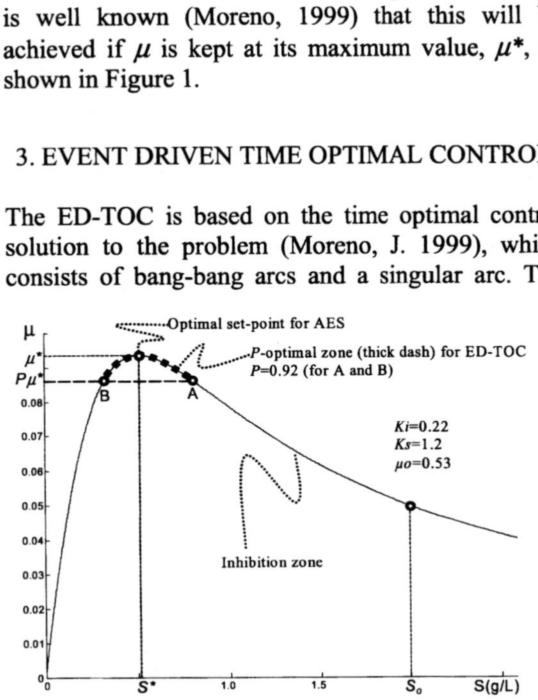

Figure 1. Haldane's type biomass specific growth rate

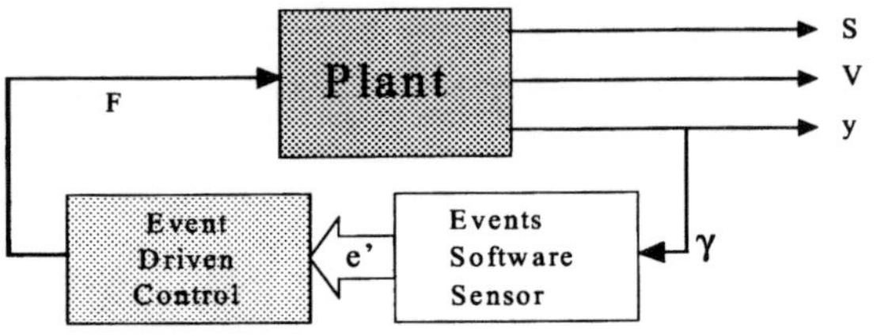

Figure 2. ED-TOC scheme

The ED-TOC estimates if μ lies inside a P-optimal zone, depicted in Figure 1, and acts accordingly to keep it inside. Figure 2 shows the ED-TOC scheme (a formal description is available in Betancur et al. 2004). The main action applied to the manipulated variable F is on/off. This bang-bang action is determined using the estimated events $\boldsymbol{e'}$. Basically the "Events Software Sensor" (ESS) generates two types of $\boldsymbol{e}$ events estimation. A parameter $P\in(0,1)$ in Figure 1 tunes the points in μ where real $\boldsymbol{e}$ are located. Event $\boldsymbol{e}=\boldsymbol{a}$ means that S is definitely above S^* and that μ is in the right-handed vicinity of point $\boldsymbol{A}$, i.e; moving away of the P-optimal zone. If the F pump is ON it must be turned OFF as shown in Figure 3. Similarly, $\boldsymbol{e}=\boldsymbol{b}$ signals the moment in which the system is leaving the left border of the P-optimal zone, near point $\boldsymbol{B}$. If F pump is OFF then it must be switched ON as shown in Figure 3. The final reaction's time t_f will be close to the optimal t_{fopt} as the established on/off cycle will keep the system as close as desired to μ^* at all possible times, by properly choosing $P\rightarrow 1$ (Table 1).

The ESS is called this way because, without measuring the variables μ or S nor knowing the parameters (S^*,μ^*), it estimates $\boldsymbol{e'}\approx\boldsymbol{e}$, using a real-time software algorithm. For doing so a γ variable must be available, such that its shape is related to the shape of μ, i.e. that as function of S its unique maximum coincides with S^*, regardless of other dependences it might have. For the case at hand γ is straightforwardly found in the gaseous product y:

$$\text{From (3):} \qquad \gamma \overset{\Delta}{=} y(S,X) = k_2 \mu(S) X \quad (6)$$

It is evident from (6) that for a given unknown $X>0$ and any unknown $k_2>0$ the shape of γ is the same as μ in Figure 1. That is, they grow monotonically from $S=0$ to $S=S^*$, and then they decrease, also monotonically, from $S=S^*$ and beyond as $S\rightarrow\infty$

But γ from (6) is useless as, except the fact they are positive, there is no available data for neither S, X nor k_2. That is why the ESS uses the time format of γ in (3) as it can be measured online and its maximum is easily found at some special times. Let's define t_k^* as the time when S just touches S^* for the k_{th} time. Then $\gamma_k^*=\gamma(t_k^*)$ is a detectable maximum. Later an k_{th} estimated event e_k' will be provided at time t_{ek} when $\gamma(t_{ek})=P\gamma_k^*$ indicating that the system is about to leave the P-optimal zone. Note that this algorithm works independently of the structure or form of μ in (5), as long as it has a unique global maximum.

It should be noted in (3) that, as $X(t)$ may change with time, a slight difference in the t_{ek} is expected compared to the situation if μ were used directly instead of γ. This estimation inaccuracy decreases when $X(t)$ varies slower than $\mu(t)$. Three cases of interest might arise. In the first, the influent pump is OFF and S decreases. It might happen that γ is increasing which means that the system is inhibited and above or in the vicinity of S^* (see left-side of Figure 3). Instead, if γ decreases, this means that the system is below S^*. The second case takes place when the pump is ON and sufficient substrate is being added to the tank as to increase S at some rate. Here, if γ is increasing, it means the system is "below" or in the vicinity of S^*. On the opposite, if γ decreases, this means that there is substrate inhibition, above S^*. The third case should be avoided. It is the limit case where S_i in F is not sufficient to guarantee that the increase in S will produce a μ with a dynamics faster than that of X. This condition is easily avoidable, even if S_i changes over time, if the inflow provides significantly more substrate than would F_{opt}, the inflow needed for the optimal case (Moreno, 1999).

Figure 3 compares the behavior of the ED-TOC for two different values of P: P=P1=0.92 corresponds to the Figure 1. The second case P=P2=0.998 provides a much better efficiency at the detriment of the switching cycle for F. Table 1 summarizes the results for different values of P. There it can be seen that the switching time is reasonable for this application and that it can be enlarged, if necessary, accepting a small loss in optimality.

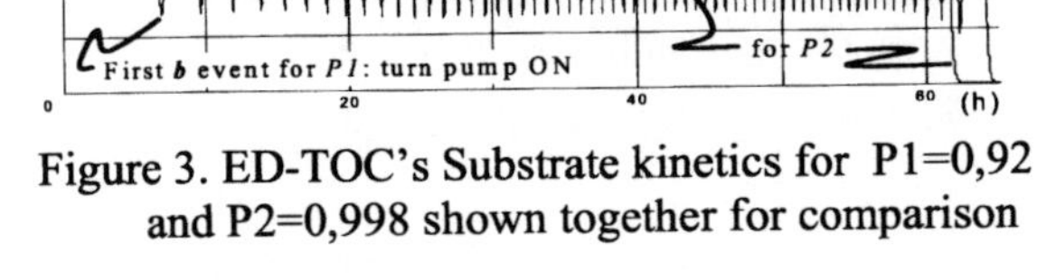

Figure 3. ED-TOC's Substrate kinetics for P1=0,92 and P2=0,998 shown together for comparison

Table 1 : ED-TOC optimality for different P

P	Optimality index t_{fopt}/t_f [%]	F pump cycles	Switching time [h]
0,998	99,9	448	0,14
0,98	99,8	151	0,41
0,96	99,2	106	0,59
0,90	96,0	62	1,0

4. ADAPTIVE EXTREMUM SEEKING

4.1. Principle and assumptions

The extremum-seeking scheme investigated here utilizes explicit structure information of the specific growth rate μ. In the case of Haldane kinetics (5) μ is maximum when the substrate concentrations equal to $S^* = \sqrt{K_S \cdot K_I}$. Then, the control objective is to keep the substrate concentration inside the reactor at the level S^*. Since the exact values of Haldane parameters are usually unknown (or at least poorly known), the adaptive extremum-seeking controller is developed to search this unknown set point such that the biomass production at the end of the reactor is maximized.

The adaptive extremum-seeking control scheme provides an adaptive control law of the dilution rate D to control the substrate S at the desired set point S^*, coupled with parameter learning laws for the Haldane model parameters. This algorithm requires the on-line knowledge of the substrate S and the gaseous outflow rate y (e.g. CO_2), as well as the related yield coefficients.

Parameter definition. Let's define:

$$\theta_\mu = \frac{\mu_0}{K_S},\ \theta_S = \frac{1}{K_S},\ \theta_I = \frac{1}{K_S \cdot K_I} \qquad (7)$$

$\theta = [\theta_\mu\ \theta_S\ \theta_I]^T$ represents the new set of kinetic parameters to be estimated on-line. The optimum for S can be re-expressed as follows: $S^* = \frac{1}{\sqrt{\theta_I}}$.

By this transformation, the optimum value is function of only one unknown parameter that has to be estimated on-line.

4.2. Estimation and Controller Design

The controller design proceeds in different steps. First, the estimation equation for S is derived from the balance model equation, then the control law and the estimation of the unknown kinetic parameters are included in a Lyapunov based derivation framework.

Estimation equation for the gaseous outflow rate y

By considering Eq. 1-3 and the kinetic parameter definition, the predicted state y is generated by:

$$\frac{d\hat{y}}{dt} = \frac{1-\hat{\theta}_i S^2}{S(1+\hat{\theta}_S S+\hat{\theta}_i S^2)}[D(S_i - S) - \theta_k y]y + \frac{\hat{\theta}_\mu Sy}{1+\hat{\theta}_S S+\hat{\theta}_i S^2} - Dy + k_y e_y \qquad (8)$$

with $k_y > 0$ and $e_y = y - \hat{y}$. $\theta_k = \frac{k_1}{k_2}$ is the ratio of the yield coefficients, assumed to be known.

Design of the adaptive extremum-seeking controller

Since the parameter θ_I is unknown, the desired set point (x) can be re-expressed as follows: $S^* = \dfrac{1}{\sqrt{\hat{\theta}_I}}$.

The controller is designed in order to drive the substrate concentration S to the estimated value of S^* (x). An excitation signal *d(t)* is designed and injected into the adaptive system such that the estimated parameter $\hat{\theta}_I$ converges to its true value. The extremum seeking control objective can be achieved when the substrate concentration S is stabilized at the optimal operating set point S^*.

Define the error control variable z_S as follows:

$$z_S = S - \frac{1}{\sqrt{\hat{\theta}_I}} - d(t) \tag{9}$$

Consider the Lyapunov candidate function as follows:

$$V = \frac{z_S^2}{2} + \frac{1}{2}\cdot\left(\frac{\tilde{\theta}_\mu^2}{\gamma_\mu} + \frac{\tilde{\theta}_S^2}{\gamma_S} + \frac{\tilde{\theta}_I^2}{\gamma_I}\right) + \frac{e_y^2}{2}(1+\theta_S\cdot S+\theta_I\cdot S^2) \tag{10}$$

where $\tilde{\theta}_\mu = \theta_\mu - \hat{\theta}_\mu$, $\tilde{\theta}_s = \theta_s - \hat{\theta}_s$ and $\tilde{\theta}_I = \theta_I - \hat{\theta}_I$, γ_μ, γ_s and γ_I are positive tuning parameters.

The adaptive extremum seeking controller algorithm is derived from the expression of the Lyapunov candidate function V, in such a way that $\dot{V}$ is negative.

After mathematical manipulations, we obtain the control law and the parameter learning laws as follows:

Control law:

$$D = \frac{1}{S_i - S}[-k_Z z_S + \theta_k y + a(t) - k_d d(t)] \tag{11}$$

Note that the resulting control law consists of two terms: the first one includes the specific rates of the biomass growth, while the second is a correcting term proportional to the error output and to the dither signal, *d(t)* defined as:

$$\dot{d}(t) = a(t) + \frac{1}{2}\cdot\hat{\theta}_I^{-\frac{3}{2}}\cdot\frac{d\hat{\theta}_I}{dt} - k_d\cdot d(t) \tag{12}$$

where $a(t)$ acts as a dither signal on the closed-loop process and k_d is a strictly positive constant.

Parameter learning laws

$$\frac{d\hat{\theta}_\mu}{dt} = \gamma_\mu e_y S y \tag{13}$$

$$\frac{d\hat{\theta}_S}{dt} = \begin{cases} \gamma_S(\Gamma_1 D+\Gamma_2), \\ \quad \text{if } \hat{\theta}_S > \varepsilon_s \text{ or } \hat{\theta}_S > \varepsilon_s \text{ and } (\Gamma_1 D+\Gamma_2)>0 \\ 0, \text{ otherwise} \end{cases} \tag{14}$$

$$\frac{d\hat{\theta}_i}{dt} = \begin{cases} \gamma_i(\Gamma_3 D+\Gamma_4), \text{if } \hat{\theta}_i > \varepsilon_i \text{ or } \hat{\theta}_i > \varepsilon_i \\ \quad \text{and } (\Gamma_3 D+\Gamma_4)>0 \\ 0, \text{ otherwise} \end{cases} \tag{15}$$

where,

$$\Gamma_1 = -\frac{e_y(1-\hat{\theta}_i S^2)(S_i - S)y}{1+\hat{\theta}_S S+\hat{\theta}_i S^2} \tag{16}$$

$$\Gamma_2 = \frac{e_y(1-\hat{\theta}_i S^2)\theta_k y^2}{1+\hat{\theta}_S S+\hat{\theta}_i S^2} - \frac{e_y \hat{\theta}_\mu S^2 y}{1+\hat{\theta}_S S+\hat{\theta}_i S^2} \tag{17}$$

$$\Gamma_3 = -e_y S(S_i - S)y + \Gamma_1 \tag{18}$$

$$\Gamma_4 = e_y S\theta_k y^2 + S\,\Gamma_2 \tag{19}$$

and with the initial condition $\hat{\theta}_S(0) \geq \varepsilon_S = \dfrac{1}{K_{S,\max}} > 0$,

and $\hat{\theta}_i(0) \geq \varepsilon_i = \dfrac{1}{K_{S,\max}K_{I,\max}} > 0$.

The update laws are projection algorithms that ensures that $\hat{\theta}_S \geq \varepsilon_S > 0$ and $\hat{\theta}_i \geq \varepsilon_i > 0$.

By using appropriate mathematical arguments (that invoke in particular LaSalle invariance principle and Barbalat's lemma), it has been shown that if the dither signal fulfils some excitation persistence condition then the parameter estimates converges to their true values and the control error z_S converges to zero (Titica et al., 2003; Marcos et al., 2004).

5. SIMULATION RESULTS

The ED-TOC and the AES where tested using Matlab simulations, under different scenarios, for the same discontinuous reactor model. Each scenario in Table 2 provides a different challenge to the controllers. The default kinetic model parameters and yield coefficients used to simulate the plant are:

$\mu_0 = 0.53h^{-1}, K_S = 1.2g/L, K_I = 0.22g/L$
$k_1 = 0.4, k_2 = 1.0$

The initial states, unless otherwise explicitly said, used during numerical simulations are:
$S_o = 2.0g/L, X_o = 7.2g/L, V_o = 1.0L$

and the substrate concentration in the inflow is $S_i = 20g/L$.

The optimality parameter for the ED-TOC controller was set to P=0.98 to calculate results in Table 3.

The design parameters for the AES controller were set to:
$\gamma_\mu = 10, \gamma_S = 200, \gamma_I = 200, k_{y,0} = 20, k_z = 0.5, k_d = 1$, $\varepsilon = 0.2$. The dither signal *a(t)* is chosen as:

$$a = \sum_{i=1}^{5} A_{1i} \sin[(0.001 + (5 - 0.001)i/4)t] + \\ + \sum_{i=1}^{5} A_{2i} \cos[(0.001 + (5 - 0.001)i/4)t] \quad (20)$$

where A_{1i} and A_{2i} are normally distributed random numbers in the interval [-0.1, 0.1]. For details about the selection of the dither signal, the reader is referred to (Titica, et al., 2003).

The control objective is to fill the tank and obtain an amount of biomass, equal to that produced by the optimal trajectory, in the smallest possible time. The final reaction time t_f has been defined as the instant when $V(t_f)=V_{max}$=40L and $S_f=S(t_f)$<0.01 g/L, where this last value has been arbitrarily assigned.

An optimality index is defined as t_{fopt}/t_f, where t_{fopt} is the optimal, theoretical minimal, finishing time. Table 2 presents the optimality index for each case. It shows that the performance of both controllers is good and close to 100%, except in Scenarios 2 and 6. In Scenario 2 bad initial conditions for the kinetic parameters are given and, additionally, in Scenario 6 a sudden change in plant's kinetics is produced. In both cases the AES performs in a less optimal way because it requires some convergence time to adapt the estimated values of the true parameters of the plant, and meanwhile the AES uses a non-optimal set point. As long as the total change in S^* is not big enough to cause big efficiency loses, the performance will not be greatly affected. This is the case if some history about the plant's behavior exists, so the initial conditions for every new batch can be given in a trustable fashion. By comparison, the performance effect in the ED-TOC is almost negligible, if any. This is so because its operational principle is of the bang-bang type, meaning that at every instant in time it will do the best possible action to get the system in the *P*-optimal zone, without the need for allowing any parameter to converge. So the time the system spends away of the defined *P*-optimal zone depends only on the actuator limits and the plant dynamics. No equilibrium or set points are searched for. Its only aim is to keep the controlled variable inside the *P*-optimal zone during the singular arc. These are the main philosophical differences in both approaches. Table 3 shows another characteristics that define the main practical differences of the two controllers.

Some complementary differences between the two methods suggest some work should be done to combine them. In the AES, information about the plant is generated, as the values of the kinetic parameters are estimated in time. More sensors are used thus generating more data, specially *S* data. Other than the dither signal, it does not produce cycling of the input. On the other hand the ED-TOC is immune to tuning problems and model uncertainties (Table 3). It performs the bang-bang arcs in an exact way, same as the theoretical optimal solution. While doing the arcs it may know the t_k* instants when $S=S^*$ but does not know $S(t_k^*)$.

Figures 4 and 5 show the simulation results for Scenario 6. In Figure 4 it is clear that the dither in *F*, produced by the ED-TOC, is only plant-dependent, and that its convergence to new operation conditions is not governed by any controller's dynamics. The amplitude and time cycle of the bang-bang control signal depends only on the plant, and during the bang-bang arcs it operates perfectly. On the other hand, the AES graciously adjust its estimated value for the set-point of the manipulated variable (see dotted line in Figure 5), using an adaptive algorithm that requires a convergence time and that depends not only on the plants behavior in the new operational zone but also on the careful selection of the tuning parameters.

Table 2 : Optimality results for different scenarios for both controllers

Test Scenario for simulation comparison	Optimality index: t_{fopt}/t_f [%]	
	ED-TOC	AES
1. Default: $\mu o=0.53$ $Ki=0{,}22$ $Ks=1{,}2$ $(S^*=0{,}5138)$, True initial conditions given.	99.8	99.7
2. Bad initial kinetics given: μo=0.108 Ki=1,645 Ks=0,01 (S* decreased 50%)	99.8	89.5
3. Bad *So* given (increased 50% respect to the real value)	99.8	99.7
4. Measurement error of +20% for *y* in *t*=(20,30)	99.8	99.8
5. Measurement error of +20% for *S* in *t*= (20,30)	99.8	99.8
6. *So*=*S**, bad initial conditions given (*S** increased 100%), and *S** changed for $t=(20,50)$: $\mu o=0.53$ $Ki=0{,}44$ $Ks=2{,}4$ (*S** increased 100%)	99.1	96.5

Table 3 : Characteristic differentiation of both control philosophies

Characteristics	ED-TOC	AES
Measured variables:	*y* (used as γ)	*S*, *y*
Controlled variables:	μ	*S*
Manipulated variables:	*F*	*D*
Tuning parameters:	-	$\gamma_\mu, \gamma_S, \gamma_i, k_z, k_{y,0}, k_d, \varepsilon, a$
Optimality parameters:	*P*	-
Initial conditions required:	-	$S(0), y(0), S_f, \theta_\mu(0), \theta_S(0), \theta_i(0)$
Convergence speed depends on:	Plant and actuator limitations only	Plant and control tuning parameters
Control type:	Hybrid (Bang-bang + continuous)	Continuous

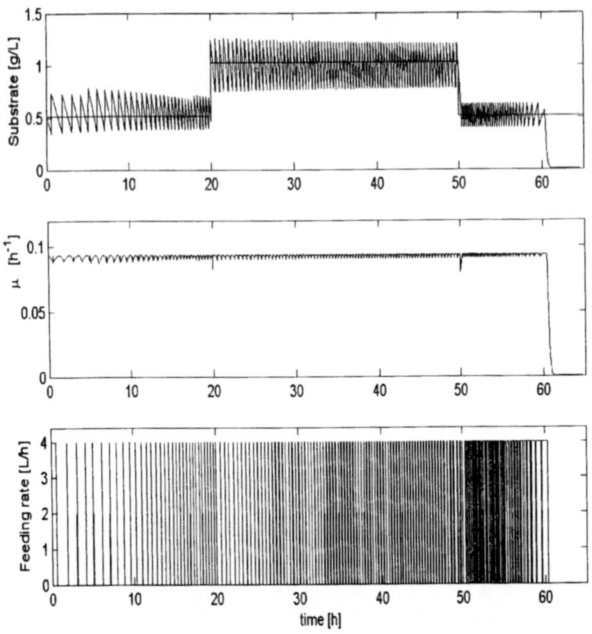

Figure 4: ED-TOC behavior using P=0.98 for Scenario 6 in Table 2. In the substrate plot S^* is drawn with a step-like line.

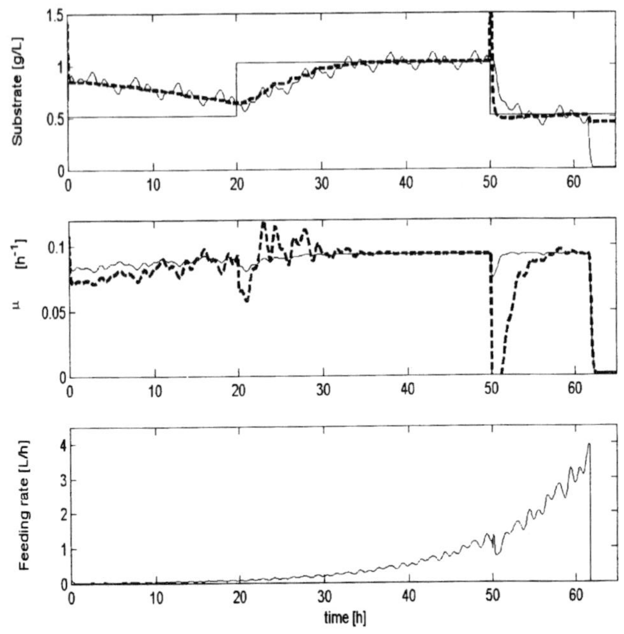

Figure 5: AES behavior for Scenario 6 in Table 2. In the substrate plot a step-like line shows S^*. Estimations are drawn with dashed lines.

6. CONCLUSIONS

Two control approaches of different philosophical nature have been compared, to understand their differences and complementarities, in a bioreactor example. The ED-TOC aims at keeping the system inside a predefined P-optimal zone. The AES aims at stabilizing the system in an optimal set point that is adapted on line. Both of them performed in a near optimal way for different test scenarios.

Advantages and disadvantages of the two philosophically different approaches were discussed, suggesting the possibility to combine them to obtain the better of each one of them. The idea is that while the system starts a batch, ED-TOC will performs in the best possible way for the singular arc. At the same time, if a S measure is available, during this bang-bang cycle a clear estimation of the P-optimal zone will be made in terms of S, including a measure for S^* at t_1^*. This value will be used to initialize the AES and, from then on, control could be made using AES policies. From time to time a bang-bang cycle could be run to test for optimality. This will reduce the on/off switching and at the same time bring valuable information about the plant model, and allow reacting to unforeseen perturbations or faults.

ACKNOWLEDGEMENTS

This paper includes results of the EOLI project that is supported by the INCO program of the European Community, Contract number ICA4-CT-2002-10012 and of the Knowledge-driven Batch Production (BatchPro) European Project HPRN-CT-2000-00039. The support of the Belgian program on Inter-University Poles of Attraction initiated by the Belgian State, Prime Minister's office for Science, Technology and Culture, is gratefully acknowledged; Thanks to CONACyT (Project 34934A) and DGAPA for its financial support; M. J. Betancur thanks the CE INCO-DEV Bursary contract ICB1-CT-2002-80006, UPB and CESAME. The scientific responsibility rests with the authors.

REFERENCES

Bastin, G., Dochain, D. (1990). On-line Estimation and Adaptive Control of Bioreactors. Elsevier, Amsterdam.

Betancur, M.J. et al. (2004). Event-Driven control for treating toxicants in aerobic sequencing batch bioreactors. *Proceedings of the 9th International Symposium of Computer Applications in Biotechnology (CAB9)*, CDROM file 1074.

Marcos, N.I., M. Guay, D. Dochain and T. Zhang (2004). Adaptive Extremum Seeking Control of a Continuous Stirred Tank Bioreactors with Haldane Kinetics, *Journal Proc. Control* , 14, pp.317-328.

Modak, J. M., H. C. Lim and Y. J. Tayeb (1986). General characteristics of optimal feed rate profiles for various fed-batch fermentation processes. *Biotechnology and Bioengineering* 28, 1396-1407.

Moreno, J. (1999). Optimal time control of bioreactors for the wastewater treatment. In: *Optimal Control Applications and Methods,* 20, pp 145-164

Titica, M., Dochain, D., Guay, M. (2003). Adaptive Extremum-Seeking Control of Fed-Batch Bioreactors, *European Journal of Control, 9, pp.614-627*

Van Impe, J.F., J. E. Claes, V. G. Ryckaert and G. Bastin (1994). Characterization of optimal feed rate profiles for fed-batch fermentation processes. *Proceedings of the 3rd IEEE Conference on Control Applications*, 1947-1952

AUTHOR INDEX

DYNAMICS AND CONTROL OF PROCESS SYSTEMS 2004
(DYCOPS-7)

A Proceedings volume from the 7th IFAC Symposium,
Cambridge, Massachusetts, USA, 5 –7 July 2004

Edited by

SIRISH L. SHAH
Department of Chemical and Materials Engineering,
University of Alberta, Edmonton, Canada

and

JOHN MacGREGOR
Department of Chemical Engineering
McMaster University, Hamilton, Canada

(In two volumes)

Volume 1

Published for the

INTERNATIONAL FEDERATION OF AUTOMATIC CONTROL

by

ELSEVIER LIMITED

ELSEVIER Ltd
The Boulevard, Langford Lane
Kidlington, Oxford OX5 1GB, UK

Elsevier Internet Homepage
http://www.elsevier.com

Consult the Elsevier Homepage for full catalogue information on all books, journals and electronic products and services.

IFAC Publications Internet Homepage
http://www.elsevier.com/locate/ifac

Consult the IFAC Publications Homepage for full details on the preparation of IFAC meeting papers, published/forthcoming IFAC books, and information about the IFAC Journals and affiliated journals.

First edition 2005

Library of Congress Cataloging in Publication Data

A catalogue record for this book is available from the Library of Congress

British Library Cataloguing in Publication Data

A catalogue record for this book is available from the British Library

ISBN 0-08-044297 8
ISSN 1474-6670

Printed in Great Britain

To Contact the Publisher

Elsevier welcomes enquiries concerning publishing proposals: books, journal special issues, conference proceedings, etc. All formats and media can be considered. Should you have a publishing proposal you wish to discuss, please contact, without obligation, the publisher responsible for Elsevier's industrial and control engineering publishing programme:

Christopher Greenwell
Publishing Editor
Elsevier Ltd
The Boulevard, Langford Lane
Kidlington, Oxford
OX5 1GB, UK

Phone: +44 1865 843230
Fax: +44 1865 843920
E.mail: c.greenwell@elsevier.com

General enquiries, including placing orders, should be directed to Elsevier's Regional Sales Offices – please access the Elsevier homepage for full contact details (homepage details at the top of this page).

7th IFAC SYMPOSIUM ON DYNAMICS AND CONTROL OF PROCESS SYSTEMS 2004

Sponsored by
International Federation of Automatic Control (IFAC)
Technical Committee on Chemical Process Control

Organized by
International Federation of Automatic Control (IFAC)
American Automatic Control Council

***International Programme Committee* (IPC)**

Shah, S. L Co-Chair
MacGregor, J. Co-Chair
Marquardt, W. IFAC Representative

Industrial Co-Chairs

Cott, B. (USA)
Duennebier, G. (Germany)

Technical Area Co-Chairs

Allgower, F. (Germany)
Bonvin, D. (Switzerland)
Braatz, R. (USA)
Doyle, F. (USA)
Dochain, D. (Belgium)
Foss, B. (Norway)
Hasebe, S. (Japan)
Huang, B. (Canada)
Jorgensen, S.B. (Denmark)
Marlin, T. (Canada)
Marquardt, W. (Germany)
Ricker, L. (USA)
Perkins, J. (UK)
Perrier, M. (Canada)
Pistikopoulos, S. (UK)
Qin, J. (USA)
Swartz, C. (Canada)
Thornhill, N. (UK)

IPC Members

Alvarez, J. (Mexico)
Arkun, Y. (Turkey)
Bakshi, B. (USA)
Biegler, L. (USA)
Boje, E. (South Africa)
Brambilla, A. (Italy)
Chen, B. (China)
Chu, J. (China)
Cinar, A. (USA)
Craig, I. (South Africa)
Daoutidis, P. (USA)
Edgar, T. (USA)
Engell, S. (Germany)
Forbes, F. (Canada)
Fujii, H. (Japan)
Jamsa-Jounela, S.-L. (Finland)
Gao, F. (HK China)
Guay, M. (Canada)
Gudi, R. (India)
Hagglund, T. (Sweden)
Hoo, K. (USA)
Huang, H.P. (Taiwan)
Isaksson, A. (Sweden)
Jacobsen, E. (Sweden)
Koivo, H. (Finland)
Kravaris, C. (Greece)
Lakshminarayanan, S. (Singapore)
Lee, J. (USA)
Lee, P. (Australia)
Lewin, D. (Israel)
Li Ping (China)
Mandler, J. (USA)
Morari, M. (Switzerland)
Narasimhan, S. (India)
Ohshima, M. (Japan)
Olivares, L.B. (Chile)
Palazoglu, A. (USA)
Park, S. (Korea)
Patwardhan, S. (India)
Preisig, H. (Norway)
Rawlings, J. (USA)
Romagnoli, J. (Australia)
Rossiter, A. (UK)
Scali, C. (Italy)
Secchi, A. (Brazil)
Tade, M.O. (Australia)
Williamson, C. (New Zealand)
Wongsri, M. (Thailand)
Yihui, J. (China)
Yoon, E.S. (Korea)
Yu, C.C. (Taiwan)

***National Organizing Committee* (NOC)**

Seider, W.D. Chair

Beaverstock, M.
Barton, P.I.
Henson, M.A.
Kothare, M.V.
McAvoy, T.J.
Muske, K.R.
Ogunnaike, B.
Soroush, M.

FOREWORD

Dycops-2004 as held in Boston was an international event. It attracted a total of 170 participants from industry and academia around the world.

Over 200 papers were submitted from which 155 were selected for presentation at this conference including 3 plenary addresses and 8 keynote talks.

Editors:

Sirish L. Shah (University of Alberta, Canada)
John MacGregor (McMaster University, Canada)

CONTENTS

VOLUME 1

PLENARY PAPER 1

MODELING AND IDENTIFICATION 1

MONITORING AND FAULT DETECTION 1

ADVANCES IN CONTROL 1

KEYNOTE PAPERS 1-4

MODELING AND IDENTIFICATION 2

DESIGN AND CONTROL

ADVANCES IN CONTROL 2

BIOPROCESS MODELING AND CONTROL 1

KEYNOTE PAPERS 5 and 6

CONTROL, OPTIMIZATION AND SCHEDULING 1

BATCH PROCESS MODELING AND CONTROL

IMPACT OF BIOLOGICAL ENIGNEERING ON PROCESS CONTROL EDUCATION

POSTER PAPERS

Bioprocess Modeling and Control

VOLUME 2

POSTER PAPERS (continued)

Monitoring and Fault Detection

Control, Optimization and Scheduling

Advances in Control

Modeling and Identification

Sensor Technologies

Industrial Applications

Batch and Semi-Batch Processes

PLENARY PAPER 3

MODELING AND IDENTIFICATION 4

INDUSTRIAL APPLICATIONS

ADVANCES IN CONTROL 3

KEYNOTE PAPERS 7 AND 8

CONTROL, OPTIMIZATION AND SCHEDULING 2

SENSOR TECHNOLOGY

BIOPROCESS MODELING AND CONTROL 2

ELSEVIER
IFAC
PUBLICATIONS
www.elsevier.com/locate/ifac

ON-LINE OPTIMIZATION VIA OFF-LINE PARAMETRIC OPTIMIZATION! - A GUIDED TOUR TO PARAMETRIC PROGRAMMING AND CONTROL

Efstratios N. Pistikopoulos [*,1] **Vivek Dua** [**,2]

* *Centre for Process Systems Engineering*
Imperial College London, SW7 2AZ, U.K.
** *Centre for Process Systems Engineering*
University College London, WC1E 7JE, U.K.

Abstract: In this paper model predictive control problems are reformulated as multi-parametric programs. The optimal value of control variables is obtained as an explicit function of the state variables. This reduces on-line optimization to simple function evaluations that need simple computational hardware.

Keywords: Real-time optimization, model predictive control, multi-parametric programming, multi-parametric dynamic optimization, hybrid control.

1. INTRODUCTION

1.1 On-line Optimization

Model Predictive Control (MPC) (Morari and Lee, 1999) has been widely adopted by industry to address on-line optimization problems with input and output constraints. MPC is based on the so called *receding horizon* philosophy: a sequence of future control actions is chosen according to a prediction of the future evolution of the system and applied to the system until new measurements are available. Then, a new sequence is determined which replaces the previous one (see Figure 1). Each sequence is evaluated by means of an optimization procedure which takes into account two objectives: optimize the tracking performance, and protect the system from possible constraint violations.

While the benefits of on-line optimization are tremendous, its application is rather restricted, considering its profit potential, primarily due to its large "on-line" computational requirements which involve a repetitive solution of an optimization problem at regular time intervals (see Figure 2). This limitation is in spite of the significant advances in the computational power of the modern computers and in the area of on-line optimization over the past many years. Thus, it is fair to state that an efficient implementation of on-line optimization tools relies on a quick and repetitive on-line computation of optimal control actions.

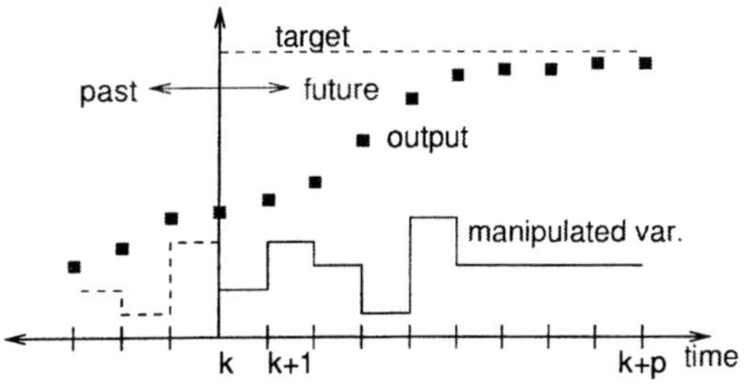

Fig. 1. Model Predictive Control

[1] e.pistikopoulos@imperial.ac.uk
[2] v.dua@ucl.ac.uk

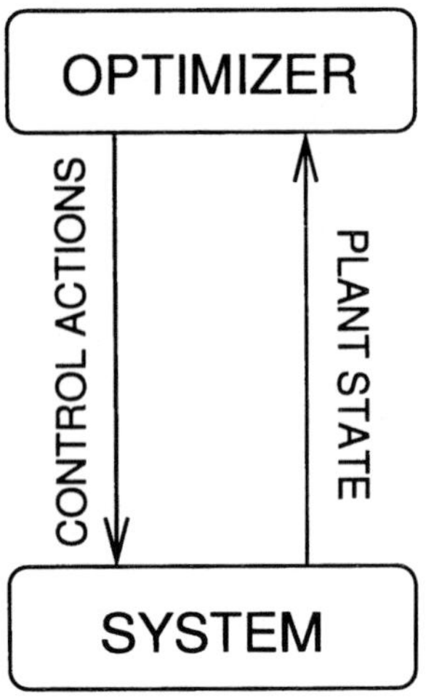

Fig. 2. On-line optimization

1.2 Multiparametric Programming

A parametric programming approach which avoids this repetitive solution is presented. In an optimization framework, where the objective is to minimize or maximize a performance criterion subject to a given set of constraints and where some of the parameters in the optimization problem are uncertain, parametric programming is a technique for obtaining (i) the objective function and the optimization variables as a function of these parameters and (ii) the regions in the space of the parameters where these functions are valid - see Figure 3. The main advantage of using the parametric programming techniques to address such problems is that for problems pertaining to plant operations, such as for process planning (Pistikopoulos and Dua, 1998) and scheduling, one obtains a complete map of all the optimal solutions and as the operating conditions fluctuate, one does not have to re-optimize for the new set of conditions since the optimal solution as a function of parameters (or the new set of conditions) is already available.

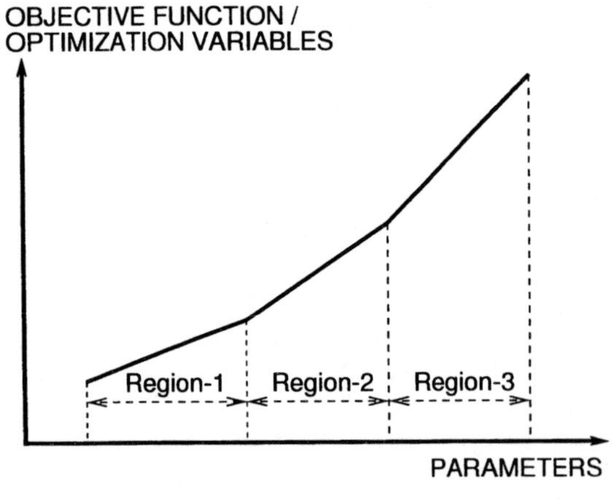

Fig. 3. Parametric Programming

By using the theory of parametric programming the control variables are obtained as a function of the state variables, and therefore on-line optimization reduces to simple function evaluations, at regular time intervals, for the given state of the plant - to compute the corresponding control actions. This results in a very small computational effort in comparison to repetitively solving an optimization problem.

The rest of the paper is structured as follows. In the next section an overview of parametric programming is provided. In section 3.1 a discrete time formulation of the on-line optimization problem is considered. While section 3.2 is concerned with continuous-time formulation, section 3.4 discusses the case when dynamics and logical decisions are simultaneously present in the problem. Concluding remarks are given in section 4.

2. PARAMETRIC PROGRAMMING AT A GLANCE

Consider the following multiparametric program:

$$\begin{aligned} & z(\theta) = \min_{x} f(x,\theta) \\ \text{s.t.} \quad & g_i(x,\theta) \leq 0, \quad \forall\ i = 1, ..., p \\ & h_j(x,\theta) = 0, \quad \forall\ j = 1, ..., q \\ & x \in X \subseteq \Re^n \\ & \theta \in \Theta \subseteq \Re^m \end{aligned} \tag{1}$$

Note that in an optimization framework f is the performance criterion to be minimized, $g \leq 0$ and $h = 0$ are the constraints, x is the vector of optimization variables and θ is the vector of parameters. The objective is to obtain the optimal $x(\theta)$, which when substituted into $f(x, \theta)$ provides the optimal objective function value, $z(\theta)$, as a function of θ. See Figure 4, where $x(\theta)$ is plotted as a function of θ - the region where a particular functional relationship between $x(\theta)$ and θ holds is known as a Critical Region (CR). Note that in the figure there are three CRs and $x(\theta)$ was obtained as a function of θ by solving only three optimization problems and therefore solving optimization problems for each and every value of θ has been avoided.

Depending upon whether f, g and h are linear, quadratic, nonlinear, convex, differentiable, or not, and also whether x is vector of continuous or mixed -continuous and integer- variables, the nature of the profiles and CRs shown in figure 4 changes (Dua and Pistikopoulos, 2000; Dua *et al.*, 2002; Dua and Pistikopoulos, 1999; Dua *et al.*, 2003; Sakizlis *et al.*, 2002*b*). For example, when f and g are convex and continuously differentiable and h is affine $z(\theta)$ is continuous and convex. Recently algorithms for the case when

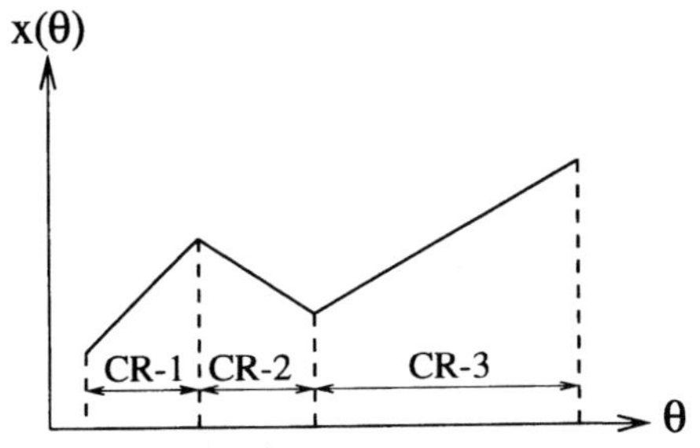

Fig. 4. Parametric Optimization

(1) involves (i) differential and algebraic equations (Sakizlis *et al.*, 2002*a*) and (ii) uncertain parameters (Sakizlis *et al.*, 2004) have been proposed.

3. PARAMETRIC CONTROL AT A GLANCE

3.1 Discrete Time Formulation

Consider the following state-space representation of a given process model (Pistikopoulos *et al.*, 2002):

$$\begin{cases} x(t+1) = Ax(t) + Bu(t) \\ y(t) = Cx(t), \end{cases} \tag{2}$$

subject to the following constraints:

$$\begin{aligned} y_{min} \le y(t) \le y_{max} \\ u_{min} \le u(t) \le u_{max}, \end{aligned} \tag{3}$$

where $x(t) \in \Re^n$, $u(t) \in \Re^m$, and $y(t) \in \Re^p$ are the state, input, and output vectors respectively, subscripts *min* and *max* denote lower and upper bounds respectively and (A, B) is stabilizable. Model Predictive Control (MPC) problems for regulating to the origin can then be posed as the following optimization problems:

$$\begin{aligned} \min_U \; & J(U, x(t)) = x'_{t+N_y|t} P x_{t+N_y|t} \\ & + \sum_{k=0}^{N_y-1} x'_{t+k|t} Q x_{t+k|t} + u'_{t+k} R u_{t+k} \\ \text{s.t.} \; & y_{min} \le y_{t+k|t} \le y_{max}, \; k = 1, \dots, N_c \\ & u_{min} \le u_{t+k} \le u_{max}, \; k = 0, 1, \dots, N_c \\ & x_{t|t} = x(t) \\ & x_{t+k+1|t} = Ax_{t+k|t} + Bu_{t+k}, \; k \ge 0 \\ & y_{t+k|t} = Cx_{t+k|t}, \; k \ge 0 \\ & u_{t+k} = Kx_{t+k|t}, \; N_u \le k \le N_y \end{aligned} \tag{4}$$

where $U \triangleq \{u_t, \dots, u_{t+N_u-1}\}$, $Q = Q' \succeq 0$, $R = R' \succ 0$, $P \succeq 0$, $N_y \ge N_u$, and K is some feedback gain. The problem (4) is solved repetitively at each time t for the current measurement $x(t)$ and the vector of predicted state variables, $x_{t+1|t}, \dots, x_{t+k|t}$ at time $t+1, \dots, t+k$ respectively and corresponding control actions $u_t, \dots, u_{t+k-1}$ is obtained.

In the following paragraphs, a parametric programming approach which avoids a repetitive solution of (4) is presented. First, we do some algebraic manipulations to recast (4) in a form suitable for using and developing some new parametric programming concepts. By making the following substitution in (4):

$$x_{t+k|t} = A^k x(t) + \sum_{j=0}^{k-1} A^j B u_{t+k-1-j} \tag{5}$$

the objective $J(U, x(t))$ can be formulated as the following Quadratic Programming (QP) problem:

$$\begin{aligned} \min_U \; & \frac{1}{2} U' H U + x'(t) F U + \frac{1}{2} x'(t) Y x(t) \\ \text{s.t.} \; & GU \le W + Ex(t) \end{aligned} \tag{6}$$

where $U \triangleq [u'_t, \dots, u'_{t+N_u-1}]' \in \Re^s$, $s \triangleq mN_u$, is the vector of optimization variables, $H = H' \succ 0$, and H, F, Y, G, W, E are obtained from Q, R and (4)–(5). The QP problem (6) can now be formulated as the following Multi-parametric Quadratic Program (mp-QP):

$$\begin{aligned} \mu(x) = \; & \min_z \frac{1}{2} z' H z \\ & \text{s.t.} \; Gz \le W + Sx(t), \end{aligned} \tag{7}$$

where $z \triangleq U + H^{-1}F'x(t), z \in \Re^s$, represents the vector of optimization variables, $S \triangleq E + GH^{-1}F'$ and x represents the vector of parameters. The main advantage of writing (4) in the form given in (7) is that z (and therefore U) can be obtained as an affine function of x for the complete feasible space of x. To derive these results, we first state the following theorem.

Theorem 1. For the problem in (7) let x_0 be a vector of parameter values and (z_0, λ_0) a KKT pair, where $\lambda_0 = \lambda(x_0)$ is a vector of nonnegative Lagrange multipliers, λ, and $z_0 = z(x_0)$ is feasible in (7). Also assume that (i) linear independence constraint qualification and (ii) strict complementary slackness conditions hold. Then,

$$\begin{bmatrix} z(x) \\ \lambda(x) \end{bmatrix} = -(M_0)^{-1} N_0 (x - x_0) + \begin{bmatrix} z_0 \\ \lambda_0 \end{bmatrix} \tag{8}$$

where,

$$M_0 = \begin{pmatrix} H & G_1^T & \cdots & G_q^T \\ -\lambda_1 G_1 & -V_1 & & \\ \vdots & & \ddots & \\ -\lambda_p G_q & & & -V_q \end{pmatrix}$$

$$N_0 = (Y, \lambda_1 S_1, \dots, \lambda_p S_p)^T$$

where G_i denotes the ith row of G, S_i denotes the ith row of S, $V_i = G_i z_0 - W_i - S_i x_0$, W_i denotes the ith row of W and Y is a null matrix of dimension $(s \times n)$.

The space of x where this solution, (8), remains optimal is defined as the Critical Region (CR^0) and can be obtained as follows. Let CR^R represent the set of inequalities obtained (i) by substituting $z(x)$ into the inequalities in (7) and (ii) from the positivity of the Lagrange multipliers, as follows:

$$CR^R = \{Gz(x) \leq W + Sx(t), \lambda(x) \geq 0\}, \quad (9)$$

then CR^0 is obtained by removing the redundant constraints from CR^R as follows:

$$CR^0 = \Delta\{CR^R\}, \quad (10)$$

where Δ is an operator which removes the redundant constraints - for a procedure to identify the redundant constraints, see Gal (1995). Since for a given space of state-variables, X, so far we have characterized only a subset of X i.e. $CR^0 \subseteq X$, in the next step the rest of the region CR^{rest}, is obtained as follows (Dua and Pistikopoulos, 2000):

$$CR^{rest} = X - CR^0. \quad (11)$$

The above steps, (8–11) are repeated and a set of $z(x), \lambda(x)$ and corresponding CR^0s is obtained. The solution procedure terminates when no more regions can be obtained, i.e. when $CR^{rest} = \emptyset$. For the regions which have the same solution and can be unified to give a convex region, such a unification is performed and a compact representation is obtained. The continuity and convexity properties of the optimal solution are summarized in the next theorem.

Theorem 2. For the mp-QP problem, (7), the set of feasible parameters $X_f \subseteq X$ is convex, the optimal solution, $z(x) : X_f \mapsto \Re^s$ is continuous and piecewise affine, and the optimal objective function $\mu(x) : X_f \mapsto \Re$ is continuous, convex and piecewise quadratic.

Based upon the above theoretical developments, an algorithm for the solution of an mp-QP of the form given in (7) to calculate U as an affine function of x and characterize X by a set of polyhedral regions, CRs, has been developed which is summarized in Table 1.

This approach provides a significant advancement in the solution and on-line implementation of MPC problems. Since its application results in a complete set of control actions as a function of state-variables (from (8)) and the corresponding regions of validity (from (10)), which are computed off-line. Therefore during on-line optimization, no optimizer needs to be called and instead for the current state of the plant, the region, CR^0, where the value of the state variables is valid, can be identified by substituting the value of these state variables into the inequalities which define the regions. Then, the corresponding control actions can be computed by using a function evaluation of the corresponding affine function (see Figure 5).

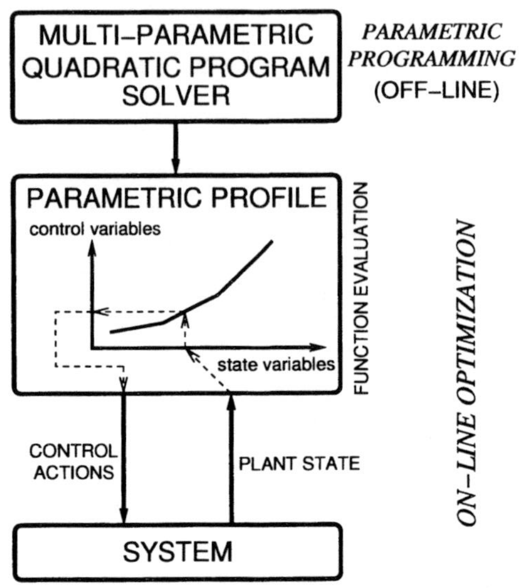

Fig. 5. On-line optimization via parametric programming

Table 1. Solution Steps of the mp-QP Algorithm

Step 1 For a given space of x solve (7) by treating x as a free variable and obtain $[x_0]$.
Step 2 In (7) fix $x = x_0$ and solve (7) to obtain $[z_0, \lambda_0]$.
Step 3 Obtain $[z(x), \lambda(x)]$ from (8).
Step 4 Define CR^R as given in (9).
Step 5 From CR^R remove redundant inequalities and define the region of optimality CR^0 as given in (10).
Step 6 Define the rest of the region, CR^{rest}, as given in (11).
Step 7 If no more regions to explore, go to the next step, otherwise go to Step 1.
Step 8 Collect all the solutions and unify a convex combination of the regions having the same solution to obtain a compact representation.

Figure 6 demonstrates how advanced controllers can be implemented on a simple hardware.

3.2 Continuous Time Formulation

Consider the following optimal control problem (Sakizlis, 2003):

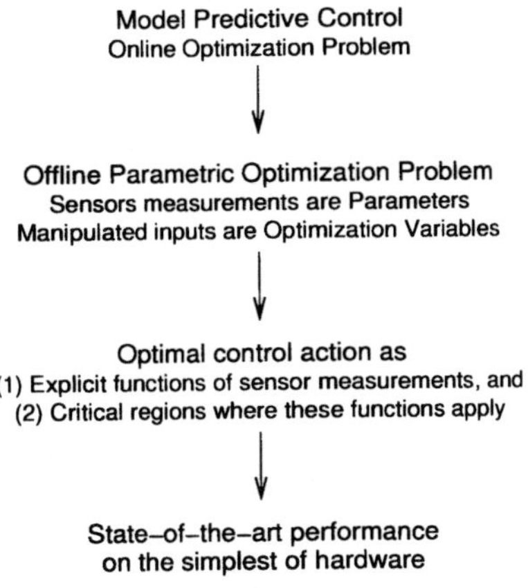

Fig. 6. Achieving state-of-the-art control performance on simple hardware

$$\hat{\phi} = \min_{x(t),v(t)} \frac{1}{2}x(t_f)^T P_1 x(t_f) + \frac{1}{2}\int_{t_o}^{t_f} [x(t)^T Q_1 x(t) + v(t)^T R_1 v(t)]dt$$

s.t.

$$\begin{aligned} \dot{x}(t) &= A_1 x(t) + A_2 v(t) \\ 0 &\geq \psi^g(x_f) = D_1 \cdot x(t_f) + b_2 \\ 0 &\geq g(x,v) = C_1 \cdot x(t) + C_2 \cdot v(t) + b_1 \\ x(t_o) &= x_o, \quad t_o \leq t \leq t_f \end{aligned} \tag{12}$$

where $x \in X \subseteq \Re^n$ are the states and $v \in U \subseteq \Re^{n_v}$ are the control manipulating inputs. Consider: $g : \Re^n \times \Re^m \times \Re^{n_v} \mapsto \Re^q$ and $\psi^g : \Re^n \times \Re^m \times \Re^{n_v} \mapsto \Re^{Q_g}$. The pair (A_1, A_2) is assumed to be stabilizable and the pair (A_1, B_1) detectable. Matrices $Q, P \succeq 0$, $R \succ 0$ constitute the quadratic performance index.

By treating the initial state conditions as parameters, problem (12) is recast as a multiparametric dynamic optimization problem (mp-DO). The solution this mp-DO problem is given by a set of expressions for the *optimal* value of the performance index and the *optimal* profiles of the control inputs and the states as a function of the initial conditions. This parametric controller has the following benefits comparing to the parametric controller for discrete-time dynamic systems: (A) The constraints are satisfied over the complete time horizon rather than at discrete time points. (B) The complexity of the control law derivation is contingent solely upon (i) the number of constraints, (ii) the system dynamics and (iii) the number of control variables.

3.3 2-state SISO illustrative example

Consider the open-loop unstable SISO plant:

$$\Sigma(s) = \frac{0.003396(s + 0.8575)}{(s-1)(s-0.6313)} \tag{13}$$

subject to the path constraint:

$$\frac{1.5s+1}{s^2 - 1.63135s + 0.6313} \leq 2.4 \tag{14}$$

A receding horizon optimal control problem is formulated as in (12) where the terminal cost P_1 is evaluated from the solution of the Riccati equation. The solution of the continuous time formulation presented in section 3.2 results in the state space partition shown in Figure 7 where the mathematical functions of the region boundaries and the comparison with the partition of the discrete-time parametric controller (parco) is also displayed.

The control policy in critical region $CR01$ is given by the function:

$$\begin{aligned} \hat{v}(t) = = -&[1.24\, e^{-10.76\,t + 10.76\,t_o}\,(1.0912\,x_{o1} + 0.93\,x_{o2}) \\ &-0.21\, e^{-0.8570\,t + 0.857\,t_o}\,(0.1329\,x_{o1} + 1.4309\,x_{o2})]; \end{aligned}$$

The execution of the derived control law is also shown in Figure 7, where the system is steered to the origin starting from a perturbed point of $x = [7 \quad -13]^T$. It is clear from Figure 7 that the discrete-time parco described in section 3.1 results in a larger number of critical polyhedral regions comparing to the continuous-time parco described in section 3.2 that generates only three regions with non-linear boundaries.

3.4 Hybrid Systems

The mathematical representation of a process system that operates in a transient mode and is subject to constraints and logical rules is considered as follows (Sakizlis *et al.*, 2002*a*):

$$\begin{aligned} \dot{x}(t) &= A_1 x(t) + A_2 v(t) + A_3 \delta(t) \\ y(t) &= B_1 x(t) + B_2 v(t) + B_3 \delta(t) \\ C_o y(t) + C_1 x(t) &+ C_2 v(t) + C_3 \delta(t) \leq 0 \end{aligned} \tag{15}$$

where $x \in \Re^n, y \in \Re^m$ are the states and the measurements of the dynamic system; $v \in \Re^q$ are the control manipulated inputs; $\delta \in \Delta \equiv \{0,1\}^b$ are binary variables that represent discrete decisions and logical conditions about the system operation.

The hybrid control problem is then formulated as follows:

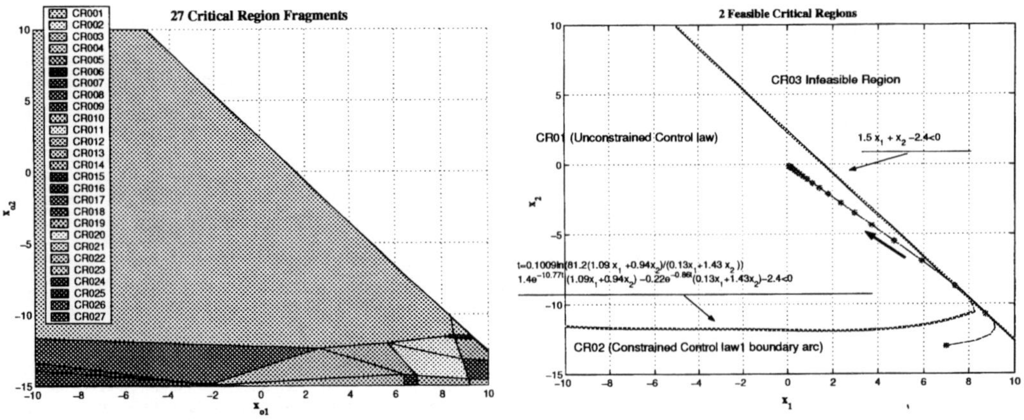

(a) Critical Regions for discrete-time parco (b) Critical Regions for continuous-time parco

Fig. 7. Comparison of state space partition of the discrete-time and continuous time parametric controller for the 2-state SISO example

$$\begin{aligned}\hat{\phi} = \min_{x,v,\delta} & \frac{1}{2}x(t_f)^T P_1 x(t_f) + ||P_2 x(t_f)|| + ||S\delta(t)|| \\ & + \int_{t_o}^{t_f} [\frac{1}{2}x(t)^T Q_1 x(t) + ||Q_2 x(t)|| \\ & + \frac{1}{2}v(t)^T R_1 v(t) + ||R_2 v(t)||]dt \\ s.t. \quad & \dot{x}(t) = A_1 x(t) + A_2 v(t) + A_3 \delta(t) \\ & y(t) = B_1 x(t) + B_2 v(t) + B_3 \delta(t) \qquad (16) \\ & \text{I. C.: } x(t_o) = x_o(t^*) \\ & C_o y(t) + C_1 x(t) + C_2 v(t) + C_3 \delta(t) \leq 0 \\ & t_o \leq t \leq t_f\end{aligned}$$

where terms $\int (1/2) \cdot x(t)^T Q_1 x(t) dt$, $(1/2) \cdot x(t_f)^T P_1 x(t_f)$ and $\int (1/2) v(t)^T R_1 v(t) dt$ correspond to the 2− norm of the state and output deviations respectively; terms $P_2 x(t_f)$, $\int Q_2 x(t)$, $\int R_2 v(t) dt$ represent either infinity $||Qx(t)||_\infty$, $||Rv(t)||_\infty$ or 1− $||Qx(t)||_1$, $||Rv(t)||_1$ norms.

After carrying out the appropriate substitutions and manipulations the optimization problem (16) is transformed to an equivalent finite dimensional problem of the following form:

$$\begin{aligned}\hat{\phi} = \min & \frac{1}{2}\{L_1 + L_2 u + L_3 x_o + L_4 \zeta + u^T L_5 u \\ & + x_o^T L_6 u + x_o^T L_7 x_o + x_o^T L_8 \zeta + u^T L_9 \zeta + \zeta^T L_{10} \zeta\} \\ & G_1 u + G_2 \zeta \leq G_3 + G_4 x_o \qquad (17) \\ & u \in \Re^{q \cdot N_q \cdot M} \quad x_o \in X_o \subset \Re^n\end{aligned}$$

where $\delta_l = \zeta_{li} \forall t \in [t_{i-1}, t_i]$, $i = 0, 1, ... N_t$, $l = 1, ... b$, $\zeta \in \{0,1\}^{b \cdot N_t}$. After removing nonlinearities from (17) by using appropriate algbraic manipulations, the problem becomes the following multi-parametric Mixed-Integer Quadratic Program (mp-MIQP):

$$\begin{aligned}\hat{\phi} = \min & \frac{1}{2}\{\bar{L}_1 + \bar{L}_2 z + \bar{L}_3 x_o + \bar{L}_4 \zeta + z^T \bar{L}_5 z + \\ & + x_o^T \bar{L}_6 x_o\} \\ & \bar{G}_1 z + \bar{G}_2 \zeta \leq \bar{G}_3 + \bar{G}_4 x_o \qquad (18) \\ & z \in \Re^{q \cdot N_q \cdot M} \quad \zeta \in \{0,1\}^{b \cdot N_t} \quad x_o \in X_o \subset \Re^n\end{aligned}$$

where $z = u + L_5^{-1} \cdot L_6^T \cdot x_o$. The procedure for solving (18) parametrically is based upon decomposing the mp-MIQP into a multi-parametric quadratic program (mp-QP) where the integers are fixed and an upper parametric bound is obtained, and a mixed-integer nonlinear program which treats the parameters x_o as free optimization variables and provides a new integer realization.

The final parametric solution provides the relation between the control and the current state of the system that ensures the optimum system regulation. This control law is proved to be (Theorem 1) piecewise affine with respect to the states and has the form:

$$\begin{aligned} & u_c(x_o) = a_c \cdot x_o + b_c; \quad \zeta_c(x_o) = d_c \\ & CR_c^1 \cdot x_o(t^*) + CR_c^2 \leq 0 \qquad (19) \\ & \text{for } c = 1, ... N_c\end{aligned}$$

where N_c is the number of regions in the state space. Matrices a_c, CR_c^1 and vectors b_c, CR_c^2, d_c are determined from the solution of the parametric programming problem and the index c designates that each region admits a different control law and different binary realizations.

4. CONCLUDING REMARKS

An overview of the parametric programming approach for the solution of on-line optimization problems has been presented. The on-line optimization problems for discrete and continuous time formulations and for hybrid systems were presented. Optimal value of the control variables is computed off-line as a function of the state variables or measurements. This is given by a set of control profiles on the space of state variables and the corresponding regions on the space of the state variables. On-line optimization is then carried out by taking measurements from the plant, identifying the region that is valid on the space of state variables and then calculating the control actions by simple function evaluations of the control profiles. The computational hardware requirements for such an implementation are very simple.

REFERENCES

Dua, V. and E. N. Pistikopoulos (1999). Algorithms for the solution of multiparametric mixed-integer nonlinear optimization problems. *Industrial and Engineering Chemistry Research* **38**, 3976–3987.

Dua, V. and E. N. Pistikopoulos (2000). An algorithm for the solution of multiparametric mixed integer linear programming problems. *Annals of Operations Research* **99**, 123–139.

Dua, V., K. P. Papalexandri and E. N. Pistikopoulos (2003). Global optimization issues in multiparametric continuous and mixed-integer optimization problems. accepted for publication in the Journal of Global Optimization.

Dua, V., N. A. Bozinis and E. N. Pistikopoulos (2002). A multiparametric programming approach for mixed-integer quadratic engineering problems. *Computers & Chemical Engineering* **26**, 715–733.

Gal, T. (1995). *Postoptimal Analyses, Parametric Programming, and Related Topics.* de Gruyter. New York.

Morari, M. and J.H. Lee (1999). Model predictive control: past, present and future. *Computers & Chemical Engineering* **23**, 667–682.

Pistikopoulos, E. N. and V. Dua (1998). Planning under uncertainty: a parametric optimization approach. In: *Proc. Third International Conference on Foundations of Computer-Aided Process Operations, pp. 164–169* (J. F. Pekny and G. E. Blau, Eds.). pp. 164–169.

Pistikopoulos, E. N., V. Dua, N. A. Bozinis, A. Bemporad and M. Morari (2002). On-line optimization via off-line parametric optimization tools. *Computers & Chemical Engineering* **26**, 175–185.

Sakizlis, V. (2003). Design of model based controllers via parametric programming. PhD thesis. Imperial College London, University of London.

Sakizlis, V., N. Kakalis, V. Dua, J. D. Perkins and E. N. Pistikopoulos (2004). Design of robust model based controllers via parametric programming. *Automatica* **40**, 189–201.

Sakizlis, V., V. Dua, J. D. Perkins and E. N. Pistikopoulos (2002*a*). The explicit control law for hybrid systems via parametric programming. Proceedings of the 2002 American Control Conference, Anchorage.

Sakizlis, V., V. Dua, J. D. Perkins and E. N. Pistikopoulos (2002*b*). The explicit model-based control law for continuous time systems via parametric programming. Proceedings of the 2002 American Control Conference, Anchorage.

ELSEVIER
IFAC
PUBLICATIONS
www.elsevier.com/locate/ifac

NOVEL PLANT TEST FOR MPC

Simone L. Kolhare * **Jorge A. Mandler** **

* *Air Products and Chemicals, Inc., Advanced Control Research Team - GEO Advanced Controls, 7201 Hamilton Blvd., Allentown, PA, 18195, USA. Mail Code: A12A2. Fax: (610) 481-8803, Phone: (610) 481-6170.*
** *Same affiliation as first author. Phone: (610)481-3413.*

Abstract: The present work is a method of plant testing for the purpose of open-loop system identification and model-based control of chemical processes, in particular, model predictive control (MPC). It is a multivariable test technique using plant-friendly and optimal amplitude binary multi-frequency signals, in conjunction with multivariable parametric modeling techniques, to achieve significant time savings in plant testing by comparison with traditional step testing and even other multivariable testing methods in the literature.

Keywords: system identification, open-loop plant test, perturbation signal design, model predictive control, model-based control, optimization

1. INTRODUCTION

The widely accepted empirical way to characterize the dynamics of a system is to apply perturbation signals at the input channels and measure the response of the system to these signals. The input and output signals are then processed to give the required estimate of the dynamics of the system. This procedure is well known as *system identification.*

In much of the literature on system identification, little attention has been paid to the perturbation signal design itself, other than to the fact that the signals should be *persistently exciting.* In the case of linear systems this means effectively that the signal should adequately span the bandwidth of the system being identified. One of the main reasons for this lack of attention has been the emphasis in the literature on identification techniques for parametric models. The main focus has been on discrete transfer function models of single-input single-output (SISO), linear, time-invariant systems. Under these circumstances, there is not usually a great deal to choose among different perturbation signal designs. However, this is quite an idealized situation in most applications and, in practice, many questions on signal design issues do in fact arise. These questions are generally associated with how to establish a suitable compromise between persistent excitation and plant friendliness during test, how to reduce the testing time without compromising the information needed, e.g., for control design, process monitoring, etc.

On the issue of system identification for model predictive control (MPC), a more recent interest has emerged in the automatic control community. However, in practice, traditional step testing is still adopted by and large in open-loop data collection for MPC modeling.

Step testing assumes that only one input channel is moved at a time. Each step move is held for a relatively long length of time. In (Boyden, 1999) it is recommended that each input move should be held for an average of half the process settling time and a series of $15 - 20$ moves should be executed for each input. In terms of their frequency characteristics, step signals tend to emphasize steady-state behavior and do not focus, therefore, on the closed-loop (faster) behavior. This means that a model with poor dynamic properties may be obtained. The plant testing time estimate for MPC, t_{test}, corresponding to this step testing technique can be expressed by the following formula:

$$t_{test} = (8 \ldots 10) \times IV \times T_{ss} \qquad (1)$$

where IV is the number of independent variables, i.e., the sum of manipulated inputs and measurable disturbances and T_{ss} is the process settling time. Naturally, t_{test} can become prohibitively long for systems with long *process settling time* and/or large *# of independent variables.* For some applications in our experience, this number can easily vary between 1 to 2 months depending on the particular process characteristics. This excessive plant testing time generally translates into practical and economical infeasibility of MPC projects.

Besides step testing, single and multivariable PRBS testing have become quite popular in the more recent literature on system identification for process control. Some relevant work includes (Cott, 1995*a*; Cott, 1995*b*; Gaikwad and Rivera, 1994; Koung and MacGregor, 1994; Ljung, 1998; Rivera *et al.*, 1990; Zhan, 1999; Zhu, 1998). In (Cott, 1995*a*) the author describes a benchmark problem proposed in the Process Identification Workshop at the 1992 Canadian Chemical Engineering Conference. Both industrial and academic participants in the workshop were challenged to identify models for the benchmark problem by using different perturbation signal designs and system identification techniques. Most of the participants utilized pseudo-random binary signals (PRBS) with different design characteristics. In (Zhu, 1998) the author proposes the use of PRBS as part of the so-called *asymptotic method* (ASYM) of system identification. In (Gaikwad and Rivera, 1994) and (Rivera *et al.*, 1990) the authors make use of PRBS as perturbation signals for system identification. In (Koung and MacGregor, 1994) the authors rotate input signals in an attempt to better capture uncertainties in the process steady-state gains. The signals used as the basis for this design are PRBS. Thus, PRBS, which started as a periodic signal generated in the time-domain through shift register circuitry, received a frequency-domain interpretation and a multivariable design which made it attractive for system identification purposes. Nonetheless, PRBS displays several weaknesses that have been reported in the literature (Cott, 1995*a*; Cott, 1995*b*; Godfrey, 1994).

Other types of time-domain periodic signals have found little application so far. These are referred to as multi-level pseudo-random signals (also known as *m-signals*) (Docter, 1999; Zierler, 1959).

Regarding frequency-domain identification, early work on signal design dates from the 1960s but most of the relevant papers are much more recent. However, they appear in journals ranging from *Nuclear Science and Engineering* to the *International Journal of Control*, from *Industrial Engineering and Chemistry* to *International Shipbuilding Progress* and in (at least) six different *IEEE Transactions.* As a result, this work is not as well known as it should be. The book edited by Keith Godfrey (Godfrey, 1994) is a relatively recent reference that collects some of this material.

In the present work, a thorough study of various perturbation signals for system identification, namely, pseudo-random binary signals, multi-level pseudo-random signals, sum-of-harmonics signals (van der Ouderaa *et al.*, 1988; Rivera, 1999) and binary multi-frequency signals (Van den Bos, 1967; Buckner and Kerlin, 1972; Harris and Mellichamp, 1980; Paehlike and Rake, 1979; Van den Bos, 1970; Van den Bos and Krol, 1979), was carried out. The findings of this study led to the selection of binary multi-frequency (BMF) signals as perturbation signals for plant test. The factors that led to this selection will be discussed later.

2. MAIN RESULTS

The main goal of this work is to considerably shorten plant testing time without compromising, and actually improving, the obtained model quality. In order to achieve this goal, the following steps are used as part of a comprehensive plant test procedure:

- Determination of Control-Relevant Plant Information
- Perturbation Signal Design
- Multivariable Plant Test Design
- Plant-Friendliness Analysis of the Plant Test
- Optimal Input Signal Amplitude Selection
- Parametric Data Analysis

Figure 1 is a flowchart of the main elements of the system identification technique proposed in this work.

2.1 Control-Relevant Plant Information

Extracting plant information which is relevant for the purpose of process control has been a topic of discussion in the literature (see, e.g., (Rivera, 1999)). The approach has been suggested as a means of enhancing controller performance. In the present work, this approach has been identified

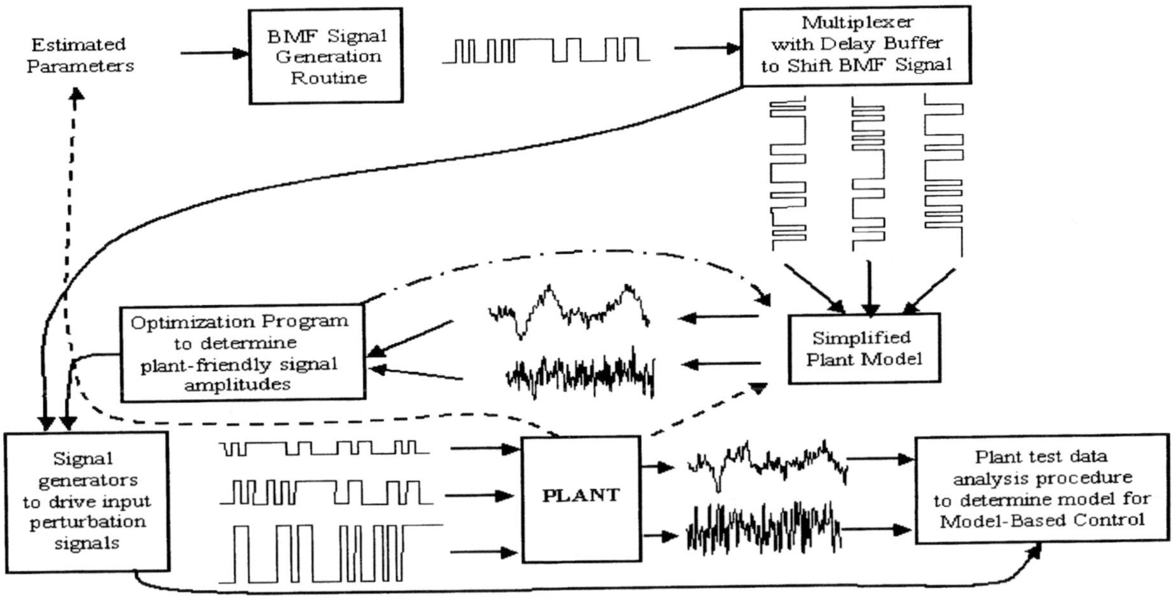

Fig. 1. Comprehensive Technique for System Identification. The dashed lines represent a priori knowledge of the plant used in designing the open-loop, non-iterative experiment. A priori knowledge may be obtained from a previously identified system of similar characteristics, operations personnel's knowledge of the system, a (dynamic) simulator, etc.

as one of the key components in reducing plant testing time.

Let us consider a first-order process with one input $u(t)$ and one output $y(t)$. If the input is excited with a unit step, the output response is given by $y(t) = K(1 - e^{-t/\tau})$, where K is the process steady-state gain and τ is the time constant. By definition, the process is considered settled when $t \geq T_{ss} \equiv 5\tau$, which implies that 99.33 % of the response has been attained. However, if $t = 4\tau$, 98.17 % of the response is still captured and if $t = 3\tau$, 95 % of the response is obtained. Thus, for a first-order process, one can design the signal so that $\omega_{low}^{process}$ is then increased to $\frac{5}{\beta T_{ss}}$, where $\beta \in [3, 5]$, and still obtain reasonable low frequency information on the process. Even if the process is not first order, most chemical processes' step responses near a given operating point resemble those of a linear system and similar arguments can be used to shorten testing time.

Now the question remains as to what should be the upper bound on the frequency range useful for control-relevant system identification. For any given linear system, a perturbation signal that spans the bandwidth of the system, i.e., $\omega \in [\frac{1}{T_{ss}}, \frac{1}{\tau_{dom}}]$ rad/unit of sampling time, where τ_{dom} is the estimated dominant time constant of the system, is adequate for the purpose of system identification if the system is to operate in open-loop. However, if the purpose of the plant test is to identify a system suitable for operation in closed-loop, the closed-loop bandwidth must be considered instead. If the closed-loop is estimated to be $\alpha > 1$ times faster than the open-loop, then the bandwidth for the system in closed-loop is given by $\omega \in [\frac{1}{T_{ss}}, \frac{\alpha}{\tau_{dom}}]$ rad/unit of sampling time. Therefore, this is the frequency range which the perturbation signal should span for control-relevant system identification.

Typically, α assumes values between 2 and 3 but it could be much higher if the controller is tuned more aggressively. Therefore, for closed-loop operation, $\omega_{high}^{process} = \frac{\alpha}{\tau_{dom}}$ rad/unit of sampling time. This frequency window will dictate the minimum period of the signal used for system identification and the speed with which this signal should vary.

2.2 Binary Multi-Frequency Signal Design

An excellent compromise between flexibility in signal power distribution, small peak factor, short plant test and ability to obtain high signal-to-noise ratio can be achieved by using the so-called *binary multi-frequency signals* (BMF) (Van den Bos, 1967; Buckner and Kerlin, 1972; Harris and Mellichamp, 1980; Paehlike and Rake, 1979; Van den Bos, 1970; Van den Bos and Krol, 1979). Like the PRBS, these are binary, discrete-interval, periodic signals with period $P = NT_{sw}$ samples. Like the sinusoids, these signals display as much power as possible in certain harmonics specified by the user. Nearly all designs are for N a power of 2, which makes it easy to ensure no spectral leakage using FFT signal processing. However, this is not a necessity and the only hard restriction on N is that it is an even number to guarantee that the signal is zero mean.

Because of the similar periodic auto- and cross-correlation properties of PRBS and BMF signals and the larger flexibility in the choice of N for the BMF, the plant testing time with BMF can always be made equivalent or shorter than that with PRBS. Thus, it has been recognized in this work that one does not pay any penalty in testing time by having a zero mean signal with a user-defined power distribution.

The *Frequency Domain Identification Toolbox* in MATLAB has two routines used to generate and improve a single binary multi-frequency signal. In these routines, the user is allowed to choose the power associated with each harmonic in a specified frequency range. The computation is an optimization where the signal with the smallest peak factor and the largest percent of useful power is computed to match, as closely as possible, the user-defined power for each harmonic. The main drawback of these routines is the lack of reproducibility of results since the resulting BMF signal will change depending on the initial random seed generated by the software. These different signals will have correspondingly different peak factor and power distribution. Therefore, the software does not necessarily provide the signal with the best compromise between peak factor and power distribution and the inexperienced user may end up with a poor signal selection. In this work these issues are addressed by improving on the existing software.

The algorithm used in these MATLAB routines is discussed in (Van den Bos and Krol, 1979) with improvement via a search technique described in (Paehlike and Rake, 1979).

Because the BMF can be designed to concentrate power in a selected number of harmonics in the control-relevant bandwidth, it is generally possible to excite the system as much with a lower amplitude BMF as with a higher amplitude PRBS.

The first design requirement on the BMF signal will be dictated by the need to capture the process long-term behavior. The lowest frequency captured by a BMF signal of period $P = NT_{sw}$ is given by $\omega_{low}^{input} = \frac{2\pi}{NT_{sw}}$ rad/unit of sampling time. Therefore, the following inequality guarantees that the signal spans the process fundamental frequency:

$$\omega_{low}^{input} \leq \omega_{low}^{process} \Longrightarrow \frac{2\pi}{NT_{sw}} \leq \frac{1}{T_{ss}}$$
$$\Longrightarrow P = NT_{sw} \geq 2\pi T_{ss} \quad (2)$$

Further reduction in testing time can be achieved if the requirement on the low frequency information is relaxed as discussed in section 2.1, i.e.:

$$P = NT_{sw} \geq 2\pi(\frac{\beta}{5})T_{ss}, \quad \text{where}\, \beta \in [3,5] \quad (3)$$

On the high-frequency side, the upper bound on the control-relevant frequency window, $\omega_{high}^{process}$, should never exceed the Nyquist frequency of the input signal which corresponds to $\omega_{high}^{input} = \omega_{Nyquist} \equiv \frac{\pi}{T_{sw}}$ rad/unit of sampling time. This condition generates the following upper bound on the switching time, T_{sw}:

$$\omega_{high}^{process} \leq \omega_{Nyquist} \Longrightarrow \frac{\alpha}{\tau_{dom}} \leq \frac{\pi}{T_{sw}}$$
$$\Longrightarrow T_{sw} \leq \frac{\pi\tau_{dom}}{\alpha} \quad (4)$$

Inequalities (3) and (4) together provide the general guidelines for the design of BMF signals for system identification. These requirements on P and T_{sw} are the same as for PRBS signals (Rivera, 1999). The fact that they extend to BMF signals is a new finding and it comes from the qualitative similarities between the periodic auto-correlation function (ACF) properties of BMF and PRBS signals.

An important additional degree of freedom in the BMF signal design is the signal power distribution over the harmonics that the signal spans. By properly designing the BMF signal, a signal with relatively small (large) T_{sw} can still invest a reasonable amount of power at lower (higher) frequencies. This allows for a very tailored design of the perturbation signal for each individual application.

2.3 Multivariable Plant Test Design

The BMF signals described in section 2.2 were used to design a new multivariable plant test. The multivariable nature of the plant test based on BMF signals proposed here is a key contributor to plant testing time reduction. As discussed in section 1, multivariable designs for perturbation signals for system identification are known for PRBS and Schröder-phased sinusoids. The multivariable statistically uncorrelated designs for PRBS and Schröder-phased sinusoids are based on time and frequency domain arguments, respectively.

In the present work, time-domain characteristics of the BMF signal, namely, the periodic auto- and cross-correlation functions (ACF and CCF), dictate the multivariable design in spite of the frequency-domain origin of the signal. This is a novel approach to multivariable plant test design using a frequency-domain based signal and generating uncorrelated copies in time-domain.

The multiple BMF signals are designed so as to ensure the smallest possible periodic unbiased

CCF between input pairs during one process settling time T_{ss}. In order to achieve this goal, the guidelines imposed by equations (3) and (4) are still relevant but may be replaced by more conservative inequalities.

In order to obtain m BMF signals with the smallest possible periodic CCFs, the initial ("mother") BMF signal is generated and $m-1$ delayed copies of this signal are created. Therefore, a new parameter appears in the MIMO design which is the *delay* between input channels.

The multivariable design, as described above, imposes the following requirements on the BMF signal period P:

$$P = NT_{sw} \geq 2\pi(\frac{\beta}{5})T_{ss}, \quad \text{for } m \leq 6$$

$$P = NT_{sw} \geq m(\frac{\beta}{5})T_{ss}, \quad \text{for } m > 6 \qquad (5)$$

with the delay between input channels, D, equal to $D = (\frac{\beta}{5})T_{ss}$.

The resulting T_{sw} in the MIMO case still satisfies inequality (4).

Since the BMF signal period in the MIMO case is given by equations (5), the minimum plant testing time is expressed by:

$$t_{test} = (2\pi + 1)(\frac{\beta}{5})T_{ss}, \quad \text{for } m \leq 6$$

$$t_{test} = (m + 1)(\frac{\beta}{5})T_{ss}, \quad \text{for } m > 6 \qquad (6)$$

Equation (6) is derived from the fact that one process settling time T_{ss} worth of data (or the modified settling time $(\frac{\beta}{5})T_{ss}$ with $\beta \in [3,5]$) needs to be discarded in the beginning of the test since it contains information related to previous input moves and not to the planned test moves.

Equation (6) is a key result of the present work. By comparing equations (6) and (1) one can infer an average $85-90$ % reduction in plant testing time with the new technique when compared to step testing.

2.4 Plant-Friendliness Analysis of the Plant Test

Once the multivariable test has been designed to tailor a particular application, it is important to predict, ahead of time, the effect that the designed perturbation signals will have on the plant. This is especially important given the multivariable nature of the test and how the different input moves interact to affect the outputs.

In this work a novel method for evaluating plant-friendliness of the designed test has been devised. Through assumption of simple linear dynamics and what is believed to be reasonable size moves on each input variable, it is possible to predict

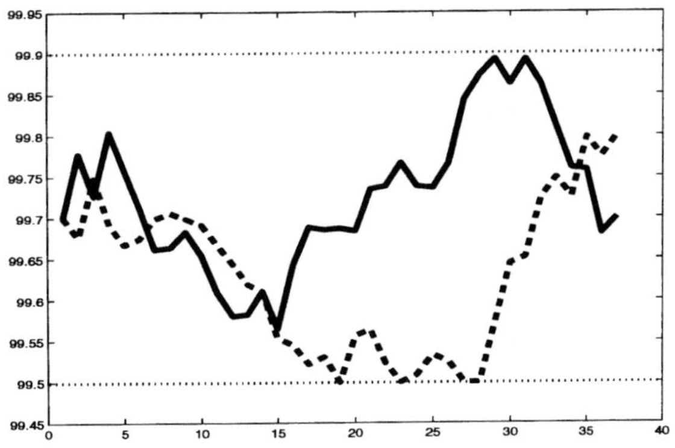

Fig. 2. Prediction of output behavior for two different plant test designs. Dashed Line: Plant-unfriendly design; Solid Line: Plant-friendly design

the plant output behavior during test. Output constraint violations can also be anticipated. This method provides a means of checking when input move combinations might lead the outputs towards constraints and what effect the input move sizes will have on the output responses.

If a design is judged to be plant-unfriendly, the first step might be to re-order the perturbation signals with respect to the input variables and re-evaluate plant-friendliness. If this does not produce satisfactory results, the test design should be re-visited.

Figure 2 illustrates how the application of this technique enables screening of a more plant-friendly test design prior to the actual test. The two curves indicate predictions of a plant output behavior using the plant-friendliness analysis tool. The dashed curve corresponds to the initial design based on the available plant information. The solid curve was obtained by simply re-ordering the BMF signals with respect to the input variables. While both designs satisfy the output constraints indicated by the horizontal dotted lines, the test design corresponding to the solid curve is believed to be more plant-friendly because it promotes output excitation above and below the initial operating point for a proportional amount of time. In the design corresponding to the dashed curve, the output tends towards the low limit constraint for the largest duration of the test. In this particular application, violation of the low limit constraint means off-spec product and, therefore, output excitation towards this constraint for a prolonged amount of time is not desirable. The same input amplitudes were considered in both cases.

2.5 Optimal Input Signal Amplitude Selection

In this work, once a plant test design is considered plant-friendly, optimal input amplitudes are computed for maximum output excitation within desired constraints.

In order to solve the dynamic optimization problem, the same model used for the predictions in the plant-friendliness analysis is utilized. For simplicity, the technique will be discussed for a first order model

$$G(t) = K\,(1 - e^{-(\frac{t}{\tau})}), \qquad (7)$$

where K is the matrix of steady-state gain estimates and $\tau \equiv \frac{T_{ss}}{5}$.

Since the input signals change at every T_{sw} samples and there are N moves in a signal period, the optimization is solved N times with $t = T_{sw}$ in formula (7).

The vector of amplitudes $\lambda_i > 0$, $i = 1, \ldots, m$, for the m inputs constitutes the set of decision variables. The optimization problem which is solved for $k = 1, \ldots, N$, is given by:

$$\begin{array}{ll} \textbf{OF:} & \min_{\lambda_1^k, \ldots, \lambda_m^k} \sum_{i=1}^{m} \ln(\frac{1}{\lambda_i^k}) \\ \textbf{IC:} & \Delta u_{min} \le \Lambda^k u^k \le \Delta u_{max} \\ \textbf{OC:} & \Delta y_{min}^k \le G(T_{sw}) \Lambda^k u^k \le \Delta y_{max}^k \\ \textbf{AC:} & \Lambda^k > 0 \end{array} \qquad (8)$$

where **OF**, **IC**, **OC** and **AC** stand for objective function, input constraints, output constraints and additional constraints, respectively. Λ^k is an $m \times m$ diagonal matrix whose elements are the input amplitudes λ_i^k, and u^k is the $m \times 1$ combination of the m input moves at the kth delay (this vector only contains elements equal to -1 or $+1$). u_{min} and u_{max} are the minimum and maximum allowed input values. $\Delta u_{min} \equiv u_{min} - u_{ss}$ and $\Delta u_{max} \equiv u_{max} - u_{ss}$ are, respectively, the minimum and maximum input deviation constraints with respect to the desired setpoint input value, u_{ss}. y_{min} and y_{max} are, respectively, the minimum and maximum output constraints which must be satisfied at all times. $\Delta y_{min}^k \equiv y_{min} - y_{ss}^k$ and $\Delta y_{max}^k \equiv y_{max} - y_{ss}^k$ are the minimum and maximum output deviation constraints with respect to the current output value y_{ss}^k. In practice, u_{ss} and y_{ss}^1 are the vectors of input and output values corresponding to the start of the plant test. Since the input deviations are always made around the initial input value u_{ss}, this value is not updated. The output values, however, vary as the inputs are implemented and the constraints have to be measured against updated values (that is why y_{ss}^k is dependent on k).

If this procedure were implemented on-line, during the plant test, as an advisor for input amplitude selection at each T_{sw} samples, it should take feedback from the plant in order to correct the outputs (instead of the update based on the assumed linear dynamics). This feedback would lead to more realistic amplitudes since it is a way to correct for the approximate dynamics in (7).

Figure 3 shows how this technique worked in a real plant test case. The figure shows the predicted output response in time obtained with optimized input amplitudes and the actual output response during plant test obtained with slightly larger amplitudes for higher output excitation. The concern with this particular critical output is that it should not go below 99.5 % since this is a product specification. The technique predicts that the test sequence will drive the output close to that limit at approximately 5250 min and again at 8300 min into the test. This indeed takes place in the actual test but the output at 5250 min actually goes below the desired limit due to the higher amplitudes used during the test. The model used for the output prediction and input amplitude optimization was a simple first order model with rough steady-state gain estimate.

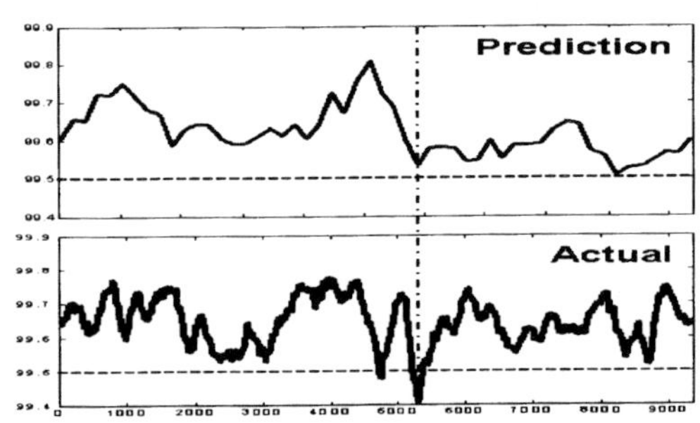

Fig. 3. Comparison of a predicted output response obtained with optimized input amplitudes with the actual response during plant test.

2.6 Parametric Data Analysis

Because the main focus of this work is to shorten testing time, this also means that less data is available for model construction. Therefore, parametric modeling methods are the only alternative if reasonable quality parameter estimates are to be obtained.

In the present work, the types of models considered were MISO or MIMO state-space or ARX models. MIMO models were utilized for highly interacting sub-systems of outputs where the output interactions must be taken into account in the model. These are the models that best capture the complex multivariable interactions in the process. This is not an issue for linear processes but, since most chemical processes are highly nonlinear, MISO and MIMO models may lead to very different responses. In this work it has been observed that MIMO identification for groups of highly interacting outputs leads to better fit of the individual output responses than MISO identification.

3. CONCLUSION

In this manuscript a method of plant testing and system identification with emphasis on test time reduction has been discussed. The plant testing time reduction is achieved via selection of binary multi-frequency (BMF) signals as input perturbation signals designed to excite the system in a control-relevant frequency range. The plant testing is a multivariable test with each input being excited with a delayed copy of the "mother" BMF signal. On an average, the minimum testing time with multivariable BMF testing is $85 - 90$ % shorter than conventional step testing. Once the test has been designed, its plant friendliness can be evaluated ahead of the plant test execution via simulation of predictive models based on a priori knowledge of the system. The amplitudes of the perturbation signals used to excite the different input channels are optimized to maximize the output variability and the signal-to-noise ratio within pre-specified input and output constraints. The resulting data from plant testing is analyzed using parametric modeling techniques.

REFERENCES

Boyden, S. (1999). *DMCplus^{TM} 1.1 - Multivariable Control Software - Training Course Slides - Reference Notes.* Aspen Technology, Inc.. Houston, TX.

Buckner, M. R. and T. W. Kerlin (1972). Optimum Binary Signals for Reactor Frequency Response Measurements. *Nucl. Sci. Eng.* **49**, 255–262.

Cott, B. J. (1995*a*). Introduction to the Process Identification Workshop at the 1992 Canadian Chemical Engineering Conference. *Journal of Process Control* **5**(2), 67–69.

Cott, B. J. (1995*b*). Summary of the Process Identification Workshop at the 1992 Canadian Chemical Engineering Conference. *Journal of Process Control* **5**(2), 109–113.

Docter, W. A. (1999). Order Reduction of Nonlinear Dynamic Models by Subspace Identification and Stepwise Regression. PhD thesis. Lehigh University. Bethlehem, PA.

Gaikwad, S. V. and D. E. Rivera (1994). Integrated Identification and Control for Model Predictive Controllers. In: *AIChE Meeting.* San Francisco, CA. Session: Model Predictive Control, paper 227a.

Godfrey, Keith, Ed.) (1994). *Perturbation Signals for System Identification.* Prentice Hall International (UK) Limited.

Harris, S. L. and D. A. Mellichamp (1980). Online Identification of Process Dynamics: Use of Multifrequency Binary Signals. *Industrial and Engineering Chemistry Process Design and Development* **19**, 166–174.

Koung, C. and J. F. MacGregor (1994). Identification for Robust Multivariable Control: the Design of Experiments. *Automatica* **30**, 1541–1554.

Ljung, L. (1998). Identification for Control - What is there to learn?. Technical Report LiTH-ISY-R-1996. Linköping University, Department of Electrical Engineering. Linköping, Sweden.

Paehlike, K. D. and H. Rake (1979). Binary Multifrequency Signals - Synthesis and Application. In: *5th IFAC Symposium on Identification and System Parameter Estimation.* Darmstadt. pp. 589–596.

Rivera, D. E. (1999). Principles of System Identification. Arizona State University, College of Engineering and Applied Sciences, Tempe, AZ. Short course notes.

Rivera, D. E., J. F. Pollard and C. E. Garcia (1990). Control-Relevant Parameter Estimation via Prediction-Error Methods: Implications for Digital PID and QDMC Control. In: *AIChE Meeting.* Chicago, IL. Session: Model Predictive Control, paper 4a.

Van den Bos, A. (1967). Construction of Binary Multifrequency Test Signals. In: *1st IFAC Symposium on Identification in Automatic Control Systems.* Prague.

Van den Bos, A. (1970). Estimation of Linear System Coefficients from Noisy Responses to Binary Multifrequency Test Signals. In: *2nd IFAC Symposium on Identification and Process Parameter Estimation.* Prague.

Van den Bos, A. and R. G. Krol (1979). Synthesis of Discrete-Interval Binary Signals with Specified Fourier Amplitude Spectra. *International Journal of Control* **30**, 871–884.

van der Ouderaa, E., J. Schoukens and J. Renneboog (1988). Peak Factor Minimization of Input and Output Signals of Linear Systems. *IEEE Trans. on Instrum. Meas.* **IM-37**, 207–212.

Zhan, Q. (1999). Multivariable Test Signal Design and Constrained Identification for the Process Industry. Technical report. Lehigh University, Department of Chemical Engineering. Bethlehem, PA. Candidacy Report.

Zhu, Y. (1998). Multivariable Process Identification for MPC: the Asymptotic Method and its Applications. *J. Proc. Com.* **6**(2), 101–113.

Zierler, N. (1959). Linear Recurring Sequences. *Journal Soc. Ind. Appl. Math.* **7**, 31–48.

ELSEVIER
IFAC
PUBLICATIONS
www.elsevier.com/locate/ifac

MULTI-OBJECTIVE INPUT SIGNAL DESIGN OF MULTI-HARMONIC SIGNALS FOR SYSTEM IDENTIFICATION

Sridharakumar Narasimhan, *
Raghunathan Rengaswamy *,[1]

* *Dept of Chemical Engineering, Clarkson University, Potsdam, NY 13699-5705, USA*

Abstract: The choice of perturbation inputs is critical in the identification and model building exercise. Accurate identification requires that the input be persistently exciting so as to excite all modes or frequencies of interest. However, there are multiple objectives that need to be considered. Multi-harmonic signals are a convenient choice for frequency domain identification. Multi-objective optimization formulations for synthesis of multi-harmonic signals are presented. *Copyright © 2004 IFAC*

Keywords: Identification, Optimal experiment design, Input signals, Multi-objective optimization

1. INTRODUCTION

One of the primary goals of system identification is to ensure that the identified model has good predictive capabilities and the model parameters are accurately estimated. It is common practice to perturb the system with specially tailored inputs and the consequential input output data are used to build the system model. The quality of the model depends strongly on the experiment design and identification and hence the input used for perturbing the system should be carefully selected.

Researchers have paid considerable attention to system identification and input signal design and a number of excellent reviews in these fields are available (Godfrey, 1993; Ljung, 1999; Pintelon and Schoukens, 2001). The theory of statistical experiment design has been applied to system identification to synthesize maximally informative inputs in the time and frequency domains (Goodwin and Payne, 1977; Kalaba and Spingarn, 1982; Mehra, 1981).

Frequency domain synthesis of the optimal input signal involves determination of the input spectrum that minimizes a scalar criterion of the uncertainty of the parameters to be estimated. Multi-frequency binary signals and multi-harmonic (sum of sines) signals are commonly used to realize the signal with the desired power spectrum. There are many advantages to using multi-sine signals with low crest factors (Guillaume *et al.*, 1991; Pintelon and Schoukens, 2001; Solomou *et al.*, 2002).

The concept of plant friendly identification that is relatively less hostile to operating conditions in process plants has received attention amongst members of the process control and identification community recently (Narasimhan *et al.*, 2003; Parker *et al.*, 2001; Rivera *et al.*, 2003). Different measures of plant friendliness have been proposed in literature.

There has been some recent application of multi-objective optimization based methods to identification and control (El-Kady *et al.*, 2003; Johansen, 1996; Johansen, 2000). However, a multi-objective approach to input design has been proposed only recently by the present authors. Illustrative formulations for optimal

[1] Corresponding author. Email: raghu@clarkson.edu

design of input signals that characterize the multi-objective nature of the design problem are presented in this contribution.

2. OBJECTIVES

2.1 Preliminaries

Consider a Linear Time Invariant system described by the Impulse Response Model of the following form

$$y(t) = \int_0^{T_s} u(t-\tau)h(\tau)d\tau + n(t) \quad (1)$$

where $y(t)$, $u(t)$, $n(t)$ are the output, input and noise respectively and $h(\tau)$ is the impulse response of the system. The frequency response function $H(j\omega)$ can be estimated as $Y(j\omega)/U(j\omega)$ where $Y(j\omega)$ and $U(j\omega)$ are the Fourier transforms of the output and input respectively. One of the primary objectives of a good system identification exercise is to accurately estimate the unknown parameters by reducing the bias and the variance of the parameter estimates. For input design purposes, it is convenient to assume that the estimator is efficient and unbiased. This implies that that the covariance of the parameter estimates can be calculated from the Cramer-Rao bound, viz., the inverse of the Fisher information matrix M. Optimal input signal design in the frequency domain thus involves determination of the optimal input spectrum that minimizes a scalar norm of M^{-1}, like $tr(M^{-1})$ or $\det M^{-1}$ or $\lambda_{\max}(M^{-1})$ where λ is an eigenvalue of M^{-1}.

Calculation of the optimal input spectrum requires *a priori* knowledge of the the noise characteristics and transfer function of the system which is supplied through an initial estimate. An analytical expression for the optimal spectrum can be calculated only for some problems. Iterative algorithms for optimization of the input power spectrum have been proposed (Pintelon and Schoukens, 2001). In general, it is difficult to realize a signal with an arbitrary spectrum and so, it is convenient to restrict the frequency space to a finite, discrete, harmonically related set of frequencies. A multi-harmonic having N_u distinct harmonics $\{\omega_u\}$ (Ref equation 2) can then be generated that has the desired amplitude spectrum.

$$u(t) = \sum_{u=1}^{N_u} a_u \cos(\omega_u t + \alpha_u) \quad (2)$$

where $\omega_u = 2\pi k_u/T$ and k_u are monotonically increasing harmonic numbers ($k_u \in \mathbb{N}$, $u = 1,2,\ldots N_u$) and $\{a_u\}$ is the specified amplitude spectrum. The advantage of using a multi-sine or multi-harmonic signal is that it is periodic and easy to generate for any given arbitrary spectrum. The steady state output of the system when subjected to the above input is

$$y(t) = \sum_{u=1}^{N_u} a_u|G(j\omega_u)|\cos(\omega_u t + \alpha_u + \phi_u) \quad (3)$$

where, $|G(j\omega_u)|$ and $\phi_u = \arg(G(j\omega_u))$ are the amplitude ratio and phase shift of the system respectively, evaluated at the frequency $\omega_u = 2\pi k_u/T$.

For a linear q input, p output system, the ith output y_i, $i = 1,2,\ldots p$ is given by

$$y_i(t) = \sum_{l=1}^{q}\sum_{u=1}^{N_u} a_{lu}|G_{li}(j\omega_{lu})|\cos(\omega_{lu}t + \alpha_{lu} + \phi_{li}(j\omega_{lu})) \quad (4)$$

where $|G_{li}(j\omega_{lu})|$ and $\phi_{li}(j\omega_{lu})$ are the gain and phase shift introduced between input l and output i at frequency $2\pi k_{lu}/T$. It can be seen that if the inputs are chosen so that the spectral lines k_{lu} do not coincide for any pair of inputs $l = 1,2,\ldots q$,

$$|Y_i(j\omega_{lu})| = |G_{li}(j\omega_{lu})|a_{lu} \quad (5)$$

Thus, ensuring that the spectral lines of the inputs do not overlap allows the MIMO system to be broken into pq SISO systems.

2.2 Crest factor minimization

Signals with a specified spectral distribution $\{a_u\}$ may be realized in different ways in the time domain. For instance in a multi-sine, the spectral characteristics are described by $\{a_u\}$ alone and the phases do not affect the spectral characteristics. However, the choice of phases can affect the time domain behaviour of the signal. It is possible that an inappropriate choice of phases results in a signal with large magnitude peaks. A more compact or compressed signal would have a small signal amplitude between the peaks. The time domain compression or compactness of a multi-sine signal $x(t)$ with a specified amplitude spectrum is described by the crest factor, (Cf), or the peak factor (Pf) (Godfrey, 1993).

$$Pf = \frac{x_{max} - x_{min}}{2\sqrt{2}x_{rms}} \quad (6)$$

$$Cf = \frac{\max|x|}{x_{rms}} \quad (7)$$

where $x_{min}, x_{max}, x_{rms}$ denote the minimum, maximum and rms values of the signal respectively. The advantage of synthesizing a minimum crest factor input signal is that large peaks are avoided. Further, it has been shown that the number of averages required to measure a signal with a specified accuracy is proportional to the crest factor (Godfrey, 1993). In addition, it has been reported that reducing the crest factor has the added advantage of reducing the effect of distortions due to certain nonlinearities (Solomou *et al.*, 2002). For a specified amplitude spectrum $\{a_u\}$, it is possible to synthesize an input signal with a low crest factor by a suitable choice of phases α_u. Different algorithms for crest factor minimization have been reported, the most efficient and common being the L_∞ algorithm based on a generalization of Polya's algorithm (Guillaume *et al.*, 1991). The crest factor is thus the ratio of the L_∞ norm and the L_2 norm where the general L_p norm of the function $x(t)$ over the interval $[0,T]$ is defined as

$$L_p(x) = \left[\frac{1}{T}\int_0^T |x(t)|^p dt\right]^{(1/p)} \tag{8}$$

and the L_∞ norm is $\max|x(t)|$. The crest factor of the output $y(t)$ can be defined in a similar manner. The output $y(t)$ is a multi-harmonic signal represented by Equation 3 for ease of calculation since the steady state output is a phase and amplitude modulated version of the multi-harmonic input signal. Thus the L_∞ algorithm can be applied to minimize the crest factor of the output if the system transfer function or an estimate is available.

2.3 Plant friendly identification

Multi-variable model based control strategies are commonly used in chemical process industries. Plant friendly identification has received the attention of researchers in recent times (Narasimhan *et al.*, 2003; Parker *et al.*, 2001; Rivera *et al.*, 2003). Identification experiments in process industries are carried out on running plants. While a persistently rich excitation with high signal to noise ratio is theoretically preferred, operational, safety, environmental and economic considerations have to be taken into account during identification.

- An input requiring aggressive and frequent movement of valves and actuators is not desirable as this can lead to equipment wear and tear.
- Identification experimentation time has to be kept to a minimum so as to minimize off-spec products and consumption of utilities. Tests using the popular Pseudo-Random Binary Signals (PRBS) usually require days to conduct (Smith, 2003).
- Output deviations should be reduced to ensure that the product quality differs as little as possible from the set point

The input spectrum is chosen either by optimization with respect to the criteria discussed above or based on the user's experience. Given an input spectrum, a convenient measure of plant friendliness (both the input and output) is the crest factor defined above. It must be noted that a minimum crest factor input signal does not necessarily imply a minimum crest factor output signal. Thus, from a practical consideration, it is necessary to use an input signal that has a low **Input Crest factor** and results in an **Output with low crest factor**.

2.4 Distortion due to nonlinearities

The linear convolution integral can be generalized to describe a general, causal, time invariant stable nonlinear system in the following manner

$$y(t) = \sum_{i=1}^{\infty} \int h_i(\tau_1,\ldots,\tau_i) \prod_{j=1}^{i} u(t-\tau_j)d\tau_j \tag{9}$$

where $h_n(\tau_1,\ldots,\tau_n)$ is the n dimensional n^{th} order Volterra kernel and can be viewed as an n dimensional impulse response.

The corresponding frequency domain representation of an nth order system is

$$Y(s_1,s_2,\ldots,s_n) = \sum_{i=1}^{n} H_i(s_1,\ldots,s_i) \prod_{j=1}^{i} U(s_j) \tag{10}$$

where $H_n(s_1,\ldots,s_n)$ is the corresponding n dimensional Laplace transform of $h_n(\tau_1,\tau_2,\ldots,\tau_n)$. When such a system is subjected to a multi-sine as described in equation 2, the output will contain frequencies other than the test or input frequencies. The frequency contributions due to the nonlinearities can be classified as Type I (arising from pairs of equal positive and negative frequencies) and Type II (which are not accounted by Type I) contributions. It is possible to suppress even order nonlinearities by considering only odd harmonics, as the nonlinear contributions of the even order effects will fall at even harmonics (Solomou *et al.*, 2002) and hence will be distinct from the linear and higher odd order contributions. Hence, for illustrative purposes, a simple cubic static nonlinearity with unit gain is considered. The nonlinear contribution of the output corresponding to a multi-sine input can be expressed in the frequency domain as

$$Y_{nl}(j\omega) = \sum_{\substack{m=-N_u \\ m\neq 0}}^{N_u} \sum_{\substack{n=-N_u \\ n\neq 0}}^{N_u} \sum_{\substack{o=-N_u \\ o\neq 0}}^{N_u} a_m a_n a_o \exp(j(\alpha_m+\alpha_n+\alpha_o)) \times\delta(\omega-(k_m+k_n+k_o)2\pi/T) \tag{11}$$

The extent of distortion due to nonlinearity at the test frequencies can be quantified by

$$E = \frac{1}{N_u}\sum_{u=1}^{N_u} |Y_{nl}(j\frac{2\pi k_u}{T})| \tag{12}$$

If the primary aim of the identification experiment is to identify the linear kernels accurately, it is possible to reduce the contribution of the nonlinearities by a suitable choice of phases.

Thus, from an optimization point of view, one could require that the phases of the input multi-sine signal be chosen so that the input crest factor is minimized and the nonlinear contributions at the test frequencies are also minimized. Again, it must be noted that minimum crest factor signals do not necessarily minimize the nonlinear contributions.

In the previous sections, two design scenarios for input signal design, each with multiple objectives were described. Since, both objectives need not be attainted simultaneously, the problem is inherently multi-objective in nature and should be treated as such. In the succeeding sections, some common techniques for solution of a multi-objective optimization problem are described.

3. MULTI-OBJECTIVE OPTIMIZATION

In traditional single objective optimization problems, the aim is to find a globally optimal solution, if it exists. Unlike single objective optimization problems, in optimization with possibly conflicting objectives, there is no unique optimal solution. System and real world design usually involves tradeoffs between different objectives and more than one decision maker. A fair amount of subjectivity and user influenced decision making are characteristics of multi-objective problems. There are several possible approaches for solving a multi-objective optimization problem (Miettinen, 1998).

Single weighted cost function: One simple and common approach to multi-objective optimization is to formulate a single weighted objective function from the individual costs J_i.

$$J = \alpha_1 J_1 + \ldots + \alpha_n J_n \tag{13}$$

such that the weights $\alpha_i \geq 0$. The resulting problem can be solved by standard methods of optimization.

Goal programming: In goal programming, a goal or aspiration value, γ_i is associated with each objective function and the weighted deviation from the target is minimized. These goals are not constraints, but only aspiration levels that may or may not be satisfied. Given an objective function J_i and goal γ_i, the deviation variable δ_i can be written as the difference of δ_i^- (negative deviation or underachievement) and δ_i^+ (positive deviation or overachievement) . It is necessary only to minimize the weighted sum of positive deviation variables δ_i^+ for minimization problems where δ_i^+ are the individual weights.

$$\text{Minimize}\{\sum w_i^+ \delta_i^+\} \ \texttt{s.t} \begin{cases} J_i + \delta_i^- - \delta_i^+ = \gamma_i \ \forall i \\ L_j \leq 0 \ \forall j \\ E_k = 0 \ \forall k \\ \delta_i^+, \ \delta_i^- \geq 0 \ \forall i \end{cases} \tag{14}$$

where E_k and L_j are equality and inequality constraints respectively.

Pareto solutions: The above techniques essentially convert a multi-objective optimization problem into a single objective problem. Unlike single objective optimization, in optimization with conflicting objectives, there is no single optimal solution. The interaction among different objectives gives rise to a set of solutions, called the Pareto optimal solutions (Ref Fig. 1). Solutions A, D, B form a Pareto optimal front and no one solution in this set can be said to be better than another in pure quantitative terms. However, solution C is dominated by solution D as solution D is better than C in both objectives. A set is called a global Pareto-optimal set, if no solution in the search space dominates any member in it. The optimization algorithm should attain two goals :- search for the global Pareto-optimal front, and maintain population diversity in the optimal front so that no bias towards any particular objective function exists. The final solution that is chosen for implementation is based on some other criteria. Evolutionary algorithms have been particularly useful in solving multi-objective optimization problems (Deb, 1999).

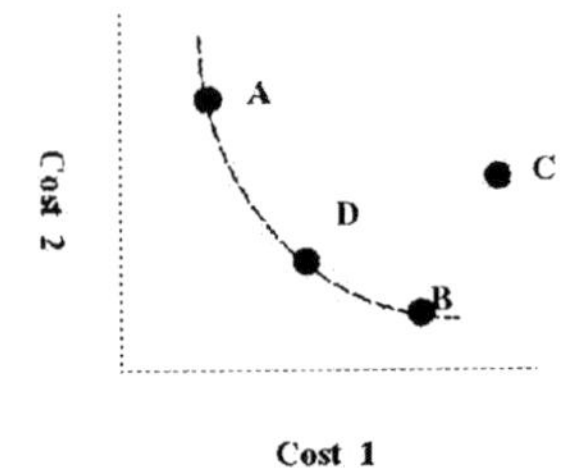

Fig. 1. Pareto optimal sets

Thus, in input signal design, there are multiple objectives, some of which cannot be satisfied simultaneously. In the subsequent section, two Pareto solutions corresponding to different cost functions are presented.

4. GENERATION OF PARETO FRONTS

4.1 Input-output crest factor minimization

Consider a 15 harmonic multisine described by equation 2 with a flat amplitude spectrum $a_u = 1$, $u = 1,2,\ldots,15$ and $k_u = 1,2,\ldots,15$, with a fundamental harmonic equal to 0.0625 rad/sec. It is required to minimize the crest factors of the input and output where the system transfer function is

$$H(s) = \frac{s}{s^3 + 2s^2 + 2s + 1} \tag{15}$$

The phases of the respective harmonics are chosen so as to minimize the input and output crest factors. As mentioned previously, use of a minimum input crest factor signal does not necessarily result in a minimum output crest factor. The L_∞ algorithm can be applied to compress the input and outputs simultaneously by defining a common norm of the input $u(t)$ and output $y(t)$ (Guillaume *et al.*, 1991). By suitably weighting the inputs and outputs, signals with different (not necessarily minimum) crest factors can be synthesized. However, the above method is not guaranteed to generate the complete Pareto front, since the solutions obtained by solving the weighted problem are weakly Pareto optimal. Another disadvantage is that it is necessary to solve a single valued objective optimization problem several times by varying the weights.

Since Genetic Algorithms (GA) work with a population of points, it is natural to use GAs for obtaining the set of non-dominated solutions simultaneously. GAs

Table 1. GA Parameters

Parameter	Crest factor	Nonlinear error
Mutation probability	0.05	0.07
Crossover probability	0.9	0.9
Population size	500	500
Number of generations	350	500
Crossover Distritution index	50	60
Mutation distribution index	50	60

do not use gradient based techniques and so, the non-differentiable nature of the L_∞ norm does not pose any analytical problems. For a specified amplitude spectrum and given system transfer function, the L_2 norm is invariant with respect to the phases, and so minimization of the crest factor is equivalent to minimizing the respective L_∞ norms. The L_∞ norm of the function $f(t), t \in [a,b]$ is approximated by determining the maximum of $|f(t)|$ over a discrete number of points in the interval $[a,b]$. The real coded variable version of the Nondominated Sorting Genetic Algorithm (NSGA-II) (Deb *et al.*, 2002) is used to generate the Pareto front of input and output crest factors. The Pareto front for the input and output crest factors at the end of 350 generations is shown in Fig. 2 (A). Parameters used for the solution are tabulated in Table 1. In order to verify that the solution is close to the true Pareto optimal solution, the following metric is defined

$$d = \sum_{i=1}^{N_p} (f_{1i}^2 + f_{2i}^2) \qquad (16)$$

where $[f_{1i}, f_{2i}]^t$ is the ith non-dominated objective vector and N_p is the number of non-dominated solutions in the current population. In this particular example, they are the input and output crest factors respectively. Fig. 3(A) shows the evolution of the distance metric d evaluated for all the non-dominated solutions in every generation. It is seen that the metric d does not increase with subsequent generations, thus confirming that the solutions are reasonably close to the true Pareto optimal solution. It must be noted that while the NSGA -II algorithm aims to preserve diversity among the population, it is possible that the front obtained is a subset of the whole front, since the true Pareto front is not known.

4.2 Reducing the effects of nonlinear distortions

Consider a muti-sine subjected to a simple nonlinear element

$$y(t) = u(t) + u(t)^3 \qquad (17)$$

Thus, the nonlinear distortion at the test frequencies is caused solely by the cubic element and can be quantified by equation 12. The objectives are to choose the phases in a 15 harmonic multi-sine with a flat amplitude spectrum $a_u = 2$, $u = 1,2,\ldots,15$ and $k_u = 1,2,\ldots,15$ so as to minimize the input crest factor and the nonlinear distortions. As above, the multi-objective optimization problem is solved using the NSGA-II algorithm (Deb *et al.*, 2002) with the parameters mentioned in Table 1. The corresponding Pareto front and the evolution of the metric d is presented in Fig. 2 (B) and Fig. 3 (B).

Thus, it is clear from the above figures that there exists a trade-off between the different objectives. It is possible to sacrifice the input friendliness to achieve a lower value of the output crest factor. Likewise, for a relatively small change in the nonlinear error, it is possible to reduce the input crest factor. Standard techniques of input signal design do not allow the user the flexibility to tailor an input signal that satisfies several objectives or consider the trade-offs. Solution of the appropriate multi-objective optimization problem results in the Pareto set of non-dominated solutions. The final choice of the input signal is made by the user. It must be noted that computing the output crest factor requires knowledge of the system transfer function. In a practical system identification setting, this is unknown or only an estimate is available. The effect of uncertainty of the transfer function on the output crest factor needs to be investigated. The effect of dynamic nonlinearities and a multi-optimization formulation to reduce the distortion of non-linearities also needs to be investigated.

5. CONCLUSION

Input signal design for system identification typically requires that multiple objectives be satisfied. Multiple objectives that arise in the design of multi-harmonic input signals were discussed and appropriate problems were formulated. A Genetic Algorithm based approach was used to solve for the Pareto front of non-dominated solutions.

ACKNOWLEDGEMENT

The authors would like to acknowledge the Kanpur Genetic Algorithms Laboratory, KANGAL [2], Indian Institute of Technology, Kanpur, 208016, India for use of the NSGA-II code.

REFERENCES

Deb, K. (1999). Evolutionary algorithms for multi-criterion optimization in engineering design. In: *Proceedings of Evolutionary Algorithms in Engineering and Computer Science* (Miettinen K., M.M. Mäkelä, P. Neittaanmäki and J. Periaux, Eds.). John Wiley and Sons.

Deb, K., A. Pratap, Agarwal S. and T. Meyarivan (2002). A fast and elitist multiobjective genetic algorithm: NSGA-II. *IEEE Trans. Evol. Comput* **6**, 182–197.

[2] `http://www.iitk.ac.in/kangal`

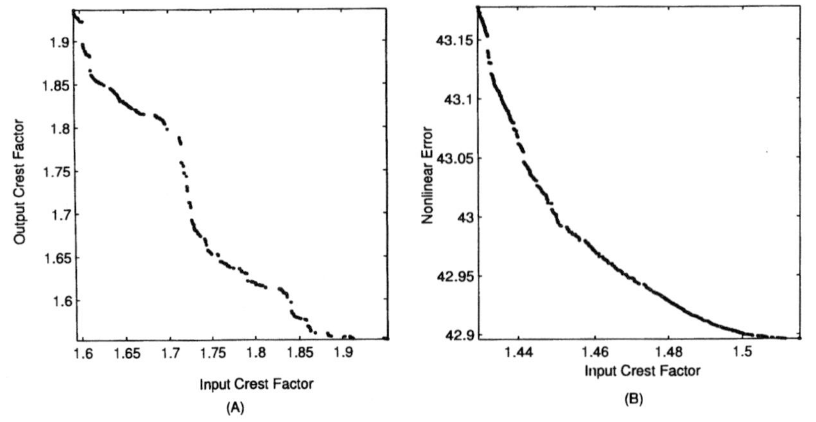

Fig. 2. Paretò optimal fronts

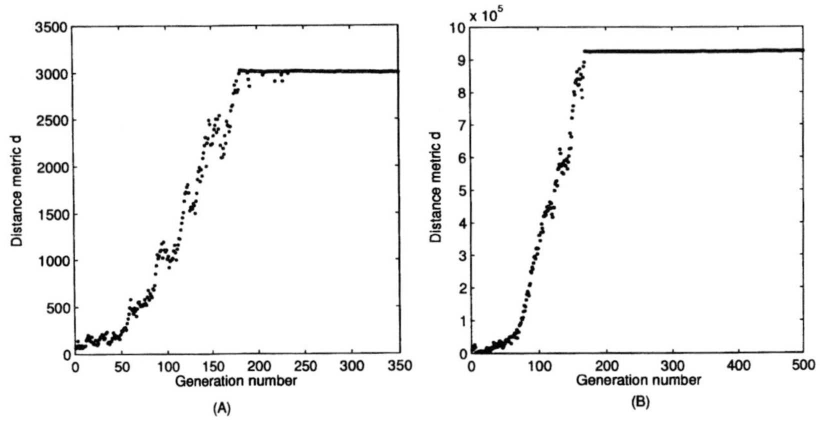

Fig. 3. Distance metric d

El-Kady, M.M., M.S. Salim and A.M. El-Sagheer (2003). Numerical treatment of multiobjective optimal control problems. *Automatica* **39**, 47–55.

Godfrey, K. (1993). *Perturbation Signals for System Identification*. Prentice Hall, USA.

Goodwin, G.C. and R.L. Payne (1977). *Dynamic System Identification: Experiment Design and Data Analysis*. Academic Press, USA.

Guillaume, P., J. Schoukens, R. Pintelon and I. Kollar (1991). Crest-factor minimization using nonlinear chebyshev approximation methods. *IEEE Trans. Instrum. and Meas.* **40**, 982–989.

Johansen, T.A. (1996). Identification of non-linear systems using empirical data and prior knowledge- an optimization approach. *Automatica* **32**, 337–356.

Johansen, T.A (2000). Multi-objective identification of fir models. In: *IFAC SYSID 2000*. Santa Barbara, USA.

Kalaba, R. and K. Spingarn (1982). *Control, Identification, and Input Optimization*. Plenum Press, New York.

Ljung, L. (1999). *System Identification, Theory for the User, 2nd Edition*. Prentice Hall, USA.

Mehra, R.K. (1981). Choice of input signals. In: *Trends and Progress in Systems Identification* (Eykhoff, Ed.). Pergamon Press, New York.

Miettinen, K.M. (1998). *Nonlinear Multibojective Optimization*. Kluwer.

Narasimhan, S., R. Srinivasan and R. Rengaswamy (2003). Multi-objective input signal design for plant-friendly identification. In: *SYSID2003*. Rotterdam, Netherlands.

Parker, R.S., D. Heemstra, F.J. Doyle III, R.K. Pearson and B.A. Ogunnaike (2001). The identification of nonlinear models for process control using tailored "plant-friendly" input sequences. *J. Process Control* **11**, 237–250.

Pintelon, R. and J. Schoukens (2001). *System Identification : A Frequency Domain Approach*. John Wiley & Sons, USA.

Rivera, D.E, H. Lee, M.W. Braun and H.D. Mittelmann (2003). Plant friendly system identification: A challenge for the process industries. In: *SYSID 2003*. Rotterdam, Netherlands.

Smith, C.L. (2003). Intelligently tune pid controllers, part 2. *Chemical Engineering* **110**, 54–59.

Solomou, M., C. Evans and D. Rees (2002). Crest factor minimization in the frequency domain. *IEEE Trans. Instrum. Meas.* **51**, 859–865.

ELSEVIER
IFAC
PUBLICATIONS
www.elsevier.com/locate/ifac

A NEW SIGNAL DESIGN TOOL FOR PROCESS MODEL IDENTIFICATION

V. C. Machado, J. O. Trierweiler, and A. R. Secchi

Group of Integration, Modeling, Simulation, Control and Optimization of Processes (GIMSCOP)
Department of Chemical Engineering, Federal - University of Rio Grande do Sul (UFRGS)
Rua Duque de Caxias, 1303 apt. 603 CEP: 90010-283 - Porto Alegre - RS - BRAZIL,
Fax: +55 51 3316 3277, Phone: +55 51 3316 4072
EMAIL: {vinicius, jorge, arge}@enq.ufrgs.br

Abstract: A new low-cost systematic procedure to obtain high-quality data for model identification is presented. Two major steps compose this methodology: first, a simplified process model is built using a PIMS (Plant Information and Management System) database, which contains limited information about the process behavior. Second, the signal amplitudes and its directions are calculated, as well as the total time of the experiment. The signal is designed to do not remove the system from its normal operating point, reducing the identification costs. The new procedure was applied to an industrial distillation unit. *Copyright © 2004 IFAC*

Keywords: Signals, Dynamic Models, Identification, Distillation Columns, Linear Control Systems

1. INTRODUTION

Model identification is the key part of modern advanced process control systems, since most process controllers are based on a model.

Generally, a model can be obtained through rigorous modeling based on the first principles or by empiric modeling and parameter estimation using plant data. In most cases, it is very difficult to create an accurate first-principles model, due to the complexity of the process and the lack of information. On the other side, the empirical modeling, although easier to obtain than the rigorous modeling, demands an intensive identification test, where all manipulated variables must be excited to produce some effects in the controlled variables. Several well-known model structures, like FIR (Finite Impulse Response), *ARX*, *OE* (Output Error), *BJ* (Box-Jenkins), (Ljung, 1999), *SUBID* (Subspace Identification) (Overschee and De Moor, 1996), etc., may be used to fit the plant data. The problem is that, normally, the identification tests remove the plant from its normal operating point and spend much time and money to be accomplished. It is important that the signal of the manipulated variables during the test be strong enough to excite the process output, producing a good signal-to-noise relation (*SNR*), given by:

$$SNR = \frac{\phi_{yy}}{\phi_{uu}} \qquad (1)$$

where ϕ_{yy} is the signal spectra of the output variable and ϕ_{uu} is the signal spectra of the input variable. High SNR guarantees a good quality of the identified model, but may drive the process away from the operating point. In this way, it is necessary to have a tool to produce good data for model identification at low cost and time.

In the literature, there is a variety of signals to dynamic model identification (Cutler, 1988; Luyben

1992). The most used signal in identification tests is the step signal, which is characterized by an abrupt level change in the value of the system inputs and is easy to implement. However, the step signal removes the plant from its normal operating point, and is time consuming, because the plant needs to achieve a new steady state after the input variation. The random binary signals and *Pseudo Random Binary Signals* (*PRBS*) (Tulleken, 1990) have the advantage to present zero mean during the identification test, maintaining the industrial unit in the operating point. However, they are also time consuming tests and are not suitable for multivariable systems as explained below. Still there are ramp signals and wave signals to identification test, but these signals are not so used in the practice.

For multivariable systems, usually, the step test is preferred in process industries, where each manipulated variable is changed individually. Clearly, this procedure is time consuming and the multivariable characteristic is not perfectly captured. Besides, the signal amplitude of each input variable is heuristically determined. The multivariable PRBS uses a lot of signal directions to excite the system. This fact is not good, because during the test a lot of data are repeated, bringing undesired correlation among all variables and the parameter estimation becomes a difficult task. These characteristics were the motivation to develop a new tool to signal design for model identification.

In the next section the proposed methodology is presented. In section 3, dynamic models of a distillation unit are identified with data generated from the proposed signal design methodology.

2. SIGNAL DESIGN METHODOLOGY

Unfortunately, most available identification tests do not use the historical plant data. Nowadays, with advent of *PIMSs* (Plant Information and Management Systems), a plenty of plant data can be stored and easily retrieved. Normally, these data are noisy and do not contain good dynamic information about the process. However, it is still possible to obtain some information about the process from these data.

In order to take into account the historical plant data in the signal design, two procedures are proposed in this work: characterization of the input and output noises and correlation analysis between inputs and outputs. The correlation analysis is used to select good plant data intervals to identify a simplified process model.

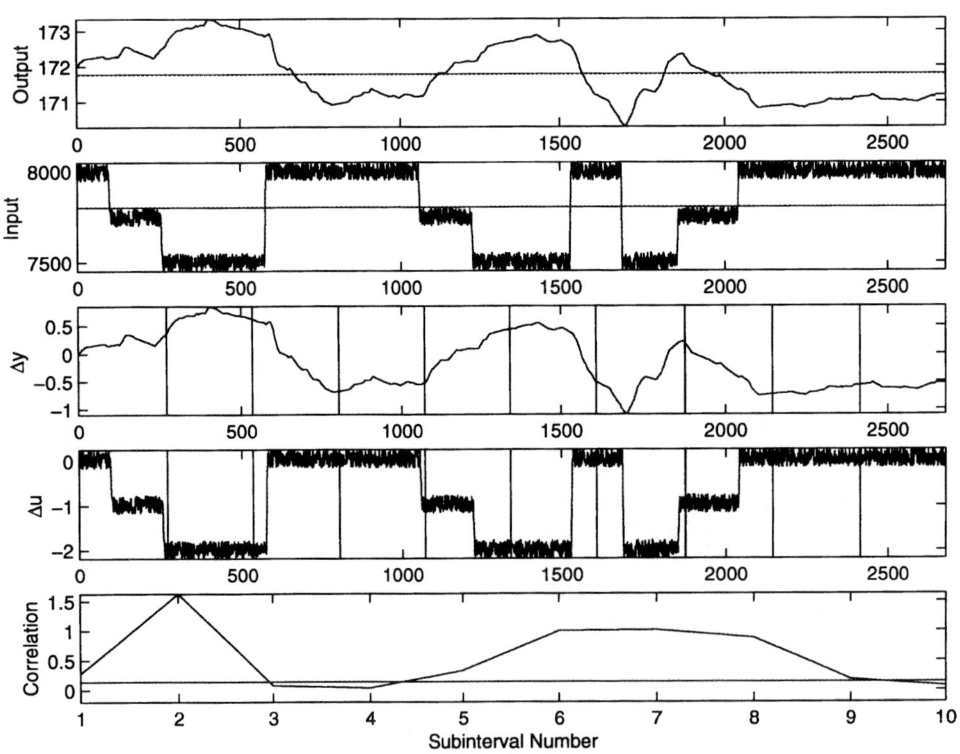

Fig. 1: Correlation analysis to select the best plant data intervals to model identification. The two first graphics show the vector *y(t)* and *u(t)* and their mean values, respectively. The third and fourth graphics show the vectors *Δu(t)* and *Δy(t)*, and each subinterval in which the correlation function was calculated. the last graphic shows the maximal correlation function values for each data subinterval and its threshold value.

2.1 Simplified process model identification

The first step of the multivariable signal design methodology is the search for plant data intervals to identify a simplified process model. It is based on a sweeping procedure to observe how inputs and outputs are correlated by calculating the correlation function, given by:

$$Corr(k) = \frac{1}{N}\sum_{t=1}^{N} \Delta u(t)\Delta y(t+k) \quad (2)$$

where k means the output shifting (in sampling intervals), t means the actual sampling interval and $t+1$ means de next sampling interval. N is the total number of points and the variables Δu and Δy are the two normalized signals being analyzed. In case of industrial plant data, Δu (related to the inputs) and Δy (related to the outputs) are calculated by :

$$\Delta y(t) = \frac{y(t) - \mu_y}{\max(|y(t) - \mu_y|)} \quad (3)$$

$$\Delta u(t) = \frac{u(t) - \mu_u}{\max(|u(t) - \mu_u|)} \quad (4)$$

where μ_y and μ_u are the output and input mean values, respectively, $y(t)$ and $u(t)$ are the original output and input vectors of plant data, respectively.

To illustrate how this correlation analysis works to select the best plant data intervals, Figure 1 presents real plant data and the result of the calculated correlation function.

It is important to note that the correlation function is calculated for all time intervals, and the best intervals are selected to compose the data set to identify a simplified process model. To decide what intervals will compose the identification data set, a threshold value, L, should be defined. If the maximal correlation value is greater than L, then the subinterval is selected. This value is related to the noise level, and it was defined as a value proportional to the geometric mean between the standard deviations of the input and output noise, given by:

$$L = 2.6\sqrt{\sigma_u \cdot \sigma_y} \quad (5)$$

where σ_u and σ_y are the standard deviation of the input and output noise, respectively. These values are obtained from the less correlated subinterval as described below.

Not only plant data time intervals need to be found, but also the input and output noises amplitudes. It is necessary to keep in mind that the input variation effects should be great than the noise amplitude to identify a good quality plant model. Although the plant data, in special the output variables, are corrupted by stochastic variations, it is convenient to affirm that the difference between the output signal (y) and its mean (μ_y), named as *ydiff* ($ydiff = y - \mu_y$), follows the normal distribution, which is represented by:

$$p(ydiff \mid \mu_y, \sigma_y) = \frac{1}{\sigma_y\sqrt{2\pi}} e^{\left(\frac{-(ydiff-\mu_y)^2}{2\cdot\sigma_y^2}\right)} \quad (6)$$

where p is the probability density of the data (*ydiff*) with deviation σ_y and mean μ_y. If the data follows a normal distribution, then 95% of the plant data must lie between $-2.6\ \sigma_y$ and $+2.6\ \sigma_y$. Considering this confidence interval, then the noise amplitude is $2.6\ \sigma_y$. Therefore, comparing the distribution of the plant output data with the normal distribution, it is possible to determine the noise level. Similar to the correlation analysis, the plant data is divided in time intervals, as shown in Figure 2, and a comparison between data distribution and the normal distribution is made in each interval. Figure 3 shows the comparison for the second time interval of Figure 2. The same comparison is done for all time intervals, and the interval which the data distribution is closer to the normal distribution is selected to evaluate the noise level. The metric used to make this selection is the absolute integral of the difference between the normal distribution and the data distribution. Observing the data of Figure 2, by visual inspection it is clear that the second time interval better represents the output noise level, which was the same result obtained by applying the proposed procedure, leading to an output noise level of 13.83 units

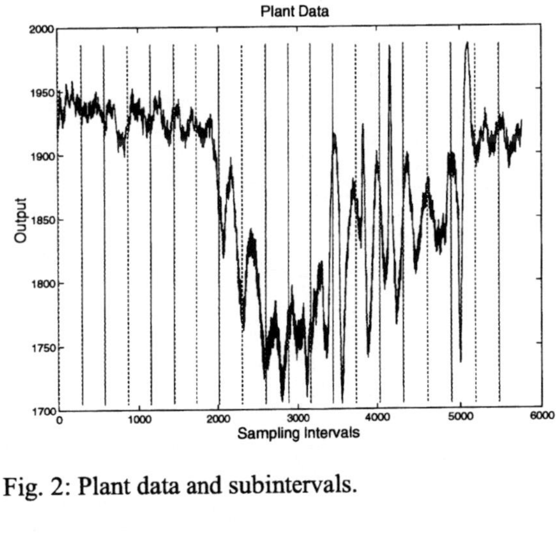

Fig. 2: Plant data and subintervals.

This procedure is repeated to the others input and output variables. As already was commented, the noise level is one of the most important information to design good signal to models identification. With the selected data, a simplified model can be identified by any identification algorithm, such as *ARX*, *OE*, *BJ*, subspace identification, etc. At this stage it is only necessary a *First Order with Dead Time* (*FODT*) model to design a good input signals, having a reasonable information about the process time constant and process gain. In the next section it is shown how to determine the manipulated variables profile to be applied during the identification test.

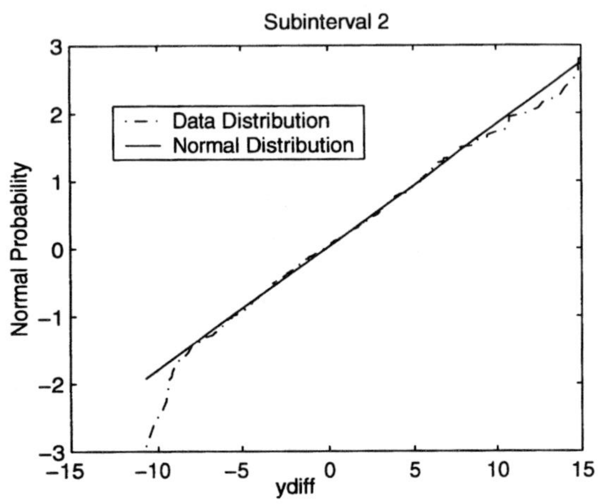

Fig. 3: Comparison between the normal distribution and the output data distribution, for the second time interval showed in Figure 2.

2.2 Input signal design

An important information in signal design is how much time is necessary to keep the system excited in order to capture the whole system response. In the proposed methodology, this information is obtained from the simplified model identified in the first step. The time constant of each channel of the transfer matrix contributes to determine the system time constant. To calculate a representative system time constant, it was chosen a geometric mean (τ_{GEO}) among all time constants present in the simplified model, calculated by:

$$\tau_{GEO} = \sqrt[N]{\prod_{i=1}^{N} \tau_i} \tag{7}$$

where N is the total number of the nonzero system time constants, and τ_i is the ith system time constant. The geometric mean gives a tradeoff between large and small time constants. A simple arithmetic mean tends to be closer to the high values of time constants. Using the representative time constant of the system, τ_{GEO}, it is defined three relevant frequencies: $\omega_1 = 0.5\tau_{GEO}^{-1}$, $\omega_2 = \tau_{GEO}^{-1}$, and $\omega_3 = 1.5\tau_{GEO}^{-1}$, to determine how much time each manipulated variable will be kept in a certain level during the identification test. These frequencies were chosen in order to design signals with high probability to capture the true system dynamics. Then, the time that each signal should be kept in a determined level is:

$$time\_of\_disturbance = \frac{\pi}{\omega_i} \tag{8}$$

For each frequency, it is necessary to determine the most important directions of the input signals. It can be done by evaluating the frequency response, $G(j\omega)$, of the simplified model G (which was determined in the first step of the methodology). However, as it is very common to observe a great difference among the system variables magnitudes, it is important to consider the scaling of $G(j\omega)$, given by:

$$G_S(j\omega) = L_S.G(j\omega).R_S \tag{9}$$

where L_S and R_S are the input and output scaling matrices, respectively. All necessary directions to design the best signals for model identification can be obtained by applying the Singular Value Decomposition (*SVD*) to $G_S(j\omega)$ yielding:

$$G_S(j\omega) = W\sum V^T \tag{10}$$

where W and V^T are the system output and input direction matrices, respectively, and Σ is diagonal matrix of the system singular values. Each column of the V^T is an input direction vector, named as U_d. The amplitudes, Δu, of the designed input signals are given by:

$$\Delta u = E_F \,.\, R_S \,.\, U_d \tag{11}$$

where E_F is an expansion factor to guarantee that all outputs will be excited, and is given by:

$$E_F = \max\left(\frac{L_S^{-1} \cdot \Delta y_{noise}}{|G_s(j\omega_i) \cdot U_d|}\right) \tag{12}$$

where Δy_{noise} is the noise amplitude (determined in the first step of the methodology). It should be clear that the above procedure has to be applied for each designed frequency ω_i. In the next section, the results of applying the proposed signals design methodology to model identification of a distillation unit.

3. INDUSTRIAL APPLICATION

The proposed methodology was applied to an industrial naphtha distillation unit. A brief description of the process: the crude petroleum feed is heated up in a heat exchanger battery, the salt is removed, and the oil goes to the first separation unit, a flash drum, to obtain a mixture of naphtha, diesel, and solvents. The mixture goes to the second and main separation unit, a distillation column, which produces heavy naphtha, diesel, and small amount of gasoline. The control objectives in this unit are to keep the high quality of diesel, observed through temperatures and volumetric flows, and to reduce the atmospheric residue, which is the fraction of raw materials that did not transform in commercial products. This residual is produced in the distillation column. To attain these control objectives, a MPC (model predictive controller) calculates the setpoints of some important variables in the system, such as the petroleum feed, the flash temperature, and the top temperature of the distillation column. The studied distillation system has 8 inputs and 14 outputs. To make a good signal design for a block diagonal controller, it is necessary to group some related input variables. In this case, three blocks were created: block 1 (input 1, petroleum feed), block 2 (inputs 2 and 6, flash drum variables) and block 3 (inputs 3,4,5,7, and 8, related to the distillation column).

The signal design to identify a dynamic model to the distillation unit starts with a generation of a simplified model, based on historic plant data. Ten historic plant data sets were analyzed, and a correlation analysis was carried out according to the methodology described in Section 2.1. The noise amplitudes were also determined, whose results are presented in Table 1.

Table 1. Noise level of the system outputs.

Variable	Noise Level	Variable	Noise Level
1	0.075	8	0.299
2	6.688	9	0.378
3	8.151	10	0.130
4	0.148	11	0.248
5	0.408	12	0.289
6	3.056	13	12.100
7	3.011	14	10.101

The designed and applied signals can be seen in Figure 4.

With the identification test data set, a dynamic model, using the discrete output error algorithm of IDENT Toolbox (Ljung, 1989), was identified. The estimated model parameters were then converted to the continuos time domain. Figure 5 shows the comparison between the identified model and the identification test data set for two output variables for each signal block. Because the system has many output variables, only two variables were chosen to show the results.

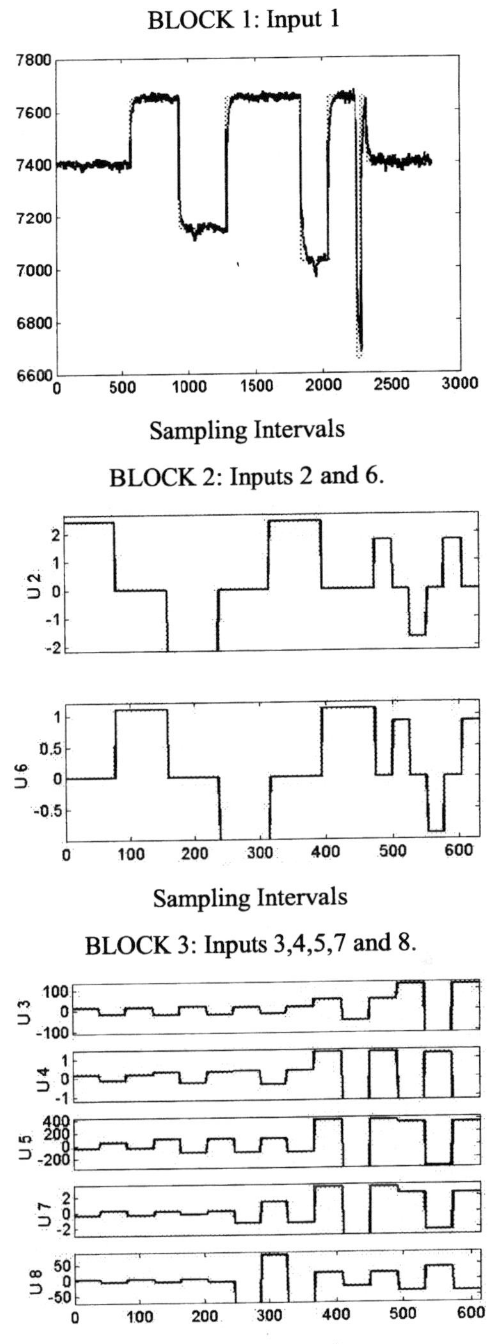

Fig. 4: Designed signals to model identification of the distillation unit.

It can be observed in Figure 5 that the obtained results were satisfactory. The identified model represents well the system behavior. It is important to mention that the process did not leave its normal operating point during the test, fact that can be verified by the plant data average and standard deviation, shown in Table 2.

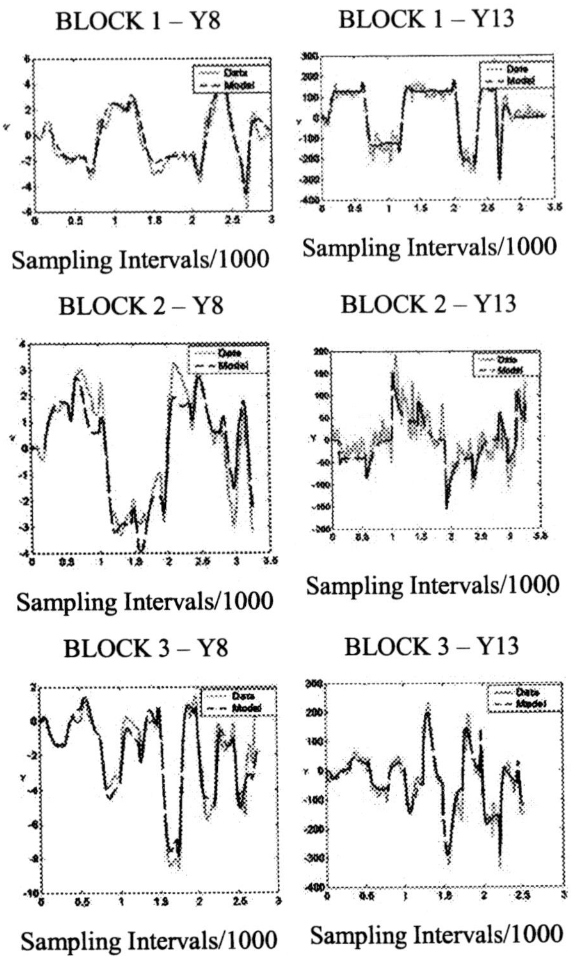

Fig. 5: Comparison between identified model and plant data for some output variables.

CONCLUSIONS

A new methodology to design signals for dynamic model identification was developed and applied to an industrial unit, generating good quality data without removes the plant from its normal operating point, at a low cost. The implemented methodology does a consistent and rational analysis of data sets contained in PIMS databases, to extract some important system information to guide a correct design of the identification signals. These signals are oriented to excite the system not only in the high gain direction but also in the low gain, to better determine the multivariable system gain.

ACKNOWLEDGMENT

The authors thank CAPES/FINEP and PETROBRAS for the financial support. In special, we thank RECAP (Capuava Refinary – São Paulo, Brazil) and its staff, by giving us the opportunity to test the proposed signal identification methodology there.

Table 2. Mean and standard deviation values of the plant data during the identification test.

Y	Mean			Deviation		
	Block 1	Block 2	Block 3	Block 1	Block 2	Block 3
1	144.07	142.64	143.80	0.55	3.01	0.51
2	758.86	753.91	783.37	17.17	62.95	12.32
3	1212.43	1252.62	1201.30	16.33	83.57	32.76
4	33.43	34.63	35.33	0.13	3.52	0.10
5	115.15	111.74	116.06	2.26	2.70	3.50
6	173.07	146.16	157.12	15.87	35.36	37.83
7	1874.53	2127.83	2227.72	106.47	118.07	92.07
8	46.35	42.53	41.86	2.06	2.85	2.39
9	249.02	235.88	227.05	4.33	3.21	4.19
10	54.03	53.03	56.48	1.92	3.91	2.56
11	-9.73	-10.05	-9.63	0.83	1.50	1.5
12	347.69	350.07	350.41	0.95	2.41	1.94
13	1839.43	1924.13	2035.30	130.03	72.47	85.50
14	3423.85	3441.15	3585.73	153.91	40.84	42.37

REFERENCES

Cutler, C.R.; Hawkins, R. B. Application of a large predictive multivariable controller to a hydrocracher second stage reactor, Proceedings of American Control Conference, 1988, 284-291.

Ljung, L. MATLAB®/IDENT-System Identification Toolbox (1989), The MathWorks.

Ljung, L. System Identification: *Theory for the User*, Prentice Hall, New Jersey. (1999)

Luyben, W. L., *Practical Distillation Control*, Van Nostrand Reinhold, New York, 1992.

Overschee, P. V., Bart D. M. *Subspace Identification for Linear Systems*. Kluwer Academic Publishers, Netherlands, 1996.

Tulleken, H. J. A. F., Generalized binary noise test-signal concept for improved identification-experiment design, Automatica, 26(1),1990, 37-49.

ELSEVIER IFAC PUBLICATIONS
www.elsevier.com/locate/ifac

IDENTIFIABILITY ANALYSIS OF A LIQUID-LIQUID PHASE-TRANSFER CATALYZED REACTION SYSTEM

A. Ben-Zvi [1], P. James McLellan, K. B. McAuley,

Department of Chemical Engineering, Queen's University, Kingston, Canada

Abstract: Identifiability is a structural property of mathematical models. A model is identifiable if the model parameters can be uniquely identified from input-output behaviour. This work focuses on the application of a novel methodology for testing differential algebraic equation systems for identifiability. In this work, a non-linear, differential algebraic model of a liquid-liquid phase-transfer reactor is tested for identifiability using a linearization-based approach. This approach is shown to be useful for the testing of identifiability even for inherently non-linear models. Furthermore, this approach allows the effect of system parameters on the non-linear system to be qualitatively analyzed. *Copyright © 2004 IFAC*

1. INTRODUCTION

Mathematical models are commonly used in almost every area of science and engineering. These models are often characterized by unknown parameters. In order to predict the behaviour of physical systems, these parameters must often be determined experimentally from the input-output behaviour of physical systems. Lack of ability to estimate parameters may be due to several reasons including noisy data, plant-model mismatch, poor choice of experiments, or poor choice of models. In this last case, the model itself is deficient and model parameters cannot be estimated regardless of the quality or quantity of input-output data obtained.

One common model deficiency is lack of identifiability. Consider a mathematical model

$$\mathcal{M}(p) : \mathcal{U} \rightarrow \mathcal{Y}$$

mapping from a set of input signals to a set of output signals and depending on a parameter vector $p \in \mathcal{P} \subseteq \mathbb{R}^s$. Note that this representation of an input-output behaviour implies the existence

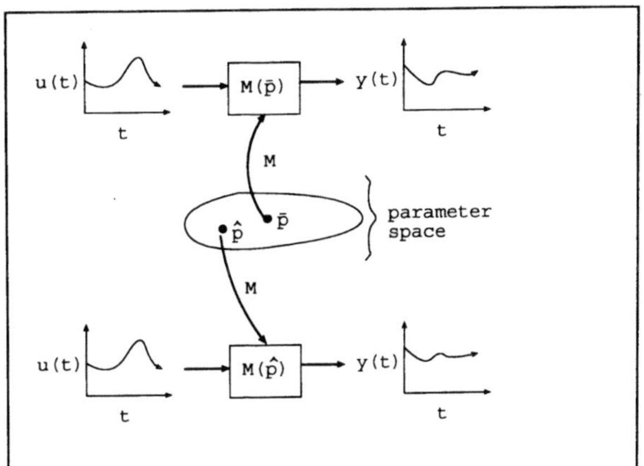

Figure 1.1: Idealized Input-Output Framework

of a mapping $\mathcal{M} : p \rightarrow \mathcal{M}(p)$ that takes a particular parameter p and maps to a particular input-output behaviour $\mathcal{M}(p)$. The situation is summarized in Figure 1.1. It is the properties of the mapping $\mathcal{M}$ that describe the effect of the system parameters on the input-output behaviour. Indeed, Walter and Pronzato (1995) have proposed the following definition for identifiability

[1] partially supported by the Government of Ontario

Definition 1. (Structural Global Identifiability). A model $\mathcal{M}$ is *structurally globally identifiable* if for almost any $\bar{p}, \hat{p} \in \mathcal{P}$,

$$\mathcal{M}(\bar{p}) = \mathcal{M}(\hat{p}) \Rightarrow \bar{p} = \hat{p}$$

where the words *almost any* imply that the identifiability condition holds at every point on a dense subset of $\mathcal{P}$.

Similarly, viewing $\mathcal{P}$ as a subset of $\mathbb{R}^s$ we can define the property of local identifiability as

Definition 2. (Structural Local Identifiability). A model $\mathcal{M}$ is *structurally locally identifiable* if for almost any $\bar{p} \in \mathcal{P}$, and $\hat{p}$ such that $\|\bar{p} - \hat{p}\| < \epsilon$ for any $\epsilon > 0$ the following holds

$$M(\bar{p}) = M(\hat{p}) \Rightarrow \bar{p} = \hat{p}$$

2. LINEAR TIME-INVARIANT DIFFERENTIAL ALGEBRAIC MODELS

Consider the following system.

$$E(p)x'(t) + M(p)x(t) + B(p)u(t) = 0 \quad \text{(1a)}$$
$$y(t) = C(p)x(t) \quad \text{(1b)}$$

where $x(t) \in \mathbb{R}^n$, $u(t) \in \mathbb{R}^\gamma$, $y(t) \in \mathbb{R}^m$ and $E(p)$, $M(p)$, $B(p)$ and $C(p)$ are appropriately sized matrices whose entries are continuously differentiable functions on $\mathcal{P}$. It has been shown by Brenan et al. (1989) that System 1 has a solution if and only if there exists a scalar $\lambda \in \mathbb{C}$ such that

$$det(E(p)\lambda + M(p)) \neq 0$$

From now on, it will be assumed that System 1 is solvable for every $p \in \mathcal{P}$.

We now turn to the problem of expressing $\mathcal{M}(p)$ for system 1 in terms of the system equations. We may take the Laplace transform of System 1, starting from the steady state solution $x(0) = 0$, $u(0) = 0$, $x'(0) = 0$ to obtain the following Laplace-domain input-output representation.

$$Y(s) = C(p)(E(p)s + M(p))^{-1}B(p)U(s)$$

where the transfer function $T(p, s) = C(p)(E(p)s + M(p))^{-1}B(p)$ can be seen as taking an input U to an output Y. For finite dimensional, multi input, multi output (MIMO) systems, $T(p, s)$ can be represented by a a matrix of rational transfer functions. Without loss of generality, we may assume that each of the transfer functions has a monic denominator, and no poles that are also zeros (i.e., all of the possible pole-zero cancellations have been carried out) . Let $s_{ij}(p) = [c_{ij1}(p), \ldots, c_{ij\alpha_{ij}}(p)]^T \in \mathbb{R}^\alpha$ be a vector whose entries are the coefficients of the ij^{th} entry in $T(p, s)$. It has been shown by Bellman and Astrom (1970) that

$$\mathcal{M}(\bar{p}) = \mathcal{M}(\hat{p}) \Leftrightarrow s_{ij}(\bar{p}) = s_{ij}(\hat{p})$$
$$i = 1 \ldots m, \quad j = 1 \ldots \gamma$$

which implies that, by definition 1, for identifiability one need only check that for every $i = 1 \ldots m$, $j = 1 \ldots \gamma$,

$$s_{ij}(\bar{p}) = s_{ij}(\hat{p}) \Rightarrow \bar{p} = \hat{p} \quad (2)$$

The above computations have the advantage that they are relatively easy to perform using computer algebra software such as $Maple^{TM}$.

3. THE PHASE-TRANSFER CATALYZED REACTOR SYSTEM

Chen et al. (1991) proposed a model to describe the behaviour of a liquid-liquid phase-transfer catalyzed reactor (PTC) system. The reactor system and kinetic scheme are shown schematically in Figure 3.1. The model proposed by Chen et al. (1991) is batch, however, in this work, the model is extended to a continuous stirred tank reactor (CSTR) model. A more comprehensive discussion of the model is available in Ben-Zvi et al. (in press 2004). The corresponding model is shown in Table 1. Equations f_1 to f_6 in Table 1 are first-order ordinary differential equations (ODEs), and equations g1 to g6 are algebraic. Equations f_1, f_2 and f_3 are organic-phase dynamic material balances for the reactant (benzyl chloride, RX_o), the phase transfer catalyst (tetrabutylammonium bromide, QY_o) and the product (benzyl bromide, RY_o), respectively. Equations f_4, f_5 and f_6 are total balances on the number of moles of bromine atoms (n_Y), chlorine atoms (n_X) and tetrabutylammonium (TBA) groups (n_Q), respectively, present in all forms, in both phases of the reaction mixture. Equations g_1, g_2 and g_3 define n_Y, n_X and n_Q, respectively, in terms of the moles of pertinent species in both phases. Equation g_4 describes the ionization equilibrium between TBA bromide (QY_w) and its ions in the aqueous phase, and g_5 describes the ionization equilibrium for TBA chloride (QX_w). g_6 describes equilibrium partitioning of TBA chloride between the two phases. Equation y defines measured outputs as RX_o, and RY_o, the organic phase concentrations of benzyl chloride and benzyl bromide, respectively.

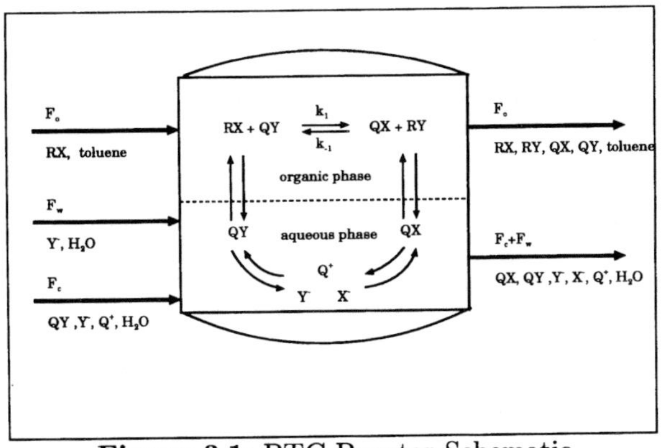

Figure 3.1: PTC Reactor Schematic

Table 1. Continuous PTC Reaction Model

$$\frac{dRX_o}{dt} = -k_1 RX_o QY_o + k_{-1} RY_o QX_o + \frac{F_o}{V_o} RX_{in} - \frac{F_o}{V_o} RX_o \qquad (f_1)$$

$$\frac{dQY_o}{dt} = -k_1 RX_o QY_o + k_{-1} RY_o QX_o + V_o^{-1} A K_{QY} (m_{QY} QY_w - QY_o) - \frac{F_o}{V_o} QY_o \qquad (f_2)$$

$$\frac{dRY_o}{dt} = k_{-1} RX_o QY_o - k_{-1} RY_o QX_o - \frac{F_o}{V_o} RY_o \qquad (f_3)$$

$$\frac{dn_Y}{dt} = F_w Y_{in} + F_c QY_{in} - (F_c + F_w)(Y_w^- + QY_w) - F_o(QY_o + RY_o) \qquad (f_4)$$

$$\frac{dn_X}{dt} = F_o RX_{in} - F_o(RX_o + QX_o) - (F_c + F_w)(X_w^- + QX_w) \qquad (f_5)$$

$$\frac{dn_Q}{dt} = F_c QY_{in} - F_o(QX_o + QY_o) - (F_c + F_w)(QX_w + QY_w + Q_w^+) \qquad (f_6)$$

$$0 = n_Y - V_o(RY_o + QY_o) - V_w(Y_w^- + QY_w) \qquad (g_1)$$

$$0 = n_X - V_o(RX_o + QX_o) - V_w(X_w^- + QX_w) \qquad (g_2)$$

$$0 = n_Q - V_o(QY_o + QX_o) - V_w(Q_w^+ + QX_w + QY_w) \qquad (g_3)$$

$$0 = Q_w^+ Y_w^- - K_{QY}^D QY_w \qquad (g_4)$$

$$0 = Q_w^+ X_w^- - K_{QX}^D QX_w \qquad (g_5)$$

$$0 = QX_o - m_{QX} QX_w \qquad (g_6)$$

$$y(t) = [RX_o \quad RY_o]^T \qquad (y)$$

The batch-reactor model originally developed by Chen et al. (1991) is a special case of this continuous model. Their batch reactor model can be obtained by setting inflow rates F_o, F_w and F_c in the model to zero, along with the accumulation terms in equations f_4 to f_6. In their batch reactor experiments, Chen et al. (1991) also observed n_Y, n_X and n_Q, the total number of moles of bromine atoms, chlorine atoms and TBA groups in the system, which remained constant over time.

Chen et al. (1991) performed a number of dynamic experiments in a liquid-liquid phase-transfer catalyzed (PTC) reactor operated in batch mode. The data from these experiments were used to estimate the kinetic and equilibrium parameters

Table 2. Parameter Values

Estimated parameters from Chen et al. (1991) and specified quantities for PTC reaction model.

Variable	Value	Units
	Parameters	
k_1	2.51×10^{-2}	$m^3/mol \cdot min$
k_{-1}	1.68×10^{-1}	$m^3/mol \cdot min$
$V_o^{-1} A K_{QY}$	1.42×10^5	$mol/m^3 \cdot min$
m_{QY}	6.96×10^{-1}	--
m_{QX}	6.60×10^{-2}	--
K_{QY}^D	6.89×10^5	m^3/mol
K_{QX}^D	8.373×10^5	m^3/mol
	Known Quantities	
V_o	7.50×10^{-5}	m^3
V_w	1.50×10^{-4}	m^3

in their model of the system. These parameters are listed in Table 2. Chen et al. (1991) had difficulties estimating the parameters in their model, so they resorted to a two-stage systematic search technique. Initial values of parameters were selected using a priori information about the magnitudes of the model parameters. Parameter estimates were refined using a DSC-Powell direct search method. Their goals were to develop a model that can predict the behaviour of the system, and to obtain "representative and physically meaningful parameters" (Chen et al., 1991). Comparison of the predictions of their model with experimental data confirmed that the model could capture the behaviour of the PTC reactor system. However, it is not known whether alternative sets of parameter values could give exactly the same model predictions. Indeed, for the parameter estimates obtained by Chen et al. (1991) to be meaningful, the parameter estimates obtained must be unique.

4. LINEARIZATION OF THE PTC MODEL

The reactor equations listed in Table 1 were linearized about the nominal parameter and state values listed in Table 3. Note that the parameter values listed in Table 3 match those estimated by Chen et al. (1991).

The resulting linear time-invariant (LTI) DAE system is given by

$$E\dot{x}(t) + M(p)x(t) + B(p)u(t) = 0 \qquad (3a)$$

$$y(t) = Cx(t) \qquad (3b)$$

where $\dot{x}(t) = \frac{dx}{dt}(t)$, $M(p) = -[\partial f/\partial x, \partial g/\partial x]_{t=0}^T$, $B(p) = -[\partial f/\partial u, \partial g/\partial u]_{t=0}^T$, and

Table 3. State and Input Values

Symbol	Definition	Initial Value	Units
States			
RX_o	$BzCl_o$	9.82×10^2	mol/m^3
QY_o	$TBABr_o$	1.40×10^{-1}	mol/m^3
RY_o	$BzBr_o$	1.02×10^3	mol/m^3
n_Y	Cl_{both}	2.99×10^{-1}	mol/m^3
n_X	BR_{both}	1.50×10^{-1}	mol/m^3
n_Q	TBA_{both}	1.41×10^{-2}	mol/m^3
QY_w	$TBABr_w$	1.40×10^{-1}	mol/m^3
QX_w	$TBACl_w$	5.69×10^{-2}	mol/m^3
QX_o	$TBACl_o$	3.76×10^{-3}	mol/m^3
Y_w^-	Br_w^-	1.49×10^3	mol/m^3
X_w^-	CL_w^-	5.09×10^2	mol/m^3
Q_w^+	TBA_w^+	9.37×10^1	mol/m^3
Inputs			
F_o	feed flow	2.08×10^{-7}	m^3/min
F_w	feed flow	3.96×10^{-7}	m^3/min
F_c	catalyst feed	2.08×10^{-8}	mol/m^3
RX_{in}	reactant in feed	2.00×10^3	mol/m^3
QY_{in}	catalyst in feed	1.88×10^3	mol/m^3
Y_{in}	reactant in feed	2.00×10^3	mol/m^3

$$x(t) = [RX_o, QY_o, RY_o, n_Y, n_X, n_Q, QY_w, Y_w^- , X_w^-, Q_w^+, QX_w, QX_o]^T \quad (4)$$

$$u(t) = [F_o, F_w, F_c, RX_{in}, QY_{in}, Y_{in}]^T$$

$$p = [k_1, k_{-1}, V_o^{-1}AK_{QY}, m_{QY}, m_{QX}, K_{QY}^D, K_{QX}^D]^T$$

$$E = \begin{bmatrix} I_6 & 0 \\ 0 & 0 \end{bmatrix} \quad C = [I_2 \quad 0] \quad (5)$$

where I_j is the j by j identity matrix. The values of $x(t)$ and $u(t)$ are the deviation in value of the states and parameters, respectively, from the values, listed in Table 3, about which the system was linearized. Note that the column-rank of the matrix $B(p)$ is five, although $B(p)$ has six columns. This is because the effect of a change in the value of one of the inputs to the aqueous phase, for example the catalyst inlet concentration Y_{in}, can be offset by varying the other three aqueous-phase inputs and consequently the effective number of input degrees of freedom is five. As a result, for the rest of this work, the value of Y_{in} will be fixed at the nominal value listed in Table 3, and we will assume, without loss of generality, that the other five inputs given by $u(t) = [F_o, F_w, F_c, RX_{in}, QY_{in}]^T$ are adjusted during experimentation.

This system is an index one LTI DAE system that is solvable when evaluated at the nominal parameter values listed in Table 2.

5. IDENTIFIABILITY ANALYSIS OF THE PTC SYSTEM

The identifiability of (the linearized) System 3 was checked using Condition 2. However, using modest computational tools, including a Toshiba Satellite laptop running at 2.66 GHz, and 512Mb of RAM memory, the computation involved in

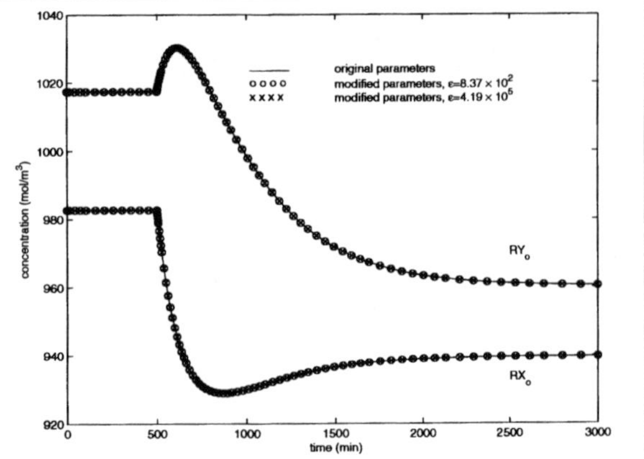

The response of nonlinear System 3 under various parameter values to a step input with $F_o = 1.88 \times 10^{-7}$, $F_w = 4.35 \times 10^{-7}$, $fc = 1.67 \times 10^{-8}$, $RX_{in} = 1.90 \times 10^3$, $QY_{in} = 1.97 \times 10^3$.

Figure 5.1: Simulation Results

checking Condition 2 became intractable. However, locally in the parameter space, Condition 2 can be checked by examining the rank of the Jacobian matrix

$$\mathcal{J}_0 = \left. \frac{\partial [s_{11}(p), \ldots, s_{m\gamma}(p)]^T}{\partial p} \right|_{p=p_0}$$

at p_0, the nominal values listed in Table 3. This computation can easily be carried out numerically even using modest computer hardware. Using Matlab version 6.5, it was found that the rank of $\mathcal{J}_0$ was six, up to the limit of machine precision. It was therefore determined that the PTC system is not identifiable.

The null-space of the matrix $\mathcal{J}_0$, which is the set of all vectors w in $\mathbb{R}^7$ such that $\mathcal{J}_0 w = 0$, was found to be spanned by the vector

$$v = -[-3.71\times10^{-12}, -5.41\times10^{-10}, -1.46\times10^{-8}, 1.90 \times 10^{-9}, 7.89 \times 10^{-8}, 2.38 \times 10^{-3}, 1.00]^T$$

which implies that, $\mathcal{M}(p_0 + \epsilon v) \rightarrow \mathcal{M}(p_0)$ as $|\epsilon|^2 \rightarrow 0$. This behaviour is illustrated in Figure 5.1 where ϵ values of 8.37×10^2 and 4.19×10^5 were used to perturb the parameters, and the resulting step response of *the original, nonlinear PTC model* was plotted. As can be seen from Figure 5.1, even for large values of ϵ, the system behaviour remained virtually identical. This fact shows that even if the system parameters were identifiable, they are not practically estimable, up to machine precision. This is especially relevant in light of the fact that the simulations were performed using no plant-model mismatch, and no added noise.

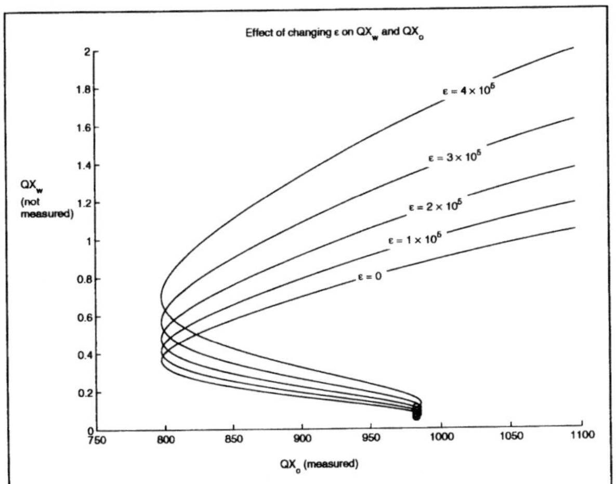

The response of nonlinear System 3 under various parameter values to a step input with $F_o = 8.33 \times 10^{-8}$, $F_w = 1.58 \times 10^{-7}$, $fc = 8.33 \times 10^{-8}$, $RX_{in} = 4.00 \times 10^3$, $QY_{in} = 2.82 \times 10^3$.

Figure 6.1: Simulation Results

6. IDENTIFIABILITY ANALYSIS OF THE ORIGINAL, NONLINEAR PTC SYSTEM

Lack of identifiability of the linearized system does not, in general imply lack of identifiability of the original, nonlinear system. Although the parameter estimates obtained by Chen et al. (1991) have been shown, by example in Figure 5.1, to be suspect, the simulation results themselves do not prove lack of identifiability. In addition, even if the original, nonlinear system is assumed to be un-identifiable, it is not clear from the results of the rank test done on the linear system what is the cause of this lack of identifiability and what can be done to correct this problem.

In order to get some insight into whether the PTC system is identifiable, we make note of the following fact: The largest entry by far in v, both in absolute terms, and relative to the size of the nominal parameter estimates, corresponds to the value of K_{QX}^D. According to Equation g_5, K_{QX}^D affects the ionization of QX_w in the aqueous phase. In addition, we note that none of the aqueous phase concentrations have been directly measured by Chen et al. (1991). In order to determine the effect of different values of ϵ on QX_w, the trajectory of the System in the $QX_w - -QX_o$ plane, in response to a step input was examined. This trajectory is shown for different values of ϵ in Figure 6.1.

The observed response of the PTC system corresponds to the horizontal axis in Figure 6.1. This implies that the experimenter would observe only the horizontal coordinate of the trajectories depicted in Figure 6.1. The trajectories shown in Figure 6.1 are identical except for a vertical translation at every point. This suggests that the trajectories of the PTC system are different for different ϵ values, but produce identical input-output behaviour because the effect of varying ϵ lies entirely in the unobserved states.

In order to test whether some alternative state trajectories can be generated using parameter values different than the ones estimated by Chen et al. (1991), we define new variables describing the unmeasured aqueous-phase concentrations $\widetilde{Q_w^+}$, $\widetilde{X_w^-}$, and $\widetilde{QX_w}$. We also define new, alternative parameter values $\widetilde{K_{QY}^D}$ and $\widetilde{m_{QX}}$. Using Equations g_4 to g_6, and insist that $\widetilde{Q_w^+} + \widetilde{QX_w} = Q_w^+ + QX_w$ and $\widetilde{X_w^-} + \widetilde{QX_w} = X_w^- + QX_w$ so that the input-output behaviour of System 3 remains the same (Ben-Zvi et al., in press 2004), we obtain the following set of five algebraic equations

$$0 = Q_w^+/K_{QY}^D - \widetilde{Q_w^+}/\widetilde{K_{QY}^D} \tag{6}$$

$$0 = \widetilde{Q_w^+}\widetilde{X_w^-} - (K_{QX}^D + \epsilon)\widetilde{QX_w} \tag{7}$$

$$0 = \widetilde{m_{QX}}\widetilde{QX_w} - m_{QX}QX_w \tag{8}$$

$$0 = \widetilde{Q_w^+} + \widetilde{QX_w} - (Q_w^+ + QX_w) \tag{9}$$

$$0 = \widetilde{X_w^-} + \widetilde{QX_w} - (X_w^- + QX_w) \tag{10}$$

in the five unknowns $\widetilde{Q_w^+}$, $\widetilde{K_{QY}^D}$, $\widetilde{X_w^-}$, $\widetilde{QX_w}$, and $\widetilde{m_{QX}}$. Solving the Equations 6 to 10 we have

$$\begin{aligned}
\widetilde{Q_w^+} &= Q_w^+ + (QX_w - \psi(\epsilon)) \\
\widetilde{K_{QY}^D} &= K_{QY}^D \frac{Q_w^+ + QX_w - \psi(\epsilon)}{Q_w^+} \\
\widetilde{X_w^-} &= X_w^- + (QX_w - \psi(\epsilon)) \\
\widetilde{QX_w} &= \psi(\epsilon) \\
\widetilde{m_{QX}} &= m_{QX} QX_w \psi(\epsilon)^{-1}
\end{aligned} \tag{11}$$

where $\psi(\epsilon)$ is the solution to

$$\begin{aligned}
0 = \psi^2 - ((K_{QX}^D + \epsilon) + Q_w^+ + 2QX_w + X_w^-)\psi \\
+ Q_w^+ QX_w + QX_w^2 + Q_w^+ X_w^- + X_w^- QX_w)
\end{aligned} \tag{12}$$

Equation 12 has two roots for each value of ϵ. For $\epsilon = 0$, these roots are $0.569\frac{mol}{m^3}$ and $8.38 \times 10^5 \frac{mol}{m^3}$. The first of these roots corresponds to the nominal value of QX_w, while the second is an un-realistic concentration. We will therefore consider only the values of Ψ which correspond to the lower root of Equation 12.

The value of $\psi(\epsilon)$ is a measure of how different the new and old variables are. Note that, given $\epsilon = 0$, the quantity $\psi(\epsilon) = QX_w$, and the old and new parameter and state values are the same. Also note that changing the value of $\psi(\epsilon)$ amounts to scaling of the unobserved state QX_w. Given small values of ϵ, Equation 12 can be solved for ψ so that new values of parameters K_{QY}^D, m_{QX}, along with $K_{QX}^D + \epsilon$ will generate the same input-output behaviour as the nominal parameter values.

Several conclusions can be drawn from the above result. First, System 3 is not strongly locally identifiable. Secondly, if one is not interested in estimating every parameter, but rather in describing the input-output behaviour of the system, then the value of $\psi(\epsilon)$ can be fixed at the nominal value of m_{QX} so that m_{QX}, which is the extractive-equilibrium constant, is effectively set to unity, and removed from the the model equations. Note that this re-parameterizations is not unique. An alternative re-parametrization is obtained by setting the value of $\psi(\epsilon)$ to $(K^D_{QY})^{-1}$ in order to eliminate the parameter K^D_{QY} from the model equations. This action corresponds to setting the dissociation constant K^D_{QY} to unity. A third alternative is to set the value of $\psi(\epsilon)$ to $\psi(K^D_{QX} - 1)$ in order to eliminate the dissociation constant K^D_{QX} by arbitrarily setting it to unity. Each of the proposed re-parameterizations will yield parameter estimates that describe the input-output behaviour, are unique to that input-output behaviour, but are, almost certainly, not close to the true values of the system parameters.

It the experimenter wishes to estimate the value of each parameter accurately then additional information about the system is needed. Specifically, if the value of the aqueous phase TBA chloride, QX_w, at some time (for example, at the initial condition) is obtained then the value of $\psi(\epsilon)$ would be known and it would be possible to determine unique estimates for all parameter values. Furthermore, given an appropriate experimental design, the parameters could be accurately estimated. Once again, the additional information required to accurately estimate the system parameters is not unique. One could also use measurements of Q^+_w, or X^-_w at some time to accurately estimate each of the model parameters.

7. CONCLUSIONS

Identifiability is a desirable property of a mathematical model. Each of the parameters in an identifiable model can be uniquely identified from input-output data. A continuous nonlinear PTC reaction model based on the batch reactor model proposed by Chen et al. (1991) was tested for identifiability using a linearization based approach. The linearized system was found to be unidentifiable. The information gathered from the linearized system coupled with additional insight into the model equations was used to show that the continuous nonlinear PTC reaction model is also un-identifiable. Simulation results were used to illustrate this lack of identifiability. This lack of identifiability implies that several parameter values produce identical input-output behaviour in the model. Furthermore, this lack of identifiability in the parameters also implies lack of identifiability of a subset of the un-observed states.

ACKNOWLEDGEMENTS

The authors gratefully acknowledge the financial support of the Province of Ontario, Queen's University , and the Natural Sciences and Engineering Research Council of Canada.

REFERENCES

R. Bellman and K. J. Astrom. On structural identifiability. *Mathematical Biosciences*, 7:329 – 339, 1970.

A. Ben-Zvi, P. J. McLellan, and K. B. McAuley. Identifiability study of a liquid-liquid phase-transfer catalyzed reaction system. *AIChE Journal*, in press 2004.

K. E. Brenan, S. L. Campbell, and L. R. Petzold. *Numerical Solutions of Initial-Value Problems in Differential-Algebraic Equations.* North-Holland, 1989.

C. T. Chen, C. Hwang, and M. Y. Yeh. Mathematical modelling of a liquid-liquid phase transfer catalyized reactor system. *Journal of Chemical Engineering of Japan*, 24(3):284 – 290, 1991.

E. Walter and L. Pronzato. Identifiabilities and nonlinearities in nonlinear systems. In A. Fossard and D. Normand-Cyrot, editors, *Modeling and Estimation*, volume 1 of *Nonlinear Systems*, pages 111–143. Chapman & Hall, London, 1995.

ELSEVIER
IFAC
PUBLICATIONS
www.elsevier.com/locate/ifac

LOW-ORDER DYNAMICS IN A LATTICE MODEL OF THIN FILM DEPOSITION, USING NONLINEAR PRINCIPAL COMPONENT ANALYSIS

Martha A. Gallivan *

* *School of Chemical and Biomolecular Engineering*
Georgia Institute of Technology, Atlanta, GA 30332
email: martha.gallivan@chbe.gatech.edu

Abstract: Linear and nonlinear principal component analysis is used to characterize the state space of a kinetic Monte Carlo simulation of thin film deposition. The film's surface is first characterized using spatial correlation functions. This high-dimensional representation is reduced using a combination of linear and nonlinear projection. When nonlinear projection is used, the dynamics of the training and test data can be captured within 2% and 7%, respectively, using three dimensions. In constrast, a three-dimensional linear reduction does not adequately describe the relationship between the training and test data. *Copyright © 2004 IFAC*

Keywords: Nonlinear principal component analysis, low-order dynamics, thin film deposition, kinetic Monte Carlo

1. INTRODUCTION

To manufacture faster computers, device size is approaching the atomic scale, and to create new materials, structure at the nanometer scale is being designed and exploited. In both cases, the manufacturing dynamics are defined by the interactions among many atoms. To optimize and control such processes, dynamic models are needed, which may take the form of molecular dynamics or Monte Carlo simulations (Nieminen, 2002; Starrost and Carter, 2002). The focus of this paper is on the fabrication of ultra-thin films, using lattice Monte Carlo simulations to describe the processing dynamics. Time-varying processes histories are sometimes beneficial in depositing very thin films (Markov *et al.*, 1991; Rosefeld *et al.*, 1995), but they are difficult to optimize due to the extremely high dimension of the dynamic models.

Model reduction provides a framework to reduce the internal dimension of these models, while preserving the map between inputs and outputs of interest. Analytical approaches to reduction have been developed for linear systems (Moore, 1981), but when the system is nonlinear, a data-driven approach is more often used. Snapshots of data are obtained from experiment or simulation, and are then reduced, yielding either a linear subspace via principal component analysis (PCA) (Holmes *et al.*, 1996), or a nonlinear manifold using nonlinear principal component analysis (NLPCA). The former method is computationally efficient, but when the relationship among the states is nonlinear, may not yield a small state dimension. Nonlinear PCA provides an alternative to linear PCA, and may be accomplished through the method of principal curves, neural net training, or a combination of both methods (Bolton *et al.*, 2003; Dong and McAvoy, 1996; Kramer, 1991). Recent application of NLPCA to high-dimensional process dynamics includes the reduction of a polymerization reaction (McLain and Henson, 2000).

Recent research has focused on the reduction of process models for thin film deposition. Straightforward application of existing methods is possible for continuum models of reacting flow (Banks *et al.*, 2002), while the surface evolution may require a non-continuum model. Principal component analysis has been applied to the surface height profile in lattice Monte Carlo simulations (Gallivan *et al.*, 2000; Raimondeau and Vlachos, 2000), but does not provide a predictive model in part due to the stochastic nature of the simulations. An alternative approach to reduction directly addresses this stochastic feature, by describing the evolution of a probability distribution (Gallivan and Murray, 2004), but requires that the set of "typical" surfaces be known. In this work, a systematic reduction method is applied to a Monte Carlo simulation of film growth, using a combination of PCA and NLPCA to identify a reduced set of state variables. This reduced state space could then be used for system identification. Section 2 describes the original high-dimensional model. The reduction method is described in Section 3, and the results of the Monte Carlo reduction are presented in Section 4. Sections 5 and 6 provide a discussion of the method and results, as well as directions for future research.

2. MODEL AND SIMULATIONS

In this work, the physical model for the processing dynamics is a lattice model. Atoms may only take spatial positions on this rigid lattice and, in this work, a cubic lattice is used. Atoms are allowed to take positions in the lattice from the gas phase, or to hop from one lattice site to an adjacent one. The rates of these transitions and there temperature dependence reflect the evolution of germanium films, as described in a previous work (Gallivan and Atwater, 2004).

The lattice model describes the evolution of a probability distribution for the occupancy of the entire lattice. However, due to its extremely high dimension, individual stochastic realizations are instead performed using the kinetic Monte Carlo (KMC) method. While Monte Carlo simulations typically describe equilibrium properties, kinetic Monte Carlo simulations capture the correct evolution in time (Bortz *et al.*, 1975). Periodic boundary conditions are used to eliminate edge effects, since the actual growth surface is much larger than the simulation domain. A KMC simulation of a 300×300 site surface is shown in Figure 1. Only a portion of the full simulation domain is shown so that the individual atoms can be resolved. Dark-colored atoms are on the edge of an atomic-height step, while light atoms have four side neighbors and comprise atomically-smooth terraces.

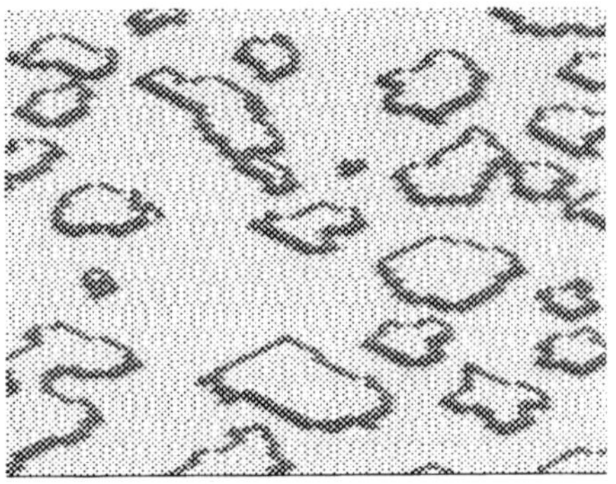

Fig. 1. Picture of a KMC simulation.

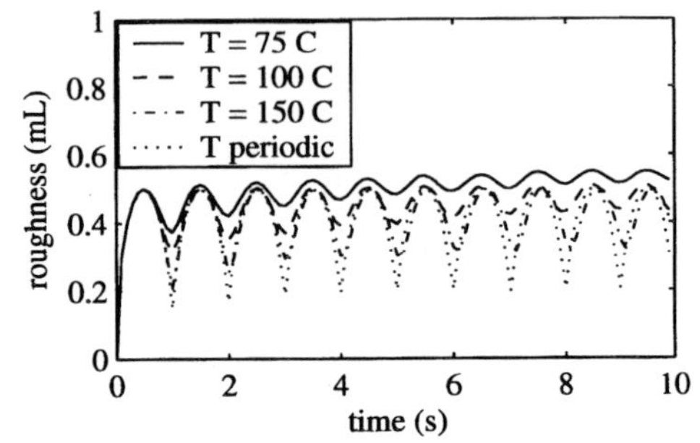

Fig. 2. Step density oscillations during epitaxial growth.

The model problem considered in this work is the description of the surface height profile, or surface morphology. As atoms adsorb onto the surface from the surrounding gas, clusters first nucleate on the surface, grow, and then coalesesce. This oscillatory behavior is illustrated by Figure 2, in which the time to deposit one monolayer (mL) of material is 1 s. When the temperature is high, the hopping rate of atoms increases, leading to fewer clusters, and a smoother surface. The oscillatory behavior decays faster at lower temperatures, because clusters nucleate on the second layer prior to full coalescence of the first layer. Periodic growth temperature can be used to achieve a smoother surface than any constant temperature in this range, as shown in Figure 2. The temperature is lowered to 75 C during the first 20% of each layer, and then raised to 150 C during the remaining 80%. As a result, many smaller clusters nucleate during the beginning of the layer, after which temperature is raised to fill in the gaps between clusters (Markov *et al.*, 1991; Rosefeld *et al.*, 1995). The goal of this paper is to generate a reduced set of coordinates that describes the dynamics of both constant and time-varying growth temperature.

While each KMC simulation generates a surface height profile, the location of individual clusters on the surface changes from realization to realization because the simulations are stochastic. Thus, alternate measures are needed to characterize each realization, and to compare simulations generated

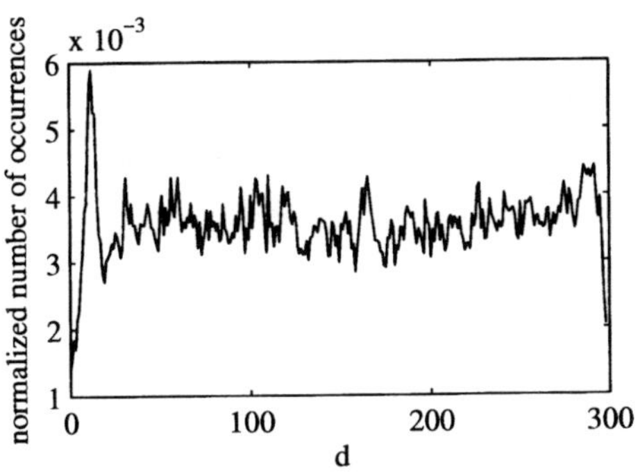

Fig. 3. Step-step correlation function (up-down) for 0.5 mL at 150 C.

under the same or different process conditions. One simple solution is to perform a Fourier transform of each surface to identify key length scales associated with cluster size and spacing. Alternatively, spatial correlation functions be used. In fact, the height-height correlation function can be computed directly from the Fourier transform (Tong and Williams, 1994). In this work the spatial correlation function for surface steps is used instead, since the films are extremely thin and are best characterized by the location and spacing of atomic-height steps. Defining the surface height to be $h(x,y)$ for site (x,y), the presence of an "up" step in the x and y directions is computed as

$$s_{ux}(x,y) = 1 \quad \text{if } h(x+1,y) > h(x,y) \tag{1}$$
$$= 0 \quad \text{otherwise} \tag{2}$$
$$s_{uy}(x,y) = 1 \quad \text{if } h(x,y+1) > h(x,y) \tag{3}$$
$$= 0 \quad \text{otherwise.} \tag{4}$$

"Down" steps are computed similarly, by reversing the inequalities. The step-step correlation function then is defined to be

$$ssc_{kl}(r) = \frac{1}{n_x\, n_y} \times \sum_{i=1}^{n_x}\sum_{j=1}^{n_y}\{s_{kx}(x_i,y_j)s_{lx}(x_i+r,y_j) + s_{kx}(x_i,y_j)s_{lx}(x_i,y_j+r)\}, \tag{5}$$

where k and l may each take values of u or d to denote the up and down directions, respectively. Due to the discrete nature of the simulations, x, y, and r are positive integers. An example of a step-step correlation function is shown in Figure 3. The peak in the correlation function represents the typical cluster size, since a cluster is composed of an up-step followed by a down-step. To characterize a surface during a KMC simulation, all four step-step correlation functions are computed, and thus represent the full state of the system. In the case of the 300×300 ($n_x = n_y = 300$) surfaces used here, the state dimension is thus 1200. The underlying physical assumption is that the evolution of two surfaces with the same initial step-step correlation function is the same, as measured by the step-step correlation function, and is justified based on physical intuition only.

3. METHOD FOR STATE SPACE REDUCTION

The method used here for reduction is analogous to other data-driven methods for state-space reduction (Holmes *et al.*, 1996; McLain and Henson, 2000). Snapshots are collected from a high-dimensional simulation, and are then projected onto a space of reduced dimension. The four steps are outlined and briefly described:

- Step 1: Collect snapshots.
  The snapshots of the surface height profile are collected at regular intervals. These snapshots *must* represent the entire range of surfaces to be considered (which is true for any black-box system identification method).
- Step 2: Express in physically meaningful coordinates.
  Earlier attempts to reduce the state dimension of KMC simulations (Gallivan *et al.*, 2000; Raimondeau and Vlachos, 2000) focused on the surface height profile. This set of coordinates changes from realization to realization because of the spatial randomness of the simulations, and does not yield sufficiently predictive modes. Here, a spatial correlation function is used instead, which is sensitive to the distribution of features, but not to their spatial locations. This provides a set of physically meaningful (although high-dimensional) coordinates to classify the state of the system.
- Step 3: Perform linear projection.
  A linear principal component analysis is first performed on the matrix of snapshots to identify a reduced dimensional linear subspace. One could theoretically skip this step and perform the nonlinear projection on the full snapshot matrix, but in practice it is advantageous to perform a linear projection first. The singular value decomposition used for PCA can easily handle large data matrices, while the nonlinear optimization used in Step 4 is more computationally demanding.
  The linear modes determined by PCA are ordered in terms of the energy associated with each mode. However, it is not clear in this example that the highest energy mode is also the most important in predicting the evolution of the state. Consequently, the PCA modes are rescaled by their singular values, so that the projection *coefficients* for each mode have a typical value of order 1.

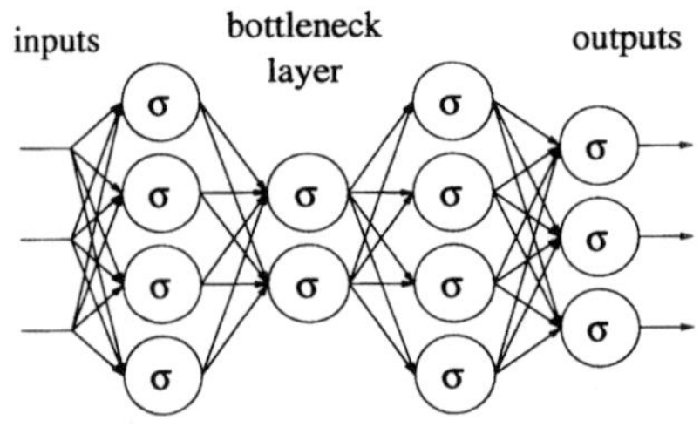

Fig. 4. Neural net with 3 inputs and outputs, 4 nodes in the mapping and demapping layers, and 2 nodes in the bottleneck layer. This net is used to reduce data from 3 dimensions to 2 dimensions.

These rescaled coefficients are then used as inputs for the nonlinear projection.

- Step 4: Perform nonlinear projection.
  The NLPCA method of Kramer (Kramer, 1991) is used to perform the nonlinear projection. This method relies on a neural net to describe the principal components, as illustrated in Figure 4. The net is trained to approximate the identity map for the data matrix, despite a hidden bottleneck layer with a reduced number of nodes. Once the net is successfully trained, the outputs of the bottleneck layer are the reduced set of coordinates.

4. RESULTS

The reduction approach presented in Section 3 is now applied to the model of Section 2. In the previous study of germanium film growth (Gallivan and Atwater, 2004), it was noted that the surfaces accessed during constant temperature growth also represented the typical surfaces seen under time-varying temperature. The benefit of the time-varying input was simply to access the smoother states after 10 s of growth, which could only be accessed under constant temperature at earlier times. Thus, constant temperature simulations were run at temperatures of T = 75, 87, 100, 125, and 150 C for 10 s of growth. They were sampled at intervals of 0.1 s, yielding 500 snapshots.

After performing the KMC simulations and collecting the snapshots, the step-step correlation function of equation (5) was computed for each snapshot. All computations associated with the reduction were performed using Matlab. The data was collected into a 1200×500 training matrix and its singular value decomposition was computed to obtain the linear principal component analysis. The corresponding singular values are shown in Figure 2. The singular values quickly decay by two orders of magnitude, and then level off. Thus, the first six linear modes are used to represent the surface structure, while the remainder are neglected and assumed to be noise. This issue will be revisited in Section 5. The error associated with this truncation is quantified as

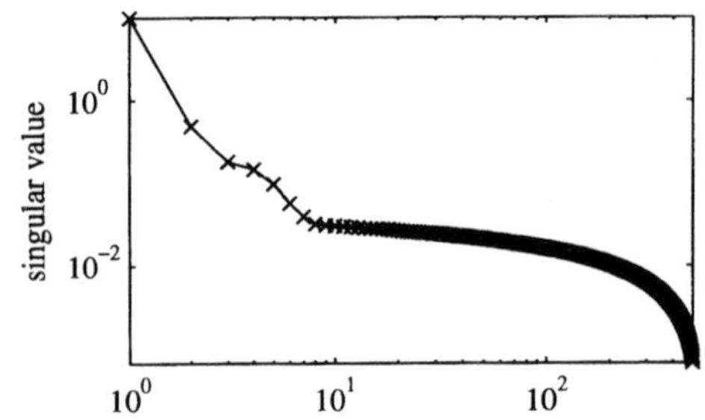

Fig. 5. Singular values of the training matrix.

$$e = \frac{\sum_{i=1}^{n_s}\sum_{j=1}^{n_i}(y(i,j) - \hat{y}(i,j))^2}{\sum_{i=1}^{n_s}\sum_{j=1}^{n_i} y(i,j)^2}, \tag{6}$$

where n_s is the number of snapshots, n_i is the number of elements in the data vector, $y(i,j)$ is the (i,j) element of a data matrix, and $\hat{y}(i,j)$ is a reconstruction of $y(i,j)$. This is the same metric used to assess the *energy* captured in PCA.

The first six linear modes retained here capture 0.9993 of the energy of the constant temperature data matrix, and also 0.9992 of the energy for the periodic input pictured in Figure 2. In the remainder of the paper, the former matrix will be referred to as the training matrix, and the latter as the test matrix. It should also be noted that the first three modes capture 0.9990 of the energy of the training matrix and the test matrix. However, it is not clear that the energy metric is the most appropriate one, since small energy features in the step-step correlation function could be critical in determinig its future evolution. Thus, before performing NLPCA, each of the six PCA modes is multiplied by its corresponding singular value, so that the projection coefficients for all six modes will be (typically) of order 1.

The nonlinear principal component analysis is performed using Matlab's Neural Net toolbox. The training matrix is used to compute a weight and a bias for a sigmoid function associated with each node, using a scaled conjugate gradient method and a maximum of 5000 iterations. Random initial guesses were used for the weights and biases. In all cases, 10 mapping and 10 demapping nodes were used for the first and third layers of the neural net, while the number of bottleneck nodes was varied. The number of inputs and outputs is six, corresponding to the number of linear principal components retained. The goal of the training is to obtain the identify map. If the identity can be well-approximated using a reduced number of bottleneck nodes, then the outputs of the bottleneck nodes may be taken as a reduced set of coordinates for the training matrix. Furthermore, if the error is also small for the test matrix, then

Table 1. Reconstruction error associated with linear and nonlinear PCA.

Data:	training		test	
Projection:	linear	nonlinear	linear	nonlinear
Dimension				
2	0.48	0.17	0.76	0.44
3	0.29	0.02	0.64	0.07
4	0.05	0.01	0.10	0.03
6	0	0.00	0	0.02

the net can be viewed as predictive and may be used to interpolate the reduced coordinates for intermediate data points not used in the training.

The error, as defined by equation (6), is shown in Table 1. The error for the two-dimensional NLPCA is high, but is greatly reduced when a third bottleneck node is added. Adding additional nodes improves the reconstruction, and should theoretically bring the error to zero when the number of bottleneck nodes equals the number of inputs (6 in this case). The error for the linear reconstruction is associated with a second PCA performed on the rescaled coefficients of the original PCA. Since each mode should contribute equally, it should not be expected that a reduced number would provide a good approximation.

Figure 6(a) shows the 3-dimensional NLPCA coordinates for the training and test data. Consistent with the small error, these coordinates reflect the physical relationship among the data points. The various temperatures are ordered from top to bottom, and the oscillatory behavior of Figure 2 is reflected in a circular arrangement of the data. As the oscillations decay, the surfaces spiral out from the center. Note that some of these cycles obscurred by the two-dimensional rendering of the data. This nonlinear projection also captures the test data, in which the system moves back and forth between the training data at the minumum and maximum temperatures.

The linear projection shown in Figure 6(b) corresponds to the errors reported in Table 1. The constant temperature training data is not clearly organized as with NLPCA, and the error is substantial (64%) for the test data. For comparison, a third plot is shown as Figure 6(c), representing the linear reconstruction of the data on the *original* three principal components, prior to the rescaling. The reconstruction does not clearly show a physically meaningful relationship among the snapshots. Perhaps more significantly, the test data does not interpolate among the training data, but instead is extrapolating out into a region not used for identification.

5. DISCUSSION AND FUTURE WORK

Nonlinear projection captures the behavior observed in the KMC simulations, using a smaller set of coordinates than is needed for linear PCA. This reduction of the state space may lead to improved understanding of the dominant dynamics of thin film morphology evolution, and also could be used for system identification. For example, the reduced state space could be discretized, with each point representing an important group of surfaces with similar properties. The transitions between groups could then be computed using KMC simulation data (Gallivan and Murray, 2004). Other identification methods could also be applied to this low-order system.

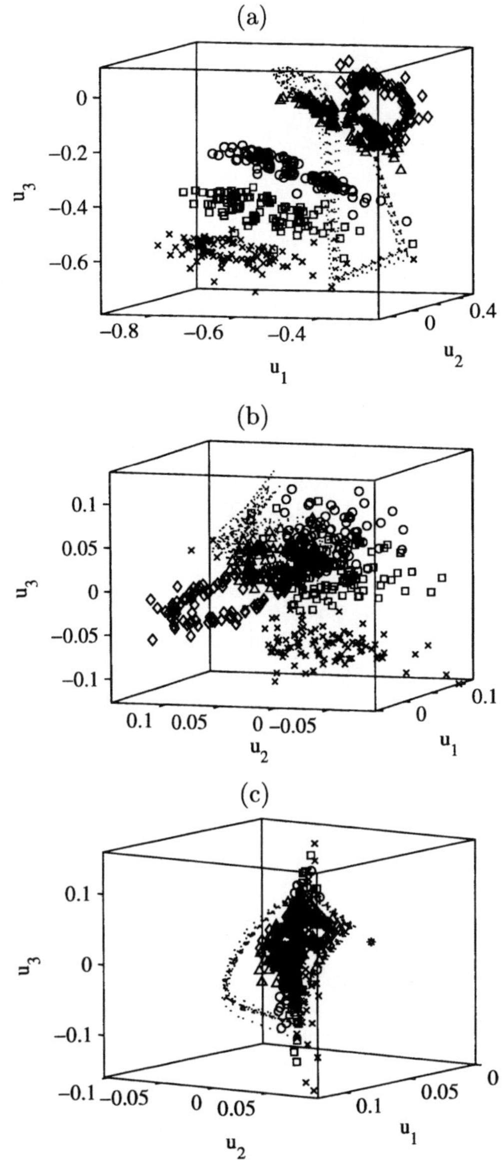

Fig. 6. 3-D projections (a) nonlinear, (b) linear with scaled coefficients, (c) linear with unscaled coefficients. The constant temperature snapshots used in the training matrix are marked by symbols—75 C (×), 87 C (square), 100 C (o), 125 C (△), and 150 C (◇). The evolution under the test data (periodic temperature) is represented by the dashed line.

This reduction method is largely automated, although some user input is required to select the number of modes in the linear projection. Open issues include the role of noise, and the importance of small energy modes which may be obscurred by noise. To reduce the noise, ensembles of KMC simulations could be collected. However, the noise can never be completely eliminated, so one will generally need to truncate modes from the PCA that have nonzero singular values. This issue remains a subject for future study.

6. CONCLUSIONS

A set of reduced-order coordinates was computed for a kinetic Monte Carlo simulation of thin film deposition. The film surface was characterized with a spatial correlation function describing the relationship between atomic-height surface steps. Principal component analysis was used to compute linear subspaces of reduced dimension for the correlation function, while nonlinear principal component analysis was accomplished using neural nets. A dimension of three provided a good approximation of the original dynamics when the nonlinear principal component analysis was used (2% error in training data, 7% in test data), while the dynamics were not adequately captured by a 3-dimensional *linear* projection. This preliminary study indicates that low-order behavior exists in the stochastic, high-dimensional Monte Carlo simulations, but it may not always be best captured using linear principal component analysis.

ACKNOWLEDGMENTS

The author thanks Georgia Tech for financial support, and I.G. Kevrekidis for fruitful discussions.

REFERENCES

Banks, H.. T., S. C. Beeler, G. M. Kepler and H. T. Tran (2002). Reduced order modeling and control of thin film growth in an HPCVD reactor. *SIAM Journal on Applied Mathematics* **62**, 1251–1280.

Bolton, R. J., D. J. Hand and A. R. Webb (2003). Projection techniques for nonlinear principal component analysis. *Statistics and Computing* **13**, 267–276.

Bortz, A. B., M. H. Kalos and J. L. Lebowitz (1975). A new algorithm for Monte Carlo simulations of Ising spin systems. *Journal of Computational Physics* **17**, 10–18.

Dong, D. and T. J. McAvoy (1996). Nonlinear principal component analysis—based on principal curves and neural networks. *Computers and Chemical Engineering* **20**, 65–78.

Gallivan, M. A. and H. A. Atwater (2004). Design of a film surface roughness-minimizing molecular beam epitaxy process by reduced-order modeling of epitaxial growth. *Journal of Applied Physics* **95**, 483–489.

Gallivan, M. A. and R. M. Murray (2004). Reduction and identification methods for markovian control systems, with application to thin film deposition. *International Journal of Robust and Nonlinear Control* **14**(2), 113–132.

Gallivan, M. A., R. M. Murray and D. G. Goodwin (2000). The dynamics of thin film growth: a modeling study. In: *CVD XV: Proceedings of the Fifteenth Symposium on Chemical Vapor Deposition* (M. D. Allendorf and M. L. Hitchman, Eds.). Vol. 616. Electrochemical Society. pp. 168–175.

Holmes, P., J. L. Lumley and G. Berkooz (1996). *Turbulence, Coherent Structures, Dynamical Systems, and Symmetry.* Cambridge University Press. Cambridge, UK.

Kramer, M. A. (1991). Nonlinear principal component analysis using autoassociate neural networks. *AIChE Journal* **37**, 233–243.

Markov, V. A., O. P. Pchelyakov, L. V. Sokolov, S. I. Stenin and S. Stoyanov (1991). Molecular beam epitaxy with synchronization of nucleation. *Surface Science* **250**, 229–234.

McLain, R. B. and M. A. Henson (2000). Principal component analysis for nonlinear model reference adaptive control. *Computers and Chemical Engineering* **2000**, 99–110.

Moore, B. C. (1981). Principal component analysis in linear systems: controllability, observability, and model reduction. *IEEE Transactions on Automatic Control* **26**, 1732.

Nieminen, R. M. (2002). From atomistic simulation toward multiscale modeling of materials. *Journal of Physics: Condensed Matter* **14**, 2859–2876.

Raimondeau, S. and D. G. Vlachos (2000). Low-dimensional approximations of multiscale epitaxial growth models for microstructure control of materials. *Journal of Computational Physics* **160**, 564–576.

Rosefeld, G., N. N. Lipkin, W. Wulfhekel, J. Kliewer, K. Morgenstern, B. Poelsema and G. Comsa (1995). New concepts for controlled epitaxy. *Applied Physics A* **61**, 455–466.

Starrost, F. and E. A. Carter (2002). Modeling the full monty: Baring the nature of surfaces across time and space. *Surface Science* **500**, 323–346.

Tong, W. M. and R. S. Williams (1994). Kinetics of surface growth—phenomenology, scaling, and mechanisms of smoothening and roughening. *Annual Review of Physical Chemistry* **45**, 401–438.

ELSEVIER
IFAC
PUBLICATIONS
www.elsevier.com/locate/ifac

PREDICTIVE CONTROL OF THIN FILM SURFACE MICROSTRUCTURE IN A COMPLEX DEPOSITION PROCESS

Dong Ni and **Panagiotis D. Christofides** [1]

Department of Chemical Engineering
University of California, Los Angeles, CA 90095-1592

Abstract: In this work, a complex deposition process including two types of molecules whose growth behaviors are very different and influenced by long range interactions is investigated. The study of this process is motivated by recent experimental results on the growth of high-κ dielectric thin films using plasma-enhanced chemical vapor deposition (PECVD). A multi-component kinetic Monte-Carlo (kMC) model is developed for the deposition. The dependence of the surface microstructure of the thin film, such as island size and surface roughness, on substrate temperature are studied. The surface morphology is found to be strongly influenced by these two factors and growth regimes governed by short and long range interactions are observed. Furthermore, a kMC model-predictive control scheme which uses the substrate temperature to control the final surface roughness of the thin film is proposed. The closed-loop simulation results demonstrate that robust deposition with controlled thin film surface roughness can be achieved under the proposed kMC model-predictive controller. *Copyright © 2004 IFAC*

1. INTRODUCTION

The industrial demands for advanced materials having certain properties, have driven the development of thin film technology. Various deposition methods have been developed and widely used to prepare thin films, however, the dependence of the thin film properties, such as uniformity, composition and microstructure, on the deposition conditions is a severe constraint on reproducing thin film's performance. Thus, real-time feedback control of thin film deposition becomes increasingly important in order to meet the stringent requirements on the quality of thin films. Significant research efforts have been made on the feedback control of thin film deposition processes with emphasis on control of film spatial uniformity in rapid thermal processing (RTP) (Theodoropoulou *et al.*, 1999; Christofides, 2001) and plasma-enhanced chemical vapor deposition (PECVD) (Armaou and Christofides, 1999).

In addition to achieving spatially uniform deposition of thin films, one would like to control film properties that characterize film quality. While deposition uniformity control can be accomplished on the basis of continuum type distributed models, precise control of film properties requires models that predict how the film state (microscopic scale) is affected by changes in the controllable process parameters (macroscopic scale). Kinetic Monte-Carlo (kMC) simulation provides a framework for modelling the effect of macroscopic process variables on the thin film microstructure and it has been widely used to simulate CVD processes (see (Battaile and Srolovitz, 2002) for a review of kMC simulation of CVD). However, the majority of these works have focused on studying the growth kinetics or interface structure while only a few works (Reese *et al.*, 2001) have addressed the computational efficiency which strongly affects the use of such kMC models in real-time feedback control systems. Recently, a methodology for feedback control of thin film growth using kMC models was developed in (Lou and Christofides, 2003*a*) and (Lou and Christofides, 2003*b*). The method was successfully applied to control surface roughness in a *GaAs* deposition process using an experimentally determined kMC process model (Lou and Christofides, 2004). Other approaches have also been developed to: (a)

* Financial support for this work from the NSF (ITR), CTS-0325246, is gratefully acknowledged.
[1] Corresponding author (email: pdc@seas.ucla.edu).

identify linear models from outputs of kinetic Monte-Carlo simulators and perform controller design by using linear control theory (Siettos *et al.*, 2003), and (b) construct reduced-order approximations of the master equation (Gallivan and Murray, 2003).

However, among these computationally attractive models, most of them consider only single component system, and long range interactions have not been modelled explicitly. In reality, more than one species participate in the film growth in most CVD processes. Moreover, direct long range interactions (Einstein, 1996) and substrate-mediated long range interactions (Merrick *et al.*, 2003) are very important in many of these processes. For example, in the PECVD ZrO_2 process, there is a large number of different species present in the gas phase during the deposition, and many of them participate in the thin film growth, particularly, zirconium hydroxide and hydrocarbon species (see (Cho *et al.*, 2002) for detailed experimental results). Moreover, recent experimental results (Cho *et al.*, 2003) have shown that, when zirconium hydroxides are the dominant species in the gas phase, the deposited ZrO_2 thin film has a very smooth surface with a roughness value less than half ZrO_2 monolayer, which suggests that the zirconium hydroxide species tend to uniformly cover the substrate surface. On the other hand, when hydrocarbons dominate the gas phase, the deposited ZrO_2 thin film has a very rough surface characterized by big islands, which suggests that the aggregation of the hydrocarbon species on the substrate surface, as a result of long range interactions, is quite significant. It is quite obvious that a single component kMC model considering only short range interactions is inadequate to describe the thin film growth in this process. Therefore, a computationally efficient kMC model of a heterogeneous deposition process in which long range interactions are accounted for is needed.

In this work, a complex deposition process including two types of molecules whose growth behaviors are very different and influenced by long range interactions is investigated. The study of this process is motivated by recent experimental results on the growth of high-κ dielectric thin films using PECVD (Cho *et al.*, 2002; Cho *et al.*, 2003; Ni *et al.*, 2003). A multi-component kMC model is developed for the deposition. The dependence of the surface microstructure of the thin film, such as island size and surface roughness, on substrate temperature are studied. The surface morphology is found to be strongly influenced by these two factors and growth regimes governed by short and long range interactions are observed. Furthermore, a kMC model-predictive control scheme which uses the substrate temperature to control the final surface roughness of the thin film is proposed. The closed-loop simulation results demonstrate that robust deposition with controlled thin film surface roughness can be achieved under the proposed kMC model-predictive controller.

2. SURFACE MICROSTRUCTURE MODEL FOR THIN FILM GROWTH

Deposition processes such as PECVD, often involve large numbers of participating species with heterogeneous growth behaviors. Here, we consider a heterogeneous deposition process in which two types of molecules, type A and type B, which behave very different from each other are considered. Type A molecule is significantly affected by long range attractions and tends to aggregate with other A molecules into clusters, i.e., favors Volmer-Weber (VW) growth mode. Hydrocarbon molecules generated from the decomposition of metal-organic (MO) precursors in a PECVD process are good exàmples of such type. Type B molecule favors surface sites of local minimum height, which usually results in Frank-van der Merwe (FM) type of film growth. Metal oxides or hydroxides originated from the MO precursors may behave similar to molecules of type B as discussed in the introduction.

The geometry of the deposition process is shown in Fig. 1. The gas flux is perpendicular to the substrate surface. Flux composition, i.e., flux of A and B, in terms of the number of molecules encountered per unit time per surface site, are taken as macro scale process parameters, that can be measured in real-time (via mass spectrometer for example). Both A and B can diffuse from the gas phase onto the substrate, however, B type molecules settle to surface sites of local minimum height simultaneously during adsorption. Surface migration and desorption processes are ignored, while hopping of A type molecules is allowed. Surface reactions are not explicitly considered in this process, however, the long range behavior of A and the local minimum adsorption behavior of B could be consequences of surface reactions (i.e., surface mediated).

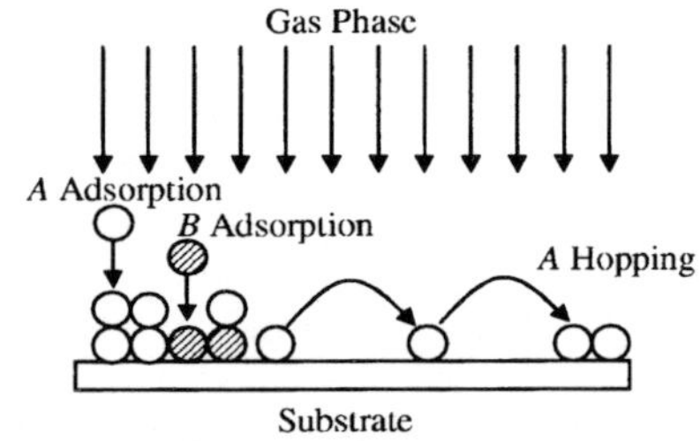

Fig. 1. Thin film growth process.

The growth micro-process model is constructed based on a standard kMC scheme (Gillespie, 1976) which assumes the growth process to be a Poisson process and its dynamics to be governed by the master equation which describes the evolution of transitions between the probabilities of the surface being in specific micro-configurations. Monte-Carlo simulation is used to obtain unbiased realizations of the master equation numerically. A simple cubic lattice structure is used and the simulated surface domain is a square grid of 100 lattice points by 100 lattice points. To improve computational efficiency, the solid-on-solid assumption is made.

Table 1. Model parameters.

Hopping Freq. Const.	k_{h0}^{A}	10^{13}	s^{-1}
Hopping Energy (bottom)	E_s^A	0.8	eV
Hopping Energy (side)	E_n^A	0.2	eV
Attraction Range	l_a	20	units

We consider multilayer growth and assumed that the adsorption rates of A (w_a^A) and B (w_a^B) can be obtained through real-time measurements as discussed above. Considering only the interactions of the first nearest neighbors (4 side neighbors and 1 bottom neighbor) to determine the hopping rate at a specific site, the hopping rate of a molecule of type A on the surface with n first nearest neighbors is given by

$$w_h^A(n) = k_{h0}^A \exp(-\frac{E_s^A + nE_n^A}{kT}) \qquad (1)$$

where k_{h0}^A is the hopping event frequency constant, E_s^A and E_n^A are the energy barriers associated with surface hopping of A for bottom and side neighbors respectively (we note that for simplicity we do not distinguish the neighboring molecules of different types).

To incorporate different growth behaviors into the kMC simulation, we set up two rules for A type molecule aggregation and B type molecule local minimum adsorption, respectively. For A type molecules, we enforce a rule on the hopping process, specifically, the hopping direction of a specific A molecule is determined by comparing the distance-weighted sum of all the A molecules in each direction and pick the one with largest value of the sum so that A molecules will aggregate from long range. Specifically, the weighted sum, for example, the weighted sum in the positive x direction of an A located at surface lattice point (x_0, y_0), $N_{h,+x}^A(x_0, y_0)$, is computed as follows:

$$N_{h,+x}^A(x_0,y_0) = \sum_{x=1}^{l_a} \sum_{y=-x}^{x} S_{(x_0+x,y_0+y)}^A (1 - \frac{\sqrt{x^2+y^2}}{l_a}) \qquad (2)$$

where l_a is the maximum range of attraction, and the value of the occupancy factor $S_{(i,j)}^A$ is unity when the surface site (i, j) is occupied by an A and zero otherwise.

For B type molecules, we enforce a rule on the adsorption process. When the target site does not have the local minimum height, a B type molecule will be adsorbed onto one of the neighboring sites that has the local minimum height. In this work, only the 4 first nearest neighbor sites and 4 second nearest neighbor sites are considered. In addition, the sticking probability of type B molecule on surface site occupied by type A molecules is considered very small (5% in this study). This is because when this sticking probability is close to unity, the surface would be smoothened by type B molecules independently of the presence of type A molecules, and thus, the dynamics of the two-component deposition would not be observable. All other sticking probabilities are considered to be unity for simplicity.

The parameters, k_{h0}^A, E_s^A, E_n^A and l_a in the model can be determined by experiments. The parameters of the process studied in this work are shown in Table 1. When the lattice is set and the rates of the three events (A adsorption, B adsorption, A hopping) are determined based on measurements or its corresponding rate expression (1), a kinetic Monte-Carlo simulation is executed following the algorithm reported in (Reese *et al.*, 2001). First, the surface A molecules are grouped into five classes based on the number of side neighbors (from 0 to 4 side neighbors); in each class, the molecules have the same hopping rates and the A and B adsorption rates are site independent. Then, a random number is generated to select an event to be run based on the rates; if the event is A hopping, the class in which the event will happen is also selected. After that, a second random number is generated to select the site where the event will be executed; if the event is A or B adsorption, the site is randomly picked from sites in the entire lattice; if the event is A hopping, the site is randomly picked from the list of the sites in the selected class. After the site is selected, the MC event is executed. If the event is adsorption, it is executed by adding one molecule on the selected site (B adsorption rule is applied if the event is B adsorption); if the event is A hopping, the A type molecule on the site is moved to the next site in the direction selected by the hopping rule. Upon an executed event, a time increment τ computed based on Eq. 2 is added to the process time t. The lattice is set to be periodic to satisfy the mass balance of the hopping molecules.

3. SIMULATION RESULTS AND DISCUSSION

We simulate the multi-component complex deposition using the proposed process model (see (Ni and Christofides, 2004) for results of single-component deposition). Both A and B types of molecules are present in the gas phase and the relative ratio of the two species is set for simplicity to be unity in the simulated case. The effect of substrate temperature (in the range of 300 K to 440K) on the surface microstructure of the deposited thin films is investigated. The root-mean-square (RMS) roughness definition is used to evaluate roughness of the thin film surface in this work.

Figs. 2 and 3 show the surface morphologies of thin films obtained by depositions at low ($T = 320$ K) and high ($T = 440$ K) substrate temperatures, respectively. The thin film deposited under low substrate temperature has a high island density but a small lateral island size, while the thin film deposited under high substrate temperature has a low island density but a large lateral island size. Such difference in surface morphology between the two thin films is similar to the single component scenario in which only A type of molecules are present in the gas phase (see (Ni and Christofides, 2004) for detailed explanation).

Fig. 4 shows the surface roughnesses of thin films deposited at different substrate temperatures. It can be clearly seen that there are two temperature regimes

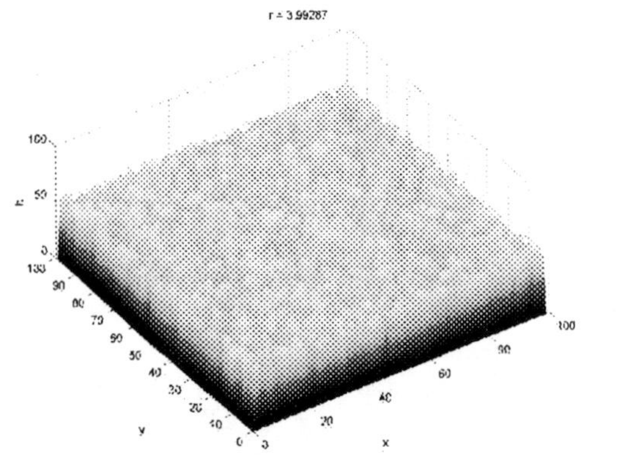

Fig. 2. Surface of a thin film deposited with $w_a^A = 0.05s^{-1}$, $w_a^B = 0.05s^{-1}$ and $T = 320K$ - $t = 900s$.

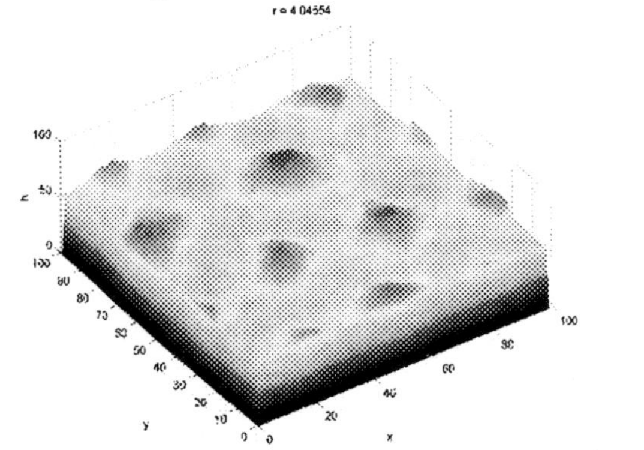

Fig. 3. Surface of a thin film deposited with $w_a^A = 0.05s^{-1}$, $w_a^B = 0.05s^{-1}$ and $T = 440K$ - $t = 900s$.

in which thin film growth is quite different. In the low temperature regime, the surface roughness drops with increasing temperature, in the high temperature regime, the surface roughness rises with increasing temperature (however, the surface roughness drops again when the substrate temperature is very high when stable surface islands start to coalesce and form islands with lateral dimension larger than the range of attraction). The transition from the low temperature regime to the high temperature regime corresponds to the change in growth process from short range interaction dominant to long range attraction dominant.

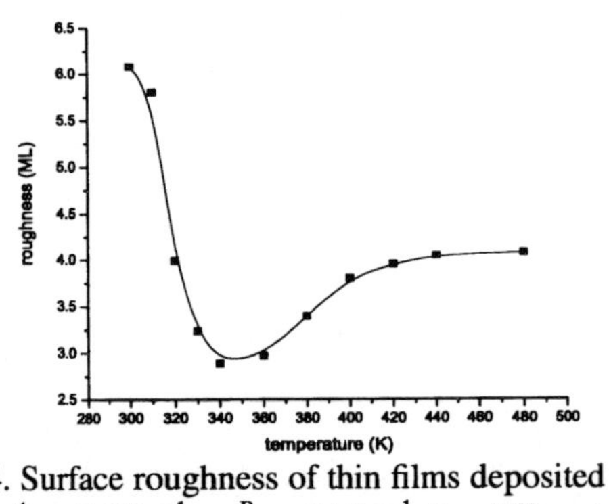

Fig. 4. Surface roughness of thin films deposited with $w_a^A = 0.05s^{-1}$, $w_a^B = 0.05s^{-1}$ for different substrate temperature - $t = 900s$.

4. FEEDBACK CONTROL

To obtain thin films of desired and reproducible surface microstructure, it is necessary to operate the deposition processes under feedback control. In this process, since there are two temperature regimes for which thin film surface morphologies are quite different, control of surface roughness in these two regimes is considered separately. Here, we only focus on the high temperature regime (see (Ni and Christofides, 2004) for results on feedback control in the low temperature regime), in which long range interactions are significant, and traditional control strategies such as proportional integral (PI) control lead to poor closed-loop performance. The final surface roughness of the thin film is selected as the controlled variable and the manipulated variable is chosen to be the substrate temperature T, which is restricted within the range of 340 K $\sim$ 420 K. The major disturbance to the deposition process is the variation of the gas phase composition.

4.1 Open-loop response

Fig. 5 shows the response profile of the surface roughness with respect to step changes in substrate temperature at time $t = 400$ s. We can see that the value of the surface roughness at the end of the deposition can be controlled by manipulating the substrate temperature, however, the responses to step changes in substrate temperature are in fact dependant on the surface microstructure of the thin film, i.e. instantaneous surface roughness. Fig. 6 shows the response profiles of surface roughness, with respect to step changes in substrate temperature at time $t = 200$ s and $t = 800$ s, when the surface of the thin film is relatively smooth and rough, respectively. It can be seen that the increase in substrate temperature corresponds to increasing final surface roughness when the surface is smooth and decreasing final surface roughness when the surface is rough. This result suggests that a more advanced controller may be needed to control the surface roughness in the high temperature regime.

Fig. 7 shows the profiles of surface roughness with respect to disturbances in the gas phase composition (i.e., gas flux pattern). It can be seen that the effect of the disturbance in the gas phase composition on the surface roughness is significant and very complex (i.e., it can not be described by a low order analytical model), and thus, the proposed growth model is needed to predict the evolution of the surface roughness.

4.2 Controller design

In order to achieve robust closed-loop operation in the high temperature regime, a kMC model-based predictive control scheme is proposed. Fig. 8 shows the block diagram of the closed-loop system. A reference trajectory of the instantaneous surface roughness of the thin film is selected based on off-line optimization, and in this work for simplicity, the profile of the surface roughness of the thin film in a ideal open-loop deposition (no disturbance during the process and the final surface roughness is just the desired value) is chosen. Using such a reference trajectory, instead of solving the receding horizon optimization problem of minimizing the difference between the final surface roughness and the desired value with multiple

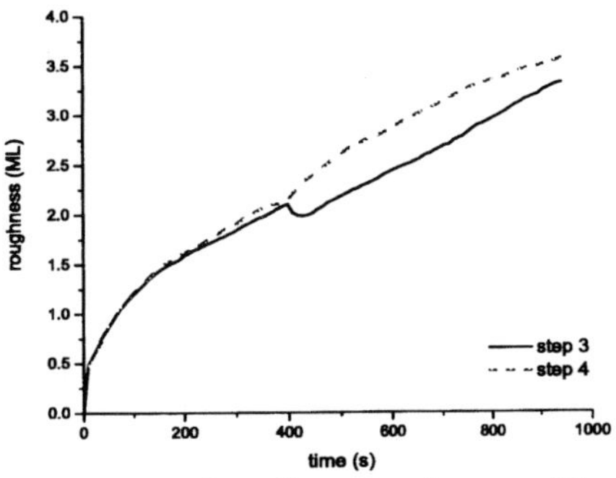

Fig. 5. Response of surface roughness with respect to step changes in substrate temperature: a. T changes from 185 K to 205 K at t=400 s (step 3); b. T changes from 185 K to 165K at t=400 s (step 4).

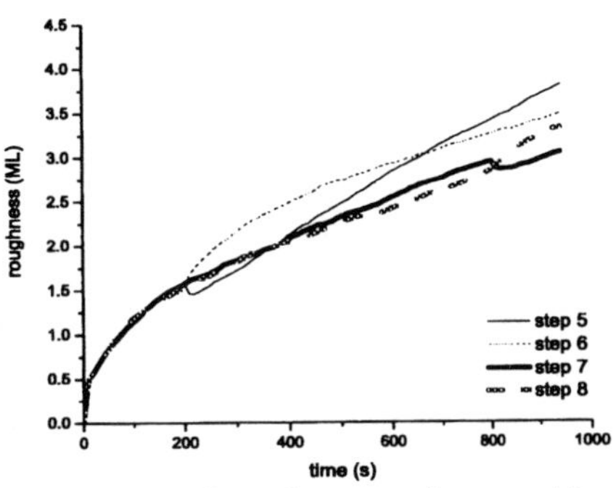

Fig. 6. Response of surface roughness with respect to step changes in substrate temperature: a. T changes from 185 K to 205 K at t=200 s (step 5); b. T changes from 185 K to 165K at t=200 s (step 6); c. T changes from 185 K to 205 K at t=800 s (step 7); d. T changes from 185 K to 165K at t=800 s (step 8).

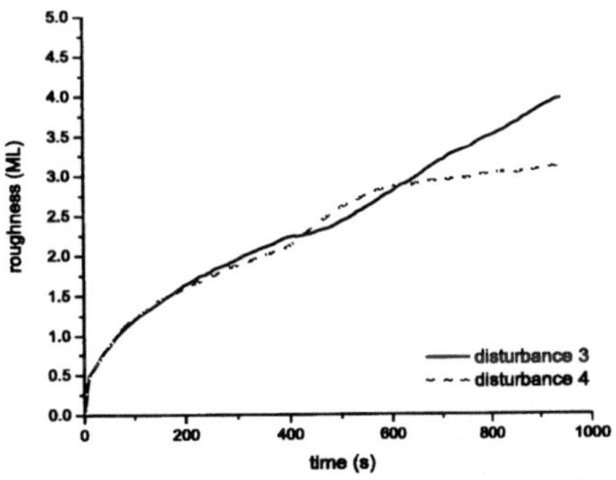

Fig. 7. Response of surface roughness with respect to step disturbance in gas phase composition: a. gas phase composition changed from $50\%A+50\%B$ to $20\%A+80\%B$ (disturbance 3); b. gas phase composition changed from $50\%A+50\%B$ to $80\%A+20\%B$ (disturbance 4).

decision variables, we only need to solve the fixed short horizon optimization problem of minimizing the difference between the instantaneous surface roughness and the reference value with a decision variable. Therefore, the computation time of each optimization is greatly reduced, since the kMC simulation duration is reduced from the scale of the total deposition time to the controller turnover time. This is very important since kMC simulation is relatively time consuming and large scale numerical optimization using kMC model is almost impossible to solve in real-time.

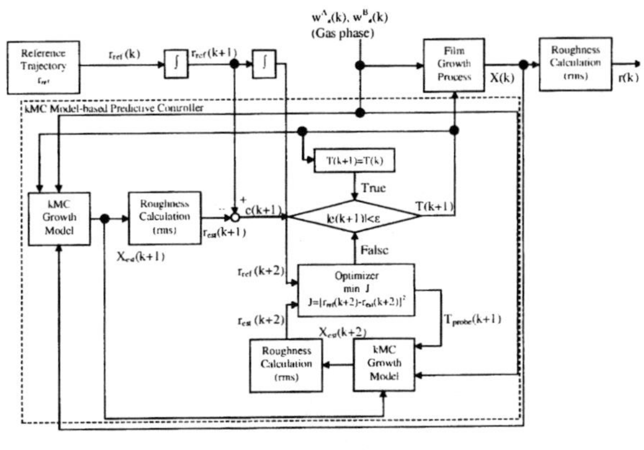

Fig. 8. Block diagram of the closed-loop system with the kMC model-predictive controller.

During each control cycle, the surface configuration $X(k)$ is first measured, an estimate of the surface configuration at the next control action time $X_{est}(k+1)$ is computed based on current process conditions using the proposed kMC model, and the estimated surface roughness value $r_{est}(k+1)$ is compared with the reference value $r_{ref}(k+1)$. If the error is less than ε ($\varepsilon = 0.05$ in this work), the next controller output $T(k+1)$ is set to be the same as the current output $T(k)$. If the error is larger than ε, the optimizer is called to compute the output value of the next control action $T(k+1)$ so that the error between the surface roughness after the next control action $r(k+2)$ and the reference value $r_{ref}(k+2)$ is minimized.

The optimizer uses direct search to find the optimal solution since the kMC model does not have a closed-form expression. The estimate of the surface roughness after the next control action $r_{est}(k+1)$ is computed using the proposed kMC model based on the estimated surface configuration before the next control action $X_{est}(k+1)$, the probe output value $T_{probe}(k+1)$ and current process conditions. The search precision specified in this work is 1 K, and since the proposed kMC model is highly computationally efficient, the optimization problem can be solved by an entry level personal computer within the controller turnover time (10 s). Furthermore, the speed and the precision of the direct search optimization algorithm can be substantially improved by parallel computing.

Fig.9 shows the profile of a closed-loop deposition with final surface roughness set-point value of $3.2ML$. It can be seen that the surface roughness value of the thin film follows the reference trajectory closely and the final surface roughness has been controlled at the desired value. Fig. 10 shows the profile of a closed-loop deposition with final surface roughness set-point value of $3.2ML$. A disturbance in the gas phase composition is introduced in this simulation in terms of a change in the adsorption rates at $t = 400s$ to $t = 500s$, specifically, w_a^A changed from $0.05s^{-1}$ to $0.1s^{-1}$ and

w_a^B changed from $0.05s^{-1}$ to $0s^{-1}$ at $t = 400s$, while at $t = 500s$, w_a^A and w_a^B both changed back to $0.05s^{-1}$. It can be seen that the surface roughness follows closely with the reference trajectory and final surface roughness has been controlled at the desired value which is 13.5% lower than open-loop. Compared to the failure of the PI controller on the same closed-loop simulation scenario (see detailed comparison with PI scheme in (Ni and Christofides, 2004)), kMC model-based predictive controller delivers substantially improved and robust closed-loop performance.

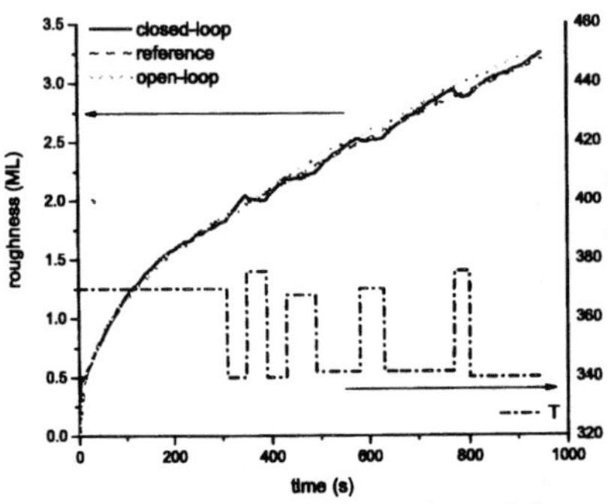

Fig. 9. Profiles of a closed-loop deposition with surface roughness set-point of $3.2ML$: a. closed-loop surface roughness (solid line, left scale); b. reference surface roughness (dashed line, left scale); c. open-loop surface roughness (dotted line, left scale); d. substrate temperature (dashed-dotted line, right scale).

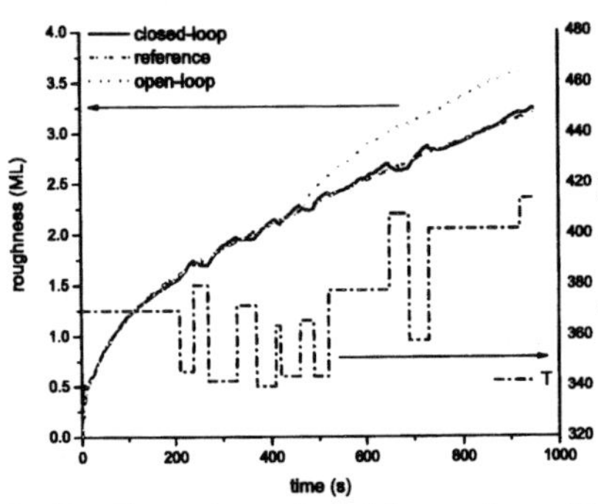

Fig. 10. Profiles of a closed-loop deposition with surface roughness set-point of $3.2ML$: a. closed-loop surface roughness (solid line, left scale); b. reference surface roughness (dashed line, left scale); c. open-loop surface roughness (dotted line, left scale); d. substrate temperature (dashed-dotted line, right scale).

REFERENCES

Armaou, A. and P. D. Christofides (1999). Plasma-enhanced chemical vapor deposition: Modeling and control. *Chem. Eng. Sci.* **54**, 3305–3314.

Battaile, C. C. and D. J. Srolovitz (2002). Kinetic-monte carlo simulation of chemical vapor deposition. *Annu. Rev. Mater. Res.* **32**, 297–319.

Cho, B., J. P. Chang, J. Min, S. H. Moon, Y. W. Kim and I. Levin (2003). Material characteristics of electrically tunable zirconium oxide thin films. *J. Appl. Phys.* **93**, 745–749.

Cho, B., J. Wang and J. P. Chang (2002). Metalorganic precursor decomposition and oxidation mechanisms in plasma-enhanced ZrO_2 deposition. *J. Appl. Phys.* **92**(8), 4238.

Christofides, P. D. (2001). *Nonlinear and Robust Control of Partial Differential Equation Systems: Methods and Applications to Transport-Reaction Processes*. Birkhäuser. Boston.

Einstein, T. L. (1996). *Handbook of Surface Science*. Chap. 11, pp. 577–650. Vol. 1. Elsevier. Amsterdam.

Gallivan, M. A. and R. M. Murray (2003). Model reduction and system identification of master equation control systems. In: *Proceedings of American Control Conference*. Denver, Colorado. pp. 3561–3566.

Gillespie, D. T. (1976). A general method for numerically simulating the stochastic time evolution of coupled chemical reactions. *J. Comp. Phys.* **22**, 403–434.

Lou, Y. and P. D. Christofides (2003*a*). Estimation and control of surface roughness in thin film growth using kinetic Monte-Carlo models. *Chem. Eng. Sci.* **58**, 3115–3129.

Lou, Y. and P. D. Christofides (2003*b*). Feedback control of growth rate and surface roughness in thin film growth. *A.I.Ch.E J.* **49**, 2099–2113.

Lou, Y. and P. D. Christofides (2004). Feedback control of surface roughness of *gaas* (001) thin films using kinetic Monte-Carlo models. *Comp. & Chem. Eng.*, to appear.

Merrick, M. L., W. Luo and K. A. Fichthorn (2003). Substrate-mediated interactions on solid surfaces: theory, experiment, and consequences for thin-film morphology. *Prog. Surf. Sci.* **72**, 117–134.

Ni, D. and P. D. Christofides (2004). Dynamics and control of thin film surface microstructure in a complex deposition process. *Chem. Eng. Sci.*, submitted.

Ni, D., Y. Lou, P. D. Christofides, L. Sha, S. Lao and J. P. Chang (2003). A method for real-time control of thin film composition using OES and XPS. *Proceedings of American Control Conference* pp. 1320–1327, To appear in *IEEE Trans. Semiconduct. Manufact.*

Reese, J. S., S. Raimondeau and D. G. Vlachos (2001). Monte Carlo algorithms for complex surface reaction mechanisms: Efficiency and accuracy. *Journal of Computational Physics* **173**, 302–321.

Siettos, C. I., A. Armaou, A. G. Makeev and I. G. Kevrekidis (2003). Microscopic/stochastic timesteppers and "coarse" control: a kMC example. *A.I.Ch.E J.* **49**, 1922–1926.

Theodoropoulou, A., R. A. Adomaitis and E. Zafiriou (1999). Inverse model based real-time control for temperature uniformity of RTCVD. *IEEE Trans. Semiconduct. Manufact.* **12**, 87.

ELSEVIER
IFAC
PUBLICATIONS
www.elsevier.com/locate/ifac

ADVANCED MONITORING OF COMPLEX AUTOCORRELATED PROCESSES

Uwe Kruger * Yiqi Zhou ** George W. Irwin *

* *Intelligent Systems and Control Research Group, Queen's University of Belfast, BT9 5AH, U.K.*
** *School of Mechanical Engineering, Shandong University, 250061 Jinan, P.R. China*

Abstract: In this work, the integration of ARMA filters into the multivariate statistical process control (MSPC) framework is presented to improve the monitoring of large-scale industrial processes. As demonstrated in the paper, such filters can remove auto-correlation from the monitored variables to avoid the production of false alarms. This is exemplified by application studies to a benchmark process from the literature.

Keywords: Statistical Process Control, ARMA Models, Auto Correlation, Large-Scale Systems, Process Models

1. INTRODUCTION

Modern industrial processes often present a large number of highly correlated process variables, which leads to huge amounts of process data to be analysed. MSPC methods are known to be effective in detecting and diagnosing abnormal behaviour in the above circumstances (MacGregor and Kourti, 1995; Wise and Gallagher, 1996). One of the most popular MSPC methods is principal component analysis (PCA), which aims to reveal linear relationships between the process variables by defining a reduced set of score variables. Using these variables, univariate statistics can be generated for on-line process monitoring.

The application of PCA, however, is based on the assumption that the process variables are stationary and normally distributed. In practice, this is rarely satisfied, as such processes are driven by random noise and disturbances. In addition, regulatory controller present feedback to the input variables, so that the impact of disturbances propagates through to both the input and output variables (MacGregor *et al.*, 1991; Kruger *et al.*, 2001). Thus, the process variables move around a steady state condition and exhibit some degree of auto-correlation.

For monitoring auto-correlated process variables, (Ku *et al.*, 1995) proposed the use of dynamic PCA (DPCA) for which time-series structures are incorporated into the PCA analysis. (Ku *et al.*, 1995) argued that: *Directly applying PCA on a data matrix actually constructs a linear static model. When the data contains dynamic information, applying PCA on the data will not reveal the exact relations between the variables, rather a linear static approximation.* It was stated further that: *The score variables will be auto-correlated and possibly cross-correlated* .

The contributions of this paper are as follows. Firstly, it is argued that if the process variables are highly correlated, the score variables can show a stronger degree of auto-correlation. Secondly, strongly auto-correlated score variables may lead to the production of false alarms that invalidate the on-line monitoring approach. Thirdly, the application of ARMA filters is proposed to remove auto-correlation from the score variables. The above findings are demonstrated using a benchmark example from the literature.

It should be noted that ARMA filters could also be applied directly to the recorded process variables. As industrial processes often present large variable sets, however, the application of ARMA filters to these variables may be practically difficult. Moreover, pre-filtering these variables may change the relationships between them, implying that PCA may be less successful in defining a much reduced set of score variables.

2. PROCESS MONITORING USING PCA

The application of PCA involves the construction of a reduced set of score variables that can describe significant variation of the process. The values of these variables can be obtained as follows:

$$\mathbf{t} = \mathbf{P}^T \mathbf{z}, \tag{1}$$

where $\mathbf{t} \in \mathbb{R}^n$ is a vector storing the values of the scores variables, $\mathbf{P} \in \mathbb{R}^{N \times n}$ is a transformation matrix, $\mathbf{z} \in \mathbb{R}^N$ is a vector in which the values of the process variables are stored and N, $n \in \mathbb{N}$ are the number of process variables and retained score variables, respectively. A more detailed analysis of PCA may be found in (Wold *et al.*, 1987).

2.1 Dynamic PCA

To accommodate auto-correlation of the process variables, (Ku *et al.*, 1995) showed that a linear time-series structure could be incorporated into the PCA analysis. This leads to an arrangement of the process variables to form an autoregressive with external input variables (ARX) model structure:

$$\mathbf{z}_k^{*^T} = \left(\mathbf{z}_k^T \cdots \mathbf{z}_{k-m}^T \right), \tag{2}$$

where $\mathbf{z}_k^{*^T}$ is an augmented set of variables, representing an ARX model structure of order m. Utilizing this "extended" set of process variables, a PCA analysis can then be carried out as described above. On the basis of the recommendations by (Zwick and Velicer, 1986), (Ku *et al.*, 1995) proposed the use of parallel analysis and a subsequent correlation analysis to determine the number of time-lagged values for the process variables and the number of retained PCs. This approach was also used here. Although more accurate relationships between the process variables can be extracted, the PCs are still to be auto-correlated, as can be seen from Equation (1).

If the process variables are auto-correlated, the score variables, which are linear combinations of the original process variables, are auto-correlated too. In addition, if a time-series structure of auto-correlated process variables is considered, the auto-correlation of the score variables will accordingly be amplified. The application study in Section 4 illustrates that strongly auto-correlated score variables may lead to the production of false alarms. Consequently, the auto-correlation of the score variables must be removed in order to prevent such alarms to occur. The next section shows how the application of ARMA filters can remove auto-correlation from the score variables.

2.2 Univariate Statistics

A univariate statistic, denoted as T^2, can be established using the retained score variables:

$$T^2 = \mathbf{t}^T \mathbf{\Lambda}^{-1} \mathbf{t} \tag{3}$$

where $\mathbf{\Lambda}$ is a diagonal matrix storing the n largest eigenvalues of the covariance matrix $\mathbf{S}_{ZZ} = \frac{1}{K-1}\mathbf{Z}^T\mathbf{Z}$ with K being the number of mean centered and appropriately scaled observations stored successively as row vectors in $\mathbf{Z} \in \mathbb{R}^{K \times N}$. The time variation in T^2 can be plotted and its confidence limit obtained as discussed by (Jackson, 1980).

Note that a second univariate statistic, referred to as Q statistic and related to the residuals of the PCA model prediction, can also be defined. However, this statistic is not considered here, since these residuals are assumed to describe measurement uncertainty which is considered to be represented by identically and independently distributed (i.i.d.) variables that are superimposed on the "true" process variables. Hence, the auto-correlation of the process variables does not affect the Q statistic.

3. APPLICATION OF ARMA FILTERS

The application of ARMA filters is now proposed to remove auto-correlations from the score variables. The general form of an ARMA filter is given by (Box *et al.*, 1994) as:

$$\frac{\mu(B)}{h(B)} = \frac{\Theta(B)}{\Phi(B)}, \tag{4}$$

where μ and h are random variables, B is the backward shift operator, i.e. $B\mu_k = \mu_{k-1}$, and Θ and Φ are polynomials in B of dimension p and q, respectively. Moreover, the polynomial $\Theta(B) = 1+\theta_1 B+\cdots+\theta_p B^p$ is the moving average (MA) operator, whilst the polynomial $\Phi(B) = 1+\phi_1 B + \cdots + \phi_q B^q$ represents the autoregressive (AR) operator. The input sequence, $h(B)$, is assumed to be a normally distributed white noise sequence with zero mean and variance σ_h^2, i.e. $h \in \mathcal{N}\left(0, \sigma_h^2\right)$, and the sequence $\mu(B)$ represents the filtered sequence of $h(B)$ using the transfer function $R_{ARMA}(B) = \frac{\Theta(B)}{\Phi(B)}$.

Such ARMA(p, q) filters can be inverted so that $\mu(B)$ represents the input sequence and $h(B)$ represents the output sequence, i.e. $h(B) = R_{ARMA}^{-1}(B)\mu(B)$. This implies that the inverse filter can be employed to filter $\mu(B)$ so that a normally distributed white noise sequence can be produced. More precisely, the calculation of h using the filter $R_{ARMA}^{-1}(B)$ at the kth time instance is given by:

$$h_k = \mu_k - \widehat{\mu}_k = \mu_k + \sum_{i=1}^{q} \phi_i \mu_{k-i} - \sum_{i=1}^{p} \theta_i h_{k-i}, \quad (5)$$

where $\widehat{\mu}_k$ represents the prediction of μ_k using the ARMA filter.

An inverse ARMA filter can be utilised to remove the auto-correlation of the score variables. More precisely, an ARMA filter can be identified for each of the score variables and the residuals of each ARMA filter can then be employed instead of the original score variables. One could also consider removing the auto-correlation of the original process variables and then establishing a PCA model. However, this would be computationally more demanding, as there are usually considerably fewer score variables, i.e. $n < N$. Furthermore, the correlation structure between the process variables might then be changed, which implies that the variable reduction, performed by PCA, might not be as efficient.

3.1 Identification of ARMA Filters

A variety of approaches can be employed for identifying the parameters of the polynomials $\Theta(B)$ and $\Phi(B)$. For example, (Liu and Hudak, 1992) proposed a maximum likelihood estimation for minimizing the variance of $h(B)$, while (Box *et al.*, 1994) suggested a three-stage iterative procedure involving the identification, the estimation and diagnostic checking of the filter obtained.

To determine the orders of the polynomials $\Theta(B)$ and $\Phi(B)$, a number of approaches have been proposed. These can be divided into information based criteria, such as the Akaike information criterion (AIC) (Akaike, 1974), the BIC criterion (Schwarz, 1978) and the minimum description length (MDL) (Rissanen, 1978), and linear algebraic approaches, such as the determinant testing algorithm (Chow, 1972), SVD based methods (Cadzow, 1983; Zhang and Zhang, 1993) and the linear, dual, decoupled procedure (LD^2-ARMA) algorithm (Ribeiro and Moura, 1991).

(Broersen, 1998) highlighted that the residuals of an ARMA process provide an objective measure for assessing the quality of different filters. In the context of this paper, however, the purpose of the filter is to remove auto-correlation from the score variables. Hence, techniques that determine the "optimal filter order", i.e. the number of AR and MA terms, on the basis of residuals may not be suitable. In this work, the mismatch between the estimated auto-correlation function (ACF) and the "ideal ACF" within a window of length l is used. This mismatch can be calculated as follows:

$$J_\mu(p, q) = \frac{1}{l} \sum_{j=0}^{l} \left(r(j) - \widehat{r}_{\mu\mu}(p, q, j)\right)^2, \quad (6)$$

where $J_\mu(p, q)$ is a cost function, $r(j)$ is the "ideal ACF", i.e. $r(j) = \delta_{0j}$ with δ being the Kronecker delta and $\widehat{r}_{\mu\mu}(p, q, j)$ is the estimated ACF including p MA and q AR terms.

Since the BIC criterion provides a consistent estimation of the number of AR and MA terms (Pauler, 1998), it is adopted in conjunction with the above cost function:

$$BIC(p, q) = \log(J_\mu(p, q)) + 2\frac{(p+q)\log(K)}{K}. \quad (7)$$

The order of both polynomials, p and q, is then selected to minimise the BIC cost function.

3.2 Identifying an ARMA based PCA Model

The first step is to record reference data from the process. Note that reference data have to be selected with care, to ensure that abnormal process behaviour is not captured. Conversely, if the size of the reference data is too small, then normal variation within the process would not be adequately represented (Kruger *et al.*, 2001). A PCA model then needs to be established. After defining the order of dynamics m using parallel analysis, (Wold *et al.*, 1987) showed that the transformation matrix $\mathbf{P}$ is constructed by the dominant eigenvectors of $\mathbf{S}_{ZZ}$, which are stored as column vectors. The number of retained score variables can also be determined by parallel analysis, including a subsequent correlation analysis to avoid that the score variables which capture process variation are not discarded. The score variables can now be determined as shown in Equation (1), and the "optimum" number of AR and MA terms then found to remove auto-correlation from these variables.

3.3 Application of ARMA based PCA Models

After the identification of an ARMA model for each score variable, on-line process monitoring can be established as follows. Given that a new observation becomes available:

(1) Obtain the values of the score variables by applying Equation (1).

(2) Apply the identified ARMA filters to each of the score variables –Equation (5).
(3) Determine the T^2 statistic –Equation (3).
(4) Update the monitoring chart.

4. APPLICATION STUDIES

This section presents an application study to a benchmark example from the literature to illustrate the influence of auto-correlated process variables on the T^2 statistic. This example involves two process input variables and two process output variables. From this process, a data set was generated which represents an ARMA process and dynamic PCA was subsequently applied.

4.1 Process Description

The process under study had the following description (Ku *et al.*, 1995):

$$\begin{pmatrix} x_k^{(1)} \\ x_k^{(2)} \end{pmatrix} = \begin{bmatrix} 0.118 & -0.191 \\ 0.847 & 0.264 \end{bmatrix} \begin{pmatrix} x_{k-1}^{(1)} \\ x_{k-1}^{(2)} \end{pmatrix} + \begin{bmatrix} 1 & 2 \\ 3 & -4 \end{bmatrix} \begin{pmatrix} v_{k-1}^{(1)} \\ v_{k-1}^{(2)} \end{pmatrix}$$
$$\begin{pmatrix} y_k^{(1)} \\ y_k^{(2)} \end{pmatrix} = \begin{pmatrix} x_k^{(1)} \\ x_k^{(2)} \end{pmatrix} + \begin{pmatrix} f_k^{(1)} \\ f_k^{(2)} \end{pmatrix}, \tag{8}$$

where $\mathbf{v}_k^T = \begin{pmatrix} v_k^{(1)} & v_k^{(2)} \end{pmatrix}$ and $\mathbf{x}_k^T = \begin{pmatrix} x_k^{(1)} & x_k^{(2)} \end{pmatrix}$ represented the process input and output variables at the kth time instance, respectively, and $\mathbf{f}_k^T = \begin{pmatrix} f_k^{(1)} & f_k^{(2)} \end{pmatrix}$ represented measurement noise superimposed to the "true" values of the output variables to form the measured output variables $\mathbf{y}_k^T = \begin{pmatrix} y_k^{(1)} & y_k^{(2)} \end{pmatrix}$. The input variables of the above process were defined as follows:

$$\begin{pmatrix} v_k^{(1)} \\ v_k^{(2)} \end{pmatrix} = \begin{bmatrix} 0.811 & -0.226 \\ 0.477 & 0.415 \end{bmatrix} \begin{pmatrix} v_{k-1}^{(1)} \\ v_{k-1}^{(2)} \end{pmatrix} + \begin{bmatrix} 0.193 & 0.689 \\ -0.320 & -0.749 \end{bmatrix} \begin{pmatrix} w_{k-1}^{(1)} \\ w_{k-1}^{(2)} \end{pmatrix}$$
$$\begin{pmatrix} u_k^{(1)} \\ u_k^{(2)} \end{pmatrix} = \begin{pmatrix} v_k^{(1)} \\ v_k^{(2)} \end{pmatrix} + \begin{pmatrix} g_k^{(1)} \\ g_k^{(2)} \end{pmatrix}, \tag{9}$$

here $\mathbf{w}_k^T = \begin{pmatrix} w_k^{(1)} & w_k^{(2)} \end{pmatrix}$ represented sequences of normally distributed values with zero mean and unit variance, i.e. $w_k^{(1)}$ and $w_k^{(2)} \in \mathcal{N}\{0,1\}$. In contrast to the simulation model by (Ku *et al.*, 1995), measurement noise was also superimposed on the "true" values of the input variables, i.e. $\mathbf{g}_k^T = \begin{pmatrix} g_k^{(1)} & g_k^{(2)} \end{pmatrix}$, to form the measured input variables $\mathbf{u}_k^T = \begin{pmatrix} u_k^{(1)} & u_k^{(2)} \end{pmatrix}$. The measurement noises were sequences of normally distributed variables with zero means and variances chosen to be 5% of the variance of each individual process variable, i.e.: $f_k^{(i)} \in \mathcal{N}\{0, 0.05 \cdot E\{x_k^{(i)^2}\}\}$ and $g_k^{(i)} \in \mathcal{N}\{0, 0.05 \cdot E\{v_k^{(i)^2}\}\}$.

The output variables were then recorded as shown in Equation (8) prior to the introduction of measurement noise to the input and output variables. Since auto-correlated process behaviour is achieved by describing the input and output variables by ARMA sequences, this process is further referred to as the ARMA process and the corresponding data set is denoted as the ARMA data set.

A data set containing a total number of 2000 samples was generated as described above. The first 500 samples were selected as the reference data and the remaining 1500 samples served as testing data to evaluate the performance of the established monitoring scheme. Given the construction of the process variables, the PCA analysis was based on the following dynamic data matrix:

$$\mathbf{Z} = \begin{bmatrix} y_2^{(1)} & y_2^{(2)} & y_1^{(1)} & y_1^{(2)} & u_1^{(1)} & u_1^{(2)} \\ y_3^{(1)} & y_3^{(2)} & y_2^{(1)} & y_2^{(2)} & u_2^{(1)} & u_2^{(2)} \\ & & \vdots & & \vdots & \\ y_{2000}^{(1)} & y_{2000}^{(2)} & y_{1999}^{(1)} & y_{1999}^{(2)} & u_{1999}^{(1)} & u_{1999}^{(2)} \end{bmatrix}. \tag{10}$$

Since the "true" values of the output variables were linear combinations of previous "true" values of the input and output variables, a total number of 4 PCs needed to be retained. The measurement noise was represented by the 2 discarded PCs. The confidence limits for the T^2 statistic was determined with a confidence of 99%.

4.2 Monitoring of the ARMA Processes

After establishing a PCA model using the first 500 samples, the auto-correlation function of the 4 retained PCs was determined. Figure (1) shows that particularly the first two score variables were strongly auto-correlated. Figure (2) shows the T^2 statistic. The number of violations of the 99% confidence limit was excessive after about 1500 samples into the data set. Furthermore, some of these violations occur in sections, i.e. more than one consecutive T^2 value violates the confidence limit. More precisely, almost 3.5% of the last 500 T^2 values violate the 99% confidence limit. This highlights the fact that the process was out-of-statistical-control.

In summary, the application of an MSPC based monitoring approach to a stationary process with auto-correlated variables may lead to false alarms

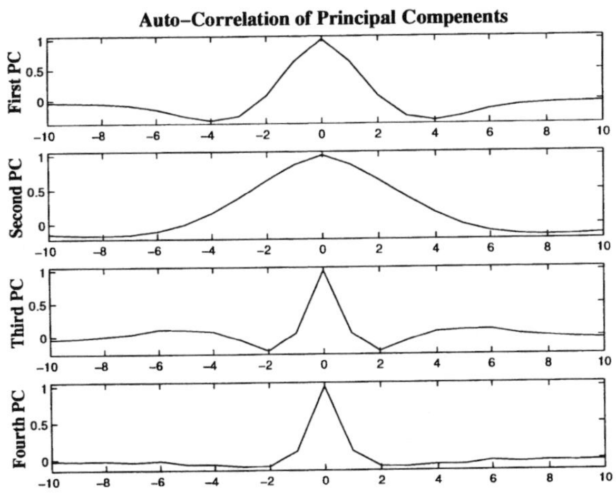

Fig. 1. Auto-correlation functions of retained PCs (original PCs)

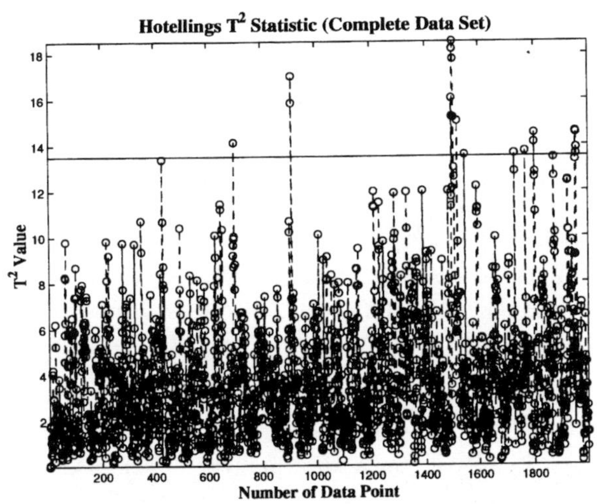

Fig. 2. Hotelling's T^2 statistic based on original score variables)

being produced although the process itself behaves normally. To circumvent such violations, the auto-correlation must be removed or filtered out as discussed in the next subsection.

4.3 Monitoring of the ARMA Process Using Box-Jenkins Filters

It is now shown that ARMA filters can remove auto-correlation from the PCs. The filters were identified using the same data set as in the previous subsection. The number of AR and MA terms, p and q, was determined by applying Equations (6) and (7). From analysing the ACF of each PC (Figure 1), the window length l was chosen as 10. Table (1) shows the selected number of AR and MA terms. The resultant ACF of each filtered PCs is shown in Figure (3) from which it can be seen that only very marginal auto-correlation remains. By comparing Figures (2) and (4), it can be seen that:

Table 1. Number of AR and MA terms for Retained PCs -ARMA Process

#PC	#AR Terms (p)	#MA Terms (q)
1	7	0
2	6	1
3	8	3
4	3	1

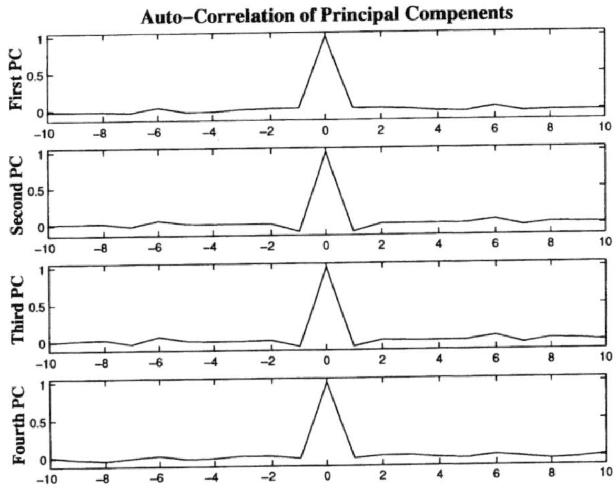

Fig. 3. Auto-correlation functions of retained PCs (filtered PCs)

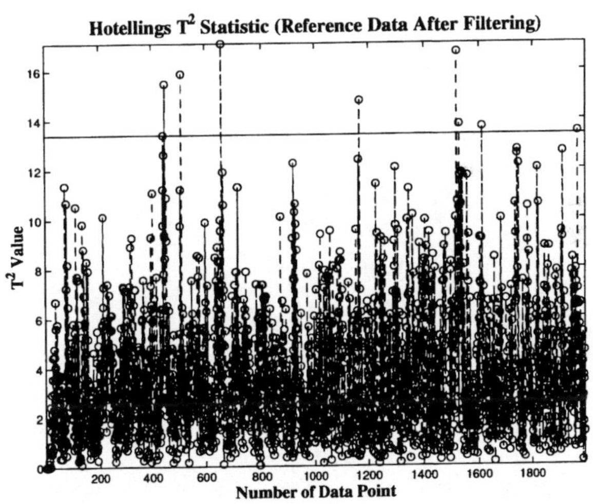

Fig. 4. Hotelling's T^2 statistic based on filtered score variables

(i) the number of violations is less than 1%; and
(ii) violations of consecutive samples, as seen in Figure (2) at around 1500 data points into the data set, were removed.

The process was therefore in-statistical-control, whereas without filtering, the T^2 statistic and the scatter diagrams suggested incorrectly that the process was out-of-statistical-control. Applying ARMA filtering of the PCs has thus prevented false alarms being produced.

5. CONCLUSIONS

This paper has studied the incorporation of ARMA filters into the MSPC framework, motivated by the fact that monitoring processes with strongly auto-correlated variables may lead to the production of false alarms. This was demonstrated using a benchmark simulation from the literature.

It was shown that the application of ARMA filters removes auto-correlation from the reduced set of PCs, thus circumventing the production of false alarms. A more robust monitoring scheme was hence established for the application study used in this work.

The benefits of filtering the PCs instead of the physical process variables are that (i) the number of ARMA filters to be established is much smaller then the number of physical process variables, (ii) the reduced variable set is then statistically independent in contrast to the large number of highly correlated process variables and (iii) auto-correlation of the PCs can be removed.

REFERENCES

Akaike, H. (1974). A new look at the statistical model identification. *IEEE Transactions on Automatic Control* **AC-19**(6), 667–674.

Box, G. E. P., G. M. Jenkins and G. C. Reinsel (1994). *Time Series Analysis*. Prentice-Hall. Englewood-Cliffs NJ.

Broersen, P. M. T. (1998). The quality of models for arma processes. *IEEE Transactions on Signal Processing* **46**(6), 1749–1752.

Cadzow, J. A. (1983). Singular-value decomposition approach to time series modeling. *Proc. Inst. Elect. Eng. F* **130**, 202–210.

Chow, J. C. (1972). On estimating the orders of an autoregressive moving average process with uncertain observations. *IEEE Transactions on Automatic Control* **17**, 707–709.

Downs, J. J. and E. F. Vogel (1993). A plant-wide industrial-process control problem. *Computers and Chemical Engineering* **17**(3), 245–255.

Jackson, J. E. (1980). Principal components and factor analysis: Part i: Principal components. *Journal of Quality Control* **12**(4), 201–213.

Kruger, U., Q. Chen, D. J. Sandoz and R. C. McFarlane (2001). Extended pls approach for enhanced condition monitoring of industrial processes. *AIChE Journal* **47**(9), 2076–2091.

Ku, W., R. H. Storer and C. Georgakis (1995). Disturbance rejection and isolation by dynamic principal component analysis. *Chemometrics and Intelligent Laboratory Systems* **30**, 179–196.

Liu, L., M. and G. B. Hudak (1992). *Forecasting and Time Series Analysis Using the SCA Statistical System.* DeKalb, Scientific Computing Associates. Illinois.

Lyman, P. R. and C. Georgakis (1995). Plant-wide control of the tennessee eastman problem. *Computers and Chemical Engineering* **19**(3), 321–331.

MacGregor, J. F. and T. Kourti (1995). Statistical process control of multivariate processes. *Control Engineering Practice* **3**(3), 403–414.

MacGregor, J. F., T. E. Marlin, J. V. Kresta and B. Skagerberg (1991). Multivariate statistical methods in process analysis and control. In: *AIChE Symposium proceedings of the 4th International Conference on Chemical Process Control.* AIChE Publication, No. P-67. New York. pp. 79–99.

Pauler, D. K. (1998). The schwarz criterion and related methods for normal linear models. *Biometrika* **85**(1), 13–27.

Ribeiro, M. I. and J. M. F. Moura (1991). Ld^2-arma identification algorithm. *IEEE Transactions on Signal Processing* **39**(8), 1822–1835.

Rissanen, J. (1978). Modelling by shortest data description. *Automatica* **14**(2), 465–471.

Schwarz, G. (1978). Estimating the model dimension. *Annals of Statistics* **6**(2), 461–464.

Wise, B. M. and N. B. Gallagher (1996). The process chemometrics approach to process monitoring and fault detection. *Journal of Process Control* **6**(6), 329–348.

Wold, S., K. Esbensen and P. Geladi (1987). Principal component analysis. *Chemometrics and Intelligent Laboratory Systems* **2**, 37–52.

Zhang, X.-D. and Y.-S. Zhang (1993). Singular value decomposition-based ma order determination of non-gaussian arma models. *IEEE Transactions on Signal Processing* **41**(8), 2657–2664.

Zwick, W. R. and W. F. Velicer (1986). *Psychol. Bull* **99**, 432–441.

ELSEVIER
IFAC
PUBLICATIONS
www.elsevier.com/locate/ifac

MONITORING OF FLOTATION PROCESSES USING MULTIRESOLUTIONAL MULTIVARIATE IMAGE ANALYSIS (MR-MIA)

J. Jay Liu and John F. MacGregor[1], Carl Duchesne[2], Gianni Bartolacci[3]

[1]*McMaster Advanced Control Consortium, Department of Chemical Engineering, McMaster University, Hamilton, ON. Canada*
[2]*Département de génie chimique, Université Laval, Québec, QC, Canada*
[3]*COREM, 1180, rue de la Minéralogie, Québec, QC, Canada*

Abstract: This paper provides a novel machine vision solution based on multiresolutional multivariate image analysis (MR-MIA) for monitoring of flotation processes. The approach based on MR-MIA is superior to the contemporary machine vision approaches in terms of efficient analysis of morphological and color information and robustness to lighting conditions. The results show that the MR-MIA based approach provides enough morphological and color information of froth needed to describe different froth status. A PCA model built from the information provided by MR-MIA can provide multivariate control charts, from which the transition as well as steady-state process status can be monitored in real time. *Copyright © 2004 IFAC*

Keywords: Flotation Froth, Machine vision, Multiresolutional Multivariate Image Analysis, Wavelet texture analysis, Wavelet size signature

1. INTRODUCTION

The flotation process is one of the important and standard techniques in the mineral processing industries to separate valuable metals from ore (Mckee, 1991). However, it impossible to monitor and control flotation process through fundamental modeling approaches due to the chaotic nature of the underlying microscopic phenomena based on surface chemistry and surface physics. In addition, state-of-the-art instrumentation technologies are not able to provide reliable on-line sensors or analyzers, measurements of which are essential for automatic monitoring and control.

It is widely known that color and morphology of the froth are closely related to mineral concentrations and process status, respectively. As a consequence, many operations are made by operators based on visual appearance of the froth together with their experience about process trends. For these reasons, over the past decade image analysis has been considered as a potentially key component to the solution to this problem.

The purpose of this paper is to provide a novel solution to the characterization and monitoring of the flotation froth based on multiresolutional multivariate image analysis (MR-MIA) (Liu and MacGregor, 2003). By combining multiresolution analysis (MRA) and multivariate image analysis (MIA), it is possible to analyze the spatial and spectral correlation of images within a single framework (Liu and MacGregor, 2003). The approach based on MR-MIA is totally different from the contemporary research in the sense that it handles morphological and color information of froth images systematically and efficiently. In addition, this approach is inherently robust to image quality or lighting conditions, contrary to most contemporary image analysis approaches.

2. VISUAL FEATURES OF FROTH AND EXTRACTION METHODS

Color and structure of flotation froth are very important visual features in the operation of flotation process. Color is strongly related with the mineral concentration carried by the froth (Bonifazi *et al.*,

1999). The structure of the froth, that is its texture or morphology (bubble's size, distribution and, shape, etc.) is known to indicate various froth characteristics such as degree of mineralization (froth load), stability, and so on (Moolman *et al.*, 1996). Operators usually determine the suitable amount of chemical reagents (i.e., control inputs or manipulated variables) based on those visual features of the froth.

Image analysis based on traditional image processing techniques has been applied to flotation processes in order to monitor the process or to infer mineral concentrations. For estimating mineral concentration, statistical features such as descriptive statistics are calculated from each RGB color channel and multiple linear regression or Partial Least Squares (PLS) are used to construct an inferential model (Bonifazi *et al.*, 1998). HSI (Hue, Saturation, Intensity) or HSV (Hue, Saturation, Value) color models other than RGB model were used in some cases in order to overcome high collinearity among R, G, and B values (Bonifazi *et al.*, 1999). In addition, bubble collapse rate, bubble mobility (speed), and various morphological features (froth bubble size, shape, etc.) were extracted via traditional image analysis methods and used in regression, but the correlation of mineral concentration with morphological features was lower than that obtained with color features (Bonifazi *et al.*, 1999).

The analysis of froth structure has frequently been carried out using segmentation methods, texture analysis methods, and Fourier transform (FT) power spectrum. Morphological features such as bubble diameter, aspect ratio, etc. were calculated for each bubble after segmentation of the froth images into bubbles and then used in estimating mineral concentration (Bonifazi *et al.*, 1999). Many other image-processing techniques were used to enhance the quality of image prior to segmentation. Statistical texture analysis methods such as gray-level co-occurrence matrix (GLCM) and its variations were used to classify the status of different froths based on froth texture in order to monitor flotation processes (Moolman *et al.*, 1995). Power spectrum from 1-D or 2-D Fourier transform (FT) was also used to extract textural feature of the froth (Niemi *et al.*, 1999).

It is clear from the literature that the correlation structure among the color in the RGB images has not been considered in most cases. Although HSI and HSV models have sometimes been used in some cases, they are rather basic approaches to handling the collinearity. Lighting condition is also crucial in analyzing color features but efforts made for removing the effect of different lighting or illumination have been so heuristic or ad-hoc that they could not be easily generalized. In this respect, MIA based on Principal Component Analysis (PCA) provides a better approach than others that have been used in the literature since PCA can easily handle the collinearity (Geladi and Grahn, 1996). Furthermore, it was experimentally verified that some features (e.g., energy and entropy) calculated after applying PCA were illumination-invariant as long as intensity saturation did not occur (Tan and Kittler, 1994).

Wavelet texture analysis (WTA) has been considered as the state of the art in texture analysis for many reasons (Liu and MacGregor, 2003). It outperforms other methods such as GLCM-based methods or FT-based methods (Bharati *et al.*, 2003) and is much more computationally efficient and robust to lighting conditions than segmentation-based methods (Bonifazi *et al.*, 1999). Therefore, MR-MIA would appear to offer an excellent alternative to contemporary image analysis approaches for monitoring and control of flotation process. MR-MIA combines advantages of MIA and MRA (Liu and MacGregor, 2003); it can handle high colinearity in RGB froth images more efficiently than HSI and HSV models and can extract structural feature of froth in a faster and more robust manner than segmentation-based or GLCM-based approaches. An overview of MR-MIA methods for extracting color and structural information from froths is discussed in the next section.

3. VISUAL FEATURE EXTRACTION USING MR-MIA

From the literature, it seems that color information and structural information in the froth are not highly correlated to each other and therefore, MR-MIA II is more preferable than MR-MIA I in this situation since spatial information and spectral information are extracted and processed separately (Liu and MacGregor, 2003); Color features will be extracted in the PCA stage of MR-MIA II and then morphological features will be calculated from WTA of the first score image, which it is almost equivalent to the grayscale version of a RGB image in most cases.

The purpose of image analysis in flotation froth is to extract features, which can indicate status of the process and thus can be used for process monitoring and control. Some of the features may be common to other flotation processes but others are not. This is because each plant can show its unique characteristics depending on the flotation cell used, the mineralogy of the ore used, etc. Furthermore, visual features are clearly subjective, and even operators in the same plant may have different criteria when interpreting the images. Therefore, we based our work on interviews about froth visual features with operators and engineers at the plant (Agnico-Eagle's Laronde plant in Quebec, Canada) where the froth image collection was performed.

3.1 CLEAR WINDOWS AND BLACK HOLES

A clear window is a watery portion of the froth, found on the top of the bubbles, that has almost no mineral content (see Figure 1). Therefore, froths with many clear windows usually have big bubbles, and are usually an indication that the degree of mineralization in the froth is very low. The color of the clear windows is much darker and blacker than other regions such as the tops and valleys of the froth. It is one of three major visual features of froth that operators at Laronde plant use to determine the state of their operations. On the other hand, one can find black holes when froths are excessively loaded with minerals (see Figure 1). A Black hole is a portion of froth surface, which is so close to the pulp that the pulp can be seen through it. Black holes usually appear when bubbles are extremely small and often provide a warning that the entire froth may collapse.

However, these features have never used for monitoring of flotation processes in the literature. Extracting clear windows and black holes can be easily done using masking in MIA since it extracts spectral (i.e., color) features independent of their spatial location. The total areas (i.e., a number pixels) of clear windows and black holes are then easily extracted. The only difficulty is that clear windows and black holes are spectrally similar (i.e., similar colors), and hence difficult to distinguish by MIA alone.

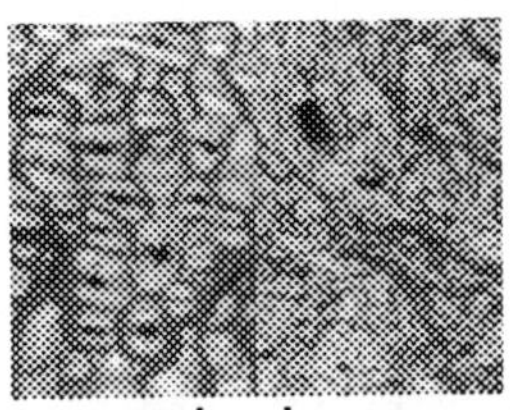

Figure 1 A composite image containing clear windows (left half) and black holes (right half)

Calculating Area of Clear Windows and Black Holes using MIA Masks. A density score plot is one of major analysis tools of MIA; Pixels having very similar spectral features lie close to one another in the score plots, regardless of their positions in the image space and such score plots can be thought as two-dimensional histograms if segmented into a number of bins. An example of a scatter score plot is shown in Figure 2 for the image in Figure 1 where the brightness of each point indicates the number of pixels falling within a bin in the score plot. We can detect different features in the original image space using a density score plot by defining a mask or a feature boundary for each feature as determined from a training image, because the coordinate of each pixel in the score plot is uniquely determined by its variable correlations (Bharati and MacGregor, 1998). The number of pixels within the feature boundaries represents the area of these features in the image space. A feature boundary is usually found by interactive inspection between such score plots and the original image. The feature boundary of clear windows and black holes for the composite image is shown in Figure 2 and pixels falling inside the boundary are highlighted in Figure 2.

3.2 FROTH BUBBLE SIZE

Bubble Size in the froth is one of the most commonly used morphological features throughout the literature and it is another major feature used by operators at the Laronde plant for characterizing process status. In fact, the measurement of froth size has been of great interest in the mining process industry as well as in the literature since it has been extensively reported that the performance of flotation process is strongly related to the size of froth (Moolman *et al.*, 1996).

We can find a size of a bubble, count the number of bubbles, and calculate a histogram of bubble size from them after segmenting each bubble. The biggest problem in using segmentation techniques in practice however is that segmentation is too sensitive to lighting conditions and segmentation techniques used (Bonifazi *et al.*, 1999) and in addition it is computationally demanding compared to other alternatives.

On the other hand, it has been also reported that the froth texture is a strong indicator of process performance (Moolman *et al.*, 1995, 1996). For this reason, statistical texture analysis techniques such as gray-level co-occurrence matrix (GLCM) and its variants have been used to classify froth images into predefined classes that correspond to different process status. Also, fractal analysis has been used in the same context (Hargrave and Hall, 1997). These approaches seem robust to lighting conditions due to the inherent robustness of statistical texture analysis techniques. However, it is difficult to characterize actual froth structure or morphology by using textural features calculated in these methods since textural features and corresponding visual features are not directly related to froth structure such as bubble size (Moolman *et al.*, 1995).

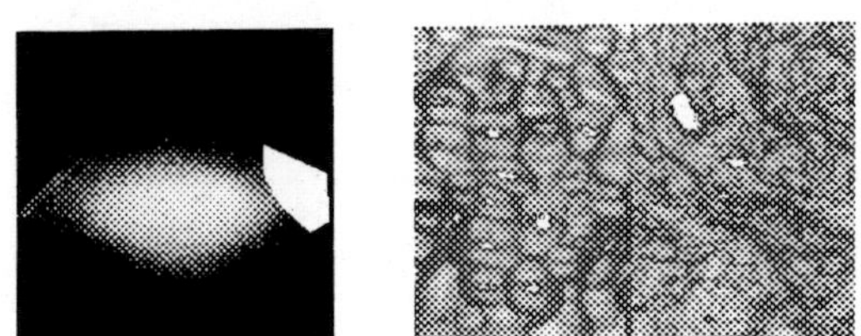

Figure 2 A t_1-t_2 score plot of the composite image with the mask for clear windows and black holes (left) and the composite image where clear windows and black holes found by the mask are highlighted in white (right)

Calculating Histogram of Froth Bubble Size using Wavelet Size Signature. A basic idea of wavelet texture analysis (WTA) is to extract a textural feature from wavelet coefficients at each resolution and assume that each texture has its unique distribution of features over all the resolutions. Therefore, different textures will have different features if the frequency spectrum is decomposed appropriately. Typical textural features in WTA are energy, entropy, or averaged l_1-norm. However, these WTA features have the same problem; they have no morphological meaning. For this reason, we develop a new feature called wavelet size signature, which can provide a histogram of froth bubble size.

Space-Frequency Representations and Uncertainty Principle. The coefficients of a 1-D discrete wavelet transform (DWT) of $f(x)$ can be computed as

$$a_{(j)}[l] = \langle f[k], \phi_{j,l}[k] \rangle \text{ and } d_{(j)}[l] = \langle f[k], \psi_{j,l}[k] \rangle, \quad (1)$$

where the $a_{(j)}$'s are expansion coefficients of the scaling function or approximation coefficients, the $d_{(j)}$'s are the wavelet coefficients or detail coefficients, and with some suitable sequence $h[k]$,

$$\phi_{j,l}[k] = 2^{j/2} h[k - 2^j l] \text{ and } \psi_{j,l}[k] = 2^{j/2} g[k - 2^j l] \quad (2)$$

where $g[k] = (-1)^k h[1-k]$. We can easily achieve a 2-D DWT by using *separable* scaling functions and wavelets (Vetterli. and Kovačević, 1995). The resulting coefficients are often called *subimages* (the wavelet coefficients are also 2-D) and at each decomposition level j, two-dimensional wavelet transform yields one approximation subimage $a_{(j)}$ and three (horizontal h, vertical v, and diagonal d) detail subimages $d^k_{(j)}$ $(k = h, v, d)$.

If we define the durations of signal $f(x)$ in space x and frequency ω by

$$\Delta_x^2 = \int_{-\infty}^{\infty} x^2 |f(x)|^2 dx \text{ and } \Delta_\omega^2 = \int_{-\infty}^{\infty} \omega^2 |F(\omega)|^2 d\omega, \quad (3)$$

respectively then one can define so called a tile in the space-frequency plane. This space-frequency tile tells us the resolutions of wavelet bases in space and frequency domains. Due to the scaling, wavelets used in the decomposition have varying space and frequency resolutions; the frequency duration goes up by 2^j and the spatial duration goes down by 2^j and vice versa. This suggests that the product of space and frequency durations of a signal, i.e., the area of a tile in the space-frequency plane of wavelet transform is a stable quantity and the Uncertainty Principle makes these statements precise, and gives a lower bound for the product.

Wavelet Size Signature. The constant tiling area in the space-frequency tiling of wavelet transform makes it ideally suited for analyzing natural signals. For identifying different size of froth using wavelet analysis froth bubbles with larger sizes will be identified by wider and lower-frequency wavelets and appear in subimages with lower frequency. Froth bubbles with smaller sizes will be identified by narrow and high-frequency wavelets and appear in subimages with higher frequency. At each subimage froth whose (vertical or horizontal) diameters fall within the corresponding width of a tile in the spatial domain will appear. Therefore, the width of a tile in spatial domain can be interpreted as a range of (vertical or horizontal) diameters of bubbles appearing at the corresponding subimage.

If we threshold a subimage then the area of the remaining parts can be thought as the total area of bubbles with sizes corresponding to the subimage. The area can be calculated simply as the fractional the number of thresholded signals (i.e., pixels) at a subimage. Let A_S's be total area of froth at a subimage S calculated from the thresholded subimages and A_T be the area of the entire scene depicted in the original image. Wavelet size signature consists of the fractional areas F_S, which is by its definition

$$F_S = A_S / A_T \quad (4)$$

The average area of a single bubble at a subimage S can be calculated as $\frac{\pi}{4} D_{H,S} D_{V,S}$ where $D_{H,S}$ and $D_{V,S}$ are horizontal and vertical average diameters of a froth calculated from the range of diameters satisfying the space-frequency tiling of the DWT and Uncertainty Principle. N_S, the total number of bubbles in the froth with size of $D_{H,S}$ and $D_{V,S}$ in a subimage S can be calculated from wavelet size signature F_S as

$$N_S = \frac{A_T F_S}{\frac{\pi}{4} D_{H,S} D_{V,S}} \quad (5)$$

Therefore, we can calculate a histogram of bubble size from wavelet size signature without requiring actual measurement of froth morphology because A_T is constant for all images. The examples of bubble size histogram calculated from wavelet size signature are shown in Figure 3 Three froth images with different morphological features in Figure 3(b) can easily be discriminated by comparing their histograms in Figure 3(a) as shown in the figure.

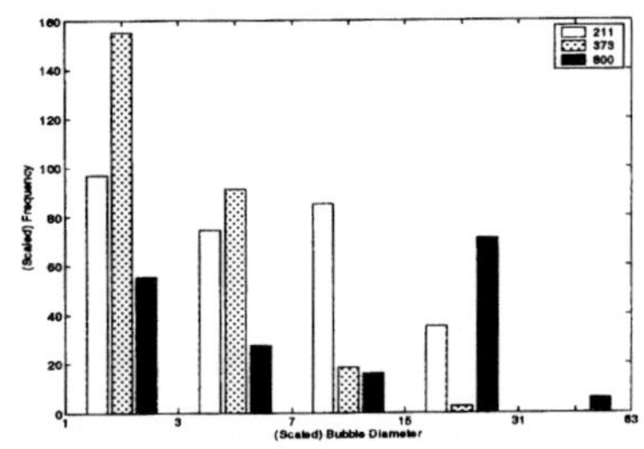

(a)

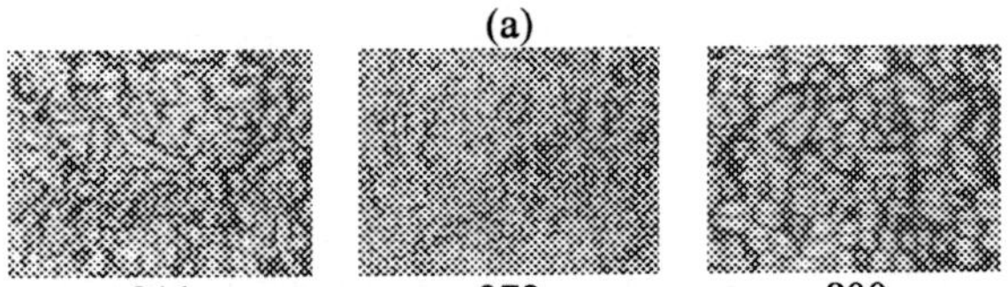

211 373 800

(b)

Figure 3 Illustration of (a) bubble size histograms calculated from wavelet size signature of (b) three different images.

4. DESCRIPTION OF THE FLOTATION PROCESS AND DATA ACQUISITION

The flotation unit in Agnico-Eagle's Laronde plant consists of 3 tank cells used for conditioning and a flotation column. The fresh feed from the grinding circuit enters the first conditioning cell, where lime and activator are added. Lime is used to adjust and control pH. The collector is added in the second conditioning cell. The last conditioning cell is used for mixing only. Air is finally added just before the pulp enters the flotation column. The pulp fresh feed rate was kept nearly constant during all the tests.

The camera was installed on top of the flotation column. It samples 24-bit, 720×480 color images at every minute. We collected images during two plant tests carried out in two different days. The duration of the step input signals was long enough to ensure the process reached new steady states.

5. FROTH FEATURE EXTRACTION USING MR-MIA II

An MIA model (i.e., loading vectors and a mask for clear windows and black holes) is developed from the composite image used in Figure 2, and used as a global model for the sequence of all images.

As mentioned earlier, clear windows and black holes are spectrally similar although they represent two independent process events. From prior knowledge about the process, the two features are correlated with very different froth morphological features and never really occur together. Therefore it is assumed that in any image, if pixels fall under the mask in Figure 2, they will be exclusively clear windows or black holes. To decide which class they represent, a classical two-class classification problem is formulated with wavelet size signature employed as feature inputs and it is solved using Fisher linear discriminant analysis (Duda *et al.*, 2000).

After calculating the first score image, WTA is applied to the first score image of each image in order to extract wavelet size signature. The decomposition level is chosen to be 6 and symlet wavelets with order 4 are used for all images. A fractional area is calculated only from an approximation subimage at each decomposition level and wavelet size signature is calculated from the difference in fractional areas between every two adjacent levels. By using only approximation subimages, the wavelet size signature of an image is a (5×1) vector. The example of the wavelet size signature is already shown in Figure 3.

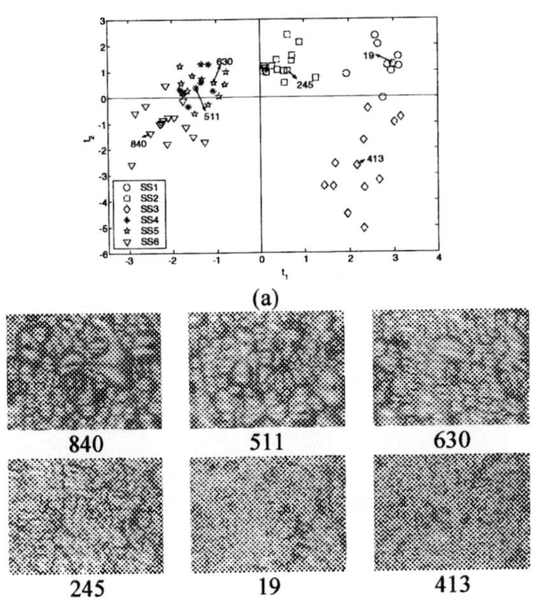

Figure 4 (a) A t_1-t_2 score plot from PCA analysis of MR-MIA II feature variables of selected steady-state images and (b) sample images from 6 steady states.

6. DEVELOPMENT OF PROCESS MONITORING CHARTS

After each (7×1) feature vector (clear windows (1×1), black holes (1×1) and a bubble size histogram (5×1)) is extracted from all images using MR-MIA II, steady state data are selected and a PCA model is built from feature vectors of the selected data. There are 6 steady states (denoted as SS1 ~ SS6 in the legend of figure) corresponding to plant step tests in the data and these steady states form 5 distinct clusters in a t_1-t_2 score plot as shown in Figure 4(a). SS4 and SS5 form one cluster and other steady states form one cluster each and this can be verified by comparing two sample images of SS4 and SS5 in Figure 4(b); they are similar to each other but completely different from 19, 245, 413, and 840 in terms of size of bubbles and area of clear windows. Starting from the fourth quadrant, froth status gradually changes counter clockwise in Figure 4(a). The sample image of SS3 has many small bubbles and some black holes, and the size of bubbles is getting bigger and bigger as one moves counter clockwise. In SS6 in the third quadrant, big bubbles and clear windows are dominant compared to the other 4 clusters. Compared to SS4 and SS5, SS6 has several bigger bubbles and larger area of clear windows. These behaviors in a score plot are verified by a corresponding p_1-p_2 loading plot.

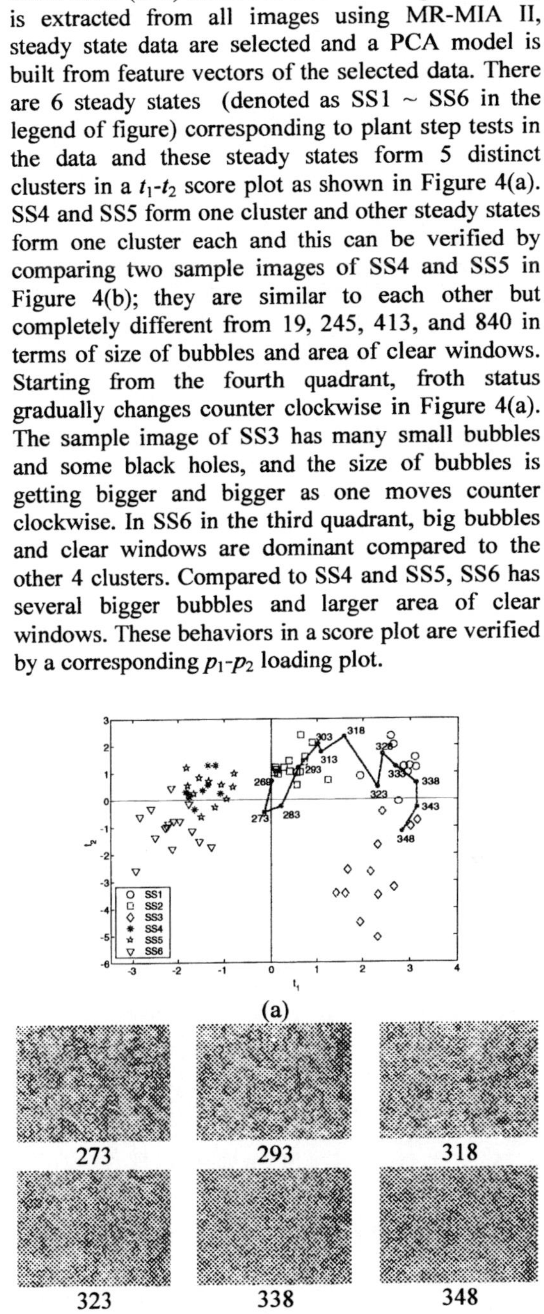

Figure 5 (a) Process transition (from SS2 to SS3) captured by the score plot in Figure 4(a) and (b) selected images during the transition

The process status during the transient states can easily be captured by the same PCA model. Predicted t_1 and t_2 score values of transient data from SS2 to SS3 are plotted over the score plot in Figure 4(a) and shown in Figure 5(a). Six selected images during the transition are shown in Figure 5(b). Starting from the

SS2 region, process passes the third quadrant (273 has several clear windows) and comes back to the SS2 region (compare 293 and 245 in Figure 4(b). They are almost identical.). Then the process moves from SS2 to SS1, passes through the SS1 region, and finally reaches to SS3. These movements can be verified by comparing 6 images in Figure 5(b) and 6 images in Figure 4(b). A process monitoring chart can be easily developed from this score plot. Usually pre-labeled classes are used in the flotation literature on monitoring and control of flotation process but those classes are not independent and discrete events as in typical classification tasks such as character recognition. In process industries, there is always continuous progression from one class to another class and flotation process is one example showing this continuous progression. Therefore, the score plot developed in this work is more suitable for monitoring the froth status in a flotation process. The score space shown in Figure 4(a) can be used as a monitoring chart when it is divided into several meaningful sub-regions based on operators experience and flotation principles. Also a residual plot can be drawn from new data and their prediction by the model in order to detect whether new data (i.e., images in this case) show abnormal behaviors. A residual plot, which corresponds to the t_1-t_2 score plot in Figure 5(a), is shown in Figure 6.

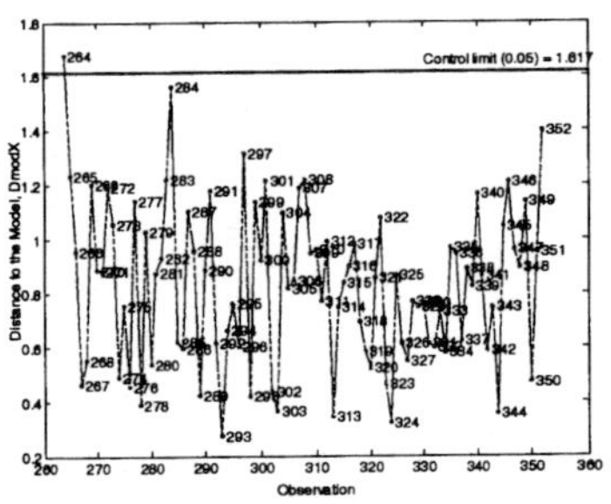

Figure 6. A residual plot corresponding to the t_1-t_2 score plot in Figure 5(a). Upper control limit is from the model built in Figure 4(a).

7. CONCLUSIONS

From the literature on image analysis of flotation process discussed in Sections 2 and 3, we infer that a good image analysis solution should (1) be able to provide rich description of froth morphology, (2) be able to handle correlation in RGB colors, (3) be robust to lighting conditions, and (4) be computationally inexpensive. The proposed approach in this article can satisfy all four requirements whereas contemporary approaches cannot; it can provide rich description of froth morphology compared to approaches based on statistical texture analysis, it is robust to lighting conditions and computationally inexpensive compared to approaches based on segmentation and statistical texture analysis, and it can handle RGB correlation better than approaches based on RGB or HSI/HSV color models.

Monitoring charts developed in this article can provide current froth status during both transient and steady-state operation. Estimation of mineral concentration using MIA (Duchesne *et al.*, 2003) can also be done within the same framework of MR-MIA II

REFERENCES

Bharati, M.H., J. Liu, and J.F. MacGregor, (2004). Image Texture Analysis: Methods and Comparisons. *Chemometrics and Intelligent Laboratory Systems*, **72**(1), 57-71.

Bharati, M.H. and J.F. MacGregor (1998). Multivariate image analysis for real-time process monitoring and control. *Ind. Eng. Chem. Res.* **37** (12), 4715-4724.

Bonifazi, G., S. Serranti, F. Volpe, and R. Zuco (1998). Flotation froth characterization by closed domain (bubbles) color analysis. *4th Int. Conf. On Quality Control by Artificial Vision*, Nov. 10-12, Takamatsu, Japan, 131-137.

Bonifazi, G., S. Serranti, F. Volpe, and R. Zuco (1999). Characterisation of flotation froth colour and structure by machine vision. *GEOVISION 99*, May. 6-7, Liege, Belgium.

Duchesne, C., G. Bartolacci, and A. Bouajila (2003). Predicting concentrate grade using multivariate image analysis of flotation froth. *CSChE*, Oct. 26 – 29, Hamilton, Ontario, Canada.

Duda, R.O., P.E. Hart, and D.G. Stork (2000). *Pattern Classification.* 2nd ed. New York: Wiley-Interscience.

Geladi, P. and H. Grahn (1996). *Multivariate Image Analysis.* John Wiley & Sons, Baffins Lane, Chichester.

Hargrave, J.M. and S.T. Hall (1997). Diagnosis of concentrate grade and mass flowrate in tin flotation from colour and surface texture analysis. *Minerals Eng.*, **10** (6), 613-621.

Liu, J. and J.F. MacGregor (2004). On the extraction of spectral and spatial information for image analysis. submitted for publication.

McKee, D.J. (1991). Automatic flotation control – a review of 20 years of effort. *Minerals Eng.*, **4** (7-11), 653-666.

Moolman, D.W., C. Aldrich, J.S.J. van Deventer, and D.J Bradshaw, (1995). Interpretation of flotation froth surfaces by using digital image analysis and neural networks. *Chemical Engineering Science*, **50** (22), 3501-3513.

Moolman, D.W., J.J. Eksteen, C. Aldrich, and J.S.J. van Deventer (1996). The significance of flotation froth appearance for machine vision control. *Int. J. Miner. Process.*, **48**, 135-158.

Niemi, A.J., H. Hyotyniemi, and R. Ylinen (1999). Image analysis and vision systems for processing plants. *Proceedings of the 2nd Int. Conf. on Intelligent Processing and Manufacturing of Materials*, **1**, 11-20.

Tan, T.S.C. and J. Kittler (1994). Colour texture analysis using colour histogram. *IEE Proc.-Vis. Image Process.*, **141** (6), 403-412.

Vetterli, M. and J. Kovačević (1995). *Wavelets and Subband Coding.* Englewood Cliffs, NJ: Prentice Hall.

ELSEVIER
IFAC
PUBLICATIONS
www.elsevier.com/locate/ifac

FAULT DETECTION AND ISOLATION IN NON-UNIFORMLY SAMPLED SYSTEMS

Weihua Li and **Sirish Shah** [1]

Department of Chemical and Materials Engineering
University of Alberta, Edmonton, AB, T6G 2G6, Canada

Abstract: This paper considers fault detection and isolation (FDI) in a non-uniformly sampled multirate system. By extending the Chow-Willsky scheme from single rate systems to multirate systems, one generates a primary residual vector (PRV) for fault detection. Further, by structuring the PRV to have different sensitivity/insensitivity to different faults, fault isolation is performed. The power of the proposed FDI scheme is illustrated via numerical examples. *Copyright ©2004 IFAC.*

Keywords: non-uniformly sampled multirate multivariate systems, fault detection and isolation, primary residual vector, structured residual vectors

1. INTRODUCTION

This work proposes a novel approach towards fault detection and isolation (FDI) in a non-uniformly sampled multirate multivariate system. A dynamic discrete-time (DT) system is referred to as *multirate* wherein variables are sampled with different rates. Further, the sampling is called *uniform/non-uniform* if a variable is sampled with *equally/non-equally* spaced intervals.

Multirate systems are common in the process industries. In many cases, sampling all variables in a process at a single rate is not possible, because of delays in sensors and laboratory analysis. Recently, research effort has shifted to FDI in multirate systems. Studies by Fadali and co-workers (Fadali and Liu, 1999; Fadali and Shabaik, 2002) consider processes where all variables are uniformly (or regularly) sampled with different rates. This paper considers FDI in a more general case: each variable in a continuous-time (CT) dynamic system is non-uniformly (or irregularly) sampled with a different rate. This represents a very general starting point. All other systems are subsets of this case. There is a lack of formal FDI methodology for this general case, to our best knowledge.

When sampling a CT system with different rates, the *lifting* technique can be used to convert the time-varying multirate DT system into a single rate time-invariant DT system (Khargonekar *et al.*, 1985). Consider a CT system represented by the state space model. We utilize the non-uniformly sampling technique (Sheng *et al.*, 2002) to *lift* the system. Then we extend the Chow-Willsky scheme (Chow and Willsky, 1984) to the lifted system to generate a primary residual vector (PRV) for fault detection. Furthermore, the PRV is transformed into a set of structured residual vectors (SRVs) for fault isolation. Eventually, the practicality and utility of the proposed FDI algorithms is illustrated by application to two numerical examples.

2. PROBLEM FORMULATION

Consider a dynamic system, which in the *fault-free* case is represented by the following CT state space equation

[1] Author to whom all correspondence should be addressed. Email: sirish.shah@ualberta.ca

$$\dot{\mathbf{x}}(t) = \mathbf{A}\mathbf{x}(t) + \mathbf{B}\tilde{\mathbf{u}}(t) + \boldsymbol{\phi}(t)$$
$$\tilde{\mathbf{y}}(t) = \mathbf{C}\mathbf{x}(t) + \mathbf{D}\tilde{\mathbf{u}}(t) \quad (1)$$

where (i) $\tilde{\mathbf{u}}(t) \in \Re^l$ and $\tilde{\mathbf{y}}(t) \in \Re^m$ are *noise-free* inputs and outputs, respectively; (ii) $\mathbf{x}(t) \in \Re^n$ is the state; (iii) $\boldsymbol{\phi}(t) \in \Re^n$ is the disturbance and can be any function of time; and (iv) $\mathbf{A}$, $\mathbf{B}$, $\mathbf{C}$ and $\mathbf{D}$ are known system matrices with appropriate dimensions.

Given a *frame period* T, throughout the paper we sample the variables in the following manners (Sheng et al., 2002):

- $\tilde{\mathbf{u}}(t)$ is sampled g times over the period $[kT, kT+T)$ at instants: $\{kT+t_1, kT+t_2, \cdots, kT+t_g\}$, where $0 = t_1 < \cdots < t_g < T$.
- $\tilde{\mathbf{y}}(t)$ is sampled p times over the same period. Within $[kT+t_i, kT+t_{i+1}]$, $\forall\ i \in [1,g]$, n_i (≥ 0) samples are taken at instants: $\{kT + t_i^1, \cdots, kT+t_i^{n_i}\}$. Similarly, $t_i \leq t_i^1 \cdots < t_i^{n_i} < t_{i+1}$, where $t_{g+1} = T$ and $p = n_1 + \cdots + n_g$.

Among the variables of the system represented by Eqn. 1, each one can be sampled differently from the others. However, for simplicity of mathematical manipulation, in the sequel we assume that (i) the inputs and the disturbances are sampled in one manner; and (ii) the m outputs are sampled in a different manner.

We consider the case of errors-in-variables (EIV). Accordingly, we denote the observed *fault-free* inputs at $kT+t_i$, $\forall\ i \in [1,g]$, by $\mathbf{u}^*(kT+t_i) = \tilde{\mathbf{u}}(kT+t_i) + \mathbf{v}(kT+t_i)$. Similarly, at $kT+t_i^j$, $\forall\ j \in [1,n_i]$, the fault-free outputs are $\mathbf{y}^*(kT+t_i^j) = \tilde{\mathbf{y}}(kT+t_i^j) + \mathbf{o}(kT+t_i^j)$. We assume $\mathbf{v}(\ )$ and $\mathbf{o}(\ )$ to be independently Gaussian distributed white noise vectors having respective covariances $\mathbf{R}_v$ and $\mathbf{R}_o$.

If the sensors are faulty, the measured outputs, $\forall\ j \in [1,n_i]$, can be represented by $\mathbf{y}(kT+t_i^j) = \mathbf{y}^*(kT+t_i^j) + \boldsymbol{\Xi}_y \mathbf{f}_y(kT+t_i^j)$, where $\boldsymbol{\Xi}_y \in \Re^{m\times d_y}$ is a matrix of fault directions; and $\mathbf{f}_y(kT+t_i^j) \in \Re^{d_y}$ are fault magnitude vectors. To represent a single sensor fault in the ith output sensor, $\forall\ i \in [1,m]$ ($d_y = 1$), $\boldsymbol{\Xi}_y \in \Re^m$ is simply the ith column of the $m \times m$ identity matrix $\mathbf{I}_m$. It must be emphasized that $\mathbf{I}_x$ always refers to an $x \times x$ identity matrix throughout the paper.

The sampled inputs are similarly represented by $\mathbf{u}(kT+t_i) = \mathbf{u}^*(kT+t_i) + \boldsymbol{\Xi}_u \mathbf{f}_u(kT+t_i)$, where $\boldsymbol{\Xi}_u \in \Re^{l\times d_u}$ and $\mathbf{f}_u(kT+t_i) \in \Re^{d_u}$ resemble $\boldsymbol{\Xi}_y$ and $\mathbf{f}_y(kT+t_i^j)$, respectively.

With $\{\mathbf{u}(kT+t_i)\}$ and $\{\mathbf{y}(kT+t_i^j)\}$, $\forall\ i \in [1,g]$, $j \in [1,n_i]$, and $k = [1,\cdots)$, the goal of FDI is to indicate when $\mathbf{f}_u(kT+t_i)$ and/or $\mathbf{f}_y(kT+t_i^j)$ are non-zero; and further identify $\boldsymbol{\Xi}_u$ and/or $\boldsymbol{\Xi}_y$.

3. NON-UNIFORMLY SAMPLED SYSTEMS

Post-multiplying Eqn. 1 by a non-singular $\mathbf{e}^{-\mathbf{A}t}$, $\forall\ t \neq 0$, leads to

$$d\left[\mathbf{e}^{-\mathbf{A}t}\mathbf{x}(t)\right]/dt = \mathbf{e}^{-\mathbf{A}t}\left[\mathbf{B}\tilde{\mathbf{u}}(t) + \boldsymbol{\phi}(t)\right] \quad (2)$$

Integrating Eqn. 2 gives

$$\mathbf{x}(k+1) = \int_{kT}^{kT+T} \mathbf{e}^{\mathbf{A}(kT+T-t)}\left[\mathbf{B}\tilde{\mathbf{u}}(t) + \boldsymbol{\phi}(t)\right]dt + \underline{\mathbf{A}}\ \mathbf{x}(k) \quad (3)$$

where $\mathbf{x}(k) \equiv \mathbf{x}(t)|_{t=kT}$, $\mathbf{x}(k+1) \equiv \mathbf{x}(t)|_{t=kT+T}$, and $\underline{\mathbf{A}} \equiv \mathbf{e}^{\mathbf{A}T}$. Consequently, the following lifted state space equation can be derived,

$$\mathbf{x}(k+1) = \underline{\mathbf{A}}\ \mathbf{x}(k) + \underline{\mathbf{B}}\ \underline{\tilde{\mathbf{u}}}(k) + \underline{\mathbf{E}}\ \underline{\boldsymbol{\phi}}(k) \quad (4)$$

where, $\forall\ i \in [1,g]$, $\underline{\mathbf{B}} = [\mathbf{B}_1\ \mathbf{B}_2\ \cdots\ \mathbf{B}_g]$, and $\underline{\mathbf{E}} = [\mathbf{E}_1\ \mathbf{E}_2\ \cdots\ \mathbf{E}_g]$ with

$$\mathbf{B}_i = \int_{T-t_{i+1}}^{T-t_i} \mathbf{e}^{\mathbf{A}t}\mathbf{B}dt,\ \mathbf{E}_i = \int_{T-t_{i+1}}^{T-t_i} \mathbf{e}^{\mathbf{A}t}dt,$$

$$\underline{\tilde{\mathbf{u}}}(k) = \begin{bmatrix} \tilde{\mathbf{u}}(kT+t_1) \\ \vdots \\ \tilde{\mathbf{u}}(kT+t_g) \end{bmatrix};\ \underline{\boldsymbol{\phi}}(k) = \begin{bmatrix} \boldsymbol{\phi}(kT+t_1) \\ \vdots \\ \boldsymbol{\phi}(kT+t_g) \end{bmatrix}.$$

It is arguable that $\underline{\boldsymbol{\phi}}(k)$ is not available. As will be shown later, $\underline{\boldsymbol{\phi}}(k)$ will be entirely removed from the PRV.

The integration of Eqn. 2 from $t \in [kT, kT+\tau]$, $\forall\ 0 < \tau < T$, can result in $\mathbf{x}(kT+\tau)$, whose substitution into Eqn. 1 leads to

$$\tilde{\mathbf{y}}(kT+\tau) = \mathbf{C}\int_{kT}^{kT+\tau} \mathbf{e}^{\mathbf{A}(kT+\tau-t)}\left[\mathbf{B}\tilde{\mathbf{u}}(t) + \boldsymbol{\phi}(t)\right]dt + \mathbf{C}\mathbf{e}^{\mathbf{A}\tau}\mathbf{x}(k) + \mathbf{D}\tilde{\mathbf{u}}(kT+\tau)$$

After defining the lifted output vector:

$$\underline{\tilde{\mathbf{y}}}(k) \equiv \begin{bmatrix} \tilde{\mathbf{y}}(kT+t_1^1) \\ \vdots \\ \tilde{\mathbf{y}}(kT+t_1^{n_1}) \\ \vdots \\ \tilde{\mathbf{y}}(kT+t_g^1) \\ \vdots \\ \tilde{\mathbf{y}}(kT+t_g^{n_g}) \end{bmatrix} \in \Re^{mp},$$

one can arrive at the following lifted output equation:

$$\tilde{\underline{\mathbf{y}}}(k) = \underline{\mathbf{C}}\ \mathbf{x}(k) + \underline{\mathbf{D}}\ \tilde{\underline{\mathbf{u}}}(k) + \underline{\mathbf{J}}\ \underline{\phi}(k) \tag{5}$$

where, $\underline{\mathbf{C}} = \left[\mathbf{C}_1'\ \mathbf{C}_2'\ \ldots \mathbf{C}_g'\right]'$ with $'$ symbolizing the transpose of the argument;

$$\underline{\mathbf{D}} = \begin{bmatrix} \mathbf{D}_{1,1} & \mathbf{0} & \ldots & \mathbf{0} \\ \mathbf{D}_{2,1} & \mathbf{D}_{2,2} & \mathbf{0} & \vdots \\ \vdots & & \ddots & \\ \mathbf{D}_{g,1} & \mathbf{D}_{g,2} & \mathbf{D}_{g,3} \ldots & \mathbf{D}_{g,g} \end{bmatrix};$$

and $\underline{\mathbf{J}} = \underline{\mathbf{D}}|_{\mathbf{D}_{i_1,i_2}=\mathbf{J}_{i_1,i_2}}$ with $i_1 \in [1,g]$ and $i_2 \in [1, i_1]$. Note that in the preceding matrices,

$$\mathbf{C}_i = \begin{bmatrix} \mathbf{C}e^{\mathbf{A}t_i^1} \\ \mathbf{C}e^{\mathbf{A}t_i^2} \\ \vdots \\ \mathbf{C}e^{\mathbf{A}t_i^{n_i}} \end{bmatrix},\ \mathbf{D}_{i,i} = \begin{bmatrix} \mathbf{C}\int_0^{t_i^1} e^{\mathbf{A}t}\mathbf{B}dt + \mathbf{D} \\ \vdots \\ \mathbf{C}\int_0^{t_i^{n_i}} e^{\mathbf{A}t}\mathbf{B}dt + \mathbf{D} \end{bmatrix},$$

$$\mathbf{D}_{i,j} = [\int_{t_i^1-t_{j+1}}^{t_i^1-t_j} (\mathbf{C}e^{\mathbf{A}t}\mathbf{B})'\,dt \cdots \int_{t_i^{n_i}-t_{j+1}}^{t_i^{n_i}-t_j} (\mathbf{C}e^{\mathbf{A}t}\mathbf{B})'\,dt]',$$

and $\mathbf{J}_{i,i} = \mathbf{D}_{i,i}|_{\mathbf{B}=\mathbf{I}_n,\mathbf{D}=\mathbf{0}}$, $\mathbf{J}_{i,j} = \mathbf{D}_{i,j}|_{\mathbf{B}=\mathbf{I}_n}$, $\forall\ j \in [1, i-1]$. The detailed derivation of Eqns. 4 and 5 can be seen in Sheng et al. (2002).

Combining Eqns. 4 and 5 eventually leads to

$$\begin{aligned} \mathbf{x}(k+1) &= \underline{\mathbf{A}}\ \mathbf{x}(k) + \underline{\mathbf{B}}\ \tilde{\underline{\mathbf{u}}}(k) + \underline{\mathbf{E}}\ \underline{\phi}(k) \\ \tilde{\underline{\mathbf{y}}}(k) &= \underline{\mathbf{C}}\ \mathbf{x}(k) + \underline{\mathbf{D}}\ \tilde{\underline{\mathbf{u}}}(k) + \underline{\mathbf{J}}\ \underline{\phi}(k) \end{aligned} \tag{6}$$

which is the lifted model of Eqn. 1 when non-uniformly sampled.

4. FAULT DETECTION

4.1 *Description of the lifted system with faults*

We define a stacked vector:

$$\tilde{\underline{\mathbf{y}}}_s(k) \equiv [\tilde{\underline{\mathbf{y}}}'(k-s)\ \tilde{\underline{\mathbf{y}}}'(k-s+1)\ \ldots \tilde{\underline{\mathbf{y}}}'(k)]',$$

where s is the order of the parity space (Chow and Willsky, 1984). We select $s = n$ for simplicity. In the sequel, any stacked vector is defined analogously to $\tilde{\underline{\mathbf{y}}}_s(k)$. Manipulating Eqn. 6 gives

$$\tilde{\underline{\mathbf{y}}}_s(k) = \underline{\mathbf{\Gamma}}_s\mathbf{x}(k-s) + \underline{\mathbf{H}}_s\tilde{\underline{\mathbf{u}}}_s(k) + \underline{\mathbf{G}}_s\underline{\phi}_s(k) \tag{7}$$

where $\tilde{\underline{\mathbf{u}}}_s(k)$ and $\underline{\phi}_s(k)$ are also stacked vectors,

$$\underline{\mathbf{\Gamma}}_s = [\underline{\mathbf{C}}'\ \ \underline{\mathbf{A}}'\underline{\mathbf{C}}'\ \ldots (\underline{\mathbf{A}}^s)'\ \underline{\mathbf{C}}']' \in \Re^{mp(s+1)\times n};$$

$$\underline{\mathbf{H}}_s = \begin{bmatrix} \underline{\mathbf{D}} & \mathbf{0} & \ldots\ \mathbf{0} \\ \underline{\mathbf{C}}\,\underline{\mathbf{B}} & \underline{\mathbf{D}} & \vdots \\ \vdots & & \ddots \\ \underline{\mathbf{C}}\,\underline{\mathbf{A}}^{s-1}\underline{\mathbf{B}} & \underline{\mathbf{C}}\,\underline{\mathbf{A}}^{s-2}\underline{\mathbf{B}} & \ldots\ \underline{\mathbf{D}} \end{bmatrix};$$

and $\underline{\mathbf{G}}_s = \underline{\mathbf{H}}_s|_{\underline{\mathbf{B}}=\underline{\mathbf{E}},\underline{\mathbf{D}}=\underline{\mathbf{J}}}$.

It follows from $\mathbf{u}(k)$ and $\mathbf{y}(k)$ that the lifted vectors of measured inputs and outputs are

$$\begin{aligned} \underline{\mathbf{u}}(k) &= \tilde{\underline{\mathbf{u}}}(k) + \underline{\mathbf{v}}(k) + \Xi_u^g\underline{\mathbf{f}}_u(k) \\ \underline{\mathbf{y}}(k) &= \tilde{\underline{\mathbf{y}}}(k) + \underline{\mathbf{o}}(k) + \Xi_y^p\underline{\mathbf{f}}_y(k) \end{aligned} \tag{8}$$

where, $\underline{\mathbf{v}}(k) = \tilde{\underline{\mathbf{u}}}(k)|_{\tilde{\mathbf{u}}()=\mathbf{v}()}$, $\underline{\mathbf{f}}_u(k) = \tilde{\underline{\mathbf{u}}}(k)|_{\tilde{\mathbf{u}}()=\mathbf{f}_u()}$, $\underline{\mathbf{o}}(k) = \tilde{\underline{\mathbf{y}}}(k)|_{\tilde{\mathbf{y}}()=\mathbf{o}()}$, and $\underline{\mathbf{f}}_y(k) = \tilde{\underline{\mathbf{u}}}(k)|_{\tilde{\mathbf{u}}()=\mathbf{f}_y()}$. In addition, $\Xi_u^g \equiv \mathbf{I}_g \otimes \Xi_u$; $\Xi_y^p \equiv \mathbf{I}_p \otimes \Xi_y$; and $\otimes$ is the Kronecker tensor product.

Stacking Eqn. 8 facilitates the relationship between the stacked vectors:

$$\begin{aligned} \underline{\mathbf{u}}_s(k) &= \tilde{\underline{\mathbf{u}}}_s(k) + \underline{\mathbf{v}}_s(k) + \Xi_{s,u}^g\underline{\mathbf{f}}_{s,u}(k) \\ \underline{\mathbf{y}}_s(k) &= \tilde{\underline{\mathbf{y}}}_s(k) + \underline{\mathbf{o}}_s(k) + \Xi_{s,y}^p\underline{\mathbf{f}}_{s,y}(k) \end{aligned} \tag{9}$$

where $\Xi_{s,u}^g = \mathbf{I}_{s+1} \otimes \Xi_u^g$, $\Xi_{s,y}^p = \mathbf{I}_{s+1} \otimes \Xi_y^p$. Moreover, using Eqn. 9 can rewrite Eqn. 7 as

$$\begin{aligned} \underline{\mathbf{y}}_s(k) - \underline{\mathbf{H}}_s\underline{\mathbf{u}}_s(k) &= \underline{\mathbf{\Gamma}}_s\mathbf{x}(k-s) - \underline{\mathbf{H}}_s\underline{\mathbf{v}}_s(k) + \underline{\mathbf{o}}_s(k) \\ -\underline{\mathbf{H}}_s\Xi_{s,u}^g\underline{\mathbf{f}}_{s,u}(k) &+ \Xi_{s,y}^p\underline{\mathbf{f}}_{s,y}(k) + \underline{\mathbf{G}}_s\underline{\phi}_s(k) \end{aligned} \tag{10}$$

4.2 *Design of the PRV for fault detection*

We select a matrix $\mathbf{W}_0$ from the *left null space* (LNS) of $\underline{\mathbf{\Gamma}}_s^G \equiv [\underline{\mathbf{\Gamma}}_s\ \underline{\mathbf{G}}_s]$. By extending the Chow-Willsky scheme (Chow and Willsky, 1984), pre-multiplying both sides of Eqn. 10 by $\mathbf{W}_0$, we obtain the PRV as follows,

$$\begin{aligned} \mathbf{e}_s(k) &\equiv \mathbf{W}_0\left[\underline{\mathbf{y}}_s(k) - \underline{\mathbf{H}}_s\underline{\mathbf{u}}_s(k)\right] \\ &= \mathbf{W}_0\left[\Xi_{s,y}^p\underline{\mathbf{f}}_{s,y}(k) - \underline{\mathbf{H}}_s\Xi_{s,u}^g\underline{\mathbf{f}}_{s,u}(k)\right] + \mathbf{e}_s^*(k) \end{aligned} \tag{11}$$

where the unknown state vector $\mathbf{x}(k-s)$ and the lifted disturbance $\underline{\phi}_s(k)$ have been completely decoupled. Besides, $\mathbf{e}_s^*(k) = \mathbf{W}_0\left[-\underline{\mathbf{H}}_s\underline{\mathbf{v}}_s(k) + \underline{\mathbf{o}}_s(k)\right]$, which is a zero-mean Gaussian distributed random vector with covariance $\mathbf{R}_{s,e}$ (Johnson and Wichern, 1998), i.e. $\mathbf{e}_s(k) \sim \aleph(\mathbf{0}, \mathbf{R}_{s,e})$.

- In the ideal case, $\mathbf{e}_s(k) = \mathbf{0}$, because $\underline{\mathbf{o}}_s(k) = \mathbf{0}$, $\underline{\mathbf{v}}_s(k) = \mathbf{0}$, $\underline{\mathbf{f}}_{s,u}(k) = \mathbf{0}$, and $\underline{\mathbf{f}}_{s,y}(k) = \mathbf{0}$.
- In the fault-free case, $\mathbf{e}_s(k) = \mathbf{e}_s^*(k)$.
- In the presence of any sensor faults,

$$\mathbf{e}_s(k) = \mathbf{e}_s^*(k) + \mathbf{e}_s^f(k) \tag{12}$$

where

$$\mathbf{e}_s^f(k) = \mathbf{W}_0\left[\Xi_{s,y}^p \underline{\mathbf{f}}_{s,y}(k) - \underline{\mathbf{H}}_s \Xi_{s,u}^g \underline{\mathbf{f}}_{s,u}(k)\right]$$

is the fault contribution. In this case, $\mathbf{e}_s(k) \sim \aleph(\mathbf{e}_s^f(k), \mathbf{R}_{s,e})$.

Therefore, fault detection is equivalent to checking if $\mathbf{e}_s(k)$ is *zero-mean*. This can be done by testing if $\eta_s(k) = \mathbf{e}_s'(k)\mathbf{R}_{s,e}^{-1}\mathbf{e}_s(k)$ follows a central chi-square distribution with degrees of freedom equal to Rank($\mathbf{W}_0$) (Johnson and Wichern, 1998). With a pre-selected level of significance α, while $\eta_s(k) < \chi_\alpha^2[\text{Rank}(\mathbf{W}_0)]$ indicates that all sensors function normally, $\eta_s(k) \geq \chi_\alpha^2[\text{Rank}(\mathbf{W}_0)]$ triggers alarming of any faulty sensors.

Finally, we investigate the calculation of $\mathbf{W}_0$. We define $\text{Rank}(\mathbf{W}_0) \equiv (mp - ng)(s+1) - n$ and will use this quantity in the sequel. It can be proved that $\mathbf{W}_0$ has Rank($\mathbf{W}_0$) independent rows and Rank($\mathbf{W}_0$) > 0 if $mp > ng$ and $s = n$. Therefore, a non-trivial solution to $\mathbf{W}_0$ can exist.

Denote $\tilde{\underline{\mathbf{H}}}_s \equiv [\mathbf{I}_{mps+mp} \mid -\underline{\mathbf{H}}_s]$. Then, it follows from Eqn. 11 that $\mathbf{e}_s(k) = \mathbf{W}_0\tilde{\underline{\mathbf{H}}}_s\left[\underline{\mathbf{y}}_s'(k)\ \ \underline{\mathbf{u}}_s'(k)\right]'$.

$\mathbf{e}_s(k)$ should have maximized sensitivity with any sensors. In line with this criterion, it comes from (Li and Shah, 2002) that

$$\mathbf{W}_0' = \text{the largest eigenvectors of } \underline{\mathbf{\Gamma}}_s^{G,\perp}\tilde{\underline{\mathbf{H}}}_s\tilde{\underline{\mathbf{H}}}_s' \text{ related to non-zero eigenvalues,}$$

where $\underline{\mathbf{\Gamma}}_s^{G,\perp} = \mathbf{I}_{mps+mp} - \underline{\mathbf{\Gamma}}_s^G\left[\left(\underline{\mathbf{\Gamma}}_s^G\right)'\underline{\mathbf{\Gamma}}_s^G\right]^{-1}\left(\underline{\mathbf{\Gamma}}_s^G\right)'$.

We consider the isolation of a *single* faulty sensor at each time. Since the considered system has l inputs and m outputs, the goal of isolation is achievable by generating $(m + l)$ SRVs, where the ith SRV is made insensitive to the ith sensor fault but most sensitive to other sensor faults, $\forall\, i \in [1, m + l]$.

Mathematically, since the ith SRV is

$$\mathbf{r}_{s,i}(k) = \mathbf{W}_i\mathbf{e}_s(k) \tag{13}$$

designing a set of SRVs is equivalent to computing $(m + l)$ transformation matrices $\mathbf{W}_i$.

Introduce a new notation $\underline{\mathbf{P}}_s \equiv \mathbf{W}_0\tilde{\underline{\mathbf{H}}}_s$. Consequently, $\mathbf{e}_s(k) = \underline{\mathbf{P}}_s\left[\underline{\mathbf{y}}_s'(k)\ \underline{\mathbf{u}}_s'(k)\right]'$, and it follows from Eqn. 13 that $\mathbf{r}_{s,i}(k) = \mathbf{W}_i\underline{\mathbf{P}}_s\left[\underline{\mathbf{y}}_s'(k)\ \underline{\mathbf{u}}_s'(k)\right]'$. Note that $\text{Rank}\left(\underline{\mathbf{P}}_s\right) = \text{Rank}\left(\mathbf{W}_0\right)$.

Denote, $\forall\, i \in [1, m]$ and $j \in [1, l]$,

$$\underline{\mathbf{y}}_s(k,i) \equiv \begin{bmatrix} \underline{\mathbf{y}}(k-s,i) \\ \vdots \\ \underline{\mathbf{y}}(k,i) \end{bmatrix}, \underline{\mathbf{u}}_s(k,j) \equiv \begin{bmatrix} \underline{\mathbf{u}}(k-s,j) \\ \vdots \\ \underline{\mathbf{u}}(k,j) \end{bmatrix},$$

where

$$\underline{\mathbf{y}}(k,i) = \begin{bmatrix} y(kT + t_1^1, i) \\ \vdots \\ y(kT + t_1^{n_1}, i) \\ \vdots \\ y(kT + t_g^1, i) \\ \vdots \\ y(kT + t_g^{n_g}, i) \end{bmatrix} \in \Re^p$$

and $\underline{\mathbf{u}}(k,j) = [u(kT + t_1, j) \cdots u(kT + t_g, j)]'$. Note that $y(kT + \tau, i)$ and $u(kT + \mu, j)$ represent the ith element of $\mathbf{y}(kT + \tau)$ for $\tau \in [t_1^1, \cdots, t_1^{n_1}, \cdots, t_g^1, \cdots, t_g^{n_g}]$, and the jth element of $\mathbf{u}(kT + \mu)$ for $\mu \in [t_1, \cdots, t_g]$, respectively.

We re-organize $\underline{\mathbf{y}}_s(k)$ and $\underline{\mathbf{u}}_s(k)$ as follows,

$$\underline{\mathbf{y}}_s(k) = \begin{bmatrix} \underline{\mathbf{y}}_s(k,1) \\ \vdots \\ \underline{\mathbf{y}}_s(k,m) \end{bmatrix}, \quad \underline{\mathbf{u}}_s(k) = \begin{bmatrix} \underline{\mathbf{u}}_s(k,1) \\ \vdots \\ \underline{\mathbf{u}}_s(k,l) \end{bmatrix}.$$

As a consequence, columns in $\underline{\mathbf{P}}_s$ must be regrouped such that the first $ps + p$ columns correspond to $\underline{\mathbf{y}}_s(k,1)$, the second $ps + p$ columns to $\underline{\mathbf{y}}_s(k,2)$, and so forth.

We use $\underline{\mathbf{P}}_{s,i}$ to represent those columns in $\underline{\mathbf{P}}_s$ associated with $\underline{\mathbf{y}}_s(k,i)$ if $i \in [1,m]$; or $\underline{\mathbf{u}}_s(k, i-m)$ if $i \in [m+1, m+l]$. Since $\mathbf{r}_{s,i}(k)$ is designed to be insensitive to $\underline{\mathbf{y}}_s(k,i)$, $\mathbf{W}_i\underline{\mathbf{P}}_{s,i} = \mathbf{0}$ must be ensured.

Moreover, each $\mathbf{W}_i$ must have a maximum covariance with $\underline{\mathbf{P}}_s$. The algorithms of calculating $\mathbf{W}_0$ can be directly applied to calculating $\mathbf{W}_i$, if replacing $\tilde{\underline{\mathbf{H}}}_s\tilde{\underline{\mathbf{H}}}_s'$ by $\underline{\mathbf{P}}_s\underline{\mathbf{P}}_s'$ and $\underline{\mathbf{\Gamma}}_s^G$ by $\underline{\mathbf{P}}_{s,i}$.

It follows from Eqns. 12 and 13 that

$$\mathbf{r}_{s,i}(k) = \mathbf{r}_{s,i}^*(k) + \mathbf{r}_{s,i}^f(k) \tag{14}$$

where $\mathbf{r}_{s,i}^f(k) = \mathbf{W}_i\mathbf{e}_s^f(k)$, and $\mathbf{r}_{s,i}^*(k) = \mathbf{W}_i\mathbf{e}_s^*(k)$. From the distribution of $\mathbf{e}_s^*(k)$, it is known that $\mathbf{r}_{s,i}^*(k) \sim \aleph(\mathbf{0}, \mathbf{R}_{s,i})$, where $\mathbf{R}_{s,i} = \mathbf{W}_i\mathbf{R}_{s,e}\mathbf{W}_i'$. In addition, $\mathbf{r}_{s,i}(k) \sim \aleph(\mathbf{0}, \mathbf{R}_{s,i})$ if the ith sensor is faulty according to the selected isolation logic; and $\mathbf{r}_{s,i}(k) \sim \aleph\left(\mathbf{r}_{s,i}^f(k), \mathbf{R}_{s,i}\right)$ once the jth $(j \neq i)$ sensor is faulty. Therefore, if $\mathbf{r}_{s,i}(k)$ is zero-mean, $\forall\ i \in [1, m+l]$, but $\mathbf{r}_{s,j}(k)$ is non-zero mean $\forall$ $\{j \neq i\} \bigcap \{j \in [1, m+l]\}$, we conclude that the ith sensor has failed.

We can also use a scalar statistic $\eta_{s,i}(k) = \mathbf{r}_{s,i}'(k)\mathbf{R}_{s,i}^{-1}\mathbf{r}_{s,i}(k)$ as a fault isolation index. We can conclude that the ith sensor is faulty, if $\eta_{s,i}(k)$ is less but $\eta_{s,j}(k)$ is larger than a pre-determined confidence limit.

5. NUMERICAL EVALUATION

A quadruple tank system (Ge and Fang, 1988), where tanks with the same height and same cross section are serially connected by outlets that have an identical cross section. In the system, the input is the water flowing into Tank 1, and the controlled variables are the levels $\mathbf{x}(t) = [x_1(t)\ x_2(t)\ x_3(t)\ x_4(t)]'$ in Tanks 1 up to 4, respectively.

The dynamics of the system can be described by the following CT state space model :

$$\begin{aligned} \dot{\mathbf{x}}(t) &= \mathbf{A}\mathbf{x}(t) + \mathbf{B}\tilde{u}(t) \\ \tilde{\mathbf{y}}(t) &= \mathbf{C}\mathbf{x}(t) \end{aligned} \quad (15)$$

where besides $\mathbf{x}(t)$, $\tilde{\mathbf{y}}(t) \in \Re^4$ is the output vector,

$$\mathbf{A} = \begin{bmatrix} -0.0457 & 0.0457 & 0 & 0 \\ 0.0457 & -0.0914 & 0.0457 & 0 \\ 0 & 0.0457 & -0.0914 & 0.0457 \\ 0 & 0 & 0.0457 & -0.0914 \end{bmatrix},$$

$$\mathbf{B} = \begin{bmatrix} 0.0020\ 0\ 0\ 0 \end{bmatrix}', \text{ and } \mathbf{C} = \mathbf{I}_4.$$

In accordance with Ge and Fang (1988), the noise-free and fault-free input to the tank system is simulated by $\tilde{u}(t) = 1 + 0.36sin(t)$.

With such an input, we use the function 'ode45' in MATLAB^{TM} to simulate Eqn. 15, generating the CT signals $\{\tilde{\mathbf{y}}(t), \tilde{u}(t)\}$.

Select a frame period $T = 0.5$ minute. For $k = [0, 1, \cdots)$, within the period $[kT, kT+T]$ we sample $\tilde{u}(t)$ at $t = kT$ and $t = kT+0.2$; and $\tilde{\mathbf{y}}(t)$ at $t = kT$ and $t = kT + 0.3$, respectively; resulting in

$$\underline{\tilde{\mathbf{u}}}(k) = \begin{bmatrix} \tilde{u}(kT) \\ \tilde{u}(kT+0.2) \end{bmatrix}, \ \underline{\tilde{\mathbf{y}}}(k) = \begin{bmatrix} \tilde{\mathbf{y}}(kT) \\ \tilde{\mathbf{y}}(kT+0.3) \end{bmatrix}.$$

Applying $\underline{\tilde{\mathbf{u}}}(k)$ and $\underline{\tilde{\mathbf{y}}}(k)$ to Eqn. 6 gives us the lifted system model, $\underline{\mathbf{A}}$, $\underline{\mathbf{B}}$, $\underline{\mathbf{C}}$, $\underline{\mathbf{D}}$, $\underline{\mathbf{E}}$, and $\underline{\mathbf{J}}$. Since there are 5 sensors in total, we generate 5 SRVs for fault isolation. We calculate the models for the PRV and 5 SRVs, respectively.

We introduce Gaussian distributed white noise to $\underline{\tilde{\mathbf{u}}}(k)$ and $\underline{\tilde{\mathbf{y}}}(k)$ to produce 1000 samples of training data that are used to estimate $\mathbf{R}_{s,e}$ and $\mathbf{R}_{s,i}$. Given $\alpha = 0.01$, the confidence limit of $\eta_s(k)$ is $\chi^2_{0.01}(36) = 58.5713$. In addition, the confidence limit for each $\eta_s^i(k)$ is $\chi^2_{0.01}(26) = 45.6420$, $\forall\ i \in [1, 5]$. We define the scaled FDI indices, $\bar{\eta}_s(k) \equiv \eta_s(k)/58.5713$ and $\bar{\eta}_s^i(k) \equiv \eta_s^i(k)/45.6420$, all of which have a confidence limit, 1, and will be plotted to show FDI results.

A drift fault simulated by $(t - t_f)/500$, where t_f is the time at which the fault begins to occur, is introduced to one sensor at $t = t_f = 501 * T = 205.5$ minutes. The FDI results are depicted in Figure 1, where in the x-axis each sample represents 0.5 minute. In addition, Fd represents $\bar{\eta}_s(k)$, and $\{Fi_1, \cdots, Fi_5\}$ represents $\{\ \bar{\eta}_s^1(k), \cdots,\ \bar{\eta}_s^5(k)\}$, respectively.

Fd is beyond 1 after the occurrence of the fault, indicating the success of fault detection. Moreover, since Fi_1 (< 1) is unaffected by the fault, while $\{Fi_2, Fi_3, Fi_4, Fi_5\}$ (> 1) have been affected by the fault, one infers that the first output sensor has a fault. Notice that there is a delay in fault detection, because a drift fault evolves with time very slowly.

5.1 *Comparison with single rate FDI schemes*

We use another quadruple tank system (Johansson, 2000) as a benchmark to conduct comparative study among the newly proposed multirate FDI scheme and the standard single rate one. Detailed description of the tank system can be found in the reference mentioned above.

The dynamics of the tank system around an operating point can be represented by the following CT state space equations (Johansson, 2000):

$$\dot{\mathbf{x}}(t) = \begin{bmatrix} -0.016 & 0 & 0.042 & 0 \\ 0 & -0.011 & 0 & 0.033 \\ 0 & 0 & -0.042 & 0 \\ 0 & 0 & 0 & -0.033 \end{bmatrix} \mathbf{x}(t)$$

$$+ \begin{bmatrix} 0.083 & 0 \\ 0 & 0.063 \\ 0 & 0.048 \\ 0.031 & 0 \end{bmatrix} \mathbf{u}(t), \ \mathbf{y}(t) = 0.5\mathbf{I}_4\mathbf{x}(t) \quad (16)$$

where in $\mathbf{x}(t)$, its ith element, $x_i(t)$ for $i \in [1, 4]$, represents the variation of the level in the ith tank. Similarly, in $\mathbf{y}(t)$, its ith element $y_i(t)$ represents the measured level variation in the ith tank. The inputs to the system are voltages applied to two pumps which provide water to the four tanks. We simulate the inputs by pseudo random binary signals with small magnitude. The frequency bands for the frequency contents of the inputs are chosen to be [0,0.03], and [0,0.05], expressed in fractions of the Nyquist frequencies.

In the second tank system, there are 6 sensors in total. Accordingly, 1 PRV and 6 SRVs are generated for FDI. Furthermore, from the PRV and SRVs, the related FDI indices are calculated and scaled to have unit confidence limit. Similarly, the ith fault isolation index is made insensitive to fault in the ith sensor but most sensitive to faults in any other sensors.

Comparative studies are conducted among (a) FDI using multirate data; and (b) FDI using a low single rate data, where different types of faults in different sensors are simulated.

A complete failure was introduced to one sensor. We sampled the two inputs with a sampling interval of 1 minute and the four outputs with a sampling interval of 2 minutes. The corresponding FDI results are shown in Figure 2, where Fd and $\{Fi_1, \cdots, Fi_6\}$ are similar to those in Figure 1. Figure 2 shows that the faulty sensor has been successfully detected and isolated.

Furthermore, we sample the inputs and outputs with an identical sampling interval of 2 minutes, and then perform FDI on the data collected with a slower rate. The relevant FDI results are demonstrated in Figure 3. Apparently, the FDI performance in this case is much worse compared with that illustrated in Figure 2. This would be expected since we are now downsampling the data and thus ignoring potentially useful information.

6. CONCLUSION

A novel approach to detection and isolation of sensor faults in non-uniformly sampled multirate dynamic systems has been proposed. This approach has been applied to a simulation example, where different types of faults including bias, drift, and precision degradation, are introduced to sensors, respectively, and are successfully detected and isolated.

Comparative studies have also been carried out via another quadruple tank system. It has been verified that multirate data-based FDI outperforms the slow single rate data-based FDI.

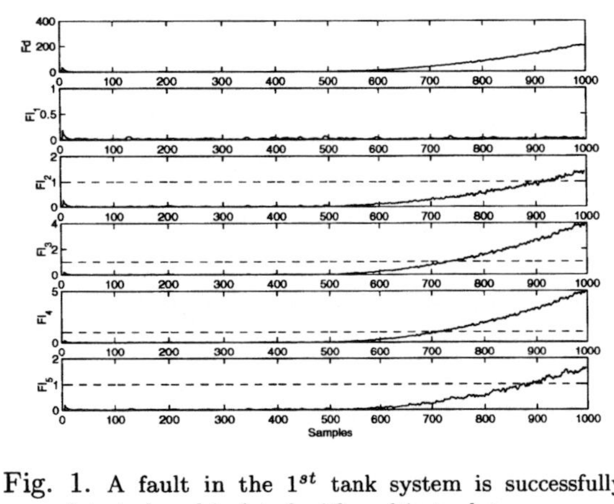

Fig. 1. A fault in the 1st tank system is successfully detected and isolated with multirate data.

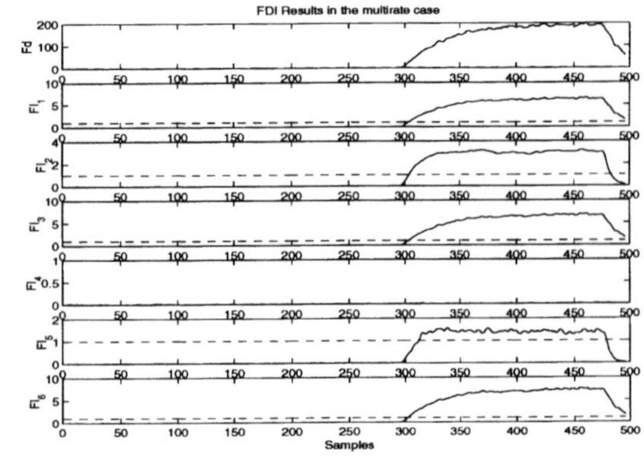

Fig. 2. A fault in the second tank system is successfully detected and isolated with multirate data.

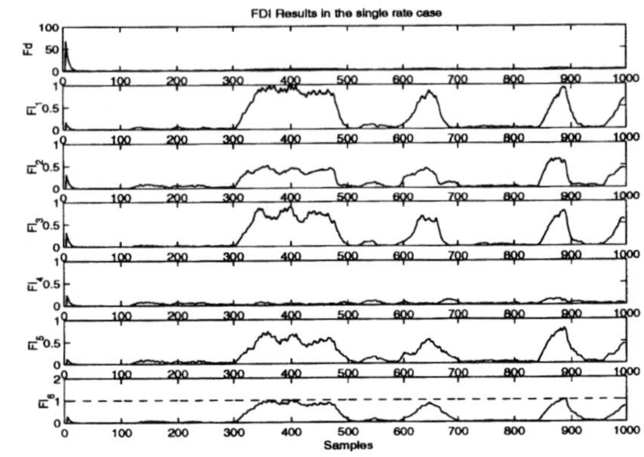

Fig. 3. Using slow single rate data fails to detect and isolate the fault in the second tank system.

7. REFERENCES

Chow, E. and A. Willsky (1984). Analytical redundancy and the design of robust failure detection systems. *IEEE Trans. Auto. Cont.*, **29**, 603–614.

Fadali, M. and W. Liu (1999). Observer-based robust fault detection for a class of multirate sampled-data linear systems. in: *Proc. of the American Control Conference*, 97–98.

Fadali, M. and H. Shabaik (2002). Timely robust detection for multirate linear systems. *Int. J. Control*, **75**, 305–313.

Ge, W. and C. Fang (1988). Detection of faulty components via robust observation. *Int. J. Control*, **47**, 581–599.

Johansson, R., (2000). The quadruple-tank process: a multivaraible laboratory process with an adjustable zero. *IEEE Trans. Cont. Sys. Tech*, **8**, 456–465.

Johnson, R., and D. Wichern (1998). *Applied Multivariate Statistical Analysis*, 4th Edition, Prentice-Hall.

Khargonekar, P., K. Poola and A. Tannenbaum (1985). Robust control of linear time-invariant plants using periodic compensation. *IEEE Trans. Auto. Cont.*, **30**, 1088–1096.

Li, W. and S. Shah (2002). Strutured residual vector-based approach to sensor fault detection and isolation. *J. of Process Control*, **12**, 429-443.

Sheng, J., T. Chen and S. Shah (2002). Generalized predictive control for non-uniformly sampled systems. *J. of Process Control*, **12**, 875–885.

ELSEVIER
IFAC
PUBLICATIONS
www.elsevier.com/locate/ifac

TO CASCADE OR NOT TO CASCADE?

Goradia D. B., M.W. Hermanto, S. Lakshminarayanan* and G.P. Rangaiah

Department of Chemical and Environmental Engineering
National University of Singapore, Singapore 117576
**Corresponding Author. Telephone: +65-68748484 Email:* chels@nus.edu.sg

Abstract: We describe an approach that is useful in deciding if significant benefits in terms of control loop performance index will be achieved by a change in control loop configuration from simple feedback (SFB) to cascade control. The problem is considered in a stochastic setting and solved using the Analysis of Variance (ANOVA) technique. The proposed methodology requires only routine operating data and knowledge of the process delay. Three simulation examples exemplify the utility of this approach. *Copyright © 2004 IFAC*

Keywords: Analysis of variance (ANOVA), Control loop performance assessment, performance monitoring, minimum variance control, cascade control

1. INTRODUCTION

Cascade control is probably the single most important performance enhancement strategy over simple feedback loops. The potential improvements in performance possible and the ease of its implementation has led to its widespread use in the chemical process industries for over five decades now. Using extra output measurement(s) (in addition to the primary controlled variable), the cascade control scheme provides timely and calculated adjustment of the manipulated variable thereby decreasing the peak error as well as the integral error for disturbances affecting the process. The efficiency of the cascade control schemes in handling disturbances entering the inner loop has been well documented in several research articles and textbooks. What is relatively less appreciated is the fact that cascade control provides better performance (as compared to the single loop case) for all types of load changes. While the improvement for disturbances entering close to the process input (i.e. secondary disturbances) can be 10 to 100 fold, the improvement in performance for disturbances entering late into the process (i.e. primary disturbances) is about 2 to 5 times (Webb, 1961; Harriott, 1984). Marlin (2000) provides an excellent review of the principles of cascade control, details the criteria for cascade design and shows several industrial examples. It is shown that the cascade scheme provides practical benefits only if the secondary process is at least three times faster than the primary process even for disturbances entering the inner loop. Krishnaswamy *et al.*, (1990) relate the benefits afforded by cascade control to the parameters of the primary and secondary process models in a deterministic setting.

Industrial control loops are designed and implemented in order to achieve specific objectives. It is important to monitor the performance of these loops periodically and make sure they provide the best possible performance. In this regard, the performance monitoring of control loops has received much attention in the last decade. Many researchers have used the minimum variance controller (MVC) performance as the benchmark – this benchmark is appropriate if the goal of control is the reduction of the variance in the controlled variable (the variance of the manipulated variables, the complexity of the MVC or its robustness is not of concern). Harris (1989) showed that the minimum variance achievable (with a MVC) can be computed from routine operating data if the process time delay is known. Since then, there has been a multitude of research articles (e.g. Desborough and Harris, 1992; Stanfelj *et al.* 1993; Huang *et al.* 1997) that consider important extensions, alternate benchmarks (Tyler and Morari, 1996; Kendra and Cinar, 1997; Swanda and Seborg, 1999), applications (Thornhill *et al.*, 1999) and industrial perspectives (e.g. Kozub, 1996; Desborough and Miller, 2001) on this topic. The user is also referred to the exceptional coverage provided by Huang and Shah (1999) and Qin (1998) to this topic. Recently, Agrawal and Lakshminarayanan (2003) described a method to determine the control loop performance achievable with PI type controllers, the optimal control settings that will yield the "best" performance and the expected robustness margins using closed loop transfer functions identified from closed loop experimental data.

Ko and Edgar (2000) established the basis of performance assessment of cascade loops. Desborough and Harris (1993) established a procedure to separate the variance contributions into components related to the controller and the disturbances by developing an analysis of variance (ANOVA) technique. Vishnubhotla *et al.* (1997) applied the ANOVA method to investigate the need for feedforward control on data sets provided by Shell, USA.

The study here is related to cascade loops. The scenario we consider is as follows: we have a process that is presently regulated by a simple feedback (SFB) controller. A control loop monitoring tool has flagged this loop as poorly performing when compared to MVC. We take a closer look at this loop and assess if the loop is performing to its full potential by taking into consideration factors such as the restricted structure of the controller – this is important because PID type controllers that are so common in the chemical process industries cannot provide minimum variance performance under many practical situations. Let us assume that such an analysis finds that the present controller is doing its best. If even better control performance is sought, the choices available are: (i) making process modifications or (ii) changing the controller structure. Two obvious enhancements to the SFB scheme are feedforward and cascade control. Feedforward control is more appropriate when measured disturbances are available. Cascade control is suited when suitable secondary measurements (the secondary measurement must be influenced by the manipulated variable; it must also have a direct impact on the primary variable) are available. Such a scenario has been mentioned in Stanfelj *et al.* (1993). Note that if the analysis had shown that the SFB loop is not performing as good as is possible, then retuning of the feedback controller would have been initiated.

In this paper, we assume that we are not satisfied with even the best performance that the SFB control system can provide and we would like to estimate the benefits that will accrue by migrating from SFB control to cascade control. Feedforward scheme is not an option due to the lack of measured disturbances. Routine operating data from the SFB control system will be utilized to perform an ANOVA decomposition of the process variance and speculate about the possible success of the cascade scheme.

This contribution is structured as follows. In the next section, we outline the basics of the performance assessment for simple feedback and cascade loops. We then discuss the ANOVA procedure as applied to a simple feedback loop and indicate the components of the variance that can be eliminated using cascade control. Examples will be used to demonstrate the utility of the proposed methodology.

2. THEORY

Consider the SFB control system shown in Figure 1. y_1 and y_2 represent the disturbance corrupted outputs of the primary and the secondary process respectively. The primary process is denoted by $T_1 = q^{-d1}\tilde{T}_1$ where d1 denotes the number of samples of time delay in the primary process and $\tilde{T}_1$ represents the delay free part of T_1. Along the same lines, the secondary process T_2 is represented as $T_2 = q^{-d2}\tilde{T}_2$. Q represents the feedback controller; N_1 and N_2 denote the disturbance transfer functions driven by zero-mean white noise sequences a_1 and a_2 respectively; disturbance a_1 is "closer" to the primary variable y_1 and disturbance a_2 is in proximity to the secondary variable y_2. 'u' represents the manipulated variable.

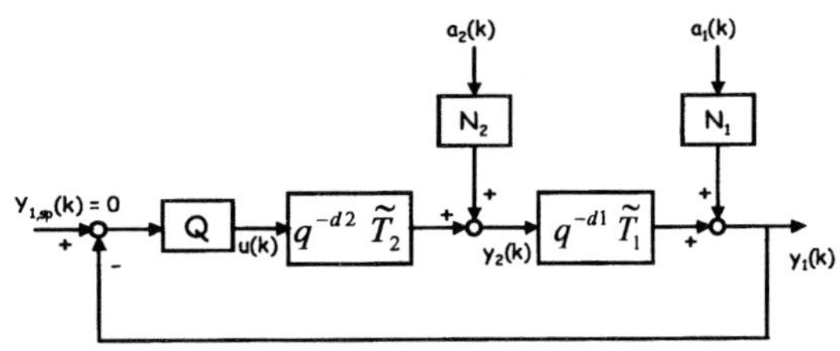

Figure 1: Simple Feedback System

Figure 2 shows a cascade system controlling the same process. In this case, Q_1 represents the primary controller and Q_2 represents the secondary controller. u_2 represents the manipulated variable that is set by the secondary controller Q_2. The setpoint for Q_2 comes from the primary controller Q_1.

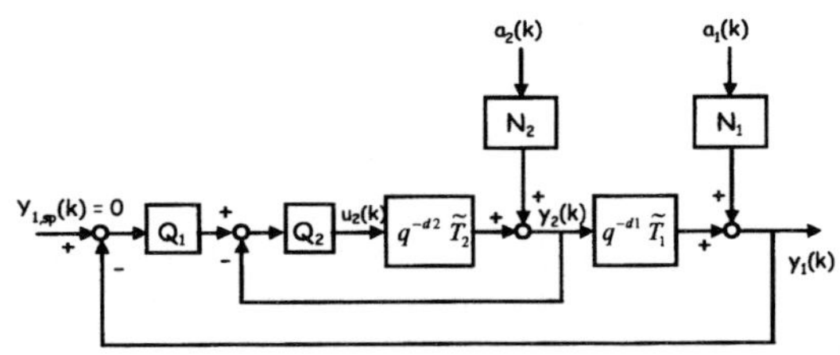

Figure 2: Cascade Control System

For a disturbance a_2 entering the system at t = 0, the output y_1 will be disturbed from 'time' d1 onwards. The controller Q in the SFB case will initiate control action at t = d1. The effect of this control action will be felt at y_1 only from (2d1+d2) onwards. y_1 will effectively be in open loop between d1 and (2d1+d2-1) samples. Under the cascade control system, Q_2 will initiate control action at t = 0, and the output y_1 will be in open loop condition only between d1 and d1+d2-1 samples. If d1 is large, the cascade scheme will provide better regulation of y_1 for the secondary disturbance a_2. Next, consider the 'primary' disturbance a_1 entering the system at t = 0. The primary controlled variable will remain in an open loop condition between t = 0 to t = d1+d2-1 for both the SFB and the cascade control system. In going from a SFB to a cascade scheme, we can hope to eliminate the effect of a_2 between t = d1+d2 and t = 2d1+d2-1. This does not mean that no more reduction in variance is possible as we change from SFB to cascade control. This aspect will be clarified later.

For the SFB, the closed loop relationship between the external signals and the output y_1 is given by:

$$y_1 = \left(\frac{N_1}{1+QT_1T_2}\right)a_1 + \left(\frac{N_2T_1}{1+QT_1T_2}\right)a_2 \tag{2.1}$$

For the cascade scheme, this relationship is modified to:

$$y_1 = \left(\frac{N_1}{1+Q_1T_1T_2^*}\right)a_1 + \left(\frac{N_2^*T_1}{1+Q_1T_1T_2^*}\right)a_2 \tag{2.2}$$

with $T_2^* = \dfrac{Q_2T_2}{1+Q_2T_2}$ and $N_2^* = \dfrac{N_2}{1+Q_2T_2}$.

For the SFB control scheme, the minimum variance is now computed. The disturbance and process transfer functions are expanded as follows:

$$N_1 = P_1 + R_1q^{-(d1+d2)} \tag{2.3}$$

$$N_2 = P_2 + R_2q^{-(d1+d2)} \tag{2.4}$$

$$P_2\tilde{T}_1 = S + Vq^{-(d1+d2)} \tag{2.5}$$

where P_1 and P_2 are monic polynomials (for N_1 and N_2 respectively) in q^{-1} of order d1+d2-1.

In equation (2.3), N_1 is expanded into two parts P_1 and $R_1q^{-(d1+d2)}$. When noise a_1 enters the process at time 0, the controller action would not have any effect on y_1 until time d1+d2-1; this makes P_1 a feedback invariant term. In equation (2.4), N_2 is expanded into two parts P_2 and $R_2q^{-(d1+d2)}$. Note that the noise a_2 entering at time 0 will upset y_2 from time 0 to d1+d2-1 irrespective of any controller action. For our purposes, the effect of a_2 on y_1 is of interest. Therefore, in equation (2.5), the product of P_2 and $\tilde{T}_1$ is expanded into S and $Vq^{-(d1+d2)}$, where S is a polynomial of order d1+ d2–1.

The closed loop transfer function shown in equation (2.1) can be divided into a feedback invariant part and feedback dependent part as shown in the following:

$$y_1 = \left(\frac{P_1 + R_1q^{-(d1+d2)}}{1+Q\tilde{T}_1\tilde{T}_2q^{-(d1+d2)}}\right)a_1 + \left(\frac{\left(P_2 + R_2q^{-(d1+d2)}\right)\tilde{T}_1q^{-d1}}{1+Q\tilde{T}_1\tilde{T}_2q^{-(d1+d2)}}\right)a_2$$

$$= \left(\frac{P_1\left[1+Q\tilde{T}_1\tilde{T}_2q^{-(d1+d2)}\right] + q^{-(d1+d2)}\left[R_1 - P_1Q\tilde{T}_1\tilde{T}_2\right]}{1+Q\tilde{T}_1\tilde{T}_2q^{-(d1+d2)}}\right)a_1$$

$$+ \left(\frac{P_2\tilde{T}_1q^{-d1}\left[1+Q\tilde{T}_1\tilde{T}_2q^{-(d1+d2)}\right] + q^{-(d1+d2)}\left[R_2 - P_2Q\tilde{T}_1\tilde{T}_2\right]\tilde{T}_1q^{-d1}}{1+Q\tilde{T}_1\tilde{T}_2q^{-(d1+d2)}}\right)a_2$$

$$= \left(P_1a_1 + Sq^{-d1}a_2\right) + \left(C_{d1+d2}q^{-(d1+d2)} + C_{d1+d2+1}q^{-(d1+d2+1)} + \ldots\right)a_1$$

$$+ \left(D_{2d1+d2}q^{-(2d1+d2)} + D_{d1+d2+1}q^{-(2d1+d2+1)} + \ldots\right)a_2 \tag{2.6}$$

From equation (2.6), the minimum variance can be written as

$$\sigma^2_{mv,SFB} = \mathrm{var}\left(P_1a_1 + Sq^{-d1}a_2\right) \tag{2.7}$$

For the cascade control system shown in Figure 2, the minimum variance can be computed as:

$$\sigma^2_{mv,CAS} = \mathrm{var}\left(P_1a_1 + S_2q^{-d1}a_2\right) \tag{2.8}$$

with polynomial P_1 as defined in equation (2.3) and S_2 is a polynomial of order d2-1 defined by equations (2.9) and (2.10).

$$N_2^* = P_2^* + R_2^*q^{-d2} \tag{2.9}$$

$$P_2^*\tilde{T}_1 = S_2 + V_2q^{-d2} \tag{2.10}$$

Remark 1: The only difference between $\sigma^2_{mv,SFB}$ and $\sigma^2_{mv,CAS}$ is in the term related to the secondary disturbance a_2.

Lemma: The d2–1 coefficients of the polynomial S_2 will be the same as the first d2-1 coefficients of the polynomial S.

Next, we seek to perform an analysis of variance for the SFB system. The variance of the primary controlled variable y_1 should be separated into an invariant component and a feedback dependent component. The result of this analysis would help in deciding if restructuring existing SFB system into cascade control system will be beneficial. In short, we are interested in predicting the *cascade achievable performance*.

The feedback invariant part for the SFB control system is given by

$$y_{1,SFB,FI} = \left(H_{10} + \ldots + H_{1(d1+d2-1)}q^{-(d1+d2-1)}\right)a_1 + \left(H_{20} + \ldots + H_{2(2d1+d2-1)}q^{-(2d1+d2-1)}\right)a_2 \tag{2.11}$$

If a cascade control system were to be established, the feedback invariant part would be:

$$y_{1,CAS,FI} = \left(H_{10} + \ldots + H_{1(d1+d2-1)}q^{-(d1+d2-1)}\right)a_1 + \left(H_{20} + \ldots + H_{2(d1+d2-1)}q^{-(d1+d2-1)}\right)a_2 \tag{2.12}$$

In equations (2.11) and (2.12), the H_1's refer to the closed loop impulse response coefficients (analytically determined or identified from routine operating data) for the primary disturbance affecting y_1. The H_2's refer to the closed loop impulse response coefficients for a_2. These coefficients are estimated by performing a multivariate autoregressive modelling using y_1 and y_2 measurements. In the SFB case, the first 2d1+d2 terms (H_{20} to $H_{2d1+d2-1}$) are used while in the cascade case only the first d1+d2 terms (H_{20} to $H_{d1+d2-1}$) are used. Keeping this difference in mind, the feedback invariant for simple feedback system can be split in to two parts as

1a) SFB and cascade invariant and
1b) Additional SFB invariant

The first part is defined as "SFB and Cascade invariant" – this variance component cannot be altered either by a simple feedback controller or a cascade control system. The second part labeled as "Additional SFB invariant" contains the variance contribution due to non-availability of the secondary controller. It is assumed that this contribution to overall variance can be reduced to zero if a *perfect* secondary controller is available. The invariant part of the SFB can hence be rearranged as follows

$$
\begin{aligned}
y_{1,SFB,FI} &= \underbrace{\left(H_{10} + \ldots + H_{1(d1+d2-1)}q^{-(d1+d2-1)}\right)a_1 + \left(H_{20} + \ldots + H_{2(d1+d2-1)}q^{-(d1+d2-1)}\right)a_2}_{\text{SFB and Cascade invariant}} \\
&+ \underbrace{\left(H_{2(d1+d2)}q^{-(d1+d2)} + \ldots + H_{2(2d1+d2-1)}q^{-(2d1+d2-1)}\right)a_2}_{\text{Additional SFB invariant}}
\end{aligned}
\tag{2.13}
$$

The variance component '1b' is given by

$$
\left(H^2_{2(d_1+d_2)} + H^2_{2(d_1+d_2+1)} + \ldots + H^2_{2(2d_1+d_2-1)}\right)\sigma^2_{a_2}
\tag{2.14}
$$

where $\sigma^2_{a_2}$ is the estimated variance of secondary noise a_2.

We are now ready to analyze the feedback dependent variance or remainder variance. The feedback-dependent part can also be separated into two distinct parts:

2a) Variance arising due to noise sequence a_1
2b) Variance arising due to noise sequence a_2.

For single feedback control system, the feedback dependent part is

$$
\begin{aligned}
y_{1,SFB,FD} &= \underbrace{\left(H_{1(d_1+d_2)}q^{-(d_1+d_2)} + H_{1(d_1+d_2+1)}q^{-(d_1+d_2+1)} + \ldots\right)a_1}_{\text{Feedback-dependent from } a_1} \\
&+ \underbrace{\left(H_{2(2d_1+d_2)}q^{-(2d_1+d_2)} + H_{2(2d_1+d_2+1)}q^{-(2d_1+d_2+1)} + \ldots\right)a_2}_{\text{Feedback-dependent from } a_2}
\end{aligned}
\tag{2.15}
$$

The individual terms in equation (2.15) can be used to determine the contributions to the variance in y_1 from the primary and secondary noise channels.

In summary, the total variance of the primary controlled variable (y_1) for a SFB system with an additional secondary output measurement y_2 can be split into four parts:

1a) The SFB and cascade invariant components
1b) Additional SFB invariant.
2a) Remainder variance due noise sequence a_1.
2b) Remainder variance due to noise sequence a_2

Out of these four parts, the cascade scheme should ideally eliminate (or reduce considerably) the variance contribution from two terms: 1b and 2b. The cascade strategy is designed specifically to reduce the overall time constant and delay to deal with a situation where the major disturbance hits the secondary process and minor stochastic disturbances hits the primary process. Hence the reason for elimination of variance contribution arising from 1b and 2b components can be easily understood. In addition, the cascade control scheme can reduce a portion of the variance attributed to component (2a). The exact amount of reduction possible with component (2a) is not easy to ascertain. We have noted in our simulations that a significant decrease (about 50% or more in all of the examples we have worked on) in the variance contribution due to a_1 (2a part) is also achieved along with practical elimination of the 1b and 2b components. The reduction in the variance contribution from the (2a) component in the cascade scheme could be due to one or more of the following reasons:

1) The severe control action applied by the primary controller along with the higher gain in secondary controller compared to SFB scheme effectively attenuates the primary disturbances (Harriott, 1984).

2) In the single loop system, the primary controlled variable (y_1) and the disturbance (a_1) are more tightly coupled than is desirable. The output y_1 will follow a load change a_1 too readily. In the cascade system, y_1 and a_1 are loosely coupled (Webb, 1961).

3) The multiple lags in the feedback path of the SFB causes the control action to be delayed. Hence the variance of y_1 remains large. These lags are greatly reduced in the cascade system so that any effect of a_1 on y_1 is greatly reduced (Webb, 1961).

3. EXAMPLES

Three simulation examples are used to demonstrate the utility of the proposed ANOVA method for predicting the possible improvement in control loop performance if cascade control is implemented and also for choosing the secondary variable (in case more than one candidate exists).

Example 1: The primary process (G_{p1}), the secondary process (G_{p2}), the primary noise transfer function (N_1), and the secondary noise dynamics (N_2) used in this example are given below. In this example, the noise dynamics affecting the primary process is purposefully kept severe compared to noise dynamics affecting the secondary process, to check the effectiveness of cascade in rejecting severe primary disturbance compared to secondary disturbance.

$$G_{P1} = \frac{z^{-2}}{1-0.9z^{-1}}; G_{P2} = \frac{z^{-1}}{1-0.5z^{-1}};$$

$$N_1 = \frac{z^{-1}}{1-0.9z^{-1}}; N_2 = \frac{z^{-1}}{1-0.5z^{-1}}$$

The PI achievable performance for the SFB is computed to be 0.29. With this "optimal" SFB control system, the variance of y_1 is 12.89; the breakup into the 1a, 1b, 2a and 2b components is 3.71, 4.03, 0.8 and 4.36 respectively. Components 1b and 2b are substantial – they make up about 65% of the variance in y_1. These are the components that can be targeted and reduced by the cascade control strategy. The analysis makes a strong case for implementing a cascade control scheme.

When a PI-P cascade scheme is implemented, the best CLPI achieved is 0.81. The improvement in performance index is about 200% and vindicates the prediction made by the ANOVA approach. With the PI-P cascade implementation, the variance in y_1 is 4.59; the breakup into the 1a, 1b, 2a and 2b components is 3.71, 0.45, 0.25 and 0.19 respectively. Note that there has been a substantial decrease (about 70%) in the 2a variance component also.

Remark 2: The proposed ANOVA approach uses only routine operating data; it cannot therefore predict the settings of the primary and secondary controller at which the optimal cascade loop performance is achieved. If suitable experimental data (collected either under open or closed loop conditions) is available and the process models are identified, the "optimal" settings of the primary and secondary controller leading to the best control loop performance can be obtained using parametric optimization.

Example 2: The transfer functions used in this example are:

$$G_{P1} = \frac{0.0038z^{-20}}{1-0.98z^{-1}}; G_{P2} = \frac{0.0278z^{-8}}{1-0.95z^{-1}};$$

$$N_1 = \frac{z^{-1}}{1-z^{-1}}; N_2 = \frac{z^{-1}}{1-z^{-1}}$$

The PI achievable performance for the SFB scheme is 0.43. At this "optimal" performance, the variance of y_1 is 55.52; the breakup into the 1a, 1b, 2a and 2b components is 23.95, 0.16, 30.86 and 0.55 respectively. The 1b and 2b components are small indicating that the benefits from a cascade control system should mainly come from the reduction of the 2a component, which accounts for about 56% of the total variance here. On the basis of our experience, we can predict that at least 50% of component 2a will be annihilated. We would expect the 2a component with the cascade scheme to be about 15 and the variance in y_1 to reduce to around 40. A more precise answer to the expected reduction in 2a component is not possible.

When a PI-P cascade scheme is implemented, the best CLPI achieved is 0.63. With the PI-P cascade implementation, the variance in y_1 is 39.71; the breakup into the 1a, 1b, 2a and 2b components is 24.99, 0.23, 13.94 and 0.55 respectively. Note that there has been a significant decrease in the 2a variance component. The overall increase in CLPI is 47%; this may be enough to justify the implementation of the cascade scheme.

Example 3: The system considered next is described by the following equations:

$$y_3 = \frac{e^{-6s}}{(5s+1)(2s+1)}u + \frac{1}{(15s+1)}a_3$$

$$y_2 = \frac{e^{-2s}}{(10s+1)}y_3 + \frac{1}{(50s+1)}a_2$$

$$y_1 = \frac{e^{-3s}}{(20s+1)}y_2 + \frac{1}{(32s+1)}a_1$$

Here, y_1 is the primary controlled variable and y_2 and y_3 represent possible secondary variables. U is the manipulated variable; a_1, a_2 and a_3 represent zero mean white noise sequences with variances $\sigma^2_{a1}, \sigma^2_{a2}$ and σ^2_{a3} respectively. The process is controlled by a PI controller with $K_c = 1$ and $\tau_I = 40$. We will examine the ANOVA results for various combinations of the noise variances and suggest the "best" secondary variable in each of those cases. In each case, 5000 samples of routine closed loop data sampled at intervals of 1 time unit were used.

Case 1: $\sigma^2_{a1} = \sigma^2_{a2} = 1$ and $\sigma^2_{a3} = 100$

If y_2 is considered as the secondary variable in the cascade scheme, the ANOVA estimates the overall variance in y_1, 1a, 1b, 2a and 2b components to be 1.38, 0.02, 0.02, 0.03 and 1.31 respectively. If y_3 were to be chosen as the secondary variable, these values are 1.40, 0.06, 0.02, 0.01 and 1.31 respectively. In this case, it does not matter whether y_2 or y_3 is chosen as the secondary variable. Since 2b component is very strong, cascade control using either y_2 or y_3 as the secondary variable will provide a vastly improved control loop performance. Between y_2 and y_3, we can choose the one that engulfs most of the disturbances as the secondary variable.

Case 2: $\sigma_{a2}^2 = \sigma_{a3}^2 = 1$ and $\sigma_{a1}^2 = 100$

If y_2 is considered as the secondary variable in the cascade scheme, the ANOVA estimates the overall variance in y_1, 1a, 1b, 2a and 2b components to be 1.83, 0.87, 0.001, 0.94 and 0.02 respectively. If y_3 were to be chosen as the secondary variable, these values are 1.83, 0.87, 0.002, 0.92 and 0.04 respectively. The 1a and 2a components are dominant in this case. Based on our experience, we conjecture that more than 50% of the 2a component will be consumed by the cascade scheme that could use either y_2 or y_3 as the secondary variable. Between y_2 and y_3, the choice will depend on their relative "location" with respect to the anticipated disturbances.

Case 3: $\sigma_{a1}^2 = \sigma_{a3}^2 = 1$ and $\sigma_{a2}^2 = 100$

If y_2 is considered as the secondary variable in the cascade scheme, the ANOVA estimates the overall variance in y_1, 1a, 1b, 2a and 2b components to be 0.445, 0.062, 0.014, 0.007 and 0.362 respectively. The 2b component is dominant here and a cascade control scheme with y_2 as the secondary variable can eliminate this variance component very effectively. If y_3 were to be chosen as the secondary variable, these values are 0.378, 0.023, 0.000, 0.342 and 0.013 respectively. Interestingly, with y_3 as the secondary variable, the 2a component is the dominant one. With cascade control we *may* not be able to eliminate this component completely (as much as we can do with the 1b or 2b components). In this case, the use of y_2 as the secondary variable seems to be a more prudent choice.

4. CONCLUSIONS

The proposed ANOVA method provides an estimate of the variance reduction possible by moving from a SFB scheme to a cascade scheme using only routine operating data and knowledge of the process time delays. Our simulations show that it is possible to achieve this predicted cascade achievable performance with a PI-P configuration or a PI-PI (when the secondary noise is non-stationary) cascade scheme.

REFERENCES

Agrawal, P. and S. Lakshminarayanan. (2003). Tuning PID controllers using Achievable Performance Index, *Ind. Eng. Chem. Res.*, **42**, 5576-5582.

Desborough, L. D. and R. M. Miller. (2001). Increasing customer value of industrial control performance monitoring – Honeywell's Experience, Proceedings of CPC VI, Tucson, USA.

Desborough, L. D. and T.J. Harris. (1992). Performance assessment measures for univariate feedback control, Can. J. Chem. Eng., **70**, 1186-1197.

Desborough, L. D. and T.J. Harris. (1993). Performance assessment measures for univariate feedforward/feedback control, Can. J. Chem. Eng., **71**, 605-616.

Harriott, P. (1984). Process Control, Tata McGraw-Hill Publishing Co. Ltd., New Delhi, India.

Harris, T. J. (1989). Assessment of control loop performance, Can. J. Chem. Eng., **67**, 856-861.

Huang, B., S. L. Shah and E. Z. Kwok. (1997). Good, bad or optimal? Performance assessment of multivariate processes, Automatica, **33**, 1175-1183.

Huang, B. and S. L. Shah. (1999), Performance assessment of control loops, Springer, London.

Ko, B. S. and T.F. Edgar. (2000). Performance Assessment of Cascade Control Loops, AIChE Journal, **46(2)**, 281-291.

Kozub, D. (1996). Controller performance monitoring and diagnosis: experiences and challenges, Proceedings of CPC V, Lake Tahoe, USA.

Krishnaswamy, P. R., G. P. Rangaiah, R. K. Jha and P. B. Deshpande. (1990). When to use cascade control, Ind. Eng. Chem. Res., **29**, 2163-2166.

Marlin, T. E. (2000). Process control: designing processes and control systems for dynamic performance, 2nd edition, McGraw-Hill International Editions, Singapore.

Qin, S. J. (1998). Control performance monitoring – a review and assessment, Comp. & Chem. Engg., **23**, 173-186.

Stanfelj, N., T. E. Marlin and J. F. MacGregor. (1993). Monitoring and diagnosing control loop performance: The single loop case, Ind. Eng. Chem. Res., **32**, 301-314.

Swanda, A. and D. E. Seborg. (1999). Controller performance assessment based on setpoint response data, Proceedings of the Amer. Cont. Conf., San Diego, USA.

Thornhill, N. F., M. Oettinger and P. Fedenczuk. (1999). Refinery-wide control loop performance assessment, J. Proc. Cont., **9**, 109-124.

Tyler, M.L. and M. Morari. (1996). Performance monitoring of control systems using likelihood methods, Automatica, **32**, 1145-1162.

Vishnubhotla, A., S. L. Shah and B. Huang. (1997). Feedback and feedforward performance analysis of the Shell industrial closed-loop data set, Proceedings of ADCHEM, Banff, Canada, 295-300.

Webb, P. U. (1961). Reducing process disturbances with cascade control, Control Engineering, August, 73-76.

ELSEVIER
IFAC
PUBLICATIONS
www.elsevier.com/locate/ifac

MULTIVARIATE FORECASTING OF BATCH EVOLUTION FOR MONITORING AND FAULT DETECTION

Salvador Garcia Muñoz, Theodora Kourti, John F. MacGregor

McMaster Advanced Control Consortium, McMaster University, Hamilton, ON, Canada.

Abstract: To monitor a batch process, which is dynamic in nature, it is necessary to consider the time varying relationship of its variables throughout the entire run. MPCA models built with batch wise unfolded data have been used extensively for batch process monitoring, these methods will not only consider the known samples to asses the ongoing batch run, but will also consider a dynamic forecast of the future unknown samples. Such forecast, implicit in the methodology, is uncovered and analyzed in this work; and proven to be a powerful feature of a batch-monitoring scheme built with MPCA and the batch-wise unfolded matrix of batch data. *Copyright © 2004 IFAC*

Keywords: Batch, Monitoring, Multivariate Statistical Process Control, Forecasting,

1. INTRODUCTION

Multivariate methods based-on multi-way PCA and PLS (Wold et al, 1987) have proven their usefulness in batch process monitoring (Kourti et al, 1995; Nomikos, 1996; Nomikos and MacGregor, 1994; Nomikos and MacGregor, 1995a; Nomikos and MacGregor, 1995b). These methods have been widely studied and extensively applied in industry (Kosanovich et al, 1996; Nomikos, 1996; Ramaker et al, 2002; Schlags and Popule, 2001; Tates et al, 1999). They are essentially the only powerful approaches for monitoring batch processes when there is no deterministic model available. Other variation of the multivariate approaches (Boque and Smilde, 1999; Wold et al, 1998) are found in literature, and have already been compared (van Sprang et al, 2002; Westerhuis et al, 1999). This work focuses on the approach proposed by Nomikos and MacGregor (Kourti et al, 1995; Nomikos, 1996; Nomikos and MacGregor, 1994; Nomikos and MacGregor, 1995a; Nomikos and MacGregor, 1995b), and in particular in the estimation of the missing trajectory information needed in the approach when monitoring a new batch.

In the Nomikos-MacGregor method the 3 dimensional batch dataset consisting of J variables, I batches and K time periods is rearranged as shown in Fig. 1. A natural problem to overcome when monitoring a new batch at time=k, where k<K is the fact that there are K-k unknown samples (Fig. 2) and therefore, the need to estimate or "fill in" these unknown samples for the ongoing batch in order to calculate a score value (note that this problem does not exist when building the PCA model, when analyzing historical data, or when monitoring only at the end of each new batch). In their work, Nomikos and MacGregor (Nomikos and MacGregor, 1994; Nomikos and MacGregor, 1995b) acknowledge this issue and propose three different ways to handle this problem: to complete the observation vector with zero deviation from the nominal trajectories (zero option: Z); complete the observation assuming that the mean centered and scaled current deviation from the nominal trajectory will be maintained throughout the rest of the batch (i.e. current deviations option: CD), or estimate the "missing" values using the PCA model (missing data option: MD).

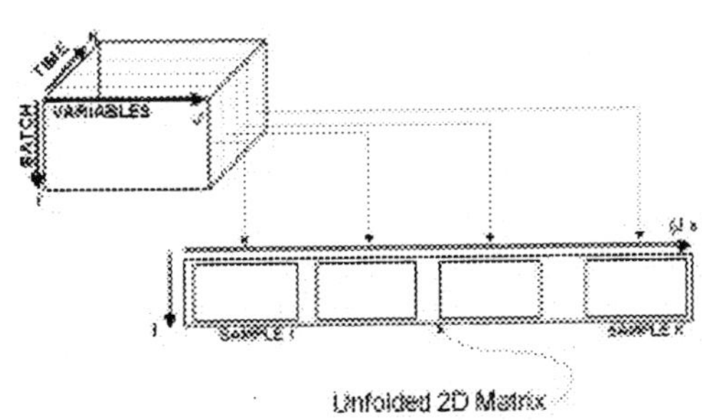

Fig. 1. Unfolding of the 3D batch data matrix.

The control limits of the resulting monitoring charts are dependent upon the used option (Nomikos and MacGregor, 1995b). This "filling in" need of the method has been criticized by several authors (Cho and Kim, 2003; Meng et al, 2003; Undey et al, 2003; van Sprang et al, 2002) without any convincing analysis to support their arguments.

In this work we demonstrate that monitoring a batch process with the Nomikos-MacGregor method, and using the MD approach to fill in the unknown samples when monitoring a new batch, is in fact a powerful model predictive monitoring scheme where the future samples of the trajectories are predicted with good accuracy even at the very beginning of the batch run.

The missing data in the future trajectory are predicted much in the same way as an adaptive multivariate time series forecast, but with optimal use of all the available measurements up to current time and detailed knowledge about the time varying correlation structure among the variables throughout all the rest of the batch.

1.1 Notation

In the following sections, a is used to represent the total number of components in a PCA model; $\mathbf{X}$ represents the original data set used to build the model; $\mathbf{T}$ represents the known scores from the original model; $\mathbf{P}$ represents the loadings of the PCA model, $\mathbf{x}$ is used to represent a vector of data from a new observation; $\hat{\mathbf{t}}$ is used to represent the estimate of the score vector corresponding to $\mathbf{x}$ in the presence of incomplete data and $\mathbf{t}$ is the value of the score vector with the complete data vector.

The dataset $\mathbf{X}$ contains I batches, each batch has J variables sampled K times throughout the batch run. $\mathbf{X}\in\Re^{I\times JK}$, $\mathbf{T}\in\Re^{I\times a}$, $\mathbf{P}\in\Re^{JK\times a}$, $\mathbf{x}\in\Re^{JK\times 1}$, $\boldsymbol{\tau}\in\Re^{a\times 1}$, $\hat{\mathbf{t}}\in\Re^{a\times 1}$.

Lower case bold letters ($\mathbf{x}$) denotes a vector; upper case bold letters ($\mathbf{X}$) denote a matrix. For a certain observation $\mathbf{x}$ with missing elements, we will group all the missing elements at the *end* of the observation, the known samples in the observation will be denoted by $\mathbf{x}^*$, and the missing samples will be denoted by $\mathbf{x}^\#$. So, for a certain observation $\mathbf{x}$ with missing elements all at the end of the vector, we denote as $\mathbf{x}^T = [\mathbf{x}^{*T}\ \mathbf{x}^{\#T}]$.

This notation is applied to other matrices or vectors as well, e.g. $\mathbf{P}^*$ refers to those rows of a loading matrix $\mathbf{P}$ that correspond to the known variables (elements) of x ($\mathbf{x}^*$); and $\mathbf{P}^\#$ refers to those rows of $\mathbf{P}$ corresponding to the missing elements of $\mathbf{x}$ ($\mathbf{x}^\#$). These definitions are illustrated in figure 2.

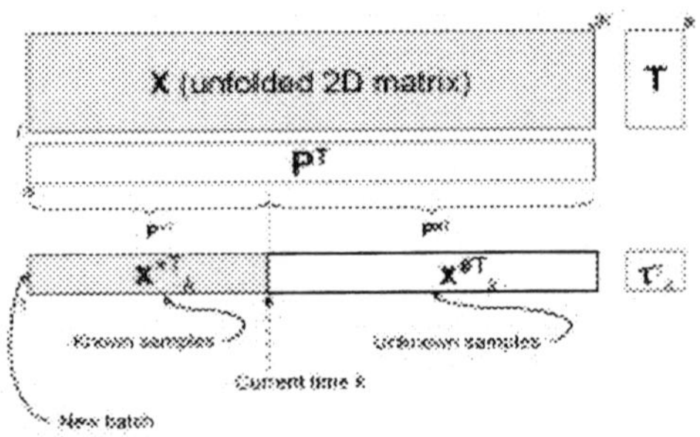

Fig 2. Notation and definitions

2. BATCH TRAJECTORY FORECASTING

2.1 Trajectory Forecasting using AR Models

To represent a batch trajectory with an AR multivariate time series we say that each batch is a collection of samples of the column vector χ such that $\mathbf{x}^T=[\,?_1^T, ?_2^T \ldots ?_K^T\,]$. At time k, the model will be able to make the forecast for the values of the samples at time $k+1$ using Eq. 1 where q is the order of the model.

$$?_{k+1} = \mathbf{f}_1 ?_k + \mathbf{f}_2 ?_{k-1} + \mathbf{f}_3 ?_{k-2} + \cdots + \mathbf{f}_n ?_{k-q} \quad (1)$$

With this fixed model, it is possible to forecast the remainder of the batch at any time, and then compute its scores. If the batch run consists of several phases (or stages) then it is desirable to fit a different model per phase.

2.2 Trajectory forecast using the PCA model

In their earlier work, Nomikos and MacGregor (Nomikos and MacGregor, 1995b) clearly state that: when using the MD approach, the missing unknown values will be *replaced* by a prediction of these done with the PCA model. In the following paragraphs this replacement mechanism is uncovered since it not done explicitly but implicitly in the method, starting from the score estimation problem.

The simplest way to estimate the score vector when an observation contains missing data is to use the single component projection method (SCP) (Nelson et al, 1996). This method will calculate each of the scores independently and sequentially as $\mathbf{f}_{ki} = \mathbf{z}^*\mathbf{p}_i^* / \mathbf{p}_i^{*T}\mathbf{p}_i^*$ where $\mathbf{z}^*$ is $\mathbf{x}^*$ at time k deflated by all the previous components. This method has proven to be the least performing of the missing data handling methods (Nelson et al, 1996), however, it has been reported to work in some cases (Lennox et al, 2000).

One natural improvement to the method is to use the forecast for the remainder of the trajectory, since it is possible to compute a prediction of the unknown measurements $\hat{\mathbf{x}}_k^{\#}$ using their corresponding loadings and the score vector estimate (Eq.2);

$$\hat{\mathbf{x}}_k^{\#} = \mathbf{P}_k^{\#}\hat{\mathbf{t}}_k^T \qquad (2)$$

Once the prediction of the unknown measurements at time k is done, a new augmented vector can be created using the known measurements and the new predicted values (Eq.3)

$$\mathbf{x}_{aug}^T = \left[\mathbf{x}^{*T}\hat{\mathbf{x}}^{\#T}\right] \qquad (3)$$

This approach was taken by Meng *et al (Meng et al, 2003)*, and Lennox *et al (Lennox et al, 2000)* to forecast the unknown part of the trajectory in their respective studies. However, this is only the first step of a more complete solution which is the *iterative imputation* (II) method proposed by Arteaga and Ferrer (Arteaga and Ferrer, 2002), where the next step to take now is to re-estimate the score vector by multiplying the augmented vector of the known measurements and the predictions, by the loadings matrix (Eq. 4)

$$\hat{\mathbf{t}}_k^T = \mathbf{x}_{aug}^T\mathbf{P} \qquad (4)$$

With this new estimate of the score it is possible to iterate by re-calculating the augmented vector (Eq. 3) and the score estimate (Eq. 4) until convergence is reached in the score estimate. At convergence, the estimate of the score is computed with a full trajectory (that includes the known measurements and the forecast for the rest of the trajectory) and the whole loading matrix.

The *iterative-imputation* (Arteaga and Ferrer, 2002) algorithm is a very illustrative way of understanding how the missing data elements of the trajectories are being replaced by the PCA forecast. This method has been proven (Arteaga and Ferrer, 2002) to be equivalent to projection to the plane method, which was used by Nomikos and MacGregor (1994, 1995a, 1995b) in their early work.

1.4 The PCA forecast model vs. a time series forecast

Since it is clear that the *MD* option uses a forecast of the trajectory to "fill in" the unknown measurements, it is desirable to analyze the properties of the forecast model embedded in the PCA projection. To better understand the PCA forecast model, it can be expressed as a time series forecast model, in order to contrast it with the properties of a conventional multivariate time series models. In the following paragraphs, the embedded one-step a head prediction model is uncovered from the PCA prediction expression to illustrate the nature of the prediction model. However, equivalent expressions for any f-step ahead forecast could have been used to illustrate the same point.

Previously, the vector $\boldsymbol{\chi}_k$ was defined as a column vector that contains the J batch variables measured at time k, such that: $\mathbf{x}^T$=[$?_1^T, ?_2^T \ldots ?_K^T$]. Using the notation defined previously, we can express the known batch samples at time k in terms of $\boldsymbol{\chi}_k$:

$$\mathbf{x}_k^{*T} = [?_1^T ?_2^T \cdots ?_k^T] \qquad (5)$$

From these equations, it is clear that the one step ahead prediction $?_{k+1}$ corresponds to the forecast of the first J elements of the $\hat{\mathbf{x}}^{\#}$ vector:

$$?_{k+1} = \hat{\mathbf{x}}_{k[1:J]}^{\#} \qquad (6)$$

Eq. 6 can be expressed as a function of the loadings matrix (Eq. 7), from Eq. (2), and taking only the first J rows of the $\mathbf{P_k}^{\#}$ matrix:

$$?_{k+1} = \hat{\mathbf{x}}_{k[1:J]}^{\#} = \mathbf{P}_{k[1:J]}^{\#}\,\hat{\mathbf{t}}_k \qquad (7)$$

At time k, and assuming that projection to the model plane (PP_{OLS}) is used to solve the missing data problem, the estimate of the score is,

$$\hat{\mathbf{t}}_k^{\mathrm{T}} = \mathbf{x}_k^{*\mathrm{T}}\mathbf{P}_k^*(\mathbf{P}_k^{*\mathrm{T}}\mathbf{P}_k^*)^{-1} \qquad (8)$$

and substituting Eq. 5 and Eq. 8 into Eq. 7:

$$?_{k+1} = \hat{\mathbf{x}}_{[1:J]}^{\#} = \mathbf{P}_{k[1:J]}^{\#}\left[\mathbf{P}_k^*(\mathbf{P}_k^{*\mathrm{T}}\mathbf{P}_k^*)^{-1}\right]^{\mathrm{T}}\begin{bmatrix} ?_1 \\ \vdots \\ ?_{k-1} \\ ?_k \end{bmatrix} \qquad (9)$$

The term $\mathbf{P}_{k[1:J]}^{\#}\left[\mathbf{P}_k^*(\mathbf{P}_k^{*\mathrm{T}}\mathbf{P}_k^*)^{-1}\right]^{\mathrm{T}}$ of Eq. 9 is a matrix of J rows and Jk columns; re-expressing each element of this matrix as ϕ_j and grouping now blocks of J columns as new square matrices $\boldsymbol{\Phi}_i$, such that each $\boldsymbol{\Phi}_i$ has J rows and J columns giving place to k number of Φ's,

$$?_{k+1} = \left[\underbrace{\begin{bmatrix} \phi_{1,1}\,\phi_{1,2}\cdots\phi_{1,J} \\ \phi_{2,1}\,\phi_{2,2}\cdots\phi_{2,J} \\ \vdots \quad \vdots \quad\quad \vdots \\ \phi_{J,1}\,\phi_{J,2}\cdots\phi_{J,J} \end{bmatrix}}_{\{\Phi_1\}}\underbrace{\begin{bmatrix} \phi_{1,J+1}\,\phi_{1,J+2}\cdots\phi_{1,2J} \\ \phi_{2,J+1}\,\phi_{2,J+2}\cdots\phi_{2,2J} \\ \vdots \quad \vdots \quad\quad \vdots \\ \phi_{J,J+1}\,\phi_{J,J+2}\cdots\phi_{J,2J} \end{bmatrix}}_{\{\Phi_2\}}\underbrace{\begin{matrix} \cdots\phi_{1,Jk} \\ \cdots\phi_{2,Jk} \\ \ddots \\ \cdots\phi_{J,Jk} \end{matrix}}_{\cdots\{\Phi_K\}}\right]\begin{bmatrix} ?_1 \\ \vdots \\ ?_{k-1} \\ ?_k \end{bmatrix} \qquad (10)$$

gives the multivariate time series form for the one-step-ahead forecast embedded in the PCA model (Eq. 11):

$$?_{k+1} = [\Phi_1\Phi_2\cdots\Phi_k]\begin{bmatrix} ?_1 \\ \vdots \\ ?_{k-1} \\ ?_k \end{bmatrix} = \Phi_1?_1 + \Phi_2?_2 + \cdots + \Phi_{k-1}?_{k-1} + \Phi_k?_k$$

(11)

This one step ahead forecast model has unique characteristics: a) The order of the PCA forecast model (Eq 15) expands, as new measurements are available (k increases); b) An even more important result is that the parameters of this model will adapt as the batch evolves since the elements in $\mathbf{P}^{\#}$ and $\mathbf{P}^{*}$ change as more samples become available.

This means that all the ϕ elements in Eq. 10 and 11 are adapted as time progresses to account for the changing auto-covariance and cross-covariance structure captured by the PCA model, this is possible because the MPCA batch model has embedded within it knowledge of the time varying covariance among all the variables over the entire time history of the batch.

This contrasts with the conventional time series model (Eq. 1) whose coefficients does not change and will always use the same number of past measurements in order to forecast the next sample. As a result, the PCA prediction equation is actually and adaptive, nonlinear function of the data available up to the prediction time k and hence provides a much more powerful prediction than any fixed multivariate time series model; c) The same form of adaptive time series prediction form of the PCA model cam be derived for any f-step ahead prediction with the same conclusions as above.

To further illustrate the adaptive nature of the PCA forecast model, consider the explicit forecast equation for the one step ahead prediction at different times. Figure 3 shows the values of the $\Phi_k^j(2,2)$ parameter for j= 5, 10, 20, 50, 80; and k=j,j-1, j-2...1; for the data set used by Nomikos and MacGregor in earlier work (Nomikos and MacGregor, 1995b).

As seen in this figure, the parameters adapt and expand, as more measurements are available. A result of the adaptive nature of this model is that the behavior of the forecast for the remaining unknown samples changes with time in the batch.

This "adaptive" behavior of the PCA predictive model is unique and results from the fact that the MPCA model built for a training set of complete batches has knowledge of how the auto and cross-covariance structure of all the variables will change over the reminder of the entire batch.

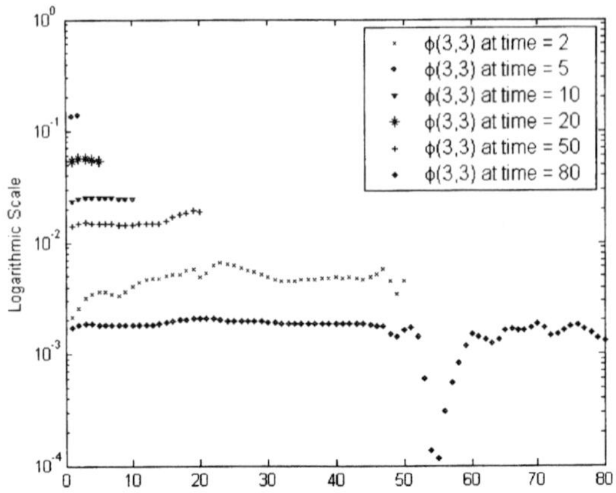

Fig. 3. Coefficient Φ_k^j (3,3) for known measurements as it adapts and expands with time

The tremendous amount of information captured by the MPCA model gives it unique forecasting capabilities, Fig. 4 shows the forecast for variable 4 in batch #7 in the industrial data set used by Nomikos and MacGregor (1995b), the forecast is done with information up to time 30, notice how accurately the prediction follows the true unknown part of the trajectory. The accuracy of the forecast done by the PCA is now contrasted among different options.

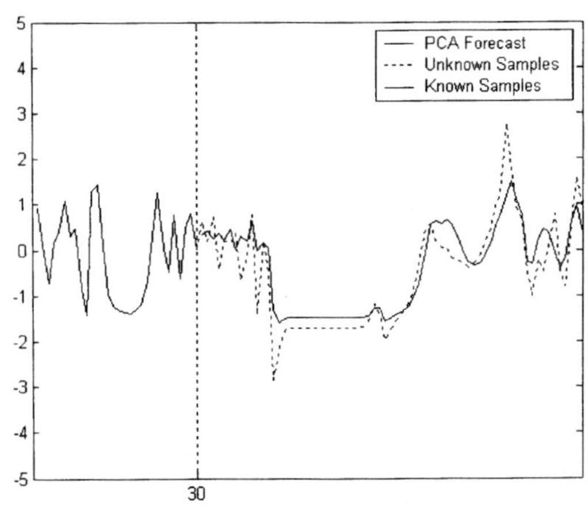

Fig. 4. Forecast done by the PCA model with samples up to time 30

3. ACCURACY COMPARISON

For batch i at time k *(given that* $k < K$*)*, the forecast for each variable j generates a vector of errors in the future prediction (Eq. 12)

$$e_k^{ij} = [e_k^{ij}\big|_1\ e_k^{ij}\big|_2\ e_k^{ij}\big|_3\ e_k^{ij}\big|_4\ e_k^{ij}\big|_5 \cdots e_k^{ij}\big|_{K-k}] \quad (12)$$

defined by Eq. 13, and illustrated in figure 5. This (K-k) vector contains the individual errors in the prediction of the future unknown trajectory of variable j, for samples $k+1$, $k+2$, $k+3$...K.

$$e_k^{ij} = x_{ijk}^{\#} - \hat{x}_{ijk}^{\#} \quad (13)$$

This e_k^{ij} vector will decrease its length as the batch evolves, because there are fewer remaining unknown measurements ($x^{\#}$).

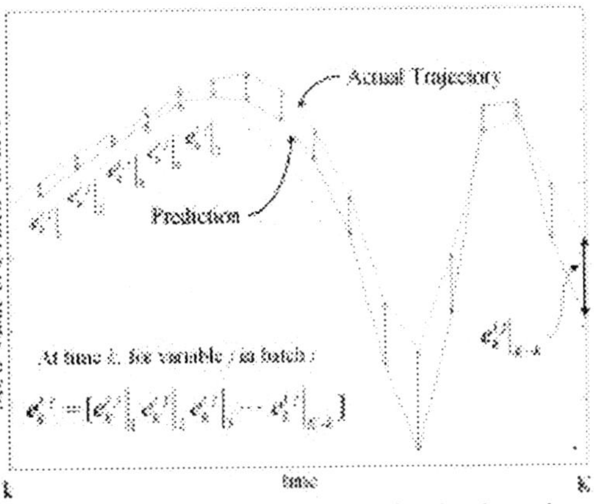

Fig. 5. Future prediction error in batch trajectory forecasting.

The original training batch data consists of I batches, each one with J variables, sampled K times throughout the batch. For each ijk sample in the original data set, we have a e_k^{ij} vector. This set of error vectors can be computed for each option available to treat the unknown part of the trajectory.

To compare all the error vectors among options, it is desirable to summarize the e_k^{ij} vector by one summary statistic in order to quantify the future prediction error as a scalar measurement of accuracy. This accuracy will vary in time and therefore can be plotted as a trajectory for each variable in each batch. To accomplish this, two different approaches are proposed:

a) To consider all individual errors ($[e_k^{ij}|_1\; e_k^{ij}|_2\; e_k^{ij}|_3 \cdots e_k^{ij}|_{K-k}]$) equally important and simply square them, and add them in to one number, the *Future Prediction Sum of Squares* (FPRESS) is then defined by Eq. 14

$$FPRESS_k^{ij} = \sum_{l=1}^{K-k} \left(e_k^{ij}\Big|_l \right)^2 \quad (14)$$

b) A second option is to weight each individual error ($e_k^{ij}|_l$) by the inverse of the distance to the current time sample, in such way that the one step ahead prediction $e_k^{ij}|_1$ becomes more important than the (k-K) step ahead prediction $e_k^{ij}|_{K-k}$; the weighted *Future Prediction Mean Square Error* (*FPMSE*) is defined in Eq. 15

$$FPMSE_k^{ij} = \frac{\sum_{l=1}^{K-k} \frac{1}{l} \left(e_k^{ij}\Big|_l \right)^2}{\sum_{l=1}^{K-k} \frac{1}{l}} \quad (16)$$

The *FPRESS* will provide a measure of global forecast accuracy, since it accounts for the error from time $k+1$ to time K, on the other hand, the *FPMSE* provides a measure of local forecast accuracy, considering the immediate forecast more important than the long term forecast.

Figure 6 shows the *mean trajectory* of the ***FPMSE*** for variable 4 in the industrial data set used by Nomikos and MacGregor (1995b), each line corresponds to a different way of handling the unknown part of the trajectory, a multivariate fixed time series model (TS), current deviations (CD), raw current deviations (CDraw), zeros (Z), and MD represents the missing data option handled with any method available, from single component projection to conditional mean replacement (Arteaga and Ferrer, 2002).

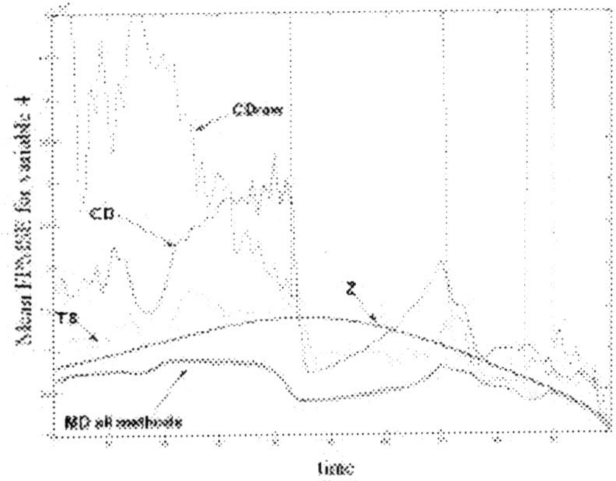

Fig. 6. Mean FPRESS (top) and mean FPMSE (bottom) for variable 1 in simulated SBR dataset.

It is very clear how, from all the methods, the MD options have the least error in the forecast at every time period over the whole trajectory, even from the very initial samples in the batch. This behavior was found to be true in general for all the variables for this data set for both of the two different future error trajectory statistics (*FPRESS* and *FPMSE*), although the CD and TS options on limited occasions showed slightly lower values at a few time intervals.

In batch process monitoring, when the missing data problem is handled correctly changes the initial suggestion of the *missing data* option being unreliable at the beginning of the batch, as illustrated in this work, and practically shown in the industrial application by Zhang, Dudzik and Vaculik (Zhang et al, 2003) where a monitoring system for start-ups is built solving the missing data problem with the PP method with excellent results even at the very beginning of the start-up – startups can be monitored by using the same modeling technique as in batch monitoring (Kourti, 2003).

4. CONCLUSIONS

From the simulation results obtained in this work, and from a careful analysis of the methods, two general conclusions arise *i)* the *filling in* mechanism is actually a powerful time-varying model predictive *forecasting* feature that uses the PCA model of the time-varying covariance structure in the data, and *ii)* the *filling in* requirement in the Nomikos-MacGregor method for batch monitoring should not be seen as a problem to overcome. The multivariate (PCA/PLS) models obtained using the Nomikos-MacGregor approach for batch systems are dynamic models that capture the time-varying relations among the variables throughout the entire batch history. Therefore, they are capable of producing forecasts of the future unknown trajectories with optimal usage of the available data and knowledge of time-varying covariance structure among the variables. These forecasts are far from being arbitrary assumptions about the trajectories.

REFERENCES

Arteaga, F. and Ferrer, A. (2002) Dealing with missing data in MSPC: several methods, different interpretations, some examples. *Journal of Chemometrics*, **16**, 408-418

Boque, R. and Smilde, A. K. (1999) Monitoring and Diagnosing Batch Processes with Multivariate Covariates Regression Models. *AICHE Journal*, **45**, 1504-1520

Cho, H. and Kim, K (2003) A Method for Predicting Future Observations in the Monitoring of a Batch Process. *Journal of Quality Technology*, **35**, 59-69

Kosanovich, K., Dahl, K. S., and Piovoso, M. J. (1996) Improved Process Understanding Using Multiway Principal Component Analysis. *Industrial & Engineering Chemistry Research*, **35**, 138-146

Kourti, T. (2003) Multivariate dynamic data modeling for analysis and statistical process control of batch processes, start-ups and grade transitions. *Journal of Chemometrics*, **17**, 93-109

Kourti, T., Nomikos, P., and MacGregor, J. F. (1995) Analysis, monitoring and fault diagnosis of batch processes using multiblock and multiway PLS. *J.Proc.Cont.*, **5**, 277-284

Lennox, B., Hiden, H. G., Montague, G. A., Kornfeld, G., and Goulding, P. R. (2000) Application of multivariate statistical process control to batch operations. *Computers and Chemical Engineering*, **24**, 291-296

Meng, X., Morris, A. J., and Martin, E. B. (2003) On-line monitoring of batch processes using a PARAFAC representation. *Journal of Chemometrics*, **17**, 65-81

Nelson, P. R. C., Taylor, P., and MacGregor, J. F. (1996) Missing Data Methods in PCA and PLS: Score calculations with incomplete observations. *Chemometrics and Intelligent Laboratory Systems*, **35**, 45-65

Nomikos, P. (1996) Detection and diagnosis of abnormal batch operations based on multi-way principal component analysis. *ISA Transactions*, **35**, 259-266

Nomikos, P. and MacGregor, J. F. (1994) Monitoring Batch Processes using Multiway Principal Components. *AICHE Journal*, **40**, 1361-1375

Nomikos, P. and MacGregor, J. F. (1995a) Multi-way partial least squares in monitoring batch processes. *Chemometrics and Intelligent Laboratory Systems*, **30**, 97-108

Nomikos, P. and MacGregor, J. F. (1995b) Multivariate SPC Charts for Monitoring Batch Processes. *Technometrics*, **37**, 41-58

Ramaker, H. J., van Sprang, E. N. M., Gurden, S. P., Westerhuis, J. A., and Smilde, A. K. (2002) Improved monitoring of batch processes by incorporating external information. *J.Process Contr.*, **12**, 569-576

Schlags, C. E. and Popule, M. P. (2001) Industrial Application of SPC to Batch Polymerization Process. *Proc.ACC (Arlington, VA)*, 25-27

Tates, A. A., Louwerse, D. J., Smilde, A. K., Koot, G. L. M., and Berndt, H. (1999) Monitoring a PVC batch process with multivariate statistical control charts. *Industrial & Engineering Chemistry Research*, **38**, 4769-4776

Undey, C., Ertunc, S., and Cinar, A. (2003) Online Batch/Fed-Batch Process Performance Monitoring, Quality Prediction, and Variable-Contribution Analysis for Diagnosis. *Industrial & Engineering Chemistry Research*, **42**, 4645-4658

van Sprang, E. N. M., Ramaker, H. J., Westerhuis, J., Gurden, S. P., and Smilde, A. K. (2002) Critical evaluation of approaches for on-line batch process monitoring. *Chemical Engineering Science*, **57**, 3979-3991

Westerhuis, J., Kourti, T., and MacGregor, J. F. (1999) Comparing Alternative Approaches for Multivariate Statistical Analysis of Batch Process Data. *Journal of Chemometrics*, **13**, 397-413

Wold, S., Geladi, P., Esbensen, K., and Ohman, J. (1987) Multi-way Principal Components and PLS-analysis. *Journal of Chemometrics*, **1**, 41-56

Wold, S., Kettaneh, N., Friden, H., and Holmberg, A. (1998) Modelling and diagnostics of batch process and analogous kinetic experiments. *Chemometrics and Intelligent Laboratory Systems*, **44**, 331-340

Zhang, Y, Dudzik, M., and Vaculik, V. (2003) Integrated Monitoring Solution to Start-up and run-time operations for continous casting. *Safeprocess2003.IFAC, Washington D.C.June 9-11.*

ELSEVIER
IFAC
PUBLICATIONS
www.elsevier.com/locate/ifac

ADVANCED PROCESS DIAGNOSIS IN COMPLEX SYSTEMS USING NONLINEAR VARIABLE RECONSTRUCTION

David Antory *,[1] **Uwe Kruger** * **George W. Irwin** *
Geoffrey McCullough *

* *Virtual Engineering Centre, Queen's University Belfast, Belfast BT9 5HN, Northern Ireland, UK*

Abstract: This paper presents a new nonlinear multivariate statistical process control technique for identifying and isolating the root cause of abnormal process behavior. The new technique is a nonlinear extension to the variables reconstruction technique by (Dunia *et al.*, 1996), based on nonlinear principal component analysis (NLPCA). This work demonstrates that the variable reconstruction (i) affects the geometry of the NLPCA model and (ii) alters the NLPCA based monitoring statistics. Incorporating such changes into the NLPCA model using reference data can address these issues. An industrial application study of a glass melter process shows that abnormal events can be identified and isolated earlier than conventional principal component analysis (PCA). *Copyright © 2004 IFAC*

Keywords: Fault diagnosis (detection, identification, isolation), monitoring, neural networks, chemical industry, nonlinear model

1. INTRODUCTION

To guarantee that complex industrial processes operate economically, safely and are environmentally friendly, it is essential to constantly monitor their performance. (MacGregor *et al.*, 1991; Kruger *et al.*, 2001) highlighted that such processes frequently produce a large set of highly correlated process variables. This has led to the evolution of a range of multivariate statistical techniques that are collectively referred to as multivariate statistical process control (MSPC), and represents an extension to more traditional univariate process control (MacGregor and Kourti, 1995).

PCA is one of the most popular MSPC techniques. PCA offers the capability to compress redundant information in the process measurements, resulting from the high degree of correlation, by retaining only essential information that can be exploited to describe the current state of the process operation (Wise and Gallagher, 1996).

(Russell *et al.*, 2000) showed that the diagnosis of anomalous process behavior entails fault detection, fault identification and fault isolation. (Jackson, 1991; Miller *et al.*, 1998) stated that the difficult issue is to determine the root cause responsible for abnormal process behavior. This is because MSPC provides monitoring statistics that only allow such behavior to be detected. (Miller *et al.*, 1998) proposed the use of contribution charts as the "missing link" in multivariate quality control. However, contribution charts only allow the identification of the potential root causes of anomalous behavior.

(Dunia *et al.*, 1996) introduced a variable reconstruction procedure to remove "fault information"

[1] Corresponding author: d.antory@qub.ac.uk

from the recorded process variables in order to provide a clearer isolation of the fault condition. This technique relies on predicting a set of process variables using the remaining set of variables and the underlying linear PCA model.

This paper presents a nonlinear extension to their linear variable reconstruction technique. This then offers a nonlinear MSPC monitoring approach that addresses the issues of fault detection, identification and isolation. The nonlinear PCA technique is based on recent work by the authors (Antory *et al.*, 2004). (Lieftucht *et al.*, 2004) showed that variable reconstruction in fact alters the underlying PCA model. Consequently, the underlying PCA model must also be "reconstructed" to incorporate the impact of this alteration. The new nonlinear variable reconstruction technique proposed here accommodates the influence of the reconstruction process.

This paper is divided into the following sections. A brief review of PCA/NLPCA is given prior to the introduction of the novel nonlinear variable reconstruction technique. A detailed analysis of the technique is included. This is followed by the application study to an industrial melter process. It is shown that a developing crack could be detected and identified using the NLPCA technique by (Antory *et al.*, 2004) and isolated using the new nonlinear variable reconstruction technique.

2. BACKGROUND TO PCA/NLPCA

In this section, a brief review of the principles of linear and nonlinear PCA is given.

2.1 PCA

For a given vector of n process measurements, $\mathbf{z} \in \mathbb{R}^n$, the application of PCA gives rise to the definition of a reduced set of "artificial" score variables.

PCA describes the measurements in $\mathbf{z}$ as follows:

$$\mathbf{z} = t_1\mathbf{p}_1 + t_2\mathbf{p}_2 + \cdots + t_k\mathbf{p}_k + \mathbf{e}, \qquad (1)$$

$$e_i = \mathbf{d}_i^T\mathbf{z} = z_i - \hat{z}_i, \qquad (2)$$

where $\mathbf{d}_i$ is the ith row of the matrix $\mathbf{D} = \left[\mathbf{I} - \mathbf{P}\mathbf{P}^T\right]$.

where t_1, t_2, ..., t_k are the score variables, $\mathbf{p}_1$, $\mathbf{p}_2$, ..., $\mathbf{p}_k$ are PCA loading vectors, $k < n$ is the number of retained principal components (PCs) and $\mathbf{e}$ represents the residuals of the PCA model. The score variables represent *significant process variation*, whilst $\mathbf{e}$ describes *insignificant and redundant variation in* $\mathbf{z}$.

2.2 NLPCA

Nonlinear PCA is a generalisation of linear PCA which describes nonlinear relationships between the recorded process variables. The work presented here relies on the application of autoassociative neural networks (AAN). The architecture of AAN relies on three hidden layers: the mapping, the bottleneck and the de-mapping layers (Kramer, 1992). The input layer receives the measured process variables, while the output layer provides the NLPCA prediction of the process variables. Consequently, this identity mapping encapsulates the important variation of the process variables in a reduced set of nonlinear scores, obtained in the bottleneck layer.

(Antory *et al.*, 2004) proposed an improved NLPCA technique based on first using PCA and subsequent application of an AAN to the linear PCA score variables. This approach has been referred to as the T2T identity mapping network. This enables the removal of linear redundant information from the recorded process data prior to the determination of the AAN architecture. Note that the score variables are (i) fewer in number and (ii) are assumed to be statistically independent (Jackson, 1991), which (iii) gives rise to a pre-conditioning for the identification of the network parameters.

The mathematical description of the T2T network is as follows:

$$\mathbf{t}_{nl} = f\left(\mathbf{t}\right) \qquad (3)$$

$$\widehat{\mathbf{t}} = g\left(\mathbf{t}_{nl}\right) \qquad (4)$$

where $\mathbf{t}_{nl} \in \mathbb{R}^{k_L}$, $k_L < k$, stores the nonlinear score variables, with k_L being the number of bottleneck nodes. $f(\circ)$ and $g(\circ)$ are nonlinear functions that represent the mapping and demapping layers of the T2T network, and $\widehat{\mathbf{t}} \in \mathbb{R}^k$ is the prediction of $\mathbf{t}$.

3. NONLINEAR VARIABLE RECONSTRUCTION

Variable reconstruction was first discussed by (Dunia *et al.*, 1996) in conjunction with linear PCA. Using this technique, a subset of process variables can be reconstructed using the remaining process variables and the identified PCA model (Dunia and Qin, 1998). Recently, (Lieftucht *et al.*, 2004) demonstrated that variable reconstruction alters the confidence limit of the Q statistic and the variance of the score variables. Further, they showed how to incorporate such changes into process monitoring.

The work presented here enhances the variable reconstruction technique by (Dunia and Qin, 1998), so that it can be applied in a nonlinear context. This nonlinear extension is applied in conjunction with NLPCA. They showed that if the Q statistic is significant, the variable reconstruction technique for linear PCA can isolate the fault signature from the recorded process variables. The Q statistic for the NLPCA model is given by:

$$\mathbf{e} = \mathbf{z} - \mathbf{P}f\left(g\left(\mathbf{P}^T\mathbf{z}\right)\right) \quad Q = \mathbf{e}^T\mathbf{e}, \tag{5}$$

with $\mathbf{P} = \left[\, \mathbf{p}_1 \; \mathbf{p}_2 \; \cdots \; \mathbf{p}_k \,\right]$, The principles of nonlinear variable reconstruction to isolate the fault signature using the NLPCA model are outlined next.

3.1 Nonlinear Reconstruction Algorithm

(Dunia *et al.*, 1996) showed that reconstruction of the jth process variable is an iterative process:

$$z_j^{new} = [\, \mathbf{z}_{-j}^T \; z_j^{old} \; \mathbf{z}_{+j}^T \,]\mathbf{c}_j \tag{6}$$

where $\mathbf{c}_j$ is the jth row vector of $\mathbf{C} = \mathbf{P}\mathbf{P}^T$, which can alternatively be formulated as follows:

$$\widetilde{z}_j = \frac{1}{1 - c_{jj}} \sum_{i=1 \neq j}^{n} c_{ji} z_i, \tag{7}$$

where $\widetilde{z}_j$ is the reconstructed value of the jth process variable.

To integrate the above linear variable reconstruction for application in conjunction with NLPCA, the iteration technique of Equation (6) has to be considered. This is because the mapping and demapping of the T2T network are based on nonlinear functions. Using linear PCA, however, these transformations are linear and involve the retained PCA loading vectors, stored in $\mathbf{P}$.

The iterative process of nonlinear reconstruction is as follows:

(i) obtain the nonlinear score variables from the current process measurements *using the mapping layer of the NLPCA model*;
(ii) predict the current process measurements *using the demapping layer of the NLPCA model*;
(iii) replace the measured value of the jth process variable by the predicted one *obtained from step (ii)*; and
(iv) go to step (ii) until the difference between two consecutive values of the predictions of the jth process variable is smaller than a given threshold, e.g. 10^{-8}.

Mathematically, the above procedure can be summarized as follows. The prediction of the current process measurements is given by:

$$\widehat{\mathbf{z}}^{(1)} = \mathbf{P}f\left(g\left(\mathbf{P}\mathbf{z}\right)\right) \tag{8}$$

Replacing the measurement of the jth process variable by its prediction leads to:

$$\mathbf{z}^* = \left(z_1 \; z_2 \; \cdots \; \widehat{z}_j \; \cdots \; z_n \right)^T \tag{9}$$

Applying Equation (7) again yields:

$$\widehat{\mathbf{z}}^{(i)} = f\left(g\left(\mathbf{z}^*\right)\right) \tag{10}$$

where i represents the actual iteration step. The iterative procedure has converged if

$$\left\| \hat{z}_j^{(i+1)} - \hat{z}_j^{(i)} \right\| < \varepsilon \tag{11}$$

More than one variable can be reconstructed, up to a maximum number equal to the number of bottleneck nodes.

4. APPLICATION STUDY: GLASS MELTER PROCESS

This section presents an application study of NLPCA, and the new nonlinear variable reconstruction technique, to an industrial melter process. The aim is to detect, identify and isolate the influence of a crack based on a historical data.

The melter process is part of the disposal procedure. Waste material is pre-processed by an evaporation treatment leading to the production of powder, which is then clad by a glass layer, provided by the melter process. The melter consists of a vessel, two exit funnels through which the melter load flows out and several induction coils.

The vessel is continuously filled with the powder while raw glass is discretely introduced in the form of glass frit. This binary composition is heated by four induction coils, which are positioned around the vessel. Because of the heating procedure, the glass is melted homogeneously.

The process of filling and heating continues until the desired height of the liquid column is reached. The molten mixture is then poured out through one of the exit funnels. After the contents of the vessel has been emptied to the height of the nozzle, the next cycle of filling and heating begins.

Measurements of 8 temperatures, the power in 4 induction coils and voltage were taken every five minutes. The filling and emptying cycles resulted in a nonlinear relationship between the temperatures, power in the induction coils and voltage.

The melter vessel is made of graphite, which is a brittle material. As a result of the strong temperature variations to which the vessel is frequently exposed, cracks may occur along the regions of high stress. These cracks not only damage the shell of the melter, but also allow the molten content to escape. It is therefore necessary to detect such cracks at an early stage.

Historical data from the melter process was available, which included normal process variation and an abnormal process situation resulting from a developing crack. The recorded data contained 1050 samples, at a sampling frequency of 5 minutes, with the last 50 points corresponded to the development of a crack in the melter vessel.

4.1 Identification of a NLPCA Monitoring Model

A PCA model was identified based on the first 1000 data points of the recorded data set. Previous work by the authors (Antory *et al.*, 2004) showed that the linear PCA model included 10 principal components. By discarding the last three PCs, i.e. $k = 10$, the reconstruction of the 13 process variables revealed that (i) only insignificant variation remained in the PCA residuals and (ii) the emptying and filling cycles were accurately described.

Furthermore, the PCs were assumed to be statistically independent, which provided favorable conditions for the subsequent identification of an AAN architecture. This network included 7 nodes in the mapping layer, 3 bottleneck nodes, i.e. $k_{n_L} = 3$, and 7 nodes in the de-mapping layer. This implies that only 3 nonlinear score variables were required to produce an accurate NLPCA model for a total of 13 process variables.

To monitor this process, the Q statistic and scatter diagrams, for which the confidence regions were obtained using a Kernel Density Estimation (KDE), were used for process monitoring. Using scatter diagrams, the influence of the developing crack was noticed almost 2 hours earlier compared to the application of conventional PCA to the same data (Chen *et al.*, 2000).

4.2 Fault Detection and Identification

Figure 1 shows the influence of the developing crack results in a statistically significant Q statistic from the 33rd data point.

The severity of the crack increases over time and led eventually to the failure of a particular temperature sensor that was closest to it. To identify potential root cases of this event, contribution charts to the Q statistic were used.

Figure 2 illustrates that the contribution chart, evaluated for the 33rd data point, identified the temperature reading of sensor 6 and and the power reading of the 3rd induction coil as the dominant contributors to this event.

This was a correct analysis of the event, since a later investigation showed that the crack indeed developed in the vicinity of the 6th temperature

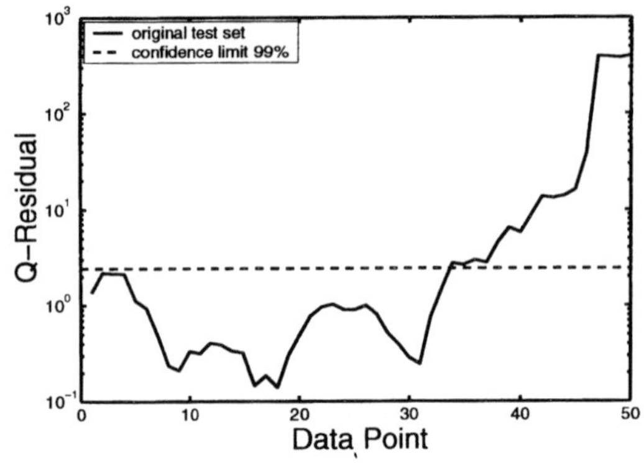

Fig. 1. The Detection of Crack in Melter Process

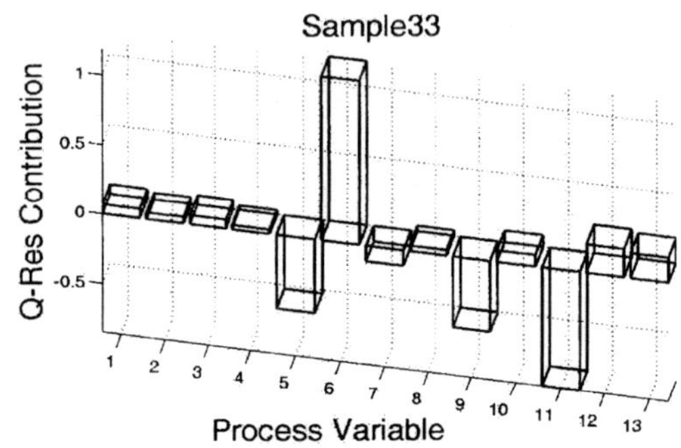

Fig. 2. Variable Contributions to the Fault

sensor. In addition, the change in the temperature profile affected the power consumed by the 3rd induction coil as a result of controller feedback.

4.3 Fault Isolation

Using the new nonlinear variable reconstruction technique, the information from the previous fault identification step is now used to isolate the signature of this event from the recorded data. Given that the 6th temperature sensor shows a significant response, the corresponding variable was reconstructed using the remaining ones.

Figure 3 compares the recorded and reconstructed sequences of the 6th temperature sensor.

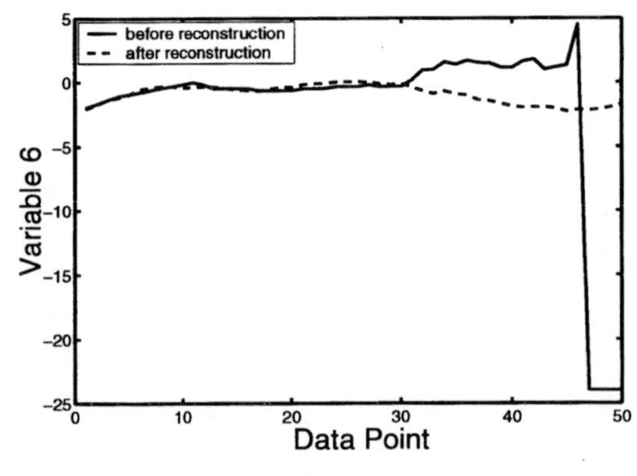

Fig. 3. Temperature Sensor 6 Before and After Reconstruction

This figure demonstrates that the nonlinear reconstruction technique successfully isolates the influence of the developing crack from the recorded sequence. In order to evaluate whether the process was operating "in-statistical-control" after the reconstruction of the 6th temperature sensor, the Q statistic was re-examined. Figure (4) highlights that no violations of the 99% confidence limit arose after the 6th temperature sensor was reconstructed. Consequently, an experienced operator would have been directed to focus attention on this sensor.

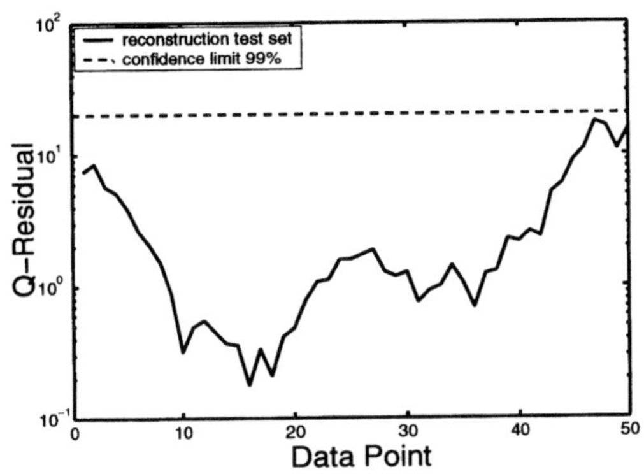

Fig. 4. Fault Detection for a New Reconstructed Test Set

As discussed by (Lieftucht *et al.*, 2004), the confidence limit for the Q statistic must be altered to accommodate the changes of the variable reconstruction on the monitoring model. This was achieved by reconstructing the readings of the 6th temperature sensor in the reference data set, i.e. the first 1000 samples, and recalculating the confidence limit. Figure 5 represents the influence of reconstructing the 6th temperature sensor on the 99% confidence limit of the Q statistic. It should be noticed that after the reconstruction, the confidence limit for the Q statistic increased, as a result of the applied reconstruction process.

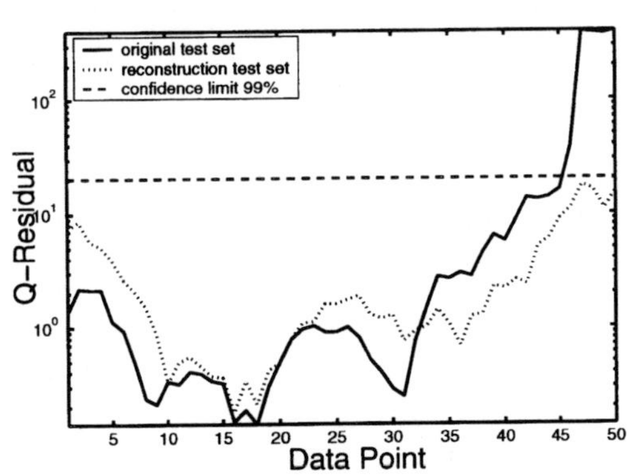

Fig. 5. Comparison of Fault Detection before and after the Reconstruction Process

4.4 Scatter Diagrams for Fault Diagnosis

To show the process variation encapsulated in the nonlinear score variables, (Antory *et al.*, 2004) proposed the use of scatter diagrams for which the confidence regions are identified using KDE. This was motivated by the fact that the nonlinear score variables may not be normally distributed and consequently, the assumptions imposed on the Hotelling's T^2 statistic may be violated.

Figures 6 to 8 show the application of scatter diagrams for fault diagnosis process.

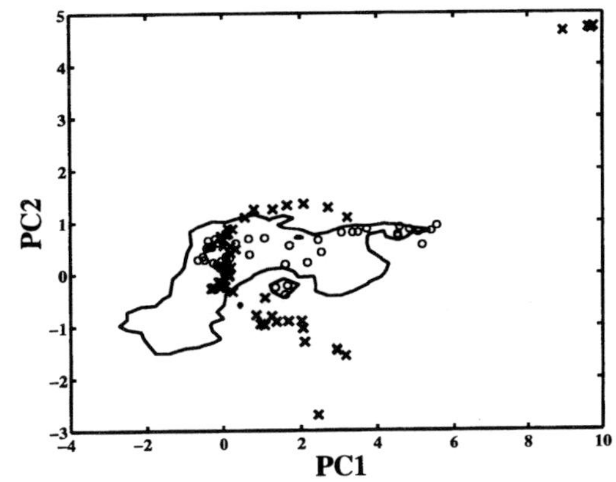

Fig. 6. Scatter Diagram: PC1 vs. PC2

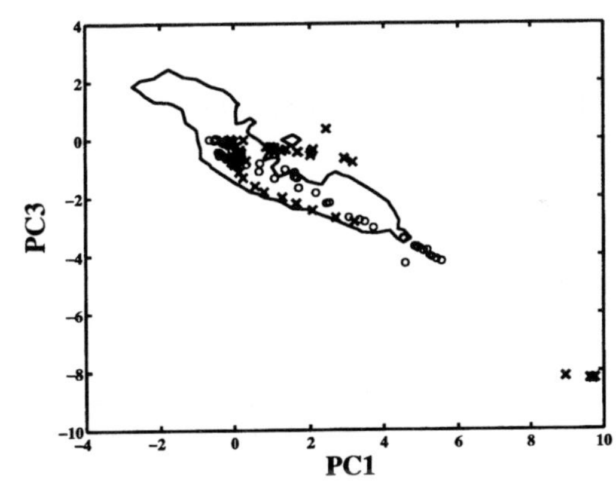

Fig. 7. Scatter Diagram: PC1 vs. PC3

The scatter points corresponding to the measured and reconstructed scatter points are respectively represented by the symbols $\times$ and $\circ$. In a similar fashion to the confidence limit of the Q statistic, the 99% confidence limits were also recalculated on the basis of the reconstruction of the 6th temperature readings in the reference data. Note that a few scatter points remain outside the confidence regions, which would indicate an "out-of-statistical-control" situation. However, these points relate to the instances 46 to 50, i.e. when the sensor failed. The failure of the 6th temperature sensor propagated back to influence to the regulatory control system and consequently adjusted the power in the induction coils. This,

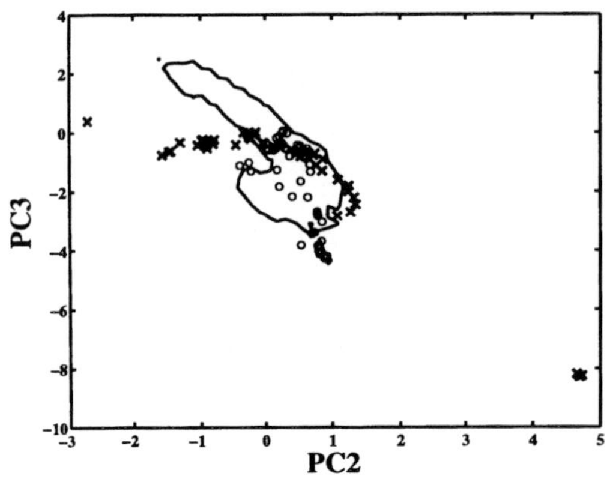

Fig. 8. Scatter Diagram: PC2 vs. PC3

in turn, caused the introduction of an excessive process variation. This excessive process variation produced the "out-of-statistical-control" situation for the last 5 instances. However, the underlying NLPCA model, after the reconstruction of the 6th temperature sensor, could accurately describe the "variation" of the melter process, since Figure 1 represents an "in-the-statistical-control" situation.

5. CONCLUSIONS

This paper presents a novel nonlinear variable reconstruction technique in conjunction with the application of an autoassociative neural network (AAN) based on nonlinear principal component analysis (NLPCA) model for process fault isolation. This reconstruction technique represents a nonlinear extension of the variable reconstruction technique by (Dunia *et al.*, 1996). Based on the findings of (Lieftucht *et al.*, 2004), the confidence limit for the Q statistic and the confidence regions for the scatter diagrams are adjusted to accommodate the changes that the reconstruction process imposes upon the PCA/NLPCA model.

The application study to an industrial melter process demonstrated that (i) a developing crack could be detected and identified using the NLPCA technique by (Antory *et al.*, 2004) and (ii) isolated using the new nonlinear variable reconstruction technique.

ACKNOWLEDGEMENT

The first author would like to acknowledge the financial support from the Virtual Engineering Centre (www.qub.vec.ac.uk), Queen's University Belfast. Valuable help and discussions with Dirk Lieftucht are greatly appreciated.

REFERENCES

Antory, D., U. Kruger, G. W. Irwin and G. McCullough (2004). Industrial process monitoring using nonlinear principal component models. *Proceedings of IEEE International Conference on Intelligent Systems (IS'04).* June 22-24, Varna, Bulgaria.

Chen, Q., R. Wayne and D. J. Sandoz (2000). The application of principal component analysis and kernel density estimation to enhance process monitoring. *Control Engineering Practice* **8**(5), 531–543.

Dunia, R. and S. J. Qin (1998). Joint diagnosis of process and sensor faults using principal component analysis. *Control Engineering Practice* **6**, 457–469.

Dunia, R., S. J. Qin, T. F. Edgar and T. J. McAvoy (1996). Identification of faulty sensors using principal component analysis. *AIChE Journal of Process Systems Engineering* **10**(10), 2797–2812.

Jackson, J. E. (1991). *A Users Guide to Principal Components.* Wiley Series in Probability and Mathematical Statistics. John Wiley. New York.

Kramer, M. A. (1992). Autoassociative neural network. *Computers and Chemical Engineering* **16**(4), 313–328.

Kruger, U., Q. Chen, D. J. Sandoz and R. C. McFarlane (2001). Extended pls approach for enhanced condition monitoring of industrial processes. *AIChE Journal* **47**(9), 2076–2091.

Lieftucht, D., U. Kruger and G. W. Irwin (2004). Improved diagnosis of sensor faults using multivariate statistics. *Proceedings of American Control Conference (ACC'04).* June 30-July 2, Boston, USA.

MacGregor, J. F. and T. Kourti (1995). Statistical process control of multivariate processes. *Control Engineering Practice* **3**(3), 403–414.

MacGregor, J. F., T. E. Marlin, J. V. Kresta and B. Skagerberg (1991). Multivariate statistical methods in process analysis and control. In: *AIChE Symposium proceedings of the 4th International Conference on Chemical Process Control.* AIChE Publication, No. P-67. New York. pp. 79–99.

Miller, P., R. E. Swanson and C. F. Heckler (1998). Contribution plots: A missing link in multivariate quality control. *Applied Mathematics and Computer Science* **8**(4), 775–792.

Russell, E. L., L. H. Chiang and R. D. Braatz (2000). *Data-Driven Techniques for Fault Detection and Diagnosis in Chemical Processes.* Advances in Industrial Control. Springer-Verlag. London, ISBN 1-85233-258-1.

Wise, B. M. and N. B. Gallagher (1996). The process chemometrics approach to process monitoring and fault detection. *Journal or Process Control* **6**(6), 329–348.

ELSEVIER
IFAC
PUBLICATIONS
www.elsevier.com/locate/ifac

GLOBALLY OPTIMAL NONLINEAR MODEL PREDICTIVE CONTROL

C. E. Long, P. K. Polisetty, and E. P. Gatzke [1]

Department of Chemical Engineering
University of South Carolina
Columbia, SC, USA 29208

Abstract

This paper presents a globally optimal nonlinear Model Predictive Control (NMPC) algorithm. Utilizing local techniques on nonlinear nonconvex problems leaves one susceptible to suboptimal solutions. In complex problems, local solver reliability is difficult to predict and often highly dependent upon the choice of initial guess. For the purpose of NMPC, local solvers can cause unexpected closed-loop results or failure of the algorithm. Stochastic attempts at global optimization of NMPC methods cannot provide rigorous bounds on the optimality of the resulting solution. Implementation of a global solution technique (Falk and Soland [1969], Horst and Tuy [1990]), which guarantees global optimality, restores the integrity of NMPC technology. Due to the combinatorial nature of nonconvex optimization, real-time considerations must be considered. The proposed algorithm's capabilities are demonstrated by the application of the controller on the benchmark control problem of the isothermal operation of a continuous stirred tank reactor (CSTR) with Van de Vusse reactions (Kremling and Allgöwer [1993]).

Keywords:
Constrained Nonlinear Control, Nonlinear Model Predictive Control, Global Optimization, Branch-and-Reduce, Convexification

INTRODUCTION

Given that a dynamic process can be approximated by a linear model near its nominal operating point, linear Model Predictive Control (MPC) methods typically provide reasonable control of a system (Morari and Lee [1991, 1997]). For these methods, a convex optimization problem is solved online at each time step for the optimal control sequence that will keep the system within a desired region of operation. However, in many processing examples, a process may not exhibit linear dynamics resulting in unstable or poor closed-loop performance using linear methods. Alternative formulations that consider nonlinear dynamics must be pursued.

For constrained nonlinear systems, a nonlinear model can be used in nonlinear Model Predictive Control (NMPC) formulations (Henson [1998]). The nonlinearity in such formulations, does however, give rise to a more complex optimization problem to be solved online. In general, convexity of the resulting online optimization problem is lost. The nonconvex nonlinear program (NLP) formulation leaves one susceptible to determination of suboptimal solutions.

With a nonlinear discrete-time state-space representation of a process, the NMPC problem can be formulated at each sampling time. The controller output will depend upon online solutions of nonconvex optimization problems which will include nonconvex constraints representing a projection of the nonlinear model states and measurements into the

[1] Author to whom correspondence should be addressed: gatzke@sc.edu

future. This work presents a new NMPC application that attempts to find the guaranteed global optimum for problems posed by general NMPC algorithms.

Deterministic methods for the global solution of nonconvex problems typically rely on convex relaxations of nonconvex functions. Numerous methods (Adjiman, Dalliwig, Floudas, and Neumaier [1998], McCormick [1976], Tawarmalani and Sahinidis [2000], Gatzke, Tolsma, and Barton [2002]) have been proposed to carry out such relaxations. For this work, the nonconvex NLP is reformulated to a convex NLP via variable transformations. A linearization strategy of (Tawarmalani and Sahinidis [2000], Gatzke et al. [2002]) is then implemented to generate a Linear Program (LP). Calculating the lower-bounds for the problem then relies on solution of these LP's, for which various methods are available (Dantzig [1963], Karmarkar [1984], ILOG [2002]). Note that nearly any algebraic function can be reformulated for convexification.

This work expands on previous contributions which considered a relative small class of nonlinear dynamic systems (Sriniwas and Arkun [1997]). Once the convex relaxation is obtained, branch-and-bound (Falk and Soland [1969]) or branch-and-reduce (Ryoo and Sahinidis [1995]) techniques can be applied to solve the modified program providing a guaranteed global optimum (within tolerances). For faster convergence to this solution, the variable space can be reduced using standard techniques such as interval analysis (Moore [1979]). Although global optimization methods are combinatorial in nature, the NMPC formulation does have a relatively restricted variable branching space. In applying this solution technique to the NMPC, other issues arise regarding the algorithms use for real-time applications.

CONTROLLER FORMULATION

Assume that a given process with n_u inputs, n_x states, and n_y outputs can be represented by a nonlinear discrete-time state space model:

$$\begin{aligned} x(k+1) &= f(x(k), u(k)) \\ y(k) &= g(x(k), u(k)) \end{aligned} \tag{1}$$

where $x(k) \in \mathbb{R}^{n_x}$ is the state vector at sample time k, $u \in \mathbb{R}^{n_u}$ is the vector of inputs, and $y \in \mathbb{R}^{n_y}$ is a vector of the predicted outputs. Note that $f : \mathbb{R}^{n_x} \times \mathbb{R}^{n_u} \rightarrow \mathbb{R}^{n_x}$ and $g : \mathbb{R}^{n_x} \times \mathbb{R}^{n_u} \rightarrow \mathbb{R}^{n_y}$. Without loss of generality, it is assumed here that g is a linear map.

The NMPC is formulated to choose a sequence of input moves over the move horizon (m) that minimizes some cost function. This cost function typically quantifies the difference between the model predicted evolution of the system and the desired setpoints over some prediction horizon (p). A 1-norm objective function may take the form:

$$\Phi = \sum_{i=1}^{p} \Gamma_y(i) e(i) + \sum_{j=0}^{m-1} \Gamma_u(j) \Delta u(j) \tag{2}$$

where $e(i)$ is the absolute value of error predicted for the $(i)^{th}$ time step into the future. The error (e) is defined and constrained as:

$$|r_y(i) - y(i)| \leq e(i) \quad \forall i = 1...p \tag{3}$$

Δu is a vector defining changes in input movements. $\Gamma_y(i)$ and $\Gamma_u(j)$ are weighting factors used to define the relative importance of each objective function term.

The optimization problem to be solved at each time step includes constraints. The predicted state and output values are constrained by the model. The model predicted output is updated to account for any plant/model mismatch. This disturbance update is defined as:

$$d(i) = y_m(0) - y_p(0) \tag{4}$$

where $y_m(0)$ and $y_p(0)$ are the measurement at the current time and the predicted value of the output at the current time, respectively. This predicted output is based on state estimates from an open-loop observer run in parallel to the process. Constraints on the input movements are implemented as

$$|u(i-1) - u(i-2)| \leq \Delta u(i-1) \qquad \forall i = 1...m \tag{5}$$

Hard constraints on the actual inputs of the process are implemented as:

$$u^l \leq u \leq u^u \tag{6}$$

The program to be solved online by the NMPC algorithm has been completely defined and can be re-written in a more compact form:

$$\begin{aligned} \min_{\mathbf{z}} \quad & C^T z \\ s.t. \quad & A_1 z \leq b \\ & h_i(z) = 0 \;\; \forall i = 1...M \\ & z^l \leq z \leq z^u \end{aligned} \tag{7}$$

where the vector $z \in \mathbb{R}^N$ is a vector of N unknowns including the desired input trajectory, the resulting state, output, and error projections, as well as the resulting Δu terms. All linear inequality constraints are represented by $A_1 z \leq b$. The M nonlinear constraints that arise from the model and the objective function are written as $h_i(z) = 0$, where $h_i : z \rightarrow \mathbb{R}^M$, $M = n_x * p + 1$. Hard constraints on the inputs are incorporated into the bounds on the unknown vector (z^l and z^u).

GLOBAL SOLUTION

Deterministic methods for global optimization depend on the generation of convex relaxations of the original nonconvex nonlinear problems. Numerous methods have been proposed for constructing such relaxations. For this work, a reformulation method (McCormick [1976]) is used which converts the original factorable nonconvex nonlinear problem into an equivalent form by the introduction of new variables and new constraints. The reformulated problem contains only linear and simple nonlinear terms for which convex relaxations can be constructed using the convex envelopes already known for such simple algebraic functions. The reformulated Nonconvex NLP is of the form:

$$\begin{aligned} \min_{\mathbf{w},\mathbf{z}} \quad & C^T z \\ s.t. \quad & A_1 z \leq b \\ & A_2 \begin{bmatrix} w \\ z \end{bmatrix} \leq 0 \\ & w = h(w, x) \\ & z^l \leq z \leq z^u \\ & w^l \leq w \leq w^u \end{aligned} \tag{8}$$

where $A_2[w\ z]^T \leq 0$ defines the new linear constraints obtained from reformulation, while $w = h(w, z)$ provides the relationship between the new and original variables. With Q new variables, $w \in \mathbb{R}^Q$ and $h : w \times z \rightarrow \mathbb{R}^Q$. Bounds on w are determined from the bounds on z. Note that h consists of simple nonlinear terms relating 2or 3 variables.

Convex relaxations of this problem are then constructed using DAEPACK (Gatzke et al. [2002], Tolsma and Barton [2000]), an automated code generation tool. The advantage of using DAEPACK tool for generating convex relaxations is it can be applied to legacy models coded in standard FORTRAN. The convex relaxations can be denoted as:

$$\check{h}(w, z, w^l, w^u, z^l, z^u) \leq w \leq \hat{h}(w, z, w^l, w^u, z^l, z^u) \tag{9}$$

where $\check{h}$ and $\hat{h}$ are the convex under and over estimates of the reformulated problem.

The linearization strategy (Tawarmalani and Sahinidis [2000], Gatzke et al. [2002]) is then used to generate an LP relaxation of the convex NLP created using DAEPACK. The resulting LP is of the form:

$$\begin{aligned} \min_{\mathbf{w},\mathbf{z}} \quad & C^T z \\ s.t. \quad & A_1 z \leq b \\ & A_2 \begin{bmatrix} w \\ z \end{bmatrix} \leq 0 \\ & A_3 \begin{bmatrix} w \\ z \end{bmatrix} \leq b_3 \\ & z^l \leq z \leq z^u \\ & w^l \leq w \leq w^u \end{aligned} \tag{10}$$

where $A_3[w\ z]^T \leq b_3$ expresses the new linear constraints resulting from the linearization process. This linearization technique is ideal because it yields an LP for which robust solvers exist (e.g., ILOG CPLEX 8.0 ILOG [2002] and the IBM OSL library I. B. M. [1997]).

Upon creation of the linear (convex) underestimates for the nonconvex nonlinear problem, the branch-and-reduce method (Ryoo and Sahinidis [1995]) is implemented. This is an extension of the traditional branch-and-bound method with bound tightening techniques for accelerating the algorithm's convergence. Within this branch-and-reduce algorithm, infeasible or suboptimal parts of the feasible region can be eliminated using range reduction techniques such as optimality based and feasibility based range reduction tests (Ryoo and Sahinidis [1995]) or interval analysis techniques (Moore [1979]). These techniques help to derive tighter variable bounds for a given partition in the search tree. Finally, the algorithm terminates when the lower bounds for all partitions either exceed or are sufficiently close (within specified tolerances) to the best upper bound. At this point, a global optimum has been found.

In general, global optimization methods are combinatorial in nature. However, for this specific application, despite the NMPC formulation involving hundreds of variables, the problem has a relatively restricted variable branching space. Only the $m\,n_u$ true decision variables are branched on during the branch and reduce algorithm. The remainder of the variable bounds are defined by the selection of the optimal input sequence bounds. This is promising, especially with regards to the feasibility of the solution approach for real-time application in the NMPC context.

A modification of standard branch-and-reduce methods was required in order to allow for rapid global solution. Typically, a node is selected from the active node list. This partition is examined for a possible new improved upper bound to the overall problem by local solution of the corresponding nonconvex optimization problem. The node is then partitioned and lower bounds are derived for the new partitions based on the convex lower bounding problems. It was found that it may be necessary for rapid convergence in this real-time application to search for an upper bound for each partition using a local search as soon as any new partition is created. This amounts to a minor reordering of events in the typical branch-and-bound search algorithm. If this upper bound local solution search is not performed as early as possible for the partition, a partition containing the global solution may be added to the active node list without a local solution and this node may not be selected from the list early on in the branch-and-bound search. This is especially problematic for problems that contain partitions with highly degenerate lower bounds, as is the case in NMPC types of problems. The lower bounding solutions may return an objective of zero for many of the early partitions. Depending on the implementation, the partition containing the global solution may not be selected from this active node list immediately in typical implementations, delaying the determination of the global solution. Performing a local searching on every node as soon as the partition is created helps determine the global solution early on, which also leads to more rapid convergence as more of the parameter space can be fathomed using the branch-and-reduce methods.

CASE STUDY

Consider the benchmark control problem of the isothermal operation of a two state continuous stirred tank reactor (CSTR) with Van de Vusse reactions. In this reactor, the Van de Vusse reactions are:

$$A \rightarrow B \rightarrow C$$
$$2A \rightarrow D$$

Material balances dictate that the system can then be described by:

$$\frac{dC_a}{dt} = (F/V)(C_{ao} - C_a) - k_1 C_a - k_3 C_a^2 \tag{11}$$

$$\frac{dC_b}{dt} = k_1 C_a - k_2 C_b - (F/V) C_b \tag{12}$$

where F is the feed flow rate of A into the reactor, V is the constant reactor volume, C_a and C_b are the reactant concentrations in the reactor, and k_i are the reaction rate constants for the three reactions. For this work, let $k_1 = 50\,h^{-1}$, $k_2 = 100\,h^{-1}$, and $k_3 = 10\,L\,gmol^{-1}\,h^{-1}$. Assume that the volume of the reactor is constant, that the feed is pure A, and that the nominal concentration of A in the feed (C_{ao}) is $10\,gmol\,L^{-1}$.

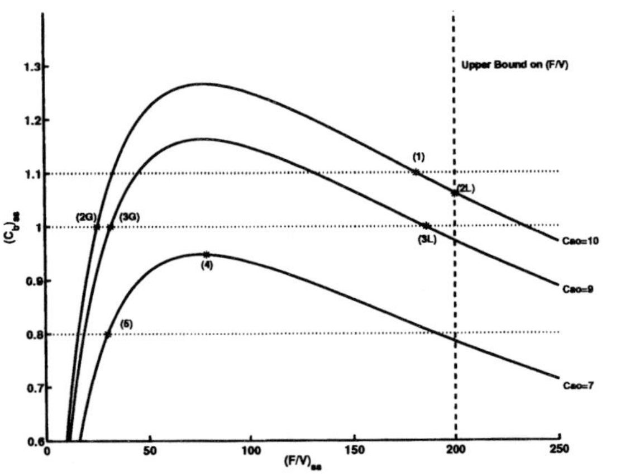

Figure 1. Steady state loci for various feed concentrations showing steady state operating points in the Local (L) and Global (G) solution methods.

For this single input single output (SISO) system, the input (u) is taken as (F/V) , the state vector (x) consists of the concentrations of A and B in the reactor, and the single measurement (y) is that of the concentration of B (i.e., $y = x_2$). An unmeasured disturbance (d) will be simulated through changes in C_{ao}. The discrete-time model necessary for the predictive aspects of this type of control is found by discretizing the nonlinear state-space model using a backward difference approximation with a sampling rate of 0.002 hours. It should be noted that this system happens to exhibit a steady-state input multiplicity. This is depicted in a plot of the steady-state loci presented in Figure 1. An upper bound on the input (F/V) is assumed to be at a value of 200 h^{-1}.

The predictive controller is tested for its abilities in both setpoint tracking and disturbance rejection. The controller is tuned with $m = 1, p = 30, \Gamma_y(p) = 100$, and $\Gamma_u = 0$. Note that $\Gamma_y(k) = 0\;\forall k \neq p$. By weighting only the pth error term in the projection, a terminal error penalty is enforced. Assume that the process is initially operated at $u = 181\,h^{-1}$ and $y = 1.1\ gmol/L$. This operating point is indicated by (1) in Figure 1. At a time of 0.1 hours, the setpoint is stepped to 1 $gmol/L$. A series of unmeasured disturbances are then introduced. At a time of 0.5 hours, the setpoint is again stepped down, this time under disturbance, to a value of 0.8 $gmol/L$.

The closed-loop results for the controller using both the local solution techniques (MINOS version 5.51) and the proposed global technique are shown in Figure 2. At the

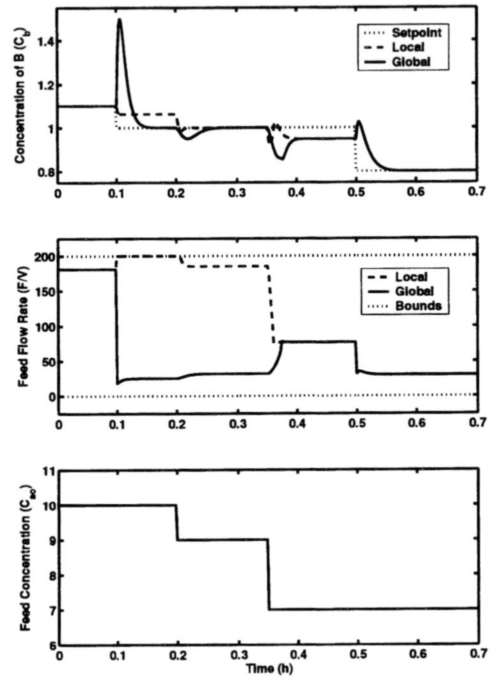

Figure 2. Closed-loop results (m=1) with setpoint changes at 0.1 and 0.5 hours and disturbance loads at 0.2 and 0.35 hours.

time of the first setpoint change, the local solver moves the system in an improving direction but ends up at a local minima (against the upper bound constraint). This is indicated by (2L) in Figure 1. The global solver is able to realize the full setpoint change by finding the global solution (2G). This example shows that the global NMPC can achieve superior performance. A sample objective function for the optimization problem encountered under this circumstance is provided in Figure 3.

At $t = 0.2$ hours, the first disturbance hits and both algorithms (using local and global solution techniques) are then able to track the setpoint under this disturbance. However, the presence of the input multiplicity allows them to do so at different operating conditions (denoted 3L and 3G in Figure 1).

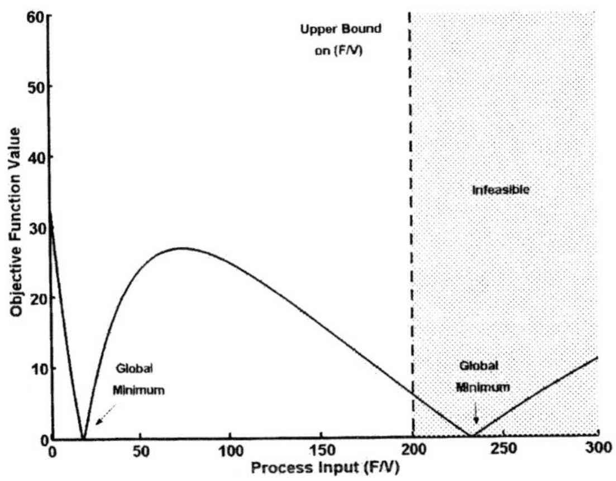

Figure 3. Sample objective function at the time of a setpoint change.

From this, it is obvious that the local solver is moving in the improving direction toward a minimum that is infeasible due to the upper bound on the input variable. At $t = 0.35$ hours, a second disturbance hits that moves the system to an operating regime in which the setpoint of $1.1\ gmol/L$ can no longer be achieved. Both the local and global solution techniques move the system to the optimal operating point denoted as (4) in Figure 1. Finally, at $t = 0.5$ hours, the second setpoint change is implemented. Both algorithms track this reference change without issue.

For a 0.7 hour simulation with a sampling rate of 0.002 hours, 350 optimization problems are solved online. The required time for finding the global solution for each problem is presented in Figure 4. In most cases using Redhat Linux 9.0 on a dual AMD 1900+ MP system, the global solver is able to guarantee global optimality sufficiently fast for real-time operation (i.e., the solver returns the global optimum within 7.2 seconds). However, at a time of 0.35 hours, the system is under a disturbance large enough that the desired setpoint can no longer be achieved. At this particular point, guaranteeing the global solution takes significantly more time than previous cases and the solver is no longer fast enough for real-time purposes. This can be attributed to a flat spot in the objective function (Figure 5). In order to account for this, the global solver is terminated at the real-time threshold and the best solution thus far is implemented. In this case, there is no degradation in the controller's performance by terminating the solve at the real-time threshold. However, the guarantee on global optimality is lost. It is likely that the global solution has indeed been found, however, some suboptimal regions of the solution space have not yet been fathomed. A measure of this optimality gap is provided in Figure 4. Convergence is dictated by bringing the lower bound within some tolerance of the upperbound. This plot shows the difference between the two upon termination.

Closed-loop results with a larger move horizon are presented in Figure 6. Again, the controller is choosing the

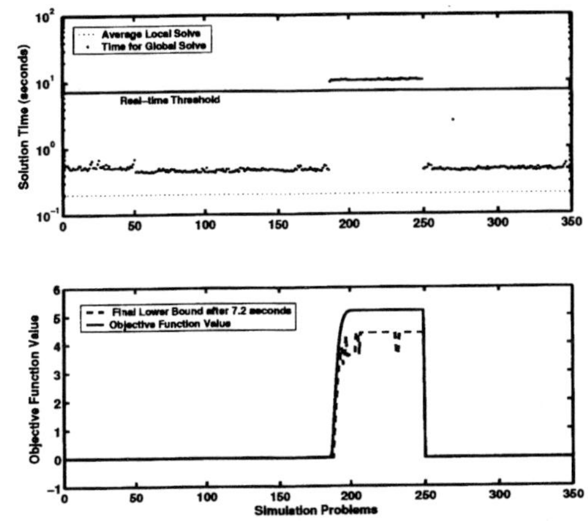

Figure 4. Solution times for the NMPC problems showing the time for the global solution, the global solution value, and the worst case lower bound upon termination for real-time application.

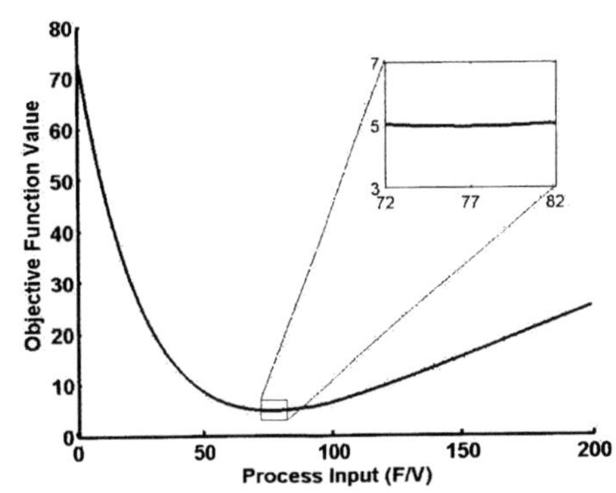

Figure 5. Sample objective function exhibiting a region of insensitivity to changes in the input.

optimal mn_u moves that minimize the objective function. The controller then implements only the first control move in that sequence. In an attempt to avoid the scenario in which the controller chooses not to move the system in this first calculated move, the objective function was modified from the previous case. An additional error term from the prediction horizon was included in the objective function to account for process dynamics. An input movement penalty was also implemented. The specific controller tunings were $m = 3$, $p = 30$, $\Gamma_y(15) = \Gamma_y(30) = 100$, and $\Gamma_u = 0.005$. Again note that $\Gamma_y(k) = 0 \quad \forall k \neq p, \frac{p}{2}$. Velocity constraints were also imposed on the system as:

$$\Delta u(i) \leqslant 70 \qquad \forall i = 1..m \tag{13}$$

The controller was subjected to the same setpoint tracking and disturbance rejection tests as in the $m = 1$ case. Interestingly enough, the NMPC using local solution techniques was able to track the reference changes and reject the disturbances without issue. In particular, when the initial setpoint change is applied, the controller based on local solution techniques is able to move the system to the setpoint instead of moving the input value to the upper bound as before. On the other hand, the NMPC algorithm using the global

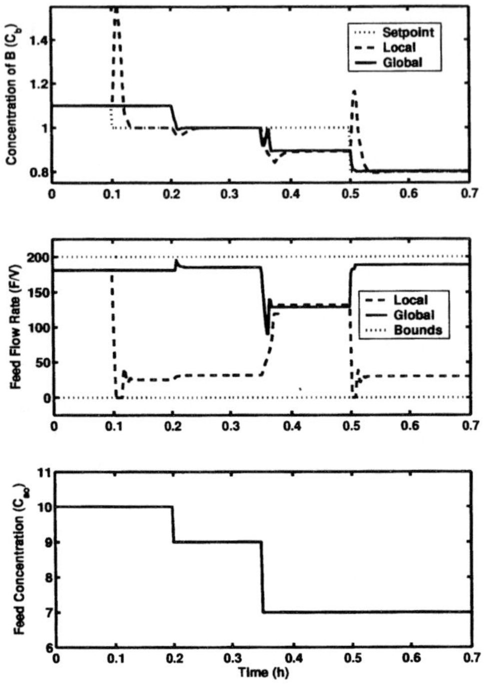

Figure 6. Closed-loop Results (m=3) with setpoint changes at 0.1 and 0.5 hours and disturbance loads at 0.2 and 0.35 hours

solution technique did not perform as well. At the time of this first setpoint transition, the controller chose not to move. The controller did indeed return the global solution of the optimization problem at this point, however, the solution dictated that it was not necessary for the controller to move the system immediately (i.e., in the first move of the chosen optimal input sequence). For the remainder of the simulation, the NMPC algorithm using the global solution technique was able to track the reference and reject the disturbances. It should be noted that it was able to do so at different input values than the local method, which is made possible by the presence of the input multiplicity.

CONCLUSIONS

A globally optimal NMPC algorithm has been proposed. A deterministic approach is used in finding the guaranteed global optimum to the nonconvex NLPs associated with the controller's operation. The global algorithm was shown to eliminate the poor performance in a simple CSTR example resulting from the suboptimal input trajectories supplied by a controller which uses local solution techniques, provided that the controller was properly tuned. In doing so, the globally optimal NMPC algorithm showed promise with regards to its real-time application. However, backup hybrid control methods should be considered to handle situations where the desired solution cannot be obtained in the allotted time.

Acknowledgments

The authors would like to acknowledge financial support from the American Chemical Society Petroleum Research Foundation grant # 38539-G9 and the National Science Foundation Early Career Development grant CTS-038663.

REFERENCES

C. S. Adjiman, S. Dalliwig, C. A. Floudas, and A. Neumaier. A Global Optimization Method, αBB, for General Twice-Differentiable Constrained NLPs - I Theoretical Advances. *Comput. Chem. Eng.*, 22(9):1137–1158, 1998.

G. B. Dantzig. *Linear Programming and Extensions*. Princeton University Press, 1963.

J. E. Falk and R. M. Soland. An Algorithm for Separable Nonconvex Programming Problems. *Management Science*, 15(9):550–569, 1969.

E. P. Gatzke, J. E. Tolsma, and P. I. Barton. Construction of Convex Function Relaxations Using Automated Code Generation Techniques. *Optimization and Engineering*, 3(3):305–326, 2002.

M. A. Henson. Nonlinear Model Predictive Control: Current Status and Future Directions. *Computers and Chemical Engineering*, (23):187–202, 1998.

R. Horst and H. Tuy. *Global Optimization: Deterministic Approaches*. Springer, Berlin, 1990.

I. B. M. . IBM Optimization Solutions and Library Linear Programming Solutions. Technical report, I. B. M., 1997.

ILOG. *ILOG CPLEX 8.0: User's Manual*. Mountain View, CA, July 2002.

N. Karmarkar. A New Polynomial-Time Algorithm for Linear Programming. *Combinatorica*, 4:373–395, 1984.

A. Kremling and F. Allgöwer. Optimaler Betrieb eines kontinuierlichen Ruhrkesselreactors mit Parallel und Folgereaction. *preprint*, 1993.

G. P. McCormick. Computability of Global Solutions to Factorable Nonconvex Programs: Part I - Convex Underestimating Problems. *Mathematical Programming*, 10:147–175, 1976.

R. E. Moore. *Methods and Applications of Interval Analysis*. SIAM, Philadelphia, 1979.

M. Morari and J. H. Lee. Model Predictive Control: The Good, the Bad, and the Ugly. In *Proceedings of the Fourth International Conference on Chemical Process Control*, pages 419–444, Padre Island, TX, 1991.

M. Morari and J. H. Lee. Model Predictive Control: Past, Present and Future. In *PSE ESCAPE-7 Symposium*, Trondheim, Norway, 1997.

H. S. Ryoo and N. V. Sahinidis. Global Optimization of Nonconvex NLPS and MINLPs with Application to Process Design. *Comput. Chem. Eng.*, 19(5):551–566, 1995.

G. R. Sriniwas and Y. Arkun. A Global Solution to the Nonlinear Model Predictive Control Algorthm Using Polynomial ARX Models. *Comput. Chem. Eng.*, 21(4):431–439, 1997.

M. Tawarmalani and N. V. Sahinidis. Global Optimization of Mixed Integer Nonlinear Programs: A Theoretical and Computational Study. Technical report, University of Illinois, 2000.

J. E. Tolsma and P. I. Barton. DAEPACK: An Open Modeling Environment for Legacy Code. *Ind. Eng. Chem. Res.*, 39(6):1826–1839, 2000.

ELSEVIER
IFAC
PUBLICATIONS
www.elsevier.com/locate/ifac

PLANT AND CONTROL-RELEVANT NONLINEARITY ANALYSIS OF A CSTR: A CASE STUDY

Y. Shastri [*,1] **T. Schweickhardt** [**] **F. Allgöwer** [**]

[*] *Department of Bioengineering, 851 S. Morgan St., University of Illinois, Chicago, IL-60607, USA*
[**] *Institute of Systems Theory in Engineering, University of Stuttgart, Pfaffenwaldring 9, 70550 Stuttgart, Germany*

Abstract: Control-relevant nonlinearity analysis has gained importance with the need to control nonlinear processes more tightly. To this effect, the recently proposed closed-loop optimal control law (OCL) nonlinearity measure quantifies the degree of nonlinearity of the optimal state feedback controller for a given control problem. The current paper attempts to gain more insight into the characteristics and behavior of the formulation through a case study using a CSTR model. The multivariate character and the correlation of plant and control-relevant nonlinearity are analyzed with particular focus. *Copyright © 2004 IFAC*

Keywords: Optimal control, Nonlinear systems, System analysis, Reactor control

1. INTRODUCTION

Nonlinear systems have been the most challenging ones for a control engineer. For many of these systems, a linear controller no longer suffices for the desired degree of performance. Regarding the complexity of nonlinear controller design, an answer is needed to the question: in which cases is the effort for nonlinear controller design justified by the resulting performance gain? The first approaches to this question considered the plant nonlinearity to be the determining factor and analysis of the degree of nonlinearity inherent to the system have been made (Guay *et al.*, 1995; Mäkilä and Partington, 2003; Sun and Hoo, 2000; Hahn and Edgar, 2001; Harris *et al.*, 2000). But the first assumption is found to be doubtful as the performance criterion plays an important role as well. This leads to the concept of control-relevant nonlinearity dealing with the assessment of the nonlinearity of a suitable controller (Stack and Doyle III, 1997).

Although an optimal state feedback controller of the form $\mathbf{u}=\mathbf{k}(\mathbf{x})$ is a good control structure for analysis, it is well-known that the derivation of such a controller is mathematically complex and not feasible for many practical systems. To circumvent these difficulties, Stack and Doyle (1997) have proposed the so called Optimal Control Structure (OCS) based on the optimal control theory and an approximation of the exact controller dynamics. Picking on the idea of analyzing the optimal controller, a more rigorous formulation, the closed-loop optimal control law nonlinearity measure has been recently proposed (Schweickhardt *et al.*, 2003) which quantifies the nonlinearity under the condition of closed-loop operation.

In this article this closed-loop optimal control law has been studied on a well-known chemical model of a CSTR. Owing to its interesting system

[1] This work was done while this author was an exchange student at the University of Stuttgart.

dynamics, the work deals with the plant as well as control-relevant nonlinearity studies for better insight. The next section first briefly introduces the general nonlinearity measure, which is the basic tool used for nonlinearity quantification, and then describes the philosophy of the closed-loop optimal control law nonlinearity measure. Section 3 deals with the CSTR case study which includes a short system description, the plant and the control-relevant nonlinearity analysis followed by the interpretation of the results. The last section draws conclusions.

2. THE CLOSED-LOOP OPTIMAL CONTROL LAW NONLINEARITY MEASURE

2.1 General nonlinearity measure

Nonlinearity measures, first proposed by Desoer and Wang (1981), rank various systems by quantifying their nonlinearity. Amongst the various formulations proposed for the nonlinearity measure in literature (Guay *et al.*, 1995; Nikolaou and Hanagandi, 1994; Sourlas and Manousiouthakis, 1992) this work uses the measure proposed by Allgöwer (1995).

Definition of the nonlinearity measure: The nonlinearity measure $\phi_N^{\mathcal{U}}$ of a nonlinear dynamic system $N : \mathcal{U} \rightarrow \mathcal{Y}$ is defined by the non-negative number

$$\phi_N^{\mathcal{U}} := \inf_{G \in \mathcal{G}} \sup_{\mathbf{u}(.) \in \mathcal{U}} \frac{\|G[\mathbf{u}] - N[\mathbf{u}]\|_{L_2}}{\|N[\mathbf{u}]\|_{L_2}} \qquad (1)$$

where $G : \mathcal{U} \rightarrow \mathcal{Y}$ is a linear dynamic operator belonging to the space of linear operators $\mathcal{G}$. $\mathcal{U}$ and $\mathcal{Y}$ are the spaces of admissible inputs and outputs, respectively and $\|.\|$ is a suitable norm in $\mathcal{Y}$. In this work the L_2-norm is used although some other norm can also be used. The measure is bounded between 0 and 1. While a value of zero (practically very close to zero) indicates that the given system is linear, a value of 1 (practically very close to 1) indicates that system is highly nonlinear. For computational purposes, the infinite dimensional min-max problem given by (1) is converted into finite dimensional constrained convex minimization problem. For this work, sinusoidal and step inputs of relevant amplitude and frequency were considered. Simulations and optimization were preformed in MATLAB using standard toolboxes.

2.2 The closed-loop optimal control law nonlinearity measure

Consider a general nonlinear system governed by the equation

$$\dot{\mathbf{x}} = \mathbf{f}(\mathbf{x}, \mathbf{u}) \qquad (2)$$

where $\mathbf{x}(t) \in R^n$ is the state vector and $\mathbf{u}(t) \in R^p$ is the control vector and the initial condition is given by $\mathbf{x}(t_0) = \mathbf{x_0}$. According to the theory of calculus of variations, for a particular objective function J, the optimal *open-loop* control policy is given by the solution of a two point boundary value problem, while the optimal *feedback law* is determined by the solution of the Hamilton-Jacobi-Bellman PDE (Kirk, 1970). While the open-loop control problem is solvable by numerical methods for practical problems, the feedback controller synthesis is not feasible for most practical problems. But note that, for a given initial condition of the plant, the solution of the open-loop control problem and the application of the optimal feedback law will lead to the same trajectories of the manipulated variable and the system states. The closed-loop optimal control law (OCL) nonlinearity measure assumes the problem to be that of infinite time horizon. In that case the optimal control law is a static state feedback law (Kirk, 1970).

Definition of the closed-loop OCL nonlinearity measure: The closed-loop optimal control law (OCL) nonlinearity measure for a control problem is defined as (Schweickhardt *et al.*, 2003)

$$\hat{\phi}_{N_{OCL}}^{\mathcal{B}} := \inf_{K \in R^{q \times n}} \sup_{\mathbf{x}_0 \in \mathcal{B}} \frac{\|N_{OCL}[\mathbf{x}_{x0}^*] - K\mathbf{x}_{x0}^*\|_{L_2}}{\|N_{OCL}[\mathbf{x}_{x0}^*]\|_{L_2}} \qquad (3)$$

with $N_{OCL}[\mathbf{x}_{x0}^*] := \mathbf{u}_{x0}^*$ being the solution of the infinite horizon open-loop optimal control problem for the initial condition x_0. The region $\mathcal{B} \subset R^n$ is the set of considered initial conditions. n is the dimension of the state space of the plant and q is the number of manipulated variables.

The definition is the application of the nonlinearity measure defined previously to the optimal control law with respect to *closed-loop trajectories* where the input signals are parameterized by their corresponding initial conditions. This measure can be efficiently computed by considering a finite set of initial conditions $\mathcal{B}$, because only optimal control *trajectories* need to be computed in order to assess the nonlinearity of the optimal control *law*. The computation is further facilitated by considering static approximations only. This approach is appropriate as the optimal control law is known to be static as well. Computations first involve the generation of optimal trajectories $\mathbf{x}_{x0}^*$ and $\mathbf{u}_{x0}^*$ through the boundary value problem solution for different x_0. These trajectories are then used in the optimization scheme to generate a linear approximation using the K matrix. It is important to note that the nonlinearity captured is the input to state nonlinearity and the output

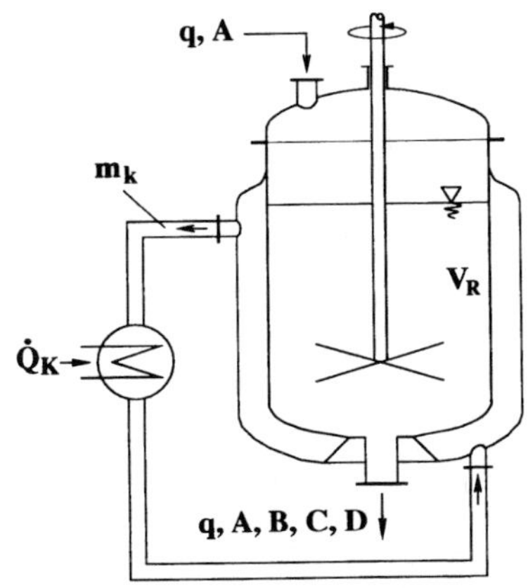

Fig. 1. Schematic of the CSTR.

nonlinearity is not accounted for. It is assumed that all the states are available. The OCL nonlinearity measure is therefore some sort of special application of the general formulation of Sec. 2.1 and is used to quantify the degree of nonlinearity of a suitable controller. The system considered in the OCL formulation is not any system but the solution to an optimal control problem that is known to be a nonlinear algebraic function and the considered class of input functions is the set of closed-loop trajectories.

3. CSTR CASE STUDY

3.1 System description

The system considered in this work for case study is a Continuous Stirred Tank Reactor (CSTR) system shown in figure 1. It consists of a CSTR with a cooling jacket carrying out the van der Vusse reaction scheme described by the following reactions:

$$A \xrightarrow{k_1} B \xrightarrow{k_2} C$$
$$2A \xrightarrow{k_3} D$$

Here B is the desired product while C and D are are the undesired byproducts. k_1, k_2 and k_3 are the reaction rate constants. The nonlinear dynamics in the reactor are governed by the following equations:

$$\frac{dc_A}{dt} = \frac{q}{V_R}(c_{A0} - c_A) - k_1(T)c_A - k_3(T)c_A^2 \quad (4)$$

$$\frac{dc_B}{dt} = -\frac{q}{V_R}c_B + k_1(T)c_A - k_2(T)c_B \quad (5)$$

$$\frac{dT}{dt} = \frac{q}{V_R}(T_0 - T) - \frac{1}{\rho C_P}(k_1(T)c_A\Delta H_{R_{AB}} + k_2(T)c_B\Delta H_{R_{BC}} + k_3(T)c_A^2\Delta H_{R_{AD}}) + \frac{k_W A_R}{\rho C_P V_R}(T_C - T) \quad (6)$$

$$\frac{dT_c}{dt} = \frac{1}{m_c C_{pc}}(\dot{Q}_K + k_W A_R(T - T_C)) \quad (7)$$

With the reaction rate constants described by

$$k_i = k_{i0}e^{(\frac{E_i}{T})}, i = 1, 2, 3. \quad (8)$$

The concentrations c_A and c_B, reactor temperature T and the coolant temperature T_C constitute the four states of the plant. The heat flow $\dot{Q}_K$ is constant. Two different stable operating points are considered. One is the optimal operating point (OP) characterized by the maximum yield with respect to the desired product B and the other one is a sub-optimal operating point (SP) with a lower yield. Further details about the process and the parameters are available in (Chen *et al.*, 1995).

3.2 Previous results on CSTR nonlinearity

The CSTR has been previously analyzed for nonlinearity at the two operating points in (Helbig *et al.*, 2000). The nonlinearity was calculated between the scaled input flow rate (q/V_R) as input for the plant and concentration of desired product B (c_B) as the output for the plant. The nonlinearity measures reported at the two operating points are:

- Optimal operating point (OP): 1.0
- Sub-optimal operating point (SP): 0.37

The difference in nonlinearity measures is quite significant and interesting. The aim of the control-relevant nonlinearity analysis of this model was thus twofold:

(1) To check the result of the closed-loop OCL nonlinearity measurement on a multi-state model
(2) To check whether the plant nonlinearity values of the CSTR get translated into the control-relevant nonlinearity

3.3 Control-relevant nonlinearity analysis of CSTR

The closed-loop OCL nonlinearity measure of the CSTR was analyzed for the two different operating points using definition (3). The performance criterion used is the quadratic objective functional given by:

$$J[u(\cdot)] = \int_0^\infty \mathbf{x}(t)^T Q \mathbf{x}(t) + \alpha u(t)^2 dt \quad (9)$$

where $\mathbf{x}(t)$ is the state vector, $u(t)$ is the input (scaled input flow rate), α is the controller weighting parameter and Q is the weighting matrix for the states.

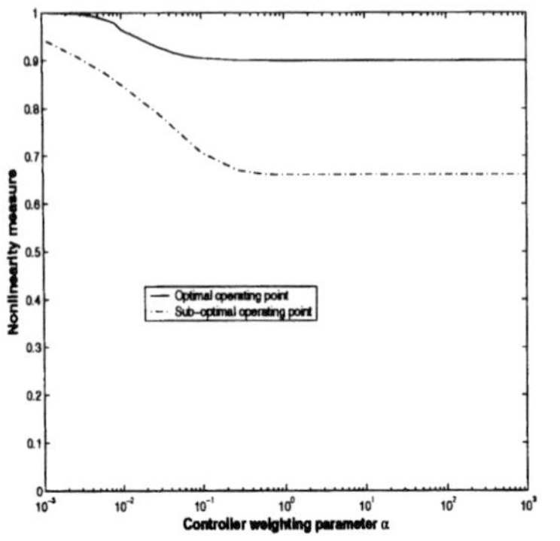

Fig. 2. Variation of OCL nonlinearity measure for the CSTR.

Since OCL nonlinearity gives input to state nonlinearity of the system, for the CSTR with four states, the OCL becomes a four input-one output (MISO) system, i.e. $n = 4$ and $q = 1$ in the definition given in (3). A higher dimension of the plant state results in a larger number of considered initial conditions and translates into an increased number of constraints in the convex optimization framework. The starting conditions x_0 should be in the range of the observed values of physical variable. For the CSTR, the range of starting conditions in terms of the deviation variables were: $c_A = \pm 0.25$, $c_B = \pm 0.10$, $T = \pm 3$ and $T_c = \pm 3$ in $[mol/l]$ and $[K]$ respectively for the sub-optimal operating point and $c_A = \pm 0.5$, $c_B = \pm 0.10$, $T = \pm 3$ and $T_c = \pm 3$ in $[mol/l]$ and $[K]$ respectively for the optimal operating point. The matrix Q in the objective function is a fourth-order identity matrix. The dependence of the OCL nonlinearity on the controller weighting parameter α was sought.

The variation at the two different operating points is plotted in figure 2. From the graph, it can be seen that the OCL nonlinearity at OP is very high which is in agreement with the high plant nonlinearity at this operating point. The nonlinearity measure is seen to vary between 0.9999 and 0.8997 for the considered range of α. But for the sub-optimal operating point (SP), the OCL nonlinearity is found to vary between 0.9454 and 0.6609 and is decreasing with increasing value of α. This result is somewhat unexpected as reported plant nonlinearity is 0.37 and so the control-relevant nonlinearity is significantly higher.

This observed difference between the plant and control-relevant nonlinearity measures was postulated to be arising from their differing analytical setups. For the CSTR plant analysis, the nonlinearity measure gives the relationship between input q/V_R and output c_B. On the other hand, the OCL nonlinearity measure, being input to state, describes a relationship between all four states of the plant and the solitary input (corresponding to four inputs and one output of the optimal control law) and thus considers dynamics between all the input-output pairs. In order to verify this hypothesis and shed more light on it, following two different kinds of analysis were carried out at both the operating points.

- Plant nonlinearity measurement for other three states as output and also considering all outputs simultaneously in a single input multi output (SIMO) setup
- OCL nonlinearity measurement using weighted states i.e. giving importance to a particular state in the objective function

3.4 Nonlinearity analysis of CSTR in a modified setup

The nonlinearity measure calculation of the CSTR model with the other three states as outputs is very similar to the previous case, with change only in the output vector considered for analysis. The measurement was carried out using sinusoidal and step inputs of amplitude up to $\pm 11hr^{-1}$ and maximum frequency (for sinusoidal inputs) of 100 Hz. The results, values of nonlinearity measures at both operating points, are reported in Table 1 where the previous values, with c_B as output are also reported for the sake of comparison.

The nonlinearity measure was also calculated considering all the states as outputs simultaneously i.e. calculation in single input multi output (SIMO) setup. The implementation of the nonlinearity measure definition for the multi output case is logically straight forward but becomes computationally more demanding due to increased dimensionality. The output, which previously was a scalar function of time, is now a vector function of time. The dynamics between different input-output combinations are approximated by different linear systems. Thus, the computational load grows linearly to the number of outputs of the plant (that are here identical to the states). Care must be taken during computation to scale the outputs to the same range so as to avoid masking of linear/nonlinear behavior due to differing ranges. The values of the nonlinearity measure for this case are also reported in table 1. The values in the table will be referred to in the next section for discussion.

3.5 OCL nonlinearity measure of CSTR in a modified setup

The previous nonlinearity analysis for the closed-loop optimal control law in Sec. 3.3 was done using an identity weighting matrix Q i.e. equal weighting given to all the four states in the objective

Table 1. Plant nonlinearity measures of the CSTR at the suboptimal operating point considering different outputs.

Output	Nonlinearity measure at OP	Nonlinearity measure at SP
c_A	0.2478	0.2524
c_B	0.9912	0.3045
T	0.9921	0.5987
T_c	0.9957	0.5988
All	0.9695	0.5877

function. As reported in the previous subsection, the (open-loop) plant nonlinearity measure depends on the input-output pairing. The aim of this section is to perform a similar analysis which will report control-relevant nonlinearity assessment focussing on single states. The optimal control law (OCL) has four states as the four inputs and the plant input q/V_R as its output. The concept of varying only one of the four inputs (states) at a time through different starting conditions does not fit in the realms of optimal control theory as the four states are dynamically highly interconnected and such a situation will never arise in actual closed loop operation. Another way to relate the results of the previous section to OCL nonlinearity measurement is to vary the weights on different states in the objective function. By doing so, different states will dominate the objective and hence the degree of nonlinearity of the optimal control law. There are thus four different cases generated by modifying the weighting matrix Q such that at any time all but one of the diagonal elements are zero, the only non-zero element being equal to one.

Using this setup, the nonlinearity measure calculation for the OCL was carried out at both operating points. Apart from the weighting scheme, all the other aspects of the computation remain the same. The plots for various cases are given in Figure 3 and Figure 4 for the optimal (OP) and suboptimal (SP) operating points respectively. The plots for the first case with diagonal weighting matrix are also included in the figure for the sake of comparison (labelled as "all states"). These plots will be used to draw conclusions in the next section.

3.6 Discussion

From Table 1, most obvious observation for both the operating points is that the nonlinearity measures with respect to the two temperatures (reactor and coolant) are significantly high whereas the nonlinearity with respect to the concentration of A, is quite low. The previously reported (Helbig *et al.*, 2000) difference in the nonlinearity with respect to the concentration of B, is also observed. In both cases, the overall nonlinearity measure

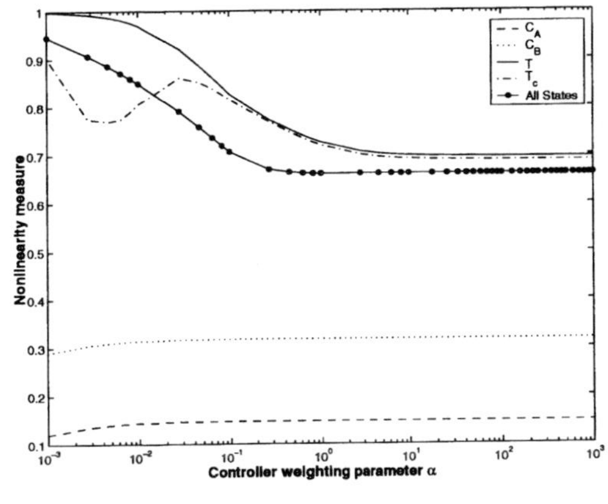

Fig. 3. Comparison of weighted OCL nonlinearities of CSTR model at sub-optimal operating point.

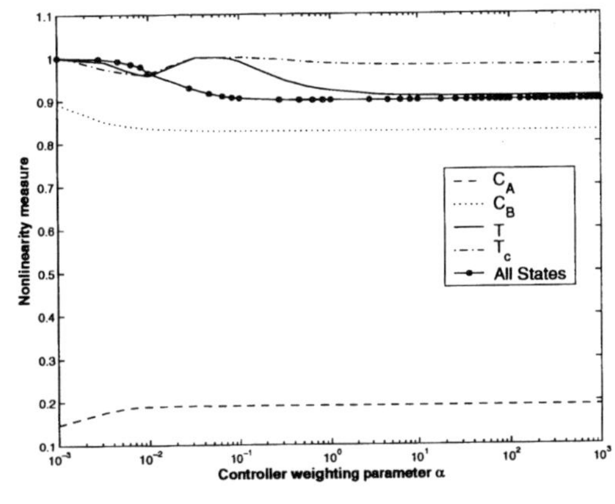

Fig. 4. Comparison of weighted OCL nonlinearities of CSTR model at optimal operating point.

is also very close to that for the temperatures indicating that the overall nonlinearity (in a multi output setup) is dominated by the temperature nonlinearities. This result leads to the conclusion that the temperature nonlinearities pose problems in the regulation of both systems (SP and OP). But the regulation of concentration of B is operating point dependent. But the results also stress the importance to consider all possible dynamic relationships within a system to draw a picture of the nonlinearity of the system as a whole.

The nonlinearity measure variation of the OCL shows a similar qualitative behavior. From the graphs it can be concluded that from control point of view also, the nonlinearity with respect to the two temperatures is always dominant. Another interesting thing to observe is that the difference in the plant nonlinearity with respect to the concentration of B is nicely reflected in the OCL nonlinearity. Not only does the qualitative behavior agree, but quantitatively also the two sets of values are close. Given these results, one can conclude that for this particular system the plant nonlinearity is getting translated into equivalent control-relevant nonlinearity for all dynamic paths of the system. The OCL nonlinearity further shows that the need for nonlinear control does not strongly depend on the cost of control action in this particular example (expressed by the weak dependence on α). But both, plant and OCL nonlinearity values show that the variable to be controlled plays an important role.

Two types of computations are involved in the determination of the OCL nonlinearity measure: first, a finite number of optimal open-loop control problems have to be solved, equivalent to two point boundary value problems. Then the best linear approximation K for the state feedback controller is determined by convex optimization. The computational demand of the optimal control problem solution certainly strongly depends on the specific plant and performance requirements at hand. For the example system, it was observed that for singular Q matrices and very low values of α (less than 0.1) the boundary value problem solution poses accuracy problems. Not surprisingly, the problem is more severe at OP where the system is highly nonlinear. The accuracy problems for the temperatures at OP could not be completely eliminated at low values of α. As a result, the somewhat fluctuating behavior at low α can possibly be due to numerical problems. The computation of the optimal K matrix poses no problem and can be completed very efficiently even for high dimensional systems with multiple inputs.

4. CONCLUSIONS

Nonlinearity analysis of an example CSTR has been done using a general and a control-relevant nonlinearity measure, the optimal control law (OCL) nonlinearity measure. Under similar conditions, the two methods show consistent results. The results suggest that at both operating points, a nonlinear controller would be necessary to control the system temperatures and a linear controller would suffice for concentration of A. The controller requirement for concentration of B though is operating point dependent. This result is in accordance to the working experience that concentration of B at sub-optimal operating point is controller by a PI controller while at the optimal operating point a linear control is not feasible.

ACKNOWLEDGMENT

The first author would like to thank DAAD (German Academic Exchange Service) for the opportunity to work at the Institute for Systems Theory in Engineering, University of Stuttgart for his Masters' project.

REFERENCES

Allgöwer, F. (1995). Definition and computation of a nonlinearity measure. In: *3rd IFAC Nonlinear Control System Design Symposium.* Lake Tahoe, CA. pp. 279–284.

Chen, H., A. Kremlin and F. Allgöwer (1995). Nonlinear predictive control of a benchmark cstr. In: *Proceedings of 3rd Europian Control Conference ECC'p5.* Rome, Italy. pp. 3247–3252.

Desoer, C.A. and Y.-T. Wang (1981). Foundations of feedback theory for nonlinear dynamical systems. In: *IEEE Trans. Circ. Syst.*. Vol. CAS-27. pp. 104–123.

Guay, M., P.J. McLellan and D.W. Bacon (1995). Measurement of nonlinearity in chemical process control systems: The steady state map. *Can. J. Chem. Eng.* **73**, 868–882.

Hahn, J. and T.F. Edgar (2001). A Gramian based approach to nonlinearity quantification and model classification. *Ind. Eng. Chem. Res.* **40**, 5724–31.

Harris, K.R., M.C. Colantonio and A. Palazoğlu (2000). On the computation of a nonlinearity measure using functional expansions. *Chem. Eng. Sci.* **55**, 2393–2400.

Helbig, A., W. Marquardt and F. Allgöwer (2000). Nonlinearity measure: definition, computation and application. *Journal of Process Control* **10**, 113–123.

Kirk, D. (1970). *Optimal Control Theory: an introduction.* Prentice-Hall, Englewood Cliffs, NJ.

Mäkilä, P.M. and J.R. Partington (2003). On linear models for nonlinear systems. *Automatica* **39**(1), 1–13.

Nikolaou, M. and V. Hanagandi (1994). The 2-norm for nonlinear processes: application to modeling and control problems. In: *Proceedings of PSE'94.* Kyongju, Korea. pp. 971–976.

Schweickhardt, T., F. Allgöwer and F.J. Doyle III (2003). Nonlinearity quantification for the optimal state feedback controller. In: *Proc. of the European Control Conference ECC'03.* Cambridge, UK.

Sourlas, D. and V. Manousiouthakis (1992). Development of linear models for nonlinear plants. In: *AIChE Annual Meeting.* Miami, FL.

Stack, A.J. and F.J. Doyle III (1997). Optimal control structure: an approach to measuring control-law nonlinearity. *Comp. Chem. Eng.* **21**(9), 1009–1019.

Sun, D. and K.A. Hoo (2000). Non-linearity measures for a class of SISO non-linear systems. *Int. J. Contr.* **73**(1), 29–37.

ELSEVIER
IFAC
PUBLICATIONS
www.elsevier.com/locate/ifac

TWO-DEGREE-OF-FREEDOM MULTIRATE CONTROLLERS FOR NONLINEAR PROCESSES

Raymond A. Wright

The Dow Chemical Company
200 Larkin Center
Midland, MI 48674
raywright@dow.com

Costas Kravaris

Department of Chemical Engineering
University of Patras
26500 Patras, Greece
kravaris@chemeng.upatras.gr

Abstract: This work focuses on developing nonlinear multirate model algorithmic controllers for nonlinear processes using a closed-loop observer. Conditions for the observer function and a gain corrector function, which is necessary to ensure that the resulting control structure has integral action, are given. The control action is calculated using the corrected states. The resulting structure is a two-degree-of-freedom multirate controller. The performance of the proposed method is evaluated through simulation of a chemical process. *Copyright © 2004 IFAC*

Keywords: nonlinear control, multirate, decoupling, multivariable control, process control

1. INTRODUCTION

Many industrial processes have measurements that are available at different rates. Temperatures, pressure, and flow rates can typically be measured with a short sampling interval. Compositions and product properties usually cannot be made available as rapidly. In the continued drive to operate chemical processes more efficiently and effectively, increasing demands are put on automated process control algorithms to make the best use of measurements made at different rates, as well as compensate for the inherent nonlinearities the systems possess. Both linear and nonlinear Model Predictive Control approaches have been proposed for multirate systems (Lee, et al, 1992, Ohshima, et al, 1994, Bequette, 1991). More recently, an approach based on Model Algorithmic Control (MAC) methods (Richalet, et al, 1978, Mehra and Rouhani, 1980, Soroush and Kravaris, 1996) was proposed for nonlinear multirate processes (Niemiec and Kravaris, 2002). Like the general MAC formulations, this work used an open-loop observer for estimates of the process states.

Methods developed recently for nonlinear single rate processes have made use of a closed-loop observer in a model-state feedback structure (Wright and Kravaris, 2000, Wright and Kravaris 2001). This extends the class of systems to which the method is applicable as well as providing an improved set of state estimates. The goal of the present work is to derive a nonlinear multirate Model Algorithmic Control structure along the lines of Niemiec and Kravaris (2002), but using the closed-loop observer as in Wright and Kravaris (2001). The resulting control algorithm is a two-degree-of-freedom control law, in the sense that the control action is not a function of the error only, but the output and set point are processed in different ways. Once developed, the proposed method will be applied to a chemical reactor system and its performance will be evaluated by simulation.

2. SYSTEM DESCRIPTION AND ASSUMPTIONS

Consider the nonlinear discrete-time MIMO system represented by:

$$\mathbf{x}(k+1) = \Phi(\mathbf{x}(k),\mathbf{u}(k)) \qquad (1)$$
$$\mathbf{y}(k) = h(\mathbf{x}(k))$$

where $\Phi: \Re^n x \Re^m \rightarrow \Re^n$, h: $\Re^n \rightarrow \Re^m$ are smooth functions, $\mathbf{x} \in \Re^n$ is the system state, $\mathbf{u} \in \Re^m$ are the system inputs, and $\mathbf{y} \in \Re^m$ are the system outputs. It is assumed that the system has locally well-defined steady-state characteristics, and the relative orders of the m controlled outputs, $r_1, \ldots, r_m$ are all finite.

With each time $k\Delta t$, where k is an integer and Δt is the time step of the model, the manipulated inputs are actuated. Of the m outputs, q are considered fast, $\mathbf{y}_f^I \in \Re^q$, meaning that they are sampled at every time instant. The remaining p = m − q outputs are considered slow, $\mathbf{y}_s^I \in \Re^p$, meaning they are sampled regularly but at a rate slower than every time instant. For each time that is a multiple of $N_i\Delta t$, where N_i is an integer constant, the output measurement $\mathbf{y}_{si}^I$ is received. N_i is defined as the ratio of the ith output sampling period to the input actuation period:

$$N_i = \frac{\text{sampling period of the input i}}{\text{period of input actuation}} \qquad (2)$$

In addition, there may be ℓ secondary outputs, $\mathbf{y}_f^{II} \in \Re^\ell$, (where $\ell \geq 0$) which are sampled at every time instant. There are no slow secondary outputs.

3. NONLINEAR MULTIRATE MODEL-ALGORITHMIC CONTROL

A main point of departure for the current work is the nonlinear multirate model-algorithmic controller (MAC) of Niemiec and Kravaris (2002). The complete structure of the controller developed in that work is shown in Figure 1.

In this structure, the vector of model states $\mathbf{x}_M$ is obtained by simulating the process model (1) online:

$$\mathbf{x}_M(k+1) = \Phi(\mathbf{x}_M(k),\mathbf{u}(k)) \qquad (3)$$

From the process model, future changes of the outputs can be predicted. The predicted changes of the output can be added to the latest available output measurement to obtain predictions of each output. For each of the outputs, a reference trajectory can be defined according to:

$$\hat{y}_i(k+r_i) = (1-\alpha_i)y_{isp} + \alpha_i \hat{y}_i(k+r_i-1) \qquad (4)$$

where sp denotes the output set point and is a tunable scalar parameter such that $0 \leq \alpha_i < 1$. Matching the output predictions with the reference trajectories

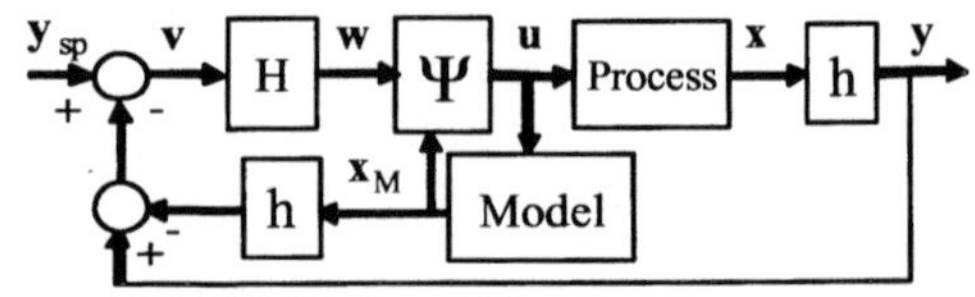

Figure 1: Structure of Nonlinear Multirate MAC

gives a set of nonlinear algebraic equations that must be solved numerically online for the input vector:

$$h_1^{r_1-1}[\Phi(\mathbf{x}_M(k),\mathbf{u}(k))] = \alpha_1 h_1^{r_1-1}[\mathbf{x}_M(k)]$$
$$+(1-\alpha_1)\left[y_{1sp} - y_1\left(\left[\frac{k}{N_1}\right]N_1\right) + h_1\left[\mathbf{x}_M\left(\left[\frac{k}{N_1}\right]N_1\right)\right]\right]$$
$$\vdots \qquad (5)$$
$$h_m^{r_m-1}[\Phi(\mathbf{x}_M(k),\mathbf{u}(k))] = \alpha_m h_m^{r_m-1}[\mathbf{x}_M(k)] + (1-\alpha_m)$$
$$\bullet\left[y_{msp} - y_m\left(\left[\frac{k}{N_m}\right]N_m\right) + h_m\left[\mathbf{x}_M\left(\left[\frac{k}{N_m}\right]N_m\right)\right]\right]$$

where $[k/N_i]$ denotes the integer part of k/N_i. The output measurement y_i is taken only at the time instants $[k/N_i]N_i$, but the manipulated inputs are actuated at every k. This equation utilizes the best available information, which is the most recent measurement at $[k/N_i]N_i$, to predict the future values of the slowly sampled output y_i. Using this structure, a controller can be derived that not only includes the necessary information for good control of outputs sampled at the lowest rates but also exhibits good disturbance rejection for outputs sampled at faster rates. For measurements available at a sampling period equal to the input actuation period (N_i=1), $[k/N_i]N_i$ will reduce to k in the notation.

The corresponding implicit algebraic function defined as the solution to (5) is given by

$$\mathbf{u}(k) = \Psi[\mathbf{x}_M(k), \mathbf{w}(k)] \qquad (6)$$

where $\Psi[*,*]$ is the same function as that in the single-rate MAC controller and

$$w_i(k) = y_{isp} - y_i\left(\left[\frac{k}{N_i}\right]N_i\right) + h_i\left[\mathbf{x}_M\left(\left[\frac{k}{N_i}\right]N_i\right)\right] \qquad (7)$$

With the model states obtained by simulating (3) online, the resulting nonlinear multirate model-algorithmic controller can be expressed as:

$$\mathbf{x}_M(k+1) = \Phi(\mathbf{x}_M(k), \Psi[\mathbf{x}_M(k),\mathbf{w}(k)])$$
$$u(k) = \Psi[\mathbf{x}_M(k),\mathbf{w}(k)] \qquad (8)$$

Note that the calculation of w(k) involves a hold operation on both $y_i(k)$ and $h[\mathbf{x}_M(k)]$:

$$w_i(k) = H\left(y_{isp} - y_i(k) + h_i[\mathbf{x}_M(k)]\right)$$

$$= y_{isp} - y_i\left(\left[\frac{k}{N_i}\right]N_i\right) + h_i\left[\mathbf{x}_M\left(\left[\frac{k}{N_i}\right]N_i\right)\right] \quad (9)$$

where H is a hold operator.

Closed-loop properties of the complete structure shown in Figure 1 are proven in Niemiec and Kravaris (2002).

4. MODEL-STATE FEEDBACK WITH CLOSED-LOOP OBSERVER

In Wright and Kravaris (2001), a model-state feedback control structure is developed for nonlinear discrete-time SISO systems using a closed-loop observer. Closed-loop observers, in general, do not preserve the steady-state gain between the manipulated inputs and process states. Incorporating only a closed-loop observer into a model-state feedback structure would result in a controller that did not possess integral action. By analogy with linear static state feedback controllers, a gain correction function is designed that restores the steady-state gain and the corrected states are used in the calculation of the manipulated input. The complete control structure is shown in Figure 2.

In this structure, the reference input v is calculated by

$$v(k) = y_{sp}(k) - y(k) + h(x_c(k)) \quad (10)$$

The manipulated inputs are calculated by a static state feedback of the form:

$$u(k) = \Psi(x_c(k), v(k)) \quad (11)$$

where $\Psi:\mathfrak{R}^n x\mathfrak{R} \to \mathfrak{R}^n$, is a smooth scalar function, with the property:

$$x = \Phi(x, \Psi(x, v)) \Rightarrow h(x) = v \quad (12)$$

which insures unity steady-state gain between v and y.

In determining the closed-loop observer and gain correction functions, it is first necessary to define what will be considered an observer for a nonlinear process (where $\rho(k) = y(k) - h(\hat{x}(k))$):

<u>Definition</u>: (Lin and Byrnes, 1995) A dynamic system of the form

$$\hat{x}(k+1) = \Pi(\hat{x}(k), u(k), \rho(k)) \quad (13)$$

is called an observer for (1) if the function $\Pi(x,u,\rho)$: $\mathfrak{R}^n x\mathfrak{R}x\mathfrak{R} \to \mathfrak{R}^n$ has the property that $\Pi(x,u,0) = \Phi(x,u)$, for all x, u

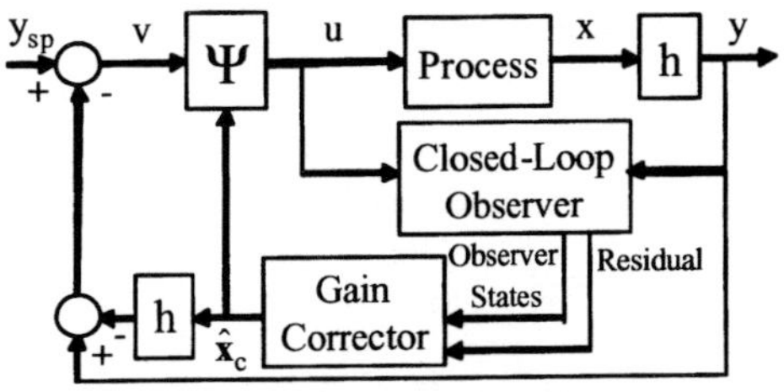

Figure 2: Model-state feedback structure with closed-loop observer

Specifications do not determine the functions $\Pi(x,u,\rho)$ and $\omega(x,\rho)$ uniquely. The Π and ω functions must satisfy the following properties:

(P1) $\Pi(x,u,0) = \Phi(x,u)$

(P2) $\Pi(x,u,\rho) = 0 \Rightarrow \Phi(\omega(x,\rho),u)=0$

(P3) the eigenvalues of $\frac{\partial\Phi}{\partial x} - \frac{\partial\Pi}{\partial\rho}\frac{\partial h}{\partial x}$ evaluated at reference conditions must assume desirable values.

A convenient functionality for the observer can be postulated which is affine in the residual with constant coefficients:

$$\Pi(x,u,\rho) = \Phi(x,u) + L\rho \quad (14)$$

This automatically satisfies property (P1). It is straightforward to show that the adjustable parameters L can be easily selected to satisfy (P3). The challenge is to find the ω-function to satisfy property (P2), which, for the above choice of the observer, becomes

$$\Phi(x,u) + L\rho = 0 \Rightarrow \Phi(\omega(x,\rho),u)=0$$

This can always be solved, but only numerically.

Alternatively, one could start by specifying the form of the correction function and then determine the observer function Π to satisfy properties (P1) and (P2). The simplest choice is the linear function:

$$\omega(x,\rho) = x - M\rho \quad (15)$$

where M is a constant vector of adjustable parameters. Then, the choice

$$\Pi(x,u,\rho) = \Phi(x-M\rho,u) + M\rho \quad (16)$$

will make properties (P1) and (P2) be satisfied. Finally, the requirement in (P3) translates into

$$\frac{\partial\Phi}{\partial x} - \left(I - \frac{\partial\Phi}{\partial x}\right)M\frac{\partial h}{\partial x} \quad (17)$$

having desirable eigenvalues.

5. TWO-DEGREE-OF-FREEDOM NONLINEAR MULTIRATE CONTROLLERS

A nonlinear multirate controller using a closed-loop observer will now be developed using results from the two previous sections. Since the observer and state corrector of the previous section were for a SISO system, the first task is to generalize the results for multiple measurements. Consider the p + ℓ fast measurements in (1) given by:

$$\mathbf{y} = \begin{bmatrix} \mathbf{y}_f^{I} \\ \mathbf{y}_f^{II} \end{bmatrix}$$

A single-rate observer and state corrector with multiple measurements can be developed driven by **y**. The observer and corrector functions must satisfy the following properties (where $?(k) = \mathbf{y}(k) - \mathbf{h}(\hat{\mathbf{x}}(k))$):

(P1) $\Pi(\mathbf{x},\mathbf{u},0) = \Phi(\mathbf{x},\mathbf{u})$

(P2) $\mathbf{x} = \Pi(\mathbf{x},\mathbf{u},\rho) \Rightarrow \omega(\mathbf{x},\rho) = \Phi(\omega(\mathbf{x},\rho),\mathbf{u})$

(P3) the eigenvalues of $\frac{\partial\Phi}{\partial x} - \frac{\partial\Pi}{\partial\rho}\frac{\partial h}{\partial x}$ evaluated at reference conditions must assume desirable values.

Using the second alternative from the previous section, a simple choice is to start by specifying a correction function of the form:

$$\omega(\mathbf{x},\rho) = \mathbf{x} - \mathbf{M}\rho \tag{18}$$

where **M** is a constant matrix of adjustable parameters. Then, the choice

$$\Pi(\mathbf{x},\mathbf{u},\rho) = \Phi(\mathbf{x}-\mathbf{M}\rho,\mathbf{u}) + \mathbf{M}\rho \tag{19}$$

will make properties (P1) and (P2) be satisfied. Finally, the requirement in (P3) translates into

$$\frac{\partial\Phi}{\partial x} - \left(I - \frac{\partial\Phi}{\partial x}\right)\mathbf{M}\frac{\partial h}{\partial x} \tag{20}$$

having desirable eigenvalues.

The resulting corrected state estimates are then used in the model-algorithmic multirate controller in place of the model values. Following the same development, fiture changes of the output can be predicted by:

$$\begin{aligned}
y_{ic}(k+1) - y_{ic}\left(\left[\frac{k}{N_i}\right]N_i\right) &= h_i^1[\hat{\mathbf{x}}_c(k)] - h_i\left[\hat{\mathbf{x}}_c\left(\left[\frac{k}{N_i}\right]N_i\right)\right] \\
y_{ic}(k+2) - y_{ic}\left(\left[\frac{k}{N_i}\right]N_i\right) &= h_i^2[\hat{\mathbf{x}}_c(k)] - h_i\left[\hat{\mathbf{x}}_c\left(\left[\frac{k}{N_i}\right]N_i\right)\right] \\
&\vdots \\
y_{ic}(k+r_i-1) - y_{ic}\left(\left[\frac{k}{N_i}\right]N_i\right) &= h_i^{r_i-1}[\hat{\mathbf{x}}_c(k)] - h_i\left[\hat{\mathbf{x}}_c\left(\left[\frac{k}{N_i}\right]N_i\right)\right] \\
y_{ic}(k+r_i) - y_{ic}\left(\left[\frac{k}{N_i}\right]N_i\right) &= h_i^{r_i-1}[\Phi(\hat{\mathbf{x}}_c(k),\mathbf{u}(k))] - h_i\left[\hat{\mathbf{x}}_c\left(\left[\frac{k}{N_i}\right]N_i\right)\right]
\end{aligned} \tag{21}$$

where y_{ic} is the ith controlled output, i.e.

$$\mathbf{y}_{ic} \in \begin{bmatrix} \mathbf{y}_f^{I} \\ \mathbf{y}_s^{I} \end{bmatrix}$$

and is calculated on the basis of the corrected state estimates. The predicted changes of the output can be added to the latest available output measurement to obtain the following predictions of the output y_i:

$$\begin{aligned}
\hat{y}_i(k+1) &= y_i\left(\left[\frac{k}{N_i}\right]N_i\right) + h_i^1[\hat{\mathbf{x}}_c(k)] - h_i\left[\hat{\mathbf{x}}_c\left(\left[\frac{k}{N_i}\right]N_i\right)\right] \\
\hat{y}_i(k+2) &= y_i\left(\left[\frac{k}{N_i}\right]N_i\right) + h_i^2[\hat{\mathbf{x}}_c(k)] - h_i\left[\hat{\mathbf{x}}_c\left(\left[\frac{k}{N_i}\right]N_i\right)\right] \\
&\vdots \\
\hat{y}_i(k+r_i-1) &= y_i\left(\left[\frac{k}{N_i}\right]N_i\right) + h_i^{r_i-1}[\hat{\mathbf{x}}_c(k)] - h_i\left[\hat{\mathbf{x}}_c\left(\left[\frac{k}{N_i}\right]N_i\right)\right] \\
\hat{y}_i(k+r_i) &= y_i\left(\left[\frac{k}{N_i}\right]N_i\right) + h_i^{r_i-1}[\Phi(\hat{\mathbf{x}}_c(k),\mathbf{u}(k))] - h_i\left[\hat{\mathbf{x}}_c\left(\left[\frac{k}{N_i}\right]N_i\right)\right]
\end{aligned} \tag{22}$$

The output measurement y_i is taken only at the time instants $[k/N_i]N_i$, but the manipulated inputs are actuated at every k. For each of the outputs, a reference trajectory can be defined according to (4).

Matching the predictions with the reference trajectories gives a set of nonlinear algebraic equations that must be solved numerically online for the input vector:

$$h_1^{r_1-1}[\Phi(\hat{\mathbf{x}}_c(k),\mathbf{u}(k))] = \alpha_1 h_1^{r_1-1}[\hat{\mathbf{x}}_c(k)] + (1-\alpha_1)\left[y_{1sp} - y_1\left(\left[\frac{k}{N_1}\right]N_1\right) + h_1\left[\hat{\mathbf{x}}_c\left(\left[\frac{k}{N_1}\right]N_1\right)\right]\right]$$

$$\vdots \qquad (23)$$

$$h_m^{r_m-1}[\Phi(\hat{\mathbf{x}}_c(k),\mathbf{u}(k))] = \alpha_m h_m^{r_m-1}[\hat{\mathbf{x}}_c(k)] + (1-\alpha_m) \bullet \left[y_{msp} - y_m\left(\left[\frac{k}{N_m}\right]N_m\right) + h_m\left[\hat{\mathbf{x}}_c\left(\left[\frac{k}{N_m}\right]N_m\right)\right]\right]$$

The corresponding implicit algebraic function defined at the solution to the previous set of equations is given by

$$\mathbf{u}(k) = \Psi[\hat{\mathbf{x}}_c(k),\mathbf{w}(k)] \qquad (24)$$

where $\Psi[*,*]$ is the same function as that in the single-rate MAC controller and

$$w_i(k) = y_{isp} - y_i\left(\left[\frac{k}{N_i}\right]N_i\right) + h_i\left[\hat{\mathbf{x}}_c\left(\left[\frac{k}{N_i}\right]N_i\right)\right] \qquad (25)$$

The structure of the complete closed-loop system is shown in Figure 3.

6. A SIMULATION EXAMPLE

The nonlinear control methodology will now be applied to a chemical process. The system consists of a CSTR, where the exothermic reaction A+B→C takes place. The reactants flow into the CSTR in separate streams. The jacket inlet temperature and the flow rate of B can be manipulated to control the outlet temperature and the outlet concentration of B. The reactor outlet temperature is measured at every time instant, but the measurement of the outlet concentration of B is measured only every N time steps. A model for this process is as follows:

$$T(k+1) = T(k) + \left[\frac{UA}{\rho C_p V}\left(T_j(k) - T(k)\right)\right]\Delta t + \left[\frac{F_A(T_{A0} - T(k)) + F_B(k)(T_{B0} - T(k))}{V}\right]\Delta t + \left[\frac{(-\Delta H)}{\rho C_p} k_0 C_A(k) C_B(k) e^{-E/RT(k)}\right]\Delta t$$

$$T_j(k+1) = T_j(k) + \left[\frac{UA}{m_j C_{pj}}\left(T(k) - T_j(k)\right)\right]\Delta t + \left[\frac{F_j \rho_j}{m_j}\left(T_{ji}(k) - T_j(k)\right)\right]\Delta t \qquad (26)$$

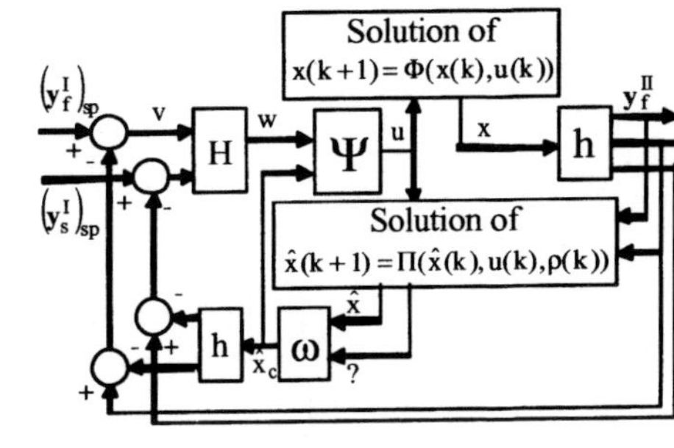

Figure 3: Structure of Nonlinear Multirate MAC with closed-loop observer

$$C_A(k+1) = C_A(k) - \left[k_0 C_A(k) C_B(k) e^{-E/RT(k)}\right]\Delta t + \left[\frac{F_A C_{A0} - (F_A + F_B(k))C_A(k)}{V}\right]\Delta t$$

$$C_B(k+1) = C_B(k) - \left[k_0 C_A(k) C_B(k) e^{-E/RT(k)}\right]\Delta t + \left[\frac{F_B(k) C_{B0} - (F_A + F_B(k))C_B(k)}{V}\right]\Delta t$$

where the symbols used are common in chemical reactor engineering. The initial steady-state values of the outputs are $T = 350$ K and $C_B = 0.781$ kmol/m^3. The corresponding steady-state values of the inputs are $T_{ji} = 358.4$ K and $F_B = 0.00013$ m^3/s. The value used for each of the parameters is given in Table 1.

The observer and corrector were derived from the model equations (26) using (18) and (19). The observer was tuned using (20) and standard pole placement routines to calculate values for the M vector. The characteristic matrix of (26) is singular. The system was extended by adding a state equation, $F_B(k+1) = \zeta(k)$, and using $\zeta(k)$ as the manipulated input. The relative orders in the extended system are equal to 2 for both outputs, the characteristic matrix is nonsingular, and the inputs can be calculated from (23).

The response of the closed-loop system to simultaneous errors in three initial conditions of the observer states are shown in Figures 4 through 6. These figures show the tracking of the observer and the rejection of the initial condition errors.

Table 1: Parameter values for the example process

$V = 1$ m^3	$-\Delta H = 60000$ kJ/kmol
$F_A = 0.000325$ m^3/s	$R = 8.345$ kJ/(kmol K)
$T_{A0} = 300$ K	$E = 82800$ kJ/kmol
$C_{A0} = 5$ kmol/m^3	$k_0 = 1.2 \times 10^9$ m^3 kmol^{-1} s^{-1}
$UA = 1.4$ kJ/(s K)	$\rho = 850$ kg/m^3
$C_p = 3.5$ kJ/(kg K)	$C_{B0} = 9$ kmol/m^3
$T_{B0} = 320$ K	$m_j = 500$ kg
Cpj = 4.2 kJ/(kg K)	$\rho_j = 100$ kg/m^3
?t = 10 seconds	N = 20

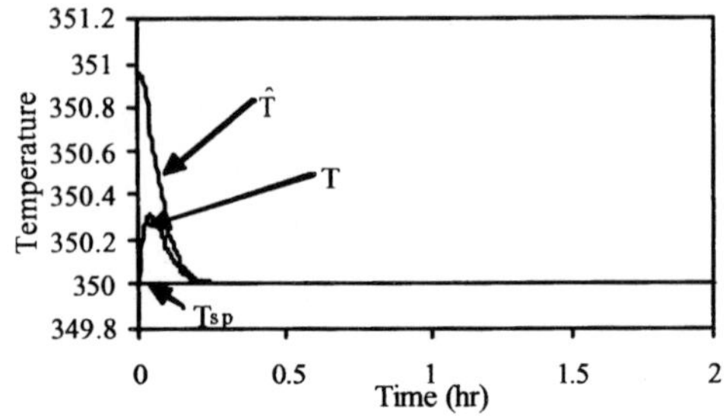

Figure 4: Response of T to a +1 K error in observer initial condition

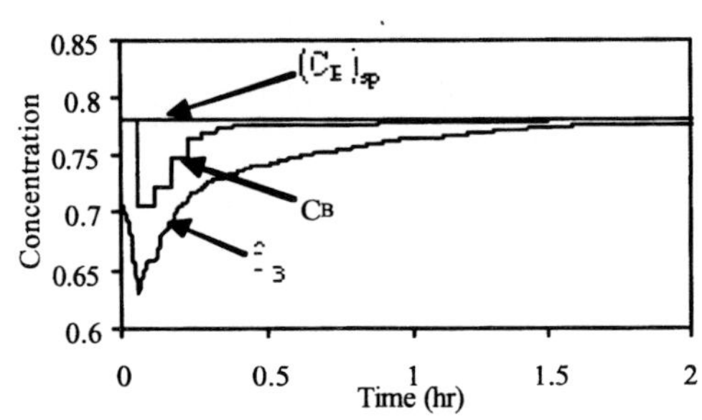

Figure 5: Response of C_B to a -10% K error in initial condition

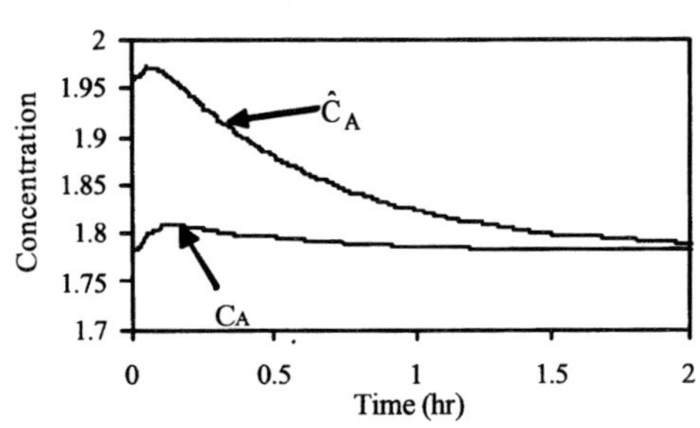

Figure 6: Response of C_A to a +10% K error in initial condition

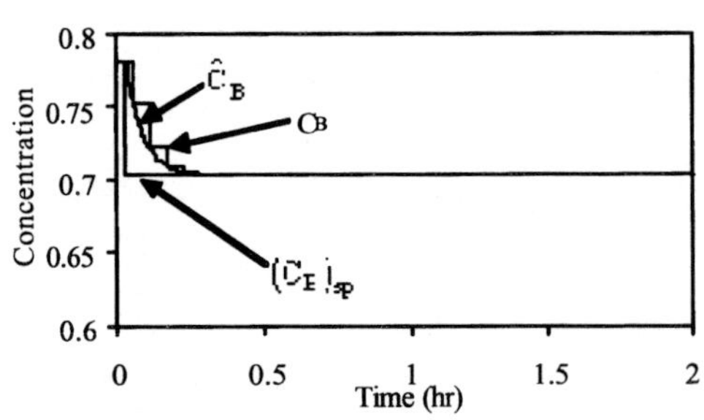

Figure 7: Response of C_B to a -10% step in C_B set point concentration

The response of the closed-loop system to a -10% step decrease in the set point of C_B concentration is shown in Figure 7. The proposed controller results in a decoupled response, there fore there is no deviation from set point in T. This figure has been omitted for brevity.

The slower rate of sampling of C_B is evident from Figure 7. Simulation results showing the ability of the proposed controller to reject unmeasured disturbances have also been omitted for brevity.

REFERENCES

Bequette, B. W. (1991). "Nonlinear Predictive Control Using Multirate Sampling," *Can. J. Chem. Eng.*, **69**, 136.

Lee, J. H.; Gelormino, M.; Morari, M. (1992). "Model Predictive Control of Multi-Rate Sampled-Data Systems: A State-Space Approach," *Int. J. Control*, **55**, 153.

Lin, W.; Byrnes, C. I. (1995). "Remarks on Linearization of Discrete-Time Autonomous Systems and Nonlinear Observer Design," *Syst. Cont. Lett.*, **25**, 31.

Mehra, R. K.; Rouhani, R. (1980). "Theoretical Considerations on Model Algorithmic Control for Nonminimum Phase Systems," *Proc. ACC* TA8-B.

Niemiec, M. P.; Kravaris, C. (2002). "Nonlinear Multirate Model Algorithmic Control. 1. Theory," *Ind. Eng. Chem. Res.*, **41**, 4054.

Ohshima, M.; Hashimoto, I.; Ohno, H.; Takeda, M.; Yoneyama, T.; Gotoh, F. (1994). "Multirate Multivariable Model Predictive Control and its Application to a Polymerization Reactor," *Int. J. Control*, **59**, 731.

Richalet, J.; Rault, A.; Testud, J. L.; Papon, J. (1978). "Model Predictive Heuristic Control: Application to Industrial Processes," *Automatica*, **14**, 413.

Soroush, M.; Kravaris, C. (1996). "Discrete-Time Nonlinear Feedback Control of Multivariable Processes," *AIChE J.*, **42**, 187.

Wright, R A.; Kravaris, C. (2000). "Two-Degree-of-Freedom Output Feedback Controllers for Nonlinear Processes," *Proc. ACC*, FM06-4.

Wright, R. A.; Kravaris, C. (2001). "Two Degree of Freedom Output Feedback Controllers for Discrete-Time Nonlinear Processes," *Proc. ACC*, TP11-2.

ELSEVIER
IFAC
PUBLICATIONS
www.elsevier.com/locate/ifac

SYSTEM'S NONLINEARITY MEASUREMENT BASED ON THE RPN CONCEPT

M. Farenzena and J. O. Trierweiler[#]

Group of Integration, Modelling, Simulation, Control and Optimization of Processes (GIMSCOP)
Department of Chemical Engineering, Federal University of Rio Grande do Sul (UFRGS)
Rua Duque de Caxias, 1303 apt. 603 CEP: 90010-283 - Porto Alegre - RS - BRAZIL,
Fax: +55 51 3316 3277, Phone: +55 51 3316 4072
E-MAIL: {farenz, jorge}@enq.ufrgs.br

Abstract: The nRPN, $nRPN_{STAT}$, and $nRPN_{DYN}$ are three novel indexes to measure system's nonlinearity. These nonlinear measurements are derived from the Robust Performance Number (RPN) concept. The total system's nonlinearity can be measured by the *nonlinear RPN* (nRPN), while the purely static nonlinearity is captured by nonlinear static RPN ($nRPN_{STAT}$) and the dynamic component by the nonlinear dynamic RPN ($nRPN_{DYN}$). The novel indexes do not require a nonlinear model, being enough a set of linear models. Therefore, they can easily be applied to quantify the nonlinearities of industrial plants and used to answer several practical important questions such as: How nonlinear is the system? Is it necessary to apply a nonlinear controller? What kind of nonlinear controller is necessary? *Copyright © 2004 IFAC*

Keywords: - nonlinearity, nonlinearity measurements, controller selection, RPN Methodology, benzene distillation column

1 INTRODUCTION

Is a nonlinear controller necessary? If the answer is yes, what kind of nonlinear controller is required? Nowadays, questions like these arise frequently. To answer these questions it is necessary to quantify the process nonlinearity degree. When the degree of nonlinearity is low, a linear controller can be applied without performance loss. Linear controllers are much simpler to tune and to maintain. Therefore, one will invest in a nonlinear controller when is really necessary. Many times, simple nonlinear gain compensation is enough to significantly improve the control performance. In other situations, the process dynamic also changes considerably. In these situations, the nonlinear controller must also compensate the system's dynamic and the nonlinear controller becomes more difficult to tune and to implement. In this work, three novel nonlinear measurements (nRPN, $nRPN_{STAT}$, and $nRPN_{DYN}$) are proposed. These nonlinear measurements are derived from the Robust Performance Number (RPN) concept. The total system's nonlinearity can be measured by the *nonlinear RPN* (nRPN), while the purely static nonlinearity is captured by nonlinear static RPN ($nRPN_{STAT}$) and the dynamic component by the nonlinear dynamic RPN ($nRPN_{DYN}$).

About nonlinear measurements, there is not too much available in the literature. For instance, Guay et al. (1997) proposed a method based on the concavity measure, while Helbig et al. (2000) quantify the degree of nonlinearity based on the best linear approximation to the nonlinear model. These approaches only measure the open-loop degree of nonlinearity and to apply these methodologies it is necessary a nonlinear phenomenological model. Therefore, the practical application is reduced and not so simple.

Trierweiler (1997) and Trierweiler and Engell (1997a) introduced the RPN for a plant set (RPPN) index to quantify the degree of nonlinearity. The

[#] Author to whom the correspondence should be addressed.

RPPN needs only a set of linear models, which can be obtained by identification or linearization of the process in different operating points. The RPPN takes the closed loop performance into account automatically. The novel indices introduced in this paper are improvement of the original RPPN.

This paper is structured as follows: in section 2 a brief description of the RPN methodology will be made. In section 3 the nonlinear RPN will be defined. The nonlinearity degree of a high purity distillation and the quadruple-tank system will be quantified in sections 4 and 5, respectively.

2 THE ROBUST PERFORMANCE NUMBER (RPN)

The RP-number is a measure of how potentially difficult it is for a given system to achieve the desired performance robustly. The RPN of a multivariable plant with transfer matrix $G(s)$ is defined as

$$\mathrm{RPN} \overset{\Delta}{=} \Gamma_{\sup}(G,T,\omega) = \sup_{\omega \in \mathrm{R}} \{\Gamma(G,T,\omega)\} \tag{1a}$$

$$\Gamma(G,T,\omega) \overset{\Delta}{=} \sqrt{\bar{\sigma}([I - T(j\omega)]T(j\omega))\left(\gamma^*(G(j\omega)) + \frac{1}{\gamma^*(G(j\omega))}\right)} \tag{1b}$$

where $\gamma^*(G(j\omega))$ is the *minimized* condition number of $G(j\omega)$ and $\bar{\sigma}([I-T]T)$ is the maximal singular value of the transfer function $[I-T]\,T$. T is the attainable desired output complementary sensitivity function that is determined for the nominal model G.

The RPN determined by two factors:

1. $\bar{\sigma}([\mathbf{I}-\mathbf{T}]\mathbf{T})$ This term acts as a weighting function that emphasizes the crossover frequency, that is the most important region for the robust stability, because feedback controller will operate in this region.
2. $\gamma^*(G)+1/\gamma^*(G)$. The origin of this term is the analysis of robust performance (RP) of inverse-based controllers (see Trierweiler, 1997).

3 NONLINEAR RPN

The nonlinear RPN (nRPN), nonlinear dynamic RPN ($\mathrm{nRPN_{DYN}}$) and nonlinear static RPN ($\mathrm{nRPN_{STAT}}$) are introduced to measure system's nonlinearity. These nonlinearity measurements are derived from the Robust Performance Number (RPN) concept. The total system's nonlinearity can be measured by the *nonlinear RPN* (nRPN), while the purely static nonlinearity is captured by nonlinear static RPN ($\mathrm{nRPN_{STAT}}$) and the dynamic component by the nonlinear dynamic RPN ($\mathrm{nRPN_{DYN}}$).

3.1 Preliminary definitions

Next, we introduce five concepts, which are used in the definition of nRPN.

***Definition 1*: Minimized condition number of a plant set P ($\gamma^\#(P)$).** Consider the polytopic system representation P consisting of NP linear models G_i

$$P \in \mathrm{Co}\{G_1,\ldots,G_{NP}\} \overset{\Delta}{=} \left\{\sum_{i=1}^{NP} \alpha_i G_i : \alpha_i \ge 0, \sum_{i=1}^{NP} \alpha_i = 1\right\}. \tag{2}$$

Then the minimized condition number of P is defined as

$$\begin{aligned}\gamma^\#(P) &\overset{\Delta}{=} \min_{L^\#,R^\#} \sup \gamma(L^\# P R^\#) \\ &= \min_{L^\#,R^\#} \sup\{\gamma_1(L^\# G_1 R^\#),\ldots,\gamma_{NP}(L^\# G_{NP} R^\#)\}\end{aligned} \tag{3}$$

where $L^\#$ and $R^\#$ are the real, diagonal, and nonsingular scaling matrices used to determine $\gamma^\#(P)$.

The scaling matrices $L^\#$ and $R^\#$ can be calculated by the solution of a generalized eigenvalue problem (Trierweiler, 1997).

***Definition 2*: Maximal difference of the minimized condition number for a plant set P ($\Delta\gamma^\#(P)$).** For the polytopic model P, the maximal difference in the condition number ($\Delta\gamma^\#(\mathrm{P})$) is defined by

$$\begin{aligned}\Delta\gamma^\#(P) = &\max(\gamma(L^\# G_1 R^\#),\ldots,\gamma(L^\# G_{NP} R^\#)) \\ &- \min(\gamma(L^\# G_1 R^\#),\ldots,\gamma(L^\# G_{NP} R^\#))\end{aligned}. \tag{4}$$

***Definition 3*: Maximal singular value ratio of a plant set P ($\sigma^\#(P)$).** For the polytopic model P, given by (2), the maximal singular value ratio of P is defined by

$$\sigma^\#(P(j\omega)) \overset{\Delta}{=} \frac{\max\{\bar{\sigma}(L^\# G_1 R^\#),\ldots,\bar{\sigma}(L^\# G_{NP} R^\#)\}}{\min\{\bar{\sigma}(L^\# G_1 R^\#),\ldots,\bar{\sigma}(L^\# G_{NP} R^\#)\}} \tag{5}$$

where $\bar{\sigma}(L^\# G_t R^\#)$ is the maximal singular value of the scaled model G_i using the scaling matrices $L^\#$ and $R^\#$.

***Definition 4*: Projection Matrix ($M^\#(P)$).** For the polytopic model P, the projection matrix $M^\#$ is a symmetric matrix with the elements $m^\#_{ij}$ given by

$$m^\#_{i,j}(P(0)) \overset{\Delta}{=} \max_{k=1..NS} \left[abs(u^H_{k,i} \cdot u_{k,j} - 1), abs(v^H_{k,i} \cdot v_{k,j} - 1)\right] \tag{6}$$

where the vectors $u_{k,i}$ and $v_{k,i}$ are the kth column vectors of the unitary matrices U_i and V_i

$$\begin{aligned} U_i &= [\bar{u}_i = u_{1,i}, u_{2,i}, \ldots, u_{no,i} = \underline{u}_i] \text{ and} \\ V_i &= [\bar{v}_i = v_{1,i}, v_{2,i}, \ldots, v_{ni,i} = \underline{v}_i] \end{aligned} \tag{7}$$

obtained by the Singular Value Decomposition (SVD) of the scaled gain matrix of the model i, i.e.,

$$\begin{aligned} L^\# G_i(0) R^\# &= U_i \Sigma_i V_i^H \\ &= \sum_{k=1}^{NS} \sigma_k (L^\# G_i(0) R^\#) u_{k,i} v^H_{k,i} \end{aligned} \tag{8}$$

and Σ_i contains a diagonal nonnegative definite

matrix Σ_{NS} of singular values arranged in descending order:

$$\Sigma = \begin{pmatrix} \Sigma_{NS} \\ 0 \end{pmatrix} \text{if } no \geq ni \text{ or } \Sigma = \begin{pmatrix} \Sigma_{NS} & 0 \end{pmatrix} \text{ if } no < ni \quad (9)$$

$$\text{and } \Sigma_{NS} = diag(\sigma_1, \sigma_2, \ldots, \sigma_{NS}\}; NS = \min\{no, ni\} \quad (10)$$

$$\text{with} \qquad \bar{\sigma} = \sigma_1 \geq \sigma_2 \geq \ldots \geq \sigma_{NS} = \underline{\sigma}. \quad (11)$$

***Definition 5*: Projection Factor ($\upsilon^{\#}(P)$).** For the polytopic model P, the projection factor is equal to 10 powered by the greatest element of the projection matrix $M^{\#}$ defined in (13), i.e.,

$$\upsilon^{\#}(P(0)) \stackrel{\Delta}{=} 10^{\wedge}\max\left(M^{\#}(P(0))\right). \quad (12)$$

The projection factor $\upsilon^{\#}(P)$ is used to detect a change of multivariable gain sign, what is captured by the projection matrix $M^{\#}$.

3.2 Nonlinear RPN curves

Based on $\Delta\gamma^{\#}(P)$, $\sigma^{\#}(P)$, and $\upsilon^{\#}(P)$ we can finally define the nonlinear RPN curves. Here, the nonlinear RPN (nRPN), the nonlinear *static* RPN ($nRPN_{STAT}$), and the nonlinear *dynamic* RPN ($nRPN_{DYN}$) curves are shown. The nRPN curve is used to capture the total process nonlinearity, while $nRPN_{STAT}$ and $nRPN_{DYN}$ are applied to quantify the static and dynamic contributions, respectively.

***Definition 6*: Nonlinear Robust Performance Number curve (nRPN curve, $\Gamma^{\#}$).** The nonlinear RPN curve is defined as

$$\Gamma^{\#}(P,T,\omega) \stackrel{\Delta}{=} \left[\sigma^{\#}(P(j\omega)) + \Delta\gamma^{\#}(P(j\omega))\right]\upsilon^{\#}(P(0))\Gamma(G,T,\omega) \quad (13)$$

where $\Gamma(G,T,\omega)$ is given by (1b). ■

The potential increase of system's directionality is captured by the term ($\Delta\gamma^{\#}(P)$), the modification of the maximal singular value is considered by $\sigma^{\#}(P)$, and the change of the multivariable sign by $\upsilon^{\#}(P)$. From the definition, the nRPN curve is always above the RPN curve.

The nominal model G used in (13) is usually the nominal operating point of the plant, where the process will mostly work. When this operating point cannot be selected among the possible plant models, then the linear model with the worst possible performance should be set as nominal model, i.e., the model with the largest RPN.

The nRPN curve is influenced by static and dynamic nonlinearities. To separate the contribution of each kind of nonlinearity, the $nRPN_{STAT}$ ($\Gamma^{\#}_{STAT}$) and $nRPN_{DYN}$ ($\Gamma^{\#}_{DYN}$) curves are defined, as follows:

$$\Gamma^{\#}_{STAT}(P,T,\omega) \stackrel{\Delta}{=} \Gamma(G,T,\omega) + \min\left(\Gamma^{\#}(P,T,\omega) - \Gamma(G,T,\omega)\right) \quad (14)$$

$$\Gamma^{\#}_{DYN}(P,T,\omega) \stackrel{\Delta}{=} \left(\Gamma^{\#}(P,T,\omega) - \Gamma^{\#}_{STAT}(P,T,\omega)\right) \quad (15)$$

Figure 1 shows a typical example of RPN, nRPN, $nRPN_{STAT}$, and $nRPN_{DYN}$ plots. The larger the difference between nRPN and RPN plots, the more nonlinear the system.

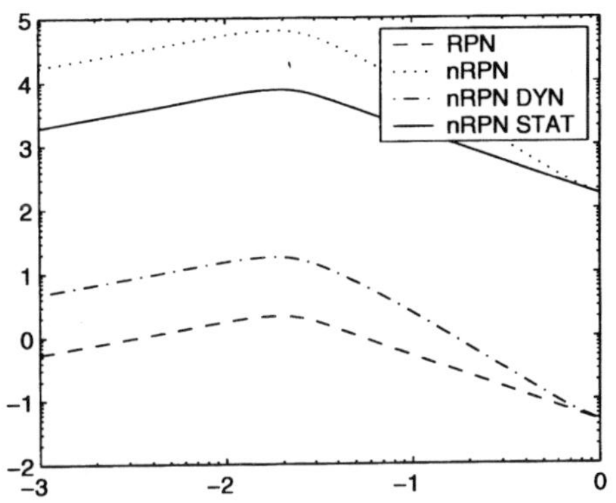

Fig. 1: RPN and nRPN curves for systems with static and dynamic of nonlinearity Note that the frequency is on a logarithmic scale so that -4 should be understood as 10^{-4}.

If the nonlinearity is purely static the difference between the nRPN and the RPN curves is constant for all frequencies, since the process models differ in the process gain only. Figure 2 illustrates this situation, which typically occurs for pH-reactors.

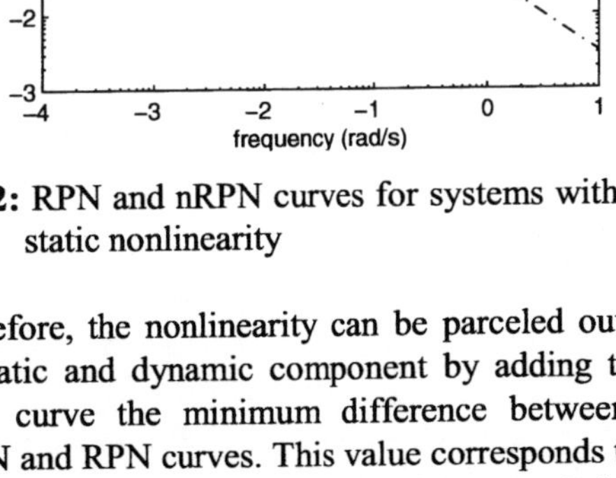

Fig. 2: RPN and nRPN curves for systems with pure static nonlinearity

Therefore, the nonlinearity can be parceled out into its static and dynamic component by adding to the RPN curve the minimum difference between the nRPN and RPN curves. This value corresponds to the nonlinear *static* RPN curve, as defined by (14). The dynamic component is given by subtracting from the total nonlinearity (nRPN curve) the static component ($nRPN_{STAT}$ curve) as in equation (15).

The analysis of the nRPN, $nRPN_{STAT}$, and $nRPN_{DYN}$ curves allow not only determining if a nonlinear controller is necessary, but also allows choosing the characteristics of the nonlinear controller that would be necessary for a given plant, what is very important for several practical applications. For example, if the $nRPN_{DYN}$ and RPN curves are closer to each other and the difference between $nRPN_{STAT}$ and RPN curves is large, a nonlinear controller with simple static gain compensation (gain scheduling controller) is recommended to improve the control performance. In this case, a controller with variable dynamic will not produce a better result. It is important to mention that nonlinear controllers with fix dynamic are much simpler to develop and maintain.

3.3 Nonlinear RPN indices

In the last section, it was shown that analyzing the nRPN curves is possible to quantify the process nonlinearities. For practical and quick analysis it is better to quantify the process nonlinearities using numbers (indices) instead of graphics. Therefore, to capture the distance between the nRPN curves we will introduce the nRPN, $nRPN_{STAT}$, and $nRPN_{DYN}$ indices. These indices are defined by the relative difference between the areas under the curves, i.e.,

$$
\begin{aligned}
nRPN &\stackrel{\Delta}{=} \log_{10}\left(\frac{A_{nRPN} - A_{RPN}}{A_{RPN}}\right) \\
nRPN_{STAT} &\stackrel{\Delta}{=} \log_{10}\left(\frac{A_{nRPN\_STAT} - A_{RPN}}{A_{RPN}}\right) \\
nRPN_{DYN} &\stackrel{\Delta}{=} \log_{10}\left(\frac{A_{nRPN\_DYN} - A_{RPN}}{A_{RPN}}\right)
\end{aligned}
\qquad (16)
$$

The areas are defined as follows:

$$
\begin{aligned}
A_{RPN} &\stackrel{\Delta}{=} \int_{\omega_{min}}^{\omega_{max}} \Gamma(G,T,\omega)\, d\log\omega \\
A_{nRPN} &\stackrel{\Delta}{=} \int_{\omega_{min}}^{\omega_{max}} \Gamma^{\#}(P,T,\omega) d\log\omega \\
A_{nRPN\_STAT} &\stackrel{\Delta}{=} \int_{\omega_{min}}^{\omega_{max}} \Gamma_{STAT}^{\#}(P,T,\omega) d\log\omega \\
A_{nRPN\_DYN} &\stackrel{\Delta}{=} \int_{\omega_{min}}^{\omega_{max}} \Gamma_{DYN}^{\#}(P,T,\omega) d\log\omega
\end{aligned}
\qquad (17)
$$

It was introduced the logarithm function in the definition of nonlinear indices to make easier their interpretation. Values smaller than 1 indicate that the performance difference between nonlinear and linear controllers is not significant, so that a linear controller is recommended. Indices greater than 2 clearly indicate that a nonlinear controller is necessary. Between 1 and 2 is a transition zone, where in many times a robust controller can stabilize all possible plants, but the performance loss can be significant.

4 Case Study: High Purity Distillation Column

Distillation columns are currently the main separation process applied in chemical and petrochemical industries (Luyben, 1992). The control of high purity distillation columns generally is not an easy task, due to the coupling between the controlled variables, susceptibility to the external unmeasured disturbances and often high non-linearity degree. Small disturbances in high purity columns are responsible for great change in the product compositions. Moreover, these columns shows high time constants and asymmetric gain variation.

In this section the nonlinearity degree of an industrial high purity distillation column used to split Benzene from Toluene from a misture Benzene-Toluene-Xilenes (BTX).

4.1 Process Description

The distillation column operates at atmospheric pressure, being fed with a mixture of BTX and small quantity of water. The column has 40 theoretical trays. The benzene is removed in a sidestream withdraw laterally at the third theoretical tray from the top. At the top, due to a subcooling the small quantity of water is separate from benzene in a two phase drum. Fig.3 shows schematically the distillation column.

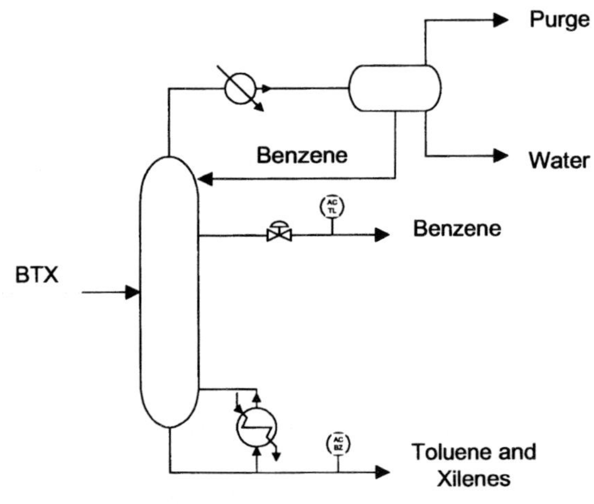

Fig. 3: Schematic representation of the Benzene column of a typical BTX unit

The controlled variables are benzene contamination in the bottom stream and toluene concentration in the sidestream. To control the compositions the corresponding manipulated variables are reboiler heat duty and sidestream flowrate.

A rigorous dynamic model of the benzene column was built in the commercial simulator Aspen Dynamics 11.1 using NRTL thermodynamic model for the liquid phase. This dynamic model can satisfactory describe the actually behavior of the real column.

4.2 Operating Regions

The operating space is divided in six operating regions defined by different compositions expressed in ppm of benzene in the bottom stream and toluene in the siedestream. Table 1 shows the operating regions analyzed in this paper. For example, the OR1 of Table 1 means that the composition of benzene and toluene are 50 ppm in the bottom and siede streams, respectively. The OR1 to OR4 are high purity operating points meaning high non-linearity degree, whilst OR5 and OR6 can be considered as moderate purity operating points.

Table 1: Definition of the Operating Regions (OR)

		Toluene (ppm) in the Siedestream	
Benzene (ppm) in the Bottom Stream		**50**	**200**
	50	OR1	OR2
	200	OR3	OR4
	100	OR5	OR6

Figures 4 and 5 respectively show the static gain and settling time for the transfer function of the channel (toluene impurity)/(siedestream flowrate). The transfer functions were obtained for several toluene composition in the siedestream with a fix benzene composition in the bottom stream equal to 50 ppm. Note that both gain and settling time abruptly change when the toluene composition is lower than 200 ppm. The gain and settling time change approx. 100 and 40 times, respectively.

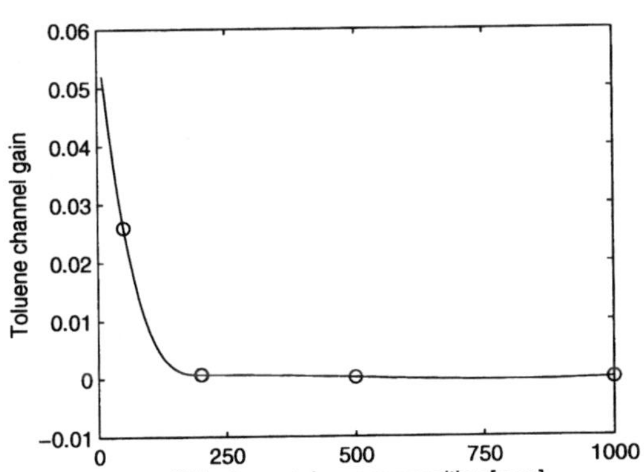

Fig. 4: Curve of the static gain for several specifications of toluene impurity with a 50 ppm benzene loss in the bottom stream.

Table 2 and 3 respectively show the gain matrix and settling time of the linearized models. These Tables indicates that the behavior presented in Figures 4 and 5 can be extended to the other channels.

The qualitative analysis of the non-linearity degree indicates that the regions of high purity (OR1 to OR4) exhibit a great variation in the static gain and dynamics. However, the regions where the specifications are moderated (OR5 and OR6), the non-linearity is less pronounced.

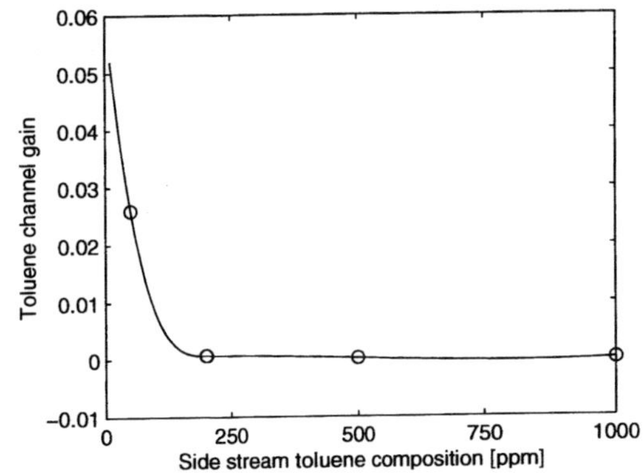

Fig. 5: Settling time for several specifications of toluene impurity and fix 50 ppm benzene loss in the bottom stream.

Table 2: Gain Matrix for the OR defined in Table 1. The output are toluene impurity and benzene loss (rows) and the manipulated variables are Sidestream flowrate and heaty duty (columns)

	50	200
50	$\begin{bmatrix} 0{,}026 & 0{,}16 \\ -0{,}13 & -0{,}86 \end{bmatrix}$	$\begin{bmatrix} 0{,}051 & 0{,}33 \\ -0{,}041 & -0{,}27 \end{bmatrix}$
200	$\begin{bmatrix} 0{,}007 & 0{,}0043 \\ -0{,}056 & -0{,}37 \end{bmatrix}$	$\begin{bmatrix} 0{,}0044 & 0{,}029 \\ -0{,}051 & -0{,}34 \end{bmatrix}$
1000	$\begin{bmatrix} 0{,}0002 & 0{,}0014 \\ -0{,}052 & -0{,}34 \end{bmatrix}$	$\begin{bmatrix} 0{,}0002 & 0{,}0014 \\ -0{,}052 & -0{,}34 \end{bmatrix}$

Table 3: Settling time (min)

	50	200
50	1900	500
200	200	100
1000	45	45

4.3 nRPN Analysis

Fig. 6 shows the nRPN curves for each OR calculated using a desired performance of 10 min rise time and 5% overshoot. The set of linear models used in each region was: the linear model determined in the nominal operating point and the models with the composition twice greater in each component (e.g. the OR1, was described by the linear models with (1) 50 and 50 ppm, (2) 100 ppm and 50 ppm, and (3) 50 ppm and 100 ppm of toluene impurity and benzene loss). Table 4 summarizes the corresponding values of the nRPN, $nRPN_{DYN}$ and $nRPN_{STAT}$ indixes.

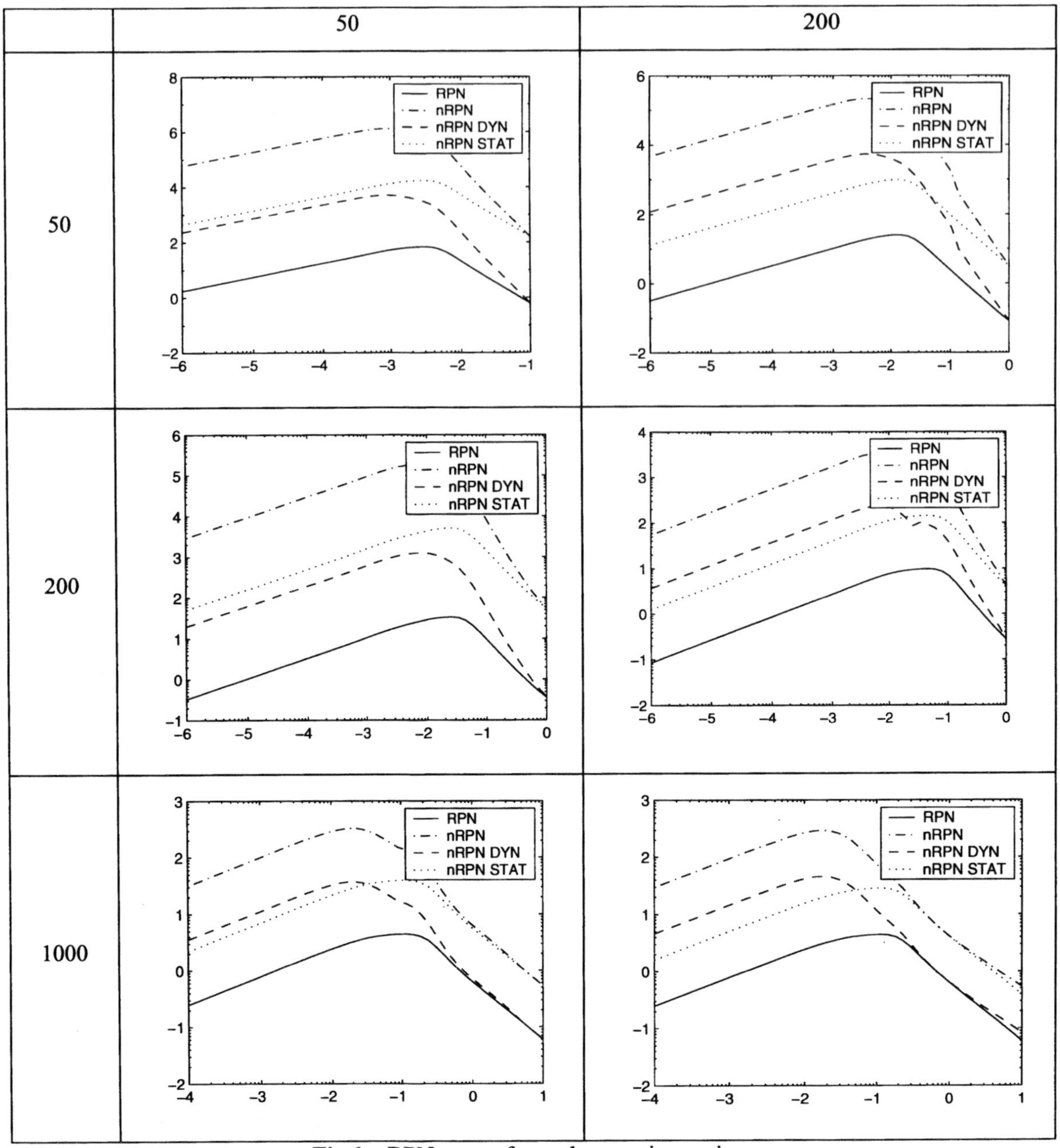

Fig.6: nRPN curves for each operating regions

Table 4: nRPN indexes corresponding to Fig. 6

	nRPN	$nRPN_{DYN}$	$nRPN_{STAT}$
OR1	3,85	1,43	2,40
OR2	3,49	1,89	1,58
OR3	3,31	1,10	2,18
OR4	1,99	0,75	1,13
OR5	1,24	0,04	0,90
OR6	1,05	-0,05	0,74

Analyzing Fig. 6 and Table 4, we can conclude that the OR1 to OR3 are the most nonlinear exhibiting both static and dynamic nonlinearities. To control the column at these regions it is necessary to apply a controller which can compensate both static and dynamic behavior. For OR4, the plant can be already controlled by a nonlinear controller with static compensation only (e.g., gain schedule controller). In this region the dynamic nonlinearity is not too pronounced. The OR5 and OR6 can be controlled by linear controllers. If tight performance is required for these OR, a simple static compensation will do the job quite well.

To show the effect of velocity of the closed loop in the nonlinear degree, the column operating at OR5 was simulated with 3 controllers with the following relations between the closed loop and open loop dynamic: (a) twice faster, (b) 6 times faster, and (c) 12 times faster. The simulations results with decentralized PI controllers which achieve these relations are show in Fig. 7 for disturbances with different magnitudes. The responses were normalized by the disturbance magnitude to make easier the analysis. Fig. 7 clearly shows that the faster the controller (closed loop), less nonlinearity degree exhibit by the plant. Note that the effect of closed loop velocity is captured automatically in the nRPN, as can be seen in Table 5. Note that for faster closed loop response the nRPN is smaller indicating a smaller nonlinearity degree.

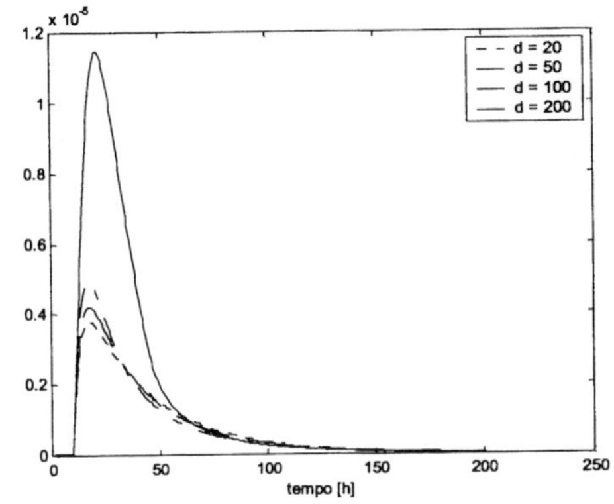

(a)

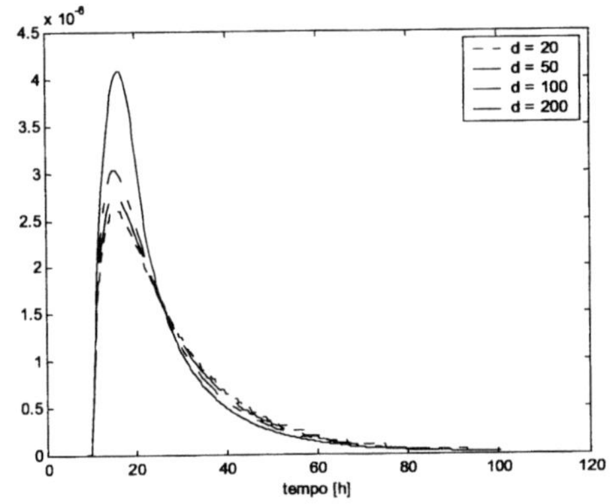

(b)

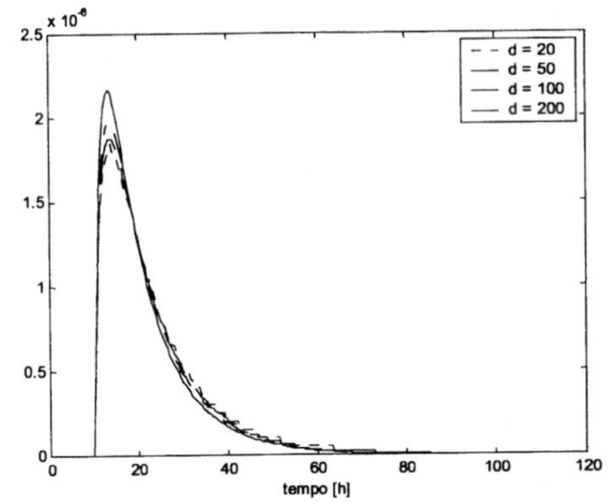

(c)

Fig.7: Simulation in closed loop with decentralized PI controllers for OR5 and different relations between the closed loop and open loop dynamic: (a) twice faster, (b) 6 times faster, and (c) 12 times faster.

Table 5: nRPN for OR5 and several rise time for the closed loop response

Rise Time [min]	nRPN
1	0.96
10	1.24
20	1.38
50	1.58
100	1.70
200	1.79

5 QUADRUPLE-TANK PLANT

5.1 Process description

The quadruple-tank process (see Fig. 8) is a laboratory process that consists of four interconnected water tanks. The linearized dynamic model of the system has a real multivariable zero, whose sign can be changed depending on operating conditions. In this way, the quadruple-tank process is ideal for illustrating many concepts in multivariable control, particularly performance limitations due to multivariable RHP zeros. The location and the direction of zero have an appealing physical interpretation. The target is to control the level in the lower two tanks (i.e., h_1 and h_2) with the inlet flowrates, F_1 and F_2.

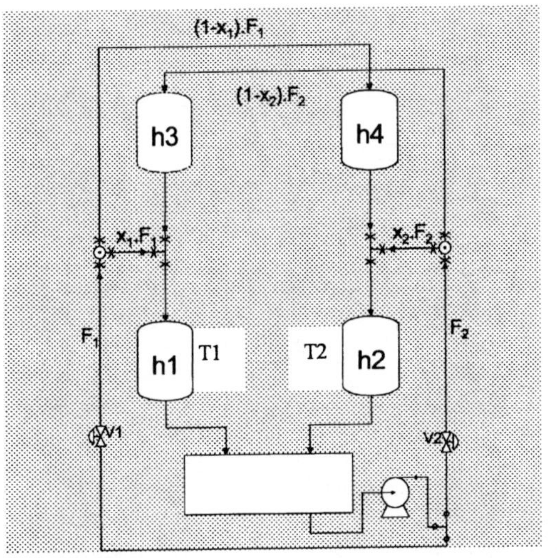

Fig. 8: Schematic representation of the quadruple-tank system

5.2 Process Model

The process model consists of the mass balance around each tank and is given by

$$
\begin{aligned}
A_1 \frac{dh_1}{dt} &= x_1 \cdot F_1 + R_3\sqrt{h_3} - R_1\sqrt{h_1} \\
A_2 \frac{dh_2}{dt} &= x_2 \cdot F_2 + R_4\sqrt{h_4} - R_2\sqrt{h_2} \\
A_3 \frac{dh_3}{dt} &= (1 - x_2) \cdot F_2 - R_3\sqrt{h_3} \\
A_4 \frac{dh_4}{dt} &= (1 - x_1) \cdot F_1 - R_4\sqrt{h_4}
\end{aligned} \tag{18}
$$

where A_i is the cross-section area of Tank i, R_i is the outlet flow coefficient of Tank i, h_i is the water level of Tank i, F_1 and F_2 are the manipulated inlet flowrates and x_1 and x_2 are the valve distribution flow factors $0 \le x_i \le 1$.

The parameters used in this work are basically the same as those in (Johansson, 2000) and are given by $A_1 = A_3 = 28\ cm^2$, $A_2 = A_4 = 32\ cm^2$, $R_1 = R_3 = 3.145\ cm^{2.5}/s$ and $R_2 = R_4 = 2.525\ cm^{2.5}/s$.

5.3 *Operating Points*

The quadruple-tank process is studied at a minimum-phase operating point (MOP) and at a nonminimum-phase operating point (NMOP), due to the presence of the RHP transmission zero. Table 6 summarizes the operating conditions of MOP and NMOP. Note that the main difference between the OPs is the valve distribution flow factors, x_1 and x_2, which are responsible for the difference in h_3 and h_4 levels. All other variables are almost the same for both OPs.

Table 6: Definition of the Operating Points

Variables	MOP	NMOP
h_1, h_2 [cm]	12.26, 12.78	12.44, 13.16
h_3, h_4 [cm]	1.63, 1.41	4.73, 4.99
F_1, F_2 [cm^3/s]	9.99 , 10.05	9.89, 10.36
x_1, x_2 [-]	0.7, 0.6	0.43, 0.34

Table 7: RHP zero and RGA

	MOP	NMOP
RHP zero	none	0.0128
RHP zero input direction	–	$u_z = \begin{bmatrix} 0.7326 \\ -0.6806 \end{bmatrix}$
RHP zero output direction	–	$yz = \begin{bmatrix} -0.7743 \\ 0.6329 \end{bmatrix}$
RGA(0)	$\begin{bmatrix} 1.4 & -0.4 \\ -0.4 & 1.4 \end{bmatrix}$	$\begin{bmatrix} -0.64 & 1.64 \\ 1.64 & -0.64 \end{bmatrix}$

5.4 *RHP Zero and RGA*

Johansson (2000) shows that the quadruple-tank system always has two transmission zeros, whose locations can be classified based on the $x_1 + x_2$ value. When $0 < x_1 + x_2 < 1$, one of the transmission zeros is located in RHP. For the case where $x_1 + x_2 = 1$, the system has a transmission zero at the origin, whereas for $1 < x_1 + x_2 < 2$ no RHP zero occurs. Table 7 shows the RHP zeros for both OPs. For NMOP, the input zero direction, u_Z, and output zero direction, y_Z, were also included in the table. The steady-state RGA (see Table 7) clearly shows that the pairing used for MOP (i.e., (F_1, h_1) and (F_2, h_2)) should not be applied to NMOP. When the system operates in NMOP the pairing is inverse, due to the transmission zero that changes its signal. A complete analysis about the controllability to the quadruple-tanks system is presented in Trierweiler and Farina (2002)

5.5 *Analysis of the Nonlinearity Degree*

To calculate the nonlinearity degree it is necessary to define a set of possible plants. Here, we will analyze the degree of nonlinearity against to inlet flow variation (i.e., F_1 and F_2) and to split ratio variation (i.e., x_1 and x_2).

Table 8: Inlet flowrates used for linearizing the nonlinear model

	Model 1	Model 2	Model 3
F_1 [cm^3/s]	8	12	10
F_2 [cm^3/s]	12	8	10

5.6 *Influence of inlet flows F_1 and F_2*

The set of models used in the analysis of inlet flow influence is defined in Table 8. The polytopic model used in the analysis is obtained by linearizing the nonlinear model at the steady state conditions corresponding to F_1 and F_2 as shown in Table 8 and for x_1=0.7 and x_2=0.6 in the case of MOP and x_1=0.43 and x_2=0.34 for NMOP.

Results for MOP

The analysis of nRPN and RPN curves indicates that the system is practically linear for inlet flowrate variations, as we can see in Fig. 9. This figure was obtained using a desired performance of 50 sec rise time and 5% overshoot.

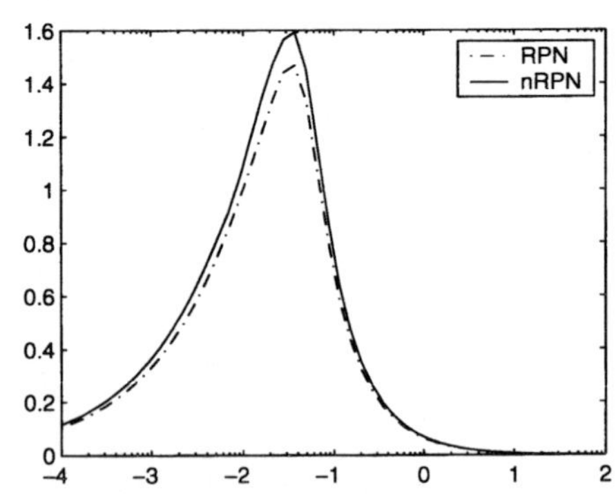

Fig. 9: nRPN and RPN plots for MOP and inlet flow variations.

Table 9 shows the corresponding nRPN indices for the influence of the feed flowrate. Note that the nonlinearity is very small and it is almost static, since

$nRPN \approx nRPN_{STAT}$. The dynamic nonlinearity is almost non-existent (i.e., $nRPN_{DYN}$= -1.84).

Table 9: nRPN indices for MOP and inlet flowrate variations

nRPN	$nRPN_{STAT}$	$NRPN_{DYN}$
-1.33	-1.84	-1.49

Fig. 10a shows the closed loop response of the nonlinear dynamic model for different setpoint changes in h_1. Two decentralized linear PI controllers were used in the simulation. In Fig. 10b the responses are normalized by the step change. This figure clearly confirms that the system is practically linear.

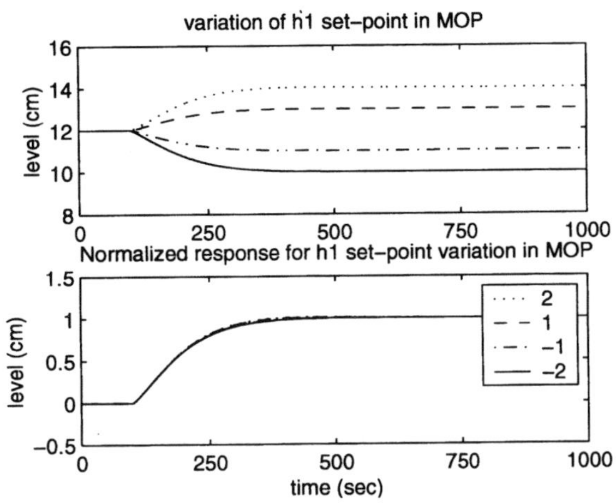

Figure 10: Closed loop step response several setpoint changes

Results for NMOP

Very similar results are obtained for the NMOP (not shown here). Therefore, we can conclude that the system is weak nonlinear against inlet flowrate variation. Simple linear controllers will succeed to control the system at least for the inlet flows listed in Table 8.

Influence of the distribution ratios x_1 and x_2

In this section, the influence of x1 and x2 in the degree of nonlinearity will be measured. x1 and x2 are responsible for the location of the transmission zeros, what can increase the nonlinearity degree.

Table 10: Distribution Ratio used to construct the polytopic model for MOP

	Model 1	Model 2
x_1 [-]	0.7	0.5
x_2 [-]	0.5	0.7

Results for MOP

The first operating region to be analyzed is around the minimum phase operating point (MOP). Table 10 shows the linear models that compose the polytopic model.

Fig. 11 shows the nRPN and RPN plots for the MOP. The desired performance was 50-sec rise time and the 5% overshoot. Table 11 presents the nRPN indices for the influence of the feed distribution ratio. Based on these results, it can be concluded that the system is weak nonlinear and a linear controller performs well.

Table 11: nRPN indices for x1 and x2 variations and MOP

NRPN	$nRPN_{STAT}$	$nRPN_{DYN}$
-0.024	-0.040	-1.73

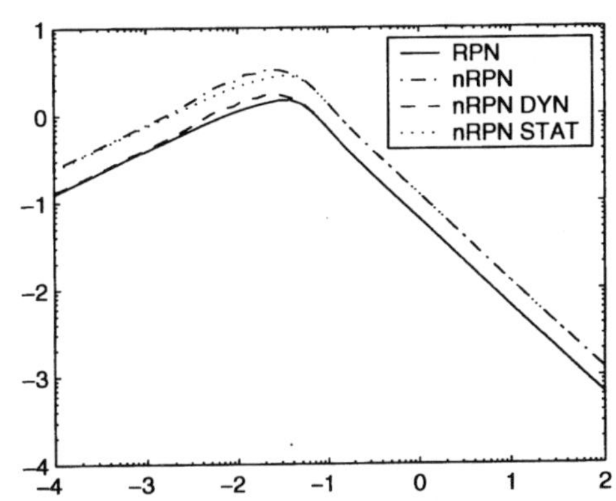

Fig. 11: nRPN and RPN plots for x1 and x2 variations and MOP (both axes are in logarithmic scale).

Fig.12 shows the closed loop response of the nonlinear dynamic model for the operating points defined by (x1,x2) = (0.8,0.7), (0.7,0.6), and (0.6, 0.7).

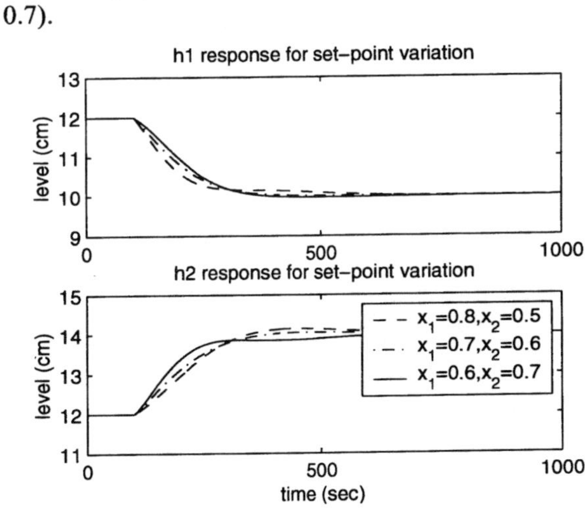

Figure 12: Closed loop step response for MOP and (x1,x2) = (0.8,0.5), (0.7,0.6), and (0.6, 0.7)

For each one of these operating points, the setpoint h1 and h2 were simultaneouly changed by -2 and 2 units, respect. A full linear PI controller was used in the simulation. This figure clearly confirms that a simple linear controller will perform well.

Results for NMOP

For the NMOP the set of models that will be analyzed is defined in Table 12. This OR shows a transmission zero located in RHP, causing inversion in the determinant of the gain matrix. The right pairing for NMOR is (F1-h2) and (F2-h1).

Table 12: Distribution Ratio used to construct the polytopic model for NMOP

	Model 1	Model 2	Model 3
x_1 [-]	0.5	0.3	0.2
x_2 [-]	0.3	0.5	0.2
RHP zero	0.0119	0.0111	0.0498

Figure 13 shows the nRPN and RPN plots for NMOR. This figure was obtained using a desired performance of 50 sec rise time and 5% overshoot. Table 13 shows the nRPN indices for the influence of the feed distribution ratio. Based on these results, it can be concluded that the NMOP is more nonlinear than the MOP, but a linear controller still performs well. nRPN lower than 1 indicates that a linear controller will perform satisfactory. Above this value, the benefits of a nonlinear controller become more evident.

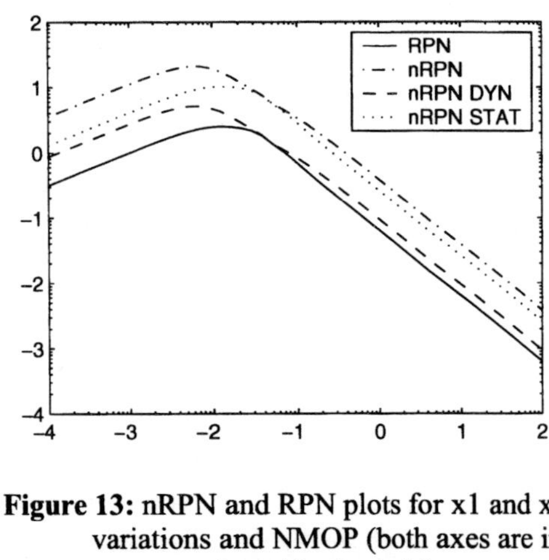

Figure 13: nRPN and RPN plots for x1 and x2 variations and NMOP (both axes are in logarithmic scale).

Table 13: nRPN indices for x1 and x2 variations and NMOR

nRPN	$nRPN_{STAT}$	$nRPN_{DYN}$
0.680	0.490	-0.380

Figure 14 shows the closed loop response of the nonlinear dynamic model for the operating points defined by (x1,x2) = (0.2,0.2), (0.5,0.3), and (0.35, 0.25). For each one of these operating points, the setpoint h1 and h2 were simultaneously changed by -2 and 2 units, respect. A full linear PI controller was used in the simulation. This figure clearly confirms that a simple linear controller will perform well. Note that for the NMOP there is a little performance degradation if compared to the MOP. If it is decided to applied a nonlinear controller, a simple gain scheduling controller is recommended, since the nonlinearity is mainly static (see Table 13).

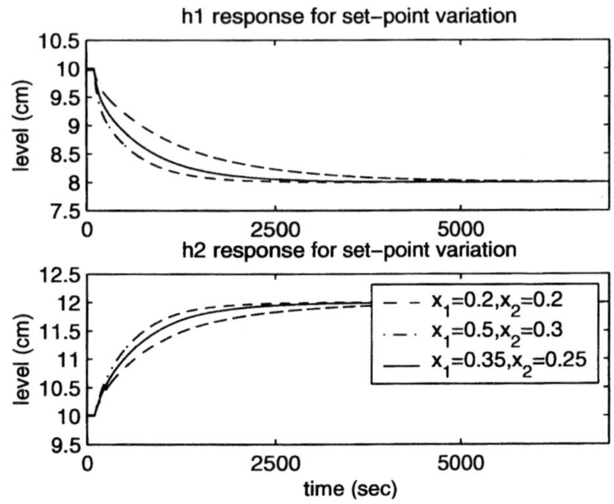

Figure 14: Closed loop step response for MOP and (x1,x2) = (0.2,0.2), (0.5,0.3), and (0.35, 0.25).

Results for both MOP and NMOP

Now, it is analyzed the case where both operating region are included. The linear models used to calculate the degree of nonlinearity are listed in Tables 10 and 12.

Figure 15 shows the nRPN and RPN plots for NMOR/MOR. This figure was obtained using a desired performance of 50-sec rise time and 5% overshoot.

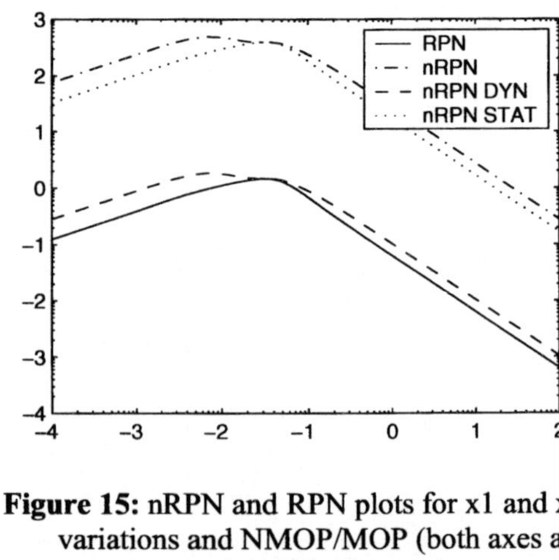

Figure 15: nRPN and RPN plots for x1 and x2 variations and NMOP/MOP (both axes are in logarithmic scale).

Table 14 shows the corresponding nRPN indices. Based on these results, it can be concluded that the NMOR/MOR is much more nonlinear than the MOR and NMOR separately. The reason is that the multivariable process gain changes its sign, so that a nonlinear controller is required.

Table 14: nRPN indices for x1 and x2 variations and NMOP/MOP

nRPN	$nRPN_{STAT}$	$nRPN_{DYN}$
2.60	2.42	-0.285

Figure 16 shows the simulation results produced by a nonlinear model predictive controller (Duraisk, 2001). With this kind of controller it is possible to work with the system in both operating regions. Equivalent simulations with linear controllers are unstable.

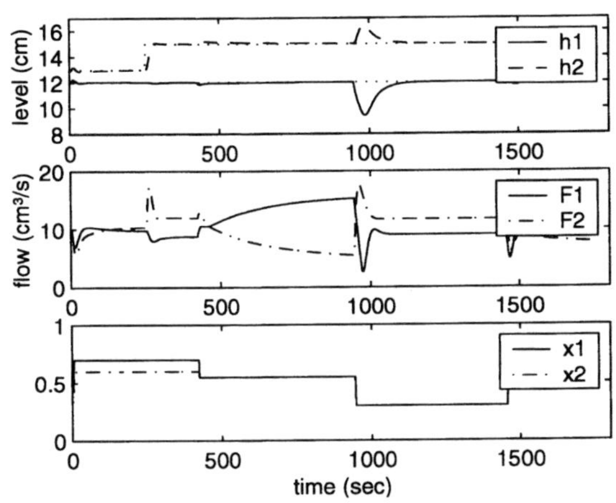

Figure 16: Set-point change and disturbance rejection capability for NMOP/MOP using NMPC (a) controlled variables and (b) manipulated variables and (c) disturbances.

The main reason for the high nonlinearity is the sign inversion of the gain matrix. If we invert the pairing for only the NMOP models, the total nonlinearity must decrease, because the inversion disappears. Figure 17 shows the nRPN and RPN plots for NMOR/MOR, where the models corresponding of NMOP have been modified to fit the (F1-h2) and (F2-h1) pairing. Table 15 shows the corresponding nRPN indices. Based on these results, we can concluded that a linear controller can stabilize the control loop if a switching mechanisms is implemented to select the correct pairing based on the operating region.

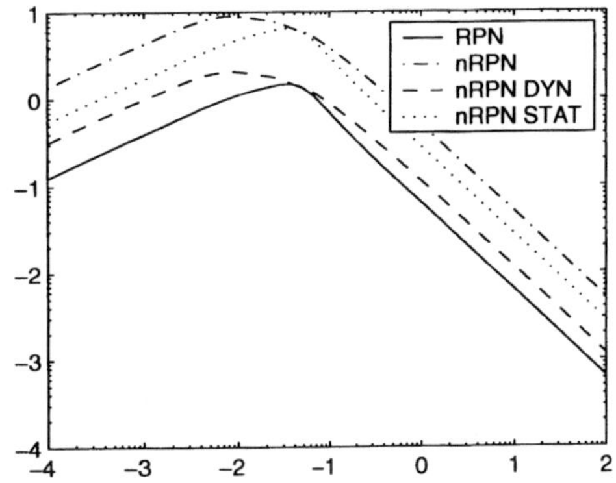

Figure 17: nRPN and RPN plots for x1 and x2 variations and NMOP/MOP (both axes are in logarithmic scale).

Table 15: nRPN indices for x1 and x2 variations and NMOP/MOP changing the paring of NMOP

$nRPN_R$	$nRPN_{STAT}$	$NRPN_{DYN}$
0.79	0.53	-0.19

6 CONCLUSIONS

Three novel indexes were introduced to measure system's nonlinearity. These nonlinear measurements are derived from the Robust Performance Number (RPN) concept. The total system's nonlinearity can be measured by the *nonlinear RPN* (nRPN), while the purely static nonlinearity is captured by nonlinear static RPN ($nRPN_{STAT}$) and the dynamic component by the nonlinear dynamic RPN ($nRPN_{DYN}$). The novel indexes do not require a nonlinear model, being enough a set of linear models. Therefore, they can easily be applied to quantify the nonlinearities of industrial plants and used to answer several practical important questions such as: how nonlinear is the system? Is it necessary to apply a nonlinear controller? What kind of nonlinear controller is necessary?

The novel indices were applied to quantify the nonlinearity degree of two case studies: an industrial high purity distillation column and the quadruple-tanks system. For the quadruple-tanks system the nonlinearity was quantified against inlet flow and distribution ratio variations. This system can perform well in almost all situations. The only condition that requires a nonlinear controller is the case where the system goes from the NMOP to MOP and vice-versa. The predictions made by nRPN indices were confirmed by nonlinear simulations with linear controllers.

ACKNOWLEDGMENT

The authors thank PETROBRAS/BRASKEM/ FINEP for supporting this work.

REFERENCES

Duraisk, R. G. (2001); Controlador Preditivo Não-Linear Utilizando Linearizações ao Longo da Tragetória; *Dissertação de Mestrado. UFRGS.*

Guay, M., McLellan, P., Bacon, D. W. (1997). Measure of Closed-Loop Nonlinearity and Interaction for Nonlinear Chemical Processes. *AiChe Journal.* Vol: 43 (9), 2261-2278.

Helbig, A., Marquardt W., Allgöwer F. (2000). Nonlinearity Measures: Definition, Computation, *Applications. Journal of Process Control* 10: 113-123.

Johansson, K. H.; The Quadruple Tank Process: A Multivariable Laboratory with a Adjustable Zero; *IEEE Transactions on Control Systems Tecnology*; Vol. 8, N°3, May 2000

Klatt, K., Engell, S (1996); Nonliner Control of Neutralization Processes by Gain-Scheduling Trajectory Control, *Ind. Eng. Chem. Res.*, 35, pp 3511-3518.

Luyben, W., L. (1992); *Practical Distillation Control*; Van Nostrand Reinhold.

Skogestad, S. and I. Postlethwaite (1996). Multivariable Feedback Control _Analysis & Design. John Wiley&Sons.

Trierweiler, J.O. and S. Engell (1997a). The Robust Performance Number: A New Tool for Control Structure Design. *Comp. chem. Eng.,* (21), *Suppl.,* pp. S409-414.

Trierweiler, J.O. (1997). A Systematic Approach to Control Structure Design. *Ph.D. Thesis,* Univ. of Dortmund.

Trierweiler, J. O. , Farina, L.A. (2002). The RPN Metodology Apllied to Quantify Process Operability and Control Design *IFAC-World Congress.*

ELSEVIER
IFAC
PUBLICATIONS
www.elsevier.com/locate/ifac

COMPLEXITY REDUCTION OF NONLINEAR SYSTEMS FOR PROCESS CONTROL

Alejandro Vargas * **Frank Allgöwer** *

* *Institute for Systems Theory in Engineering, University of Stuttgart, Pfaffenwaldring 9, D-70550 Germany, e-mail: {vargas,allgower}@ist.uni-stuttgart.de*

Abstract: While models of high complexity are very useful for simulation and prediction purposes in the process industry, they may not be adequate for controller design. This paper proposes a procedure for model complexity reduction for this specific purpose by combining several nonlinear identification and optimization techniques. The available complex simulation model is used as surrogate plant to generate control relevant input and output trajectories and both an optimization of the input and an iterative nonlinear identification procedure are used. The results are illustrated on a nonlinear exothermic copolymerization reactor example.

Keywords: nonlinear identification, model reduction, process control, orthogonal basis function, Volterra representation

1. INTRODUCTION

In the process sector, detailed models of very high fidelity are almost always needed during the design phase. They are usually first principles models based on physical and/or chemical considerations (Ogunnaike and Ray, 1996). Such simulation models give a good approximation of the real process behavior, but tend to be too complex to be used for advanced process controller design, mainly because they are not explicitly developed for this purpose.

Assuming a given fully characterized first principles model, this paper proposes an iterative procedure for model complexity reduction from dedicated simulations (experiments) on the complex model. The novelty in the approach lies in combining several black-box empirical modeling techniques (Ljung, 1999; Nelles, 2001; Pintelon and Schoukens, 2001) with a certain knowledge of the plant, which permits optimizing the input design phase and building the model step by step in an iterative manner. This might be unthinkable in the usual identification setting, where the number of experiments is a main limitation and furthermore all signals are corrupted by noise.

The main assumption made on the system is that it admits a Volterra representation (Schetzen, 1980). It is a sufficiently general system class to fit the input/output behavior of many system processes and furthermore, at least several model based controllers can easily be designed based on this representation (Doyle III *et al.*, 2002).

Three known results from system identification are used by the proposed procedure: frequency domain identification in the presence of nonlinearities (Schoukens *et al.*, 1998), the theory of generalized orthonormal bases for system representation (Ninness *et al.*, 1999), and the forward selection orthogonal least squares methods (Chen *et al.*, 1989; Chen *et al.*, 2003). The procedure is additionally iterative, and on each step the pole of a basis function is chosen and the input excitation for identification is optimized.

In the next section the assumptions on the input/output behavior of the system are given, which concern mainly the Volterra representation and its parametrization with respect to basis functions. Section 3 gives an overview of the linear and nonlinear identification tools used, namely frequency domain identification and orthogonal least squares techniques. The proposed procedure for model complexity reduction is fully explained in Section 4, while in Section 5 the procedure is tested on the model of a copolymerization reactor. Finally, some conclusions are presented. Throughout the text, $\mathbb{R}$ and $\mathbb{C}$ denote the real and complex sets, respectively, $\langle\cdot,\cdot\rangle$ denotes scalar product, and $\bar{\xi}$ denotes the complex conjugate of ξ.

2. ASSUMPTIONS ON THE PROCESS

Given an input time sequence $u(t)$, the output $y(t)$ can be obtained through simulation of the given complex model $\mathcal{S}$, i.e. $y(t) = \mathcal{S}[u(t)]$. It is assumed that initial conditions are always the nominal operating conditions. The aim is thus to obtain a (possibly nonlinear) model $\mathcal{M}$,

$$\hat{y}(t) = \mathcal{M}[u(t)] \approx \mathcal{S}[u(t)] = y(t), \tag{1}$$

that is useful for process control design. This paper limits the reduced model $\mathcal{M}$ to be discrete and, without loss of generality, t is taken as normalized discrete time, i.e. $\tau = tT_s$, where T_s is the sampling period and τ is real time.

Most control systems satisfy the assumption that their input/output behavior $\mathcal{S}$ can be sufficiently well approximated by a Volterra series (Lesiak and Krener, 1978). The finite discrete Volterra series is given by (Doyle III *et al.*, 2002)

$$\hat{y}(t) = y_0 + \sum_{j=1}^{n} v_m^j(t), \tag{2}$$

with $v_m^j(t) = v_m^j\left(u(t-1), \ldots, u(t-m)\right)$ the j-th order terms with memory length m, i.e.

$$v_m^j(t) = \sum_{i_1=0}^{m} \cdots \sum_{i_j=0}^{m} \alpha_j(i_1, \ldots, i_j) \prod_{k=1}^{j} u(t - i_k). \tag{3}$$

The coefficients $\alpha_j(i_1, \ldots, i_j)$ are called the j-th order kernels. It is a well known fact that *fading memory systems* (Boyd and Chua, 1985) may be approximated uniformly on bounded input sets by a finite Volterra system with n and m sufficiently large. Notice, however, that the number of terms can become quite large.

Finite Volterra models are BIBO stable, have periodic responses to periodic inputs without subharmonic generation, and can exhibit input multiplicities (same steady state output y_s for distinct steady state inputs u_s), but no output multiplicities (one u_s leads to various y_s). They also preserve asymptotic constancy, i.e. if $u(t) \to u_s$ as $t \to \infty$, then $y(t)$ also approaches a constant limit y_s.

If the memory m is allowed to be infinite, then the Volterra class may encompass more complex nonlinear behaviors, but this is impossible to implement directly. Therefore the following parametrization in terms of *generalized orthonormal basis functions* (GOBFs) is considered (Ninness and Gustafsson, 1997).

Given a set of poles $\{\xi_1, \xi_2, \ldots,\}$ satisfying

$$\sum_{k=0}^{\infty} (1 - |\xi_k|) = \infty, \qquad |\xi_k| < 1, \\ \xi_k \in \mathbb{C} \implies \xi_{k+1} = \overline{\xi_k}, \tag{4}$$

the k-th GOBF is given by

$$B_k(z) = \frac{\sqrt{1 - |\xi_k|^2}}{z - \xi_k} \prod_{i=1}^{k-1} \frac{1 - \bar{\xi}_i z}{z - \xi_i}. \tag{5}$$

The basis $\{B_1(z), B_2(z), \ldots\}$ is complete and the GOBFs are orthonormal, i.e.

$$\langle B_i, B_j \rangle = \frac{1}{2\pi} \int_{-\pi}^{\pi} B_i(e^{j\omega}) \overline{B_j(e^{j\omega})} d\omega = 0 \tag{6}$$

for $i \neq j$ and $\langle B_i, B_i \rangle = 1$. This set of basis functions generalizes the Laguerre basis ($\xi_k = \alpha$ for all k), the Kautz basis ($\xi_{k,k+1} = \gamma \exp(\pm j\phi)$) and the finite impulse response basis ($\xi_k = 0$).

If the Volterra kernels $\alpha_j(i_1, \ldots, i_j)$ vary regularly as functions of their arguments, then they can be expanded as

$$\alpha_j(i_1, \ldots, i_j) = \\ \sum_{r_1=1}^{\infty} \cdots \sum_{r_j=r_{j-1}}^{\infty} \gamma_j(r_1, \ldots, r_j) \prod_{k=1}^{j} b_{r_k}(i_k). \tag{7}$$

where $\gamma_j(\cdot)$ are kernels (i.e. parameters) and $b_i(t)$ is the inverse z-transform of $B_i(z)$, i.e.

$$b_i(t) = \mathcal{Z}^{-1}\left\{B_i(z)\right\}. \tag{8}$$

Such an expansion is exact if the j-th order kernel is *stably separable* and *strictly proper* (Seretis and Zafiriou, 1997), which basically means that it can be expressed as the product of j first order kernels and $\alpha_j(i_1, \ldots, i_j) = 0$ for $i_1 \cdots i_j = 0$. A direct substitution of (7) in (3) with $m = \infty$ and regrouping leads to

$$v_\infty^j(t) = \sum_{r_1=1}^{\infty} \cdots \sum_{r_j=r_{j-1}}^{\infty} \gamma_j(r_1, \ldots, r_j) \times \\ \times \prod_{k=1}^{j} \left(\sum_{i=0}^{\infty} b_{r_k}(i) u(t - i) \right). \tag{9}$$

Defining

$$\psi_r(t) = \sum_{i=0}^{\infty} b_r(i) u(t - i) \tag{10}$$

and truncating the sums up to q, the finite GOBF-Volterra series is finally given by:

$$\hat{y}(t) = y_0 + \qquad (11)$$
$$+\sum_{j=1}^{n}\sum_{r_1=1}^{q}\cdots\sum_{r_j=r_{j-1}}^{q}\gamma_j(r_1,\ldots,r_j)\prod_{k=1}^{j}\psi_{r_k}(t).$$

Compared to the finite Volterra series (2), this parametrization usually requires only $q \ll m$ basis functions for comparable accuracy, provided the poles are chosen adequately. Notice (10) is the response of the (linear) r-th GOBF to the input $u(t)$, so it quite easy to generate all required responses $\psi_r(t)$ by implementing a realization of the linear system with transfer function matrix

$$\boldsymbol{B}(z) = \left[B_1(z) \cdots B_q(z)\right]^T. \qquad (12)$$

Having these responses, identification of the parameters $\gamma(\cdot)$ is simplified substantially, since (11) is linear in the parameters and usual least squares techniques may be used (Nelles, 2001). The difficulty lies in adequately selecting the poles and the number of GOBFs used. For linear systems it is known that poles close to the dominating frequencies will yield a more compact representation (Gómez, 1998). The proposed procedure iteratively selects these poles based on frequency response data, using a *best* linear identification of the error system.

3. SYSTEM IDENTIFICATION TOOLS

Linear system identification theory is a very mature subject (Ljung, 1999). Its main objective is to obtain a system model under restrictive experimental conditions, which basically pertain to measurement and generator noise. Under the framework studied here these problems do not appear, but the complexity (e.g. nonlinearity) of the system $\mathcal{S}$ makes the task no less difficult.

In the iterative procedure to be outlined in the next section, two identification techniques are used. First the identification of a *best linear* system is performed, and then an orthogonal least squares technique for structure selection and parameter estimation is implemented. The first identification has as objective to select a set of candidate GOBFs based on the significant poles of the linear approximation to the *difference* between the complex model and the current reduced model. It is based on the following result (Schoukens *et al.*, 1998).

Assume a nonlinear system with a convergent Volterra representation (2) is excited with a *random multisine*, i.e. a periodic signal with fixed amplitude spectrum, but random phase spectrum. Then the measured frequency response function (FRF) will have the following decomposition under ideal measurements

$$G(j\omega_k) = G_0(j\omega_k) + G_B(j\omega_k) + G_S(j\omega_k). \qquad (13)$$

The measured FRF is the quotient $Y(k)/U(k)$ of the measured output spectrum and the measured input spectrum, whereby k is an index for the excited frequency lines. The term $G_0(j\omega_k)$ in (13) corresponds to the underlying linear system and is independent of the spectra of the input signal. The second term corresponds to the *systematic* nonlinear contributions, which depend only on the (deterministic) amplitude spectrum but not on the random phases. The third term corresponds to the *stochastic* nonlinear contributions, which are completely dependent on the (random) input spectrum, so that $Y_S(k) = G_S(j\omega_k)U(k)$ behaves like noise, uncorrelated with the input and with zero mean. For this reason, the response of the nonlinear system (2) to a random multisine can be regarded as the response through a *related linear dynamical system* $G_R(j\omega_k) = G_0(j\omega_k) + G_B(j\omega_k)$ plus some *nonlinear noise* due to the stochastic contributions.

There are a host of techniques available for system identification in the frequency domain (see e.g. the book by Pintelon and Schoukens (2001) and the references therein), given either the measured data $Y(k), U(k)$ or the measured FRF $G(jw_k)$ and certain noise assumptions. *Subspace algorithms* implemented in the frequency domain (McKelvey *et al.*, 1996) are used in the present approach. These produce a state space realization $(\hat{A}, \hat{B}, \hat{C}, \hat{D})$ for the fit $\hat{G}(z)$ to the FRF $G(j\omega_k)$. On one hand, a consistent estimate of $G_R(j\omega_k)$ with respect to the nonlinear noise is guaranteed as the number of realizations of the same random multisine tends to infinity. On the other hand, being based on principal component analysis, the dynamic order must not be fixed *a priori*. This is important in the proposed procedure, since only significant candidate poles should be tested and these are directly given by the eigenvalues of $\hat{A}$.

The other nonlinear identification scheme used in the procedure pertains structure selection and parameter estimation using orthogonal least squares techniques (Chen *et al.*, 1989). The assumed structure for the system is linear in the parameters, i.e.

$$y(t) = \boldsymbol{\varphi}^T(t)\boldsymbol{\theta} + e(t) \qquad (14)$$

where $\boldsymbol{\theta} \in \mathbb{R}^r$ is the parameter vector, $\boldsymbol{\varphi} \in \mathbb{R}^r$ is the *regressor* vector, and $e(t)$ stands for the modeling error. The regressors must be functions of the measured data (e.g. filtered inputs). From (11) it is clear that grouping all possible $\gamma(\cdot)$ in $\boldsymbol{\theta}$, the regressor vector is composed of products of the type $\prod_i \psi_{r_i}(t)$. Notice that the number of regressors for (11) can be quite large, namely $r = \binom{n+q}{q}$, so estimating $\boldsymbol{\theta}$ as the least squares solution may be numerically difficult. Furthermore, it would contradict the reduced complexity goal.

The identification scheme considers

$$\boldsymbol{y} = \boldsymbol{\Phi}\boldsymbol{\theta} + \boldsymbol{e} = \boldsymbol{W}\boldsymbol{\vartheta} + \boldsymbol{e} \tag{15}$$

where the data is given as N time domain samples:

$$\boldsymbol{y} = [y(1), y(2), \ldots, y(N)]^T, \tag{16}$$

and the t-th row of the $N \times r$ matrix $\boldsymbol{\Phi}$ corresponds to the regressor $\boldsymbol{\varphi}^T(t)$. Assume $\boldsymbol{\Phi}$ has been orthogonally decomposed, i.e.

$$\boldsymbol{\Phi} = \boldsymbol{W}\boldsymbol{A}, \qquad \boldsymbol{\vartheta} = \boldsymbol{A}\boldsymbol{\theta} \tag{17}$$

where $\boldsymbol{A} \in \mathbb{R}^{r \times r}$ is an upper triangular matrix with ones in the main diagonal and $\boldsymbol{W} = [\boldsymbol{w}_1, \ldots, \boldsymbol{w}_r]$ is a matrix with orthogonal columns.

The main aspect about orthogonal least squares techniques is that each orthogonal regressor $\boldsymbol{w}_i$ contributes independently to minimizing the quadratic cost function of the output error and $\hat{\vartheta}_i = \hat{\vartheta}_i(\boldsymbol{w}_i, \boldsymbol{y})$. Assume $\boldsymbol{w}_1$ is the most significant regressor in this sense and $\boldsymbol{w}_r$ is the least significant one. To obtain a good balance between model complexity —measured by the number of regressors— and good prediction capabilities (a small error), consider only the first, say ρ, orthogonal regressors. In essence most structure selection algorithms iteratively reorder the columns of the matrix $\boldsymbol{\Phi}$ so that the reduced model is given by

$$\boldsymbol{y} \approx \widetilde{\boldsymbol{W}}\tilde{\boldsymbol{\vartheta}} \quad \text{and} \quad \tilde{\boldsymbol{\vartheta}} = \widetilde{\boldsymbol{A}}\tilde{\boldsymbol{\theta}}, \tag{18}$$

where $\widetilde{\boldsymbol{W}}$ comprises the first ρ columns of $\boldsymbol{W}$ and $\widetilde{\boldsymbol{A}}$ comprises the upper left $\rho \times \rho$ sub-matrix of $\boldsymbol{A}$.

Most orthogonal least squares algorithms select in a forward manner the most significant regressors while following an orthogonalization procedure, e.g. Gram-Schmidt (Chen *et al.*, 1989). A recent paper (Chen *et al.*, 2003) proposes to combine the modified Gram-Schmidt procedure with local regularization, together with a D-optimality design in order to obtain a sparse model. The local regularization aims at minimizing over-fitting the model with spurious regressors, while the D-optimality criterion aims at maximizing the information content of the selected regressors, much as it is done with Akaike's (1974) well known information criterion (AIC).

4. ITERATIVE PROCEDURE FOR MODEL COMPLEXITY REDUCTION

The rationale behind the proposed procedure is the following: GOBF poles are selected based on measuring the FRF of the error system, the identification input is then optimally selected to *invalidate* the current model, and finally an orthogonal least squares procedure is used to select a reduced number of regressors.

The proposed methodology starts with $\kappa = 1$ and the candidate nonlinear model $\mathcal{M}^{(0)} = 0$ (iterations are denoted by parenthesized indices).

Step 1. Measure the frequency response function $G_{\mathcal{E}}(j\omega_k)$ of the error between simulation $\mathcal{S}$ and current model $\mathcal{M}^{(\kappa-1)}$ and fit a linear model $\hat{G}_{\mathcal{E}}(z)$ to it using a subspace algorithm.

Step 2. For each pole of $\hat{G}_{\mathcal{E}}(z)$ build a set of candidate filters $\boldsymbol{B}_i^{(\kappa-1)}(z)$ by adding a generalized basis function to the current filter $\boldsymbol{B}^{(\kappa-1)}(z)$ of already selected GOBFs (see (12)).

Step 3. Design an optimized control relevant input $u^{(\kappa)}(t)$ to *invalidate* the current model $\mathcal{M}^{(\kappa-1)}$ with respect to $\mathcal{S}$. Build the signals $y^{(\kappa-1)}(t)$ and $\boldsymbol{\psi}_i^{(\kappa-1)}(t)$ as the outputs of $\mathcal{S}$ and each $\boldsymbol{B}_i^{(\kappa-1)}(z)$ to this input.

Step 4. For each i construct the columns of the regressor matrix $\boldsymbol{\Phi}^{(\kappa)}$, considering a certain nonlinear order n in (11). Then perform a structure selection and parameter estimation using orthogonal least squares techniques to end up with a set of candidate models $\mathcal{M}_i^{(\kappa)}$.

Step 5. Cross-validate each candidate model and select the "best" one as $\mathcal{M}^{(\kappa)}$; also update $\boldsymbol{B}^{(\kappa)}$. If the model is adequate enough, then **stop.** Otherwise return to **Step 1** (add another GOBF) or **Step 4** (increase nonlinear degree).

Regarding **Step 1**, the error is defined as

$$\begin{aligned} e^{(\kappa)}(t) &= \mathcal{E}^{(\kappa)}[\bar{u}^{(\kappa)}(t)] \\ &= \mathcal{S}[\bar{u}^{(\kappa)}(t)] - \mathcal{M}^{(\kappa-1)}[\bar{u}^{(\kappa)}(t)] \end{aligned} \tag{19}$$

Since $\mathcal{M}^{(\kappa-1)}$ is a Volterra model and $\mathcal{S}$ is assumed to have a Volterra representation, then so does $\mathcal{E}^{(\kappa)}$ and (13) is valid for random multisines. It is recommended to design $\bar{u}^{(\kappa)}(t)$ as a random multisine with odd frequency components in order to minimize the effect of even nonlinearities (Schoukens *et al.*, 1998). Furthermore, to minimize the effect of "nonlinear noise" several realizations of $\bar{u}^{(\kappa)}(t)$ should be considered and the FRF averaged over these measurements.

With respect to **Step 2**, minimal state space realizations for a set of basis functions are easy to derive (Gómez, 1998). Furthermore, adding a candidate basis function $B_i^{(\kappa-1)}(z)$ to $\boldsymbol{B}^{(\kappa-1)}(z)$ in order to build $\boldsymbol{B}_i^{(\kappa-1)}$ is straightforward. However, care should be taken when pairs of complex conjugate poles are considered to avoid complex parameters (Ninness and Gustafsson, 1997).

Step 3 comprises one of the main features of this procedure. The previous model $\mathcal{M}^{(\kappa-1)}$ is the best possible model for a GOBF-Volterra structure of some nonlinear order and some basis functions $\boldsymbol{B}^{(\kappa-1)}(z)$. From (11) it can be seen that adding a basis function $B_i^{(\kappa)}(z)$ simply adds more terms to the structure of $\mathcal{M}^{(\kappa-1)}$, which by themselves may compose a model for $\mathcal{E}^{(\kappa)}$. By designing an input that somehow *maximizes* the output of such an error system, the current nonlinear model $\mathcal{M}^{(\kappa-1)}$ is being invalidated. To carry on this input design, multisines are again considered, given their reported success for nonlinear Volterra system identification (Evans *et al.*, 1996).

Having chosen a frequency grid with N lines, the input signal $u^{(\kappa)}(t)$ is parametrized by its (complex) input spectrum $U(k)$. The optimization problem to solve is

$$U^*(k) = \arg\sup_{U(k)\in\mathbb{U}} \left\| \left(\mathcal{S} - \mathcal{M}^{(\kappa-1)}\right)[u(t)] \right\|^2 \quad (20)$$

whereby $\mathbb{U}$ refers to the set where the input spectrum is to be constrained, e.g. by fixing a maximal RMS value of the signal, and $\|\cdot\|$ is some norm. To solve this constrained optimization problem only a finite number of elements in the set $\mathbb{U}$ is considered and the sup-operator is replaced by a max-operator. The solution may be need to be found by exhaustive search or using global optimization strategies (Nelles, 2001).

Several candidate models $\mathcal{M}_i^{(\kappa)}$ with $\rho_i^{(\kappa)}$ parameters/regressors are obtained in **Step 4** using the same input/output data. In **Step 5** a validation is performed to select the model that has better prediction capabilities. Each model is excited with L realizations $u_\ell^{(\kappa)}(t)$ of a random multisine with the fixed amplitude spectrum of $U^*(k)$. The following performance index is then calculated:

$$\begin{aligned} J(\mathcal{M}_i^{(\kappa)}) &= g(\rho_i^{(\kappa)}) + \\ &+ \frac{1}{L}\sum_{\ell=1}^{L}\sum_{t=1}^{N} \left\| \mathcal{S}[u_\ell^{(\kappa)}(t)] - \mathcal{M}_i^{(\kappa)}[u_\ell^{(\kappa)}(t)] \right\|^2 \quad (21) \end{aligned}$$

where $g\colon \mathbb{N} \to \mathbb{R}_+$ is a monotonically increasing function that penalizes a large number of parameters $\rho_i^{(\kappa)}$. The "best" model is chosen as $\mathcal{M}^{(\kappa)} = \mathcal{M}_{i^*}^{(\kappa)}$ with $i^* = \arg\min_i J(\mathcal{M}_i^{(\kappa)})$.

5. EXAMPLE

The procedure is illustrated on the model of a continuous copolymerization reactor with exothermic kinetics. The model has 15 states, 6 inputs, and 7 outputs (Bindlish, 1999). The inputs are the feed rates of the inhibitor (m-dinitrobenzene), the transfer agent (acetaldehyde), and the two monomers (vinyl acetate and methyl methacrylate), plus the coolant and feed temperatures. The outputs are the effluent compositions of the two monomers and the solvent (benzene) in the separator, the copolymer composition, its intrinsic viscosity, the reactor temperature, and the polymer production rate.

Although the proposed procedure is extensible to MIMO systems, for simplicity it was applied in this example to an equivalent SISO system. The system was set to certain operating conditions and only one input was made time-varying, while only one output was measured. The main nonlinearity of the system is due to the temperature dependent kinetics of Arrhenius type, so to make the problem more interesting, the coolant temperature was chosen as input. The output chosen was the copolymer production rate.

The sampling time was set to $T_s = 2.5$ min. For **Step 1** the input used was an odd random multisine with maximum frequency $f_{\max} = 255 f_0$ and base frequency $f_0 \approx 3.9 \times 10^{-4}\text{min}^{-1}$. Input deviations of ± 15K were considered. Two periods were measured, but only the last one was considered for identification to account for transients.

The procedure was tested several times and consistently only two iterations were sufficient to obtain a satisfactory reduced complexity model. The GOBFs resulted with the following ordered poles:

$$\{0.8546 \pm j0.1934, 0.8955 \pm j0.2451\}, \quad (22)$$

which leads to a fourth order system. Before the nonlinear identification procedure was carried out, the gain of the GOBF filters was adjusted so that $\psi_i(t)$, $i = 1, \ldots, 4$, was normalized for the optimal input generated in **Step 3**. The reduced resulting structure for the system has only five terms:

$$\begin{aligned} y(t) &= \theta_1\psi_1(t) + \theta_2\psi_2(t) + \theta_3\psi_3(t) + \\ &+ \theta_4\psi_1(t)\psi_2(t) + \theta_5\psi_1(t)\psi_4(t) \quad (23) \end{aligned}$$

with parameter vector

$$\boldsymbol{\theta} = \left[0.148, 0.0149, 0.0513, -0.0801, -0.0627\right]^T. \quad (24)$$

After the identification, the resulting model was tested on a completely different validation input, namely a random step input (Pearson, 1998). The output of both the complex simulation model and the model (23) is shown in Fig. 1, whereas the error between both signals is shown in Fig. 2.

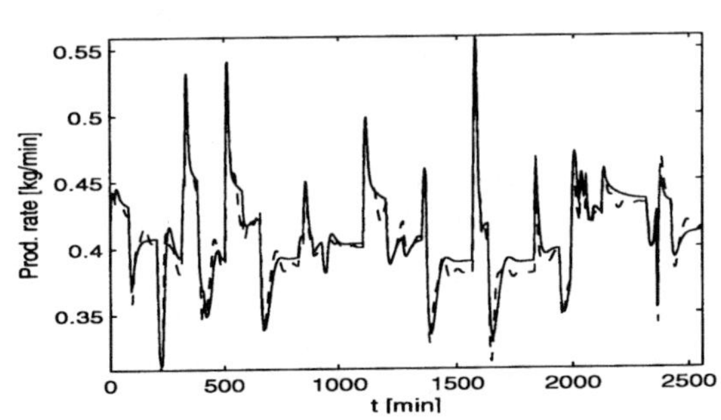

Fig. 1. Output of complex simulation model (solid) and reduced complexity model (dashed) to a random step input.

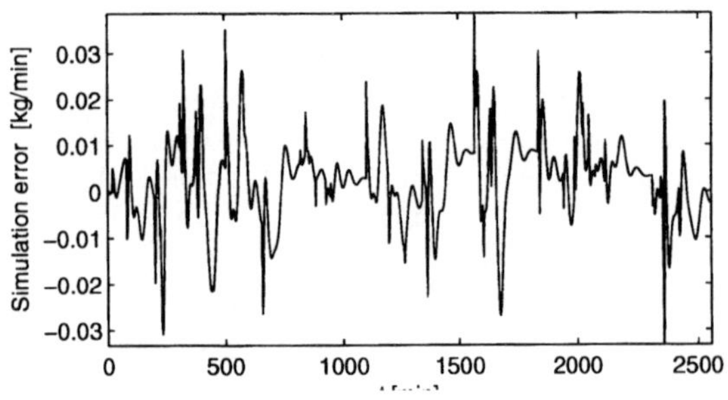

Fig. 2. Error between complex and reduced complexity models.

6. CONCLUSIONS

A procedure for complexity reduction combining frequency domain linear identification, constrained input optimization techniques, and substructure selection and parameter estimation methods has been proposed. Testing the procedure on a realistic example shows promising results and further investigations are under way to refine it.

ACKNOWLEDGEMENTS

This work was partly funded by the German National Science Foundation (DFG) in course of the SFB412. The authors thank Z.K. Nagy for kindly providing the simulation model of the reactor and P. Wolfrum for performing some simulations.

REFERENCES

Akaike, H. (1974). A new look at the statistical model identification. *IEEE Trans. Automatic Control* **19**(6), 716–723.

Bindlish, Rahul (1999). Modeling and Control of Polymerization Processes. PhD thesis. University of Wisconsin. Madison.

Boyd, S. and L.O. Chua (1985). Fading memory and the problem of approximating nonlinear operators with Volterra series. *IEEE Trans. Cir. Sys.* **32**(11), 1150–1161.

Chen, S., S.A. Billings and W. Luo (1989). Orthogonal least squares methods and their application to nonlinear identification. *Int. J. Control* **50**(5), 1873–1896.

Chen, S., X. Hong and C.J. Harris (2003). Sparse kernel regression modelling using combined locally regularized orthogonal least squares and D-optimality experimental design. *IEEE Trans. Automatic Control* **48**(6), 1029–1036.

Doyle III, F.J., R.K. Pearson and B.A. Ogunnaike (2002). *Identification and Control Using Volterra Models.* Springer. London.

Evans, C., D. Rees, L. Jones and M. Weiss (1996). Periodic signals for measuring nonlinear Volterra kernels. *IEEE Trans. Instrum. Meas.* **45**(2), 362–371.

Gómez, J.C. (1998). Analysis of Dynamic System Identification using Rational Orthonormal Bases. PhD thesis. The University of Newcastle. Australia. Available at www.fceia.unr.edu.ar/~jcgomez.

Lesiak, C. and A.J. Krener (1978). The existence and uniqueness of Volterra series for nonlinear systems. *IEEE Trans. Automatic Control* **23**(6), 1090–1095.

Ljung, L. (1999). *System Identification. Theory for the user.* 2nd ed.. Prentice–Hall. Upper Saddle River.

McKelvey, T., H. Akçay and L. Ljung (1996). Subspace-based multivariable system identification from frequency response data. *IEEE Trans. Automatic Control* **41**(7), 960–979.

Nelles, O. (2001). *Nonlinear System Identification.* Springer. Berlin.

Ninness, B. and F. Gustafsson (1997). A unifying construction of orthonormal bases for system identification. *IEEE Trans. Automatic Control* **42**(4), 515–521.

Ninness, B., H. Hjalmarsson and F. Gustafsson (1999). The fundamental role of general orthonormal bases in system identification. *IEEE Trans. Automatic Control* **44**(7), 1384–1406.

Ogunnaike, B.A. and W.H. Ray (1996). *Process Dynamics, Modeling, and Control.* Oxford University Press. New York.

Pearson, R.K. (1998). Input sequences for nonlinear modeling. In: *Nonlinear Model Based Process Control* (R. Berber and C. Kravaris, Eds.). pp. 599–621. Number 353 In: *NATO ASI.* Kluwer Academic Publishers. Dordrecht.

Pintelon, R. and J. Schoukens (2001). *System Identification. A frequency domain approach.* IEEE Press. Piscataway.

Schetzen, M. (1980). *The Volterra and Wiener Theories of Nonlinear Systems.* John Wiley & Sons. New York.

Schoukens, J., T. Dobrowiecki and R. Pintelon (1998). Parametric and nonparametric identification of linear systems in the presence of nonlinear distortions —A frequency domain approach. *IEEE Trans. Automatic Control* **43**(2), 176–190.

Seretis, C. and E. Zafiriou (1997). Nonlinear dynamical system identification using reduced Volterra models with generalized orthonormal basis functions. In: *Proc. American Control Conference.* Albuquerque, New Mexico. pp. 3042–3046.

ELSEVIER
IFAC
PUBLICATIONS
www.elsevier.com/locate/ifac

INTERACTION BETWEEN CONTROL AND STATE ESTIMATION IN NONLINEAR MPC

Morten Hovd [*,1] **Robert R. Bitmead** [**]

[*] *Department of Engineering Cybernetics, Norwegian University of Technology and Science, N-7491 Trondheim, Norway*
[**] *Department of Mechanical and Aerospace Engineering, University of California San Diego, 9500 Gilman Drive, La Jolla, CA 92093-0411, USA*

Abstract: This paper shows how to take the quality of the state estimation into account in the formulation of the optimization criterion for model predictive control (MPC). This is relevant for the control of nonlinear plants, for which the separation principle in general does not apply. The method is illustrated on an example which is locally weakly unobservable at the reference state.

Keywords: Nonlinear MPC, state estimation

1. INTRODUCTION

Model predictive control (MPC) has found widespread application, particularly in the chemicals processing industry and refineries. Originally, MPC was proposed using linear plant models, with constraints on inputs and states (or outputs) being the only type of nonlinearity that was addressed. However, in the last two decades, there has been a number of publications on MPC using nonlinear process models (Allgöwer *et al.*, 1999) and a number of actual industrial applications using such nonlinear models (Qin and Badgwell, 2003).

The typical academic publication on nonlinear MPC focuses on developing stability guarantees in the face of plant nonlinearities, based on the assumption that the plant states are available either through direct measurements or through the use of a state estimator.

In the processing industries, which has seen the bulk of MPC applications so far, one seldom enjoys the luxury of being able to measure all states directly. One therefore has to resort to some sort of state estimation, the Extended Kalman Filter (EKF) or Receding Horizon Estimation (RHE) being the estimators most commonly proposed.

Although the control methodology (nonlinear MPC) and state estimation methodology (EKF / RHE) are powerful, this approach totally ignores the fact that the separation principle in general does not apply to nonlinear plants. In (Imsland *et al.*, 2003) and (Adetola and Guay, 2003) it is shown how the performance of an output feedback NMPC controller can achieve the performance and region of attraction of a state feedback NMPC controller. However, these results are restricted to highly specific nonlinear model structures and requires very high observer gains.

This paper shows how to take into account the quality of the state estimate in the formulation of the optimization problem for the MPC. In previous work, Yan and Bitmead (2002) have shown how to include the quality of the state estimate into the constraint formulation for MPC. Here we will in include the effects of manipulated variable moves on the state estimate covariance in both the constraints and the optimization criterion. This

[1] To whom correspondence should be addressed. e-mail: morten.hovd@itk.ntnu.no

will of course result in control moves that differ from those that would be obtained if the estimation accuracy was ignored, i.e., it will 'degrade' the control to obtain more accurate state estimates. However, insisting on accurate control based on inaccurate state estimates is not very meaningful. The proposed modification is clearly only relevant for nonlinear plants, since for linear plants the quality of the state estimation is independent of the control action.

In section 2 the model predictive control formulation is presented, and the state estimation approaches (EKF and RHE) are presented briefly. The proposed control and estimation method is illustrated on a simple example in section 3. The properties of the proposed MPC formulation are addressed in section 4. The final section presents conclusions and points out directions for further work.

2. MPC AND STATE ESTIMATION FORMULATIONS

2.1 *MPC formulation*

The MPC formulation is based on a discrete nonlinear state space model of the type

$$x_{k+1} = f(x_k, u_k) + w_k \quad (1)$$

$$y_k = g(x_k, u_{k-1}) + v_k \quad (2)$$

where x is the state vector, u is the input vector, y is the measurement vector, whereas w and v are state excitation noise and measurement noise, respectively. The noise sequences are assumed to be Gaussian white noise. The subscript is a sampling time index. It is assumed that the measurement is obtained before calculating the new value for the input for a given sample interval, therefore y_k may depend on u_{k-1} but not on u_k.

The proposed MPC may be formulated as

$$\min_{\left[u_k^T, u_{k+1}^T, \cdots, u_{k+n_p-1}^T\right]^T} \quad (3)$$

$$\sum_{i=1}^{n_p} (x_{k+i} - x_{k+i}^r)^T Q_{k+i} (x_{k+i} - x_{k+i}^r)$$

$$+ \sum_{i=1}^{n_p} (u_{k+i-1} - u_{k+i-1}^r)^T R_{k+i} (u_{k+i} - u_{k+i}^r)$$

$$+ \sum_{i=1}^{n_p} \sum_{m=1}^{n_v} \rho_{i,m} v_m^T P_{k+i} v_m$$

such that

$$H_c x_{k+i} + G_c u_{k+i} + L_c \sigma_{c,k+i} \leq b_c$$

$$x_{k+i} = f(x_{k+i-1}, u_{k+i-1})$$

The matrix P_{k+i} is the *à posteriori* EKF state covariance matrix at time $k+i$. This will sometimes be denoted $P_{k+i|k+i}$ to clearly distinguish it from the *à priori* covariance matrix $P_{k+i|k+i-1}$. Although the EKF state covariance matrix is not an exact representation of the state probability distribution for nonlinear systems, it is a measure of state estimation accuracy that can be obtained with relative ease. Note also that P_{k+i} is easily predicted along a future state trajectory. The vectors v_m are typically chosen to be orthonormal, and allow using different weights on the state covariance for different directions of the state space. In the constraint equations, $\sigma_{c,k+i}$ represents the predicted standard deviation of the state estimate in the direction of relevance for constraint c. No constraint can be guaranteed to hold in the future if the plant is subject to stochastic inputs (at least if these stochastic inputs are unbounded, as is the case for Gaussian noise). However, by including the standard deviation of the state estimate in the constraints, one has a handle for tuning the controller to ensure an acceptable probability for constraint violation.

The vectors x^r and u^r are reference vectors for the states and manipulated inputs, respectively. These reference vectors are assumed to be known *à priori.*

To enable efficient computation in on-line applications, there is a practical requirement that the optimization problem must be solvable within one sampling period. Therefore, MPC optimization problems are in general formulated as LP or QP problems. Problems with significant nonlinearities are typically approximated by QP problems, with line search along the solution vector to improve both the accuracy and the rate of convergence to the solution (Biegler, 1998). Further improvements in accuracy may be obtained by using the result of one QP with line search as the starting point for a new optimization, if computing power and available time allow.

In Eq. (3), it is the non-linear plant model which makes the optimization problem differ from a standard QP problem. For linear plant models (subject to Gaussian noise) the state estimate covariance matrix and corresponding state estimate standard deviation would be independent of the future input and state trajectories.

In order to approximate Eq. (3) with a QP problem, we will linearize the plant model around given future state and input trajectories, denoted by x_{k+i}^0 and u_{k+i}^0, respectively. Typically, these future state and input trajectories are obtained from by simulating of the non-linear plant model n_p timesteps into the future, starting from the current state estimate x_k, and applying the 'tail' of the vector of manipulated variables obtained at the previous time step (assuming zero state excitation noise w_{k+i}). In addition to providing future trajectories for the states and manipulated variables, such a simulation can also provide nominal future values for the EKF state covariance matrix P^0_{k+i}. Thus, the nonlinear plant model, Eqs. (1,2), are replaced in the approximate QP problem by the time varying linear model

$$x_{d,k+i+1} = A_{k+i}x_{d,k+i} + B_{k+i}u_{d,k+i} \quad (4)$$

$$y_{d,k+i} = C_{k+i}x_{d,k+i} + D_{k+i}u_{k+i-1} \quad (5)$$

The subscript d on the measurements, states, and manipulated variables indicate that these are here the *deviations* from the values around which the linearization is performed.

Having thus 'removed' the non-linearity in the constraints that are caused directly by the nonlinear plant model, we must next do the same for the predicted values of the state covariance matrix in the optimization criterion, and the state standard deviation in the constraints. This is done by a Taylor series expansion around the nominal future state and manipulated variable trajectories. We will henceforth adopt the more economic notation

$$\chi = [x^T_{d,k+1}, \cdots, x^T_{d,k+n_p}]^T$$

$$\nu = [u^T_{d,k}, \cdots, u^T_{d,k+n_p-1}]^T$$

$$\chi_0 = [(x^0_{k+1})^T, \cdots, (x^0_{k+n_p})^T]^T$$

$$\nu_0 = [(u^0_k)^T, \cdots, (u^0_{k+n_p-1})^T]^T$$

Thus, we get

$$v^T_m P_{k+i} v_m \approx v^T_m P^0_{k+i} v_m + dP_{\chi,m,k+i}\chi \quad (6)$$
$$+ dP_{\nu,m,k+i}\nu + \frac{1}{2}[\chi^T \nu^T] ddP_{m,k+i} \begin{bmatrix} \chi \\ \nu \end{bmatrix}$$

where $dP_{\chi,m,k+i} = \frac{\partial(v^T_m P_{k+i} v_m)}{\partial\chi}$, $dP_{\nu,m,k+i} = \frac{\partial(v^T_m P_{k+i} v_m)}{\partial\chi}$, and $ddP_{m,k+i}$ is the corresponding matrix of second partial derivatives

$$ddP_{m,k+i} = \begin{bmatrix} \frac{\partial^2(v^T_m P_{k+i} v_m)}{\partial\chi^2} & \frac{\partial^2(v^T_m P_{k+i} v_m)}{\partial\chi\partial\nu} \\ \frac{\partial^2(v^T_m P_{k+i} v_m)}{\partial\nu\partial\chi} & \frac{\partial^2(v^T_m P_{k+i} v_m)}{\partial\nu^2} \end{bmatrix} \quad (7)$$

The vectors v_m are independent of states and manipulated variables, and the evaluation of the partial derivative matrices defined above therefore requires the evaluation of the partial derivatives of P_{k+i}. To do this, we need to start from the familiar EKF covariance equations

$$P_{k+i|k+i-1} = A_{k+i-1}P_{k+i-1|k+i-1}A^T_{k+i-1} + W \quad (8)$$

$$P_{k+i|k+i} = P_{k+i|k+i-1} - P_{k+i|k+i-1}C^T_{k+i}MC_{k+i}P_{k+i|k+i-1} \quad (9)$$

where $M = (V + C_{k+i}P_{k+i|k+i-1}C^T_{k+i})^{-1}$. The next step is to differentiate the linearized model equations around the predicted future state and input trajectories. Note that the linearized model equations will only depend on the state and input values at the same timestep. The effects of states and inputs at previous timesteps are accounted for through the model equations themselves, whereas future states and inputs obviously cannot have any effect. On the other hand, the EKF state covariance matrix will depend on deviations from the nominal state and input trajectories at previous time steps, through the EFK covariance equations above.

Thus, we have

$$A^{x_j}_{k+i} = \left.\frac{\partial A_{k+i}}{\partial x_{j,k+i}}\right|_{x^0_{k+i},u^0_{k+i}} \quad (10)$$

$$A^{u_j}_{k+i} = \left.\frac{\partial A_{k+i}}{\partial u_{j,k+i}}\right|_{x^0_{k+i},u^0_{k+i}} \quad (11)$$

$$C^{x_j}_{k+i} = \left.\frac{\partial C_{k+i}}{\partial x_{j,k+i}}\right|_{x^0_{k+i},u^0_{k+i-1}} \quad (12)$$

$$C^{u_j}_{k+i} = \left.\frac{\partial C_{k+i}}{\partial u_{j,k+i-1}}\right|_{x^0_{k+i},u^0_{k+i-1}} \quad (13)$$

Here the index j refers to a particular element of the state or manipulated variable vector at time $k+i$. Note the time index on u in Eq. (13) above. This reflects the assumed order of execution of tasks at each sampling instant:

(1) Obtain measurements.
(2) Update state estimates.
(3) Calculate new manipulated variables.

Thus, the manipulated variable that are applied at the time when y_{k+i} is obtained, is actually u_{k+i-1}.

Tedious, but straight forward application of matrix calculus will then produce the needed partial derivatives $dP_{\chi,m,k+i}$, $dP_{\nu,m,k+i}$, and $ddP_{m,k+i}$. Due to space limitations, this derivation is omitted from the paper.

What remains before the MPC problem can be formulated is to linearize the state standard deviation in the constraints in Eq. (3). Thus, we use the approximation

$$\sigma_{c,k+i} \approx \sigma^0_{c,k+i} + \begin{bmatrix} \frac{\partial\sigma_{c,k+i}}{\partial\chi} & \frac{\partial\sigma_{c,k+i}}{\partial\nu} \end{bmatrix} \begin{bmatrix} \chi \\ \nu \end{bmatrix} \quad (14)$$

Let h^r_c be row r of the matrix H_c in Eq. (3). The standard deviation of the states in the direction perpendicular to the constraint boundary is then

$$\sigma^r_{c,k+i} = \sqrt{h^r_c P_{k+i} (h^r_c)^T} \tag{15}$$

Thus,

$$\begin{bmatrix} \frac{\partial \sigma_{c,k+i}}{\partial \chi} & \frac{\partial \sigma_{c,k+i}}{\partial \nu} \end{bmatrix} = \frac{1}{\sqrt{h^r_c P_{k+i} (h^r_c)^T}} \times$$

$$\begin{bmatrix} \frac{\partial (h^r_c P_{k+i} (h^r_c)^T)}{\partial \chi} & \frac{\partial (h^r_c P_{k+i} (h^r_c)^T)}{\partial \nu} \end{bmatrix} \tag{16}$$

Hence, we will also here require partial derivatives of P_{k+i} w.r.t. states and manipulated variables.

Everything is now ready for formulating an approximation to Eq. (3) as a QP problem. The required manipulations are fairly standard, and are not repeated here. The MPC computations at each timestep are organized as follows:

(1) The 'tail' of the manipulated variable vector from the previous timestep is used as an initial guess for the solution.
(2) This manipulated variable sequence is applied to the nonlinear model in Eq. (1), and the model is linearized around the manipulated variable and the state trajectories.
(3) The QP problem is formulated and solved.
(4) A line search along the solution vector from the QP problem is performed. In the line search, the actual nonlinear model and nonlinear predictions of the EKF covariance matrices are used.

2.2 *State estimation*

Two related approaches are used for state estimation, the Extended Kalman Filter (EKF) and Receding Horizon Estimation (RHE).

For the EKF, standard formulae are used, using local linearizations of the system dynamics. The EKF can be found in many textbooks on advanced control.

For RHE, a formulation similar to the constrained linear Moving Horizon Estimator of Rao (e.g. (Rao, 2000)), is used. That is, a QP formulation is used to approximate the observed system outputs over the estimation horizon, with the linearized equations describing the system dynamics as constraints. The free variables in the optimization are measurement and state excitation noise sequences, and the state at the beginning of the estimation horizon. The arrival cost (i.e., weight on the state at the beginning of the estimation horizon) is based on the EKF covariance, as described by Rao (2000).

The QP problem gives us a state excitation noise sequence, as well as the change from the *à priori* to the *à posteriori* state estimate at the start of the estimation horizon. For any multiple of these QP outputs, the system may be simulated using the nonlinear system model, and the corresponding measurement noise sequence required to obtain the observed measurements can be calculated. This is used in a line search procedure after solving the QP problem, in much the same way as line search is commonly used in nonlinear MPC.

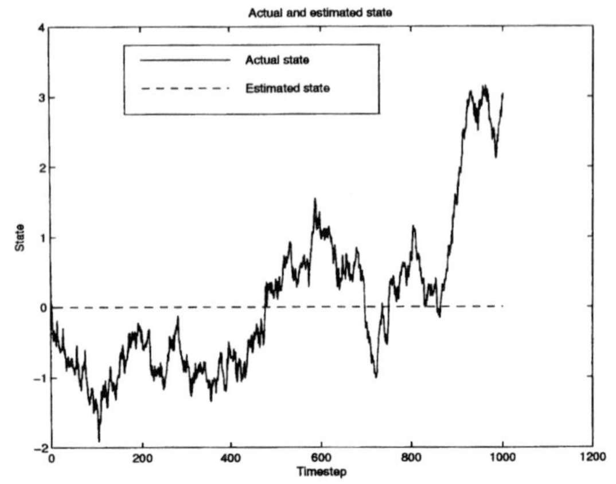

Fig. 1. Results when using no weight on the EKF state covariance estimate .

3. EXAMPLE

The method is applied to a simple example, given by

$$x_{k+1} = x_k + 0.1 x_k u_k + u_k + w_k \tag{17}$$

$$y_k = x_k^3 + v_k \tag{18}$$

The state excitation noise w is assumed to be Gaussian with variance 0.01, whereas the measurement noise v has variance 0.1. The MPC has prediction horizon $n_p = 10$, state weight $Q_{k+i} = 2$ for $i = 1, \cdots, 9$, $Q_{k+10} = 10$, and $R_{k+i} = 1$. All simulations are started from an initial state $x_0 = 0.5$ and a small initial state covariance $P_0 = 0.01$. Figure 1 shows the results when not taking the covariance of the state estimate into account ($\rho_i = 0$ in Eq. (3)). The state estimation is performed with an Extended Kalman Filter. We see that the controller quickly controls the estimated state to zero. Once the estimated state is zero, the system is no longer observable, and the EKF can no longer correct for the observed offset between actual and predicted measurement. Consequently, the state starts to drift. In Fig. 2 we see the corresponding results when a value of $\rho_i = 0.3$ is used. The manipulated variable starts moving when the state covariance starts becoming large. This moves the state estimate into a region where the system is observable, and the EKF can correct the state. It may not be immediately obvious that the performance achieved in Fig. 2 is better than that achieved in Fig. 1. However, over

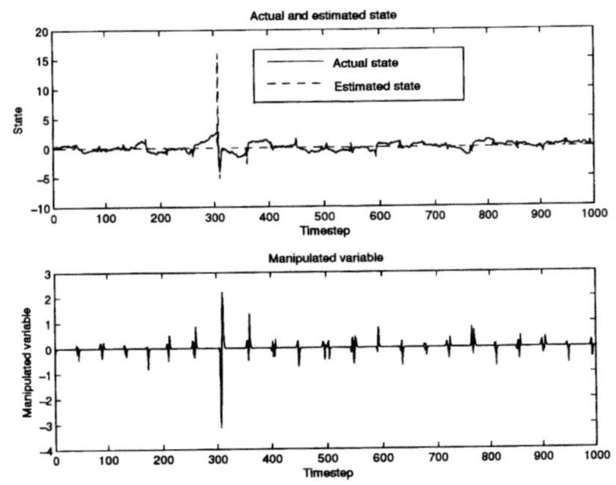

Fig. 2. Results when using the weight $\rho_i = 0.3$ on the EKF state covariance estimate .

a sufficiently long operating period, the actual state in Fig. 1 may drift arbitrarily far from the reference, whereas this does not happen in Fig. 2.

If the state estimate in Fig. 2 has drifted any significant distance from the reference (at which the system is unobserable), the gain from state to measurement will be much larger at the actual state value than at the estimated state. The EKF, using a model linearized around the estimated state, therefore modifies the state estimate excessively. This is particularly evident just after 300 timesteps in Fig. 2. Figure 3 shows the results obtained when using Receding Horizon Estimation with an estimation horizon of 10. The main difference between the performance of the EKF and the RHE is here that a line search along the solution vector is used in the RHE. Since the nonlinear model is used in the line search, the overcorrection that was evident in the EKF is significantly reduced. On the other hand, a somewhat more 'noisy' state estimate is obtained, leading to more active use of the manipulated variables. Nevertheless, Fig. 3 clearly shows the best performance among the three cases shown here. Note that since the measurement function is the state cubed, the overshoot shown in Fig. 2 would appear much more dramatic if the measurement and not the state were shown.

4. PROPERTIES OF THE PROPOSED METHOD

In this paper, a method is proposed for taking the quality of the state estimate into account in the optimization problem in MPC. This consideration is only of relevance for nonlinear systems; It is well known that the control and estimation problems can be separated for linear systems. The theoretical properties of this type of MPC controller has not been extensively studied. However, some relevant issues are discussed below.

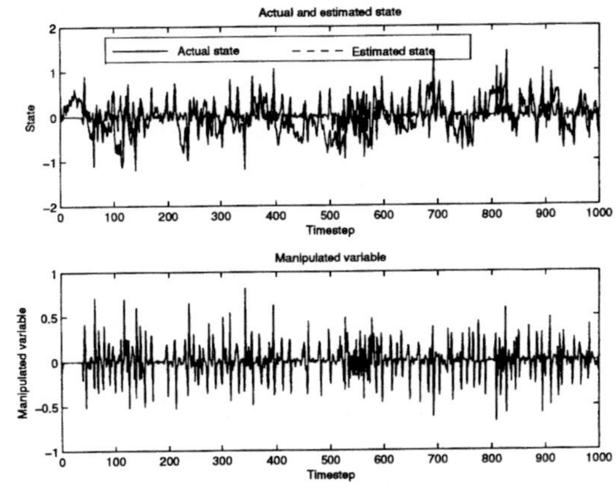

Fig. 3. Results when using the weight $\rho_i = 0.3$ on the EKF state covariance estimate in the MPC, while RHE is used for state estimation.

4.1 *Asymptotic stability*

A formal stability analysis for the type of MPC controller presented here is likely to be hard. Clearly, this type of controller will not result in an *asymptotically* stable closed loop system, if the system is not observable at the reference state x^r (and corresponding input u^r). This is not a shortcoming of the proposed control method, though, it merely means that the system needs to be excited (i.e., moved away from the reference state) in order to update the state estimate. For acceptable control to be achievable, it must be possible to obtain acceptable state estimates in a region not too far from the reference state.

4.2 *Computational issues*

4.2.1. *Boundedness of solutions* Obviously, it is a practical requirement that the solution to the optimization problem should be bounded. For convex QP problems this is automatically ensured by the shape of the objective function itself. For a non-convex problem, this must be ensured by the constraints. Applying maximum and minimum value constraints to the manipulated variables will automatically ensure this. For most problems, such constraints are imposed by the physics of the problem anyway.

4.2.2. *Required computation time* Typical MPC formulations are formulated as LP or convex QP problems. For the method proposed here, it is possible that the problem can loose convexity if the EKF state covariance estimate grows so large as to dominate the cost of control error and manipulated variable moves. Non-convex QP problems are generally 'np-hard', meaning that the required time to guarantee that the optimal solution is found will in general grow exponentially with problem size. Thus, it may take unacceptably long to find the global optimum even for problems of moderate size. However, this does not imply that it will necessarily take long to find an acceptable solution. The constraints used in this MPC formulation are linear, and thus finding a *feasible* starting point is no harder than with traditional MPC formulations. Any gradient-based optimization code can then be used to find a local optimum – what is hard is to ascertain whether such a local optimum is also globally optimal. Even if a solution that is only locally optimal is used in the control, the resulting control performance may well be acceptable.

4.2.3. *The need for line search* If the weight on the EKF covariance matrix is set to zero, the objective function is by construction convex, i.e., it resembles a 'bowl' in shape. Consider a case where the plant is locally weakly unobservable at the reference state, like in the example above. Keeping the (estimated) state at the reference will then necessarily mean that the state covariance grows with time. With a non-zero weight on the covariance matrix, this will first flatten the bottom of the bowl described by the objective function, and then introduce a 'hill' around the reference state. Formulating the optimization problem as a QP problem means that we are locally fitting a quadratic function to approximate the true objective function. Fitting the quadratic function to the top of the 'hill' in the bottom of the 'bowl' can clearly lead to gross errors when the solution is based on the fitted quadratic function only. Performing a line search along the direction specified by the solution vector is the standard way to guard against this type of problems in nonlinear optimization.

5. CONCLUSIONS AND FURTHER WORK

This work presents a novel way of including state estimate quality into the MPC formulation. It represents a significant extension of the work of Yan and Bitmead (2002). A topic for further research is to extend this approach to simultaneous state and parameter estimation, which would a step towards a dual MPC controller, since it has the capability to generate probing signals to better identify states and parameters. This will also introduce back-off from constraints due to uncertain model parameters, while the controller will actively probe the control plant in order to identify the parameters. Note that in this extension of our work, the probing will only be active when parameter estimates are uncertain, and the probing will only excite directions affected by the uncertain parameters. This is in contrast to the Model Predictive Control and Identification (MPCI) formulation proposed in (Shouche *et al.*, 2002), which imposes a persistence of excitation constraint, and hence excites all model directions continuously.

The MPC formulation proposed herein does require heavy on-line computation. Further research is needed to analyze how the resulting optimization problem can be solved efficiently. Emerging results on explicit MPC formulations for nonlinear systems (see e.g. Johansen (2002)) appears to be a promising direction for further research. This approach may also provide more control on the degree of sub-optimality of the implemented controller.

6. REFERENCES

Adetola, V. and M. Guay (2003). Nonlinear output feedback receding horizon control of sampled data systems. In: *Proc. of the American Control Conference*. Denver, Colorado. pp. 4914–4919.

Allgöwer, F., T. A. Badgewell, J. S. Qin, J. B. Rawlings and S. J. Wright (1999). Nonlinear model predictive control and moving horizon estimation - an introductory overview. In: *Advances in Control. Highlights of the ECC'99.* Springer.

Biegler, L. T. (1998). Advances in nonlinear programming concepts for process control. *Journal of Process Control* **8**, 301–311.

Imsland, L., R. Findeisen, E. Bullinger, F. Allgöwer and B.A. Foss (2003). A note on stability, robustness and performance of output feedback nonlinear model predictive control. *Journal of Process Control* **13**, 633–644.

Johansen, T. A. (2002). On multi-parametric nonlinear programming and explicit nonlinear model predictive control. In: *IEEE Conf. Decision and Control.* Las Vegas, NV, USA. pp. 2768–2773.

Qin, S. J. and T. A. Badgwell (2003). A survey of industrial model predictive control technology. *Control Engineering Practice* pp. 733–764.

Rao, C. V. (2000). Moving Horizon Strategies for the Constrained Monitoring and Control of Nonlinear Discrete-Time Systems. PhD thesis. Department of Chemical Engineering, University of Wisconsin - Madison.

Shouche, M. S., H. Genceli and M. Nikolaou (2002). Effect of on-line optimization techniques on model predictive control and identification (MPCI). *Computers & Chemical Engineering* **26**, 1241–1252.

Yan, J. and R. R. Bitmead (2002). Model predictive control and state estimation: A network example. In: *IFAC World Congress.* Barcelona, Spain.

ELSEVIER
IFAC
PUBLICATIONS
www.elsevier.com/locate/ifac

CONTROL AND MONITORING OF SEMICONDUCTOR MANUFACTURING PROCESSES: CHALLENGES AND OPPORTUNITIES

S. Joe Qin, Gregory Cherry, Richard Good, Jin Wang, and Christopher A. Harrison

Department of Chemical Engineering
The University of Texas at Austin
Austin,. TX 78712, USA

Abstract:
The semiconductor industry is going through a technology transition from 200mm to 300mm wafers to improve manufacturing efficiency and reduce manufacturing cost per chip. These technological changes present a unique opportunity to optimally design the process control systems for the next generation fabs. In this paper we first propose a hierarchical fab-wide control framework with the integration of 300mm equipment and metrology tools and highly automated material handling system. Relevant existing run-to-run technology is reviewed and analyzed in the fab-wide control context, process and metrology data monitoring are discussed with an example, and missing components are pointed out as opportunities for future research and development. Concluding remarks are given at the end of the paper.

Keywords:
semiconductor manufacturing, fab-wide control, electrical parameter control, run to run control, fault detection and classification, metrology data monitoring

1. INTRODUCTION

The semiconductor industry is going through a technology transition from 200mm to 300mm wafers to improve manufacturing efficiency and reduce manufacturing cost per chip. Along with this transition is the doubling of capital expenditure in a 300mm fab versus a 200mm fab. Other technological changes include:

- Single wafer processing capability;
- Fully automated material handling systems (AMHS) with inter-bay and intra-bay transportation;
- Integrated metrology that allows for timely quality control; and
- Highly automated process control and fault diagnosis

Owing to the capital intensity of the new generation fabs, it is critical to maintain highly efficient operations, minimize downtime of equipment, and optimize the yield of high quality products. The International Technology Roadmap (Semiconductor Industry Association, 2003) clearly identifies that factory information and control systems are a critical enabling technology to reduce cycle-time and improve yield. These technological changes present a unique opportunity to optimally design the process control systems for the new generation fabs.

A persistent challenge in semiconductor manufacturing control is the lack of critical in-situ sensors to provide real time information of the wafer

[1] Corresponding author. e-mail: qin@che.utexas.edu.

status for feedback control and optimization. Fortunately, recent advance in metrology technology provides an opportunity for improving the timeliness and usefulness of the measurement data. Typically a modern fab has the following measurement data available for analysis and control:

(1) Real time trace data at the tool level which reflect the equipment health condition and provide feedback for real time control;
(2) Integrated metrology and in-line metrology data available for geometric dimensions after a major processing step, with small to moderate metrology delay;
(3) Sample and final electrical test (E-test) data available for electrical properties with medium or long time delay, but they have the most important information about the manufacturing effectiveness.

Advanced control and optimization methodology should maximize the use of all the information in an integrated hierarchy for highly efficient manufacturing and tight product quality control.

Monitoring and control of semiconductor manufacturing processes have been investigated at a number of U.S. universities and industrial research laboratories. Representative work includes U.C. Berkeley (Lee and Spanos, 1995; May *et al.*, 1991) on statistical modeling and control of plasma etchers, Michigan on real-time and run to run multivariable control (Hamby *et al.*, 1998), as well as MIT on different sensor and control technologies (Boning *et al.*, 1995; Boning *et al.*, 1996). Due to lack of in-situ sensors much of the control work is developed from the run-to-run (R2R) control strategy (Butler and Stefani, 1994; Sachs *et al.*, 1995). Research groups at University of Maryland contributed in the area of run to run control (Adivikolanu and Zafiriou, 2000; Baras and Patel, 1996; Zafiriou *et al.*, 1995). Sematech, a consortium of leading semiconductor manufacturers, posted several benchmark problems on plasma equipment fault detection and diagnosis (Bakshi, 1997). Adaptive and nonlinear control for R2R operations is proposed by (Del Castillo and Yeh, 1998). Model predictive control is applied to R2R control as well which has additional capability in handling constraints explicitly (Edgar *et al.*, 1999). At UT-Austin we have developed (i) stability conditions and tuning guidelines for multivariable EWMA and double EWMA control with metrology delays (Good and Qin, 2002; Good and Qin, 2003), (ii) multivariate statistical monitoring of RTA and etchers (Yue *et al.*, 2000; Yue *et al.*, 2001), and (iii) multivariate statistical control of CD metrology data from lithography (Cherry *et al.*, 2002). Other new development and applications of control and fault detection are reported at recent SPIE conferences and AEC/APC Symposia organized by Sematech and summarized in Del Castillo and Hurwitz (Del Castillo and Hurwitz, 1997) and Moyne et al. (Moyne *et al.*, 2001). Manufacturing companies like AMD, Intel, Motorola, and TI and vendors like Applied Materials, Brooks-PRI Automation, and Yield Dynamics are leaders in deploying APC technologies at the manufacturing lines.

In this paper we draw the analogy between semiconductor manufacturing fabs and chemical plants and propose a hierarchical optimization and control system for semiconductor fab control. A schematic diagram is shown in Figure 1 for this analogy. The equipment level control involves automatic feedback control of tool parameters and small scale run-to-run control using integrated metrology. The next level run-to-run control involves sharing information from multiple steps to achieve feedforward and predictive control. Since there are multiple tools in each module and each of them are different in terms of manufacturing effectiveness, a module level optimization is needed to make sure that each wafer is processed by the best combination of tools (or threads). The top level of the hierarchy is the fab-wide control which is the highest level optimization to achieve desired electrical properties by recalculating the optimal geometric targets and dosage for the lower level.

The organization of the paper is given as follows. We first propose a hierarchical fab-wide control strategy with the integration of 300mm equipment and metrology tools and highly automated material handling system. Relevant existing run-to-run technology is reviewed and analyzed in the fab-wide control context, process and metrology data monitoring are discussed with an example, and missing components are pointed out as opportunities for future research and development. Concluding remarks are given at the end of the paper.

2. A FRAMEWORK FOR FAB-WIDE CONTROL

Almost all existing development is on R2R control which adjusts recipes of a step based on metrology data at the equipment level. These are known as islands of control as illustrated in the lower part of Figure 1. None of the existing control strategies examine the coordination of multiple manufacturing steps to improve the overall product quality in terms of electrical parameters. The R2R controllers compensate for equipment drifts through metrology feedback, but they cannot compensate for metrology drifts and uncertainties. The direct control of electrical parameters proposed here can compensate for metrology drifts and systematic errors in the geometric measurements. It is believed that the control and optimization of electrical parameters represent the next generation of semiconductor manufacturing control system as it directly controls the electrical properties to a desired product profile by manipulating the operation requirements for lower level R2R controllers. The electrical parametric control and optimization will maximize the yield of high-grade products or reduce operational cost when a demand profile is specified by market orders.

The fab-wide control framework in Figure 1 provides optimization and coordination from step to step to reduce variability, reworks, and scraps, thus improving the overall equipment effectiveness and reducing manufacturing cost. This framework was first presented by Qin and Sonderman (Qin and Sonderman, 2002) after having deployed many R2R controllers at AMD and analyzed the need for a higher level control. The equipment level control involves automatic feedback control of tool parameters. The next level is run-to-run control using integrated or in-line metrology to achieve a specified target. The third level is a module level control that shares information from multiple steps to perform feedforward and feedback control and tool performance matching. The top level of the hierarchy, which is one of the focuses of the proposal, is electrical parametric control (EPC) to achieve desired electrical properties by recalculating the optimal targets for the lower levels. Equipment drifts, metrology drifts, and material variations are compensated by feedback at the EPC level, leading to improved process and metrology availability and reduced use of calibration and test wafers.

This multiple level control framework resembles the hierarchical control framework that has been successful in the refinery industry (Qin and Badgwell, 1997), but significant differences exist: (i) the lowest level control is mostly batch operations; (ii) the middle level R2R control has virtually no R2R process dynamics except for disturbance dynamics; and (iii) the top level EPC is a multi-step operation control that aims to compensate for errors made in prior steps, regardless of the nature of the errors as long as step-wise metrology measurement is available. This makes it different from model predictive control (MPC) of batch processes with shrinking horizons.

The E-test data are used to update the device model parameters based on mismatch between the E-test data and the model. After parameter estimation is performed, the estimated parameters are sent to a fab-wide optimizer, which distributes targets to lower-level controllers that regulate steps within the manufacturing process. The model updated with the new set of model parameters is used for EPC control.

The multi-step EPC control minimizes an objective function that penalizes the difference between the desired electrical properties and the updated model output subject to constraints. At the beginning before the first step is processed, all geometric input parameters are used to minimize the objective function. After the first step is processed for a wafer or a lot, metrology data for the first step is available as u_1, which will be different from the target $\hat{u}_1$ calculated before implementing the first step. To compensate for the manufacturing error in the first step the objective function is re-optimized over the remaining inputs with the first input fixed at u_1. Then the second processing step is implemented with the newly optimized target. This procedure is repeated until all processing steps are implemented for the processed wafer or lot. The objective function at each step is given as follows:

$$J = \|y_{EPC} - f(\hat{u}, u)\|^2 + \lambda \|\hat{u} - u_{nom}\|^2 \quad (1)$$

subject to

$$u_{min} \leq \hat{u} \leq u_{max}$$
$$y_{min} \leq f(\hat{u}, u) \leq y_{max}$$

where u contains implemented inputs at the step, $\hat{u}$ contains the remaining inputs that can still be optimized, and u_{nom} is some nominal values for u that could be the values from previous runs. λ is a weight parameter that balances between the output target error and the input deviation from the nominal values. In the case that the output target is exactly feasible within the constraints, the optimization gives a solution that has minimal change from the nominal values or the values from previous runs. The feature is desirable to generate stable targets for the lower level R2R controllers. The constraints are derived from product specifications and requirements.

The vector y_{EPC} can have multiple entries containing multiple electrical parameters to be controlled simultaneously. Some of the electrical parameters have a target value based on the product specifications; others have only upper and lower constraints to make sure that these parameters meet the specifications. The EPC objective can be adapted to represent several modes of operations. For logic products one can choose to optimize the oscillation frequency subject to constraints that all major electrical parameters meet the specifications. The target y_{EPC} in this case is set at the highest desirable value for the product. In another mode of operation the target y_{EPC} is set based on the demand profile. In this case the target y_{EPC} can be easily achievable; the optimization generates a target that has smallest deviation from the nominal values, thus minimizing variability at lower levels due to target adjustment.

The optimization algorithm can be fulfilled using a nonlinear programming solution when the device model is nonlinear. There is no perceived difficulty in the implementation of such an algorithm. A case study of a flash memory EPC is presented in (Harrison *et al.*, 2003).

3. RUN TO RUN CONTROL ALGORITHMS

3.1 *Run to Run Control*

In recent years, run-to-run (R2R) control technology has received tremendous interest in semiconductor manufacturing. Moyne and Hurwitz (2001) (Moyne *et al.*, 2001) define the run-to-run control as "a form of discrete process and machine control in which the product recipe with respect to a particular process is modified *ex situ*, i.e., between machine 'runs', so as to minimize process drift, shift, and variability". In order to modify the recipe to address the process drift, shift and other variability, the current tool and wafer states need to be estimated. One class of widely used run-to-run controllers is based on the exponentially weighted moving average (EWMA) statistics to estimate process disturbances.

The EWMA has been used for a long time for quality monitoring purposes (Box and Jenkins, 1963). Its use as a basis for run-to-run control is relatively recent (Sachs *et al.*, 1991). For a time series of measurement $\{x[n], x[n-1], \cdots\}$, where n denotes the run number, the EWMA is given in the following recursive formula:

$$\hat{x}[n] = \omega\hat{x}[n-1] + (1-\omega)x[n] \qquad (2)$$

where $\hat{x}$ is the EWMA estimate of x, ω is the EWMA weight, and $x[n]$ is the measurement of the process disturbance or parameter to be estimated. For a linear process model

$$y[n] = bu[n] + x[n] \qquad (3)$$

the disturbance

$$x[n] = y[n] - bu[n]$$

After the EWMA filter has estimated the process offset, a control law is used to determine the control input (or recipe) for the following run. In the unconstrained SISO case, the recipe is determined through simple model inversion,

$$u[n+1] = \frac{T - \hat{x}[n+1]}{b} \qquad (4)$$

where T is the process target.

The MIMO control law is somewhat more complicated as b may be non-square so that an inverse is not attainable. The MIMO control law therefore can take several different forms depending on the objective function of the optimization problem. A few of the commonly seen unconstrained MIMO control laws are listed below.

(1) Minimize the sum of the manipulated variables squared subject to the model hitting the process target:

$$\min_{u[n+1]} J = u[n+1]^T u[n+1]$$
$$s.t. \quad T = bu[n+1] + \hat{x}[n+1]$$

$$u[n+1] = b^T(bb^T)^{-1}(T - \hat{x}[n+1]) \qquad (5)$$

(2) Minimize the sum of the change in the manipulated variables squared subject to the the model hitting the process target (Tseng *et al.*, 2002):

$$\min_{u[n+1]} J = \delta u[n+1]^T \delta u[n+1]$$
$$s.t. \quad T = bu[n+1] + \hat{x}[n+1]$$

$$u[n+1] = b^T(bb^T)^{-1}(T - \hat{x}[n+1]) + (b^T(bb^T)^{-1}b - I)u[n] \qquad (6)$$

(3) Minimize the sum of squares deviation from target (Del Castillo and Rajagopal, 2002):

$$\min_{u[n+1]} J = \hat{y}[n+1]^T \hat{y}[n+1]$$
$$s.t. \quad \hat{y}[n+1] = bu[n+1] + \hat{x}[n+1]$$

$$u[n+1] = (b^T b)^{-1} b^T (T - \hat{x}[n+1]) \qquad (7)$$

(4) Model predictive control formulation:

$$\min_{u[n+1]} J = (T - \hat{y}[n+1])^T Q (T - \hat{y}[n+1]) + u[n+1]^T R u[n+1] + \Delta u[n+1]^T S \Delta u[n+1]$$
$$s.t. \quad \hat{y}[n+1] = bu[n+1] + \hat{x}[n+1]$$

$$u[n+1] = (b^T Q b + R + S)^{-1}(Su[n] + b^T Q (T - \hat{x}[n+1])) \qquad (8)$$

The first two control laws are used when the number of outputs exceeds the number of inputs. In this case there are an infinite number of control inputs that will bring the process to the expected target, T. An objective function is defined to establish a criteria to choose the 'best' controller input. The first objective function is to minimize the sum of squared controller input. The second objective function is to minimize the sum of squared *change* in the controller input. The third control law is used when the number of outputs exceeds the number of inputs. In this case there exists no controller inputs that will bring the process to the expected target. The objective function in this case is to minimize the sum of squared deviation of the expected output from the target. The final control law is a more general controller as the objective is to find the optimal balance between missing the process target, the absolute controller input, and the change in the controller input from the previous run. The control law in (8) can be made to return equivalent results as the other three control laws by selecting the appropriate values of Q, R, and S.

Ingolfson and Sachs (Ingolfsson and Sachs, 1993) show that the EWMA controller is a discrete integral controller, which explains why it is able to compensate for process shifts and offsets. Butler and Stefani (Butler and Stefani, 1994) noticed that for processes with severe drifts, the EWMA controller is insufficient even when large weights are used. This problem becomes more severe when there is a measurement delay, which is almost inevitable in semiconductor manufacturing.

In order to control drifting processes, a predictor-corrector controller (PCC) (Butler and Stefani, 1994) and a double EWMA (dEWMA) controller (Chen and Guo, 2001) have been developed. The PCC algorithm uses two parameters, ω_1 and ω_2 to weight noise and drift respectively. The double-EWMA is very similar to PCC, as can be seen from the following equations:

$$\begin{aligned} a[n] &= \omega_1 x[n] + (1-\omega_1)(a[n-1]+p[n-1]) \\ p[n] &= \omega_2(x[n]-a[n-1]) + (1-\omega_2)p[n-1] \\ \hat{x}[n] &= a[n]+p[n] \end{aligned}$$

The only difference between double-EWMA and PCC is the estimate of intercept term a_i. Chen and Guo (Chen and Guo, 2001) show that both PCC and double-EWMA controller are in effect Integral-double-Integral (I-II) controllers, which are able to control drifting processes. However, since offset is often coupled with the noise of the process, the second filter may add variability to the control action in the presence of significant noise (Bode, 2001). In addition, tuning the second filter is not as intuitive as a single EWMA filter. Therefore, PCC or double-EWMA controller is not as widely used as EWMA controllers.

Like all feedback controllers, run-to-run control is subject to closed loop instability. The first study on the conditions for stability of the single input-single output (SISO) EWMA controller was published shortly after Ingolfsson and Sachs's first work on run-to-run control (Ingolfsson and Sachs, 1993). They noted that a process will become unstable when the input-output relationship between the tool recipes and quality measurements are not accurately estimated. Their work showed the allowable range of model mismatch that a process can have and still maintain asymptotic stability. This work was later extended by Tseng *et al* to show the stable region of a particular formulation of the multiple input-multiple output (MIMO) EWMA controller (Tseng *et al.*, 2002) . In addition, the effect of metrology delay on the stability of the SISO and multiple input-single output (MISO) EWMA controller was studied by Adivikolanu and Zafiriou (Adivikolanu and Zafiriou, 2000). They utilized an internal model control approach to derive the stability region of an EWMA controller with a delay of one run. A numerical method was then introduced for determining the stability region of the EWMA controller for processes with longer metrology delays. Good and Qin (Good and Qin, 2002; Good and Qin, 2003) extend the work of Tseng *et al* and Ingolffson and Sachs by deriving the stability conditions of the MIMO EWMA controller with metrology delay.

3.2 *Simplified dEWMA with RLS*

To simplify the double EWMA control Wang et al. (Wang *et al.*, 2004) consider the double EWMA control in the recursive least squares (RLS) framework and reduce the tuning parameters to a forgetting factor. Consider the process model as

$$y[n] = g(u[n]) + x[n] \tag{9}$$

where g is the nonlinear input-output model and the disturbance x has the following polynomial form,

$$\begin{aligned} x[n+i] &= \theta_0 + \theta_1 i + \theta_2 \frac{i^2}{2!} + \cdots + \theta_k \frac{i^k}{k!} + \epsilon[n+i] \\ &= \sum_{j=0}^{k} \theta_j \frac{i^j}{j!} + \epsilon[n+i] \end{aligned} \tag{10}$$

where $\theta = \begin{bmatrix} \theta_0 & \theta_1 & \cdots & \theta_k \end{bmatrix}^T$ are the parameters of the model to be determined, i is the time index, and ϵ is the a sequence of uncorrelated errors with variance σ^2. k denotes the model order, which can be determined by cross-validation. The model adequacy can be checked by calculating the sample autocorrelations of the residuals (Abraham and Ledolter, 1983).

The model (10) assumes that the model parameters are constant over all time periods. However, in many instances, the assumption of a time invariant model is restrictive and a locally constant model would be more reasonable. By applying the forgetting factor in the least squares criterion, more weight is given to more recent observations and past observations are discounted. Letting $\varphi[n+i] = \begin{bmatrix} 1 & i & \cdots & \frac{i^k}{k!} \end{bmatrix}^T$, the model parameter estimates are determined by minimizing the following loss function,

$$V(\theta, n) = \frac{1}{2}\sum_{i=1}^{n} \lambda^{n-i}(x[i] - \theta^T \varphi[i])^2 \tag{11}$$

where λ $(0 < \lambda \leq 1)$ is the forgetting factor that gives more weight to recent prediction error $(x[i]-$

$\theta^T\varphi[i])$. The recursive algorithms to estimate the model parameters is (Åström and Wittenmark, 1995; Ljung, 1999):

$$\hat{\theta}[n] = \hat{\theta}[n-1] + K[n](x[n] - \varphi^T[n]\hat{\theta}[n-1])$$
$$K[n] = P[n-1]\varphi[n](\lambda + \varphi^T[n]P[n-1]\varphi[n])^{-1}$$
$$P[n] = (I_m - K[n]\varphi^T[n])P[n-1]/\lambda$$

where $K[n]$ is an $(k+1) \times 1$ vector of gains and $P[n]$ is an $(k+1) \times (k+1)$ matrix proportional to the covariance matrix of the estimated parameter. In the above recursive algorithm, the initial estimates $\hat{\theta}[0]$ and $P[0]$ can be obtained from a priori knowledge. The better the initial estimates, the smaller the effect of the transient behavior (Del Castillo and Hurwitz, 1997).

For $k = 0$, model (10) becomes the constant mean model,

$$x[n+i] = \theta_0 + \epsilon[n+i] \quad (12)$$

In this case the RLS gives a form that is very similar to EWMA except that $K[n]$ depends on the number data points n. Wang et al. (2004) show that the RLS converges exactly to EWMA with $\omega = 1 - \lambda$ when n approaches infinity. For finite n, $K[n]$ varies with n which is similar to the unsteady state Kalman filter.

For $k = 1$, model (10) becomes a linear trend model,

$$x[n+i] = \theta_0 + \theta_1 i + \epsilon[n+i] \quad (13)$$

In this case the RLS gives a form that is very similar to double EWMA except that $K[n]$ depends on the number data points n. Wang et al. (2004) show that the RLS converges exactly to double EWMA when n approaches infinity. The equivalence is achieved with

$$\omega_1 = 1 - \lambda^2 \quad (14)$$
$$\omega_2 = (1-\lambda)^2 \quad (15)$$

This result relates the double EWMA tuning parameters to the forgetting factor in RLS. The number of tuning parameters reduces from two to one which has the physical meaning of a forgetting factor.

3.3 *Kalman Filter Implementations*

Realizing (9) and 10 in a state space form,

$$x[n+1] = Ax[n] + w[n] \quad (16)$$
$$y[n] = Cx[n] + g(u[n]) + v[n] \quad (17)$$

where w and v are process and measurement noise, the estimated disturbance is

$$\hat{x}[n+1] = A\hat{x}[n] + K[n](y[n] - g(u[n]) - C\hat{x}[n])$$
$$= (A - K[n]C)\hat{x}[n] + K[n](y[n] - g(u[n]))$$

For a constant disturbance model, $A = 1, C = 1$. In this case the Kalman filter is equivalent to EWMA with $K[n] = \omega$ as n approaches infinity. For a linear trend model $A = \begin{bmatrix} 1 & 1 \\ 0 & 1 \end{bmatrix}$ and $C = \begin{bmatrix} 1 & 1 \end{bmatrix}$. In this case the Kalman filter is equivalent to the double EWMA with $K[n] = \begin{bmatrix} \omega_1 & \omega_2 \end{bmatrix}^T$ as n approaches infinity. In the Kalman filter interpretation both tuning constants are necessary and are determined by the process and measurement noise.

3.4 *Time Control as R2R Control*

One of the most effective manipulated variables in R2R control is the processing time within a processing step such as etch time, exposure time, and planarization time. The controlled variables in the case are typically the extent to which the process develops under the processing time, such as depth of etch and critical dimensions. Given the processing rate as $r(t,n)$ for Run n, the process model can be described as

$$y[n] = \int_0^{T_f} r(t,n)dt \quad (18)$$

The processing rate $r(t,n)$ typically varies with run and time in a run. In the rare case that the rate can be measured in real time, the control problem is trivial. If the rate is not measurable in real time but some kind of end pointing mechanism or indirect measurement is available, the monitoring of the indirect measurements are effective, such as optical emission spectra in dielectric etch (Yue *et al.*, 2001). In the most typical case no real time measurement is available. In this case end of the step metrology measurement can be used for R2R control, with or without metrology delay (Wang *et al.*, 2004). Only the average rate can be estimated from previous processing steps and (18) becomes

$$y[n] = \bar{r}[n]T_f[n]$$

This multiplicative model does not fit into the typical linear state space model presented earlier, but it can be converted to the linear state space model with process and measurement noise by simply taking the logarithm. Therefore, all the control algorithms presented earlier in this are applicable to time control.

4. FAULT DETECTION AND MONITORING

Processing tool data such as temperatures, pressures, and gas flow rates will be used to monitor recipes applied to single wafers or batches of wafers. Some typical processing operations include plasma etching, thin film deposition, rapid thermal annealing, ion implantation, and chemical mechanical planarization. At most processing steps, sensors collect data for each wafer or batch of wafers that are processed on the tool. This data can be in the form of real-time traces for a recipe, summary statistics available at the end of each run, or data from more advanced sensor platforms such as optical emission spectroscopy.

While the batch nature of semiconductor manufacturing provides plenty of opportunities for applying multi-way process monitoring (Nomikos and MacGregor, 1995), many forms of semiconductor metrology data are naturally organized in three dimensions. One such case is CD metrology, where the three dimensions are wafer, site, and parameter. Batch data are also commonly available from processing tools, which exhibits the dimensions of batch, time, and parameter (Figure 2).

Multi-way PCA has been successfully applied for batch process monitoring across many different industries. In the field of semiconductor manufacturing, Yue *et al.* (Yue *et al.*, 2000) demonstrated the concept of unfolding data by applying multi-way PCA to optical emission spectra for plasma etchers. For metrology and processing tool monitoring, the data can be unfolded by site or time (every row represents one site on a wafer or time instant in a batch) or by wafer (every row represents one wafer). In this work, wafer level fault detection and identification is desired, so the latter has been chosen as the more appropriate unfolding method (Figure 4). However, as will be discussed later, the advantages of analyzing data by site or time can be realized simply by implementing the multiblock approach.

4.1 *Metrology Monitoring*

While processing operations build the structures, metrology operations characterize them. Some examples of metrology measurements include development inspection critical dimension (DICD), final inspection critical dimension (FICD), and film thickness. Metrology measurements are normally taken at several locations on the semiconductor wafers, oftentimes for multiple features at the same site (i.e., top and bottom DICD). Fault detection and identification applied to site-level metrology data is intended to validate whether the structures built on the semiconductor wafers hit their targets and do so uniformly across the wafer surfaces.

As an example, we will use PCA to perform fault detection and identification on DICD data from Advanced Micro Devices' Fab25 in Austin, Texas. The DICD is the width of the pattern in the photoresist after the photoresist has been developed and before the next trim step. As shown in Figure 3, isotropy in development results in a small difference between the top of the photoresist and the bottom. The data set consists of 700 wafers, where both the top and bottom of the resist are measured at nine sites on each wafer.

An initial PCA model is built using the first 100 wafers and the VRE method (Qin and Dunia, 2000) selects 10 components as the optimal number for reconstruction. While this may seem high, keep in mind that the goal of the VRE method is the reconstruction of variables. After the PCA model is formulated, it is applied to the remaining 600 wafers, with the fault detection indices, SPE, T^2, and the combined index ϕ (Yue and Qin, 2001) provided in Figure 5. Note that the statistics have been scaled against their respective confidence limits and logarithms have been taken. The result is that the value of 1 represents the 99% threshold for each of SPE_r, T_r^2, and φ_r.

The indices charted in Figure 5 demonstrate that there is some behavior within those 600 wafers that is not consistent with the initial set of 100 wafers. In order to identify the cause of the excursion, multiblock contributions to the combined index are calculated. For the case of site-level metrology, the most logical blocking of the 18 variables are by parameter (top and bottom) and by site (sites 1 through 9). These multiblock indices are tracked by wafer in Figures 6 and 7.

Inspection of Figure 6 allows one to quickly reach the conclusion that the general trend is caused by drifts in both the bottom and top DICDs as time goes on. On the other hand, the larger outliers are more pronounced in either the top or bottom, but not both. Because the two CDs are highly correlated, the presence of such extreme outliers is likely to be caused by inaccurate values provided by the CD metrology tool, rather than problems with the physical structures on the wafers. While sensor faults are of interest, the more important issue is the process drift, which is first experienced in the bottom CD around wafer 200 and then propagates to the top 100 wafers later, where the signal becomes even more pronounced.

While Figure 6 grouped all 9 sites together for each of the two parameters, the Figure 7 contributions take into account both parameters at each individual site. These nine plots make it easy to identify problems based on wafer location. For the data set provided, the excursion appears strongest on sites 2, 3, and 4, while it is hardly noticeable at sites 6, 8, and 9. With knowledge of each site's location on the wafer it would be possible to use these plots to troubleshoot possible tilt or focus issues with the masking tool.

While tracking the block contributions is good practice for identifying excursions that influence a large number of wafers, one must also consider the case where a problem is identified on a single wafer, and the cause needs to be identified. To demonstrate this functionality, contribution plots have been generated for wafers 395 and 450 (indicated with arrows in Figure 5.

The contribution plot for wafer 395 is provided in Figure 8. It is easy to see that measurements 4 (*Bottom−Site* 4), 12 (*Top−Site* 3), and 13 (*Top−Site* 4) are suspect. Problems are also indicated in both *Bottom* and *Top* parameters, along with *Site* 4 as the only extreme site contribution. A logical interpretation of these plots are that there was a significant issue with *Site* 4 and further investigation into that location on the wafer may be warranted to explore any issues that may affect product yield or performance. Although the top dimension on *Site* 3 was also singled out, the overall site contribution was normal when both the top and bottom were considered collectively.

The contribution plot for wafer 450 is provided in Figure 9. As previously shown in the φ_r plot (Figure 5), this was a less extreme fault than 395, but it did lie within the set of wafers experiencing some form of process drift. Although none of the contributions are as extreme as those for wafer 395, the plots suggest that four of the nine sites were faulty for both the bottom and top CDs. This fault signature appears to be quite typical of many wafers during the same processing time span, as corroborated in the variable and site contribution plots tracked for the entire sequence. For the case of wafer 450, the contribution plot showed that all faults indicated in the measurement contribution also propagated themselves to their corresponding multiblock contributions for both parameters and sites.

5. CHALLENGES AND OPPORTUNITIES

5.1 *Modeling of Electrical Parameters*

To implement the fab-wide control it is important to develop a physics-based device model that maps from geometric dimensions to electrical parameters such as oscillation frequency, erase time of flash memories, and sheet resistance. This model is different from process models used in R2R controllers that describe the relation between process operation conditions to geometric parameters such as critical dimensions, depth, or thickness. Since optimization is involved in EPC, which could use the model in fairly wide operation regions, nonlinear physics based models are chosen to accomplish this task. The models suitable for EPC must be implementable in real-time, making it different from simulation and design models. To illustrate the modeling task we use a flash memory cell to demonstrate the steps needed for these tasks.

5.2 *Model Update with Long Delay*

The developed models can work well for one operating condition. As the process metrology and material change over time it is important to adapt the model from real data using parameter estimation. In order to estimate a set of parameters, a nonlinear least-squares method is employed for the nonlinear physical model. The least squares objective is a dual of the EPC objective that minimizes the difference between the E-test data and the model output of finished wafers or lots subject to possible constraints. An effective parameter for model update is the intercept term or constant disturbance model. A challenging task in model updating is the long measurement delay in E-test data. The updating mechanism should respond only to long term persistent changes, not short-lived temporary errors. The integrated learning control and real time feedback control framework by (Chin et al., 2003) is a possible solution. The updated model is then used by the EPC controller to generate input targets for the next incoming lot.

5.3 *Integration of FDC and R2R*

As illustrated in Figure 1 each step in the fab wide control framework has a R2R controller and an FDC module for the step. The FDC are designed to monitor deviation from normal situations based on historical data analysis. One of the approaches is the multi-way PCA approach to equipment monitoring. The co-existence of the FDC and R2R control presents a challenge for their integration in two ways. First, FDC methods usually assumes repeatable batch profiles with similar or equal batch lengths. On the other hand, the R2R module is designed to adjust the recipe, e.g., process time to minimize variability due to normal process drifts. The FDC module, if not properly designed, could consider normal R2R adjustments as deviation from normal situations and signal a false alarm. Another challenge is the impact of R2R feedback on the FDC module. Because of tool control feedback the root cause of the fault could be transferred from one variable to another due to the existence of feedback. The use of feedback invariant subspace for fault diagnosis by (McNabb and Qin, 2004) is a possible solution.

5.4 *Integrated Metrology for Control*

The transition from 200mm to 300mm technology and beyond makes it possible to replace inline metrology with integrated metrology into the tools. The integrated metrology reduces the time delay in the measurement and provides the possibility of wafer to wafer (W2W) control which can reduce the variability further. Another possibility is within wafer (WiW) control that allows one to control from die to die (Sonderman and Bode, 2004).

6. CONCLUDING REMARKS

The semiconductor industry is becoming one of the most capital-intensive industries with a high ratio of capital investment to revenue. On the other hand, the optimization and control of manufacturing operations have received significant attentions only recently and shown to be a necessary competitive advantage. A well designed fab-wide control framework provides improved competitiveness of the semiconductor manufacturers as they transition to 300mm technology and 450mm technology in the foreseeable future. The automated material handling systems and automated R2R control capabilities provide the necessary foundation for implementing fab-wide control and fault detection in all levels of the hierarchy. It is envisioned by the leading manufacturers that most of the routine operations will move from clean rooms to a centralized control room in the future. This transition provides exciting challenges and opportunities to process control researchers and engineers to develop a new standard for this vigorously growing industry.

ACKNOWLEDGMENTS

Financial support from National Science Foundation under CTS-9985074 and a Faculty Research Assignment grant from University of Texas is gratefully acknowledged.

7. REFERENCES

Abraham, B. and J. Ledolter (1983). *Statistical Methods for Forecasting.* John Wiley and Sons.

Adivikolanu, S. and E. Zafiriou (2000). Extensions and performance/robustness tradeoffs of the EWMA run-to-run controller by using the internal model control structure.. *IEEE Transactions on Electronics Packaging Manufacturing* **23**, 56–68.

Åström, Karl J. and Björn Wittenmark (1995). *Adaptive Control.* Prentice-Hall. Addison-Wesley Publishing Company, Inc.

Bakshi, V. (1997). Benchmarking of commercial software for fault detection and classification (fdc) of plasma etchers for semiconductor manufacturing equipment. In: *Proceedings of the American Control Conference.* Albuquerque, New Mexico.

Baras, J.S. and N.S. Patel (1996). Designing response surface model-based run-by-run controllers: A worst case approach. *IEEE Transactions on Components, Packaging, and Manufacturing Technology - part C* **19**, 98–104.

Bode, C.A. (2001). Run-to-Run Control of Overlay and Linewidth in Semiconductor Manufacturing. PhD thesis. The University of Texas at Austin.

Boning, D., W. Moyne and T. Smith (1995). Run by run control of chemical-mechanical polishing. In: *1995 IEEE/CPMT International Electronics Manufacturing Technology Symposium.* pp. 81–87.

Boning, D.S., W.P. Moyne, T.H. Smith, J. Moyne, R. Telfeyan, A. Hurwitz, S. Shellman and J. Taylor (1996). Run by run control of chemical-mechanical polishing. *IEEE Transactions on Semiconductor Manufacturing* **9**, 307–314.

Box, G.E.P. and G.M. Jenkins (1963). Further contributions to adaptive quality control: Simultaneous estimation of dynamics: Nonzero costs. *Bulletin of the International Statistical Institute* **34**, 943–974.

Butler, S.W. and J.A. Stefani (1994). Supervisory run-to-run control of a polysilicon gate etch using in situ ellipsometry. *IEEE Transactions on Semiconductor Manufacturing* **7**, 193–201.

Chen, A. and R.S. Guo (2001). Age-based double EWMA controller and its application to CMP processes. *IEEE Transactions on Semiconductor Manufacturing* **14**, 11–19.

Cherry, G., R. Good, and S.J. Qin (2002). Semiconductor process monitoring and fault detection with recursive multiway pca based on a combined index. In: *AEC/APC Symposium XIV.* Salt Lake City, Utah.

Del Castillo, E. and A. Hurwitz (1997). Run-to-run process control: Literature review and extensions. *Journal of Quality Technology* **29**, 184–196.

Del Castillo, E. and J.Y. Yeh (1998). An adaptive run-to-run optimizing controller for linear and nonlinear semiconductor processes. *IEEE Transactions on Semiconductor Manufacturing* **11**, 285–295.

Del Castillo, E. and R. Rajagopal (2002). A multivariate double EWMA process adjustment scheme for drifting processes. *IIE Transactions* **34**, 1055–1068.

Edgar, T.F., W.J. Campbell and C. Bode (1999). Model based control in microelectronics manufacturing. In: *Proceedings of the Conference on Decision and Control.* Vol. 38. pp. 4185–4191.

Good, R. and S.J. Qin (2002). Stability analysis of double EWMA run-to-run control with metrology delay. In: *Proceedings of the American Control Conference.* Anchorage, AK. pp. 2156–2161.

Good, R. and S.J. Qin (2003). On the stability of MIMO EWMA run-to-run controllers with metrology delay. *IEEE Transactions on Semiconductor Manufacturing.* submitted.

Hamby, E.S., P.T. Kabamba and P.P. Khargonekar (1998). A probabilistic approach to run-to-run control. *IEEE Transactions on Semiconductor Manufacturing* **11(4)**, 654–669.

Harrison, Christopher A., Richard Good, Daniel Kadosh and S. Joe Qin (2003). Multi-step supervisory control of flash memory device production via a simple first-principles model. In: *AEC/APC Symposium XV.* Colorado Spring, Denver.

Ingolfsson, A. and E. Sachs (1993). Stability and sensitivity of an EWMA controller. *Journal of Quality Technology* **25 (4)**, 271–287.

Lee, S.F. and C.J. Spanos (1995). Prediction of wafer state after plasma processing using real-time tool data. *IEEE Transactions on Semiconductor Manufacturing.*

Ljung, L. (1999). *System Identification: Theory for the User.* Prentice-Hall, Inc.. Englewood Cliffs, New Jersey.

May, G.S., J. Huang and C.J. Spanos (1991). Statistical experimental design in plasma etch modeling. *IEEE Transactions on Semiconductor Manufacturing* **4**, 83–98.

McNabb, C. A. and S. Joe Qin (2004). Fault diagnosis in the control invariant subspace of closed-loop systems. *Ind. Eng. Chem. Res.* to be submitted.

Moyne, J., E. del Castillo and A. M. Hurwitz (2001). *Run-to-Run Control in Semiconductor Manufacturing.* CRC Press.

Nomikos, P. and J.F. MacGregor (1995). Multivariate SPC charts for monitoring batch processes. *Technometrics* **37**(1), 41–59.

Qin, S. J. and R. Dunia (2000). Determining the number of principal components for best reconstruction. *J. Proc. Cont.* **10**, 245–250.

Qin, S. J. and T. Sonderman (2002). From chemical process control to semiconductor manufacturing control. In: *Keynote at the AEC/APC Symposium XIV.* Salt Lake City, Utah.

Qin, S. Joe and T.A. Badgwell (1997). An overview of industrial model predictive control technology. In: *Chemical Process Control - V* (J. Kantor, C. Garcia and B. Carnahan, Eds.). Fifth International Conference on Chemical Process Control. CACHE and AIChE. pp. 232–256.

Sachs, E., A. Hu and A. Ingolfsson (1991). Modeling and control of a epitaxial silicon deposition process with step disturbance. In: *Proceedings of 1991 IEEE/SEMI Advanced Semiconductor Manufacturing Conference.* pp. 104–107.

Sachs, E., A. Hu and A. Ingolfsson (1995). Run by run process control: Combining SPC and feedback control. *IEEE Transactions on Semiconductor Manufacturing* **8**, 26–43.

Semiconductor Industry Association (2003). The international technology roadmap for semiconductors. San Jose, CA.

Sonderman, T. and C. Bode (2004). Automated precision manufacturing at AMD. In: *Spring Meeting of Texas-Wisconsin Modeling and Control Consortium.* http://www.che.utexas.edu/twmcc/.

Tseng, S.T., R.J. Chou and S.P. Lee (2002). A study on a multivariate EWMA controller. *IIE Transactions* **34**, 541–549.

Wang, Jin, Q. Peter He, S. Joe Qin, Christopher A. Bode and Matthew A. Purdy (2004). Recursive least squares estimation for run-to-run control with metrology delay and its application to an STI etch process. *IEEE Trans. on Semiconductor Manufacturing.* revised for publication.

Yue, H. and S. Joe Qin (2001). Reconstruction based fault identification using a combined index. *Ind. Eng. Chem. Res.* **40**, 4403–4414.

Yue, H., S.J. Qin, J. Wiseman and A. Toprac (2001). Plasma etching endpoint detection using multiple wavelengths for small open-area wafers. *J. of Vacuum Science and Technology A* **19**, 66–75.

Yue, H., S.J. Qin, R. Markle, C. Nauert and M. Gatto (2000). Fault detection of plasma etchers using optical emission spectra. *IEEE Trans. on Semiconductor Manufacturing* **13**, 374–385.

Zafiriou, Evanghelos, Hung-Wen Chiou and R.A. Adomaitis (1995). Nonlinear model-based run-to-run control for rapid thermal processing with unmeasured variable estimation. In: *187th Electrochemical Society Meeting.*

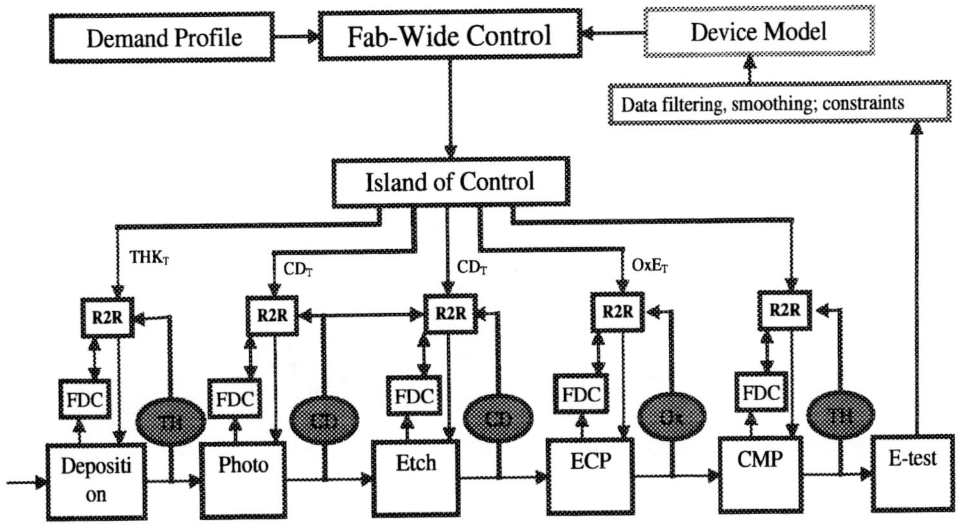

Fig. 1. Electrical parameter control as the top level of fab-wide control

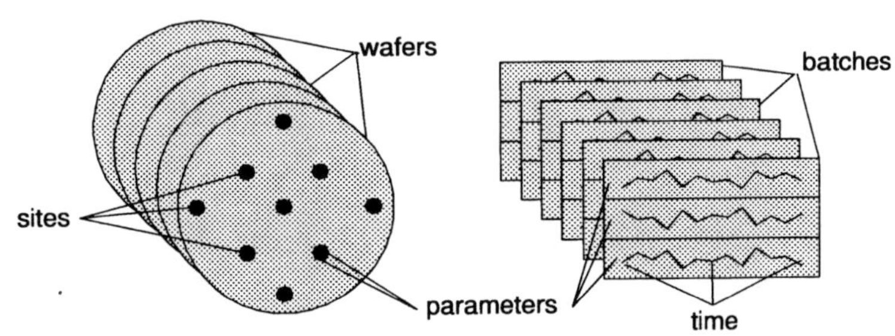

Fig. 2. Organization of site-level and batch data.

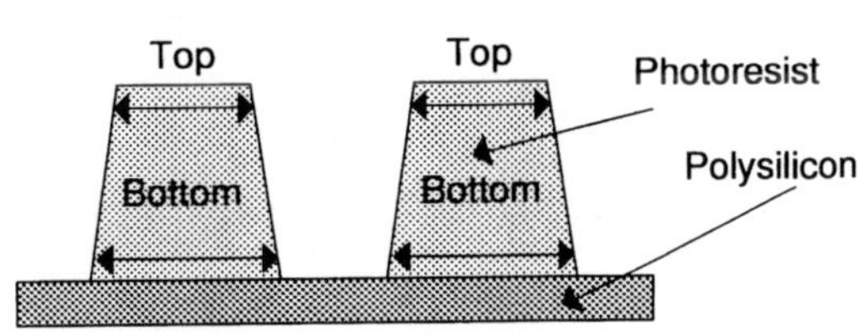

Fig. 3. Development inspection critical dimension (DICD).

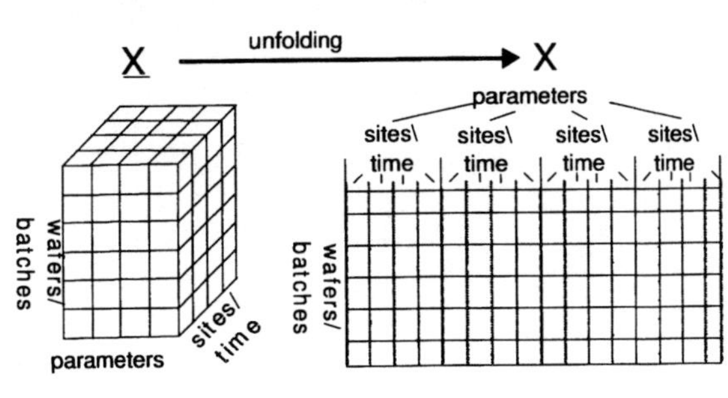

Fig. 4. Unfolding of site-level and batch data.

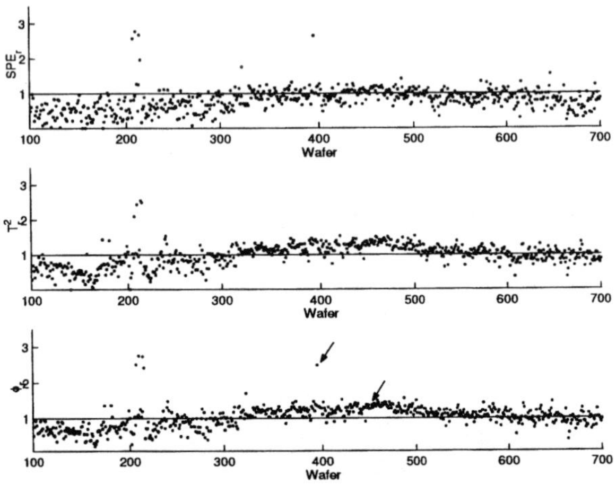

Fig. 5. DICD fault detection using SPE_r, T_r^2, and φ_r.

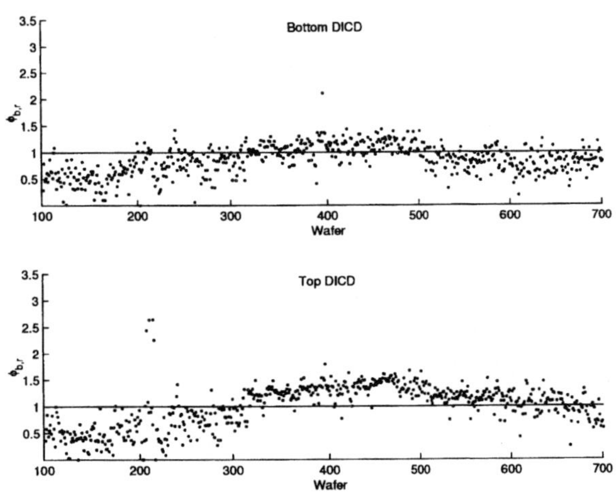

Fig. 6. DICD fault identification using parameter contributions.

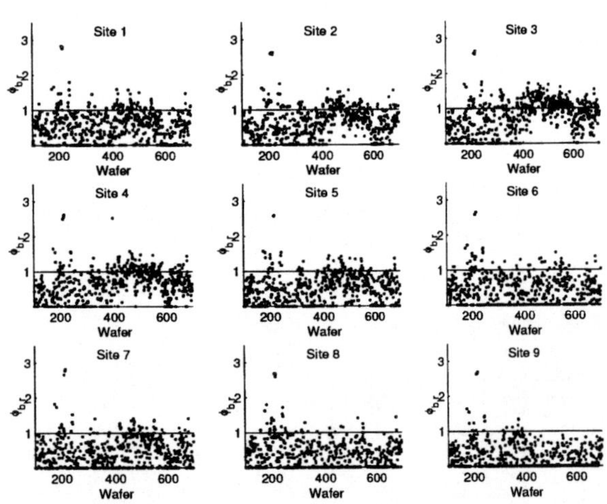

Fig. 7. DICD fault identification using site contributions.

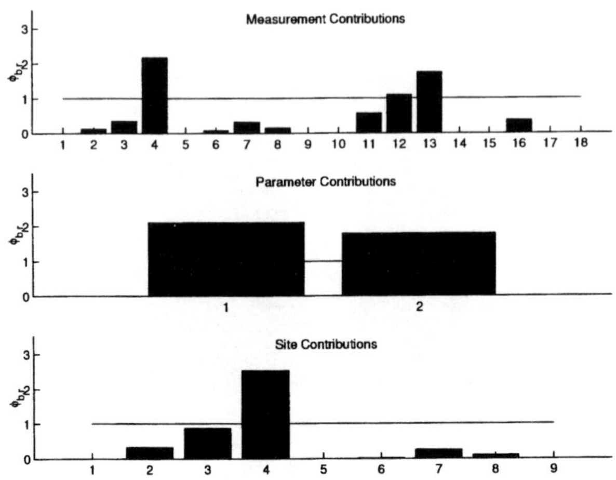

Fig. 8. Wafer 395 - DICD fault identification using measurement, parameter, and site contributions.

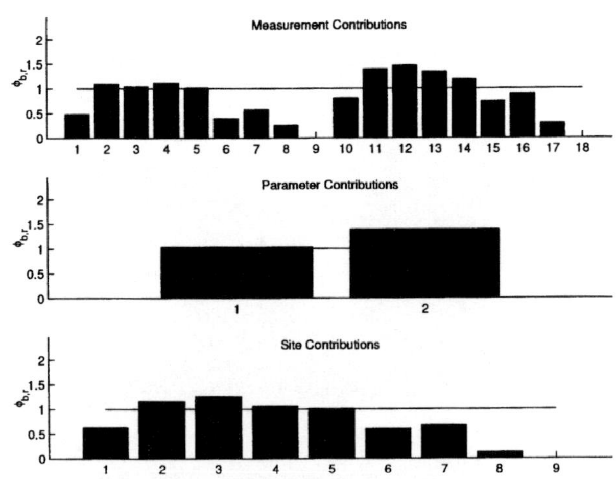

Fig. 9. Wafer 450 - DICD fault identification using measurement, parameter, and site contributions.

ELSEVIER
IFAC
PUBLICATIONS
www.elsevier.com/locate/ifac

CHALLENGES IN POLYMERIZATION REACTOR MODELING AND OPTIMIZATION: A POPULATION BALANCE PERSPECTIVE

Costas Kiparissides

Department of Chemical Engineering, Aristotle University of Thessaloniki & Chemical Process Engineering Research Institute, P.O. Box 472, 54006, Thessaloniki, Greece

Abstract: In the present article, recent developments on modeling, monitoring, optimization and control of particulate polymerization processes are reviewed. A unified population balance approach is described to follow the time evolution of molecular and morphological polymer properties in batch and continuous polymerization reactors. Recent advances on the online monitoring of "polymer quality" are discussed in the context of available hardware and software sensors. The problem of real-time optimization of polymerization processes under parametric uncertainty is also examined along with its online implementation. Finally, new issues related with the modeling, numerical solution and control of multidimensional PBEs are conferred. *Copyright © 2004 IFAC*

Keywords: Polymerization, Particulate processing, Product quality, Distributions, Computational methods, Estimators, Optimization, Control

1. INTRODUCTION

Process modeling, optimization and control can have a significant impact on polymer plant operability and economics. Polymer manufacturers face increasing pressures for production cost reductions and more stringent quality requirements. However, product quality in polymer manufacturing is a much more complex issue than in more conventional simple chemical systems. Thus, a major objective of polymerization reactor modeling is to understand how the reaction mechanism, the physical transport phenomena (e.g., mass and heat transfer), mixing, reactor type and operating conditions affect the "polymer quality". The last term includes all the polymer molecular properties (e.g., molecular weight distribution (MWD), copolymer composition distribution (CCD), sequence length distribution (SLD), long and short chain branching distributions (SCB, LCB), stereoregularity, etc.) as well as the morphological properties of the product (e.g., particle size distribution, (PSD), pore size distribution, bulk density, etc.). One of the most difficult issues in polymer reactor optimization and control is to determine the quantitative relationships between the product physical, thermal, mechanical, rheological, chemical properties and the molecular/morphological polymer properties. Table 1 summarizes both types of quality control measures.

Since the end-use properties of the produced polymers are directly linked with the molecular characteristics of the polymer chains, control of the polymer chain microstructure is of profound interest to the polymer manufacturing industry. This presupposes a thorough knowledge of the polymerization kinetics and the availability of advanced mathematical models to quantify the effects of process operating conditions on the molecular and morphological polymer properties. A great number of modeling studies on the prediction of MWD, CCD, PSD, etc. for linear and branched polymers have been published. Most of the models follow the general framework of population balances but having as internal coordinates the chain length, the chain composition, etc. of the polymer molecules or the particle volume.

Synthetic polymers are produced via a multitude of reaction mechanisms (e.g., free-radical, catalytic, ionic, etc.) and processes (e.g., bulk, solution, suspension, etc.). In general, polymerization

Table 1. Some Measures of Polymer Product Quality

MOLECULAR AND MORPHOLOGICAL PROPERTIES	END-USE PROPERTIES
• Average Molecular Weights and Molecular Weight Distribution (MWD) • Copolymer Composition Distribution (CCD) • Sequence Length Distribution (SLD) • Long-Chain Branching Distribution (LCBD) • Stereoregularity (Tacticity) • Particle Size Distribution (PSD) • Particle Porosity and Surface Area, etc.	• Physical and thermal properties (e.g., density, clarity, melting point, temperature stability, swellability, plasticizer uptake, etc.) • Chemical (e.g., corrosion resistance, etc.) • Mechanical properties (e.g., strength, toughness, stress crack resistance, abrasion resistance, impact resistance, etc.) • Rheological properties (e.g., flow properties, shear viscosity and elasticity, melt index, extensional viscosity, extrudate swell, etc.)

processes can be broadly classified into homogeneous and heterogeneous ones. The first class comprises polymerizations which are carried our in a single phase. The latter involves the presence of several phases (i.e., solid, liquid and gas). In heterogeneous polymerization, the final product is obtained in the form of a distribution of polymer particles in the sub-micron or micron size range. Table 2 presents the main characteristics of typical particulate polymerization processes employed for the production of some well-known commodity polymers (Kiparissides, 1996).

The analysis of particulate polymerization processes is a rather complex problem due to the highly coupled kinetics, thermodynamics, heat and mass transfer phenomena, taking place in a heterogeneous process. Furthermore, the physical properties (e.g., viscosity, density, transport coefficients, interfacial tension, etc.) of the various phases typically vary by several orders of magnitude in the course of polymerization. Thus, the framework of population balances is ideally suited to the description of the complex dynamics of a wide range of particulate polymerization processes.

Efficient polymer reactor process control requires correct information on the state of the polymerization process. Therefore, the availability of on-line measurements is an essential requirement for the implementation of high-level process and polymer quality optimization schemes. On-line monitoring devices have to be able to measure continuously (or at least semi-continuously) and often condition the sampled material (and sometimes even chemically process it); they must have a satisfactory analysis and response time behaviour; and, finally, they have to supply correct results over long periods of time in environments that are often physically and chemically aggressive. In a review article (Kammona, *et al.*, 1999), a comprehensive survey on recent advances on hardware and software sensors for on-line monitoring of polymer quality was presented.

The operating objectives in batch and continuous polymerization processes must satisfy complex molecular and morphological property requirements for the final product and simultaneously achieve the greatest economic potential for the plant. Basically, two types of optimization problems are encountered in polymerization processes. The first deals with the selection of the best (optimal) time invariant controls so that, in the absence of process disturbances, the final product attains some desired molecular and morphological properties. The second, the time optimal control problem, deals with the calculation of the time-optimal control trajectories to ensure the satisfaction of the "polymer quality" requirements and the operational process constraints. These optimal control policies can be calculated offline and implemented as set-point changes of the regulatory process controllers. However, the operation of a polymerization process is heavily affected by disturbances (e.g., process-related sources or inherent model parameter uncertainties) that cause the process to drift away from the offline

Table 2. Typical Particulate Polymerization Processes

Process:	Suspension	Emulsion	Precipitation	Liquid Slurry	Gas-Solid Polymerization
Particle Size (μm)	50-500	0.1-1.0	100-10^4	10-10^4	10-10^4
Reactor Type	Batch	Batch, CSTR	Batch, CSTR	Batch, Loop	Fluid Bed, Stirred Bed
Kinetic Mechanism	FR, I	FR, I	All	ZN	ZN
Polymers	PMMA PS, PVC	PS, PMMA, PVA, PTFE	PVC, PTFE	HDPE, PP, EP	HDPE, PP, LLDPE, EP

Kinetic Mechanisms: FR – Free Radical, I- Ionic, ZN- Ziegler-Natta
Polymers: PS – Polystyrene, PMMA – Poly(methyl methacrylate), PVA – Poly(vinyl acetate), PVC – Poly(vinyl cloride), HDPE – High density polyethylene, LLDPE – Linear low density polyethylene, PTFE – Poly(tetrafluorethylene), PP – Polypropylene, EP – Ethylene- Propylene copolymers

calculated optimal trajectories. To alleviate the above problems, an online estimation-optimization approach needs to be implemented (Kiparissides, *et al.*, 2002, Chatzidoukas, *et al.*, 2003). The updating of the time optimal control policies is carried out in two steps: in the first step, a state/parameter estimator utilizes the available process measurements to obtain reliable estimates of the state variables and the time-varying model parameters. In the optimization step, the time optimal trajectories are periodically reevaluated based on the most recent information about the process. The calculation of the time optimal control policy is obtained via dynamic programming techniques (including sequential and simultaneous discretization methods).

The present paper deals with the modeling and optimization of particulate polymerization processes using a population balance approach. In particular, the paper addresses the following topics. In section two, a unified population balance approach is presented to follow the molecular weight (e.g., MWD, LCB, etc.) and morphological (e.g., PSD) changes in a particulate polymerization process. The solution of the general PBE is reviewed in relation to available numerical methods (e.g., M-I and M-II approaches, finite differences, collocation, finite elements, wavelets, etc.). Subsequently, the general population balance approach is applied to typical polymerization processes to predict the dynamic evolution of MWD, PSD, etc. in batch and continuous systems. In section three, the calculation of the time optimal control policies under structural and parametric uncertainty is reviewed along with their on-line implementation to polymerization processes. Finally, in section four, future directions related to the modeling, monitoring, optimization and control of particulate polymerization reactors are conferred.

2. THE POPULATION BALANCE EQUATION

The development of the general population balance equation (PBE) follows the works of Hulburt and Katz (1964) and Ramkrishna (1985). Let us consider a distribution of countable entities (e.g., polymer chains, monomer droplets, polymer particles, etc.) $n(x, y, z, p_1, p_2, \ldots p_k, t)$ where x, y and z represent ordinary spatial coordinates, t is the time, and p_i represents the ith property of the entity. For example, p_i could denote the degree of polymerization, or the comonomer composition or the volume of a polymer particle. Actually, the term $n\,dxdydzdp_1 \ldots dp_k$ represents the fraction of entities in the geometric volume element $\delta V = \delta x \delta y \delta z$, with property values in the range $\delta p_1, \delta p_2, \ldots \delta p_k$. Note that the distribution function n will satisfy the following condition:

$$\int n(x, y, z, p_1, p_2, \ldots p_k)dxdydz\,dp_1 dp_2 \ldots dp_k = 1 \quad (1)$$

In the above formulation, there are (3 + k) independent variables plus time involved that can be viewed as a (3 + k) dimensional space. Thus, for an ordinary small volume element in this space, Ω, the general population balance becomes (Himmelblau and Bischoff, 1968):

$$\frac{d}{dt}\int_{\Omega} n d\Omega = \int_{\Omega}(B - D)d\Omega \quad (2)$$

where

$$d\Omega = dxdydzdp_1 dp_2 \ldots dp_k \;;\quad \int_{\Omega} = \int_x\int_y\int_z\int_{p_1}\ldots\int_{p_k} \quad (3)$$

B and D represent the corresponding rate of "birth" and "death" of entities per unit time, unit geometric volume and unit property change. The left-hand side of eq (2) can be modified by using the general form of Leibnitz's rule for differentiating definite integrals. Thus, following the general developments of Himmelblau and Bischoff, the differential form of eq (2) can be obtained:

$$\begin{aligned}&\frac{\partial n}{\partial t} + \frac{\partial}{\partial x}(u_x n) + \frac{\partial}{\partial y}(u_y n) + \frac{\partial}{\partial z}(u_z n) + \\ &+ \sum_{i=1}^{k}\frac{\partial}{\partial p_i}(u_i n) + D - B = 0\end{aligned} \quad (4)$$

where $u_x = dx/dt$, $u_y = dy/dt$ and $u_z = dz/dt$ denote the usual geometric velocities in the x, y and z spatial coordinates, respectively. $u_i = dp_i/dt$ is the rate of change of the property p_i. Very often, the average values of the properties in the entire reactor volume, V, are only required. Accordingly, one can define the following geometrically averaged distribution function, $\bar{n}$.

$$\bar{n} = \frac{1}{V}\int_V n dV \quad (5)$$

By integrating eq (4) over the geometric volume, V, and using the Gauss' divergence theorem, the following macroscopic population balance equation is obtained:

$$\begin{aligned}&\frac{1}{V}\frac{\partial}{\partial t}(V\bar{n}) + \sum_{i=1}^{k}\frac{\partial}{\partial p_i}(u_i\bar{n}) + \bar{D} - \bar{B} = \\ &= \frac{1}{V}(q_{in}\bar{n}_{in} - q_{out}\bar{n}_{out})\end{aligned} \quad (6)$$

where q_{in} and q_{out} are the inlet and outlet volumetric flow rates, respectively.

2.1 Calculation of Molecular Weight Distribution

Let us assume that the kinetic mechanism of the chemically initiated free-radical polymerization of a vinyl monomer comprises the following elementary reactions:

Initiator decomposition:

$$I \xrightarrow{k_d} 2R^{\bullet}$$

Chain initiation:

$$R^{\bullet} + M \xrightarrow{k_I} R_1$$

Propagation:

$$R_x + M \xrightarrow{k_p} R_{x+1}$$

Chain transfer to monomer:

$$R_x + M \xrightarrow{k_{fm}} P_x + R_1$$

Chain transfer to solvent (CTA):

$$R_x + S \xrightarrow{k_{fs}} P_x + R_1$$

Termination by disproportionation:

$$R_x + R_y \xrightarrow{k_{td}} P_x + P_y$$

Termination by combination:

$$R_x + R_y \xrightarrow{k_{tc}} P_{x+y}$$

Chain transfer to polymer:

$$R_x + P_y \xrightarrow{k_{fp}} R_y + P_x$$

Reaction with terminal double bond:

$$R_x + P_y^{=} \xrightarrow{k_{db}} R_{x+y}$$

The subscripts x and y denote the number of monomer units in a polymer chain. The above kinetic mechanism includes propagation and termination reactions by both disproportionation and combination, molecular weight control reactions via transfer to monomer and to chain transfer agent, long-chain branching formation by transfer to polymer and reaction with terminal double bonds.

Based on the above kinetic mechanism, we can derive the following population balance equations for the "live", $R(x,t)$, and "dead", $P(x,t)$, polymer chains of length x. Assuming that the degree of polymerization, x, can only take discrete values $(x:1,2,\ldots N_f)$, the inlet and outlet terms in eq (6) are equal to zero (i.e., batch reactor) and the rate of change of property x is equal to zero, one can easily derive the following macroscopic population balances for the entities $R(x,t)$ and $P(x,t)$.

$$\begin{aligned}
\frac{1}{V}\frac{d}{dt}(VR) = &\left\{k_I[R^\bullet][M] + (k_{fm}[M] + k_{fs}[S])\sum_{y=1}^{\infty} R(x,t)\right\}\delta_{x,1} + \\
&+ k_p\{R(x-1,t) - R(x,t)\}[M] - \{k_{fm}[M] + k_{fs}[S]\}R(x,t) + \\
&+ k_{fp}\{x\,D(x,t)\}\sum_{y=1}^{\infty} R(y,t) - k_{fp}R(x,t)\sum_{y=2}^{\infty} yD(x-y,t) - \\
&- k_{td}R(x,t)\sum_{y=1}^{\infty} R(y,t) - k_{tc}R(x,t)\sum_{y=1}^{\infty} R(y,t) - \\
&- k_{db}R(x,t)\sum_{y=2}^{\infty} D(x-y,t) + k_{db}\sum_{y=1}^{x-1}\{R(x-y,t)D(x-y,t)\}
\end{aligned} \tag{7}$$

$$\begin{aligned}
\frac{1}{V}\frac{d}{dt}(VP) = &\{k_{fm}[M] + k_{fs}[S]\}R(x,t) - \\
&- k_{fp}\{x\,D(x,t)\}\sum_{y=1}^{\infty} R(y,t) + k_{fp}R(x,t)\sum_{y=2}^{\infty}\{yD(x-y,t)\} + \\
&+ k_{td}R(x,t)\sum_{y=1}^{\infty} R(y,t) + \frac{1}{2}k_{tc}\sum_{y=1}^{x-1}\{R(y,t)R(x-y,t)\} - \\
&- k_{db}D(x,t)\sum_{y=1}^{\infty} R(y,t)
\end{aligned} \tag{8}$$

where $\delta_{x,1}$ is the Kronecker's delta function (i.e., $\delta_{x,1} = 1$ if $x = 1$ and $\delta_{x,1} = 0$ if $x \neq 1$).

Note that the right-hand side terms in eqs (7-8) represent the net production rates of "live" and "dead" polymer chains. The actual number of rate equations for the "live" and "dead" polymer chains will depend on the total degree of polymerization, N_f, that may be of the order of hundreds or/and thousands monomer units. Consequently, the computational effort associated with the solution of the complete set of differential equations becomes prohibitively high for most cases of interest and makes the on-line application of such a model unrealistic. To deal with the above high-dimensionality problem, several methods have been proposed to reduce the infinite system of differential equations into a low-order system of DAEs. These can be broadly classified into kinetic lumping methods, continuous variable approximations, z-transforms, method of moments, discrete weighted Galerking, orthogonal collocation on finite elements and so forth.

In the kinetic lumping methods (Crowley and Choi, 1997a,b, 1998), the polymer chains population is divided into a specified number of "chain lump" domains and the resulting population balance equations are solved numerically. A uniform concentration is assumed for all polymer chains that are members of a "chain lump" domain. However, the calculation of MWD for branched polymers (e.g., systems involving transfer to polymer and terminal double bond reactors) suffers from significant inaccuracies. Continuous variable approximation and z-transform methods, besides their elegant mathematical formulation, lack the ability to handle complex polymerization systems efficiently. Polynomial expansion methods of the MWD require the calculation of high-order moments and may lead to slow convergence (Tobita and Ito, 1993). Pladis and Kiparissides (1998) proposed a polymer chain fractionation approach and used the method of moments to reconstruct the joint molecular weight – long-chain branching distribution for branched polymers. The total population of the polymer chains was divided into a number of classes with respect to the number of long-chain branches. However, in addition to the well-known problem of closure of the "higher-order" moments, the reconstruction of the overall MWD at high monomer conversions and high LCB content requires a very large number of classes to reduce the approximation errors for high molecular weight fractions. Monte-Carlo simulations (Tobita, 1993, 1995) are straightforward techniques that can generally handle complex kinetic mechanisms but usually require significant computational effort for the determination of the MWD. Discrete weighted Galerkin formulation (Deuflhard and Wulkow, 1989, Wulkow, 1992), even though are computationally demanding, provide a powerful tool for the prediction of the MWD in complex polymerization systems (Iedema et al., 2000). However, the approximation of the infinite summation terms (e.g., resulting from termination by combination reactions) requires special treatment.

Method of Moments. The method of moments is based on the statistical representation of the average molecular properties of the polymer (e.g., number average, M_n, and weight average, M_w, molecular weights) in terms of the leading moments of the number chain length distributions (NCLDs) of "live" and "dead" polymer chains, defined by the following equations:

$$\lambda_k = \sum_{x=1}^{\infty} x^k R_x \; , \; \mu_k = \sum_{x=1}^{\infty} x^k P_x \; , \; k = 0, 1, 2 \quad (9)$$

The relevant rate functions for the moments of the NCLDs of the "live" and "dead" polymer chains can be obtained by multiplying each term of eqs (7-8) by x^k and summing the resulting expressions over the total degree of polymerization, N_∞. The resulting rate equations for the leading moments are (Pladis and Kiparissides, 1998):

$$\begin{aligned}\frac{d\lambda_k}{dt} &= k_I[R^\bullet][M] + k_{fm}[M]\lambda_0 + k_{fs}[S]\lambda_0 + \\ &+ k_p[M]\sum_{i=0}^{k}\binom{k}{i}\lambda_i - k_p[M]\lambda_k - k_{fm}[M]\lambda_k - \\ &- k_{fs}[S]\lambda_k - k_{fp}\lambda_k\mu_1 + k_{fp}\lambda_0\mu_{k+1} - k_{db}\lambda_k\mu_0 + \\ &+ k_{db}\sum_{i=0}^{k}\lambda_i\mu_{k-i} - k_t\lambda_0\lambda_k \end{aligned} \quad (10)$$

$$\begin{aligned}\frac{d\mu_k}{dt} &= k_{fm}[M]\lambda_k + k_{fs}[S]\lambda_k + k_{fp}\lambda_k\mu_1 - \\ &- k_{fp}\lambda_0\mu_{k+1} - k_{db}\lambda_0\mu_k + \frac{1}{2}k_{tc}\sum_{i=0}^{k}\binom{k}{i}\lambda_i\lambda_{k-i} + k_{td}\lambda_0\lambda_k\end{aligned} \quad (11)$$

where $[R^\bullet]$, $[M]$ and $[S]$ denote the concentrations of "live" radicals, monomer and chain transfer agent, respectively. Usually, one needs to know the leading moments (i.e., λ_0, λ_1, λ_2 and μ_0, μ_1, μ_2 of the "live" and "dead" polymer distributions) to calculate the values of the number-average, M_n, and weight-average, M_w, molecular weights.

Orthogonal Collocation on Finite Elements. A key characteristic of the OCFE method is the treatment of the discrete polymer chain length domain as a continuous one. Hence, the concentrations of "live" and "dead" polymer chains are handled as continuous variables. Accordingly, the chain length domain is divided into a number of finite elements, NE, with element boundaries at the points: $\zeta_0=1$, ζ_1, $\zeta_2, \ldots, \zeta_{NE-1}$, $\zeta_{NE} = N_f$, where N_f is the final degree of polymerization. For each element, a number of n interior collocation points, $[s_1, s_2, \ldots, s_n]$ are specified. The concentrations of the "live" and "dead" polymer chains are then approximated by continuous low-order polynomial functions within each finite element. In the present study, Lagrange interpolation polynomials were used to approximate the concentrations of the "live" and "dead" polymer chains:

$$\tilde{R}(s) = \sum_{i=0}^{n} W_{i,j}^R(s)\tilde{R}(s_{i,j}) \quad \zeta_{j-1} \le s \le \zeta_j \quad j = 1,\ldots,NE \quad (12)$$

$$\tilde{P}(s) = \sum_{i=1}^{n} W_{i,j}^P(s)\tilde{P}(s_{i,j}) \quad \zeta_{j-1} \le s \le \zeta_j \quad j = 1,\ldots,NE \quad (13)$$

The tilde denotes approximation variables. The functions, $W_{i,j}^R(s)$ and $W_{i,j}^P(s)$ are Lagrange interpolation polynomials of order n+1 and n, respectively, given by the expressions:

$$W_{i,j}^R(s) = \prod_{\substack{k=0\\k\neq i}}^{n}\frac{s - s_{k,j}}{s_{i,j} - s_{k,j}} \quad \zeta_{j-1} \le s \le \zeta_j \qquad i = 0,\ldots,n \quad j = 1,\ldots,NE \quad (14)$$

$$W_{i,j}^P(s) = \prod_{\substack{k=1\\k\neq i}}^{n}\frac{s - s_{k,j}}{s_{i,j} - s_{k,j}} \quad \zeta_{j-1} \le s \le \zeta_j \qquad i = 1,\ldots,n \quad j = 1,\ldots,NE \quad (15)$$

The Lagrange polynomials $W_{i,j}^R(s)$ and $W_{i,j}^P(s)$ are equal to zero at the collocation points $s_{i,j}$, for k≠i and equal to unity for k=i. To take into account the concentration of "live" polymer chains at chain length x=1, the left boundary point of the first element was included as an interpolation point.

A set of residual equations are then derived by substituting eqs (12-13) into eqs (7-8). The main requirement of the OCFE formulation forces the residual balances to vanish at the selected collocation points, $s_{i,j}$. The number of the finite element controls the density of the collocation points in the overall chain length domain. In general, a high density of collocation points is required in chain length regions where steep changes in the concentrations of "live" and "dead" polymer chains are foreseen. Extensive simulations showed that low-order interpolation polynomials predict the overall MWD more accurately, compared to high-order polynomials for the same total number of collocation points. This was attributed to the oscillatory behavior of the high-order Langrange polynomials.

Fixed Pivot Method. The inherent limitations of the discretized PBE approach, resulting from the discretization of the chain length domain, can be avoided in the more general formulations (i.e., fixed and moving pivot techniques) of Kumar and Ramkrishna (1996a,b) and (1997). The last methods guarantee the correct calculation of any two moments of the distribution and are applicable to any type of discretization of the chain length domain.
The fixed pivot technique is a very efficient method for the calculation of the number chain length distribution (NCLD). It assumes that the overall polymer chain population can be assigned to selected discrete chain lengths. Specific reaction steps (e.g., termination, propagation, etc.) leading to the formation of polymer chains other than the selected ones, are incorporated in the set of discrete equations in such a way that specified properties of the NCLD (i.e., total number, mass of polymer chains, etc., corresponding to any two moments of the NCLD) are exactly preserved.

Simulation results. The outlined methods (Moments, OCFE and FPM) were applied to the free-radical batch polymerization of methyl-methacrylate (MMA). The values of the kinetic rate constants as well as the gel and glass effect models were taken from the original paper of Chiu, *et al.*, (1983). The polymerization was carried out isothermally at 60°C. For the fixed pivot method, the chain length domain

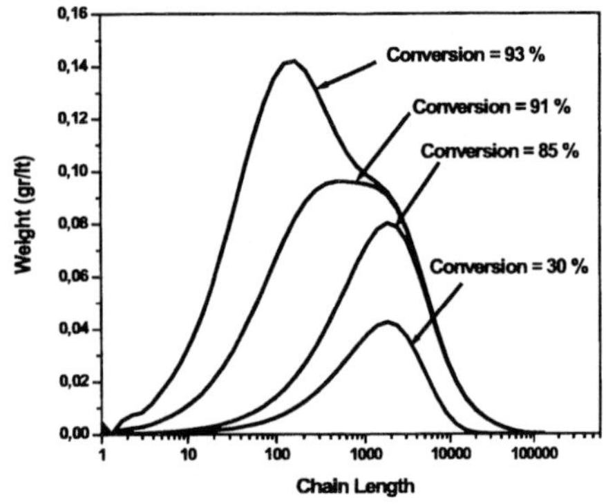

Fig. 1. Time evolution of the MWD calculated by the FP method.

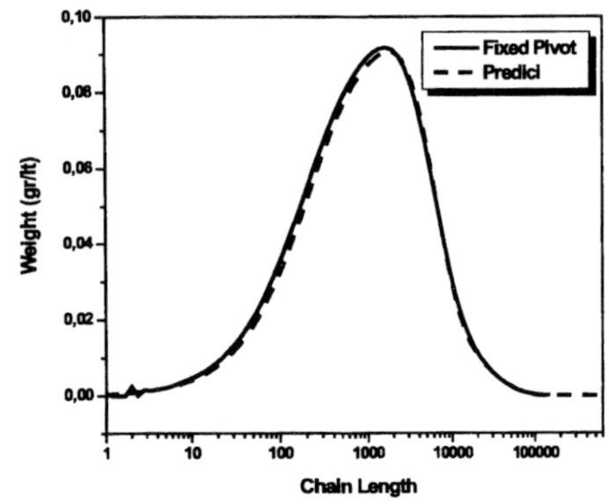

Fig. 2. Comparison of MWDs calculated by the FP and discrete Galerkin methods.

was partitioned into 70 finite elements leading to a number of 2×NE+2 = 142 residual differential equations.

Figure 1 depicts the time evolution of the molecular weight distribution of PMMA as calculated by the fixed pivot method. As can be seen, as the monomer conversion increases the MWD becomes broader. At very high monomer conversions (e.g., > 90%), a bimodal distribution is obtained. In Figure 2, a comparison of the MWDs calculated by the FP method and the discrete Galerkin h-p method (PREDICI simulation package) is shown at 90% monomer conversion. It is evident that the calculated distributions are in excellent agreement, verifying the ability of the FP method to accurately predict the MWD.

2.2 Calculation of Particle Size Distribution

To follow the dynamic evolution of the PSD in a particulate process, a population balance approach is commonly employed. The distribution of the particulates (e.g., solid particles, liquid droplets, etc.) is considered to be continuous over the volume variable and is commonly described by a number density function. Thus, the term n(V,t)dV represents the number of particles per unit volume in the differential volume size range (V to V+dV). For a dynamic particulate system undergoing simultaneous particle nucleation, growth and aggregation, the rate of change of the number density function with respect to time and volume is given by the following nonlinear integro-differential population balance equation:

$$\frac{\partial n(V,t)}{\partial t}+\frac{\partial\big(G_V(V)n(V,t)\big)}{\partial V}=\frac{q_{in}\,n_{in}(V,t)-q_{out}\,n(V,t)}{V}-$$
$$-\frac{n(V,t)}{V}\frac{\partial V}{\partial t}+\int_{V_{min}}^{V/2}\beta(U,V-U)\,n(U,t)\,n(V-U,t)\,dU-$$
$$-n(V,t)\int_{V_{min}}^{V_{max}}\beta(V,U)\,n(U,t)\,dU+S_v(V,t) \qquad (16)$$

where q_{in} and q_{out} are the inlet and outlet volumetric flow rates, respectively. $n_{in}(V,t)$ is the inflow number density function, V the total volume of the reactor, G_V is a particle volume growth rate function and β(V,U) is a particle aggregation rate kernel for particles of volumes V and U. V_{min} and V_{max} denote the corresponding minimum and maximum size of particles present in the system. Finally, $S_v(V,t)$ is the particle nucleation rate. In general, eq 16 will satisfy the following initial condition:

$$n(V,0)=n_0(V)\ , \quad \text{at} \quad t=0 \qquad (17)$$

The numerical solution of the dynamic PBE for a particulate system, especially for a reactive one, is a notably difficult problem due to both numerical complexities and model uncertainties regarding the particle nucleation, growth, aggregation and breakage mechanisms that are often poorly understood. Usually, the numerical solution of the PBE requires the discretization of the particle volume domain into a number of discrete elements that results in a system of stiff, nonlinear differential or algebraic/differential equations that is solved numerically. In the open literature, several numerical methods have been developed for solving the steady-state or dynamic PBE. These include the full discrete method (Hidy, 1965), the method of classes (Marchal, *et al.*, 1988; Chatzi and Kiparissides, 1992), the discretized PBE (Batterham, *et al.*, 1981; Hounslow, *et al.*, 1988), the fixed and moving pivot discretized PBE methods (Kumar and Ramkrishna, 1996a,b), the high order discretized PBE methods (Landgrebe and Pratsinis, 1990; Bleck, 1970; Gelbard and Seinfeld, 1980), the orthogonal collocation on finite elements (Gelbard and Seinfeld, 1979), the Galerkin method (Nicmanis and Hounslow, 1998), and the wavelet-Galerkin method (Chen, *et al.*,1996). In the reviews of Ramkrishna (1985), Dafniotis (1996), and Kumar and Ramkrishna (1996a,b), the various numerical methods available for solving the PBE are described in detail. Moreover, in three publications by Kostoglou and Karabelas (1994, 1995), and Nicmanis and Hounslow (1996), comparative studies on the different numerical methods are presented. Based on the conclusions of these studies, the discretized PBE method of Litster, *et al.*, (1995) the pivot methods of Kumar and Ramkrishna (1996a,b), the Galerkin and the orthogonal collocation on finite element methods were found to be the most accurate and stable numerical techniques.

In Table 3, selective publications on the numerical solution of the general PBE are reported.

Table 3. Numerical Methods for the Solution of the PBE

Methods	Author	Test Cases	Comments
M-I	Hiddy (1965)	➢ Aggregation	• Full discrete method.
	Batterham et al. (1981)	➢ Aggregation	• Geometric discretization of the form $V_{i+1}=2V_i$.
	Marchal et al. (1988)	➢ Aggregation / Growth	• Method of classes.
	Hounslow et al. (1988)	➢ Aggregation / Growth / Nucleation	• Fixed discretization of the form $V_{i+1}=2V_i$.
	Litster et al. (1995)	➢ Aggregation / Growth (Batch) ➢ Aggregation / Growth / Nucleation (CSTR)	• Discretization based on the rule $V_{i+1}=2V_i^{1/q}$, $q \geq 2$
	Kumar and Ramkrishna (1996[a,b]; 1997)	➢ Aggregation / Growth / Nucleation	• Arbitrary discretization.
	Vanni (2000)	➢ Aggregation /Breakage	• Comparison of DPBE methods based on a very coarse discretization
M-II	Bleck (1970)	➢ Aggregation	• Use of two different types of discretization. (i.e . $V_{i+1}=2^{1/2} V_i$ and $V_{i+1}=2^{1/3} V_i$).
	Gelbard et al (1980)	➢ Aggregation	• Generalization for the multi-component PBE case (Gelbard and Seinfeld, 1980).
	Satry and Gaschignard (1981)	➢ Aggregation	• Different set of equations for every moment of the distribution.
	Landgrebe and Pratsinis (1990)	➢ Aggregation	• Comparison of the numerical results only with respect to the moments of the distribution.
WR	Sigh and Ramkrishna (1975)	➢ Aggregation	• Employment of problem-specific polynomials as basis functions.
	Sampson and Ramkrishna (1985)	➢ Aggregation	• Employment of root-shifted problem-specific polynomials as basis functions.Improved accuracy based on the correct choice of the collocation points
FE	Gelbard and Seinfeld (1978)	➢ Aggregation / Growth / Nucleation	• Comparison of two collocation methods (i.e., spline collocation and orthogonal collocation on finite elements).
	Pillinis (1990)	➢ Aggregation / Growth	• Galerkin on finite elements method. Simulations of multi-component particulate systems.
	Erasmus et al. (1994)	➢ Aggregation / Growth	• Galerkin on finite elements method with b-splines.
	Nicmanis and Hounslow (1996 and 1998)	➢ Aggregation / Growth /Nucleation / Breakage (CSTR)	• Steady state formulation of the PBE. Collocation and Galerkin techniques with quadratic basis functions.
	Mahoney and Ramkrishna (2002)	➢ Aggregation / Growth / Nucleation	• Galerkin on finite elements with linear basis functions. Special treatment of moving discontinuities that arise in problems with particle growth.
	Rigopoulos and Jones (2003)	➢ Aggregation / Growth /Nucleation / Breakage (Batch, CSTR)	• Galerkin on finite elements with linear basis functions.
	Sandu and Borden (2003)	➢ Aggregation / Growth	• General framework for the solution of PBE with finite element techniques. Several formulations are considered with respect to the choice of basis functions, co-ordinates etc.

Common problems related to the numerical solution of the PBE include the inaccurate calculation of the PSD for highly aggregating processes, numerical instabilities for growth-dominated processes, increased stiffness of the system of DAEs for processes involving rapid particle nucleation, and domain errors for high-order aggregation kernels (Kumar and Ramkrishna, 1996a,b ; Dafniotis, 1996). In two recent publications (Alexopoulos, et al., 2004; Alexopoulos and Kiparissides, 2004), a comprehensive investigation on the solution of the dynamic population balance equation, in the presence of particle growth, aggregation and nucleation, is presented and the numerical difficulties and limitations of two very commonly employed methods (i.e., the orthogonal collocation on finite elements and the discretized PBE approach) are critically examined.

The Discretized Population Balance Equation. In the discretized PBE approach, it is assumed that the number density function, n(V,t), remains constant in the discrete volume interval (V_i to V_{i+1}).

Accordingly, a particle number distribution, $N_i(t)$, corresponding to the "i" element is defined:

$$N_i(t) = \int_{V_i}^{V_{i+1}} n(V,t)\, dV = \bar{n}_i(t)\left(V_{i+1} - V_i\right) \qquad (18)$$

where $\bar{n}_i(t)$ is the average value of n(V,t) in the element "i".

Following the original developments of Litster, *et al.*, (1995) the total volume domain (V_{min} to V_{max}) is divided into a number of elements using the fractional geometric discretization rule, $V_{i+1} = 2^{1/q} V_i$, where q is an integer, positive number. As the value of q increases the total number of volume elements increases (i.e., a finer grid is generated) and so does the computational effort for the calculation of the PSD. Accordingly, the discretized PBE for a batch particulate system undergoing simultaneous growth, aggregation and nucleation becomes:

$$\begin{aligned}
\frac{dN_i(t)}{dt} &= S_0(t)\sum_{j=1}^{3}\delta(i-j)f_i + \frac{2G_V}{3(1+r)V_i}\left\{\frac{r\,N_{i-1}}{r^2-1} + N_i - \frac{N_{i+1}}{r^2-r}\right\} + \\
&+ \sum_{k=2}^{q}\ \sum_{j=i-S(q-k+2)-k+1}^{i-S(q-k+1)-k} \beta_{i-k,j} N_{i-k} N_j \frac{2^{(j-i+1)/q} - 1 + 2^{-(k-1)/q}}{2^{1/q}-1} + \\
&+ \frac{1}{2}\beta_{i-q,i-q} N_{i-q}^2 - \sum_{j=1}^{i-S(q)} \beta_{i,j} N_i N_j \frac{2^{(j-i)/q}}{2^{1/q}-1} + \\
&+ \sum_{k=2}^{q}\ \sum_{j=i-S(q-k+2)-k+2}^{i-S(q-k+1)-k+1} \beta_{i-k+1,j} N_{i-k+1} N_j \frac{2^{1/q} - 2^{(j-i)/q} - 2^{-(k-1)/q}}{2^{1/q}-1} - \\
&- \sum_{j=i-S(q)+1}^{\infty} \beta_{i,j} N_i N_j + \sum_{j=1}^{i-S(q)-1} \frac{2^{(j-i+1)/q}}{2^{1/q}-1}\beta_{i-1,j} N_{i-1} N_j
\end{aligned} \qquad (19)$$

where $S(q) = q(q+1)/2$. "ne" is the total number of elements, $\beta_{i,j} = \beta(V_i,V_j)$ is the equivalent discrete kernel for particle aggregation and r is equal to the ratio D_{i+1}/D_i.

Finite Element Techniques. The continuous form of the PBE (eq 16) can be solved using the finite element method (FEM). Recently, Nicmanis and Hounslow, (1998), employed a collocation and Galerkin FEM to determine the steady state PSD for a continuous particulate process undergoing combined particle growth and aggregation.

In the OCFE, the particle volume domain is first divided into "ne" elements based on an appropriately selected volume discretization rule. Then, "nc" internal collocation points are specified within each element. Accordingly, the unknown number density function is approximated at the internal and boundary collocation points of each element, "e", in terms of Lagrange basis functions, φ_j,:

$$n(V,t) = \sum_{j=0}^{nc+1} n_j^e(t)\, \varphi_j(V) \tag{20}$$

where n_j^e denotes the value of n(V, t) at the "j" internal or boundary collocation point. The above approximation, results in a total number of (ne(nc+1)+1) unknown values of the number density function, n_j^e.

Following the general developments of Finlayson (1980), eq 16, for a batch process, can be recast into a system of (ne•nc) residual equations, corresponding to all the internal points of the "ne" volume elements.

$$\dot{n}_i^e(t) = -\sum_{j=0}^{nc+1}\left(\frac{d\varphi_j}{d\xi}\right)_i n_j^e(t)\,|J|^e\, G_V(V_i) - \left(\frac{dG_V}{dV}\right)_i n_i^e(t) -$$
$$-n_i^e(t)\sum_{f=1}^{ne}\sum_{k=1}^{nc} w_k^G\,|J|^f\,\beta\left(V_i^e, V_k^f\right) n_k^f(t) +$$
$$+\sum_{f=1}^{g-1}\sum_{j=1}^{nc}\left[w_j^G\,|J|^f\,\beta(V_i^e - V_j^f, V_j^f)\, n_j^f n(V_i^e - V_j^f)\right] +$$
$$+\int_{V_1^g}^{V_i^e/2}\beta(V_i^e - V, V)\, n(V)\, n(V_i^e - V)\,dV \tag{21}$$

According to the standard finite element formulation, the global volume domain of each element "e" is linearly transformed into a local domain [-1,1]. The index "g" denotes the element containing the $(V_i^e/2)$ discrete point. $|J|^f$ is the Jacobean of the volume transformation and w_k^G are the weights of the Gauss-Legendre quadrature integration rule. More details regarding the derivation of eq 21 can be found in Alexopoulos, *et al.*, (2004).

At the boundary points between the various elements, the number density function and its first derivative are forced to be continuous. Thus, the following (ne-1) continuity conditions between all the adjacent pair of elements (e and e+1) are written:

$$\sum_{j=1}^{nc}\left(\frac{d\varphi_j}{d\xi}\right)_{\xi_{nc+1}} n_j^e(t)\,|J|^e = \sum_{j=1}^{nc}\left(\frac{d\varphi_j}{d\xi}\right)_{\xi_0} n_j^{e+1}(t)\,|J|^{e+1} \tag{22}$$

Since the total number of unknown nodal values of n_j^e (i.e., ne (nc + 1) + 1), is more than the total number of residual equations and continuity conditions (i.e., ne (nc + 1) − 1), two additional equations are needed to produce a closed system of DAEs. These equations correspond to the values of n_j^e at the minimum V_{min} (e = 1, j = 0) and maximum V_{max} (e = ne, j = nc+1) value of the volume integration domain. At $V = V_{max}$, a residual equation similar to eq 21 can be written. On the other hand, at $V = V_{min}$ the zero boundary condition, $n(V_{min},t) = 0$, can be employed.

2.3 Application of PBE to Particulate Polymerizations

The Suspension Polymerization Process. In suspension polymerization, the monomer is initially dispersed in the continuous aqueous phase by the combined action of surface active agents (e.g., inorganic or/and water-soluble polymers) and agitation. All the reactants (i.e., monomer, initiator(s), etc.) reside in the organic or "oil" phase. The polymerization occurs in the monomer droplets that are progressively transformed into sticky, viscous monomer-polymer particles and finally into rigid, spherical polymer particles of size 50-500 μm. The polymer solids content in the fully converted suspension is typically 30-50% w/w. Large quantities of polymers, including poly(vinyl chloride) and polystyrene, are produced by the suspension polymerization process.

One of the most important issues in the operation of a suspension polymerization reactor is the control of the final particle size distribution (Yuan, *et al.*, 1991). The initial monomer droplet size distribution as well as the final polymer particle size distribution in general depend on the type and concentration of the surface active agents, the quality of agitation and the physical properties (e.g., density, viscosity and interfacial tension) of the continuous and dispersed phases. The transient droplet / particle size distribution is controlled by two dynamic processes, namely, the drop breakage and drop coalescence rates. The former mainly occurs in regions of high shear stresses (i.e., near the agitator blades). The latter is either increased or decreased with the intensity of the turbulent flow field and can be assumed to be negligible for dilute dispersions at sufficiently high concentrations of surface active agents (Chatzi, *et al.*, 1989).

The dynamic behavior of a liquid-liquid dispersion in a stirred vessel and, therefore, the drop-size distribution are determined by the droplet breakage and coalescence rates. In general, the drop-size distribution in a batch system can be predicted by the solution of the following population balance equation:

$$\frac{\partial(n(V,t))}{\partial t} = \int_V^{V_{max}} k(U,V)\, u(U)\, g(U)\, n(U,t)\, dU +$$
$$+\int_{V_{min}}^{V/2}\beta(V-U,U)\, n(V-U,t)\, n(U,t)dU - \tag{23}$$
$$- n(V,t)\, g(V) - n(V,t)\int_{V_{min}}^{V_{max}}\beta(V,U)\, n(U,t)\, dU$$

The left-hand side term in eq 23 represents the rate of change of the number density function, n(V,t). The first term on the right-hand side of eq 23 represents the generation of droplets in the size range (V , V+dV) due to drop breakage. k(U,V) is a daughter drop probability function, accounting for the probability that a drop of volume V to be formed by the breakage of a drop of volume U. u(U) is the number of droplets formed by the breakage of a drop of volume U and g(U) is the breakage rate of a droplet of volume U. The second term represents the

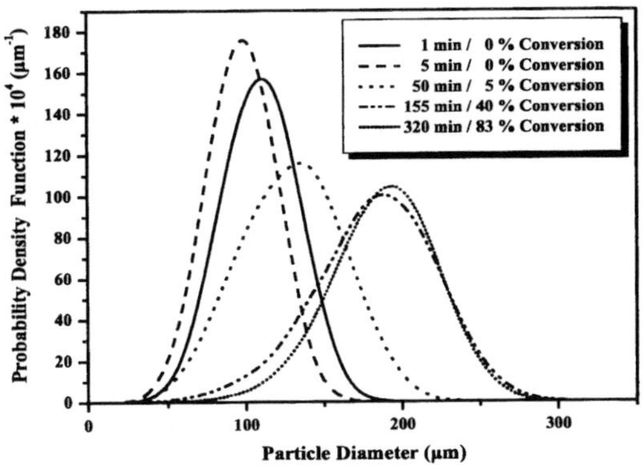

Fig. 3. Dynamic evolution of the volume probability density function with respect to polymerization time for the VCM suspension polymerization (Polymerization temperature: 56.5 °C, impeller speed: 330 rpm, dispersed phase volume fraction: 40%).

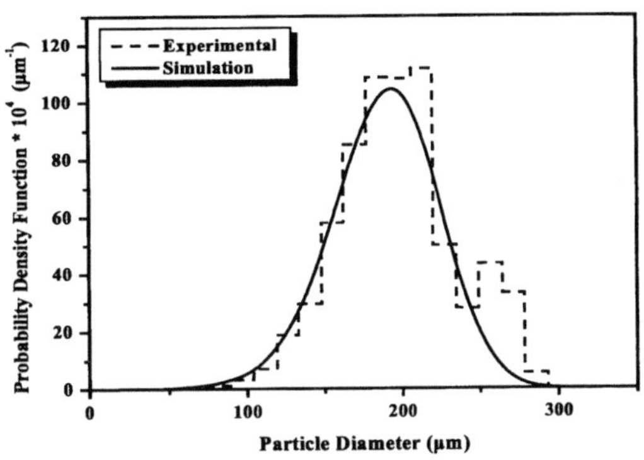

Fig. 4. Predicted and experimental PSDs for the VCM suspension polymerization

generation of drops in the size range (V , V+dV) due to the coalescence of two smaller droplets. β(V, U) is the coalescence rate between two drops of volume V and U. Finally, the third and fourth terms represent the disappearance rates of drops of volume V due to breakage and coalescence, respectively.

The predicting capabilities of the present model were demonstrated by a direct comparison of model predictions with experimental data on the drop/particle size distribution for a VCM suspension polymerization system. The experiments were carried out in a 30 lt batch reactor, using 40 % v/v VCM in water. The polymerization temperature was set equal to 56.5 °C while the agitation speed remained constant at 330 rpm. Detailed experimental measurements on polymerization rate, monomer conversion and PSD were provided by ATOFINA.

Figure 3 illustrates the dynamic evolution of the calculated PSD at different monomer conversions. More specifically, the volume probability density function (defined as $d_v^3 n/V_{tot}$) is plotted with respect to the particle diameter. One can easily distinguish the three stages that the suspension polymerization undergoes (Maggioris, *et al.*, 2000). During the initial low-viscosity period (e.g., up to 5% monomer conversion), drop breakage is the dominant mechanism and the distribution shifts to smaller sizes. Afterwards, during the sticky stage of the polymerization (e.g., 5-75 % monomer conversion), the drop breakage rate decreases while the drop coalescence becomes the dominant mechanism. In Figure 4, the predicted final PSD (continuous line) is compared with the experimental one (discrete line). As can be seen, the simulation results are in very good agreement with the experimental measurements.

Emulsion Polymerization. In emulsion polymerization, an aqueous dispersion of monomer(s) is converted by free-radical polymerization into a stable dispersion of polymer particles of 0.1-1 μm in diameter. A typical emulsion recipe comprises the dispersing medium (e.g., water), monomer(s), a water soluble initiator and an emulsifying agent (e.g., sodium and potassium salts of saturated long-chain acids). The excess of surfactant forms micelles (ca. 5-10 nm in diameter). A very small part of the monomer can be found in solution, some is solubilized by the micelles but most stays in the monomer droplets (ca. 10μm in diameter). Usually, the concentrations of micelles and monomer droplets are 10^{18} micelles/cm^3 and 10^{10}-10^{11} droplets/cm^3, respectively. Polymerization occurs either in the interior of the micelles (micellar nucleation) or in the continuous phase (homogeneous nucleation) when the monomer is at least slightly soluble. Typical emulsion processes include copolymerization of styrene and butadiene (SBR rubber) and polymerization of chloroprene (neoprene rubber). Also vinyl acetate and several acrylic monomers are polymerized in emulsion in the manufacture of latex paints.

The mathematical modeling of emulsion polymerization reactors has been the subject of a great number of publications (Saldivar and Ray, 1997a,b; Richards, *et al.*, 1989; Storti, *et al.*, 1989; Napper and Gilbert, 1989; Dougherty, 1986). Nevertheless, the development of a comprehensive mathematical model, capable of predicting the molecular (e.g., molecular weight distribution, copolymer composition, etc.) and morphological (e.g., particle size distribution) properties of the latex is not a trivial task, for there is a number of physical and chemical phenomena, which cannot be experimentally assessed and, thus, be easily modelled.

In a recent study by Kammona, *et al.*, (2003), a detailed experimental and theoretical study on the Styrene / 2-Ethylhexyl Acrylate emulsion copolymerization was reported. Kinetic experiments were carried out in a laboratory batch reactor to assess the effect of key process variables (e.g., concentrations of initiator and nonionic surfactant, temperature and initial monomer molar ratio) on the polymerization rate, monomer conversion, copolymer composition and particle size distribution (PSD). To predict the evolution of molecular and morphological properties, a comprehensive mathematical model was developed in terms of a detailed emulsion copolymerization kinetic mechanism, the dynamic molar species balances in the aqueous and polymer phases and a population balance equation (PBE) describing the time evolution of PSD in the reactor. A generalized model

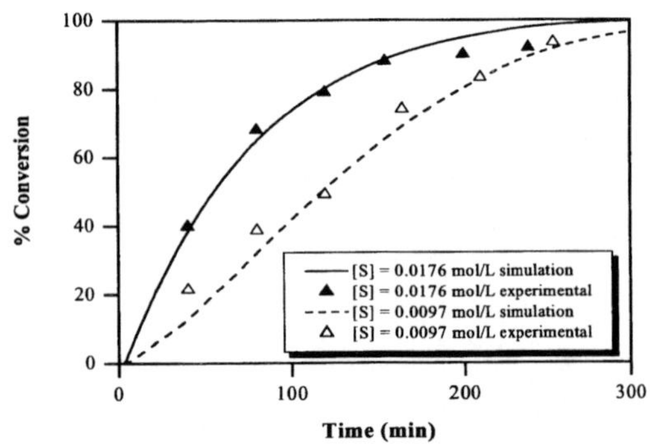

Fig. 5. Comparison of model predictions and experimental measurements on overall monomer conversion. Effect of surfactant concentration on the overall monomer conversion.

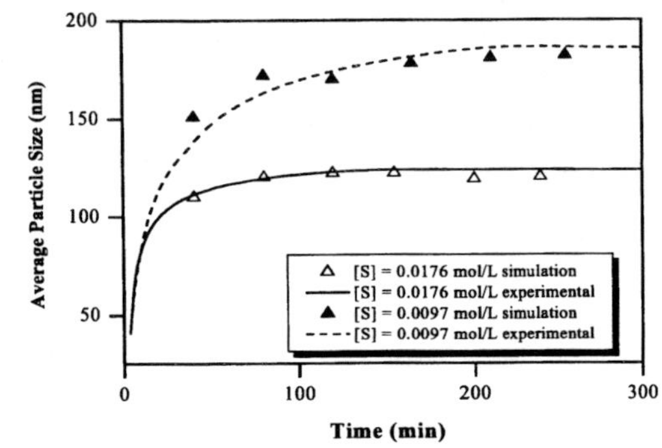

Fig. 6. Comparison of model predictions and experimental measurements on average particle size. Effect of surfactant concentration on the average particle size.

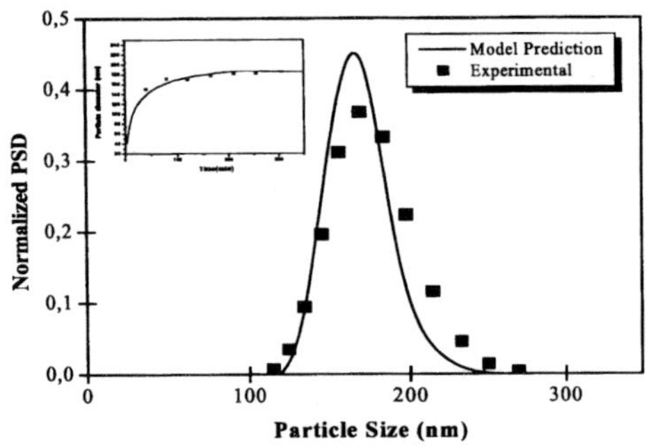

Fig. 7. Predicted and experimental PSDs (f_{10} = 0.81, [APS]= 0.0095 mol/L and [NP-30] = 0.0097 mol/L).

(Lazaridis, *et al.*, 1999) was employed for the calculation of the total interaction potential of sterically stabilized particles, in order to describe particle stability during the polymerization process. The predictive capabilities of the proposed model were thoroughly assessed by a direct comparison of theoretical predictions with experimental data on monomer conversion, average particle size and PSD.

In Figures 5 and 6, model predictions are compared with experimental measurements on overall monomer conversion and average particle diameter, respectively, for an initial styrene mol fraction of f_{10} = 0.81 at two different concentrations of the nonionic surfactant (NP-30) (e.g., 0.0097, 0.0176 mol/L). As can be seen, model predictions are in very good agreement with experimental measurements. As expected, an increase in emulsifier concentration results in an increase of the number of generated particles which leads to a decrease of the average particle size (see Figure 6) and to a respective increase of the polymerization rate (see Figure 5). In Figure 7, the predicted final PSD is compared with the experimentally measured one. As can be seen, model predictions are in very good agreement with the experimentally measured PSD. The small inserted graph shows the time evolution of the average diameter.

3. REAL-TIME OPTIMIZATION AND CONTROL

Process optimization and control can have a significant strategic impact on polymer plant operability and economics. The main goals in operating a polymer reactor (e.g., high yield, better product quality and safe operation) are very difficult, if not impossible, to achieve without efficient and reliable polymer characterization techniques. Although the weakest link in polymer reactor control is undoubtedly the on-line instrumentation, lack of understanding of the process dynamics, the highly sensitive and nonlinear behavior of polymer reactors and the lack of well structured control strategies all contribute to the impairment of competitiveness. Thus, the main problems to be addressed in the development of an optimization/control policy for a polymer process are:

- Better understanding of process dynamics (via mechanistic/data-based modelling and experimentation).
- Ability to measure and characterize a whole range of polymer quality variables (via on-line sensors and nonlinear estimation algorithms).
- Development of nonlinear model-based predictive controllers with emphasis on achieving superior performance and constraint handling.

The key to building a reliable model is to mathematically describe the chemical and physical phenomena involved in a process and to derive the necessary material, energy and momentum balances. These mathematical descriptions involve nonlinear algebraic and differential equations. Needless to say that the rigour and accuracy of a simulation model are strongly affected by the prediction of the thermophysical properties of the polymerization system.

Hardware and Software Sensors. In Table 4, selective references on the online monitoring of "polymer quality" are reported. In several occasions, the measurements become available with delay and poor signal to noise ratio. On the other hand, mathematical models based on fundamental material, energy and population balances are capable of calculating the process states accurately. However, the predictions from mathematical models frequently diverge from the real process output variables due to the presence of time varying model parameters and unknown process disturbances or/and the inherent model-plant mismatch. The first problem to be addressed

Table 4. On-line Hardware Sensors for the Monitoring of Polymer Quality

Measured Property	Sensing Technique
Monomer Conversion Zeaiter et al.(2002), Mikes et al.(2002), Rudschuck et al.(1999), Van Den Brink et al. (2001), Sasic et al.(2002)	*Densimetry, Gas Chromatography, Calorimetry/Reactor Energy Balances, Ultrasound, Raman Spectroscopy, Near Infrared Spectroscopy, Mid-range Infrared Spectroscopy, UV reflection spectroscopy, Refractive Index Measurements*
Copolymer Composition Stavropoulos et al.(2001), Chatzi et al. (1997), Asua et al.(1995), Gugliotta et al.(1995)	*Gas Chromatography, Near Infrared Spectroscopy, Mid-range Infrared Spectroscopy, Calorimetry/ Reactor Energy Balances*
Molecular Weight and MWD Vega et al. (2001), Vicente et al.(2001), Cherfi et al. (2002)	*Viscometry, Gel Permeation Chromatography, Calorimetry, Near-Infrared Spectroscopy*
Particle Size Distribution Thomas and Dimonie (1990), Gossen et al. (1993), Brandolin and Garcia-Rubio (1991), DosRamos and Silebi (1991)	*Dynamic Light Scattering, Turbidimetry, Near-Infrared Spectroscopy, Packed Column Chromatography, Capillary Hydrodynamic Chromatography*

in a real-time optimization or/and control application is the design of an efficient nonlinear state estimator to establish an information feedback from the process to the model that enables the correction of modelling errors. Apart from the nonlinear control applications, estimators can play a key role in process monitoring for an early detection of undesired reactor conditions (e.g., violation of safety regulations, environmental standards, operating constraints and product quality specifications).

Soroush (1998) categorized the design methods for nonlinear estimators into three classes. The first class includes methods that are based on the extension of the Luenberger observer and Kalman filter estimation using a linearized process model. The second class refers to methods that incorporate the estimation problem within an optimization framework, utilizing the full nonlinear process model. The estimated values for the state variables result from the minimization of the difference beween the measured and predicted responses over a specified time horizon. The third class is based on model linearization through coordinate transformation derived from results in nonlinear control theory.

MacGregor, *et al.*, (1986) illustrated the successful application of an Extended Kalman Filter (EKF) to batch and continuous styrene emulsion polymerization reactors for the online tracking of conversion and latex particle size. De Valliere and Bonvin, (1989), Gagnon and MacGregor, (1991) and Kozub and MacGregor (1992) demonstrated through simulation examples the need to augment the state vector with additional stochastic states to account for the non-stationary variations and eliminate the bias from the state estimates. Ellis, *et al.*, (1994) formulated a multi-rate sampled data EKF incorporating infrequently online GPC measurements of the "dead" polymer concentrations along with temperature and conversion that successfully traced the evolution of the molecular weight distribution in a methyl methacrylate polymerization batch reactor. Muhta, *et al.*, (1997) proposed an algorithm that can handle several measurements with different time delays. Dimitratos, *et al.*, (1989) and (1991) developed several different nonlinear EKFs and experimentally applied them to a semi-batch vinyl acetate/butyl acrylate copolymerization reactor with excellent state and parameter tracking performance. Crowley and Choi (1998) and Ahn, *et al.*, (1999) implemented EKF algorithms for the optimal control of the molecular weight distribution.

Jang and coworkers (1987) implemented an optimization-based estimator to obtain estimates of the unknown initial state vector. Such an approach was shown to outperform the EKF with increased robustness with respect to measurement noise. Later, Jang, *et al.*, (1987) used similar ideas to identify unmeasured disturbances and to estimate the states and the model parameters for control purposes of a simple CSTR. Muske, *et al.*, (1995) studied the convergence properties and the performance of the receding horizon state estimation technique. Vos, *et al.*, (1997) suggested a receding horizon parameter estimation technique for process monitoring that uses past information for the model parameter behavior in order to enhance the robustness of the procedure. Nonlinear state observers based on coordinate transformations (Soroush, 1997 ; Fevotte, *et al.*, 1998a,b) is another alternative for reconstructuring the state vector of dynamic systems from process measurements but suffers in practical implementation when complex process models are involved.

In Table 5, recent publications on monitoring, estimation, optimization and control of particulate polymerization processes are reported.

3.1 Optimization under Parametric Uncertainty

The real-time optimization of a polymerization process, compared to its off-line counterpart, has as a distinctive merit the inherent capability of exploiting the information springing continually from the process. As a result, it can more easily cope with the process variability and redefine in context the optimal operating conditions. In a recent paper by Kiparissides, *et al.*, (2002), an on-line, multi-level estimation/optimization procedure was described to control the MWD in batch polymerization reactors. The proposed estimation/optimization scheme combined a state/parameter estimator with an on-line optimizer that calculated periodically the time optimal control policies based on the most recent information about the process. The state/parameter estimator included an extended Kalman filter (EKF) for estimating the state variables and the time-varying kinetic rate constants. Subsequently, the time optimal trajectories that drive the process to the desired final property specifications were determined using a dynamic programming approach.

Figure 8 shows the schematic of the information flowsheet for the on-line optimizing controller. The measured output variables, $\mathbf{Y}_{\mathbf{meas}}$, are compared to the model predictions, $\mathbf{Y}_{\mathbf{m}}$. The anticipated error incorporates the unmeasured process disturbances, **d**, the measurement noise and the process-model mismatch. The state/parameter estimation block uses the error term to retrieve the entire state vector of the system, $\mathbf{Y}_{\mathbf{est}}$, as well as the values for the stochastic time varying model parameters. Subsequently, the updated values for the state variables and the model

Table 5. Selective Publications on Monitoring, Estimation, Optimization and Control of Molecular and Morphological Polymer Properties

Author	Test Case	Comments
Catalgil-Giz, *et al.* (2002)	S : MMA / St	Monitoring : MWD / CCD
Cerrillo & MacGregor, (2002)	E : St	Control : PSD
Chang & Hung, (2002)	S : MMA	Optimization : MWD
Chatzidoukas, *et al.* (2003)	C: Olefin	Optimization : MWD / CCD
Christofides, (2001)	E	Control : PSD
Congalidis, *et al.* (1989)	S : MMA / VAc	Control: MWD
Crowley, *et al.* (2000)	E : St	Control : PSD
Crowley & Choi (1997a)	S : MMA	Optimization / Control : MWD
Crowley & Choi (1998)	S : MMA	Optimization / Control : MWD
Crowley & Choi (1999)	So : St /MMA	Optimization / Control : MWD / CCD
Dimitratos, *et al.* (1994)	E : Acrylics / Vinyl monomer/ St	Control : MWD / CCD / PSD
Doyle III, *et al.* (2003)	E : St	Control : PSD
Ellis, *et al.* (1988)	S : MMA	Estimation : MWD
Embirucu, *et al.* (1996)	E / S / C	Modelling / Estimation / Optimization / Control : MMW / CCD
Fevotte, *et al.* (1998b)	E : MMA / VAc	Modelling / Estimation : PSD
Immanuel, *et al.* (2002)	E : VAc	Estimation : PSD
Immanuel & Doyle (2002)	E : VAc / BuA	Optimization / Control : PSD
Kiparissides, *et al.* (2002)	S : MMA	Optimization / Control : MWD
Ohmura, *et al.* (1998)	E : VAc	Control : PSD
Park, *et al.* (2002)	So : MMA /MA	Modelling / Estimation : MWD / CCD
Pringle & MacGregor, (1998)	So : St	Optimization / MWD
Saldivar & Ray (1997a)	E : MMA / VAc	Control : MWD / CCD
Santos, *et al.* (2000)	E : MMA / VAc	Modelling / Estimation : CCD
Sayer, *et al.* (2000)	E : MMA / BuA	Control : MWD / CCD
Sayer, *et al.* (2001)	E : MMA / BuA	Optimization : MWD / CCD
Semino & Ray (1995)	E	Control : PSD
Stavropoulos, *et al.* (2001)	E : St / 2-EHA	Modelling : CCD
Valappil and Georgakis (2002)	E : St	Control : MWD / CCD
Vicente, *et al.* (2001)	E : St / BuA	Control : MWD / CCD
Yabuki & MacGregor (1997)	E : St	Control : MWD
Yoo, *et al.* (1999)	So : St	Control : MWD

Process. C : Catalytic, E : Emulsion, S : Suspension, So : Solution

parameters, $\mathbf{Y_{est}}$, enter the optimizer block. A set of target values associated with the molecular weight properties, $\mathbf{Y_{q,sp}}$, and a performance index that specifies the overall process control objectives in a hierarchical order conclude the input stream to the optimization block. The optimizer computes a series of optimal set-point changes for the controlled variables that would ensure the satisfaction of the overall control objectives in an optimal sense. The time optimal sequence of set-point changes, $\mathbf{Y_{opt,sp}}$, are then passed to the regulatory controller that forces the process to follow the optimal trajectory as close as possible. The execution of the optimization task is performed periodically, depending on the rate of change of the model parameters, the frequency of the process disturbances and the dynamic behavior of the process in tracking the optimal control trajectory.

Optimizing Control of a Batch Polymerization Reactor. The control scheme outlined in Figure 8 was applied to a simulated model of a free-radical methyl-methacrylate (MMA) batch polymerization reactor. The polymerization temperature was controlled by a cascade control system consisting of a primary PID and two secondary PI controllers. The polymerization exhibits a strong acceleration in reaction rate due to gel-effect (e.g., the termination rate constant decreases with conversion). Usually,

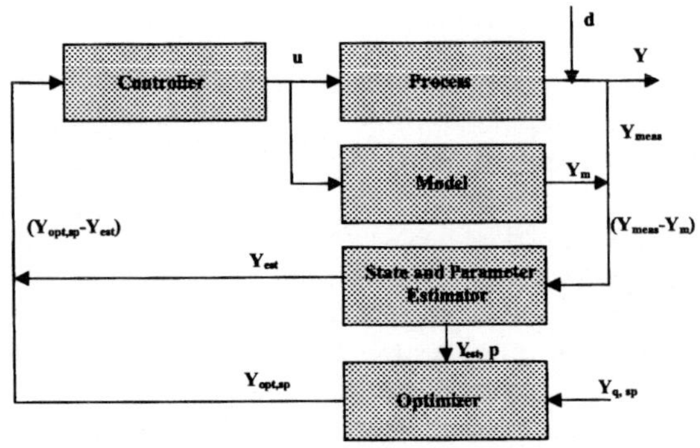

Fig. 8. Schematic representation of the control loop

the gel-effect contribution on the termination rate constant can be expressed in terms of an intrinsic chemical rate constant, k_{t0}, and a diffusion-controlled gel-effect function g_t, accounting for the observed decrease of k_t with monomer conversion. Because of the uncertainty in determining the value of g_t, a correction term, $g_{t,corr}$, was employed to account for the imprecise knowledge of g_t.

$$k_t = k_{t0}\, g_t\, g_{t,corr} \tag{24}$$

An EKF was employed to provide estimates for the process state variables and the time varying parameter $g_{t,corr}$ that was assumed to follow a random walk model:

$$g^s_{t,corr} = w^s \tag{25}$$

where $\mathbf{w}^s$ is a zero mean, white Gaussian noise.

At the optimization level, the time optimal sequence of temperature set-point changes was calculated by minimizing the following performance index:

$$J = w_1\left(\frac{X_f}{X_d} - 1.0\right)^2 + w_2\left(\frac{t_f}{t_d}\right)^2 + w_3\left(\frac{M_{n,f}}{M_{n,d}} - 1.0\right)^2 + w_4\left(\frac{M_{w,f}}{M_{w,d}} - 1.0\right)^2 + w_5\sum_{i=1}^{NC}\left(\frac{WCLD_i}{WCLD_{i,d}} - 1.0\right)^2 \tag{26}$$

where X_f, t_f, $M_{n,f}$, $M_{w,f}$, and $WCLD_{i,f}$ are the monomer conversion, batch time, number and weight average molecular weights and the values of the WCLD (evaluated at N_C collocation points), respectively. X_d, t_d, $M_{n,d}$, $M_{w,d}$, and $WCLD_{i,d}$ are the corresponding desired values of the variables. The weighting factors, w_1-w_5 in eq 26 specify the relative importance of the corresponding competing terms.

The polymerization temperature was selected as the control variable because of its apparent effect on the reaction kinetics and consequently, on the final molecular weight properties and monomer conversion. The minimization of the performance index in eq 26 subject to the polymerization model equations was carried out using a simultaneous discretization approach. An OCFE technique was employed to discretize the set of model differential equations.

The time optimal polymerization temperature that minimized the objective function (eq 26) was calculated for a desired polymer grade ($M_{n,d}$=4.0·10^5 kg/kmol, $M_{w,d}$=2.0·10^6 kg/kmol, X_d=0.85). The state variables and the time varying gel-effect parameter, $g^s_{t,corr}$, were estimated every two minutes. A new optimal temperature trajectory was evaluated every 20 minutes using the most recent value of the termination rate constant. In the simulated process-model mismatch case, the value of $g_{t,corr}$ term in eq 24 was assumed to increase linearly with time (i.e., $g_{t,corr}$=1+ct).

Figure 9 depicts the time optimal temperature profiles at different time instances, calculated through successive state/parameter estimation and optimization steps. Notice that the updated time optimal temperature profiles differ from the polymerization temperature profile calculated at the beginning of the batch both in shape and duration due to the process-model mismatch caused by the variation in the termination rate constant. In Figure 10 the average molecular weights corresponding to the uncorrected and updated temperature profiles are plotted with respect to batch time. Finally, Figure 11 shows the time evolution of the WCLD. It can be seen that the final WCLD coincides with the desired one. The results of Figures 10 and 11 indicate that the updated temperature trajectories result in a definite improvement of the final product properties and successfully compensate for the effect of the time varying termination rate constant, k_t.

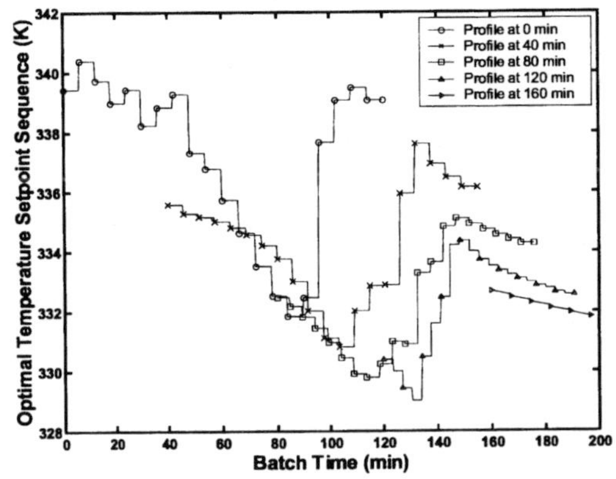

Fig. 9. Updated time optimal temperature trajectories for a positive disturbance scenario in k_t ($g_{t,corr}$=1+8.0×10^{-3} t).

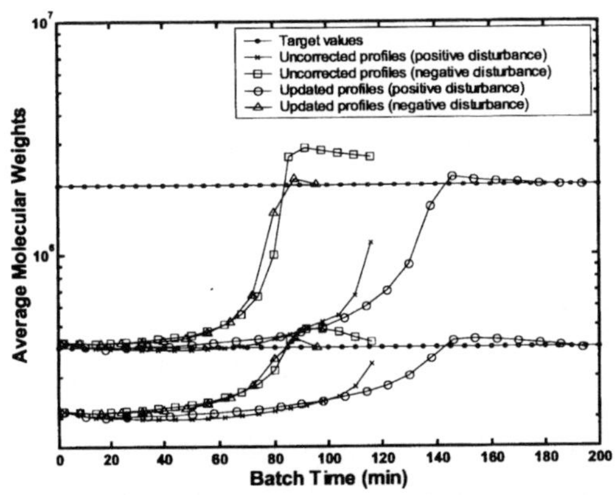

Fig. 10. Uncorrected and updated number and weight average molecular weight profiles.

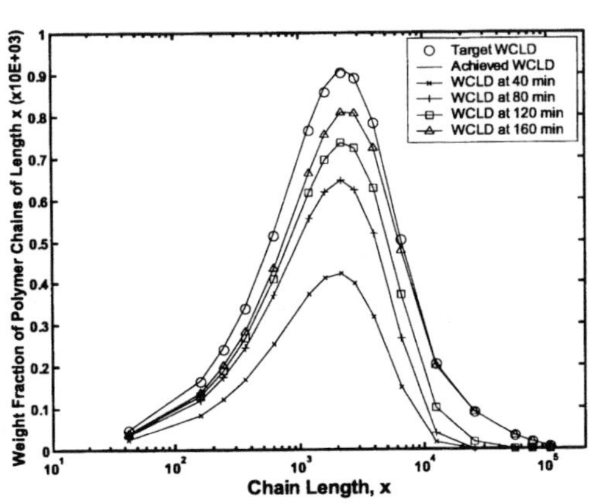

Fig. 11. Time evolution of WCLD.

4. FUTURE DIRECTIONS

New Issues on Population Balance Models. In the general form of PBE (eq. 4), it is recognized that the distribution function $n(x,y,z,p_1,\ldots,p_k,t)$ has some "external" and some "internal" coordinates. The external coordinates (x,y,z) express the change in the population in a particular volume element due to the particle velocity, $\underset{\sim}{u}$. The summation term expresses the change in the population due to the particle properties, p_i. This term always includes the particle size as an internal coordinate but the summation sign recognizes the fact that there may be more than one property of interest. As discussed in the recent paper of Verkoeijen, *et al.* (2002), the internal particle parameter that determines changes in the population should be the particle volume. As a result, the basic PBE should be expressed in terms of the mass distribution, or the volume distribution if the density is constant, so the mass in the system is automatically conserved.

Traditionally, one-dimensional PBEs have been employed to describe the evolution of a population. However, in many cases this assumption does not hold true. Thus, to follow the molecular and compositional developments in a copolymerization reactor one needs, for example, to introduce tri-variate distributions for the "live" and "dead" polymer chains. Accordingly, the notation P(n,m,b,t) denotes the concentration of "dead" polymer chains having n units of monomer M_1, m units of monomer M_2, and b long chain branches (LCB) per molecule. The general rate functions for such a polymerization system are given in Achilias and Kiparissides (1992). It should be pointed out that the calculation of the tri-variate molecular weight – copolymer composition – long chain branching distribution is an extremely difficult problem. Moreover, in many particulate polymerization processes, 2-D or 3-D population balances are required to describe the evolution of the PSD in the reactor. Thus, in addition to the standard particle volume property one needs to include the particle age or catalyst activity to follow the evolution of PSD in a gas-solid Ziegler-Natta olefin polymerization reactor. On the other hand, in emulsion polymerization, it is often necessary to introduce a bivariate distribution n(V,i,t) to describe the time-evolution of PSD with respect to the particle volume, V, and the number of radicals, i, in the particle. The numerical solution of the resulting 2-D or 3-D dynamic PBE in batch and continuous polymerization reactors is a very demanding numerical problem. In the review article of Ramkrishna and Mahoney (2002), present challenges and requirements regarding the formulation and numerical solution of multi-dimensional PBEs are discussed. Needless to say that the identification of the functional forms of particle nucleation, growth, "death" and "birth" rate terms appearing in a multi-dimensional PBE is by no means a trivial task.

In most publications on the modeling of particulate polymerization processes, it is assumed that the multiphase reaction system can be approximated by a pseudo-homogeneous medium, (i.e., the population does not vary with respect to the external (spatial) coordinates. However, this is rarely the case in industrial polymerization processes. For example, in industrial semi-batch emulsion polymerization, the mixing of the monomer feed stream with the reactor content and, thus, the transfer of monomer to the particulate polymer phase may vary with respect to the spatial coordinates. Moreover, particulate polymerization processes are characterized by a particle volume increase during polymerization that strongly influences the rheological properties and, thus, mixing of the dispersion. Consequently, a range of different problems can arise during the course of a particulate polymerization, including the appearance of a secondary particle size distribution, catastrophic particle agglomeration, particle settling, hot spots, etc. In gas-solid polyolefin fluidized bed reactors, the distribution of solids along the bed-height can be far from uniform. Thus, particle segregation in the bed should be accounted for in any relevant PSD modeling study. In a recent publication by Alexopoulos, *et al.* (2002), a computational fluid dynamic (CFD) approach is described for calculating the spatial distribution of energy dissipation rate in a batch suspension polymerization reactor. A two-compartment model is derived to calculate the time evolution of the monomer droplet size distribution by taking into account the large spatial variations of the turbulent kinetic energy in the reactor.

Such complex multiphase polymerization processes face challenges both on issues of model formulation as well as computation. In fact, the mutual coupling of the particulate and continuous fluid phases in regard to the transport of momentum, mass and energy as well as the quantification of particle nucleation, growth and aggregation mechanisms in terms of spatial coordinates increase significantly the modeling and computational efforts for calculating the time evolution of PSD.

New Issues on Optimization and Control. Despite the rich literature on the dynamics of population balance systems, there are limited industrial applications on the dynamic optimization and control of particulate polymerization processes. There are several valid reasons for the lack in closed-loop "quality" control applications for particulate polymerization processes. More specifically,

1. Particulate polymerization processes are highly nonlinear, containing a large number of time-varying kinetic and transport parameters.
2. Modeling of particulate polymerization systems typically leads to nonlinear integro-differential equations. Their infinite-dimensional nature does not allow their direct use for the design of nonlinear controllers that can be easily implemented in real-time.
3. On-line measurement of MWD, CCD, LCB, PSD, etc. is very difficult and the available secondary measurements are often insufficient for the accurate inference of the entire molecular weight or/and particle size distribution.
4. Formulation of a meaningful objective function in terms of distributed molecular and morphological properties (e.g., MWD, PSD, etc.) is not easy, especially when the end-use properties of the product need to be controlled. Furthermore, the controlled variables react in opposite directions to changes in the manipulated variables.

However, recent advances in online measurements (Kammona, et al. 1999) and in control theory (Christofides, 2001 ; Chiu and Christofides, 1999 ; Crowley, *et al.* 2000 ; Valappil and Georgakis 2002 and 2003) have made possible the nonlinear optimization and control of molecular weight and particle size distributions in particulate polymerization processes. Key to the successful application of optimal control to a particulate polymerization processes is the availability of appropriate mathematical models that need to range from detailed fundamental mechanistic forms to database forms such as those provided by statistical (Yabuki, *et al.* 2002) and neural network approaches.

Needless to say that the diverse requirements imposed by various particulate polymerization processes require equally diverse control strategies to address the specific process goals. In particular, for semi-batch processes different control strategies, including batch-to-batch, on-line model-based control and mid-course correction, have been proposed by Lee, *et al.* (2002), Pringle and MacGregor (1998), Yabuki and MacGregor (1997) and Doyle, *et al.* (2003) to control the MWD and PSD developments in a polymerization reactor. On the other hand, for continuous particulate polymerization processes, on-line "polymer quality" soft-sensing, optimal grade changeover and scheduling are some of the issues to be addressed for maximizing productivity, improving product quality while keeping safety regulations (Chatzidoukas, *et al.* 2003 ; BenAmor, *et al.* 2004).

5. REFERENCES

Achilias, D. and Kiparissides, C. (1992). Toward the Development of a General Framework for Modeling Molecular Weight and Compositional Changes in Free Radical Copolymerization Reactions, *J.M.S.- Rev. Macromol. Chem. Phys.* **C32**, 183-234.

Ahn, S., Park, M.J. and Rhee, H.K. (1999). Extended Kalman Filter-Based Nonlinear Model Predictive Control for a Continuous MMA Polymerization Reactor. *Ind. Eng. Chem. Res.*, **38**, 3942-3949.

Alexopoulos, A.H., Maggioris, D. and Kiparissides, C. (2002). CFD Analysis of Turbulence Non-Homogeneity in Mixing Vessels a Two-compartment Model, *Chem Eng Sci,* **57(10)**, 1735-1752.

Alexopoulos A.H., Roussos, A.I. and Kiparissides, C. (2004). Part I: Dynamic Evolution of the Particle Size Distribution in Particulate Processes Undergoing Combined Particle Growth and Aggregation. *Chem. Eng. Sci. (submitted)*

Alexopoulos A.H. and Kiparissides, C. (2004) Part II: Dynamic Evolution of the Particle Size Distribution in Particulate Processes: Nucleation Combined with Particle Growth and Aggregation. *Chem. Eng. Sci. (submitted)*

Asua, J.M., Saenz de Buruaga, I., Arotcarena, M., Urretabizkaia, A., Armitage, P.D., Guiglotta,

L.M. and Leiza, J.R. (1995). On-line Control of Emulsion Polymerization Reactors. *DECHEMA Monographs, VCH Verlagsgesellschaft*, **131**, 655.

Batterham, R.J., Hall, J.S., Barton, G. and Pelletizing (1981). Kinetics and Simulation for Full-Scale Balling Circuits, in: *Proc.3rd.Int. Symp. on Aggregation,* Nurnberg, W. Germany, p. A136.

BenAmor, S., Doyle III, F.J. and McFarlane, R. (2004). Polymer Grade Transition Control Using Advanced Real-time Optimization Software. *J. Proc. Contr.*, **14**, 349-364

Bleck, R. (1970). A Fast, Approximate Method for Integrating the Stochastic Coalescence Equation. *J. Geophysical Res.* **75**, 5165-5171.

Brandolin, A. and Garcia-Rubio, L.H. (1991). On-line Particle Size Distribution Measurements for Latex Reactors, in *Particle Size Distribution II – Assessment and Characterization*, Th. Provder, Ed., ACS Symposium Series, ACS Washington DC, Vol. **472**, p. 64.

Catalgil-Giz, H., Giz, A., Alb, A.M., Koc, O. and Reed W. F. (2002). Online Monitoring of Composition, Sequence Length, and Molecular Weight Distributions during Free Radical Copolymerization, and Subsequent Determination of Reactivity Ratios. *Macromolecules,* **35**, 6557 – 6571

Cerrillo, J.F. and MacGregor, J.F. (2002). Control of Particle Size Distributions in Emulsion Semi-batch Polymerization using Mid - Course Correction Policies. *Ind. Eng. Chem. Res.*, **41**, 1805 – 1814.

Chang, J.S. and Hung, B.C. (2002). Optimization of batch polymerization reactors using neural-network rate-function models. *Ind. Eng. Chem. Res.*, **41**, 2716 – 2727

Chatzi, E.G., Gavrielides, A.D. and Kiparissides, C. (1989). Generalized Model for Prediction of the Steady-State Drop Size Distributions in Batch Stirred Vessels. *Ind. Eng. Chem. Res.* **28**, 1704.

Chatzi, E.G. and Kiparissides, C. (1992). Dynamic Simulations of Bimodal Drop Size Distributions in Low-Coalescence Batch Dispersion Systems. *Chem. Eng. Sci.,* **47**(2), 445.

Chatzi, E.G., Kammona, O. and Kiparissides, C. (1997). Use of a Mid-Range Infrared Optical-Fiber Probe for the On-line Monitoring of 2-Ethylhexyl Acrylate/Styrene Emulsion Copolymerization, *J. Appl. Pol. Sci.*, **63**, 799.

Chatzidoukas, C., Perkins, J.D., Pistikopoulos, E.N. and Kiparissides, C. (2003). Optimal Grade Transition and Selection of Feedback Control Loops in a Gas-Phase Polyolefin FBR using a Mixed Integer Dynamic Optimisation Approach. *Chem Eng Sci*, **58**, 3643-3685.

Chen, M.-Q., Hwang, C. and Shih, Y.-P. (1996). A Wavelet-Galerkin Method for Solving Population Balance Equations. *Comp. Chem. Eng.*, **20**(2), 131-145.

Cherfi, A., Fevotte, G. and Novat, C. (2002). Robust On-line Measurement of Conversion and Molecular Weight using NIR Spectroscopy during Solution Polymerization. *J. Appl. Pol. Sci.*, **85**, 2510.

Chiu, W. Y., Carratt, G. M., and Soong, D. S. (1983). A Computer Model for the Gel Effect in Free-Radical Polymerization. *Macromolecules*, **16**, 348-357

Chiu, W. Y. and Christofides P. (1999). Nonlinear Control of Particulate Processes. *AIChE J.*, **45**, 1279.

Christofides, P.D. (2001). Control of Nonlinear Distributed Process Systems: Recent Developments and Challenges. *AIChE J.*, **47**, 514 – 518.

Congalidis, J.P., Richards, J.R. and Ray, W.H. (1989). Feedforward and Feedback Control of a Solution Copolymerization Rector. *AICHE J.*, **35**, 891 – 907

Crowley, T.J. and Choi K.Y. (1997a). Discrete Optimal Control of Molecular Weight Distribution in a Batch Free Radical Polymerization Process. *Ind. Eng. Chem. Res.*, **36**, 3676 – 3684

Crowley, T.J. and Choi, K.Y. (1997b). Calculation of Molecular Weight Distribution From Molecular Weight Moments in Free-Radical Polymerization. *Ind. Eng. Chem. Res.*, **36**, 1419-1423.

Crowley, T. J. and Choi, K. Y. (1998). Experimental Studies on Optimal Molecular Weight Distribution Control in a Batch Free-Radical Polymerization Process. *Chem. Eng. Sci.*, **53**(15), 2769-2790.

Crowley, T.J. and Choi K.Y. (1999). Control of Copolymer Hydrodynamic Volume Distribution in a Semibatch Free Radical Copolymerization Process. *Comput Chem Eng,* **23**, 1153 – 1165

Crowley, T.J. Meadows, E.S., Kostoulas, E. and Doyle III, F.J. (2000). Control of Particle Size Distribution described by a Population Balance Model of Semi-Batch Emulsion Polymerization. *J. Process Control*, **10**, 419 – 432

Dafniotis P. (1996) Modeling of Emulsion Copolymerization Reactors Operating below the Critical Micelle Concentration. *Ph.D. Thesis, University of Wisconsin-Madison.*

De Valiere, P. and Bonvin, D. (1989). Application of Estimation Techniques to Batch Reactors-II. Experimental Studies in State and Parameter Estimation. *Comput. Chem. Eng.*, **13**, 11-20.

Deuflhard, P. and Wulkow, M. (1989). Computational Treatment of Polyreactions Kinetics by Orthogonal Polynomials of a Discrete Variable. *Impact of Computing in Science and Engineering*, **1**, 269-301.

Dimitratos, J., Georgakis, C., El-Aasser, M.S. and Klein, A. (1989). Dynamic Modeling and State Estimation for an Emulsion Copolymerization Reactor. *Comput. Chem. Eng.*, **13**, 21.

Dimitratos, J., Georgakis, C., El-Aasser, M.S. and Klein, A. (1991). An Experimental Study of Adaptive Kalman Filtering in Emulsion Copolymerization. *Chem. Eng. Sci.*, **46**, 3203-3218.

Dimitratos, J., Elicabe, G. and Georgakis, C. (1994). Control of Polymerization Reactors. *AIChE J.* **40**, 1993 – 2021

DosRamos, J.G. and Silebi, C.A. (1991). Size Analysis of Simple and Complex Mixtures of Colloids in the Submicrometer Range using Capillary Hydrodynamic Fractionation, in

Particle Size Distribution II – Assessment and Characterization, Th. Provder, Ed., ACS Symposium Series, ACS Washington DC, Vol. **472**, p. 292.

Dougherty, E.P., (1986). The SCOPE Dynamic Model for Emulsion Polymerization I. Theory. *J. Appl. Pol. Sci.*, **32**(1), 3051.

Doyle III, F.J., Harrison, C.A. and Crowley, T.J. (2003). Hybrid Model-Based approach to Batch-to-Batch Control of Particle Size Distribution in Emulsion Polymerization. *Comp. Chem. Eng.*, **27**, 1153 – 1163

Ellis, M.F., Taylor, T.W., Gonzalez, V. and Jensen K. (1988). Estimation of the Molecular Weight Distribution in Batch Polymerization. *AIChE J.*, **34**, 1341 – 1353

Ellis, M.F., Taylor, T.W. and Jensen K.F. (1994). Online Molecular Weight Distribution Estimation and Control in Batch Polymerization. *AIChE J.*, **40**, 445-462.

Embirucu, M., Lima, E.L. and Pinto, J.C. (1996). A Survey of Advanced Control of Polymerization Reactors. *Polymer Engineering Science*, **36**, 433 – 447

Erasmus, L.D., Eyre, D. and Everson, R.C. (1994). Numerical Treatment of the Population Balance Equation Using a Spline-Galerkin method, *Comput Chem Eng*, **18**(9), 775-783

Fèvotte, G., McKenna, T.F., Othman, S. and Hammouri, H. (1998a). Nonlinear Tracking of Glass Transition Temperatures for Free Radical Emulsion Copolymers. *Chem. Eng. Sci.*, **53**, 773-786.

Fèvotte, G., McKenna, T.F., Othman, S. and Santos, A.M. (1998b). A Combined Hardware/Software Sensing approach for On-Line Control of Emulsion Polymerization Process. *Comp. Chem. Eng.*, **22**, 443 – 449

Finlayson, B.A. (1980). *Nonlinear Analysis in Chemical Engineering.* McGraw-Hill, NY, New York

Gagnon, L. and MacGregor, J.F. (1991). State Estimation for Continuous Emulsion Polymerization. *Can. J. Chem. Eng.*, **69**, 648-656.

Gelbard, F. and Seinfeld, J.H. (1978). Numerical solution of the dynamical equation for particulate systems, *Jf Comput Phys*, ***28***, 357-375

Gelbard, F. and Seinfeld, J.H. (1979). Exact Solution of the General Dynamic Equation for Aerosol Growth by Condensation. *J. Colloid Interface Sci.*, **68**(1), 173.

Gelbard, F. and Seinfeld, J.H. (1980). Simulation of Multicomponent Aerosol Dynamics. *J. Colloid Interface Sci.*, **78**(2), 485.

Gelbard F., Tambour Y. and Seinfeld J.H. (1980) Sectional Representation for Simulating Aerosol Dynamics. *J. Colloid Interface Science*, **76**(2), 541-556

Gossen, P.D., MacGregor, J.F. and Pelton, R.H. (1993). Composition and Particle Diameter for Styrene/Methyl Methacrylate Copolymer Latex using UV and NIR Spectroscopy. *Appl. Spectr.*, **47**(11), 928.

Guiglotta, L.M., Arotcarena, M., Leiza, J.R. and Asua, J.M. (1995). Estimation of Conversion and copolymer Composition in Semicontinuous Emulsion Polymerization using Calorimetric Data. *Polymer*, **36**(10), 2019.

Hidy, G.M. (1965). On the Theory of the Coagulation of Noninteracting Particles in Brownian Motion. *J. Colloid Sci.*, **20**, 123.

Himmelblau, D.M. and Bischoff, K.B. (1968). *Process Analysis and Simulation. Deterministic Systems.* John Wiley & Sons, Inc

Hounslow M.J., Ryall, R.L. and Marshall, V.R. (1988). Discretized Population Balance for Nucleation, Growth, and Aggregation. *AIChE J.*, **34**(11), 1821-1832.

Hulburt, H.M. and Katz, S. (1964). Some Problems in Particle technology. A Statistical Mechanical Formulation. *Chem. Eng. Sci.*, **19**, 555

Iedema, P. D., Wulkow M. and Hoefsloot, C. J. (2000). Modeling Molecular Weight and Degree of Branching Distribution of Low-Density Polyethylene. *Macromolecules*, **33**(19), 7173-7184.

Immanuel, C.D., Cordeiro, C.F., Sundaram, S.S., Meadows, E.S., Crowley, T.J. and Doyle III, F.J. (2002). Modeling of Particle Size Distribution in Emulsion Co-Polymerization: Comparison with Experimental Data and parametric Sensitivity Studies. *Comp. Chem. Eng.*, **26**, 1133 – 1152

Immanuel, C.D. and Doyle III, F.J. (2002). Open-Loop Control of Particle Size Distribution in Semibatch Emulsion Copolymerization using a Genetic Algorithm. *Chem. Eng. Sci.*, **57**, 4415 – 4427

Jang, S.S., Joseph, B. and Mukai, H. (1987). Control of Constrained Multivariable Nonlinear Process using a Two-Phase Approach. *Ind. Eng. Chem. Res.*, **26**, 2106-2114.

Kammona, O., Chatzi, E.G. and Kiparissides, C. (1999). Recent Developments in Hardware Sensors for the On-line Monitoring of Polymerization Reactions. *J. Macromol Sci., Rev Macromol Chem Phys*, **C39**(1), 57-134

Kammona, O., Pladis, P., Frantzikinakis, C. and Kiparissides, C. (2003). A Comprehensive Experimental and Theoretical Investigation of the Styrene/2-Ethylhexyl Acrylate Emulsion Copolymerization. *Macromol.Chem.Phys.*, **204**, 983-999

Kiparissides, C. (1996). Polymerization reactor modeling: A Review of Recent Developments and Future Directions. *Chem. Eng. Sci.*, **51**, 1637

Kiparissides, C., Seferlis, P., Mourikas, G. and Morris, A.J. (2002). Online Optimizing Control of Molecular Weight Properties in Batch Free-Radical Polymerization Reactors. *Ind. Eng. Chem. Res.*, **41**, 6120 – 6131

Kostoglou M. and Karabelas A.J. (1994). Evaluation of Zero Order Methods for Simulating Particle Coagulation. *J. Colloid Interface Sci.*, **163**, 420.

Kostoglou M. and Karabelas A.J. (1995). Evaluation of Numerical Methods for Simulating an Evolving Particle Size Distribution in Growth Processes. *Chem. Eng. Comm.*, **136**, 177.

Kozub, D.J. and MacGregor, J.F. (1992). State Estimation for Semi-Batch Polymerization Reactors. *Chem. Eng. Sci.*, **47**, 1047-1062.

Kumar, S. and Ramkrishna, D. (1996a). On the Solution of Population Balance Equations by

Discretization-I. A Fixed Pivot Technique. *Chem. Eng. Sci.*, **51**(8), 1311-1332.

Kumar, S. and Ramkrishna, D. (1996b). On the Solution of Population Balance Equations by Discretization-II. A Moving Pivot Technique. *Chem. Eng. Sci.*, **51**(8), 1333-1342.

Kumar, S. and Ramkrishna, D. (1997). On The Solution of Population Balance Equations by Discretization III. Simultaneous Nucleation, Growth and Aggregation. *Chem. Eng. Sci., 52*, 4659-4679.

Landgrebe, J.D. and Pratsinis, S.E. (1990). A Discrete-Sectional Model for Powder Production by Gas-Phase Chemical reaction and Aerosol Coagulation in the Free-Molecular Regime. *J. Colloid Interface Sci.*, **139**(1), 63-86.

Lazaridis, N., Alexopoulos, A.H., Chatzi, E.G. and Kiparissides, C. (1999). Steric Stabilization in Emulsion Polymerization Using Oligomeric Nonionic Surfactants. *Chem. Eng. Sci.*, **54**, 3251-3261.

Lee, K., Lee, J.H., Yang, D.R. and Mahoney, A.W. (2002). Integrated Run-to Run and On-line Model-Based Control of Particle Size Distribution for a Semi-Batch Precipitation Reactor, *Comp. Chem. Eng.*, **26**, 1117-1131.

Litster, J.D., Smit, D.J. and Hounslow, M.J. (1995). Adjustable Discretized Population Balance for Growth and Aggregation. *AIChE J.*, **41**, 591-603.

MacGregor, J.F., Kozub, D.J., Penlidis, A. and Hamielec, A.E. (1986). State Estimation for Polymerization Reactors. *IFAC Symp. On Dynamic and Control of Chemical Reactors and Distillation Columns*, Bournemouth, UK, Pergamon Press.

Maggioris, D., Goulas, A., Alexopoulos, A.H., Chatzi, E.G. and Kiparissides C. (2000). Prediction of Particle Size Distribution in Suspension Polymerization Reactors: Effect of Turbulence Nonhomogeneity. *Chem. Eng. Sci.*, **55**, 4611.

Mahoney, A.W. and Ramkrishna, D. (2002). Efficient Solution of Population Balance Equations with Discontinuities by Finite Elements. *Chem Eng Sci*, **57**, 1107-1119

Marchal P., David, R., Klein, J.P. and Villermaux, J. (1988). Crystallization and Precipitation Engineering-I. An Efficient Method for Solving Population Balance in Crystallization with Agglomeration. *Chem. Eng. Sci.*, **43**(1), 59-67.

Mikes, F., Gonzalez-Benito, B., Serrano, B., Bravo, J. and Baselga, J., (2002). Fluorescence Monitoring of Polymerization Reaction. A New Method for Treating Fluorescence Experimental Data. *Polymer*, **43**, 4331.

Muske, K.R., Rawlings, J.B. and Lee, J.H. (1995). Receding Horizon Recursive State Estimation. In R. Berber (ed), *Methods of Model Based Process Control*, 349-365, Kluwer, Dordrecht.

Mutha, R.K., Cluett, W.R. and Penlidis, A. (1997). A New Multirate-Measurement-Based Estimator: Emulsion Copolymerization Batch Reactor Case Study. *Ind. Eng. Chem. Res.*, **36**, 1036-1047.

Napper, D. H and Gilbert, R.G. (1989). Polymerization in Emulsions, in: *Comprehensive Polymer Science*, I. Pirma, Eds., Pergamon, Oxford, Vol. IV, p.171.

Nicmanis, M. and Hounslow, M.J. (1996). A Finite Element Analysis of the Steady State Population Balance Equation for Particulate Systems: Aggregation and Growth. *Comp. Chem. Eng.*, **20**, S261.

Nicmanis, M. and Hounslow, M.J. (1998). Finite-Element Methods for Steady-State Population Balance Equations. *AIChE J.*, **44**, 2258.

Ohmura, N., Kataoka, K., Watanabe, S. and Okubo, M. (1998). Controlling Particle Size by Self-Sustained Oscillations in Continuous Emulsion Polymerization of Vinyl Acetate. *Chem. Eng. Sci.*, **53**, 2129 – 2135

Park, M.J., Hur, S.M. and Rhee, H.K. (2002). Online Estimation and Control of Polymer Quality in a Polymerization Reactor. *AIChE J.*, **48**, 1013 – 1021

Pilinis C. (1990) Derivation and Numerical Solution of the Species Mass Distribution Equation for Multicomponent Particulate Systems, *Atmos. Environ. A*, **24** (7), 1923–1928

Pladis, P. and Kiparissides, C. (1998). A Comprehensive Model for the Calculation of Molecular Weight-Long-Chain Branching Distribution in Free-Radical Polymerizations. *Chem. Eng. Sci.*, **53**(18), 3315-3333.

Pringle, T.L.C., and MacGregor, J. F. (1998). Optimization of Molecular Weight Distribution using Batch-to-Batch Adjustments. *Ind. Eng. Chem. Res.*, **37**, 3660 – 3669

Ramkrishna, D. (1985). The Status of Population Balances. *Reviews Chem. Eng.*, **3**(1), 49.

Ramkrishna, D. and Mahoney, A.W. (2002). Population Balance Modeling. Promise for the Future. *Chem Eng Sci*, **57**, 595-606

Richards, J.R., Congalidis, J.P. and Gilbert, R., (1989). Mathematical Modeling of Emulsion Copolymerization Reactors. *J. Appl. Pol. Sci.*, **37**, 2727.

Rigopoulos S., and Jones A.G. (2003). Finite-Element Scheme for Solution of the Dynamic Population Balance Equation. *AIChE J.*, **49**(5), 1127-1139

Rudschuck, S., Adams, J., Fuhrmann, J., (1999). Online Observation of Emulsion Polymerization by Fluorescence Technique. *Nuclear Instruments and Methods in Physics Research B*, **151**, 341.

Saldivar, E. and Ray, W.H., (1997a). Control of Semicontinuous Emulsion Copolymerization Reactor. *AIChE J.*, **43**, 2021 – 2033

Saldivar, E. and Ray, W.H., (1997b). Mathematical Modeling of Emulsion Copolymerization Reactors: Experimental Validation and Application to Complex Systems. *Ind. Eng. Chem. Res.*, **36**(4), 1322

Sampson K.J., and Ramkrishna D. (1985) A New Solution to the Brownian Coagulation Equation through the Use of Root-Shifted Problem-Specific Polynomials, *J. Coll Int. Sci.*, **103**(1), 245-254.

Santos, A. M., Fevotte, G., Othman, N. and McKenna, T.F. (2000). On-line monitoring of methyl-methacrylate-vinyl acetate emulsion copolymerization. *J. Appl. Polym. Sci.*, **75**, 1667 – 1683

Sandu, A. and Borden, C. (2003). A Framework for the Numerical Treatment of Aerosol Dynamics. *Appl. Num. Math.*, **45**, 475-497.

Sasic, S., Ozaki, Y., Olinga, A. and Siesler, H.W., (2002). Comparison of Various Chemometric Evaluation approaches for On-line FT-NIR Transmission and FT-MIR/ATR Spectroscopic Data of Methyl Methacrylate Solution Polymerization. *Anal. Chim. Acta*, **452**, 265.

Sastry K.V.S, and Gaschignard P., '*Discretization Procedure for the Coalescence of Particulate Processes*', Ind. Eng. Chem. Fundam., **20**, 355-361 (1981).Sayer, C., Lima, E.L., Pinto, J.C., Arzamendi, G. and Asua, J.M. (2000). Molecular Weight Distribution in Composition controlled Emulsion Polymerization. *J. Appl. Polym. Sci.*, **38**, 1100 – 1109

Sayer, C., Arzamendi, G., Asua, J.M., Lima, E.L. and Pinto, J.C. (2001). Dynamic Optimization of Semicontinuous Emulsion Copolymerization Reactions: Composition and Molecular Weight Distribution. *Comp. Chem. Eng.*, **25**, 839 – 849

Semino, D., and Ray, W.H. (1995). Control of Systems described by Population Balance Equations – I. Controllability Analysis. *Chem. Eng. Sci.*, **50**, 1805 – 1824

Singh P.N. and Ramkrishna D. (1975). Transient Solution of the Brownian Coagulation Equation by Problem-Specific Polynomials, *J. Coll Int. Sci.*, **53**(2), 214-223

Soroush, M. (1997). Nonlinear State-Observer Design with Application to Reactors. *Chem. Eng. Sci.*, **52**, 387-404.

Soroush, M. (1998). State and Parameter Estimations and their Applications in Process Control. *Comput. Chem. Eng.*, **23**, 229-245.

Stavropoulos, Y., Kammona, O., Chatzi, E.G. and Kiparissides, C. (2001). Multivariate Statistical Analysis of Mid-Infrared Spectra for the Online Monitoring of 2-Ethylhexyl Acrylate / Styrene Emulsion Copolymerization. *J. Appl. Polym. Sci.*, **82**, 1776 – 1787

Storti, G., Morbidelli, M. and Carra, S. (1989). Detailed Modeling of Multicomponent Emulsion Polymerization Systems", in: *Computer Applications in Applied Polymer Science II: Automation, Modelling and Simulation*, p.379, T. Provder, Eds., ACS Symposium. Series, 404, Washington DC,.

Tobita, H. (1993). Molecular Weight Distribution in Free-Radical Polymerization with Long-Chain Branching. *J. Polymer Sci.*, **31**, 1363.

Tobita, H. (1995). Molecular Weight Distribution in Free-Radical Polymerization with Chain-Length-Dependent Bimolecular Termination. 1. Bulk Polymerization. *Macromolecules*, **28**, 5119.

Tobita, H. and Ito K. (1993). On The Calculation of Molecular Weight Distribution from the Moments Using Laguerre Polynomials. *Polymer Reaction Eng.*, **1**(3), 407-425.

Thomas, J. and Dimonie, V. (1990). Fiber Optic Dynamic Light Scattering from Concentrated Dispersions. 3: Particle Sizing in Concentrates. *Appl. Optics*, **29**(36), 5332.

Valappil, J. and Georgakis, C. (2002). Nonlinear Model Predictive Control of End-Use Properties in Batch Reactors. *AIChE J.*, **48**, 2006 – 2021

Valappil, J. and Georgakis, C. (2003). Accounting for Batch Reactor Uncertainty in the Nonlinear MPC of End-Use Properties. *AIChE J.*, **49**, 1178-1192

Van Den Brink, M., Hansen, J.-F., De Peinder, P., Van Herk, A.M. and German, A.L., (2001). Measurement of Partial Conversions During the Solution Copolymerization of Styrene and Butyl Acrylate Using On-line Raman Spectroscopy. *J. Appl. Pol. Sci.*, **79**, 426.

Vanni M., (2000). Approximate Population Balance Equations for Aggregation-Breakage Processes. *J. Coll Int. Sci.*, **221**, 143-160.

Vega, M.P., Lima, E.L. and Pinto, J.C. (2001). In-line Monitoring of Weight Average Molecular Weight in Solution Polymerizations using Intrinsic Viscosity Measurements. *Polymer*, **42**, 3009.

Verkoeijen, D., Pouw, G.A., Meesters, G.M.H and Scarlett, B. (2002). Population Balances for Particulate Processes – A Volume Approach. *Chem. Eng. Sci.*, **57**, 2287-2303

Vicente, M., Leiza, J.R. and Asua, J.M. (2001). Simultaneous Control of Copolymer Composition and MWD in Emulsion Copolymerization. *AIChE J.*, **47**, 1594 – 1606

Vos, B.J., Grievink, J., de Jong, P.J. and Verheijen P.J.T. (1997). A Robust Mode-Based Approach for Online Monitoring of Dynamic Process Performance. *Paper 194d*, Annual AIChE Meeting, Los Angeles.

Wulkow, M. (1992). Adaptive Treatment of Polyreactions in Weighted Sequence Spaces. *Impact of Computing in Science and Engineering*, **4**, 153-193

Yabuki, Y., and MacGregor, J.F. (1997). Product Quality Control in Semibatch Reactor using Midcourse Correction policies. *Ind. Eng. Chem. Res.*, **36**, 1268 – 1275

Yabuki, Y., Nagasawa, T. and MacGregor, J.F. (2002). Industrial Experiences with Product Quality in Semi-Batch Processes. *Comp. Chem. Eng.*, **26**, 205 – 212

Yoo, K.Y., Jeong, G. and Rhee, H.K. (1999). Molecular Weight distribution Control in a Batch Polymerization Reactor using the On-line Two-Step Method. *Ind. Eng. Chem. Res.*, **38**, 4805 – 4814

Yuan, H.G., Kalfas, G. and Ray W.H. (1991). Suspension polymerization. *JMS – Rev. Macromol. Chem. Phys.*, **C31**, 215.

Zeaiter, J., Gomes, V.G., Romagnoli, J.A. and Barton, G.W., (2002). Inferential Conversion Monitoring and Control in Emulsion Polymerization through Calorimetric Measurements. *Chem. Eng. J.*, **89**, 37.

ELSEVIER
IFAC
PUBLICATIONS
www.elsevier.com/locate/ifac

PERSPECTIVES ON THE DESIGN AND CONTROL OF MULTISCALE SYSTEMS

R. D. Braatz, R. C. Alkire, E. Seebauer, E. Rusli, R. Gunawan, T. O. Drews, X. Li, Y. He

Department of Chemical and Biomolecular Engineering, University of Illinois at Urbana-Champaign, 600 South Mathews Avenue, Urbana, Illinois 61801

Abstract: New applications in materials, medicine, and computers are being discovered where the control of events at the molecular and nanoscopic scales is critical to product quality, although the primary manipulation of these events during processing occurs at macroscopic length scales. This motivates the creation of tools for the design and control of multiscale systems that have length scales ranging from the atomistic to the macroscopic. This paper describes a systematic approach that consists of stochastic parameter sensitivity analysis, Bayesian parameter estimation applied to *ab initio* calculations and experimental data, model-based experimental design, hypothesis mechanism selection, and multistep optimization. *Copyright © 2004 IFAC*

Keywords: nonlinear systems, distributed parameter systems, complex systems, stochastic systems, uncertain dynamic systems, optimal control, system sensitivity.

1. INTRODUCTION

New applications in materials, medicine, and computers are being discovered where the control of events at the molecular and nanoscopic length scales is critical to product quality, although the primary manipulation of these events during processing occurs at macroscopic length scales (e.g. temperature of the system, valves on flows into and out of the system, applied potential between two electrodes). These applications include nanobiological devices, micromachines, nanoelectronic devices, and protein microarrays and chips (Drexler, 1992; Hoummady and Fujita, 1999; Lee *et al.*, 2003; Prokop, 2001; Sematech, 2002; Tsukagoshi, 2002). While many of these devices are designed using highly simplified models or trial-and-error experimentation, recent advances in computer speed and memory, numerical algorithms, and sensor technologies suggest that a more systematic approach to the design and control of these multiscale systems is possible.

The potential applications motivate the creation of tools for the design and control of multiscale systems that have length scales ranging from the macroscopic to the atomistic. This paper describes the challenges to building such multiscale systems tools, which include uncertainties in the physicochemical mechanisms as well as the values of thermodynamic and kinetic parameters, complexities in the simulation of model equations that can span a wide range of time and length scales, lack of manipulated variables and direct measurements of most properties at the nanoscale during processing, and the inapplicability of most existing systems tools to address systems described by noncontinuum and dynamically coupled continuum-noncontinuum models. These challenges specify the requirements for multiscale systems tools.

This paper describes how these requirements can be satisfied by a systematic approach to the design and control of multiscale systems that consists of stochastic parameter sensitivity analysis, Bayesian parameter estimation applied to *ab initio* calculations and experimental data, model-based experimental design, hypothesis mechanism selection, and multi-step optimization. This enables multiscale systems to be designed and controlled based on the numerical algorithms that are most appropriate for simulating each of the length scales of the process.

2. CHALLENGES AND REQUIREMENTS

The challenges associated with the design and control of multiscale systems specify the requirements for multiscale systems tools. To make the description of the challenges as concrete as possible, the manufacture of on-chip copper interconnections in electronic devices is used as an illustrative example. In this process, an applied potential is used to electrodeposit copper on surfaces and in trenches and vias. The product quality of the deposit is a function of nucleation at the atomic scale, surface morphology at the nanoscale, shape evolution at the nano- to micro-length scales, and deposit uniformity over the wafer surface. This electrodeposition process involves phenomena that are simultaneously important over ten orders of magnitude in time and length scales (Alkire and Verhoff, 1995; 1998). According to the International Technology Roadmap for Semiconductors (Sematech, 2002), the manufacture of next-generation interconnects will require design and control of all of these length scales.

Figure 1 is a schematic of the electrodeposition of copper into a trench, in which Cu^{2+} ions in solution diffuse and migrate to the surface in response to a potential applied between the reference and working electrodes. Although the introduction of organic chemical additive cocktails to the solution to produce void-free copper deposits in sub-100 nm trenches is well established (Andricacos, 1999; Andricacos *et al.*, 1998), the precise physicochemical mechanisms of the interactions of these additives with the copper surface are not well understood (Datta and Landolt, 2000; Kondo *et al.*, 2004; Moffet *et al.*, 2000; 2001; 2004; Tan and Harb, 2003; West, 2000), making it difficult to design new additive cocktails able to produce void-free deposits in smaller features. A challenge in applying systems principles to these and other multiscale systems is that the underlying mechanisms, as well as the thermodynamic and kinetic parameters associated with the steps in these mechanisms, are uncertain. *Multiscale systems tools are needed that can handle uncertain mechanisms, as well as uncertain parameters.*

Another challenge to engineering multiscale systems is that the codes used to simulate these systems are computationally expensive. For example, consider Figure 2, which is a schematic of a multiscale simulation model for the electrodeposition of copper into trenches. Chemical reactions and the diffusion and migration of species in the solution boundary layer are described by a system of partial differential-algebraic equations, which are typically simulated using the finite volume or finite element method. The height of the boundary layer is typically ~50 μm and a typical time step for such a code is ~1 ms. The nucleation, surface chemistry, and roughness evolution of the trench surface are most accurately simulated using noncontinuum methods such as kinetic Monte Carlo (KMC) simulation (Battaile *et al.*, 1997; Drews *et al.*, 2004a; Henkelman and Jónsson, 2001; Levi and Kotrla, 1994; Yang *et al.*, 2002). KMC methods are used to simulate structural properties of matter that cannot be represented by a macroscopic continuum description. A KMC simulation is a realization of the Master equation (Fichthorn and Weinberg, 1991):

$$\frac{\partial P(\sigma,t)}{\partial t} = \sum_{\sigma'} W(\sigma',\sigma)P(\sigma,t) - \sum_{\sigma'} W(\sigma,\sigma')P(\sigma,t) \quad (1)$$

where σ and σ' are successive states of the system, $P(\sigma, t)$ is the probability that the system is in state σ at time t, and $W(\sigma',\sigma)$ is the probability per unit time that the system will undergo a transition from state σ' to σ. For a particular system being studied, the KMC code chooses randomly among the possible transitions of the system and accepts particular transitions with appropriate probabilities. After each accepted or attempted transition, the time variable is incremented by one Monte Carlo time step (typically the time step is ~1 ns), and the process is repeated. By selecting the probabilities to satisfy certain conditions, the real time variable t corresponding to the number of Monte Carlo time steps can be computed.

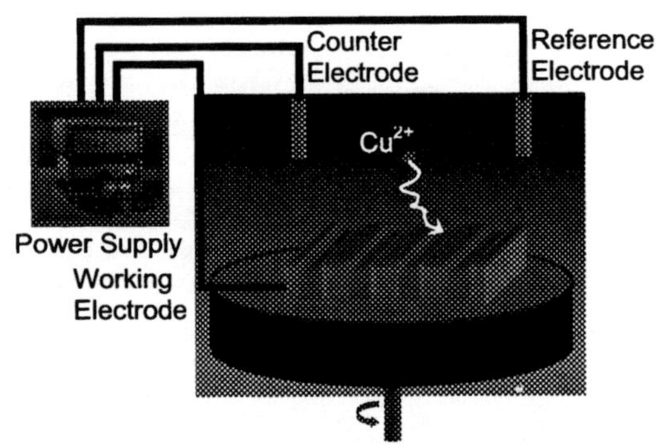

Fig. 1. Electrochemical process for manufacturing on-chip copper interconnects, in which a rotating disk creates a boundary layer above the wafer surface (not drawn to scale).

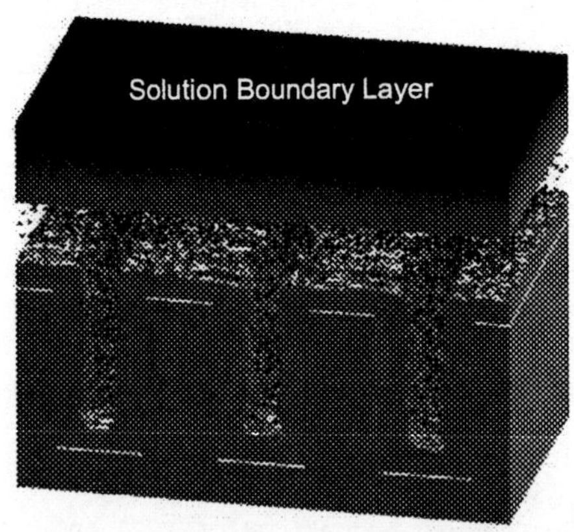

Fig. 2. Multiscale simulation of the electrochemical process for manufacturing on-chip copper interconnects (not drawn to scale). The dots represent Cu^{2+} ions in solution, with the film on the surface being metallic copper.

The bottom part of Figure 2 shows the infill of several trenches simulated using a three-dimensional (3D) KMC code, that tracks adsorption, desorption, bulk and surface diffusion, and chemical reactions (Drews *et al.*, 2004bc). To reduce the computational load, the 3D KMC code was coarse-grained (Gear *et al.*, 2003; Ismail *et al.*, 2003ab; Jönsson *et al.*, 2003; Lopez *et al.*, 2002; Shelley and Shelley, 2000), such that clusters of molecules were tracked instead of individual molecules (Drews *et al.*, 2003b; Katsoulaokis *et al.*, 2003; Katsoulakis and Vlachos, 2003; Pricer *et al.*, 2002ab), and periodic boundary conditions were used at all sides of the simulation domain. Even with these simplifications, it takes ~1 day to perform one simulation run. This greatly limits the number of simulation runs that a systems tool is allowed to make in a coupled simulation-optimization algorithm such as used in control vector parameterization (Ray, 1981). Further, systems techniques that write the simulation code as an algebraic system of equations to be embedded into a structured nonlinear program (Jockenhovel *et al.*, 2003), are not computationally feasible for multiscale systems, as there would be $>10^{16}$ algebraic equations in the structured nonlinear program. *Multiscale systems tools must be much more computationally efficient than most existing systems tools.* Note that the state dimension of KMC codes is very high, while the numbers of simulation inputs (e.g., applied potential) and outputs (e.g., surface roughness, fraction of voids) are much lower. *This motivates the creation of multiscale systems tools that act directly on simulation inputs and outputs*, to keep the computational cost low.

As a further complication, the codes in Figure 2 must be dynamically coupled when dilute additives are included in the simulation, as the surface chemistry and transport determines the amount of depletion of additives in the boundary layer, and the boundary layer influences the rate that chemical species reach the surface. *Multiscale systems tools are needed that can handle models described by dynamically coupled continuum and noncontinuum codes.*

Another characteristic of noncontinuum codes is that their outputs typically have significant stochastic fluctuations, which can be non-Gaussian. For example, Figure 3 is the current density response from a dynamically coupled KMC-finite difference simulation of copper electrodeposition in response to a staircase function of the applied potential (Rusli *et al.*, 2003). The current density only takes on discrete values, which are associated with electron transfer at the copper surface (e.g., as a Cu^{2+} ion gains two electrons to form copper metal). This response is in sharp contrast to the typical step and staircase responses reported in the controls literature (Levine, 1996; Ray, 1981). Although most existing systems tools are applicable to stochastic models with Gaussian fluctuations, these tools always assume that the deterministic part of the model is known. *Multiscale systems tools must be able to address models with large amounts of non-Gaussian noise, for which a deterministic model is unavailable.*

Another challenge in multiscale systems is the lack of key measurements during processing at industrially relevant operating conditions. For example, the only on-line measured variables for the copper electrodeposition process are temperature and current. There are no concentration measurements at the surface, where the uncertain chemical mechanisms and most of the uncertain parameters are located. The key measurement data, which are atomic force microscopy images, are only available at the end of the process (see Figure 4). *Multiscale systems tools must include experimental design methods that maximize the information from sensors, to create models that are predictive.*

Another characteristic of multiscale systems is a lack of manipulated variables at the molecular and nanoscopic length scales during processing. For example, the only variable manipulated during the electrodeposition process in Figure 1 is the applied potential, which does not provide enough degrees of freedom to produce void-free copper in the 0.13 μm trenches used in modern microelectronic devices. This is why industrial practice is to introduce additional degrees of freedom in the initial conditions, which is done through the selection and concentrations of organic chemicals added to the solution. In general, *most multiscale systems require that molecular and nanoscale manipulation be treated as a design focus, to exploit self-assembly during processing.*

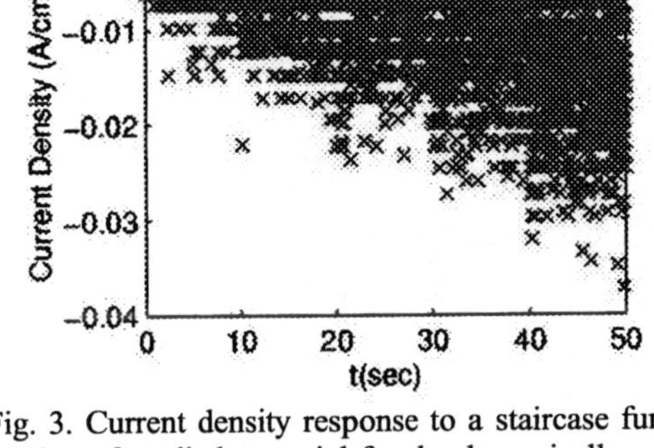

Fig. 3. Current density response to a staircase function of applied potential for the dynamically coupled simulation of the electrochemical process for manufacturing on-chip copper interconnects (each step of the staircase was 10 s long).

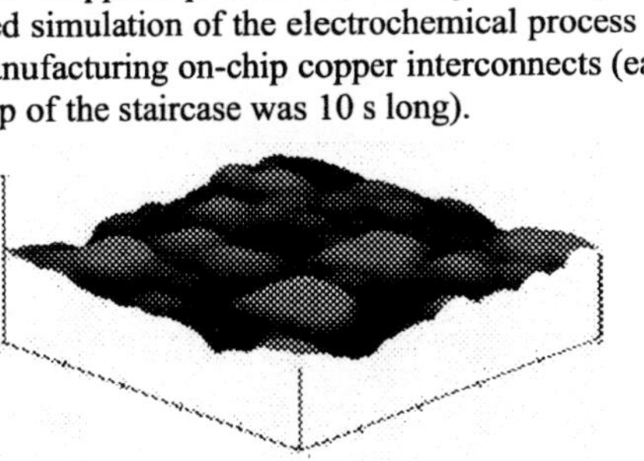

Fig. 4. Atomic force microscopy image of an electrodeposited copper surface.

3. COMPLEX SYSTEMS TOOLS

Before looking at multiscale systems tools, it is useful to review some systems tools developed for complex continuum models that address most, but not all, of the challenges of multiscale systems (see Figure 5). The first step is the identification of the model, which is an iterative procedure. The first experiment is designed using engineering judgment on how to excite the dynamics of the system, or is computed using initial estimates of the model parameters and some experimental design objective such as minimizing the uncertainties in the parameters (Atkinson and Donev, 1992; Bard, 1974; Beck and Arnold, 1977). Improved estimates of the parameters and an associated uncertainty description are computed from the dynamic data collected from the experiment. This can be augmented with *ab initio* computational chemistry calculations (such as density functional theory) using Bayesian parameter estimation techniques (Gunawan *et al.*, 2003b). When several hypothesized physicochemical mechanisms are available, model discrimination techniques are used to select which mechanism is most consistent with the experimental data techniques (e.g., Burke *et al.*, 1997; Gunawan *et al.*, 2002b; Reilly *et al.*, 1974). The model parameters and uncertainty description are used to design the next laboratory experiment, which can be constructed to further reduce the model uncertainties or to maximize the ability to distinguish among the multiple hypothesized mechanisms (Atkinson and Donev, 1992). Parameter estimates obtained from this iterative procedure can be many orders-of-magnitude more accurate than estimates obtained from data collected from trial-and-error experimentation. Once the model parameters are accurate enough, the simulation model is incorporated into an optimization algorithm to compute the physical design variables, initial conditions, startup procedures, setpoint trajectories, and the feedback control system.

These complex systems tools are well established for continuum models described by general integro-partial differential algebraic equations (IPDAEs), and have been applied to the manufacture of photographic chemicals (Matthews and Rawlings, 1998), pharmaceuticals (Braatz, 2002; Fujiwara *et al.*, 2004), and semiconductors (Gunawan *et al.*, 2004). These applications have been to a wide range of real physical systems at Merck, Eastman Kodak, and Sematech (an international consortium of semiconductor companies), indicating the wide applicability of these tools. The key to the generality of these tools is that they act directly on the inputs and outputs of simulation codes, rather than requiring a particular form for the equations or the numerical algorithm used to simulate the equations. Also key to providing generality is that the numerical algorithms implemented in the steps in Figure 5 incorporate parameter sensitivity methods based on finite differences (e.g., Beck and Arnold, 1977; Caracotsios and Stewart, 1985; Kamrunnahar *et al.*, 2004) or automatic differentiation (Cao *et al.*, 2002; 2003; Feehery *et al.*, 1997; Galan *et al.*, 1999; Li *et al.*, 2000). The parameter sensitivities improve the numerical conditioning

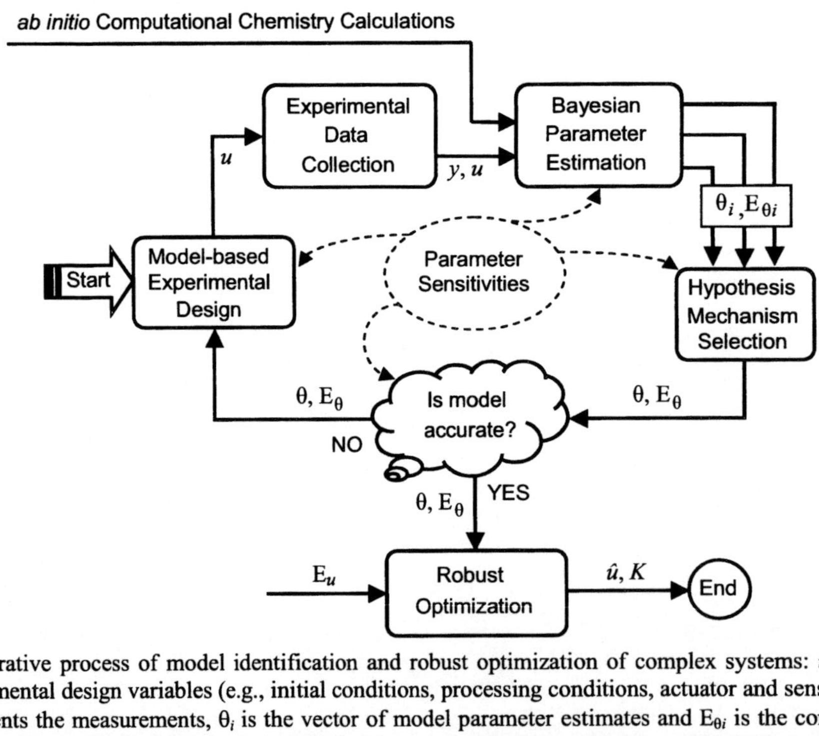

Fig. 5. Iterative process of model identification and robust optimization of complex systems: u represents all experimental design variables (e.g., initial conditions, processing conditions, actuator and sensor locations), y represents the measurements, θ_i is the vector of model parameter estimates and $E_{\theta i}$ is the confidence region for the ith hypothesized mechanism, θ is the vector of parameter estimates and E_θ is the confidence region for the most likely mechanism, $\hat{u}$ is the optimal control policy (including initial conditions, equipment specifications, operating conditions, actuator and sensor locations, etc.), K is the optimal controller, and E_u is the uncertainty in the implementation of the optimal control policy, including the effects of disturbances.

of the systems tools while focusing the modeling effort towards only the key parameters whose values must be known with high certainty to produce predictive models. This is critical when the number of parameters is large, as is typical for semiconductor processes (Gunawan *et al.*, 2003; Maroudas, 2000).

In recent years fast techniques have been developed to quantify the effects of model uncertainties on all states and outputs for the batch and semibatch processes used to manufacture the most complex systems (Ayyub, 1998; Elishakoff, 1999; Ma *et al.*, 1999; Matthews *et al.*, 1996), and to integrate robustness into all design variables. Some of these techniques have little or no restrictions on the zero dynamics, the integro-differential structure of the equations, the simulation algorithms, or the form of the uncertainties (Pan *et al.*, 1998; Ma and Braatz, 2001; Nagy and Braatz, 2003b; 2004; Phenix, 1998; Tatang, 1997). These analysis algorithms have been incorporated into optimization algorithms to produce designs and controllers that are robust to the uncertainties (Nagy and Braatz, 2003a; 2004).

To illustrate the complexity of systems that can be addressed by these systems tools, consider their application to the manufacture of ultrashallow junctions. The current technology for the formation of ultrashallow junctions in microelectronic logic devices relies almost exclusively on ion implantation to introduce dopants into the substrate (see Figure 6). Although junctions can be made shallower by reducing the implant energy, the effectiveness of this approach has been limited by the need to anneal the resulting structure to over 1000°C both to activate the dopant electrically and to eliminate implant-induced defects in the crystal structure (see Figure 7 for an example temperature trajectory). Defects mediate unwanted diffusion of dopants during the anneal process, which leads to a significant undesired increase in the junction depth. The aforementioned systems tools have been applied to the post-implant annealing process, to construct a simulation model and to minimize the junction deepening while maximizing dopant activation.

The simulation model includes the coupled mass balance equations for interstitial atoms, interstitial clusters, and related defects. These equations have the general form for species i:

$$\frac{\partial N_i}{\partial t} = -\frac{\partial J_i}{\partial x} + G_i \qquad (2)$$

where N_i denotes concentration and G_i is the net generation rate. The flux J_i incorporates terms due to diffusion and drift in response to electric fields. The model also includes Poisson's equation describing the electric field generated by spatial imbalance of the charge density. The simulation model consisted of ~25 partial differential equations which were nonuniformly spatially discretized using between 200 and 800 points in the depth direction, resulting in up to 20,000 extremely stiff ordinary differential equations that were solved using the public domain software FLOOPS (Law and Cea, 1998), which integrates the equations using a combination of the one-step trapezoidal rule and the multiscape backward differentiation formula (Bank *et al.*, 1985).

The activation energies in the expressions for G_i and J_i were obtained by Bayesian parameter estimation, which incorporated information from density functional theory (DFT) calculations, past experimental studies, and boron secondary ion mass spectroscopy data from the International Sematech consortium of semiconductor companies (Gunawan *et al.*, 2003b). A combination of parameter sensitivity analysis and kinetic insights was used to select the physical mechanism (Gunawan *et al.*, 2002a; 2003a), in which the most important part was the specification of the network of chemical reactions for the clusters. Parameter sensitivity analysis was a necessity in the construction of the physical mechanism, as the number of kinetic parameters was large, including 18 activation energies associated with the interstitial diffusion, cluster association, and cluster dissociation reactions.

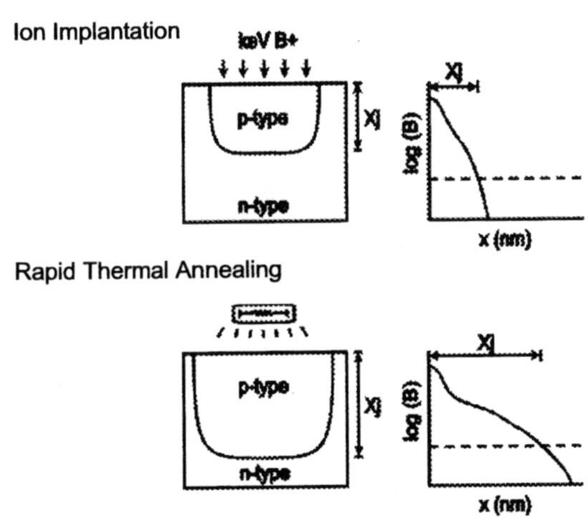

Fig. 6: Schematic diagram of the junction increase that takes place during rapid thermal annealing after ion implantation of dopant.

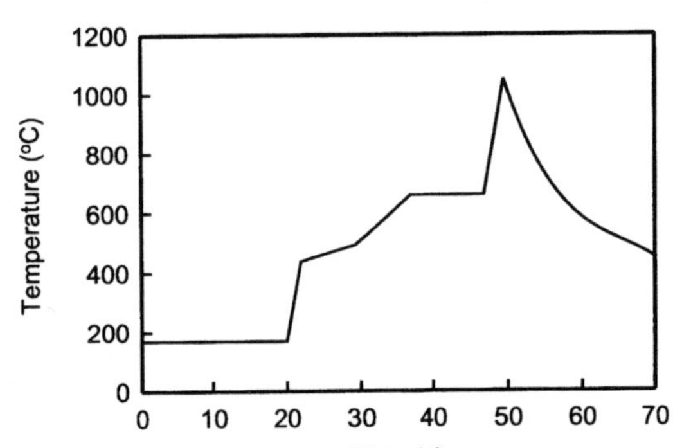

Fig. 7. A typical rapid thermal anneal temperature program, which consists of a stabilization step and a spike-anneal (*i.e.*, a fast linear heating step followed by a natural cool down step).

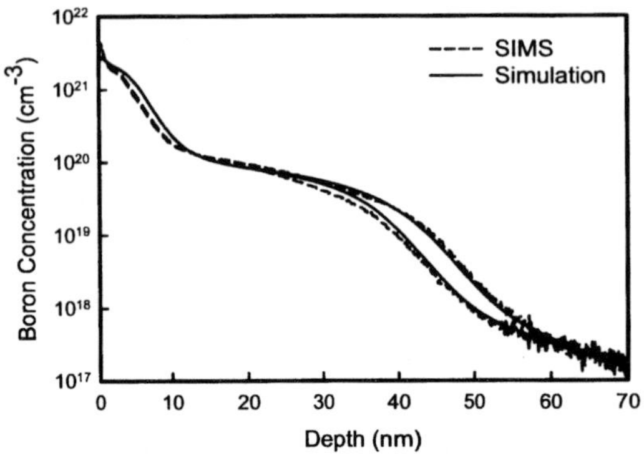

Fig. 8. Experimental and simulated boron dopant profiles, for two batch operating recipes (Gunawan *et al.*, 2003b).

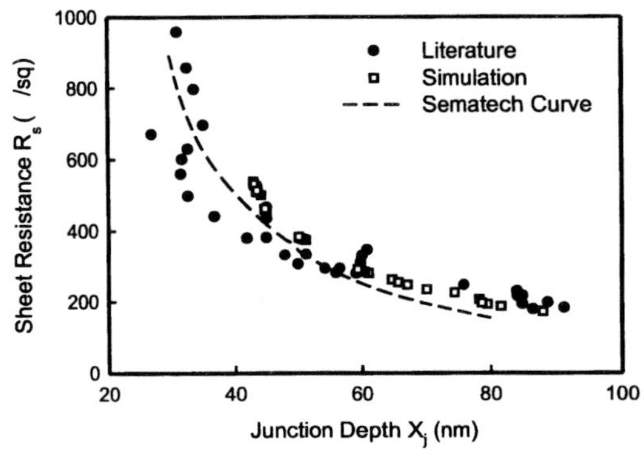

Fig. 9. Comparison of junction depth–sheet resistance pairs from various published experimental papers and TED simulations employing various heating and cooling rates, and annealing temperatures. The Sematech curve summarizes the sheet resistance and junction depth data in experimental studies performed by International Sematech (Gunawan *et al.*, 2003b).

The complex systems tools in Figure 5 permitted the construction of a clear picture of the fundamental kinetic processes that govern diffusion and electrical activation of dopant (Jung *et al.*, 2003; 2004a). The agreement between the simulated and experimental boron dopant profiles was within 2 nm for the entire junction (see Figure 8). To provide further validation of the simulation model, the junction depth and sheet resistance (this is a measure of dopant activation) were computed for a wide range of temperature profiles, and compared to a large number of experimental values reported in the literature and to the "Sematech curve," which summarizes additional experimental values (see Figure 9). The predictions of the simulation model are highly consistent with reported experimental values.

The simulation model was incorporated into an optimal control problem to compute an annealing temperature trajectory that minimized the junction depth while maintaining a desirable level of boron activation (Gunawan *et al.*, 2004). Robustness analysis as shown in Figure 5 was applied to rigorously quantify the performance degradation from uncertainties in the feedback control implementation and the model parameters. The analysis indicated that limited improvement in product quality is achievable using existing metrology and rapid thermal processing controllers.

As discussed in Section 2, the limited actuation available at macroscopic length scales motivates the application of *molecular design*, and it was argued that the potential impact of such design optimization can be much greater than the potential benefit of improved feedback control. As it is highly advantageous in terms of electronic device properties to restrict the chemistry to dopant and silicon molecules, we have been keeping the atomic species unchanged, but using the simulation model to change the bond structure at the silicon surface (Jung *et al.*, 2004). We have shown that the effects of the structure of bonds at the silicon surface have a substantial effect on the junction depth, due to a change in the effective surface boundary condition for interstitials. These simulation studies have motivated discrete changes in processing conditions, which are being evaluated in a systematic experimental study (Dev *et al.*, 2004).

These complex systems tools address most of the requirements for application to multiscale systems, in that non-Gaussian stochastic behavior and uncertain mechanisms and parameters are taken into account; the tools are computationally efficient, general purpose, and act directly on simulation inputs and outputs; experimental design methods are included that maximize the information from sensors to create predictive models; and that design and control are optimized simultaneously, which enables molecular and nanoscale manipulation in the design problem to be considered jointly with the manipulation of the online variables by feedback controllers during processing. The complex systems tools; however, do not consider all of the issues particular to models described by noncontinuum and coupled continuum and noncontinuum simulation codes, namely, the lack of an underlying deterministic model for the noncontinuum simulation codes, and the additional numerical stability issues that can arise when codes are linked. The next section discusses efforts to extend these complex systems tools to address these two additional requirements of multiscale systems.

4. NONCONTINUUM AND COUPLED CONTINUUM/NONCONTINUUM CODES

For the noncontinuum models that describe complex systems, the optimizations that occur in the model-based experimental design, Bayesian parameter estimation, hypothesis mechanism selection, and optimal design and control steps in Figure 5 are stochastic with no closed form expression for the underlying deterministic system. For KMC codes, it has been proposed to construct reduced-order models by truncating unlikely configurations and grouping probabilities that evolve together (Gallivan and Atwater,

2004; Gallivan and Murray, 2003; 2004), using smaller lattices (Lou and Christofides, 2003ab), or applying least-squares systems identification methods (Gallivan, 2003; Gallivan *et al.*, 2000). Similar reduced-order models have been defined for dynamically coupled continuum/noncontinuum codes (Raimondeau and Vlachos, 2000; Rusli *et al.*, 2003). Although such approaches are acceptable for controller design, one of the main points in Section 2 is that molecular and nanoscale manipulation should be treated as a *design* problem, to exploit self-assembly during processing – *not* primarily as feedback controller design. For design purposes, the reduced-order models must be a function of the physicochemical parameters in the simulation codes to be manipulated in the design optimization, which can be changed through modifying the chemistry of the system, rather than as a function of macroscopic manipulated variables such as temperature. Developing reduced-order models that are globally applicable and a function of both the physicochemical parameters and the macroscopic manipulated variables is a challenging problem, considering the high computational cost of the simulation codes.

Another consideration is that simulation codes which include molecular-scale phenomena are stochastic, which implies that averaging techniques (Rusli *et al.*, 2003) or stochastic optimization algorithms such as simulated annealing are appropriate (Aarts and Korst, 1989). We have shown that even the finite difference calculation of parameter sensitivities typically used in systems engineering algorithms must be formulated as a stochastic optimization, to obtain estimates of the highest accuracy (Drews *et al.*, 2003a). More specifically, the finite difference expressions for sensitivities reported in textbooks and papers are derived assuming that the underlying system can be described as a deterministic Taylor series expansion, whereas this assumption is invalid for simulations that include noncontinuum models, whose outputs are stochastic. A much more accurate and appropriate formulation is to include a stochastic term in the series expansion, and determine the finite difference expression by solving an optimization problem whose objective is to compute either the minimum variance or maximum likelihood estimate of the parameter sensitivity. As in continuum models, these parameter sensitivities are a key step needed to reduce the complexity and improve the numerical conditioning of the stochastic optimizations that define the systems tasks in Figure 5.

Recently we incorporated our stochastic parameter sensitivity algorithm into a multi-step optimization algorithm (Raimondeau *et al.*, 2003), that uses sensitivity analysis to determine the key parameters, followed by solution mapping to parameterize the responses of the simulation model as low-degree polynomials of the key parameters, and simulated annealing to optimize the key parameters. The low-order parameterization is used to reduce the number of runs of the computationally expensive stochastic simulation code required to converge to the parameter estimates. This revised algorithm (see Figure 10) has been used to estimate kinetic parameters associated with copper electrodeposition from measurements of the applied potential and the surface using atomic force microscopy images (see Figure 4), where the simulation model consisted of a coarse-grained KMC code dynamically coupled to a finite difference continuum code (Drews *et al.*, 2003c).

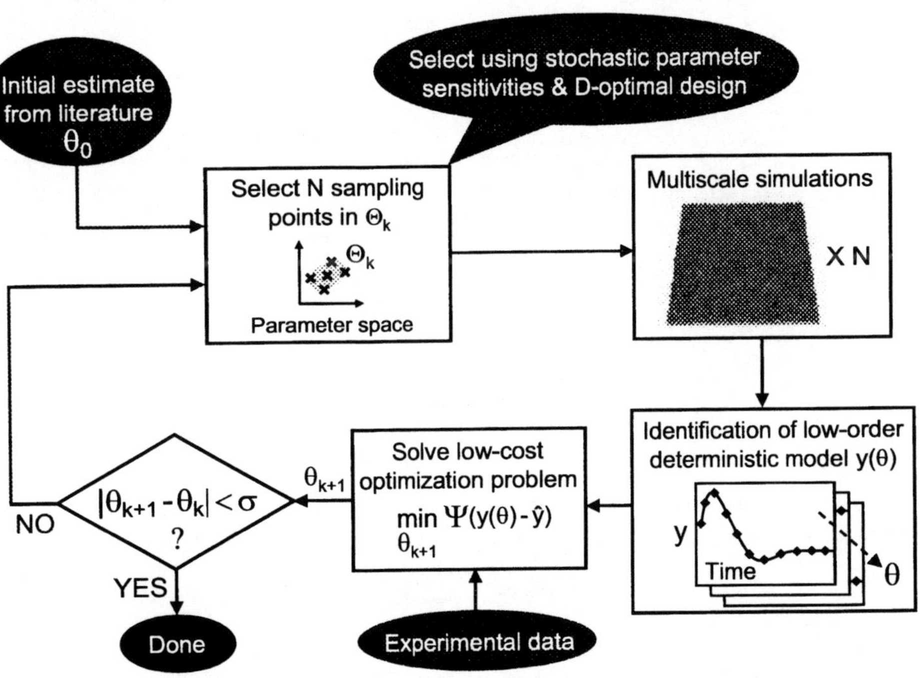

Fig. 10. A simplified schematic of the multi-step optimization algorithm for parameter estimation in stochastic simulation codes. Kinetics from *ab initio* calculations such as DFT are included using a Bayesian formulation (Gunawan *et al.*, 2003b).

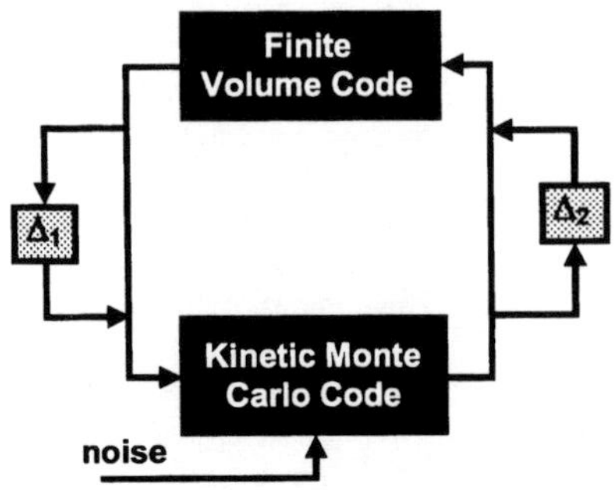

Fig. 11. Time- and length-scale mismatches at interfaces between simulation codes can be modeled as perturbations Δ_i on the information passed between simulation codes. The effects of these mismatches and uncertainties in kinetic parameters in the individual simulation codes on the simulation outputs can be analyzed using nonlinear systems theory and uncertainty-based simulation techniques.

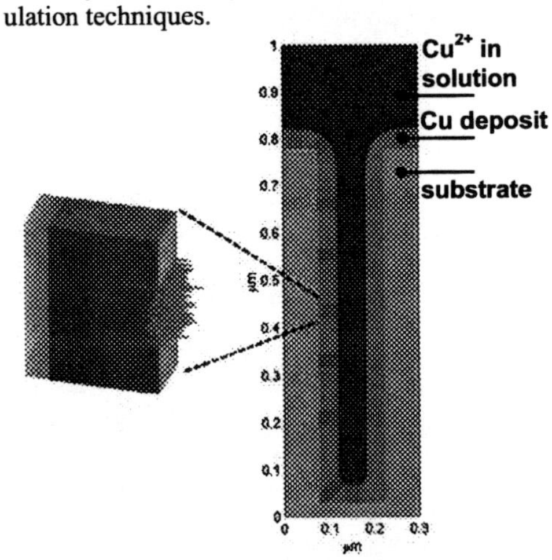

Fig. 12. A coupled simulation with "master-worker" computational paradigm. The stripes in the copper deposit denote the location of KMC simulations. The zoom-in picture of the copper deposit shows the distinction of the actual surface morphology and the surface seen by the MB code in the simulation.

This application coarse-grained the KMC code to further reduce the computational expense; other acceleration methods such as the tau-leaping (Gillespie, 2001; Gillespie and Petzold, 2003; Rathinam *et al.*, 2003) and gap-tooth methods (Gear, *et al.*, 2003) could be applied, in isolation or in addition to coarse-graining. Recall that coupling simulation codes induce an additional systems issue – linkage instabilities. While numerically stable codes are available for simulating each length scale, numerical instabilities can be induced in the coupling of such codes by temporal and spatial mismatches at the interfaces between the codes. Control systems theory has been used to design numerical linkage algorithms that modify the dynamic information passed between the simulation codes to numerically stabilize their coupling, and to increase the numerical accuracy of the simulation results (Drews *et al.*, 2004c, Rusli *et al.*, 2004ab). In this approach, the simulation codes are represented by deterministic or stochastic discrete-time nonlinear operators, with mismatches at the interfaces between simulation codes modeled as norm-bounded perturbations, as is commonly done in robust control theory (see Figure 11). Further, results from nonlinear systems theory have been applied to produce a constructive procedure for testing whether an arbitrary interconnection of simulation codes is well-posed (Rusli *et al.*, 2004ab), and a sufficient condition for the numerical stability of dynamically coupled simulation codes.

More recently we have been using nonlinear systems theory to guide the design of much more complex dynamic coupling of simulation codes than shown in Figure 11. An improved simulation model for the electrochemical process used to manufacture copper interconnects (shown in Figures 1 and 2) is shown in Figure 12. The multiscale simulation model couples multiple instances of a solid-on-solid KMC simulation code (Drews *et al.*, 2004a) to an internally coupled moving boundary (MB) finite-volume/level set continuum simulation code (Chang *et al.*, 1996; Li *et al.*, 1999; Wheeler *et al.*, 2003) to simulate the filling of on-chip features (trenches) by electrodeposition in the presence of additives. The KMC and MB simulations dynamically pass interface conditions during the simulations. The MB code sends surface concentrations and the solution potential to each KMC code, which computes reaction rates from the simulation of the chemistry and physics that occur at the electrode surface. The KMC codes send species fluxes to the MB code which are used as surface boundary conditions. The MB code advances the copper-solution interface using the level-set method, and simulates the chemistry and physics in the electrolyte in and above the trench using the finite volume method.

It is useful to relate the systematic approach of this paper to projective methods for constructing bifurcation maps (Koronaki *et al.*, 2003; Schroff and Keller, 1993; Siettos *et al.*, 2003a; Tuckerman and Barkley, 2000). Projective methods satisfy one of the key requirements for multiscale systems tools described in Section 2, in that they act directly on the inputs and outputs of simulation codes. Although originally developed for continuum codes, these methods can be applied to noncontinuum codes (Gear *et al.*, 2002; Gear and Kevrekidis, 2004; Hummer and Kevrekidis, 2003; Makeev *et al.*, 2002ab; Siettos *et al.*, 2003a). It is straightforward to show that these methods apply to dynamically coupled continuum/noncontinuum codes as well. The main assumption underlying the projective methods is the requirement of a clear separation of time scales (Gear *et al.*, 2002; Runborg *et al.*, 2002), which holds for some but not all physical systems (note that the aforementioned methods in Sections 3 and 4 do not require a clear separation in time scales). For systems in which this assumption holds, projective methods enable the analysis of the nonlinear dynamical behavior of multiscale systems, which can be used for an initial assessment of whether the qualitative dynamic behavior of a hypothesized mechanism is consistent with experimental observations. Hypothesized mechanisms with qualitative dynamics that are consistent with experiments are removed from further consideration in the mechanism selection step in Figure 5.

5. CONCLUSIONS

This paper describes the characteristics of multiscale systems, using the manufacture of on-chip copper interconnections as a specific example to illustrate the key points. These characteristics specify the requirements for multiscale systems tools. One of the key points was that limitations in manipulations available during processing for most multiscale systems imply that *a much larger impact on product quality can be achieved from molecular design* rather than on designing better feedback controllers. Systems tools applicable to complex continuum models were summarized that satisfy the requirements for multiscale systems, except for issues specific to dealing with noncontinuum and coupled noncontinuum-continuum codes. This was illustrated in an application to the manufacture of ultrashallow junctions. An extension of the complex systems tools to deal with the requirements of multiscale systems was described, that incorporates stochastic sensitivity analysis within a multistep optimization algorithm.

Although the systems principles are the same for multiscale systems as for macroscopic systems, the problem formulations and the numerical algorithms designed to solve these formulations are different. Further, a new issue arises in multiscale systems, which is how to address numerical instabilities that can arise during the linkage of individual simulation codes. Some results in applying nonlinear control theory to design numerical linkage algorithms were described. More theory is needed to provide a systematic methodology for the numerical stabilization of multiscale simulation codes, and on algorithms for Bayesian parameter estimation, model-based experimental design, hypothesis mechanism selection, and robust optimization.

ACKNOWLEDGEMENTS

This material is based upon work supported by the National Science Foundation under Grant Numbers CTS-0135621, CISE-9619019, and PACI NRAC-MCA01S022. Any opinions, findings, and conclusions or recommendations expressed in this paper are those of the authors and do not necessarily reflect the views of the National Science Foundation.

REFERENCES

Aarts, E.H.L., and J.H.M. Korst (1989). *Simulated Annealing and Boltzmann Machines: A Stochastic Approach to Combinatorial Optimization and Neural Computing*. Wiley, New York.

Alkire, R., and M. Verhoff (1995). Electrochemical reaction engineering in materials processing. *Chem. Eng. Sci.*, **49**, 4085-4093.

Alkire, R., and M. Verhoff (1998). The bridge from nanoscale phenomena to macroscale processes. *Electrochim. Acta*, **43**, 2733-2741.

Andricacos, P.C. (1999). Copper on-chip interconnections – A breakthrough in electrodeposition to make better chips. *The Electrochemical Society Interface*, **8**, 32-37.

Andricacos, P.C., C. Uzoh, J.O. Dukovic, J. Horkans and H. Deligianni (1998). Damascene copper electroplating for chip interconnections. *IBM J. of Res. and Dev.*, **42**, 567-574.

Atkinson, A.C., and A.N. Donev (1992). *Optimum Experimental Designs*. Clarenden Press, Oxford, United Kingdom.

Ayyub, B.M., editor (1998). *Uncertainty Modeling and Analysis in Civil Engineering*, CRC Press, Boca Raton, Louisiana, U.S.A.

Bank, R.E., W.M. Coughran, Jr., W. Fichtner, E.H. Grosse, D.J. Rose and R.K. Smith (1985). Transient simulation of silicon devices and circuits. *IEEE Trans. on Computer-Aided Design of Integrated Circuits and Systems*, **4**, 436-451.

Bard, Y. (1974). *Nonlinear Parameter Estimation*. Academic Press, New York, U.S.A.

Battaile, C.C., D.J. Srolovitz and J.E. Butler (1997). A kinetic Monte Carlo method for the atomic-scale simulation of chemical vapor deposition: application to diamond. *J. Appl. Phys.*, **82**, 6293-6300.

Beck, J.V., and K.J. Arnold (1977). *Parameter Estimation in Engineering and Science*. Wiley, New York, U.S.A.

Braatz, R.D. (2002). Advanced control of crystallization processes. *Annual Reviews in Control*, **26**, 87-99.

Burke, A.L., T.A. Duever and A. Penlidis (1997). Choosing the right model: Case studies on the use of statistical model discrimination experiments. *Can. J. Chem. Eng.*, **75**, 422-436.

Cao, Y., S. Li, and L.R. Petzold (2002). Adjoint sensitivity analysis for differential-algebraic equations: Algorithms and software. *J. Comp. Appl. Math.*, **149**, 171-192.

Cao, Y., S. Li, L.R. Petzold and R. Serban (2003). Adjoint sensitivity analysis for differential-algebraic equations: The adjoint DAE system and its numerical solution. *SIAM J. Sci. Comput.*, **24**, 1076-1089.

Caracotsios, M., and W.E. Stewart (1985). Sensitivity analysis of initial value problems with mixed ODEs and algebraic equations. *Comp. Chem. Eng.*, **9**, 359-365.

Chang, Y.C., T.Y. Hou, B. Merriman and S. Osher (1996). A level set formulation of Eulerian interface capturing methods for incompressible fluid flows. *J. Comp. Phys.*, **124**, 449-464.

Datta, M., and D. Landolt (2000). Fundamental aspects and applications of electrochemical microfabrication. *Electrochim. Acta*, **45**, 2535-2558.

Dev, K., M.Y.L. Jung, R. Gunawan, R.D. Braatz and E.G. Seebauer (2003). New mechanism for coupling between properties of interfaces and bulk semiconductors. *Phys. Rev. B*, **68**, 195311.

Drews, T.O., R.D. Braatz and R.C. Alkire (2003a). Parameter sensitivity analysis of Monte Carlo simulations of copper electrodeposition with multiple additives. *J. Electrochem. Soc.*, **150**, C807-C812.

Drews, T.O., J.C. Ganley and R.C. Alkire (2003b). Evolution of surface roughness during copper electrodeposition in the presence of additives: Comparison of experiments and Monte Carlo simulations. *J. Electrochem. Soc.*, **150**, C325-C334.

Drews, T.O., F. Xue, X. Li, H. Deligianni, P. Vereecken, E. Cooper, P. Andricacos, R.D. Braatz and R.C. Alkire (2003c). Parameter estimation of a copper electrodeposition additive mechanism using data obtained from a D-optimal experimental design. In: *Proc. of the Topical Conf. on Electrodeposition Processes*, Paper 189b, *AIChE Annual Meeting*, San Francisco, U.S.A.

Drews, T.O., R.D. Braatz and R.C. Alkire (2004a). Coarse-grained kinetic Monte Carlo simulation of copper electrodeposition with additives. *Int. J. Multiscale Computational Engineering*, in press.

Drews, T.O., S. Krishnan, J. Alameda, D. Gannon, R.D. Braatz and R.C. Alkire (2004b). Multiscale simulations of copper electrodeposition onto a resistive substrate. *IBM J. Res. & Dev.*, in press.

Drews, T.O., E.G. Webb, D.L. Ma, J. Alameda, R.D. Braatz and R.C. Alkire (2004c). Coupled mesoscale-continuum simulations of copper electrodeposition in a trench. *AIChE J.*, **50**, 226-240.

Drexler, K.E. (1992). *Nanosystems: Molecular Machinery, Manufacturing, and Computation.* Wiley Interscience, New York, U.S.A.

Elishakoff, I., editor (1999). *Whys and Hows in Uncertainty Modelling: Probability, Fuzziness, and Anti-optimization*, Springer-Verlag, Wien, Germany.

Feehery, W.F., J.E. Tolsma and P.I. Barton (1997). Efficient sensitivity analysis of large-scale differential-algebraic equations. *Appl. Numer. Math.*, **25**, 41-54.

Fujiwara, M., Z.K. Nagy, J.W. Chew and R.D. Braatz (2004). First-principles and direct design approaches for the control of pharmaceutical crystallization. *J. of Process Control*, accepted.

Galan, S., W.F. Feehery and P.I. Barton (1999). Parametric sensitivity functions for hybrid discrete/continuous systems. *Appl. Numer. Math.*, **31**, 17-47.

Gallivan, M.A. (2003). Optimization, estimation, and control for kinetic Monte Carlo simulations of thin film deposition. In: *Proc. of the Conference on Decision and Control*, IEEE Press, Piscataway, New Jersey, U.S.A., 3437-3442.

Gallivan, M.A., and H.A. Atwater (2004). Design of a film surface roughness-minimizing molecular beam epitaxy process by reduced-order modeling of epitaxial growth. *J. Appl. Phys.*, **95**, 483-489.

Gallivan, M.A., and R.M. Murray (2003). Model reduction and system identification for master equation control systems. In: *Proc. of the American Control Conference*, IEEE Press, Piscataway, New Jersey, U.S.A., 3561-3566.

Gallivan, M.A., and R.M. Murray (2004). Reduction and identification methods for Markovian control systems, with application to thin film deposition. *Int. J. of Robust and Nonlinear Control*, **14**, 113-132.

Gallivan, M.A., R.M. Murray and D.G. Goodwin (2000). The dynamics of thin film growth: a modeling study. In: *CVD XV: Proc. of the Fifteenth Symposium on Chemical Vapor Deposition* (M.D. Allendorf and M.L. Hitchman, eds.), The Electrochemical Society, **616**, 168-175.

Gear, C.W., J. Li and I.G. Kevrekidis (2003). The gap-tooth method in particle simulations. *Phys. Lett. A*, **316**, 190-195.

Gear, C.W., and I.G. Kevrekidis (2004). Computing in the past with forward integration. *Phys. Lett. A*, **321**, 335-343.

Gear, C.W., I.G. Kevrekidis, and C. Theodoropoulos (2002). 'Coarse' integration/bifurcation analysis via microscopic simulators: Micro-Galerkin methods, *Comp. Chem. Eng.*, **26**, 941-963.

Gillespie, D.T. (2001). Approximate accelerated stochastic simulation of chemically reacting systems. *J. Chem. Phys.*, **115**, 1716-1733.

Gillespie, D.T. and L.R. Petzold (2003). Improved leap-size selection for accelerated stochastic simulation. *J. Chem. Phys.*, **119**, 8229-8234.

Gunawan, R., M.Y.L. Jung, R.D. Braatz and E.G. Seebauer (2003a). Parameter sensitivity analysis of boron activation and transient enhanced diffusion in silicon. *J. Electrochem. Soc.*, **150**, G758-G765.

Gunawan, R., M.Y.L. Jung, R.D. Braatz and E.G. Seebauer (2002a). Systems analysis applied to modeling dopant activation and TED in rapid thermal annealing. In: *Proc. of the 10th IEEE International Conference on Advanced Thermal Processing of Semiconductors*, IEEE Press, Piscataway, New Jersey, U.S.A., 107-110.

Gunawan, R., M.Y.L. Jung, E.G. Seebauer and R.D. Braatz (2003b). Maximum *a posteriori* estimation of transient enhanced diffusion energetics, *AIChE J.*, **49**, 2114-2123.

Gunawan, R., M.Y.L. Jung, E.G. Seebauer and R.D. Braatz (2004). Optimal control of transient enhanced diffusion in a semiconductor process. *J. of Process Control*, **14**, 423-430.

Gunawan, R., D.L. Ma, M. Fujiwara and R.D. Braatz (2002b). Identification of kinetic parameters in a multidimensional crystallization process. *Int. J. of Modern Physics B*, **16**, 367-374.

Henkelman, G., and H. Jónsson (2001). Long time scale kinetic Monte Carlo simulations without lattice approximation and predefined event table. *J. Chem. Phys.*, **115**, 9657-9666.

Hoummady, M., and H. Fujita (1999). Micro-machines for nanoscale science and technology. *Nanotechnology*, **10**, 29-33.

Hummer, G., and I.G. Kevrekidis (2003). Coarse molecular dynamics of a peptide fragment: Free energy, kinetics, and long-time dynamics computations. *J. Chem Phys.*, **118**, 10762-10773.

Ismail, A.E., G.C. Rutledge and G. Stephanopoulos (2003a). Multiresolution analysis in statistical mechanics – I. Using wavelets to calculate thermodynamic properties. *J. Chem. Phys.*, **118**, 4414-4423.

Ismail, A.E., G.C. Rutledge and G. Stephanopoulos (2003b). Multiresolution analysis in statistical mechanics – II. The wavelet transform as a basis for Monte Carlo simulations on lattices. *J. Chem. Phys.*, **118**, 4424-4431.

Jockenhovel, T., L.T. Biegler and A. Wachter (2003). Dynamic optimization of the Tennessee Eastman process using the OptControlCentre. *Comp. Chem. Eng.*, **27**, 1513-1531.

Jönsson, M., M. Skepö, F. Tjerneld and P. Linse (2003). Effect of spatially distributed surface residues on protein-polymer association. *J. Phys. Chem. B*, **107**, 5511-5518.

Jung, M.Y.L., R. Gunawan, R.D. Braatz and E.G. Seebauer (2003). Ramp-rate effects in transient enhanced diffusion and dopant activation. *J. Electrochem. Soc.*, **150**, G838-G842.

Jung, M.Y.L., R. Gunawan, R.D. Braatz and E.G. Seebauer (2004a). A simplified picture for transient enhanced diffusion of boron in silicon. *J. Electrochem. Soc.*, **151**, G1-G7.

Jung, M.Y.L., R. Gunawan, R.D. Braatz and E.G. Seebauer (2004b). Effect of near-surface band bending on dopant profiles in ion-implanted silicon. *J. Appl. Phys.*, **95**, 1134-1139.

Kamrunnahar, M., R.D. Braatz and R.C. Alkire (2004). Parameter sensitivity analysis of pit initiation at single sulfide inclusions in stainless steel. *J. Electrochem. Soc.*, **151**, B90-B97.

Katsoulaokis, M.A., A.J. Majda and D.G. Vlachos (2003). Course-grained stochastic processes and Monte Carlo simulations in lattice systems. *J. Comp. Phys.*, **186**, 250-278.

Katsoulakis, M.A., and D.G. Vlachos (2003). Coarse-grained stochastic processes and kinetic Monte Carlo simulation for the diffusion of interacting particles. *J. Chem. Phys.*, **119**, 9412-9427.

Kondo, K., T. Matsumoto and K. Watanabe (2004). Role of additives for copper damascene electrodeposition: Experimental study on inhibition and acceleration effects. *J. Electrochem. Soc.*, **151**, C250-C255.

Koronaki, E.D., A.G. Boudouvis and I.G. Kevrekidis (2003). Enabling stability analysis of tubular reactor models using PDE/PDAE integrators. Comp. & Chem. Eng., **27**, 951-964.

Law, M.E., and S.M. Cea (1998). Continuum based modeling of silicon integrated circuit processing: An object oriented approach. *Comput. Mat. Sci.*, **12**, 289-308.

Lee, S.Y., S.J. Lee and H.T. Jung (2003). Protein microarrays and chips. *J. Ind. Eng. Chem.*, **9**, 9-15.

Levi, A.C., and M. Kotrla (1994). Theory and simulation of crystal growth. *J. Phys. Condens. Matter*, **9**, 299-344.

Levine, W.S., editor (1996). *The Control Handbook.* IEEE Press, Boca Raton, Florida, U.S.A.

Li, S., L.R. Petzold and W. Zhu (2000). Sensitivity analysis of differential-algebraic equations: A comparison of methods on a special problem. *Appl. Numer. Math.*, **32**, 161-174.

Li, Z., H. Zhao and H. Gao (1999). A numerical study of electro-migration voiding by evolving level set functions on a fixed Cartesian grid. *J. Comp. Phys.*, **152**, 281-304.

Lopez, C.F., P.B. Moore, J.C. Shelley, M.Y. Shelley and M.L. Klein (2002). Computer simulation studies of biomembranes using a coarse grain model. *Comp. Phys. Comm.*, **147**, 1-6.

Lou, Y.M., and P.D. Christofides (2003a). Estimation and control of surface roughness in thin film growth using kinetic Monte-Carlo models. *Chem. Eng. Sci.*, **58**, 3115-3129.

Lou, Y.M., and P.D. Christofides (2003b). Feedback control of growth rate and surface roughness in thin film growth, *AIChE J.*, **49**, 2099-2113.

Ma, D.L., and R.D. Braatz (2001). Worst-case analysis of finite-time control policies. *IEEE Trans. on Control Syst. Tech.*, **9**, 766-774.

Ma, D.L., S.H. Chung and R.D. Braatz (1999). Worst-case performance analysis of optimal batch control trajectories. *AIChE J.*, **45**, 1469-1476.

Makeev, A.G., D. Maroudas and I.G. Kevrekidis (2002a). Course stability and bifurcation analysis using stochastic simulators: Kinetic Monte Carlo examples. *J. Chem. Phys.*, **116**, 10083-10091.

Makeev, A.G., D. Maroudas, A.Z. Panagiotopoulos and I.G. Kevrekidis (2002b). Coarse bifurcation analysis of kinetic Monte Carlo simulations: A lattice-gas model with lateral interactions, *J. Chem. Phys.*, **117**, 8229-8240.

Maroudas, D. (2000). Multiscale modeling of hard materials: Challenges and opportunities for chemical engineering, *AIChE J.*, **46**, 878-882.

Matthews, H.B., S.M. Miller and J.B. Rawlings (1996). Model identification for crystallization: Theory and experimental verification. *Powder Technology*, **88**, 227-235.

Matthews, H.B., and J.B. Rawlings (1998). Batch crystallization of a photochemical: Modeling, control and filtration. *AIChE J.*, **44**, 1119-1127.

Moffat, T.P., J.E. Bonevich, W.H. Huber, A. Stanishevsky, D.R. Kelly, G.R. Stafford and D. Josell (2000). Superconformal electrodeposition of copper in 500-90 nm features. *J. Electrochem. Soc.*, **147**, 4524-4535.

Moffat, T.P., D. Wheeler, W.H. Huber and D. Josell (2001). Superconformal electrodeposition of copper. *Electrochemical and Solid State Letters*, **4**, C26-C29 (2001).

Moffat, T.P., D. Wheeler and D. Josell (2004). Electrodeposition of copper in the SPS-PEG-Cl additive system. *J. Electrochem. Soc.*, **151**, C262-C271.

Nagy, Z.K., and R.D. Braatz (2003a). Robust nonlinear model predictive control of batch processes. *AIChE J.*, **49**, 1776-1786.

Nagy, Z.K., and R.D. Braatz (2003b). Worst-case and distributional robustness analysis of finite-time control trajectories for nonlinear distributed parameter systems. *IEEE Trans. on Control Syst. Tech.*, **11**, 494-504.

Nagy, Z.K., and R.D. Braatz (2004). Open-loop and closed-loop robust optimal control of batch processes using distributional worst-case analysis. *J. of Process Control*, **14**, 411-422.

Pan, W.W., M.A. Tatang, G.J. McRae and R.G. Prinn (1998). Uncertainty analysis of indirect radiative forcing by anthropogenic sulfate aerosols, *J. of Geophysical Research - Atmospheres*, **103**, 3815-3823.

Phenix, B.D., J.L. Dinaro, M.A. Tatang, J.W. Tester, J.B. Howard and G.J. McRae (1998). Incorporation of parametric uncertainty into complex kinetic mechanisms: Application to hydrogen oxidation in supercritical water. *Combustion and Flame*, **112**, 132-146.

Pricer, T.J., M.J. Kushner and R.C. Alkire (2002a). Monte Carlo simulation of the electrodeposition of copper - I. Additive-free acidic sulfate solution. *J. Electrochem. Soc.*, **149**, C396-C405.

Pricer, T.J., M.J. Kushner and R.C. Alkire (2002b). Monte Carlo simulation of the electrodeposition of copper - II. Acid sulfate solution with blocking additive. *J. Electrochem. Soc.*, **149**, C406-C412.

Prokop, A. (2001). Bioartificial organs in the twenty-first century - Nanobiological devices. *Bioartificial Organs III: Tissue Sourcing, Immunoisolation, and Clinical Trials, Annals of the New York Academy of Sciences*, **944**, 472-490.

Raimondeau, S., P. Aghalayam, A.B. Mhadeshwar and D.G. Vlachos (2003). Parameter optimization of molecular models: Application to surface kinetics. *Ind. Eng. Chem. Res.*, **42**, 1174-1183.

Raimondeau, S., and D.G. Vlachos (2000). Low-dimensional approximations of multiscale epitaxial growth models for microstructure control of materials, *J. Comp. Phys.*, **160**, 564-576.

Rathinam, M., L. Petzold, D. Gillespie (2003). Stiffness in stochastic chemically reacting systems: The implicit tau-leaping method. *J. Chem. Phys.*, **119**, 12784-12794.

Ray, W.H. (1981). *Advanced Process Control.* McGraw Hill, New York, U.S.A.

Reilly, P.M., and G.E. Blau (1974). The use of statistical methods to build mathematical models of chemical reacting systems. *Can. J. Chem. Eng.*, **52**, 289-299.

Runborg, O., C. Theodoropoulos and I.G. Kevrekidis (2002). Effective bifurcation analysis: A time-stepper-based approach. *Nonlinearity*, **15**, 491-511.

Rusli, E., T.O. Drews, D.L. Ma, R.C. Alkire and R.D. Braatz (2003). Nonlinear feedback control of a coupled kinetic Monte Carlo-finite difference simulation. In: *Proc. of the IFAC Symposium on Advanced Control of Chemical Processes,* Hong Kong.

Rusli, E., T.O. Drews and R.D. Braatz (2004a). Systems analysis and design of dynamically coupled multiscale reactor simulation codes. *18th International Symposium on Chemical Reaction Engineering*, Chicago, Illinois, June 6-9. Paper 24.

Rusli, E., T.O. Drews and R.D. Braatz (2004b). Control systems analysis of a multiscale simulation code for copper electrodeposition. In: *Proc. of the American Control Conference*, IEEE Press, Piscataway, New Jersey, U.S.A., in press.

Sematech (2002). International Technology Roadmap for Semiconductors, International Sematech.

Shelley, J.C., and M.Y. Shelley (2000). Computer simulation of surfactant solutions. *Cur. Opin. in Col. and Surf. Sci.*, **5**, 101-110.

Siettos, C.I., M.D. Graham and I.G. Kevrekidis (2003a). Coarse Brownian dynamics for nematic liquid crystals: Bifurcation, projective integration, and control via stochastic simulation. *J. Chem. Phys.*, **118**, 10149-10156.

Siettos C.I., C.C. Pantelides and I.G. Kevrekidis (2003b). Enabling dynamic process simulators to perform alternative tasks: A time-stepper-based toolkit for computer-aided analysis. *Ind. Eng. Chem. Res.*, **42**, 6795-6801.

Taephaisitphongse, P., Y. Cao and A.C. West (2001). Electrochemical and fill studies of a multicomponent additive package for copper deposition. *J. Electrochem. Soc.*, **148**, C492-C497.

Tan, M., and J.N. Harb (2003). Additive behavior during copper electrodeposition in solutions containing Cl^-, PEG, and SPS. *J. Electrochem. Soc.*, **150**, C420-C425.

Tatang, M.A., W.W. Pan, R.G. Prinn and G.J. McRae (1997). An efficient method for parametric uncertainty analysis of numerical geophysical models. *J. Geophysical Res.*, **102**, 21952-21932.

Tsukagoshi, K., N. Yoneya, S. Uryu, Y. Aoyagi, A. Kanda, Y. Ootuka and B.W. Alphenaar (2002). Carbon nanotube devices for nanoelectronics. *Physica B-Condensed Matter*, **323**, 107-114.

West, A.C. (2000). Theory of filling of high-aspect ratio trenches and vias in presence of additives. *J. Electrochem. Soc.*, **147**, 227-232.

Wheeler, D., D. Josell and T.P. Moffat (2003). Modeling superconformal electrodeposition using the level set method. *J. Electrochem. Soc.*, **150**, C302-C310.

Yang, Y.G., R.A. Johnson and H.N.G. Wadley (2002). Kinetic Monte Carlo simulation of heterometal epitaxial deposition. *Surf. Sci.*, **499**, 141-151.

ELSEVIER
IFAC
PUBLICATIONS
www.elsevier.com/locate/ifac

MODEL REDUCTION AND CONTROL IN REACTOR–HEAT EXCHANGER NETWORKS

Michael Bâldea * **Prodromos Daoutidis** *

* *Department of Chemical Engineering and Materials Science, University of Minnesota, Minneapolis, MN 55455*
Telephone: (612) 625 8818, email: daoutidi@cems.umn.edu

Abstract: This paper focuses on energetic aspects of the dynamics and control of process networks with high material recycle. Two prototype systems are analyzed, for which the large material recycle acts as an energy carrier, leading to a high energy throughput for the entire network. Using singular perturbation arguments, we show that the variables in the energy balance of these networks evolve in the fast time scale, while the terms in the material balance equations can exhibit both fast and slow transients. We present a procedure for deriving reduced-order, non-stiff models for the fast and slow dynamics, the latter typically of low order, and a framework for rational control system design, that accounts for the time scale separation exhibited by the system dynamics. The theoretical results are illustrated in a reactor-external heat exchanger network example. *Copyright © 2004 IFAC*

Keywords: model reduction, nonlinear control, energy recycle, singular perturbations

1. INTRODUCTION

Integrated process networks, consisting of individual units interconnected through material and energy recycle, are the rule, rather than the exception in the process industries. The dynamics and control of such networks present distinct challenges, since, in addition to the nonlinear behavior of the individual units, the feedback interactions among these units, induced by the recycle, typically give rise to more complex overall network dynamics (*e.g.* (Morud and Skogestad, 1994; Mizsey and Kalmar, 1996; Jacobsen and Berezowski, 1998; Bildea and Dimian, 1998)).

At the same time, the efficient transient operation of such networks is of critical importance, as the current economic environment imposes frequent changes in operating conditions and objectives (*e.g.* changes in product grade and feed switching), requiring tighter coordination of the plant-wide optimization and advanced control levels (Marquardt, 2000; Kulhavy et al., 2000). A major bottleneck towards analyzing, optimizing and improving the control of process networks is the often overwhelming size and complexity of their dynamic models, which make dynamic simulation computationally intensive and the design of fully centralized nonlinear controllers on the basis of entire network models impractical (such controllers are almost invariably difficult to tune, expensive to implement and maintain, and sensitive to measurement errors and noise). Indeed, the majority of studies on control of networks with material (*e.g.* (Luyben, 1993; Yi and Luyben, 1997)) and energy recycle (*e.g.* (Luyben, 1998; Reyes and Luyben, 2000; Chen and Yu, 2003)) are within a multi-loop linear control framework. The strong coupling between the control loops in different process units in a process network has consequently been recognized as a major issue that must be addressed in a plant-wide control setting (Price

[1] Partial support for this work by ACS-PRF, grant 38114-AC9 and NSF-CTS, grant 0234440 is gratefully acknowledged.

and Georgakis, 1993; Luyben et al., 1997; Ng and Stephanopoulos, 1998), and several strategies have been proposed to this end. (Georgakis, 1986) suggested the use of empirically identified extensive fast and slow variables for the synthesis of nonlinear controllers for a process network. Using the concept of passivity, (Farschman et al., 1998; Hangos et al., 1999) introduced a formal framework for stability analysis and stabilization of process networks using distributed control, subject to thermodynamic and equipment constraints. Finally, (Tyreus, 1999) proposes an analysis method for identifying a small number of dominant variables of a network, that are subsequently used in a partial control framework (Kothare et al., 2000).

In our previous work (Kumar and Daoutidis, 2002; Kumar and Daoutidis, 2003), we considered networks and staged processes with high internal flows compared to the throughput. Within the framework of singular perturbations, we established that the large recycle induces a time scale separation, with the dynamics of individual processes evolving in a fast time scale with weak interactions, and the dynamics of the overall system evolving in a slow time scale where these interactions become significant; this slow dynamics is usually nonlinear and of low order. Motivated by this, we proposed a method for deriving nonlinear low-order models of the slow dynamics, and a controller design framework comprising of properly coordinated controllers designed separately in the fast and slow time scales.

In (Baldea et al., 2004), we focused on process networks with recycle, in which small quantities of inert components are present and a small purge stream is used for their removal. Adopting again a singular perturbation perspective, we established the presence of a slow dynamics associated with the inert, derived explicit descriptions of this dynamics and outlined a framework for rationally addressing the control of inert levels in the network.

In the present paper we focus on the energetic aspects of process networks with high material recycle. We analyze two prototype systems for which the large material recycle acts as an energy carrier, leading to a high energy throughput for the entire network. First, we consider the case of a reactor where a highly exothermic set of reactions takes place, interconnected with an external heat exchanger through a large material recycle stream for more effective heat removal. Using singular perturbation arguments, we show that the dynamics of such networks typically exhibit two time scales, with the energy dynamics evolving in the fast time scale, and the material balance dynamics of the entire network evolving in the slow time scale. We describe a method for deriving approximate, non–stiff, reduced–order models for the dynamics in each time scale, and propose a controller design framework that accounts for this time scale separation. Subsequently, we turn our attention to high purity distillation with large internal recycle, and show that the energy throughput in the column is also high. We also show that the energy dynamics of the column are fast, while the variables related to the material balance exhibit both fast and slow transients in their evolution, and, finally, we outline a procedure for deriving reduced-order, non-stiff, nonlinear models for the fast and slow dynamics of the column.

Throughout our derivations, we use the standard order of magnitude notation $\mathcal{O}(.)$.

2. DYNAMICS OF REACTOR–EXTERNAL HEAT EXCHANGER NETWORKS

In processes in which reactions with significant thermal effects are present, adiabatic reactor operation is not possible and direct heating/cooling for isothermal operation is often impractical or infeasible. In such cases, large material recycle streams are frequently used as heat carriers, connecting the reaction unit to an external heat exchange system (Seider et al., 1999). This configuration allows for more efficient heat exchange, due to the high flowrates of the recycle and the heating/cooling medium. The efficiency of the external heat exchanger can be increased further by increasing the heat capacity of the recycle stream, either by using excess quantities of a reactant or by introducing an inert diluent in the recycle loop, along with a separation unit. Such configurations can be used in both batch and continuous processes, and are quite common in processes featuring fast, highly exothermic reactions (*e.g.* polymerization).

Existing literature on the control of reactor–external heat exchanger networks is relatively scarce, concerning mostly the implementation of linear (Alhumaizi, 2000; Henderson and Cornejo, 1989) and nonlinear (Dadebo et al., 1997) control structures on specific processes. These studies report several control challenges, including difficult tuning of PID and model–based controllers due to the ill–conditioning of the process model.

2.1 *Modeling of Reactor–External Heat Exchanger Networks*

We consider a process network, comprising of a reactor and a heat exchanger. For the case in which an inert heat transfer medium is used, a unit separating the medium from the reaction mass could also be considered. Let M denote the reactor holdup, M_R the holdup in the tube side of the heat exchanger and M_C the holdup in the shell side. Let F_o be the

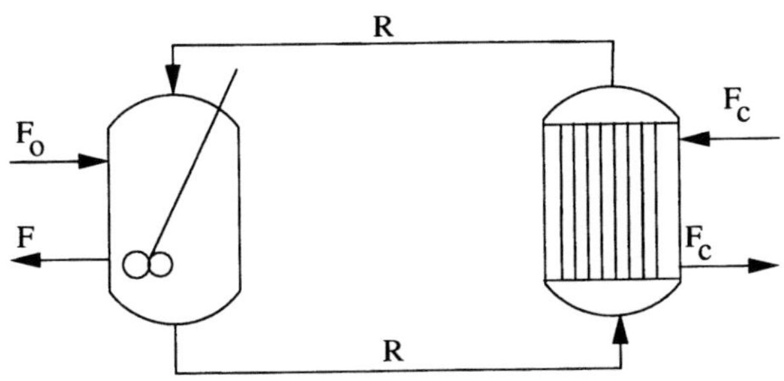

Fig. 1. Schematic diagram of a process network with external heat exchanger

feed flowrate to the reactor, F the effluent flowrate from the network, F_c the coolant flowrate and R the recycle flowrate. Let T_o be the temperature of the feed stream, T the reactor temperature, T_R the temperature of the reaction mass in the tube-side of the heat exchanger, T_{co} and T_c the inlet and outlet temperature of the cooling medium, respectively. $\mathcal{C}$ components are present in the network and participate in $\mathcal{R}$ stoichiometrically independent reactions, with reaction rate r_i, $i = 1, \ldots, \mathcal{R}$ and stoichiometric matrix $\underline{\underline{S}} \in \mathbb{R}^{\mathcal{C}\times\mathcal{R}}$. We denote the heat of reaction vector by $\underline{\Delta H} = [\Delta H_1, \ldots, \Delta H_{\mathcal{R}}]^T$. We assume that the thermal effect of the reactions is very high and that the adiabatic operation of the reactor is not possible. In order to control the reactor temperature, the reaction mass is recycled at a high rate (compared to the feed) through the heat exchanger. For simplicity, we consider the density and heat capacity of the reactants and products (ρ and C_p) and of the cooling medium used in the heat exchanger (ρ_c and C_{pc}) to be constant, and C_p and C_{pc} to be of comparable magnitude, *i.e.* $C_p/C_{pc} = k_{cp} = \mathcal{O}(1)$. Assuming that all units are modeled as lumped parameter systems, and a linear approximation for the temperature gradient in the heat exchanger, the model of the CSTR-external heat exchanger network becomes:

$$
\begin{aligned}
\dot{M} &= F_o - F \qquad (1)\\
\underline{\dot{C}} &= \underline{\underline{S}}\underline{r} + \frac{F_o}{M}(\underline{C_o} - \underline{C})\\
\dot{T} &= -\frac{1}{C_p}\underline{\Delta H}^T\underline{r} + \frac{F_o}{M}(T_o - T) + \frac{R}{M}(T_R - T)\\
\dot{T}_R &= \frac{R}{M_R}(T - T_R) - \frac{UA}{C_pM_R}(T_R - T_C)\\
\dot{T}_C &= \frac{Fc}{M_C}(T_{Co} - T_C) + \frac{UA}{C_{pc}M_C}(T_R - T_C)
\end{aligned}
$$

where U denotes the overall heat transfer coefficient in the heat exchanger and A the heat transfer area.

Let us now define:

$$\varepsilon = \frac{F_{os}}{R_s} \qquad (2)$$

where the subscript s denotes steady-state values. Since the recycle flowrate R_s is much larger than the reactor feed F_{os}, $\varepsilon \ll 1$. Also, we define the scaled (potentially manipulated) inputs $u_o = F_o/F_{os}$, $u_F = F/F_s$, $u_R = R/R_s$ and $u_c = F_c/F_{cs}$, and the $\mathcal{O}(1)$ quantity $k_f = F_s/F_{os}$.

The model of Eq. 1 thus becomes:

$$
\begin{aligned}
\dot{M} &= F_{os}(u_o - k_F u_f) \qquad (3)\\
\underline{\dot{C}} &= \underline{\underline{S}}\underline{r} + \frac{F_{os}}{M}u_o(\underline{C_o} - \underline{C})\\
\dot{T} &= -\frac{1}{C_p}\underline{\Delta H}^T\underline{r} + \frac{F_{os}}{M}u_o(T_o - T) + \frac{1}{\varepsilon}\frac{F_{os}}{M}u_R(T_R - T)\\
\dot{T}_R &= \frac{1}{\varepsilon}\frac{F_{os}}{M_R}u_R(T - T_R) - \frac{UA}{C_pM_R}(T_R - T_C)\\
\dot{T}_C &= \frac{Fc}{M_C}(T_{Co} - T_C) + \frac{UA}{C_{pc}M_C}(T_R - T_C)
\end{aligned}
$$

In practical applications, for useful energy removal or recovery, the flowrates of external cooling utility streams for the heat exchanger will be in direct relationship with the reaction mass throughput, *i.e.* a high recycle rate will require a high coolant flowrate. Hence, we can assume that $F_{cs}/R_s = k_r = \mathcal{O}(1)$ and consequently $F_s/F_{cs} = \mathcal{O}(\varepsilon)$. Also, we assume that $\frac{UA}{C_pM_C}$ is sufficiently large so that the cross-stream heat transfer rate in the heat exchanger is of the same order of magnitude as the net rate at which heat is input to the heat exchanger by the recycle stream R:

$$\frac{(UA(T_R - T_C))_s}{(RC_p(T - T_R))_s} = \mathcal{O}(1)$$

or, that the time constants for heat transfer and mass transport are of the same order of magnitude, *i.e.*

$$\frac{\frac{UA}{C_pM_R}}{\frac{R_s}{M_R}} = k_h = \mathcal{O}(1)$$

or, using Eq. 2,

$$\frac{UA}{C_pM_R} = k_h\frac{F_{os}}{\varepsilon M_R}$$

With the above notation, the dynamic model of the process network in Fig. 1 can be written as:

$$
\begin{aligned}
\dot{M} &= F_{os}(u_o - k_F u_f) \qquad (4)\\
\underline{\dot{C}} &= \underline{\underline{S}}\underline{r} + \frac{F_{os}}{M}u_o(\underline{C_o} - \underline{C})\\
\dot{T} &= -\frac{1}{C_p}\underline{\Delta H}^T\underline{r} + \frac{F_{os}}{M}u_o(T_o - T) + \frac{1}{\varepsilon}\frac{F_{os}}{M}u_R(T_R - T)\\
\dot{T}_R &= \frac{1}{\varepsilon}\frac{F_{os}}{M_R}u_R(T - T_R) - \frac{1}{\varepsilon}\frac{k_hF_{os}}{M_R}(T_R - T_C)\\
\dot{T}_C &= \frac{1}{\varepsilon}\frac{k_rF_{os}}{M_C}u_c(T_{Co} - T_C) + \frac{1}{\varepsilon}\frac{k_hk_{cp}F_{os}}{M_C}(T_R - T_C)
\end{aligned}
$$

Due to the presence of flowrates of different magnitudes and of fast heat transfer, the above model is stiff, its stiffness being captured by the small singular perturbation parameter ε.

In the following section we show, via a singular perturbation analysis, that the dynamics of the network (1) exhibits two time scales, and obtain non–stiff, reduced–order models of the dynamics in each time scale.

2.2 *Model reduction and Control*

We proceed with our analysis starting from the fast time scale. To this end, we define the "stretched", fast time scale $\tau = t/\varepsilon$ in which Eq. 4 becomes:

$$\begin{aligned}
\frac{dM}{d\tau} &= \varepsilon F_{os}(u_o - k_F u_f) \\
\frac{d\underline{C}}{d\tau} &= \varepsilon\left[\underline{\underline{S}}\underline{r} + \frac{F_{os}}{M}u_o(\underline{C_o} - \underline{C})\right] \\
\frac{dT}{d\tau} &= -\varepsilon\left[\frac{1}{C_p}\underline{\Delta H}^T\underline{r} + \frac{F_{os}}{M}u_o(T_o - T)\right] + \\
&\quad \frac{F_{os}}{M}u_R(T_R - T) \\
\frac{T_R}{d\tau} &= \frac{F_{os}}{M_R}u_R(T - T_R) - \frac{k_h F_{os}}{M_R}(T_R - T_C) \\
\frac{T_C}{d\tau} &= \frac{k_r F_{os}}{M_C}u_c(T_{Co} - T_C) + \frac{k_h k_{cp} F_{os}}{M_C}(T_R - T_C)
\end{aligned} \tag{5}$$

Then, we consider the limit $\varepsilon \to 0$, corresponding to infinitely large recycle and cooling medium flowrates and infinitely fast heat transfer in the heat exchanger. In this limit, we obtain the following description of the process network dynamics in the fast time scale:

$$\begin{aligned}
\frac{dT}{d\tau} &= \frac{F_{os}}{M}u_R(T_R - T) \\
\frac{T_R}{d\tau} &= \frac{F_{os}}{M_R}u_R(T - T_R) - \frac{k_h F_{os}}{M_R}(T_R - T_C) \\
\frac{T_C}{d\tau} &= \frac{k_r F_{os}}{M_C}u_c(T_{Co} - T_C) + \frac{k_h k_{cp} F_{os}}{M_C}(T_R - T_C)
\end{aligned} \tag{6}$$

Notice that the large recycle and coolant flowrates u_R and u_C are the only manipulated inputs available in this fast time scale, and can be used to address temperature stabilization and regulation objectives.

Turning to the slow dynamics, multiplying Eq. 4 by ε and considering the limit $\varepsilon \to 0$, we obtain the following quasi-steady state constraints:

$$\begin{aligned}
0 &= \frac{F_{os}}{M}u_R(T_R - T) \\
0 &= \frac{F_{os}}{M_R}u_R(T - T_R) - \frac{k_h F_{os}}{M_R}(T_R - T_C) \\
0 &= \frac{k_r F_{os}}{M_C}u_c(T_{Co} - T_C) + \frac{k_h k_{cp} F_{os}}{M_C}(T_R - T_C)
\end{aligned} \tag{7}$$

or, equivalently,

$$\begin{aligned}
0 &= T_R - T \\
0 &= T_R - T_C \\
0 &= T_{Co} - T_C
\end{aligned} \tag{8}$$

The constraints in Eq. 8 are linearly independent and hence they can be solved for the quasi-steady state values $\underline{\Theta}^\star = [T^\star, T_R^\star, T_C^\star]$ of the variables $\underline{\Theta} = [T, T_R, T_C]$, *i.e.*

$$T^\star = T_R^\star = T_C^\star = T_{Co}$$

Substituting the value for $T^\star$, we then obtain:

$$\begin{aligned}
\dot{M} &= F_s(k_f u_o - u_F) \\
\dot{\underline{C}} &= \underline{\underline{S}}\underline{r}(T^\star) + \frac{k_f F_s}{M}u_o(\underline{C_o} - \underline{C})
\end{aligned} \tag{9}$$

which represents the model of the slow dynamics of the process network. Note that only the small feed and effluent flowrates u_o and u_F are available as manipulated inputs in this slow time scale.

Remark 1. Due to the independence of the constraints (8), the system in Eq. 5 is in a *standard* singularly perturbed form (Kumar and Daoutidis, 1999), whereby one can distinguish between the fast variables $\underline{\Theta}$ (Eq. 6), and the slow ones, M and $\underline{C}$ (Eq. 9). Equivalently, the energy-related variables and the variables in the material balance of the process network in Figure 1 evolve in different time scales, with the former being faster than the latter.

Remark 2. Notice that the rates of heat generation from the $\mathcal{R}$ reactions are the product of two terms, ΔH_i and r_i, corresponding to the heat of reaction and reaction rate, respectively. Consequently, a high rate of heat generation by reaction could occur both in reacting systems in which fast reactions with moderate reaction enthalpies are present, and in reacting systems in which the reactions have moderate rates and a high heat of reaction. In the former case, the material balance of Eq. 9 will itself be in a nonstandard singularly perturbed form (Vora and Daoutidis, 2001), and further reduction steps will be necessary in order to obtain non-stiff descriptions of the intermediate and slow dynamics.

Remark 3. The analysis framework we presented is also applicable if an inert component is used as a heat carrier. In this case, the the model (1) would be augmented by the equations corresponding to the model of the separation unit. However, the stoichiometric matrix $\underline{\underline{S}}$ and and reaction rates $\underline{r}$ would remained unchanged, as the inert component does not partake in any reaction.

Remark 4. The arguments presented above indicate that the control objectives related to the energy-balance related fast variables T and T_R should be addressed using the large flowrates u_R and u_C, whereas the control objectives involving the slow variables in the material balance (such as the reactor holdup and product purity or distribution) should be addressed using u_o and u_F.

2.3 *Illustrative example*

Consider the process network in Fig. 2. It comprises of a CSTR, of volume M connected to an external heat exchanger by a recycle stream R. Two parallel

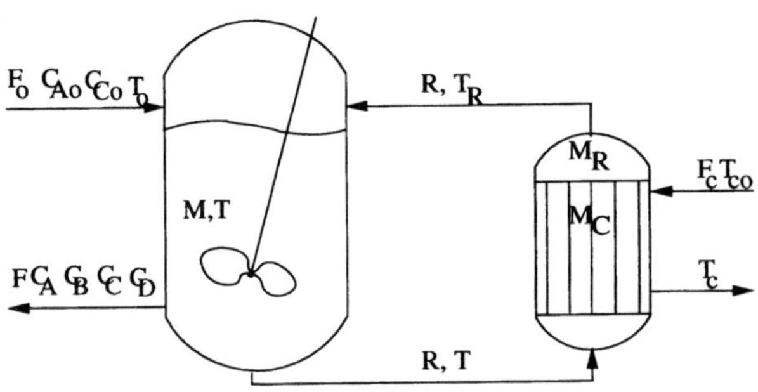

Fig. 2. CSTR with external heat exchanger

reactions take place in the reactor:

$$A \xrightarrow{k_1} 2B \tag{10}$$

$$B + C \xrightarrow{k_2} D \tag{11}$$

The feed stream F_o contains the reactants A and C and its composition C_{Ao}, C_{Co} is assumed to be constant. The intermediate B and the product D, as well as the unreacted A and C are removed at a rate F. We assume that the amount of heat generated in the two reactions is high, and that the reactor cannot be run adiabatically. In order to control the reactor temperature and prevent reaction runaway, the reaction mass is recycled through the tube side of the heat exchanger, at a (constant) high rate R. The coolant, of heat capacity C_{pc}, initially at temperature T_{co} is circulated though the heat exchanger shell at a rate F_c. The holdups in the heat exchanger shell and tube sides, respectively M_R and M_C are considered to be constant. We denote the heat transfer coefficient in the heat exchanger by U and let A be the transfer area. All units are modeled as lumped parameter systems, with a linear approximation for the temperature gradient in the heat exchanger.

The objectives for this process are control of the reactor temperature T, together with the control of the total holdup of the network (in this case, equivalent to the control of the reactor holdup M) and the product purity C_D.

With the assumptions stated above, the model of the process in Fig. 2 has the following form:

$$\begin{aligned}
\dot{M} &= F_0 - F \qquad (12)\\
\dot{C}_A &= \frac{F_o}{M}(C_{Ao} - C_A) - k_{10}e^{-\frac{Ea_1}{RT}}C_A\\
\dot{C}_B &= -\frac{F_o}{M}C_B + 2k_{10}e^{-\frac{Ea_1}{RT}}C_A - k_{20}e^{-\frac{Ea_2}{RT}}C_BC_C\\
\dot{C}_C &= \frac{F_o}{M}(C_{Co} - C_C) - k_{20}e^{-\frac{Ea_2}{RT}}C_BC_C\\
\dot{C}_D &= -\frac{F_o}{M}C_D + k_{20}e^{-\frac{Ea_2}{RT}}C_BC_C\\
\dot{T} &= \frac{F_o}{M}(T_o - T) + \frac{R}{M}(T_R - T)\\
&\quad -\frac{\Delta H_1}{C_p}k_{10}e^{-\frac{Ea_1}{RT}}C_A\\
&\quad -\frac{\Delta H_2}{C_p}k_{20}e^{-\frac{Ea_2}{RT}}C_BC_C\\
\dot{T}_R &= \frac{R}{M_R}(T - T_R) - \frac{UA}{C_pM_R}(T_R - T_C)\\
\dot{T}_c &= \frac{Fc}{M_C}(T_{Co} - T_C) + \frac{UA}{C_{pc}M_C}(T_R - T_C)
\end{aligned}$$

which, by defining the small singular perturbation parameter

$$\varepsilon = \frac{F_{os}}{R_s}$$

the $\mathcal{O}(1)$ quantities

$$k_r = \frac{F_{cs}}{R_s}$$

$$k_f = \frac{F_s}{F_{os}}$$

$$k_{cp} = \frac{C_p}{C_{pc}}$$

$$k_h = \frac{UA}{R_sC_p}$$

and the manipulated inputs $u_o = F_o/F_{os}$, $u_F = F/F_s$, $u_R = R/R_s$ and $u_c = F_c/F_{cs}$, can be rewritten as:

$$\begin{aligned}
\dot{M} &= F_{os}(u_o - k_fu_F) \qquad (13)\\
\dot{C}_A &= \frac{F_{os}}{M}u_o(C_{Ao} - C_A) - k_{10}e^{-\frac{Ea_1}{RT}}C_A\\
\dot{C}_B &= -\frac{F_{os}}{M}u_oC_B + 2k_{10}e^{-\frac{Ea_1}{RT}}C_A\\
&\quad -k_{20}e^{-\frac{Ea_2}{RT}}C_BC_C\\
\dot{C}_C &= \frac{F_{os}}{M}u_o(C_{Co} - C_C) - k_{20}e^{-\frac{Ea_2}{RT}}C_BC_C\\
\dot{C}_D &= -\frac{F_{os}}{M}u_oC_D + k_{20}e^{-\frac{Ea_2}{RT}}C_BC_C\\
\dot{T} &= \frac{F_{os}}{M}u_o(T_o - T) + \frac{1}{\varepsilon}\frac{F_{os}}{M}u_R(T_R - T)
\end{aligned}$$

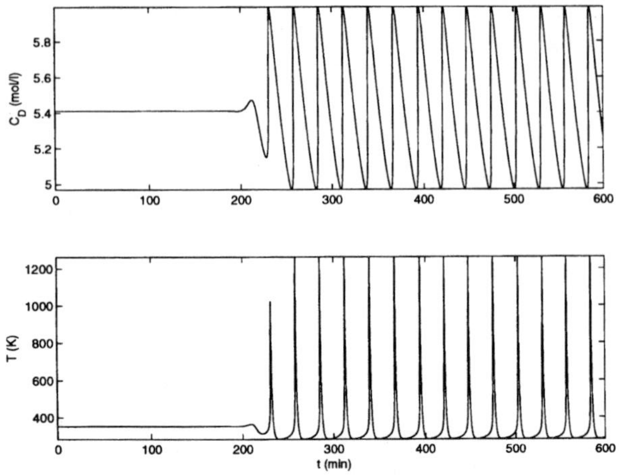

Fig. 3. Oscillatory open loop behavior of the reactor-external feed exchanger network (12). Temperature control is switched off at $t = 100$ min

$$
\begin{aligned}
&-\frac{\Delta H_1}{C_p} k_{10} e^{-\frac{Ea_1}{RT}} C_A \\
&-\frac{\Delta H_2}{C_p} k_{20} e^{-\frac{Ea_2}{RT}} C_B C_C \\
\dot{T}_R &= \frac{1}{\varepsilon}\frac{F_{os}}{M_R} u_R (T - T_R) - \frac{1}{\varepsilon}\frac{k_h F_{os}}{M_R} (T_R - T_C) \\
\dot{T}_C &= \frac{1}{\varepsilon}\frac{k_r F_{os}}{M_C} u_c (T_{Co} - T_C) + \frac{1}{\varepsilon}\frac{k_h k_{cp} F_s}{M_C} (T_R - T_C)
\end{aligned}
$$

We now apply the model reduction framework outlined in Section 2.2. In order to obtain a description of the fast dynamics, we define the fast time scale $\tau = t/\varepsilon$, and, in the limit of the recycle and coolant flow rate and the heat transfer coefficient in the heat exchanger becoming infinite *i.e.* $\varepsilon \to 0$, we obtain a description of the fast dynamics of the reactor-heat exchanger process network (13):

$$
\begin{aligned}
\frac{dT}{d\tau} &= \frac{F_{os}}{M} u_R (T_R - T) \qquad (14) \\
\frac{dT_R}{d\tau} &= \frac{F_{os}}{M_R} u_R (T - T_R) - \frac{k_h F_{os}}{M_R} (T_R - T_C) \\
\frac{dT_c}{d\tau} &= \frac{k_r F_{os}}{M_C} u_c (T_{Co} - T_C) + \frac{k_h k_{cp} F_s}{M_C} (T_R - T_C)
\end{aligned}
$$

According to the analysis in Section 2.2, we address the control of the reactor temperature T in the fast time scale, using the large flowrate u_C as a manipulated input, and the proportional feedback law:

$$u_C = 1 + K_C (T - T_{sp}) \qquad (15)$$

Eq. (14) yields the quasi–steady state value $T^\star = T_{Co}$. Substituting $T^\star$ in Eq. 13, we obtain a description of the slow dynamics of the reactor-heat exchanger network:

$$\dot{M} = F_o - F \qquad (16)$$

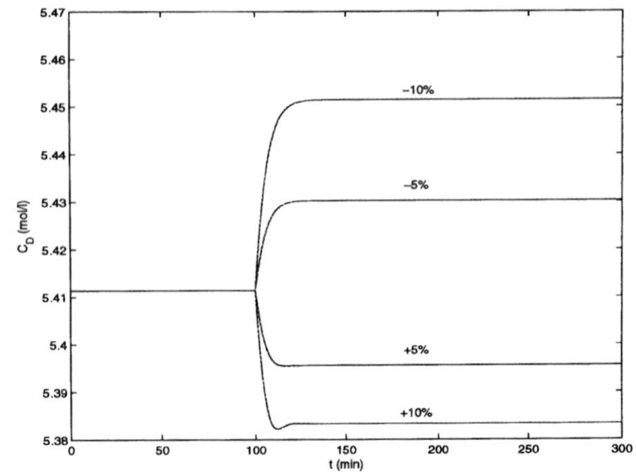

Fig. 4. Nonlinear behavior of the slow dynamics of the reactor-external heat exchanger network: change in C_D for different step changes in the feed flowrate F_o, from the nominal value F_{os}

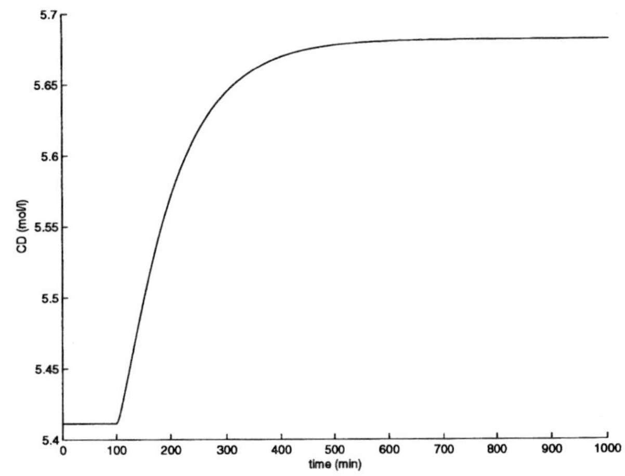

Fig. 5. Closed loop evolution of the product purity

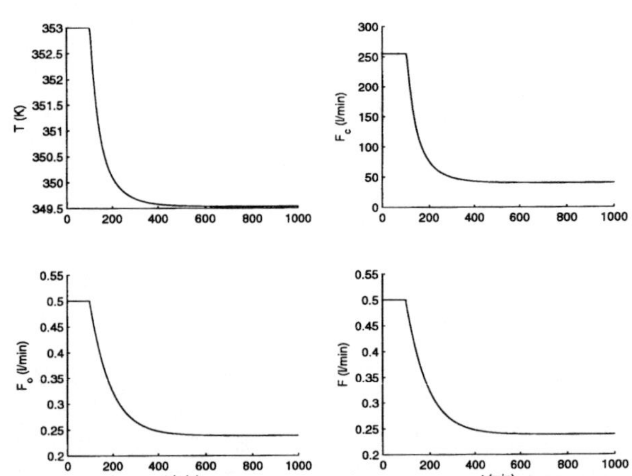

Fig. 6. Closed loop evolution of the reactor temperature and of the feed, product and coolant flowrates

$$\dot{C}_A = \frac{F_o}{M}(C_{A0} - C_A) - k_{10} e^{-\frac{Ea_1}{RT_{Co}}} C_A$$

$$\dot{C}_B = -\frac{F_o}{M}C_B + 2k_{10}e^{-\frac{Ea_1}{RT_{Co}}}C_A - k_{20}e^{-\frac{Ea_2}{RT_{Co}}}C_BC_C$$
$$\dot{C}_C = \frac{F_o}{M}(C_{C0} - C_C) - k_{20}e^{-\frac{Ea_2}{RT_{Co}}}C_BC_C$$
$$\dot{C}_D = -\frac{F_o}{M}C_D + k_{20}e^{-\frac{Ea_2}{RT_{Co}}}C_BC_C$$

We carried out numerical simulations, using the nominal values in Table 1. Initially, we simulated the system switching the temperature controller (15) off at t = 100 min, and recording the subsequent behavior. Without temperature control, the behavior of the reactor-heat exchanger network is oscillatory (Fig. 3), with very large deviations in the reactor temperature T. In the simulation presented in Fig. 3, the proportional controller in Eq. 15 (with K_C = 0.9) stabilizes the fast dynamics of the network.

According to the analysis in Section 2.2, after setting the reactor temperature with the control law (14), the control of the total holdup M and of the product purity C_D should be addressed in the slow time scale using the small flowrates F_0 and F. Figure 4 shows the profiles for the product purity C_D for different step changes in F_0, in the system with the proportional controller of Eq. 14. The responses indicate a nonlinear behavior of the slow dynamics of the reactor-external heat exchanger network.

Based on the reduced order model (16) we designed a multivariable input–output linearizing feedback controller with integral action (Daoutidis and Kravaris, 1994) for the product purity and for the total holdup (that acts as an integrator), using respectively u_o and u_F as manipulated inputs, requesting a first–order decoupled response:

$$C_D + \beta_1\frac{dC_D}{dt} = C_{D,sp} \qquad (17)$$
$$M + \beta_2\frac{dM}{dt} = M_{sp}$$

with β_1 = 30 min and β_2 = 30 min.

Figures 5–6 present the closed–loop behavior of the reactor–heat exchanger system for a 5% increase in the product purity setpoint at t = 100 min in the nominal case. Clearly, the proposed controller yields the desired performance, imposing the requested first–order response. Figures 5–6 show the corresponding profiles in the case of a modeling error (a −16% error in the heat transfer coefficient U and a $+5K$ unmeasured disturbance in the coolant initial temperature T_{Co}). The controller performance is very robust with respect to these disturbances.

3. DYNAMICS OF HIGH PURITY DISTILLATION COLUMNS

High purity distillation columns rely on a high internal recycle for increasing the purity of the

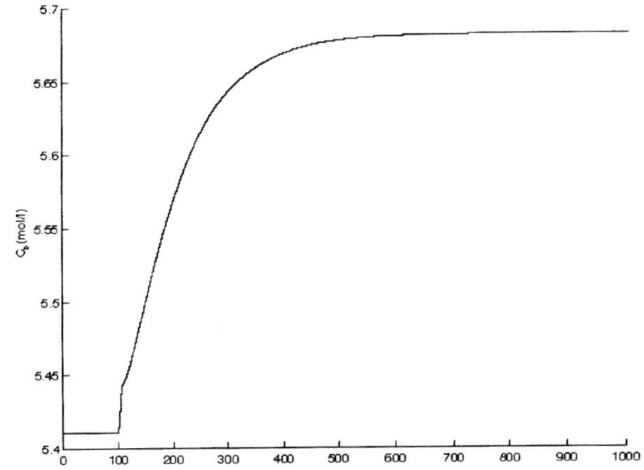

Fig. 7. Closed loop evolution of the product purity in the presence of modeling errors

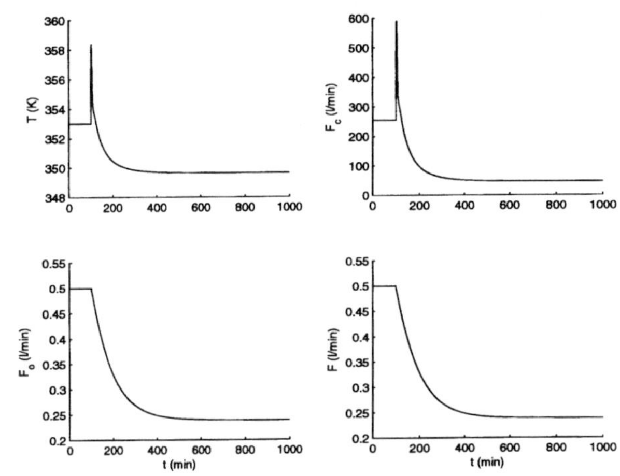

Fig. 8. Closed loop evolution of the reactor temperature and of the feed, product and coolant flowrates in the presence of modeling errors

distillate/bottoms streams. In our previous work (Kumar and Daoutidis, 2003), we have provided a rigorous justification of the two time scale behavior exhibited by such columns, and developed an explicit nonlinear low-order model of the slow input/output dynamics that is suitable for analysis and controller design. In the present work, we analyze the energy dynamics of high purity distillation columns. Using singular perturbation arguments, we show that the

Table 1. Nominal values for the process parameters

F_0	0.5 l/min	T_o	298 K
F	0.5 l/min	T_{co}	278 K
R	60 l/min	T	353 K
F_C	255.1 l/min	T_R	296 K
M	50 l	k_{10}	129.32 min^{-1}
M_R	40 l	k_{20}	70.50 $l/(mol\ min)$
M_C	40 l	ΔH_1	-92000 J/mol
C_{A0}	3 mol/l	ΔH_2	-229000 J/mol
C_{C0}	10 mol/l	Ea_1	20300 J/mol
UA	120000 $JK^{-1}min^{-1}$	Ea_2	30000 J/mol
K_C	0.9	Cp_c	229.9 $Jl^{-1}K^{-1}$
		Cp	220 $Jl^{-1}K^{-1}$

presence of the large internal flowrates leads to a high energy *throughput* (compared to the energy content of the product streams) and that the energy dynamics of the column evolves in the fast time scale.

3.1 *Modeling of a high purity distillation column*

We consider a distillation column with N trays (numbered from top to bottom), to which a saturated liquid containing a mixture of three components with mole fractions x_{1f}, x_{2f} of components 1 and 2 respectively, is fed at (molar) flowrate F_0 on tray N_f. The heavy component 3 which is the desired product is removed at the bottom from the reboiler at a flowrate B, while the lighter components 1 and 2 are removed at the top from the condenser at a flowrate D. In this column, large (compared to the feed, distillate and bottom product flowrates) vapor boilup V_B and liquid recycle R are used to attain a high purity of the desired component 3 in the bottom product. For simplicity, we model the heat transfer in the reboiler and condenser by using heat duties and denote by Q_R and Q_C the heat duties in the reboiler and the condenser, respectively. We assume that the relative volatilities of the three components are constant, and hence, that the phase equilibrium relationships are given by :

$$y_{1,i} = \frac{\alpha_1 x_{1,i}}{1 + (\alpha_1 - 1)x_{1,i} + (\alpha_3 - 1)x_{3,i}} \tag{18}$$

$$y_{3,i} = \frac{\alpha_3 x_{3,i}}{1 + (\alpha_1 - 1)x_{3,i} + (\alpha_3 - 1)x_{3,i}}$$

We consider that the heat capacities $C_{p,l}$ and $C_{p,v}$ of the liquid and vapor phases are constant. Under the above assumptions, a standard dynamic model of the column is obtained:

$$\dot{M}_C = V_B - R - D \tag{19}$$

$$\dot{x}_{1,D} = \frac{V_B}{M_C}(y_{1,1} - x_{1,D})$$

$$\dot{x}_{3,D} = \frac{V_B}{M_C}(y_{3,1} - x_{3,D})$$

$$\dot{T}_C = \frac{1}{M_C C_{p,l}}[V_B(Cp_V T_1 + \sum_{j=1}^{3} y_{j,1}\lambda_j - C_{p,l}T_C) - Q_C]$$

$$\vdots$$

$$\dot{x}_{1,i} = \frac{1}{M_i}[V_B(y_{1,i+1} - y_{1,i}) + R(x_{1,i-1} - x_{1,i})]$$

$$\dot{x}_{3,i} = \frac{1}{M_i}[V_B(y_{3,i+1} - y_{3,i}) + R(x_{3,i-1} - x_{3,i})]$$

$$\vdots$$

$$\dot{x}_{1,N_f} = \frac{1}{M_{N_f}}[V_B(y_{1,N_f+1} - y_{1,N_f}) + R(x_{1,N_f-1} - x_{1,N_f}) + F(x_{1,N_f-1} - x_{1,N_f})]$$

$$\dot{x}_{3,N_f} = \frac{1}{M_{N_f}}[V_B(y_{3,N_f+1} - y_{3,N_f}) + R(x_{3,N_f-1} - x_{3,N_f}) + F(x_{3,N_f-1} - x_{3,N_f})]$$

$$\vdots$$

$$\dot{x}_{1,i} = \frac{1}{M_i}[V_B(y_{1,i+1} - y_{1,i}) + R(x_{1,i-1} - x_{1,i}) + F(x_{1,i-1} - x_{1,i})]$$

$$\dot{x}_{3,i} = \frac{1}{M_i}[V_B(y_{3,i+1} - y_{3,i}) + R(x_{3,i-1} - x_{3,i}) + F(x_{3,i-1} - x_{3,i})]$$

$$\vdots$$

$$\dot{M}_R = R - V_B + F - B$$

$$\dot{x}_{1,B} = \frac{1}{M_R}[R(x_{1,N} - x_{1,B}) - V_B(y_{1,B} - x_{1,B}) + F(x_{1,N} - x_{1,B})]$$

$$\dot{x}_{3,B} = \frac{1}{M_R}[R(x_{3,N} - x_{3,B}) - V_B(y_{3,B} - x_{3,B}) + F(x_{3,N} - x_{3,B})]$$

$$\dot{T}_R = \frac{1}{M_R C_{p,l}}[RC_{p,l}(T_n - T_R) + FC_{p,l}(T_n - T_R) - V_B(C_{p,v} - C_{p,l})T_R - V_B\sum_{j=1}^{3} y_{j,B}\lambda_j + Q_B]$$

In Eq. 19, $M_C, x_{1,D}, x_{3,D}, y_{1,D}, y_{3,D}$ and T_D are the molar liquid holdup, liquid mole fractions, vapor mole fractions of components 1 and 3 and the temperature in the condenser, $M_i, x_{1,i}, x_{3,i}, y_{1,i}, y_{3,i}$ and T_i are the molar liquid holdup, liquid mole fractions, vapor mole fractions of components 1 and 3 and (constant) temperature on tray i and $M_R, x_{1,B}, x_{3,B}, y_{1,B}, y_{3,B}$ and T_R are the corresponding holdup, liquid mole fractions, vapor mole fractions and temperature in the reboiler.

3.2 *Model reduction*

In the case of the high purity distillation column, the presence of a large liquid recycle R implies an equally large liquid boilup V_B at the nominal steady state. On the other hand, the feed flow rate F, the distillate flow rate F and the bottom product flow rate B are of the same order of magnitude. Therefore, we can define $\varepsilon = (D_s/R_s) \ll 1$ and $k_1 = V_{B,s}/R_s = \mathcal{O}(1)$, where the subscript s refers to the nominal steady state.

Note that the energy content of the large internal flowrates R and V_B is much larger than that of the feed, distillate and bottom product, *i.e.* :

$$\frac{(DC_{p,l}T_C)_s}{(RC_{p,l}T_C)_s} \triangleq \frac{H_{Ds}}{H_{Rs}} = \varepsilon \tag{20}$$

and, similarly, $H_{Ds}/H_{V_B s} = \mathcal{O}(\varepsilon)$. In order to obtain a large vapor boilup V_B, the nominal heat duty of the reboiler $Q_{R,s}$ will be equally large, *i.e.* $Q_{Rs}/H_{V_B s} = \mathcal{O}(1)$ and, consequently, $H_{Ds}/Q_{Rs} = \varepsilon/k_R$, where $k_R = \mathcal{O}(1)$. Note also that Q_C and Q_R will be of comparable magnitude, as a large condenser heat duty is required when a large vapor boilup is present. Hence, we can write $H_{Ds}/Q_{Cs} = \varepsilon/k_C$, $k_C = \mathcal{O}(1)$.

With the above notation, the model of the high purity distillation column can be rewritten in the form:

$$
\begin{aligned}
\dot{M}_C &= \frac{D_s}{\varepsilon}(\bar{V}_B - \bar{R}) - D \qquad (21)\\
\dot{x}_{1,D} &= \frac{D_s k_1 \bar{V}_B}{\varepsilon M_C}(y_{1,1} - x_{1,D})\\
\dot{x}_{3,D} &= \frac{D_s k_1 \bar{V}_B}{\varepsilon M_C}(y_{3,1} - x_{3,D})\\
\dot{T_C} &= \frac{D_s}{\varepsilon M_C C_{p,l}}[k_1 \bar{V}_B (Cp_V T_1 + \\
&\sum_{j=1}^{3} y_{j,1}\lambda_j - C_{p,l}T_C) - k_C \frac{H_{Ds}}{D_s}\bar{Q}_C]\\
&\vdots\\
\dot{x}_{1,i} &= \frac{D_s}{\varepsilon M_i}[k_1 \bar{V}_B (y_{1,i+1} - y_{1,i}) + \bar{R}(x_{1,i-1} - x_{1,i})]\\
\dot{x}_{3,i} &= \frac{D_s}{\varepsilon M_i}[k_1 \bar{V}_B (y_{3,i+1} - y_{3,i}) + \bar{R}(x_{3,i-1} - x_{3,i})]\\
&\vdots\\
\dot{x}_{1,N_f} &= \frac{D_s}{\varepsilon M_{N_f}}[k_1 \bar{V}_B (y_{1,N_f+1} - y_{1,i}) + \\
&\bar{R}(x_{1,N_f-1} - x_{1,N_f})] + \frac{F}{M_{N_f}}(x_{1,N_f-1} - x_{1,N_f})\\
\dot{x}_{3,N_f} &= \frac{D_s}{\varepsilon M_{N_f}}[k_1 \bar{V}_B (y_{3,N_f+1} - y_{3,N_f}) + \\
&\bar{R}(x_{3,N_f-1} - x_{3,N_f})] + \frac{F}{M_{N_f}}(x_{3,N_f-1} - x_{3,N_f})]\\
&\vdots\\
\dot{x}_{1,i} &= \frac{D_s}{\varepsilon M_i}[k_1 \bar{V}_B (y_{1,i+1} - y_{1,i}) + \bar{R}(x_{1,i-1} - x_{1,i})] + \\
&\frac{F}{M_i}(x_{1,i-1} - x_{1,i})\\
\dot{x}_{3,i} &= \frac{D_s}{\varepsilon M_i}[k_1 \bar{V}_B (y_{3,i+1} - y_{3,i}) + \bar{R}(x_{3,i-1} - x_{3,i})] + \\
&\frac{F}{M_i}(x_{3,i-1} - x_{3,i})]\\
&\vdots\\
\dot{M}_R &= \frac{D_s}{\varepsilon}(\bar{R} - k_1 \bar{V}_B) + F - B\\
\dot{x}_{1,B} &= \frac{D_s}{\varepsilon M_R}[\bar{R}(x_{1,N} - x_{1,B}) - \\
&k_1 \bar{V}_B (y_{1,B} - x_{1,B})] + \frac{F}{M_R}(x_{1,N} - x_{1,B})\\
\dot{x}_{3,B} &= \frac{D_s}{\varepsilon M_R}[\bar{R}(x_{3,N} - x_{3,B}) - \\
&k_1 \bar{V}_B (y_{3,B} - x_{3,B})] + \frac{F}{M_R}(x_{3,N} - x_{3,B})]\\
\dot{T_R} &= \frac{D_s}{\varepsilon M_R C_{p,l}}[\bar{R} C_{p,l}(T_n - T_R) - k_1 \bar{V}_B ((C_{p,v} - C_{p,l})T_R - \\
&\sum_{j=1}^{3} y_{j,B}\lambda_j) + k_B \frac{H_{Ds}}{D_s}\bar{Q}_B] + \frac{F}{M_R}C_{p,l}(T_n - T_R)
\end{aligned}
$$

In the representation (21), $\bar{R}$, $\bar{Q}_C$ and $\bar{Q}_R$ denote scaled (and potentially manipulated) input variables:

$$\bar{R} = \frac{R}{R_s}, \quad \bar{Q}_C = \frac{Q_C}{Q_{C,s}}, \quad \bar{Q}_R = \frac{Q_R}{Q_{R,s}}$$

and

$$\bar{V}_B = \frac{V_B}{V_{B,s}}$$

Thus, the process model has the general form:

$$\dot{\underline{x}} = \underline{f}(\underline{x}) + \underline{\underline{g}}^s(\underline{x})\underline{u}^s + \underline{\underline{g}}^l(\underline{x})\underline{u}^l \qquad (22)$$

where

$$x = [X\ \Theta]^T = [M_C \quad x_{1,D} \quad x_{3,D} \cdots x_{1,i} \quad x_{3,i} \cdots \\ M_R \quad x_{1,B} \quad x_{3,B} \quad T_C\ T_R]^T$$

is the vector of state variables ($\underline{\Theta} = [T_C\ T_R]^T$), $u^s = [D\ B]^T \in \mathbb{R}^2$ is the vector of manipulated inputs corresponding to small flowrates and $u^l = [\bar{R}\ \bar{Q}_C\ \bar{Q}_R]^T \in \mathbb{R}^3$ is the vector of manipulated inputs corresponding to large material and energy flows, while $\underline{f}(\underline{x})$ is a smooth vector field and $\underline{\underline{g}}^s(\underline{x})$ and $\underline{\underline{g}}^l(\underline{x})$ are smooth matrices. To account for the presence of both material and energy balance equations, we rewrite $\underline{\underline{g}}^l(\underline{x})$ as:

$$\underline{\underline{g}}^l(\underline{x}) = \begin{bmatrix} \underline{\underline{g}}^{lx}(\underline{x}) \\ \underline{\underline{g}}^{lt}(\underline{x}) \end{bmatrix} \qquad (23)$$

with $\underline{\underline{g}}^{lx}(\underline{x}) \in \mathbb{R}^{(n-2)\times 3}$ and $\underline{\underline{g}}^{lt}(\underline{x}) \in \mathbb{R}^{(2)\times 3}$.

Note that the model (22) exhibits stiffness owing to the presence of the small parameter ε. In what follows, we analyze the time scale multiplicity induced by the model stiffness in the dynamic behavior of the high-purity column. We begin with a derivation of a description of the fast dynamics of the process. To this end, let us define the fast time scale $\tau = t/\varepsilon$. In this time scale, the system in Eq. 22, becomes:

$$\frac{d\underline{x}}{d\tau} = \varepsilon \underline{f}(\underline{x}) + \varepsilon \underline{\underline{g}}^s(\underline{x})\underline{u}^s + \underline{\underline{g}}^l(\underline{x})\underline{u}^l \qquad (24)$$

Considering the limit $\varepsilon \rightarrow 0$, we obtain the following description of the fast dynamics of the system:

$$\frac{d\underline{x}}{d\tau} = \underline{\underline{g}}^{l}(\underline{x})\underline{u}^{l} \tag{25}$$

Note that the inputs $\underline{u}^{s}$ have no effect on the fast dynamics; only the inputs $\underline{u}^{l}$, corresponding to the large flow rates and large reboiler and condenser heat duties have an effect, and can be used for control in this fast time scale.

We now turn to the slow time scale t, in order to obtain a description of the slow dynamics. In particular, multiplying Eq. 22 by ε, and considering the limit $\varepsilon \rightarrow 0$, we obtain the following algebraic constraints, that need to be satisfied in the slow time scale:

$$\begin{bmatrix} -\bar{R} + k_1\bar{V}_B \\ k_1\bar{V}_B(y_{1,1} - x_{1,D}) \\ k_1\bar{V}_B(y_{3,1} - x_{3,D}) \\ \vdots \\ \bar{R}(x_{1,i-1} - x_{1,i}) + k_1\bar{V}_B(y_{1,i+1} - y_{1,i}) \\ \bar{R}(x_{3,i-1} - x_{3,i}) + k_1\bar{V}_B(y_{3,i+1} - y_{3,i}) \\ \vdots \\ \bar{R} - k_1\bar{V}_B \\ \bar{R}(x_{1,N} - x_{1,B}) - k_1\bar{V}_B(y_{1,B} - x_{1,B}) \\ \bar{R}(x_{3,N} - x_{3,B}) - k_1\bar{V}_B(y_{3,B} - x_{3,B}) \\ [k_1\bar{V}_B(Cp_V T_1 + \sum_{j=1}^{3} y_{j,1}\lambda_j - C_{p,l}T_C) - \\ k_C\bar{Q}_C\frac{H_{D_S}}{D_s}] \\ [\bar{R}(T_n - T_R) - \\ k_1\bar{V}_B\frac{((C_{p,v} - C_{p,l})T_R - \sum_{j=1}^{3} y_{j,B}\lambda_j)}{C_{p,l}} + \\ k_B\bar{Q}_R\frac{H_{D_S}}{D_s C_{p,l}}] \end{bmatrix} = \underline{0} \tag{26}$$

or, in a general form:

$$\begin{bmatrix} \underline{\underline{g}}^{lx}(\underline{x})\underline{u}^{l} \\ \underline{\underline{g}}^{lt}(\underline{x})\underline{u}^{l} \end{bmatrix} = \underline{0} \tag{27}$$

The constraints (26) denote the quasi-steady state condition for the fast dynamics described in Eq. 25. Once the large manipulated inputs $\bar{R}$, $\bar{Q}_C$ and $\bar{Q}_R$ are specified with appropriate control laws, the Jacobian matrix $\partial\underline{\underline{g}}^{lt}(\underline{x})\underline{u}^{l}/\partial\underline{\Theta}$, is non-singular, and hence the last two constraints can be solved for the quasi-steady state values $\underline{\Theta}^{\star} = [T_C^{\star}\ T_R^{\star}]^T$ of the condenser and reboiler temperatures. This implies that the temperatures $\underline{\Theta}$ only exhibit *fast* dynamics.

With the above observation, referring back to the system description in Eq. 22 and considering the limit $\varepsilon \rightarrow 0$ in the slow time scale t, it follows that the terms $\underline{\underline{g}}^{lx}(\underline{\chi}, \underline{\Theta}^{\star})\underline{u}^{l}/\varepsilon$ are indeterminate. Defining these finite, but unknown terms as the additional variables $\underline{z} = \lim_{\varepsilon \rightarrow 0} \underline{\underline{g}}^{lx}(\underline{\chi}, \underline{\Theta}^{\star})\underline{u}^{l}/\varepsilon$, we obtain a description (28) of the slow dynamics of the system of Eq. 22.

$$\begin{aligned} \dot{\chi} &= \underline{f}(\underline{\chi}, \underline{\Theta}^{\star}) + \underline{\underline{g}}^{s}(\underline{\chi}, \underline{\Theta}^{\star}) + \underline{\underline{b}}(\underline{\chi}, \underline{\Theta}^{\star})\underline{z} \\ \underline{0} &= \underline{\underline{g}}^{lt}(\underline{x})\underline{u}^{l} \end{aligned} \tag{28}$$

where $\underline{\underline{b}}(\underline{\chi}, \underline{\Theta}^{\star})$ is the diagonal matrix:

$$\underline{\underline{b}}(\underline{\chi}, \underline{\Theta}^{\star}) = diag \begin{bmatrix} D_s \\ \frac{D_s}{M_C} \\ \frac{D_s}{M_C} \\ \vdots \\ \frac{D_s}{M_i} \\ \frac{D_s}{M_i} \\ \vdots \\ D_s \\ \frac{D_s}{M_R} \\ \frac{D_s}{M_R} \end{bmatrix} \tag{29}$$

Eq. (28) is a high-index differential algebraic equation (DAE) system, and the derivation of the corresponding ODE representation should be addressed using *e.g.* the methods presented in (Kumar and Daoutidis, 1999). Also, note that only the small flowrates $\underline{u}^{l}$ are present in the model (28), and are available (potentially in conjunction with the *set-points* of the fast controllers) for addressing control objectives in the slow time scale.

4. CONCLUSIONS

In this paper, we analyzed the energy dynamics of two prototype process networks with large recycle, a reactor with external heat exchanger and a high purity distillation column. In these networks, the large material recycle acts as an energy carrier, leading to a high energy throughput for the entire network. The presence of the large recycle stream causes stiffness in the network models, their dynamics exhibiting a time-scale separation. Using singular perturbation arguments, we showed that the variables in the energy balance of these networks evolve in the fast time scale, while the terms in the material balance equations can exhibit both fast and slow transients. Also within the framework of singular perturbations, we proposed a procedure for deriving reduced-order, non-stiff models for the fast and slow dynamics, the latter typically of low order.

Furthermore, our approach allowed for a rational separation of the available material flow rates and/or heat duties into two distinct sets of manipulated inputs, that act and can be used to address control objectives in the two time scales. Specifically, the large flowrates and heat duties only act upon the fast dynamics, while the small ones act in the slow time scale.

Finally, the application of the proposed analysis and model reduction procedure was illustrated through numerical simulations in a reactor-external heat exchanger network example.

5. REFERENCES

Alhumaizi, K.L. (2000). Stability analysis of the ethylene dimerization reactor for the selective production of butene-1. Trans IChemE **78**, 492–498.

Baldea, M., P. Daoutidis and A. Kumar (2004). Dynamics of process networks with recycle and purge: time scale separation and model decomposition. In: Preprints of ADCHEM'03. Hong Kong. pp. 645–650.

Bildea, C.S. and A.C. Dimian (1998). Stability and multiplicity approach to the design of heat integrated PFR. AIChE J. **44**(12), 2703–2712.

Chen, Y.H. and C.C. Yu (2003). Design and control of heat integrated reactors. Ind. Eng. Chem. Res. **42**, 2791–2808.

Dadebo, S.A., M.L. Bell, P.J. McLellan and K.B. McAuley (1997). Temperature control of industrial gas phase polyethylene reactors. J. Proc. Contr. **7**(2), 83–95.

Daoutidis, P. and C. Kravaris (1994). Dynamic output feedback control of minimum-phase multivariable nonlinear processes. Chem. Eng. Sci. **49**, 433–447.

Farschman, C. A., K. P. Viswanath and B. E. Ydstie (1998). Process systems and inventory control. AIChE J. **44**(8), 1841–1857.

Georgakis, C (1986). On the use of extensive variables in process dynamics can control. Chem. Eng. Sci. **41**, 1471–1484.

Hangos, K. M., A. A. Alonso, J. D. Perkins and B. E. Ydstie (1999). Thermodynamic approach to the structural stability of process plants. AIChE J. **45**(4), 802–816.

Henderson, L.S. and R.A. Cornejo (1989). Temperature control of continuous, bulk styrene polymerization reactors and the influence of viscosity: An analytical study. Ind. Eng. Chem. Res. **28**, 1644–1653.

Jacobsen, E. and M. Berezowski (1998). Chaotic dynamics in homogeneous tubular reactors with recycle. Chem. Eng. Sci. **23**, 4023–4029.

Kothare, M. V., R. Shinnar, I. Rinard and M. Morari (2000). On defining the partial control problem: Concepts and examples. AIChE J. **46**, 2456–2474.

Kulhavy, R., J. Lu and T. Samad (2000). Emerging technologies for enterprise optimization in the process industries. In: Preprints of Chemical Process Control - 6. Tucson, Arizona. pp. 411–422.

Kumar, A. and P. Daoutidis (1999). Control of Nonlinear Differential Equation Systems. Vol. 397 of Research Notes in Mathematics Series. Chapman & Hall/CRC.

Kumar, A. and P. Daoutidis (2002). Dynamics and control of process networks with recycle. J. Proc. Contr. **12**, 475–484.

Kumar, A. and P. Daoutidis (2003). Nonlinear model reduction and control for high-purity distillation columns. IECres **42**, 4495–4505.

Luyben, M. L., B. D. Tyreus and W. L. Luyben (1997). Plantwide control design procedure. AIChE J. **43**(12), 3161–3174.

Luyben, W. L. (1993). Dynamics and control of recycle systems. parts 1-4.. Ind. Eng. Chem. Res. **32**, 466–486,1142–1162.

Luyben, W.L. (1998). External versus internal open-loop unstable processes. Ind. Eng. Chem. Res. **37**, 2713–2720.

Marquardt, W. (2000). Fundamental modeling and model reduction for optimization based control of transient processes. In: Preprints of Chemical Process Control - 6. Tucson, Arizona. pp. 30–60.

Mizsey, P and I. Kalmar (1996). Effects of recycle on control of chemical processes. Comput. chem. Engng. **20**, S883–S888.

Morud, J. and S. Skogestad (1994). Effects of recycle on dynamics and control of chemical processing plants. Comput. chem. Engng. **18**, S529–S534.

Ng, C. and G. Stephanopoulos (1998). Plant-wide control structures and strategies. In: Proc. of DYCOPS-5. Corfu, Greece. pp. 1–16.

Price, R. M. and C. Georgakis (1993). Plantwide regulatory control design procedure using a tiered framework. Ind. Eng. Chem. Res. **32**, 2693.

Reyes, F. and W.L. Luyben (2000). Steady-state and dynamic effects of design alternatives in heat-exchanger/ furnace/ reactor processes. Ind. Eng. Chem. Res. **39**, 3335–3346.

Seider, W.D., J.D. Seader and D.R. Lewin (1999). Process Design Principles. John Wiley and Sons, Inc.. New York.

Tyreus, B.D. (1999). Dominant variables for partial control. 1. A thermodynamic method for their identification. Ind. Eng. Chem. Res. **38**, 1432—1443.

Vora, N.P. and P. Daoutidis (2001). Nonlinear model reduction of chemical reaction systems. AIChE J. **47**, 2320–2332.

Yi, C. K. and W. L. Luyben (1997). Design and control of coupled reactor/column systems - parts 1-3.. Comput. chem. Engng. **21**(1), 25–68.

ELSEVIER
IFAC
PUBLICATIONS
www.elsevier.com/locate/ifac

CLOSED-LOOP TIME DELAY ESTIMATION OF SISO PROCESSES FOR CONTROL PERFORMANCE MONITORING

Christopher A. Harrison and S. Joe Qin [1]

Department of Chemical Engineering
The University of Texas at Austin
Austin, TX 78712, USA

Abstract: A novel time delay estimation method for SISO control loop monitoring has been introduced. The proposed method requires a process regulated by a SISO controller during routine operation to be temporarily switched to relay control. The input and output data gathered from the process under relay control can be used to produce a time delay estimate. *Copyright © 2004 IFAC*

Keywords: performance monitoring, time delay estimation, closed-loop, SISO process

1. INTRODUCTION

Many important contributions have been made in the area of control performance monitoring in recent years. These include a method for the computation of a minimum output variance benchmark from closed-loop data of a SISO process (Harris, 1989). From this benchmark, engineers can evaluate the performance of feedback controllers of the SISO process. This algorithm assumes an *a priori* knowledge of the process time delay. Because the process time delay is a limiting factor for feedback controller performance, the computed minimum variance benchmark is sensitive to the estimated time delay. Reliable and unobtrusive techniques for estimating the process time delay in the closed-loop are necessary to utilize the existing performance assessment techniques.

To address this issue, Lynch and Dumont (1996) make use of a closed-loop time delay estimation technique presented by Elnaggar (1990) known as fixed model variable regressor estimation (FMVRE). The FMVRE method determines the delay by minimizing the expectation of the model prediction error in a least squares square sense. Candidate models for FMVRE have a known ARMA structure, but an unknown delay. The time delay and model parameters are decoupled, and any parameter estimation technique can be used to estimate these quantities. Although this method is straightforward to implement and converges quickly, it has some limitations. One limitation of this method is the ARMA model structure must be known before the time delay and model parameters can be estimated. Elnaggar (1990) demonstrates FMVRE for a first-order plus dead time model. A flexible implementation which does not require process knowledge would make time delay estimation more widely applicable. Another limitation in employing FMVRE for time delay estimation is the process input must be sufficiently rich at high frequencies. This requirement means another design factor for engineers implementing time delay estimation; it is desirable to have a method which requires minimal user input.

[1] Author to whom all correspondence should be addressed. Email: qin@che.utexas.edu. Fax: (512)471-7060.

Owing to the fact that relay-based auto-tuning is very popular in practice, we propose a new delay estimation method based on a relay (Yu, 1999). The new estimation method addresses the limitations of the previous methods. It requires the process, which is regulated by a feedback controller during routine operation, to be temporarily regulated by a relay controller. The input and output data collected from the process under relay control can be used to obtain a process time delay estimate. As a byproduct of this approach, an estimate and confidence limits for the minimum variance benchmark can be established.

2. CLOSED-LOOP TIME DELAY ESTIMATION

Since the closed-loop impulse response coefficients before the time delay are feedback-invariant, they should remain unchanged even if the controller is changed dramatically. Therefore, we can calculate the impulse responses from two different controllers, one from the current controller and another from the relay controller. The delay can be determined by examining the impulse response coefficients from both controllers. The proposed method for time delay estimation consists of the following steps:

(1) Collect output history from the SISO process under routine feedback control.
(2) Form several windows of output history, fit an ARMA model from the output history in each window.
(3) Calculate an impulse response from each of the ARMA models identified in Step 2.
(4) Calculate the mean and confidence interval for each coefficient of the impulse responses calculated in Step 3.
(5) Switch control of the process to a relay controller. Collect input and output history of the process under relay control.
(6) Form several windows of input and output history, fit an ARMAX model to obtain the process transfer function from each window.
(7) Extract the average time-delay estimate from the models in Step 6.
(8) Determine the number of the feedback-invariant impulse response coefficients in the mean impulse response found in Step 4 by comparing coefficients with the mean impulse response found in Step 6. This calculation will provide another time delay estimate, which will confirm the value found in Step 7.

2.1 Computation of a Mean Impulse Response for a Closed-loop Process Using Routine Operation Data

The closed-loop SISO system is described by the following equations:

$$y(t) = H(q^{-1})u(t) + N(q^{-1})a(t) \quad \text{(1a)}$$

$$u(t) = -K(q^{-1})y(t) \quad \text{(1b)}$$

where K is the linear transfer function for a PID controller:

$$K(q^{-1}) = \frac{\kappa_0 + \kappa_1 q^{-1} + \kappa_2 q^{-2}}{1 - q^{-1}} \quad (2)$$

It is assumed H is stable, N can be either stable or with integrating elements, and *a(t)* is white noise. The closed-loop transfer function, which is determined by applying feedback to the linear process, is given by the following expression:

$$y(t) = \frac{N(q^{-1})}{1 + K(q^{-1})H^*(q^{-1})q^{-1-f}}a(t) \quad (3)$$

where $b=1+f$ is the time delay. As shown by Harris (1989), the structure and parameters of the closed-loop transfer function are found by fitting a time-series model, typically a model of an ARMA process, to the closed-loop output data. The impulse response coefficients of the closed-loop process can be determined by performing long division on the closed-loop transfer function. The result will have the following form:

$$y(t) = \left[\underbrace{(1 + \psi_1 q^{-1} + ... + \psi_f q^{-f})}_{\text{Feedback - Invariant}} + \underbrace{(\psi_{f+1} q^{-f-1} + \psi_{f+2} q^{-f-2}) + ...}_{\text{Feedback - Varying}}\right] a(t) \quad (4)$$

The first $f+1$ coefficients of the impulse response from the closed-loop transfer function are feedback-invariant. They are identical to the first $f+1$ impulse response coefficients that determined by performing long division on the disturbance transfer function $N(q^{-1})$ (Harris, 1989).

We desire to obtain an estimate of the impulse response of closed-loop process under PID control from the available output history. The following assumptions are made to make this task more tractable:

(1) The plant dynamics and time delay are approximately constant over the time period over which the closed-loop data is collected.

(2) The disturbance dynamics are approximately constant over the time period in which the closed-loop data is collected.

To obtain an estimate, we first use many sample sets of the output history to obtain estimates of the closed-loop transfer function of the process. A moving window approach is applied to obtain the sample sets of output history. From these closed-loop transfer functions, samples of the impulse response are collected, and a mean impulse response and its confidence limits are determined. As more samples of the impulse response are collected, the estimation error in the mean impulse response and its confidence limits will decrease. Knowledge of the confidence limits will allow for the uncertainty in the minimum variance benchmark to be determined once the time delay is known.

2.2 Estimation of Time Delay by Switching to a Relay Controller

After a mean impulse response and its confidence-limits have been established for the closed-loop process under PID control, it is necessary to switch the process controller to an on-off relay controller. Implementation of this controller requires that a relay of magnitude h be used as the input u in the feedback control system. Initially, the input is at 0. If the output is increased above its setpoint (y=0) by a disturbance, then the input u is switched to -h. As the output decreases below its setpoint, the input is switched to h. At some time after the time delay b has elapsed, the output will begin to increase. As the output crosses zero, the input is switched back to -h, and the cycle continues in this manner (Yu, 1999). Since the output lags behind to input by at least $-\pi$ radians, a limit cycle will be present in the output. In the limit cycle, the output period is the ultimate period. The time delay of the process along with the process time constant(s) will determine the length of the ultimate period according to the standard phase angle equation (Yu, 1999). If we neglect the contribution of the time constant(s) in the standard phase angle equation, half of the period will provide an upper bound on the time delay for the SISO process:

$$Delay \leq \frac{\pi}{\omega} = \frac{1}{2}Period \tag{5}$$

The on-off relay controller is a nonlinear controller. This characteristic of the relay controller will result in there being no linear correlation between the output and input used to provide feedback. For modeling purposes, we treat the closed-loop data as though it were collected from an open-loop process. Thus, an average time delay of the process can be extracted by fitting time series models to the closed-loop input and output history of the process under relay control.

Another means of determining the time delay involves the calculation the mean impulse response of the disturbance model of the transfer function of the closed-loop process identified under relay control. The impulse response coefficients from the disturbance model can be superimposed on the mean impulse response and its confidence limits identified from the process under PID control. Before the time delay has elapsed, the impulse response coefficients not affected by feedback should be the same for both impulse responses. An estimate of the time delay can be determined by observing where the impulse response identified under relay control has deviated two standard deviations from the mean impulse response identified under PID control. This analysis is useful for verifying the average time delay estimate extracted from fitting closed-loop data obtained relay control to time series models.

2.3 Calculation of an Estimate of the Minimum Variance Benchmark

At this point, a time delay estimate b has been obtained from the process in a non-invasive manner. According to Harris (1989), the performance of the minimum variance controller can be calculated from the first *f+1* coefficients:

$$\text{var}\{y(t)\} = \left(1 + \psi_1^2 + ... + \psi_f^2\right)\sigma_a^2 \tag{6}$$

In addition, the uncertainty in the minimum variance benchmark can be obtained by using the mean impulse response and confidence limits found under PID control.

3. SIMULATION RESULTS

3.1 Closed-loop Time Delay Estimation for a Stable Plant with an Added Stationary Disturbance

The proposed approach for closed-loop time delay estimation was applied to a process with a stable plant model and time delay and a stationary disturbance driven by normally distributed, white noise. Assume the process model with delay is given by

$$H(q^{-1}) = H^*\left(q^{-1}\right)q^{-1-f} = \left[\frac{0.1}{1-0.5q^{-1}}\right]q^{-6} \tag{7}$$

Further, assume the stationary disturbance is given by

$$N(q^{-1}) = \frac{1}{(1 - 0.5q^{-1})}, \; \sigma_a^2 = (0.01)^2 \quad (8)$$

In routine closed-loop operation, it is assumed that the process is regulated by a PID controller. The tuning constants for the PID controller are $(\kappa_0, \kappa_1, \kappa_2) = (2, -2, 0.2)$, and are chosen so as to give closed-loop poles that stabilize the system. Upon implementing the chosen PID controller, the process is an ARMA process. In order to estimate the impulse response and confidence limits of the closed-loop transfer function, we fit multiple ARMA models to a time series of output data collected in the closed-loop. To accomplish this task, a moving window of 1000 discrete time steps is used over 3000 time steps of the process to collect the process output history. After the first 1000 steps of output history have been collected, an ARMA model is fit to this data, and the corresponding estimate of the impulse response is calculated. The window of data is shifted forward 100 time steps each time 100 additional time steps have been collected, so that 21 time series are collected. In each window, an ARMA model is fit to the data and, and the corresponding estimate for the impulse response is calculated.

A common approach to selecting the time series model structure is to use Akaike's final prediction error (FPE) as criterion. This criterion, which consists of a loss function quantifying the candidate model fit to the training data and a penalty for a candidate model of higher complexity, is minimized to determine model structure (Ljung, 1999). A linear search over the range of possible orders of polynomials A, B, and C to find the minimum FPE for the ARMAX model (Ljung, 1995):

$$A(q^{-1})y(t) = B(q^{-1})u(t-k) + C(q^{-1})a(t) \quad (9)$$

For the particular case in which the process is known to be an ARMA process, the order of polynomial B is assigned to be 0, and the orders of A and C are varied in the minimization of the FPE. The output, white noise, and input sequences for the process under PID control are shown in Figure 1.

A moving window is applied to obtain sets of time series by which to determine the mean impulse response. To simplify the ARMAX model identification, the polynomial orders are assumed to be equal, so that it is only necessary to search for two parameters. In each window, the polynomial orders of A and C were varied from 1 to 6, and the time delay k was varied from 1 through 12. The mean impulse response and its confidence limits are displayed in Figure 2.

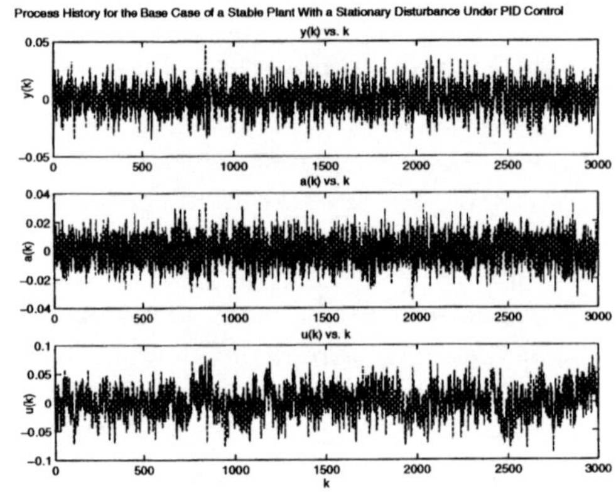

Fig. 1. Output, white noise, and input sequences for the stable plant with stationary disturbance under PID control

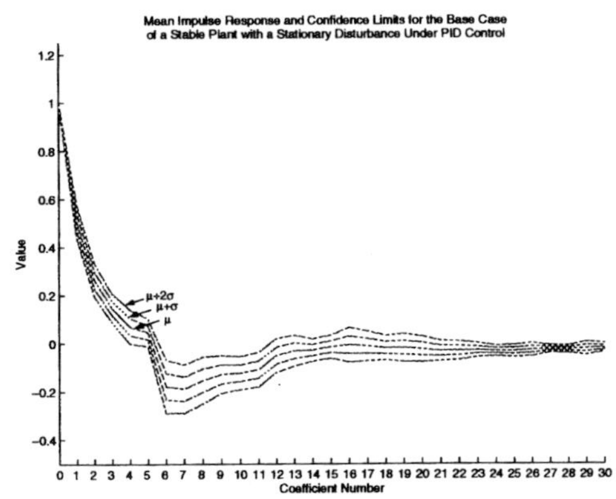

Fig. 2. Mean impulse response and its σ and 2σ confidence limits for the stable plant with stationary disturbance under PID control

In order to obtain the time delay estimate, the controller of the process is switched to an on-off relay controller. The control amplitude must chosen large enough so that the process output exhibits a limit cycle. In this simulation, the control amplitude is chosen to be 0.8 in the relay controller. The output and input history of the process under relay control, featured in Figure 3, is collected for 200 time steps.

A moving window of 100 time steps is used to obtain the mean impulse response of the disturbance model from this closed loop data. The moving window is slid every 10 samples after 100 samples have been collected, so that 11 estimates of the disturbance model are obtained. The process history within each window is fit to an ARMAX model, where the orders of polynomials A, B, C and the time delay are determined by minimizing the FPE. The polynomial orders of A, B, and C were varied from 1 through 6, and the time delay k was varied from 1 through 12. The mean impulse response of the estimated disturbance model is superimposed on the impulse responses obtained under PID control in Figure 5.

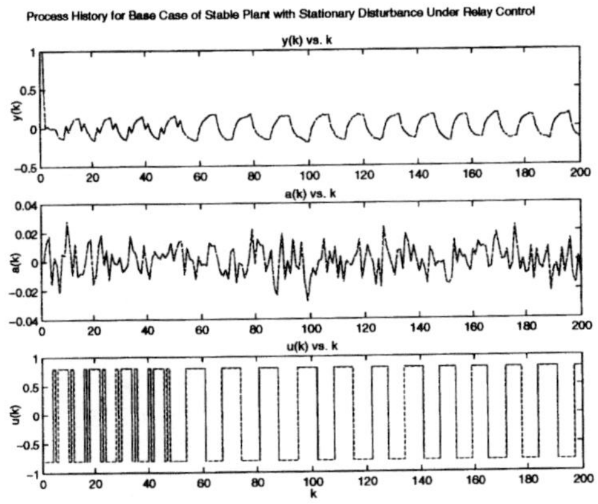

Fig. 3. Output and input history for base case of stable plant with stationary disturbance under relay control

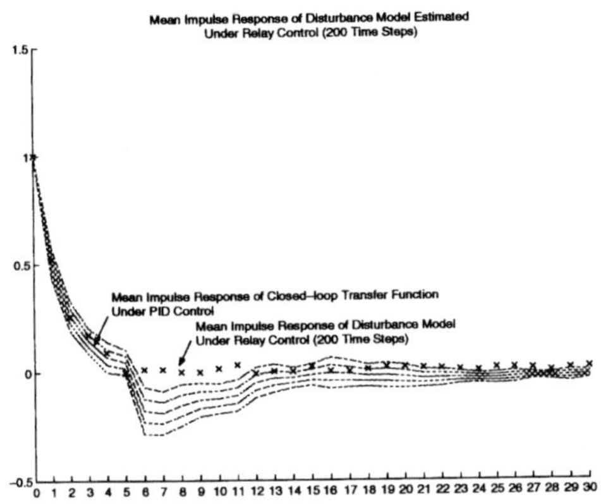

Fig. 4. Impulse response of the disturbance model estimated under relay control (200 time steps)

The 2σ confidence limit is used as the criterion for deciding which coefficients of the impulse response identified relay control are feedback-invariant. From Figure 5, it can be seen that this criterion indicates the first 5 to 6 coefficients are feedback-invariant. The average of the time delay given by the ARMAX models fit in each of the windows is found to be 5.7273.

To determine the effect of lowering the estimation error of the mean impulse response of the disturbance model on the time delay estimate, 2000 time steps of the base case for the process were simulated under relay control. A moving window of 1000 steps was implemented with the window being slid forward every 100 time steps collected after 1000. The mean impulse response of the estimated disturbance model is superimposed on the impulse responses obtained under PID control in Figure 6.

From Figure 6, it is observed that the more accurate impulse response estimate for the disturbance model yields a time delay estimate between 5 and 6. The average time delay estimated by the ARMAX models fit to process data is approximately

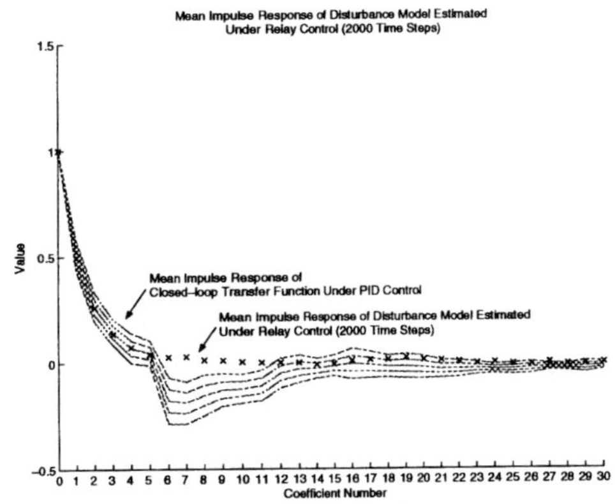

Fig. 5. Impulse response of the disturbance model estimated under relay control (2000 time steps)

6 if 200 time steps are used and 5 if 2000 time steps are used. The time delay estimates for this process are summarized in Table 1. The minimum variance benchmarks for time delays of 5 and 6 are shown in Table 2.

Table 1. Time Delay Estimate for Base Case

Amount of Data (Time Steps)	Feedback-Invariant Coeff. Approach	Average ARMAX Model Estimate
200	5-6	5.7273
2000	5-6	5.3636

Table 2. Output Variance Summary

	MV Benchmark (b=5)	MV Benchmark (b=6)
$\mu+2\sigma$	0.001368	0.001657
μ	0.001184	0.001406
$\mu-2\sigma$	0.001015	0.001179
	Output Variance of Current PID control:	0.002305

If a time delay estimate of 5 is considered, the current output variance is much greater than the upper 2σ confidence limit. This result indicates the current control strategy can likely be improved by retuning the PID control or by using a different controller. For the more conservative time delay estimate of 6, the current output variance is still much greater than the upper 2σ confidence limit, which provides further support to the claim that controller performance can be improved. The time delay estimates obtained from averaging the time delays extracted from the identified ARMAX models of the process underestimate the actual delay in this case. This result suggests the next highest integer delay is the better choice for the delay estimate.

3.2 Effect of Process Time Constant on Time Delay Estimate

We now consider a more general representation for the stable, minimum phase plants in the process:

$$H(q^{-1}) = \left[\frac{0.1}{1-\alpha q^{-1}}\right] q^{-6} \qquad (10)$$

If $\alpha = 0.01$ is used to simulate the process, the plant dynamics will be fast as compared to the disturbance dynamics. As a result, the disturbance dynamics will dominate the output response. If $\alpha = 0.99$ is used to simulate the process, the plant dynamics will be slow as compared to the disturbance dynamics. As a result, the plant dynamics will dominate the output response. The time delay estimates obtained by applying the proposed method to these two processes are featured in Tables 3 and 4.

Table 3. Time Delay Estimate for Case of $\alpha = 0.01$

Amount of Data (Time Steps)	Feedback-Invariant Coeff. Approach	Average ARMAX Model Estimate
200	5-6	5.7273
2000	5-6	5.2727

Table 4. Time Delay Estimate for Case of $\alpha = 0.99$

Amount of Data (Time Steps)	Feedback-Invariant Coeff. Approach	Average ARMAX Model Estimate
200	7-8	5.0000
2000	8-9	4.6364

The results of Table 3 indicate the time delay estimation method performs well when the plant dynamics are faster than the disturbance dynamics. In Table 4, we observe the accuracy of the time delay estimates are poorer when the plant dynamics are slower than the disturbance dynamics. In general, the average time delay estimate obtained from the ARMAX models is robust to processes with varying plant dynamics.

3.3 Upper Limit of Time Delay Estimate from Standard Phase Angle Equation

As mentioned previously, the standard phase angle equation gives an upper limit on the time delay as half of the period of the output response for the process under relay control. The upper limit on the time delay for the three simulated process is shown in Table 5. The period is an average over five oscillations of the output response.

Table 5. Upper Limit of Time Delay from Standard Phase Angle Equation

$\alpha = 0.01$	$\alpha = 0.5$	$\alpha = 0.99$
6.0	6.7	10.8

As observed in Table 5, the upper limit on the time delay approaches the time delay if the plant dynamics are much slower than the disturbance dynamics. In this situation, the upper limit provides a convenient technique for estimating the process time delay.

4. CONCLUSIONS

A novel time delay estimation method for SISO control loops has been introduced. The proposed method requires a process regulated by a PID controller during routine operation to be temporarily switched to relay control. The input and output data gathered from the process under relay control can be used to produce a time delay estimate.

The proposed time delay estimation method yields an average time delay estimate from the fitting of time series models to the input and output data. A time delay estimate can also be gleaned from application of the feedback-invariant property of the impulse response of the closed-loop transfer function. The latter time delay estimate is less reliable than the former estimate, but is useful for verifying the first estimate. An upper limit to time delay is obtained by calculated the period of the output response for the process under relay control.

Future studies should be focused on determining the robustness of the time delay estimates for processes with higher order plant models, higher order disturbance models, nonstationary disturbances, and higher white noise variances.

5. ACKNOWLEDGEMENTS

Financial support from the National Science Foundation Graduate Research Fellowship Program and the National Defense Science and Engineering Fellowship Program is gratefully acknowledged.

REFERENCES

Elnaggar, A. (1990). New method for delay estimation. In: *Proc. 29th Conf. on Decision and Control.* Honolulu, HI. pp. 1629–1630.

Harris, T.J. (1989). Assessment of control loop performance. *Can. J. Chem. Eng.* **67**(10), 856–861.

Ljung, L. (1995). *System Identification Toolbox User's Guide.* The MathWorks, Inc. Natick, MA.

Ljung, L. (1999). *System Identification: Theory for the User.* Prentice-Hall, Inc. Englewood Cliffs, NJ.

Lynch, C.B. and G.A. Dumont (1996). Control loop performance monitoring. *IEEE Trans. Cont. Sys. Tech.* **4**(2), 185–192.

Yu, C. (1999). *Autotuning of PID Controllers.* Springer-Verlager London Limited. London, UK.

ELSEVIER
IFAC
PUBLICATIONS
www.elsevier.com/locate/ifac

DIRECTIONAL LEAKAGE AND PARAMETER DRIFT

Morten Hovd *,[1] **Robert R. Bitmead** **

* *Department of Engineering Cybernetics, Norwegian University of Technology and Science, N-7491 Trondheim, Norway*
** *Department of Mechanical and Aerospace Engineering, University of California San Diego, 9500 Gilman Drive, La Jolla, CA 92093-0411, USA*

Abstract: A new method for eliminating parameter drift in parameter estimation problems is proposed. Existing methods for eliminating parameter drift work either on a limited time horizon, restrict the parameter estimates to a range that has to be determined *à priori*, or introduce bias in the parameter estimates which will degrade steady state performance. The idea of the new method is to apply leakage only in the directions in parameter space in which the exciting signal is not informative. This avoids the problem of parameter bias associated with conventional leakage.

Keywords: Parameter estimation, parameter drift, leakage

1. INTRODUCTION

Parameter drift is a well known problem in parameter estimation problems where the exciting signal is not sufficiently rich. The parameters estimated will typically converge to a manifold in parameter space on which the observed relationships between the input and output data are well explained. Noise in the observed data will then cause the parameter estimates to drift along this manifold in parameter space.

When the parameter estimates are used in an adaptive controller, the parameter drift will be unproblematic until parameter values are approached for which the model becomes uncontrollable. When this happens, bursting occurs. Bursting involves violent moves in the input as the controller attempts to control a plant which it erroneously thinks is close to uncontrollable, with corresponding large moves in the controlled output. The bursting typically supplies the parameter estimation algorithm with informative data, causing the parameter estimates to move to more reasonable values, from which the parameter drift starts anew.

There is a substantial literature on parameter drift and avoidance of bursting. Common modifications are the use of a deadzone in the parameter update, or the use of leakage. With a deadzone, the parameter updates are stopped as long as the model follows the observed system behaviour with reasonable accuracy. This will reduce the parameter drift, which can only occur when the parameter update is active, but will also reduce the accuracy with which the observed system behaviour can be approximated by the model.

With leakage, on the other hand, one adds an extra term in the parameter update law which drives the parameters towards a particular choice of reference values. Sufficiently strong leakage can eliminate bursting, but will bias the parameter values toward the chosen reference values. A good portion of luck is required for the reference values for the parameters to accurately describe the

[1] To whom correspondence should be addressed. e-mail: morten.hovd@itk.ntnu.no

observed input-output behaviour of the system, and hence leakage typically degrades steady state performance.

In this work, the properties of the exciting signal and the structure of the model is used such that the leakage is applied only in the directions in parameter space in which the input-output data is not informative. Hence, steady state performance is not degraded by the applied leakage.

1.1 *Preliminaries*

This section defines notation and states some established results on parameter estimation. Much of the material in this section is extracted from Ioannou and Sun (1996), where a more comprehensive treatment of parameter estimation and adaptive control can be found. We consider the estimation of model parameters for linear, time-invariant plants. The plant model is expressed as an nth-order difference equation given by

$$y_{k+n} + a_{n-1}y_{k+n-1} + \cdots + a_0 y_k = b_{n-1}u_{n-1} + b_{n-2}u_{n-2} + \cdots + b_0 u \quad (1)$$

Obviously, this corresponds to the discrete transfer function representation

$$y(z) = \frac{b_{n-1}z^{n-1} + b_{n-2}z^{n-2} + \cdots + b_0}{z^n + a_{n-1}z^{n-1} + z_{n-2}z^{n-2} + \cdots + a_0}u(z) = g(z)u(z) \quad (2)$$

The unknown parameters are lumped in a parameter vector

$$\theta = [b_0, b_1, \cdots, b_{n-1}, a_0, a_1, \cdots, a_{n-1}]^T \quad (3)$$

and all input-output signals are collected in the signal vector

$$\phi_{k-1} = [u_{k-n}, u_{k-n+1}, \cdots, u_{k-1}, -y_{k-n}, -y_{k-n+1}, \cdots, -y_{k-1}]^T \quad (4)$$

Thus, Eq. (1) can be compactly expressed as

$$y_k = \phi_{k-1}^T \theta \quad (5)$$

In a practical situation, measurements will be corrupted by noise. Therefore it is commonplace to filter both y_k and ϕ_{k-1} in the equation above with a common low pass filter. However, in order to keep the presentation simple, we will ignore such filtering in most of the analysis to follow.

Let $\hat{\theta}$ denote an estimate of the parameter vector θ, and $\tilde{\theta} = \theta - \hat{\theta}$ denote the corresponding parameter error vector. The observed model error is denoted e, and is given by

$$e_k = y_k - \phi_{k-1}^T \hat{\theta}_{k-1} = \tilde{\theta}_{k-1}^T \phi_{k-1} \quad (6)$$

1.2 *The gradient method*

We choose an instantaneous optimization criterion of the form

$$J(\hat{\theta}_k) = \frac{1}{2}(y_k - \phi_{k-1}^T \hat{\theta}_k)^2 + \frac{c}{2}(\hat{\theta}_k - \hat{\theta}_{k-1})^T(\hat{\theta}_k - \hat{\theta}_{k-1}) \quad (7)$$

Differentiating with respect to $\hat{\theta}_k$ we obtain

$$\left(\frac{dJ}{d\hat{\theta}_k}\right) = -(y_k - \phi_{k-1}^T \hat{\theta}_k)\phi_{k-1} + c(\hat{\theta}_k - \hat{\theta}_{k-1}) = -e_k\phi_{k-1} + \phi_{k-1}^T(\hat{\theta}_k - \hat{\theta}_{k-1})\phi_{k-1} + c(\hat{\theta}_k - \hat{\theta}_{k-1}) \quad (8)$$

Setting the gradient to zero, and noting that $(cI + \phi_{k-1}\phi_{k-1}^T)^{-1}\phi_{k-1} = \frac{1}{c+\phi_{k-1}^T\phi_{k-1}}\phi_{k-1}$, we obtain the standard equation error parameter estimation algorithm

$$\hat{\theta}_k = \hat{\theta}_{k-1} + \frac{1}{c + \phi_{k-1}^T \phi_{k-1}}\phi_{k-1}\left(y_k - \phi_{k-1}^T \hat{\theta}_{k-1}\right) \quad (9)$$

We have from Eq. (5) that Eq. (9) in the absence of measurement noise may be expressed as

$$\tilde{\theta}_k = \tilde{\theta}_{k-1} + \frac{1}{c + \phi_{k-1}^T \phi_{k-1}}\phi_{k-1}\phi_{k-1}^T \tilde{\theta}_{k-1} \quad (10)$$

The matrix $\phi\phi^T$ is clearly singular (of rank 1) at any one time instant. Equation (10), on the other hand, contains $2n$ integrators. It might therefore appear as if there is only one integrator (or linear combination of integrators) that is stabilized by feedback, and that the remaining integrators are left to integrate the noise in the measurement signal. However, if the systematic variation in ϕ is such that the sum of $\phi\phi^T$ over any time interval $[k, k+t]$ is positive definite, there is actually negative feedback around all the integrators in Eq. (10). This results in an asymptotically stable parameter estimation, with the estimated parameters in the absence of measurement noise converging to the true values at steady state.

The requirement that the sum of $\phi\phi^T$ should be positive definite is the familiar *persistent excitation* requirement. The problem of parameter drift and the bursting that results from it arise when the persistent excitation requirement is not met.

1.3 *Least squares parameter estimation algorithm*

The least squares parameter estimation algorithm with exponential data weighting results from recursively minimizing the optimization criterion

$$J_L(\hat{\theta}_k) = \frac{1}{2}\sum_{t=1}^{k}\beta^{k-t}(y_t - \phi_{t-1}^T\hat{\theta}_k)^2 + \frac{1}{2}\beta^k(\hat{\theta}_k - \hat{\theta}_0)\Pi_0^{-1}(\hat{\theta}_k - \hat{\theta}_0) \quad (11)$$

starting from a given initial parameter guess $\hat{\theta}_0$ and a positive definite Π_0. Here β $(0 < \beta \leq 1)$ is known as the exponential forgetting factor, and is normally chosen to be slightly less than 1. It is shown in, e.g., Goodwin and Sin (1984) that minimizing Eq. (11) results in the recursive scheme

$$\hat{\theta}_k = \hat{\theta}_{k-1} + \frac{\Pi_{k-2}\phi_{k-1}}{\beta + \phi_{k-1}^T\Pi_{k-2}\phi_{k-1}}\left[y_k - \phi_{k-1}^T\hat{\theta}_{k-1}\right] \quad (12)$$

$$\Pi_{k-1} = \frac{1}{\beta}\left[\Pi_{k-2} - \frac{\Pi_{k-2}\phi_{k-1}\phi_{k-1}^T\Pi_{k-2}}{\beta + \phi_{k-1}^T\Pi_{k-2}\phi_{k-1}}\right] \quad (13)$$

Choosing $\beta = 1$ results in no forgetting of old data. As a result, Π_k will approach zero, and the parameter estimation will essentially turn itself off. A consequence will be that the parameter estimation will be unable to track even slow parameter variations. To avoid this problem, one may choose β slightly smaller than 1. Whereas this makes the tracking of slow parameter variations possible, it also gives room for parameter drift in cases where the exciting signal is not sufficiently rich. It is well know that an excitation signal containing m sinusoids is *sufficiently rich of order* $2m$, i.e., will allow accurate estimation of $2m$ parameters (or $2m$ linear combinations of parameters).

2. DIRECTIONAL LEAKAGE

In this section we will propose a method for performing parameter updates in the subspace of parameter space for which the I/O data is informative, while preventing parameter drift by implementing leakage in the orthogonal subspace. We will first consider the gradient method for parameter estimation, and thereafter consider the least squares method. For both parameter estimation methods, stability is analyzed using averaging, in a manner similar to that of (Bitmead and Johnson, 1987).

2.1 *Directional leakage with the gradient method*

After selection of a basis for the informative subspace, and collecting the basis vectors in a matrix Υ, this is achieved simply by modifying the parameter update equation Eq. (9) as follows

$$\hat{\theta}_k = \hat{\theta}_{k-1} + P\frac{1}{c + \phi_{k-1}^T\phi_{k-1}}\phi_{k-1}\phi_{k-1}^T\left(\theta - \hat{\theta}_{k-1}\right) + rP^{\perp}\left(\theta_{ref} - \hat{\theta}_{k-1}\right) \quad (14)$$

where $0 < r < 1$, $P = \Upsilon(\Upsilon^T\Upsilon)^{-1}\Upsilon^T$ is a projection matrix and $P^{\perp} = I - P$ its orthogonal complement (and $P^{\perp}$ is itself a projection matrix).

We can decompose the dynamics of the parameter estimation described by Eq. (14) into directions driven by input-output data and directions driven by the leakage, by premultiplying both sides of Eq. (14) by P and $P^{\perp}$, respectively.

$$P\hat{\theta}_k = P\left(I - \frac{1}{c + \phi_{k-1}^T\phi_{k-1}}\phi_{k-1}\phi_{k-1}^T\right)\hat{\theta}_{k-1} + P\frac{1}{c + \phi_{k-1}^T\phi_{k-1}}\theta \quad (15)$$
$$= PH_{k-1}\hat{\theta}_{k-1} + PG_{k-1}\theta$$

$$P^{\perp}\hat{\theta}_k = P^{\perp}(1-r)\hat{\theta}_{k-1} + rP^{\perp}\theta_{ref} \quad (16)$$

The dynamics in the directions described by Eq. (16) are trivially asymptotically stable, and have a steady state described by $P^{\perp}(\theta_{ref} - \theta_{\infty}) = 0$. The dynamics described by Eq. (15) require more careful analysis.

Assuming that $\phi_{k-1}\phi_{k-1}^T$ is in the range of P, and that P is of rank $2m$ (where $m < n$), PH_{k-1} has $2(n-m)$ eigenvalues at the origin (the singular directions of P), one eigenvalue inside the unit disk, and the remaining eigenvalues at $+1$. One therefore has to study the evolution of the parameters over several timesteps to conclude about stability. Considering an entire oscillation period for the input signal, i.e. $s = \tau/T$ timesteps, one obtains

$$P\hat{\theta}_{k+s} = \prod_{i=k}^{k+s-1} PH_i\hat{\theta}_k + \sum_{i=k}^{k+s-1}\left(\prod_{l=i+1}^{k+s-1} PH_l\right)PG_i\theta \quad (17)$$

The system signals are repeated every s timesteps, and therefore the stability of the parameter estimation will depend on $\prod_{i=k}^{k+s-1} PH_i$. To simplify notation, denote $\frac{1}{c+\phi_{i-1}^T\phi_{i-1}}$ by ϵ_i and $\phi_{i-1}\phi_{i-1}^T$ by X_i. For sufficiently large c, we have $\epsilon_i \approx \epsilon << 1 \forall i$. Tedious, but straightforward calculation[2] then results in

$$\prod_{i=k}^{k+s-1} H_i = P - P(\epsilon_k X_k + \epsilon_{k+1}X_{k+1} + \cdots + \epsilon_{k+s-1}X_{k+s-1})P + O(\epsilon^2\frac{s(s-1)}{2}) \quad (18)$$

[2] We will require the technical assumption $PX_k = X_kP$, which will be fulfilled by our subsequent choice of P.

Provided $\epsilon^{\frac{s(s-1)}{2}} << 1$, the linear term in Eq. (18) will dominate. Thus, the parameter estimation will be asymptotically stable provided the rank of $P(X_k + X_{k+1} + \cdots + X_{k+s-1})P$ equals the rank of P. Thus, we have to determine which basis to use as the columns of Υ, to ensure that this rank condition holds. Classical results on sufficient richness of signals provide us with the dimension of the informative subspace, i.e., the number of columns in Υ. A straightforward approach is to sum $\phi^T\phi$ over some window of past data with window length of some multiple of s timesteps. The singular vectors corresponding to the $2m$ largest singular values of the resulting matrix can then be chosen as the range space of P. In the ideal case (in the absence of noise) this would lead to a time invariant projection matrix P once the effects of initial conditions have died out.

Example 1. The idea of using directional leakage is illustrated using a simple example with $g(z) = \frac{b}{z+a}$. The input is a unit step signal (corresponding to $\omega = 0 \Rightarrow z = 1$), and is thus sufficiently rich to estimate only one parameter. It is well known that in this situation only the steady state gain, $g(1) = \frac{b}{1+a}$ can be estimated. The true parameters in this example are given by $a = 0.7$, $b = 0.7$. The estimation is simulated in Simulink, and the measurement is corrupted with white noise with power 0.1. The estimation is simulated both with no leakage, and with directional leakage. Initial parameter estimates are $a = 2$, $b = 6$ in both simulations. In the simulations, both input and output are low pass filtered through identical low pass filters $f(z) = \frac{1}{z+0.8}$.

The results without leakage are shown in Fig. 1. The estimation quickly converges to parameter values giving the correct steady state gain (the straight line in the figure), and then starts to drift while staying close to the correct steady state gain.

The results obtained when using directional leakage are shown in Fig. 2. The reference values θ_{ref} are chosen equal to the initial parameters. The parameter estimates converge to a point on the line indicating the correct steady state gain, and no significant parameter drift can be found.

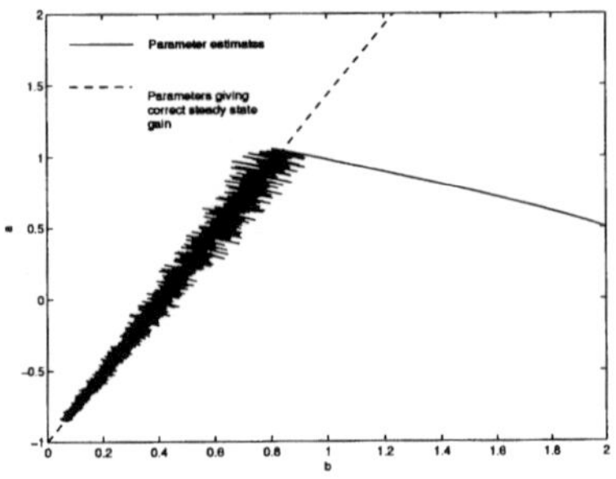

Fig. 1. Results for example 1 when using no leakage.

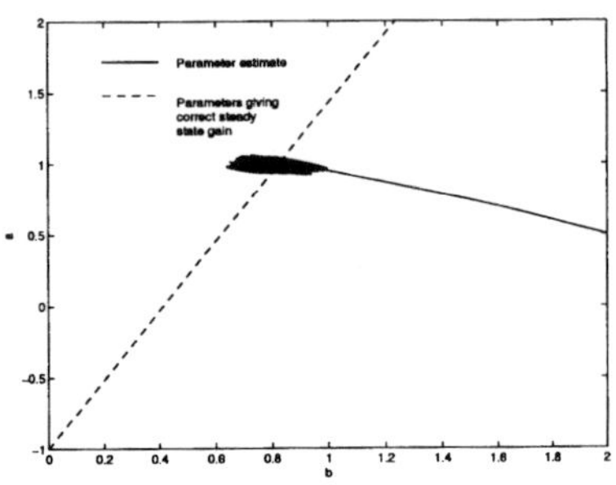

Fig. 2. Results for example 1 when using directional leakage.

2.2 *Directional leakage with the least squares parameter estimation method*

It was noted above that without exponential forgetting the least squares parameter estimation method will eventually turn itself off. Thus, parameter drift and leakage (whether directional or not) only become relevant when $\beta < 1$.

Introducing directional leakage in Eq. (12) we obtain

$$\hat{\theta}_k = \hat{\theta}_{k-1} + \left\{ P \frac{\Pi_{k-2}\phi_{k-1}}{\beta + \phi_{k-1}^T \Pi_{k-2}\phi_{k-1}} \times \left[y_k - \phi_{k-1}^T \hat{\theta}_{k-1} \right] \right\} + P^{\perp}(\theta_{ref} - \theta_{k-1}) \quad (19)$$

whereas Eq. (13) in principle remains unchanged. However, it is clear from Eq. (19) that the matrix Π_{k-2} affects the gain of the parameter estimation. With exponential forgetting, Π_{k-2} may become very large in directions that are poorly excited - and even grow without bounds in directions that are not excited at all. We will therefore impose an upper limit on each of the singular values of Π. Other (and less computationally demanding) methods for keeping Π bounded are discussed in (Goodwin and Sin, 1984).

Proceeding with the analysis for the exponentially weighted least squares, we arrive at an equation mirroring Eq. (17) above. However, we here get that $H_i = I - \frac{1}{\beta + \phi_{k-1}^T \Pi_{k-2}\phi_{k-1}} \Pi_{k-2}\phi_{k-1}\phi_{k-1}^T$. In evaluating $\prod_{i=k}^{k+s-1} PH_i$, we then face the problem that in general Π_{k+m} and $\phi_{k+n}\phi_{k+n}^T$ do not commute. We are therefore unable to arrive at an equally simple expression as in Eq. (18). Instead, we obtain

$$\prod_{i=k}^{k+s-1} H_i = P - P(\epsilon_k \Pi_k X_k + \epsilon_{k+1}\Pi_{k+1}X_{k+1} + \cdots + \epsilon_{k+s-1}\Pi_{k+s-1}X_{k+s-1}) + O((\bar{\epsilon}\bar{\Pi})^2 \frac{s(s-1)}{2}) \quad (20)$$

where here $\epsilon_i = \frac{1}{\beta+\phi_{i-1}^T \Pi_{i-2}\phi_{i-1}}$ whereas $X_i = \phi_{i-1}\phi_{i-1}^T$ as before. The symbols $\bar{\epsilon}$ and $\bar{\Pi}$ represent average values of ϵ_i and Π_i, respectively. We find that for sufficiently small $\bar{\Pi}$, the linear term in Eq. (20) will dominate, here $\bar{\epsilon}$ may well be close to 1. A sufficiently small $\bar{\Pi}$ is obtained by choosing a forgetting factor β close to 1, *and* effectively constraining growth of Π in directions that are insufficiently excited. It should be noted that in the directions in the input-output data that are poorly but persistently excited, Π may grow large unless β is very close to 1 or the singular values of Π are properly constrained. Still, we can conclude from Eq. (20) that given a sufficiently small $\bar{\Pi}$, the parameter estimation will remain asymptotically stable provided $rank(P) = rank(P(\epsilon_k \Pi_k X_k + \epsilon_{k+1}\Pi_{k+1}X_{k+1} + \cdots + \epsilon_{k+s-1}\Pi_{k+s-1}X_{k+s-1})$ Since Π_i is positive definite for all i, this rank condition is essentially equivalent to the rank condition found for the gradient method.

Example 2. The use of directional leakage will next be illustrated used on a problem in adaptive control. The plant is given by

$$g(z) = \frac{2z - 1.8}{z^3 - 2.78z^2 + 2.5715z - 0.79135} \quad (21)$$

The control objective is to track the reference signal

$$r(t) = \sin(\frac{2\pi t}{360}) + \sin(\frac{2\pi t}{45}) \quad (22)$$

Thus, we have five unknown parameters, whereas the reference signal would only be sufficiently rich to identify four parameters. There is also an additive measurement noise, modelled as a normally distributed, zero mean random variable with standard deviation 0.04. The control design is accomplished using adaptive pole placement, using a discrete version of the method described in (Ioannou and Sun, 1996). A sampling interval of 1 is used, and all poles of the closed loop characteristic polynomial are placed at $z = 0.7$. For the parameter estimation, state variable filters for both inputs and outputs equal to $f(z) = \frac{1}{(z-0.8)^3}$ are used. The estimation is done with the least squares method with a forgetting factor $\beta = 0.998$. The singular values of Π are constrained to be no larger than $1e-4$. The initial parameter estimates correspond to the transfer function.

$$\hat{g}(z, \hat{\theta}) = \frac{2(z - 0.91)}{(z - 0.99)(z - 0.93)(z - 0.86)} \quad (23)$$

Although these initial parameter values would appear to be quite good, initial control performance is appalling. The performance quickly improves, and good performance is obtained after a short initial transient. Luckily, this paper addresses long term rather than initial behavior, and parameter estimation rather than control. Similar control performance is obtained for the simulated time period both without leakage and with directional leakage. Figure 3 may represent the typical control performance for both cases. The results (Figs. 4 - 8) show that the parameter estimates drift when no leakage is applied. When directional leakage is applied, the noise does affect the parameter estimates, but no parameter drift can be seen. The effect of the noise that is seen in the parameter estimates appear have no effect on the control performance. However, the effect of noise may be reduced by reducing the maximum singular value bound for the covariance matrix, and/or using a larger forgetting factor. The forgetting factor of $\beta = 0.998$ corresponds to a time constant for forgetting of 500 samples. This may appear to be somewhat too fast when the signal to be tracked has a sinusoid of period 360.

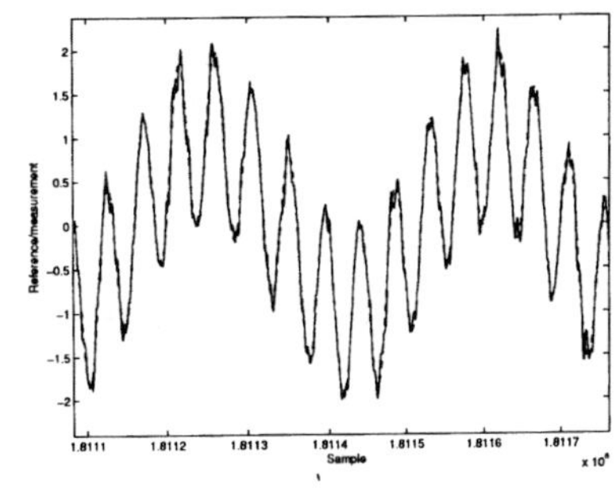

Fig. 3. Typical control performance for Example 2. Solid: noise-corrupted measurement, dashed: reference signal.

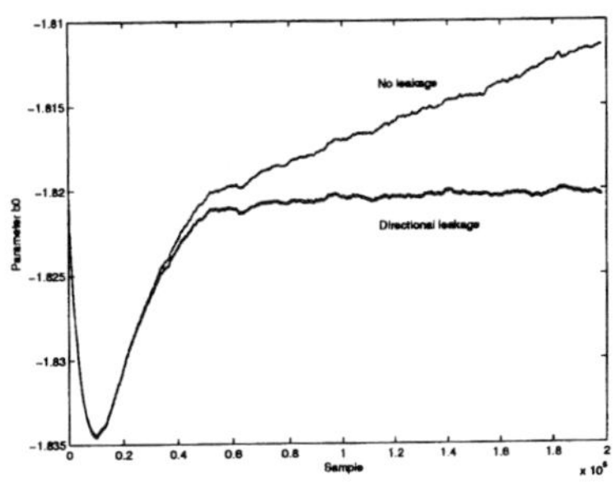

Fig. 4. Estimates of parameter b_0 in Example 2.

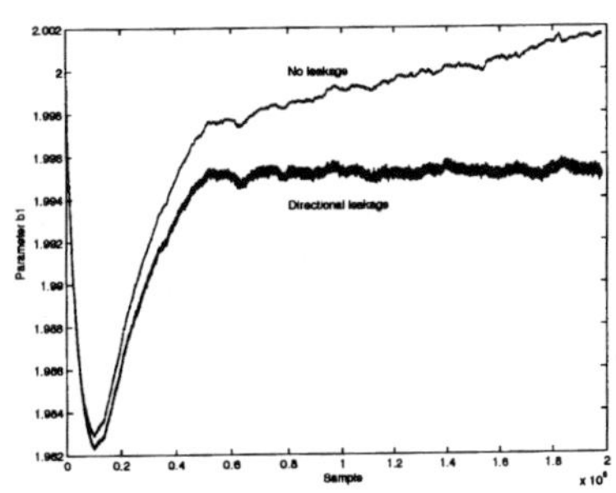

Fig. 5. Estimates of parameter b_1 in Example 2.

3. DISCUSSION AND CONCLUSIONS

This paper has focused on the gradient and least squares methods for parameter estimation. Note

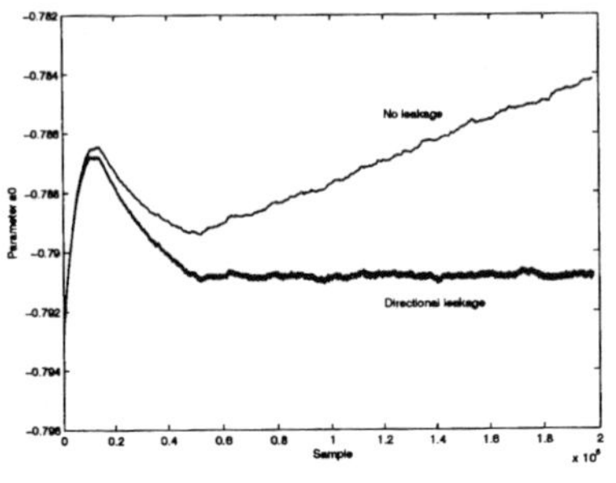

Fig. 6. Estimates of parameter a_0 in Example 2.

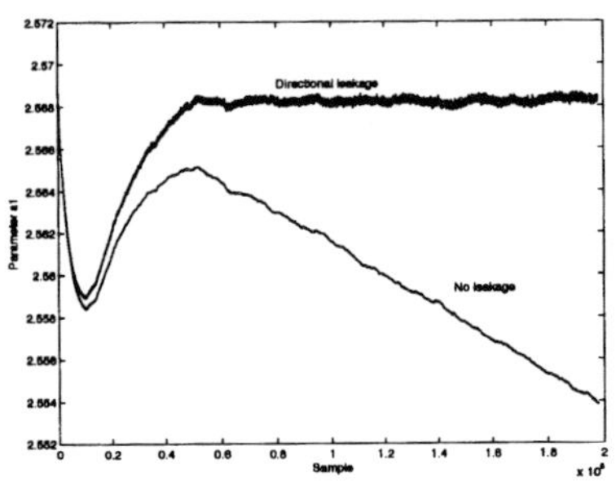

Fig. 7. Estimates of parameter a_1 in Example 2.

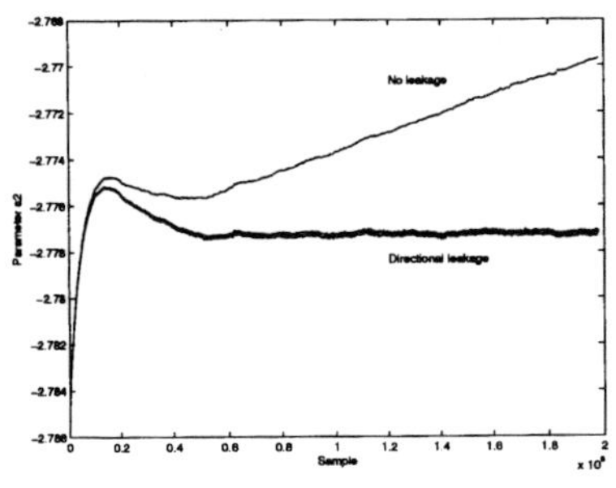

Fig. 8. Estimates of parameter a_2 in Example 2.

that other parameter estimation algorithms, including parameter estimation using the Extended Kalman Filter, suffer from the same parameter drift problems whenever the system is not sufficiently excited and tracking of time-varying parameter values is desired. 'Conventional' leakage is a simple method for eliminating drift, but unfortunately conventional leakage results in bias in the parameter estimates. This paper shows how leakage can be applied only in the directions in parameter space where the plant is not properly excited. This stabilizes the parameter estimation (stops parameter drift) without degrading parameter estimation accuracy in the directions where the input-output data is informative.

It was noted previously that bursting will occur if the parameter estimates approach a hypersurface in parameter space where the model is uncontrollable. Although directional leakage, when properly tuned, makes the parameter estimation asymptotically stable, that does not rule out the possibility that during the initial transient the parameters may encounter such an uncontrollable hypersurface. For problems where this is a serious concern, directional leakage may easily be combined with so-called *parameter projection* (Ioannou and Sun, 1996), which constrains the parameter estimates to stay within an *à priori* defined region of the parameter space where the true model is assumed to lie, and which is also assumed to be free of uncontrollable hypersurfaces.

Thus, despite the advantages of directional leakage noted above, it does not totally eliminate the problem of bursting. Note, however, that such bursting during initial transients may also occur when the exciting signals are persistently exciting, if the initial parameter estimates are ill-chosen. Therefore, any failure to prevent bursting during transient conditions should not be considered a shortcoming of (directional) leakage - it is simply not the problem it is intended to prevent.

The use of a dead-zone (see, e.g., (Egardt, 1979)) in the parameter estimation is a common way of reducing parameter drift. The dead-zone reduces parameter drift by stopping the parameter update when model predictions are close to the physical measurement. Directional leakage, on the other hand, does not stop parameter estimation when the model output is close to the measurement. Since leakage is applied only in the directions in parameter space where the signals are not informative, the asymptotic model accuracy is not affected. Directional leakage should therefore be an attractive alternative to the use of dead-zones.

Acknowledgements The first author got the idea for this paper while attending Miroslav Krstic's Adaptive Control class during a sabbatical at UCSD, and wishes to thank prof. Krstic for his inspiring teaching.

4. REFERENCES

Bitmead, R. R. and C. R. Johnson (1987). *Discrete Averaging Principles and Robust Adaptive Identification.* pp. 237–271. Vol. 25 of *Advances in Control and Dynamic Systems.* Academic Press. New York.

Egardt, B. (1979). *Stability of Adaptive Controllers.* Vol. 20 of *Lecture Notes in Control and Information Sciences.* Springer-Verlag. Berlin, Germany.

Goodwin, C. G. and K. S. Sin (1984). *Adaptive Filtering Prediction and Control.* Prentice-Hall. Englewood Cliffs, New Jersey, USA.

Ioannou, P. A. and J. Sun (1996). *Robust Adaptive Control.* Prentice-Hall. Upper Saddle River, New Jersey, USA.

ELSEVIER
IFAC
PUBLICATIONS
www.elsevier.com/locate/ifac

ROBUST AND EFFICIENT JOINT DATA RECONCILIATION – PARAMETER ESTIMATION USING A GENERALIZED OBJECTIVE FUNCTION

Yen Yen Joe[i), iii)] [1], David Wang[ii)], Jose Romagnoli[ii)] and Arthur Tay[i)]

[i)] *Dept. of Electrical & Computer Engineering, The National University of Singapore, 10 Kent Ridge Cres., Singapore 119260*

[ii)] *Dept. of Chemical Engineering, The University of Sydney, NSW 2006, Australia*

[iii)] *Institute of Chemical and Engineering Sciences, Ayer Rajah Cres., Block 28, Unit #02-08, Singapore 139959*

Abstract: This paper proposes the use of a generalized distribution, namely the Generalized T (GT) distribution in the joint estimation of process states and model parameters. The desirable properties of the GT-based estimator are its robustness, simplicity, flexibility and efficiency for the wide range of commonly encountered distributions (including Box-Tiao and t-distributions) which belong to the GT distribution family. To achieve the efficiency, the parameters of the GT distribution are adapted from the data through preliminary estimation. The strategy is applied to the virtual version of a practical chemical engineering plant. *Copyright © 2004 IFAC*

Keywords: Parameter estimation, data processing, robust estimation, robustness, estimators, error correction

1. INTRODUCTION

Data reconciliation (DR) and parameter estimation (PE) are important components in process modelling, control, optimization and other applications that require reliable process data and accurate process model. Due to measurement errors, measurements of variables are usually equilibriated according to some conservation equations prior to their use in parameter estimation. In obtaining the model parameter estimates, however, the measurement data are further adjusted such that the model parameters and the adjusted data satisfy the process model equations. The inefficiency of the two-step DR and PE is hence noted. A remedy will be to couple the two procedures, thereby subjecting the parameter estimates and the measurement data to both the conservation equations and the process model, yielding estimates that are consistent to both sets of equations. Such estimates are expected to be more accurate. This is confirmed by the findings of MacDonald and Howat (1988), who examined both the sequential and coupled DR and PE and concluded that more reliable estimates are obtained using the coupled method.

The coupled DR and PE (DRPE) can also be viewed as the error-in-all-variables-measured (EVM) strategy, although DRPE is a more general formulation due to the inclusion of DR constraints that do not include unknown parameters. Mathematically speaking, the two strategies are similar in terms of issues that might arise in obtaining reliable estimates. Several aspects that have been discussed in the literature are: the general algorithm for the solution strategy (Valko and Vadja, 1987), the optimization strategy (Kim et al, 1990; Tjoa and Biegler, 1991) and the robustness of the estimation (Albuquerque and Biegler, 1996; Arora

[1] Author to whom correspondence should be addressed: g0202653@nus.edu.sg

and Biegler, 2001). Our proposed strategy will mainly deal with the robustness of the estimation.

The most prevalent form of estimator is the weighted least squares (WLS) formulation, where estimates minimizing the squared error in the measurements, normalized by the measurement covariance, are sought. Then again, it is also well known that the optimality of the weighted least squares is conditional upon the normality of the error distribution. Departure from normality, which could stem from the presence of gross errors and the empirical character of the error, often renders the weighted least square estimates unreliable. Attempts to treat these deviations include the a priori assignment of distributions with thicker tails, such as the contaminated Gaussian distribution (Tjoa and Biegler, 1991), or other particular forms of estimators to the same effect, i.e. ones that down-weight measurements with large residuals (e.g. the Fair Function by Albuquerque and Biegler, 1996; the redescending estimator in Arora and Biegler, 2001). An argument against pre-determined distributions is that the efficiency of the estimates may not be optimal, as the actual error distribution may not be close to the one assigned. If the parameters of these pre-determined forms of estimator are estimated posteriori, partially adaptive estimator results, and efficiency may be improved, although it will still be limited by the forms spanned by the possible values of the tuned parameters. A more flexible alternative is the non-parametric approach, where the estimator does not assume any fixed form and can fully adapt to the error distribution. The drawback of this approach is the relatively large number of unknown tuning parameters, and the possibility that it might not perform well for sample sizes frequently encountered in practice (Butler et al, 1990).

Our proposed approach is the robust DRPE strategy using a generalized distribution function, i.e. the Generalized T (GT) distribution. We will estimate the distribution parameters, and as the GT covers a wide range of commonly encountered distributions, it is expected to be more efficient than other partially adaptive estimators mentioned above. It is also simpler than the fully adaptive approach, as there are only a few tuning parameters. The GT-based estimator has been applied to parameter estimation (Butler et al, 1990) and data reconciliation (Wang and Romagnoli, 2003). In this paper, we extend the application to the more general and efficient DRPE strategy. The organization of the paper is as follows. The next section discusses about the robustness of M-estimators. Our proposed GT-based estimator is introduced here and its robustness properties discussed, along with that of two other estimators that we use for comparison in the case study, namely the weighted least square and the contaminated Normal estimator. Next, Section 3 discusses the partial adaptiveness and the efficiency of the proposed strategy. In Section 4, the strategy is applied to an application case study and the results presented and discussed. Finally, Section 5 concludes the paper.

2. THE M-ESTIMATOR AND ITS ROBUSTNESS

Our proposed GT-based estimator falls under the category of M-estimator, which is the generalization of the maximum likelihood estimation and can be expressed as:

$$z = \arg\max f(u) = \arg\min -\log(f(u)) = \arg\min \rho(u) \quad \text{s.t. constraints} \qquad (1)$$

where u = y-x, y=measurements, x=variables estimate, z=[x, p], p=model parameters to be estimated, f(u) is the probability density function (PDF) of u and $\rho(u) = -\log(f(u))$. It is of paramount importance, then, that the measurement error distribution indeed follows the PDF f(u), otherwise it will be forced to yield estimates that maximize f(u) when actually it does not resemble f(u), leading to inaccurate estimates.

The weighted least squares (WLS) is the most widely used estimator. Within the context of M-estimation, it corresponds to f(u) being the normal distribution, i.e. the error is assumed to be normally distributed.

To overcome the shortcomings of the WLS estimator in the event of gross errors, various robust estimators have been developed. Huber (1981) and Hampel et al (1986) provide a unified theoretical framework for statistical robustness, by means of which the various robust estimators can be analysed and compared. The analysis based on Hampel et al's Influence Function (IF) will be adopted in our analysis. To simplify presentation, the derivation of formulae will be omitted here; the interested reader could refer to Hampel et al (1986) for details. The IF aims to describe the behaviour of an estimator in the neighbourhood of the parametric distribution assumed. If the residual u is drawn from a PDF f(u) and if T[f(u)] is the unbiased estimate corresponding to u, then the IF of a residual u_0 is given by:

$$IF = \psi(u_0) = \lim_{t\to 0} \frac{T[(1-t)f + t\delta(u-u_0)] - T[f]}{t} \qquad (2)$$

where $\delta(u-u_0)$ is the delta function centred about u_0. Thus, the IF quantifies the amount of influence that a residual u_0 has on the estimation. For M-estimators, the influence function IF is proportional to the derivative of $\rho(u)$, i.e.:

$$\mathrm{IF} = \psi(u) \propto \partial\rho/\partial u = \partial \ln(f(u))/\partial u\,. \qquad (3)$$

In order for the estimator to be robust, the IF should be bounded such that a single large residual cannot dominate and distort the estimation. Additional desirable properties for the IF would be for it to be descending to very small values as the residual gets large, and to be continuous such that the estimator is well-behaved.

We will now use IF to analyse the robustness of a few estimators, starting from the weighted least squares (WLS). For WLS, $\rho(u) = u^{\mathrm{T}}\Phi^{-1}u$, where

Φ is the covariance matrix of the measurements and represents the weight. The IF of the WLS is then

$$\psi(u) = \ln \rho(u) = \ln(u^T \Phi^{-1} u) \propto u \tag{4}$$

which is a straight line and clearly unbounded for large values of residuals u (Figure 1). The amount of influence is proportional to the residual magnitude, allowing large residues to distort the estimation.

A robust alternative proposed by Tjoa and Biegler (1991) combines the normal PDF with another normal PDF with much larger variance to account for outliers. This results in the contaminated Normal distribution with PDF:

$$f(u)=(1-p)\frac{1}{\sqrt{2\pi}\alpha}\exp\left(-\frac{u^2}{2\alpha^2}\right)+p\frac{1}{\sqrt{2\pi}\alpha b}\exp\left(-\frac{u^2}{2\alpha^2 b^2}\right) \tag{5}$$

The PDF has two parameters: p is the probability of the gross error occurrence, and b is the ratio of the standard deviation of the larger Normal PDF to that of the smaller one. The IF of this estimator is shown in Figure 1. The improvement in robustness can be observed: the IF behaves like that of WLS for small residuals, but starts to descend after a certain point where the residuals are considered too large.

Our proposed GT distribution has the following mathematical form:

$$f_{GT}(u;\sigma,p,q)=\frac{p}{2\sigma q^{1/p}B(1/p,q)\left(1+\frac{|u|^p}{q\sigma^p}\right)^{q+1/p}};\ -\infty<u<\infty \tag{6}$$

It is symmetric about zero and unimodal. It is characterised by the distribution parameters $\{p,q,\sigma\}$: p and q determine the shape of the distribution, while σ determine the scale of the distribution. The GT defines a very general family of density function and combines two general forms that include as special cases most of the stochastic specifications encountered in practice. Some of the more commonly known special cases, along with the particular values of $\{p,q,\sigma\}$ for each case, are depicted in Figure 2.

The IF of the GT-based estimator can be obtained as:

$$\psi_{GT}(u;\sigma,p,q)=\frac{(pq+1)sign(u)|u|^{p-1}}{q\sigma^{1/p}+|u|^p} \tag{7}$$

Figure 3 shows the IF with several different sets of values of $\{p,q,\sigma\}$. It shows that generally the IF is bounded and actually descending when the residuals get large. However, we also observe that as p increases, the IF for large residuals increases and as q increases, the IF becomes less bounded. In fact, we notice that one special case where $q \to \infty$, with $p=2, \sigma=\alpha\sqrt{2}$ (α is the standard deviation) is none other than the Normal distribution. To ensure that the GT-based estimator is insensitive to large residuals, therefore, bounds must be imposed on the values that p and q can take. In our work, $1 \le p \le 5, 0.5 \le q \le 50$.

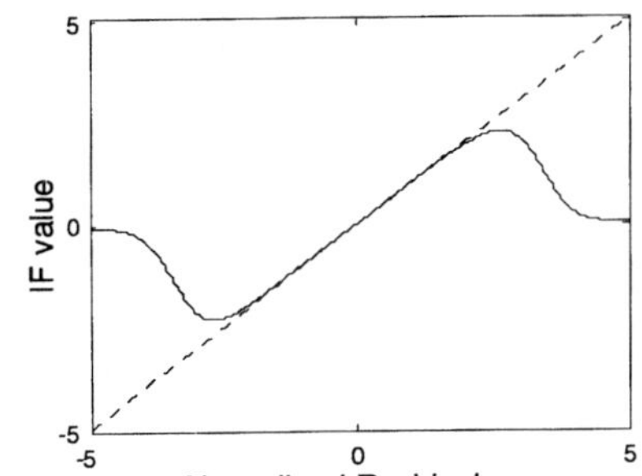

Figure 1. Plots of Influence Function for Weighted Least Square Estimator (dashed line) and the Robust Estimator based on Contaminated Normal Distribution (solid line)

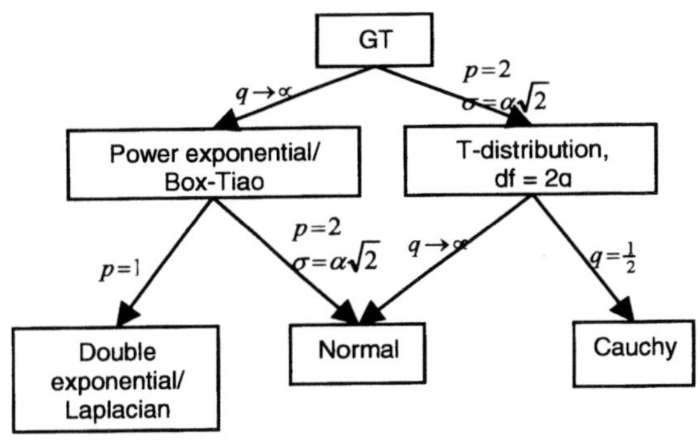

Figure 2. GT Distribution Family Tree, Depicting the Relationships among Some Special Cases of the GT Distribution

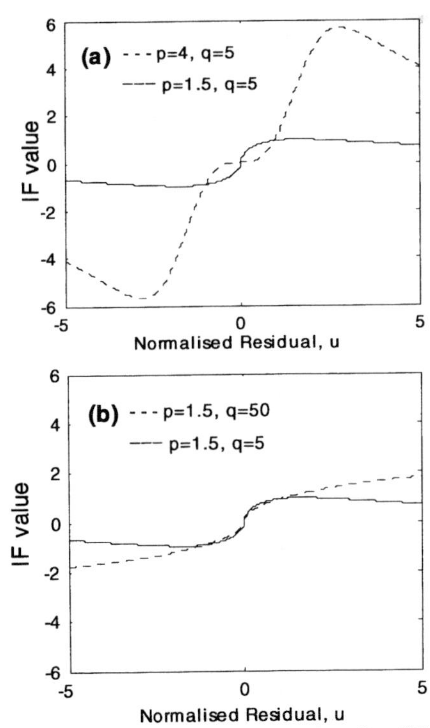

Figure 3. Plots of Influence Function for GT-based Estimator with different parameter settings

We will estimate the distributional parameters $\{p,q,\sigma\}$ using residuals obtained by performing a preliminary estimation on historical data. These residuals should have the same characteristics as the actual error, provided that the preliminary estimation is robust enough. As such, using the estimated $\{p,q,\sigma\}$ will ensure that the estimator characterises the error better, and in turn, yields more efficient estimates. The efficiency of this partially adaptive estimator will be further discussed in the next section.

3. PARTIALLY ADAPTIVE ESTIMATION AND EFFICIENCY

The motivation for partially adaptive estimation is to include the information about the error characteristics into the estimation. This inclusion of prior information corresponds to the formulation of posterior density which is the Bayes estimation. For our GT-based estimator, the distribution parameters $\{p,q,\sigma\}$ are estimated from the residuals of preliminary estimates. The residuals should be descriptive of the underlying error distribution, hence ensuring that the estimates of $\{p,q,\sigma\}$ result in the GT shape that best characterizes the error. It is therefore important for the preliminary estimator to be sufficiently robust so that the preliminary estimates are unbiased. If the preliminary estimator is not robust enough, iterating the estimation of the distribution parameters can result in better distribution parameter estimates. The amount of historical data must also be sufficient to represent most of the data that are likely to be obtained. However, since there are only three distributional parameters to be tuned, the tuning is comparably simpler than that of non-parametric estimators.

For the GT-based approach, the distributional parameters $\{p,q,\sigma\}$ can be estimated using the maximum likelihood estimator (McDonald and Newey, 1988):

$$\{p,q,\sigma\}=\arg\max\sum\log f_{GT}(u_0;\{p,q,\sigma\}) \tag{8}$$

where u_0 is the residual from the preliminary estimates. The values of $\{p,q,\sigma\}$ are then obtained as the parameters of a GT member from which the data are most likely sampled. The GT estimator using such estimated values of $\{p,q,\sigma\}$ has been shown to be asymptotically efficient among all estimators, when the error distribution is within the GT family. Rigorous mathematical proof can be found in McDonald and Newey (1988). Nothing can be said about the efficiency in the case of non-GT distributed errors, and some loss of efficiency is possible. However, since the GT family includes a wide range of commonly encountered distributions, the fact that it is asymptotically efficient for these distributions make its application very appealing.

In our approach, we use the maximum likelihood estimator in (8) to estimate $\{p,q,\sigma\}$. To obtain the residuals u_0 in (8), we take the median of the data set as the estimated values. Taking the median as the estimate corresponds to the use of robust L-estimator (Albuquerque and Biegler, 1996; Hampel et al., 1986). It is feasible in our case as we assume steady state measurements, i.e. the actual values are assumed to be constant over the time horizon considered. This method of estimating the distribution parameters is simple, robust and computationally more convenient, as compared to performing a full preliminary DRPE. Since the asymptotic distribution of the estimates depend only on the limit of the distribution parameters, and not on the particular way by which they are estimated (McDonald and Newey, 1988), our approach is justifiable. In cases where the measured variables are not constant over the time horizon considered, a full preliminary DRPE must be performed to obtain the preliminary residuals.

4. APPLICATION CASE STUDY

The proposed GT-based DRPE strategy is applied to a case study of the pilot-scale setting of a general purpose plant containing two CSTRs, a mixer and a number of heat exchangers (Figure 4). Material feed from the feed tank is heated before being fed to the first reactor and the mixer. The effluent from first reactor is then mixed with the material feed in the mixer, and then fed to the second reactor. The effluent from the second reactor is, in turn, fed back to the feed tank and the cycle continues.

Steady-state analysis of the system structure results in three redundant equations involving seven redundant variables. The model parameters estimated are the product of the heat transfer coefficient with the effective heat transfer area of the cooling coil of the first reactor. For parameter estimation, two more model equations with four redundant variables and one non-redundant variable are included.

Associated with the pilot-scale plant, a virtual environment that mimics the actual plant behaviour has been developed within the Matlab/Simulink framework and will be used in this paper while the plant is being commissioned. Simulation data are generated with several different distributions: Normal, Laplacian and t distribution. A data set having Normal distribution and with large random shifts as gross error is also generated.

Five DRPE methods are compared, i.e. **(i)** weighted least squares (WLS); **(ii)** contaminated normal with distribution parameters fixed at p = 0.05, b = 10, and σ = standard deviation of the normal noise; this configuration is suggested in Tjoa and Biegler (1991) if the number of outliers are not known apriori, which is the condition we assume here; **(iii)** GT method with distribution parameters fixed at p = 1.5, q = 5, and σ = standard deviation of the normal noise multiplied by square root of two; this configuration is assigned based on our analysis on the influence function with different sets of parameter values, where the analysis shows that the current configuration will be sufficiently robust; **(iv)** partially adaptive contaminated normal with distribution parameters p, b and σ estimated from preliminary residuals; we obtain the preliminary residuals and estimate the distribution parameters in a similar way as we do for the partially adaptive GT method (as described in Section 3); and **(v)** partially adaptive GT method, with distribution parameters p, q and σ estimated from preliminary residuals.

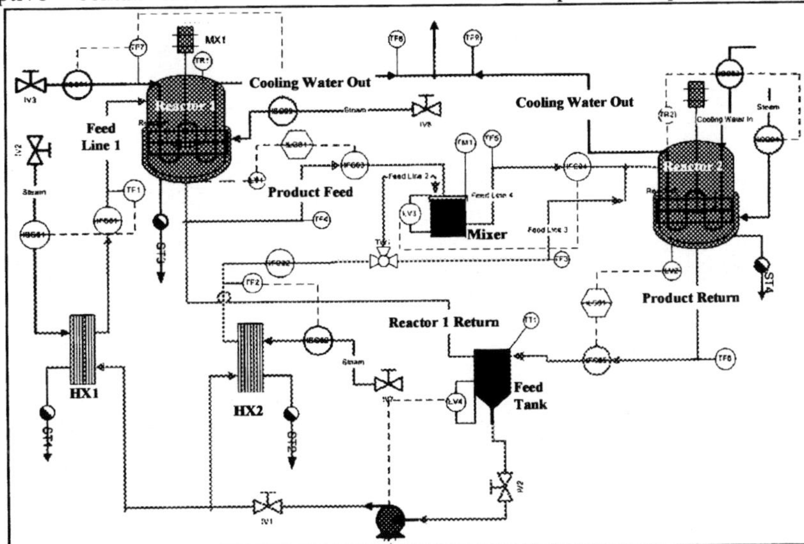

Figure 4. Flow Diagram of General Purpose Plant for Application Case Study

Two measures are used to quantify the efficiency of the DRPE methods, i.e. the mean square error (MSE) and the percentage of model parameter estimation accuracy (%-discrepancy). The MSE is calculated as:

$$\text{MSE} = \frac{1}{\text{mK}} \sum_{j=1}^{\text{K}} \sum_{i=1}^{m} \frac{(\hat{x}_{i,j} - x_{i,j})^2}{\sigma_i^{\,2}} \tag{9}$$

where m is the number of measured variables and K is the number of data sets used for the DRPE (m=12, K=24 in this study). $\hat{x}_{i,j}$ and $x_{i,j}$ are the estimates of the reconciled data and the actual value of the variable, respectively, while σ_i is the standard deviation of the Gaussian noise on sensor *i*. The %-accuracy is calculated as:

$$\%\,\text{discrepancy} = 100\% \times \frac{|\,\text{parameter estimate - actual value}|}{\text{actual value}} \tag{10}$$

Figure 5 and 6 chart the MSE and %-discrepancy, respectively. The partially adaptive GT method performs most efficiently for all types of outliers considered, i.e. it has the smallest overall estimation errors for both parameter and states. The advantage of partially adaptive scheme is apparent by the lower estimation errors achieved by GT and contaminated Normal methods with distribution parameters estimated from the data, as compared to their counterparts with distribution parameters assigned and fixed at certain values. However, although the partially adaptive contaminated Normal method performs very well, it is seen that the partially adaptive GT method is still more efficient for Laplacian and t noise, which are within the GT family of distributions.

Comparing the non-partially adaptive GT and contaminated Normal methods, it is seen that in this particular case study, GT generally performs better. However, the choice of distribution parameters greatly influences the estimation results, and since we have only one instance of all possible settings of distribution parameters, the results here cannot be generalized to the conclusion that all non-partially adaptive GT estimators perform better than all non-partially adaptive contaminated methods.

The results of the preliminary estimation for the GT distribution parameters for some variables when the error follows Cauchy and t distributions are listed in Table 2. Referring back to the distribution tree in Figure 2, we can see that the values of the estimated p and q are close to the ideal p and q for the respective distributions. For example, for Laplacian noise, where the ideal p=1 and $q \to \infty$, the estimated p are close to one, while q are large or close to the upper bound, i.e. q=50.

To visualise how well the estimators characterize the data, the probability density functions of the distributions assumed by each estimator are plotted in Figure 7 (for the variable feed flow to reactor 1 in Table 1, in the case of t noise), along with the relative frequency histogram of the noise. The histogram represents the actual empirical error distribution. The PDFs are plotted based on the distribution parameter values assumed by the estimators, i.e. normal $N(0,\sigma^2)$, where σ =standard deviation of the measurement (in this case = 0.5) for WLS; partially adaptive contaminated Normal with parameters p=0.0546, b=5.0000, and σ =1.3240 for the contaminated Normal estimator; and GT with $\{p,q,\sigma\}$= {1.5343, 2.6729, 1.4498} for the GT-based estimator. It is seen that the both GT and contaminated Normal estimator characterizes the data very well. However, the GT shape resembles the data more, as it can assume the different shapes of distribution within the GT family, while the contaminated Normal shape is limited to the shape of two normal PDFs combined.

5. CONCLUSION

The DRPE strategy based on GT distribution strikes a balance between the simplicity of strict parametric approaches where a certain distribution is assumed and the flexibility of non-parametric approaches which involves many unknown parameter approaches and may not perform well for common sample sizes in practice. The GT-based estimator is robust for a range of its distribution parameter values, by adapting these distribution parameters to the data, it is efficient particularly when the underlying error distribution is within the GT family. Within the range of its robust distribution parameter values, the GT family encompasses a wide variety of important statistical distributions, and thus it is a worthy candidate for DRPE as it can handle a wide range of error distributions with simple adaptations of its distribution parameters.

6. REFERENCES

Albuquerque, J. S., Biegler, L.T. (1996). Data Reconciliation and Gross-Error Detection for Dynamic Systems. *AIChE J.*, **Vol. 42**, No. 10, pp. 2841-2856.

Arora, N., Biegler, L.T. (2001). Redescending Estimators for Data Reconciliation and Parameter Estimation. Comp. Chem. Eng., **Vol. 25**, pp. 1585-1599.

Butler, R.J., McDonald, J.B., Nelson, R.D., White, S.B. (1990). Robust and Partially Adaptive Estimation of Regression Models. *Rev. Econ. Stat.*, **Vol 72**, No. 2, pp.321-327.

Hampel, F.R., Ronchetti, E.M., Rousseeuw, P.J., Stahel, W.A. (1986). *Robust Statistics: The Approach Based on Influence Function.* Wiley, New York.

Huber, P.J. (1981). *Robust Statistics.* Wiley, New York.

Kim, I.W., Liebman, M.J., and Edgar, T.F. (1990). Robust Error-in-Variable Estimation Using Non-linear Programming techniques. *AIChE J*, **Vol. 36**, pp. 985-993.

MacDonald, R.J., Howat, C.S. (1988). *AIChe J*, **Vol. 34**, 1.

McDonald, J.B., Newey, W.K. (1988). Partially Adaptive Estimation of Regression Models via the Generalized T Distribution. *Econometric Theory*, **Vol. 4**, pp.428-457.

Tjoa, I.B., Biegler, L.T. (1991). Simultaneous Strategies for Data Reconciliation and Gross Error Detection of Nonlinear Systems. *Comp. Chem. Eng.*, **Vol. 15**, No. 10, pp. 679-690.

Valko, P., Vadja, S. (1987), An extended Marquardt-type Procedure for Fitting Error-in-Variable Models. *Comp. Chem. Eng.*, **Vol. 11**, pp. 37-43.

Wang, D., Romagnoli, J.A. (2003). A Framework for Robust Data Reconciliation Based on a Generalized Objective Function. *Ind.Eng.Chem.Res.*, **Vol.42**, No.13, pp. 3075-3084.

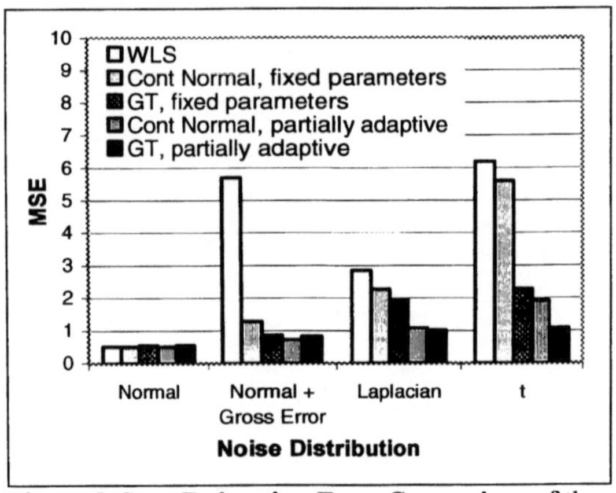

Figure 5. State Estimation Error Comparison of the WLS, Contaminated Normal and GT-based Estimators for Different Noise Profiles

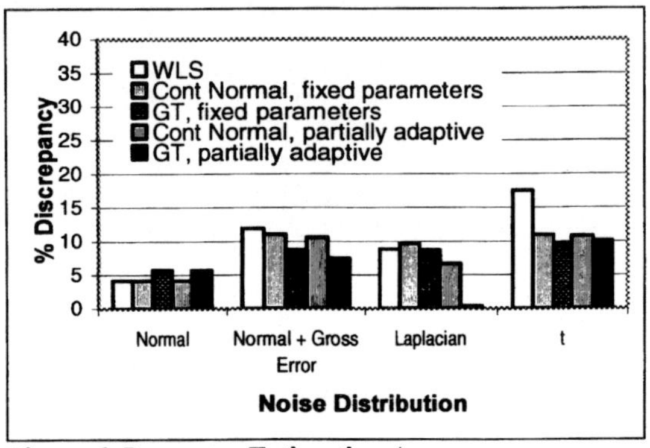

Figure 6. Parameter Estimation Accuracy Comparison of the WLS, Contaminated Normal and GT-based Estimators for Different Noise Profiles

Table 1. GT Distribution Parameter Estimates

Variable	Laplacian (p=1,q→inf)		t with 2 dof (p=2,q=1)	
	p	q	p	q
Feed Flow to Reactor 1	1.32	50.00	1.53	2.67
Feed Flow to Reactor 2	1.01	46.75	3.52	0.50
Cooling Water Flow of Reactor 1	1.15	47.71	3.13	0.62

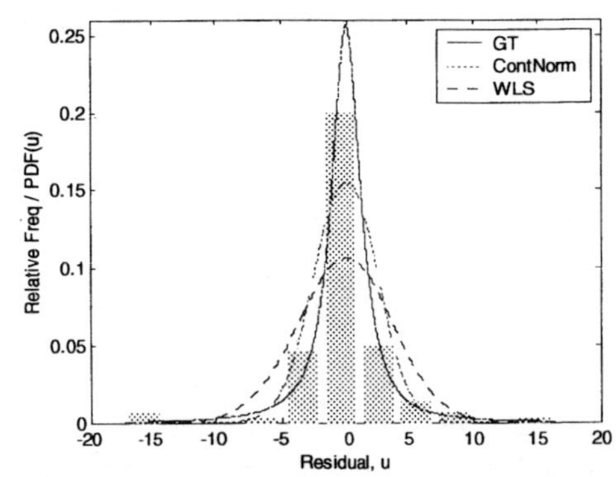

Figure 7. Plots of Relative Frequency Histogram and the Various Error Distributions Assumed by the Different Estimators

ELSEVIER
IFAC
PUBLICATIONS
www.elsevier.com/locate/ifac

MISSING DATA TREATMENT USING ITERATIVE PCA AND DATA RECONCILIATION

S.A. Imtiaz * S. L. Shah *,[1] S. Narasimhan **

* *Department of Chemical and Materials Engineering University of Alberta, Edmonton AB, Canada, T6G 2G6*
** *Department of Chemical Engineering IIT, Madras*

Abstract: Two methods, one based on Iterative Principal Components Analysis(IPCA) and the other based on Data Reconciliation have been developed for estimating a model from a data matrix containing missing data. These algorithms are iterative in nature and analogous to the method based on PCA for treating missing data. The methods incorporate information about the measurement errors to develop the models and are optimal in a maximum likelihood sense. The close connection of the methods with the Expectation Maximization (EM) algorithm is also established. Simulated data from a Flow Network system with a variety of error structures and missing data is used to evaluate the performance of the proposed methods. In all cases, models estimated by the proposed methods were superior to those obtained by the classical PCA-based missing data treatment algorithms for nonuniform error.

Keywords: Missing Data, PCA, IPCA, Data Reconciliation, Expectation Maximization (EM)

1. INTRODUCTION

Process industries are now using multivariate statistical methods frequently along side with univariate methods. As a result treatment of missing data has become more important from statistical process control (SPC) perspective. Unless we have an appropriate method for dealing with missing measurements, one may end deleting an entire row of data even if a single measurement in that row is missing. When a process is operating at quasi-steady state or in dynamic mode the samples have time stamps and deletion of rows due to missing or bad data leads to irregular sampling intervals. Most of the popular model building techniques such as Principal Components Analysis (PCA), Partial Least Squares (PLS) were originally developed for dealing with uniformly sampled data without any missing value. Modifications are necessary to accommodate missing values in the data matrix. In this paper our focus is on Principal Components Analysis only. PCA is a widely used method to build models from a data matrix of large number of highly correlated variables. Models developed using PCA are subsequently used to monitor process operating performance and detect impending faults. In dealing with data matrices containing missing values, PCA faces difficulties at two stages. Firstly, in building model from data sets with missing values and, secondly in monitoring phase when model is available from previous analysis but some of the monitored variables have missing values. Both problems have received considerable attention from researchers. Historically, NIPALS algorithm was used originally for build-

[1] author to whom all correspondence should be addressed. E-mail: sirish.shah@ualberta.ca

ing principal component models. Christofferson (1970) extended NIPALS algorithm for finding first and second principal component in presence of missing values in the data matrix (Grung and Manne, 1998). Now it has become more customary to use Singular Value Decomposition (SVD) algorithms for finding the PCs. In the presence of missing values in the data matrix, an iterative imputation approach is used to fill the missing values. Initially, the missing values are filled with the unconditional mean of the respective variable and SVD is done on the complete data set. PCA solution is iteratively improved by updating the missing values with the estimated values of the current model until the imputed values stabilize (Grung and Manne, 1998; Troyanskaya *et al.*, 2001; Walczak and Massart, 2001). Assuming that the model is available from a previously obtained data without any missing value Nelson et al.(1996) and Arteaga et al.(2001) developed and analyzed several methods for calculating scores from a data set with missing values. These methods such as, Trimmed Score (TRI), Single Component Projection (SCP), Conditional Mean Replacement (CMR), Projection to Model Plane (PMR) are useful from a monitoring point of view. However all these methods use PCA in its natural form (Nelson and MacGregor, 1996; Artega and Ferrer, 2002). Though PCA is a robust and widely used tool for model building but it has some limitations too. As such, algorithms based on PCA for building models from data matrix with missing values also suffer from these limitations. In this paper we will point out two such restrictions and subsequently propose two algorithms which outperform currently used PCA-based missing data treatment techniques in these situations.

Firstly, PCA is often not the optimal method for model estimation in Maximum Likelihood sense. For example, ordinary least squares is a maximum likelihood method when measurement errors in the regressor variables are negligible and error in the dependent variable are independent and identically distributed. Likewise, PCA can be considered maximum likelihood if all the measurement errors are uncorrelated and of equal variance (i.e. independent and identically distributed, iid). This is a very restrictive assumption as process measurements come from a wide variety of sensors (i.e. thermocouple, gas chromatograph etc.) and these different types of sensors have different measurement accuracy. Sometimes instead of using these raw variables, derived variables are used in the data matrix, for example enthalpy which is a derived variable obtained by multiplying flowrate with temperature and heat capacity. Thus, there is a possibility that errors may be cross correlated as well. Wentzell et al.(1997) developed Maximum Likelihood Principal Component Analysis (MLPCA) which relaxed these assumptions and gives ML estimate of the model using alternating regression method. MLPCA requires a priori knowledge of the true values of the error covariance matrix (Wentzell *et al.*, 1997). In chemometrics applications often replicates of the measurements are taken and an estimate of the error covariance is available. Whereas, in process industries replicates of measurements are rarely taken. Recently, Narasimhan and Shah (2004) developed an iterative technique, Iterative Principal Components Analysis (IPCA) which estimates both the model and the error covariance matrix simultaneously (Narasimhan and Shah, 2004). In this paper we will show the use of IPCA in missing data context.

The other difficulty associated with PCA is to decide on the number of principal components to be included in the model. PCA relies on ad hoc. methods such as, scree test, cross validation etc. These methods are not backed by theory and they often lead to inaccurate model order selection. The problem becomes even more serious in the presence of missing values in the data matrix. IPCA provides an accurate way of determining the number of independent variables by looking at the singular values.

Two algorithms, one based on IPCA and the other based on IPCA and Data Reconciliation have been proposed in this paper for treating missing data. Both methods are iterative imputation type and henceforth labelled as IPCA Imputation Algorithm (IPCAIA) and IPCA Data Reconciliation (IPCADR). The algorithms are optimal in the Maximum Likelihood Sense for steady state and quasi-steady state processes. Several assumptions are inherent in the development of the methods. First, a true underlying model of lower dimension than the total number of variables does exist, Second, deviations of the measurements are only due to random measurement errors, Third, measurement errors are normally distributed and Finally, the methods are only valid for treating missing data which are Missing Completely At Random (MCAR) or Missing At Random (MAR). Treatment of missing data with Non Ignorable (NI) mechanism is beyond the scope of the present paper (Little and Rubin, 2002).

The paper is organized as follows, in section 2 IPCA algorithm and theory behind the algorithm have been described. Missing data treatment algorithm using IPCA is introduced in section 3. Section 4 gives a brief overview of data reconciliation and shows its application in combination with IPCA algorithm for handling missing data. In section 5 improved performance of these newly developed algorithms has been demonstrated by a flow network simulation example.

2. ITERATIVE PRINCIPAL COMPONENT ANALYSIS (IPCA)

The details of IPCA can be found in Narasimhan and Shah(2004). Here only the main steps of IPCA have been discussed for completeness. IPCA has two main steps, first, Optimally Scaled Principal Components Analysis (OSPCA) and second, Estimation of the error covariance matrix. IPCA iteratively alternates between these two steps until convergence.

An optimal scaling strategy has been used in IPCA. It was shown that PCA is scaling invariant with this scaling scheme. Let us consider a case when all the measurements contain random errors and are described by

$$Y = X + \varepsilon$$

where $\varepsilon \sim N(0, \Sigma_\varepsilon)$ and the underlying model is such that,

$$AX = 0$$

Scaling factor L is defined by,

$$LL^T = \Sigma_\varepsilon$$

After scaling the measurements with the scaling matrix the transformed measurements are given by,

$$Y_s = L^{-1}X + L^{-1}\varepsilon$$

Covariance of Y_s is given by,

$$\Sigma_{y_s} = S_{x_s} + I$$

This is an important result as it provides a convenient way to select the order of the model. From the *eigenvalue* shift theorem, eigenvectors of noise corrupted covariance matrix, Σ_{y_s} is equal to the eigenvectors of the noise free covariance matrix, S_{x_s}. So there is no distortion in the eigenvectors due to the presence of noise in the signals. On the other hand, eigenvalues of the noise corrupted covariance matrix are shifted by unity from the noise free eigenvalues. For example, if the rank of the data matrix X_s is m, then the last n-m eigenvalues of S_{x_s} will be exactly zero as the data matrix is noise free and the last n-$m$eigenvalues of Σ_{y_s} will be unity. The eigenvectors corresponding to these unity eigenvalues define the basis vectors of the residual space, which is the constraint model in scaled domain, A_s. Therefore, the model order selection is not arbitrary. Rather it provides a definitive way of selecting the model order. The constraint model in the original domain is simply given by,

$$A = A_s L$$

The only unknown parameter is Σ_ε. To estimate Σ_ε an iterative approach is taken. An initial estimate of the constraint model, A^0 is obtained by ordinary PCA on the unscaled data matrix Y. Using this initial estimate residuals at each instant are calculated as, $r(t) = A^0 y(t)$

If the estimated model is exact, then the residuals will be independent normally distributed variables with zero mean. Thus the joint density function of $r(1)\ r(2)\ r(3)\dots.r(N)$ is easily obtained and the maximum likelihood estimate of Σ_ε is estimated by maximizing the log likelihood function of $r(1)\ r(2)\ r(3)\dots.r(N)$, which is equivalent to minimizing the following function:

$$\min_{\Sigma_\varepsilon} N\log\left|\hat{A}^0\Sigma_\varepsilon\left(\hat{A}^0\right)^T\right| + \sum_{i=1}^{N}\left(r_i^T(t)\left(\hat{A}^0\Sigma_\varepsilon\left(\hat{A}^0\right)^T\right)^{-1} r_i(t)\right) \quad (1)$$

The maximum number of elements in Σ_{y_s} that can be estimated is restricted by the number of constraints or the rank of A. If rank of A is m then the maximum number of elements that can be estimated is less than or equal to $m(m+1)/2$. The method only allows error covariance in the spatial direction.

3. IPCA IMPUTATION ALGORITHM (IPCAIA)

IPCA algorithm has been discussed in the previous section. In this section we will describe the main computational steps of IPCAIA and show its link to the much celebrated Expectation Maximization (EM) Algorithm.

Let $Y^{n\times N}$ be the data matrix containing the noise corrupted measurements which include both dependent and independent variables, where n is the number of variables and N is the total number of samples. If Y_{obs} and Y_{mis} denote the observed and the missing values respectively then the data matrix, $Y = \{Y_{obs}, Y_{mis}\}$.

$$Y = X + \varepsilon$$

where $X$is the noise free variables and ε is measurement noise. There is no assumption on the distribution of the variables Y or X. Only assumption is that, ε follows a multivariate normal distribution with mean zero and covariance Σ_ε, and the underlying model is such that,

$$AX = 0$$

The main steps of IPCAIA are given below:

(1) The missing values of the data matrix are filled with the unconditional mean of the variables. For example, the missing values of the data matrix are filled by the row averages of Y_{obs}.
(2) The filled data matrix Y is supplied to the IPCA algorithm. IPCA automatically determines the number of significant principal components and gives an estimate of the constraint matrix A, scaling matrix L and predicts the noise free variables in scaled domain, $\hat{X}_s$.

(3) The estimated noise free variables $\hat{X}_s$ are converted to $\hat{X}$ in the original domain, $\hat{X} = L\hat{X}_s$. Missing values in the data matrix are filled with these predicted $\hat{X}$ values.
(4) Return to step (2) and repeat these steps until convergence.

The link between IPCA and Expectation Maximization (EM) algorithm is clear from the iterative nature of the two algorithms. Each iteration of EM consists of an Expectation step and a Maximization step. The E-step calculates the conditional expectation of the sufficient statistics given the observed data and the current estimated parameters. Because of filling the missing values with the conditional expectation the variances are underestimated and the main feature of the E-step is adding corrections to the variances. Since error variances are estimated in IPCAIA this can be done implicitly by adding a scaled error term to each of the estimates of the missing value. In step 3 of the above algorithm, instead of filling the missing values with $\hat{X}$ missing values may be filled with $\hat{Y}$ where,

$$\hat{Y} = \hat{X} + L\nu$$
$$\nu \sim N(0, I)$$

The estimated conditional expectation of the sufficient statistics from the E-step is substituted in the log likelihood function and parameters of the model are estimated by maximizing the function. This is the M-step of EM algorithm. Important fact is that parameter estimation has to be maximum likelihood estimate for the observed data set(Little and Rubin, 2002).

In IPCAIA parameter estimation using IPCA is equivalent to the .maximization step of the EM algorithm. The parameters of the model are obtained by minimizing the objective function,

$$\underbrace{\min}_{A_s, X_s} \sum_{i=1}^{n} (Y_{s,i} - X_{s,i})(Y_{s,i} - X_{s,i})'$$

Because of the optimal scaling strategy, in the scaled domain the error covariance matrix is iid. Thus, IPCA gives an ML estimate of the model parameters under this optimal scaling scheme. Since the missing values are filled by values from the distribution of the data, the method remains optimal in the presence of missing data.

4. IPCA DATA RECONCILIATION (IPCADR)

The IPCA Data Reconciliation algorithm for treating missing data is a combination of two methods: IPCA and Data Reconciliation. Before describing the actual algorithm a brief introduction on Data Reconciliation is necessary. In presence of measurement error, the balance equations are not satisfied exactly. The objective of data reconciliation is to estimate the underlying noise free variables so that they satisfy the balance equations. In accordance with the earlier notation,

$$y(t) = x(t) + \varepsilon(t)$$

where $y(t)$ is a $nx1$ vector of observations, $x(t)$ is a $nx1$ vector of the underlying variables and $\varepsilon(t)$ normally distributed measurement error. If the time invariant constraint model is A then data reconciliation solves the following weighted least squares problem:

$$\begin{aligned} &\min_x (y - x)^T \Sigma_\varepsilon^{-1} (y - x) \\ &s.t. Ax = 0 \end{aligned} \tag{2}$$

If all the measurements are not available then the x values corresponding to these unmeasured variables are estimated using the constraints and the available measurements. As the number of unmeasured values increases the degree of freedom also increases, so the estimation becomes poor. In extreme cases when there is not enough redundancy between the measurements and the model, data reconciliation breaks down(Sanchez and Romagnoli, 1996). This is important with regard to missing data treatment because the degree of redundancy changes with missing measurement. So, while using data reconciliation for filling missing data we expect that the method will perform better for small number of missing values in the data matrix.

The main steps of the IPCADR algorithm are as follows:

(1) The missing values of the data matrix are filled with the unconditional mean of the variables. For example, the missing values of the data matrix are filled by the row averages of Y_{obs}.
(2) The filled data matrix Y is supplied to IPCA algorithm. IPCA automatically determines the number of significant principal components and gives an estimate of the constraint matrix A and the error covariance matrix Σ_ε.
(3) The estimated constraint matrix and the error covariance matrix are used in equation (2) to estimate the underlying variables $x(t)$ at each instant, where t=*1. . . .N*. If enough redundancy is not present in the data then the missing values are filled by the unconditional mean.
(4) Return to step(2) and repeat these steps until convergence.

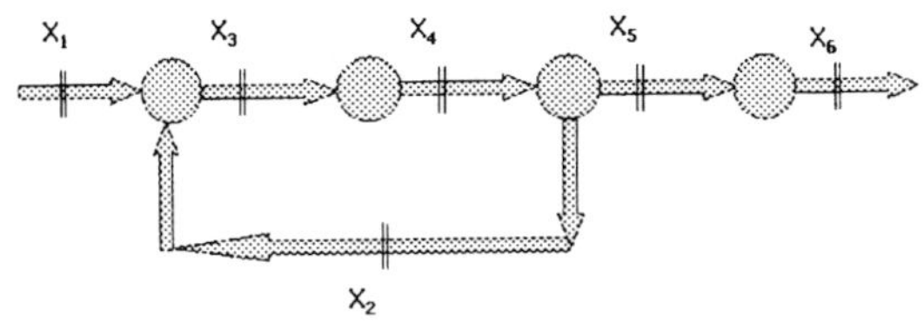

Fig. 1. *Schematic Diagram of the Flow Network*

Table 1. Transfer Functions of the Deterministic Signals

X_1	X_2
$\frac{1}{0.9s+1}$	$\frac{1}{0.8s+1}$

5. SIMULATION RESULTS AND DISCUSSIONS

A flow network process, shown in figure 1, has been chosen to demonstrate the strength of the newly proposed methods. It is assumed that the fluid flowing through the network is incompressible and there is no time delay in the process. The constraint model, A of the process can be obtained easily from the mass balance equation at the junctions. Four mass balance equations can be written for this flownetwork system, so the model order of the constraint matrix is four. In the above example X_1 and X_2 were chosen as independent variables. These are two deterministic signals with slow dynamics. Transfer functions used to simulate these two flows are given in table 1. The rest of the flow rates, X_3 to X_6 were estimated from the mass balance equations. In accordance with the previous notations these are the noise free variables and satisfy the model,

$$AX = 0$$

The observed flow rates are corrupted by measurement noise only,

$$Y = X + \varepsilon$$

Measurement noise were assumed uncorrelated but with unequal variances $\left(i.e.\varepsilon \sim N\left(0, \sigma_i^2 I\right)\right)$. Each measurement vector has 200 data points.

Two indicators were used to assess the performance of the proposed methods. First, Total Sum of Squared Error (TSE), which is defined as,

$$TSE = \sum_{j=1}^{N} \sum_{i=1}^{n} \left(X_{ij}^{mis} - \hat{X}_{ij}^{mis}\right)^2$$

Secondly, Subspace Angle(θ), this is the angle between the two subspaces specified by their columns. If the true constraint model is A^{6x4} and the estimated constraint model is $\hat{A}^{6x4}$, then the angle between each column is defined by,

$$\theta_i = \left\| A_{.i} - A_{.i}\hat{A}^T \left(\hat{A}\hat{A}^T\right)^{-1} \hat{A} \right\|$$

$$\theta = \sum \theta_i$$

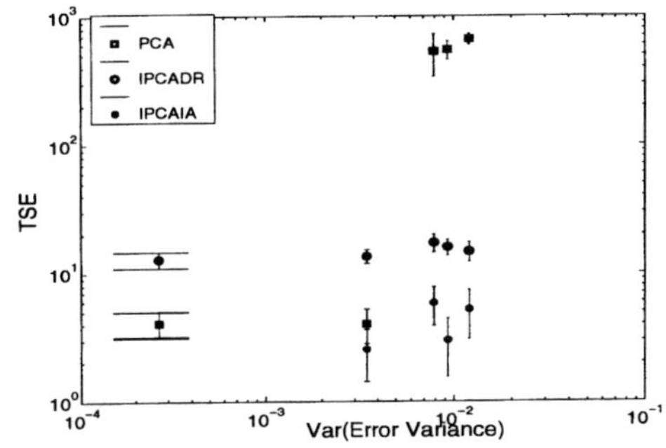

Fig. 2. *Prediction Capability Comparison. Total Sum Square Error vs. variance of error variances*

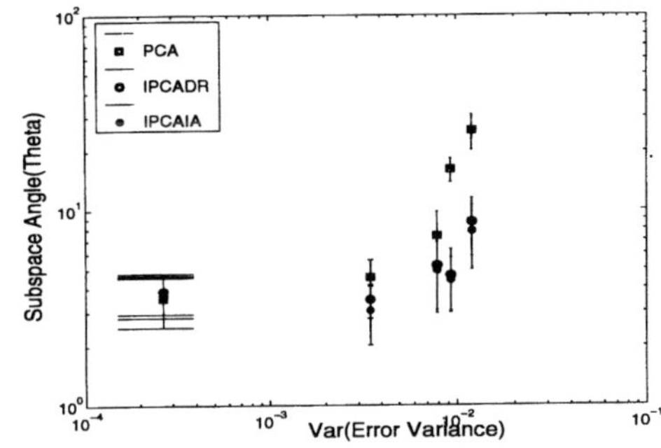

Fig. 3. *Model Quality Comparison. Subspace Angle between the estimated model and actual model vs. variance of error variances*

For a two dimensional space this is the angle at the intersection of two planes. If the angle is small, the two matrices will be nearly linearly dependent.

It was claimed in the previous sections that the proposed methods perform better than PCA based imputation algorithms in two situations, first, model estimation when the errors are not iid $\left(i.e.\Sigma_\varepsilon \neq \sigma^2 I\right)$ and second, model order selection in the presence of missing data. To demonstrate the first point, the Total Sum Square Error (TSE) and the Subspace Angle (θ) estimated by all three methods are plotted in figure 2 and 3. The deviations of the noise variance from iid was quantified by var(error variance). In both cases 7.3% of the total values were missing. TSE gives a measure of the prediction capability of the algorithms. When the noise covariance is close to iid performance of all three methods are practically indistinguishable. But as it deviates further from the iid assumption the performance of the PCA based algorithm deteriorates sharply. The subspace angle is also smaller for the models estimated by IPCAIA and IPCADR algorithms compared to the PCA based algorithm.

The results on the robustness of the algorithms are shown in figure 4. In this case the error covariance matrix deviates from iid only moderately; var(error variance) is 0.0035. As the missing data increases above 10% of the total data, the difference between the estimated models becomes apparent. Estimated model using IPCAIA algorithm

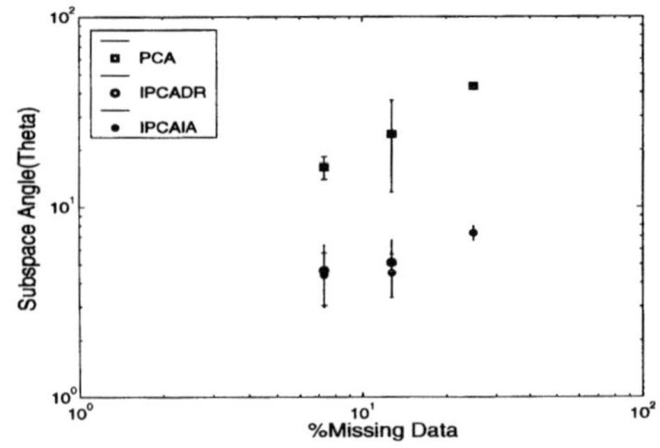

Fig. 4. *Robustness Comparison Plot. Subspace Angle between the estimated model and actual model vs. % missing data*

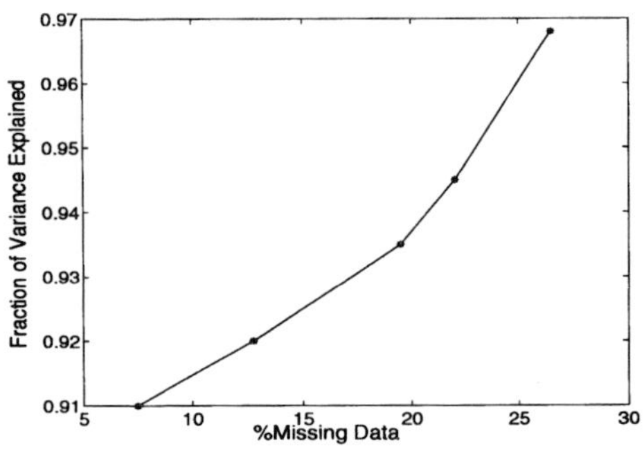

Fig. 5. *Fraction of Variance Explained vs. %Missing Value*

is very good for missing values as high as 25%, while the PCA-based algorithm performs poorly for this large amount of missing data. On the other hand, IPCADR algorithm performs well up to moderate range of missing data. But at a high range the method breaks down because of lack of redundancy.

The improved model order selection criteria in the presence of missing data are evident from figures 5 and 6. Figure 5 shows that as more values are missing in the data matrix, total variance explained by first two principal components increases (figure 5). So, methods used in PCA for selecting model order, such as Scree plots suffers even more in the presence of missing data. On the other hand, the last four eigenvalues from IPCAIA and IPCADR do not deviate significantly from unity (figure 6). Therefore, the model order selection is precise for these two methods even in the presence of missing data.

6. CONCLUSIONS

In this work two limitations of PCA in connection with missing data treatment have been pointed out: 1) model order selection in the presence of missing data, 2) model building from data with unequal noise variances. The proposed algorithms IPCAIA and IPCADR provide a precise way of model order selection by looking at the eigenvalues. Error covariance is incorporated in the

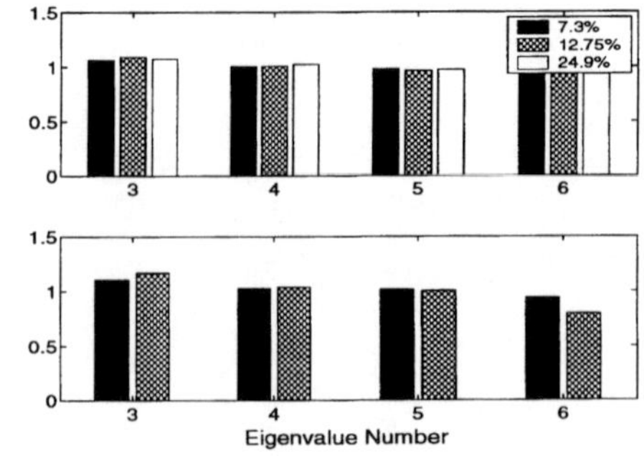

Fig. 6. *Last four eigenvalues estimated by IPCAIA(top) and IPCADR (bottom)*

model estimation procedure. Thus, a maximum likelihood estimate of the model parameters can be obtained. From the simulation study it was evident that IPCAIA was able to estimate good models up to a fairly significant amount of missing data (25%), while IPCADR showed good performance up to moderate range (13%). Both methods outperformed PCA-based missing data treatment algorithm for the unequal noise situation.

REFERENCES

Artega, R. and A. Ferrer (2002). Dealing with missing data in MSPCS: several methods, different interpretations, some examples. *Journal of Chemometrics* **16**, 408–418.

Grung, B. and R. Manne (1998). Missing values in principal component analysis. *Chemometrics and Intelligent Laboratory Systems* **42**, 125–139.

Little, R.J.A. and D.B. Rubin (2002). *Statistical Analysis with Missing Data.* Vol. 2. John Willy and Sons.

Narasimhan, S. and S.L. Shah (2004). Model identification and error covariance matrix estimation from noisy data using PCA. ADCHEM.

Nelson, P.R.C., Taylor P.A. and J.F. MacGregor (1996). Missing data methods in PCA and PLS: Score calculations with incomplete observations. *Chemometrics and Intelligent Laboratory Systems* **35**, 45–65.

Sanchez, M. and J. Romagnoli (1996). Use of orthogonal transformations in data classification reconciliation. *Computers and Chemical Engineering* **20**(5), 483–493.

Troyanskaya, O., M. Cantor, G. Sherlock, P. Brown, T. Hastie, R. Tibshirani, D. Botstein and R.B. Altman (2001). Missing value estimation methods for DNA microarrays. *Bioinfomatics* **17**(6), 520–525.

Walczak, B. and D.L. Massart (2001). Dealing with missing data. *Chemometrics and Intelligent Laboratory Systems* **58**, 15–27.

Wentzell, P.D., E.T. Andrews, A.D. Hamilton, K. Faber and B.R. Kowalski (1997). Maximum likelihood principal component analysis. *Journal of Chemometrics* **11**, 339–366.

ELSEVIER
IFAC
PUBLICATIONS
www.elsevier.com/locate/ifac

SPECIFYING THE DIRECTIONALITY OF FAULT PROPAGATION PATHS USING TRANSFER ENTROPY

Margret Bauer*, Nina F. Thornhill*, Adrian Meaburn+

**Department of Electronic and Electrical Engineering, Torrington Place, University College London, London WC1E 7JE, UK.*
+BP Chemicals, Salt End, Hull, HU12 8DS, UK.

Abstract: In continuous chemical processes, variations of process variables usually travel along propagation paths in the direction of flow. The aim of this study was to find a data-driven method for identifying the direction of variation propagation using historical process data. Transfer entropy is a recently proposed method based on the probability density function (PDF) that measures directionality of variation with respect to time. An industrial case study illustrates the method which detects the influence of a temperature controller on downstream temperature measurements. A reversal of directionality was noted during a disturbance and a physical explanation offered.

Keywords: Chemical industry; fault diagnosis; plant-wide; probability density function; propagation; statistical process control.

1. INTRODUCTION

Plant-wide disturbances occur if a fault distributes along a fault propagation path and affects several process variables. The detection and diagnosis of plant-wide disturbances is a major problem in the process industry (Qin, 1998). It is therefore important to understand the mechanism of propagation.

A description of the propagation mechanism in causal qualitative models can identify the order of occurrence of events and specify the paths of fault propagation. Thus, the root cause of the disturbance can be retraced along the propagation path.

This paper describes work in progress on a data-driven method that serves as a basis for the automatic construction of digraphs. The entropy related measure of transfer entropy is chosen for determining the direction of fault propagation and to answer questions of

- Directionality: Does a measurement X influence a second measurement Y more than measurement Y influences X?
- Consistency: Does the directionality change if a fault is present?

Digraph based models usually express the relationship between faults and symptoms and define the propagation paths by incorporating process knowledge of experts (Nam et al., 1996). A drawback is that extracting expert knowledge is very time consuming and the knowledge not always available. If a mathematical representation of the process exists then digraphs can be developed systematically from differential and algebraic equations (Maurya et al., 2003a,b). However, complete descriptions are rarely available for complex chemical processes and large plants.

An incorporation of the data-driven approach in a causal map consisting of digraphs was presented by Chiang and Braatz (2003). They derive causal maps from expert knowledge and from the measurements affected by a fault. Broken dependencies along the propagation paths are identified using data-driven, entropy related measures. The direction of propagation is not automatically detected but derived from expert knowledge.

The contribution of the work reported in this paper explores a measure for the directionality of variation propagation in historical process data. The result can then serve as a basis for qualitative model generation in form of digraphs.

The next section explains the concept of transfer entropy and its use for fault propagation. To illustrate the concept of transfer entropy it is then applied to an industrial case study in Section 3. Results of directionality and consistency detected by transfer entropy are discussed.

2. DIRECTIONALITY ANALYSIS

Entropy related measures regard the measurement of a physical variable as the outcome of a random process. Information contained in the outcome is expressed by its probability density function (PDF). Assuming that the process is ergodic over a defined timeframe the PDF can be estimated by the relative frequency. This is achieved by dividing the amplitude axis into discrete amplitude bins and assigning every sample data value to a bin accordingly. The number of data points per bin accumulates to the relative frequency.

Entropy measures the information, also referred to as uncertainty or randomness, of a process by logarithmical summing over all bins. When a fault is present the PDF and consequentially the entropy can change significantly (Chiang and Braatz, 2003). Thus, entropy can be used for detecting faults as it is insensitive to outliers or temporary biases.

2.1 Transfer Entropy

A focus on PDF destroys the time trend of the sequence. To show the significance of this problem, two signals are considered: the first is a random signal whose amplitude is uniformly distributed between -1 and +1, the second a triangular wave oscillating regularly between -1 and +1. The PDFs of these two signals are identical yet they are obviously

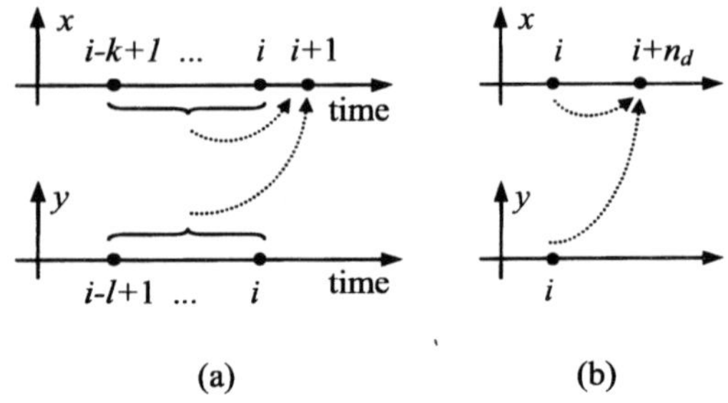

Fig. 1. Time samples involved in the computation of the transfer entropy, general case (a) and special case for k=l=1 with time delay n_d (b).

not the same signal. Thus, PDF and entropy cannot capture time dependency.

Transfer entropy has been recently proposed by Schreiber (2000) to measure dependency in time and thus which of two variables causes the other. As an example, the dependency between heart and breath rate was investigated, finding that the heart rate influences the breath rate rather than vice versa. Transfer entropy relates k previous samples of variable X and l previous samples of variable Y to predict the next value of X and thus incorporates the time dependency. This relationship is illustrated in Figure 1(a). Transfer entropy is then calculated by summing the joint and conditional PDF of two time sequences logarithmically:

$$T_{Y \to X} = \sum_{x_{i+1}} \sum_{\mathbf{x}_i^k} \sum_{\mathbf{y}_i^l} p(x_{i+1}, \mathbf{x}_i^k, \mathbf{y}_i^l) \log \frac{p(x_{i+1} \mid \mathbf{x}_i^k, \mathbf{y}_i^l)}{p(x_{i+1} \mid \mathbf{x}_i^k)} \quad (1)$$

where the sums are over all amplitude bins. The joint PDF $p(x_{i+1}, \mathbf{x}_i^k, \mathbf{y}_i^l)$ is the probability that the combination of x_{i+1}, $\mathbf{x}_i^k$ and $\mathbf{y}_i^l$ have particular values. The conditional PDF $p(x_{i+1} | \mathbf{x}_i^k, \mathbf{y}_i^l)$ is the probability that x_{i+1} has a particular value when the value of previous samples $\mathbf{x}_i^k$ and $\mathbf{y}_i^l$ are known.

The index $Y \to X$ indicates that the influence of sequence Y on sequence X is measured. The reverse dependency $X \to Y$ is calculated by exchanging x and y of the joint and conditional PDFs. To specify whether X influences Y more than Y influences X a measure for the directionality is introduced:

$$t_{X \to Y} = (T_{X \to Y} - T_{Y \to X}) / \min\{T_{X \to Y}, T_{Y \to X}\} . \quad (2)$$

If $t_{X \to Y}$ is greater than zero then X influences Y. Scaling the difference by the minimum of both values provides a standardized index.

2.2 Time Delays

Transfer entropy uses the presence of time delays to identify which variable influences the other. In continuous processes time delays occur as the product streams along the propagation path. Thus, disturbances and variations can be observed with a time delay at two sequential measuring points. In most cases, the time delay of the stream is unknown and varies with throughput over time. This issue has to be considered when adjusting the parameters for the transfer entropy.

The distance n_d between two consecutive samples, as shown in Figure 1(b) for k=l=1, is an important parameter which captures the time delay t_d. The transfer entropy and also $t_{X\rightarrow Y}$ become a function of n_d. Eq. 1 reveals that no symmetry with respect to n_d holds but it can be noted that $T_{X\rightarrow Y}(n_d) = T_{Y\rightarrow X}(-n_d)$.

Ideally, the parameter n_d should equal the actual time delay t_d. Since t_d is usually not known, a weighted sum of the directionality measure t_{XY} for a number of N of values of n_d can be introduced:

$$t'_{X\rightarrow Y}(N) = \sum_{n_d=1}^{N} \frac{1}{n_d} t_{X\rightarrow Y}(n_d)\,. \tag{3}$$

The number of samples N has to be sufficiently large to ensure that the time delay t_d is captured. The directionality for $n_d = t_d$ should be much larger than the average value for the directionality the sum in Eq. 3 therefore will be dominated by the $n_d = t_d$ term and will capture the dependency. The factor $1/n_d$ originates from the assumption that small time delays are generally more relevant than large time delays. This intuitive assumption is based on the principle of Markov processes.

2.3 Challenges

A drawback of transfer entropy is that it depends strongly on the bin size of the PDF. A large number of time samples are required to find a suitable representation of the PDF with a sufficient large number S of amplitude bins. Additionally, the number of bins increases exponentially with larger k and l as the total number of the combination of bins is S_{tot}=$S^{k+1}\cdot S^{l}$.

The obstacle of poor resolution or too large bin size of the PDF can be overcome by using more sophisticated estimation than binning. Fewer time samples are needed to estimate the PDF when using Kernel functions (Silverman, 1986) which are centred around the sample points and thus accumulate to the PDF. Alternatively, a uniform or linear distribution can be assumed and approximated by integrating over each bin (Hukkanen and Braatz, 2003). As this research is still ongoing, Kernel functions will be implemented in a next step.

3. THE INDUSTRIAL PROCESS

The process schematic of the industrial case study is shown in Figure 2. The process is the solvent recycle path in a gas purification system. Measurements of the temperature are taken at seven points along the recycle stream. The level of the liquid at the bottom of both distillation columns and the differential pressure within are also captured.

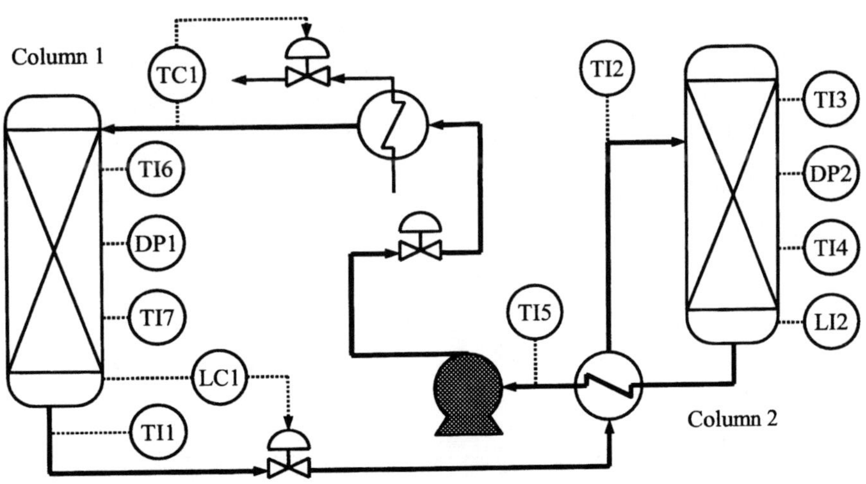

Fig. 2. Process schematic of the industrial case study with process variable indicators and controllers for temperature (T), level (L) and differential pressure (DP).

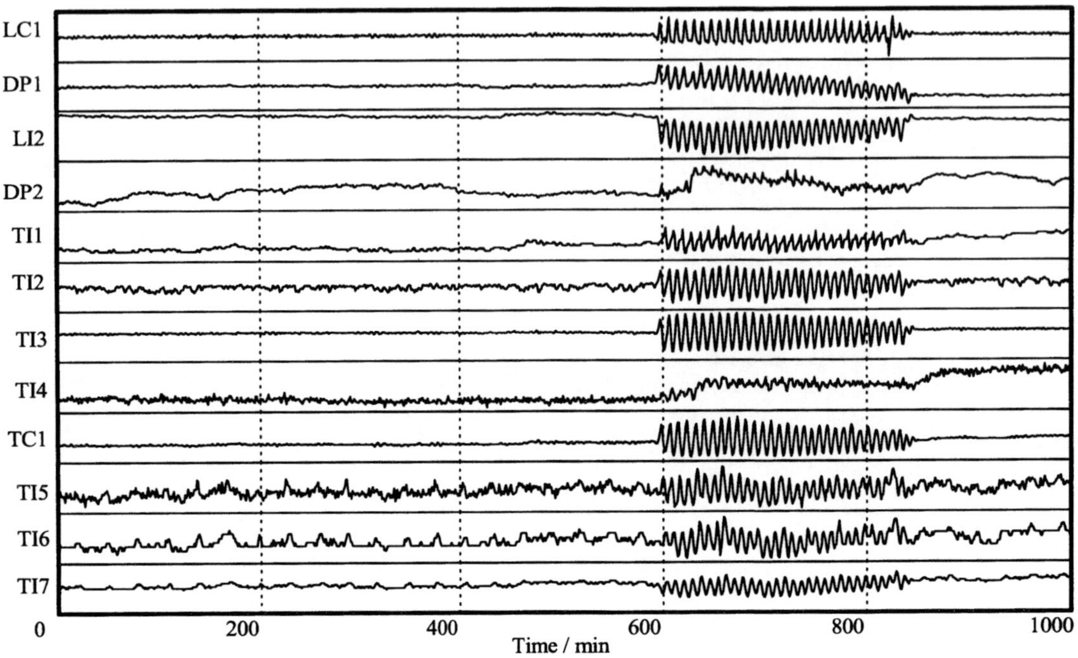

Fig. 3. Time trends of process variables.

The purpose of applying transfer entropy to case study is to provide a measure for the direction of propagation and to identify directionalities that change during the presence of a disturbance.

3.1 The Data Set

The full data set consists of 1000 samples taken at the twelve measurement points at a sample rate of once per minute. Figure 3 shows the mean centred and normalized time trend of the process variables. From visual inspection an oscillatory disturbance can be identified from sample 600 to 840. The identical period of oscillation and the coinciding start and ending of the disturbance for all measurements suggest a plant-wide disturbance that affects more or less all process variables.

3.2 Direction of Propagation

Five sequential temperature measurements are investigated for applying the directionality measure: indicator TI5 next to the recycle heat exchanger, the process variable of controller TC1, two indicators at distillation column 1 (TI6 and TI7), and the measurement at the exit of column 1, TI1.

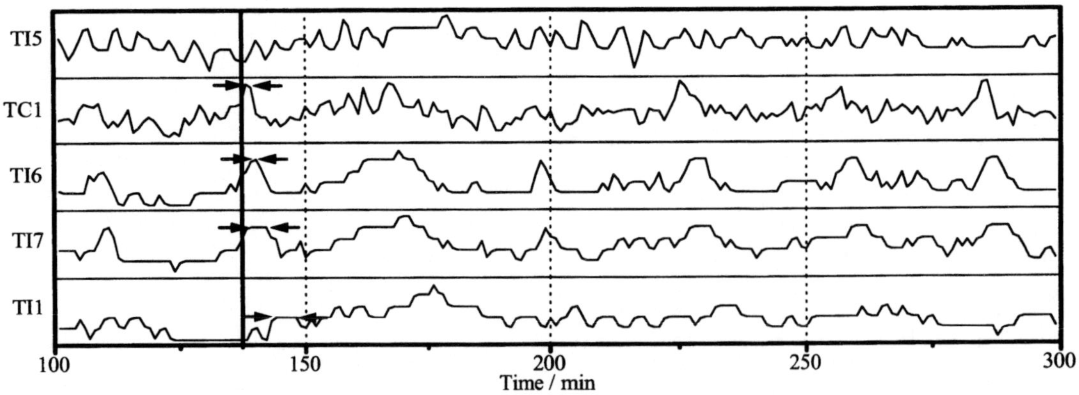

Fig. 4. Close up of temperature measurements.

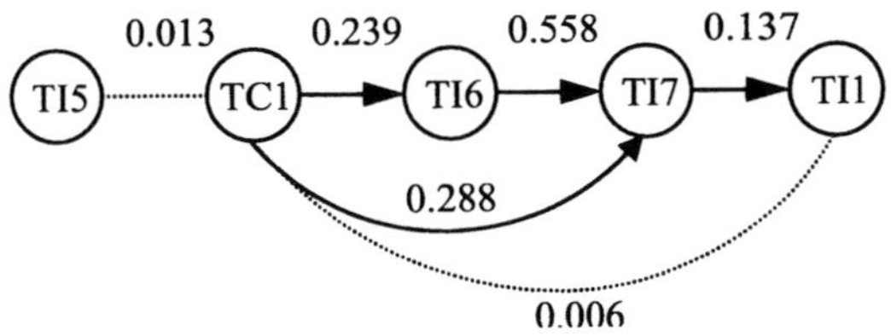

Fig. 5. Directionality $t'_{X \to Y}$ for sequential temperature measurements of the industrial process (k=2, l=1).

A close-up of the time trends of these measurements is shown in Figure 4. Distinct features can be observed which occur with a time delay in all variables. A large peak just before 200 minutes provides an example. It occurs first in TC1 and moves to TI6, TI7 and TI1. A requirement for the new transfer entropy method is that it should match this visual observation.

Transfer entropy was able to measure the directionality of sequential variables. The variables and the direction represented by arcs are shown in Figure 5. The values assigned with the arcs are values for the directionality measure $t'_{X \to Y}$ (K=5). Large values indicate a strong directionality.

As expected, the two temperature indicators at column 1 are strongly dependent ($t'_{TI6 \to TI7}$ = 0.558) and TI6 influences TI7. As a result, TI7 is approximately influenced by TC1 as much as TI6. However, the directionality between TC1 and TI1 is almost zero ($t'_{TC1 \to TI1}$ = 0.006), while TI7 influences TI1 distinctively ($t'_{TI7 \to TI1}$ = 0.137). The directionality measure proves to give a representation of the actual dependencies in the process because it follows the direction of material flow. In particular, Figure 5 shows an influence of TC1 on TI7 as much as on TI6. This means that the structure is correctly represented even if the measurement of TI6 is not available. It can be concluded correctly that TC1 influences all temperature measurements in the column.

3.3 Consistency of Propagation

Consistency is a second issue to be solved by transfer entropy since the direction of signature propagation might change when a disturbance is present. Transfer entropy showed that during normal operation, differential pressure DP1 influences level LC1 while during the disturbance LC1 influences DP1. Physical experiments suggest that the disturbance is caused by foaming in column 1. If that is so then a possible explanation for the observed change in directionality is as follows.

During normal operation (up to 600 min in Figure 3), any pressure increase also increases the liquid hold-up in the column and decreases the flow rate of the liquid. Thus, the level of liquid fluctuates. It is known that foaming affects the level measurement on this column. This in turn will cause LC1 to move the output and this will alter both the liquid flow and liquid hold-up because of the resulting movement of the control valve. The changes in hold-up and flow rates will in turn cause the differential pressure to change.

Figure 6 shows transfer entropy $T_{LC1 \to DP1}$ and $T_{DP1 \to LC1}$ over a number of values for the time delay n_d. The directionality measure t' is the weighted sum over all time delays such that during normal operation $t'_{DP1 \to LC1}$ = 0.805 and during the disturbance $t'_{LC1 \to DP1}$ = 0.850. In the upper plot, representing the undisturbed period, DP1 influences LC1. In the lower plot, representing the period when the disturbance is present, LC1 influences DP1. This result suggests that transfer entropy can capture consistency in fault propagation by finding directionality changes when a fault is present.

4. CONCLUSIONS

The concept of transfer entropy (Schreiber, 2000) has been proposed for finding the direction of flow during normal operation and to identify changes when a fault is present.

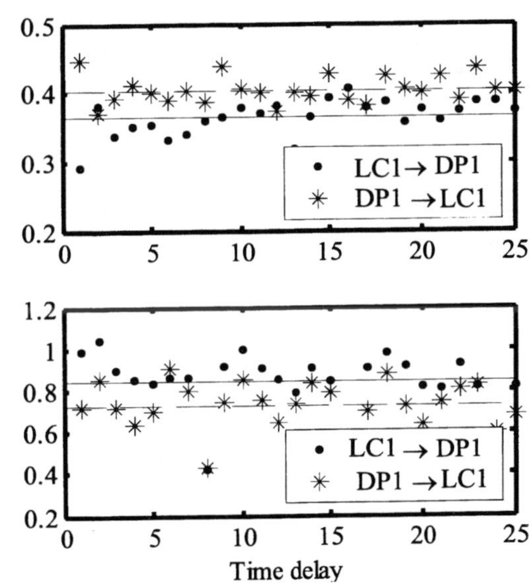

Fig. 6. Transfer entropy of level (LC1) and differential pressure (DP1) during normal operation (upper plot) and when a disturbance is present (lower plot).

The measure gives a basis for creation of digraph models using process data rather than expert knowledge. In an industrial case study, the direction of variables was identified from the process data and an inverse direction was detected during a period of disturbance. The directionality measure is showing exciting results even though the unknown time lag can only be approximated and not exactly captured. It appears to be robust within bounds; however, the next step is to optimize the parameters for the number of time delays N and the number of samples of joint and conditional probability.

ACKNOWLEDGEMENTS

The principal author gratefully acknowledges the financial support of the University College London Graduate School. The UCL authors appreciate the support of BP Chemicals, Hull.

REFERENCES

Chiang, L.H. and Braatz, R.D. (2003). Process monitoring using causal map and multivariate statistics: fault detection and identification. *Chemometrics and Intelligent Laboratory Systems*, **65**, 159-178.

Hukkanen, E. J. and Braatz, R.D. (2003). Measurement of particle size distribution in suspension polymerization using in situ laser backscattering. *Sensors and Actuators B: Chemical*, **96**, 451-459.

Maurya, M.R., Rengaswamy, R. and Venkatasubramanian, V., 2003. A systematic framework for the development and analysis of signed digraphs for chemical processes. 1. Algorithms and analysis. *Industrial and Engineering Chemical Research*, **42**, 4789-4810.

Maurya, M.R., Rengaswamy, R. and Venkatasubramanian, V., 2003. A systematic framework for the development and analysis of signed digraphs for chemical processes. 2. Control loops and flowsheet analysis. *Industrial and Engineering Chemical Research*, **42**, 4811-4827.

Nam, D.S., Han, C., Jeong, C.W. and Yoon, E.S. (1996). Automatic construction of extended symptom-fault associations from the signed digraph. *Computers and Chemical Engineering*, **20**, S605-S610.

Qin, S.J. (1998). Control performance monitoring – a review and assessment. *Computers and Chemical Engineering*, **23**, 173-186.

Schreiber, T. (2000). Measuring information transfer. *Physical Review Letters*, **85**, 461-464.

Silverman, B.W. (1986). *Density estimation for statistics and data analysis*. Chapman & Hall, London.

ELSEVIER
IFAC
PUBLICATIONS
www.elsevier.com/locate/ifac

ROBUST STABLE NONLINEAR CONTROL AND DESIGN OF A CSTR IN A LARGE OPERATING RANGE

Johannes Gerhard, Martin Mönnigmann, Wolfgang Marquardt

*Lehrstuhl für Prozesstechnik, RWTH Aachen
Turmstr. 46, D-52064 Aachen, Germany*

Abstract: In this work an approach based on nonlinear dynamics is used for the integrated design and controller tuning of a CSTR. The approach enables integrated design optimization and robust tuning of a linearizing feedback controller. The controller setting found by the approach guarantees robust stability of the process over a large range of set point variations even in the presence of parameter uncertainty. The method enforces robust stability by introducing lower bounds on the parametric distance of the operating point to critical boundaries in the space of process and controller parameters. For stabilization of a large range of operations, a lower bound on the distance to a nontransversal Hopf bifurcation has to be considered in this particular case. *Copyright ©2004 IFAC*

1. INTRODUCTION

Chemical processes generally exhibit nonlinear behavior. In process and control design it is of particular importance to guarantee stable behavior of the closed-loop process. Ideally, the operating point should be robust against uncertainties in the underlying process model. For nonlinear chemical processes with a frequently changing set point it is furthermore desirable to guarantee robust stability over a wide range of operating conditions.

In this paper the issue of robust stability over a wide range of operating conditions is discussed for a temperature controlled CSTR (continuously stirred tank reactor). Control is realized by linearizing feedback control. A bifurcation analysis of the CSTR performed by Hahn *et al.* (2003) shows that model uncertainties may lead to unstable operating points for a particular range of set points. The analysis also reveals that a lower bound of the control parameter exists above which the unstable region vanishes completely. This behavior can be attributed to a so-called nontransversal Hopf bifurcation.

In this work a novel design method based on nonlinear dynamics and bifurcation theory is adopted and used for an integrated process design and controller tuning of a CSTR. The method has been reported recently by Mönnigmann and Marquardt (2002a, 2003). It enables the optimization of process systems with respect to some profit function guaranteeing robust stability in the presence of parametric uncertainty. The method is applicable to both open and closed-loop systems. In case of the CSTR considered, stability is ensured by maintaining a specified lower bound on the distance between the operating point and the critical boundary formed by the set of all nontransversal Hopf points. Results of the approach are presented for two illustrating design scenarios.

2. PROCESS MODEL AND LINEARIZING FEEDBACK CONTROLLER

We consider a cooled CSTR with an exothermic first order reaction $A \rightarrow B$. Assuming perfect level control, the CSTR model consists of the material and energy balances. It reads

$$
\dot{c_A} = \frac{q}{V}(c_{Af} - c_A) - \Delta k_0 \exp\left(-\frac{E}{RT}\right) c_A\,,
$$

$$
\dot{T} = \frac{q}{V}(T_f - T) - \frac{\Delta H k_0}{\rho C_p} \exp\left(-\frac{E}{RT}\right) c_A + \frac{UA}{V\rho C_p}(T_c - T)\,. \quad (1)
$$

The temperature of the cooling fluid T_c is the manipulated variable and the reactor temperature T is the measured and controlled variable (Uppal *et al.*, 1974).

Control is realized by means of linearizing feedback including integral action. The control law reads as (Hahn *et al.*, 2003)

$$
u = \frac{-\frac{q}{V}(T_f - T) + \frac{\Delta H}{\rho C_p} k_0 \exp(-\frac{E}{RT}) c_A + \frac{UA}{V\rho C_p} T}{\frac{UA}{V\rho C_p}} + \frac{\frac{2}{\epsilon}(T_{sp} - T) + \frac{1}{\epsilon^2}\int_0^t (T_{sp} - T) d\tau}{\frac{UA}{V\rho C_p}}\,. \quad (2)
$$

The tuning parameter ϵ corresponds to the time constant of the closed loop dynamics, i.e. the smaller ϵ the faster is the dynamics of the closed loop. The integral action $1/\epsilon^2 \int_0^t (T_{sp} - T) d\tau$ is included to compensate for potential set point offsets.

3. BIFURCATION ANALYSIS

In a previous article (Hahn *et al.*, 2003) the closed loop process model (1), (2) has been analyzed by parameter continuation and bifurcation theory in the presence of unmodeled dynamics and parametric uncertainties. Unmodeled dynamics have been introduced into the model by an overdamped second order process between the controller output u and the manipulated variable T_c according to

$$
\begin{aligned}
\epsilon_v \dot{z} &= -u + z \\
\epsilon_v \dot{T_c} &= -T_c + z\,,
\end{aligned} \quad (3)
$$

where ϵ_v represents the unknown time constant of the unmodeled dynamics. The controller (2) remains unchanged. The assumed value of ϵ_v and the other process parameters are summarized in Table 1.

Table 1. Parameters used for analysis

Parameter	Value	Parameter	Value
ΔH	$-5 \cdot 10^4 \frac{\text{J}}{\text{mol}}$	$\frac{E}{R}$	8750 K
c_{Af}	$1 \frac{\text{mol}}{\text{l}}$	k_0	$7.2 \cdot 10^{10} \frac{1}{\text{min}}$
T_f	350 K	UA	$833.3 \frac{\text{W}}{\text{K}}$
V	100 L	C_p	$0.239 \frac{\text{J}}{\text{gK}}$
ρ	$1000 \frac{\text{g}}{\text{l}}$	q	$100 \frac{\text{mol}}{\text{min}}$
ϵ_v	0.05 min		

The results of the analysis are briefly summarized here in order to assert the existence of a particular critical point, a so-called nontransversal Hopf (NTH) bifurcation, which fully controls the stability loss in this case. Thus the design approach presented in Section 4 can solely be based on this NTH bifurcation.

Continuation of the equilibria from a stable operating point with varying set point temperature T_{sp} yields two Hopf bifurcations. The Hopf

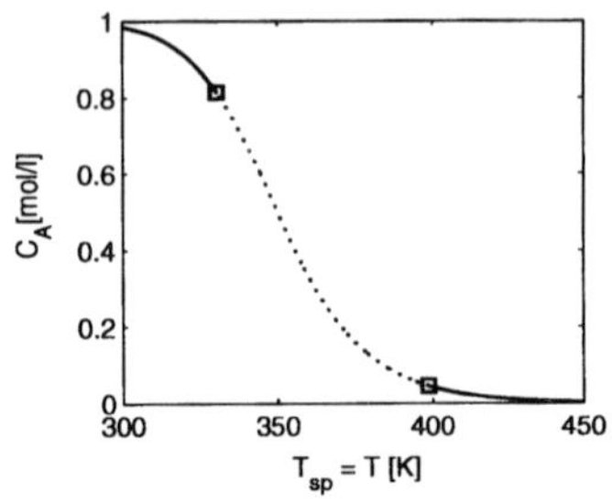

Fig. 1. Curve of steady states for variations in T_{sp}. Two Hopf bifurcations (□) divide the curve into stable states (solid line) and unstable states (dotted line).

points separate the curve of steady states shown in Fig. 1 into two stable parts, and an unstable part between the two stable parts. Introducing the control parameter ϵ as a second free parameter, a curve of Hopf points can be evaluated. This

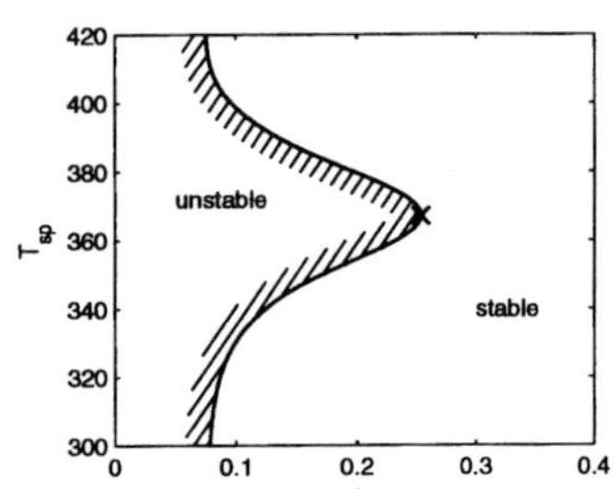

Fig. 2. Curve of Hopf points and nontransversal Hopf point (×) at $\epsilon = 0.25$ min.

curve of Hopf points shown in Fig. 2 separates the parameter space (ϵ, T_{sp}) into a stable and an unstable region. The distance between the two Hopf points diminishes with increasing values of ϵ. At $\epsilon = 0.25$min, $T_{sp} = 367$K, they merge and form an extremum with respect to the control parameter ϵ.

The bifurcation at the extremum of the Hopf curve is the already mentioned NTH bifurcation. The NTH point represents a lower boundary for the robust tuning of the controller. For values of ϵ larger than the extremal value, stability can be guaranteed for *all* values of T_{sp}. Hence the process is stable over the entire range of operating conditions of interest.

4. CONTROL SYNTHESIS FOR STABILITY

This section gives a short introduction into the basic idea of the design method. For a rigorous derivation of the method the reader is referred to Mönnigmann and Marquardt (2002a). The method is applied to the NTH bifurcation which was found by means of continuation analysis of the process. The dynamic equations of the process model (1), the controller (2) and the unmodeled dynamics (3) are denoted by

$$\dot{x} = f(x, \alpha, p), \quad x(t_0) = x_0 . \tag{4}$$

In this notation, $x \in \mathbb{R}^{n_x}$ are the state variables, $\alpha \in \mathbb{R}^{n_\alpha}$ the uncertain process and controller parameters and $p \in \mathbb{R}^{n_p}$ the known parameters.

The robustness of the operating point has to be guaranteed only for the uncertain parameters of the process model. It is assumed that there are known upper and lower uncertainty boundaries $\pm\Delta\alpha$ for all α, i. e.

$$\alpha_i \in \left[\alpha_i^{(0)} - \Delta\alpha_i, \alpha_i^{(0)} + \Delta\alpha_i\right], i = 1, \ldots, n_\alpha .$$

These boundaries are defined e. g. by the accuracy of measurements. The scaling of the parameters

$$\alpha_i \rightarrow \frac{\alpha_i}{\Delta\alpha_i}, \alpha_i^{(0)} \rightarrow \frac{\alpha_i^{(0)}}{\Delta\alpha_i}$$

gives the dimensionless parameters

$$\alpha_i \in \left[\alpha_i^{(0)} - 1, \alpha_i^{(0)} + 1\right] . \tag{5}$$

Robustness against the uncertainties is achieved by enforcing a minimal distance between the critical manifold and the operating point in the space of uncertain parameters. The minimal distance is equal to the radius $\sqrt{n_\alpha}$ of the n_α-dimensional ball enclosing the n_α-dimensional cube of sidelength 2 defined by eq. (5). In Fig. 3, a robust operating point is shown for $n_\alpha = 2$. The ball with a radius

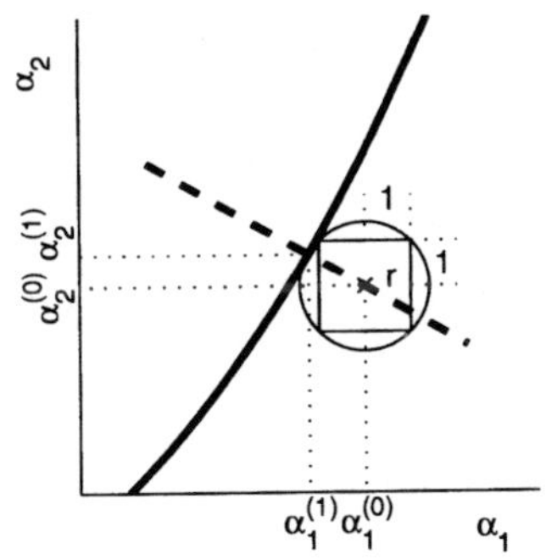

Fig. 3. Robust operating point $\alpha^{(0)}$ with stability boundary (thick line). The critical point $\alpha^{(1)}$ nearest to the operating point is in the direction of the normal vector r (dashed thick line) at a distance of $\sqrt{2}$ from $\alpha^{(0)}$.

of $\sqrt{2}$ touches the critical boundary tangentially guaranteeing stability for the nominal operating point $\alpha_i^{(0)}$ and for all variations of the parameters within their range of uncertainty. The minimal distance $l = \sqrt{n_\alpha}$ between the critical manifold and the operating point occurs along the direction of the normal vector r to the critical manifold. Thus the constraints for the robustness are given by

$$\alpha^{(0)} = \alpha^{(1)} + l\frac{r}{\|r\|} \tag{6}$$

$$l \geq \sqrt{n_\alpha} . \tag{7}$$

A scheme for the derivation of the system of equations defining r for a generalized critical point has been derived by Mönnigmann and Marquardt (2002a). The NTH point that is found in the bifurcation analysis of the reactor model is a degenerated Hopf point. Thus the normal vector system of the NTH point can be derived from the normal vector system of a regular Hopf point. According to Mönnigmann and Marquardt (2002a) the augmented system for the normal vector r on a Hopf point is given by

$$0 = f(x, p, \alpha) \tag{8a}$$

$$\begin{aligned} 0 &= f_x w^{(1)} + \omega w^{(2)} \\ 0 &= f_x w^{(2)} - \omega w^{(1)} \\ 0 &= w^{(1)T} w^{(1)} + w^{(2)T} w^{(2)} - 1 \\ 0 &= w^{(1)T} w^{(2)} \end{aligned} \tag{8b}$$

$$\begin{aligned} 0 &= f_x^T v^{(1)} - \omega v^{(2)} + \gamma_1 w^{(1)} - \gamma_2 w^{(2)} \\ 0 &= f_x^T v^{(2)} + \omega v^{(1)} + \gamma_1 w^{(2)} + \gamma_2 w^{(1)} \\ 0 &= v^{(1)T} w^{(1)} + v^{(2)T} w^{(2)} - 1 \\ 0 &= v^{(1)T} w^{(2)} - v^{(2)T} w^{(1)} \\ 0 &= f_x^T u + v^{(1)T} f_{xx} w^{(1)} + v^{(2)T} f_{xx} w^{(2)} \end{aligned} \tag{8c}$$

$$\begin{aligned} 0 &= r - f_\alpha^T u - v^{(1)T} f_{x\alpha} w^{(1)} \\ &\quad - v^{(2)T} f_{x\alpha} w^{(2)} . \end{aligned} \tag{8d}$$

The system of equations (8) is divided into four parts. Eqns. (8a) define the steady state of the process. Eqns. (8b) denote the augmented system for Hopf bifurcations, with f_x being the Jacobian of the model equations, $i\omega$ the eigenvalue of the Jacobian with vanishing real part and $w^{(1)} + iw^{(2)}$ the corresponding complex eigenvector. Eqns. (8c) introduce the complex eigenvector $v^{(1)} + iv^{(2)}$ corresponding to f_x^T and the eigenvalue $-i\omega$, and the auxiliary variables u, γ_1 and γ_2 for the calculation of the normal vector r given by eqns. (8d).

It is easy to derive the normal vector system for the NTH point by means of a geometric argument. From the curve of Hopf points in Fig. 2 it is obvious that the component of the normal vector in direction of T_{sp}, $r_{T_{sp}}$, is equal to zero at the extremum of the curve. The system that defines the normal vector of the NTH point is therefore identical to the set of equations (8) except for the subsystem (8d) that contains $r_{T_{sp}}$. Without restrictions it is assumed that $T_{sp} = \alpha_1$. Then the eqns. (8d) have to be replaced by

$$0 = f_{\alpha_1}^T u + v^{(1)T} f_{x\alpha_1} w^{(1)} + v^{(2)T} f_{x\alpha_1} w^{(2)}$$

$$0 = r - f_{\alpha_i}^T u - v^{(1)T} f_{x\alpha_i} w^{(1)} - v^{(2)T} f_{x\alpha_i} w^{(2)}, \quad i = 2, \ldots, n_\alpha . \tag{9}$$

The equations defining the normal vector on the NTH bifurcation (8a)–(8c), (9) are summarized by

$$G(x, \tilde{x}, p, \alpha, r) = 0, \tag{10}$$

where $\tilde{x}$ denotes the auxiliary variables (e.g. w, v, u). Together with the eqns. (6) and (7) that define the minimal distance between an operating point $(x^{(0)}, \alpha^{(0)}, p^{(0)})$ and a critical point $(x^{(1)}, \alpha^{(1)}, p^{(1)})$, eqns. (10) represent the necessary constraints that must hold in order to guarantee robust stability for a large range of operating conditions. Thus the robust optimum of the process model (4) with respect to a cost function ϕ can be found by solving the following constrained nonlinear programm

$$\min_{x^{(0)}, \alpha^{(0)}, p^{(0)}} \quad \phi(x^{(0)}, \alpha^{(0)}, p^{(0)}) \tag{11a}$$

$$\text{s. t.} \quad 0 = f(x^{(0)}, \alpha^{(0)}, p^{(0)}), \tag{11b}$$

$$\begin{aligned} 0 &= G(x^{(1)}, \tilde{x}^{(1)}, \alpha^{(1)}, p^{(1)}, r), \\ 0 &= \alpha^{(1)} - \alpha^{(0)} + l \frac{r}{\|r\|}, \\ 0 &\leq l - \sqrt{n_\alpha}. \end{aligned} \tag{11c}$$

Eqns. (11b) ensure that the optimal operating point is a steady state of the process model. The constraints (11c) enforce the minimal distance of the optimum to the critical points. It is important to note that certain restrictions (Mönnigmann and Marquardt, 2003) must hold for the system (4) in order to apply the normal vector approach. In particular it has to be assumed that variations of the uncertain parameters are slow compared to the time scales of the process, i. e. all α may only vary quasistatically with respect to the system dynamics.

5. ILLUSTRATIVE EXAMPLES

In this section the theory described above will be illustrated by two exemplary integrated design and control problems associated with the CSTR. In the first example, the design problem is the maximization of the reactor yield. The control problem consists of finding a robust controller setting that not only stabilizes the optimal operating point but furthermore allows set point variations without loss of stability. The robustness against set point variations is crucial for a reactor with frequently varying demands on the product quality. The design problem in the second example addresses the minimization of the investment costs attributed to the heat transfer area for a fixed rate of production. The control design problem is again to find a robust controller setting for the optimal reactor design that allows for setpoint variations without loss of stability. The constrained optimization problem (11) is used for the solution of both integrated problems. From a technical point of view, (11) guarantees the robust stability for all values of the parameter T_{sp} by ensuring a specified distance to the manifold of NTH bifurcations. For both examples, results of the optimization without the robustness constraints (11c) are included for reference.

5.1 Optimization of the reactor yield

Here the CSTR is optimized with respect to the amount of B produced per unit time. The production of B is equivalent to the amount of converted A. The function to be maximized is therefore chosen to be

$$\phi = q\,(c_{Af} - c_A) \tag{12}$$

with a constant concentration $c_{Af} = 1.0\,\text{mol/l}$ of the feed. The objective function may be increased by increasing the feed rate q or decreasing c_A. A decrease of c_A can be achieved by setting the reactor temperature $T = T_{sp}$ to higher values. To ensure the thermal stability of the product it is therefore necessary to impose an upper bound on T_{sp}. Assuming that water is used as the cooling medium a lower bound has to be set on T_c. For the optimization problem these bounds represent additional constraints, e. g.

$$\begin{aligned} T_{\max} &\leq 400\text{K}, \\ T_{c,\min} &\geq 300\text{K}. \end{aligned} \tag{13}$$

The time constant ϵ_v of the unmodeled dynamics and the feed rate q are assumed to be the uncertain parameters ($n_\alpha = 2$). The uncertainties are set to

$$\begin{aligned} \Delta q &= 10\,\text{mol/min}, \\ \Delta \epsilon_v &= 0.01\text{min}, \end{aligned} \tag{14}$$

which correspond to 10% and 20% of the nominal values given in Table 1, respectively. While the nominal value of q may vary, the estimated nominal value of $\epsilon_v = 0.05\,\text{min}$ is assumed to be constant. T_{sp} and the controller parameter ϵ are known parameters ($n_p = 2$) of the optimization problem (11).

For reference, the results for the optimization without any normal vector constraints are shown in Fig. 4. For this optimization run the control parameter ϵ is fixed at the value $\epsilon = 0.25\,\text{min}$ corresponding to the NTH bifurcation found in the analysis given in Section 3. The figure shows that the operating point is on the wrong side of the robustness boundary. The optimal operating point itself is stable, but continuation analysis reveals that variations of T_{sp} result in unstable operating points between $352.0\,\text{K} < T_{sp} < 388.3\,\text{K}$.

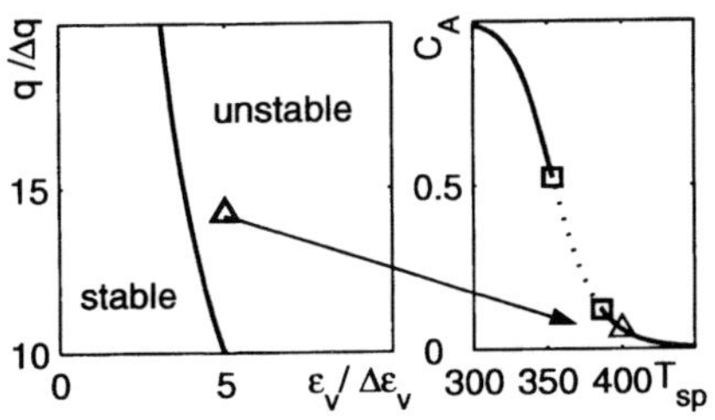

Fig. 4. Left: Nonrobust optimum (△) at maximum yield. Right: Continuation analysis reveals two Hopf bifurcations (□) and an unstable region (dotted).

To achieve the required stability over the entire range of T_{sp} an optimization including the normal vector constraints (11c) has to be carried out. Fig. 5 illustrates the obtained result. The robustness circle with the radius $r = \sqrt{n_\alpha} = \sqrt{2}$ around the optimal operating point touches the critical boundary. Variations of T_{sp} confirm that the setting of the control parameter found, $\epsilon = 2.28\,\text{min}$, guarantees stability for the entire region of operating conditions.

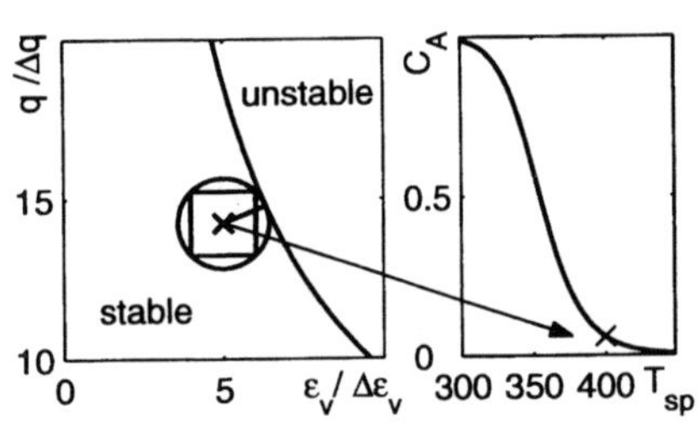

Fig. 5. Left: Robust optimum (×) at maximum reactor yield. Right: Continuation analysis shows stability for all T_{sp}.

The values of the objective function are the same for both optimizations. The two operating points only differ in the value of the controller parameter ϵ. The cost function (12) obviously defines the reactor parameters and the normal vector constraints (11c) define the controller parameter in the optimization problem. Comparing both stability boundaries it becomes obvious that the higher value of ϵ enlarges the region of stable parameter settings. Loosely speaking the normal vector constraints shift the critical boundary by increasing ϵ until the optimal operating point is found on the robust side of the boundary. The constraints guarantee that no NTH bifurcation occurs for any variations of the uncertain parameters within their specified uncertainties. The drawback for the stabilization of the entire range of operating points is the conservative setting of ϵ. If only robust stabilization of the optimal set point is required much tighter control becomes feasible.

The optimal robust parameter settings are summarized in Table 2. The temperature constraints (13) are both active at the optimum.

Table 2. Robust optimal operating point

Parameter	Value	Parameter	Value
q	$142.4\,\frac{1}{\text{min}}$	ϵ	2.28 min
T_c	300.0 K	$T_{sp} = T$	400.0 K
c_A	$0.06\,\frac{\text{mol}}{\text{l}}$	ϕ	$134.0\,\frac{\text{mol}}{\text{min}}$

5.2 Minimization of the heat transfer area

The aim of the optimization is to find an optimal reactor design that produces a specified amount of B with minimized investment cost and a controller setting that allows for variations of the set point temperature without loss of stability. The objective function is given by the investment cost attributed to the heat transfer area. According to Douglas (1990) they can be estimated to be

$$\phi = 2285 A^{0.65} [\$] \,, \tag{15}$$

with A in m^2. Here a constant heat transfer coefficient $U = 100\,\text{W/m}^2\text{K}$ is assumed. The value of (15) is only influenced by UA which becomes an additional parameter for the optimization problem in this example. However, it is not possible to set UA to arbitrary small values since the constraints on the reactor temperature and cooling water temperature (13) apply to this optimization problem as well. The desired production rate of B is arbitrarily set to be

$$q(c_{Af} - c_A) = 85\,\text{mol/min}\,. \tag{16}$$

UA is assumed to be uncertain. The uncertainty is set to

$$\Delta UA = 83.3\,\text{W/K}\,, \tag{17}$$

which corresponds to 10% of the nominal value given in Table 1. Together with the uncertainties Δq and $\Delta\epsilon_v$ specified in eq. (14), there are three uncertain parameters ($n_\alpha = 3$) and two known parameters (ϵ and T_{sp}, $n_p = 2$) in this optimization problem. The desired production rate (16) constitutes an additional equality constraint.

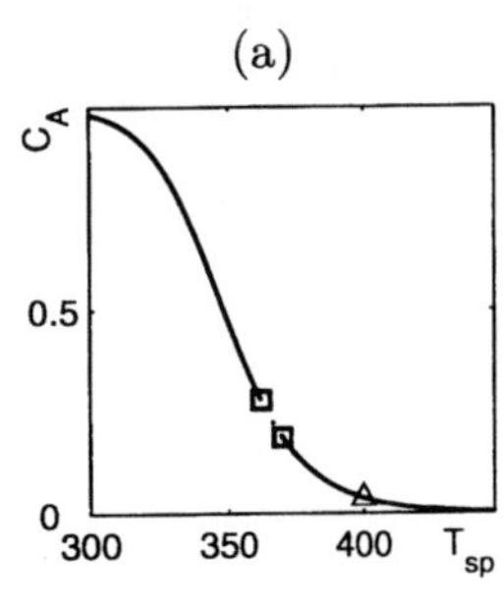

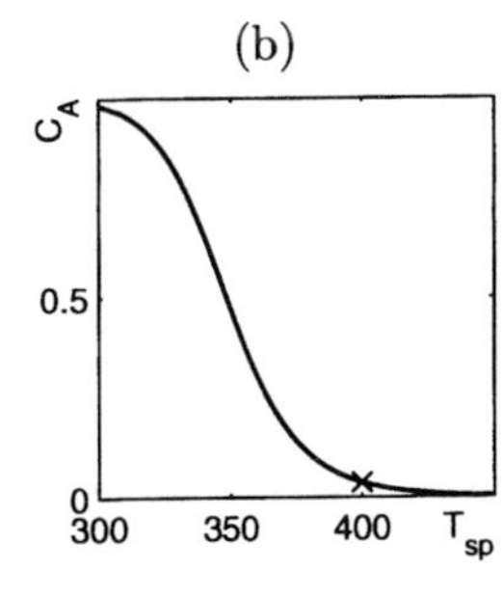

Fig. 6. (a) Continuation of the non robust optimum (△) with two Hopf bifurcations (□) and an unstable region (dotted). (b) Continuation of the robust optimum (×) yields set of stable operating points.

For reference, results without normal vector constraints are shown in Fig. 6a. The controller parameter is set to $\epsilon = 0.25\,\text{min}$. As in the previous example this optimization results in a stable operating point. Variations of T_{sp}, however, again give rise to Hopf bifurcations and a region of unstable operating points at $360.5\,\mathrm{K} < T_{sp} < 371.3\,\mathrm{K}$.

The required stability for all values of T_{sp} at the optimal reactor design is obtained by including the normal vector constraints in the optimization problem.

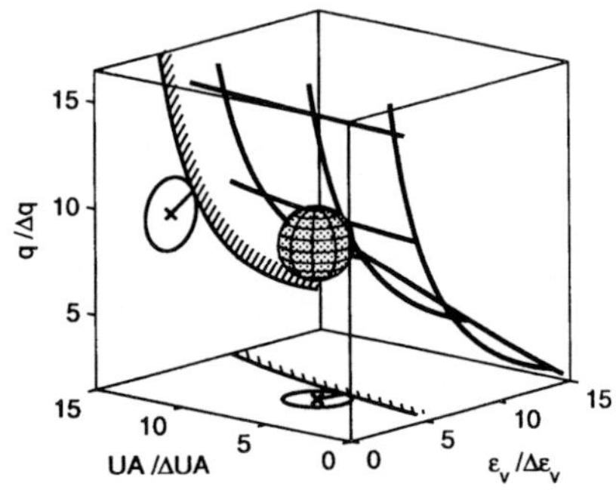

Fig. 7. Robust optimization with cost function (15); the operating point is surrounded by its robustness sphere; the critical manifold is indicated by thick lines; the optimal operation point, the robustness sphere and the critical manifold are projected onto the planes (ϵ_v, q) and (ϵ_v, UA).

Fig. 7 shows the robust operating point with the controller setting of $\epsilon = 0.63\,\mathrm{min}$ in relation to the critical boundary spanned by the uncertain parameters ϵ_v, q and UA. The robustness sphere with the radius $r = \sqrt{3}$ around the operating point touches the critical manifold. Thus it is guaranteed that no NTH bifurcation occurs within the specified range of uncertainties. As in the previous example, the value of the objective function (15) corresponds to the value obtained without the robustness constraints. Robust stability is achieved by increasing ϵ. The continuation analysis in Fig. 6b shows the desired stability over the entire range of T_{sp}. The results of the optimized operating point are summarized in Table 3. The determined reactor design admits a faster control in comparison with the control of the robust maximum of the reactor yield. Relaxing the robustness requirements to solely stabilizing the optimal operating point, however, would admit even tighter control.

Table 3. Robust optimal operating point

Parameter	Value	Parameter	Value
q	$88.3\,\frac{\mathrm{l}}{\mathrm{min}}$	c_A	$0.04\,\frac{\mathrm{mol}}{\mathrm{l}}$
$q(c_{Af} - c_A)$	$85\,\frac{\mathrm{mol}}{\mathrm{min}}$	ϵ	0.63 min
$T_{sp} = T$	400.0 K	T_c	300.0 K
UA	$532.5\,\frac{\mathrm{W}}{\mathrm{K}}$	ϕ	6776 \$

6. CONCLUSIONS

In this paper, an integrated method of robust design optimization and controller tuning for a temperature controlled CSTR is presented. The controller is realized by means of linearizing feedback. The integrated method guarantees stability of the optimal operating point for variations of uncertain parameters within specified bounds and for all variations of the set point. Robust stability is achieved by constraints that ensure a specified distance in the direction of the normal vector from the operating point to a critical boundary of NTH bifurcations. The critical boundary separates the parameter space into one region where set point variations always result in a stable operating point and another region where variations of the set point give rise to instabilities. For an extension of the method to constraints that guarantee not only stability but also a specified performance of the closed loop system the reader is referred to Mönnigmann and Marquardt (2002b). In this work critical manifolds of steady states are considered at which the real part of the leading eigenvalue attains a user specified value $\sigma_0 < 0$.

Clearly the presented integrated design method is neither restricted to the simple examples presented nor to linearizing feedback control. It can be applied to any process that exhibits a NTH bifurcation controlling the stability loss, if stability has to be guaranteed in the presence of uncertainties even in a large region of operating conditions.

REFERENCES

Douglas, J.M. (1988). *Conceptual Design of Chemical Processes.* 1994. 2nd ed.. McGraw-Hill. New York.

Hahn, J. M. Mönnigmann and W. Marquardt (2003). Robust tuning of feedback linearizing controllers via bifurcation analysis. In: *Proceedings of ADCHEM 2003, Hong Kong, China.* pp. 525–530.

Mönnigmann, M. and W. Marquardt (2002*a*). Normal vectors on manifolds of critical points for parametric robustness of equilibrium solutions of ODE systems. *Journal of Nonlinear Science* **12**, 85–112.

Mönnigmann, M. and W. Marquardt (2002*b*). Parametrically robust control-integrated design of nonlinear systems. In: *Proceedings of the American Control Conference 2002 (6), IEEE.* pp. 4321–4326.

Mönnigmann, M. and W. Marquardt (2003). Steady state process optimization with guaranteed robust stability and robust feasibility. *AIChE J.* **49**, 3110–3126.

Uppal, A. W.H. Ray and A.B. Poore (1974). On the dynamic behavior of continuous stirred tank reactors. *Chem. Eng. Sci.* **29**, 967–985.

ELSEVIER
IFAC
PUBLICATIONS
www.elsevier.com/locate/ifac

LEXICOGRAPHIC OPTIMIZATION BASED SENSOR NETWORK DESIGN FOR ROBUST FAULT DIAGNOSIS

Mani Bhushan * **Raghunathan Rengaswamy** **

* *Chemical and Materials Engineering Department, University of Alberta, Edmonton, AB, Canada*
** *Department of Chemical Engineering, Clarkson University, Potsdam, NY, USA*

Abstract: An appropriately designed sensor network is crucial for the success of any fault diagnostic strategy. In previous works (Bhushan and Rengaswamy, 2002a,b) strategies for optimally locating sensors based on criteria of reliability maximization and cost minimization have been devised. These opposing objectives were treated in a lexicographic (a special type of multiobjective optimization) manner. Signed digraph (SDG) based process models were used to generate the cause-effect information and fault occurrence and sensor failure probabilities were used to calculate a measure of reliability of a sensor network. While reliability and cost were considered in selecting the optimal sensor network, the robustness of the selected network with respect to uncertainties/errors in the underlying signed directed graph models and the available probability data was not considered explicitly. In this article, lexicographic formulations which incorporate some robustness enhancing criteria while designing cost-optimal sensor network for reliable fault diagnosis are presented. Some robustness to modeling errors in the SDG can be incorporated by choosing a distributed sensor network. Robustness to available probability data can be incorporated by maximizing reliability of the faults involving uncertain probability data. Integer linear programming (ILP) formulations incorporating these criteria in a lexicographic manner along with overall reliability-maximization and cost-minimization objectives are presented. The utility of the proposed approach is demonstrated through application to the Tennessee Eastman case study.

Keywords: Sensor, Fault Diagnosis, Multiobjective Optimization

1. INTRODUCTION

For safe and optimal operation of a chemical plant, it is essential to quickly detect and identify faults when they occur. Hence an efficient fault diagnosis methodology is very useful for modern day chemical plants. The efficiency of any diagnostic system depends critically on the location of the sensors monitoring important process variables. With hundreds of process variables available for measurement in a typical chemical plant, selection of crucial and optimum sensor positions is an important problem with possibly significant implications for various fault diagnosis strategies.

While the area of sensor location in general has been popular among researchers for quite sometime now, the area of sensor location from a fault diagnosis perspective has not received adequate attention. Even though related, there are significant differences in these two areas. The aim of sensor location problem as posed and solved in literature has been to identify variables to be measured such that various requirements on observability (ability to estimate values of all/important variables), estimation accuracy for key variables/parameters, and reliability (the probability of being able to estimate a variable) amongst others, are met, while the aim of sensor location from fault diagnostic perspective is to identify variables to be measured such that various faults when they occur can be detected, and diagnosed. Criteria such as resolution (ability to differentiate between different faults) achieved, cost and reliability of the sensor network, speed of detection and diagnosis, and false alarm rate are some of the criteria that can be incorporated in this approach.

In previous work, optimization formulations for sensor network design for reliable fault diagnosis have been presented (Bhushan and Rengaswamy, 2002*a*). These formulations incorporate various constraints on the sensor network design problem, as well as utilize quantitative information such as fault occurrence and sensor failure probabilities, and cost of sensors. In this article, formulations for incorporating robustness to the data (fault occurrence probability and sensor failure probability) used in the problem, are presented. Robustness to modeling errors/uncertainties is also considered. These formulations are presented in a lexicographic optimization framework. The application of the proposed formulations is demonstrated on the Tennessee Eastman case study.

2. RELATED PREVIOUS WORK

Bhushan and Rengaswamy (2002a) posed sensor location problem as a constrained lexicographic optimization problem based on the concept of unobservability minimization. The unobservability of a fault is defined as the probability of the fault occurring and remaining undetected and can be calculated as:

$$U_i = f_i \prod_{j=1}^{n} (s_j)^{(d_{ij} x_j)} \tag{1}$$

Here index j is for the variables which can be potentially measured, x_j is the number of sensors placed to measure variable j with probability of failure of each sensor being s_j, index i is for the faults with f_i being the probability of occurrence of fault i, U_i is the unobservability of fault i, constant $d_{ij} = 1$ if fault i affects variable j and is zero otherwise. Construction of matrix D (with elements d_{ij}) is based on fault simulation either in the process DG, the process SDG, or the process SDG with gains (Bhushan and Rengaswamy, 2002*b*). For the case study presented later in this article, SDG with gains is used to generate the D matrix.

A good sensor network should lead to low unobservability values for the faults. The measure of process unobservability chosen (Bhushan and Rengaswamy, 2002*a*) is the maximum unobservability among all faults. Apart from unobservability minimization, cost minimization was also considered, and these two objectives were combined in a lexicographic manner. The sensor location design problem was posed as an integer linear programming (ILP) problem as follows:

Formulation I: One-Step Optimization considering unobservability and cost

$$\min_{(x_j)} [U - \alpha\ x_s] \tag{2}$$

subject to,

$$U \geq \log U_i, \quad i = 1, ..., m \tag{3}$$

$$\sum_{j=1}^{n} c_j x_j + x_s = C^* \tag{4}$$

$$x_j \in \mathbb{Z}^+, \quad j = 1, ..., n \tag{5}$$

$$x_s \in \mathbb{R}^+ \tag{6}$$

In the above problem, the first term in the objective function U is the maximum unobservability among all faults (this is captured by constraints 3), C^* is the available cost (resources) for performing sensor location, and c_j is the cost of installing one sensor on variable j . The problem is linear in the decision variables (x_j) since instead of unobservability U_i (as given by equation 1) for a fault, the log of unobservability is being used. Hardware redundancy is incorporated in the above formulation by not restricting the decision variables to be binary, but rather allowing them to take any non-negative integer value. The variable x_s is the slack in the cost constraint, which takes non-zero real values. The higher the value of the variable x_s, lower is the cost used for sensor location. α is a positive constant which has to be chosen such that the primary objective (minimizing unobservability) still attains its earlier optimal value. For such a case, negative contribution of the second term $(\alpha\ x_s)$ in the objective function will ensure that among all solutions which yield minimum unobservability, the one which has highest x_s will be chosen. Thus, if the constant α is appropriately chosen, the solution to Formulation I will give a sensor network, which will have the least cost among all the networks which yield the minimum unobservability. A sufficient range for the value of α was given as: $0 < \alpha < 1/(aC^*)$, where a is a constant that can be calculated based on the given fault and sensor probabilities. The optimization problem as presented above is known as lexicographic optimization , where the idea is to arrange multiple objectives in a lexicographic order. This ordering means that the more important objective is infinitely more important than a less important objective (Miettinen, 1998). Sherali (1982) has presented an algorithm for selecting suitable weights for combining multiple objectives in a lexicographic manner. The selection of constant α as presented above by Bhushan and Rengaswamy (2002a) is similar to the approach used by Sherali (1982). Bhushan and Rengaswamy (2002a) have also pointed out that the sensor network design problem as presented above can be solved not only for the basic fault observability case, but also for other cases, such as single or multiple fault resolution. The basic idea is to convert any given scenario into a suitable fault-observability case by generating appropriate pseudo-faults. The same holds true for formulations to be presented next in this article. Also, similar to formulation I, all the formulations presented in the next section are ILP problems.

3. ROBUST SENSOR NETWORK DESIGN

The formulation I for sensor location requires fault occurrence and sensor failure probability data, and assumes that this data is accurately known. In practice, this may not always be the case. Also, it is possible that the underlying fault-models (SDG with gains for this article) are not accurate and have errors/uncertainties in them. In this section, some criteria to incorporate robustness to these factors will be considered.

3.1 Robustness to available probability data

For a given process, it is quite possible that some fault occurrence probabilities are accurately known (based on experience or past plant data), and for others only approximate values are available. Sensor failure probability data would be comparatively easier to obtain as there are several possible sources of such data, such as vendors, instrumentation handbooks, or lab/pilot scale experiments. However, similar to fault occurrence probability data, it is quite possible that depending on the source/reliability of information, values for some sensor failure probabilities are only approximately known (for example, the values calculated using lab experiments may not be exactly same as that obtained during the actual process operation). Hence it is desirable to incorporate some robustness to such data mismatch in the sensor location formulations. In this section, two scenarios are considered: (i) inaccuracies in occurrence probabilities of some faults (all sensor probabilities accurately known) and (ii) inaccuracies in failure probabilities of some sensors (all fault occurrence probabilities accurately known). In the rest of the article, to simplify the notation, a fault whose unobservability calculation involves inaccurately known data, will be referred to as "inaccurate" fault. A fault can be inaccurate for either of the two scenarios.

The central idea is again based on the concept of lexicographic optimization. The primary objective still is to minimize system unobservability (maximum unobservability amongst all faults) calculated using nominal values for fault occurrence and sensor failure probabilities. The secondary objective is to try to maximize the slack in the unobservability constraints (3) of inaccurate faults. The idea here is that since the system unobservability characterizes only the maximum unobservability amongst all faults, a process with two different sensor networks can have the same system unobservability with both the networks, and yet have different unobservabilities for some faults. Amongst these two networks, one would like to choose the network which yields lower unobservability for the inaccurate fault. For a given overall system unobservability, the higher the slack in the unobservability constraint of a fault, the lower is its individual unobservability. In general, there can be more than one inaccurate fault. The secondary objective is then to maximize the minimum slack amongst all the slack variables in the unobservability constraints of these faults. With optimization of these two objective functions (overall system unobservability minimization and slack maximization for inaccurate faults), it is still possible to obtain multiple optimal solutions with possibly different cost required for each solution. The third objective is then to minimize the cost used. The three objective functions are combined in a single-weighted objective function using appropriately computed weighting constants.

The formulations for the two scenarios are presented next. Even though the basic idea for the two scenarios is the same, there are some differences.

Formulation II: Robustness to Inaccurate Fault Occurrence Probability Data

$$\min_{(x_j)} [\lambda_1 U - \lambda_2 \phi - x_s] \tag{7}$$

subject to,

$$\sum_{j=1}^{n} c_j x_j + x_s = C^* \tag{8}$$

$$U \geq \log(U_i), \quad i = 1, ..., m_1 \tag{9}$$

$$U = \log(U_i) + \phi_i, \quad i = m_1 + 1, ..., m \tag{10}$$

$$\phi \leq \phi_i, \quad i = m_1 + 1, ..., m \tag{11}$$

$$x_j \in \mathbb{Z}^+, \quad (x_s, \phi_i, \phi) \in \mathbb{R}^+ \tag{12}$$

In the above formulation, ϕ_i is the (positive) slack variable of the unobservability constraint for the ith fault. This slack variable is added to the unobservability constraints of those faults (faults $m_1+1,...,m$) whose occurrence probabilities are not accurately known. The secondary objective is to then maximize the minimum slack (ϕ) amongst all the faults with inaccurately known probabilities. Constraints 11 and maximization of ϕ in the objective function ensures that ϕ is equal to the minimum value amongst all individual ϕ_i values. Positive constants λ_1 and λ_2 ensure that the solution to Problem II solves a lexicographic optimization problem with unobservability minimization, slack maximization and cost minimization as the objectives in decreasing order of preference. The values for λ_1 and λ_2 can be derived using the idea presented by Sherali (Sherali, 1982) and are:

$$\lambda_1 = (1+M)(1+C^*) \tag{13}$$

$$\lambda_2 = (1+C^*) \tag{14}$$

where the constant

$$M = \max_{(i=m_1+1,...,m)} (\phi_i^*) \tag{15}$$

The constants ϕ_i^* are defined in the next paragraph. For the sake of brevity, the procedure followed in deriving the above values is not discussed in this article.

Formulation II as presented has one small problem. It can lead to solutions where extra cost is spent in increasing ϕ to meaningless high values. To understand this, consider a process with one inaccurate fault with nominal fault occurrence probability of 10^{-2}. In the worst case, the actual occurrence probability of this fault will tend to $10^0 = 1$. Considering constraint 10 for this fault, it can be seen that if ϕ_i for this fault is equal to 2, then even in the worst case, the system unobservability will not be higher than the value calculated using nominal fault occurrence probability data. Hence, spending extra cost to increase ϕ above 2 for this case is not required. This is achieved by replacing constraints 11 by the following set of constraints

$$\phi \leq M \tag{16}$$

$$\phi \leq \phi_i + My_i, \quad i = m_1+1,...,m \tag{17}$$

$$Py_i \geq \phi_i - \phi_i^*, \quad i = m_1+1,...,m \tag{18}$$

$$P(y_i - 1) \leq \phi_i - \phi_i^*, \quad i = m_1+1,...,m \tag{19}$$

$$y_i \in \{0,1\}, \quad i = m_1+1,...,m \tag{20}$$

In the above constraints, P is a large positive constant (such as 100), y_i is a binary variable, and ϕ_i^* is the maximum meaningful value of the slack variable for the ith inaccurate fault (2 in the above example). The basic idea behind introducing variable y_i is that if for a fault i, the slack variable $\phi_i > \phi_i^*$, then the value ϕ is not restricted by value of ϕ_i, and increasing ϕ_i further will not increase ϕ. Constraints 18 and 19 ensure that $y_i = 1$ for this case, and this ensures that ϕ_i (constraint 17) does not limit the upper value of ϕ. For the case when $\phi_i < \phi_i^*$, constraints 18 and 19 force y_i to be 0, and this reduces constraint 17 for fault i to $\phi \leq \phi_i$ as it should be. For the case when $\phi_i = \phi_i^*$, constraints 18 and 19 allow y_i to be either 0 or 1, but maximization of ϕ in the objective function ensures that $y_i = 1$, if the maximum value of ϕ is being restricted by ϕ_i (constraint 17). Constraint 16 ensures that the maximum value for ϕ is not more than the maximum meaningful value among all the slack variables.

To summarize, the formulation that is proposed for incorporating robustness to fault occurrence probability data is: Formulation II with constraints 16-20 used in place of constraints 11. Formulation to incorporate robustness to sensor failure probability data is considered next.

Formulation III: Robustness to Inaccurate Sensor Failure Probability Data

$$\min_{(x_j)} [\lambda_1 U - \lambda_2 \phi - x_s] \tag{21}$$

subject to,

$$\sum_{j=1}^{n} c_j x_j + x_s = C^* \tag{22}$$

$$U = \log(U_i) + \phi_i, \quad i = 1,...,m \tag{23}$$

$$\phi_i^* = -\sum_{j \in S_u} d_{ij} (\log s_j) x_j, \quad i = 1,...,m \tag{24}$$

$$Py_i \geq \phi_i - \phi_i^*, \quad i = 1,...,m \tag{25}$$

$$P(y_i - 1) \leq \phi_i - \phi_i^*, \quad i = 1,...,m \tag{26}$$

$$\phi \leq \phi_i + My_i, \quad i = 1,...,m \tag{27}$$

$$\phi \leq M \tag{28}$$

$$x_j \in \mathbb{Z}^+, \quad (x_s, \phi_i, \phi) \in \mathbb{R}^+, y_i \in \{0,1\} \tag{29}$$

In the above formulation, S_u is the set of sensor with uncertain failure probabilities, constant M is the maximum meaningful value for ϕ for a given problem, and P is a large positive constant (for example, 100). The constant M is now given as

$$M = \max_i \sum_{j \in S_u} d_{ij} (\log s_j) x_j^* \tag{30}$$

where x_j^* is an upper bound on the maximum number of jth sensors that can be selected, and can be calculated as quotient of C^*/c_j. Constraint 24 is written for each fault, and it calculates the maximum meaningful value of slack required in a fault unobservability constraint, based on the chosen sensors. For example, suppose that the unobservability calculation of a fault does not depend on any sensor with inaccurate failure probability. Then no slack is required for this fault constraint. As can be seen from equation 24, ϕ_i^* for this fault will be 0. Constraints 25-28 are same as the constraints used in Formulation II.

Formulation II incorporated robustness to inaccurate fault occurrence probabilities (assuming all sensor failure probabilities were exactly known), and formulation III considered robustness to inaccurate sensor failure probabilities (assuming all fault occurrence probabilities were exactly known). A formulation where both types of uncertainties are considered can also be formulated as a combination of the two formulations. This formulation is not presented in this article.

3.2 Robustness to Modeling Errors

Formulations to incorporate robustness to inaccurate probability data were presented above. It can also be the case that the effects of faults on the measurable variables (matrix D with elements d_{ij} in our formulations) as used in the formulation are not accurate. This may happen since the process may not be exactly known and/or the simulation procedure to obtain the matrix D may make some restrictive assumptions (such as steady state gains for SDG with gain model). Also, since the effects of process/sensor noise, fault magnitude, and sampling times for sensors have not been included in the formulations, the effects of various faults predicted on the variables may not match exactly with those measured in the plant. The efficacy of a selected sensor network will then be affected by these unknown factors. In order to incorporate some robustness to such modeling errors, it is proposed that a distributed sensor network should be preferred, since for a given fault, even if effects on some variables are wrongly modeled, the chances of correctly diagnosing that fault would be higher since some other variables being effected by that fault are also being measured. Once again, this is done in a lexicographic framework, and the formulation that is proposed to achieve this is as follows:

Formulation IV: Lexicographic formulation for achieving a distributed sensor network when some fault occurrence probabilities are inaccurately known

$$\min_{(x_j)} [\alpha_1 U - \alpha_2 \phi - \alpha_3 x_s - N] \tag{31}$$

subject to

$$\sum_{j=1}^{n} c_j x_j + x_s = C^* \tag{32}$$

$$U = \log(U_i) + \phi_i, \quad i = 1, ..., m \tag{33}$$

$$\phi \leq M \tag{34}$$

$$\phi \leq \phi_i + M y_i, \quad i = m_1 + 1, ..., m \tag{35}$$

$$P y_i \geq \phi_i - \phi_i^*, \quad i = m_1 + 1, ..., m \tag{36}$$

$$P(y_i - 1) \leq \phi_i - \phi_i^*, \quad i = m_1 + 1, ..., m \tag{37}$$

$$n_j \leq x_j, \quad j = 1, ..., n \tag{38}$$

$$N = \sum_{j=1}^{n} n_j \tag{39}$$

$$N, x_j \in \mathbb{Z}^+, \quad j = 1, ..., n \tag{40}$$

$$x_s, \phi_i, \phi \in \mathbb{R}^+ \tag{41}$$

$$n_j \in \{0, 1\}, \quad j = 1, ..., n \tag{42}$$

$$y_i \in \{0, 1\}, \quad i = m_1 + 1, ..., m \tag{43}$$

In the above formulation, N is the total number of variables measured in the process (different from total number of sensors used since a variable may have more than one sensor), and characterizes the sensor distribution in the process. Variable n_j is a binary variable, which is 1 if variable j is measured (irrespective of the number of sensors used to measure variable j), and is 0 if variable j is not measured. This is ensured by constraints 38, 39 and maximization of N in the objective function. Constraints 33-37 are the same as those used in Formulation II. α_1,α_2 and α_3 are positive constants which ensure that solution to Problem IV is optimal in the lexicographic sense with unobservability minimization, slack maximization, cost minimization, and network distribution maximization being the four objectives in decreasing order of priority. These constants can be calculated using the algorithm by Sherali (1982) as:

$$\alpha_1 = (1 + N^*)(1 + C^*)(1 + M) \tag{44}$$

$$\alpha_2 = (1 + N^*)(1 + C^*) \tag{45}$$

$$\alpha_3 = (1 + N^*) \tag{46}$$

where N^* is an upper bound on the maximum number of different sensors that can be selected in the process for a given available cost C^*, and can be calculated based on the given data. Even though theoretically sound, depending on the values of constants N^*, C^*, M, in the case study section, for some cases scaling problems were encountered, which were handled by appropriately scaling some variables (such as N). Alternatively, for this case, one can obtain the lexicographic solution by solving more than one problem in sequence (Sherali, 1982).

Formulation IV is posed assuming that the some fault probability data is inaccurately known. Incase, some sensor failure probabilities are not accurately known, a formulation similar to Formulation IV can be posed, with constraints borrowed from Formulation III. This formulation is referred to as formulation V in the case study section.

4. CASE STUDY: TENNESSEE EASTMAN PROCESS

To illustrate the utility of the ideas presented in this article, the Tennessee Eastman (TE) process (Downs and Vogel, 1993) is used as a case study.

Table 1. Results for inaccurate fault occurrence probability data

Form.	C^*	Cost Used	U	ϕ	N	Sensors Selected
I	2900	2100	-5	0	5	2,8,9,13,58
	5400	4200	-8	0	5	2(2),8(2),9(2) 13(2),55(2)
II	2900	2200	-5	2	6	1,2,8,9,13,58
	5400	4300	-8	2	8	1,2(2),5,7,8, 9,13(2),57(2)
IV	2900	2200	-5	2	6	1,2,8,9,13,54
	5400	4300	-8	2	9	1,2(2),5,7,8, 9,13(2),57,58

Based on a reduced model of this process proposed by (Ricker and Lee, 1995), Bhushan and Rengaswamy (2002b) developed a signed digraph with gains for this process, and used it to generate the cause-effect information required for sensor network design. This cause-effect information is used in this section to design sensor networks for problems posed in this article. In this process, 61 potential measurements and 15 faults are considered. These possible locations of sensors, sensor failure probabilities, sensor costs, faults considered and probabilities of their occurrences and the diagraph based cause-effect model used, are taken from (Bhushan and Rengaswamy, 2002*b*) and are not presented here. All the results presented in this section are only for the single fault resolution (Bhushan and Rengaswamy, 2002*a*) case and were obtained by solving the formulations in the integer linear programming package LINDO.

Case I Robust design in the presence of inaccuracies in fault occurrence probability data: It is now assumed that the occurrence probabilities for faults $F14$ and $F15$ (with nominal values of 10^{-2} and 10^{-1} respectively) are not accurately known. Sensor location is performed for different cases and the some typical results are summarized in Table 1. Comparing the results of formulation I and II for available cost $C^* = 2900$ units, it is seen that for formulation II (where slack maximization for inaccurate faults is considered), the selected sensor network ensures that even if the true probability of the inaccurately known faults ($F14, F15$) were to be higher by 2 orders of magnitude (than their nominal values used in the optimization), the overall system unobservability will not change. The same cannot be said about the sensor location obtained with Formulation I. Comparison of results for $C^* = 5400$ for formulations II and IV indicate that the network is more distributed for formulation IV than for II, thereby incorporating robustness to some amount of modeling error as compared to Formulation II.

Case II Robust design in the presence of inaccuracies in sensor failure probability data: Results for cases when failure probabilities of sensors $S13$ and $S60$ (with nominal values of 10^{-3} each) are inaccurately known are presented in Table 2. Conclusions regarding utility of various formulations similar to Table 1 can be derived based on this Table also.

Table 2. Results for inaccurate sensor failure probability data

Form.	C^*	Cost Used	U	ϕ	N	Sensors Selected
III	2900	2700	-5	6	5	2,8,9,13(3),55
	5400	5400	-8	12	5	2(2),8(2),9(2), 13(6),58(2)
V	2900	2700	-5	6	5	2,8,9,13(3),57
	5400	5400	-8	12	8	2(2),5,7,8, 9,13(6),55,57

5. CONCLUSIONS

A solution to the problem of designing robust sensor networks for fault diagnosis has been attempted in this article. Depending on the requirements of the fault monitoring system, appropriate optimization problems have been posed. The proposed formulations incorporate various criteria, such as reliability (in terms of unobservability), cost, and robustness to fault/sensor probability data and process-model mismatch, in a lexicographic optimization framework. In the case study section, the results of the application of the various formulations to the Tennessee Eastman process are compared. These demonstrate the utility of the various optimization objectives presented in this article.

REFERENCES

Bhushan, M. and R. Rengaswamy (2002*a*). Comprehensive design of a sensor network for chemical plants based on various diagnosability and reliability criteria: I. framework. *Ind. Eng. Chem. Res.* **41**, 1826–1839.

Bhushan, M. and R. Rengaswamy (2002*b*). Comprehensive design of a sensor network for chemical plants based on various diagnosability and reliability criteria: II. application. *Ind. Eng. Chem. Res.* **41**, 1840–1860.

Downs, J. J. and E. F. Vogel (1993). A plant-wide industrial process control problem. *Comput. Chem. Engng.* **17**(3), 245.

Miettinen, K. M. (1998). *Nonlinear Multiobjective Optimization.* Kluwer Academic Publishers: Boston.

Ricker, N. L. and J. H. Lee (1995). Nonlinear modeling and state estimation for the tennessee eastman challenge process. *Comput. Chem. Engng.* **19**(9), 983.

Sherali, H. D. (1982). Equivalent weights for lexicographic multi-objective programs: Characterizations and computations. *European Journal of Operational Research* **11**, 367.

ELSEVIER
IFAC
PUBLICATIONS
www.elsevier.com/locate/ifac

PROCESS DESIGN FOR REDUCED DISTURBANCE SENSITIVITY OF INTEGRATED PLANTS

Hong Cui Carlemalm and Elling W. Jacobsen [1]

S3-Automatic Control, Royal Institute of Technology - KTH
S-100 44 Stockholm, Sweden

Abstract: The disturbance sensitivity of an integrated plant is to a large extent caused by interactions between the process units, imposed by material and energy recycle flows. In this paper we show how a plug flow (delay) tank integrated in the recycle path can be used to modify these interactions such as to reduce the disturbance sensitivity in a given frequency range. The principle idea is to modify the physical feedback present in the process such that it provides disturbance damping, rather than amplification, in the frequency range where feedback controllers can not be made effective. As we show, the required capacity for a given level of disturbance attenuation with the proposed method can be severalfolds smaller than that required by a traditional cascaded buffer system. The results are illustrated by application to a reactor-separator system with recycle of unconverted material. *Copyright © 2004 IFAC*

1. INTRODUCTION

Storage capacities, in the form of buffer and surge tanks, are extensively used in plants in the process industry. The purpose is partly to facilitate operatios, such as startup, and partly to improve control performance by damping disturbances. Due to both economic considerations and safety aspects, it is however desirable to keep the use of buffer tanks at a minimum.

For attenuation of disturbances, a control system is usually the most effective solution. However, all plants contain inherent properties that limit the achieveable control performance. If these limitations are in conflict with the performance requirements, then acceptable performance can only be achieved by modifying the process design. In this case, adding buffer or surge tanks to dampen disturbances often offers the least expensive solution. To keep the buffer size at a minimum, the buffer system should be designed so that it primarily act as a complement to the control system (Faanes and Skogestad, 2003).

Traditional buffers are typically cascaded next to some process unit with the aim of damping the disturbance magnitude, i.e., acting as low-pass filters. However, the disturbance sensitivity of integrated plants is to a large extent determined by interactions between the various process units. Thus, rather than using a capacity to directly dampen disturbances, it may be more efficient to utilize it to modify the disturbance amplification provided by the process unit interactions. To achieve this, it is necessary to first understand how the interactions affect the disturbance sensitivity, and how the interactions should be modified in order to reduce the disturbance sensitivity.

We start the paper with a brief introduction to controllability analysis, since this forms a basis for designing buffers that complement control systems. We then use linear systems theory to analyze the impact of process unit interactions, in plants with recycle flows, on the overall plant disturbance sensitivity. Based on these results, we show how a plug flow tank integrated in a recycle path can be systematically designed to achieve a desired reduction in the plant disturbance sensitivity. We compare the results with those obtained with traditional cascaded buffer tanks as considered in Faanes and Skogestad (2003) and Mahajanam and Zheng (2002).

2. INPUT-OUTPUT CONTROLLABILITY

Consider a linear dynamic model of a process

$$y(s) = G(s)u(s) + G_d(s)d(s) \tag{1}$$

where y is the output to be controlled, u is the control input and d is the disturbance, all in deviations from their nominal steady-state values. We

[1] jacobsen@s3.kth.se

assume all signals to be scalar, but extensions to multivariable systems is straightforward. We assume the variables are scaled such that acceptable performance corresponds to all variables having magnitude less than 1.

Consider now a frequency response analysis, and define ω_d as the frequency where the scaled disturbance gain is unity, i.e.

$$|G_d(j\omega_d)| = 1 \tag{2}$$

Assume that this frequency is unique. For $\omega < \omega_d$, we then have $|y|/|d| > 1$ and hence the disturbance sensitivity needs to be reduced in the frequency range $\omega \in [0, \omega_d]$. This can be achieved either through feedback control, a modification of the process design, or a combination of the two. Usually, feedback control is the least expensive solution and thus a modification of the process design should only be considered when acceptable performance can not be achieved through feedback control alone.

Applying the feedback control law $u(s) = -C(s)y(s)$, the system can be described as

$$y(s) = S(s)G_d(s)d(s) \tag{3}$$

where $S(s) = 1/(1 + G(s)C(s))$ is the closed-loop sensitivity function. Acceptable disturbance sensitivity is obtained when $|SG_d(j\omega)| < 1$, $\forall\omega$. Define the bandwidth ω_B as the frequency for which $|S(j\omega_B)| = 1$. Then, for acceptable disturbance rejection, we get the bandwidth requirement $\omega_B > \omega_d$, i.e. the control must be effective at least up to the frequency ω_d.

A plant always has fundamental limitations which restrict the highest bandwidth ω_B that the feedback control system can achieve, even with the best possible controller. Fundamental limitations come from the process itself, e.g. in the form of time delays, unstable zero dynamics and input constraints. If the achieveable ω_B is smaller than ω_d, acceptable disturbance sensitivity can not be achieved using feedback control alone, and some modification of the process design is required in order to either increase the attainable bandwidth ω_B, or reduce the disturbance sensitivity of the process in the frequency range $\omega \in [\omega_B, \omega_d]$. In this paper we consider the latter option, and base our design modification on addition of capacities, or tanks, to the process.

3. EXAMPLE PROCESS: REACTOR-SEPARATOR PLANT

Chemical reaction followed by separation and recycle of unconverted reactant is a common process arrangement in the process industry. Figure 1 shows the simple reactor-separator system we consider here. The problem considered is that of

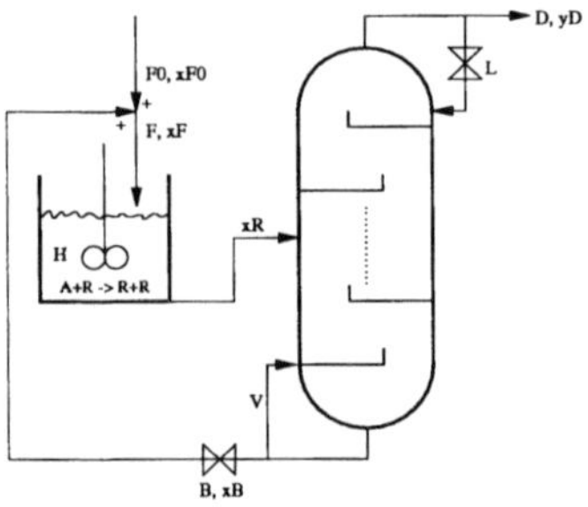

Figure 1. Reactor-separator system.

keeping the product composition y_D within prespecified bounds in the presence of feed composition disturbances.

The individual model for the reactor is

$$dx_R = \underbrace{\frac{0.42}{2s+1}}_{G_r}(dx_{F0} + dx_B) \tag{4}$$

where x_R is the fraction of the reaction product in the reactor, and x_{F0} and x_B are the corresponding fractions in the two feed flows F_0 and B, respectively. By "individual" we here imply that the input x_B is assumed to be an independent variable when deriving the model, while in the overall plant x_B is determined by the separation unit.

For the separator we employ the individual model

$$dy_D = \underbrace{\frac{0.09}{10s+1}}_{G_{c1}} dx_R; \qquad dx_B = \underbrace{\frac{2}{10s+1}}_{G_{c2}} dx_R \tag{5}$$

where y_D and x_B are the product fractions in the two product flows of the separator.

The recycle flow is assumed to be equal to the fresh feed flow, i.e. $B = F_0$. Combining the two models (4) and (5), i.e. by eliminating x_R and x_B, we obtain the overall transfer-function from the disturbance x_{F0} to the product composition y_D

$$y_D(s) = \frac{0.236}{(73s+1)(1.7s+1)} x_{F0}(s) \tag{6}$$

We assume that the feed composition x_{F0} at most can change by 20% and that acceptable deviations in the product composition y_D is ± 0.005. This implies that the disturbance model (6) should be scaled by a factor 40, such that the scaled disturbance model becomes

$$G_d(s) = \frac{9.45}{(73s+1)(1.7s+1)} \tag{7}$$

Figure 2 shows the frequency response of G_d, from which we find that the disturbance sensitivity exceeds 1, and hence needs to be reduced, up to $\omega_d = 0.125$. Using feedback control only, this requires a corresponding bandwidth $\omega_B = \omega_d$, or a closed-loop time-constant of approximately $1/\omega_B = 8$ *min*. If the achievable bandwidth is less than this, then the process design must be modified to reduce the disturbance sensitivity in the range $\omega \in [\omega_B, \omega_d]$.

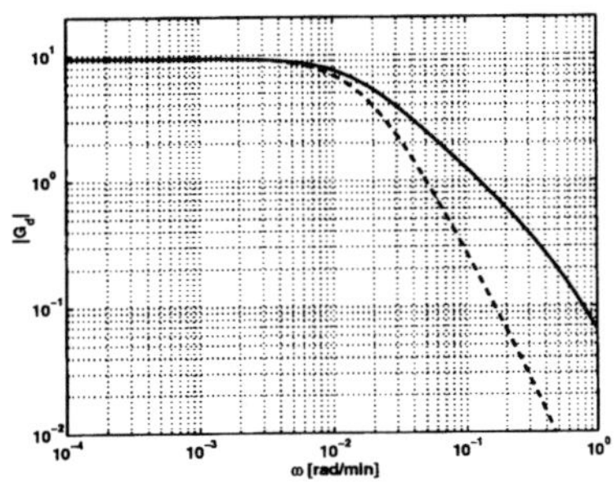

Figure 2. Reactor-separator system. Frequency response of distillate composition y_D to the disturbance feed composition x_{F0}. Dashed line shows response after addition of cascaded buffer with residence time $\tau_B = 45.4\ min$.

Assume that measurement delays and actuator dynamics prevents a bandwidth above $\omega_B = 0.05$, corresponding to a closed-loop time-constant of approximately 20 min. The process disturbance sensitivity must then be reduced by a factor $k_d = 2.5$ at $\omega = 0.05$. The required size of a cascaded mixed buffer system that achieves this can be calculated using the results in Faanes and Skogestad (2003), from which we find a single tank with residence time $\tau_B = 45.4\ min$. The resulting modified disturbance sensitivity is shown in Figure 2.

We next consider the potential improvement obtained by integrating the buffer, so as to modify the process unit interactions.

4. DISTURBANCE SENSITIVITY IN PLANTS WITH RECYCLE

A block diagram representation of a typical recycle process is shown in Figure 3. Note that we for simplicity assume all variables to be scalar and that the disturbances act on the output via variables in the recycle loop only. To relax these assumptions, see Carlemalm (2003). Here $G_f(s)$

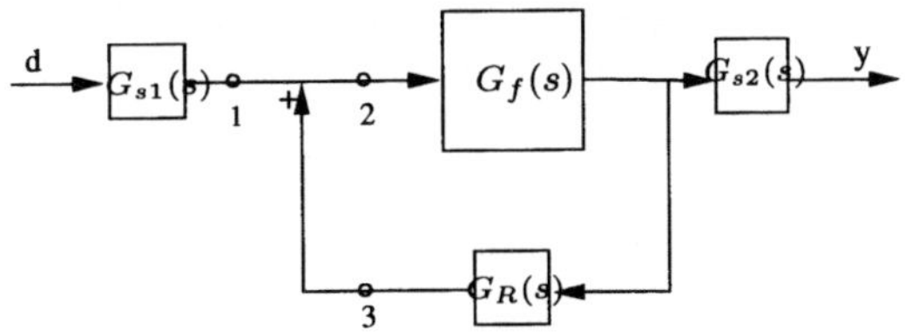

Figure 3. Block diagram representation of a typical recycle system. The numbers 1, 2 and 3 denote possible locations for a buffer tank.

is the transfer function of the forward path of the recycle loop, $G_r(s)$ represents the recycle path transfer function, and $G_{s1}(s)$, $G_{s2}(s)$ are transfer functions representing process units placed outside the recycle loop. For instance, for the reactor-separator in Figure 1, with x_{F0} as input and y_D as output, the reactor is in the forward loop, the separator bottoms in the recycle path, while the effect of reactor composition x_R on the top of the column y_D is external to the loop. Note that we here include variable scaling, as discussed above, in $G_{s1}(s)$ and $G_{s2}(s)$, such that $|d| \leq 1$, $|y| \leq 1$ correspond to acceptable performance. Also note that all transfer-functions in Figure 3 are derived for the unit operations individually, i.e. with the recycle flow replaced by an equivalent fixed flow.

Introducing

$$G_0 = G_f G_R \ ; \quad G_k = \frac{G_{s1}G_{s2}}{G_R} \tag{8}$$

the overall scaled disturbance sensitivity becomes

$$G_d(s) = \frac{y(s)}{d(s)} = \frac{G_k(s)G_0(s)}{1 - G_0(s)} \tag{9}$$

Based on (9) we are in a position to divide the overall disturbance sensitivity into one part caused by the individual units, i.e. the forward part $G_k(s)G_0(s)$, and one part caused by the unit interactions, i.e. the feedback part

$$S_p(s) = \frac{1}{1 - G_0(s)} \tag{10}$$

The function S_p, which describes the feedback amplification of disturbances, is known as the *sensitivity function* in feedback control theory, and we here correspondingly label it the *process sensitivity function* to emphasize that it describes the relative change in disturbance sensitivity due to the physical feedback present in the process itself.

In feedback control systems, the sensitivity is typically made small at low frequencies by making the loop-gain $|G_0|$ large and the phase $\angle G_0$ close to $-180°$ (negative feedback). However, a typical process unit has a positive steady-state gain, and hence $\angle G_0(0) = 0$. If we also assume stability of the individual process units as well as of the integrated plant, then it follows from Bodes stability theorem that $G_0(0) < 1$. With these assumptions we get from (10) that

$$S_p(0) > 1 \tag{11}$$

Thus, for most processes the physical feedback imposed by recycling will serve to increase the steady-state disturbance sensitivity. This is a well known fact, see e.g. (Gilliland *et al.*, 1964).

However, for higher frequencies the gain $|G_0(j\omega)|$ and phase lag $\angle G_0(j\omega)$ of the process units will change and the hence the feedback properties will also change with frequency. For feedback systems which are open-loop stable, i.e. individual process units stable, and for which there are at least two more poles than zeros in $G_0(s)$, the Bode Sensitivity Integral applies

$$\int_0^\infty ln|S_p(j\omega)|d\omega = 0 \tag{12}$$

See e.g. (Skogestad and Postlethwaite, 1996). According to (12), if $|S_p| > 1$ for some frequencies, then we must have $|S_p| < 1$ for some other frequencies. Thus, while recycling serves to increase the disturbance sensitivity at some frequencies, it will effectively reduce the disturbance sensitivity at other frequencies.

Figure 4 shows the process sensitivity function

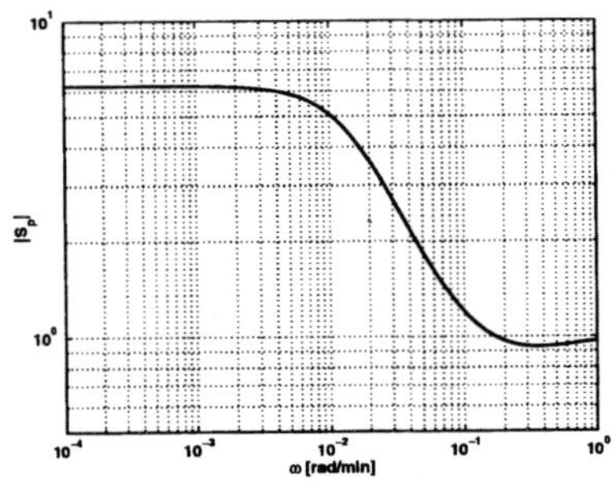

Figure 4. Process sensitivity function S_p as a function of frequency for reactor-separator.

$|S_p|$ for the reactor-separator process. As can be seen, the recycle serves to increase the disturbance sensitivity by a factor more than 6 at low frequencies, while it serves to slightly attenuate disturbances for frequencies above 0.15 rad/s.

Thus, we can conclude that while recycling typically provides disturbances amplification at low frequencies, it will always provide disturbance attenuation in some frequency range. This suggests that a design modification aimed at reduced disturbance sensitivity should, at least partially, aim at modifying the properties of the loop units such that the recycling provides disturbance attenuation, rather than amplification, in the frequency range where feedback controllers can not be made effective, i.e. around the bandwidth ω_B.

To better understand the conditions for when the feedback imposed by recycling will provide disturbance damping, consider Figure 5. According

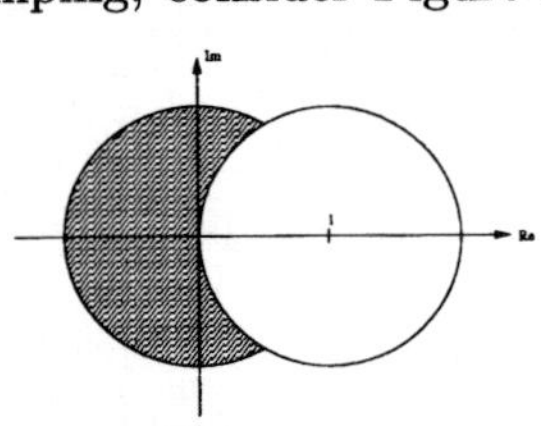

Figure 5. Region (scratched) of complex plane where the frequency response of the recycle loop $G_0(j\omega)$ should be for recycle feedback to provide disturbance damping.

to (10) the feedback will provide sensitivity reduction when $|1 - G_0(j\omega)| > 1$, i.e. when the distance from the point 1 to the frequency response $G_0(j\omega)$ in the complex plane exceeds unity. Thus, $|S_p| < 1$ when $G_0(j\omega)$ is outside a circle with centre at 1 and radius 1. See Figure 5. If we also assume that $|G_0(j\omega)| < 1 \; \forall\omega$, corresponding to assuming that the process units has low-pass properties and that recycling does not affect stability, which applies to most process units, then we get that the feedback imposed by recycling will provide disturbance damping when the frequency response $G_0(j\omega)$ is within the scratched region of Figure 5.

The polar form of the open loop frequency response is

$$G_0(j\omega) = re^{j\theta} \tag{13}$$

where $r = |G_0(j\omega)|$ and $\theta = \angle G_0(j\omega)$. Then, the region outside the white circle in Figure 5, i.e. for which $|1 - G_0| > 1$ corresponding to disturbance damping from recycling, can be shown to correspond to

$$cos(\theta) < \frac{r}{2} \tag{14}$$

With the additional condition $0 < r < 1$, a sufficient condition for $|S_p| < 1$ is that $\theta \in [-3\pi/2, -\pi/2]$ while $\theta \in [-\pi/3, \pi/3]$ is a sufficient condition for $|S_p| > 1$. For other values of θ the conclusion will depend on the size of r. The minimum sensitivity is achieved for $\theta = -\pi$ and $r = 1$ for which $|S_p| = 0.5$.

By integrating a buffer in the recycle loop, we will modify both the loop-gain r as well as the phase-lag θ and can therefore expect to significantly modify the feedback properties. In particular, from the above discussion we see that the phase lag of the loop is crucial to obtain reduced disturbance sensitivity.

4.1 *Integrated Mixed Buffer Tank*

We consider adding a mixed buffer with transfer-function

$$G_B(s) = \frac{1}{\tau_B s + 1} \tag{15}$$

inside the recycle loop of a process system, with the aim of modifying the recycle feedback properties. We consider placement in the recycle path only. The disturbance sensitivity of the modified process becomes

$$G_{d1}(s) = \frac{y(s)}{d(s)} = \frac{G_k(s)G_0(s)}{1 - G_0(s)G_B(s)} \tag{16}$$

where the residence time $\tau_B = V/q_3$ and q_3 is the recycle flow, i.e. the flow at position 3 in Fig. 3.

The aim of the buffer design is to determine τ_B such that

$$|G_{d1}(\tau_B, j\omega_B)| = 1 \tag{17}$$

On polar form

$$G_0(j\omega_B) = re^{j\theta} \; ; \quad \frac{1}{G_B}(j\omega_B) = \tau_B\omega_B i + 1 \tag{18}$$

The required buffer residence time can now be determined by inserting the frequency responses into (16) and solving for $|G_{d1}(j\omega_B)| = 1$. This gives the required residence time of the buffer as

$$\tau_B = \frac{r sin\theta \pm \sqrt{r^2 sin^2\theta + (k_d^2 - 1)(1 - k^2r^2)(kr/k_d)^2}}{(1 - k^2r^2)\,\omega_B} \quad (19)$$

where $k = |G_k(j\omega_B)|$.

Note that all the values of τ_B, as computed from (19), may be physically unrealizable, being either complex or negative, in which case the required buffering effect can not be achieved with a mixed buffer placed in the recycle path. When there are two positive real roots τ_{B1} and τ_{B2} of (19), the smallest value gives the optimal residence time.

Example revisited: for the reactor-separator problem we have

$$G_0(s) = G_r(s)G_{c2}(s)\ ; \quad G_k(s) = 40\frac{G_{c1}(s)}{G_{c2}(s)} \quad (20)$$

From (19) we find that the smallest frequency for which the integrated buffer can reduce the disturbance sensitivity $|G_{d1}(j\omega)|$ to 1 is $\omega = 0.09$, with $k_d = 1.4$. For this bandwidth an integrated buffer with residence time $\tau_B = 8.3\ min$ is sufficient, while a cascaded buffer cascaded requires $\tau_B = 10.9\ min$. If we assume that the control system bandwidth is limited to $\omega_B = 0.05$, then an integrated buffer is not sufficient. However, Figure 6 shows the overall disturbance sensitivity as a function of the buffer residence time at $\omega = 0.05$, with an integrated and cascaded buffer, respectively. As can be seen, the integrated buffer is significantly more effective in reducing the disturbance sensitivity at low τ_B, but is not able to reduce the disturbance sensitivity below 1.25. and in fact increases the sensitivity if τ_B is increased further. This latter effect is explained by the fact that increasing the buffer size further will start to increase the process sensitivity function $|S_p|$ due to phase-lag effects.

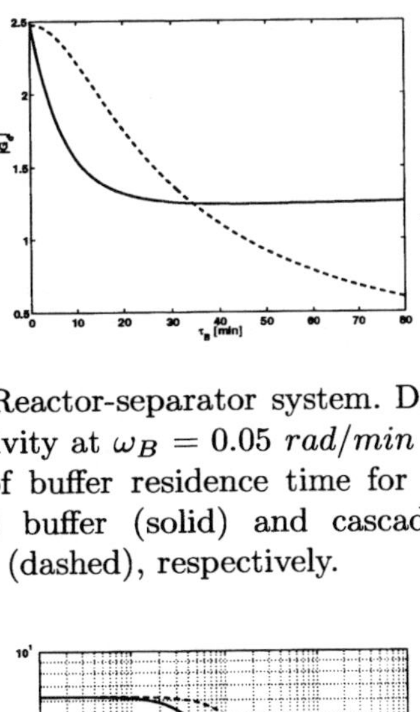

Figure 6. Reactor-separator system. Disturbance sensitivity at $\omega_B = 0.05\ rad/min$ as a function of buffer residence time for integrated mixed buffer (solid) and cascaded mixed buffer (dashed), respectively.

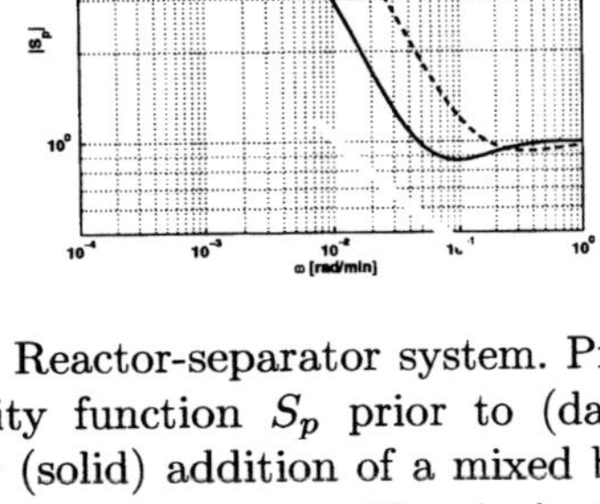

Figure 7. Reactor-separator system. Process sensitivity function S_p prior to (dashed) and after (solid) addition of a mixed buffer with residence time $\tau_B = 20\ min$ in the recycle path.

The effect of an integrated buffer on the process sensitivity is illustrated in Figure 7, showing the process sensitivity of the reactor-separator process before and after addition of a buffer with residence time $\tau_B = 20$ min. As can be seen, the buffer significantly reduces the sensitivity in the frequency range $\omega \in [0.002, 0.2]$, while it increases the sensitivity slightly for $\omega > 0.2$.

4.2 *Integrated Delay Tank*

According to (10), the effect of process unit interactions on the disturbance sensitivity is reduced by increasing $|1 - G_0G_B(j\omega_B)|$. The geometrical interpretation is that the buffer should be used to increase the distance of the loop-gain $G_0G_B(j\omega_B)$ from the point 1 in the complex plane. See Fig. 5.

Consider the integrated buffer that minimizes $|S_p|$ at ω_B. On polar form

$$G_0(j\omega_B) = re^{j\theta}\ ; \quad G_B(j\omega_B) = r_Be^{j\theta_B} \quad (21)$$

As before we assume $r < 1$ and $r_B \leq 1$. Then, the maximum of $|1 - rr_Be^{j(\theta+\theta_B)}|$ is achieved for $r_B = 1$ and $\theta_B = -\theta - (2m+1)\pi$, with m an integer, resulting in

$$\min_{G_B} |S_p(j\omega_B)| = \frac{1}{1+r} \quad (22)$$

Thus, the buffer $G_B(s)$ which yields the minimum process sensitivity function has unit magnitude and a non-zero phase-lag. This corresponds to a pure delay

$$G_B(s) = e^{-\tau_B s} \quad (23)$$

A delay process can be realized by a plug-flow tank with residence time τ_B.

The optimal solution to the buffer design problem, i.e. to make the disturbance sensitivity $|G_{d1}(j\omega_B)| = 1$ for the smallest total buffer volume, does not necessarily correspond to making $|S_p(j\omega_B)|$ minimal. However, we find that for most values of $G_0(j\omega_B)$, the distance $|1 - G_0G_B(j\omega_B)|$ is indeed maximized for a given τ_B by employing $G_B(s) = e^{-\tau_B s}$, i.e. a pure delay.

By integrating a delay $G_B = e^{-\tau_B s}$ in the recycle path, the transfer-function for the modified disturbance sensitivity becomes

$$G_{d2} = \frac{G_k G_0}{1 - G_0 e^{-\tau_B s}} \quad (24)$$

and the frequency response at $\omega = \omega_B$

$$|G_{d2}(j\omega_B)| = \frac{kr}{|1 - re^{j(\theta - \tau_B \omega_B)}|} \quad (25)$$

where r and θ are defined by (21) and $k = |G_k(j\omega_B)|$. The requirement $|G_{d2}(j\omega_B)| = 1$ then yields

$$\tau_B = (\theta + \underbrace{cos^{-1}(\frac{1 + (1-k^2)r^2}{2r})}_{E_1})/\omega_B \quad (26)$$

Note that the addition of a delay will not always be sufficient to reduce the overall disturbance sensitivity $|G_{d2}|$ to 1 at ω_B. This will occur when the minimum process sensitivity function $|S_p| = 1/(1+r)$ is not sufficient. in which case (26) does not have a solution.

Reactor-separator: With $\omega_B = 0.05$ we find from (26) that a delay tank with residence time $\tau_B = 23.5$ *min* will reduce the overall disturbance sensitivity to 1 at ω_B. The resulting disturbance and process sensitivity are shown in Figure 8. As can be seen, the delay reduces the process sensitivity function at ω_B from about 2 to 0.7, i.e., almost by a factor 3. Note that the delay introduces a resonance in the system such that the sensitivity is increased at higher frequencies. However, the total disturbance sensitivity $|G_{d2}|$ stays less than 1 for all frequencies above the bandwidth. The required delay tank size is almost half of the optimal mixed cascaded buffer which requires a residence time of $\tau_B = 45.4$ *min*. The effective filtering effect

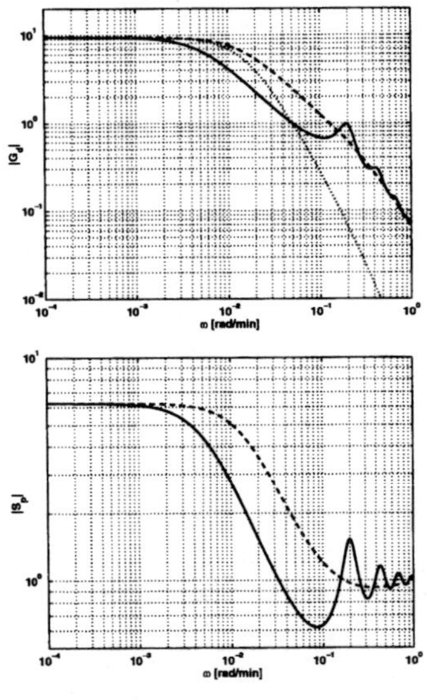

Figure 8. Reactor-separator system with integrated delay tank, $\tau_B = 23.5$ *min*. Top: disturbance sensitivity G_d. Bottom: process sensitivity function S_p. Solid line: delay buffer. Dashed line: original process. dotted line: cascaded buffer with $\tau_B = 45.4$ *min*.

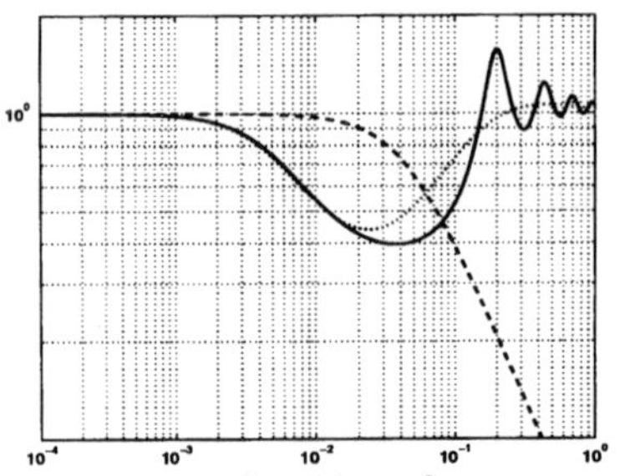

Figure 9. Reactor-separator system. Effective disturbance filtering of integrated delay (solid), cascaded (dashed) and integrated (dotted) mixed tank, $\tau_B = 23.5$ *min*.

obtained by an integrated delay with residence time $\tau_B = 23.5$ and a corresponding mixed buffer cascaded and integrated, respectively, is shown in Figure 9. As can be seen, the filtering effect of the delay tank is more than twice of the cascaded mixed tank, and about 30% more than the integrated mixed tank, at the frequency $\omega = 0.05$. As noted above, integrating a delay in the recycle path can give rise to resonance peaks in the disturbance sensitivity. Due to space limitations we refer to Carlemalm (2003) for results on how to deal with such resonances.

5. CONCLUSIONS

The disturbance sensitivity of an integrated plant is to a large extent determined by feedback interactions between the process units, induced by recycling of material and energy. We considered the use of storage capacities to modify the interactions such as to obtain favorable feedback properties in the frequency range where disturbance attenuation is needed, but infeasible with feedback control. As shown, adding a delay in the recycle path is the most effective means of reducing the process sensitivity in a given frequency range. We only considered monovariable recycle here, but extensions to multivariable cases are discussed in Carlemalm (2003).

6. REFERENCES

Faanes, A. and S. Skogestad (2003). Buffer tank design for acceptable control performance. *Ind. Eng. Chem. Res.*, **42**(10), 2198 – 2208.

Gilliland, E. R., L. A. Gould and T. J. Boule (1964). Dynamic Effects of Material Recycle. In: *Preprints JACC*. Stanford. pp. 140–146.

Mahajanam, R. V. and A. Zheng (2002). Evaluation of cost of dynamic controllability and optimal surge capacity without dynamic models. *Ind. Eng. Chem. Res.*, **41**(16), 3877–3883.

Skogestad, S. and I. Postlethwaite (1996). *Multivariable Feedback Control.* John Wiley & Sons.

ELSEVIER
IFAC
PUBLICATIONS
www.elsevier.com/locate/ifac

PROCESS DESIGN AND CONTROL OF A REACTIVE DISTILLATION SYSTEM

Pinky Panjwani, Myrian Schenk[1], Michael C. Georgiadis and Efstratios N. Pistikopoulos

Imperial College London,
Department of Chemical Engineering, Centre for Process Systems Engineering, London SW7 2AZ, UK

Abstract: A Mixed-Integer Dynamic Optimization (MIDO) model and solution strategy to study the interactions between process design and process control/operability of an ethyl acetate reactive distillation system is presented. It is shown that the resulting design and control scheme can guarantee feasible operation under bounded uncertainty at a minimum total average cost, representing 17 to 20% savings over the original design. *Copyright © 2004 IFAC*

Keywords: Process design; Control; Reactive distillation; Mixed integer optimization.

1. INTRODUCTION

In recent years chemical process industries have shown an increasing interest in the development of reactive distillation processes combining reaction and separation mechanisms into a single, integrated unit. Such processes offer several important advantages such as reduction in total costs and energy consumption, overcoming of thermodynamic limitations, (e.g. azeotropes) and increased reaction yield and selectivity. However, reactive distillation is not extensively used in the chemical process industries since its operation is more difficult and poses higher requirements on the quality of the design and control than conventional flowsheets in which a reactor is typically followed by a train of distillation columns (Engell and Fernholz, 2003).

Modelling of reactive distillation has received considerable attention over the last 10 years and several key contributions have appeared in the literature (see the book of Doherty and Malone, 2001 and the overview of Noeres *et al.*, 2003). Pilavachi *et al.* (1997) presented an extensive discussion of several important aspects that affect the accurate modelling of reactive distillation processes. Schenk *et al.* (1999) described in considerable detail a hybrid-modelling environment in which a reactive distillation process can be simulated using a combination of equilibrium and mass transfer models, both in steady state and dynamic modes.

The control of reactive distillation has received some attention in recent years. Sneesby *et al.* (1999) discussed the advantages of a combined composition and conversion scheme for an ETBE column. Monroy-Loperena *et al.* (2000) studied the control problem of an ethyl glycol reactive distillation system and proposed a robust PI control configuration. Sneesby *et al.* (2000) proposed an integrated control scheme for an ETBE column by permitting the control objectives to be changed on-line in order to reflect changing economic contracts. Vora and Daoutidis (2001) studied the dynamics and control of an ethyl acetate reactive distillation system and proposed a new feed configuration for the two reactants that allows higher conversion and purity than the conventional configuration, which involves feeding in a single tray. Al-Arfaj and Luyben (2002) compared the control of an ideal reactive distillation column with that of a similar real chemical system, the production of methyl acetate. A number of control structures were evaluated for both systems. Gruner *et al.* (2003) proposed a non-linear control scheme for an industrial reactive distillation column operated by Bayer AG. Bisowarno *et al.* (2003) investigated model gain scheduling for an ETBE reactive distillation column. It was illustrated that the proposed control strategy outperforms the standard PI control.

The work in the area of integration of design and control in reactive distillation systems is very limited. Heath *et al.* (2000) studied the interactions of design

---

[1] Present address, Air Products PLC, Molesey Road, Walton-on-Thames, Surrey, U.K.

and control in an ethyl glycol reactive distillation system assuming that the process structure is fixed (e.g. number of trays, feed tray location, etc). Schenk (1999) and Georgiadis *et al.* (2002) compared sequential and simultaneous approaches to design and control of a reactive distillation system, using advanced optimization techniques. Again the process and control structure were assumed fixed.

The aim of this work is to explore the synergistic benefits between process design, process control and operability in a reactive distillation column for the production of ethyl acetate. The process structure (number of trays, feed tray location), the process design (column diameter, reboiler and condenser surface areas) and control structure/design (pairings of measurements and manipulated variables, controller tuning parameters) are all simultaneously optimized. The problem is posed as a Mixed-Integer Dynamic Optimization (MIDO) problem and solved using recent algorithmic advances developed in our group at Imperial College (Bansal, *et al.*, 2003).

2. PROBLEM DEFINITION

The reactive system of ethyl acetate was considered in this work, since experimental data is openly available in the literature. This enables to validate the modelling results as well as the application of the proposed optimization framework.

The main problems encountered in achieving high purity products in the ethyl acetate reactive distillation system are summarized below (Bock *et al.* (1997); Chang and Seader (1988)):

- ✓ unfavourable reactant conversion;
- ✓ similar K-values of ethanol, water and ethyl acetate;
- ✓ temperature profile in the column;
- ✓ the system is strongly non-ideal due to the presence of ethanol, acetic acid and water. The separation of pure components is very difficult due to the existence of five normal azeotropes, namely, ethanol-water; water-acetic acid; ethyl acetate-ethanol; ethyl acetate-water; ethanol-ethyl acetate-water.

Full details of this system as well as a rigorous dynamic model together with the thermodynamic method employed are presented in Georgiadis *et al.* (2002).

Figure 1 depicts the ethyl acetate column superstructure considered where a single feed configuration is assumed. The objective is then to design the column and its required PI control scheme at minimum total annualised cost (capital cost and operating cost), which can guarantee feasible operation over a finite time horizon, subject to a set of constraints (Georgiadis *et al.*, 2002) such as:

- ✓ model equations;
- ✓ high-frequency sinusoidal disturbance in the acetic acid inlet composition;
- ✓ slow-moving disturbance in the cooling water inlet temperature representing diurnal, ambient variations;
- ✓ product quality specifications;
- ✓ minimum column diameter to avoid flooding;
- ✓ temperature driving forces in the reboiler and condenser;
- ✓ fractional entrainment limits;
- ✓ control structure specifications.

The solution of this problem would determine:

(a) the optimal process design, i.e., number of trays and feed location (discrete decisions), the diameter of the column, and the surfaces areas of reboiler and condenser (continuous decisions);

(b) the optimal PI control design, i.e., the pairings of manipulated and controlled variables (discrete decisions) and the tuning parameters (gain, integral time) of the PI controllers (continuous decisions).

Note that an original design of diameter 5.2 m, reboiler area 195 m^2 and condenser area 495 m^2 is considered as a base case for comparison of the results obtained in this paper.

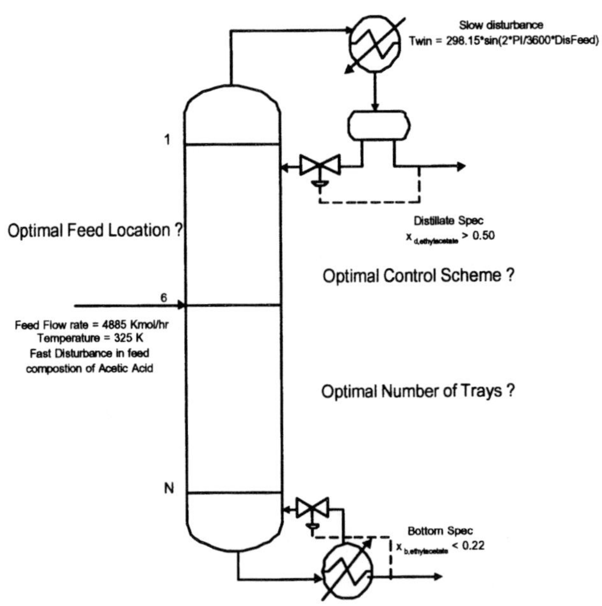

Fig. 1 Ethyl acetate column superstructure

3. SOLUTION STRATEGY

The methodology employed in this work relies on a systematic mixed-integer dynamic optimization (MIDO) framework exploring simultaneously the interactions between process and control system design (Bansal *et al.*, 2003). Two strategies were employed. In the first strategy, the control scheme is fixed, while the design of the column is optimized. The column design includes both discrete decisions such as number of trays and feed location, as well as continuous decisions (column diameter, exchanger areas, controller tuning parameters). The second strategy additionally incorporates the control structure selection (discrete decision). In this way, the economic and operability benefits of simultaneously optimizing the process and control design can be revealed and further assessed. Full details and files can be obtained from the authors by request.

3.1 MIDO Strategy with Fixed Control Structure

The ethyl acetate column superstructure consisted of 13 trays with feed located on the 6th tray. Some restrictions were imposed for the solution of this large-scale problem. Constraints like feed and reflux entering in only one tray; and, feed entering below the reflux were included in the mathematical framework. The fixed control scheme incorporated had two proportional integral controllers, as normally used in distillation systems, namely:

- PI1: Reflux rate (R) – Top ethyl acetate composition ($x_{4,d}$)
- PI2: Steam flow rate (F_s) – Bottom ethyl acetate composition ($x_{4,r}$)

Since the period of the slow-moving disturbance in the cooling water temperature is 3600 hours, it is deemed that this time is adequate as time horizon while rendering the large-scale dynamic optimization problem tractable. The objective function is defined as follows,

$$(Expected\ )Cost = \min_{d} \left\{ C_{cap} + \sum_{j=1}^{ns} w_j \cdot C_{op,j} \right\} \quad (1)$$

For this particular case, the resulting MIDO problem consists of 1652 model equations (81 differential and 1571 algebraic) with 26 binary (0-1) variables and 36 optimization parameters (9 inequality constraints and 27 equality constraints). The MIDO problem is decomposed into primal problem and master problem. The dynamic optimization primal problems (NLPs) were solved using gPROMS/gOPT (Process Systems Enterprise Ltd., 1999) and the master problems (MILPs) were solved using GAMS/CPLEX. (Brooke *et al.*, 1992).

Final convergence, i.e. the solution of the problem was obtained in two iterations. Table 1 presents the full details of the primal and master solutions for both iterations. The optimal column configuration has 12 trays with the feed located on the 6th tray and the corresponding final total cost is 536125 \$/yr.

3.2 Overall MIDO Strategy

The overall MIDO strategy considers the same column superstructure, objective function and optimization variables as the previous strategy but now also includes the control structure selection as additional optimization decisions.

Some process knowledge restrictions were imposed for the solution. In the master problems, based on the fact that a very long time response is undesirable between the action of a manipulated variable and its effect on a controlled variable, pairings such as Reflux rate – Bottom ethyl acetate composition, Distillate flowrate – Bottom ethyl acetate composition, Steam flowrate – Top ethyl acetate composition, Bottoms flowrate – Top ethyl acetate composition, are not considered from the search space; and, due to the available degrees of freedom in the system, only two controllers are allowed.

Table 1 Progress of Iterations for the MIDO strategy with fixed control structure

Iteration No	1	2	3
Primal Solutions			
Discrete Decisions			
No of trays	13	12	11
Feed location	6	6	5
Process Design			
Diameter of column	5.00	4.95	
Reboiler surface area	170.00	168.00	--Inf-- (Infeasible point)
Condenser surface area	480.00	478.00	
Controller gains			
PI1	25	25	--Inf--
PI2	5726.00	5725.66	
Controller Reset times			
PI1	60	60	--Inf-
PI2	4936.00	4935.54	
Controller Bias			
PI1			
PI2	3065.00	3074.54	-Inf--
	59000.0	58950.0	
Capital Cost	297000	286500	
Operating Cost	249280	249625	--Inf--
Final Cost	546280	536125	
UB	546280	536125	
Master Solutions			
No of trays	12	11	
Feed tray	6	5	
LB	534330	535890	
UB-LB<Tolerance	NO	NO	STOP

The solution to this problem is summarised in Table 2. The optimal column configuration consists of 12 trays with optimal feed located on 7th tray. The final total cost obtained is 517000 \$/yr. The optimal control scheme obtained is pairing reflux flowrate with top temperature of the column (T_c) and steam flowrate with feed tray temperature (T_7). The noticeable fact, from this result, is that in order to obtain better purity and maximum conversion, in the case of reactive distillation, it is important to control the temperature of the column, both feed tray as well as product (top column temperature, for this particular case). This somehow contrasts with the conventional control scheme in distillation columns, where top and bottom compositions (or temperatures) are controlled in order to attain better performance. Moreover, for this particular case the composition of ethyl acetate is reduced in the bottom of the column by almost 10%, indicating less product losses and better separation when using this control scheme.

Figs. 2 and 3 show the top and bottom product composition at the optimal solution.

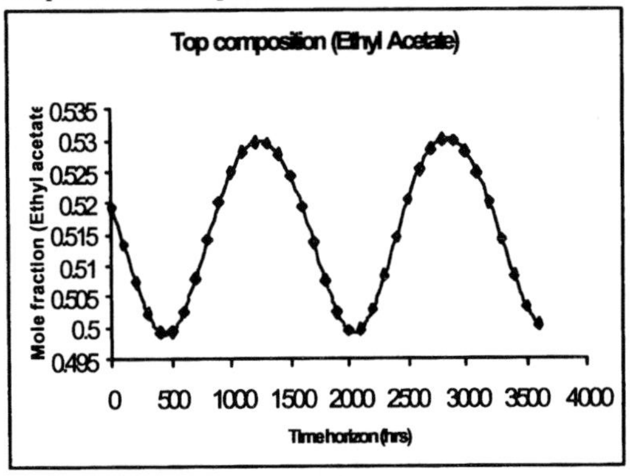

Fig. 2 Ethyl acetate top composition

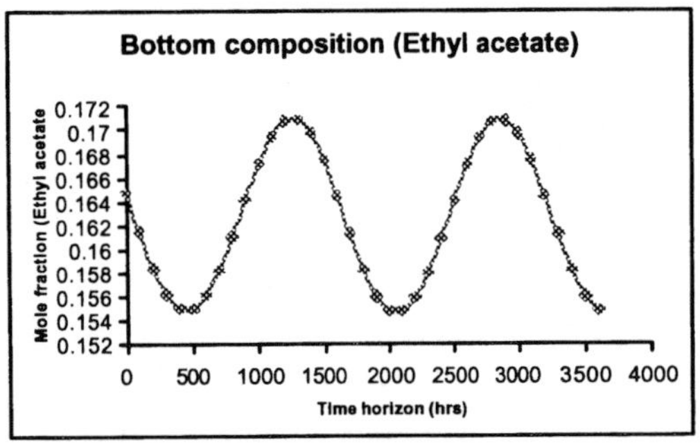

Fig. 3 Ethyl acetate bottom composition

Both figures illustrate that the ethyl acetate composition satisfies the product quality constraints over the entire time horizon (at least 50% of ethyl acetate at the top; and, at the most 20% at the bottom of the column).

The temperature profile of a reactive distillation column is closely related to the extent of reaction or reaction rate. Figure 4 depicts the reaction rate throughout the column on the solution of the MIDO problem. It is observed that most of the reaction is actually taking place at the bottom part of the column. Since reaction takes place mainly in the bottom section of the column with tray 7 (feed tray) having the maximum extent of reaction, a new column configuration could be considered where top trays are considered as non-reactive and bottom trays as reactive. By introducing the binary variables to model reactive and non-reactive trays in the superstructure, it is expected that the performance of the design obtained will be better than that of the fully reactive distillation column.

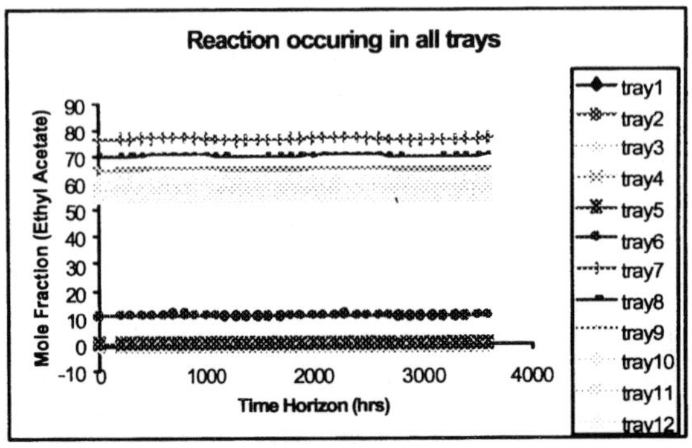

Fig. 4 Reaction profile in all the trays

3.3 Discussion of Results

The results obtained by using integrated MIDO approach are summarised as follows:

- The simultaneous consideration of process design and control results in economic savings of around 20% compared to the original design thus highlighting the importance of addressing the interactions and synergistic benefits that exist between process design and process control in the early design stage. For the case when only process design is considered the savings are around 17%.
- It was also found that by determining the design and control structure simultaneously, savings of almost 3.7% are obtained compared to the case where the control structure is fixed. The reduction in the capital cost and the operating cost are approximately 1.5% and 6% respectively. There is also a considerable reduction in the reboiler duty and the condenser duty, in the order of 10%. The reboiler and the condenser areas are significantly reduced. Moreover, the reflux and steam flow rates are reduced. This strongly portrays the impact that the interactions between design and control have on the operability.
- Simultaneous design and control strategy results in the reduced composition of ethyl acetate in the bottom stream, which implies less product losses, than with fixed control structure strategy.
- For this reactive distillation system, it is required to control the temperature of the reactive tray in which maximum extent of reaction takes place.
- Controlling feed tray temperature emphasizes the fact that the behaviour of reactive distillation column is different than the conventional distillation, where normally the composition of products (or top and bottom temperatures) are controlled to obtain better purity. The temperature in the reactive section needs to be controlled to obtain maximum conversion and better purity of the product.

Table 2 Progress of Iterations for the overall MIDO strategy

Iteration No	1	2	3
Primal Solutions *Discrete Decisions*			
No of trays	13	12	11
Feed location	6	7	3
1st controller	$R\text{-}x_{4,d}$	$R\text{-}T_c$	$R\text{-}x_{4,d}$
2nd controller	$F_s\text{-}x_{4r}$	$F_s\text{-}T_7$	$F_s\text{-}T_c$
Process Design			
Diameter of column *(m)*	5.00	5.10	
Reboiler surface area *(m2)*	170.00	140.00	--Inf--
Condenser surface area *(m2)*	480.00	470.00	
Controller gains			
1st	25	42.52	--Inf--
2nd	5726.00	5174.00	
Controller Reset times			
1st	60	28.00	--Inf--
2nd	4936.00	4966.00	
Controller Bias			
1st	3065.00	2790.00	
2nd	59000.0	55830.0	--Inf--
Capital Cost	297000	282000	
Operating Cost	249280	235000	--Inf--
Final Cost	546280	517000	
UB	546280	517000	
Master Solutions			
No of trays	12	11	
Feed tray	7	3	
1st controller	$R\text{-}T_c$	$R\text{-}x_{4,d}$	
2nd controller	$F_s\text{-}T_7$	$F_s\text{-}T_c$	
LB	445135	470000	
UB-LB<Tolerance	NO	NO	STOP

4. CONCLUSIONS

This work considers an advanced MIDO framework for studying the interaction between process design, process control and operability in a reactive distillation system. The results demonstrate the benefits from pursuing a simultaneous approach in process and control design rather than the traditional sequential approach. Economic advantages of the order of 3.7% are obtained with feasible dynamic performance, production specifications and operational restrictions despite the presence of rapidly varying disturbances.

Although this work has focused on the design of traditional PI controllers, more advanced control schemes (e.g. model-based parametric controllers) could be incorporated based on recent advances in our group (Sakizlis *et al.*, 2003; Sakizlis, 2003). It is thus expected that even more inexpensive designs could be achieved with proven dynamic operation and economics over the simultaneous process and control design featuring conventional controllers. Furthermore, the importance of new binary variables in the MIDO framework for the automatic identification of the type of tray (reactive or non-reactive) is expected to lead to more economic benefits since the cost of catalyst would be reduced.

ACKNOWLEDGEMENT

The authors would like to thank financial support from the European Union, Shell Chemicals and Air Products and Chemicals. P. Panjwani would also like to acknowledge financial support from the British Chevening scholarship shared with Imperial College.

REFERENCES

Al-Arfaj and W.L. Luyben (2002). Comparative control study of ideal and methyl acetate reactive distillation. *Chemical Engineering Science*, **57**, 5039-5050.

Bansal, V., V. Sakizlis, R. Ross, J.D. Perkins and E.N. Pistikopoulos (2003). New algorithms for Mixed-Integer Dynamic Optimization, *Computers and Chemical Engineering*, **27**, 647-668.

Bisowarno B.H., Y.C. Tian and M.O. Tade (2003). Model gain scheduling control of an ethyl tert-butyl ether reactive distillation column, *Industrial and Engineering Chemistry Research*, **42**, 3584-3591.

Bock, H., Jimoh, M. and G. Wozni (1997). Analysis of reactive distillation using the esterification of acetic acid as an example. *Chemical Engineering Technology*, **20**, 182-191.

Brooke, A., D. Kendrick and A. Meeraus (1992) GAMS Release 2.25: A Users's Guide. The Scientific Press. San Francisco.

Chang, Y. A. and J.D. Seader (1988). Simulation of continuous reactive distillation by a homotopy-continuation method. *Computers and Chemical Engineering*, **12**, 1243-1257.

Doherty, M.F., and M.F. Malone (2001). Conceptual design of distillation systems. McGraw-Hill Chemical Engineering Series. New York.

Engell, S., and G. Fernholz (2003). Control of a reactive distillation process. *Chemical Engineering and Processing*, **42**, 201-210.

Georgiadis, M.C., M. Schenk, E.N. Pistikopoulos and R. Gani (2002). The interactions of design, control and operability in reactive distillation systems, *Computers and Chemical Engineering*, **26**, 735-746.

gPROMS/gOPT (1999) – User Guide. Process Systems Enterprise Ltd..

Gruner, S., K.D. Mohl, A. Kienle, E.D. Gilles, G. Fernholz and M. Friedrich (2003). Non-linear control of a reactive distillation column, *Control Engineering Practice*, **11**, 915-925.

Heath, J.A, I. Kookos and J.D. Perkins (2000). Process control structure selection based on economics, *American Institute of Chemical Engineers Journal*, **46**, 1998-2016.

Monroy-Lopera, R., E. Perez-Cisneros and J. Alvarez-Ramirez (2000). A robust PI control configuration for a high purity ethylene glycol reactive distillation column, *Chemical Engineering Science*, **55**, 4925-4937.

Noeres, C., E.Y. Kenig and A. Gorak, (2003). Modelling of reactive separation processes: reactive absorption and reactive distillation, *Chemical Engineering and Processing*, **42**, 157-178.

Pilavachi, P., M. Schenk, E. Perez-Cisneros and R. Gani (1997). Modelling and Simulation of Reactive Distillation Operations, *Industrial and Engineering Chemistry Research*, **36**, 3188-3197.

Sakizlis, V., J.D. Perkins and E.N. Pistikopoulos (2003). Parametric controllers in simultaneous process and control design optimization, *Industrial and Engineering Chemistry Research*, 42, 4545-4563.

Sakizlis, V., (2003) Design of Parametric Controllers via Parametric Programming. PhD Thesis. Imperial College London, University of London.

Schenk, M.A., (1999), Design of operable reactive distillation columns. PhD Thesis. University of London.

Schenk, M., R. Gani, D. Bogle and E.N. Pistikopoulos (1999). A Hybrid Modelling Approach for Separation Systems Involving Distillation, *Chemical Engineering Research and Design*, Part A, **77**,519-534.

Sneesby, M.G., O.M. Tade and T.N. Smith (1999). Two-point control of a reactive distillation column for composition and conversion, *Journal of Process Control*, **9**, 19-31.

Sneesby, M.G., M.O. Tade, T.N. Smith (2000). A multi-objective control scheme for an ETBE reactive distillation column, *Chemical Engineering Research and Design*, **78** (A2), 283-292.

Vora, N. and P. Daoutidis (2001). Dynamics and control of an ethyl acetate reactive distillation column, *Industrial and Engineering Chemistry Research*, **40**, 833

ELSEVIER IFAC PUBLICATIONS
www.elsevier.com/locate/ifac

DESIGN AND CONTROL FOR ENERGY INTEGRATION IN A BIO-PROCESS

Michael L. Luyben

DuPont Engineering Technology
1007 Market St. – B7434
Wilmington, DE 19898 USA

Abstract: This paper presents an industrial viewpoint on designing for energy integration in processes with substantial water removal (e.g. bio-processes) to ensure dynamic controllability. The use of evaporation and distillation for water removal and product purification in a bio-process is energy intensive, since the concentration of desired product leaving the fermenter is often quite low. The nature of a fermentation process causes inherent variation in the feed composition to downstream units. Potential opportunities for energy integration can be exploited, but must be done in a way that ensures the process is controllable. An example is presented of a multi-effect evaporator system combined with two distillation columns, describing how the design of energy integration between them requires some mechanism or manipulator to prevent disturbances from propagating around the material and energy loops.

Keywords: Distillation columns, integrated plant control .

1. INTRODUCTION

Many researchers and practitioners have recognized that the design of a chemical process (unit operations, process conditions and parameters, equipment layout, available measured and manipulated variables, etc.) plays a dominating role in determining how controllable the process is. Furthermore, it is economically advantageous to consider operability at the design stage rather than wait until after the process is built. Van Schijndel and Pistikopoulos (1999) provide a comprehensive summary of work on the integration of design, control, and operation.

The viewpoint in this paper (as in Downs and Doss, 1991) argues that controllability and control strategy design should not be simply an afterthought of the process design. Processes designed with a focus on inherently eliminating variability can often be controlled by relatively simple, straightforward, decentralized control strategies (rather than a focus on the development of complicated control algorithms). This paper uses an example of separations in a bio-based process to show how controllability can be designed into the flowsheet.

One area of new technology development within the chemical industry focuses on the use of renewable feedstocks (as opposed to those that use raw materials from petroleum) in biologically-based processes that produce chemicals and other materials. Such new processes will take some carbohydrate source (e.g. corn syrup) and convert it with microorganisms via fermentation into a desired product (Vanhoek and Reeder, 2003). The Cargill Dow process to make lactic acid (for use in production of polylactide polymer) is one example. The DuPont process to make 1,3-propanediol (for use in production of polytrimethylene terephthalate or 3GT polymer) is another.

One consequence of using microorganisms in a bio-process is that the desired product may leave the fermenter in a large amount of water. For example, product titer of 50 g/L translates into roughly 95 wt% water, depending on steps taken after fermentation. Many different separation technologies are practiced on the fermentation broth to produce the desired product with the required product quality and consistency. These include filtration (rotary drum, micro, nano, ultra, etc.), ion exchange beds, carbon beds, chromatography, evaporation, extraction, crystallization, distillation, and others. Two references, among many others, include Belter *et al.* (1988) and Subramanian (1998). The focus of this paper is on evaporation and distillation for water removal and product purification and how to design with energy integration to ensure the integrated process is controllable.

2. EVAPORATION

The technology of evaporation has existed for millennia (chapter 11 in Perry and Green, 1999, provides some general background). It is widely used in the food, beverage, sugar, dairy, mining, mineral, pulp, paper, pharmaceutical, health care, chemical, polymer, salt, seawater, and many other industries. The purpose of this paper is not to present specific details on evaporator design and operation. Much of that is available from vendors or the literature (e.g. Nisenfeld, 1985). There are many different types of evaporators including natural circulation, rising and falling film tubular, forced circulation, wiped film, plate, etc. The choice depends upon the heat sensitivity of the process materials, the viscosity or corrosivity, the scale and capacity, and the relative costs of steam and electricity.

Since evaporation is energy intensive, many techniques have been developed to conserve energy. One is the use of a multi-effect system, where the vapor produced in one effect becomes the heat source in the second effect at lower pressure (Figure 1). Another is the use of mechanical vapor recompression, where a fan or compressor increases the pressure of the evaporated vapor product which becomes the heat source. One important factor in evaporation is the boiling point rise, which is the difference between the boiling point of the solution and the boiling point of water at the same pressure. This governs many of the process conditions and evaporator types.

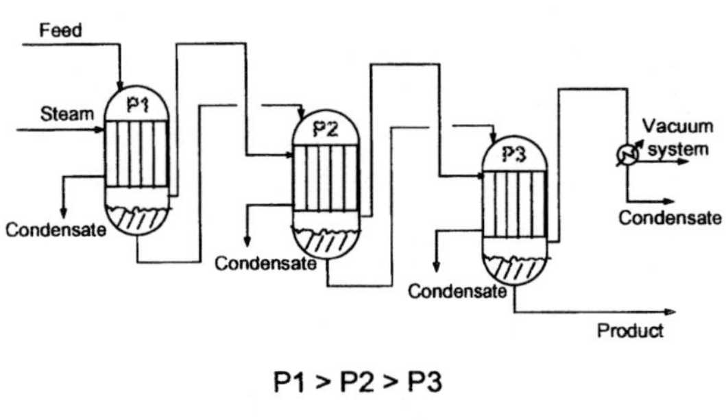

Fig. 1. Multi-effect evaporation system

3. BIO-PROCESS EXAMPLE

A generic bio-process shown in Figure 2 contains downstream steps for the removal of water and other impurities. The desired product from the fermenter may be more or less volatile than water. Some alcohols, for example, may be more volatile then water. The process design would be different from what is presented here, but the same controllability issues and principles apply. In this particular example, water is assumed to be the lightest component. Three other components A, B, and C are present (in order of decreasing volatility). Component B is the desired product. The feed stream of 150 kmol/h contains 0.97 mol frac water, 0.007 mol frac A, 0.02 mol frac B, and 0.003 mol frac C. The design for this process needs to consider equipment sizes, energy costs, and process constraints to arrive at an economic optimum, but those details are not included here due to conserve space. The equipment sizing and costing are standard.

For purposes of this example, a three-stage multi-effect evaporator is used for water removal (assuming the evaporators to be falling film). The vapor generated by steam in the first stage is used as the heating medium in the second evaporator stage (operating at lower pressure). Liquid from the first evaporator is the feed to the second. Liquid and vapor flow similarly from the second to the third evaporator. The flow of liquid and vapor is co-current so that the lowest pressure and temperature are in the final stage. The liquid from the third evaporator feeds the first distillation column that separates A in the distillate from B and C in the bottoms. The second column separates B in the distillate from C in the bottoms.

Table 1 contains process conditions for the evaporators in this example, while Table 2 contains process data for the distillation columns.

Table 1 Process conditions in evaporators

Parameter	Value	Parameter	Value
E1 duty	584 kW	E3 liquid out	4.9 kmol/h
E1 liquid out	104 kmol/h	E3 temperature	101°C
E1 temperature	122°C	E3 pressure	0.15 atm
E1 pressure	2 atm	E3exit H2O	0.08 mol frac
E2 liquid out	54 kmol/h	E3 exit A	0.21 mol frac
E2 temperature	110°C	E3 exit B	0.62 mol frac
E2 pressure	1.3 atm	E3 exit C	0.09 mol frac

4. ENERGY INTEGRATION

In this design, steam is used both in the column reboilers and in the first multi-effect evaporator and cooling water is used in the two column condensers. However, observation shows, and a pinch analysis (e.g. Seider *et al.*, 1999) confirms, that process conditions have been set so that the top column temperatures are both higher than the temperature in the first evaporator. Also, the sum of the two condenser duties exactly matches the evaporator duty (584 kW).

Table 2 Process conditions in columns

Parameter	Value	Parameter	Value
C1 cond duty	330 kW	C2 cond duty	254 kW
C1 th stages	8	C2 th stages	8
C1 feed	4 from bs	C2 feed	4 from bs
C1 pressure	0.8 atm	C2 pressure	0.15 atm
C1 top temperature	142°C	C2 top temperature	159°C
C1 dist	1.9 kmol/h	C2 dist	2.5 kmol/h
C1 dist mol frac	H2O=0.2 A=0.54 B=0.26	C2 dist mol frac	A=0.001 B=0.999
C1 base	3 kmol/h	C2 base	0.5 kmol/h
C1 bs mol frac	A=0.001 B=0.85 C=0.15	C2 bs mol frac	B=0.01 C=0.99

It therefore is feasible to integrate the design further by using the overhead vapor streams from both columns as the heating medium in the first evaporator (Figure 3). The liquid condensate must be kept separate by partitioning the evaporator chest so that the liquid condensed can go to separate reflux drums for each column. This integrated design reduces both steam and cooling water consumption and reduces capital by eliminating the need for individual column condensers. For companies that seek to minimize capital investment and reduce operating costs, this integrated design offers significant opportunities. The questions then focus on whether the design can run in actual operation.

At steady-state, the heat duties can be balanced exactly between the column condensers and the first evaporator. However, from the viewpoint of operability, important problems must be identified and solved: (1) can the system be started up and shut down safely, (2) can it produce consistent product quality, and (3) can it deal with changes in operating conditions (water content in crude feed, column reflux flows, etc.)? Westphalen *et al.* (2003) summarize some of the basic principles about the control of heat exchanger networks and ways to assess their controllability. Unless such questions are addressed prior to a finalized design, the process may never be able to run as designed.

In examining the plantwide control of any integrated process design, one of the key objectives is to provide mechanisms or manipulators that prevent disturbances from propagating around the material and energy loops. If necessary, the process design needs to be changed to make it more operable. Such changes need to be justified in terms of economics.

For example, one expected disturbance is a change in the water content of the crude evaporator feed. If the crude feed water concentration increases, then the liquid and vapor flows from the evaporators will change. The amount of water in the feed to the distillation columns will increase since there is not enough energy coming from the column overhead vapor streams. The disturbance will take much longer to disappear than with no energy integration as shown in Figure 4, where the evaporator is run open-loop with no adjustment to maintain a certain water concentration in the third stage (while the two columns are run in closed-loop) compared with the design with no energy integration.

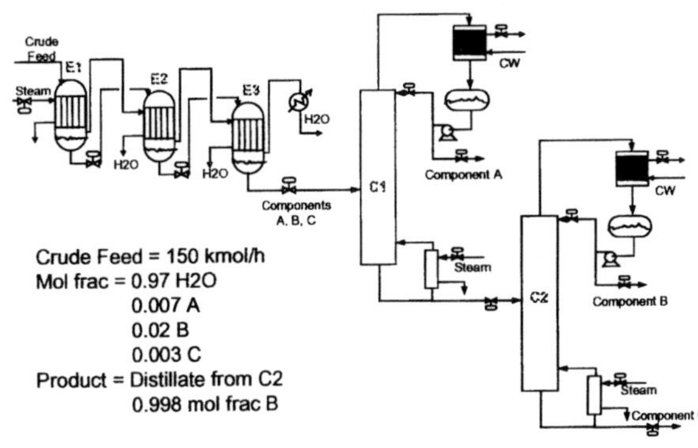

Fig. 2. Bio-process downstream example

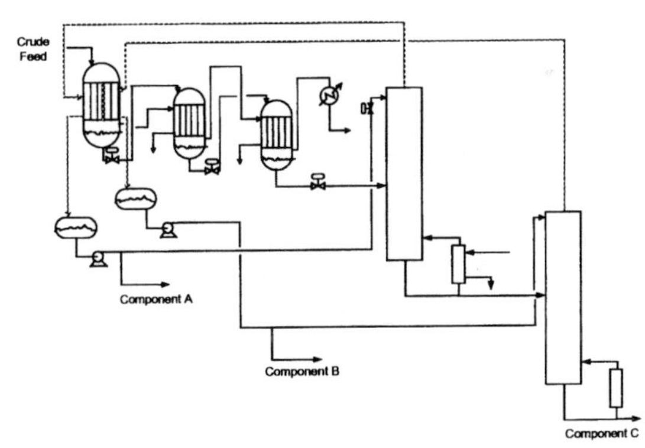

Fig. 3. Design with energy integration

5. CONTROL OF EVAPORATOR EXIT WATER CONTENT

The main goal for this system then becomes the control of the exit water content (as measured by temperature) from the third evaporator as the way to isolate disturbances. Heath *et al.* (2000) provide some discussion on the control of a single evaporator. The main alternative control strategies are:

(1) Bypass some of the crude feed to the third evaporator assuming that more energy comes from the column overhead streams than is required to remove the water content. If the water content increases in the exit stream, the bypass flow is decreased.
(2) Manipulate the crude feed flow and other liquid flows (going backwards). If the water content increases in the exit stream, the liquid flow from stage 2 to 3 is decreased, and so on.
(3) Underdesign the evaporators so that extra steam flow must always be added to the third evaporator. If the water content increases in the exit stream, the supplemental steam flow is increased.

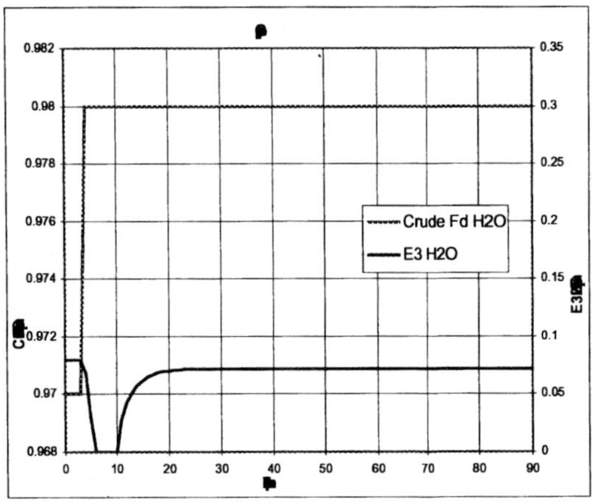

Fig. 4a. Open loop response to increase in feed water content with no energy integration

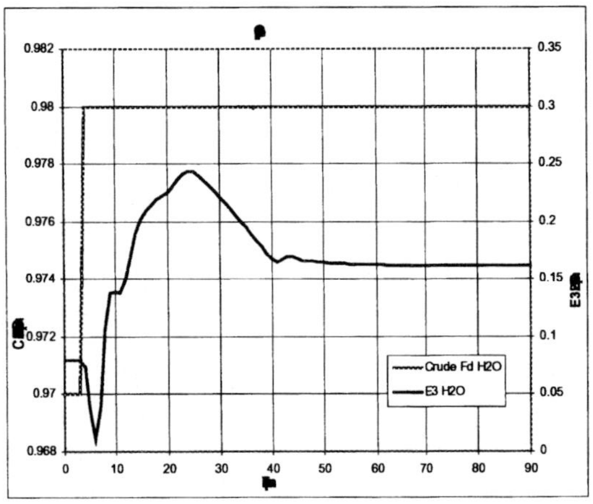

Fig. 4b. Open loop response to increase in feed water content with no energy integration

Each of these alternatives has some potential advantages and disadvantages that must be evaluated via dynamic analysis and simulation. A purely steady-state evaluation will not provide the complete picture. The details of this dynamic analysis cannot be presented here completely. One of the main responsibilities of a control engineer working in process development is to generate ideas for alternative control strategies and process designs and then to evaluate them based on dynamic performance and economics. The ideas come from a fundamental understanding and a quantitative assessment of the process (not from purely mathematical algorithms).

It turns out that the process design must have a manipulator to serve as the key break point for avoiding interaction between disturbances that affect the evaporators and columns. The nature of a fermentation process inherently causes variability in the feed composition to downstream units, adding to the importance of designing for control

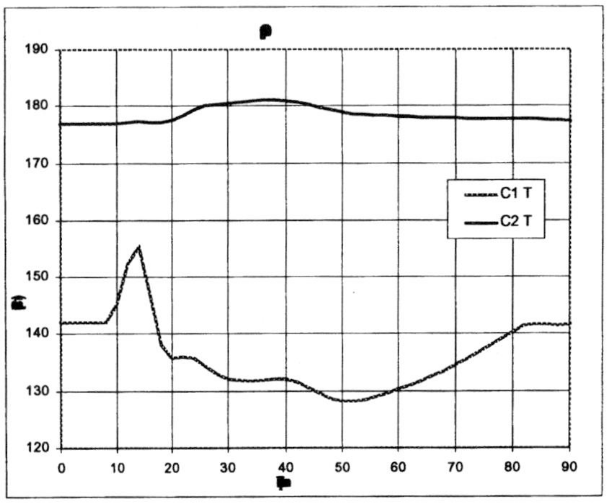

Fig. 4c. Open loop response to increase in feed water content with energy integration

6. DESIGN AND CONTROL

During abnormal situations, such as the loss of crude feed flow, the integrated design would create significant problems. Since the columns rely on the evaporators to condense their overhead streams, a loss of evaporator crude feed flow or other parts of the evaporator system would require an immediate shut down of the two distillation columns to prevent over-pressurization. This would cause the column contents to dump and means that the columns could not be run under total reflux conditions. Start up of the integrated system would be lengthened since crude feed is needed to bring up the columns, which would produce off-specification material until the desired temperature and composition profiles are achieved.

There are several alternative design changes that might potentially achieve some degree of greater independence between the evaporators and columns. One is to include auxiliary column condensers in parallel to the first evaporator stage, so that the columns could always be run on total reflux. This requires capital, which it was hoped could be avoided by using energy integration in the first place.

A second is to install the ability to feed make-up water into the first evaporator as a way to provide an energy sink. The water vapor generated in this way still must be removed and condensed, which means enough water must be fed to go through all three effects. This requires some additional capital, but much less than auxiliary condensers. It does, though, still rely on operation of the evaporation.

A third is to avoid the direct integration completely by generating steam with the column overhead streams. The steam would go into a header and supply the first evaporator stage. Excess steam would be vented or used elsewhere in the process and any deficiency in steam could be supplied from the plant steam source. The trade-off is the additional capital costs of the condensers and boilers with twice the heat transfer area (for the two temperature differences).

A fourth design alternative is to explore the value of a buffer tank between the third evaporator stage and the first column. A tank that is large enough would in principle filter out any possible changes in water concentration or in flow. This too requires extra capital and introduces inventory in a piece of equipment that adds no inherent value to the process.

Figure 5 shows the integrated system with one possible control strategy and design. Here the third stage evaporator temperature (water content) is controlled with the make-up steam flow to the third evaporator. Make-up water to the first evaporator is added in case the crude feed flow is lost. This could be done by override pressure controllers on the two columns. If the pressure in either column rises too much (indicating a loss of evaporator crude feed), then makeup water is added to the first evaporator stage so that the column overhead vapor streams can be condensed. The rest of the control strategy is also shown with the two column temperature controllers to maintain final product purity. Only a single temperature needs to be controlled in either column to maintain the desired composition profile and product purities. Ratios of reflux to feed flow can be used if more direct control of composition is needed.

Figure 6 shows the closed-loop dynamic response to a change in crude feed water content with this control strategy. These results come from a rigorous nonlinear dynamic simulation of the entire evaporator and distillation system. The increase in the feed water concentration is handled by adjusting the steam flow to the third evaporator. This results in a fairly minor disturbance for the column temperature controllers and disappears more quickly than the case where the evaporator was run open-loop with no control of exit water content. Similar analysis can be done with other design and control alternatives.

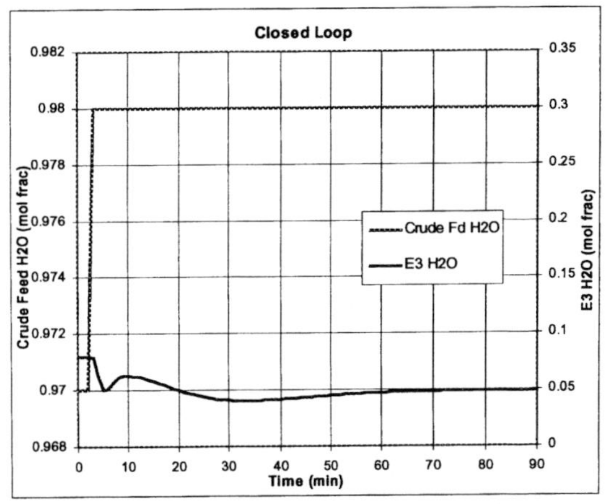

Fig. 6a. Closed loop response to increase in feed water content with control of exit evaporator water

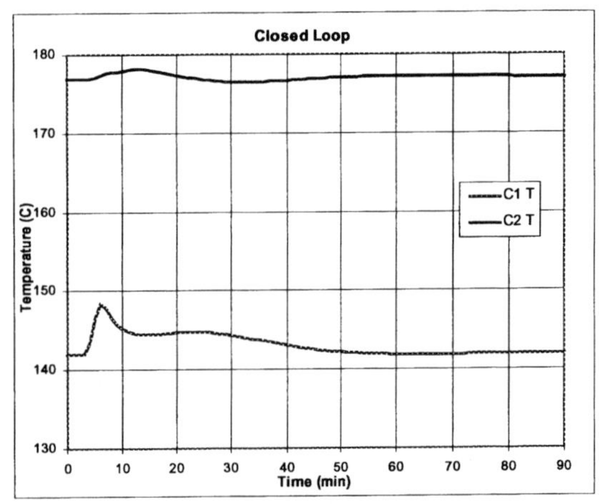

Fig. 6b. Closed loop response to increase in feed water content with control of exit evaporator water

7. CONCLUSION

This paper has presented an industrial view on the design and control of energy integration for bio-processes where evaporation and distillation are used for water removal and product purification. Such challenges will arise as new chemical processes are developed and commercialized using renewable feedstocks to manufacture chemicals and other materials. Because a fermentation process will inherently create disturbances in the feed composition to downstream units, controllability must be examined at the design stage, particularly with increased energy integration. This also highlights the large incentive to develop alternative separation processes that require less energy than evaporation or distillation.

The main objective for control of systems that interact with both energy and material flows is to provide mechanisms or manipulators to prevent disturbances from propagating. In the example discussed in this paper, this meant controlling the water concentration from the last evaporator stage before it reaches the distillation columns. If necessary, changes to the process design must be made to ensure the system is controllable.

REFERENCES

Belter, P.A., E.L Cussler, and W.S. Hu (1988). *Bioseparations – Downstream Processing for Biotechnology*. J. Wiley, New York.

Downs, J.J. and J.E. Doss (1991). *Present Status and Future Needs – A View from North American Industry*. In Proceedings of the Chemical Process Control IV Conference. 53.

Heath, J.A., I.K. Kookos, and J.D. Perkins (2000). Process Control Structure Selection Based on Economics. *AIChE J.*, **46**, 1998.

Nisenfeld, A.E. (1985). *Industrial Evaporators: Principles of Operation and Control*. ISA, Research Triangle Park.

Perry, R.H. and D.W. Green, eds. (1999). *Perry's Chemical Engineers' Handbook*. McGraw-Hill, New York.

Seider, W.D., J.D. Seader, and D.R. Lewin (1999). *Process Design Principles*. J. Wiley, New York.

Subramanian, G., ed. (1998). *Bioseparation and Bioprocessing: A Handbook*. J. Wiley, New York.

Vanhoek, P., D. Reeder (2003). *From Corn to Plastic*. In American Institute of Chemical Engineers Annual Meeting. San Francisco, Paper 201a.

Van Schijndel, J. and E.N. Pistikopoulos (1999). *Towards the Integration of Process Design, Process Control, and Process Operability: Current Status and Future Trends*. In Proceedings of Foundations of Computer-Aided Process Design. 99.

Westphalen, D.L., B.R. Young, and W.Y. Svrcek (2003). A Controllability Index for Heat Exchanger Networks. *Ind. Eng. Chem. Res.*, **42**, 4659.

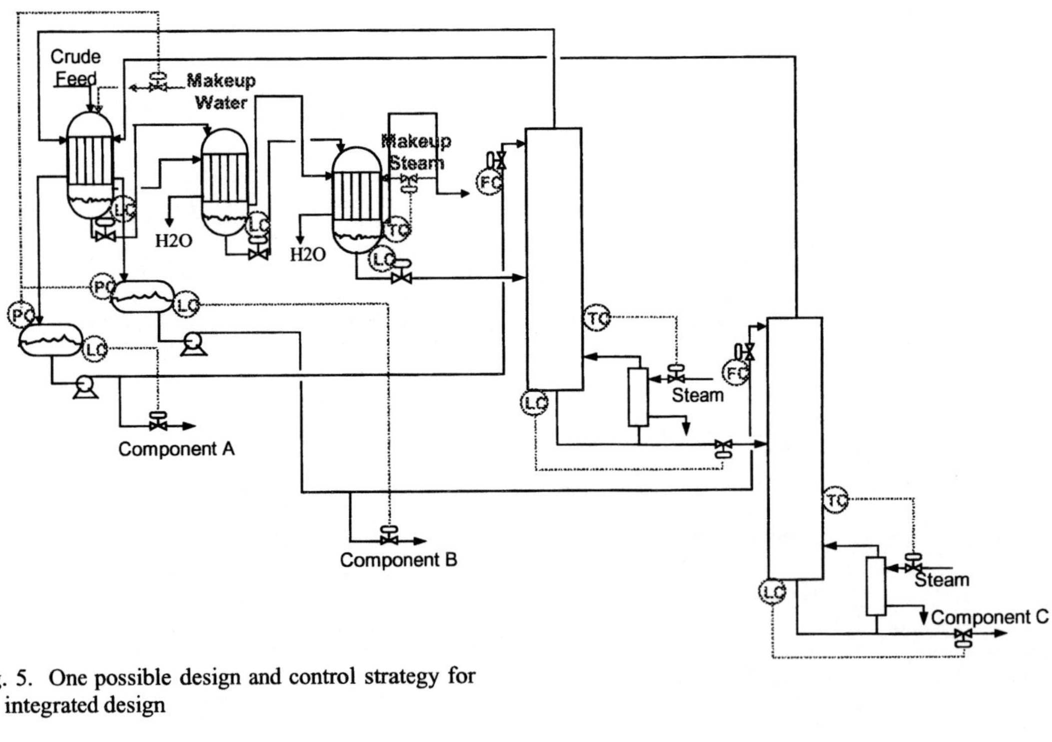

Fig. 5. One possible design and control strategy for integrated design

ELSEVIER
IFAC
PUBLICATIONS
www.elsevier.com/locate/ifac

A DESIGN OF PID CONTROLLERS FUSED CMACS WITH NEURAL NETWORKS

Kenji Takao [*] **Ryota Kurozumi** [**]
Toru Yamamoto [**] **Takao Hinamoto** [*]

[*] *Graduate School of Engineering, Hiroshima University, 1-4-1 Kagamiyama, Higashi-hiroshima, Hiroshima, 739-8527, Japan*
[**] *Graduate School of Education, Hiroshima University 1-1-1 Kagamiyama, Higashi-hiroshima, Hiroshima, 739-8524, Japan*

Abstract: Several neural-net based PID controllers have been proposed for nonlinear process systems. However, they have been not so widely used in process industries due to the considerably computational cost. This paper presents a new intelligent PID tuning scheme, whose PID tuner is constructed by the fusional structure of a cerebellar model articulation controller and a neural network. This PID tuner gives us the higher learning efficiency which has not been realized by the conventional neural-net based controllers, and it enables us to tune PID gains in an on-line manner. The behaviour of the proposed scheme is examined by a simulation example for a chemical reactor model. *Copyright © 2004 IFAC*

Keywords: PID Control, CMAC, Neural Network, Nonlinear Control, Process Control

1. INTRODUCTION

Neural-net based controllers have been proposed for nonlinear systems. The reason why the neural network(NN)(K.S.Narendra and K.Parthasarathy, 1990; C.-Y.Seong and B.Widrow, 2001) is employed for such systems is that they have the capability of highly approximating nonlinear properties. The back propagation(BP) scheme is usually employed in updating weights included in neural networks. According to the neural-net based controllers using the BP scheme, however, the problem is pointed out which considerably computational time is required until the good control performance is obtained. Therefore, the neural-net based controllers are not necessarily suitable in the real-time control.

On the other hand, PID control(J.G.Zieglar and N.B.Nichols, 1942; K.L.Chien and J.B.Reswick, 1972; K.J. Åström, 1988) schemes still continue to be widely used for most industrial control systems, particularly in the chemical process industry. This is mainly because PID controllers have simple control structures, and are simple to maintain and tune. Therefore, it is still attractive to design discrete-time control systems with PID control structures. Especially, some neural-net based PID controllers(S.Omatu *et al.*, 1995) have been considered for nonlinear systems in the reason that most real process systems have nonlinear properties. As mentioned above, however, since these controllers require much computational cost to training the neural networks, they have been hardly utilized in real process systems.

By the way, a cerebellar model articulation controller(CMAC) has been proposed by Albus (J.S.Albus, 1975*b*; J.S.Albus, 1975*a*), which is a kind of artificial neural networks, and especially applied to some robot control systems and chemical process systems. The weights included in the CMAC are trained based on a learning scheme that has the common memory reference structure. Therefore, since the learning effect effectively spreads around the learning point, it has the feature that learning cost is drastically reduced in comparison with the conventional BP method. However, it has also the following problems.

(1) Since the conventional CMAC directly generates the control input, it is difficult to explicitly grasp the nonlinear properties of the controlled object. It is a very important factor to know characteristics of the controlled object in operating equipments in real process systems.

(2) Since input signals for the CMAC are quantized, it is difficult to obtain accurate control performance even if it has the rapidity in the initial stage of training the CMAC. This problem can be solved by increasing the number of weight tables included in the CMAC. However, since the capacity of the computer memories becomes considerably large, it is not useful in the practical systems.

The objective of this paper is to propose a new design method of neural-net based intelligent PID controllers with high learning efficiency. The proposed controller has a fusional structure of the CMAC and the NN. The new scheme to generate PID parameters using the CMAC, is firstly proposed. As mentioned above, although it is impossible to know the properties of the controlled object using the conventional CMAC, the operators to deal with real systems can roughly grasp them through PID gains. To the best of our knowledge, such a controller that generates PID gains using the CMAC, has been not reported until now. Next, the fusional structure is considered, which is constructed using the CMAC and the NN. This structure can complement each problem in the CMAC and the NN mutually. In other words, the CMAC is suitable to work in the initial stage of learning, because it demonstrates effect in the static mapping. On the other hand, the accurate learning performance can be finally obtained using the NN, although the NN requires much learning time. Therefore, the CMAC is firstly worked in tuning PID gains, and the PID tuner which has been roughly trained, is switched from the CMAC to the NN. Then, the NN works so that the control performance is further improved. This structure gives us the higher learning efficiency which has not been realized by the conventional neural-net based controllers.

This paper is organized as follows. The intelligent PID controller with the fusional structure of the CMAC and the NN, is proposed. Next, the CMAC-based PID tuner firstly is explained, followed by the NN-based PID tuner. Furthermore, a criterion in order to switch from the CMAC to the NN is considered. Finally, the effectiveness of the proposed scheme is numerically evaluated on a simulation example based on a chemical reactor model.

2. PID CONTROLLER DESIGN

2.1 Outline of controller

The velocity-type PID control law which has variable gains, is described as

$$\Delta u(t) = K_I(t)e(t) - K_P(t)\Delta y(t) - K_D(t)\Delta^2 y(t) \quad (1)$$

$$\mathbf{K}(t) = [K_P(t), K_I(t), K_D(t)], \quad (2)$$

where $u(t)$ and $y(t)$ denote the control input and the corresponding system output, respectively. And also, $e(t)$ denotes the control error, which is defined as follows:

$$e(t) := r(t) - y(t), \quad (3)$$

where $r(t)$ denotes the reference signal, and $K_P(t)$, $K_I(t)$ and $K_D(t)$ are respectively the proportional gain, the integral gain and the derivative gain. Furthermore, Δ denotes the differencing operator defined as $\Delta := 1 - z^{-1}$.

The control performance strongly depends on PID gains in (1). Especially, PID gains must be carefully determined for nonlinear systems. Several neural-net based PID controllers have been proposed for such systems, in which PID gains adequately change corresponding to the operating points. However, it is pointed out that these schemes require much computational time until the good control performance is obtained.

On the other hand, the CMAC which is one of artificial neural networks, can roughly approximate the nonlinear components with quite small learning time, although the CMAC does not give an accurate control performance such as the NN.

Therefore, this paper newly presents an intelligent PID tuner which has the fusional structure of the CMAC and the NN as shown in Fig.1. Here, the **S.W** in this figure denotes the switching from the CMAC to the NN, and two **S.W**s work together. That is, the CMAC firstly works in the initial stage in order to roughly approximate the nonlinear components, and the PID tuner is switched from the CMAC to the NN in order to further improve the control performance. This structure

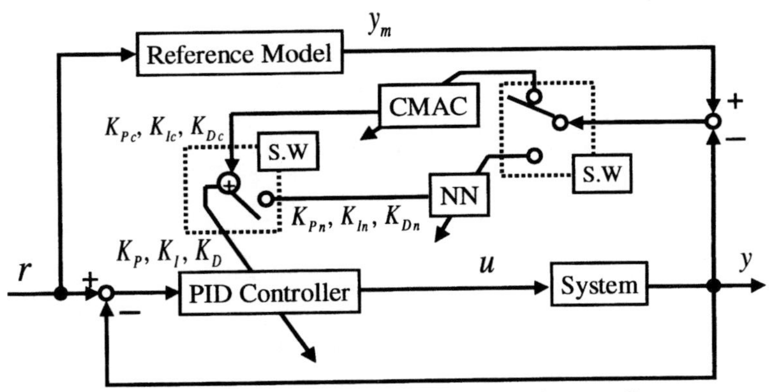

Fig. 1. Block diagram of the proposed controller.

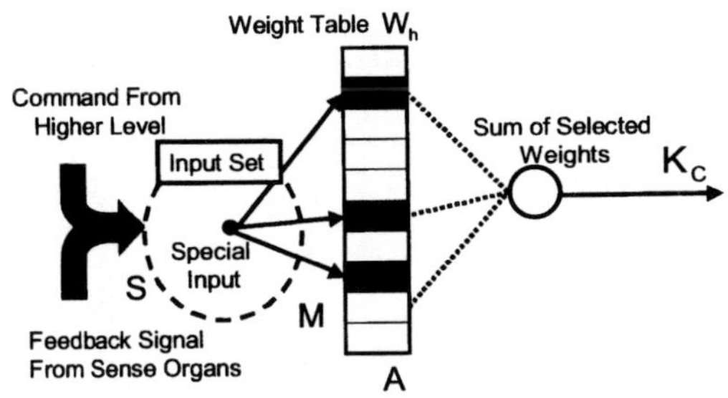

Fig. 2. Conceptual diagram of a CMAC.

gives us the high learning efficiency, because a good control performance can be obtained with quite small learning time. The detailed tuning scheme is discussed below.

2.2 PID parameter tuning by CMAC

The CMAC is one of artificial neural networks, and is a controller for the body motion system. The conceptual figure of the CMAC is shown in Fig.2. The CMAC is defined as the following mapping relation:

$$S \to M \to A \to \mathbf{K_c}, \tag{4}$$

where S, M, A and $\mathbf{K_c}$ denote the input vectors, the intermediate variables to code S, the association cell vectors with M and the output from the CMAC, respectively. The output from the CMAC, $\mathbf{K_c}$ consists of PID gains defined by

$$\mathbf{K_c}(t) := [K_{Pc}(t), K_{Ic}(t), K_{Dc}(t)]. \tag{5}$$

These mapping can be explained in detail as follows. Firstly, the CMAC composes the label space M for the input vector S ($S \to M$). Next, the weights are selected from the weight table for the label group M ($M \to A$). Finally, the output $\mathbf{K_c}$ is obtained by summing the selected weights($\mathbf{W_h}(t)$) ($A \to \mathbf{K_c}$). Thus, $\mathbf{K_c}(t)$ is generated by the following equation:

$$\mathbf{K_c}(t) = \sum_{h=1}^{k} \mathbf{W_h}(t), \tag{6}$$

where k denotes the number of weights which are extracted from the weight table.

In this paper, the inputs for the CMAC consist of the reference signal $r(t)$, $e(t)$ and $\Delta e(t)$. Here, $\Delta e(t)$ is defined as

$$e(t) := r(t) - y(t). \tag{7}$$

Furthermore, the only CMAC is worked until the error criterion is satisfied. Moreover, the learning rule to update the weights in the CMAC is as follows:

$$\mathbf{W_h}^{new}(t) = \mathbf{W_h}^{old}(t) - g(t)\frac{1}{k}\frac{\partial J(t+1)}{\partial \mathbf{K_c}(t)} \tag{8}$$

$$J(t+1) = \frac{1}{2}\varepsilon^2(t+1) \tag{9}$$

$$\varepsilon(t) := \frac{1}{2}\{y_m(t) - y(t)\}^2, \tag{10}$$

where $g(t)$ is given by

$$g(t) = \frac{1}{\beta_3 + \beta_1 \cdot \exp(-\beta_2|\varepsilon(t)|)}. \tag{11}$$

Here, β_1, β_2 and β_3 are the user-specified parameters. Moreover, each partial differentiation of Eq.(8) is developed as follows:

$$\left.\begin{aligned}
\frac{\partial J(t+1)}{\partial K_{Pc}(t)} &= \frac{\partial J}{\partial \varepsilon}\frac{\partial \varepsilon}{\partial y}\frac{\partial y}{\partial u}\frac{\partial u}{\partial K_{Pc}} \\
&= -\varepsilon(t+1)\Theta_1(t)\frac{\partial y(t+1)}{\partial u(t)} \\
\frac{\partial J(t+1)}{\partial K_{Ic}(t)} &= \frac{\partial J}{\partial \varepsilon}\frac{\partial \varepsilon}{\partial y}\frac{\partial y}{\partial u}\frac{\partial u}{\partial K_{Ic}} \\
&= -\varepsilon(t+1)\Theta_2(t)\frac{\partial y(t+1)}{\partial u(t)} \\
\frac{\partial J(t+1)}{\partial K_{Dc}(t)} &= \frac{\partial J}{\partial \varepsilon}\frac{\partial \varepsilon}{\partial y}\frac{\partial y}{\partial u}\frac{\partial u}{\partial K_{Dc}} \\
&= -\varepsilon(t+1)\Theta_3(t)\frac{\partial y(t+1)}{\partial u(t)},
\end{aligned}\right\} \tag{12}$$

where

$$\Theta_k(t) = \begin{cases} -\Delta y(t) & : k = 1 \\ e(t) & : k = 2 \\ -\Delta^2 y(t) & : k = 3. \end{cases} \tag{13}$$

However, the system Jacobian $\partial y(t+1)/\partial u(t)$ is required in order to calculate Eq.(12). Here, if this relation $x = |x|\mathrm{sign}(x)$ is used, the system Jacobian is given as follows:

$$\frac{\partial y(t+1)}{\partial u(t)} = \left|\frac{\partial y(t+1)}{\partial u(t)}\right| \mathrm{sign}\left(\frac{\partial y(t+1)}{\partial u(t)}\right), \quad (14)$$

where $\mathrm{sign}(x) = 1(x > 0), -1(x < 0)$. Now, if $|\partial y(t+1)/\partial u(t)|$ can be included in $g(t)$, it is enough to know only the sign of the system gradient in advance(S.Omatu and T.Yamamoto, 1996). Therefore, in this paper, it is assumed that this sign of the system gradient is known.

$y_m(t)$ denotes the output of the reference model, which is designed as follows:

$$y_m(t) = \frac{z^{-1}T(1)}{T(z^{-1})} r(t), \quad (15)$$

where $T(z^{-1})$ is the desired polynomial and is defined by

$$T(z^{-1}) := 1 + t_1 z^{-1} + t_2 z^{-2}. \quad (16)$$

Furthermore, $T(1)$ denotes the static gain of $T(z^{-1})$, and $T(z^{-1})$ is designed based on a literature (T.Yamamoto and S.L.Shah, 1998).

Next, the error criterion to switch from the CMAC to the NN is determined as follows:

$$\eta(epoc) := \frac{1}{N}\sum_{t=1}^{N}\left\{\frac{\varepsilon(t)}{r(t)}\right\}^2 \quad (17)$$

$$\Delta\eta(epoc) \le \bar{\eta}, \quad (18)$$

where N and $\eta(epoc)$ denote the number of steps per 1[epoc] and the scaling error, respectively. Here, since the design of $\bar{\eta}$ influences the control performance greatly, $\bar{\eta}$ is set as 0.01 as a measure in this paper, which is determined by examining several cases.

2.3 PID parameter tuning by NN

After the PID tuner is switched from the CMAC to the NN, PID gains are tuned using the NN whose detailed tuning rule is discussed below. The multilayered NN shown in Fig.3 is used in this paper. From Fig.3, the input layer for the NN is given as

$$\mathbf{I}(t) = [y(t-1), \cdots, y(t-n), u(t-1), \cdots, u(t-m), u_c(t-1), r(t), I_o], \quad (19)$$

where I_o denotes the threshold of the input layer, and $u_c(t)$ is given by

$$\Delta u_c(t) = K_{Ic}e(t) - K_{Pc}\Delta y(t) - K_{Dc}\Delta^2 y(t). \quad (20)$$

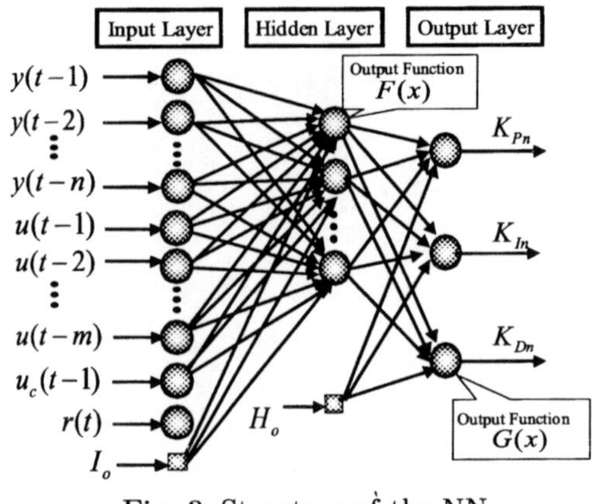

Fig. 3. Structure of the NN.

Furthermore, output functions of the hidden layer and the output layer are respectively defined as the following sigmoidal functions:

$$F(x) = 2\left\{\frac{1}{1+\exp(-ax)} - \frac{1}{2}\right\} \quad (21)$$

$$G(x) = c\left\{\frac{1}{1+\exp(-bx)} - \frac{1}{2}\right\}. \quad (22)$$

The output of the NN, $\mathbf{K_n} := [K_{Pn}, K_{In}, K_{Dn}]$, is calculated using the following equations:

$$T_j(t) = \sum_{i=0}^{p} V_{ij}(t) I_i(t) \quad (23)$$

$$H_j(t) = 2\left\{\frac{1}{1+\exp(-aT_j(t))} - \frac{1}{2}\right\} \quad (24)$$

$$S_k(t) = \sum_{j=0}^{q} W_{jk}(t) H_j(t) \quad (25)$$

$$O_k(t) = c\left\{\frac{1}{1+\exp(-bS_k(t))} - \frac{1}{2}\right\} \quad (26)$$

$$\mathbf{K_n}(t) = [O_1(t), O_2(t), O_3(t)], \quad (27)$$

where a, b and c denote the user-specified parameters, and p and q denote the number of neurons included in the input layer and the hidden layer, respectively. Furthermore, V_{ij} and W_{jk} respectively denote the weights of the input layer and the hidden layer. And H_o is the threshold of the hidden layer. Here, the PID gains $\mathbf{K}$ can be obtained by summing $\mathbf{K_c}$ and $\mathbf{K_n}$.

$$\mathbf{K}(t) = \mathbf{K_c}(t) + \mathbf{K_n}(t). \quad (28)$$

Next, the weights of the NN (V_{ij}, W_{jk}) are updated based on the BP method as follows:

$$W_{jk}(t+1) = W_{jk}(t) - \alpha_1 \frac{\partial J(t+1)}{\partial W_{jk}(t)}$$

$$= W_{jk}(t) + \alpha_1 \Theta_k(t)\varepsilon(t)\Big(\frac{c}{2} - O_k(t)\Big)\Big(\frac{c}{2} + O_k(t)\Big) \cdot \frac{b}{c} H_j(t) \frac{\partial y(t+1)}{\partial u_n(t)} \quad (29)$$

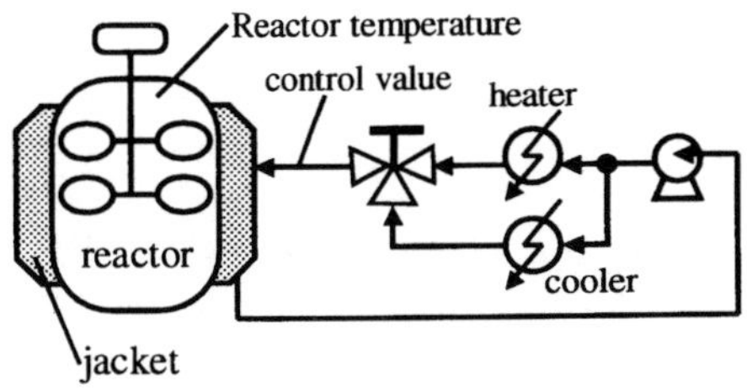

Fig. 4. The schematic figure of the polystyrene reactor model.

$$V_{ij}(t+1) = V_{ij}(t) - \alpha_2 \frac{\partial J(t)}{\partial V_{ij}(t)}$$

$$= V_{ij}(t) + \alpha_2 \Theta_k(t)\varepsilon(t)\frac{b}{c}\left(\frac{c}{2} - O_k(t)\right)\left(\frac{c}{2} + O_k(t)\right)$$

$$\cdot\frac{a}{2}(1-H_j(t))(1+H_j(t))I_i(t)W_{jk}(t)\frac{\partial y(t+1)}{\partial u_n(t)}, \quad (30)$$

where $\alpha_1 (> 0)$ and $\alpha_2 (> 0)$ are the learning coefficients. Furthermore, if the relation $\eta(\text{epoc}) \leq \eta_0$ is satisfied, the NN learning is stopped.

3. SIMULATION EXAMPLES

In order to evaluate the effectiveness of the newly proposed scheme, a simulation example for a nonlinear system is considered. As the nonlinear system, the following polystyrene reactor model is discussed. The schematic figure of the system is shown in Fig.4. The relation between the jacket temperature $u(t)$ and the reaction temperature $y(t)$ is described as follows(E.Nakanisi and Y.Hanakuma, 1992):

$$\begin{aligned} y(t) &= 0.804y(t-1) + 5.739 \times 10^{15} \\ &\cdot \exp\{-E_a/R(y(t-1)+273)\} \\ &+0.148u(t-1) + \xi(t), \end{aligned} \quad (31)$$

where $E_a = 240$, $R = 0.01986$, and $\xi(t)$ denotes the white Gaussian noise with zero mean and variance 0.1^2. Nonlinear properties of this system become strong in the case of $y(t) \geq 75$. The reference signal $r(t)$ is given as follows:

$$r(t) = \begin{cases} 60(0 \leq t < 100) \\ 70(100 \leq t < 200) \\ 80(200 \leq t < 300) \\ 75(300 \leq t \leq 400). \end{cases} \quad (32)$$

Furthermore, the desired polynomial $T(z^{-1})$ was designed as follows:

$$T(z^{-1}) = 1 - 1.558z^{-1} + 0.449z^{-2}. \quad (33)$$

And, the user-specified parameters included in the CMAC and the NN were determined as shown in Table 1.

First, for the purpose of comparison with the conventional schemes, the fixed PID control scheme

Table 1. User-specified parameters on the CMAC and the NN

(a) **CMAC**

Width of quantization	10
Number of weight tables	10
User-specified parameters included in $g(t)$	β_1=100 β_2=0.01 β_3=10^5

(b) **NN**

Number of units in input layer	$p = 9$
Number of units in hidden layer	$q = 20$
User-specified parameters included in sigmoidal functions	$a = 1$ $b = 1$ $c = 2$
Learning rate (hidden layer)	α_1=0.0001
Learning rate (output layer)	α_2=0.0001
Threshold in input layer	I_o=1
Threshold in hidden layer	H_o=1
Permissible error to finish training	η_0=1.0×10^{-4}

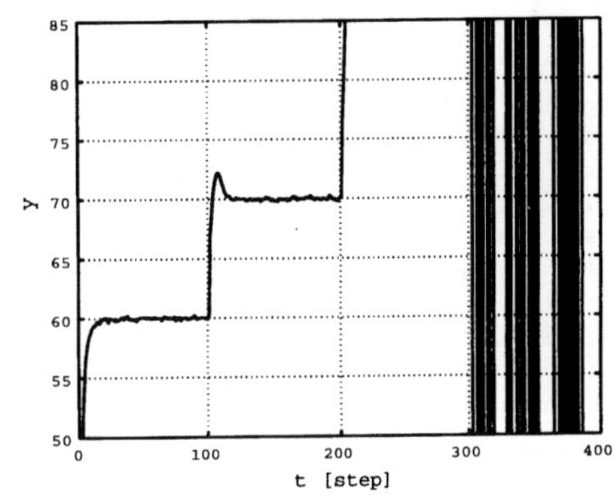

Fig. 5. Control result using the fixed PID controller.

widely used in industrial processes is employed, whose PID gains are tuned by using Chien, Hrones & Reswick (CHR) method (K.L.Chien and J.B.Reswick, 1972). Then, PID gains are as follows:

$$K_P = 6.5422, \quad K_I = 1.5522, \quad K_D = 1.8856. \quad (34)$$

The control result is shown in Fig.5. From Fig.5, owing to the nonlinearities of the controlled object, the control result by the fixed PID controller is oscillatory after 300[step].

Next, the control result using the proposed control scheme is shown in Fig.6, and then trajectories of PID gains are shown in Fig.7. From Fig.6 and Fig.7, the good control result can be obtained using the proposed method, because PID gains are adjusted adequately. Furthermore, error behaviors corresponding to the proposed method, the NN-PID scheme and the CMAC-PID scheme are shown in Fig.8. From Fig.8, it is clear that the learning speed using the NN-PID scheme is slow, and learning speed using the CMAC-PID scheme is very fast in the initial stage, but the learning speed is very slow subsequently. Here, the number of the first learning by the CMAC was 18[epoc] and the number of the additional learning by the

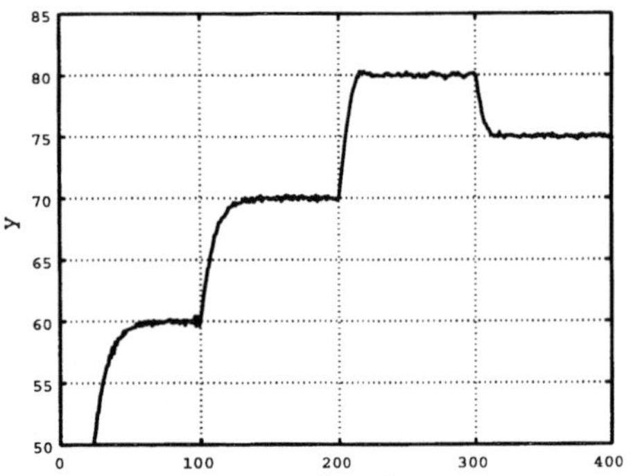

Fig. 6. Control result using the newly proposed control scheme.

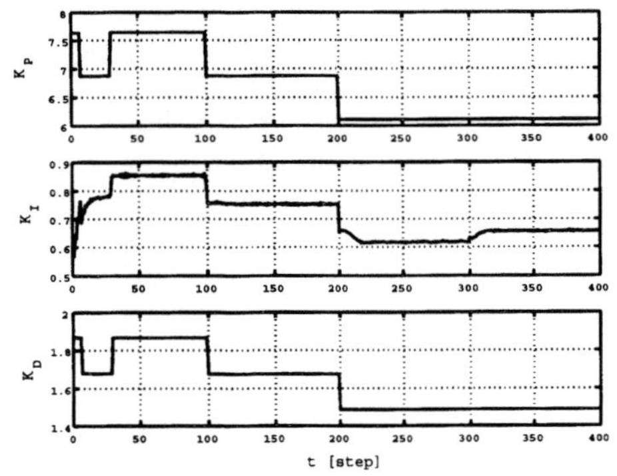

Fig. 7. Trajectories of PID gains corresponding to Fig.6.

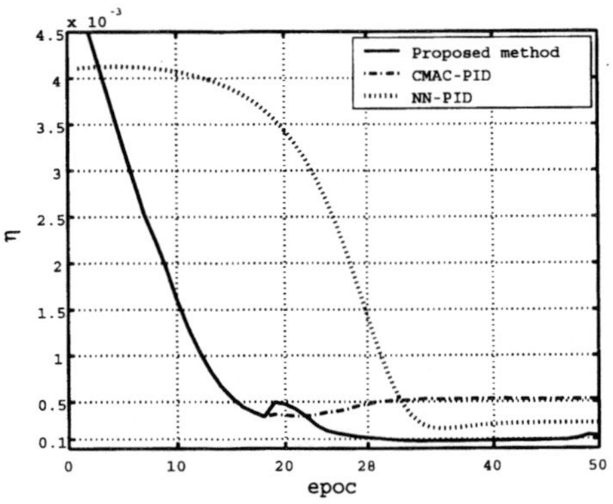

Fig. 8. Error behaviors where solid line, dashed and dotted line and dotted line are respectively corresponding to the proposed scheme, CMAC-PID scheme and NN-PID scheme.

NN was 10[epoc], that is, the total number of the learning was 28[epoc] till this relation $\eta \leq \eta_0$ was satisfied. On the other hand, control results could not be obtained so that the criterion $\eta \leq \eta_0$ was satisfied in the case where the only CMAC or the only NN was employed.

From these results, it is clear that the newly proposed scheme works well. The adaptability for non-trained reference signals has been examined. The control results are omitted due to the page limit.

4. CONCLUSIONS

In this paper, a new intelligent PID tuner has been proposed, which has a fusional structure of the CMAC and the NN. According to the newly proposed scheme, the roles in learning the PID tuner are divided into the CMAC and the NN. That is, the CMAC firstly work in the initial stage in order to roughly approximate the nonlinear components, and then the NN works to further improve the control performance. This structure enables us to drastically reduce the computational burden. The effectiveness of the newly proposed scheme has been verified on a simulation example of the chemical reactor model. PID gains have changed adequately corresponding to the operating points (reference signals), and suitable control results have been obtained. The application study of the proposed scheme is our future work.

REFERENCES

C.-Y.Seong and B.Widrow (2001). Neural dynamic optimization for control systems. *IEEE Trans. on Systems, Man, and Cybernetics* **31**, 482–513.

E.Nakanisi and Y.Hanakuma (1992). *The foundation and the practice of a process control(in Japanese).* Asakura. Tokyo.

J.G.Zieglar and N.B.Nichols (1942). Optimum settings for automatic controllers. *Trans. ASME* **64**, 759–768.

J.S.Albus (1975*a*). Data storage in the control cerebellar model articulation control (cmac). *Trans. on ASME, J. of Dynamic Systems, Measurement, and Control* **9**, 228–233.

J.S.Albus (1975*b*). A new approach to manipulator control cerebellar model articulation control (cmac). *Trans. on ASME, J. of Dynamic Systems, Measurement, and Control* **9**, 220–227.

K.J. Åström, T. Hägglund (1988). *Automatic Tuning of PID Controllers.* Instrument Society of America. America.

K.L.Chien, J.A.Hrones and J.B.Reswick (1972). On the automatic control of generalized passive systems. *Trans. ASME* **74**, 175–185.

K.S.Narendra and K.Parthasarathy (1990). Identification and control of dynamical systems using neural networks. *IEEE Trans. on Neural Networks* **1**, 1–27.

S.Omatu and T.Yamamoto (1996). *Self-Turning Control.* SICE Learned Book. Japan.

S.Omatu, K.Marzuki and Y.Rubiyah (1995). *Neuro-Control and Its Applications.* Springer-Verlag. London.

T.Yamamoto and S.L.Shah (1998). Design and experimental evaluation of a multivariable self-tuning pid controller. *Proc. of IEEE Conference on Control Application* pp. 1230–1234.

ELSEVIER
IFAC
PUBLICATIONS
www.elsevier.com/locate/ifac

BOUNDING LINEAR TIME VARYING HYBRID SYSTEMS WITH TIME EVENTS [1]

Cha Kun Lee * **Paul I. Barton** *

* *Department of Chemical Engineering*
Massachusetts Institute of Technology
77 Massachusetts Avenue, Cambridge, MA 02139, USA

Abstract: The problem of bounding the states of a linear time varying hybrid system at fixed transition times where the mode sequence is allowed to vary is considered. It is shown via an illustrative example that a simple decomposition algorithm produces weak bounds which get worse as the number of epochs increases. To address this issue, a novel algorithm is proposed, based on solving families of relaxed linear programming problems, which allows the incorporation of additional constraints derived from physical insight. This provides tighter bounds while avoiding explicit enumeration of all possible mode trajectories.

Keywords: varying mode sequence, exact state bounds, differential equations, optimization problems

1. INTRODUCTION

Continuous time hybrid systems have become the modeling framework of choice for a wide variety of applications that require detailed dynamic models with embedded discontinuities. In general, these time dependent, nonlinear models exhibit model switching and state jumps as a consequence of both time and state dependent events (Barton and Lee, 2002). Hybrid system models are important in many areas of science and engineering, including the analysis of digital circuits, signaling and decision making mechanisms in (biological) cells, robotic systems, air and ground traffic management systems, sequential operations, safety interlock systems, and embedded systems. Economic and safety considerations in these applications, such as the automated design of safe operating procedures and the formal verification of embedded systems, strongly motivate the development of algorithms and tools for the global optimization of hybrid systems.

Many modern, general methods for deterministic global optimization in Euclidean spaces rely on the notion of a convex relaxation of a nonconvex function (McCormick, 1976). This is a convex function which underestimates a nonconvex function on the set of interest. The convex programs that result from convex relaxation of all nonconvex objective and constraint functions in a problem formulation can (in principle) be solved to guaranteed global optimality, which, for example, can be used to generate rigorous lower bounds on the nonconvex problem for a branch and bound (B&B) algorithm (Horst and Tuy, 1993).

Recently, a convexity theory has been developed that enables well known symbolic convex relaxations on Euclidean spaces (McCormick, 1976; Adjiman *et al.*, 1998) to be harnessed in the construction of convex relaxations of general, nonconvex Bolza type functionals subject to an embedded linear time varying (LTV) hybrid system

[1] This material is based upon work supported by the National Science Foundation under Grant No. CCR-0208956.

where the transition times are fixed, and the sequence of modes, T_μ, is known (Lee *et al.*, 2004). The construction of a set enclosing the image of the parameter space under the solution of the hybrid system, $\mathcal{X}^{(i)}$, is critical for obtaining tight (accurate) convex relaxations of the participating functionals. Note that $\mathcal{X}^{(i)}$ represents the state bounds for the embedded hybrid system for all values of the parameters. The better the estimate of $\mathcal{X}^{(i)}$ is, the tighter the relaxations obtained. Consequently, tighter lower bounds are obtained, increasing the efficiency of the global optimization algorithm.

In (Barton and Lee, 2003), a mixed-integer reformulation is proposed to deal with the problem when T_μ is allowed to vary and becomes an optimization parameter. This results in a nonconvex mixed-integer nonlinear programming (MINLP) problem. In particular, auxiliary optimization parameters, $\mathbf{Z}$, are introduced to represent the initial conditions for each epoch. In this case, $\mathcal{X}^{(i)}$ becomes a larger set containing the image of the parameter space under the solution of the hybrid system for all possible T_μ, and the bounds on $\mathbf{Z}$ are obtained from estimating $\mathcal{X}^{(i)}$ at the time events, i.e., the state bounds at the beginning of each epoch.

For analogous reasons to those presented above, it is very desirable to obtain tight bounds for $\mathbf{Z}$, as these bounds are needed to construct a convex relaxation of the nonconvex MINLP. Although the exact state bounds at the beginning of each epoch can be obtained with an explicit enumeration of all possible T_μ, the cost of doing so clearly becomes prohibitive (increases exponentially) as the number of epochs increases. To deal with this problem, a decomposition algorithm for estimating valid state bounds was proposed in (Barton and Lee, 2003).

In this paper, we formally present the decomposition algorithm, and show that it produces weak bounds which deteriorate as the number of epochs increases. To address this issue, a novel algorithm is proposed to obtain tighter state bounds, based on solving a family of relaxations of mixed-integer linear programming (MILP) problems as LP problems.

2. LTV HYBRID SYSTEMS

The modeling framework of (Barton and Lee, 2002) is used to define the LTV hybrid system of interest. The time horizon is partitioned into contiguous intervals called *epochs*. We define a *hybrid time trajectory*, T_τ, as a finite sequence of epochs $\{I_i\}$ terminating with epoch I_{n_e}, where n_e is fixed, and is the total number of epochs. Each epoch is a closed time interval $I_i = [\sigma_i, \tau_i] \subset \mathbb{R}$, $\sigma_i = \tau_{i-1}$ for $i = 2, \ldots, n_e$, $\sigma_1 \leq \tau_1$, and $\tau_{i-1} \leq \tau_i$ for all $i = 2, \ldots, n_e$. For epoch I_i, the system evolves continuously in time if $\sigma_i < \tau_i$, and it evolves discretely by making an instantaneous transition if $\sigma_i = \tau_i$. The continuous state subsystems are called *modes* and the corresponding sequence of modes for T_τ is called the *hybrid mode trajectory*, T_μ. At the end of epoch I_i, a transition is made from the *predecessor* mode in I_i to a *successor* mode in epoch I_{i+1}, also called an *event*, when the *transition condition* is satisfied.

Definition 1. The LTV ODE hybrid system of interest is defined by the following.

1. An index set of modes potentially visited along T_μ, $M = \{1, \ldots, n_m\}$, and a fixed T_τ with given time events (i.e., explicit transition times) $\sigma_1, \tau_1, \tau_2, \ldots, \tau_{n_e}$. It is clear that $T_\mu = \{m_i\}$, where $m_i \in M$. Henceforth, the superscript (m) will refer to any mode in M, while superscript (m_i) will refer to the active mode in epoch I_i;
2. An invariant structure system where the number of continuous state variables is constant between modes, $V = (\mathbf{x}(\mathbf{p}, T_\mu, t), \mathbf{p})$, where $\mathbf{p} \in P \subset \mathbb{R}^{n_p}$, and $\mathbf{x}(\mathbf{p}, T_\mu, t) \in \mathbb{R}^{n_x}$ for all $(\mathbf{p}, T_\mu, t) \in P \times M^{n_e} \times I_i$, $i = 1, \ldots, n_e$;
3. The LTV ODE system for each mode $m \in M$, which is given by
$$\dot{\mathbf{x}}(\mathbf{p}, T_\mu, t) = \mathbf{A}^{(m)}(t)\mathbf{x}(\mathbf{p}, T_\mu, t) + \mathbf{B}^{(m)}(t)\mathbf{p} + \mathbf{q}^{(m)}(t), \quad (1)$$
where $\mathbf{A}^{(m)}(t)$ is continuous on $[\sigma_1, \tau_{n_e}]$, $\mathbf{B}^{(m)}(t)$ and $\mathbf{q}^{(m)}(t)$ are piecewise continuous on $[\sigma_1, \tau_{n_e}]$ and defined at any point of discontinuity, for all $m \in M$;
4. The transition conditions for the transitions between epochs I_i and I_{i+1}, $i = 1, \ldots, n_e - 1$, which are explicit time events:
$$L^{(m_i)} := (t = \tau_i), \quad (2)$$
indicating the transition from mode m_i in epoch I_i to mode m_{i+1} in epoch I_{i+1} at time τ_i;
5. The system of transition functions, which is given by
$$\mathbf{x}(\mathbf{p}, T_\mu, \sigma_{i+1}) = \mathbf{D}_i\mathbf{x}(\mathbf{p}, T_\mu, \tau_i) + \mathbf{E}_i\mathbf{p} + \mathbf{k}_i, \quad \forall\, i = 1, \ldots, n_e - 1, \quad (3)$$
for the transition from mode m_i in epoch I_i to mode m_{i+1} in epoch I_{i+1}; and
6. A given initial condition for mode m_1,
$$\mathbf{x}(\mathbf{p}, T_\mu, \sigma_1) = \mathbf{E}_0\mathbf{p} + \mathbf{k}_0. \quad (4)$$

Definition 2. Let P be a nonempty compact convex subset of $\mathbb{R}^{n_p}$. Define the following sets for all $i = 1, \ldots, n_e$ where $\underline{t}$ denotes fixed t:
$$\mathcal{X}^{(i)}(\underline{t}) \equiv \{\mathbf{x}(\mathbf{p}, T_\mu, \underline{t}) \mid \mathbf{p} \in P, T_\mu \in M^{n_e}, \underline{t} \in I_i\}, \quad (5)$$

$$\mathcal{X}^{(i)} \equiv \bigcup_{\underline{t} \in I_i} \mathcal{X}^{(i)}(\underline{t}). \quad (6)$$

3. THE DECOMPOSITION ALGORITHM

Consider the following dynamic system,

$$\dot{\mathbf{x}}(\mathbf{p}, \tilde{\mathbf{z}}, t) = \mathbf{A}^{(m)}(t)\mathbf{x}(\mathbf{p}, \tilde{\mathbf{z}}, t) + \mathbf{B}^{(m)}(t)\mathbf{p} + \mathbf{q}^{(m)}(t), \quad (7)$$

$$\mathbf{x}(\mathbf{p}, \tilde{\mathbf{z}}, \sigma) = \tilde{\mathbf{z}}, \quad (8)$$

for some $m \in M$, where $\sigma < \tau, t \in T \equiv [\sigma, \tau]$, $\mathbf{p} \in P \subset \mathbb{R}^{n_p}$, $\tilde{\mathbf{z}} \in \tilde{Z} \subseteq \mathbb{R}^{n_x}$. Define the following set:

$$\mathcal{X}^a(\underline{t}) \equiv \{\mathbf{x}(\mathbf{p}, \tilde{\mathbf{z}}, \underline{t}) \mid \mathbf{p} \in P, \tilde{\mathbf{z}} \in \tilde{Z}, \underline{t} \in T\}. \quad (9)$$

Theorem 3. (Singer and Barton, 2003) Given $P \equiv [\mathbf{p}^L, \mathbf{p}^U]$ and $\tilde{Z} \equiv [\tilde{\mathbf{z}}^L, \tilde{\mathbf{z}}^U]$, the set $\mathcal{X}^a(\underline{t}) \equiv [\mathbf{x}^L(\underline{t}), \mathbf{x}^U(\underline{t})]$ for $\underline{t} \in T$ can be calculated pointwise in time from the following interval equation,

$$[\mathbf{x}](\underline{t}) = \mathbf{M}(\underline{t})[\mathbf{w}] + \mathbf{n}(\underline{t}), \quad (10)$$

where $\mathbf{w} = (\mathbf{p}, \tilde{\mathbf{z}})$, $\mathbf{w} \in W \equiv [\mathbf{w}^L, \mathbf{w}^U]$, $\mathbf{w}^L = (\mathbf{p}^L, \tilde{\mathbf{z}}^L)$, $\mathbf{w}^U = (\mathbf{p}^U, \tilde{\mathbf{z}}^U)$, and $\mathbf{M}(t)$ and $\mathbf{n}(t)$ are given by the solution of the following LTV system,

$$\dot{\mathbf{M}}(t) = \mathbf{A}^{(m)}(t)\mathbf{M}(t) + \mathbf{H}^{(m)}(t), \quad (11)$$

$$\dot{\mathbf{n}}(t) = \mathbf{A}^{(m)}(t)\mathbf{n}(t) + \mathbf{q}^{(m)}(t), \quad (12)$$

$$\mathbf{M}(\sigma) = \mathbf{L}, \quad (13)$$

$$\mathbf{n}(\sigma) = \mathbf{0}, \quad (14)$$

where $\mathbf{H}^{(m)}(t) = [\mathbf{B}^{(m)}(t)\ \mathbf{0}]$, $\mathbf{L} = [\mathbf{0}\ \mathbf{I}]$, and $\mathbf{I}$ is the identity matrix of rank n_x.

Remark 4. The functional form of the solution of the LTV system is affine in the parameters $\mathbf{w}$,

$$\mathbf{x}(\mathbf{w}, t) = \mathbf{M}(t)\mathbf{w} + \mathbf{n}(t). \quad (15)$$

The entries in $\mathbf{M}(t)$ are clearly the parametric sensitivities of the dynamic system, $\frac{\partial \mathbf{x}}{\partial \mathbf{w}}(t)$. Hence, (11) and (13) are simply the forward sensitivity equations of the embedded dynamic system in (7) and (8). We note that for problems where the number of parameters is much greater than the state variables, it might be more attractive to employ adjoint methods to calculate the required parametric sensitivities at the specified final time, i.e., calculating $\mathbf{M}(\tau)$ to construct $\mathcal{X}^a(\tau)$. However, it is beyond the scope of this paper to elaborate on this issue.

Remark 5. The bounds $\mathbf{x}^L(t)$ and $\mathbf{x}^U(t)$ from (10) are *exact* in the following sense. For any $i \in \{1, \ldots, n_x\}$, and any $\underline{t} \in T$, the following relationship holds,

$$x_i(\mathbf{w}^*, \underline{t}) = x_i^L(\underline{t}) \leq x_i(\mathbf{w}, \underline{t}) \leq x_i^U(\underline{t}) = x_i(\mathbf{w}^\dagger, \underline{t}), \ \forall\ \mathbf{w} \in W, \quad (16)$$

for some $\mathbf{w}^*, \mathbf{w}^\dagger \in W$.

From Remark 5, we know that exact bounds for $\mathbf{x}(\tau)$ can be constructed for each subproblem in the mixed-integer reformulation presented in (Barton and Lee, 2003) once the bounds for $\tilde{\mathbf{z}}$ are known. This suggests the following decomposition algorithm for estimating the bounds on $\mathbf{Z} \in Z$, where $\mathbf{z}_i$ represents the initial conditions for epoch I_i.

Algorithm 1. (A1).

(1) Initialize i=1.

(2) For $m = 1$ to n_m do:

  (a) Integrate the following system from σ_1 to τ_1, and store $\mathbf{M}_1^{(m)}(\tau_1)$ and $\mathbf{n}_1^{(m)}(\tau_1)$.

$$\dot{\mathbf{M}}_1^{(m)}(t) = \mathbf{A}^{(m)}(t)\mathbf{M}_1^{(m)}(t) + \mathbf{B}^{(m)}(t), \quad (17)$$

$$\dot{\mathbf{n}}_1^{(m)}(t) = \mathbf{A}^{(m)}(t)\mathbf{n}_1^{(m)}(t) + \mathbf{q}^{(m)}(t), \quad (18)$$

$$\mathbf{M}_1^{(m)}(\sigma_1) = \mathbf{E}_0, \quad (19)$$

$$\mathbf{n}_1^{(m)}(\sigma_1) = \mathbf{k}_0. \quad (20)$$

  (b) Calculate and store $[{\mathbf{x}^{(m)}}^L(\sigma_2), {\mathbf{x}^{(m)}}^U(\sigma_2)]$ from

$$[\mathbf{x}^{(m)}](\sigma_2) = \left(\mathbf{D}_1\mathbf{M}_1^{(m)}(\tau_1) + \mathbf{E}_1\right)[\mathbf{p}] + \mathbf{D}_1\mathbf{n}_1^{(m)}(\tau_1) + \mathbf{k}_1. \quad (21)$$

(3) For $j = 1$ to n_x do:

  (a) Calculate and store the jth element of $[\mathbf{z}_{i+1}^L, \mathbf{z}_{i+1}^U]$ from

$$(\mathbf{z}_{i+1}^L)_j = \min_{m \in M} {x_j^{(m)}}^L(\sigma_{i+1}), \quad (22)$$

$$(\mathbf{z}_{i+1}^U)_j = \max_{m \in M} {x_j^{(m)}}^U(\sigma_{i+1}). \quad (23)$$

(4) For $i = 2$ to $(n_e - 1)$ do:

  (a) For $m = 1$ to n_m do:

    (i) Integrate the system (11), (12), (13) and (14) from $\sigma = \sigma_i$ to $\tau = \tau_i$, and store $\mathbf{M}_i^{(m)}(\tau_i) \leftarrow \mathbf{M}(\tau)$ and $\mathbf{n}_i^{(m)}(\tau_i) \leftarrow \mathbf{n}(\tau)$.

    (ii) Calculate and store $[{\mathbf{x}^{(m)}}^L(\sigma_{i+1}), {\mathbf{x}^{(m)}}^U(\sigma_{i+1})]$ from

$$[\mathbf{x}^{(m)}](\sigma_{i+1}) = \left(\mathbf{D}_i\mathbf{M}_i^{(m)}(\tau_i) + \mathbf{L}_i\right)[\mathbf{w}] + \mathbf{D}_i\mathbf{n}_i^{(m)}(\tau_i) + \mathbf{k}_i. \quad (24)$$

where $\mathbf{w}^L = (\mathbf{p}^L, \mathbf{z}_i^L)$, $\mathbf{w}^U = (\mathbf{p}^U, \mathbf{z}_i^U)$, and $\mathbf{L}_i = [\mathbf{E}_i\ \mathbf{0}]$.

  (b) Calculate and store $[\mathbf{z}_{i+1}^L, \mathbf{z}_{i+1}^U]$ as in Step (3) above.

Remark 6. The staggered corrector method can be used for efficient integration of the dynamic systems (Feehery *et al.*, 1997).

Remark 7. The system (11), (12), (13) and (14) is independent of the parameters $\mathbf{w}$, hence the values of $\mathbf{M}(\tau)$ and $\mathbf{n}(\tau)$ are also independent of $\mathbf{w}$. Hence, if the epochs are of equal duration, i.e., $\tau_i - \sigma_i$ is constant for all i, and we have a linear time invariant hybrid system, step (4ai) only needs to be executed once for $i = 2$.

Remark 8. Note that $\mathbf{M}_1$ is a $n_x \times n_p$ matrix, while $\mathbf{M}_{i\neq 1}$ is a $n_x \times (n_p + n_x)$ matrix.

Although Theorem 3 guarantees exact bounds for the system (7) and (8), the bounds obtained from implementing (A1) have no guarantee of being exact for $\mathbf{Z}$ past the first epoch. This arises because bounds for different elements of $[\mathbf{z}_i^L, \mathbf{z}_i^U]$, $i > 2$, could come from different predecessor modes I_{i-1}, and this is illustrated in the example presented below.

One way to obtain exact bounds is to solve the bounding equations (see (Lee *et al.*, 2004) for obtaining the exact bounds for fixed T_μ) for all possible combinations of T_μ. This method clearly suffers from exponential complexity in the number of epochs, and an alternative algorithm for computing tighter bounds for $\mathbf{Z}$ is needed.

4. THE RELAXED LP ALGORITHM

Consider the following problem.

Problem 9. (P1(α,β)).

$$\min_{\mathbf{p}\in P, \mathbf{Y}\in Y^b, \mathbf{Z}} \mathbf{e}_\beta^{\mathrm{T}} \mathbf{z}_{\alpha+1} \tag{25}$$

$$\text{s.t. } \sum_{m=1}^{n_m} y_{mi} = 1, \ \forall\, i = 1,\ldots,\alpha, \tag{26}$$

$$\mathbf{z}_{i+1} = \sum_{m=1}^{n_m} y_{mi}\Big(\mathbf{D}_i \mathbf{x}_{mi}(\mathbf{p},\mathbf{Z},\tau_i) + \mathbf{E}_i\mathbf{p} + \mathbf{k}_i\Big), \quad \forall\, i = 1,\ldots,\alpha, \tag{27}$$

$$\mathbf{z}_1 = \mathbf{E}_0\mathbf{p} + \mathbf{k}_0, \tag{28}$$

where $Y^b \equiv \{0,1\}^{n_m\times\alpha} \subset Y \equiv [0,1]^{n_m\times\alpha}$, $\mathbf{Z} \in \mathbb{R}^{n_x\times(\alpha+1)}$, and the unit vector $\mathbf{e}_\beta$ is the βth column of the rank n_x identity matrix; $\mathbf{x}_{mi}(\mathbf{p},\mathbf{Z},t)$ are given by the solution of the following embedded LTV ODE systems for all $m \in M$, $i = 1,\ldots,\alpha$,

$$\dot{\mathbf{x}}_{mi}(\mathbf{p},\mathbf{Z},t) = \mathbf{A}^{(m)}(t)\mathbf{x}_{mi}(\mathbf{p},\mathbf{Z},t) + \mathbf{B}^{(m)}(t)\mathbf{p} + \mathbf{q}^{(m)}(t), \ \forall\, t \in I_i, \tag{29}$$

$$\mathbf{x}_{mi}(\mathbf{p},\mathbf{Z},\sigma_i) = \mathbf{z}_i. \tag{30}$$

Problem (P1) determines the exact lower bound for the βth component of $\mathbf{x}(\mathbf{p}, T_\mu, \sigma_{\alpha+1}) = \mathbf{z}_{\alpha+1}$. We can construct a convex relaxation for (P1) by treating the bilinear terms in (27) using the exact linearizations in (Glover, 1975). We can then formulate the following, equivalent, MILP.

Problem 10. (P2(α,β)).

$$\min_{\mathbf{p},\mathbf{Y},\mathbf{Z},\mathbf{V},\mathbf{W},\mathbf{S}} \mathbf{e}_\beta^{\mathrm{T}} \mathbf{z}_{\alpha+1} \tag{31}$$

$$\text{s.t. } \sum_{m=1}^{n_m} y_{mi} = 1, \ \forall\, i = 1,\ldots,\alpha, \tag{32}$$

$$\mathbf{z}_{i+1} = \sum_{m=1}^{n_m} \mathbf{s}_{mi}, \ \forall\, i = 1,\ldots,\alpha, \tag{33}$$

$$\mathbf{z}_1 = \mathbf{E}_0\mathbf{p} + \mathbf{k}_0, \tag{34}$$

$$\mathbf{v}_{mi}^U(y_{mi} - 1) + \mathbf{v}_{mi} \le \mathbf{s}_{mi} \le \mathbf{v}_{mi}^L(y_{mi} - 1) + \mathbf{v}_{mi}, \ \forall\, m \in M, i = 1,\ldots,\alpha, \tag{35}$$

$$\mathbf{v}_{mi}^L y_{mi} \le \mathbf{s}_{mi} \le \mathbf{v}_{mi}^U y_{mi}, \ \forall\, m \in M, i = 1,\ldots,\alpha, \tag{36}$$

$$\mathbf{v}_{m1} = \Big(\mathbf{D}_1\mathbf{M}_1^{(m)}(\tau_1) + \mathbf{E}_1\Big)\mathbf{p} + \mathbf{D}_1\mathbf{n}_1^{(m)}(\tau_1) + \mathbf{k}_1, \ \forall\, m \in M, \tag{37}$$

$$\mathbf{v}_{mi} = \Big(\mathbf{D}_i\mathbf{M}_i^{(m)}(\tau_i) + \mathbf{L}_i\Big)\mathbf{w}_i + \mathbf{D}_i\mathbf{n}_i^{(m)}(\tau_i) + \mathbf{k}_i, \ \forall\, m \in M, i = 2,\ldots,\alpha, \tag{38}$$

$$\mathbf{w}_i = (\mathbf{p}, \mathbf{z}_i), \ \forall\, i = 2,\ldots,\alpha, \tag{39}$$

$$\mathbf{w}_1 = \mathbf{0}, \tag{40}$$

where $\mathbf{Y} \in Y^b \equiv \{0,1\}^{n_m\times\alpha} \subset Y \equiv [0,1]^{n_m\times\alpha}$, $\mathbf{Z} \in \mathbb{R}^{n_x\times(\alpha+1)}$, $\mathbf{V} \in V \subset \mathbb{R}^{n_x\times n_m\times\alpha}$, $\mathbf{W} \in \mathbb{R}^{(n_p+n_x)\times\alpha}$, $\mathbf{S} \in \mathbb{R}^{n_x\times n_m\times\alpha}$, and the unit vector $\mathbf{e}_\beta$ is the βth column of a rank n_x identity matrix; $\mathbf{L}_i = [\mathbf{E}_i\ \mathbf{0}]$; $\mathbf{M}_1^{(m)}(\tau_1)$ and $\mathbf{n}_1^{(m)}(\tau_1)$ are given by the solution of the system (17), (18), (19) and (20) from σ_1 to τ_1 for $m \in M$; and $\mathbf{M}_i^{(m)}(\tau_i)$ and $\mathbf{n}_i^{(m)}(\tau_i)$ are given by the solution of the system (11), (12), (13) and (14) from $\sigma = \sigma_i$ to $\tau = \tau_i$, for $m \in M$, $i = 2,\ldots,\alpha$.

The required bounds on the auxiliary variables $\mathbf{V}$ (see (35) and (36)) constitute the set V, and can be determined sequentially for each epoch (see algorithm below). The variables $\mathbf{Z}$, $\mathbf{W}$, $\mathbf{S}$ are left as free or unrestricted variables. While it is impractical to solve a family of MILPs (P2) to obtain the tightest bounds for $\mathbf{Z}$, it is much cheaper to solve (P2) on the relaxed space $\mathbf{Y} \in Y$, resulting in solving a family of relaxed LPs to provide valid (but not exact) bounds for $\mathbf{Z}$. This constitutes the following algorithm.

Algorithm 2. (A2).

(1) Execute steps (1), (2) and (3) in (A1).
(2) For $i = 2$ to $(n_e - 1)$ do:
  (a) For $m = 1$ to n_m do:
    (i) Integrate the system (11), (12), (13) and (14) from $\sigma = \sigma_i$ to $\tau = \tau_i$, and store $\mathbf{M}_i^{(m)}(\tau_i) \leftarrow \mathbf{M}(\tau)$ and $\mathbf{n}_i^{(m)}(\tau_i) \leftarrow \mathbf{n}(\tau)$.
    (ii) Calculate and store $[\hat{\mathbf{x}}^{(m)^L}(\sigma_{i+1}), \hat{\mathbf{x}}^{(m)^U}(\sigma_{i+1})]$ from

$$[\hat{\mathbf{x}}^{(m)}](\sigma_{i+1}) = \left(\mathbf{D}_i\mathbf{M}_i^{(m)}(\tau_i) + \mathbf{L}_i\right)[\mathbf{w}] + \mathbf{D}_i\mathbf{n}_i^{(m)}(\tau_i) + \mathbf{k}_i. \quad (41)$$

where $\mathbf{w}^L = (\mathbf{p}^L, \mathbf{z}_i^L)$, $\mathbf{w}^U = (\mathbf{p}^U, \mathbf{z}_i^U)$, and $\mathbf{L}_i = [\mathbf{E}_i \ \mathbf{0}]$.

  (b) For $m = 1$ to n_m do:
    (i) For $j = 1$ to n_x do:
      (A) Solve (P2(i,j)), with $[\mathbf{v}_{\lambda\theta}] = [\mathbf{x}^{(\lambda)}](\sigma_{\theta+1})$, $\theta = 1, \ldots, i-1$, and $[\mathbf{v}_{\lambda i}] = [\hat{\mathbf{x}}^{(\lambda)}](\sigma_{i+1})$, for all $\lambda \in M$, on the relaxed space Y, with the following constraint,

$$y_{mi} = 1, \quad (42)$$

and store $x_j^{(m)^L}(\sigma_{i+1}) \leftarrow$ objective.
      (B) Repeat step (A) as a maximization problem, and store $x_j^{(m)^U}(\sigma_{i+1}) \leftarrow$ objective.

  (c) For $j = 1$ to n_x do:
    (i) Calculate and store the jth element of $[\mathbf{z}_{i+1}^L, \mathbf{z}_{i+1}^U]$ from

$$(\mathbf{z}_{i+1}^L)_j = \min_{m \in M} x_j^{(m)^L}(\sigma_{i+1}), \quad (43)$$

$$(\mathbf{z}_{i+1}^U)_j = \max_{m \in M} x_j^{(m)^U}(\sigma_{i+1}). \quad (44)$$

5. AN ILLUSTRATIVE EXAMPLE

Example 11. Consider an isothermal plug flow reactor (PFR) operating at steady state, and 3 possible choices of catalyst. The reaction scheme, initial conditions and associated rate constants are shown in Fig. 1, where x_i represents the molar concentration of component i (kmol m^{-3}) and k_j represents the rate constant of reaction j (h^{-1}). The PFR has a uniform cross-sectional area of 1 m^2, and a constant volumetric flow rate of 1 m^3 h^{-1}. In this example, the independent variable t is the length, l, of the reactor. Determine the bounds on the concentration of the reactant and products at the beginning of each reactor section.

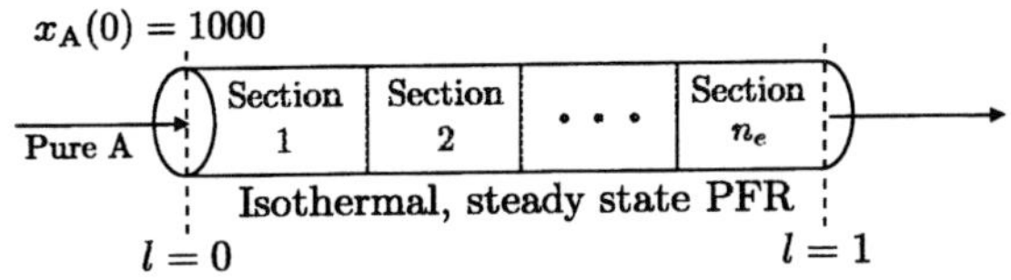

Reaction scheme

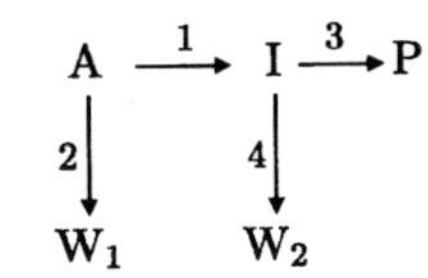

Catalyst

	k_1	k_2	k_3	k_4
1	2.098	1.317	0.021	0.033
2	29.53	110.2	0.295	0.079
3	182.6	2325	1.826	0.143

Kinetics

$\dot{x}_A = -(k_1 + k_2)x_A$ $\quad \dot{x}_I = k_1 x_A - (k_3 + k_4)x_I$

$\dot{x}_{W_1} = k_2 x_A$ $\quad \dot{x}_{W_2} = k_4 x_I$

$\dot{x}_P = k_3 x_I$

Fig. 1. Chemical reaction scheme and kinetics for PFR example

Note that the choice of catalyst corresponds to the choice of the sequence of modes in a linear hybrid system with 3 modes (each mode corresponds to the choice of a different catalyst) and n_e epochs (each epoch corresponds to a section of the reactor), with state continuity at the transitions.

Tables 1 and 2 show the bounds obtained for $\mathbf{z}_8$ and $\mathbf{z}_{15}$ when $n_e = 15$ when explicit enumeration (EE) (which obtains the exact bounds), (A1) and (A2) are used. As can be seen, (A2) produces tighter bounds than (A1). When physical infor-

Table 1. Bounds for $\mathbf{z}_8$ where $n_e = 15$

Species	(EE) $\mathbf{z}_8^L$	(EE) $\mathbf{z}_8^U$	(A1) $\mathbf{z}_8^L$	(A1) $\mathbf{z}_8^U$
A	0.00	203.18	0.00	203.18
W$_1$	307.30	927.18	78.52	3628.49
I	29.08	493.23	29.08	735.13
W$_2$	1.48	12.68	0.76	29.59
P	2.98	139.37	0.49	373.51
Species	**(A2) $\mathbf{z}_8^L$**	**(A2) $\mathbf{z}_8^U$**	**(A2) with (45) $\mathbf{z}_8^L$**	**(A2) with (45) $\mathbf{z}_8^U$**
A	0.00	203.18	0.00	203.18
W$_1$	230.54	1734.62	307.30	959.02
I	29.08	493.23	29.08	493.23
W$_2$	1.19	16.16	1.19	16.16
P	1.26	190.19	1.28	180.97

Table 2. Bounds for $\mathbf{z}_{15}$ where $n_e = 15$

Species	(EE) $\mathbf{z}_{15}^L$	(EE) $\mathbf{z}_{15}^U$	(A1) $\mathbf{z}_{15}^L$	(A1) $\mathbf{z}_{15}^U$
A	0.00	41.28	0.00	41.28
W$_1$	369.73	927.18	78.52	4365.72
I	11.60	567.77	11.60	815.88
W$_2$	2.43	27.87	1.08	79.06
P	7.34	293.77	0.69	1005.21
Species	**(A2) $\mathbf{z}_{15}^L$**	**(A2) $\mathbf{z}_{15}^U$**	**(A2) with (45) $\mathbf{z}_{15}^L$**	**(A2) with (45) $\mathbf{z}_{15}^U$**
A	0.00	41.28	0.00	41.28
W$_1$	230.54	2030.39	321.03	981.59
I	11.60	567.77	11.60	567.77
W$_2$	1.54	44.77	1.54	44.77
P	1.51	544.39	1.53	451.06

mation from the problem can be used, e.g., conservation of molar species, we can add the following additional constraints to (P2),

$$\sum_{j=1}^{n_x}(\mathbf{s}_{mi})_j = 1000y_{mi},\ \forall\ m \in M, i = 1,\ldots,\alpha. \tag{45}$$

When this physical insight is employed, it can be seen that the bounds obtained from (A2) with (45) produces tighter bounds than using (A2) alone. The reason why (A2) itself does not produce bounds which obey this conservation law is that the linearizations of the bilinear terms in (27) are only exact on the space Y^b, and not on the space Y. Hence, we have to enforce the law with (45). Note that there is no way to incorporate additional constraints within (A1). For further illustration, the upper bound computed for species W_1 at the beginning of each section when $n_e = 10$ is shown in Table 3.

Table 4 shows the bounds obtained for W_1 when the algorithms are trivially extended to calculate the bounds at $l = 1$. It can be seen that the bounds obtained from (A1) and (A2) deteriorate significantly from the exact bounds as n_e increases. When physical insight (45) is employed in conjunction with (A2), much tighter bounds are obtained.

All calculations were performed on an AMD 1.2 GHz, 1 GB RAM machine using CPLEX 7.5 as the LP solver. All LPs were started cold, and we note that the computational times for (A2) would improve if the LPs were warm started where possible. The computation times for the algorithms are shown in Table 5, from which the exponential explosion of (EE) is clear.

6. CONCLUSION

The deterministic solution of global dynamic optimization problems with LTV hybrid systems embedded depends on the construction of convex relaxations of the participating functions. The quality of these relaxations in turn relies on the computation of tight state bounds. While the exact state bounds can be computed via explicit enumeration, this approach quickly becomes impractical as the number of epochs is increased. We show through an example that a simple and efficient decomposition approach based on calculating the exact state bounds for the subproblems of a mixed-integer reformulated problem produces weak bounds which deteriorate as the number of epochs increases. A novel algorithm is proposed based on the solution of families of MILPs as relaxed LPs. This algorithm is able to incorporate physical insight as additional constraints in the LPs, and produces significantly tighter bounds than the decomposition algorithm.

Table 3. Upper bound for W_1 ($n_e = 10$)

Section	(EE)	(A1)	(A2)	(A2) with (45)
2	927.18	927.18	927.18	927.18
3	927.18	1586.13	927.18	927.18
4	927.18	2054.45	1161.34	932.50
5	927.18	2387.29	1245.91	945.55
6	927.18	2623.83	1356.05	954.30
7	927.18	2791.95	1401.19	961.80
8	927.18	2911.43	1461.41	967.78
9	927.18	2996.34	1482.13	972.70
10	927.18	3056.69	1516.14	976.73

Table 4. Upper bound for W_1 at $l = 1$

n_e	(EE)	(A1)	(A2)	(A2) with (45)
5	927.18	1811.88	1094.46	967.02
10	927.18	3099.58	1523.92	980.04
15	927.18	4404.00	2052.69	983.43
20	927.18	5712.63	2605.73	984.89

Table 5. CPU times (s)

n_e	(EE)	(A1)	(A2)	(A2) with (45)
5	0.04	0.04	1.6	2.3
10	0.4	0.04	8.3	12.8
15	135	0.04	27.1	45.7
20	44227	0.04	50.7	86.7

REFERENCES

Adjiman, C. S., S. Dallwig, C. A. Floudas and A. Neumaier (1998). A global optimization method, αBB, for general twice-differentiable constrained NLPs - I. Theoretical advances. *Comput. Chem. Eng.* **22**(9), 1137–1158.

Barton, P. I. and C. K. Lee (2002). Modeling, simulation, sensitivity analysis and optimization of hybrid systems. *ACM Trans. Model. Comput. Simul.* **12**(4), 256–289.

Barton, P. I. and C. K. Lee (2003). Design of process operations using hybrid dynamic optimization. In press, *Comput. Chem. Eng.*

Feehery, W. F., J. E. Tolsma and P. I. Barton (1997). Efficient sensitivity analysis of large-scale differential-algebraic systems. *Appl. Numer. Math.* **25**(1), 41–54.

Glover, F. (1975). Improved linear integer programming formulations of nonlinear integer programs. *Management Sci.* **22**(4), 455–460.

Horst, R. and H. Tuy (1993). *Global Optimization.* Springer-Verlag. Berlin.

Lee, C. K., A. B. Singer and P. I. Barton (2004). Global optimization of linear hybrid systems with explicit transitions. *Syst. Control Lett.* **51**(5), 363–375.

McCormick, G. P. (1976). Computability of global solutions to factorable nonconvex programs: Part I - convex underestimating problems. *Math. Programming* **10**, 147–175.

Singer, A. B. and P. I. Barton (2003). Global solution of linear dynamic embedded optimization problems. In press, *J. Optim. Theor. Appl.*

ELSEVIER
IFAC
PUBLICATIONS
www.elsevier.com/locate/ifac

A NOVEL TOOL FOR MULTI-MODEL PID CONTROLLER DESIGN

Flávio Faccin and Jorge O. Trierweiler

Group of Integration, Modeling, Simulation, Control and Optimization of Processes (GIMSCOP)
Department of Chemical Engineering, Federal University of Rio Grande do Sul (UFRGS)
Rua Duque de Caxias, 1303 apt. 603. CEP: 90010-283 - Porto Alegre - RS - BRAZIL
Fax: +55 51 3316 3277, Phone: +55 51 3316 4072
E-MAIL: {faccin, jorge}@enq.ufrgs.br

Abstract: In this paper, it is shown how a robust performance PID controller can be synthesized for a set of possible linear models. The method is carried out by solving a min-max optimization problem formulated in frequency domain. As the synthesis of a low-order controller for a high-order plant is in general nonconvex, the original optimization problem is divided into two convex optimization subproblems, which are solved iteratively until to achieve the complete convergence. These subproblems are constructed using a general two degree-of-freedom control configuration, in which all commercial industrial PID controllers can be converted. *Copyright © 2004 IFAC*

Keywords: PID controllers, frequency response methods, multi-objective optimization, nonlinear control systems, uncertain linear systems.

1. INTRODUCTION

There is no doubt that the PID algorithm is the most widely used controller in the industrial process control systems of the whole world. It is a robust easily understood algorithm that can provide good control performance despite the varied dynamic characteristics of process plants. Since the creation of PID controllers, roughly sixty years ago, several PID tuning methods were and are being proposed. These methods can be classified in empirical (e.g., Ziegler-Nichols), analytical (e.g., direct synthesis), or based on some kind of optimization (e.g., ITAE optimization criterion). But all of them have at least one gap or a deficiency. Usually, they work for only some special conditions producing an unsatisfactory result for some classes of plant behavior.

In order to fill these gaps, this paper presents a novel tool, which intends to help the PID design. It works with a multi-model system (different linear models representing a nonlinear system, or a set of parameter uncertainties derived from a fixed linear model), and designs the parameters (for any kind of PID) based on an optimization problem to bring all the closed loop systems as close as possible to the same attainable closed loop performance.

2. MYTHS AND FACTS ABOUT PID CONTROLLERS

According Dorf and Bishop (1998), "the popularity of PID controllers can be attributed partly to their robust performance in a wide range of operating conditions and partly to their functional simplicity, which allows engineers to operate them in a simple, straightforward manner". Skogestad (2003), however, believes that "although the proportional-integral-derivative (PID) controller has only three parameters, it is not easy, without a systematic procedure, to find good values (settings) for them. In fact, a visit to a process plant will usually show that a large number of the PID controllers are poorly tuned".

In Åström and Hägglund (1995), an interesting and amazing study was reported, which concludes, "a typical mill has more than 2000 control loops and that 97% use PI control. Only 20% of the control loops were found to work well and decrease process variability. Reasons for poor performance were poor tuning (30%) and valve problems (30%). The remaining 20% of the controllers functioned poorly for a variety of reasons such as: sensor problems, bad choice of sampling rates, and anti-aliasing filters".

In another study also reported by the same authors it was claimed that "30% of installed process controllers operate in manual, that 20% of the loops use 'factory tuning', i.e., default parameters set by the controller manufacturer, and that 30% of the loops function poorly because of equipment problems in valves and sensors".

Due to the enormous amount of PID controllers do not working in a suitable and satisfactory manner, some control systems have packages for automatic controller performance assessment. This study area is relatively young, but very important, since even the most good tunings, with the time, present a deteriorated performance, that occurs because natural and continuously variations in the process dynamics, as discussed by Kempf (2003) and Harris et al. (1999). Although this powerful technique can warn when a loop shows a poor performance, it must be combined to a design method, which is used to retune the control loop.

After this review, one question needs to be answered: how this so simple, so studied, so old and the most used controller in the world is *not* working satisfactorily in the majority of the cases? Regardless of technical trouble (as sensor and valves), which is an instrumentation matter, the most important factor is how to tune the controller in a suitable manner, despite of the enormous amount of design methods, proposed throughout the last six decades. All these methods are compiled in special books about PID controllers, such as O'Dwyer (2003), Åström and Hägglund (1995) or in any textbooks about process control.

Actually, the PID tuning is only so easy for simple dynamics, as sluggish systems, when just a PI controller is already enough. To more complicated systems, the derivative action is needed to improve the closed loop performance, and the process of tuning becomes much more difficult. According Piazzi and Visioli (2002), "for many processes the derivative term of the controller is not useful and it is often difficult to tune, so that practitioners prefer to avoid its use". Luyben (2001) and Ingimundarson and Hägglund (2002) also agree with this idea. Isaksson and Graebe (2000) have including attracted attention to the fact that "there is an industrial myth that derivative action does not work".

In general, a control engineer working in a process plant is very busy and he does not want to have more work to do, so at first he tries to tune a provisional PI with simple memorized rules. If the control loop has a fairly good performance, he does not spend more time with this loop and the provisional PI becomes a permanent controller. This is one reason for 20% of the controllers use 'factory tuning', and the majority of them are PI. In fact, the derivative action tuning is normally much more complicated and implicates in the use of additionally filters, when the system noise is significant.

In addition, due to the existence of many different types of parameterization (specially in commercial control systems) when a design method is used, the results obtained may not be compatible with the industrial controller. Goodwin et al. (2001) have already pointed out "caution must be exercised when applying PID tuning rules, as there are a number of possible parameterizations". If a practitioner tries to use a design method, which is not compatible with the existing in the control system, the result may not be satisfactory, and then he gives up using the derivative action and even thinks that the method is not good. It helps to keep the 'myth'.

Another important factor is that in spite of the advanced techniques in PID tuning, the nearly whole amount of these methods were done for *ideal parameterizations*, without derivative action filter and passing the derivative action through the error signal. In practice, the industrial use of these parameterizations is not conceivable, due to the strong control action at high frequency produced by measurement noise and suddenly setpoint changes. Some of them consider the derivative filter action, but pretty few use the derivative action passing just at the process variable signal, even the most modern optimization techniques. This means that the results of these methods are not completely satisfactory for industrial uses. One exception is presented in Carotenuto et al. (2002), which uses a PID controller with derivative action on the filtered output and a global optimization method, but it is not able to work with multi-model systems, and it claims that the method convergence for a PID is *more difficult* to attain, due to the complexity of the method.

The problems above quoted are only related to the PID algorithm. Up to now the process worries were not taken into account, but of course they need to. A common tuning method is normally model-based, or in some cases, based in some measured characteristics of the system (such as: ultimate gain, Ku, and period, Pu), but in all of them only one operating region is considered. It is well known that the industrial processes are nonlinear at some extension. In spite of the intrinsic process nonlinearities, usually the dynamic behavior can be satisfactory approximated by a linear model at each operating point. Therefore, if the process works in several operating points, a set of linear models can be constructed to represent the system behavior. Even if the process is linear, during the identification procedure, usually more then one linear model is identified, or an uncertain bound is given for the identified parameters, what again produce a set of linear models. Another source for multi-model representation is time variant systems, what often occurs in the process industry, since during the process operation the dynamic behavior can change due to the equipment fouling, impurities level in the row materials, etc. All these situations can be well and easily described by a multi-model approach (i.e., a set of possible linear models).

3. MULTI-MODEL APPROACH (MMA) PID CONTROLLER DESIGN

In this section, it is shown how a robust performance PID controller can be synthesized for a set of possible linear models. The method is carried out by solving a min-max optimization problem formulated in frequency domain. As the synthesis of a low-order controller for a high-order plant is in general nonconvex (Safonov et al., 1994), the original optimization problem is divided into two convex optimization subproblems, which are solved iteratively until to achieve the complete convergence.

3.1 Control Configuration

The optimization subproblems are constructed using a general two degree-of-freedom (2DOF) feedback control configuration shown in Fig. 1, where y and y_{SET} are respectively, the control variable and the setpoint, G is the plant model, and the C_{SP}, C_{PI} and C_{PV} blocks constitute together the PID controller. The C_{PI} block is a PI controller whose structure is always fixed and given by (1), whilst C_{SP} and C_{PV} are dependent on the PID controller parameterization.

$$C_{PI} = K_C\left(1+\frac{1}{T_I s}\right) \tag{1}$$

The advantages to use the 2DOF control configuration are threefold: (a) It divides a typical nonconvex optimization problem (when the standard configuration is used) into two convex problems, which can be solved with the Sequential Iterative Optimization Method (SIOM) proposed by Faccin and Trierweiler (2004). (b) It consists in a common base, in which all possible industrial PID parameterization can be converted. In Faccin (2004) it is shown this conversion for several industrial PID parameterizations. (c) The controller order can be easily increased and implemented in modern DCS. For example, process filters for noise averting can be synthezed and incorporated into C_{PV}.

The controller used in this article is the ISA standard form (Åström and Hägglund, 1995) with $b = 1$, $c = 0$, $N = 10$ and has three adjustable parameters: K_C, T_I and T_D, given by (2). When it is put into the general configuration of Fig. 1, the controller blocks C_{SP} and C_{PV} are given by (3) and (4).

$$C = K_C\left[\left(1+\frac{1}{T_I\cdot s}\right)Y_{SET} - \left(1+\frac{1}{T_I\cdot s}+\frac{T_D\cdot s}{1+0.1T_D\cdot s}\right)Y\right] \tag{2}$$

$$C_{SP} = 1 \tag{3}$$

$$C_{PV} = \frac{1.1\cdot T_I\cdot T_D\cdot s^2+(T_I+0.1\cdot T_D)\cdot s+1}{0.1\cdot T_I\cdot T_D\cdot s^2+(T_I+0.1\cdot T_D)\cdot s+1} \tag{4}$$

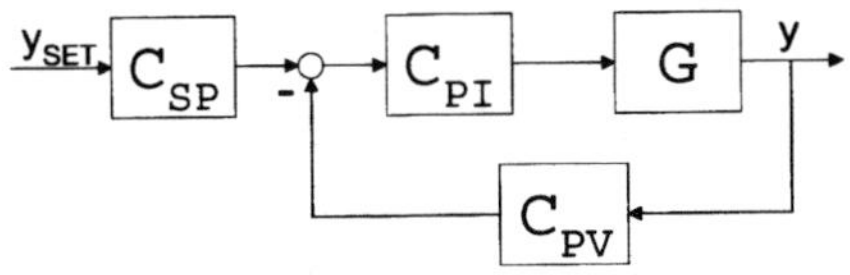

Fig. 1. The 2DOF feedback control configuration.

3.2 Optimization Problem

To design and convert the PID parameters, it is formulated a multiobjective optimization problem in frequency domain:

$$\begin{aligned} &\min_{\gamma, x\in R^+} \ \gamma \\ &subject\ \ to: \ F_n(x) - w_n\cdot\gamma \le 0; \quad n = 1..N \end{aligned} \tag{5}$$

Where N is the total number of plant models, w_n is the weight given to the model n (as default, all weights are equal to 1), γ is an auxiliary optimization variable, x is the vector of the decision variables (the controller parameters which are optimized), and $F_n(x)$ is given by:

$$F_n(x) = \sum_{s=j\omega_0}^{j\omega_1}\left(\left[T_n(s,x)-T_0(s)\right]\frac{1}{s}\right)^2 \tag{6}$$

It consists of minimizing the difference between the closed loop transfer function of each model (T_n) and the desired attainable performance (T_O) on the frequency domain ($s = j\omega$) for a logspaced frequency range $\omega = [\omega_0\ \omega_1]$. The term ($1/s$) is used to emphasize that both response should be as close as possible for stepwise setpoint changing. Considering the control configuration of Fig. 1, the closed loop transfer function is given by:

$$T_n(s,x) = \frac{G_n(s)\cdot C_{PI}(s,x)\cdot C_{SP}(s,x)}{1+G_n(s)\cdot C_{PI}(s,x)\cdot C_{PV}(s,x)} \tag{7}$$

In (7), G_n is the model n and the C_{PI}, C_{SP} and C_{PV} are given respectively by (1), (3) and (4). When the parameters of these blocks are simultaneously optimized, the optimization problem is nonconvex. To overcome the nonconvexity, a sequential iterative procedure is applied, where the C_{PI} block is separately optimized to the other two blocks in two steps. At the first iteration, the following approximation is used to calculate the start point for the PI parameters (i.e., K_C and T_i):

$$F_n(x) = \sum_{s=j\omega_0}^{j\omega_1}\left\{\begin{bmatrix} G_n(s)\cdot C_{PI}(s,x)\cdot\left(1-T_0(s)\right)^2 \\ -T_0(s)\cdot\left(1-T_0(s)\right)\end{bmatrix}\frac{1}{s}\right\}^2 \tag{8}$$

With these parameters, the C_{PI} block is fixed and a new optimization using (6) is done to calculate C_{PV} and C_{SP}. With the converged C_{PV} and C_{SP} a new C_{PI} is determined with a fixed C_{PI} obtained in the last iteration in the denominator of (7) and then with the new C_{PI}, the blocks C_{PV} and C_{SP} can be updated, and thus for ahead until all three blocks converge. The default stop criterion is 1% of relative error (from two sequential iterations) for all the parameters.

Particularly, the selected controller parameterization used in this paper, the ISA standard form given by (2), has a fixed C_{SP} block (no optimization parameter available). In spite of the MMA method is able to work with 2DOF controllers, to make the comparison with another tuning method, it was just designed this 1DOF controller.

In the optimization problem (5), another additional constraint can be considered (such as: maximal sensitivity or control action boundaries), but in the manner as the optimization problem was built, it is not normally necessary, since all these characteristics can be considered selecting an appropriated attainable performance.

3.3 Attainable Performance

The attainable performance is a very important factor in the MMA procedure. It must be a function with all desired characteristics (e.g., stable, unit gain and fast convergence to the setpoint, without a significant overshoot and good robustness characteristics). It could be a second order time delay model, with specified overshoot and rise time. However, to obtain better results, the order of this function should be compatible with the corresponding closed loop system order. In Dorf and Bishop (1998), transfer functions for several orders with the optimal coefficients considering an ITAE criterion for step response setpoint changes are presented. These functions are properly to be used as attainable performance, since they have unit gain, small overshoot, the smallest ITAE criteria and only one adjustable parameter (ω_n). The coefficients of these functions' denominator are shown in Table 1.

When a model with dead-time is used, the attainable performance must consider this constraint as a limitation. It could be done in two ways. In the first one, a higher order model without dead time approximates the dead time model. Large poles (that do not influence in the system dynamics, and which sum is exactly equivalent to the dead time) are added in the model. In this case, a higher order attainable performance is required to ensure a compatibility with the models. In the second way, the higher dead time needs to be included in the attainable performance. In this case, the solution becomes more difficult. Therefore, the first way was used in this article.

3.4 User-Friend Interface

An interface with the whole procedure was made in Matlab® platform, which turns easier the understanding and the obtaining of satisfactory results. This interface lets the practitioner to choose a suitable attainable performance by varying its order and the parameter ω_n. A graph with a step response of this function and all models together helps this chosen. Based on this function, an appropriated frequency vector is calculated automatically.

Table 1 Optimum coefficients of $T_O(s)$ based on the ITAE Criterion for step response

$s+\omega_n$
$s^2+1.4\omega_n\cdot s+\omega_n^2$
$s^3+1.75\omega_n\cdot s^2+2.15\omega_n^2\cdot s+\omega_n^3$
$s^4+2.1\omega_n\cdot s^3+3.4\omega_n^2\cdot s^2+2.7\omega_n^3\cdot s+\omega_n^4$
$s^5+2.8\omega_n\cdot s^4+5.0\omega_n^2\cdot s^3+5.5\omega_n^3\cdot s^2+3.4\omega_n^4\cdot s+\omega_n^5$

Several types of PID parameterizations are available and can be easily selected. All results are shown in a table where F_n values for each model and each iteration step are shown. Therefore, it is easy to visualize the convergence of the optimization method. The converged controller is then simulated for load disturbance and setpoint changes. In addition, robustness and performance criteria are calculated for each model.

4. CASE STUDY: SPHERICAL TANK LEVEL CONTROL

A nonlinear spherical tank system (Fig. 2) was studied aiming to show the benefits of the proposed procedure. The objective is to control the fluid level (h) through varying the inlet flow (Fin). This system is related to storage tanks, however it is not so common in the process industries. Cylindrical tanks are normally used but the spherical ones have a more nonlinear behavior, and because that they are more difficult to control.

4.1 Process Model and Transfer Function

Through a material balance for the system shown in Fig. 2, where the outlet flow ($Fout$) is dependent of the fluid level considering turbulent flow, equation (9) is obtained.

$$\frac{dh}{dt}=\frac{Fin-K\cdot\sqrt{h}}{\pi\cdot h\cdot(D-h)} \tag{9}$$

In (9), D is the tank diameter and K is the outlet flow capacity coefficient. For each inlet flow, a transfer function can be obtained, which is given by:

$$G=\frac{\Delta h(s)}{\Delta Fin(s)}=\frac{K_P}{\tau\cdot s+1}\cdot e^{-\theta\cdot s} \tag{10}$$

$$K_P=\frac{2\cdot Fin}{K^2} \tag{11}$$

$$\tau=\frac{2\cdot\pi\cdot Fin^3\cdot\left(D\cdot K^2-Fin^2\right)}{K^6} \tag{12}$$

A small pure time delay θ was included to consider the delays produced by valve and sensor dynamics. The stationary tank heigth is $h=(Fin/K)^2$. Here, an operating inlet flow range of 0 to 10 L/min, D = 25 cm, and K = 2 L.min^{-1}.cm$^{0.5}$ are used as parameters of the model. Fig. 3 shows the static gain, the time constant and the stationary tank height as a function of the inlet flowrate.

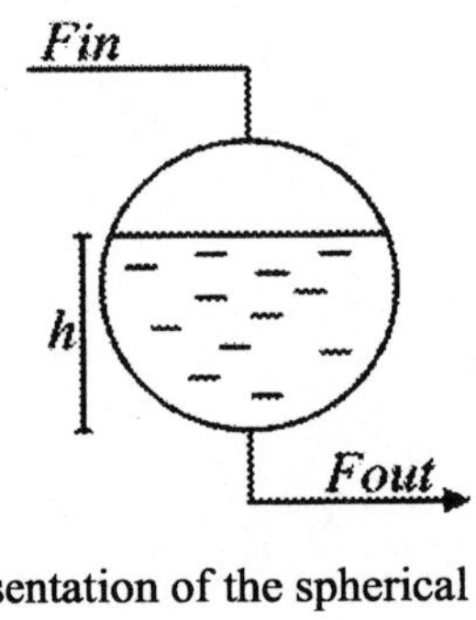

Fig. 2. Representation of the spherical tank system.

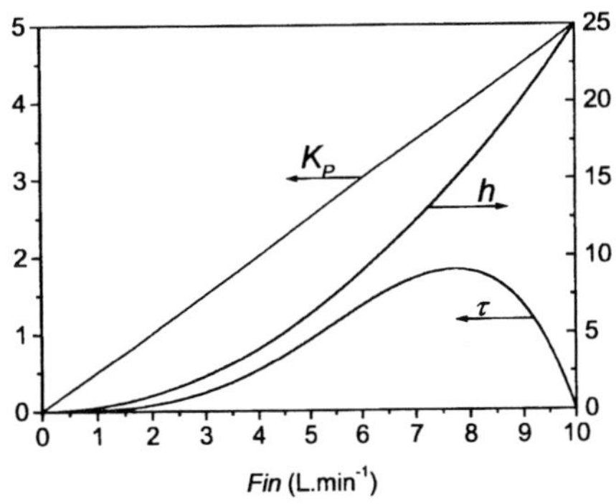

Fig. 3. Static gain (K_P, cm.min.L^{-1}), time constant model (τ, min) and stationary height (h, cm), for the whole inlet flowrate (*Fin*) range.

There is a significant nonlinearity in the model parameters, mainly in the time constant τ, which should be intuitively higher at the middle than at the top or bottom, however the maximum point is not at the middle, but a little higher up, at exactly 60% of height (or 15 cm). This asymmetry is a consequence of the nonlinear relation between *Fout* and *h*.

Assuming the nominal operating point as $h = 12.5$ cm and the normal operating range between 2 and 23 cm, three models representing these operating points where selected to describe the plant behavior. The transfer function G_1 is based on the minimum operating point (Fin = 2.83 L.min^{-1}), G_2 is the nominal one (Fin = 7.07 L.min^{-1}) and G_3 is based on the maximum operating point (Fin = 9.59 L.min^{-1}). A pure time delay of 0.1 min is used in all cases producing the following final expressions:

$$G_1 = \frac{1.41 \cdot e^{-0.1 \cdot s}}{0.204 \cdot s + 1}; \quad G_2 = \frac{3.54 \cdot e^{-0.1 \cdot s}}{1.736 \cdot s + 1}; \quad G_3 = \frac{4.80 \cdot e^{-0.1 \cdot s}}{0.693 \cdot s + 1}$$

4.2 Controller Design

A third order attainable performance with $\omega_n = 8$ was chosen because it is feasible to be achievable for all models. Fig. 4 shows the step responses for all models normalized by the respective gain and the desired attainable performance T_0. For the MMA procedure, the frequency range was divided in 200 points and the stop criterion was 0.1% of relative error. As the G_2 is the slowest model, the nominal one, and it represents the region where the system will operate more frequently, its weight (w_2) is 0.5, while the other both 1.

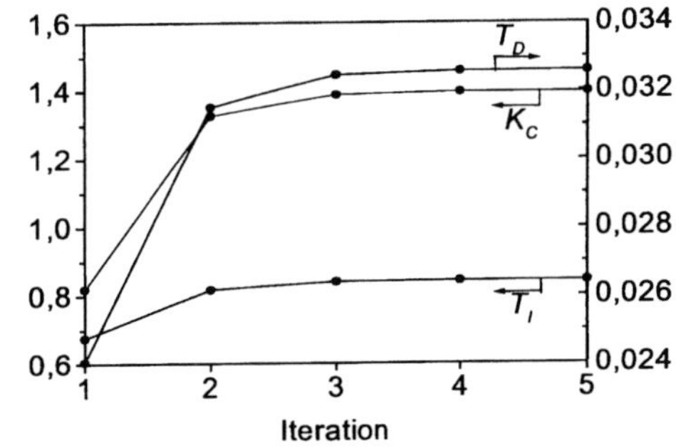

Fig. 5. Parameters convergence (K_C, T_I and T_D), for the iterations of the MMA optimization.

With these settings the optimization process converged in 5 iterations and the solution obtained was: K_C = 1.396; T_I = 0.843; T_D = 0.033. Fig. 5 shows the convergence of these parameters throughout the iterations. As it can be seen, the initial estimates are very similar to the final solution, and the procedure converges quickly to the optimal values. The corresponding step response to setpoint change for all linear models is shown in Fig. 6.

Normally, when a multi-model system needs to be controlled by only one set of PID parameters, the controller is designed for the limiting model. If the controller can control this model, it will also control the other ones, of course with a loss of performance. Nevertheless, determine the worst-case is not an easy task, mainly when there are differences in the gain and in the dynamic. When there are differences in the dead time or in the model order, it becomes even more difficult and can be strongly dependent on the desired closed loop response.

Therefore, to apply a simple tuning rule for this system, first it is necessary to find the worst-case. With an analysis on the bode diagram (not shown) one will conclude that it is the G_3 model. In this way, a Simple IMC (SIMC) tuning rule proposed by Skogestad (2003) was also employed to compare the performances. The SIMC tuning rules have for first order time-delay models only PI setting, which applied to the G_3 gives K_C = 0.722 and T_I = 0.693. Through a comparison between these parameters and the ones produced by the proposed multi-model approach (MMA), it can be seen that the MMA has a small derivative action, which is fundamental to the system's robustness. Fig. 7 shows the closed loop responses for the SIMC controller.

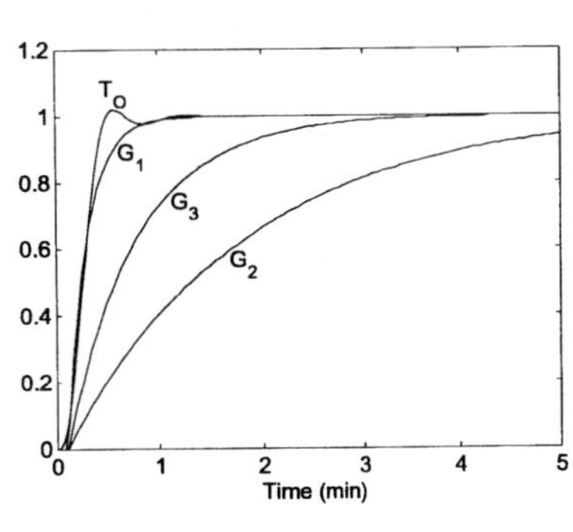

Fig. 4. Step responses for T_O and for G_1, G_2 and G_3 models normalized by the static gain.

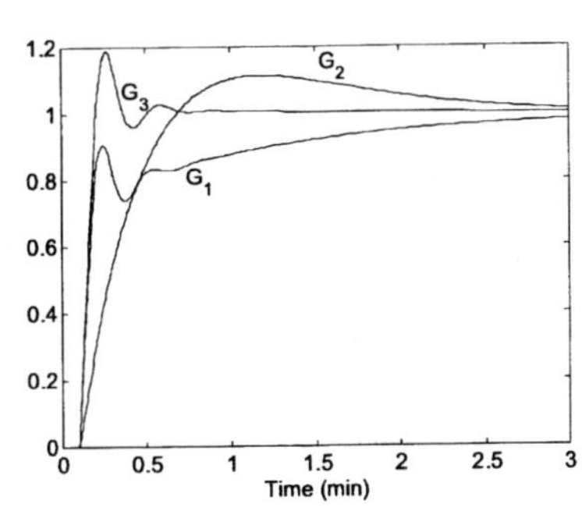

Fig. 6. Step response to setpoint change for G_1, G_2 and G_3 models with MMA controller.

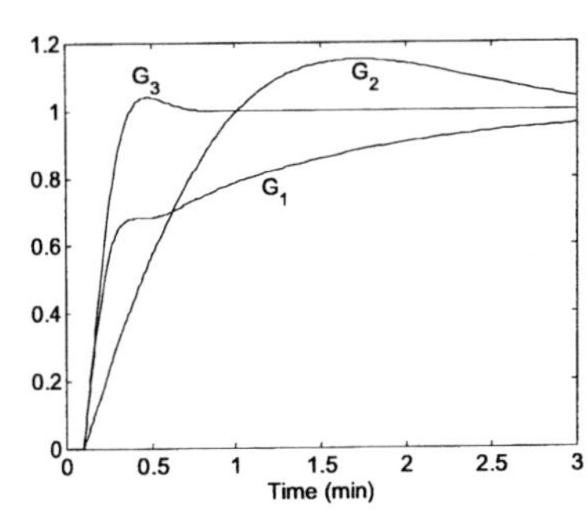

Fig. 7. Step response to setpoint change for G_1, G_2 and G_3 models with SIMC controller.

Table 2 Comparison criteria between MMA and SIMC tuning for all closed loop models

	G_1		G_2		G_3	
	MMA	SIMC	MMA	SIMC	MMA	SIMC
MS	1.27	1.07	1.04	1.02	1.21	1.06
GM	1.95	3.65	6.12	10.24	1.83	3.14
PM	79.2	107.2	67.2	58.7	53.3	61.3
IAE	0.404	0.631	0.455	0.678	0.185	0.217
ITAE	0.287	0.516	0.277	0.538	0.031	0.029
ST	3.06	3.88	2.62	3.25	0.64	0.61
OV	-	-	11.4	15.0	18.9	4.1
MD	0.293	0.376	0.238	0.355	0.379	0.549

Now, the performances of these two controllers are compared at Table 2 using maximal sensitivity (MS), gain margin (GM) and phase margin (PM) as robustness criteria and integral of the absolute value of the error (IAE), integral of product of time and the absolute value of the error (ITAE), 2% settling time (ST, in min), maximal overshoot (OV, in % of the setpoint), and the maximal deviation to the setpoint, when a load disturbance of 0.5 $L.min^{-1}$ on the inlet flowrate is done (MD, in cm) as performance criteria.

Several remarks can be taken from Table 2. The SIMC tuning produced more robust response for all models, of course, because it was designed for the worst-case, but it has a significant performance loss for models G_1 and G_2 (higher ST values). The MMA tuning has shown responses with better performance (smaller values for IAE and ITAE), a little aggressive for G_3 model (high OV), but with suitable robustness criteria. For load disturbance it seems clear that MMA tuning is much better (the MD values for SIMC tuning is 40% higher in average), since with a more aggressive tuning, it can reject more easily the disturbances.

5. CONCLUSIONS

A new powerful multi-model approach (MMA) for PID controller tuning was presented in details. The MMA produces a robust controller with a trade-off among the performance of all considered models and it is flexible to be used for any parameterization type. The desired attainable performance is the main tuning parameter, which can be selected to produce better performance or robustness. The paper showed how the desired attainable performance could be easily set. The proposed tool considerably extends the set of process where a PID controller will produce good results, since a low order controller with robust performance can be easily and fastly synthesize.

The proposed method can work with models of different orders and structures, i.e., it is not based on a fixed model type (such as first order plus time delay, used by several methods). This characteristic together with the multi-model approach are responsible for the high flexibility and suitability of the method for industrial applications, where classical tuning methods do not always give satisfactory results.

ACKNOWLEDGMENT

The authors would like to thank Prof. M.Sc. Vinícius C. Machado to suggest the spherical tank system as an illustrative example for this work.

REFERENCES

Åström, K.J. and T. Hägglund (1995). *PID Controllers: Theory, Design, and Tuning.* Instrument Society of America, Research Triangle Park.

Carotenuto, L., P. Pugliese and Y.D. Sergeyev (2002). Maximizing Performance and Robustness of PI and PID Controllers by Global Optimization. *15th IFAC World Congress*, Barcelona.

Dorf, R.C. and Bishop R.H. (1998). *Modern Control Systems*. Addison-Wesley, Menlo Park.

Faccin, F. (2004). *Development and Application of a Novel Tool for PID Controller Design* (In Portuguese). Master Thesis. Federal University of Rio Grande do Sul, Chemical Eng. Department, Brazil.

Faccin, F. and J.O. Trierweiler (2004). Performance Limitations of Some Industrial PID Controllers. Submitted to *14th ESCAPE Symposium*, Lisboa.

Goodwin, G.C., S.F. Graebe and M.E. Salgado (2001). *Control System Design.* Prentice-Hall, New Jersey.

Harris, T.J., C.T. Seppale and L.D. Desborough (1999). A review of performance monitoring and assessment techniques for univariate and multivariate control systems. *Journal of Process control*, **9**, 1-17.

Ingimundarson, A. and T. Hägglund (2002). Performance comparison between PID and dead-time compensating controllers. *Journal of Process Control*, **12**, 887-895.

Isaksson, A.J. and S.F. Graebe (2000). The derivative filter is an integral part of PID design. Presented at *Control Systems 2000.* Victoria, British Columbia.

Kempf, A.O. (2003). *Control Loops Performance Assessment* (In Portuguese). Master Thesis. Federal University of Rio Grande do Sul, Chemical Engineering Department, Brazil.

Luyben, W.L. (2001). Effect of Derivative Algorithm and Tuning Selection on the PID Control of Dead-Time Process. *Ind. Eng. Chem. Res.*, **40**, 3605-3611.

O'Dwyer, A. (2003). *Handbook of PI and PID Controller Tuning Rules.* World Scientific Pub Co, London.

Piazzi, A. and A. Visioli (2002). Improved PI Control via Dynamic Inversion. *15th IFAC World Congress*, Barcelona.

Safonov, M.G., K.C. Goh and J.H. Ly (1994). Control System Synthesis via Bilinear Matrix Inequalities. *Proc. of American Control Conference*, pp. 45-49, Baltimore.

Skogestad, S. (2003). Simple analytic rules for model reduction and PID controller tuning. *Journal of Process Control*, **13**, 291-309.

ELSEVIER
IFAC
PUBLICATIONS
www.elsevier.com/locate/ifac

PREDICTIVE CONTROL OF SWITCHED NONLINEAR PROCESSES WITH SCHEDULED MODE TRANSITIONS [1]

Prashant Mhaskar, Nael H. El-Farra and Panagiotis D. Christofides [2]

Department of Chemical Engineering
University of California, Los Angeles, CA 90095-1592

Abstract: A predictive control framework is proposed for the constrained stabilization of switched nonlinear processes. The problem of stabilization of a switched process, transiting between its constituent modes at prescribed transition times, is considered. The main idea is to design Lyapunov–based predictive controllers for each of the constituent modes in which the switched process operates, and incorporate constraints in the predictive controller design of the individual modes which ensure that transitions between the modes at pre-determined switching times occur in a way so as to guarantee stabilization of the switched closed–loop process. This is achieved as follows: For each constituent mode, a Lyapunov-based predictive controller (MPC) is designed. Then, constraints are incorporated in the MPC design which ensure that: (1) the states of the closed–loop process, at the time of the transition, reside in the stability region of the mode that the process switches into, and (2) the value of the Lyapunov function of the process for each of the modes is nonincreasing whenever the mode is active, thereby guaranteeing stabilization. The proposed control method is demonstrated through application to a chemical process example. *Copyright © 2004 IFAC*

Keywords: Switched Systems, Process Schedule, Nonlinear Processes, Hybrid control, Input constraints, Predictive Control, Bounded Control.

1. INTRODUCTION

Process operation often involves controlled, discrete transitions between multiple, continuous dynamical modes of operation, in order to handle, for example, changes in raw materials, energy sources, product specifications and market demands, giving rise to an overall process behavior that is more appropriately viewed as a hybrid system, i.e., intervals of piecewise continuous behavior interspersed by discrete transitions. Compared to purely continuous processes, the hybrid nature of these systems and their changing dynamics makes them more difficult to describe, analyze, or control.

A class of hybrid systems that has attracted significant attention recently, because it can model several practical control problems that involve integration of supervisory logic-based control schemes and feedback control algorithms, is the class of switched (or multimodal) processes. For this class, results have been developed for stability analysis using the tools of multiple Lyapunov functions (MLFs) for linear (Peleties and DeCarlo, 1991) and nonlinear (Branicky, 1998) systems, and the concept of dwell time (Hespanha and Morse, 1999); the reader may refer to (Liberzon and Morse, 1999) for a survey of results in this area. These results have motivated the development of methods for control of various classes of switched processes (see, e.g., (Hu *et al.*, 1999)). Despite this progress, however, significant research remains to be done in the direction of control of switched nonlinear processes, especially

[1] Financial support from NSF, CTS-0129571, is gratefully acknowledged
[2] Corresponding author (e-mail:pdc@seas.ucla.edu)

since the majority of practical switched processes exhibit inherently nonlinear dynamics.

Motivated by this, in (El-Farra and Christofides, 2003*a*), a framework for coordinating feedback and switching for control of hybrid nonlinear processes was developed. The key feature of the proposed control methodology is its hierarchical structure involving the integrated synthesis of: (1) a family of lower-level bounded nonlinear controllers that stabilize the continuous dynamical modes, and provide an explicit characterization of the stability region associated with each mode, and (2) upper-level switching laws that ensure stability of the overall switched closed-loop process. While this approach serves to *determine* whether or not a switch can be made at a time without loss of stability guarantees, the controller design does not address the problem of *ensuring* that such a switch can be made safely at a pre-decided time.

Guiding the process through a prescribed trajectory requires a control algorithm that can achieve optimal closed–loop trajectory behavior in the presence of constraints. A practical controller design, for handling of state and input constrains, in an optimal control setting is model predictive control (MPC). One of the important issues that arise in the practical implementation of predictive control policies for the purpose of stabilization, however, is the difficulty they typically encounter in identifying, a priori (i.e., before controller implementation), the set of initial conditions starting from where feasibility and closed-loop stability are guaranteed. This typically results in the need for extensive closed-loop simulations to search over the whole set of possible initial conditions, thus adding to the overall computational load. The fallout of this problem is more pronounced when considering MPC of hybrid processes that involve switching between multiple constrained modes. Re-tuning the parameters of each predictive controller (e.g., horizon length) on-line, or running extensive closed-loop simulations in the midst of mode transitions, to determine the feasibility of switching, becomes computationally intractable, especially if the hybrid process involves a large number of modes with frequent switches.

In this work, a predictive control framework is proposed for the constrained stabilization of switched nonlinear processes. The problem of stabilization of a switched process, transiting between its constituent modes at prescribed transition times, is considered. The main idea is to design Lyapunov–based predictive controllers for each of the constituent modes in which the switched process operates, and incorporate constraints in the predictive controller design of the individual modes which ensure that transitions between the modes at pre-determined switching times occur in a way so as to guarantee stabilization of the switched closed–loop process. This is achieved as follows: For each constituent mode, a Lyapunov-based predictive controller (MPC) is designed. Then, constraints are incorporated in the MPC design which ensure that: (1) the states of the closed–loop process, at the time of the transition, reside in the stability region of the mode that the process switches into, and (2) the value of the Lyapunov function of the process for each of the modes decreases whenever the mode is active, thereby guaranteeing stabilization. The proposed control method is demonstrated through application to a chemical process example.

2. PRELIMINARIES

We consider the class of switched nonlinear processes represented by the following state-space description:

$$\begin{aligned} \dot{x}(t) &= f_{\sigma(t)}(x(t)) + G_{\sigma(t)}(x(t))u_{\sigma(t)} \\ u_{\sigma(t)} &\in \mathcal{U}_\sigma \\ \sigma(t) &\in \mathcal{K} := \{1, \cdots, p\} \end{aligned} \tag{1}$$

where $x(t) \in \mathbb{R}^n$ denotes the vector of continuous-time state variables, $u_\sigma(t) = [u_\sigma^1(t) \cdots u_\sigma^m(t)]^T \in \mathcal{U}_\sigma \subset \mathbb{R}^m$ denotes the vector of constrained manipulated inputs taking values in a nonempty compact convex set $\mathcal{U}_\sigma := \{u_\sigma \in \mathbb{R}^m : \|u_\sigma\| \leq u_\sigma^{max}\}$, where the notation $\|\cdot\|$ denotes the euclidian norm, $u_\sigma^{max} > 0$ is the magnitude of the constraints, $\sigma : [0, \infty) \to \mathcal{K}$ is the switching signal which is assumed to be a piecewise continuous (from the right) function of time, i.e., $\sigma(t_k) = \lim_{t \to t_k^+} \sigma(t)$ for all k, implying that only a finite number of switches is allowed on any finite interval of time. p is the number of modes of the switched process, $\sigma(t)$, which takes different values in the finite index set $\mathcal{K}$, represents a discrete state that indexes the vector field $f(\cdot)$, the matrix $G(\cdot)$, and the control input $u(\cdot)$, which altogether determine $\dot{x}$. Throughout the paper, we use the notations $t_{k_r^{in}}$ and $t_{k_r^{out}}$ to denote the time at which, for the r-th time, the k-th subsystem is switched in and out, respectively, i.e., $\sigma(t_{k_r^{in}}^+) = \sigma(t_{k_r^{out}}^-) = k$. With this notation, it is understood that the continuous state evolves according to $\dot{x} = f_k(x) + G_k(x)u_k$ for $t_{k_r^{in}} \leq t < t_{k_r^{out}}$. It is assumed that all entries of the vector functions $f_k(x)$, and the $n \times m$ matrices $G_k(x)$, are sufficiently smooth on $\mathbb{R}^m$ and that $f_k(0) = 0$ for all $k \in \mathcal{K}$. Throughout the paper, the notation $L_f\bar{h}$ denotes the standard Lie derivative of a scalar function $\bar{h}(x)$ with respect to the vector function $f(x)$, $L_f\bar{h}(x) = (\partial\bar{h}/\partial x)f(x)$.

In this work, we consider the problem of stabilization of continuous-time nonlinear processes where mode transitions are decided and executed at pre-determined times. In order to provide the necessary background for our main results in section 3, we will briefly review in the remainder of this section the design procedure for, and the stability properties of a bounded controller design, stability properties of which are then exploited in the design of a Lyapunov–based model predictive controller that guarantees stability for an explicitly characterized set of initial conditions. For simplicity, we focus on the state feedback problem assuming measurements of $x(t)$ to be available for all t.

2.1 *Bounded Lyapunov-based control*

Consider the system of Eq.1, for a fixed $\sigma(t) = k$, where $\mathcal{U}_k = \{u \in \mathbb{R}^m : \|u\| \leq u_k^{max}\}$, and a control Lyapunov function V_k for which, we construct, using the results in (Lin and Sontag, 1991) (see also (El-Farra and Christofides, 2001; El-Farra and Christofides, 2003*b*)), the following continuous bounded control law

$$u_k(x) = -k_k(x)(L_{G_k}V_k)^T(x) \equiv b_k(x) \quad (2)$$

where $k_k(\cdot) =$

$$\frac{L_fV_k + \sqrt{L_fV_k^2 + (u_k^{max}\|(L_{G_k}V_k)^T\|)^4}}{\|(L_{G_k}V_k)^T\|^2\left[1 + \sqrt{1 + (u_k^{max}\|(L_{G_k}V_k)^T\|)^2}\right]} \quad (3)$$

and $L_{G_k}V_k(x) = [L_{g_k^1}V_k \cdots L_{g_k^m}V_k]$ is a row vector where g_k^i is the ith column of G_k. For the above controller, one can show, using standard Lyapunov arguments, that whenever the closed–loop state trajectory, x, evolves within the region described by the set $\Phi_k(u_k^{max}) =$

$$\{x \in \mathbb{R}^n : L_fV_k(x) < u_k^{max}\|(L_{G_k}V_k)^T(x)\|\} \quad (4)$$

then the controller satisfies the constraints, and the time-derivative of the Lyapunov function is negative-definite. An estimate of the stability region can be obtained by using the level sets of V_k, i.e.,

$$\Omega_k(u_k^{max}) = \{x \in \mathbb{R}^n : V_k(x) \leq c_k^{max}\} \quad (5)$$

where $c_k^{max} > 0$ is the largest number for which $\Phi_k(u_k^{max}) \supset \Omega_k(u_k^{max})\backslash\{0\}$. $\Omega_k(u_k^{max})$ then provides an estimate of the stability region, starting from where the origin of the constrained closed–loop process for the mode k, under the control law of Eqs.2-3, is guaranteed to be asyimptotically stable.

The bounded controller of Eqs.2-3 possesses a robustness property with respect to measurement errors, that preserves closed–loop stability when the control action is implemented in a discrete (sample and hold) fashion with a sufficiently small hold time, Δ. Specifically, given the size of a neighborhood of the origin (characterized by δ'), one can compute a positive real number Δ^*, such that if Δ is less than Δ^*, then for all initial conditions in Ω_k, the closed–loop state remains in Ω_k and eventually converges to the specified neighborhood of the origin (for proof, see (Mhaskar *et al.*, 2004*b*)). This robustness property will be exploited in the Lyapunov-based predictive controller design of section 2.2.

2.2 *Model predictive control*

In this section, we consider model predictive control of the system of Eq.1, for a fixed $\sigma(t) = k$ for some $k \in \mathcal{K}$. We present here a Lyapunov–based design of MPC that guarantees feasibility of the optimization problem and hence constrained stabilization of the closed–loop process from an explicitly characterized set of initial conditions. For this MPC design, the control action at state x and time t is obtained by solving, on-line, a finite horizon optimal control problem of the form:

$$P(x,t) : \min\{J(x,t,u_k(\cdot))|u_k(\cdot) \in S_k\} \quad (6)$$

$$s.t. \ \dot{x} = f_k(x) + G_k(x)u_k \quad (7)$$

$$V_k(x(t+\Delta)) < V_k(x(t)) \ if \ V_k(x(t)) > \delta_k'(8)$$

$$V_k(x(t+\Delta)) \leq V_k(x(t)) \ if \ V_k(x(t)) \leq \delta_k'(9)$$

where $S_k = S_k(t,T)$ is the family of piecewise continuous functions (functions continuous from the right), with period Δ less than Δ^*, mapping $[t, t+T]$ into $\mathcal{U}_k$ and T is the specified horizon. A control $u_k(\cdot)$ in S_k is characterized by the sequence $\{u_k[j]\}$ where $u_k[j] := u_k(j\Delta)$ and satisfies $u_k(t) = u_k[j]$ for all $t \in [j\Delta, (j+1)\Delta)$. The performance index is given by

$$J(x,t,u_k(\cdot)) = \int_t^{t+T} \left[\|x^u(s;x,t)\|_Q^2 + \|u_k(s)\|_R^2\right] ds$$

where R and Q are strictly positive definite, symmetric matrices and $x^u(s;x,t)$ denotes the solution of Eq.1, due to control u, with initial state x at time t. The minimizing control $u_k^0(\cdot) \in S_k$ is then applied to the plant over the interval $[j\Delta, (j+1)\Delta)$ and the procedure is repeated indefinitely. This defines an implicit model predictive control law

$$M_k(x) := argmin(J(x,t,u_k(\cdot))) = u_k^0(t;x,t) \quad (10)$$

The predictive controller formulation of Eqs.6–10 requires that the value of the Lyapunov function decrease after the first step only. Since the optimization problem is guaranteed to be initially and successively feasible for all initial conditions in Ω_k, every control move that is implemented, enforces a decay in the value of the Lyapunov function, leading to stability (for a proof and further details, see (Mhaskar *et al.*, 2004*b*)).

3. PREDICTIVE CONTROL OF SWITCHED NONLINEAR PROCESSES

Consider now the nonlinear switched process of Eq.1. The control problem is formulated as the one of designing a Lyapunov-based predictive controller that guides the closed–loop process trajectory in a way that the schedule, defined by the sets $\mathcal{T}_{k,in}$ and $\mathcal{T}_{k,out}$, for all $k \in \mathcal{K}$, is followed while also, stability of the closed–loop process is achieved. The implementation of the predictive controller that addresses this problem is described algorithmically below (for a proof and further details, see (Mhaskar *et al.*, 2004*b*)). Without loss of generality, it is assumed that the switching schedule dictates initializing the process in mode 1:

(1) Given the process model of Eq.1 and the constraints on the input, design the bounded controller of Eqs.2–3 for each mode and compute

the stability region estimate, $\Omega_k(u_k^{max})$, for the bounded controller using Eq.4-5.

(2) Given the size of the neighborhood, that the state is required to converge to, compute Δ^* such that the predictive controller for each of the individual modes can drive the process to the given neighborhood of the origin for discretization times less than Δ^*, and use a $\Delta \in (0, \Delta^*]$ for the purpose of MPC implementation.

(3) Consider the time $t_{k_r^{in}}$, which designates the time that the closed–loop process is in the k-th mode for the r-th time, and the state belongs to the stability region of the k-th mode (for the purpose of initialization, i.e., at $t = 0$, this would correspond to $t_{1_1^{in}}$, and the state belonging to the stability region $\Omega_1(u_1^{max})$).

(4) If at any time, the switching sequence is aborted, go to step 7; else continue to step 5.

(5) From the set of prescribed switching times, pick $t_{m_j^{in}} = t_{k_r^{out}}$ ($t_{m_j^{in}}$, therefore, is the time that the next switch takes place, and that the process, upon exiting from the current mode enters mode m for the j-th time).

(6) Implement the predictive controller of Eqs.6-10, together with the transition constraint $V_m(t_{m_j^{in}}) < V_m(t_{m_{j-1}^{in}})$ that requires that when the closed–loop process enters the mode m, the value of V_m is less than what it was at the time that the process last entered mode m (this is a version of the multiple Lyapunov function stability condition, see (Branicky, 1998)). If the process has never entered mode m before, i.e., for $j = 1$, set $V_m(t_{m_{j-1}^{in}}) = c_m^{max}$ (this requires that the first time the process enters mode m, it should enter in a way that the state belongs to the stability region corresponding to mode m). If the closed–loop state has already entered the desired ball around the origin, implement $V_m(t_{m_j^{in}}) < \delta_m^{'}$, that ensures that the state stays within the ball. Implement the predictive controller up-to time $t_{k_r^{out}}$, i.e., until the time that the process switches into mode m. Go back to step 2 to proceed with the rest of the switching sequence.

(7) Implement the Lyapunov-based predictive controller of Eqs.6-10 for the current mode to stabilize the closed–loop process.

Remark 1: The closed–loop state evolves in the stability region of the current mode (the mode in which the switched process is operating at a given time) in such a way that at the transition times the state resides in the stability region of the target mode. If at any time, the switching sequence is aborted, the predictive control algorithm is able to stabilize the closed-loop process, because by virtue of the constraint of Eqs.8-9, the state evolves such that it remains in the stability region of the current mode also. The requirement that the Lyapunov function value of the mode, upon entry, be less than that when the process last exited this mode, serves to ensure stabilization of the closed-loop process when the switching sequence is infinite (see (Branicky, 1998) for details). Additionally, the transition constraint serves to fulfil the requirement that the state belong to the stability region of the target mode at the time of the switch.

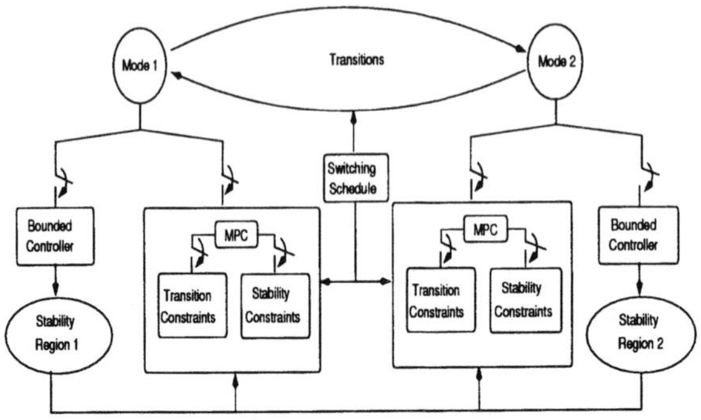

Fig. 1. A schematic representation of the predictive control structure comprising of the predictive and bounded controllers for the individual modes, together with transition constraints.

Remark 2: The constraints of Eqs.8-9 require that the closed–loop state evolves so that the value of the Lyapunov function for the current mode continues to decay, and therefore that the closed–loop state trajectory evolves in a way that it enters the stability region of the target mode (due to the transition constraint), while continuing to evolve in the stability region of the current mode. This condition may be restrictive, and hamper the feasibility of the optimization problem, especially if the stability regions of the constituent modes do not have a significant overlap. When dealing with a finite switching sequence, these constraints can be relaxed for all times before the final switch takes place. For example, a relaxed version of the transition constraint that may be used up to the time of the terminal switch, requires that $V_m(t_{m_j^{in}}) < c_m^{max}$ (i.e., the closed–loop state resides in the stability region of the target mode).

Remark 3: For linear processes, the problem of control of hybrid processes can be handled by formulating it in the form of a mixed logical dynamical system and solving a mixed-integer linear program to come up with an optimal switching sequence and switching times (see (Bemporad and Morari, 1999)). For nonlinear processes, while one can, in principle, still set up the mixed integer nonlinear programming problem, the solution to such a problem is hard to compute and online implementation of such a control algorithm is not possible. The proposed approach provides a systematic framework for control of switched nonlinear processes by decoupling the problem through appropriate constraints on the control implemented on the control action in each of the individual modes. Note also, that if it is possible to solve the mixed integer optimization problem off-line, it can be used to provide a set of optimal switching times, and an associated switching sequence, which can be then implemented online using the proposed approach.

Remark 4: For purely continuous processes, the problem of implementing MPC with guaranteed stability regions was recently addressed for linear processes under state (El-Farra *et al.*, 2004*b*) and output feedback (Mhaskar *et al.*, 2004*a*) and in (El-Farra *et al.*, 2004*a*) for nonlinear processes, by means of a hybrid control structure that unites bounded control and MPC. The hybrid control structure was used to embed the implementation of MPC within the stability region of a Lyapunov-based bounded controller which serves as a fall-back component that can be switched to in the event of infeasibility or instability of the predictive controller. In this work, the bounded controller design is used for the purpose of providing an estimate of the stability region for the Lyapunov-based predictive controller, and feasible initial guesses for the control moves (the decision variables in the optimization problem). In the event, however, that the MPC solver malfunctions, and fails to yield a solution (due, for example, to numerical problems), the switching schedule needs to be aborted and the bounded controller can be switched in to preserve stability.

4. APPLICATION TO A CHEMICAL PROCESS EXAMPLE

We consider a continuous stirred tank reactor where an irreversible, first-order exothermic reaction of the form $A \xrightarrow{k} B$ takes place. The operation schedule requires switching between two available inlet streams consisting of pure A at flow rates F_1, F_2, concentrations C_{A1}, C_{A2}, and temperatures T_{A1}, T_{A2}, respectively. For each mode of operation, the mathematical model for the process takes the form:

$$\begin{aligned} \dot{C}_A &= \frac{F_\sigma}{V}(C_{A\sigma} - C_A) - k_0 e^{\frac{-E}{RT_R}} C_A \\ \dot{T}_R &= \frac{F_\sigma}{V}(T_{A\sigma} - T_R) + \frac{(-\Delta H)}{\rho c_p} k_0 e^{\frac{-E}{RT_R}} C_A \\ &+ \frac{Q_\sigma}{\rho c_p V} \end{aligned} \tag{11}$$

where C_A denotes the concentration of the species A, T_R denotes the temperature of the reactor, Q_σ is the heat removed from the reactor, V is the volume of the reactor, k_0, E, ΔH are the pre-exponential constant, the activation energy, and the enthalpy of the reaction, c_p and ρ, are the heat capacity and fluid density in the reactor and $\sigma(t) \in \{1, \ 2\}$ is the discrete variable. The values of all process parameters can be found in Table 1. The control objective is to stabilize the reactor at the unstable equilibrium point $(C_A^s, \ T_R^s) = (0.57 \ mol/l, 395.3 \ K)$ using the rate of heat input, Q_σ, and change in inlet concentration of species A, $\Delta C_{A\sigma} = C_{A\sigma} - C_{A\sigma_s}$ as manipulated inputs with constraints: $|Q_\sigma| \leq 1KJ/hr$ and $|\Delta C_{A\sigma}| \leq 1 \ mol/l$, $\sigma = 1, 2$. For each mode, we construct a bounded controller and compute its stability region estimate, Ω_1 and Ω_2, shown in Fig.2. The parameters in the objective function of Eq.2.2 are chosen as $Q = qI$, with $q = 1$, and $R = rI$, with $r = 1.0$. The constrained nonlinear optimization problem is solved using the MATLAB subroutine fmincon, and the set of ODEs is integrated using the MATLAB solver ODE45. We first demonstrate the implementation of the Lyapunov-based predictive controller to a single mode operation of the chemical reactor, i.e., one in which the process is operated for all times in mode 1. To this end, we consider an initial condition that belongs to the stability region of the predictive controller for mode 1. As shown by the solid line in Fig.2, starting from the initial condition $(C_A, \ T_R) = (0.14 \ mol/l, \ 404.9 \ K)$, which belongs to the stability region of the predictive controller for mode 1, successful stabilization of the closed–loop process is achieved. The corresponding state and input profiles are shown in Fig.3.

Table 1. Process parameters and steady–state values.

V	=	0.1	m^3
R	=	8.314	$kJ/kmol \cdot K$
C_{A1_s}	=	0.79	$kmol/m^3$
C_{A2_s}	=	1.0	$kmol/m^3$
T_{A1}	=	352.6	K
T_{A2}	=	310.0	K
Q_{1_s}	=	0.0	KJ/hr
Q_{2_s}	=	0.0	KJ/hr
ΔH	=	-4.78×10^4	$kJ/kmol$
k_0	=	1.2×10^9	s^{-1}
E	=	8.314×10^4	$kJ/kmol$
c_p	=	0.239	$kJ/kg \cdot K$
ρ	=	1000.0	kg/m^3
F_1	=	3.34×10^{-3}	m^3/s
F_2	=	1.67×10^{-3}	m^3/s
T_{Rs}	=	395.33	K
C_{As}	=	0.57	$kmol/m^3$

To demonstrate the need to implement the algorithm proposed in section 3 for stabilization when switching is involved, we choose a schedule involving a switch from inlet stream 1 (mode 1) to inlet stream 2 (mode 2) at time $t = 0.1 \ hr$. Once again the process is initialized within the stability region of mode 1, and the predictive controller for mode 1 is implemented. Up until $t = 0.1 \ hr$, the state of the closed–loop process moves towards the desired steady state (as seen from the dashed–lines in Fig.2); however, when the process switches to mode 2, the MPC controller, designed for the stabilization of the process in mode 2, does not yield a feasible solution. If the bounded controller for mode 2 is implemented, the resulting control action is not able to stabilize the closed–loop process (dashed lines in Figs.2–3). This happens because at the time of the transition, the state of the closed–loop process (marked by the *o* in Fig.2) does not belong to the stability region of mode 2. Note also that while the predictive formulation of Eqs.6-10 guarantees stabilization for all initial conditions belonging to the stability region of mode 1, it does not incorporate constraints which enable or ensure a safe transition to mode 2.

Finally, the predictive control algorithm described in section 3 (which incorporates constraints that account for switching) is implemented (dash-dotted lines in Figs.2–3). The MPC controller of mode 1 is designed to drive the state of the closed–loop process such that the state belongs to the stability region of mode 2 at the switching time. Consequently, when the process switches to mode 2 at $t = 0.1\ hr$, the closed–loop process state at the switching time (marked by the $\diamond$ in Fig.2) belongs to the stability region of the MPC designed for mode 2. At this time, when the process switches to mode 2 and the corresponding predictive controller is implemented, closed–loop stability is achieved.

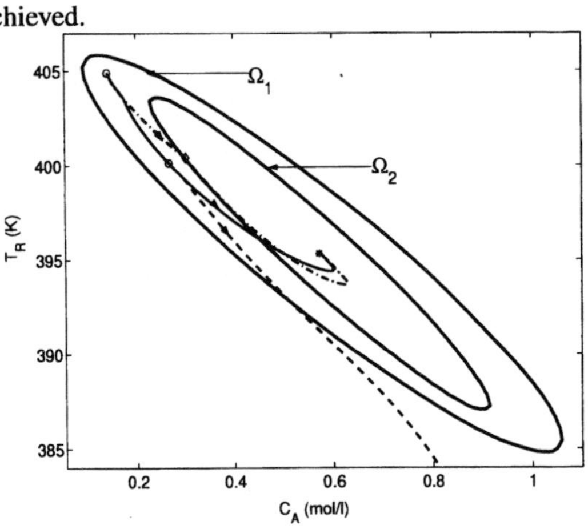

Fig. 2. Closed–loop state trajectory when the reactor is operated in mode 1 for all times under the stabilizing MPC formulation of Eqs.6-10 (solid line), when the reactor operation involves switching from mode 1 to mode 2 at $t = 0.1\ hr$, under the predictive controller design of Eqs.6-10 (dashed line), and when the reactor operation involves switching from mode 1 to mode 2 at $t = 0.1\ hr$, under the predictive controller described in section 3 (dashed–dotted line).

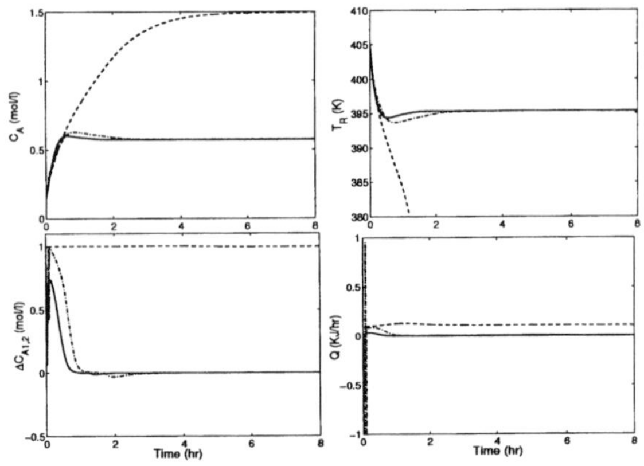

Fig. 3. Closed-loop state (top plots) and manipulated input (bottom plots) profiles: when the reactor is operated in mode 1 for all times under the stabilizing MPC formulation of Eqs.6-10 (solid line), when the reactor operation involves switching from mode 1 to mode 2 at $t = 0.1\ hr$, under the predictive controller design of Eqs.6-10 (dashed line), and when the reactor operation involves switching from mode 1 to mode 2 at $t = 0.1\ hr$, under the predictive controller described in section 3 (dashed–dotted line).

5. REFERENCES

Bemporad, A. and M. Morari (1999). Control of systems integrating logic, dynamics and constraints. *Automatica* **35**, 407–427.

Branicky, M. S. (1998). Multiple Lyapunov functions and other analysis tools for switched and hybrid systems. *IEEE Trans. Automat. Contr.* **43**, 475–482.

El-Farra, N. and P. D. Christofides (2003*a*). Coordinated feedback and switching for control of hybrid nonlinear processes. *AIChE J.* **49**, 2079–2098.

El-Farra, N. H. and P. D. Christofides (2001). Integrating robustness, optimality and constraints in control of nonlinear processes. *Chem. Eng. Sci.* **56**, 1841–1868.

El-Farra, N. H. and P. D. Christofides (2003*b*). Bounded robust control of constrained multivariabe nonlinear processes. *Chem. Eng. Sci.* **58**, 3025–3047.

El-Farra, N. H., P. Mhaskar and P. D. Christofides (2004*a*). Hybrid predictive control of nonlinear systems: Method and applications to chemical processes. *Int. J. Rob. & Non. Contr.* **4**, 199–225.

El-Farra, N. H., P. Mhaskar and P. D. Christofides (2004*b*). Uniting bounded control and MPC for stabilization of constrained linear systems. *Automatica* **40**, 101–110.

Hespanha, J. P. and A. S. Morse (1999). Stability of switched systems with average dwell time. In: *Proceedings of 38th IEEE Conference on Decision and Control*. Phoenix, AZ. pp. 2655–2660.

Hu, B., X. Xu, P. J. Antsaklis and A. N. Michel (1999). Robust stabilizing control law for a class of second-order switched systems. *Syst. Contr. Lett.* **38**, 197–207.

Liberzon, D. and A. S. Morse (1999). Basic problems in stability and design of switched systems. *IEEE Control Systems Magazine* **19**, 59–70.

Lin, Y. and E. D. Sontag (1991). A universal formula for stabilization with bounded controls. *Syst. & Contr. Lett.* **16**, 393–397.

Mhaskar, P., N. H. El-Farra and P. D. Christofides (2004*a*). Hybrid predictive control of process systems. *AIChE J., in Press.*

Mhaskar, P., N. H. El-Farra and P. D. Christofides (2004*b*). Predictive control of switched nonlinear systems with scheduled mode transitions. *IEEE Trans. Automat. Contr., submitted.*

Peleties, P. and R. DeCarlo (1991). Asymptotic stability of m-switched systems using Lyapunov-like functions. In: *Proceedings of American Control Conference*. Boston, MA. pp. 1679–1684.

ELSEVIER
IFAC
PUBLICATIONS
www.elsevier.com/locate/ifac

FEEDBACK CONTROL OF SURFACE ROUGHNESS IN A DEPOSITION PROCESS USING A STOCHASTIC PDE ⋆

Yiming Lou and Panagiotis D. Christofides [1]

Department of Chemical Engineering
University of California, Los Angeles, CA 90095-1592

Abstract: This paper focuses on feedback control of surface roughness in a deposition process, based on a stochastic partial differential equation (PDE) which describes the fluctuation of surface height in the spatial domain. Specifically, we focus on control of surface roughness in a deposition process on a 1-dimensional lattice, whose fluctuation of surface height can be described by the Edwards-Wilkinson equation, a second-order stochastic PDE. We initially reformulate the stochastic PDE into a system of infinite stochastic ordinary differential equations by using modal decomposition. A finite-dimensional approximation of the Edwards-Wilkinson equation is then derived that captures the dominant mode contribution to the surface roughness. A state feedback controller is designed based on the finite-dimensional approximation to control the surface roughness. Analysis of the closed-loop system shows that the controller can drive the surface roughness governed by the infinite-dimensional system to desired levels. The effectiveness of the proposed method is demonstrated by numerical simulations.

Keywords: stochastic partial differential equations, feedback control, deposition processes, roughness control

1. INTRODUCTION

The surface roughness of thin films deposited from gas phase precursors is an important variable to control because it strongly affects the quality of such films. In a thin film growth process, the film is directly formed from microscopic random processes (e.g., particle adsorption, desorption, migration and surface reaction). Therefore, the stochastic nature of thin film growth processes must be fully considered in the modeling and control of the surface roughness of thin films. The desire to understand and control the thin film micro-structure has motivated extensive research on fundamental mathematical models describing the deposition processes, which include 1) kinetic Monte-Carlo methods (e.g., (Gillespie, 1976; Fichthorn and Weinberg, 1991; Lam and Vlachos, 2001)), and 2) stochastic partial differential equations (SPDEs) (e.g., (Edwards and Wilkinson, 1982; Villain, 1991)).

The kinetic Monte-Carlo simulation method can be used to predict properties of the thin film, by explicitly accounting for the micro-processes that directly shape thin film microstructure. Recently, a methodology for feedback control of thin film growth using kinetic Monte-Carlo models has been developed in (Lou and Christofides, 2003*a*; Lou and Christofides, 2003*b*). The methodology leads to the design of (a) real-time roughness estimators by using multiple small lattice kinetic Monte-Carlo simulators, adaptive filters and measurement error compensators and (b) feedback controllers based on the real-time roughness estimates. The method was successfully applied to control surface roughness in a *GaAs* deposition process using an experimentally determined kinetic Monte-Carlo process model (Lou and Christofides, 2004*b*).

⋆ Financial support for this work from the NSF (ITR), CTS-0325246, is gratefully acknowledged.
[1] Corresponding author (email: pdc@seas.ucla.edu).

Other approaches have also been developed to: (a) identify linear models from outputs of kinetic Monte-Carlo simulators and perform controller design by using linear control theory (Siettos *et al.*, 2003), and (b) construct reduced-order approximations of the master equation (Gallivan and Murray, 2003).

However, the fact that kinetic Monte-Carlo models are not available in closed-form makes it very difficult to perform model-based controller design directly on the basis of kinetic Monte-Carlo models. To achieve better closed-loop performance, it is desirable to design feedback controllers on the basis of deposition process models. This motivates research on feedback control of deposition processes based on stochastic PDE models of thin film growth.

This paper focuses on feedback control of surface roughness in a deposition process, based on a stochastic partial differential equation (PDE) which describes the fluctuation of surface height in the spatial domain. Specifically, we focus on control of surface roughness in a deposition process on a 1-dimensional lattice, whose fluctuation of surface height can be described by the Edwards-Wilkinson equation, a second-order stochastic PDE. We initially reformulate the stochastic PDE into a system of infinite stochastic ordinary differential equations by using modal decomposition. A finite-dimensional approximation of the Edwards-Wilkinson equation is then derived that captures the dominant mode contribution to the surface roughness. A state feedback controller is designed based on the finite-dimensional approximation to control the surface roughness. Analysis of the closed-loop system shows that the controller can drive the surface roughness governed by the infinite-dimensional system to desired levels. The effectiveness of the proposed method is demonstrated by numerical simulations. Due to space limitations, we will present in this paper the main results of this research; further results and analysis can be found in (Lou and Christofides, 2004*c*).

2. PRELIMINARIES

We consider a deposition process on a 1-dimensional lattice. In this process, particles land on the surface at a rate, r_a. The rules for the deposition are as follows: a site, l, is first randomly picked among the sites of the whole lattice and the deposition site is determined according to the following rules: 1) if the height of this site is lower than or equal to that of both each nearest neighbors, this site is picked as the deposition site; 2) if the height of only one of the two nearest neighbor sites is lower than that of the original site, deposition is on that site; 3) if the height of each one of the nearest neighbor sites is lower than that of the original site, the deposition site is randomly picked with equal probability between the two nearest neighbor sites. A schematic of the rules of the deposition is shown in Fig.1. There is no particle migration and desorption taking place on this process (see (Lou and Christofides, 2003*a*; Lou and Christofides, 2003*b*) for film growth processes that involve these phenomena).

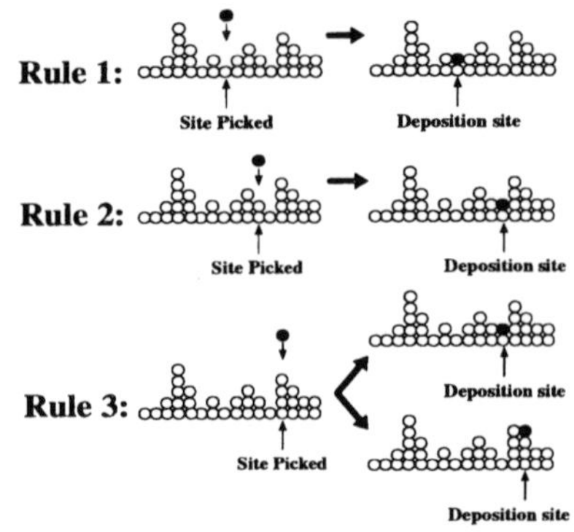

Fig. 1. Schematic of the rules of the deposition.

The deposition process is a stochastic process. Kinetic Monte-Carlo simulation can be used to predict the evolution of the surface configuration in this process. The kinetic Monte-Carlo model of the deposition process is a first-principle model in the sense that the deposition rules can be explicitly considered in the model. However, kinetic Monte-Carlo models are not available in closed-form, which prohibits their use for model-based control design. As an alternative, closed-form stochastic PDE models can be derived based on the deposition rules to describe the evolution of the surface configuration, which is consistent to that predicted by kinetic Monte-Carlo models. In this work, we focus on model-based feedback control design for surface roughness control using a stochastic PDE model of the deposition process described in Fig.1.

The equation for the height fluctuations of the surface in this deposition process was first developed by (Edwards and Wilkinson, 1982). Recently, the same equation was derived directly from the microscopic transition rules of the process (Vvedensky, 2003). Specifically, the height fluctuation of the surface is described by the following stochastic partial differential equation:

$$\frac{\partial h}{\partial t} = a^2 r_a \frac{\partial^2 h}{\partial x^2} + \xi(x,t) \tag{1}$$

where $x \in [-\pi, \pi]$ is the spatial coordinate, t is the time, $h(x,t)$ is the height of the surface at position x and time t, a is the lattice size and $\xi(x,t)$ is a Gaussian noise with zero mean and covariance:

$$\langle \xi(x,t)\xi(x',t') \rangle = \varsigma^2 \delta(x-x')\delta(t-t') \tag{2}$$

where $\varsigma^2 = a^3 r_a$, $\delta(\cdot)$ is the dirac function, and $\langle \cdot\cdot \rangle$ denotes the average. Note that the noise covariance depends on both space x and time t.

The surface roughness, r, is given by the following expression:

$$r(t) = \sqrt{\frac{1}{2\pi} \int_{-\pi}^{\pi} [h(x,t) - \bar{h}(t)]^2 dx} \tag{3}$$

where $\bar{h}(t) = \frac{1}{2\pi}\int_{-\pi}^{\pi} h(x,t)dx$ is the average surface height.

Our objective is to control the surface roughness of the deposition process described by Fig.1. The controller design is based on the SPDE model of the process (Eqs.1 and 2). To do this, we formulate a distributed control problem in the spatial domain $[-\pi,\pi]$. The control problem is described by the following stochastic partial differential equation:

$$\frac{\partial h}{\partial t} = \nu\frac{\partial^2 h}{\partial x^2} + \sum_{i=1}^{p} b_i(x)u_i(t) + \xi(x,t) \tag{4}$$

subject to periodic boundary conditions:

$$\frac{\partial^j h}{\partial x^j}(-\pi,t) = \frac{\partial^j h}{\partial x^j}(\pi,t),\ j=0,1 \tag{5}$$

and the initial condition:

$$h(x,0) = h_0(x) \tag{6}$$

where $\nu = a^2 r_a$, u_i is the ith manipulated input, p is the number of manipulated inputs and b_i is the ith actuator distribution function (i.e., b_i determines how the control action computed by the ith control actuator, u_i, is distributed (e.g., point or distributed actuation) in the spatial interval $[-\pi,\pi]$)

To study the dynamics of Eq.4, we initially consider the eigenvalue problem of the linear operator of Eq.4, which takes the form:

$$\begin{aligned} A\bar{\phi}_n(x) &= \nu\frac{d^2\bar{\phi}_n(x)}{dx^2} = \lambda_n\bar{\phi}_n(x),\ n=1,\cdots,\infty, \\ \frac{d^j\bar{\phi}_n}{dx^j}(-\pi) &= \frac{d^j\bar{\phi}_n}{dx^j}(\pi), \qquad j=0,1 \end{aligned} \tag{7}$$

where λ_n denotes an eigenvalue and $\bar{\phi}_n$ denotes an eigenfunction. A direct computation of the solution of the above eigenvalue problem yields $\lambda_0 = 0$ with $\psi_0 = 1/\sqrt{2\pi}$, and $\lambda_n = -\nu n^2$ (λ_n is an eigenvalues of multiplicity two) with eigenfunctions $\phi_n = (1/\sqrt{\pi})\sin(nx)$ and $\psi_n = (1/\sqrt{\pi})\cos(nx)$ for $n=1,\cdots,\infty$. From the solution of the eigenvalue problem shown in Eq.7, it follows that for a fixed value of $\nu > 0$ the distance between two consecutive eigenvalues (i.e., λ_n and λ_{n+1}) increases as n increases. Furthermore, the eigenspectrum of operator A in Eq.7, $\sigma(A)$ can be partitioned as $\sigma(A) = \sigma_1(A)\bigcup\sigma_2(A)$, where $\sigma_1(A)$ contains the first m (with m finite) eigenvalues (i.e. $\sigma_1(A) = \{\lambda_1,\ldots,\lambda_m\}$) and $\sigma_2(A)$ contains the remaining eigenvalues (i.e., $\sigma_2(A) = \{\lambda_{m+1},\ldots,\}$.

To present the method that we use to control the stochastic PDE of Eq.4, we first derive stochastic ODE approximations of Eq.4 using modal decomposition. To this end, we first expand the solution of Eq.4 in an infinite series in terms of the eigenfunctions of the operator of Eq.7 as follows:

$$h(x,t) = \sum_{n=1}^{\infty}\alpha_n(t)\phi_n(x) + \sum_{n=0}^{\infty}\beta_n(t)\psi_n(x) \tag{8}$$

where $\alpha_n(t)$, $\beta_n(t)$ are time-varying coefficients. Substituting the above expansion for the solution, $h(x,t)$, into Eq.4 and taking the inner product with the adjoint eigenfunctions, $\phi_n^*(z) = (1/\sqrt{\pi})\sin(nz)$ and $\psi_n^*(z) = (1/\sqrt{\pi})\cos(nz)$, the following system of infinite stochastic ODEs is obtained:

$$\begin{aligned} \frac{d\alpha_n}{dt} &= -\nu n^2\alpha_n + \sum_{i=1}^{p} b_{i\alpha_n}u_i(t) + \xi_\alpha^n(t) \\ \frac{d\beta_n}{dt} &= -\nu n^2\beta_n + \sum_{i=1}^{p} b_{i\beta_n}u_i(t) + \xi_\beta^n(t);\ n=1,\ldots,\infty \end{aligned} \tag{9}$$

where $b_{i\alpha_n} = \int_{-\pi}^{\pi}\phi_n(x)b_i(x)dx$, $b_{i\beta_n} = \int_{-\pi}^{\pi}\psi_n(x)b_i(x)dx$, $\xi_\alpha^n(t) = \int_{-\pi}^{\pi}\xi(x,t)\phi_n(x)dx$, $\xi_\beta^n(t) = \int_{-\pi}^{\pi}\xi(x,t)\psi_n(x)dx$.

The covariances of $\xi_\alpha^n(t)$ and $\xi_\beta^n(t)$ can be computed by using the following result (Åström, 1970):

Result 1: If (1) $f(x)$ is a deterministic function, (2) $\eta(x)$ is a random variable with $\langle\eta(x)\rangle = 0$ and covariance $\langle\eta(x)\eta(x')\rangle = \sigma^2\delta(x-x')$, and (3) $\varepsilon = \int_a^b f(x)\eta(x)dx$, then ε is a random number with $\langle\varepsilon\rangle = 0$ and covariance $\langle\varepsilon^2\rangle = \sigma^2\int_a^b f^2(x)dx$.

Using Result 1, we obtain $\langle\xi_\alpha^n(t)\xi_\alpha^n(t')\rangle = \varsigma^2\delta(t-t')$ and $\langle\xi_\beta^n(t)\xi_\beta^n(t')\rangle = \varsigma^2\delta(t-t')$.

In this work, the controlled variable is the expected value of surface roughness, $\sqrt{\langle r^2\rangle}$. According to Eq.8, we have $\bar{h}(t) = \beta_0(t)\psi_0$. Therefore, $\sqrt{\langle r^2\rangle}$ can be rewritten in terms of α_n and β_n as follows:

$$\begin{aligned} \sqrt{\langle r^2\rangle} &= \sqrt{\frac{1}{2\pi}\langle\int_{-\pi}^{\pi}(h(x,t)-\bar{h}(t))^2dx\rangle} \\ &= \sqrt{\frac{1}{2\pi}\sum_{i=1}^{\infty}[\langle\alpha_i^2\rangle + \langle\beta_i^2\rangle]} \end{aligned} \tag{10}$$

Therefore, the surface roughness control problem for the stochastic PDE system of Eq.4 is formulated as the one of controlling the covariance of the states α_n and β_n in the stochastic ODE system of Eq.9.

3. FEEDBACK CONTROL

In this section, we design a linear state feedback controller for the system of Eq.9 so that the surface roughness defined in Eq.10 can be controlled to a desired level.

3.1 Model reduction

Owing to its infinite-dimensional nature, the system of Eq.9 cannot be directly used for the design of controllers that can be implemented in practice (i.e., the practical implementation of controllers which are designed on the basis of this system will require the computation of infinite sums which cannot be done by a computer). Instead, we will base the controller design on finite-dimensional approximations of this system. Subsequently, we will show that the resulting controller will enforce the desired control objective in the closed-loop infinite-dimensional system.

Specifically, we rewrite the system of Eq.9 as follows:

$$\begin{aligned}\frac{dx_s}{dt} &= \Lambda_s x_s + B_s u + \xi_s \\ \frac{dx_f}{dt} &= \Lambda_f x_f + B_f u + \xi_f\end{aligned} \tag{11}$$

where $x_s = [\,\alpha_1 \cdots \alpha_m\ \beta_1 \cdots \beta_m\,]^T$, x_f=$[\alpha_{m+1}\ \beta_{m+1} \cdots]^T$, $\Lambda_s = diag[\,-\nu \cdots -m^2\nu\ -\nu \cdots -m^2\nu\,]$, $\Lambda_f = diag[-(m+1)^2\nu\ -(m+1)^2\nu \cdots]$, $u = [\,u_1 \cdots u_p\,]$, $\xi_s = [\,\xi_\alpha^1 \cdots \xi_\alpha^m\ \xi_\beta^1 \cdots \xi_\beta^m\,]$, and $\xi_f = [\xi_\alpha^{m+1}\ \xi_\beta^{m+1} \cdots]$.

$$B_s = \begin{bmatrix} b_{1\alpha_1} & \cdots & b_{p\alpha_1} \\ \vdots & \ddots & \vdots \\ b_{1\alpha_m} & \cdots & b_{p\alpha_m} \\ b_{1\beta_1} & \cdots & b_{p\beta_1} \\ \vdots & \ddots & \vdots \\ b_{1\beta_m} & \cdots & b_{p\beta_m} \end{bmatrix} \quad B_f = \begin{bmatrix} b_{1\alpha_{m+1}} & \cdots & b_{p\alpha_{m+1}} \\ b_{1\beta_{m+1}} & \cdots & b_{p\beta_{m+1}} \\ b_{1\alpha_{m+2}} & \cdots & b_{p\alpha_{m+2}} \\ b_{1\beta_{m+2}} & \cdots & b_{p\beta_{m+2}} \\ \vdots & \vdots & \vdots \end{bmatrix} \tag{12}$$

We note that the subsystem x_f in Eq.11 is infinite-dimensional.

Neglecting the x_f subsystem, the following $2m$-dimensional system is obtained:

$$\frac{d\tilde{x}_s}{dt} = \Lambda_s \tilde{x}_s + B_s u + \xi_s \tag{13}$$

where the tilde symbol in $\tilde{x}_s$ denotes that this state variable is associated with a finite-dimensional system.

3.2 Feedback control design

We design the state feedback controller on the basis of the finite-dimensional system of Eq.13. To simplify our development, we assume that $p = 2m$ and pick the actuator distribution functions such that B_s^{-1} exists. The state feedback control law then takes the form:

$$u = B_s^{-1}(\Lambda_{cs} - \Lambda_s)\tilde{x}_s \tag{14}$$

where the matrix Λ_{cs} contains the desired poles of the closed-loop system; $\Lambda_{cs} = diag[\lambda_{c\alpha 1} \cdots \lambda_{c\alpha m}\ \lambda_{c\beta 1} \cdots \lambda_{c\beta m}]$, $\lambda_{c\alpha i}$ and $\lambda_{c\beta i}$ $(1 \le i \le m)$ are desired poles of the closed-loop finite-dimensional system, which can be computed from the desired closed-loop surface roughness level.

We first analyze the dependence of the covariances of the states α_n and β_n $(n = 1,\cdots,m)$ on the poles of the finite-dimensional system of Eq.13. Then, we will show in subsection 3.3 that the surface roughness of the infinite-dimensional system of Eq.9 can be controlled to the desired level by using the state feedback controller of Eq.14, which only uses a finite number of actuators.

By applying the controller of Eq.14 to the system of Eq.13, the closed-loop system takes the form:

$$\frac{d\tilde{x}_s}{dt} = \Lambda_{cs}\tilde{x}_s + \xi_s(t) \tag{15}$$

To analyze the effect of the feedback controller on the covariance of the state $\tilde{x}_s$, we discretize Eq.15 in the time domain, using Δt as time step, as follows:

$$X_s(k+1) = G_{cs}X_s(k) + \zeta_s(k);\ k = 0,\cdots,\infty \tag{16}$$

where $X_s(k) = \tilde{x}_s(k\Delta t)$, $G_{cs} = e^{\Lambda_{cs}\Delta t}$, $\zeta_s(k) = \int_{k\Delta t}^{(k+1)\Delta t} e^{\Lambda_{cs}(t-k\Delta t)}\xi_s(t)dt$. According to (Åström, 1970, Chapter 3), if all eigenvalues of G_{cs} are within the unit circle on the complex plane, the covariance matrix of $X_s(k)$, $P(k) = \langle X_s(k)X_s(k)^T\rangle$ converges to $P(\infty)$, which is the solution of the following equation:

$$P(\infty) = G_{cs}P(\infty)G_{cs}^T + R_1 \tag{17}$$

where $R_1 = \langle \zeta_s\zeta_s^T\rangle$. Eq.17 can not be solved, in general, analytically. However, for the specific deposition system considered in this work, the analytical solution for $P(\infty)$ can be obtained as follows:

$$P(\infty) = \begin{bmatrix} P_\alpha(\infty) & 0 \\ 0 & P_\beta(\infty) \end{bmatrix} \tag{18}$$

where $P_\alpha(\infty) = diag[\,\langle\alpha_1(\infty)^2\rangle \cdots \langle\alpha_m(\infty)^2\rangle\,]$, $P_\beta(\infty) = diag[\,\langle\beta_1(\infty)^2\rangle \cdots \langle\beta_m(\infty)^2\rangle\,]$. Using Result 1, $\langle\alpha_n(\infty)^2\rangle$ and $\langle\beta_n(\infty)^2\rangle$ $(n = 1,\cdots,m)$ can be computed by using the following expressions:

$$\langle\alpha_n(\infty)^2\rangle = -\frac{\varsigma^2}{2\lambda_{c\alpha_n}};\ \langle\beta_n(\infty)^2\rangle = -\frac{\varsigma^2}{2\lambda_{c\beta_n}} \tag{19}$$

From Eq.19, we can see that by assigning the closed-loop poles $\lambda_{c\alpha_n}$ and $\lambda_{c\beta_n}$ $(n = 1,\cdots,m)$ at desired locations, the covariances of the states α_n and β_n $(n = 1,\cdots,m)$ can be controlled to desired levels. Therefore, according to Eq.10, the contribution to the surface roughness from the finite-dimensional system of Eq.13 can be controlled to the desired level.

3.3 Analysis of the closed-loop infinite-dimensional system

In this subsection, we show that when the state feedback controller of Eq.14 is used to manipulate the poles of the finite-dimensional system of Eq.13, the contribution to the surface roughness from the α_f and β_f subsystem of the system of Eq.11 is bounded and can be made arbitrarily small by increasing the dimension of the x_s subsystem.

By applying the feedback controller of Eq.14 into the infinite-dimensional system of Eq.11, we obtain the following closed-loop system:

$$\begin{aligned}\frac{dx_s}{dt} &= \Lambda_{cs}x_s + \xi_s \\ \frac{dx_f}{dt} &= \Lambda_\varepsilon x_s + \Lambda_f x_f + \xi_f\end{aligned} \tag{20}$$

where $\Lambda_\varepsilon = B_f B_s^{-1}(\Lambda_{cs} - \Lambda_s)$.

The boundedness of the state of the above system follows directly from the stability of the matrices Λ_{cs} and Λ_f and the structure of the system, where the x_s subsystem is independent of the x_f state (see

(Christofides, 2001) for results and techniques for analyzing the stability properties of such systems).

Due to the structure of the eigenspectrum of operator A (Section 2), the effect of the control action computed from Eq.14 to the poles of the x_f subsystem can be reduced by increasing m. Therefore, by picking m sufficiently large, the $\Lambda_\varepsilon x_s$ can be made very small compared to $\Lambda_f x_f$ and thus, the closed-loop system of Eq.20 can be adequately described by the following system:

$$\begin{aligned} \frac{dx_s}{dt} &= \Lambda_{cs} x_s + \xi_s \\ \frac{dx_f}{dt} &= \Lambda_f x_f + \xi_f \end{aligned} \tag{21}$$

On the basis of the above system, it can be shown that the covariance of the state of the x_f subsystem converges to $[\ \langle \alpha_{m+1}(\infty)^2 \rangle\ \langle \beta_{m+1}(\infty)^2 \rangle\ \cdots\ \cdots]$, where

$$\langle \alpha_n(\infty)^2 \rangle = \frac{\varsigma^2}{2n^2 \nu};\ \langle \beta_n(\infty)^2 \rangle = \frac{\varsigma^2}{2n^2 \nu};\ n > m \tag{22}$$

Therefore, the overall contribution to the surface roughness from the x_f subsystem in Eq.11 can be computed as follows:

$$\frac{\varsigma}{\sqrt{2\pi(m+1)\nu}} < \sqrt{\frac{1}{2\pi} \sum_{n=m+1}^{\infty} \left[\frac{\varsigma^2}{\nu n^2}\right]} < \frac{\varsigma}{\sqrt{2\pi m \nu}} \tag{23}$$

Clearly, as $m \longrightarrow \infty$, the contribution to the surface roughness from the α_f and β_f subsystem goes to zero.

In summary, under the controller of Eq.14, the closed-loop surface roughness, for m sufficiently large, can be adequately described by the following expression:

$$\sqrt{\langle r^2 \rangle} = \varsigma \sqrt{\frac{1}{2\pi} [\lambda^* + \sum_{n=m+1}^{\infty} \frac{1}{\nu n^2}]} \tag{24}$$

where

$$\lambda^* = \sum_{i=1}^{m} \left(-\frac{1}{2\lambda_{c\alpha_i}} - \frac{1}{2\lambda_{c\beta_i}}\right) \tag{25}$$

Remark 1. Note that to control the closed-loop surface roughness to $\sqrt{\langle r_d^2 \rangle}$, we need to design a controller to assign the poles of the finite-dimensional system of Eq.15 to appropriate values. The controller which assigns the poles of the system of Eq.15 to satisfy Eq.25 is not unique. Consequently, for a fixed number of actuators, p, the controller that can drive the closed-loop surface roughness to a desired level is also not unique.

4. SIMULATION RESULTS

In this section, we present an application of the proposed state feedback controller to the deposition process described in Fig.1 to regulate the surface roughness to a desired level. Specifically, the deposition occurs on a lattice containing 1000 sites. Therefore, $a =$ 0.00628. The deposition rate for each site is $r_a = 1s^{-1}$. A $600th$ order stochastic ordinary differential equation approximation of the system of Eq.4 is used to simulate the process (the use of higher-order approximations led to identical numerical results, thereby implying that the following simulation runs are independent of the discretization). The δ function involved in the covariances of ξ_α^n and ξ_β^n is approximated by $\frac{1}{\delta t}$.

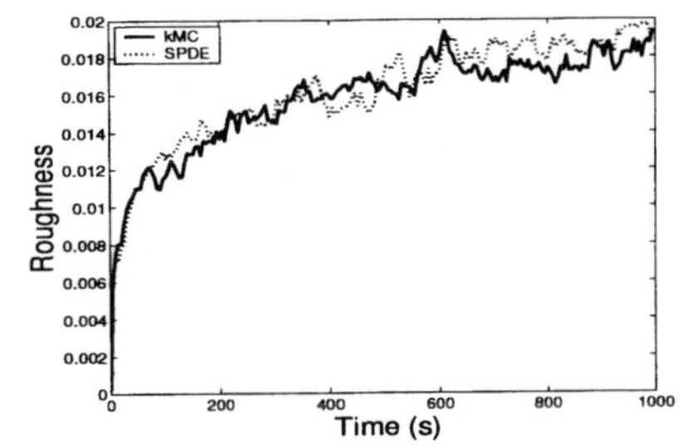

Fig. 2. Comparison of the open-loop profile of surface roughness from the kinetic Monte-Carlo simulator (solid line) and that from the solution of the SPDE using 300 modes (dotted line).

In the first simulation, we compare the open-loop surface roughness profile of the deposition process from the solution of the stochastic PDE model of Eq.1 to that from a kinetic Monte-Carlo simulation. We use the kinetic Monte-Carlo algorithm developed in (Vlachos, 1997) to simulate the process. First, a random number is generated to pick a site among all the sites on the 1-D lattice. If the height of this site is lower than or equal to that of both each nearest neighbors, this site is picked as the deposition site and the height of this site increases by a; if the height of only one of the two nearest neighbor sites is lower than that of the original site, deposition is on that site and the height of that site increases by a; if the height of each one of the nearest neighbor sites is lower than that of the original site, a second random number is generated to randomly pick one of the two nearest neighbors with equal probability and the height of the picked site increases by a. Upon an executed event, a time increment, dt, is computed by $dt = \frac{-\ln \zeta}{N \times r_a}$, where ζ is a random number in the $(0, 1)$ interval and N is the total number of sites on the lattice.

The profiles of surface roughness from the kinetic Monte-Carlo simulation and the solution of the stochastic PDE model are shown in Fig.2. The two profiles are very close, which means that the stochastic PDE model of Eq.4 can adequately (see Remark 2 below for a discussion on this issue) simulate the evolution of surface roughness of the deposition process described in Fig.1.

Subsequently, we design a state feedback controller based on a $60th$ order stochastic ODE approximation constructed by using the first 60 eigenmodes of the system of Eq.9. 60 control actuators are used to control the system. The ith actuator distribution function is taken to be:

$$b_i(z) = \frac{1}{\sqrt{\pi}}[\sin(iz) + \cos(iz)]; \ i = 1, \cdots, 60 \quad (26)$$

Our desired roughness is around 0.0058. Using Eq.24, we design the state-feedback controller such that $\lambda_{c\alpha_i} = \lambda_{c\beta_i} = -3.55$, for $i = 1, \cdots, 30$. Under this state feedback controller, the closed-loop surface roughness is in the range $[0.0057, 0.0058]$. Fig.3 shows a comparison between the closed-loop roughness profile (solid line) and the open-loop roughness profile (dotted line). We can see the controller successfully drives the surface roughness to the desired level, which is much lower than that corresponding to open-loop operation ($u_i(t) = 0, i = 1, \ldots, 60$).

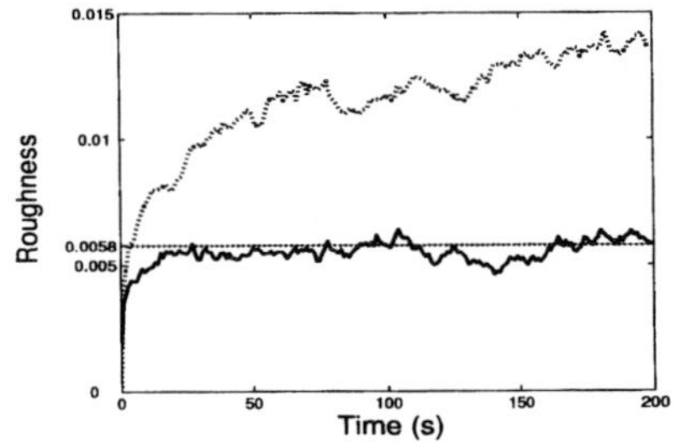

Fig. 3. Closed-loop roughness profile (solid line) when the controller is designed based on the first 60 modes vs. the open-loop roughness profile (dotted line).

Remark 2. Note that all simulation results of roughness profiles shown in this section are realizations of a stochastic process. Therefore, fluctuations can be observed in all simulation results. Our control objective is to drive the expected roughness to the desired level, so roughness profiles from each simulation fluctuate around this desired level due to the stochastic nature of the process. Conceptually, if we run infinite number of simulations with the same parameters, and average the roughness profiles obtained from each simulation run, the desired expected roughness profile can be obtained.

Remark 3. Note that the controller design method developed in this work can be applied to other processes described by stochastic PDE models. For example, it can be used to control the stochastic Kuramoto-Sivashinsky equation (Lou and Christofides, 2004*a*), which describes evolution of surface microstructure in a variety of physical and chemical processes including ZrO_2 thin film growth by reactive ion beam sputtering (Qi *et al.*, 2003).

REFERENCES

Åström, K. J. (1970). *Introduction to Stochastic Control Theory*. Academic Press. New York.

Christofides, P. D. (2001). *Nonlinear and Robust Control of PDE Systems: Methods and Applications to Transport-Reaction Processes*. Birkhäuser. Boston.

Edwards, S. F. and D. R. Wilkinson (1982). The surface statistics of a granular aggregate. *Proc. R. Soc. Lond. A* **381**, 17–31.

Fichthorn, K. A. and W. H. Weinberg (1991). Theoretical foundations of dynamical Monte Carlo simulations. *Journal of Chemical Physics* **95**, 1090–1096.

Gallivan, M. A. and R. M. Murray (2003). Model reduction and system identification of master equation control systems. In: *Proceedings of American Control Conference*. Denver, Colorado. pp. 3561–3566.

Gillespie, D. T. (1976). A general method for numerically simulating the stochastic time evolution of coupled chemical reactions. *Journal of Computational Physics* **22**, 403–434.

Lam, R. and D. G. Vlachos (2001). Multiscale model for epitaxial growth of films: Growth mode transition. *Physical Review B* **64**, 035401.

Lou, Y. and P. D. Christofides (2003*a*). Estimation and control of surface roughness in thin film growth using kinetic Monte-Carlo models. *Chemical Engineering Science* **58**, 3115–3129.

Lou, Y. and P. D. Christofides (2003*b*). Feedback control of growth rate and surface roughness in thin film growth. *A.I.Ch.E Journal* **49**, 2099–2113.

Lou, Y. and P. D. Christofides (2004*a*). Feedback control of surface roughness in a sputtering process using the stochastic Kuramoto-Sivashinsky equation. *Submitted to 43rd IEEE Conference on Decision and Control.*

Lou, Y. and P. D. Christofides (2004*b*). Feedback control of surface roughness of GaAs (001) thin films using kinetic Monte-Carlo models. *Computers & Chemical Engineering,* in press.

Lou, Y. and P. D. Christofides (2004*c*). Feedback control of surface roughness using stochastic PDEs. *A.I.Ch.E. Journal;* in press.

Qi, H. J., L. H. Huang, Z. S. Tang, C. F. Cheng, J. D. Shao and Z. X. Fan (2003). Roughness evolution of ZrO_2 thin films grown by reactive ion beam sputtering. *Thin Solid Films* **444**, 146–152.

Siettos, C. I., A. Armaou, A. G. Makeev and I. G. Kevrekidis (2003). Microscopic/stochastic timesteppers and "coarse" control: a kMC example. *A.I.Ch.E J.* **49**, 1922–1926.

Villain, J. (1991). Continuum models of crystal growth from atomic beams with and without desorption. *J. Phys. I* **1**, 19–42.

Vlachos, D. G. (1997). Multiscale integration hybrid algorithms for homogeneous-heterogeneous reactors. *A.I.Ch.E Journal* **43**, 3031–3041.

Vvedensky, D. D. (2003). Edwards-wilkinson equation from lattice transition rules. *Physical Review E* **67**, 0251021–0251024.

ELSEVIER
IFAC
PUBLICATIONS
www.elsevier.com/locate/ifac

ON-LINE INDUSTRIAL IMPLEMENTATION OF PROCESS MONITORING / CONTROL APPLICATIONS USING MULTIVARIATE STATISTICAL TECHNOLOGIES: CHALLENGES AND OPPORTUNITIES

Michael Dudzic*, **Yale Zhang**

Process Automation, Dofasco Inc.
Box 2460, Hamilton, Ontario, Canada, L8N 3J5

Abstract: The global steel industry is striving to improve product quality through excellence in operation. To support this, significant investments have been made in upgrading instrumentation, data acquisition and computing infrastructures. The expectation is that with more process and product data readily available, useful information and better process knowledge can be gained in a timely fashion. The problem that has developed is that with the large volumes of data available, the associated data analysis and modeling have become increasingly complex. As a result, much of the data is either not used or summarized / heavily compressed. This means that a significant amount of the information and knowledge resident in the data is lost, diminishing the returns from the investment made in the information technology infrastructure. A class of technologies that Dofasco has used to meet this data challenge is multivariate statistics (MVS), with a primary focus on Principal Components Analysis (PCA) and Projection to Latent Structures (PLS). These methods have been successfully applied to analyze data for a variety of purposes, which includes the development of on-line predictive models and process monitoring systems. Since 1993, Dofasco has been involved with over 70 off-line / on-line applications of this technology at our steel facility in Hamilton, Ontario, Canada. Through these applications, significant financial returns to the company have been generated. *Copyright © 2004 IFAC*

Keywords: multivariate statistics; model-based control; process monitoring; continuous caster; desulphurization

1. INTRODUCTION

As the ability to collect industrial data increases and improves, the key challenge is to exploit the useful information so as to maximize process understanding and knowledge. The concern then becomes the appropriate selection of data analysis techniques that can be used with large data sets in order to provide production engineers and operators with useful and easy-to-understand results.

From an automation engineer's perspective, improved data analysis and statistical modeling can be used to: (1) maximize information and knowledge generation to aid in both business and process related decisions, (2) develop predictive models for both inference and control, (3) facilitate useful statistical process control (SPC) schemes, (4) provide the ability to supplement first principle modeling activities, and (5) provide reliable results under difficult industrial conditions. The obstacles are that typical industrial data include a large number of highly correlated, noisy and possibly missing measurements. For such data sets, the use of traditional data analysis and modeling technologies can be both time consuming and produce results that do not capture the essential and useful information resident in the data. For any on-line application, robustness and long-term sustainability are key requirements for success.

In this paper, we will present some our industrial experiences in the use of multivariate statistics and the benefits from it. A brief introduction to the MVS technologies will be presented from an industrial perspective. This will be followed by a description of two on-line applications that continue to provide significant value to the company. The first example is an on-line process monitoring system that is used to observe the complete operation of the casting process in the mould area of Dofasco's #2 Continuous Slab Caster. The second example is a predictive model used at Dofasco's Desulphurization facility for determining the optimal amount of reagent needed to accurately remove sulphur from pig iron. The paper

* Author to whom correspondence should be addressed. Tel.: +1(905)548-7200 ext. 6986. E-mail: mike_dudzic@dofasco.ca.

concludes with the following two sections; our thoughts for future applied research in this area that would be of value to industry in general and some reflections on the industrial application of MVS.

2. OVERVIEW OF MULTIVARIATE STATISTICAL TECHNOLOGIES

The practical value of PCA and PLS modeling methods is that these techniques allow for the systematic examination and interpretation of large amounts of highly correlated data. Examination of MVS model outputs can provide insight into the operation of an industrial process during monitoring and quality assurance activities. With PCA and PLS, the systematic interpretation of dominant patterns in the data, and the isolation of the most important contributors to these patterns are also possible. This allows the classification of data relationships according to normal and abnormal operation. Some of the numerous advantages of PCA and PLS latent variable modeling methods have over traditional monitoring and prediction technologies are: (1) provision for data dimension reduction, (2) robustness to highly correlated, noisy and missing data, (3) the meaningful graphical display of model outputs, and (4) applicability to both continuous and batch processes.

The following section describes some concepts of multivariate statistical modeling from an industrial viewpoint. Further excellent theoretical reviews of applying MVS in industrial applications can be found in Kourti (2002), and Kourti and MacGregor (1996).

2.1. An Industrial Perspective on Multivariate Statistical Modeling

In both SPC and empirical modeling applications, the sample covariance matrix, $\mathbf{S_{xx}}=(n-1)^{-1}\mathbf{X}^T\mathbf{X}$, and its inverse, often form the basis of the statistics being calculated, where $\mathbf{X}$ is an n × p matrix of n observations of p variables. When developing linear predictive models, and in the calculation of the Hotelling T^2 statistic in monitoring applications, the inversion of an ill-conditioned $\mathbf{S_{xx}}$ matrix can be problematic. It is precisely when the $\mathbf{S_{xx}}$ matrix is ill-conditioned that multivariate statistics become particularly useful. Our focus on the use of MVS as the basis for SPC and modeling applications in industrial settings, is because in such applications, it is very common to have large data sets that include many variables that are highly correlated, and hence result in $\mathbf{S_{xx}}$ matrices that are ill-conditioned.

The Hotelling T^2 statistic is defined in Eq. (1) as the multidimensional extension of the t^2 statistic.

$$T^2=(\mathbf{x}-\boldsymbol{\mu}_\mathbf{x})^T\mathbf{S_{xx}}^{-1}(\mathbf{x}-\boldsymbol{\mu}_\mathbf{x}) \qquad (1)$$

where $\mathbf{x}$ is a p × 1 vector of variables whose covariance structure is described by $\mathbf{S_{xx}}$, and $\boldsymbol{\mu}_\mathbf{x}$ is a p × 1 vector of the associated sample means of the variables in $\mathbf{x}$. The Hotelling T^2 statistic is often used as the basis for a control chart in applications where there are many variables to be monitored, although there are other options such as the multivariate exponentially weighted moving average (MEWMA) or the multivariate cumulative sum (MCUSUM) statistics. However, due to the correlated nature of many process data, the matrix inversion in Eq. (1) can often produce poor numerical results due to ill-conditioning. To remedy this, the $\mathbf{S_{xx}}$ matrix may be decomposed using linear projection based methods such as principal component analysis (PCA). PCA projects the original process data ($\mathbf{x}$) into a lower dimensional space spanned by the first A eigenvectors of $\mathbf{S_{xx}}$. Once projected into this space, the resulting transformed vectors ($\mathbf{t}$) are orthogonal and have better numerical properties that allow for fault monitoring using the T^2 statistic given in Eq. (1) with $\mathbf{X}$ and $\mathbf{x}$ replaced by $\mathbf{T}$ and $\mathbf{t}$ respectively, where $\mathbf{T}$ is an n × A matrix of projected data ('scores'), and $\mathbf{t}$ is an A × 1 vector of scores for a single observation. The control chart is used by evaluating the value of T^2 at prescribed intervals and comparing this value to a control limit. Control charts can also be constructed to monitor the vector of $\mathbf{t}$ values individually.

The PCA projection of data is accompanied by some loss of information since the dimension of the space spanned by the first A eigenvectors of $\mathbf{S_{xx}}$ is smaller than the dimension of $\mathbf{S_{xx}}$ (i.e. $A < p$). To measure this, a companion statistic called the 'squared prediction error' (SPE) is also monitored for fault detection when projections are used. This statistic measures the squared orthogonal distance from the un-projected data point ($\mathbf{x}$) to the space spanned by the eigenvectors of $\mathbf{S_{xx}}$ and is also compared to a threshold limit in fault detection. This distance captures the information in the space spanned by the last p-A eigenvectors of $\mathbf{S_{xx}}$.

The definition of these statistics, their distributional properties, and their interpretation can be found in several references including Morrison (1990), Mason and Young (2002), Jackson (1991). Kourti and MacGregor (1996) have advocated a means of rewriting the T^2 and SPE statistics in terms of the elements that are used in the summation equations that define these terms. These elements, when separated and scaled in a particular way, can be related to the un-projected X-data and are said to be 'contributors' to the T^2 and SPE statistics. These contributors can be used to investigate faults in a Pareto fashion by focusing effort on the large contributors.

In many industrial applications, it is necessary to use the information in measured process variables to predict values of output variables that are often related to product quality and performance, or to define values for input variables such as controller setpoints. Approaches to modeling for prediction include the development of first principles modeling, and empirical modeling. While first-principals models offer more promise in developing theoretical understanding of process behaviour, they are often difficult to formulate and costly to develop. In some applications, these models can be too computationally intense to solve for direct on-line use. In comparison, many empirical models, including those based on MVS, offer less theoretical insight but can be developed and solved more rapidly than first-principles models. Furthermore, in contrast to first-principles models, empirical models can include variables that theoretical models do not account for, are generally only valid in the region spanned by the data used to build the model, and should not be expected to produce reliable results when extrapolating. Nevertheless, empirical models are often useful in industrial settings where speed of implementation and robustness to practical issues such as missing data are important.

One useful class of empirical models are based on linear predictions. These predictive models have the following form:

$$\mathbf{y}_p = \boldsymbol{\beta}\, \mathbf{x} \qquad (2)$$

where the output variable $\mathbf{y}_p$ is a $k \times 1$ vector of predicted quantities that result from the combination of the input vector $\mathbf{x}$. Predictions can be made when the elements of the $k \times p$ matrix $\boldsymbol{\beta}$ have been computed. There are various approaches to determining values for $\boldsymbol{\beta}$, many of which rely on the inversion of the $\mathbf{X}^T\mathbf{X}$ (i.e. (n-1) $\mathbf{S}_{xx}$) matrix. If the variables forming the columns of the data matrix $\mathbf{X}$ are nearly collinear, as is often the case, the inversion of $\mathbf{X}^T\mathbf{X}$ can become numerically unstable and can produce poor predictions that are of little practical use. This phenomenon is well known and remedies such as PLS are available.

PLS is a projection-based method that explicitly employs the covariance structures between the elements of $\mathbf{x}$ to those of $\mathbf{y}$, and those of $\mathbf{x}$ to $\mathbf{x}$, and $\mathbf{y}$ to $\mathbf{y}$, in model development. The differences between this and the objectives in modeling with other methods were analyzed by Burnham et. al. (1999), and the numerical methods associated with PLS are summarized in Nelson et al. (1996). In general, PLS and other projection-based methods can produce estimates that are more stable because the correlation among the variables being modeled has been taken into account.

From a robustness point-of-view, the eigenvalue-eigenvector model structure used in PCA can be used to treat missing data in many ways during both the model estimation phase and during on-line use of the model. One approach to missing data relies on the model estimation algorithm. During the PCA model estimation phase, the application of the Non-linear Iterative Partial Least Squares (NIPALS) algorithm extracts eigenvectors from the **X** data sequentially via the application of an iterative series of linear regressions. When data in any of the columns of the **X** matrix are missing, the regressions are preformed using the data that are present and the missing points are ignored. As long as the number of variables in any row or column is greater than the number of principle components being estimated, the NIPALS algorithm can find a solution. A detailed description of various missing data handling methods for both PCA and PLS is given by Nelson *et. al.* (1996).

3. ON-LINE MONITORING APPLICATION – #2 CONTINUOUS CASTER STABLE OPERATION MONITORING SYSTEM

In this section is a description of selected features of the integrated on-line monitoring system at Dofasco's #2 Continuous Caster. This will highlight the use of PCA in a multivariable SPC monitoring application.

3.1. Continuous Steel Caster Process Description

In the manufacture of steel, the conversion of liquid steel into solid steel slabs is commonly achieved through a process known as continuous casting, which is illustrated in Figure 1.

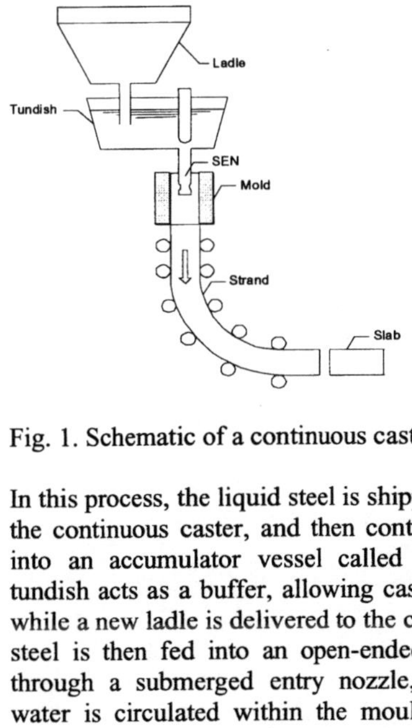

Fig. 1. Schematic of a continuous caster.

In this process, the liquid steel is shipped by a ladle to the continuous caster, and then continuously poured into an accumulator vessel called a tundish. The tundish acts as a buffer, allowing casting to continue while a new ladle is delivered to the caster. The liquid steel is then fed into an open-ended copper mould through a submerged entry nozzle, SEN. Cooling water is circulated within the mould walls so that liquid steel in contact with the copper mould solidifies, forming a steel strand with a solid shell and liquid center. The strand is continuously withdrawn from the mould into additional cooling chambers where the internal liquid steel solidifies under controlled cooling conditions. Once the strand has fully solidified, it is cut into slabs according to customer requirements.

In a continuous caster, the typical operating sequence consists of a brief start-up transition (called a start-cast), followed by a prolonged continuous, run-time operation, and then a shutdown operation. Within the run-time operation, there are occasional process transitions that occur to accommodate equipment or product specification changes.

3.2. Motivation for Improved Monitoring

The focus for improvement was to detect the onset and to prevent the occurrence of caster strand breakouts, with a secondary focus on providing early detection of other abnormal operating situations. A breakout is a catastrophic failure in the steel casting process. It occurs when the solidifying shell of the cast steel strand does not form uniformly on leaving the bottomless mould and ruptures, allowing molten steel to escape. This spillage of molten steel can be potentially hazardous and can cause equipment damage and costly process delays. The onset of a caster strand breakout is known to be related to changes in mould thermocouple signals, cooling water flows, mould level and other process variables that are all interrelated.

Experience shows that breakouts occur during start-cast, run-time or under process transition operation. Breakouts can be avoided by reducing the casting speed thereby giving more residence time in the mould for the steel to solidify. To avoid the occurrence of a breakout, it is critical to detect improper solidification of the steel shell in advance with enough lead-time to appropriately slow down the casting.

Monitoring continuous casting processes and predicting the onset of breakouts in advance has been attracting considerable industrial interest for several decades. To date, commercially available monitoring systems only address those types of breakouts in run-time cast operations that have definite signatures in the process data. Our challenge was to develop an improved monitoring scheme that included provisions for: (1) the ability to detect many different classes of breakouts, (2) advanced warning, as early as possible, to the high probability onset of a breakout under all applicable operating conditions, and (3) easy-to-access diagnostics for real-time troubleshooting of processing problems that could potentially result in a breakout.

3.3. PCA-based Multivariate SPC Monitoring Solution

In the following subsections, are descriptions of selected features of the PCA-based monitoring system under run-time, start-cast and process transition operating conditions.

3.3.1. Run-time Operations Monitoring

PCA has been applied to monitor the continuous run-time cast operations, where a principal component model was developed based on selected historical data from over half a year of normal run-time operations. The model building task involved the selection of significant input measurements, the incorporation of variable lagging to capture process dynamics, and the appropriate pre-treatment of the data to ensure a representative data set. The resulting PCA model is a projection of 323 inputs into a 10 dimensional score space. Two control charts, the Hotelling T^2 (HT) statistic and the squared prediction error (SPE), then monitor the real-time output of this model.

For this application, 6 PCA models were required to cover all the different operating scenarios at the caster. Model selection is automatic so that the operator only views the correct model output. During run-time operation, the system uses the appropriate model to compare current inputs to their historical benchmark in real-time at a sub-second frequency. The resulting HT and SPE control charts provide a continuous measurement of the overall stability of the casting process. In addition, variable contributions are calculated and ranked in order of magnitude to provide direction to operators in troubleshooting their process.

HMI (Human Machine Interface) screens have been custom designed to supply sufficient information for operators to monitor caster operations in real-time and perform further diagnosis if abnormal situations are detected. As shown in Figure 2, the information contained in the main screen of the integrated caster monitoring system includes: (1) general casting information such as heat number, steel grade, casting speed, etc..., (2) SPE and HT control charts indicate the stability of the casting operation, and (3) a graphical representation of the mould and its surrounding thermocouples. The SPE and HT are scaled to [0,1] with respect to the control limits, respectively. There is a warning limit set at 0.8 and an audible alarm is generated if the scaled SPE or HT statistics exceed the value of one. Other screens include contribution plots (see Figure 3), customized trending, etc..., which operators can access via navigation buttons.

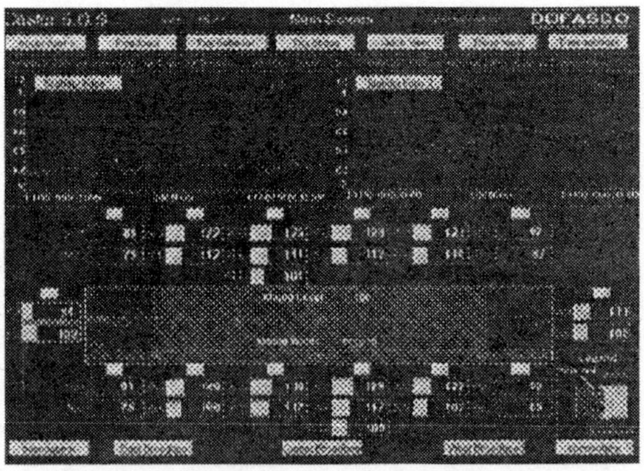

Fig. 2. Main screen of integrated monitoring system for #2 Continuous Caster

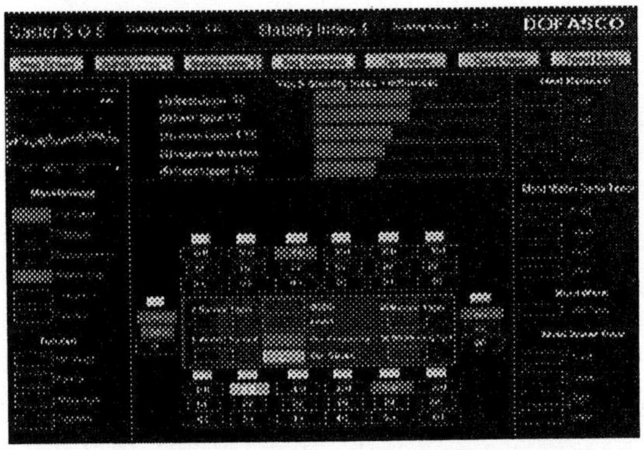

Fig. 3. Contribution plot screen of integrated monitoring system for #2 Continuous Caster

Further details on run-time MVS caster monitoring can be found in Vaculik, *et. al.* (2001).

3.3.2. Start-cast Operations Monitoring

The start-cast is essentially a dynamic process in which most process variables (such as thermocouple temperatures, heat transfer flux through each side of the mould, etc...) show non-linear trajectories versus time and they are highly auto-correlated. In continuous processes, it is expected that the relationship among the variables remains stationary through time. In order to capture the non-stationary, non-linearity and the dynamic nature of the start-cast process, it is necessary to include data through time for all variables.

In approaching this problem, a start-cast is considered a batch process over a pre-defined, finite duration. For our case, the start cast duration begins with the time the casting speed starts to increase, and ends when the strand length exceeds a specific value (as determined by process knowledge). The multivariate statistics technology applicable for this application is multi-way PCA (MPCA). MPCA is an extension of the PCA algorithm, as was demonstrated by Nomikos and MacGregor (1994 and 1995) for batch process monitoring. The MPCA algorithm organizes the data such that each sample of process measurements at different sampling intervals is considered a new variable in the MPCA model. This allows the characterization of the changing relationships among the variables.

Some key issues in applying MPCA to monitoring start-casts are briefly described here. Further details on start-cast MVS caster monitoring can be found in Zhang, *et. al.* (2003). For theoretical details on the use of MVS in dynamic data modeling for batch processes, see Kourti (2003).

To construct the start-cast data, large quantities of historical data for normal start-cast operations are required. The data can be shaped into a 3-dimensional data block in which the 3 dimensions are: a number of normal start-cast *operations* (must cover the full region of operation); the set of process *variables*; and the samples through time (*observations*) over the start-cast duration. In the model development, 147 start-cast *operations,* with 60 process *variables,* over 800 *observations,* were used.

In the actual operation, the start-cast duration is variable. This does not comply with one of underlying requirements in the MPCA algorithm that, all trajectories must be synchronized with the evolution of the batches. To solve this problem, the indicator variable approach (Nomikos and MacGregor, 1994) was used to synchronize the process trajectories, where the strand length acts as the indicator variable since it progresses monotonically in time and has the same starting and ending value for each start-cast. All process trajectories of start-casts are synchronized using interpolation based on a set of predefined scales in the strand length. The synchronization scales are determined by a quadratic function of time such that it possesses small intervals at the beginning of the start-cast duration and large intervals at the end. Such non-uniform scales provide a better opportunity for early detection of abnormal situations.

The core concept of the MPCA algorithm is to handle the resulting 3-dimensional, synchronized dynamic data. The algorithm first unfolds the 3-dimensional data block to a 2-dimensional data matrix to preserve the direction of operations (see Figure 4).

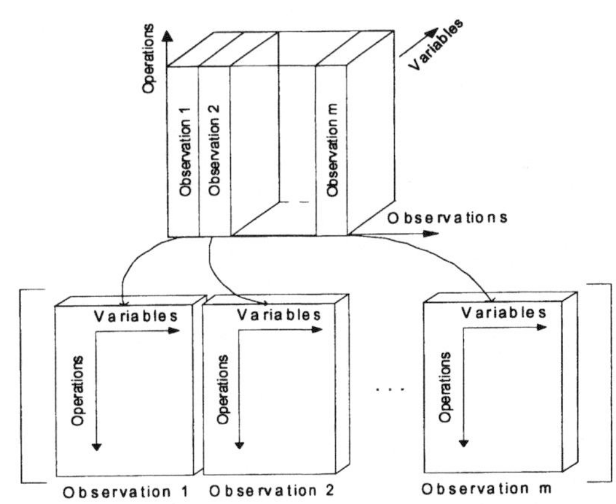

Fig. 4. 3-Dimensional data unfolding

This is accomplished by the following two steps. First, the data block is sliced vertically along the observation direction; next the slices are juxtaposed to build the data matrix with a large column dimension such that each row corresponds to a start-cast. A standard PCA algorithm is then applied to this unfolded data matrix to build a statistical model for the deviation of each process variable from its average trajectory over the start-cast duration, benchmarking the operation-to-operation variance existing in the historical data of normal start-casts. Similar to the run-time operation monitoring, two statistics, SPE and HT, are defined at each observation over the start-cast duration in such a way that they are able to describe how each start-cast is compared with normal operation as described by the model. As a new start-cast evolves, its deviation from the average trajectories is examined by comparing it to the model. If the new start-cast is statistically different from normal operation, then an alarm is generated to indicate the abnormal situation. Once an alarm is generated, the variables of the new start-cast that contribute most to the process deviation are identified in the contribution plots.

3.3.3. Process Transition Operation Monitoring

The remaining operating condition to monitor for involves process transitions. There are some well defined process transitions involved with the continuous casting process related to equipment or product specification changes where abnormal operation resulting in a breakout can occur. Our solution to monitoring one important process transition, the changing of the submerged entry nozzle, is discussed in detail in the companion paper at this conference titled: "Industrial Experience on Process Transition Monitoring for Continuous Steel Casting Operation".

3.4. Integrated Monitoring System Architecture

The schematic of the fully integrated on-line monitoring system for start-cast, process transitions and the continuous, run-time cast operations is depicted in Figure 5. Real-time measurements of various sensors are collected on-line by a data acquisition module, and then sent to an on-line process monitoring module as well as a historical database for data archiving purposes. Once the monitoring module receives the real-time process measurements, a series of computations are performed based on a selected multivariable statistical model from the model set.

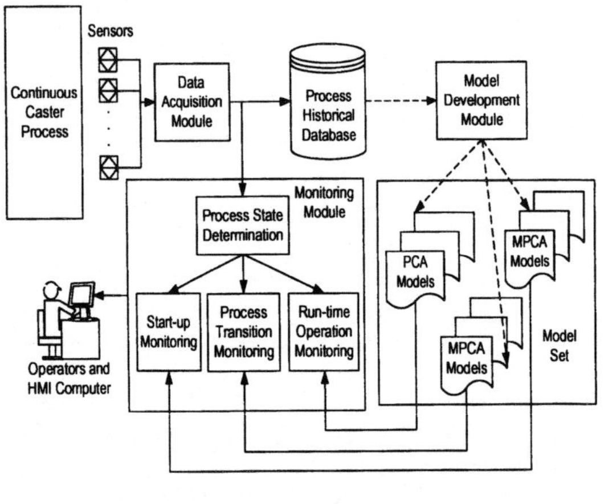

Fig. 5. Schematic of the integrated on-line caster monitoring system

The monitoring module consists of 4 computation functions, i.e., start-up monitoring, process transition monitoring, run-time operation monitoring and process state determination. The latter is able to determine the process state based on the current process conditions (described by casting speed, strand length, etc...), and automatically selects the appropriate model and monitoring function. Five general process states are defined: shut-down, start-up, specific process transition, run-time and idle (which is designed to handle some special operating conditions). Based on current process conditions, a state is determined and the corresponding model results are presented to the operator. For presentation simplicity and ease-of-use for the operators, the HMI screens are identical for all of the operating states and transition seamlessly between them.

3.5. Operational Results and Benefits

Since the deployment of the caster monitoring system at #1 Continuous Caster in April 1997, Dofasco has enjoyed its highest productivity levels with an approximate 50% reduction in breakouts over the last 7 years from its previous best year. The system has also aided in early detection of other critical operational situations such as mould level sensor problems. In general, this system has helped operators have greater confidence in running the caster at higher production rates.

The run-time monitoring application at #2 Continuous Caster was implemented on-line in February 2002. That year, there were 4 high probability cases of breakouts that were prevented through the aid of this new monitoring system. A significant reduction in breakouts was also observed in 2003, as compared to previous years. A portion of the start-cast monitoring feature was implemented in May 2003 and fully commissioned January 2004. In 2003, there was one high probability start-cast breakout case that was prevented through the aid of this new monitoring feature. The process transition monitoring feature is still under development.

The run-time monitoring application is patented with the start-cast and process transition monitoring features patent pending.

4. ON-LINE PREDICTIVE CONTROL APPLICATION – DESULPHURIZATION REAGENT CONTROL (DRC) SYSTEM

4.1. Torpedo Car Desulphurization Process

Dofasco uses a Torpedo Car Desulphurization facility to remove sulphur from most batches of hot metal in order for the steel to meet the specified quality standard. A schematic of the facility is illustrated in Figure 6. Desulphurization refers to the process of injecting the hot metal with chemical agents in order to bind, and thereby remove, a portion of the sulphur in the metal. At this facility, 2 different reagents are utilized.

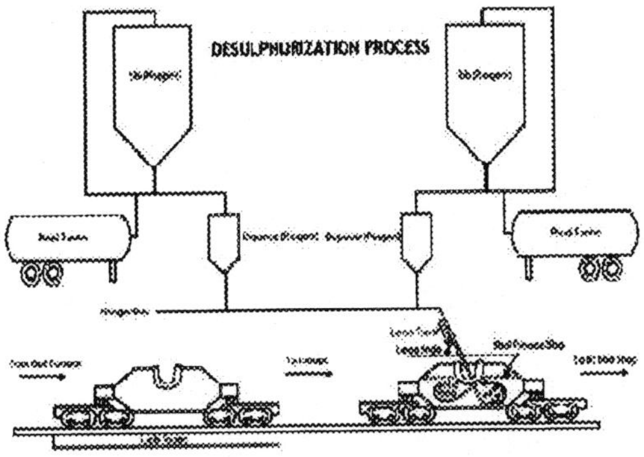

Fig. 6. Torpedo car desulphurization process

4.2. Motivation for Controller Change

The harsh environment at this facility makes it challenging to obtain reliable, high-quality process data. Because of this, only very simple regression models, utilizing a fraction of the available data, were robust enough for the original on-line model-based control system. An additional concern was that the maintenance of these models was a time-consuming, cumbersome task. The operations staff felt there was room to improve the performance of this control system in order to better control the reagent usage and accurately meet downstream sulphur targets.

4.3. Adaptive PLS Control Strategy

An emprirical modeling methodology was required in this application because the mechanisms of the chemical and mechanical effects are not fully understood. As an approach, PLS was favoured because it enabled us to most effectively use the multitude of highly correlated data available for this operation. Also, PLS is able to robustly handle situations where real-time data are missing, a reality in this application. In the desulphurization application, it is known that the process does shift and drift due to changes in the characteristics of the purchased chemical reactants. Therefore, it is important that the modeling technology have an adaptation feature.

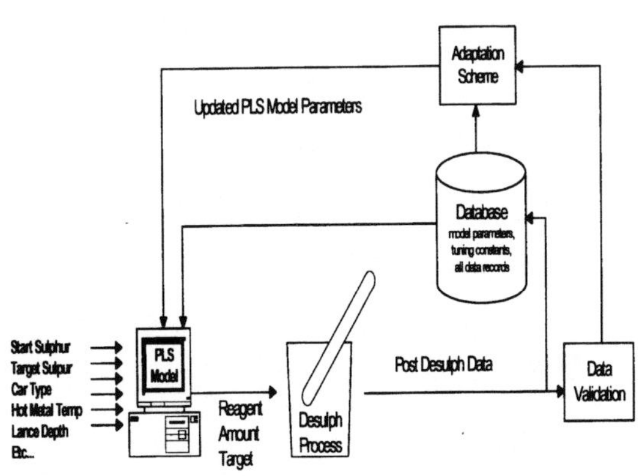

Fig. 7. Adaptive PLS desuphurization reagent control logic

The control structure developed is illustrated in Figure 7. The algorithm used is a modified version of the Adaptive Modified Kernel Algorithm described in Dayal and MacGregor (1997a,b). The PLS models at the Desulph Station are defined in terms of:

1. 28 β values (from Eq. 2) – 14 β values for each of the two reagents, where each value of β is the coefficient associated with one of the 14 input variables (which include initial and target sulphur values, mass of hot metal, all available chemistry data, and process variables such as the type of torpedo car in use).
2. 14 xm values – these are the weighted means for each of the transformed input variables (x).
3. 2 ym values – these are analogous to the xm values but represent the weighted means of the transformed output variables (y)
4. 14 xw values – these are the weighted standard deviations for each of the transformed x variables
5. 2 yw values – these are analogous to the xw values but represent the weighted standard deviations of the transformed y variables

Together, these values fully specify the model and allow the calculation of the required amounts of the two reagents. The adaptation algorithm updates the values of all of these parameters when an appropriate new set of data is available. The model updating is done as follows.

The xm and ym values are updated to reflect the new data according to:

$$\begin{aligned} xm_{j,i} &= (1-\alpha_1)xm_{j,i-1} + \alpha_1 \bar{x}_{j,i} \\ ym_{j,i} &= (1-\alpha_1)ym_{j,i-1} + \alpha_1 \bar{y}_{j,i} \end{aligned} \quad (3)$$

where $xm_{j,i}$ is the new weighted average value for the jth x variable, $xm_{j,i-1}$ is the weighted average from the previous iteration of the adaptation scheme, $\bar{x}_{j,i}$ is the mean of the transformed values for the jth x-variable in the new dataset, and α_1 is a tuning parameter that can take on values between zero and one and determines how quickly old data are discounted in the adaptation routine. The y variables are defined in a similar way.

The xw and yw values are updated to reflect changes in the standard deviations over time according to:

$$xw_{j,i} = \sqrt{(1-\alpha_2)\sigma^2_{j,i-1} + \alpha_2 \breve{x}^2_j} \quad (4)$$

where $\sigma^2_{j,i-1}$ is the weighted variance for the jth x-type variable from the previous time the adaptation algorithm was executed, $\breve{x}_{j,i}$ is variance for the transformed j[th] x-variables in the new dataset, $xw_{j,i}$ is the new value of the weighting factor for the jth x-variable, and α_2 is the discounting factor that determines how much weight is given to previous values of the variances of the variables. The yw values are updated similarly.

The values of the β's are determined based on the kernel PLS algorithm using $\mathbf{X^TX}$ and $\mathbf{X^TY}$. It is the $\mathbf{X^TX}$ and $\mathbf{X^TY}$ matrices that are updated using the new data. These matrices are used to update the "old" covariance structures. This updating is done using a standard moving average scheme as follows.

$$\left(\mathbf{X^TX}\right)_{updated}=\alpha\left(\mathbf{X^TX}\right)_{current}+(1-\alpha)\left(\mathbf{X^TX}\right)_{new} \quad (6)$$

$$\left(\mathbf{X^TY}\right)_{updated}=\alpha\left(\mathbf{X^TY}\right)_{current}+(1-\alpha)\left(\mathbf{X^TY}\right)_{new} \quad (7)$$

The updated correlation matrices are then used to fit a new PLS model. Finally, the kernel PLS algorithm is updated in the following manner:

$$\beta_{updated} = f\left(\left(\mathbf{X^TX}\right)_{updated}, \left(\mathbf{X^TY}\right)_{updated}\right) \quad (8)$$

where $\beta_{updated}$ represents the new set of all β values. Note that together with the parameters of the model, the most recent values for $\mathbf{X^TX}$ and $\mathbf{X^TY}$ must be stored for use in the next call to the adaptation algorithm. One advantage of the kernel algorithm is that only these matrices need to be stored and not the original data matrices $\mathbf{X}$ and $\mathbf{Y}$.

4.4. On-line Implementation

The initial model was fitted using approximately three months worth of historical data, which amounts to a few thousand observations. The initial version of the on-line PLS application was implemented in July 1995. It was integrated into the operation as a PC based control system with an easy-to-use Visual Basic HMI. In 2000, changes to operating practices required a new model algorithm to handle multiple (co-injected) reagents. The updated PLS control strategy, which included the adaptation feature, was implemented in September 2000.

In demonstrating the new control system performance, Figure 8 shows the "before" and "after" histograms for the distribution of the final sulphur values for one aim. Approximately three months worth of data for both periods of operation were used to generate the frequency values for each distribution. The figure shows that the new model provided more precise estimates of the amounts of reagent required (narrowed the distribution) and this allowed Dofasco to move its target sulphur values closer to the constraint.

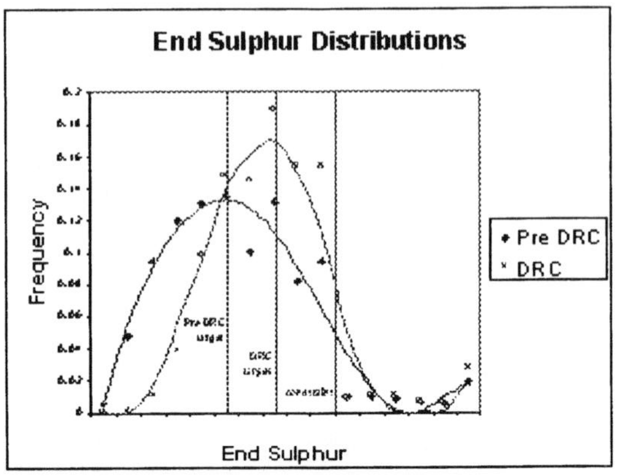

Fig. 8. Distributions of End Sulphur values before and after the implementation of the adaptive PLS model

In demonstrating the adaptation capability, Figure 9 shows a plot of the predicted quantities of reagent required to achieve a desired desulphurization (based on the PLS model) versus the amounts of reagent actually used to achieve the result. If the models were perfect and there was no noise in the data, this plot would be a straight line with a slope of one. There are two groups of data shown on this plot. The red diamonds represent data based on a PLS model with no adaptation. The black stars are points that were generated using an adaptive PLS model. Both groups of data show remarkably good clustering close to the diagonal line indicating that reliable predictions are possible despite noise in the raw data. However, the static results show a slight upwards offset from the adaptive results. This represents a small bias in the static model such that it is consistently predicting slightly too much reagent required. Even a small bias like this can result in substantial over-consumption of reagent over the long term. Model adaptation eliminates the offset and provides assurance that the model is reliable and up-to-date.

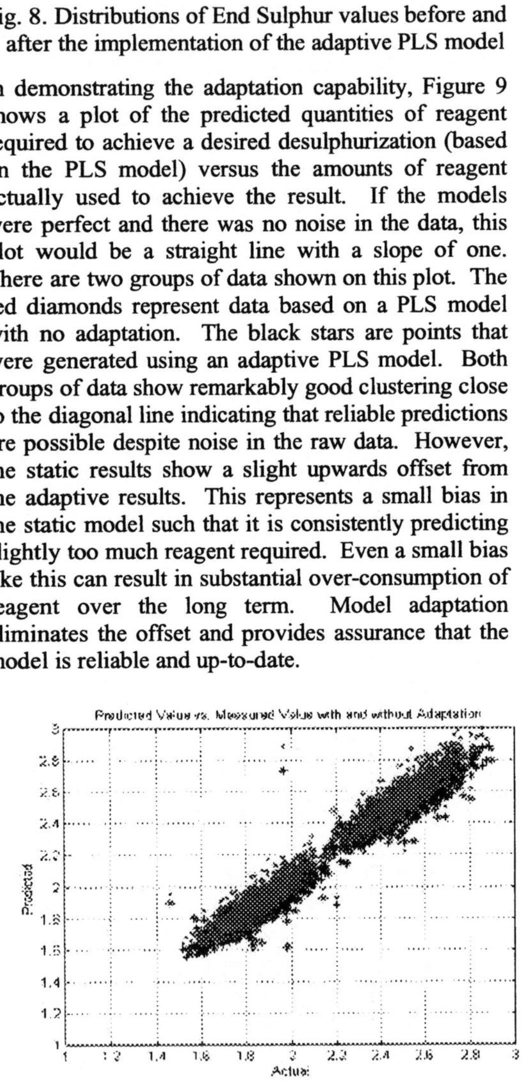

Fig. 9. Predicted versus actual reagent quantities. (KEY: red diamonds - without adaptation; black stars - with adaptation)

4.5. Operational Results

There was immediate positive feedback from operations on the performance of the new PLS reagent control system as compared to the previous model-based control system. At the end of 1995, a formal audit of the new control system was established. The results were: (1) the root mean square (RMS) error between the aim and final sulphur was reduced by 50%, (2) the re-desulphurization rate (i.e. the times where a second reagent addition is necessary to meet target) was reduced by 70%, (3) the amount of reagent purchased was reduced by 8.5%, and (4) the process iron yield was improved by 0.25% due to less iron tied up in slag. The performance of the enhanced adaptive PLS algorithm was also excellent, with minimal model maintenance effort being required. Since the introduction of the co-injection operating strategy, improvement initiatives (i.e. the adaptive PLS model, automation upgrades and new operating practices) have achieved another 17% reduction in reagent consumption (Quinn et al., 2002), while accurately meeting sulphur targets. Improved process understanding and process control have helped in unforeseen ways. For example, our relationships with our supplies have improved because our combined expertise has been maximized - Dofasco is the process expert, while the suppliers are the product experts.

The adaptive PLS desulphurization reagent control system is patented.

5. CONSIDERATIONS FOR FURTHER MVS BASED INVESTIGATIONS

Through the experiences of implementing on-line MVS applications at Dofasco, much has been learned about what it takes to successfully commission such applications. In our opinion, there are several challenges that industrial practitioners will have to overcome to be able to effectively utilize the full potential of this technology in on-line applications. In general, what is needed is a flexible, industrially-focussed, methodology for applying MVS technology. The success of this technology in industry will be directly related to the availability of MVS systems that are useable and understandable by people of varying technical backgrounds. In our experience, MVS solutions in industry need to pay special attention to issues that can be categorized under the following headings: (1) identifying and defining appropriate application scope, (2) data gathering / preprocessing, (3) off-line modeling to on-line solution, and (4) long-term maintenance strategies. These are discussed further below. In identifying these issues, we have also tried to provide suggestions on how to address them; however, we believe that many of these issues remain open areas for research and we hope this paper may open new academic and industrial work in these areas.

5.1. Identifying and Defining an Appropriate Scope

In an industrial setting, it is necessary to be able state the economic justification. In this regard, the practitioner must be able to determine if the capability and applicability of the MVS technology makes sense over alternate technologies. Similarly, one must define the appropriate modeling strategy necessary to handle many complex operating situations such as: many different products produced by a process, large feed rate changes, recycle processes, and more complicated batch processing. We believe that academic studies on applying MVS in these cases would be beneficial, and that results could be compelling in defining project scope for future industrial MVS work. We stress this because we have found that appropriately defining scope for the application greatly improves its chances of long-term success. For example, it is important to determine how small or how big a modeling application should be, and whether it is better to have many small applications versus one large application. Defining these matters at the outset of a project will impact many issues from long-term maintenance to data acquisition and processing.

5.2. Data Gathering and Data Preprocessing

We have found that a key to success in any multivariate statistical application is the early involvement of process experts. Although multivariate statistical models are empirical models, having access to expert advice on which variables are "known" to be important and what transformations or combinations of variables may be important, can affect not only the speed of development of a system but also its ultimate success. Furthermore, process experts are usually helpful in constructing appropriate training and validation data sets if historical data is to be used. Should there prove to be a need for designed experimentation, process experts can help in that too. Ultimately, process experts are required in determining the depth and breadth of data to cover the operating window of the process. In most cases we have encountered, it was important that data pre-processing take into account operation-specific needs which may include: non steady-state process dynamics, non-stationary process conditions, non-linear process conditions, and large process dead-times. We suggest more in-depth collaboration between industry and academia in developing an understanding of industrial data sets and, data presentation and visualization needs.

5.3. Application of Models On-line

Typically, the first step in putting an MVS model on-line is justifying the proposal to do so. This involves quantifying the benefits in terms of cost, revenue, safety and/or quality. Therefore, meaningful measures of the applicable benefits should be developed and computed. What these criteria are will depend on the application, but typically they are related to the performance of the process or expected changes in the performance as opposed to being related to model performance such as the percentage of variability explained by the model. For example, in monitoring and fault detection applications, tuning of the alarm levels is very important. Although some academics advocate setting limits based on a choice of a probability (tail) value, tuning limits based on achieving a desired average run length (ARL) may be more useful in some instances, especially relative to economic costs incurred while using control charts where process adjustments are made based on both T^2 and Q^2. We suggest more academic work be preformed on statistical topics such as this, perhaps following Box and Jenkins (1963), Box and Kramer (1992), and Lorenzen and Vance (1986).

5.4. Long-term Maintenance Strategies

Having long-term maintenance strategies in place is important in ensuring success. Typically, the sophistication of these strategies is determined by the rate at which the process is expected to change (i.e. how often it is expected that the models or alarm limits will need tuning). In cases where the process is known to shift or drift over relatively short time periods, it may be necessary to consider on-line adaptation, among other things. Especially in the area of maintenance strategies, we believe that there is still research to be done. The following are areas we feel require further work: on-line adaptation of alarm limits in monitoring applications, monitoring the performance of MVS systems to determine when tuning / remodeling is required, methods for expediting the remodeling process when a major change is made to the process, and simple and efficient ways to minimize the number of models and limits that need to be maintained. Dofasco has developed internal methodologies to address some of these needs, however, there is a significant opportunity to further address these for the benefit of all process industries.

6. REFLECTIONS, SUMMARY, CONCLUSIONS AND FUTURE DIRECTIONS

This paper has described two important on-line MVS-based applications at Dofasco that continue to provide significant value to the company. Since our first off-line analysis in 1993, we have utilized MVS-based technologies on over 70 applications. These areas can be categorized in the following groups: (1) off-line process analysis / troubleshooting, (2) on-line process monitoring, and (3) on-line model-based control. To date, Dofasco has implemented 9 on-line MVS-based applications, 6 in the area of process monitoring and 3 in model-based control strategies. The remainder involve off-line process analysis / troubleshooting. Off-line process analysis is a prerequisite proof-of-concept activity that provides justification for moving forward with on-line applications. Off-line troubleshooting applications typically involve discovering specific insight hidden within a large dataset to support a team effort to fix a problem. The applicability of this technology over a wide range of problems / opportunities that we deal with make it a very valuable tool for the company.

Multivariate Statistical technologies have been successful at Dofasco because of the support provided at all levels of management. The MVS applications developed at Dofasco stem from a structured approach in developing applications, and have delivered valuable business results that can be related to company profits. The MVS applications have proven to be robust to many practical matters such as missing data, and have proven to be amenable to various maintenance and updating strategies. MVS provides models and control chart schemes that are readily understood by production personnel once training is provided. Although MVS methods use many process variables simultaneously for modeling and fault detection, the methods provide a built-in means of relating SPC chart, and model calculations to the individual variables used for modeling. This has facilitated troubleshooting and process diagnosis on the part of shop floor personnel, and has helped in building acceptance of the MVS methods.

Perhaps more than anything else, operator (final user) acceptance is critical to the long-term success of an on-line MVS system. Early involvement of the operators in the design of the user interfaces has proven beneficial in our experience. Also, seeking advice from the operators early, and throughout the process has helped in acceptance of the system, and has helped alleviate feelings of being intimidated by the technology. Operators often have very good suggestions on how to integrate the new technology into existing control room infrastructure and / or how to incorporate signals and information from other systems into the new MVS system.

This paper has addressed many technical and operational issues in the on-line application of MVS technology. We note here that there are also other corporate issues that need to be considered. The most critical and perhaps the most difficult to quantify or influence is that of having a corporate culture to support this type of development. For Dofasco, this has meant that a willingness to perform MVS work and to manage it in the form of an engineering project was necessary. When coupled with project management methods, periodic project evaluation, and the on-going development of computer infrastructure, the framework we presented has helped to create success and has led to changes in the way analysis work is done in general. The success of the MVS, and other, often much larger, data-driven projects has led to more use of data by shop floor personnel during daily activities like production meetings, more discussion on how to improve quality, and most importantly, more ideas on how to use available data to improve production methods and products.

Dofasco's vision to the future for MVS technologies is to "institutionalize" it throughout the corporation by: (1) maximizing the understanding and benefits of this technology through technical awareness and training, (2) integrating, where appropriate, the technology into our corporate quality initiatives, and (3) continuing to build technical breadth and depth while leveraging leading-edge aspects of the technology with the academic experts (especially with linkages at McMaster University). To facilitate the propagation of this technology throughout the corporation, Dofasco has developed an in-house generic software platform for on-line MVS-based solutions. This platform is architected with all the latest open system features for application on various computing platforms that are resident in the company. This also facilitates and simplifies the long-term maintenance of a growing application base.

In general, Dofasco will continue to leverage multivariate statistical methods, with other advanced automation technologies, to develop new, high-valued on-line control and monitoring applications.

7. REFERENCES

Box, G. E. P., G. M. Jenkins (1963), Further Contributions to Adaptive Quality Control: Simultaneous Estimation of Dynamics: Non-Zero Costs, Bulletin of the International Statistical Institute, Proceedings of the 34-th Session, Ottawa, Canada.

Box, G. E. P., T. Kramer (1992), Statistical Process Monitoring and Feedback Adjustment. A Discussion, Technometrics, **34** (3), pp.251-285.

Burnham, A., J.F. MacGregor, R. Viveros (1999), Latent Variable Multivariate Regression Modeling, Chemometircs and Intelligent Laboratory Systems, **48**, pp.167-180.

Dayal, B., J. F. MacGregor (1997a), Improved PLS Algorithms, Journal of Chemometrics, **11**, pp.73-85.

Dayal, B., J. F. MacGregor (1997b), Recursive Exponentially Weighted PLS and its Application to Adaptive control and Prediction, Journal of Process Control, **7** (3), pp.169-179.

Dudzic M., V. Vaculik and I. Miletic (2000), "On-line applications of multivariate statistics at Dofasco", in *Proc.* IFAC Workshop on Future Trends in Automation in Mineral and Metal Processing, Heisinki, Finland, pp.438-443.

Jackson, J. E. (1991), A User's Guide to Principal Components. John Wiley and Sons, New York.

Kourti, T., J.F. MacGregor (1996), Recent Developments in Multivariate SPC Methods for Monitoring and Diagnosing Process and Product Performance, Journal of Quality Technology, **28** (4), pp.409-428.

Kourti, T., J. Lee, J.F. MacGregor (1996), Experiences with Industrial Applications of Projection Methods for Multivariate Statistical Process Control, Computers in Chemical Engineering, **20**, S745-S750.

Kourti, T. (2002), Process Analysis and Abnormal Situation Detection: From Theory to Practice, IEEE Control Systems, **22** (5), pp.10-25.

Kourti T. (2003), Multivariate dynamic data modeling for analysis and statistical process control of batch processes, start-ups and grade transitions, *J. Chemometrics*, **17**, pp.93-109

Lorenzen, T. J., L. C., Vance (1986), The Economic Design of Control Charts: A Unified Approach, Technometrics, **28** (1), pp.3-10.

Morrison, D. F. (1990), "Multivariate Statistical Methods, Third Edition", McGraw-Hill, New York.

Nelson P.R.C., P.A. Taylor and J.F. MacGregor (1996), Missing data methods in PCA and PLS: score calculations with incomplete observations, *Chemometric Intell. Lab. Syst.*, **35**, pp.45-65

Nomikos P and J.F. MacGregor (1994), Monitoring of batch processes using multi-way principal component analysis, *AICHE J.*, **40**, pp.1361-1375

Nomikos P and J.F. MacGregor (1995), Multivariate SPC charts for monitoring batch processes, *Technometrics*, **37**, pp.41-59

Quinn, S.L., V. Vaculik (2002), Improving the Desulfurization Process Using Adaptive Multivariate Statistical Modeling, AISE Steel Technology, **79** (10) pp.37-41.

Vaculik, V., B. MacCuish, M. Dudzic, I. Miletic, M. Hough, K. Smith (2001), "Innovative monitoring technology improves Caster operation at Dofasco", Iron and Steelmaker, **28**, No. 9, pp. 23-28.

Zhang Y. M. Dudzic and V. Vaculik (2003), Integrated monitoring solution to start-up and run-time operations for continuous casting, *Annual Reviews in Control,* **27**, pp.141-149

ELSEVIER
IFAC
PUBLICATIONS
www.elsevier.com/locate/ifac

CLOSED-LOOP SUBSPACE IDENTIFICATION: AN ORTHOGONAL PROJECTION APPROACH

B. Huang [*,1] **S.X. Ding** [**] **S.J. Qin** [***]

[*] *Dept. of Chemical and Materials Engg., Uinv. of Alberta, Edmonton, AB, Canada, T6G 2G6*
[**] *Univ. of Duisburg-Essen, Inst. Auto. Cont. and Comp. Sys./Faculty 5, Bismarckstrasse 81, BB511, 47048 Duisburg, Germany*
[***] *Dept. of Chemical Engineering, Univ. of Texas at Austin, 1 University Station, C0400 Austin, TX 78712, USA*

Abstract: In this paper, a closed-loop subspace identification approach through an orthogonal projection and subsequent singular value decomposition is proposed. As a by-product of this development, it explains why some existing subspace methods may deliver a bias in the presence of the feedback control and suggests a remedy to eliminate the bias. Furthermore, as the proposed method is a projection based method, it can simultaneously provide extended observability matrix, lower triangular block-Toeplitz matrix, and Kalman filtered state sequences. Therefore, using this method, the system state space matrices can be recovered either from the extended observability matrix/the block-Toeplitz matrix or from the Kalman filter state sequences.

Keywords: Subspace identification, closed-loop identification, projection, instrument variable method, PCA, subspace PCA, singular value decomposition.

1. INTRODUCTION

Identification of the subspace matrices from closed loop data has received an increasing attention by a number of researchers. It is found that the regular open-loop subspace identification algorithm yields a biased estimate when applied to closed-loop data(Ljung and McKelvey, 1996). Several modified algorithms have been proposed (Overschee and Moor, 1996; Ljung and McKelvey, 1996; Ljung and McKelvey, 1996). Another class of subspace system identification is called the instrument variable methods (Chou and Verhaegen, 1997; Wang and Qin, 2002). In the class of the instrument variable methods, the effect of disturbances is eliminated by appropriate selection of the instrument variables that are independent of disturbances. Under the framework of MOESP, (Chou and Verhaegen, 1997) developed an instrument variable subspace identification algorithm. They claimed that the algorithm worked for closed-loop systems provided there is at least one sample time delay in the *controller*, which may be restrictive in practice. Aiming at solving the open-loop error in variable (EIV) identification problem, (Wang and Qin, 2002) developed an instrument variable subspace identification method via principal components analysis, which may be applicable to closed-loop identification but only under certain conditions.

[1] Corresponding author.

In this paper we develop a closed-loop subspace identification algorithm through an orthogonal subspace projection, and then use either the extended observability matrix/lower block-Toeplitz matrix or the Kalman filter states resulting from the projection to extract system models (Due to the space limit, we will not discuss the procedure to recover the model from the Kalman filter state sequence). Since the model is obtained using projections of subspace, as is the case in most other subspace identification algorithms, it has certain additional properties compared to the instrument methods. It is further shown that the existing instrument subspace identification via PCA, although works well for open-loop EIV systems, yields a biased solution in closed-loop under certain conditions. A remedy to eliminate this bias is also proposed in this paper.

The remainder of this paper is organized as follows. Several subspace notations adopted throughout this paper are revisited in Section 2. Our main results, the proposed subspace closed-loop identification approach and a solution to an existing instrument subspace identification method for closed-loop identification, are discussed in Section 3. Simulation results are presented in Section 4, followed by concluding remarks in Section 5.

2. SUBSPACE IDENTIFICATION

Consider a state space model in the innovation form

$$x_{t+1} = Ax_t + Bu_t + Ke_t \tag{1}$$
$$y_t = Cx_t + Du_t + e_t \tag{2}$$

where $x_t \in \mathbb{R}^n$, $u_t \in \mathbb{R}^p$, $y_t \in \mathbb{R}^q$, and $e_t \in \mathbb{R}^q$ is white noise innovation sequence with covariance Σ_e.

Following the standard subspace notation, one can derive, through the iterative substitution of Eqs(1) and (2), the subspace matrix equations as

$$Y_f = \Gamma_i X_f + H_i^d U_f + H_i^s E_f \tag{3}$$
$$Y_p = \Gamma_i X_p + H_i^d U_p + H_i^s E_p \tag{4}$$
$$X_f = A^i X_p + \Delta_i^d U_p + \Delta_i^s E_p \tag{5}$$

where subscript p stands for the "past" and f for the "future". All notations used in Eqs(3)-(5) are standard and can be found in most of subspace identification literatures. Γ_i is extended observability matrix, Δ_i^d and Δ_i^s are reversed extended controllability matrices for process input and disturbance input respectively, H_i^d and H_i^s are the lower triangular block-Toeplitz matrices for process input and disturbance input respectively. The past and future input block-Hankel matrices are defined as

$$U_p \triangleq U_{0|i-1} = \begin{pmatrix} u_0 & u_1 & \cdots & u_{j-1} \\ u_1 & u_2 & \cdots & u_j \\ \cdots & \cdots & \cdots & \cdots \\ u_{i-1} & u_i & \cdots & u_{i+j-2} \end{pmatrix} \tag{6}$$

$$U_f \triangleq U_{i|2i-1} = \begin{pmatrix} u_i & u_{i+1} & \cdots & u_{i+j-1} \\ u_{i+1} & u_{i+2} & \cdots & u_{i+j} \\ \cdots & \cdots & \cdots & \cdots \\ u_{2i-1} & u_{2i} & \cdots & u_{2i+j-2} \end{pmatrix} \tag{7}$$

where $U_p, U_f \in \mathbb{R}^{pi\times j}$. Note that, in Eqs(6) and (7), the row dimension of U_f is allowed to be different from that of U_p, which would provide an extra freedom to tune the identification algorithm.

The output and innovation block-Hankel matrices $Y_p, Y_f \in \mathbb{R}^{qi\times j}$, $E_p, E_f \in \mathbb{R}^{qi\times j}$, respectively, are defined conformably with U_p, U_f.

The states are defined as

$$X_p \triangleq X_0 = \begin{pmatrix} x_0 & x_1 & \cdots & x_{j-1} \end{pmatrix} \tag{8}$$
$$X_f \triangleq X_i = \begin{pmatrix} x_i & x_{i+1} & \cdots & x_{i+j-1} \end{pmatrix} \tag{9}$$

where $X_p, X_f \in \mathbb{R}^{n\times j}$.

In subspace identification literature, the following short-hand notation is often used:

$$W_p = \begin{pmatrix} Y_p \\ U_p \end{pmatrix}$$

The following two projections are frequently used throughout this paper.

Orthogonal projection: The orthogonal projection of the row space of A onto the row space of B is denoted by A/B and can be calculated through

$$A/B = AB^+B$$

where B^+ is the pseudo inverse of B.

Oblique projection: The oblique projection of the row space of $A \in \mathbb{R}^{p\times j}$ along the row space of $B \in \mathbb{R}^{q\times j}$ on the row space of $C \in \mathbb{R}^{r\times j}$ are defined as $A/_B C$ and can be calculated via (using MATLAB matrix index notation)

$$A/_B C = A \begin{pmatrix} C \\ B \end{pmatrix}^+ (:, 1:r)C$$

Two important properties of the oblique projection are often used and they are

$$A/_A C = 0 \tag{10}$$
$$A/_B A = A \tag{11}$$

These two properties are straightforward results by the definition of the oblique projection.

3. CLOSED-LOOP IDENTIFICATION

3.1 A preliminary solution

Let's now revisit Eq(3)

$$Y_f = \Gamma_i X_f + H_i^d U_f + H_i^s E_f \tag{12}$$

The essential system information is contained in the extended observability matrix Γ_i or in the state X_f. That is, the first term on the right hand side of Eq(12) deserves our main attention. To calculate, for example, Γ_i from Eq(12), one has to get rid of the terms containing U_f and E_f. If E_f is independent of past input U_p, past output Y_p (or equivalently their combination W_p), and future input U_f, then one can easily achieve the above objective by performing an oblique projection of Eq(12) along the row space U_f onto the row space of W_p, i.e.

$$Y_f/_{U_f}W_p = \Gamma_i X_f/_{U_f}W_p + H_i^d U_f/_{U_f}W_p + H_i^s E_f/_{U_f}W_p \tag{13}$$

It is easy to see that the last two terms of Eq(13) are zero, $U_f/_{U_f}W_p = 0$ by the property of the oblique projection, Eq(10); $E_f/_{U_f}W_p = 0$ and this is based on the assumption that future disturbance is independent of past input/output and future input. This assumption holds under the open-loop condition. Thus Eq(13) can be simplified to

$$Y_f/_{U_f}W_p = \Gamma_i X_f/_{U_f}W_p \tag{14}$$

This result indicates that the column space of Γ_i is the same as the column space of $Y_f/_{U_f}W_p$, which can be calculated by the SVD decomposition of $Y_f/_{U_f}W_p$. Similarly, the row space of $Y_f/_{U_f}W_p$ is the same as the row space of $X_f/_{U_f}W_p$, which is Kalman filter state solution with $X_p/_{U_f}W_p$ as its initial condition (Favoreel, 1999). Therefore, the Kalman state sequence can also be calculated from the SVD decomposition of $Y_f/_{U_f}W_p$. Subsequently, the system state space matrices can be recovered either from the extended observability matrix or from the Kalman state sequence. This is the solution to open-loop subspace system identification.

The situation becomes more complex for closed-loop identification where the future disturbance E_f is no longer independent of the future input U_f due to the feedback. The implication of this dependency is that the oblique projection of E_f along U_f onto W_p is no longer zero although the orthogonal projection of E_f onto W_p is zero.

To solve this problem, by adopting the EIV structure, we move the term related to U_f into the left hand side of Eq(12) as it would be a troublesome term if left in the right hand side of the equation. This yields a new equation with both input and output variables in the same side of the equation.

$$\begin{bmatrix} I & -H_i^d \end{bmatrix} \begin{pmatrix} Y_f \\ U_f \end{pmatrix} = \Gamma_i X_f + H_i^s E_f \tag{15}$$

Using the short-hand notation

$$W_f = \begin{pmatrix} Y_f \\ U_f \end{pmatrix}$$

Eq(15) can be simplified to

$$\begin{bmatrix} I & -H_i^d \end{bmatrix} W_f = \Gamma_i X_f + H_i^s E_f \tag{16}$$

Performing an *orthogonal projection* of Eq(16) onto the row space of W_p yields

$$\begin{bmatrix} I & -H_i^d \end{bmatrix} W_f/W_p = \Gamma_i X_f/W_p + H_i^s E_f/W_p \tag{17}$$

The last term of Eq(17) is an orthogonal projection of the future disturbance (white noise) onto the row space of past input and output matrix W_p, which is zero. Therefore, Eq(17) can be simplified to

$$\begin{bmatrix} I & -H_i^d \end{bmatrix} W_f/W_p = \Gamma_i X_f/W_p = \Gamma_i \hat{X}_f \tag{18}$$

Remark 1. Eq(18) is a natural result through the projection as is often done in subspace system identification literature. The orthogonal projection of Eq(16) onto the row space of W_p results in Eq(18), which includes a multiplication term between the extended observability matrix Γ_i and non-steady state Kalman state $\hat{X}_f$. On the other hand, (Wang and Qin, 2002) used an instrument variable method to arrive an equation as

$$\begin{bmatrix} I & -H_i^d \end{bmatrix} W_f W_p^T = \Gamma_i X_f W_p^T \tag{19}$$

where the instrument variable is the past input and output W_p. This equation is derived by multiplying Eq(16) by W_p^T and noticing the independency between W_p and E_f.

The projection method may also be mathematically regarded as an instrument variable method where the instrument is $W_p^T(W_pW_p^T)^{-1}W_p$. However, by using the projection, it results in a multiplication term between the extended observability matrix and the Kalman filtered state, which provides some additional feature and performance. We shall call the proposed projection method as the subspace orthogonal projection identification method, abbreviated SOPIM, while the subspace identification method via PCA of (Wang and Qin, 2002) is abbreviated as SIMPCA.

Now multiplying both sides of Eq(18) by the orthogonal column space of Γ_i, denoted by $\Gamma_i^{\perp}$, yields

$$(\Gamma_i^{\perp})^T \begin{bmatrix} I & -H_i^d \end{bmatrix} W_f/W_p = 0 \tag{20}$$

Denoting $Z = W_f/W_p$, the problem is transferred to finding the orthogonal column space of Z, which should equal to the column space of $\left((\Gamma_i^{\perp})^T \begin{bmatrix} I & -H_i^d \end{bmatrix} \right)^T$.

Perform SVD decomposition of Z as

$$Z = \begin{pmatrix} U_1 & U_2 \end{pmatrix} \begin{pmatrix} \Sigma_1 & \\ & 0 \end{pmatrix} \begin{pmatrix} V_1^T \\ V_2^T \end{pmatrix} \tag{21}$$

where, in practice, Z is not singular and one has to determine its rank by checking its singular values. In theory, its rank should be $pi + n$(Wang and Qin, 2002) assuming that the external excitation is persistent excitation. The rank determination is equivalent to the determination of system orders.

With Eq(21), one can easily find the orthogonal column space of Z, which is U_2. Therefore

$$\left((\Gamma_i^{\perp})^T \begin{bmatrix} I & -H_i^d \end{bmatrix} \right)^T = U_2 M \tag{22}$$

where M is any constant matrix. Partition

$$U_2 M = \begin{pmatrix} P_1 \\ P_2 \end{pmatrix}$$

Then Eq(22) can be written as

$$\begin{pmatrix} (\Gamma_i^{\perp}) \\ -(H_i^d)^T \Gamma_i^{\perp} \end{pmatrix} = \begin{pmatrix} P_1 \\ P_2 \end{pmatrix} \tag{23}$$

Therefore,

$$\Gamma_i = P_1^{\perp} \tag{24}$$

$$-(P_1)^T H_i^d = P_2^T \tag{25}$$

The remaining problem is to solve for Γ_i and H_i^d, and then to extract the system matrices A, B, C, D from Γ_i and H_i^d. Many methods are available for this purpose, for example, (Wang and Qin, 2002).

Can SOPIM and SIMPCA work under closed-loop conditions? Up to now, it appears that there should be no question for both to work under closed-loop conditions as the input U_f, the correlation of which with the future disturbance E_f is the key problem in closed-loop identification, is not involved in the projection or in the instrument variable. Both equations (18) and (19) appear to be able to uniquely determine the process model irrespective of open or closed loop. However, simulation results indicate that this is not the case although both work for open-loop systems.

3.2 The problem and the solution

The problem must have resulted from the controller since they both work well under the open-loop condition. Without presenting a full version of analytical derivation due to the space limit, in this section, we will make a heuristic analysis of the problem and then present our solution.

To find the problem, let's consider that the controller is described by the following state space model:

$$x_{t+1}^c = A_c x_t^c + B_c(r_t - y_t) \tag{26}$$

$$u_t = C_c x_t^c + D_c(r_t - y_t) \tag{27}$$

where r is the setpoint excitation. Using subspace notations, we should have the controller expressed as

$$U_f = \Gamma_i^c X_f^c + H_i^c(R_f - Y_f) \tag{28}$$

$$U_p = \Gamma_i^c X_p^c + H_i^c(R_p - Y_p) \tag{29}$$

where R_p and R_f are data Hankel matrices of the setpoint, X_p^c and X_f^c are the state matrices of the controller, Γ_i^c is the extended observability matrix, and H_i^c is the lower triangular block-Toeplitz matrix, of the controller.

Eq(28) can be re-arranged to give

$$\begin{bmatrix} H_i^c & I \end{bmatrix} W_f = \Gamma_i^c X_f^c + H_i^c R_f \tag{30}$$

Projecting Eq(30) to W_p yields

$$\begin{bmatrix} H_i^c & I \end{bmatrix} W_f / W_p = \Gamma_i^c X_f^c / W_p + H_i^c R_f / W_p \tag{31}$$

Comparing Eq(31) with Eq(18), one can find the overlap of two subspaces represented by the two equations, one for process model subspace and the other for controller model subspace, if $H_i^c R_f / W_p \rightarrow 0$ or $R_f W_p^T \rightarrow 0$ (in the case of SIMPCA). This situation can occur, for example, when R_f is white noise excitation.

Our solution to this problem (omitting details due to space limit) is: for closed-loop identification, one should replace W_p by W_{pr} according to Eq(32)

$$W_{pr} \triangleq \begin{pmatrix} R_f \\ W_p \end{pmatrix} \tag{32}$$

This will guarantee $H_i^c R_f / W_p \neq 0$ or $R_f W_p^T \neq 0$. With the modification, all the computation procedures discussed in the last section are valid for closed-loop identification after replacing W_p by W_{pr}, and then both SOPIM and SIMPCA can be truly applied to closed-loop data. We shall call the modified algorithms as closed-loop SOPIM (abbreviated as CSOPIM) and closed-loop SIMPCA (abbreviated as CSIMPCA), respectively.

Remark 2. From the above discussion, one can see that the undesired effect of the feedback control on the identifiability of open-loop instrument and/or projection subspace methods may also be alleviated if the setpoint r_t "stays away" from whiteness, i.e. if they are (strongly) autocorrelated or colored. But this "non-witeness" is ambiguous and there is no *a priori* indication on how R_f/W_p or $R_f W_p^T$ will be different from zero even if r_t is not white. In addition, the effect of whiteness or non-whiteness on the identifiability also depends on the controller in the feedback and the disturbances that are affecting the process. For example, if the output can not follow the setpoint closely due to large disturbances, then W_p can have a little correlation with R_f even if the setpoint is not white, resulting $R_f W_p^T \to 0$ in SIMPCA or $R_f / W_p \to 0$ in SOPIM. Consequently, W_p may not be suitable to be an instrument in this case even though the external excitation is non-white.

4. SIMULATION

In this section, we will use a benchmark problem to evaluate the proposed approach and compare it with other existing subspace identification algorithms. We will apply the following representative subspace algorithms in the literature: closed-loop algorithm by (Overschee and Moor, 1996), closed-loop algorithm by (Verhaegen, 1993), closed-loop algorithm by (Ljung and McKelvey, 1996), and the two classical subspace algorithms, N4SID, MOESP and two versions of CVA, i.e. MATLAB N4SID with CVA weighting and CVA according to (Larimore, 1990). To comply with the standard practice in subspace identification literature (Overschee and Moor, 1996), we will perform Monte-Carlo simulations and the averaged Bode magnitude plot from the Monte-Carlo simulation will be used to represent bias error while the scatter plot of estimated poles will be used to represent the variance error of the estimation.

The system to be considered(Verhaegen, 1993; Overschee and Moor, 1996), expressed in the innovation state space form (Overschee and Moor, 1996), is given by Eqs(1) and (2) with the following numerical values

$$A = \begin{pmatrix} 4.40 & 1 & 0 & 0 & 0 \\ -8.09 & 0 & 1 & 0 & 0 \\ 7.83 & 0 & 0 & 1 & 0 \\ -4.00 & 0 & 0 & 0 & 1 \\ 0.86 & 0 & 0 & 0 & 0 \end{pmatrix}, B = \begin{pmatrix} 0.00098 \\ 0.01299 \\ 0.01859 \\ 0.0033 \\ -0.00002 \end{pmatrix},$$

$$C^T = \begin{pmatrix} 1 \\ 0 \\ 0 \\ 0 \\ 0 \end{pmatrix}, K = \begin{pmatrix} 2.3 \\ -6.64 \\ 7.515 \\ -4.0146 \\ 0.86336 \end{pmatrix}$$

The state space model of the feedback control has the following values:

$$A_c = \begin{pmatrix} 2.65 & -3.11 & 1.75 & -0.39 \\ 1 & 0 & 0 & 0 \\ 0 & 1 & 0 & 0 \\ 0 & 0 & 1 & 0 \end{pmatrix}, B_c = \begin{pmatrix} 1 \\ 0 \\ 0 \\ 0 \end{pmatrix}$$

$$C = \begin{pmatrix} -0.4135 & 0.8629 & -0.7625 & 0.2521 \end{pmatrix}$$

with $D_c = 0.61$. The simulation conditions exactly resemble those used by (Overschee and Moor, 1996): e_t is a Gaussian white noise sequence with variance 1/9; the reference signal r_t is a Gaussian white noise sequence with variance 1, injected after the controller and before the plant. Each simulation run generates 1200 data points. We generate 100 data sets, each time with the same reference input r_t but with a different noise sequence e_t.

(Overschee and Moor, 1996) used this example to compare several closed-loop identification algorithms. The conclusion from their simulations was that algorithm 1 of Van Overschee and De Moor is superior to other algorithms in terms of performance. However, the algorithms proposed in (Overschee and Moor, 1996) required the precise information of at least first i Markove parameters of the controller model, while others did not.

In this simulation, we will reproduce the results of (Overschee and Moor, 1996) and compare them with the most representative algorithm proposed in this paper, CSOPIM. The simulation results are shown in Fig.1. From this figure, we can see that the proposed algorithm CSOPIM has an almost identical performance as that of algorithm Van Overschee and De Moor in both bias and variance aspects. However, the proposed algorithm has the advantage over algorithm Van Overschee and De Moor in the sense that the proposed algorithm does not need any knowledge about the controller model while algorithm Van Overschee and De Moor does. To see the further advantage of the proposed algorithm, we consider EIV case by adding white noises to measurements of both u_t and y_t. We do two Monte-Carlo simulations for the EIV case, one with measurement noise variance 0.2 and the other 0.5. The comparison results are shown in Fig.2. From this figure, one can see that the proposed algorithm performs better than algorithm Van Overschee and De Moor in the presence of measurement noises.

To appreciate closed-loop subspace identification algorithms, we have also applied classical subspace algorithms, N4SID, MOESP, two CVAs (CVA according to (Larimore, 1990) and MATLAB N4SID with CVA weighting) to closed-loop data. Our results indicate that N4SID, MOESP and MATLAB based CVA deliver essentially the same performance and all are biased in the presence of feedback control. The CVA programmed according to (Larimore, 1990) gives some improved performance compared to the MATLAB N4SID with CVA weighting but the bias error remains.

5. CONCLUSION

In this paper, by adopting the EIV model structure, a subspace orthogonal projection identification method (SOPIM) is proposed. It, however, yields a bias for closed-loop identification

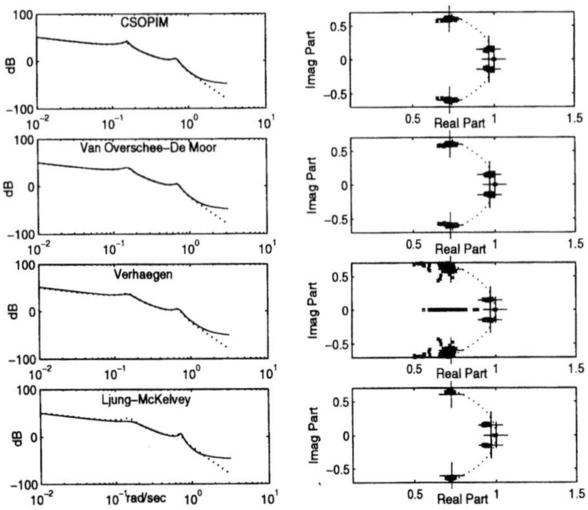

Fig. 1. Closed-loop Monte-Carlo simulations. The left column is Bode magnitude plots; the dotted lines are the true values and the solid lines are the estimated values averaged from 100 runs. The right column is the scatter plots of the eigenvalues of the estimated A matrix.

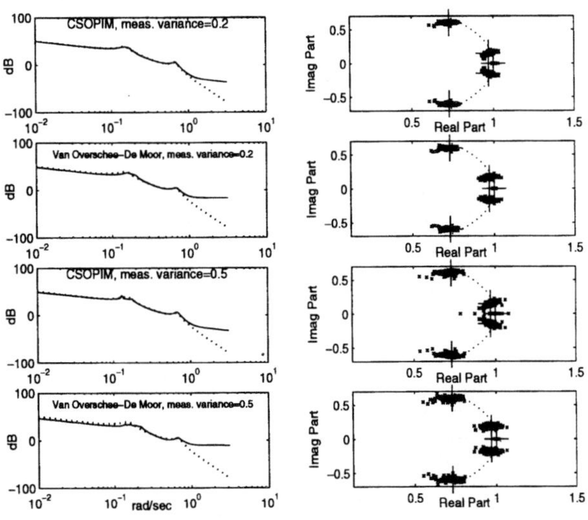

Fig. 2. Closed-loop Monte-Carlo simulations for the EIV system. The left column is Bode magnitude plots; the dotted lines are the true values and the solid lines are the estimated values averaged from 100 runs. The right column is the scatter plots of the eigenvalues of the estimated A matrix.

at least under the condition that the external excitation is white. Through analysis of the bias error of SOPIM under closed-loop conditions, it is discovered that other existing instrument subspace methods in the literature may also suffer from the same bias error and therefore, these methods only yield a partial solution to closed-loop identification. Motivated by this discovery, a remedy is derived to eliminate bias error for SOPIM as well as for one of the existing instrument subspace identification algorithms, SIMPCA, for the sake of closed-loop identification. As a result two new subspace closed-loop identification algorithms CSOPIM and CSIMPCA, named after SOPIM and SIMPCA, respectively, are developed. In addition, the orthogonal projection method proposed in this paper provides both extended observability matrix and Kalman filter state sequence. Therefore, system models may also be recovered from the estimated Kalman state sequence. Simulations based on a benchmark problem compare the performance of the proposed algorithms with a number of well-known subspace identification algorithms and verify the feasibility and closed-loop applicability of the proposed algorithms.

ACKNOWLEDGEMENTS

This work is supported by Alexander von Humboldt Research Fellowship of Germany and this support is greatly acknowledged by BH and SXD. SJQ also acknowledges the support from Natural Sciences Foundation of China in the form of an Outstanding Young Investigator Award for Overseas (60228001).

REFERENCES

Chou, C.T. and M. Verhaegen (1997). Subspace algorithms for the identification of multivariable dynamic errors-in-variables models. *Automatica* **33(10)**, 1857–1869.

Favoreel, W. (1999). Subspace methods for identification and control of linear and bilinear systems. PhD thesis. Katholieke Universiteit Leuven.

Larimore, W.E. (1990). Canonical variate analysis in identification, filtering, and adaptive control. In: *Proceedings of the 29th Conference on Decision and Control*. pp. 596–604.

Ljung, L. and T. McKelvey (1996). Subspace identification from closed loop data. *Signal Processing* **52**, 209–215.

Overschee, P. Van and B. De Moor (1996). Closed-loop subspace system identification. Technical Report ESAT-SISTA/TR 1996-52I. Katholieke Universiteit Leuven.

Verhaegen, Michel (1993). Application of a subspace model identification technique to identify LTI systems operating on closed-loop. *Automatica* **29(4)**, 1027–1040.

Wang, J. and J. Qin (2002). A new subspace identification approach based on principal component analysis. *Journal of Process Control* **12**, 841–855.

ELSEVIER
IFAC
PUBLICATIONS
www.elsevier.com/locate/ifac

SUBSPACE IDENTIFICATION USING THE PARITY SPACE

Jin Wang ** **S. Joe Qin** *,[1]

* *Department of Chemical Engineering*
The University of Texas at Austin, Texas, USA

** *Advanced Process Control*
Advanced Micro Devices, Inc. Texas, USA

Abstract: It is known that most subspace identification algorithms give biased estimates for closed-loop data due to a projection performed in the algorithms. In this work, consistency analysis of SIMPCA is given and the exact input requirement is formulated. The effect of column weighting in subspace identification algorithms is discussed and the column weighting for SIMPCA is designed which gives consistent estimates of state-space models from both open loop and closed-loop data. A simulation example is given to demonstrate the performance of the proposed algorithm. *Copyright © 2004 IFAC*

Keywords: subspace identification, instrumental variables, closed-loop identification, principal component analysis, consistency analysis, parity space

1. INTRODUCTION

Subspace identification algorithms have been developed in the past two decades. Based on numerically robust singular value decomposition (SVD), these algorithms are simple and completely bypass the need for the estimation of structure index, which causes no additional difficulty when handling MIMO systems (Gevers, 2003). The asymptotic properties of these subspace algorithms also have been investigated in the past decade and consistency conditions of the estimates have been identified (Deistler *et al.*, 1995; Peternell *et al.*, 1996; Jansson and Wahlberg, 1998; Bauer *et al.*, 1999; Knudsen, 2001).

Because most subspace identification algorithms perform a projection of the future output onto the orthogonal complement of future input, which requires future input to be uncorrelated to past noise, the application of these algorithms to closed-loop data typically gives biased estimates even though the data satisfy identifiability conditions for prediction error methods. To address this aspect, algorithms for closed-loop identification have been developed in the last a few years (Chou and Verhaegen, 1997; Ljung and McKelvey, 1996; Forssell and Ljung, 1999; Wang and Qin, 2002; Huang *et al.*, 2003).

The aim of this paper is to present a modification of our recently developed algorithm — SIMPCA (subspace identification method via principal component analysis), which gives consistent estimates of state-space models from both open loop and closed-loop data, under some input excitation conditions. The remaining part of the paper is organized as follows. Section 2 gives the problem formulation and assumptions. Section 3 reviews the original SIMPCA algorithm. The exact input requirement and consistency analysis are

[1] Supported by NSF under CTS-9985074 and an Overseas Young Investigator Award (60228001) from NSF China

given in this section. Section 4 presents the modified algorithm, SIMPCA with column weighting (SIMPCA-Wc), and discusses the effect of column weighting. Section 5 gives the simulation example. The final section gives conclusions to the paper.

2. PROBLEM FORMULATION AND ASSUMPTIONS

The original SIMPCA algorithm (Wang and Qin, 2002) was developed to address errors-in-variables case, and it is applicable to innovation formulation. Consider the linear time-invariant system in its innovation representation:

$$\mathbf{x}(k+1) = \mathbf{A}\mathbf{x}(k) + \mathbf{B}\mathbf{u}(k) + \mathbf{K}\mathbf{e}(k) \quad (1)$$
$$\mathbf{y}(k) = \mathbf{C}\mathbf{x}(k) + \mathbf{D}\mathbf{u}(k) + \mathbf{e}(k) \quad (2)$$

Here $\mathbf{x}(k) \in \Re^n$ is the state vector, $\mathbf{u}(k) \in \Re^l$ and $\mathbf{y}(k) \in \Re^m$ are the measured input and output signals. $\mathbf{e}(k) \in \Re^m$ is the innovation process.

2.1 Assumptions

We introduce the following assumptions:

A1: The system is asymptotically stable.

A2: $(\mathbf{A},\ \mathbf{C})$ is observable.

A3: $(\mathbf{A},\ [\mathbf{B}\ \ \mathbf{K}])$ is controllable.

A4: The input $\mathbf{u}$ and innovation $\mathbf{e}$ are jointly stationary and one-way uncorrelated, i.e.,

$$\bar{\mathbf{E}}[\mathbf{e}(k)\mathbf{e}(l)^T] = \mathbf{R}_e\delta_{kl} \quad (3)$$
$$\bar{\mathbf{E}}[\mathbf{e}(k)\mathbf{u}(l)^T] = \mathbf{0}, \quad k > l \quad (4)$$

where $\bar{\mathbf{E}}$ is defined as (Ljung, 1999)

$$\bar{\mathbf{E}}\{\bullet\} = \lim_{N\to\infty}\frac{1}{N}\sum_{k=1}^{N}\mathbf{E}\{\bullet\} \quad (5)$$

Eqn (4) allows us to include closed-loop identification in the proposed method. In order to get consistent identification results, a persistent excitation condition on the input is introduced later.

2.2 Preliminaries

An extended state-space model can be formulated,

$$\mathbf{Y}_f = \mathbf{\Gamma}_f\mathbf{X}_k + \mathbf{H}_f\mathbf{U}_f + \mathbf{G}_f\mathbf{E}_f \quad (6)$$
$$\mathbf{Y}_p = \mathbf{\Gamma}_p\mathbf{X}_{k-p} + \mathbf{H}_p\mathbf{U}_p + \mathbf{G}_p\mathbf{E}_p \quad (7)$$

where $\mathbf{\Gamma}_f$ is the extended observability matrix with rank n, $\mathbf{H}_f$ and $\mathbf{G}_f$ are two triangular Toeplitz matrices as follows,

$$\mathbf{\Gamma}_f = \begin{bmatrix}\mathbf{C}\\ \mathbf{CA}\\ \vdots\\ \mathbf{CA}^{f-1}\end{bmatrix} \in \Re^{mf\times n} \quad (8)$$

$$\mathbf{H}_f = \begin{bmatrix}\mathbf{D} & \mathbf{0} & \cdots & \mathbf{0}\\ \mathbf{CB} & \mathbf{D} & \cdots & \mathbf{0}\\ \vdots & \vdots & \ddots & \vdots\\ \mathbf{CA}^{f-2}\mathbf{B} & \mathbf{CA}^{f-3}\mathbf{B} & \cdots & \mathbf{D}\end{bmatrix} \in \Re^{mf\times lf} \quad (9)$$

$$\mathbf{G}_f = \begin{bmatrix}\mathbf{I} & \mathbf{0} & \cdots & \mathbf{0}\\ \mathbf{CK} & \mathbf{I} & \cdots & \mathbf{0}\\ \vdots & \vdots & \ddots & \vdots\\ \mathbf{CA}^{f-2}\mathbf{K} & \mathbf{CA}^{f-3}\mathbf{K} & \cdots & \mathbf{I}\end{bmatrix} \in \Re^{mf\times mf} \quad (10)$$

The future and past data are arranged in the following Hankel form:

$$\mathbf{Y}_f = \begin{bmatrix}\mathbf{y}(k) & \mathbf{y}(k+1) & \cdots & \mathbf{y}(k+N-1)\\ \mathbf{y}(k+1) & \mathbf{y}(k+2) & \cdots & \mathbf{y}(k+N)\\ \mathbf{y}(k+f-1) & \mathbf{y}(k+f) & \cdots & \mathbf{y}(k+f+N-2)\end{bmatrix}$$
$$\equiv \left[\mathbf{y}_f(k)\ \mathbf{y}_f(k+1)\ \cdots\ \mathbf{y}_f(k+N-1)\right] \in \Re^{mf\times N}$$

$$\mathbf{Y}_p = \begin{bmatrix}\mathbf{y}(k-p) & \mathbf{y}(k-p+1) & \cdots & \mathbf{y}(k-p+N-1)\\ \mathbf{y}(k-p+1) & \mathbf{y}(k-p+2) & \cdots & \mathbf{y}(k-p+N)\\ \mathbf{y}(k-1) & \mathbf{y}(k) & \cdots & \mathbf{y}(k+N-2)\end{bmatrix}$$
$$\equiv \left[\mathbf{y}_p(k)\ \mathbf{y}_pf(k+1)\ \cdots\ \mathbf{y}_p(k+N-1)\right] \in \Re^{mp\times N}$$

$\mathbf{U}_f$, $\mathbf{U}_p$, $\mathbf{E}_f$ and $\mathbf{E}_p$ are formulated similarly.

Eqn (6) is the basis for many subspace identification methods (SIM).

3. THE ORIGINAL SIMPCA ALGORITHM

SIMPCA includes two steps: i) identifying the extended observability matrix $\mathbf{\Gamma}_f$ and the block triangular Toeplitz matrix $\mathbf{H}_f$; ii) estimating system matrices $\mathbf{A}$, $\mathbf{B}$, $\mathbf{C}$ and $\mathbf{D}$.

We begin with the extended state-space model of Eqn (6). Instead of focusing on the observable subspace, SIMPCA algorithm focuses on the parity subspace, which is commonly used in fault detection (Chow and Willsky, 1984). Pre-multiplying Eqn (6) with $(\mathbf{\Gamma}_f^{\perp})^T$, the orthogonal complement of $\mathbf{\Gamma}_f$ with *full* column rank, and moving the input term to the left hand side, Eqn (6) becomes

$$\left(\mathbf{\Gamma}_f^{\perp}\right)^T[\mathbf{I}\mid -\mathbf{H}_f]\,\mathbf{Z}_f = \left(\mathbf{\Gamma}_f^{\perp}\right)^T\mathbf{G}_f\mathbf{E}_f \quad (11)$$

where $\mathbf{Z}_f \equiv \begin{bmatrix}\mathbf{Y}_f\\ \mathbf{U}_f\end{bmatrix}$.

In order to achieve the consistent estimates of $\mathbf{\Gamma}_f$ and $\mathbf{H}_f$ under the noise-corrupted condition, we use instrumental variables to remove the noise. An important property of the instrumental variables is that they should be uncorrelated with the noise, and sufficiently correlated with the informative part of the data. Other work (Verhaegen and Dewilde, 1992; Verhaegen, 1994) shows that past input and past output together is a good choice for instrumental variables. It is easy to see that

$$\lim_{N\to\infty}\frac{1}{N}\mathbf{E}_f\mathbf{Z}_p^T = \lim_{N\to\infty}\frac{1}{N}\mathbf{E}_f[\mathbf{Y}_p^T \quad \mathbf{U}_p^T] = \mathbf{0} \quad (12)$$

because of the Assumption A4.

In order to obtain consistent estimation under the noise-corrupted condition, the input excitation condition is needed, which is stated in the following lemma.

Lemma 1. Given general Assumptions A1 $\sim$ A4 and $p \geq f > n$, $\frac{1}{N}\mathbf{Z}_f\mathbf{Z}_p^T$ has $mf - n$ zero singular values when $N \to \infty$, if $\bar{\mathbf{E}}\left\{\begin{bmatrix}\mathbf{x}(k)\\ \mathbf{u}_f(k)\end{bmatrix}\mathbf{z}_p(k)^T\right\}$ has full row rank.

Proof. See Appendix A.

A similar condition ($\bar{\mathbf{E}}\left\{\begin{bmatrix}\mathbf{u}_f(k)\\ \mathbf{x}(k)\end{bmatrix}\begin{bmatrix}\mathbf{u}_f(k)\\ \mathbf{z}_p(k)\end{bmatrix}^T\right\}$ has full row rank) was derived in Viberg *et al.* (1997) for the IV-4SID method.

Remark 1 Although regarded as fully excited for other subspace identification methods, white noise input is not a suitable input for SIMPCA. If the input is white noise, $\lim_{N\to\infty}\frac{1}{N}\mathbf{U}_f\mathbf{Z}_p^T = 0$ and the input excitation condition is not satisfied for SIMPCA. Similarly, for closed-loop identification, the input perturbation needs to be correlated in order to meet the input excitation condition.

Based on lemma 1, we formulate the following theorem for the combined determinate-stochastic realization.

Theorem 1. Given the general assumptions A1 $\sim$ A4, the persistent excitation condition in Lemma 1, and the following PCA decomposition,

$$\frac{1}{N}\mathbf{Z}_f\mathbf{Z}_p^T = \mathbf{P}\mathbf{T}^T + \tilde{\mathbf{P}}\tilde{\mathbf{T}}^T \quad (13)$$

when $N \to \infty$, we have

$$\begin{bmatrix}\mathbf{\Gamma}_f^{\perp}\\ -\mathbf{H}_f^T\mathbf{\Gamma}_f^{\perp}\end{bmatrix} = \tilde{\mathbf{P}}\mathbf{M} \quad (14)$$

where $\mathbf{M} \in \Re^{(mf-n)\times(mf-n)}$ is a non-singular matrix.

Proof. See Appendix B.

Remark 2 Because projecting out the future input is avoided in SIMPCA, we only require that the future noise is independent of past input, which is true for closed-loop operation. Therefore, SIMPCA is applicable to closed-loop identification provided that the input excitation condition is satisfied.

Remark 3 SIMPCA is applicable to errors-in-variables situation, which has been discussed in Wang and Qin (2002). Besides, SIMPCA is applicable to the colored noise that is finitely correlated. We only need to separate $\mathbf{Z}_f$ and $\mathbf{Z}_p$ farther beyond the correlation window. Colored noise is also treated in Stoica *et al.* (1995) and Li and Qin (2001).

After the parity space is identified, $\mathbf{\Gamma}_f$, $\mathbf{H}_f$ and matrices $\mathbf{A}$, $\mathbf{B}$, $\mathbf{C}$ and $\mathbf{D}$ can be estimated following Wang and Qin (2002).

4. SIMPCA WITH COLUMN WEIGHTING

Like other subspace-based system identification methods, SIMPCA also obtains an estimate of the extended observability matrix first, that is, the column space of the extended observability matrix. It is thus of great interest to optimize the estimate of the observability matrix. One special direction is to apply certain weighting matrix to improve the estimate. Several contributions in the literature have appeared in this area (Jansson and Wahlberg, 1996; Viberg *et al.*, 1997; Gustafsson, 2002). Although different approaches of analysis were adopted, it is interesting to note that the column weightings in Jansson and Wahlberg (1996) and Verhaegen (1994) all correspond to the CVA weighting of Larimore (1983).

In Gustafsson (2002), an asymptotic analysis of the estimated observability matrix is presented, and an optimal column weighting matrix $\mathbf{W}_c^o$ is derived. It is also interesting to note that an approximation of $\mathbf{W}_c^o$ given in Gustafsson (2002) is the same as the column weighting employed by CVA and MOESP, as pointed out in Viberg *et al.* (1997).

Define $\mathbf{R}_{\mathbf{xy}}(\tau)$ as the cross-covariance matrix of $\mathbf{x}$ and $\mathbf{y}$

$$\mathbf{R}_{\mathbf{xy}}(\tau) = \bar{\mathbf{E}}\{\mathbf{x}(t+\tau)\mathbf{y}^T(t)\} \quad (15)$$

and let $\mathbf{R}_{\mathbf{xy}} = \mathbf{R}_{\mathbf{xy}}(0)$, the approximate optimal weighting in Gustufsson (2002) is

$$\begin{aligned}\mathbf{W}_c^a &= (\mathbf{R}_{\mathbf{z}_p\mathbf{z}_p} - \mathbf{R}_{\mathbf{u}_f\mathbf{z}_p}^T\mathbf{R}_{\mathbf{u}_f\mathbf{u}_f}^{-1}\mathbf{R}_{\mathbf{u}_f\mathbf{z}_p})^{-\frac{1}{2}}\\ &= (\frac{1}{N}\mathbf{Z}_p\mathbf{\Pi}_{\mathbf{U}_f}^{\perp}\mathbf{Z}_p^T)^{-\frac{1}{2}} \quad (16)\end{aligned}$$

where $\mathbf{\Pi}_{\mathbf{U}_f}^{\perp} = \mathbf{I} - \mathbf{U}_f^T(\mathbf{U}_f\mathbf{U}_f^T)^{-1}\mathbf{U}_f$.

In the framework studied in Gustafsson (2002), an instrumental variable approach reduces Eqn (6) to the following relation,

$$\lim_{N\to\infty}\frac{1}{N}\mathbf{Y}_f\mathbf{\Pi}_{\mathbf{U}_f}^{\perp}\mathbf{Z}_p^T = \lim_{N\to\infty}\frac{1}{N}\mathbf{\Gamma}_f\mathbf{X}_k\mathbf{\Pi}_{\mathbf{U}_f}^{\perp}\mathbf{Z}_p^T \quad (17)$$

Post-multiplying the column weighting in Eqn (16) leads to:

$$\lim_{N\to\infty}\frac{1}{N}\mathbf{Y}_f\mathbf{\Pi}_{\mathbf{U}_f}^{\perp}\mathbf{Z}_p^T(\mathbf{Z}_p\mathbf{\Pi}_{\mathbf{U}_f}^{\perp}\mathbf{Z}_p^T)^{-\frac{1}{2}}$$

$$= \lim_{N\to\infty} \frac{1}{N}\mathbf{\Gamma}_f \mathbf{X}_k \mathbf{\Pi}_{\mathbf{U}_f}^{\perp} \mathbf{Z}_p^T (\mathbf{Z}_p \mathbf{\Pi}_{\mathbf{U}_f}^{\perp} \mathbf{Z}_p^T)^{-\frac{1}{2}} \quad (18)$$

Denoting $\mathbf{Z}_{\mathbf{U}^\perp} \equiv \mathbf{Z}_p \mathbf{\Pi}_{\mathbf{U}_f}^{\perp}$ as the generalized instrumental variables, the left hand side of the above equation reduces to

$$\mathbf{Y}_f \mathbf{Z}_{\mathbf{U}^\perp}^T (\mathbf{Z}_{\mathbf{U}^\perp} \mathbf{Z}_{\mathbf{U}^\perp}^T)^{-\frac{1}{2}} = \mathbf{Y}_f \mathbf{Z}_{\mathbf{U}^\perp}^T \mathbf{W}_c^a \quad (19)$$

Therefore, the approximate optimal column weighting can be interpreted as scaling the generalized instrumental variable to unit variance since

$$(\mathbf{Z}_{\mathbf{U}^\perp}^T \mathbf{W}_c^a)^T (\mathbf{Z}_{\mathbf{U}^\perp}^T \mathbf{W}_c^a) = \mathbf{W}_c^a \mathbf{Z}_{\mathbf{U}^\perp} \mathbf{Z}_{\mathbf{U}^\perp}^T \mathbf{W}_c^a = \mathbf{I} \quad (20)$$

Because SIMPCA also picks the subspace by performing SVD (PCA) on process data and instrumental variables are applied to remove noise, it would be beneficial to scale the instrumental variables to unit variance. In SIMPCA, since projecting out of future input is not needed, the generalized instrumental variables is just past data $\mathbf{Z}_p$. In this case, $\mathbf{W}_c^a$ is simply $\left(\mathbf{Z}_p \mathbf{Z}_p^T\right)^{-\frac{1}{2}}$, which is the same weighting applied in IV-4SID (Viberg, 1995).

In summary, SIMPCA with column weighting (SIMPCA-Wc) is based on the following equation,

$$\frac{1}{N}\left(\mathbf{\Gamma}_f^{\perp}\right)^T [\mathbf{I} \mid -\mathbf{H}_f] \mathbf{Z}_f \mathbf{Z}_p^T \left(\mathbf{Z}_p \mathbf{Z}_p^T\right)^{-\frac{1}{2}}$$

$$= \frac{1}{N}\left(\mathbf{\Gamma}_f^{\perp}\right)^T \mathbf{G}_f \mathbf{E}_f \mathbf{Z}_p^T \left(\mathbf{Z}_p \mathbf{Z}_p^T\right)^{-\frac{1}{2}} \quad (21)$$

The right hand side of Eqn (21) goes to zero as $N \to \infty$ due to Assumption A4. Therefore, performing PCA on $\mathbf{Z}_f \mathbf{Z}_p^T \left(\mathbf{Z}_p \mathbf{Z}_p^T\right)^{-\frac{1}{2}}$ can extract the residual subspace which contains the system information. The remaining steps in SIMPCA-Wc to estimate $\mathbf{A}$, $\mathbf{B}$, $\mathbf{C}$, $\mathbf{D}$ are the same as in SIMPCA. Simulation results show that the column weighting matrix $\left(\mathbf{Z}_p \mathbf{Z}_p^T\right)^{-\frac{1}{2}}$ can significantly improve the variance of estimates compared to SIMPCA, as shown in the next section. It is interesting that applying the CVA column weighting can achieve the similar performance as the SIMPCA column weighting.

In Huang et al. (2003) a subspace orthogonal projection identification method (SOPIM) for closed-loop identification is proposed based on the SIMPCA structure. In the proposed SOPIM, the parity space is extracted through an orthogonal projection of the future input and output on the row space of past input and output, i.e.,

$$\left(\mathbf{\Gamma}_f^{\perp}\right)^T \left[\mathbf{I} \; -\mathbf{H}_f\right] \mathbf{Z}_f \mathbf{\Pi}_{\mathbf{Z}_p} = \left(\mathbf{\Gamma}_f^{\perp}\right)^T \mathbf{G}_f \mathbf{E}_f \mathbf{\Pi}_{\mathbf{Z}_p} \quad (22)$$

where the right hand side goes to zero when the number of observations goes to infinity. It is interesting to notice that the projection of future data on the row space of past data is

$$\mathbf{Z}_f \mathbf{\Pi}_{\mathbf{Z}_p} = \mathbf{Z}_f \mathbf{Z}_p^T (\mathbf{Z}_p \mathbf{Z}_p^T)^{-1} \mathbf{Z}_p \quad (23)$$

which is closely related to SIMPCA-Wc, as shown in the following,

$$\mathbf{Z}_f \mathbf{Z}_p^T \mathbf{W}_c = \mathbf{Z}_f \mathbf{Z}_p^T (\mathbf{Z}_p \mathbf{Z}_p^T)^{-\frac{1}{2}} \quad (24)$$

it is clear that $\mathbf{\Pi}_{\mathbf{Z}_p} = \left(\mathbf{Z}_p^T \mathbf{W}_c\right)\left(\mathbf{Z}_p^T \mathbf{W}_c\right)^T$ also has unit variance.

5. SIMULATION STUDY

In this section a simulation study is presented to demonstrate the performance of SIMPCA algorithm. Results from SIMPCA, N4SID with CVA weighting in the Matlab System Identification Toolbox (version 5.0) (CVA via N4SID), SIMPCA with column weighting (SIMPCA-Wc) and MOESP-PO are presented for comparison. The simulation example is a first order SISO system, where both open-loop and closed-loop operations are examined. The system order is given.

The system under test is given by the following difference equation:

$$y(k) - 0.9y(k-1) = u(k-1) + e(k) + 0.9e(k-1) \quad (25)$$

The input to the system has the following structure

$$u(k) = -\lambda y(k) + r(k) \quad (26)$$

where λ can be set to different values to achieve both open loop and closed-loop operations. The open loop input excitation signal $\mathbf{r}(k)$ is a moving average process:

$$r(k) = (1 + 0.8\, q^{-1} + 0.6\, q^{-2}) r_0(k) \quad (27)$$

where $r_0(k)$ is zero mean white noise sequence with unit variance.

5.1 Open-loop case

In this case $\lambda = 0$. Process noise $\mathbf{e}(k)$ is added to the system, where $\mathbf{e}(k)$ is zero mean white noise sequence with standard deviation $\sigma_e = 1.2$. Monte-Carlo experiments are conducted and 50 runs are performed. For each run, 7000 data points are collected for identification. Fig. 1 shows the Bode plot of the identified system from different algorithms. In order to compare the relative efficiency of different methods, the asymptotical performance is shown in Fig. 2, where the mean squared error of the pole estimation given different number of samples is plotted. It shows that SIMPCA-Wc performs similarly to CVA(via N4SID) and MOESP, while SIMPCA is worse than the other three methods.

5.2 Closed-loop case

In this case $\lambda = -0.6$. Fig. 3 shows the Bode plot under combined deterministic-stochastic case where the number of data point is 10^5. Apparently CVA(via N4SID) and MOESP give biased estimates, while estimates from SIMPCA and SIMPCA-Wc are unbiased. We also plot the asymptotic performance of different methods under closed-loop in Fig. 4, which confirms that SIMPCA and SIMPCA-Wc give unbiased estimates. Compared to more efficient SIMPCA-Wc, SIMPCA requires more data to achieve same level of estimate variance.

As we discussed in Remark 1, white input perturbation is not ideal for SIMPCA because it does not satisfy the input excitation condition. Huang et al. (2003) point out that if the input perturbation is white noise, the controller dynamic will also fall into the left null space of $\mathbf{Z}_f\mathbf{Z}_p^T$, which will make the estimate from SIMPCA biased. A remedy is also proposed in Huang et al. (2003). By including the input perturbation as part of the instrumental variable, it is guaranteed that the left null space of $\mathbf{Z}_f\mathbf{Z}_p^T$ contains only the process dynamics.

6. CONCLUSION

In this paper, consistency analysis of SIMPCA with focus on the deterministic part is given. Because SIMPCA makes use of parity space to estimate system model and avoids projecting out the future input, it is applicable to closed-loop data, provided that the input excitation condition is satisfied. A modification of original SIMPCA (SIMPCA-Wc) is presented. By scaling the instrumental variables to unit variance through column weighting, SIMPCA-Wc can significantly improve the estimate efficiency. A simulation example shows that the designed column weighting can significantly reduce the variance of the estimates.

REFERENCES

Bauer, D., M. Deistler and W. Scherrer (1999). Consistency and asymptotic normality of some subspace algorithms for systems without observed inputs. *Automatica* **35**, 1243–1254.

Chou, C.T. and M. Verhaegen (1997). Subspace algorithms for the identification of multivariable dynamic errors-in-variables models. *Automatica* **33**(10), 1857–1869.

Chow, E.Y. and A.S. Willsky (1984). Analytical redundancy and the design of robust failure detection systems. *IEEE Trans. Auto. Cont.* **AC-29**(7), 603–614.

Deistler, M., K. Peternell and Scherrer (1995). Consistency and relative efficiency of subspace methods. *Automatica* **31**, 1865–1875.

Forssell, U. and L. Ljung (1999). Closed-loop identification revisited. *Automatica* **35**, 1215–1241.

Gevers, M. (2003). A personal view on the development of system identification. In: *Proceedings of 13th IFAC symposium on System Identification.* Rotterdam, Netherlands. pp. 773–784.

Gustafsson, T. (2002). Suspace-based system identification: weighting and pre-filtering of instruments. *Automatica* **38**, 433–443.

Huang, B., S. Ding and S.J. Qin (2003). Closed-loop subspace identificaiton: an orthogonal projection approach. Submitted to Journal of Process Control.

Jansson, M. and B. Wahlberg (1996). A linear regression approach to state-space subspace system identification. *Signal Processing* **52**, 103–129.

Jansson, M. and B. Wahlberg (1998). On consistency of subspace methods for system identification. *Automatica* **34**(12), 1507–1519.

Knudsen, T. (2001). Consistency analysis of subspace idenfification methods based on a linear regression approach. *Automatica* **37**, 81–89.

Larimore, W.E. (1983). System identification, reduced-order filtering and modeling via canonical variate analysis. In: *Proc. 1983 American Control Conf., H.S. Rao and T. Dorato (Ed), IEEE.* New York.

Li, W. and S.J. Qin (2001). Consistent dynamic PCA based on errors-in-variables subspace identification. *J. Process Control* **11(6)**, 661–678.

Ljung, L. (1999). *System Identification: Theory for the User.* Prentice-Hall, Inc.. Englewood Cliffs, New Jersey.

Ljung, L. and T. McKelvey (1996). Subspace identification from closed loop data. *Signal Processing* **52**, 209–215.

Peternell, K., W. Scherrer and M. Deistler (1996). Statistical analysis of novel subspace identification methods. *Signal Processing* **52**, 161–177.

Stoica, P., M. Cedervall and A. Eriksson (1995). Combined instrumental variable and subspace fitting approach to parameter estimation of noisy input-output systems. *IEEE Transactions on Signal Processing* **43(10)**, 2386–2397.

Verhaegen, M. (1994). Identification of the deterministic part of MIMO state space models given in innovations form from input-output data. *Automatica* **30**, 61–74.

Verhaegen, M. and P. Dewilde (1992). Subspace model identification. part i: the output-error state-space model identification class of algorithms. *International Journal of Control* **56**, 1187–1210.

Viberg, M. (1995). Subspaced-based methods for the identification of linear time-invariant systems. *Automatica* **31(12)**, 1835–1851.

Viberg, M., B. Wahlberg and B. Ottersten (1997). Analysis of state space system identification methods based on instrumental variables and subspace fitting. *Automatica* **33**, 1603–1616.

Wang, J. and S.J. Qin (2002). A new subspace identification approach based on principal component analysis. *Journal of Process Control* **12(8)**, 841–855.

Appendix A. PROOF OF LEMMA 1

From the extended state-space model Eqn (6), we have

$$\mathbf{Z}_f = \begin{bmatrix} \mathbf{Y}_f \\ \mathbf{U}_f \end{bmatrix} = \begin{bmatrix} \mathbf{\Gamma}_f & \mathbf{H}_f \\ \mathbf{0} & \mathbf{I}_{lf} \end{bmatrix} \begin{bmatrix} \mathbf{X}_k \\ \mathbf{U}_f \end{bmatrix} + \begin{bmatrix} \mathbf{G}_f \\ \mathbf{0} \end{bmatrix} \mathbf{E}_f \tag{A.1}$$

Post-multiply the instrumental variable $\mathbf{Z}_p^T$ to remove noise,

$$\lim_{N\to\infty} \frac{1}{N}\mathbf{Z}_f\mathbf{Z}_p^T = \begin{bmatrix} \mathbf{\Gamma}_f & \mathbf{H}_f \\ \mathbf{0} & \mathbf{I}_{lf} \end{bmatrix} \begin{bmatrix} \mathbf{X}_k \\ \mathbf{U}_f \end{bmatrix} \mathbf{Z}_p^T \tag{A.2}$$

Because $f > n$, $rank(\mathbf{\Gamma}_f) = n$, and $\begin{bmatrix} \mathbf{\Gamma}_f & \mathbf{H}_f \\ \mathbf{0} & \mathbf{I}_{lf} \end{bmatrix}$ has full column rank, which is $lf + n$. Therefore, if $\bar{\mathbf{E}}\left(\begin{bmatrix} \mathbf{x}(k) \\ \mathbf{u}_f(k) \end{bmatrix} \mathbf{z}_p(k)^T\right)$ has full row rank, $rank\left(\frac{1}{N}\mathbf{Z}_f\mathbf{Z}_p^T\right) = lf + n$ as $N \to \infty$.

Appendix B. PROOF OF THEOREM 1

Eqn (6) can be rearranged as

$$[\,\mathbf{I} \mid -\mathbf{H}_f\,]\,\mathbf{Z}_f = \mathbf{\Gamma}_f\mathbf{X}_k + \mathbf{G}_f\mathbf{E}_f \tag{B.1}$$

In order to remove the noise term, multiply $\frac{1}{N}\mathbf{Z}_p^T$ to both sides of the above equation,

$$\frac{1}{N}[\,\mathbf{I} \mid -\mathbf{H}_f\,]\,\mathbf{Z}_f\mathbf{Z}_p^T = \frac{1}{N}\mathbf{\Gamma}_f\mathbf{X}_k\mathbf{Z}_p^T + \frac{1}{N}\mathbf{G}_f\mathbf{E}_f\mathbf{Z}_p^T \tag{B.2}$$

while

$$\lim_{N\to\infty} \frac{1}{N}\mathbf{E}_f\mathbf{Z}_p^T = 0 \tag{B.3}$$

when $N \to \infty$

$$\frac{1}{N}[\,\mathbf{I} \mid -\mathbf{H}_f\,]\,\mathbf{Z}_f\mathbf{Z}_p^T = \frac{1}{N}\mathbf{\Gamma}_f\mathbf{X}_k\mathbf{Z}_p^T \tag{B.4}$$

pre-multiply the above equation with $\left(\mathbf{\Gamma}_f^{\perp}\right)^T$ to remove state,

$$\frac{1}{N}\left(\mathbf{\Gamma}_f^{\perp}\right)^T [\,\mathbf{I} \mid -\mathbf{H}_f\,]\,\mathbf{Z}_f\mathbf{Z}_p^T = 0 \tag{B.5}$$

Following the same procedure as in proof of theorem 1, it is straight forward to show that

$$\begin{bmatrix} \mathbf{\Gamma}_f^{\perp} \\ -\mathbf{H}_f^T\mathbf{\Gamma}_f^{\perp} \end{bmatrix} = \tilde{\mathbf{P}}\mathbf{M} \tag{B.6}$$

where $\mathbf{M} \in \Re^{(mf-n)\times(mf-n)}$ is a non-singular matrix.

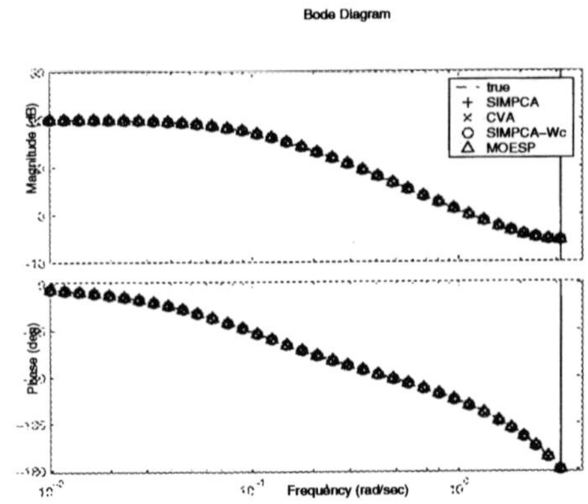

Fig. 1. Bode plot of the identified system for the open loop case

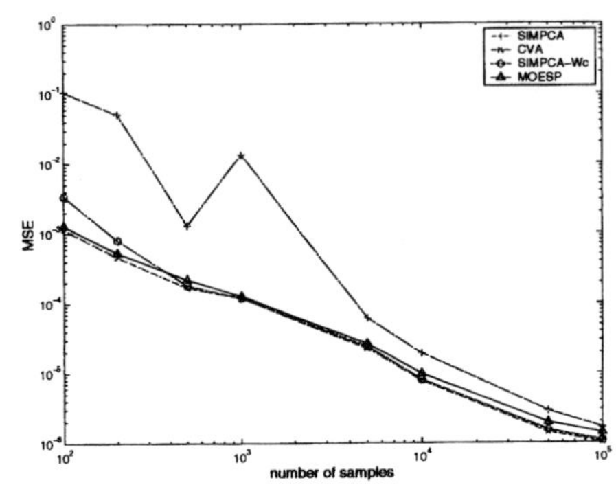

Fig. 2. Mean squared error of pole estimate versus number of samples for the open loop case

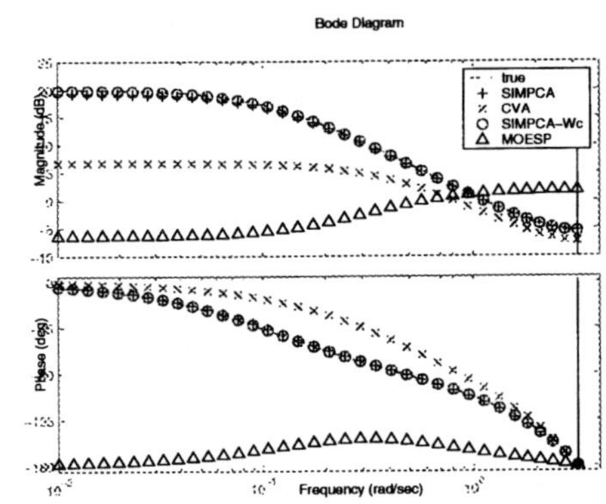

Fig. 3. Bode plot of the identified system for the closed-loop case

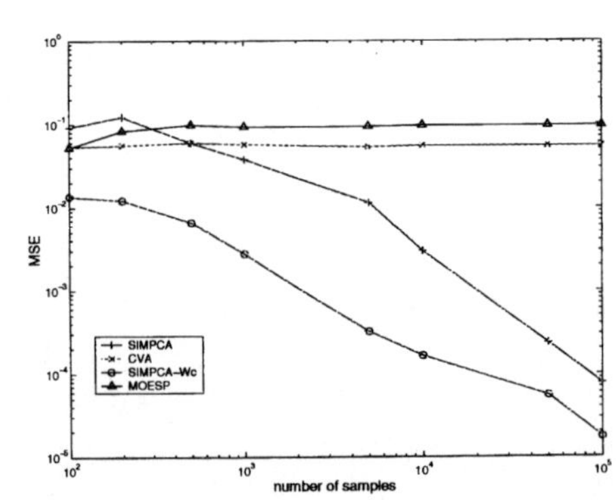

Fig. 4. Mean squared error of pole estimate versus number of samples for the closed-loop case

ELSEVIER
IFAC
PUBLICATIONS
www.elsevier.com/locate/ifac

LARGE SAMPLE EFFICIENCY FOR ADAPTX SUBSPACE SYSTEM IDENTIFICATION WITH UNKNOWN FEEDBACK

Wallace E. Larimore

Adaptics, Inc, 1717 Briar Ridge Road, McLean, VA 22101 USA
Phone: 703 532-0062, Fax: 703 536-3319, Email: `larimore@adaptics.com`

Abstract. Over the last two decades, the canonical variate analysis method for subspace system identification has been widely applied. A number of these applications have demonstrated near maximum likelihood accuracy of the `adaptx` CVA subspace algorithm in large samples with unknown feedback. The critical step in the algorithm is the use of an ARX model estimated by conditional maximum likelihood to remove the effects of future inputs on future outputs. It is shown that the subspace estimates can be considered as restrictions on the ML ARX estimates to a subspace of the parameters obtained by projection methods. As a result, the errors between the models are orthogonal to the subspace model, and the subspace parameter estimates are asymptotically ML. A critical step in showing this orthogonality is use of the multistep form of the likelihood function. *Copyright©2004 IFAC*

Key Words: Large sample, Subspace identification, Canonical variate analysis, Closed-loop, Feedback, Multistep likelihood.

1. OVERVIEW

The purpose of this paper is to give an outline of the large sample efficiency of `adaptx` for the case of unknown feedback. A more detailed technical development will appear elsewhere. Asymptotic efficiency means the parameter estimation error approaches the minimum variance bound for large sample size.

Over the past two decades, the computational methods, statistical theory, and applications of canonical variate analysis (CVA) have been developed considerably. The basic algorithm (Larimore, 1983) has been significantly improved with model order selection (Larimore, 1990a; 1990b), confidence bands on spectral functions such as frequency response and power spectrum (Larimore, 1993), monitoring and fault detection (Larimore, 1997a; Wang et al, 1997; Juricek et al, 2004; Conner et al, 2004), and delay estimation (Larimore, 2003).

There were early empirical demonstrations of near optimal estimation approaching the Cramer-Rao lower bound (Larimore et. al., 1984), with more detailed simulations to follow (Deistler et al, 1995; Larimore, 1996a, 1996b; Peternell et al, 1996). In the case of no inputs, this was followed by considerable effort on the asymptotic theory, as the sample size becomes large, showing the optimal properties of asymptotic normality and minimum variance (Bauer, 1998; 2004).

A much discussed aspect in the literature has been the behavior of subspace system identification for the case of colored inputs perhaps with feedback. The fundamental problem is the necessity to compute and remove the effects of future inputs on future outputs before the CVA is done to determine the system state. But it appears that the CVA solution itself is required to compute these effects on future outputs. In Larimore (1996a, 1996b), simulation results were presented that strongly suggest such efficiency for that simulation model. The algorithm used in those simulations, and incorporated in the first release of the ADAPTx^{TM} software (Larimore, 1992) as well as all subsequent releases, is as follows:

- *Fit ARX.* Using conditional maximum likelihood (ML), fit ARX models recursively on order and evaluate the AIC_C statistic to determine the optimal number ℓ of delayed inputs and outputs to use in the CVA computation.
- *Remove effects of future inputs q_t on future outputs f_t.* Compute the multistep predictor matrix Ω using the ARX model, and compute the corrected future $f_t|q_t = f_t - \Omega q_t$.
- *CVA.* Do a CVA between the past p_t and corrected future $f_t|q_t$ to determine the states ordered by their associated canonical correlation.

- *Select State Order k.* Compute the estimated one-step prediction error covariance matrix for each state order from 0 to order $\ell\text{Dim}(y_t)$, compute the associated AIC_C for each order, and select the minimum AIC_C.
- *Estimate Model.* Compute estimates of the state space matrices and the one-step error covariance in the state equations by regression.
- *Alternate Model Forms.* Solve the Riccati equation and compute the innovations, overlapping parameterization, and ARMAX models.

It may seem surprising that the use of the ARX model to remove the effects of future inputs from future outputs results in an optimal procedure with asymptotic efficiency. Questions that come to mind are the well known issues:

- *High Order ARX.* The ARX model can have far more parameters to obtain a reasonable approximation to the process than the state space model especially for a process with moving average terms in the noise requiring a high AR order.
- *ARX Model Error.* Such a high order ARX model will have modeling error proportional to the number of estimated parameters so the modeling error for the ARX could be much larger than that potentially achievable using a SS model.
- *SS Model Error.* Thus using the ARX model to remove the effect of future inputs on future outputs could result in additional error in the future outputs, and consequently increase the error in fitting the SS model in subsequent steps.

While these issues are well founded concerns, it will be shown that there is much additional structure to the problem that effectively projects these additional errors to zero.

The `adaptx` algorithm is discussed below in terms of a number of statistical concepts and how they impact the estimation problem. It has long been noted in the literature (Larimore, 1990a) that the difficulty is the presence of future inputs that introduce errors in the prediction of the future outputs from the past, and this introduces errors in the CVA step to determine the state. The use of the ARX model avoids this problem for a number of reasons that will become more evident in later sections. The basic concept is given in Cox and Hinkley (1974, pp. 307, 321-4) concerning nested models, projection, and sufficiency. The use of the ARX model to remove future inputs from future outputs has the following advantages:

- *Linear Computation.* Fitting of the ARX model permits the approximate maximum likelihood identification of a model using efficient and non-iterative linear computations that are needed also to determine the number of lags ℓ of the past to use in the CVA calculation.
- *Order-recursive Computation.* A process can be approximated arbitrarily closely by an ARX process, and recent methods permit the use of an efficient (order ℓ^3 verses ℓ^4 multiplications) order-recursive computation that is highly accurate with no error accumulation (Larimore, 1990b, 2002).
- *ML is Immune to Colored Inputs and Feedback.* The ARX procedure is asymptotically ML and as such the estimates of the plant model from input and output data do not depend on knowledge of the spectrum of the inputs or feedback system, i.e. there is no bias in the estimates (Larimore, 1997b; Gustavsson et al, 1977)
- *Nested Model.* The ARX model class contains the state space model that is fitted by regression so that the subspace model is nested in the ARX model. Specifically, the state space model parameters lie in a subspace of the ARX model.
- *Projection to Low Dimension.* Because of the nested model structure, fitting of the SS model by regression projects the ARX model onto the low dimensional state subspace of the ARX space of delayed inputs and outputs.
- *Decomposition of the ARX Model.* The ARX model decomposes into two pieces, the low-dimensional SS model and the part of the ARX model orthogonal to the SS model. This orthogonal piece projects to zero, i.e. errors in this part of the ARX model go to zero when projecting on the SS model.
- *ARX Model is Sufficient for SS Model.* From model nesting, all of the information in the sample for inference about the SS model is contained in the ARX model parameter estimates.
- *Multistep Likelihood Function.* The equivalence of the onestep and multistep likelihood functions plays a key role in the technical details to demonstrate orthogonality.

While there have been a number of recent papers on new subspace algorithms to handle colored inputs and feedback, there has been very little discussion concerning the asymptotic efficiency of these subspace methods. An exception is Peternell et al (1996) who discuss two algorithms, one imposing a block shift structure on the model involving future inputs, and the other using an iteration to refit the previous model for removing the effects of inputs. By simulation, the first method was shown not to be efficient, and the second appeared to be. But the iterative method appears not to have been pursued, presumably because a major advantage of CVA is the lack of any iteration.

A method was developed by Ljung and McKelvey (1996) using ARX models to remove the effect of future inputs on future outputs. However, the ARX model is used in a completely different way to predict the future outputs that are then used in place of the measurements. A major disadvantage is that such a procedure will lead to biased estimates of the noise covariance matrix. They mention the potential illconditioning in fitting high order ARX models. Illconditioning is avoided in the `adaptx` algorithm by using the order-recursive factorization algorithm (Larimore 1990b, 2002, 2003) that has been demonstrated to be accurate to machine precision even in the case of highly rank deficient data (Larimore, 2002).

Shi (2001) and Shi and MacGregor (2001) discuss several algorithms and consider the use of the ARX model to remove the effects of future inputs on future outputs and show it gives unbiased estimates in the presence of unknown feedback. There is no discussion of the efficiency of the procedure.

2. ML ESTIMATION UNDER FEEDBACK

In this section, the maximum likelihood estimation of input-output models under the effect of unknown feedback is discussed. ML estimation of models has the considerable advantage of being immune to the presence of colored inputs or feedback.

An easy way to see the immunity of ML estimation to feedback is based on simple conditional probability relationships, as shown in Larimore (1997b). The following notation will be used in the development, $Y_1^N = (y_N, \ldots, y_1)$ and similarly for U_1^N. Also let p_t denote the inputs and outputs in the strict past of t. The joint likelihood function of the outputs Y_1^N and the inputs U_1^N conditional on the initial state expressed by the past p_1 at time $t = 1$ and as a function of the unknown parameters θ can be expressed

$$p(Y_1^N, U_1^N | p_1; \theta) = [\prod_{t=1}^{N} p(y_t | u_t, p_t; \theta)][\prod_{t=1}^{N} p(u_t | p_t; \theta)] \tag{1}$$

The probability densities above involve the conditional random variable $y_t|(u_t, p_t)$ that is the usual output innovations process of the plant input-output model. The conditional random variable $u_t|p_t$ is the innovation of the feedback system with a required delay of one time step between y_t and u_t. The joint likelihood function of (Y_1^N, U_1^N) is expressed as the product of two terms that are thus independently distributed. Each of these terms is the product of probabilities of independently distributed innovations processes.

The above factoring of the likelihood function into two terms as in (1) always holds and is the consequence of simple conditional probability rules. The real usefulness comes, however, when the plant and feedback pieces of the system can be parameterized separately. Suppose that the parameter vector can be written as $\theta = (\theta_p, \theta_f)$ where the two subvectors respectively parameterize the plant and feedback parts of the systems. In this case, the maximum of the likelihood function is the product of the maxima of each of the two pieces. Thus under the hypothesis that the process is in a plant-feedback form with the only relationships between them appearing in the plant inputs and outputs, then ML estimation of the plant does not depend upon the presence or absence of feedback. The actual computation of the ML estimates for the ARX model and other details are discussed in the next section.

3. PROJECTION IN ARX AND MARKOV MODELS

The fitting of ARX models using conditional ML and the fitting of state space models using CVA involve the use of regression. Projection is a very useful concept in regression that greatly clarifies some fundamental orthogonality relationships among the identified parameters. The result of this is the elimination of the effect of future inputs on future outputs even in the presence of unknown feedback in the system.

Consider the multivariate ARX model

$$y(t) = \sum_{s=1}^{\ell} \alpha(s) y(t-s) + \sum_{s=0}^{\ell} \beta(s) u(t-s) + e(t) \tag{2}$$

for $t = \ell + 1, \ldots, N$, and where ℓ is the AR and X orders and the error e_t is normally distributed with covariance matrix Σ and independently for different t. The $\alpha(s)$ are the autoregressive (AR) coefficients and the $\beta(s)$ are the exogenous (X) input coefficients.

In fitting the ARX model using least squares (LS), also called conditional maximum likelihood (ML), the equations (2) are used for $t = \ell + 1, \ldots, N$, and are transposed and stacked up to give

$$Y = Z\Theta + E \tag{3}$$

where $Y^T = [y_{\ell+1}, \ldots, y_N]$ with the first ℓ observations of the output not used in the regression so it is conditional on the first ℓ observations. Also denote $\Theta^T = [\alpha_1, \ldots, \alpha_\ell, \beta_0, \ldots, \beta_\ell]$ and

$$Z^T = \begin{bmatrix} y_\ell & \cdots & y_1 & u_{\ell+1} & \cdots & u_1 \\ \vdots & & & & & \vdots \\ y_{N-1} & \cdots & y_{N-\ell} & u_N & \cdots & u_{N-\ell} \end{bmatrix}$$

The linear model (3) applies to much more general processes than ARX models, that will be denoted by Θ_A when needed. The LS and conditional ML estimates are given as

$$\hat{\Theta} = (Z^T Z)^{-1} Z^T Y$$

$$\hat{\Sigma} = Y^T Y - \hat{\Theta}^T Z^T Z \hat{\Theta}$$

The model for y_t is the right hand side of (2) without the noise e_t, which is the conditional expectation of y_t given the past p_t and present input u_t. This is the systematic part of the model for Y. The ML estimates $\hat{\Theta}$ minimize the error $E = Y - \hat{Y}$ with

$$\hat{Y} = Z\hat{\Theta} = Z_1 \hat{\Theta}_{1*} + \cdots + Z_m \hat{\Theta}_{m*}$$

where Z_i is the i-th column of Z and $\hat{\Theta}_{i*}$ is the i-th row of $\hat{\Theta}$.

A subspace projection interpretation clarifies the nesting of parameter spaces. Primarily the univariate case is discussed for conceptual simplicity (see Schaffe, 1959, pp. 43, for a detailed discussion), but it extends to the multivariate case (Anderson, 1984, pp. 295).

In the case that Y is a vector so that $\hat{\Theta}$ is a vector of parameters, then $\hat{\Theta}$ is the linear combination of the columns of Z that gives the model $\hat{Y}$ for Y. Thus the model $Z\hat{\Theta}$ is an $N - \ell$ dimensional vector that lies in the m-dimensional subspace generated by the m columns of Z, denoted $\mathcal{S}(Z)$. Also the parameters $\hat{\Theta}_i$ can be associated with the basis vectors Z_i, respectively, and are coordinates for the subspace. A change of coordinates can be used to define a different parameterization of the subspace. In the multivariate case that Y is a matrix, then the above interpretation applies to each column Y_i of Y using the corresponding column

$\hat{\Theta}_{*i}$ of $\hat{\Theta}$ so that the model for the i-th components Y_i of the observations is

$$\hat{Y}_i = Z\hat{\Theta}_{*i} = Z_1\hat{\Theta}_{1i} + \cdots + Z_m\hat{\Theta}_{mi} \tag{4}$$

This has the following projection interpretation. The estimated model $\hat{Y} = Z\hat{\Theta} = Z(Z^TZ)^{-1}Z^TY$ involves the orthogonal projection operator $Z(Z^TZ)^{-1}Z^T$. The error $Y - \hat{Y}$ is orthogonal to the estimate $\hat{Y}$ since substituting the above for $\hat{Y}$ reduces $\hat{Y}^T(Y - \hat{Y})$ to zero. So $\hat{Y}$ is the orthogonal projection of columns of Y onto the subspace $S(Z)$ span by the columns of Z with the projections defined by the linear combinations (4) specified by the columns of $\hat{\Theta}$.

Substituting $Y = Z\Theta + E$ into $\hat{Y} = Z\hat{\Theta} = Z(Z^TZ)^{-1}Z^TY$ gives

$$\hat{Y} = Z\Theta + Z(Z^TZ)^{-1}Z^TE \tag{5}$$

Thus, under the hypotheses that the true process lies in a lower dimensional subspace, the first observation is that except for the noise, the estimate $\hat{Y}$ is equal to the true noiseless value $Z\Theta$ plus noise. The second observation is that projecting the data on a lower dimensional subspace reduces the degrees of freedom of the noise to the dimension of the subspace. This is a major concept in obtaining asymptotic efficiency.

In the case of static regression where the regressors Z are not random variables but fixed known values, parameter estimates are unbiased since

$$\begin{aligned} \mathcal{E}[\hat{\Theta} - \Theta] &= \mathcal{E}[(Z^TZ)^{-1}Z^T)Y - \Theta] \\ &= \mathcal{E}[(Z^TZ)^{-1}Z^T(Z\Theta + E) - \Theta] = 0 \end{aligned}$$

and the parameter estimation error between any two columns $\hat{\Theta}_i$ and $\hat{\Theta}_j$ of $\hat{\Theta}$ is

$$\begin{aligned} \text{Cov}(\hat{\Theta}_i, \hat{\Theta}_j) &= \mathcal{E}(Z^TZ)^{-1}Z^TE_{i.}E_{j.}^TZ(Z^TZ)^{-1} \\ &= (Z^TZ)^{-1}Z^T\sigma_{ij}Z(Z^TZ)^{-1} = \sigma_{ij}(Z^TZ)^{-1} \end{aligned}$$

Suppose the space $S(Z)$ decomposes into two subspaces that are orthogonal so $Z_a = (Z_1, \ldots, Z_r)$ and $Z_b = (Z_{r+1}, \ldots, Z_m)$ with $Z = (Z_a\ Z_b)$ and the orthogonality condition $Z_a^TZ_b = 0$. Then the corresponding decomposition of the parameters $\hat{\Theta} = (\hat{\Theta}_a; \hat{\Theta}_b)$ have diagonal covariance matrix with

$$\text{Cov}(\hat{\Theta}_i, \hat{\Theta}_j) = \sigma_{ij}\text{diag}((Z_a^TZ_a)^{-1}, (Z_b^TZ_b)^{-1})$$

so parameter estimates $\hat{\Theta}_a$ and $\hat{\Theta}_b$ are uncorrelated. The converse is also true; if $\hat{\Theta}_a$ and $\hat{\Theta}_b$ are uncorrelated, then Z_a and Z_b are orthogonal.

Now given a subspace $S(Z_S)$ of a larger space $S(Z_A)$, the orthogonal compliment Z_{A-S} can always be constructed by orthonormalization, that in turn defines orthogonal parameter estimates Θ_S and Θ_{A-S}. The Z_S and Θ_S are said to be *nested* respectively in Z_A and Θ_A. Denoting the restricted model as $\hat{Y}_S = Z_S\hat{\Theta}_S$ in such a nested structure, the error $\hat{Y}_A - \hat{Y}_S$ is orthogonal to the estimate $\hat{Y}_S$ as illustrated in Fig. 1 and the parameter estimates $\hat{\Theta}_S$ and $\hat{\Theta}_{A-S}$ are uncorrelated.

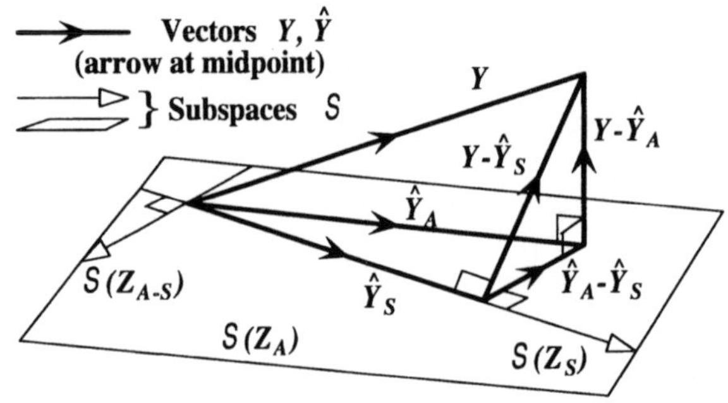

Figure 1: Nested Subspaces and Orthogonality Relationships.

In the case of estimating an ARX time series with Z random rather than a static regression, the above properties also hold asymptotically for large sample under appropriate assumptions (Lütkepohl, 1993).

Now, consider any finite dimensional multivariable Markov process with vector input u_t and output y_t of the form

$$x_{t+1} = \Phi x_t + Gu_t + w_t \tag{6}$$

$$y_t = Hx_t + Au_t + Bw_t + v_t \tag{7}$$

where x_t is a k-order Markov state and w_t and v_t are white noise processes that are independent with covariance matrices Q and R respectively. An alternative representation is the innovations form where the noise terms w_t and $Bw_t + v_t$ are replaced, respectively, with Kv_t and the output innovation v_t, where K is the Kalman gain obtained from solving the Riccati equation. The state expressed as $x_t = J_k^\infty p_t^\infty$ in terms of the infinite past p_t^∞ is

$$x_t = \sum_{i=1}^{\infty}(\Phi - KH)^{i-1}[(G - KA)u_{t-i} + Ky_{t-i}] \tag{8}$$

that results from recursively substituting (6) for x_t in (6). Eq. (8) is equivalent to (6) provided that $J_k^\infty(\Theta)$ is parameterized as in (6) and (7). By truncating, the approximation $x_t = J_kp_t$ is obtained. The approximation error decreases as $(\Phi - KH)^\ell$ that is exponential in the length ℓ of the past p_t so it can be ignored asymptotically. Since (7) with $x_t = J_kp_t$ is in the ARX form (2) with additional restrictions on the parameters, the Markov model (6) and (7) is nested within the ARX model class, asymptotically.

In the `adaptx` subspace algorithm, the fitting of the Markov model is done in two steps. First, a reduced-rank regression is done to estimate $\hat{J}_k$ of fixed rank in $x_t = \hat{J}_kp_t$ and with no parametric constraints on $\hat{J}_k$ so it is not parameterized as in (8). The reduced-rank regression is performed using a canonical variate analysis between past and future as developed in Larimore (1997a) for the case of no inputs. The case of inputs with feedback is developed in the next section. In the second step, the constraints are then introduced by regression using (6) and (7) with the state given by $x_t = \hat{J}_kp_t$. In particular, let X^+ denote X with the time index t replaced by $t+1$, and project $(X^+\ Y)$ on

$S(X\ U)$ to obtain $(\Phi\ G; H\ A)$. Implicit in this regression are additional constraints among the parameters that lead to the various state space canonical forms (Candy et al, 1979).

This can be viewed as a succession of restrictions on ML models starting with the ARX, then the reduced rank regression using CVA, and finally the state space regression using (6) and (7). The latter two involve nonlinear parameterizations, and the projection methods developed above are generalized in the next section to such nonlinear parameterizations.

4. SUBSPACE DETERMINATION BY CVA

In this section, the CVA procedure for the determination of the state space subspace is developed. The main difficulty in the case of feedback is to remove the effects of future inputs on future outputs. This is done using the estimated ARX model and a multistep form of the likelihood function.

The future $f_t = (y_t^T, \ldots, y_{t+\ell}^T)^T$ of the process is related to the past p_t through the state x_t and the future inputs $q_t = (u_t^T, \ldots, u_{t+\ell}^T)^T$ in the form

$$f_t = \Psi^T x_t + \Omega^T q_t + e_t \tag{9}$$

where x_t lies in some fixed subspace of p_t, $\Psi^T = (H; H\Phi; \ldots; H\Phi^{\ell-1})$ and the i, j-th block of Ω is $H\Phi^{j-i}G$. The presence of the future inputs q_t creates a major problem in determining the state space subspace from the observed past and future. If the term $\Omega^T q_t$ could be removed from the above equation, then the state space subspace could be estimated accurately. The method used in the adaptx algorithm is to fit an ARX model and compute an estimate $\hat{\Psi}$ of Ψ based on the estimated ARX parameters. Note that an ARX process can be expressed in state space form with state $x_t = p_t$ so that the above expressions for Ω and Ψ in terms of the state space model can be used as well for the ARX model. Then the ARX state space parameters (Φ, G, H, A) and Ψ and Ω are themselves functions of the ARX model parameters $\hat{\Theta}_A$.

In Larimore (1997a), the determination of the state space subspace by CVA was developed from basic principles and the maximum likelihood method with a rank constraint. This leads to a CVA between the past p_t and $f_t - \Omega^T q_t$, the future outputs with the effects of future inputs removed. To justify the CVA for a correlated time series, a multistep ahead form of the likelihood function was derived. The log likelihood function of the outputs $Y_{\ell+1}^N$ conditional on the inputs U_1^N and the past $p_{\ell+1}$ at time $\ell + 1$ is of the form

$$\log p(Y_{\ell+1}^N | p_{\ell+1}, Q, \Theta, \Sigma) = \frac{1}{\ell} \sum_{t=\ell+1}^{N} \log p((f_t - \Omega^T(\Theta) q_t) | p_t, \Theta, \Sigma)) \tag{10}$$

To be exact, a maximized likelihood function should be satisfied by the ML estimates $\hat{\Theta}_S$ for the SS model. But of course these estimates are not available for computing $\Omega(\hat{\Theta}_S)$ at the point of trying to determine the state space subspace, so the corrected future cannot be computed.

Now consider removing the effect of future inputs from future outputs using the ARX parameter estimates $\hat{\Theta}_A$. The first point is that the multistep likelihood function (10) is asymptotically equivalent to the one-step prediction likelihood (see Bauer, 2004). Thus parameter estimates from the two are asymptotically equivalent and can be used interchangeably. The estimates of the ARX parameters Θ_A do not involve the estimates of the covariance matrix Σ_A so Θ_A can be estimated separately.

Consider the model (9) of the multistep output f_t with $\Psi(\Theta)$ and $\Omega(\Theta)$ nonlinear in the parameters Θ that appear in the multistep likelihood function (10). The notation $F = (f_{\ell+1}, \ldots, f_{N-\ell})^T$ is used where each column of F is one component of the future vector f_t. Then for an estimate $\hat{\Theta}$, the model for F is

$$\hat{F} = X\Psi(\hat{\Theta}) + Q\Omega(\hat{\Theta}) = W\Pi(\hat{\Theta}) \tag{11}$$

with $\Pi(\hat{\Theta}) = (\Psi(\hat{\Theta})\ \Omega(\hat{\Theta}))$. As in the linear regression case, $\hat{F}$ lies in the subspace span by the columns of $W = (X\ Q)$.

Suppose the Θ_A and Θ_S are two nested models with $\Theta_S \subset \Theta_A$. As in the linear regression case, consider the refinement of the ML parameter estimate $\hat{\Theta}_A$ by restriction to the model $\hat{\Theta}_S$ that is assumed to be true. The orthogonality condition

$$F(\hat{\Theta}_S)[F(\hat{\Theta}_A) - F(\hat{\Theta}_S)] = 0$$

holds asymptotically for large sample in the nonlinear case. This is proven using results from Magnus and Neudecker (1988) where it is shown that the first differential of the likelihood function leads to the above orthogonality condition in the case of univariate nonlinear regression, and is easily extended to multivariate nonlinear regression. Thus as in the linear regression case, the error $F(\hat{\Theta}_A) - F(\hat{\Theta}_S)$ in the estimate is orthogonal to the estimate $F(\hat{\Theta}_S)$ asymptotically for large sample. For any two columns $\hat{\Pi}_i = \Pi_i(\hat{\Theta})$ and $\hat{\Pi}_j = \Pi_j(\hat{\Theta})$ and where σ_{ij} applies to e_t of (9), it can be shown that asymptotically

$$\sigma_{ij}(W^T W)^{-1} = \mathrm{Cov}(\hat{\Pi}_i, \hat{\Pi}_j)) = \frac{\partial \Pi_i}{\partial \Theta^T} \mathrm{Cov}(\hat{\Theta}, \hat{\Theta}) \frac{\partial \Pi_j^T}{\partial \Theta} \tag{12}$$

In the last section, the parameter space Θ_A is reparameterized by Θ_S and the difference Θ_{A-S} that are uncorrelated. Let $\Pi(\Theta_S)$ or $\Pi(\Theta_{A-S})$ denote Π as a function with Θ restricted to the respective subspaces Θ_S and Θ_{A-S}, and let W_S and W_{A-S} be the corresponding subspaces of W. Using (12), it can be shown that asymptotically W_S and W_{A-S} are orthogonal if and only if $\hat{\Theta}_A$ and $\hat{\Theta}_{A-S}$ are uncorrelated.

If $\Omega(\hat{\Theta}_S)$ was known, then a canonical variate analysis between $f_t - \Omega(\hat{\Theta}_S)^T q_t$ and the past p_t would estimate the state space subspace. If the ARX estimate $\hat{\Theta}_A$ is used instead of $\hat{\Theta}_S$, then from (9)

$$f_t - \Omega(\hat{\Theta}_A)^T q_t = \Psi(\hat{\Theta}_S)^T x_t - [\Omega(\hat{\Theta}_A) - \Omega(\hat{\Theta}_S)]^T q_t$$

$$+[\Psi(\Theta_S) - \Psi(\hat{\Theta}_S)]^T x_t + [\Omega(\Theta_S) - \Omega(\hat{\Theta}_S)]^T q_t + v_t$$

where the first term is the projection on the ML estimated state $\hat{x}_t$ and the second is error between the ML ARX and SS estimates in removing future input effects, with the first term orthogonal to the second. Remaining terms are due to ML estimation errors that are minimum and the error v_t. From the above, it can be shown that the CVA using $f_t - \Omega(\hat{\Theta}_A)^T q_t$, the future corrected with the ARX rather than the SS ML parameter estimates, and the past p_t to determine the state space subspace introduces no bias or added variance asymptotically. This reduces the case of inputs and feedback to the no inputs case, that achieves the ML lower bound asymptotically (Bauer, 2004).

REFERENCES

Anderson, T.W. (1984), *An Introduction to Multivariate Statistical Analysis*, New York: Wiley.

Bauer, D. (1998). *Some asymptotic theory for the estimation of linear systems using maximum likelihood methods or subspace algorithms*. Ph.D. thesis, Technischen Universitat Wien, Austria

Bauer, D. (2004) "Comparing the CCA subspace method to pseudo maximum likelihood methods in the case of no exogenous inputs", submitted to *J. Time Series Analysis*.

Candy, J.V., Bullock, T.E., and Warren, M.E. (1979), "Invariant Description of the Stochastic Realization," *Automatica*, Vol. 15, pp. 493-5.

Conner, J.S., D.E. Seborg, and W.E. Larimore (2004), "A Theoretical Analysis of the DeltaAIC Statistic for Optimal Detection of Small Changes", to appear *2004 American Control Conference*, to be held June 30 - July 2, Boston MA.

Cox, D.R., and D.V. Hinkley (1974), *Theoretical Statistics*, New York: Chapman and Hall.

Deistler, M., K. Peternell and W. Scherrer (1995), "Consistency and Relative Efficiency of Subspace Methods," *Automatica*, Vol. 31, pp. 1865-1875.

Gustavsson, G., L. Ljung, and T. Soderström (1977), "Identification of Processes in Closed Loop – Identifiability and Accuracy Aspects," *Auotmatica*, Vol. 13, No. 1, pp. 59-75.

Juricek, B.C., D.E. Seborg, and W.E. Larimore (2004), "Fault Detection Using Canonical Variate Analysis," *Ind. Eng. Chem. Res.*, Vol. 43, pp. 458-474.

Larimore, W.E. (1983). "System Identification, Reduced-Order Filtering and Modeling Via Canonical Variate Analysis", *Proc. 1983 American Control Conference*, H.S. Rao and T. Dorato, Eds., pp. 445-51. New York: IEEE.

Larimore, W.E., S. Mahmood and R.K. Mehra (1984), "Multivariable Adaptive Model Algorithmic Control," *Proc. Conference on Decision and Control*, Eds. A.H. Haddad and M. Polis, Vol. 2, pp. 675-80. Held December 12-14, 1984, Las Vegas, NV. New York: IEEE.

Larimore, W.E. (1990a), "Canonical Variate Analysis for System Identification, Filtering, and Adaptive Control, " *Proc. 29th IEEE Conference on Decision and Control*, Honolulu, Hawaii, December, Vol. 1, pp. 635-9.

Larimore, W.E. (1990b), "Order-Recursive Factorization of the Pseudoinverse of a Covariance Matrix", *IEEE Trans. of Automatic Control*, Vol. 35, pp. 1299-1303.

Larimore, W.E. (1992), *ADAPT$_X$ Automated System Identification Software Users Manual*, Adaptics, Inc, 40 Fairchild Drive, Reading, MA 01867.

Larimore, W.E. (1993), "Accuracy Confidence Bands Including the Bias of Model Under-fitting," *Proc. 1993 American Control Conference*, San Francisco, CA, June 2-4, 1993, Vol. 2, pp. 1995-9.

Larimore, W.E. (1996a), "Statistical Optimality and Canonical Variate Analysis System Identification," *Signal Processing*, Vol. 52, pp. 131-144.

Larimore, W.E. (1996b), "Optimal Order Selection and Efficiency of Canonical Variate Analysis System Identification," *Proc. 13th IFAC World Congress*, Vol. I, San Francisco, July 1-5, 1996, Vol. I, pp. 151-156.

Larimore, W.E. (1997a), "Optimal Reduced Rank Modeling, Prediction, Monitoring, and Control Using Canonical Variate Analysis," *IFAC Internat. Symp. on Advanced Control of Chemical Processes*, held June 9-11, Banff, Canada, pp. 61-6.

Larimore, W.E. (1997b), "System Identification of Feedback and 'Causality' Structure using Canonical Variate Analysis," *Preprints 11th IFAC Symposium on system Identification*, held July 8-11, 1997, Fukuoka, Japan, Vol. 3, pp. 1101-6.

Larimore, W.E. (2002). Reply to 'Comment on 'Order-recursive factorization of the pseudoinverse of a covariance matrix' '. *IEEE Trans. Automat. Contr.*, Vol. 47, pp. 1953-7.

Larimore, W.E. (2003). "Inferring Multivariable Delay and Seasonal Structure for Subspace Modeling". *Preprints 13th IFAC Symposium on System Identification*, held August 27-29, 2003, Rotterdam, Netherlands.

Ljung, L and T. McKelvey (1996), "Subspace identification from closed loop data", *Signal Processing*, Vol. 52, pp. 209-215.

Lütkepohl, H. (1993), *Introduction to Multiple Time Series Analysis*, Second Edition, New York: Springer-Verlag.

Magnus, J.R., and H. Neudecker (1988), *Matrix Differential Calculus with Applications in Statistics and Econometrics*, New York: Wiley.

Peternell, K., W. Scherrer, and M. Deistler (1996), "Statistical analysis of novel subspace identification methods", *Signal Processing*, Vol. 52, pp. 161-177.

Schaffé, H. (1959), *The Analysis of Variance*, New York: Wiley.

Shi, R., Ph.D thesis, McMaster University, ON, Canada, 2001

Shi, R., and J.F. MacGregor (2001), "A Framework for Subspace Identification Methods," *Proc. American Control Conference*, Arlington, VA, June 25-7, pp. 3678-83.

Wang, Y., D.E. Seborg and Larimore (1996), "Process Monitoring based on Canonical Variate Analysis," *IFAC 1997 Int. Symp. on Advanced Control of Chemical Processes* held June 9-11, 1997, Banff, Canada, pp. 523-8.

ELSEVIER
IFAC
PUBLICATIONS
www.elsevier.com/locate/ifac

IDENTIFICATION FOR DECENTRALIZED MPC

Gudi, R.D.[1] , Rawlings, J.B. [*2], Venkat, A[2]. and Jabbar, N.[3]

[1]Department of Chemical Engineering, Indian Institute of Technology, Bombay, Mumbai 400 076, India.

[*]Department of Chemical and Biological Engineering, University of Wisconsin, Madison 53705, USA.

[3] Jordan Institute of Science and Technology, Irbid 22110, Jordan.

Abstract: This paper addresses the problem of identifying interaction dynamics that exist between units operating in a decentralized control scheme. Identification of such interaction relationships is crucial to the deployment of coordinated decentralized control schemes. The proposed methodology is based on (i) the use of partial correlation analysis to identify the interacting channels in closed loop, and (ii) closed loop identification of the concerned dynamics using data obtained from suitable dithering. Alternative identification schemes relevant for this scenario are briefly analyzed in this paper and validation studies on a representative system is presented.. *Copyright © 2004 IFAC*

Keywords: Closed loop identification, interaction estimation, partial correlation analysis.

1. INTRODUCTION

Centralized control schemes that are based on a complete description of the cause and effect relationships are known to yield optimized control performance for multivariable systems. However, for optimization and control of large-scale systems, partitioning and decentralized control schemes have been eminently recommended over centralized approaches (Siljak, 1996). This choice is due to a number of factors. Firstly, the computational intensity for control is known to grow quire rapidly than size. Therefore, it is practical to design and implement control schemes for smaller sub-systems by looking at their local cause and effect relationships and the effect of interactions from other subsystems. The second and perhaps more important limitation for centralized control of large-scale systems is from a modeling and identification perspective. It is widely recognized that identification of cause-effect relationships in large-scale systems is a relatively difficult task. While such relationships are identified easily at a local level, for example at a unit level in a chemical or power generation plant, the interaction between such levels is usually associated with a lot of uncertainty. Often times, such interactions are also not perceivable during direct modeling, but manifest when the local control loops are closed. Identification of such interaction dynamics is a critical requirement for implementing coordinated decentralized schemes, which are known to yield closed loop performance that approaches that of a centralized control scheme.

As indicated earlier, the interacting dynamics are often not easily characterized, unless all available control inputs for the large scale system are perturbed for identification. Therefore, such dynamics are best identified under closed loop conditions, wherein the large-scale system is partitioned into smaller sub-systems with local controllers. Having identified these interactions, co-ordinated and decentralized control can then be achieved with these local controllers by implementing a higher-level co-ordinator or a peer-level communication (Mesarovic *et al.* 1970, Katebi and Johnson (1997)). These co-ordination mechanisms thus require knowledge of the interaction to achieve the desired level of co-ordination. This interaction thus needs to be identified under controlled conditions using closed-loop identification methodologies.

Closed-loop identification strategies have been extensively proposed in the literature and an excellent review of the state of the art can be found in Forssell and Ljung (1999). The primary motivation in these strategies has been towards identifying direct models between the inputs and outputs using closed loop data, when for example, the plant is open loop unstable or there are inherent feedback mechanisms implemented for safety reasons. Another important reason for identifying such direct models has been towards obtaining reduced order models with a view to achieve better control(Gevers and Ljung, 1986). Broadly, three different approaches to closed loop identification of the direct dynamics, viz (i) the direct, (ii) the indirect and (iii) the joint input-output methods, have been proposed. These methods differ in terms of apriori knowledge assumed about the nature of the controller and the assumptions made regarding the noise models. The relative merits of these strategies, in terms of consistency of estimates and the applicability of these methods (depending on the accuracy of the noise models) are discussed in Ljung (1999), Forssell and Ljung (1999).

The identification problem considered in this paper involves the characterization of the interaction dynamics between decentralized control loops. The challenges encountered are somewhat similar to what one would expect for identification of the direct plant in closed loop, discussed in the earlier paragraph. However, some key differences exist. Firstly, as is the case with closed loop identification, lack of informative data for identification is a key problem. This problem is overcome in regular closed loop identification, via the use of a dither signal applied either at the controller output or at the setpoint. However, for the case of identification of interaction involving several decentralized loops, dithering of each of the controller outputs is not practical. A more systematic method to prune down this set of variables is necessary. Secondly, in regular closed loop identification, the direct method (i.e. ignoring the presence of feedback) has been recommended in view of its optimal statistical properties. For the case of interaction identification, the direct method is not applicable because (as will be shown later) the inter-relationships between any two interacting decentralized loops also involves the individual loop sensitivities. Thus, one has to resort to other closed loop identification methods to first explicitly estimate the individual loop properties and then factor them out. Finally, a recognized drawback of the indirect method of closed loop identification is that it requires knowledge of the feedback controller mechanisms. In the identification problem considered in this paper, each of the individual decentralized controllers is multivariable and perhaps constrained in nature, and therefore the use of the indirect method is precluded. Also, in view of the multivariable nature of the concerned loops and the interacting dynamics, the use of subspace based state-space identification methods (Verdult and Verhaegen (2002)) greatly facilitates the identification procedure.

This paper analyzes the above mentioned problems encountered in identifying the interaction dynamics in large scale systems, with a view to achieving co-ordinated decentralized control. The approach proposed here to identify the interaction is based on using closed loop data. As mentioned before, dithering all the control inputs regardless of whether they contribute to interaction is perhaps not practical. The interaction between loops in large scale systems is generally sparse. Therefore, towards isolating the channels that contribute to interaction and minimize the number of control inputs that need to be dithered, we propose the use of partial correlation analysis. Having dithered a smaller set of relevant manipulative variables and generated sufficiently rich closed loop data, the problem of identifying the interacting dynamics needs to be addressed. Towards this end, we present and analyze three methods for identification of the interacting dynamics from the viewpoints of apriori knowledge necessary as well as their applicability in the constrained controller case. The proposed identification strategies have been validated on representative examples taken from the literature.

2. PROBLEM DEFINITION

The identification problem that we seek to address in the paper is shown in Figure 1. For purposes of explanation, we consider the case of two decentralized loops and seek to identify the interacting dynamics $\mathbf{G_{dj}(q)}$ between them.

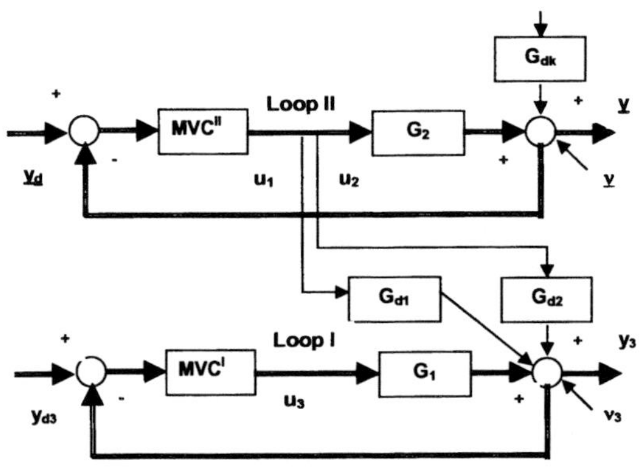

Figure 1: Schematic of interacting multivariable controllers

Each of the individual loop outputs is assumed to be affected by noise and unmeasured disturbances **v**. Each of the individual controllers are assumed to be multivariable in nature. For purposes of explanation again, we assume that the controllers involved with loop I and loop II are multivariable and of size n x n. These controllers are designed based on the local dynamics $\mathbf{G_1(q)}$ and $\mathbf{G_2(q)}$, which are assumed to be known. We further assume that the interaction dynamics $\mathbf{G_d}$ is sparse. This is a realistic assumption because if the structure was full, one could deploy a centralized (rather than decentralized) scheme that is based on the complete enumeration of all the cause and effect relationships. In general, no other apriori knowledge is assumed about the interactions and the channels in which they could exist. We seek to estimate the interacting dynamics $\mathbf{G_d(q)}$ under closed loop conditions. The intent is to then use knowledge of the interacting dynamics in co-ordinated control schemes so as to achieve centralized performance but using decentralized control structures as shown above.

In general, closed loop data is known to be less informative from an identification viewpoint. Hence a dither signal **d** at the controller output is commonly employed. Figure 2 shows the block diagram of a single decentralized loop with the introduction of the dither signal

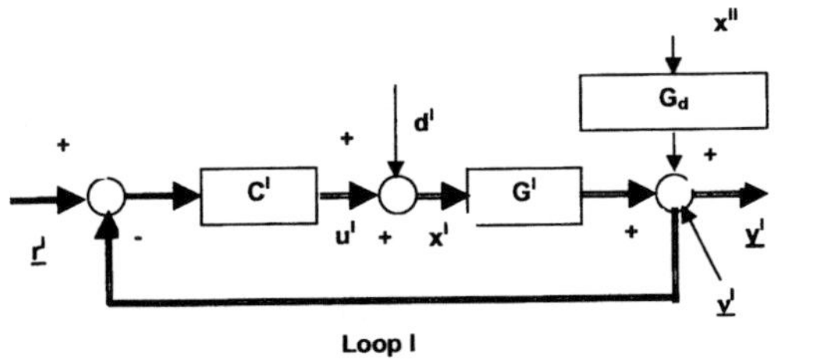

Figure 2: Signals associated with a single loop.

3. PARTIAL CORRELATION ANALYSIS

Consider the block diagram shown in Figure 2. During normal plant operation, the dither signal $\mathbf{d}^I$ is zero and therefore the signals in the vectors **x** and **u** are the same. Routine operating data is collected on each of the variables in **u** and **y** . We seek to use this routine operating data to assess the existence of interaction, i.e. we test the correlation between any signal p in $\mathbf{x}^I$ and a set of signals q, s and t (say) in $\mathbf{x}^{II}$. Since we would like to test this correlation in a dynamic sense, (i.e. the cause and effect relationships are dynamic in nature, rather than static, with at least a single lag due to zero order hold), we consider a dynamic map of these signals. Let Y denote the vector of measured observations of p and X denote the matrix of measured and lagged observations of signals q,s and t, with n being the number of lags in each of the latter signals. In other words, the kth row of Y would then consist of $p(k)$ and that of X would contain [$q(k\text{-}1)$ ... $q(k\text{-}n)$, $s(k\text{-}1)$, ...$s(k\text{-}n)$, $t(k\text{-}1)$...$t(k\text{-}n)$]. A regular correlation analysis based on the evaluation of the Spearman correlation coefficient (Draper and Smith(1981), would essentially look at a matrix Z=[Y X] and evaluate the correlation coefficients in terms of the covariance of Z . These coefficients would then be tested for significance using the t-test. However, due to the possibility of the variables in X being correlated, we propose to use the partial correlation analysis. The partial correlation coefficient analysis can be used to suppress the effect of correlation and clearly discriminate the influence of correlated causal variables in X on the effect variable in Y.

Let us assume that the X matrix has dependent columns and therefore a reduced rank. Consider the problem of assessing the correlation between any column variable x_i with the variable Y. Let matrix X_R contain all variable x_j ($j \neq i$). We first find a vector e_{x_i} that contains the negation of the effect of all other variables on x_i.. To do this, we first consider the regression problem between x_i and X_R. Assuming a linear relationship, the regression can be written as,

$$x_i \quad X_R\theta_1 \quad E_1 \quad (1)$$

This regression problem can and can be solved for θ_1 using the simple Moore-Penrose inverse as,

$$\theta_1 = (X_R{}^T X_R)^{-1} X_R{}^T x_i \quad (2)$$

We now evaluate the prediction errors for the above model as,

$$e_{x_i} = x_i - X_R(X_R{}^T X_R)^{-1} X_R{}^T x_i \quad (3)$$

In the next step, the influence of X_R on Y is again estimated using a regression model. The relevant equations can be written as,

$$Y = X_R\theta_2 + E_2 \quad (4)$$

from which θ_2 can be estimated as,

$$\theta_2 = (X_R{}^T X_R)^{-1} X_R{}^T Y \quad (5)$$

and the prediction errors for y can be estimated as,

$$e_y = y - X_R(X_R{}^T X_R)^{-1} X_R{}^T Y \quad (6)$$

In the above equations, e_{x_i} and e_y can be assumed to consist of those components of x_1 and Y that are independent of X_R. These quantities can be used to check for the partial correlation of x_1 with y. This partial correlation can therefore be written as a regular correlation between e_{x_i} and e_y as,

$$pr_{x_i,y} = \frac{e_{x_i}^T e_y}{(m-1)\sigma_{e_{x_i}}\sigma_{e_y}} \quad (7)$$

It can be seen that, if X is full rank, i.e. if x_i is completely uncorrelated with all columns in X_R; then the predictions in Equations 3 will be white and so e_{x_i} will be equal to x_1 itself. Similarly, e_y will contain only the effect of x_1 on y (i.e. the effect of X_R on y will be suppressed). The partial correlation can thus give an assessment of the existence of a linear dependence between the cause and effect variables in the presence of correlated cause variables.

Remark 1: It must be mentioned since we use lagged variables in X, the resulting indices would indicate the existence of a linear, dynamic relationship between the signals in $\mathbf{x}^{II}$ (i.e. q, s and t) and p in $\mathbf{x}^I$. For the variables amongst q,s and t that are correlated with p, one would obtain symmetric blocks of non-zero correlation indices of size n, which would indicate the evidence of a non-zero interacting dynamics in $\mathbf{G_d}$ for the assumed pair of signals.

Remark 2: It is possible that the reduced signals e_{x_i} and e_y might also reflect the effect of unmeasured disturbances that routinely affect the loop. These effects would corrupt the interpretation from the above analysis. However, if there is any evidence of such disturbances affecting the loop, then the regression problems posed in Equation (1) and (4) may be solved using more sophisticated methods such as the PEM or the instrumental variables (IV) method, rather than the simple Moore-Penrose method as indicated above.

Remark 3: If the matrix X contains perfectly correlated variables (*i.e.* if the signals q, s and t are perfectly correlated), the above method would still not be able to discriminate between signals that have a dependent relationship from those that do not. However, perfect correlation is difficult to exist/ obtain in practice. Even an uncorrelatedness in the measurement noise associated with the signals q, s and t could be adequate to discriminate between these signals.

4. CLOSED LOOP IDENTIFICATION

Having isolated the channels in which interaction exists, we now focus on the problem of identifying the interacting dynamics. For purposes of explanation we consider that only one signal p in $\mathbf{x}^{II}$ affects the loop I (see Figure 2) through some disturbance dynamics, which needs to be estimated using closed loop identification methods. It is now assumed that the signal p can be perturbed as a dither signal at the output of controller II (MVC_2) (see Figure 1), in a way that is uncorrelated with loop I set-point changes. As well, through regular target specifications by the upper LP layer in a typical MPC setting (Shah *et al.* (2002)), it is assumed that sufficient excitation exists at the set-point for Loop I.

For loop I, the relationship between the two independent, deterministic signals p and r^I and the dependent signals can be written $\mathbf{x}^I$ and $\mathbf{y}^I$ as,

$$x^I = (I + C^I G^I)^{-1} C^I r^I + (I + C^I G^I)^{-1} C^I G_d p + (I + C^I G^I)^{-1} d^I \quad (8)$$

$$y^I = (I + G^I C^I)^{-1} G^I C^I r^I + (I + G^I C^I)^{-1} G_d p + (I + G^I C^I)^{-1} G^I d^I \quad (9)$$

In the above equations, the effect of stochastic and unknown noise elements can also be added in the above equation but are omitted for brevity. Also, it is important to note that the loop sensitivities associated with the loop are different for the general multivariable case and are related by the equation,

$$(I + C^I G^I)^{-1} C^I = C^I (I + G^I C^I)^{-1} \quad (10)$$

As mentioned before, in regular closed loop identification using the direct method, the effect of feedback is ignored and the signals $\mathbf{y}^I$ and $\mathbf{u}^I$ are used to estimate the transfer function $\mathbf{G}^I$.using Equation 9. In the above problem, we seek to estimate G_d and in this case the direct method of closed loop identification is not really an option. This is because the direct method based on the signals p and y (or $\mathbf{x}^I$) will only yield the product of G_d and one of the loop sensitivities; it is therefore necessary to explicitly estimate and factor out the sensitivity dynamics to generate an estimate of G_d. In earlier approaches (Van den Hof and Schrama (1993), Huang and Shah (1997)), an alternative two step, indirect method of closed loop identification of $\mathbf{G}^I$ was proposed with some of its theoretical properties being reviewed subsequently by Gevers *et al.* (2001). Here we propose to use both single step as well as two step methods for the identification of the interacting dynamics G_d assuming that the plant dynamics $\mathbf{G}^I$ is known. This latter assumption is valid considering that cause and effect relationships can more easily be characterized in smaller decomposed sub-systems. The methods for closed loop identification can be presented as follows:

Method 1: *Single Step Method*: Consider that the closed loop system is simultaneously perturbed both at the set-point $\mathbf{r}^I$ and at the output of C^{II} for signal p in an uncorrelated fashion. The dither input d^I is assumed to be zero in Figure 2. Using Equation (8), it can be seen that this results in a multi-input single output, open-loop identification problem in the variables $\mathbf{r}^I$ and p as inputs and $\mathbf{x}^I$ as the output. This identification problem can be solved in the presence of stochastic disturbances affecting loop I, using the prediction error method (PEM) to generate unbiased estimates of the dynamics $G_{cl}=(I+C^IG^I)^{-1}C^I$ and the $G_{d,cl}=(I+C^IG^I)^{-1}C^IG_d$. Having obtained an estimate of these dynamics, the interacting dynamics can be obtained as,

$$G_d = (G_{cl})^{-1} G_{d,cl} \quad (11)$$

An important aspect that must be recognized towards realization of G_d as above is that Loops I and II are multivariable in nature and could involve a good number of manipulated and controlled variables. Towards ease of parameterization of the dynamics in G_{cl} and $G_{d,cl}$ during identification, it is more suitable to use appropriate order state space models, identify them using suitable sub-space methods and then realize the transfer function representations of G_d, if necessary as indicated by Equation 11 above. However, as will be seen in the next section, inverses are also easily realizable in state space forms.

Method 2: *Two step method*
Equation 8 is again the starting point for the two step method. Again considering that the dither d^I is zero, routine plant data involving the variables r^I and x^I can be collected. This routine plant data would contain the effect of interactions from Loop II through signal p as well as effect of other stochastic disturbances. In the first step of the two step identification method, the dynamics G_{cl} can be estimated by considering the signals r^I and x^I and by using the output-error method (OEM) which is a variant of the PEM method. The OEM is a special case of the PEM method with noise parameterized as unity, and it has been shown (Soderstrom and Stoica (1989)) that (i) if the true dynamics belongs to the model set, and (ii) the driving signals are orthogonal to the noise, unbiased estimates of the concerned dynamics can be obtained. Thus an unbiased estimate of the dynamics in state space form of G_{cl} can be obtained in the first step, using a suitable sub-space method. Some theoretical considerations related to analysis of bias in the estimates of G_{cl} in the presence of interacting dynamics G_d is presented in Gudi and Rawlings (2003).

In the second step, the dither at the set-point r^I is set to zero and the signal p is dithered. The signals x^I and

p can now be used to estimate G_d. Towards this end, a new signal x^I_f is generated as a filtered version of x^I as given by,

$$x^I{}_f = \{(I + C^I G^I)^{-1} C^I\}^{-1} x^I \quad (12)$$

This filtered signal x^I_f is then used along with the measurements of p to generate an estimate of the interacting dynamics G_d using the OEM or the PEM methodology. A critical step in the above filtering is the realization of the inverse of $G_{cl}=(I+C^I G^I)^{-1} C^I$ as required in Equation 12. In this context as well, parameterization of state space models for capturing the dynamics is more desirable, as state space representations are relatively easier to invert than transfer function representations. It can be shown that the relationship between the state space matrices A,B,C,D of any forward model are related to the state space matrices Φ,Γ, C_i and D_i of the inverse model as given by the relationships,

$$\Phi = A - BD^{-1}C; \Gamma = BD^{-1}$$
$$C_i = -D^{-1}C; D_i = D^{-1}$$

It is also possible to show that for the identification of the loop sensitivities, the inverse of D always exists and so these transformations are realizable.

Method 3: *Direct Method: Constrained controller case*

In the presence of constraints, it is well-known (Zafiriou (1990)) that the controller dynamics and hence the loop sensitivities are no longer linear. In such a case, it is inappropriate to use the methods proposed in the earlier sub-sections, which inherently are based on the assumption of linearity. In such a case, it is important to exploit the fact that the plant dynamics $\mathbf{G}^I$ is well characterized. Defining the prediction error, $\varepsilon = y^I - G^I_m x^I$, where $\mathbf{G}_m{}^I$ is the model of the plant $\mathbf{G}^I$, one can write the closed loop relationships between ε and the independent deterministic loop inputs as,

$$\varepsilon = (S_0 G^I C^I - G_m{}^I R_0{}^i) r + (S_0 + G_m{}^I R_0{}^i) G_d p \quad (13)$$

In the above equation, $S_0 = (I + G^I C^I)^{-1}$, $S_0^i = (I + C^I G^I)^{-1}$ and $R_0{}^i = (I + C^I G^I)^{-1} C^I$ and it can be shown that for $\mathbf{G}_m{}^I = \mathbf{G}^I$, the deterministic relationship between ε and p can be written as,

$$\varepsilon = G_d p \quad (14)$$

Thus, in the presence of stochastic disturbances affecting at the output, a prediction error method based on the measured signals ε and p can be used to estimate the dynamics G_d.

A relative assessment of the above three methods of identifying the interacting dynamics can be presented as follows: In the presence of a perfect model $\mathbf{G}_m{}^I$, one would always prefer Method 3 as it is a single step, direct method that accommodates the most general constrained controller case. It does not require any other knowledge of the controller mechanisms or the loop sensitivities and as such is a parallel counterpart to the Direct Identification scheme proposed for identifying direct models in closed loop. However, it is based on the key assumption of a perfect plant model. If this assumption cannot be justified, one would then choose amongst the first two methods. The merits of the first method is that it is a one step method that just uses closed loop data to simultaneously identify the concerned dynamics in a MISO framework. However, the drawback is that sufficiently strong and uncorrelated dither has to be applied simultaneously at both the set-point and at the manipulated variables, to generate unbiased models. This may pose problems in terms of the acceptance of this strategy at the plant level. The second method on the other hand is a two step method which involves the identification of the loop sensitivities in the first step. This identification may be done without strong dithering at the set-point. Regular set-point updating by the upper LP layers could be sufficient to provide the necessary excitation. The identification could be performed using data collected over extended periods of time; for the closed loop case, availability of such data is not a constraint. After this first step, regular dithering of the manipulated variables of the other controllers over a short period of time is adequate to generate the excitation necessary of identification of the interacting dynamics. Thus, the second method would be preferable when dithering at the setpoint is not permissible.

5. CASE STUDY

We consider the following 3 x 3 transfer function and choose to partition the system for decentralized control as shown below:

$$G(s) = \begin{bmatrix} \frac{12.8}{16.7s+1} & \frac{-18.9}{21s+1} & \frac{-1}{3.3s+1} \\ \frac{6.6}{10.9s+1} & \frac{-19.4}{14.4s+1} & \frac{2}{5s+1} \\ \frac{-1.333*K_1}{3.33s+1} & \frac{-0.875*K_2}{1.25s+1} & \frac{21*K_3}{10s+1} \end{bmatrix}$$

The input-ouput pair $\{y_1,y_2 - u_1,u_2\}$ forms a decentralized sub-system which is controlled by two PI controllers (K_c=0.47 and τ_I = 3.26 for y_1-u_1 pairing and K_c=-0.09 and τ_I = 9.35 for the y_2-u_2 pairing) with decouplers. The pair y_3-u_3 forms another decentralized sub-system that is also regulated by a PI controller with K_c=0.09 and τ_I = 10. The two decentralized loops were assumed to be corrupted at the output with random noise having variance of 0.1. It was proposed to first evaluate the partial correlation analysis for identifying the interactions for zero and non-zero values of the parameter K_i (see the y_i-u_3 dynamics). For closed loop identification, the set-points and the controller outputs were dithered using random signals having variance 0.3. These were designed to be uncorrelated whenever necessary, as for example in Method 1. For method 2, simple step like set-point changes were introduced in the first step to estimate the sensitivities. For method 3, the controller outputs were clipped using an appropriate saturation block, to simulate constrained controller action. The canonical variate analysis (CVA) based sub-space method was used for the identification of the state space models using closed loop data. The model order for the state space models were selected based on the Akaike Information Criteria and the models identified were cross-validated using closed loop data.

Isolating interacting channels
The partial correlation analysis was used to assess the existence of interaction between the two sub-systems. Towards this end, K_2 was deliberately set to zero in the plant and K_1 and K_3 were set to 1. Closed loop data for u_3 and u_1, u_2 was used along with partial correlation analysis to infer the presence of interaction. A total of 10 sets of data consisting of 1000 points (sampling time of one time unit) was used to perform both the regular and partial correlation analysis.

It was found that the Spearman correlation coefficient based on the regular correlation analysis was not able to differentiate the two channels u_1-u_3 (interaction exists) and u_2-u_3 (absence of interaction) in the closed loop for a majority of the sets. This was obviously because both u_1 and u_2 were correlated amongst themselves. In fact, the Spearman correlation coefficient for the pair u_1 and u_2 was consistently above 0.8 for most of the sets indicating a strong correlation between these variables. However, partial correlation analysis was able to remove the correlation between these variables and consistently assess the underlying closed loop correlation between u_3 and u_1,u_2.

Estimating interacting dynamics
Both the single step and two step methods were able to accurately identify the interacting dynamics between u_1 and y_3. For method 2, the slower set-point excitation in the first step did require a larger data set to provide consistent estimates of the sensitivities. As mentioned before, this requirement of a larger data set is not a drawback in closed loop. However, in both the methods, the identified loop sensitivities as well as the interaction dynamics matched quite closely with the true values. The direct method (Method 3) was a simple open loop identification problem because of the assumption of a perfect model. The interaction model between u_1 and y_3 was also accurately estimated as expected. However, in terms of applicability this requirement of a perfect model is a constraint.

6. CONCLUSIONS

The focus of this paper was on the identification of interacting dynamics in large decentralized multivariable controllers. Partial correlation analysis was proposed to isolate the interacting channels and three methods of closed loop identification was proposed. A representative case study was presented to highlight the practicality of the proposed methodology.

REFERENCES

Siljak, D.D, "Decentralized Control and Computations: Status and Prospects", Annual Reviews in Control, **20**, 131-142, (1996)

Mesarovic, M.D., Macko D. and Takahara, "Two coordination principles and their application in large scale systems control", Automatica, **6**, 2647-2473, (1999).

Katebi, M.R and Johnson, M.A., " Predictive Control Design for Large Scale Systems", Automatica, **33** (3), pp. 421-425, (1997).

Forssell, U. and Ljung, L., " Closed Loop Identification revisited", Automatica, 35(7), pp. 1215-1241, (1999).

Gevers, M. and Ljung, L. "Optimal Experiment Designs with respect to intended model application", Automatica,5, 543-554, (1986).

Ljung. L., " System Identification: Theory for the user", Prentice Hall, (1999).

Verdult, V. and Verhaegen, M., Subspace identification of multivariable linear parameter-varying systems, Automatica, 38(5), pp. 805-814, (2002).

Draper, N. R., and Smith, H. "Applied Regression Analysis," 2nd ed., New York, John Wiley and Sons, 1981.

Shah, S.L., Patwardhan, R. and Huang, B. "Multivariable Controller Performance Analysis" , CPC – VI, Proceedings of sixth international conference on Chemical Process Control, Rawlings, J.B., Ogunnaike B.A., Eaton, J.W.(*Eds*)., AIChE symposium series, Vol. 98, 2002.

Van den Hof, P.M.J, and Schrama, R.J.P. " An indirect method for transfer function estimation using closed loop data", Automatica, 29(6), 1523-1527 (1993).

Huang, B. and Shah, S.L., "Closed Loop Identification – A two step approach", Journal of Process Control, 7, pp. 425-438, (1997).

Gevers M., Ljung, L and Van den Hof, P., " Asymptotic variance expressions for closed loop identification", Automatica, **37**, 781-786, (2001).

Soderstrom, T. and Stoica, P., "System Identification", Prentice-Hall, (1989).

Gudi, R.D. and Rawlings, J.B, " Issues in identification of interacting dynamics in large decentralized multivariable systems", (In preparation) (2003).

ELSEVIER
IFAC
PUBLICATIONS
www.elsevier.com/locate/ifac

MODEL-BASED AUTOTUNING SYSTEM USING ANN AND RELAY FEEDBACK TEST

Hsiao-Ping Huang [1] **Jyh-Cheng Jeng**
Feng-Yi Lin

Department of Chemical Engineering
National Taiwan University
Taipei 10617, Taiwan

Abstract: A model-based autotuning system is presented for PI/PID controllers. A conventional relay feedback test is used to generate the response for estimations of parameters such as: process gain, ultimate gain, ultimate frequency, and apparent deadtime. Two ANNs are built based on the data generated from standard SOPDT processes to enhance the estimation of apparent deadtime. Upon having the estimated results, the dynamics of the process is classified into one of two groups. In one group, PI/PID controller will be tuned based on the FOPDT parameterization. In the other group, only PID controller will be considered for dynamic control, and the controller will be tuned according to the SOPDT parameterization. The tuning rules are given in terms of ultimate gain, ultimate frequency, and the apparent deadtime. Since the tuning rules are derived from model-based control, they are superior to the Z-N rules in performance. *Copyright © 2004 IFAC*

Keywords: relay feedback, autotuning, ANN, FOPDT, SOPDT

1. INTRODUCTION

In 1984, Åström and Hägglund (1984) presented a relay feedback system to generate sustained oscillation as an alternative to the conventional continuous cycling technique for controller tuning. This relay feedback test was soon referred as autotune variation (ATV) (Luyben, 1987). As shown in Fig. 1(a) is the block diagram of the ATV loop. Figure 1(b) illustrates the typical response curves from the ATV system. The test provides ultimate gain (k_{cu}) and ultimate period (P_u) of the following to apply Ziegler-Nichols method for controller tuning.

$$k_{cu} = \frac{4h}{\pi A}\ ; \qquad \omega_u = \frac{2\pi}{P_u} \tag{1}$$

Controller tuning via the above mentioned ATV test is attractive, because it is operated under closed-loop and no *a priori* knowledge of system is need. But, the control performance thus obtained is, in general, inferior to those tuned by model-based, such as IMC and other related methods. This fact is most obvious when the processes have significantly underdamped and second-order dynamics. To employ model-based tuning, an adequate and reduced order dynamic model (usually, FOPDT or SOPDT) is required. In literature, there are reported efforts (Luyben, 1987; Li *et al.*, 1991; Chang *et al.*, 1992; Wang *et al.*, 1997; Huang *et al.*, 2000; Kaya and Atherton, 2001; Huang and Jeng, 2003) trying to develop such models from the ATV tests. Nevertheless, according to those reports in literature, two major difficulties are encountered. One of difficulties is that estimation of ultimate gain and frequency according to Eq.(1) is subject to errors, sometimes as high as 20 percent for ultimate gain (Li *et al.*, 1991). The second difficulty is due to the fact that one simple ATV test, in general, does not provide sufficient data for identifying a parametric model aforementioned. Improvements to give better accuracy and efficiency by using saturation relay (Shen *et al.*, 1996) or by reducing high-order harmonic terms using the Fourier analysis (Lee *et al.*, 1995; Sung and Lee, 1997; Wang *et al.*, 1997) have also been reported. But, these does not alter the encountered second difficulty above mentioned. To overcome the second difficulty, more than one ATV test is usually required (Li *et al.*, 1991; Scali *et al.*, 1999). In order to have the superior performance that the model-based controller used to have, while having the simple experiment like the conventional autotuning via ATV, a new autotuning method is presented.

[1] Corresponding author. E-mail:huanghpc@ntu.edu.tw

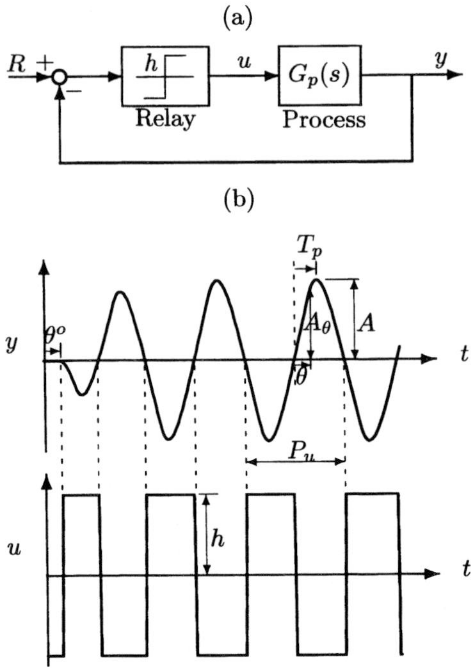

Fig. 1. (a) Block diagram of a relay feedback system (b) Response curves in relay feedback test

In the following, it is assumed that zero offset is a common specification of control. For zero offset, in general, PI or PID controller is considered for control. Usage of either type of the controllers mentioned depends not only on the application occasions, but also on their dynamic characteristics. Thus, an autotuning system should provide the best flexibility to cope with the application demand and the dynamics. For this purpose, model-based tuning rules including PI and PID controllers are considered, and, classification based on the dynamics of the processes into two groups for controller tuning is presented. The first group of processes are suitable for either PI or PID controllers for dynamic compensations, and, the tuning formula are derived based on an FOPDT model. The second group of processes, on the other hand, are considered good only for PID controllers which are derived based on an underdamped SOPDT model. Identification for classification of these two groups is thus presented. Since the apparent deadtime of process is crucial for applying the proposed classification, a novel approach enhanced by artificial neural network for estimation of the apparent deadtime is presented. The autotuning procedure starts with an ATV test. The steady-state gain, k_p, and k_{cu} are estimated along with the experiment until constant cycling occurs. The constant cycles at the output are then used to estimate the apparent deadtime, and the process is classified for tuning. After classifying to either of the two groups, model-based tuning rules are assigned and the controller parameters are computed accordingly. Several examples are used to illustrate this proposed method.

2. MODEL-BASED CONTROLLERS DESIGN

In general, models with dynamics up to second order plus deadtime are adequate for designing PI/PID controllers. For higher order processes, dynamics are usually represented by reduced order models of the following:

- FOPDT $\quad G_p(s) = \dfrac{k_p\, e^{-\theta s}}{\tau s + 1} \qquad (2)$
- SOPDT $\quad G_p(s) = \dfrac{k_p\, e^{-\theta s}}{\tau^2 s^2 + 2\tau\zeta s + 1} \qquad (3)$

2.1 Tuning PI/PID based on FOPDT model

A direct synthesis approach is used to synthesize PI/PID controller using an FOPDT parameterization. In case of demanding for a PI controller, the controller is devised so as to compensate the process dynamics to have a loop transfer function (LTF) of the following standard form:

$$G_{loop}(s) = \frac{k_o}{\theta}\frac{e^{-\theta s}}{s} \qquad (4)$$

In those cases that demand for a PID controller, LTF is chosen to be compensated the process to become:

$$G_{loop}(s) = \frac{k_o}{\theta}\frac{(1 + a\theta s)}{s\,(1 + \tau_f s)}\; e^{-\theta s} \qquad (5)$$

The use of loop transfer function in the form of Eq.(4) or Eq.(5) has been explained in a late work of Huang and Jeng (2002). According to the standard forms, parameters of the PI/PID controllers can be derived as following:

$$k_c^{'} = \frac{k_o\,\tau}{k_p\,\theta}\;;\quad \tau_R^{'} = \tau\;;\quad \tau_D^{'} = a\,\theta \qquad (6)$$

Notice that the PID controller used here is the actual, or series, form. In this tuning, k_o and a are parameters for performance or robustness specification, and the filter time constant, τ_f, is taken arbitrarily small (e.g. $0.05\tau_D^{'}$). The defaulted values of k_o and a are suggested as 0.65 and 0.4, respectively, for the PID controller. On the other hand, the defaulted values of k_o for the PI controller are 0.5. These defaulted values will provide about 3 and 60° as gain and phase margin of the system, respectively.

Using ultimate gain and ultimate frequency, the tuning formula of Eq.(6) that are originally in terms of the FOPDT parameterization can be re-written as:

$$\begin{aligned} k_c^{'} &= \frac{k_o\sqrt{K_u^{\,2} - 1}}{k_p[\pi - \tan^{-1}(\sqrt{K_u^{\,2} - 1})]} \\ \tau_R^{'} &= \frac{\sqrt{K_u^{\,2} - 1}}{\omega_u} \\ \tau_D^{'} &= a\left[\frac{\pi - \tan^{-1}(\sqrt{K_u^{\,2} - 1})}{\omega_u}\right] \end{aligned} \qquad (7)$$

where $K_u = k_p k_{cu}$.

2.2 Tuning PID based on SOPDT model

The PID controller in terms of SOPDT parameterization is derived in the similar way to give a loop transfer function of Eq.(4). The resulting controller parameters are given as:

$$k_c = \frac{k_o\ (2\tau\zeta)}{k_p\ \theta}\ ;\quad \tau_R = 2\tau\zeta\ ;\quad \tau_D = \frac{\tau}{2\zeta} \tag{8}$$

In this case, the controller is an ideal PID. As has been mentioned, the defaulted value of k_o is taken as 0.5. As a result, in terms of the ultimate gain and ultimate frequency, the PID parameters of Eq.(8) can be written as:

$$\begin{aligned} k_c &= \frac{k_o\ K_u \sin(\omega_u \theta)}{k_p\ \omega_u\ \theta} \\ \tau_R &= \frac{K_u \sin(\omega_u \theta)}{\omega_u} \\ \tau_D &= \frac{1 + K_u \cos(\omega_u \theta)}{\omega_u K_u \sin(\omega_u \theta)} \end{aligned} \tag{9}$$

If k_{cu} and ω_u of the given process are obtained from ATV test, the parameters of the PI or PID controllers can be computed based on either Eq.(7) or Eq.(9). However, as one can seen in Eq.(9), the tuning formula need to know the values of k_p and the apparent deadtime θ in advance. The estimation for k_p and θ will be given in the next section.

3. PROCESS IDENTIFICATION USING ATV TEST

To apply the model-based controllers derived in the previous section, reduced order dynamic models parameterized as FOPDT of Eq.(2), and, as SOPDT of Eq.(3) are required. An ATV response is thus used to develop such parametric models. As shown in Fig. 1(a) is a ATV system which consists of process, G_p, and a relay controller. The relay controller provides output at $+h$ or $-h$ only as on-off control. The controlled process output consists of transient oscillations after a pure deadtime, θ^o, and develops constant cycling with magnitude, A, and period, P_u. Notice that A and P_u are used in the autotuning system of Åström and Hägglund (1984) to apply Z-N method to compute the PI/PID controller parameters. In the meantime, T_p, θ and an associated height, A_θ, within one of the constant cycles are also indicated in Fig. 1(b). Here, θ is used to designate an apparent deadtime in the FOPDT or the SOPDT model, which is used to represent the dynamics of higher order processes. All these quantities measured from an ATV response are governed by the dynamic model and the relay. In other words, they can be expressed as:

$$\frac{A}{k_p h} = f_1(\bar{\tau}) \quad \text{or} \quad f_1(\bar{\tau}, \zeta) \tag{10}$$

$$\frac{P_u}{\theta} = f_2(\bar{\tau}) \quad \text{or} \quad f_2(\bar{\tau}, \zeta) \tag{11}$$

$$\frac{A_\theta}{A} = f_3(\bar{\tau}) \quad \text{or} \quad f_3(\bar{\tau}, \zeta) \tag{12}$$

$$\frac{\theta}{T_p} = f_4(\bar{\tau}) \quad \text{or} \quad f_4(\bar{\tau}, \zeta) \tag{13}$$

where $\bar{\tau}$ represents τ/θ. The equations given above describe the dynamic features of the ATV response in time domain. Two equations which correspond to Eqs.(10) and (11) derived from frequency domain are:

$$\omega_u \theta + \tan^{-1}\left(\frac{2\ \omega_u \tau \zeta}{1 - {\omega_u}^2 \tau^2}\right) = \pi \tag{14}$$

$$\frac{k_p}{\sqrt{(1 - {\omega_u}^2 \tau^2)^2 + 4\ {\omega_u}^2 \tau^2 \zeta^2}} = \frac{1}{k_{cu}} \tag{15}$$

where ω_u is taken as $2\pi/P_u$.

Theoretically, provided that k_p and k_{cu} are given, the identification problem to find an reduced order SOPDT model can be solved by finding τ, ζ, and θ that fit Eqs.(10)-(13) or Eqs.(12)-(15) in the sense of least-squares. The explicit functional forms for Eqs.(10) -(13) are not available, and numerical method to solve the above equations will not be convenient. Since estimation of apparent deadtime θ as well as k_p and k_{cu} is required for this proposed autotuning method, in the following, a simplified algorithm will be presented to estimate these required data from ATV test.

3.1 Estimation of k_p and k_{cu}

To estimate k_p, the experimental ATV test is started with a temporal disturbance to either the set-point or the process input (i.e. u) for a short period of time and restored back to its origin. The disturbance introduced has two main purposes. One is to initialize the relay feedback control, the other one is to generate data for computing the steady-state process gain, k_p. The estimation is made along the ATV test in one run as the following.

Let y^I and u^I designate the integrations of y and u from the very beginning of the experiment in one run. That is:

$$y^I(t) = \int_0^t y(\tau)\ d\tau\ ;\quad u^I(t) = \int_0^t u(\tau)\ d\tau \tag{16}$$

For some $t > T$ when y in an ATV test starts to oscillate with constant period and amplitude, y^I and u^I will have similar cycling responses. Let y^I_{av} and u^I_{av} designate the average heights of constant cycles of y^I and u^I, respectively. The value of k_p can be estimated as:

$$k_p = \frac{y^I_{av}}{u^I_{av}} \tag{17}$$

On the other hand, due to the use of describing function for estimation, the ultimate gain computed from

Eq.(1) is subjected to error, which may, sometimes, be as high as 20%. For tuning purpose, estimation of the ultimate gain k_{cu} with improved accuracy is desirable. Since $u(t)$ and $y(t)$ are periodic with period P_u, they can be expanded into Fourier series. If the first harmonics are extracted, their coefficients give one point of process frequency response at ultimate frequency ω_u via the following equation (Wang *et al.*, 1997):

$$G_p(j\omega_u) = \frac{\int_{t_0}^{t_0+P_u} y(t)\, e^{-j\omega_u t}\, dt}{\int_{t_0}^{t_0+P_u} u(t)\, e^{-j\omega_u t}\, dt} \tag{18}$$

where t_0 is taken as any time instant in a constant cycle. Then, the ultimate gain can be computed exactly as:

$$k_{cu} = \frac{1}{|G_p(j\omega_u)|} \tag{19}$$

3.2 Estimation of apparent deadtime

The apparent deadtime is the deadtime appearing in an FOPDT or SOPDT model that approximates best the higher order process. As a result, this apparent deadtime differs, in general, from its true deadtime which is designated as θ^o and can be detected at the very beginning of the test. Features in the cycling response can be used to distinguish the FOPDT from SOPDT dynamics. For example, in case of FOPDT process, T_p equals θ^o or θ. The same equality does not apply to the SOPDT case. In an ATV test, two measured quantities, A_θ/A and θ/T_p, are used to characterize the effect of the apparent deadtime. These two quantities, as mentioned earlier, are functions of $\bar{\tau}$ and ζ. To explore their functional relations, simulations of ATV tests on the standard SOPDT processes covering wide range of $\bar{\tau}$ (in dimensionless form) and ζ are carried out. Results of A_θ/A and θ/T_p for underdamped SOPDT processes are plotted as a graph as shown in Fig. 2(a), using $\bar{\tau}$ and ζ as parameters. Each pair of the two measured quantities corresponding to a point in the graph, where a specific pair of values for $\bar{\tau}$ and ζ can be found. As shown in the figure, it is difficult to read $\bar{\tau}$ and ζ from the figure. In order to make each curve in Fig. 2(a) more readable, the coordinate is rotated through the following transformation.

$$\begin{aligned} X &= \frac{\theta}{T_p}\cos(\pi/3) - \frac{A_\theta}{A}\sin(\pi/3) \\ Y &= \frac{\theta}{T_p}\sin(\pi/3) + \frac{A_\theta}{A}\cos(\pi/3) \end{aligned} \tag{20}$$

The results are plotted in Fig. 2(b). Moreover, for applying these data efficiently, two artificial neural networks (ANN) are constructed. The network architecture is as shown in Fig. 3. These two neural networks are fed with X and Y, and compute $\bar{\tau}$ and ζ, respectively. Each network consists of feedforward net with one input layer, one hidden layer, and one

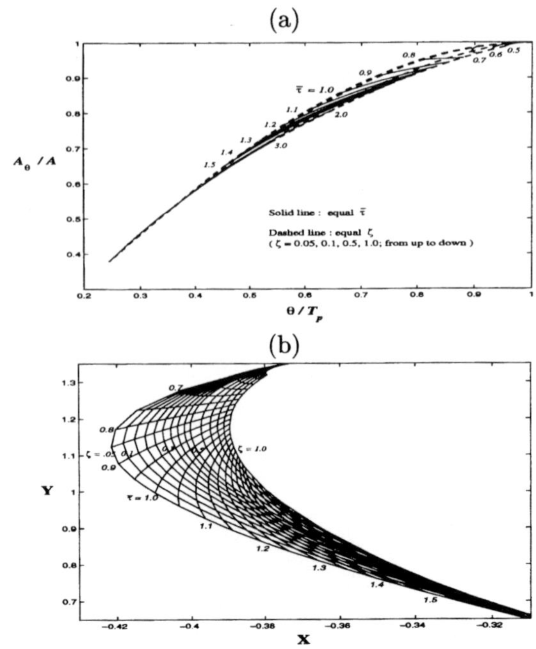

Fig. 2. Results of A_θ/A and θ/T_p in the ATV test for SOPDT processes (a) original (b) transformed

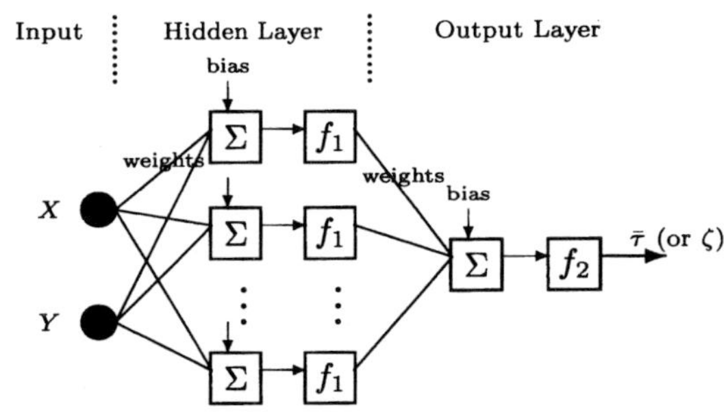

Fig. 3. Neural network architecture

output layer. The sigmoid function f is used in an error backpropagation technique to minimize the error between prediction and target values.

With these two networks, estimation of the apparent deadtime can be proceeded. The estimation makes uses of the phase criterion in Eq.(14). Notice that three unknowns are required to satisfy at least four functional relations (i.e. Eqs.(10)-(13) or Eqs.(12)-(15)), and iterative check is thus necessary to find the solution in a least-squares sense. By solving Eqs.(12)-(14), a unique solution for τ, ζ, and θ can be found. But, the resulting solution may not necessarily satisfy Eq.(15). The final solution needs further iterative procedures. To satisfy the extra Eq.(15), it is found that manifold in the space of τ and ζ results from the relation of the following:

$$\tan^{-1}\left(\frac{2\omega_u\tau\zeta}{1-{\omega_u}^2\tau^2}\right) = \pi - \omega_u\theta \tag{21}$$

With ω_u and θ being fixed, there are many pairs of τ and ζ that satisfy Eq.(14), and one of these pairs would make Eq.(15) satisfied. However, in this proposed autotuning system, only the apparent deadtime

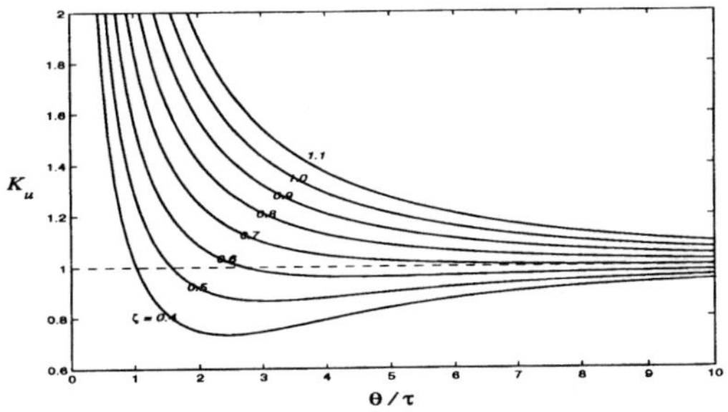

Fig. 4. K_u for different SOPDT processes

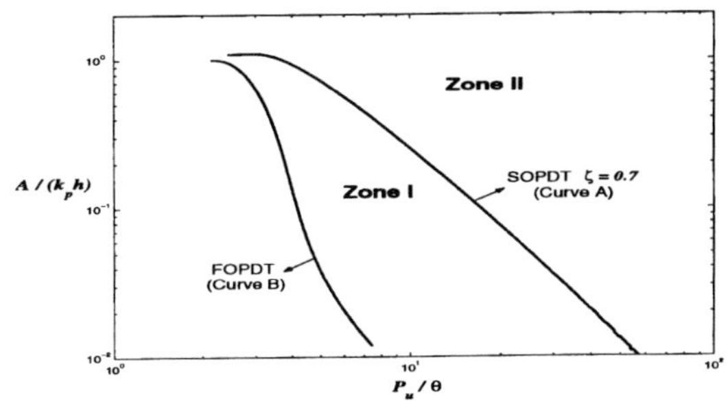

Fig. 5. Curves for process classification

of the process is required, estimation of the exact values of τ and ζ can be skipped except a complete process model is desired.

With the theory presented above, the algorithm for the estimation of apparent deadtime can be made in a simpler way as the following:

1) Starting from a guessed value of θ, which is initially taken as θ^o.
2) The values of X and Y are calculated and fed into two networks. Then, parameters $\bar{\tau}$ and ζ are computed, and check if Eq.(14) holds. The procedure proceeds iteratively by increasing the guess value of θ until θ equals T_p.
3) When, at certain value of guessed θ, Eq.(14) holds true, the resulting guess value θ is taken as the estimated value of apparent deadtime.
4) If, until the guessed value of θ exceeds T_p and no candidate solution is found, the underdamped SOPDT model is not good for describing the dynamics of the process. As a result, model of FOPDT or overdamped SOPDT should be used.

3.3 Classification for controller tuning

In general, processes with SOPDT dynamics, which are overdamped or slightly underdamped, are sometimes identified with FOPDT models for controller design without significant difference in performance. It is thus curious to know under what condition can an SOPDT process has controller parameters in terms of an FOPDT parameterization. The result in Eq.(7) indicates that it is necessary with $K_u > 1$. As shown in Fig. 4, K_u of SOPDT processes is plotted against θ/τ using ζ as a parameter. It is found that $K_u > 1$ happens when $\zeta \geq 0.7$. Thus, the uses of Eq.(7) and of Eq.(9) for controller tuning are discriminated using $\zeta = 0.7$ as a boundary. As mentioned earlier, from the ATV test, it provides $A/(k_p h)$ and P_u/θ which are functions of $\bar{\tau}$ and ζ. As shown in Fig. 5, there are two curves that correspond to SOPDT processes with $\zeta = 0.7$ (curve A) and FOPDT processes (curve B). Curve A and curve B are found to be represented, respectively, by the following equations:

$$\Omega = 0.465\Lambda^3 - 2.002\Lambda^2 + 0.958\Lambda - 0.007 \qquad (22)$$

$$\Omega = 30.93\Lambda^3 - 56.78\Lambda^2 + 29.03\Lambda - 4.51 \qquad (23)$$

where $\Omega = \log(A/k_p h)$ and $\Lambda = \log(P_u/\theta)$. These two curves divide the graph of Fig. 5 into two zones. One is between the curve A and curve B (i.e. Zone I) that represents the feasible region for using parameterization of FOPDT due to $\zeta \geq 0.7$, and the other region above curve A (i.e. Zone II) that represents the feasible region to use parameterization of underdamped SOPDT. Thus, once k_p and the apparent deadtime θ are obtained, the normalized value of $A/(k_p h)$ and P_u/θ can be calculated. Then, compare the resulting value of $\log(A/k_p h)$ with the one calculated from Eq.(22). If the former is smaller, then this process is classified into the group (*Group I*) where Eq.(7) applies to tune PI/PID controllers. Otherwise, the process is classified into the other group (*Group II*) where Eq.(9) will be used for tuning PID controllers.

4. AUTOTUNING PROCEDURES WITH ILLUSTRATED EXAMPLES

To perform the autotuning, the relay feedback of the ATV test is initialized by a short period of manual disturbance. After the response has constant cycling, A and P_u are recorded, and the following procedure steps are taken:

1) Compute k_p from Eq.(17) and k_{cu} from Eq.(19).
2) Estimate the apparent deadtime θ.
3) Having the values of k_p and θ, the process is classified by comparing the value of $A/(k_p h)$ from ATV test with the one calculated by Eq.(22).
4) If the process belongs to *Group I*, Eq.(7) is used for tuning PI or PID controllers. Otherwise, Eq.(9) is used to tune a PID controller.

Notice that, in case of significantly overdamped dynamics, the trajectory of the $(\theta/T_p, A_\theta/A)$ will be crowded at the lower edge of the graph in Fig. 2(a). In that case, the process can be classified into *Group I* directly without estimating the apparent deadtime.

For practical application, the issue to be considered is the measurement noise. When the measured signals are blurred with noise, data used in the proposed method are taken as the average values of measurements from several constant cycles to ensure the success of this autotuning.

In order to illustrate the above autotuning procedures, a few examples are used for simulation. The results of applying proposed autotuning method are

Table 1. Results of autotuning for simulated processes

Example	Process	ATV test		Estimated parameters			Classification	Controller tuning		
		A	P_u	k_p	k_{cu}	θ		k_c'	τ_R'	τ_D'
1	$\frac{e^{-1.5s}}{(s+1)^5}$	0.72	12.14	1.01	1.81	2.91	*Group I*	0.46	2.96	1.66
2	$\frac{(0.5s+1)\ e^{-s}}{(s+1)^2(2s+1)}$	0.35	6.56	1.0	3.74	1.32	*Group I*	1.29	3.78	0.77
								k_c	τ_R	τ_D
3	$\frac{e^{-2s}}{(9s^2+2.4s+1)(s+1)}$	1.17	15.90	1.0	1.11	2.72	*Group II*	0.45	2.47	3.96

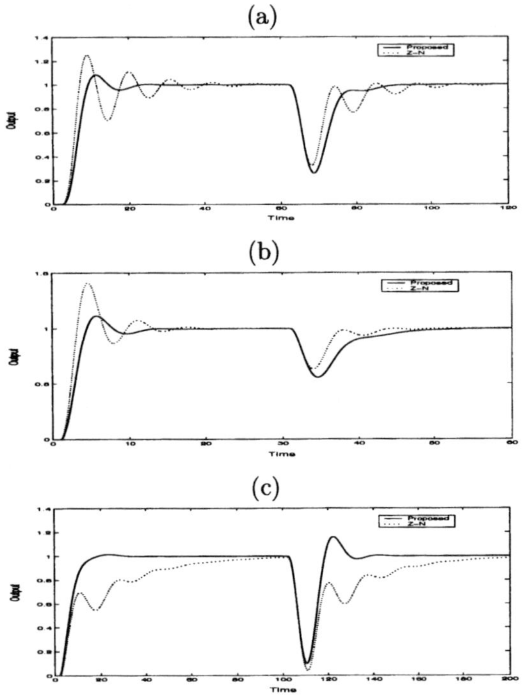

Fig. 6. Control responses of (a) Example 1 (b) Example 2 (c) Example 3

summarized in Table 1. The control results using proposed model-based method and Z-N method are shown in Fig. 6. Notice that the Z-N settings are obtained from the same set of experiment data. These results show that the proposed method can give more satisfactory control performance than conventional autotuning system.

5. CONCLUSIONS

In this paper, a systematic procedure is used to perform the autotuning of PI/PID controllers using one ATV experiment. The usage of PI or PID controller depends on the control application and on the dynamic characteristics of the process. The autotuning system considered the effective damping factor for classifying the process into one of two groups. In either case, besides parameters k_p and θ, the tuning formula are given in terms of ultimate gain and ultimate frequency obtained from a constant cycle. These parameters can be estimated within one ATV experiment. Two ANNs are constructed to enhance the estimation of apparent deadtime. The simulation results have shown that this proposed autotuning system is efficient and self-contained.

ACKNOWLEDGMENT

The authors thank R. C. Panda (CLRI/CSIR, India) for the help in constructing the ANNs.

This work is supported by the National Science Council of Taiwan under Grant no. NSC 92-2214-E-002-013.

REFERENCES

Åström, K. J. and T. Hägglund (1984). Automatic tuning of simple regulators with specifications on phase and amplitude margins. *Automatica* **20**, 645–651.

Chang, R. C., S. H. Shen and C. C. Yu (1992). Derivation of transfer function from relay feedback systems. *Ind. Eng. Chem. Res.* **31**, 855–860.

Huang, H. P. and J. C. Jeng (2002). Monitoring and assessment of control performance for single loop systems. *Ind. Eng. Chem. Res.* **41**, 1297–1309.

Huang, H. P. and J. C. Jeng (2003). Identification for monitoring and autotuning of pid controllers. *J. Chem. Eng. Japan* **36**, 284–296.

Huang, H. P., M. W. Lee and I. L. Chien (2000). Identification of transfer function models from the relay feedback test. *Chem. Eng. Comm.* **180**, 231–253.

Kaya, I. and D. P. Atherton (2001). Parameter estimation from relay autotuning with asymmetric limit cycle data. *J. Process Control* **11**, 429–439.

Lee, T. H., Q. G. Wang and K. K. Tan (1995). A modified relay-based technique for improved critical point in process control. *IEEE Trans. Cont. Sys. Tech.* **3**, 330–337.

Li, W., E. Eskinat and W. L. Luyben (1991). An improved autotune identification method. *Ind. Eng. Chem. Res.* **30**, 1530–1541.

Luyben, W. L. (1987). Derivation of transfer functions for highly nonlinear distillation columns. *Ind. Eng. Chem. Res.* **26**, 2490–2495.

Scali, C., G. Marchetti and D. Semino (1999). Relay with additional delay for identification and autotuning of completely unknown processes. *Ind. Eng. Chem. Res.* **38**, 1987–1997.

Shen, S. H., H. D. Yu and C. C. Yu (1996). Use of saturation-relay feedback method for autotune identification. *Chem. Eng. Sci.* **51**, 1187–1198.

Sung, S. W. and I. B. Lee (1997). Enhanced relay feedback method. *Ind. Eng. Chem. Res.* **36**, 5526–5530.

Wang, Q. G., C. C. Hang and B. Zou (1997). Low-order modeling from relay feedback. *Ind. Eng. Chem. Res.* **36**, 375–381.

ELSEVIER
IFAC
PUBLICATIONS
www.elsevier.com/locate/ifac

IDENTIFICATON OF ALBEBRAIC AND STATE SPACE MODELS USING GENETIC PROGRAMMING

Kyaw Tun and S. Lakshminarayanan*

Department of Chemical and Environmental Engineering
National University of Singapore, Singapore 117576
**Corresponding Author. Telephone: +65-68748484 Email: chels@nus.edu.sg*

Abstract: We describe the development of an automated modelling system that uses the genetic programming paradigm. The Genetic Modelling System (GeMS) is capable of generating appropriate algebraic (linear and nonlinear) and ODE systems from experimental data. The utility of GeMS is illustrated through several examples involving laboratory systems and simulated kinetic data. *Copyright © 2004 IFAC*

Keywords: System identification, Nonlinear models, State space models

1. INTRODUCTION

The present day chemical industry is characterized as "data rich and information poor" – there is a need for tools and methods that are capable of extracting useful process information from the terabytes of data that are archived in plant databases. A plethora of tools such as multivariate statistics, neural networks, Bayesian networks, self-organizing maps, CART (Classification and Regression Trees) etc. are available at present to convert data into useful process knowledge (models, rules etc.). A relatively recent addition to this family is the genetic programming (GP) that has its origins in genetic algorithms (GA) (Holland, 1975). GP is a biologically inspired, domain-independent method that genetically breeds populations (of computer programs, mathematical formulae etc.) to solve problems. The first runs of genetic programming were made in 1987 but the first detailed description was available in a textbook published several years later (Koza, 1992). When accompanied by robust parameter estimation methods, GP has the ability to determine the "optimal" structure of the model that "best" describes the given data set. By its very nature, GP also provides a family of comparable solutions from which the informed user can select a model that is commensurate with the physics/chemistry of the system as well as the intended end use. The GP method is computationally intensive but can come quite handy in situations where little or nothing is known about the structure of the process model.

GP has been applied in many areas ranging from engineering and finance to medicine. We primarily focus on applications in the field of process systems engineering here. McKay et al. (1997) apply GP to the steady state modelling of chemical processes and consider a tutorial example followed by applications to two typical processes - a vacuum distillation column and a chemical reactor system. Willis et al., (1997) extend the GP methodology to the development of dynamic input-output models and demonstrate its workability by the development of inferential estimation models for a vacuum distillation column and a twin screw cooking extruder. Marenbach (1998) use GP for the construction and refinement of models for a simulated bioprocess. This work employs block diagrams (such as those available in SIMULINK® under MATLAB®) for representing the models – this is in contrast to other works that are equation oriented. Greeff and Aldrich (1998) illustrate GP-based empirical modeling by means of three examples, two of which are based on data pertaining to leaching experiments. Lakshminarayanan et al. (2000) use a composite GP-PCA (genetic programming combined with a multivariate statistical technique called principal components analysis (PCA)) approach to generate nonlinear models for industrial product design applications. Gao and Loney (2001) combine GP with neural net to evolve a polymorphic neural network and apply it to predicting pH in a simulated CSTR. Grosman and Lewin (2002) employ GP for nonlinear model predictive control (NMPC). In their work, the nonlinear model predictive control uses predictions provided by the GP generated model and this is shown to improve the control performance on two multivariable simulated processes: a mixing tank and a Karr liquid-liquid extraction column.

In this paper, we wish to describe the features of a Genetic Modelling System (GeMS) that has been developed in our laboratory. GeMS uses the genetic programming paradigm to identify models ranging form algebraic to state space models from experimental data. The paper is organized as follows. In section 2, we provide a quick overview of GP based modelling and detail some of the innovative concepts implemented in GeMS. Section 3 provides some interesting simulation and experimental case studies of modelling using GeMS. One of the key features of GeMS is the automated assembly of a nonlinear ordinary differential equation (ODE) system to represent the behaviour of process systems using experimentally observed data. This will be demonstrated in Section 4. Conclusions and planned future work will be spelt out in the final section.

2. GENETIC PROGRAMMING

Since a lot of information is available in the literature about genetic operators etc. we will only point out the unique features of GeMS here.

2.1 Data structure for gene and chromosome

The most important thing is how the population data is stored. We need to store all the chromosomes from initial to final population; it is easy to see that the storage requirements can be very high. Whenever a new chromosome is created, it is compared with all the members that were ever "created" to ensure that unnecessary calculations are avoided. It is easy to understand that this comparison must be performed very fast.

GeMS uses a novel approach to store the genes and chromosomes – this provides fast access and makes it easily extensible. GeMS employs indexing to the gene library. It is similar to using pointers in C language. In contrast to using string to represent chromosomes, the use of indexing requires less memory to store and facilitates faster comparison and manipulation. Since MATLAB is optimized for matrix manipulations, the entire population is stored in a single three-dimensional matrix. The last (third) dimension denotes the chromosome number. Let X denote the whole population. The second chromosome or chromosome number 2 is referred as X(:,:,2) - this is a two dimensional matrix. The first row contains the index of the gene library and the remaining rows represent properties or states of gene of respective column. For a typical run having population size of 200, 20 generations and with 30 as the maximum chromosome size, the storage requirement is roughly 5 MB. The size is reduced by a factor of 3 to 5 when saved to disk using sparse matrix format.

The main reason behind this flexibility is the implementation of the representation scheme using a pointer or index to the gene library. Addition of a new gene to the gene library can be accomplished without hard coding of the program simply by pointing the gene functions file (m-file) to gene's evaluation function handle. Single information (a pointer) is not enough to fully represent a gene in most cases. Additional information such as number of parameters, number of input arguments, state properties (for ODE systems) and so on must also be stored along with the gene. To address these issues, a novel representation is used. A two-dimensional matrix represents the chromosome. Each column represents a gene. Usual genetic operators manipulate along the column of this matrix. Some special genetic operators such as the adaptation operator manipulate the third row or state of the gene.

The advantages of using a matrix to represent a chromosome are: (i) Flexibility, generality and ease of implementing novel genes (ii) it requires less memory to store the information. This is particularly important since GeMS stores all chromosomes that are generated. (iii) it permits fast searching and manipulation of chromosome and (iv) it allows easy implementation of genetic operators. The unique encoding of chromosomes allows us to handle different domains without altering the main GP algorithm.

GeMS also exploits any available *a priori* knowledge in order to limit the search space or to transform the model assembled by GP into more effective models. GeMS assembles models randomly, so sometimes it may assemble absurd model structures such as $\exp(\exp(\exp(u_1)))$. In such cases, we may limit how many operands one function may take. GeMS also determines the optimal parameter position in the model. For example, the model $k_1u_1 * k_2u_2$ is reduced to $k_1u_1u_2$. In order to accommodate varied *a priori* knowledge, a new concept called evolution policy is implemented. The user can define the evolution policy using keywords for action to be taken (inclusion, exclusion, parameter reduction, etc) for conditions such as repetition of a specific gene, existence of a specific gene, etc.

2.2 Parameter Estimation

GP is used only for structure generation according to the well-known and some newly developed (superposition crossover and adaptation) genetic principles. Proven global and local optimization methods combined with a user-defined rule-based parameter pruning technique are used for parameter estimation in GeMS.

A run of GP spends most of the time in parameter estimation than in the genetic operations itself. The model structure may be nonlinear and has several local minima. Choosing right optimizer is critical. Torni *et al.* (1999) mention that choosing the right optimizer is based on features of problem, which is postulated depending on 1) region of attraction of global minimum 2) cost of function evaluations 3) embedded or isolated global minimum and 4) number of local minima. GeMS uses stochastic "global" optimization methods such as simulated annealing, genetic algorithms and differential evolution in conjunction with local optimization techniques like the Gauss

Newton method to determine the optimum values of the parameters.

2.3 Fitness measure

Fitness is a numeric value assigned to each member of the population to provide a measure of the appropriateness of the individual as a solution to the problem in question. The definition of fitness measure has a direct and significant bearing on the resulting solution. Fitness measure is defined such that the most suited model will have the highest (or lowest) score whereas the poorest model will have the lowest (or highest) score and all other models lie in the continuum between these extremes.

The goal of having a fitness evaluation is to give continuous feedback to the evolutionary algorithm regarding which individuals should have a higher probability of being allowed to multiply and reproduce and which individuals should have a higher probability of being removed form the population. The fitness function is calculated on the validation data set or the combination of both training and validation data sets.

GeMS uses well-known model selection criteria employed for system identification such as Minimum Description Length (MDL) or Akaike's Information Criterion (AIC). Generally, such criteria include two terms: (a) a term that accounts for mismatch between the experimental data and model predictions and (b) penalty for number of parameters. The first term generally decreases with increasing model complexity and the second term increases with increasing model complexity.

GeMS also incorporates a modified fitness function that basically adds a heuristically defined complexity index to the MDL or AIC criterion. This strategy is used to generate and nurture less complex models and hence the resulting models turn out to be much simpler.

GeMS uses MATLAB ODE solvers such as RK45 and ODE15S to determine the model responses for each model generated by the GP and parameter estimation sub-modules.

3. CASE STUDIES

Several researchers have employed GP for the steady state and dynamic modeling of chemical process (e.g. McKay et al., 1997). The dynamic modeling is achieved by including the lagged values of the inputs and/or the output variables thereby converting the problem into an algebraic modeling exercise. Interestingly, even some simple algebraic systems cannot be identified if suitable transformations are not enforced on the dependent variables (y). Usually, the programs attempt to construct models of the form $y = f(\underline{x},\underline{k}) + \varepsilon$ where $f(\underline{x},\underline{k})$ is a functional form (to be identified) that involves the input variables $\underline{x}$ and the parameter vector $\underline{k}$. ε denotes the noise component. However, this form is not always adequate. We have therefore made provisions in GeMS to also search for implicit models of the form $g(y,x) = f(\underline{x},\underline{k}) + \varepsilon$ where $g(y,x)$ is a functional form that also needs to be identified along with $f(x,k)$. For such cases, a chromosome consists of two trees: the first tree for function f and last tree for function g. Standard genetic operations are then used to identify the model.

Example 1: Nonlinear Algebraic Equation

The data for this example was generated from the equation $x^2 + y^2 = 1$. The output 'y' was corrupted with IID noise. The [x, y] data was presented to both the explicit and implicit algebraic equation identification modules. All runs of the explicit algebraic identification resulted in the model y = -0.02, which is nothing but the mean value of y in the data set. Eight of the ten runs with the implicit model identification procedure identified the correct model as (y*y)=(1+(-0.992*(x*x))). The data, true model and the identified model are shown in Figure 1.

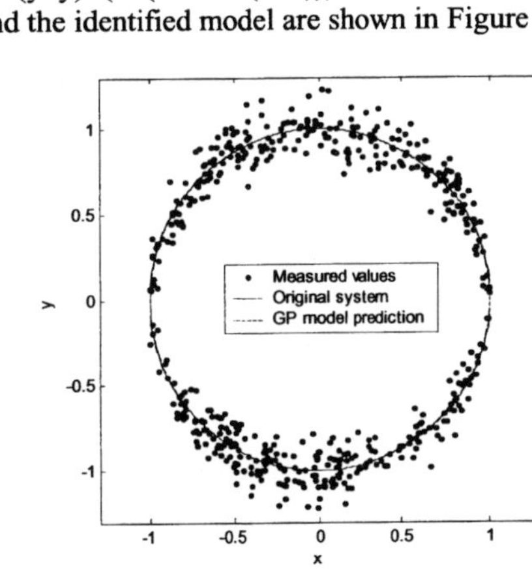

Fig 1. Data vs. model prediction for example 1

Example 2: Identification of a dynamic model for a laboratory process

An experimental process system (see schematic in Figure 2) was designed and installed recently in our research laboratory. This equipment facilitates experimental studies related to the dynamics and control of linear as well as nonlinear multivariable processes. The system is fitted with industrial grade sensors and valves; the data acquisition and control operations are performed using MATLAB/SIMULINK running on a personal computer. Industrial standard OPC communication is used. Besides such "remote control", the process can also be controlled using local devices. The sensors (thermocouples) are located such that the process dynamics and control studies can be conducted over a wide range of time delays.

The focus of our current application is the exit temperature of water in Tank 1. The inlet water that comes in at about 27°C is heated using an electric heater that can supply up to 9 kW power. The level of water in Tank 1 is kept constant by manipulating the inlet water flow into Tank 1. The water exiting from tank 1 flows to Tank 2 through a long-winded tube. The temperature of water in Tank 2 is the controlled variable (TT5; y); the input variables are the heating power applied to Tank 1 (HT1; u1) and the temperature of the water entering Tank 1 (TT8; u2). These measurements are available at 40 seconds interval. The objective, here, is to build a model that makes a one step ahead prediction of TT5 given past values of all the variables. Two sets of data are collected – one for model construction and the other for pure validation.

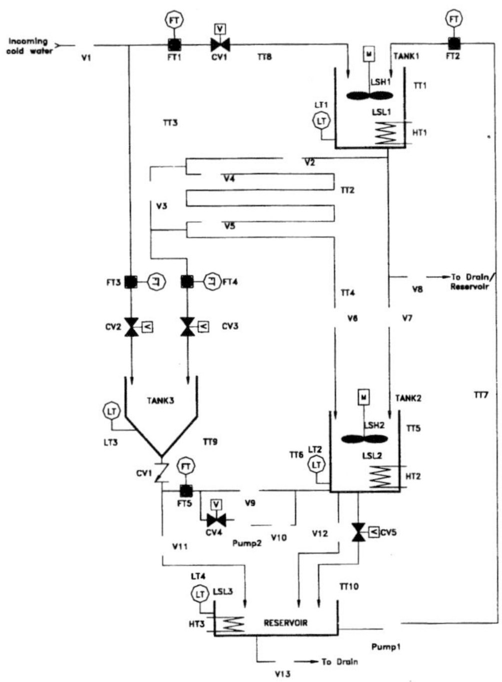

Fig 2. Experimental equipment setup (Example 2)

Ten runs were conducted. The best model obtained from GeMS is given below:

$$y_k = 26.39 + 0.711\, y_{k-1} - 2.11\, y_{k-2} + 0.1\, y_{k-1}\, y_{k-2} - 0.027\, y_{k-1}^2 - 0.00075\, y_{k-2}^2\, y_{k-1} + 0.0021\, u1_{k-3} + 0.0042\, u1_{k-4}$$

This model has a RMSE of 0.056 in the modelling data set and a RMSE of 0.066 during its "pure" validation on the test data set. The model validation is shown in Figure 3. The low RMSE value for the prediction set as well as the results shown in Figure 3 points to the fact that the process characteristic is captured well by the model reported above. Note that the variable u2 (temperature of the water entering Tank 1) does not feature in the model. This is due to the lack of excitation in u2 and is not due to the physical characteristics of the system. The GP procedure has demonstrated that it is able to distinguish between the important and unimportant variables.

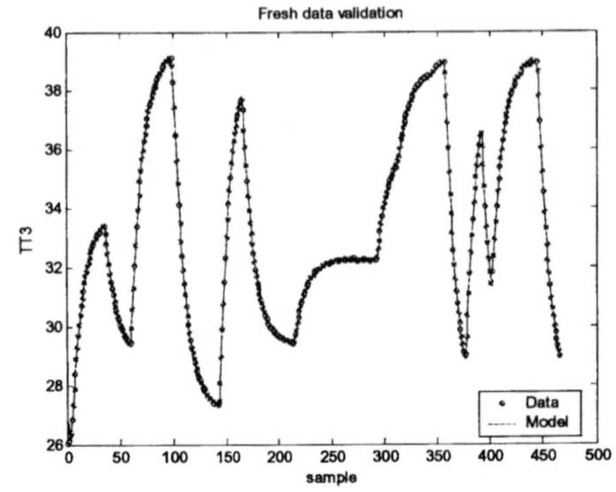

Fig 3. Model validation for Example 2

Example 3: Identification of a dynamic model for an experimental heat exchanger

For this case study, we consider the modelling of data obtained from an experimental heat exchanger system. The data set was provided by Dr. Eskinat and is also considered in the work by Eskinat *et al.* (1991) in the context of identification of a block oriented nonlinear model (hammerstein model: the model is characterized by a nonlinear static element followed by a linear dynamic element). The details of the experimental system can be found in the above reference. The nonlinearity in the system is caused by the presence of two distinct operating regions corresponding to the high and low process water flow rates. A dataset containing 334 input-output samples was available. Sixty-five percent of this data set was used to obtain the model using DACS-GP. The remaining data was used for model validation.

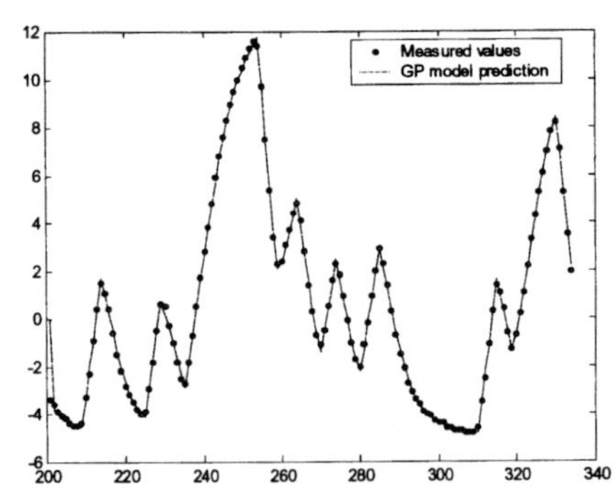

Fig 4. Model validation for Example 3

The best model obtained from GeMS is:

$$y_k = 0.84\, y_{k-1} - 6.17\, u_{k-1} + 7.55\, u_{k-1}^2 + u_{k-4}$$

The validation of the model on the test portion of the data is shown in Figure 4. The model provides an excellent approximation to the true system behavior. It is interesting to note that the identified model is very close to the hammerstein model (except for the u_{k-4}

term) structure. In fact, the above model provided a slightly better validation RMSE as compared to that obtained by methods (Eskinat *et al.* 1991; Lakshminarayanan *et al.* 1995) that assumed the hammerstein model structure and estimated its parameters from the given data.

4. IDENTIFICATION OF STATE SPACE MODELS

The genetic programming technique has been applied to a wide range of domain including finance, medicine and engineering. However, most of the research articles (at least in the engineering literature) focus on the identification of algebraic models – even dynamic models are converted to algebraic models through the incorporation of lagged variables (e.g. Rodriguez and Fleming, 1998; Hinchliffe and Willis, 2003). To the best of our knowledge, we have seen only one published work on the identification of an ordinary differential equation (ODE) system using the GP paradigm (Cao *et al.* 1999). Their models are shown to have good prediction capability. However, the ODE systems they generate are often found to be non-autonomous. The identification of ODE systems pose some interesting issues: firstly, the GP generated models might turn out to be very stiff and integrating them could pose run time problems (stalling). The choice of a suitable integrator becomes crucial; secondly, the computational cost can be exorbitant during the parameter estimation and fitness function evaluation steps. The parameter estimation for systems of known structure is itself a challenging task (Esposito and Floudas, 2000). In our case, the GP could come out with complex structures involving many parameters thereby making the parameter estimation even more difficult.

GeMS uses a multi-tree scheme to represent several simultaneous ODE's. It used one tree for each state equation. Apart from this modification in representation, no other changes in genetic operations are required. In this module, crossover is allowed across different states or trees.

The efficiency of GeMS in modeling batch data from three chemical reaction systems is demonstrated below. These examples have been taken from the parameter estimation literature in which the structure of the ODE system is assumed to be fixed and known.

Example 4: Cracking of gas oil

This model represents the catalytic cracking of gas oil (A) to gasoline (Q) and other side products (S). Only the concentrations of A and Q were measured; therefore, the concentration of S does not appear in the model for estimation. Tjoa and Biegler (1991) estimated the parameters of a proposed differential equation system model using experimental data from this system. Owing to space constraints, we do not report the data here – the reader is referred to Table 23 of Esposito and Floudas (2000) for the data set. Here, we assume no knowledge of the chemistry or the reaction mechanism. We only provide GeMS with the measurements of the two states A and Q. We specify that the terminal gene should include state variable z_1, z_2 and constant. The arithmetic operators that were included are '+', '–', '*'and '/'.

The best models from the ten independent runs conducted are summarized in the Table 1.

Table 1: Best models of GP runs for example 4

Fitness	RMSE	Model
-93.1	0.0083	(-14.81*(z1*z1)), ((12.32*(z1*z1))+(-7.859*z2))
-93.1	0.0083	(-14.81*(z1*z1)), ((12.32*(z1*z1))+(-7.859*z2))
-91.6	0.0083	(-14.81*(z1*(z1-(z1-z1)))), ((-7.859*z2)+ (12.32*(z1*z1)))
-93.1	0.0083	(-14.81*(z1*z1)), ((12.32*(z1*z1))+(-7.859*z2))
-93.1	0.0083	(-14.81*(z1*z1)), ((12.32*(z1*z1))+(-7.859*z2))

For comparison, the "true" model that was used to generate this data set is:

$$\frac{dz_1}{dt} = -14\, z_1^2 \; ; \frac{dz_2}{dt} = 12\, z_1^2 - 8\, z_2$$

The parameters in the model generated by GeMS are different from the true parameters owing to the noise added to the measurements.

Example 5: Lotka-Volterra system

This problem has been studied by Luus (1998). This model is a representation of the predator-prey model used in ecology. The system is described by two differential equations:

$$\frac{dz_1}{dt} = k_1 z_1 (1 - z_2)$$

$$\frac{dz_2}{dt} = k_2 z_2 (z_1 - 1)$$

with the initial conditions z_0 = [1.2 1.1]. The experimental data is available for t = [1,10].

In this model, z_1 represents the population of the prey and z_2 the population of the predator. The solutions to these equations point to the periodic nature of the predator and prey populations. The data used in the study were generated using values for the parameters as $k_1 = 3$ and $k_2 = 1$ with a small amount of normally distributed random error with σ = 0.02 and zero mean added to the observations. The data set is available in Table 26 of Esposito and Floudas (2000). The best models obtained from six runs are shown in Table 2. The fourth model has the best fitness value. It is clear that the best identified model matches with the true model very closely. With this model, a plot of the measured data vs. model prediction appears as shown in Figure 5.

Table 2: Best models of GP runs for example 5

Fitness	RMSE	Model
-31.92	0.024	(-11.099*(z1-(z2*z2))), (-10.859*(z1-z2))
-31.76	0.025	(4.6951*((z2*z2)-z1)), (4.4575*(z2-z1))
-31.68	0.025	(9.8078*(z1-(z1*z2))), (0.34589*(z1-z2))
-51.09	0.0031	(2.9812*((z1+(-0.0047458*z1))-(z2*z1))), ((z2*z1)-z2)
-33.17	0.019	((8.1585*(z2-(z2*z2)))+(0.26568*z1)), (0.3633*(z1-z2))
-43.46	0.0075	(-7.4976*(z1-(z2*z1))), (2.9281*(z1-(z1*z1)))

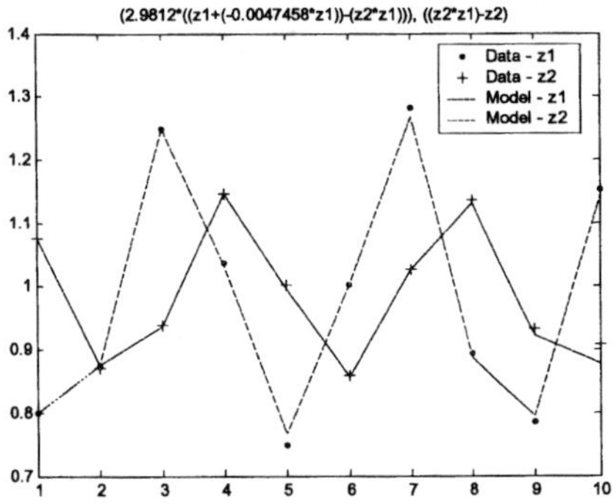

Fig 5. Model fit obtained for Example 5

5. CONCLUSIONS

We have described and demonstrated the features of a GP-based modelling system called GeMS. The capabilities include the identification of nonlinear models for static and dynamical systems. GeMS is able to identify explicit and implicit form of equations. A unique feature of GeMS is the implementation of a robust module that can identify nonlinear ODE's directly from measured data. We have also integrated some nonparametric and exploratory data analysis techniques into GeMS (not explained here) so that its search space can be quite focused.

ACKNOWLEDGEMENTS

Financial support from the Academic Research Fund (ARF), National University of Singapore is gratefully acknowledged.

REFERENCES

Cao, H., Yu, J., Kang, L., Chen, Y., and Chen, Y. (1999). The kinetic evolutionary modeling of complex systems of chemical reactions. *Computers & Chemistry,* **23**(2), 143-152.

Eskinat, E., Johnson, S. H. and Luyben, W. L. (1991). Use of Hammerstein models in identification of nonlinear system. *AICh.E J.*, **37**(2), 255-268.

Esposito, W. R. and Floudas, C. A. (2000). Global Optimization for the Parameter Estimation of Differential-Algebraic Systems. *Ind. Eng. Chem. Res.*, **39**, 1291-1310.

Gao, L. and Loney, N.W., (2001). Evolutionary Polymorphic Neural Network in Chemical Process Modelling, *Comp. & Chem. Engg.*, **25**, 1403-1410.

Greeff, D. J. and Aldrich, C. (1998). Empirical Modelling of Chemical Process Systems with Evolutionary Programming. *Comp. & Chem. Engg.*, **22**(7-8), 995-1005.

Grosman, B. and Lewin, D.R., (2002). Automated nonlinear model predictive control using genetic programming. *Comp. & Chem. Engg.*, **26**, 631–640.

Hinchliffe, M.P and Willis, M.J. (2003) Dynamic systems modeling using genetic programming. *Comp. & Chem. Engg.*, **27**, 1841-1854.

Holland, J. H., (1975). *Adaptation in natural and artificial systems: an introductory analysis with applications to biology, control, and artificial intelligence.* MIT Press, Cambridge, MA.

Koza, J. R. (1992). *Genetic programming: on the programming of computers by means of natural selection.* Cambridge, Mass., MIT Press.

Lakshminarayanan, S., Shah. S. L. and Nandakumar K. (1995). Identification of Hammerstein models using multivariate statistical tools. *Chem. Engg. Sci.*, **50** (22), 3599-3613.

Lakshminarayanan, S., Fujii, H., Grosman, B., Dassau, E., and Lewin, D.R., (2000). New product design via analysis of historical databases. *Comp. & Chem. Engg.,* **24**, 671-676.

Luus, R., (1998). Parameter Estimation of Lotka-Volterra Problem by Direct Search Optimization. *Hung. J. Ind. Chem.,* **26**, 287.

McKay, B., Willis, M. and Barton, G., (1997). Steady-state modeling of chemical processing system using genetic programming. *Computer Chem. Engg,*. **21** (9), 981-996.

Marenbach P. (1998). Using prior knowledge and obtaining process insight in data based modeling of bioprocesses. *SAMS,* **31**. 39-59.

Rodriguez-Vazquez, K. and Fleming, P.J. (1998). Multiple-objective genetic programming for nonlinear system identification, Electronics Letters, 34(9), 930-931.

Tjoa, T. B., Biegler, L. T. (1991). Simultaneous Solution and Optimization Strategies for Parameter Estimation of Differential-Algebraic Equation Systems. *Ind. Eng. Chem. Res.*, **30**, 376.

Törni, A., Ali, M.M., and Viitanen, S. (1999). Stochastic Global Optimization: Problem Classes and Solution Techniques. *Journal of Global Optimization,* **14**, 437–447.

Willis, M., Hiden, H., Hinchliffe, M., McKay, B. and Barton, G. W. (1997) Systems modeling using genetic programming. *Comp. & Chem. Engg.*, **21**, S1161-S1166.

ELSEVIER
IFAC
PUBLICATIONS
www.elsevier.com/locate/ifac

ROBUST CONSTRAINED ESTIMATION VIA UNSCENTED TRANSFORMATION

Pramod Vachhani[a], **Shankar Narasimhan**[b] and **Raghunathan Rengaswamy** [a 1]

[a]*Department of Chemical Engineering, Clarkson University, Potsdam, NY -13699, USA.*
[b]*Department of Chemical Engineering, IIT-Madras, Chennai - 600036, India.*

Abstract: The task of improving the quality of the data so that it is consistent with material and energy balances is called reconciliation. Since chemical processes often operate dynamically in nonlinear regimes, techniques like Extended Kalman Filter (EKF) and Nonlinear Dynamic Data Reconciliation (NDDR) have been developed. There are various issues that arise with the use of either of these techniques: EKF cannot handle inequality or equality constraints, while the NDDR has high computational cost.
In this paper, first, a recursive nonlinear dynamic data reconciliation (RNDDR) formulation is discussed. The RNDDR formulation extends the capability of the EKF by allowing for incorporation of algebraic constraints and bounds during correction. The covariance calculations arising in the RNDDR are same as EKF, i.e., both, nonlinearity and constraints are neglected during covariance propagation and calculation of uncertainty in filtered estimates. The use of Unscented Transformation with the RNDDR gives the Unscented Recursive Nonlinear Dynamic Data Reconciliation (URNDDR) formulation, which addresses all the aspects of nonlinearity and constraints in a recursive estimation framework, thus proving to be an efficient tool for real-time estimation.

Keywords: State and Parameter Estimation, Nonlinearity, Algebraic Constraints

1. INTRODUCTION

The quality of process data in a chemical process significantly affects the performance and benefits gained from activities like performance monitoring, online optimization and control. Processes are invariably subject to random disturbances and process measurement is corrupt with random errors. In order to ameliorate the effect of these random errors, estimation methods can be used to obtain accurate estimates of the process states and parameters.

Several different estimation methods have been proposed in the literature depending on the assumptions made. For linear dynamic systems, the Kalman Filter (KF) gives optimal estimates in presence of measurement and state uncertainties (Gelb, 1988). For nonlinear systems, Extended Kalman Filters (EKF) have been developed, which are based on linearising the nonlinear equations and applying the Kalman filter update equations to the linearised system. The advantages of the KF, the EKF and their variants lie in their predictive-corrective form and the recursive nature of estimation. The recursive form of these estimation methods allows for rapid estimation in real-time, which is extremely important for online deployment. A major disadvantage of the KF and all its variants is that they cannot take into account bounds on process variables or other algebraic constraints.

[1] Corresponding Author: raghu@clarkson.edu

In this paper, first, a Recursive Nonlinear Dynamic Data Reconciliation (RNDDR) formulation is presented. The RNDDR formulation extends the capability of EKF by allowing for incorporation of algebraic constraints and bounds. The covariance calculations arising in the RNDDR formulation are similar to the EKF, which is, linearized model based covariance propagation and the calculation of uncertainty in filtered estimates by using the Kalman gain i.e. by assuming an unconstrained correction. This formulation has been extended with the aid of Unscented Transformation (Julier and Uhlmann, 1997), to give the Unscented Recursive Nonlinear Dynamic Data Reconciliation (URNDDR) formulation which addresses nonlinearity and constraints arising in both the propagation and correction steps.

2. RECURSIVE NONLINEAR DYNAMIC DATA RECONCILIATION (RNDDR)

In Kalman filter and its variants, the estimation procedure at each sampling instant can be regarded as composed of two steps (as described in Vachhani *et al.* (2003)). In the first step, the state estimates from the previous time instant are propagated through the process dynamic equations (along with its error covariance matrix), while in the second step, the predicted estimates are corrected using the measurements available at the current time instant. It is known that the optimal updated estimates for KF (as also its variants) are obtained by solving an unconstrained optimization problem for which the objective function is given in Equation A.5. In the absence of any constraints, the solution of this optimization problem is given by the standard Kalman filter update equation for the state estimates given in Equations A.6 - A.7. The Kalman filter can be extended to obtain simultaneous state and parameter estimates, by treating the parameters to be estimated as augmented states (Jazwinski, 1970).

If algebraic constraints or bound constraints have to be imposed on the state estimates, these can be conveniently included in this optimization problem. In this case, the solution of the optimization problem has to be obtained numerically. This forms the basis for the RNDDR method.

$$x_{k+1} = x_k + \int_{k\Delta t}^{(k+1)\Delta t} f(x(\tau), u_k, p)d\tau + w_k$$
$$y_{k+1} = g(x_{k+1}) + v_{k+1} \quad (1)$$

Consider the system given by Equation 1 with bounds and algebraic constraints imposed on the states and parameters. Let $\hat{x}_{k|k}$, $\hat{p}_{k|k}$, and $P_{k|k,a}$ (a in the subscript here refers to the augmented system) be given at time instant 'k'. The predicted state estimates $\hat{x}_{k+1|k}$ is determined by integration for the given parameter estimate $\hat{p}_{k|k}$ (same as $\hat{p}_{k+1|k}$) and the variance of uncertainty in the predicted estimates is calculated by covariance propagation as with EKF.

For covariance propagation, the nonlinear state space model is linearized around $[\hat{x}_{k|k}, \hat{p}_{k|k}]$,

$$\begin{bmatrix} \dot{x} \\ \dot{p} \end{bmatrix} = \underbrace{\begin{bmatrix} A_k & \Gamma_{pk} \\ 0 & 0 \end{bmatrix}}_{A_{k,a}} \begin{bmatrix} x \\ p \end{bmatrix}$$

and the state transition matrix is approximated assuming an equivalent LTI system, $\bar{A}_{k,a} = exp(A_{k,a}\Delta t)$. Using this linearized approximation, the covariance matrix of estimation errors is propagated as

$$P_{k+1|k,a} = [\bar{A}_{k,a}]P_{k|k,a}[\bar{A}_{k,a}]^T + \begin{bmatrix} Q_k & 0 \\ 0 & Q_{pk} \end{bmatrix} \quad (2)$$

In order to obtain the updated state estimates, the following optimization problem is solved.

$$\min_{x_{k+1}, p_{k+1}} \quad (y_{k+1} - g(x_{k+1}))^T R_{k+1}^{-1} (y_{k+1} - g(x_{k+1})) +$$
$$\left(\begin{bmatrix} x_{k+1} \\ p_{k+1} \end{bmatrix} - \begin{bmatrix} \hat{x}_{k+1|k} \\ \hat{p}_{k+1|k} \end{bmatrix}\right)^T (P_{k+1|k,a})^{-1} \left(\begin{bmatrix} x_{k+1} \\ p_{k+1} \end{bmatrix} - \begin{bmatrix} \hat{x}_{k+1|k} \\ \hat{p}_{k+1|k} \end{bmatrix}\right)$$

subject to the following constraints,

$$x_L \leq x_{k+1} \leq x_U$$
$$p_L \leq p_{k+1} \leq p_U$$
$$h(x_{k+1}, p_{k+1}) \leq 0$$
$$e(x_{k+1}, p_{k+1}) = 0$$

The optimal solution to this problem (x_{k+1}^*, p_{k+1}^*) provides the corrected state and parameter estimates. The solution obtained using EKF is used as an initial guess for solving the above optimization problem. It should be noted that if the measurement model is linear, and none of the inequality constraints are active in the optimal solution, then the solution for the updated state estimates obtained will be the same as the one computed using Equation A.6. The covariance matrix of the error in the updated state and parameter estimates is computed using Equations (A.7 and A.8). It should also be noted that while equations (A.6, A.7 and A.8) are for state estimation, the Kalman filter can be used for simultaneous state and parameter estimation by augmenting the system with parameters as additional state variables and applying the Kalman filter equations to the augmented system. By using these equations, the effect of the constraints on the covariance matrix of estimation errors is neglected. This limitation is addressed in the proposed URNDDR formulation.

3. UNSCENTED TRANSFORMATION

The use of unscented transformation in estimation theory was first studied by Julier and Co-workers (Julier and Uhlmann, 1997; Julier *et al.*, 2000). Traditionally, the covariance calculation in the Extended Kalman filter for nonlinear systems have used linearized approximation of the process for propagation, see Equation 2. This approximation has two drawbacks:

(i) The linearized model may prove to be coarse, and
(ii) The analytical expression for the Jacobian matrix

may not be easy for large systems, requiring the use of finite difference methods. This would involve further computational errors.

The motivation for the unscented transformation is that *it is easier to approximate a probability distribution than it is to approximate an arbitrary nonlinear function or transformation* (Julier and Uhlmann, 1997). In the unscented transformation a set of weighed points are chosen to parameterize the mean and covariance of the probability distribution. The set of points undergo the nonlinear transformation at hand, and the statistical property of the result is calculated from the transformed data set. The exact procedure of the unscented transformation is presented below.

A n-dimensional random variable $\mathbf{x}$ with mean $\bar{\mathbf{x}}$ and covariance $\mathbf{P_{xx}}$ is approximated by $2n+1$ weighed points around the mean of distribution. The set of points $\mathscr{X}$ is also referred to as the sigma set. The first point chosen is the mean of the distribution,

$$\mathscr{X}_0 = \bar{\mathbf{x}} \quad W_0 = \frac{\kappa}{n+\kappa}$$

The rest of the $2n$ translated sigma points are chosen symmetrically around the mean as,

$$\mathscr{X}_i = \bar{\mathbf{x}} + (\sqrt{(n+\kappa)\mathbf{P_{xx}}})_i \quad W_i = \frac{1}{2(n+\kappa)}$$

and

$$\mathscr{X}_{i+n} = \bar{\mathbf{x}} - (\sqrt{(n+\kappa)\mathbf{P_{xx}}})_i \quad W_{i+n} = \frac{1}{2(n+\kappa)}$$

where $(\sqrt{(n+\kappa)\mathbf{P_{xx}}})_i$ is the ith column of the matrix square root of $((n+\kappa)\mathbf{P_{xx}})$ and W_i is the weight associated with the corresponding point. Here κ is a tuning parameter and the heuristic $\kappa+n=3$ has been suggested for normal distribution (Julier and Uhlmann, 1997).

The transformed set of sigma points are evaluated by $\mathscr{Y}_i = g(\mathscr{X}_i)$ and the predicted mean is computed as

$$\bar{\mathbf{y}} = \sum_{i=0}^{2n} W_i \mathscr{Y}_i \tag{3}$$

The predicted covariance is calculated from the transformed set and the predicted mean

$$\mathbf{P_{yy}} = \sum_{i=0}^{2n} W_i [\mathscr{Y}_i - \bar{\mathbf{y}}][\mathscr{Y}_i - \bar{\mathbf{y}}]^T \tag{4}$$

Thus, the unscented transformation can be used to better model the changes in the statistical properties of a random variable through a nonlinear transformation. In the unscented transformation for more than three dimensional variables, κ can take a negative value and in such a case there is a possibility that the predicted covariance becomes non-positive semi-definite. In this situation, it is possible to modify the prediction algorithm (Julier and Uhlmann, 1997).

Inequality constraints like bounds, for example $x \leq x_U$, take the following form

$$\text{if } \mathscr{X}_i > x_U, \quad \text{then } \mathscr{X}_i = x_U$$

This briefly describes the unscented transformation. Detailed discussion on Unscented Kalman Filter can be found in Wan and van der Merwe (2000). In the next section, the use of unscented transformation is proposed towards addressing constrained nonlinear transformations arising in the RNDDR formulation.

4. UNSCENTED RECURSIVE NONLINEAR DYNAMIC DATA RECONCILIATION (URNDDR)

In the previous section, the unscented transformation was explained. The advantage of the unscented transformation is that it can handle nonlinearity and constraints in the transformation. Here, the algorithmic implementation of the unscented recursive nonlinear dynamic data reconciliation formulation is presented.

Consider the system given by Equation 1 with bounds and algebraic constraints imposed on the states and parameters. Let $\hat{x}_{k|k}$, $\hat{p}_{k|k}$, and $\mathbf{P_{k|k,a}}$ be given at time instant 'k'. In the first step of prediction, the unscented transformation is used to determine the mean and covariance of uncertainty in the propagated states. A set of sigma points are selected for propagation, involving uncertain states and uncertainty in the equations (Wan and van der Merwe, 2000). The selection follows the approach outlined in the section on unscented transformation (section 3). Bounds are imposed on the above set of points.

The constrained sigma points are propagated through the nonlinear differential equations governing the process to arrive at the set of translated sigma points. The bounds are imposed on the set of translated sigma points, to compute the propagated sigma points. Once the propagated set of data has been determined, the mean $\hat{x}_{k+1|k,a}$ and the variance $\mathbf{P_{k+1|k,a}}$ are calculated using the unscented transformation. Notice that in the propagation step, the equality and inequality constraints are not taken into account. These constraints are conveniently handled in the correction step which is solved as an optimization problem. Further, imposing bounds and constraints on the sigma points (in the correction and propagation steps), might not preserve the original properties of the unscented transformation. This has not been addressed in this paper.

In the correction step of the URNDDR, there is an uncertainty associated with the n_a (number of states + number of parameters) propagated states $\hat{x}_{k+1|k,a}$ ($\mathbf{P_{k+1|k,a}}$). Therefore, $2(n_a)+1$ sigma points are used for the correction step. The selection of the sigma points is,

$$\left[\hat{\mathscr{X}}_{0,k+1|k,a}\right] = \left[\hat{x}_{k+1|k,a}\right], W_0 = \frac{\kappa}{(n_a+\kappa)}$$

$$\left[\hat{\mathscr{X}}_{i,k+1|k,a}\right] = \left[\hat{x}_{k+1|k,a}\right] + \left(\sqrt{(n_a+\kappa)\left[\mathbf{P_{k+1|k,a}}\right]}\right)_i$$

$$W_i = \frac{1}{2(n_a+\kappa)}$$

$$\left[\hat{\mathscr{X}}_{i+n_a,k+1|k,a}\right] = \left[\hat{x}_{k+1|k,a}\right] - \left(\sqrt{(n_a+\kappa)\left[\mathbf{P_{k+1|k,a}}\right]}\right)_i$$

$$W_{i+n_a} = \frac{1}{2(n_a+\kappa)}$$

and $2(n_a)+1$ optimization problems are solved.

$$\min_{\mathscr{X}_{j,k+1,a}} \; (y_{k+1} - g(\mathscr{X}_{j,k+1}))^T \mathbf{R}_{k+1}^{-1} (y_{k+1} - g(\mathscr{X}_{j,k+1})) + \left(\mathscr{X}_{j,k+1,a} - \hat{\mathscr{X}}_{j,k+1|k,a}\right)^T (\mathbf{P_{k+1|k,a}})^{-1} \left(\mathscr{X}_{j,k+1,a} - \hat{\mathscr{X}}_{j,k+1|k,a}\right)$$

subject to the following constraints,

$$x_L \leq \mathscr{X}_{j,k+1} \leq x_U$$

$$p_L \leq \mathscr{P}_{j,k+1} \leq p_U$$

$$h(\mathscr{X}_{j,k+1}, \mathscr{P}_{j,k+1}) \leq 0$$

$$e(\mathscr{X}_{j,k+1}, \mathscr{P}_{j,k+1}) = 0$$

The optimal solutions to the $2(n_a)+1$ optimization problems are the corrected sigma point set of $\hat{\mathscr{X}}_{j,k+1|k+1,a}$. The mean $\hat{x}_{k+1|k+1,a}$ and the covariance $\mathbf{P_{k+1|k+1,a}}$ of the corrected estimates are calculated using the unscented transformation.

For comparison purposes, the Unscented Kalman filter (UKF) is also implemented in a similar fashion, with the Kalman gain being used in the correction step and constraints are not applied on the translated sigma points during propagation.

5. STATE ESTIMATION PROBLEM

The first case study selected for the URNDDR is the state estimation in a gas phase reversible reaction in an isothermal batch reactor (Haseltine and Rawlings, 2003)

$$2A \xrightarrow{\bar{k}} B \quad \bar{k} = 0.16$$

where reaction rate $r = \bar{k}P_A^2$. The partial pressures are the state variables and the total pressure is measured.

$$x = \begin{bmatrix} P_A \\ P_B \end{bmatrix}, \quad y_k = [1 \ \ 1]x_k$$

The initial state is $x_0 = [3 \ \ 1]^T$. For state estimation problem, the following parameters have been used:

$$\Delta t = 0.1, \; P_0 = \begin{bmatrix} 36 & 0 \\ 0 & 36 \end{bmatrix}, Q = \begin{bmatrix} 10^{-6} & 0 \\ 0 & 10^{-6} \end{bmatrix}$$

$$R = 0.01, \hat{x}_0 = [0.1 \ \ 4.5]^T, \; x_L = [0 \ \ 0]^T, \\ x_U = [100 \ \ 100]^T$$

First the EKF result is presented in Figure 1 for the above initialization scheme. As it can be seen without the information about the constraints the EKF could not converge to the actual dynamics and gives unrealistic estimates for partial pressure of component 'A'. The same was the case with UKF. With the added information of constraints the RNDDR result is presented in Figure 2.

The RNDDR did converge to the actual dynamics of the reaction system but it was stuck at the lower bound for few sample instants before it could respond. This can be explained by the fact that the covariance calculation did not take into account the presence of constraint and the correction step relies on the reliability of the covariance calculation. In case of the URNDDR, which takes the effects of constraints as well as process nonlinearity during covariance calculation, it would be expected that it tracks the actual dynamics better than RNDDR. This is verified by the results presented in Figure 3 for the URNDDR. The URNDDR converged to the actual dynamics within a few sample instants. This result is comparable with the MHE results presented by Haseltine and Rawlings (Haseltine and Rawlings, 2003) which is computationally a more expensive procedure.

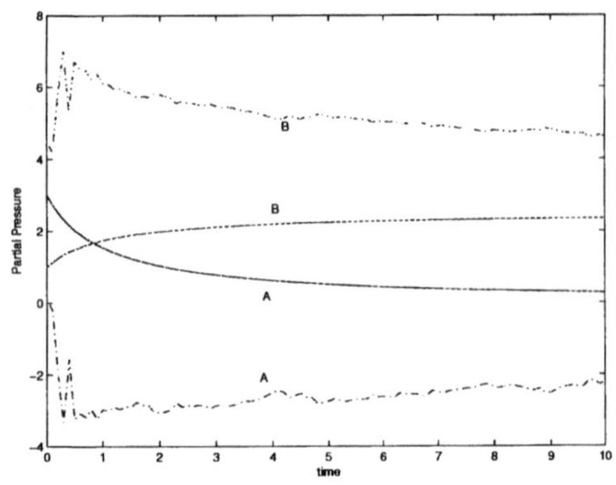

Fig. 1. EKF Case 1 State Estimates. Line: Actual, Dash-Dot: Estimated

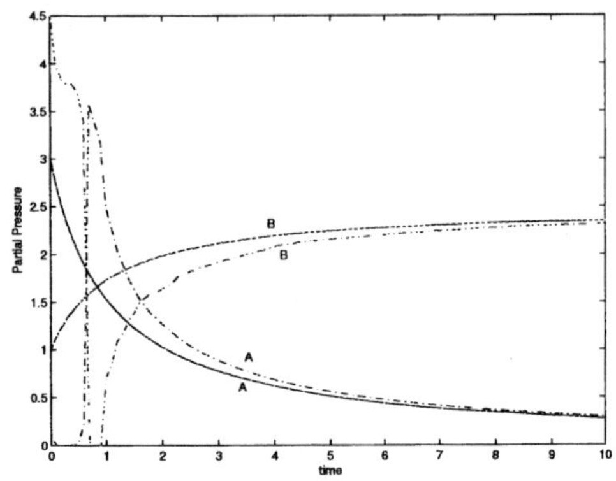

Fig. 2. RNDDR Case 1 State Estimates. Line: Actual, Dash-Dot: Estimated

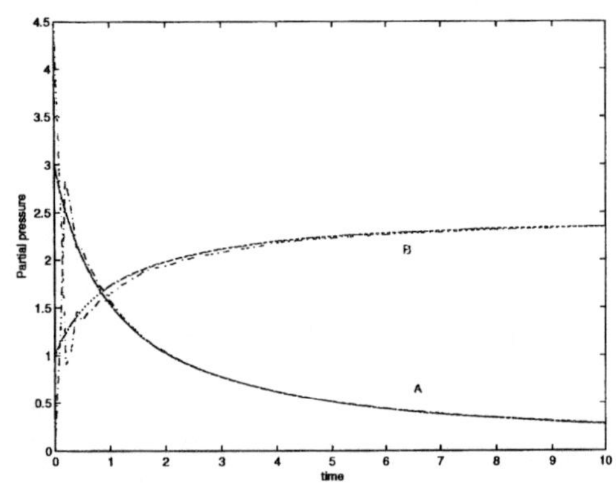

Fig. 3. URNDDR Case 1 State Estimates. Line: Actual, Dash-Dot: Estimated

6. CSTR CASE STUDY

In the previous section, a state estimation example was chosen to show the efficacy of the proposed approach. In this section, the proposed URNDDR formulation is rigorously compared against the EKF, Unscented Kalman Filter (UKF) and the RNDDR formulations for simultaneous state and parameter estimation in a Continuous Stirred Tank Reactor (CSTR). The simulation result presented is the case when the catalyst activity undergoes a change, and it is required to simultaneously estimate the states and the catalyst activity parameter. The implementation details of the case study can be found in Vachhani *et al.* (2003).

Three different step changes in catalyst activity of 1 to 0.7, 1 to 0.8 and 1 to 0.9 at sampling instant of 20 are introduced. The process is simulated for 100 sampling instants with sampling time of 0.05 hour, and the four estimators namely: EKF, RNDDR, UKF and URNDDR are implemented. The Root Mean Square (RMS) errors in the measurements (difference between measured and true values) and the RMS errors in the estimates (difference between estimated and true values) of both measured variables and parameters are computed and averaged over the three simulation runs for each of the methods, the results of which are presented in Table 1. As it can be seen the performance of the URNDDR is better than RNDDR in terms of accuracy of results. The RMS errors in the UKF result are marginally better than the URNDDR result but at the trade-off of unrealistic estimates of catalyst activity greater than one, as shown in Figure 4. Hence, UKF is not a valid estimator for this problem. A comparison of RMS errors reflects the accuracy of the URNDDR formulation. It can be seen that the URNDDR performs relatively better than the RNDDR formulation. The computational cost for this problem correspondingly increases approximately from 0.2 seconds for RNDDR to 5 seconds for URNDDR formulation. The RMS errors in the UKF result are similar to URNDDR formulation, but at the trade-off of unrealistic estimates of catalyst activity greater than one, as shown in Figure 4. Hence, UKF is not a valid estimator for this problem.

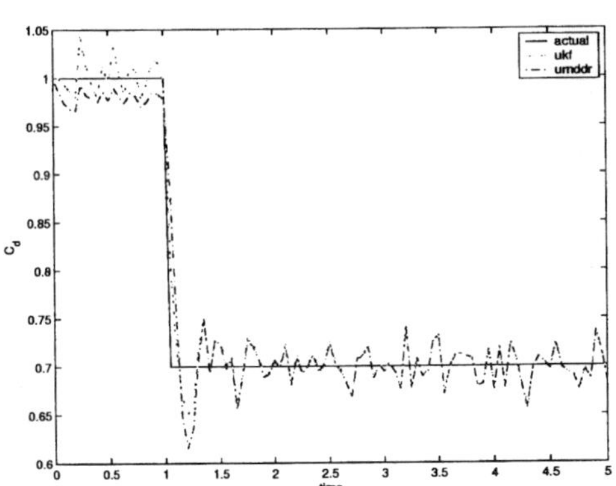

Fig. 4. Catalyst activity estimate - UKF and URNDDR

In Figure 5, the results for all the discussed formulations are presented for a ramp change in the catalyst activity. On comparing the URNDDR with the RNDDR result, it can be seen that with the use of the unscented transformation, realistic estimates which are consistently away from the constraints are achieved. The result of the URNDDR is same as the UKF if the constraints are not activated, this can be seen in Figure 5 when catalyst activity has decreased to less than one.

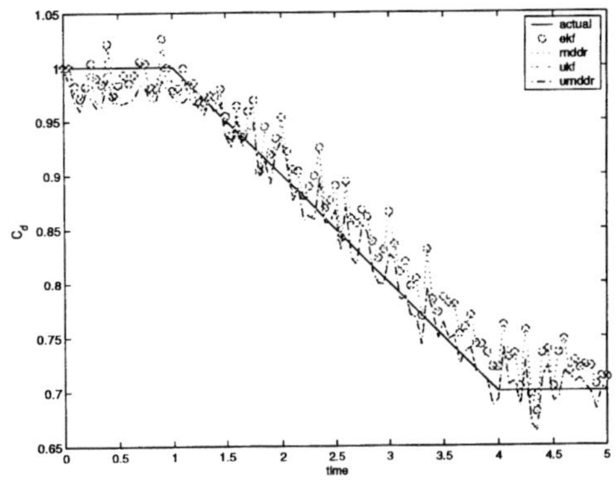

Fig. 5. Catalyst activity estimate - EKF, RNDDR, UKF and URNDDR

7. CONCLUSIONS

The Unscented RNDDR is motivated by the limitation of the RNDDR in taking into account effect of nonlinearity and algebraic constraints on the covariance matrix of estimation errors. In the RNDDR, the basis of covariance propagation is on a linearized representation of the process and the covariance calculation arising in the correction step is based on an equivalent unconstrained minimization solution. With the use of the unscented transformation, the effect of nonlinearity and constraints is taken into account in both propagation and correction steps of the URNDDR formulation.

The proposed method is an extension of the unscented Kalman filter, without sacrificing the essential recursive computation advantage of the Kalman filter. The results of simulation studies indicate that the proposed method provides more accurate estimates of states and parameters, than previous approaches without significant increase in computation effort. The method can be deployed online for real time state and parameter estimation in nonlinear processes.

Appendix A. KALMAN FILTER

Let the discrete linear stochastic state-space system be of the form,

$$x_{k+1} = \bar{A}_k x_k + w_k$$

$$y_{k+1} = \bar{C}_{k+1} x_{k+1} + v_{k+1} \tag{A.1}$$

where w_k and v_{k+1} are independent normally distributed random variables with covariance Q_k and R_{k+1} respectively.

Assume unbiased estimates of the state at time instant 'k' ($\hat{x}_{k|k}$) and the measurement at time instant '$k+1$' (y_{k+1}) are available. The state estimate for time

Table 1. Catalyst deactivation

RMS Errors	V	P	T	C_A	T_c	Par Est
Data	0.4892	1.9373	0.6913	0.0050	0.6553	-
EKF	0.1629	2.0122	0.4419	0.0020	0.4128	0.0297
UKF	0.1740	1.6352	0.1981	0.0006	0.1659	0.0207
RNDDR	0.1528	1.9023	0.4056	0.0018	0.3768	0.0271
URNDDR	0.1452	1.8350	0.2277	0.0010	0.1830	0.0248

instant '$k+1$' ($\hat{x}_{k+1|k+1}$) can be expressed as a linear combination of the two,

$$\hat{x}_{k+1|k+1} = K'_{k+1}\hat{x}_{k|k} + K_{k+1}y_{k+1} \tag{A.2}$$

For $\hat{x}_{k+1|k+1}$ to be an unbiased estimate of x_{k+1}, we require $K'_{k+1} = (I - K_{k+1}\bar{C}_{k+1})\bar{A}_k$. Thus the recursive estimator can be rewritten in two parts, first for prediction and the second for correction.

$$\begin{aligned} \hat{x}_{k+1|k} &= \bar{A}_k\hat{x}_{k|k} \\ \hat{x}_{k+1|k+1} &= \hat{x}_{k+1|k} + K_{k+1}\left(y_{k+1} - \bar{C}_{k+1}\hat{x}_{k+1|k}\right) \end{aligned} \tag{A.3}$$

We also assume $P_{k|k}$, defined by the covariance matrix of errors in the state estimates $\hat{x}_{k|k}$, is known. Therefore, the uncertainty in $\hat{x}_{k+1|k}$ can be calculated as,

$$P_{k+1|k} = \bar{A}_k P_{k|k} \bar{A}_k^T + Q_k \tag{A.4}$$

The Kalman gain matrix is arrived at by solving the following unconstrained optimization problem

$$\begin{aligned} \min_{x_{k+1}} \; & (y_{k+1} - \bar{C}_{k+1}x_{k+1})^T (R_{k+1})^{-1} (y_{k+1} - \bar{C}_{k+1}x_{k+1}) \\ & + (x_{k+1} - \hat{x}_{k+1|k})^T (P_{k+1|k})^{-1} (x_{k+1} - \hat{x}_{k+1|k}) \end{aligned} \tag{A.5}$$

giving the filtered state estimate as,

$$\hat{x}_{k+1|k+1} = \hat{x}_{k+1|k} + K_{k+1}(y_{k+1} - \bar{C}_{k+1}\hat{x}_{k+1|k}) \tag{A.6}$$

$$K_{k+1} = P_{k+1|k}\bar{C}_{k+1}^T(R_{k+1} + \bar{C}_{k+1}P_{k+1|k}\bar{C}_{k+1}^T)^{-1} \tag{A.7}$$

The covariance matrix of errors in the filtered state estimates is given by

$$P_{k+1|k+1} = (I - K_{k+1}\bar{C}_{k+1})P_{k+1|k} \tag{A.8}$$

REFERENCES

Gelb, A. (1988). *Applied optimal estimation.* M.I.T. Press, Cambridge.

Haseltine, E. and J. Rawlings (2003). A critical evaluation of extended kalman filtering and moving horizon estimation. Technical report. Texas-Wisconsin Modeling and Control Consortium. Available at *http://www.che.wisc.edu/jbr-group/tech-reports/twmcc-2002-03.pdf.*

Jazwinski, A. H. (1970). *Stochastic processes and filtering theory*. Academic Press, New York.

Julier, S. and J. Uhlmann (1997). A new extension of the kalman filter to nonlinear systems. In: *International Symposium Aerospace/Defense Sensing, Simulation and Controls.*

Julier, S., J. Uhlmann and H. Whyte (2000). A new method for the nonlinear transformation of means and covariances in filters and estimators. *IEEE Transactions on Automatic Control* **45**, 477–482.

Vachhani, P., V. Gangwal, S. Narasimhan and R. Rengaswamy (2003). Recursive techniques for state and parameter estimation in nonlinear dynamical processes. *submitted to AIChE J.*

Wan, E. A. and R. van der Merwe (2000). The unscented kalman filter for nonlinear estimation. In: *Proceedings of IEEE Symposium 2000 (AS-SPCC), Lake Louise, Alberta Canada.*

ELSEVIER
IFAC
PUBLICATIONS
www.elsevier.com/locate/ifac

DETECTION AND DIAGNOSIS OF DATA RECONCILIATION PROBLEMS IN AN INDUSTRIAL CHEMICAL INVENTORY SYSTEM

Zhengang Han * **Sirish Shah** *,[1]
Shankar Narasimhan [2] **Hasna Zaknoun** [3]

* *Department of Chemical and Materials Engineering, University of Alberta, Edmonton, Canada*

Abstract: According to one industrial engineer, "reconciliation of mass balances to monitor chemical tank inventories is a struggle even at the best of times". This paper is concerned with the design of an offline and online fault detection and diagnosis monitoring system for the caustic tank inventory process at a pulp and paper company. Even with limited instrument redundancy, the offline monitoring analysis was able to correctly detect and diagnose sensor calibration errors. The scheme is now undergoing online implementation tests. *Copyright © 2004 IFAC*

Keywords: Data reconciliation, gross error detection, process monitoring, sensor calibration error, fault detection and diagnosis

1. INTRODUCTION

Modern chemical plants are usually equipped with hundreds of instruments and many complex processing units. Therefore, even if only one single element of the whole plant does not function normally, the whole process performance and operation can degrade. In order to avoid this and run the whole process more efficiently, a scheme which can monitor the "health" of all the instruments and processing units is necessary from operation point of view.

The basic idea in process monitoring is to timely check the newly measured process variables and to see if they are consistent with the process models under the fault free situation. If there is a large deviation between these two then a fault is said to be detected. This is usually referred as "fault detection". Further, by manipulating the deviation, the location of the fault can be obtained by comparing it with the preset fault signatures. This is known as "fault isolation".

Two elements are essential for fault detection and isolation (FDI): process models and process measurement. For the purpose of robust and yet accurate FDI, one has to obtain appropriate process models. Once the process measurements are found to be inconsistent with respect to these models, one can then infer that the instruments or the process equipment are abnormal without bothering to consider the possibility of model inaccuracy. For example physical relationships, such as mass and energy balances can be used as a process model for FDI. For fault diagnosis, process measurements need to have a level of redundancy or the models need have extra degrees of freedom, i.e. each measured process variable can be inferred from the other measured variables as well as process

[1] Author to whom correspondence should be addressed. Email: Sirish.Shah@UAlberta.ca, Tel: 1-780-4925162
[2] Department of Chemical Engineering, Indian Institute of Technology Madras, Chennai, India
[3] Millar Western Forest Products Ltd., Whitecourt, Canada

models. This type of redundancy is also referred as "analytical redundancy" (Chow and Willsky, 1984).

Data reconciliation (DR), which has been well studied in recent years (Mah 1990, Madron 1992, Crowe 1996, Bagajewicz 2000, Narasimhan and Jordache 2000), is concerned with the validation of process variables that satisfy physical constraints of the process. However, to obtain accurate estimates, some action has to be taken to eliminate the influence of gross errors. Hypothesis testing has been extensively used for detection, isolation and identification of gross errors. Therefore, gross error detection (GED) is closely related to DR. Literature surveys of GED can be found in Narasimhan and Mah (1987), Mah (1990), Madron (1992), Sánchez and Romagnoli (2000) and Narasimhan and Jordache (2000).

This paper is organized as follows: section 2 provides the theoretical background on offline and online process monitoring system design; the process of interest is introduced in section 3; the detailed description of offline and online analysis and results are presented in section 4; and the paper ends with concluding remarks in section 5.

2. SENSOR CALIBRATION ERROR DETECTION AND ISOLATION

Suppose the fault-free process measurement vector $\mathbf{y}(k) \in \Re^m$ can be described by:

$$\mathbf{y}(k) = \mathbf{x}(k) + \boldsymbol{\varepsilon}(k) \tag{1}$$

where $\mathbf{x}(k) \in \Re^m$ is the vector of true values at time k and $\boldsymbol{\varepsilon}(k) \in \Re^m$ is a zero mean normal distributed random vector representing the measurement noise, i.e. $\boldsymbol{\varepsilon}(k) \sim \mathcal{N}(\mathbf{0}, \boldsymbol{\Sigma})$ where $\boldsymbol{\Sigma} \in \Re^{m\times m}$ is the covariance matrix of noise $\boldsymbol{\varepsilon}(k)$. In most cases, the noise vector is mutually independent, i.e. $\boldsymbol{\Sigma}$ is a diagonal matrix.

Assume that the process model is given by

$$\mathbf{A}\mathbf{x}(k) = \mathbf{0} \tag{2}$$

where $\mathbf{A} \in \Re^{q\times m}$ is a known matrix and q is the number of constraints. If sensor(s) are suffering from calibration problems, their measurements may be different from $\mathbf{x}$. Different types of calibration errors as illustrated in Fig. 1 are possible. In this paper, the most general case - a bias plus slope error - is considered and described mathematically as:

$$y_i(k) = \alpha_i x_i(k) + \beta_i + \varepsilon_i(k) \tag{3}$$

where α_i and β_i are calibration parameters associated with slope and bias errors respectively, and x_i

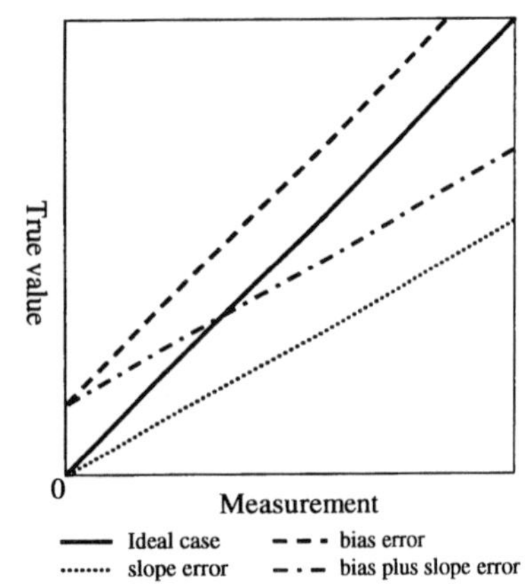

Fig. 1. Illustration of different types of calibration errors

and y_i denote the ith element of vector $\mathbf{x}$ and $\mathbf{y}$. In the fault diagnosis literature (Gertler, 1988), the slope error is classified as a multiplicative fault, while the bias error is referred to as an additive fault. In the fault-free case, $\alpha_i = 1$ and $\beta_i = 0$.

To detect the sensor calibration error, one can directly project the constraint matrix $\mathbf{A}$ on the measured variables $\mathbf{y}(k)$ and check the residuals $\mathbf{r}(k) = \mathbf{A}\mathbf{y}(k)$. Obviously, if the sensors do not have any calibration error, the residuals will be only related to the measurement noise, i.e. $\mathbf{r}(k) = \mathbf{A}\boldsymbol{\varepsilon}(k)$. Therefore, the residuals will be zero-mean, i.e. $\mathbf{r}(k) \sim \mathcal{N}(0, \Sigma_r)$. However, if some of the sensors suffer from calibration errors, the residuals will not be zero-mean. Thus, one can easily detect possible sensor calibration error(s) by checking the mean of residuals.

Once a calibration problem is detected, the next step is to proceed to the diagnosis stage, i.e. to determine the exact location and type of the sensor calibration error. In this step, one has to categorize the process variables into two sets: suspected $\mathbf{x}^s(k) \in \Re^{m_1}$ and unsuspected $\mathbf{x}^u(k) \in \Re^{m_2}$. Thus, the process model $\mathbf{A}$ can be rearranged as $\mathbf{A} = [\mathbf{A}^s \;\; \mathbf{A}^u]$ where $\mathbf{A}^s$ and $\mathbf{A}^u$ correspond to $\mathbf{x}^s$ and $\mathbf{x}^u$ respectively. Therefore, Eqn. 2 can be re-written as:

$$\mathbf{A}^s\mathbf{x}^s(k) + \mathbf{A}^u\mathbf{x}^u(k) = \mathbf{0} \tag{4}$$

with the assumption of

$$\mathbf{y}^s(k) = \Lambda\mathbf{x}^s(k) + \Delta + \varepsilon^s(k)$$
$$\mathbf{y}^u(k) = \mathbf{x}^u(k) + \varepsilon^u(k)$$

where $\Lambda \in \Re^{m_1\times m_1}$ is an unknown diagonal matrix and $\Delta \in \Re^{m_1}$ is an unknown vector.

When one applies the above assumption on the process constraint Eqn. 4, the following can be obtained:

Table 1. Measured variables of chemical tank inventory system

Names	Descriptions	Units
L	Tank level	%
F1	Flow rate in outflow #1	L/min
F2	Flow rate in outflow #2	kg/min
F3	Flow rate in outflow #3	kg/min
F4	Flow rate in outflow #4	L/min
F5	Flow rate in outflow #5	kg/min
F6	Flow rate in outflow #6	kg/min
P1	Valve position of outflow #1	%
P2	Valve position of outflow #2	%
P3	Valve position of outflow #3	%
P4	Valve position of outflow #4	%
P5	Valve position of outflow #5	%
P6	Valve position of outflow #6	%

$$\begin{aligned}\mathbf{r}(k) &= \mathbf{A}^u\mathbf{y}^u(k) + \mathbf{A}^s\Lambda^{-1}\mathbf{y}^s(k) - \mathbf{A}^s\Lambda^{-1}\Delta \\ &= \mathbf{A}^s\Lambda^{-1}\varepsilon^s(k) + \mathbf{A}^u\varepsilon^u(k) \qquad (5)\end{aligned}$$

Then, the problem can be formulated in the form of a *least-squares optimization* problem as follows:

$$J = \min_{\Lambda,\Delta} \mathbf{r}^T(k)\mathbf{r}(k) \qquad (6)$$

Please note that only $2m_1$ parameters need to be determined by the optimization routine as Λ is a diagonal matrix. By observing the estimated value of the diagonal elements of Λ and Δ, one can determine which sensor(s) are mostly likely to suffer from calibration errors by comparing the estimates with the fault-free case.

3. PROCESS DESCRIPTION

The process of interest is a caustic tank inventory system at Millar Western Forest Products Ltd., Whitecourt, Canada. The process serves as a buffer to supply the caustic usage for the whole plant. The simplified schematics of the tank inventory system is illustrated in Fig. 2.

In the process, caustic is delivered by transportation trucks at periodic intervals depending on the plant chemical usage. Therefore, the inlet flow rate is not continuous and hence not measured. There are 6 independent outlet flow control loops. Each pipeline supplies caustic to different parts of the plant and the flow rates in the outlet pipes are controlled independently.

The level of the tank is not controlled but measured. The tank is a vertical cylinder tank with a diameter of 6.8 meters and a height of 6 meters. Thus the nominal volume is 218 cubic meters. The level sensor is located 0.45 meters above the tank bottom. Therefore, the measurable volume of the tank is about 200 cubic meters. All the related measured variables are listed in Table 1.

Notice from Table 1, that the six flow rate sensors do not use the same measurement units. Two of them (F1 and F4) are volume flow sensors in L/min, while the other four are mass flow sensors in kg/min. In this paper, it is assumed the delivered caustic having a constant density: $\rho = 1.52kg/L$.

Furthermore, all six valves are equal percentage valves, which are essentially nonlinear valves. The characteristic of equal percentage valve are described in the literature (Johnson 1988). All flow rate, level and valve position data are sampled every 1 minute, i.e. $T_s = 1$ min.

4. MONITORING SYSTEM DESIGN AND DIAGNOSIS RESULTS

4.1 Process model description

4.1.1. Mass Balance The relationship between the volume of caustic in the tank and the inlet and outlet flow rates can be represented by the following equation:

$$\rho\frac{dV(t)}{dt} = \rho F_{in}(t) - \rho F_{out}(t) \qquad (7)$$

where $V(t)$ is the chemical volume in the tank, and $F_{in}(t)$ and $F_{out}(t)$ are the inlet and outlet volume flow rates respectively.

In discrete time domain, the above equation can be written as:

$$\begin{aligned}r(k) &= V(k) - V(k-1) - (F_{in}(k) - F_{out}(k))\,T_s \\ &= 0 \qquad (8)\end{aligned}$$

where $r(k)$ is defined as the "mass balance residual". This discrete-time representation implies that the inlet and outlet flow rates do not change in between samples. In this process, since the sampling interval is relatively large, this assumption does not hold most of the time. Therefore, when this discrete-time model is used, small deviation of the residuals from zero can be expected due to this systematic model-plant mismatch (MPM).

$F_{out}(k)$ can be easily calculated by summing all the outlet flow rates. $V(k)$ can also be obtained from the tank level sensor L and *a priori* knowledge of the total volume of the storage tank. However, $F_{in}(k)$ is not measured. Therefore, according to the essential elements of process monitoring mentioned in the introduction section, the process measurements are not "redundant".

Nevertheless, this does not mean that one can not design a monitoring scheme on the current system. Once we take a closer look at the current system, the operation can be categorized into two stages - "during delivery" and "zero delivery" - with different F_{in}s as below:

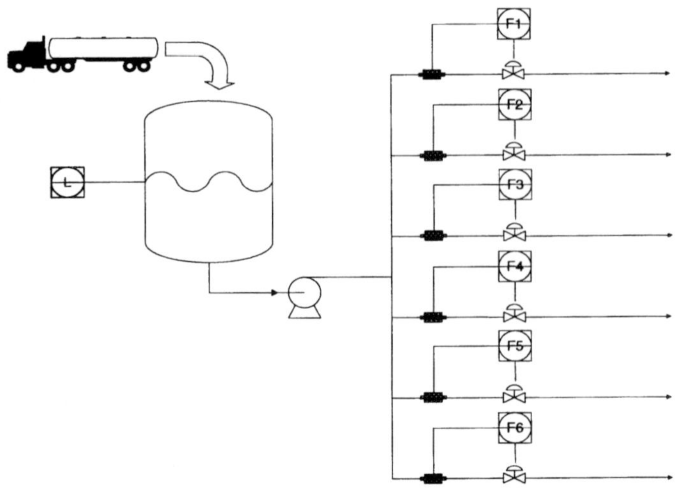

Fig. 2. The schematics of chemical tank inventory process at Millar Western Forest Products Ltd.

$$F_{in}(k) = \begin{cases} 0, & \text{zero delivery} \\ \text{unknown}, & \text{during delivery} \end{cases} \quad (9)$$

Thus, during the period of "zero delivery", one exactly knows F_{in} even though there is no measurement. Therefore, when there is no delivery, the "redundancy" condition can be satisfied. One can make use of this "redundancy" for the purpose of process monitoring. The details are shown later in this paper.

4.1.2. Valve characteristics All control valves in this process are equal percentage valves. Theoretically, the flow rate and valve position should obey the following relationship (Johnson 1988):

$$F = F_{min} R^{S/S_{max}} \quad (10)$$

where $R = \dfrac{F_{max}}{F_{min}}$ is referred to as *rangeability*, F_{max} and F_{min} are the maximum and minimum flow rates respectively, F is flow rate passing through the valve, and S and S_{max} are the stem position and maximum stem position.

By taking logarithm of Eqn. 10, we obtain:

$$\log F = \frac{S}{S_{max}} \cdot \log R + \log F_{min} \quad (11)$$

By doing this, the nonlinear equal percentage valve characteristic has been transformed into a linear framework. Both the flow rate F and the valve position S/S_{max} are measured in this system. Therefore, there are some measurement redundancies with respect to the valve characteristic. The estimates of the unknown parameters $\log R$ and $\log F_{min}$ can be obtained through *least squares* estimation by using archived process data of flow and valve position measurements, provided these measurement do not contain any bias. Once the parameter estimates have been obtained, the valve equations provide the necessary redundancy for fault diagnosis even during delivery periods. Due to limited space, the details of the parameter estimates are omitted.

However, the models of control valves are different from the mass balance model in the following aspects. Firstly, unlike mass balance, valve characteristic cannot be guaranteed to be valid for ever. Due to wear and tear, the parameters of the valve model may change with time. This can be treated as a parametric fault and distinguished from sensor biases using the available data. The diagnosis problem becomes more complex, but it is solvable. Secondly, the mass balance equation is known exactly. But for the valve characteristics, the unknown parameters, such as $\log R$ and $\log F_{min}$, need to be determined from the data. Because of these differences, these two types of models need to be treated differently in the later analysis.

4.2 Offline analysis for detection and diagnosis of sensor calibration errors

After selecting the data portion corresponding to the "zero delivery" case, Eqn. 8 becomes:

$$r(k) = V(k) - V(k-1) + F_{out}(k)T_s \quad (12)$$

where $F_{out}(k) = \sum_{i=1,4} F_i(k) + \frac{1}{\rho} \sum_{i=2,3,5,6} F_i(k)$ and $V(k) = 200,000 \times L(k)$.

Thus, the mass balance residual at each time instant can be obtained. The mass balance residual for a period of 40,000 data points is shown in Fig. 3. From this figure, note that the mean of the residual is negative, and it is not stationary throughout the range. After consulting with the process engineer and confirming that no leakage is taking place, we suspected that the problem may

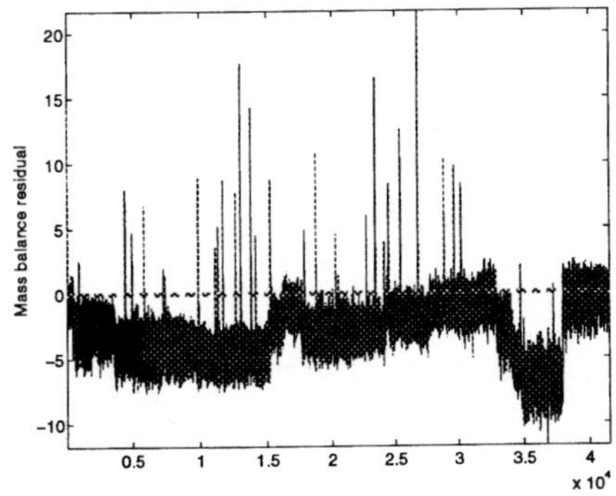

Fig. 3. Mass balance residual before calibration correction

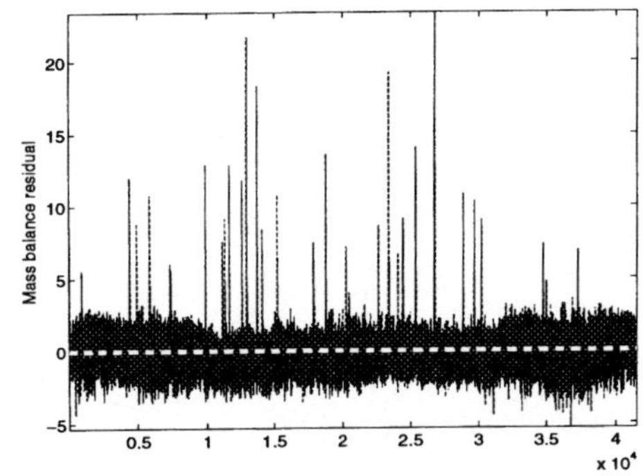

Fig. 4. Mass balance residual after calibration correction

possibly be due to flow sensor calibration error. The target was then to find out which sensor(s) have the calibration problem and what is the magnitude of this calibration problem.

In order to confirm the suspicion, it is assumed that the level sensor is accurate, while all the flow sensors may suffer from bias plus slope calibration errors, i.e. $F_i^f(k) = \alpha_i F_i^o(k) + \beta_i + \varepsilon_i(k)$, $\forall\, i = 1, \cdots, 6$, where $F_i^f(k)$ and $F_i^o(k)$ denote the ith measured and true flow rates respectively, α_i and β_i are constant coefficients, and $\varepsilon_i(k)$ are zero mean normally distributed measurement noise. If the ith flow rate sensor works fine, then ideally $\alpha_i = 1$ and $\beta_i = 0$. Therefore, the mass balance equation can be further written as

$$r(k) = V(k) - V(k-1) + \sum_{i=1}^{6} \frac{T_s}{\alpha_i} F_i^f(k) - \sum_{i=1}^{6} \frac{\beta_i \cdot T_s}{\alpha_i} = \sum_{i=1}^{6} \frac{T_s}{\alpha_i} \varepsilon_i(k) \qquad (13)$$

Since all the β_i are included in the same summation operator, one cannot distinguish the β_is from each other. Thus, define $\beta = \sum_{i=1}^{6} \frac{\beta_i}{\alpha_i}$. From the objective function shown in Eqn. 6, it is simply a least-squares problem. Then, from Eqn. 13 through the least-squares estimation, the estimated coefficients and their confidence limits are shown in Table 2.

Since the estimated value of β is small in comparison with the flows, it can be deduced that there are no additive biases in any of the flow sensors (in the absence of bias cancellations). Also from the estimated values of αs for all sensors except F2, it can be inferred that the corresponding flow sensors do not have a slope error. The estimated value of α_2 which is significantly greater than unity, indicates a slope error in the corresponding flow sensor.

The afore-mentioned analysis results were further confirmed by the process control engineer. The problem was traced to inconsistency in the sensor ranges in the DCS system and the sensor. It was determined that the flow rate sensor F2 was changed and re-calibrated to the range of 0 - 15 L/M. Whereas, the conversion range in the DCS system still remained at the original 0 - 25 L/M. Therefore, this gives the scaling factor to be $25/15 = 1.67$, which is close to the estimate 1.73. The residual after calibration correction shown in Fig. 4 is more or less zero mean.

4.3 Online monitoring design and implementation

In order to provide online monitoring scheme for the caustic tank inventory process, the possible abnormal situations that are considered in this paper are listed below:

- *Sensor faults*: the whole process is equipped with 13 sensors. All of these sensors could potentially malfunction during the operation.
- *Process faults*: this type of fault include any malfunction of the process equipment other than instrument errors, e.g. tank leakage.

To correctly detect and further isolate the afore-mentioned faults, analytical redundancy has to be used. In this project, there are totally 7 equations representing various relationships among measured variables. They consist of one mass balance and six valve characteristic equations. For detection, as long as one of these equations becomes invalid, one can infer that a fault has occurred, i.e. a fault is detected. For isolation, since different types of faults affect different equations, one can determine the root cause by observing which equations are affected. Table 3 shows how different faults affect different equations. The following remarks regard to Table 3:

- The Y-axis denotes possible faults.
- The X-axis denotes different equations - MB stands for mass balance and V_i represents the valve characteristic of valve i.
- '×' denotes the intersection where a fault affects the corresponding equation.

Table 2. Estimated calibration coefficients

	α_1	α_2	α_3	α_4	α_5	α_6	β
Estimates	1.0410	1.7354	1.0437	1.0532	1.0084	1.0478	-0.2708
Upper bound	1.2808	2.0551	1.2129	1.3327	1.1965	1.3356	1.2430
Lower bound	0.8011	1.4157	0.8745	0.7737	0.8203	0.7600	-1.7846

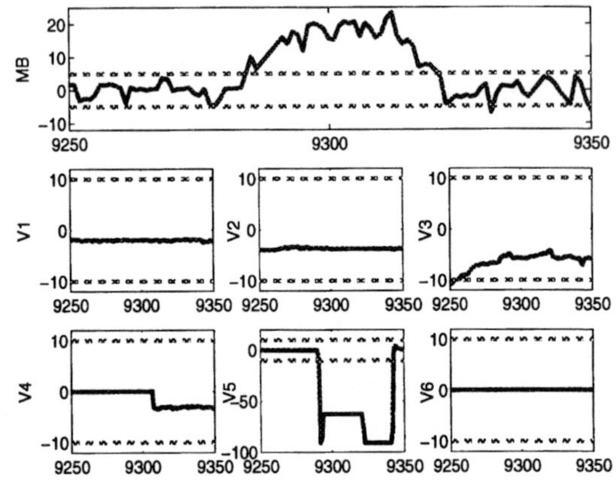

Fig. 5. Residuals when an alarm was generated ·

Table 3. Fault isolation logic

	MB	V_1	V_2	V_3	V_4	V_5	V_6
L	×						
F1	×	×					
F2	×		×				
F3	×			×			
F4	×				×		
F5	×					×	
F6	×						×
P1		×					
P2			×				
P3				×			
P4					×		
P5						×	
P6							×
Process	×						

- If an intersection is blank then there is no fault affecting the corresponding equation.
- Different faults affect different combinations of equations. Thus, if one or more equations are not valid, one can look up Table 3 to find out the root cause of the fault.
- One cannot distinguish the difference between level sensor and process faults because they affect the process model in the same way due to the limited degrees of freedom.

On May 29 2003, one alarm was generated by the afore mentioned monitoring scheme. Fig. 5 shows the 7 residuals during that period of time. From Fig. 5, the valve characteristic equation of flow loop #5 was found to be invalid as well as the mass balance residual, while the other valve characteristic equations are still valid. Therefore, according to the fault isolation logic presented in Table 3, the flow sensor F5 was suspected to have measurement problem during that period of time. This online analysis result was also confirmed by process control engineer.

5. CONCLUDING REMARKS

In this paper, the detection and diagnosis of sensor calibration error has been proposed and experimentally evaluated on an industrial caustic tank inventory monitoring process at Millar Western Forest Products Ltd. The proposed method uses a first order model to represent the calibration error and applied optimization routine to perform diagnosis of this problem. The effectiveness of this method has been proved by successfully isolating a mis-calibrated sensor. It is shown that even if additive faults cannot be isolated with the given level of redundancy in a system, it is possible to isolate multiplicative faults, by suitably exploiting the nonlinear effect of these faults. In addition, an online monitoring scheme has been designed and implemented by making use of the process model structure presented in Table 3. The effectiveness of this online monitoring scheme has also been verified.

REFERENCES

Bagajewicz, M. (2000). *Process Plant Instrumentation: Design and Upgrade.* CRC.

Chow, E. and A. Willsky (1984). Analytical redundancy and the design of robust failure detection systems. *IEEE Trans. Auto. Cont.* **29**, 603–614.

Crowe, C. (1996). Data reconciliation - progress and challenges. *J. Proc. Cont.* **6**, 89–98.

Gertler, J. (1988). Survey of model-based failure detection and isolation in complex plants. *IEEE Cont. Sys. Mag.* **12**, 3–11.

Johnson, C. (1988). *Process control instrumentation technology.* 3rd ed.. John Wiley and Sons

Madron, F. (1992). *Process Plant Performance, Measurement Data Processing for Optimization and Retrofits.* Ellis Horwood. West Sussex, England.

Mah, R. (1990). *Chemical Process Structures and Information Flows.* Butterworths. Stoneham, MA, USA.

Narasimhan, S. and C. Jordache (2000). *Data Reconciliation and Gross Error Detection. An Intelligent Use of Process Data.* Gulf Publishing Company.

Narasimhan, S. and R. Mah (1987). Generalized likelihood ratio method for gross error identification. *AIChE J.* **33**, 1514–1521.

Sánchez, M. and J. Romagnoli (2000). *Processing and Reconciliation for Chemical Process Operations.* Academic Press.

ELSEVIER
IFAC
PUBLICATIONS
www.elsevier.com/locate/ifac

A Systematic and Fully Automated Procedure for Water and Element Balancing over Pulp and Paper Mills A Case Study at Visy Tumut Mill

T.D. Nguyen, M.J.T. Bennett, J.A. Romagnoli
Laboratory for Process Systems Engineering
The University of Sydney
Sydney, NSW, 2006, Australia

ABSTRACT

A general strategy to perform water and element balances was developed and implemented for a pulp and paper mill to identify the flow profiles of water and key elements. The materials covered in the balance were water, sodium (Na), sulfur (S), calcium (Ca), potassium (K), and chlorine (Cl). A sampling strategy for the whole plant was first developed and then a number of analytical techniques were chosen, taking into account the special/difficult characteristics of the samples found in a typical Kraft pulp and paper mill. To perform the balances, boundaries around each section of the plant were selected and additional information was created by incorporating extra knowledge/relationships within each section, thus providing redundant information. Error analysis and data reconciliation were performed to evaluate the reliability of the results and correct them, to finally obtain an adjusted set of data consistent with the balance equations. The reconciled data obtained, as expected, showed better consistency in regards to the balance equations and reduced uncertainties. The final mass balance simulation toolkit developed enhanced the calculation and data processing activities, allowing material balances to be performed continuously to provide up-to-date results to serve for "on time" monitoring and "on time" decision making by mill staff.

I. INTRODUCTION

Environmental and economical sustainability is driving the pulp and paper industry towards ever increasing closure of the water and chemical cycles within the pulp and paper making process. A high degree of closure means that non-process elements, such as chlorine and potassium that enter in trace amounts in the wood and makeup chemicals, tend to accumulate within a mill. This can cause a number of problems such as corrosion, scaling, recovery boiler fouling and plugging [1], and deficient product quality. Further, restrictions on water usage and the utilization of treated process effluent for irrigation, means that water usage and effluent contaminant levels must also be carefully monitored. Therefore, a material balance study is often considered as a practical tool to provide water and chemical flow profiles for monitoring the closure in pulp and paper mills.

A material balance study was conducted over the Visy Tumut Pulp and Paper Mill (VTPPM), allowing tracking of flow profiles, and continuous monitoring of the mill's operational performance. The materials covered in the balance were water, and the key elements: sodium (Na), sulfur (S), calcium (Ca), potassium (K), and chlorine (Cl). The scope of this project encompassed a balance over all main functional areas, including those responsible for providing service utilities, such as cooling water.

The VTPPM employs the Kraft pulping process to produce strong paper suitable for linerboard packaging. The mill is a highly integrated process consisting of three main flow cycles: fiber processing, chemical recovery, and cooling water (Figure 1).

To complete the balance study, firstly a systematic procedure for sampling and sample analysis was developed to determine the amount of water, and key elements within the process streams of the mill. Secondly, a series of mass balances were constructed over all the areas previously mentioned. Thirdly, error analysis and data reconciliation techniques were applied, to evaluate the reliability of the results and compensate for experimental uncertainty, and obtain an adjusted data set more consistent with the balance equations. Finally, a material balance toolkit was developed to fully automate the overall procedure.

II. SAMPLE COLLECTION

In order to conduct the required mass balances it was necessary to collect and analyze the critical process materials within the mill.

The greatest possible accuracy to the mass balance could have been obtained by sampling every material within the mill. This however was not practical or necessary, and sampling was limited to materials within the mill containing significant and unknown quantities of the balance components. Very dilute streams and purchased chemicals were excluded.

With the materials requiring sampling identified, it was necessary to collect samples over a sufficient length of time, and with a sufficient frequency, to capture an accurate picture of the water and key elemental concentrations within the mill. The ideal period and frequency should provide sufficient resolution to capture small process fluctuations, such as those caused by the action of a control value maintaining a set point, and should do this over a sufficient length of time to capture gradual changes, such as those caused by changes in the source region for the mill's wood. In choosing the period and frequency it is was also necessary to consider the project time constraints (6 months) and resources available in addition to these factors, as sample analysis is a time consuming and costly procedure.

The ideal period and frequency is unfortunately best determined from experience, as each mill is different and there are many factors affecting concentrations fluctuations. It was generally accepted within the mill that a change in the liquor cycle takes three days to propagate, and that two weeks of continuous sampling is required to close a mass balance over an individual process area. Using this knowledge, samples were collected three times a week, for a period of four weeks, to provide a sufficient margin of error to capture major concentration changes in the mill, and provide enough time to close the balances.

This sampling program still provided a significant number of samples. To reduced the number within the project time constraints and resources, and yet still maintain an accurate representation, each weekly sample set was composited, to give four weekly sample sets.

III. SAMPLE ANALYSIS

The use of Inductively Coupled Plasma Emission Spectrometry (ICP) was chosen as the main means of analysis. ICP was chosen because of its ability to simultaneously analyze for a wide range of elements, including sulfur that is too electronegative to be analyzed by other methods [3]. ICP also offers relative freedom from chemical interference due to the long residence times and extreme temperatures experienced by a sample, resulting in almost complete atomization [4]. To increase accuracy, additional analysis of some samples containing calcium in very high or very low concentrations relative to the other elements, such as in the grits and dregs, was performed using Atomic Absorption Spectroscopy (AAS). AAS was used over ICP, as it is able to detect calcium over a much wider concentration range, and is a cheap and rapid way to determine the concentration of a single element.

Chlorine is too electronegative to be detected by any spectrographic method, and thus it was necessary to use an alternate method of analysis. A review of the analytical methods for chlorine analysis [4] showed that the most rapid and practical method for determining chlorine concentration is the use of an ion selective electrode (ISE). This method used to analyze the samples for total chlorine.

All of the analytical methods used require the sample to be in the form of an aqueous solution. For accurate analysis the samples must ideally consists of free ions in solution, unhindered by organic and colloidal material; this is essential for use of an ISE. With the exception of some water soluble samples, most samples collected required digestion to put them in this form. A number of methods and reagents are available to perform sample digestion, most of which involve heating a sample with a mixture of strong acids and oxidizing agents.

The need to analyze for chlorine and sulfur limited both the choice of method and reagents available, as both of these elements are the basis of many commonly used strong acids, and in addition these elements from volatile compounds during digestion. These limitations were overcome by performing the digestions using a mixture of nitric acid and hydrogen peroxide in microwave digester, which due to the use of sealed pressure vessels minimizes volatile loss. Typically 5mls of liquid samples or 0.5g of solid samples were digested with nitric acid (5ml, 69%) and hydrogen peroxide (1ml, 30%) using a mild digestion program (250W – 1min, 0W – 2min, 250W – 5min, 400W – 5min, 600W – 5min).

IV. MASS BALANCE CONSTRUCTION

Mass balances were performed over all parts of the mill, using a combination of the sample analysis results, historical process data from the mill's distributed control system, and chemical inventory records.

To perform these balances, it was necessary to choose boundaries over which to perform the input and output flow calculations. These boundaries can be made to arbitrarily contain any fraction of the total process, but generally the smaller each boundary, the greater the available information about the process and difficulty of the balance. A boundary was used for each individual processing area within the mill, and it was found that this provided sufficiently detailed information about the process, while allowing the balances to be solved with the process information available, and a minimal degree of difficulty.

The VTPPM features one of the world's most advanced distributed control systems (DCS). This DCS measures and records most of the important material flows and state variables within the mill. With the exception of some of the chemical inputs, the DCS was sufficiently dimensioned that any unmeasured material flows could be determined from other measured flows. To determine the unmeasured chemical inputs, chemical inventory records were used. For the purposes of a mass balance the inclusion of such an advanced control system is extremely valuable. Process data can be rapidly obtained for the entire operating history of the mill, making very representative averages obtainable.

To perform the balances, process data was collected from the DCS corresponding to the sampling period. This data was processed by calculating the numerical mean and the uncertainty within this mean from the variance within the data set. Typically, raw process data will contain outliers caused by process upsets and breakdowns, which in large numbers can skew the mean. The standard method to remove these outliers is to apply a confidence interval to the data, which relies on the data conforming to a known distribution. It was found that the data from the VTPPM did not conform to any standard distribution, and thus without rigorous statistical processing, confidence intervals could not be applied. This was acceptable considering the project aim was to create a mass balance representing the normal operation of the mill, which will include process upsets and breakdowns. When using variance to calculate the uncertainty in the mean of a measured process variable, it is necessary to assume that any uncertainty contributed from instrumental error is insignificant when compared to that from process fluctuations, which is reasonable given that typically the difference is several orders of magnitude.

V. RESULTS

The results of the water balance showed that water was mainly circulated within the paper machine, before being used for counter-current washing within the fibreline and then the digester.

Sodium, potassium and sulfur all behaved similarly as these elements where almost entirely circulated within the chemical recovery loop (i.e. from the white liquor to the weak black liquor and vice versa). The loss of sodium and sulfur (cooking chemicals) were accounted for in the amount of these elements remaining in the paper product, as well as in the losses from the chemical recovery process. The amount of these elements remaining in the paper is determined by the efficiency of the pulp washing process in the fibreline. The calcium balance results revealed that calcium, both dissolved and suspended, was carried from the paper machine into fibreline by the white water used for pulp washing in a counter-current flow fashion from the machine.

The uncertainties in the averages and analysis results were propagated through all of the mass balance calculations, allowing the results of the mass balances to be evaluated.

It was generally found that the discrepancies (residuals) within the component balances (water, Na, S, Ca, K, Cl) were close to or smaller than the uncertainty within the discrepancies, indicating that the discrepancies could have been negligible within the bounds of experimental uncertainty. This verified the accuracy of the balances in accounting for all of the material flows within the mill, and of the procedures for sample collection and analysis.

The only exception was the chlorine component balances. These showed discrepancies much larger than the uncertainties within the discrepancies, suggesting that a large systematic error had occurred within the balance procedures. Given that this systematic error was absent from the other component balances, the error could be traced back to method used to analyze the samples for chlorine. Chloride ion selective electrodes are very sensitive to sulfide ions, which at high concentrations cause the surface of the electrode to become passivated. At low concentrations the ions cause interference, and inaccurate readings. Even though measures were taken to destroy the sulfide in all samples known to contain sulfide (acidification and oxidization), these measures proved to be insufficient, as shown by the results of the chlorine component balances.

Samples of the results obtained from the balances can be found in Figure 2, Figure 3, and Table 1.

VI. DATA RECONCILIATION

It was found that the data measurements, both laboratory analysis measurements and process measurements, all contained a small but significant degree of uncertainty. These uncertainties will lead to some degree of inaccuracy when the balance results are used to monitor, control, or make any decision regarding the process. Hence, advanced data reconciliation (DR) was used to compensate for the uncertainties associated with the data measurements.

In the context of this balance study, the key results are the elemental flows obtained from the material balance equations which were a linear combination of the elemental flows. Therefore, a linear DR technique was utilized, and a batch processing method [2] was used to solve the DR problem. The data reconciliation problem was formulated to minimize the weighted least-squares estimation:

$$\underset{x}{Min}\ J = (y-x)^T\ \Psi^{-1}\ (y-x)$$
$$s.t.\ \ Ax = 0$$

Where **A** is a matrix (m × g) of known constants, and **x** is the vector of variables containing the measured process variables. In this case, all variables are redundant. The solution to this problem is obtained using the Lagrange multipliers method. The new estimate of the process variable ($\hat{x}$) can be obtained as:

$$\hat{x} = y - \Psi A_1^T (A_1 \Psi A_1^T)^{-1} A_1 y$$

The Microsoft Excel Spreadsheet Package was used along with its Visual Basic Macro Language to perform all calculations. The reconciled balance results showed that for the flow values, all the balances were now satisfied (i.e. balance discrepancies were equal or close to zero) (Figure 3 and Table 1). The reconciled results also showed that the associated uncertainties were reduced/minimized.

VII. MATERIAL BALANCE TOOLKIT

To make any decision on time, the mill requires up-to-date results. Therefore, the time taken for acquiring and processing the data to provide useful and reliable results needs to be reduced, which can be achieved with the aid of a fully automated procedure. This section will discuss the main steps involved in the construction of this mass balance toolkit.

At the VTPPM, process variables are measured by various measurement devices linked to the Distributed Control System (DCS), processed, and finally stored in the Information Management System (IMS). Each measured variable has a unique tag, consistent in both the DCS and the IMS. To retrieve the flow data from the IMS for a certain period of time, Simple Query Language (SQL) was used to communicate with the IMS and extract the required data for a specified tag and time period. The data retrieved was then imported into an Excel Spreadsheet with the aid of Visual Basic. Error analysis functions and data reconciliation formulation were also constructed in Visual Basic to process the propagated uncertainty calculations. A historical data base was also built into the toolkit to record and store key variables before each new update. This database is useful for monitoring the mill's operation, checking process consistency, and makes an excellent trouble shooting tool, as the source of process changes and upsets can be tracked.

The results obtained are reported using three main techniques which allow staff and management to obtain an informative overview of the mill's current operation. Reporting techniques include: (1) Graphical Interfaces (Figure 4), (2) Graphical Trends, and (3) Tabulation. The material balance simulation toolkit has made the mass balance fast and repeatable in a standard and systematical procedure. It also allows the mill staff to check the consistency of the mill's performance, as well as to perform trouble shooting by accessing the historical database.

VIII. CONCLUSIONS

The material balance strategy developed to determine the flow profiles of water and key elements within the VTPPM is repeatable, simple, labor saving, and rational, without compromising the level of accuracy. This procedure allows ongoing monitoring and management of water and chemicals within the VTPPM.

IX. REFERENCES

[1] Minday M, Douglas B., Brown C. (1997). "An Overview of Various Strategies for Balancing Saltcake, Chloride and Potassium Levels in an ECF Kraft Mill". TAPPI Minimum Effluent Symposium, San Francisco, October 23-24, 1997.

[2] Romagnoli, J.A, and Sanchez, M.C. (2000), "Data Processing and Reconciliation for Chemical Processes Operations", Academic Press.

[3] Siren, K (1998), "Analytical Methods for Determination of Metals, Silicon and Phosphorous in Kraft Mill Samples", TAPPI Journal, Vol. 3, p 1113-1119.

[4] Skoog D. A., West, D. M., and Holler, F. J. (1996), "Fundamentals of Analytical Chemistry", 7th edition, Sounders College Publishing,

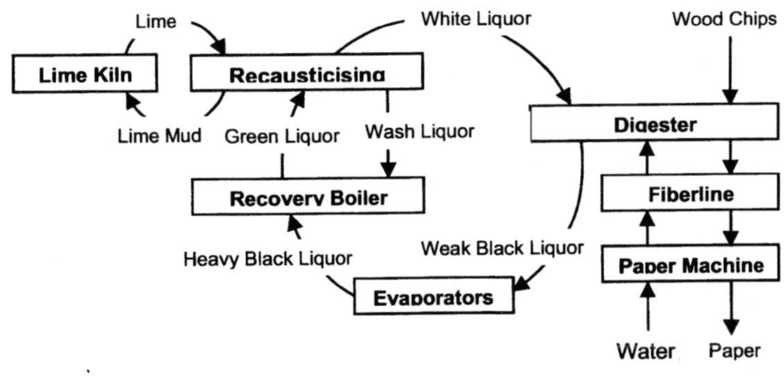

Figure 1: Schematic representation of the Kraft process within the mill.

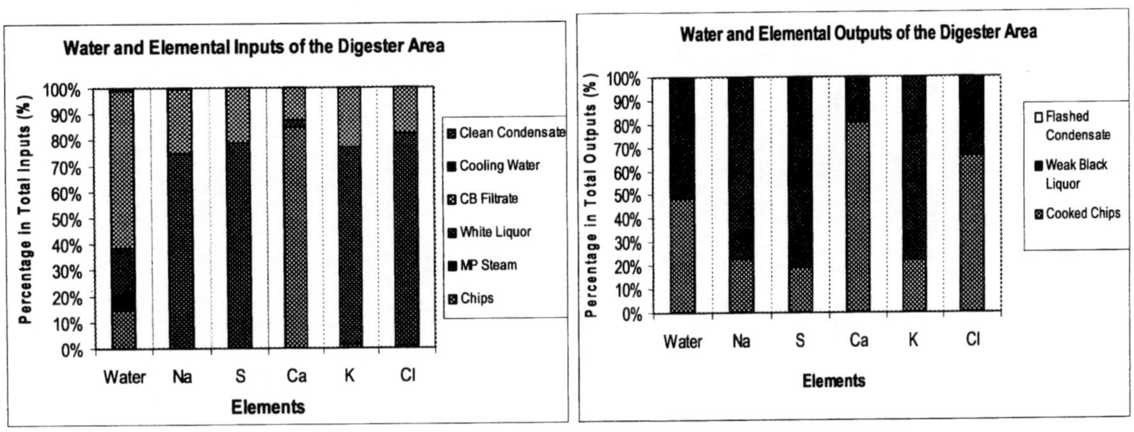

Figure 2: Percentages of water and elements input and output flows in the Digester Area.

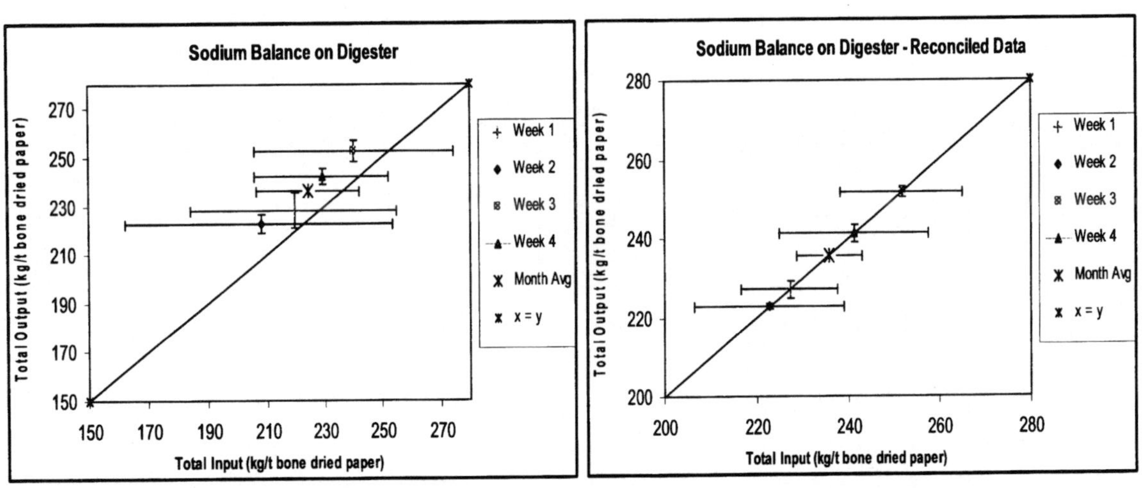

Figure 3: Elemental sodium balance results before and after data reconciliation in the Digester Area.

Table 1: Values of Na flows in different streams before and after the data reconciliation within the Fibre-Paper processing area.

Streams	Na Before		Na After	
	Average	Uncertainties	Average	Uncertainties
Chips	0.02	0.00	0.0226	0.0009
MP Steam	0.00	0.00	0.000	0.000
White Liquor	170	20	182	7
Cold Blow Filtrate	55	5	52	1
Cooling Water	0.004	0.006	0.005	0.001
Clean Condensate	1.18	0.00	1	0.00
Cooked Chips	53	1	53.9	0.3
Weak Black Liquor	183	2	182	1
Condensate to Cond Mill	0.00	0.00	0.000	0.00
Back Water	3.8	0.4	3.6	0.1
Seal Water	0.00	0.00	0.002	0.00
Pulp	5.2	0.5	5.5	0.1
Steam to HE	0.00	0.00	0.000	0.00
MP Steam to Steam Sys	0.00	0.00	0.000	0.00
Clean Condensate	0.01	0.00	0.015	0.004
White Water Condensate	0.03	0.00	0.034	0.001
Fresh Water	0.00	0.00	0.0024	0.0003
Recycle Paper	0.29	0.05	0.39	0.05
H_2SO_4	0.00	0.00	0.000	0.00
Alum	0.00	0.00	0.000	0.00
Paper	2.33	0.03	2.29	0.02
Condensate to Boiler	0.00	0.00	0.000	0.00
Steam Losses	0.000	0.00	0.000	0.00

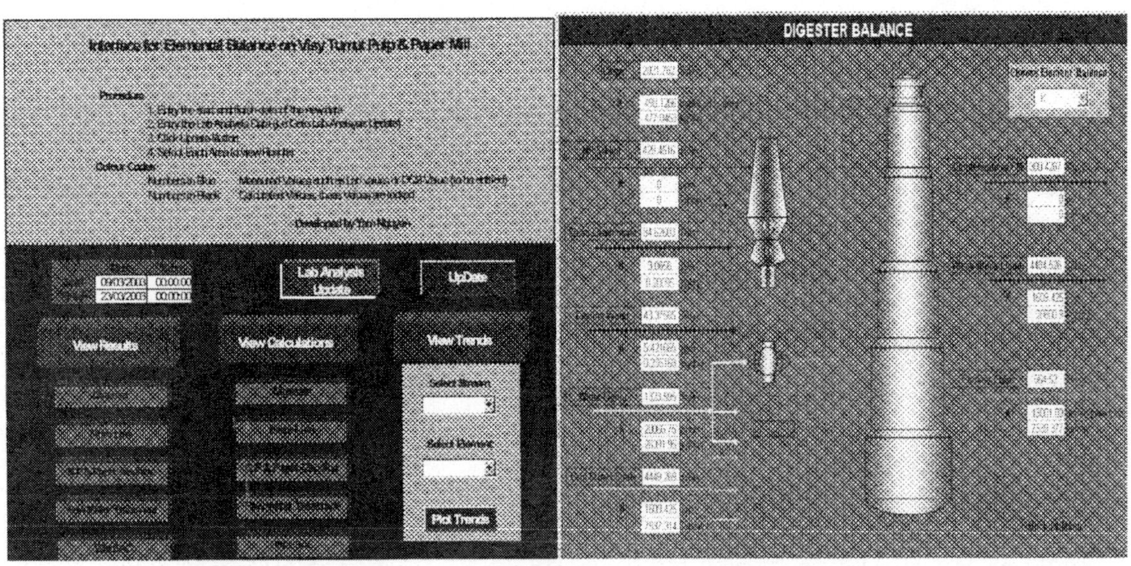

Figure 4: Graphical Interface of the Menu.

ELSEVIER
IFAC
PUBLICATIONS
www.elsevier.com/locate/ifac

PRACTICAL SOLUTIONS TO MULTIVARIATE FEEDBACK CONTROL PERFORMANCE ASSESSMENT

B. Huang *,[1] **S.X. Ding** **

* *Dept. of Chemical and Materials Engg., Uinv. of Alberta, Edmonton, AB, Canada, T6G 2G6*
** *Univ. of Duisburg-Essen, Inst. Auto. Cont. and Comp. Sys./Faculty 5, Bismarckstrasse 81, BB511, 47048 Duisburg, Germany*

Abstract: Performance assessment of multivariate feedback control with minimum variance control as the benchmark requires an interactor matrix to filter the closed-loop output. This is to transfer the coordinate of the original variables into a new one in order to separate the control invariant disturbance dynamics from the first few terms of the closed-loop output Markov parameters. There has been a great deal of interest to simplify this approach, in particular, to find methods that do not need the general interactor matrix. With this motivation, this paper discusses practical solutions to multivariate control performance assessment without relying on the general interactor matrices. In particular, we will consider two practical scenarios, 1) no *a priori* knowledge about the process model at all and 2) known time delays between each pairs of inputs and outputs. Solutions to these two scenarios are proposed. Several examples illustrate the feasibility of the proposed approaches. *Copyright © 2004 IFAC*

Keywords: Performance monitoring, performance assessment, control monitoring, multivariate systems.

1. INTRODUCTION

In recent years, there are growing research interests in reducing the complexity of the *a priori* knowledge requirement for multivariate control performance assessment, such as (Ko and Edgar, 2001; Kadali and Huang, 2004; McNabb and Qin, 2001). Although these attempts have reduced the complexity of the *a priori* knowledge requirement to some extent, they all require certain information that is computationally simpler but fundamentally equivalent to the interactor matrices, for example, the open-loop process Markov parameter matrices, the lower triangular Toeplitz matrix, or the multivariate time delay (MTD) matrix. That is, they all require *a priori* knowledge that is beyond the pure time delays between each pairs of the inputs and outputs.

In the univariate case, one interprets output variance under minimum variance control as the variance of the optimal prediction error for the given time delay of a process. One can imagine that if a closed-loop output is highly predictable, then one should be able to do better, i.e. to compensate the predictable content by a well designed controller. Should a better controller be implemented, then the closed-loop output would have been less predictable. Therefore, the high predictability of

[1] Corresponding author. This work is supported in part by Humboldt Research Fellowship of Germany.

a closed-loop output implies the potential to improve its performance by control re-tuning and/or re-design, or in other word, the existing controller may not have been satisfactory.

The actual process often has time delays, which prevent the complete compensation of the predictable content of the output. For example, if a univariate process has two sample time delays, then the compensation control action will not take effect on the output until two steps later and the one step ahead prediction will not be useful for its compensation. In this case, the best a controller can do is to compensate the predicted content according to the two step optimal prediction (multistep optimal prediction) and the minimum control error will coincide with the two-step prediction error. Therefore, the two-step optimal prediction error is the lower bound of the output error that can be achieved by a feedback controller. This lower bound is also known as the minimum variance that is often used for control loop performance assessment (Harris, 1989).

Although the same rationale can not be exactly carried over to multivariate processes due to the relatively complex delay structure for multivariate processes, multi step optimal predictions provide useful information about the control performance. We will show that the multi-step optimal prediction error is analogous to closed-loop step response of the univariate process from the white noise to the output. The analogy provides an interesting interpretation of the multi-step optimal prediction error for multivariate processes and results in a new method for performance assessment.

While the prediction error based approach is general, if certain process model knowledge such as time delays between each pairs of inputs and outputs are available, one may be able to do more about control performance assessment than the prediction error based approach. It is known that the diagonal form of the interactor matrix only depends on the pair-wise time delays (delays between each pairs of the inputs and outputs). If one can determine that the process has the diagonal form of the interactor, the computation for the performance assessment can be greatly simplified (Huang and Shah, 1999). Thus there is a need to determine whether a process has a diagonal interactor matrix from the given pair-wise time delays.

Motivated by the above discussions, this paper dedicates to (1) development and analysis of the practical performance assessment approach based on optimal predictions, (2) development of methods to determine whether the process has a simple or diagonal form of interactor matrices. The remainder of this paper is organized as follows. In section 2, results on multivariate feedback control performance assessment without relying on any *a priori* process model knowledge are presented. The performance assessment problem, when the time delays between each pairs of inputs and outputs are known, is discussed in Section 3, followed by concluding remarks in section 4.

2. ASSESSMENT OF MULTIVARIATE CONTROL PERFORMANCE WITHOUT ANY A PRIORI KNOWLEDGE OF PROCESS MODEL

2.1 Performance assessment using optimal multi step prediction errors

The following multivariate process will be considered in this paper:

$$Y_t = TU_t + Na_t \tag{1}$$

where $T$and N are proper (causal), rational transfer function matrices in the backshift operator q^{-1}; Y_t, U_t and a_t are output, input and noise vectors of appropriate dimensions. a_t is further assumed to be white noise with zero mean and $Var(a_t) = \Sigma_a = I$. However, if $\Sigma_a \neq I$, one can always normalize N such that $N \leftarrow N\Sigma_a^{1/2}$ and $a_t \leftarrow \Sigma_a^{-1/2} a_t$, and then the new a_t will satisfy $\Sigma_a = I$. Furthermore, we assume that T does not have nonmimimum phase zeros in multivariable sense (Huang and Shah, 1999). The interactor matrix of T is denoted as D with order d and the interactor matrix has three possible forms: simple, diagonal and general. For details on the three forms of the interactor matrix, readers are referred to (Huang and Shah, 1999).

If the process (1) is controlled by a linear feedback controller, the closed-loop model from a_t to Y_t can be represented by a moving average or a Markov parameter form:

$$Y_t = F_0 a_t + \cdots + F_{i-1} a_{t-(i-1)} + F_i a_{t-i} + \cdots \tag{2}$$

This moving average model can be estimated from routine operating data without any *a priori* knowledge about the process.

Since a_t is white noise, the optimal ith step prediction is given by

$$Y_{t|t-i} = F_i a_{t-i} + F_{i+1} a_{t-i-1} + \cdots \tag{3}$$

and the prediction error $e_{t|t-i} = Y_t - Y_{t|t-i}$ is given by

$$e_{t|t-i} = F_0 a_t + F_1 a_{t-1} + \cdots + F_{i-1} a_{t-(i-1)} \tag{4}$$

The covariance of the prediction error can be calculated as

$$Cov(e_{t|t-i}) = F_0 F_0^T + F_1 F_1^T + \cdots + F_{i-1} F_{i-1}^T$$

and its scalar measure

$$s_i \triangleq tr[Cov(e_{t|t-i})] = tr(F_0F_0^T + \cdots + F_{i-1}F_{i-1}^T)$$

The incremental of the prediction error can be calculated as

$$r_i \triangleq tr[Cov(e_{t|t-i}) - Cov(e_{t|t-(i-1)})] = tr(F_{i-1}F_{i-1}^T)$$

If we plot s_i versus i, then the plot reflects how the prediction error increases with the prediction horizon. Note that as $i \to \infty$, $Cov(e_{t|t-i}) \to Cov(Y_t)$. This fact can be seen by comparing eqn(2) and (4).

A plot of r_i versus i indicates how the incremental of the prediction error changes with the prediction horizon i. These two plots (r_i and s_i) will be shown to be useful for the interpretation of the control performance.

To understand s_i and r_i, a deterministic interpretation of eqn(2) is helpful. The moving average model of eqn(2) is applicable to both stochastic and deterministic input. Considering that a_t is a unit impulse, then the coefficient matrices $F_0, F_1, \cdots, F_i, \cdots$ are the impulse response coefficients of the closed-loop system to the impulse disturbance. Since $r_i = tr(F_iF_i^T)$, r_i is a 2-norm measure of the impulse response coefficients and is analogous to the squared impulse response coefficients of a univariate process. Therefore a plot of r_i versus i reflects how the disturbance is regulated by the controller, i.e. how the control error is reduced versus time. Each point of the plot, for example r_i, represents the squared error of the closed-loop response at time i due to an impulse disturbance. Therefore, this plot of r_i versus i, is a good indication of closed-loop performance of a multivariate controller, which has been suggested in the literature for multivariate control performance monitoring (Shah *et al.*, 2002). However, we will show that a s_i plot is more useful.

Since $s_i = tr(F_0F_0^T + F_1F_1^T + \cdots + F_iF_i^T)$, s_i is nothing but the sum of squared error of the closed-loop response to an impulse disturbance up to time i. If we plot s_i versus i, each point of the plot, for example s_i, represents the sum of squared error (SSE) of the closed-loop response up to time i. If, after time i, the disturbance can be completely controlled (by a deadbeat control for example), then the total error of the response will remain to be s_i, which is the i step optimal prediction error from stochastic view point. There are two characters in a s_i plot worth noting. 1) It is monotonically increasing or non-decreasing curve. 2) Its steady state is $tr[Cov(Y_t)]$, the actual variance of the output. Since s_i is the integration of the squared closed-loop impulse response, it is analogous to the step response and can be used to determine dynamic information such as the settling time of the closed-loop response to the disturbance.

Motivated by the interpretation of s_i, we define the closed-loop potential p_i as

$$p_i \triangleq \frac{s_\infty - s_i}{s_\infty} \tag{5}$$

Since s_i is monotonically increasing with i, p_i is monotonically decreasing. Since $s_0 = tr[Cov(Y_t - Y_{t|t})] = 0$, $p_0 = 1$. Therefore, p_i starts from 1 at $i = 0$ and monotonically decreases to 0 and $0 \leq p_i \leq 1$. p_i can be interpreted as follows: If a deadbeat control action can be applied from time i, then the process output SSE can be reduced by $100 \times p_i$ percent. From stochastic view point, if i is greater than the interactor order d (Huang and Shah, 1999), there exists a control such that the variance of the multivariate output can be reduced by $100 \times p_i$ percent of the current variance (see Lemma 1 next for this control law). Since the order of the actual interactor matrix is usually unknown, one would look for the trajectory of the closed-loop potential versus a range of possible d. Potential plots such as those illustrated in Fig. 2 are useful. Faster the potential decays to zero, less the possibility to improve control performance. Due to the monotonically decreasing nature of the potentials and fixed starting and ending values of the potentials, the area below the potential plot well reflects the rate of its decaying. Therefore, it is possible to define a scalar index to monitor the change of the closed-loop potential. This index is called relative potential index and can be calculated as

$$\eta_p = \frac{\int p_i^{(2)}}{\int p_i^{(1)}} - 1 \tag{6}$$

where $p_i^{(1)}$ is a reference potential calculated, for example, from the data sampled before control tuning, and $p_i^{(2)}$ is calculated from data sampled after the tuning. The value of η_p gives the percent change of the closed-loop potential with the positive sign indicating an increased potential and the negative sign indicating a decreased potential. Note that an increase of the potential implies a deteriorated tuning while a decrease of the potential implies an improved tuning.

Lemma 1. If there is a controller with $Q^* = Q_0^*$ expressed in the IMC framework such that its closed-loop output can be written as

$$Y_t = \underbrace{F_0 + \cdots + F_{i-1}q^{-(i-1)}}_{F} + q^{-i}R \tag{7}$$

where $i \geq d$ and R is a proper rational transfer function matrix, then there exists another physically realizable controller with $Q^* = Q_i^*$ such that its closed-loop output can be written as

$$Y_t = \underbrace{F_0 + \cdots + F_{i-1}q^{-(i-1)}}_{F} \tag{8}$$

and the controller can be written as

$$Q_i^* = Q_0^* + \tilde{T}^{-1}(q^{-i}D)RN^{-1} \tag{9}$$

where, the relation between the IMC control, Q^*, and the actual feedback control, Q, is given by

$$Q = Q^*(I - TQ^*)^{-1} \tag{10}$$

PROOF. Omitted due to space limit.

This lemma tells us that s_i, the i-step prediction error or the sum squared error, is achievable by a physically implementable controller if $i \geq d$.

2.2 Example

Example 2. Consider a 2×2 multivariable process with the open-loop transfer function matrix T and disturbance transfer function matrix N given by

$$T = \begin{bmatrix} \dfrac{q^{-1}}{1-0.4q^{-1}} & \dfrac{0.5q^{-2}}{1-0.1q^{-1}} \\ \dfrac{0.3q^{-1}}{1-0.4q^{-1}} & \dfrac{q^{-2}}{1-0.8q^{-1}} \end{bmatrix}$$

$$N = \begin{bmatrix} \dfrac{1}{1-0.5q^{-1}} & \dfrac{-q^{-1}}{1-0.6q^{-1}} \\ \dfrac{q^{-1}}{1-0.7q^{-1}} & \dfrac{1.0}{1-0.8q^{-1}} \end{bmatrix}$$

The white noise excitation, a_t, is a two-dimensional normally-distributed white noise sequence with $\Sigma_a = I$.

Consider that the following multiloop controller is implemented in the process:

$$Q = k \begin{bmatrix} \dfrac{0.5-0.20q^{-1}}{1-0.5q^{-1}} & 0 \\ 0 & \dfrac{0.25-0.200q^{-1}}{(1-0.5q^{-1})(1+0.5q^{-1})} \end{bmatrix}$$

In this example, three control gains, $k = 2.8, 3, 3.2$ respectively, are considered. s_i and r_i for $i = 1, 2, \cdots, 10$ are calculated and plotted in Fig.1. The s_i plot (top sub-figure) indicates that the closed-loop settling time increases with the increasing of the control gain, so does the SSE. For example, the settling time for $k = 2.8$ is about 5 samples while the settling time for $k = 3.2$ is more than 10 samples. The r_i plot shown in the bottom sub-figure of Fig.1 presents the similar information as s_i such as the information about the

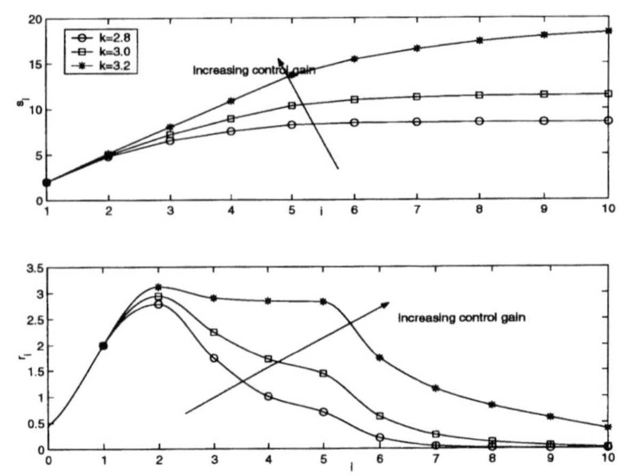

Fig. 1. s_i and r_i plots.

settling time. However, unlike the s_i plot which is monotonically increasing, the r_i plot has a more complicated and hard-to-interpret shape. We will therefore recommend to use the s_i plot and the p_i plot(to be discussed next).

The potential plot of p_i shown in Fig.2 is possibly more useful in the interpretation of control performance. For example, s_i for $k = 3.2$ has a slowest rate to approach its steady state and thus its potential decreases to zero at the slowest rate. For a considerable range of the process delays (expressed by interactor order d for example), its potential is significantly different from zero. For example, for an interactor order up to 5 samples, the potential is larger than 0.3, i.e. 30% reduction of variance is possible for the interactor order up to 5. On the other hand, for the tuning of $k = 2.8$, the potential dies to zero quickly. In this case, there is not much potential left after the interactor order is greater than 3.

For control tuning of multivariate systems or control upgrading from multiloop control to multivariable control such as MPC, one is interested in whether control performance is indeed improved. If an existing control gain is $k = 3$, assume that the gain is tuned to 2.8 or 3.2 and representative closed-loop data are sampled before and after the tuning. Then the scalar measures of the relative closed-loop potentials calculated from the data are -0.21 or 0.39 for tunings $k = 2.8$ or 3.2 with $k = 3.0$ as the reference. These results indicate that 1) if the controller gain increases to 3.2, then the resulting system has increased closed-loop potential by 39%, indicating a deteriorated performance; 2) if the controller gain decreases to 2.8, then the resulting system has reduced closed-loop potential by 21%, indicating an improved performance.

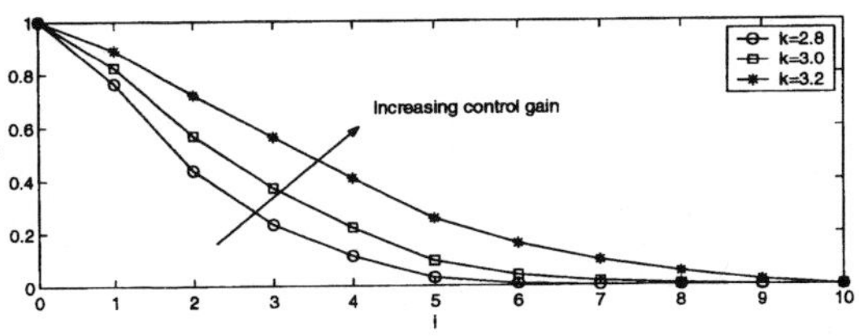

Fig. 2. Potential p_i plot.

3. ASSESSMENT OF MULTIVARIATE CONTROL PERFORMANCE WITH KNOWN PAIR-WISE TIME DELAYS

If the time delays between each pair of inputs and outputs are known *a priori*, we should search for a possible simple or diagonal structure of the interactor matrix, which can greatly simplify the computation of the multivariate minimum variance. Both the simple and the diagonal interactor matrices can be calculated from the time delays between each pairs of inputs and outputs of the process. Once the interactor matrix is known, multivariate control performance assessment problem is readily solved following the procedure of (Harris *et al.*, 1996; Huang and Shah, 1999). One may surprisingly find that the simple and diagonal interactor matrices are not uncommon, particularly in industrial process, where the sparse structure of the transfer function matrix is often observed. The sparse structure also facilitates the determination of the interactor structure.

Consider a multivariable transfer function matrix of dimension $n \times m$ given by

$$T = \begin{pmatrix} T_{11}q^{-d_{11}} & T_{12}q^{-d_{12}} & \cdots & T_{1m}q^{-d_{1m}} \\ T_{21}q^{-d_{21}} & T_{22}q^{-d_{22}} & \cdots & T_{2m}q^{-d_{2m}} \\ \cdots & \cdots & \cdots & \cdots \\ T_{n1}q^{-d_{n1}} & T_{n2}q^{-d_{n2}} & \cdots & T_{nm}q^{-d_{nm}} \end{pmatrix} \quad (11)$$

where T_{ij} is a scalar transfer function from the jth input to the ith output. Define a delay matrix

$$\Psi = \begin{pmatrix} t_{11}q^{-d_{11}} & t_{12}q^{-d_{12}} & \cdots & t_{1m}q^{-d_{1m}} \\ t_{21}q^{-d_{21}} & t_{22}q^{-d_{22}} & \cdots & t_{2m}q^{-d_{2m}} \\ \cdots & \cdots & \cdots & \cdots \\ t_{n1}q^{-d_{n1}} & t_{n2}q^{-d_{n2}} & \cdots & t_{nm}q^{-d_{nm}} \end{pmatrix} \quad (12)$$

where d_{ij}'s are time delays that are assumed known; t_{ij} is the first non-zero impulse response coefficient from the jth input to the ith output, which is typically unknown. From Ψ, we can arrive a diagonal matrix

$$\Theta = \begin{pmatrix} q^{d_1} & & & \\ & q^{d_2} & & \\ & & \ddots & \\ & & & q^{d_n} \end{pmatrix} \quad (13)$$

where $d_i = min\{d_{ij} : j = 1, \cdots, m\}$. Then the following lemma is true:

Lemma 3. If T has a diagonal interactor matrix D then $D = \Theta$, where T and Θ are given by eqn(11), and (13), respectively.

PROOF. Omitted due to space limit.

The following lemma provides a useful criterion to determine whether a process T has a diagonal interactor matrix:

Lemma 4. If $K = \lim_{q^{-1}\to 0} \Theta\Psi$ is of full rank for all $t_{ij} \neq 0$, then 1) the interactor matrix is diagonal and 2) $D = \Theta$.

PROOF. The first part can be proved by noticing the fact that $\lim_{q^{-1}\to 0} \Theta T = \lim_{q^{-1}\to 0} \Theta\Psi = K$. By the definition of the interactor matrix, if K is of full rank, Θ must be an interactor matrix of T. Based on the result of the first part, the second part of the proof directly follows from Lemma 3.

Lemma 4 provides a sufficient condition for the determination of the interactor structure. In practice, one can relax this condition by checking the determinant of K (if K is not a square matrix, one has to use the singular values instead), to determine conditions for the singularity. MATLAB symbolic toolbox is useful for such an application. The procedure is as follows: 1) calculate the determinant or singular values, 2) find condition for the determinant to be zero or the singular values to be zero, and 3) check whether these conditions hold. Next, we will demonstrate the method of determining the structure of the interactor matrix using an industrial example.

Example 5. Consider an industrial 6×6 process presented in (Gao *et al.*, 2003), which has the following pair-wise delay matrix:

$$\Psi = \begin{pmatrix} q^{-1} & 0 & 0 & 0 & 0 & 0 \\ q^{-1} & q^{-1} & q^{-1} & 0 & q^{-1} & 0 \\ q^{-1} & q^{-1} & q^{-1} & 0 & q^{-1} & 0 \\ 0 & 0 & 0 & 0 & q^{-1} & 0 \\ q^{-2} & q^{-2} & 0 & q^{-2} & q^{-2} & q^{-2} \\ 0 & 0 & 0 & q^{-1} & q^{-1} & 0 \end{pmatrix} \quad (14)$$

An unitary interactor matrix was calculated in (Gao *et al.*, 2003) using *complete knowledge* of the process transfer function matrix. The result was

$$D = \begin{pmatrix} 0.059q & 0.72q & 0.4q & -0.56 & 0 & 0 \\ 0.006q & 0.42q & -0.90q & -0.09q & 0 & 0 \\ -0.023q & -0.31q & -0.098q & -0.47q & 0 & 0.82q \\ -0.033q & -0.44q & -0.14q & -0.67q & 0 & -0.57q \\ -0.997q & 0.068q & 0.025q & 0 & 0 & 0 \\ 0 & 0 & 0 & 0 & q^2 & 0 \end{pmatrix} \quad (15)$$

Using the method discussed in this section, we would get

$$\Theta = \begin{pmatrix} q & & & & & \\ & q & & & & \\ & & q & & & \\ & & & q & & \\ & & & & q^2 & \\ & & & & & q \end{pmatrix} \quad (16)$$

and K matrix

$$K = \begin{pmatrix} t_{11} & 0 & 0 & 0 & 0 & 0 \\ t_{21} & t_{22} & t_{23} & 0 & t_{25} & 0 \\ t_{31} & t_{32} & t_{33} & 0 & t_{35} & 0 \\ 0 & 0 & 0 & 0 & t_{45} & 0 \\ t_{51} & t_{52} & 0 & t_{54} & t_{55} & t_{56} \\ 0 & 0 & 0 & t_{64} & t_{65} & 0 \end{pmatrix} \quad (17)$$

The determinant of K can be calculated as $det(K) = t_{25}t_{64}t_{56}(t_{22}t_{33} - t_{32}t_{23})$. Due to the sparse structure of industrial processes, the determinants often have such a simple structure. Since $t_{ij} \neq 0$ by the definition, the condition for the determinant to be zero is $\frac{t_{22}}{t_{32}} = \frac{t_{23}}{t_{33}}$. Whether this condition holds can be easily determined by checking the variables of the process. As discussed in (Gao *et al.*, 2003), CV_2 is temperature, CV_3 internal reflux ratio, MV_2 reboiler stem flow, and MV_3 internal reflux flow. The chance for the condition to be true has the probability $\rightarrow 0$. Therefore, the interactor matrix should have a diagonal structure and the complete knowledge of the multivariate process, which was assumed in (Gao *et al.*, 2003), is not necessary. In the worst case, if one is not able to determine whether the condition is true, one would at most need to find out the first non-zero impulse response coefficients of four sub-transfer functions, a significant reduction of the *a priori* knowledge than the complete transfer function matrices.

Is this result in a contradiction to that of (Gao *et al.*, 2003) shown in eqn(15)? It has been shown in (Huang and Shah, 1999) that the two unitary interactor matrices are equivalent if

$$\bar{D} = \Gamma D \quad (18)$$

where Γ is a unitary constant matrix. It can be shown by using QR decomposition that if

$$\Gamma = \begin{pmatrix} -0.0592 & -0.0065 & 0.0236 & 0.0330 & 0.9974 & 0 \\ -0.7234 & -0.4251 & 0.3103 & 0.4417 & -0.0679 & 0 \\ 0.4000 & -0.9003 & -0.0976 & -0.1389 & 0.0249 & 0 \\ -0.5596 & -0.0934 & -0.4734 & -0.6738 & 0 & 0 \\ 0 & 0 & 0 & 0 & 0 & 1 \\ 0 & 0 & 0.8183 & -0.5748 & 0 & 0 \end{pmatrix} \quad (19)$$

pre-multiplying eqn(15) by Γ results in eqn(16), a diagonal interactor matrix! Therefore, difficult to see or not, eqn(15) is in fact a diagonal interactor matrix. The thumb of rule is that higher the dimension, higher the chance to have a diagonal interactor structure.

Once the diagonal structure of the ineractor matrix is determined, the multivariate minimum variance can be determined by following the approaches discussed in (Harris *et al.*, 1996; Huang and Shah, 1999) in addition to the multi-step optimal prediction based approach for performance assessment.

4. CONCLUSION

In this paper, we have discussed two practical approaches for multivariate feedback control performance assessment. The first approach assumes that there is no any *a priori* knowledge about the process model. The solution is based on the multi-step optimal prediction error. This approach is general and applicable to processes with or without *a priori* model information. The second approach assumes that time delays between each pair of inputs and outputs are available. Then the structure of the interactor matrices may be determined from the given time delays. The simulation examples have shown the feasibility of the proposed algorithms.

REFERENCES

Gao, J., R. Patwardhan, K. Akamatsu, Y. Hashimoto, G. Emonto, S.L. Shah and B. Huang (2003). Performance evaluation of two industrial mpc controllers. *Control Engineering Practice* **11**, 1371–1387.

Harris, T. (1989). Assessment of closed loop performance. *Can. J. Chem. Eng.* **67**, 856–861.

Harris, T.J., F. Boudreau and J.F. MacGregor (1996). Performance assessment of multivariable feedback controllers. *Automatica* **32**(11), 1505–1518.

Huang, B. and S.L. Shah (1999). *Control Loop Performance Assessment: Theory and Applications.* Springer Verlag, London.

Kadali, R. and B. Huang (2004). Multivariate control performance assessment without interactor matrix. In: *Preprints IFAC Advanced Control of Chemical Processes.*

Ko, B.S. and T.F. Edgar (2001). Performance assessment of multivariate feedback control systems. *Automatica* **37**(5), 899–905.

McNabb, C.A. and S.J. Qin (2001). Projection based mimo control performance monitoring: I - covariance monitoring in state space. *Journal of Process Control* **13**, 739–757.

Shah, S.L., R. Patwardhan and B. Huang (2002). Multivariate controller performance assessment: methods, applications and challenges. In: *Chemical Process Control - CPC VI, CACHE.* pp. 190–207.

ELSEVIER
IFAC
PUBLICATIONS
www.elsevier.com/locate/ifac

BAYESIAN ESTIMATION BY SEQUENTIAL MONTE CARLO SAMPLING: APPLICATION TO HIGH-DIMENSIONAL NONLINEAR DYNAMIC SYSTEMS

Wen-shiang Chen [*] **Bhavik R. Bakshi** [*]
Prem K. Goel [**] **Sridhar Ungarala** [***]

[*] *Dept. Chem. Eng., The Ohio State University Columbus, OH 43210, USA*
[**] *Dept. Statistics, The Ohio State University Columbus, OH 43210, USA*
[***] *Dept. Chem. Eng., Cleveland State University Cleveland, OH 44115, USA*

Abstract: Modern industrial enterprises have invested significant resources for collecting and distributing data, with the expectation that it will enhance profitability via better decision making. Due to the complexity of these problems, existing approaches tend to make convenient, but invalid assumptions so that tractable solution may be found. For example, for estimation in nonlinear dynamic systems, extended Kalman filtering (EKF) relies on Gaussian approximation and local linearization to find a closed-form solution. Moving horizon based least-squares estimation (MHE) also relies on Gaussian approximation, but the use of nonlinear models and constraints eliminates most of the computational benefits of this approximation, but can provide more accurate estimates than EKF. Unfortunately, in most practical nonlinear dynamic systems, the posterior distributions are often far from Gaussian, and continually change their shape.
Our previous work has developed rigorous Bayesian methods for estimation in nonlinear dynamic systems with constraints. These methods rely on recent theoretical developments in Sequential Monte Carlo Sampling (SMC). It does *not* rely on assumptions about the shape of the distributions, or nature of the models. Furthermore, this approach is expected to be computationally more efficient due to its recursive formulation that does not rely on nonlinear programming. These claims have been supported via applications to relatively small scale CSTR case studies. However, there are no illustrations or theoretical proofs to indicate how SMC performs for high-dimensional systems.
This paper applies our previous work on Bayesian rectification by SMC to large scale nonlinear dynamic systems and compares the computational efficiency and accuracy with MHE and EKF. The model of the selected polymerization reactor contains eight variables, exhibit significant linear and nonlinear dynamics under different operating conditions, and requires satisfaction of process constraints. Results indicate that the accuracy and computational benefits of SMC are significant even for such high-dimensional systems. *Copyright © 2004 IFAC*

Keywords: Bayesian estimation, Sequential Monte Carlo sampling

1. INTRODUCTION

Efficient operation of chemical and manufacturing processes relies on cleaning or rectification of measured data and estimation of unknown quantities. Data rectification and estimation form the foundation for process operation tasks such as process control, fault detection and diagnosis, real-time estimation, process monitoring, and process scale-up. Since most processes are nonlinear and subject to constraints, significant efforts have already focused on methods for estimation of nonlinear dynamic systems (Kramer and Mah, 1994; Robertson *et al.*, 1996).

In our previous work, we have developed rigorous Bayesian methods for estimation in nonlinear dynamic systems. This approach relies on recent theoretical developments at the interface of statistical physics and Bayesian statistics. Unlike existing methods, this approach of sequential Monte Carlo sampling (SMC) does not rely on assumptions about the shape of the distributions, or nature of the models. Instead, it allows the distributions to adopt any shape according to system dynamics or constraints. Furthermore, the proposed approach is expected to be computationally efficient due to its recursive formulation. Previous results have illustrated that SMC shows significant improvement in estimation accuracy over existing methods, and in computation time over moving horizon based least-squares estimation (MHE) for cases with or without constraints (Chen *et al.*, 2004; Chen *et al.*, 2003*b*). However, these illustrations were based on relatively low-dimensional examples, and were not enough to address skepticism about the feasibility and advantages of SMC based Bayesian estimation for higher dimensional problems.

Such skepticism is well-founded for data-based methods due to the "curse of dimensionality", and may also apply to sampling based methods. Traditionally, Bayesian methods have not been feasible for high-dimensional systems since these methods usually require integration in high-dimensional space over a relatively fine grid, which leads to formidable requirement of computation. This experience leads to the perception and expectation that the computation load of sampling based methods might also increase exponentially thus making them impractical for most large scale problems.

This paper demonstrates that Bayesian estimation by SMC is feasible even for high-dimensional systems via application to a polymerization reactor. The selected process model is nonlinear, dynamic and involves eight state variables. Non-negativity constraints are also imposed on some variables. The estimation accuracy and computational efficiency of EKF, MHE and SMC are studied under a variety of conditions. These studies show that SMC can be significantly faster than MHE for similar estimation accuracy. Since the posterior distributions for the selected operating conditions may be approximated quite well as Gaussian distributions, the accuracy of SMC and MHE can be comparable. The case study indicates that Bayesian estimation via SMC may not be subject to the curse of dimensionality, and can retain its theoretical rigor and practical benefits for high-dimensional problems. The theoretical basis for this observation is also discussed in this paper.

In the following sections, a review of Bayesian estimation is first introduced. After that, the algorithm for SMC is provided. Performance of the proposed approach is compared with that of existing approaches in the case study section.

2. BAYESIAN ESTIMATION FOR DYNAMIC SYSTEMS

In general, the goal of estimation may be expressed as follows. Given measurements $y_{1:k} = \{y_1, y_2, \ldots, y_k\}$, process models, and the distribution of the initial condition $p(x_0)$, determine the current state, x_k. Process models may be expressed as follows,

$$x_k = f_{k-1}(x_{k-1}, \omega_{k-1}) \tag{1}$$

$$y_k = h_k(x_k, \nu_k) \tag{2}$$

where $x_k \in \Re^{n_x}$ is the state vector and $f_k : \Re^{n_x} \times \Re^{n_\omega} \rightarrow \Re^{n_x}$ is the system equation. Measurements, $y_k \in \Re^{n_y}$, are related to the state vector through the measurement equation, $h_k : \Re^{n_x} \times \Re^{n_\nu} \rightarrow \Re^{n_y}$.

Bayesian estimation provides a rigorous way of maximizing the use of all available information and can handle all types of errors, models and constraints. All variables are considered to be stochastic, and previous belief can be combined with the current measurements via Bayes rule.

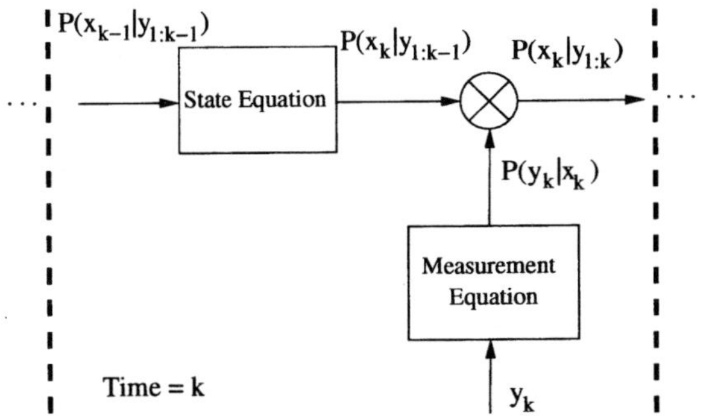

Fig. 1. Algorithm of recursive Bayesian estimation.

The recursive formulation of Bayesian estimation has been proposed decades ago and may be represented as follows (Ho and Lee, 1964),

$$p(x_k|y_{1:k}) = \frac{p(y_k|x_k)\ p(x_k|y_{1:k-1})}{p(y_k|y_{1:k-1})}, \quad (3)$$

where $p(x_k|y_{1:k})$ is the posterior. $p(x_k|y_{1:k-1})$ may be found by solving the following two equations.

$$p(x_k|y_{1:k-1}) = \int p(x_k|x_{k-1})\ p(x_{k-1}|y_{1:k-1})\ dx_{k-1} \quad (4)$$

$$p(x_k|x_{k-1}) = \int \delta(x_k - f_{k-1}(x_{k-1}, \omega_{k-1}))\ p(\omega_{k-1})\ d\omega_{k-1} \quad (5)$$

Likelihood distribution may be found as follows.

$$p(y_k|x_k) = \int \delta(y_k - h_k(x_k, \nu_k))\ p(\nu_k)\ d\nu_k \quad (6)$$

Further detail of the derivation of these equations can be found in Chen *et al.* (2004).

The algorithm for recursive Bayesian estimation may be visualized as in Figure 1. Information in previous measurements up to time $k-1$ is captured by the prior, $p(x_{k-1}|y_{1:k-1})$. Prediction of distribution of the current state is implemented by utilizing Equations (4) and (5). Information in current measurement is represented as the likelihood function based on Equation (6). The posterior can then be found by combining previous and current information by Equation (3).

After the posterior is available, a point estimate and its uncertainty may be obtained via,

$$E[\mathcal{L}(x_k)] = \int \mathcal{L}(x_k)\ p(x_k|y_{1:k})\ dx_k, \quad (7)$$

where $\mathcal{L}(\cdot)$ is a chosen loss or criterion function (Jazwinski, 1970). Mean, mode or median of the posterior are popular choices for the point estimate.

In general, there is no closed-form solution for Equations (4) to (6). Direct integration is computationally expensive and may not be practical for high-dimensional systems. Most estimation methods address these computational challenges by making simplifying assumptions about the nature of the model and/or posterior distributions at the cost of accuracy and computational efficiency. However, recent theoretical advances coupled with fast computation are providing the foundation of building a feasible Bayesian approach even for large scale systems. This computationally efficient algorithm is based on sequential Monte Carlo sampling and will be discussed in Section 3.

2.1 *Bayesian view of existing methods*

Existing methods may be described as special cases of Bayesian estimation. EKF assumes Gaussian prior and additive Gaussian noise and relies on local linearization to have closed-form solution (Jazwinski, 1970). MHE also relies on the assumption of Gaussian prior and additive Gaussian noise to have the least-squares formulation (Robertson *et al.*, 1996). MHE may be more accurate results than EKF by keeping the nonlinearity and being able to handle constraints. However, a closed-form solution is not available anymore. Instead, MHE needs to solve a constrained nonlinear programming problem over each moving window and lacks a recursive formulation. This makes MHE a computationally demanding algorithm. In the presence of constraints, MHE implicitly uses truncated Gaussian prior. Modifications via approximating truncated Gaussian distributions with combination of Gaussian or other fixed-shape distributions have also been suggested (Robertson and Lee, 2002). However, a general algorithm to decide the optimal number of Gaussian distributions for approximation is not easy to find. Further discussion of EKF and MHE with SMC can be found in Chen *et al.* (2004).

3. BAYESIAN ESTIMATION BY SEQUENTIAL MONTE CARLO SAMPLING

Monte Carlo sampling based Bayesian approach has been an active area of research for a few years (Gordon *et al.*, 1993; Doucet *et al.*, 2000; Andrieu *et al.*, 2003). Existing methods may be categorized into two groups, sequential Monte Carlo sampling (SMC) and Markov chain Monte Carlo sampling (MCMC). Both SMC and MCMC use Monte Carlo sampling for its convenience in computing the properties of distributions from available samples. MCMC employs iterative algorithm for generating samples, while SMC draws samples from an importance function and adjusts samples' importance with weight. In this paper, SMC is favored for its computational efficiency.

The algorithm for SMC may be represented in pseudo-code as follows (Chen *et al.*, 2003*b*):

- FOR times $k = 1, 2, 3, \dots$
  - · FOR samples $i = 1, 2, 3, \dots, N$
    - Draw sample, $x_k(i)$ from an importance function, $\pi(x_k|x_{k-1}(i), y_k)$
    - Enforce constraints by acceptance and rejection algorithm
    - Assign a weight to $x_k(i)$, $q_k^*(i)$
  - · END FOR
  - · Normalize $q_k^*(i)$ to find $q_k(i)$
  - · Implement resampling when necessary
  - · Find $E[\phi_k(x_k)] \approx \sum_{l=1}^{N} \phi_k(x_k(l))q_k(l)$
- END FOR

where

$$q_k^*(i) = q_{k-1}(i)\frac{p(y_k|x_k(i))\ p(x_k(i)|x_{k-1}(i))}{\pi(x_k(i)|x_{k-1}(i), y_k)} \quad (8)$$

A convenient choice of importance function is to use samples of prior as the importance function (Gordon *et al.*, 1993),

$$\pi(x_k(i)|x_{k-1}(i), y_k) = p(x_k(i)|x_{k-1}(i)) \quad (9)$$

This choice simplifies Equation (8) to

$$q_k^*(i) = q_{k-1}(i)\ p(y_k|x_k(i)) \quad (10)$$

More sophisticated choice of importance functions is expected to improve the robustness of SMC (Doucet *et al.*, 2000).

Various issues of applying SMC have been found and practical solutions are available. SMC may develop degeneracy after several time steps when most samples have insignificant importance to the distribution. Degeneracy can be easily improved by performing resampling Chen *et al.* (2004). SMC may also encounter slow initial convergence when poor initial guess is used. Slow initial convergence can be significantly improved via empirically estimating the initial state (Chen *et al.*, 2004).

3.1 SMC of High-Dimensional Problems

The computational challenges faced by direct computation of Bayesian integrals (Equations (4) to (6)) are formidable for low-dimensional problems, and could become intractable for high-dimensional cases. As discussed earlier, this computational challenge has motivated the application of Monte Carlo sampling for Bayesian estimation. However, the feasibility of SMC for high-dimensional problems is not obvious since the number of samples to maintain equal density or coverage in the solution space increases exponentially with increasing dimensionality. Furthermore, since more samples are required to fill the solution space, it is possible that SMC faces more frequent and serious encounters with various practical challenges like, degeneracy or/and slow initial convergence. It is not clear whether Monte Carlo sampling based approaches suffer more from the curse of dimensionality than other competing approaches like EKF and MHE. Recent theoretical research has been addressing these and other related issues. The key relevant findings are summarized in this subsection based on the survey of Crisan and Doucet (2002).

Theoretical studies indicate that the convergence of SMC in terms of mean-squares error toward zero is almost sure when the importance weights, $\{q_k(i)\}$, are upper bounded, and when a standard resampling scheme is applied. Under these slightly restrictive conditions, the convergence rate is inversely proportional to the number of samples, and is independent of the dimensionality of the problem. Even though more studies are needed to explore the existence of a general convergence property, the number of samples required for SMC in high-dimensional systems may not have to be exponentially increasing. Thus the computation load of SMC may not have to be increased dramatically to obtain accurate estimates. Furthermore, the convergence of SMC may be independent or weakly dependent on the dimensionality.

Another benefit of SMC is that it can be readily implemented in parallel. This is because each sample can be computed independently for most of the computation algorithm. Parts of the algorithm that can not be incorporated into the parallel computation include the occasional implementation of resampling, and the inference from the posterior. Thus, a parallel implementation can retain the computational efficiency even when large number of samples become necessary.

Recent theoretical development shows that the convergence of SMC may be independent of the dimension of the problem, and is inversely proportional to the number of samples. In addition, the recursive formulation also reduces the complexity of the problem compared to a moving horizon formulation. The computational efficiency of SMC may actually be practical for high-dimensional cases, and our results confirm the expectation.

4. CASE STUDY

Start-up of a polymerization process in a continuous stirred tank reactor (CSTR) is studied (Tatiraju *et al.*, 1999). Monomer styrene, initiator, and solvent are continuously fed into the

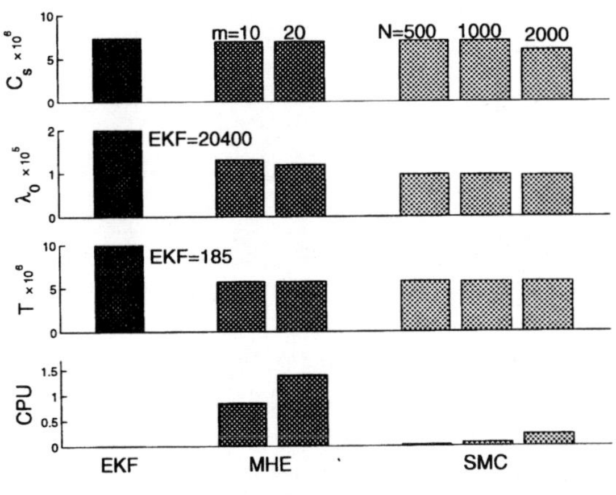

Fig. 2. Performance comparison of Polymerization case study. Top three plots illustrate MSE results, while the bottom one shows CPU load for each method. Not all results are presented for clarity.

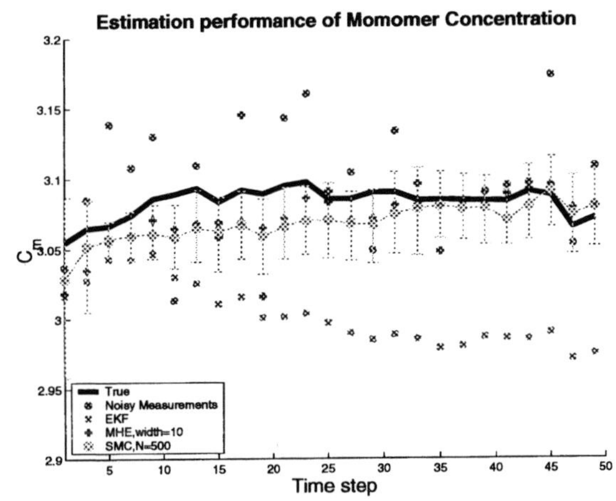

Fig. 3. Estimation results of polymerization case. Estimates provided by SMC are bounded with 95% error bar.

reactor with constant flow rates and compositions. The system has eight state variables, including concentrations of initiator, solvent, and monomer in the CSTR, temperatures in the CSTR and in the cooling device, and first three moments of the molecular weight distribution. Finite difference is used to simulate the true states. Nonlinear process model and simulation conditions and parameters can be found in Chen *et al.* (2003*a*).

All system states are assumed to be measurable and both states and measurements are contaminated with independent and identically distributed Gaussian noise. Estimation performance compares three estimation methods, EKF, MHE and SMC, on the accuracy and efficiency of estimation with noisy measurements. Non-negative constraints on all state variables are enforced. Results provided in Figure 2 and Table 1 are based on 100 realizations of simulation by Octave on a personal computer with Pentium 2.0 GHz and 512MB RAM. 50 measurements are rectified in each realization. Accuracy is compared based on mean-squares error (MSE). For comparison purpose, point estimate is chosen for SMC. Mean is chosen simply for computational convenience. MHE is based on the codes by James Rawlings' group at University of Wisconsin. No systematic way of finding the proper window size of MHE exists, and usually the number of state variables is used as the horizon width. CPU time required is also studied (in units of CPU seconds per time step) with SMC by uncompiled code, while MHE is based on compiled code.

The proposed approach, SMC, exhibits significant efficiency in data rectification of a high-dimensional problem. SMC easily outperforms EKF in MSE since no linearization is required. SMC and MHE has competitive performance on MSE since the posterior exhibits Gaussian like features. However, SMC requires less than 7% of computation than MHE even when SMC is based on uncompiled code. This result indicates that methods based on Gaussian approximation need not be computationally more efficient than methods based on other distributions.

Bayesian estimation can readily provide uncertainty information. Figure 3 shows SMC estimate with 95% error bar. Estimating the error bounds or any moment of the posterior is straightforward and can be implemented by choosing proper loss function, $\mathcal{L}(\cdot)$, in Equation (7).

5. CONCLUSION AND DISCUSSIONS

This paper extends our previous work on Bayesian estimation by sequential Monte Carlo sampling (SMC) to high-dimensional systems. The formulation is general and has shown significantly better performance for a variety of systems including, linear and nonlinear, Gaussian and non-Gaussian, constrained and non-constrained systems. To date, most applications were to relatively small scale problems (Chen *et al.*, 2004; Chen *et al.*, 2003*b*). Application to high-dimensional systems is essential to identify and address any new challenges, and to support claims about the practicality of the proposed approach.

CPU results indicate that SMC may not suffer severely from the curse of dimensionality as commonly expected. Rectification performance on an eight state variables nonlinear dynamic systems shows that SMC requires less than 7% of computation load than MHE while providing competitive estimation accuracy. The accuracy of SMC can be much better than MHE and EKF for systems with highly non-Gaussian posterior distributions. In this example, estimates of MHE and SMC have

Table 1. Mean-Squares Error and CPU time for Polymerization Case Study. Not All Results Are Presented for Clarity.

	$MSEC_s$	$MSE\lambda_0$	$MSEC_m$	CPU	Parameter
EKF	$7.3 \times 10^{-6} \pm 6.7 \times 10^{-6}$	$2.0 \times 10^{-1} \pm 2.7 \times 10^{-3}$	$1.8 \times 10^{-4} \pm 5.5 \times 10^{-5}$	0.003	N/A
MHE	$6.9 \times 10^{-6} \pm 6.6 \times 10^{-6}$	$1.3 \times 10^{-5} \pm 9.8 \times 10^{-6}$	$5.7 \times 10^{-6} \pm 6.6 \times 10^{-6}$	0.85	$m = 10$
	$6.9 \times 10^{-6} \pm 6.6 \times 10^{-6}$	$1.2 \times 10^{-5} \pm 9.6 \times 10^{-6}$	$5.7 \times 10^{-6} \pm 6.6 \times 10^{-6}$	1.39	$m = 20$
SMC	$7.0 \times 10^{-6} \pm 6.6 \times 10^{-6}$	$9.6 \times 10^{-6} \pm 1.0 \times 10^{-5}$	$5.7 \times 10^{-6} \pm 6.5 \times 10^{-6}$	0.03	$N = 500$
	$7.0 \times 10^{-6} \pm 6.6 \times 10^{-6}$	$9.5 \times 10^{-6} \pm 1.0 \times 10^{-5}$	$5.7 \times 10^{-6} \pm 6.3 \times 10^{-6}$	0.07	$N = 1000$
	$6.9 \times 10^{-6} \pm 6.6 \times 10^{-6}$	$9.4 \times 10^{-6} \pm 1.0 \times 10^{-5}$	$5.7 \times 10^{-6} \pm 6.7 \times 10^{-6}$	0.23	$N = 2000$

comparable accuracy which results from the Gaussian like priors of the system under the operating condition studied. SMC has shown significant improvement over MHE in situations when non-Gaussian priors exist with or without constraints (Chen *et al.*, 2004; Chen *et al.*, 2003*b*). Other operating conditions with non-Gaussian distributions are currently being explored.

The results in this paper compare point estimates of SMC with other methods. Since details about the posterior distribution at each time point are available in SMC, choice of the loss function can have a significant effect on the accuracy of the point estimate. The use of the mean value, as in current work may not be appropriate, particularly for multi-modal distributions. Such a point estimate may not utilize all the benefits of Bayesian estimation, and its use for performance comparison with other methods that rely on Gaussian approximation may be biased against SMC. Research on appropriate loss functions for non-Gaussian distributions is on-going for a better and fairer portrayal of the benefits of SMC-based Bayesian estimation.

6. ACKNOWLEDGMENTS

Special thanks to Eric L. Haseltine and Dr. James Rawlings at the University of Wisconsin for providing MHE code and assistance in running the program. Financial support from the National Science Foundation (CTS-9733627 and CTS-0321911) is also gratefully acknowledged.

REFERENCES

Andrieu, Christophe, Nando de Freitas, Arnaud Doucet and Michael Jordan (2003). An introduction to MCMC for machine learning. *Machine Learning* **50**(1-2), 5–43.

Chen, Wen-shiang, Bhavik R. Bakshi, Prem K. Goel and Sridhar Ungarala (2003*a*). Bayesian estimation by sequential Monte Carlo sampling: Application to large-scale nonlinear dynamic systems. Technical report. Department of Chemical Engineering, The Ohio State University.

Chen, Wen-shiang, Bhavik R. Bakshi, Prem K. Goel and Sridhar Ungarala (2003*b*). Bayesian estimation of nonlinear dynamic systems - dealing with constraints. Technical report. Department of Chemical Engineering, The Ohio State University.

Chen, Wen-shiang, Bhavik R. Bakshi, Prem K. Goel and Sridhar Ungarala (2004). Bayesian estimation of unconstrained nonlinear dynamic systems via sequential Monte Carlo sampling. *Industrial & Engineering Chemistry Research.* Submitted for publication.

Crisan, Dan and Arnaud Doucet (2002). A survey of convergence results on particle filtering methods for practitioners. *IEEE Transactions on Signal Processing* **50**(3), 736–746.

Doucet, Arnaud, Simon Godsill and Christophe Andrieu (2000). On sequential Monte Carlo sampling methods for Bayesian filtering. *Statistics and Computing* **10**, 197–208.

Gordon, N. J., D. J. Salmond and A. F. M. Smith (1993). Novel approach to nonlinear/non-Gaussian Bayesian state estimation. *IEE Proceedings-F* **140**(2), 107–113.

Ho, Y. C. and R. C. K. Lee (1964). A Bayesian approach to problems in stochastic estimation and control. *IEEE Transactions on Automatic Control* pp. 333–339.

Jazwinski, Andrew H. (1970). *Stochastic Processes and filtering theory.* Academic Press. New York.

Kramer, M. A. and R. S. H. Mah (1994). Model-based monitoring. In: *Proceedings of the International Conference on Foundations of Computer Aided Process Operations* (D. Rippin, J. Hale and J. Davis, Eds.). CACHE. Austin, TX. pp. 45–68.

Robertson, Douglas G. and Jay H. Lee (2002). On the use of constraints in least squares estimation and control. *Automatica* **38**, 1113–1123.

Robertson, Douglas G., Jay H. Lee and James B. Rawlings (1996). A moving horizon-based approach for least-squares estimation. *AIChE Journal* **42**(8), 2209–2224.

Tatiraju, Srinivas, Masoud Soroush and Babatunde Ogunnaike (1999). Multirate nonlinear state estimation with application to a polymerization reactor. *AIChE Journal* **45**(4), 769–780.

ELSEVIER
IFAC
PUBLICATIONS
www.elsevier.com/locate/ifac

CORRELATION DIMENSION AND LYAPUNOV EXPONENT BASED ISOLATION OF PLANT-WIDE OSCILLATIONS

Xiaoyun Zang, John Howell

University of Glasgow, Centre for Systems and Control,
Department of Mechanical Engineering, G12 8QQ, Glasgow, UK
j.howell@mech.gla.ac.uk

Abstract: The correlation dimension and maximal Lyapunov exponent have a direct relationship with the harmonic complexity contained in non-linearity induced oscillations in chemical process plants. When combined with knowledge of the harmonic propagation of plant-wide oscillations, they can be applied to locate the source of the non-linearity. The method is demonstrated on both simulated data and real industrial data.

Keywords: correlation dimension, Lyapunov exponent, non-linearity, oscillations, process control

1. INTRODUCTION

Oscillations are a common type of plant-wide disturbance whose detection and diagnosis have generated considerable interest in recent years (Thornhill, *et al.*, 2001; Thornhill, *et al.*, 2003a; Thornhill and Hagglund 1997; Xia, *et al.*, 2003). A key issue is the determination of the root cause of a plant-wide oscillation (Qin 1998). That is an oscillation that is observed in a number of neighbouring process units and control loops. The most likely sources of oscillations are either due to valve non-linearity such as stiction or hysteresis or due to a badly tuned controller (Wallen 1997). Thornhill, *et al.*, (2001) have focused on isolating a problem non-linear element in a control loop; the most likely loop is selected by finding that measurement record that has the maximum distortion factor (D- factor), which is a measure of the harmonic content of a record. In the same paper and in (Thornhill, *et al.*, 2003b), the authors focus on an alternative measure, on what they call *non-linearity*. They recommend the application of a non-linearity statistic (the N measure), and hypothesize that the non-linearity in a measurement record obtained near the source will be stronger than one that it is recorded further away from the source. Both methods are based on the fact that process plants are usually 'low-pass' in nature, so filter out high frequency components in the oscillations.

In this paper, the correlation dimension and Lyapunov exponents are found to have some relationship with the harmonic content in the time series recorded from the oscillating loops. Nonlinearity induced oscillations in chemical process plants usually have asymmetric waves which contain a fundamental, plus even and odd harmonics (Zang and Howell, 2003). Hence correlation dimension and Lyapunov exponents can provide additional information towards the location of the source of nonlinearity induced, plant-wide oscillations.

The correlation dimension and Lyapunov exponents are recently developed descriptions which provide quantities for the characterisation of nonlinear, deterministic and chaotic data (Kantz and Schreiber 1997). These methods have been widely used to analyse non-linear time series in mechanical systems (Wang and Lin 2003), in medical science (Muller, *et al.*, 2003), and in chemical engineering (Guo, *et al.*, 2003). Sometimes the correlation dimension and/or the maximal Lyapunov exponent indicate non-linear or deterministic chaotic behaviour; whilst other references conclude that a larger correlation dimension or a larger maximal Lyapunov exponent corresponds to richer dynamics (Boltezar, *et al.*, 1999) or more system complexity (Burioka, *et al.*, 2001). Few have related the correlation dimension and maximal Lyapunov exponent of a time series to its harmonic content because chaotic time series, although usually periodic and oscillatory, do not contain distinct fundamentals and harmonics.

The definitions and characteristics of the correlation dimension and maximal Lyapunov exponent will be introduced first. A simple example will then be described to demonstrate their relationships with the harmonic content of an oscillating time series. A method of applying these measures to find the root cause of the oscillation will then be given. Finally three case studies will be outlined, one is simulated and two industrial.

2. TECHNIQUES

Both techniques are based on a discrete, multi-dimensional phase space representation of the data. Given the time series $x_1, x_2, x_3, \cdots, x_N$ then the multi-dimensional phase space is formed from:

$$\mathbf{X}_i = (x_i, x_{i+T}, x_{i+2T}, \cdots x_{i+(m-1)T}),\ i = 1, 2, \cdots, N \quad (1)$$

where T is the time interval and m is known as the embedding dimension; $\mathbf{X}_i$ is called an embedding vector (point) of m-dimension.

2.1 Correlation dimension: definition and estimation

The correlation dimension provides a tool to quantify self-similarity. A larger correlation dimension corresponds to a larger degree of complexity and less self-similarity. Stochastic signals are infinite-dimensional. The most frequently used procedure to estimate the correlation dimension was introduced by Grassberger and Procaccia (1983a, 1983b). They defined the correlation sum for a collection of points $\mathbf{X}_i$ $(i=1,2,...,N)$ in some phase space to be the fraction of all possible pairs of points which are closer than a given distance ε in a particular norm:

$$C(m,\varepsilon) = \frac{2}{(N-m)(N-m-1)} \sum_{i=m}^{N} \sum_{j=i+1}^{N} \Theta(\varepsilon - \|\mathbf{X}_i - \mathbf{X}_j\|) \quad (2)$$

where Θ is the Heaviside step function, $\Theta(x) = 0$ if $x \le 0$ and $\Theta(x) = 1$ for $x > 0$. Thus Equation 2 counts the pairs ($\mathbf{X}_i$, $\mathbf{X}_j$) whose distance is smaller than ε. To eliminate the temporal correlation, the Theiler window can be used to exclude those pairs of points that are too close in time (Kantz and Schreiber 1997).

It has been shown by Sauer & Yorke (1993) that in the limit of an infinite amount of data (i.e. $N \to \infty$) and for small ε, C scales like a power law, $C(\varepsilon) \propto \varepsilon^{D_2}$, where D_2 is known as the correlation dimension. Thus D_2 is defined by

$$D_2 = \lim_{\varepsilon \to 0} \lim_{N \to \infty} \frac{\partial \ln(C(m,\varepsilon))}{\partial \ln(\varepsilon)}. \quad (3)$$

Convergence to a finite correlation dimension can be checked by plotting "effective dimensions" versus scale (ε) for various embeddings (*m*). The easiest way to proceed is to compute (numerically) the derivative of $\ln C(m,\varepsilon)$ with respect to $\ln \varepsilon$, for example by fitting straight lines to the log-log plot of $C(\varepsilon)$. If the scale is large, then self-similarity should be minimal, whereas if it is sufficiently small one observes a relationship with the embedding dimension i.e. D_2 is not totally independent. This effect is due to noise. Only on the intermediate scales can one see the desired *plateau* where the results are approximately independent of *m* and ε i.e. where D_2 is *invariant*. The region where the scale rule holds, not just the range selected for straight line fitting, is called the *scaling range*. Takens-Theiler have developed an alternative estimate of the correlation dimension:

$$D_{TT}(\varepsilon) = \frac{C(\varepsilon)}{\int_0^{\varepsilon} \frac{C(\varepsilon')}{\varepsilon'} d\varepsilon'} \quad (4)$$

2.2 Lyapunov exponents: definition and estimation

Lyapunov exponents measure the exponential divergence (positive exponents: chaotic motion) or convergence (negative exponents: regular motion) of two initially neighbouring trajectories in a phase space. In other words, they measure the degree of unpredictability of the future. There are many different Lyapunov exponents for a dynamical system. The most important is known as the *maximal Lyapunov exponent* (λ_1) (Kantz and Schreiber 1997). Let $\mathbf{X}_i$ and $\mathbf{X}_j$ be two points in a phase space with a distance $\|\mathbf{X}_i - \mathbf{X}_j\| = \delta_0 \ll 1$ between them. Let δ_t denote the distance obtained between the two trajectories *t* units of time later i.e. $\delta_t = \|\mathbf{X}_{i+t} - \mathbf{X}_{j+t}\|$. Then λ_1 is defined by $\delta_t = \delta_0 e^{\lambda_1 t}$, $\delta_t \ll 1$. A positive λ_1 means that there is an exponential divergence of these trajectories, i.e. chaos; a negative λ_1 implies the existence of a stable fixed point; if the motion settles down onto a limit cycle, λ_1 is zero.

Based on this understanding, a robust, consistent and unbiased estimator for the maximal Lyapunov exponent was proposed by Kantz and Schreiber (1997). They compute

$$S(\varepsilon,m,t) = \frac{1}{N} \sum_{i=1}^{N} \ln\left(\frac{1}{|U(\mathbf{X}_i)|} \sum_{\mathbf{X}_j \in U(\mathbf{X}_i)} |\mathbf{X}_{i+t} - \mathbf{X}_{j+t}|\right) \quad (5)$$

where the reference points $\mathbf{X}_i$ are embedding vectors and $U(\mathbf{X}_i)$ is known as the neighbourhood of $\mathbf{X}_i$ with diameter ε. If $S(\varepsilon,m,t)$ exhibits a linear increase with fixed slope for all m larger than some m_0 and for a reasonable range of ε, then the slope of S is deemed to be *invariant* over this range. This slope can be taken as an estimate of the maximal exponent λ_1.

The correlation sum, correlation dimension, and the maximal Lyapunov exponent estimation in this paper are all based on the TISEAN software package (Hegger, *et al.*, 1999). Throughout the paper, $D_{TT}(\varepsilon)$ and $S(\varepsilon,m,t)$ are formed in 1-15 dimensions, D_2 and λ_1 are then estimated from the values of $D_{TT}(\varepsilon)$'s and slopes of $S(\varepsilon,m,t)$'s in 2-15 dimensions where the *plateau* or *linear increase* is well established, giving the mean and variance of these values.

2.3 Relationship between these invariants and harmonic content

Intuitively, a time series with high harmonic content will be more complex and unpredictable than one with a low harmonic content, and hence will have a larger correlation dimension and a larger maximal Lyapunov exponent. Here a simple example will be used to reinforce this intuition. Three 4096-sample time series were analysed to examine the effect of increasing the harmonic content:

$$\begin{aligned} &s1(k)=\sin(2\pi fk),\\ &s2(k)=\sin(2\pi fk)+0.2\sin(4\pi fk),\\ &s3(k)=\sin(2\pi fk)+0.2\sin(4\pi fk)+0.05\sin(6\pi fk),\\ &k=1,\cdots,4096 \end{aligned} \tag{6}$$

where $f=0.01$ is the fundamental frequency. The correlation dimensions and maximal Lyapunov exponents of the three time series are listed in Table 1.

It can be seen that the correlation dimension and maximal Lyapunov exponent both increase with harmonic content. This supports the claims that the correlation dimension reflects the self-similarity and the maximal Lyapunov exponent reflects the predictability. The future of time series with more harmonics should be more unpredictable and hence should have a larger invariant.

Table 1. Invariant of sinusoidal signals

Time series	Correlation dimension (D_2)	Maximal Lyapunov Exponent (λ_1)
s1(k)	1.04±0.04	2.21e-5±1.44e-5
s2(k)	1.07±0.04	2.10e-4±2.36e-5
s3(k)	1.08±0.03	8.12e-4±2.11e-4

2.4 Invariant based location of the source of the oscillations

In general the dynamic behaviour of physical processes are inherently low-pass, so time series that are recorded close to the root cause of a plant-wide oscillation will have a high harmonic content, and those recorded further away from the root cause will become more sinusoidal. However the first few harmonics of an oscillation could be amplified if the fundamental is sufficiently low to locate the first few harmonics on the positive slope of the disturbance transfer function bode plot. In this case, and since the first few harmonics dominate, the time records further away from the source could be more complex (with large signal-to-noise ratio).

Based on this understanding of how harmonic content is affected as an oscillation propagates through a plant, the correlation dimension and maximal Lyapunov exponent can provide features to locate the source of plant-wide oscillations. More specifically, the time series pertaining to the root cause should have the largest correlation dimension and maximal Lyapunov exponent in cases where the plant is inherently low pass in nature over the frequency range of concern, because the harmonics have been attenuated through the propagation and also because the time series become less complex and more predictable further away from the source; and the correlation dimension and maximal Lyapunov exponent pertaining to the root cause will be the smallest in low-frequency cases where there is amplification of the first few harmonics.

3. EXAMPLES

3.1 Simulated time series

A simulated time series was generated by simulating a simple distillation column model due to Seborg, *et al.*, (1989), which was adapted by incorporating a valve model into one of its loops. Details pertaining to the revised model are given in Fig. 1; stiction in the valve (in Loop 1) was increased until the model oscillated. The static friction model that was used is that due to Horch and Isaksson (1998). Both loops oscillated. Their controlled process variables (*PVs*) were recorded, resulting in two 4096-sample time series (Fig. 2). The correlation dimension plots, D_{TT} versus ε, and the plots of $S(\varepsilon,m,t)$ versus t are given in Fig. 3 and Fig. 4 respectively.

This is a low-frequency case, because the fundamental oscillating frequency is much lower than the cut-off frequency of the disturbance transfer function bode plot of Loop2. So the correlation dimension and maximal Lyapunov exponent of the source should be smaller than the disturbed loop. The results obtained (Table 2) support this view.

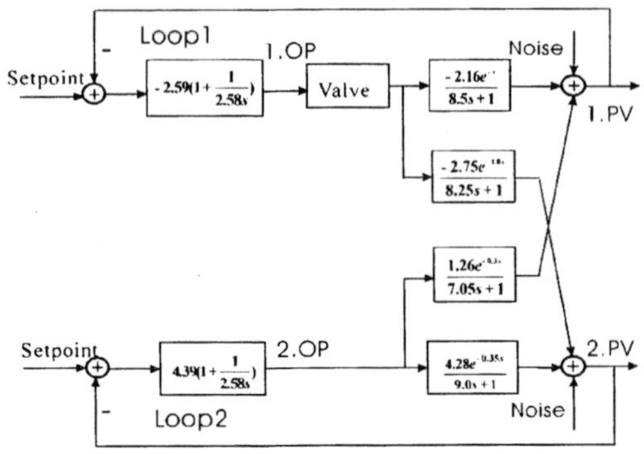

Fig. 1 The simulated plant schematic

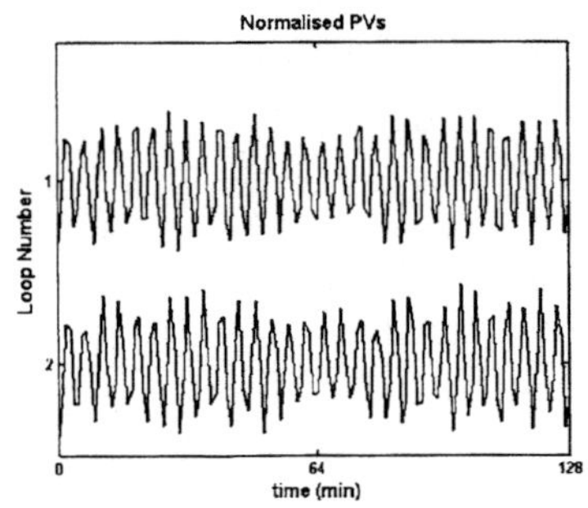

Fig. 2 Process variable time series for the two simulated loops

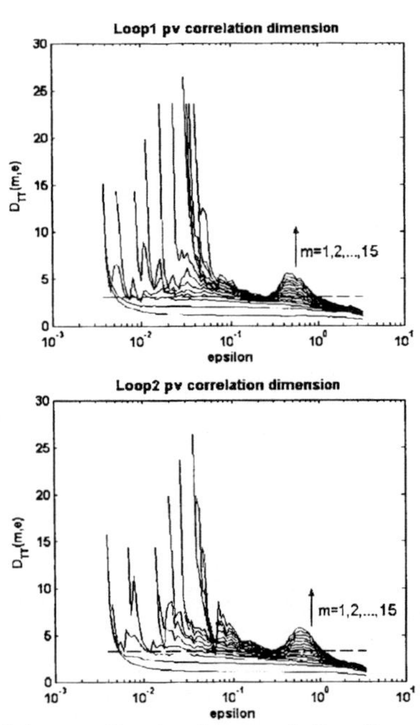

Fig. 3 Takens estimator for correlation dimension of the two loops

Table 2. Invariant estimation for the simulated case

Loop Number	Correlation Dimension (D_2)	Maximal Lyapunov exponent (λ_1)
1	3.00±0.13	0.30±0.01
2	3.28±0.13	0.31±0.01

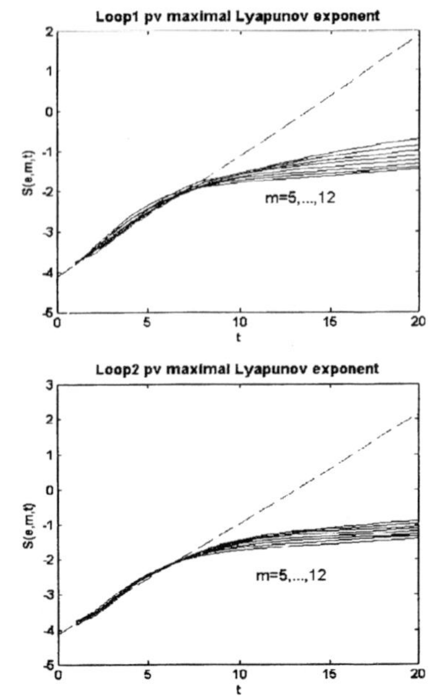

Fig. 4 Maximal Lyapunov exponents of the two loops

3.2 Case study 1 (low-frequency case)

Case study 1 pertains to industrial data, provided by the Eastman Chemical Company, which was recorded whilst a plant was oscillating. A plant assessment found a sticking valve in the loop with Tag22 (Thornhill, *et al.*, 2002; Xia, *et al.*, 2003). Figure 5 gives the 4096-sample time series for each of the six dominant loops that showed these plant-wide oscillations. The correlation dimensions and maximal Lyapunov exponents of each time series are plotted and estimated. The results are listed in Table 3.

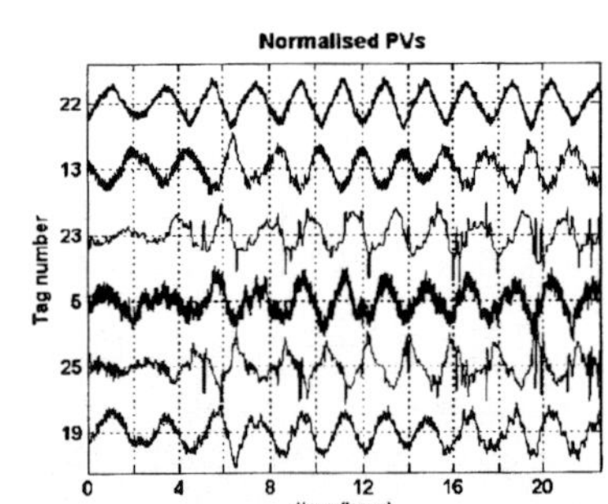

Fig. 5 Time series plots of the six loops of interest

Table 3. Invariant estimation for Easterman case study

Tag Number	Correlation Dimension (D_2)	Maximal Lyapunov exponent (λ_1)
22	1.52±0.03	0.004±0.0001
13	2.33±0.01	0.022±0.004
23	-	-
5	-	-
25	3.86±0.01	0.029±0.007
19	2.29±0.05	0.007±0.0004

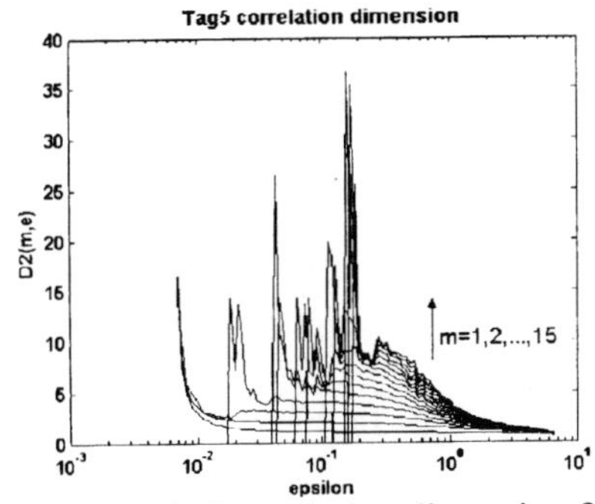

Fig. 6 (a) No typical correlation dimension found in Tag5

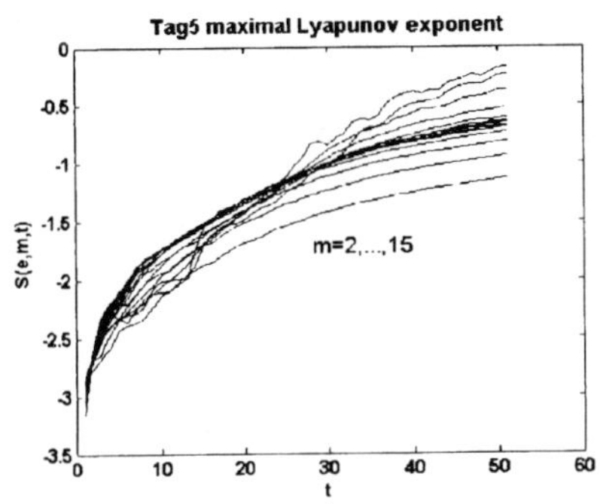

Fig. 6 (b) No exponential divergence of Tag5

Once again, this is a low-frequency case study. Tag22 has the smallest correlation dimension and maximal Lyapunov exponent inferring that it should be the nearest to the root cause. Interestingly Fig. 6 (a) shows the correlation dimension plot of Tag5, where no typical correlation dimensions can be found. All curves from different embedding dimensions behave different and there is no common behaviour. Fig. 6 (b) shows that the time series pertaining to Tag5 exhibit no linear increase, reflecting the lack of exponential divergence of nearby trajectories. And the time series pertaining to Tag23 also has no typical dimension and maximal Lypunov exponent. A possible explanation for this is that the stochastic content (for example, noise) in the time series dominates. These observations also match the propagation schematic provided by (Thornhill, *et al.*, 2002), where Tag23 and Tag5 are both relatively far away from Tag22. Although Tag19 is farther away than Tag13, it has a smaller correlation dimension and maximal Lyapunov exponent. This is because the intrinsic low-pass filtering attenuates the higher harmonics and simplifies the time series pertaining to Tag19.

3.3 Case Study 2 (high-frequency case)

A set of refinery data (courtesy of a SE Asian refinery) was examined. Previously Thornhill, *et al.*, (2001) have performed spectral PCA on the data and suggested that 12 loops were associated with a plant-wide oscillation observed in the data. They then determined the distortion factor and N- measure of each of these 12 loops and concluded that the source of non-linearity was located in the loop associated with one of Tags 13, 33 or 34. Figure 7 shows the time trends of the 12 loops. The estimated correlation dimensions and maximal Lyapunov exponents for the 12 loops are listed in Table 4.

This is a high-frequency case study. The source of such plant-wide oscillations should have the largest correlation dimension and maximal Lyapunov exponent. From Table 4, it can be seen that the time series pertaining to Tag13 and Tag33 have larger correlation dimensions and maximal Lyapunov exponents than those pertaining to other loops, which agrees with the conclusions made by Thornhill, *et al.*, (2001). Note that the time series pertaining to Tag11 also has a large correlation dimension and maximal Lyapunov exponent. It is suspected that this loop contained harmonics which were not multiples of the fundamental pertaining to the oscillation (0.06 min^{-1}). These harmonics may contribute to the large correlation dimension and maximal Lyapunov exponent. Note that the time series pertaining to Tags 24, 25, 4 and 19 have no typical correlation dimensions and maximal Lyapunov exponents, because these loops are contaminated by noise.

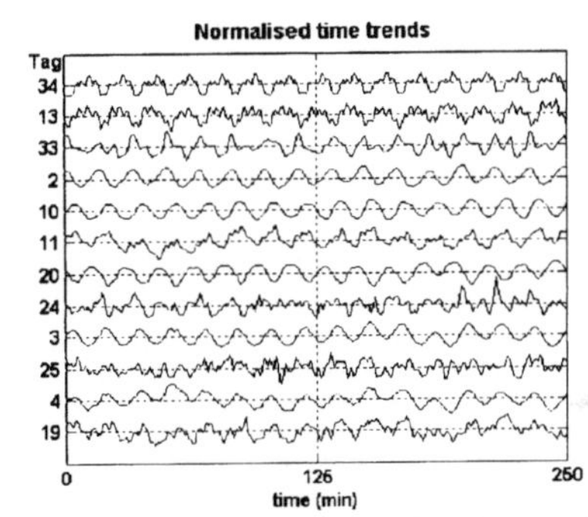

Fig. 7 Time series plots of the twelve loops of interest

Table 4. Invariant estimation for SE Asian case study

Tag Number	Correlation Dimension (D_2)	Maximal Lyapunov exponent (λ_1)
34	1.24±0.01	0.011±0.0003
13	2.78±0.06	0.057±0.0008
33	2.16±0.12	0.030±0.004
2	1.88±0.08	0.010±0.0002
10	1.79±0.10	0.011±0.001
11	2.87±0.04	0.027±0.0007
20	1.87±0.05	0.019±0.001
24	-	-
3	2.05±0.02	0.029±0.001
25	-	-
4	-	-
19	-	-

4. CONCLUSION

The correlation dimension and maximal Lyapunov exponent of a time series are demonstrated to be associated with its harmonic content. Based on the assessment of the propagation of harmonic content of plant-wide oscillations, it is possible to find the source of non-linearity by analysing invariants of the time series pertaining to the oscillating loops. Compared to the N-measure proposed by Thornhill, *et al.*, (2001), these methods do not involve surrogate data, and are hence more time efficient.

Second-order Volterra models that are capable of exhibiting asymmetric responses to symmetric input changes (Doyle *et al.*, 1995) might provide additional information.

ACKNOWLEDGEMENTS

Xiaoyun Zang & John Howell gratefully acknowledge John Cox & Michael Paulonis from the *Eastman Chemical Company, Kingsport, TN, USA*, for the supply of real plant data.

REFERENCES

Boltezar, M., N. Jaksic, I. Simonovski and A. Kuhelj (1999). Dynamical behaviour of the planar non-linear mechanical system -- PART II: Experiment. *Journal of Sound and Vibration*, **226**(5), 941-953.

Doyle, F. J., B. A. Ogunnaike and R. K. Pearson (1995). Nonlinear model-based control using second-order Volterra models. *Automatica*, **31**(5), 697-714.

Grassberger, P. and I. Procaccia (1983a). Characterisation of strange attractors. *Physical Review Letters,* **50**, 346-349.

Grassberger, P. and I. Procaccia (1983b). Measuring the strangeness of strange attractors. *Physica D: Nonlinear Phenomena*, **9**(1-2), 189-208.

Guo, Q., G. Yue, T. Suda and J. Sato (2003). Flow characteristics in a bubbling fluidized bed at elevated temperature. *Chemical Engineering and Processing*, **42**(6), 439-447.

Hegger, R., H. Kantz and T. Schreiber (1999). Practical implementation of nonlinear time series methods: the TISEAN package. *Chaos*, **9**, 413.

Horch, A. and A. J. Isaksson (1998). A method for detection of stiction in control valves. In: *Proceedings of the IFAC Workshop on on-line fault detection and supervision in the chemical process industry*. Session 4B. Lyon, France.

Kantz, H. and T. Schreiber (1997). *Nonlinear time series analysis*. Cambridge University Press, Cambridge.

Muller, V., W. Lutzenberger, H. Preil, F. Pulvermuller and N. Birbaumer (2003). Complexity of visual stimuli and non-linear EEG dynamics in humans. *Cognitive Brain Research*, **16**(1), 104-110.

Qin, S. J. (1998). Control performance monitoring -- a review and assessment. *Computers & Chemical Engineering*, **23**(2), 173-186.

Sauer, T. and J. Yorke (1993). How many delay coordinates do you need? *Int.J.Bifurcation and Chaos*, **3**, 737.

Seborg, D. E., T. F. Edgar and D. A. Mellichamp (1989). *Process Dynamics and Control*. John Wiley & Sons, Inc.

Thornhill, N. F. and T. Hagglund (1997). Detection and diagnosis of oscillation in control loops. *Control Engineering Practice*, **5**(10), 1343-1354.

Thornhill, N. F., S. L. Shah and B. Huang (2001). Detection of distributed oscillations and root-cause diagnosis. In: *Proceedings of the CHEMFAS-4*. Cheju, Korea. pp. 167-172.

Thornhill, N. F., C. Xia, J. Howell, J. Cox and M. Paulonis (2002). Analysis of plant-wide disturbances through data-driven techniques and process understanding. In: *Proceedings of the IFAC World Congress*. Barcelona, Spain.

Thornhill, N. F., B. Huang and H. Zhang (2003a). Detection of multiple oscillations in control loops. *Journal of Process Control*, **13**(1), 91-100.

Thornhill, N. F., J. W. Cox and M. A. Paulonis (2003b). Diagnosis of plant-wide oscillation through data-driven analysis and process understanding. *Control Engineering Practice,* **11**(12), 1481-1490.

Wallén, A. (1997). Valve diagnositics and automatic tuning. In: *Proceedings of 1997 American Control Conference*. Albuquerque, New Mexico. pp. 2930-2934.

Wang, W. J. and R. M. Lin (2003). The application of pseudo-phase portrait in machine condition monitoring. *Journal of Sound and Vibration*, **259**(1), 1-16.

Xia, C. and J. Howell (2003). Isolating multiple sources of plant-wide oscillations via independent component analysis. In: *Proceedings of the 5th IFAC Symposium on Fault Detection, Supervision and Safety for Technical Processes (SAFEPROCESS'03)*. Washington DC, USA.

Zang, X. and J. Howell (2003). Discrimination between bad tuning and non-linearity induced oscillations through bispectral analysis. In: *SICE Annual Conference 2003 in Fukui proceedings*. Fukui, Japan.

ELSEVIER
IFAC
PUBLICATIONS
www.elsevier.com/locate/ifac

CONTINUOUS-DISCRETE OBSERVER DESIGN FOR A CHO-K1 CELL CULTURE IN SUSPENSION

Jens E. Haag * **Alain Vande Wouwer** *,[1]

* *Service d'Automatique, Faculté Polytechnique de Mons, Boulevard Dolez 31, B-7000 Mons, Belgium*

Abstract: Bioprocess models are often uncertain due to the lack of experimental data for structure selection and parameter estimation. Maximum-likelihood parameter estimation techniques, which take the measurement errors into account, allow confidence intervals in the identified parameters to be evaluated. In turn, this information can be advantageously exploited in the design of (more) robust state estimators (or software sensors) for process monitoring and control. In this paper, the formulation of continuous-discrete Kalman filters and receding horizon observers are extended to include *a posteriori* knowledge on model parameter uncertainties. These extended observers/filters are then successfully applied to a real-case study, i.e., animal cell cultures in perfusion mode.

Keywords: state estimation, parameter estimation, Kalman filter, receding-horizon observer, biotechnology

1. INTRODUCTION

State estimation techniques are particularly important for monitoring bioprocess operation. Indeed, biomass and component concentrations are difficult to measure on-line and measurements are often corrupted by relatively large errors. As on-line measurements are collected at discrete times only, and with relatively low sampling rates (a few hours to 1-2 days), continuous (model)–discrete (measurements) software sensors taking the stochastic nature of the measurement signals into account are the first choices in bioprocess applications (Bogaerts and Vande Wouwer, 2003).

Stochastic continuous-discrete software sensors include the Kalman filter (Gelb, 1989) and the receding horizon observer (Allgöwer *et al.*, 1999; Bogaerts and Hanus, 2001). In the former, uncertainties are introduced in the form of uncorrelated white noises corrupting the state and measurement equations. The selection of the covariance matrices of these noises is often based on some *a priori* knowledge and some form of tuning. In the latter, measurement errors are introduced in a maximum-likelihood or Gauss-Markov cost function, which is repeatedly minimized in order to estimate most-likely model initial conditions.

The main objective of this paper is to show how *a posteriori* information on model parameter uncertainties, which results from the model parameter identification procedure, can be incorporated in the observer/filter design. Indeed, maximum-likelihood parameter identification yields estimates of the parameter covariance matrices which are very useful for model analysis and validation purposes, but which are seldom exploited when considering the state estimation problem.

As a result, extended forms of the Kalman filter and receding-horizon observer are developed. These software sensors are then applied to a real-case study, i.e., CHO-K1 cell cultures in a bioreactor operated in perfusion mode.

This paper is organized as follows. The next section deals with the extended Kalman filter whereas section 3 presents the full- or receding-horizon observer.

[1] Corresponding author, fax: +32 65 37 4136, e-mail: Alain.VandeWouwer@FPMs.ac.be

Section 4 describes the experimental set-up and shows results of the application of both state estimation techniques. Finally, section 5 is devoted to some concluding remarks.

2. EXTENDED KALMAN FILTER

The Kalman Filter is one of the most popular state estimation algorithms, which takes the stochastic nature of the process and measurement noises into account.

Assume that the state equation for $x \in \mathbb{R}n_x$ and the output $y \in \mathbb{R}n_y$ are corrupted by normally distributed white noise with zero mean according to

$$\dot{x}(t) = f(x(t), u(t), t) + w(t); \quad w(t) \sim \mathcal{N}(0, Q(t)) \quad (1)$$
$$y_i = h(x(t_i)) + v_i; \quad v_i \sim \mathcal{N}(0, R_i), \quad (2)$$

where $Q \in \mathbb{R}^{n_x \times n_x}$ and $R \in \mathbb{R}^{n_y \times n_y}$ are the state and output error covariance matrices, respectively.

The continuous-discrete extended Kalman Filter (Gelb, 1989) makes use of a linearisation along the state estimate trajectory and corresponds to continuous-time prediction and discrete-time measurements:

- initialization $(i = 0, 1, 2, \dots)$

$$\hat{x}(t_i) = \hat{x}_i \quad (3)$$
$$P(t_i) = P_i \quad (4)$$

- continuous prediction step for $t_i \leq t < t_{i+1}$

$$\dot{\hat{x}}(t) = f(\hat{x}(t), u(t), t) \quad (5)$$
$$\dot{P}(t) = F(\hat{x}(t), u(t), t)P(t) + P(t)F^{\mathrm{T}}(\hat{x}(t), u(t), t) + Q(t) \quad (6)$$

- discrete correction step at $t = t_i$ $(i = 1, 2, \dots)$

$$\hat{x}_i = \hat{x}(t_i^-) + K_i\left(y_i - h_i(\hat{x}(t_i^-))\right) \quad (7)$$
$$P_i = \left(I - K_i H_i(\hat{x}(t_i^-))\right) P(t_i^-) \quad (8)$$
$$K_i = P(t_i^-)H_i^{\mathrm{T}}(\hat{x}(t_i^-)) \cdot \left(H_i(\hat{x}(t_i^-))P(t_i^-)H_i^{\mathrm{T}}(\hat{x}(t_i^-)) + R_i\right)^{-1} \quad (9)$$

- linearisations

$$F(\hat{x}(t), u(t), t) = \left.\frac{\partial f(x(t), u(t), t)}{\partial x^{\mathrm{T}}}\right|_{x(t)=\hat{x}(t)} \quad (10)$$
$$H_i(\hat{x}(t \rightarrow t_i)) = \left.\frac{\partial h(x(t))}{\partial x^{\mathrm{T}}}\right|_{x(t)=\hat{x}(t \rightarrow t_i)} \quad (11)$$

The matrices P_0, Q and R_i play the following roles:

- The initial state covariance matrix P_0 represents the confidence paid to the initial guess for the state vector $\hat{x}_0$. If the latter is arbitrarily chosen, the elements of P_0 should be set to high values (so that these estimates are almost replaced by the measurements at the first correction instant, where possible).
- The state covariance matrix Q is often used to keep a certain level of uncertainty on the state estimates preventing that the correction matrix K_i vanishes for $i \rightarrow \infty$, which would result in the disappearance of the correction and therefore in a pure simulation.
- The measurement covariance matrix R contains information on the measurement noise. Large values for the diagonal elements are equivalent to significant measurement noise in the output signal and non-zero off-diagonal elements indicate correlations between the respective signals. Less confidence is therefore paid to the measurements in the correction step.

A further extension of the Kalman filter is derived in the following, taking the information about the modelling errors into account. This information can be quantified with the error covariance matrices obtained at the model identification stage.

Consider the system

$$\dot{x}(t) = f(x(t), u(t), p); \quad x(t_0) = x_0 \quad (12)$$
$$y_i = h(x(t_i)) + v_i; \quad v_i \sim \mathcal{N}(0, R_i). \quad (13)$$

Compared to (1), the disturbance of the state equation does not appear explicitly, and the model parameter vector $p \in \mathbb{R}^{n_p}$ is considered instead.

Consider an analogous system for the state estimate,

$$\dot{\hat{x}}(t) = f(\hat{x}(t), u(t), \hat{p}); \quad \hat{x}(t_0) = \hat{x}_0 \quad (14)$$

with the estimated initial state $\hat{x}_0$ and the estimated parameter vector $\hat{p}$. Their error covariance matrices are denoted P_0 and Q_p, respectively, and are assumed to be known, e.g., from a preliminary parameter estimation.

The estimation error $\tilde{x} = \hat{x} - x$ is due to the errors on the initial state and the parameters, i.e. in first-order approximation

$$\tilde{x} \approx S_p(x, u, p)\tilde{p} + S_0(x, u, p)\tilde{x}_0 \quad (15)$$

where the sensitivity matrices are defined as

$$S_p = \frac{\partial x}{\partial p}; \quad S_0 = \frac{\partial x}{\partial x_0} \quad (16)$$

The covariance matrix of the estimation error is then written

$$P = S_p Q_p S_p^{\mathrm{T}} + S_0 P_0 S_0^{\mathrm{T}} \quad (17)$$

The equations for the sensitivities of the state with respect to the parameter vector and the initial state are obtained through differentiation of the state equation with respect to p and x_0, respectively:

$$\dot{S}_p = F_x(x, u, p)S_p + F_p(x, u, p); \quad S_p(t_0) = 0 \quad (18)$$
$$\dot{S}_0 = F_x(x, u, p)S_0; \quad S_0(t_0) = I \quad (19)$$

with the Jacobian matrices

$$F_x(x, u, p) = \frac{\partial f(x, u, p)}{\partial x} \quad (20)$$
$$F_p(x, u, p) = \frac{\partial f(x, u, p)}{\partial p}. \quad (21)$$

The dynamic equation for the propagation of P is derived by differentiation of (17) with respect to time:

$$\dot{P} = \dot{S}_p Q_p S_p^{\mathrm{T}} + S_p Q_p \dot{S}_p^{\mathrm{T}} + \dot{S}_0 P_0 S_0^{\mathrm{T}} + S_0 P_0 \dot{S}_0^{\mathrm{T}}; \quad P(t_0) = P_0 \quad (22)$$

Substitution of the time derivatives $\dot{S}_p$ and $\dot{S}_0$ with (18) and (19), respectively,

$$\dot{P} = F_x S_p Q_p S_p^{\mathrm{T}} + F_p Q_p S_p^{\mathrm{T}} + S_p Q_p S_p^{\mathrm{T}} F_x^{\mathrm{T}} + S_p Q_p F_p^{\mathrm{T}} + F_x S_0 P_0 S_0^{\mathrm{T}} + S_0 P_0 S_0^{\mathrm{T}} F_x^{\mathrm{T}}; \quad P(t_0) = P_0 \quad (23)$$

leads with (17) to

$$\dot{P} = F_x P + P F_x^{\mathrm{T}} + F_p Q_p S_p^{\mathrm{T}} + S_p Q_p F_p^{\mathrm{T}}; \quad P(t_0) = P_0. \quad (24)$$

The difference bewtween equations (24) and (6) is the dynamic definition of the uncertainty in the state equation

$$Q(t) = F_p(\hat{x}(t), u(t), \hat{p}) Q_p S_p^{\mathrm{T}}(t) + S_p(t) Q_p F_p^{\mathrm{T}}(\hat{x}(t), u(t), \hat{p}) \quad (25)$$

$$\dot{S}_p(t) = F_x(\hat{x}(t), u(t), \hat{p}) S_p(t) + F_p(\hat{x}(t), u(t), \hat{p}); \quad S_p(t_0) = 0 \quad (26)$$

3. FULL-HORIZON AND RECEDING-HORIZON OBSERVERS

The so-called Full-Horizon observer is very much alike an on-line parameter estimator, since it uses an optimization algorithm to determine the most likely initial conditions of the bioprocess model (Bogaerts, 1999; Bogaerts and Hanus, 2001), which is equivalent to a cross-validation of the model. If, in addition, some parameters are uncertain and re-estimated in the course of state observation, the solution of the estimation problem is equivalent to a direct model validation. A major difference lies the reduced number of on-line signals, as compared to a classical off-line parameter estimation procedure.

Consider the non-linear system (12, 13). The principle of the Full-Horizon observer is to estimate the best initial condition in a maximum-likelihood sense and to predict the current state through simulation of the system:

$$\dot{\hat{x}} = f(\hat{x}, u, \hat{p}); \quad \hat{x}(t_0) = \hat{x}_{0|i} \quad (27)$$

with

$$[\hat{x}_{0|i}, \hat{p}_i] = \arg\min_{x_0, p} j(x_0, p) \quad (28)$$

$$j(x_0, p) = \sum_{k=0}^{i} (y_k - h(x(t_k, x_0, p)))^{\mathrm{T}} R_k^{-1} \cdot (y_k - h(x(t_k, x_0, p))) + (x_0 - \hat{x}_{0|0})^{\mathrm{T}} P_0^{-1} (x_0 - \hat{x}_{0|0}) + (p - \hat{p}_0)^{\mathrm{T}} Q_p^{-1} (p - \hat{p}_0) \quad (29)$$

The first term in the cost function (28) represents the classical sum of squared deviations between the experimental data y_k and model-predicted output $h(x(t_k))$, weighted by the measurement covariance matrix R_k. The last two terms are introduced here to take the *a priori* knowledge on the initial condition and the parameter vector with their respective covariance matrices P_0 and Q_p, e.g., calculated in a preliminary parameter estimation procedure, into account.

The current estimate of the initial condition serves as an initial guess for the following optimization step, since it is supposed to be the best estimate based on the available data.

The uncertainty associated with the estimated variable vector $\theta_i^{\mathrm{T}} = \left[x_{0|i}^{\mathrm{T}} \; p^{\mathrm{T}}\right]$, containing the initial condition and the re-estimated model parameter vector, which is obtained from data between t_0 and t_i, can be evaluated via the Fisher information matrix F_i:

$$F_i(\hat{x}_{0|i}, \hat{p}_i) = \sum_{k=0}^{i} \left(\left.\frac{\partial h(x(t,\theta))}{\partial \theta}\right|_{t_k} \right)^{\mathrm{T}} R_k^{-1} \left(\left.\frac{\partial h(x(t,\theta))}{\partial \theta}\right|_{t_k} \right) = \sum_{k=0}^{i} \left(\left.\frac{\partial h}{\partial x}\right|_{\hat{x}(t)} \cdot \left[S_0(t_k) \; S_p(t_k)\right] \right)^{\mathrm{T}} R_k^{-1} \cdot \left(\left.\frac{\partial h}{\partial x}\right|_{\hat{x}(t)} \cdot \left[S_0(t_k) \; S_p(t_k)\right] \right) \quad (30)$$

with the sensitivity matrices S_0 and S_p defined in (16). In order to calculate the actual joint covariance matrix Q_i of the estimated initial condition $\hat{x}_{0|i}$ and the re-estimated parameters $\hat{p}_i$ the expression for the Cramer-Rao inequality is extended as follows:

$$Q_i(\hat{x}_{0|i}, \hat{p}_i) \geq \left(F_i(\hat{x}_{0|i}, \hat{p}_i) + \begin{bmatrix} P_0 & 0 \\ 0 & Q_p \end{bmatrix}^{-1} \right)^{-1} \quad (31)$$

assuming that the initial estimates $\hat{x}_{0|0}$ and $\hat{p}_0$ are not correlated. The current state covariance matrix $P(t)$ is obtained from equation (17)

$$P(t) = \left[S_0(t) \; S_p(t)\right] Q_i \left[S_0(t) \; S_p(t)\right]^{\mathrm{T}} \quad (32)$$

and the matrix differential equation system for the sensitivity matrices, respectively,

$$\dot{S}_0 = F_x(\hat{x}, u, \hat{p}) S_0; \quad S_0(t_0) = I \quad (33)$$

$$\dot{S}_p = F_x(\hat{x}, u, \hat{p}) S_p + F_p(\hat{x}, u, \hat{p}); \quad S_p(t_0) = 0 \quad (34)$$

with the Jacobian matrices (20,21).

Thus, the diagonal elements of $P(t)$ represent the variances of the current state estimates $\hat{x}(t)$, and their respective approximate 95% confidence intervals

$$\mathbb{X}(t) = \hat{x}(t) \pm \Delta x_{0.95}(t) \quad (35)$$

can be determined with:

$$\Delta x_{0.95}(t) \approx 1.96 \sqrt{P_{jj}(t)}; \quad j = 1, \ldots, n_x \quad (36)$$

When the system under consideration is operated in fed-batch or continuous mode and the influence of the initial condition decreases as time evolves, a possible strategy is to define a limited-range horizon into the past, and to consider only the data within this time window. As time evolves, the horizon of the observer moves forward leading to a Receding-Horizon observer,

$$\dot{\hat{x}} = f(\hat{x}, u, \hat{p}_0); \quad \hat{x}(t_0) = \hat{x}_{i-l+1|i} \tag{37}$$

with the new initial condition of the considered time window obtained by

$$\hat{x}_{i-l+1|i} = \arg\min_{x_{i-l+1}} \Big\{ \sum_{k=i-l+1}^{i} (y_k - h(x(t_k)))^{\mathrm{T}} R_k^{-1} \cdot (y_k - h(x(t_k))) \Big\} \tag{38}$$

with a fixed horizon of l preceding output instants.

An adaptive version of the Receding-Horizon observer results, if the horizon l is adjusted such that a minimum confidence level in the state estimate can be guaranteed. Based on the current estimate $x_{i-l+1|i}$ resulting from the optimization (38), the confidence in this estimate is approximated by

$$P_{i-l+1|i} \geq \left(\sum_{k-l+1}^{i} \left(\frac{\partial h}{\partial x}\bigg|_{\hat{x}(t)} \frac{\partial \hat{x}}{\partial \hat{x}_0}\bigg|_{t_k} \right)^{\mathrm{T}} R_k^{-1} \cdot \left(\frac{\partial h}{\partial x}\bigg|_{\hat{x}(t)} \frac{\partial \hat{x}}{\partial \hat{x}_0}\bigg|_{t_k} \right) \right)^{-1} . \tag{39}$$

An approximation of the covariance matrices for horizons of different lengths, assuming that the trajectory remains the same, i.e. $\hat{x}_{k|i} \approx \hat{x}(t_k, \hat{x}_{i-l+1|i}, \hat{p}_i)$, is then written analogously:

$$P_{k|i} \approx \left(\sum_{\tilde{k}=k}^{i} \left(\frac{\partial h}{\partial x}\bigg|_{\hat{x}(t)} \frac{\partial \hat{x}}{\partial \hat{x}_0}\bigg|_{t_{\tilde{k}}} \right)^{\mathrm{T}} R_{\tilde{k}}^{-1} \cdot \left(\frac{\partial h}{\partial x}\bigg|_{\hat{x}(t)} \frac{\partial \hat{x}}{\partial \hat{x}_0}\bigg|_{t_{\tilde{k}}} \right) \right)^{-1} . \tag{40}$$

Since the matrices fulfil the inequality $P_{k_1|i} \leq P_{k_2|i}$, for $k_1 < k_2$, it is straightforward to determine the minimum horizon $l \in \{1, \ldots, i\}$ in $k = i - l + 1$ that fulfils the confidence condition $P_{k|i} \leq P_{\max}$, i.e.,

$$l_i = \min \left\{ \tilde{l} \mid P_{i-\tilde{l}+1|i} \leq P_{\max} \right\} \tag{41}$$

4. APPLICATION TO A CHO-K1 CELL CULTURE

Both state estimation techniques are now applied to cultures of animal cells, i.e., Chinese Hamster Ovary (CHO-K1) cells, in a bioreactor operated in perfusion mode. In this mode, the culture growth medium (glucose, amino-acids, and metabolites such as ammonium and lactate) is continuously replaced, whereas biomass is retained within the bioreactor thanks to a filtering device.

A dynamic model of this bioprocess has been developed by Haag *et al.* (2004), which consists of the following mass balance equations for the medium concentrations $c \in \mathbb{R}^{n_c}$:

$$\frac{dc}{dt} = \Upsilon r(c) + (c_{\text{in}} - c)\frac{\dot{V}_{\text{in}}}{V} + (c - c_{\text{out}}(c))\frac{\dot{V}_{\text{out}}}{V}; \quad c(t_0) = c_0, \tag{42}$$

where the first term on the right-hand side represents production or consumption according to the considered reaction scheme, the second term represents dilution due to the feed flow rate and the third term represents a correction term in the case where the harvesting concentration differs from the reactor concentration, i.e., due to the presence of a filtering device. The reaction term consists of the constant stoichiometric matrix $\Upsilon \in \mathbb{R}^{n_c \times n_r}$ containing the yield coefficients and a reaction rate vector $r \in \mathbb{R}^{n_r}$ non-linearly dependent on the medium concentrations c.

A specific structure of the reaction kinetics is assumed (Haag *et al.*, 2003). Therein, each component can have a modulating effect – from limitation to inhibition – in each considered reaction. The reaction rate of reaction j is therefore formulated as follows:

$$r_j(c) = r_{j,\max} \prod_{i=1}^{n_c} a_{ji} \tag{43}$$

with the modular functions $0 \leq a_{ji} \leq 1$ defined as

$$a_{ji} = \frac{c_i^{\operatorname{sign}\{K_{ji}^*\}}}{c_i^{\operatorname{sign}\{K_{ji}^*\}} + K_{ji}^{*2}} \tag{44}$$

$$= \begin{cases} \dfrac{c_i}{c_i + K_{ji}^{*2}} & \text{(limitation), if } K_{ji}^* > 0; \\ 1 & \text{(neutral), if } K_{ji}^* = 0; \\ \dfrac{K_{ji}^{*-2}}{c_i + K_{ji}^{*-2}} & \text{(inhibition), if } K_{ji}^* < 0. \end{cases} \tag{45}$$

Four experiments were carried out. Among them, three are used for parameter estimation and one is left for model cross-validation. The dynamical model involves 47 model parameters, which have been identified using a systematic procedure described by Haag *et al.* (2004). The resulting model is briefly summarized in Table 1 representing a pseudo-reaction scheme with the respective actual maximum specific reaction rates.

4.1 Extended Kalman Filter

An Extended Kalman Filter allows the information on the output and modelling errors to be taken into account with the covariance matrices R and Q_p, which are a direct result of the identification procedure. The measurement covariance matrix R is diagonal and of dimension 7×7 with the variances given in Table 2. The symmetric parameter covariance matrix is calculated from the 47×47 Fisher information matrix (19 estimated stoichiometric coefficients in Υ and 28

Table 1. Simplified reaction scheme obtained by model identification. Stoichiometric coefficients are normalized with respect to the viable biomass in units of $^{mmoles}/_{10^9 cells}$.

$$25.0\,Glc + 4.0\,Gln + 1.6\,Lac \xrightarrow{\max\mu_1 \approx 0.2\,1/d} X_v + 0.9\,Glu + 28.9\,NH_3$$
$$7.8\,Glc + 0.1\,Gln + 2.7\,Glu \xrightarrow{\max\mu_2 \approx 0.2\,1/d} X_v + 27.8\,Lac + 23.9\,NH_3$$
$$5.6\,Glc + 1.0\,Gln + 4.5\,NH_3 \xrightarrow{\max\mu_3 \approx 0.6\,1/d} X_v + 4.4\,Lac + 0.5\,Glu$$
$$X_v \xrightarrow{\max k_d \approx 0.2\,1/d} X_d$$

estimated modulation coefficients $K_{ij}^*, i = 1,\ldots,4, j = 1,\ldots,7$).

Table 2. Output errors estimated by model identification.

component	unit	$\hat{\sigma}_y$
viable cells	10^6 cells/ml	2.62
non-viable cells	10^6 cells/ml	0.96
glucose	g/l	0.30
lactate	g/l	0.14
glutamate	mg/l	20.1
glutamine	mg/l	41.5
ammonia	mg/l	14.4

The initial condition of the observer is chosen according to the mean value of the experimental data that have served for model identification, i.e., the measurements of Experiments 2 to 4. The confidence in this initial condition (expressed as the covariance matrix P_0) is given by the variances of these data.

Consider the case, where all states are measured at discrete sampling times. Figure 1 shows the graphical results of state estimation based on the measurement data of Experiment 1, which have not been used for identification. In this case, the benefit from state estimation is the continuous-time reconstruction and prediction of all the states.

The graphical results demonstrate the efficiency of the state estimation, which takes all available information on the parameter uncertainties in the model and the measurement errors into account. The measurements mostly remain within the estimated confidence intervals of the predicted states.

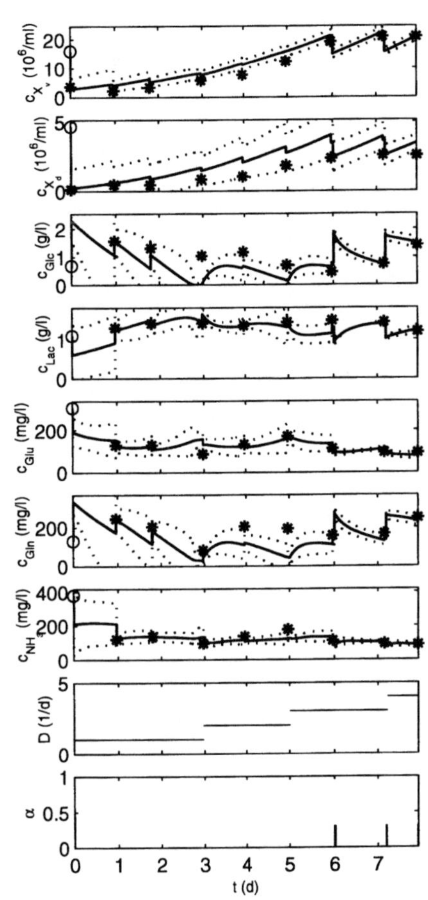

Fig. 1. Extended Kalman Filter for CHO-K1 cell culture applied on Experiment 1. Stars: Measurement outputs; solid line: continuous-time state estimation; dotted line: estimated 95% state confidence interval; circle: initial state estimate.

4.2 Full- and Receding-Horizon observer

The graphical results of the Full-Horizon observer are shown in Figure 2.

Apparently, these results are not as good as those of the Extended Kalman filter, especially towards the end of the observation time.

In fact, the problem lies in the estimation of the initial conditions of a system, which is continuously diluted. In some sense, the history of the process, and particularly the initial condition, is washed out, and the sensitivity of the current state with respect to this initial condition vanishes. Since the dilution rate $D = \dot{V}_{in}/V$ increases from $1\,^1/_d$ to $4\,^1/_d$ in the considered experiments, i.e., the bioreactor volume is exchanged several times a day, this washing out effect is increased.

An appealing alternative is therefore the Receding-Horizon observer, which considers a limited number of measurements into the past (in this specific case, the five most recent time instants) in order to estimate the initial condition of the current time window through a Gauss-Markov criterion.

The graphical results given in Figure 3 show indeed much better results than the Full-Horizon version and are comparable with the performance of the Extended Kalman Filter.

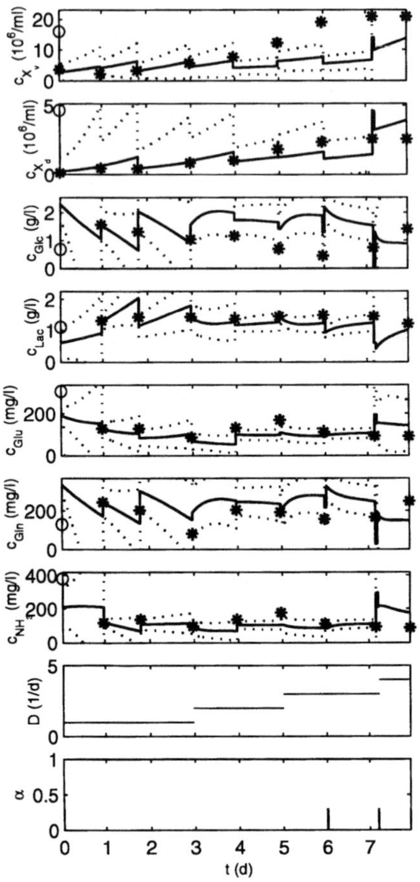

Fig. 2. Full-Horizon observer for CHO-K1 cell culture applied on Experiment 1. Stars: Measurement outputs; solid line: continuous-time state estimation; dotted line: estimated 95% state confidence interval; circle: initial state estimate.

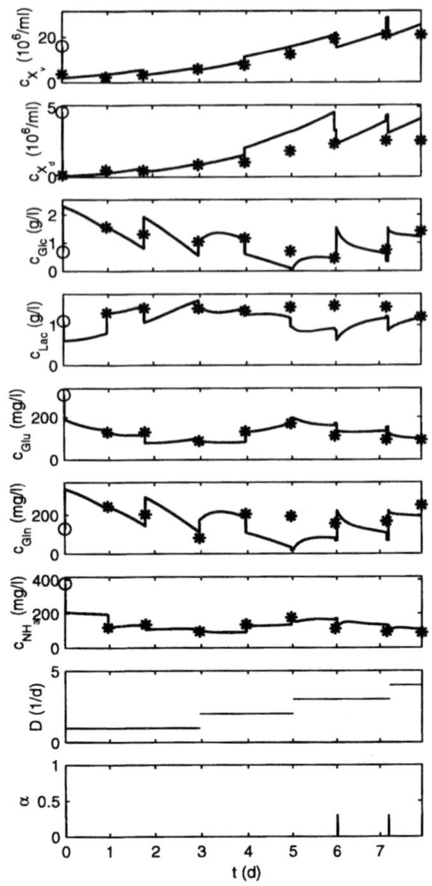

Fig. 3. Receding-Horizon observer for CHO-K1 cell culture applied on Experiment 1. Stars: Measurement outputs; solid line: continuous-time state estimation; circle: initial state estimate.

5. CONCLUSION

In this paper, extensions of the Kalman filter and receding-horizon observer are described, which take advantage of the information on the covariance matrices of the model parameter errors, the measurement errors and the initial state estimate errors, which results from the model parameter identification procedure. An application of both techniques is shown on a real experimental case study, namely animal cell cultures in a perfused continuous bioreactor.

ACKNOWLEDGEMENTS

The financial support of Henogen s.a., Charleroi (Belgium) is gratefully acknowledged. The authors also wish to thank Prof. A. O. A. Miller (CIL Biotech, Mons, Belgium) for providing the experimental data.

REFERENCES

Allgöwer, F., T.A. Badgwell, J.S. Qin, J.B. Rawlings and S.J. Wright (1999). Nonlinear predictive control and moving horizon estimation - an introduction overview. In: *Advances in Control (Highlights of ECC'99)* (P. M. Frank, Ed.). Springer-Verlag. London. pp. 391–449.

Bogaerts, Ph. (1999). Contribution À la Modélisation Mathématique Pour la Simulation et l'Observation D'États Des Bioprocédés. PhD thesis. Université Libre de Bruxelles. Brussels.

Bogaerts, Ph. and A. Vande Wouwer (2003). Software sensors for bioprocesses. *ISA Transactions* **42**, 547–558.

Bogaerts, Ph. and R. Hanus (2001). On-line state estimation of bioprocesses with full horizon observers. *Mathematics and Computers in Simulation* **56**, 425–441.

Gelb, A., Ed.) (1989). *Applied optimal estimation.* MIT Press. Cambridge.

Haag, J. E., A. Vande Wouwer and M. Remy (2003). A general model of reaction kinetics in biological systems. In: *European Control Conference.* Cambridge, UK.

Haag, J. E., A. Vande Wouwer, M. Remy and Ph. Bogaerts (2004). Systematic model identification of complex bioprocesses: Application to a cho-k1 cell culture. In: *9th International Conference on Computer Applications in Biotechnology*. IFAC. Nancy. accepted for publication.

ELSEVIER
IFAC
PUBLICATIONS
www.elsevier.com/locate/ifac

DESIGN AND PRACTICAL USE OF PROBABILISTIC OBSERVERS FOR MASS-BALANCE BASED BIOPROCESS MODELS

Benoît Chachuat [*,**] **Olivier Bernard** [*]

[*] *INRIA Comore, BP 93, 06902 Sophia-Antipolis, France*
fax: +33 492 387 858
email: `Olivier.Bernard@inria.fr`
[**] *Currently at: MIT, Dpt. of Chemical Engineering, Cambridge, MA 02139-4307, USA*
email: `bchachua@mit.edu`

Abstract: In this paper, the design of probabilistic observers for mass-balance based bioprocess models is investigated. It is assumed that the probability density function of every uncertain parameter, input and/or initial state is known *a priori*. Then, the probability density functions of the state variables are obtained, at any time, by considering the image of this initial probability density function by the flow of the dynamical system. In comparison to classical open-loop interval observers, the method provides information on the confidence level of the estimates rather than simple upper and lower bounds. The numerical implementation of the method is closely considered and an application to an industrial anaerobic digester is detailed.

Keywords: non-linear estimation, unknown inputs, uncertain parameters, mass-balance models, bioprocesses, anaerobic digestion.

1. INTRODUCTION

One of the main difficulties in the monitoring and control of biological reactors lies in the absence, in most applications, of cheap and reliable sensors capable of providing direct, on-line measurements of the biological state variables. The design of *observers* for online monitoring of the state variables which are not measurable in real time has thus received much attention in the literature, and has given rise to numerous practical applications [Bastin and van Impe, 1995].

Bioprocess models generally consist of two parts: (*i*) a part based on mass-balance considerations which requires few phenomenological knowledge; (*ii*) another part that describes the biological reactions (kinetics) and therefore includes a large part of phenomenological knowledge. This later part introduces large uncertainties in the models, and it is well-known that the classical observers such as the extended Kalmann filter and the high gain observer [Gauthier et al., 1992, Bernard et al., 1998] may exhibit a poor convergence rate or even fail to converge. Taking advantage of the special model structure, *asymptotic observers* have been proposed which do not require the knowledge of the kinetics [Bastin and Dochain, 1990]. More recently, *interval observers* have been designed in order to deal with these uncertainties [Rapaport and Gouzé, 2003]. The idea is to estimate rigorous intervals for the state variables by considering conservative bounds on the uncertainties (such as imprecisions in the influent measurements or model parameters). This requires to weaken the observation principle in the sense that the observation error is no longer expected to asymptotically converge to zero. Several applications have been successfully considered for biological systems [*e.g.* Gouzé et al., 2000]. But even if interval observers are more robust than the aforementioned asymptotic observers, poor information is obtained on the estimates since "only" guaranteed upper and lower bounds are derived.

In many situations however, more information than simple upper and lower bounds is available on the uncertainties, *e.g.*, their probability density functions (p.d.f.). For the uncertain parameters in particular, confidence intervals can be easily obtained from the application of a parameter identification procedure. In this work, we extend the concept of interval observers by developing *probabilistic observers* which estimate the probability density functions of the unmeasured states. Major differences between the proposed observers and the extended Kalman filter are that no assumption is stated here for the kinetic expressions and that aribitrary p.d.f.s can be selected for the uncertain parameters.

The paper is organised as follow: some results on both asymptotic and interval observers are first recalled in section 2; probabilistic observers are considered in section 3 for general mass-balance bioprocess models, and their numerical implementation is closely considered; the method is illustrated in section 4 by a practical application of a probabilistic observer to estimate the COD concentration in an industrial anaerobic digester.

2. ASYMPTOTIC AND INTERVAL OBSERVERS FOR BIOPROCESSES

A bioprocess operated in a stirred tank bioreactor is often described by means of general mass-balance based models of the following form [Bastin and Dochain, 1990]:

$$\dot{\boldsymbol{\xi}} = \mathcal{K}\boldsymbol{r}(\boldsymbol{\xi}) + D(\boldsymbol{\xi}^{in} - \boldsymbol{\xi}) - \boldsymbol{q}(\boldsymbol{\xi}) \tag{1}$$

where $\boldsymbol{\xi} \in \mathbb{R}^{n_\xi}$ (resp. $\boldsymbol{\xi}^{in}$) denotes the concentration vector in the liquid phase (resp. in the influent), $\mathcal{K} \in \mathbb{R}^{n_\xi \times n_r}$ the yield coefficient matrix, $\boldsymbol{r} \in \mathbb{R}^{n_r}$ the reaction rate vector, $\boldsymbol{q} \in \mathbb{R}^{n_\xi}$ the gaseous exchange vector, and $D \in \mathbb{R}$ the dilution rate.

Throughout the paper, it is assumed that $(n_\xi - n_b)$ components of the state variables are measured on-line with $n_\xi - n_b \geq n_r$, and we denote $\boldsymbol{\xi}_a \in \mathbb{R}^{n_\xi - n_b}$, $\boldsymbol{\xi}_b \in \mathbb{R}^{n_b}$ the measured and unmeasured state variables respectively. It is also assumed that the gaseous flow rates $\boldsymbol{q}$ are measured on-line. It follows that Eq. (1) can be rephrased as:

$$\dot{\boldsymbol{\xi}}_a = \mathcal{K}_a\boldsymbol{r}(\boldsymbol{\xi}) + D(\boldsymbol{\xi}_a^{in} - \boldsymbol{\xi}_a) - \boldsymbol{q}_a(\boldsymbol{\xi}) \tag{2}$$
$$\dot{\boldsymbol{\xi}}_b = \mathcal{K}_b\boldsymbol{r}(\boldsymbol{\xi}) + D(\boldsymbol{\xi}_b^{in} - \boldsymbol{\xi}_b) - \boldsymbol{q}_b(\boldsymbol{\xi}) \tag{3}$$

with $(\mathcal{K}_a, \mathcal{K}_b)$, $(\boldsymbol{\xi}_a^{in}, \boldsymbol{\xi}_b^{in})$ and $(\boldsymbol{q}_a, \boldsymbol{q}_b)$ being the induced partition of $\mathcal{K}$, $\boldsymbol{\xi}^{in}$ and $\boldsymbol{q}$ respectively.

We state the following hypothesis:

Hypothesis 1. Matrix $\mathcal{K}_a \in \mathbb{R}^{(n_\xi - n_b)\times n_r}$ has full rank.

From hypothesis 1 and a linear change of variables, the following n_b-dimensional auxiliary system can be derived:

$$\dot{\boldsymbol{z}} = -D(\boldsymbol{z} - \boldsymbol{z}^{in}) - \mathcal{P}\boldsymbol{q}_a - \boldsymbol{q}_b \tag{4}$$
$$\boldsymbol{\xi}_b = \boldsymbol{z} - \mathcal{P}\boldsymbol{\xi}_a \tag{5}$$
$$\text{where:} \quad \mathcal{P} \triangleq -\mathcal{K}_b\mathcal{K}_a^{\dagger}$$

and $\mathcal{K}_a^{\dagger}$ is a left inverse of $\mathcal{K}_a$. Note in particular that, in the auxiliary system (4,5), the measured variable $\boldsymbol{\xi}_a$ is used as an input.

The following result holds.

Lemma 1. Under Hypothesis 1 (and provided D is persistently exiting), the solution $\hat{\boldsymbol{\xi}}_b$ of the following open-loop observer converges asymptotically towards the solution $\boldsymbol{\xi}_b$ of the reduced system (3):

$$\dot{\hat{\boldsymbol{z}}} = -D(\hat{\boldsymbol{z}} - \boldsymbol{z}^{in}) - \mathcal{P}\boldsymbol{q}_a - \boldsymbol{q}_b \tag{6}$$
$$\hat{\boldsymbol{\xi}}_b = \hat{\boldsymbol{z}} - \mathcal{P}\boldsymbol{\xi}_a \tag{7}$$

Proof. see, *e.g.*, Bastin and Dochain [1990]. □

Such observers have however several defects due to their open-loop nature. In particular, it is implicitly assumed that the mass-balance part of the model is perfectly known and that neither the measurements, nor the values of the feeding inputs, nor the estimates of the yield coefficients (matrix $\mathcal{K}$) are biased. Otherwise, the predictions of the mass-balance based observer (6,7) will be corrupted and might provide poor estimates of the unmeasured concentrations $\boldsymbol{\xi}_b$.

In the case of large uncertainties, it is no longer possible to build exact observers guarantying that the observation errors converges to zero. Therefore, the observation principle must be revisited and the results must be weakened. A complementary approach, called interval observers, provides guaranteed enclosures on the estimated states, whenever upper and lower bounds are known on the uncertain inputs and parameters. Such observers consist in coupling two estimators providing each an over-estimate $\boldsymbol{x}^+(t)$ and an under-estimate $\boldsymbol{x}^-(t)$ of the unknown state variables $\boldsymbol{x}(t)$ at any time. Details on interval observers can be found, *e.g.*, in Gouzé et al. [2000], Rapaport and Gouzé [2003].

The following proposition applies the concept of interval observers to the auxiliary system (4) derived from the general mass-balance based model (1).

Proposition 1. The following pair of systems is an interval observer for the variables $z(t)$ solution of (4):

$$\dot{z}^+ = -Dz^+(t) + Dz^{in^+}(t) - \mathcal{P}^-(p^+,p^-)\, q_a(t) + q_b(t)$$

$$\dot{z}^- = -Dz^-(t) + Dz^{in^-}(t) - \mathcal{P}^+(p^+,p^-)\, q_a(t) + q_b(t)$$

$$\text{with: } z^+(0) = \xi_b^+(0) + \mathcal{P}^+(p^+,p^-)\, \xi_a(t)$$

$$z^-(0) = \xi_b^-(0) + \mathcal{P}^-(p^+,p^-)\, \xi_a(t)$$

where the unknown parameters $p \in \mathbb{R}^{n_p}$ and inputs $z^{in} \in \mathbb{R}^{n_b}$ are characterised by their upper and lower bounds:

$$p_j^- \le p_j \le p_j^+ \ , \ \forall j = 1 \dots n_p$$

$$z_i^{in^-}(t) \le z_i^{in}(t) \le z_i^{in^+}(t) \ , \ \forall t \ , \ \forall i = 1 \dots n_b$$

and the upper and lower bound matrices $\mathcal{P}^+$, $\mathcal{P}^-$ are defined componentwise:

$$\mathcal{P}_{i,j}^-(p^+,p^-) \le \mathcal{P}_{i,j}(p) \le \mathcal{P}_{i,j}^+(p^+,p^-) \ ,$$

$$\forall p \in [p^-,p^+] \ , \ \forall i = 1 \dots n_b \ , \ j = 1 \dots n_\xi - n_b$$

Proof. See, *e.g.*, Rapaport and Gouzé [2003]. □

Interval observers provide guaranteed bounds on the estimation of the unmeasured variables, given rigorous bounds on the parameters and the inputs. However, the resulting intervals might be large if the uncertainty is high. In this case, it could be interesting to build subintervals on the estimates corresponding to different confidence levels on the uncertainty. Tools to derive such observers are discussed in the next section.

3. PROBABILISTIC OBSERVERS

As mentioned before, it is not rare that the influent concentrations and/or model parameters are not known precisely for biological processes. For parameters in particular, estimates as well as confidence intervals can be obtained under the application of a parameter identification procedure. More knowledge than simple upper and lower bounds can obviously be obtained on the unmeasured state variables, *e.g.* the probability distribution of the estimates.

3.1 Mathematical background

In mass-balance based bioprocess models of the form (1), the uncertainties may correspond to the yield coefficients in matrix $\mathcal{K}$, to the influent concentrations ξ^{in}, and/or to the unmeasured state variables at initial time ξ_{b0}. In addition, the measurements of the state variables ξ_a and the gaseous flow rates q_a, q_b might be noisy. In order to derive the p.d.f.s of the unmeasured state variables ξ_b, we consider the auxiliary dynamical system (4,5) defined in section 2.

Hypothesis 2. The auxiliary dynamical system (4,5) falls into the following class of dynamical systems:

$$\begin{cases} \dot{z}(t) = \mathcal{A}(p,t)\, z(t) + b(p,t) \\ \xi_b(t) = z(t) + c(p,t) \\ z(0) = \xi_b(0) - c(p,0) \end{cases} \qquad (\mathcal{S})$$

where $p = (p_1, \dots, p_{n_p})^t \in \mathbb{R}^{n_p}$ denotes the vector of the uncertain parameters, and the components of $\mathcal{A} \in \mathbb{R}^{n_b \times n_b}$, $b \in \mathbb{R}^{n_b}$ and $c \in \mathbb{R}^{n_b}$ are C^1 with respect to $p_1, \dots, p_{n_p}$. In addition, the uncertain parameters p and initial conditions z_0 have *independent* p.d.f.s.

Remark 1. For sake of simplicity, we consider that parameters $p_1, \dots, p_{n_p}$ are constants with known p.d.f..

Notations 1. In the sequel, the following notations are used: $\chi_0 = (\xi_{b0}^t, p^t)$, $\mathcal{Z}_0 = (z_0^t, p^t)$, $\chi_t = (\xi_{bt}^t, p^t)$ and $\mathcal{Z}_t = (z_t^t, p^t)$. In addition, we denote f_{χ_0}, $f_{\mathcal{Z}_0}$, f_{χ_t} and $f_{\mathcal{Z}_t}$ their respective p.d.f..

The estimation of the p.d.f. of the unmeasured state variables ξ_b at a given time t can be decomposed into three successive steps:

1) estimate the p.d.f. of the random variable $\mathcal{Z}_0$ at initial time, from the individual p.d.f. of the unmeasured state variables ξ_{b0} and the uncertain parameters p;
2) compute the image $f_{\mathcal{Z}_t}$ at time t of the p.d.f. of $f_{\mathcal{Z}_0}$, by the flow of the auxiliary dynamical system (4);
3) estimate the p.d.f. of the unmeasured state variables ξ_{bt} at time t, from the p.d.f. of $\mathcal{Z}_t$.

The following Theorem [see, *e.g.*, Grimmett and Stirzaker, 2001] provides a general framework to calculate the image of a given p.d.f. by a C^1-diffeomorphism for each aforementioned step.

Theorem 1. Let $\mathcal{U}$, $\mathcal{V}$ be open subsets of $\mathbb{R}^n$, K be a compact subset of U, and $\phi : \mathcal{U} \longmapsto \mathcal{V}$ be a C^1 diffeomorphism. If z is a random variable with a p.d.f. f_z, then the random variable $\omega = \phi(z)$ has a p.d.f. f_ω given by:

$$f_\omega(\omega) = \begin{cases} f_z(\phi^{-1}(\omega)) \det(J_{\phi^{-1}}(\omega)) & \text{if } \omega \in \phi(K) \\ 0 \quad \text{otherwise} \end{cases}$$

where J_ϕ denotes the Jacobian matrix of ϕ.

For steps 1) and 3), let the mapping ψ_t be defined as:

$$\psi_t : \mathcal{U}_t' \longrightarrow \mathcal{U}_t = \psi_t(\mathcal{U}_t')$$

$$\mathcal{Z}_t \longmapsto \chi_t = (\xi_{bt}^t, p^t)^t$$

where $\mathcal{U}_t'$ is an open subset of $\mathbb{R}^{n_b+n_p}$. From Hypothesis 2, the following property is immediate.

Property 1. ψ_t is a C^1-diffeomorphism.

For step 2), consider the mapping φ_0^t defined as:

$$\begin{aligned}\varphi_0^t : \mathcal{U}_0' &\longrightarrow \mathcal{U}_t' = \varphi_0^t(\mathcal{U}_0)\\ \mathcal{Z}_0 &\longmapsto \mathcal{Z}_t = \left(z_t^t, p^t\right)^t\end{aligned}$$

with $\mathcal{U}_0'$ being an open subset of $\mathbb{R}^{n_b+n_p}$. The mapping φ_0^t can be explicitly defined. Indeed, from linear systems theory, a general solution of system $(\mathcal{S})$ can be obtained from:

$$z_t = \Phi(0,t,p)\, z_0 + \int_0^t \Phi(\tau,t,p)\, b(\tau,p)\, d\tau \quad (8)$$

where $\Phi(\cdot,\cdot,p) \in \mathbb{R}^{n_b \times n_b}$ is the transition matrix, *i.e.* the solution of the matrix differential equation:

$$\begin{aligned}\frac{d}{dt}\Phi(\tau,t,p) &= \mathcal{A}(t,p)\,\Phi(\tau,t,p)\ ,\ \forall t\\ \Phi(\tau,\tau,p) &= I_{n_b \times n_b}\end{aligned}$$

Based on the analytical solution of system $(\mathcal{S})$ and Hypothesis 2, it can be shown that the following property holds:

Property 2. φ_0^t is a C^1-diffeomorphism.

Proposition 2. The p.d.f. f_{χ_t} corresponding to the image $\chi_t = \left(\xi_{b_t}^t, p^t\right)^t$ of the random variable $\chi_0 \in K_0 \subset \mathcal{U}_0$ is given by:

$$f_{\chi_t}(\chi_t) = \begin{cases} \dfrac{f_{\chi_0}\left(\psi_0 \circ \varphi_0^{t\,-1} \circ \psi_t^{-1}(\chi_t)\right)}{\det\,\Phi(0,t,p)} \\ \quad \text{if} \quad \chi_t \in K_t = \varphi_0^t(K_0) \\ 0 \quad \text{otherwise}\end{cases}$$

Proof. From Properties 1 and 2, the transformations in steps 1), 2) and 3) are C^1-diffeomorphisms. The overall transformation $\psi_0 \circ \varphi_0^{t\,-1} \circ \psi_t^{-1}$ is therefore a C^1-diffeomorphism (chain rule of differentiation), and Theorem 1 applies. Since,

$$\det\,J_{\psi_0} = \det\,J_{\psi_t} = I_{n_b+n_p}$$

then,

$$\det\,J_{\left(\psi_0 \circ \varphi_0^{t\,-1} \circ \psi_t^{-1}\right)} = \det\,J_{\varphi_0^{t\,-1}}$$

In addition,

$$\det J_{\varphi_0^t} = \det\begin{pmatrix}\Phi(0,t,p) & \star\\ 0_{n_p \times n_b} & I_{n_p \times n_p}\end{pmatrix} = \det\Phi(0,t,p)$$

which completes the proof. □

Finally, individual p.d.f.s for the unmeasured state variables ξ_b can be obtained, at any time t, by integrating the joint state/parameter p.d.f. with respect to the uncertain parameters p. Also note that the aforementionned procedure does not require any particular assumption for the initial state/parameter p.d.f.s, and is therefore not restricted to unimodal p.d.f.s.

3.2 Numerical implementation and practical application

In the sequel, we restrict ourselves to the subclass of dynamical systems $(\mathcal{S})$ for which the matrix $\mathcal{A}$ is Hurwitz. Otherwise the dynamical system would be unstable, which is unrealistic here since $(\mathcal{S})$ is intended to be used as an auxiliary system/observer of the stable BIBO class of models (1).

The numerical implementation of the previous framework can be decomposed into two successive steps:

1) Compute the image of the joint p.d.f. by the diffeomorphism $\psi_0 \circ \varphi_0^{t\,-1} \circ \psi_t^{-1}$. This can be done, *e.g.*, by defining a grid in the joint state/parameter space. Note that due to the linearity of the dynamical system in the state variables, it is only necessary to integrate the trajectories of the dynamical system $(\mathcal{S})$ at the edges of the state variation range; for fixed values of the parameters, evenly spaced points in the initial grid indeed remain evenly spaced at any time (see Eq. (8)).
2) Integrate the joint state/parameter p.d.f. at a given time t, with respect to the parameters, in order to obtain individual p.d.f.s for the state variables.

It should be stressed that step 2) is by far the most complex and time consuming step in the procedure. As the influence of the initial conditions z_0 vanish with increasing time, the joint p.d.f. tends to degeneracy (it becomes infinitely thin), and eventually results in inaccurate p.d.f.s. This can however be easily detected in pratice, *e.g.*, by monitoring the error $\varepsilon = \left|1 - \int_{K_t} f_{\chi_t} d\chi_t\right|$. The degeneracy of the joint p.d.f. is illustrated in the example below.

ILLUSTRATIVE EXAMPLE – To illustrate the practical difficulties encountered when attempting to implement the probabilistic observer numerically, we consider the following simple dynamic system

$$\begin{cases}\dot z(t) = -z(t) + p\ ,\ \forall t > 0\\ z(0) = z_0\end{cases} \qquad (\mathcal{T})$$

and assume that either p.d.f. of z_0 and p are normal distributions, denoted $\mathcal{N}(\overline{z_0}, \sigma_{z_0})$ and $\mathcal{N}(\overline{p}, \sigma_p)$ respectively. Then applying Proposition 2 allows us to compute the joint p.d.f. of $\mathcal{Z}_t = (z_t, p)^t$ at a given time $t > 0$ as

$$f_{\mathcal{Z}_t}(\mathcal{Z}) = \frac{f_{\mathcal{Z}_0}\left(\varphi_0^{t\,-1}(\mathcal{Z})\right)}{\det\,\Phi(0,t)}$$

where

$$\Phi(0,t) = \mathrm{e}^{t-t_0}$$

$$f_{\mathcal{Z}_0}(\mathcal{Z}) = \frac{1}{2\pi\sigma_{z_0}\sigma_p}\mathrm{e}^{-\frac{1}{2}\left[\left(\frac{z-\overline{z_0}}{\sigma_{z_0}}\right)^2 + \left(\frac{p-\overline{p}}{\sigma_p}\right)^2\right]}$$

$$\varphi_0^{t\,-1}(\mathcal{Z}) = \begin{pmatrix}\mathrm{e}^{t-t_0}z + \left[1 - \mathrm{e}^{t-t_0}\right]p\\ p\end{pmatrix}$$

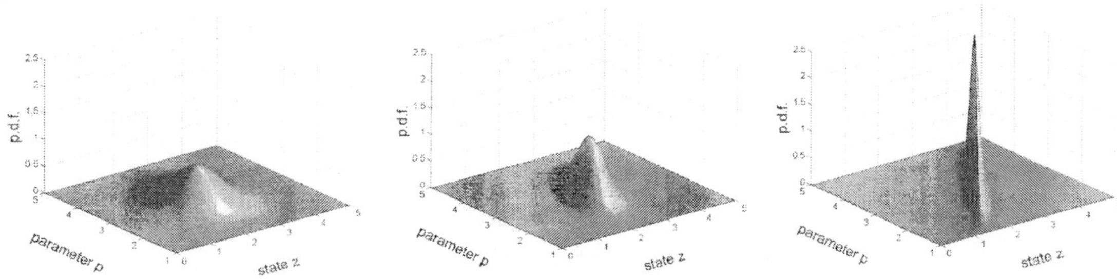

Figure 1. Evolution of the joint state/parameter p.d.f. *Left plot*: $t = 0$; *Middle plot*: $t = 0.5$; *Right plot*: $t = 1.5$.

for any $\mathcal{Z} = (z, p)^t$. By combining these expressions and letting t tend to infinity, it can be proved that

$$f_{Z_t}(\mathcal{Z}) \xrightarrow{t \longrightarrow +\infty} +\infty \ , \text{ if } z = p$$
$$f_{Z_t}(\mathcal{Z}) \xrightarrow{t \longrightarrow +\infty} 0 \quad , \text{ otherwise}$$

This is illustrated in Fig. 1. □

It is frequent, from a monitoring point of view, that the key state variables are measured at a low frequency from off-line analysis in order to check for proper process operation (*e.g.*, once a week in anaerobic digesters for COD and VSS). Accordingly, the approach used in this work to avoid the aforementionned numerical problems consists in repeatedly reseting the observer from a periodic process sampling. For practical applications where the observer cannot be reseted, it is beleived that the numerical difficulties could be circumvented by adding an artificial term acting as a *diffusion* term [see, *e.g.*, Zeeman, 1988]; this will be the topic of future research.

4. CASE STUDY

A simple model of the anaerobic digestion process is considered by accounting for a single substrate (COD) and a single bacterial population, denoted s and x respectively. The biological reaction is represented as

$$k\, s \xrightarrow{r(\xi) = \mu(\cdot)\, x} x$$

where k is a yield coefficient, and $\mu(\cdot)$ represents the bacterial growth rate. By assuming perfect mixing in the digester, the following 2-dimensional dynamical model is obtained

$$\dot{x}(t) = -D(t)\, x(t) + \mu(\cdot)\, x(t) \qquad (9)$$
$$\dot{s}(t) = -D(t)\left(s(t) - s^{in}(t)\right) - k\, \mu(\cdot)\, x(t)$$

where $s^{in}(t)$ stands for the concentration of substrate in the feeding stream. In addition, the methane flow rate is defined as $q_{CH_4}(t) = k'\, \mu(\cdot)\, x(t)$ with k' being the yield coefficient associated to methane production.

Based on model (9), the following auxiliary system can be derived for the biomass concentration provided that the methane gaseous flow rate as well as the influent COD are measured on-line

$$\dot{\hat{s}}(t) = -D(t)\, \hat{s}(t) + \left[D(t)\, s^{in} - \frac{k}{k'}\, q_{CH_4}(t)\right] \qquad (10)$$

Experimental data from a $2,000\,\mathrm{m}^3$ industrial digester located at Agralco Ltd. in Estella, Spain, are considered. The digester is fed with wine distillery vinasses, and the measurements consist of the dilution rate D and the methane outflow rate q_{CH_4} at a high sampling frequency over more than 140 days. In our objective to design a probabilistic observer for estimating the COD concentration based on Eq. (10), we consider the uncertainties on (i) the parameter $\gamma = \frac{k}{k'}$, (ii) the influent concentration s^{in} and, (iii) the COD concentration s_0 at initial time.

- The value of γ was estimated off-line based on mass-balance considerations along with its standard deviation: $\gamma = 5.30\ 10^{-2} \pm 0.98\%$. It is assumed that the p.d.f. for this parameter is Gaussian.
- The uncertainty on s^{in} is handled by considering a multiplicative noise ε defined as $s^{in}(t) = \overline{s^{in}}\,(1 + \varepsilon(t)), \forall t.$ It was found from experimental data that $\overline{s^{in}} = 35.9\,\mathrm{g.L}^{-1}$ and that the p.d.f. of ε can be assimilated to a normal distribution $\mathcal{N}(0, \sigma_\varepsilon)$ with $\sigma_\varepsilon = 0.124$.
- In accordance to the discussion in subsection 3.2, the observer is reinitialised every one week by considering off-line COD measurements; the p.d.f. of these initial concentrations $z_k = z^{\exp}(t_k)$ are assumed to be uniform in the range $z_k \pm 20\%$.

Also note that the experimental data have been filtered and, for the sake of simplicity, no measurement error is taken into account here for D and q_{CH_4}.

A probabilistic observer can be derived for the COD concentration from Eq. (10) by applying Proposition 2. In this case, note that $\psi_t = \psi_0^{-1} = I$ and $\Phi(\tau, t) = \exp\left[-\int_\tau^t D(s)\, ds\right]$. The resulting p.d.f. of the estimate $\hat{s}$ are plotted in Fig. 2 (upper plot) at different time instants. Experimental points are also displayed for the sake of comparison. Note that the variation range of $\hat{s}$ (corresponding to the vertical segments) is rather conservative, but a significant part of the range has a null probability. Also note that the experimental points are correctly enclosed within

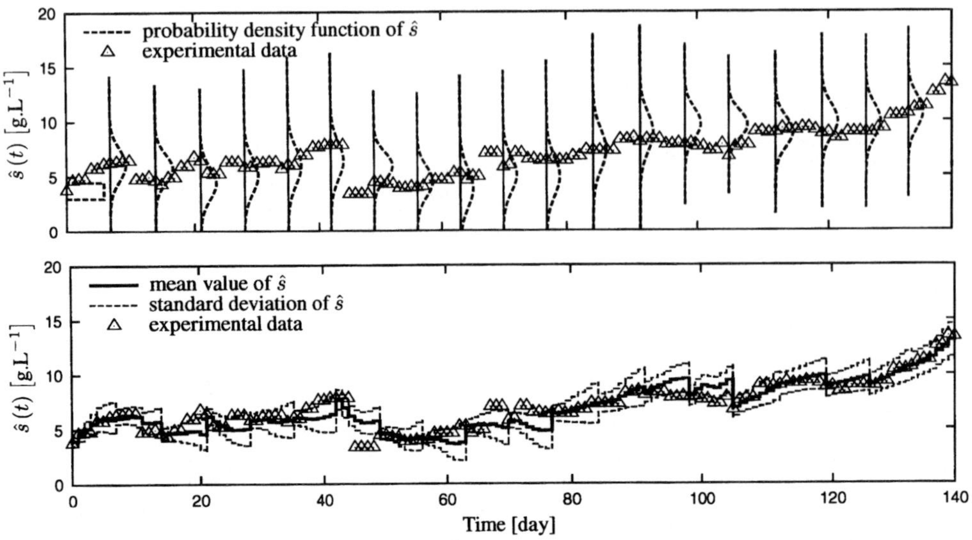

Figure 2. Estimated COD concentration in the digester. *Upper plot*: Probability density function. *Lower plot*: Mean value and standard deviation.

the non-null probability part, and that the maximum likelihood estimate is close to the experimental points.

A straightforward interpretation for the p.d.f. can be obtained by extracting the mean and standard deviation values of $\hat{s}$. These values are pictured in Fig. 2 (lower plot). It can be seen in particular that the estimates are close to the experimental measurements, and the interval corresponding to the standard deviation of $\hat{s}$ mainly encloses these measurements. From a sensitivity point of view, the quality of the standard-deviation based bounds is strongly affected by the p.d.f. of parameter γ, while the influence of the initial p.d.f. after the observer is reseted rapidely vanishes and therefore remains limited.

5. CONCLUSIONS

In this paper, the design of probabilistic observers was investigated for general mass-balance based bioprocess models. It was shown how the probability density functions of the unmeasured state variables can be derived from the knowledge of the probability density functions of the uncertain parameters and the measurement noises. An application of the method to the estimation of the COD concentration in an industrial anaerobic digester was then presented, which demonstrate the tractability and good performances of the approach.

Although only an asymptotic rate of convergence is guaranteed in this paper, it is worth pointing out that this rate can be chosen in some cases [Rapaport and Gouzé, 2003]. Such probabilistic observers can therefore be used as tools to improve bioprocess monitoring. In addition, they could also be applied in the scope of bioprocess supervision and fault detection, *e.g.*, by defining a given tolerance expressed in term of a threshold confidence level.

Acknowledgements: This work has been carried out with the support provided by the European Commission, Information Society Technologies program, Key action I Systems & Services for the Citizen, contract TELEMAC number IST-2000-28256. This information is provided under the sole responsibility of the author and does not necessarily represent the opinion of the European Commission, which is not responsible for any use that might be made of it.

REFERENCES

G. Bastin and D. Dochain. *On-line estimation and adaptive control of bioreactors*. Elsevier, Amsterdam, The Netherlands, 1990.

G. Bastin and J. van Impe. Nonlinear and adaptive control in biotechnology: a tutorial. *European Journal of Control*, 1(1): 1–37, 1995.

O. Bernard, G. Sallet, and A. Sciandra. Nonlinear observers for a class of biological systems. Application to validation of a phytoplanktonic growth model. *IEEE Trans. Automat. Control.*, 43:1056–1065, 1998.

J.-P. Gauthier, H. Hammouri, and F. Celle. A simple observer for nonlinear systems – Applications to bioreactors. *IEEE Trans. Automat. Control.*, 37:875–880, 1992.

J.-L. Gouzé, A. Rapaport, and M. Z. Hadj-Sadok. Interval observers for uncertain biological systems. *Ecological Modelling*, 133:45–56, 2000.

G. R. Grimmett and D.R. Stirzaker. *Probability and Random Processes*. Oxford University Press Inc., New York, 3rd edition edition, 2001.

A. Rapaport and J.-L. Gouzé. Parallelotopic and practical observers for non-linear uncertain systems. *International Journal of Control*, 76(3):237–251, 2003.

E.C. Zeeman. Stability of dynamical systems. *Nonlinearity*, 1:115–155, 1988.

ELSEVIER
IFAC
PUBLICATIONS
www.elsevier.com/locate/ifac

Population Balance Model for Cellular Processes in Biological Systems: Biochemical and Biomedical Applications

Charles David Immanuel
Dept. of Chemical Engineering and Chemical Technology, Imperial College, London SW7 2AZ, UK
c.immanuel@imperial.ac.uk; Phone: +44 (0)20 7594 5594; Fax: +44 (0)20 7594 6606

Abstract

In this paper, the generic problem of the development of a population balance model for those biological systems involving cellular processes is discussed. Such models are of interest for bioreactor optimisation and control, as well as for different biomedical problems including the study of cancer. The generic model and its features are presented, serving partly to consolidate the studies that have been reported in the literature already. The major challenges and issues with this problem are highlighted. One of the challenges is with respect to the solution technique for these complex models, which issue is addressed here. The algorithm proposed for solving these complex models is based on a hierarchical two-tier solution strategy that has been proposed previously (Immanuel and Doyle, III, Chem. Eng. Sci., 2003). The two-tier strategy involves the calculation of the rates of sub-processes that contribute to the population dynamics in the first tier, with the population itself being updated in the second tier. This in turn enables performing a major portion of the calculations off-line (once at the start of the simulation), thereby achieving a major reduction in the computational load. The algorithm is implemented on a six-dimensional population balance bioreactor model.

Introduction

The interest in the development of detailed mathematical models for biological processes is undergoing a tremendous boost in recent years among the Chemical Engineers. This is because of the various purposes that a mathematical model of biological processes can serve [16]. The primary purpose is the growing interest in a biochemical route to the manufacture of commodity chemicals, pharmaceuticals, food products, and agricultural products. Apart from the bioreactors, the models will be very versatile tools for the analysis of biomedical problems and in therapeutical studies. Very detailed studies on the various individual aspects of biological systems is carried out by the biologists, biochemists, and the researchers in the medical fields. These individual aspects need to be consolidated to form a comprehensive picture, to study the effect of these individual mechanisms on the overall process, and hence to bring these individual studies to further fruition. A comprehensive model will provide a framework to do this, and will in most cases be the only tool available to explain the various intricate experimental observations. The mathematical model can aid in the study of the underlying processes themselves, by suggesting appropriate and effective experimentation. The net result is a symbiotic improvement in the process understanding and model development. The resultant model will be in a form suitable for the different end applications, be it to perform design, optimisation and control of bioreactors or for therapeutical studies, as the case may be.

The focus of this study is on biological systems that involve cellular processes. The most common and simple methodology for the mathematical modelling of such systems is based on the so-called unstructured models, which do not distinguish the individual cells in the aqueous medium. However, the complexity of the bioprocesses warrents the introduction of structure into the model, to obtain a more rigorous representation of the processes. *A more rigorous modelling approach will account for the heterogenous nature of the reaction medium, the non-uniformity of the cell populations with respect to their mass and internal metabolites, the cell cycles (the processes of growth, division, death, and in certain cases even recombination), and the interaction of the cell cycle with the environment. This can be naturally done employing the population balance framework.*

Although the advantages of pursuing a population balance approach for bioreactor modelling has been recognised since the 1960's by Fredrickson and Tsuchiya [5], the actual application of these models in reactor optimisation and control has not been pursued until much recently [17, 10]. This is partly attributed to a lack of sufficient depth of understanding of the underlying mechanisms. The complexity of the metabolic pathway, and hence of the mathematical model, has also been a major deterent in this problem. Although these deterents have been sufficiently overcome in recent years, the control studies reported are based on total-mass-structured cell population balance models, with a single internal coordinate. However, the complexity of the process, with the large number of metabolites and reactions involved, warrants and even necessitates the introduction of further structure into the models. The cell cycle phases, and the growth and division rates, are dependent not only on the total mass of the cells, but more so on the mass of the different metabolites within the cells. Thus, there is a need to develop higher dimensional population balance models, by accounting for the distribution of different properties (and not just the total cell mass) among the cell populations.

On a different level of time scales and to a certain extent the model complexity, the modelling of cell cycles assumes importance for characterising the physiological conditions and understanding several medical problems. For instance, the cells that constitute the various organs in the human body are involved in the processes of growth, division and death, and their interaction with the surroundings and the other cells. Alterations to the cell cycles, due to perturbations either internally within the cells or externally in the surrounding, lead to cancerous growth of the cells and to tumours. Even though modelling studies on cancer have appeared in the literature (such as the Marchuk's Model for general immune response), these studies do not account for the cell cycle phases. Instead, they mainly account for the interaction between normal cells, cancer cells and the antigens, resulting in relatively simple (three state) models (similar in structure to the first models on HIV [12]). Since the major disease factor lies in the cell cycle phase, the best model for cancer has to account for the cell cycles. Thus, again, a population balance model becomes the most appropriate modelling tool for cancer. In a similar vein, *population balance models of microbial populations also assume importance for the study of diseases caused by bacterial populations*, which co-exist and interact with each other as well as with bacteriophages. In this case again, unlike in the case of the relatively short-spanned bioreactors, the interactions between mutation and recombination (crossover) will assume a critical factor.

Measurement of the cell populations and their distribution will be a critical factor in model development, to ensure that the right level of detail is present in the model to provide it with the predictive capabilities. *Flow cytometry is a boon for the development of cell population balance models.* Although it stems from the age-old concept of gram-staining of the cells, it has been considerably developed over the years to be amenable to current requirements [15]. Flow cytometers can also be adapted for on-line measurement of cell populations, with sample times of 18 minutes being reported [1]. These developments are very positive aspects for model development. The instrument can be used to monitor the distribution of select individual metabilites, or to monitor the total mass distribution of the population.

In this paper is presented a general, yet realistic formulation of the population balance-based models for the different biological problems highlighted above. Further, the major challenges in implementing this approach are identified. *Preliminary results that overcome some of these challenges are presented, particularly with respect to the numerical solution of bioreactor models.*

Opportunities

Cell cycle and regulation

As indicated in the introduction, a realistic model of the cellular biological processes has to account for the cell cycles in the heterogeneous medium. The cells in the population undergo a cycle of changes through their life time. The individual cells synthesise metabolites including proteins, RNA and DNA, and thereby grow in size. Upon the duplication of the genetic carriers – the DNAs, the cells undergo cell division into two usually identical daughter cells, contributing new and younger members to the population. The cell cycle is divided into two regimes – the interphase and the mitosis phase (M-phase). The interphase is that regime of the cell cycle that characterises the synthesis of metabolites, cell growth and DNA replication. The M-phase characterises the cell division. In the case of the Eukaryotes, the interphase is further divided into the G1 phase, the S phase, and the G2 phase. The G1 phase accounts for the synthesis of the metabolites required for DNA replication, the S phase accounts for DNA synthesis, the G2 phase accounts for the gap between DNA synthesis and cell division that serves to ensure proper replication of the DNA. There could also be a G0 phase, which represents an additional gap during which the cells are at rest. The cell division (the M phase) is itself divided into several phases (Prophase, Metaphase, Anaphase and Telophase) that represent the processes of separation of the DNA, cytokinesis and formation of the cell membrane dividing the daughter cells. The cells usually cease to divide (senescent state) after a certain number of generations of reproduction. Cell senescence usually sets in after 60-80 generations. The cells also undergo apoptosis and necrosis (cell death). The duration of the cell cycle depends upon the organism, and in the human body, the cell cycle duration also depends upon the organs that the cells constitute.

The G1 and S phases are characterised by the complex metabolic reaction scheme that occurs within the cells. The amount of information available on the various pathways (glycolysis pathway, TCA cycle pathway *etc.*) far exceeds the amount that can be modelled reasonably with the available computational resources. Thus, in developing models for bioreactors and for biomedical applications, the metabolic pathways that are the most relevant will need to be adopted. Techniques such as the cybernetic modelling concepts [8] can be exploited to account for both the simplification that is introduced into the metabolic pathways, and more importantly to account for the selectivity and regulations that are inherent within the cells.

The passage from one phase to the other is dictated by several regulatory mechanisms (check points), some of which are triggered by external impulses. The most important regulation is the RAS signalling, which regulates growth and proliferation as well as survival of the cell populations. The RAS signalling operates through several pathways such as the RAF pathway and the Mitogen-Activated Protein Kinase (MAPK) pathway, which determine the transition of the cells into the synthesis phase of the cell cycle [3]. The protein p53, a type of cyclin activated kinase inhibitor, is crucial to maintain normal cell cycles by supressing cell proliferation and regulating cell growth. Mutants of p53 and the perturbations that appear in these regulatory mechanisms are the major reasons behind the occurence of cancer and malignancy, as has been borne out by several studies [11, 13]. In addition, the cell senescence is contingent on telomere shortening (telomeres being precursors to replicated DNAs), which is perturbed by the expression of the enzyme telomerase that leads to 'immortality' of cells (return from the senescent state back to a proliferative state) and hence to cancer [11]. Similarly, cell apoptosis is regulated by the transforming growth factor TGF-β [14]. Population balance provides a natural framework to model these various mechanisms.

Model formulation

The general population balance model as applicable to biological systems is given in Equation (1). In this equation, $F^n(\mathbf{x},\mathbf{t})$ accounts for the population density of the cells at generation n (after n divisions), $\mathbf{x}$ accounts for the internal coordinates (the concentration of the metabolites within the cells), and N is the number of internal coordinates considered in the problem. $\frac{dx_i}{dt}$ accounts for the rate of change of the metabolites due to kinetic pathways within the cells. $\Re_{div}^{n-1}(\mathbf{x},\mathbf{t})$ accounts for the formation of cells in generation n by cell division in the previous generation $n-1$. Likewise, $\Re_{div}^{n}(\mathbf{x},\mathbf{t})$ accounts for the depletion of cells in the current generation n due to cell division. $\Re_{div}^{n}(\mathbf{x},\mathbf{t}) = \mathbf{0}$ for n beyond the threshold generation for cell senescence to occur. In bioreactor models, cell senescence is usually not of significance as the residence time of the cells is much shorter. But this assumes importance for cancer modelling. $\Re_{death}(\mathbf{x},\mathbf{t})$ accounts for cell death and $\Re_{cell-cell}(\mathbf{x},\mathbf{t})$ can be employed to account for cell-cell interactions such as the recombination aspects in bacterial populations.

$$\frac{\partial}{\partial t}F^n(\mathbf{x},\mathbf{t}) + \Sigma_{\mathbf{i=1}}^{\mathbf{N}} \frac{\partial}{\partial \mathbf{x_i}}(\mathbf{F^n}(\mathbf{x},\mathbf{t})\frac{\mathbf{dx_i}}{\mathbf{dt}}) = \Re_{\mathbf{div}}^{\mathbf{n-1}}(\mathbf{x},\mathbf{t}) - \Re_{div}^{n}(\mathbf{x},\mathbf{t}) - \Re_{\mathbf{death}}(\mathbf{x},\mathbf{t}) + \Re_{\mathbf{cell-cell}}(\mathbf{x},\mathbf{t}) \quad (1)$$

Employing the assumption of a deterministic process, the various phases are mutually exclusive in terms of the concentration of the cell metabolites, as clearly explained in a recent article [4]. Based on this argument, one need not introduce further distinction into the population to distinguish among the various phases of the cells. However, one might need different growth kernels ($\frac{dx_i}{dt}$) for the different cell cycle phases. In the above cited paper, in the final phase (M-phase), the age of the cells in this phase is also taken into account in order to account for the variable time delay associated with the processes of cytokinesis and the development of the cell partition before the cell division. In the model presented in equation (1), this distinction is not made to keep the model simple, and to maintain the right level of detail.

On the other hand, one also needs to model the effect of the environment on the cell cycle phases, such as through regulatory mechanisms. As indicated previously, the modelling of the regulation will be critical mainly for medical problems. In modelling the regulatory processes, one might adopt the models for these mechanisms that have been published in the literature. For example, the MAPK pathway has been recently modelled [2]. While a strategy similar to the one proposed in that study can be employed in our population balance formulation, a strategy based on the extension of the cybernetic approach will also be a very viable alternative (both in terms of simplifying the model and in terms of avoiding difficulties in representing the regulatory aspects of the pathway).

Modelling for Bioreactor control

Due to the relatively small life time of the cells within the reactors, cell senescence aspects will not be important. Thus, the model presented above can be simplified substantially by not distinguish between the different generations of the cells. Also, the cell-cell interaction effects can be ignored. In effect, the PBE reduces to the form shown in Equation (2).

$$\frac{\partial}{\partial t}F(\mathbf{x},\mathbf{t}) + \Sigma_{\mathbf{i}=1}^{\mathbf{N}}\frac{\partial}{\partial \mathbf{x_i}}(\mathbf{F}(\mathbf{x},\mathbf{t})\frac{\mathbf{dx_i}}{\mathbf{dt}}) = \Re_{\mathbf{div}}(\mathbf{x},\mathbf{t}) - \Re'_{div}(\mathbf{x},\mathbf{t}) - \Re_{\mathbf{death}}(\mathbf{x},\mathbf{t}) \quad (2)$$

The cell division terms are modelled as in Equations (3) and (4). The extracellular metabolites are modelled as in Equation (5), accounting for the diffusion out of the cells and any deactivation in the aqueous medium.

$$\Re'_{div}(\mathbf{x},\mathbf{t}) = \mathbf{\Gamma}(\mathbf{x})\mathbf{F}(\mathbf{x},\mathbf{t}) \quad (3)$$

$$\Re_{div}(\mathbf{x},\mathbf{t}) = \int_{\mathbf{x_1}}^{\mathbf{x_{1,max}}}\int_{\mathbf{x_2}}^{\mathbf{x_{2,max}}}\int_{\mathbf{x_3}}^{\mathbf{x_{3,max}}}\int_{\mathbf{x_4}}^{\mathbf{x_{4,max}}}\int_{\mathbf{x_5}}^{\mathbf{x_{5,max}}}\int_{x_6}^{x_{6,max}}\Gamma(x)F(x,t)dx_1dx_2dx_3dx_4dx_5dx_6 \quad (4)$$

$$\frac{dx_{ex}}{dt} = \int_{x_1}\int_{x_2}\int_{x_3}\int_{x_4}\int_{x_5}\int_{x_6}\kappa(x - x_{ex}) F(x,t)dx_1dx_2dx_3dx_4dx_5dx_6 - k_w x_{ex} \quad (5)$$

Modelling for Cancer Studies

A more detailed population balance model becomes important for this application, in view of the larger 'residence time' of the process. In particular, the model will need a detailed account of the regulatory mechanisms (check points). One may need to account for a larger number of internal metabolites in this case, and moreover, the generation number of the cell populations will need to be monitored.

The model developed will aid in the study of the sensitivity of the different regulatory mechanisms, and hence to identify the most sensitive mechanism that one should target for treatment. For example, the effect that the RAS signalling and the other individual pathway have on the overall process of tumour growth can be analysed using the mathematical model (even if only on a qualitative level). These sensitivity informations can then be used to direct further experimentation and research along the most promising path for therapeutic studies. For therapeutic studies, the antibody dynamics and its interaction with both the normal cells and the cancerous cells will need to be incorporated into the model.

Challenges

The challenges in the development of population balance models for biological systems can be categorised into three areas. The first category pertains to the complexity of the process, and the challenges in identifying an appropriate level of detail for the various applications at hand and a suitable mathematical formulation for each mechanism. The knowledge base with regard to biological processes has reached such a stage that one has to be eclectic in choosing the aspects and details that one needs to incorporate into the model. This issue was illustrated briefly above in selecting the models for bioreactors. In this case, there is much less need to incorporate the regulatory mechanisms. Also, the portion of the intracellular metabolic pathway that is relevant to the process can be focussed upon. On the other hand, for biomedical applications, one needs to have a more comprehensive metabolic pathway, and also the regulatory aspects assume great importance. Extensions of the cybernetic approach can be employed at different stages of the model.

The second category pertains to the identification of the appropriate kernels for the various portions of the model. While the development of the growth kernels might be an easy task, the same cannot be said about the division kernel. Also, the partition functions for non-symmetric divisions might be another challenge to obtain. However, the ability to measure and monitor individual portions of the process through flow cytometry provides great promise for the identification of the appropriate kernels.

The third category pertains to the solution of the models. The need is a generic yet realistic and efficient numerical solution technique to solve these complex models. Such a technique will segregate the effect of solution capabilities on the level of details in the models, and enable attaining the full potential in the model that reflects current process knowledge.

Representative model simulation

In this paper, a representative model for a bioreactor is presented. The yeast system described in Henson *et al.* [6] is chosen as the model system of study. A six-dimensional population balance model is developed, as recommended there. The six internal coordinates chosen are the intra-cellular concentrations of glucose; glyceraldehyde 6-phosphate; 1,3 biphospho glycerate; pyruvate; NADH; and ATP. The intra-cellular metabolic pathways and the associated reaction kinetics are modelled employing a variation of the Monod kinetics [6]. These kinetics will form the basis for modelling the cell growth phenomenon in Equations (1) and (2). The nutrient glucose is a continuous input to the cells, while pyruvate is a metabolite produced by the cells that is excreted to the aqueous medium (with mass transfer coefficient κ).

Applying traditional solution techniques such as the discretisation methods or the methods of weighted residuals results in a large system of equations, even with a modest choice of discretisation parameters. This aspect, combined with the stiffness of the model equations, leads to very large computational requirements. The largest system for which a method of weighted residuals has been applied is a 3-dimensional population balance model [9], with much simplified kinetics. For a six-dimensional system, Henson *et al.* [6] report a constant-N Monte Carlo-like solution method. This particular method is based on the superposition of multiple single cell models rather than on the solution of the population balance equation itself. Although this is a viable starting point, these techniques do not allow a rigorous characterisation of the population dynamics. *In the present study, a realistic, feasible and efficient solution technique is employed for the population balance models.* The hierarchical two-tier technique that was developed for single-dimensional models [7] is extended to this 6-D case. In this technique, the population balance equation is explicitly formulated in terms of the individual phenomena of the nucleation, growth, division and death. As indicated earlier, the continuous growth (or shrinkage) process is modelled using partial differentials, while the discrete processes that include the aggregation (coagulation) and breakage processes are modelled in terms of complex integrals. The evaluation of these integrals require very high computational loads by themselves. In the hierarchical two-tier solution technique, the individual rates of the nucleation, growth and death phenomena are computed in the first tier of the algorithm, and the population is updated in the second tier. To reduce the high computational demand of modelling the aggregation/breakage integrals, an off-line solution strategy is employed. In this strategy, the complex quadratures are solved semi-analytically in terms of the (unknown) population densities *a priori*, thereby substantially reducing the on-line computational load.

For bioreactor problems, cell aggregation is not important. However, one must account for cell division. The derivation of the *off-line semi-analytical solutions for the integral terms representing the cell division phenomenon* is relatively easier than the corresponding aggregation integral. Cell division is assumed to be connected with the concentration of glucose within the cell, with the cells dividing upon reaching a threshold glucose concentration. Also, the division is assumed to be binary, with a uniform split of each of the metabolites.

The algorithm is based on a finite-volume discretisation of the distribution domains. The particle population is assumed to be uniform within each of the finite volumes. The growth phenomenon is modelled based

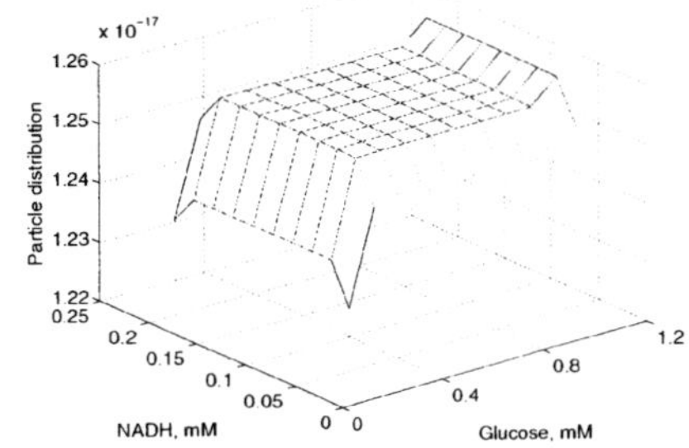

(a) t=6 $\times$ 10^{-3} s

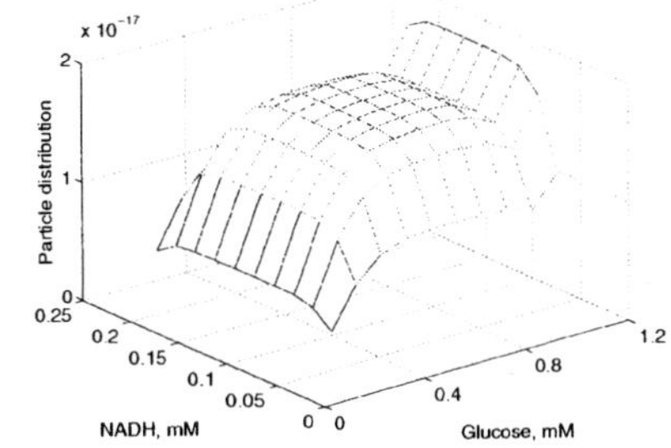

(b) t=0.6 s

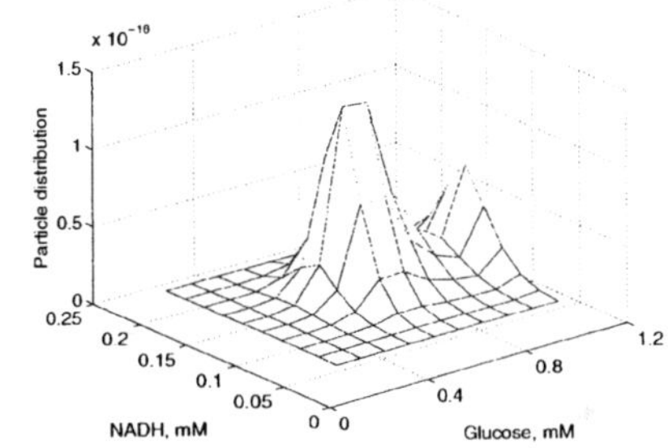

(c) t=6 s

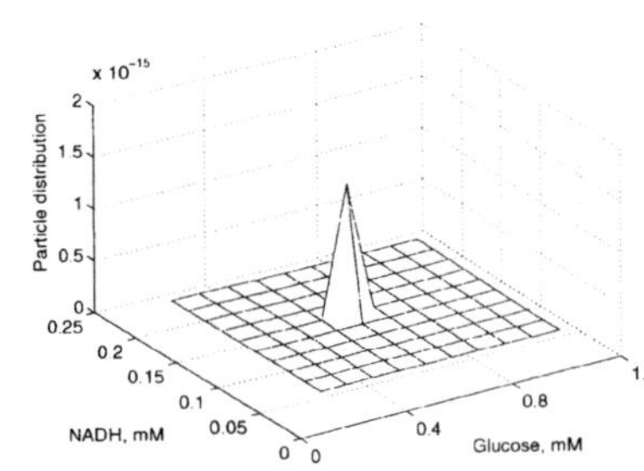

(d) t=1800 s

Figure 1: Simulated evolution of the cell populations, depicted with respect to two of the internal co-ordinates. The internal coordinates are the intra-cellular concentrations of glucose and nicotinamide adenide dinucleotide hydrogen (NADH). As the time progresses, a synchronisation of the population is observed.

on the metabolic reaction rates at the six-dimensional boundaries of the finite volumes. As explained earlier, the 6-dimensional integrals that characterise cell division are reduced to simpler multiplication and addition of terms, which while being accurate as analytical solutions, substantially reduces the on-line computational load. Space constraints prevent the inclusion of the detailed algorithm and the model equations here, but will be elaborated elsewhere.

Since a rigorous population balance formulation is adopted here, the modelling of the interaction of the cells with the aqueous medium can be done in a straightforward manner, without having to resort to unnecessary assumptions and simplifications that would be necessary in the statistical methods. In the later stages of the project, the *few* macroscopic parameters in the model will be determined by fitting the simulation results to experimental data from flow cytometry. Preliminary simulation results are presented in Figures 1. This indicates the synchronisation of the cell populations, which effect has been observed in other studies [6]. Parametric sensitivity studies also indicate an ability to capture complex dynamics such as the oscillatory and limit cycle behaviour that have been observed experimentally.

Conclusion

A general and realistic framework was presented for the modelling of cellular processes in biological systems, employing the population balance ideas. The model will have ramifications for biochemical studies (design, optimisation and control of bioreactors) as well as for biomedical studies (in particular, to understand the occurence of cancer and to identify the most sensitive regimes of the cellular processes to target for therapeutic investigation). The major challenges that confront model development were highlighted. One of these is the challenge in the solution of these complex models in an efficient manner. It is imperative to have an efficient solution technique, so as to remove the computational burden from being a determining factor for the model features and details. A hierarchical solution strategy is presented, which enables performing a major portion of the computations *a priori*. Despite being based on a rigorous six-dimensional population balance formulation, the technique gives solution times of about an hour for a 1 hour run. This will be adequate for design and optimisation of bioreactors, and also for analysis of medical problems. For on-line control purposes, one can resort to simpler, one-dimensional population balance models or the faster cell ensemble solution technique [6]

References

[1] N. Abu-absi, A. Zamamiri, J. Kacmar, S. J. Balogh, and F. Srienc. Automated Flow Cytometry for Acquisition of Time-Dependent Population Data. *Cytometry Part A*, 51A:87–96, 2003.

[2] A. R. Asthagiri and D. A. Lauffenburger. A Computational Study of Feedback Effects on Signal Dynamics in a Mitogen-Activated Protein Kinase (MAPK) Pathway Model. *Biotechnol. Prog.*, 17:227–239, 2001.

[3] J. Downward. Targetting RAS Signalling Pathways in Cancer Therapy. *Nature Reviews: Cancer*, 3:11–22, 2003.

[4] A. G. Fredrickson. Population Balance Equations for Cell and Microbial Cultures Revisited. *AIChE J.*, 49:1050–1059, 2003.

[5] A. G. Fredrickson and H. M. Tsuchiya. Continuous Propagation of Microorganisms. *AIChE J.*, 9(4):459–468, 1963.

[6] M. A. Henson, D. Muller, and M. Reuss. Cell population modelling of yeast glycolytic oscillations. *Biochem J.*, 368:433–446, 2002.

[7] C. D. Immanuel and F. J. Doyle III. Computationally-Efficient Solution of Population Balance Models Incorporating Nucleation, Growth and Coagulation. *Chem. Eng. Sci.*, 58(16):3681–3698, 2003.

[8] D. S. Kompala, D. Ramkrishna, and G. T. Tsao. Cybernetic Modeling of Microbial Growth on Multiple Substrates. *Biotech. Bioeng.*, 26:1272–1281, 1984.

[9] N. V. Mantzaris, P. Daoutidis, and F. Srienc. Numerical solution of multi-variable cell population balance models: III. Finite elements methods. *Comput. Chem. Eng.*, 25:1463–1481, 2001.

[10] N. V. Mantzaris, F. Srienc, and P. Daoutidis. Nonlinear productivity control using a multi-staged cell population balance model. *Chem. Eng. Sci.*, 57:1–14, 2002.

[11] N. F. Mathon and A. C. Lloyd. Cell Senescence and Cancer. *Nature Reviews: Cancer*, 3:203–213, 2001.

[12] J. E. Mittler, B. Sulzer, A. U. Neumann, and A. S. Perelson. Influence of Delayed Viral Production on Viral Dynamics in HIV-1 Infected Patients. *Math. Biosci.*, 152:143–163, 1998.

[13] D. Ruggero and P. P Pandolfi. Does the Ribosome Translate Cancer? *Nature Reviews: Cancer*, 3:179–192, 2003.

[14] P. M. Siegel and J. Massague. Cytostatic and Apoptotic Actions of TGF-beta in Homeostasis and Cancer. *Nature Reviews: Cancer*, 3:807–820, 2003.

[15] F. Srienc. Cytometric data as the basis for rigorous models of cell population dynamics. *J. Biotechnol.*, 71:233–238, 1999.

[16] G. Stephanopoulos. Invited comment: Chemical and Biological Engineering. *Chem. Eng. Sci.*, 58:4931–4933, 2003.

[17] G. Zhu, A. Zamamiri, M. A. Henson, and M. A. Hjortso. Model predictive control of continuous yeast bioreactor using cell population balance models. *Chem. Eng. Sci.*, 55:6155–6167, 2000.

ELSEVIER
IFAC
PUBLICATIONS
www.elsevier.com/locate/ifac

PREFERENTIAL ESTIMATION VIA TUNING OF THE KALMAN FILTER

L. Bodizs, B. Srinivasan and D. Bonvin

Laboratoire d'Automatique
École Polytechnique Fédérale de Lausanne
CH-1015 Lausanne, Switzerland

Abstract: Estimation problems have been traditionally formulated so as to minimize the estimation error of the full state vector. However, in applications that involve the tracking of only a few unmeasured variables, it is sufficient to limit the attention along certain directions in state space. This way, it is hoped that better accuracy can be obtained along the desired directions, possibly at the cost of poorer estimates along the other directions. This problem, termed preferential estimation, is formally formulated in this paper as a least-squares minimization problem. Using calibration measurements of the preferred variables, the above mentioned problem is solved numerically via tuning of the Kalman filter. The approach is illustrated in simulation on the optimization of a penicillin fermentation process, where preferential estimation is used successfully to reduce the error in tracking a single unmeasured variable, the substrate concentration.

Keywords: State estimation, Kalman filter, Preferential estimation, Tuning via optimization, Penicillin fermentation.

1. INTRODUCTION

State estimation is an important topic within the theory of dynamic systems (Maybeck, 1979; Brown and Hwang, 1992). State estimation is a necessary component of sophisticated monitoring and control techniques since they typically require information that is too expensive or impossible to obtain from direct measurements. Estimation attempts to reconstruct the needed information from the available measurements and prior knowledge in the form of a dynamic model (Dochain, 2003; Doyle, 1998).

Traditional estimation techniques estimate the full state vector, and the main issues are how good these estimates are in terms of bias and variance (Chui and Chen, 1987). The reason for estimating all the states is due to the close link between the estimation theory and the full-state feedback literature. However, information on all the states is not always needed, especially when the objective is to track only a few unmeasured variables.

For cases where only some of the state variables have to be estimated, reduced-order estimators have been proposed in the literature. The order of such estimators typically corresponds to the difference between the system order and the number of measurements (Luenberger, 1979). These techniques have been applied succesfully to chemical reactors (Soroush, 1997) and utilized for inferential control purposes (Doyle, 1998). The main drawback is that the reduction in the number of variables to estimate is generally not significant, due to the small number of measured variables.

More important reduction in the order of the estimator can be achieved through projection methods (Marquardt, 2001). These methods rely on the fact that the most important characteristics (for a specific purpose) of the dynamics of a complex system can be described by the dominant subspace. Thus, the system can be described by a dynamic model whose dimension is equal to that of the dominant subspace. Significant reduction has been reported for various applications (Hahn and Edgar, 2002). However, the projection methods often need to be tailor made to an application, and there are too few guidelines for selecting the appropriate method for a particular application.

Hence, new alternatives for estimating accurately a subset of the states are welcome, and this paper aims at proposing such an alternative. The idea of the proposed method is to focus attention only along certain preferred directions, where a better accuracy is required, at the cost of possibly poorer estimation along the other directions. This problem, termed *preferential estimation*, is formulated formally in this paper. Increasing the accuracy along the preferred directions is done using one-time calibration measurements. Thus, preferential estimation draws a strong parallel with the calibration approach, and many concepts therein can be used (Martens and Naes, 1989).

The main question in preferential estimation is whether such a thing is at all possible. Can the error along certain preferred directions be reduced by increasing the errors in other directions ? This will be shown with a bioreactor example, where the parameters of a Kalman filter used to reconstruct a single state will be tuned via numerical optimization. It turns out that preferential estimation is indeed possible and can be used advantageously for tracking a desired substrate profile.

The paper is organized as follows. In Section 2, the standard full-state estimation problem and the Kalman filter are briefly reviewed. Section 3 describes the concept of preferential estimation, while Section 4 deals with the tuning of the estimator via numerical optimization. The application of preferential estimation is illustrated on a simulated batch fermentation process in Section 5. Section 6 discusses some features of preferential estimation, and Section 7 concludes the paper.

2. STANDARD ESTIMATION

Consider the following discrete-time system:

$$x_{k+1} = G(x_k, u_k) + w_k, \quad x_0 = x_o \tag{1}$$

$$y_k = H(x_k) + v_k \tag{2}$$

where u are the inputs, x the states, x_o the initial values of states, w the process noise, y the measurements, and v the measurement noise. G are the nonlinear functions describing the system dynamics, and H the nonlinear measurement functions.

The standard estimation problem consists of determining the best state estimates, $\hat{x}$, given the measurements y. One of the most successful full state estimators used in practice is the Extended Kalman Filter (EKF) (Kalman, 1960; Jazwinski, 1970; Grewal and Andrews, 1993). It determines the Kalman gain matrix K_k that minimizes the a posteriori estimation error covariance P_k:

$$\min_{K_k} \quad P_k = E\langle (x_k - \hat{x}_k)(x_k - \hat{x}_k)^T \rangle \tag{3}$$

$$s.t. \; \hat{x}_k^- = G(\hat{x}_{k-1}, u_{k-1}), \quad \hat{x}_0 = E\langle x_o \rangle \tag{4}$$

$$\hat{x}_k = \hat{x}_k^- + K_k(y_k - H(\hat{x}_k^-)) \tag{5}$$

where $\hat{x}_k^-$ is the a priori estimate, and $\hat{x}_k$ the a posteriori estimate.

The solution of Problem (3)-(5) is:

$$\begin{aligned} P_k^- &= A_k P_{k-1} A_k^T + Q, \quad P_0 = E\langle x_o x_o^T \rangle \\ P_k &= (I - K_k C_k) P_k^- \\ K_k &= P_k^- C_k^T (C_k P_k^- C_k^T + R)^{-1} \end{aligned} \tag{6}$$

where P_k^- is the a priori estimation error covariance, Q and R the process and measurement noise covariances, A_k the Jacobian of G with respect to $\hat{x}_k$, and C_k the Jacobian of H with respect to $\hat{x}_k^-$.

3. CONCEPT OF PREFERENTIAL ESTIMATION

In many applications, the main interest is on estimating rather accurately the values of z, a vector of much smaller dimension than x. Consider the following extension to System (1)-(2):

$$\begin{aligned} x_{k+1} &= G(x_k, u_k) + w_k, \quad x_0 = x_o \\ y_k &= H(x_k) + v_k \\ z_k &= L(x_k), \qquad \dim(z_k) \ll \dim(x_k) \end{aligned} \tag{7}$$

where z_k are the quantities to be estimated, and L a nonlinear function.

One way of obtaining an estimate of z is to estimate $\hat{x}$ with standard techniques and then use the function L to get $\hat{z}_k = L(\hat{x}_k)$. The drawback of this approach is that the problem formulation (3) concentrates on the accuracy of $\hat{x}$ rather than that of $\hat{z}$. In other words, the full state $\hat{x}$ is computed as accurately as possible and then used to compute $\hat{z}$. Consequently, the resulting $\hat{z}$ will inherit the accuracy of $\hat{x}$, though it could probably be estimated more accurately if the attention were placed exclusively on its estimation.

In contrast, the objective of this paper is not to maximize the accuracy of $\hat{x}$ (standard estimation,

SE), but to address directly the problem of maximizing the accuracy of $\hat{z}$ (preferential estimation, PE) regardless of the accuracy of $\hat{x}$.

Thus, the objective of PE is to focus on estimating $\hat{z}$ as accurately as possible. An optimization problem can be formulated as follows:

$$\min_{K_k} \quad J_k = E\langle (z_k - \hat{z}_k)^T (z_k - \hat{z}_k) \rangle \tag{8}$$

$$\begin{aligned} s.t. \ \hat{x}_k^- &= G(\hat{x}_{k-1}, u_{k-1}), \quad \hat{x}_0 = E\langle x_o \rangle \\ \hat{x}_k &= \hat{x}_k^- + K_k(y_k - H(\hat{x}_k^-)) \\ \hat{z}_k &= L(\hat{x}_k) \end{aligned} \tag{9}$$

where J_k is the mean-squared estimation error in z_k. There are two main differences between the formulations (3) and (8):

- The former deals with the states x, while the latter concerns only the preferred directions z, whose dimension is much smaller than x
- The former minimizes a matrix that represents the estimation error covariance of x, while the latter deals with a scalar that corresponds to the mean-squared estimation error in z.

The reason for choosing a scalar cost function in (8) rather than a matrix as in (3) is that, except for the special case of linear systems without uncertainty, there exists no unique choice of the decision variables that minimizes every element of the matrix. Thus, a weighted sum of the various elements of the matrix is necessary in order to define a solution. The mean-squared estimation error, $E\langle (z_k - \hat{z}_k)^T (z_k - \hat{z}_k) \rangle$, which is the trace of the matrix $E\langle (z_k - \hat{z}_k)(z_k - \hat{z}_k)^T \rangle$, represents one possible weighting of the covariance matrix.

Since Problem (8) has no analytical solution, a numerical approach is needed. However, this approach requires the knowledge of z_k in order to evaluate the objective function J_k. Thus, z_k has to be measured once, during the tuning process, in order to obtain the preferential estimator. But once the estimator is tuned, measurements of z_k are no longer needed. The computation of the expectation in (8) requires numerous realizations (obtained via Monte-Carlo simulation) with known output variables y_k and desired variables z_k. Thus, a set of calibration measurements (y_k, z_k) is necessary for preferential estimation.

Hence, a parallel to calibration can be drawn, where the calibration model is tuned based on experimental data and then used for prediction (Martens and Naes, 1989). The main difference between calibration and preferential estimation are: (i) Traditional calibration methods are static, while preferential estimation deals with dynamic systems, (ii) Dynamic calibration methods such as dynamic PLS (Lakshminarayanan *et al.*, 1997), work with black-box dynamic prediction models built just on the available calibration data. On the contrary, preferential estimation a mechanistic, first-principles or other system models (built for other purposes from a different information) along with the available calibration measurements.

The numerical solution is computationally expensive since it has to test each choice of the estimator parameters K_k over numerous realizations. Though many ideas (such as analytically computing an approximation of the cost function without taking an expectation using Monte-Carlo realizations) can be used to ease this numerical burden, a brute-force approach will be used in this paper for tuning an extended Kalman filter.

4. PREFERENTIAL ESTIMATION VIA TUNING OF EXTENDED KALMAN FILTER

In the ideal case, i.e. linear process model and white noise of known covariances (Q and R are chosen equal to the true process and noise covariances that are assumed to be known), the Kalman filter is the best unbiased estimator. In this case, the P_k obtained upon convergence reflects the true estimation error covariance, and the expectation with infinitely many realizations is the true value, i.e. $E\langle \hat{x}_k \rangle = x_k$.

However, this correspondence is lost in other cases (nonlinearities and/or model mismatch and/or non-white noise), and the Q and R matrices are simply used as tuning parameters, for which tuning rules are available (Grewal and Andrews, 1993; Chui and Chen, 1987). The objective here is typically to minimize the estimation error of $\hat{x}$ (standard estimation). However, Q and R could also be tuned so as to minimize the estimation error of $\hat{z}$ (preferential estimation).

Thus, in the context of the extended Kalman filter with Q and R being considered as tuning parameters, SE and PE can be formulated as the following optimization problems involving Q and R as decision variables:

- *SE* - minimizes the estimation error $\bar{J}_k = E\langle (x_k - \hat{x}_k)^T (x_k - \hat{x}_k) \rangle = \text{tr}(P_k)$

$$\min_{Q,R} \quad \bar{J}_k = E\langle (x_k - \hat{x}_k)^T (x_k - \hat{x}_k) \rangle \tag{10}$$
$$s.t. \ (4) - (6)$$

- *PE* - minimizes the estimation error $J_k = E\langle (z_k - \hat{z}_k)^T (z_k - \hat{z}_k)$

$$\min_{Q,R} \quad J_k = E\langle (z_k - \hat{z}_k)^T (z_k - \hat{z}_k) \rangle \tag{11}$$
$$s.t. \ (4) - (6), (9)$$

The data points x_k and z_k are obtained from a simulated reality. Problems (10) and (11) can be solved using numerical optimization where the expectation is computed via Monte-Carlo simulations. An additional difficulty is that these optimization problems exhibit many local minima and, thus, a search for the global minimum is necessary.

In this paper, the two-step numerical optimization method proposed in (Nagy *et al.*, 2001) is used:

(1) A global search method (Genetic Algorithm, GA (Houck *et al.*, 1995)) is used as pre-optimization to find the neighborhood of the global minimum, and
(2) A gradient-based local search algorithm ('fmincon' from Matlab) is used to find the global minimum.

This combined method is computationally very expensive, especially in combination with Monte-Carlo simulations. However, all these efforts are off-line and so, for the moment, the speed of solving (10) and (11) has not been given much importance.

5. PENICILLIN FERMENTATION - IMPLEMENTATION OF PREFERENTIAL ESTIMATION

The proposed estimation techniques are tested on the fed-batch penicillin fermentation process (Srinivasan *et al.*, 2003). The reactions taking place are:

$$S \to X, \quad S \xrightarrow{X} P$$

where S, X and P are the substrate, biomass and product, respectively. The specific rate of the first reaction is described by a Haldane type expression, while that of the second reaction is constant. The third-order dynamic model of the system reads:

$$\begin{aligned}
\dot{c}_X &= \mu(c_S)c_X - \frac{F}{V}c_X,\ c_X(0) = c_{Xo} \\
\dot{c}_S &= -\frac{\mu(c_S)c_X}{Y_X} - \frac{\nu c_X}{Y_P} + \frac{F}{V}(c_{Sin} - c_S),\ c_S(0) = c_{So} \\
\dot{V} &= F,\ V(0) = V_o \qquad (12) \\
c_P &= \frac{c_{Po}V_o}{V} + c_{Sin}(V - V_o)\frac{Y_P}{V} - (c_S V - \\
&\quad - c_{So}V_o)\frac{Y_p}{V} - (c_X V - c_{Xo}V_o)\frac{Y_P}{Y_X V} \\
\mu(c_S) &= \frac{\mu_m c_S}{K_m + c_S + \frac{c_S^2}{K_i}}
\end{aligned}$$

where c_S, c_X and c_P are the substrate, biomass and product concentrations, respectively, V the volume, F the feed flow rate, c_{Sin} the inlet substrate concentration, ν, μ_m, K_m and K_i kinetic parameters. The numerical values are given in Table 1.

Table 1. Model parameters, initial conditions and operating bounds

μ_m	0.53	L/h
K_m	1.2	g/L
K_i	22	g/L
Y_X	0.4	
Y_P	1	
ν	0.5	L/h
c_{Sin}	20	g/L
c_{Xo}	1	g/L
c_{So}	5.138	g/L
c_{Po}	0	g/L
V_o	2	L
F_{min}	0	L/h
F_{max}	1	L/h
c_{Xmax}	3	g/L
t_f	8	h

Consider the problem of maximizing the amount of product at the given final time t_f by adjusting the feed flow rate:

$$\begin{aligned}
&\max_{F(t)} \ c_P(t_f) \qquad (13) \\
&s.t.\ (12) \\
&\qquad c_X(t) \le c_{Xmax}, \quad F_{min} \le F(t) \le F_{max}
\end{aligned}$$

The solution of (13) corresponds to maximizing the rate of the first reaction $\mu(c_S)$ (Srinivasan *et al.*, 2003). This can be achieved by maintaining c_S at $c_S^* = \sqrt{(K_m K_i)} = 5.131$ g/L, the maximum of the Haldane-type expression. Therefore, the capability of tracking c_S^* accurately is vital for optimal process operation.

On-line measurement of c_S is typically not available or is too expensive, while that of c_P is mostly available. So, the problem of state estimation arises. Also, since optimal process operation requires the tracking of a single state, it is of interest to formulate the estimation problem as that of minimizing the estimation error in c_S. Thus, using the notations of the earlier sections, $x_k = [c_{Xk}, c_{Sk}, V_k]^T$; $u_k = F_k$; $y_k = c_{Pk}$; $z_k = c_{Sk}$; G - the discretized version of (12) using the sampling time $\Delta t = 0.01s$. This model is used as the simulated reality, and zero-mean measurement noise (v_k) with 5% variance is considered.

The model used for estimation has the same structure as G, but with errors in the parameters and initial conditions: $\mu_m = 0.5$ L/h; $K_m = 1$ g/L; $K_i = 20$ g/L; $c_{Xo} = 0.6$ g/L; $c_{So} = 5.4$ g/L.

The estimation problem is solved in both the standard and the preferential contexts by solving numerically Problems (10) and (11). The performance criterion, the bias, and the standard deviation are presented in Table 2. The error distribution plots are given in Figure 1, while the time evolution of one realization is shown in Figure 2.

Table 2. Comparison of SE and PE over 100 realizations ($\bar{J}$ - estimation error for x; J - estimation error for z; $E\langle x-\hat{x}\rangle$ - bias; σ_x - standard deviation)

	SE	PE
$\bar{J}$	3.7171	4.7727
J	0.7850	**0.3378**
$E\langle x-\hat{x}\rangle$	$\begin{bmatrix}0.0120\\0.0135\\-0.0343\end{bmatrix}$	$\begin{bmatrix}0.0040\\\mathbf{0.0010}\\-0.0463\end{bmatrix}$
σ_x	$\begin{bmatrix}0.0077\\0.0081\\0.0022\end{bmatrix}$	$\begin{bmatrix}0.0092\\0.0084\\0.0066\end{bmatrix}$

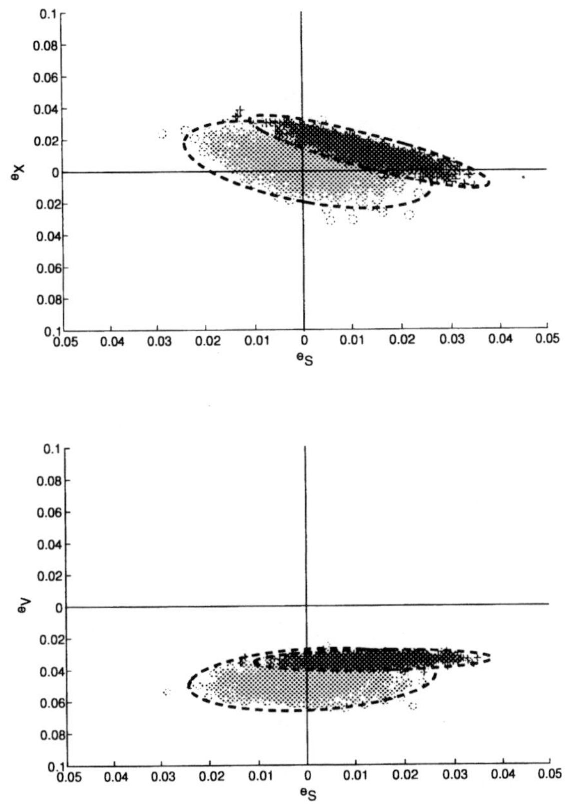

Fig. 1. Error distribution plots of $[e_X, e_S, e_V]^T = x_k - \hat{x}_k$ for $1h \leq t \leq 4h$ over 100 realizations (+ (dark gray) - SE; o (light gray) - PE).

The estimation error of $\hat{z} = \hat{c}_S$ is reduced considerably in PE by reducing the bias in $\hat{c}_S$ (Table 2, Figure 2). This is also visible on the distribution plots (Figure 1), where e_S is centered around 0 in PE. The price to pay, however, is a greater bias in V and an increased variance in all three states, which is seen both in Table 2 and Figure 1.

Due to the reduced bias in $\hat{c}_S$, the tracking of c_S^* based on PE is more accurate, as illustrated on the time evolution of one realization in Figure 3. Hence, a slightly higher amount of product at final time can be obtained: $c_P(t_f)_{PE} = 2.428$ g/L $> c_P(t_f)_{SE} = 2.419$ g/L.

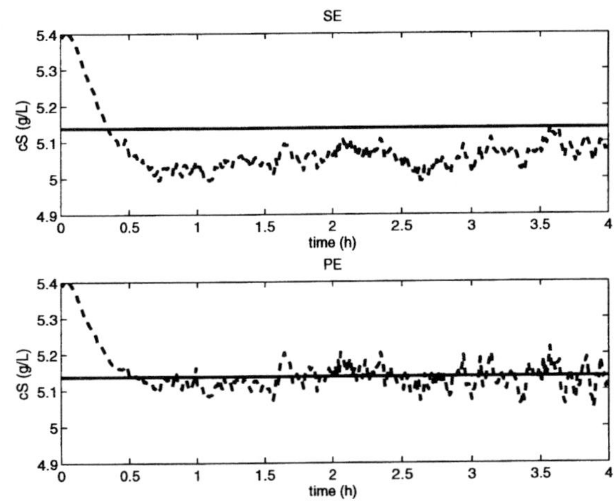

Fig. 2. Estimation of c_S using SE and PE (dashed - $\hat{c}_S$; continuous - c_S)

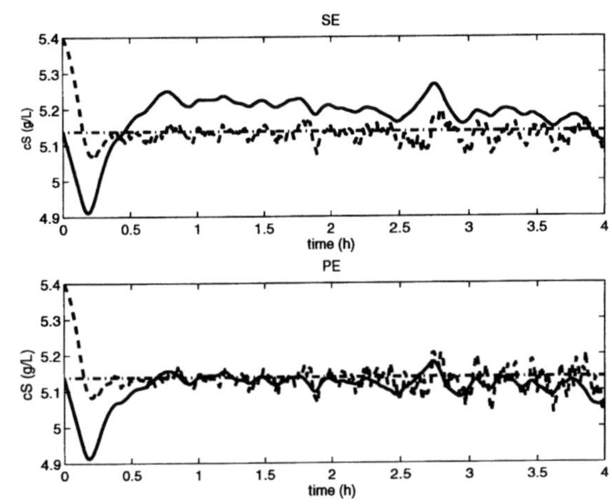

Fig. 3. Tracking of c_S^* using a P controller and $\hat{c}_S$ based on SE and PE (dashed-dot - c_S^*; dashed - $\hat{c}_S$; continuous - c_S)

6. DISCUSSION

The previous section has shown that it is possible to reduce the mean-squared estimation error in z, the preferred direction. The mean-squared estimation error has two parts:

$$\mathrm{tr}\ \mathrm{E}\langle \mathrm{e}_{z_k}\mathrm{e}_{z_k}^{\mathrm{T}}\rangle = \mathrm{tr}\ \mathrm{E}\langle \bar{\mathrm{e}}_{z_k}\bar{\mathrm{e}}_{z_k}^{\mathrm{T}}\rangle + \mathrm{tr}\ \mathrm{E}\langle (\mathrm{e}_{z_k} - \bar{\mathrm{e}}_{z_k})(\mathrm{e}_{z_k} - \bar{\mathrm{e}}_{z_k})^{\mathrm{T}}\rangle \quad (14)$$

where $e_{z_k} = z_k - \hat{z}_k$, and $\bar{e}_{z_k}$ is its mean over various time instants and realizations. tr $E\langle \bar{e}_{z_k}\bar{e}_{z_k}^T\rangle$ is the bias and $E\langle (e_{z_k} - \bar{e}_{z_k})(e_{z_k} - \bar{e}_{z_k})^T\rangle$ is the covariance.

It is possible to reduce the mean-squared estimation error by a tradeoff between the bias and variance parts of (14). In the bioreactor example, the bias part of $\hat{z}$ was reduced and the variance slightly increased to achieve the desired goal. In some sense, PE uses the bias-variance tradeoff that is well-known in statistical literature (Hoerl and Kennard, 1970). Note that the tradeoff used here is the opposite of that used in partial least-squares techniques, where the bias is increased to reduce the variance (Martens and Naes, 1989).

In addition, tradeoffs between the different directions can be observed. For example, the bias along V is increased in order to reduce the bias along c_S. Thus, in the context of PE, bias-bias tradeoffs between different directions can also exist.

In the ideal case (linear model, no model mismatch and zero-mean gaussian white noise), the bias is zero. In this case, the Kalman filter is the best unbiased estimator (Section 4), and all the elements of $E\langle(x_k - \hat{x}_k)(x_k - \hat{x}_k)^T\rangle$ are minimal. However, it should be noted that, in certain cases, a biased estimator may result in a smaller mean-squared estimation error. This issue has been extensively studied in the context of static systems and calibration (principle component regression, partial least squares, etc.), while there has been little study on biased estimation techniques for dynamic systems. This is one of the issues preferential estimation addresses, but it clearly calls for a different estimator structure, i.e. simply tuning the parameters of a Kalman filter will not do it.

In other cases (nonlinear model and/or model mismatch and/or non-white noise), the bias is nonzero. This provides much more potential for preferential estimation as was seen in the bioreactor example, and various tradeoffs can be exploited in this context. In minimizing the mean-squared estimation error, the tradeoff between the bias and the variance of a given variable (direction) naturally arises. Either the variance can be reduced by increasing the bias (as in usual bias-variance tradeoff) or the bias can be reduced by increasing the variance (as in the example of Section 5). Which option to choose depends upon the relative magnitudes of the bias and variance. In the example considered in this paper, the bias was more important (modeling error, errors in initial conditions) and, consequently, the bias in $\hat{c}_S$ was reduced and its variance increased.

7. CONCLUSION

This paper has formulated the problem of preferential estimation, where the issue is more accurate estimation along a few preferred directions in state space. PE was implemented on a penicillin fermentation example by tuning a Kalman filter via numerical optimization. It was shown that PE exploits the bias-variance and bias-bias tradeoffs. Searching for a systematic way of influencing these tradeoffs, instead of the numerical optimization method used in this paper, constitutes a promising future research direction.

REFERENCES

Brown, R. G. and P. Y. C. Hwang (1992). *Introduction to Random Signals and Applied Kalman Filtering.* 2nd ed.. John Wiley.

Chui, C. K. and G. Chen (1987). *Kalman Filtering with Real-Time Applications.* Springer-Verlag.

Dochain, D. (2003). State and parameter estimation in chemical and biochemical processes : a survey. *To appear in J. Process Control.*

Doyle, F. J. (1998). Nonlinear inferential control for process applications. *J. Process Contr.* **8**, 339–353.

Grewal, M. S. and A. P. Andrews (1993). *Kalman Filtering. Theory and Practice.* Prentice Hall.

Hahn, J. and T. F. Edgar (2002). An improved method for nonlinear model reduction using balancing of empirical gramians. *Comp. Chem. Eng.* **26**, 1379–1397.

Hoerl, A. E. and R. W. Kennard (1970). Ridge regression: biased estimation for nonorthogonal problems. *Technometrics* **12**(1), 55–67.

Houck, C., J. Joines and M. Kay (1995). A Genetic Algorithm for Function Optimization: A Matlab Implementation. Technical Report NCSU-IE-TR-95-09. North Carolina State University, Raleigh, NC.

Jazwinski, A. H. (1970). *Stochastic Processes and Filtering Theory.* Academic Press.

Kalman, R.E. (1960). A new approach to linear filtering and prediction problems. *Transaction of the ASME-Journal of Basic Engineering* pp. 35–45.

Lakshminarayanan, S., S. L. Shah and K. Nandakumar (1997). Modeling and control of multivariable processes: a dynamic PLS approach. *AIChE J.* **43**, 2307–2322.

Luenberger, D. G. (1979). *Introduction to Dynamic Systems.* John Wiley, New York.

Marquardt, W. (2001). Fundamental modeling and model reduction for optimization based control of transient processes. In: *Chemical Process Control-6.* Tucson, Arizona.

Martens, H. and T. Naes (1989). *Multivariate Calibration.* John Wiley.

Maybeck, P. S. (1979). *Stochastic Models, Estimation and Control.* Vol. 1. Academic Press.

Nagy, Z., S. Agachi, F. Allgower, R. Findeisen, M. Diehl, H. G. Bock and J. P. Schloder (2001). Using genetic algorithm in robust nonlinear model predictive control. In: *ESCAPE-11.* Kolding, Denmark.

Soroush, M. (1997). Nonlinear state-observer design with application to reactors. *Chem. Eng. Science* **52**, 387–404.

Srinivasan, B., S. Palanki and D. Bonvin (2003). Dynamic optimization of batch processes: I. Characterization of the nominal solution. *Comp. Chem. Eng.* **27**, 1–26.

ELSEVIER
IFAC
PUBLICATIONS
www.elsevier.com/locate/ifac

EFFECTS OF CELL POPULATION HETEROGENEITY ON THE DYNAMICS OF CELL POPULATIONS

Nikos V. Mantzaris

Department of Chemical Engineering, Rice University, Houston, TX 77005

Abstract: A novel methodological and computational framework for the quantitative assessment of the effects of cell population heterogeneity on the dynamics of cell populations is presented. We focus on populations of cells carrying an artificial genetic network, consisting of a system of two promoter-repressor pairs. Detailed numerical simulations indicate that taking into account cell population heterogeneity leads to agreement with experimental data, while neglecting it results in significant qualitative and quantitative differences both transiently and at balanced growth. *Copyright © 2004 IFAC*

Keywords: cell population heterogeneity, genetic networks, cell population balance models, cell-to-cell variability,.

1. INTRODUCTION

Biological systems are undoubtedly complex. Their complexity originates from a variety of sources. First, the DNA of organisms is comprised of a large number of genes, which, depending on the intracellular state, maybe on or off or have intermediate expression levels. This, in turn gives rise to a huge number of possible gene states. In addition, cells contain a large variety of chemical components, including ribonucleic acids, lipids, amino-acids, proteins, and metabolites of many different chemical compositions. These cellular components participate in many different processes, including signal transduction, DNA transcription, DNA replication, translation of mRNA into proteins, transport between different cellular compartments or between the cell and the extracellular space, as well as transformation of chemical compounds into metabolic products. Furthermore, products of one set of processes typically affect (inhibit or enhance) the rates of another set of processes, leading to highly nonlinear interactions. Finally, intracellular processes occur at multiple, vastly different time scales. For example, cell proliferation may occur at the time scale of minutes or hours or days depending on the strain or cell type, the media, and the environmental conditions. However, regulation molecules typically exert their influence in the time scale of seconds.

All of the aforementioned sources of complexity are related to events at the *single-cell level*. However, the majority of the powerful experimental techniques that are available today (e.g. DNA arrays, 2-D gels, LC-MS etc), collect measurements from entire *cell populations*, instead of individual cells. In addition, the objective of most biotechnological applications is to maximize the productivity of products formed by a population of cells instead of maximizing the production in an individual cell. Since, in systems engineering, the definition of a system depends on the purposes and specific objectives of the study, these considerations lead us to define a biological system as a system consisting of a cell population rather than one comprised of an individual cell and its components.

Such a biotechnologically meaningful definition necessitates the consideration of an extra level of complexity originating from the fact that cell populations are *heterogeneous* systems in the sense that cellular properties are unevenly distributed amongst the cells of a population. Thus, at any given point in time, cells of an isogenic cell population contain different amounts of DNA, mRNA, proteins, metabolites etc. In short, cell population heterogeneity can be defined as phenotypic variability amongst the cells of an isogenic cell population. This biological fact has been known for a long time. As early as 1945, Delbrück showed significant variations in phage burst sizes (Delbrück, 1945). Moreover, cell population heterogeneity has been observed in cell division times (Powell, 1956), the lysogenic states of phage-infected cells (Ptashne, 1987), the tumbling and smooth-swimming states of flagellated bacteria (Spudich and Koshland, 1976), flagellar phases (Stocker, 1949), induction or repression states of bacterial differentiation (Russo-Marie, et al., 1993), sporulating cultures of *B. subtilis* containing fusions between sporulation genes and *lac*Z (Chung and Stephanopoulos, 1995), while population heterogeneity in β-galactosidase activities of cell populations expressing the lac operon genes has been demonstrated in various systems (Novick and Weiner, 1957). Recently, through an elegant set of experiments, the inherently heterogeneous nature of various isogenic *E. coli* strains was also established using two different fluorescent proteins as reporters (Elowitz, et al., 2002).

The phenotype of each cell depends on the type and number of genes that are expressed at any given point in time, as well as on the subsequent, numerous intracellular reactions comprising what is simply known as metabolism. The outcome of these cellular processes is greatly influenced by the extracellular environmental conditions, while, intracellularly, these processes are tightly controlled by *regulatory* molecules. Due to the instrumental role that regulatory molecules play in the determination of single-cell phenotype, phenotypic variability among the cells of a population is tightly related to regulation of gene expression and, hence, the architecture and dynamics of single-cell regulatory networks. The astonishing recent progress in genomic research has offered the exciting capability of synthesizing artificial genetic networks with pre-specified, tunable functions. This, in turn, presents the revolutionary opportunity to engineer desirable cellular properties by building the organism's gene-regulatory architecture from appropriate, interconnected genetic blocks with well-defined regulatory functions. However, before embarking into the exciting challenge of building the organism's regulatory architecture (Guet, et al., 2002), it is of fundamental importance to understand how each specific gene-regulatory module affects the phenotype of the entire cell population.

Based on these considerations, the natural, fundamental question for the Systems Biologist becomes: "How does one accurately describe the dynamics of cell populations?" The traditional approach consists of formulating chemically structured continuum models. These are ordinary differential equation (ODE) models describing the dynamics of *population average* intracellular concentrations and their coupling with the dynamics of the nutrients and extracellular environmental conditions, in general. Models of this type contain a significant amount of biological detail on processes that occur at the single-cell level. Moreover, they are rather easy to formulate and simulate. In addition, there exist well-developed, general, theoretical and computational tools, which utilize ODE models for process design, control and optimization. Finally, there exists a variety of advanced parameter estimation techniques allowing the routine determination of parameters that appear in such models from appropriate experimental data. However, chemically-structured continuum models are characterized by fundamental limitations. First, by construction, they do not describe how moments of the cell property distribution higher than the first evolve with time, hence providing only a limited view of the biological system of interest. One might argue that in many instances, knowledge of just the population average dynamics is sufficient. However, in order to obtain closure in deriving a set of equations describing just the first moments of the cell property distribution, continuum models inherently assume that a cell population is a homogeneous system. More specifically, it is assumed that a cell population behaves like a lumped biophase, where all cells behave exactly like the average population cell. Such an assumption obviously neglects the effects of cell population heterogeneity and might lead to inaccurate predictions of the dynamics of the average.

Such a restrictive assumption is not present in the cell population balance (CPB) modeling approach (Ramkrishna, 2000). These models describe the dynamics of the entire cell property distribution, thus naturally accounting for the heterogeneous nature of cellular processes, nutrient uptake and product formation. Furthermore, they can accommodate all the information contained in continuum models on the reaction kinetics occurring at the single-cell level. In addition, they explicitly incorporate in their formulation the division process, as well as the mechanism by which dividing cells partition their material into two daughter cells. Thus, one of the fundamental sources of cell population heterogeneity is explicitly accounted for in the mathematical formulation. However, CPB models consist of first-order, hyperbolic, partial integro-differential equations describing the time evolution of the cell property distribution, which are typically, nonlinearly coupled with integro-ordinary differential equations describing the dynamics of the extracellular environment. Such formulations are characterized by significant mathematical complexity and their numerical simulation represents a challenging task. This mathematical complexity in turn results in additional difficulties in developing control and optimization strategies for biological systems described by CPB models.

In this work we present a novel methodological framework that isolates the quantitative effects of cell population heterogeneity for a given set of biomolecular reactions occurring at the single-cell level by combining the chemically-structured continuum with the cell population balance modeling approaches. In section 2 we present the derivation of this framework, and we sketch the general computational algorithm that was developed in order to quantitatively assess the effects of cell population heterogeneity on the dynamics of cell populations. Finally, in section 3 we apply the presented mathematical/computational framework to a well-known artificial gene switching network, known as the genetic toggle (Gardner et al., 2000).

2. METHODOLOGICAL FRAMEWORK

Motivated by the natural way that CPB models account for population heterogeneity, we developed a general modeling and computational framework aiming at: a) quantitatively assessing the effects of population heterogeneity on the dynamics of cell populations for a given single-cell gene-regulatory network and b) accurately predicting the entire cell property distribution dynamics.

To achieve our aim, we observe that the fundamental difference between chemically-structured continuum models and CPB models is that the former models assume that all cells in the population behave identically to the average cell, while the latter are more general and do not make this restrictive assumption. We also note that, by construction, continuum models can only predict the dynamics of average cellular properties, and none of the other cell distribution characteristics. Therefore, for a given reaction network, in order to isolate the effects of extrinsic population heterogeneity, it suffices to develop a framework for comparing the predictions for the average cellular property dynamics made by a CPB model with those made by its corresponding continuum model. The question that naturally arises is the following: *Given a CPB model, what is its corresponding continuum formulation*?

The starting point of our approach for addressing this question is the generalized CPB model subject to regularity boundary conditions (Ramkrishna, 2000):

$$\frac{\partial N(\mathbf{x},t)}{\partial t}+\nabla_{\mathbf{x}}\cdot\left[\mathbf{r}(\mathbf{x},\mathbf{S})N(\mathbf{x},t)\right]+\Gamma(\mathbf{x},\mathbf{S})N(\mathbf{x},t) = 2\int_G \Gamma(\mathbf{y},\mathbf{S})p(\mathbf{x},\mathbf{y},\mathbf{S})N(\mathbf{y},t)d\mathbf{y}-DN(\mathbf{x},t) \quad (1)$$

$$\int_G \nabla_{\mathbf{x}}\cdot\left[\mathbf{r}(\mathbf{x},\mathbf{S})N(\mathbf{x},t)\right]d\mathbf{x}=0 \quad (2)$$

where **x** is an r-dimensional vector describing the cellular state; G is the space of admissible states; S is a p-dimensional vector with the extracellular substrate concentrations; D is the dilution rate in continuous bioreactors; N(**x**,t)d**x** is the number of cells which at time t, have a state between **x** and **x**+d**x**; **r**(**x**,**S**), Γ(**x**,**S**), and p(**x**,**y**,**S**) describe single-cell, state dependent processes. They represent, respectively, the r-dimensional vector with the net reaction rates, the scalar division rate and the partition probability density function. The latter describes the mechanism by which a mother cell of state **y** partitions its cellular material into two daughter cells of state **x** and **y**-**x**. We note that the integro-partial differential equation (1) is, in general, nonlinearly coupled with a set of integro-ordinary differential equations describing the dynamics of the p substrates **S** (not shown). Moreover, in the case of a batch reactor with excess substrates and in the absence of cell death, the *state distribution function* N(**x**,t) does not reach a steady state as the cells continue to grow in the exponential phase. Manipulation of eqs. (1) and (2), as well as properties of p(**x**,**y**,**S**), leads to the equation describing the dynamics of the *number density function* n(**x**,t), which is defined as the ratio of N(**x**,t) over the cell density $N_t(t)=\int_G N(\mathbf{x},t)d\mathbf{x}$:

$$\frac{\partial n(\mathbf{x},t)}{\partial t}+\nabla_{\mathbf{x}}\cdot\left[r(\mathbf{x},\mathbf{S})n(\mathbf{x},t)\right]+\Gamma(\mathbf{x},\mathbf{S})n(\mathbf{x},t) = 2\int_G \Gamma(\mathbf{y},\mathbf{S})p(\mathbf{x},\mathbf{y},\mathbf{S})n(\mathbf{y},t)d\mathbf{y} - n(\mathbf{x},t)\int_G \Gamma(\mathbf{x},\mathbf{S})n(\mathbf{x},t)d\mathbf{x} \quad (3)$$

Due to the presence of the extra nonlinear term accounting for dilution due to cell proliferation, n(**x**,t) reaches a steady state, known as the *state of balanced growth*. At this state, N(**x**,t) reaches a time-invariant shape, while cells continue to proliferate. Taking the first moment of eq. (3) and applying conservation of mass for each cellular component at cell division, leads to a set of equations describing the dynamics of the *average* intracellular concentrations ($\bar{x}_i$, i =1,...,r):

$$\frac{d\bar{x}_i}{dt}=\int_G r_i(\mathbf{x},\mathbf{S})n(\mathbf{x},t)d\mathbf{x}-\bar{x}_i\int_G \Gamma(\mathbf{x},\mathbf{S})n(\mathbf{x},t)d\mathbf{x} \quad (4)$$

Notice that the dynamics of the average intracellular concentrations are coupled with those of the number density function (eq. (3)). To derive the corresponding continuum model, it suffices to apply to eq. (4), the fundamental assumption of continuum models, namely, that all cells in the population behave exactly like the average cell. This can be mathematically captured by representing n(**x**,t) as a product of delta functions centered around the average intracellular states:

$$n(\mathbf{x},t)=\prod_{i=1}^{r}\delta\left(x_i-\bar{x}_i(t)\right) \quad (5)$$

Substitution of (5) into (4) yields the continuum model, which corresponds to the CPB model:

$$\frac{d\bar{x}_i}{dt}=r_i\left(\bar{\mathbf{x}},\mathbf{S}\right)-\Gamma\left(\bar{\mathbf{x}},\mathbf{S}\right)\bar{x}_i \quad (i=1,...,r) \quad (6)$$

The difference in the predictions of equations (4) and (6) constitutes the basis for quantitatively assessing the effects of population heterogeneity for a given biomolecular network. Notice that both sets contain the same parameters. The only difference is that eqs. (4) account for the heterogeneous nature of cell growth, while eqs. (6) do not. A careful inspection of eqs (4) and (6) reveals that the effects of population heterogeneity can safely be neglected when the single-cell division and reaction rates are either state-independent or depend linearly on the cellular state. In all other cases, differences between the two approaches are expected. Finally, notice that the proposed framework is bidirectional: Given a CPB one can derive the corresponding continuum model as described. Similarly, given a continuum model describing the dynamics of a certain network, the corresponding CPB model can be derived, by relaxing the homogeneity assumption captured in eq. (5). One also needs to model the partitioning mechanism. However, p(**x**,**y**,**S**) affects the dynamics of the average only indirectly, since it does not explicitly appear in the equations describing the population average dynamics (eqs. (4)).

2.1 *Numerical Algorithm*

The computational challenge in applying this theoretical framework lies in the necessity to solve the cell population balance model. To this end, significant progress has been made in recent years (Mantzaris, et al., 1999; Mantzaris et al., 2001a,b,c).

The most efficient algorithms suggested in the literature share one common characteristic: they discretize the cell population balance equation in a fixed domain. However, the upper bounds of the physiological state space are typically known only initially. Moreover, large differences in the time scales involved in the time evolution of the cell property distribution may lead even the most efficient, fixed-domain algorithms to significant inaccuracies. Motivated by the above considerations, we developed a moving boundary algorithm for the solution of population balance models. The number density function in the moving physiological state domain was expanded using a Galerkin spectral method with sinusoidal basis functions, since this set of basis functions requires a low number of modes in order to accurately capture the dynamics of the distribution. For the time integration of the system of ODEs resulting after expansion of the distribution function using a finite number of basis functions, the one-step, time explicit Runge-Kutta 4th order algorithm was found to be the most efficient (Mantzaris, et al., 2001b). This algorithm results in a more accurate and efficient simulation of the population balance model, since at each time step, only the part of the physiological state space where cells exist is discretized. It simulated very efficiently population balance equations describing vastly different (cellular and non-cellular) processes. The details of the derivation and the performance tests are presented elsewhere (Mantzaris, 2003).

3. THE GENETIC TOGGLE SYSTEM

Motivated by the potential of artificial genetic networks to tailor the phenotype of various strains according to specific biotechnological needs, we focus our attention on a specific artificial genetic network, known as the pTAK genetic toggle. It was constructed by Gardner and coworkers (Gardner et al., 2000) and consists of two promoter-repressor pairs. In particular, the *lac*I repressor gene is under the control of the *clts* (phage λ repressor) - repressed P_Ls1con promoter, while the λ repressor gene is under the control of the *lac*I-repressed, IPTG-inducible Ptrc-2 promoter. The gfpmut3 gene expressing fluorescent GFP serves as the reporter and is placed after the *clts* gene. The population dynamics of cells carrying the genetic toggle were well-characterized by Gardner and co-workers using flow cytometry (Gardner, et al., 2000).

To predict the induction dynamics, the authors also developed a two-state, chemically-structured continuum model, which, as stated earlier assumes *a homogeneous population behavior.* The model describes the expression dynamics of the two genes according to the dimensionless equations:

$$\frac{dx_1}{dt} = \frac{a_1}{1+x_2^{\beta}} - x_1 \qquad (7)$$

$$\frac{dx_2}{dt} = \frac{a_2}{1+\left(\dfrac{x_1}{f([\mathrm{IPTG}])}\right)^{\gamma}} - x_2 \qquad (8)$$

where a_1, a_2 are the effective rates of synthesis of the two genes; β, γ are Hill coefficients describing cooperative inhibition of the expression of each gene by the gene product of the other; f([IPTG]) is a function describing the induction process of the Ptrc-2 promoter (controlling the expression of the λ repressor) by the extracellular inducer isopropyl-β-D-thiogalactopyranoside (IPTG):

$$f([\mathrm{IPTG}]) = \left(1+\frac{[\mathrm{IPTG}]}{K}\right)^{\eta} \qquad (9)$$

The parameter values that were used in all simulations are given in Table 1 and are identical to those used in the original work of Gardner and co-workers (Gardner, et al., 2000).

Equations (7) and (8) constitute the continuum model, which was used as the basis for the development of the corresponding CPB model through the application of the theoretical framework presented earlier. We assumed that the division rate depends on the λ repressor expression according to an expression of the form:

$$\Gamma(x_1,x_2) = \frac{x_2^m}{\int_0^{x_{1,max}}\int_0^{x_{2,max}} x_2^m n(x_1,x_2,t)dx_2} \qquad (10)$$

where m is a parameter. Notice that the average division rate obtained from eq. (10) is equal to 1, in agreement with the average division rate given by the continuum model. Finally, we assumed that the partitioning function is equal with respect to the lac repressor and a symmetric beta distribution with respect to the λ repressor:

$$p(x_1,x_2,z_1,z_2) = \delta\left(x_1-\frac{z_1}{2}\right)\cdot\left(\frac{x_2}{z_2}\right)^q\cdot\left(1-\frac{x_2}{z_2}\right)^q \qquad (11)$$

where δ is the Dirac function; x_1,x_2 and z_1,z_2 are the expression levels of the two genes in the daughter and mother cells respectively, and q is a parameter.

Table 1: Parameter values used in simulations

Parameter	Value
a_1	156.25
a_2	15.6
β	2.5
γ	1.0
K	2.9618x10^{-5}M
η	2.0015
m	3
q	10

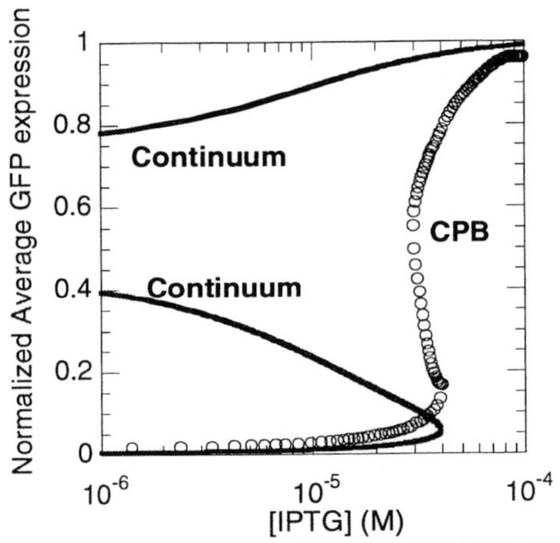

Figure 1: Average, normalized expression levels at balanced growth. Solid line: continuum model; Open circles: Cell population balance model

3.1. Results

The theoretical framework discussed in section 2 was applied in order to understand the interplay between the single-cell dynamics of the genetic toggle system and those of the entire cell population. The natural bifurcation parameter was the extracellular IPTG concentration. The comparison between the predictions of the two modeling approaches for the normalized, average balanced growth GFP expression levels (quantifying the expression levels of the λ repressor) is shown in Figure 1. Notice that both modeling approaches predict similar balanced growth average expression levels at high induction levels. Notice also the agreement between the modeling approaches in the prediction of the bifurcation point around 40 μM IPTG. However, the continuum model predicts a huge range of IPTG concentrations where the system exhibits steady-state multiplicity, with the upper and low states being stable, while the intermediate is unstable. On the contrary, taking into account population heterogeneity leads to a much narrower region of multiple steady states. The predictions of the CPB model are in excellent agreement with the experimental measurements of Gardner and co-workers (figure 5a, Gardner, et al., 2000). In particular, they showed that the population displays a much sharper transition from the low to the high state with a narrow region of steady-state multiplicity around the transition point of 40 μM IPTG.

Analogous similarities and differences are observed transiently. In particular, figure 2 shows two sets of simulations for low-intermediate induction levels ([IPTG]=20μM). In the first set, the initial average expression level is lower than in the second set. Notice that vastly different phenotypes are predicted by neglecting population heterogeneity. Low initial average expression level leads to an uninduced state, while for the higher initial condition the system rests at a highly induced state. On the contrary, after some initial dynamics and for both initial conditions, the inclusion of population heterogeneity leads the system to an uninduced state, which is higher than the one predicted by the continuum model.

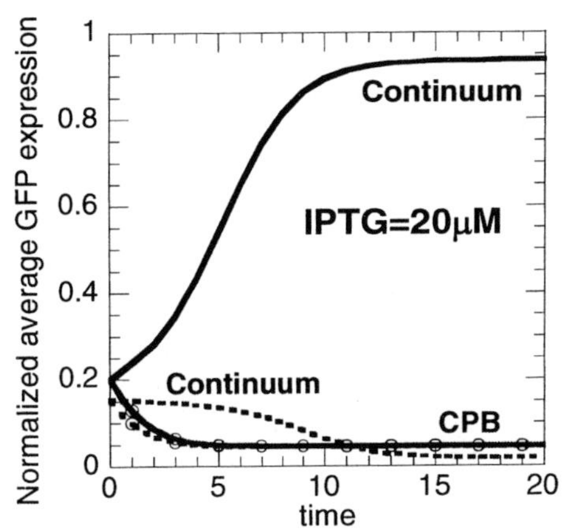

Figure 2: Transient simulations for the average, normalized expression levels for high and low initial average expression level at [IPTG]=20μM. Solid line: continuum model-high initial average; Dashed line: continuum model-low initial average; Solid line-open circles: CPB-high initial average; Dashed line-open circles: CPB-low initial average.

To better understand the source of the discrepancy between the predictions of the two models, we show the dynamics of the entire number density function for the same IPTG concentration (figure 3a). The system initially exhibits some transient bimodal distributions before reaching a unimodal distribution with a low mean value. These bimodal transient distributions have also been experimentally observed (see Figure 6 in (Gardner et al., 2000)). Despite the low average value, there exist some cells with at relatively high expression state, accounting for the fact that the average expression level at [IPTG]=20μM predicted by the CPB model is higher than the one predicted by the continuum model, (see Figure 2).

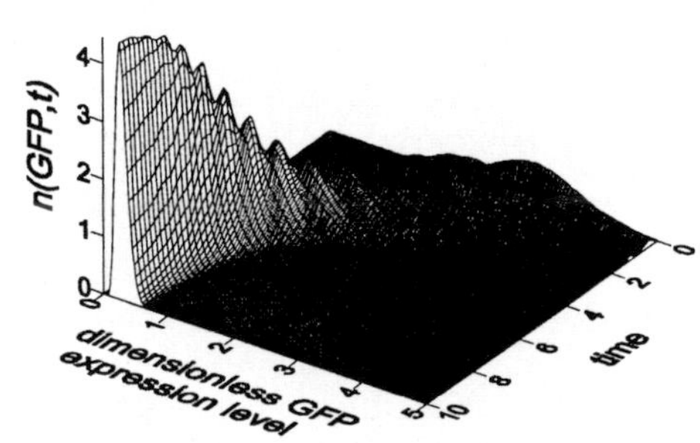

Figure 3: The dynamics of the GFP (or λ repressor) number density function for [IPTG]=20μM as predicted by the CPB model.

The effects of population heterogeneity are less pronounced at high induction levels as figure 4 shows. In this case, most cells exist above the threshold separating the low and high state and consequently the phenotype of the cell population is not significantly affected by the uneven distribution of expression levels. This result can also be explained mathematically: at high induction levels, the expression rate of the λ repressor is almost state-independent and, as explained earlier, linear or constant rate expressions result in less significant quantitative effects of cell population heterogeneity. The small differences in the predictions of the two modeling approaches are mainly attributed to the nonlinearity of the division rate.

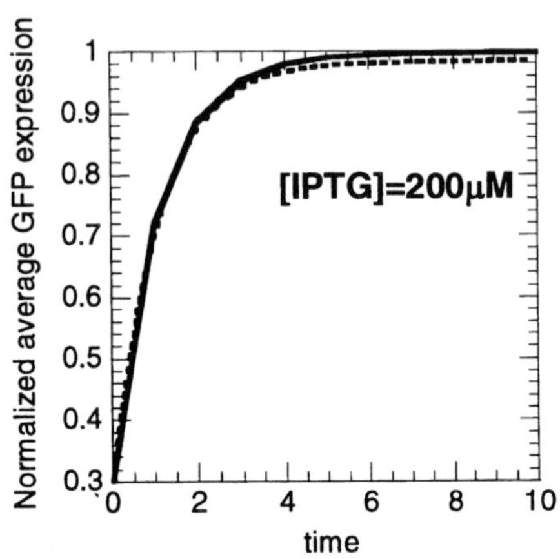

Figure 4: Transient simulations for the average, normalized expression levels at [IPTG]=200μM. Solid line: continuum; Dashed line: CPB model.

4. CONCLUSIONS

We have developed a theoretical/computational framework for isolating the quantitative effects of cell population heterogeneity for a given set of intracellular processes. The framework was applied to study the population dynamics of cells carrying an artificial genetic network, consisting of two promoter-repressor pairs. Our studies clearly showed that neglectimg cell population heterogeneity, may lead to significant qualitative and quantitative errors in the predictions of average population behavior, both transiently and at steady state. Thus, even in cases where only the prediction of the average is of interest, cell population heterogeneity needs to be accounted for. The effects of cell population heterogeneity were shown to be more pronounced at low-intermediate induction levels and smaller at high induction levels. This behavior is in qualitative agreement with experimental data (Elowitz, et al., 2002). The cell population balance modeling approach which accounts for cell-to-cell variability predicted a narrower region of multiple steady states, in agreement with experimental data (Gardner et al., 2000). This qualitative behavior was found to be insensitive to the exact division and partitioning mechanisms (results not shown). Thus, accounting for the heterogeneous nature of cellular processes leads to predictions with less rich dynamics. This, in turn, is an indication that cell-to-cell variability may be responsible for the robustness of cell populations as well as their adaptation ability to various environmental perturbations.

5. ACKNOWLEDGMENTS

Financial support by NSF/BES-0331324 is gratefully acknowledged.

6. REFERENCES

Chung, J.D. and G. Stephanopoulos (1995). "Studies of transcriptional state heterogeneity in sporulating cultures of *Bacillus subtilis*", *Biotechnol. Bioeng.*, **47**, 234-242.

Delbrück, M. (1945). "The burst size distribution in the growth of bacterial viruses (bacteriophages)", *J.Bacteriol.*, **50**, 131-135.

Elowitz, M.B., A.J. Levine, E.D. Siggia, and P.S. Swain (2002). "Stochastic gene expression in a single cell", *Science*, **297**, 1183-1186.

Gardner, T.S., C.R., Cantor and J.J. Collins (2000). "Construction of a genetic toggle switch in *Escherichia coli*", *Nature*, **403**, 339-342.

Guet, C.C., M.B., Elowitz, H. Weihong and S. Leibler (2002). "Combinatorial synthesis of genetic networks", *Science*, **296**, 1466-1470.

Mantzaris, N.V., J-J Liou, P. Daoutidis and F. Srienc (1999). "Numerical solution of a mass-structured cell population balance model in an environment of changing substrate concentration", *J. Biotechnol.*, **71** (1-3), 157-174.

Mantzaris, N.V., P. Daoutidis and F. Srienc (2001a). "Numerical solution of multivariable cell population balance models. I: Finite difference methods", *Comp. & Chem. Eng.*, **25** (11-12), 1411-1440.

Mantzaris, N.V., P. Daoutidis and F. Srienc (2001b). "Numerical solution of multivariable cell population balance models. II: spectral methods", *Comp. & Chem. Eng.*, **25** (11-12), 1441-1462.

Mantzaris, N.V., P. Daoutidis and F. Srienc (2001c). "Numerical solution of multivariable cell population balance models. III: Finite element methods", *Comp. & Chem. Eng.*, **25** (11-12), 1463-1481.

Mantzaris, N.V. "A moving boundary algorithm for the numerical solution of population balance models", (in preparation).

Novick, A. and M. Weiner (1957). "Enzyme induction as an all-or-none phenomenon", *Proc. Natl. Acad. Sci. USA*, **43**, 553-566, (1957).

Powell, E.O. (1956). "Growth rate and generation time of bacteria, with special reference to continuous culture", *J. Gen. Microbiol*, **15**, 492-511.

Ptashne, M. (1987). "*A genetic switch: Gene control and phage lamda*", Cambridge, MA: Cell Press, and Palo Alto, CA: Blackwell Science.

Ramkrishna, D. (2000). *Population balances: theory and applications to particulate systems in engineering*, Academic Press, San Diego, CA.

Russo-Marie, F., M. Roederer, B. Sager, L.A. Herzenberg and D. Kaiser (1993). "β-galactosidase activity in single differentiating bacterial cells", *Proc. Natl. Acad. Sci. USA*, **90**, 8194-8198.

Spudich, J.L. and D.E Koshland (1976). "Non-genetic individuality: chance in the single cell", *Nature*, **262**, 467-476.

Stocker, B.A.D. (1949). "Measurements of rate of mutation of flagellar antigenic phase in *Salmonella typhimurium*", *J. Hyg. Camb.*, **47**, 398-413.

ELSEVIER
IFAC
PUBLICATIONS
www.elsevier.com/locate/ifac

DYNAMIC MODELING OF FILAMENTOUS BULKING IN LAB-SCALE ACTIVATED SLUDGE PROCESSES

E.N. Banadda, I.Y. Smets, R. Jenné, and J.F. Van Impe*

Chemical Engineering Department
Katholieke Universiteit Leuven, B-3001 Leuven (Belgium)
Fax: +32-16-32.29.91
e-mail: jan.vanimpe@cit.kuleuven.ac.be

*Corresponding author

Abstract: Activated sludge is a complex ecosystem constituted mainly of bacteria and protozoa. Filamentous bulking, a phenomenon when the filamentous organisms dominate the activated sludge is still a widespread problem in the operation of activated sludge processes. Image analysis offers promising perspectives for early detection of filamentous bulking because the morphology parameters of the activated sludge react very fast to changing process conditions. This paper is aimed at identifying dynamic ARX and state space models as a function of organic loading and digital image analysis information (such as the total filament length per image and some representative mean floc shape parameters) to describe the evolution of the Sludge Volume Index (SVI). Their performances are compared based on an adequate quality performance criterion. *Copyright © 2004 IFAC*

Keywords: Environmental engineering, image analysis, modeling, monitoring.

1. INTRODUCTION

Activated sludge systems encompass biodegradation and sedimentation processes which take place in the aeration and sedimentation tanks, respectively. The performance of the activated sludge process is, however, to a large extent dictated by the ability of the sedimentation tank to separate and concentrate the biomass from the treated effluent. Since the effluent from the secondary clarifier is most often not treated any further, a good separation in the settler is critical for the whole plant to meet the effluent standards. Apart from bad operating strategies or poorly designed clarifiers, settling failures can be mainly attributed to *filamentous bulking*. As a general guideline, filamentous bulking is said to occur when the Sludge Volume Index [1] (SVI) is greater than $150 \mathrm{mLg}^{-1}$, regardless of its cause. Apart from the physico-chemical parameters, biological parameters are also increasingly consulted through microscopic observation. However, microscopic observation is not only subjective (i.e., operator dependent) but also time consuming, often leaving insufficient time for corrective interventions. Image analysis, a computer assisted procedure through which analysis of digital images is performed, could provide an objective means to assist the operator's decision making. Furthermore, mathematical models, capable of predicting the evolution of, e.g., the SVI value, will be indispensable for the development of early warning and detection tools. Although filamentous bulking has been studied intensively during the last decades (Jenkins *et al.*, 1993), the phenomenon is so complex (i.e., influenced by so many different factors) that a first principles model is still lacking. The identification of a model to predict in real-time - with reasonable accuracy - the appearance of sludge bulking is therefore of great importance, in view of the potential for improvement in plant efficiency and cost saving (Novotny *et al.*, 1990). This paper aspires to identify ARX and state space type models which can predict the settling characteristic SVI as a function of either organic loading (Q) and/or digital image analysis information as will be explained later on. The model performances are compared based on a quality performance criterion. Since static (instantaneous) correlation models, as reported by, e.g., da Motta and coworkers (da Motta *et al.*, 2002) cannot provide future SVI predictions nor take into account the system's delay, we will focus on *dynamic* models. In other words, we focus on how the past input data, i.e., in this case organic loading and/or digital image analysis information, affect(s) the future

[1] volume in milliliters occupied by 1 g of a suspension after 30 minutes settling

output data, i.e., the SVI. In a first approach *linear* models (ARX and state space models) have been preferred over *nonlinear* ones in modeling the filamentous bulking phenomena although the latter could be characterized by nonlinear behavior (since living organisms are involved). Linear model identification is however a much more *mature* field.

2. MATERIALS AND METHODS

This study relies on results obtained from two laboratory wastewater treatment experiments. Both systems are operated under more or less similar conditions so as to mimic a large-scale continuous system. However, it is worth mentioning that for both experiments, conditions were chosen to favor filamentous bulking outbreaks (Jenné *et al.*, 2003; Banadda *et al.*, 2003).

2.1 Lab-scale activated sludge system

The laboratory set-up is a continuous type activated sludge system in a classic configuration: an aeration tank (5.5 L) followed by a sedimentation tank (3 L) and sludge recycle (Figure 1).

Fig. 1. Left: lab-scale activated sludge system. Right: microscope (Olympus BX51) and video camera (Sony DXC-950P) used for daily capturing of activated sludge images.

This system was inoculated with activated sludge from a domestic wastewater treatment plant at Huldenberg (Belgium). The influent was synthetic wastewater, nevertheless, different feed regimes and carbon sources were administered (Figure 2), i.e., the use of glucose in the second experiment as opposed to sodium acetate (first experiment) aimed at inducing earlier filamentous bulking outbreaks. Quantitative information which includes analytical results of the effluent quality are recorded: organic matter, measured as Chemical Oxygen Demand (COD), the settleability measured as Sludge Volume Index (SVI), the activated sludge concentration measured as the Mixed Liquor Suspended Solids (MLSS) and the effluent Suspended Solids (SS). Also, digital image analysis is performed on a daily basis on 50 images per sample as reported in previous studies (Jenné *et al.*, 2002). The first experiment lasted 70 days while the second lasted 38 days (see Figure 2).

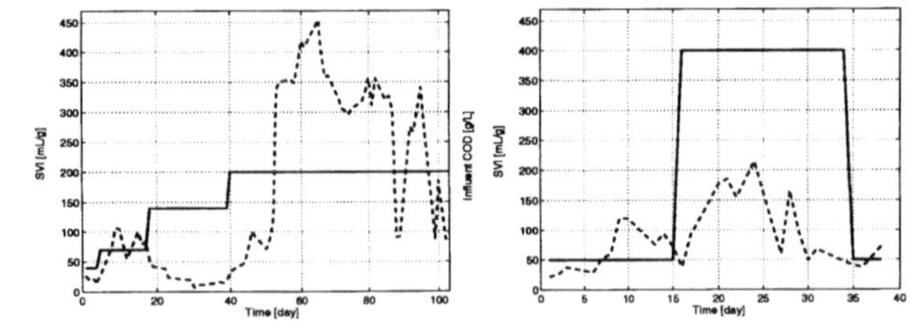

(a) (b)

Fig. 2. Evolution of the SVI (- - -) and influent COD feed regime (—) for (a) the first experiment and (b) the second experiment.

2.2 Image analysis procedure

Images of activated sludge samples were recorded by means of a light microscope (Olympus BX51) with phase contrast equipped with a 3CCD color video camera (Sony DXC-950P), shown in Figure 1. The magnification of the microscope objective used was 10 × 10 times. A fully automatic image analysis method for recognition and characterization of flocs and filaments in an activated sludge sample has been developed in previous work (Jenné *et al.*, 2002; Jenné *et al.*, 2003), and is applied to all colored phase contrast images.

2.3 Measurements

The principal aim of the image processing procedure is the extraction of certain information from the digital images. Once the objects in the image (i.e., the flocs and the filaments) are distinguished from the background, several size and shape related parameters can be computed.

Filament measurement. The filament length per image (F) describes the total length of the filament skeleton.

Floc size measurement. The size of the sludge flocs is an important parameter with respect to the settling properties (Ganczarczyk, 1994). The size of the flocs is expressed as the equivalent circle diameter D_{eq}, calculated from the real projected area A:

$$D_{eq} = 2\sqrt{A/\pi} \qquad (1)$$

Floc shape measurements. It is mentioned in the literature (Eriksson and Hardin, 1984) that the shape of sludge flocs is related to the settling properties. Many shape quantifying parameters can be measured by means of image analysis. Four parameters are considered in this study.

- The form factor (FF) is particularly sensitive to the *roughness* of the boundaries. A circle has an FF value equal to one.

$$\text{FF} = 4\pi \frac{\text{area}}{\text{perimeter}^2} \qquad (2)$$

- The aspect ratio (AR) is mainly influenced by the elongation of an object. It varies between 1 and infinity. A circle has an AR value equal to one.

$$\mathrm{AR} = 1 + \frac{4 \cdot (\text{Length-width})}{\pi.\text{width}} \quad (3)$$

- The roundness (R) is also mainly influenced by the elongation of an object. It varies between 0 and 1. A circle has an R value equal to one.

$$\mathrm{R} = \frac{4 \cdot \text{area}}{\pi.\text{length}^2} \quad (4)$$

- The reduced radius of gyration (RG) is also influenced by the elongation of an object. A more elongated floc will have a larger RG. A circle has an RG value equal to $\frac{\sqrt{2}}{2}$.

$$\mathrm{RG} = \frac{\sqrt{M_{2x} + M_{2y}}}{\frac{D_{eq}}{2}} \quad (5)$$

M_{2x} and M_{2y} are second order moments.

3. EXPERIMENTAL RESULTS AND DISCUSSION

Previous studies by Jenné *et al* (2003), have shown that a reasonably good correlation exists between the digital image analysis information and the SVI. In this paper, new model inputs in addition to those mentioned in (Banadda *et al.*, 2003), are investigated. Data collected during the two experimental periods are used to identify ARX and state space models that describe the SVI evolution, i.e., the model output, based on either (*i*) information gathered with the image analysis procedure (i.e., F, FF, D_{eq}, R, RG) and/or (*ii*) the organic loading (Q) profile as the model inputs. Afterwards the identified models from one experiment are cross-validated on the data of the other experiment.

3.1 Optimization criterion

The criterion to be maximized is the *R-squared adjusted* (R^2_{adj}) value, which is often expressed in percent (Equation (6)). The values obtained from the criterion reflect the percentage of output variation explained by the model (i.e., $y_h(t)$). Moreover, the R^2_{adj} criterion takes into account both the number of data points N and the model parameters (degrees of freedom) DF.

$$R^2_{adj} = 100 \cdot \left(1 - \frac{(N-1) \cdot \sum_{t=1}^{N} (y(t) - y_h(t))^2}{(N-DF) \cdot \sum_{t=1}^{N} (y(t) - mean(y(t)))^2}\right) \quad (6)$$

with $y(t)$ the measured output at discrete time t, $y_h(t)$ the model output at discrete time t.

3.2 ARX models

Procedure. ARX models relate the current output $y(t)$ to a finite number of past outputs $y(t-k)$ and inputs $u(t-k)$.

$$y(t) + a_1 y(t-1) + (\cdots) + a_{na} y(t-na) = b_1 u(t-nk) + b_2 u(t-nk-1) + (\cdots) + b_{nb} u(t-nk-nb+1) + e(t) \quad (7)$$

with $y(t)$ equal to the output response at discrete time t, $u(t)$ the input at discrete time t, na the number of poles, nb the number of zeros, nk the pure time-delay (the dead-time) in the system and $e(t)$ a white noise signal. a_i and b_j are model parameters, with $i = 1 \ldots na$ and $j = 1 \ldots nb$. The model structure is entirely defined by the three integers na, nb, and nk.

The purpose of this section is to identify which image analysis parameters characterize the SVI evolution via *dynamic* ARX models. This is done by simulating the ARX models with organic loading (Q) and/or digital image analysis information parameters and analyzing prediction system performance based on the chosen criterion, R^2_{adj}. ARX models are identified by using the **arx** command in the System Identification Toolbox 5.0.1 in MATLAB (The Mathworks, Inc., Natick), which allows to specify a specific *focus* during model identification. Three different options are available, i.e., a focus on *prediction*, *simulation* or *stability*. *Prediction* means that the model is determined by minimizing the prediction errors. With focus on *simulation*, the model approximation is such that the model will produce as good simulations as possible, when applied to inputs with the same spectra as used for the estimation. A stable model is guaranteed. Finally, a *stability* focus implies that the algorithm is modified so that a stable model is guaranteed, but the weighting still corresponds to prediction. Previous studies (Banadda *et al.*, 2003), have shown that the performance of the optimal models with focus on *prediction* in combination with our own stability check is as good and sometimes even better than that of the models resulting from the *stability* focus. Therefore, identification results of the latter are omitted.

Space does not allow to show all the identification and validation results but both Tables 1 and 2 give a representative view on the identification and validation outcomes for ARX and state space models, respectively. Optimal combinations (organic loading and image analysis information) were sought in the range of 1 to 5 for the number of poles (na) and zeros (nb) (with nb smaller than or equal to na) at fixed delays nk of 0, 1 and 2. The maximum value for na is deliberately kept low to avoid overfitting.

Identification and validation results.

One input case. F as a single input performs best in both identification and validation although other inputs such as Q can be reasonable in identification (e.g., for the first experiment) but validate badly. This result encourages the use of image analysis as an activated sludge monitoring tool.

Two input case. The input F in combination with another input is reasonable in identification and validation (except the validation with [F Q] in the first experiment). Excluding F as an input has a severe (negative) impact on the validation quality of the model. Overall, a combination of F and D_{eq} is a satisfying input combination.

Remarkably, the nk value is always equal to zero for F in the first experiment. For the second experiment, this nk value can deviate from zero but only in the multi-input case in combination with the input Q. These models are only a tenth of a percent better in identification but validate obviously rather bad on the first experiment. Therefore, in the search for optimal models for the data of the second experiment, the nk value corresponding to F is restricted to zero.

Three input case. No positive validation results are recorded except in the case of the [F D_{eq} Q] input combination for the second experiment model on the first experiment data. Relatively, good validation results are noted when F is one of the inputs. In general, [F D_{eq} Q] is a well performing input combination. Figure 3 shows the performance of the ARX models in identification (R^2_{adj} = 78.87%, second experiment) and validation (R^2_{adj} = 85.00%, first experiment) as compared to the measured SVI.

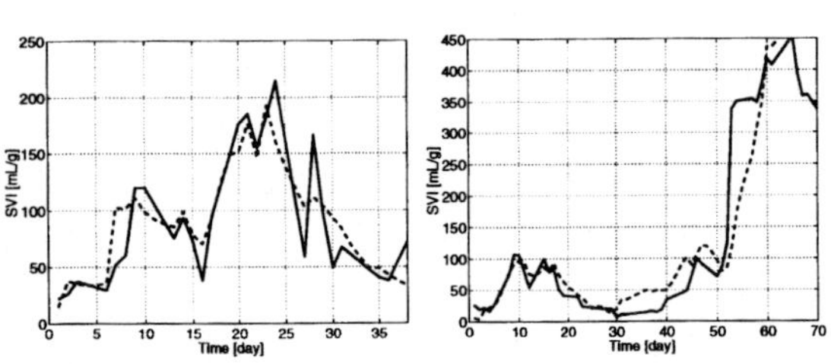

(a) (b)

Fig. 3. Measured SVI (—) and modeled SVI (- - -) with the optimal 1st-order ARX model and 3 model inputs [F D_{eq} Q] during (a) identification and (b) validation.

3.3 State space models

Procedure. Apart from ARX models, also state space models have been explored. In discrete time, a state space model has the following form

$$\begin{aligned} x(t+1) &= Ax(t) + Bu(t) + Ke(t) \\ y(t) &= Cx(t) + Du(t) + e(t) \end{aligned} \tag{8}$$

with $x(t)$ equal to the state vector at discrete time t, $y(t)$ equal to the output response, $u(t)$ the input, $e(t)$ a white noise signal, and A, B, C, D and K the state space matrices of appropriate dimensions.

In this section, state space models are identified by means of the `n4sid` command in the System Identification Toolbox 5.0.1 in MATLAB (The Mathworks, Inc., Natick), which allows, similar to the ARX command, to choose a specific *focus* during model identification. `n4sid` is a subspace-based method that does not use iterative search. As for ARX modeling, optimal combinations of the organic loading and image analysis information (state space model inputs) for both the first and the second experiment were sought in the order range of 1 to 4 and their performance evaluated based on the R^2_{adj} criterion (Equations (6)).

Identification and validation results.

One input case. As for ARX modeling, again F as a single input performs best in identification and validation while other inputs such as Q show relatively lower identification values and validate badly.

Two input case. For the first experiment, no positive validation results are obtained. As for the second experiment, F must be in it; [F D_{eq}] or [F Q] are appropriate input combinations.

Three input case. For the first experiment, no positive validation results are noted. For the second experiment, F must be in it and this in combination with Q or D_{eq}, with [F D_{eq} Q] giving the highest validation results on the first experiment. Figure 4 elucidates the performance of the state space models in identification (R^2_{adj} = 69.56%, second experiment) and validation (R^2_{adj} = 80.29%, first experiment) as compared to the measured SVI.

3.4 Discussion

Several interesting *general* remarks remain to be stated. They are listed below.

- The validation performance of the single input model with F is as good as, or at times better than that of some given multi-input models but the latter are most often better in identification. However, from a *prediction*

Table 1. Optimal ARX model output with 1, 2 or 3 model input(s) for the first experiment and the second experiment.

Focus on	Model inputs	na	nb	nk	R^2_{adj} (%) ARX^1	R^2_{adj} (%) ARX^{1-2}	na	nb	nk	R^2_{adj} (%) ARX^2	R^2_{adj} (%) ARX^{2-1}
Prediction	F	5	1	0	79.89	54.49	1	1	0	67.85	84.29
	Q	1	1	2	62.04	-51.28e+01	5	1	0	12.55	-56.52e-01
	F, Q	5	[1 5]	[0 2]	82.66	-70.93	1	[1 1]	[0 2]	72.27	86.00
	F, AR	5	[1 1]	[0 2]	79.04	52.02	1	[1 1]	[0 0]	71.55	75.15
	F, D_{eq}	5	[1 1]	[0 1]	79.57	53.09	1	[1 1]	[0 0]	71.81	76.20
	Q, D_{eq}	5	[2 5]	[1 2]	78.34	-86.93e+01	5	[2 1]	[0 2]	32.33	-18.43
	AR, D_{eq}	5	[1 5]	[2 1]	63.92	-18.31e+02	3	[1 1]	[0 0]	48.87	79.50e-01
	F, D_{eq}, Q	5	[3 5 2]	[0 2 1]	91.23	-61.06e+01	1	[1 1 1]	[0 0 2]	78.87	85.00
	F, FF, R	5	[3 5 2]	[0 2 0]	88.80	-11.78e+02	3	[3 1 1]	[0 0 0]	78.84	-79.71
	F, FF, RG	4	[3 2 1]	[0 1 2]	87.25	-77.57e+01	2	[2 1 1]	[0 0 0]	74.06	-10.94
	F, FF, AR	5	[3 5 2]	[0 2 1]	88.42	-15.63e+02	4	[1 3 2]	[0 1 2]	77.08	-80.86e-01
Simulation	F	4	4	0	83.83	57.59	2	1	0	67.48	82.36
	Q	1	1	2	64.58	-42.83e+01	2	1	0	26.30	38.25e-01
	F, Q	4	[3 2]	[0 0]	92.57	-12.30e+01	5	[1 4]	[0 1]	75.63	78.76
	F, AR	3	[3 2]	[0 2]	85.03	58.74	1	[1 1]	[0 0]	73.52	33.89
	F, D_{eq}	3	[3 1]	[0 2]	83.60	46.04	1	[1 1]	[0 0]	71.99	66.58
	Q, D_{eq}	3	[3 3]	[2 2]	90.58	-10.80e+02	5	[1 2]	[0 2]	74.79	-14.58
	AR, D_{eq}	4	[1 2]	[2 0]	86.77	-61.42e+02	1	[1 1]	[0 1]	55.17	16.87
	F, D_{eq}, Q	3	[3 3 2]	[0 2 0]	94.74	-45.62e+01	1	[1 1 1]	[0 0 2]	79.47	81.63
	F, FF, R	4	[4 4 4]	[0 2 2]	95.85	-89.23e+01	3	[3 1 1]	[0 0 0]	79.50	-85.03
	F, FF, RG	5	[3 5 5]	[0 2 2]	96.16	-65.39e+01	3	[1 2 2]	[0 0 0]	76.65	-82.60
	F, FF, AR	5	[4 5 5]	[0 2 2]	96.14	-13.59e+02	4	[1 2 3]	[0 2 1]	77.21	-52.06
	FF, R, AR	2	[2 1 2]	[2 1 2]	94.21	-42.21e+02	3	[1 3 2]	[0 0 1]	81.68	-16.50+01

ARX^{1-2}: R^2_{adj} values of ARX models identified from the first experiment (ARX^1) and validated on the second experiment.
ARX^{2-1}: R^2_{adj} values of ARX models identified from the second experiment (ARX^2) and validated on the first experiment.

Table 2. Optimal state space model output with 1, 2 or 3 model input(s) for the first experiment and the second experiment.

Focus on	Model inputs	Optimal order	R^2_{adj} (%) ss^1	R^2_{adj} (%) ss^{1-2}	Optimal order	R^2_{adj} (%) ss^2	R^2_{adj} (%) ss^{2-1}
Prediction	F	3	73.05	40.81	1	59.75	79.68
	Q	2	59.19	-704.47	1	0.20	1.34
	F, Q	2	65.41	-48.43e+01	1	65.60	76.47
	F, AR	2	55.25	-13.02	1	57.22	76.87
	F, D_{eq}	4	71.08	-20.63e–01	1	60.04	79.28
	Q, D_{eq}	1	74.92	-32.67e+01	3	33.25	-30.09
	AR, D_{eq}	2	73.26	-13.89e+02	3	54.82	15.07e-02
	F, D_{eq}, Q	2	75.41	-33.52e+01	1	69.56	80.29
	F, FF, R	1	62.67	-64.26e+01	1	64.33	98.16e-01
	F, FF, RG	1	63.89	-45.15e+01	1	60.36	21.91
	F, FF, AR	1	66.49	-59.13e+01	1	64.55	-25.35
	FF, R, AR	1	59.50	-54.32e+01	1	64.07	-11.44e+01
Simulation	F	2	83.14	-12.64	1	61.47	80.03
	Q	2	67.09	-2.41e+03	2	25.31	-13.08
	F, Q	4	88.12	-24.11e+02	1	66.65	79.58
	F, AR	3	83.35	-60.24e+01	1	61.91	73.17
	F, D_{eq}	2	83.01	-18.32	1	63.46	74.87
	Q, D_{eq}	3	92.18	-15.62e+02	3	48.35	-28.56
	AR, D_{eq}	2	83.37	-17.76e+02	3	57.26	26.54e-01
	F, D_{eq}, Q	2	89.03	-34.46e+01	1	70.54	79.95
	F, FF, R	4	87.22	-32.47e+01	1	64.71	30.22e-01
	F, FF, RG	4	88.78	-39.37e+01	1	62.61	51.46
	F, FF, AR	3	88.11	-25.86e+01	1	65.85	-49.56e-01
	FF, R, AR	2	94.30	-13.95e+02	2	66.71	-17.11e+01

ss^{1-2}: R^2_{adj} values of state space models identified from the first experiment (ss^1) and validated on the second experiment.
ss^{2-1}: R^2_{adj} values of state space models identified from the second experiment (ss^2) and validated on the first experiment.

point of view, the filament length input is less suitable since the n_k value is equal to zero, indicating a change in the parameter *at the same time* as the change in the SVI.

- Validation of the models identified on the data of the first experiment is bad or non existing. This could be due to the fact that too much effort has been dedicated in identifying the SVI jump, i.e., values in the range of 250mLg^{-1} to 450mLg^{-1} during the first experiment, a jump that is not present in the second experiment.
- If the second experiment models validate well on data of the first experiment, the R^2_{adj} criterion value is for yet unexplainable reasons higher in validation than in identification. This is also true for the R^2 criterion that does not compensate for the number of data points and the number of degrees of freedom (results not shown).
- Most often, combinations of F, FF, Q and D_{eq} with the floc elongation related parameters R, RG or AR give rise to more or less similar identification values, implying that

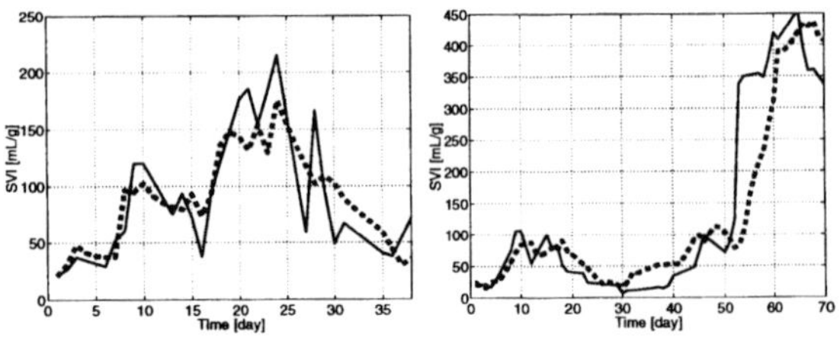

(a) (b)

Fig. 4. Measured SVI (—) and modeled SVI (- - -) with the optimal 1st-order state space model and 3 model inputs [F D_{eq} Q] during (a) identification and (b) validation.

probably only one of the roundness related floc characteristics will be sufficient.

- A final remark is related to the MATLAB implementation. Although not needed, we did test the model stability when implementing the *simulation* focus. We then noticed that the System Identification Toolbox is rather generous in assigning the stable label to a model. Models, whose poles lie within a circle of a radius of 1.01, are regarded as stable by MATLAB.

4. CONCLUSIONS

In the quest for an early forecast tool for filamentous bulking in activated sludge wastewater treatment systems, organic loading and/or digital image analysis information have been exploited into dynamic black box models to predict the Sludge Volume Index evolution based on a quality performance criterion, i.e., R^2_{adj}. If the floc and filament characteristics change *before* the SVI reaches its critical limit, a *dynamic* model, taking into account previous input and output values, could aid in predicting filamentous bulking before it really happens, which is not feasible with static, instantaneous models. Therefore, ARX and state space models were identified from two individual data sets and afterwards cross-validated on the data of the other experiment.

ARX models are better (higher criterion values) in identification than state space models and as good in validation. Previous studies have earmarked image analysis information as a promising activated sludge monitoring tool (Jenné *et al.*, 2003). In this paper this conclusion is confirmed. An interesting observation is, e.g., that the total filament length per image (F) could either be used as a single input or in combination with other (floc related) inputs, especially if good validation quality is to be attained. The validation exercise with a single input shows F as a strong candidate to model the SVI behavior as compared to Q. [F D_{eq}] and [F D_{eq} Q] come out as well performing multi-input combinations though the validation results with a single input F are equally good. The drawback of the filament length as an input is the lack of *predictive* power since the characteristic changes occur at the same time as the change in the SVI values. As additional experimental data and other image analysis based model inputs become available, re-identification and further validation of ARX models will be performed.

ACKNOWLEDGEMENTS

Work supported by Projects OT/99/24, OT/03/30 and IDO/00/008 of the KULeuven Research Council, and the Belgian Program on Interuniversity Poles of Attraction, initiated by the Belgian Federal Science Policy Office. Ilse Smets is a postdoctoral fellow with the fund for Scientific Research Flanders. The scientific responsibility is assumed by its authors.

REFERENCES

Banadda, E.N., R. Jenné, I.Y. Smets and J. F. Van Impe (2003). Predicting the onset of filamentous bulking in biological wastewater treatment systems by exploiting image analysis information. European Control Conference ECC2003 CD ROM 6p, University of Cambridge, UK.

da Motta, M., M.N. Pons and N. Roche (2002). Study of filamentous bacteria by image analysis and relation with settleability. *Water Science and Technology* **46**, 363–369.

Eriksson, L. and A.M. Hardin (1984). Settling properties of activated sludge related to floc structure. *Water Science and Technology* **16**(10-11), 55–68.

Ganczarczyk, J.J. (1994). Microbial aggregates in wastewater treatment. *Water Science and Technology* **30**(8), 87–95.

Jenkins, D., M. G. Richard and G. T. Daigger (1993). *Manual on the causes and control of activated sludge bulking and foaming.* 2nd edition. Lewis. London (UK).

Jenné, R., C. Cenens, A.H. Geeraerd and Van Impe J.F. (2002). Towards on-line quantification of flocs and filaments by image analysis. *Biotechnology Letters* **24**(11), 931–935.

Jenné, R., E.N. Banadda, N. Philips and J. F. Van Impe (2003). Image analysis as a monitoring tool for activated sludge properties in lab-scale installations. *Environmental Science and Health Part A* **38**(10), 2009–2018.

Novotny, V., H. Jones, X. Feng and A.G. Capodaglio (1990). Time series analysis models of activated sludge plants. *Water Science and Technology* **23**, 1107–16.

ELSEVIER
IFAC
PUBLICATIONS
www.elsevier.com/locate/ifac

MARKET-ORIENTED SCHEDULING, ECONOMIC OPTIMIZATION AND STOCHASTIC CONSTRAINED CONTROL OF CONTINUOUS MULTI-GRADE CHEMICAL PROCESSES

O.H. Bosgra * R.L. Tousain *,[1] D.H. van Hessem *,[2]

* *Delft Centre for Systems and Control, Delft University of Technology, Mekelweg 2, 2628CD Delft, The Netherlands*

Abstract: In this paper an approach for flexible production scheduling for continuous multi-grade chemical processes is proposed. The approach integrates the economics of production and of company-market interaction for single-machine multi-grade continuous processes. The resulting grade transitions are realized using a newly developed closed-loop stochastic MPC framework, that decomposes this task into a deterministic feedforward constrained trajectory optimization task and a stochastic feedback disturbance suppression task. The back-off used in the former is provided by the latter. The approach is demonstrated on a gas phase HDPE manufacturing plant. *Copyright © 2004 IFAC*

Keywords: process optimal operation, continuous scheduling, grade transitions, closed-loop model predictive control, stochastic MPC, process control

1. INTRODUCTION

During the past decades the chemical industry has been faced with a major change into a *globally competing* and *demand driven* mode of operation. Companies are required to respond quickly to changing market situations and must meet customer-specified product specifications. A main challenge and a key to demand-driven operation, is the allocation of production resources to comply with orders and physical constraints such as plant capacity and storage facilities. Such problems are generally referred to as *scheduling* problems, and their role in the internal supply chain is broadly acknowledged. An overview can be found in (Reklaitis, 1992). Although most of the scheduling literature focuses on *batch* operations, there is recent work on continuous process scheduling see (Mendez and Cerda, 2002), (Giannelos and Georgiadis, 2002). The approach here will consider continuous, single-machine, multi-grade chemical plants, and differs from the work cited in several aspects. Our way of including the effect of process transitions on the material flows appears new. Further, most scheduling studies assume the order base to be fixed in advance and strive for 'minimum makespan' or 'minimum lateness'. In our approach, the negotiation of sales orders and purchases is an integral part of the decision making that is supported by the scheduler, and to this end the scheduler selects a set of appealing purchase and sales transactions from a larger set of possible transactions (denoted *opportunities*). The optimal scheduling approach considered here includes a formulation of the grade change problem that includes a truly economic objective. To

[1] Present affiliation: Philips Centre for Industrial Technology, P.O. Box 218, 5600 MD Eindhoven, The Netherlands
[2] Present affiliation: Shell International Chemicals B.V. P.O. Box 38000, 1030 BN Amsterdam, The Netherlands

implement grade changes under practical process conditions, the off-line computed process trajectories will be transferred into reality by a real-time MPC algorithm. We will extend current MPC capabilities (Rawlings, 2000) for this purpose to deal with back-off under stochastic disturbances, and to obtain guaranteed control performance with respect to disturbance suppression. Lee and Ricker (Lee and Ricker, 1994) have proposed to decompose the stochastic MPC problem into an optimal Gaussian estimation and a deterministic prediction problem, to be solved as separate optimization problems in a receding horizon implementation. This view has become a major paradigm for MPC research, in which stochastics are considered in the past but not in the future. Three limitations of this paradigm are (1) there is no back-off to the constraints, (2) there is no possibility to tune the implied feedback properties and (3) the underlying assumption of the validity of the certainty equivalence property may be questionable in the case of active inequality constraints. The approach taken here suggests to decompose the predictive control formulation into a feedforward trajectory and a feedback controller. By creating a back-off to the constraints, the feedback controller retains its linear behaviour, enabling feedback properties to be assigned using an appropriate disturbance suppression objective. This in turn provides an estimate for the back-off to be used in the feedforward trajectory optimization, as in a bootstrap technique. Consequently, optimal plant transitions can be realized using feedforward whereas the inequality constraints are guaranteed not to be violated under the closed-loop control. The paper presents a survey of work available in detail in (Tousain, 2002) and (van Hessem, 2004).

2. DEMAND DRIVEN OPERATION

The scheduling task concerns the timing of feedstock and grade changes. Most of today's multigrade plants are still operated according to a predetermined sequence of product grades, called a *product slate* or a *product wheel*, see figure 1 (Sinclair, 1987). The sequence is constructed such that the necessary grade changes are relatively easy, safe and well-known by the operating staff. In order to operate market-responsive, short-term production scheduling should consider the internal supply chain in interaction with the markets involved. The simple model is represented schematically in figure 2. It provides a new, broader scope for *process control* from a perspective of *intentional dynamics*, (Marquardt *et al.*, 1998). It captures some of the main mechanisms in practical chemical supply chains: company-market interactions, inventory control, process

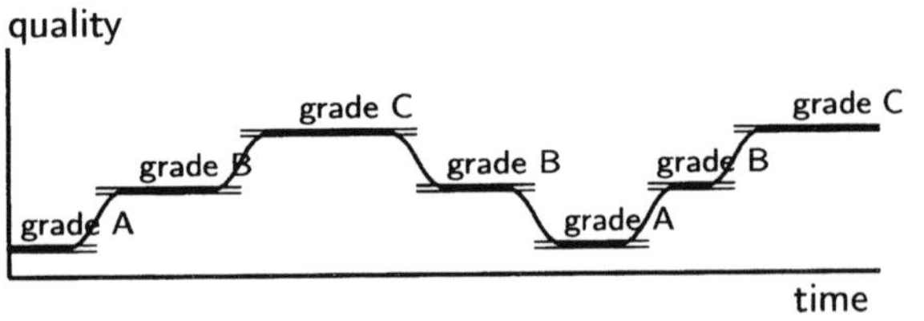

Fig. 1. A product grade slate A-B-C-B-A with on-spec-ranges

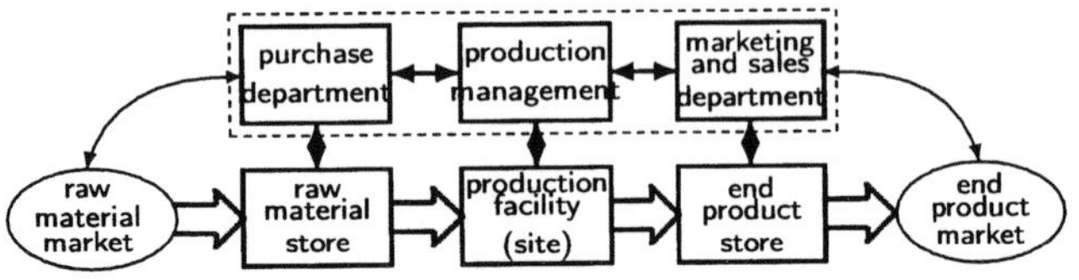

Fig. 2. A simple supply chain model for a continuous chemical manufacturing site

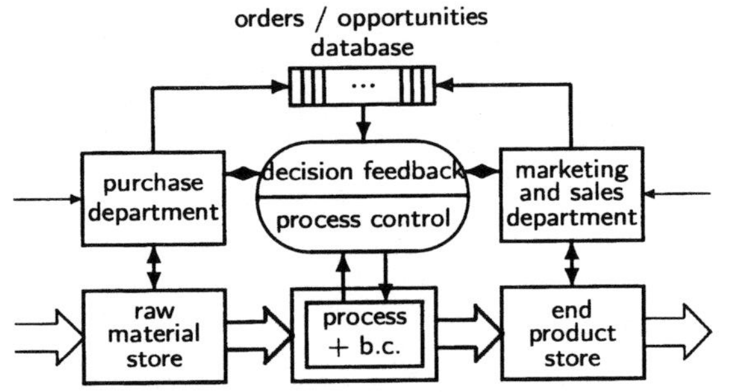

Fig. 3. A single level approach to supply-chain-conscious production management and process control

dynamics such as transitions, load changes, and internal supply chain organization. Although dynamic supply chain phenomena such as demand amplification (Towill, 1996) are important, here we focus on production management decisions for multi-grade processes. These involve scheduling, plant optimal operation and control tasks. The integration of these tasks needs care to avoid inconsistencies. In (Bassett *et al.*, 1996), three types of model-based integration are distinguished: *single-level control*, *multi-level control* and *conceptual decomposition*. A *single-level control strategy* to the problem of supply-chain-conscious process operations implies to control the process while supporting the purchasing and sales decision making tasks, as represented schematically in Figure 3. This necessitates to include all relevant time scales of the operations problem (i.e. weeks/months to capture market changes, and seconds/minutes to describe plant dynamics which should be controlled) into one control problem. This will generally lead to combinatorial control problems which are of enormous dimension and hence computationally infeasible. The main decomposition that we propose is hence to deal separately with the

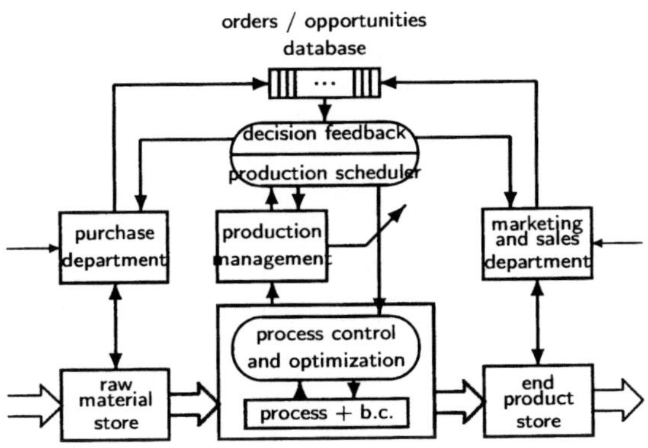

Fig. 4. Decomposition of supply-chain-conscious production management and process control

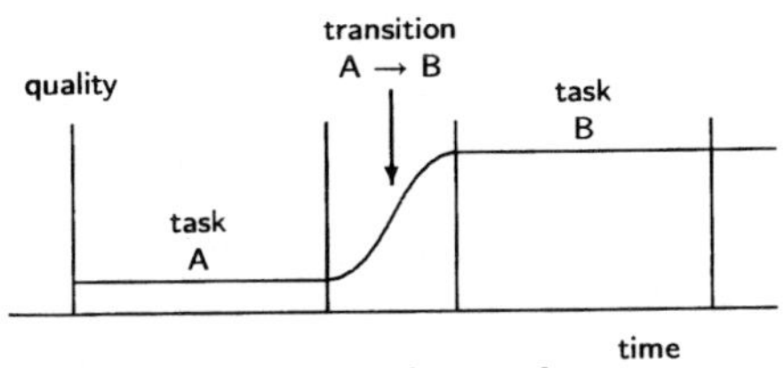

Fig. 5. Operation of a multi-grade process considering static production tasks and transition tasks

two questions 1: what products/feedstocks will be produced/processed when? and 2: how will the production be realized? We split up the production management problem into a *production scheduling problem* (section 3), and a *production control problem*, (section 5), see figure 4. Both solutions utilize the same dynamic model, which is an important prerequisite for the consistency of the decomposition approach.

3. SHORT TERM SCHEDULING DESIGN

To solve the production scheduling problem, we concentrate on a description of the plant in terms of operating tasks with accompanying precedence rules. The use of quasi-static and dynamic tasks in modeling the behavior of a continuous multi-grade plant is illustrated in figure 5. Let a model of the plant, *including the basic control system*, be given by the following set of DAE's

$$\begin{aligned} \dot{x} &= f(x,u,y) \\ 0 &= g(x,u,y) \\ z &= C_x x + C_u u, \end{aligned} \tag{1}$$

where $x(t) \in \mathbb{R}^{n_x}$, $u(t) \in \mathbb{R}^{n_u}$, and $y(t) \in \mathbb{R}^{n_y}$ are respectively the state, input and algebraic variables of the model. For ease of notation, the dependency of these variables on time will be omitted in the remainder. $z \in \mathbb{R}^{n_z}$ contains the so-called performance channels, i.e. all variables that are required for the performance evaluation of the plant. The performance computation involves a so-called objective function as well as the violation of constraints. These operating constraints are expressed as

$$h(z) < 0. \tag{2}$$

All feasible steady state operating conditions are given by the set

$$\mathcal{F} = \{x, u \mid \exists\, y, z \text{ such that } f(x,u,y) = 0, \\ g(x,u,y) = 0,\ z = C_x x + C_u u,\ h(z) < 0\} \tag{3}$$

In general we are only interested in a limited number of interesting scenarios, e.g. a finite number of product grades or feedstock conditions. The production grade conditions $\mathcal{G}_g$ are by definition given by specific sets of conditions (z) that obey the corresponding constraints:

$$\mathcal{G}_g = \{(x,u) \in \mathcal{F} \mid \exists z = C_x x + C_u u, \text{ such that} \\ g_g(z) < 0\} \tag{4}$$

where g_g defines the quality bounds of grade g. For example, in a distillation plant, these would be lower and upper limits on the purity. The different production grades are connected via process transitions. We define a transition T_g^h from an element $(x^g, u^g) \in \mathcal{G}_g$ to an element $(x^h, u^h) \in \mathcal{G}_h$ as a quadruple (x, u, y, z) satisfying

plant	init. cond.	end cond.	constraints
$\dot{x} = f(x,u,y)$	$x(0) = x^g$,	$x(T) = x^h$,	$h(z) < 0$
$0 = g(x,u,y)$	$u(0) = u^g$	$u(T) = u^h$	
$z = C_x x + C_u u$			

(5)

for some $T > 0$. Let $\mathcal{T}_g^h$ be the set of all these transitions. The sets $\mathcal{G}_g$ and $\mathcal{G}_h$ are said to be compatible if there exists a transition T_g^h from any $(x^g, u^g) \in \mathcal{G}_g$ to any $(x^h, u^h) \in \mathcal{G}_h$. Finally, all sequences of transitions are feasible if all pairs $[g, h]$ are compatible. The verification of such conditions is not straightforward, only for specific cases there may exist a computationally feasible approach. We assume that during the static production tasks, the operating conditions are determined according to the *maximization* of an economic criterion. As the basic economic criterion we will use the *added value*. Let C_r and Y_e denote the consumption of raw materials and utilities r, and the production of end product e, respectively. We assume these are given as functions of the performance variables: $C_r = C_r(z)$, $Y_e = Y_e(z)$. Then, in case of a static production task, the added value rate is given by

$$L(z) := -\sum_r p^{C,r} C_r(z) + \sum_e p^{Y,e} Y_e(z) \tag{6}$$

where $p^{C,r}$ and $p^{Y,e}$ are the instantaneous prices of respectively raw material r and the product e. Using this expression for the added value and disregarding the effect of noise and disturbances, the optimal operating conditions for grade g are found by solving the following, static optimization problem

$$(\bar{x}^g, \bar{u}^g) = \text{argmin}\,\{-L(z) \mid \exists z = C_x x + C_u u \text{ such that } (x,u) \in \mathcal{G}_g\} \quad (7)$$

and the corresponding raw material and product flows are given by $C_r^g = C_r(\bar{z}^g)$ and $Y_e^g = Y_e(\bar{z}^g)$, where $\bar{z}^g = C_x\bar{x}^g + C_u\bar{u}^g$. The dynamic transition tasks are designed to be the solution of a dynamic optimization with an economic criterion as performance index. We also require that the process control system governing the implementation of process transitions is based on dynamic optimization involving the same criterion. A general formulation of the corresponding trajectory optimization problem is given as follows

$$\min_{T,u\in\mathcal{U}} \left\{ \int_0^T L^d(z)dt \,\middle|\, \exists x, y, z, \text{ s.t. } (x,u,y,z) \in \mathcal{T}_{\bar{g}}^{\bar{h}}, \; x(0) = \bar{x}^g, \, x(T) = \bar{x}^h \right\} \quad (8)$$

where L^d is chosen so as to represent the economics of the transition. In case of a **demanding market**, the main incentive is to maximize production. Then transition times should be limited to the minimum. Transition costs are of secondary importance. In case of **low demand** and moderately filled order books, there is no incentive to minimize the transition times. Instead, the difference between the transition revenues and the transition costs should be maximized. This leads to a truly economic transition optimization, where $L^d(z)$ is selected to represent the negative added value rate during the transition. The grade change optimization problem is hard to solve using standard gradient-based numerical optimization techniques due to the fact that the production rate of a certain grade depends discontinuously on the quality variables. One possible approach uses a smooth approximation of the definition of the grade region and exploits the structure of the problem in the definition of a Nonlinear Programming (NLP) based inner loop optimization to compute accurate search directions. A second approach uses integer variables to describe the grade regions and solves a sequence of MILP's to converge to a solution. The grade change problem is non convex and both approaches can be expected to converge to local minima only. However they are believed to do so much faster than the conventional control parametrization method.

The **company-market interaction** is represented here by a transactions-based framework. The sales actions are **(1) orders** originating from long term sales contracts or short term commitments to customers; the main order attributes are the quantity, the price and the time span during which the order can be delivered; **(2) opportunities** arising from predictions of the market-

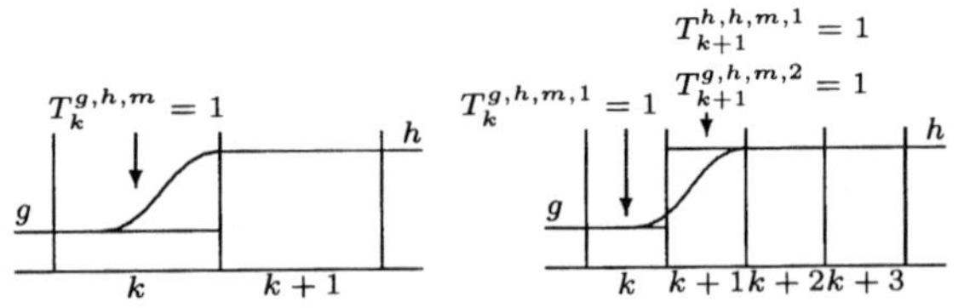

Fig. 6. Representation of transition tasks in UDM framework in case (left) transition time is smaller than τ, and (right) transition time is smaller than 2τ

company interaction made by the sales decision makers. In opportunities they express their estimate of future sales deals. The attributes of the opportunities are the same as those of the orders. The representation of the time axis is discussed by (Zentner *et al.*, 1998). Here we adopt the *Uniform Discretization of time Modeling* (UDM) framework, in which the horizon is divided into a finite number of time slots of uniform length. Only at the beginning of each interval changes may occur. The choice of the interval length, denoted τ in the sequel, is a trade-off between solution resolution and computational tractability. Transition tasks can occur within a single time interval or within more than one, as shown in figure 6. Here we will only consider the first case. We introduce decision variables G_k^g, $g = 1,\dots,n_g$ as the *production decision* variables, where $G_k^g = 1$ means that production task g is started at the beginning of the k-th time span and executed during this time span. Only one task can be performed at the same time:

$$\sum_g G_k^g = 1. \quad (9)$$

To model the transitions we introduce variables $T_k^{g,h,m}$ which, if equal to 1, indicate that a transition from grade g to h and of mode[3] m is executed at the end of time span k. $T_k^{g,h,m}$ relates to G_k^g and G_{k+1}^h in the following manner.

$$\sum_g \sum_h \sum_m T_k^{g,h,m} = 1, \; \sum_m T_k^{g,h,m} \le G_k^g, \; \sum_m T_k^{g,h,m} \le G_{k+1}^h. \quad (10)$$

Further, let TM_k^m be one if transition mode m is executed in time span k and zero otherwise. Then the following holds:

$$T_k^{g,h,m} < TM_k^m. \quad (11)$$

Because all transition times are shorter than τ we need to take the quasi-static production preceding the transition into account to arrive at the appropriate production attributes (raw material consumption and end product yield). Let the transition time for a mode m transition from grade g to grade h be denoted $\tau^{g,h,m}$. Then, the

---

[3] The transition mode m refers to the market scenario for which the transition characteristics are computed.

consumption of raw material r and the production of end product e for the corresponding transition are computed as

$$TC_r^{g,h,m} = (\tau - \tau^{g,h,m})C_r^g + C_r^{g,h,m}, \quad (12)$$

$$TY_e^{g,h,m} = (\tau - \tau^{g,h,m})Y_e^g + Y_e^{g,h,m}. \quad (13)$$

For 'transitions' $T_k^{g,g,m}$ the transition time is zero which yields the corresponding steady state production figures: $TC_r^{g,g,m} = \tau C_r^g$ and $TY_e^{g,g,m} = \tau Y_e^g$. C_r^g, Y_e^g, $C_r^{g,h,m}$, and $Y_e^{g,h,m}$ are the consumption and yield variables as defined in the previous section. To formulate the *transaction-based market description*, a binary decision variable $S_k^{e,s} \in \{0,1\}$ is used to indicate whether sales order/opportunity s for end product e is executed in time span k (in which case $S_k^{e,s} = 1$) or not ($S_k^{e,s} = 0$). Further, for each order/opportunity we introduce a set of time spans $\Omega^{e,s}$ outside which it may not be executed. Attached to each sales order/opportunity the market database stores the amount of product, $SA^{e,s}$ and the unit price offered, $S\$^{e,s}$. By definition of $\Omega^{e,s}$ we have:

$$S_k^{e,s} = 0, \quad \forall k \notin \Omega^{e,s}. \quad (14)$$

Further, each order must be executed exactly once and each opportunity at most once. This gives rise to the following constraints

$$SO^{e,s} \leq \sum_{k \in \Omega^{e,s}} S_k^{e,s} \leq 1, \quad (15)$$

where $SO^{e,s}$ is zero if s is an opportunity and one if s is an order. Similar reasoning for purchasing orders and opportunities leads to the introduction of binary decision variables $P_k^{r,p}$ and the constraints

$$P_k^{r,p} = 0, \quad \forall k \notin \Omega^{r,p}, \quad (16)$$

$$PO^{r,p} \leq \sum_{k \in \Omega^{r,p}} P_k^{r,p} \leq 1, \quad (17)$$

where $\Omega^{r,p}$ is defined as the set of time spans k outside which purchase ord./opp. p may not be executed and $PO^{r,p}$ is zero if transaction p for raw material r is a purchase opportunity and 1 if it is an order. The purchase attributes are $PA_k^{r,p}$ and $\$P_k^{r,p}$: the amount of feedstock r and its unit price for purchase ord./opp. p. Let us introduce ES_k^e as the storage level of end-product e at the beginning of time span k and RS_k^r as the storage level of raw material r at the beginning of time span k. The *material balances* for the raw material storage and the end-product storage are defined as follows

$$ES_{k+1}^e = ES_k^e + \sum_g \sum_h \sum_m T_k^{g,h,m} TY_e^{g,h,m} - \sum_s S_{k+1}^{e,s} SA^{e,s}, \quad (18)$$

$$RS_{k+1}^r = RS_k^r - \sum_g \sum_h \sum_m T_k^{g,h,m} TC_r^{g,h,m} + \sum_p P_k^{r,p} PA^{p,r} \quad (19)$$

with initial conditions $ES_1^e = ES_{initial}^e$ and $RS_1^r = RS_{initial}^r$. Note that the recursive formulation is conservative. Minimum and maximum constraints on the *storage capacity* can be imposed as follows

$$ES_l^e \leq ES_k^e \leq ES_u^e, \quad (20)$$

$$RS_l^r \leq RS_k^r \leq RS_u^r. \quad (21)$$

In a simple form, the cumulative added value (CAV) is the difference between all revenues through sales of end products and all expenses on raw materials over a certain period of time. Here, the objective is defined as the CAV extended by interest on the capital balance, and can be formulated as follows

$$V_{k+1} = (1+\gamma)V_k + \sum_e \sum_s S_{k+1}^{e,s} SA_S^{e,s} \$^{e,s} - \sum_r \sum_p P_{k+1}^{r,p} PA^{r,p} P\$^{r,p} \quad (22)$$

$$J = V_H + \sum_r RS_{r,H} RV_r + \sum_e ES_{e,H} EV_e \quad (23)$$

where γ is the fractional interest rate and H is the horizon length. The last two terms in (23) account for end-storage appreciation which is necessary because we solve a finite horizon approximation of an infinite (or much longer)-horizon problem. The number of binary variables changes with the choice of the transition modeling. A formulation with multiple, distinct transition modes leads to the introduction of $H(n+m)$ binary variables *only* for the production modeling. Additional binary variables arise from the modeling of the purchase and sales actions. Their number can be limited through an appropriate choice of the validity sets $\Omega^{e,s}$ and $\Omega^{r,p}$.

Because the objective as well as all constraints are linear and the total set of variables contains binary as well as continuous variables, the scheduling problem is a MILP:

maximize the objective J (23) subject to the objective recursion (22), the storage constraints (20, 21), the inventory recursion (18, 19), the purchase constraints (16, 17), the sales constraints (14, 15), the transition constraints (10, 11), and the grade constraints (9).

This MILP is, like most scheduling problems, NP-hard. This means that no polynomial time algorithm has been found for solving the problem. For NP-hard problems one can in general only hope that an acceptable solution be found in a reasonable time.

4. CASE STUDY: GAS PHASE HDPE PLANT

We consider a nonlinear model of an industrial high density poly-ethylene (HDPE) fluidized bed

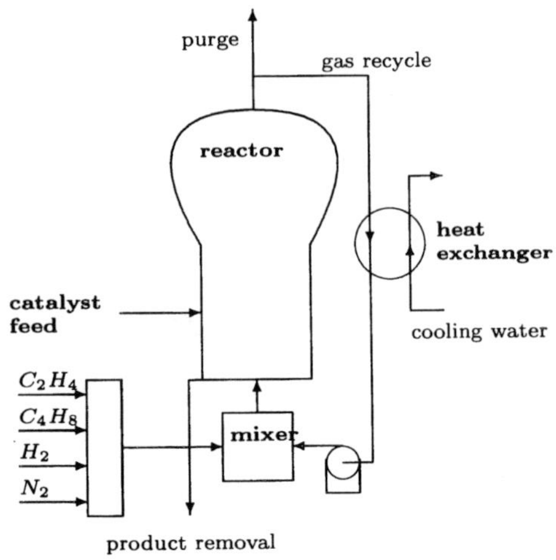

performance channels (z)		inputs (u)	
1	polymer density	1	butylene feed flow
2	polymer melt index	2	hydrogen feed flow
3	production flow	3	catalyst feed flow
4	cooling water flow	4	setpoint pressure control
5	ethylene feed flow		
6	butylene feed flow		
7	hydrogen feed flow		
8	catalyst feed flow		
9	setpoint pressure control		

Fig. 7. Schematic process flow sheet of the gas phase HDPE polymerization process

Table 1. The five production grades for the HDPE reactor.

g	constraints	constraints
1	$932.8 < z^1 < 933.2$	$0.8 < z^2 < 1.2$
2	$937.8 < z^1 < 938.2$	$0.8 < z^2 < 1.2$
3	$937.8 < z^1 < 938.2$	$2.8 < z^2 < 3.2$
4	$932.8 < z^1 < 933.2$	$2.8 < z^2 < 3.2$
5	$942.8 < z^1 < 943.2$	$0.8 < z^2 < 1.2$

reactor, schematically described in figure 4. Here we consider 5 different product grades. The product grades are defined by bounds on the density (z^1) and the natural logarithm of the melt index (z^2) of the polymer as given in Table 1. The model is based on (Choi and Ray, 1985) and (McAuley, 1991) and further refined by Tousain and Van Brempt [4]. It contains about 1000 variables amongst which 40 differential variables. The model has the following properties:

- The fluidized bed consists of a bubble phase, an emulsion gas phase and an emulsion polymer phase. Both emulsion phases are assumed perfectly mixed. Mass transfer between the bubble phase and the emulsion gas phase is modeled using Fick's law.
- Co-polymerization reactions (butylene comonomer) occur on the surface of Ziegler-Natta catalyst particles in the emulsion polymer phase. The following reactions are modeled according to Ziegler-Natta kinetics (Dotson *et al.*, 1996): 1. catalyst activation, 2. chain initiation, 3. chain propagation, 4. chain transfer, 5. catalyst deactivation.
- On top of the reactor a wide gas cap is modeled as being ideally mixed.
- The counter current heat exchanger is modeled using a multi-compartment model.
- A purge outlet flow makes it possible to remove nitrogen (and other gases) from the reactor.
- Four low-level PI controllers stabilize the process
- A flow driven representation is used.

[4] IMPACT (Eureka project number E! 2063), IPCOS Technology, Leuven, Belgium

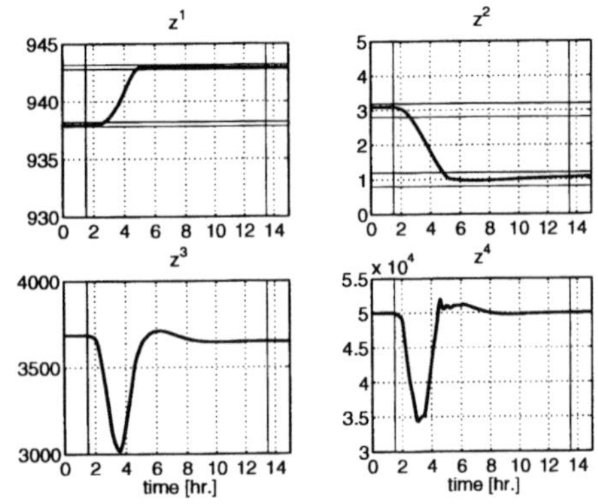

Fig. 8. (SSQP) Optimal trajectories of the density (z^1), melt index (z^2), production (z^3) and cooling water flow (z^4) for a transition from grade 3 to grade 5

The general mathematical formulation of the grade change problem is given by (8). For the HDPE process, the economic objective is defined as

$$V_{ec} = \int_0^T \left(\sum_{r=1}^{3} p_F^r C_r(z) - \sum_{e \in \{g,h,6\}} p_P^e Y_e(z) \right) dt, \tag{24}$$

where the feed flows are given by $C_1(z) = z^5$, $C_2(z) = z^6$, and $C_3(z) = z^7$. Y_g and Y_h refer to the production flows of the departure grade and the target grade, respectively. Y_6 refers to the production of off-spec material. It is assumed that during a transition, all material that is not within the specifications of either one of the grades connected by the transition is *off-spec* material. With the product prices, chosen according to a representative market scenario, 20 grade transitions have been optimized. These are used to construct the production database for the scheduler. A typical optimized transition result is shown in Figure 8.

A uniform discretization of time is chosen with production intervals of 12 hours. All transition times being smaller than 12 hours, we can use the single-interval transition model using production data which is computed on the basis of the static optimization of the grades and the dynamic optimization of grade transitions using (12,13). In

Table 2. Sales order and opportunity database for the HDPE production facility.

Sales of grade 1

s	$SA^{1,s}$	$S\$^{1,s}$	$\Omega_{1,s}$	$SO^{1,s}$
1	300,000	0.7	{8}	1
2	150,000	1.55	{15,22}	0
3	200,000	0.71	{15,22}	1
4	250,000	0.72	{43,50}	1
5	180,000	1.20	{71,78}	0

Sales of grade 2

s	$SA^{2,s}$	$S\$^{2,s}$	$\Omega_{2,s}$	$SO^{2,s}$
1	250,000	0.70	{8,15}	1
2	200,000	1.31	{22,29}	0
3	250,000	1.60	{28,35}	0
4	150,000	1.20	{43,50,57}	0
5	200,000	0.65	{64,71}	1

Sales of grade 3

s	$SA^{3,s}$	$S\$^{3,s}$	$\Omega_{3,s}$	$SO^{3,s}$
1	250,000	1.60	{8,15}	0
2	300,000	0.70	{22,29}	1
3	200,000	1.65	{50,57}	0
4	150,000	1.30	{64,65}	0

Sales of grade 4

s	$SA^{4,s}$	$S\$^{4,s}$	$\Omega_{4,s}$	$SO^{4,s}$
1	250,000	0.70	{8}	1
2	200,000	1.15	{22,29,36}	0
3	300,000	0.71	{36,43,50}	1
4	150,000	1.20	{57,64,71}	0

Sales of grade 5

s	$SA^{5,s}$	$S\$^{5,s}$	$\Omega_{5,s}$	$SO^{5,s}$
1	200,000	0.60	{8,15}	1
2	250,000	1.35	{15,22,29}	0
3	200,000	1.50	{29,36,43}	0
4	220,000	0.73	{50,57,64}	1
5	150,000	1.25	{71,78}	0

Sales of off-spec material

s	$SA^{6,s}$	$S\$^{6,s}$	$\Omega_{6,s}$	$SO^{6,s}$
1	100,000	0.32	{8,..,36}	0
2	100,000	0.33	{43,..,78}	0

Table 3. Raw material storage attributes for the HDPE plant.

r	RS^r_l	RS^r_u	RS^r_1	$RS^r_{H,l}$	$R\$^r$
1	0	10,000,000	1,100,000	600,000	0.45
2	0	450,000	60,000	35,000	0.60
3	0	2,000	250	100	3.30

addition, a fictional order/opportunity database has been constructed such that about 80% of the production capacity is used for production orders, the remaining is to be used for attractive sales opportunities. The sales order/opportunity database is given in Table 2. The sets of time instances at which transactions can take place are chosen sparse in order to limit the number of binary decision variables. The order/opportunity database should be seen as being the result of previous interactions of sales managers with the market. Initial inventory levels, the storage constraints and the end-storage appreciation of respectively the raw materials and the end products are finally given in tables 3 and 4. In traditional slate scheduling, the slate is going back and forth through a sequence of grades. The sequence is determined in such a way that the overall grade change effort, which can be characterized for example by grade change time or off-spec production, is minimized. Here, we exploit that further improvements can be achieved if, by means of advanced process control technology, all transitions are enabled, leading to a flexible scheduling approach. In this approach the production is allowed to switch to any other grade at each production interval, providing much improved flexibility to deal with changes in the market situation and to react to attractive sales opportunities. A solution to the flexible scheduling problem was found in several hours and to a guaranteed optimality of slightly more than 2.3 %. The objective value was equal to 3,339,845 which is over 16.8 % better than

Table 4. End product storage attributes for the HDPE plant.

e	ES^e_l	ES^e_u	ES^e_1	$ES^e_{H,l}$	$E\$^e$
1	0	5,000,000	325,000	300,000	1.02
2	0	5,000,000	315,000	300,000	0.98
3	0	5,000,000	270,000	300,000	1.06
4	0	5,000,000	270,000	300,000	1.05
5	0	5,000,000	306,000	300,000	1.03
6	0	5,000,000	22,000	0	0.33

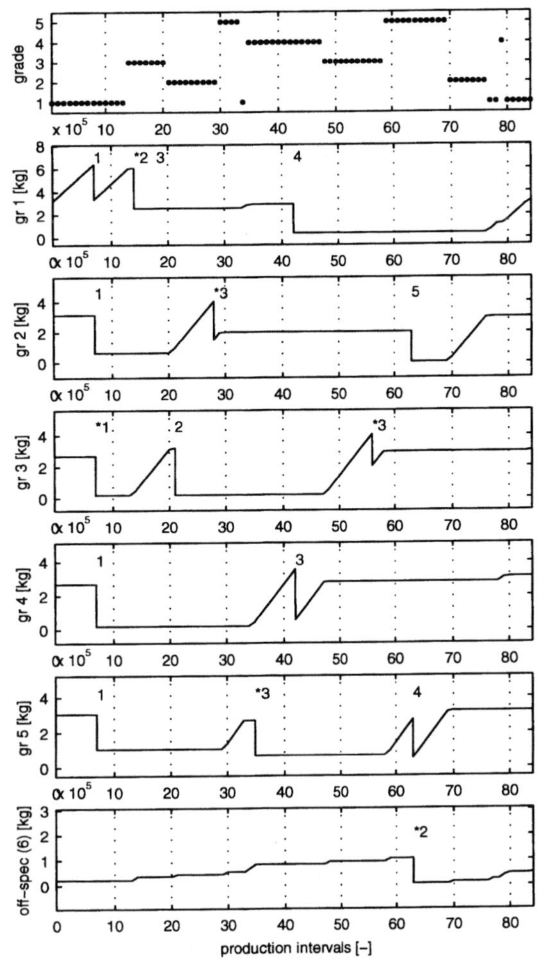

Fig. 9. Flexible schedule for HDPE production. Grades (top) and storage levels of the end-products (bottom)

the performance achieved with a fixed-duration slate. The result is displayed in figure 9.

5. CLOSED-LOOP STOCHASTIC MPC

In this section a closed-loop stochastic model-predictive control formulation will be discussed. The formulation is model-based and predictive, with constraints on the process variables. The approach is developed using the following observations:

- The smooth dynamic model (1) is assumed to be available, enabling to determine numerically a trajectory along which a time-varying linearized model can be computed. The trajectory is initially given by the scheduler and is subsequently improved in an iterative implementation of the closed-loop MPC formulation (CLMPC)
- CLMPC exploits the availability of future measurements, i.e. future feedback, and derives desirable feedback properties from this structure
- CLMPC considers stochastic disturbances, here assumed as Gaussian
- The constraints are taken into consideration with a backoff resulting from the effect of the stochastic disturbances
- In (van Hessem and Bosgra, 2003) we have shown that for any controller of the CLMPC there exists an equivalent finite horizon LQG controller (FHLQG) having the same performance and which can be implemented in a receding horizon mode
- The receding horizon CLMPC controller implementation inherits the stability properties of the optimal CLMPC solution
- In (van Hessem and Bosgra, 2002*a*),(van Hessem and Bosgra, 2002*b*) it has been shown how a Youla parametrization that renders CLMPC convex can be replaced by an innovations feedback controller structure which allows the solution to be put in a receding horizon formulation. It has an observer-based structure as shown in figure 11.
- For its actual solution, the CLMPC problem is decomposed into a deterministic constrained optimization problem and a minimum variance problem
- The solution of the latter provides the back-off used in the former
- The idea is shown in figure 10 based on a strong concept of classical LQG control (Athans, 1971). The deterministic optimization using the back-off is implemented as feedforward, the minimum variance controller provides the feedback. In fact any (robust) feedback controller with desirable disturbance-suppressing properties can be used here.

Consider a linear discrete time-varying stochastic system derived from the model (1):

$$\begin{pmatrix} x_{k+1}(\xi) \\ z_k(\xi) \\ y_k(\xi) \end{pmatrix} = \begin{pmatrix} A_k & G_k & B_k \\ C_k^z & O & D_k^z \\ C_k & F_k & O \end{pmatrix} \begin{pmatrix} x_k(\xi) \\ w_k(\xi) \\ u_k(\xi) \end{pmatrix} \tag{25}$$

where ξ is a Gaussian random process, $w_k(\xi)$ is a white noise sequence with variance matrix W_k and with the property $G_k W_k F_k^T = 0$ (process and measurement noise are independent). In 'lifted' form (see (Furuta and Wongsaisuwan, 1993)) over a time horizon of n samples, the vector-stochastic processes are

$$\mathbf{y}_k(\xi) = \begin{pmatrix} y_k(\xi) \\ \vdots \\ y_{k+n}(\xi) \end{pmatrix}, \mathbf{z}_k(\xi) = \begin{pmatrix} z_k(\xi) \\ \vdots \\ z_{k+n}(\xi) \end{pmatrix} \tag{26}$$

and so on, representing each signal $\mathbf{y}, \mathbf{z}$ from sample k to sample $k+n$. The two stochastic processes $\mathbf{y}, \mathbf{z}$ are the measured and performance output process, respectively; $\mathbf{z}$ contains variables appearing in the objective function or in the constraints, or in the inputs.

$$\mathbf{y}_k(\xi) = G_{yx} x_k(\xi) + G_{yu} \mathbf{u}_k(\xi) + G_{yw} \mathbf{w}_k(\xi)$$
$$\mathbf{z}_k(\xi) = G_{zx} x_k(\xi) + G_{zu} \mathbf{u}_k(\xi) + G_{zw} \mathbf{w}_k(\xi)$$

Define a finite horizon observer, having time varying-feedback gains

$$\begin{pmatrix} \hat{x}_0(\xi) \\ \hat{x}_1(\xi) \\ \vdots \\ \hat{x}_n(\xi) \end{pmatrix} = \begin{pmatrix} I \\ \Phi^e_{1,0} \\ \vdots \\ \Phi^e_{n,0} \end{pmatrix} \hat{x}_0(\xi) \tag{27}$$
$$+ \begin{pmatrix} O & O & \cdots & O \\ B_0 & O & \cdots & O \\ \vdots & \vdots & \ddots & \vdots \\ \Phi^e_{n,1}B_0 & \Phi^e_{n,2}B_1 & \cdots & O \end{pmatrix} \begin{pmatrix} u_0(\xi) \\ u_1(\xi) \\ \vdots \\ u_n(\xi) \end{pmatrix}$$
$$+ \begin{pmatrix} O & O & \cdots & O \\ N_0 & O & \cdots & O \\ \vdots & \vdots & \ddots & \vdots \\ \Phi^e_{n,1}N_0 & \Phi^e_{n,2}N_1 & \cdots & O \end{pmatrix} \begin{pmatrix} y_0(\xi) \\ y_1(\xi) \\ \vdots \\ y_n(\xi) \end{pmatrix}$$

where the transition matrix $\Phi^e_{k,j}$ for the observer system mapping $\hat{x}_j$ to $\hat{x}_k$ is given for $k > j$ by

$$\Phi^e_{k,j} = A^e_{k-1}A^e_{k-2}\cdots A^e_j, \quad \Phi^e_{j,j} = I, \quad A^e_k = A_k - N_k C_k$$

and N_k is the Kalman predictor gain. Define

$$e_k(\xi) := x_k(\xi) - \hat{x}_k(\xi) \tag{28}$$

which recursively satisfies

$$e_{k+1}(\xi) = \underbrace{(A_k - N_k C_k)}_{A^e_k} e_k(\xi) + \underbrace{(G_k - N_k F_k)}_{G^e_k} w_k(\xi) \tag{29}$$

In lifted form: $\mathbf{e}_0(\xi) = G_{ee} e_0(\xi) + G_{ew} \mathbf{w}_0(\xi)$ and in terms of $(\Phi^e_{k,l}, G^e_k)$:

$$\begin{pmatrix} e_0(\xi) \\ e_1(\xi) \\ \vdots \\ e_n(\xi) \end{pmatrix} = \begin{pmatrix} I \\ \Phi^e_{1,0} \\ \vdots \\ \Phi^e_{n,0} \end{pmatrix} e_0(\xi) \tag{30}$$

$$+ \begin{pmatrix} O & O & \cdots & O \\ G^e_0 & O & \cdots & O \\ \vdots & \vdots & \ddots & \vdots \\ \Phi^e_{n,1}G^e_0 & \Phi^e_{n,2}G^e_1 & \cdots & O \end{pmatrix} \begin{pmatrix} w_0(\xi) \\ w_1(\xi) \\ \vdots \\ w_n(\xi) \end{pmatrix}$$

The corresponding innovation sequence (Kailath, 1968), is given by

$$v_k(\xi) := y_k(\xi) - \hat{y}_k(\xi) = C_k e_k(\xi) + F_k w_k(\xi)$$

In lifted form

$$\mathbf{v}(\xi) = G_{ve} e_0(\xi) + G_{vw}\mathbf{w}(\xi) \tag{31}$$

and in terms of the matrices $(\Phi^e_{k,l}, G^e_k, C, F)$

$$\begin{pmatrix} v_0(\xi) \\ v_1(\xi) \\ \vdots \\ v_n(\xi) \end{pmatrix} = \begin{pmatrix} C_0 \\ C_1\Phi^e_{1,0} \\ \vdots \\ C_n\Phi^e_{n,0} \end{pmatrix} e_0(\xi) \tag{32}$$

$$+ \begin{pmatrix} F_0 & O & \cdots & O \\ C_1G^e_0 & O & \cdots & O \\ \vdots & \vdots & \ddots & \vdots \\ C_n\Phi^e_{n,1}G^e_0 & C_n\Phi^e_{n,2}G^e_1 & \cdots & F_n \end{pmatrix} \begin{pmatrix} w_0(\xi) \\ w_1(\xi) \\ \vdots \\ w_n(\xi) \end{pmatrix}$$

In (van Hessem and Bosgra, 2003) it has been shown that for the optimal CLMPC problem, an optimal output feedback controller can be equivalently replaced by an optimal innovations feedback controller. Let this be the law (assuming zero reference trajectories w.l.o.g.)

$$\mathbf{u}_0(\xi) = K_0\mathbf{v}_0(\xi) \tag{33}$$

where $K_0 \in$ is a non-anticipative controller, i.e.

$$K_0 = \begin{pmatrix} K^{11} & O & \cdots & O \\ K^{21} & K^{22} & \cdots & O \\ \vdots & \vdots & \ddots & \vdots \\ K^{n1} & K^{n2} & \cdots & K^{nn} \end{pmatrix}.$$

Then, at any time instant in the future, there exist matrices L_k and K_k of appropriate dimension, such that the control law

$$\mathbf{u}_k(\xi) = L_k\hat{x}_k(\xi) + K_k\mathbf{v}_k(\xi) \tag{34}$$

generates the optimal control sequence for the remainder of the horizon, provided the matrices are appropriately parametrized. This is a solution to a receding horizon implementation, using a recursive expression for the variance matrices and the feedback law. Observe that the actual controls applied at each instant are given by L^1_k and K^{11}_k which are the first n_u rows of the control law L_k and K_k respectively:

$$u_k = L^1_k\hat{x}_k + K^{11}_k v_k$$

From the preceding analysis it follows that $\mathbf{z}_k(\xi)$ in closed-loop is some function of $\hat{x}_k(\xi), e_k(\xi), \mathbf{w}_k(\xi)$. Since $\hat{x}_k(\xi), e_k(\xi)$ are independent of $\mathbf{w}_k(\xi)$ it follows that we only need to keep track of the joint

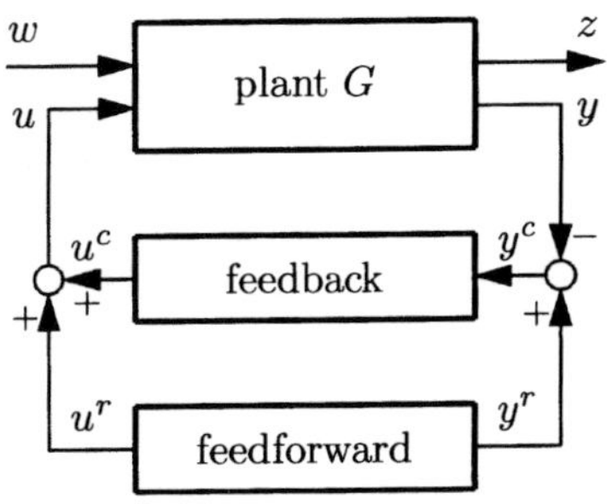

Fig. 10. Generalized plant for MPC

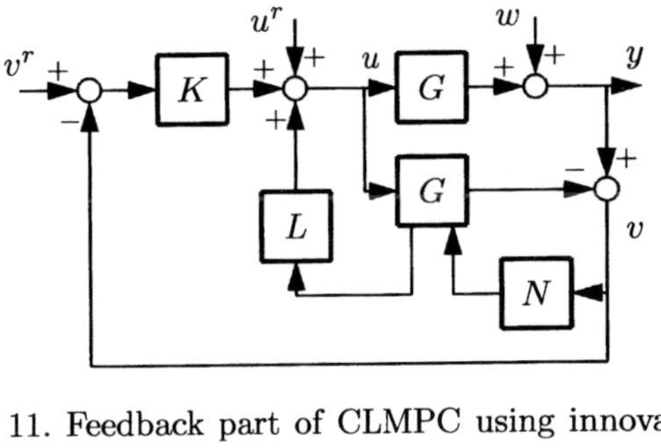

Fig. 11. Feedback part of CLMPC using innovations and state feedback

variance matrix V_k of $\hat{x}_k(\xi)$ and $e_k(\xi)$, generated by

$$\begin{pmatrix} \hat{x}_{k+1}(\xi) \\ e_{k+1}(\xi) \end{pmatrix} = \begin{pmatrix} A_k + B_kL^1_k & (N_k + B_kK^{11}_k)C_k \\ O & A_k - N_kC_k \end{pmatrix} \begin{pmatrix} \hat{x}_k(\xi) \\ e_k(\xi) \end{pmatrix} + \begin{pmatrix} (N_k + B_kK^{11}_k)F_k \\ G_k - N_kF_k \end{pmatrix} w_k(\xi)$$

and recursively given by

$$V_{k+1} = \begin{pmatrix} A_k + B_kL^1_k & (N_k + B_kK^{11}_k)C_k \\ O & A_k - N_kC_k \end{pmatrix} V_k \begin{pmatrix} \star & \star \\ \star & \star \end{pmatrix}^T + \begin{pmatrix} (N_k + B_kK^{11}_k)F_k \\ G_k - N_kF_k \end{pmatrix} W_k \begin{pmatrix} \star \\ \star \end{pmatrix}^T, \quad V_0 = \begin{pmatrix} O & O \\ O & P_0 \end{pmatrix}$$

As $e_k(\xi)\perp\hat{x}_k(\xi)$ for any k, the joint variance matrix V_k is block-diagonal for each k by construction. Thus the variance matrices follow from the Riccati recursions for the estimation error

$$P^e_{k+1} = A_kP^e_kA^T_k - N_k(C_kP^e_kC^T_k + F_kW_kF^T_k)N^T_k + G_kW_kG^T_k$$

having boundary condition $P^0_e = P_0$. The recursion for the state-estimate

$$P^{\hat{x}}_{k+1} = (A_k + B_kL^1_k)P^{\hat{x}}_k(A_k + B_kL^1_k)^T + (N_k + B_kK^{11}_k)(C_kP^e_kC^T_k + F_kW_kF^T_k)(N_k + B_kK^{11}_k)^T$$

having boundary condition $P^{\hat{x}}_0 = O$. L^1_k and K^{11}_k are given externally in every cycle by the solution of the control problem. The Kalman gain is

$$N_k = A_kP^e_kC^T_k(C_kP^e_kC^T_k + F_kW_kF^T_k)^{-1}$$

The factored variance matrix of the initial condition and disturbances are

$$P^x_k = Ex_k(\xi)x_k(\xi)^T = P^{\hat{x}}_k + P^e_k = F_xF^T_x, \quad W_k = F_wF^T_w.$$

On the basis of the previous development, the closed-loop MPC problem is defined as

$$\textbf{(CLMPC)} \quad \begin{matrix} \min\limits_{\mathbf{z}^r_k \in \mathbb{R}^{n_u}, K \in \mathbf{K}_0} & f(\mathbf{z}^r_k) \\ r\sqrt{h^T_j Z_k h_j} + h^T_j \mathbf{z}^r \leq g_j & \end{matrix} \tag{35}$$

where $Z = E(\mathbf{z}_k(\xi) - \mathbf{z}_k^r)(\mathbf{z}_k(\xi) - \mathbf{z}_k^r)^T$. and f is a convex function. The constraints above follow from the nominal inequality constraints which trajectories must obey:

$$h_j^T \mathbf{z}^r \leq g_j, \quad j = 1 \ldots n_z$$

The extra term $r\sqrt{h_j^T Z_k h_j}$ in CLMPC is added as back-off to the constraints using an ellipsoidal relaxation, with the intention to prevent violation of constraints under stochastic disturbances. This problem is a convex optimization problem that can be solved for the global optimum using a solver for second-order cone problems, (Lobo *et al.*, 1998) . However, it is a computationally expensive problem in its present format for medium- and large-sized problems with many constraints and long prediction horizons, due to the present state of development of state-of-the-art cone programming solvers. These solvers utilize a vectorization of the problem data and given the large number of parameters in the controller K, the size of the problem tends to grow. This especially holds for the number of Lagrange multipliers in primal-dual interior-point methods. One solution to this situation is to derive new algorithms exploiting the ellipsoidal structure of the **(CLMPC)**, see for example (Dabbene *et al.*, 2003). However, there exists a different approach. Given the quadratic objective function

$$E\sum_{k=1}^{n}(z_k(\xi) - z_k^r)^T Q_k (z_k(\xi) - z_k^r) + \Delta u_k(\xi)^T R_k \Delta u_k(\xi) \tag{36}$$

where $\Delta u_k = u_k - u_{k-1}$, and given the fact that the optimal solution to **(CLMPC)** is a finite horizon LQG controller. Then in the absence of any inequality constraints, (36) can be separated into a deterministic optimal control problem and a stochastic optimal feedback problem by the separation property of LQG optimal control. The fact that the inequality constraints are treated with a back-off implies that the unconstrained case is effective for the stochastic problem. This motivates to separate (36) in a stochastic feedback problem and a deterministic feedforward problem. This implies a considerable simplification in the computational load of the problem, and makes real-time implementation a realistic opportunity. Assume that the weighting matrix $R = R^T > 0$ has a full rank Cholesky factorization $F_R F_R^T$. R and S are weighting parameters in the quadratic performance index in an LQG problem. The first step is to solve for given F_R, S, s the minimal variance problem (subproblem $\mathbf{CFHLQG}^A$):

$$\min_{K \in \mathbf{K}} \ \text{tr } S F_R F_Z F_Z^T F_R^T S^T \tag{37}$$

for

$$F_Z = \begin{pmatrix} G_{zx}F_x & G_{zw}F_u \end{pmatrix} + G_{zu}K \begin{pmatrix} G_{vx}F_x & G_{vw}F_u \end{pmatrix}$$

Suppose one has solved this problem for the optimal $Q^\star$, then the optimal variance matrix $Z^\star$ is also known, and the back-off terms using the ellipsoidal relaxation are readily computed:

$$\nu_j^\star = r\sqrt{h_j^T Z^\star h_j} \tag{38}$$

In the second step (subproblem $\mathbf{CFHLQG}^B$) one solves the optimal transition, in which the terms 38 are used to keep back-off to the constraints:

$$\begin{aligned} &\min_{\mathbf{u}^r} \quad (S\hat{\mathbf{z}} - s)^T R (S\hat{\mathbf{z}} - s) \\ &\nu_j^\star + h_j^T \mathbf{z}^r \leq g_j, j = 1, \ldots, m \\ &\mathbf{z}^r = G_{zx}x_0^r + G_{zu}\mathbf{u}^r + G_{zw}\mathbf{w}^r \end{aligned} \tag{39}$$

where the performance weighting matrix S and vector s have to be chosen. Note that the separation into two subproblems also implies that separate tuning of the feedback part and of the feedforward transition part has been realized. This enables especially to modify the criterion in the second step to include a strong component representing the economic performance of a transition.

6. CASE STUDY (CONTINUED)

Application of a grade change from grade 5 to grade 3 has been investigated for the CLMPC approach. We require the grade change to be done under two bias disturbances: +0.0025 kg/h on the catalyst flow, and −0.05 kg/h on the hydrogen flow, results of imprecise valve positions and errors in flow measurements. The hydrogen bias influences the melt-index and a negative value of this disturbance is chosen to counteract the transition from grade 5 to grade 3. The controller must increase the hydrogen feed to compensate for this bias possibly violating the feed limitation. A positive bias in the catalyst feed leads to a significant increase in energy hold-up. The PI loop in the basic control loop increases the cool water flow threatening to saturate the flow constraint. Additionally, measurement noise is active on all output measurements, such that controller parameters cannot grow unbounded (high gain feedback) to compensate the persistent disturbances, which adds robustness to the control design. The performance index has appropriately chosen quadratic weights on performance variables 1, 2 and 4 and on inputs 1, 2 and 3. The constraints are given in table 5. The computed open-loop performance

Table 5. Constraints on process variables

Variable		lo	up
z_4	[10^4kg/h]	2.5	5.2
z_5	[10^2kg/h]	0.0	1.5
z_6	[kg/h]	0.1	1.0
z_7	[kg/h]	0.2	2.0
z_9	[bar]	0.0	1.0

trajectories of density, melt index, coolwater flow and production rate are shown in figure 12. Without feedback a serious offset results from the computed optimal reference trajectories, due to the active disturbances which are not taken into consideration in the optimization. In figure 13, the

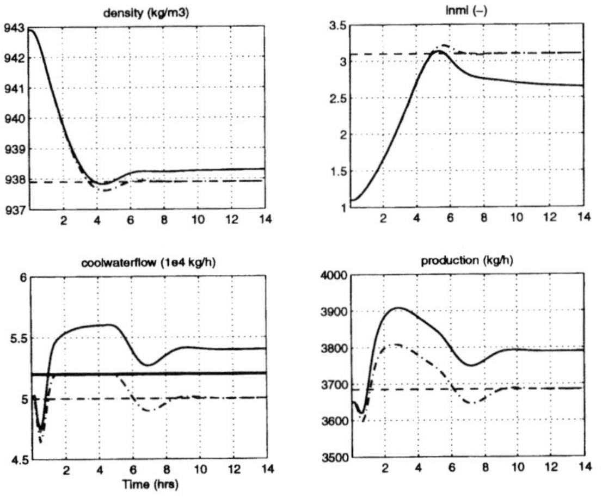

Fig. 12. Open loop: real (solid), reference (dash-dot), target (dash)

performance trajectories of the same performance variables are shown under closed-loop CLMPC feedback. We see no violation of the feedwater flow constraint as a consequence of the appropriate back-off in the computation of the reference trajectories. Figures 14 and 15 shows a close-up

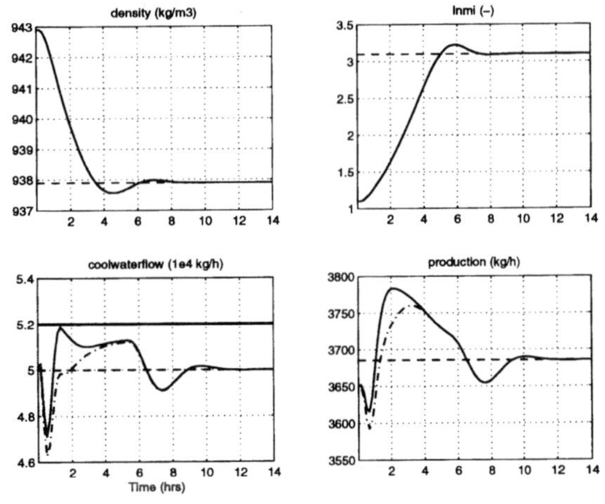

Fig. 13. Closed-loop: real (solid), reference (dash-dot), target (dash)

of the reference trajectories of the cool water and hydrogen feed flows, respectively, where the back-off is visualized via the 1-dimensional confidence intervals (projected ellipsoids). Due to the back-off, inequality constraints play no role in the part of the control moves related to disturbance rejection. Consequently, the control updates needed to remove the biases on the feeds are very smooth. Apart from the real actual trajectories, the error bars in the figures show the evolution of the variance disturbances which are expected (and can be accomodated) by the CLMPC controller.

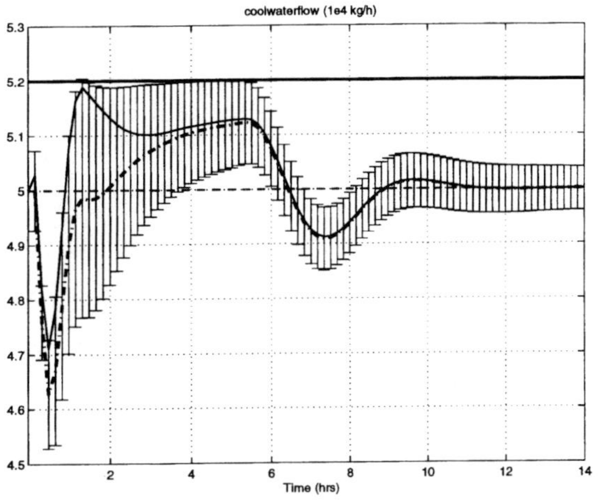

Fig. 14. Closed-loop cool water flow: real (solid), reference (dash-dot), target (dash)

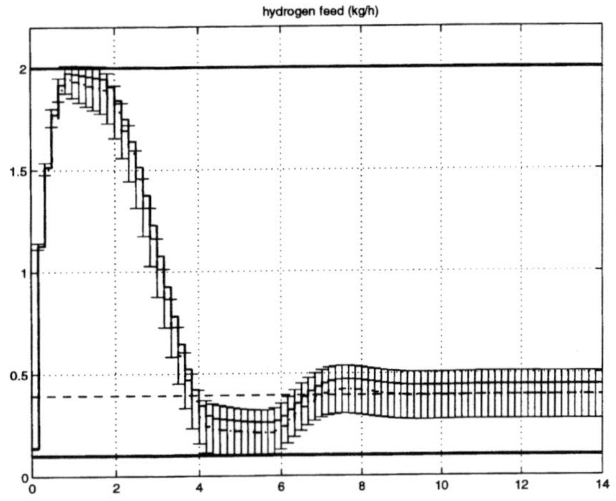

Fig. 15. Closed-loop hydrogen feed flow: real (solid), reference (dash-dot), target (dash)

7. CONCLUSIONS

In this paper a scheduling approach has been presented that enables the production management of a multi-grade chemical plant to operate the plant according to a market-responsive and economics-based objective. For single-machine multi-grade processes, such a scheduler can be designed by capturing the purchasing, production and sales decisions and their effect on the company's objective into a MILP which can be solved using standard software. Process transitions and their effect on material flows are included in the formulation without the need to introduce additional binary variables. Reasonable solution times were encountered for the HDPE production test problem that was considered in this paper. The application of the flexible short term scheduling approach to a gas phase HDPE manufacturing plant for a fictional market situation demonstrates a significant increase in added value compared to the traditional fixed and variable duration slate scheduling.

For the actual realization of grade transitions the paper has presented a new closed-loop MPC formulation that separates a stochastic MPC problem into a deterministic constrained feedforward trajectory optimization solution and a stochastic minimum-variance feedback solution. The stochastic solution provides the back-off needed in the feedforward part. Conversely, the implementation of the back-off allows the feedback control to realise a proper disturbance suppression task. The combined scheduling and closed-loop MPC solution provides a coherent integration of process control tasks covering the days-to-minutes range of time scales.

REFERENCES

Athans, M. (1971). The role and use of the stochastic linear-quadratic-Gaussian problem in control system design. *IEEE Transactions on Automatic Control* **16**, 529–552.

Bassett, M.H., P. Dave, F.J. Doyle III, G.K. Kudva, J.F. Pekny, G.V. Reklaitis, S. Subrahmanyam, D.L. Miller and M.G. Zentner (1996). Perspectives on model based integration of process operations. *Computers and Chemical Engineering* **20**, 821–844.

Choi, K.-Y. and W.H. Ray (1985). The dynamic behaviour of fluidized bed reactors for solid catalysed gas phase olefin polymerization. *Chemical Engineering Science* **40**, 2261–2279.

Dabbene, F., P. Gay and B.T. Polyak (2003). Recursive algorithms for inner ellipsoidal approximation of convex polytopes. *Automatica* **39**, 1773–1781.

Dotson, N.A., R. Galvan, R.L. Laurence and M. Tirrell (1996). *Polymerization Process Modeling.* Wiley - VCH. New York, NY.

Furuta, K. and M. Wongsaisuwan (1993). Closed-form solutions to discrete-time LQ optimal control and disturbance attenuation. *Systems and Control Letters* **20**, 427–437.

Giannelos, N.F. and M.C. Georgiadis (2002). A novel event-driven formulation for short-term scheduling of multipurpose continuous processes. *Industrial and Engineering Chemistry Research* **41**, 2431–2439.

Kailath, T. (1968). An innovations approach to least-squares estimation. I. Linear filtering in additive white noise. *IEEE Transactions on Automatic Control* **13**, 646–655.

Lee, J.H. and N.L. Ricker (1994). Extended Kalman filter based nonlinear model predictive control. *Industrial and Engineering Chemistry Research* **33**, 1530–1541.

Lobo, M.S., L. Vandenberghe and S. Boyd (1998). Applications of second-order cone programming. *Linear Algebra and its Applications* **284**, 193–228.

Marquardt, W., O.H. Bosgra and T. Backx (1998). Towards intentional dynamics in supply chain conscious process operations. In: *Proceedings of the 3rd International Conference on Foundations of Computer Aided Process Operations, Snowbird, Utah, USA.*

McAuley, K.B. (1991). *Modelling, estimation, and control of product properties in a gas phase polypropylene reactor. PhD thesis, McMaster University.* Hamilton, Ontario.

Mendez, C.A. and J. Cerda (2002). An efficient MILP continuous-time formulation for short-term scheduling of multiproduct continuous facilities. *Computers and Chemical Engineering* **26**, 687–695.

Rawlings, J.B. (2000). Tutorial overview of model predictive control. *IEEE Control Systems Magazine* **20**, 38–52.

Reklaitis, G.V. (1992). Overview of scheduling and planning of batch process operations. In: *Proceedings of the Nato ASI on Batch Processing Systems Engineering, Antalya, Turkey.* pp. 660–705.

Sinclair, K.B. (1987). Grade change flexibility - defined, determined, compared. In: *Proceedings of the 5th International Conference on Polyolefins, Houston, Texas, USA.* pp. 2–19.

Tousain, R.L. (2002). *Dynamic optimization in business-wide process control. PhD thesis, Delft University of Technology.* Delft University Press. Delft, The Netherlands.

Towill, D.R. (1996). Industrial dynamics modelling of supply chains. *Int. J. of Physical Distribution and Logistics Management* **26**, 23–42.

van Hessem, D.H. (2004). *Stochastic Inequality constrained closed-loop model predictive control with application to chemical process operation. PhD thesis, Delft University of Technology.* to appear.

van Hessem, D.H. and O.H. Bosgra (2002*a*). Closed-loop stochastic dynamic process optimization under input and state constraints. In: *Proc. Amer. Contr. Conf.*. pp. 2023–2028.

van Hessem, D.H. and O.H. Bosgra (2002*b*). A conic reformulation of Model Predictive Control including bounded and stochastic disturbances under state and input constraints. In: *Proc. IEEE Conf. Dec. Contr.*. pp. 4643–4648.

van Hessem, D.H. and O.H. Bosgra (2003). A full solution to the constrained stochastic closed-loop MPC problem via state and innovation feedback, and its receding horizon implementation. In: *Proc. IEEE Conf. Dec. Contr.*. pp. 929–934.

Zentner, M.G., A. Elkamel and J.F. Pekny (1998). A language for describing process scheduling problems. *Computers and Chemical Engineering* **22**, 125–145.

ELSEVIER
IFAC
PUBLICATIONS
www.elsevier.com/locate/ifac

IDENTIFICATION FOR CONTROL: ACHIEVEMENTS AND OPEN PROBLEMS [1]

Michel Gevers *

* *Center for Systems Engineering and Applied Mechanics (CESAME)*
Université Catholique de Louvain
Bâtiment Euler, 4 Av. Georges Lemaître,
B-1348 Louvain-la-Neuve, Belgium
`Phone: +32-10-472590`
`Email: gevers@csam.ucl.ac.be`

Abstract: This paper presents the author's views on the development of identification for control. The paper reviews the emergence of this subject as a specific topic within the larger discipline of system identification, the major progress accomplished over the last 15 years and its practical impact, as well as the remaining open problems.

Keywords: Identification, identification for control, robust control, optimal design.

1. INTRODUCTION

Identification for control! Identification and control have initially emerged as two separate disciplines, developed by two distinct communities.

Identification has for a long time been the territory of mathematicians, statisticians, time-series analysts and econometricians. The history of system identification goes as far back as the work of Gauss and Legendre in the late 18th and early 19th century. An excellent presentation of the history of system identification can be found in (Deistler, 2002). Control theory has always been the territory of engineers. Until about 1960, most of control design was based on model-free methods, using Bode, Nyquist, or Ziegler-Nichols plots, combined with graphical design techniques. The introduction of state-space models in 1960, together with the solution of optimal control and optimal filtering problems in a Linear Quadratic Gaussian framework (Kalman, 1960*a*; Kalman, 1960*b*), gave birth to a tremendous development of model-based control design methods. Successful applications abounded, particularly in aerospace, where accurate models were readily available.

1965 saw the start of identification activity in the control community. The paper (Ho and Kalman, 1965) set the stage for state space realization theory which, 25 years later, became the founding block for what is now called subspace identification. The paper (Åström and Bohlin, 1965) introduced into the control community the Maximum Likelihood framework for the identification of input-output models. This gave rise to the celebrated Prediction Error (PE) framework (Ljung, 1999) that has since proven so successful. Undoubtedly the advent of identification theory was spurred by a desire to extend the applicability of model-based control design to broader and broader fields of applications, for which no reliable models could be obtained from first principles.

From 1965 to the late eighties, model-based control was applied to ever growing classes of dynamical systems and processes, with models obtained via the newly emerging identification techniques.

[1] This paper presents research results of the Belgian Programme on Interuniversity Attraction Poles, initiated by the Belgian Federal Science Policy Office. The scientific responsibility rests with its author.

The prevailing habit at that time was to separate the identification step and the control design step. A model was identified first using the best available techniques; subsequently, a model-based control design was performed based on the "certainty equivalence principle", i.e. the model was treated as if it represented the true system. Dual control and adaptive control were two early attempts to address the issue of parametric uncertainty and model-based control design in a synergistic way. In dual control the parameter estimation and the control design mechanism are obtained as the result of a single but complex optimization problem. In adaptive control, the parameter adjustment scheme is subsidiary to the control objective. Both schemes were developed for the case where the structure of the true system is assumed to be known, which severely limits their practical applicability. The solution of the dual control problem proved to be computationally intractable, even in the simplest cases. As for adaptive control, the major difficulty is that the parameters of the feedback control system change at every sampling instant, making the closed-loop dynamics nonlinear and their stability analysis extremely complex. After convergence mechanisms had been devised for the ideal case where the system is in the model set, attempts were made to robustify the adaptive control algorithms in order to take account of some modest degree of uncertainty. These attempts essentially consisted of introducing cautionary safeguards in the computation of the gain of the parameter adjustment scheme; see e.g. (Anderson *et al.*, 1986).

One of the main contributions of the control community to system identification theory was to consider identification as an exercise in estimating the *best possible approximate model* within some model set, rather than as a search for the *true system*. Together with this effort came the characterization of the approximate model in terms of bias error and variance error on the estimated transfer functions.

If the model of a system is exact, it is optimal for all applications. However, if the model is only an approximation of the "true system", then the quality of the model should be dependent on the intended application. It thus makes sense to tune the identification towards the objective for which the model is to be used, i.e. to ensure that the distribution of model error is such that it does not deteriorate the objective too much. This gave rise to the paradigm of *goal-oriented identification* and it led one to view identification as a *design problem*. Identification for control has been the major outlet for this new paradigm. The reasons for this are many: (i) control is very often the main motivation for model building; (ii) high performance control can often be achieved with very simple models, provided some basic dynamical features of the system are accurately reflected; (iii) a powerful robust control theory, based on nominal models and uncertainty sets, had been developed all through the eighties, but these models and uncertainty sets were not data-based for lack of a proper theory; (iv) identification for control research led to iterative model and controller tuning tools that were intuitive, practical and easy to implement by the process engineers.

One early approach to optimal identification design for control, that established a direct link between experimental conditions and controller performance, was obtained in (Gevers and Ljung, 1986) by considering variance errors only, i.e. the system was assumed to be in the model set. This approach consists, for a given certainty equivalence control design procedure, of computing the experimental conditions of the identification that minimize the average performance degradation that results from the fact that the controller is computed from an estimated (and hence random) model rather than from the exact true system. Like with all optimal experiment design methods, the optimal experiment depends on the unknown system: see Section 4. Hence, even though such results give useful insights, they do not provide an operational design method for the identification for control problem. In addition, they are based on a certainty equivalence controller design mechanism, rather than on a robust control design.

Except for the methods already mentioned (dual control, adaptive control, control-oriented optimal design), which are restricted to the case where the true system is in the model set, the first contributions in which identification and control design with restricted complexity models were looked upon as a combined design problem appeared only around 1990. The plenary (Gevers, 1991) at the 1991 IFAC Symposium on System Identification addressed many of the key issues; however, it was more an agenda for research than a presentation of solutions. Indeed, there was very little understanding at that time about the interplay between system identification and robust control. The two theories had been developed by separate communities with very little interaction. In the nineties, the activity in identification for control surged: in the plenary (Hjalmarsson, 2003), the author estimated that about 1,500 papers have appeared on the topic of identification for control after the plenary (Gevers, 1991).

When the application of a model is the design of a controller, then what really matters is the performance achieved by this model-based controller on the "true system", not the intrinsic quality of the model. We illustrate this idea with a very simple example inspired by (Skelton, 1989). Let a 'true system' be represented by $G_0(s) = \frac{1}{s+1}$. Then the 'model' $\hat{G}(s) = \frac{1}{s}$ would clearly be deemed to be absurd as a model for G_0. However, if the objective is to design a high gain static output feedback controller, then $\hat{G}$ would be a perfectly acceptable model. Indeed, with a high gain static output feedback $u = -Ky$, the closed loop transfer functions $\frac{K}{s+1+K}$ and $\frac{K}{s+K}$ become indistinguishable. Thus, whether or not a model is appropriate for control design depends as much on the controller that will be implemented as it depends on the plant/model mismatch.

In practice, the true system is unknown, the model is unknown at the identification design stage, and the controller that will be implemented is unknown because it depends on that model. What is typically known in control-oriented identification is the control performance objective. Some prior knowledge about the true system may also be available. Ideally, the design of a control-oriented identification procedure could then be formulated as follows: *Given a control performance objective, design the identification in such a way that the performance achieved by the model-based controller on the true system is as high as possible.* Identification design includes many choices: input data, feedback configuration (possibly), data length, model structure, identification criterion, validation criterion, etc. The identification design problem is impossible to solve in such generality. In order to get a handle on the problem, it has been customary to fix some of the choices; the number of data is usually taken to be fixed, and so is the model structure.

In the first half of the nineties, the research focused on the bias error distribution, assuming that low complexity models were being used for the design of the controller. It produced a string of results on the design of control-oriented nominal models (Schrama and Bosgra, 1993; Lee *et al.*, 1993; Åström and Nilsson, 1994; Zang *et al.*, 1995; de Callafon and Van den Hof, 1997). The first and rather obvious result was to establish that a model is good for control design if the closed loop system obtained by the feedback connection of that model with the designed controller is close to the system obtained by the feedback connection of the true system with that same controller. Since the 'to be designed controller' is not available at the identification stage, this led to the necessity of using an iterative scheme of model updates and controller updates.

A second important result that emerged from the understanding of control-oriented nominal models was that the experimental conditions of the identification should resemble as much as possible the experimental conditions obtained when the designed controller is applied to the true system (Anderson and Gevers, 1998). This led to the observation that, for most control performance objectives, identification should be performed in closed loop (Hjalmarsson *et al.*, 1996; Forssell and Ljung, 2000). This observation triggered a revival of interest for closed loop identification, and the emergence of new identification methods specifically designed for this situation: see e.g. (Hansen *et al.*, 1989; Van den Hof and Schrama, 1993).

The iterative identification and control design schemes do not necessarily converge to a stationary point, corresponding to a stable closed loop system. Thus, they must be applied with caution, and a lot of work has gone into developing tools for safe model and controller updates. Even so, these iterative schemes have had a remarkably fast transfer into the world of applications. There are two main reasons for this:

- whereas much of the industrial world was still living with the belief that one should 'open the loop' to perform a valid identification experiment, here was a theory that showed the benefits of closed-loop identification; this came as welcome news to process control engineers who had never really liked the idea of opening the loop;
- in the process industry, thousands of measurements flow into the computer; here was a theory that showed how these data could be used for the design of a better controller.

As stated above, the work on control-oriented nominal models focused on the bias error distribution of the identified model, with the controller computed from the model in a certainty equivalence framework. Thus, that work did not incorporate the robust control concepts developed during the eighties. It focused on the design of identification criteria that minimize a (control-oriented) measure of the model error.

The second half of the nineties saw a shift towards the definition and estimation of *control-oriented uncertainty sets* (Kosut and Anderson, 1994; Mäkilä *et al.*, 1995; de Vries and Van den Hof, 1995; Bombois *et al.*, 2000) in order to put the control design into the framework of robust control design. The focus turned to shaping the distribution of the variance error of the identified models, i.e. on manipulating the shape of the uncertainty set. Indeed, the paradigm of robust control design is to compute a controller that achieves the best possible worst case performance, i.e. the best possible performance over all models in an uncertainty set. Such best worst case performance depends as much on the controller as it depends on the uncertainty set, and this set is directly dependent on the experimental conditions under which the identification is performed. The study of the interplay between experiment conditions of the identification and properties of the robust controller has been split up into two questions:

(1) what is the connection between a model uncertainty set and the properties of robust controllers computed from that set and, consequently, how should one define a control-oriented uncertainty set?
(2) how should one design the identification experiment in such a way that the uncertainty set around the identified model has such 'control-oriented' property?

Even though many new insights have been gained on the first question, there is at this point no clear view as to the most operational definition of a 'control-oriented uncertainty set'. We shall come back to this in Section 7.

To summarize this brief historical account, the work on identification of the last 15 years has been essentially developed in three directions: optimal control-oriented experiment design for identification, the definition and computation of control-oriented nominal models, and the connection between data-based uncertainty sets and the properties of robust controllers resulting from such sets. In addition, the demands of this highly visible field have generated lots of parallel work on estimation and validation of uncertainty sets from data, and on closed-loop identification. Some of the concepts that have emerged from these 15 years of work, such as the idea of improving the performance of an existing controller on the basis of closed loop data collected with the presently operating controller, have immediately found their way into practice. However, many questions remain unsolved, particularly on the optimal tuning of uncertainty sets, and we are still far from an automatic procedure that would go from the collection of experimental data to the design of a robust controller via a model and its uncertainty set.

In the rest of this paper, we explain in some more detail the specific achievements and the remaining unsolved questions in the different subtopics that we have sketched: what is optimal control-oriented experiment design? what is a control-oriented nominal model? why iterative design? what is a control-oriented uncertainty set? how can we match experiment design and uncertainty set? We have chosen to do the presentation of these ideas in a Prediction Error (PE) identification framework, because PE identification is by far the most successful and widely used identification method. In the next section, we first present the bare essentials of PE identifcation that are necessary to understand the remaining issues.

2. THE BARE ESSENTIALS OF PREDICTION ERROR IDENTIFICATION

The purpose of this paper is to present concepts rather than technicalities. Thus, to simplify the presentation, we assume that the unknown true system can be represented by a single-input single-output linear time-invariant system:

$$\mathcal{S}: y_t = G_0(z)u_t + v_t = G_0(z)u_t + H_0(z)e_t, \quad (1)$$

where $G_0(z)$ is a linear time-invariant causal operator, y is the measured output, u is the control input, and v is noise, assumed to be quasistationary, modelled as the output of a model $v_t = H_0(z)e_t$, where e is white noise. One considers a parametrized model set: $\mathcal{M} = \{(G(z,\theta), H(z,\theta), \ \theta \in D_\theta \subset \mathbb{R}^d\}$ where $G(z,\theta)$ and $H(z,\theta)$ are typically rational transfer functions, and D_θ is a subset of admissible values for the parameter vector θ. To every θ corresponds a one-step ahead predictor:

$$\begin{aligned}\hat{y}_{t|t-1}(\theta) = & H^{-1}(z,\theta)G(z,\theta)u_t \\ & +[1 - H^{-1}(z,\theta)]y_t, \end{aligned} \quad (2)$$

and hence a one-step ahead prediction error:

$$\begin{aligned}\varepsilon_t(\theta) &\triangleq y_t - \hat{y}_{t|t-1}(\theta) \\ &= H^{-1}(z,\theta)[(G_0(z) - G(z,\theta))u_t + v_t]\end{aligned} \quad (3)$$

These prediction errors can, possibly, be filtered by a data filter $D(z)$, thus defining the filtered prediction errors $\varepsilon_t^f(\theta) = D(z)\varepsilon_t(\theta)$. The least squares PE estimate $\hat{\theta}_N$ based on N input-output data is then defined as

$$\hat{\theta}_N = \arg\min_{\theta \in D_\theta} V_N(\theta), \quad (4)$$

where the PE criterion is defined as

$$V_N(\theta) = \frac{1}{N}\sum_{t=1}^{N}[\varepsilon_t^f(\theta)]^2. \quad (5)$$

The estimate $\hat{\theta}_N$ defines the model $\hat{G} = G(z,\hat{\theta}_N)$, $\hat{H} = H(z,\hat{\theta}_N)$. Under reasonable conditions, $\hat{\theta}_N \overset{N\to\infty}{\longrightarrow} \theta^*$, where $\theta^* \triangleq \arg\min_{\theta\in D_\theta} \bar{V}(\theta)$, with $\bar{V}(\theta) \triangleq E[\varepsilon_t^f(\theta)]^2$ (Ljung, 1999) .

Two different situations need to be considered. The first one is when the model structure $\mathcal{M}$ has been chosen sufficiently complex that the true system belongs to the model set. This is denoted $\mathcal{S} \in \mathcal{M}$, and means that there exists a value $\theta_0 \in D_\theta$ such that $G(z,\theta_0) = G_0(z)$ and $H(z,\theta_0) = H_0(z)$. In such case, under reasonable conditions, $\theta^* = \theta_0$, which means that the PE estimates of the transfer functions converge to the true transfer functions: $G(z,\hat{\theta}_N) \overset{N\to\infty}{\longrightarrow} G_0(z)$, $H(z,\hat{\theta}_N) \overset{N\to\infty}{\longrightarrow} H_0(z)$. When $\mathcal{S} \in \mathcal{M}$, the parameter error converges to a Gaussian random variable:

$$\sqrt{N}(\hat{\theta}_N - \theta_0) \overset{N\to\infty}{\longrightarrow} N(0, P_\theta). \quad (6)$$

The asymptotic parameter covariance P_θ can be estimated from the data, and the true parameter vector θ_0 belongs to an ellipsoid:

$$U_\theta = \{\theta | N(\theta - \hat{\theta}_N)^T P_\theta^{-1}(\theta - \hat{\theta}_N) < \chi^2\} \quad (7)$$

with probability $\alpha(d, \chi^2) = Pr(\chi^2(d) \leq \chi^2)$, where $\chi^2(d)$ denotes the χ^2 distribution with d degrees of freedom. Thus, when the system is in the model set, PE identification delivers a nominal model $G(z, \hat{\theta}_N), H(z, \hat{\theta}_N)$, together with an ellipsoidal confidence region in parameter space. This, in turn, defines an uncertainty region in the space of transfer functions:

$$\mathcal{D} = \{G(z, \theta) \mid \theta \in U_\theta\}. \quad (8)$$

In the more general situation where the system is not in the model set, $\lim_{N\to\infty} \hat{\theta}_N = \theta^* \neq \theta_0$. In such case, the transfer function error, $G_0(e^{j\omega}) - G(e^{j\omega}, \hat{\theta}_N)$, at a given frequency ω, can be decomposed as:

$$G_0(e^{j\omega}) - G(e^{j\omega}, \hat{\theta}_N) = \quad (9)$$
$$\underbrace{G_0(e^{j\omega}) - G(e^{j\omega}, \theta^*)}_{\text{bias error}} + \underbrace{G(e^{j\omega}, \theta^*) - G(e^{j\omega}, \hat{\theta}_N)}_{\text{variance error}}.$$

The bias error arises when the model structure is unable to represent the true system. The variance error is caused by the noise and the finiteness of the data set; by definition of θ^*, it goes to zero asymptotically. In the mid-eighties, Ljung produced some important formulas for the characterization of bias and variance errors of identified transfer functions (Ljung, 1985; Wahlberg and Ljung, 1986). The bias was characterized implicitly by representing θ^* as the minimizing argument of a frequency integral. A variance error estimate for the estimated transfer functions was obtained under an assumption of model order going to infinity. More recent work has produced formulas for the estimation of an uncertainty set $\mathcal{D}$ around $\hat{G}$, with the property that $G_0 \in \mathcal{D}$ with probability α, where α is any desired level close to 1 (e.g. $\alpha = 0.95$) even in the case where the system is not in the model set, and for finite model orders: see Section 6. Finally, we note that the results described in this section are valid for both open loop and closed loop identification.

3. THE GAME AND THE PLAYERS

In identification for control, a typical situation is that we can perform experiments on the true system (1) with the purpose of designing a feedback controller. The system may already be under feedback control, in which case the task is to replace the present controller by one that achieves better performance. This situation is representative of many practical industrial situations. We then denote the present controller by C_{id} and the reference signal, if any, by r_t:

$$u_t = C_{id}(z)[r_t - y_t]. \quad (10)$$

Using N data collected on the system, in open loop or in closed loop, we can compute a model $\hat{G}$ of the unknown G_0, and possibly also a noise model $\hat{H}$ of H_0 by PE identification. Since the complexity of a model-based controller is of the same order as that of the model, one often performs the identification with a low order model.

The traditional scenario in model-based robust control design was: *First estimate a model $\hat{G}$ and an uncertainty set $\mathcal{D}$, then design a new controller $C(z)$ that achieves closed-loop stability and meets the required performance with all models in $\mathcal{D}$, and hence with the unknown true system G_0.* For this scenario to be successful, a very accurate model $\hat{G}$ was typically required.

The objective in *identification for control* is to replace that traditional scenario by the following. *On the basis of the required performance, and of any knowledge of the unknown system, design a control-oriented identification experiment that produces a model $\hat{G}$ and an uncertainty set $\mathcal{D}$; then design a new controller C that achieves closed-loop stability and meets the required performance with all models in $\mathcal{D}$, and hence with the unknown true system G_0.* If necessary, repeat this design procedure, possibly with a more demanding performance criterion. In some scenarios, one first computes a class $\mathcal{C}(\hat{G}, \mathcal{D})$ of controllers, each of which achieves the required performance with all models in $\mathcal{D}$; the controller C is then chosen within this class in such a way as to have some additional nice features (e.g. low complexity).

The goal of the new scenario is to achieve the same or better performance based on models (and hence controllers) of lower complexity. The class $\mathcal{C}$ of controllers that achieve the required performance is larger if the model uncertainty set $\mathcal{D}$ can be tuned towards that aim.

The players within this (iterative) identification and robust control design scenario are therefore:

- the unknown plant G_0
- the optimal controller C^{opt} for G_0
- the present controller C_{id} (if any)
- the present model $\hat{G}_{init}$ (if any)
- the identified model $\hat{G}$
- the uncertainty set of models $\mathcal{D}$ around $\hat{G}$
- the set $\mathcal{C}$ of controllers that achieve the prescribed performance
- the new model-based controller $C \in \mathcal{C}$

Except for the unknown plant G_0 and its corresponding optimal controller C_0, the designer has a handle on all other players. It is the complexity of the interplay between these players that makes the problem challenging and interesting. To illustrate the interplay, it is important to understand that one deals with five different feedback loops, which impact on one another: see Figures 1 to 5.

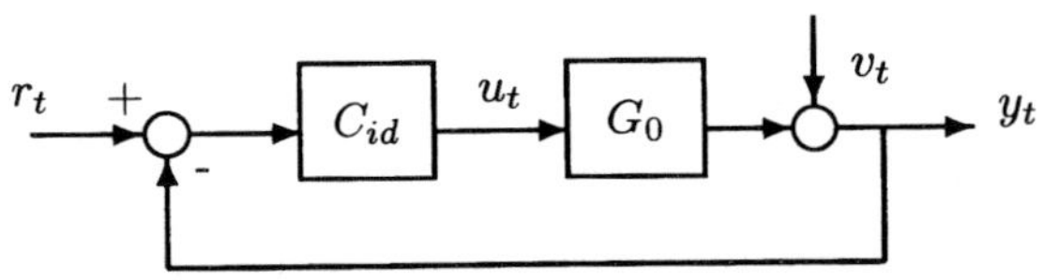

Fig. 1 : Experimental loop

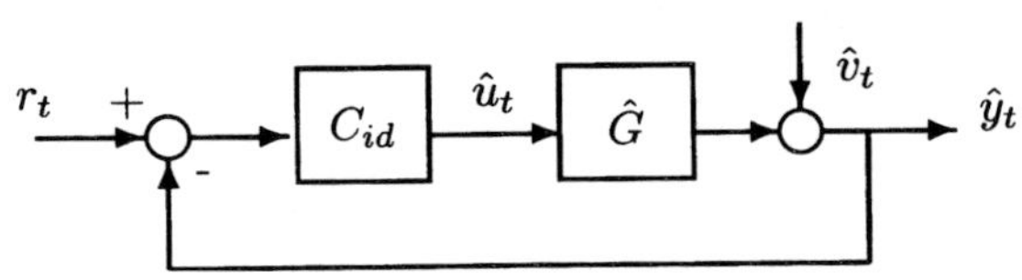

Fig. 2 : Identified system

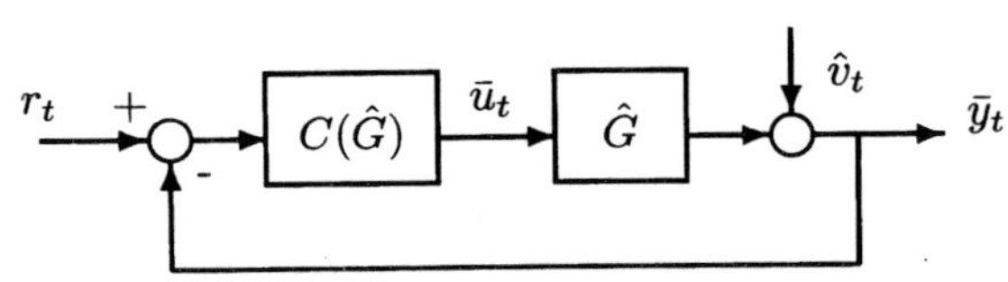

Fig. 3 : Design loop

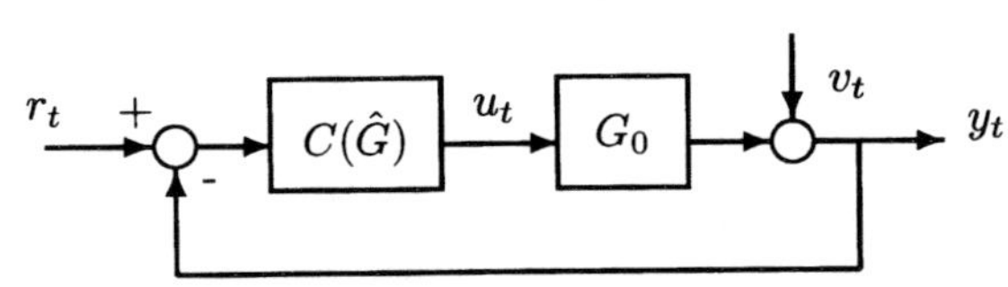

Fig. 4 : Achieved loop

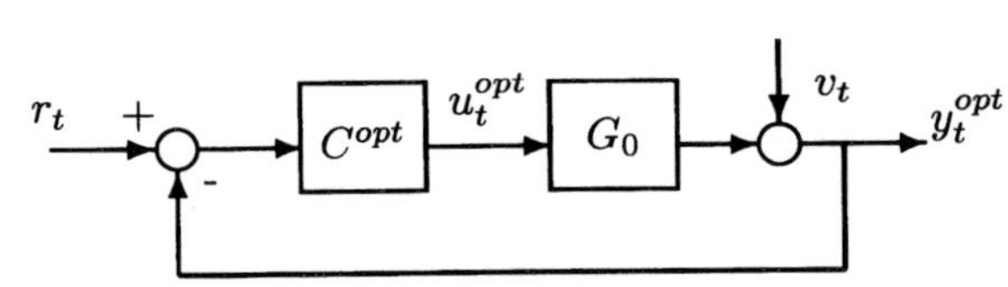

Fig. 5 : Optimal loop

In identification for control, the designer collects data on the experimental loop of Figure 1, and estimates a model $\hat{G}$ such that the closed loop system of Figure 2 is "as close as possible" to the actual system of Figure 1. Sometimes, one of his design choices is the choice of a controller C_{id} and of a reference signal r_t in the experimental loop. On the basis of the identified model $\hat{G}$, possibly with an estimated uncertainty set $\mathcal{D}$, the designer then computes the new controller $C(\hat{G})$, or $C(\hat{G}, \mathcal{D})$: this generates the designed loop of Figure 3. However, what really matters is the performance achieved by this controller with the real system, i.e. the performance of the achieved loop of Figure 4. Thus, if the identifier performs a good job, then the identified loop will be close to the experimental loop; but what is really desired is that the achieved performance on the loop of Figure 4 is close to the designed performance on the loop of Figure 3. If the experimental setup was such that C_{id} was identical (or at least close to) the "to-be-designed controller" C, then closeness of the loops of Figures 1 and 2 would entail closeness of the loops of Figures 3 and 4. This explains why it pays to have experimental conditions that closely match the conditions in which the "to-be-designed controller" will operate.

The first identification for control results, mentioned earlier, were based on a different approach. The aim was to find the optimal experiment conditions such that the output of the achieved loop of Figure 4 is as close as possible to the output of the optimal loop of Figure 5 that would be obtained if the true system were known exactly. Because of its historical precedence, and its intuitive appeal, we first present this approach in the next section.

4. OPTIMAL CONTROL-ORIENTED IDENTIFICATION DESIGN

In this approach an identification experiment design is called "optimal" if the controller computed from the estimated model is one that minimizes the average performance degradation vis-à-vis the performance that would be achieved with the ideal controller. The ideal controller is the controller that would be computed if the true system were known. We now explain this in some more detail.

We denote by $J(G, H, C)$ the control design criterion, and by $C = c(G, H)$ the certainty equivalence mapping that maps a model (G, H) into the corresponding optimal controller. In particular,

$$C^{opt} = c(G_0, H_0) = \arg\min_C J(G_0, H_0, C). \quad (11)$$

We consider PE identification of a parametric model from N data, and assume that $\mathcal{S} \in \mathcal{M}$. The control design mapping then defines a controller $\hat{C}_N = c(\hat{G}_N, \hat{H}_N)$ for each model $(\hat{G}_N, \hat{H}_N)$.[2] The controller $\hat{C}_N$ is a random variable, because the estimated parameter vector $\hat{\theta}_N$ is random, and hence also the model. Applying $\hat{C}_N$ (rather than C^{opt}) to the true system results in an achieved cost $J(G_0, H_0, \hat{C}_N) \geq J(G_0, H_0, C^{opt})$. This results in a "performance degradation" $J^{deg} = J(G_0, H_0, \hat{C}_N) - J(G_0, H_0, C^{opt})$, which is again a random variable.

The problem statement of optimal identification design for control is then phrased as follows: "Find the experimental conditions $\mathcal{X}$ that minimize the

[2] Here $\hat{G}_N$ is a shorthand notation for $G(z, \hat{\theta}_N)$, and similarly for $\hat{H}_N$.

average performance degradation"[3]. In view of what precedes, this can be formulated as:

$$\min_{\mathcal{X}} EJ(G_0, H_0, c(\hat{G}_N, \hat{H}_N)). \quad (12)$$

The expected value is taken with respect to the noise, which affects the model estimate, and hence the controller estimate.

In the context of certainty equivalence control design, this is probably the most logical (and ideal) problem formulation for an *optimal identification for control design.* However, there are several difficulties with this formulation (see Chapter 9 in (Albertos and Sala, 2002)), the main one being that the optimal experiment $\mathcal{X}$ defined by (12) necessarily depends on the unknown system (G_0, H_0). This is specific to all experiment design problems. It does not mean that such results are meaningless: they give useful guidelines for the identification design, and they may lead to iterative schemes that converge to the optimal experiment design: see e.g. (Hjalmarsson *et al.*, 1996).

The first application of this optimal experiment design concept to control-oriented identification was in (Gevers and Ljung, 1986), where an application to Minimum Variance control was treated. The results showed that the optimal experiment consists of performing closed-loop identification with the unknown optimal Minimum Variance controller in the loop. These results were later extended to other control performance criteria in (Hjalmarsson *et al.*, 1996; Forssell and Ljung, 2000).

5. ITERATIVE DESIGN FOR THE NOMINAL MODEL

In this section we discuss the control-oriented identification design of the nominal model, and we show why the pursuit of a control-oriented objective leads to iterative model and controller updates. This observation was made independently in the early nineties by several research teams, who were using different combinations of identification method and control design criterion. To understand the need for iterative design, consider the closed loop systems of Figures 3 and 4, and assume for simplicity that there is no noise ($v_t = 0$ and $\hat{v}_t = 0$), i.e. the control objective is a tracking performance objective.

[3] $\mathcal{X}$ denotes the set of all admissible experimental conditions that have an effect on the quality of the model estimates $(\hat{G}_N, \hat{H}_N)$, such as use of open-loop or closed-loop data, choice of input spectrum distribution, of regulator in the case of closed-loop identification, etc. By 'admissible' experimental conditions, we refer to conditions that obey possible constraints on signal powers or signal energies.

The controller is designed on the basis of the model $\hat{G}$ and then applied to the system G_0. The achieved performance will therefore be close to the desired performance if the two closed loop transfer functions, or a weighted version of these, are close to one another. Thus, we want the following error to be small[4]:

$$\frac{G_0 C}{1 + G_0 C} - \frac{\hat{G}C}{1 + \hat{G}C} = (G_0 - \hat{G})CS_0\hat{S} \quad (13)$$

where $S_0 = \frac{1}{1+G_0C}$ and $\hat{S} = \frac{1}{1+\hat{G}C}$. Now, closed loop PE identification with a given model structure $\mathcal{M} = \{G(z,\theta) \mid \theta \in D_\theta\}$, and with a controller C_{id} in the loop, will asymptotically deliver a model $G(z, \theta^*)$, where $\theta^* = \arg\min_{\theta \in D_\theta} V(\theta)$, with $V(\theta)$ given by

$$V(\theta) = \int_{\pi}^{\pi} |G_0 - \hat{G}(\theta)|^2 |C_{id}S_0|^2 |D|^2 \Phi_r d\omega. (14)$$

Here $D(z)$ is the data filter that can be freely chosen by the user. Observe that, if $C_{id} = C$, and if the data filter is chosen such that $|D(e^{j\omega})|^2\Phi_r(\omega) = |\hat{S}(e^{j\omega})|^2$, then the model $G(z, \theta^*)$ obtained asymptotically by such closed loop PE identification will make the error (13) small in an H_2 sense. By such design, the identification criterion would be matched to the control performance criterion. However, there are two difficulties with such design: (i) the controller C in the design loop of Figure 3 is a function of the identified model, $C = C(\hat{G})$, and it is therefore impossible to choose $C_{id} = C(\hat{G})$ at the identification design stage; (ii) the sensitivity function $\hat{S}$ is also a function of the estimated model, $\hat{S} = \hat{S}(\hat{G})$. These observations have led to the concept of iterative design, where successive steps of closed loop identification and model-based controller design are performed. Thus, at iteration k, where a model $\hat{G}_k$ has been obtained, select $C_{id,k} = C(\hat{G}_k)$, and $|D_k|^2\Phi_r = |\hat{S}_k|^2$, where $\hat{S}_k = \frac{1}{1+\hat{G}_kC(\hat{G}_k)}$.

The research work of the early nineties on the definition and computation of control-oriented nominal models led to several important conclusions, that can be summarized as follows.

- The identification criterion for the nominal model should aim at minimizing the distance between the achieved and the designed loop, where this distance is measured in a norm determined by the control performance criterion. An application of this principle to LQG control can be found in (Zang *et al.*, 1995), to H_∞ control in (Schrama and Bosgra, 1993) and to GPC in (Shook *et al.*, 1992), the latter result being based on dual control ideas.

[4] For simplicity of notation, we omit all ω-dependent arguments whenever there is no risk of confusion.

- The identification must be performed in closed loop, with a specific data filter.
- The data filter is model-dependent (i.e. θ-dependent). Thus, one has to resort to iterative model/controller updates for the practical implementation of this design.

In summary, in identification for control, the control performance objective shapes the bias error distribution of the nominal model. This means that the nominal model has a bias error that is small in the frequency areas where it needs to be small for the design of a better controller, typically around the present cross-over frequency.

Iterative identification and control schemes flourished in the nineties, with various combinations of control criteria and identification criteria. The reader is referred to (Gevers, 1993; Bitmead, 1993; Van den Hof and Schrama, 1995) for details and for a survey on such iterative schemes. Unfortunately, it was found (Hjalmarsson *et al.*, 1995) that these iterative schemes do not generically converge to the achievable minimum (within the model/controller set) of the control performance cost.

Despite this, the concept of iterative identification and control design was rapidly adopted in process control applications: see e.g. (Partanen and Bitmead, 1995; Schrama and Bosgra, 1993; de Callafon *et al.*, 1993; Holmberg *et al.*, 2000; Cooley and Lee, 2001). One reason is that it is typical in such applications that large numbers of closed loop data are flowing into the control computer, and it then makes sense to use these data to replace the existing controller by one that achieves better performance. The practical impact of iterative model and controller redesign has been assessed in (Landau, 1999), where some interesting observations are made on the distinction between this batch-like mode of operation and the more classical methods of adaptive control.

6. MODEL UNCERTAINTY SETS AND THE ROBUST CONTROL PARADIGM

We have shown how control-oriented nominal models are obtained by minimization of an identification criterion that is determined by the overall control performance criterion. In establishing this result, one has assumed that the controller is computed from the estimated model $\hat{G}$ using the certainty equivalence paradigm. At no point has the model uncertainty due to the noise been taken into account. Conversely, the control design step that has led to the iterative design schemes developed above is not based on robust control principles.

The paradigm of modern [5] robust control design can be briefly summarized as follows.

One wants to design a controller for an unknown system G_0. Some prior knowledge about G_0 allows one to assume that G_0 belongs to some *model uncertainty set* $\mathcal{D}$. Most often, a nominal model $\hat{G}$ of G_0 is available, typically the centre of $\mathcal{D}$. Some performance objective is often given in the form of a criterion $J(G, C)$, to be minimized. If the true system were perfectly known, the optimal controller would then be defined as $C^{opt} = \arg\min J(G_0, C)$, where the minimization is performed over some predefined set of admissible controllers $\mathcal{C}$. In the robust control paradigm, the true system is unknown, but the information $(\hat{G}, \mathcal{D})$ is available. One then seeks a *robust controller* C with the following properties:

(1) C must stabilize all models in $\mathcal{D}$;
(2) the worst performance achieved by C on any model in $\mathcal{D}$ must be as high as possible.

The robust controller C is often selected as one that stablizes all models in $\mathcal{D}$ and achieves the best *worst-case performance* over $\mathcal{D}$, i.e. C is computed as

$$C = \arg\min_{C \in \mathcal{C}} \sup_{G \in \mathcal{D}} J(G, C). \tag{15}$$

The uncertainty model set $\mathcal{D}$ plays a central role in this design strategy. In the robust control theory developed through the eighties and nineties, this set is god-given in that it is based on so-called "prior assumptions" about the model and its uncertainty. Many different descriptions of model uncertainty have been considered; most often they are expressed as frequency domain sets, containing both structured and unstructured components: see e.g. (Zhou *et al.*, 1995; Skogestad and Postlethwaite, 1996; Morari and Zafiriou, 1989). The following are representative examples of commonly used uncertainty sets; for simplicity, we consider scalar transfer functions only.

Additive uncertainty set:

$$\mathcal{D}_A = \{G_\Delta(z) \mid G_\Delta(z) = G(z) + \Delta(z), \quad \text{with } |\Delta(e^{j\omega})| < W(e^{j\omega}) \ \forall \omega\},$$

where $G(z)$ is a "nominal model" and $W(e^{j\omega})$ is a frequency weighting function.

Youla-Kucera uncertainty set:

$$\mathcal{D}_{YK} = \{G_\Delta(z) \mid G_\Delta(z) = \frac{N_x + D_c\Delta}{D_x - N_c\Delta}, \quad \text{with } |\Delta(e^{j\omega})| < W(e^{j\omega}) \ \forall \omega\},$$

[5] We refer to modern robust control design as the theory developed in the eighties, that is based on model uncertainty sets; in contrast, the classical robust design theory relies on robustness margins expressed in Bode, Nyquist or Nichols plots.

where $N_x, D_x, N_c, D_c, \Delta$ are stable, rational, proper transfer functions, $\hat{G} = \frac{N_x}{D_x}$ is a "nominal model", and $C = \frac{N_c}{D_c}$ is any stabilizing controller of G_0. This uncertainty set, dual of the Youla-Kucera set of all stabilizing controllers of a given system, was introduced for the description of model sets in (Hansen *et al.*, 1989).

The new approach, initiated around 1990, was to consider the estimation of uncertainty sets from data. Unfortunately, the available PE identification theory had rather little to offer to the existing robust control theory, for two reasons: (1) there were no adequate expressions for the estimation of the total error on an identified transfer function, or some upper bound on this error; the main difficulty was the estimation of the bias error, which could only be characterized by implicit integral expressions; (2) the available uncertainty descriptions were not given as frequency domain sets. As a result, a wide range of new identification techniques were developed. New model assumptions, noise assumptions and identification criteria were introduced whose main merit was to deliver computable error bounds on the estimated models. The survey paper (Ninness and Goodwin, 1995) is probably still one of the best presentations of these alternative methods. Some of these methods have been shown to lead to very conservative upper bounds on the uncertainty set. But clearly, this work has led to important advances in the characterization of bounds on transfer function error estimates.

One uncertainty set that was available in PE identification theory was the set (8) defined via the ellipsoidal set (7) in parameter space. For open loop identification, this set can be described as follows.

Prediction Error uncertainty set:

$$\mathcal{D}_{PE} = \{G(z,\theta) \mid G(z,\theta) = \frac{N(z,\theta)}{D(z,\theta)} \text{ with } (\theta - \hat{\theta}_N)^T R(\theta - \hat{\theta}_N) < 1\}$$

Here N and D are polynomials parametrized by θ, and R is proportional to the inverse of the estimated covariance matrix of $\hat{\theta}_N$. The difficulty with this PE uncertainty set is that it could be computed easily only in the case where the system is in the model set, and it did not connect with the robust control theory and design tools that were available in the early nineties. Thus, all through the nineties, a lot of research was produced by the identification community to

(1) extend the use of uncertainty sets in order to also include the bias error; this was achieved by either embedding the bias error in a stochastic framework (Goodwin *et al.*, 1992; Hakvoort and Van den Hof, 1997), or by estimating the bias error through a validation step that uses a full order model (Ljung, 2000; Gevers *et al.*, 2003).
(2) develop a robust control stability and performance theory for PE uncertainty sets characterized by ellipsoids in parameter space (Bombois *et al.*, 2001).

Many important advances have thus been made in the characterization of bounds on transfer function error estimates, but this subject will undoubtedly remain an object of intense debate and activity for years to come.

7. TOWARDS CONTROL-ORIENTED UNCERTAINTY SETS?

In the second half of the nineties, one began to seriously study the interplay between model uncertainty sets and robust control objectives, in order to address the question of building *control-oriented uncertainty sets*. The motivation for this is based on the following two observations.

- It follows from the properties that define a robust controller (see the previous section) that satisfaction of these two properties hinges as much on the choice of the controller C as it does on the uncertainty set $\mathcal{D}$.
- The shape of a data-based uncertainty set depends very much on the experimental conditions under which it is estimated.

Combining these two observations leads to the idea of constructing control-oriented uncertainty sets by proper choice of experimental conditions.

To illustrate the connection between experiment conditions and model-based control properties, we consider a very simple gedankenexample, presented in (Gevers *et al.*, 1998), where we focus attention only on robust stability. Consider a "true system" G_0 described by the following simple ARX model:

$$(1-1.4z^{-1}+0.45z^{-2})y_t = z^{-1}(1+0.25z^{-1})u_t+e_t,$$

where e is a unit variance white noise. With a constant gain feedback law $u_t = r_t - Cy_t$, the closed loop system is stable for $C < C_{max} = 2.2$. Consider now that we estimate the parameters of this ARX model by PE identification, using the exact structure. We can thus estimate an ellipsoid in parameter space, to which the true parameters belong with probability 95%, say. We can then compute the proportional output feedback controller with the highest gain, $\hat{C}_{max}$, that stabilizes all models $\hat{G}$ whose parameters lie in that 95% uncertainty set. We call this controller the "optimal robust controller". In order to show the effect of the experimental setup of the identification on the optimal robust controller, we have compared, by Monte Carlo simulations, two experimental conditions:

(1) open loop identification with a unit variance white noise as input signal u;
(2) closed loop identification with a controller $u_t = r_t - y_t$ in the loop during data collection, and a white noise reference signal r with variance 18.38; this choice yields the same output variance as in the open loop experiment.

Each of these two experiments was run 100 times, each time with 1000 input-output data. For each run, the 95% confidence ellipsoid was computed in parameter space, and the gain $\hat{C}^i_{max}$ of the i-th run was computed as the largest gain that would stabilize all models in the corresponding model set. On the basis of 100 runs, we then computed the average $\hat{C}_{max}$ and its variance for each of the two experimental conditions. The following results were obtained for open loop (O.L.), respectively closed loop (C.L.), identification:

$$\text{O.L.}: \hat{C}_{max} = 1.36,\ \sigma^2_{\hat{C}max} = 0.12$$
$$\text{C.L.}: \hat{C}_{max} = 2.04,\ \sigma^2_{\hat{C}max} = 0.02$$

Remember that, for the true system G_0, we have $C_{max} = 2.2$. This example clearly shows that the experimental conditions used in the feedback experiment are more control-oriented than those in the open loop experiment, since they lead to a much less conservative estimate of the limit gain for the robust controller. This example serves to illustrate the effect of the identification experiment on the set of admissible controllers, via the model uncertainty set. In a more realistic setup, the interplay between the design of the identification experiment, the corresponding model uncertainty set, and the set of admissible controllers, is a lot harder to understand and analyze.

In order to relate the identification design with the properties of a robust controller, one has resorted to splitting up the problem into its two components:

- understanding the effect of experimental conditions on model quality, or more precisely on the properties of the uncertainty set;
- understanding the interplay between the uncertainty set and the properties of the ensuing robust controller.

While the first problem is reasonably easy and has already yielded many results, the second problem is much harder and there is, as of today, no consensus on the best way to address it. We now elaborate on these two subproblems.

In PE identification, there has been a continuing activity aimed at characterizing the bias, variance, and total mean square error on the estimation of parametric transfer functions, dating back to the mid-eighties. The variance formulas of (Ljung, 1985) were based on an assumption of model order going to infinity; more accurate approximations have recently been obtained for the variance formulas of finite order transfer function estimates (Ninness and Hjalmarsson, 2003). The bias formulas of (Wahlberg and Ljung, 1986) have given way to explicit expressions based on a stochastic embedding of the bias error, or on its estimation based on validation with full order models, as mentioned in the previous section.

All these formulas for the estimation of the bias, the variance or the total mean square error of identified transfer functions explicitly contain the effect of the experimental conditions (e.g. number of data, input spectrum, noise spectrum, feedback configuration, etc) on the error measure. This yields the possibility of optimizing over some relevant experimental design variable in order to minimize a particular measure of this error. In the 1970's, optimal input design for system identification was an active area of research, with different quality measures of the identified model being used for this optimal design (Zarrop, 1979; Goodwin and Payne, 1977). Thus, given a clear view as to what constitutes a *control-oriented uncertainty set*, one could certainly extend these optimal experiment design ideas to the construction of such sets.

So, what constitutes a *control-oriented uncertainty set*? There is no clear consensus yet on a good definition. One possible view is to say that a model uncertainty set $\mathcal{D}$ is "control-oriented" if the corresponding set of admissible controllers, $\mathcal{C}$, is large. Such track has been pursued in e.g. (Douma *et al.*, 2003) where different uncertainty structures are compared, and in (Gevers *et al.*, 2003) where the worst-case ν-gap has been proposed as a control-oriented measure of size of $\mathcal{D}$, because it is related to the size of the corresponding set $\mathcal{C}$ of stabilizing controllers. In (Hildebrand and Gevers, 2003) the corresponding optimal input design problem for the minimization of this worst-case ν-gap measure has been solved.

However, one could also define a control-oriented uncertainty set in a very different way as follows. Consider an uncertainty set $\mathcal{D}$ with center $\hat{G}$ containing the true G_0, and a controller $C = C(\hat{G}, \mathcal{D})$ with nominal stability margin $b(\hat{G}, C)$ and nominal performance $J(\hat{G}, C)$. Then $\mathcal{D}$ could be called control oriented if C stabilizes all models in $\mathcal{D}$, if the worst case stability margin $\sup_{G\in\mathcal{D}} b(G, C)$ of C with all models in $\mathcal{D}$ is close to the nominal margin $b(\hat{G}, C)$, and if the worst case performance $\sup_{G\in\mathcal{D}} J(G, C)$ is close to the nominal performance $J(\hat{G}, C)$.

Thus, much work remains to be done on the definition of control-oriented uncertainty sets, and on the computation of the corresponding quality measures. This is certainly one area where one can expect a lot of research activity in the coming years.

8. CONCLUSIONS

We have attempted to explain the major issues related to the problem of identification for control, to present the successive solutions that have been brought to the overall problem, and to display the major remaining open problems. Most certainly, the major impact so far in terms of transfer of technology to the industrial world has been the iterative schemes of model and controller updates, which have provided a methodological background for iterative data-based controller performance enhancement. As for the synergy of robust control concepts and identification design concepts, we believe that most challenges are still ahead of us. The main reason is that a full understanding of the interplay between identification design and robust control analysis and synthesis cannot bypass the role of the uncertainty set, and that the analysis of this problem is difficult and involves techniques from several different subdisciplines.

9. REFERENCES

Albertos, P. and A. Sala (2002). *Editors, Iterative Identification and Control.* Springer Verlag.

Anderson, B.D.O. and M. Gevers (1998). Fundamental problems in adaptive control. *Perspectives in Control, a Tribute to Ioan Doré Landau* pp. 9–21.

Anderson, B.D.O., R.R. Bitmead, C.R. Johnson Jr, P.V. Kokotovic, R.L. Kosut, I.M.Y. Mareels, L. Praly and B.D. Riedle (1986). *Stability of Adaptive Systems: Passivity and Averaging Analysis.* MIT Press. USA.

Åström, K.J. and J. Nilsson (1994). Analysis of a scheme for iterated identification and control. In: *Proc. IFAC Symp. on Identification.* Copenhagen, Denmark. pp. 171–176.

Åström, K.J. and T. Bohlin (1965). Numerical identification of linear dynamic systems from normal operating records. In: *Proc. IFAC Symposium on Self-Adaptive Systems.* Teddington, UK.

Bitmead, R.R. (1993). Iterative control design approaches. In: *Proc. 12th IFAC World Congress.* Vol. 9. Sydney, Australia. pp. 381–384.

Bombois, X., M. Gevers and G. Scorletti (2000). A measure of robust stability for an identified set of parametrized transfer functions. *IEEE Transactions on Automatic Control* **45**(11), 2141–2145.

Bombois, X., M. Gevers, G. Scorletti and B.D.O. Anderson (2001). Robustness analysis tools for an uncertainty set obtained by prediction error identification. *Automatica* **37**(10), 1629–1636.

Cooley, Brian L. and Jay H. Lee (2001). Control-relevant experiment design for multivariable systems described by expansions in othonormal bases. *Automatica* **37**(2), 273–281.

de Callafon, R.A. and P.M.J. Van den Hof (1997). Suboptimal feedback control by a scheme of iterative identification and control design. *Mathematical Modelling of Systems* **3**(1), 77–101.

de Callafon, R.A., P.M.J. Van den Hof and M. Steinbuch (1993). Control-relevant identification of a compact disc pick-up mechanism. In: *Proc. 1993 IEEE Conf. on Decision and Control.* San Antonio, Texas. pp. 2050–2055.

de Vries, D.K. and P.M.J. Van den Hof (1995). Quantification of uncertainty in transfer function estimation: a mixed probabilistic - worst-case approach. *Automatica* **31**, 543–558.

Deistler, M. (2002). System identification and time series analysis: Past, present, and future. In: *'Stochastic Theory and Control', Festschrift for Tyrone Duncan.* Springer. Kansas, USA. pp. 97–108.

Douma, S.G., P.M.J. Van den Hof and O.H. Bosgra (2003). Controller tuning freedom under plant identification uncertainty: double youla beats gap in robust stability. *Automatica* **39**, 325–333.

Forssell, U. and L. Ljung (2000). Some results on optimal experiment design. *Automatica* **36**, 749–756.

Gevers, M. (1991). Connecting identification and robust control: A new challenge. In: *Proc. IFAC/IFORS Symposium on Identification and System Parameter Estimation.* Budapest, Hungary. pp. 1–10.

Gevers, M. (1993). Towards a joint design of identification and control?. In: *Essays on Control: Perspectives in the Theory and its Applications, Birkhauser, New York* (H.L. Trentelman and J.C. Willems, Eds.). Birkhauser. New York. pp. 111–151.

Gevers, M. and L. Ljung (1986). Optimal experiment designs with respect to the intended model application. *Automatica* **22**, 543–554.

Gevers, M., B. D.O. Anderson and B. Codrons (1998). Issues in modeling for control. In: *Proc. American Control Conference.* Philadelphia, USA. pp. 1615–1619.

Gevers, M., X. Bombois, B. Codrons, G. Scorletti and B.D.O. Anderson (2003). Model validation for control and controller validation in a prediction error identification framework - Part I: theory. *Automatica* **39**(3), 403–415.

Goodwin, G. and R.L. Payne (1977). *Dynamic System Identification: Experiment Design and Data Analysis.* Academic Press. New York.

Goodwin, G.C., M. Gevers and B. Ninness (1992). Quantifying the error in estimated transfer functions with application to model order selection. *IEEE Trans. Automatic Control* **37**, 913–928.

Hakvoort, R.G. and P.M.J. Van den Hof (1997). Identification of probabilistic system uncertainty regions by explicit evaluation of bias and variance errors. *IEEE Trans. Automatic Control* **42**(11), 1516–1528.

Hansen, F., G. Franklin and R. Kosut (1989). Closed-loop identification via the fractional representation: Experiment design. *Proc. American Control Conference* pp. 1422–1427.

Hildebrand, R. and M. Gevers (2003). Identification for control: optimal input design with respect to a worst-case ν-gap cost function. *SIAM Journal on Control and Optimization* **41**(5), 1586–1608.

Hjalmarsson, H. (2003). From experiments to closed-loop control (plenary lecture). In: *CD-ROM of 13th IFAC Symposium on System Identification.* Rotterdam, The Netherlands.

Hjalmarsson, H., M. Gevers and F. De Bruyne (1996). For model-based control design, closed-loop identification gives better performance. *Automatica* **32**, 1659–1673.

Hjalmarsson, H., S. Gunnarsson and M. Gevers (1995). Optimality and sub-optimality of iterative identification and control design schemes. In: *Proc. American Control Conference 95.* Vol. 4. Seattle, Washington. pp. 2559–2563.

Ho, B.L. and R.E. Kalman (1965). Effective construction of linear state-variable models from input-output functions. *Regelungstechnik* **12**, 545–548.

Holmberg, U., S. Valentinotti and D. Bonvin (2000). An identification-for-control procedure with robust performance. *Control Engineering Practice* **8**, 1107–1117.

Kalman, R.E. (1960*a*). Contributions to the theory of optimal control. *Boletin de la Sociedad Matematica Mexicana* **5**, 102–119.

Kalman, R.E. (1960*b*). A new approach to linear filtering and prediction problems. *Transactions ASME, Series D, J. Basic Eng.* **82**, 34–45.

Kosut, R.L. and B.D.O. Anderson (1994). Least-squares parameter set estimation for robust control design. In: *Proc. American Control Conference.* Baltimore,Maryland. pp. 3002–3006.

Landau, I.D. (1999). From robust control to adaptive control. *Control Engineering Practice* **7**(9), 1113–1124.

Lee, W.S., B.D.O. Anderson, R.L. Kosut and I.M.Y Mareels (1993). A new approach to adaptive robust control. *Int. Journal of Adaptive Control and Signal Processing* **7**, 183–211.

Ljung, L. (1985). Asymptotic variance expressions for identified black-box transfer function models. *IEEE Trans. Automatic Control* **AC-30**, 834–844.

Ljung, L. (1999). *System Identification: Theory for the User, 2nd Edition.* Prentice-Hall. Englewood Cliffs, NJ.

Ljung, L. (2000). Model error modeling and control design. In: *CD-Rom Proc. of IFAC Symposium on System Identification, paper WeAM1-3.* Santa Barbara, California.

Mäkilä, P.M, J.R. Partington and T.K. Gustafsson (1995). Worst-case control-relevant identification. *Automatica* **31**(12), 1799–1819.

Morari, M. and E. Zafiriou (1989). *Robust Process Control.* Prentice-Hall. Englewood Cliffs, NJ.

Ninness, B. and G.C. Goodwin (1995). Estimation of model quality. *Automatica* **31**(12), 32–74.

Ninness, B. and H. Hjalmarsson (2003). Variance error quantifications that are exact for finite model order. *submitted for publication.*

Partanen, A.G. and R.R. Bitmead (1995). The application of an iterative identification and controller design to a sugar cane crushing mill. *Automatica* **31**, 1547–1563.

Schrama, R.J.P. and O.H. Bosgra (1993). Adaptive performance enhancement by iterative identification and control design. *Int. Journal of Adaptive Control and Signal Processing* **7**(5), 475–487.

Shook, D.S., C. Mohtadi and S.L. Shah (1992). A control-relevant identification strategy for gpc. *IEEE Trans. Automatic Control* **AC-37**(7), 975–980.

Skelton, R.E. (1989). Model error concepts in control design. *Int. Journal of Control* **49**(5), 1725–1753.

Skogestad, S. and I. Postlethwaite (1996). *Multivariabel Feedback Control: Analysis and Design.* J. Wiley. Chichester, U.K.

Van den Hof, P.M.J. and R.J.P. Schrama (1993). An indirect method for transfer function estimation from closed loop data. *Automatica* **29**, 1523–1527.

Van den Hof, P.M.J. and R.J.P. Schrama (1995). Identification and control - closed-loop issues. *Automatica* **31**, 1751–1770.

Wahlberg, B. and L. Ljung (1986). Design variables for bias distribution in transfer function estimation. *IEEE Trans. Automatic Control* **AC-31**, 134–144.

Zang, Z., R.R. Bitmead and M. Gevers (1995). Iterative weighted least-squares identification and weighted LQG control design. *Automatica* **31**(11), 1577–1594.

Zarrop, M. (1979). *Optimal experiment design for dynamic system identification.* Lecture Notes in Control and Information Sciences, Vol. 21, Springer Verlag. Berlin, New York.

Zhou, K., J.C. Doyle and K. Glover (1995). *Robust and Optimal Control.* Prentice Hall, New Jersey.

ELSEVIER
IFAC
PUBLICATIONS
www.elsevier.com/locate/ifac

OPTIMAL CONTROL OF EMULSION COPOLYMERIZATION: APPLICATION TO A PILOT-SCALE REACTOR UNDER A DCS ENVIRONMENT

B. Alhamad, R.Willis, J. A. Romagnoli*, V. G. Gomes

Laboratory for Process Systems Engineering (PSE)
Department of Chemical Engineering,
The University of Sydney
NSW 2006, Australia

ABSTRACT

A detailed model was developed for the emulsion copolymerization of styrene and MMA to predict the evolution of the particle size distribution (PSD) and molecular weight distribution (MWD) over the entire range of monomer conversion. A system exhibiting zero-one kinetics was employed. The model was used to optimise the emulsion copolymerization process in order to maximize particle size polydispersity index (PSPI) and Molecular weight polydispersity index (MWPI). Five variables were used as manipulated variables, styrene monomer feed rate, MMA monomer feed rate, surfactant feed rate, initiator feed rate, and the temperature of the reaction by controlling the jacket temperature. The results from optimization were validated against experimental and found to show good agreement. The whole strategy has been developed within a gPROM-Excel/OPC-DCS environment allowing for direct transfer of the technology to a general industrial application. *Copyright © 2004 IFAC*

Keywords: Optimization, Copolymerization, PSPI, PSD, MWPI, MWD

1. INTRODUCTION

Emulsion copolymerisation is widely used in industry to produce products ranging from paints, adhesives through to tyres and wet-suits. To improve the efficiency and safety of the process, model development becomes necessary to ensure optimized processes and product quality. However, the modelling of emulsion copolymerisation is a challenging task, since it involves complex physico-chemical sub-processes, some of which are not well-understood. Especially, the particle formation mechanism and inclusion of two or monomers increases the complexity. Moreover, it is a multi-phase environment consisting of an aqueous phase, surfactants, initiators, buffers, monomer droplets and polymer particles. The reaction mechanisms are yet to be fully elucidated under various reaction conditions. The goal is to develop a practical tool to predict polymer production rate and key product attributes for ultimate application in industry.

There is relatively little literature on the control of the full PSD, and it is even littler for the copolymerization process. In contrast, there is a considerable number of studies reported in the literature on the control of lumped properties. (Semino and Ray, 1995b, Semino and Ray, 1995a) addressed the very pertinent issue of the controllability of population balance systems. They found in their system the controllability of the distributions for the unconstrained case is ensured by employing the feed concentration of surfactant, initiator and inhibitor as manipulated variables. (Kozub and Macgregor, 1992) applied the idea of a two-tier control strategy-open loop feed forward generation of an optimal recipe, which is then recomputed online based on feedback from process measurements to a semibatch emulsion polymerization system, for the multivariable control of composition and average molecular weight. (Saldivar and Ray, 1997) studied the control of copolymer composition and averaged molecular weight for semi-continuous emulsion polymerization. (Clarke-Pringle and MacGregor, 1998) presented a batch to batch adjustment strategy for the control of molecular weight distribution (MWD) to reoptimize the inputs for the next batch based on end-point measurement of the MWD. (Crowley et al., 2001) used a hybrid modelling strategy in batch to batch optimization for PSD control. (Crowley et al., 2000) are among the first researchers to address the control of the full particle size distribution. (Immanuel and Doyle, 2002) presented an open loop optimization study for the control of the full PSD in the emulsion copolymerization of vinyl acetate and butyl acrylate. (Zeaiter et al., 2002) developed a model for emulsion polymerization of styrene to optimise and control the PSD.

A comprehensive dynamic model for a copolymerization reactor was developed by (Alhamad et al., 2003, Alhamad et al., 2004), which allows the prediction of key polymer properties such as: average particle size, conversion, PSD, MWD, M_n, and M_w. A modified zero-one kinetic model is used in the formulation allowing prediction of secondary nucleation under starving monomer conditions. In this work, based on this comprehensive model, a multi-layer model-based framework was developed and implemented within an industrial distributed control system environment, to optimize the Particle Size Polydispersity Index (PSPI) and Molecular Weight Polydispersity Index (MWPI).

2. EXPERIMENTAL SETUP

2.1 Data Acquisition and Control Environment

The polymer lab consists of a 5 litre jacketed stirred reactor; a Julabo heating circulator to provide heat to the reactor through the jacket; 4 solenoid dosing pumps for providing the monomers, surfactant and initiator to the reactor; 3 RTDs for monitoring temperature; and 4 precision balances to determine the quantities of the reactants used. The polymer lab is controlled by a Honeywell C200 controller on a Honeywell Plantscape DCS. The uploading and configuration of the control schemes to the controller is done using the Control Builder from the Honeywell Plantscape r500.1. The polymer lab I/O consists of 3 RTDs, 3 analogue inputs, 5 analogue outputs, 4 digital outputs and a 4 port serial to Ethernet converter. The RTDs are the jacket inlet temperature, the jacket outlet temperature, and the internal reactor temperature. The analogue inputs are the impeller speed within the reactor; the Julabo heating circulator operating temperature; and the user specified setpoint temperature from the Julabo. The analogue outputs are setpoint temperature for the Julabo heating circulator, and the dosing frequencies or stroke rates for each of the dosing pumps. The digital outputs are used to stop or resume the dosing pumps from pumping. The 4 port serial to Ethernet converter is used in conjunction with software developed within the PSE group to send flowrates or differential weight readings from the balances to the controller. All the manipulated variables profiles are put into the Control Builder to be operated automatically.

Data Acquisition and control of the polymerization reactor was performed using Honeywell's Plantscape R300 software residing on a server. This developed set-up allows the configuration and implementation of multilayer control scheme for advanced operation and control of the process and a client station is used for operator manipulations. The lower level control consists of a series of conventional PID controllers to control the monomer feeds, surfactant, and initiator as well as to control the temperature. The inputs to the PID controllers are the set-point (upper-layer) provided from either one of the following three sources: 1) *Manual set-point:* this is used when the key polymerization variables are required to be set at a constant temperature (used during model validation), 2) *Rampsoak set-point:* this is a set-point varying with time and it changes according to an operator built-in profile (used during off-line optimization) and 3) *External set-point:* this is used when an external program is used to provide the set-point (this is the case in our environment when implementing the real-time optimizing control using an MPC strategy).
The model developed is used within the proposed strategy to provide the optimal set-points (off-line) in case 2 and to provide optimal trajectories and as a real-time soft sensor to control PSD and MWD in case 3.

An intelligent control hierarchy is formulated for such distributed parameter system incorporating three different levels; offline optimization, on-line DMC and regulatory control successively. The novelty of this approach is the incorporation of a validated high order dynamic model as a soft sensor for on-line feedback of the PSD and MWD. Figure 1 shows the schematic diagram of the overall control hierarchy.

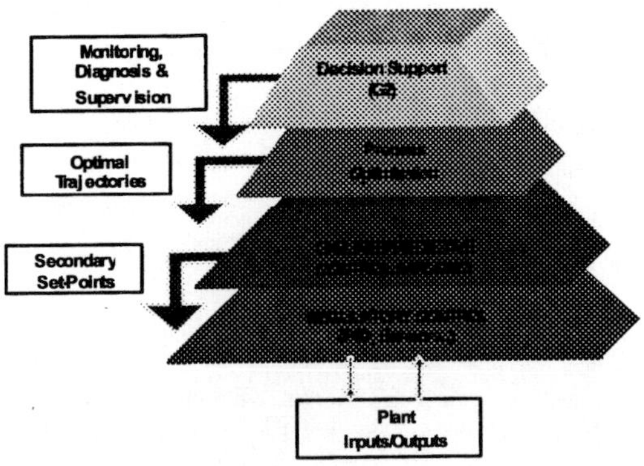

Figure 1: Overall Control Hierarchy

2.2 Sample Preparation and Characterization

Styrene (99% purity, inhibited with 4-tert-butylcatechol) and MMA (99% purity, inhibited with hydroquinone) monomers were obtained from Fluka, while the water used was purified to a Milli-Q standard. Surfactant (sodium dodecyl sulfate) and initiator (potassium persulfate) were both obtained from Sigma-Aldrich. Semi-batch emulsion copolymerisations of styrene and MMA were carried out at 70, 75, 80 and 85 ℃ under slight nitrogen pressure. Note that the monomers were purifed using an inhibitor column, which is also obtained from Sigma Aldrich. Monomer conversion was gravimetrically determined off-line by taking samples from the reactor. The MWD of the polymer samples was determined off-line by GPC, while most of the PSDs were obtained using the zetasizer and the CHDF.

3. Optimal Control Studies

3.1. Model Development

The population balance equations (PBE), based on particle formation through both homogeneous and micellar nucleation, provide estimates for the PSD. The PBEs for the three types of particles: those containing no radicals (type n_0 particles), those containing one monomeric radical (type n_1^M particles), and those containing one polymeric radical (type n_1^P particles), are given by:

$$\frac{\partial n_1^P(V,t)}{\partial t}=k_p^1 C_P n_1^M+\rho_{init}n_0-\rho n_1^P-k_{tr}C_P n_1^P-\frac{\partial(Kn_1^P)}{\partial V}$$
$$+\delta(V-V_o)\left[C_{micelle}\sum_{i=z}^{j_{crit}-1}\left[k_{e,micelle}^{i}[IM_i]\right]-k_{p,aq}^{j_{crit}-1}C_W[IM_{j_{crit}-1}]\right]$$
$$+\int_0^\infty B(V,V-V')[n_0(V')n_1^P(V-V')+n_1^P(V')n_0(V-V')]dV'$$
$$-n_1^P(V)\int_0^\infty B(V,V')[n_0(V')+n_1^P(V')]dV' \qquad (1)$$

$$\frac{\partial n_0(V,t)}{\partial t}=\rho[n_1^P+n_1^M-n_0]+k_{dM}n_1^M+\int_0^\infty B(V,V-V')[n_0(V')n_0(V-V')$$
$$+n_1^P(V')n_1^P(V-V')]dV'-n_0(V)\int_0^\infty B(V,V')[n_0(V')+n_1^P(V')]dV' \qquad (2)$$

$$\frac{\partial n_1^M(V,t)}{\partial t}=k_{tr}C_P n_1^P+k_{eE}[E]n_0-(k_p^1 C_P+k_{dM}+\rho)n_1^M \qquad (3)$$

Total number particles are given by:

$$n(V,t)=n_1^P(V,t)+n_0(V,t)+n_1^M(V,t) \qquad (4)$$

The term $C_{micelle}$ is the micelle concentration, which is determined by the rate of surfactant consumption. The terms k_p, k_e, k_{tr}, are the kinetic rate coefficients for the propagation, entry and transfer respectively.

The transient monomer molar balance for semi-batch reactor operation is given by:

$$\frac{dN_{m,A}}{dt}=F_{mA,in}-R_{pAp}V_p-R_{pAw}V_w \qquad (5)$$

where $R_{p,A,p}$ and $R_{p,A,w}$ are the rate of reactions of reactions of the polymer for monomer A and B in the particle and water phase, respectively. The rate of constant are defined as follows:

$$R_{pAp}=\frac{\left(kp_{AA}\cdot kp_{BB}\cdot(r_A\cdot C_{pA}^{\ 2}+C_{pA}\cdot C_{pB})\right)}{k_{pBB}\cdot r_A\cdot C_{pA}+k_{pAA}\cdot r_B\cdot C_{pB}}\bullet\frac{\bar{n}\cdot N_{tot}}{N_a\cdot V_p} \qquad (6)$$

$$R_{pAw}=\frac{\left(kp_{AA}\cdot kp_{BB}\cdot(r_A\cdot C_{wA}^{\ 2}+C_{wA}\cdot C_{wB})\right)}{k_{pBB}\cdot r_A\cdot C_{wA}+k_{pAA}\cdot r_B\cdot C_{wB}}\bullet T \qquad (7)$$

where k_p is the propagation coefficient, N_{tot} is the total number of particles, N_a is the Avogadro's number, r_A is the reactivity ratio, T is the total number of radicals, and V_p is the volumer of the polymer phase.

The volume of polymer particles is defined for two separate regimes: when monomer droplets exit (Vp) and when there are no longer any droplets. So the copolymer particle volume is calculat ed by two different equations in two stages. The first regime is calculated by:

$$\frac{dV_p}{dt}=\frac{M_{w,A}\cdot(R_{p,A,p}\cdot V_p+R_{p,A,w}\cdot V_w)+M_{w,B}\cdot(R_{p,B,p}\cdot V_p+R_{p,B,w}\cdot V_w)}{d_p} \qquad (8)$$

whereas the second regime is calculated by the following:

$$\frac{dV_p}{dt}=\frac{M_{w,A}\cdot(R_{p,A,p}\cdot V_p+R_{p,A,w}\cdot V_w)+M_{w,B}\cdot(R_{p,B,p}\cdot V_p+R_{p,B,w}\cdot V_w)}{d_p} \qquad (9)$$
$$-\left[M_{w,A}\cdot R_{p,A,p}\cdot V_p\cdot\left(\frac{1}{d_{m,A}}-\frac{1}{d_{p,A}}\right)+M_{w,B}\cdot R_{p,B,p}\cdot V_p\cdot\left(\frac{1}{d_{m,B}}-\frac{1}{d_{p,B}}\right)\right]$$

P_A and P_B are the time-averaged probabilities of finding a free radical with ultimate unit of type A and B, respectively. They are calculated as follows:

$$P_A=\frac{k_{pBA}\cdot C_{p,A}}{k_{pBA}\cdot C_{p,A}+k_{pAB}\cdot C_{p,B}} \qquad (10)$$

$$P_B=1-P_A \qquad (11)$$

where C_{pA} and C_{pB} are the concentrations of monomers A and B in the particle phase. They are obtained by using the partition coefficients (K) between the three phases, water (w), monomer (m) and particles (p).

$$C_{p,A}=MIN\left(C_{p,A}^{sat},\frac{N_{m,A}}{K_{A,mw}\cdot K_{A,wp}\cdot V_m+K_{A,wp}\cdot V_w+V_p}\right) \qquad (12)$$

where N_m is the number of moles and V is the volume.

3.2 Optimization Strategy Setup

As stated previously, the main objective in this study is to investigate the use of a detailed (validated) mechanistic model (Alhamad et al., 2003, Alhamad et al., 2004) to develop an advanced control strategy for the optimal operation of the reactor. The proposed strategy will provide set-point trajectories for the manipulated variables (for example, monomer feed rate and reactor temperature) so as to ensure the production of a copolymer with a defined PSD and MWD in the minimum reaction time.

Several objective functions were studied, in terms of both PS and MW. For the purpose of obtaining the desired broad PSD, a polydispersity index was chosen as an objective function to be maximized. For MW both MWPI as well as simply the molecular weight were investigated as the objective functions to be optimized. The particle concentration densities, particle diameter, temperature shift, copolymer composition and the total amount of monomer to be added to the reactor were included as constraints along with the reaction time.

The objective functions are optimized by using the five manipulated variables, styrene feed rate, MMA feed rate, surfactant feed rate, and initiator feed rate and temperature. The temperature was used as a manipulated variable for maximizing MWPI, since the molecular weight is largely affected by the temperature, rather than the PSPI, which is not significantly affected by the temperature. The MWPI was also maximized without the temperature as a manipulated variable investigate the actual effect of temperature.

A polydispersity index indicates the spread of the distribution. The particle size polydispersity index (PSPI) is estimated as follows:

$$PSPI = \frac{<r^2>}{<r>^2} \tag{13}$$

The number average radius was computed from

$$\langle r \rangle = \frac{\sum (r_{uns} \cdot n)}{\sum n} \tag{14}$$

The objective functions to be optimized were defined as: Max [PSPI(r, t_{final})] and Max[MWPI(r, m, t_{final}]. Where: m is the molecular weight and t_{final} is the processing time. For operational reasons, the monomers, surfactant, initiator feed rates and temperature (for molecular weight optimisation) were specified within the following upper and lower bounds:

$$0.12815\ g/s \le F_{m,A} \le 0.2\ g/s \tag{15}$$

$$0.12815\ g/s \le F_{m,B} \le 0.2\ g/s \tag{16}$$

$$0.12815\ g/s \le F_S \le 0.2\ g/s \tag{17}$$

$$0.12815\ g/s \le F_I \le 0.2 g/s \tag{18}$$

$$343\ ^oC \le T_{reac} \le 358\ ^oC \tag{19}$$

The final PSD shape was included in this optimisation in the form of end point inequality constraints formulated in terms of the final molar concentration density of particles

$$n_{\min} \le n(r, t_{final}) \le n_{\max} \tag{20}$$

Above, n_{min} and n_{max} denote the lower and upper limits, respectively, and were specified to match the required distribution. The final copolymer composition required was 50/50, so a condition was specified as follows:

$$0.49 \le \overline{F} \le 0.51 \tag{21}$$

Since this is a semibatch process with monomer fed to the reactor, the maximization of PSPI and MWPI must also be subjected to additional constraints to account for the total amount of monomer in the recipe ($N_{m,T}$), and the total reaction time. These constraints are defined as follows:

$$N_{m,T} = 8 \text{ mol and } t_{\min} \le t \le t_{\max} \tag{22}$$

The solution to this constrained optimal control problem was obtained via an interface to the gOPT dynamic optimization code. Five bounded time intervals (of initially equal duration) were specified for the manipulated variables, $F_{m,A}$, $F_{m,B}$, F_S, F_I and T_{reac}, such that it remained constant over a certain time interval, before moving discretely to the value for the next interval. A violation over 0.1°C is put for the temperature, so that the temperature when jumping from one temperature to the other, the temperature rises gradually.

3.3 Results and Discussions

As discussed, the optimal control strategy involved the computation of a monomer feed trajectory which would drive the process to give a PSD with a specified broad distribution. The required PSD was generated by fixing the reactor temperature at 75 °C, and allowing a 25 min (batch) pre-period to allow for initial particle nucleation, before any further addition. Implementation of the optimal control strategy for the reminder of the run was fully automatically achieved through the control and data acquisition strategy developed, using the Rampsoak set-point option from operator built-in profiles. The final shape of the experimental PSD, MWPI and MW was in good agreement with the simulation results. Figure 2 shows the results obtained to maximize PSPI. Figure 3 shows the results obtained in order to maximize MWPI using only the flowrates, that is, temperature was not used as a manipulated variable. Figure 4was used to maximize MWPI using the additional variable, which is the temperature. The MWD, average particle size, PSD and conversion were used as a validation to the optimization process. It can be seen that on maximizing PSPI, a bimodal distribution was produced, since the requirement was to have a very broad distribution. Another observation regarding the molecular weight is that the molecular weight produced with the temperature as a variable was much larger, and that shows the great effect that temperature has on molecular weight. It has also been observed that maximizing the MWPI would result in the maximizing of the M_n

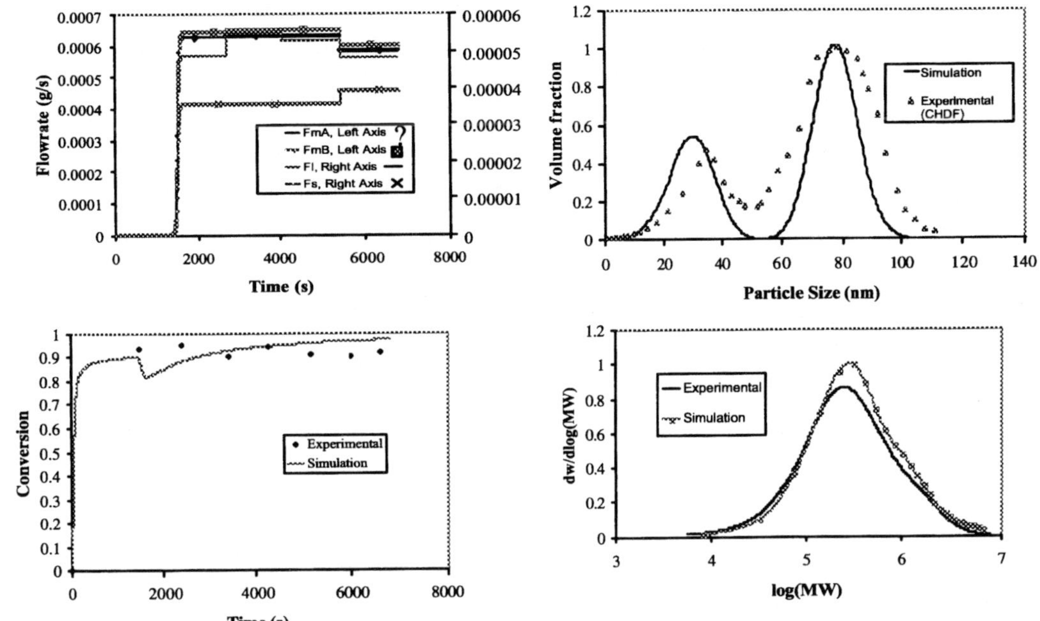

Figure 2. Validation of optimal trajectories to maximize .(a) manipulated variables optimal profile; (b) PSD; (c) Conversion; (d) MWD

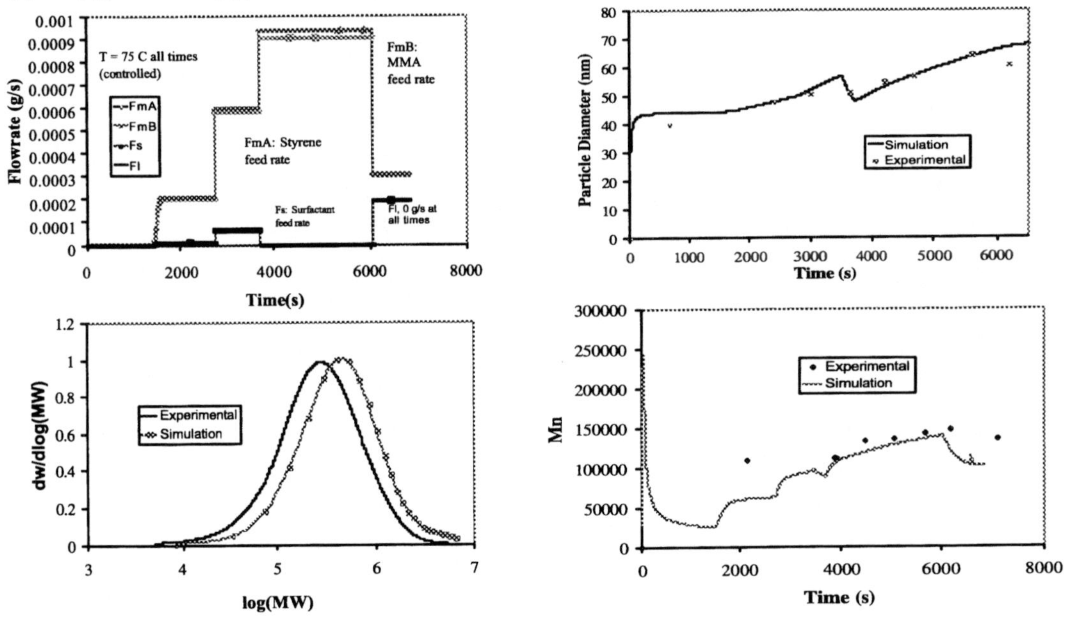

Figure 3: Validation of optimal trajectories to maximize MWPI. (a) manipulated variables optimal profile; (b) Particle diameter; (c) MWD; (d) M_n

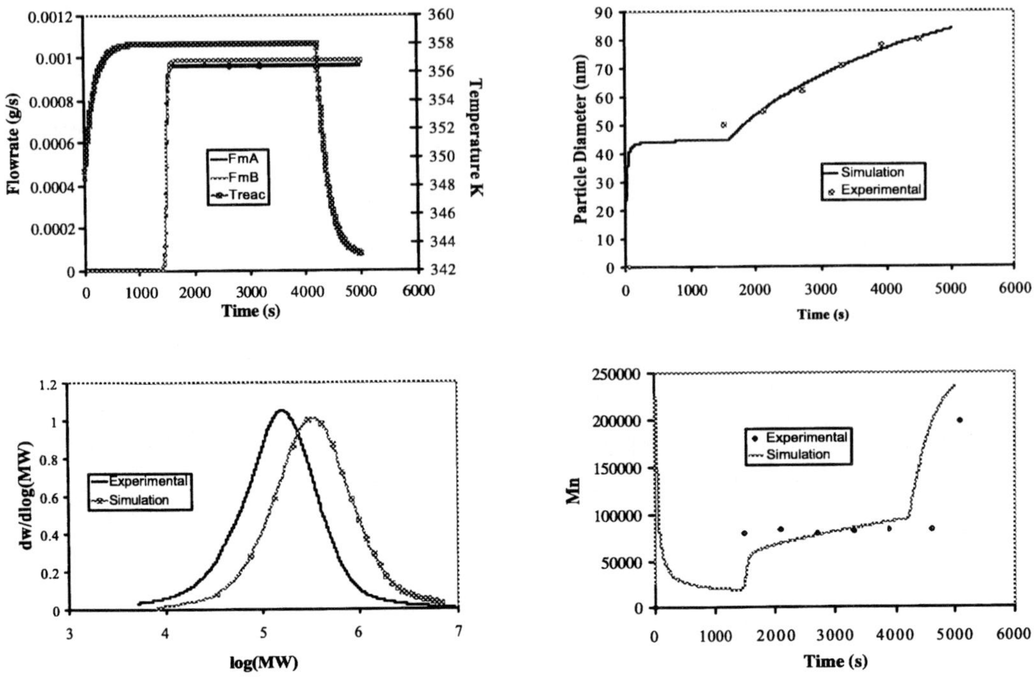

Figure 4: Validation of optimal trajectories to maximize M_n. (a) manipulated variables optimal profile; (b) Particle Diameter; (c) MWD (d) M_n

CONCLUSIONS

A model-based framework to optimally control key copolymerization variables has been developed and implemented within a pilot-scale environment under DCS. Validation results show very good agreement between model prediction and experimental runs. Work is currently underway to implement a simultaneous MWD/PSD control strategy in an on-line environment within a model predictive control scheme.

REFERENCES

Alhamad, B., Romagnoli, J. A. and Gomes, V. G. (2003) In *31st Australasian Chemical Engineering Conference CHEMECA,* Adelaide, South Australia, pp. 138.

Alhamad, B., Willis, R., Romagnoli, J. A. and Gomes, V. G. (2004) In *ESCAPE-14,*Lisbon.

Clarke-Pringle, T. L. and MacGregor, J. F. (1998) *Industrial & Engineering Chemistry Research,* **37,** 3660-3669.

Crowley, T. J., Harrison, C. A. and Doyle III, F. J. (2001) In *2001 American Control Conference, Jun 25-27 2001,* Vol. 2 Institute of Electrical and Electronics Engineers Inc., Arlington, VA, pp. 981-986.

Crowley, T. J., Meadows, E. S., Kostoulas, E. and Doyle, F. J. I. (2000) *Journal of Process Control 14th IFAC World Congress, Jul 5-Jul 9 1999,* **10,** 419-432.

Immanuel, C. D. and Doyle, F. J. I. (2002) *Chemical Engineering Science,* **57,** 4415-4427.

Kozub, D. and Macgregor, J. F. (1992) *Chemical Engineering Science,* **47,** 929-942.

Saldivar, E. and Ray, W. H. (1997) *Aiche Journal,* **43,** 2021-2033.

Semino, D. and Ray, W. H. (1995a) *Chemical Engineering Science,* **50,** 1805-1824.

Semino, D. and Ray, W. H. (1995b) Chemical Engineering Science, 50, 1825-1839.

Zeaiter, J., Romagnoli, J. A., Barton, G. W., Gomes, V. G., Hawkett, B. S. and Gilbert, R. G. (2002) In Chemical Engineering Science, Vol. 57 Elsevier Science Ltd, pp. 2955-2969.

ELSEVIER
IFAC
PUBLICATIONS
www.elsevier.com/locate/ifac

DIRECT SEQUENTIAL DYNAMIC OPTIMIZATION WITH AUTOMATIC SWITCHING STRUCTURE DETECTION

Martin Schlegel, Wolfgang Marquardt [1]

Lehrstuhl für Prozesstechnik, RWTH Aachen University
D–52056 Aachen, Germany

Abstract: In this paper we present a novel method for the numerical solution of dynamic optimization problems. After obtaining a first solution at a coarse resolution of the control profiles with a direct sequential approach, the structure of the control profiles is analyzed for possible switching times and arcs. Subsequently, the problem is reformulated and solved as a multi-stage problem, with each stage corresponding to a potential arc. Order and resolution of the control parameterization are adapted to the type of the particular arc. With a case study we show that accurate solutions with only few degrees of freedom can be obtained.

Keywords: optimal control, dynamic optimization, sequential approach, solution structure, switching times, arcs, multi-stage problem

1. INTRODUCTION

The optimization of the operation of batch processes or transient phases in continuous processes for either off-line or on-line applications requires the solution of dynamic optimization problems. It is still a challenge to obtain a high-quality solution for such problems efficiently, especially when the problem formulation contains large-scale models, e.g. those stemming from industrial applications. However, even for problems with just small models where the computation time is not an issue, the solution quality which can be obtained by numerical methods is not always satisfactory.

One important reason for this is that the analytical solution of a dynamic optimization problem consists of one or more intervals, the so-called *arcs* (Bryson and Ho, 1975). The control variables to be optimized are continuous and differentiable within each interval, but can jump from one interval to the next at the so-called *switching times*. This may pose problems to numerical solution methods, because the quality of the solution depends on the chosen parameterization order and resolution of the control variables. The solution quality can be insufficient, if the parameterization does not properly reflect the switching structure. However, the sequence and nature of the arcs is typically not known beforehand.

For practical applications, these problems are usually treated by either a) accepting the sometimes limited accuracy of the numerical solution, or by b) trying to interpret the numerical solution and to subsequently improve the solution. The latter approach typically involves human interaction such as visual inspection of the numerical solutions for finding the type and sequence of the arcs and often also requires the use of physical insight.

To the authors' knowledge there is only one attempt to try to determine the sequence of arcs in the solution structure automatically. Winderl and Büskens (2002) mention an algorithm for this purpose, though no details are presented. Some

[1] Corresponding author: marquardt@lpt.rwth-aachen.de

solution methods like the one of Vassiliadis *et al.* (1994) allow free interval lengths for control profile discretization, but the switching structure is not explicitly considered there.

In this paper we present the combination of a direct single-shooting approach with an automatic structure detection method. This procedure allows to solve a dynamic optimization problem with a control variable parameterization specifically tailored to reflect the sequence and nature of the different arcs, leading to a high solution quality involving only relatively few degrees of freedom. This is accomplished by a reformulation and subsequent solution of the problem as a *multi-stage problem*, where each stage corresponds to an arc of the solution structure as determined by the detection algorithm. This way a robust and efficient solution technique can be obtained.

2. PRELIMINARIES

2.1 Problem formulation

We consider an optimal control problem of the following form:

$$\min_{\boldsymbol{u}(t),t_f} \Phi\left(\boldsymbol{x}\left(t_f\right)\right) \tag{P1}$$

$$\text{s.t.} \quad \dot{\boldsymbol{x}} = \boldsymbol{f}(\boldsymbol{x},\boldsymbol{u},t), \quad t \in [t_0,t_f], \tag{1}$$

$$\boldsymbol{0} = \boldsymbol{x}(t_0) - \boldsymbol{x}_0, \tag{2}$$

$$0 \geq \boldsymbol{h}^x(\boldsymbol{x},t), \quad t \in [t_0,t_f], \tag{3}$$

$$0 \geq \boldsymbol{h}^u(\boldsymbol{u},t), \quad t \in [t_0,t_f], \tag{4}$$

$$0 \geq \boldsymbol{e}\left(\boldsymbol{x}\left(t_f\right)\right). \tag{5}$$

In this formulation, $\boldsymbol{x}(t) \in \mathbb{R}^{n_x}$ denotes the vector of state variables with the given initial conditions $\boldsymbol{x}_0$. The process model (1) is formulated in form of the vector function $\boldsymbol{f}$. The time-dependent control variables $\boldsymbol{u}(t) \in \mathbb{R}^{n_u}$ are the degrees of freedom for the optimization. The final time t_f can be either fixed or an unknown decision variable, as well. The objective function Φ is formulated as a terminal cost criterion for simplicity. Note that the more general formulation of an integral cost term can be easily converted into the above formulation. Furthermore, path constraints on the states (3) and control variables (4) and endpoint constraints on the state variables (5) can be employed. We assume that each constraint is formulated in terms of simple (lower and upper) bounds on the specific variables. More complex constraints and combined state and control path constraints can be converted into this formulation by adding additional equations to the model.

2.2 Necessary conditions of optimality

By employing Pontryagin's Minimum Principle (Bryson and Ho, 1975), problem (P1) can be reformulated by introducing the *Hamiltonian* function $H(t)$ as

$$\min_{\boldsymbol{u}(t),t_f} H(t) = \boldsymbol{\lambda}^T\boldsymbol{f}(\boldsymbol{x},\boldsymbol{u}) + \boldsymbol{\mu}^T\boldsymbol{h}(\boldsymbol{x},\boldsymbol{u}) \tag{P2}$$

$$\text{s.t.} \quad \dot{\boldsymbol{x}} = \boldsymbol{f}(\boldsymbol{x},\boldsymbol{u},t), \quad t \in [t_0,t_f], \tag{6}$$

$$\boldsymbol{0} = \boldsymbol{x}(t_0) - \boldsymbol{x}_0, \tag{7}$$

$$\dot{\boldsymbol{\lambda}}^T = -\frac{\partial H}{\partial \boldsymbol{x}}, \tag{8}$$

$$\boldsymbol{0} = \boldsymbol{\lambda}^T(t_f) - \left.\frac{\partial \Phi}{\partial \boldsymbol{x}}\right|_{t_f} - \boldsymbol{\nu}^T\left.\left(\frac{\partial \boldsymbol{e}}{\partial \boldsymbol{x}}\right)\right|_{t_f}, \tag{9}$$

$$0 = \boldsymbol{\mu}^T\boldsymbol{h}(\boldsymbol{x},\boldsymbol{u},t), \tag{10}$$

$$0 = \boldsymbol{\nu}^T\boldsymbol{e}(\boldsymbol{x}(t_f)). \tag{11}$$

Here, $\boldsymbol{\lambda}(t) \neq \boldsymbol{0}$ represents the vector of adjoint variables, $\boldsymbol{\mu}(t) \geq \boldsymbol{0}$ the vector of Lagrange multipliers for the path constraints and $\boldsymbol{\nu} \geq \boldsymbol{0}$ the vector of Lagrange multipliers for the terminal constraints. Note that the path constraints on states and control variables are combined into the vector function $\boldsymbol{h} = [\boldsymbol{h}^{xT}, \boldsymbol{h}^{uT}]^T$. An optimal solution of problem (P2) fulfills the necessary conditions of optimality:

$$\frac{\partial H(t)}{\partial \boldsymbol{u}} = \boldsymbol{\lambda}^T\frac{\partial \boldsymbol{f}}{\partial \boldsymbol{u}} + \boldsymbol{\mu}^T\frac{\partial \boldsymbol{h}}{\partial \boldsymbol{u}} = \boldsymbol{0}, \tag{12}$$

$$0 = \boldsymbol{\mu}^T\boldsymbol{h}(\boldsymbol{x},\boldsymbol{u},t), \tag{13}$$

$$0 = \boldsymbol{\nu}^T\boldsymbol{e}(\boldsymbol{x}(t_f)), \tag{14}$$

$$\mu_i = 0, \text{ if } h_i < 0; \ \mu_i > 0, \text{ if } h_i = 0, \tag{15}$$

$$\nu_i = 0, \text{ if } e_i < 0; \ \nu_i > 0, \text{ if } e_i = 0. \tag{16}$$

If a free final time is allowed, an additional transversality condition has to be also satisfied. The complementary conditions (15), (16) can be interpreted in a way that a specific Lagrange multiplier is positive if the corresponding constraint is active and zero otherwise.

2.3 Solution structure

An optimal control profile $\boldsymbol{u}(t)$ consists of one or more arcs in such a way, that the control is continuous and differentiable within each arc, but may jump at the switching points. Conclusions about the possible solution structure can be derived from the necessary conditions (12)-(16). Equation (12) can be written separately for each control $u_i(t)$ as

$$\frac{\partial H(t)}{\partial u_i} = \boldsymbol{\lambda}^T\frac{\partial \boldsymbol{f}}{\partial u_i} + \boldsymbol{\mu}^T\frac{\partial \boldsymbol{h}}{\partial u_i} = 0. \tag{17}$$

The expression consists of two parts, a system dependent part $\boldsymbol{\lambda}^T\frac{\partial \boldsymbol{f}}{\partial u_i}$ and a constraint dependent part $\boldsymbol{\mu}^T\frac{\partial \boldsymbol{h}}{\partial u_i}$. Assuming that we look at the solution structure in a given interval, we can distinguish two cases depending on the value of $\boldsymbol{\lambda}^T\frac{\partial \boldsymbol{f}}{\partial u_i}$:

(1) If $\boldsymbol{\lambda}^T\frac{\partial \boldsymbol{f}}{\partial u_i} \neq 0$, then it follows that $\boldsymbol{\mu}(t) \neq 0$ in order to fulfill (17). This implies that at least one of the path constraints $\boldsymbol{h}(\boldsymbol{x},\boldsymbol{u},t)$ must be active.

(2) If $\lambda^T \frac{\partial f}{\partial u_i} = 0$, then $\mu^T \frac{\partial h}{\partial u_i} = 0$ must hold. Together with (15) this indicates that no path constraint can be active in that interval. The solution therefore lies inside the feasible region; it is not determined by the constraints. Such an interval is termed a *singular arc*.

The first case can be further distinguished in active control constraints or active state constraints. This leads to four different possibilities for the profile $u_i(t)$ in a specific interval, namely

(1) $u_i = u_{i,min}$: the control is at its upper bound,
(2) $u_i = u_{i,max}$: the control is at its lower bound,
(3) $u_i = u_{i,path}$: the control is determined by an active state path constraint,
(4) $u_i = u_{i,sing}$: the control is singular.

This distinction is useful for an automatic detection of the control structure.

3. NUMERICAL SOLUTION

Solution strategies for dynamic optimization problems of form (P1) can be classified into indirect methods, which use the first-order necessary conditions from Pontryagin's Minimum Principle (see Section 2.2) for a reformulation of the problem as a multi-point boundary value problem, and direct methods, which solve problem (P1) directly. We refer to Srinivasan *et al.* (2003) for a review of the various variants of these methods.

In this work, we use the control vector parameterization approach, also referred to as single-shooting or sequential approach, a direct method which solves the problem by conversion into a nonlinear programming problem (NLP) through discretization of the control variables $\boldsymbol{u}(t)$. It is important to note that our solution approach does not require an explicit derivation of the necessary conditions (12)-(16).They are just used for a theoretical justification of the method proposed later.

3.1 Control vector parameterization

In the control vector parameterization approach (Kraft, 1985) only the control variables $\boldsymbol{u}(t)$ are discretized explicitly. The discretization parameters are the degrees of freedom for the optimization. The profiles for the state variables $\boldsymbol{x}(t)$ are obtained by forward numerical integration of the model (1) for a given input. For the parameterization of the control profiles $u_i(t)$ often piecewise polynomial approximations are applied. We use a B-spline representation for this purpose where the discretized control variables can be written as

$$u_i(t) = \sum_{j=1}^{n_i^u} \hat{u}_{i,j}\, \varphi_j^{(m)}\,, \tag{18}$$

with n_i^u denoting the number of parameterization functions for the control variable u_i. Depending on the choice of the order m of the B-spline function $\varphi^{(m)}$ different orders can be realized. Our solution framework uses piecewise constant ($m = 1$) and piecewise linear ($m = 2$) parameterizations. Let Δ_i denote the set of discretization time points for each control variable $u_i(t)$.

3.2 Reformulation as NLP problem

Once the control profiles have been discretized, problem (P1) can be reformulated as an NLP:

$$\min_{\hat{\boldsymbol{u}},t_f} \Phi = \Phi\left(\boldsymbol{x}\left(\hat{\boldsymbol{u}}, t_f\right)\right) \tag{P3}$$

$$\text{s.t.} \quad 0 \geq \boldsymbol{h}(\boldsymbol{x}, \hat{\boldsymbol{u}}, t_i), \quad \forall t_i \in \Delta\,, \tag{19}$$

$$0 \geq \boldsymbol{e}\left(\boldsymbol{x}\left(t_f\right)\right)\,. \tag{20}$$

The vector $\hat{\boldsymbol{u}}$ contains all discretization parameters of the control variables. For evaluation of the state path constraints in (19) we cannot directly use the continuous formulation of (P1), since the NLP requires a finite number of constraints. Therefore, all state path constraints are evaluated point-wise, for example on the unified mesh of all control variables $\Delta := \bigcup_{i=1}^{n_u} \Delta_i$. The number of mesh points contained in Δ is denoted with n_Δ.

3.3 Solution of the NLP problem

Once problem (P3) has been formulated, it can be solved by a suitable NLP solver for a given $\hat{\boldsymbol{u}}$. Typically, a sequential quadratic programming (SQP) method (Nocedal and Wright, 1999) is used for this purpose. A detailed discussion of NLP theory is beyond the scope of this paper. We just mention the key aspects which are required later.

An optimal solution of the NLP problem (P3) fulfills the Karush-Kuhn-Tucker conditions of optimality (Nocedal and Wright, 1999). These are formulated based on the *Lagrangian* function, which is defined as

$$L(\hat{\boldsymbol{u}}, t_f, \hat{\boldsymbol{\mu}}, \hat{\boldsymbol{\nu}}) = \Phi(\boldsymbol{x}(\hat{\boldsymbol{u}}, t_f)) - \hat{\boldsymbol{\mu}}^T \boldsymbol{h}(\hat{\boldsymbol{u}}, \boldsymbol{x}, t) - \hat{\boldsymbol{\nu}}^T \boldsymbol{e}(\boldsymbol{x}(t_f))\,. \tag{21}$$

Each discrete constraint has an associated discrete Lagrange multiplier, $\hat{\mu}_i$ or $\hat{\nu}_i$, respectively. They are related to the Lagrange multipliers $\boldsymbol{\mu}(t)$ and $\boldsymbol{\nu}$ of the continuous problem (P2). The value of each of the discrete multipliers provides information about the status of the particular constraint at the optimal solution. Analogously to the continuous case, they fulfill complementary conditions. In the following we will make use of the convention, that $\hat{\mu}_i = 0$, if the corresponding constraint is not active, $\hat{\mu}_i > 0$, if it is at the upper bound and $\hat{\mu}_i < 0$, if it is at the lower bound.

3.4 Resolution of the control discretization

Obviously, it is desirable to obtain a solution of the discretized problem (P3) which is close to the true solution of the optimal control problem (P1). However, a very fine discretization mesh Δ_i is not a favorable option for practical problems mainly because of three reasons: a) numerical accuracy, b) robustness and c) computational efficiency. If a control profile is represented by a very fine discretization of the form (18), the NLP solution algorithm has to deal with a large number of decision variables. However, as indicated before, typical control profiles may exhibit regions where there are no significant differences between the values of neighboring control vector parameters $\hat{u}_{i,j}$, for example in those parts of the control profile where it is governed by an active control path constraint. In these regions, a fine discretization would not be required to reflect the true solution. But also on singular arcs, a too fine parameterization can lead to numerical problems. Also, the computational effort required for the solution of a dynamic optimization problem with the sequential method is strongly correlated to the number of parameterization functions and renders a fine discretization unattractive from an efficiency perspective. If a-priori knowledge about the solution structure is available, it can be considered while setting up a possibly non-equidistant discretization mesh. The method suggested in the following does not require such an a-priori information. Instead, an appropriate parameterization is determined automatically.

4. AUTOMATIC STRUCTURE DETECTION

The proposed method consists of three main steps:

(1) a (probably coarse) solution of problem (P3),
(2) a detection of the arcs in the solution structure,
(3) a reformulation and solution of a multi-stage problem.

The multi-stage problem in step 3 treats each arc determined in step 2 as a separate stage, where the parameterization order (constant or linear) is chosen according to the type of the arc. For example, a stage corresponding to an arc with $u = u_{max}$ will be parameterized by piecewise constant functions, whereas a piecewise linear parameterization is used on a singular arc.

4.1 Structure detection algorithm

The detection of the arcs and switching times in the control profiles obtained as a solution of problem (P3) works with the help of the concepts of optimal control theory which have been outlined in Sections 2.2 and 2.3. The solution of the discretized problem (P3) including the Lagrange parameters for the path constraints can be used for this purpose. In the following, the corresponding discrete Lagrange parameters for h_i^x and h_i^u are denoted with $\hat{\mu}_i^x$ and $\hat{\mu}_i^u$, respectively. For brevity we restrict ourselves to the case of a single control variable and a single state path constraint. Then, a simplified algorithm looks as follows:

Algorithm 1 Structure detection

```
for i = 1, n_Δ do
  if μ̂^u(i) < 0 then
    utype(i) ← 'min'
  else if μ̂^u(i) > 0 then
    utype(i) ← 'max'
  else
    if μ̂^x(i) ≠ 0 then
      utype(i) ← 'path'
    else
      utype(i) ← 'sing'
    end if
  end if
end for
j ← 1, atype(1) ← utype(1)
for i = 1, n_Δ − 1 do
  if utype(i+1) ≠ utype(i) then
    j ← j + 1
    atype(j) ← utype(i)
  end if
end for
L ← j
```

As a result, we obtain the number of arcs L and their type in the vector *atype*. *utype* denotes the type (u_{max}, u_{sing}, etc.) of the control variable in each discretization interval. Also, initial values for the lengths of the intervals and therefore the switching times can be extracted from the single-stage solution. This information is used for a subsequent reformulation as a multi-stage problem.

4.2 Multi-stage problem formulation

The model $\boldsymbol{f}$ in problem (P1) is continuous and only comprises a single set of state variables $\boldsymbol{x}$ and controls $\boldsymbol{u}$. Such a model describes a process in a single mode or a single discrete state and can therefore be called a *single-stage model.* The corresponding optimization problem is a *single-stage problem.* In contrast, *multi-stage models* are used to describe situations, where the process consists of a sequence of modes or discrete states, with switching from one mode to the next at some time instant t_k (Vassiliadis *et al.*, 1994).

In the context of this paper we explicitly only treat problems which only have one physical stage, i.e. single-stage problems. However, we can use the multi-stage formalism in order to obtain an alternative formulation and solution method for these problems by using the results of the structure detection algorithm.

We define a set $K = \{1, \ldots, L\}$ comprising the indices of all stages. A multi-stage reformulation of the original problem then can be written as

$$\min_{\boldsymbol{u}_k(t), \boldsymbol{x}_k(t_k), t_k} \quad \Phi_L(\boldsymbol{x}_k(t_k)) \tag{P4}$$

$$\dot{\boldsymbol{x}}_k = \boldsymbol{f}_k(\boldsymbol{x}_k, \boldsymbol{u}_k, t),\ t \in [t_{k-1}, t_k], \forall k \in K, \tag{22}$$

$$\boldsymbol{0} = \boldsymbol{x}_1(t_0) - \boldsymbol{x}_0\,, \tag{23}$$

$$\boldsymbol{0} = \boldsymbol{x}_{k+1}(t_k) - \boldsymbol{x}_k(t_k)\,, \forall k \in K \tag{24}$$

$$0 \geq \boldsymbol{h}_k(\boldsymbol{x}_k, \boldsymbol{u}_k, t),\ t \in [t_{k-1}, t_k], \forall k \in K, \tag{25}$$

$$0 \geq \boldsymbol{e_L}(\boldsymbol{x_L}(t_L))\ . \tag{26}$$

The index k denotes quantities belonging to stage k. The time indices are chosen such that e.g. the time horizon of stage 1 runs from t_0 to t_1. The final time of the last stage t_L corresponds to the final time t_f of the single-stage problem. The models $\boldsymbol{f}_k$, $k \in K$, are all identical copies of model (1). The initial conditions of the first stage ($k = 1$) are set to $\boldsymbol{x}_0$ by equation (23) and are therefore also the same as in equation (2). In addition, the multi-stage problem requires so-called stage transition or mapping conditions (24), which map the state variable values $\boldsymbol{x}_k$ across the stage boundaries. The objective function of a multi-stage problem is often obtained by summing up individual costs $\Phi_k(\boldsymbol{x}_k(t_k))$ formulated for each model stage $k \in K$. Here, just the objective function value at the endpoint of the last stage $\Phi_L(\boldsymbol{x}_L(t_L))$ has to be minimized. The influence of the other stages is propagating into the final stage via the initial condition $\boldsymbol{x}_L(t_{L-1})$.

The path constraints obviously should be fulfilled in all stages and are therefore also copied as in equation (25). Similarly to the objective function the endpoint constraints should be only evaluated at the final time point t_L. Therefore, they are formulated only for stage L (equation (26)).

In the single-stage problem, the final time t_f might be a degree of freedom. In the reformulated problem each stage corresponds to an arc of the solution structure. Since we know that the solution of the discretized problem (P3) is not likely to exactly match the switching times of the true solution, we would like to introduce those switching times as additional degrees of freedom. The multi-stage formulation provides a well-suited framework for this purpose, because the length of the arcs (which can be easily converted into the switching times) are just the parameters t_k. Therefore, these are included into the set of optimization degrees of freedom in problem (P4). Note that the times t_k are even free in those cases where t_f is fixed, though in principle one of them then can be determined from the others.

The control variables $\boldsymbol{u}(t)$ in the original problem are now also present in each model stage as $\boldsymbol{u}_k(t)$. In order to be able to fulfill the mapping conditions at the stage boundaries, the multi-stage optimization problem requires additional degrees of freedom: the initial values of the state variables $\boldsymbol{x}_k$ of all stages but the first. Those have to be determined by the optimization algorithm such that at the optimal point the conditions (24) are satisfied. As indicated earlier, the order m of the B-splines used for discretization is adapted to the type of arc. More precisely, we use $m = 1$ and one interval, if atype = 'min' or 'max', and $m = 2$ and two discretization intervals for atype = 'path' or 'sing'.

The maybe coarse solution of the single-stage problem allows to derive good initial values for the optimization variables of (P4). A proper initialization of $\boldsymbol{u}_k(t)$ is available through interpolation of the single-stage solution in the particular region of the time horizon belonging to arc (or stage) k. The multi-stage problem is then solved in the same way as the single-stage problem, i.e. through conversion into an NLP problem as described above.

5. ILLUSTRATIVE EXAMPLE

The numerical concepts presented above have been implemented into the software tool DyOS (DyOS, 2002). The implementation allows optimization of any model (1) compliant to the Cape-Open ESO interface standard (Keeping and Pantelides, 2000), e.g. through the modeling and simulation package gPROMS (gPROMS, 2002). The user just has to specify the problem in form of (P1). Switching structure detection, reformulation and solution of the multi-stage problem are carried out automatically. The sample problem has been solved by using SNOPT (Gill *et al.*, 1998) as NLP solver and a modified version of LIMEX (Schlegel *et al.*, 2004) as numerical integrator.

As an example we consider the optimization of a fed-batch bioreactor with inhibition and a biomass constraint. Details of the problem can be found elsewhere (Srinivasan *et al.*, 2003). The problem has been solved for two different cases: starting from a single-stage solution with $n_\Delta = 8$ and $n_\Delta = 32$ equidistant discretization intervals. As the solution plots in Figure 1 show, the optimal profile of the substrate feed rate exhibits a complex switching structure with $L = 4$ arcs. The arcs of the true solution have been detected correctly. The reformulated multi-stage solution is of higher quality, because a) the switching times have been captured properly (which was not the case in the single-stage solutions) and b) the piecewise-linear parameterization in the singular and path-constrained regions provide good approximations of the true solution with very few degrees of freedom. Note that the multi-stage solutions derived from the two different single-stage solutions are essentially identical.

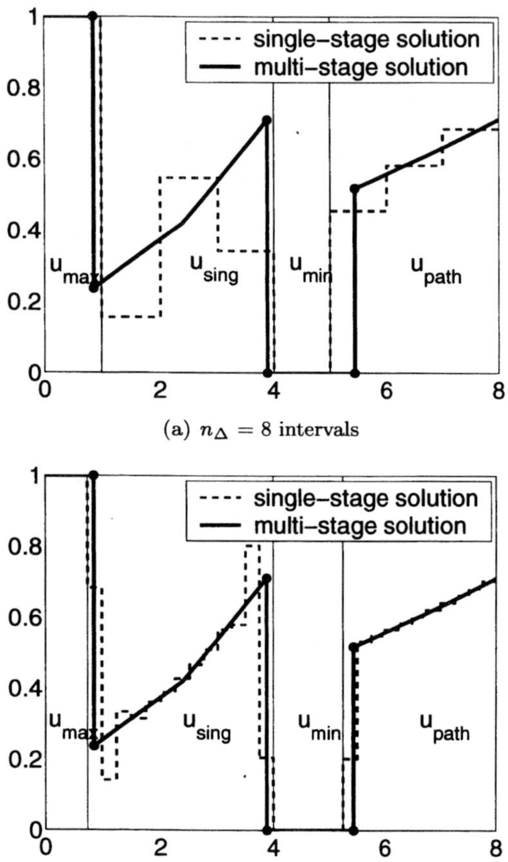

(a) n_{Δ} = 8 intervals

(b) n_{Δ} = 32 intervals

Fig. 1. Optimal solution profiles for u.

6. CONCLUSIONS

We have presented a direct method for the numerical solution of dynamic optimization problems, which determines and explicitly considers the structure of the optimal control profiles. The method yields a high solution accuracy with a small number of degrees of freedom. In contrast to indirect methods the method can be easily applied to large-scale problems. No a-priori knowledge about the solution structure is required.

The multi-stage solution can be useful in on-line applications like moving-horizon optimization, because taking the switching structure into account can add robustness and computational efficiency to the numerical solution method.

As presented here, the arcs in the solution structure can be only captured properly, if they are already present in the single-stage solution. This assumption might be violated in some cases, e.g. if the single-stage discretization is too coarse. A highly resolved solution of the initial problem would most likely contain all arcs, but cannot be solved efficiently. Current work is therefore aiming at including grid point adaptation techniques on both, the single and the multi-stage level in order to capture the correct solution structure efficiently also in complex problem. Recent extensions of the method allow to handle problems with more than one control variable, as well as enable an efficient treatment of large-scale process models.

REFERENCES

Bryson, A.E. and Y.-C. Ho (1975). *Applied Optimal Control.* Taylor & Francis. Bristol, PA.

DyOS (2002). *DyOS User Manual, Release 2.1.* Lehrstuhl für Prozesstechnik, RWTH Aachen University. Aachen, Germany.

Gill, P.E., W. Murray and M.A. Saunders (1998). SNOPT: An SQP algorithm for large-scale constrained optimization. Technical report. Stanford University. Stanford, USA.

gPROMS (2002). *gPROMS User Guide (Release 2.1.1).* Process Systems Enterprise Ltd.. London, UK.

Keeping, B. and C. Pantelides (2000). Numerical Solvers Open Interface Specification Draft. Technical Report CO-NUMR-EL-03. CPSE, Imperial College. London.

Kraft, D. (1985). On converting optimal control problems into nonlinear programming problems. *Comput. Math. Prog.* **15**, 261–280.

Nocedal, J. and S.J. Wright (1999). *Numerical Optimization.* Springer-Verlag.

Schlegel, M., W. Marquardt, R. Ehrig and U. Nowak (2004). Sensitivity analysis of linearly-implicit differential-algebraic systems by one-step extrapolation. *Appl. Num. Math.* **48**(1), 83–102.

Srinivasan, B., S. Palanki and D. Bonvin (2003). Dynamic optimization of batch processes I. Characterization of the nominal solution. *Comp. Chem. Eng.* **27**, 1–26.

Vassiliadis, V.S., R.W.H. Sargent and C.C. Pantelides (1994). Solution of a class of multistage dynamic optimization problems: 1. Problems without path constraints. *Ind. Eng. Chem. Res.* **33**(9), 2111–2122.

Winderl, S. and C. Büskens (2002). Exploiting the special structure for real-time control of optimal control problems with linear controls: Numerical experience. In: *Proc. Appl. Math. Mech.*. Vol. 1. Wiley. pp. 484–485.

ELSEVIER
IFAC
PUBLICATIONS
www.elsevier.com/locate/ifac

AN AGENT-BASED FRAMEWORK FOR CONTROL OF REACTOR NETWORKS WITH AUTOCATALYTIC REPLICATORS

Eric Tatara, Fouad Teymour, and Ali Cinar

Department of Chemical and Environmental Engineering,
Illinois Institute of Technology, Chicago, IL 60616

Abstract: Several recent studies on autocatalytic reactions in single and coupled continuous stirred-tank reactor (CSTR) networks have demonstrated a rich spectrum of complex behavior. From a control systems perspective, the operating regime of a CSTR network can be manipulated by changing the flow rates between the reactors. Systems of more than one CSTR require multiple controllers and may need transients through several operating regimes to achieve the desired operation. This may require a hierarchical control structure whereby local control objectives can change dynamically in order to achieve the global control objective of the system. An agent-based control system is used to observe and control various aspects of a CSTR network. The primary focus of analyzing complex emergent behavior is demonstrating methods or combinations of methods that influence the behavior and capabilities of the agent-based control system. Simulation studies illustrate scenarios of interest, especially conditions that lead to emergent behavior.

Keywords: Intelligent Agents, Agent-Based Control, Reactor Networks

1. INTRODUCTION

The behavior of continuous stirred tank reactors (CSTRs) has been studied extensively over the past several decades. Fascinating static and dynamic phenomena have been observed for many classes of reactions including autothermal reactions, autocatalytic reactions and polymerization reactions. The analysis of Uppal and Ray (1974) on the behavior of a CSTR with a non-isothermal reaction was one of the first thorough investigations of complex behavior in chemical reaction systems, extending the initial ideas presented by Bilous and Amundson (1955). Complex dynamic behavior has been identified through analytical solutions of the state equations and their stability characteristics. Further investigation of the non-isothermal system (Uppal *et al.*, 1976; Farr and Aris, 1986) using the reactor feed flow rate as a bifurcation parameter yield even more types of complex behavior, such as mushrooms and isolas.

The work of Lin (1981) and Gray and Scott (1983) on the isothermal autocatalytic reaction $A+nB \rightarrow (n+1)B$ with decay $B \rightarrow C$ show that cubic ($n = 2$) and quadratic ($n = 1$) autocatalytic reactions exhibit similar behavior to non-isothermal reactions. The solution multiplicities include isolas, mushrooms and limit cycles. Further work has determined how the various model parameters affect the stability criteria and conditions for multiplicity as well as more sophisticated numerical analysis of the periodic solutions (Gray and Scott, 1984). Birol and Teymour (2000) have studied isothermal autocatalysis when two competing species are introduced to the system. It was shown that with N species, although there may be as many as $2(2^N - 1)$ steady states, only one species can exist stably in the reactor as $t \rightarrow \infty$.

If several identical CSTRs are connected, such that material is exchanged between them, the system becomes spatially heterogeneous compared with a single CSTR. The degree of heterogeneity can be varied by manipulating the interconnections between reactors. Taylor and Kevrekidis (1993) have extensively studied the effects of reactor coupling via numerical bifurcation techniques on the non-isothermal $A \rightarrow B$ system. Oscillatory states tend to synchronize when the frequency of the oscillations in each reactor are not too different. Weak coupling of the reactors is usually used, since strong coupling of a reactor network causes the network to behave as one large reactor.

Recent work on multiple reactor configurations with cubic autocatalytic reactions has demonstrated that spatial heterogeneity enlarges the boundaries of species survival (Birol *et al.*, 2002). Furthermore, detailed analysis has shown that networks of reactors with autocatalytic replicators produce highly complex bifurcation structures and that the number of steady states increases exponentially with size of the system (Tatara *et al.*, 2003). With the autocatalytic reaction scheme, larger networks permit more steady states and spatial combinations thereof than smaller networks. Although much of the bifurcation diagram is dominated by unstable steady states, there exists a number of stable steady states for a large range of feed flow rates.

In the case of a single CSTR with competing autocatalytic species, a nonlinear control scheme is necessary to first remove the invading species and then return the host species to the original steady state (Chaivorapoj *et al.*, 2003). Systems of more than one CSTR require multiple controllers and may require transients through several operating regimes to achieve the desired operation. Furthermore, when the system contains multiple reactors, situations arise such that global control objective can be satisfied by several different combinations of local control objectives. This leads to the requirement of a hierarchical control structure whereby local control objectives can change dynamically in order to achieve the global control objective of the system.

2. REACTOR NETWORK MODEL

A network of I interconnected CSTRs is modeled by specifying the mass balance for an individual reactor at position i in the network, where $i = 1..I$. Figure 1 shows the schematic for a system of four CSTRs. The cubic autocatalytic reaction for a single autocatalytic species is

$$R + 2P \xrightarrow{k} 3P \tag{1}$$

$$P \xrightarrow{k_d} D \tag{2}$$

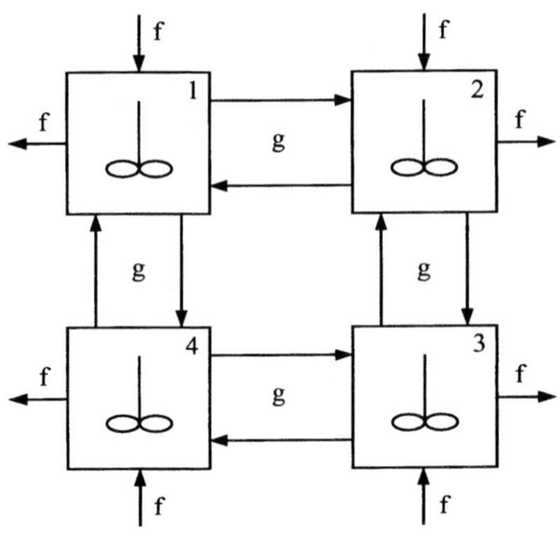

Fig. 1. 4-CSTR network schematic

R is the resource concentration, P is the species concentration, D is a dead (inert) species, k is the species growth rate constant, and k_d is the species death rate constant.

The production rates of the resource and species concentrations for a network of size I are

$$V\frac{dR_i}{dt'} = -VkR_iP_i^2 + F(R_0 - R_i) + G(R_{i-1} + R_{i+1} - 2R_i) \tag{3}$$

$$V\frac{dP_i}{dt'} = VkR_iP_i^2 - P_i(F + Vk_d) + G(P_{i-1} + P_{i+1} - 2P_i) \tag{4}$$

where R_0 is the resource concentration in the feed, R_i is the resource concentration in reactor i , P_i is the species concentration in reactor i, F is the feed flow rate, G is the interconnection flow rate and V is the reactor volume. The feed stream contains only resource. The state equations can be written in dimensionless form as

$$\frac{dr_i}{dt} = -kr_ip_i^2 + f(1-r_i) + g(r_{i-1} + r_{i+1} - 2r_i) \tag{5}$$

$$\frac{dp_i}{dt} = kr_ip_i^2 - p_i(f+d) + g(p_{i-1} + p_{i+1} - 2p_i) \tag{6}$$

by redefining the variables as $r_i = R_i/R_0$, $p_i = P_i/R_0$, $f = F/(VR_0^2)$, $g = G/(VR_0^2)$, $d = k_d/R_0^2$, and $t = R_0^2t'$. For $i > 3$, analytical solutions become intractable, although it should be noted that a single trivial steady state ($r_i = 1$, $p_i = 0$) exists $\forall i$ for every combination of model parameter values. This state represents total extinction in the system. The feed flow rates and interconnection flow rates are treated as manipulated variables. Constraints on the reactor flow rates ensure that material is conserved.

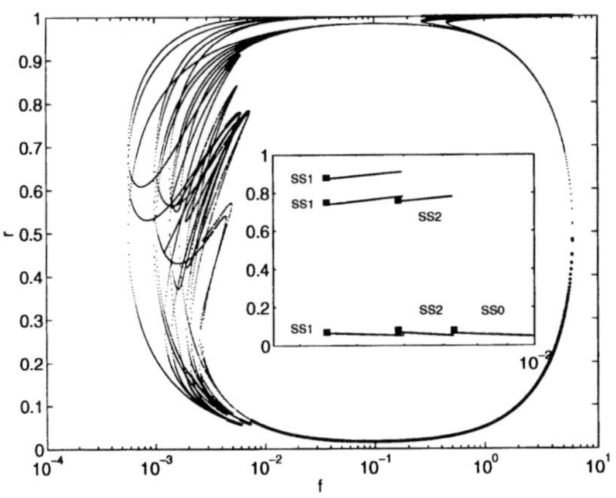

Fig. 2. Bifurcation diagram of r vs f for a 4-CSTR bidirectional network with $k = 25$, $d = 0.1$, and $g = 0.002$. Unstable steady states are indicated with a light line and stable steady states are indicated with a dark line. Inset: stable steady state detail. The ■ symbol represents a Hopf bifurcation point.

3. STEADY STATE BIFURCATION STRUCTURE

The nonlinear solver KINSOL (Hindmarsh and Taylor, 1998) is used to map the bifurcation structure of reactor networks. The steady state solutions are found by sweeping the bifurcation parameter through the region of interest and using a large number of randomized initial guesses in a manner similar to that of Ourique *et al.* (2002). Although this method is more computationally demanding than traditional numerical bifurcation techniques, it removes the constraint of a priori knowledge of the steady state branches. Traditional bifurcation analysis will miss isolated solution branches, unless their location is known beforehand. Stability information is obtained by examining the eigenvalues of the Jacobian matrix.

The steady state bifurcation diagram of r versus f for a single species in 4-CSTR bidirectional ring network is shown in Figure 2. The bifurcation diagram is constructed for all possible steady state values, thus the omission of the reactor number subscript, i. The resource concentration steady states in the bifurcation diagram represent those values that *can* occur in an individual reactor, but not the spatial arrangement of the steady states in the network. The 4-CSTR bidirectional network has two combinations of stable steady states at low feed flow rates in addition to the steady state of the single CSTR. The first combination (SS1) displays three distinct steady states. The spatial pattern consists of one high resource reactor surrounded by two identical lower resource reactors, followed by the lowest resource reactor. Reducing the feed flow rate from SS1 results in the system moving to a periodic regime via a Hopf bifurcation. When the feed flow rate is increased from the SS1, the network displays quasi-periodicity and chaos. Increasing the feed flow rate further causes the network to move to the single CSTR periodic regime in which all reactors synchronize to single CSTR limit cycle. A further increase in the feed flow rate results in the nontrivial single CSTR stable steady state (SS0).

The second reactor configuration (SS2) in the 4-CSTR bidirectional network mimics a 2-CSTR setup through locking neighboring CSTRs in pairs. Note that, though two stable steady states are possible in four reactors, the configuration is always a pair of pairs, as opposed to alternating high and low concentrations. Reducing the feed flow rate from SS2 causes the system to pass through a Hopf bifurcation point and move into a periodic regime. If the feed flow rate is increased from SS2, the system collapses to the nontrivial single CSTR stable steady state because the stability region of the single CSTR SS0 begins just as the stability region of 4-CSTR network terminates. Note that, while for the previous spatial pattern of states in the 4-CSTR network, the system could be moved to an oscillatory regime by increasing the flow rate, in this case, the ending and starting points of the stability region meet, and this does not allow periodic behavior to occur.

4. CONTROL SYSTEM ARCHITECTURE

From a control perspective, reactor networks pose a tough challenge because several different control strategies are necessary to achieve the desired operational goal. Intelligent supervisory control systems (Kendra *et al.*, 1994) adapt to changing process operating conditions, thereby facilitating the control of such processes. Furthermore, the operation of highly nonlinear systems like autocatalytic replicator networks benefit from *evolutionary control* because the optimal operating regime may not be known a priori. Agent-based control systems (Jennings and Bussmann, 2003) provide the capability for localized and global control strategies that are both reactive in controlling disturbances and proactive in searching for more better operational solutions.

Software agents are an extension of object oriented programming in that both agents and objects encapsulate information. The difference between agents and software objects is that agents are semi-autonomous units that employ a form of reasoning to negotiate with other agents - either for self interest or that of the collective (Jennings, 2000). Multi-agent systems have several properties that make them particularly attractive for use with large, complex systems (Lesser, 1999). The first, and usually most impor-

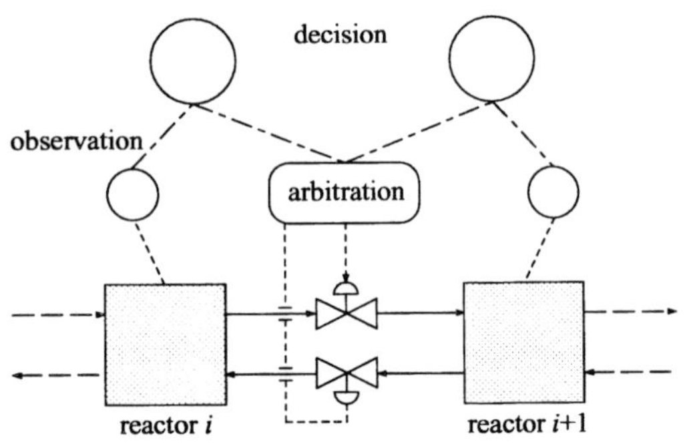

Fig. 3. Control agent architecture.

tant in critical systems, is a high level of reliability. Modularity and scalability also play an important role in multi-agent systems. Software agents often produce different solutions to the same problem. *Solution multiplicity* arises when several agents, using completely independent methods, arrive at different conclusions based on the presented data. Negotiation between agents is therefore required to resolve the situation.

The agent-based control system architecture consists of several sub-systems, each of which are highly modularized (Figure 3). At the process level, network elements such as reactors and valves interface with the higher level agents via low level agents. Each reactor is monitored by an *observation* agent that is responsible for sampling data requested by other agents as well as storing the data in a history for some specified time. Data collection occurs asynchronously with data requests from other agents. Upon instantiation, the frequency at which an observation agent samples data from the process is set by a superior agent. The observation agent maintains the process history and provides data to superior agents when queried. The interconnection flow rates are manipulated by *actuation* agents (not shown) that receive commands from superior control agents.

The next layer in the control hierarchy is the *local decision* layer. Local decision agents are responsible for monitoring control functions and proactively improving the overall performance of the network based on the control objectives of the individual agents and the network as a whole. Due to the number of control responsibilities of decision agents, each agent may use sub-agents. For example, the local control decision agent requires information regarding the state of the process. A sub-agent is therefore tasked with checking whether the system is at steady state, on a limit cycle, or a quasi-periodic regime.

During network operation, local decision agents attempt to satisfy their individual control objectives. An objective may be to maximize a specific species concentration for example. However, in most cases, a decision agent can never fully reach it's desired objective due to conflicts with other agent's control objectives. If an agent desires to modify the interconnection flow rate between a reactor and its neighbor in order to meet a control objective, the adjacent reactor will be affected as well. Naturally, disputes will arise as to the value of the interconnection flow rates between neighboring reactors.

Arbitration agents serve both as a communication channel between decision agents as well as a means to resolve disputes between agents. The arbitration agents receive requested operational procedures from the local decision agents and then presents a solution to them. For example, a decision agent must maximize the species concentration in a reactor and attempts to do so by manipulating the interconnection flow rate. The decision agent sends a set of acceptable values for the manipulated variables to the arbitration agent which then tries to match the desired operational values for neighboring decision agents. If the arbitration agent cannot find an acceptable solution to the control requests from two or more decision agents, then the control move is not permitted. Finally, *supervision* agents function as the topmost layer in the control system hierarchy. This layer is responsible for setting the desired global operating conditions for the entire network.

Considering the nonlinearity of reactor networks, it is difficult to predict how the bifurcation structure of the system changes when the system parameters are manipulated. Consequently, one cannot easily predict how to change operating conditions of the network by manipulating the flow rates. Decision agents are given the task of increasing the concentration of the autocatalytic species in their reactor by manipulating the interconnection flow rates between neighboring reactors. Various methods may be used to guide the decision agents in planning their control strategies including dynamic exploration of the parameter space, rule-based heuristic models, or first-principles based models.

Decision agents may exploit a model of the reactor network, say by knowing the precise location of stable branches and oscillatory regimes. For example, the complete bifurcation structure shown in Figure 2 would be quite valuable to the decision agents in formulating a control strategy. However, since the number of steady states increases exponentially with the size of the system, this method is not scalable to larger systems. An effective solution is provided via rule-based heuristic models coupled with dynamic exploration techniques.

A heuristic model of the reactor networks consists of rules that describe how the manipulated variables effect the system behavior. For example, in the four reactor network bifurcation diagram

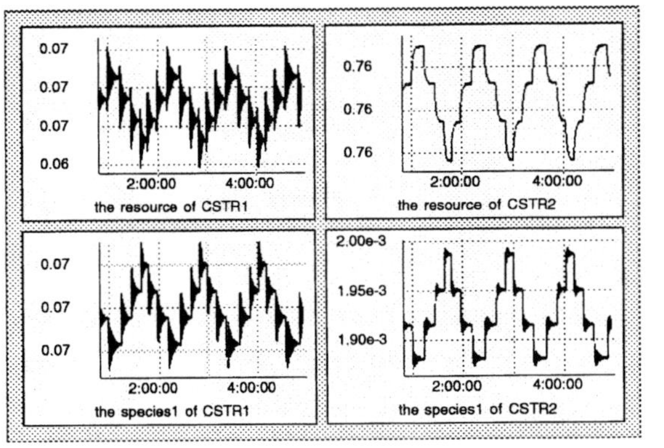

Fig. 4. Non-cooperative agent control strategy

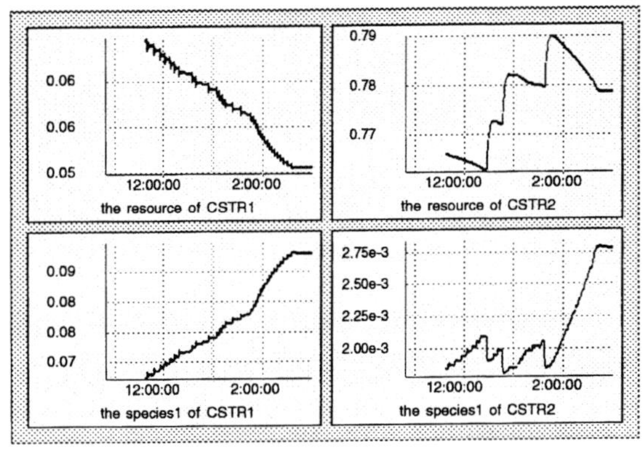

Fig. 5. Cooperative control strategy performance

shown in Figure 2, the stable steady states occupy only certain portions of the diagram. This information is provided to the decision agents in the form of rules to guide their control actions. Furthermore, the decision agents are allowed to "probe" the system by making small, temporary changes to the manipulated variables and observing the resulting system behavior. This dynamic exploration provides additional flexibility to the decision agents when the generalized heuristic model cannot explain system behavior.

The software agents have been developed in G2 (Gensym, 2003), which is a graphical knowledge base development environment for creating intelligent real-time applications. G2 provides an excellent platform for the development of agent-based monitoring and control systems. The ordinary differential equations that describe the autocatalytic reactions in each CSTR are solved numerically using the CVODE solver (Cohen and Hindmarsh, 1994). The solver code is written in C and linked to the G2 agent-based system via a custom software bridge. The reactor class definition in G2 is designed such that the reactor objects have the same attributes as defined in the ODE model. For example, the feed rate to any particular CSTR object is mapped to the specific array location in the ODE solver. When the simulator is initialized, the user may specify the size of the reactor network, initial conditions and inputs. The appropriate number of reactor objects are automatically created by the agent-based system and the reactor objects are modified to include the initial conditions. The CVODE solver simply requires the initial states and parameters to describe the system. The solver then dynamically creates the appropriate equations internally and returns the output to G2, where the states are then mapped back to the reactor objects.

5. EFFECTS OF CONTROL STRATEGY

The performance of two different control strategies is examined in case studies of agent interactions between neighboring reactors. The reactor network is operated initially on a stable steady state. The control objective for the two neighboring decision agents is to maximize the concentration of autocatalytic species in their respective reactors. A competitive strategy is defined by allowing the decision agents to make selfish control moves without regard for its neighbor. In contrast, a cooperative strategy permits the neighboring agents to coordinate their control decisions by allowing a temporary degradation in performance in return for a long-term performance gain.

The result of using a purely competitive strategy on neighboring reactor concentrations is shown in the time series charts of Figure 4. Initially the network is located on steady state SS2 (Figure 2). The competitive control strategy used by the local decision agents allows each to agree on a new operating regime only if both agents see a improvement in their reactor. As shown in Figure 4, the agents are able to improve the operating conditions (increase in species concentration) only slightly before the performance begins to degenerate once again. The control strategy gets stuck in the parameter space and will oscillate indefinitely unless halted by the supervision agents.

A more robust approach to this problem is to permit the local decision agents to suffer some performance loss, but only for a small number of control moves. Figure 5 shows the concentrations of neighboring reactors when the decision agents are designed to cooperatively optimize the network performance. The rules governing the arbitration agent permit a decision agent to make a control move that is detrimental to its neighbor for only one iteration, otherwise it must return to its previous state. This strategy proves to be very effective at improving the network performance. Although the performance of CSTR 2 occasionally suffers while the performance of CSTR 1 improves continuously, this loss is only relative to the previous move and ultimately results in a net gain in performance for the whole network.

6. CONCLUSIONS

Interconnected networks of CSTRs with an autocatalytic reaction mechanism produces highly complex static and dynamic behavior. The extent of the complexities in the bifurcation structure of the system cannot be understood simply by examining a few regions of the parameter space and network configurations. These uncertainties naturally result in complications in controlling the system, either in disturbance rejection or changing the operational regimes of the system.

The concept of agent-based control has been successfully applied to a reactor network to improve the overall performance of the system. The multiagent control system is able to explore the parameter space of the network and intelligently manipulate the network flow rates such that the specified goal is achieved.

Furthermore, it was shown that cooperative relationships between the decision agents provide a more effective paradigm for improving the network performance than selfish relationships. The overall network performance is improved, despite the fact that individual reactors may see a temporary decrease in performance.

Acknowledgment: This material is based upon work supported by the National Science Foundation under Grant No. CTS-0325378.

REFERENCES

Bilous, O. and N.R. Amundson (1955). Chemical reactor stability and sensitivity. *AIChE Journal* **1**(4), 513–521.

Birol, I. and F. Teymour (2000). Statics and dynamics of multiple cubic autocatalytic reactions. *Physica D* **144**, 279–297.

Birol, I., S.J. Parulekar and F. Teymour (2002). Effect of environment partitioning on the survival and coexistence of autocatalytic replicators. *Physical Review E* **66**, 051916.

Chaivorapoj, W., I. Birol, A. Cinar and F. Teymour (2003). Feedback control of a continuous-flow stirred tank reactor with competing autocatalators. *Industrial and Engineering Chemistry Research* **42**, 3765–3785.

Cohen, S.D. and A.C. Hindmarsh (1994). *CVODE User Guide.* LLNL Report UCRL-MA-118618.

Farr, W.W. and R. Aris (1986). Yet who would have thought the old man to have had so much blood in him?-reflections on the multiplicity of steady states of the stirred tank reactor. *Chemical Engineering Science* **41**(6), 1385–1402.

Gensym (2003). *G2 Reference Manual.* Gensym Corporation, Boston, MA.

Gray, P. and S.K. Scott (1983). Absence of restrictions on permissible paths for changing the efficiencies of operation of flow reactors. *Journal of Physical Chemistry* **87**(11), 1835–1838.

Gray, P. and S.K. Scott (1984). Autocatalytic reactions in the isothermal, continuous stirred tank reactor: Oscillations and instabilities in the system $A + 2B \rightarrow 3B; B \rightarrow C$. *Chemical Engineering Science* **39**(6), 1087–1097.

Hindmarsh, A.C. and A.G. Taylor (1998). Pvode and kinsol: Parallel software for differential and nonlinear systems. *Lawrence Livermore National Laboratory technical report UCRL-ID-129739.*

Jennings, N.R. (2000). On agent-based software engineering. *Artificial Intelligence* **117**, 277–296.

Jennings, N.R. and S Bussmann (2003). Agent-based control systems. *IEEE Control Systems* **23**, 61–73.

Kendra, S.J., M.R. Basila and A. Cinar (1994). Intelligent process control with supervisory knowledge-based systems. *IEEE Control Systems* **14**, 37–47.

Lesser, V.R. (1999). Cooperative multiagent systems: A personal view of the state of the art. *IEEE Transactions on Knowledge and Data Engineering* **11**(1), 133–142.

Lin, K.F. (1981). Multiplicity, stability and dynamics for isothermal autocatalytic reactions in cstr. *Chemical Engineering Science* **36**(9), 1447–1452.

Ourique, C.O., E.C. Biscaia and J.C. Pinto (2002). The use of particle swarm optimization for dynamical analysis in chemical processes. *Computers and Chemical Engineering* **26**, 1783–1793.

Tatara, E., I. Birol, F. Teymour and A. Cinar (2003). Static and dynamic behavior of autocatalytic replicators in heterogeneous reactor networks (paper 457f). *AIChE Annual Meeting.* San Francisco, CA, November 16-21.

Taylor, M.A. and I.G. Kevrekidis (1993). Couple, double, toil and trouble: A computer assisted study of two coupled CSTRs. *Chemical Engineering Science* **48**(11), 2129–2149.

Uppal, A. and W.H. Ray (1974). On the dynamic behavior of continuous stirred tank reactors. *Chemical Engineering Science* **30**, 967–985.

Uppal, A., W.H. Ray and A.B. Poore (1976). The classification of the dynamic behavior of continuous stirred tank reactors-influence of reactor residence time. *Chemical Engineering Science* **31**, 205–214.

ELSEVIER
IFAC
PUBLICATIONS
www.elsevier.com/locate/ifac

INCLUSION OF ACTUATOR SATURATION AS COMPLEMENTARITY CONSTRAINTS IN INTEGRATED DESIGN AND CONTROL

Rhoda Baker and Christopher L. E. Swartz

Department of Chemical Engineering, McMaster University, Hamilton, Ontario, L8S 4L7, Canada

Abstract: The design of a process can significantly affect its ability to be satisfactorily controlled, which motivates the need to design the process and its associated control system simultaneously. However, since all physical systems have constraints, it is necessary to include these within the model as well. A bilinear formulation for rigorous saturation handling is able to correctly describe the logical conditions which accompany saturation of the manipulated variable and is suitable for incorporation within a simultaneous optimization framework. This paper explores a number of issues. First, a bilinear formulation is used to examine the effect of including anti-reset windup within an integrated design. Second, the performance of a solver specifically tailored to solve optimization problems with complementarity constraints is compared to other nonlinear solvers for the solution of a simple nonlinear case study. Finally, its performance on a more complex nonlinear CSTR case study is discussed.

Keywords: saturation, anti-reset windup, complementarity, integrated design

1. INTRODUCTION

It has long been recognized that the design of a process can significantly affect its inherent ability to be controlled using feedback. This has motivated the need for techniques which allow the control system to be designed in conjunction with the process design. One means of achieving this is through an optimization-based approach, which allows the simultaneous consideration of both the steady state economics as well as the dynamic operation of the process under feedback control.

Applying such optimization techniques requires a mathematical model of the process as well as its associated control system for inclusion within an optimization problem. However, all physical systems are subject to constraints and, in order to model processes correctly, these constraints should be included within the model formulation as well. These include, in particular, constraints on manipulated inputs where a valve, for example, cannot be more than 100% open. However, when creating a model of a process for the purpose of optimization based integrated control and design, merely including these limits as upper or lower bounds may be insufficient to correctly describe the behavior of the actual system.

In particular, when dealing with a control system, a disturbance of sufficiently large magnitude may in reality cause the manipulated variable to saturate, whereas the above constraint specification would merely allow the modelled manipulated variable to touch the bound and then move away as illustrated in Figure 1. This is because input saturation essentially corresponds to a set of logical conditions, and requires special formulation in order to be included within a conventional optimization framework. A bilinear or mixed-integer formulation for rigorous saturation handling can be used to correctly describe this phenomenon (Young *et al.*, 2004).

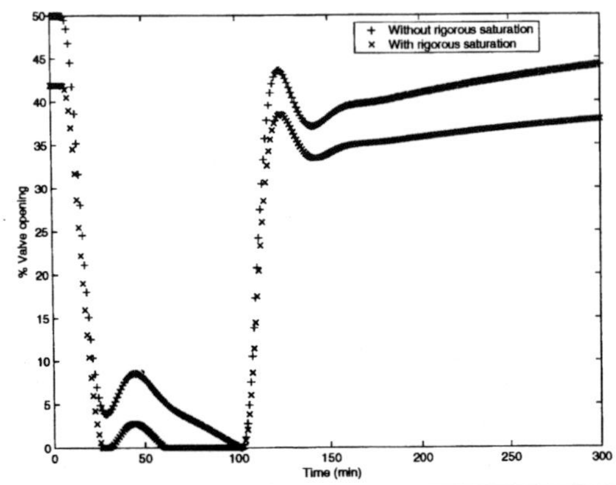

Fig. 1. The effect of including rigorous saturation handling in an integrated design and control optimization framework.

Briefly, in the case of the bilinear formulation, the inclusion of the following constraints in an integrated control and design problem will correctly account for the difference between the actuator, u_a, and the controller output generated by the control law, u_c, by means of the slack variables S_L and S_U at time step k. The complementary terms (1b) and (1c) ensure that a discrepancy occurs only when the actuator is at either the lower or upper bound, u_L and u_U respectively, and that the rest of the time u_a is equal to u_c.

$$
\begin{aligned}
u_c(k) &= u_a(k) - S_L(k) + S_U(k) && \text{(1a)}\\
0 &= S_L(k)(u_a(k) - u_L) && \text{(1b)}\\
0 &= S_U(k)(u_a(k) - u_U) && \text{(1c)}\\
u_L &\le u_a(k) \le u_U && \text{(1d)}\\
0 &\le S_L(k) && \text{(1e)}\\
0 &\le S_U(k) && \text{(1f)}
\end{aligned}
$$

Recent work shows that failure to handle the actuator constraints rigorously within an integrated design and control optimization framework can lead to a suboptimal design (Baker, 2000).

On a related note, depending on the implementation of the controller, saturation of the manipulated variable can lead to the phenomenon of reset windup. Reset windup occurs in controllers with reset or integral action when the integral mode continues to integrate the error when it does not affect the output. In the case of input saturation, this occurs because the manipulated variable is unable to respond to changes in the controller output signal because it is at a bound. However, reset windup can also occur in processes with selective control systems, which have a manipulated variable that can be controlled by one of many controllers at different times.

There exist a number of approaches for dealing with reset windup. Simply limiting the controller output signal does not eliminate reset windup because there is still a discrepancy between the calculated controller output and the controller output as seen by the process. One approach is to freeze the integral term if windup is detected, and reactivate it once the cause for windup has been removed. Alternatively, if the digital control algorithm is implemented using the velocity form of the controller, then if the calculated value of the input change is added to the actual value of the manipulated variable, no windup will occur (Khandheria and Luyben, 1976).

In this paper we specifically focus on the effects of including anti-reset windup measures within an integrated control and design problem in processes that experience input saturation.

As mentioned earlier, previous work had considered two formulations for modelling input saturation within an optimization framework; namely the bilinear and mixed-integer formulations. When the bilinear problem was solved using a standard nonlinear solver it would occasionally find a suboptimal solution, or sometimes not find a solution at all depending on the initial starting guess.

The inclusion of the complementarity constraints of the bilinear formulation within the optimization framework leads to a problem within the class of mathematical programs with equilibrium constraints (MPECs). The general form of the MPEC with complementarity constraints is:

$$
\begin{aligned}
\min_{x,y,z} \quad & f(x,y,z)\\
\text{s.t.} \quad & g(x,y,z) \ge 0\\
& c(x,y,z) = 0\\
& 0 \le F(x,y,z) \perp y \ge 0
\end{aligned}
\qquad (2)
$$

where $x \in \Re^n$, $y \in \Re^m$, $z \in \Re^l$, $f : \Re^{n+m+l} \to \Re$ is the objective function, and the constraints are $g : \Re^{n+m+l} \to \Re^p$, $c : \Re^{n+m+l} \to \Re^q$ and $F : \Re^{n+m+l} \to \Re^m$. The last constraint can be interpreted as an "either/or" condition which can be rewritten as the following set of complementarity constraints:

$$
\left.\begin{aligned}
F_i(x,y,z)\, y_i &= 0\\
F_i(x,y,z), y_i &\ge 0
\end{aligned}\right\} \quad \forall \quad i = 1,\ldots,m \qquad (3)
$$

Using simple examples, it was shown that these problems may be nonconvex or have feasible solution sets that may be discontinuous or non-closed (Luo *et al.*, 1997), which might explain some of the symptoms described previously. Unfortunately, this class of problems has only recently come under scrutiny and solvers for MPECs are consequently not widely available. Examples of algorithms and software developed to specifically solve MPECs are the Penalty Interior Point Algorithm (Luo *et al.*, 1997), and IPOPT, which is a general interior point based NLP solver that has recently been modified to solve MPECs (Raghunathan and Biegler, 2003). There is also work which provides encouraging results indicating that certain general NLP solvers are able to perform well in comparison to specialized algorithms when solving MPECs (Fletcher and Leyffer, 2002).

The approach IPOPT uses in its MPEC solving mode differs from the approach a conventional NLP solver would take in that it handles the complementarity constraints differently from the manner in which it deals with conventional equality and inequality constraints. In particular, the algorithm relaxes the complementarity constraints by adding a nonnegative slack variable which is then also included in the log barrier function and the entire problem solved using an interior point method (Raghunathan and Biegler, 2003).

In this study, the IPOPT algorithm is applied to the optimization-based integrated design and control of three mixing tanks in series as well as a stirred tank reactor, both with rigorous input saturation handling provided by including the bilinear formulation for saturation handling. There is also a brief comparison of the performance of the IPOPT solver against general nonlinear solvers for a problem of this type.

2. CASE STUDY I

The system under consideration is a set of three mixing tanks in series depicted in Figure 2 (Marlin, 2000). The controlled variable is the percentage of component A exiting the final tank, while the manipulated variable is the percentage valve opening for the flow of stream A. The disturbance entering the system is a change in the concentration of A in stream B. The aim of the control system is to maintain the concentration of A in the final tank as close to the set point as possible in the face of disturbances and set point changes and to this end, the objective used in this case is the integral absolute error.

The transfer functions relating the concentration of A exiting the final tank to the valve position and to the disturbance are respectively:

$$G_p(s) = \frac{0.039}{(5.0s+1)^3}$$

$$G_d(s) = \frac{1.0}{(5.0s+1)^3}.$$

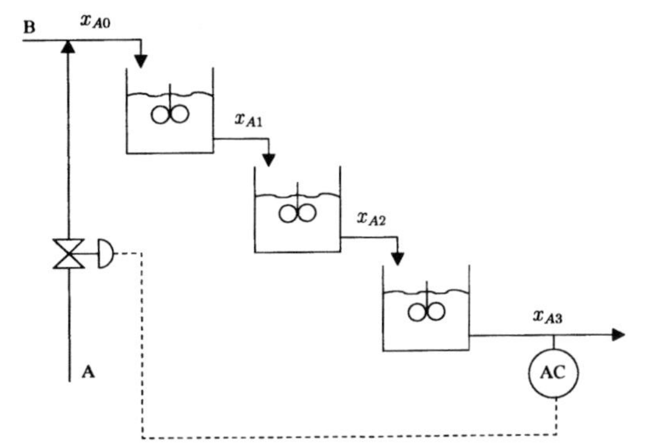

Fig. 2. Schematic of 3 mixing tanks in series.

The disturbance being rejected in this case study is a pulse of 4% in the concentration of A in feed stream B of duration $t = 5$ min to $t = 100$ min. The size of the disturbance increase is such that the actuator has to saturate in order to reject the disturbance. The set point for the outlet concentration of A, and therefore the initial percentage valve opening, as well as the controller reset time are fixed.

Under investigation is the effect that the inclusion of anti-reset windup has on the objective function and on the design of the controller.

The process was modelled in AMPL by means of a finite impulse response model with the saturation logic handled by including the set of equations (1).

In the first example the controller parameters are fixed, and since there are no other design degrees of freedom, the optimization problem reduces essentially to a simulation. This will demonstrate that the bilinear formulation is able to model mathematically the logical conditions associated with input saturation. It will also demonstrate the effect of including anti-reset windup within an optimization-based integrated control and design problem.

The results for the two cases with and without anti-reset windup are presented in Table 1. The resulting trajectories are displayed in Figure 3.

Table 1. Results of scenarios with and without anti-reset windup.

	Objective
Without anti-reset windup	369.68
With anti-reset windup	233.78

As expected, the example without anti-reset windup in the controller has a larger integral absolute error. This demonstrates worse control performance in comparison to the case with anti-reset windup for these particular parameters.

In the second example, the velocity form of the digital controller will be used for both scenarios:

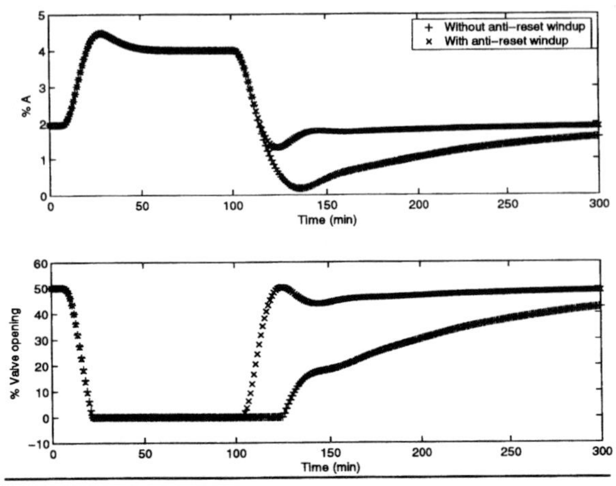

Fig. 3. Results of scenarios with and without anti-reset windup, control parameters fixed.

$$\Delta u_c(i) = K_c \left[e(i) - e(i-1) + \frac{\Delta t}{\tau_I} e(i) \right] \quad (4)$$

The key difference between the two scenarios is that, in the first, the change in the controller output is calculated based on the previous predicted controller position, which may differ from the actual value entering the process.

$$\Delta u_c(i) = u_c(i) - u_c(i-1) \quad (5)$$

In the second case, reset windup is handled correctly and the true actuator value is used instead:

$$\Delta u_c(i) = u_c(i) - u_a(i-1) \quad (6)$$

In this example, the gain is allowed to vary and we compare the results of a controller with and without correct anti-reset windup. It is required that the optimization scheme choose the best gain to reject this particular disturbance and so the controller gain will be included as a design variable within this problem.

The results of the two formulations are reported in Table 2. The case without correct anti-reset windup has a larger integral absolute error in comparison to the case where the correct actuator position was fed back to the controller. It can also be seen in Figure 4 that reset windup occurs in the case without anti-reset windup and results in poor control performance, since there is a delay before the manipulated variable reacts to the termination of the disturbance pulse at $t = 100$ min. The optimal gain obtained for equation (5) is also smaller than that for equation (6) because a larger gain would exacerbate the effect of the reset windup. It can also be seen that at the end of the time horizon the controlled variable trajectory has still not achieved steady state at the set point in the case of equation (5), because the controller gain is less aggressive.

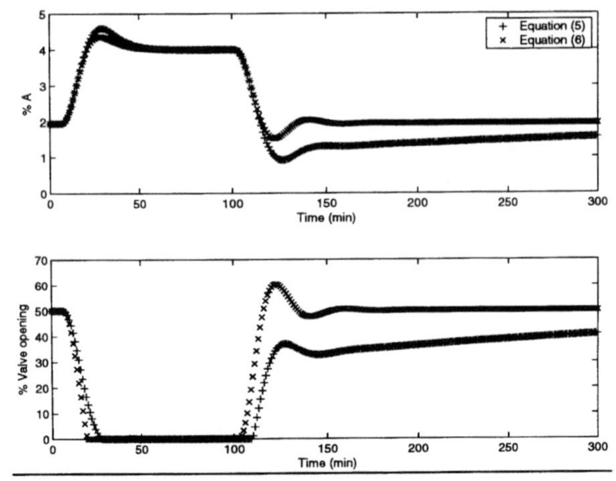

Fig. 4. Results using equations (5) and (6) with the gain allowed to vary.

Table 2. Results using equations (5) and (6) with the gain allowed to vary.

	Objective	Kc
Equation 5	310.25	17.29
Equation 6	206.99	25.69

3. COMPARISON OF IPOPT AND NONLINEAR SOLVERS

As mentioned previously, IPOPT/MPEC uses a different approach to solving problems with complementarity constraints than the conventional general nonlinear solver. In this section, the reliability of the IPOPT/MPEC solver is compared to IPOPT as a nonlinear solver (IPOPT/NLP), MINOS and CONOPT2 running under GAMS. The case study in question is the three mixing tanks in series example from the previous section with anti-reset windup provided by equation (6).

The nonlinearity of the case study arises from two sources: the complementarity constraints of the bilinear formulation for rigorous saturation handling and the fact that the controller gain is included as a decision variable.

The methodology of the comparison is as follows. The results for the case study of the previous example are taken as the base case. From the base case, the values of all the variables are extracted. Uniformly distributed random factors are then added to these nominal values and the resulting values are provided as an initial guesses to each of the solvers. Thus each solver uses the same initial guess. Each random factor is chosen from a "ball" of radius α around the nominal value of the variable. This was then repeated ten times for each particular value of α.

The IPOPT solver was compiled with the MA27, MA28, and the MA47 routines from the Harwell Science Library. The performance of the augmented Lagrangian line search was found to be superior to that of the default filter method, and therefore used in all the studies. All of the solvers were limited to a maximum of 1000 iterations.

It should be noted that since the case study under investigation is a nonlinear problem for which the convexity has not yet been characterized, it is possible that the resulting solution is not the global optimum. For the sake of notational brevity however, we will refer to the lowest known feasible solution as the "optimal" solution.

The solvers were compared based on whether they found the optimal solution, converged to a point that was not the optimal value, or failed and were unable to to converge to a solution within a given number of iterations. This was done for three different values of α (0.5, 1 and 10) and the combined results are reported in Table 3.

Table 3. Summary of results of comparitive study

	Failure	Suboptimal	Optimum
IPOPT/MPEC	6.7%	0%	93.3%
IPOPT/NLP	100%	0%	0 %
MINOS	90%	10 %	0 %
CONOPT2	6.7%	0%	93.3%

The IPOPT/NLP solver performed the least satisfactorily, although it is possible that a different set of algorithm parameters might have resulted in better performance. MINOS also performs fairly poorly. IPOPT/MPEC and CONOPT2 were the only solvers that found the known optimum. For some initial guesses IPOPT would result in line search errors terminating the solver prematurely. At other times the objective function would come fairly close to the known optimum, but the solver would require over a 1000 iterations. Once again, it should be borne in mind that a different set of algorithmic parameters might have yielded different results.

4. CASE STUDY II

The following case study is an example of the integrated design and control of a single continuous stirred tank reactor. The first-principles model described in (Schweiger and Floudas, 1998) was used.

Briefly, the reaction taking place is a first order, exothermic irreversible reaction ($A \rightarrow B$). The controlled variable is the temperature in the tank, while the flow rate of water through the cooling jacket is the manipulated variable. The disturbance is a step increase of 18.3 °C in the temperature of the feed to the tank. The design variables of this example are: the height and diameter of the tank; the nominal operating temperature in the CSTR and the reactor jacket; the nominal concentration of reactant A exiting the tank; and the gain and reset time of the associated controller. The disturbance for this CSTR example is such that the system exhibits saturation behavior when the input saturation model formulations are included in the integrated design and control optimization framework, and the lower and upper limits of the cooling water flow rate were set as 0 and 12.88 m^3/h respectively. The objective function for the problem is the total cost of the process operating over 4 years, which includes the capital cost of the CSTR as well as the utility cost or cost of operation over the period. The cost of off-spec product is included as part of the objective function by means of a penalty on the integral square error of the concentration of reactant A in the tank. Furthermore, the anti-reset windup formulation described in equation (6) is used.

Table 4. Results of CSTR integrated control and design case study

	No saturation	Saturation
Total Cost, $	572.7 $\times 10^3$	502 $\times 10^3$
Capital Cost, $	559.5 $\times 10^3$	488.7 $\times 10^3$
Diameter, m	8.57	7.97
Height, m	4.29	3.98
Controller gain	-68.74	-100
Reset time	2	1.83

Previous work attempted to solve this problem using the mixed-integer mathematical formulation of input saturation, but found that the time to solve the optimization problem with currently available MINLP solvers could become prohibitive as the number of variables increased (Baker, 2000). For this reason, the interior point based solver IPOPT/MPEC will be applied to the CSTR example instead to assess whether it is able to successfully solve the optimization problem within a reasonable amount of time.

IPOPT/MPEC was successfully able to solve the problem with results shown in Table 4 and Figure 5. It can also be seen that the case in which the system is allowed saturate has a lower cost than a comparative case in which the actuator behavior is assumed to be strictly linear.

Using results for the system without saturation handling as an initial guess, IPOPT/MPEC was able to solve the rigorous saturation problem in around 1 minute of CPU time, converging in 255 iterations.

5. CONCLUSION

This paper has discussed the use of complementarity constraints to describe actuator saturation effects in integrated design and control type problems. Neglecting to include anti-reset windup was shown to lead to a suboptimal design. In addition, a comparison of the IPOPT/MPEC

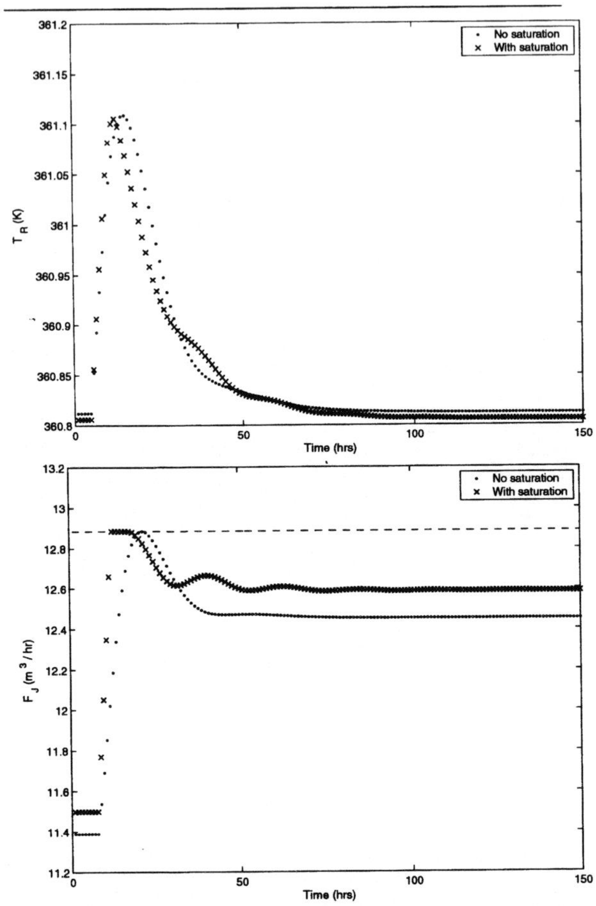

Fig. 5. Rigorous saturation handling for a CSTR.

solver against certain general nonlinear solvers on a problem with rigorous saturation handling was conducted, and showed IPOPT/MPEC and CONOPT2 to have superior performance. Finally, IPOPT/MPEC was successfully used to solve a large scale nonlinear integrated design and control problem with rigorous saturation handling.

REFERENCES

Baker, R. (2000). Inclusion of input saturation in the design of dynamically operable plants. Master's thesis. University of Cape Town.

Fletcher, R. and S. Leyffer (2002). Numerical experience with solving MPECs as NLPs. Technical report. Department of Mathematics, University fo Dundee, Dundee, UK.

Khandheria, J. and W.L. Luyben (1976). Experimental Evaluation of Digital Algorithms for Antireset Windup. *Industrial and Engineering Chemistry, Process Design and Development* **15**(2), 278–285.

Luo, Z.-Q., J.S. Pang and D. Ralph (1997). *Mathematical Programs with Equilibrium Constraints.* Cambridge University Press.

Marlin, T.E. (2000). *Process Control: Designing Processes and Control Systems for Dynamic Performance.* 2nd ed.. McGraw-Hill.

Raghunathan, A.U. and L.T. Biegler (2003). Mathematical Programs with Equlibrium Constraints (MPECs) in Process Engineering. *Computers and Chemical Engineering* **27**(10), 1381–1392.

Schweiger, C.A. and C.A. Floudas (1998). Interaction of design and control: Optimization with dynamic models. In: *Optimal Control: Theory, Algorithms, and Applications* (W.W. Hager and P.M. Pardalos, Eds.). pp. 388–435. Kluwer Academic Publishers.

Young, J.C.C., R. Baker and C.L.E. Swartz (2004). Input Saturation Effects in Optmizing Control - Inclusion within a Simultaneous Optimization Framework. *Computers and Chemical Engineering, in press.*

ELSEVIER
IFAC
PUBLICATIONS
www.elsevier.com/locate/ifac

ON-LINE PARTICLE SIZE DISTRIBUTION CONTROL STRATEGY IN AN EMULSION COPOLYMERIZATION REACTOR

Myung-June Park, Mustafa T. Dokucu,
Francis J. Doyle III [1]

Department of Chemical Engineering, University of California, Santa Barbara, CA 93106, U. S. A.

Abstract: Model order reduction is applied to a population balance equation (PBE) describing the particle size distribution in an emulsion copolymerization reactor. The reduced order model is used in linear model predictive control. The performance of controllers with different controlled outputs is compared in the presence of two classes of disturbances, one related to the nucleation mechanism and the other affecting the growth rate, to determine the most robust control strategy for the regulation of the distribution at the final time. *Copyright © 2004 IFAC*

Keywords: Reference trajectory tracking, on-line feedback-based control, model order reduction, particle size distribution.

1. INTRODUCTION

Several control strategies for polymer properties, including molecular weight distribution, copolymer composition, and particle size distribution, have been developed to improve the end-use properties of a polymer product. Some of strategies are based on the tracking of trajectories calculated off-line using optimization methods (Choi and Butala, 1989; Saldivar and Ray, 1997; Crowley *et al.*, 2000) and on-line feedback algorithms have been applied to overcome the limitations of open-loop optimal trajectory tracking (Kozub and MacGregor, 1992; Echevarria *et al.*, 1998; Vicente *et al.*, 2001). However, on-line feedback-based control of particle size distribution (PSD) in particulate systems is rarely reported as the process is highly nonlinear and numerical solution techniques for the population balance equation lead to a high dimension problem. Therefore, most research work has focused on the control of lumped properties (Zeaiter *et al.*, 2002; Chiu and Christofides, 1999).

In this study, model order reduction based on principal component analysis (PCA) is applied to the first-principles model, and a control strategy is developed for tracking the reference trajectory in an effort to control the final PSD in an emulsion copolymerization system. This modeling and control is more difficult than the homopolymerization case because the two monomers influence the stability of the particle in different manners.

2. SEMIBATCH VAC/BUA EMULSION COPOLYMERIZATION REACTOR

A mechanistic model of a semibatch VAc/BuA emulsion copolymerization reactor developed by Immanuel *et al.* (2002) is summarized in Table 1. This model has several advantages in that the size-dependent growth rate is taken into account and the average number of radicals per particle is calculated via a balance between the rates of entry, desorption and termination of

[1] To whom all correspondence should be addressed. Tel: +1 (805) 893-8133. Fax: +1 (805) 893-4731. E-mail: doyle@engineering.ucsb.edu

Table 1. Model equations for a semibatch emulsion copolymerization system

Mass balance equations	
Oxidizer	$d([I_w]V_{aq})/dt = -k_{d1}[I_w][Y_1^r] + v_{I_w}$
Reducer	$d([Y_2]V_{aq})/dt = -r_I k_{d1}[I_w][Y_1^r] + v_{Y_2}$
Initiator radical	$d([R_w]V_{aq})/dt = k_{d1}[I_w][Y_1^r] - V_{aq}\sum_{i=1}^{2} k_{ri}[R_w][M_i]_w - V_{aq}k_{tav}^w[R_w]\left(\sum_{l=0}^{j_{cr}-1}[P_w]^l + [R_w]\right)$
Monomers	$\frac{dM_j}{dt} = v_{M_j} - \sum_{i=1}^{2}(k_{pij}^w + k_{trij}^w)p_{wi}[P_w][M_j]_w V_{aq} - \sum_{i=1}^{2}(k_{pij} + k_{trij})p_i[M_j]_p \int_{r_{nuc}}^{r_{\max}} \bar{n}(r,t)F(r,t)dr$
Population balance equation	
Density function	$\frac{\partial}{\partial t}F(r,t) + \frac{\partial}{\partial r}\left(F(r,t)R_{growth}(r,t)\right) = R_{nuc}(r,t) + R_{coag}(r,t)$
Growth rate	$R_{growth}(r,t) = \frac{dr}{dt} = \frac{3}{4\pi r^2\rho_p}\sum_{i=1}^{2}\sum_{j=1}^{2} k_{pij}p_i\frac{\bar{n}(r,t)}{N_A}[M_j]_p MW_j$
Nucleation rate	$R_{nuc} = R_{micellar} + R_{homo} = \sum_{l=0}^{j_{cr}-1}\sum_{i=1}^{2} e_{i,micelle}^l p_{wi}[P_w]^l C_{micelle}V_{aq} + k_{pav}^w[P_w^{j_{cr}-1}]V_{aq}$
Coagulation rate	$R_{coag}(r,t) = H(r_{uppper} - r)R_{formation}(r,t) - H(r_{cutoff} - r)R_{depletion}$
	$R_{formation}(r,t) = \frac{1}{V_{aq}}\int \beta(r',r'')F(r',t)F(r'',t)\frac{r^2}{(r^3-(r')^3)^{2/3}}dr'$
	$R_{depletion}(r,t) = \frac{1}{V_{aq}}\int_{r_{nuc}}^{r_{\max}} \beta(r,r')F(r,t)F(r',t)dr'$ where $\beta(r,r') = c_1 4\pi D_0(r+r')/W_s$
Output functions	
Weight averaged PSD	$W(r_i,t) = r_i^3F_i/\sum_i r_i^3F_i$ where $F_i = \int_{r_{b,i-1}}^{r_{b,i}} F(r,t)dr$
Total particle number	$N_p = N_A\sum_i F_i$ Solid content $S_c = N_A\left(\sum_i 4\pi r_i^3\rho_p F_i/3\right)/m_T$

radicals. The kinetic rate constants and physical constants were drawn from the literature (see Immanuel *et al.*, 2002 and references) and Immanuel *et al.* (2002) investigated the parametric sensitivity on the distribution at the final time of the batch.

The population balance equation (PBE) is solved numerically using the solution technique proposed by Immanuel and Doyle III (2003*a*), in which the continuous PSD is divided into a finite number of sections within which an integral quantity of the distribution is defined. To discretize the population balance equation for the particle density function (*cf.* Table 1), 250 elements (or grids) with a width of 2 nm are used and ordinary differential equations for mass balance equations of two monomers, aqueous phase volume, surfactant, oxidizer, reducer and initial radicals are calculated using the ODE solver (ode45) in MATLAB.

3. MODEL ORDER REDUCTION

The finite discretization of the governing PBE leads to very high order, ill-conditioned and uncontrollable dynamical systems, thus motivating the application of model order reduction (MOR) methods to the mechanistic model for semibatch VAc/BuA copolymerization. In general, MOR methods involve the projection of the full order states onto a state space with a suitable reduced order through a linear transformation:

$$x = \mathbf{P}z \tag{1}$$

where z is a q-th order projection of state $x \in R^N$ in the reduced order state space, and $\mathbf{P}$ represents an orthonormal matrix for a transformation from the reduced one to the original state space. There are a number of methods to select the transformation matrix $\mathbf{P}$. In the present study, principal component analysis (PCA) (Sharaf *et al.*, 1986) is applied. PCA is one of several multivariate statistical projection techniques in which the original number of (possibly) correlated variables are transformed into a (smaller) number of uncorrelated variables known as *principal components.* This technique has several advantages in that it addresses limitations due to measurement noise, correlated variables and unknown variables, and the data set dimensionality problem can be managed.

Some key issues in applying the transformation matrix to a nonlinear system are associated with storage and evaluation (Park and Doyle III, 2004). To solve this problem, the nonlinear model is linearized at every sample time using a nominal point and the zero-order hold is used to calculate discrete system matrices ($\mathbf{A}_k$, $\mathbf{B}_k$, and $\mathbf{H}_k$) under the assumption that the state between sample times is not changed significantly (Garcia, 1984; Gattu and Zafiriou, 1992). The transformation matrix $\mathbf{P}$ in Eq. (1) is applied to the linearized model and consequently the following reduced order model is obtained:

$$\bar{z}_{k+1} = \mathbf{A}_k^r\bar{z}_k + \mathbf{B}_k^r\bar{u}_k \text{ and } \bar{y}_k = \mathbf{H}_k^r\bar{z}_k \tag{2}$$

where $\mathbf{A}_k^r = \mathbf{P}^T\mathbf{A}_k\mathbf{P}$, $\mathbf{B}_k^r = \mathbf{P}^T\mathbf{B}_k$ and $\mathbf{H}_k^r = \mathbf{H}_k\mathbf{P}$, respectively. The over-bar represents the deviation from the nominal value.

4. RESULTS AND DISCUSSION

In order to generate the database for the calculation of the linear transformation matrices, pseudo random multi-level signals are introduced to the plant (the first principles model) with the assumption that this signal produces representative responses over the region of interest. At every sample time, the input is changed with the probability of 7% and the value is chosen

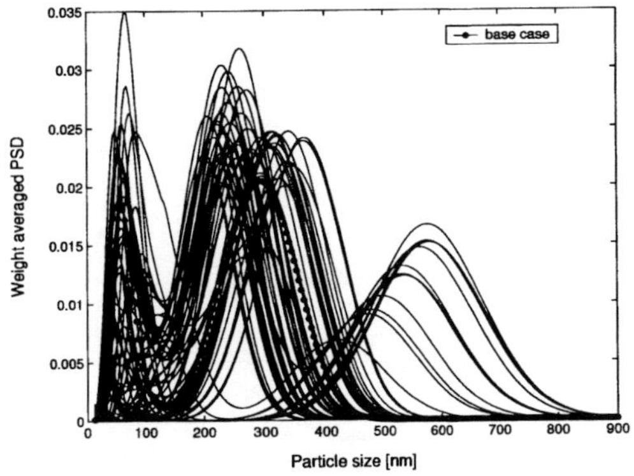

Fig. 1. Weight averaged particle size distribution at the final time generated by pseudo random 4-level signals (60 batches).

among 4-level values (0, 3.5×10^{-4}, 7.0×10^{-4}, 1.05×10^{-3} for u_1 and 0, 1.66×10^{-4}, 3.32×10^{-4}, 4.97×10^{-3} for u_2). The corresponding responses, including states and outputs, are saved in the database at every sample time (1 min). Experimental conditions and additional details are listed in Table 2.

Table 2. Conditions for semibatch VAc/BuA copolymerization system

Initial reactor charge	
Water	1.0L
VAc	52g
Ferrous ammonium sulphate	0.1g
Sodium benzoate	1.12g
Reactor temperature	67.5°C
Flow rates of feed solutions [mol/s]	
BuA feed	$2.81 \times 10^{-4} \rightarrow 0$ at 94 min
Oxidizer feed	$5.4 \times 10^{-4} \rightarrow 9.0 \times 10^{-4}$ at 107 min
Reducer feed	$4.86 \times 10^{-4} \rightarrow 8.09 \times 10^{-4}$ at 107 min
Operating range of inputs [mol/s]	
VAc flow rate (u_1)	$0 \sim 1.05 \times 10^{-3}$
surfactant flow rate (u_2)	$0 \sim 4.97 \times 10^{-3}$ mol/s

Figure 1 shows the results of weight averaged particle size distribution at the final time for 60 batches. Since the inputs are specified between upper and lower bounds, this figure accounts for the reachable region of the system under the current operation condition. Due to the slow supply of surfactant, the amount of free surfactant in the reactor is below the critical micelle concentration (cmc), thus only homogeneous nucleation takes place in the early stage while micellar nucleation becomes dominant after the surfactant concentration exceeds the cmc (Immanuel *et al.*, 2002). Hence the PSD at the final time shows a bimodal distribution; the primary peak in right-hand side by homogeneous nucleation and the secondary peak in left-hand side by micellar nucleation.

To simplify the calculations of the linear time varying transformation matrices, all the data are mean-centered and scaled to unit variance. For the principal component analysis, the Statistics Toolbox for MATLAB is used. The number of principal components (PCs) is determined so that the transformation matrix constitutes more than 99.9% of the total variance because the PCs for 90.0% fail to approximate the entire distribution (Park and Doyle III, 2004). In general, the PCs for 80% or 90% variance predict the behavior of original system in most of the results reported in the literature, while the PCs for 99.9% are used in this system. It is worth emphasizing that PCA approach described in this paper employs the nonlinear fundamental model directly, hence the unusually large variance is reasonable. If, on the other hand, data were employed to generate the PCs, a lower variance would be more practical to avoid capturing noise effects (Park and Doyle III, 2004). The number of PCs for the particle density function (F) is chosen to be 16 at every sample time while the number of PCs for the weight averaged PSD that predicts more than 99.9% cumulative variance is determined to be 15. The state variables for the mass balance equations, such as monomer concentration, are not projected to the latent variable space but are normalized as $x_{normalized} = (x - x_{\min})/(x_{\max} - x_{\min})$. After the transformation matrices are obtained, the reduced order model is used in the prediction equation of linear model predictive control (MPC).

Immanuel and Doyle III (2003*b*) suggested a control strategy where time varying reference trajectories for the total number of particles (N_p) and solid content (S_c) can be used for the control of the nucleation rate and the growth rate, respectively. They applied the Nondominated Sorting Genetic Algorithm (NSGA) (Bhaskar *et al.*, 2000) to calculate the optimal input trajectories to track the reference trajectories for N_p and S_c. Meanwhile, Park and Doyle III (2004) introduced a control strategy that drives the weight averaged particle size distribution directly to its time varying reference trajectory using a model predictive control based on the reduced order model. In this study, the performance of these strategies is evaluated for the case when a disturbance or a model mismatch exists in the system.

Three control strategies are compared in this study: (*strategy* 1) the utilization of N_p and S_c time varying reference trajectories; (*strategy* 2) the utilization of weight averaged PSD time varying reference trajectory; and (*strategy* 3) the utilization of all of the reference trajectories. For all cases, it is assumed that the state variables are available by feedback and the 250 original state variables in the particles density function (F) are projected onto the latent variable space. The total number of particles is scaled using the normalization method with high and low bounds as 1×10^{18} and 0, respectively, and the weight averaged PSD is reduced using PCA.

The controller is tuned to exhibit acceptable performance in rejecting a disturbance in the surfactant feed concentration (the concentration of surfactant feed in the plant is less than that in the model by 10%), and the weighting matrices are determined by trial and error (listed in Table 3). State variables are assumed to be available by state feedback. For all strategies, the prediction and control horizons are 25 and 15, respec-

Table 3. Weighting matrices for each control strategy

	weights on the control error (Λ^y)
Strategy 1	$81 \times \mathbf{I}_{15\times15}$
Strategy 2	$diag([5.5 \times 10^3, 1.0 \times 10^3])$
Strategy 3	$diag([94 \times \mathbf{I}_{15\times15}, 5.5 \times 10^3, 1.0 \times 10^3])$
	weights on the control input (Λ^u)
Strategy 1	$diag([0.1, 0.51])$
Strategy 2	$diag([0.1, 0.12])$
Strategy 3	$diag([0.1, 0.22])$

tively, and the sampling time is 1 min. Constraints on the input magnitude are specified with the operating range given in Table 2, and the rate of input change is constrained as follows:

Constraints on the rate of input change

$$\begin{array}{ll} -4.0 \times 10^{-4} \leq \Delta u_1 \leq 4.0 \times 10^{-4} & [\text{mol/s}] \\ -2.0 \times 10^{-3} \leq \Delta u_2 \leq 2.0 \times 10^{-3} & [\text{mol/s}] \end{array} \quad (3)$$

Figures 2–4 illustrate the controlled distribution at the final time of the batch and the corresponding input trajectories determined by each strategy. As the results in Figure 2 indicate, the direct feedback of variables that are sensitive to the disturbances leads to the use of both inputs to eliminate the disturbance. The controller in strategy 2 uses the flow rate of surfactant feed solution primarily to drive the total number of particles to its time varying reference trajectory as the disturbance in the surfactant feed solution mainly effects N_p. As a result, different levels of control performance are observed between strategies 1 and 2, in that strategy 1 shows better performance in the secondary peak than the primary peak, while strategy 2 results in satisfactory performance in both peaks. Considering the characteristics of the controller as well as the system, the different results of the controllers are attributed to the limitations of each controller; strategy 1 has a disadvantage in that different control weights cannot be used for different peaks because the distribution is projected to a latent variable space, and the shortcoming of strategy 2 is that, even after the deviated trajectories of N_p and S_c due to the disturbance are converged to their respective reference trajectories, the effect of deviations in the early stage still remains in the weight averaged particle size distribution at the final time.

To compensate for the disadvantages of each controller in strategies 1 and 2, all the outputs (the distribution, N_p, and S_c) are utilized in the strategy 3 and the result is indicated in Figure 4. It is noted that the same control weights as those in the previous strategy are used for N_p and S_c, but the VAc feed flow rate between 20 and 40 min shows higher values while the surfactant feed flow rate becomes smaller compared to the input trajectories in strategy 2. This feature indicates that the utilization of the distribution as a controlled output compensates for the error in strategy 2 and consequently, the controlled distribution in strategy 3 shows a better level of performance than those of strategies 1 or 2.

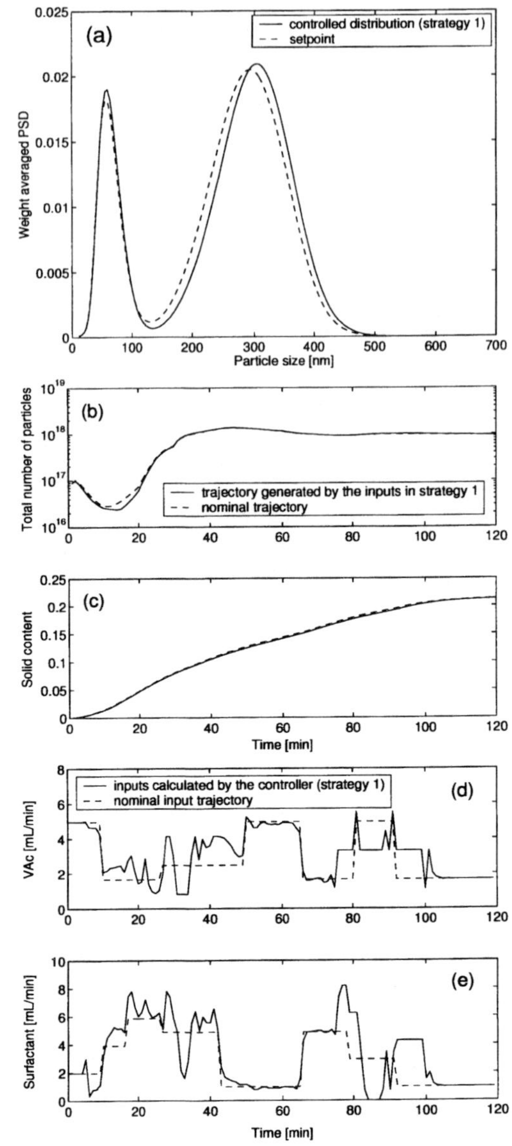

Fig. 2. The performance of the controller in strategy 1 (a) the weight averaged particle size distribution (b) the total number of particles (c) the solid content (d) the input trajectory (VAc) (e) the input trajectory (surfactant).

To evaluate robustness, the performance of the controllers is compared when disturbance magnitudes are varied (controller tuning is fixed). The following relative error is used for comparison

$$\frac{\|W_i(t_f) - W_{i,\text{desired}}\|_2^2}{\|W_{i,\text{desired}}\|_2^2} \quad (4)$$

where W_i denotes the sampled version of the weight averaged PSD and the result is presented in Figure 5 (surfactant disturbance). When the disturbance is less than 0.9, the performance of strategy 1 becomes worse and the magnitude of the difference in relative errors between strategies 1 and 2 is increased. Therefore, utilization of the distribution hinders the performance of strategy 3 compared to that of strategy 2. However,

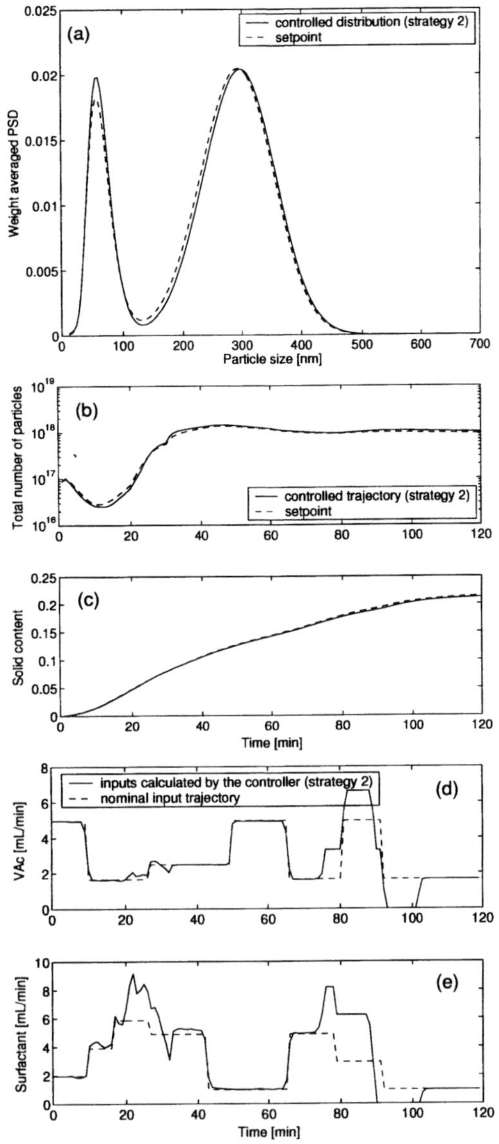

Fig. 3. The performance of the controller in strategy 2 (a) the weight averaged particle size distribution (b) the total number of particles (c) the solid content (d) the input trajectory (VAc) (e) the input trajectory (surfactant).

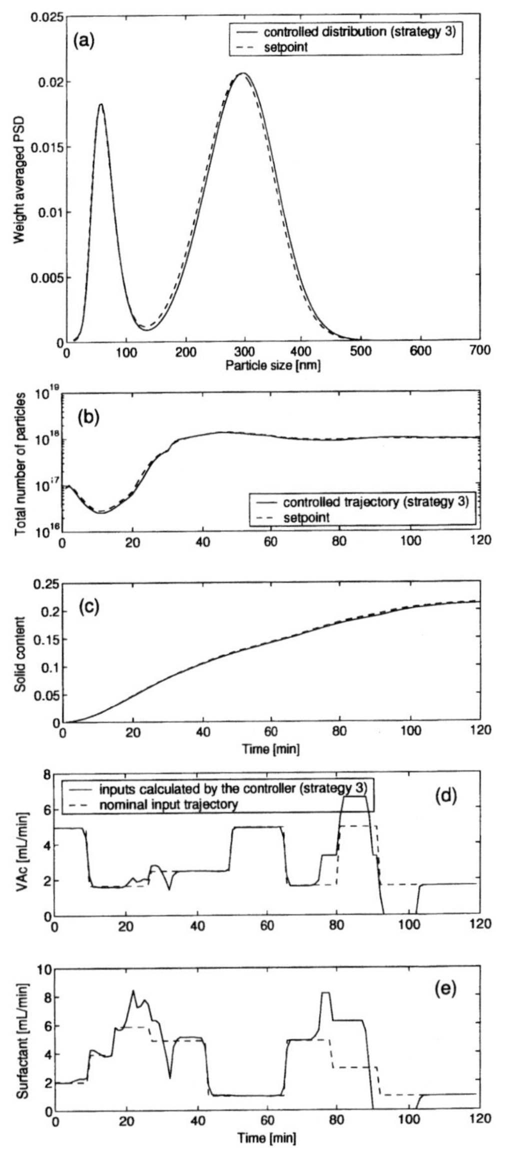

Fig. 4. The performance of the controller in strategy 3 (a) the weight averaged particle size distribution (b) the total number of particles (c) the solid content (d) the input trajectory (VAc) (e) the input trajectory (surfactant).

in the region of disturbance larger than 1.0 (nominal case), the performance of strategy 1 becomes best whereas strategy 2 as well as 3 (including N_p and S_c) performs worse. As the magnitude of the disturbance is increased, the output trajectories for N_p and S_c deviate from the reference trajectories and then converge to them again. However, the deviation of lumped parameters in the middle of the reaction causes a deviation of the PSD at the final time. Therefore, the utilization of lumped parameters in the controlled outputs results in worse performance compared to the direct tracking of the PSD reference trajectory. From the results, it is inferred that the direct use of the distribution in the controller is robust to the different magnitudes of disturbances.

Figure 6 shows the performance of strategies when a different disturbance from the previous case is im-

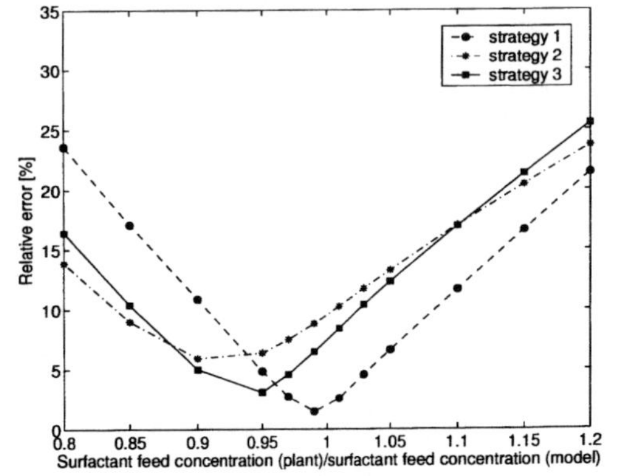

Fig. 5. Comparison of performance between strategies in the presence of a disturbance in the surfactant feed.

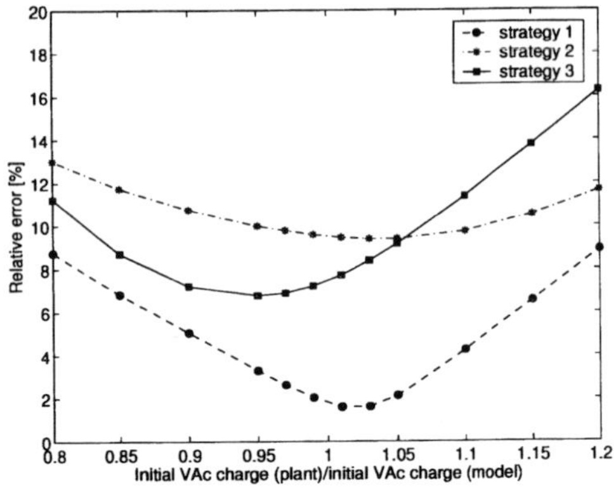

Fig. 6. Comparison of performance between strategies in the presence of a disturbance in the initial VAc charge.

posed on the system. In this case, a disturbance is introduced in the initial charge of VAc. This type of disturbance brings about an error in the growth rate while the disturbance in the previous case is related to the nucleation rate. In this case, the result also shows the effectiveness of strategy 1 since it utilizes most effectively both control inputs. The presence of the disturbance in the early stage of the reaction is also responsible for the inferior performance of strategies 2 and 3 because the tracking of N_p and S_c trajectories produces a deviation in the initial stage.

5. SUMMARY

A model order reduction method based on principal component analysis (PCA) is applied to a nonlinear mechanistic model for a semibatch VAc/BuA emulsion copolymerization reactor. The resulting model is used to design a model predictive controller. Strategies using different controlled outputs are compared in the presence of disturbances on the basis of particle size distribution at the final time of the reaction. The results show that the performance of the controller utilizing lumped parameters is satisfactory in the region of "optimal" tuning. However, compared to the strategy utilizing the direct measurement of particle size distribution, a lumped parameter-based method leads to inferior performance for various disturbances (a practical situation for the application of the controller to industrial processes). In summary, the strategies using the reduced order model and trajectory tracking for the distribution are proven to be most effective for the generation of a desired distribution at the final time.

REFERENCES

Bhaskar, V., S. K. Gupta and A. K. Ray (2000). Multiobjective optimization of an industrial wiped-film pet reactor. *AIChE J.* **46**, 1046–1058.

Chiu, T. and P. D. Christofides (1999). Nonlinear control of particulate processes. *AIChE J.* **45**, 1279–1297.

Choi, K. Y. and D. N. Butala (1989). Synthesis of open-loop controls for semibatch copolymerization reactors by inverse feedback control method. *Automatica* **25**, 917–923.

Crowley, T., E. Meadows, E. Kostoulas and F. J. Doyle III (2000). Control of particle size distribution described by a population balance model of semibatch emulsion polymerization. *J. Proc. Control* **10**, 419–432.

Echevarria, A., J. R. Leiza, J. C. Cal and J. M. Asua (1998). Molecular weight distribution control in emulsion polymerization. *AIChE J.* **44**, 1667–1679.

Garcia, C. E. (1984). Quadratic/dynamic matrix control of nonlinear processes: An application to a batch reaction process. AIChE Annual Meeting, San Francisco, CA.

Gattu, G. and E. Zafiriou (1992). Nonlinear quadratic dynamic matrix control with state estimation. *Ind. Eng. Chem. Res.* **31**, 1096–1104.

Immanuel, C. D. and F. J. Doyle III (2003*a*). Computationally-efficient solution of population balance models incorporating nucleation, growth and coagulation: Application to emulsion polymerization. *Chem. Eng. Sci.* **58**, 3681–3698.

Immanuel, C. D. and F. J. Doyle III (2003*b*). Hierarchical multiobjective strategy for particle-size distribution control. *AIChE J.* **49**, 2383–2399.

Immanuel, C. D., C. F. Cordeiro, S. S. Sundaram, E. S. Meadows, T. J. Crowley and F. J. Doyle III (2002). Modeling of particle size distribution in emulsion co-polymerization: Comparison with experimental data and parametric sensitivity studies. *Comp. Chem. Eng.* **26**, 1133–1152.

Kozub, D. J. and J. F. MacGregor (1992). Feedback control of polymer quality in semi-batch copolymerization reactors. *Chem. Eng. Sci.* **47**, 929–942.

Park, M.-J. and F. J. Doyle III (2004). Nonlinear model order reduction and control of particle size distribution in a semibatch vinyl acetate/butyl acrylate emulsion copolymerization reactor. *Korean J. Chem. Eng.* submitted **21**, 168–176.

Saldivar, E. and W. H. Ray (1997). Control of semi-continuous emulsion copolymerization reactors. *AIChE J.* **43**, 2021–2033.

Sharaf, M. A., D. L. Illman and B. R. Kowalski (1986). *Chemometrics*. John Wiley & Sons. New York.

Vicente, M., J. R. Leiza and J. M. Asua (2001). Simultaneous control of copolymer composition and mwd in emulsion copolymerization. *AIChE J.* **47**, 1594–1606.

Zeaiter, J., J. A. Romagnoli, G. W. Barton, V. G. Gomes, B. S. Hawkett and R. G. Gilbert (2002). Operation of semi-batch emulsion polymerisation reactors: Modeling, validation and effect of operating conditions. *Chem. Eng. Sci.* **57**, 2955–2969.

ELSEVIER
IFAC
PUBLICATIONS
www.elsevier.com/locate/ifac

MULTIVARIATE ANALYSIS FOR QUALITY IMPROVEMENT OF AN INDUSTRIAL FERMENTATION PROCESS

Leo Chiang[*a], Arthur Kordon[a], Lawrence Chew[b], Duncan Coffey[a], Robert Waldron[a], Torben Bruck[b], Keith Haney[b], Annika Jenings[b], and Hank Talbot[b]

[a] *The Dow Chemical Company, Corporate R&D, 2301 Brazosport Blvd., Freeport, TX 77541*
[b] *The Dow Chemical Company, Biotechnology, 5501 Oberlin Dr., San Diego, CA 92121*

Abstract: The results of a successful implementation of multivariate analysis on multiple fermentations are shown in the paper. To minimize batch-to-batch variability of an industrial fermentation process, multi-way partial least squares (PLS) was used. Eighteen baseline batches from different fermentors were analyzed. After applying diagnostics tools and contribution charts, three batches were clearly identified to be abnormal, and even among the remaining batches, inconsistencies were found. The root causes of the variability were determined by a combination of variable importance in the projection plot and fermentation knowledge. A new fermentation procedure was applied and the quality improvement was demonstrated on 20 new batches. These batches were more consistent as evidenced by the improvement in the model fit (R^2X = 0.829 for the new batches versus R^2X = 0.681 for the baseline batches) and in the percent of out-of-specification (3.3% for the new batches versus 20.4% for the baseline batches). The on-line multi-way PLS model was shown to detect bad batches promptly and to determine abnormal variables accurately. The success of this implementation demonstrates the value of applying multivariate analysis to large-scale industrial batch bioprocesses.

Keywords: Multivariable systems, fault detection, batch control, fermentation processes, and biotechnology

1. INTRODUCTION

To achieve consistent product quality from a batch process, minimizing batch-to-batch variability is important. Multivariate statistical techniques such as principal component analysis (PCA) or partial least squares (PLS) are useful for quality improvement (Chiang *et al.*, 2001; Beebe *et al.*, 1998; Zhang and Lennox, 2004). In off-line applications, these techniques can identify and pinpoint the root causes of batch-to-batch variability. In on-line applications, PCA/PLS models are used to monitor batch conditions. The objective is to identify and correct abnormal conditions early enough to avoid out-of-specification product.

In this paper multi-way PLS analysis was used for quality improvement of an industrial fermentation process at the San Diego biotech facility of The Dow Chemical Company. This application illustrates the use of multivariate analysis for solving typical industrial problems where root causes are unknown This paper also addresses practical issues of industrial data analysis and outlines the steps needed to minimize variability in the final product quality.

2. METHODS

2.1 Multi-way PLS

Applying multi-way analysis for batch data is a two-step process. The first step is to unfold the three-way batch data into two-way data (see Figure 1) and the second step is to apply regular PLS/PCA analysis.

In this mode of unfolding, each column contains a particular variable over all time periods for all batch runs and each row contains all variables at a particular time for a particular batch. After unfolding, regular PLS is applied with the X block (independent variables) as the unfolded matrix and the Y block (dependent variable) as the maturity variable (Wold *et al.*, 1998). A monotonically increasing variable that is related to percent completion of a batch is used as maturity variable. It is preferable to use a quality variable. However, for the case when the uncertainty in the quality variable is too high, elapsed time can be used as maturity variable instead.

This way of unfolding is attractive for on-line implementations because the score calculation and maturity prediction can be computed directly based

* To whom all correspondence should be addressed. E-mail: hchiang@dow.com

on on-line measurements. Contribution charts can be easily computed and interpreted. Similar to regular PLS, the T^2 and Q statistics are used to monitor the process, while the predicted Y variable gives information about the maturity of a batch. Because of these advantages, this mode of unfolding was used.

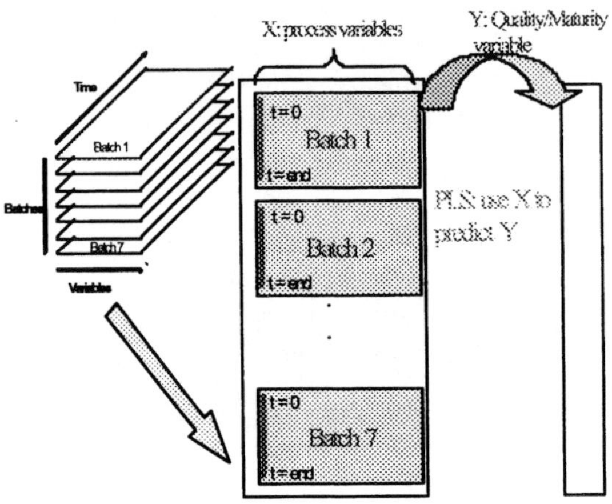

Fig. 1. Illustration of multi-way PLS for batch data.

2.2 Fermentation process

The multivariable analysis was performed on experimental batches from the Dow Chemical San Diego facility. The scope of the investigation included six 20 L fermentors. The key objective was to eliminate problematic fermentation issues that prevent 100% fermentation success rate. The results from this project will be leveraged to other large-scale bioreactors.

Fermentation consists of two phases: growth and production. To enhance the sensitivity on analyzing batch data, separate PLS models have been applied for each phase using the SIMCA-P software (Umetrics, 2003). In SIMCA-P, non-linear iterative partial least squares (NIPALS) is used in PLS computation and leave 1/7 out cross validation is used to determine the optimal number of PLS components. In the growth phase, X-block contains on-line process variables of interest and Y-block contains the optical density (OD), which measures the dynamics and the maturity of cell growth. In the production phase, X-block contains the same variables as in the growth phase, but Y-block contains relative activity/protein yield (referred to as activity in this paper). Quality decisions are made based on OD and activity. Therefore, it is more meaningful to use these two variables, instead of elapsed time, in the Y block.

The purposes of applying PLS for analyzing the batch data are:

- Identify batch-to-batch consistency
- Identify key variables that correlate with the OD and activity
- Understand the root causes of bad batches

Eighteen batches were collected during a month period to identify the baseline performance of the process. Process data were measured on-line every minute, while quality variables (OD, activity, etc.) were measured on an hourly basis. Because OD and activity were monotonically increasing, all missing values between sample points for the Y-block were linearly interpolated using SIMCA-P.

3. RESULTS AND DISCUSSION

3.1 Consistency for the baseline batches

Figure 2 plots the predicted against the observed OD for the 18 baseline batches in the growth phase. The model fit (R^2X = 0.816 and R^2Y = 0.980 for a PLS model with 6 components) is reasonably good, indicating that some process variables correlate well with the OD. This is in spite of the fact that OD and elapsed time for each batch vary at the end of the growth phase.

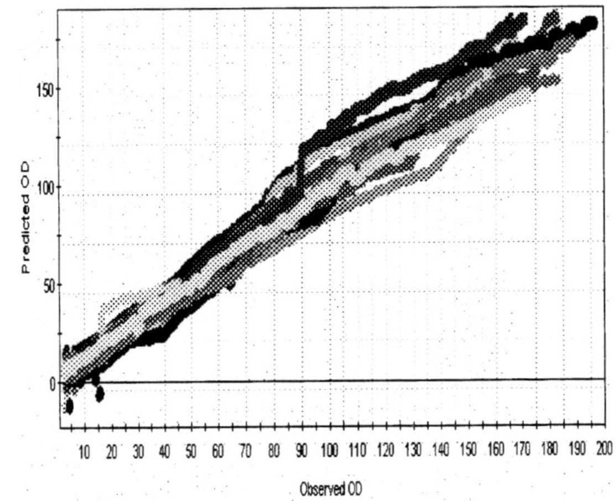

Fig. 2. Predicted OD versus observed OD for the baseline batches in the growth phase.

From a control standpoint, it is important to determine the key variables that are related to the OD. To accomplish this, the variable importance in the projection (VIP) plot is used (Eriksson *et al.*, 2001). The VIP plot, computed based on the PLS weights and the variability explained, ranks the process variables in terms of their relative contribution in predicting OD.

As shown in Figure 3, the top variables in the VIP plot have higher correlation with the OD. On the average, variables with magnitude greater than one in the plot are more relevant in predicting OD. Poor control in these variables will definitely result in a large inconsistency in the OD. At the same time, variables with low correlation to OD must not be ignored. If these variables are related to the key process parameters (e.g., through control loop), poor control of these variables will affect the key variables. This will have a cascade effect on the consistency of the OD. The VIP plot prioritizes the effort to determine the root causes of batch-to-batch inconsistency.

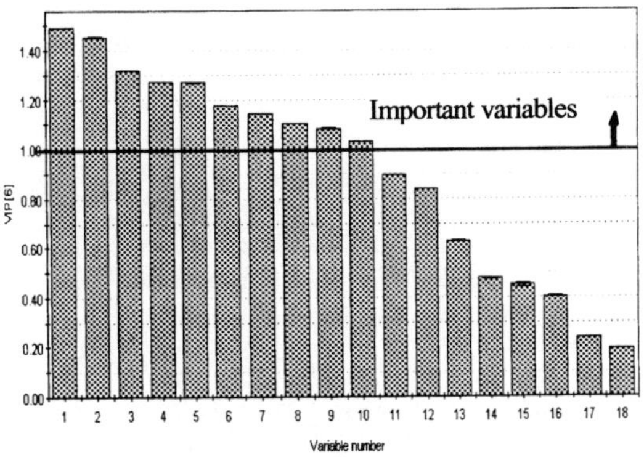

Fig. 3. The VIP plot for the baseline batches in the growth phase.

Figure 4 plots the predicted against the observed activity for the production phase. Comparison between production and growth phases shows that the model fit is worse in production phase ($R^2X = 0.642$ and $R^2Y = 0.848$ for a PLS model with 6 components). This indicates that correlation between the process variables and activity is weak for some batches and the correlation is not consistent from batch to batch. Note also that there are some spikes in the predicted activity. This means that some process measurements are noisy.

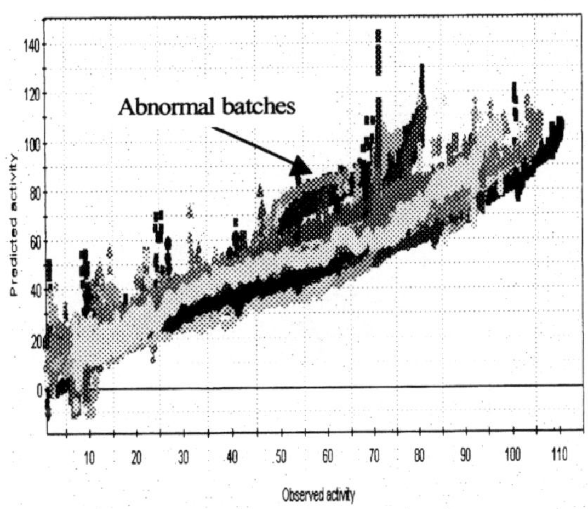

Fig. 4. Predicted activity versus observed activity for the baseline batches in the production phase.

Model disagreement is especially large for some of the batches at the end of production. For two batches, the predicted activity is higher than the observed activity for the entire production phase. This is a clear indication of abnormal batches.

3.2 Identification of bad batches

To understand the consistency of all the variables in the growth phase, a number of diagnostics tools can be used. One of them is the plot of Y-block latent variable versus X-block latent variable. If nonlinear relationships are observed for all batches, then it will be more appropriate to apply non-linear PLS (Qin and McAvoy, 1992). However, if linear relationships are observed for most batches, but nonlinear relationships are observed for others, this indicates the presence of bad batches.

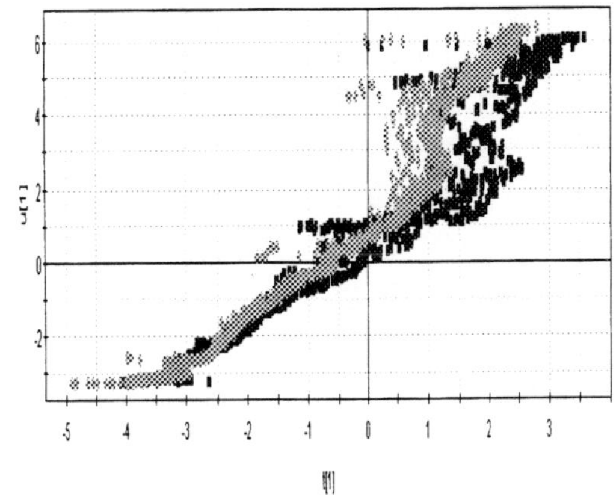

Fig. 5. Y-block latent variable versus X-block latent variable for two abnormal batches

As shown in Figure 5, nonlinear relationship is observed for two batches in the growth phase. With the use of the contribution charts (Miller and Swanson, 1998), many abnormal variables were revealed. As an example, one of these variables is plotted against OD for all batches in Figure 6. Three batches have lower values in this variable than the rest of the batches.

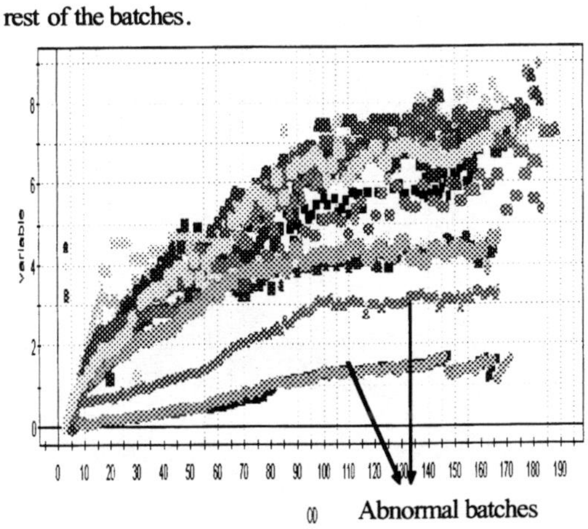

Fig. 6. Abnormal variable versus OD in the growth phase for all batches.

Diagnostic tools were also used to identify bad batches in the production phase. Two of the abnormal batches in the growth phase were also identified in the production phase. Operators confirmed that they encountered operating problems during these 3 batches. After removal of the inconsistent batches, the model fit for both phases improved. An additional result supporting the removal was that now the Y-block latent variable versus X-block latent variable is linear for the rest of 15 batches.

3.3 Root cause determination

Root cause analysis was based on the assumption that the top variables in the VIP plot explained most of the batch-to-batch variability. Using this information and fermentation knowledge, a new procedure with several corrective actions was applied to the process. Experiments were performed to validate improvements in the key parameter. It is shown in Figure 7 that variability in this key parameter becomes significantly smaller after the improvement (standard deviation decreases from to 0.328 to 0.0732).

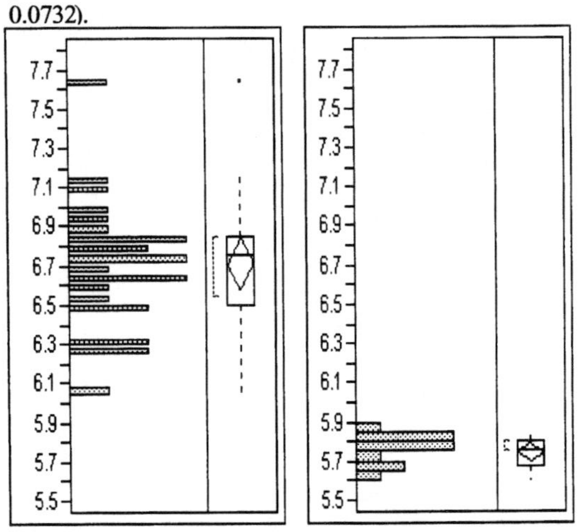

Fig. 7. Variability of the key variable before (left) and after (right) the implementation of new procedure.

3.4 Consistency for the batches under new procedure

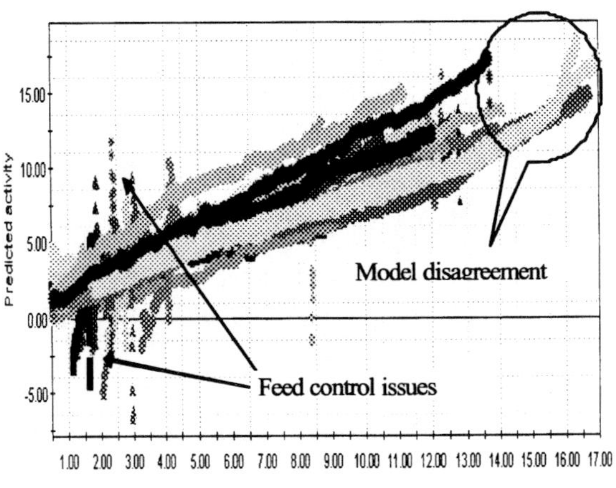

Fig. 8. Predicted activity versus observed activity for the new batches in the production phase.

Twenty batches were collected in order to validate the improvements. Of the 20 batches, seven batches are golden (no main problems were encountered and process data correlate well with the quality variables). One batch is known to be bad. The other 12 batches are questionable in a sense that some difficulties were encountered during the batch runs.

PLS models were applied here to identify the consistency of the batches. Improvement in the growth phase was validated. In the production phase the improvement is less clear on the first glance of Figure 8, which plots the observed and predicted activity ($R^2X = 0.635$ and $R^2Y = 0.866$ for a PLS model with 4 components). Six batches encountered feed control issues for two hours during the production phase. This is illustrated in Figure 8.

Figure 8 also illustrates another effect that is typical in real industrial applications. As it is shown, there are large model disagreements for seven batches. Without careful examination of the data and detailed discussion with the process engineers, one may incorrectly conclude that improvement has not taken place. However, a closer examination of the data shows that activity measurements for these seven batches are biased high. Large model disagreement means that a consistent model cannot be obtained to relate process measurements to the activity accurately for all 20 batches. This does not necessary mean that process measurements are inconsistent.

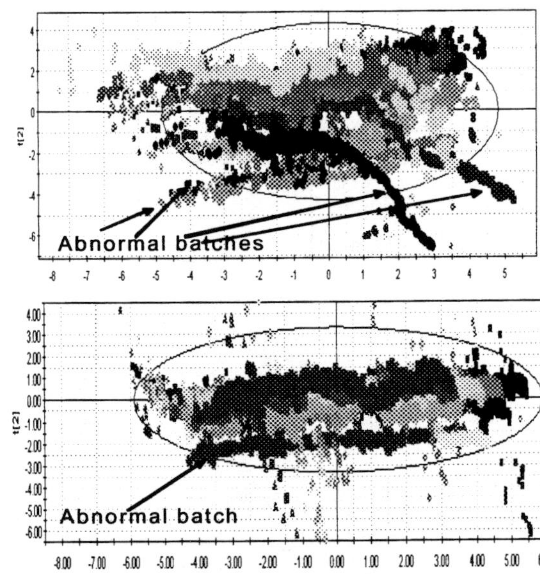

Fig. 9. Score plot consistency comparison between the baseline batches (top) and the new improved batches (bottom).

To understand the root cause, it is desired to examine the consistency of the process measurements using the score plot (see Figure 9). Each color in the plot represents a batch trajectory. Consistency in the batch (process data) is reflected in the consistency of the batch trajectories. In other words, if the process data are consistent among all batches, their batch trajectories will overlap. Of the 18 baseline batches (top of Figure 9), four batches (22%) clearly have different trajectories. This indicates that some process measurements are inconsistent from the rest of the batches. For the 20 new improved batches (bottom of Figure 9), only one batch (5%) has a different trajectory (this batch is known to be bad). The spikes in the score plot are results of known feed control issues. Process data were more consistent for the new batches, which validated the improvements that were implemented.

From the score plot, it is clear that the process measurements are consistent for the new batches. Because the same fermentation procedure was applied for these seven batches, activity measurements were expected to be similar. This posted a question of the activity measurements for these batches.

After discussion with the process engineers, it became apparent that the deviation in activity was due to a change of analyst. The measurements were precise but not accurate. Because of this, it is not meaningful to interpret the model fit and to compare the results to the baseline batches. Model fit for the new batches would improve if the hardware problems did not occur and activity measurements were not biased. Although the model fit cannot be compared directly, an examination between Figure 4 and Figure 8 shows that the activity prediction for the new batches is less noisy (spikes in the predicted activity disappear in Figure 8). This is an indication of improvement in the measurement system.

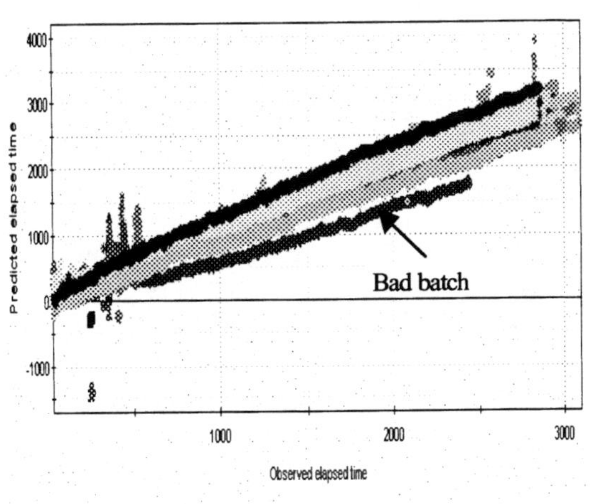

Fig. 10. Predicted elapsed time versus observed elapsed time for the new batches in the production phase.

To validate that the process data are indeed more consistent for the new batches than the baseline batches and that activity measurement are biased for some batches, additional PLS models were built with elapsed time, rather than activity, as the maturity variable. For these models, we are exploring the correlation between the process data and elapsed time for all batches. As shown in Figure 10, the model fit for the new batches (R^2X =0.829; R^2Y = 0.957 for a PLS model with 9 components) is better than the model fit in the baseline batches (R^2X =0.681; R^2Y = 0.955 for the model with 6 components) elapsed time as the maturity variable). This indicates that the process data are correlated with elapsed time consistently for most of the new batches.

For passing quality testing for a batch, six specifications have to be met. There are 18 baseline batches, which translates to 108 opportunities for failure. The baseline performance is 22 defects out of 108 opportunities (20.4%). For the 20 new batches, there are 4 defects out of 120 opportunities (3.3%) Therefore, quality improvement is clearly demonstrated.

4. ON-LINE MONITORING

To monitor performance during a batch run, it is effective to use on-line multivariate control charts. To illustrate this concept, golden batch data were used to build multi-way PLS models. The models capture normal variability of the batches that are acceptable to the users. Unlike Figure 9, the 7 golden batches overlap in the score plot (see Figure 11), which confirms that the process data are consistent for these batches.

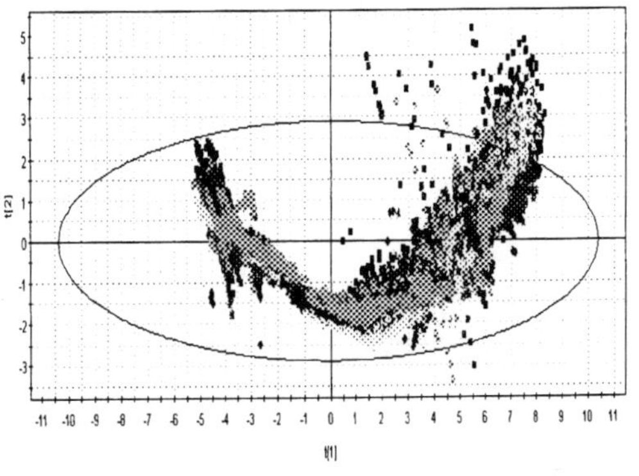

Fig. 11. Score plot for the 7 golden batches in the growth phase.

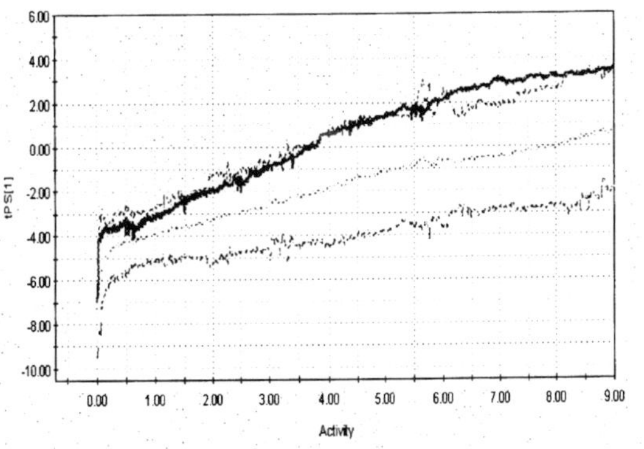

Fig. 12. Detection of an abnormal batch (solid line) using the first latent variable on-line. Dash lines and dotted line represent the critical limits and the average of the 7 golden batches, respectively.

After the models are built off-line, on-line monitoring takes place. Figure 12 shows the first latent variable for monitoring abnormal behavior during the growth phase of one bad batch. As process data are collected, on-line predictions of the latent variables and the quality variables are computed instantly. The critical limits (dashed lines) are defined based on the 3 standard deviation limit of the golden batches. The statistic was above the limit in the beginning for this batch, and then it went below the limit, and finally the statistic gradually increased above the limit again. Because no corrective actions were done during the batch run, the batch remained out of control. To identify abnormality during this run, contribution charts were used.

One of the variables that was identified in the contribution chart is plotted in Figure 13, which clearly shows that variable 10 (solid line) is above the average (dotted line) of the 7 golden batches. If on-line monitoring model was implemented during the batch run, problems could have been identified promptly, which could in turn avoid the production of out-of-specification product for this batch.

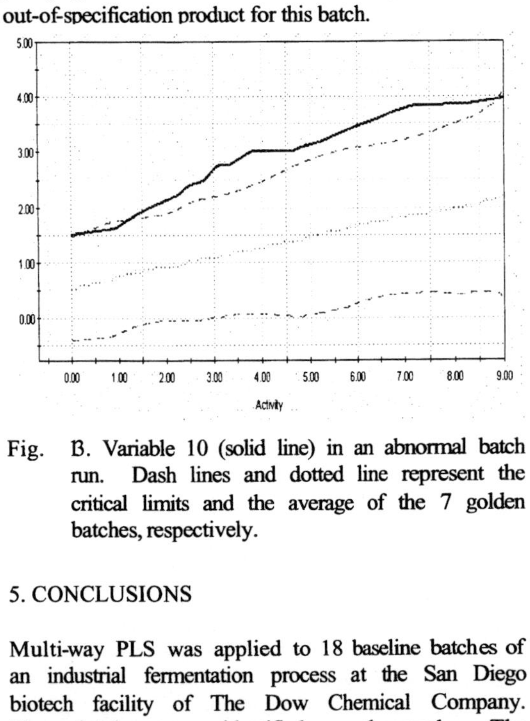

Fig. 13. Variable 10 (solid line) in an abnormal batch run. Dash lines and dotted line represent the critical limits and the average of the 7 golden batches, respectively.

5. CONCLUSIONS

Multi-way PLS was applied to 18 baseline batches of an industrial fermentation process at the San Diego biotech facility of The Dow Chemical Company. Three batches were identified as abnormal. The variability between the remaining batches was more significant in the production phase than the growth phase. The top variable in the VIP plots explained most of the variability. A new procedure was carried out to control this key parameter and twenty new batches were made with the new procedure. Multi-way PLS analysis showed that the new batches were more consistent in both phases as evidenced in the improvement in the model fit (R^2X = 0.829 for the new batches versus R^2X = 0.681 for the baseline batches) and in the percent of out-of-specification (3.3% for the new batches versus 20.4% for the baseline batches).

Seven golden batches were used to build a PLS model to capture normal operation conditions. Bad batches were detected promptly and abnormal variables were determined accurately.

The use of multivariate analysis plays an important role at The Dow Chemical Company. Value has been realized in many applications including batch quality improvement (this paper), scale-up of new batch agrochemicals (Schnelle and Armstrong, 2003), variable selection for multivariate calibration (Leardi *et al.*, 2002), and soft sensor development (Kordon *et al.*, 2002). Implementation of on-line batch monitoring in the future is expected to bring further value.

ACKNOWLEDGMENT

Karl Schnelle and Mary Beth Seasholtz are thanked for their inputs on the manuscript.

REFERENCES

Beebe, K. R., R. J. Pell and M. B. Seasholtz (1998). *Chemometrics: A Practical Guide*, John Wiley & Sons.

Chiang, L. H., E. L. Russell and R. D. Braatz (2001). *Fault Detection and Diagnosis in Industrial Systems*, Springer-Verlag.

Eriksson, L., E. Johansson, N. Kettaneh-Wold and S. Wold (2001). *Multi- and Megavariate Data Analysis*, Umetrics Academy.

Kordon, A. K., G. F. Smits, E. Jordaan and E. Rightor (2002). Robust soft sensors based on integration of genetic programming, analytical neural networks, and support vector machines, *Proceedings of WCCI*, 896 - 901.

Leardi, R., M. B. Seasholtz and R. J. Pell (2002). Variable selection for multivariate calibration using a genetic algorithm: prediction of additive concentrations in polymer films from Fourier transform-infrared spectral data. *Anal. Chim. Acta*, **461**, 189-200.

Miller, P. and R. E. Swanson (1998). Contribution plots: a missing link in multivariate quality control. *Appl. Math. and Comp. Sci.*, **8**, 775-792.

Qin, S. J. and T. J. McAvoy (1992). Nonlinear PLS modeling using neural networks, *Comput. & Chem. Engr.*, **16**, 379-391.

Schnelle, K. and K. Armstrong (2003). Statistical analysis of large pilot plant datasets, *Proceedings Foundations of Computer-Aided Process Operations (FOCAPO)*, 459-462.

Umetrics, Inc. (2003). SIMCA-P+, version 10, Multivariate process modeling software, www.umetrics.com.

Wold, S., N. Kettaneh, H. Friden and A. Holmberg (1998). Modelling and diagnostics of batch processes and analogous kinetic experiments. *Chemom. Intell. Lab. Syst.*, **44**, 331-340.

Zhang, H. and B. Lennox (2004). Integrated condition monitoring and control of fed-batch fermentation processes. *J. of Process Control*, **14**, 41-50.

ELSEVIER
IFAC
PUBLICATIONS
www.elsevier.com/locate/ifac

MULTI-PARAMETRIC NONLINEAR PROGRAMMING AND THE EVALUATION OF IMPLICIT OPTIMIZATION MODEL ADEQUACY

Elaine T. Hale *,[1] **S. Joe Qin** *

* *The University of Texas at Austin*
Department of Chemical Engineering
Austin, Texas, USA

Abstract: An algorithmic framework for numerically approximating multi-parametric nonlinear programming (mp-NLP) solutions is given, along with a method that uses mp-NLP for evaluating the adequacy of the nominal model used in Implicit Optimization. The mp-NLP algorithm builds on numerical methods for single parameter nonlinear programming and for the approximation of implicit manifolds. An example problem is presented. *Copyright © 2004 IFAC*

Keywords: parametric optimization, implicit optimization, real-time optimization, dynamic optimization, implicit manifolds

1. INTRODUCTION

A multi-parametric nonlinear program (mp-NLP) is a nonlinear program whose solution is a function of some finite number of uncertain model parameters. In particular, let $u \in \mathbb{R}^n$ be the vector of variables to be minimized over and $\alpha \in \mathbb{R}^d$ be the set of uncertain parameters. Then the mp-NLP problem is to find

$$u^*(\alpha) = \arg \left\{ \begin{array}{c} \min\limits_u f(u, \alpha) \\ s.t. \\ h(u, \alpha) = 0 \\ g(u, \alpha) \leq 0 \end{array} \right\}, \qquad (1)$$

where

$$\begin{aligned} &f \in C^2 : \mathbb{R}^n \times \mathbb{R}^d \to \mathbb{R}, \\ &h \in C^2 : \mathbb{R}^n \times \mathbb{R}^d \to \mathbb{R}^q, \\ &\text{and } g \in C^2 : \mathbb{R}^n \times \mathbb{R}^d \to \mathbb{R}^p. \end{aligned} \qquad (2)$$

No numerical methods for solving this problem have been given in the literature, however, methods for solving various simplifications of this problem have, see for example (Gal and Nedoma, 1972; Tøndel *et al.*, 2003; Guddat *et al.*, 1990). This work extends the predictor-corrector methods given for single parameter nonlinear programming (Lundberg and Poore, 1993; Guddat *et al.*, 1990) to mp-NLP using the multi-dimensional predictor-corrector algorithm for general parameterized nonlinear equations described by Rheinboldt and Brodzik (Rheinboldt, 1988; Brodzik, 1996).

There are many applications for which having a solution to the problem given in Equation 1 would be quite useful. For example, extending Tøndel's work, nonlinear MPC problems can be thought of in this light, with the initial state as the set of uncertain parameters. Then an mp-NLP algorithm would make it possible to do the bulk of the calculations necessary to implement nonlinear MPC offline, and in a manner that is more likely to be able to deal with the complications introduced by

[1] Supported by the American Association of University Women, the University of Texas at Austin, and the National Science Foundation.

non-convexity. Also, process synthesis problems often contain highly uncertain parameters such that one might want to look at the whole space of possible solutions, and multi-objective optimization problems can be considered by introducing parameters that indicate the relative importance of the various objectives. This work, however, concentrates on the application of mp-NLP to the question of whether or not a process model is accurate enough for use in *Implicit Optimization.*

Implicit Optimization is a method for optimizing batch processes proposed by Srinivasan, Bonvin, and their colleagues in a recent paper (Srinivasan *et al.*, 2003). The assumption that must be satisfied in order for this method to be successful is that the optimum of the nominal model of the process, as compared to the actual process optimum, must be qualitatively correct. Specifically, a *qualitative optimum* is defined by parsing the nominal optimum into a sequence of sets of active constraints, and is implemented in the plant via control loops that are designed to follow these constraints and/or set-points, switching at predetermined times which are updated using run-to-run control. The method is appealing in its online simplicity and potential to be robust to plant-model mismatch. However, one question looms: how does one know if the nominal model is close enough to the actual process to produce the correct qualitative optimum?

A three step process for answering this question is proposed. First, the nominal model is augmented with a) any structural elements that are not in the nominal model, but are suspected of being present and significant in the actual process and b) a list of uncertain model parameters. The parameter list should include parameters directly related to whether or not each of the structural elements in a) are present. Secondly, the parametric programming problem that results from using the augmented model in the original optimization problem is solved via discretization of any ODE's and the algorithm discussed in this paper. Thirdly, the region in the parameter space where the nominal qualitative optimum is valid is identified by inspection, and the likelihood of the process being in this region is assessed using process knowledge and data.

Sections 2 and 3 of this paper provide some background and outline a method for solving multi-parametric nonlinear programming (mp-NLP) problems. The fourth section then discusses the three part process for validating an Implicit Optimization result using an example problem. Finally, conclusions and future work are given.

2. MATHEMATICAL FOUNDATIONS

2.1 mp-NLP Problem Formulation

This work solves the mp-NLP problem using the Fritz-John first-order necessary conditions combined with an active set strategy (Lundberg and Poore, 1993; Guddat *et al.*, 1990). In particular, once the set, A, of active inequality constraint indices is defined:

$$A = \{i : g_i(u, \alpha) = 0\}, \tag{3}$$

solving the mp-NLP problem (Equation 1) is equivalent to solving the following set of parameterized nonlinear equations combined with appropriate active set switching:

$$F_A(z, \alpha) = \begin{bmatrix} \nabla_u \mathcal{L}(z, \alpha) \\ h(u, \alpha) \\ g_i(u, \alpha), \ \forall i \in A \\ \lambda^T\lambda + \mu^T\mu + \lambda_0^2 - \beta_0^2 \end{bmatrix} \tag{4}$$

where $\mathcal{L}$ is the Lagrangian (Mangasarian and Fromovitz, 1967):

$$\mathcal{L}(z, \alpha) = \lambda_0 f(u, \alpha) + \lambda^T h(u, \alpha) + \mu^T g(u, \alpha) \tag{5}$$

and

$$z = (u, \lambda, \mu, \lambda_0). \tag{6}$$

2.2 Implicit Manifolds

Parameterized nonlinear equations of the form

$$\begin{aligned} &F(z, \alpha) = 0 \\ &F : \mathbb{R}^N \times \mathbb{R}^d \to \mathbb{R}^N \end{aligned} \tag{7}$$

have been studied quite extensively in their own right. In this setting, a point

$$x = (z, \alpha) \tag{8}$$

is *regular* if $DF(x)$ $(= (\nabla F(x))^T)$ has maximal rank N. Then, given sufficient smoothness of F, the set

$$M = \{x : x \text{ is regular and } F(x) = 0\} \tag{9}$$

is an *implicit, d-dimensional differentiable manifold* (Brodzik, 1996). Also, Sard's theorem tells us that if $F \in C^k, k > d$ then almost all (in the sense of Lebesgue measure) points are regular (Guddat *et al.*, 1990); therefore M encompasses almost all solutions of $F(x) = 0$.

Numerical methods for approximating the manifold M are based on local parameterizations and tangent spaces. Given a regular point $x_0 \in M$, the tangent space of x_0, $T_{x_0}M$, is a d-dimensional linear subspace, is naturally associated with the function F:

$$\begin{aligned} T_{x_0}M &= \ker DF(x_0), \\ &= \{y \in \mathbb{R}^{N+d} : DF(x_0)\, y = 0\}, \end{aligned} \tag{10}$$

and intuitively represents the directions one can travel from x_0 and expect to remain close to the solution manifold M. The tangent space of x_0 also induces a local parameterization once an orthonormal basis for $T_{x_0}M$ has been introduced. Let U be a $(N+d) \times d$ matrix whose columns form such an orthonormal basis. Then, within an open neighborhood of x_0, $\mathcal{U}_2$, all points on M are uniquely identified with coordinates in this basis, t, that are within a given open neighborhood of the origin, $\mathcal{U}_1$:

$$M \cap \mathcal{U}_2 = \left\{x : x = x_0 + Ut + (DF(x_0))^T \Phi(t) \right. \left. , t \in \mathcal{U}_1\right\} \tag{11}$$

where

$$\Phi : \mathbb{R}^d \supset \mathcal{U}_1 \rightarrow \mathbb{R}^N \tag{12}$$

is a unique, smooth function (Fink and Rheinboldt, 1987).

2.3 Triangulations

An approximation of M can be obtained using the above ideas by considering a neighborhood of the origin of $T_{x_c}M$, which has coordinates in $\mathbb{R}^d$ and known point x_c at its origin. New points, t_i's, are predicted in this space, and then projected onto M using Equation 11 and a nonlinear equation solver to approximate $(DF(x_0))^T \Phi(t)$. The overall structure of M is easier to analyze, however, if connectivity is defined between these approximating points, such that a piecewise linear approximation of M is obtained. In this work, that connectivity is defined for the t_i's in relation to the origin of $T_{x_c}M$ and other relevant points using a *triangulation*.

A triangulation is a set of simplices. In $\mathbb{R}^d$, a *simplex* is the convex hull of $d+1$ points, which are called *vertices* or *nodes* of the simplex. Therefore, if $d = 1$, a simplex is a line segment, if $d = 2$, a simplex is a triangle, if $d = 3$, a simplex is a tetrahedron, etc. Also, the convex hull of any number of vertices of a simplex is a *face* and the convex hull of d vertices is a *facet*.

Besides being a set of simplices, a triangulation of some set of distinct points, $\mathcal{P}$, also satisfies several additional properties (Lawson, 1986):

(1) All vertices of the simplices are members of $\mathcal{P}$.
(2) The interiors of the simplices are pairwise disjoint.
(3) Each facet of a simplex is either on the boundary of the convex hull of $\mathcal{P}$ or is a common facet of exactly two simplices.
(4) Each simplex contains no points of $\mathcal{P}$ other than its vertices.
(5) The union of the simplices in the triangulation is the convex hull of $\mathcal{P}$.

In this work, local triangulations in $T_{x_c}M$ will satisfy all of these properties, but the d-dimensional triangular approximation of M only satisfies the first four properties and only if "convex hull of $\mathcal{P}$" in Property 3 is replaced by "current approximation of M".

2.4 Singular Points and Qualitative Solutions

The developments in the previous two subsections provide the basis for numerically approximating M (Equation 9) as defined by F_A (Equation 4). However, our real desire is to elucidate important qualitative features of this manifold. With respect to our optimization problem, this means identifying when the sought optimum is no longer near the neighborhood of M being considered, which happens when the need to change active sets is detected or the classification of the current optimum (minimum, maximum, or saddle) changes. Regions within such boundaries are considered to have the same *qualitative solution*, and the adjacent qualitative solutions are found by either changing A (and therefore F_A) or by looking at another region on M which has the correct classification and values of α. Poore et al. have shown and exploited the connection between the boundaries of qualitative solutions and properties of solutions of $F_A(x) = 0$ with active set switching, however, to present the main theorem that demonstrates this connection, some additional notation is needed.

First of all, Equation 4 can be written without using the active set strategy:

$$F(z,\alpha) = \begin{bmatrix} \nabla_u \mathcal{L}(z,\alpha) \\ h(u,\alpha) \\ \mu_i g_i(u,\alpha), \ \forall i \\ \lambda^T\lambda + \mu^T\mu + \lambda_0^2 - \beta_0^2 \end{bmatrix} = 0. \tag{13}$$

This is the original form of the Fritz-John necessary conditions (still leaving off the sign conditions on μ, g and λ_0), but includes the non-smooth complementarity conditions $\mu_i g_i = 0$, which are difficult to solve numerically.

Secondly, we will use the concept of the eigenvalues of a matrix restricted to a subspace. In particular, let $B \in \mathbb{R}^{n \times n}$ be a square matrix, and V be a k-dimensional subspace of $\mathbb{R}^n$. Then if $Z \in \mathbb{R}^{n \times k}$ is an orthonormal basis for V, the eigenvalues of B restricted to V, $\lambda(B|_V)$ are just the eigenvalues of Z^TBZ, and $B|_V$ is singular if and only if zero is an eigenvalue of Z^TBZ.

Now we present the theorem of interest without proof (Lundberg and Poore, 1993):

Theorem 1. Let $(z_0, \alpha_0) = (u_0, \lambda_0, \mu_0, \lambda_{00}, \alpha_0)$ be a solution to Equation 13. Assume that f, g and $h \in C^2$, the original definition of the set A

(Equation 3) holds, and define a tangent space T as:

$$T = \{y \in \mathbb{R}^n : D_u h_i (u_0, \alpha_0)\, y = 0, \forall i \\ D_u g_j (u_0, \alpha_0)\, y = 0, \forall j \in A\} \quad (14)$$

Then $D_z F(z_0, \alpha_0)$ is nonsingular *if and only if* the following three conditions hold:

(i) $\forall j \in A$, $\mu_j \neq 0$;

(ii) $S := \{\nabla_u h_i (u_0, \alpha_0)\}_{\forall i} \cup \{\nabla_u g_j (u_0, \alpha_0)\}_{\forall j \in A}$ is a linearly independent collection of $q + |A|$ vectors;

(iii) The Hessian of the Lagrangian restricted to tangent space T, $\nabla_u^2 \mathcal{L}|_T (z_0, \alpha_0)$, is nonsingular.

Therefore, detecting the transition from one qualitative solution to another is equivalent to monitoring the signs of μ and g (the active set), and detecting singularities in $D_z F_A$, since the characterization of non-singular points is completely based on the active set, the signs of the active μ_i's, and the signs of the eigenvalues of $\nabla_u^2 \mathcal{L}|_T (z_0, \alpha_0)$ (Guddat *et al.*, 1990).

3. A MULTI-PARAMETRIC NONLINEAR PROGRAMMING ALGORITHM

An algorithmic framework for solving mp-NLP problems is now given. The basic structure is taken from Brodzik's work on approximating implicitly defined manifolds (of any finite dimension) (Brodzik, 1996), the contributions of this work being the customization of Brodzik's algortihm to the mp-NLP problem (largely using the results of Lundberg and Poore), and the addition of auto-scaling, adaptive step-size adjustment, and parameter bounds following. The algorithm currently does not switch active sets, but that feature is forthcoming. (Switching active sets is related to singularity detection, approximation, and following, as indicated by Theorem 1. The current algorithm only detects singularities.) I will now give an overview of the entire algorithm. Details are excluded due to space constraints.

The proposed algorithm is best described as a multi-dimensional predictor-corrector. Given a starting point x_0 (which is a non-singular solution of Equation 13), the active set A of x_0 is determined, and the algorithm seeks to approximate the manifold

$$M = \{x : x \text{ is regular and } F_A(x) = 0\}. \quad (15)$$

To initialize the algorithm, a reference triangulation of the origin in $\mathbb{R}^d$, $\mathcal{T}_{ref}$, is calculated using triangulations by reflections (Allgower and Georg, 1978), x_0 is placed at its origin, and for each other vertex in the triangulation a new point is predicted. This new point is considered to be in $T_{x_0}M$ and is represented in the full space by

$$x_i^0 = x_0 + h_{x_0} U t_i, \quad (16)$$

where U is an orthonormal basis for $T_{x_0}M$, h_{x_0} is an initial step size, and t_i is a point in $\mathcal{T}_{ref}$. These points, x_i^0, are then projected onto the solution manifold, M, using a nonlinear equation solver to solve the augmented equations

$$G(x_i) = \begin{bmatrix} F_A(x_i) \\ U^T (x_i - x_i^0) \end{bmatrix} = 0, \quad (17)$$

characterize the new point with regards to the optimization problem, and detect any singularities in $D_z F$ (Lundberg and Poore, 1993). Finally, x_0, the new points and the triangulation structure specified by $\mathcal{T}_{ref}$ are stored in the database structure described in (Brodzik, 1998).

Once initialized, the algorithm enters the main loop. The first step is to find, if it exists, the next node in the database which is on the boundary of the current approximation of M, is non-singular, and is feasible. This node is labeled x_c (the center point), and analogous to the algorithm initialization, x_c becomes the origin of its tangent space in $\mathbb{R}^d$. Any points in the database which are already in a simplex with x_c are projected onto this tangent space:

$$y_j = U^T (x_j - x_c), \quad (18)$$

with special note being made of points that are on an *open facet*, which is a facet with only one adjacent simplex. Then $\mathcal{T}_{ref}$ is rotated such that the direction of one of its vertices (from the origin) lines up with one of the open facet vertices' directions, and any points in $\mathcal{T}_{ref}$ lying between the open facets and not too close to any of them are used to predict new points in the approximation. These new points, the y_i's on the open facets, and the origin are then triangulated using Joe's algorithm (Joe, 1993). Finally, the new points are projected onto the solution manifold (using the same procedure as the initialization phase), and the algorithm checks to see if any of the new simplices overlap with any simplices that are already in the database. If there is no overlap, the new points, simplices and facets are added to the database. Otherwise, this iteration of the loop is ignored and x_c is marked as temporarily unusable as a center point. The main loop is repeated as long as potential center nodes exist.

4. IMPLICIT OPTIMIZATION EXAMPLE

In order to demonstrate the three part process for establishing Implicit Optimization model adequacy discussed in the introduction, a hypothetical scenario is considered. Imagine that an engineer is charged with starting up a new reaction process based on the following chemical reaction:

$$A \to B,\ r_1 = k_1 c_A,\ k_1 = 0.05s^{-1}. \tag{19}$$

Since this is the reaction that produces the product, B, the chemists who studied it were thorough and are sure that the reaction rate constant, k_1, is very accurate. However, they did not estimate the rate constant of the following known side reaction as precisely:

$$B \to C,\ r_2 = k_2 c_B,\ k_2 = \alpha_1 k_1,\ \alpha_1 \approx 0.5, \tag{20}$$

and suspect a second side reaction, but are not sure if it is significant:

$$A \to C,\ r_3 = k_3 c_A,\ k_3 = \alpha_2 k_1,\ \alpha_2 \approx 0. \tag{21}$$

The reactions are to take place in a batch reactor with capacity $V_{max} = 1000L$ that is equipped with an automatic feed and emptying system, with minimum and maximum flowrates $F_{min} = 0.01L/s$ and $F_{max} = 1L/s$. The process objective is to maximize the amount of product B produced per unit time, where the total time of a batch is the amount of time it takes to fill the reactor to its initial volume, V_0, plus the duration of the batch, T, plus the amount of time it takes to empty the reactor at the end of the batch. In order to simplify the problem, we will assume that the reactor will be operated in batch or semi-batch mode. More specifically, the feed to the reactor is pure A at a concentration of $c_{A0} = 2M$, and the reaction is either started with the reactor full of feed ($V_0 = V_{max}$), which is batch mode, or it is started only partially full ($V_0 < V_{max}$), and is filled the rest of the way during the first portion of the batch, using a constant flowrate, F, until time τ, when $V(\tau) = V_{max}$ and the flow is shut off, which is semi-batch mode. This process can be modeled by two sets of differential equations, one that is valid for $t < \tau$ ($\tau = 0$ for batch mode and $\tau = \frac{V_{max} - V_0}{F}$ for semi-batch mode):

$$\begin{aligned}
\frac{d\hat{n}_A}{dt} &= c_{A0}F - (1+\alpha_2)\,k_1\hat{n}_A, & \hat{n}_A(0) &= c_{A0}V_0\\
\frac{d\hat{n}_B}{dt} &= k_1\hat{n}_A - \alpha_1 k_1 \hat{n}_B, & \hat{n}_B(0) &= 0\\
\frac{d\hat{V}}{dt} &= F, & \hat{V}(0) &= V_0
\end{aligned} \tag{22}$$

and another that is valid for $t \geq \tau$:

$$\begin{aligned}
\frac{dn_A}{dt} &= -(1+\alpha_2)\,k_1 n_A, & n_A(\tau) &= \hat{n}_A(\tau)\\
\frac{dn_B}{dt} &= k_1 n_A - \alpha_1 k_1 n_B, & n_B(\tau) &= \hat{n}_B(\tau)\\
\frac{dV}{dt} &= 0, & V(\tau) &= V_{max}.
\end{aligned} \tag{23}$$

Let us assume that the Implicit Optimization method is to be used to optimize this process. The first step is to maximize the objective function using the nominal model ($\alpha_1 = 0.5$ and $\alpha_2 = 0$), and to characterize the resulting solution. That optimization problem can be written as follows, noting that the above differential equations are simple enough that we can obtain an analytic expression for the number of moles of B present at the end of the batch, $n_B(V_0, F, T)$:

$$\begin{aligned}
&\max_{V_0,F,T} \frac{n_B(V_0,F,T)}{\frac{V_0}{F_{max}} + T + \frac{V_{max}}{F_{max}}}\\
&\text{s.t.}\\
&\quad 0 \leq V \leq V_{max}\\
&\quad F_{min} \leq F \leq F_{max}\\
&\quad \tau \leq T.
\end{aligned} \tag{24}$$

(In a more complex problem, the discretized ODE's would be equality constraints in the NLP.) The solution to this problem is $V_0 = 978.7L$, $F = 1L/s$ and $T = 27.7s$, which is characterized by noting that V_0 and T are at unconstrained values and that F is at its upper bound. If we implemented this Implicit Optimization solution directly we would use the given values as a starting point and adjust V_0 and T online using run-to-run control. However, it may very well be that this qualitative optimum is incorrect since α_1 and α_2 are so uncertain. Therefore, the three-step method outlined in the introduction should be used to evaluate the likelihood that this qualitative optimum is correct.

Step one (augmenting the nominal model with suspected structural elements and identifying uncertain model parameters) was done implicitly during the presentation of the problem. Specifically, the nominal model does not include the third reaction, Equation 21, and the uncertain parameters are α_1 and α_2. We will allow fairly wide bounds on these parameters for the purpose of solving the mp-NLP, namely

$$0 \leq \alpha_1, \alpha_2 \leq 5.0. \tag{25}$$

The second step is to solve the parametric optimization problem

$$(V_0,F,T)^*(\alpha_1,\alpha_2) = \arg\left\{\begin{aligned}
&\max_{V_0,F,T} \frac{n_B(V_0,F,T,\alpha_1,\alpha_2)}{\frac{V_0}{F_{\max}} + T + \frac{V_{\max}}{F_{\max}}}\\
&s.t.\\
&\quad 0 \leq V \leq V_{\max}\\
&\quad F_{\min} \leq F \leq F_{\max}\\
&\quad \tau \leq T
\end{aligned}\right\} \tag{26}$$

using the method outlined in the second and third sections of this paper. The corresponding numerical results are shown in Figure 4, which depicts the approximation of M projected onto $\alpha_1 \times \alpha_2 \times V_0$. As you can see, the results show no qualitative change in the solution in the region about x_0, as no singularities were detected, however, the algorithm cannot yet fill in gaps that are left in order to avoid overlap or follow singular paths. The addition of these features should both speed up the algorithm and allow more global exploration of the solution manifold. As the results stand now, there is no reason to believe that Implicit Optimization will not be successful for values of α near nominal, but the more global search forthcoming in future work may find that a different solution is the global optimum for some values of α near enough to the nominal values that the alternative solution(s) should be considered further.

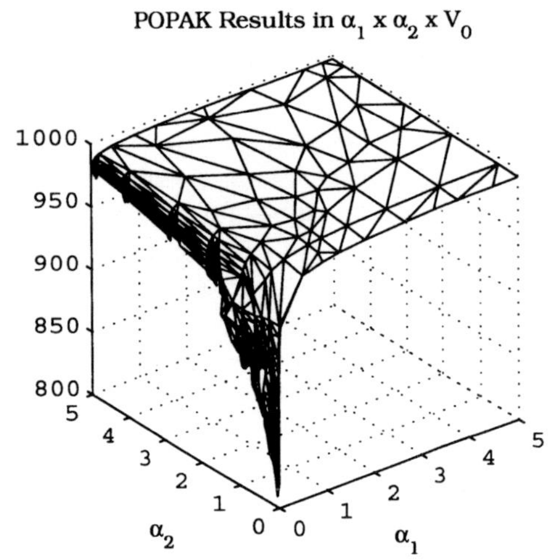

Fig. 1. Numerical results for the Implicit Optimization example. All of the points shown are feasible minimums.

5. CONCLUSIONS

Two novel methods have been presented: an algorithm for solving multi-parametric nonlinear programming (mp-NLP) problems, and a method for evaluating the adequacy of the nominal model used in Implicit Optimization (Srinivasan *et al.*, 2003), which relies heavily on solving an mp-NLP problem. The mp-NLP algorithm is designed to provide qualitative sensitivity information about the underlying NLP problem, particularly by defining regions in the parameter space where a single qualitative optimum holds. This emphasis is particularly useful for evaluating Implicit Optimization model adequacy since the main crux of this method is its dependance on the nominal optimum being qualitatively correct. This paper presents preliminary results for both of these methods, demonstrating their potential. Improvements to the mp-NLP algorithm that will allow more global exploration of the solution space are forthcoming in future work.

REFERENCES

Allgower, E. L. and K. Georg (1978). Triangulations by reflections with applications to approximation. In: *Numerische Methoden der Approximationstheorie* (L. Collatz, G. Meinardus and H. Werner, Eds.). Vol. 4. Birkhäuser Verlag. Basel, Switzerland.

Brodzik, M. L. (1996). Numerical Approximation of Manifolds and Applications. PhD thesis. University of Pittsburgh.

Brodzik, M. L. (1998). The computation of simplical approximations of implicitly defined p-dimensional manifolds. *Comput. Math. Appl.* **36**(6), 93–113.

Fink, J. P. and W. C. Rheinboldt (1987). A geometric framework for the numerical study of singular points. *SIAM J. Numer. Anal.* **24**(3), 618–633.

Gal, T. and J. Nedoma (1972). Multiparametric linear programming. *Manage. Sci.* **18**, 406.

Guddat, J., F. Guerra Vazquez and H. Th. Jongen (1990). *Parametric Optimization: Singularities, Pathfollowing and Jumps.* B. G. Teubner. Stuttgart.

Joe, B. (1993). Construction of k-dimensional delaunay triangulations using local transformations. *SIAM J. Sci. Comput.* **17**(6), 1415–1436.

Lawson, C. L. (1986). Properties of n-dimensional triangulations. *Computer Aided Geometric Design* **3**, 231–246.

Lundberg, B. N. and A. B. Poore (1993). Numerical continuation and singularity detection methods for parametric nonlinear programming. *SIAM J. Opt.* **3**(1), 134–154.

Mangasarian, O. L. and S. Fromovitz (1967). The Fritz John necessary optimality conditions in the presence of equality and inequality constraints. *J. Math. Anal. Appl.* **17**, 37–47.

Rheinboldt, W. C. (1988). On the computation of multi-dimensional solution manifolds of parameterized equations. *Numer. Math.* **53**, 165–181.

Srinivasan, B., D. Bonvin, E. Visser and S. Palanki (2003). Dynamic optimization in batch processes: II. Role of measurements in handling uncertainty. *Comput. Chem. Eng.* **27**(1), 27–44.

Tøndel, P., T. A. Johansen and A. Bemporad (2003). An algorithm for mulit-parametric quadratic programming and explicit MPC solutions. *Automatica* **39**, 489–497.

ELSEVIER
IFAC
PUBLICATIONS
www.elsevier.com/locate/ifac

ITERATIVE LEARNING CONTROL WITH INPUT SHIFT

C. Welz, B. Srinivasan and D. Bonvin

Laboratoire d'Automatique, École Polytechnique Fédérale de Lausanne
CH-1015 Lausanne, Switzerland

Abstract: Iterative learning control (ILC) is a technique to realize system inversion in a run-to-run manner. Though most of the techniques presented in the literature consider zero tracking error between the desired and achieved outputs, perfect inversion is often not feasible and in many cases not even desirable. Approximate inversion with good convergence and robustness properties (at the cost of a nonzero tracking error) has been proposed by using a forgetting factor on the input of the previous run. In this paper, approximate inversion is achieved by shifting the input of the previous run backwards in time. In addition, anticipatory ILC and current cycle feedback are used. The advantage of input shift over the use of a forgetting factor is that, when the reference trajectory is constant and the system stable, the tracking error decreases with run time. The proposed scheme is illustrated in simulation on a batch distillation system.

Keywords: Batch processes, Iterative learning control, Run-to-run control, Batch distillation.

1. INTRODUCTION

Iterative learning control (ILC) has been developed in order to improve the tracking performance of repetitive processes (Arimoto *et al.*, 1984). This is accomplished by utilizing the tracking error of the previous run to update the input of the current run. ILC has been successfully applied in robotics (Arimoto *et al.*, 1984; Kuc *et al.*, 1991; Wang, 2000) and in batch chemical processing (Lee *et al.*, 1996; Choi *et al.*, 1996; Lee *et al.*, 1999; Xu *et al.*, 2001).

The various ILC methods that have been reported in the literature can be classified into two main categories: (i) schemes with zero tracking error (Kuc *et al.*, 1991; Xu *et al.*, 1995), and (ii) those with nonzero tracking error (Arimoto *et al.*, 1990; Chien and Liu, 1994). System inversion is realized in the former, while the inversion is only approximated in the latter. Though the nonzero error techniques have the disadvantage of a residual error, they typically show better convergence and robustness properties (Arimoto *et al.*, 1990).

The standard technique for approximate inversion is to have a forgetting factor in the input update (Arimoto *et al.*, 1990). This causes the tracking error to be nonzero over the entire interval. However, the main difficulty with the feasibility of system inversion arises during the first part of the trajectory due to unmatched or varying initial conditions (Lee and Bien, 1991; Heinzinger *et al.*, 1992). After a certain catch-up time, trajectory following is relatively easy. Thus, the idea is to allow a nonzero tracking error early in the run and have the error decrease progressively with the run time t. This can be achieved with an *input* shift. The idea of a shift comes from anticipatory ILC techniques that use a shift in the *error* of the previous run (Park *et al.*, 1998; Wang, 2000).

The first ILC approaches used only the error from the previous run and thus could only handle repetitive disturbances. The addition of current cycle feedback has been proposed to handle non-repetitive disturbances (Xu *et al.*, 1995). The general ILC scheme proposed in this paper will include the following aspects: (i) input shift, (ii) error shift, and (iii) current cycle feedback.

The paper is organized as follows. Section 2 reviews ILC schemes and studies convergence conditions and the residual tracking error. In Section 3, a novel ILC scheme that shifts both the error and the feedforward trajectory of the previous cycle is proposed. A batch distillation system is presented as an application in Section 4, and conclusions are drawn in Section 5.

2. GENERAL ILC FORMULATION

2.1 Problem Formulation

In this section, a general ILC law is formulated in order to classify the common ILC schemes. For simplicity of notation, the following single input, single output system is considered in operator notation:

$$y_k = Gu_k + \bar{y} \tag{1}$$

where u_k and y_k represent the input and output trajectories in the run k, G is the operator representing the system, and $\bar{y}$ the response to initial conditions. Note that the input trajectory u_k and the output trajectory y_k are defined for the run time $t \in [0, t_f]$, where t_f is the final time of the run.

The idea of ILC is to improve trajectory tracking for repetitive processes by utilizing the previous cycle tracking error. The ILC update law for the input u_{k+1} in the next run is given as

$$u_{k+1} = Au_k + Be_k, \qquad e_k = y_{ref} - y_k \tag{2}$$

where y_{ref} is the reference trajectory to be tracked and A and B are operators applied to the previous cycle input and tracking error trajectories, respectively.

2.2 Convergence

An ILC law is convergent if the following limits exist:

$$\lim_{k\to\infty} y_k = y_\infty \quad \text{and} \quad \lim_{k\to\infty} u_k = u_\infty \tag{3}$$

Since the converged output trajectory y_∞ is not necessarily equal to the reference trajectory y_{ref}, the residual error can be nonzero.

One way of ensuring convergence is for the relation between u_k and u_{k+1} to be a contraction mapping in some appropriate norm $|\cdot|$:

$$|u_{k+1}| \le \rho |u_k|, \qquad 0 \le \rho < 1 \tag{4}$$

Using (1) in (2) gives:

$$u_{k+1} = (A - BG)\, u_k + B(y_{ref} - \bar{y}) \tag{5}$$

Note that $(y_{ref} - \bar{y})$ does not represent an error term as in (2), but the difference between two exogeneous signals, the reference trajectory and the response to initial conditions. Also, contrary to (2) where e_k depends on u_k, $(y_{ref} - \bar{y})$ is independent of u_k. Since the convergence of the algorithm depends on the homogeneous part of (5), the condition for convergence is:

$$|A - BG| < 1 \tag{6}$$

2.3 Residual Error

If the input trajectory converges to $u_k = u_{k+1} = u_\infty$, (5) gives:

$$u_\infty = (I - A + BG)^{-1} B(y_{ref} - \bar{y}) \tag{7}$$

The residual tracking error is then given by:

$$\begin{aligned} e_\infty &= y_{ref} - (Gu_\infty + \bar{y}) \\ &= \left(I - G(I - A + BG)^{-1} B\right)(y_{ref} - \bar{y}) \end{aligned} \tag{8}$$

Multiplying both sides with B and rearranging, this equation can be rewritten as:

$$Be_\infty = (I - A)(I - A + BG)^{-1} B(y_{ref} - \bar{y}) \tag{9}$$

Note that the error term Be_∞ is zero when $A = I$ and nonzero otherwise. When B is a derivative operator and $A = I$, $\dot{e}_\infty$ will be zero, but not e_∞. In other words, integral action along the run index k is needed for zero error. Zero error implies that system inversion has been achieved iteratively. However, methods with nonzero error might show better convergence properties.

2.4 Current Cycle Feedback

In conventional ILC schemes, only the previous cycle tracking error is used to adjust the input in the current cycle. To reject perturbations during a run, the error occuring during the current cycle need to be used as well. From the point of view of convergence and error analysis, these modified schemes simply correspond to different choices of the operators A and B in (2) as will be shown next. Let

A	B	ILC type	Literature
I	kD	derivative	(Arimoto *et al.*, 1984; Heinzinger *et al.*, 1992)
I	k	proportional	(Saab, 1994)
I	$k + \bar{k}D$	proportional-derivative	(Kuc *et al.*, 1991)
I	$k\Delta$	anticipatory	(Park *et al.*, 1998; Wang, 2000)
$(I+CG)^{-1}(\bar{A}+CG)$	$(I+CG)^{-1}(\bar{B}+(I-\bar{A})C)$	current cycle feedback	(Xu *et al.*, 1995; Jang *et al.*, 1995)
I	$(G^TQG+R)^{-1}G^TQ$	quadratic criterion based	(Amann *et al.*, 1996; Lee *et al.*, 2000)
βI	k	forgetting factor	(Arimoto *et al.*, 1990; Chien and Liu, 1994)
$(I+CG)^{-1}(\Delta+CG)$	$(I+CG)^{-1}(k\Delta+(I-\Delta)C)$	input shift, anticipatory, current cycle feedback	present paper

Table 1. ILC schemes with selected references (k, $\bar{k}$: proportional controllers, D: differentiation operator, Δ: shift operator, Q, R: weighting matrices, $0 \leq \beta < 1$: forgetting factor).

$$u_{k+1} = u_{k+1}^{ff} + u_{k+1}^{fb} \quad (10)$$

$$u_{k+1}^{ff} = \bar{A}u_k^{ff} + \bar{B}e_k, \quad u_{k+1}^{fb} = Ce_{k+1} \quad (11)$$

where the superscripts $(.)^{ff}$ and $(.)^{fb}$ represent the feedforward and feedback parts of the input, and $\bar{A}$, $\bar{B}$ and C are operators. Expressions (10) and (11) can be combined with (1)-(2) to give:

$$u_{k+1} = (I+CG)^{-1}(\bar{A}(I+CG) - \bar{B}G)u_k + (I+CG)^{-1}(\bar{B} + (I-\bar{A})C)(y_{ref} - \bar{y}) \quad (12)$$

which is of the form (5) with $A = (I+CG)^{-1}(\bar{A}+CG)$ and $B = (I+CG)^{-1}(\bar{B} + (I-\bar{A})C)$. Thus, the convergence condition and residual error can be analyzed using (6) and (8), respectively. Note that A and B are now functions of G due to the current cycle feedback.

2.5 Overview of ILC schemes

The various ILC schemes reported in the literature can be viewed as special cases of the general scheme presented in (5). They correspond to different choices of the operators A and B, as this was done in (12) for the current cycle feedback. Table 1 gives an overview of some ILC schemes and corresponding references.

The ILC approach first described by (Arimoto *et al.*, 1984) uses the error derivative in the ILC law, i.e. $A = I$, $B = kD$, where k is a proportional controller and D the differentiation operator. Since numerical differentiation leads to noise amplification, proportional ILC was proposed, i.e. $B = k$. However, the convergence of proportional ILC was shown for a more restrictive class than for the derivative counterpart (Saab, 1994). Also, proportional derivative controllers were proposed, $B = k+\bar{k}D$ (Kuc *et al.*, 1991). Another possibility that has been explored is a shift in the error trajectory $B = k\Delta$, where Δ is the shift operator (Wang, 2000). Since a time shift in discrete time corresponds to a differentiation, the anticipatory ILC scheme combines the robustness of derivative ILC with the low noise amplification of proportional ILC.

As discussed earlier, schemes with current cycle feedback can be cast in the above framework as well. In particular, the schemes proposed in (Xu *et al.*, 1995) and (Jang *et al.*, 1995) correspond to the choice $\bar{A} = I$, and arbitrary $\bar{B}$ and C. Note that the choice $\bar{A} = I$ results in $A = I$. Current cycle feedback is also used in ILC based on a quadratic performance criterion (Amann *et al.*, 1996).

A simple way of having a nonzero tracking error, and thus enforcing only approximate inversion, is to use a forgetting factor (Arimoto *et al.*, 1990), $A = \beta I$ with $\beta < 1$. Such a scheme has been shown to provide more robustness.

The idea of this paper is to use a non-identity operator for A so as to accept approximate inversion but improve convergence and robustness. At the same time, it is desirable to have the residual error, which could be large at the beginning of the run, get small towards the end of the run. Such a situation can be created by using $A = \Delta$ as will be shown in the next section.

3. ANTICIPATORY ILC WITH INPUT SHIFT AND CURRENT CYCLE FEEDBACK

Anticipatory ILC applies a shift to the error from the previous cycle. In the following, the use of an additional time shift of the feedforward input in order to increase robustness will be investigated. This idea was first used in (Welz *et al.*, 2004), with the same shift for the error and the input. The goal of this paper is to extend this idea and use different shifts for the error and the input.

The proposed ILC approach has three components: (i) shift of the feedforward part of the input, (ii) shift of the previous cycle error, and (iii) current cycle feedback. The iterative update law can be written as:

$$u_{k+1}(t) = u_{k+1}^{ff}(t) + k^{fb}e_{k+1}(t) \quad (13)$$

where k^{fb} is the proportional controller used in the current cycle feedback.

The feedforward part of the current input consists of shifted versions of the feedforward part of the previous input and the previous cycle tracking error:

$$u_{k+1}^{ff}(t) = u_k^{ff}(t + \delta_u) + k^{ff} e_k(t + \delta_e) \quad (14)$$

where k^{ff} is the proportional feedforward controller, δ_u the time shift of the feedforward trajectory and δ_e the time shift of the previous run error trajectory. This defines the signals $u_k^{ff}(t)$ and $e_k^{ff}(t)$ up to the times $(t_f - \delta_u)$ and $(t_f - \delta_e)$, respectively. The remaining parts of these signals are kept constant at the values $u_k^{ff}(t_f - \delta_u)$ and $e_k(t_f - \delta_e)$. In operator notation, the introduced shifts are expressed with the shift operators Δ_u and Δ_e:

$$u_{k+1}^{ff} = \Delta_u u_k^{ff} + k^{ff} \Delta_e e_k \quad (15)$$

Thus, $\bar{A} = \Delta_u$, $\bar{B} = k^{ff}\Delta_e$ and $C = k^{fb}$, and from (12), $A = (I + k^{fb}G)^{-1}(\Delta_u + k^{fb}G)$ and $B = (I + k^{fb}G)^{-1}(k^{ff}\Delta_e + (I - \Delta_u)k^{fb})$. Hence, the convergence condition follows from (6):

$$\left|(I + k^{fb}G)^{-1}\left(\Delta_u(I + k^{fb}G) - k^{ff}\Delta_e G\right)\right| < 1 \quad (16)$$

and the error term $(I - k^{fb}G)Be_\infty$ from (9):

$$\begin{aligned}(k^{ff}\Delta_e + (I - \Delta_u)k^{fb})e_\infty = (I - \Delta_u) \\ \left((I - \Delta_u)(I + k^{fb}G) + k^{ff}\Delta_e G\right)^{-1} \\ (k^{ff}\Delta_e + (I - \Delta_u)k^{fb})(y_{ref} - \bar{y}) \quad (17)\end{aligned}$$

It is seen that the residual error is zero if $\Delta_u = I$ (no shift of the feedforward input). However, the main interest for using an input shift, instead of a scalar $\beta < 1$, is the possibility to make $e_\infty(t_f)$ close to zero. In other words, the error can be shaped within the run.

The error term has the form $Se_\infty = (I - \Delta_u)T(y_{ref} - \bar{y})$, where S and T are appropriate operators. The term $(I - \Delta_u)$ corresponds to a differentiation since it approximates the derivative using forward difference. If the operator T is stable, then the trajectory $T(y_{ref} - \bar{y})$ will approach a constant value with increasing run time t. Therefore, its differentiation gives a zero error. However, since the batch time t_f is finite, it cannot be guaranteed that the error will indeed be zero at t_f.

Such an error shaping is particularly interesting when the initial conditions do not correspond to the desired reference trajectory. Without direct transmission from input to output in System (1), the initial error cannot be reduced by proportional ILC. Instead, the initial error add up from run to run, and convergence cannot be guaranteed. Direct transmission can be obtained by derivative ILC (Heinzinger *et al.*, 1992), so that the initial error is reduced from run to run. For anticipatory ILC, Chen and Wen (1999) have proposed initial state learning to compensate the effect of differences in initial conditions. On the other hand, Saab (1994) has shown convergence with the use of a forgetting factor. With the input shift presented in this paper, an initial error can be tolerated since time is provided for the system to catch up with the desired trajectory.

4. APPLICATION TO A BATCH DISTILLATION SYSTEM

4.1 Problem formulation

A binary batch distillation system is utilized to illustrate in simulation the developments of the previous sections. This example has been described in Welz *et al.* (2004), but the main features are recalled here for completeness.

Under typical assumptions for simple distillation models, a column with p equilibrium stages is considered. Writing molar balance equations for the holdup in the reboiler and for the liquid on the various stages and in the condenser, the following model of order $(p + 2)$ is obtained:

$$\dot{M}_1 = -(1 - r)V \quad (18)$$

$$\dot{x}_1 = \frac{V}{M_1}(x_1 - y_1 + rx_2) \quad (19)$$

$$\dot{x}_i = \frac{V}{M_i}(y_{i-1} - y_i + r(x_{i+1} - x_i)) \quad (20)$$

$$\dot{x}_c = \frac{V}{M_c}(y_p - x_c) \quad (21)$$

with $i = 2, \ldots, p$, x_i the molar liquid fraction, y_i the molar vapor fraction, M_i the molar holdup on Stage i, V the boilup rate and D the distillate flowrate. Stage 1 refers to the reboiler and Stage p to the top of the column. M_c is the holdup in the condenser. The internal reflux ratio $r = \frac{V-D}{V}$, is considered as the manipulated variable. The vapor-liquid equilibrium relationship is:

$$y_i = \frac{\alpha x_i}{1 + (\alpha - 1)x_i}, \quad i = 1, \cdots, p \quad (22)$$

where α is the relative volatility. The model parameters and the initial conditions are given in Table 2.

The composition of the accumulated distillate, x_d, which is measured with the sampling time $T_s = 0.1\ h$, is given by:

$$x_d(t) = \frac{\sum_{i=1}^{p} x_i(t)M_i(t) - x_i(0)M_i(0)}{M_1(t) - M_1(0)} \quad (23)$$

p	10		$x_{d,des}$	0.9	kmol/kmol
α	1.6		$M_1(0)$	100	kmol
M_i	0.2	kmol	$x_1(0)$	0.5	kmol/kmol
M_c	2	kmol	$x_i(0)$	0.5	kmol/kmol
V	15	kmol/h	$x_c(0)$	0.5	kmol/kmol

Table 2. Model parameters and initial conditions, $i = 2, \cdots, p$

A batch is divided into 2 intervals of operation:

(1) Start-up phase with full reflux, $r = 1$, $t = [0, t_s]$, $t_s = 1.415\ h$.
(2) Distillation phase, $r \in (0,1)$, $t = [t_s, t_f]$, $t_f = 10\ h$.

The objective is to track a reference trajectory for the distillate purity $x_d(t)$ in Interval 2. This trajectory ends up at the desired distillate purity at final time, $x_{d,des}$. Meeting the purity constraint can then be realized by closely tracking this reference. The reference trajectory is chosen to be linear with $x_{d,ref}(t_s) = 0.925\ kmol/kmol$ and $x_{d,ref}(t_f) = 0.9\ kmol/kmol$. The accumulated distillate purity $x_d(t)$ is assumed to be measurable once some distillate has been collected in Interval 2. The initial feedforward trajectory for iterative learning schemes is also linear with $r(t_s) = 0.898$ and $r(t_f) = 0.877$.

In order to obtain a realistic test scenario, the following uncertainty is considered:

- Perturbation: Boilup rate equally distributed in the range $V = [13\ 17]\ kmol/h$, changed every $0.5\ h$.
- Measurement noises: Product composition x_d with 5% multiplicative gaussian noise.

The values of the squared tracking error $\sum_{t=t_s}^{t_f} e(t)^2$ and the final tracking error $e(t_f)$ are averaged over 20 realizations of the perturbation and measurement noise. Also, the variance $v_{e(t_f)}$ of the residual error is calculated from 20 realizations.

4.2 Implementation of trajectory tracking schemes

The tracking error cannot be reduced to zero because of a nonzero initial tracking error ($x_d(t_s) \neq x_{d,ref}(t_s)$) arising from uncertainties in the start-up phase. As a consequence, the methods assuming zero initial tracking error do not converge in this example. Instead, a nonzero tracking error has to be allowed in order to guarantee convergence. This can be accomplished by applying a forgetting factor or a time shift of the feedforward input.

4.2.1. Iterative learning control without on-line feedback (ILC): Three ILC schemes without on-line feedback are considered:

- (ILC $\beta = 0.999$): A forgetting factor β is applied to the feedforward input trajectory in (11) with $\bar{A} = \beta$ and $\bar{B} = k^{ff}$.
- (ILC $\delta_u = 1\ h, \delta_e = 1\ h$): The same large shift is applied to the feedforward input and the error in (14).
- (ILC $\delta_u = 0.25\ h, \delta_e = 0\ h$): A smaller shift of the feedforward input trajectory is imposed and the error trajectory is not shifted in (14).

The run-to-run gain $k^{ff} = 0.1\ kmol/kmol$ was determined as a compromise between robustness and performance. ILC with forgetting factor converges after 30 runs, while the methods with time shift of the trajectories converge after 25 runs (Figure 1). The residual error is slightly smaller with the latter methods, especially when the time shift of the feedforward trajectory is reduced to $\delta_u = 0.25\ h$ (Table 3).

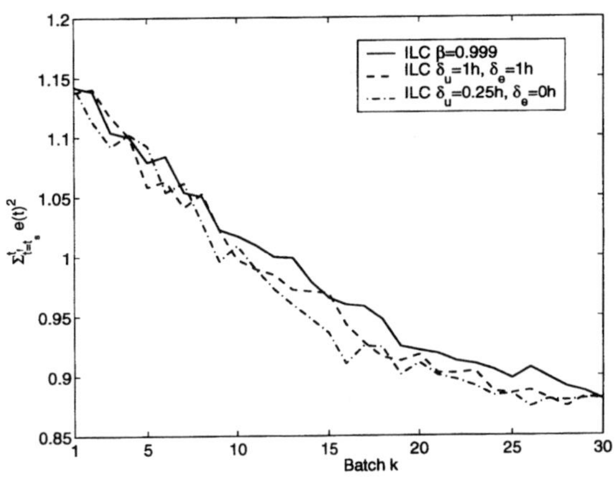

Fig. 1. Evolution of the squared tracking error for ILC methods without current cycle feedback.

Strategy	$\sum_{t=t_s}^{t_f} e(t)^2$ $[kmol^2/kmol^2]$	$e(t_f)$ $[kmol/kmol]$	$v_e(t_f)$ $[kmol^2/kmol^2]$
ILC $\beta = 0.999$	0.874	+0.0065	$4.1 \cdot 10^{-05}$
ILC $\delta_u = 1h$, $\delta_e = 1h$	0.883	-0.0022	$3.5 \cdot 10^{-05}$
ILC $\delta_u = 0.25h$, $\delta_e = 0h$	0.874	-0.0005	$3.0 \cdot 10^{-05}$
FB	0.869	-0.0069	$1.5 \cdot 10^{-05}$
FB+ILC $\beta = 0.999$	0.865	-0.0053	$3.0 \cdot 10^{-05}$
FB+ILC $\delta_u = 1h, \delta_e = 1h$	0.865	-0.0048	$2.7 \cdot 10^{-05}$
FB+ILC $\delta_u = 0.25h, \delta_e = 0h$	0.864	-0.0036	$2.0 \cdot 10^{-05}$

Table 3. Comparison of the various tracking schemes after the 30th run in terms of squared tracking error $\sum_{t=t_s}^{t_f} e(t)^2$, tracking error at final time $e(t_f)$ and its variance $v_e(t_f)$.

4.2.2. ILC with current cycle feedback (FB+ILC): The on-line feedback utilized is the PI controller:

$$r(t) = r^{ff}(t) - k_p \left(e(t) + k_i \int_{t_s}^{t} e(\tau) d\tau \right) \quad (24)$$

where r^{ff} is the feedforward term. The parameters $k_p = 8\ kmol/kmol$ and $k_i = 0.02\ 1/h$ were tuned manually. The error does not reduce to zero since integral action is not sufficient to drive it to zero within the finite time of the batch. Also, increasing the gains for faster error reduction causes instability.

Three ILC schemes are compared as in Section 4.2.1, but now with the additional current cycle feedback. Though the squared tracking errors are similar for all on-line schemes, the final error can be reduced using the time shift $\delta_u = 0.25\ h$ (Table 3). The residual error with current cycle feedback and $\delta u = 0.25$ is larger than that without feedback, but its variance is smaller, because within-run perturbations can be compensated.

5. CONCLUSION

This paper has provided an overview of ILC schemes with general conditions for convergence and an analysis of the residual tracking error. A new scheme that consists of anticipatory ILC with a shift for both the feedforward part of the input and the error trajectory and current cycle feedback has been presented. By allowing a nonzero residual error, this scheme provides good robustness properties. In contrast to ILC with a forgetting factor, which also allows a nonzero residual error, the proposed scheme has been shown to reduce the error as a function of run time for a constant input trajectory. This property is useful for batch processes, where often the initial conditions are uncertain and a constraint has to be met at final time. The proposed scheme has been applied to a simulated batch distillation system in the presence of uncertainties, where tracking of a distillate purity reference is utilized to meet the terminal constraint on distillate purity. It has been shown that shifting the input trajectory results in a smaller residual error than using a forgetting factor.

REFERENCES

Amann, N., D.H. Owens and E. Rogers (1996). Iterative learning control for discrete-time systems with exponential rate of convergence. *IEE Proc.-Control Theory Appl.* **143**(2), 217–224.

Arimoto, S., S. Kawamura and F. Miyazaki (1984). Bettering operation of robots by learning. *J. Robotic Syst.* **1**(2), 123–140.

Arimoto, S., T. Naniwa and H. Suzuki (1990). Robustness of P-type learning control with a forgetting factor for robotic motion. In: *Proceedings of the 29th Conference on Decision and Control.* Honolulu, Hawaii. pp. 2640–2645.

Chen, Y. and C. Wen (1999). *Iterative Learning Control.* Lecture Notes in Control and Information Sciences. Springer, London.

Chien, C.J. and J.S. Liu (1994). A P-type iterative learning controller for robust output tracking of nonlinear time-varying systems. In: *Proceedings of the American Control Conference.* Baltimore.

Choi, J.W., H.G. Choi, K.S. Lee and W.H. Lee (1996). Control of ethanol concentration in a fed-batch cultivation of Acinetobacter calcoaceticus RAG-1 using a feedback-assisted iterative learning algorithm. *Journal of Biotechnology* **49**, 29–43.

Heinzinger, G., D. Fenwick, B. Paden and F. Miyazaki (1992). Stability of learning control with disturbances and uncertain initial conditions. *IEEE Trans. Autom. Contr.* **37**, 110–114.

Jang, T.J., C.H. Choi and H.S. Ahn (1995). Iterative learning control in feedback systems. *Automatica* **31**(2), 243–248.

Kuc, T.Y., K. Nam and J.S. Lee (1991). An iterative learning control of robot manipulators. *IEEE Trans. Rob. Autom.* **7**(6), 835–842.

Lee, J.H., K.S. Lee and W.C. Kim (2000). Model-based iterative learning control with a quadratic criterion for time-varying linear systems. *Automatica* **36**(5), 641–657.

Lee, K.H. and Z. Bien (1991). Initial condition problem of learning control. *IEE Proceedings-D* **138**(6), 525–528.

Lee, K.S., I.S. Chin and H.J. Lee (1999). Model predictive control technique combined with iterative learning for batch processes. *AIChE J.* **45**(10), 2175–2187.

Lee, K.S., S.H. Bang, S.Yi, J.S Son and S.C. Yoon (1996). Iterative learning control of heat-up phase for a batch polymerization reactor. *J.Proces.Contr.* **6**, 255–262.

Park, K.H., Z. Bien and D.H. Hwang (1998). Design of an iterative learning controller for a class of linear dynamic systems with time delay. *IEE Proc.-Control Theory Appl.* **145**(6), 507–512.

Saab, S.S. (1994). On the P-type learning control. *IEEE Trans. Autom. Contr.* **39**, 2298–2302.

Wang, D. (2000). On D-type and P-type ILC designs and anticipatory approach. *Int. J. Control* **73**(10), 890–901.

Welz, C., B. Srinivasan and D. Bonvin (2004). Combined on-line and run-to-run optimization of batch processes with terminal constraints. In: *ADCHEM'03.* Hong-Kong.

Xu, J.X., Q. Hu, T.H. Lee and S. Yamamoto (2001). Iterative learning control with Smith time delay compensator for batch processes. *J.Proces.Contr.* **11**, 321–328.

Xu, J.X., X.W. Wang and L.T. Heng (1995). Analysis of continuous iterative learning control systems using current cycle feedback. In: *Proceedings of the American Control Conference.* Seattle, Washington. pp. 4221–4225.

ELSEVIER
IFAC
PUBLICATIONS
www.elsevier.com/locate/ifac

FRONTIERS OF CHEMICAL ENGINEERING:

THE SYSTEMS APPROACH

T.F. Edgar[1] and J.B. Rawlings[2]

[1]*University of Texas, Department of Chemical Engineering, Austin, Texas 78712*

[2]*University of Wisconsin, Department of Chemical and Biological Engineering,*

Madison, WI, 53706

Abstract: A dramatic shift in chemical engineering undergraduate education has been proposed based on NSF-funded, discipline-wide workshops that took place in 2003. This has led to a broad consensus regarding foundations for chemical engineering undergraduate education in the future based on the organizing principles of (1) molecular transformations; (2) multiscale analysis, and (3) the systems approach. The third area is the focus of this paper, as it relates to systems and control and the increasing importance of biological systems in chemical engineering as well as the development of new educational materials. *Copyright © 2004 IFAC*

Keywords: biological systems, process control, chemical engineering curriculum, educational materials.

1. INTRODUCTION

Chemical engineering skills are critical to such sectors as microelectronics, medicine, biotechnology, and new materials. While demands for well-trained engineers in these areas have continued to grow, undergraduate educational programs have not kept pace, especially in these new areas. The curriculum will need to change if chemical engineers are to play a leading role in meeting the workforce needs in emerging technologies based on molecular transformations. Current teaching materials and textbooks do not reflect recent advances that have taken place in the underlying sciences of chemistry and biology.

In 2003, three NSF-funded workshops were held to assess the chemical engineering curriculum. Faculty from more than 53 universities and industry representatives from 5 companies reached strong consensus that there is a need for significant change. The participants reached broad consensus on basic principles for chemical engineering undergraduate education in the future; these principles address fundamental knowledge, skills and attributes of graduates, and methods of engagement with students. See http://web.mit.edu/che-curriculum for full proceedings of these workshops. A key driver for curriculum reform was to address the increasingly central role of biology in the traditional industries that hire chemical engineers and the need to prepare students for versatile, multifaceted careers.

Workshop participants defined the scope of what chemical engineers do and described the elements of an undergraduate chemical engineering education without relying on the conventional categories of the traditional curriculum. Three organizing principles emerged. First, chemical engineers seek to understand, manipulate, and control the molecular basis of matter, and the molecular-level processes – physical, chemical, and biological – that underlie observed phenomena in nature and technology. Molecular transformation is a unified treatment of phenomena at this level.

Second, chemical engineers are effective because they combine macroscopic engineering tools with molecular understanding. Multiscale analysis covers the tools appropriate to a given length scale (molecular dynamics, continuum equations, macroscopic averages), an appreciation for the ways in which phenomena occur at different scales (e.g., in a packed bed reactor, ranging from kinetic mechanism to heat duty), and an understanding of how phenomena at one scale affect another (e.g., molecular structure affects macroscopic properties). This also deals with connection of transient and steady-state processes.

Third, realistic chemical engineering problems (that is, the dynamic behavior of batch and continuous processes and systems in nature, technology, and society) feature multiple interacting components and draw important information from fields outside chemical engineering. The analysis of such problems depends on coordinating a variety of tools. Systems analysis and synthesis is the organizing principle that leads to the ability to manipulate systems to achieve desired behavior or performance. Furthermore, chemical engineers design and create products and processes, so that there is a strong component of synthesis, as well.

In short, the chemical engineer leverages knowledge of molecular processes across multiple length scales to synthesize and manipulate complex systems that encompass both processes and products. In this paper we focus on the systems approach and its strong connection to education of students in the subjects of process modeling and control.

1.1. Desirable Attributes of Graduates and the Systems Approach

Engineers are fundamentally problem solvers, seeking to achieve some objective of design or performance among technical, social, economic, regulatory, and environmental constraints. The chemical engineer brings particular insight to problems in which the molecular nature of matter is important. Educators cannot teach students everything that might be encountered; instead the aim is to equip graduates so they grasp fundamentals and engineering tools, enabling them to specialize or diversify as opportunity and initiative allow. The systems component of the curriculum should:

(1) Cultivate professional attributes, such as willingness to make estimates and assumptions, readiness to face open-ended problems and noisy data, and ability to visualize the solution.
(2) Hone the professional skills of problem solving; estimating uncertainty; using computational tools; economic analysis; and the ability to plan, execute, and interpret experiments.
(3) Integrate knowledge and information to aid in solution of chemical engineering problems.

2. THE SYSTEMS APPROACH IN THE CHEMICAL ENGINEERING CURRICULUM

The systems component of the curriculum ensures that chemical engineering graduates should be able to:

(1) create and understand mathematical descriptions of physical phenomena,
(2) scale variables and perform order-of-magnitude analysis,
(3) structure and solve complex problems,
(4) manage large amounts of messy data, including missing data and information,
(5) resolve complex and sometimes contradictory issues of process design: sensitivity of solutions to assumptions, uncertainty in data, what if questions, process optimization.

The systems approach is a fundamental concept that is explicitly addressed only in a few chemical engineering courses. The concept of analyzing a collection of components and processes as an overall system, rather than as individual components, is critical for frontier areas of chemical and biological engineering, as well as traditional areas.

The systems component of the curriculum is the part that trains the students in the tools for synthesis, analysis and design of chemical and biological processes, units and collections thereof. Systems education teaches the students how to convert scientific facts and principles of chemical and biological systems into engineering decisions. The knowledge base of systems consists of methods for dynamic and steady-state simulation at multiple length and time scales, statistical analysis of data, sensitivity analysis, optimization, parameter estimation and system identification, design and analysis of feedback, methods for online monitoring and diagnosis, and methods for design of products and processes.

New educational materials that enable instructors to integrate the systems concepts into the curriculum at each stage of the undergraduate educational program are needed. This integration is a new and essential component. As the students learn new scientific concepts, the systems tools that enable specific scientific knowledge to be harnessed for engineering purposes can be presented in parallel. This purposeful and tight integration marks one significant change to the traditional curriculum. The changing scientific principles of interest, which include both newly emerging concepts in molecular biochemistry and cellular biology as well as the expanding tools of molecular modeling, require a concomitant change and expansion of the systems tools currently used. The four years of the curriculum can be structured as shown below.

The Freshman Systems Experience

- Plant-wide and product viewpoints
- Exposure to multi-faceted, real-world problems
- Degrees-of-freedom analysis
- Computer programming concepts and simple computer applications

The Engineering Systems Experience (Sophomore Year)

- Conservation laws for simple dynamic and steady-state systems
- Simple models for an experimental dynamic system (chemically reacting system)
- Acquisition and analysis of noisy, complex, dynamic laboratory data
- Numerical simulation for simple models (single ODE)
- Parameter estimation for simple models (one or two parameters estimated from one or two dynamic sensor measurements)
- Equipment construction and sensor design

Methods for Molecular Systems (Junior Year)

- Random variables, probability and statistics
- Stochastic systems and molecular level reactions as systems
- Stochastic kinetic models and Monte Carlo models
- Simulation as an enabling technology
- Optimization principles for design, parameter estimation, and decision making
- General principles of experimental design for static and dynamic systems
- Use of models in predicting and understanding system behavior (analysis) and subsequent use of models in shaping system behavior (synthesis)
- Systems biology: sequence to function: in metabolic networks, gene expression networks, integrated gene-metabolic networks
- Examples from microelectronics, catalysis, systems biology

Systems Integration and the Marketplace (Senior Year)

- Multi-scale systems, separation and resolution of time and length scales
- Design and analysis of feedback control systems
- Frequency response and analysis of spectroscopic data
- Monitoring and fault detection
- Energy and mass integration, design for environment and process efficiency, network targeting concepts
- Process operations: planning, scheduling, and the supply chain
- The design experience: economics and business skills, safety, marketing, environmental impact, life cycle analysis, ethics, intellectual property, globalization, social and national needs

A significant portion of the new teaching materials can be developed as case studies and modules that can be integrated into each year of the curriculum. Eventually new textbooks will be produced, but other forms of dissemination are also effective. For example, modules in the systems area can be distributed electronically and take advantage of computer simulation and animation to illustrate the concepts.

Case studies and modules supporting the concepts could include the following real-world engineering applications, some in non-traditional areas of chemical engineering:

- Desalination of sea water
- Hydrogen from biomass
- Synthesis of specialty polymers
- Global climate change
- Regional air quality analysis
- Production, separation, and purification of natural products and recombinant proteins
- Reconstruction of cellular networks. Design of cells and biomolecules.
- Insulin regulation
- Pharmacokinetic and pharmacodynamic models
- Biomedical control systems
- Drug patch design
- The human body as a chemical process

- Environmental cycles (water, carbon, nitrogen, etc.)
- Chemical plant design and operation (refining, plastic, power plant, pharmaceutical, etc.)
- Cell design (human, animal, plant, etc.), DNA replication, RNA transcription, translation, regulation of metabolic pathways
- Product design (need, conception, product engineering, process manufacturing, disposal, life cycle analysis)
- Batch processing for semiconductor manufacturing
- Particulate processes (cell dynamics, aerosols in drug inhalers, smog, crystallization, dispersions, emulsions)

3. IMPLEMENTATION OF THE NEW CURRICULUM

The preceding view that the systems approach is a critical component in chemical engienering is not shared by all faculty. At the first Frontiers Workshop, after the first breakout sessions, different groups reported a long and varied list of proposed changes to the curriculum. Notably, a few groups reported that the subject of process control was high on the "hit list". While process control was not perceived as the only option for elimination, it was disconcerting to see it in such an egregious position. Apparently the value of process control to academic chemical engineers is apparently not as high as many in the control community believe it should be.

In a related article, Cussler et al. (2002) declared that a number of fields like thermodynamics, reaction engineering, transport, and control can be relegated to the scrap heap of "mature technologies" that will not have much future impact in the gain in knowledge. They proposed dropping courses on control and optimization but added several disclaimers: "First, we accept without question the importance of process optimization to commodity chemicals. Secondly, we recognize that process control has a key role in ensuring the success of those other cornerstones of competitive advantage in specialty product manufacture; safety, consistency and quality. Our third hesitation stems from our unwillingness to sacrifice any of our technical core to less-quantitative business ideas. Still, we recognize that a large part of our future is going to be in areas where different skills are needed". This article has sparked spirited discussions in many departments who are considering curriculum change.

Industrial chemical engineers seem to have little doubt that process control is important to keeping modern chemical plants operating, a view that was articulated by several industrial attendees at the Frontiers Workshops. Perhaps the disconnect between the two positions is how the typical faculty member views where employment opportunities will reside in the future, say in 2020. If chemical engineers are not involved in making value-added "stuff" at a desired quality level, the contribution of chemical engineering to the national economy will undoubtedly be greatly reduced.

Perhaps the current emphasis of the typical process control course that heavily focuses on Laplace transforms, analytical solutions to linear differential equations, linear algebra, frequency response, and multiple methods to tune a PID controller needs to change. Computer simulation should not take a back seat to theoretical analysis. In fact the availability (since 1998) of computer-based tools such as Simulink in MATLAB or Control Station (www.controlstation.com) has completely changed the way in which process control can be taught.

3.1 New Educational Materials in Process Control

It is clear that a system viewpoint is very important for dealing with biological processes and biotechnology. Unsteady-state behavior and feedback control are important concepts in living systems, and any organism that is at steady-state is dead (arguing for an understanding of dynamic process models).

The topics I cover when I teach process control include dynamic behavior (with about one lecture on Laplace transforms and analytical solutions to ODEs), physical and empirical modeling, computer simulation, measurement and control hardware technology, basic feedback and feedforward control concepts, and advanced control strategies. Many of these topics can be presented to reflect applications in biochemical or materials engineering. More emphasis could be placed on modeling, optimization, and data analysis/statistics in a revised course. Unfortunately existing textbooks (including one I have co-authored) mostly use examples from continuous processes in petrochemical plants (vs. batch specialty products). Therefore it is valuable to introduce different examples to emphasize principles that are relevant to non-traditional applications.

Three process control textbooks contain a number of examples and exercises that illustrate applications of process control to biological engineering as follows:

(1) Ogunnaike and Ray (1995): drug delivery, blood pressure control, bioreactors
(2) Bequette (2003): biochemical reactor, pharmacokinetic models, drug delivery, blood glucose control, blood pressure control
(3) Seborg, Edgar, and Mellichamp (2004): fed-batch bioreactor, drug delivery

Below is an example of an open-ended homework problem recently assigned in the process control course at the University of Texas that helps the student understand the difficulty of control in a biomedical application. The problem statement is as follows:
Diabetes Mellitus is characterized by insufficiency of the pancreas to produce enough insulin to regulate the blood sugar level. In Type I Diabetes the pancreas produces no insulin, and the patient is totally dependent on insulin from an external source to be infused at a rate to maintain blood sugar levels at normal levels. Hyperglycemia occurs when blood glucose level rises much higher than the norm (>8 mmol/L) for prolonged periods of time; hypoglycemia occurs when the blood sugar level falls below values of 3 mmol/L. Both situations can be deleterious to the individual's health. Hyperglycemia can lead to blindness, kidney failure, and other complications on a long-term basis. The effects of hypoglycemia are more critical on a short-term basis, leading to loss of consciousness and coma within a few hours. The normal range of blood sugar in a healthy patient should staty between 3.8-5.6 mmol/L, the target range for a controller regulating blood sugar.

A Type I Diabetic needs your help to maintain her blood sugar within an acceptable range (3 mmol/L < glucose < 8 mmol/L). She has just eaten a large meal (a disturbance) that you estimate will release glucose according to $D(t) = 0.5\ e^{-0.05t}$, where t is in minutes and D(t) is in mmol/L-min. She has a subcutaneous insulin pump that can release insulin up to 115 mU/min (mU = 10^{-3} Unit of Insulin). The flowrate of insulin is the manipulated variable.

A model of her blood glucose level is given by (Lynch and Bequette, 2002; Bequette, 2002):

dG/dt = -P1 * (G - G_basal) - (X - X_basal) * G + D;
dX/dt = -P2 * (X - X_basal) + P3 * (I - I_basal);
dI/dt = -n * I + U / V1;

where P1, P2, P3, V1, n are constants. G, X, and I are values for glucose concentration in the blood (mmol/L, insulin concentration in the body (mU/L), and plasma insulin concentration, respectively. Basal values refer to the initial values for each one. G_basal = 4.5 mmol/L and X_basal = I_basal = 15 mU/L. D is the rate of glucose release into the blood (mmol/L-min) as the disturbance. U is the flowrate of insulin (mU/min) as the manipulated variable.

This model is implemented in a MATLAB function 'blood_glucose.m'. You can calculate the glucose response by calling an integrator in MATLAB, such as ode45. Appropriately label any graphs that you produce.

(a) What will happen to her blood glucose level if the pump is shut off initially?
See Figure 1 for the solution.

(b) What will happen to her blood glucose level if the pump injects at a constant rate of 15 mU/min?
See Figure 2 for the solution.

(c) Is there a constant infusion rate of insulin that will help her stay within an acceptable glucose range (3mmol/L < G < 8mmol/L) for the next 400 minutes?

By trial and error, the students can determine that a constant rate strategy will not work, which leads to the need to develop more elaborate control strategies. Student reaction to this example was quite positive, especially because some of them had diabetic relatives. About one-third of the class is taking a "bio"-emphasis in their course selection for the B.S. Ch.E. degree, but even for students in the traditional curriculum this type of problem is quite understandable. Developing new educational materials such as proposed in the NSF workshops will certainly add to the base of interesting control applications. In addition, a CACHE task force on Biosystems, chaired by Frank Doyle, will be developing educational materials in this area.

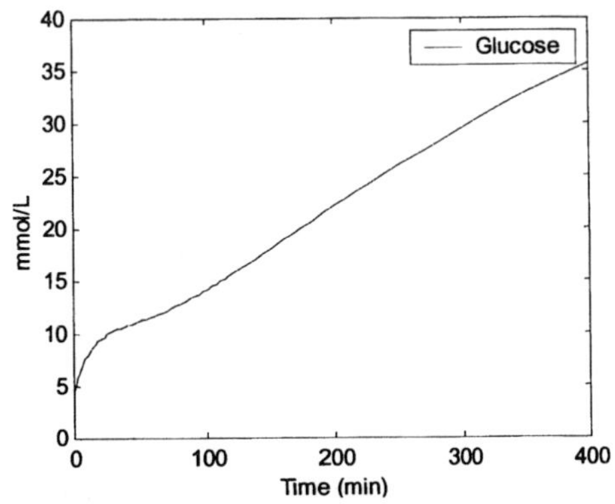

Fig. 1. Effect of No Insulin Flow on Glucose After Meal Disturbance.

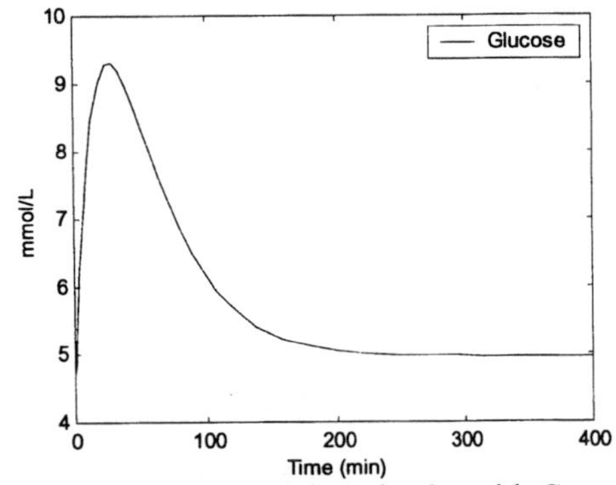

Fig 2. Glucose Dynamic Behavior with Constant Insulin Flow After Meal Disturbance.

CONCLUSIONS

Most educators agree that a systems viewpoint is valuable for chemical engineering graduates. Dynamics, feedback, and stability are intellectual underpinnings required for understanding many new and complex systems of interest to chemical engineers. Control, like design, can be taught in a way so that students must, integrate knowledge from other core ChE courses in process modeling and analysis of process behavior. There are not many courses in the curriculum that fulfill these needs. New educational materials should be developed to augment the limited number of examples and exercises in existing textbooks.

REFERENCES

Bequette, B.W. (2003). *Process Control.* Prentice-Hall, Upper Saddle River, NJ.

Cussler, E.L., D.W. Savage, A.P.J. Middleburg, and M. Kind (2002). Refocusing Chemical Engineering, *Chem. Engr. Prog.*, p. 265.

Lynch, S.M., and B.W. Bequette (2002). Model Predictive Control of Blood Glucose in Type I Diabetics using Subcutaneous Glucose Measurements, *Proc. ACC*, p. 4039.

Ogunnaike, B.A., and W.H. Ray (1995). *Process Dynamics, Modeling, and Control.* Oxford University Press, London.

Seborg, D.E., T.F. Edgar, and D.A. Mellichamp (2004). *Process Dynamics and Control, 2nd edition*, Wiley, New York.

ELSEVIER
IFAC
PUBLICATIONS
www.elsevier.com/locate/ifac

INTEGRATION OF BIOLOGICAL SYSTEMS CONTENT INTO THE PROCESS DYNAMICS AND CONTROL CURRICULUM

Robert S. Parker *,[1] **Francis J. Doyle III** ** **Michael A. Henson** ***

* *Department of Chemical and Petroleum Engineering, University of Pittsburgh, Pittsburgh, PA*
** *Department of Chemical Engineering, University of California at Santa Barbara, Santa Barbara, CA*
*** *Department of Chemical Engineering, University of Massachusetts at Amherst, Amherst, MA*

Abstract: The evolution of the chemical engineering discipline motivates a reevaluation of the process dynamics and control curriculum. A key requirement of future courses is the introduction of theoretical concepts and application examples relevant to emerging areas such as complex biological systems. We outline the critical concepts required to integrate biological content within the traditional framework of process dynamics and control. Our initial experience with three courses is outlined to demonstrate alternative approaches to achieve this integration. The paper concludes with a discussion of open issues which require further attention from the process control and biological systems communities.

Keywords: Process control, process dynamics, education, biological systems, biomedical systems, biochemical systems, systems biology.

1. INTRODUCTION

The discipline of chemical engineering is evolving, as evidenced by the recent wave of departmental name changes. These have primarily focused on the addition of "bio", such as "biomolecular" (Cornell, Illinois, Notre Dame, Pennsylvania), "biological" (Colorado, Northwestern, Rensselaer Polytechnic Institute, Wisconsin), and "biochemical" (Rutgers). Such name changes reflect both the increasing number of chemical engineering faculty involved in research on biology-oriented topics, and the fact that the percentage of chemical engineering undergraduates going to work for companies in the biotechnology and biomedical sectors has increased from 4.6% in 1998 to 10.3% in 2001–02 (AIChE, 2002).

A series of MIT-organized workshops, supported by the National Science Foundation, examined the current state of undergraduate chemical engineering education and recommended a sweeping set of content and delivery changes (Massachusetts Institute of Technology, 2003). Foremost among the proposed changes were the introduction of *biology as a core science*, the importance of *addressing complexity*, and the expanded use of the *systems approach*. The present discussion focuses on these three elements within the context of the traditional process dynamics and control curriculum.

The dynamics and control course usually taught in the late junior year or senior year is a natural point for including biological systems content along with chemical process material. Due to the focus on key principles rather than specific case studies, biological systems can be integrated seamlessly without detracting from the coverage of more tradi-

[1] To whom correspondence should be addressed: rparker@pitt.edu; +1-412-624-7364; 1249 Benedum Hall, Pittsburgh, PA 15261 USA

tional applications. The study of complex biological systems often necessitates the integration of concepts from other undergraduate courses such as mass and energy balances, heat and mass transfer and reaction engineering. As a result, the integration of biological systems content has the potential to further expand the role of the dynamics and control course as a capstone experience in the chemical engineering curriculum. Due to their inherent complexity, biological systems offer a rich set of dynamic problems to which well trained chemical engineers can make a significant contribution.

This expanded vision of the system dynamics and control curriculum requires the following difficult issues to be addressed: (1) how can these complex systems be introduced in a meaningful way to undergraduate chemical engineers with little background in biology?; and (2) what changes are required to include biological content without sacrificing the traditional core of process dynamics and control? The objective of this paper is to provide some answers to these questions through the discussion of three courses taught at our respective institutions. The first two examples illustrate the introduction of biological content into the traditional process control course, while the third example focuses on the development of a new course in which the systems approach is applied to a diverse set of biological problems.

2. INTEGRATION OF BIOLOGICAL SYSTEMS CONTENT

As with most other courses in the undergraduate chemical engineering curriculum, the process dynamics and control course contains a large span of new material that is covered at a rather brisk pace. To produce students who can apply their knowledge of single- and multi-variable analysis and control in an industrial context is a formidable challenge even when the focus is purely on chemical process systems. The addition of biological content along with the requisite modeling and analysis techniques requires a carefully crafted course to avoid overwhelming students. A possible structure for a semester long system dynamics and control course is illustrated by the syllabus in Table 1 where NL is the number of lectures allotted to the specific topics listed in ALL CAPS. **Bold** entries represent new topics specific to biological systems. *Italicized* entries are theoretical topics often considered optional in a traditional control course but which are viewed as important for a biologically oriented course.

The introduction of state-space models and associated analysis tools is essential for the treatment of biological systems due to their complexity (*e.g.*, high order, multivariable, highly nonlinear) often precludes simple Laplace-based analysis. Because students often retain little of the mathematics "learned" in their lower level courses, a few lectures on matrix algebra and linear state-space systems are required to review core

Table 1. Proposed syllabus for a biologically oriented dynamics and control course.

NL	Topics
4	DYNAMIC MODELING Principles of fundamental modeling; chemical and **biological process examples**; introduction to empirical modeling
7	LINEAR AND NONLINEAR SYSTEMS ANALYSIS *Matrix algebra and linear state-space systems*; linear systems theory; introduction to nonlinear systems theory; dynamic simulation; chemical and **biological process examples**; introduction to the Laplace transform
7	FEEDBACK SYSTEMS Basis principles of feedback; **physiological control systems**; **homeostasis as a setpoint-free feedback system**; **feedback in biochemical reaction networks**; closed-loop response analysis; servo vs. load behavior; feedback control of chemical process systems; **closed-loop drug delivery**
8	FEEDBACK CONTROL SYNTHESIS Basic principles of model-based controller design; PID controller design & tuning; advanced single-variable control techniques; multivariable control techniques; *model predictive control*; chemical and **biological process examples**
4	ADVANCED TOPICS *Large-scale systems & plant-wide control*; *parameter estimation and experimental design*; *state estimation*; **introduction to systems biology**

material and to ensure that students with deficient backgrounds understand the basic concepts. When combined with the linear systems analysis lecture, this material allows the calculation of eigenvalues to determine stability and matrix rank for the analysis of controllability and observability. The nonlinear systems theory lecture includes the traditional topic of Jacobian linearization as well as introductory coverage of phase plane analysis, multiplicity and bifurcations.

Feedback is a concept easily introduced in the context of biological system examples. Consider the healthy human body versus the insulin-dependent diabetic as a case study. Circulating glucose concentrations are sensed by pancreatic β-cells. Based on these measurements, β-cells release an appropriate amount of insulin to maintain the glucose concentration at a physiological reference value. Increasing glucose concentrations lead to increased insulin release and *vice versa*. The representation of biological control systems using various elements of the traditional block diagram is an appealing framework for presenting biological content. However, this approach can fail to reveal the complexity of the underlying biological processes. In the case of circulating glucose control, a limited degree of complexity can be incorporated by utilizing a nonlinear proportional-derivative controller to represent the pancreatic response to glucose concentration (Nomura *et al.*, 1984). Additional complexity can be added as necessary to allow the application of increasingly advanced closed-loop analysis tools.

Table 2. Possible case studies for the process dynamics and control course.

Chemical Processes
Continuous and/or fed-batch polymerization reactors; distillation column; continuous pulp digester; paper machine; simple plantwide example (*e.g.*, reactor & separator); semiconductor process (*e.g.*, lithography); photovoltaic fi lm processing; fuel cell
Biotechnological Systems
Continuous and/or fed-batch fermentors; yeast energy metabolism; cell stress response (*e.g.*, heat shock); eukaryotic cell cycle; bacterical chemotaxis
Biomedical Systems
Baroreceptor vagal reflex (blood pressure control system); insulin-dependent diabetic patient (glucose-insulin metabolism/control); circadian rhythm gene regulatory network; anesthesia control; drug delivery for HIV treatment; drug delivery for cancer treatment

Throughout the topic sequence in Table 1, a number of examples serve to highlight the breadth of opportunities for application of the theoretical concepts presented in the course. Table 2 provides a list of case studies that could be integrated into a typical dynamics and control class. For each topic where examples are listed in the syllabus, two chemical process and two biological system examples could be used to develop lecture material, in-class exercises and recitation problems. Ideally the biological problems are divided equally between the biotechnology and biomedical lists. However, some material may be illustrated more clearly by selecting both biological examples from the same list. For example, homeostasis is most easily discussed within the context of biomedical processes.

One similarity between the traditional chemical and biological examples listed in Table 2 is the scale on which these systems can be analyzed. Polymerization reactor models can be developed using input-output representations (Parker *et al.*, 2001*a*) or based on detailed descriptions of the individual polymer particles and their interactions (Immanuel *et al.*, 2003). Analogous models can be developed for microbial fermentors where lumped descriptions of cellular process are provided by unsegregated models (Henson and Seborg, 1992) and detailed descriptions of the individual cells are provided by cell population models (Henson, 2003). A major conclusion of the MIT-organized education workshops was that multi-scale phenomena should be incorporated throughout the undergraduate chemical engineering curriculum (Massachusetts Institute of Technology, 2003). Several of the biological examples listed in Table 2 are well suited for studying system level phenomena characterized by events at small length scales. For example, the baroreceptor vagal reflex can be used to illustrate how activity at the single cell level affects systemic blood pressure control (Doyle III *et al.*, 1997). From the controller design perspective, different types of biological complexity can be examined with the goal of determining the model structure required to synthesize effective controllers for specific bioprocesses. While the introduction of biological systems content is not necessarily required to illustrate these concepts, we feel that an integrated program of chemical and biological examples will reinforce key concepts and demonstrate that these diverse examples are conceptually similar.

3. SYSTEM DYNAMICS AND CONTROL

The process control course at the University of Massachusetts has traditionally focused on Laplace transform analysis and chemical process applications. This course usually represents the only extensive exposure to dynamic modeling and feedback control in the undergraduate curriculum. While classical material must be covered for students to gain a working knowledge of the field, advanced control techniques and emerging applications also must be emphasized to reflect current industrial trends. Biological systems were chosen as an appropriate vehicle for introducing the key elements of biological transformations, multi-scale phenomenon and systems level analysis identified in the MIT-sponsored education workshops (Massachusetts Institute of Technology, 2003). Rather than completely change the existing course content, a more conservative approach based on the integration of biological systems and the requisite analysis techniques is being pursued. The short-term objective is to utilize biological system dynamics and feedback to demonstrate the wide applicability of the analytical and computational techniques covered in the course.

The current syllabus for the UMass process control course (ChE 446) is shown in Table 3 where new topics introduced in the past year are *italicized.* The first few weeks are focused on fundamental modeling because undergraduate students typically have little experience formulating dynamic balance equations. A case study approach with traditional chemical process examples and biochemical system examples focusing on yeast metabolism is utilized. A continuous yeast fermentor model is introduced and revisited in lectures and homeworks throughout the semester. Both time domain and Laplace domain analysis techniques receive extensive coverage. An introduction to matrix algebra is necessary since this material is not covered in the required mathematics courses. A major focus is the formulation and stability analysis of linear state-space models.

Engineered and natural feedback systems are introduced in parallel to highlights their common features and unique properties. Multi-scale characteristics of biological control systems are covered by discussing feedback in a yeast metabolic network and the architecture of the baroreceptor reflex. Closed-loop stability is analyzed in both the time and Laplace domains. While most of the material on single-loop controller synthesis is traditional, an introduction to

Table 3. Syllabus for UMass Course ChE 446: Process Control.

NL	Topics
5	FUNDAMENTAL MODELING Basic principles; chemical process examples (non-isothermal chemical reactor; *binary flash unit; binary distillation column*); *biochemical system examples (continuous fermentor model; metabolically structured yeast cell model)*
7	DYNAMIC SYSTEM ANALYSIS *Linear algebra (solution of matrix equations, state-space models; eigenvalues & eigenvectors); time domain analysis (basic stability concepts, linearization of nonlinear models, linear stability analysis, continuous fermentor example)*; Laplace transforms; transfer functions models; empirical models
6	FEEDBACK SYSTEMS Process control systems; *biological feedback systems (engineered versus natural feedback systems, yeast sulfate assimilation pathway, baroreceptor vagal reflex)*; closed-loop transfer functions; closed-loop stability
7	FEEDBACK CONTROL SYNTHESIS PID controller tuning; internal model control; *time domain controller design (state feedback, pole placement, model matching, continuous fermentor example)*; feedforward control; cascade control
5	MULTIVARIABLE CONTROL Control loop interactions; decentralized control; *discrete-time models (discretization of continuous-time models; convolution models, prediction models); model predictive control (controller design & tuning, constraint handling, real-time optimization; continuous fermentor example)*

time domain controller design techniques is provided to parallel the Laplace domain methods. The final few weeks are focused on the analysis and design of multivariable control systems. The main emphasis is linear model predictive control since many students entering the oil refining, petrochemical and chemical industries will encounter this technology. The continuous yeast fermentor model is used to illustrate the controller design techniques introduced throughout the course.

To accommodate new material on biological systems and time domain analysis and design techniques, some material typically covered in the process control course had to be deemphasized or virtually eliminated. Topics which received reduced coverage included transfer function models, Laplace domain analysis and design techniques, advanced single loop control strategies and traditional chemical process examples. Frequency domain analysis and design techniques received very limited coverage. While these topics are admittedly valuable, a broader view of dynamic systems and feedback control was deemed to be more important given current trends in the chemical engineering profession.

As part of the ABET mandated evaluation process, students are requested to complete an assessment concerning the course objectives. Four new objectives pertaining to the biological systems content were included in the assessment for the fall of 2003. Each student provided a score ranging from "5" if they strongly agreed the objective was achieved to "1" if they strongly disagreed the course objective was achieved. Results obtained from the twenty one respondents are summarized in Table 4. The average scores are similar to those obtained for the other course objectives, thereby indicating that the biological content was successfully integrated into the course. Continued improvements in student learning and satisfaction are anticipated as the course content is further developed in the spirit of Table 1.

Table 4. Student Responses to Biological Systems Content in the UMass Process Control Course.

Score	Question
3.83	I can construct a dynamic model of a biological system.
3.78	I can perform dynamic system analysis and controller design in the time domain.
3.89	I can apply dynamic system analysis techniques to biological systems to evaluate properties such as stability.
3.83	I can describe the relevance of feedback control theory to biological systems.

4. SYSTEMS–LEVEL BIOMEDICINE

The University of Pittsburgh Medical Center is a research active hospital system, and the medical doctors collaborate actively with faculty from engineering, business, and the sciences. The biology component in the dynamics and control course (ChE 1034) at Pittsburgh focuses on the analysis of, and controller synthesis for, biomedical systems at the whole-organism level. By integrating the research activities in modeling and control of diabetic and cancer case studies within the undergraduate class, students are exposed to a component of chemical engineering they might otherwise overlook. This format has resulted in a steady flow of undergraduates interested in undergraduate research, and an increased interest in graduate study.

ChE 1034 is approached from a model-based perspective; approximately half of the course is focused on modeling systems using both fundamental to empirical approaches, in both continuous and sampled-data (discrete) domains. From the fundamental modeling perspective, the students are taught to distinguish pharmacokinetics (the time profile of a drug) from pharmacodynamics (the disease dynamics, effect of the drug on the disease, and toxicity) in much the same way valve dynamics and process output response are captured by separate blocks in a block diagram. In an input-output framework, these two physiologically-motivated effects are indistinguishable, as the pharmacodynamic effect is the observed response. The remainder of the course focuses on the model-based synthesis and analysis of classical and advanced control systems, as in Table 1.

As a case study, consider the insulin-dependent diabetic patient depicted in Figure 1. The fundamental model usually covers a recitation session, introducing the students to the key variables of the diabetic patient problem and demonstrating the utility of skills developed elsewhere in the curriculum (*e.g.* dynamic mass balance with reaction, transport resistance) in the modeling of biomedical problems. As homework or an in-class exercise, the students are asked to reduce this model to a low-order equivalent and justify any assumptions made. The usual result is a model structure containing two differential equations, one each for glucose and insulin, which is conceptually similar to that published in (Bolie, 1961).

The dynamics and control course material is often challenging to the students, who have had a limited introduction to dynamical systems prior to the course. One classical approach to address the challenge of depth of treatment versus limited classroom time is the case study method (Mustoe and Croft, 1999). To provide the students with a consistent example, a collection of case study examples that correspond to the various sections of the curriculum outlined in Table 1 is required. To this end, selections from the insulin-dependent diabetic patient literature have been mapped onto the course outline, as shown in Table 5.

For an expert in the field of diabetic patient modeling and control, this table is fairly straightforward to construct. For the faculty member less familiar with biology, or the diabetic patient problem, the map provides a guide to focused literature reading to quickly bring biomedically motivated problems into the classroom. This aspect is most useful for faculty who are not dynamics and control experts, but who are responsible for teaching the course. This course is challenging for non-experts to teach, and is often rated

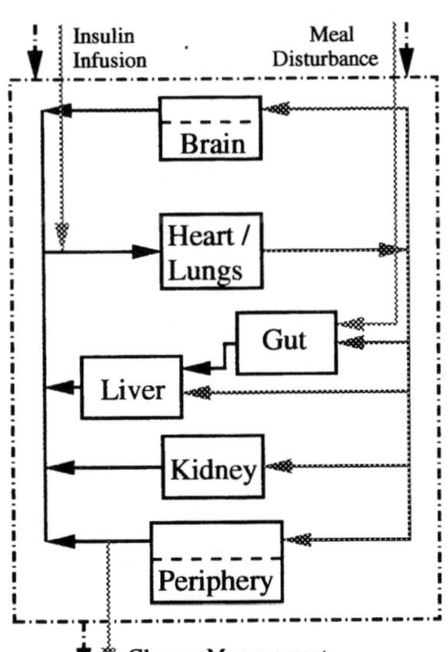

Fig. 1. Open-loop schematic of the diabetic patient. Small solid blocks represent the fundamental model, with manipulated input insulin delivery rate, meal disturbance, and glucose concentration measurement.

Table 5. Integration of sample case study (insulin-dependent diabetic patient) with course outline topics.

DATA-DRIVEN MODELING Sorensen FOTD (1985), Bolie two-state linear (1961), Bergman "Minimal" Model (1981)
FIRST PRINCIPLES MODELING Physiologically–based Pharmacokinetic / Pharmacodynamic (Sorensen, 1985; Parker *et al.*, 2000*b*)
LINEAR SYSTEMS ANALYSIS Bolie two-state linear ODEs (1961)
LINEAR SYSTEMS ANALYSIS w/ LINEARIZATION Linearize and analyze Bergman "Minimal" (1981)
DYNAMIC SIMULATION All models, including AIDA as a different performance classification (Sorensen, 1985; Bolie, 1961; Bergman *et al.*, 1981; Parker *et al.*, 2000*b*; Lehmann *et al.*, 1994; Agar *et al.*, 2002)
FEEDBACK SYSTEMS Glucose-insulin interactions (Bolie, 1961); nonlinear feedback response (Bergman *et al.*, 1981), healthy pancreas response (Sorensen, 1985; Nomura *et al.*, 1984)
CLOSED-LOOP ANALYSIS Sorensen healthy patient (Sorensen, 1985)
PID CONTROL Controller design from FOTD (Sorensen, 1985), low-order ODEs (Bolie, 1961), and linearized systems and/or effects of nonlinearity (Bergman *et al.*, 1981)
ADVANCED CONTROL Feedforward for meal disturbances (Lehmann and Deutsch, 1992) and exercise (Lenart and Parker, 2002), with simple (Bolie, 1961; Bergman *et al.*, 1981) or complex (Sorensen, 1985) case studies
MULTIVARIABLE CONTROL MISO (glucose and insulin inputs; G, I and exercise inputs) (Parker *et al.*, 2000*a*) or MIMO (glucose and insulin control) for a variety of systems (Sorensen, 1985; Bolie, 1961; Bergman *et al.*, 1981)
MODEL PREDICTIVE CONTROL Linear MPC in analytical (Bolie, 1961) or data-driven (Parker *et al.*, 1999) forms; MPC with a linearized model (Sorensen, 1985; Bergman *et al.*, 1981; Parker *et al.*, 2001*b*; Lenart and Parker, 2001); nonlinear MPC if desired (Sorensen, 1985; Bergman *et al.*, 1981; Parker *et al.*, 2001*b*; Lenart and Parker, 2001; Rubb and Parker, 2003)

the "least desired teaching assignment" (based on a three-university non-randomized sample of chemical engineering educators). A collection of these paper-topic maps, for traditional and biological case studies, would provide those teaching the dynamics and control course a variety of examples tailored to each section of the course.

5. SYSTEMS BIOLOGY

In addition to the required dynamics and control course, described in Section 2, there is a demand in many chemical engineering programs for elective courses that facilitate specialization in either systems engineering or biotechnology/biomedical engineering. At UCSB, a new course has been offered in the Spring 2004 quarter, entitled "Engineering Approaches to Systems Biology". The course is taught at a dual-level (seniors and new graduate students), and fulfills the track requirement for both systems and biology emphases in the undergraduate chemical engineering program. In addition, it will be cross-listed, in its subsequent offering, with the biomolecular science and engineering (BMSE) program. The current syllabus is listed in Table 6, detailing the topics for a single quarter course (20 lectures of duration 75 minutes).

The course focuses on the emerging problems in systems biology and computational biology. There is a substantial level of effort being invested in these areas in both academia and industry, and the demand for the training of students has increased in proportion. These advances have been facilitated by developments in both computational modeling and high throughput biology – enabling a systematic approach to analyzing complexity in biophysical networks, which was previously untenable. Advances in molecular biology over the past decade have made it possible to probe experimentally the causal relationships between microbiological processes initiated by individual molecules within a cell, and their macroscopic phenotypic effects on cells and organisms. These studies provide increasingly detailed insights into the underlying networks, circuits, and pathways responsible for the basic functionality and robustness of biological systems and create new and exciting opportunities for the development of quantitative and predictive modeling and simulation tools. Model development involves translating identified biological processes into coupled dynamical equations which are amenable to numerical simulation and analysis. These equations describe the interactions between various constituents and the environment, and involve multiple feedback loops, responsible for system regulation, and noise attenuation and amplification.

The discipline of *Systems Biology* has emerged in response to these challenges (Kitano, 2002), and combines approaches and methods from systems engineering, computational biology, statistics, genomics, molecular biology, biophysics, and other fields. The recurring themes include: (i) integrative viewpoints

Table 6. Syllabus for UCSB Course: ChE 154 – Engineering Approaches to Systems Biology.

NL	Topics
6	CELLULAR REGULATION Central dogma; genome sequences; genome expression; genomic circuits; protein, metabolic, signaling networks; high throughout biological data; biological databases
6	MATH MODELING AND SYSTEMS ANALYSIS TOOLS Modeling strategies; boolean models; nonlinear ODE models; discrete stochastic models; systems biology modeling packages; network analysis – robustness, identifi ability; design of experiment issues
6	BIOSYSTEMS CASE STUDIES Bacterial chemotaxis; lambda phage; circadian rhythm; caspase signaling cascade; pap pili phase variation
2	COURSE PROJECTS Midterm progress reports; fi nal presentations

towards unraveling complex dynamical systems, and (ii) tight iterations between experiments, modeling, and hypothesis generation.

In response, there have been a number of courses introduced in a variety of departments across the country that address elements of systems biology and computational biology. These have been targeted at both undergraduate and graduate audiences, and in some cases involve continuing education participants from industry. The balance of topics in the syllabus in Table 6 is approximately one third on basic cellular regulation, one third on applications of systems engineering tools to biological problems, and one third on detailed case studies to illustrate current methodologies and future challenges. Although the UCSB curriculum is based on quarters, the same general template could be extended to a semester-long course, without significant modification.

Assignments for this course consist of short homework problems, primarily at the beginning of the course, and a major course project. The project entails a midterm progress report, a final presentation, and a written report. The case study offers a mechanism to tailor the course to a diverse student population – seniors work on teams with a reduced scope, while graduate students work as individuals on a more detailed project.

6. OPEN ISSUES IN UPDATING THE PROCESS DYNAMICS AND CONTROL COURSE

6.1 Reducing the Emphasis on the Laplace Domain

Traditional process control courses emphasize the Laplace transform for solution of dynamic models as well as for the analysis of open- and closed-loop systems. In the coverage of multivariable systems, a common approach is to construct single-input, single-output (SISO) transfer function models for each input-output pair and then combine the SISO models into a transfer function matrix. While classical analysis may be facilitated by using Laplace domain system representations, inherent limitations (*e.g.* linear system, deviation variables, "small" input-output dimension) severely restrict the utility of this approach for complex systems commonly encountered in biological applications. Biological systems are inherently nonlinear with phenomena ranging from protein interactions in gene regulatory networks to adaptation in systemic reflexes. Furthermore, modeling of biological systems at resolutions below the macroscopic scale often leads to high state dimension (Henson, 2003; Parker *et al.*, 2000*b*).

>From this perspective, complex dynamic systems are most effectively addressed in the time domain. Nonlinear analysis techniques can be introduced explicitly in the time domain (see Section 3), thereby exposing students to theoretical concepts and analysis

tools with wider applicability than Laplace domain methods. Moreover, the formulation of large-scale system models (in terms of state and/or input-output dimensions) is more readily performed in the time domain via conservation equations. Consequently, the syllabus in Table 1 focuses on the derivation and analysis of linear and nonlinear state-space systems. Connections with the corresponding Laplace domain concepts can be introduced as necessary (*e.g.* stability via eigenvalues vs. poles).

As is evident from Table 1, frequency domain analysis and design techniques have been effectively removed from the proposed curriculum. While we do not dispute the potential value of these methods, our experience is that this material is difficult for most students to understand and apply. Because frequency domain analysis often starts with the substitution of $s = j\omega$ into the transfer function model, the removal of frequency domain material is consistent with the reduced emphasis on the Laplace transform. On the other hand, the syllabus in Table 1 is sufficiently flexible that limited coverage of frequency domain methods at the expense of other topics is possible. For instructors interested in including these tools, brief coverage of the fast Fourier transform (FFT) and pulse testing may be represent a suitable compromise.

The Laplace transform is particularly useful for the analysis of systems with time delay and/or zero dynamics. While numerical simulation of these systems is straightforward, analysis in the time domain is more difficult. Computational tools such as the FFT can be employed, but such methods are not well suited for use in conventional written exams. Analytical treatment of zeros in the time domain is more involved than the corresponding Laplace domain methods for SISO linear systems.

Time domain analysis of transportation and measurement delays is most conveniently performed using a discretized framework. Linear state-space models without delay can be constructed by performing state augmentation and using shift matrices. While this approach can lead to potentially large state dimensions, properly formulated exam problems can partially mitigate this problem. Multiple delays and delays between blocks in a closed-loop system produce more complicated state-space models that are problematic in the context of written exams. Consequently, evaluating student understanding of this material can be challenging. Two possible approaches are: (i) prepare challenging problems and conduct exams in computational classrooms where students have access to the tools necessary to complete the analysis; and (2) construct two-part problems with the first part focuses on problem formulation and the second part addresses the analysis problem on a related but simplified problem.

6.2 Multivariable Control

While most traditional courses treat multivariable systems as a straightforward extension of SISO systems, a more comprehensive approach which addresses the unique challenges of multivariable controller design is required. A formal introduction to decentralized control would support the "systems viewpoint" when performing control relevant analysis of multivariable processes – a set of optimal SISO feedback loops generally does not result in overall system optimality. While coverage of this material is relatively straightforward, the vast majority of multivariable systems are suboptimally controlled when single-loop controller design techniques are employed. Furthermore, students working in the chemical process industries will invariably encounter model predictive control applications. While comprehensive treatment of model predictive control is beyond the scope of this course, students need to develop a working knowledge of this high performance control technology. As outlined in the course syllabus in Table 3, the introduction of this topic necessitates limited coverage of discrete-time models and real-time optimization.

6.3 Robustness

A critical topic in the analysis of both process control systems and biological regulation is robustness. The remarkable levels of robust performance attained in nature is enviable from an engineering perspective and is not a well understood issue in biology. The importance of robustness in understanding disease states, as well as evolution and development, motivate its incorporation in the dynamics and control curriculum. Of course, a detailed theoretical treatment, such as that provided in (Skogestad and Postlethwaite, 1996), is beyond the scope of a typical undergraduate course. However, key concepts of robustness can be emphasized using simple tools such as sensitivity analysis — effectively capturing the gains from uncertain elements in a system to the controlled output or performance measure. Analysis can include the closed-loop strategies that have been adopted in nature to deal with robustness, such as redundancy, feedback, filtering and modular protocols.

6.4 Teaching Control for Non-Expert Faculty

Our experience indicates that the process dynamics and control class is not a popular choice as a teaching assignment among non-experts in the field. The lack of interest is due to a variety of issues, including the mathematical complexity of the material and the significant focus on feedback controller synthesis. An additional concern is that the material is challenging to students, who have had a limited introduction to dynamical systems prior to this course. The syllabus in Table 1 represents a significant departure from the traditional controller synthesis dominated course to a more balanced presentation of system dynamics and feedback.

A notable benefit of the proposed syllabus is the degree of potential customization. While our focus has been on the introduction of biological systems content, the treatment of other application areas such as advanced materials or microelectronics can be accomplished in a similar manner. This flexibility provides an excellent opportunity for instructors to integrate their research interests into the course. The three courses described in Sections 3–5 were heavily influenced by the work performed in the corresponding research groups. Possible benefits of such integration include: (i) increasing the diversity of application examples by encouraging non-experts to teach the course; and (ii) introducing students to cutting edge engineering research which influences their perception of the field and may affect their future career directions.

Experts in process dynamics and control can contribute to the development of instructional materials in a variety of ways. The construction of extended case studies such as Table 5 for different applications (*e.g.* electronic materials) would ease the burden on non-experts to incorporate novel examples into the curriculum. Software tools such as the Process Control Modules (Doyle III *et al.*, 2000) and Java-based Control Modules (Yang and Lee, 2002) are well suited for introducing traditional concepts and applications. New software tools are needed to increase the exposure of chemical engineering undergraduates to biological complexity and to allow the application of the theoretical concepts introduced in the course to representative biological systems. Ongoing efforts, such as those organized by MIT and the CACHE organization, are focused on the development and refinement of biologically-relevant systems courses. A task force headed by the second author of this paper is currently working on course revisions as well as software module design as a means to integrate biological content throughout the chemical engineering curriculum. More details will be made available at `www.cache.org`.

REFERENCES

Agar, B. U., G. Birol and A. Cinar (2002). Virtual experiments for controlling blood glucose level in type I diabetes. In: *Proc. Second Joint EMBS/BMES Conf.*. p. 2609.

AIChE (2002). 2001-2002 initial placement of chemical engineering graduates. http://www.aiche.org/careerservices/trends/placement.htm.

Bergman, R.N., L.S. Phillips and C. Cobelli (1981). Physiologic evaluation of factors controlling glucose tolerance in man. *J. Clin. Invest.* **68**, 1456–1467.

Bolie, V. W. (1961). Coeffi cients of normal blood glucose regulation. *J. Appl. Physiol.* **16**, 783–788.

Doyle III, F. J., M. A. Henson, B. A. Ogunnaike, J. S. Schwaber and I. Rybak (1997). Neuronal modeling of the baroreceptor reflex with applications in process modeling and control. In: *Neural Networks for Control* (D.L. Elliott, Ed.). Springer Verlag.

Doyle III, F. J., R. S. Parker and E. P. Gatzke (2000). *Process Control Modules: A Software Laboratory for Control Design.* Prentice Hall International Series in the Physical and Chemical Engineering Sciences. PH PTR. Upper Saddle River, NJ 07458.

Henson, M. A. (2003). Dynamic modeling of microbial cell populations. *Current Opinion in Biotechnology* **14**, 460–467.

Henson, M. A. and D. E. Seborg (1992). Nonlinear control strategies for continuous fermentors. *Chem. Eng. Sci.* **47**, 821–835.

Immanuel, C. D., C. F. Cordeiro, S. S. Sundaram, E. S. Meadows, T. J. Crowley and F. J. Doyle III (2003). Modeling of particle size distribution in emulsion co-polymerization: Comparison with experimental data and parametric sensitivity studies. *Comp. Chem. Eng.* **26**, 1133–1152.

Kitano, H. (2002). Systems biology: A brief overview. *Science* **295**, 1662–1664.

Lehmann, E. D., T. Deutsch, E. R. Carson and P. H. Sonksen (1994). AIDA: An interactive diabetes advisor. *Comp. Meth. Prog. Biomed.* **41**, 183–203.

Lehmann, E.D. and T. Deutsch (1992). A physiological model of glucose-insulin interaction in type 1 diabetes mellitus. *J. Biomed. Eng.* **14**, 235–242.

Lenart, P. J. and R. S. Parker (2001). Glucose control during exercise in type I diabetic patients. Proceedings of the Topical Conference on Bioinformatics and Genomics. AIChE Annual Meeting.

Lenart, P. J. and R. S. Parker (2002). Modeling exercise effects in type I diabetic patients. Proceedings of the 15th IFAC World Congress on Automatic Control. Barcelona, Spain.

Massachusetts Institute of Technology (2003). Frontiers in chemical engineering education initiative. http://web.mit.edu/che-curriculum/.

Mustoe, L. R. and A. C. Croft (1999). Motivating engineering students by using modern case studies. *International Journal of Engineering Education* **15**, 469–476.

Nomura, M., M. Shichiri, R. Kawamori, Y. Yamasaki, N. Iwama and H. Abe (1984). A mathematical insulin-secretion model and its validation in isolated rat pancreatic islets perfusion. *Comput. Biomed. Res.* **17**, 570–579.

Parker, R. S., D. Heemstra, F. J. Doyle III, R. K. Pearson and B. A. Ogunnaike (2001*a*). The identifi cation of nonlinear models for process control using tailored "plant–friendly" input sequences. *J. Proc. Control* **11, Sp. Issue SI**, 237–250.

Parker, R. S., E. P. Gatzke and F. J. Doyle III (2000*a*). Advanced model predictive control (MPC) for type I diabetic patient blood glucose control. In: *Proc. American Control Conf.*. Vol. 5. pp. 3483–3487.

Parker, R. S., F. J. Doyle III and N. A. Peppas (2001*b*). The intravenous route to blood glucose control. *IEEE Eng. Med. Biol.* **20**, 65–73.

Parker, R. S., F. J. Doyle III and N. A. Peppas (1999). A model-based algorithm for blood glucose control in type I diabetic patients. *IEEE Trans. Biomed. Eng.* **46**(2), 148–157.

Parker, R. S., J. H. Ward, N. A. Peppas and F. J. Doyle III (2000*b*). Robust H_∞ glucose control in diabetes using a physiological model. *AIChE Journal* **46**, 2537–2549.

Rubb, J. D. and R. S. Parker (2003). Glucose control in type I diabetic patients: A Volterra model-based approach. In: *Proc. ADCHEM 2003*. Vol. in press.

Skogestad, S. and I. Postlethwaite (1996). *Multivariable Feedback Control.* John Wiley & Sons, New York, NY.

Sorensen, J.T. (1985). A Physiologic Model of Glucose Metabolism in Man and its Use to Design and Assess Improved Insulin Therapies for Diabetes. PhD thesis. Department of Chemical Engineering, MIT.

Yang, D. R. and J. H. Lee (2002). Process control education software using Java applet. AIChE Annual Meeting. http://dot.che.gatech.edu/Information/research/issicl/che4400/javamodule.html.

ELSEVIER
IFAC
PUBLICATIONS
www.elsevier.com/locate/ifac

WAVELET-BASED MODEL REDUCTION OF BREAKAGE PROCESSES

Y. Liu[a] and M. O. Tadé[b,*]

[a] *Department of Electronic, Electrical and Computer Engineering,
The University of Birmingham, Birmingham, B15 2TT, U.K.*

[b] *Department of Chemical Engineering, Curtin University of Technology,
GPO Box U 1987, Perth, WA 6845, Australia*

Abstract: Mathematical models of breakage processes are composed of partial integro-differential equations. For process optimisation and controller synthesis, a low-order model is preferred. A wavelet-based model reduction procedure is proposed. A multi-resolution (wavelet) approach is employed to achieve a set of high-order ordinary differential equations. A procedure which is motivated by the theory of singular perturbation is then proposed for the construction of approximate inertial manifold. The dominant behaviour is found from a multi-scale decomposition. The proposed technique has been tested for two typical case studies. In both cases, the predicted and analytical particle size distributions are in excellent agreement. *Copyright © 2004 IFAC*

Keywords: Model reduction; singular perturbation method; multi-resolution method; breakage processes; wavelets

1. INTRODUCTION

Breakage processes are very common in engineering practice. For efficient operations and product effectiveness, control of a breakage process is an important problem to industries. Examples of these systems include the grinding of coal in ring-roller mills (Sato, *et al.*, 1996; Shoji, *et al.*, 1998), the wet stirred ball attrition of alumina and synthetic diamond (Shinohara, *et al.*, 1999), the attrition of alumina hydrate in a tumbling mill (Frances and Laguerie, 1998), the dry stirred ball attrition of quartz (Ma, *et al.*, 1998) and the wet stirred bead attrition of carbon (Varinot, *et al.*, 1999). Other examples in chemical engineering are polymer engineering and liquid-liquid dispersions (Kostoglou and Karabelas, 1998). These processes are usually described by the population balance formulation and especially the breakage equations. Since the processes have distributed properties, it is not surprising that there are an infinite number of spatially dependent state variables.

With advances in scientific computation, sophisticated spatial-dependent models developed using first principle approaches can be used as effective tools to provide realistic predictions or estimates of the system behaviour. However, one of the main problems associated with this approach is that the dimension of the dynamical system is very large and not practical for the design and implementation of effective controllers. Moreover, on-line simulation of these high-order, complex models for open-loop optimisation or parameter estimation is computationally costly.

The breakage equation is a linear partial integro-differential equation and its solution is not trivial. For this reason, the field of development of new solution method for the breakage equation is very active and a lot of new approaches have been proposed during the last decade as briefly discussed below.

The recent method of generations can be considered as an extension of the procedure used for the analytical solution to arbitrary breakage functions (Liou, *et al.*, 1997). Discretization (sectional) methods are the most important class for numerical solution of breakage equations. Several methods of

* Corresponding author. Fax: +61 8 9266 2681
Email: tadem@che.curtin.edu.au (M. O. Tadé)

this class have been recently proposed by Hill and Ng (1995), Kumar and Ramkrishna (1996), Kostoglou and Karabelas (2002), and Vanni (1999). Another class of the methods employs the approximation of the unknown function with higher order polynomials. However, all the methods are characterised by computational inefficiency in tackling complicated spatially dependent problems. Cheng *et al.* (1996) applied wavelets in a class of breakage problems.

The control of a breakage process is usually accomplished by transforming the breakage equations to a system of ordinary differential equations (ODEs) that permit the synthesis of a controller. High order solutions of the systems of breakage equations almost always result in an impossible synthesis due to poor robustness.

A natural framework to address the problem of deriving low-dimensional ODE systems that accurately reproduce the solution of dissipative systems is based on the concept of inertial manifold (IM). An IM is a positive, invariant, finite-dimensional Lipschitz manifold, which attracts every trajectory exponentially. However, the explicit derivation of the inertial form requires the computation of the analytical form of the IM. In order to overcome the problems associated with the existence and construction of IMs, the concept of approximate inertial manifold (AIM) has been introduced (Foias, *et al.*, 1988; 1989) and used for the derivation of ODE systems whose dynamic behaviour approximates those of the inertial form.

In the area of modelling and simulation of breakage processes, few papers have appeared in the literature dealing with the model reduction. Consequently, this paper makes an original contribution to this area using a multi-resolution (wavelet) approach to arrive at a high-order ODE model. A procedure which is motivated by the theory of singular perturbation is then proposed for the construction of AIMs for breakage systems. The AIMs are used for the derivation of ODE systems with dimension equal to the number of slow modes. The dominant behaviour was found from a multi-scale decomposition using Daubechies family of wavelets. The solutions of the low-order dominant ODE systems are close to those of original breakage systems for almost all times. The dimensions of reduced model are less than 10% of those of original high order models, which may be used to design model-based controllers.

2. PRELIMINARIES

2.1 Breakage equations

The breakage process can be described in general by the following linear partial integro-differential equation:

$$\frac{\partial m(x,\tau)}{\partial \tau} = \int_x^\infty p(x,y)b(y)m(y,\tau)dy - b(x)m(x,\tau) \quad (1)$$

where τ is time, x is particle volume, $m(x,\tau)$ is particle size distribution, $b(x)$ is breakage rate and $p(x,y)$ is probability distribution of a particle of volume x resulting from the breakup of a particle of volume y.

Let $m_0(x) = m(x,0)$ be the initial particle size distribution. The total mass, total number and mean size of the initial distribution are as follows:

$$M_0 = \int_0^\infty x m_0(x)dx \quad (2)$$

$$N_0 = \int_0^\infty m_0(x)dx \quad (3)$$

$$x_0 = \frac{M_0}{N_0} \quad (4)$$

The function and variables already introduced can be expressed in a dimensionless form as follows (Kostoglou and Karabelas, 2002):

$$r = \frac{x}{x_0};\ \bar{r} = \frac{y}{x_0};\ t = b(x_0)\tau;\ n(r,t) = \frac{x_0 m(x,t)}{N_0};$$

$$s(r) = \frac{b(x)}{b(x_0)};\ \beta(r,\bar{r}) = \frac{p(x,y)}{x_0}$$

and equation (1) can be rewritten as a dimensionless form:

$$\frac{\partial n(r,t)}{\partial t} = \int_r^\infty \beta(r,\bar{r})s(\bar{r})n(\bar{r},t)d\bar{r} - s(r)n(r,t) \quad (5)$$

2.2 Daubechies orthonormal wavelets

Wavelet transform involves representing general functions in terms of simple, fixed building blocks at different scales and positions. These building blocks, which are actually a family of wavelets, are generated from a single fixed function called "mother wavelet" by translation and dilation (scaling) operations. In contrast to the traditional trigonometric basis functions which have infinite support, wavelets have compact support. Therefore wavelets are able to approximate a function not by cancellation, but through placement of the right wavelets at appropriate locations. The multi-resolution analysis properties of wavelets render them attractive candidates for representing functions in differential and integral equations which are being solved numerically (Bertoluzza and Naldi, 1996).

In the basic work of Daubechies (1988), a family of compactly supported orthonormal wavelets is constructed. The family includes members from highly localised to highly smooth. Each wavelet number is governed by a set of L (an even integer) coefficients $\{p_k : k = 0,1,\cdots,L-1\}$ through the two-scale relation:

$$\phi(x) = \sum_{k=0}^{L-1} p_k \phi(2x-k) \quad (6)$$

where function $\phi(x)$ is called the scaling function. The fundamental support of the scaling function $\phi(x)$ is in the interval $[0, L-1]$. The coefficients p_k

appearing in the two-scale relation (6) are called wavelet filter coefficients. For more details see Daubechies' paper (1988).

Denote $L^2(R)$ the space of square integrable functions on the real line. Let V_j be the subspace as the L^2-closure of the linear combination of:

$$\phi_{jk}(x) = 2^{j/2}\phi(2^j x - k) \tag{7}$$

for $k \in Z = \{\cdots, -1, 0, 1, \cdots\}$. A function $f(x) \in V_j$ can be represented by the wavelet series:

$$f(x) = \sum_{k \in Z} f_{jk}\phi_{jk}(x) \tag{8}$$

Wavelets automatically provide a hierarchical organization of bases which can be taken advantage by using multigrid-like methods.

2.3 Wavelet interpolation and treatment of the boundary conditions

In the wavelet collocation method (Bertoluzza and Naldi, 1996), trial functions are a class of interpolating functions generated by autocorrelation of the compactly supported Daubechies scaling functions. Such an autocorrelation function $\theta(\cdot)$ verifies trivially the equality $\theta(n) = \delta_{0n}$, and generates a multi-resolution analysis. The approximate solution of the problem will be a function $u_j(\cdot)$ that will be written in terms of its values in the dyadic points:

$$u_j(x) = \sum_n u_j(2^{-j}n)\theta(2^j x - n) \tag{9}$$

and such a function is exact at the dyadic points.

Consider the function:

$$\theta(x) = \int_{-\infty}^{+\infty}\phi(y)\phi(y-x)dy \tag{10}$$

Denote V_j the linear span of the set $\{\theta(2^j x - k), k \in Z\}$. It is possible to prove that V_j forms a multi-resolution analysis where $\theta(\cdot)$ plays the role of a scaling function (nonorthonormal). The autocorrelation function has the following properties:

$$\theta(x) = \int_{-\infty}^{+\infty}\phi(y-x)\phi(y)dy = \int_0^{L-1}\phi(y-x)\phi(y)dy$$

$$\theta(x) = \theta(2x) + \frac{1}{2}\sum_{l=1}^{L/2} a_{2l-1}[\theta(2x-2l+1) + \theta(2x+2l-1)]$$

$$\begin{cases} a_{2k} = 0, \quad k = 1, \cdots, L/2-1 \\ a_k = 2\sum_{l=0}^{L-1-k} p_l p_{l+k}, \quad k = 1, 3, \cdots, L-1 \end{cases}$$

The following integrals can be calculated exactly using the above properties through an iterative strategy:

$$I(x) = \int_{-\infty}^{x}\theta(y)dy \tag{11}$$

$$H_k(x) = \int_{-\infty}^{x}\theta(y-k)\theta(y)dy \tag{12}$$

Integrals of autocorrelation functions on intervals can be similarly computed according to the method proposed by Bertoluzza *et al.* (2000). The data calculated from the algorithm can be used to compute the operational matrices for the integral term in a breakage equation.

In this paper, some modified interpolating functions, which were introduced by Bertoluzza and Naldi (1996) are constructed in order to achieve an interpolating operator on the interval with the same accuracy as the counterpart on the line. However, they were employed to solve breakage equations rather than partial differential equations in this paper.

Consider a function $u(\cdot)$ on [0, 1]. The modified interpolating functions are defined as follows:

$$\theta^l_{jk} = \theta_{jk} + \sum_{n=-L+1}^{-1} a_{nk}\theta_{jn}, \quad k = 0, \cdots, 2M-1$$

$$\theta^r_{jk} = \theta_{jk} + \sum_{n=2^j+1}^{2^j+L-1} b_{nk}\theta_{jn}, \quad k = 2^j - M + 1, \cdots, 2^j$$

where $a_{nk} = l^1_k(x_n)$, $b_{nk} = l^2_k(x_n)$, $l^1_k(\cdot)$ and $l^2_k(\cdot)$ are the Lagrange polynomials related to the two (2*M*-1)-tuples of interpolation points $\{x_0, \cdots, x_{2M-1}\}$ and $\{x_{2^j-2M+1}, \cdots, x_{2^j}\}$, $x_k = k/2^j$, $k = 0, 1, \cdots, 2^j$, $\theta_{jk} = \theta(2^j x - k)$. Therefore:

$$\mathrm{I}_j u = \sum_{k=0}^{2M-1} u(x_k)\theta^l_{jk} + \sum_{k=2M}^{2^j-2M} u(x_k)\theta_{jk} + \sum_{k=2^j-2M+1}^{2^j} u(x_k)\theta^r_{jk} \tag{13}$$

The approximate solution $I_j u$ is developed using $2^j + 1$ basis functions expressed in equation (13). In this paper, $L = 6$ and $M = 3$ are chosen for compromise of computational time and the smooth of functions.

There is also another simple method by introducing the following functions:

$$\widetilde{\theta}_{j0} = \sum_{k=-L+1}^{0}\theta_{jk}, \quad \widetilde{\theta}_{j2^j} = \sum_{k=2^j}^{2^j+L+1}\theta_{jk}$$

where $I_j u$ has the form:

$$\mathrm{I}_j u = u(0)\widetilde{\theta}_{j0} + \sum_{k=1}^{2^j-1} u(x_k)\theta_{jk} + u(1)\widetilde{\theta}_{j2^j} \tag{14}$$

Considering the breakage processes, usually there is a pulse function on the right boundary. Therefore the simple procedure is used on the right boundary to avoid introducing the pulse function to other collocation points. On the left boundary, the Lagrange polynomial method is employed.

2.4 Wavelet-collocation algorithms for general breakage equations

The general breakage equation (5) must be rescaled to a finite domain prior to numerical solution. Choosing N_a and N_b as the lower and upper limits of the particle value r and the linear transformation

$v=(r-N_a)/(N_b-N_a)$, the breakage equation becomes:

$$\frac{\partial\bar{n}(v,t)}{\partial t}=(N_b-N_a)\int_v^1 Q(v,\bar{v})\bar{n}(\bar{v},t)d\bar{v}-\bar{s}(v)\bar{n}(v,t) \quad (15)$$

where $Q(v,\bar{v})=\bar{\beta}(v,\bar{v})\bar{s}(\bar{v})$. Let $Q(v,\bar{v})$ and $\bar{n}(v,t)$ be approximated by the J-level wavelet series:

$$Q(v,\bar{v})\approx Q_J(v,\bar{v})= \quad (16)$$
$$=\sum_{k=0}^{2M-1}Q(v,\bar{v}_k)\theta^l_{Jk}(\bar{v})+\sum_{k=2M}^{2^J-2M}Q(v,\bar{v}_k)\theta_{Jk}(\bar{v})+\sum_{k=2^J-2M+1}^{2^J}Q(v,\bar{v}_k)\theta^r_{Jk}(\bar{v})$$

$$\bar{n}(v,t)\approx\bar{n}_J(v,t)= \quad (17)$$
$$=\sum_{i=0}^{2M-1}\bar{n}_J(v_i,t)\theta^l_{Ji}(v)+\sum_{i=2M}^{2^J-1}\bar{n}_J(v_i,t)\theta_{Ji}(v)+\bar{n}_J(v_{2^J},t)\tilde{\theta}_{J2^J}(v)$$

where $\bar{n}_J(v_i,t), v_i=i/2^J, i=0,\cdots,2^J$, are 2^J+1 unknown wavelet series coefficients to be determined. Substituting equations (16) and (17) in equation (15) and applying the collocation discretization scheme, the following ODEs on the uniform grid are obtained:

$$\frac{d\bar{n}_J(v_i,t)}{dt}=\bar{N}\sum_{k=0}^{2^J}M^{op}_{i,k}\bar{n}_J(v_k,t)-\bar{s}(v_i)\bar{n}_J(v_i,t), \quad (18)$$
$$i=0,1,2,\cdots,2^J$$

with initial conditions, $\bar{N}=N_b-N_a$ and operational matrix on the uniform grid for the breakage operator:

$$M^{op}=\frac{1}{2^J}\left[\sum_{l=0}^{2^J}Q(v_i,v_l)\bar{H}_{k-l}(k)\right]_{(2^J+1)\times(2^J+1)} \quad (19)$$

where the definition of $\bar{H}_j(k)$ is the same as equation (12), however, the modified interpolating functions in equations (16) and (17) are employed.

3. PROPOSED METHODOLOGY

Using wavelet-based collocation method above, the breakage equation (15) is converted to the following ODEs:

$$\frac{dC(t)}{dt}=M\,C(t) \quad (20)$$

where $C(t)$ is the coefficient vector with respect to the wavelet basis; M is the matrix derived from the equation (18).

A new set of basis functions is formed by multiplying the previous wavelet basis using eigenvectors of the matrix M. This new basis set is then divided into two sets $\chi_s(v)$ and $\chi_f(v)$ corresponding to the magnitude of eigenvalues that are Λ_{big} for large eigenvalue and Λ_{small} for small eigenvalues, respectively.

This provides a new approximation for the solution of breakage equations:

$$\bar{n}(v,t)\approx\sum_{k=0}^{n_s}\chi_{s_k}(v)c_{s_k}(t)+\sum_{j=0}^{n_f}\chi_{f_j}(v)c_{f_j}(t)$$

and $n=n_s+n_f$, where n is the number of basis functions used in the original basis set, n_s is the number of elements in the slow set, and n_f is the number of elements in the fast set.

Applying the new sets of basis functions to breakage equation (15) gives the following system of ODEs:

$$\begin{cases}\dfrac{dC_s(t)}{dt}=M_{ss}C_s(t)+M_{sf}C_f(t)\\[2mm] \dfrac{dC_f(t)}{dt}=M_{fs}C_s(t)+M_{ff}C_f(t)\end{cases} \quad (21)$$

C_s and C_f are the corresponding states associate large eigenvalue set Λ_{big} and small eigenvalue set Λ_{small}, respectively. Since the magnitude of the large eigenvalue set Λ_{big} is much more than that of the small eigenvalue set Λ_{small}, the equation (21) can be approximated using a singular perturbation framework and the equations become a pseudo-steady-state equation. With this assumption, the above system of equations can be rewritten as:

$$\begin{cases}\dfrac{dC_s(t)}{dt}=M_{ss}C_s(t)+M_{sf}C_f(t)\\[2mm] M_{fs}C_s(t)+M_{ff}C_f(t)=0\end{cases} \quad (22)$$

An approximate inertial manifold can be found using the following equation:

$$C_f(t)=-M_{ff}^{-1}M_{fs}C_s(t) \quad (23)$$

The result can be substituted into equation (21) so that a new set of equations is obtained:

$$\frac{dC_s(t)}{dt}=\left(M_{ss}-M_{sf}M_{ff}^{-1}M_{fs}\right)C_s(t) \quad (24)$$

It is found that the dimension of C_s is small (less than 10% of original dimension). Hence the resulting low-order approximation is of the following form:

$$\bar{n}(v,t)\approx C_s(t)\chi_s(v)-M_{ff}^{-1}M_{fs}C_s(t)\chi_s(v)$$

where the coefficients C_s follow the dynamics given in equation (24).

4. CASE STUDY

We consider two examples to illustrate the usefulness of the technique developed here, uniform binary breakup and multiple breakup. The numerical results presented in this section are obtained for a monodispersed initial condition. The results are compared with analytical solutions and numerical solutions from a conventional discretized approach. All the results presented were calculated on a 1GHz Pentium III running Windows 2000 Professional.

4.1 Uniform binary breakup

For the case, the dimensionless probability distribution $\beta(r,\bar{r})=2/\bar{r}$ and breakage rate $s(r)=r^2$, Ziff and McGrady (1985) provided analytical solution for it. Choose $N_a=10^{-5}$, $N_b=1$ based on the interesting size range. ODE15s of

MATLAB was chosen as ODE solvers. The resolution levels of $J = 8, 9$ and different reduced model orders were tested. Figure 1 shows the comparisons of numerical and analytical results for dimensionless particle size distributions obtained at the dimensionless time corresponding to total particle number 250. The figure shows that the predictions are in good agreement with the analytical results for large values of total number. The results are much better than the predictions from the descritized population balance (DPB) method with order 20 (Hill and Ng, 1995).

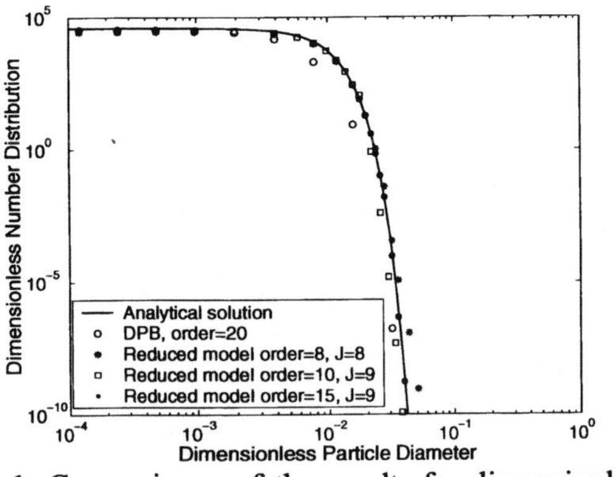

Fig. 1. Comparisons of the results for dimensionless particle size distributions of uniform binary breakage

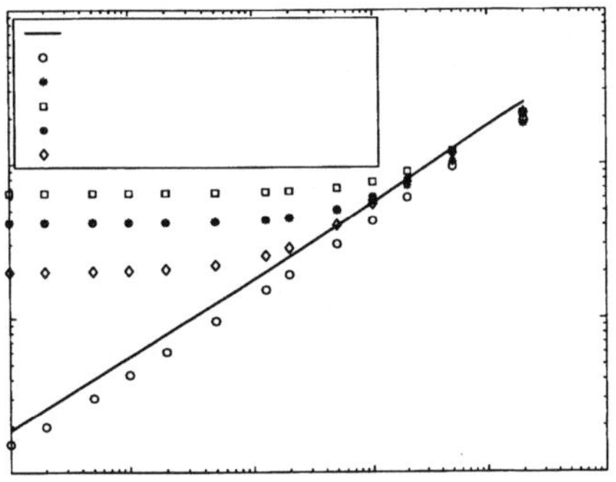

Fig. 2. Comparisons of the results for dimensionless total numbers of particles of uniform binary breakage

Figure 2 shows that the numerical solution for total number is in very good agreement with the analytical results if the response time is large enough (about 200 dimensionless time). In this case, desired accuracy can be obtained for an order reduction rate 3% compared with the order of ODEs resulted from standard wavelet-based collocation method.

4.2 Multiple breakup

A similar test has been carried out for multiple breakup with dimensionless probability distribution $\beta(r,\bar{r}) = \dfrac{33}{32\bar{r}}\left(\dfrac{r}{\bar{r}}\right)^{-\frac{31}{32}}$ and breakage rate $s(r) = r^{\frac{3}{2}}$. Ziff (1991) provided analytical solution for it. In this case, 33 daughter particles are produced due to the breakage of a single parent particle and the number distribution approaches infinity as the volume of particle goes to zero. Figure 3 show that even for this case, the particle size distributions are predicted very well for long time with the resolution levels $J = 9, 10$ and different reduced model orders.

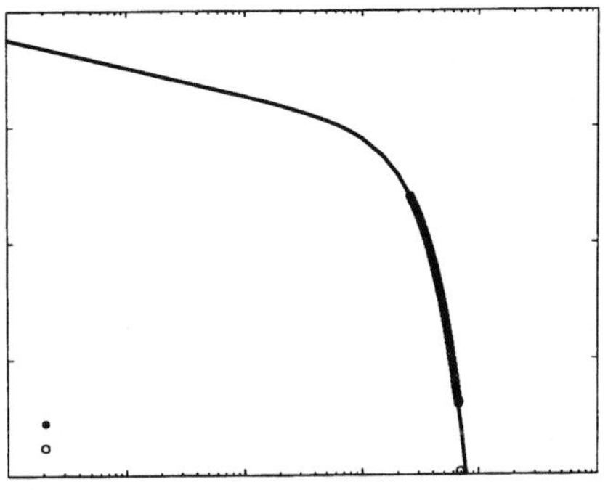

Fig. 3. Comparison of the results for dimensionless particle size distributions of multiple breakup

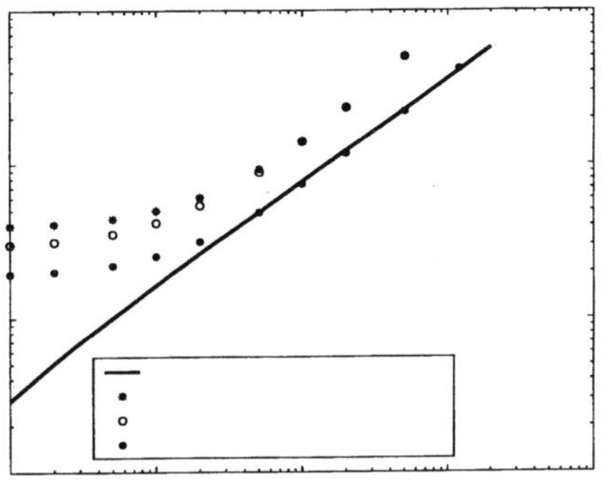

Fig. 4. A comparison of the results for dimensionless total numbers of particles of multiple breakup

Figure 4 shows that the numerical solution for total number is in very good agreement with the analytical results if the response time is large enough (about 20 dimensionless time). In this case, desired accuracy can be obtained for an order reduction rate 6% compared with the order of ODEs resulted from standard wavelet-based collocation method.

5. CONCLUSIONS

In this paper, a wavelet-base general procedure for model reduction was presented for breakage systems. It produces low-dimension solutions that capture the dominant modes of the system, which uses wavelet-base collocation method to arrive at the reduced order solutions. To get the accurate solutions, the order reduction rates depend on the characteristics of the systems, i.e. the gradients of the solutions. If the gradient is sharp, the order of the approximation may be large due to large resolution level. These results are very promising, Online control systems such as case 1 are possible because the computing time for an 8th order approximation is only about 0.01- 0.02 CPU time of a 1GHz Pentium III computer. For a 20th order DPB approximation it is about 0.1 CPU time with worse accuracy.

REFERENCES

Bertoluzza, S. and G. Naldi (1996). A wavelet collocation method for the numerical solution of partial differential equations. *Applied and Computational Harmonic Analysis*, ***3***, 1-9.

Bertoluzza, S., C. Canuto and K. Urban (2000). On the adaptive computation of integrals of wavelets. *Applied Numerical Mathematics*, ***34***, 13-38.

Chen, M. Q., Hwang, C. and Y. P. Shih (1996). A wavelet-Galerkin method for solving population balance equations. *Computer & Chemical Engineering 20,* 131-145.

Daubechies, I. (1988). Orthonormal bases of compactly supported wavelets. *Communications on Pure and Applied Mathematics*, ***41***, 909-996.

Foias, C., M. S. Jolly, I. G. Kevrekidis, G. R. Sell and E. S. Titi (1988). On the computation of inertial manifolds. *Physical Letters A*, ***131***, 433-437.

Foias, C., G. R. Sell and E. S. Titi (1989). Exponential tracking and approximation of inertial manifolds for dissipative equations. *Journal of Dynamic Differential Equations*, ***1***, 199-244.

Frances, C. and C. Laguerie (1998). Fine wet grinding of an alumina hydrate in a ball mill. *Powder Technology*, ***99***, 147-153.

Hill, P. J. and K. M. Ng (1995). New discretization procedure for the breakage equation. *AIChE Journal*, ***42***, 1600-1611.

Kostoglous, M. and A. J. Karabelas (1998). On the attainment of steady state in turbulent pipe flow of dilute dispersion. *Chemical Engineering Science*, ***53***, 505-513.

Kostoglou, K. and A. J. Karabelas (2002). An assessment of low-order method for solving the breakage equation. *Powder Technology*, ***127***, 116-127.

Kumar, S. and D. Ramkrishna (1996). On the solution of population balance equations by discretization -- I. A fixed pivot technique. *Chemical Engineering Science 51,* 1311-1332.

Liou, J., F. Srienc and A. G. Frerikson (1997). Solution of population balance models based on successive generations approach. *Chemical Engineering Science*, ***52***, 1529-1540.

Ma, Z., S. Hu, S. Zhang and X. Pan (1998). Breakage behaviour of quartz in a laboratory stirred ball mill. *Powder Technology*, ***100***, 69-73.

Sato, K., N. Meguri, , K. Shoji, H. Kanemoto, T. Hasegawa, and T. Maruyama (1996). Breakage of coals in ring-roller mills: Part I. The breakage properties of various coals and simulation model to predict steady-state mill performance. *Powder Technology*, ***86***, 275-283.

Shinohara, K., B. Golman, T. Uchiyama, and M. Otani (1999). Fine-grinding characteristics of hard materials by attrition mill. *Powder Technology*, ***103***, 292-296.

Shoji, K., N. Meguri, K. Sato, H. Kanemoto, T. Hasegawa, and T. Maruyama (1998). Breakage of coals in ring-roller mills: Part II. An unsteady-state simulation model. *Powder Technology*, ***99***, 46-52.

Vanni, M. (1999). Discretization procedure for the breakage equation. *AIChE Journal*, ***45***, 916-919.

Varinot, C., H. Berthiaux and J. Dodds (1999). Prediction of the product size distribution in associations of attired bead mills. *Powder Technology*, ***105***, 228-236.

Ziff, R.M. and E.D. McGrady (1985). The kinetics of cluster fragmentation and deploymerisation. *Journal of Physics A: Mathematical and General*, ***18***, 3027-3037.

Ziff, R. M. (1991). New solutions to the fragmentation equation. *Journal of Physics A: Mathematical and General*, ***24***, 2821-2828.

NOMENCLATURE

a_{nk}, b_{nk}	coefficients of polynomials; $b(x)$ breakage rate
$C(t)$	coefficient vector with respect to the wavelet basis
$C_f(t)$	coefficient vector corresponding to small eigenvalue
$C_s(t)$	coefficient vector corresponding to large eigenvalue
d_k^j	*j*-level wavelet coefficients at position *k*
$f(x)$	arbitrary function;
$I(x)$	integral of autocorrelation function
I_j	*j*-level interpolation operator
$H_k(x)$	two-term integral of autocorrelation function
$\overline{H}_k(x)$	two-term integral of modified interpolating function
k	position of resolution; l_k^1, l_k^2 Lagrange polynomials
L	Daubechies number; $m(x,\tau)$ particle size distribution
$m_0(x)$	initial particle size distribution
M	degree of polynomial or matrix
M^{op}	operational matrix for integral term
M_0	initial total mass of particles
$n(r,t)$	dimensionless particle size distribution
$\overline{n}(v,t)$	normalised dimensionless particle size distribution
$N(t)$	total number of particles at dimensionless time t
N_0	initial total number of particles
N_a	lower limit of the particle volume
N_b	upper limit of particle volume
$\overline{N}$	range of particle volume;
$p(x,y)$	possibility distribution
p_k	wavelet filter coefficient;
r	dimensionless particle volume
$s(r)$	dimensionless breakage rate
$\overline{s}(v)$	normalised dimensionless breakage rate
t	dimensionless time; $u(x)$ arbitrary function
$u_0(x)$	initial function; u_j *j*-level resolution
v	normalized dimensionless particle volume
V_j	*j*-level linear span of autocorrelation functions
x	particle volume; x_k *k*-th collocation point
x_k^j	*k*-th collocation point of *J*-level resolution
Z	set of integers

Greek letters

τ	time;
$\beta(r,\overline{r})$	dimensionless probability distribution
$\overline{\beta}(v,\overline{v})$	normalised dimensionless probability distribution
$\chi_s(v)$	sub-basis corresponding big eigenvalue
$\chi_f(v)$	sub-basis corresponding small eigenvalue
δ_{mn}	*Dirac-delta* function; $\phi(x)$ scaling function
$\phi_{jk}(x)$	*j*-level scaling function at position *k*
$\theta(x)$	autocorrelation function of scaling function
$\theta_{jk}(x)$	*j*-level autocorrelation function at position *k*
$\theta_{jk}^l(x)$	left *j*-level modified interpolating function at position *k*
$\theta_{jk}^r(x)$	right *j*-level modified interpolating function at position *k*
Λ_{big}	large eigenvalue set; Λ_{small} small eigenvalue set

ELSEVIER
IFAC
PUBLICATIONS
www.elsevier.com/locate/ifac

MATHEMATICAL MODELING FOR ADIPIC ACID CRYSTALLIZATION PROCESS

Caliane B. B. Costa, Aline C. Costa, Rubens Maciel Filho

Chemical Engineering School, State University of Campinas (UNICAMP), CP 6066, CEP; 13081-970, Phone: +551937883971, Fax: +551937883965 Campinas, SP, Brazil

Abstract: This work is part of a study aiming to improve crystallization processes. The batch crystallization of adipic acid is chosen as a case study, a process in which the supersaturation necessary for the crystals to appear and grow is generated by the cooling of the solution. This part of the study involves the process modeling and an analysis of the influence of the process variables on the final crystal size distribution (CSD) and on the quantity of solids. The results have shown that the developed model is a good representation of the process, being able to reproduce literature results. *Copyright © 2004 IFAC*

Keywords: Mathematical models, Solids processing, Process models, Dynamic Behaviour, Temperature profiles

1. INTRODUCTION

Crystallization is an ancient unit operation and is widely used, since solids of high purity can be obtained. Batch crystallization is specially used in the production of high value added products, mainly because it offers flexible and simple processing steps for plants with frequently changing recipes and product lines (Zhang and Rohani, 2003; Rawlings *et al.*, 1993). For this kind of material, product purity and the crystal size distribution (CSD) are of prime importance. Furthermore, the crystals produced through a crystallization process have a decisive influence on the downstream processing and, therefore, the CSD should be reproducible in each operation and regular as much as possible (Ma *et al.*, 2002). Therefore, the variables that affect the crystallization process must be known and controlled so as to be kept within an acceptable range, in order to satisfy the requirements concerning the final product quality and the production process (Rawlings *et al.*, 1993).

This work is part of a study aiming to improve the performance of batch crystallization processes. The batch cooling crystallization of adipic acid is chosen as a case study. In a well-mixed batch crystallizer, the final crystal product is determined by the supersaturation profile, the initial seed mass, and the seed crystal size distribution (Ma *et al.*, 2002). The supersaturation evolution during batch crystallization processes determines the magnitude of the many kinetic phenomena that comprise the process. In order to perform a successful study of the process, the accurate modelling of the process is crucial. It turns possible the calculation of further on-line optimal operating policies, as well as real time applications (Zhang and Rohani, 2003). In order to analyse the performance of suitable control structures and to perform control calculations, an accurate modelling is necessary to be developed.

Seeding is frequently applied to avoid a supersaturation peak at the beginning of the process (Giulietti *et al.*, 2001). This occurs because the supersaturation values sufficient for crystal growth are lower than the necessary ones for spontaneous nucleation. So, if the system could be kept at a region in the metastable zone with not so high supersaturation values, and with addition of seeds, just seeds growth will occur (Mullin, 1993; Rawlings *et al.*, 1992).

In this work, the developed mathematical modeling of the adipic acid crystallization process is presented. The role of the several process variables in the final product is evaluated as part of the study of ways to improve the process.

2. MATHEMATICAL MODELING

In order to completely model a crystallization process, mass balance, energy balance and a description of the crystal size distribution are necessary. This description is necessary because the process produce a mass of particles of many sizes, whose description of size distribution is crucial to its characterization. The models used in this description are called population balance models. According to Puel *et al.* (2003), the population balance modeling is firmly established as a basic theoretical framework for all particle processes.

Two phenomena dominate the crystallization kinetics: nucleation and crystal growth. Both phenomena consume the mass of the desired material during the crystallization process and are, therefore, competing mechanisms. They are different in the manner they consume material: nucleation involves the formation of new crystals while in crystal growth the crystals become larger with the deposition of material onto the existing crystals (Lang *et al.*, 1999). Apart from nucleation and growth, other phenomena, such as agglomeration and breakage, may occur during the process.

2.1 Population Balance – The Method of Classes.

Population Balance equation (PBE) is a hyperbolic partial differential equation, which involves all the crystallization kinetic phenomena. Being strongly nonlinear, the PBE do not possess an analytical solution in most cases, requiring the development and adaptation of numerical techniques. According to Wulkow *et al.* (2001), in the present state of PBE studies, no standard numerical method has been established for population balance models and none of the presently available methods produce an efficient and accurate solution for a broad class of models. Discretization sizing techniques appear to be robust. Marchal *et al.* (1988) developed the Method of Classes, a method that transforms the partial differential equation into an ordinary differential equations system by discretizing the range of variation of the variable L, related to the crystal size. The obtained differential equations are then no longer written with population density functions but with absolute numbers of crystals in each class (Nallet *et al.*, 1998). In a recent work, Puel *et al.* (2003) also used the Method of Classes to solve the population balance equation, but they extended the method to consider two characteristic dimensions.

In the Method of Classes, the particles sizes are defined as $L_0, L_1, L_2, \ldots, L_N$, where L_0 is the size of the nuclei and L_N is the size of the largest crystals. This sizes determine the existence of N granulometric classes C_i, whose widths are defined by $\Delta C_i = L_i - L_{i-1}$ and the characteristic size of the class C_i is $S_i = (L_{i-1}+L_i)/2$ (see Figure 1).

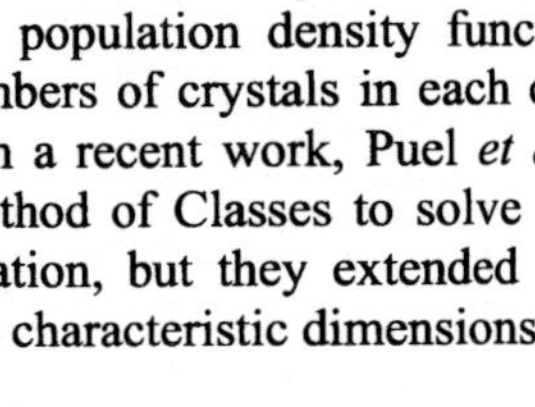

Fig. 1. Division of Classes.

The population of crystals is described by the number density function Ψ. Ψ(L,t)dL is the number of crystals of size between L and L+dL per unit volume at time t (Nallet *et al.*, 1998). Ni(t), the number of crystals per unit of volume in the *ith* class (C_i) at time t, is given by:

$$N_i(t) = \int_{L_{i-1}}^{L_i} \Psi(L,t)\,dL \qquad (1)$$

The model assumes that the number density function is constant at each class, turning possible the transformation of the population balance equation into a system of ordinary differential equations:

$$\begin{cases} \dfrac{dN_1}{dt} + \dfrac{1}{V_{susp}}\dfrac{dV_{susp}}{dt}N_1 + \dfrac{Q_s N_1 - Q_e N_{1e}}{V_{susp}} + \dfrac{G(L_1)}{2\Delta C_2}N_2 + \dfrac{G(L_1)}{2\Delta C_1}N_1 = r_N + R_{A,1} - R_{B,1} \\ \dfrac{dN_i}{dt} + \dfrac{1}{V_{susp}}\dfrac{dV_{susp}}{dt}N_i + \dfrac{Q_s N_i - Q_e N_{ie}}{V_{susp}} + \dfrac{G(L_i)}{2\Delta C_{i+1}}N_{i+1} + \dfrac{G(L_i) - G(L_{i-1})}{2\Delta C_i}N_i \\ \quad - \dfrac{G(L_{i-1})}{2\Delta C_{i-1}}N_{i-1} = R_{A,i} - R_{B,i} \\ \dfrac{dN_N}{dt} + \dfrac{1}{V_{susp}}\dfrac{dV_{susp}}{dt}N_N + \dfrac{Q_s N_N - Q_e N_{Ne}}{V_{susp}} + \dfrac{-G(L_{N-1})}{2\Delta C_N}N_N \\ \quad - \dfrac{G(L_{N-1})}{2\Delta C_{N-1}}N_{N-1} = R_{A,N} - R_{B,N} \end{cases} \qquad (2)$$

In the development of the modelling of the batch cooling crystallization process, the Method of Classes was chosen, because it is an efficient method widely used in literature, including in modelling of growth-dominated processes. Puel *et al.* (2003) stress that this numerical technique reliability is confirmed by three different papers dealing with bi-dimensional balance equations where discretization methods were used, despite differences in the formalism and the application area. However, the number of classes considered is the major key to guarantee the accurate solution, as was pointed out by Puel *et al* (2003). This was also observed in the present work.

Growth Mechanism. The expression for the growth rate developed by Marchal *et al.* (1988) is based on the film model and may be written as:

$$G = \frac{dL}{dt} = \frac{k_a MM\, k_c}{3\rho_c k_v}\eta_r (c - c^*)^{j'} \qquad (3)$$

The effectiveness factor, η_r, which expresses the diffusional limitations in crystal growth, causing a distribution of the growth rates among the classes, is found by the solution of equation (4). The mass transfer coefficient, k_d, is evaluated through the expression for Sherwood (equation (5)).

$$\left[\frac{k_c}{k_d}(c - c^*)^{j'-1}\right]\eta_r + \eta_r^{1/j'} - 1 = 0 \qquad (4)$$

$$Sh = \frac{k_d L}{D} = 2{,}0 + 0{,}47\left[\frac{L^{4/3}\varepsilon^{1/3}}{\nu}\right]^{0,62}\left[\frac{Diam}{Diam_T}\right]^{0,17}\left[\frac{\nu}{D}\right]^{0,36} \qquad (5)$$

From the equation (5), it is easy to note that each granulometric class has a value for the mass transfer coefficient, what means that the growth rate is size dependent.

The data of solubility of adipic acid in water were extracted from Postnikov and Nalivaiko (1971).

Nucleation Mechanism. Both primary and secondary nucleation are considered. Their expressions are respectively given by equations (6) and (7):

$$r_{N1} = A \exp\left[-\frac{B}{\ln^2\left[\frac{(HR)}{(HR)^*}\right]} \right] \tag{6}$$

$$r_{N2} = k'_N \left[(HR) - (HR)^*\right]^{i'} C_s^{k'} \tag{7}$$

Agglomeration Mechanism. Still in the work of Marchal *et al.* (1988), a model for the agglomeration mechanism is proposed, considering only the agglomeration of two particles. The agglomeration is considered as chemical reaction of particle *m* and particle *n*, leading to the formation of a crystal of size *q*. Considering that the pseudo-chemical reaction describing the agglomeration can be schematically represented by equation (9), its stoichiometric coefficient may be found by $\nu_{est} = (S_m^3+S_n^3)/S_q^3$. The class q, where the agglomerate fits, is found by the relation: $L_{q-1} < (S_m^3+S_n^3)^{1/3} \leq L_q$.

$$(m) + (n) \longrightarrow \nu (q) \tag{8}$$

All possible agglomerations between two particles (*m* and *n*, $n \geq m$) can be arranged in a series, assuming a rank $l_{m,n}$ that represents its position in this series. For N granulometric classes, N(N+1)/2 different binary agglomerations are present and the series are represented by (1,1), (1,2) ... (1,N), (2,2), (2,3) .... (m,n) .... (N,N). The position $l_{m,n}$ of the agglomeration of a particle *m* with a particle *n* is found by relation (9):

$$l_{m,n} = N(m-1) - \frac{m(m-1)}{2} + n \tag{9}$$

This agglomeration will affect class *i*, only if *i* is equal to *m*, *n* (in both cases, $\nu_{est} = -1$) or *q* (with $\nu_{est} = (S_m^3+S_n^3)/S_q^3$). In this way, an overall stoichiometric coefficient of class i, with respect to agglomeration of rank *l* can be computed:

$$\nu_{est\,l,i} = \left(\frac{S_m^3 + S_n^3}{S_q^3}\right)\delta_{i,q} - \left(\delta_{i,m} + \delta_{i,n}\right) \tag{10}$$

where $\delta_{ij} = 1$ if $i = j$ and $\delta_{ij} = 0$ if $i \neq j$.

The net rate of particle production by agglomeration in the *i*th class is calculated by equation (11), where r(*l*) is the intrinsic rate of agglomeration of rank *l*, being a function of the number of collisions per unit volume per time and of the supersaturation.

$$R_{A,i} = \sum_{l=1}^{\frac{N(N+1)}{2}} \nu_{est\,l,i}\, r(l) \tag{11}$$

David *et al.* (1991) proposed an expression for the calculation of the intrinsic rate of agglomeration (equation 12), based on phenomenological and fluid mechanical considerations, taking into account the concentration of particles, the supersaturation, the power dissipation per mass unit, the crystallizer size and of the crystals being agglomerated. The agglomeration contribution in the developed software is computed through equation (12).

$$r = k'_A S_m \left(1 + \frac{S_n}{S_m}\right)^2 N \; Diam \; f\left(\frac{S_n}{S_m}\right)\left[1 - \frac{(S_n + S_m)^2}{\lambda_e^2}\right] k'_d (c - c^*) \times N_n N_m (S_m - \delta) \; H(\lambda_e - S_n - S_m) \tag{12}$$

where H(*x*) = 1 for $x \geq 0$ and H(*x*) = 0 for $x < 0$. The coefficient k_d' is calculated by the same expression for k_d (Eq. 5), but the size of crystal considered must be the size of the agglomerate ($S_m + S_n$). The f function represents a relative shape function of both crystals. Considering both particles as spheres:

$$f\left(\frac{S_n}{S_m}\right) = \frac{4\left(1 + \frac{S_n}{S_m} - \sqrt{\left(\frac{S_n}{S_m}\right)^2 - 1}\right)}{\frac{1}{3} + \frac{S_n}{S_m} - \sqrt{\left(\frac{S_n}{S_m}\right)^2 - 1} - \left(\frac{S_n}{S_m} - \sqrt{\left(\frac{S_n}{S_m}\right)^2 - 1}\right)^2 \left(\frac{2\frac{S_n}{S_m}}{3} + \frac{\sqrt{\left(\frac{S_n}{S_m}\right)^2 - 1}}{3}\right)} \tag{13}$$

λ_e is defined as Lagrangian microscale, which is taken as having the same magnitude as the Taylor microscale, calculated by:

$$\lambda_e = 0{,}3\,\pi\, N\, Diam \left(\frac{60\nu}{10\varepsilon}\right)^{1/2} \tag{14}$$

The calculation of further on-line optimal operation policies takes an advantage with the inclusion of agglomeration in the model, compared to most batch optimization studies encountered in literature. Most of them neglect agglomeration and considers just nucleation and growth, which may be a poor consideration, as is the case with adipic acid, a classic-known substance that tends to agglomerate during crystallization processes.

2.2 *Mass Balance – Batch Crystallizer.*

The material balance of the solute is made based on the fact that a change in the solution concentration results in a change of the mass of crystals per unit volume.

Since the simulated crystallizer is a batch one, the solute present in the solution in the beginning of the batch is the whole mass of adipic acid available for crystallization. The mass balance is made using the dissociation constant of adipic acid, considered as a monoacid: the concentration of solid in the suspension can be connected with the concentration of protons in the solution (Marchal *et al.*, 1988):

$$V_o C_o = \frac{(H^+)^2}{K}\left[1 + \frac{K}{(H^+)}\right] V_o + \frac{V_o}{1 - \frac{MM}{\rho} C_s} C_s \tag{15}$$

2.3 *Energy Balance.*

The energy balance must take into account the differences in enthalpy of the streams in and out, the heat of crystallization and the heat removed by the cooling system. In the case study, batch crystallization, the energy balance equation is:

$$\rho C_p V \frac{dT}{dt} = -\Delta H_c 3\rho_c k_v V_{susp} \int_0^{\infty} nL^2 G dL - UA_c (T - T_c) \qquad (16)$$

Using the Method of Classes, the integral present in the previous equation is substituted by a summation over all granulometric classes.

The heat of crystallization of adipic acid was extracted from Postinikov and Nalivaiko (1971).

The specific heat of the slurry is considered to be not so different from the water, since the process considered is the crystallization from solution, which is characterized to have not so high concentrations of solids.

3. MODEL VALIDATION

In order to test the validity of the constructed model, results from the literature were reproduced. The results obtained by Marchal *et al.* (1988) are compared with the ones generated with the developed model with the same data. To illustrate this point, Figure 2 depicts a comparison between the results for evolution of the relative supersaturation during the crystallization. The results generated by the developed software fit pretty well to the results obtained by Marchal *et al.* (1988), both for crystallization considering agglomeration or not. Considering the availability of data, the model is validated by the similarities of plots, not only for supersaturation evolution, but also for solids concentration evolution, CSD in number and in mass at the end of the batch and effectiveness factor for three granulometric classes during the batch.

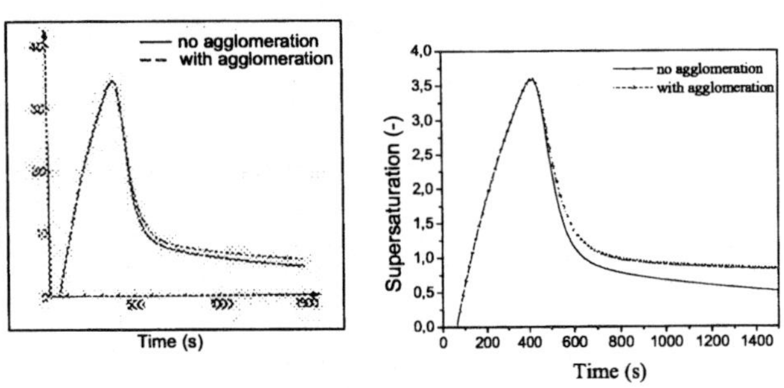

Fig.2: Comparison between Marchal *et al.* (1988) result (left) and model predictions (right)

4. SIMULATIONS – PROCESS VARIABLES INFLUENCE

Some simulations were made in order to evaluate the influence of several process variables. The main results are presented here. They are concerned to seeding, cooling rate, influence of the global heat transfer coefficient and of the power dissipation. It should be pointed that the effect of model uncertainties (i.e., model parameters) on the accuracy of the model predictions should be analysed. It was evaluated, through analysis of the effect of model parameters variations (±15%) on the final response through experimental design. The study showed that the model uncertainties do not affect the analysis that can be extracted from the influence of the process variables: the same qualitative results illustrated in Figures 3 – 5 are obtained.

The first result is the influence of the seeding in the process. In order to analyze the influence of seeding, the plots on Figure 3 are presented. It shows the results of operation trajectories, solution temperature evolution and supersaturation evolution for two batches that differ each other only in the seeding: the first one (non-seeded) has insignificant amount of seeds, while the second one has a considerable mass of crystals working as seeds. At these two simulations, a parabolic profile of coolant temperature, with negative second derivative, was employed.

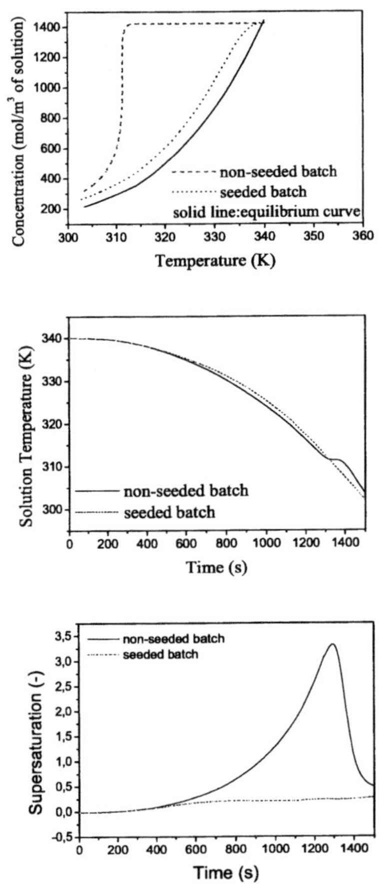

Fig. 3: Operation curve, evolution of the solution temperature and evolution of the supersaturation for a non-seeded and a seeded batch crystallization

The first plot in Figure 3 illustrates the operation curves (trajectories) for the two simulations. The operation trajectory is formed by the pairs of values of solution concentration and temperature followed during the batch time. The difference in the path followed during the process in the two cases is extremely visible. When no seeds are added to the crystallization process, the solution is cooled with constant concentration until the metastable zone limit is probably reached. An enormous quantity of crystals is generated by nucleation, releasing great amount of heat of crystallization (exothermic reaction), making the solution temperature to rise (second plot of Fig. 3) – what means that the cooling system was not able to remove all the heat released. As a consequence of the operation curve, a great peak of supersaturation is noted (last plot of Fig. 3).

On the other hand, if the crystallization is conducted the same way, except by the addition of seeds, the cooling of the solution happens softly, while the concentration of adipic acid in solution decreases (first plot of Fig. 3), probably keeping the operation far from the metastable zone limit. It is kept at a reasonable distance from the equilibrium line (Lang *et al.*, 1999). Not so high values of supersaturation are reached in the process (last plot of Fig. 3) and the nucleation is disfavoured. There is no great release of heat of crystallization and so there is no increase of solution temperature (second plot of Fig. 3).

Apart from the influence of the seeding, other important factor in the crystallization performance is the rate of cooling during all the process. According to Mullin (1993), the use of natural cooling is not the best choice. By natural cooling, one understands that the coolant is passed through the exchange device at constant both temperature and flow rate. If natural cooling is used, the temperature inside the crystallizer decreases exponentially, making the supersaturation to increase very fast at the first moments of the process, favouring nucleation. As a result, at the end of the batch small crystals and a large CSD are obtained. But, if the cooling profile is characterized to have a soft decrease at the early stages and a more pronounced one at the end of the process, the operation is favoured (Mullin, 1993).

Two seeded simulations that differ in the coolant temperature profile illustrate appropriately this feature. In the first one, natural cooling is simulated (coolant temperature and flow rate are both constant). In the next simulation, the same amount of seed is used, but a parabolic profile of coolant temperature, with negative second derivative, is employed. This last cooling rate is characterized to have a soft decrease at the early stages and a more pronounced one at the end of the process. The results of the two simulations can be observed in Fig. 4. When natural cooling is employed, a great number of fines is obtained (last plot in Fig. 4). A great peak of supersaturation occurs at the early stages of the crystallization process, what causes the appearance of thousands of nuclei. The concentration of solids in the suspension increases too fast in the first moments (first plot of Fig. 4). Except for the disadvantage of the great number of fines, this kind of cooling has the advantage of being able of extracting a great amount (mass) of solute of the solution, bringing them into crystals.

When a negative parabolic temperature profile is employed, much less number of fines is obtained and the solid concentration evolves softly (see Fig. 4), with seeds growth and not so many fines being generated. This is a consequence of the evolution of the supersaturation during the batch, showing that it is desirable to avoid peak supersaturation in order to favour growth. This confirms the results from literature (Lang et al, 1999), indicating that the optimum temperature profile would produce no peak in supersaturation at all, being almost flat. However, the slow evolution of the process when the supersaturation is kept low implies in less extraction of solute from the solution (less mass is obtained compared to natural cooling).

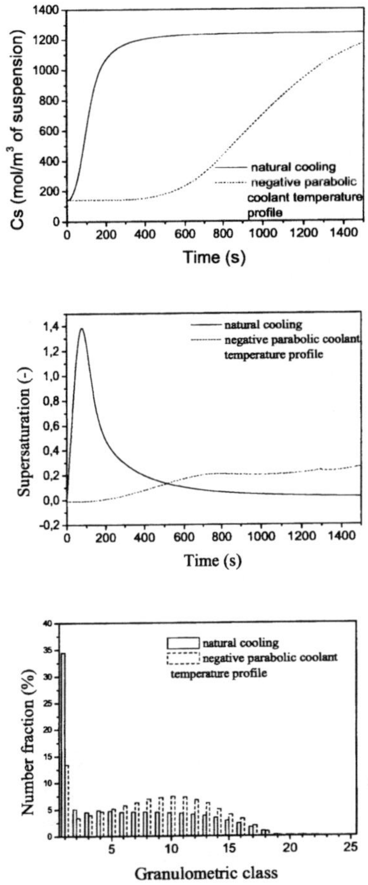

Fig. 4: Evolutions of solid concentration and supersaturation, as well as the CSD in number at the end of the batch in a seeded crystallization with natural cooling and with a negative parabolic coolant temperature profile.

At the study of the influence of the global heat transfer coefficient, its variation would express a variation in the coolant flow rate: the grater the value of the global heat transfer coefficient, the greater the coolant flow rate. The study showed that an increase in the fluid flow rate would be advantageous only in cases where the coolant temperature profile decreases slowly in early stages and sharper latter in time, what means to pass the coolant fluid with lower temperature and higher flow rate as the batch time passes. In this way, more mass is obtained in the process.

The role of the power dissipation is evaluated in the present work and this variable has shown itself to be important in the distribution of crystals in the various classes (Fig. 5). It is an expected result, since the power dissipation is present in the expression of the mass transfer coefficient, which influences the growth rate in each class, as well as the intrinsic rate of agglomeration. It has shown to be not so important in the amount (mass) of crystals obtained.

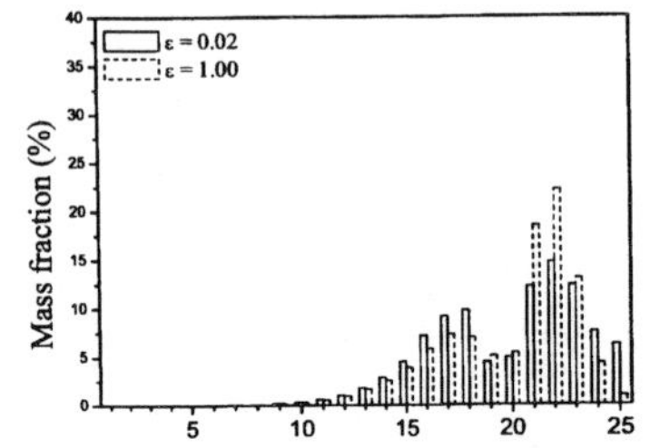

Fig. 5: CSD in mass at the end of the batch in two simulations (ε = 0.02 and ε = 1.00).

5. BATCH OPTIMIZATION STUDY

In order to control batch cooling crystallization processes, it is necessary to have an off-line optimized cooling profile (set point profile), which is calculated through optimal control theory. The accurate model of the process is essential in obtaining this set point profile, because it is obtained through the solution of a non-linear optimization problem. A general non-linear optimization problem subjected to equality and inequality constraints h (x) and g (x), respectively, may be defined as:

$$\begin{aligned} &\text{Min} \quad f(x) \\ &\text{s. t.:} \quad h_j(x) = 0, j = 1, 2, ..., m \\ &\qquad g_k(x) \leq 0, k = 1, 2, ..., p \end{aligned} \tag{17}$$

In the batch cooling crystallization optimization, the function to be minimized is set as a function of the final CSD and the problem constraints are the model equations, as well as physical constraints.

On-line batch optimal operating policies are calculated through the use of a model-based controller, which uses the process model to predict future answers of the process and take actions in order to obtain the best product at the end of the batch.

The optimal operating profiles may be successfully obtained by the proposed model together with the optimal control theory.

6. CONCLUSIONS

The developed model proved to be a good one, expressing representative results of the process. Literature results were satisfactory reproduced from the available information. The predictions generated through the model allow analyzing the impact of the process variables in the system behaviour. The proposed model may be used in a non-linear optimization problem, in order to find out the optimal cooling profile.

7. ACKNOWLEDGEMENTS

Financial support from FAPESP is gratefully acknowledged.

REFERENCES

David, R., P. Marchal, J.P. Klein and J. Villermaux (1991). Crystallization and Precipitation Engineering – III. A Discrete Formulation of the Agglomeration Rate of Crystals in a Crystallization Process. *Chemical Engineering Science*, **46**, 205-213.

Giulietti, M., M.M. Seckler, S. Derenzo, M.I. Ré and E. Cekinski (2001). Industrial Crystallization and Precipitation from Solutions: State of the Technique. *Brazilian Journal of Chemical Engineering*, **18**, 423-440.

Lang, Y.D., A.M. Cervantes and L.T. Biegler (1999). Dynamic Optimization of a Batch Cooling Crystallization Process. *Industrial &. Engineering Chemistry Research*, **38**, 1469-1477.

Ma, D.L., D.K. Tafti and R.D. Braatz (2002). Optimal control and simulation of multidimensional crystallization processes. *Computers and Chemical Engineering*, **26**, 1103-1116.

Marchal, P., R. David, J.P. Klein and J. Villermaux (1988). Crystallization and Precipitation Engineering – I. An Efficient Method for Solving Population Balance in Crystallization with Agglomeration. *Chemical Engineering Science*, **43**, 59-67.

Mullin, J.W. (1993). *Crystallization*, Butterworth-Heinemann, Oxford.

Nallet, V., D. Mangin and J.P. Klein (1998). Model identification of batch precipitations: application to salicylic acid. *Computers and Chemical Engineering*, **22**, S649-652.

Postnikov, V.A. and A.S. Nalivaiko (1971). Solubility and Heat of Crystallisation of Adipic Acid. *Russian Journal of Physical Chemistry*, **45**, 1356.

Puel, F., G. Févotte and J.P. Klein (2003). Simulation and analysis of industrial crystallization processes through multidimensional population balance equations. Part 1: a resolution algorithm based on the method of classes. *Chemical Engineering Science*, **58**, 3715-3727.

Rawlings, J.B., W.R. Witkowski and J.W. Eaton (1992). Modeling and control of crystallizers. *Powder Technology*, **69**, 3-9.

Rawlings, J.B., S.M. Miller and W.R. Witkowski (1993). Model Identification and control of Solution Crystallization Processes: A Review. *Industrial &. Engineering Chemistry Research*, **32**, 1275-1296.

Wulkow, M., A. Gerstlauer and U. Nieken (2001). Modeling and simulation of crystallization processes using parsival. *Chemical Engineering Science*, **56**, 2575-2588.

Zhang, G.P. and S. Rohani (2003). On-line optimal control of a seeded batch cooling crystallizer. *Chemical Engineering. Science.*, **58**, 1887-1896.

ELSEVIER
IFAC
PUBLICATIONS
www.elsevier.com/locate/ifac

ENHANCED IMC FOR GLUCOSE CONTROL IN TYPE I DIABETIC PATIENTS

Y. Ramprasad, G.P. Rangaiah*, and S. Lakshminarayanan

Department of Chemical and Biomolecular Engineering
National University of Singapore, Singapore 117576
**Corresponding Author. Email: chegpr@nus.edu.sg*

Abstract: Maintaining the glucose concentration in normoglycemic range in Type I diabetic patients is challenging. In this study, the Enhanced Internal Model Controller (EIMC) is developed for a diabetic, due to its simple structure, better disturbance attenuation and uncertainty reduction capabilities. A first order plus time delay (FOTD) model approximation of nominal patient model is used for developing the EIMC. The controller performance is assessed based on its ability to track normoglycemic set point (81.1 mg/dl) in response to a 50 g meal disturbance. In the nominal case, the controller maintains the glucose concentration within ± 6.5 mg/dl of set point. The robustness of the controller is studied by application to a fairly "diverse" group of "patients", and compared with that of IMC controller. Copyright © 2004 IFAC.

Keywords: Blood glucose control, Glycaemia control, Enhanced IMC, Robust control

1. INTRODUCTION

Diabetes mellitus is a major chronic disease affecting millions of people worldwide, Type I or insulin-dependent diabetes mellitus (IDDM) is characterized by the insufficient secretion of insulin from the β-cells of pancreas, resulting in glucose concentrations elevated beyond the normoglycemic range (70-100 mg/dl) (Ashcroft and Ashcroft, 1992). The prevalence of sustained hyperglycaemia (arterial glucose concentration > 120 mg/dl) is the reason for long-term complications associated with diabetes such as nephropathy and retinopathy. The current treatment methods of insulin therapy are largely "open loop" in nature. Intuitively, it can be said that "closed loop" control approaches can result in good glycaemia control over extended periods of time. Such a strategy would involve three major components – a mechanical pump, *in-vivo* glucose sensor and a mathematical algorithm to regulate the pump and set appropriate insulin dosage based on a sensor measurement. These aspects have been active areas of research in the diabetes literature. Present work focuses on the automatic control aspects of blood glucose regulation.

Early modelling studies of the diabetic condition (Bolie, 1961; Ackerman et al., 1965) established a precedent for mathematical analysis of insulin-glucose interactions. Later studies utilized the more complicated nonlinear models such as Bergman's "Minimal Model" (Bergman et al., 1981) and incorporated physiological system knowledge in the model structure (Cobelli and Mari, 1983; Sorensen, 1985). Various model based optimal control algorithms have been developed utilizing these models in either an explicit or implicit fashion (Sorensen, 1985; Ollerton, 1989; Fischer 1991; Parker et al., 1999). In the synthesis of these controllers, however, the inherent uncertainty in the model has not been properly addressed. Such control strategies could lead to significant performance degradation, in the presence of inevitable patient-model mismatch.

Significant inter- and intra-patient variability has been documented in literature (Bremer and Gough, 1999; Pucket and Lightfoot, 1995; Simon et al., 1987; Steil et al., 1994), resulting in the controller to be retuned for each patient. The control algorithm employed in the insulin delivery device must be able to compensate for the uncertainty that exists between the internal model and the actual patient. This clearly motivates the synthesis of a controller that can handle the model-patient mismatch. Parker et al. (2000) used the H_∞ framework to explicitly treat the model uncertainty. They employed a detailed and fairly complex physiological patient model (19 state nonlinear dynamic model) because it allows uncertainty characterization on particular tissues or effects that are responsible for the insulin or glucose variability. This detailed model is used in this study too but the focus is on an Enhanced Internal Model Controller (EIMC) scheme of Zhu et al. (1995). owing to its simple structure and better uncertainty and disturbance attenuation. Despite these attractive features, EIMC has not been applied for blood glucose control so far. Its robustness to typical inter- and intra-patient variability will be demonstrated.

2. THE MATHEMATICAL MODEL

A nonlinear pharmacokinetic/pharmacodynamic compartmental model of the diabetic patient has been constructed previously (Guyton et al., 1978; Sorensen, 1985; Parker, 1999), and is detailed in Parker et al. (2000). A few typographical errors in the differential equations provided there are corrected and listed in the appendix of this article. The meal disturbance model of Lehmann and Deutsch (1992) was included in the model of Parker et al. (2000). The specific operating conditions for the diabetic patient model used in testing the robust controller algorithms are described in Parker (1999). The diabetic patient model had two inputs and one measured output. Insulin delivery rate, represented as deviation from its 22.3 mU/min nominal delivery rate, was the manipulated variable (represented as $\overline{u}$). The meal disturbance had a nominal value of 0 mg/min (absorption into the blood stream), and its signal was denoted as $\overline{m}_d$. The measured variable $\overline{Y}$ represented the deviation in blood glucose concentration from the nominal value of 81.1 mg/dl. All these three variables were scaled as mentioned in Parker et al. (2000).

$$m_d = \frac{1}{360}\overline{m}_d, u = \frac{1}{33.125}\overline{u}, Y = \frac{1}{20}\overline{Y} \quad (1)$$

Here, the disturbance scaling was determined by its maximum value, and the output scaling by maximum allowable deviation in glucose concentration. Note that the measured value, Y is (blood) plasma glucose concentration, obtained by multiplying the arterial glucose concentration (3rd state in the model of Parker et al., 2000) by a correcting factor of 0.925.

The EIMC controller synthesis requires a linear model of the system to be controlled. A 19th order linear model was obtained by linearizing the nonlinear model of the diabetic around the nominal plasma glucose concentration of 81.1 mg/dl. Subsequently, the 19th order model was reduced to a 3rd order model using "balanced realization" technique. The patient model was also subjected to ±5% step change in the manipulated variable (i.e. insulin) and the resulting step response was used to identify a first order plus time delay (FOTD) model using the method of Sundaresan and Krishnaswamy (1977). From the step responses of the above-mentioned models (Fig 1), it appears that a better fit of the patient dynamics is obtained with the FOTD model.

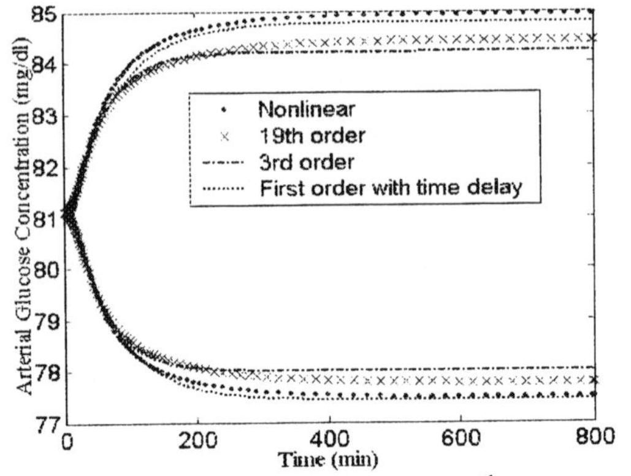

Fig. 1. Response of nonlinear (dot), 19th order linear (cross), 3rd order linear (dash-dot), first order with time delay (dotted) models to ±5% step changes in insulin from the nominal 22.3 mU/min.

3. UNCERTAINTY DESCRIPTION

Uncertainty due to the differences between an actual patient and the patient model could be translated to variations in the model parameters. The glucose and insulin dynamics were found to be most sensitive to variations in the metabolic parameters (Parker et al., 1998). In the patient model, glucose metabolism is mathematically described by threshold functions with the following structure

$$\Gamma_e = E_r\{A_r - B_r \tanh[C_r(x_i + D_r)] \quad (2)$$

The subscript i in equation (2) is the state vector element involved in the metabolic effect and the e subscript denotes specific effects within the model: the effect of glucose on hepatic glucose production (EGHGP), the effect of glucose on hepatic glucose uptake (EGHGU) or the effect of insulin on peripheral glucose uptake (EIPGU). Inter- or intra-patient uncertainty could be classified physiologically as either receptor (D_r parameter) or post-receptor (E_r parameter) defect, and this was modelled mathematically by adjusting the inflection point of the hyperbolic tangent function or the maximum value of Γ_e respectively. Differences in insulin clearance (metabolism) between patients also exist, and could be modelled as deviations in the fraction of clearance from a given compartment such as the fraction of hepatic clearance (FHIC) or the fraction of peripheral insulin clearance (FPIC). This uncertainty formulation essentially focused on the liver (variability in five parameters) and the peripheral (muscle/fat) tissues (variability in three parameters) as these were considered to be more relevant to the control study (Parker et al., 2000).

In the absence of physical data from which to identify ranges for parametric variations, Parker et al. (2000) assumed ±40% parametric variability in each parameter represented a broad range of potential patients. The exception was FHIC, which was limited to ±20% to guarantee non-negative glucose concentrations. From these eight parameters, sets of three parameters were chosen. Each of these three parameters were tested at three levels (nominal, low and high) yielding a total of $^8C_3 \times 3^3 = 1512$ "patients". Patients with similar values for all the eight parameters were removed and this resulted in a set of 577 unique patients. These patients are assumed to capture all the inter- and intra-patient variability among Type I diabetics. Each of these patients was subjected to a 50 g meal disturbance at t = 0 under closed-loop to test the robustness and disturbance attenuating capabilities of the controller designed.

Bounds on open-loop responses of patient models to step change in insulin (from the value required to maintain the output at 81.1 mg/dl) to 0 mU/min, are shown in Fig 2. These responses are very similar to those reported in Parker (1999), confirming the validity of patient models employed in this study. They also show the broad range of patient dynamics. Note that the glucose profiles shown in Fig. 2 are for ±50% variation in parameters where as only ±40% variation in parameters is considered to test controller robustness.

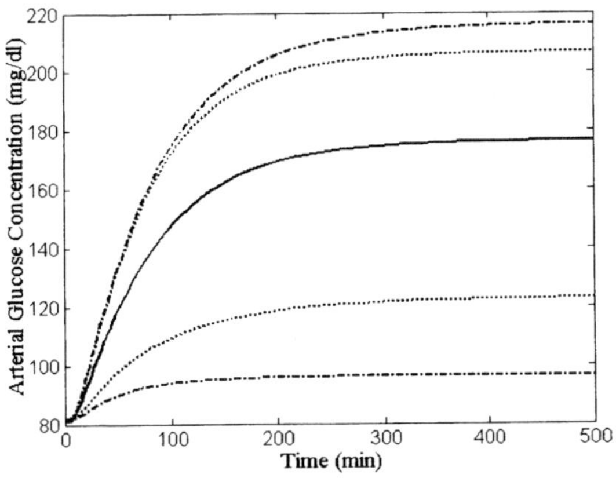

Fig. 2. Response of some patient models to the step change in insulin to 0 mU/min: solid - nominal patient model; dashed - response bounds for ± 50% variations in EGHGP-E_Γ; dash-dot - response bounds for the simultaneous ± 50% variations in EGHGP-E_Γ and EIPGU-D_Γ.

4. SYNTHESIS OF THE EIMC

The conventional IMC structure is shown in Fig 3, where G_p is the plant to be controlled, G_m is a model of the plant, and C is the IMC controller, R is the reference input to the control system, Y is the system output and D is the equivalent external disturbance. The IMC controller design involves two steps

Step1: The process model $G_m(s)$ is factorized as

$$G_m(s) = G_m^+(s)\, G_m^-(s) \tag{3}$$

where $G_m^+(s)$ contains any time delays and right-half plane zeros. It is specified so that its steady-state gain is 1. And, $G_m^-(s)$ contains the remaining invertible dynamics of $G_m(s)$.

Step 2: The IMC controller is specified as

$$C(s) = \frac{1}{G_m^-(s)} f(s) \tag{4}$$

where f (s) is a low-pass filter with steady-state gain of 1. This filter typically has the form

$$f(s) = \frac{1}{(\lambda s + 1)^r} \tag{5}$$

where λ is the tuning parameter to be selected by the control engineer so as to meet the robustness and performance requirements of the control system. The parameter 'r' is a positive integer that is selected so that either C(s) is a proper transfer function or the order of its numerator exceeds the order of the denominator by 1, if ideal derivative action is allowed.

From Fig 3, the closed loop servo transfer function G^R and the closed loop disturbance transfer function G^D of the conventional IMC system can be derived as follows.

$$G^R = [I + (C^{-1} - G_m)\, G_P^{-1}]^{-1} \tag{6}$$

$$G^D = [I + G_P (C^{-1} - G_m)^{-1}]^{-1} \tag{7}$$

The EIMC structure (Zhu et al., 1995) is shown in Fig 4. It can be seen that an additional path is appended to the IMC system within the plant-model parallel structure. Here, in addition to the IMC controller, C_1 (same as C in Fig. 3), C_2 is the 'complementary' IMC compensator and K is pure gain. By inserting this additional path into the original IMC system, a complementary control signal generated from the plant-model error is injected into the output of the original IMC controller. This leads to the EIMC structure whose performance was shown to be superior to IMC in the presence of modelling errors (Zhu et al., 1995).

From Fig 4, the closed loop servo transfer function (G_1^R) and the closed loop disturbance transfer function (G_1^D) of the EIMC structure can be derived as follows.

$$G_1^R = [I + (C_1^{-1} - G_m)\, M]^{-1} \tag{8}$$

and

$$G_1^D = (I - G_m C_1)\,[I + (G_p - G_m)C_1 + G_p C_2 K]^{-1} \tag{9}$$

Here,

$$C_1(s) = \frac{1}{G_m^-(s)} f_1(s); \qquad C_2(s) = \frac{1}{G_m^-(s)} f_2(s);$$

$$f_1(s) = \frac{1}{(\lambda_1 s + 1)^{r_1}}, \qquad f_2(s) = \frac{1}{(\lambda_2 s + 1)^{r_2}}$$

and $\quad M = [G_p (I + C_2 K G_p)^{-1} (I + C_2 K G_m)]^{-1}$

The disturbance attenuating capability of the EIMC is improved with large K and a smaller λ_2 as the denominator in the equation (9) becomes high. The values of both r_1 and r_2 are 1 in the present study.

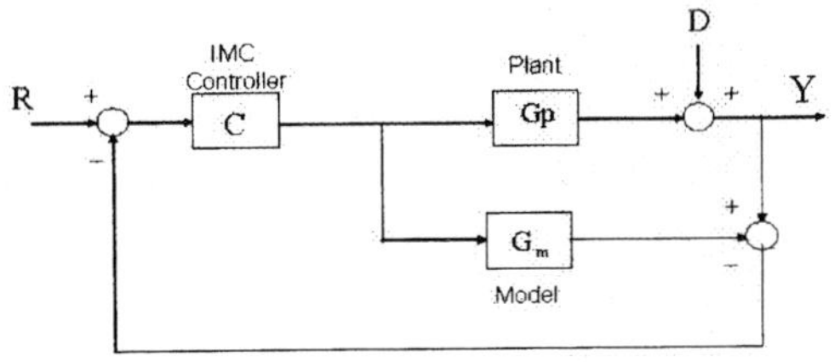

Fig. 3. Conventional IMC system

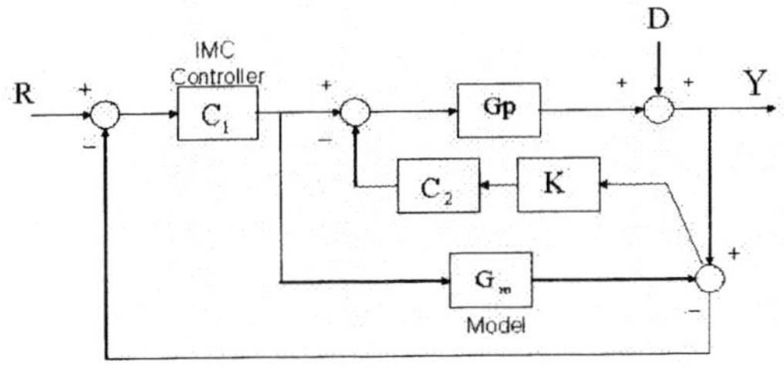

Fig. 4. Enhanced IMC structure

5. RESULTS AND DISCUSSION

Internal Model Controller (IMC) was developed for regulating the glucose level in diabetics. The filter constant, λ of 25 was initially chosen based on robust performance and stability analysis. Then, the IMC performance with various values of λ was tested by subjecting the nominal patient to 50 g meal disturbance (see the results in Fig. 5 for K = 0 and λ = 25). Disturbance rejection of the IMC improved with decrease in λ. Robustness of the IMC was then tested on all 577 patients and the results are summarized in Table 1. Number of patients whose blood glucose concentration was maintained by IMC within normoglycemic range (70-100 mg/dl), is given in the second column. Glucose concentration below 60 mg/dl is the dangerous hypoglycaemic region. Hence, number of patients whose glucose concentration was maintained by the IMC within 60-100 mg/dl, is also presented in Table 1. Also, average IAE and standard deviation of IAE based on all 577 patients are reported in this table. Results in Table 1 show that decreasing λ has increased number of patients whose glucose concentration is maintained between 70-100 mg/dl and 60-100 mg/dl while rejecting 50 g meal disturbance, and also the average IAE and standard deviation of IAE decreased. These and transient responses (not shown here) indicate that decreasing λ has improved the performance of IMC. But the IMC even with λ =2 is not able to maintain the glucose concentration within normoglycemic range in 46% of the patients. This led us to investigate if the EIMC strategy can deliver better disturbance rejection even in the presence of model uncertainty as compared to the conventional IMC.

For EIMC, conservative values of λ_1 = 25 and λ_2 = 20 were selected. The performance of the EIMC for various values of K, in rejecting 50 g meal disturbance acting on the nominal patient at t = 0, is shown in Fig. 5 Overshoot, undershoot and IAE of the closed-loop response are summarized in Table 2. As expected from equation (9), the disturbance attenuating capability of the EIMC improves with increasing K; however, oscillations in the glucose concentration also increase with K. Table 2 shows that increase of K from 0 to 8 reduces overshoot, undershoot and IAE, and increasing K further to 12 increases overshoot, undershoot and IAE indicating possible stability problems at higher K.

Table 1: Performance of IMC with various values of filter constant, for all 577 patients subjected to 50 g meal disturbance at time t = 0.

λ	No. of Patients with Glucose Conc. (mg/dl)		Average IAE for all 577 patients	Standard Deviation of IAE
	70-100	60-100		
25	113	198	318.9	185.9
15	163	294	262.5	170.4
5	250	412	193.5	145.5
2	310	451	169.4	134.7

Note: In this and subsequent tables, IAE values for a simulation time of 800 min for each patient.

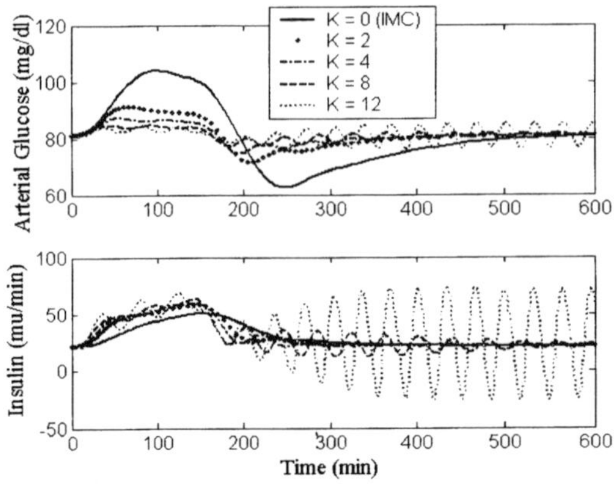

Fig. 5. Disturbance rejection of the EIMC with λ_1 = 25 and λ_2 = 20 for various values of K: nominal patient model subjected to 50 g meal disturbance.

Table 2. Effect of K on rejecting the 50 g meal disturbance by the EIMC: nominal patient model.

K	Overshoot (mg/dl)	Undershoot (mg/dl)	IAE
0	22.77	18.38	272.6
2	9.99	9.63	111.5
4	6.33	6.49	69.4
8	3.54	3.85	42.4
12	3.99	4.90	94.6

From the results in Fig. 5 and Table 2, K = 4 is considered to be most suitable in the EIMC in order to obtain good performance and robustness, and will be used in the subsequent simulations. The effect of λ_1 and λ_2 on the performance of the EIMC was tested while keeping K = 4. As before, 577 patients were subjected to 50 g meal disturbance and the results are summarized in Table 3. Reducing λ_1 and λ_2 does not provide much improvement in performance, and the marginal increase in the average IAE and standard deviation of IAE is due to oscillations in the response at lower λ_1 and λ_2. The parameters K = 4, λ_1 = 20 and λ_2 = 20 will be used in the EIMC for subsequent simulations. For these parameter values, only 75 patients (13%) have entered the dangerous hypoglycaemic region (i.e., glucose concentration < 60 mg/dl). In all these 75 patients, one parameter, namely, FPIC is at 0.21 (a +40% variation from nominal). This number is comparable to 72 patients mentioned by Parker et al. (2000) when H_∞ controller was employed.

Table 3: Effect of λ_1 and λ_2 on EIMC (with K = 4) performance for all 577 patients, when each of them is subjected to a 50 g meal disturbance.

λ_1	λ_2	No. of Patients with Glucose Conc. (mg/dl)		Average IAE for all 577 patients	Std. Dev. of IAE
		70-100	60-100		
25	20	461	502	102	95.6
20	20	462	502	93.2	89.2
20	15	473	505	95.3	119

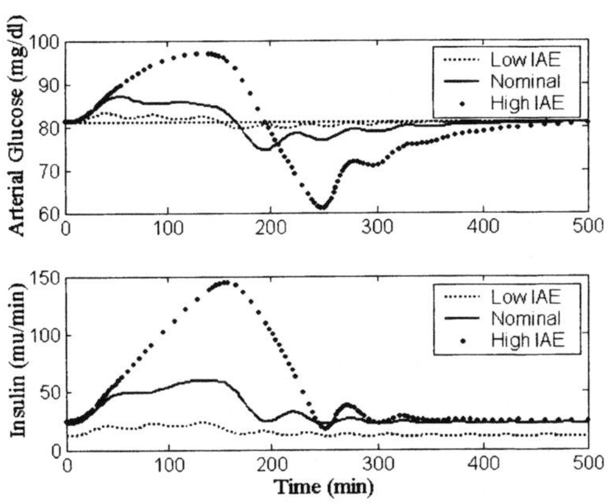

Fig. 6. Transient response of the EIMC for two perturbed patients (giving the lowest and highest IAE) and the nominal case, in attenuating 50 g meal disturbance at time t = 0.

Performance of the EIMC on two perturbed patients and nominal patient are depicted in Fig 6. The two perturbed patients considered are those having the lowest and highest IAE value among the 502 patients whose glucose concentration was controlled within 60-100 mg/dl. The parameters of the patient with the lowest IAE are EGHGU-D_Γ=-0.88; FHIC=0.32; FPIC =0.09, while those of the patient with the highest IAE are EIPGU-E_Γ= 1.4; EGHGU-E_Γ=1.4; FPIC=0.21. Rest of the parameters are at their nominal values for both cases. Responses of about 250 perturbed patients with parameters different from those of the nominal patient, when subjected to 50 g meal disturbance, are similar, and one of these responses is shown in Fig 7. The transients in Figs. 6 and 7 have just a few oscillations and are acceptable.

Next, the EIMC controller is tested on a patient whose parameters are different from the 577 patients considered in this study. This case is specified as the worst-case performance of the continuous-time H_∞ controller, including uncertainty weighting and parametric uncertainty in Parker et al., (2000). Disturbance rejection by the H_∞ controller for this case was shown in Figure 9 of Parker et al. (2000). The response by the EIMC for this patient model, illustrated in Fig 8, is better than that by the H_∞ controller.

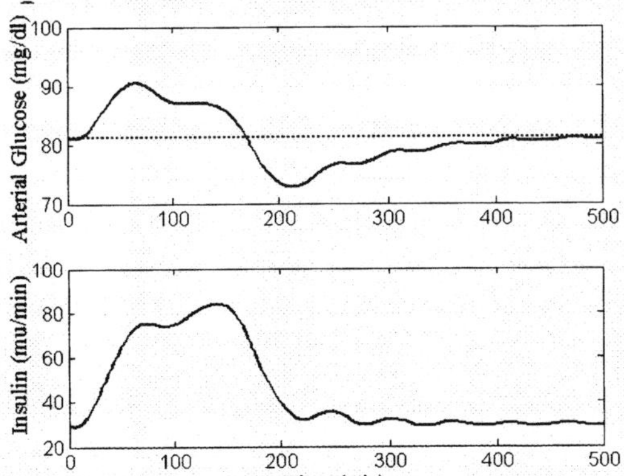

Fig. 7. Performance of EIMC on a perturbed patient model with parameters EGHGP-E_Γ = 1.4, EIPGU-D_Γ = -8.15, EGHGP-D_Γ = -0.6957. The remaining five parameters are at their nominal values.

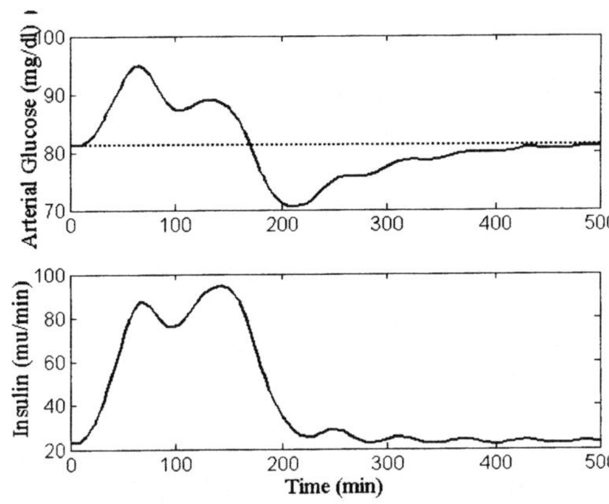

Fig. 8. Performance of EIMC on a perturbed patient model with parameters: EIPGU-D_Γ = -8.15, EGHGU-D_Γ = -2.072, FHIC = 0.36. The remaining five parameters are at their nominal values.

6. CONCLUSIONS

The EIMC structure utilizes a simple FOTD model and is shown to provide a very good regulation of blood glucose concentration of diabetics. It is able to maintain the glucose concentration above the dangerous hypoglycaemic range (< 60 mg/dl) in 87% of 577 patient models tested. For these patients, average IAE and standard deviation of IAE are reduced by nearly half by the EIMC scheme compared to the IMC structure. Further, the performance of the former is comparable to that of H_∞ controller. Thus, the EIMC is very attractive for glycaemia control owing to its simple structure and design as well as good robustness.

REFERENCES

Ackerman, E., L.C. Gatewood., J.W. Rosevear, and G.D. Molnar, (1965), Model Studies of Blood-Glucose Regulation, *The Bulletin Of Mathematical Biophysics*, **27**(Suppl.), 21.

Ashcroft, F.M. and S.J.H. Ashcroft, (1992), Insulin: Molecular Biology to Pathology, Oxford University Press, New York.

Bergman, R.N., L.S. Phillips and C. Cobelli, (1981), Physiologic Evaluation of Factors Controlling Glucose Tolerance in Man, *Journal of clinical investigation*, **68**, 1456.

Bolie, V.W. (1961), Coefficients of Normal Blood Glucose Regulation, *Journal of Applied Physiology*, **16**, 783.

Bremer, T. and D.A. Gough, (1999), Is Blood Glucose Predictable From Previous Values? A solicitation of Data, *Diabetes*, **48**, 445.

Cobelli, C. and A. Mari, (1983), Validation of Mathematical Models of Complex Endocrine-Metabolic Systems. A Case Study on a Model of Glucose Regulation, *Medical & biological engineering & computing*, **21**, 390.

Fischer, M.E. (1991), A semiclosed-loop Algorithm for the Control of Blood Glucose Levels in Diabetics, *IEEE Transactions on Biomedical Engineering*, **38**, 57.

Guyton, J.R., R.O. Foster, J.S. Soeldner, M.H. Tan, C.B. Kahn, L. Koncz, and R.E. Gleason, (1978), A model of Glucose-Insulin Home-ostasis in Man that Incorporates the Heterogeneous Fast Pool Theory of Pancreatic Insulin Release, *Diabetes*, **27**, 1027.

Lehmann, E.D. and T. Deutsch, (1992), A Physiological Model of Glucose-Insulin Interaction in Type 1 diabetes Mellitus, *Journal of Biomedical Engineering*, **14**, 235.

Ollerton, R.L. (1989), Application of Optimal Control Theory to Diabetes Mellitus, *International Journal of Control*, **50**, 2503.

Parker, R.S. (1999), *Model-Based Analysis and Control for Biosystems*, PhD thesis, Dept. of Chemical Engineering, University of Delaware.

Parker, R.S. (2002), Personal correspondence.

Parker, R.S., F.J. Doyle III, and N. A. Peppas, (1999), A Model-Based algorithm for Blood Glucose Control in Type I Diabetic Patients, *IEEE Transactions on Biomedical Engineering*, **46**, 148.

Parker, R.S., F.J. Doyle III., J.H. Ward and Nicholas A. Peppas, (2000), Robust H_∞ Glucose control in Diabetes Using a Physiological Model, *AIChE J*, **46**, 2537-2549.

Puckett, W.R. and E.N. Lightfoot, (1995), A Model for Multiple Subcutaneous Insulin Injections Developed from Individual Diabetic Patient Data, *American Journal of Physiology*, **269** (Endocrinology metabolism 32), E1115.

Simon, G., G. Brandenberger and M. Follenius, (1987), Ultradian Oscillations of Plasma Glucose, Insulin, and C-Peptide in Man During continuous Enteral Nutrition, *Journal of Clinical Endocrinology and Metabolism*, **64**, 669.

Sorensen, J.T. (1985), *A physiologic Model of Glucose Metabolism in Man and its use to Design and Assess Improved Insulin Therapies for Diabetes*, PhD Thesis, Dept. of Chemical Engineering, MIT.

Steil, G.M., J. Murray, R.N. Bergman and T.A. Buchanan, (1994), Repeatability of Insulin Sensitivity and Glucose Effectiveness from the Minimal Model – Implications for Study Design, *Diabetes*, **43**, 1365.

Sundaresan, K.R. and P.R. Krishnaswamy, (1977), Estimation of time delay, time constant parameters in time, frequency, and Laplace domains, *Canadian Journal of Chemical Engineering*, **56**, 257.

Zhu, H.A., C.L. Teo., A.N. Poo and G.S. Hong, (1995), An Enhanced Internal Model Structure, *Control Theory and Advanced Technology*, Vol **10**, No 4, Part 2, pp. 1115-1127.

APPENDIX

Equations A13, A20, A25 and A29 in Parker et al. (2000), corrected for typographical errors, are listed below. The parameter $F_{KC} = 0.3$ is missing in Table A1 of Parker et al. (2000). All these were confirmed by Parker (2002).

$$\dot{A}_{IHGP} = \frac{1}{25}[1.2088 - 1.138 \quad (\tanh(1.669(\frac{I_L^C}{21.43} - 0.8885))) - A_{IHGP}] \tag{A13}$$

$$\dot{I}_H^C = (I_B^C Q_B + I_L^C Q_L + I_K^C Q_K + I_P^C Q_P - I_H^C Q_H + \Gamma_{IVI})\frac{1}{V_H^C} \tag{A20}$$

$$\dot{I}_P^T = (I_P^C - I_P^T)\frac{1}{T_P^I} + \frac{\Gamma_{SIA} - \Gamma_{PC}}{V_P^T} \tag{A25}$$

$$\Gamma_{PC} = \frac{I_P^T}{\frac{1 - F_{PC}}{F_{PC}}\frac{1}{Q_P} - \frac{T_P^I}{V_P^T}} \tag{A29}$$

ELSEVIER
IFAC
PUBLICATIONS
www.elsevier.com/locate/ifac

A PROTOTYPE MODEL FOR INDOLE-3-ACETIC ACID (IAA) PRODUCTION BY *AZOSPIRILLUM BRASILENSE* SP245

Ilse Y. Smets [*] **, Kristel Bernaerts** [*] **, Astrid Cappuyns** [*]
Ositadinma Ona [**] **, Jos Vanderleyden** [**] **, Els Prinsen** [***]
and Jan F. Van Impe [*,1]

* *BioTeC–Katholieke Universiteit Leuven,*
W. de Croylaan 46, B-3001 Leuven (Belgium)
Fax: +32-16-32.29.91
e-mail: jan.vanimpe@cit.kuleuven.ac.be
** *CMPG–Katholieke Universiteit Leuven,*
Kasteelpark Arenberg 20, B-3001 Leuven (Belgium)
*** *Department of Biology, University of Antwerp,*
Universiteitsplein 1, B-2610 Antwerpen (Belgium)

Abstract: Given the vital role of nitrogen for plant growth on the one hand, versus the severe impact on the environment of extensive use of inorganic nitrogen fertilizers on the other hand, the need for alternative fertilization becomes apparent. In this respect, the nitrogen fixing bacteria of the genus *Azospirillum* offer interesting perspectives since inoculation of the plant roots with these bacteria positively affects plant growth. Studies have shown that the growth promoting power of the bacteria can probably be attributed more to the production of the auxin indole-3-acetic acid (IAA) than to the nitrogen fixation qualities of the genus. The aim of this study is therefore to characterize and quantify the IAA production of *Azospirillum brasilense* Sp245 on a macroscopic level as to exploit the resulting model in optimal experiment design studies. The data obtained from these optimal experiments should be rich enough to aid in further unravelling the studied IAA production mechanism.

Keywords: *Azospirillum brasilense*, IAA production, bioreactor model identification

1. INTRODUCTION

Given the fact that nitrogen is undoubtedly one of the most important plant nutrients (together with inorganic phosphate), the production of inorganic nitrogen fertilizers has been of immense benefit for the agricultural sector. Unfortunately, the intensive application of fertilizers gives rise to severe ecological problems (Wright and Black, 1979). Alternative means have been sought to increase crop production without compromising the environment. In this context, the nitrogen fixing bacteria of the genus *Azospirillum* present a challenging alternative for, or a supplement to chemical fertilization (Baldani *et al.*, 1983; Dobbelaere *et al.*, 2001) since they belong to the class of *plant growth promoting rhizobacteria.* Indeed, after inoculation of *Azospirillum* spp., a significant change in root morphology can be detected. The key factor in this root number and root surface increase after *Azospirillum* inoculation is generally believed to be the auxin indole-3-acetic acid (IAA) (Okon and Vanderleyden, 1997). Several recent studies have shown significant enhancement in crop yields under glasshouse and field conditions in response to inoculation with *Azospirillum* spp. (Dobbelaere *et al.*, 2001; Dobbelaere *et al.*, 2002). In addition, it has been shown that *Azospirillum* inoculants neither cause any environmental hazards (Fages, 1992) nor present any health prob-

[1] Corresponding author

lems in plants (Okon and Vanderleyden, 1985). Unfortunately, not all field results are consistent.

Since the (factors that influence the) IAA production mechanisms by *Azospirillum* are not fully understood yet, the aim of this paper is to quantitatively assess the IAA production by *Azospirillum brasilense* Sp245 based on a set of strictly controlled bioreactor experiments. The resulting model can then be used to design future, *information rich* experiments to rule out certain hypotheses and to extend as such our knowledge on this production mechanism. The structure of the paper is as follows. First the material and methods are introduced after which the growth of the biomass on the carbon source is modelled (Section 3). In a second stage (Section 4), the IAA production process is quantified. Afterwards, it is outlined how the obtained preliminary model can be exploited to design future experiments (Section 5). Finally, Section 6 summarizes the conclusions.

2. MATERIALS AND METHODS

Bacterial strain. *Azospirillum brasilense* Sp245 was collected from the stock held in the CMPG laboratory at the K.U.Leuven. The preculture used to inoculate the bioreactor was initiated by inoculating a 250 mL Erlenmeyer flask containing 150 mL of a minimal malate medium (MMAB) (Vanstockem *et al.*, 1987) with a loop of cells taken from an overnight plate culture. The culture was incubated in an incubator rotary shaker (New Brunswick Scientific, USA) at 200 rpm and 30°C until an optical density (at 600 mm) of about 1.8 was reached.

Experimental conditions. Four batch cultures were conducted in the computer controlled BioFlo 3000 benchtop fermentor (New Brunswick Scientific, USA) with an autoclavable vessel of 1.25 to 5 L working volume. 100 mL of the preculture was transferred to the fermentor vessel containing 3.0 L MMAB medium. PID cascade controllers ensured that the fermentation temperature was kept at 30°C and the dissolved oxygen (DO) at 3% (micro-aerobic range) by changing the agitation rate while keeping the aeration rate at 0.1 L/min. L-malate is provided as carbon source. To test whether pH has a significant effect on growth and/or production, in the first and third experiment the pH is kept at 6.3 while in the second and fourth experiment the pH is kept at 6.8. Exogenous tryptophane (Trp) is only added at the beginning of the batch culture (5 μg/mL) of the first and the second experiment.

Measurements. Culture media samples were removed at regular intervals and each sample divided into subsamples. Optical density at 600 nm was measured with a Perking Elmer Lamda 2 UV/VIS spectrophotometer. L-malate concentrations were determined using test kits from Roche. Indole-3-acetic acid and tryptophane concentrations in the fermentation medium were determined by Gas Chromatography/Mass spectrophotometry (GC/MS) (Prinsen *et al.*, 2000). All data shown are the average value of at least two replicates.

Mathematical tools. The implemented identification routine for model parameter estimation is the **e04UCF** routine from the NAG library (Numerical Algorithms Group) in Fortran. Apart from Fortran, Matlab 6.1 (The Mathworks Inc., Natick) was used as simulation software. It must be mentioned that the absolute and relative tolerances of the integration routines had to be set to a very small value (i.e., 10^{-8}) to ensure correct simulations.

3. GROWTH MODELLING

Since there is no evidence that malate consumption or biomass growth is influenced by the presence of Trp or IAA in the medium, the growth process is first tackled as a separate phenomenon. When focussing on (*macroscopic*) mass balance equations for batch reactors, the only term that needs to be specified is the reaction term since no transport has to be taken into account.

For the modelling of the specific growth rate μ, a simple monotonically increasing function, i.e., the Monod equation is proposed. The link between the specific growth rate and the substrate consumption is the so-called *linear law* in which, for the time being, the maintenance term is neglected. All substrate consumed is therefore assumed to be built in as new biomass with a certain efficiency or yield factor $Y_{X/S}$ [OD L /g]. The evolution in time of the substrate concentration (i.e., malate) C_S [g/L] and the biomass concentration C_X [OD] is then described by following system of mass balance equations:

$$\begin{aligned} \frac{dC_S}{dt} &= -\frac{\mu}{Y_{X/S}} \cdot C_X \\ \frac{dC_X}{dt} &= \mu \cdot C_X \end{aligned} \tag{1}$$

in which

$$\mu = \mu_{max} \frac{C_S}{C_S + K_M}. \tag{2}$$

In this specific growth rate expression, μ_{max} [1/h] is the maximum specific growth rate and K_M [g/L] the half saturation constant. As can be seen, only the influence of the carbon source is taken into account at this stage.

Inspection of Equations (1) and (2), reveals that 5 parameters have to be identified, i.e., the yield factor $Y_{X/S}$, the Monod growth constants μ_{max} and K_S and the initial values of the substrate and biomass concentration $C_S(0)$ and $C_X(0)$ (to account for measurement errors these latter initial conditions are also considered as degrees of freedom during parameter identifcation). Hereto, following sum of squared errors (SSE) between the model predicted and experimentally obtained values for substrate and biomass concentration is minimized:

$$\begin{aligned} SSE_G = & \sum_{k=1}^{n_{C_S}} (C_S(k) - C_{S,exp}(k))^2 \\ & + \sum_{k=1}^{n_{C_X}} (C_X(k) - C_{X,exp}(k))^2 \end{aligned} \tag{3}$$

	Growth parameters				
	Initial[a]	pH6.3+	pH6.8+	pH6.3-	pH6.8-
$C_S(0)$	exp[b]	2.426	2.477	2.427	2.595
$C_X(0)$	exp	0.079	0.099	0.088	0.195
μ_{max}	1.000	0.421	0.612	0.414	0.253
K_S	1.000	0.439	1.425	0.473	$1{\cdot}10^{-7}$
$Y_{X/S}$	0.800	0.777	0.756	0.745	0.636
SSE_G		0.090	0.088	0.045	0.036

[a]the initial values to start the identification procedure
[b]the first experimental data point

Table 1. Growth parameter estimates for the 4 experiments. The + and - sign denote addition of exogenous Trp or not.

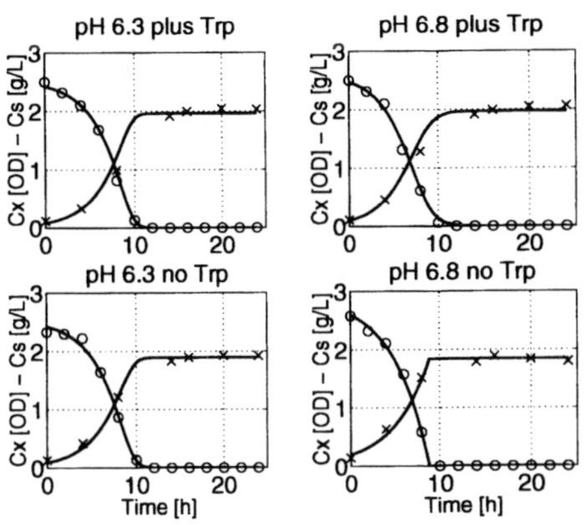

Fig. 1. Biomass and malate concentration profiles at different pH values and with or without extra Trp: experimental values (x,o) and model predictions (–) (Equations (1) and (2) and Table 1).

with n_{C_S} and n_{C_X} the number of experimental data points for the substrate and biomass concentration, respectively. Since the measurements are in the same order of magnitude, no rescaling of the individual errors is performed.

The obtained growth parameter sets for each experiment are listed in Table 1 together with the corresponding SSE_G value. As can be seen from Figure 1, these parameter values give rise to excellent model fits.

However, correct identification of the parameters is not a trivial task since (*i*) the experimental data points are scarce and (*ii*) batch experiments are known as not the most optimal setup for estimation of both Monod constants at once (Holmberg and Ranta, 1982). When plotting the value of the SSE_G for a whole range of μ_{max} and K_S values as is done in the contour plot of Figure 2 for the first experiment, it is apparent that their is a whole *valley* in which several local minima can appear. Indeed, the final values of the parameter estimation routine are largely affected by (*i*) the initialization of the parameters and (*ii*) the quality and positioning of the data points. Since there are relatively few data points at the zones of significant transition in dynamics, a small difference in the measurements for C_S and C_X will

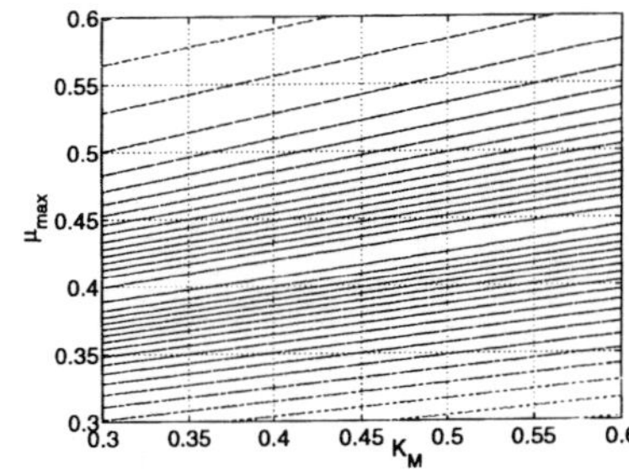

Fig. 2. Contour plot for the SSE_G value obtained for a wide range of μ_{max} and K_S values.

have a large influence on the final value, probably explaining the difference in the K_S values for both pH setpoints. Nevertheless, the estimated parameter sets from one experiment can be used to describe another data set, i.e., at another pH value (results not shown due to space limitation) and a significant pH effect on growth cannot be inferred from the data at hand.

Section 5 discusses how the identifiability problem related to the Monod kinetics constants can be overcome.

4. IAA PRODUCTION MODELLING

While for the modelling of the biomass growth and substrate consumption one could rely on intuitive microbiological knowledge, the quantification by a mass balance type model of the IAA production by *Azospirillum* has, to the authors knowledge, not been published yet. Evidently, some hypotheses have been formulated with respect to the influencing factors of this production process. One of these hypotheses, namely, the need for additional, exogenous tryptophan as a precursor (Zimmer *et al.*, 1991), is explored in this paper. The experimental results for IAA in the four data sets are illustrated in Figure 3. When no exogenous Trp is added (upper plots), very limited (in fact hardly detectible) amounts of IAA are produced. If, however, an initial amount of 5 μg/mL of Trp is added to the batch bioreactor, the IAA production soon becomes significant, i.e., already during the exponential growth phase (lower plots).

Remains the question whether the production of IAA is growth related or not, i.e., whether the growth limiting substrate (or the energy derived from it) is directly needed in the IAA production pathway or not. Two model types for the IAA production are therefore proposed. In the first model hypothesis, IAA is formed due to the presence of Trp only but the biomass is needed as (bio)catalyst. In the second hypothesis, production of IAA is also growth related. Following sets of additional mass balance equations are proposed for the evolution in time of the IAA concentration [μg/mL] and Trp concentration [μg/mL].

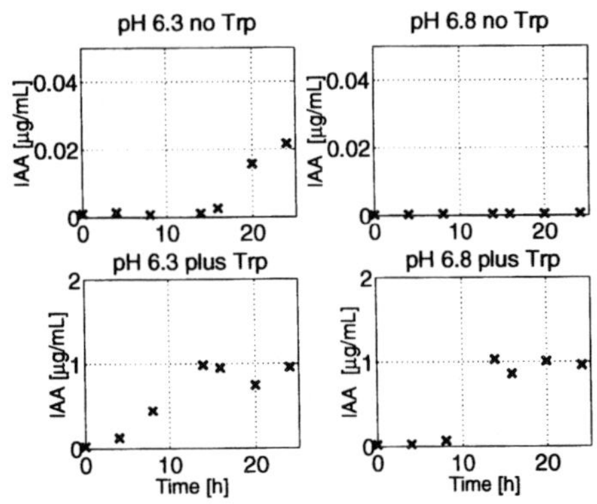

Fig. 3. Experimental IAA profiles at different pH values and with or without extra Trp.

Production model hypothesis 1 (H1)

$$\frac{dTrp}{dt} = -A_T \frac{Trp}{Trp + K_T} \cdot C_X$$
$$\frac{dIAA}{dt} = Y_{P/X} \cdot \frac{Trp}{Trp + K_T} \cdot C_X \quad (4)$$

Production model hypothesis 2 (H2)

$$\frac{dTrp}{dt} = -A_T \frac{Trp}{Trp + K_T} \cdot C_X$$
$$\frac{dIAA}{dt} = Y_{P/X} \cdot \mu \cdot \frac{Trp}{Trp + K_T} \cdot C_X \quad (5)$$

with μ [1/h] the Monod expression of Equation (2), A_T [μg/h] and K_T [μg/mL] Monod kinetics constants for the Trp consumption and $Y_{P/X}$ [μg/h] for H1 and [μg] for H2 a yield like coefficient of IAA on biomass.

In the first model hypothesis, the left hand side of the balance equation for IAA reflects the need for Trp via a Monod like *switching* function: if the amount of Trp is very low, no IAA will be produced but once Trp is present (to some extent) IAA production can occur. In the second model hypothesis this Trp dependent term is multiplied with the specific growth rate μ to reflect the presumed dependency on the growth.

The global identification cost SSE_T is equal to the sum of the SSE for growth (Equation (3)) and an SSE term for the production related concentrations:

$$SSE_T = SSE_G + SSE_P \quad (6)$$

with

$$SSE_P = \sum_{k=1}^{n_{Trp}} (Trp(k) - Trp_{exp}(k))^2 + \sum_{k=1}^{n_{IAA}} (IAA(k) - IAA_{exp}(k))^2 \quad (7)$$

Production parameters					
		Hypothesis 1		Hypothesis 2	
	Initial	pH6.3+	pH6.8+	pH6.3+	pH6.8+
$Trp(0)$	exp	4.024	4.082	4.022	4.080
$IAA(0)$	exp	0.109	$1{\cdot}10^{-4}$	$1{\cdot}10^{-4}$	$1{\cdot}10^{-4}$
A_T	1.000	0.276	0.246	0.275	0.245
K_T	1.000	$1{\cdot}10^{-4}$	$1{\cdot}10^{-4}$	$1{\cdot}10^{-4}$	$1{\cdot}10^{-4}$
$Y_{P/X}$	1.000	0.056	0.058	0.476	0.457
SSE_T		0.197	0.542	0.176	0.878

Table 2. The identified production parameter sets for the experiments with exogenous Trp (with the identified growth parameters of Table 1) for both hypothesis.

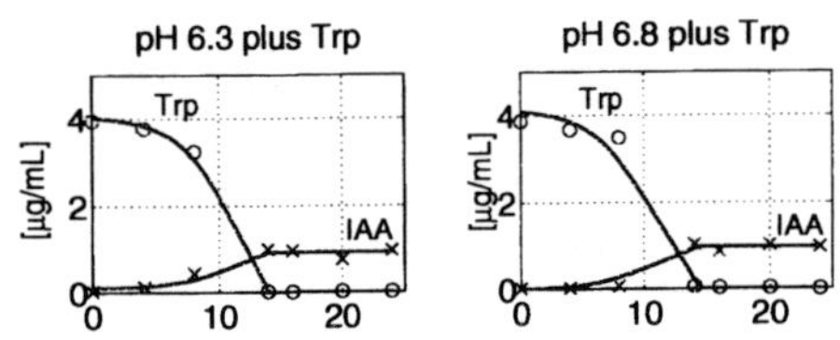

Fig. 4. IAA and Trp profiles at different pH values and with exogenous Trp: experimental values (x,o) and model predictions (–) following H1, i.e., Equations (1), (2) and (4).

with n_{Trp} and n_{IAA} the number of experimental data points for the Trp and IAA concentration, respectively.

To avoid an overload of parameters to be identified, the growth related parameters were first fixed to their optimal values (as listed in Table 1). The identified values for the remaining parameters (for both hypotheses) can be found in Table 2. Re-identification of all 10 parameters afterwards revealed that the identified values hardly changed. The difference between the identified values for $Y_{P/X}$ in both hypotheses, being an order of a magnitude smaller for H1 can probably be attributed to the different units for this parameter.

Obviously, no production related parameters have been identified for the experiments to which no Trp is added, and in which accordingly hardly any IAA is produced. Both model hypotheses will predict zero IAA production which is in accordance with the experimental data since the recorded values, which are close to the detection limit, are questionable.

Figures 4 and 5 illustrate the model predictions following the first and second hypothesis for the experiments to which exogenous Trp was added. As can be seen, for H1, the IAA production ceases when the Trp concentration goes to zero, while for H2, the IAA production levels already when the malate concentration becomes depleted. However, the experimental data at hand do not enable discrimination between the proposed hypotheses, since both figures illustrate a positive match between the model predictions and the experimental data.

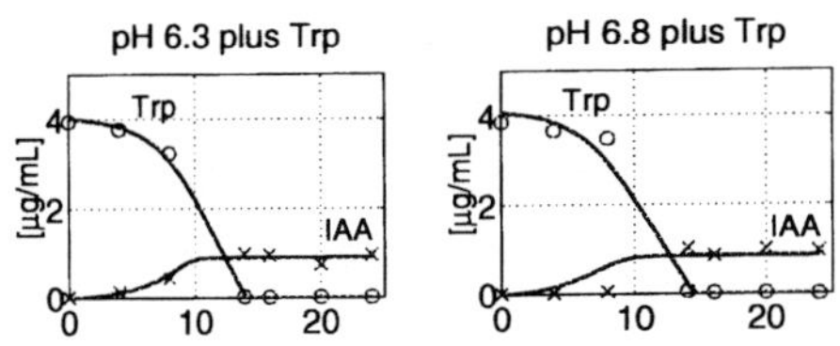

Fig. 5. IAA and Trp profiles at different pH values and with exogenous Trp: experimental values (x,o) and model predictions (–) following H2, i.e., Equations (1), (2) and (5).

	Set 1	Set 2	Set 3
$C_S(0)$	2.390	2.364	2.263
$C_X(0)$	0.056	0.011	0.010
μ_{max}	0.476	41.078	0.589
K_S	0.475	136.002	0.010
$Y_{X/S}$	0.802	0.828	0.875
$Trp(0)$	3.983	3.933	4.036
$IAA(0)$	0.005	0.013	0.022
A_T	0.281	0.295	416.975
K_T	$1{\cdot}10^{-4}$	$1{\cdot}10^{-4}$	2243.264
$Y_{P/X}$	0.467	0.453	330.529
SSE_T	0.161	0.237	0.598

Table 3. Different parameter sets for the first experiment, obtained by starting from different initial parameter estimates.

5. OPTIMAL EXPERIMENT DESIGN

To refine (and/or validate) the proposed model structure, following necessary steps are proposed. First of all, the uniqueness problem of the identified parameter values has to tackled. Afterwards, the alleged growth dependency of the production process must be investigated. Finally, it must be examined whether IAA production is persistent with prolonged availability of malate and tryptophan.

5.1 *Uniqueness of the parameter values*

If Monod type kinetics have to be identified from batch experiments, it is known that the parameters will be highly correlated and a unique identification will be hard since the cost surface is littered with local minima. As an illustration, Table 3 lists three sets of parameters that exhibit more or less the same identification performance (reflected by similar SSE_T values). The first two sets differ at the level of growth constants while the latter has different Trp consumption rate constants.
Intuitively, one could propose a *fed-batch* experiment with a very low feed rate in which malate is present. This more controlled feeding strategy enables to discriminate between the different parameter sets as illustrated in Figure 6. A more rigorous approach would be based on optimal experiment design. Through the optimization of a scalar function of the Fisher Information Matrix an optimal input (i.e., feeding profile) can be designed, that ensures high information content

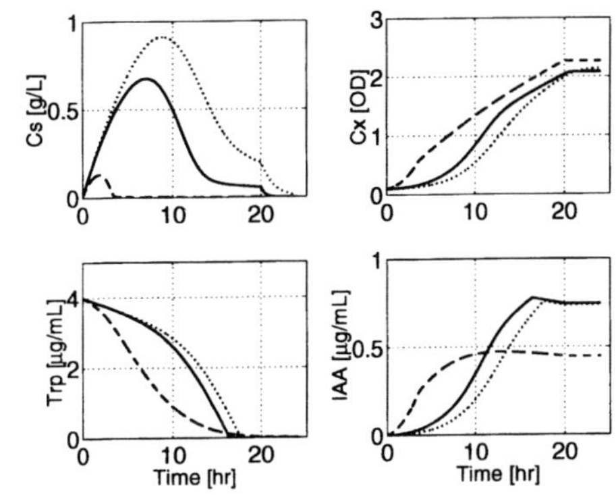

Fig. 6. Proposed fed-batch experiment to discriminate between different parameter sets. Set 1 (–), set 2 (...) and set 3 (--) as listed in Table 3. Fin = 0.0625 L/h, t_{feed}=20 h, $C_{S,in}$ = 10 g/L, $C_X(0)$ = 0.1 OD, $Trp(0)$ = 4 µg/mL, $C_S(0)$=$IAA(0)$=0 g/L.

of the data with respect to uncorrelated, unique parameter estimation (Versyck *et al.*, 1997). This is the subject of ongoing research.

5.2 *IAA production: growth related or not?*

To overrule or validate the hypothesis that the IAA production is growth related, the time instant at which the malate becomes depleted must be significantly different from the instant at which Trp becomes exhausted. Although chemostat cultures are often cited in this respect, as a first indicative experiment, a simple batch experiment with another initial malate and Trp concentration can rapidly elucidate the problem. As illustrated in Figure 7 (solid line for H1, dotted line for H2), it should then be possible to infer from the experimental data, whether IAA production stops when malate is depleted or continues until Trp is depleted (since both hypotheses are only different at the level of the IAA production mass balance, the predictions for C_S, C_X and Trp are the same). A high sampling rate during significant changes in the culture dynamics is recommended. Also for this type of *model structure discrimination*, techniques of optimal experiment design exist, some of which will be tested in future (Munack, 1992).

5.3 *Persistence of IAA production*

With the obtained model (and its parameters) a fed-batch experiment can be designed that maintains the substrate and Trp concentration in the reactor at a (low) constant level. As such, depletion of the growth/production limiting factors can be avoided. It is then possible to check whether *Azospirillum* IAA production is persistent or that other (limiting) factors or product degradation have to be taken into account.

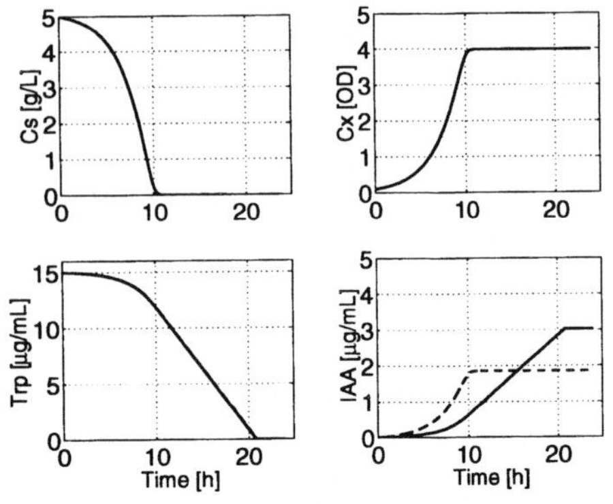

Fig. 7. Proposed batch experiment to discriminate between the two production hypotheses H1 (–) and H2 (- -). $C_S(0)$=5 g/L, $Trp(0) = 15$ μ/mL, $C_X(0) = 0.1$ OD, $IAA(0)$=0 μg/mL.

6. CONCLUSIONS

In this paper, a prototype model for the production of IAA by *Azospirillum brasilense* Sp245 is proposed. IAA is an auxine that could play an important role with respect to soil fertilization since it is shown that the inoculation of plants with the IAA producing *Azospirillum* species, has a positive effect on plant growth and crop yield. Based on the experimental batch data at hand, a simple, Monod law based model for the growth of the bacteria on malate is deduced. pH was not observed to have a significant effect on growth nor on production. The unique identification of the kinetic parameters μ_{max} and K_S asks however for additional fed-batch experiments. For the IAA production process, no distinction between the two hypotheses (only Trp or Trp and growth dependency) could be made on the basis of the available data. Future research will focus on accurate parameter estimation and model structure discrimination by means of optimal experiment design and the realization of the optimal proposed experiments.

ACKNOWLEDGMENTS

Ilse Smets and Kristel Bernaerts are postdoctoral fellows with the Fund for Scientific Research Flanders (FWO). Work supported in part by Projects FWO-G.0085.03, OT/99/24 and OT/03/30 of the Research Council of the Katholieke Universiteit Leuven and the Belgian Program on Interuniversity Poles of Attraction, initiated by the Belgian State, Prime Minister's Office for Science, Technology and Culture. The scientific responsibility is assumed by its authors.

7. REFERENCES

Baldani, V.L.D., J.I. Baldani and J. Döbereiner (1983). Effects of *Azospirillum* inoculation on root infection and nitrogen incorporation in wheat. *Canadian Journal of Microbiology* **29**, 924–929.

Dobbelaere, S., A. Croonenborghs, A. Thys, D. Ptacek, J. Vanderleyden, P. Dutto, C. Labandera Gonzalez, J. Caballero-Mellado, J.F. Aguirre, Y. Kapulnik, S. Brener, S. Burdman, D. Kadouri, S. Sarig and Y. Okon (2001). Response of agronomically important crops to inoculation with *Azospirillum*. *Australian Journal of Plant Fysiology* **28**, 871–879.

Dobbelaere, S., A. Croonenborghs, A. Thys, D. Ptacek, Y. Okon and J. Vanderleyden (2002). Effect of inoculation with wild type *A. brasilense* and *A. irakense* strains on development and n-content of spring wheat and grain maize. *Biology and Fertility of Soils* **36**, 284–297.

Fages, J. (1992). An industrial view of *Azospirillum* inocullants: production and application in technology can stimulate plants. *Symbiosis* **13**, 15–26.

Holmberg, A. and R. Ranta (1982). Procedures for parameter and state estimation of microbial growth process models. *Automatica* **18**(2), 281–193.

Munack, A. (1992). Some improvements in the identification of bioprocesses. In: *Modeling and Control of Biotechnical Processes* (M.N. Karim and G. Stephanopoulos, Eds.). pp. 89–94. Pergamon Press, Oxford.

Okon, Y. and J. Vanderleyden (1985). *Azospirillum* as a potential inoculant for agriculture. *Trends in Biotechnology* **3**, 223–228.

Okon, Y. and J. Vanderleyden (1997). Root associated *Azospirillum* can stimulate plants. *ASM News* **63**, 366–370.

Prinsen, E., S. Van Laer, S. Öden and H. Van Onckelen (2000). Plant hormone protocols. In: *Auxin analysis* (G.A. Tucker, Ed.). pp. 49–56. Humana Press, Totowa, N.J.

Vanstockem, M., K. Michiels, J. Vanderleyden and A. Van Gool (1987). Transposon mutagenesis of *Azospirillum brasilense* and *Azospirillum lipoferum*: physical analysis of Tn5 and Tn5-mob insertion mutants. *Applied and Environmental Microbiology* **53**, 410–415.

Versyck, K.J., J.F. Claes and J.F. Van Impe (1997). Practical identification of unstructured growth kinetics by application of optimal experimental design. *Biotechnology Progress* **13**(5), 524–531.

Wright, J.R. and A.L. Black (1979). Range fertilization: plant response and water use. *Journal of Water Range Management* **32**, 345–349.

Zimmer, W., C. Aparicio and C. Elmerich (1991). Relationship between tryptophan biosynthesis and indole-3-acetic acid production in *Azospirillum*: identification and sequencing of a trpgdc cluster. *Molecular and General Genetics* **229**, 41–51.

ELSEVIER
IFAC
PUBLICATIONS
www.elsevier.com/locate/ifac

MONITORING AND CONTROL BASED ON A FIA-BIOSENSOR SYSTEM WITH AUTOMATIC CORRECTION

L. S. Ferreira[1*], J. O. Trierweiler[1] and M. B. De Souza Jr.[2]

1 Group of Integration, Modelling, Simulation, Control, and Optimization of Process, Chemical Engineering Department, Federal University of Rio Grande do Sul - UFRGS. R. Luiz Englert, s/n°. Campus Centro. CEP.: 90040-040. Porto Alegre. RS. Brazil.
E-mails: {luciane,jorge}@enq.ufrgs.br
2 Chemical Engineering Department, Chemistry School, Federal University of Rio de Janeiro - UFRJ. Bloco E - C - Ilha do Fundão. CP 68542. CEP.: 21949-900. Rio de Janeiro. RJ. Brazil
E-mail: mbsj@eq.ufrj.br

Abstract: In this work, an enzymatic FIA-Biosensor system for lactose is used to on-line monitoring and control of a cultivation process to produce the enzyme lactase. Temperature effects on the output signal of the sensor are also studied. The whole system is modeled and, with this model, an automatic compensation is proposed to deal with temperature variations. *Copyright © 2004 IFAC*

Keywords: Biotechnology; Compensation; Control applications; Bio control; Sampling systems; Dynamic modeling.

1. INTRODUCTION

During the last decades, the application of biotechnological processes has grown in several sectors, such as pharmaceuticals/medicine, agrochemicals, food technology, sustainable development etc. As operating parameters - quality of substrate and microorganisms - may change, an automatic control of the substrate supply is mandatory for the vast majority of the continuous and fed-batch processes, in order to keep high productivity levels. It will be more efficiently reached if appropriate measuring devices are available to provide on-line information about state variables. The development of specific biosensors interfaced with adequate sampling systems appeared as an important contribution to maximize the productivity of the bioprocesses. Biosensors are composed of a biological element connected to a transducer system, that can transform a biological signal into a measurable electrical signal. Several kinds of biosensors may be found in the literature, applied to a broad range of areas (Vega *et al.*,1998; Gill *et al.*, 1998; Adányi *et al.*, 1999; Heim *et al.*, 1999; Rasooly and Rasooly, 1999; Qin *et al.*, 2000; D'Souza, 2001; Schroth *et al.*, 2001; Berney *et al.*, 2000).

Sampling systems (e.g., Flow Injection Analysis) are necessary for the on-line utilization of the biosensor, as its in-situ use is restricted by the fact that the sterilization procedures do not preserve the biological integrity of the biosensor. Integrated FIA-Biosensor systems have been extensively reported in the literature (Adányi *et al.*, 1999; Milardovic *et al.*, 1999;, Min *et al.*, 1999; Folly *et al.*, 1996; Weigel *et al.*, 1996), however only a very small number was effectively employed for automated control purposes (Ferreira *et al.*, 2001a; Hitzmann *et al.*, 2000).

The output signal of the FIA-Biosensor system is influenced by several parameters including the carrier rate and concentration, pH, temperature and composition of the sample. For that reason, a supervisory control, able to systematically analyze the data and detect instrument failures, should also be available (Olsson *et al.*, 1998, Linko, 1998, Lohn and Hitzmann, 1999). This work investigates the utilization of a lactose biosensor, integrated to a FIA system, for the on-line monitoring and control of a cultivation process. A dynamical model able to describe the whole *measurement system* is also formulated. This model allows the automated correction of the measurements for several parameters.

2. FIA-BIOSENSOR SYSTEM

The technology of the biosensor employed in this work was developed by Weigel (1995). It is composed by two enzymes, β-galactosidase (β-gal) and glucose oxidase (GOD), which catalyze the following enzymatic reactions:

$$\text{Lactose} + H_2O \xrightarrow{\beta\text{-gal}} \text{D-glucose} + \text{D-Galactose} \quad (1)$$

$$\text{D-glucose} + O_2 \xrightarrow{GOD} \text{Gluconic Acid} + H_2O_2 \quad (2)$$

The oxygen consumption was measured and related to the lactose concentration in the sample. Fig. 1 shows the calibration curve to each microreactor. The responses are in volts as the amperometric electrode was coupled with a voltage converter (Wheatstone Bridge). The lowest concentration measured was 1 g/L.

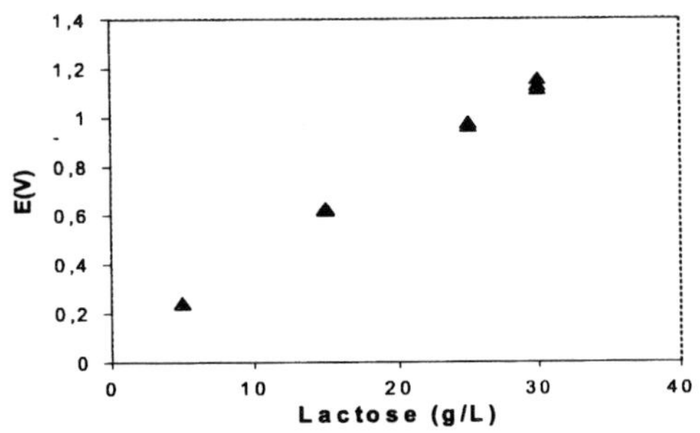

Fig. 1. Calibration curve for lactose measurements.

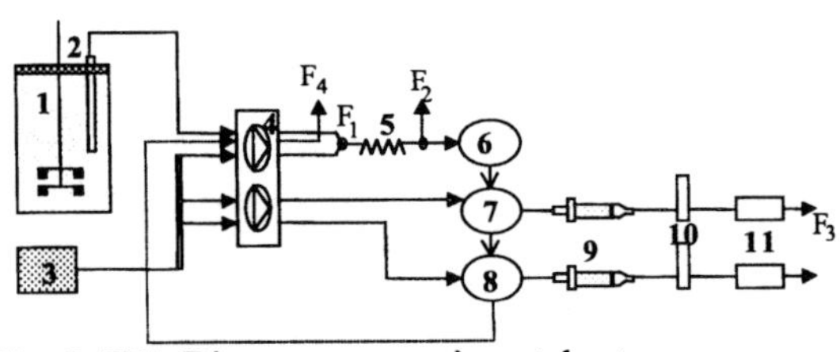

Fig. 2. FIA-Biosensor experimental set-up.

The experimental set-up is shown in Fig. 2. A sample is withdrawn from the bioreactor(1) through a micro-filtration membrane(2), being diluted immediately after the pump(4). After passing by a mixer(5), part of the mixture is discarded and the remainder follows to a selection valve(6). The calibration solutions (glucose and lactose) are connected to it, allowing to make recalibrations during the cultivation. From the selection valve, the sample goes to the injection valves(7 and 8). In this valve, two positions are possible: **L**(load) or **I** (injection). On **L** position, only the sample passes through the injection loop and the buffer(3) is directly sent to the lines of the biosensors(10) and amplifiers(11). On **I** position, the buffer goes through the injection loop, carrying the sample to the biosensors.

3. "FIA-BIOSENSOR" SYSTEM TO ON-LINE MONITORING AND CONTROL

3.1 Materials and Methods

Cultivations were conducted in BIOSTAT B (B. Braun) bioreactors, in a 5 L vessel, with 600 rpm agitation and 8 L/min input air flow rate. Temperature and pH were respectively controlled at 37°C and 5.5. These operating conditions were determined by Rech (1998). *Kluyveromyces marxianus* CBS 6556 yeast was the microorganism used for the cultivations. The culture media was composed of reconstituted powdered cheese whey (7% w/v) and yeast extract (0.5% w/v). The feed stream of the fed-batch was composed of cheese whey (28% w/v) and yeast extract (2% w/v). Cells were measured by optical density and lactose is measured by the DNS Method (Chaplin and Kennedy, 1994). The CAFCA software (ANASYSCON) completely automated the sampling and analysis measurement.

3.2 Controlled Fed-batch Cultivation.

The FIA-biosensor system provided signals related to the lactose concentration in the bioreactor, which were used to control the fed-batch cultivation manipulating the substrate feed rate. Effective integration of biosensors in control schemes are still very rare in the literature. To the knowledge of the authors, only one previous work (Hitzmann *et al.*, 2000) effectively used biosensors in a closed control loop. In the cited work (Hitzmann *et al.*, 2000), glucose concentration was controlled during a fed-batch cultivation, using a glucose biosensor and also a Kalman Filter to estimate cell concentration and the parameter μ_{max} of the Monod model.

In the present contribution, the measurements from the FIA-biosensor system were used by a PI controller implemented in MATLAB (gain K_c=0.025 g h/L^2 , reset time τ_i=2.8 h) to control the lactose concentration at 2 g/L (set-point). This value was chosen because it is known (Rech, 1998) that, in this cultivation, high enzyme activity rates are reached at low substrate concentrations. The results for the controlled variable are shown in Fig. 3. It can be seen that the lactose concentration was lead to the set-point in an oscillating fashion. The following reasons may have contributed for this behavior: (1) the PI was tuned by trial and error based on simulations of a model (Secchi *et al.*, 2001), which did not provide a good representation for fed-batch operation (because the model was originally fitted for batch operation mode); (2) the system's intrinsic noise disturbed the controller.

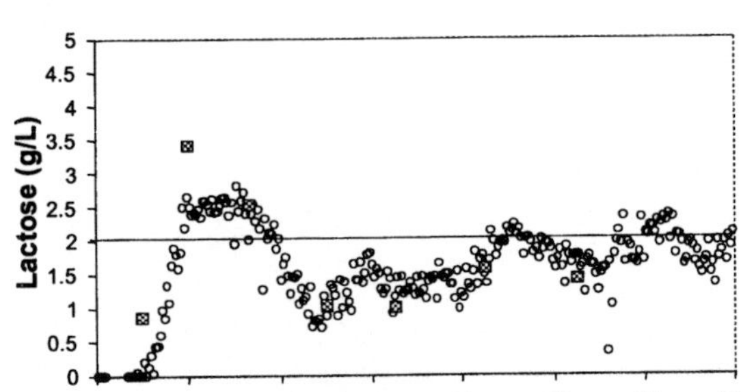

Fig. 3. Measurements of lactose concentration, on and off-line, during the control phase.

In one of the cultivations, at the beginning of a batch operation, the biosensor erroneously indicated that the lactose concentration was increasing, instead of decreasing. These results should not have been accounted for, as the temperature increased gradually during the first hours of the experiment. When the room temperature was kept constant and the system was recalibrated, the results coherently agreed with the off-line measurements. This fact clearly indicates that the temperature has a strong influence on the biosensor measurements. Hence, either the FIA-biosensor system should be thermally isolated or a compensation should be sought for the temperature effects. The second option is addressed here as presented in the sequel.

4. MODELING THE FIA-BIOSENSOR SYSTEM IN THE PRESENCE OF TEMPERATURE EFFECTS

Fig. 4 shows the behavior of the output signal at different temperatures. The data plotted are the media and standard deviation of three experiments, to each temperature and concentration.

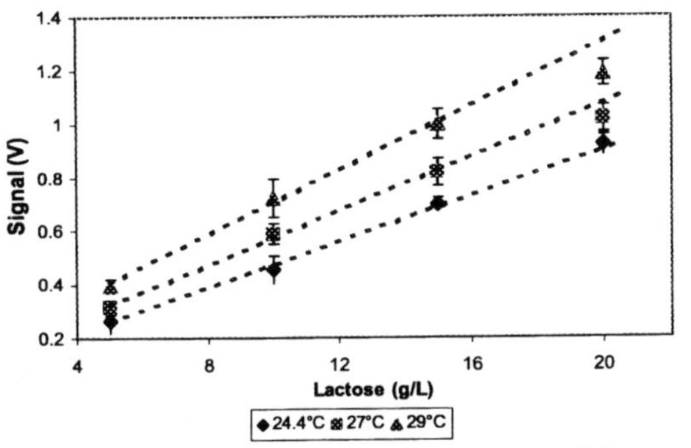

Fig. 4. Temperature influence on biosensor measurements.

It can be depicted from the figure that the temperature significantly affects the signal, imposing a nonlinear behavior which is stronger at highest temperatures and concentrations. If this effect is not considered, the sensor measurement will not be precise nor reliable. A compensation for the temperature effect was studied previously by the authors of this paper, using empirical correlations based on nonlinear regression and neural networks (Ferreira *et al.*, 2001b). In the present contribution, physical and chemical phenomena are considered in the development of the model allowing to include temperature effects. It must be emphasized that this approach is innovative, being not found in the present literature.

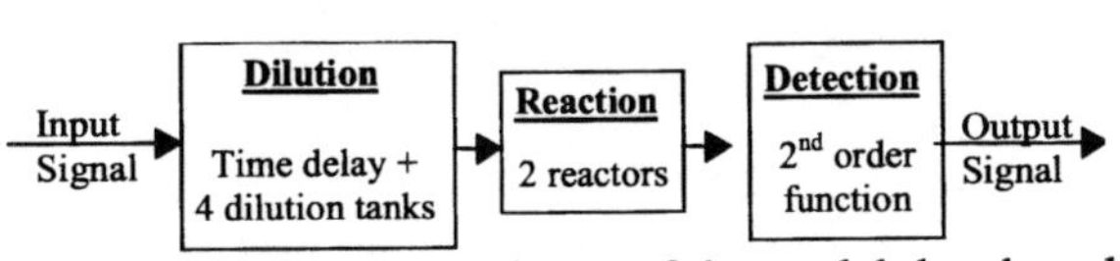

Fig. 5. Representative scheme of the model developed for the 'FIA-Biosensor' system.

The system was modeled taking the following elements into account (see Fig. 5): (i) dilution of the sample; (ii) enzymatic reaction in the biosensor; and, (iii) detection by the oxygen electrode.

4.1 Representation of the input signal

Even though the working characteristics of the injection valve make the input signals have a pulse shape, as the duration of the injection (0,0194 min) is very short, the input signal could be represented by an impulse function.

4.2 Modeling of the dilution line

The dilution line, that carries the sample to the biosensor, induces the concentration gradient responsible for the characteristic shape of the FIA's peaks (Fig. 8): the delay in the response of the sensor and the axial diffusion. In order to impart these characteristics to the model of the dilution line, n tanks were considered in series with a pure time delay element, as illustrated by Fig. 6.

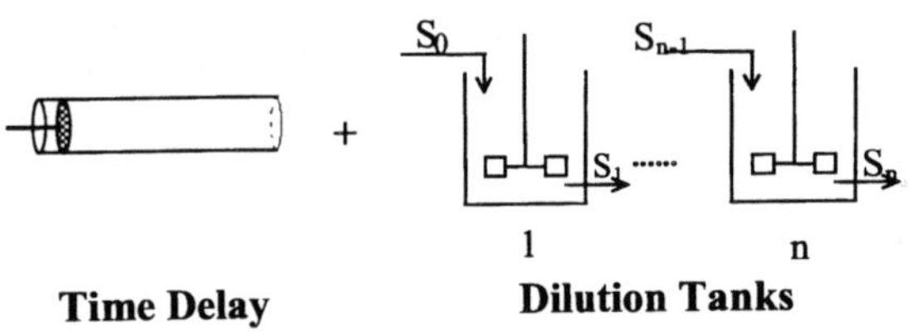

Fig. 6. Schematic representation of the dilution line, where n is the number of tanks.

The transfer function relating the output lactose concentration (S_n) to the input lactose concentration (S_0), in the dilution line, is:

$$\frac{S_n(s)}{S_0(s)} = e^{-t_d s} / (\tau s + 1)^n \qquad (3)$$

where s is the complex variable in the Laplace domain.

This approach allows to develop an analytical model to the dilution line by the Inverse Laplace Transform. According to the experimental data, the time delay for temperatures 27 and 29°C is 50 sec. For 24.4°C, this value is smaller, 45 sec. The dilution line is represented by 4 tanks in series, with a time constant of 0.117 min.

4.3 Modeling the enzymatic microreactor

As the biosensor is an enzymatic reactor, models describing the hydrolysis of lactose by immobilized enzymes were reviewed. In Ladero *et al.* (2001b) several enzymatic kinetic models were studied and a model based on the Michaelis-Menten equation, with competitive inhibition by galactose, proved to be most suitable. This model is described by the scheme and equations below:

$$E + S \rightarrow ES \rightarrow E + P$$
$$E + P \rightarrow EP$$

$$r_S = -kC_E S \Big/ \left(K_m \left(1 + \frac{P}{K_i} \right) + S \right) \qquad (4)$$

where:
S [g/L]: lactose concentration ; P [g/L]: product that inhibits the main reaction (glucose and/or galactose); C_E: concentration of the immobilized enzyme (mg of enzyme/g of support); k, K_m and K_i parameters of the Michaelis-Menten Model, including inhibition.

Ladero *et al.* (2000) defined temperature dependent parameters according to Arrhenius equation and estimated their values using experimental hydrolysis data. The case studied in this paper is similar to the one of Ladero *et al.*(2000), with the difference that, beyond β-galactosidase, glucose oxidase is also present. However, in the development of the present model it was considered that glucose was rapidly consumed. Hence, only the lactose hydrolysis by the β-galactosidase enzyme is taken into consideration.

Because of the characteristics of FIA, the sample injected to the system reaches the biosensor diluted approximately by a factor of 36 (D=36 min^{-1}). Consequently, the amount of galactose produced by the 1st reaction is very small (<< K_i) and competitive inhibitory effects were discarded from the model. Each reactor was modeled as a CSTR, according to equation 5.

$$\frac{dS_R}{dt} = D_R(S_n - S_R) - kC_E S_R / K_m + S_R \qquad (5)$$

where S_R and S_n are the lactose concentrations in the reactor and in the feed, respectively; D_R is the dilution rate of the reactor.

After linearization and application of Laplace Transform to the previous equation, a transfer function is obtained for each reactor (eq.6).

$$\frac{S_R'(s)}{S_n'(s)} = \frac{D_R}{\left(s + D_R + \dfrac{kC_E K_m}{(K_m + S_e)^2} \right)} \qquad (6)$$

where S_e is the lactose concentration at initial steady state.

The reaction was represented by two microreactors in series, each one having a volume of 0.05 mL. The concentration of the immobilized enzyme is 37.17 mg of enzyme/g of support. At the beggining, only the buffer (carrier) reaches the sensing element, then S_e was considered equal to zero. The kinetic parameters were estimated for each studied temperature(see.Table 1) and are represented by the following equations.

$$\ln k = \frac{-4384.3}{T} + 19.636 \qquad (7)$$

$$K_m = -3.378T + 1032.94 \qquad (8)$$

$$\ln K_i = \frac{-4521{,}7}{T} + 19.885 \qquad (9)$$

Table 1. Parameters to the reactor kinetic model.

T (°C)	24.4	27	29
k [min^{-1}]	135.5	150	170
K_m [g/L]	28	18.8	12.5
K_i [g/L]	108.91	123.44	137.35

4.4 Modeling of the oxygen electrode

An YSI oxygen electrode was used to measure the oxygen consumption. It was mentioned in the manufacturer's manual that steady state should be reached in 30 seconds and that 90 % of the response should be completed in 10 seconds. A transfer function describing the dynamics of the electrode was built based on these data, as shown in Fig. 7. The term 1/B describes the gain and converts the concentration input to a signal, in volts. The resulting function G can be modeled by transfer function in parallel, as shown in Fig. 7, where G_1 and G_2 exhibit slow and fast dynamics, respectively.

$$G = G_1 + G_2 = (8s+1)/(20s^2 + 12s + 1) \qquad (10)$$

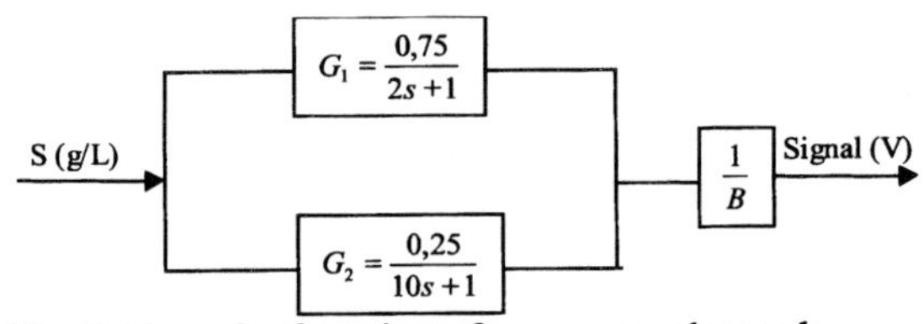

Fig. 7. Transfer functions for oxygen electrode.

4.5 Model Simulation

The final transfer function model can be derived from the product of the transfer functions of all components of the FIA-biosensor system:

$$G_{total} = G_{dilution}\, G_{reactor}\, G_{electrode} \qquad (11)$$

G_{total} is an 8th order transfer function. Fig. 8 shows the simulation of the model for the temperature of 24.4°C.

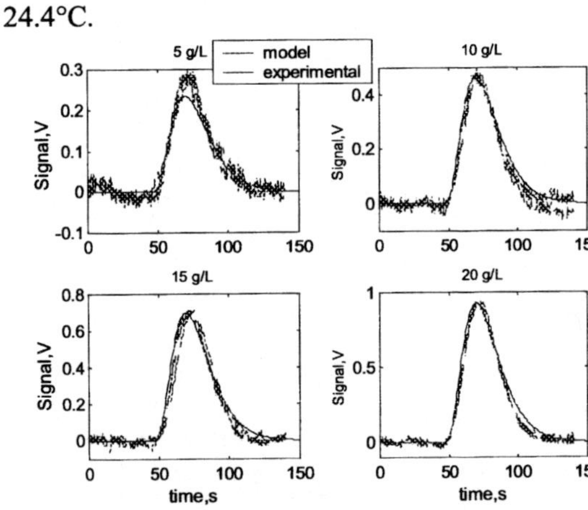

Fig. 8. Results of model simulation to T = 24.4°C.

The results of the simulation to the other temperatures are similar to ones shown in Fig. 8. However, a minor deviation in the model prediction at the end of the FIA cycle appears due to the increase of the non-linearity at high temperatures and concentrations.

4.6 *Innovative approach of temperature compensation*

The analytical solution of the model can be derived by the application of the inverse Laplace transform of eq. 11, which explicitly relates the input signal (S_0), to the temperature and the output signal (S). The literal analytical solution is an equation composed of eight terms (Ferreira, 2002). The analytical solutionis described in the Appendix. This approach allows the implementation of an automatic temperature compensation as schematized in Fig. 9. The output signal of the FIA-Biosensor system, represented by **S**, is analyzed (by the height of the peak, in this study). If a disturbance is detected in the temperature (which must be measured on-line), the module of compensation is fired; on the opposite case, only the calibration curve is used. The term $f(t,p(T),\tau)$ represents the exponential terms in the equation of the Appendix. According to the methodology explained, the experimental values were supplied as inputs (**S**) to the module of compensation and the results obtained (**S_0**) can be seen in Table 2.

Table 2. Results of temperature compensation.

S_0 (g/L)	24.4°C	27 °C	29°C
5	5.81	5.59	5.92
10	9.87	10.4	10.8
15	15.1	14.5	14.9
20	20.0	18.0	17.8

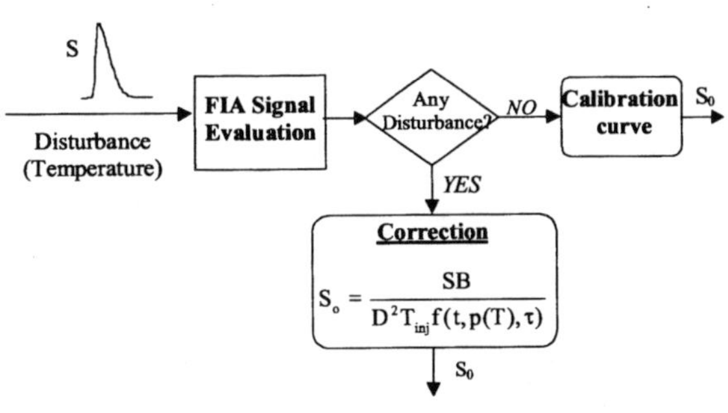

Fig. 9. Implementation proposal to compensate on-line temperature effects.

The following aspects of the proposed phenomenological model should be emphasized: (i) the use of the module of compensation can be extended to any variable considered in the model; for example carrier rate, time delay or injection time. Hence, any disturbance in these variables can be compensated; and, (ii) the dynamic behavior of the FIA-Biosensor system is completely modeled, what is important from the control point of view.

5. CONCLUSION

The following experimental characteristics of the biosensors were evidenced in this investigation:

- Short time for a complete analysis (about 2 minutes);
- Stability of the biosensor after almost 7000 measurements.

These features indicate biosensors as suitable devices for on-line monitoring of bioprocesses. Even so, based on the experimental experience acquired, the authors suggest for the case of industrial applications: (1) to follow the viability of the membrane that covers the oxygen electrode; (2) to measure the disturbances of the system and compensate them using models of the system; (3) to proceed frequent and automated recalibrations of the system (to avoid systematic errors caused by unmeasured disturbances); and, (4) to provide – in addition to recalibration – a failure detection scheme for the sensor, mainly when the system is being used for control. On-line and precise measurements make the implementation of the control substrate viable, allowing better conditions for the microorganisms to produce lactase. A first attempt to control this process is presented here. The initial results show that it is possible to use the biosensor for that purpose. However, additional studies should be endeavored to improve the performance of the controller.

ACKNOWLEDGMENTS

The authors are particularly grateful to Profs. Marco Antônio Z. Ayub and Rosane Rech, from Instituto de Ciência e Tecnologia de Alimentos of UFRGS, and also to Prof. Bernd Hitzmann and graduate student Olaf Broxtermann for technical help and discussions. To CNPq for the research grant. To FAPERGS for financial support and to the PROBRAL 199/00 Program (CAPES/DAAD) for the visiting fellowship during this work.

REFERENCES

Adányi N. , E.E. Szabó and M. Váradi (1999). Multi-enzyme biosensors with amperometric detection for determination of lactose in milk and dairy products. *Eur. Food Res. Technol.*, **209**, pp. 220 – 226.

Berney, H., J. West, E. Haefele, J. Alderman, W. Lane and J.K. Collins (2000). A DNA diagnostic biosensor: development, characterization and performance, *Sensors and Actuators B*, **68**, pp. 100 – 108.

Chaplin M.F. and J.F. Kennedy (1994). In: *Carbohydrate Analysis: A Practical Approach.* 2nd edition, IRL Press, Oxford.

D'souza, S.F. (2001). Microbial Biosensors, *Biosensors and Bioelectronics*,**16**, pp. 337 – 353.

Ferreira L.S., M.B. De Souza Jr. and R.O.M. Folly (2001a). Development of an alcohol fermentation control system based on biosensor measurements interpreted by neural networks, *Sensors and Actuators B*, **75**, pp. 166 – 171.

Ferreira L.S., M.B. De Souza Jr., J.O. Trierweiler, O. Broxtermann, R.O.M. Folly, B. Valdman and B. Hitzmann (2001b). *Aspects concerning the use of biosensors for process control: experimental and simulation investigations, Annals of Pan-American Workshop CEPAC, Guarujá, São Paulo, Brazil, Sept. 19-21.* .

Ferreira, L.S (2002). In: *Investigação de um sistema FIA-Biossensor para monitoração e controle.* Rio de Janeiro: EQ/UFRJ, Dissertation (in portuguese).

Folly R.O.M., B. Valdman, F. Valero and C. Solá (1996). Potentiometric sensor for on line glucose determination, *Biotechnology Techniques*, **10**, pp. 867 – 870.

Gill A., D.G. Bracewell, C.H. Maule, P.A. Lowe and M. Hoare (1998). Bioprocess monitoring: an optical biosensor for rapid bioproduct analysis, *Journal of Biotechnology*, **65**, pp. 69 – 80.

Heim S., I. Schnieder, D. Binz, A. Vogel and U. Bilitewski (1999). Development of an automated microbial sensor system, *Biosensors and Bioelectronics*, **14**, pp. 187 – 193.

Hitzmann B., O. Broxtermann, Y.-L. Cha, O. Sobieh, E. Stärk and T. Scheper (2000). The control of glucose concentration during yeast fed-batch cultivation, *Bioprocess Engineering*, **23(4)**, pp. 337-341.

Olsson, L., U. Schulze and J. Nielsen (1998). On-line bioprocess monitoring - an academic discipline or an industrial tool?, *Trends in Analytical Chemistry*, **17(2)**, pp. 88 – 95.

Ladero, M., A. Santos and F. Garcia-Ochoa (2000). Kinetic modeling of lactose hydrolysis with an immobilized beta-galactosidase from *Kluyveromyces fragilis*, *Enzyme and Microbial Technology*, **27 (8)**, pp. 583-592.

Linko, S. (1998). Expert systems - what can they do for the food industry?, *Trends in Food Science & Technology*, **9**, pp. 3 – 12.

Löhn, A. and B. Hitzmann (1999). A knowledge-based system for real-time validation of calibrations and measurements, *Chemometrics and Intelligent Laboratory Systems*, **46**, pp. 57 – 66.

Milardovic, S., I. Kruhak and B.S. Grabaric (1999) Urea determination in FIA mode by a newly designed urease-based biosensor, *Laboratory Robotics and Automation*, **11(5)**, pp. 266 – 271.

Min, R.W., V. Rajendran, N. Larsson, L. Gorton, J. Planas and B.H. Hägerdal (1998). Simultaneous monitoring of glucose and L-lactic acid during a fermentation process in an aqueous two-phase system by on-line FIA with microdialysis sampling and dual biosensor detection, *Analytica Chimica Acta*, **366**, pp. 127 – 135.

Qin W., A. Zhang and Y. Peng (2000). Plant tissue-based chemiluminescence flow biosensor for urea, *Analytica Chimica Acta*,**407**, pp. 81 – 86.

Rasooly L. and A. Rasooly (1999). Real time biosensor analysis of Staphylococcal enterotoxin A in food, *International Journal of Food Microbiology*, **49,** pp. 119 – 127.

Rech R (1998). *Aproveitamento do soro de queijo para a produção de lactase por Kluyveromyces marxianus.* Porto Alegre: DEQUI/UFRGS. Dissertation (in portuguese).

Schroth, P., M.J. Schoring, H. Luth, B. Weissbecker, H.E. Hummel and S. Schutz (2001). Extending the capabilities of an antenna/chip biosensor by employing various insect species, *Analytica Chimica Acta,* **78**, pp. 1 – 5.

Secchi, A.R., L.G.S. Longhi, P.R. Barrera, J.O. Trierweiler, R. Rech, M.A.Z. Ayub, L.S. Ferreira and M.B. de Souza Jr.*(2001). Lactase Production From Whey Cheese By Kluyveromyces Marxianus Yeast. Part II: Kinetic Modeling, Annals of 3rd Mercosur Congress in Process Engineering, Santa Fe, Argentine, Sept. 16-20.*

Vega, F.A., C.G. Núnez, B. Weigel, B. Hitzmann and J.C.D. Ricci (1998). On-line monitoring of galactoside conjugates and glycerol by Flow Injection Analysis, *Analytica Chimica Acta*, **373**, pp. 57 – 62.

Weigel B (1995). *Entwicklung von Enzym-Kartuschen und FIA-Methoden zur On-line-Analytik in Biomedien.* Hannover: Uni-Hannover/IFTC. Thesis (in german).

Weigel B., B. Hitzmann, G. Kretzmer, K. Schugerl, A. Huwig and F. Giffhom (1996). Analysis of various sugars by means of immobilized enzyme coupled flow injection analysis, *Journal of Biotechnology*, **50, pp.** 93 - 106.

APPENDIX

Analytical solution equation of the phenomenological model.

$$S(t)=\frac{D_R{}^2\cdot S_0\cdot T_{inj.}}{B}\left\{\left[\frac{\frac{1}{6}(t-t_d)^3}{\tau^2(p\tau-1)^2}-\frac{(t-t_d)^2}{\tau(p\tau-1)^3}+\frac{3\tau}{(p\tau-1)^4}-\frac{4\tau}{(p\tau-1)^5}\right]e^{\frac{-(t-t_d)}{\tau}}\right\}+\left\{\frac{(t-t_d)}{(p\tau-1)^4}+\frac{4\tau}{(p\tau-1)^5}\right\}e^{-p(t-t_d)} \quad \text{(A1)}$$

where: $p=D_R+\frac{kK_m}{(K_m+S_e)^2}$; t, time; t_d, time delay; τ, time constant of dilution tanks; S_0, sample concentration of lactose; D, dilution rate.

ELSEVIER
IFAC
PUBLICATIONS
www.elsevier.com/locate/ifac

A FAST COMPUTATIONAL PROCEDURE FOR THE PREDETERMINATION OF PARAMETERS IN NON-LINEAR BIOPROCESS MODELS

Jens E. Haag * **Alain Vande Wouwer** *,[1]

* *Service d'Automatique, Faculté Polytechnique de Mons, Boulevard Dolez 31, B-7000 Mons, Belgium*

Abstract: In this study, a semi-analytical computational procedure, which allows bioprocess model parameters to be quickly evaluated from experimental data, is developed and illustrated with an application example. The approximate model can be used to investigate the qualitative behavior of key components of interest and to check modeling assumptions. The approximate model parameter values can also be used as starting points for more rigorous identification methods. *Copyright © 2004 IFAC*

Keywords: modeling, parameter estimation, biotechnology

1. INTRODUCTION

Modeling and parameter identification of bioprocesses are notoriously difficult tasks, due to the following issues:

- the lack of experimental data, i.e. the number of samples and measurements is often restricted by time- and money-consuming laboratory analysis;
- the difficulty to perform experiments in a reproductible way, due to the large number of factors influencing cell growth;
- the lack of a priori knowledge on the biological reaction scheme, in terms of yield coefficients and kinetics;
- the potential model complexity, i.e. the potentially large number of involved components and number of reactions, and nonlinearity (mostly involved in the kinetics).

Once a model structure and parametrisation has been selected, it is required to infer the numerical values of the model parameters from experimental data. Recently, a systematic identification procedure has been proposed in (Bogaerts and Hanus, 2000), which is based on the state transformation originally introduced in (Bastin and Dochain, 1990), and which allows the yield coefficients to be estimated independently from the kinetic parameters. This procedure takes all the measurement errors into account through the formulation of maximum-likelihood criteria. Even though very efficient, this procedure can sometimes be time-consuming, and there is still a need for fast data evaluation and modeling procedure.

In this connection, this study aims at developing a semi-analytical procedure which allows a fast predetermination of model parameters. To allow analytical developments to be performed, some simplifying assumptions are required, which makes the problem solution approximate. However, a first evaluation of the model parameters can be very useful to check basic modeling assumptions, and to investigate the qualitative behavior of the key components of interest. In addition, the estimated parameter values can serve as starting points for nonlinear parameter identification approaches as those proposed in (Bogaerts and Hanus, 2000).

This paper is organized as follows. The next section deals with the devolopment of a computational procedure for the predetermination of parameters in

[1] Corresponding author, fax: +32 65 37 4136, e-mail: Alain.VandeWouwer@FPMs.ac.be

bioprocess models whereas section 3 summarizes the results. Section 4 presents a simple numerical application. Finally, section 5 is devoted to some concluding remarks.

2. PREDETERMINATION OF MODEL PARAMETERS

In order to develop an analytical procedure, we will make the assumption that the specific production/consumption, growth and death rates are time-invariant in a finite time interval, e.g. the first sampling interval $[t_0,t_1]$ (in bioprocess applications, this time interval can be relatively long, e.g. a few hours). This assumption allows the following dynamic model to be written

$$\dot{c} = \bar{\nu} x_v + D(c_f - c); \qquad c(t_0) = c_0 \tag{1}$$

$$\dot{x}_v = (\bar{\mu} - \bar{k}_d)\, x_v - D_x x_v; \quad x_v(t_0) = x_{v0} \tag{2}$$

$$\dot{x}_d = \bar{k}_d x_v - D_x x_d; \qquad x_d(t_0) = x_{d0} \tag{3}$$

In these equations, the components c are produced or consumed with an average specific rate $\bar{\nu}$ by the viable cells x_v. The latter grow at an average specific growth rate $\bar{\mu}$ and viable cells turn into dead cells x_d with a mean mortality rate $\bar{k}_d$.

D and D_x are dilution rates of the dissolved components c and the biomass x, respectively. In order to be able to describe various operating modes – batch, fed-batch, continuous and perfusion – we assume also that the reactor can be fed with fresh growth medium and that material is withdrawn from the reactor in several ways. Particularly, in a perfusion mode biorecator, the biomass can be retained inside the bioreactor thanks to a filtering device, while the growth medium can be continuously replaced. The feed flow rate and the withdrawal rates of filtered and unfiltered culture medium are denoted Q_{in}, $Q_{out,c}$ and $Q_{out,x}$, respectively. The dilution rates are then defined as

$$D(t) = \frac{Q_{in}(t)}{V(t)} \tag{4}$$

$$D_x(t) = \frac{Q_{in}(t) - Q_{out,c}}{V(t)}. \tag{5}$$

The general solution of the system (1)–(3) is written:

$$c(t) = \left(c_0 + \int_{t_0}^{t} (\bar{\nu} x_v(\tau) + D(\tau) c_f(\tau))\, e^{\int_{t_0}^{\tau} D(\tilde{\tau}) d\tilde{\tau}} d\tau \right) \cdot e^{-\int_{t_0}^{t} D(\tau) d\tau} \tag{6}$$

$$x_v(t) = x_{v,0} e^{(\bar{\mu} - \bar{k}_d)(t - t_0) - \int_{t_0}^{t} D_x(\tau) d\tau} \tag{7}$$

$$x_d(t) = \left(x_{d,0} + \bar{k}_d \int_{t_0}^{t} x_v(\tau) e^{\int_{t_0}^{\tau} D_x(\tilde{\tau}) d\tilde{\tau}} d\tau \right) \cdot e^{-\int_{t_0}^{t} D_x(\tau) d\tau}. \tag{8}$$

Fitting the data triplet $\{c_1, x_{v,1}, x_{d,1}\}$ at the instant t_1 yields the following specific mean rates:

$$\bar{\mu}^* = \frac{\ln\left\{ \frac{x_{v,1}}{x_{v,0}} \right\} + \int_{t_0}^{t_1} D_x(\tau) d\tau}{t_1 - t_0} \tag{9}$$

$$\bar{k}_d = \frac{x_{d,1} \exp\left\{ \int_{t_0}^{t_1} D_x(\tau) d\tau \right\} - x_{d,0}}{\int_{t_0}^{t_1} x_v(\tau) \exp\left\{ \int_{t_0}^{\tau} D_x(\tilde{\tau}) d\tilde{\tau} \right\} d\tau} = \frac{x_{d,1} \exp\left\{ \int_{t_0}^{t_1} D_x(\tau) d\tau \right\} - x_{d,0}}{x_{v,1} - x_{v,0}} \bar{\mu}^* \tag{10}$$

$$\bar{\mu} = \bar{\mu}^* + \bar{k}_d \tag{11}$$

$$\bar{\nu} = \frac{c_1 \exp\left\{ \int_{t_0}^{t_1} D(\tau) d\tau \right\} - c_0}{\int_{t_0}^{t_1} x_v(\tau) \exp\left\{ \int_{t_0}^{\tau} D(\tilde{\tau}) d\tilde{\tau} \right\} d\tau} - \frac{\int_{t_0}^{t_1} D(\tau) c_f(\tau) \exp\left\{ \int_{t_0}^{\tau} D(\tilde{\tau}) d\tilde{\tau} \right\} d\tau}{\int_{t_0}^{t_1} x_v(\tau) \exp\left\{ \int_{t_0}^{\tau} D(\tilde{\tau}) d\tilde{\tau} \right\} d\tau} \tag{12}$$

These calculations can be applied to each pair of consecutive experimental data points, for each time interval $[t_{i-1}, t_i]$, $i = 1, \ldots n_t$. Therefore, an approximate time evolution of the consumption and production rates can be determined.

The problem formulation becomes more complicated, if the reactor medium is diluted at discrete times with fresh medium. This corresponds however to common practice, where a certain volume $V_{out,i}$ is withdrawn from the reactor at the time instant t_i, and a volume $V_{in,i}$ of fresh medium with the concentration $c_{in,i}$ is added to the remaining volume in the reactor.

The algebraic relationships between the volume and the concentrations before and after the dilution are written:

$$V_r(t_i) = V_{r,i} - V_{out,i} + V_{in,i} \tag{13}$$

$$c(t_i) = \frac{(V_{r,i} - V_{out,i})\, c_i + V_{in,i} c_{in,i}}{V_r(t_i)} = \alpha_i c(t_i) + (1 - \alpha_i c_{in,i}) \tag{14}$$

$$x_v(t_i) = \frac{(V_{r,i} - V_{out,i})\, x_{v,i}}{V_r(t_i)} = \alpha_i x_{v,i} \tag{15}$$

$$x_d(t_i) = \frac{(V_{r,i} - V_{out,i})\, x_{d,i}}{V_r(t_i)} = \alpha_i x_{d,i}, \tag{16}$$

where $V_{r,i}$, c_i, $x_{v,i}$ and $x_{d,i}$ are the reactor volume and concentrations just before the dilution.

To take these discrete-time dilutions into account, we reformulate our time notations and consider that $[t_0, t_n]$ represents a sampling interval with the first measurement c_0 being available at t_0 and the second measurement c_n being available at t_n. In between, there are n dilutions according to (14) at the discrete times instants $t_0, \ldots, t_{n-1}$ (it is therefore assumed that the sample c_n is taken just before the dilution at t_n); see figure 1.

The evolution of the concentration on the interval $[t_{i-1}, t_i]$ is written according to (7):

$$c_i = \Bigg(c(t_{i-1}) + \int_{t_{i-1}}^{t_i} (\bar{\nu} x_{\mathrm{v}}(\tau) + D(\tau) c_{\mathrm{f}}(\tau))\, e^{\int_{t_{i-1}}^{\tau} D(\tilde{\tau})\mathrm{d}\tilde{\tau}} \mathrm{d}\tau \Bigg) \cdot e^{-\int_{t_{i-1}}^{t_i} D(\tau)\mathrm{d}\tau}$$
$$= \Bigg(\alpha_{i-1} c_{i-1} + (1-\alpha_{i-1})\, c_{\mathrm{in},i-1} + \int_{t_{i-1}}^{t_i} (\bar{\nu} x_{\mathrm{v}}(\tau) + D(\tau) c_{\mathrm{f}}(\tau))\, e^{\int_{t_{i-1}}^{\tau} D(\tilde{\tau})\mathrm{d}\tilde{\tau}} \mathrm{d}\tau \Bigg) \cdot e^{-\int_{t_{i-1}}^{t_i} D(\tau)\mathrm{d}\tau} \quad (17)$$

for $i = 1, \ldots, n$, and the final concentration c_n is consequently determined recursively from the initial concentration c_0 of the sampling time interval. Using the substitutions

$$E_D(t_i, t_{i-1}) = \exp\left\{ -\int_{t_{i-1}}^{t_i} D(\tau)\mathrm{d}\tau \right\} \quad (18)$$
$$I_{x,c}(t_i, t_{i-1}) = \int_{t_{i-1}}^{t_i} x_{\mathrm{v}}(\tau) \exp\left\{ \int_{t_{i-1}}^{\tau} D(\tilde{\tau})\mathrm{d}\tilde{\tau} \right\} \mathrm{d}\tau \quad (19)$$
$$I_u(t_i, t_{i-1}) = \int_{t_{i-1}}^{t_i} D(\tau) c_{\mathrm{f}}(\tau) \exp\left\{ \int_{t_{i-1}}^{\tau} D(\tilde{\tau})\mathrm{d}\tilde{\tau} \right\} \mathrm{d}\tau, \quad (20)$$

the recursion is written

$$\begin{aligned}
c_n =& \big(\alpha_{n-1} c_{n-1} + (1-\alpha_{n-1})\, c_{\mathrm{in},n-1} \\
&+ \bar{\nu} I_x(t_n, t_{n-1}) + I_u(t_n, t_{n-1})\big) E_D(t_n, t_{n-1}) \\
=& \Big(\alpha_{n-1} \big(\alpha_{n-2} c_{n-2} + (1-\alpha_{n-2} c_{\mathrm{in},n-2}) \\
&+ \bar{\nu} I_x(t_{n-1}, t_{n-2}) + I_u(t_{n-1}, t_{n-2})\big) \\
&\cdot E_D(t_{n-1}, t_{n-2}) \\
&+ (1-\alpha_{n-1} c_{\mathrm{in},n-1}) \\
&+ \bar{\nu} I_x(t_n, t_{n-1}) + I_u(t_i, t_{i-1}) \Big) \\
&\cdot E_D(t_n, t_{n-1}) \\
=& \alpha_{n-1}\alpha_{n-2} c_{n-2} E_D(t_{n-1}, t_{n-2}) E_D(t_n, t_{n-1}) \\
&+ \alpha_{n-1}(1-\alpha_{n-2})\, c_{\mathrm{in},n-2} E_D(t_{n-1}, t_{n-2}) \\
&\cdot E_D(t_n, t_{n-1}) + (1-\alpha_{n-1})\, c_{\mathrm{in},n-1} E_D(t_n, t_{n-1}) \\
&+ \bar{\nu}\big(\alpha_{n-1} I_x(t_{n-1}, t_{n-2}) E_D(t_{n-1}, t_{n-2}) E_D(t_n, t_{n-1}) \\
&+ I_x(t_n, t_{n-1}) E_D(t_n, t_{n-1})\big) \\
&+ \alpha_{n-1} I_u(t_{n-1}, t_{n-2}) E_D(t_{n-1}, t_{n-2}) E_D(t_n, t_{n-1}) \\
&+ I_u(t_n, t_{n-1}) E_D(t_n, t_{n-1}) \\
&\ldots
\end{aligned}$$

and after some further manipulation and recursion

$$\begin{aligned}
c_n =& \prod_{j=0}^{n-1} \alpha_j E_D(t_{j+1}, t_j) c_0 \\
&+ \sum_{j=1}^{n-1} \prod_{k=j}^{n-1} \alpha_k (1-\alpha_{j-1})\, c_{\mathrm{in},j-1} \prod_{k=j-1}^{n-1} E_D(t_{k+1}, t_k) \\
&+ (1-\alpha_{n-1})\, c_{\mathrm{in},n-1} E_D(t_n, t_{n-1}) \\
&+ \bar{\nu} \Bigg(\sum_{j=1}^{n-1} \prod_{k=j}^{n-1} \alpha_k I_{x,c}(t_j, t_{j-1}) \prod_{k=j-1}^{n-1} E_D(t_{k+1}, t_k) \\
&+ I_{x,c}(t_n, t_{n-1}) E_D(t_n, t_{n-1}) \Bigg) \\
&+ \sum_{j=1}^{n-1} \prod_{k=j}^{n-1} \alpha_k I_u(t_j, t_{j-1}) \prod_{k=j-1}^{n-1} E_D(t_{k+1}, t_k) \\
&+ I_u(t_n, t_{n-1}) E_D(t_n, t_{n-1})
\end{aligned} \quad (21)$$

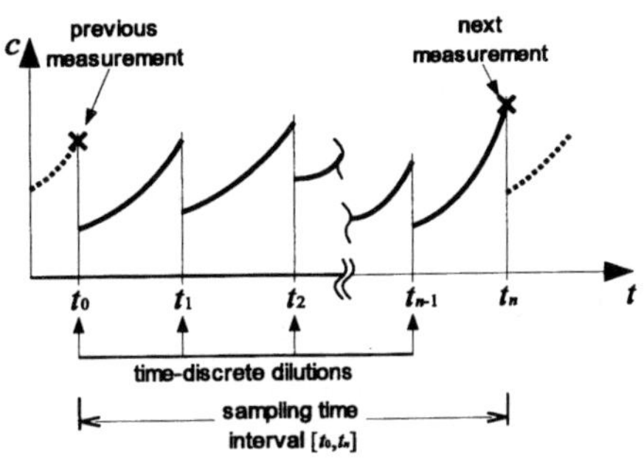

Fig. 1. Several time-discontinuous dilutions in one sampling interval $[t_0, t_n]$.

Analogously, the evolution of the biomass is given by

$$\begin{aligned}
x_{\mathrm{v},n} =& \alpha_{n-1} x_{\mathrm{v},n-1} e^{\bar{\mu}^*(t_n - t_{n-1}) - \int_{t_{n-1}}^{t_n} D_x(\tau)\mathrm{d}\tau} \\
=& \alpha_{n-1}\alpha_{n-2} x_{\mathrm{v},n-2} e^{\bar{\mu}^*(t_{n-1} - t_{n-2}) - \int_{t_{n-2}}^{t_{n-1}} D_x(\tau)\mathrm{d}\tau} \\
&\cdot e^{\bar{\mu}^*(t_n - t_{n-1}) - \int_{t_{n-1}}^{t_n} D_x(\tau)\mathrm{d}\tau} \\
=& \alpha_{n-1}\alpha_{n-2} x_{\mathrm{v},n-2} e^{\bar{\mu}^*(t_n - t_{n-2}) - \int_{t_{n-2}}^{t_n} D_x(\tau)\mathrm{d}\tau} \\
=& \ldots.
\end{aligned}$$

Further manipulation leads to the final equation

$$x_{\mathrm{v},n} = \prod_{j=0}^{n-1} \alpha_j x_{\mathrm{v},0} \exp\left\{ \bar{\mu}^* (t_n - t_0) - \int_{t_0}^{t_n} D_x(\tau)\mathrm{d}\tau \right\}. \quad (22)$$

The evolution equation for the dead biomass x_{d} is structurally identical to the evolution equation (21) for c with $D = D_x$, $\bar{\nu} = \bar{k}_{\mathrm{d}}$ and $c_{\mathrm{f}} = c_{\mathrm{in}} = 0$ instead.

In the case of discrete-time dilution, the more general set of equations for the determination of the mean specific rates in a time interval $[t_0, t_n]$ are therefore written as follows

$$\bar{\mu}^* = \frac{\ln x_{\mathrm{v},n} - \ln x_{\mathrm{v},0} - \sum_{j=0}^{n-1} \ln \alpha_j + \int_{t_0}^{t_n} D_x(\tau)\mathrm{d}\tau}{t_n - t_0} \quad (23)$$
$$\bar{\nu} = \frac{c_n - P_\alpha c_0 - S_u}{S_{x,c}} \quad (24)$$
$$\bar{k}_{\mathrm{d}} = \frac{x_{\mathrm{d},n} - P_\alpha x_{\mathrm{d},0}}{S_{x,x}} \quad (25)$$

with the respective substitutions

$$P_\alpha = \prod_{j=0}^{n-1} \alpha_j E_D(t_n, t_0) \tag{26}$$

$$\begin{aligned} S_u = & \sum_{j=1}^{n-1} \prod_{k=j}^{n-1} \alpha_k \left(I_u(t_j, t_{j-1}) + (1-\alpha_{j-1})\, c_{\text{in},j-1}\right) \\ & \cdot E_D(t_n, t_{j-1}) \\ & + \left(I_u(t_n, t_{n-1}) + (1-\alpha_n)\, c_{\text{in},n-1}\right) \\ & \cdot E_D(t_n, t_{n-1}) \end{aligned} \tag{27}$$

$$\begin{aligned} S_{x,c} = & \sum_{j=1}^{n-1} \prod_{k=j}^{n-1} \alpha_k I_{x,c}(t_j, t_{j-1}) E_D(t_n, t_{j-1}) \\ & + I_{x,c}(t_n, t_{n-1}) E_D(t_n, t_{n-1}) \end{aligned} \tag{28}$$

$$\begin{aligned} S_{x,x} = & \sum_{j=1}^{n-1} \prod_{k=j}^{n-1} \alpha_k I_{x,x}(t_j, t_{j-1}) E_{D_x}(t_n, t_{j-1}) \\ & + I_{x,x}(t_n, t_{n-1}) E_{D_x}(t_n, t_{n-1}). \end{aligned} \tag{29}$$

The equations developed so far are general and can be applied to bioreactors in various operating modes. In perfusion mode, with constant feed concentration c_f, constant perfusion rate D, constant reactor volume V and complete cell retention ($D_x \equiv 0$), the following closed analytical expressions are obtained:

$$\bar{\mu}^* = \frac{\ln x_{v1} - \ln x_{v0}}{t_1 - t_0} \tag{30}$$

$$\bar{k}_d = \frac{x_{d1} - x_{d0}}{x_{v1} - x_{v0}} \bar{\mu}^* \tag{31}$$

$$\bar{\mu} = \bar{\mu}^* + \bar{k}_d \tag{32}$$

$$\bar{\nu} = \frac{(c_1 - c_f)\, e^{Dt_1} - (c_0 - c_f)\, e^{Dt_0}}{x_{v1} e^{Dt_1} - x_{v0} e^{Dt_0}} \left(\bar{\mu}^* + D\right). \tag{33}$$

These mean consumption/production, growth and mortality rates can now be used to make a preliminary estimation of the complete stoichiometry of the reaction system and the respective reaction rates. Thus far, the specific rates for each component are determined independently of each other. However, macroscopic reaction schemes relate the consumption of some reactants with the formation of some products according to the stoichiometry. In other words, the stoichiometric matrix $\Upsilon \in \mathbb{R}^{n_q \times n_r}$ maps the reaction rate vector $\rho \in \mathbb{R}^{n_r}$ into a consumption/production rate vector $q \in \mathbb{R}^{n_q}$ of each component by linear combination of the individual reaction rates of the reaction scheme

$$q = \Upsilon \rho \tag{34}$$

with q and ρ respectively defined as

$$q = \begin{bmatrix} \nu_1 \\ \vdots \\ \nu_{n_q-2} \\ \mu^* \\ k_d \end{bmatrix}; \qquad \rho = \begin{bmatrix} \rho_1 \\ \vdots \\ \rho_{n_r-1} \\ k_d \end{bmatrix}. \tag{35}$$

The consumption/production rates have been determined approximately for each sampling time interval, while the constant stoichiometric matrix and the reaction rates in each sampling time interval are unknown.

To determine these unknown values, a necessary condition is therefore $n_r < n_q$, i.e. the number of reactions in the macroscopic reaction scheme must be lower than the number of components. More precisely, if n_c different concentrations are measured, then $n_t n_c$ data points are available for the determination of n_r reaction rates at each of the n_t sampling intervals and the stoichiometric coefficients.

If the stoichiometric matrix Υ is constructed as follows:

$$\Upsilon = \left[\begin{array}{ccc|c} & Y & & 0 \\ \hline 1 & \cdots & 1 & -1 \\ 0 & \cdots & 0 & 1 \end{array}\right] \tag{36}$$

with the yield matrix $Y \in \mathbb{R}^{(n_q-2)(n_r-1)}$, the total number of unknowns is $n_r n_c + (n_q - 2)(n_r - 1)$. In this expression, it is assumed that the dead biomass is exclusively formed through the mortality reaction, and the stoeciometry is normalized with respect to the viable biomass.

3. A PRACTICAL RECIPE

In the following, a recipe summarizing the results obtained so far is given.

(1) Set up a time vector $t = \left[t_0 \ \cdots \ t_{n_t}\right]^{\mathrm{T}}$ containing all discrete time instants corresponding to sampling and/or dilutions.

(2) Determine the specific mean growth rate $\bar{\mu}^*$ for each time interval $[t_i, t_{i+1}]$, $i = 0, \ldots, n_t - 1$, according to equation (23).

(3) Determine the unmeasured viable cell concentrations at all time instants of t according to

$$x_{v,i+1} = x_{v,i} \alpha_i \exp\left\{ \bar{\mu}^* (t_{i+1} - t_i) - \int_{t_i}^{t_{i+1}} D_x(\tau) d\tau \right\} \quad i = 1, \ldots, n_t - 1, \tag{37}$$

which is easily deduced from (22).

(4) Determine the exponential $E_D(t_{i+1}, t_i)$by equation (18) for all time intervals.

(5) Determine the integrals $I_u(t_{i+1}, t_i)$ and $I_{x,c}(t_{i+1}, t_i)$ by equations (20) and (19) for all time intervals.

(6) Determine the dilution factor $\alpha_i = \frac{V_{r,i} - V_{\text{out},i}}{V_{r,i} - V_{\text{out},i} + V_{\text{in},i}}$ according to its definition in equation (14).

(7) Calculate the specific mean rates $\bar{\nu}_i$ and $\bar{k}_{d,i}$ by equations (24) and (25) and $\bar{\mu}$ by equation (11).

(8) The mean production/consumption, growth and mortality rates can serve for the preliminary estimation of the yield matrix $\hat{Y}$ and the specific reaction rates $\hat{\rho}_i$ subject to equation (34), e.g. by optimization of a weighted least-squares criterion

$$[\hat{Y},\hat{\rho}] = \arg\min_{Y,\rho} \sum_{i=0}^{n_t-1} (\bar{q} - \Upsilon(Y)\rho_i)^T \Omega^{-1} \cdot (\bar{q} - \Upsilon(Y)\rho_i) \quad (38)$$

with a weighting matrix Ω^{-1}, e.g. chosen according to the order of magnitude, $\Omega = \operatorname{diag}\left\{\max_i q_{ji}^2\right\}_j$.

In a second step, the estimated specific reaction rates $\hat{\rho}_i$ for each time interval $[t_i, t_{i+1}]$ can be used to estimate the kinetic parameters. At this stage, the states have to be known, whereas they are available at some discrete time instants only. In order to get a representative value for each interval, a mean value of the initial and final values is evaluated:

$$\bar{c}_i = \frac{c_{i+1} + \alpha_i c_i + (1-\alpha_i)\, c_{\text{in},i}}{2} \quad (39)$$

$$\bar{x}_{v,i} = \frac{x_{v,i+1} + \alpha_i x_{v,i}}{2} \quad (40)$$

$$\bar{x}_{d,i} = \frac{x_{d,i+1} + \alpha_i x_{d,i}}{2}. \quad (41)$$

In this manner, approximate points representing the dependence of the kinetics on the state vector can be collected for all time intervals $[t_{i-1}, t_i]$, $i = 1, \ldots, n_t$, and fit to an assumed structure of the kinetics. This structure is generally written in the following form:

$$\rho(c, x_v, x_d, p_{\text{kin}}) = \begin{bmatrix} \nu(c, x_v, x_d, p_{\text{kin}}) \\ \mu(c, x_v, x_d, p_{\text{kin}}) \\ k_d(c, x_v, x_d, p_{\text{kin}}) \end{bmatrix} \quad (42)$$

Based on this reaction vector, the kinetic parameter vector p_{kin} can be estimated using the corresponding $\hat{\rho}_i$ that has been previously determined for all intervals $i = 1, \ldots, n_t - 1$. For instance, the weighted least-squares criterion

$$\hat{p}_{\text{kin}} = \arg\min_{p_{\text{kin}}} \left\{ \sum_{i=0}^{n_t-1} (\hat{\rho}_i - \rho(\bar{c}, \bar{x}_v, \bar{x}_d, p_{\text{kin}}))^T \Omega^{-1} \cdot (\hat{\rho}_i - \rho(\bar{c}, \bar{x}_v, \bar{x}_d, p_{\text{kin}})) \right\} \quad (43)$$

could be used.

Finally, a complete optimization can be performed with respect to the yield matrix $Y(p_{\text{st}})$ containing the stoichiometric parameters p_{st} as unknown elements and the kinetic parameters p_{kin}, given the specific mean rates $\bar{q}$ and subject to

$$q = \Upsilon(p_{\text{st}})\rho(c, x_v, x_d, p_{\text{kin}}), \quad (44)$$

e.g. using again the weighted least-squares criterion

$$[\hat{p}_{\text{st}}, \hat{p}_{\text{kin}}] = \arg\min_{p_{\text{st}}, p_{\text{kin}}} \left\{ \sum_{i=0}^{n_t-1} (\bar{q} - \Upsilon(p_{\text{st}})\rho(\bar{c}, \bar{x}_v, \bar{x}_d, p_{\text{kin}}))^T \cdot \Omega^{-1} (\bar{q} - \Upsilon(p_{\text{st}})\rho(\bar{c}, \bar{x}_v, \bar{x}_d, p_{\text{kin}})) \right\} \quad (45)$$

Note that relatively small measurement errors directly influence the results of the preliminary data evaluation, and that sufficiently small sampling time intervals are needed in order to avoid a too strong deviation between the real reaction rates and their mean on each time interval (which is basic assumption of the parameter redetermination).

4. A SIMPLE APPLICATION

Consider the following reaction system consisting of two metabolic reactions and one mortality reaction

$$S_1 \xrightarrow{\rho_1} X_v + P \quad (46)$$

$$S_2 \xrightarrow{\rho_2} X_v \quad (47)$$

$$X_v \xrightarrow{\rho_3} X_d \quad (48)$$

involving five components: the substrates S_1 and S_2, the product P and the viable and non-viable biomasses, X_v and X_d, respectively. The specific growth rates are defined by Monod kinetics

$$\rho_1(s_1) = \mu_{1,\max}\frac{s_1}{s_1 + K_1} \quad (49)$$

$$\rho_2(s_2) = \mu_{2,\max}\frac{s_2}{s_2 + K_2} \quad (50)$$

with $\mu_{1,\max} = 0.3$, $K_1 = 1$, $\mu_{2,\max} = 0.2$ and $K_2 = 2$, while the specific mortality rate is constant, $\rho_3 = 0.1$.

The culture is performed in a stirred-tank reactor, first operated in batch mode, $D = 0$, for $0 \le t < 5$, with the initial condition

$$\left[s_{1,0}\ s_{2,0}\ p_0\ x_{v,0}\ x_{d,0}\right]^T = \left[2\ 3\ 0\ 1\ 0\right], \quad (51)$$

and then operated in continuous mode with a perfusion rate, $D = 1$, for $5 \le t \le 10$, with the following feed concentrations:

$$\left[s_{1,f}\ s_{2,f}\ p_f\ x_{v,f}\ x_{d,f}\right]^T = \left[2\ 2\ 0\ 0\ 0\right] \quad (52)$$

Samples are taken every $\Delta t = 1$ time units, i.e. 11 samples are taken during the experiment and a total of 55 measurements are available for identification purposes (see Figure 2). They serve for a preliminary calculation of the consumption and production rates in each of the 10 sampling time intervals (Table 1).

Table 1. Estimated average production and consumption rates in each time interval.

	[0,1]	[1,2]	[2,3]	[3,4]	[4,5]
S_1	−0.196	−0.187	−0.174	−0.155	−0.130
S_2	−0.119	−0.117	−0.113	−0.109	−0.105
P	0.393	0.374	0.347	0.311	0.260
X_v	0.415	0.404	0.387	0.365	0.335
X_d	0.100	0.100	0.100	0.100	0.100

	[5,6]	[6,7]	[7,8]	[8,9]	[9,10]
S_1	−0.153	−0.172	−0.174	−0.171	−0.166
S_2	−0.098	−0.095	−0.093	−0.091	−0.089
P	0.305	0.343	0.347	0.342	0.332
X_v	0.346	0.366	0.366	0.362	0.356
X_d	0.100	0.100	0.100	0.100	0.100

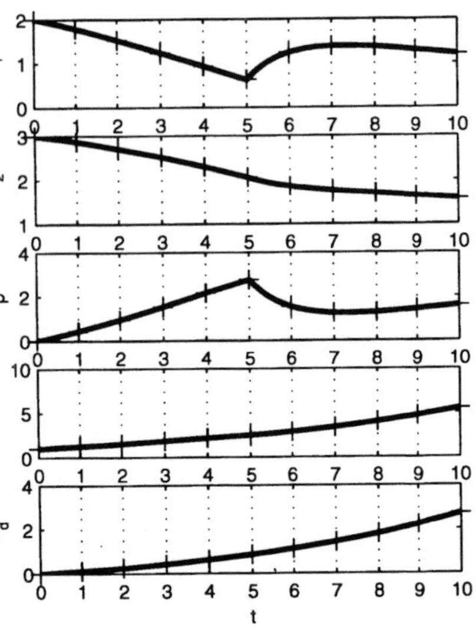

Fig. 2. Measurements in simulated experiment.

Assuming a reaction scheme with two reactions, as it is the case for the reference system, the stoichiometric matrix $\hat{\Upsilon}_0$ (Table 2) is obtained from the identification of the linear system (34) as well as the corresponding matrix of specific reaction rates ρ at all time intervals. The resulting predicted specific production and consumption rates are shown in Figure 3 together with the average rates from Table 1 and the true continuous time evolution.

Table 2. Stoichiometric matrices of the reference and approximate systems.

	reference			estimation		
rcn. #	1	2	3	1	2	3
S_1	−1	0	0	−1.260	−0.107	0
S_2	0	−1	0	0.275	−0.906	0
P	2	0	0	2.521	0.213	0
X_v	1	1	−1	1	1	−1
X_d	0	0	1	0	0	1

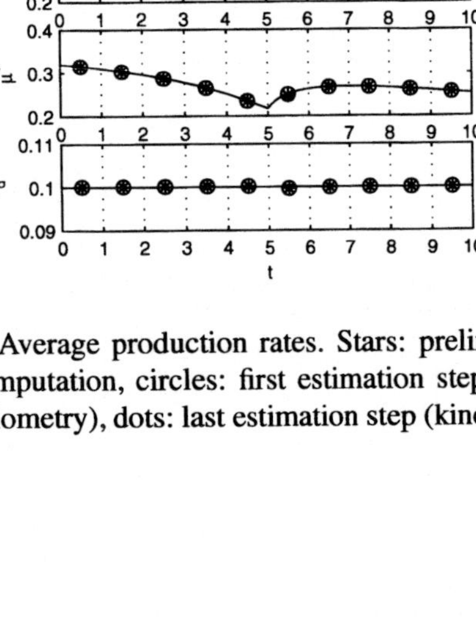

Fig. 3. Average production rates. Stars: preliminary computation, circles: first estimation step (stoichiometry), dots: last estimation step (kinetics).

The reaction rates ρ are then used for a preliminary estimation of the kinetic coefficients $\hat{\rho}_{j,\max,0}$ and $\hat{K}^*_{ij,0}$, $i = 1,\ldots,5$ and $j = 1,2$, according to the general kinetic model structure introduced by Haag *et al.* (2003). These kinetic coefficients can then be used together with the stoichiometric matrix Υ_0 as initial estimates for a final estimation of the kinetic model in combination with the stoichiometry. The result is displayed in table 3. The resulting stoichiometric matrix remains almost unchanged, $\hat{\Upsilon} \approx \hat{\Upsilon}_0$.

Table 3. Kinetic parameters ρ_{max} and K^* for the reference and identified systems.

	reference			estimation		
rcn. #	1	2	3	1	2	3
$\rho_{\max}$	0.3	0.2	0.1	0.270	0.261	0.106
$K^*_{S_1}$	1	0	0	0.944	0.299	−0.161
$K^*_{S_2}$	0	1.41	0	0.300	1.101	0.113
K^*_P	0	0	0	−0.169	−0.143	0.116
$K^*_{X_v}$	0	0	0	−0.141	0.091	0.063
$K^*_{X_d}$	0	0	0	0.000	0.011	0.006

As is shown in Tables 2 and 3, these preliminary results are already very close to the real model parameters, despite the strong assumption of constant reaction rates in each sampling interval.

5. CONCLUSION

In this study, a procedure for the fast prototyping of bioprocess models is proposed. In a first step, mean production/consumption, growth and mortality rates are obtained, which can serve for a preliminary evaluation of the stoichiometric matrix and specific reaction rates. In a second step, approximate values of the parameters of an assumed kinetic model structure can be evaluated. Finally, in a third step, all the stoichiometric and kinetic parameters can be re-estimated more accurately. The proposed procedure allows experimental data to be assessed and modeling assumptions to be checked. Finally, the estimated parameter values can be used as starting points for more rigorous identification methods.

REFERENCES

Bastin, G. and D. Dochain (1990). *On-Line Estimation and Adaptive Control of Bioreactors*. Vol. 1 of *Process Measurement and Control*. Elsevier. Amsterdam.

Bogaerts, Ph. and R. Hanus (2000). Macroscopic modelling of bioprocesses with a view to engineering applications. In: *Focus on Biotechnology, vol. IV (Engineering and Manufacturing for Biotechnology)* (Ph. Thonart and M. Hofman, Eds.). Kluwer Academic Publishers.

Haag, J. E., A. Vande Wouwer and M. Remy (2003). A general model of reaction kinetics in biological systems. In: *European Control Conference*. Cambridge, UK.

ELSEVIER
IFAC
PUBLICATIONS
www.elsevier.com/locate/ifac

MULTIOBJECTIVE OPTIMIZATION IN *Aspergillus niger* FERMENTATIONS FOR SELECTIVE PRODUCT ENHANCEMENTS

Mandal Chaitali, G. K. Suraishkumar[1] and R. D. Gudi

Department of Chemical Engineering
Indian Institute of Technology, Bombay
Mumbai-400076, India

A more comprehensive formulation of optimal control problem to enhance a desired product in a multiproduct fermentation is proposed in this work. Optimal control profiles for multiple substrate additions in fed-batch fermentation of *Aspergillus niger* to maximize either catalase or protease have been developed and experimentally evaluated. The liquid phase oxygen supply strategy (LPOS) has been used to overcome limitations of aeration due to high gas-liquid transport resistance. The optimal control problem involves linearly appearing control variables and the decision space is constrained by state and end-point constraints. The proposed multiobjective optimization is solved by Differential Evolution algorithm, a relatively superior population based stochastic optimization strategy.

Keywords: Bioprocess, Bio-control, Multiobjective optimisation, Stochastic optimisation, Multisubstrate optimisation, *Aspergillus niger*

1. INTRODUCTION

Optimal operation of bioprocesses involving multiple products is required to selectively maximize one product in the space of the minimization of the other products. The application of this multi-objective optimisation in fermentation has not been extensively explored so far, but has been useful in the context of the chemical process application (Kasat et al, 2002). Design of optimal control strategies may involve multiple substrate feeding strategies with state-path and end point constraints.

Differential Evolution (DE), (Storn and Price, 1996) is a stochastic optimisation algorithm and a parallel direct search method, that operates on a population of potential solutions by applying the principle of the survival of the fittest, to produce an optimal solution. It has been used in recent past to solve non-conventional optimization problems involving general constraints (Chiou and Wang, 1999). However, the Differential Evolution strategy has not been applied to solve a multi-objective and multiple substrates feed rate optimization problem. Further, the experimental evaluations of the theoretical strategies are not widely reported.

This work is concerned with the development of optimal control profiles for multiple substrates to maximize one selective product in the space of the minimization of the other product in a multiproduct fed-batch fermentation. The system of focus is the fed-batch fermentation for catalase and protease enzyme production by *Aspergillus niger*. A novel method of oxygen supply through liquid phase H_2O_2 addition has been used for this system with a view to overcome limitations of aeration due to high gas-liquid transport resistance (Sriram et al., 1998).

Email: [1]gksuresh@che.iitb.ac.in

Other substrates that are crucial to the fermentation are the sucrose and nitrogen levels. The problem is bound with the end point constraints to reflect realistic constraints on the final levels of product concentrations. The problem is also typically (upper) bounded by the H_2O_2 level in the fermentation, which is treated in our work as a path constraint. We additionally impose end point constraints to minimize residual levels of substrates at the end of the fermentation. To solve this complex problem we propose the use of the Differential Evolution algorithm. We demonstrate the applicability of the proposed methodology through simulations and experimental validations involving fed batch fermentation of *Aspergillus*. niger.

2. PROCESS MODEL

The production of catalase and protease enzyme by *Aspergillus niger* , with liquid phase H_2O_2 addition as an alternative source of oxygen, has been chosen as a model system. The models developed are discussed below.

A mass balance for H_2O_2 in the medium, considering that the flux of H_2O_2 in the cell is proportional to the difference of H_2O_2 concentration between the medium and cell, can be written as,

$$\frac{dH}{dt}=\frac{F_H H_{feed}}{V}-\frac{(F_H+F_C+F_N)H}{V}-K_H\left(\frac{4XV}{d\rho}+\frac{\pi d^2}{2}\right)\frac{(H-H_i)}{V} - K_d P_{cat} H \quad \text{(1)}$$

A balance on the H_2O_2 concentration, considering the cell as the system, yields. the following dynamic balance in terms of the intracellular hydrogen peroxide concentration, Hi,

$$\frac{dH_i}{dt} = K_H\left(\frac{4XV}{d\rho} + \frac{\pi d^2}{2}\right)\frac{(H - H_i)}{V}\frac{\rho}{XV} - \frac{(F_H + F_C + F_N)H_i}{V}$$
$$- KE_i H_i \rho - H_i \mu_m \frac{K_i}{(K_i + H)}\frac{N}{(K_N + N)} \quad (2)$$

The dynamic equation describing cell growth can be written as

$$\frac{dX}{dt} = \mu_m X \frac{K_i}{(K_i + H)}\frac{N}{(K_N + N)} - \frac{(F_H + F_C + F_N)X}{V} \quad (3)$$

The dynamic equation for the catalase production, a growth associated product, is written as

$$\frac{dP_{cat}}{dt} = q_P \mu_m X \frac{K_i}{(K_i + H)}\frac{N}{(K_N + N)}\frac{H_i}{(K_{H_i} + H_i)}\frac{C}{(K_C + C)} -$$
$$\frac{(F_H + F_C + F_N)P_{cat}}{V} \quad (4)$$

The dynamic equation for the protease formation, a non-growth associated product is developed as Luedeking-Piret model

$$\frac{dP_{prot}}{dt} = (a_P \mu_m X \frac{K_i}{(K_i + H)}\frac{N}{(K_N + N)} + b_P X)\frac{C}{(K_C + C)}$$
$$- \frac{(F_H + F_C + F_N)P_{prot}}{V} \quad (5)$$

The mass balances on sucrose and nitrogen concentrations, are represented as:

$$\frac{dC}{dt} = \frac{F_C C_{feed}}{V} - \frac{(F_H + F_C + F_N)C}{V} - Y_{\frac{C}{X}} \mu_m X \frac{K_i}{(K_i + H)}\frac{N}{(K_N + N)}$$
$$- Y_{\frac{C}{cat}} q_P \mu_m X \frac{K_i}{(K_i + H)}\frac{N}{(K_N + N)}\frac{H_i}{(K_{H_i} + H_i)}\frac{C}{(K_C + C)}$$
$$- Y_{\frac{C}{prot}} (a_P \mu_m X \frac{K_i}{(K_i + H)}\frac{N}{(K_N + N)} + b_P X)\frac{C}{(K_C + C)} \quad (6)$$

$$\frac{dN}{dt} = \frac{F_N N_{feed}}{V} - \frac{(F_H + F_C + F_N)N}{V}$$
$$- Y_{\frac{N}{X}} \mu_m X \frac{K_i}{(K_i + H)}\frac{N}{(K_N + N)} \quad (7)$$

$$\frac{dV}{dt} = (F_H + F_C + F_N) \quad (8)$$

Towards obtaining minimal concentration of the substrate and a specific product at the end of fermentation, end point constraints are

$$C(t_f) - C_f \leq 0 \quad (9)$$

$$N(t_f) - N_f \leq 0 \quad (10)$$

$$P(t_f) - P_f \leq 0 \quad (11)$$

End point constraint on the final batch volume

$$V_t - V_f \leq 0 \quad (12)$$

The sucrose and nitrogen concentrations in the feed and the initial batch volume are treated as decision variables. The lower and upper bound on sucrose and nitrogen source feed are imposed as:

$$C_{feed_{min}} \leq C_{feed} \leq C_{feed_{max}} \quad (13)$$

$$N_{feed_{min}} \leq N_{feed} \leq N_{feed_{max}} \quad (14)$$

$$V_{min} \leq V \leq V_f \quad 15)$$

The variable path is also bounded by a state-path constraint on the level of H_2O_2 in the broth as

$$\frac{H}{X} < 0.005 \, mol / g \, of \, cell \quad (16)$$

3. OPTIMAL CONTROL FORMULATION

The objective function is to maximize the final product concentration. Therefore it can be posed as

$$\text{Max } J = P(t_f) \quad (17)$$

Over an assumed value of t_f, N_d discretizations in the time duration $[0, t_f]$ are made and each individual nutrient addition rate is assumed to be approximated by a piecewise constant value. The DE method is a parallel direct search method and it utilizes N_P vectors of decision parameters of dimension ($N_d \times 3$) as a population for each generation, W. The initial population Z_i ($i = 1, \ldots N_P$) is randomly selected as

$$Z_i = Z_{min} + \rho i (Z_{max} - Z_{min}) \quad (18)$$

where ρ_i denotes a random number generated from a uniform distribution.

The three main operations in DE are mutation, crossover and evaluation. In mutation, the essential step is generation of the difference vector, D_{jk}, between any two individuals of the N_P vectors. The weighted difference vector is then added to a third randomly selected individual or best performing individual to have a perturbed individual (Storn and Price, 1996) as

$$D_{jk} = (Z_j^{W-1} - Z_k^{W-1}) \quad (19)$$

$$Z_i^W = Z_P^{W-1} + f(Z_i^{W-1} - Z_k^{W-1}) \quad (20)$$

where $f \in (0, 1.2)$ is a scaling factor.

During crossover, the resulting perturbed individual, Z_i^W from Equation (20) and a target individual in current generation are selected by a binomial distribution to perform crossover operation to generate, an off-spring. In the crossover operation the jth gene of the ith individual at the next generation is produced from the perturbed individual $Z_i^W=[Z_1....Z_{Nd}]_i^W$ and current individual as $Z_i^{W-1}=[Z_1....Z_{Nd}]_i^{W-1}$

$$Z_{ji} = Z_{ji}^{W-1} \quad \text{if a random number} > C_R$$
$$Z_{ji}^{W} \quad \text{otherwise} \qquad (21)$$

Where $j = 1,...N_d$, $i = 1,...N_P$ and the crossover factor $C_R \in [0,1]$ is a user specified real number.

In DE the evaluation function of an offspring competes one to one with that of its parent (target individual in current generation) and the parent is replaced by its offspring in the next generation only if it is superior than its parent.

In these methods, the penalty terms associated with the violation of constraints are added to the objective function.

$$J[\underline{u}(t)] = P(t_f) + \sum_{l=i}^{l} \varepsilon_l^2 \qquad (22)$$

4. RESULTS AND DISCUSSIONS

In all the simulations, the fed-batch fermentation was assumed to start with an initial condition on cell mass (0.75 g l^{-1}), nitrogen source (1.5 g l^{-1}), sucrose (6.0 g l^{-1}), catalase and protease (0.0 g l^{-1}), and H_2O_2 concentration both in the medium and in the cell (0.0 moles l^{-1}). The overall bioreactor run was assumed to be of 30 hours duration and the total run time was divided into 15 intervals of 2 hours duration each, i.e. the value of N_d was assumed to be 15. Thus, in this problem a total number of 49 decision variables were used consisting of 45 decision variables generated from each of the glucose, nitrogen source and H_2O_2 additions, 2 variables for glucose and nitrogen concentration in the feed and 2 variables for initial batch volume and fermentation time. A population of 100 vectors of decision variables in one generation was used i.e., N_P is 100. The initial value of the decision and control variables were determined by using equation (18). A crossover factor (C_R) of 0.5 was used and the program was run for 10,000 generations.

4.1 *Optimum run for Catalase Maximization*

The results of the catalase maximization are shown in Fig. 1-2. The optimizer predicted that the fed-batch must start from an optimum initial batch volume of 850 ml and optimum nitrogen source and sucrose concentrations in the feed to be 295 g l^{-1} and 300 g l^{-1}, respectively. From this Figure 1, it is clear that the nitrogen source feed additions occur almost in one initial pulse. Whereas, sucrose feed addition is predicted as two initial pulse followed by two end pulse. For H_2O_2 the additions occur over the entire time duration mostly with an increasing profile. As is evident from the results, these profiles are predicted by the optimizer to balance the three effects, viz. those resulting from constraints on final batch volume, the end point residual nutrient and

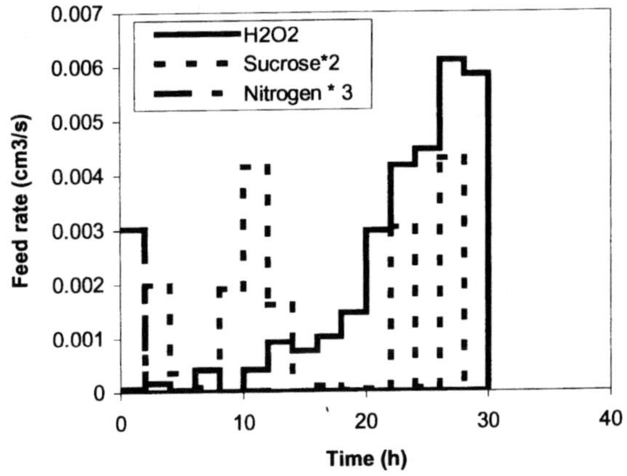

Fig. 1 Optimal feed rate profiles for H_2O_2, sucrose and nitrogen source in catalase maximization run

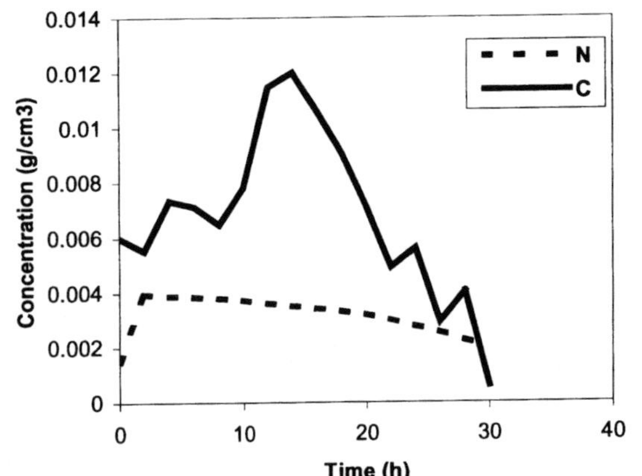

Fig. 2: Sucrose and Nitrogen profiles for Catalase maximization run

protease concentration as well as the nutrient requirements for growth and product formation. A final cell and catalase concentrations of 7.77 g l^{-1} and 4.24 × 10^{-7} M respectively were obtained. The final protease concentration in the catalase maximization run of 5.5 u $^{-1}$l was obtained.

4.2 *Optimum run for Protease Maximization*

The results of the protease maximization are shown in Fig. 3. In this second case, the optimizer predicted that the fed-batch must start from an optimum initial batch volume of 1250 ml and optimum nitrogen source and sucrose concentrations in the feed to be 300 g l^{-1} and 295 g l^{-1}, respectively. From Fig. 3, it is clear that the nitrogen source feed additions are almost same as catalase run; in one initial pulse. Whereas, the sucrose feed additions are different from that of catalase run; a small initial feed addition followed by an increasing feed rate after 18 hrs of fermentation, to support the non-growth associated protease production, is seen in Fig. 3.

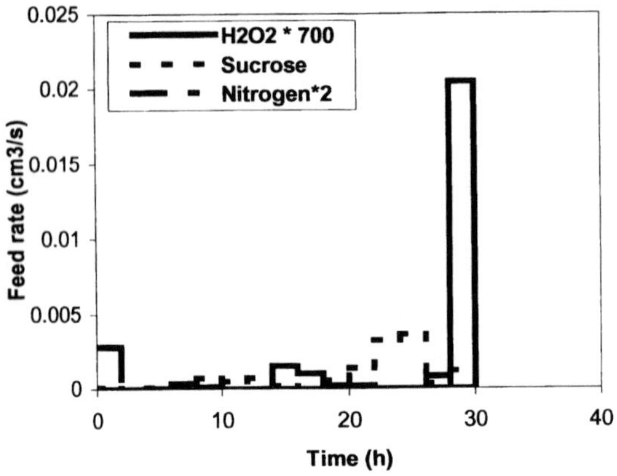

Fig. 3. Optimal feed rate profiles for H_2O_2, sucrose and nitrogen source in protease maximization run

For H_2O_2, the additions occur in a small pulses over the entire time duration with a comparatively large pulse at the end. A final cell concentrations of 12.58 g l^{-1} and a final protease of 7.79 u l^{-1} were obtained. The final catalase concentration in protease maximization run of 3.18×10^{-9} M was obtained.

4.3 Analysis of the optimum runs

The optimal profiles obtained from both catalase and protease maximization are different in nature. The cell mass concentration was found to be higher in case of protease than catalase. The non-growth associated model for protease requires higher cell growth to maximize the final protease concentration. The optimum initial batch volume is lower in case of catalase than protease. In the catalase run there is also a higher dilution due to the continuous addition of H_2O_2. Therefore optimizer recognizes this requirement and predicts a lower initial batch volume for catalase to allow the higher dilution to meet the constraints on the maximum batch volume. In case of protease run the prediction of higher batch volume does not violate the constraint on maximum batch volume. The continuous feed addition profile for H_2O_2 enhanced the catalase production and the pulse addition of H_2O_2 is sufficient to supply the oxygen for cell growth to maximize the protease. The initial and end pulse addition of sucrose is able to maintain an increasing trend of sucrose concentration in the broth, shown in Fig. 2. to support the growth associated catalase production, followed by a decrease in sucrose concentration to meet the minimum residual concentration at the end of the fermentation. On the other hand there is initially small addition of sucrose followed by an increasing feed addition to support the non-growth associated protease production.

4.4 Analysis of the other simulation results

Different runs

1. Optimal run for catalase maximization with end point constraint on sucrose and nitrogen.
2. Catalase Max. without any end point constraint
3. Catalase Max with end protease upwards than the optimal threshold value
4. Catalase Max with end protease down wards than the optimal threshold value
5. Optimal run for protease maximization
6. Protease Max. without end point constraint
7. Protease Max with end catalase upwards than the optimal threshold value
8. Protease Max with end catalase down wards than the optimal threshold value

The results obtained from different simulations are presented in Table 1.

Table 1: Final concentration obtained from different simulations

Run	X (g/l)	Catalase (mole/l)	Protease (units/l)	Nitrogen (g/l)	Sucrose (g/l)
1	7.8	4.24×10^{-7}	5.53	1.99	0.6
2	8.2	4.48×10^{-7}	5.78	8.2	0.3
3	8.5	4.11×10^{-7}	6.0	1.99	0.6
4	4.1	3.04×10^{-7}	2.8	1.99	0.6
5	12.6	3.18×10^{-9}	7.79	1.99	0.9
6	13.2	1.63×10^{-11}	8.21	8.50	2.9
7	12.6	3.52×10^{-7}	4.92	1.99	0.9
8	12.6	1.47×10^{-9}	7.77	1.99	0.6

Effect of end protease concentration in catalase maximization

In catalase maximization, Run 3, the specification of a higher final protease concentration than its threshold optimal value decreases the final catalase concentration from its optimal value. This is because the enhancement of one product in a multi-product fermentation, simultaneously decrease the concentration of the other product to maintain the material balance. In Run 4, the decrease in protease concentration than its threshold optimal value should increase the final optimal catalase concentration. But Run 4 shows a lower value of final catalase than its optimal value. The reason for this behaviour is that, the end point constraint to lower down the final protease concentration from its optimal threshold value directed the optimizer towards the low cell mass production. Therefore catalase production being a growth associated product decreased from its optimal value by this low cell concentration.

Effect of end catalase concentration in protease maximization

In case of protease maximization, Run 8, the specification of a lower final catalase concentration from its optimal value directed the optimizer towards a decrease in the cell concentration. As an effect of this decrease in the cell concentration, the final protease concentration also decreases from its optimal value, though the change in the final concentration is almost negligible. On the other hand the end point constraint to up wards the final catalase concentration, Run 7, from its optimal value decreases the protease concentration from its optimal value. Therefore we can say that the enhancement of one product reduced the concentration of the other product to maintain the mass balance in the competitive substrate utilization system.

End Point Constraints on the Final Residual Nutrient Concentrations

One of the other interesting aspects of this work is the reduction of the final residual nutrient concentrations by applying end point constraints on them; this has been proposed here to ease separation during downstream processing. The optimization problems without imposing any end point constraints on the final residual nutrient concentration, Run 2 and Run 6, yields a high residual nutrient concentration at the end of the fermentation that could interfere with the downstream processing. It shows a high residual nitrogen concentration of 8.2 gl^{-1} (run 2) and 8.5 gl^{-1} (run 6) at the end of the fermentation when there is no end point constraint imposed on residual nitrogen. On the other hand in case of the optimum (Run 1 and 5) a very low residual nitrogen concentration of 1.99 gl^{-1} is obtained for both the run. This indicates a trade-off between high final concentrations of the product and the residual nutrient levels. Relaxation of the end-point constraints could yield higher final product concentrations but at the expense of higher residual nutrient levels which translate to higher processing/downstream costs.

State-Path Constraints on Nutrient Additions

As stated above in equation (16), a state-path constraint is used in this optimization problem to bound the toxic level of H_2O_2 concentration on the cell growth. The results obtained are presented in Fig. 4. It shows that the H/X value was maintained below the threshold value of 0.005 mol/g-cell over the entire duration of fermentation.

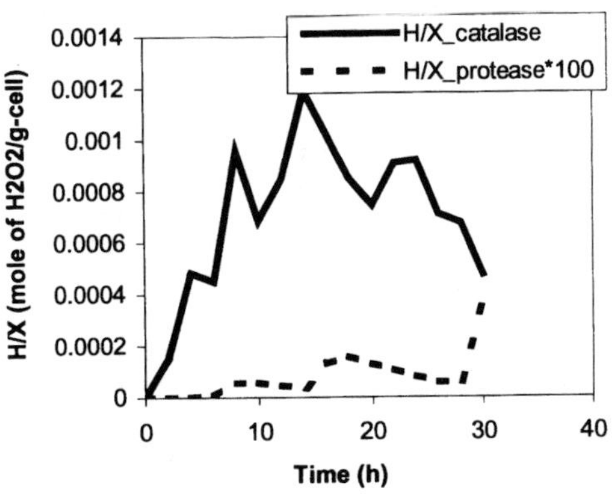

Fig 4: Profiles of the constraint, H/X

4.5 Experimental Evaluation of the Simulation Results

Catalase Maximization

The catalase and the cell concentration profiles obtained from the simulation results have been validated here through experiments shown in Fig 5 and 6. It was observed that the experimental results were broadly in agreement with the simulation results. A final cell concentration of 7.28 g l^{-1} obtained from the experiment compares reasonably well with 7.77 g l^{-1} from simulations. Also the catalase concentration of 2.65 $\times$ 10^{-7} M from experiments, at 28 hours of fermentation, is also comparable with than of 3.61 $\times 10^{-7}$ M form simulation, at 28 hours. A 70% higher catalase than batch cultivation with constant feeding of H_2O_2 at 12 ml h^{-1} was predicted by the optimiser and realised experimentally.

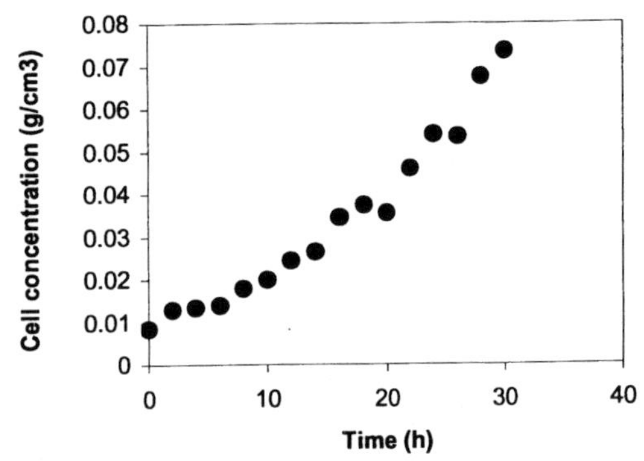

Fig 5: Cell concentration profiles from experimental results for catalase maximization

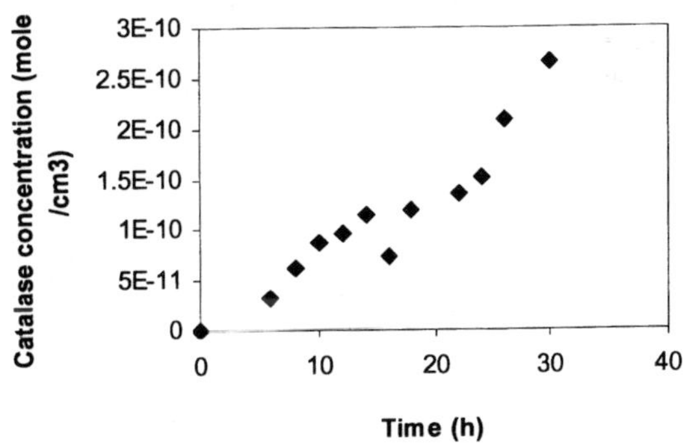

Fig 6: Catalase profiles from experimental results for catalase maximization

Protease Maximization

The product and the cell concentration profiles obtained from the simulation result have been validated here through experiments shown in Fig. 7 and 8. It was observed that the experimental results were broadly in agreement with the simulation results upto 17 hours of fermentation. After that the

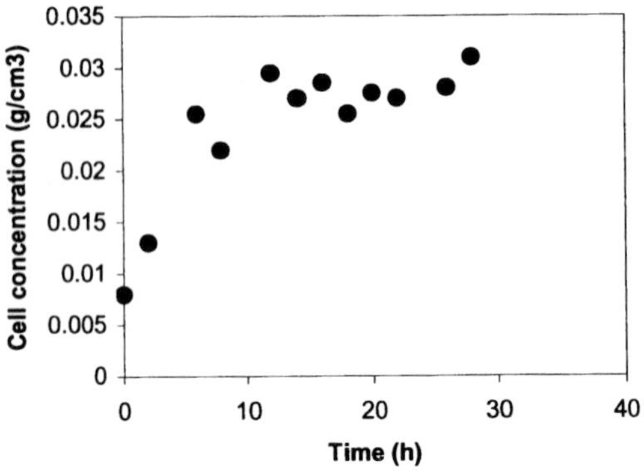

Fig 7: Cell concentration profiles from experimental results for protease maximization

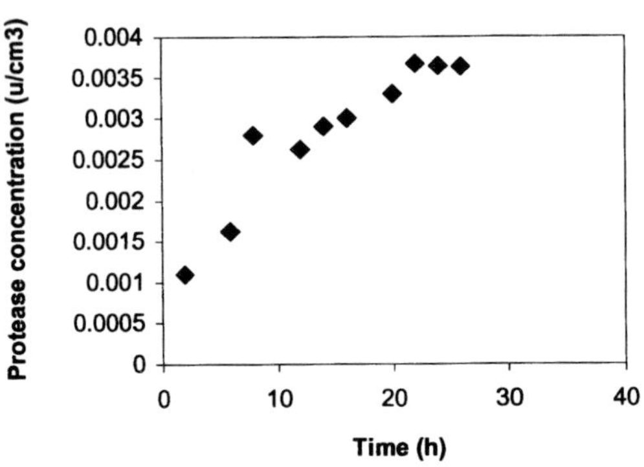

Fig. 8: Protease profiles from experimental results for protease maximization

experimental growth profile follows the stationary phase whereas the simulation result follow the exponential growth due to exponential growth model. A prolonged stationary phase of 18 hours and a final cell concentration of approx 3.1 g/l were obtained from the experiment. A final protease concentration of 3.7 u l^{-1} was obtained. A 31% higher protease than batch cultivation with constant feeding of H_2O_2 at 12 ml h^{-1} was obtained.

5. CONCLUSIONS

This article proposed a more complete formulation of the optimal control problem for multiobjective optimisation in fed-batch fermentations. The formulation proposed an inclusion of end point and state-path constraints to pose the fed-batch fermentation requirements more realistically. A robust stochastic optimization algorithm based on differential evolution was found to solve the resulting optimization problem. The results obtained from simulations were verified experimentally and showed an increase of about 70% and 31% final product concentration compared to conventional batch cultivation.

6. NOMENCLATURE

E_i	intracellular enzyme (2.23×10^{-13} mole cm^{-3} g^{-1})
F_H	hydrogen peroxide feed rate ($cm^3 s^{-1}$)
F_C	sucrose feed rate ($cm^3 s^{-1}$)
F_N	nitrogen source feed rate ($cm^3 s^{-1}$)
K	rate constant (7500000 (mole $cm^{-3})^{-1} s^{-1}$)
K_H	H_2O_2 permeability across the cell (0.812×10^{-7} cm s^{-1})
K_{Hi}	Monod type constant (5×10^{-7} mole cm^{-3})
K_N	Monod type constant for nitrogen (1.4×10^{-4} g cm^{-3})
K_C	Monod type constant for sucrose (0.3×10^{-4} g cm^{-3})
K_i	inhibition constant for H_2O_2 (0.24×10^{-4} mole cm^{-3})
q_P	catalase yield (mol catalase (13.26×10^{-9} g mass)$^{-1}$)
a_P	protease yield (0.044 unit protease (g mass)$^{-1}$)
b_P	protease yield with respect to maintenance (5.43×10^{-7} unit protease (g mass)$^{-1}$ s^{-1})
d	diameter of the hyphae (13.43 cm)
ρ	density of the hyphae (1.02 g cm^{-3})
$Y_{N/X}$	inverse of cell yield with respect to nitrogen (0.015)
$Y_{C/X}$	inverse of cell yield with respect to sucrose (0.68)
$Y_{C/Pcat}$	inverse of catalase yield with respect to sucrose (1.01×10^8)
$Y_{C/Pprot}$	inverse of protease yield with respect to sucrose (8.54)
V_f	max. working reactor volume (1.4 l)

Greek Letters

μ_m	maximum specific growth rate (2.77×10^{-5} s^{-1})
ε	penalty factor

REFERENCES

Storn, R. and K. Price (1996). Minimizing the real functions of the ICEC'96 contest by differential evolution. *IEEE Conference on Evolutionary Computation*, **Nagoya**, 842-844.

Sriram, G, Y. ManjulaRao, A.K Suresh and G.K Sureshkumar (1998). Oxygen supply without gas-liquid film resistance to *Xanthomonas campestris* cultivation. *Biotechnol. Bioeng.*, **59**: 714-723.

Chiou, J.P and F. S. Wang (1999). Hybrid method of evolutionary algorithms for static and dynamic optimization problems with application to a fed-batch fermentation Process. *Comput. Chem. Eng.*, *23*, 1277-1291.

Kasat, R. B, D. Kunzru, D. N. Saraf and S. K. Gupta (2002). Multiobjective optimisation of Industrial FCC Units Using Elitist Nondominated Sorting Genetic Algorithm . *Ind. Eng. Chem. Res.*, **41**, 4765-4776.

ELSEVIER
IFAC
PUBLICATIONS
www.elsevier.com/locate/ifac

REFERENCE TRAJECTORY TRACKING OF SUPERFICIAL TEMPERATURE IN FOOD DECONTAMINATION

Ruben Zuñiga Ulloa, Lionel Boillereaux*, Michel Havet and Olivier Rouaud
GEPEA-ENITIAA, UMR CNRS 6144
Rue de la Géraudière, BP 82225, 44322 Nantes cedex 3, France

Abstract

This paper proposes a preliminary study, for the tracking control of the superficial temperature in rapid decontamination of foods surface. The data are issued from a heat and mass transfer simulator, taking into account all the specificities of the process and of the product. This study, supported by a on-going European project, aims at imposing an increasing and decreasing ramp-type temperature profile to the surface of foods, to achieve a complete or partial destruction of microorganisms. A complete model including water activity, food desiccation and heat transfer is proposed to describe the phenomena induced by the process. A simplified model, based only on heat transfer, is finally proposed to design a tracking control of a superficial temperature reference profile. The robustness of the solution is discussed through different conditions of use. *Copyright © 2004 IFAC*

Keywords

Heat transfer, mass transfer, decontamination, tracking control, trajectory planning.

Nomenclature

a_w	Water activity	%
C_v	Concentration of water vapor	kg. m^{-3}
Cp	Specific heat	J. kg^{-1}. K^{-1}
D	Diffusion coefficient	m^2. s^{-1}
h	Heat transfer coefficient	W. m^{-2}. K^{-1}
k_m	Mass transfer coefficient	m. s^{-1}
M	Moisture content	kg w. kg dm^{-1}
M_w	Molecular weight of water	kg. kmole^{-1}
t	Time	s
T	Temperature	°C
z	Vertical coordinate	m
α	Thermal diffusivity	m^2. s^{-1}
Δz	thickness of each finite volume	m
λ	Thermal conductivity	W. m^{-1}. K^{-1}
ρ	Density	kg. m^{-3}
R_u	Universal gas constant	J.kmole^{-1}. K^{-1}
P	Water vapor pressure	Pa

a-dimensional numbers

Le	Lewis number
Pr	Prandtl number
Sc	Schmidt number

Subscripts

a	Air
dm	Dry matter
m	melting
inf	Bottom surface
n	Top surface
p	Product
v	Vapor
w	Water

Introduction

The surface of food products may be contaminated by microorganisms which could threaten the health and even the life of consumers. Decontamination of the surface of meat products is necessary and very important. It is proved that heat treatments are efficient to decrease significantly this microbial contamination. However, the determination of the level of microbial destruction that a thermal treatment induces in a product requires the understanding of the amount of heat delivered to every portion of the food product. A too intensive treatment should be avoided as it may cook a thin layer of product and affect the surface color. For these reasons, the treatment consists, in a first stage, in heating the product at the desired temperature and, in a second stage, in cooling it towards a low temperature which will prevent recontamination. A on-going European project (EU QLK1-CT-2001-01415, BUGDEATH) coordinated by the Food Refrigeration and Process Engineering Research Centre (University of Bristol, UK) aims, in part, to develop models and control strategies that predict and optimize the effect of thermal treatments of decontamination of food products surfaces. The key parameter of decontamination processes is the surface temperature which depends on the heating/cooling medium and on the thermal properties of the food. In this work, we consider the decontamination treatment of lean beef by hot/cold air at high velocity (20m/s). The physical phenomena close to the surface are rather complex because the heat is transferred by convection and evaporation during the heating stage and by convection and condensation in the cooling phase. The evaporation is crucial because it is governed by the water activity of the food which evolves with time. The evaporation modifies the water content of the product and leads to coupled heat and mass transfer. An accurate modeling of this process requires the knowledge of all parameters and a robust numerical algorithm.

To follow a ramp-type (heating then cooling) set point profile for the superficial temperature, we propose to design a simple but robust reference tracking control, computed from a simplified model of the phenomena. The control law obtained is applied on a simulator derived from a complete modeling of the phenomena, that is, taking into account water activity, mass diffusion inside the food and evaporation at the surface The follow-up of the internal temperature could thus be used to reconsider the procession parameters, as instance to avoid non expected cooking of the internal layers and to dimension heating and cooling systems.

This paper is organized as follows:

- The first part is dedicated to the process description,
- In a second part, a complete model of heat transfer inside the food is developed, taking into account the evaporation phenomena at the exchange surface. This model constitutes the simulator on which the control algorithm will be tested in simulation.
- The third part deals with the tracking control of the superficial temperature, presenting a particularly efficient and simple way to implement closed-loop approach.

Process description.

The decontamination rig used for the decontamination process can heat the surface of a food sample from 5 to 100°C in a time span of up to 900 seconds, and a cooling system allows quickly decreasing the temperature to prevent recontamination. The apparatus is fully computer controlled and the duration of heat treatment, holding and cooling phase is entered into a graphical user interface (GUI) written in Microsoft® Visual Basic, constituting the reference to track.

For the process, the food sample is first placed in the sample holder of the machine, a small dish 18 mm deep with a diameter of 50 mm. By pressing an initial button from the GUI, a pneumatic actuator moves the sample into the chamber. The sample is then subjected to a very high velocity air stream heated by a 3.3 kW heater, allowing a rapid increasing of the superficial temperature of the products. Then, the temperature is hold at a constant set superficial temperature during a variable time, and at last a rapid cooling is provoked with air at –20°C. Hot and cold air jets are blown through 2 separate nozzles placed inside the chamber 35 mm away from the leading edge of the sample in such a way that each airflow is parallel to the surface (no impingement). The rotation of the sample using a motor ensures a uniform temperature of the food surface during treatment. So heat and mass transfer can be assumed to be 1D. The surface temperature is measured using an infra red (IR) sensor attached to the top of the chamber 80 mm above the surface of the sample. During treatment, a control status box on the computer display informs the user of what is currently happening and the time remaining of the different tasks. Especially, a control accuracy bar displays the surface temperature measured by the IR sensor and the desired temperature. Today, the surface temperature control is limited to a basic split-range threshold controller, acting either on the heater or on the pneumatic valves of the cooling system. It is obvious that such a control strategy is particularly insufficient to warrant the microbiological specifications.

Process modeling.

heat – mass transfer model.

In the considered process, heat and mass transfer must be taken into account together, so the evaporation phenomena are important with elevated temperature and convection coefficient. To develop an efficient control of the superficial temperature during decontamination, a preliminary study must be undertaken. We propose beforehand to develop an accurate model of the heat and mass transfers, which will be used as a simulator for the control law development.

Let us now consider the following assumptions:

- sample is an infinite slab;
- the product is isotropic and homogeneous;
- no internal heat generation and uniform initial conditions are considered;
- the physical and thermal properties are constant;
- water vapor flux in the product during drying is negligible.

The flow of moisture within the product occurs by diffusion and is governed by the diffusion coefficient according to Fick's second law (Braud et al., 2001):

$$\frac{\partial M}{\partial t} = D_w \cdot \frac{\partial^2 M}{\partial z^2} \tag{1}$$

Heat transfer within the product is driven by conduction as a temperature gradient develops along its thickness according to Fourier's second law:

$$(\rho \cdot Cp)_p \cdot \frac{\partial T}{\partial t} = \lambda_p \cdot \frac{\partial^2 T}{\partial z^2} \tag{2}$$

The bottom surface is insulated, and the top one receives heat by convection:

$$(\rho \cdot Cp)_p \left.\frac{\partial T}{\partial t}\right|_{z=0} = h(T_a - T(0,t)) \tag{3}$$

For the boundary condition at z=0 (top surface), the liquid flux to the surface plus vapor flux away from the surface into the external drying media must be equal to the instantaneous variation of water mass content in the solid:

$$\rho_{dm} \cdot D_w \cdot \frac{\partial M}{\partial z} + k_m \cdot (C_{va} - C_{vn}) = \rho_{dm} \cdot \left(\frac{\Delta z}{2}\right) \cdot \frac{\partial M}{\partial t} \tag{4}$$

The corresponding boundary condition for heat transfer is:

$$\lambda_p \cdot \frac{\partial T}{\partial z} + h \cdot (T_a - T_n) + \rho_{dm} \cdot D_w \cdot \frac{\partial M}{\partial z} \cdot Cp_w \cdot T_n + k_m \cdot (C_{va} - C_{vn}) \cdot Cp_v \cdot T_n = (\rho \cdot Cp)_p \cdot \left(\frac{\Delta z}{2}\right) \cdot \frac{\partial T}{\partial t} \tag{5}$$

Concerning the boundary condition at z=L (bottom surface), it is assumed that there are no mass and heat transfers across that surface of the drying material:

$$D_w \cdot \frac{\partial M}{\partial z} = \left(\frac{\Delta z}{2}\right) \cdot \frac{\partial M}{\partial t} \tag{6}$$

$$\lambda_p \cdot \frac{\partial T}{\partial z} + \rho_{dm} \cdot D_w \cdot \frac{\partial M}{\partial z} \cdot Cp_w \cdot T_{inf} = (\rho \cdot Cp)_p \cdot \left(\frac{\Delta z}{2}\right) \cdot \frac{\partial T}{\partial t} \tag{7}$$

Parameters used in the model.

The thermal properties of the product (Rahman, 1995), air and water (Wark, 1996 and Holman, 1997) were taken from the dedicated literature.

Assuming water vapor as ideal gas (Wang and Brennan, 1995), the concentrations of water vapour is equal to

$$C_v = \frac{P_v \cdot M_w}{R_u \cdot T} \tag{8}$$

The mass transfer coefficient can be related to convective heat transfer coefficient (Kuitche et al., 1996) using the Lewis relation:

$$k_m = \frac{h}{\rho_a \cdot Cp_a \cdot Le^{2/3}} \tag{9}$$

The water activity is approximated by the ratio of the partial pressure of water vapor at the surface of the product to the vapor pressure of pure water at the product temperature (Lind and Rask, 1991)

$$a_w = \frac{P_{vn}}{P_{vsat}} \tag{10}$$

The equilibrium moisture content and the water activity of a food at given temperature and pressure are related by the sorption isotherms graphs. The sorption isotherm can also be described mathematically, and several mathematical models have been suggested (Delgado and Sun, 2002).

Tracking control of reference superficial temperature.

Model simplification

The solutions to track a reference trajectory are numerous, especially in the literature dedicated to robots (Ider et al., 2002), but they can be found process engineering as well (Sakamoto et al., 1998, Boillereaux et al., 1999). The tracking problem can be seen two ways, either by following a reference trajectory function, or by specifying position, velocity and acceleration (if necessary) at each time.

Simple PID controllers can be employed to achieve such control strategies, but it is well-known that their robustness in regards with the modeling errors is limited. Some authors (Alonso et al., 1998) propose to use adaptive control strategies that are more robust since they are, e.g., coupled with an IMC strategy. In this work, we propose a simple and robust strategy based on an approximated model of the process, assuming that the requested state variables are available (measured or estimated).

A complete model of the phenomena has been proposed and discussed in the previous section, and comprises a certain number of uncertainties, as for instance the water activity evolution, the mass diffusion coefficient and the convective exchange parameter. Moreover, the measurement of the mass transfer is very difficult even impossible to get on-line. Due to these drawbacks, the control law will be established considering a simplified model of the phenomena, but will be applied on the complete simulator.

In the simplified model, we propose to neglect the mass transfer. By neglecting the evaporation term, the evolution of the superficial temperature, issuing from an energy assessment at the superficial volume, is:

$$\frac{dT_n}{dt}=\frac{2h}{(\rho Cp)_p\Delta z}(T_a-T_n)+\frac{2\lambda_p}{(\rho Cp)_p\Delta z^2}(T_{n-1}-T_n) \quad (11)$$

For all the internal volumes, it can be written that, following the same reasoning of simplification, the temperatures evolve as expressed in the following equation:

$$\frac{dT_i}{dt}=\frac{\lambda_p}{(\rho Cp)_p\Delta z^2}(T_{i+1}-2T_i+T_{i-1}) \quad \forall i\in[2,n-1] \quad (12)$$

$$\frac{dT_1}{dt}=\frac{2\lambda_p}{(\rho Cp)_p\Delta z^2}(T_{i+1}-T_i) \quad (13)$$

Control law design

Consider the ambient temperature as the manipulated variable, denoted *u*(*t*), and the surface temperature measurement, which is the control one, denoted *y*(*t*). The previous system can be written in the following matrix form, leading to a linear state space representation:

$$\begin{aligned}\dot{x} &= Ax+Bu\\ y &= Cx\end{aligned} \quad (14)$$

$x=[T_n \;\; T_{n-1} \;\; \cdots \;\; T_1]^T$ represents the vector of the temperatures of each finite volume,

$$A=\frac{\alpha}{\Delta z^2}\begin{bmatrix} -2-{}^{2h\Delta z}/_{\lambda} & 2 & 0 & \cdots & \cdots & \cdots & 0\\ 1 & -2 & 1 & 0 & \cdots & \cdots & 0\\ 0 & 1 & -2 & 1 & 0 & & \vdots\\ \vdots & \ddots & \ddots & \ddots & \ddots & & \vdots\\ \vdots & & \ddots & \ddots & \ddots & \ddots & 0\\ \vdots & & & \ddots & 1 & -2 & 1\\ 0 & \cdots & \cdots & \cdots & 0 & 2 & -2 \end{bmatrix}$$ is a 3-diagonal matrix,

$C=[1 \;\; 0 \;\; \cdots \;\; 0]$ and $B=\frac{h}{\rho C_p\Delta z}[1 \;\; 0 \;\; \cdots \;\; 0]^T$ are respectively the output and input vectors.

By replacing y in Eq. (11), the following expression is straightforwardly obtained:

$$u=\frac{(\rho Cp)_p\Delta z}{2h}\left(\dot{y}-\frac{2\lambda_p}{(\rho Cp)_p\Delta z^2}(T_{n-1}-y)\right)+y \quad (15)$$

Let us denote y_{sp} the set point profile of the superficial temperature. The expression of the input to apply to the process, in an open-loop scheme, is obtained by replacing $\dot{y}$ and y respectively by $\dot{y}_{sp}$ and y_{sp} in Eq. (15).

However, it appears that the computation of the control variable requires the knowledge of the temperature in the second volume of the meshing, denoted T_{n-1}. As mentioned previously, all the state variables requested are assumed available, either by direct measurement or estimation.

Let us now design the closed-loop solution. Denote $\varepsilon=y_{sp}-y$ the output error. This error must respect the following differential equation:

$$\varepsilon^{(p)}(t)=k_0\varepsilon(t)+k_1\dot{\varepsilon}(t)+k_2\ddot{\varepsilon}(t)+\cdots+k_{p-1}\varepsilon^{(p-1)}(t) \quad (16)$$

where the scalars $k_0,\cdots,k_{p-1}$ are chosen to warrant the stability of Eq. (16). p is the relative degree of the considered output. In our case, it can be easily verified that:

$$p=\min_{k\in N}\left\{\frac{\partial x_n^{(k)}(x,u)}{\partial u}\right\}=1 \quad (17)$$

The closed-loop control is directly obtained from Eq. (15) by replacing $\dot{y}$ by $\dot{y}_{sp}+k_0(y_{sp}-y)$, according to Eq. (16):

$$u=\frac{(\rho Cp)_p\Delta z}{2h}\left(\dot{y}_{sp}+k_0\varepsilon-\frac{2\lambda_p}{(\rho Cp)_p\Delta z^2}(T_{n-1}-y)\right)+y \quad (18)$$

Application in simulation

The control approach has been carried out on a simulator built from the model expressions given in the section dedicated to the process modeling, using an implicit finite volumes schemes to solve the problem.

The food under consideration is lean beef, and the thermal parameters used in the simulator and in the control law are assumed to be well known.

The control objective is to track a superficial set point temperature composed first of a 1 $K.s^{-1}$ ramp-type profile, and second of a -0.5 $K.s^{-1}$ ramp-type profile.

In this first example, the evaporation term is omitted, as if the product were packed. Figure 1a represents the reference to track (solid line), the superficial temperatures with $k_0 = -0.5$ (dashed lines) and $k_0 = -3$ (dotted lines), while Figure 1b represents the corresponding control temperature evolution.

In a second example, we consider that the food is exposed directly to the air jet, and an important evaporation appears. We propose here to test the model robustness of our control approach. The control approach is used with the same parameters, that is, $k_0 = -0.5$ and $k_0 = -3$ (see Figure 2a and Figure 2b).

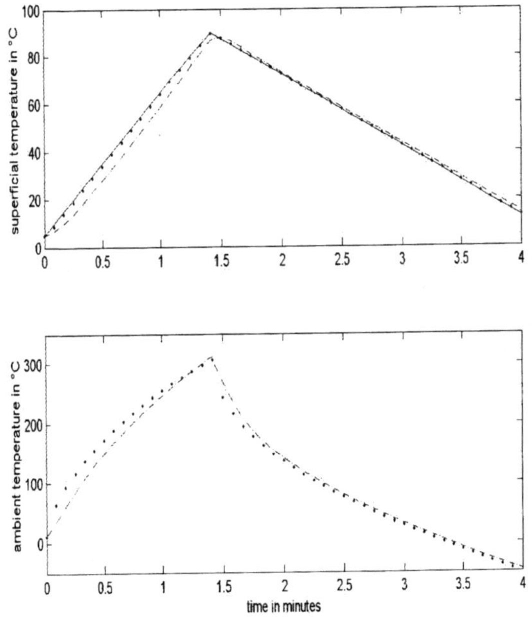

Figure 1.a- Reference tracking without evaporation.
b- Control variable evolution

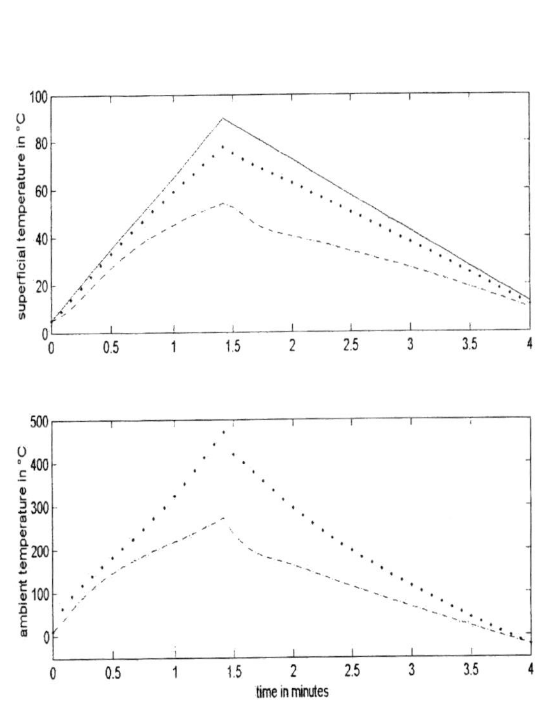

Figure 2.a- Reference tracking with evaporation.
b- Control variable evolution

It can be noticed that the robustness is limited, especially when the corrective gain k_0 is small. Of course, the temperature decrease due to the evaporation leads to an important increasing of the control temperature, as expected.

Addition of an integral correction

To reduce the effects of model uncertainties, an efficient improvement consists in adding an integral action to the correction. With this modification, the expression of the input becomes:

$$u^* = \frac{(\rho Cp)_p \Delta z}{2h}\left(\dot{y}_{sp} + k_0\varepsilon + k_I\int_0^t \varepsilon(\tau)d\tau\right) - \frac{\lambda_p}{h\cdot\Delta z}(T_{n-1} - y) + y \tag{19}$$

On Figure 3a, the tracking error is represented with (k_0, k_I) equal to $(-3, -1)$, while the corresponding control variable evolution is given on Figure 3b.

The addition of an integral term leads to a increasing of energy consumption of about 13% in a first approximation (proportionally to $\int u^*(t)dt$ on the whole experiment), with $k_0 = -3$, but the robustness is considerably improved.

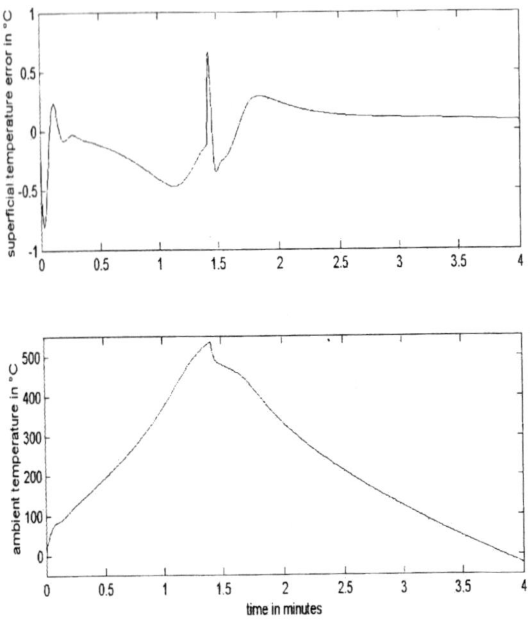

Figure 3.a- Superficial temperature tracking errors.
b- Control variable evolution

Conclusion.

The aim of our investigations about food preservation consists in imposing a specific temperature profile to the surface of a food during decontamination. The superficial temperature profiles presented in the previous section are recommended by the microbiologists taking part in the project, but must be adapted to each type of food. However, to achieve this objective, the profile is composed of ramp-type set point temperature profiles, in heating then cooling modes, with or without an intermediary plateau. We have thus proposed a tracking control of the reference of the superficial temperature, based on a simplified model of the transfer phenomena, totally neglecting the evaporation phenomena. The control law has been carried out in different cases, on a simulator resulting from a complete modeling of the phenomena, including water activity, weight loss due to evaporation and thermal properties evolution with mass transfer.

In this paper, it has been shown that the robustness of the control is limited, when the evaporation phenomena is considerable. To improve the results, a corrective term based on the integral of the tracking error has been added, leading to an efficient superficial temperature planning.

However, it is important to notice that the control law requires the knowledge of the temperature located in volume 2, that is, immediately under the superficial layer. In further investigations, this measurement (here the simulator) will be removed and replaced by an observer, based on the surface temperature measurement.

The next stage of this work, of course after experimental validation on the rig, presented in a next work, is to develop a tracking control strategy considering some constraints on state variables, as instance limited time-temperature couples to avoid partial cooking of the product during decontamination.

Acknowledgments

The authors thank the European Community for the financial support of this project, and the associated partners: the FRPERC of Bristol, for the pilot development, the INRA of Theix for the modeling contribution.

References

Alonso, A.A., Banga, J.R., Perez-Martin, R. (1998). Modeling and adaptive control for batch sterilization. *Computers and Chemical Engineering, 22(3),* 445-458.

Boillereaux, L., Chourot, J.M., Havet, M. (1999). Nonlinear trajectory control of high pressure thawing. *Journal of Process Control, 9(4)*, 351-356.

Braud, L. M., Moreira, R. G., Castell-Perez, M. E. (2001). Mathematical modeling of impingement drying of corn tortillas. *Journal of Food Engineering, 50*, 121-128.

Delgado, A. E., Sun, D. (2002). Desorption isotherms for cooked and cured beef and pork. *Journal of Food Engineering, 51*, 163-170.

Holman, J. P. (1997). Heat transfer. McGraw Hill Book Company, Inc.

Ider, S.K., Özgören, M.K., Ay, V.(2002). Trajectory tracking control of robots with flexible links. *Mechanims and Machine Theory, 37*, 1377-1394.

Kuitche, A., Daudin, J. D., Letang, G. (1996). Modelling of temperature and weight loss kinetics during meat chilling for time-variable conditions using an analytical-based method. I. The model and its sensitivity to certain parameters. *Journal of Food Engineering, 28*, 55-84.

Lind, I., Rask, C. (1991). Sorption isotherms of mixed minced meat, dough, and bread crust. *Journal of Food Engineering, 14*, 303-315.

Rahman, S. (1995). Food properties Handbook. *CRC Press*, Boca Raton.

Sakamoto, N., Masubuchi, M., Kawata, S.(1998). Variable structure control using the idea of fictituous set point temperature tracking control of a batch reactor process. *Control Engineering Practice, 6*, 1151-1158.

Wark, K. (1996). Thermodynamics. McGraw Hill Book Company, Inc.

Wang, N. and Brennan, J. G. (1995). A mathematical model of simultaneous heat and moisture transfer during drying of potato. *Journal of Food Engineering, 24*, 47-60.